Vol. 3

Frontiers in Nanobiomedical Research

HANDBOOK OF NANOBIOMEDICAL RESEARCH

Fundamentals, Applications and Recent Developments

❹ Biology, Safety and Novel Concepts in Nanomedicine

Frontiers in Nanobiomedical Research

ISSN: 2251-3965

Series Editors: Martin L. Yarmush *(Harvard Medical School, USA)*
Donglu Shi *(University of Cincinnati, USA)*

Published

Vol. 1: Handbook of Immunological Properties of Engineered Nanomaterials
edited by Marina A. Dobrovolskaia and Scott E. McNeil (SAIC-Frederick, Inc., USA)

Vol. 2: Tissue Regeneration: Where Nano-Structure Meets Biology
edited by Qing Liu (3D Biotek, USA and Tongji University, China) and Hongjun Wang (Stevens Institute of Technology, USA)

Vol. 3: Handbook of Nanobiomedical Research: Fundamentals, Applications and Recent Developments (In 4 Volumes)
edited by Vladimir Torchilin (Northeastern University, USA)

Forthcoming titles

Cancer Therapeutics and Imaging: Molecular and Cellular Engineering and Nanobiomedicine
edited by Kaushal Rege (Arizona State University, USA)

Nano Vaccines
edited by Balaji Narasimhan (Iowa State University, USA)

Nano Pharmaceuticals
edited by Rajesh N. Dave (New Jersey Institute of Technology, USA)

Thermal Aspects in Nanobiomedical Systems and Devices
by Dong Cai (Boston College, USA)

Nano Mechanochemistry in Biology
edited by Jeffrey Ruberti (Northeastern University, USA)

Nanomaterial Probes of Biological Processes and Systems
edited by David Mast (University of Cincinnati, USA)

Handbook of Biomaterials
edited by Donglu Shi (University of Cincinnati, USA) and Xuejun Wen (Clemson University, USA)

HANDBOOK OF NANOBIOMEDICAL RESEARCH

Fundamentals, Applications and Recent Developments

❹ Biology, Safety and Novel Concepts in Nanomedicine

Vol. 3

Frontiers in Nanobiomedical Research

editor

Vladimir Torchilin

Northeastern University, USA

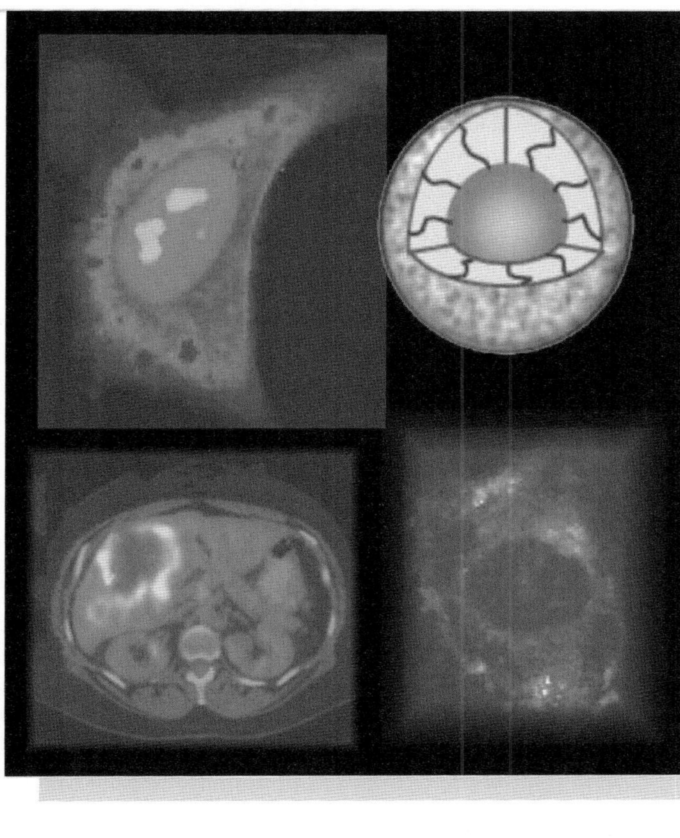

World Scientific

NEW JERSEY • LONDON • SINGAPORE • BEIJING • SHANGHAI • HONG KONG • TAIPEI • CHENNAI

Published by

World Scientific Publishing Co. Pte. Ltd.
5 Toh Tuck Link, Singapore 596224
USA office: 27 Warren Street, Suite 401-402, Hackensack, NJ 07601
UK office: 57 Shelton Street, Covent Garden, London WC2H 9HE

Library of Congress Cataloging-in-Publication Data
Handbook of nanobiomedical research : fundamentals, applications, and recent developments / editor, Vladimir Torchilin.
p. ; cm. -- (Frontiers in nanobiomedical research ; vol. 3)
Includes bibliographical references and index.
ISBN 978-9814520645 (set : alk. paper) -- ISBN 978-9814520676 (volume 1 : alk. paper) --
ISBN 978-9814520683 (volume 2 : alk. paper) -- ISBN 978-9814520690 (volume 3 : alk. paper) --
ISBN 978-9814520706 (volume 4 : alk. paper)
I. Torchilin, V. P., editor. II. Series: Frontiers in nanobiomedical research ; v. 3. 2251-3965
[DNLM: 1. Nanomedicine. 2. Nanostructures. 3. Nanotechnology. QT 36.5]
R857.N34
610.28--dc23

2014024090

British Library Cataloguing-in-Publication Data
A catalogue record for this book is available from the British Library.

Typeset by Stallion Press
Email: enquiries@stallionpress.com

Printed in Singapore

Contents

A:
Biology and Safety

Chapter A1

Biological Requirement and Safety Assessment of Nanomedicines

*Banu Zolnik, Simona Bancos, Jae Seung Kim and Nakissa Sadrieh**

Director, Cosmetics Staff
Office of Cosmetics and Colors
U.S. Food and Drug Administration
4300 River Road
Room 1034 (HFS-125)
College Park, MD 20740
** Nakissa.Sadrieh@fda.hhs.gov*

Abstract

The findings and conclusions in this chapter have not been formally disseminated by the Food and Drug Administration and should not be construed to represent any Agency determination or policy. The mention of commercial products, their sources, or their use in connection with material reported herein is not to be construed as either an actual or implied endorsement of such products by the Department of Health and Human Services.

The word "nanotechnology" confers images of extremely small objects that are very precisely engineered to perform targeted and complex tasks, by virtue of their small size. While this notion might be accurate for electronics and sensor nanotechnology applications, the progress of nanomedicine development is not as advanced as in other fields. Although, there are, at present, examples of marketed drug products containing nanoscale materials, one cannot say that drug development has been revolutionized by the advent of nanotechnology. However, a number of steps have been taken and we now know much more than we did a decade ago, and we expect to learn even more in the coming years, to advance nanomedicine development.

At the heart of the promise of nanotechnology, lies the concept of "size", which was and continues to be the central theme in many academic, regulatory and marketing/advertising discussions. The National Nanotechnology Initiative (NNI) developed a definition for nanotechnology, with a size limit of 100 nm, which has been accepted by certain organizations such as the Environmental Protection Agency (EPA), other organizations, such as the FDA, have not yet adopted a formal definition.[1] Alternatively, the FDA has identified criteria that can help to better review applications that might contain nanotechnology, despite the lack of adaptation of nanotechnology with definite size limits.

This chapter focuses on some of the reported factors related to the biological requirements and characterization of nanoparticles (NPs). In the following section, we will take a look at what is the biological basis for size considerations, because the relative size of the drug particles in relation to the physiological and anatomical barriers may play a critical role in the biological properties of the NPs in the body. Following that, we will focus on the size consideration that can impact the passage of NPs across various barriers, after which we will look at the safety assessment of nanomedicines, with considerations for toxicokinetic and toxicology studies, and finally we will focus on FDA's approach to the regulation of nanomedicines with a chronologic look at some of the agency milestones in nanotechnology policy development.

1. Biological Factors that Underscore Size Considerations

Following intravenous (*i.v.*) administration, NPs must overcome a series of barriers, such as the reticuloendothelial system (RES), the endothelial walls, extravascular spaces before they reach their target sites.[2,3] Most vascular walls are generally continuous with the endothelial cells and basal membranes. The gap junctions between endothelial cells are tight (1–5 nm), thus preventing NPs with sizes larger than the gap limits from exiting the circulation.[4] Liver endothelial cells have fenestrated structures that allow NPs to pass from the blood capillaries to the liver cells.[5] Table 1 shows examples of critical size barriers which might be relevant to the biodistribution and clearance of NPs in the body, and which will be discussed in the following section.

2. Size Considerations for Liver Transport

Liver fenestrae act as a sieve which can control the passage of NPs based on their size, allowing only those particles smaller than the fenestrae to reach the

Table 1. Biological responses with reported size constraints.

Biological responses	Size limits	Comments
Glomerular filtration	5–10 nm	Physiologic upper limit by renal clearance
NPs transport through liver sinusoid	300 nm	Liver fenestrae act as a sieve plate and can control the passage of NPs, allowing only those particles smaller than the fenestrae to reach the liver cells. Size distribution of sinusoidal fenestrae among different species; 75–300 nm in rats, 45–255 nm in rabbits, 55–320 nm in mice and 50–300 in humans. Particle shapes and rigidity may further affect the size limit, for example, liposome sizes of up to 400 nm were able to cross the liver fenestrae.
Enhanced permeability and retention (EPR)	300–600 nm	EPR effects may vary depending on the species, tumor implantation sites and tumor types. Pore size may also be different in xenografts than in natural tumors. Nanoscale particles up to 300 nm in diameter extravasate through leaky tumor vasculature and selectively accumulate in tumor tissue via EPR effect. In the study involving liposomes, sizes up to 600 nm in diameter were able to permeate through the tissues. Thus, average pore sizes may be approximately assumed to be in 300–600 nm range.
Reticuloendothelial system (RES) uptake	300 nm	NPs are cleared by size-dependent phagocytosis by the cells in RES. Extended blood residence time for surface modified NPs are up to 200–300 nm.

parenchymal cells. In the liver, the size of fenestrae plays a critical role in the function of the liver. The fenestrations can affect the extraction of nanoscale materials from the circulation.[6,7] The reported size distribution of sinusoidal fenestrae among different species are similar; 75–300 nm in rats,[5] 45–255 nm in rabbits, 55–320 nm in mice and 50–300 in humans.[7,8] The FDA presented internal data at an advisory committee meeting in August of 2012,[9] indicating that nearly 90% of drugs under review were under 300 nm, based on at least one size parameter assessed by pharmaceutical manufacturers. A recent publication reported on the state of investigational and approved nanomedicine and the authors also selected 300 nm as the size limit that would encompass the unique physiologic behavior that occurs on the nanoscale.[10] The authors' reason for selecting 300 nm as the cutoff for nanomedicine is because much of nanomedicines utilize feature sizes at or below 100 nm, and this size cutoff would exclude many applications of their significant clinical benefit. The observed liver uptake of NPs with particle size larger than 300 nm is reported in the literature. This could be because many particles are not spherical and rigid, resulting in their mechanical deformation in the vasculature, which may allow the passage of bigger particles across the liver sinusoids. Therefore, based on published literature, a 300 nm size limit may be a reasonable estimate for the delivery of NPs to the liver.

3. Size Implication for Neovasculature

Cancer endothelium is leaky compared to normal vasculature and is devoid of lymphatic drainage, therefore, allowing NPs in the systemic circulation to accumulate in tumor tissue. In fact, it is this exact property which is capitalized upon for the delivery of nanoscale drugs to tumors. Enhanced Permeability and Retention (EPR) refers to the porous endothelial surface and leaky lymphatic drainage due to abnormal vascular formations in cancer cells.[11–16] EPR effects may vary depending on the species,[17] tumor implantation sites and tumor types.[18,19] Pore size may also be different in xenografts than in natural tumors.[20] Generally, average pore sizes may be assumed in the 200–400 nm range. Nanoscale particles less than about 300 nm in diameter extravasate through leaky tumor vasculature and selectively accumulate in tumor tissue via EPR effect.[21] A study involving liposomes found them to be permeable up to 400 nm in diameter, suggesting that the cutoff size of the pores is approximately 200–400 nm.[11]

4. Size Considerations for NP Drug Delivery Systems

NP drug delivery systems may extend the blood residence time of drugs when engineered in a way to avoid elimination from the blood compartment. Taking advantage of the long residence time of certain NP constructs allows NPs the possibility to stay in the circulation before they can reach the target tissues and deliver the therapeutic loads.

Renal clearance is one of the primary routes for nanomaterials to be eliminated from the blood.[22] Particle size thresholds for the renal filtration is approximately 5–10 nm.[4,22] Particles that exceed the renal filtration limit are taken up in the cells of reticuloendothelial system (RES). Macrophages in RES can uptake the particles according to size.[23] However, many factors such as size, composition and surface charge can influence RES uptake of particles. Particles with unmodified surfaces are rapidly taken up by the RES. With suitable surface coatings, particles can remain in the blood based on their size and are gradually eliminated. Data in animal studies have shown that surface modified particles in the range of up to 200–300 nm exhibited extended blood half-lives compared to their unmodified counterparts.[24–37] Particle sizes in 200–300 nm range with suitable surface modification may maintain the extended blood half-life and increase their chances to reach the target tissues and deliver the therapeutic loads. It should be recognized, however, that these particle size cutoffs are not definitive and can only be considered as trends because cross-study comparisons are difficult to assess because of differences in techniques, and study materials used in each study.

Having discussed some of the anatomical and biological barriers in the above section, the next section will focus on the physiological barriers affecting the oral absorption of NPs, as well as some of the physiological targets for the design of nanomedicines. Other routes of administration such as parenteral, dermal and inhalation will not be discussed in this chapter; however, one should recognize that there are unique considerations for each route of administration. We will end the chapter with a focus on some of the milestones that CDER and FDA have reached, during the journey that began in 2006, when the FDA nanotechnology Task Force was established, after a brief survey of toxicokinetics and toxicology assessment considerations.

5. Oral Delivery Route for the NPs

The oral route of drug delivery is usually the most desirable route of administration due to patient compliance and ease of administration. Because of various anatomical and functional barriers, such as immunological responses, the oral delivery of marketed NPs is so far limited to nanocrystals. However, there are other NPs such as solid lipid, polymeric micelles, and dendrimers that are reported in the literature, which are being explored for oral delivery.[38–41] In order to maximize the bioavailability of NPs following the oral route, we will examine the physiological barriers of the gastrointestinal (GI) tract. The primary function of the intestinal barriers is to allow nutrients to reach the systemic circulation while defending the body from the unwanted molecules. Therefore, for development of effective nanomedicine, one must capitalize on the prevailing pathways for absorption while avoiding the immune system by custom designing NPs.

The GI tract presents these barriers which include physical barriers and immunological barriers. These barriers will be described in this section and included in Sections a, b and c of this chapter. The acidic environment of the stomach (approximately pH 1.7 fasting and pH 6.7 fed state)[42] and the presence of digestive enzymes can present a challenge to the stability of NPs in the GI lumen. Briefly, drugs and nutrients go through different compartments of the GI tract, where they encounter varying pH environments and digestive enzymes before being absorbed into the blood and excreted. The absorption site for most drugs is the duodenum and the upper region of the small intestine, where the pH is around 6 to 6.5. Pancreatic enzymes which are secreted from the bile duct into the duodenum, as well as enzymes such as trypsin, chymotrypsin, carboxypeptidase, amylase, lactase and lipase, are responsible for digestion of proteins, carbohydrates and fats. In addition to affecting the absorption of drugs, these enzymes also play a role in the stability of drugs in the GI tract. Unique anatomical features of the duodenum include a high surface area created by structures such as the folds of Kerkring, villi and microvilli, which allow drugs to be passively absorbed into the bloodstream. Once past the duodenum, the GI content reaches the jejunum, which is the middle portion of the small intestine, where further digestion of proteins and carbohydrates take place. Upon reaching the ileum or the terminal part of the small intestine, the pH increases from 6–6.5 to around 7 due to the presence of bicarbonate secretion, which increases the solubility of the acidic drugs in this region of the GI tract. The presence of bile secretion also aids hydrophobic drugs to solubilize in this region.

It is reported that bioadhesive materials with varying compositions have been designed to react differentially with the three regions of the small intestine, and this tissue specific approach to anatomical features could be utilized in the future for effective oral delivery of nanomedicines.[43] The colon is the terminal region of the GI tract, where limited absorption takes place due to the viscous and semisolid nature of the GI content combined with the low pH of this region (around 5.5–7), as well as to the presence of aerobic and anaerobic microorganisms which have the capacity to metabolize some drugs.

Relevant to oral absorption of nanomedicines, the next section will examine each of the barriers of the GI tract.

6. Dissolution, Solubility and Stability

The solubility of a drug molecule that is taken orally, depends on its physicochemical properties in the pH environment of the GI tract. Some nanocarriers are insoluble in any environment, such as gold (Au) NPs, and their route of absorption will not be affected by the pH of the GI tract, rather it will be mediated by passive or active transport. However, if a drug molecule can be released from a nanocarrier platform, then its absorption is governed by the solubility of the drug when it is released from the nanocarrier.

7. Barriers in the Gut

The first barrier is the gut microbiota, which prevents the adhesion of pathogens and toxins and hinders paracellular diffusion of such compounds. The next barriers are the mucus layer and the monolayer of the intestinal epithelial cells below the mucus layer, which not only provide an anatomical barrier but also contain immune cells. Immune cells are organized in a specialized manner known as the gut associated lymphoid tissue (GALT). Dynamic "communication" exists between these layers and the immune cells to protect the body from foreign compounds. Immune cells such as dendritic cells and M-cells in Peyer's patches are able to internalize the foreign compounds and present antigens to T-lymphocytes which are responsible for the production of cytokines and the elicitation of the immune response. Interaction between NPs and the immune system is reviewed later in this chapter.

7.1. *Gut microbiota*

Gut microbiota consist of several hundreds of different species of bacteria, fungi and yeast. While their functions are complex, they present a barrier, metabolize drugs, toxins and nutrients and even play a role in synthesis of vitamins and hormones to protect the body.[44]

7.2. *Mucus layer*

The mucus barrier acts as a lubricant and prevents the adhesion of foreign compounds. It consists of two layers; one layer is 50 μm thick and dense and firmly adherent to the epithelial cells and the second layer is unattached and is approximately 100 μm thick. The mucus layer is wrapped by a layer of highly glycosylated MUC2 mucin, which forms a net like structure consisting of a highly hydrated polymer, where the O-glycans on the mucin can act as an attachment site. It has been reported that positively charged NPs are transported easily because of the electrostatic interactions between the negatively charged mucin in the mucus layer. Pegylated NPs have been reported to exhibit muco-penetrating properties and improved bioavailability compared to NPs without any coating.[45,46] One study reports the preclinical oral relative bioavailability of 180 nm pegylated NPs loaded with paclitaxel to be 85%,[45] while another report examined the influence of the type of surface conformation (PEG brush versus loop) on the interaction with the mucosa, and predicted that brush PEG corona may facilitate the penetration and adhesive properties of the NPs, whereas loop conformation may increase the residence time of the adhered NPs in the mucosa.[46]

7.3. *Epithelial cells*

The gut epithelium is made up of a 20 μm thick monolayer of cells consisting of five different cell types, including enterocytes, M-cells, enteroendocrine cells producing GI hormones or peptides, mucus producing goblet cells, and antimicrobial peptides and protein secreting Paneth cells.[47] Uptake and transcellular transport of NPs by the enterocytes is an energy-dependent process and may occur via pinocytosis (macropinocytosis) and endocytosis (clathrin-mediated or caveolae-mediated or other independent mechanism). Details on the uptake and transcellular mechanisms for NPs are widely studied and reviewed in the literature,[48] therefore we will not review these processes in this chapter. However, we will summarize the requirements from the biological perspective.

Macropinocytosis, which is not a receptor mediated transport system, includes formation of large vesicles at the cell surface and is involved in the uptake of particles with sizes greater than 0.2 μm.[49] Studies indicate that receptor mediated transport mechanisms such as clathrin-coated pits, result in the formation of 120 nm size vesicles. On the other hand, for caveolae-mediated endocystosis, which is associated with the cholesterol and sphingolipid rich domains of the cell membranes, vesicle size is estimated to be between 50–80 nm. The vesicles fuse with early endosomes and then become acidified and form late endosomes and finally lysosomes. Inside the lysosomes, the NPs may be destroyed by the enzymes or be exocytosed to the lymphatic system or to the bloodstream. It has been shown that cationic NPs can disrupt the endosomes, which may explain some of the reported toxicities associated with the cationic NPs.

Tight junctions (TJs) are located between enterocytes, and their function is to regulate the paracellular passage of the solutes, ions and water. There are two types of junctions: adherent junctions (AJs) and TJs, which are formed by transmembrane proteins such as, occludin, claudins, junctional adhesion molecule (JAM) and tricellulin.[50–52] It was reported that pore diameter of the TJs are in the order of 4–14 Å.[53] The passage through TJs is very limited for the NPs unless there exists a disease state, or there are surface modification of the NPs with surfactants, which can play a role in opening up the TJs. Alteration of these junctions can be a target of nanomedicines, in order to improve drug delivery and to improve intestinal permeability. However, surface modification of NPs may also result in activation of immune cells and trigger an inflammatory response. M-cells are specialized cells of the mucosa associated lymphoid tissue (O-MALT) in the Peyer's patches, and they are responsible for the uptake and transcytosis of macromolecules and micro-organisms to the immune cells in the lymphoid cells beneath the epithelium. M-cells are capable of uptake of particles up to 1 μm and may occur with different mechanisms such as fluid phase endocytosis, adsorptive endocytosis and phagocytosis. M-cell specific delivery has been studied extensively with specific ligands such as RGD peptide, IgA, ganglioside attached to the NPs. NPs may be designed to bind to the apical M-cell membrane, and this interaction may either be non-specific or integrin and lectin specific interaction. However, the development of M-cell specific delivery by nanomedicines may be limited because M-cells only make up 1% of the epithelial cells, and NPs end up being trapped in the local lymph nodes following their transcytosis through M-cells. A survey of the literature indicates that NPs with a particle size less than 300–400 nm size and with optimal surface chemistry and charge may be the most promising candidates for the future of nanomedicine drug delivery.

Following the discussion of the mechanisms of absorption of NPs, from the oral route, we will discuss some of the considerations that may be important in the safety assessment of NPs.

8. Safety Assessment of Nanomedicines

Safety assessment is an evaluation and characterization of the possible undesirable effects caused by drugs, which for the purpose of this chapter will be limited to nanomedicines. Safety assessment normally consists of a preclinical and a clinical phase. So far, in this chapter, we have focused on discussions pertaining to the physico-chemical properties, stability, size and factors associated with the uptake of NPs following oral delivery. In the following section, we will focus on the preclinical safety assessment of nanomedicines, using insoluble gold NPs as an example. It should be noted that some of this considerations may or may not apply to soluble NPs or NPs with certain surfaces chemistries.

8.1. *Nanomaterial toxico kinetics in animal models*

Intravenous administration of gold NPs (Au NPs; 10–250 nm) has shown that while smaller NPs had a wider distribution to multiple organs (blood, liver, spleen, kidney, thymus, heart, lung, brain), all Au NPs, independent of their size, were localized in blood, liver and spleen.[54] Au NPs accumulating in the

RES organs, independent of their size, was also demonstrated by using primary particles, aggregates or agglomerates of Au NPs.[55] Conversely, a study showed that although accumulation of Au NPs in the liver is unquestionable, the amount accumulated depends on the NPs charge.[56] For example negatively charged Au NPs accumulated more (81%) in the liver compared to Au NPs of the same size that accumulated in the spleen (72%). Table 2 provides a summary of nanomaterial distribution in animal models discussed in this chapter.

Many of the studies reported in the literature use a single dose of NPs. However, most drugs are not given as a single administration, and one should consider that nanomedicines are likely to be given as multiple doses. This means that safety assessment studies should be adequately designed to include multiple-dose studies, using relevant animal models. For instance, repetitive administration of Au NPs leads to a dose-dependent increase in liver weight.[57] Similar studies conducted with silver (Ag) NPs showed that while one dose of Ag NPs determined accumulation in the liver or spleen (depending on Ag NPs size), repeat doses of Ag NPs determined increased accumulation in RES organs (liver, spleen and lung).[58]

Surface coating of drugs has long been associated with a prolonged circulation time, and in the case of NPs, surface coatings can also be taken advantage of to prevent NPs agglomeration. Studies have shown that intraperitoneal (*i.p.*) administration of polyethylene glycol (PEG)-coated Au NPs accumulated in the liver, spleen, blood and bone-marrow or preferentially in the blood cells.[59] Other studies showed that more than half of administered PEG-coated Au NPs accumulate in the liver (45%) and spleen (20%), while the rest was recovered from kidney, lung, brain, testis.[60] Distribution of Au NPs with citrate, 11-mercaptoundecanoid acid (11-MUA) and three pentapeptides was also evaluated. Citrate-Au NPs was rapidly removed from the circulation with the highest accumulation in the liver and spleen; MUA-coated Au NPs presented a similar distribution however liver exhibited higher Au NPs concentration.[61]

While the majority of the safety assessment studies are conducted in healthy animals, absorption, distribution, metabolism, excretion (ADME) profiles can change depending on the animal model used. PEG-coated Au nanoshells (NSs) and Au nanorods (NRs) administration to mice with subcutaneous tumor xenografts showed that Au NRs accumulated more in the tumor compared to Au NSs. Higher doses of PEG–Au NSs and PEG–Au NRs determined accumulation in the liver.[62] In tumor-bearing rabbit model Au NPs mostly accumulated in the spleen, liver and to a lesser extent in tumor.[63]

We need to keep in mind that nanomedicines are complex structures that are designed to deliver their load to the target tissues and the presence of the drug or biologic that is delivered can affect the toxicokinetics profile of the nanomedicine. CYT-6091 NP system consisting of a TNF-α–coated Au NPs was used in a tumor-bearing mouse model. While TNF-α was delivered to the tumor, it was shown that after TNF-α degradation Au NPs were taken up preferentially by the liver.[64]

The toxicokinetics profile of orally administered nanomaterials is less studied. In a repeat 28-day oral dosing toxicology study, Ag NPs were shown to accumulate in intestine, liver and kidney. Oral administration of chitosan and PEG coated Poly, DL, lactic-co-glycolic acid (PLGA) NPs showed that they are taken up by the peritoneal macrophages.[65] Other studies showed no significant accumulation in any of the organs or tissues. For example, rats orally administered Au NPs of different sizes and different charges showed that while there was some retention of Au NPs in the liver, there was no clear trend detected regarding Au NPs distribution.[66] Oral administration of TiO2 NP and Zn NPs showed differences in ADME in a rat model.[67] TiO2 NPs distribution to liver, spleen, kidney and brain was minimal due to the minimal absorption rate, while high concentration of TiO2 NPs was detected in feces. ZnO NPs on the other hand, showed high distribution and accumulation in liver and kidney.

Safety assessments of inhaled nanomaterials are primarily done for environmental toxicity studies. In the case of inhaled nanomedicine products, there are fewer toxicokinetic studies available. After inhalation, nanomaterials deposit on the pulmonary surfactant layer which can cause nanomaterial agglomeration and translocation of nanomaterials to the blood. However, it has been shown that the majority of inhaled Au NPs are associated with the lung tissue and the amount of Au NPs cleared from the lung was low at early time-points.[68] Au NPs that accumulated in the lungs after inhalation are 90% associated with the lung macrophages and decreased to ~60% one week later.[69] Comparison of bare Au NPs and PEG-coated Au NPs in inhalation studies in rats showed that the majority of both types of Au NPs accumulated in the lung and only a small portion of PEG–Au NPs was found in the blood circulation.[70] Nanomaterial translocation from lung to blood circulation is not significantly increased in the case of repeat dose inhalation. For example, repeat instillation of Au NPs determined significant increase of Au NPs accumulation in lung macrophages and only a very small fraction of Au NPs was found in the liver.[71]

Clearance of nanomaterials such as Au NPs can take place through the kidneys, bile or feces. However, accumulation in the kidneys can impair glomerular basement membrane permeability which in turn can impact Au NPs elimination via the urine.[72] Excretion of Au NPs might be size-dependent.

Table 2. Summary of nanomaterial distribution in animal models.

Core	Shape	Size	Coating	Single/ Repeat	Animal model	Route of administration	Targeted	Major site of accumulation	References
Au	NP	10 nm; 50 nm; 100 nm; 250 nm	—	Single	rat	i.v	No	blood, liver, spleen	dejong, 2008
Au	NP	30–200 nm aggregates; 50–2000 nm agglomerates	—	Single	mouse	i.v	No	liver, spleen, lung	Keene, 2011
Ag	NP	20 nm; 80 nm; 110 nm	—	Single	—	—	No	liver, spleen	Lankveld, 2010
Ag	NP	20 nm; 80 nm; 110 nm	—	Repeat	rat	i.v	No	liver, spleen, lung	Lankveld, 2010
Au	NP	12.5 nm	—	Repeat	mouse	i.p	No	liver	Lasagna-Reeves, 2010
Au	NP	5 nm; 10 nm; 30 nm; 60 nm	PEG	Single	mouse	i.p	No	liver, spleen, bone marrow, blood	Zhang, 2011
Au	NP	13 nm	PEG	Single	mouse	i.v	No	liver, spleen, kidney, lung, brain, testis	Cho, 2009
Au	NP	20 nm; 80 nm; 110 nm	Citrate; 11-MUA; 3 pentap eptides	Single	rat	i.v	No	liver, spleen	Morais, 2012

(*Continued*)

Table 2. (*Continued*)

Core	Shape	Size	Coating	Single/ Repeat	Animal model	Route of administration	Targeted	Major site of accumulation	References
Au	NS; NR	120 nm; 24 × 7 nm	PEG	Repeat	mouse tumor xenograf	i.v	No	tumor	Puvanakrishnan, 2012
Au	NP	5 nm; 25 nm	—	Single	rabbit (liver tumor)	i.v	No	liver, spleen, tumor	Glazer, 2011
Au	NP	27 nm	PEG	Single	mouse (tumor)	i.v	Yes	tumor, liver, spleen	Goel, 2009
PLGA	NP	50–150 nm; 200 nm	PEG	Single	mouse	oral; i.p	No	peritoneal macrophages	Semete, 2007
TiO_2	NP	20 nm	—	Repeat	rat	oral	No	feces	Cho, 2013
Zn	NP	40 nm	—	Repeat	rat	oral	No	liver, kidney	Cho, 2013
Au	NP	2 nm; 40 nm; 100 nm	—	Repeat	mouse	i.t	No	lung	Sadauskas, 2009

The level of Au NPs (20 nm) was higher in the kidney and intestine compared to Au NPs (80 nm) suggesting that smaller size NPs might be excreted easier.[73] Conversely, others[70] showed that excretion is not size- but surface coating-dependent. For example, Au NPs coated with PEG with a hydrodynamic diameter of 31 nm were present at higher level in the feces than PEG-coated Au NPs with a hydrodynamic diameter of 21 nm.

Although some clearance of nanomaterial occurs, biopersistence has been reported for a wide range of NPs.[74–76] Exposure to the NPs may last a long time and potentially span the lifetime of the patient. This exposure is especially important in the treatment of chronic diseases where therapies are administered repeatedly to the affected population.

After administration, nanomaterials reach the blood circulation; as mentioned earlier, the circulation time is influenced by the presence of surface coating and surface charges. Although variations exist between studies, and it is difficult to draw a comprehensive conclusion on the existing evidence, data presented so far indicate that nanomaterials distribute and accumulate mostly in the liver, spleen and lungs. The accumulation in RES is independent of size, charge and core (e.g. Au NPs versus Ag NPs), mode of administration, animal model or targeted delivery. The nanomaterials that were not excreted normally remains accumulated in macrophages of RES. Therefore, the interaction of nanomaterials with the immune system constitutes a very important part of the safety assessment of nanomedicine.

8.2. *Nanomedicines toxicology*

Toxicology studies to support drug safety are conducted both *in vitro* and *in vivo*. A desired aim is to develop *in vitro* tests that identify toxicities and that are in some way predictive of *in vivo* behavior. *In vitro* tests are fast and often less expensive than whole animal studies, and they can be useful for screening, identification of toxicities and determination of mechanisms of action. However, there are currently no validated *in vitro* tests for nanomaterials that help predict *in vivo* behavior. In the following section, we will focus on some of the *in vitro* and *in vivo* studies that have been described in the literature to have some utility in predicting and discriminating the safety of nanomedicines. The Nanotechnology Characterization Laboratory (NCL), NIH focused exclusively on the development of many of the existing predictive safety assessment tests, as well as protocols for *in vitro* and *in vivo* assay cascades and characterization methodologies, all of which can be found at http://ncl.cancer.gov/.

8.2.1. *In vitro* toxicology

The first body compartment that is encountered by nanomaterials after systemic administration is the blood. Nanomaterial coating prolongs blood circulation and reduces nanomaterial uptake by the immune system cells, however prolonged blood circulation of nanomaterials can increase the risk of hemotoxicity. Blood constituents that are relevant to nanomaterials are the plasma proteins (~3,000 in plasma including immunoglobulins and complement system) and the circulating blood cells. While interaction of nanomaterials with plasma proteins and formation of a protein "corona" can influence the biodistribution and clearance of nanomaterials,[77] interaction of nanomaterials with erythrocytes and platelets, as well as with the complement system, can predict nanomaterial hemotoxicity, thrombosis and complement activation.[78] Below, we will present a few hemotoxicity in vitro studies that identified factors predictive of hemotoxicity, clotting and complement system activation.

Au NPs-induced hemolysis (damage of erythrocytes measured by hemoglobin release) was shown to take place in a concentration-dependent manner.[79] Positively charged Au NPs seemed to result in higher levels of hemolysis, versus negatively charged Au NPs. However, studies conducted with Au NPs of the same size, charge and synthesis method showed that the difference in hemolysis patterns was attributed to differences in protein adsorption.[79] When the assays are conducted in plasma, hemolysis did not take place regardless of the surface charge due to the formation of a protein corona which inhibits damage to erythrocytes.[80] Differences in hemolytic capacity were also detected between different NPs. For example Ag NPs exhibited a higher hemolytic capacity compared to Au NPs or Pt NPs.[81]

Nanomaterials were shown to interact with and influence the plasma coagulation system (PCS). PCS, composed of platelets, endothelial cells and plasma proteins, is responsible for hemostasis: thrombus formation in the case of blood vessel injury or inhibition of clotting in physiological conditions. Most of the *in vitro* studies that address the influence of nanomaterials on thrombosis, focus on leukocyte procoagulant activity (PCA) and platelet function. It was shown that anionic and neutral polyamidoamine (PAMAM) dendrimers did not induce leukocyte PCA *in vitro*, independent of their size and surface charge.[82] However, large and small cationic and large neutral PAMAM dendrimers induce platelet aggregation *in vitro*,[83] altered platelet morphology and determined adherence to surfaces.[84]

The complement system is the part of the innate immunity that helps the immune cells (such as macrophages, T- and B-cells) to clear the invading pathogens from the body. There are three main pathways of activation: classical, alternative and lectin. The complement system consists of a number of proteins (C3, C5, C6, etc.) that induce a cascade reaction and complement the activity of the immune cells by: opsonization (increases phagocytosis), chemotaxis (recruitment of neutrophils and macrophages to the site of inflammation) and cell lysis of foreign cells. While the morphology of a nanomaterial core seems to have no effect on complement system, the surface chemistry is important for complement activation. It was shown that the length of surface coating and surface architecture can influence the complement activation cascade. Nanocapsules with different PEG chain length and density induced C3 activation, but the activation occurred to a lesser extent in the case of nanocapsules with longer PEG chains and higher density.[85] PEG-coated nanocapsules were weaker activators of the complement system compared to PLA [poly(D,L-lactide)]-coated nanocapsules or nanospheres.[85] The influence of surface coating and configuration was evaluated using Poloxamine 908-coated NPs with polyethylene oxide (PEO) surface chains displaying mushroom, mushroom–brush and brush configurations. Alteration of surface architecture from mushroom to brush switched complement activation from C1q-dependent classical to the lectin pathway.[86]

8.2.2. *In vivo* toxicity

Due to the limitation of the *in vitro* assays, *in vivo* assays are often used for adequate risk assessment of nanomedicine products. General toxicity studies evaluate hemato-, hepato-, nephro- and cardio-toxicity as well as immunogenicity and signs of inflammation. These findings are analyzed in conjunction with pharmacokinetic (PK) profiles. PK profiles of the nanomedicine products are influenced by several factors such as: physio-chemical characteristics (such as core, size, charge, coating, stability, shape, etc.), dose, route of administration, animal model (healthy versus tumor-bearing).[87] As described above, nanomaterials tend to accumulate in the RES organs, namely in the macrophages. Therefore, herein we will focus on how the interaction between mononuclear phagocyte system (MPS) and nanomaterials affects the pharmacokinetics of certain nanomedicines.

As soon as nanomaterials enter the blood circulation, they undergo opsonization (adsorption of proteins such as immunoglobulins and complement system proteins) which in turn affects nanomaterial clearance by facilitating nanomaterial phagocytosis by MPS. Opsonization is believed to be the major biological barrier for delivery of polymeric NPs.[88] Serum proteins are adsorbed rapidly on hydrophobic particles therefore opsonization of hydrophobic particles occurs more quickly than in the case of hydrophilic particles.[89] One of the most frequently used methods to decrease NP opsonization is pegylation. It is postulated that PEG chains are compressed by opsonizing proteins, which condense the PEG chains, and create a repulsive force between opsonins and nanomaterial surface.[90] Administration of poly (epsilon-caprolactone) (PCL) and pegylated PCL showed that the naked PCL was present in MPS tissues. However, PEG hydrophilic chains not only decreased C3 cleavage and complement activation, but also suppressed opsonization and phagocytosis, resulting in increased blood retention of PEG–PCL versus naked PCL.[88] Poly methoxypolyethyleneglycol (MePEG) cyanoacrylate-co-n-hexadecyl cyanoacrylate (PEG-PHDCA) NPs containing recombinant human tumor necrosis factor-alpha (rHuTNF-alpha) were prepared using different MePEG molecular weights, and their biodistribution and PK values were determined. Serum absorption and phagocytosis was decreased for high molecular weight PEG–PHDCA NPs, extending the half-life of rHU-TNF-alpha in S-180 tumor-bearing mice by 24-fold.[37]

After opsonization takes place, the cells of the MPS attach themselves to the opsonized nanomaterial. Attachment occurs through adherence of macrophages to the opsonized nanomaterial or through the interaction of the macrophage receptors with the opsonins.[90] Nanomaterials can be incorporated by macrophages through different mechanisms (phagocytosis, receptor-mediated endocytosis or pinocytosis) that are well-described elsewhere.[91] Although the majority of nanomaterials accumulate in the macrophages of the MPS, some studies indicate that nanomaterials are preferentially taken up by either liver or spleen macrophages depending on size and surface characteristics. These observations might be useful in cases of tissue-targeting.

Monocytes differentiate in the bone marrow and circulate in the blood. Monocytes migrate into the tissues where they differentiate into macrophages that perform phagocytosis, incorporating cell debris, microorganisms or nanomaterials. Although there are several articles describing the phagocytosis of nanomaterials by macrophages, there is little information on where the phagocytosis occurs. Namely, are tissue resident macrophages incorporating nanomaterials from the blood passing through the MPS organs or are the circulating monocytes incorporating the nanomaterial ("hijacking") in the blood after which monocytes migrate to the tissues and differentiate into macrophages? The second theory has been postulated with results from clinical trials of S-CKD602 (a pegylated liposomal formulation of CKD-602) showing that monocytes were more sensitive to the effect S-CKD602 compared to neutrophils.[92] Monocytopenia is believed to be caused by incorporation of S-CKD602 in monocytes, release of CKD-602 which induces monocyte toxicity.[92]

In cases where nanomaterials are taken up by tissue macrophages, nanomaterials have to pass through the blood vessels into the tissues. The vascular endothelium and the underlying basement membrane limit passage of materials to the tissues. However, as mentioned in an earlier section, liver and spleen endothelium is fenestrated, allowing nanomaterials (up to 300 nm for liver and over 100 nm for spleen) to pass from the vascular endothelium to the parenchymal cells.[93] On the other hand, in the case of pathological conditions (e.g. tumors), the endothelium becomes leaky leading to EPR (which was discussed earlier in the chapter). This EPR feature is capitalized upon, when designing tumor targeted nanomedicine drugs, in order to increase nanomedicine accumulation in tumors.[94,95]

The most frequently used animal models are rodents (mice and rats) while other models (pigs) are less studied. This aspect is important in the biodistribution context. Nanomaterial accumulation in spleen was shown to take place in mice in the marginal zone macrophages of the spleen whereas the accumulation occurred in the red pulp macrophages of rat.[96] Compared to rodents where foreign materials are cleared in spleen or liver, in calves, sheep, goats, cats and pigs foreign materials are cleared in the lung.[97] Surprisingly the vascular clearance rate of 198Au colloid was similar in 13 different animal species (sheep, calf, pig, cat, monkey, hyrax, rabbit, guinea pig, rat, mouse and chicken). However, a predominant liver accumulation was determined to occur for monkey, hyrax, rabbit, guinea pig, rat, mouse and chicken whereas cat, pig, calf and sheep showed lung accumulation by the lung macrophages.[98]

From the data presented so far it can be concluded that in order for a successful nanomedicine delivery to take place, the nanomaterial should not interact with or activate the blood components, should not be incorporated by the MPS, and most importantly, should reach the target tissue. However, the majority of nanomedicine products that are approved or in clinical phase trials, show some type of interaction with the immune system[48] (the formulations that were intended to have immunomodulatory effects should be excluded since they were designed to affect the immune system). An important aspect to consider is whether or not accumulation of nanomaterials in MPS has an effect on macrophage/monocyte function. As presented above, monocytopenia induced by S-CKD602 was due to released drug into monocytes and was not caused by the liposomes. However, there is no reported data on the interaction of hard-core, non-degradable nanomaterials with monocytes/macrophages. Therefore, the questions related to hard-core nanomaterial accumulation, excretion, blood levels reached after single or repeat dose remains to be addressed. *In vitro* and *in vivo* methods require additional consideration depending on factors such as the type of nanomaterial used, the physico-chemical characteristics, the dose used and the animal model chosen for conducting safety studies. The safety assessment of nanomedicines is an area that will in the near future undergo further development and standardization.

Having focused on some of size related characteristics of biological barriers in the body, as well as the factors involved in the absorption, uptake and safety assessment of nanomaterials, we will end the chapter with a brief focus on FDA's approach to the regulation of nanotechnology products.

9. FDA's Approach to the Regulation of Nanotechnology Products

The application of nanotechnology may result in product attributes that differ from those of conventionally-manufactured products. Therefore, evaluations of safety or effectiveness of FDA-regulated products that include nanomaterials or otherwise involve the application of nanotechnology should consider the unique properties and behaviors that nanomaterials may exhibit. However, at present, FDA does not categorically judge all products containing nanomaterials or otherwise involving the application of nanotechnology as intrinsically benign or harmful. FDA continues to regulate nanotechnology products under existing statutory authorities, in accordance with the specific legal standards applicable to each type of product under its jurisdiction.[1]

FDA's approach respects variations in legal standards for different product-classes. Drugs are evaluated not only on the basis of their risk profile but also their predicted benefit. Where premarket review authority exists, attention to nanomaterials is being incorporated into standing procedures. For example, new drugs, new animal drugs, biologics, food additives, color additives, certain human devices and certain new dietary ingredients in dietary supplements are subject to premarket review requirements. Premarket review processes for these products require applicants to submit data to answer questions related to the safety, effectiveness (where applicable), or regulatory status of the product. Individual premarket review procedures include attention to whether the use of nanomaterials suggests the need for additional data on safety or effectiveness, as applicable.[99]

Where statutory authority does not provide for premarket review, consultation is encouraged to reduce the risk of unintended harm to human or animal health. Some FDA regulated products, such as dietary supplements (except certain new dietary ingredients), cosmetics (except color additives), and food (except food or color additives) are not subject to mandatory pre-market review. In these cases, FDA relies on publicly available or voluntarily submitted information, adverse event reporting (where applicable), and on post-market surveillance activities, to provide oversight. However, FDA will continue post-market monitoring for such products. FDA will continue to monitor the marketplace for all products regardless of whether it contains nanomaterials and will take actions, as needed, to protect consumers.

Since 2004, FDA has been working towards a better understanding of nanotechnology-related science issues. The following represents a chronologic view of the activities related to nanotechnology, undertaken by the U.S. Food and Drug Administration.

April 2004: Nanotechnology Awareness topic presented at FDA Advisory Committee for Pharmaceutical Sciences.

March 2006: The U.S. FDA included nanotechnology in its Critical Path Initiative and formed a Nanotechnology Task Force to determine the regulatory approaches for the development of innovative, safe and effective products which contains nanoscale material.

July 2007: FDA's first nanotechnology report highlighted scientific and regulatory issues involving nanoscale material containing products.

July 2008: The Office of Pharmaceutical Science (OPS) within the Center for Drug Evaluation and Research (CDER) convened an advisory committee meeting focusing on Nanotechnology. FDA has convened four advisory committee meetings since 2004, on the topic of nanotechnology, with the most recent meeting in August of 2012.

September 2008: Following the publication of the report issued by the FDA Nanotechnology Task Force, the agency held its first Public Meeting where the public was invited to present its experience with nanomaterials in FDA-regulated products.

July 2010: CDER issued a Manual of Policies and Procedures (MaPP) entitled Reporting Format for Nanotechnology-Related Information in Chemistry, Manufacturing, Controls (CMC) Review in 2010. The goal of this MaPP was to identify and track nanomaterial-containing drug applications (IND, NDA or ANDA) submitted to FDA. Since the issuance of the MaPP, CDER has followed up with the establishment of a risk assessment working group, which published its first paper in the AAPS Journal in 2013.[100]

June 2011: FDA issued a draft guidance a Draft Guidance for Industry entitled, "Considering Whether an FDA-Regulated Product Involves the Application of Nanotechnology" to present its thinking on considerations related to nanotechnology.[101] In that draft guidance, FDA indicated that the Agency would issue product-specific guidances in the future, as appropriate.

April 2012: FDA issued two Draft Guidances for Industry.

— Safety of nanomaterials in cosmetic products.
— Assessing the effects of significant manufacturing process changes, including emerging technologies, on the safety and regulatory status of food ingredients and food contact substances, including food ingredients that are color additives.[102]

Therefore, since 2004, when CDER presented nanotechnology as an awareness topic at advisory committee meeting for the first time, the FDA has taken a number of steps towards better understanding the science of nanotechnology and towards providing better clarity regarding the regulatory status of products containing nanotechnology. Throughout this period, the FDA has also developed a robust research program in nanotechnology and there have been a number of papers published by FDA scientists in the area of nanotechnology.

In this chapter, we covered some of the biological processes that are relevant to nanomedicine uptake and clearance and we discussed the size considerations that might influence some of these processes. We also discussed factors associated with the absorption of nanomedicines given by the oral route of administration, focusing on the characteristics of nanotechnology-related drugs that might influence their absorption processes and biodistribution, such as surface modification and charge. Once in the body, it is important to have methods in order to assess the safety of nanomedicines, therefore, we also discussed in this chapter, some of the testing approaches that would be predictive of nanomaterial safety. Specifically, we looked at nanomaterial toxicokinetics testing, *in vitro* and *in vivo* testing approaches. Finally, we focused on the recent FDA activities, such as public meeting and policy development activities, which have helped to shape FDA's approach to the regulation of nanomaterials.

Acknowledgements

The authors would like to thank Roxane Modares for her valuable help in editing and proofreading of the chapter, as well as for her help with the preparation of the list of references.

References

1. FDA. Nanotechnology, http://www.fda.gov/ScienceResearch/SpecialTopics/Nanotechnology/default.htm.
2. M. D. Schoenberg, P. A. Gilman, V. Mumaw and R. D. Moore, Proliferation of the reticuloendothelial system and phagocytosis. *Exp. Mol. Pathol.* **2,** 126–143 (1963).
3. S. Stolnik, L. Illum and S. S. Davis, Long circulating microparticulate drug carriers. *Adv. Drug Deliv. Rev.* **64**, Supplement,: 290–301 (2012).
4. H. Sarin, Physiologic upper limits of pore size of different blood capillary types and another perspective on the dual pore theory of microvascular permeability. *J. Angiogenes Res.* **2,** 2040–2384 (2010).
5. E. Wisse, R. B. De Zanger, K. Charels, P. Van Der Smissen and R. S. McCuskey, The liver sieve: considerations concerning the structure and function of endothelial fenestrae, the sinusoidal wall and the space of Disse. *Hepatology* **5,** 683–692 (1985).
6. E. Wisse, F. Jacobs, B. Topal, P. Frederik and B. De Geest, The size of endothelial fenestrae in human liver sinusoids: implications for hepatocyte-directed gene transfer. *Gene Ther.* **15,** 1193–1199 (2008).
7. J. Snoeys, J. Lievens, E. Wisse, F. Jacobs, H. Duimel, D. Collen, P. Frederik and B. De Geest, Species differences in transgene DNA uptake in hepatocytes after adenoviral transfer correlate with the size of endothelial fenestrae. *Gene Ther.* **14,** 604–612 (2007).
8. F. Braet and E. Wisse, Structural and functional aspects of liver sinusoidal endothelial cell fenestrae: a review. *Comp. Hepatol.* **1,** 1 (2002).
9. A. Commitee, *Advisory Committee for Pharmaceutical Science and Clinical Pharmacology*, Food and Drug Administration (2012).
10. M. L. Etheridge, S. A. Campbell, A. G. Erdman, C. L. Haynes, S. M. Wolf and J. McCullough, The big picture on nanomedicine: the state of investigational and approved nanomedicine products. *Nanomedicine* **9,** 1–14 (2013).
11. F. Yuan, M. Dellian, D. Fukumura, M. Leunig, D. A. Berk, V. P. Torchilin and R. K. Jain, Vascular permeability in a human tumor xenograft: molecular size dependence and cutoff size. *Cancer Res.* **55,** 3752–3756 (1995).
12. D. M. McDonald and P. Baluk, Significance of blood vessel leakiness in cancer. *Cancer Res.* **62**(18), 5381–5285 (2002).
13. S. R. MacEwan, D. J. Callahan and A. Chilkoti, Stimulus-responsive macromolecules and nanoparticles for cancer drug delivery. *Nanomedicine* **5,** 793–806 (2010).
14. D. Lu, M. G. Wientjes, Z. Lu and J. L. Au, Tumor priming enhances delivery and efficacy of nanomedicines. *J. Pharmacol. Exp. Ther.* **322,** 80–88 (2007).
15. R. K. Jain, Normalizing tumor vasculature with anti-angiogenic therapy: a new paradigm for combination therapy. *Nat. Med.* 7, 987–989 (2001).
16. H. Hashizume, P. Baluk, S. Morikawa, J. W. McLean, G. Thurston, S. Roberge, R.K. Jain and D. M. McDonald, Openings between defective endothelial cells explain tumor vessel leakiness. *Am. J. Pathol.* **156,** 1363–1380 (2000).
17. P. P. Adiseshaiah, J. B. Hall and S. E. McNeil, Nanomaterial standards for efficacy and toxicity assessment. *Wiley Interdiscip. Rev. Nanomed. Nanobiotechnol.* **2,** 99–112 (2010).
18. M. Dellian, B. P. Witwer, H. A. Salehi, F. Yuan and R. K. Jain, Quantization and physiological characterization of angiogenic vessels in mice: effect of basic fibroblast growth

factor, vascular endothelial growth factor/vascular permeability factor, and host microenvironment. *Am. J. Pathol.* **149**, 59–71 (1996).
19. S. K. Hobbs, W. L. Monsky, F. Yuan, W. G. Roberts, L. Griffith, V. P. Torchilin and R. K. Jain, Regulation of transport pathways in tumor vessels: role of tumor type and microenvironment. *Proc. Natl. Acad. Sci. USA* **95**, 4607–4612 (1998).
20. S. Lauk, A. Zietman, S. Skates, R. Fabian and H. D. Suit, Comparative morphometric study of tumor vasculature in human squamous cell carcinomas and their xenotransplants in athymic nude mice. *Cancer Res.* **49**, 4557–4561 (1989).
21. H. Maeda and Y. Matsumura, Tumoritropic and lymphotropic principles of macromolecular drugs. *Crit. Rev. Ther. Drug Carrier Syst.* **6**, 193–210 (1989).
22. H. S. Choi, W. Liu, P. Misra, E. Tanaka, J. P. Zimmer, B. Itty Ipe, M. G. Bawendi and J. V. Frangioni, Renal clearance of quantum dots. *Nat. Biotechnol.* **25**, 1165–1170 (2007).
23. G. Schuhmann-Giampieri, Liver contrast media for magnetic resonance imaging. Interrelations between pharmacokinetics and imaging. *Invest Radiol.* **28**, 753–761 (1993).
24. B. H. Simon, H. Y. Ando and P. K. Gupta, Circulation time and body distribution of 14C-labeled amino-modified polystyrene nanoparticles in mice. *J. Pharm. Sci.* **84**, 1249–1253 (1995).
25. K. Ogawara, M. Yoshida, K. Higaki, T. Kimura, K. Shiraishi, M. Nishikawa, Y. Takakura and M. Hashida, Hepatic uptake of polystyrene microspheres in rats: effect of particle size on intrahepatic distribution. *J. Control Release* **59**, 15–22 (1999).
26. Z. Panagi, A. Beletsi, G. Evangelatos, E. Livaniou, D. S. Ithakissios and K. Avgoustakis, Effect of dose on the biodistribution and pharmacokinetics of PLGA and PLGA-mPEG nanoparticles. *Int. J. Pharm.* **221**, 143–152 (2001).
27. P. Wunderbaldinger, L. Josephson and R. Weissleder, Tat peptide directs enhanced clearance and hepatic permeability of magnetic nanoparticles. *Bioconjug. Chem.* **13**, 264–268 (2002).
28. S. E. Gratton, P. D. Pohlhaus, J. Lee, J. Guo, M. J. Cho and J. M. Desimone, Nanofabricated particles for engineered drug therapies: a preliminary biodistribution study of PRINT nanoparticles. *J. Control Release* **121**, 10–18 (2007).
29. J. Zheng, J. Liu, M. Dunne, D. A. Jaffray and C. Allen, In vivo performance of a liposomal vascular contrast agent for CT and MR-based image guidance applications. *Pharm. Res.* **24**, 1193–1201 (2007).
30. M. Liu, H. Li, G. Luo, Q. Liu and Y. Wang, Pharmacokinetics and biodistribution of surface modification polymeric nanoparticles. *Arch. Pharm. Res.* **31**, 547–554 (2008).
31. G. Gaucher, K. Asahina, J. Wang and J. C. Leroux, Effect of poly(N-vinyl-pyrrolidone)-block-poly(D,L-lactide) as coating agent on the opsonization, phagocytosis, and pharmacokinetics of biodegradable nanoparticles. *Biomacromol.* **10**, 408–416 (2009).
32. Z. Yang, J. Leon, M. Martin, J. W. Harder, R. Zhang, D. Liang, W. Lu, M. Tian, J.G. Gelovani, A. Qiao and C. Li, Pharmacokinetics and biodistribution of near-infrared fluorescence polymeric nanoparticles. *Nanotechnology* **20**, 0957–4484 (2009).
33. C. H. Choi, J. E. Zuckerman, P. Webster and M. E. Davis, Targeting kidney mesangium by nanoparticles of defined size. *Proc. Natl. Acad. Sci. USA* **108**, 6656–6661 (2011).
34. D. P. Lankveld, R. G. Rayavarapu, P. Krystek, A. G. Oomen, H. W. Verharen, T. G. van Leeuwen, W. H. De Jong and S. Manohar, Blood clearance and tissue distribution of

PEGylated and non-PEGylated gold nanorods after intravenous administration in rats. *Nanomedicine* **6**, 339–349 (2011).

35. Y. Wenger, R. J. Schneider, 2nd, G. R. Reddy, R. Kopelman, O. Jolliet and M. A. Philbert, Tissue distribution and pharmacokinetics of stable polyacrylamide nanoparticles following intravenous injection in the rat. *Toxicol. Appl. Pharmacol.* **251**, 181–190 (2011).
36. K. S. Yadav, K. Chuttani, A. K. Mishra and K. K. Sawant, Effect of Size on the Biodistribution and Blood Clearance of Etoposide-Loaded PLGA Nanoparticles. *PDA J. Pharm. Sci. Technol.* **65**, 131–139 (2011).
37. C. Fang, B. Shi, Y. Y. Pei, M. H. Hong, J. Wu and H. Z. Chen, In vivo tumor targeting of tumor necrosis factor-alpha-loaded stealth nanoparticles: effect of MePEG molecular weight and particle size. *Eur. J. Pharm. Sci.* **27**, 27–36 (2006).
38. W. Ke, Y. Zhao, R. Huang, C. Jiang and Y. Pei, Enhanced oral bioavailability of doxorubicin in a dendrimer drug delivery system. *J. Pharm. Sci.* **97**, 2208–2216 (2008).
39. S. H. Kwon, S. Y. Kim, K. W. Ha, M. J. Kang, J. S. Huh, T. J. Im, Y. M. Kim, Y. M. Park, K. H. Kang, S. Lee, J. Y. Chang, J. Lee and Y. W. Choi, Pharmaceutical evaluation of genistein-loaded pluronic micelles for oral delivery. *Arch. Pharm. Res.* **30**, 1138–1143 (2007).
40. D. R. Kalaria, G. Sharma, V. Beniwal and M. N. Ravi Kumar, Design of biodegradable nanoparticles for oral delivery of doxorubicin: in vivo pharmacokinetics and toxicity studies in rats. *Pharm. Res.* **26**, 492–501 (2009).
41. K. Manjunath and V. Venkateswarlu, Pharmacokinetics, tissue distribution and bioavailability of clozapine solid lipid nanoparticles after intravenous and intraduodenal administration. *J. Control Release* **107**, 215–228 (2005).
42. J. B. Dressman, R. R. Berardi, L. C. Dermentzoglou, T. L. Russell, S. P. Schmaltz, J. L. Barnett and K. M. Jarvenpaa, Upper gastrointestinal (GI) pH in young, healthy men and women. *Pharm Res.* **7**, 756–761 (1990).
43. N. Oliva, S. Shitreet, E. Abraham, B. Stanley, E. R. Edelman and N. Artzi, Natural tissue microenvironmental conditions modulate adhesive material performance. *Langmuir* **28**, 15402–15409 (2012).
44. M. Rajilic-Stojanovic, Function of the microbiota. *Best Pract. Res. Clin. Gastroenterol.* **27**, 5–16 (2013).
45. V. Zabaleta, P. Calleja, S. Espuelas, L. Corrales, R. Pio, M. Agueros and J. M. Irache. [Mucopenetrating nanoparticles: vehicles for the oral administration of paclitaxel]. *Ann. Pharm. Fr.* **71**, 109–118 (2013).
46. K. Yoncheva, S. Gomez, M. A. Campanero, C. Gamazo and J. M. Irache, Bioadhesive properties of pegylated nanoparticles. *Expert. Opin. Drug Deliv.* **2**, 205–218 (2005).
47. H. C. Clevers and C. L. Bevins, Paneth cells: maestros of the small intestinal crypts. *Annu. Rev. Physiol.* **75**, 289–311 (2013).
48. M. A. Dobrovolskaia and S. E. McNeil, Immunological properties of engineered nanomaterials. *Nat. Nanotechnol.* **2**, 469–478 (2007).
49. J. P. Lim and P. A. Gleeson, Macropinocytosis: an endocytic pathway for internalising large gulps. *Immunol. Cell Biol.* **89**, 836–843 (2011).
50. I. Martin-Padura, S. Lostaglio, M. Schneemann, L. Williams, M. Romano, P. Fruscella, C. Panzeri, A. Stoppacciaro, L. Ruco, A. Villa, D. Simmons and E. Dejana, Junctional adhesion molecule, a novel member of the immunoglobulin superfamily that distributes

at intercellular junctions and modulates monocyte transmigration. *J. Cell Biol.* **142**, 117–127 (1998).

51. M. Furuse, K. Fujita, T. Hiiragi, K. Fujimoto and S. Tsukita, Claudin-1 and -2: novel integral membrane proteins localizing at tight junctions with no sequence similarity to occludin. *J. Cell Biol.* **141**, 1539–1550 (1998).
52. M. Furuse, T. Hirase, M. Itoh, A. Nagafuchi, S. Yonemura and S. Tsukita, Occludin: a novel integral membrane protein localizing at tight junctions. *J. Cell Biol.* **123**, 1777–1788 (1993).
53. D. Gunzel and A. S. Yu, Claudins and the modulation of tight junction permeability. *Physiol. Rev.* **93**, 525–569 (2013).
54. W. H. De Jong, W. I. Hagens, P. Krystek, M. C. Burger, A. J. Sips and R. E. Geertsma, Particle size-dependent organ distribution of gold nanoparticles after intravenous administration. *Biomaterials* **29**, 1912–1919 (2008).
55. A. M. Keene, D. Peters, R. Rouse, S. Stewart, E. T. Rosen and K. M. Tyner, Tissue and cellular distribution of gold nanoparticles varies based on aggregation/agglomeration status. *Nanomedicine* 7, 199–209 (2012).
56. S. Hirn, M. Semmler-Behnke, C. Schleh, A. Wenk, J. Lipka, M. Schaffler, S. Takenaka, W. Moller, G. Schmid, U. Simon and W. G. Kreyling, Particle size-dependent and surface charge-dependent biodistribution of gold nanoparticles after intravenous administration. *Eur. J. Pharm. Biopharm.* 77, 407–416 (2011).
57. C. Lasagna-Reeves, D. Gonzalez-Romero, M. A. Barria, I. Olmedo, A. Clos, V. M. Sadagopa Ramanujam, A. Urayama, L. Vergara, M. J. Kogan and C. Soto, Bioaccumulation and toxicity of gold nanoparticles after repeated administration in mice. *Biochem. Biophys. Res. Commun.* **393**, 649–655 (2010).
58. D. P. Lankveld, A. G. Oomen, P. Krystek, A. Neigh, A. Troost-de Jong, C. W. Noorlander, J. C. Van Eijkeren, R. E. Geertsma and W. H. De Jong, The kinetics of the tissue distribution of silver nanoparticles of different sizes. *Biomaterials* **31**, 8350–8361 (2010).
59. X. D. Zhang, D. Wu, X. Shen, P. X. Liu, N. Yang, B. Zhao, H. Zhang, Y. M. Sun, L. A. Zhang and F. Y. Fan, Size-dependent in vivo toxicity of PEG-coated gold nanoparticles. *Int. J. Nanomedicine* **6**, 2071–2081 (2011).
60. W. S. Cho, M. Cho, J. Jeong, M. Choi, H. Y. Cho, B. S. Han, S. H. Kim, H. O. Kim, Y. T. Lim and B. H. Chung, Acute toxicity and pharmacokinetics of 13 nm-sized PEG-coated gold nanoparticles. *Toxicol. Appl. Pharmacol.* **236**, 16–24 (2009).
61. T. Morais, M. E. Soares, J. A. Duarte, L. Soares, S. Maia, P. Gomes, E. Pereira, S. Fraga, H. Carmo and L. Bastos Mde, Effect of surface coating on the biodistribution profile of gold nanoparticles in the rat. *Eur. J. Pharm. Biopharm.* **80**, 185–193 (2012).
62. P. Puvanakrishnan, J. Park, D. Chatterjee, S. Krishnan and J. W. Tunnell, In vivo tumor targeting of gold nanoparticles: effect of particle type and dosing strategy. *Int. J. Nanomedicine* 7, 1251–1258 (2012).
63. E. S. Glazer, C. Zhu, A. N. Hamir, A. Borne, C. S. Thompson and S. A. Curley, Biodistribution and acute toxicity of naked gold nanoparticles in a rabbit hepatic tumor model. *Nanotoxicology* **5**, 459–468 (2011).
64. R. Goel, N. Shah, R. Visaria, G. F. Paciotti and J. C. Bischof, Biodistribution of TNF-alpha-coated gold nanoparticles in an in vivo model system. *Nanomedicine* **4**, 401–410 (2009).

65. B. Semete, L. I. Booysen, L. Kalombo, J. D. Venter, L. Katata, B. Ramalapa, J. A. Verschoor, and H. Swai, In vivo uptake and acute immune response to orally administered chitosan and PEG coated PLGA nanoparticles. *Toxicol. Appl. Pharmacol.* **249**, 158–165 (2010).
66. C. Schleh, M. Semmler-Behnke, J. Lipka, A. Wenk, S. Hirn, M. Schaffler, G. Schmid, U. Simon and W. G. Kreyling, Size and surface charge of gold nanoparticles determine absorption across intestinal barriers and accumulation in secondary target organs after oral administration. *Nanotoxicology* **6**, 36–46 (2012).
67. W. S. Cho, B. C. Kang, J. K. Lee, J. Jeong, J. H. Che and S. H. Seok, Comparative absorption, distribution, and excretion of titanium dioxide and zinc oxide nanoparticles after repeated oral administration. *Part Fibre Toxicol.* **10**, 1743–8977 (2013).
68. C. Schleh, U. Holzwarth, S. Hirn, A. Wenk, F. Simonelli, M. Schaffler, W. Moller, N. Gibson and W. G. Kreyling, Biodistribution of inhaled gold nanoparticles in mice and the influence of surfactant protein D. *J. Aerosol. Med. Pulm. Drug. Deliv.* **26**, 24–30 (2013).
69. S. Takenaka, W. Moller, M. Semmler-Behnke, E. Karg, A. Wenk, O. Schmid, T. Stoeger, L. Jennen, M. Aichler, A. Walch, S. Pokhrel, L. Madler, O. Eickelberg and W. G. Kreyling, Efficient internalization and intracellular translocation of inhaled gold nanoparticles in rat alveolar macrophages. *Nanomedicine* 7, 855–865 (2012).
70. J. Lipka, M. Semmler-Behnke, R. A. Sperling, A. Wenk, S. Takenaka, C. Schleh, T. Kissel, W. J. Parak and W. G. Kreyling, Biodistribution of PEG-modified gold nanoparticles following intratracheal instillation and intravenous injection. *Biomaterials* **31**, 6574–6581 (2010).
71. E. Sadauskas, N. R. Jacobsen, G. Danscher, M. Stoltenberg, U. Vogel, A. Larsen, W. Kreyling and H. Wallin, Biodistribution of gold nanoparticles in mouse lung following intratracheal instillation. *Chem. Cent. J.* **3,** 3–16 (2009).
72. S. K. Balasubramanian, J. Jittiwat, J. Manikandan, C. N. Ong, L. E. Yu and W. Y. Ong, Biodistribution of gold nanoparticles and gene expression changes in the liver and spleen after intravenous administration in rats. *Biomaterials* **31**, 2034–2042 (2010).
73. G. Zhang, Z. Yang, W. Lu, R. Zhang, Q. Huang, M. Tian, L. Li, D. Liang and C. Li, Influence of anchoring ligands and particle size on the colloidal stability and in vivo biodistribution of polyethylene glycol-coated gold nanoparticles in tumor-xenografted mice. *Biomaterials* **30**, 1928–1936 (2009).
74. T. Liu, L. Li, X. Teng, X. Huang, H. Liu, D. Chen, J. Ren, J. He and F. Tang, Single and repeated dose toxicity of mesoporous hollow silica nanoparticles in intravenously exposed mice. *Biomaterials* **32**, 1657–1668 (2011).
75. E. Sadauskas, G. Danscher, M. Stoltenberg, U. Vogel, A. Larsen and H. Wallin, Protracted elimination of gold nanoparticles from mouse liver. *Nanomedicine* **5**, 162–169 (2009).
76. S. Sadekar, A. Ray, M. Janat-Amsbury, C. M. Peterson and H. Ghandehari, Comparative biodistribution of PAMAM dendrimers and HPMA copolymers in ovarian-tumor-bearing mice. *Biomacromolecules* **12**, 88–96 (2011).
77. M. Lundqvist, J. Stigler, G. Elia, I. Lynch, T. Cedervall and K. A. Dawson, Nanoparticle size and surface properties determine the protein corona with possible implications for biological impacts. *Proc. Natl. Acad. Sci. USA* **105**, 14265–14270 (2008).

78. M. A. Dobrovolskaia, P. Aggarwal, J. B. Hall and S. E. McNeil, Preclinical studies to understand nanoparticle interaction with the immune system and its potential effects on nanoparticle biodistribution. *Mol. Pharm.* **5**, 487–495 (2008).
79. S. A. Love, J. W. Thompson and C. L. Haynes, Development of screening assays for nanoparticle toxicity assessment in human blood: preliminary studies with charged Au nanoparticles. *Nanomedicine* 7, 1355–1364 (2012).
80. A. J. Paula, D. S. T. Martinez, R. T. Araujo Júnior, A. G. Souza Filho and O. L. Alves, Suppression of the hemolytic effect of mesoporous silica nanoparticles after protein corona interaction: independence of the surface microchemical environment. *J. Brazilian Chem. Soc.* **23**, 1807–1814 (2012).
81. P. V. Asharani, S. Sethu, S. Vadukumpully, S. Zhong, C. T. Lim, M. P. Hande and S. Valiyaveettil, Investigations on the Structural Damage in Human Erythrocytes Exposed to Silver, Gold, and Platinum Nanoparticles. *Adv. Funct. Mat.* **20**, 1233–1242 (2010).
82. M. A. Dobrovolskaia, A. K. Patri, T. M. Potter, J. C. Rodriguez, J. B. Hall and S. E. McNeil, Dendrimer-induced leukocyte procoagulant activity depends on particle size and surface charge. *Nanomedicine* 7, 245–256 (2012).
83. M. A. Dobrovolskaia, A. K. Patri, J. Simak, J. B. Hall, J. Semberova, S. H. De Paoli Lacerda and S. E. McNeil, Nanoparticle size and surface charge determine effects of PAMAM dendrimers on human platelets in vitro. *Mol. Pharm.* **9**, 382–393 (2012).
84. C. F. Jones, R. A. Campbell, Z. Franks, C. C. Gibson, G. Thiagarajan, A. Vieira-de-Abreu, S. Sukavaneshvar, S. F. Mohammad, D. Y. Li, H. Ghandehari, A. S. Weyrich, B. D. Brooks and D. W. Grainger, Cationic PAMAM dendrimers disrupt key platelet functions. *Mol. Pharm.* **9**, 1599–1611 (2012).
85. V. C. Mosqueira, P. Legrand, A. Gulik, O. Bourdon, R. Gref, D. Labarre and G. Barratt, Relationship between complement activation, cellular uptake and surface physicochemical aspects of novel PEG-modified nanocapsules. *Biomaterials* **22**, 2967–2979 (2001).
86. I. Hamad, O. Al-Hanbali, A. C. Hunter, K. J. Rutt, T. L. Andresen and S. M. Moghimi, Distinct polymer architecture mediates switching of complement activation pathways at the nanosphere-serum interface: implications for stealth nanoparticle engineering. *ACS Nano* **4**, 6629–6638 (2010).
87. W. C. Zamboni, V. Torchilin, A. K. Patri, J. Hrkach, S. Stern, R. Lee, A. Nel, N. J. Panaro and P. Grodzinski, Best practices in cancer nanotechnology: perspective from NCI nanotechnology alliance. *Clin. Cancer Res.* **18**, 3229–3241 (2012).
88. X. Shan, Y. Yuan, C. Liu, X. Tao, Y. Sheng and F. Xu, Influence of PEG chain on the complement activation suppression and longevity in vivo prolongation of the PCL biomedical nanoparticles. *Biomed. Microdevices.* **11**, 1187–1194 (2009).
89. R. H. Müller, K. H. Wallis, S. D. Tröster and J. Kreuter, In vitro characterization of poly(methyl-methaerylate) nanoparticles and correlation to their in vivo fate. *J. Control. Release* **20**, 237–246 (1992).
90. D. E. Owens, 3rd and N. A. Peppas, Opsonization, biodistribution, and pharmacokinetics of polymeric nanoparticles. *Int. J. Pharm.* **307**, 93–102 (2006).
91. A. Vonarbourg, C. Passirani, P. Saulnier and J. P. Benoit, Parameters influencing the stealthiness of colloidal drug delivery systems. *Biomaterials* **27**, 4356–4373 (2006).
92. W. C. Zamboni, L. J. Maruca, S. Strychor, B. A. Zamboni, S. Ramalingam, R. P. Edwards, J. Kim, Y. Bang, H. Lee, D. M. Friedland, R. G. Stoller, C. P. Belani and

R. K. Ramanathan, Bidirectional pharmacodynamic interaction between pegylated liposomal CKD-602 (S-CKD602) and monocytes in patients with refractory solid tumors. *J. Liposome Res.* **21**, 158–165 (2011).

93. M. C. Garnett and P. Kallinteri, Nanomedicines and nanotoxicology: some physiological principles. *Occup. Med.* **56**, 307–311 (2006).
94. N. Desai, Challenges in development of nanoparticle-based therapeutics. *Aaps J.* **14**, 282–295 (2012).
95. H. B. Frieboes, M. Wu, J. Lowengrub, P. Decuzzi and V. Cristini, A computational model for predicting nanoparticle accumulation in tumor vasculature. *PLoS One* **8**, 28 (2013).
96. M. Demoy, J. P. Andreux, C. Weingarten, B. Gouritin, V. Guilloux and P. Couvreur, Spleen capture of nanoparticles: influence of animal species and surface characteristics. *Pharm. Res.* **16**, 37–41 (1999).
97. G. C. Winkler, Pulmonary intravascular macrophages in domestic animal species: review of structural and functional properties. *Am. J. Anat.* **181**, 217–234 (1988).
98. J. D. Brain, R. M. Molina, M. M. DeCamp and A. E. Warner, Pulmonary intravascular macrophages: their contribution to the mononuclear phagocyte system in 13 species. *Am. J. Physiol.* **276**, L146–L154 (1999).
99. FDA. FDA's Approach to Regulation of Nanotechnology Products, http://www.fda.gov/ScienceResearch/SpecialTopics/Nanotechnology/ucm301114.htm.
100. C. N. Cruz, K. M. Tyner, L. Velazquez, K. C. Hyams, A. Jacobs, A. B. Shaw, W. Jiang, R. Lionberger, P. Hinderling, Y. Kong, P.C. Brown, T. Ghosh, C. Strasinger, S. Suarez-Sharp, D. Henry, M. Van Uitert, N. Sadrieh and E. Morefield. CDER risk assessment exercise to evaluate potential risks from the use of nanomaterials in drug products. *Aaps J.* **15**, 623–628 (2013).
101. FDA. Considering Whether an FDA-Regulated Product Involves the Application of Nanotechnology, http://www.fda.gov/regulatoryinformation/guidances/ucm257698.htm.
102. FDA. Assessing the Effects of Significant Manufacturing Process Changes, Including Emerging Technologies, on the Safety and Regulatory Status of Food Ingredients and Food Contact Substances, Including Food Ingredients that are Color Additives, http://www.fda.gov/Food/GuidanceRegulation/GuidanceDocumentsRegulatoryInformation/IngredientsAdditivesGRASPackaging/ucm300661.htm.

Chapter A2

Anatomical and Physicochemical Factors Controlling Nanoparticle Pharmacokinetics

S. M. Moghimi[*,†,‡] *and Z. S. Farhangrazi*[†]

[*] *Nanomedicine Research Group, Centre for Pharmaceutical Nanotechnology and Nanotoxicology, Faculty of Health and Medical Sciences, University of Copenhagen, DK-2100 Copenhagen Ø, Denmark*

[†] *NanoScience Centre, Faculty of Science, University of Copenhagen, DK-2100 Copenhagen Ø, Denmark*

Biotrends International, 4582 S. Ulster Street, Denver Technological Center, CO 80237–2639, USA

[‡] *imedicine.moghimi@outlook.com*

[‡] *moien.moghimi@sund.ku.dk*

Abstract

The blood circulation times and tissue distribution of intravenously injected particulate drug carriers are controlled by a complex and inter-related biological and physicochemical processes. This chapter highlights important anatomical features (macrophage locations, capillary types and structures, and escape routes from the vasculature) and physicochemical aspects of carrier engineering (size, surface properties and shape) that regulate nanoparticle pharmacokinetics.

1. Introduction

Particulate systems such as liposomes, polymeric nanospheres and composite nanoshells have long been established as carriers of therapeutic agents.[1] Drug encapsulation in these carriers offers a number of pharmaceutical and clinical advantages. This may afford protection against drug degradation or inactivation *en route* to the target site and it may reduce the amount of active agent needed to obtain a beneficial effect thus effectively reducing drug-induced toxicity. In addition to these, encapsulation technologies could provide a viable strategy for selection and administration of highly potent but discarded drugs on the grounds of high toxicity, instability in the systemic circulation and difficulty in making them soluble for intravenous (*i.v.*) injection.[2] Drug entrapped in a particulate carrier is not bioavailable unless is released, where the rate of drug release from the carrier will determine the therapeutic response; the latter in turn will depend on drug intensity, the duration of effect within the therapeutic window and the area under the curve for drug within the therapeutic window. Accordingly, pharmacokinetics of encapsulated drugs will be controlled by the biological performance of drug carriers. The pharmacokinetics of particulate carriers is controlled by a complex array of inter-related physicochemical (e.g. chemical composition, morphology, size distribution, surface properties, and interface forces) and biological factors (e.g. stability in biological fluids, host defense responses, anatomical barriers and integrity, and dosing schedule).[3–7] These concepts have been critically reviewed, recently.[2] Here, we provide a brief integrated analysis of nanoparticle pharmacokinetics on *i.v.* route of injection with particular emphasis on anatomical considerations and barriers as well as the physicochemical concepts of nanoparticle engineering.

2. Anatomy and the Vasculature "Highway Code"

2.1. *Capillary structures and escape routes from the vasculature*

Three systemic veins, the coronary sinus (the main vein of the heart, which receives almost all venous blood from the myocardium), superior vena cava (which drains the head, neck, chest, and upper limbs) and inferior vena cava (which drains the abdomen, pelvic, and lower limbs), return deoxygenated blood to the right atrium. The right and left pulmonary arteries are the only arteries that carry deoxygenated blood from the right ventricle to alveoli within the lungs, where the branches divide and subdivide to form capillaries around the air sacs. The pulmonary capillaries unite to form venules and pulmonary veins that carry the oxygenated blood to the left atrium. Aorta carries oxygenated blood from the left ventricle to systemic capillaries via a number of divisions and branches (ascending aorta, arch of aorta, thoracic aorta, and abdominal aorta). In a resting person, the circulation time (the time taken by a drop of blood to travel from the right atrium, through the pulmonary circulation, back to the left atrium, through the systemic circulation and back again to the right atrium) is approximately one minute.

In accordance with the circulatory routes, intravenously injected particles, depending on their size, shape, deformability, and adhesion properties, may rapidly become trapped in lung capillaries.[2,8] Trapping in lung capillaries usually requires rigid particles greater than 7 μm in size.[8] Nanoparticles may also get trapped in lung capillaries during structural transformation in the blood. One example is the formation of large particle aggregates in the blood as a consequence of protein adsorption and protein–protein interactions.[2] A recent attempt has further assessed and optimized particle delivery to the lung through simultaneous active targeting of lung vasculature (e.g. through surface attachment of intracellular adhesion molecule 1) and fabrication of both particle size and shape (e.g. discoid and spherical geometries), while paying attention to the dislodging forces under flow conditions.[9]

In different healthy tissues, the structure of the microcirculation varies to meet specific functional requirements. Particle escape from microcirculation, therefore, depends on capillary structure variables,[1] where there are three main types. The first type is continuous capillaries, the type found in most tissues where the endothelial cells form an uninterrupted lining. Here, the capillary endothelium is supported by a thin basement membrane and adjacent collagen fibrils. Pericytes may further embrace the capillary and they are also supported by their own basement membrane. Particle extravasation from the blood to the underlying parenchyma of continuous capillaries is a rare event, unless there are pathologically mediated anatomical and architectural openings. However, particle translocation across the continuous capillaries can be modulated through selected receptor-mediated processes as in the case for apoE-coated entities with the ability to translocate across cerebral capillary endothelial cells via ligation with lipoprotein receptors.[10]

On the other hand, fenestrated capillaries (Fig. 1), where the endothelial interface contains numerous large pores or fenestrations (50–100 nm in diameter), are found in selected tissues such as the kidney, small intestine and endocrine glands. In these capillaries, the fenestrations are usually traversed by a thin electron-dense diaphragm (Fig. 1(a)), but a diaphragm is not seen across the fenestrations of the glomerular capillaries of the kidney (Fig. 1(b)). Like continuous endothelium type capillaries, all fenestrated capillaries are surrounded by a basement membrane, which is continuous across the fenestrations as well as being thick (200–400 nm). Pericytes are rarely found in fenestrated capillaries. The capillaries of the glomerular filter in kidney are further lined by processes of podocytes, giving rise to pedicels that interdigitate with primary processes. The openings between pedicels are in the order

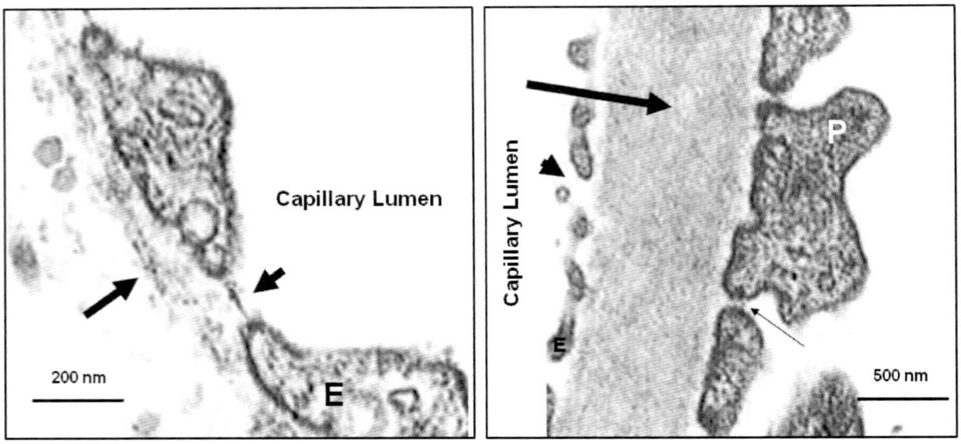

(a) (b)

Fig. 1. Ultrastructure of fenestrated capillaries. Fenestrations across endothelial cells (*E*) are traversed by a thin-electron dense line (small arrow), which may constitute a diaphragm, panel (a). A diaphragm is not seen across the fenestrations of the glomerular capillaries, arrowhead in panel (b). Fenestrated capillaries are supported by a basement membrane (large thick arrows in panels (a) & (b)), which is continuous across the fenestrations. Panel (b) further shows podocyte secondary foot processes (*P*) in the glomerular capillaries, separated by filtration slits of almost uniform width and bridged by electron-dense diaphragms (thin arrow).

of 25 nm in width and bridged by a diaphragm with a thickness of 4 nm (Fig. 1(b)). These morphological characteristics also provide torturous routes for nanoparticle translocation. However, recent studies suggest that particle shape and size may play a significant role in renal clearance.[11–13] For instance, intravenously injected amino-functionalized single-walled carbon nanotubes with average lengths of 200–300 nm that by-pass efficient macrophage sequestration, undergo rapid renal clearance by glomerular filtration.[11] This process involves partial tubular reabsorption and transient translocation into the proximal tubular cell nuclei, but does not involve active secretion via specific transporters.[12] Spherical nanoparticles of 5 nm or smaller in size (e.g. quantum dots) are also filtered by kidneys and may undergo urinary excretion.[13] Finally, cationic dendrimers may accumulate in kidneys, a process that is believed to arise from electrostatic interaction with the negatively charged capillary endothelium of the golmerular filter and the outer surface of podocytes.[14] Podocytes further possess phagocytic function.

Sinusoids are capillaries of wide diameter and are found in the liver (Fig. 2), spleen (Fig. 3), lymph nodes, and bone marrow. The endothelium of sinusoids may be continuous, fenestrated or discontinuous. The latter type is only restricted to the liver sinusoids, where the size of fenestrae can be as

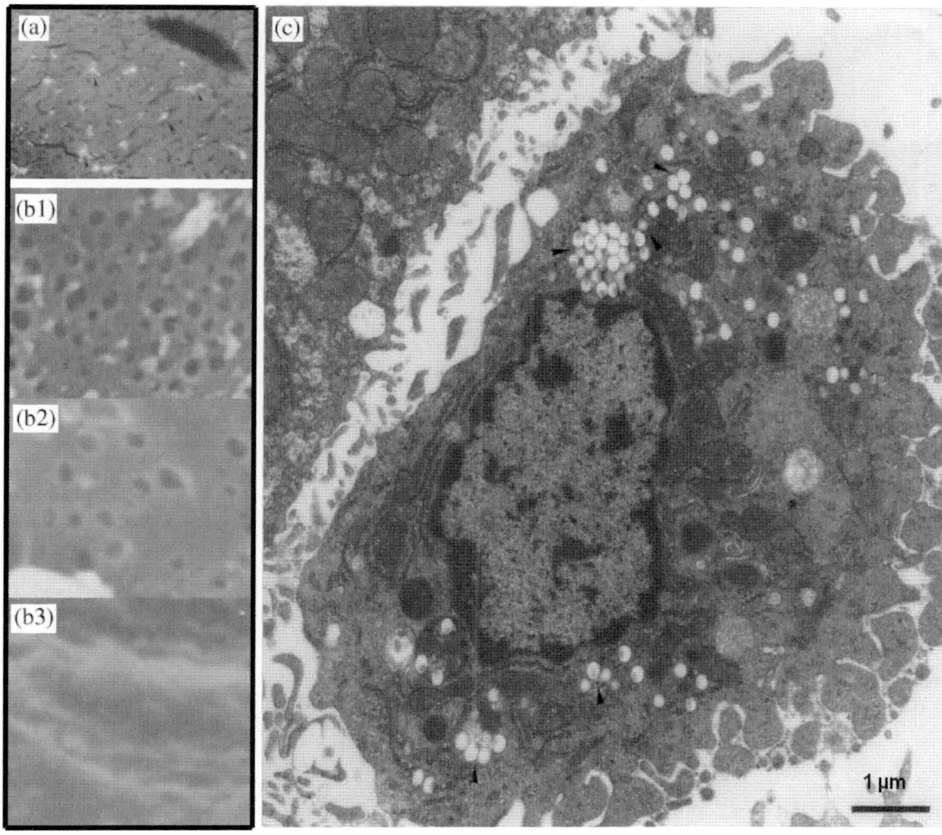

Fig. 2. Ultrastructure of liver sinus capillaries. Panel (a) represent fluorescent light microscopy of a rat liver after administration of fluorescently labeled polystyrene nanoparticles (60 nm in size). The fluorescence representing nanoparticles is localized in large quantities in Kupffer cells (arrows). Panel (b) shows sinusoidal wall of a normal human liver with fenestrae arranged in sieve plates (b1), sinusoidal lining wall of the liver from a chronic alcoholic with a reduced number of fenestrae, and total defenestration in the liver of chronic alcoholic subject. Panel (c) shows an electron micrograph of a rat Kupffer cell with ingested polystyrene nanoparticles (200 nm).

large as 150 nm. Some small particles (100 nm or smaller) as well as deformable particles of 150 nm or larger in size may get access to the hepatic parenchyma through the space of Disse by "endothelial massaging".[1,3] This exposes particles to hepatocytes and hepatic stellate cells, but particle internalization by these cells may depend on receptor-mediated processes. The role of particle shape in "endothelial massaging" has not received much attention.

In sinusoidal spleens (e.g. human, rat), the circulating blood enters the lumen of venous sinuses through the slits between endothelial cells that are

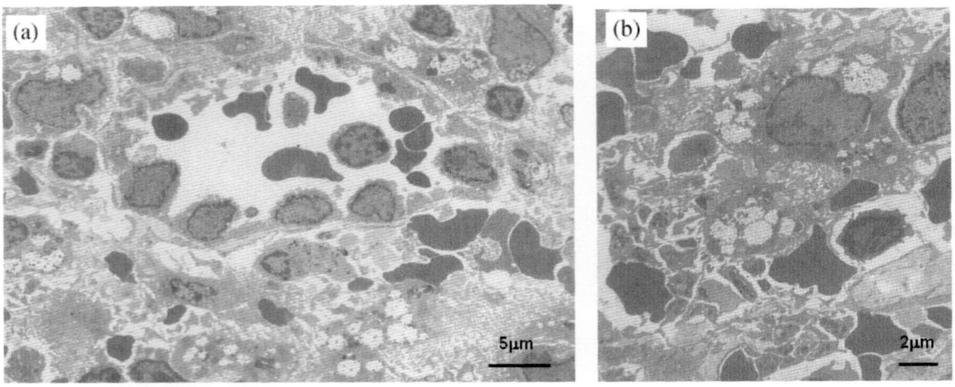

Fig. 3. Ultrastructure of a rat red-pulp splenic venous sinus. Panel (a) shows accumulation of poloxamine 908-coated nanospheres (220 nm in size) in macrophages. These nanoparticles are resistant to ingestion by Kupffer cells. Particles are filtered first at inter-endothelial cell slits before macrophage engulfment (nanoparticle filtration still proceed after destruction of red-pulp macrophages by clodronate-containing liposomes). Panel (b) is an enlarged view of a red-pulp macrophage with ingested poloxamine 908-coated nanoparticles. Modified with permission from Ref. 42.

termed the interendothelial cell slits (IES) (Fig. 3), which then drain into pulp, trabecular and capsular veins. Approximately, 20% of the total number of IES present anatomically allows passage of erythrocytes during any 5 min period.[15] However, the slit rarely exceeds 200 nm in width, even with an erythrocyte in passage.[16] Therefore, retention of blood cells and particulate agents at IES depend on their deformability/rigidity.[17,18] These slits are the sites where erythrocytes containing rigid inclusions (e.g. malarial plasmodia) are believed to be "pitted" of their inclusions, which are eventually cleared by the nearby resident macrophages.[18] Pulp venules which receive flow freely through the open-ended fenestrations in their walls lack IES. The size of fenestrations is mostly large enough to allow unimpeded entry of erythrocytes into the venule. Finally, the splenic filtration beds do not normally show a high level of filtration since 90% of the arterial inflow travels through fast pathways, which bypasses these sites. However, contractile fibroblastic cells (also known as "barrier-forming cells"), which fuses with one another in the filtration beds of the spleen to form syncytia, in coordination with macrophages and other stromal cells regulate normal basal splenic filtration at red-pulp and marginal zones, blood flow, cell homing and migration.[19,20] Barrier cells occur in small numbers in the normal spleen but are present in large numbers in murine (a non-sinusoidal spleen) and human spleens under some pathological conditions.[19,20]

The medullary sinusoidal system of the bone marrow is formed from an uninterrupted layer of endothelial cells, an outer layer of adventitial cells that covers a variable proportion of the abluminal surface of the endothelium, and a discontinuous basal lamina.[21] Three types of membrane structures have been identified in bone marrow sinus endothelium. These are diaphragmed pits, diaphragmed channels, and diaphragmed fenestrae.[21] The former structures are similar to other coated or non-coated pits but are spanned by a diaphragm. Diaphragmed channels arise from intraendothelial vesicles spanning across the endothelium, facing both luminal and abluminal surfaces of the endothelium, and present diaphragmed apertures at both sides of the endothelium. Finally, portions of the endothelial cells lining the marrow sinuses may be markedly attenuated, forming fenestrae of 85–100 nm width, which are spanned by thin diaphragms, thus maintaining endothelial continuity. Non-diaphragmed channels across the bone marrow sinusoidal wall do not exist. The bone marrow sinus endothelium is capable of removing particles by both transcellular and intercellular routes. The former route occurs through the diaphragmed fenestrae of endothelial walls and the latter route is associated with the formation of bristle-coated pits on the luminal surface of endothelium, where internalized particles are often directed to dense bodies.[1,3,21,22] The endothelium of other myeloid vessels is not capable of efficient uptake and transport of particles. An exception to this is in the guinea pig.[23] The single layer of endothelium, which constitute the wall of the vein resemble the sinusoidal endothelium not only in structure, but also in particle uptake and transport properties.

2.2. *Macrophages in contact with blood*

Intravenously injected particulate drug carriers are predominantly intercepted by hepatic macrophages (Kupffer cells) (Figs. 1(a) and 1(c)), which constitute the largest population of macrophages in direct contact with the blood in human , mice and rats, and macrophages in the marginal zone and the red-pulp regions of the sinusoidal spleens (Fig. 3).[1–3,24] Blood monocytes (and other leukocytes) are also capable of internalizing particulate matters,[24] but due to their low numbers relative to Kupffer cells, their role in particle extraction remains marginal. Among the hepatic macrophages, periportal Kupffer cells (approximately comprising 45% of total liver macrophages) are the main particle scavengers; this is also reflected in their larger size and higher lysosomal enzyme activity (on a per cell basis), compared with Kupffer cells located in the hepatic midzonal and perivenous regions.[25] However, pulmonary intravascular macrophages are the main scavengers in calves, pigs, cats, goats, and sheep.[26] These macrophages are junctionally adherent to the capillary endothelium of lungs and morphologically similar to the Kupffer cells. Differentiated pulmonary intravascular macrophages are rare in newborn pigs, whereas monocytes are closely associated with the pulmonary capillary endothelium. Additionally, persinal macrophages in the bone marrow, which are also in direct, open contact with the bloodstream may further participate in particle extraction; this process, again, is species-dependent.[21,24,27,28] On the other hand, access to stromal or hematopoietic macrophages of the bone marrow is typically through nanoparticle passage across the bone marrow sinus endothelium.[21]

The *i.v.* route also offers a gateway for particle access (depending on their size and surface characteristics) to interstitial macrophages at locations where the endothelial barrier is perturbed by inflammatory processes or by dysregulated angiogenesis (e.g. arthritis, atherosclerotic lesions and solid tumors).[1–3,29] However, there is marked variability in endothelial permeability at pathological sites that may control the extent of particle extravasation and hence access to local macrophages. For example, variability in endothelial function and permeability has been widely documented in different tumor types, different vessels within the same tumor and during tumor growth, regression, and relapse.[30,31] Pore frequency and cut-off size (200–1200 nm) in tumors may be controlled by microenvironmental factors and may increase with the histological grade and malignant potential of tumors.

3. Physicochemical Aspects

3.1. *Particle surface characteristics and size*

The physicochemical parameters of particulate drug carriers are often changed in contact with the blood.[1,2,8,32,33] These alterations may further impose significant effects on macrophage/monocyte recognition and clearance rate of particulate matters.[24] For instance, an important consequence of particle contact with blood is adsorption and deposition of a range of plasma proteins on the particle surface.[1,2,8,32,33] The composition of the plasma protein coat differs considerably in amount and heterogeneity among different particles and even within the same population of nanoparticles due to surface heterogeneity (in terms of surface charge and ligand density, surface defects, etc.).[2,32–37] Some of these proteins act as opsonic factors such as various subclasses of antibodies that are capable of mediating particle recognition by different macrophage Fc receptors.[32] Other example of opsonic molecules includes complement activation products C3b and iC3b, which also prime the particle surface for interaction with a plethora of macrophage complement receptors.[8,32,35–40] The binding of different opsonic proteins and/or dynamic changes in surface opsonization processes may determine which subpopulation of macrophages will host the particles.[24] Therefore, dynamic opsonization processes could indicate a recognition hierarchy phenomenon, wherein a specific macrophage receptor (or population) might recognize the earliest surface changes and other receptors (or other subpopulations) might engage later. Understanding of differential opsonization processes may further allow means for design of drug carriers with high specificity toward only one macrophage subset in a typical organ.

There have been rapid developments in stealth technologies, resulting in engineering of particulate delivery systems that remain resistant to rapid clearance by macrophages in contact with the blood.[2,8] Stealth characteristics were initially achieved through surface adsorption of polyoxypropylene–polyoxyethylene linear or star-shaped block copolymers (e.g. poloxamers and poloxamines) on hydrophobic nanoparticles and later by covalent attachment of poly(ethylene glycol)-2000 (PEG_{2000}) molecules to the surface of vesicles (liposomes) and nanoparticles.[1,2,8,41] Today, there are a large library of polymers and other constructs that can confer longevity to nanoparticles following surface grafting or adsorption.[2,3,32] Typically, stealth particles are below 150 nm in size where the surface projected PEG or polyethylene oxide chains provide stability to the particle suspension by repulsion through a steric mechanism of stabilization involving both enthalpic and entropic contributions.[32] Due to its hydrophilicity and close association with water molecules, the steric barrier, and the structured-water surrounding the barrier, dramatically suppresses protein adsorption, but this may not necessarily prevent opsonization processes.[39,41] For instance, complement activation and fixation still proceeds with some surface engineered particles where the extent of complement activation and fixation depends on surface density and conformation of the projected polymers.[34,36–39] The surface polymer conformational changes have further been shown to switch complement activation from one pathway to another.[34,37] The steric barrier on the nanoparticle surface, however, can interfere with the binding of C3b and/or iC3b to their corresponding macrophage receptors.[34] Stealth nanoparticles have widely been used for passive targeting of certain pathological sites where anatomical openings exists (e.g. mostly sarcomas and to some extent certain carcinomas).[1–3]

Understanding of anatomical organization of splenic microcirculation has further allowed rational strategies for development of splenotropic nanoparticles with minimum sequestration by Kupffer cells.[17,18] Morphologically, splenotropic nanoparticles are rigid non-deformable entities of 220 nm to

250 nm in diameter exhibiting stealth characteristic to the hepatic Kupffer cell recognition (e.g. through surface coating with poloxamine 908). As a result of poor recognition and clearance by Kupffer cells, nanoparticle concentration in the blood remains high. Therefore, nanoparticles are gradually exposed to the spleen (the rate of blood flow to the spleen is substantially lower than the liver). It is the combined particle size and non-deformable characteristics that contribute to particle filtration at IES in splenic venous sinuses (the width of IES slits is below 220 nm). Remarkably, macrophages in the vicinity of cell slits eventually take up filtered nanoparticles, but the mechanism of recognition remains unknown, Fig. 3.[42] We have eliminated the role for the splenic macrophage complement receptors in sequestration of splenotropic nanoparticles. Splenotropic particles may have important applications with respect to examining splenic microcirculatory pathways in addition to selective delivery of antimicrobials and other therapeutic molecules to marginal zone and red-pulp macrophages.

3.2. *Particle shape*

Recent evidence attests that particles of variable shapes are capable of initiating phagocytosis in at least one orientation.[43] The particle shape at the point of initial contact could control whether macrophages proceed with phagocytosis or rest over particles, but the particle size may primarily affect the completion of phagocytosis in cases where particle volume exceed the cell volume. Similarly, shape-switching particles (engineered from stimulus-responsive polymers) also exhibit unique interactions with macrophages; elliptical disk-shaped particles that are not phagocytosed by macrophages were made to internalize through shape switch into spherical particles.[44] Particle shape can further modulate the specificity of antibody-displaying nanoparticles. For example, antibody-coated rod-shaped particles were shown to exhibit higher specific uptake, and lower non-specific uptake, in cells compared with spherical counterparts.[45] Accordingly, shape-induced enhancement of antibody specificity and avidity may dramatically improve therapeutic efficacy.

Long entities such as filomicelles (2–18 μm) were shown to exhibit prolonged circulation half-lives, often in the order of days.[6] Hydrodynamic shears tend to stretch and flow-align the cylindrical filomicelles in most blood vessels, which minimizes filomicelle interactions with macrophages by pulling off the filomicelles as they come into contact with the phagocytes. However, following cell contact, a "nanofragment" of a filomicelle might break off and be taken up by macrophages in the liver and the spleen.[6]

4. Conclusions

The pharmacokinetics of intravenously injected nanoparticles is controlled by a complex, but inter-related anatomical, pathophysiological, immunological and physicochemical factors. Although many studies have attempted to map out the key parameters that control nanoparticle pharmacokinetics, more concerted actions are necessary to outline inter-related core and interfacial physicochemical and biological factors that control pharmacokinetics in relation to type, developmental stage and location of the disease. For instance, only limited attention has been paid to geometric parameters that regulate nanoparticle flow properties, vascular adhesion, escape from vasculature, and cellular internalization processes. The bulk of current available data is mostly restricted to spherical entities.

Sophistication in design and fabrication technologies may yield generation of more heterogenous populations of nanoparticles. These developments may cause more difficulties in interpretation of pharmacokinetic parameters following *i.v.* injection. Development of technologies that yield more homogenous populations of complex nanoparticles in terms of surface chemistry, architecture and morophology can have a major impact clinically.[46,47] One such development is Particle Replication In Non-wetting Templates (PRINT).[48] This process is based on the exploitation of the low surface energy of novel fluoropolymeric molds to produce a variety of organic particles with well-defined controllable physicochemical parameters.[48] Indeed, PRINT allows for the precise control over particle size (from tens of nanometers to several microns), geometry (spheres, cylinders, discs and toroids), porosity and texture and yield monodisperse particles. Additionally, the concentration of drug cargos in the particles can be exactly chosen to meet specific needs since this technology does not rely on kinetic drug trapping as in the case with liposomes and micelles. Finally, detailed attention must also be paid to drug bioavailability on particulate drug carrier administration in relation to dosing schedule and dosing intensities in order to avoid unexpected toxicity and altered biodistribution.[2,46]

Acknowledgments

Financial support by the Danish Agency for Science, Technology and Innovation (Det Strategiske Forskningsråd), reference 2106–080–081, and the European Union 7th Framework Programme under the grant agreement number 310337 (CosmoPHOS-nano) is gratefully acknowledged.

References

1. S. M. Moghimi, A. C. Hunter and J. C. Murray, Nanomedicine: Current progress and future prospects. *FASEB J.* **19**, 311–330 (2005).
2. S. M. Moghimi, A. C. Hunter and T. L. Andresen, Factors controlling nanoparticle pharmacokinetics: An integrated approach and perspective. *Ann. Rev. Pharmacol. Toxicol.* **52**, 481–503 (2012).
3. S. M. Moghimi, Recent developments in polymeric nanopartricle engineering and their applications in experimental and clinical oncology. *Anti-Cancer Agent Med. Chem.* **6**, 553–561 (2006).
4. P. Decuzzi and M. Ferrari, Design maps for nanoparticles targeting the diseased microvasculature. *Biomaterials* **29**, 377–384 (2008).
5. P. Decuzzi *et al.*, Size and shape effect in the biodistribution of intravenously injected particles. *J. Control. Release* **141**, 320–327 (2010).
6. Y, Geng *et al.*, Shape effects of filaments versus spherical particles in flow and drug delivery. *Nat. Nanotechnol.* **2**, 249–255 (2007).
7. R. A. Petros and J. M. DeSimone, Strategies in the design of nanoparticles for therapeutic applications. *Nat. Rev. Drug Discov.* **9**, 615–627 (2010).
8. S. M. Moghimi, A.C. Hunter and J. C. Murray, Long-circulating and target-specific nanoparticles: Theory to practice. *Pharmacol. Rev.* **53**, 283–318 (2001).
9. S. Muro *et al.*, Control of endothelial targeting and intracellular delivery of therapeutic enzymes by modulating the size and shape of ICAM-1-targeted carriers. *Mol. Ther.* **16**, 1450–1458 (2008).
10. H. R. Kim *et al.*, Translocation of poly(ethylene glycol-co-hexadecyl)cyanoacrylate nanoparticles into rat brain endothelial cells: Role of apolipoproteins in receptor-mediated endocytosis. *Biomacromolecules* **8**, 793–799 (2007).
11. L. Lacerda *et al.*, Carbon-nanotube shape and individualization critical for renal excretion. *Small* **4**, 130–132 (2008).
12. A. Ruggiero *et al.*, Paradoxical glomerular filtration of carbon nanotubes. *Proc. Natl. Acad. Sci. USA* **107**, 12369–12374 (2010).
13. H. S. Choi *et al.*, Renal clearance of quantum dots. *Nat. Biotechnol.* **25**, 1165–1170 (2007).
14. J. C. Roberts, M. K. Bhalgat and R. T. Zera, Preliminary biological evaluation of polyamidoamine (PAMAM) dendrimers. *J. Biomed. Mat. Res.* **30**, 53–65 (1996).
15. I. C. MacDonald, D. M. Ragan, E. E. Schmidt and A. C. Groom, Kinetics of red cell passage through interendothelial slits into venous sinuses in rat spleen, analyzed by *in vivo* microscopy. *Microvasc. Res.* **33**, 118–134 (1987).
16. L. T. Chen and L. Weiss, The role of the sinus wall in the passage of erythrocytes through the spleen. *Blood* **41**, 529–537 (1973).
17. S. M. Moghimi, C. J. H. Porter, I. S. Muir, L. Illum and S. S. Davis. Non-phagocytic uptake of intravenously injected microspheres in rat spleen: Influence of particle size and hydrophilic coating. *Biochem. Biophys. Res. Commun.* **177**, 861–866 (1991).
18. S. M. Moghimi, Mechanisms of splenic clearance of blood cells and particles: Towards development of new splenotropic agents. *Adv. Drug Deliv. Rev.* **17**, 103–115 (1995).
19. L. Weiss, R. Powell and F. J. Schiffman, Mechanisms of splenic control of murine malaria: Reticular cell activation and the development of a blood-spleen barrier. *Am. J. Anat.* **176**, 251–285 (1986).

20. L. Weiss, Barrier cells. *Immunol. Today* **12**, 24–29 (1991).
21. S. M. Moghimi, Exploiting bone marrow microvascular structure for drug delivery and future therapies. *Adv. Drug Deliv. Rev.* **17**, 61–73 (1995).
22. C. J. H. Porter, S. M. Moghimi, L. Illum and S. S. Davis, The polyoxyethylene/polyoxypropylene block co-polymer poloxamer-407 selectively redirects intravenously injected microspheres to sinusoidal endothelial cells of rabbit bone marrow. *FEBS Lett.* **305**, 62–66 (1992).
23. G. Hudson and J. M. Yoffey, Reticuloendothelial cells in the bone marrow of guinea pig. *J. Anat.* **97**, 409–416 (1963).
24. S. M. Moghimi *et al.*, Particulate systems for targeting of macrophages: Basic and therapeutic concepts. *J. Innate Immun.* **4**, 509–528 (2012).
25. E. C. Sleyster and D. L. Knook, Relation between localization and function of rat liver Kupffer cells. *Lab. Invest.* **47**, 484–490 (1982).
26. G. C. Winkler, Pulmonary intravascular macrophages in domestic animal species: Review of structural and functional properties. *Am. J. Anat.* 1988; **181**:217–234.
27. K. Sou, B. Goins, S. Takeoka, E. Tsuchida and W. T. Phillips, Selective uptake of surface-modified vesicles by bone marrow macrophages *in vivo*. *Biomaterials* **28**, 2655–2666 (2006).
28. K. Sou, B. Goins, B. O. Oyajobi, B. Travi and W. T. Phillips, Bone marrow-targeted liposomal carriers. *Exp. Opin. Drug Deliv.* **8**, 317–328 (2011).
29. J. Hamzah *et al.*, Specific penetration and accumulation of a homing peptide with atherosclerotic plaques of apolipoprotein E-deficient mice. *Proc. Natl. Acad. Sci. USA* **108**, 7154–7159 (2011).
30. S. K. Hobbs *et al.*, Regulation of transport pathways in tumor vessels: Role of tumor type and microenvironment. *Proc. Natl. Acad. Sci. USA* **95**, 4607–4612 (1998).
31. L. L. Munn, Aberrant vascular architecture in tumors and its importance in drug-based therapies. *Drug Discov. Today* **8**, 396–403 (2003).
32. S. M. Moghimi and J. Szebeni, Stealth liposomes and long circulating nanoparticles: Critical issues in pharmacokinetics, opsonization and protein-binding properties. *Prog. Lipid Res.* **42**, 463–478 (2003).
33. M. P. Monopoli *et al.*, Physical–chemical aspects of protein corona: Relevance to *in vitro* and *in vivo* biological impacts of nanoparticles. *J. Am. Chem. Soc.* **133**, 2525–2434 (2011).
34. I. Hamad *et al.*, Distinct polymer architecture mediates switching of complement activation pathways at the nanosphere-serum interface: Implications for stealth nanoparticle engineering. *ACS Nano* **4**, 6629–6638 (2010).
35. S. M. Moghimi and A. C. Hunter, Recognition by macrophages and liver cells of opsonised phospholipid vesicles and phospholipid headgroups. *Pharm. Res.* **18**, 1–8 (2000).
36. A. J. Andersen *et al.*, Complement activation by PEG-functionalized MWCNTs is independent of PEG molecular mass and surface density. *Nanomed. Nanotechnol. Biol. Med.* **9**, 469–473 (2013).
37. A. J. Andersen *et al.*, Single-walled carbon nanotubes surface control of complement sensing and activation. *ACS Nano* **7**, 1108–1119 (2013).
38. S. M. Moghimi *et al.*, Material properties in complement activation. *Adv. Drug Deliv. Rev.* **63**, 1000–1007 (2011).
39. S. M. Moghimi, I. Hamad, T. L. Andresen, K. Jørgensen and J. Szebeni, Methylation of the phosphate oxygen moiety of phospholipid-methoxy(polyethylene glycol) conjugate

prevents PEGylated liposome-mediated complement activation and anaphylatoxin production. *FASEB J.* **20**, 2591–2593 (2006).
40. D. Brambilla *et al.*, PEGylated nanoparticles bind to and alter amyloid-beta peptide conformation: Towards engineering of functional nanomedicines for Alzheimer's disease. *ACS Nano* **6**, 5897–5908 (2012).
41. S. M. Moghimi and A. C. Hunter, Capture of stealth nanoparticles by the body's defences. *Crit. Rev. Ther. Drug Carr. Syst.* **18**, 527–550 (2001).
42. S. M. Moghimi, I. S. Hedeman, I. S. Muir, L. Illum and S. S. Davis, An investigation of the filtration capacity and the fate of large filtered sterically-stabilized microspheres in rat spleen. *Biochim. Biophys. Acta. -Mol. Cell Res.* **1157**, 233–240 (1993).
43. J. A. Champion and S. Mitragotri, Role of target geometry in phagocytosis. *Proc. Natl. Acad. Sci. USA* **103**, 4930–4934 (2006).
44. J. W. Yoo and S. Mitragotri, Polymer particles that switch shape in response to a stimulus. *Proc. Natl. Acad. Sci. USA* **107**, 11205–11210 (2010).
45. S. Barua *et al.*, Particle shape enhances specificity of antibody-displaying nanoparticles. *Proc. Natl. Acad. Sci. USA* **110**, 3270–3275 (2013).
46. S. M. Moghimi, D. Peer and R. Langer, Re-shaping the future of nanopharmaceuticals: Ad iudicium. *ACS Nano* **5**, 8454–8458 (2011).
47. S. M. Moghimi and Z. S. Farhangrazi, Nanomedicine and complement paradigm. *Nanomed. Nanotechnol. Biol. Med.* **9**, 458–460 (2013).
48. S. E. A. Gratton *et al.*, Nanofabricated particles for engineered drug therapies: A preliminary biodistribution study of PRINT™ nanoparticles. *J. Control. Release* **121**, 10–18 (2007).

Chapter A3

Immunological Issues with Nanomedicines: Immunogenicity, Hypersensitivity, Accelerated Clearance and Immune Suppression

Janos Szebeni and Wim Jiskoot*†

**Nanomedicine Research and Education Center, Semmelweis University, Budapest and Department of Nanobiotechnology and Regenerative Medicine, Miskolc University, Miskolc, Hungary*
jszebeni2@gmail.com

†*Division of Drug Delivery Technology, Leiden Academic Centre for Drug Research, Leiden University, P. O. Box 9502, 2300 RA Leiden, The Netherlands*
w.jiskoot@lacdr.leidenuniv.nl

Abstract

Some nanomedicines are recognized by the immune system as foreign, and their systemic (*i.v.*) use may entail adverse effects afflicting the host, the drug, or both. This review focuses on four such adverse effects: (i) formation of antibodies against the drug (immunogenicity), (ii) rise of hypersensitivity reactions (HSRs), (iii) accelerated blood clearance (ABC) of the drug and (iv) immunosuppression. In addition to providing essential information on these processes, it was our goal to highlight the inter-relations among their progress, pointing out the initial, key role of complement (C) activation in a vicious cycle of immune harm. The review outlines mechanistic details of immunogenicity, C activation-related pseudoallergy (CARPA) and the ABC phenomenon, and draws attention to the benefits of immune suppression, at least in the therapeutic effect of liposomal doxorubicin (Doxil). Our main message is that immune side effects still present a major barrier to the clinical use of many state-of-art nanomedicines, so success in nanodrug development is inconceivable without understanding and close control of immune responses.

1. Introduction

Nanomedicines, because of their relatively large size and complex structure (in comparison with traditional small-molecular drugs), carry an inherent risk of raising an immune response. An important reason is that our immune system has evolved to recognize particulate matter, such as viruses, as foreign and many nanomedicines are very similar to viruses in size, shape and certain surface properties. In fact, most of our vaccines are based on nanoparticles, such as live or inactivated viruses and virus like particles, exploiting the exceptional sensitivity of immune surveyance in this dimension. Furthermore, in case of vaccines based on recombinant antigens, a common way to enhance their often poor immunogenicity is to either adsorb them to particulate adjuvants, such as colloidal aluminium salts, or encapsulate them in particulate, nanosized systems, such as liposomes, polymeric nanoparticles, ISCOMs or nano-emulsions.[1,2]

This chapter aims to address four major adverse immune effects of nanomedicines; immunogenicity, complement (C) activation and consequent hypersensitivity reactions (HSRs), called C activation-related pseudoallergy (CARPA), accelerated blood clearance (ABC) phenomenon and immune suppression. As shown in Fig. 1, C activation, immunogenicity, anti-drug antibody (ADA) formation and its binding to nanomedicines feed a vicious cycle, whose adverse impacts include CARPA and the ABC phenomenon.

Since most of the references on nanoparticle-triggered immunological events concern liposomal drugs, our discussion is liposome-focused to some

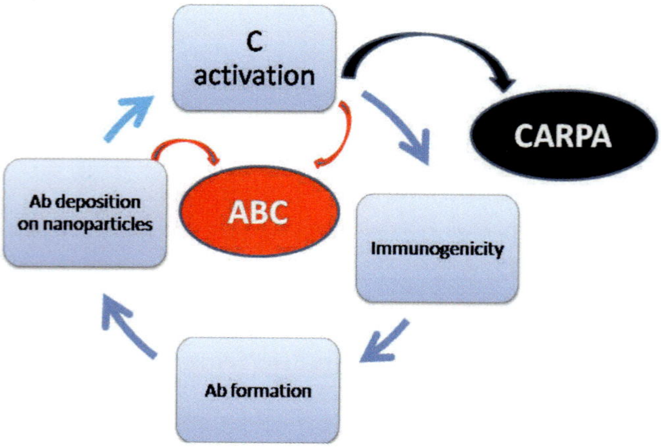

Fig. 1. The vicious cycle of C activation, immunogenicity, Ab formation and Ab deposition on nanoparticles, with spin-off adverse effects: CARPA and ABC.

degree, however, many of the examples and concepts discussed will also apply to other nanomedicines. As for the definition, properties, therapeutic and diagnostic use and non-immune adverse effects of different kinds of nanomedicines, we refer to other chapters in this volume, as well as a great number of books and reviews in this rapidly advancing field of experimental medicine.

It should be emphasized that by focusing on adverse immune effects we certainly do not intend to discourage scientist colleagues to work with nanoparticle systems, but rather to help improving their awareness of immunological events that could happen *in vivo*, upon clinical application of nanomedicines. Although absolute absence of immune reactivity may be difficult to achieve, the balance between immunological risk and therapeutic benefit will ultimately determine clinical success. Therefore, all immunological risks should be taken into account and tested rather sooner than later in the development of nanomedicines.

The sections below address the three mentioned issues, potential adverse immune effects of nanomedicines.

2. Immunogenicity

In this chapter, we refer to immunogenicity as the capability of a drug or agent to induce a specific immune response, which can be antibody-mediated (humoral) or cell-mediated (cellular), or both. Thus, immunogenicity is assessed by measuring the production of specific antibodies, or specific T cell responses. It should be pointed out as an interesting controversy, that many nanomedicines, such as liposomes and polymeric nanoparticles, are widely studied for applications both as vaccines and as therapeutic drug delivery systems, yet immunogenicity is dealt with almost exclusively in relation to vaccine applications. Obviously, this is due to the fact that vaccines are aimed to elicit immunogenicity and the success of vaccine delivery concepts is measured in terms of the strength and type of immunological responses, whereas for drug delivery, a scientist's primary focus is to prove therapeutic efficacy rather than immune complications. It might be for this reason that the number of research laboratories focusing on immunogenicity of therapeutic or diagnostic nanomedicines is limited at this time, probably to a few groups in the world.

Where immunogenicity gets real attention, in addition to vaccine research, is the field of biopharmaceuticals, whereupon monoclonal antibodies and other proteins are used by following repeated or chronic administration protocols. Some of the patients treated this way mount an immune response against these drugs, mostly via production of ADAs. These antibodies can change the pharmacokinetics of the drug after the first or second administration, and thus interfere with the drug's therapeutic effect. In addition, ADAs can lead to adverse immune effects, sometimes severe toxicities. Examples of such severe adverse responses to protein therapies include anti-erythropoietin ADA-induced pure red cell aplasia and anti-thrombopoietin ADA-induced idiopathic thrombocytopenic purpura ITP.[3–5] Since the impact of immunogenicity can be quite severe, regulatory agencies are developing risk-based guidelines for immunogenicity screening.[6] This illustrates that immunogenicity remains a major concern with the therapeutic use of biologicals, whose study provides a rich and relevant source of information on yet a further field where immunogenicity is a critical issue: nanomedicine.

As with biologicals, the rush to deliver novel nanomedicines to market may lead to overlooking of their immunogenicity, which represents an intrinsic risk associated with the use of nanoparticulate medicines. To our knowledge, fortunately, no catastrophic immunogenicity-related side effect has been reported to date for any nanomedicines introduced in the clinics, nevertheless administration of liposomes and drug-polymer conjugates were shown to lead to the formation of antibodies against these agents.[7] Thus, the risk of adverse clinical effects similar to those observed with protein therapeutics does exist with liposomes and drug-polymer conjugates and most likely many more nanomedicines as well.

2.1. *Mechanisms of immunogenicity*

2.1.1. T cell independent activation of B cells through repetitive antigen arrays

B cells can be directly activated by antigens (i.e. molecules capable of interacting with components of the immune system). In principle, any molecule or structure can function as an antigen, but direct B cell activation by antigens only happens when they are presented to these cells in a repetitive manner. Direct activation of B cells via repetitive epitopes is the basis for the mode of action of capsular polysaccharide vaccines against, e.g. meningococcal and pneumococcal polysaccharide vaccines.[8] Although these vaccines do not generate memory and therefore are protective for at most a year or two, they have been widely used until the introduction of polysaccharide-protein conjugate vaccines, which (owing to the protein carrier) can induce T cell dependent class switching, affinity maturation and therefore much stronger protective anti-polysaccharide antibody responses, and immunological memory.

Dintzis and co-workers were among the first ones to show that even small molecules (also called haptens), such as dinitrophenol, can be rendered immunogenic by conjugating them with polymeric carriers.[9] Analogous to the plain polysaccharides mentioned above, these conjugates lack T-helper epitopes but do result in antibody production through direct B cell activation, resulting from cross-linking of the B cell receptor by an array of identical B cell epitopes, i.e. the haptens.[10] This principle has been confirmed by several groups for other hapten-polymer conjugates and studies have indicated that key characteristics for direct B cell activation are a sufficiently close spacing (5–10 nm) of the B cell epitopes (Fig. 2) and a minimally required number of repeat units (10–20), as reviewed elsewhere.[7]

So, if closely spaced repetitive structures are present on the surface of a nanomedicine, B cells, recognizing those via their B cell receptors, can become activated resulting in their proliferation to antibody-producing plasma cells. If there are no further signaling molecules present, this typically leads to a transient IgM response. It is most probably via this mechanism that administration of liposomes can result in the formation of antibodies against constituents of the lipid bilayer that are exposed at the liposomal surface, such as phospholipids and cholesterol.[11–14]

Intravenously administered PEGylated liposomes have been reported to elicit IgM antibodies.[7] These antibodies have been suggested to be directed against the PEG chains, but a recent study indicates that they may be directed against the linker between the PEG moiety and the liposomal bilayer.[15]

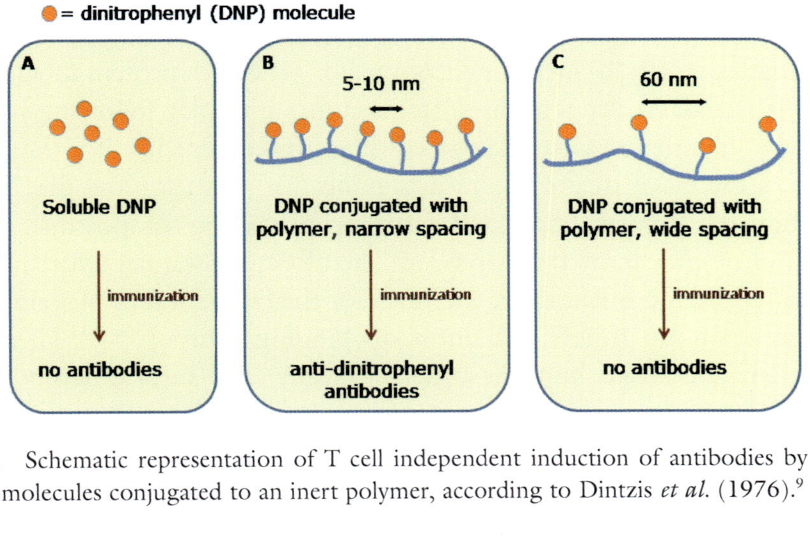

Fig. 2. Schematic representation of T cell independent induction of antibodies by closely spaced molecules conjugated to an inert polymer, according to Dintzis *et al.* (1976).[9]

Although IgM responses lack memory and are relatively weak (compared to what would be needed for a potent vaccine), such antibodies typically lead to the ABC phenomenon of nanomedicines upon repeated administration.[16–25] Recently, it was shown that this can also occur after subcutaneous administration of PEGylated nanoparticles.[26]

2.1.2. T cell dependent antibody responses

For conversion of an early IgM response into a robust IgG response, T cell help is required. T cells are most likely involved in this immune response when IgG class ADAs are observed. This implies that the risk of a potent immune response will be considerably increased when the nanomedicine contains protein-based materials (as carrier, drug, or targeting ligand) that contain T cell epitopes. In that case dendritic cells (DCs) most probably play a key role. DCs are professional antigen-presenting cells (APCs) that can internalize particulate matter, process proteins derived from it, and present processed peptide epitopes in MHC class II molecules to $CD4^+$ helper T cells, which in turn can activate B cells. From the vaccine delivery literature we know that the extent to which this happens depend on many factors, such as particle properties (e.g. size, surface charge), presence of co-stimulatory molecules, dose and dosing schedule and route of administration.[27–29] Furthermore, the fact that virus like particles, containing highly repetitive structures on their surface and viral proteins carrying T cell epitopes, induce potent immune responses show that direct B cell activation and DC-mediated B cell activation likely act synergistically.[30] This explains why immunoliposomes carrying IgG molecules as targeting moieties on their surface can induce IgG antibodies directed against the targeting ligand.[31,32]

Immunogenicity, as mentioned, is a major issue with protein therapeutics, when the antigenic character of proteins comes into play following repeated or chronic treatment. Quantitation of the immunogenicity of a therapeutic protein is therefore of great clinical significance, both for the diagnosis of ADA-induced abnormalities, and also for prediction of possible adverse effects during preclinical development. Therefore, substantial effort has been devoted to develop quantitative methods for the evaluation of protein immunogenicity, among others, via antibody binding assays.[4,5,33–38] These tests include enzyme-linked immunosorbent (ELISA) and surface plasmon resonance (Biacore) assays, which measure the binding of ADAs to the protein drug immobilized on a surface (ELISA plates, Biacore chips). Usually, an initial screening for bound ADAs needs to be followed by specificity confirmation and additional tests to determine whether the formed and bound

Table 1. Shortcomings of immunogenicity assays used for protein therapeutics.

- low sensitivity (particularly in the detection of low-affinitiy antibodies)
- lack of ability to detect immunoglobulins with a wide range of affinities to the drug molecule
- low signal-to-noise ratio due to non-specific binding and possible presence of preexisting (natural) antibodies
- arbitrary choices of background and outlier limits
- subjective use of fixed, floating or dynamic cut-points distinguishing positive from negative samples
- inconsistent use of statistics, e.g. parametric or non-parametric
- high inter-assay variability
- low "assay drug tolerance" (i.e. high likelihood of interference of the test drug with the assay
- potential presence of matrix effects

antibodies are functionally active or not, i.e. whether they neutralize the effect of the drug or not (neutralizing assays). The antibody assays mentioned above and developed for protein therapeutics may be usable for nanomedicines as well, making it worth to remind of the difficulties and flaws of these assays (Table 1).

Apart from these assay limitations the most arduous problem in this field is that immunogenicity of proteins or nanomedicines in humans can only be definitively assessed in human studies, and even human studies will not fully predict the consequences of ADA generation on the individual level. At this time the available ADA tests are diagnostic, and there is no *in vitro* or animal model that could be used to predict whether or not antibody formation against a protein drug or nanomedicine will be inconsequential, or will have a major clinical impact.[39,40] These facts taken together suggest that instead of further validation and optimization of available assays, the field would benefit from introducing novel technologies for predicting immunogenicity already in the preclinical state of drug development.

2.2. *Approaches to avoid antibody responses against nanomedicines*

Our knowledge of structure-immunogenicity relationships of nanomedicines are curtailed due to the lack of systematic studies. However, from the vaccine delivery literature we do know that surface characteristics are important.

For instance, cationic liposomes interact more strongly with APCs than negatively charged liposomes.[41] From a theoretical perspective it makes sense to avoid high surface densities of targeting ligands. In the early work by Dintzis *et al.* it was shown that an epitope spacing >20 nm minimizes the risk of T cell independent B cell activation,[9] whereas virus like particles with antigen spacings of about 5 nm are highly immunogenic.[30] Furthermore, in the choice of carrier materials one should consider potential immune-stimulating activity. For instance, trimethylated chitosan, which has been widely studied as carrier for the delivery of both therapeutic proteins and vaccines,[42] has been shown to have instrinsic immune-stimulatory activity.[43] While this makes it an interesting polymer for inclusion in vaccines, it should raise concerns when used as protein drug delivery carrier.[44]

Furthermore, one may consider avoiding the use of proteinaceous targeting ligands. It is well-known that small amounts of aggregated or particulate proteins in therapeutic protein formulations increase the risk of an ADA response.[45,46] So, deliberately making nanoparticulate protein-containing medicines could be seen as "asking for problems". One possible approach to overcome immunogenicity problems with such systems is the "deimmunization" of proteins by removing T cell epitopes from the protein molecule.[40,47] However, the general clinical success of this approach is yet to be determined.

3. Hypersensitivity Reactions

3.1. *C activation underlying HSRs to nanomedicines*

3.1.1. The C system and its activation

The C system is a protein network consisting of some 30 glycoproteins in blood and cell surfaces. Along with natural antibodies, it represents the humoral, non-specific arm of the immune system, a main provider of innate immunity. Although it was discovered nearly 120 years ago, the discovery of novel physiological functions of different C proteins, mostly C3, keeps going on.[48] Figure 3 illustrates that C activation is a complex proteolytic cascade triggered via three different pathways called classical, alternative and lectin.[49–56] Each pathway has its own sensor molecules which non-specifically recognize molecules on the surface of foreign particles, namely, IgG, IgM, C1q and C reactive protein (CRP) in the classical, C3 and properdin in the alternative and mannose binding lectin (MBL) and ficolin in the lectin pathways. These molecules usually recognize repetitive surface structures, such as polymer coatings, protein or peptide ligands, charged surface groups or sugar

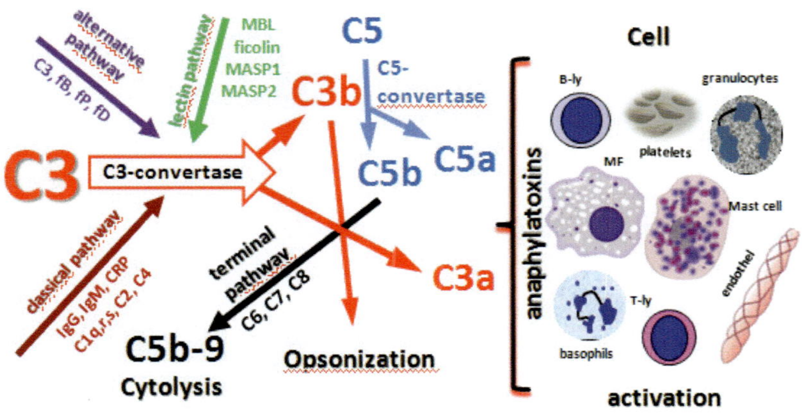

Fig. 3. The C cascade and C-responsive blood cells. The figure visualizes the main routes of C activation and the cells stimulated in blood as a consequence of anaphylatoxin liberation.

residues on foreign as well as host membranes, although in the latter case the presence of C inhibitor proteins prevents C activation.[57]

C activation via different pathways may be independent or simultaneous, redundant and activation of the alternative pathway not only represents an independent activation mechanism, but it is also a positive feedback loop triggered by activation of the other two pathways, thus it provides effective amplification of the activation process. In addition to these amplification steps, the multiple functions of C3 and C5-derived anaphylatoxins (stimulating almost all blood cells as well and the endothelial cells), and the deposition of C3-derived opsonins (mainly C3b) on foreign or dying cells (labeling them for elimination by a variety of clearance mechanisms) all act in a concerted effort to eliminate foreign particles.

Among a large number of textbook and internet illustrations of the C cascade, Fig. 3 is unique in that it combines all protein and cellular elements involved in *in vivo* C activation. Taken together with the large number of C split products and natural inhibitors that are not shown, the figure illustrates the huge complexity, and, hence, substantial individual variation of the changes and symptoms of C activation.

3.1.2. Nanomedicines causing HSRs

Table 2 lists the clinically used drugs and agents that have been reported to cause acute HSRs. Among them, liposomes and micelles carrying water insoluble drugs are typical and unquestionable nanomedicines according to

Table 2. Examples of clinically used drugs and agents that cause CARPA.

Liposomal drugs	Micellar drugs	Monoclonal antibodies	Contrast agents	Enzymes	Miscellaneous
Ambisome	Etoposide	Avastin	Iodixanol	Abbokinase	Copaxone
Amphocyl	Fasturec	Campath	Iohexol	Actimmune	Eloxatin
Amphotec	Elitec	Erbitux	Iopromide	Activase	Orencia
DaunoXome	Taxol	Herceptin	Iothalamate	Aldurazyme	Salicilates
Doxil	Taxotere	Mylotarg	Ioversol	Avonex	
Myocet	Vumon	Remicade	Ioxaglate	Fasturtec	
Visudyne		Rituxan	Ioxilan	Neupogen	
		Vectibix	Lopamidol	Plenaxis	
		Xolair	Magnevist	Zevalin	
			Odixanol		
			SonoVue		

the definition of nanomedicines (nanoscale drugs produced via nanoengineering). However, the nanomedicine status" of monoclonal antibodies and other protein-based pharmaceuticals is ambiguous, in as much as they are in the nano range (8–10 nm), however, the technology involved in their production is bio-, rather than nanotechnology. The supramolecular organization of radiocontrast media and the other listed inducers of HSRs in blood is not clear, implying that not all C reactogenic drugs are nanomedicines, but all nanomedicines has the potential to induce C activation and related HSRs.

In the case of commercially available liposomes,[55,58–63] micellar drugs and solvents[64–67] and radiocontrast media[59,68–70] there is ample evidence for C activation both *in vitro* and *in vivo*, just as for many other nanomedicines in research or preclinical development[71–75] providing basis to assume that the HSRs they cause represent CARPA, at least in part.

3.1.3. Symptoms and characteristics of nanomedicine-induced HSRs

The most common symptoms of CARPA are rash, back pain, dyspnea (shortness of breath), chest pain, hypotension, hypertension, chill and tachycardia. Nevertheless, almost any systemic manifestation of allergy can occur, with sudden death in < 0,5–0,01% of patients. Table 3 shows the distribution of different symptoms in the case of Doxil-induced HSRs.

Table 3. Relative frequency of the various symptoms of HSRs caused by Doxil.[a]

Symptom	%	Symptom	%
Rash	31,3	flashing	8,4
back pain	29,8	arrhythmia	6,8
Dyspnea	29,5	lung infiltration	6,3
chest pain	26,5	bradycardia	6,1
Hypotension	20,6	anaphylaxis	6,0
Hypertension	19,4	multi-organ failure	5,7
Chill	16,6	mental status changes	5,7
Tachycardia	16,4	anaphylactic shock	4,6
Shock	15,3	confusion	3,4
Fibrillation	14,6	delirium	2,9
cardiac arrest	13,5	apnoe	2,3
myocardial infarction	12,5	ventricular fibrillation	1,3
cardiac failure	10,3	sudden death	0,4
Pruritus	10,2		

[a] Percentage estimates are based on data in the Doxil adverse events reports web site (http://www.patientsville.com/medication/doxil18_side_effects.htm#reports). Symptoms are ordered in decreasing frequency.

There are major differences between true and pseudoallergy in several respects, most importantly CARPA arises upon the first exposure to the medicine, and the strength of the reaction decreases or the reaction completely disappears following repeated administration of the drug. The diagnosis, special features, assessment and management of drug-induced acute HSRs are discussed in more detail in other reviews and papers on the subject.[61–63,76–81]

3.1.4. Causes of C activation by nanomedicines

There are several reasons for C activation by nanomedicines. The relatively large size of nanoparticles, making them indistinguishable from viruses by the immune system, has been mentioned in the introduction. Further to this point, it should be mentioned that visual comparison of pathogenic viruses and liposomes give clear evidence for the similarity, with the resemblance between Doxil and HIV being particularly striking.[61,82] As it was pointed out earlier,[61] the resemblance may not be simple coincidence but rather the consequence of those common physicochemical rules and forces that force bilayer membranes to form vesicles in test tubes, just like in living systems. For this reason liposomes are considered as the archetype bilayer membrane that enabled cellular evolution over millions of years.[83–85]

In addition, liposomes and other nanomedicines lack all those C inhibitory molecules that normally protect self-cells against C attack, including C receptor type I (CR1), membrane cofactor protein (MCP), decay accelerating factor (DAF) and CD59.[54,57,86] These highly conserved C binding proteins, also called C control proteins (CCP), usually include repetitive amino acid sequences called short consensus repeats, or "Sushi" domains.[87] They accelerate the decay and enzymatic digestion of membrane-deposited C proteins, most importantly the C3 convertase and C3b. Their positioning and relative sizes on membranes were illustrated earlier.[61,88]

Yet a further, important contributor to C activation by nanomedicines is the presence of molecules and residues on their surface that are known to bind C proteins. These include certain lipids, carbohydrates, polymers and proteins. The phospholipids and cholesterol of liposomes can bind natural antibodies, a process that was shown to underlie liposomal C activation and associated leakage of entrapped compounds.[11–13,89–91] Negative or positive charges on any nanoparticle surface, including liposomes[92] and dendrimers[93] bind C1q directly as well as via antibodies or CRP.[58] Polyethyleneglycol (PEG), also on liposomes, binds anti-PEG antibodies[18,19,22,23] while carbon nanotubes (CNTs) were shown to bind MBL and ficolin, among other proteins.[73,94,95]

3.1.5. Factors influencing C activation by nanomedicines

Among the structural factors of nanomedicines that directly influence C activation, size, shape, curvature and the presence of aggregates have been shown to play key roles in different systems, under different conditions. The effects of these factors can easily be rationalized with the surface-dependent propagation of C activation on particles by way of successive depositions of classical, and then alternative pathway C3 convertases. Considering the MW and molecular dimensions of these initially bimolecular (C4bC2b and C3bBb, 2–300 kDa, ≥ 8–10 nm), later trimolecular complexes, their sequential deposition is not conceivable on high-curvature nanoparticles with d ≤ 20–30 nm. Thus, according to the existing paradigm of surface-enabled C activation, small nanoparticles, such as dendrimers (3–5 nm) and micelles (8–20 nm) can only activate C in solution if these particles form aggregates to provide a sufficiently large "landing field" for C3 and C5 convertases. On the same grounds, the explanation of C activation by liposomes and other particles with $d \geq 80$–100 nm is relatively straightforward in light of the relatively large surface area available for C deposition. Nevertheless, C activation by dendrimers, which was attributed to the high cationic surface charge of 3–4 nm particles,[93] just as those of C-activating PEG-PEI polymers,[96] cannot provide sufficient surface for C3 and C5 convertase formation (unless they would aggregate), leaving the mechanism of C activation in dark.

3.1.6. The CARPA concept and its clinical significance

Adverse drug reactions (ADRs) can be classified as allergic or non-allergic, also referred to as type A and B reactions.[78] Type A reactions are dose dependent and predictable, while type B reactions are not dose dependent and are unpredictable. There are many other classifications and categorizations for ADRs,[78,97–101] among which one, the "CARPA concept", was conceived following the recognition that C activation may explain many non-allergic (pseudoallergic or type B) reactions, for which the liposome-induced cardiopulmonary distress of pigs provides a sensitive model.[102,103] Later CARPA was proposed[104] to represent a subcategory of Gell and Coombs' Type I reactions[105] thereby alleviating the unsolved paradox that many HSRs and ADRs could not find place in Gell and Coombs' system.[106] In short, the CARPA concept states that a large fraction of non-allergic, type B ADRs and HSRs arise through a process wherein C activation is a trigger, or co-trigger, or it plays a potentiating role.[104] These pseudoallergies can be distinguished from other pseudoallergies as "C activation-related", the term being vague enough not to implicate direct or indirect causality, yet highlighting the role of the C system. By turning the attention to this system, the innovation of taxonomy might result in innovation in therapy as well, for example if a C inhibitor turns out to be more useful in preventing these reactions than any anti-allergy medication.

3.1.7. The role of C activation in CARPA

The exact role of C activation in CARPA was further refined in the clinical study by Chanan-Khan *et al.*,[107] wherein C activation was measured in cancer patients following Doxil infusion. The number of patients showing clinical symptoms (clinical reactors) was much lower than the number of patients in whom the blood level of a C marker, SC5b-9 significantly rose (i.e. laboratory reactors), implying that C activation contributed to the reactions, but it was not rate limiting to HSRs. Obviously, there are many more co-factors and controlling steps in the complex pathomechanism of CARPA. Further discrepant facts regarding the CARPA concept include the finding that some liposomes and other nanoparticles, that do not activate C in (all) human serum, still can cause hemodynamic changes in pigs, which are symptoms of mild CARPA. More precisely, we find dose dependence and clear correlation between *in vitro* C activation and reactogenicity in pigs only in the case of strong C activator liposomes[63,108,109]; weak C activator substances, such as PEGylated polyethylene-imines (PEI) are inconsistent activators of porcine CARPA.[96] These facts, taken together with the phenomenon of tachyphylaxis (i.e. self-induced tolerance) in pigs observed with PEGylated liposomes (Doxil)[62] suggest that CARPA may actually be more complex than a simple anaphylatoxin effect, and may involve a second "hit", namely, the binding of reactogenic particles to receptors on mast cells and/or macrophages, which mediate identical or similar intracellular signaling that leads to release reaction of masts cells and/or macrophages. This theory was described very recently[62,103]; here we just acknowledge that the new term CARPA may soon require revision perhaps to "DIARPA" (= dual innate immunity activation-related pseudoallergy).

3.1.8. The clinical significance of CARPA

Regarding the clinical relevance of CARPA, it is known that ADRs affect 10–20% of hospitalized people, and HSRs represent about 25–33% of ADRs.[78,110] Further statistics on the portion of non-IgE-mediated HSRs within all ADRs (10–17%)[78,110,111] suggests the incidence of mild to severe CARPA to be several hundred thousand per year in the USA.[78,104,110,111] Nevertheless, the statistics on ADRs and HSRs show wild variation among different authors.[78]

The clinical evaluation and judgment on CARPA depends on the indication area of the reactogenic drug in question. For example, in oncology, the risk of CARPA is far outweighed by the benefit of a life-saving drug therapy. On the other hand, in non-fatal illnesses, and, particularly, in diagnostic applications of reactogenic drugs, the tolerance for potentially lethal adverse reactions is much less. A rational approach to minimize the risk of adverse events is to exclude those people from therapy who are at increased risk to develop fatal reactions, such as patients with advanced cardiovascular disease or severe allergy. The prediction of CARPA will soon be an important procedure in "personalized" pharmacotherapy relying more and more on targetable nanomedicines and biologicals.

3.1.9. Prediction and prevention of CARPA

CARPA is a "capricious" adverse immune effect since its rise, strength and exact symptoms are not predictable by the standard allergy tests[112] or any other test that is currently in medical practice. This makes it very difficult or impossible at this time to specifically prevent the reaction on an individual basis, either by avoiding its cause, or by selectively pre-treating only those subjects who are susceptible for allergic reactions in general. The currently used approaches for prevention include non-specific pre-medication with steroids, antihistamines and non-steroidal anti-inflammatory drugs, along with application of slow and/or interrupted and/or gradually escalating infusion.[77,79,80,113]

3.1.10. Mechanism of CARPA: the double hit hypothesis

As for the molecular mechanism of CARPA, it was mentioned that the original view that the actions of anaphylatoxins on mast cells and basophils might account for all physiological changes is probably overly simplistic.[70,104,114] The anaphylatoxin effects may be just a "pathway" of CARPA as the "classical pathway" is just one pathway of C activation. As delineated above, a theoretical "alternative" pathway of CARPA may be the binding of reactogenic particles and drugs to mast cell and/or macrophage surface receptors, whose intracellular signaling might feed into, or act in synergy with the signaling apparatus of G-protein coupled anaphylatoxin receptors (i.e. C3a/C5a/C5L2 receptors).

The two processes acting together may lead to super-threshold triggering of release reactions (Fig. 4). Subsequent release of secondary vasoactive mediators, including histamine, tryptase, PAF, leukotrienes (LTB2, LTB4, LTC4, LTD4, LTE4, TXA2, PGD2 and TXD4) exert their effects on smooth muscle and endothelial cells in the lung, heart, gut and many other organs in the body, causing the symptoms of CARPA. Some of these "allergomedins" (e.g. PAF, histamine, tryptase and TXA2) are preformed and are liberated from the cells immediately upon activation, while others are *de novo* synthesized and are liberated slower.

As for the identity of mast cell and/or macrophage surface receptors, pattern recognition receptors (PRRs) were considered,[61,62,103] since particle-bound polymers, including PEG, may under certain conditions act like pathogen-associated molecular patterns (PAMPs).[61,73,115] That PEG on the surface of liposomes (i.e. stealthness) may in fact be a critical factor in reaction induction and tachyphylaxis is also supported by our earlier finding that the

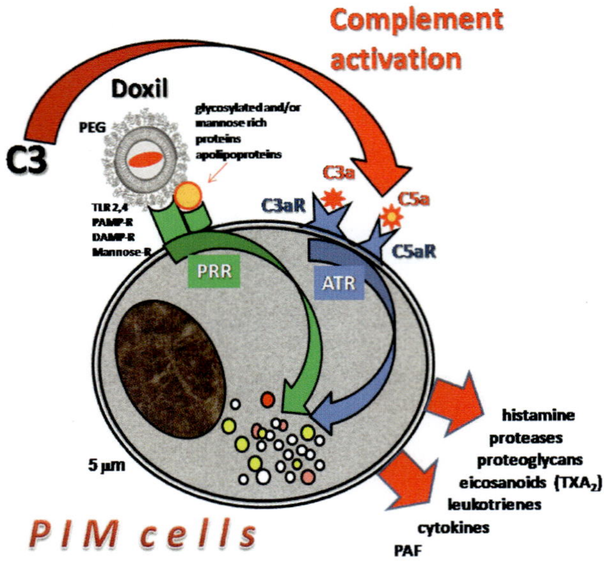

Fig. 4. The double-hit hypothesis of CARPA. C is activated by nanoparticles (illustrated by Doxil in the figure), leading to the liberation of anaphylatoxins. The binding of anaphylatoxins to their respective receptors (anaphylatoxin receptors, ATR) on mast cells (or pulmonary intravascular macrophages, or PIM cells in pigs) lead to release reaction, i.e. secretion of vasoactive secondary mediators. The particles also bind to the cells via pattern recognition receptors (PRR), possibly toll-like receptors (TLR 2,4), pathogen-associated molecular pattern receptors (PAMP-R), danger-associated molecular pattern receptors (DAMP-R) or mannose receptors (mannose-R). In addition to binding to PRR, apolipoproteins and glycosylated and/or mannose rich proteins can mediate the binding of particles to mast cells. Figure modified from Ref. 103.

reactogenicity of some PEGylated PEI polymers are tachyphylactic,[96] while PEG-free Ambisome, which is similar to Doxil in size, and which causes similar first reactions in pigs as Doxil, does not lose reactogenicity upon repeated injections (its effect is non-tachyphylactic) [Szebeni 2012 #3725].

3.1.11. The porcine model of CARPA

Pigs represent a highly sensitive animal model of human CARPA, as these animals closely mimic the symptoms of human HSRs with relatively small individual variation. As discussed in recent reviews in detail,[62,103] the unique reactivity of pigs to liposomes and other particulate reaction inducers is most likely due to the presence of PIM cells in the lung of pigs.[116] Besides alveolar, interstitial, pleural and free vascular macrophages, PIMs residing in the pulmonary capillaries are part of the macrophage system of the lung. Like Kupffer cells in the liver, their main task is to rapidly and directly remove blood-born particles, such as cellular debris, bacteria, viruses, endotoxins, etc. With their extended, glycocalyx-covered pseudopods and spikes, these cells display high phagocytic efficacy and immediate and significant secretory reactions Brain, 1992 #3791; Atwal, 1992 #4001], which is in keeping the rapid response to *i.v.* administered liposomes and other particulate matter.

4. Accelerated Clearance of Nanomedicines

4.1. *Enhancement of adaptive immune responses by C activation*

Table 4 lists some of the major findings attesting to a key role of C in the development of specific immune responses. In fact, C promotes almost all processes in adaptive immunity, including antigen presentation, antibody

Table 4. C effects promoting specific (adaptive) immunity.

Effect of C via C3 ligands[a]	Target cell(s)	Reference
Facilitation of the cooperation between APC and T cells	T cells APC	117 and 118
Stimulation involving CD21/CD19/CD81 signaling Stimulation involving CD21/CD35 signaling	B cells follicular dentritic cells	118–121
Increasing cell survival via CD21/CD19/Tapa-1 signaling	B cells	122–125
Stimulation of germinal center formation in lymph nodes	B cells, T cells	126–128
Increased cytotoxicity against CR2 (CD21)+ target cells via surface-bound C3b	CD8 T cells	129
Acceleration of antibody affinity maturation via ligation to CR2	B cells	130
Induction of IL-1 and IL10 production	monocytes	131 and 132
Suppression of IL-6, IL-12, TNFα	Monocytes, macrophages, DCs, PBMCs	128, 132–134

[a]C3 ligands, C3b, C3c, C3dg, C3d. Abbreviations: APCs, antigen-presenting cells; DCs, dendritic cells; PBMCs, peripheral blood mononuclear cells.

production and helper and cytotoxic T cell responses via anaphylatoxins, opsonins and C3b targeting (i.e. using C3b and its derivatives as targeting ligands for T cell activities).

4.2. *The ABC phenomenon*

An example of non-standard, partial immunogenicity is the so-called "ABC phenomenon", i.e. ABC of PEGylated liposomes, a phenomenon that has great clinical relevance. It has been discovered by Ishida and colleagues[16,18–23,25,135] that repeated injection of PEGylated liposomes in mice and rats causes rapid clearance of liposomes from the bloodstream, due to the formation of anti-PEG IgM in the spleen. Importantly, the phenomenon is absent with PEGylated liposomes encapsulating doxorubicin, which is consistent with the lack of ABC in cancer patients treated with Doxil. On the other hand, free doxorubicin given in doses that correspond to the amount given in Doxil, restores ABC. These data indicate that immune cells responsible for the ABC phenomenon might be selectively affected by doxorubicin encapsulated in PEGylated liposomes, as detailed in the section on liposome-induced immune suppression below. Since the ABC phenomenon was also observed in BALB/c nu/nu mice, but not in BALB/c SCID mice, it was suggested that antibody production represents a T cell-independent B cell response, and that PEGylated liposomes might be recognized by B cells as a thymus-independent type 2 antigen.[25] Based on the array theory, as delineated above, it can be hypothesized that PEGylated liposomes trigger B cells via some of their TLRs, for PEG looks to B cells like viral spikes.

5. Immune Suppression by Liposomes

Because liposomes are taken up mainly by cells of the reticuloendothelial system (RES) in the liver, spleen, bone marrow and elsewhere, the possibilities of macrophage saturation and secondary immunosuppression have always been a concern ever since liposomes are in human therapy.[136] Nevertheless, there is ample evidence that clinically applied doses of non-cytotoxic liposomes generally do not cause immune suppression, at least not major, clinically important blockage of macrophage function. However, the situation is different with anticancer liposomes loaded with cytotoxic drugs, which may cause different levels of immunosuppression.

Administration of Doxil in mice, for example, was shown to interfere with the clearance of bacteria from blood, which was explained by macrophage suppression.[137] Further evidence, in mice, of partial RES blockage by liposomal doxorubicin was the dose-dependent pharmacokinetics of Doxil, resulting in slower clearance and disproportional increase of tumor uptake at higher doses (in the 2.5 to 20 mg/kg range).[138]

Although interference with innate antibacterial defense was not found in Doxil-treated patients to date (by now more humans were injected with Doxil than mice), clinical signs of partial macrophage suppression do seem to be present in cancer patients treated with Doxil. Evidence for this includes the rise of circulation half-life of Doxil after repeated administrations,[139] and the inhibition of HSRs to carboplatin by co-administered Doxil.[140] The implications of partial immune suppression by Doxil remain to be evaluated, including two potential clinical benefits. One is that tumor uptake of Doxil might increase upon repeated doses (as seen with dose escalation in mice[138]), resulting in increased therapeutic efficacy by the same dose, or allowing dose reduction without loss of efficacy. The second is that Doxil may provide therapeutic advantage over paclitaxel or gemcitabine in combination chemotherapies with carboplatin, as HSRs to carboplatin have become dose-limiting during treatment of platinum sensitive recurrent cancer.[140]

6. Conclusions and Outlook

Immunosafety may become a key issue in current and future R&D of liposomes and other nanomedicines. Its significance is emphasized by examples of postmarketing drug recalls, industry guidelines on immunotoxicity testing and recent (FP7) calls for collaborative research in the fields of nanotoxicology and immunogenicity.[141–143] Clearly, the currently used immunotoxicology endpoints (e.g. lymhoid organ weight and histology, lymphocyte proliferation, cytotoxicity and antibody production tests, cytokine assays, skin irritation test) have little or no predictive value in assessing the risk of HSRs and the outcome of immunogenicity of complex liposomes and other nanomedicines. An almost insurmountable difficulty in this regard is the high complexity and species and individual variability of the immune processes involved in the latter phenomena.

As for assessing the risk of HSRs mediated by C, to our knowledge, the only direct approaches are the *in vitro* C activation measurements in normal human sera (NHS), the *in vivo* porcine and canine CARPA tests and the basophil leukocyte activation-based allergenicity assays, provided as service by a few CROs in the world. These tests can predict in a rough, semi-quantitative way the likely presence, severity and frequency of CARPA in patients. The unique benefit of a large animal (pig and dog) model of CARPA is that the endpoints, i.e. cardiopulmonary and hematological changes, when they occur, not only reveal the potential reactogenicity of the drug, but also model the physiological changes occurring in hypersensitive individuals. As mentioned in the text above liposome-induced CARPA in pigs and dogs provide a model of the infusion reactions seen in hypersensitive individuals. Focusing on liposomes, the C and CARPA tests are useful in assessing their risk of causing HSRs, enabling early abandonment of highly reactogenic formulations. In lack of valid animal models for assessing the immunogenicity of human or humanized protein-containing liposomes, securing of immunosafety of these products seems to be an even greater challenge than solving the HSR problem.

What gives hope in this challenging "terra nova" is that interest will never cease in improving the therapeutic efficacy of available drugs, or creating new drugs whose molecular buildup gets increasingly complex. Hopefully it is not too far in the future that we can equip liposomes and other drug carrier nanosystems with immune evasive capabilities, and/or "teach" the immune system to distinguish these nanotech marvels from harmful pathogens or agents.

Acknowledgments

JS gratefully acknowledges the financial supports by the National Office for Research and Technology (NKTH), Budapest (NANOMEDI and TÁMOP-4.2.1.B- 09/1/KMR-2010-0001 and 4.1.2.B-10/2/KONV-2010-0001 and Gedeon Richter Pharmaceuticals in Budapest, Hungary

References

1. M. Singh, A. Chakrapani and D. O'Hagan, Nanoparticles and microparticles as vaccine-delivery systems. *Expert Rev. Vaccines* **6**, 797–808 (2007).
2. J.-P. Amorij, S. M. Bal, J. A. Bouwstra, W. L. J. Hinrichs, A. Huckriede, G. F. A. Kersten, V. Saluja, B. Slütter, W. F. Tonnis and W. Jiskoot, Towards tailored vaccine delivery: needs, challenges and perspectives. *J. Control. Release* **161**, 363–376 (2012).
3. J. Rossert, N. Casadevall and K. U. Eckhardt, Anti-erythropoietin antibodies and pure red cell aplasia. *Am. Soc. Nephrol.* **15**, 398–406 (2004).
4. E. Koren, L. A. Zuckerman and A. R. Mire-Sluis, Immune responses to therapeutic proteins in humans-clinical significance, assessment and prediction. *Curr. Pharm. Biotechnol.* 3349–3360 (2002).
5. E. Koren, H. W. Smith, E. Shores, G. Shankar, D. Finco-Kent, B. Rup, Y. C. Barrett, V. Devanarayan, B. Gorovits, S. Gupta, T. Parish, V. Quarmby, M. Moxness, S. J. Swanson, G. Taniguchi, L. A. Zuckerman, C. C. Stebbins and A. Mire-Sluis, Recommendations on risk-based strategies for detection and characterization of antibodies against biotechnology products. *J. Immunol. Methods* 3331–3339 (2008).
6. FDA, Guidance for Industry Assay Development for Immunogenicity Testing of Therapeutic Proteins. *Federal Register* **74**, 63758 (2009).

7. W. Jiskoot, R. M. F. van Schie, M. G. Carstens and H. Schellekens, Immunological risk of injectable drug delivery systems. *Pharm. Res.* **26**, 1303–1314 (2009).
8. W. Jiskoot, G. F. A. Kersten and E. Mastrobattista, Vaccines, in *Pharmaceutical Biotechnology*, D. J. A. Crommelin, R. D. Sindelar and B. Meibohm (Eds.), 3rd ed. London: Informa Healthcare (2007), pp. 405–427.
9. H. M. Dintzis, R. Z. Dintzis and B. Vogelstein, Molecular determinants of immunogenicity: the immunon model of immune response. *Proc. Natl. Acad. Sci. USA* **73**, 3671–3675 (1976).
10. Q. Vos, A. Lees, Z. Q. Wu, C. M. Snapper and J. J. Mond, B cell activation by T cell-independent type 2 antigens as an integral part of the humoral immune response to pathogenic microorganisms. *Immunol. Rev.* **176**, 154–170 (2000).
11. C. R. Alving, R. L. Richards and A. A. Guirguis, Cholesterol-dependent human complement activation resulting in damage to liposomal model membranes. *J. Immunol.* **118**, 342–347 (1977).
12. C. R. Alving, Natural antibodies against phospholipids and liposomes in humans. *Biochem. Soc. Trans.* **12**, 342–344 (1984).
13. C. R. Alving and N. R. Wassef, Naturally occurring antibodies to cholesterol: a new theory of LDL cholesterol metabolism, *Immunol. Today* (1999).
14. C. R. Alving, Antibodies to lipids and liposomes: immunology and safety. *J. Liposome Res.* **16**, 157–166 (2006).
15. K. Shiraishi, M. Hamano, H. Ma, K. Kawano, Y. Maitani, T. Aoshi, K. Ishii and M. Yokoyama, Hydrophobic blocks of PEG-conjugates play a significant role in the accelerated blood clearance (ABC) phenomenon. *J. Control. Release* **165**, 183–190 (2013).
16. T. Ishihara, M. Takeda, H. Sakamoto, A. Kimoto, C. Kobayashi, N. Takasaki, K. Yuki, K. Tanaka, M. Takenaga, R. Igarashi, T. Maeda, N. Yamakawa, Y. Okamoto, M. Otsuka, T. Ishida, H. Kiwada, Y. Mizushima and T. Mizushima, Accelerated blood clearance phenomenon upon repeated injection of PEG-modified PLA-nanoparticles, *Pharm. Res.* **10**, 2270–2279 (2009).
17. T. Ishida and H. Kiwada, Accelerated blood clearance (ABC) phenomenon upon repeated injection of PEGylated liposomes. *Int. J. Pharm.* **354**, 56–62 (2008).
18. T. Ishida, M. Ichihara, X. Wang, K. Yamamoto, J. Kimura, E. Majima and H. Kiwada, Injection of PEGylated liposomes in rats elicits PEG-specific IgM, which is responsible for rapid elimination of a second dose of PEGylated liposomes. *J. Control. Release* **112**, 15–25 (2006).
19. T. Ishida, M. Ichihara, X. Wang and H. Kiwada, Spleen plays an important role in the induction of accelerated blood clearance of PEGylated liposomes. *J. Control. Release* **115**, 243–250 (2006).
20. T. Ishida, M. Harada, X. Y. Wang, M. Ichihara, K. Irimura and H. Kiwada, Accelerated blood clearance of PEGylated liposomes following preceding liposome injection: effects of lipid dose and PEG surface-density and chain length of the first-dose liposomes. *J. Control. Release* **105**, 305–317 (2005).
21. T. Ishida, T. Ichikawa, M. Ichihara, Y. Sadzuka and H. Kiwada, Effect of the physicochemical properties of initially injected liposomes on the clearance of subsequently injected PEGylated liposomes in mice. *J. Control. Release* **95**, 403–412 (2004).
22. T. Ishida, R. Maeda, M. Ichihara, K. Irimura and H. Kiwada, Accelerated clearance of PEGylated liposomes in rats after repeated injections, *J. Control. Release* 35–42 (2003).

23. T. Ishida, K. Masuda, T. Ichikawa, M. Ichihara, K. Irimura and H. Kiwada, Accelerated clearance of a second injection of PEGylated liposomes in mice. *Int. J. Pharm.* **255**, 167–174 (2003).
24. T. Tagami, K. Nakamura, T. Shimizu, T. Ishida and H. Kiwada, Effect of siRNA in PEG-coated siRNA-lipoplex on anti-PEG IgM production. *J. Control. Release* **137**, 234–240 (2009).
25. H. Koide, T. Asai, K. Hatanaka, Akai S, Ishii T, E. Kenjo, T. Ishida, H. Kiwada, H. Tsukada and N. Oku, T cell-independent B cell response is responsible for ABC phenomenon induced by repeated injection of PEGylated liposomes. *Int. J. Pharm.* **392**, 218–223 (2010).
26. Y. Zhao, L. Wang, M. Yan, Y. Ma, G. Zang, Z. She and Y. Deng, Repeated injection of PEGylated solid lipid nanoparticles induces accelerated blood clearance in mice and beagles. *Int. J. Nanomedicine* 7, 2891–900 (2012).
27. D. Mohanan, B. Slütter, M. Henriksen-Lacey, W. Jiskoot, J. A. Bouwstra, Y. Perrie, T. M. Kündig, B. Gander and P. Johansen, Administration routes affect the quality of immune responses: a cross-sectional evaluation of particulate antigen-delivery systems. *J. Control. Release* **147**, 342–349 (2010).
28. B. Slütter, S. M. Bal, Z. Ding, W. Jiskoot and J. A. Bouwstra, Adjuvant effect of cationic liposomes and CpG depends on administration route. *J. Control. Release* **154**, 123–130 (2011).
29. S. M. Bal, B. Slütter, R. J. Verheul, J. A. Bouwstra and W. Jiskoot, Adjuvanted, antigen loaded N-trimethyl chitosan nanoparticles for nasal and intradermal vaccination: adjuvant- and site-dependent immunogenicity in mice. *Eur. J. Pharm. Sci.* **45**, 475–481 (2012).
30. M. F. Bachmann and G. T. Jennings, Vaccine delivery: a matter of size, geometry, kinetics and molecular patterns. *Nat. Rev. Immunol.* **10**, 787–796 (2010).
31. N. C. Phillips and J. Dahman, Immunogenicity of immunoliposomes: reactivity against species-specific IgG and liposomal phospholipids. *Immunol. Lett.* **45**, 149–152 (1995).
32. J. A. Harding, C. M. Engbers, M. S. Newman, N. I. Goldstein and S. Zalipsky, Immunogenicity and pharmacokinetic attributes of poly (ethylene glycole)-grafted immunoliposomes. *Biochim. Biophys. Acta.* **1327**, 181–192 (1997).
33. A. R. Mire-Sluis, Y. C. Barrett, V. Devanarayan, E. Koren, H. Liu, M. Maia, T. Parish, t. G. Scot, G. Shankar, E. Shores, S. J. Swanson, G. Taniguchi, D. Wierda and L. A. Zuckerman, Recommendations for the design and optimization of immunoassays used in the detection of host antibodies against biotechnology products. *J. Immunol. Methods* **289**, 1–16 (2004).
34. CHMP, Committee for Medicinal Products: Concept paper on immunogenicity assessment of therapeutic proteins, *EMEA/CHMP/BMWP/246511/2005*, Vol. 2006, www.fda.gov/downloads/Drugs/GuidanceComplianceRegulatoryInformation/Guidances/UCM192750.pdf.
35. J. A. Lofgren, S. Dhandapani, J. J. Pennucci, C. M. Abbott, D. T. Mytych, A. Kaliyaperumal, S. J. Swanson and M. C. Mullenixl, Comparing ELISA and surface plasmon resonance for assessing clinical immunogenicity of panitumumab. *J. Immunol.* **178**, 7467–7472 (2007).
36. M. Liang, S. L. Klakamp, C. Funelas, H. Lu, B. Lam, C. Herl, A. Umble, A. W. Drake, P. Min, N. Ageyeva, R. Pasumarthi and L. K. Roskos, Detection of high- and

low-affinity antibodies against a human monoclonal antibody using various technology platforms. *Assay Drug Dev. Technol.* 655–662 (2007).

37. G. Shankar, V. Devanarayan, Y.-C. Barrett, R. Bowsher, D. Finco-Kent, M. Fiscella, B. Gorovits, S. Kirschner, M. Moxness, T. Parish, V. Quarmby, E. Shores, H. Smith, W. Smith, J. Zhong, L. Zuckerman and E. Koren, Recommendations for the validation of immunoassays used for detection of host antibodies against biotechnology products. *J. Pharm. Biomed. Anal.* **48**, 1267–1281 (2008).
38. B. Gorovits, Antidrug antibody assay validation: Industry survey results. *AAPS J.* **11**, 133–138 (2009).
39. V. Brinks, W. Jiskoot and H. Schellekens, Immunogenicity of therapeutic proteins: the use of animal models. *Pharm. Res.* **28**, 2379–2385 (2011).
40. R. J. Johnson and W. Jiskoot, Models for the evaluation of relative immunogenic potential of protein particles in biopharmaceutical formulations. *J. Pharm. Sci.* **101**, 3586–3592 (2012).
41. C. Foged, C. Arigita, A. Sundblad, W. Jiskoot, G. Storm and S. Frokjaer, Interaction of dendritic cells with antigen-containing liposomes: effect of bilayer composition. *Vaccine* **22**, 1903–1913 (2004).
42. M. Amidi, E. Mastrobattista, W. Jiskoot and W. Hennink, Chitosan-based delivery systems for protein therapeutics and antigens. *Adv. Drug Deliv. Rev.* **62**, 59–82 (2010).
43. S. Bal, B. Slütter, E. van Riet, A. Kruithof, Z. Ding, W. Jiskoot and J. A. Bouwstra, Efficient induction of immune responses through intradermal vaccination with N-trimethyl chitosan containing antigen formulations. *J. Control. Release* **142**, 374–383 (2010).
44. W. Jiskoot, T. W. Randolph, D. B. Volkin, C. R. Middaugh, C. Schöneich, G. Winter, W. Friess, D. J. A. Crommelin, F. John and J. F. Carpenter, Protein instability and immunogenicity: Roadblocks to clinical application of injectable protein delivery systems for sustained release. *J. Pharm. Sci.* **101**, 946–954 (2012).
45. S. Hermeling, D. J. A. Crommelin, H. Schellekens and W. Jiskoot, Structure-immunogenicity relationships of therapeutic proteins. *Pharm. Res.* **21**, 897–903 (2004).
46. M. M. C. Van Beers and M. Bardor, Minimizing immunogenicity of biopharmaceuticals by controlling critical quality attributes of proteins. *Biotechnol. J.* 7, 1473–1484 (2012).
47. S. Kontos and J. A. Hubbell, Drug development: longer-lived proteins. *Chem. Soc. Rev.* **41**, 2686–2695 (2012).
48. M. Leslie, The new view of complement. *Science* **337**, 1034–1037 (2012).
49. E. Fosse, T. E. Mollnes, A. O. Aasen, J. H. Trumpy and T. Stokke, Complement activation following multiple injuries. *Acta. Chir. Scand.* **153**, 325–330 (1987).
50. H. W. Lim, The complement system. Activation, modulation, and clinical relevance. *Dermatol. Clin.* **8**, 609–18 (1990).
51. V. Videm and T. E. Mollnes, Human complement activation by polygeline and dextran 70. *Scand. J. Immunol.* **39**, 314–320 (1994).
52. C. D. Krohn, O. Reikers, T. E. Mollnes and A. O. Aasen, Complement activation and release of interleukin-6 and tumour necrosis factor-alpha in drained and systemic blood after major orthopaedic surgery. *Eur. J. Surg.* **164**, 103–108 (1998).
53. E. Fosse, J. Pillgram-Larsen, J. L. Svennevig, C. Nordby, A. Skulberg, T. E. Mollnes and M. Abdelnoor, Complement activation in injured patients occurs immediately and is dependent on the severity of the trauma. *Injury* **29**, 509–514 (1998).

54. T. E. Mollnes, Therapeutic manipulation of the complement system. In *The Complement System: Novel Roles in Health and Disease*, J. Szebeni (Ed.), Boston: Kluwer (2004), pp. 483–516.
55. J. Szebeni, In *The Complement System: Novel Roles in Health and Disease*. Boston: Kluwer (2004), p. 512.
56. M. Basta, Ambivalent effect of immunoglobulins on the complement system: activation versus inhibition. *Mol. Immunol.* **45**, 4073–4079 (2008).
57. C. L. Harris and B. P. Morgan, The many faces of the membrane regulators of complement. In *The Complement System: Novel Roles in Health and Disease*, J. Szebeni (Ed.), Boston: Kluwer (2004).
58. J. Szebeni, The interaction of liposomes with the complement system, *Crit. Rev. Ther. Drug Carrier Syst.* **15**, 57–88 (1998).
59. J. Szebeni, Complement activation-related pseudoallergy: Mechanism of anaphylactoid reactions to drug carriers and radiocontrast agents. In *The Complement System: Novel Roles in Health and Disease*, J. Szebeni (Ed.), Boston: Kluwer (2004), pp. 399–440.
60. J. Szebeni, R. Bunger, L. Baranyi, P. Bedocs, M. Toth, L. Rosivall, Y. Barenholz and C. R. Alving, Animal models of complement-mediated hypersensitivity reactions to liposomes and other lipid-based nanoparticles, *J. Liposome Res.* **17**, 107–117 (2007).
61. J. Szebeni and Y. Barenholz, Complement activation, immunogenicity, and immune suppression as potential side effects of liposomes. In *Handbook of Harnessing Biomaterials in Nanomedicine: Preparation, Toxicity, and Applications*, D. Peer (Ed.), Singapore: Pan Standford Publishing (2012), pp. 309–334.
62. J. Szebeni, P. Bedőcs, R. Urbanics, R. Bunger, L. Rosivall, M. Tóth and Y. Barenholz, Prevention of infusion reactions to PEGylated liposomal doxorubicin via tachyphylaxis induction by placebo vesicles: a porcine model. *J. Control. Release* **160**, 382–387 (2012).
63. J. Szebeni, P. Bedőcs, Z. Rozsnyay, Z. Weiszhár, R. Urbanics, L. Rosivall, R. Cohen, O. Garbuzenko, G. Báthori, M. Tóth, R. Bünger and Y. Barenholz, Liposome-induced complement activation and related cardiopulmonary distress in pigs: factors promoting reactogenicity of Doxil and AmBisome, *Nanomedicine NBM* **8**, 176–184 (2012).
64. J. Szebeni, C. R. Alving, S. Savay, Y. Barenholz, A. Priev, D. Danino and Y. Talmon, Formation of complement-activating particles in aqueous solutions of Taxol: Possible role in hypersensitivity reactions , *Int. Immunopharmacol.* **1**, 721–735 (2001).
65. J. Szebeni, F. M. Muggia and C. R. Alving, Complement activation by Cremophor EL as a possible contributor to hypersensitivity to paclitaxel: an *in vitro* study. *J. Natl. Cancer Inst.* **90**, 300–306 (1998).
66. Z. Weiszhár, J. Czúcz, C. Révész, L. Rosivall, J. Szebeni and Z. Rozsnyay, Complement activation by polyethoxylated pharmaceutical surfactants: Cremophor-EL, Tween-80 and Tween-20. *Eur. J. Pharm. Pharmaceutics* **45**, 492–498 (2012).
67. S. Qiu, Z. Liu, L. Hou, Y. Li, J. Wang, H. Wang, W. Du, W. Wang, Y. Qin and Z. Liu, Complement activation associated with polysorbate 80 in beagle dogs. *Int. Immunopharmacol.* **15**, 144–149 (2013).
68. P. Lieberman, Anaphylactoid reactions to radiocontrast material. *Clin. Rev. Allergy* **9**, 319–338 (1991).
69. J. Szebeni, Complement activation-related pseudoallergy caused by liposomes, micellar carriers of intravenous drugs and radiocontrast agents. *Crit. Rev. Ther. Drug Carrier. Syst.* **18**, 567–606 (2001).

70. J. Szebeni, Hypersensitivity reactions to radiocontrast media: the role of complement activation. *Curr. Allergy Asthma Rep.* **4**, 25–30 (2004).
71. S. M. Moghimi and J. Szebeni, Stealth liposomes and long circulating nanoparticles: critical issues in pharmacokinetics, opsonization and protein-binding properties. *Prog. Lipid Res.* **42**, 463–478 (2003).
72. S. M. Moghimi and I. Hamad, Liposome-mediated triggering of complement cascade. *J. Liposome Res.* **18**, 195–209 (2008).
73. S. M. Moghimi, A. J. Andersen, S. H. Hashem, B. Lettiero, D. Ahmadvand, A. C. Hunter, T. L. Andresen, I. Hamad and J. Szebeni, Complement activation cascade triggered by PEG-PL engineered nanomedicines and carbon nanotubes: the challenges ahead. *J. Control. Release* **146**, 175–181 (2010).
74. A. J. Andersen, S. H. Hashemil, T. L. Andresen, A. C. Hunter and S. M. Moghimi, Complement: alive and kicking nanomedicines. *J. Biomed. Nanotechnol.* **5**, 364–372 (2009).
75. A. J. Andersen, J. T. Robinson, H. Dai, A. C. Hunter, T. L. Andresen and S. M. Moghimi, Single-Walled Carbon Nanotube Surface Control of Complement Recognition and Activation. *ACS Nano* **2013** (2013).
76. D. J. Naisbitt, M. Pirmohamed and B. K. Park, Immunopharmacology of hypersensitivity reactions to drugs. *Curr. Allergy and Asthma Rep.* **3**, 22–29 (2003).
77. A. Cheifetz and L. Mayer, Monoclonal antibodies, immunogenicity, and associated infusion reactions. *Mt. Sinai. J. Med.* **72**, 250–256 (2005).
78. E. R. Gomes and P. Demoly, Epidemiology of hypersensitivity drug reactions. *Curr. Opin. Allergy Clin. Immunol.* **5**, 309–316 (2005).
79. H. J. Lenz, Management and preparedness for infusion and hypersensitivity reactions. *Oncologist* **12**, 601–609 (2007).
80. W. H. Vogel, Infusion reactions: diagnosis, assessment, and management, *Clinical J. Oncol. Nurs.* **14**, E10–E21 (2010).
81. P. A. Demoor, Y. Matusov, C. Kelly, S. Kolan, L. Barnachea and L. A. Bazhenova, A retrospective review of the frequency and nature of acute hypersensitivity reactions at a medium-sized infusion center: comparison to reported values and inconsistencies found in literature. *J. Cancer* **10**, 153–164 (2011).
82. J. Szebeni, Hemocompatibility testing for nanomedicines and biologicals: predictive assays for complement mediated infusion reactions. *Eur. J. Nanomed.* **5**, 33–53 (2012).
83. A. D. Bangham, Liposomes: the Babraham connection. *Chem. Phys. Lipids* **64**, 275–285 (1993).
84. D. W. Deamer, Origins of life: How leaky were primitive cells? *Nature* **454**(7200), 37–38 (2008).
85. D. Deamer, On the origin of systems: Systems biology, synthetic biology and the origin of life. *EMBO Rep.* **10**(Suppl 1), S1–S4 (2009).
86. T. E. Mollnes and M. Kirschfink, Strategies of therapeutic complement inhibition. *Mol. Immunol.* **43**, 107–121 (2006).
87. I. D. Campbell, M. Baron, A. J. Day, R. B. Sim, D. G. Norman and P. N. Barlow, Three-dimensional structure of a complement control protein module in solution. *J. Mol. Biol.* **219**, 717–725 (1991).
88. J. Szebeni, F. Muggia, G. Gabizon and Y. Barenholz, Activation of complement by therapeutic liposomes and other lipid excipient-based therapeutic products: Prediction and prevention. *Adv. Drug Deliv. Rev.* **63**, 1020–1030 (2011).

89. C. R. Alving, S. C. Kinsky, J. A. Haxby and C. B. Kinsky, Antibody binding and complement fixation by a liposomal model membrane. *Biochemistry* **8**, 1582–1587 (1969).
90. C. R. Alving and G. M. Swartz, Antibodies to cholesterol, cholesterol conjugates, and liposomes: implications for atherosclerosis and autoimmunity. *Crit. Rev. Immunol.* **10**, 441–453 (1991).
91. Y. Sherer and Y. Shoenfeld, Antiphospholipid syndrome, antiphospholipid antibodies, and atherosclerosis. *Curr. Atheroscler. Rep.* 328–333 (2001).
92. A. Chonn, P. R. Cullis and D. V. Devine, The role of surface charge in the activation of the classical and alternative pathways of complement by liposomes. *J. Immunol.* **146**, 4234–4241 (1991).
93. C. Plank, M. X. Tang, A. R. Wolfe and F. C. J. Szoka, Branched cationic peptides for gene delivery: role of type and number of cationic residues in formation and *in vitro* activity of DNA polyplexes. *Hum. Gene Ther.* **10**, 319–332 (1999).
94. I. Hamad, A. C. Hunter, K. J. Rutt, Z. Liu, H. Dai and S. M. Moghimi, Complement activation by PEGylated single-walled carbon nanotubes is independent of C1q and alternative pathway turnover. *Mol. Immunol.* **45**, 3797–3803 (2008).
95. I. Hamad, O. Al-Hanbali, A. C. Hunter, K. J. Rutt, T. L. Andresen and S. M. Moghimi, Distinct polymer architecture mediates switching of complement activation pathways at the nanosphere-serum interface: implications for stealth nanoparticle engineering. *ACS Nano* **4**, 6629–6638 (2010).
96. O. Merkel, R. Urbanics, P. Bedőcs, Z. Rozsnyay, L. Rosivall, M. Toth, T. Kissel and J. Szebeni, *In vitro* and *in vivo* complement activation and related anaphylactic effects associated with polyethylenimine and polyethylenimine-graft-poly(ethylene glycol) block copolymers. *Biomaterials* **32**, 4936–4942 (2011).
97. J. K. Aronson and R. E. Ferner, Joining the DoTS: new approach to classifying adverse drug reactions, *Br. Med. J.* **327**, 1222–1225 (2003).
98. T. Zuberbier, Pseudoallergy or nonallergic hypersensitivity. *J. Allergy* **54**, 397–398 (1999).
99. P. Demoly, R. Kropf, A. Bircher and W. J. Pichler, Drug hypersensitivity: questionnaire. EAACI interest group on drug hypersensitivity. *Allergy* **54**, 999–1003 (1999).
100. S. G. O. Johansson, T. Bieber, R. Dahl, P. S. Friedmann, B. Q. Lanier, R. F. Lockey, C. Motala, J. A. Ortega Martell, T. Platts-Mills, J. Ring, F. P. V. C. Thien and H. C. Williams, Revised nomenclature for allergy for global use: Report of the Nomenclature Review Committee of the World Allergy Organization. *J. Allergy Clin. Immunol.* **113**, 832–836 (2003).
101. W. J. Pichler, D. J. Naisbitt and B. K. Park, Immune pathomechanism of drug hypersensitivity reactions. *J. Allergy Clin. Immunol.* **127**(Suppl), S74–S78 (2011).
102. J. Szebeni, J. L. Fontana, N. M. Wassef, P. D. Mongan, D. S. Morse, D. E. Dobbins, G. L. Stahl, R. Bünger and C. R. Alving, Hemodynamic changes induced by liposomes and liposome-encapsulated hemoglobin in pigs: a model for pseudo-allergic cardiopulmonary reactions to liposomes. Role of complement and inhibition by soluble CR1 and anti-C5a antibody. *Circulation* **99**, 2302–2309 (1999).
103. J. Szebeni, P. Bedőcs, D. Csukás, L. Rosivall, R. Bunger and R. Urbanics, A porcine model of complement-mediated infusion reactions to drug carrier nanosystems and other medicines. *Adv. Drug Deliv. Rev.* **64**, 1706–1716 (2012).
104. J. Szebeni, Complement activation-related pseudoallergy: a new class of drug-induced immune toxicity. *Toxicology* **216**, 106–121 (2005).

105. R. R. A. Coombs and P. G. H. Gell, Classification of allergic reactions responsible for drug hypersensitivity reactions. In *Clinical Aspects of Immunology*, R. R. A. Coombs and P. G. H. Gell (Eds.), 2nd edn. Philadelphia: F. A. Davis Company, (1968), pp. 575–596.
106. J. Descotes and G. Choquet-Kastylevsky, Gell and Coombs's classification: is it still valid? *Toxicology* **158**, 43–49 (2001).
107. A. Chanan-Khan, J. Szebeni, S. Savay, L. Liebes, N. M. Rafique, C. R. Alving and F. M. Muggia, Complement activation following first exposure to pegylated liposomal doxorubicin (Doxil): possible role in hypersensitivity reactions. *Ann. Oncol.* **14**, 1430–1437 (2003).
108. J. Szebeni, L. Baranyi, S. Sávay, M. Bodó, J. Milosevits, C. R. Alving and R. Bünger, Complement activation-related cardiac anaphylaxis in pigs: role of C5a anaphylatoxin and adenosine in liposome-induced abnormalities in ECG and heart function. *Am. J. Physiol. Heart Circ. Physiol.* **290**, H1050–H1058 (2006).
109. J. Szebeni, B. Baranyi, S. Savay, M. Bodo, D. S. Morse, M. Basta, G. L. Stahl, R. Bunger and C. R. Alving, Liposome-induced pulmonary hypertension: Properties and mechanism of a complement-mediated pseudoallergic reaction. *Am. J. Physiol. Heart Circ. Physiol.* **279**, H1319–H1328 (2000).
110. J. Lazarou, B. H. Pomeranz and P. N. Corey, Incidence of adverse drug reactions in hospitalized patients. A meta-analysis of prospective studies. *JAMA* **279**, 1200–1205 (1998).
111. P. Demoly, R. Kropf, A. Bircher and W. J. Pichler, Drug hypersensitivity: questionnaire. EAACI interest group on drug hypersensitivity. *Allergy* **54**, 999–1003 (1999).
112. C. Svensson, Commentary: Drug Hypersensitivity — Where Do We Stand? *AAPS J.* **8**, E236 (2006).
113. A. A. Gabizon and F. M. Muggia, Initial clinical evaluation of Pegylated Liposomal Doxorubicin in solid tumors. In *Long-Circulating Liposomes: Old Drugs, New Therapeutics*, M. Woodle and G. Storm (Eds.), Austin: Landes Bioscience, (1998), pp. 165–174.
114. J. Szebeni, Complement activation-related pseudoallergy: Mechanism of anaphylactoid reactions to drug carriers and radiocontrast agents. In *The Complement System: Novel Roles in Health and Disease*, J. Szebeni (Ed.), Boston: Kluwer, (2004), pp. 399–440.
115. I. Hamad, A. C. Hunter, J. Szebeni and S. M. Moghimi, Poly(ethylene glycol)s generate complement activation products in human serum through increased alternative pathway turnover and a MASP-2-dependent process, *Mol Immunol.* **46**, 225–232 (2008).
116. G. C. Winkler, Pulmonary intravascular macrophages in domestic animal species: review of structural and functional properties. *Am. J. Anat.* **181**, 217–234 (1988).
117. A. Erdei, V. Köhler, H. Schéfer and R. Burger, Macrophage-bound C3 fragments as adhesion molecules modulate presentation of exogenous antigens. *Immunobiology* **185**, 314–326 (1992).
118. Y. Ara, T. Saito, T. Takagi, E. Hagiwara, Y. Miyagi, M. Sugiyama, S. Kawamoto, N. Ishii, T. Yoshida, D. Hanashi, T. Koshino, H. Okada and K. Okuda , Zymosan enhances the immune response to DNA vaccine for human immunodeficiency virus type-1 through the activation of complement system. *Immunology* **103**, 98–105 (2001).
119. L. Clemenza, F. Dieli, M. Cicardi and A. Salerno, Research on complement: old issues revisited and a novel sphere of influence. *Trends Immunol.* **24**, 292–296 (2003).

120. M. L. Roberts, A. T. Luxembourg and N. R. Cooper, Epstein-Barr virus binding to CD21, the virus receptor, activates resting B cells via an intracellular pathway that is linked to B cell infection. *J. Gen. Virol.* 77(Pt 12), 3077–3085 (1996).
121. T. Roberts and E. C. Snow, Cutting edge: recruitment of the CD19/CD21 coreceptor to B cell antigen receptor is required for antigen-mediated expression of Bcl-2 by resting and cycling hen egg lysozyme transgenic B cells. *J. Immunol.* **162**, 4377–4380 (1999).
122. M. Carroll, Role of complement receptors CD21/CD35 in B lymphocyte activation and survival. *Curr. Top Microbiol. Immunol.* **246**, 63–69 (1999).
123. M. C. Carroll, CD21/CD35 in B cell activation. *Semin. Immunol.* **10**, 279–286 (1998).
124. M. C. Carroll, The role of complement and complement receptors in induction and regulation of immunity. *Annu. Rev. Immunol.* **16**, 545–568 (1998).
125. M. C. Carroll and A. P. Prodeus, Linkages of innate and adaptive immunity. *Curr. Opin. Immunol.* **10**, 36–40 (1998).
126. H. Molina, V. M. Holers, B. Li, Y. Fung, S. Mariathasan, J. Goellner, J. Strauss-Schoenberger, R. W. Karr and D. D. Chaplin, Markedly impaired humoral immune response in mice deficient in complement receptors 1 and 2. *Proc. Natl. Acad. Sci. USA* **93**, 3357–3361 (1996).
127. J. M. Ahearn, M. B. Fischer, D. Croix, S. Goerg, M. Ma, J. Xia, X. Zhou, R. G. Howard, T. L. Rothstein and M. C. Carroll, Disruption of the Cr2 locus results in a reduction in B-1a cells and in an impaired B cell response to T-dependent antigen. *Immunity* **4**, 251–262 (1996).
128. A. Erdei, E. Molnár, E. Csomor, Z. Bajtay and J. Prechl, Coordination of Adaptive Immune Responses by C3. In *The Complement System: Novel Roles in Health and Disease*, J. Szebeni (Ed.), Boston: Kluwer, (2004), pp. 77–96.
129. O. F. Ramos, I. Algarra, G. Sarmay, E. Yefenof, J. Gergely and E. Klein, Lymphocytes stimulated by allogeneic B cell lines cleave the third component of complement and fix C3 fragments. Their nonspecific lytic capacity is elevated against complement receptor type 2-carrying targets. *J. Immunol.* **142**, 217–223 (1989).
130. V. M. l. Holers, Complement receptors and the shaping of natural antibody repertoire. *Springer Semin. Immunopathol.* **26**, 405–423 (2005).
131. F. Bacle, N. Haeffner-Cavaillon, M. Laude, C. Couturier and M. D. Kazatchkine, Induction of IL-1 release through stimulation of the C3b/C4b complement receptor type one (CR1, CD35) on human monocytes. *J. Immunol.* **144**, 147–152 (1990).
132. Y. Yoshida, K. Kang, M. Berger, G. Chen, A. C. Gilliam, A. Moser, L. Wu, C. Hammerberg and K. D. Cooper, Monocyte induction of IL-10 and down-regulation of IL-12 by iC3b deposited in ultraviolet-exposed human skin. *J. Immunol.* **161**, 5873–5879 (1998).
133. T. Marth and B. L. Kelsall, Regulation of interleukin-12 by complement receptor 3 signaling. *J. Exp. Med.* **185**, 1987–1995 (1997).
134. S. M. Yentis, R. P. Gooding and P. G. Riches, The effects of IgG and immune complexes on the endotoxin-induced cytokine response. *Cytokine* **6**, 247–254 (1994).
135. H. Koide, T. Asai, K. Hatanaka, T. Urakami, T. Ishii, E. Kenjo, M. Nishihara, M. Yokoyama, T. Ishida, H. Kiwada and N. Oku, Particle size-dependent triggering of accelerated blood clearance phenomenon. *Int. J. Pharm.* **362**, 197–200 (2008).

136. P. E. Belchetz, J. C. Crawley, I. P. Braidman and G. Gregoriadis, Treatment of Gaucher's disease with liposome-entrapped glucocerebroside: beta-glucosidase. *Lancet* **8029**, 116–117 (1977).
137. G. Storm, M. T. ten Kate, P. K. Working and A. Bakker-Woudenberg, Doxorubicin entrapped in sterically stabilized liposomes: effects on bacterial blood clearance capacity of the mononuclear phagocyte system. *Clin. Cancer Res.* **4**, 111 (1998).
138. A. Gabizon, D. Tzemach, L. Mak, M. Bronstein and A. T. Horowitz, Dose dependency of pharmacokinetics and therapeutic efficacy of pegylated liposomal Doxorubicin (DOXIL) in murine models. *J. Drug Target.* **10**, 539–548 (2002).
139. A. Gabizon, R. Isacson, O. Rosengarten, D. Tzemach, H. Shmeeda and R. Sapir, An open-label study to evaluate dose and cycle dependence of the pharmacokinetics of pegylated liposomal doxorubicin. *Cancer Chemother. Pharmacol.* **61**, 695–702 (2007).
140. D. S. Alberts, P. Y. liu, S. P. Wilczynski, M. C. Clouser, M. A. Lopez, D. P. Michelin, V. J. Lanzotti and M. Markman, Randomized trial of pegylated liposomal doxorubicin (PLD) plus carboplatin versus carboplatin in platinum-sensitive (PS) patients with recurrent ovarian or peritoneal carcinoma after failure of initial platinum-based chemotherapy (South West Oncology group protocol SO200). *Gynecol. Oncol.* **108**, 90–94 (2008).
141. EMEA, Note for guidance on repeated dose toxicity, *CPMP/SWP/1042/99,* Vol. 2000, pp.
142. K. L. Hastings, Implications of the new FDA/CDER immunotoxicology guidance for drugs. *Int. Immunopharmacol.* **11**, 1613–1618 (2002).
143. I. H. Dujmovic, Comparison of two guideleines on immunotoxicology testing of medicinal products. *Arh. Hig. Rada Toksikol.* **56**, 265–268 (2005).

Chapter A4

Toxicology of Mesoporous Silica Particles and Their Uses in Nanomedicine

Xin Xia, Natalia Kupferschmidt† and Alfonso E. Garcia-Bennett*,‡*

**Department of Materials and Environmental Chemistry, Arrhenius Laboratory, Stockholm University, 10691 Stockholm, Sweden*

†Department of Engineering Sciences, Uppsala University SE-751 21 Uppsala, Sweden

‡alfonso@mmk.su.se

Abstract

Nanomaterials are being explored intensely for applications in pharmaceutical sciences and nanomedicine. The same properties of these materials that make them interesting for the development of novel excipients, therapies and diagnostics are those which often make them potentially harmful. A thorough investigation of the interaction of any new nanomaterial with the immune system, coupled with both generic and specific toxicological information must hence be conducted. However, in many instances common toxicological assays and endpoints are not applicable and may need revising. Here, we highlight some of these barriers using mesoporous silica materials as an example and suggest some important material properties to be taken into account when performing a full toxicological appraisal.

1. Introduction

Nanomedicine is the application of nanotechnology via the use of synthetic nanoscale or nanostructured materials, specifically for medical applications or research.[1–3] Whilst the synthesis and design of nanoparticles has been an ongoing pursuit of research and industrial laboratories for at least 100 years, the dedicated and deliberate use of nanomaterials within medicine is a relatively new field and as a result there are so far few standard procedures to evaluate the toxicological and immune-modulatory properties of emerging nanomaterials with potential applications in medicine.[4–6]

It is important to note that nanomaterials are in the same size range as many biological structures and hence the properties which make nanomaterials interesting for biomedical applications, as for example their high surface to volume ratio, increased reactivity or agglomeration, may also make them potentially harmful. For example, synthetic nanoparticles with various compositions (e.g. fullerenes, carbon nanotubes (CNT) and quantum dots) have been shown to generate reactive oxygen species (ROS) and cause oxidative stress,[7] damaging cells by peroxidizing lipids, altering proteins, disrupting DNA, interfering with signaling functions and modulating gene transcription.[8,9] As commercial interest in industrial applications requiring large-scale production of nanomaterials increases, as for example in the preparation of petrochemical catalysts through sol–gel routes, so does the risk of exposure both for workers in production plants and to handlers of the products. Maynard *et al.* performed laboratory exposure studies concluding that after sufficient agitation, unrefined CNT released as fine particles into the air, even though concentrations generated while handling material were very low (below 53 $\mu g/m^3$).[10] A large effort is hence being devoted to analyze the exposure and risk of exposure to biological systems, very often *de novo,* as new variants of nanoparticles are being discovered, engineered or mass produced. These studies must include their toxicological properties as well as their mode of action under a variety of general contexts and specific administration routes. The interaction with the immune system which is specialized in the recognition and defense against foreign entities such as microorganisms and particles, is central in the efforts to characterize nanomaterials for their toxicological understanding and for their use nanomedicine.[11] Organic based liposomes, dendrimers or other polymeric nanomaterials are being designed as therapeutics agents within drug development, as drug delivery vehicles, as functional excipients and as adjuvants.[12–15] The possibilities of extending the current *state-of-the-art* in diagnostic devices by incorporating both therapeutic and diagnostic schemes within a single nanoparticle platform have been reported and exciting.[16–19]

Amongst inorganic nanoparticles mesoporous silicon dioxide particles with amorphous (non-crystalline) compositions offer some general advantages of both organic and inorganic systems, as will be discussed in the following sections. The number of reported publications using mesoporous

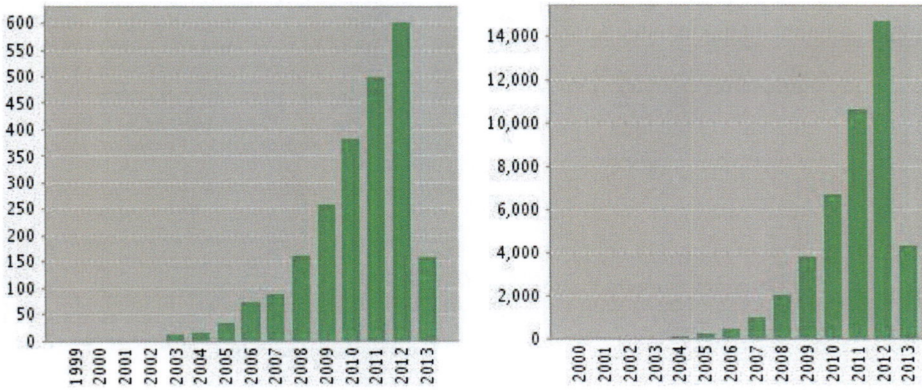

Fig. 1. Number of reported publications and citations for studies of mesoporous materials in the field of drug delivery.
Source: Thomson Reuters, ISI Web of Knowledge, search parameter: Mesoporous, refined for Drug Delivery.

materials as a platform for pharmaceutical drug delivery has increased from 2 per year in 2002 to 600 per year in 2012 (Fig. 1). Mesoporous materials (with pores in the mesoscale, or 2–50 nm) were discovered early during 1990s and most early research focused on their synthesis and catalytic properties.[20–24] Whilst mesoporous silica materials are deemed useful catalyst supports for a variety of fine chemical process they are characterized by a poor hydrothermal stability arising from their easily hydrolyzed amorphous walls, preventing their use on high temperature process such as hydrocracking or steam cracking petrochemical process.[25] The specificity arising from their sharp pore size distributions was hence investigated for other industrial applications such as gas adsorption, scavenging of reaction impurities in catalysis, chromatographic separation, as sensors; and eventually in the pharmaceutical industry as drug delivery vehicles.[26–30] Colloidal non-porous silicon dioxide has a long standing history as a pharmaceutical excipient. It is industrially used even in nanoparticle form, as an anticaking agent in pharmaceutical, food and cosmetic formulations.[31,32] Mesoporous silica, although possessing very similar compositions as materials described in the pharmacopeia, demonstrate different fundamental properties than their non-porous counterparts, such as their extremely high surfaces areas (almost an order of magnitude higher and often above 1000 m^2/g) or controllable and sharp pore size distributions. This alone poses many unavoidable questions, for example: how will these interact with the immune system in comparison to conventional colloidal silica particles? ; how will systemic adsorption and opsonization of the particles be affected by their large number of surface silanol (Si-OH) groups? For any particular family of nanomaterials a large collection of physico-chemical properties can be designed and manufactured, as is the case of mesoporous silica particles. The functionalization of their large surfaces via the use of silylating agents has been widely reported through the use of organosilanes.[33] Hydrophobic mesoporous silica as well as those presenting –COOH, –SH, $-NH_2$, or other functionalities have been prepared and allow to fine tune the properties of mesoporous silica particles for particular applications. Hence, how to tackle with the complexity of studying a myriad of surface properties? These are common themes in nanoparticle toxicology.[35]

Common knowledge in toxicology states that *there is no risk without exposure* and hence understanding the routes of nanoparticle exposure to humans becomes evidently important. Clearly the exposure and related potential risk of a nanoparticle will differ depending on the administration route and the potential use. If applied orally and not degraded by the acidic conditions of the stomach, particles will need to pass through several cell layers of the intestinal epithelium before reaching the blood stream and becoming systemic. Whilst particles injected into the bloodstream will start interacting directly with the organism affecting blood cells, blood vessel walls, organs, etc. The relation between exposed dose and toxicological implications is additionally often ignored when quoting research results, primarily by the non-scientific media. For example, titanium dioxide particles are used in numerous products as UV-protecting agents in sun screens. Its uptake properties via the skin have been studied in depth, in various large project consortia, concluding a lack of dermal penetration through passive diffusion.[36] Titanium dioxide micro- and nanoscale particles are approved food additives (E-171) and concerns regarding the potential harmful effects if they are internalized into the body via the gastrointestinal tract have been unfounded.[37] Studies on relevant gastrointestinal scenarios showed no important acute effects after oral administration of TiO_2 to mice, showing slight adverse effects at single doses of 5 g/kg body weight. These doses are comparable to those of common table salt, with an LD50 value for oral administration of 4 g/kg in mice or a fatal dose for humans between 0.75–3 g/kg (45–420 g for a 60 kg person).[38] All of this data has not prevented an online debate on the safety of TiO_2 particles, as a simple search will show.

This chapter attempts to use mesoporous silica particles, as a suitable example to highlight the ongoing effort to apply these versatile materials within nanomedicine applications by academic, industrial and regulatory actors. We start by putting the novelty of these materials in the context of their wider family of silica inorganic solids, describing some of the differences and why they are of interest to nanomedicine. We highlight some of the seminal publications and toxicological studies conducted on the materials followed by some suggestions on how the field may progress in order to get closer to real life clinical applications in therapeutic, excipient or diagnostic use of these nanostructured materials.

2. Silicon Dioxide

Silicon is the second most abundant element in the earth's crust, after oxygen. The oxide of silicon is water-soluble and is naturally occurring in many forms with different properties and characteristics, for example natural zeolites and diatomaceous earths. Some forms of silica are essential nutrients, obtained from dietary sources such as drinking water, cereals and plant fiber.[39] Silicon dioxide, hereon termed silica, is often found in hair, skin and nail supplements and a correlation between the maintenance of these tissues and the intake of dietary forms of silica has been found.[40] Silica is essential for collagen formation and vital for healthy bones, cartilage and other tissue (including ligaments, tendons, arteries, the aorta and trachea).[41] It is well-understood that low levels of the elemental silicon in old age contribute to conditions such as wear and tear on joints, osteoporosis, wrinkled skin and hardening of the arteries.

Silica displays a wide range of properties depending on its crystalline structure, composition, degree of impurities and its preparation. It is widely used in large volumes many industries; as a catalyst in the form of crystalline porous silica in the petrochemical industry (as synthetic and natural zeolites); as a paint additive in colloidal amorphous form (non-crystalline form) to promote the viscosity of paints; as a bleaching agent in the paper and pulp

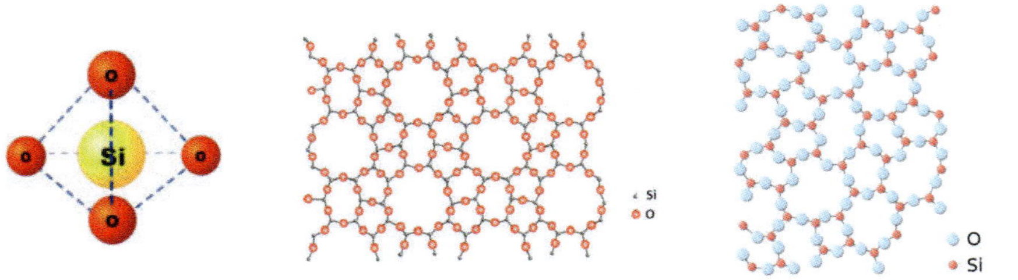

Silicon Oxide tetrahedral unit | *The structure of a crystalline zeolite* | *Amorphous silica, note that there is no long range ordering of silicon atoms*

Fig. 2. Some of the basic structural arrangements of silica. Despite all having SiO_4^{4-} building units (right), silica may display crystalline (middle) and amorphous, atomic scale ordering (left).

industry; as a moisture adsorbent in packaging; and as a colorant/pigment in many health-care and cosmetic products.[42] Silica in the form of doped silicates is additionally the main component of glass, a material central to society and whose optical and thermal properties have been exploited since 3500 BC.[43]

2.1. *Basic classification of nanostructured silica*

Silica owes its many properties from hard and robust industrial solid to a soft biocompatible material, to its ability to form flexible tetrahedral SiO_4^{4-} units which polymerize to form a varied array of crystalline and amorphous structures (Fig. 1). One may classify silica and silicates (including doped mineral silicates) as crystalline or amorphous, i.e. displaying long-range order at the atomic scale.[44] For the purpose of their applications in nanomedicine one may additionally classify silica into porous and non-porous forms. Porous amorphous silica is here defined as a material and capable of absorbing a certain amount of nitrogen gas above a level of 50 cm^3/g (as measurement by nitrogen sorption experiments). Amorphous silica materials absorbing nitrogen below 50 cm^3/g can be considered to be non-porous though they may still possess relatively high surface areas, and may be otherwise referred to as colloidal silicas. Other terms typically used for non-porous colloidal silica include; precipitated silicon dioxide, silica gel, hydrous silica, hydrated silicic acid, polysilicic acid gel and E-551 (food additive classification).[31] They are generally described under Chemical Abstract Service (CAS) numbers: 7631-86-9 (silica), 112945-52-5 (silica, amorphous, fumed, crystalline-free) and 112926-00-8 (Silica gel and precipitated silica, crystalline-free). Unfortunately, at the moment the CAS classification does not take into account the textural properties of these materials and does not help in identifying which properties, or synthesis routes the different types of silica materials have undergone. Hence mesoporous silica particles may be classed as CAS number 7631-86-9.

2.2. *Silica in medicine*

There is 1–2 g of silicon present in the body. Although there are previous reports of silicon and silica as useful nutrients it was not until the confirmation of silicon as an essential element for bone formation that it was considered important as a dietary supplement.[45] During the 1970s, animal studies showed that dietary silicon deficiency produces defects in connective and skeletal tissues and that silicates are at the mineralization front of growing bone structures. Subsequent reports made the connection between inadequate dietary intake of silicon and atherosclerosis and hypertension, as well as some bone disorders and the ageing process.[46,47]

The use of silica in the form of Bioglass® was pioneered in the 1970s from collaborative research between medical surgeons and material scientist, for applications in bone regeneration and implants. The first devices cleared were used to treat conductive hearing loss by replacing the bones of the middle ear.[48] This development led to a rapid interest in the use of silica materials and ceramics in medicine and surgery, providing a cellular and molecular basis for development of third-generation biomaterials for *in situ* tissue regeneration and repair, with minimally invasive surgery, as well as an example of a successful cross-disciplinary collaborations between material scientists and medical professionals.

The worldwide production of silica in 1992 was estimated at ~100,000 metric tons (for pyrogenic amorphous silica), ~800,000 metric tons (for precipitated amorphous silica), ~115,000 metric tons (silica gels and sols). The European consumption in the year 2000 (including imports and excluding exports) was ~408,500 metric tons with 368,000 metric tons constituting synthetic amorphous silica (SAS) and 40,450 metric tons for other silicates.[49] In 2002, the overall consumption amounted to ~481,050 metric tons. In consumer products silica is used to thicken pastes and ointments, to maintain flow properties in powder products, as a carrier for fragrances or flavors and is present in cosmetics (especially toothpaste). Pharmaceuticals and food, account for most of this production. They are also used in animal feed, rubber and silicones (as fillers), paints (as pigments), lacquers (as flattening agents) and plastics (to prevent plastic films from sticking together and for thixotropic control).[49] Amorphous colloidal silica is registered as a biocide in the EU and United States has been successfully employed against juvenile and adult store-product pests, predominantly exerting lethal activity on juvenile and adult forms by sorption of the cuticular lipid layer, thus causing dehydration of the insects. Silicon dioxide was registered as a pesticide as far back as 1956.[50]

2.3. *Brief history of mesoporous silica*

Mesoporous silica particles are porous amorphous materials that are distinguished by having pore size distributions in the order between 2 nm and 50 nm (the mesoscale range). They are high surface area materials and in general the term refers to those porous amorphous silica materials that are prepared through sol–gel precipitation of silica precursors (very often tetraethyl orthosilicate, TEOS) in the presence of surfactant micelles as the template. By control of the synthesis conditions, such as pH, surfactant concentration, temperature and addition of co-solvents; a favorable charge matching interaction can be achieved between the surfactant head-group and the hydrolyzed silica species. This allows for the growth silica particles with *ordered* mesoscale periodicity. The resulting amorphous silica material may be calcined or solvent extracted in order to remove the surfactant micelles resulting in white colored particles possessing well-defined ordered arrays of pores. The discovery of ordered mesoporous materials was reported in 1992 by researchers at the Mobil Corporation, in the United States,[20] with the preparation of MCM-41 material (Mobil Composition of Matter) templated by the cationic amphiphilic surfactant cetyltrimethyl ammonium bromide, CTAB (Fig. 2). This was done in an alkaline solution, promoting the interaction of the cationic head-group with the negatively charged silica species. After removal of the surfactant through calcination the particles produced were compositionally the same as other SAS, but with hexagonally ordered and narrow pore sizes distributions with large specific surface areas and pore volumes (typically above 0.5 cm^3/g) as measured by nitrogen sorption experiments.[51,52]

Although the Mobil research group of are generally acknowledged in the scientific community as the first to synthesize ordered mesoporous silica particles, another group led by Kuroda *et al.* (Waseda University, Japan) synthesized a regularly ordered pore system composed of amorphous walls and their publication date is prior to the first Mobil Corporation patent filing date.[52–54] The mechanism proposed by Kuroda *et al.* is difficult to generalize in order

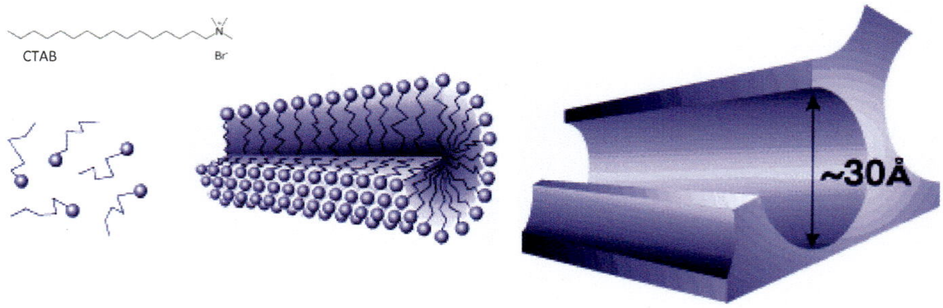

Fig. 3. The templating of MCM-41 through silica condensation around aggregations of organic surfactant molecules and the resulting pore. Surfactant molecules (e.g. CTAB, left) self-assembled in aqueous conditions to form hexagonal rod type micelles (middle). Addition of a silica source results in a charge matching interaction between the head-groups of the surfactant and silica oligomers, which create a wall around the organic micelle. Calcination or solvent extraction of the micelles after silica condensation results in a porous replica of the micelle (right).

to prepare other meso-structures, as the procedure relies on the intercalation of silicate sheets by surfactants known as *folded sheet materials* (FSM-*n*).[55] Instead Kresge *et al.* described a general pathway that has been extended to other material compositions and mesopore structures, showing the versatility of their surfactant templating approach. Incidentally, there is a prior patent, filed in 1971, disclosing the synthesis of a material comparable to mesoporous MCM-41, but whose researchers did not foresee the full context of the material produced.[56]

Since the 1990s, numerous synthesis protocols have been reported for the preparation of mesoporous silica with varying structures; cubic, hexagonal and tetragonal pore arrangements; two-dimensional and three-dimensional (3d) pore connectivities; caged-type and cylindrical pores. Literally all family of surfactants have been used as templates for mesoporous silica materials, including neutral polymeric, anionic and zwitterionic surfactants.[22,57,58] In 2008, the first preparation of ordered mesoporous silica materials without the use of surfactants as templates was reported.[59]

The synthesis relies on the use of folic acid as a self-assembling organic template, and other similar approaches relying in non-surfactant assemblies have followed.[60] These new approaches offer the possibility of utilizing functional molecules as templates; folic acid is a well-known dietary supplement and a tumor targeting molecule.[61] Furthermore, new properties of materials may be derived by their encapsulation within mesopores. The fundamental properties of mesoporous materials can in general be summarized as: high surface areas, large pore volumes, sharp pore size distributions, chemically

Fig. 4. Scanning electron microscopy (SEM) images of mesoporous material STA-11 (left),[62] and transmission electron microscopy (TEM) images recorded along the 110, 111 and 311 orientations (from left to right) of its 3d-bicontinous cubic structure. Inset shows a unit cell model of the bicontinuous cylindrical meso-structure.

and thermally stable compositions and tailored functionalized surfaces with controllable particle sizes and shapes.

2.4. *Application of mesoporous silica in nanomedicine*

It is surprising that the link between pharmaceutical excipients and mesoporous silica was not made until 10 years after their discovery, an indication of the slow transfer of knowledge between research fields. Furthermore, drug delivery and formulation problems are very much *the order of the day* in the pharmaceutical industry, as it struggles to meet formulation challenges associated with drug discovery and repositioning of pharmaceutical actives.[63,64] Poor bioavailability from poorly soluble compounds, which make up a large proportion of lead compounds in pharmaceutical companies, and even their poor dissolution in suitable media for simple toxicology studies are major reasons for the rejection of lead compounds even before clinical trials.[65] The improvement of bioavailability and subsequent increases in potency or selectivity, the mitigation of toxicity, improving patient compliance through more patient friendly administration routes or reduced number of doses, improving drug stability or reducing first pass effects are some of the important reasons that drive research within new drug delivery vehicles. Vallet-Regi *et al.* made this connection in 2001, and pioneered the field of pharmaceutical drug release using mesoporous materials.[26,27,66] As functional excipients for the

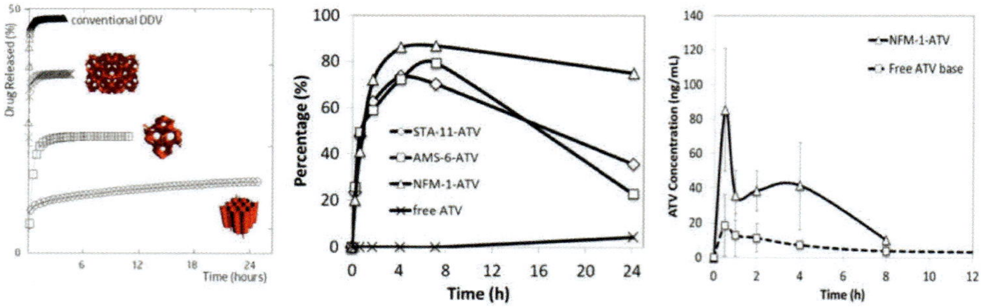

Fig. 5. Comparison of drug release properties for a model drug in three different types of mesoporous structure (left), adapted from Ref. 68. Dissolution curves of free HIV drug atazanavir (ATV) and loaded ATV into mesoporous materials with different pore size in simulated intestine fluid (SIF) performed under sink conditions (middle) adapted from Ref. 70. Pharmacokinetic profile for ATV in whole blood samples for mesoporous material NFM-1-ATV and free ATV co-administered with Omeaprazole to Sprague Dawley rats (right), adapted from Ref. 70.

pharmaceutical industry ordered mesoporous silica particles have been shown *in vitro* and *in vivo* to: load large pay loads of single or multiple active molecules;[67] tailor the pharmacokinetic release profiles through diffusion or other mechanisms;[68] target the release of pharmaceutical products to specific cell types;[69] increase the bioavailability of pH sensitive drug candidates;[70] enhance the solubility of hydrophobic pharmaceutical actives with high partition constants;[71] act as adjuvants in immunotherapies;[72] act as a diagnostic and theranostic particles; and enhance the growth of apatite layers in tissue generation and bone implants.[73]

Several reliable strategies have emerged to tailor the release kinetics of pharmaceutical drugs, DNA, siRNA, nutraceuticals and growth factors from the internal pore space of mesoporous materials.[74] These include structural variations, variations in the pore diameter, electrostatic or covalent interactions between the wall and the drug, functionalization of the internal pore space, particle pegylation and external functionalization of the pore entrances.[74] A change in pore structure alone from 2d- to 3d-pore connectivity has been shown to increase the pharmacokinetic properties of model drugs from the internal pore space by several orders of magnitude. The desire to tailor release properties via an external (e.g. optical response or magnetic field) or internal stimuli (e.g. gastric changes in pH) has led to several interesting pore gatekeeping strategies. These include the use of photo-switching molecules that release the contents of the pores upon the use of certain wavelengths (e.g. coumarin at 310 nm). Furthermore, intelligent drug delivery vehicles based on mesoporous silica particles that respond *in situ* to changes in blood plasma composition of certain disease markers have already been achieved and represent the next wave of drug delivery technologies. Lin *et al.* developed glucose-triggered release of both gluconic acid-modified insulin (G-Ins) and cyclic adenosine monophosphate (cAMP) from mesoporous particles by immobilizing G-Ins on the exterior surface of the particles. The immobilized insulin served as a gatekeeper to encapsulate cAMP inside the mesopores.[75] With a similar goal Zhao *et al.* have achieved triggered release of insulin based on sequential coatings of enzymatic layers cross-linked with glutaraldehyde which were shown to act as a valve to control the release of insulin in response to external stimuli over mesoporous silica.[76]

3. Toxicology of Mesoporous Silica Particles

Current guidelines for the approval of new pharmaceutical excipients, such as mesoporous silica, are complex and by definition uncharted.[77] In some cases it may be necessary for novel excipient to be tested as a new drug compound or in combination with the drug to be used in the final formulation for the excipient. Pre-clinical toxicity studies for drug compounds must include: acute toxicity, subchronical and chronical assays, genotoxicity, reproductive toxicity, pharmacokinetics and pharmacology for absorption, distribution, metabolism and excretion.[77] As mesoporous silica are potential excipients with some function (e.g. controlling drug dose, as adjuvants or in diagnostics) their toxicity needs to be carefully assessed. This is problematic for a variety of nanoparticles being proposed for clinical use, and there is a clear need for a unification of toxicological assessment techniques and standards, which may facilitate the comparison of studies and endpoints. This must involve first a careful and accurate characterization of material properties, such as their: (i) "*in vivo*" particle size (i.e. in the relevant media to be used in assays or clinical context), (ii) surface chemistry, (iii) composition, (iv) porosity (surface area, pore volume and pore size), (v) degree of opsonization and (vi) potential for the formation of ROS.

With respect to particle size, several comparison studies have already been performed on mesoporous silica materials. Vallhov *et al.* studied the effect of mesoporous silica particle size on particle uptake by human dendritic cells.[72] Mesoporous particles AMS-6 (270 nm in particle size) and AMS-8 (2.5 μm) were used in the study. Smaller particles affected the viability and cytokine production of DCs to a lesser extent than the larger AMS-8 particles. None of the particles were taken up at 4°C, indicating that both types are taken up via an active mechanism such as endocytosis. AMS-6 particles were found within membrane-enclosed compartments while AMS-8 were found directly in the cytosol indicating that they escape from the lysosome compartments. The interaction of SBA-15 and MCM-41 mesoporous silica particles with human red blood cells was investigated by Zhao *et al.*[78] Larger SBA-15 particles (~531 nm, SEM) where taken up by red blood cells to a greater extent than MCM-41 particles (~122 nm, SEM). MCM-41 particles were adsorbed onto the cell surface without disturbing the cell membrane while the adsorption of SBA-15 particles to the cell surface led to greater local membrane deformation, and internalization and subsequent rupture of the red blood cells (hemolysis).

Surface chemistry can influence the reactivity, solubility, interaction and agglomeration of particles in different environments, as well as their accumulation in organs and tissues. Surface chemistry will also strongly influence the nature of adsorbed protein corona and their interaction strength in biological media. The inner "hard" corona layer contain proteins that are strongly adsorbed to the particle's surface whereas the outer "soft" corona layer is composed of proteins with shorter residence times and lower affinities to the surface. Whilst many studies have focused on surface chemistry *vis-a-vis* of conjugation to fluorescent dyes or bio-molecules (drugs, proteins, etc.) no through study with mesoporous material has so far been conducted in order to evaluate the potential to control the adsorbed protein corona and its effects on cell uptake or cell viability.

Analysis methods should be chosen to avoid their interactions with tested particles. The suitability of using the MTT assay to evaluate biocompatibility *in vitro* has been put into question for several nanoparticles including mesoporous silica.[6,79] Material-dye interactions and dye adsorption may interfere with the experimental outcome. When assessing the effect of mesoporous silica on cell viability, the MTT colorimetric assay overestimates cell death compared to other techniques such as flow cytometry due to an enhanced formazan exocytosis.[80] Moreover, formazan can form crystals within the pores of CNT limiting its solubilization which result in a decreased color and absorbance in the analyzed solution and an underestimation of the cell viability. For this reason other techniques than MTT, such as Trypan blue exclusion assay, PI (propidium iodide) and Annexin-V staining, should be used in conjunction.[81]

Toxicological and biocompatibility evaluations of mesoporous particles have been performed in several *in vitro* and *in vivo* systems via several administration routes. We follow by summarizing some of the major findings.

3.1. *Oral administration*

Oral administration offers several advantages over other administrations *vis-a-vis* of patient compliance and hence it is one of the more desired routes in pharmaceutical development. Mesoporous materials are not likely to become systemic through the passage of the gastrointestinal tract making, lowering some of the toxicological implications of their administration.

We recently reported an *in vivo* oral toxicology study with two different types of mesoporous particles, AMS-6 and NFM-1 (with 300 nm and 1.5 μm particle size, 3 nm and 5 nm pore size, respectively), which were administrated by oral gavage to rats.[82] A fixed-dose procedure was used to analyze the acute toxicity of the particles, as this is considered a more humane method than the LD_{50} test for acute toxicity and is recommended in OECD guidelines.[83] In the dose range finding test (DRF), the animals (two males and two females) were administrated with a fixed and single dose and then monitored

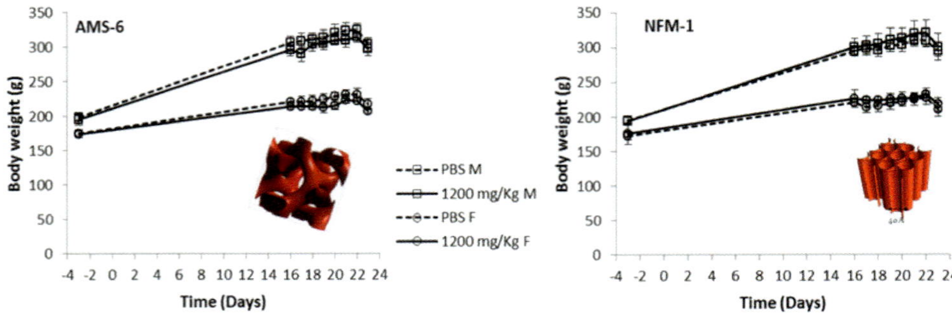

Fig. 6. The graphs show the evolution of the body weight in Sprague Dawley rats after oral administration of two mesoporous materials with varying particle size and pore structure. No difference in body weight evolution was observed in the animal receiving particle formulations compared to animals receiving vehicle alone (PBS) during seven consecutive days at doses of 2000 mg/kg for NFM-1 (2d-hexagonal, 27Å pore size, 2 μm particle size) and of 1200 mg/kg for AMS-6 (3d-cubic, 40 Å pore size, 230 nm particle size). On day 23: Overnight fasted animals (only food was withheld not water). Adapted from Ref. 82.

in order to detect any signs of discomfort or toxicity. If any negative signs were observed, the next group of animals was administrated with the next higher fixed dose. The maximum tolerated dose (MTD) in the DRF was not reached, i.e. no signs of toxicity were observed in response to the particles at any of the doses tested, between 30 mg/kg to 2000 mg/kg for NFM-1 and 40 mg/kg to 1200 mg/kg for AMS-6. The dispersion properties of the particles in PBS, in particular NFM-1, limited the formulation preparation needed for their administration at higher doses. Hence, a study of even higher doses was not possible. In the subsequent confirmatory test (CT) the highest doses utilized were 1200 mg/kg and 2000 mg/kg for AMS-6 and NFM-1, respectively (Fig. 5). The CT involved daily administrations for seven consecutive days. The results confirm the high tolerance of the particles after oral administration, and a lack of systemic adsorption of mesoporous silica particles of sizes above 300 nm. All animals showed normal weight development and there were no differences in the hematology or the biochemistry parameters from animals treated with NFM-1 or AMS-6 compared to the control group. Breathing difficulties that could be associated with toxic effects exerted by the particle formulation reaching the lungs were observed at high particle doses with some animals. However, reflux may occur in rats after administration by oral gavage and can lead to respiratory difficulties and mortality.[84] We concluded that this was the cause of the observed effect, and not an effect by the particles. In general, results showed that the tested particles are highly tolerated in rats after oral administration.

3.2. *Intravenous administration*

Intravenous (*i.v.*) administration results in systemic distribution and offers the possibility of targeting via passive or active targeting. This is particularly important in the treatment of cancer with nanoparticles. Passive targeting occurs as a result of the tumor's leaky vasculature and reduced lymphatic drainage system. Particles of up to ~300 nm in diameter accumulate in the tumor via the enhanced permeability and retention (EPR) effect.[85] Active targeting uses specific ligands attached on the particle surface to bind to specific or overexpressed surface molecules on the target tissues. Increased accumulation in the target can improve the effectiveness of toxic drugs and vaccines. This may result in the use of lower quantities of toxic drug compounds and decreased adverse effects. Mesoporous silica particles have been shown to be well-tolerated when delivering efficient doses of anticancer drugs to tumors in human cancer xenographt mice after intraperitoneal (*i.p.*) and *i.v.* injection.

Lu *et al.* have found that 10 daily doses of >100 mg/kg of ~100–130 nm particles administered by *i.p.* injection or *i.v.* were well-tolerated by mice.[69] In a recent publication the MTD after *i.v.* administration of mesoporous particles (spheres of 120 nm and rods of 36 x 1028 nm) and non-porous silica particles (150 nm) was respectively 30–65 mg/kg and 450 mg/kg.[86] For comparison the MTD of amorphous non-porous silica particles after *i.v.* administration (15 nm and 55 nm) and crystalline silica particles (400 nm, quartz) has been determined in rats to be respectively; ~50, ~125 and 100 mg/kg.[87]

Hudson *et al.* reported that *i.p.* and *i.v.* injections of ~1.2 g/kg doses of mesoporous silca materials (MCM-41, SBA-15 and MCF particles with particle size of 29–140 nm, 800 nm and 4 μm and pore sizes of 3, 7, 16 nm respectively), were lethal or lead to euthanasia of mice. Dose reduction to 40 mg/kg was safe.[88]

The silica content in different organs has been shown to decrease over time indicating that the particles can be biodegraded, after injection. However, low toxicity by other mesoporous silica particles after *i.v.* injection has been described in several publications.[89] An entire particle clearance from the organism after ~4 weeks with no signs of toxicity after a single dose (500 mg/kg) and daily doses (14 days, 80 mg/kg) has been demonstrated.[90]

4. Immuno-Modulatory Properties of Mesoporous Silica Particles

There is a dual reason for studying the immune-modulatory properties of nanoparticles. First, defensive mechanisms which have developed to protect against invading external pathogens, such as bacteria, viruses, fungi and parasites, as well as against internal harmful components such as cancer and damaged cells. The skin and mucous membranes, lining the internal surfaces of the respiratory, gastrointestinal and urogenital tracts form a physical, chemical and antimicrobial barrier preventing the entrance of most pathogens to the body. In this context nanoparticles must be considered as a potential pathogen, and may interact with immune competent cells — once clearing these barriers — causing an immune response. It is important hence to study these interactions from a toxicological point of view.

Second, there is a need for development of new adjuvants in vaccine delivery. An ideal vaccine is a simple and stable formulation that can be easily administrated and provide life-long immunity against a specific pathogen. Adjuvants are compounds that are added to vaccine formulations in order to improve the immune response against the vaccine antigen.[91] Adjuvants mimic infection, stimulate innate immunity and direct the type of the adaptive (acquired) immunity induced by the vaccine.[10] This leads to increased effectiveness and dose reductions with implications in the production of antigens.[92] Important additional issues required for the development of new adjuvants are: stability, production cost, suitability to a broad range of vaccines and minimal adverse effects. The only adjuvants approved for human use are aluminum salts, known as alum. Alum has a good safety record but induces mainly Th2 type immune responses. Alum is therefore not optimal in the treatment of allergies and in inducing IgA antibody mucosal responses.[93] Additionally, the development of vaccines with controlled antigen delivery systems may avoid the use of booster doses, controlling the spatial and temporal presentation of antigens to the immune system.

Immunological studies performed on mesoporous silica with different particle size and pore sizes, showed that particle uptake by macrophage cells (immune competent cells) did not impair their functions or affect their cell viability. Smaller (AMS-6, 300 nm particle size, 4.5 nm pore size) particles were encapsulated into vesicular compartments while the larger (AMS-8, 3 μm particle size, 3.0 nm pore size) particles were found directly in the cytosol within human monocyte derived dendritic cells (MDDC).[72] Witasp *et al.* showed that the uptake of AMS-6 and AMS-8 by MDDCs occurs through an energy dependent (active) mechanism since none of the particles were taken up when cells were incubated at 4°C, which inhibits active internalization.[94] Endosomal entrapment has also been shown for MCM-41 (150 nm particle size) in studies with HeLa cells.[95]

In a recent study, ovalbumin (OVA) sensitized mice were treated with SBA-15-OVA or alum-OVA, before challenging with OVA. The treatment with the particles was shown to skew the existing allergic inflammation (Th2-type response) toward a more Th-1 like response. This was indicated by higher IgG2a serum levels and INF-γ production, measured in splenocytes after *ex vivo* stimulation with OVA, compared to alum adjuvant. Moreover, the OVA-specific IgE levels were lower than those in the alum-OVA treated group.[96] These results make mesoporous silica particles suitable adjuvant candidates for vaccine/immunotherapy when a more Th1-like immune response is preferred over Th2. No symptoms of local toxicity or granuloma formation at the site of administration were observed in accordance with previous studies using SBA-15 as adjuvant.[97,98]

5. Regulatory Status of Mesoporous Silica Particles

Although numerous pre-clinical animal studies have been reported, to the best of the authors knowledge only one clinical study in humans (Phase I) has been performed using mesoporous silica particles.[99] This was the first-ever clinical study to demonstrate the bioavailability enhancing properties of silica in humans. In this study, the bioavailability profile of fenofibrate formulated with mesoporous silica was compared to the marketed micronized formulation (Lipanthyl$^{(R)}$). The study results show a 54% higher bioavailability for the silica formulation than the marketed formulation, reproducing in humans what has been shown previously *in vitro* and *in vivo*. Currently there appears to be at least three companies actively developing mesoporous materials for commercial applications in the field of pharmaceutical sciences and nanomedicine, and regulatory engagement in order to approve mesoporous materials for clinical use is ongoing.[99] At least one Drug master file (DMF) type IV can be found in the FDA's registry for new excipients, which attests to a high pore volume porous silica material.[100]

6. Conclusions and Outlook

The large number of research reports and the versatility of the mesoporous silica particles as a platform for drug delivery and nanomedicine demonstrate the importance for an accurate and thorough evaluation of their toxicological properties, *de novo.* Whilst colloidal non-porous silica particles are already included in consumer products and generally regarded as safe, their porous counterparts are clearly different both in terms of their material properties and biologically. A recent study in our group, shows that mesoporous particles above a certain pore size may absorb gastrointestinal enzymes if administered with their pores open.[101] Whilst such a result shows promise in combating disease such as obesity, on the other hand it may cause undesired effects if the materials are to be used as drug delivery vehicles in an oral administration context, in particular if co-administered with other drug formulations. Hence a full toxicological appraisal, for individual type of mesoporous materials must be completed but also with a focus on how the particles may affect co-administrations of other compounds. As has been mentioned elsewhere it is not feasible screen all nanomaterials using animal models, hence *in vitro* systems that reliably predict nanotoxicology markers, must be developed. Finding toxicology markers specific to mesoporous materials, as for example their potential to cause thrombosis if administered *i.v.*, or their potential to reduce the efficacy of co-administered drug compounds if administered orally; is ongoing.

Whilst mesoporous silica materials are only 10 years into their adventure in nanomedicine, they have already shown to be useful in a general sense as excipients in drug delivery, but also as specific drug delivery vehicles for cancer treatment. More complex schemes are being realized, such as glucose concentration dependent release of insulin, which show that the large pore volumes, external and internal surfaces can be designed independently to provide multifunctional devices. Like for many other nanoparticles all of these schemes are ahead of regulatory and industrial acceptance of the materials however, and these must now go *hand in hand* if the particles are to succeed in entering the clinical phase.

Acknowledgments

AEGB acknowledges support from the Swedish Research Council and the Seventh Framework Programme of the European Commission: EC-FP-7-ORCHID (grant no. 261378). AEGB is grateful to colleagues at Karolinska Institute (Stockholm, Sweden): Professor Bengt Fadeel, Professor Annika Scheynus and Dr. Susanne Gabrielsson, for many helpful discussions and collaboration over recent years.

References

1. L. Fatehi *et al.*,. *J. Law Med. Ethics* **40**, 716–750 (2012).
2. V. K. Varadan, L. Chen and J. Xie, *Nanomedicine: Design and Applications of Magnetic Nanomaterials, Nonsensors and Nanosystems.* UK: John Wiley and Sons (2008).
3. B. Fadeel *et al.*, *WIREs Nanomed. Nanobiotech.* **5**, 111–129 (2013).
4. N. Sanvicens and M. P. Marco, *Trends Biotechnol.* **26**(8), 425–433 (2008).
5. T. R. Sathe *et al.*, Mesoporous silica beads embedded with semiconductor quantum dots and iron oxide nanocrystals: dual function microcarriers for optical encoding and magnetic separation. *Anal. Chem.* **78,** 5627–5632 (2006).
6. B. Fadeel and A. E. Garcia-Bennett, *Adv. Drug Deliv. Rev.* **62**(3), 362–374 (2010).
7. A. Nel, T. Xia, L. Madler and N. Li, *Science* **311**, 622–627 (2006).
8. G. Oberdörster, E. Oberdörster and J. Oberdörster, *Environ. Health Perspect.* **113**, 823–839 (2005).
9. M. D. Brown, K. Donaldson, J. P. Borm, R. P. Schins, M. Dehnhardt, P. Gilmour, L. Jimenez and V. Stone, *Am. J. Physiol. Lung Cell. Mol. Physiol.* **286**, L344–L353 (2004).
10. A. D. Maynard, P. A. Baron, M. Foley, A. Shvedova, E. R. Kisin and V. Castranova, *J. Toxicol. Environ. Health* **67**, 87–107 (2004).
11. W. K. Oh, S. Kim, M. Choi, C. Kim, Y. S. Jeong, B. R. Cho, J. S. Hahn and J. Jang, *ACS Nano* **4**, 5301–5313 (2010).
12. C. M. Sayes, J. D. Fortner, W. Guo, Lyon, R. A. Petros and J. M. DeSimone, *Nat. Rev. Drug Discov.* **9**(8), 615–627 (2010).
13. L. D. Leserman, J. Barbet, F. Kourilsky and J. N. Weinstein, *Nature* **288**, 602–604 (1980).
14. O. C. Farokhzad and R. Langer, *Adv. Drug Deliv. Rev.* **58**(14), 1456–1459 (2006).
15. A. Z. Wang *et al.*, *Expert Opin. Biol. Ther.* **8**(8), 1063–1070 (2008).
16. L. Zhang *et al.*, *ACS Nano* **2**(8), 1696–1702 (2008).
17. D. A. Tomalia, L. A. Reyna and S. Stevenson, *Biochem. Soc.Transac.* **35**(1), 61–67 (2007).
18. G. von Maltzahn *et al.*, *Nat. Mater.* **10**, 545–552 (2010).
19. W. T. Al-Jamal and K. Kostarelos, *Acc. Chem. Res.* **44**(10), 1094–1104 (2011).
20. C. T. Kresge, M. E. Leonowicz, J. Roth, J. C. Vartuli and J. S. Beck, *Nature* **359**, 710–712 (1992).
21. K. Kuroda, *J. Porous Mat.* **3**(2), 107–114 (1996).
22. Y. Wan and D. Y. Zhao, *Chem. Rev.* **107**, 2821–2860 (2007).
23. A. Taguchi and F. Schüth, *Microporous Mesoporous Mat.* 77, 1–45 (2004).
24. S. Che, A. E. Garcia-Bennett, T. Yokoi, K. Sakamoto, H. Kunieda, O. Terasaki and T. Tatsumi, *Nature Mat.* **2**, 801–805 (2003).
25. X. Yongde and R. Mokaya, *J. Phys. Chem. B* **107**(29), 6954–6960 (2003).
26. M. Vallet-Regi, A. Ramila, R. P. Real and J. Perez-Pariente, *Chem. Mat.* **13**, 308–312 (2001).
27. M. Vallet-Regi, F. Balas and D. Arcos, *Angew. Chem. Int. Ed.* **46**, 7548–7558 (2007).
28. B. Trewyn et al., *Acc. Chem. Res.* **40**(40), 846–853 (2007).
29. A. E. Garcia-Bennett, *Nanomedicine* **6**(5), 867–877 (2011).
30. F. Tang, L. Li and D. Chen, *Adv. Mater.* **24**, 1504–1534 (2012).

31. European Parlament and Council Directive No 95/2/EC of 20 February 1995 on food additives other than colours and sweeteners (OJ No L 61, 18. 3. 1995, p. 1).
32. British Pharmacopoeia Volume I & II Monographs: Medicinal and Pharmaceutical Substances Colloidal Anhydrous Silica (2009).
33. S. Huh, J.W. Wiench, J.-C. Yoo, M. Pruski and V. S.-Y. Lin, *Chem. Mater.* **15**, 4247–4256 (2003).
34. D. Brühwiler, *Nanoscale* **2**, 887–892 (2010).
35. B. Fadeel, *Swiss Med. Wkly.* **142**, 13609 (2012).
36. G. J. Nohynek, E.K. Dufour and M. S. Roberts, *Skin Pharmacol. Physiol.* **21**(3), 136–49 (2008).
37. J. Wang *et al.*, *Toxicol Lett.* **168**(2), 176–85 (2007).
38. R. C. Dart, *Medical toxicology*: 3rd ed. Philadelphia: Lippincott, Williams & Wilkins (2004).
39. D. Napierska *et al.*, *Particle Fibre Toxicol.* **7**(39), (2010).
40. T. K. Barik, B. Sahu and V. Swain, *Parasitol Res.* **103**, 253–258 (2008).
41. E. T. Degens, *Top. Curr. Chem.* **64**, 1–112 (1976).
42. M. Schmidt and F. Schwertfeger, *J. Non-Crystall. Sol.* **225,** 364–368 (1998).
43. P. J. Heaney, *Rev. Mineral. Geochem.* **29**(1), 1–40 (1994).
44. C. J. Brinker and G. W. Scherer, *Sol-Gel Science: The Physics and Chemistry of Sol-Gel Processing* London: Academic Press (1990).
45. R. R. Wickett *et al.*, *Arch Dermatol Res.* **299**(10), 499–505 (2007).
46. C. Seaborn and F. H. Nielsen, *Nutrion Today* July/August, 13, (1973).
47. A. Lassus, *J. Int. Med. Res.* **25**(4), 206–209 (1993).
48. L. L. Hench, *J. Mater. Sci.: Mater. Med.* **17**, 967–978 (2006).
49. SIDS Initial Assessment Report, Synthetic amorphous silica and silicates, *SIAM* **19**, 19–22 (2004).
50. United States Pesticides And Environmental Protection Agency, *Toxic Substances*, 738-F-91–107 (1991).
51. D. Y. Zhao *et al.*, *Science* **279**(5350), 548–552 (1998).
52. M. Kruk and M. Jaroniec, *Chem. Mater.* **13**(10), 3169–3183 (2001).
53. T. Shimizu, T. Yanagisawa, K. Kuroda and C. Kato, Abstract No. 1XII D42(I-761), Annual Meeting of the Chemical Society of Japan (1988).
54. K. Kuroda, T. Yanagisawa, T. Shimizu and C. Kato, *Abstracts of the 9th Int. Clay Conf. Strasbourg*, **222**(1989).
55. T. Yanagisawa, T. Shimizu, K. Kuroda and C. Kato, *Bull. Chem. Soc. Jpn.* **63**, 988 (1990).
56. V. Chiola, J. E. Ritsko, C. D. Vanderpool, *Process for producing low-bulk density silica* Application No. US 3556725DA, filed on 26-Feb-1969, Publication No. US 3556725A published on 19-Jan-1971.
57. F. Kleitz, *Ordered Mesoporous Materials*. In *Handbook of Heterogeneous Catalysis* (2008), pp. 178–219.
58. S. Che, A. E. Garcia-Bennett, T. Yokoi, K. Sakamoto, H. Kunieda, O. Terasaki and T. Tatsumi, *Nat. Mat.* **2**, 801–805 (2003).
59. R. Atluri, N. Hedin and A. E. Garcia-Bennett, *J. Am. Chem. Soc.* **131**(9), 3189–3191 (2009).
60. H. Qiu, J. Xie and S. Che, *Chem. Commun.* **47**, 2607–2609 (2011).

61. A. E. Garcia-Bennett, M. Nees and B. Fadeel, *Biochem. Pharmacol.* **81**(8), 976–984 (2011).
62. R. P. Hodgkins, A. E. Garcia-Bennett and P. A. Wright, *Microporous Mesoporous Mat.* **79**, 241–252 (2005)
63. P. Cuatrecasas, *J. Clin. Invest.* **116**(11), (2006).
64. A. J. J. Wood, *N. Engl. J. Med.* **355**, 618–623 (2006).
65. T. Takagi *et al.*, *Mol. Pharm.* **3**(6), 631–643 (2006).
66. P. Horcajada, A. Rámila, J. Pérez-Pariente and M. Vallet-Regı, *Microporous and Mesoporous Mat.* **68**(1&3), 105–109 (2004).
67. R. Mellaerts, R. Mols, J. A. Jammaer, C. A. Aerts, P. Annaert, J. V. Humbeeck, G. Van den Mooter, P. Augustijns and J. A. Martens, *Eur. J. Pharm. Biopharm.* **69**, 223–230 (2008).
68. U. Brohede, R. Atluri, A. E. Garcia-Bennett and M. Stromme, *Curr. Drug Deliv.* **5**(3), 177–185 (2008).
69. J. Lu, M. Liong, Z. Li, J. I. Zink and F. Tamanoi, *Small* **6**(16), 1794–1805 (2010).
70. X. Xia, C. Zhou, L. Ballel and A. E. Garcia-Bennett, *Chem. Med. Chem.* **1**, 43–48 (2012).
71. M. Van Speybroeck *et al., Eur. J. Pharm. Biopharm.* **21**, 623–630 (2010).
72. H. Vallhov, S. Gabrielsson, M. Stromme, A. Scheynus and A. E. Garcia-Bennett, *Nano Lett.* 7(12), 3576–3582 (2007).
73. C. Wu and J. Chang, *Interface Focus* **2**(3), 292–306 (2012).
74. J. Rosenholm, C. Sahlgren and M. Linden, *Nanoscale* **2**, 1870–188 (2010).
75. Y. Zhao, B. G. Trewyn, I. I. Slowing and V. S. Y. Lin, *J. Am. Chem. Soc.* **131**, 8398–8401 (2009).
76. W. Zhao *et al.*, *Chem. Commun.* **47**, 9459–9461 (2011).
77. K. Kolter, *Pharm. Technol. Eur.* **23**(10), (2011).
78. Y. Zhao *et al.*, *ACS Nano* **5**(2), 1366–1375 (2011).
79. T. Laaksonen *et al., Chem. Res. Toxicol.* **20**(12), 1913–1918 (2007).
80. N. A. Monteiro-Riviere, A.O. Inman and L.W. Zhang, *Toxicol. Appl. Pharm.* **234**(2), 222–235 (2009).
81. J. M. Worle-Knirsch, K. Pulskamp and H. F. Krug, *Nano Lett.* **6**(6), 1261–1268 (2006).
82. N. Kupferschmidt, X. Xia, R. Hanoi, R. Atluri, L. Ballel and A. E. Garcia-Bennett, *Nanomedicine* **8**(1), 57–64 (2013).
83. The OECD, *Guideline for Testing of Chemicals,* **420**, (2001).
84. S. Damsch *et al, Toxicol. Pathol.* **39**(2), 348–360 (2011).
85. P. P. Adiseshaiah, J. B. Hall and S. E. McNeil, *Wiley Interdiscip. Rev. Nanomed. Nanobiotechnol.* **2**(1), 99–112 (2010).
86. T. Yu *et al.*, *ACS Nano* **6**(3), 2289–301 (2012).
87. T. R. Downs *et al., Mutat. Res.* **745**(1&2), 38–50 (2012).
88. S. P. Hudson *et al.,* **29**(30), 4045–4055 (2008).
89. X. Huang *et al.*, *ACS Nano* **5**(7), 5390–5399 (2011).
90. T. Liu *et al.*, *Biomaterials* **32**(6), 1657–1668 (2011).
91. J. P. Amorij *et al.*, *J. Control. Release* **161**(2), 363–76 (2012).
92. J. A. Hubbell, S. N. Thomas and M. A. Swartz, *Nature* **462**, 449–460 (2009).
93. J. M. Brewer, *Immunol. Lett.* **102**(1), 10–15 (2006).
94. E. Witasp *et al.*, *Toxicol. Appl. Pharmacol.* **239**, 306–319 (2009).

95. I. Slowing, B. G. Trewyn and V. S. Y. Lin, *J. Am. Chem. Soc.* **128**(46), 14792–14793 (2006).
96. Kupferschmidt *et al., Nanomedicine,* under revision (2013).
97. L. P. Mercuri *et al., Small* **2**(2), 254–256 (2006).
98. R. S. Ferreira Jr. *et al., J. Toxicol. Environ. Health A,*73(13&14), 926–933 (2010).
99. Available at: http://www.grace.com/media/NewsItem.aspx?id=1685777 .
100. An up-to-date list of submitted DMFs can be found here: http://www.fda.gov/Drugs/DevelopmentApprovalProcess/FormsSubmissionRequirements/DrugMasterFilesDMFs/default.htm .
101. Kupferschmidt *et al., Nanomedicine*, under revision (2013).

Chapter A5

Cellular Responses Induced by Nanoparticles

Masanori Horie

Institute of Industrial Ecological Sciences,
University of Occupational and Environmental Health,
Kitakyushu, Fukuoka 807-8555, Japan
masa-horie@med.uoeh-u.ac.jp

Abstract

Understanding of cellular responses induced by nanoparticles is essential to its application for life science. There are three important factors in cellular influences of nanoparticles; metal ion release, protein adsorption and cellular uptake. Subsequently, induction of oxidative stress is important in nanotoxicology. Generation of intracellular reactive oxygen species (ROS) is a trigger for the induction of oxidative stress by nanoparticles. Many kinds of nanoparticles cause the induction of intracellular ROS level. Cellular uptake of nanoparticles and intracellular metal ion release are important in intracellular ROS generation. Intracellular ROS is generated by the interaction of nanoparticles and cells. An increase of intracellular ROS levels by nanoparticles induces stress response enzyme proteins such as HO-1. If oxidative stress is strong or continuous, lipid peroxidation and oxidative DNA damage occur. However, not all kinds of nanoparticles induce oxidant stress. Intracellular ROS generation by nanoparticles depends on their physical and chemical properties, such as crystalline phase, adsorption ability and solubility.

1. Introduction

Nanomaterial is defined as a material whose diameter in one dimension is a minimum of 1–100 nm (ISO/TS 27687: 2008, Nanotechnologies — Terminology and definitions for nano-objects — nanoparticle, nanofiber and nanoplate). Among the various nanomaterials available, many types of nanoparticles with a diameter of 1–100 nm are industrially produced. Some nanoparticles such as titanium dioxide (TiO_2), zinc oxide (ZnO), silicon dioxide (SiO_2), calcium carbonate ($CaCO_3$), and carbon black are bulk-produced and used on a daily basis in products such as paint, pigment, sunscreen and cosmetics.

Nanoparticles are applied also to life sciences besides the industrial application. For instance, there are a number of investigations regarding use of nanoparticles as plasmid DNA vectors in cells for gene therapy.[1–5] There are also reports that modified nanoparticles showed selective growth inhibition of cancer cells.[6–8] However, in recent years, many studies have described the biological influences of manufactured nanoparticles, including toxicity, and risk assessment of nanoparticles is therefore attracting increasing attention. Many *in vivo* and *in vitro* investigations have been made, enhancing our knowledge of the biological influences of nanoparticles. A number of benefits of using nanoparticles in the life sciences industries are known, e.g. in a previous study, nanoparticles enhanced the fibrillation of β_2-microglobullin.[9] In order to apply nanoparticles to the life sciences industries, we have to understand the cellular influences caused by nanoparticles. In their industrial use, workers may unintentionally be exposed to nanoparticles; hence, in many cases, skin- and lung-derived cells are examined for determining nanoparticle cytotoxicity. However, for their use in the life sciences industries, nanoparticles need to be modified with various functional groups or proteins. Such nanoparticle modification is necessary for biological functions such as target cell uptake; it is therefore necessary to carefully examine the cellular influences of the source nanoparticles.

Nanoparticles can cause various cellular influences such as cytotoxicity (in this paper, a biomolecule response excluding death observed by exposure to nanomaterials is called a "cellular influence" [*in vitro*] or "biological influence" [*in vivo*]). The chemical composition of any nanoparticle is the same as its submicron or microscale particles. For instance, titanium dioxide is "TiO_2" regardless of the particle size. Novel biological activities of nanoparticles are attributable to their novel physical and chemical properties. The term "nanoparticle" includes many types of materials and does not have a determined chemical formula. Nanoparticles are defined only by the particle size, and include both extremely cytotoxic materials and non-cytotoxic materials. Illustrating this, Cr_2O_3 nanoparticles exhibit the induction of strong oxidative stress and cytotoxicity on cultured cells.[10] Conversely, platinum nanoparticles and nanodiamonds did not induce oxidative stress, and their cellular influences were small.[11,12] In some cases, even if the chemical component is identical, the cellular influences of individual nanoparticles are different. For instance, it has been reported that CeO_2 nanoparticles both induce oxidative stress in cultured cells[13,14] and have antioxidative effects.[15,16] These features complicate the risk assessment of nanoparticles.

The cellular influences of nanocarbons, often used in the field of life sciences[17] are generally lower than that of metal or metal oxide nanoparticles. However, nanocarbons do exert some form of cellular influence. For example, although fullerene C_{60} does not affect cell viability, it induces the generation of intracellular reactive oxygen species (ROS).[18] Similarly, single wall carbon nanotubes (SWCNT) induces the generation of intracellular ROS levels but does not cause cell death.[12]

The physical and chemical properties of nanoparticles are deeply related to their cellular influences. Some types of nanoparticles show greater solubility than fine-scale particles, and thus, a large amount of metal ion is released into the dispersion medium.[19,20] Increase in solubility is one of the important features of nanoparticles, and metal ion release is a major cellular influence factor. There are 3 key components in the cellular influence of nanoparticles: metal ion release, cellular uptake, and protein adsorption.

Many physical and chemical properties are involved in the cellular influences of nanoparticles, including not only the chemical components but also the particle size, specific surface area, aggregation state, adsorption ability and solubility. Cellular influences are a result of interactions of these physical and chemical properties.

In many cases, cellular influences of nanoparticles are triggered by the induction of intracellular ROS generation. Numerous types of nanoparticles can cause cellular oxidative stress. Oxidative stress is defined as "a disturbance in the pro-oxidant/antioxidant system in favor of the pro-oxidant."[21] ROS, which is one of the major causes of oxidative stress, is generated by mitochondrial respiration in healthy cells. Phagocytes such as macrophages and neutrophils also generate ROS in order to digest phagocytized organic substances. In healthy cells, the intracellular oxidative state is maintained by the antioxidant system. However, excess ROS generation induces dysfunction of the cellular antioxidative system, causing accumulation of oxidative biomolecules.

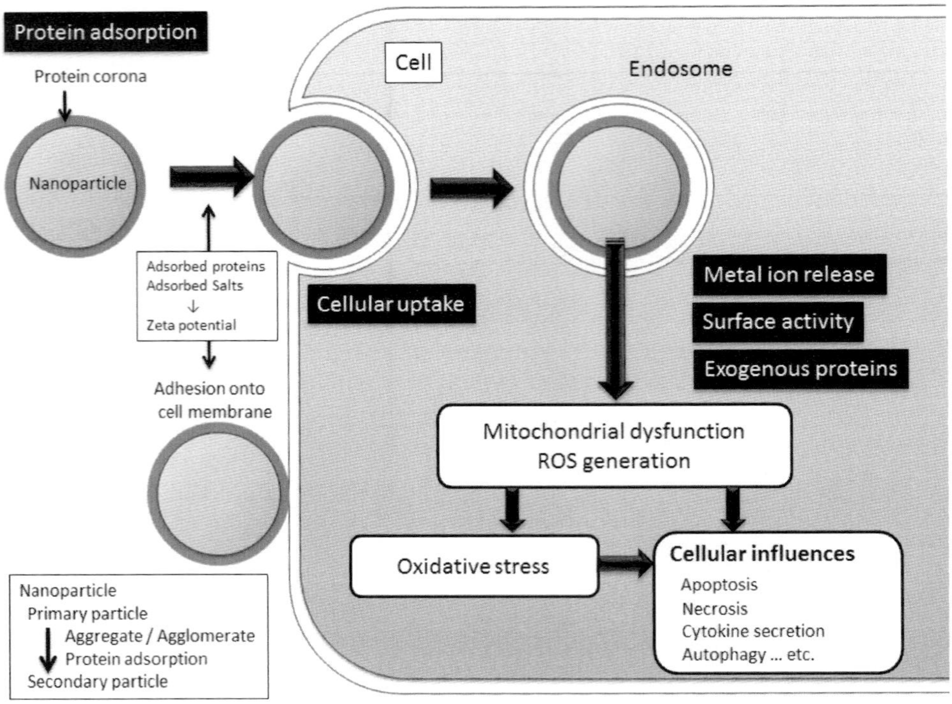

Fig. 1. Outline of cellular influences caused by nanoparticles.

ROS are also involved in cell signaling. Therefore, induction of oxidative stress should be considered for the application of nanoparticles, particularly for applications in life sciences industries such as nanomedicine. In this chapter, cellular responses induced by nanoparticles are described by three important cellular influence factors: protein adsorption, cellular uptake and metal ion release. Additionally, the most important cellular response of nanoparticles, the induction of oxidative stress, is comprehensively explained.

An outline of cellular influences caused by nanoparticles is shown in Fig. 1.

2. Metal Ion Release

Many metal oxides and metal nanoparticles exhibit higher solubility than their corresponding fine particles. Even if a fine particle is classified as insoluble, its nanoparticles often release metal ions into the dispersion medium. The high solubility may be due to the presence of many atoms on the surface of the nanoparticle as compared with the fine particle. For example, nickel(II) oxide (NiO) is generally classified as a water-insoluble material. In fact, the water solubility of NiO fine particles is negligible and their cellular influences are weak, whereas NiO nanoparticles are soluble in water.[19] Additionally, the nanoparticles showed a larger solubility in the cell culture medium than in water.[22] Common cell culture media such as Minimum Essential Medium (MEM), Dulbecco's Modified Eagle Medium (DMEM) and RPMI 1640 contain salts, amino acids, vitamins and sugars. The composition of the culture medium is similar to the composition of the biological fluid-like serum. Therefore, it is possible that the nanoparticles release metal in biological fluids such as serum, saliva and sweat. Released metal ions cause various cellular influences such as the induction of oxidative stress, cytokine release and apoptosis.[20,23–25] It has been reported that intracellular zinc levels are positively correlated with intracellular ROS levels.[26] However, the precise relationship between released metal ions from nanoparticles and cellular influences is unclear because of the difficulties during quantification of the intracellular released metals from nanoparticles.

3. Protein Adsorption Onto Nanoparticles

Many nanoparticles exhibit larger adsorption abilities than microscale particles regardless of the material type. Among the adsorbed nanoparticle materials, protein is the most important. Proteins adsorbed on the nanoparticles affect the stability of their dispersion and cellular uptake *in vitro*. *In vivo*, nanoparticles may adsorb proteins from the saliva, pulmonary surfactants, etc. The adsorbed proteins may also change the biological activity of the nanoparticles.

Dawson *et al.*[27] compared the adsorbed proteins on the surface of nanoparticles to a solar corona, and named it "protein corona". The type of adsorbed protein depends on the environment, and the adsorbed protein content may change according to the environment. Adsorbed proteins on the nanoparticles are in chemical equilibrium. If any specific protein could be adsorbed to "harmless" nanoparticles, these nanoparticles may be promising as medicinal carriers.

4. Cellular Uptake of Nanoparticles

Cellular uptake of nanoparticle is an important factor of cellular influences. However, cellular uptake itself is not a nanoparticle-specific property. Both nanoparticles and microscale particles are taken up into the cell.[13,19,28] The combination of cellular uptake, metal ion release and protein adsorption is very important in terms of cellular influences of nanoparticles. Namely, internalized nanoparticles release metal ions from within the cell, and exogenous proteins are taken up into the cells with nanoparticles. Soluble metal compounds such as metal chloride attach to the outside of cells. The extracellular soluble metal cannot easily pass the cell membrane. Typically, the intracellular metal concentration is closely controlled by ion channels on the cell membrane, and intracellular release of metals from internalized nanoparticles directly affects the subcellular organelles and cell signaling pathways. Generally, nanoparticles behave as "particles" and are taken up into cells by endocytosis. The internalized nanoparticles then behave as a source of metal ions, with the nanoparticles having features of both the particle and the metal.

Nanocarbons, essentially only carbon, may include metal impurities such as iron and nickel, and this may cause the uptake and release of these metal impurities with the nanocarbons inside the cells. It is suggested that the efficiency of endocytosis of the nanoparticles is affected by adsorbed proteins, zeta potentials and modified functional groups.[29] For nanomedicine applications, nanoparticles are often modified with functional groups for cellular uptake by design. Cellular uptake of polyethylene glycol (PEG) hydrogel-coated magnetic iron oxide nanoparticles, which were modified with specific amino acid sequences (arginine-glycine-aspartic acid-serine [RGDS]) in HeLa cells, was increased 17-fold over the unmodified nanoparticles.[30] The RGD sequence is found in extracellular matrix proteins such as fibronectin, and it binds to integrins on the cell membrane.[31] The RGD sequence is also involved in neutrophil phagocytosis.[32] The efficiency of cellular uptake of $-NH_2$ added quantum dot (QD) and –COOH added QD to macrophage J774.A1 cells were compared.[33] Eventually, both $-NH_2$ added QD and the –COOH added QD were taken up into cells; however, the –COOH added QD was taken up more rapidly than the $-NH_2$ added QD. Internalized QDs were localized at the lysosome and mitochondria. Localization of the internalized QDs at the mitochondria suggests the possibility of mitochondrial dysfunction and subsequent oxidative stress induction, as does the heat inactivation of serum-affected protein corona composition and cellular uptake of polystyrene nanoparticles. On the other hand, it is also reported that the internalized SiO_2 nanoparticles was not found at mitochondria in HeLa cells. They were localized at the endosome and lysosome.[34] When the effect of the adsorbed proteins was excluded by using serum-free medium, carboxyl-modified polystyrene nanoparticles exhibited adhesion and cellular uptake in melanoma cells.[35] Efficiency of the cellular uptake of nanoparticles is enhanced by modification of nanoparticles with high-affinity functional groups to the cell membrane. When adsorbed proteins are not involved in the cellular uptake of the nanoparticles, modified functional groups and zeta potential are important for cellular uptake. Adsorbed proteins, functional groups on the surface and the zeta potential are important factors for the cellular uptake of nanoparticles.

4.1. *Induction of oxidative stress by nanoparticles*

Oxidative stress is a key factor in cellular influences of nanoparticles.[36] Oxidative stress is a failure of the balance between oxidation and antioxidative potency. ROS, one of the major causes of oxidative stress, are generated by mitochondria respiration.[37] Phagocytes such as macrophages and neutrophils also generate ROS and radicals in order to digest organic substances. In other words, all of us undergo oxidative stress on a daily basis. Generally, oxidative stress is suppressed by the antioxidative system. However, if we are exposed to some kinds of nanoparticles, excess oxidative stress occurs, and this excess oxidative stress leads to the oxidation of biological molecules, lipids, proteins and nucleic acids. The oxidized biological molecule loses its biological function, and accumulated oxidized biological molecules due to continuous oxidative stress leads to various diseases. Furthermore, an appropriate amount of oxidative stress leads to an adaptive response by activation of the antioxidative system.[38] As a result, we acquire tolerance against oxidative stress. Therefore, evaluation of oxidative stress is essential for risk assessment of nanoparticles.

We can determine the level of oxidative stress by the measurement of various biomolecules and chemicals called "oxidative stress markers." Generally, oxidative stress markers are divided into three categories: (1) active species such as ROS, reactive nitrogen species (RNS) and other radicals which trigger oxidative stress; (2) oxidized biomolecules and metabolic substances such as malondialdehyde (MDA), 4-hydroxy-2-nonenal (4-HNE) and 8-OHdG; and (3) antioxidants such as glutathione (GSH), ascorbic acid and antioxidation-related enzymes such as heme oxygenase-1 (HO-1) and catalase.

4.2 *Intracellular ROS generation by nanoparticles*

Many reports showed increased intracellular ROS levels in cells exposed to nanoparticles. The measurement of intracellular ROS production is simple, compared to that in animals. 2′,7′-dichlorodihydrofluorescin (DCFH) is a fluorogenic probe commonly used for the measurement of radicals. Generally, 2′,7′-dichlorodihydrofluorescin diacetate (DCFH-DA), an acetylated form of DCFH, is employed to measure intracellular ROS levels. We can determine the ROS levels by measuring the fluorescence of 2′,7′-dichlorofluorescein (DCF), which is an oxidized product of DCFH. However, the type of ROS cannot be determined by DCFH, and the reactivity of DCFH with H_2O_2 is lower compared with that of radicals.[39] The use of electron spin resonance (ESR) is effective for the identification of the type of ROS generated from nanoparticles.[40–42] Using these methods, many nanoparticles cause an induction of intracellular ROS levels. Particularly, some metal oxide nanoparticles such as NiO, ZnO, CuO and Cr_2O_3 showed remarkable increases in intracellular ROS levels. These nanoparticles also showed lipid peroxidation, depletion of GSH and cell death.[43] Measurement of cells exposed to TiO_2 nanoparticles with UV irradiation by ESR showed generation of hydroxyl radicals (*OH), carboxyl radical anions (CO_2^-), and superoxide radical anions (O_2^-).[41,42] Moreover, TiO_2 also generated free radicals such as O_2 and *OH by reaction with oxygen, water or H_2O_2 without photoactivation.[40]

In many cases, soluble metal oxide nanoparticles and metal nanoparticles induced an increase in intracellular ROS levels. Metal ion release is an important factor in intracellular ROS generation by nanoparticles.[20] For instance, although microscale NiO particles are generally classified as an insoluble material, NiO nanoparticles release a large amount of Ni^{2+}. NiO nanoparticles induced an increase of intracellular ROS and cytotoxicity.[19,44] CuO, ZnO, Cr_2O_3 and MoO_3 nanoparticles also showed metal ion release and strong cytotoxicity with an increase in the intracellular ROS levels.[10,20,45] Intracellular ROS generation by soluble metal-related nanoparticles is explained by Fenton-like reactions.[46,47] Although ROS such as H_2O_2 and O_2^- are generated in mitochondria, they are removed by antioxidant defenses such as superoxide dismutase, catalase or glutathione, under normal conditions.[37,48–50] The intracellular metal ion that is released from internalized nanoparticles leads to mitochondrial dysfunction and ROS generation by a Fenton-like reaction. As a result, the redox balance fails, and lipids, proteins and DNA are oxidized in the cell. ZnO nanoparticles induce ROS generation, formamidopyrimidine-DNA-glycosylase (fpg)-sensitive DNA damage and mitochondria-mediated apoptosis.[51] Mitochondrial dysfunction is important in ROS generation by metal-related nanoparticles. Details of the mechanism of ROS generation from metals are described in reviews by Valko *et al.* and Leonard *et al.*[46,52]

Generation of intracellular ROS and the cytotoxicity of insoluble nanoparticles, in contrast, are typically small. However, these insoluble nanomaterials also showed induction of intracellular ROS. Y_2O_3, Gd_2O_3 and SnO_2 nanoparticles caused an induction of intracellular ROS levels, although these metal oxide nanoparticles barely dissolved in the culture medium.[20]

TiO_2 nanoparticles also generate ROS in cells, as mentioned above. TiO_2 nanoparticles, especially anatase, have strong photocatalytic activity.

The generation of hydroxyl radicals from H_2O by photoactivated TiO_2 nanoparticles is well-known. TiO_2 has a 3.0 eV bandgap; hence, it is activated by irradiation of wavelengths shorter than 415 nm.[53] UV-irradiated TiO_2 nanoparticles generated hydroxyl radicals and induced DNA injury and cell death.[41,42,54] The cytotoxicity induced by photoactivated TiO_2 nanoparticles was decreased by catalase and tryptophan, which remove *OH.[55] These results suggest that the cytotoxicity of photoactivated TiO_2 nanoparticles occurs due to extracellular ROS generation from TiO_2 nanoparticles. However, there is likely no possibility of photoactivation of inhaled TiO_2 nanoparticles in the lung, and even if the TiO_2 nanoparticles are irradiated by an excitation wavelength on the skin, the effect of the generated radicals will be limited by the cutaneous barrier.[56] Cytotoxicity of TiO_2 nanoparticles without the excitation wavelength is lower, compared with photoactivated TiO_2 nanoparticles.[54] Moreover, ROS generation in a cell-free system does not always reflect the intracellular ROS generation; although TiO_2 nanoparticles generated radicals in a cell-free system, no cellular influence was observed.[57]

Intracellular ROS levels of cells exposed to TiO_2 nanoparticles also increased without UV irradiation.[40] The induction of oxidative stress by TiO_2 was weaker than that by CuO and ZnO nanoparticles, but TiO_2 nanoparticles also induced significant oxidative stress. Anatase-type TiO_2 nanoparticles induced an increase in the intracellular ROS levels in HaCaT cells.[28] Intracellular ROS level in cells exposed to TiO_2 nanoparticles for 24 h at a concentration of 30 μg/mL was approximately 1.5–2 times higher than those in unexposed cells. Gurr *et al.*[58] reported that TiO_2 nanoparticles induced an increase of intracellular ROS levels, lipid peroxidation and oxidative DNA damage in BEAS-2B cells. Mouse macrophage-like RAW264.7 cells showed transient ROS induction in the early stage of TiO_2 exposure. On the other hand, ZnO nanoparticles induced continuous strong induction of intracellular ROS levels in RAW264.7 cells[22] and decreased cell viability. In contrast, the influence of TiO_2 nanoparticles on cell viability was lower than that of ZnO nanoparticles.[20,43,59,60]

Some nanocarbons also induced an increase in intracellular ROS levels. An increase in intracellular oxidative stress was observed in cells exposed to fullerene C_{60} and carbon black.[18,61] Carbon nanotube and gold nanoparticles also induced an increase in intracellular ROS levels.[12,62–64]

The mechanisms of ROS generation from insoluble nanoparticles are not completely understood. In many cases, mitochondria seem to be important in ROS induction by nanoparticles; however, further investigations are required.

4.3. *Activation of antioxidative stress system by nanoparticles*

ROS generated via nanoparticles and lipid radicals, a secondary product of oxidative stress, activate the antioxidative system. The major antioxidative molecule in cells is GSH. GSH contains a thiol group and prevents the oxidation of intracellular molecules caused by free radicals or ROS. Under oxidative stress conditions, intracellular GSH is converted to glutathione disulfide (GSSG), which is an oxidized form of GSH. Additionally, the expression of glutathione reductase is induced. Glutathione reductase converts GSSG to GSH. We can determine the oxidative stress level by measuring the intracellular GSH level and the expression of redox-related enzymes such as superoxide dismutase (SOD) and catalase. Yang *et al.*[65] compared the cellular influences of SiO_2 and ZnO nanoparticles in primary mouse embryonic fibroblasts. Both SiO_2 and ZnO nanoparticles induced an increase in intracellular ROS levels, a decrease in GSH and SOD levels and an increase in lipid peroxidation. The induction of oxidative stress by ZnO nanoparticles was markedly stronger than that of SiO_2 nanoparticles. GSH depletion was also observed in cells exposed to Cr_2O_3 nanoparticles.[10] Hydroxyapatite nanoparticles showed an induction of intracellular ROS levels and a decrease in SOD in rat glioma C6 cells.[66] Exposure of airway epithelial HEp-2 cells to CuO nanoparticles induced the activation of GPx and inhibited the activity of glutathione reductase and catalase. Inhibition of the catalase activity was also observed upon exposure to Fe_2O_3 nanoparticles.[67] Carbon black nanoparticles induced a decrease in GSH but did not affect HO-1 expression.[61]

HO-1, a major stress response enzyme, is also induced by oxidative stress due to nanoparticle exposure. HO-1 catalyzes the reaction that converts heme to biliverdin, CO, and Fe, and bilirubin, which has strong antioxidative activity, is produced as the final product. Thus, the induction of HO-1 reduces oxidative stress levels. Many types of nanoparticles induce HO-1 expression. TiO_2 nanoparticles induced intracellular ROS generation and the expression of HO-1 in alveolar macrophages.[68] In addition, exposure to iron-doped amorphous SiO_2 nanoparticles, silver and NiO nanoparticles induced HO-1 expression.[44,69,70]

In vivo activation of the antioxidative system by nanoparticles has also been reported. Intratracheal instillation of NiO nanoparticles induced HO-1 gene expression in the mouse lung. However, TiO_2 nanoparticles did not induce HO-1 expression.[71] GSH concentration in mouse bronchoalveolar lavage fluid (BALF) was decreased by intratracheal instillation of iron oxide.[72]

Nanoparticles drive an oxidative stress-related signal pathway. Exposure of CeO_2 nanoparticles to human bronchus BEAS-2B cells caused the induction of intracellular ROS levels and expression of HO-1. At the same time, phosphorylation of p38MAP kinase and nuclear localization of Nrf2 were observed.[14] SiO_2 nanoparticles showed induction of intracellular ROS levels, HO-1 expression, phosphorylation of extracellular signal-regulating kinase and Nrf2 nuclear translocation.[73] Nano-WC-Co particles drove the Nrf2 signaling pathway and induced the expression of glutathione S-transferase (GST) in JB6 cells.[74] It has also been reported that silver nanoparticles induced HO-1 expression that was dependent on Nrf2. The phosphoinositide 3-kinase and p38MAP kinase signaling cascades were involved in the induction of HO-1 by silver nanoparticles.[75] SWCNT, a major nanofiber, also showed intracellular ROS induction and activation of p38 in human mesothelial cells.[76] These results suggest that p38MAP kinase and Nrf2 are involved in nanoparticle-induced HO-1 expression. Some antioxidative enzymes may be induced by oxidative stress due to nanoparticles via the Nrf2-MAP kinase pathway. Nrf2 is the oxidative stress response transcription factor that regulates gene expression of antioxidative stress enzymes, including HO-1.[77] Under normal conditions, Nrf2 is localized in the cytosol as a complex with Keap1. Keap1 releases Nrf2 by ROS and electrophile species, and Nrf2 travels into the nucleus where it enhances the transcription of antistress response genes such as HO-1 and GST.

In many cases, nanoparticles induced the activation of oxidative stress response proteins with an increase in intracellular ROS levels. When the nanoparticle-derived oxidative stress is weak or transient, the effect of the oxidative stress on cells and the body is limited. However, if the oxidative stress is strong or continuous, biomolecules are oxidized.

4.4. *Lipid oxidation and DNA damage by nanoparticles*

If the antioxidative system fails to protect the cells and body from nanoparticle-derived oxidative stress, biomolecules such as proteins, lipids and nucleic acids are oxidized. Among the biomolecules, fatty acids and, in particular, polyunsaturated fatty acids, are easily oxidized,[78] and lipid peroxide is accumulated by the lipid peroxidation chain reaction.[79] Thus, the oxidized fatty acids and their metabolic products are employed as oxidative stress markers. MDA, which is the end product of lipid peroxidation, is commonly used as a marker of lipid peroxidation. MDA is normally measured by the thiobarbituric acid reactive substances (TBARS) assay. Lipid peroxidation in cells can be measured with diphenyl-1-pyrenylphosphine (DPPP).[80] According to these measurement methods, lipid oxidation was observed when nanoparticles were used, particularly when cells were exposed to metal oxide and metal nanoparticles, with the induction of intracellular ROS levels.

Increases in lipid peroxide and 8-iso-prostaglandin (8-iso-PG), an oxidized product of arachidonic acid, were observed in cells exposed to silver nanoparticles.[81] NiO nanoparticles also induced an increase in lipid peroxidation and hydroxyoctadecadienoic acid (HODE), an oxidized product of linoleic acid, in A549 cells. An increase in HODE was also observed *in vivo*; intratracheal instillation of NiO nanoparticles to rat induced an increase of HODE in BALF at 3–7 days after instillation.[44] With these nanoparticles, induction of intracellular ROS generation and depletion of GSH were also observed. CuO nanoparticles also caused the induction of intracellular ROS levels, depletion of GSH and increased lipid peroxidation in HEp-2 cells.[67] CuO-induced lipid peroxidation was reduced by co-treatment of cells with resveratrol. Many oxidative products of lipids are potentially toxic. MDA has a high reactivity and form DNA adducts.[82] It is suggested that oxidative products of lipids due to nanoparticles such as 4-HNE and MDA induce further oxidative stress.[61]

Oxidative DNA damage was also observed in cells exposed to nanoparticles. In many cases, oxidative DNA damage was detected by comet assay with fpg. Fpg repairs DNA injury from oxidative stress, and it removes oxidized purines such as formamidopyrimidine and 8-oxoguanine from damaged DNA.[83] 8-Hydroxy-2′-deoxyguanosine (8-OHdG) is also used as an oxidative stress-related DNA damage marker.[84] A few reports suggest that TiO_2 and SiO_2 nanoparticles induced the formation of 8-OHdG.[85,86] DNA damage with the induction of intracellular ROS levels was also observed with ZnO, CuO and Cr_2O_3 nanoparticles.[10,45,51,87,88] Karlsson *et al.*[45] examined the cellular influences induced by various metal oxide nanoparticles. When cells were exposed to CuO, TiO_2, ZnO, $CuZnFe_2O_4$, Fe_3O_4 and Fe_2O_3 nanoparticles, intracellular ROS levels were increased only in the CuO-exposed cells, whereas oxidative DNA damage was observed in Fe_3O_4-, $CuZnFe_2O_4$-, ZnO- and CuO-exposed cells. The oxidative DNA damage induced by CuO nanoparticles was stronger than that of CuO fine particles. CuO fine particles did not induce significant DNA damage.[89] Karlsson *et al.*[89] explained that the cellular influences induced by CuO nanoparticles were due to released Cu^{2+}. These soluble nanoparticles induce not only DNA damage but also lipid peroxidation and cell death. Oxidative DNA damage induced by soluble nanoparticles tends to be greater than that of insoluble nanoparticles. However, oxidative DNA damage was also induced by some insoluble nanoparticles. Fullerene C_{60} also caused the induction of intracellular ROS levels and DNA damage, but did not show lipid peroxidation or cell death.[18]

Possibly, the mechanism of oxidative stress induction by fullerene C_{60} differs from that by metal oxide and the metal nanoparticles. The mechanism of oxidative stress induced by C_{60} is presently not known. Some studies have concluded that TiO_2 nanoparticles induced oxidative DNA damage,[85,90] although others showed that TiO_2 nanoparticles did not induce oxidative DNA damage.[91,92] These conflicting results may be due to different physical properties such as the crystalline phase and photocatalytic activity. The oxidative DNA damage induced by Fe_2O_3 and Fe_3O_4 nanoparticles was small to negligible.[45,85]

The synergistic oxidative effect of nanoparticles has been reported. Both Fe_2O_3 and carbon black nanoparticles alone did not induce oxidative stress in cells. However, co-exposure of Fe_2O_3 and carbon black induced protein and lipid oxidation.[93,94] At the same time, released Fe^{3+} from Fe_2O_3 nanoparticles was reduced to Fe^{2+} by carbon black. This result suggests that a redox status of released metal ion from nanoparticles is important in the induction of oxidative stress.

4.5. *Oxidative stress-related apoptosis by nanoparticles*

Some nanoparticles induce apoptosis, such as SiO_2 nanoparticles that induce apoptosis in human lung fibroblasts.[95] Concomitantly, we observed enhancement of p53 expression, release of cytochrome c from mitochondria, and activation of caspase-9. Silver nanoparticles also caused the induction of intracellular ROS levels, release of cytochrome c, and the activation of caspase-9 in human Chang liver cells, which then died due to apoptosis.[81] Additionally, silver nanoparticles induced mitochondrial-related apoptosis in *Candida albicans* with increased hydroxyl radical levels.[96] TiO_2 and Cr_2O_3 nanoparticles induced apoptosis with the activation of caspase-3 and caspase-9.[10,28] When cells are exposed to a low level of ROS, apoptosis is induced.[18,97] For example, 250 μM H_2O_2 induced apoptosis in A549 and HaCaT cells.[18] These results suggest that oxidative stress is involved in the induction of apoptosis by nanoparticles. In many cases, nanoparticles induced apoptosis with the generation of intracellular ROS and activation of caspase-9.

4.6. *Mechanism of oxidative stress in nanotoxicology*

In both *in vitro* and *in vivo* investigations, many but not all types of nanoparticles induce oxidative stress. Compared with microscale particles, nanoparticles have features such as increased solubility, large surface adsorption and surface reactivity. The induction of oxidative stress by nanoparticles is attributable to these physical and chemical properties. Therefore, insoluble and inactive nanoparticles do not induce oxidative stress. For example, nanodiamond did not induce any oxidative stress on cultured cells even at a high concentration of 1.0 mg/mL.[12] Similarly, insoluble platinum nanoparticles did not induce oxidative stress in cells.[11] The size of the particle does not directly relate to the oxidative stress caused by nanoparticles, and the specific surface area itself is not directly related to the oxidative stress induced by nanoparticles. However, increases in metal ion release or surface reactivity caused by an increase in the specific surface area are more important in the induction of oxidative stress.

In many cases, an increase in intracellular ROS levels is observed in cells exposed to nanoparticles. However, although ZnO nanoparticles induced strong oxidative stress in cultured cells, ZnO nanoparticles did not generate ROS in a cell-free system. Song *et al.* reported that ZnO nanoparticles caused the induction of intracellular ROS levels in mouse macrophage Ana-1 cells, but ESR measurements did not indicate that ROS were generated from ZnO nanoparticles.[98] The induction of intracellular ROS levels by nanoparticles may be caused by mitochondrial dysfunction rather than by the generation of ROS on the surface of the nanoparticles. Intracellular ROS is likely generated by the interaction between the nanoparticles and the cell.

A major factor in oxidative stress induction by nanoparticles is released metal ion. ROS generation by the interaction of cells with released metals ions triggers further oxidative stress. Increase in solubility is one of the characteristic features of nanoparticles. In some cases, even if the particle is classified as insoluble at the microscale, it releases metal ions at the nanoscale.[10,19] The cytotoxicity induced by nanoparticles is metal toxicity. Although the induction of intracellular ROS levels is due to Fenton-like reactions by the released metal ions, the metal ion is provided by internalized nanoparticles. Generally, extracellular metal ions cannot pass easily through the cell membrane. Intracellular metal ion concentration is strictly controlled by ion channels and metal-binding proteins.[99] Nanoparticles are taken up into cells by endocytosis.[10–13,19,28,100] The internalized nanoparticles will release metal ions inside the cell and induce oxidative stress. Soluble nanoparticles have features of both the particle and the metal ion. Cellular uptake and metal ion release are major factors in nanoparticle-mediated induction of intracellular ROS generation and the subsequent oxidative stress.

Nanoparticles also induce pulmonary oxidative stress *in vivo* by inhalation, intratracheal instillation or aspiration of pharynges. The induction of intracellular ROS generation is important in the *in vivo* toxicity of nanoparticles as a primary oxidative stress; however, measurement of *in vivo* ROS generation is difficult. Oxidized biomolecules and the antioxidative system indicate the presence of oxidative stress in lung. *In vivo*, it is necessary to consider secondary oxidative stress from the accumulation of neutrophils and macrophages in addition to the nanoparticle-induced ROS generation. Inhaled nanoparticles may induce ROS generation in the lung, and the subsequently generated ROS activate the antioxidation system. If a large amount of ROS is generated, lipids, proteins and DNA are oxidized. ROS or lipid peroxide enhances the gene expression of oxidative stress response proteins such as HO-1 via the Nrf2-MAP kinase pathway. Additionally, immunostaining for p65 (RELA) showed NF-κB nuclear translocation after exposure of ZnO and CdS nanoparticles to the proximal tubule HK-2 cells.[43] Nuclear translocation of NF-κB was not observed in cells exposed to TiO_2 nanoparticles. ROS generation and activation of NF-κB were also observed in cells exposed to SWCNT.[100] NF-κB enhances the expression of interleukins such as IL-1, IL-6 and IL-8, and inducible nitric oxide synthases (iNOS).[101] The secretion of these cytokines induces neutrophil and macrophage recruitment. Accumulated macrophages and alveolar macrophages take in nanoparticles by phagocytosis and generate ROS.[102] Neutrophils generate hypochlorous acid (HClO) form H_2O_2 and Cl^- by the catalysis of myeloperoxidase (MPO). Neutrophils also generate tyrosyl radicals by catalysis of MPO.[103] Gene expression of iNOS was increased in TiO_2 nanoparticle-exposed alveolar macrophages.[104] iNOS catalyzes the formation of RNS from NO and ROS.[105] As a result of the reaction, ROS such as peroxynitrite ($ONOO^-$) are generated in cells. These active species and radicals generated by macrophages and neutrophils induce further oxidative stress or "secondary oxidative stress" in the lung. The continuation of oxidative stress in the lung may depend on the clearance of nanoparticles from the lung. Continued inflammation accompanies continued oxidative stress and causes various diseases. In the lung, it is possible that continuous oxidative stress and inflammation leads to lung fibrosis via activation of the NF-κB signal pathway[63] and subsequent induction of cancer. In many cases, nanoparticles that induce strong oxidative stress tend to have high solubility. These highly soluble particles may clear rapidly from the lung. For instance, ZnO nanoparticles induce strong oxidative stress in the lung but they have less persistence.[106] However, compared with soluble nanoparticles, the clearance of insoluble nanomaterials such as CNT from the lung is slow.[107] Although oxidative stress induced by insoluble nanoparticles is weaker than that induced by soluble nanoparticles, oxidative stress by inhaled nanoparticles in the lung may be more persistent. However, details of a nanoparticle-induced oxidative stress related "disease" remain to be elucidated. Further investigations into the relationship between clearance and diseases such as fibrosis and cancer by nanoparticles are necessary.

5. Summary

Currently, a number of applications of nanoparticles for life sciences industries, such as drug delivery systems and gene therapy, are being investigated. For this purpose, nanoparticles are often directly injected into body. Therefore, understanding the characteristic cellular influences of the nanoparticles is essential. There are three important components for the cellular influences of the nanoparticles: (1) metal ion release (increase of solubility), (2) large adsorption ability (particularly protein adsorption) and (3) cellular uptake. These features of the nanoparticles are also important for life sciences applications. Oxidative stress often results from the influence of these factors in cells, and cellular oxidative stress leads to general oxidative stress.

An outline of oxidative stress in nanotoxicology is shown in-Fig. 2. The induction of oxidative stress is important in cellular influences of nanoparticles.

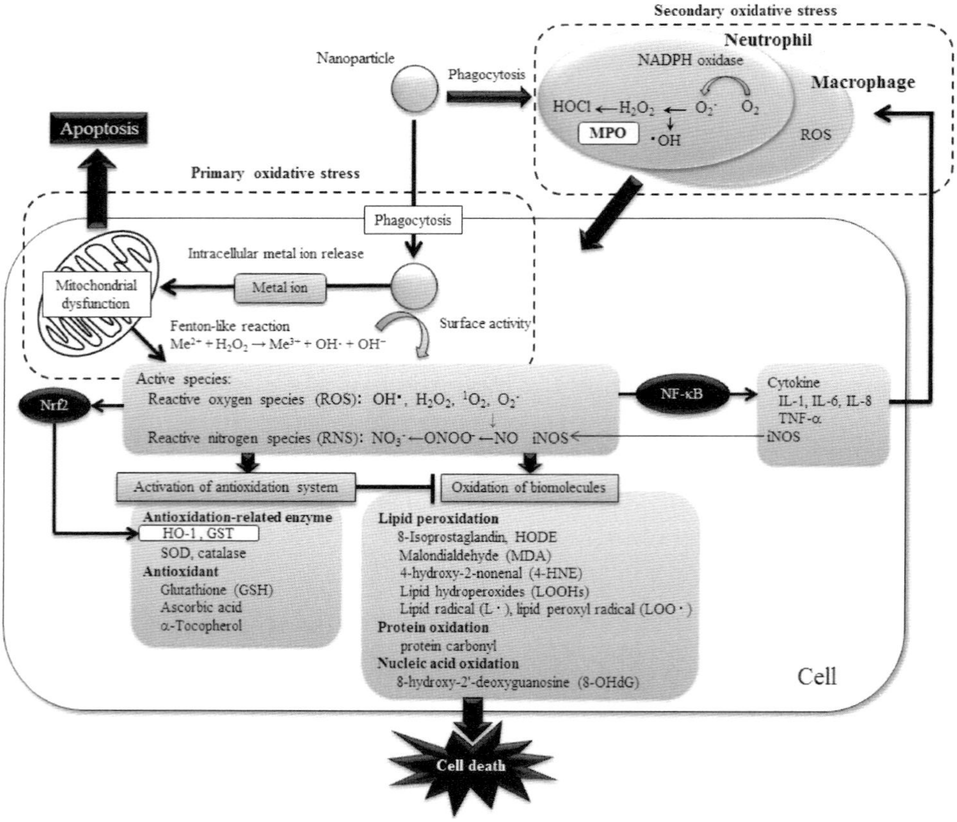

Fig. 2. Outline of nanoparticles-related oxidative stress.

In vitro investigations revealed that nanoparticles have the potential for intracellular ROS induction. In cells, accumulated oxidative stress leads to cell death, including apoptosis. However, *in vivo*, the primary oxidative stress, ROS generation, leads to the recruitment of phagocytes, and following this, secondary oxidative stress occurs. Primary and secondary oxidative stresses *in vivo* are indistinguishable. Nonetheless, an understanding of the mechanism of oxidative stress generation from nanoparticles is essential for an understanding of nanotoxicology.

According to *in vitro* investigations, intracellular ROS generation serves as a trigger for nanoparticle-induced oxidative stress. Cellular uptake and intracellular metal ion release are important factors in intracellular ROS generation. The solubility of nanoparticles is one of the most important factors in intracellular ROS generation. The oxidative stress depends on the kinds of metals,[20] but the load of the oxidative stress by nanoparticles cannot completely be explained by metal ion release alone.

Insoluble nanoparticles such as TiO_2 and C_{60} also induce intracellular ROS generation. It can be said without fear of misunderstanding that oxidative stress by nanoparticles is different in individual cases. In some cases, although the chemical component is the same, the strength of the oxidative stress differs. For example, the strength of oxidative stress by TiO_2 nanoparticles is differs with the case.[28,58,41,108] This potential of nanoparticles to induce different oxidative stresses depends on their physical and chemical properties such as crystalline phase, protein adsorption ability, and zeta potential. For example, intracellular ROS induction by nanodiamonds depends on its zeta potential.[12] The intracellular ROS generation mechanism is still not completely understood. There are some difficulties faced in the induction of oxidative stress in nanotoxicology; i.e. the interaction between oxidative stress and *in vivo* toxicity, and intracellular ROS generation from insoluble nanoparticles.

References

1. R. J. Fields, C. J. Cheng, E. Quijano, C. Weller, N. Kristofik, N. Duong, C. Hoimes, M. E. Egan and W. M. Saltzman, *J. Control. Release* **164**, 41 (2012).
2. T. Y. Cheang, B. Tang, A. W. Xu, G. Q. Chang, Z. J. Hu, W. L. He, Z. H. Xing, J. B. Xu, M. Wang and S. M. Wang, *Int. J. Nanomedicine* 7, 1061 (2012).
3. Y. Niut, A. Popatt, M. Yu, S. Karmakar, W. Gu and C. Yu. *Ther. Deliv.* **3**, 1217 (2012).
4. T. N. Do, W. H. Lee, C. Y. Loo, A. V. Zavgorodniy and R. Rohanizadeh, *Ther. Deliv.* **3**, 623 (2012).
5. C. Liu and N. Zhang, Prog. *Mol. Biol. Transl. Sci.* **104**, 509 (2011).

6. M. Jeyaraj, M. Rajesh, R. Arun, D. Mubarakali, G. Sathishkumar, G. Sivanandhan, G. K. Dev, M. Manickavasagam, K. Premkumar, N. Thajuddin and A. Ganapathi, *Colloids Surf. B Biointerfaces* **102**, 708 (2013).
7. M. R. Cheng, Q. Li, T. Wan, B. He, J. Han, H. X. Chen, F. X. Yang, W. Wang, H. Z. Xu, T. Ye and B. B. Zha, *World J. Gastroenterol* **18**, 6076 (2012).
8. J. Meng, J. Xing, X. Ma, W. Cao, J. Lu, Y. Wang, X. Gao, B. Sun, X. Liang and Y. Zhao, *Nanomedicine (Lond)*, in press.
9. S. Linse, C. Cabaleiro-Lago, W. F. Xue, I. Lynch, S. Lindman, E. Thulin, S. E. Radford and K. A. Dawson, *Proc. Natl. Acad. Sci. USA* **104**, 8691 (2007).
10. M. Horie, K. Nishio, S. Endoh, H. Kato, K. Fujita, A. Miyauchi, A. Nakamura, S. Kinugasa, K. Yamamoto, E. Niki, Y. Yoshida and H. Iwahashi, *Environ. Toxicol.*, in press, doi: 10.1002/tox.20695.
11. M. Horie, H. Kato, S. Endoh, K. Fujita, K. Nishio, L. K. Komaba, H. Fukui, A. Nakamura, A. Miyauchi, T. Nakazato, S. Kinugasa, Y. Yoshida, Y. Hagihara, Y. Morimoto and H. Iwahashi, *Metallomics* **3**, 1244 (2011).
12. M. Horie, L. K. Komaba, H. Kato, A. Nakamura, K. Yamamoto, S. Endoh, K. Fujita, S. Kinugasa, K. Mizuno, Y. Hagihara, Y. Yoshida and H. Iwahashi, *Diamond Relat. Mater.* **24**, 15 (2012).
13. M. Horie, K. Nishio, H. Kato, K. Fujita, S. Endoh, A. Nakamura, A. Miyauchi, S. Kinugasa, K. Yamamoto, E. Niki, Y. Yoshida, Y. Hagihara and H. Iwahashi, *J. Biochem.* **150**, 461 (2011).
14. H. J. Eom and J. Choi, *Toxicol. Lett.* **187**, 77 (2009).
15. A. Y. Estevez, S. Pritchard, K. Harper, J. W. Aston, A. Lynch, J. J. Lucky, J. S. Ludington, P. Chatani, W. P. Mosenthal, J. C. Leiter, S. Andreescu and J. S. Erlichman, *Free Radic. Biol. Med.* **51**, 1155 (2011).
16. S. M. Hirst, A. S. Karakoti, R. D. Tyler, N. Sriranganathan, S. Seal and C. M. Reilly, *Small* **5**, 2848 (2009).
17. A. Nunes, K. Al-Jamal, T. Nakajima, M. Hariz and K. Kostarelos, *Arch. Toxicol.* **86**, 1009 (2012).
18. M. Horie, K. Nishio, H. Kato, N. Shinohara, A. Nakamura, K. Fujita, S. Kinugasa, S. Endoh, K. Yamamoto, O. Yamamoto, E. Niki, Y. Yoshida and H. Iwahashi, *J. Biochem.* **148**, 289 (2010).
19. M. Horie, K. Nishio, K. Fujita, H. Kato, A. Nakamura, S. Kinugasa, S. Endoh, A. Miyauchi, K. Yamamoto, H. Murayama, E. Niki, H. Iwahashi, Y. Yoshida and J. Nakanishi, *Chem. Res. Toxicol.* **22**, 1415 (2009).
20. M. Horie, K. Fujita, H. Kato, S. Endoh, K. Nishio, L. K. Komaba, A. Nakamura, A. Miyauchi, S. Kinugasa, Y. Hagihara, E. Niki, Y. Yoshida and H. Iwahashi, *Metallomics* **4**, 350 (2012).
21. H. Sies, *Am. J. Med.* **91**, 31S (1991).
22. T. Xia, M. Kovochich, M. Liong, L. Mädler, B. Gilbert, H. Shi, J. I. Yeh, J. I. Zink and A. E. Nel, *ACS Nano* **2**, 2121 (2008).
23. D. Cohen, Y. Soroka, Z. Ma'or, M. Oron, M. Portugal-Cohen, F. M. Brégégère, D. Berhanu, E. Valsami-Jones, N. Hai and Y. Milner, *Toxicol. In Vitro.* **27**, 292 (2013).
24. Z. Wang, N. Li, J. Zhao, J. C. White, P. Qu and B. Xing, *Chem. Res. Toxicol.* **25**, 1512 (2012).

25. G. Cai, J. Zhu, C. Shen, Y. Cui, J. Du and X. Chen, *Biol. Trace Elem. Res.* **150**, 200 (2012).
26. H. Fukui, M. Horie, S. Endoh, H. Kato, K. Fujita, K. Nishio, L. K. Komaba, J. Maru, A. Miyauhi, A. Nakamura, S. Kinugasa, Y. Yoshida, Y. Hagihara and H. Iwahashi, *Chem. Biol. Interact.* **198**, 29 (2012).
27. T. Cedervall, I. Lynch, S. Lindman, T. Berggård, E. Thulin, H. Nilsson, K. A. Dawson and S. Linse, *Proc. Natl. Acad. Sci. USA.* **104**, 2050 (2007).
28. M. Horie, K. Nishio, K. Fujita, H. Kato, S. Endoh, M. Suzuki, A. Nakamura, A. Miyauchi, S. Kinugasa, K. Yamamoto, H. Iwahashi, H. Murayama, E. Niki and Y. Yoshida, *Toxicol. In Vitro.* **24**, 1629 (2010).
29. M. Lundqvist, J. Stigler, T. Cedervall, T. Berggård, M. B. Flanagan, I. Lynch, G. Elia and K. Dawson, *ACS Nano* **5**, 7503 (2011).
30. C. Nazli, T. I. Ergenc, Y. Yar, H. Y. Acar and S. Kizilel, *Int. J. Nanomedicine.* **7**, 1903 (2012).
31. E. Ruoslahti and M. D. Pierschbacher, *Science* **238**, 491 (1987).
32. E. J. Brown and J. L. Goodwin, *J. Exp. Med.* **167**, 777 (1988).
33. M. J. Clift, C. Brandenberger, B. Rothen-Rutishauser, D. M. Brown and V. Stone, *Toxicology.* **286**, 58 (2011).
34. M. Al-Rawi, S. Diabaté and C. Weiss, *Arch. Toxicol.* **85**, 813 (2011).
35. H. Shinto, Y. Ohta and T. Fukasawa, *Adv. Powder Technol.* **23**, 693 (2012).
36. H. Sies, *Am. J. Med.* **91**, S31 (1991).
37. S. Lee, E. Tak, J. Lee, M. A. Rashid, M. P. Murphy, J. Ha and S. S. Kim, *Cell Res.* **21**, 817 (2011).
38. Z. H. Chen, Y. Yoshida, Y. Saito and E. Niki, *Neurosci. Lett.* **383**, 256 (2005).
39. Y. Tampo, S. Kotamraju, C. R. Chitambar, S. V. Kalivendi, A. Keszler, J. Joseph and B. Kalyanaraman, *Circ. Res.* **92**, 56 (2003).
40. I. Fenoglio, G. Greco, S. Livraghi and B. Fubini, *Chemistry* **15**, 4614 (2009).
41. J. F. Reeves, S. J. Davies, N. J. Dodd and A. N. Jha, Hydroxyl radicals (*OH) are associated with titanium dioxide (TiO_2) nanoparticle-induced cytotoxicity and oxidative DNA damage in fish cells. *Mutat. Res.* **640**, 113 (2008).
42. N. J. Dodd and A. N. Jha, *Mutat. Res.* **660**, 79 (2009).
43. I. Pujalté, I. Passagne, B. Brouillaud, M. Tréguer, E. Durand, C. Ohayon-Courtès and B. L'Azou, *Part. Fibre. Toxicol.* **8**, 10 (2011).
44. M. Horie, H. Fukui, K. Nishio, S. Endoh, H. Kato, K. Fujita, A. Miyauchi, A. Nakamura, M. Shichiri, N. Ishida, S. Kinugasa, Y. Morimoto, E. Niki, Y. Yoshida and H. Iwahashi, *J. Occup. Health* **53**, 64 (2011).
45. H. L. Karlsson, P. Cronholm, J. Gustafsson and L. Möller, *Chem. Res. Toxicol.* **21**, 1726 (2008).
46. M. Valko, H. Morris and M. T. Cronin, Metals, toxicity and oxidative stress. *Curr. Med. Chem.* **12**, 1161 (2005).
47. A. Dhakshinamoorthy, S. Navalon, M. Alvaro and H. Garcia, *ChemSusChem.* **5**, 46 (2012).
48. J. M. Suski, M. Lebiedzinska, M. Bonora, P. Pinton, J. Duszynski and M. R. Wieckowski, *Methods Mol. Biol.* **810**, 183 (2012).
49. G. R. Buettner, *Anticancer Agents Med. Chem.* **11**, 341 (2011).

50. S. Marchi, C. Giorgi, J. M. Suski, C. Agnoletto, A. Bononi, M. Bonora, E. De Marchi, S. Missiroli, S. Patergnani, F. Poletti, A. Rimessi, J. Duszynski, M. R. Wieckowski and P. Pinton, *J. Signal Transduct.* **2012**, 329635 (2012).
51. V. Sharma, D. Anderson and A. Dhawan, *Apoptosis* **17**, 852 (2012).
52. S. S. Leonard, G. K. Harris and X. Shi, *Free Radic. Biol. Med.* **37**, 1921 (2004).
53. A. Fujishima and K. Honda, *Nature* **238**, 37 (1972).
54. C. M. Sayes, R. Wahi, P. A. Kurian, Y. Liu, J. L. West, K. D. Ausman, D. B. Warheit and V. L. Colvin, *Toxicol. Sci.* **92**, 174 (2006).
55. R. Cai, Y. Kubota, T. Shuin, H. Sakai, K. Hashimoto and A. Fujishima, *Cancer Res.* **52**, 2346 (1992).
56. K. Adachi, N. Yamada, K. Yamamoto, Y. Yoshida and O. Yamamoto, *Nanotoxicology* **4**, 296 (2010).
57. T. Xia, M. Kovochich, J. Brant, M. Hotze, J. Sempf, T. Oberley, C. Sioutas, J. I. Yeh, M. R. Wiesner and A. E. Nel, *Nano Lett.* **6**, 1794 (2006).
58. J. R. Gurr, A. S. Wang, C. H. Chen and K. Y. Jan, *Toxicology* **213**, 66 (2005).
59. I. S. Kim, M. Baek and S. J. Choi, *J. Nanosci. Nanotechnol.* **10**, 3453 (2010).
60. C. C. Huang, R. S. Aronstam, D. R. Chen and Y. W. Huang, *Toxicol. In Vitro.* **24**, 45 (2010).
61. L. Foucaud, S. Goulaouic, A. Bennasroune, P. Laval-Gilly, D. Brown, V. Stone and J. Falla, *Toxicol. In Vitro.* **24**, 1512 (2010).
62. M. Pacurari, Y. Qian, W. Fu, D. Schwegler-Berry, M. Ding, V. Castranova and N. L. Guo, *J. Toxicol. Environ. Health A* **75**, 129 (2012).
63. X. He, S. H. Young, D. Schwegler-Berry, W. P. Chisholm, J. E. Fernback and Q. Ma, *Chem. Res. Toxicol.* **24**, 2237 (2011).
64. W. Gao, K. Xu, L. Ji and B. Tang, *Toxicol. Lett.* **205**, 86 (2011).
65. H. Yang, C. Liu, D. Yang, H. Zhang and Z. Xi, *J. Appl. Toxicol.* **29**, 69 (2009).
66. J. Xu, P. Xu, Z. Li, J. Huang and Z. Yang, *J. Biomed. Mater. Res. A* **100**, 738 (2012).
67. B. Fahmy and S. A. Cormier, *Toxicol. In Vitro.* **23**, 1365 (2009).
68. A. M. Scherbart, J. Langer, A. Bushmelev, D. van Berlo, P. Haberzettl, F. J. van Schooten, A. M. Schmidt, C. R. Rose, R. P. Schins and C. Albrecht, *Part. Fibre Toxicol.* **8**, 31 (2011).
69. D. Napierska, V. Rabolli, L. C. Thomassen, D. Dinsdale, C. Princen, L. Gonzalez, K. L. Poels, M. Kirsch-Volders, D. Lison, J. A. Martens and P. H. Hoet, *Chem. Res. Toxicol.* **25**, 828 (2012).
70. T. H. Kim, M. Kim, H. S. Park, U. S. Shin, M. S. Gong and H. W. Kim, *J. Biomed. Mater. Res. A* **100**, 1033 (2012).
71. M. Horie, H. Fukui, S. Endoh, J. Maru, A. Miyauchi, M. Shichiri, K. Fujita, E. Niki, Y. Hagihara, Y. Yoshida, Y. Morimoto and H. Iwahashi, *Inhal. Toxicol.* **24**, 391 (2012).
72. E. J. Park, H. Kim, Y. Kim, J. Yi, K. Choi and K. Park, *Toxicology* **275**, 65 (2010).
73. H. J. Eom and J. Choi, *Environ. Health. Toxicol.* **26**, e2011013 (2011).
74. X. D. Zhang, J. Zhao, L. Bowman, X. Shi, V. Castranova and M. Ding, *J. Environ. Pathol. Toxicol. Oncol.* **29**, 31 (2010).
75. S. J. Kang, I. G. Ryoo, Y. J. Lee and M. K. Kwak, *Toxicol. Appl. Pharmacol.* **258**, 89 (2012).
76. M. Pacurari, X. J. Yin, J. Zhao, M. Ding, S. S. Leonard, D. Schwegler-Berry, B. S. Ducatman, D. Sbarra, M. D. Hoover, V. Castranova and V. Vallyathan, *Environ. Health Perspect.* **116**, 1211 (2008).

77. T. Ishii, K. Itoh, H. Sato and S. Bannai, *Free Radic. Res.* **31**, 351 (1999).
78. E. N. Frankel, *Prog. Lipid Res.* **23**, 197 (1984).
79. C. Mylonas and D. Kouretas, *In Vivo.* **13**, 295 (1999).
80. Y. Okimoto, E. Warabi, Y. Wada, E, Niki, T. Kodama and N. Noguchi, *Free Radic. Biol. Med.* **35**, 576 (2003).
81. M. J. Piao, K. A. Kang, I. K. Lee, H. S. Kim, S. Kim, J. Y. Choi, J. Choi and J. W. Hyun, *Toxicol. Lett.* **201**, 92 (2011).
82. L. J. Marnett, *Mutat. Res.* **424**, 83 (1999).
83. L. Serre, K. Pereira de Jésus, S. Boiteux, C. Zelwer and B. Castaing, *EMBO J.* **21**, 2854 (2002).
84. Y. Matsumoto, Y. Ogawa, R. Yoshida, A. Shimamori, H. Kasai and H. Ohta, *J. Occup. Health* **50**, 366 (2008).
85. K. Bhattacharya, M. Davoren, J. Boertz, R. P. Schins, E. Hoffmann and E. Dopp, *Part. Fibre Toxicol.* **6**, 17 (2009).
86. C. Gong, G. Tao, L. Yang, J. Liu, H. He and Z. Zhuang, *Mol. Biol. Rep.* **39**, 4915 (2012).
87. V. Sharma, P. Singh, A. K. Pandey and A. Dhawan, *Mutat. Res.* **745**, 84 (2012).
88. M. Ahamed, M. A. Siddiqui, M. J. Akhtar, I. Ahmad, A. B. Pant and H. A. Alhadlaq, *Biochem. Biophys. Res. Commun.* **396**, 578 (2010).
89. H. L. Karlsson, J. Gustafsson, P. Cronholm and L. Möller, *Toxicol. Lett.* **188**, 112 (2009).
90. Z. Magdolenova, D. Bilaničová, G. Pojana, L. M. Fjellsbø, A. Hudecova, K. Hasplova, A. Marcomini and M. Dusinska, *J. Environ. Monit.* **14**, 455 (2012).
91. K. Gerloff, I. Fenoglio, E. Carella, J. Kolling, C. Albrecht, A. W. Boots, I. Förster and R. P. Schins, *Chem. Res. Toxicol.* **25**, 646 (2012).
92. H. Du, X. Zhu, C. Fan, S. Xu, Y. Wang and Y. Zhou, *Environ. Toxicol.* **27**, 590 (2012).
93. J. M. Berg, S. Ho, W. Hwang, R. Zebda, K. Cummins, M. P. Soriaga, R. Taylor, B. Guo and C. M. Sayes, *Chem. Res. Toxicol.* **23**, 1874 (2010).
94. B. Guo, R. Zebda, S. J. Drake and C. M. Sayes, *Part. Fibre Toxicol.* **6**, 4 (2009).
95. Z. Xu, L. Chou and J. Sun, *J. Appl. Toxicol.* **32**, 358 (2012).
96. I. S. Hwang, J. Lee, J. H. Hwang, K. J. Kim and D. G. Lee, *FEBS J.* **279**, 1327 (2012).
97. Y. Saito, K. Nishio, Y. Ogawa, J. Kimata, T. Kinumi, Y. Yoshida, N. Noguchi and E. Niki, *Free Radic. Res.* **40**, 619 (2006).
98. W. Song, J. Zhang, J. Guo, J. Zhang, F. Ding, L. Li and Z. Sun, *Toxicol. Lett.* **199**, 389 (2010).
99. G. Kovacs, T. Danko, M. J. Bergeron, B. Balazs, Y. Suzuki, A. Zsembery and M. A. Hediger, *Cell Calcium.* **49**, 43 (2011).
100. M. Pacurari, X. J. Yin, J. Zhao, M. Ding, S. S. Leonard, D. Schwegler-Berry, B. S. Ducatman, D. Sbarra, M. D. Hoover, V. Castranova and V. Vallyathan, *Environ. Health Perspect.* **116**, 1211 (2008).
101. A. Panwalkar, S. Verstovsek and F. Giles, *Cancer* **100**, 1578 (2004).
102. H. J. Forman and M. Torres, *IUBMB Life* **51**, 365 (2001).
103. J. W. Heinecke, W. Li, G. A. Francis and J. A. Goldstein, *J. Clin. Invest.* **91**, 2866 (1993).
104. A. M. Scherbart, J. Langer, A. Bushmelev, D. van Berlo, P. Haberzettl, F. J. van Schooten, A. M. Schmidt, C. R. Rose, R. P. Schins and C. Albrecht, *Part. Fibre. Toxicol.* **8**, 31 (2011).

105. R. Domenico, *Curr. Pharm. Des.* **10**, 1667 (2004).
106. R. G. Cooper, *Indian J. Occup. Environ. Med.* 12, 10 (2008).
107. M. J. Osmond-McLeod, C. A. Poland, F. Murphy, L. Waddington, H. Morris, S. C. Hawkins, S. Clark, R. Aitken, M. J. McCall and K. Donaldson, *Part. Fibre. Toxicol.* **8**, 15 (2011).
108. M. Horie, L. K. Komaba, H. Kato, S. Endoh, K. Fujita, K. Nishio, A. Nakamura, A. Miyauchi, S. Kinugasa, Y. Hagihara, Y. Yoshida and H. Iwahashi, *Nano Biomedicine* **2**, 182 (2010).

Chapter A6

Regulatory Aspects of Pharmaceutical Nanopreparations

Gerard M. Jensen and Linda Soo Hoo*

Gilead Sciences, 650 Cliffside Drive, San Dimas California 91773, USA
**gjensen@gilead.com*

Abstract

The regulation of nanotechnology derived medicines, epitomized by liposomal nanopreparations in the past, and featuring highly diverse chemical and physical modalities in the present and future, is far from settled. We explore the factors motivating the need for class specific regulation, and update the current status of efforts to provide a regulatory framework that fosters innovation yet protects the public.

1. Introduction

In 1959, Richard Feynman[1] delivered a lecture that is widely credited with launching the field of nanotechnology — doing work and making things on nanometer scales, and issuing a challenge to the scientific community to solve the problems of scale — with many suggestions for approaches; essentially taking what had been in the realm of science fiction and making it begin to seem achievable.

Nanomedicine in the past decade has grown into a much hyped area of research. With three-quarters of new drug research being related to nanomedicines, and more than half of US patents focused on drug delivery systems, there is little doubt that nanotechnology related regulations will have a significant impact on the future commercial development of such emerging technologies and the subsequent ability to treat life threatening and/or debilitating diseases.[2] In 2011, the global nanomedicine market reportedly reached $72.8 billion and is projected to grow to over $130 billion by 2016.[3] What are the current trajectories of government regulation and oversight, and fundamentally what scope — and what definitions — are emerging as those that will be utilized?

Widespread public acceptance and use of nanomedicines may falter ultimately, if public perception of all things "nano" is grounded in fiction rather than good science. The level of public knowledge of nanotechnology remains consistently low each year. In 2004, nearly 80–85% of Americans who were surveyed reported to have limited or no familiarity with nanotechnology.[4] By 2010, the number decreased to approximately 50%, which compared similarly to surveys of Japanese and Europeans.[5,6] Naturally, the lack of knowledge and adequate communication leads to doubt, fear and mistrust. The risks associated with nanotechnology may even be amplified to reach "hysteria" type levels akin to those experienced when genetically modified foods (GMO) were first introduced in Europe. Take for example the first "nanoscare" involving the aerosol Magic-Nano products introduced in Germany in 2006.[7] Roughly 100 people reported adverse side effects including coughing, headaches and vomiting, along with a small number hospitalized for the treatment of pulmonary edema. The incident eventually resulted in a national recall in that same year. Ironically, the culprit behind the first "nanoscare" did not involve any nanoparticles *per se*. The investigation found that the symptoms were caused by the *absence* of silicon nanoparticles owing to the particles precipitating out of the solution due to an inadvertent change in the pH of the aerosols.[7]

Regardless of the cause, the cat is out of the bag. Perception is reality; and a general negativity may be sufficient to hamper public willingness to accept nanotechnology as applied in medicine. The unintended consequences of an ill-informed public include challenges to the future research funding in nanotechnology and potential barriers to commercialization of such promising innovative therapies. Given the lack of public familiarity with nanotechnology and misunderstandings concerning the potential toxicity risks associated, examining the actual scientific evidence available in the literature on safety risks in pre-clinical models and clinical experience associated with its use is pertinent. This of course is the primary element driving the appropriate level of regulation.

Should nanomedicines be presumed safe? Of course not; there are evidence based concerns, just as there are with any new class of medicines. Given the complexity and diversity of nanoparticles and the body of evidence documented thus far, the most meaningful conclusion one can derive based on the data is that toxicity for any given nanoparticle based therapeutic must largely be judged on a case-by-case basis. We note here that complexity *per se* is not uniquely the province of nanosystems: If we look at camptothecin and its many analogs,[8] there are variants that impact binding to albumin, and thence pharmacokinetics, that impact function via the lactone ring opening equilibrium, that form elements of pro-drugs activated *in vivo*, and wide ranges of hydropathy leading to instrinsic differences in biofate and in how the drugs must be formulated. Lessons have been learned from the class that have been applied intelligently, but each drug has a unique developmental and regulatory history.

Nearly 50 years after Feynman's talk, the US FDA and other world wide agencies are trying to tackle the issue of appropriate regulatory guidance for nanomedicines. Some have balked at the merits of this initiative;[9] The Biotechnology Industry Association despairs the lack of any useful definition being available; and that a definition might be sweeping and add regulatory burden to many existing and future products. From its onset, the FDA has had a dual mission. Per the agency mission statement, the "FDA is responsible for protecting the public health by assuring the safety, efficacy and security of human ... drugs, biological products, medical devices,...". Unlike a more purely safety focused agency like the Federal Aviation Administration, the "FDA is also responsible for advancing the public health by helping to speed innovations that make medicines more effective, safer and more affordable ....." Thus, the agency's challenge is to regulate, to achieve and assure safety but also to foster utilization of the technology where it can benefit patients. The effort for nanopharmaceutical preparations, however, nearly fails at takeoff; the first question — what is a nanomedicine? — is not a trivial one. Any definition does indeed risk being overly inclusive, or overly restrictive.

As the experience with nanotechnology is still developing, regulatory health agencies such as the FDA thus far have been reluctant to adopt any regulatory definition of "nanoscale materials" or "nanotechnology" officially.[10,11] In the absence of a clear definition, various health regulatory agencies around the world disagree on its definition and, moreover, are reluctant to establish one definitively until scientific experience is sufficiently acquired. At present, the definition proposed by the US National Nanotechnology Initiative that often is used proposes that "nanotechnology is the understanding and control of matter at dimensions between approximately 1 nm and 100 nm, where unique phenomena enable novel applications".[12]

Based on this definition to be sure, many of today's consumer products would be considered "nano" products. Some of these products include cosmetics, sunscreens, bio-sensors, fuel cells, paints, food ingredients, consumer electronics, cleansers and car tires to name a few — all of which utilize nanomaterials.[13] A considerable number of these products including cosmetics and medicines have even been on the market for a long time. For years, about one-third of cosmetic products are sunscreen lotions, where most of which contain titanium dioxide and zinc oxide nanoparticles. In medicine, nanoparticles such as superparamagnetic iron oxide, used as a contrasting agent for magnetic resonance imaging, and liposomes, typically sub-100 nm vesicles that carry either antineoplastic or antifungal agents, have been approved by the FDA since the mid-1990s and used in patients for over 20 years.

Such a size-based definition has been explicitly called out in the recent FDA draft guidance issued to industry that aims to aid manufacturers in determining whether their products involve nanotechnology.[14] While the FDA acknowledges the nanoscale definition provided by the NNI, the FDA's Center for Drug Evaluation and Research (CDER) has published internal procedures directing their CMC reviewers to consider "any materials with at least one dimension smaller than 1000 nm" under the purview of the definition for "nanomaterial."[15] By a similar token, the European Commission in 2011 further refined the nanomaterial definition that implicitly acknowledges the median size value of a particle which strongly depends on particle mass fractions:[16]

> *"A natural, incidental or manufactured material containing particles, in an unbound state or as an aggregate or as an agglomerate and where, for 50% or more of the particles in the number size distribution, one or more external dimensions is in the size range 1–100 nm. In specific cases and where warranted by concerns for the environment, health, safety or competitiveness the number size distribution threshold of 50% may be replaced by a threshold between 1% and 50%. By derogation from the above, fullerenes, graphene flakes and single wall carbon nanotubes with one or more external dimensions below 1 nm should be considered as nanomaterials."*

Alternatively, other experts in the field such as Batwa have proposed a more practical definition — one not restricted by an arbitrary size limitation — that captures the complexities of nanotechnology-based applications:[17] "The design, characterization, production, and application of structures, devices, and systems by controlled manipulation of size and shape at the nanometer scale (atomic, molecular and macromolecular scale) that produces structures, devices and systems with at least one novel/superior characteristic or property."

This lack of an agreement on a well-established definition for nanotechnology also carries over to the field of "nanomedicine". While the NNI emphasizes the sub-100 nm scale in the definition of a nanomedicine, the European Science Foundation (ESF) and the European Technology Platform on Nanomedicine emphasizes the practical applications of nanotechnology in improving human health. ESF defines nanomedicine as "...the science and technology of diagnosing, treating and preventing disease and traumatic injury or relieving pain, and of preserving and improving human health, using molecular tools and molecular knowledge of the human body".[18] Similarly, the European Technology Platform refers to nanomedicine as "the application of nanotechnology to health. It exploits the improved and often novel physical, chemical and biological properties of materials at the nanometric scale. Nanomedicine has potential impact on the prevention, early and reliable diagnosis and treatment of diseases".[19]

Defining what precisely qualifies as a nanomaterial is rather difficult in part because the myriad of molecules that are studied in biological systems exist at nanoscale dimensions. Peptides are similar in size to quantum dots; some viruses, recombinant protein-based products such as monoclonal antibodies and enzymes — all fall below the 100 nm size cut off.[20] Strict interpretation of NNI's definition would even include existing technologies currently in clinical trials or on the market such as short RNA interference (siRNA) and the RNA antisense drug, Kynmaro, recently approved by the FDA in 2013.[21] Despite the fact that much of biology exists in the nanoscale range, no one has suggested recategorizing these existing technologies as "nano". Moreover, the size constraint that the NNI definition imposes excludes many materials and devices that fall outside of their nanoscale definition including, for example, carbon nanotubes and fullerenes that typically exist at dimensions below 1 nm. Other nanoparticles, including nanocrystals can range from 100 nm to 1000 nm in particle size and block copolymer micelles typically hover between 50 nm to 200 nm.[22] The first FDA approved protein-based commercialized nanoparticle, Abraxane®, has a 130 nm mean particle size.[23] Furthermore, there are increasingly more complex next-generation nanomedicines that are highly integrated combination products with different modes of action encompassing properties characteristic of a drug, a biologic and/or device. Such so-called "theranostic" nanoparticles can easily surpass the 100 nm upper size limit. (The 100 nm size cutoff is of course not new; in the development of liposomal nanomedicines this size — usually quoted between 100 nm and 150 nm — is the key size limit in avoidance or targeting of reticuloendothelial system uptake, and in achieving uptake through fenestrations in tumor or other poorly formed vasculature).[24]

What classifies a material as "nano", and any resulting regulatory conclusions, is further complicated by the physicochemical properties that such materials exhibit at that scale. At those dimensions, it is well-known that nanomaterials are sensitive to different environmental conditions. Owing to their small size and large surface area, nanoparticles can flocculate during production or storage and can form agglomerates (i.e. weakly-bound particles held by forces such as van der Waals) and even aggregates (i.e. strongly- or covalently-bonded particles).[25] Since aggregates and agglomerates of nanoparticles may grow beyond the sub-100 nm range, dissecting out the constituent particles by mechanical means may not be physically possible. Often times microscopic methods such as transmission electron microscopy (TEM) or cryo-TEM, already plagued with concerns about the statistical relevance of sample size, are unable to resolve the discrete dimensions of the individual constituent particles within an aggregate or aggolomerate species.[25] Where weakly-bound agglomerates typically may be dispersed by ultrasonic based methods, breaking apart strongly-bound aggregates may likely lead to destroying the original nanoparticle and in turn generate artefacts. Additionally, size distribution for nanoparticles under physiological conditions (e.g. pH and ionic strength) may differ significantly from the distribution *in vitro*.[26,27] In the bloodstream, many nanoparticles are well-known to bind to plasma proteins. The size of such a protein-nanoparticle aggregate may be a more relevant determinant to *in vivo* performance than that of a free-particle measured in an aqueous solution or in powdered form. Unfortunately, few of the characterization methods available today can reliably distinguish an aggregate or agglomerate from a single particle.

Added to this complication is the fact that nanoparticles are irregularly shaped and are often naturally polydisperse. While it is possible to achieve similar particle size measurements using different methods when particles are monodisperse and spherically shaped, it is often not possible when either or both aspects deviate. Even for monodisperse and spherical species, different measurement methods may produce significantly different particle size values. The sources of variability can be attributed to the type of instrument or algorithm used to determine the particle size distribution.[28] For instance, dynamic light scattering (DLS) instruments can produce skewed measurements when a small fraction of large particles are present in the sample. This size bias is due to the fact that the intensity of light scattered from a moving particle in a liquid suspension is inversely proportional to the sixth power of the radius of the nanoparticle.[25] Different size distribution measurement can stem from sampling or sample preparation errors. Even the same size distribution can produce different interpretation of the particle size if different algorithms are used for normalization. According to the 2011 report, the Joint Research Center essentially concluded that implementing the nanomaterial definition put forth by the European Commission will be significantly challenging given these analytical technical limitations and that no single method currently available is able to completely determine whether a material fulfills the definition of a nanomaterial.[25] It is also important to emphasize that none of the particle sizing methods available are currently validated for general purpose usage. And since the common reality is that the technical limitations with today's sizing methods at best lead to incomplete information about a particle's size and will undoubtedly result in miscategorizing some NCEs as nanomedicines based on an overly prescriptive size-based definition.[27]

As more elaborate and complex structures are generated for the next-generation nanoparticles such as pegylated or polymer drug and protein conjugates, the ability to size correctly may become more acutely challenging.[29] Thus, determining which external dimensions to consider as part of the "nanoparticle" is not as seemingly straightforward as it first appears, and the issue of size, morphology and sub-populations of different size or morphology are the key technical challenges.

2. Nanotechnology and Medicine — Reasons for Concern and Regulatory Approach

In the introduction to the popular novel *Prey* by science-in-fiction author Michael Crichton[30] the author notes the explosion in nanotechnology applications, and looks forward to further developments including nanomachines and even self-replicating nanotechnology systems. He notes that even a major proponent of nanotechnology, K. Eric Drexler[31] has expressed concerns about hidden dangers in the technology and the ability of society to appropriately control and regulate the field while enjoying its benefits.

Though we have not yet created the fictional and deadly nanobeasties of *Prey*, and much of public concern may not be on a firm basis, there are precedents for concern in the setting where nanomaterials interact with living bodies. Is the concern over the potential safety risks of nanomedicines justified? Let us examine the evidence.

The potential toxicity for nanomedicine stems primarily from its unique physicochemical properties associated with the nanometer size range, shape, surface characteristics (including coatings and chemically conjugated molecules) and chemical composition.[32,33] As a function of its size, nanomedicines have a large surface area-to-volume ratio compared to the same corresponding bulk or micrometer sized particles composed of the same material. This characteristic enables nanoparticles to cross membrane barriers more efficiently, avoid some clearance mechanisms, and penetrate much deeper into tissues than their larger size counterparts.[34] Such size-dependent behavior of nanomedicines may likely be true in tumor capillaries that make up a substantial fraction of blood vessels in the solid tumor microvascular network. On the other hand, many nanoparticle based drug delivery carriers become sequestered in the bloodstream, thereby thwarting their extravasations through the vascular membrane wall. This phenomenon has potential toxic implications associated with increased residency in the circulatory system, including decreased safety margins and possibly non-linear kinetics.[35]

If given by inhalation, the large surface-to-volume property can have a particularly strong influence on the biodistribution behavior and potent reactivity. Much of the research related to nanotech associated toxicity that is reported in the literature originates from ultrafine particles stemming from pollution-related causes such as internal combustion.[36] For example, numerous toxicology and epidemiological studies from inhaled nanoparticles have shown that lung injury is strongly associated with the size of the particles.[33] When inhaled, the ultrafine particles (airborne particulates < 100 nm) induce greater pulmonary inflammatory toxicity, oxidative stress and distal organ deposition.[37] Inhaled ultrafine particles have previously been reported to travel to the brain via an olfactory neuronal pathway that was not observed from earlier studies of larger molecules.[38] The most famous of inhaled nanoparticles, that has fueled and will likely continue to fuel consumer concern and perceived risks associated with nanomedicines disproportionately, is asbestos, the widely commercially-used silicate fibrous crystals that are strongly linked to fibrosis and carcinogenesis such as mesothelioma. Since it falls within the <100 nm size range, it has been suggested by the Royal Society and Royal Academy of Engineering that asbestos is considered a nanomaterial.[33] In a similar manner, inorganic nanoparticles such as gold and silver nanoparticles, cadmium-based quantum dots, titanium dioxide and mesoporous silica all can induce cytotoxicity, inflammation, fibrosis and oxidative stress-induced injury.[39] Size-dependent cytotoxicity has been reported for gold nanoparticles. Cell line studies have demonstrated that 1.4 nm gold nanoparticles caused cell death in contrast to slightly larger particles (3.7 nm) and in a dose-dependent manner.[40] Nanomedicines can be antigenic as a function of their physicochemical properties. Toxicities associated with nanomedicines currently on the market including liposomes, nanoparticles, and micelles themselves have reportedly elicited side effects ranging from proinflammatory responses, myelosuppression, to cytotoxicity and carcinogenesis.[41] For organic nanoparticles, injections of carbon-based particles such as C60 fullerenes or single walled carbon nanotubes, have been shown to be immunogenic in mice.[42,43]

In the past two decades, much of the knowledge on the toxicological effects of nanomedicines has revolved around studies carried out using cell culture based- and animal based-model systems exposed to disparate levels of the nanomaterials tested. Cell culture studies account for the majority of all *in vitro* nanotoxicity experiments conducted. Assessing the toxicity of nanoparticles *in vivo* has not always correlated positively with those observed in cell culture studies. For example, various cadmium-based quantum dots have shown minimal toxicity at low dosages in mouse models but were quite toxic in cell culture studies for the same particles.[44] Often times, the doses or dosage frequency are staggeringly high in *in vitro* and *in vivo* model systems too, and thus there may be limited value in extrapolating the results to humans. To date, there is no clear consensus on the toxicological effects of nanomaterials in biological systems and thus it is unclear to what extent, if any, peculiar adverse effects intrinsic to nanomaterials could impact humans.

Furthermore, the toxicities associated with surface modifications of nanomedicines with plasma proteins, when injected into the blood stream, have been shown to be an important determining factor. Other important factors to consider in regards to the toxicity of engineered nanomedicines are the biodistribution behavior and biodegradation of nanoparticles. Adverse effects can potentially arise from abnormally high concentrations of nanomedicines that concentrate in different organs and tissues and are not readily eliminated.

Toxicities of nanomedicines stem not only from physicochemical properties and the chemical composition of the nanoparticles themselves but also from the complex manufacturing process of assembling them. Variations in manufacturing over time (from GLP stages to scale-up during commercialization to factory technology transfers) can lead to alterations in toxicity profile.

In the next section, we examine two formulations of paclitaxel, one of which would fall into most definitions of nanomedicine, and one which would not. The nanomedicine formulation carries specific manufacturing process technology elements, and is a paradigm for Chemistry Manufacturing Controls (CMC) regulations as well, as the story behind the commercial development of Abraxane echoes some of the same challenges faced by first-generation nanomedicines during scale-up.[45] These are also the challenges to be anticipated in the development and approval of generic equivalents of these products.

Abraxane® (Celgene Corporation) is a well-known example of a re-formulation where a proven drug, Taxol® (Bristol-Myers Squibb Co.), containing the active ingredient paclitaxel, is refashioned into an albumin-bound nanoparticle (*nab*-paclitaxel) with an average size of 130 nm (Abraxane package insert). In the original formulation, paclitaxel, which is highly hydrophobic is solubilized in a mixture of Cremophor-EL (CrEL, a non-ionic surfactant poly-oxy-ethylated caster oil and ethanol manufactured by BASF). Side effects that limit the drug therapeutic index for Taxol and a biodistribution profile that appears to decrease drug distribution into the tumor are directly attributed to the presence of the Cremophor-EL solvent.[46] Toxicities commonly seen related to this solvent include potentially severe hypersensitivity, neutropenia and prolonged peripheral neuropathy.[43] Following infusion, Cremophor-EL paclitaxel in the plasma can form large micelles that sequester the drug from reaching its peak concentration at the tumor target site resulting in non-linear pharmacokinetics.[47] This altered pharmacokinetic behavior can lead to reduced drug exposure time and inadvertently lead to the use of higher doses which are associated with higher toxicities. In contrast, clinical studies of Abraxane have demonstrated that the pharmacokinetics of *nab*-paclitaxel behaved in a linear fashion across a dose rage of 135–300 mg/m^2.[48] By avoiding the use of toxic solvents, *nab*-paclitaxel affords an improved safety profile wherein clinical experiences of infusion-related acute hypersensitivities are remarkably non-existent. Thus, premedication with histamine blockers and corticosteroids is not needed prior to infusion with Abraxane. Additional distinct advantages with using *nab*-paclitaxel include a shorter infusion time (30 min versus 3 h), a higher tolerated dose of paclitaxel in Abraxane compared to Taxol (260 mg/m^2 versus 175 mg/m^2, respectively), and an improved overall response rate to tumor progression.[49] A fascinating aspect of this is that infusion related hypersensitivity reactions — including complement mediated reactions — are notoriously associated with nanomedicines. Here a definitive nanomedicine — Abraxane — appears bereft of this aspect, and a definitive "free drug" system — Taxol — has these side effects to what was originally a lethal level. A regulatory system wherein Abraxane had been required to assume a class related toxicity warning about infusion reactions would clearly have missed the mark here — either owing to its paclitaxel or nanomedicine heritages. (Acute toxicities of reasonable percentage occurrence should always be evaluable in clinical trials, and warnings from class effects of this type ought not be applied without some other factor).

To manufacture Abraxane, paclitaxel is added to an aqueous solution of albumin which then is homogenized under high-pressure to achieve the appropriate nanometer size range.[50] In *nab*-paclitaxel, human albumin is associated with the drug via hydrophobic interactions and acts as a natural carrier protein delivering the nanomedicine across the blood endothelium. Albumin enhances uptake into the cell by binding to a 60 kDa glycoprotein (gp60) on an endothelial cell surface which in turn activates caveolin-1, a scaffold protein decorated on the plasma membrane surface. Activated caveolin-1 induces the albumin-bound gp60 complexes to cluster, resulting in the formation of small vesicles called caveolae. It is thought that internalized caveolae bearing gp60/*nab*-paclitaxel complexes are transported from the apical to the basal membrane side in a process called transcytosis and subsequently released to the extracellular space, whereby the drug can accumulate in the tumor interstitium. Due to presence of albumin, there is some evidence suggesting that SPARC (Secreted Protein, Acidic and Rich in Cysteine), a glycoprotein secreted by tumor cells, plays a possible role in binding to the *nab*-paclitaxel complexes and further facilitating its preferential accumulation in the intratumoral space.[46,51]

The regulatory approval pathway to drug approval for nanotechnology based drugs is presently judged on a case-by-case basis. For drugs where an existing commercially approved drug is reformulated using a nanotechnology platform, the pathway to approval may be less burdensome. A good example of this is FDA's approval of Abraxane which obtained FDA approval in 2005 using the 505(b)(2) pathway. With this pathway, pre-clinical genetic toxicology and comparative PK studies on Taxol proved to be valuable in supporting Abraxane approval.

3. The Regulatory Landscape — Liposomes

Well ahead of other efforts, the US FDA began work on guidance for liposomal therapeutics, eventually taking on aspects of CMC and bioequivalence in light of manufacturing changes. Though patent expiry was at this time still somewhat in the distance, the agency already anticipated generics of liposomal therapeutics, and sought a somewhat carrot and stick approach to industry: if requirements for generics are set too high, perhaps requirements for manufacturing changes also need to be onerous. The goal was to get reasonable feedback from industry, but in taking this approach, the agency may be effectively equating the knowledge basis of innovator and generic producer; that is not likely to be a good assumption. We will discuss the bioequivalence/generic issue further below. The FDA utilized as input material both academic and industry advisors, and used the existing product NDAs. These were all liposomal re-formulations of proven active molecules — AmBisome (liposomal amphotericin B injection), Doxil (liposomal doxorubicin HCl injection), Myocet (another liposomal doxorubicin not yet approved), Depocyte (liposomal cytarabine) and DaunoXome (liposomal daunorubicin citrate injection).

In July 2001, an open hearing was held before the Advisory Committee for Pharmaceutical Science.[52] (It is worth noting that the liposome field never approached FDA in a concerted manner when the technology was first being brought to medical application in the 1980s; at around the same time, the developers of second-generation powder inhaled therapies banded together to influence the FDA and teach the agency how such products are best characterized and regulated; there resulted from this relatively harmonized approaches to things like particle size characterization. In contrast the various liposome technology companies were instead involved in litigation, and in any case, never sought to influence regulatory authorities as a group.)

At the 20 July 2001 meeting, Christine Swenson from Élan presented information on the drug Abelcet — a multi-micron sized lipid complex formulation of amphotericin. Structural aspects of the drug material were emphasized, but the product itself was excluded from the final guidance because it was not considered a liposome (liposomes are phospholipid bilayer structures that maintain at least one entrapped aqueous space).

Mei-Ling Chen from FDA introduced the Agency's approach. The liposome program was part of a larger effort around complex drugs, including cyclodextrins and proteins. The scope was guidance for CMC and bioavailability/bioequivalence. Arthur Shaw of the FDA office of new drug chemistry reviewed the agency's conclusions around CMC. A comprehensive characterization gave key elements to be considered. The approach focused on placing various liposome variants into taxonomic classifications; and introducing the concept of "functional excipients". Kofi Kumi presented the agency's analysis of the bioavailability/bioequivalence issue. At this time, injectable drugs were still formally assumed to be 100% bioavailable, but the concept of drug release from a carrier was of course appreciated. The key elements were determined to be released after injection — in essence, stability in the plasma or blood — and being available to carry out therapeutic effect at the target site of action. Again a taxonomic approach was taken: for a given liposome, is there or is there not reticuloendothelial system uptake? For an RES avoiding system, what is the stability on injection? For a very stable system, when conducting pharmacokinetic studies, total drug in the blood should be measured, but for an unstable one, perhaps total and free. Of course, many systems are not purely one way or the other.

Frank Martin from J&J Alza presented information on Doxil. At this time, the issue for Doxil was not the issue of generics, but the issue of a competing liposomal doxorubicin Myocet then in Phase III. The two formulations are fundamentally different, with the long circulating, RES avoiding pegylated liposome Doxil, with very low drug release level, contrasted with

Mycocet, which is taken up actively by the RES and liver, and wherein drug release is relatively rapid. The presentation emphasized the taxonomic approach and, perhaps in hind sight, focused too much on basic design and formulation. As critical as formulation (composition) and basic design are, they are relatively easy to regulate. Gerard Jensen of Gilead provided perhaps an opposite focus: the importance of production process, raw material quality and testing, and having sufficient analytical capability to test what, after years of experience, still requires every batch testing. This talk focused on a given liposome of a given formulation compared to a liposome of the same formulation but a varying manufacturing process, in contrast to the J&J Alza presentation that was more than half devoted to Doxil versus Myocet. Examples were given of varying half-life for liposomes of the same composition; the need for additional particle size measurement technology looking for contaminants (aggregates/flocculants) due to the insufficiency of DLS, with examples of batches that are identical in median size but varying in contaminating particle populations and aggregation rate owing to process differences; examples of mixed lipid intermediate structural differences; and differences in drug entrapment leading directly to measurable toxicity differences — again for the same formulation but different process conditions. Especially in a case where small amounts of free drug are toxic, simple pharmacokinetic studies would never see the key differences in product batches. An updated version of this last presentation has appeared in the literature.[53]

The guidance "Liposomal Drug Products: Chemistry, Manufacturing, and Controls; Human Pharmacokinetics and Bioavailability; and Labeling Documentation" was issued by CDER in August of 2002.[54] It remains "Draft Guidance" more than a decade later, though this is not so uncommon. In addition to the elements already noted, the guidance contains a substantial emphasis on accelerated model development for product storage — useful if successful but difficult as few liposomes obey Arrhenius thermal stability and in many cases changes seen at higher or lower temperatures cannot be obtained at storage conditions. There is emphasis on proper characterization of size, size distribution, surface charge, aggregation patterns and states. Uncommon to much of commercial drug manufacturing is the availability of second tier methods — chemical, physical or biological — to use for larger or less certain changes in manufacturing or raw materials or process validation, methods the innovator presumably developed in an expansive early development program and during years of commercial manufacturing and scale-up. The "same-ness" of GLP batches — often studies conducted before final manufacturing processes exist — may be key if they are used to address toxicity and/or pharmacokinetic assumptions. Study of morphology of the liposomes, volume of entrapment and protein binding are deemed important, though we would in our experience not see these as a batch to batch item during routine production. Validation of the sterilizing filter, if such is used, must use the real drug product. The document goes into pharmacokinetics, but not enough on the functional consequence of the liposome system — including what could be called pharmacodynamics. The entrapment/encapsulation assay methods assume a long half-life of the liposome and very rapid clearance of the drug (relatively). Mass balance of drug including free drug is discussed, but this is easier said than done (many values of "below limit of detection", e.g. in urine, can lead to a lot of missing – or extra - drug). There is discussion about a warning label — the consequences of substituting a liposome for a free drug formula or a different liposome — but often the warning would best be applied to the free drug formula, which for the set of molecules under consideration at the time would all have been older generic drug products.

With respect to the notion of "functional excipients" — there are two areas of emphasis. The first is: lipid to lipid and/or lipid to drug ratios; these should be set within manufacturing experience and/or knowledge of functional consequence. The second is chemical characterization of the lipid raw material — to say that they must be characterized " at the same level of detail expected for drug substance" is probably not appropriate in most cases; known degradants or impurities of functional consequences should be considered. For example, the rate of hydrolysis of lipids is well-known or easily discernable and the impact of such hydrolysis on pharmacokinetics, drug entrapment, etc. should be understood. It is very fortunate that high purity and high quality lipids and cholesterol are available with tight specification control from the innovators of those products.

In the area of bioequivalence, great emphasis is placed on the development and use of an *in vitro* release test. An example is provided by Xu *et al.* (2012)[55] In many cases, such a test can be very useful, but as with many other aspects of liposome characterization not every case is the same. The examples of Xu *et al.* afford full release of drug, whereas, for example, a small unilamellar liposome of stable composition but with a basic drug encapsulated by a relatively small chemical potential gradient[53] would come to equilibrium rather than full drug release. While the method of Xu *et al.* demonstrates sensitivity, it is to change in lipid composition — thus useful in formulation development but likely not in identifying small changes batch to batch. Again, the *in vitro* release assay may or may not be a solid and useful assay to grade variability in a liposomal setting. If one can establish such, the regulatory pathway in the US at least is simplified in regards to determining the equivalence of different manufacturing situations.

In the guidance, the definition of a liposome is not fully clear, however "lipid complexes" are explicitly excluded. One grey area is the application to fast release liposomes — cases where the lipid system, liposomal or not, is used just as a solvent for a hydrophobic active that is essentially instantly released on injection (see a liposomal propofol example[56]). The presence of a regulatory guidance can be of great help in dictating what choices to make in a complex setting — but it can also be a disincentive from pursuit of a given technology. In our experience, an internal hydrophobic drug candidate was slated for a phase I study where an injectable quick release formulation was needed. To avoid any use of this liposome guidance document, a simple SoyPC formulation was passed over in favor of a Cyclodextran one — only because the latter had no governing guidance.

At the time of this publication, a final guidance has not appeared, but some product specific documents have emerged, discussed in a later section.

Recently, the EMA has issued, for comment, a reflection paper on the data requirements for intravenous (*i.v.*) liposomal products developed with reference to an innovator liposomal product. The comment period ended on 31 January 2012. The document "Reflection Paper on the Data Requirements for Intravenous Liposomal Products Developed with Reference to an Innovator Liposomal Product"[57] acknowledges and assumes that innovators have information about manufacturing processes, raw materials and testing that are not available to a generic manufacturer. The innovator can develop a meaningful and data informed failure mode analysis, but the generic manufacturer cannot, without sufficient investment in development and manufacturing experience of their own.

The EMA document has a specific focus — generics. There is less emphasis on morphology and other elements and more emphasis on purpose — pharmacokinetic/pharmacodynamic goals. The document is a guideline of principles, not specifics, and up front acknowledges that the systems are too complex to avoid non-clinical/clinical data. A great emphasis is placed on effects at the tissue level — for example, effects of dose escalation, toxicity in the tissues and not just drug release are emphasized. Given the inability to do work in healthy volunteers for many of the existing liposomal drugs, additional complexity in clinical design of comparison trials is warranted. Any chemistry difference — an example cited as a change in PEG linkage in a PEG modified liposome — would also warrant increased clinical scope in the analysis. Specific focus is given to hypersensitivity reactions — acknowledging that small differences in products can lead to potentially large differences in the occurrence and severity of these responses. Animal models[58] for this phenomenon do exist, such that the biology of the phenomenon can be readily studied, but are likely too sensitive to predict human rates and severity. One has to find out in Man, and find out for each product. (An anecdote may be of value here to point out that even this may not be so simple; in a phase I trial in healthy volunteers of a liposomal amikacin product, itching was noted. A possible complement mediated effect, informed consent was modified to include direct disclosure of this risk to the enrolling patients. Several additional subjects in the study experienced itching. However, as this study grew beyond the one site where this was noted, no other case of itching was ever found, and none were found in later phase II studies.) A direct and formal link is made in this document to the biosimilar issues around copies of protein based therapeutics, the issues being in many ways parallel. CMC comments are similar to those provided by FDA, but are less prescriptive.

As a first step in forging a regulatory path toward generics, the US FDA has begun to publish draft product-specific guidances for determining the bioequivalence of nanoparticles. Non-binding FDA draft guidances on determining the bioequivalence of doxorubicin pegylated liposomes and nanoparticle albumin-bound (nab)-paclitaxel injectable formulations were published in 2010 and 2012, respectively.[59,60]

In February 2013, the FDA approved for the first time a parenterally administered generic PEGylated liposome. The approval of the generic version of Doxil/Caelyx (called Lipodox) by Sun Pharmaceuticals/Caraco Pharmaceuticals marks a major milestone that ushers in the era of generic nanomedicines. A drug shortage of the innovator product may or may not have influenced this process.

To demonstrate bioequivalency, the agency recommended that generic manufacturers conduct at least one small clinical single-dose, crossover *in vivo* standard PK study and one particle size distribution *in vitro* study. For generic nanoparticles, the bar for sameness to the pioneer product is set. The generic nanoparticle would be required to contain the exact same active ingredient, the exact same lipids that are derived from the same source (natural or synthetic) and in the exact same formulation as that of the innovator product. Based on the FDA guidance for liposomal doxorubicin hydrochloride, generic nanoparticles are required to match the reference physicochemical properties including lipid composition and orientation, drug encapsulation by ammonium sulfate gradient, nanoparticle surface morphology, surface charge, size distribution, internal environment and drug leakage under *in vitro* conditions. The guidance also stipulates that the clinical studies use a test product produced at the proposed commercial scale manufacturing. In the case of the nanoparticle paclitaxel, the guidance identifies some critical physicochemical attributes such as nanoparticle size, morphology, surface charge, palcitaxel crystallinity, free versus bound paclitaxel or albumin in suspension, the bond between paclitaxel and albumin, and *in vitro* drug release kinetics. For both generic *nab*-paclitaxel and Doxil, bioequivalence of generic nanoparticles to the reference product is ostensibly based on analysis between nanoparticle-encapsulated and free drug conducted in clinical studies using cancer patients. There is currently a debate as to whether these guidances adequately address the critical physicochemical properties of nanoparticles that impact the safety and efficacy *in vivo*.

The development of generic copies of pegylated liposomal doxorubin (PLD) (Doxil®/Caelyx®) illustrate well the challenges faced today in testing nanomedicines with similar pharmacokinetic properties.[61] Two previous studies found that slight modifications in liposome formulations can produce profound differences in acute toxicity and therapeutic index.[62,63] One study compared two different sized liposomal formulations with different (but not very different in terms of chemical potential) ammonium sulfate concentrations (PLD-75 at 300mM and PLD-100 at 250mM) with identical lipid composition and lipid/drug ratio in both normal and tumor xenograft murine models. These slight changes in formulation resulted in faster drug release rates with PLD-75 than with PLD-100. Correspondingly, PLD-75 exhibited higher levels of toxicity expressed as greater decrease in body weight. In regards to antitumor efficacy, PLD-75 was therapeutically more effective than PLD-100 at the same doxorubicin concentration level showing a reduction in tumor size of 1.9-fold. This is despite higher concentrations of PLD-100 observed at the tumor site than with PLD-75. Based on these findings, the authors conclude that the standard approach of measuring PK in the blood or plasma is not a sufficiently adequate yardstick by which to determine bioequivalency between generics and the innovator drug.

Another study[63] that compared Doxil with six different PLD formulations showed that minor changes in the particle size, lipid or internal buffer composition also resulted in flawed outcomes for toxicity and antitumor efficacy that did not positively correlate with plasma levels of doxorubicin. While changing the ammonium counterion to dextran sulfate did not alter PK, it nevertheless resulted in increased levels of markers for liver, kidney, heart, cardiac and bone marrow toxicities.[63] Even formulations that yielded lower systemic exposure levels of doxorubicin produced comparably higher levels of hematological toxicities along with elevated liver and bone toxicities. Taken together, these studies call into question whether plasma systemic exposure of the drug is suitable as the *de facto* method of choice for measuring the bioequivalency of two similarly formed PLD products.

AmBisome is a liposomal formulation of the fungicidal drug amphotericin B. The active ingredient is entrapped in the lipid bilayer of small unilamellar liposomes. It has been produced and marketed by Gilead Sciences since the 1990s. The liposomal formulation of amphotericin B reduces the toxicity of the active substance, providing both high blood levels for extended periods and direct fungicidal interaction of AmBisome liposomes with fungi. AmBisome has been shown to be safer than, and at least as effective as conventional amphotericin B in clinical trials. The primary enhancement of therapeutic index of the formulation over the conventional "free drug" formulation (the deoxycholate suspension Fungizone) is a lowering of the toxic effects of the amphotericin B active drug by stable entrapment of the drug in the liposome carrier vehicle.[64,53]

This achievement of a stable drug entrapment in the liposomal formulation is achieved by: the choice of formula and component chemicals; quality control measures utilized on these component materials; from rigorous chemical and physical testing from the raw materials through the manufacturing process; and through precision control of the manufacturing process itself.

Gilead has developed an *in vitro* assay to measure the liberation of intracellular potassium from rat red blood cells upon incubation with AmBisome. In this test, amphotericin B formulations at various concentrations are incubated with red cells and the concentration of released potassium ions is measured. The potassium ions are released when amphotericin B forms ion channels in the red cell membrane and the test thus provides an assessment of the extent to which the drug is "available" to form these pores. The reported parameter in this test is the concentration of amphotericin B leading to release of 50% of the potassium in the red cells (K50) after fixed hours of incubation. During development of the potassium release assay, experiments were performed comparing AmBisome with the conventional "free drug" formulation, Fungizone. As deoxycholate amphotericin B essentially fully partitions immediately into red cell membranes, a comparison of AmBisome and Fungizone samples requires only a short incubation time to assess the extent to which the drug is "freely available" from AmBisome liposomes. For 4 h incubations, the K50 values for Fungizone and AmBisome were 0.4 mg/mL and ~400 mg/mL, respectively. Therefore, ~1000-fold more drug was required to achieve potassium release for AmBisome relative to Fungizone, indicating very small amounts of immediately, freely available amphotericin B in AmBisome (approximately 0.1%). The potassium release test and K50 parameter is an *in vitro* indication and predictor of the toxicity profile of a given amphotericin B formulation and has been shown to correlate with *in vivo* toxicity.[64]

As noted above, examples have been provided from within Gilead's in house experience of substantially different formulation performance for batches of identical composition but different manufacturing process elements, which demonstrate that product composition is not all that matters in these complex dosage forms. The potential for increased toxicity owing to these manufacturing process-related differences was demonstrated following testing by Gilead of the Argentinean sourced liposomal amphotericin B formulation Anfogen, which is ostensibly a copy of AmBisome. It was shown that the Anfogen product had substantially different physical properties and that these manifested tangibly in a substantially worse toxicity profile relative to AmBisome.[65] A second example came to light in the form of another liposomal amphotericin B product, Amphiprol, which is licensed though not marketed in Greece, and we provide data here to provide predictions about the toxicity profile of this product.

Only a single vial of Amphiprol (PROEL Pharmaceuticals) was available to be tested. Results indicate that, unlike AmBisome, reconstituted Amphiprol yields a very turbid suspension with significant populations of both small (~80 nm) and large (>3 μm) sized particles, indicating a very different particle size distribution. Amphiprol also exhibits a very poor degree of drug encapsulation; indicated by a poor result in potassium release testing against red cells (K50 = 1.3 μg/mL for a 12 h incubation).

The very low K50 value places the predicted toxicity profile into the range of the Argentine product Anfogen[65] or the free drug formulation Fungizone. The observation of significant populations of large particles present in Amphiprol, with a substantially worse profile than the Argentine product Anfogen, where one should instead see particle sizes well below 100 nm, would suggest the likelihood of a substantially different biodistribution on injection, and with the expectation of significant loading of drug in the liver and spleen, relative to AmBisome. In essence, Amphiprol fails to reach the levels of encapsulation and the dispersion quality of AmBisome. Overall, this physical/chemical profile for Amphiprol is at the same level or worse than that of the Argentine product Anfogen, which was studied in established animal models.[65] In those studies, Anfogen was found to be significantly more toxic than AmBisome in mice with 0% 14-Day survival at 20 mg/kg, 60% survival at 10 mg/kg dosing, and 100% survival at 5 mg/kg. AmBisome by contrast showed 100% survival for mice at all doses between 5 mg/kg to 100 mg/kg. These results were consistent with the relationship of lethal dose to K50 response established previously.[64] In addition, the murine study of the

Argentine product Anfogen showed evidence of substantially enhanced kidney toxicity, and even attenuated efficacy in an Aspergillosis model.[65] It is reasonable to assume that similar toxicities could be seen for Amphiprol in mice, and in turn to manifest in patient injury should Amphiprol ever be dosed in humans. An additional concern pertains to a significant number of very large particles present in reconstituted Amphiprol i.e. 44% by volume are in the 3.8 μm size range. These large particles may be filtered out with the co-packaged 5 μm filter, possibly leading to an under-dosing of drug administered to the patient.

Chemically similar liposome systems can exhibit very different colloidal or toxicity profiles as previously demonstrated. The overall test results generated by Gilead reflect a product made with inadequate manufacturing procedures and process controls, which cannot be said to be identical or even similar to AmBisome, and which could potentially pose harm to patients. A recent article by Balesegaram *et al.* (2012) highlights this issue, exacerbated by the high doses now being used in some treatment settings, which makes a reliable efficacy and toxicity profile critically important.[66]

A comprehensive pharmacokinetics/pharmacodynamics characterization is needed to fully assure that a generic will afford the therapeutic index proffered by the innovator drug. Elements that can be adduced from non-clinical animal testing (see Adler-Moore *et al.*, 2006; Olsen *et al.*, 2006; Zolnik and Sadrieh, 2009; and Lestner *et al.*, 2010)[67–70] but as noted by the EMA human data is almost certainly needed. Success to meet this challenge requires an enhanced, specialized and science based development program (see e.g. Lionberger, 2008; Burgess *et al.*, 2010; and Schellekens *et al.*, 2011)[71,72,73] not only at the onset of drug launch and production but throughout drug production and drug usage life.

4. The Regulatory Landscape — Non-Liposomal Nanomedicines

There is currently no regulatory guidance in place that is designed specifically for non-liposome nanomedicines. As such, the FDA, EMA and other regulatory agencies formally regard nanomedicines to pose no more a risk as a class than conventional medicines. (We note, however, that agencies are beginning to ask, for example, the application for authorization from Swissmedic asks if the drug contains "synthetic nanoparticles" and defines such as "...particles [that] have at least one dimension on the nanoscale (1–1000 nm) plus a function and/or mode of action based on nanotechnology characteristics.")

First-generation polymer therapeutics, including polymer-drug and polymer-protein conjugates, are judged on a case-by-case basis using the regulatory approval standards prevailing at the time to gain successful market entry. For example, the pegylated aptamer Macugen was approved under an NDA without any specific polymer — or polynucleotide — based guidance. The active pegaptanib sodium was simply treated as a New Chemical Entity. Polymer therapeutics is comprised of a tripartite structure that includes the polymer, linker and the active pharmaceutical ingredient. From a CMC standpoint, some of the key issues to consider for polymer therapeutics that impact toxicity, biodistribution and efficacy, include polydispersity, size, shape and the surface chemistry of the polymer.

Tyner and Sadrieh (2011)[74] provide a guide, in the absence of formal guidance, for elements to consider in development and regulatory approval pathways for nanomedicines. They highlight a thorough physicochemical characterization, including structure, composition, particle size/distribution, shape/morphology, surface chemistry and propensity to aggregate/flocculate. Functional demonstration of desired nanotechnology function is fundamental — some nanotechnology formulations are for simple solubilization, others for sophisticated targeting. Focus should be applied to the state of the nanotechnology system in the mode of use — actual in-use conditions, finished dosage form, and on methods of study to assure the methods themselves do not alter the state of the nanoparticle in question. Also critical is understanding the biofate of the particles: decoration by proteins, biodistribution, impact on and by tissues. Animal toxicity studies must address conventional concerns and the aspects of the nanotechnology system that may be of relevance — either to model potential nanotechnology related toxicities, or any influence or alteration of normal animal models to the different dosage forms. Analytical assays typically require adaptation (see Jensen *et al.*, 2010[75] for a clinical analytical example). They may also involve technologies not normally seen in pharmaceutical settings.

Other articles of interest in the regulation of nanomedicines to which the reader is referred are to be found in Refs. 76–96.

In the end, the innovator of a given nanotechnology derived therapeutic will best understand the critical parameters for CMC, non-clinical and clinical study, if they have properly developed the drug. Complete characterization, factoring in the unique and general aspects brought by the nanotechnology and the specific use thereof in a given system, will reduce the burden on government regulators to try to find general guidance that are overly prescriptive.

Acknowledgments

The authors thank all at Gilead for valuable discussions and support.

References

1. Feynman, *Caltech Engineering and Science*, 30(5), 22 (1960); lecture given December 29, 1959, also printed in *J. Microelectromecahnical Systems* **1**(1), 60 (1992).
2. Editorial, *Nature Materials* **6**(4), 249 (2007).
3. BCC Research, *Market research report "Nanotechnology in Medical Applications: The Global Market"* (January 20, 2012). http://www.prnewswire.com/news-releases/nanotechnology-in-medical-applications-the-global-market-137803583.html
4. M. D. Cobb, J. Macoubrie, *J. Nanoparticle Research*, **6**(4), 395 (2004).
5. T. Satterfield, M. Kandlikar, C. E. H. Beaudrie, J. Conti and B. H. Harthorn. *Nature Nanotechnologies*, **4**, 752 (2009).
6. J. Besley, *Emerging Heath Threats Journal*, 3, **e8** (2010).
7. H. Wolinsky, *EMBO Reports*, 7(9), 858 (2006).
8. V. R. Adams and T. G. Burke, Eds, *Camptotecins in Cancer Therapy*, (Humana Press, Totowa New Jersey, 2005) 474 pp.
9. N. Morrision, *In_Pharma Technology.com*, 22 November 2011.
10. FDA Science & Research. Nanotechnology Task Force Report 2007. http://www.fda.gov/downloads/ScienceResearch/SpecialTopics/Nanotechnology/ucm110856.pdf.
11. M. A. Hamburg, *Science*, **336**, 299 (2012).
12. National Science and Technology council, Committee on Technology, Subcommittee on Nanoscale Science, Engineering, and Technology: The National Nanotechnology Initiative Strategic Plan. http://www.nano.gov/
13. JRC Reference Reports: Considerations on a definition of nanomaterial for regulatory purposes, (2010).http://ihcp.jrc.ec.europa.eu/our_activities/nanotechnology/report-definition-nanomaterial/at_multi_download/file?name=jrc_reference_report_201007_nanomaterials.pdf
14. FDA Draft Guidance for Industry: Considering whether an FDA-regulated product involves the application of nanotechnology, (June 2011). http://www.fda.gov/regulatoryinformation/guidances/ucm257698.htm
15. FDA, MAPP 5015.9, Reporting format for nanotechnology-related information in CMC Review (June 3, 2010). http://www.fda.gov/downloads/AboutFDA/CentersOffices/CDER/ManualofPoliciesProcedures/UCM214304.pdf
16. European Commission, Commission recommendation of 18 October 2011 on the definition of nanomaterial. (2011/696/EU), (October 25, 2011). http://ec.europa.eu/environment/chemicals/nanotech/pdf/commission_recommendation.pdf
17. R. Batwa, *Current Drug Delivery*, **8**, 227 (2011).
18. European Science Foundation's Forward look Nanomedicine: An EMRC Consensus Opinion, (2005). http://www.esf.org
19. European Technology Platform on Nanomedicine, Nanotechnology for Health: Vision paper and basis for a strategic research agenda for nanomedicine, (September 2005), EC Publication office.

20. B. Y. S. Kim, J. T. Rutka and W. C. W. Chan, *New England Jouranl of Medicine*, **363**(25), 2434 (2010).
21. A. Pollack, *New York Times* (January 29, 2013). http://www.nytimes.com/2013/01/30/business/fda-approves-genetic-drug-to-treat-rare-disease.html
22. R. Duncan R. and R. Baspar, *Molecular Pharmaceutics*, **8**, 2101 (2011).
23. See Abraxis Bioscience Inc., Oncologic drugs advisory committee meeting briefing package: Abraxane 5 (2006). http://www.fda.gov/ohrms/dockets/ac/06/briefing/20064-235b20-10-1abraxisbioscience-background.pdf
24. G. M. Jensen and T. Bunch, *J. Liposome Research*, **17**(3), 121 (2007).
25. T. P. J. Linsinger, G. Roebben, D. Gilliland, L. Calzolai, F. Rossi, N. Gibson, and C. Klein, Joint Research Center (JRC): Requirements on the measurements for the implementation of the European Commission definition of the term 'nanomaterial', (2012). http://publications.jrc.ec.europa.eu/repository/bitstream/111111111/26399/2/irmm_nanomaterials%20%28online%29.pdf
26. J. B. Hall, M. A. Dobrovolskaia, A. K. Patri and S. E. McNeil, *Nanomedicine*, **2**(6), 789 (2007).
27. S. K. Singh, N. Afonina, M. Awwad, K. Bechtold-Peters, J. T. Blue, D. Chou, M. Cromwell, H.-J. Krause, H.-C. Mahler, B. K. Meyer, L. Narhi, D. P. Nesta and T. Spitznagel, *J. Pharm. Sci.* **99**(8), 3302 (2010).
28. D. J. Burgess, D. J. A. Crommelin, A. S. Hussain, M.-L. Chen, *AAPS Pharm Sci*, **6**(1, Article 11), 1 (2004).
29. R. Gaspar and R. Duncan, *Advanced Drug Deliver Reviews*, **61**, 1220 (2009).
30. M. Crichton, *Prey* (HarperCollins Publishers, 2002).
31. K. E. Drexler, in *Prospects in Nanotechnology: Towards Molecular Manufacturing*, Ed. M. Krummenacker and J. Lewis (Wiley & Sons, New York, 1995), p. 21.
32. L. Bimbo, L. Peltonen, J. Hirvonen and H. A. Santos, *Current Drug Metabolism,* **13**, 1068 (2012).
33. E. Fröhlich, *Intl. J. of Nanomedicine*, **7**, 5577 (2012).
34. A. Nel , T. Xia, L. Mädler, N. Li, *Science*, **311**, 622 (2006).
35. K. Namdee, A. J. Thompson, P. Charoenphol and O. Eniola-Adefeso, *Langmuir*, DOI: 10.1021/la304746p (2013).
36. G. Oberdörster, V. Stone and K. Donaldson, *Nanotoxicology*, **1**(1), 2 (2007).
37. S. Bakand, A. Hayes and F. Dechsakulthorn, *Inhalation Technology*, **24**(2), 125 (2012).
38. G. Oberdörster, Z. Sharp, V. Atudorei, A. Elder, R. Gelein, W. Kreyling and C. Cox. *Inhaled Toxicology*, **16**(6–7), 437 (2004).
39. B. Fadeel B and A. E. Garcia-Bennett. *Adv. Drug Delivery Reviews*, **62**(3), 362 (2010).
40. A. Kunzmann, B. Andersson, T. Thurnherr, H. Krug, A. Scheynius and B. Fadeel, *Biochimica et Biophysica Acta*, **1810**, 361 (2011).
41. F. Jian, Y. Zhang, J. Wang, K. Ba, R. Mao, W. Lai and Y. Lin, Current Drug Metabolism, **13**, 440 (2012).
42. B.-X. Chen, S. R. Wilson, M. Das, D. J. Coughlin and B. F. Erlanger, *PNAS*, **95**, 10809 (1998).
43. B. F. Erlanger, B.-X. Chen, M. Zhu and L. Brus, *Nano Letters*, **1**(9), 465 (2005).
44. K. T. Yong, W. C. Law, R. Hu, L. Ye, L. Liu, M. T. Swihart and P. N. Prasad, *Chemical Society Review*, **42**, 1236 (2013).
45. N. Desai, *The AAPS Journal*, **14**(2), p. 282 (2012).

46. E. Miele, G. P. Spinelli, E. Miele, F. Tomao and S. Tomao, *Intl. J. of Nanomedicine*, **4**, 99 (2009).
47. A. Sparreboom, L. van Zuylen, E. Brouwer, W. J. Loos, P. de Bruijn, H. Gelderblom, M. Pillay, K. Nooter, G. Stoter and J. Verweij, *Cancer Research*, **59**, 1454 (1999).
48. P. Vishnu and V. Roy, *Breast Cancer: Basic and Clinical Research*, **5**, 53 (2011).
49. W. J. Gradishar, S. Tjulandin, N. Davidson, H. Shaw, N. Desai, P. Bhar, M. Hawkins and J. O'Shaughnessy, *J. of Clinical Oncology*, **23**(31), 7794 (April 2005).
50. S. C. Hock, Y. M. Ying and C. L. Wah, *PDA J. Pharmacology Science and Technology*, **65**(2), 177 (March-April 2011).
51. W. J. Gradishar, *Expert Opinion Pharmacotherpy*, 7, 1041 (2006).
52. FDA Advisory Committee for Pharmaceutical Science meeting presentations (July 20, 2001). http://www.fda.gov/ohrms/dockets/ac/01/slides/3763s2.htm
53. G. M. Jensen, T. H. Bunch, N. Hu and C. G. S. Eley, in *Liposome Technology Third Edition Volume 1: Liposome Preparation and Related Techniques*, Ed. G. Gregoriadis (Informa Healthcare, New York, London, 2006), p. 297.
54. FDA Guidance for Industry: Liposome Drug Products: Chemistry, Manufacturing, Controls; Human Pharmacokinetcs and Bioavailability; and labeling Documentation. http://www.fda.gov/downloads/Drugs/GuidanceComplianceRegulatoryInformation/Guidances/ucm070570.pdf
55. X. Xu, M. A. Khan, D. J. Burgess, *J. Int. J. Pharmaceutics*, **426**, 211 (2012).
56. G. M. Jensen, C. S. Ashvar, S. W. Bunte, C. D. Barzak, T. H. Bunch, R. L. Fahrner, N. Hu, J. Kennavane, H. Pham, C. Skenes and S. Yang *Theor. Chem. Account*, **119**, 291 (2008).
57. EMA reflection paper: "Reflection Paper on the Data Requirements for Intravenous Liposomal Products Developed with Reference to an Innovator Liposomal Product". http://www.ema.europa.eu/docs/en_GB/document_library/Scientific_guideline/2011/07/WC500109479.pdf
58. J. Szebeni, P. Bedöcs, Z. Rozsnyay, Z. Weiszhár, R. Urbanics, L. Rosivall, R. Cohen, O. Garbuzenko, G. Báthori, M. Tóth, R. Bűnger and Y. Barenholz, *Nanomedicine, Nanotechnology, Biology, and Medicine*, **8**, p 176 (2012). We note in this paper that AmBisome has a listed mean size of 141 nm; if correct, this may imply some preparation issue with the product used, as this size is much larger than typical for AmBisome (<80 nm).
59. US FDA. Draft guidance on doxorubicin hydrochloride (February 2010). Available at http://www.fda.gov/downloads/Drugs/.../Guidances/UCM199635.pdf.
60. US FDA Draft guidance on Paclitaxel (September 2012). Available at http://www.fda.gov/downloads/Drugs/GuidanceComplianceRegulatoryInformation/Guidances/UCM320015.pdf
61. Y. Barenholz, *J. Controlled Release*, **160**, 117 (2012).
62. J. Cui, C. Li, W. Guo, Y. Li, C. Wang, L. Zhang, L. Zhang, Y. Hao and Y. Wang, *J. of Controlled Release*, **118**, 204 (2007).
63. Rao, N. V. S. Mamidi, S. Weng, S. Stellar, C. Wang, N. Yu, T. Huang, A. P. Tonelli, M. F. Kelley, A. Angiuoli and M.-C. Fung, *Cancer Chemotherarpy Pharmacology*, **66**, 1173 (2010).
64. G. M. Jensen, C. R. Skenes, T. H. Bunch, C. A. Weissman, N. Amirghahari, A. Satorius, K. L. Moynihan and C. G. S. Eley, *Drug Delivery*, **6**(2), 1 (1999).
65. J. A. Olson, J. P. Adler-Moore, G. M. Jensen, J. Schwartz, M. C. Dignani and R. T. Proffitt, *Advances in Antimicrobial Chemotherapy*, **52**(1), 259 (2008).
66. M. Balasegaram, K. Ritmeijer, M. A. Lima, S. Burza, G. O. Genovese, B. Milani, S. Gaspani, J. Potet and F. Chappuis, *Expert Opin. Emerging Drugs*, **17**(4), 493 (2012).

67. J. P. Adler-Moore, G. M. Jensen, R. T. Proffitt, in *Liposome Technology Third Edition Volume III: Interactions of Liposomes with the Biological Milieu*, Ed. G. Gregoriadis (Informa Healthcare, New York, London, 2006) p. 405.
68. J. A. Olson, J. P. Adler-Moore, J. Schwartz, G. M. Jensen, R. T. Proffitt, *Antimicrobial Agents and Chemotherapy*, **50**(6), 2122 (2006).
69. B. S. Zolnik, N. Sadrieh, *Adanced Drug Delivery Reviews*, **61**, 422 (2009).
70. J. M. Lestner, S. J. Howard, J. Goodwin, L. Gregson, J. Majithiya, T. J. Walsh, G. M. Jensen and W. W. Hope, *Antimicrobial Agents and Chemotherapy*, **54**(8), 3432 (2010).
71. R. A. Lionberger, *The AAPS Journal*, **10**(1), 103 (2008).
72. P. Burgess, P. B. Hutt, O. C. Farokhzad, R. Langer, S. Minick, S. Zale, *Nature Biotechnology*, **28**(12), 1267 (2010).
73. H. Schellekens, E. Klinger, S. Muhlebach, J.-F. Brin, G. Storm and D. J. A. Crommelin, *Regulatory Toxicology and Pharmacology*, **59**, p. 176 (2011).
74. K. Tyner, N. Sadrieh, in *Characterization of Nanoparticles Intended for Drug Delivery, Methods in Molecular Biology Volume 687*, Ed. S. E. McNeil (Springer Science + Business Media, 2011), p. 17.
75. G. M. Jensen, T. H. Bunch, S. Wolf and S. Laybourne, *Clinica Chimica Acta*, **411**, 1900 (2010).
76. V. Morigi, A. Tocchio, C. B. Pellegrini, J. H. Sakamoto, M. Arnone and E. Tasciotti, *J. Drug Delivery*, 2012, 1 (2012).
77. R. Bosetti and L. Vereeck, *Nanomedicine*, **6**(4), 747 (2011).
78. N. Ross, *Regulatory Focus*, 8 (2011).
79. V. S. W. Chan, *Regulatory Toxicology and Pharmacology*, **46**, 218 (2006).
80. B. J. Culliton, *Health Affairs*, **27**(4) p. w315 (2008).
81. T. A. Faunce, *Medical Journal of Australia*, **186**(4), 189 (2007).
82. N. Chowdhury, *Nanomedicine* **5**(1), 135 (2010).
83. L. F. Hogle, *Journal of Law, Medicine, and Ethics*, p. 749 (2009).
84. J. Moore, *J. Law Med.*, **19**(1), 112 (2011).
85. N. Sadrieh and K. M. Tyner, *Therapeutic Delivery*, **1**(1), 83 (2010).
86. C. L. Ventola, *Pharmacy & Therapeutics*, **37**(11), 631 (2012).
87. T. A. Faunce, *Australian Health Review*, **33**(2), 258 (2009).
88. T. Faunce, K. Shats, *Journal of Law* & *Medicine*, **15**, 128 (2007).
89. J. Paradise, *Saint Louis University Law Journal*, **56**, 465 (2012).
90. J. Miller, *The Columbia Science and Technology Law Review*, IV, 1 (2003).
91. H. Johnston, G. Pojana, S. Zuin, N. R. Jacobsen, P. Møller, S. Loft, M. Semmler-Behnke, C. McGuiness, D. Balharry, A. Marcomini, H. Wallin, W. Kreyling, K. Donaldson, L. Tran, and V. Stone, *Critical Reviews in Toxicology*, **43**(1), 1 (2013).
92. P. Boisseau, B. Loubaton, *Comptes Rendus de l'Académie des Sciences*, 1 (2011).
93. European Commission Directorate-General for Health & Consumers, Scientific Committee on Emerging and Newly Identified Health Risks (SCENIHR), Risk Assessment of Products of Nanotechnologies (January 19, 2009) http://ec.europa.eu/health/ph_risk/committees/04_scenihr/docs/scenihr_o_023.pdf
94. D. Farrell, K. Ptak, N. J. Panaro and P. Grodzinski, *Pharmaceutical Research*, **28**, 273 (2011).
95. T. X. Viegas and F. M. Veronese, in *PEGylated Protein Drugs: Basic Science and Clinical Applications*, Ed. F. M. Veronese (Birkhäuser Verlag, Switzerland, 2009), p. 273.
96. R. Nijhara, K. Balakrishnan, *Nanomedicine: Nanotechnology, Biology, and Medicine*, **2**, 127 (2006).

Chapter A7

A Short Course on the Chemistry Manufacturing and Control of Liposome-based Pharmaceutical Products

*Simon Eastman and Tom Redelmeier**

Northern Lipids Inc., 8855 Northbrook Court,
Burnaby, BC, V5J 5J1, Canada
**tomredelmeier@northernlipids.com*

Abstract

Liposome technology has progressed significantly in the last 35 years since the first products entered clinical trials. The technology supports a broad range of products as diverse as vaccines, gene therapy and the formulation of small molecules for the treatment of infectious diseases and cancer. Over this time, there has been a growing recognition by the pharmaceutical community of how liposome technology can assist scientists in developing pharmaceutical products.

In parallel to these developments, and in particular, in response to more than 150 products that have entered clinical development, has been an advancement in our understanding of the Chemistry Manufacturing and Controls (CMC) of liposome formulations. The application of a systematic approach to the identification of high quality excipients, optimization of manufacturing processes and the development of suitable analytical methods accelerates commercialization of liposomal pharmaceuticals. This chapter discusses recent trends in this area, and in particular, advances in the large-scale manufacture of liposome products, including their characterization.

1. Introduction

This chapter discusses the Chemistry Manufacturing and Controls (CMC) of liposome-based pharmaceuticals with an emphasis on studies that are appropriate for drug products undergoing preclinical or clinical evaluation. It is general in nature and attempts to provide a broad background that is applicable to most parenterally administered liposome-based pharmaceuticals. However, given the broad application of liposome technology in the formulation of vaccines, small molecules for the treatment of infectious diseases and oncology, and gene therapy, all of the examples provided are not necessarily relevant to all applications. It is the nature of the drug product, its material properties, and specifically the physical-chemical properties of the Active Pharmaceutical Ingredient (API) that drive the selection and optimization of the manufacturing process, and the development of appropriate Quality Systems. This chapter does not discuss CMC issues that relate specifically to the production and Quality Control (QC) of the drug substance.

Table 1 lists the liposome-based pharmaceutical products that have received marketing approval in China, Europe, India, Japan and North America. This list does not include drug products that have been introduced as generics or used for topical or oral administration. From the date of approval, and the evident commercial success of several of these products (annual sales of greater than US$500 M for DOXIL®, AmBisome® and

Table 1. Liposome-based drug products that have received marketing approval.

Product	Active ingredient	Year approved
AmBisome®	Amphotericn B	1990
DOXIL®	Doxorubicin HCl	1995
DaunoXome®	Doxorubicin HCl	1996
Epaxal®	Inactivated HAV	1997
Inflexal®	Influenza antigens	1998
Visudyne®	Verteporfrin	2000
Definity®	Octafluoropropane	2001
Myocet®	Doxorubicin HCl	2002
Depocyte®	Cytarabine	2002
DepoDure®	Morphine	2004
Lipusu®	Paclitaxel	2007
Marqibo®	Vincritsine	2012

Visudyne®), we can safely conclude that liposome-based pharmaceuticals can be successfully commercialized. Though this assertion may seem obvious now, it nonetheless contradicts earlier comments made during the 1980s that liposome formulations do not have suitable pharmaceutical properties (such as chemical and physical stability, compatibility with vials, closures, and infusion sets). Indeed at that time, some technology leaders bemoaned the lack of high quality excipients, quality standards, and availability of equipment and processes that are suitable for commercial manufacturing under cGMP.[1]

Today, the development of over 135 products that have entered clinical trials in the last two decades and the industry's successes in developing the products listed in Table 1, lead us to conclude that liposome technology has matured. This is not to say however, that the industry does not continue to optimize equipment, manufacturing processes or develop new analytical tools to characterize formulations. Rather, it recognizes that many of the lessons that we have learned in the past can be applied to the current manufacturing challenges that we face today. Further, and as importantly, the infrastructure necessary to the development of a liposome product is now available, and there are a number of Contract Research Organizations (CRO), Contract Manufacturing Organizations (CMO) and Excipient Suppliers that have specialized experience in supporting the development of liposome-based pharmaceuticals. In short, many of the practical obstacles to the development of liposome products that existed 10 or 20 years ago, have disappeared and been replaced by an infrastructure that is capable of supporting the next generation of liposome-based products.

An understanding of how liposome technology can assist scientists in developing pharmaceutical products has grown in parallel with the development of liposome-based therapeutics. Liposome technology can be viewed as a tool set that can be used to address formulation challenges that arise during the development of New Chemical Entities (NCEs). In Sec. 1.1, we discuss some of the uses of liposome technology, as it applies to applications that involve parenteral administration of pharmaceutical products. Our discussion does not cover applications in the area of topical, transdermal, oral, cosmetic or nutraceutical sciences.

1.1. *Formulation of sparingly soluble compounds*

Many NCEs developed today are sparingly soluble in water and exhibit poor bioavailability. The formulation of sparingly soluble compounds in liposomes significantly increases their solubility. These formulations are safe, stable to dilution, do not leach materials from plastics, and are compatible with administration by aerosols or subcutaneous, intramuscular or *i.v.* injection. Visudyne®, a liposome formulation of verteporfin that is approved for the treatment of wet stage macular degeneration, is an example of such a commercial liposome product. In our hands, we have found that sparingly soluble compounds often have a higher solubility (expressed as a drug/lipid ratio) in a lipid bilayer than the hydrophobic core of a micelle or an emulsion. This is presumably because sparingly soluble NCEs have polar groups that orient the molecule in the bilayer to take advantage of the polarity of the head group and hydrophobicity of the acyl chain region.

The attractiveness of using liposome formulations to dissolve sparingly soluble compounds stems also from the proven safety of lipid excipients. Studies conducted in the mid-1990s by Esperion established that daily *i.v.* administration of up to 40 gm of 1-palmitoyl-2-oleoyl-phosphatidylcholine (POPC) was well tolerated in human volunteers. Given that even the most uncooperative NCEs ("brick–dust") can typically be formulated in liposomes at a drug/lipid ratio of 0.01 wt/wt, this safety margin would support the daily administration of up to 400 mg of an NCE. Liposomes represent a relatively risk-free, safe and rapid approach to the formulation of NCEs for first-in-man studies.

Though it appears intuitively obvious that liposomes may be used to dissolve sparingly soluble compounds, we note also that sparingly soluble compounds can be "remote loaded" into liposomes with some surprising (non-intuitive) results. The process of remote loading involves the use of transmembrane ion gradients to "capture" drugs in the vesicle interior[2] and is described in more detail in Sec. 3 of this chapter. Membrane permeable weak acids and bases redistribute across the liposome bilayer in response to transmembrane pH gradients, and accumulate in the vesicle interior, often reaching levels equivalent to a supersaturated solution. It is generally applicable only to compounds that: (i) can diffuse across the liposomal membrane; and (ii) are simple weak acids or bases or (iii) have a relatively high affinity for complexing agents such as transition metals or polyanions/cations.

The advantage of using remote loading to formulate sparingly soluble compounds is illustrated in Fig. 1. In this study, two strategies were used to formulate a sparingly soluble NCE into liposomes at the indicated drug/lipid ratio. In the first approach (termed passively loaded), the drug and lipid were dissolved in t-butanol and a powder was generated following lyophilization. The lipid/drug powder was hydrated and vesicles were extruded through polycarbonate membranes. The passively loaded liposomes were further

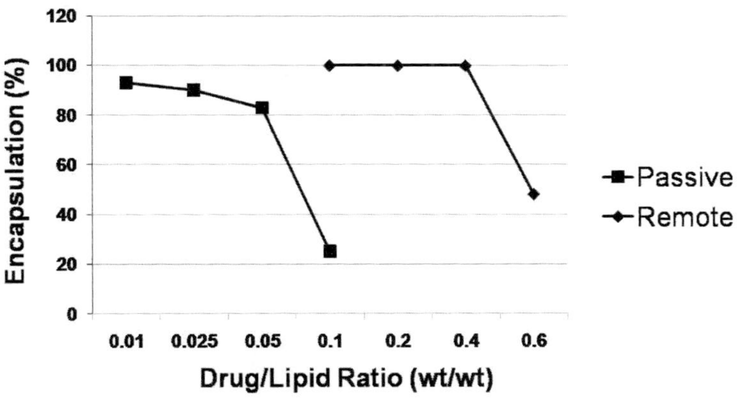

Fig. 1. A comparison of the encapsulation of sparingly soluble compounds by passive and remote loading strategies.

isolated by gel filtration chromatography and the drug/lipid ratio was determined. The encapsulation efficiency was observed to be inversely correlated to the drug/lipid ratio. In contrast, when a remote loading strategy was applied to encapsulate the NCE, a higher drug/lipid ratio could be achieved. We have found this to be the case in a number of "challenging-to-formulate" NCEs that had simple acid/base properties. An attractive feature of this approach is that formulation of up to 50 mg/mL API can be developed for compounds that have a nominal concentration of less than 0.1 mg/mL.

1.2. *Reduction in vesicant properties*

It is perhaps not surprising that the administration of drugs is often accompanied with localized irritation or necrosis, especially for sparingly soluble compounds. This may arise from drug precipitation, or a high concentration of the drug at the site of injection, leading to partitioning of the compound into cells and concomitant cellular damage and inflammation. One of the advantages of liposome formulations is that they invariably reduce the vesicant properties of the API. This was observed in the development of Myocet®, a liposome formulation of doxorubicin.[3] In this example, the liposome formulation of doxorubicin reduced the vesicant properties observed upon extravasation presumably because remote loading of the drug encapsulates it in the vesicle interior. However, we have also observed this reduction in vesicant properties of NCEs for passively loaded formulations and believe that this is a general property of liposome formulations.

1.3. *Improvement in the safety profile*

The most successful commercial liposome products have derived their success, in part, due to a reduction in toxic side effects, which enable the administration of a higher drug dose (AmBisome®, Myocet®, Marqibo®) or supports further cycles of treatment (Myocet®, DOXIL®). The reduction in toxicity observed for AmBisome® is attributed to altered biodistribution wherein less amphotericin B is delivered to the kidney, the site of dose limiting toxicity. In the case of the doxorubicin formulations, advantages stem from reduced exposure to the heart, whereas for Marqibo®, a liposome formulation of vincristine, the reduction in toxicity is attributed to reduced delivery to the central nervous system. In our hands, we have noted that the development of controlled release liposome formulations often reduces side-effects attributed to an elevated Cmax.

1.4. *An enhancement in the therapeutic activity through changes in biodistribution or pharmacokinetics*

The primary rationale for the development of liposome formulations is often the improvement in therapeutic activity that may be expected upon significant changes in the biodistribution of the compound, or an altered pharmacokinetic profile. There is ongoing interest in the use of liposomes to passively-target sites of inflammation, a process that has been termed the "enhanced permeation retention" (EPR) effect.[4] Liposomes can be designed to have "long-circulating" properties such that a significant fraction of the injected dose persists in the plasma 24 h after administration. This is achieved by appropriate choice of liposome composition, particle size, material properties of the liposome surface and, most importantly, by choosing a formulation wherein the drug is slowly released from the carrier (e.g. $t_{1/2}>4$ h). Drug formulations may have clearance rates with half-lives that extend from hours to several days (e.g. DOXIL®, SPI77, CPX-351), and clearance of the drug from the plasma compartment can be tailored to meet the requirements of the drug product. There has been a great deal of work in animal models that has established long-circulating liposomes accumulate in sites of inflammation or solid tumors, observations that are consistent with clinical studies in man.[5] Improvements in therapeutic outcome may well be expected to derive from formulations that enhance the drug delivery to disease sites.

Additional benefits may also derive from prolonged exposure of diseased tissue to the drug. In fact, it is often challenging to identify whether the improved therapeutic effect of a liposome formulation derives from altered biodistribution or from prolonged exposure. As illustrated in Fig. 2, the impact of liposome formulation on the exposure of a tissue to a drug can be very dramatic. In this example, the exposure of lungs to a drug that has been instilled in water or a liposome formulation is compared. The results establish that liposomes act as a controlled release system wherein the liposome formulation increases the tissue AUC by several orders of magnitude. This observation has important implications for the control of site specific delivery.

Though liposomes may have dramatic, order of magnitude, impact on drug pharmacokinetics, not all formulations are modified release dosage forms. In fact, many liposome formulations designed to increase drug solubility or reduce vesicant properties can be classified as "immediate" release. An example of this is presented in Fig. 3, where three formulations that were

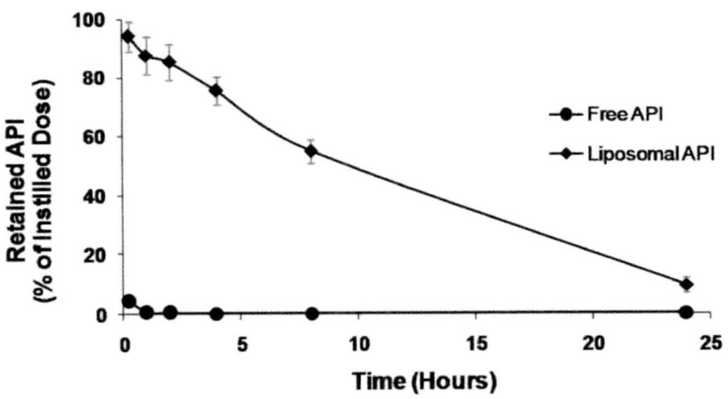

Fig. 2. Controlled release of sparingly soluble compounds following pulmonary administration.

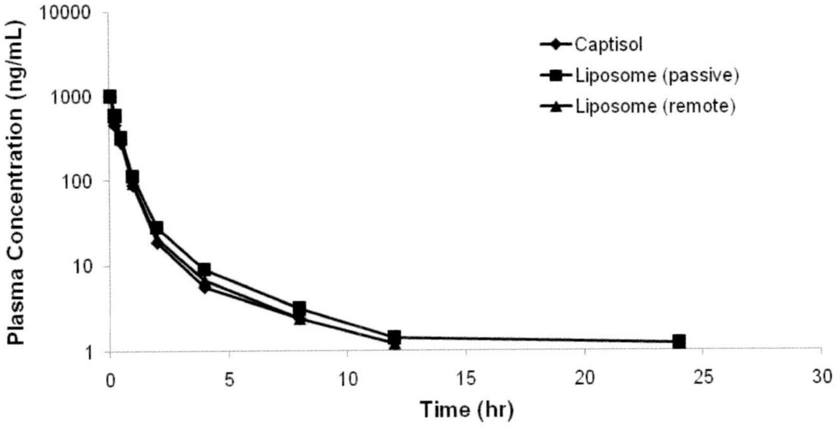

Fig. 3. Comparison of the Pharmacokinetic profile of two liposome formulations with Captisol.

developed to reduce the vesicant properties of a NCE are compared. Interestingly, all formulations had similar plasma pharmacokinetics despite the different formulation strategies and excipient compositions.

1.5. *Passive targeting to specific cell types*

Underlying the development of liposome-based vaccines or adjuvants is the understanding that liposomes can be used to formulate antigens and influence their delivery to professional antigen presenting cells, such as macrophages and dendritic cells. There are a number of examples of liposomal adjuvants (e.g. muramyl tripeptide, MPLA) that have been extensively studied in non-clinical studies,[6,7] reached clinical development, or are marketed as adjuvants (mifurmatide).

1.6. *Presentation of antigens or haptens on surfaces*

Targeting liposome formulations of APIs to cells remains an area of sustained interest, and several targeted liposome products (MBP-426, MM-302) are under clinical development. Though it is generally accepted that targeting does not impact the extravasation of liposomes at sites of inflammation or necrosis, it has been postulated that targeting can enhance intracellular delivery of agents to cancer cells.

1.7. *Intracellular delivery*

A prime interest of scientists engaged in the development of novel drug delivery vehicles is to identify strategies that efficiently introduce drugs into the cytosol and/or nucleus of cells.[8] This technology would profoundly impact our ability to control gene expression and could potentially be used to synthesize proteins (either through the delivery of DNA or mRNA), or alternatively down regulate gene expression (through the introduction of antisense or siRNA). Targeting remains an important facet of strategies to enhance intracellular delivery of nucleic acids,[9] though optimal delivery also requires efficient fusion of the carriers with cellular membranes. Recent promising results from Phases 1 and 2a clinical studies of the first and second generations of liposomal formulations of siRNA are recent encouraging developments that suggest liposome formulations can be designed to efficiently deliver nominally impermeable agents into the cytosol of targeted cells.

2. Characterization of Liposome Products

Analytical assays are critical to the development of liposome-based therapeutics. They are used to: (i) characterize the drug product; (ii) establish the identity and quality of the raw materials; (iii) demonstrate and support control of the manufacturing process; and (iv) release and establish the stability of the drug product. Assays that are frequently used to characterize the drug product are summarized in Tables 2–4, and some are discussed below. We note that this list is not exhaustive, and does not include; for example, assays

Table 2. Assays that may be used to characterize drug products.

Characterization assays
CryoElectron Microscopy
Freeze-fracture Electron Microscopy
Encapsulated Volume
Internal pH
Internal and External Ion Concentration
Differential Scanning Calorimetry (DSC)
Circular Dichroism
X-ray Diffraction (SAXS)
^{1}H-NMR
^{31}P-NMR
Leakage tests upon chemical, thermal or physical stress
Protein binding

Table 3. Assays that may be used to support control of the manufacturing process.

Assay	Method
API Content	HPLC/UV
API Encapsulation	Physical separation and HPLC
Lipid Content	HPLC/ELSD, HPLC/CAD
Particle Size	Dynamic Light Scattering
Zeta Potential	
pH	USP/EP
Osmolality	USP/EP
Residual Solvents	USP/EP

Table 4. Assays that may be used to release liposome-based drug products.

Assay	Method
Appearance	Visual inspection
API Identity	HPLC
API Content	HPLC
API Related Compounds	HPLC
API Encapsulation	Physical separation and HPLC
IVR	Physical separation and HPLC
Identity of Lipids	HPLC/ELSD, HPLC/CAD, TLC
Content of Lipids	HPLC/ELSD, HPLC/CAD
Lipid Related Compounds	HPLC/ELSD, HPLC/CAD, TLC
Ligand Identity	Various
Ligand Content	Various
Ligand Binding	Various
Particle Size	Dynamic Light Scattering
Zeta Potential	Various
pH	USP/EP
Osmolality	USP/EP
Residual Solvents	USP/EP
Particulates	USP/EP
Endotoxins	USP/EP
Sterility	USP/EP

that may be used to characterize secondary structure of protein targeted systems or liposome formulation of nucleic acids. A discussion of assays used to characterize raw materials is also beyond the scope of this chapter.

2.1. *API content and related compounds*

The availability of robust assays that accurately and precisely quantitate the API and related compounds is critical to the timely development of manufacturing processes. The development of these assays should be considered a priority at the earliest practical opportunity.

HPLC with a UV/VIS, PDA (photodiode) or fluorescence detector are the principal methods used to quantitate APIs in liposome formulations. The assays are specific to the chemical properties of the API and typically developed at an early stage of the program to support production of the drug substance. The strategies to validate these assays for their intended purpose are well recognized in the pharmaceutical industry (ICH Guidelines Validation of Analytical Methods Procedures: Q2[R1]) and readily applied to liposome formulations. We note that liposomes have to be either "broken open" to release encapsulated drug or, more commonly, dissolved in an appropriate diluent. Diluents are typically apolar and care must be taken to ensure that the diluent does not interfere with reverse phase methods that are developed for the drug substance. It is possible to develop strategies to remove excipients by extraction. However, a diluent that dissolves both the API and the excipients, and allows for chromatographic resolution is more frequently selected.

In establishing the suitability of the analytical method(s) for identifying related compounds, we recommend the evaluation of thermally stressed samples of the drug product or prototype. This is especially important for assays that use low wavelength (<210 nm) spectroscopy to quantitate the API or related compounds, since lipid excipients and/or their degradation products (lysolipids, fatty acids) may interfere with the quantitation process. Assays that identify and preferably quantitate related compounds can be used to determine whether processing of the API impacts chemical stability.

2.2. *API encapsulation*

API encapsulation is a fundamental property of liposome products and is typically used as a Drug Product Specification for release and stability. It is perhaps less necessary to have this as a specification for "immediate release" products or for sparingly soluble compounds that have a maximum solubility in the buffer of less than 1% of the nominal concentration of the API. For these products, it may be more appropriate to determine whether the liposomes are contaminated with drug precipitates.

API encapsulation can be determined by physical separation of the liposome and the unencapsulated drug. This may be achieved by gel filtration chromatography, ion exchange chromatography, filtration through membranes to remove drug precipitates or alternatively filtration through membranes which retain the liposome and allow free drug to pass through the membrane. In a QC environment, methods should be selected that are straightforward, robust, have appropriate requirements for accuracy and precision and enable fast turnaround time. Consideration should also be given to whether the method separates encapsulated drug or drug that is simply bound (i.e. dissolved) in the external leaflet of the membrane. The results

Table 5. "Free" drug recovery in the eluent following separation on an Amicon filtration device.

Conditions	Recovery (%)
20 μg/mL "free" API	44
20 μg/mL "free" API + Liposomes	3
200 μg/mL "free" API	89
200 μg/mL "free" API + Liposomes	30

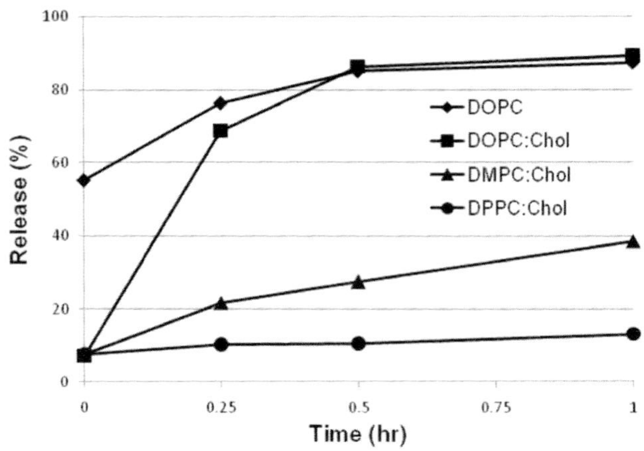

Fig. 4. Effect of lipid composition on the in vitro drug release.

presented in Table 5 illustrate that a nominally water soluble compound may still have a tendency to bind with liposomes, and this can impact separation on an Amicon filtration device. In this example, recovery of "free" drug in the eluant was influenced by the API concentration, the presence of lipid (control liposomes without a transmembrane ion gradient), and also the drug/lipid ratio (not shown).

2.3. *In vitro release*

In Vitro Release (IVR) assays are often developed at the initiation of the R&D program to aid in the selection and design of liposome prototypes. Figure 4 provides an example where IVR studies were used to establish the influence of lipid composition on the *in vitro* drug release from the vesicles. Assays may be used in this manner to rank order prototypes and identify whether lipid composition, buffer composition, particle size, and drug/lipid ratio impact the rate of drug release from the carrier. IVR assays may not be relevant to the development of "immediate release" formulations since the material properties of the carriers do not impact the bioavailability of the drug.

IVR assays are also developed for QC purposes. For example, they can be used to establish whether storage of the drug product impacts release. They are also frequently used to establish equivalence if changes are made in the composition or manufacturing process during commercialization (changes in venue or scale). IVR assays should be simple, straightforward, robust and designed to be completed within a standard work day. There is not one standard method that works for all products, since the mechanism and rate of release varies widely between different drug products.

IVR assays are typically applied to liposome formulations of small molecules. They are not routinely applied to vaccines, liposome formulations of nucleic acids, or immediate release formulations. Since many liposome formulations of small molecules are designed to release over a prolonged period, even after exposure to serum proteins, it is common to use IVR assays that contain components or conditions that accelerate the drug release from the carrier. Suitable perturbants include: (i) temperature; (ii) buffers; (iii) serum; (iv) detergents; and (v) organic solvents. The ideal perturbant would accelerate, but not change the mechanism of release.

In developing IVR methods, we advise following these guidelines:

i. Conditions should be established that result in release of more than 70% over a 2–4 h incubation period; or
ii. 4 time points should be taken, including a "zero" time point, a time point where intermediate levels of release are observed and a time point where the maximum level of release is observed;
iii. The solubility of the drug in the vehicle should not impact the rate of release;
iv. Ion gradients should be completely dissipated;
v. Small variations in temperature, pH and drug concentration do not lead to significant changes in the release profile.

The results presented in Fig. 5 provide an example of a stability study conducted under accelerated conditions where the chemical stability of the lipid excipients and IVR was followed for a product stored for 6 months at 5°C or 25°C. At 5°C, there was no observable change in the lipid content of the formulation or the rate of release. Under accelerated storage conditions

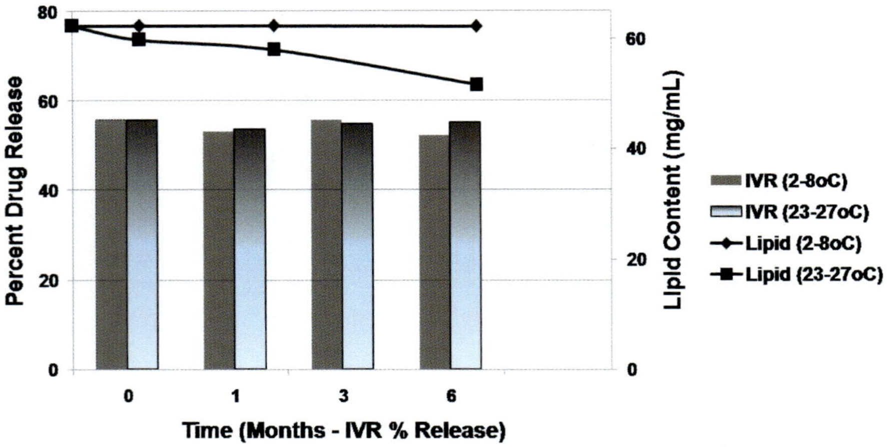

Fig. 5. Effect of lipid degradation on the in vitro drug release.

(25°C), lipid degradation was observed with approximately 25% decrease in total lipid after storage for 6 months. Interestingly, the rate of drug release did not change for these samples, providing evidence of the robustness of the formulation.

2.4. *Lipid content and related compounds*

Regulatory guidances recommend that the lipid content of liposome formulations be fully characterized. In practice, assays should be available to quantitate the principal lipid components, as well as their putative degradation products. An example of such an assay is provided in Fig. 6. Since most lipids are not readily detected in the UV/VIS range, many assays use HPLC with an Evaporative Light Scattering Detector (ELSD) or a Charged Aerosol Detector (CAD). Both the ELSD and CAD rely upon nebulization of the column eluent, followed by detection of the particles using light scattering (ELSD or CAD). The ELSD is considered a non-linear detector and has a lower sensitivity and dynamic range than UV/VIS or CAD detectors. These detectors, are often considered near "universal", though in practice are not suitable for the quantitation of volatile compounds that would include lipid degradants with short acyl chains (myristic acid, glycerol). Assay development may be challenging as responses can be sensitive to peak shape.

Attention is directed toward the quantitation of the principal lipid excipients in the early stages of program development. However, multiple assays may need to be developed to quantitate all of the lipid excipients and putative lipid degradants. Assays that are used to quantitate these related compounds

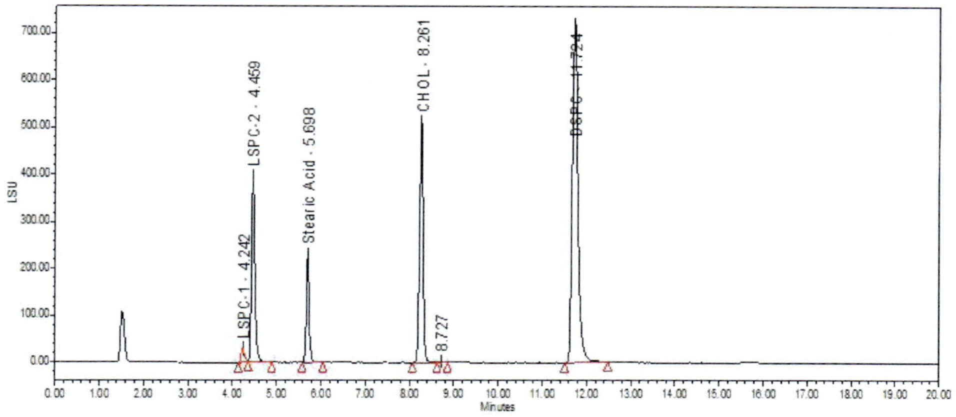

Fig. 6. HPLC/ELSD analysis of lipids.

are often not introduced for routine QC activities at an early stage of product development unless the formulation exhibits instability upon routine storage conditions. The availability of assays that sensitively monitor the level of lipid degradants facilitates manufacturing process development, and the establishment of hold times for manufacturing processes.

2.5. *Particle size*

The particle size of a liposome is a critical material property that may impact pharmacokinetics of the carrier, as well as drug release. Particle size, and in particular, the homogeneity of the liposomes invariably impact the ease of sterile filtration. The primary tool to determine particle size is Dynamic Light Scattering (DLS), also known as Quasi Elastic Light Scattering (QELS). Cryo-electron microscopy may also be used to determine particle size, but is not used routinely for this purpose.

The data presented in Fig. 7 illustrate both the strength and challenge associated with particle size analysis by DLS. In this study, a liposome formulation with a particle size of approximately 400 nm was spiked into a "homogenous" vesicle preparation with a nominal particle size of 80 nm. When DLS is performed at a scattering angle of 90°, the analysis is generally insensitive to the presence of contaminants. For example, the average diameter for samples with 10% (wt/wt) of putative contaminants is observed to increase by less than 20%. DLS size measurement using a 90° scattering angle remains the predominant method to provide accurate and precise estimations of the diameter of liposome preparations, but does not provide a sensitive indication of contamination.

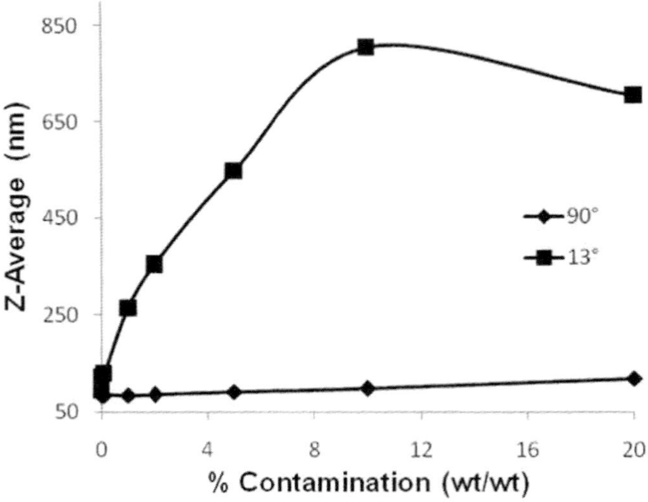

Fig. 7. The sensitivity of DLS at 90° and 12.8° to particulate impurities.

In contrast, DLS performed at a scattering angle of 12.8° is a sensitive marker of contamination producing observable differences when less than 0.1% (wt/wt) of contaminant is introduced into the homogenous liposome formulations. Though DLS measured at 12.8° provides additional information, the data is qualitative and not an accurate measurement of the particle size. For example, the particle size of a preparation containing 99% (wt/wt) of 80 nm vesicles and 1% (wt/wt) of 400 nm vesicles had an apparent diameter of 265 nm with a polydispersity index (PDI) of greater than 0.4. Moreover, DLS when measured at a 12.8° scattering angle is so sensitive to particle contamination that it is likely to skew the analysis of some formulations.

Direct comparisons to light scattering are not presented here, however we note that light scattering also provides qualitative information on particle size heterogeneity and is used as a QC measurement for AmBisome®.

3. Manufacturing Strategies

The optimal manufacturing strategy for a given liposomal therapeutic is dictated by a number of considerations ranging from the material properties of the drug product (e.g. chemical stability of the API, lipid composition, particle size, particle size homogeneity), robustness of the manufacturing processes, and commercial considerations such as batch size and equipment capacity. Though it is true that several different strategies can be applied to manufacture a given liposome product, there are some processes that provide an opportunity for improved manufacturing control and these should be applied at an early stage of preclincal development, as they are likely to greatly reduce the complexity of process scale-up and validation. We recommend that the manufacturing process be reviewed at the initiation of preclinical studies and that studies are conducted to establish the suitability of the manufacturing strategy for its stated purpose.

On the basis of the properties of the API, the preparation of liposome formulations can be subdivided into four distinct strategies: (i) partitioning of sparingly soluble compounds in the membrane; (ii) passive encapsulation of water soluble compounds into the vesicle interior; (iii) formation of ion pairs with charged lipids; and (iv) remote loading of weak acids and bases into liposomes in response to transmembrane ion gradients. We briefly discuss the common approaches to manufacturing these products in Sec. 3.1.

3.1. *Sparingly soluble compounds*

Sparingly soluble compounds will partition into liposome membranes. The solubility of the drug in the liposome bilayer is dictated by the material properties of the liposomes and the physiochemical properties of the compound. Hydrophobic compounds that have a log P between 2 and 4 and a MW less than 400 Daltons appear to be good candidates. However, it is possible to have compounds which are exceptions to this rule.

Several approaches can be taken to manufacture the liposome formulation.

Partitioning:

i. Dissolve the lipid in t-butanol;
ii. Lyophilize;
iii. Hydrate the lipid and extrude to form vesicles of desired particle size;
iv. Incubate with API to dissolve drug;
v. Dilute to label claim, and sterile filter.

Preparation of a lipid powder:

i. Dissolve the lipid and API in t-butanol;
ii. Lyophilize;
iii. Hydrate the powder and extrude to form vesicles of desired particle size;
iv. Dilute to label claim, and sterile filter.

Preparation by Solvent Dilution:

i. Dissolve the lipid and API in an organic solvent;
ii. Dilute to form vesicles;
iii. Extrude to achieve desired particle size and remove drug precipitates;
iv. Diafilter to remove organic solvent and unincorporated API;
v. Ultrafilter (concentrate) to label claim;
vi. Dilute to label claim, and sterile filter.

Sparingly soluble compounds can also be manufactured by remote loading, which will be discussed elsewhere.

3.2. *Water soluble compounds that do not diffuse across membranes*

Water soluble compounds that are zwitterionic, have MW greater than 800 Daltons or are very polar do not diffuse readily across liposome membranes. They may be prepared by the following processes.

Preparation by Solvent Dilution:

i. Dissolve the lipid in an organic solvent and API in aqueous solvent;
ii. Mix organic and aqueous solvents to form vesicles;
iii. Extrude to achieve desired particle size;
iv. Diafilter to remove organic solvent and unincorporated API;
v. Ultrafilter (concentrate) to label claim;
vi. Dilute to label claim, and sterile filter.

3.3. *Water soluble compounds that form ion pairs with charged lipids*

The use of ion pairs to formulate nucleic acids is well established. Though the original approach involved mixing of charged liposomes with the nucleic acids, the following general strategy is used for many of the products undergoing preclinical and clinical testing.

Preparation by Solvent Dilution:

i. Dissolve the lipid in an organic solvent and API in aqueous solvent;
ii. Mix organic and aqueous solvents to form vesicles;
iii. Extrude (if necessary) to achieve desired particle size;

iv. Diafilter to remove organic solvent and unincorporated API;
v. Ultrafilter (concentrate) to label claim;
vi. Dilute to label claim, and sterile filter.

3.4. *Weak acids and bases that can be remote loaded*

Remote loading refers to the spontaneous encapsulation of weak acids and bases into liposomes in response to transmembrane ion gradients. Originally described for biological amines,[10] the strategy requires that a transmembrane ion gradient be established. Loading of the API into the vesicles may be accomplished using ion gradients,[11] a pH gradient,[12] a permeable weak base,[13] a permeable weak acid,[14] or a transition metal.[15]

Following strategy is generally used:

i. Dissolve the lipid in an organic solvent;
ii. Mix organic and aqueous solvents to form vesicles;
iii. Extrude to achieve desired particle size;
iv. Diafilter to remove organic solvent and create transmembrane ion gradient;
v. Load drug into liposomes;
vi. If necessary, diafilter to remove unencapsulated drug and ultrafilter (concentrate) to label claim;
vii. Dilute to label claim, and sterile filter.

4. Manufacturing Processes

In the following discussion, we shall provide guidance in choosing appropriate manufacturing processes and examples of the type of study that is used to identify or optimize it. It is worthwhile at the onset of this discussion to indicate that our preference is always to simplify the manufacturing process as much as practical. It is our experience that research & development studies are often carried out using processes that are unnecessarily complex, though often convenient and available. For example, though it is possible to achieve the desired particle size after extrusion through filters of varying pore size, it is preferable to identify a process that achieves the desired particle size and homogeneity after a single pass through multiple stacked filters of a single pore size. Not only does the simplification of this process improve the prospects of scaling, it also decreases the exposure of the lipid (and possibly API) to shear and (often) elevated temperatures.

4.1. *Lipid dissolution*

Lipids and, in some cases, APIs are dissolved in organic solvents prior to formation of the liposomes. This is an essential first step taken to ensure that the composition of individual liposomes reflects the composition of the formulation. For large-scale manufacture under cGMPs, attention should be directed toward choosing solvents with appropriate toxicity profiles (ICH Topic Q3C (R4) Impurities: Guideline for Residual Solvents). Preferred solvents include those classified as Class 3 e.g. ethanol, butanol and DMSO. Chlorinated solvents such as chloroform and methylene chloride are generally to be avoided since the handling of large quantities of these solvents in commercial parenteral manufacturing facilities is often discouraged. The use of Class 3 solvents also reduces concerns related to solvent removal of Class 2 (or 1) solvents in the finished drug product.

The choice of solvent is often dictated by the solubility of the lipids, and in turn this is impacted by the lipid composition. Though lipids, by definition, are apolar by nature, considerable differences in solubility in organic solvents are observed as a function of the polar head group, acyl chain length and degree of acyl chain saturation. Relatively polar lipids, such as phosphatidylglycerols and cardiolipin require more polar solvents which may be achieved by adding water to the organic solvent (e.g. ethanol/water mixtures). In contrast, apolar lipids such as cholesterol have limited solubility in polar solvents and often precipitate if the polarity is too high.

There are often advantages in working with relatively high concentrations of lipid e.g. 50% lipid by weight. Under these conditions, the lipids themselves may play a role in dissolving other excipients, including cholesterol, and may well also impact the solubility of the API. For example, the maximum solubility of cholesterol in ethanol is approximately 20 mg/mL, whereas significantly higher solubility can be achieved in the presence of other lipid excipients (e.g. phosphatidylcholines). These phenomena may also apply to the solubility of APIs in a particular solvent, where high concentrations of lipids may act as a cosolvent.

The dissolution of the lipid in the solvent is a critical process that has the potential to impact downstream manufacturing processes, including extrusion, diafiltration and/or sterile filtration. The choice of lipid concentration impacts the dissolution conditions, notably the mixing rate, temperature and time of mixing. A lipid concentration that is too high or too close to its maximum solubility may be overtly sensitive to the water content of the solvents, and may require prolonged heating and mixing to achieve complete dissolution. Though our experience primarily revolves around the use of Class 3

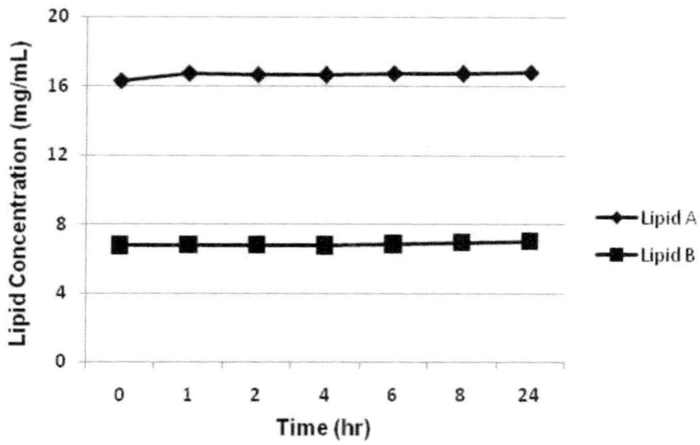

Fig. 8. The chemical stability of lipid excipients dissolved in an organic solvent.

solvents, the solubility of lipids and excipients in chloroform is also sensitive to water content.

Lipids that have high (>40°C) gel-to-liquid phase transitions may require heating to achieve dissolution. We advise that process development studies be conducted to determine the chemical stability of the lipids (and API) over prolonged periods at temperatures that are representative of the manufacturing conditions. The results provided in Fig. 8 illustrate the type of study that we have previously conducted to support manufacturing process development, and in particular for setting allowable guidelines for holding intermediates in the manufacturing process. As a general rule, we have found that lipid excipients dissolved in organic solvents show good chemical stability, though recommended that this be established during the development of the manufacturing process.

4.2. *Solvent removal*

Under most circumstances the organic solvents are subsequently removed from the formulation to meet ICH Guidelines for Residual Solvents (Q3C[R3]). We discuss here the strategies that are used to remove the solvents prior to further processing, though note that it is far more common to process the liposomes and remove the solvents in a subsequent diafiltration step (see Sec. 4.4).

In R&D studies, solvent evaporation is often conducted using a rotoevaporator. This facilitates the evaporation of methylene chloride (or chloroform) and co-solvents such as methanol to form a dry thin film of lipid which coats the surface of the flask. In practice, this approach is best suited for research purposes and is not an optimal process for large-scale manufacture under cGMP. Control of solvent removal is challenging and compounds with significantly different solubility in the solvent system may form precipitates that interfere with downstream processing (e.g. extrusion, sterile filtration). The use of type 2 organic solvents such as chloroform, methylene chloride and methanol in the production of pharmaceuticals should be avoided, if possible.

Spray drying may be used in place of a rotoevaporator, though in practice the requirement to have specialized equipment especially for applications that involve handling of flammable solvents and potent compounds has limited the use of this strategy. An alternative strategy has been to replace chloroform (or methylene chloride) with ethanol and subsequently evaporate the solvent under reduced pressure. While circumventing the use of Class 2 solvent, this strategy does not circumvent the possible generation of particulates that can potentially impact particle size reduction by extrusion, and/or sterile filtration.

One approach to remove solvents that has been scaled to support commercial manufacture is the production of stable plurilamellar vesicles.[16] In this case an emulsion is formed with an immiscible organic solvent (e.g. chloroform) before removing the solvent under reduced pressure in the presence of an aqueous buffer. The resulting vesicles may then be further processed using extrusion.

We frequently use lyophilization from t-butanol as a strategy for production of lipid powders. In general, the quality of the powders is very good, and this is often a suitable process for making lipid pre-mixes and/or incorporating hydrophobic (sparingly soluble) drugs into liposomes. The process scales easily from R&D to commercial production.

Regardless of the strategy selected to remove the organic solvent, it is worthwhile to remember that it may impact downstream processing such as the level of drug incorporation into the vesicles, the ease of extrusion, particle size homogeneity and sterile filtration.

4.3. *Particle size reduction*

The particle size of most liposome-based products is sub-micron, typically less than 120 nm. This facilitates the sterile filtration of the formulations by ensuring that the formulation can pass through 0.22 μm filters, a critical manufacturing step for most liposome formulations. It is also consistent with most formulations that are designed for "long-circulation" and controlled release.

Liposome formulations, such as DepoCyte, that have a larger particle size cannot be sterilized by filtration and are manufactured aseptically. It may be argued that vesicle size and in particular the homogeneity of the particle size is not a critical material property for immediate release formulations since drug bioavailability is unlikely to be dependent upon small changes in these material properties. In contrast, the particle size of modified release formulations is considered a critical material property since it will impact: (i) the trap volume; (ii) concentration of the drug and excipients in the liposome formulation; (iii) the interaction of the carriers with cells; and (iv) the clearance of the liposomes from the plasma compartment. All of these considerations point to the importance of maintaining tight control of particle size and homogeneity during the manufacturing process.

Three different strategies can be used to prepare liposomes formulations of the appropriate particle size: (i) solvent dilution; (ii) homogenization; and (iii) extrusion. There are strengths and weaknesses to each approach. The particle size of liposomes prepared by solvent dilution is sensitive to the: (i) lipid composition; (ii) lipid concentration; (iii) presence of the API; (iv) solvent composition; (v) solvent concentration; (vi) equipment design; (vii) flow rate; and (viii) manufacturing scale. The process requires optimization for a specific formulation, but can be used under some circumstances for select formulations. The process has not, to our knowledge, been used for liposome products that have been commercialized, though it has been used for several products that have entered clinical development. The process may well be most suited for the preparation of liposome formulations where the API forms an ion-pair with a charged lipid. However, control of the trapped volume, a critical parameter for remote-loaded formulations, may be challenging to achieve.

Homogenization, a technology that was originally used in the food industry, has been successfully applied to the production of liposomes for commercial applications. However, in our experience, homogenization does not generate homogenous particles. This is attributed to several aspects of the process: (i) inadequate temperature control; (ii) the requirement for multiple passes to achieve appropriate particle size; (iii) the use of high pressures (typically greater than 25,000 psi) which in turn leads to the generation of limit-sized vesicles; and (iv) the possibility of contamination related to lipid coated drug precipitates. Commercial products that have been manufactured by homogenization are used either in the cosmetic/nutraceutical field or for formulation where particle size, homogeneity and trap volume may not be as critical to the safety and therapeutic activity of the drug product.

The process that we recommend for the preparation of homogenous liposome formulations relies upon medium-pressure extrusion through polycarbonate filters of a defined pore size.[17] Equipment specifically designed for the preparation of liposome formulations by extrusion was introduced by LIPEX® Biomembranes in 1984. Early applications of this technology focused on the extrusion of hydrated lipid powders, typically through two stacked filters with a nominal pore size of 100 nm. However, this was later extended to polycarbonate filters of sizes varying from 50–400 nm. In practice, vesicles passed through membranes of greater than 100 nm are multilamellar, with size distributions that are lower than the nominal pore size of the filters. In contrast, vesicles passed through polycarbonate filters that are less than 100 nm are predominantly, if not exclusively, unilamellar with average particle size that is close to the pore size of the membranes.

Extrusion can be used to prepare uniform dispersions of liposomes following the hydration of lipid powders without any further processing. Parameters that should be controlled and/or varied to optimize particle size homogeneity include: (i) filter size; (ii) number of stacked filters; (iii) lipid concentration; (iv) temperature and (v) number of passes. We recommend that particle size and particle size homogeneity should be monitored using DLS at 12.8° and 90°. We note that particle size reduction is often sensitive to process parameters that are "upstream" of the extrusion processes. For example, the quality of the raw materials (excipients), method of preparation of the dry lipid powder and hydration conditions.

Though one of the original advantages of extrusion was that it could be used to prepare liposomes in the absence of an organic solvent, extrusion is often applied to the preparation of the liposomes in the presence of a solvent such as ethanol. In this approach, lipids are dissolved in an organic solvent and then subsequently diluted into a buffer to form multilamellar vesicles. The API can be dissolved in the organic solvent (for sparingly soluble compounds), in the buffer (for water soluble compounds) or subsequently remote loaded (for weak acids and bases). Following dilution, the vesicles are passed through polycarbonate filters to achieve the target particle size and level of homogeneity.

A picture of a LIPEX™ Extruder is presented in Fig. 9. The equipment is custom designed and consists of two heat-jacketed pumps that operate in parallel to mix lipids (dissolved in solvent) with buffer. Liposomes are subsequently passed through two 293 mm filter holders. The filter holder can be designed to operate in parallel and can be practically operated with up to four (145 mm or 293 mm) filter holders.

Fig. 9. The footprint of a LIPEX® extruder.

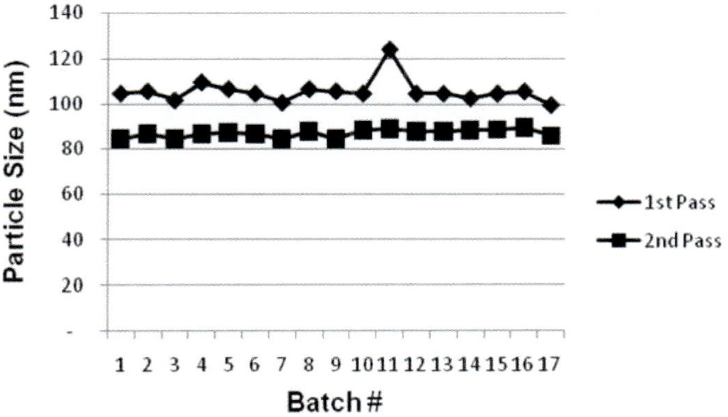

Fig. 10. Particle size analysis following single pass extrusion.

Extrusion conditions can be optimized to generate homogeneous particles of the desired particle size after 1–3 passes through the filters. The data presented in Fig. 10 provides an example of the particle size reduction. For a series of seventeen 30L batches, particle size was monitored after 1 or 2 passes through stacked polycarbonate filters. Under these conditions, the target particle size (85–90 nm) was achieved after a second pass with an average diameter of 87 nm (Std. Dev. 2.2 nm).

The results provided in Fig. 11 are an example of an experiment designed to optimize the solvent concentration at the time of extrusion. In this experiment, the solvent concentration was varied between 6 and 14% and the

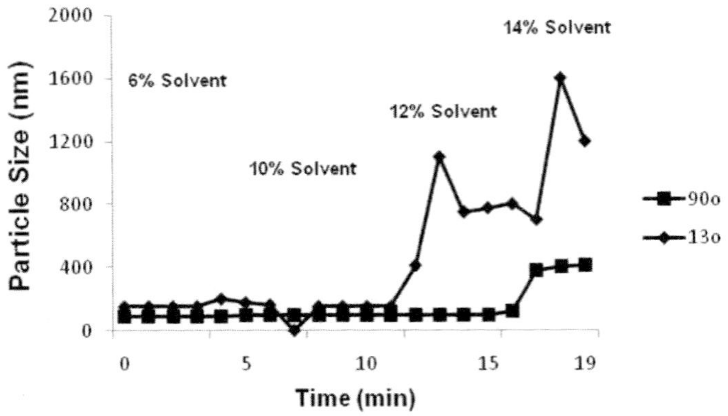

Fig. 11. Influence of solvent concentration on particle size and homogeneity.

particle size monitored by DLS at 12.8° and 90°. Interestingly, DLS at 12.8° identified vesicle size heterogeneity at a solvent concentration of 12%, which was not detected by comparable DLS analysis at 90°.

4.4. *Solvent and solute removal and exchange of buffers*

Following particle size reduction, liposome preparations may be subjected to tangential flow filtration to remove organic solvents, unencapsulated API, exchange buffers and/or to concentrate the product. The typical process involves passing the liposomes through hollow fiber membranes, which act to retain the liposome in the fibers and allow the passage of solutes and water through the pores. The process exposes the liposomes to moderate shear forces which can lead to particle size growth under some circumstances. The results presented in Table 6 provide an example of a series of experiments that were conducted to optimize diafiltration conditions to avoid the generation of particulates that could impact downstream processes such as sterile filtration.

One of the most common applications of diafiltration is the exchange of buffers to create ion gradients. This is the basis of remote loading, as described in Sec. 3.4. The results presented in Fig. 12 provide an example where the establishment of a pH gradient is monitored as a function of buffer wash volumes. The results also demonstrate the stability of the pH gradient following up to 20-wash volumes.

4.5. *Drug loading*

Remote loading of drugs into liposomes is typically performed at elevated temperatures, and for sufficient time to achieve maximal loading. Care must

Table 6. Identification of diafiltration conditions which preserve particle size.

Manufacturing Step	90° (nm)	PDI	12.8°	PDI
Before Diafiltration	86	0.044	87	0.083
Diafiltration #1	88	0.043	181	0.401
Diafiltration #2	88	0.027	200	0.324
Diafiltration #3	88	0.032	123	0.218
Diafiltration #4	90	0.009	104	0.076

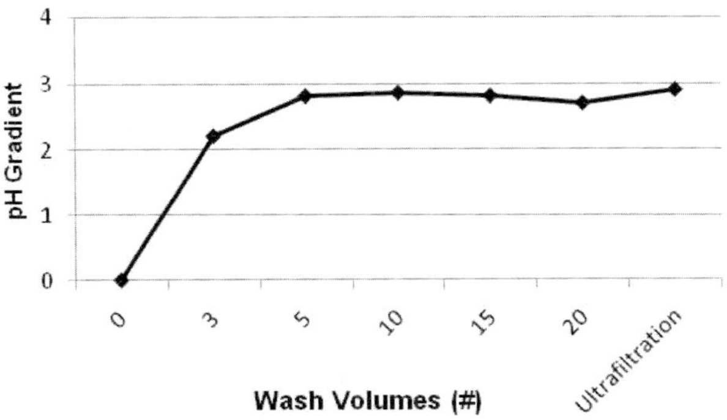

Fig. 12. Monitoring the establishment of an ion gradient.

be taken to ensure appropriate mixing of the drug with the liposomes, and temperature control including control of heating and cooling rates. Conditions should be established that result in reliable encapsulation of the API while maintaining an adequate gradient that will serve to maintain encapsulation during storage of the drug product.

4.6. *Sterile filtration*

Redundant sterile filtration of liposomes through two 0.22 μm sterilizing grade filters in series is often challenging, even though the nominal size of the vesicles is often less than 100 nm (0.1 μm). In practice, liposomes that have phase transition temperatures that are greater than 40°C are particularly problematic, presumably because they deform while passing through the filters. We often find that the ease of sterile filtration is impacted by upstream processing, including the quality of the raw materials, the homogeneity of the vesicles after extrusion and the stability of the particles to moderate shear forces during processing. At an early stage of development, we optimize

Table 7. Process optimization facilitates filter sterilization.

Manufacturing Step	90° (nm)	PDI	13° (nm)	PDI	Capacity (mL/cm^2)
Process #1	89	0.07	204	0.32	≈ 0.3
Process #2	87	0.045	94	0.17	≥ 2.0

manufacturing process parameters on the basis of the flow rate (and capacity) of the drug product through sterilizing grade filters. An example of the impact on appropriate selection of process parameters is illustrated in Table 7 where a five-fold increase in capacity was observed upon optimization of the manufacturing process.

5. Conclusion

Liposome technology has progressed significantly in the last 35 years since the first products entered clinical trials. The technology itself is broad and supports applications as diverse as vaccines, gene therapy and the formulation of small molecules for the treatment of infectious diseases and cancer. Delivery systems can be designed for immediate, intermediate and prolonged release, as well as for systemic or localized therapies. Carriers can be designed to be cleared rapidly from circulation, or exhibit prolonged half-lives. Recent clinical studies demonstrate that liposomes can be designed to efficiently deliver their payload to the cytosol of targeted cells to obtain a desired therapeutic response. There is growing recognition by the pharmaceutical community that liposome technology can assist scientists in developing pharmaceutical products.

Advancement in our understanding of the CMC of liposome formulations has occurred in parallel with the emergence of more than 150 products that have been commercialized or entered clinical development. Manufacturing strategies exist for most liposome applications, though require optimization to meet the physiochemical properties of the API and commercial requirements of the program. Analytical methods that support the characterization of the drug product should be developed to support preclinical and clinical studies. Though these assays will be refined during clinical development, and may only be validated during the later stages of the clinical program, they should nonetheless be available at an early stage in the program to support measurements of equivalence during manufacturing process scale-up and venue changes. The application of a systematic approach to the identification of high quality excipients, optimization of manufacturing processes and the development of suitable analytical methods accelerates commercialization of liposomal pharmaceuticals.

References

1. G. Poste, In: *Liposome Letters*, A. D. Bangham (Ed.). London: Academic Press (1984), p. 375.
2. T. D. Madden, P. R. Harrigan, L. C. Tai, M. B. Bally, L. D. Mayer, T. E. Redelmeier, H. C. Loughrey, C. P. Tilcock, L. W. Reinish and P. R. Cullis, *Chem. Phys. Lipids* **53**, 37 (1990).
3. J. Balazsovits, L. D. Mayer, M. B. Bally, P. R. Cullis, R. S. Ginsberg and R. Falk, *Cancer Chemother. Pharmacol.* **23**, 81 (1989).
4. Y. Matsumura and H. Maeda, *Cancer Res.* **46**, 6387 (1986).
5. K. J. Harrington, S. Mohammadtaghi, P. S. Uster, D. Glass, A. M. Peters, R. G. Vile and J. S. Stewart, *Clin. Cancer Res.* 7, 243 (2001).
6. D. T. Hickman, M. P. Lopez-Deber, D. Mlaki, A. B. Silva, D. Nand, M. Pihlgren, V. Giriens, R. Madani, A. St-Pierre, H. Karasteneva, L. Nagel-Steger, D. Willibold, D. Riesner, C. Nicolau, M. Baldus, A. Pfeifer and A. Muhs, *J. Biol. Chem.* **286**, 13966 (2011).
7. C. Theunis, N. Crespo-Biel, V. Gafner, M. Pihlgren, P. Lopez-Deber, P. Reis, D. Hickman, O. Adolfsson, N. Chuard, D. Mlaki Ndao, P. Borghgraef, H. Devijver, F. Van Leuven, A. Pfeifer and A. Muhs, *PLOS one* **8**, e72301 (2013).
8. M. Jayaraman, S. M. Ansell, B. L. Mui, Y. K. Tam, J. Chen, X. Du, D. Butler, L. Eltepu, S. Matsuda, J. K. Narayanannair, K. G. Rajeev, I. M. Hafez, A. Akinc, M. A. Maier, M. A. Tracy, P. R. Cullis, T. D. Madden, M. Manoharan and M. J. Hope, *Angew. Chem. In. Ed. Engl.* **51**, 8529 (2012).
9. A. Akinc, W. Querbes, S. De, J. Qin, M. Frank-Kamenetsky, K. N. Jayaprakash, M. Jayaraman, K. G. Rajeev, W. L. Cantley, J. R. Dorkin, J. S. Butler, L. Qin, T. Racie, A. Sprague, E. Fava, A. Zeigerer, M. J. Hope, M Zerial, D. W. Sah, K. Fitzgerald, M. A. Tracy, M. Manoharan, V. Koteliansky, Ad Fougerolles and M. A. Maier, *Mol. Ther.* **18**, 1357 (2010).
10. J. W. Nichols and D. W. Deamer, *Biochim. Biophys. Acta* **455**, 269 (1976).
11. L. D. Mayer, M. B. Bally, M. J. Hope and P. R. Cullis, *Biochim. Biophys. Acta* **816**, 294 (1985).
12. L. D. Mayer, M. B. Bally and P. R. Cullis, *Biochim. Biophys. Acta* **857**, 123 (1986).
13. D. D. Lasic, P. M. Frederik, M. C. Stuart, Y. Barenholz and T. J. McIntosh, *FEBS Lett.* **312**, 255 (1992).
14. S. Clerc and Y. Barenholz, *Biochim Biophys Acta* **1240**, 257 (1995).
15. B. C. Cheung, T. H. Sun, J. M. Leenhouts and P. R. Cullis, *Biochim. Biophys. Acta* **1414**, 205 (1998).
16. S. M. Gruner, R. P. Lenk, A. S. Janoff and M. J. Ostro, *Biochemistry* **24**, 2833 (1985).
17. M. J. Hope, M. B. Bally, G. Webb and P. R. Cullis, *Biochim. Biophys. Acta* **812**, 55 (1985).

Chapter A8

Preclinical to Clinical Development of Nanomedicines

*Sonke Svenson**

Drug Delivery Solutions LLC
16 Temple Street, Arlington, MA 02476, USA
**ssvenson@drugdeliverysolution.com*

Abstract

Nanomedicines on the market and in various stages of clinical translation are being reviewed, grouped around carrier platforms such as liposomes/polymersomes, polymeric micelles and nanoparticles, polymeric prodrugs, emulsions, dendrimers, and metal ion-containing complexes used in imaging applications. The goal is to provide a realistic overview of the field in order to understand the importance of nanomedicines for the treatment of patients, mainly those suffering from cancer. It was found that more than 30 nanomedicines are on the market and at least 50 nanomedicines are at different stages in clinical translation, clearly demonstrating that there is no reason for unbecoming hype or pessimism in the field. The next steps to further improve nanomedicines have been discussed: answering the question about ligand-driven versus EPR-driven targeting, and extending nanomedicines to theranostics by combining therapy and imaging/diagnostics into one treatment, which would allow personalizing the treatment and thus improving efficacy and reducing side effects.

1. Introduction

Nanomedicine is the application of nanotechnology to medicine, *i.e.* the use of nanocarriers to deliver therapeutic and/or imaging agents. Nanocarriers are generally below 100 nm in size in at least one dimension and may be constructed from a wide range of organic and inorganic materials, forming liposomes and polymersomes, polymeric micelles and nanoparticles, emulsions, dendrimers, silica-based carriers, carbon nanotubes (CNT), magnetic nanoparticles, and quantum dots (QDs) to name the main platforms. These nanocarriers are being used to encapsulate or solubilize therapeutic agents for improved drug delivery *in vivo* or to provide unique optical, magnetic and electrical properties for imaging and therapy. Several functional nanoparticles are being evaluated in preclinical and clinical studies and some nanoparticles have reached the patient, including clinically approved liposome drug formulations and metallic imaging agents. Current research in nanoparticles is focused on the consolidation of functions into multifunctional devices, which may ultimately facilitate the realization of personalized therapy. These nanoparticles may be capable of: (1) improved delivery of poorly water-soluble drugs; (2) identifying malignant cells via molecular detection; (3) visualizing their location in the body by providing enhanced contrast in medical imaging techniques; (4) fostering transcytosis of drugs across tight epithelial and endothelial barriers; (5) targeting and killing disease cells with minimal side effects through selective cell or tissue targeting; (6) temporal control of drug release, (7) monitoring treatment in real-time; and (8) co-delivery of multiple drugs for combination therapy.[1–3]

A major focus of nanomedicine development is on oncology. Most drugs utilized in cancer chemotherapy are cytotoxic and cause severe side effects. In addition, these drugs are often poorly soluble in biological fluids, quickly recognized by the mononuclear phagocyte system (MPS) and cleared from the body. Some anticancer drugs, especially the newer generation of biomacromolecules such as short interfering RNA (siRNA) and peptides, are instable in systemic circulation due to changes in local pH or the presence of enzymes. Encapsulating these drugs into nanocarriers serves two purposes: the body is protected against off-site toxicities, and the drug is protected against the body's defense system. Another benefit of nanocarriers is their improved pharmacokinetic (PK) profile compared to small molecule drugs. Small drugs distribute evenly throughout the body, resulting in very large volume of distribution (V_d). This distribution profile requires the use of high doses to achieve the therapeutic drug concentration at the tumor site, leading to severe side effects including bone marrow suppression, cardiac and kidney toxicity, hair loss, and mucositis. Nanocarriers, on the other hand, can take advantage of the enhanced permeability and retention (EPR) effect of leaky vasculature in tumorous and inflamed tissue.[4] Poorly aligned endothelial cells in the fast growing tumor vasculature with fenestrations larger than 100 nm in size, and reduced lymphatic drainage in tumor tissue result in preferred accumulation of nanocarriers in these tissues over healthy tissue, reducing V_d while improving the PK profile, and consequently, the efficacy of the anticancer drug. The combination of reduced toxicity and enhanced efficacy greatly improves the therapeutic window of the drug and is a main driver for the development of nanomedicines.[5,6] Besides taking advantage of EPR-driven targeting, targeting ligands attached to the surface of nanocarriers are being employed to increase target specificity and reduce V_d through interaction of these ligands with complementary receptors on the cell surface. However, these ligands can alter the surface composition of nanocarriers and result in reduced systemic circulation and accelerated clearance by the MPS. Careful selection of type and surface density of targeting ligands are therefore imperative to find the beneficial balance between selectivity and circulation time/clearance.[7,8] Regardless of the therapeutic benefit, the introduction of targeting ligands adds yet another layer of complexity to scale-up and regulatory approval of nanocarrier formulations. Furthermore, the number of extracellular targets that are overexpressed uniformly in a patient population is limited. Not surprising, therefore, that from a total of 41 recently reviewed nanoparticle-based formulations on the market or under investigation at different levels of clinical development, 37 nanomedicines relied on targeting through the EPR effect while only four nanoparticle formulations employed ligand-driven targeting.[9]

Most nanocarrier-based delivery systems rely on physical entrapment of the drug, causing two major challenges: (1) The hydrophilic or hydrophobic nature of the drug has to match that of the carrier to achieve sufficiently high drug loading; and (2) release of physically entrapped drugs generally shows bi-phasic behavior, where an initial burst release of the drug is followed by a first-order, sustained release. The bi-phasic behavior is a consequence of drug loading by diffusion, a reversible process. All drug molecules on or close to the nanocarrier surface will be released instantaneously under the sink condition of the bloodstream, followed by those drug molecules that first need to diffuse to the carrier surface or are released during hydrolysis of the carrier matrix. Generally only drug molecules following the sustained, first-order release reach the tumor site and are efficacious, while drug molecules released in the bloodstream contribute to toxic side effects. Alternatively, drug molecules can be chemically conjugated to the nanocarrier either directly or through selected linker molecules. Drug conjugation has several advantages over physical entrapment: (1) The initial burst release of drugs in the bloodstream can be greatly reduced or avoided; (2) the ability to design linker molecules between drug and polymer chains that can release drugs with varying kinetics in response to changes in pH or enzyme levels provides an additional tool to control release kinetics of drug molecules besides diffusion and degradation of the nanocarrier matrix; and (3) conjugation of drugs to nanocarrier constituents allows nanoparticle formation regardless of the drugs' hydrophilic or hydrophobic character because hydrophobic interactions between the nanocarrier constituents and not the compatibility between drug and nanocarrier material determine the stability of the nanocarriers. Finally, adding an imaging or diagnostic feature to nanocarriers allows to more closely follow the fate of these "theranostic" nanocarriers in the body and, potentially, opens the door to personalized medicine.[10]

In this chapter, the preclinical to clinical translation of nanomedicines will be reviewed, divided by platforms of nanocarriers (Fig. 1).

2. Platforms of Nanomedicines

2.1. *Liposomes and polymersomes*

Liposomes are spherical, closed structures composed of one or several concentric lipid bilayers surrounding an aqueous core. Depending on their membrane structure they are divided into unilamellar or multilamellar liposomes. Surface-modification with poly (ethylene glycol) (PEG) polymers provides them with stealth or long circulation properties. Liposomes are biocompatible and can carry hydrophilic drugs in their aqueous core and hydrophobic drugs within the lipid membrane. Liposomes are the oldest class with regard to clinical translation. Doxil®, approved in the U.S. in 1995 for refractory Karposi's sarcoma, is the PEGylated liposomal formulation of doxorubicin.[6,11,12] One disadvantage

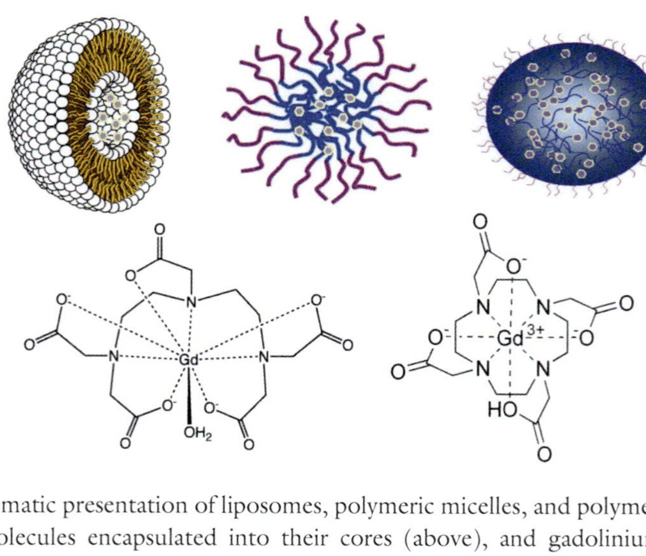

Fig. 1. Schematic presentation of liposomes, polymeric micelles, and polymeric nanoparticles with drug molecules encapsulated into their cores (above), and gadolinium MRI contrast agents Gd-DTPA and Gd-DTOA (below). Schemes are not to scale.

of liposomes is the mobility of the constituent lipids within and between the liposome layers ("flip-flop" mechanism), potentially resulting in leakiness and instability of the carrier.[13] Connecting the inner and outer membrane of a liposome and using higher molecular weight constituents such as block copolymers leads to the formation of polymersomes. Compared with liposomes, polymersomes have many distinct properties. Their membrane thickness, which can be controlled by the molecular weight of the hydrophobic block of the copolymer, determines polymersome properties such as elasticity, permeability, mechanical stability, and loading capacity for hydrophobic drugs.[14,15]

2.2. *Liposome/polymersome-based nanomedicines*

The observation that phospholipids in aqueous systems can form closed bilayer structures was described in 1965.[16] Not surprising, therefore, that liposomes are the oldest class of nanomedicines, with Doxil® receiving approval by the U.S. Food and Drug Administration (FDA) in 1995. Doxil is based on three principles: (1) prolonged drug circulation time and avoidance of the MPS due to the PEGylated ("stealth") surface; (2) high and stable remote loading of doxorubicin driven by a transmembrane ammonium sulfate gradient; and (3) a lipid bilayer membrane in a "liquid ordered" phase composed of phosphatidylcholine and cholesterol.[17] Doxisome® from Taiwan Liposome Company, Ltd. (TLC) is described as being equivalent to Doxil in terms of anticancer activity, PKs, and toxicity, potentially qualifying it as the generic form of Doxil. Lipo-Dox®, another liposomal doxorubicin formulation from TLC, was successfully launched in Taiwan and approved by the Taiwanese Department of Health in 2001 for AIDS-Kaposi sarcoma, breast cancer, and ovarian cancer. FDA approval of non-PEGylated liposome formulations DaunoXome®, AmBisome®, DepoCyt®, and Visudyne® was given in 1996, 1997, 1999, and 2000, respectively. More recent liposomal nanomedicines are Depodur®, FDA approved in 2004, and Myocet®, which is approved in Europe and has FDA approval pending as of March 2013. LEP-ETU is a liposomal delivery system of paclitaxel developed by NeoPharm, Inc. It is currently in a Phase 2 trial for the treatment of metastatic breast cancer, sponsored by Insys Therapeutics, Inc. Similarly, LE-DT, a liposomal formulation of docetaxel developed by NeoPharm and sponsored by Insys, is in Phase 2 trial to treat locally advanced or metastatic pancreatic cancer. Protein stabilized liposomes (PSL™) have been developed by Azaya Therapeutics, Inc. for the delivery of water-insoluble drugs, embedded into the liposomal membrane. Two products are in clinical trials, ATI-1123, the PSL formulation of docetaxel using human serum albumin (HSA) as the stabilizing protein, and

ATI-0918, the PSL formulation of doxorubicin. ATI-0918 is being evaluated as generic version against Doxil in a recruiting Phase 1 study.[18]

Another promising liposomal clinical candidate employing doxorubicin is ThermoDox®, developed by Celsion Corp. However, ThermoDox very recently failed to reach its endpoint in clinical trial. The liposome formulation was evaluated in Phase 3 for the treatment of hepatocellular carcinoma (HCC or primary liver cancer), using lysolipid thermally sensitive liposomes (LTSL) as the carrier. LTSL become leaky and release the drug at 42°C, and therefore, the liposomes are designed to be used in combination with hyperthermic treatments such as radiofrequency thermal ablation (RFA). Heat treatment enhances the leakiness of tumor vasculature and increases the number of liposomes that enter tumor tissue. The goal of the ThermoDox approach is to capture micrometastases which are most commonly responsible for post-treatment disease recurrence.[19] The Phase 3 study was designed to show a 33% improvement in progression-free survival (PFS) with 80% RFA power and a p-value = 0.05; however, the study did not meet the goal of demonstrating evidence of clinical effectiveness that could form the basis for regulatory approval in the population chosen for the study.[20]

The neoadjuvant EndoTAG®-1 from Medigene AG is a novel composition of the established drug paclitaxel combined with neutral and positive lipids. The positively charged lipids interact with newly developing, negatively charged endothelial cells, which are primarily required for the growth of tumor blood vessels. The EndoTAG-1 paclitaxel component attacks the endothelial cells as they divide, thus targeting the blood supply to tumors without affecting the supply to healthy tissue. Therefore, EndoTAG-1 is expected to prevent the formation of new tumor blood vessels and to inhibit tumor growth. Medigene assumes that this new treatment can be applied to those tumors that have developed resistance to conventional paclitaxel therapy. This hypothesis is being verified in a Phase 2 clinical trial on HER2-negative breast cancer. Previous Phase 2 trials have tested EndoTAG-1 alone and with paclitaxel as a rescue therapy for patients with relapsed or metastatic triple receptor negative breast cancer and combined with gemcitabine versus gemcitabine monotherapy in pancreatic cancer.[21]

Besides taxanes and anthracyclines, camptothecin and its derivatives irinotecan (CPT-11) and topotecan have been employed in the development of liposomal nanomedicines. IHL-305, a liposomal formulation of irinotecan, is in Phase 1 for the treatment of advanced solid tumors, sponsored by Yakult Honsha Co., Ltd. Another irinotecan formulation, NL CPT-11 (nanoliposomal CPT-11) is in Phase 1 for the treatment of recurrent high-grade gliomas, sponsored by University of California, San Francisco (last update January 2013).

Brakiva™ (or TLI, topotecan liposomes injection), the liposomal formulation of topotecan, has been developed by Talon Therapeutics to replace standard topotecan therapies. A multicenter, open-label, dose-escalation Phase 1 study to establish the safety, tolerability, and schedule of TLI in adults with advanced solid tumors that have relapsed, are refractory to standard therapy, or for whom there is no standard therapy available is ongoing but not recruiting. The liposome platform used in the TLI formulation, sphingomyelin-based liposomes called Optisome™, is also applied in the formulations of Alocrest™ (Vinorelbine liposomes injection (VLI)) and Marqibo®. VLI is listed in an ongoing but not recruiting Phase 1 study that will determine the safety, tolerability, and PKs of VLI in patients with advanced solid tumors, non-Hodgkin's lymphoma, or Hodgkin's disease. The clinically most advanced formulation, Marqibo (liposomal vincristine sulfate) is in recruiting a Phase 3 study involving patients 60 years old or older with newly diagnosed acute lymphoblastic leukemia (ALL) and a recruiting Phase 2 study to learn if Marqibo can help to control metastatic uveal melanoma. However, in January 2013, Talon has decided to review strategic alternatives that may lead to a possible transaction, including the merger, business combination, or sale of the company, and therefore, the fate of these liposomal formulations is unknown.[22]

One of the few non-cancer therapies in clinical trials is Arikace® from Insmed Inc., the liposomal formulation of amikacin, a FDA-approved aminoglycoside antibiotic, for inhalation treatment of cystic fibrosis patients with *Pseudomonas* lung infections. Arikace is in Phase 3 clinical trial in Europe and Canada, and in Phase 2 for nontuberculous mycobacteria (NTM) lung infections in the U.S.[23] Another liposomal inhalation formulation, developed by PARI Pharma GmbH, contains the immunosuppressant cyclosporin A (L-CsA). L-CsA is being evaluated in a recruiting Phase 3 study for long-term safety in the prevention of bronchiolitis obliterans syndrome (BOS) following lung transplantation. BOS is an incurable disease of the small airways affecting approximately 60,000 patients worldwide after lung transplantation.[24]

In an extension of the liposome platform, liposomes with targeting ligands have been formulated with anticancer drugs. MBP-426 developed by Mebiopharm Co., Ltd., is a liposome composed of N-glutarylphosphatidylethanolamine (NGPE), conjugated with transferrin (Tf) ligands and carrying oxaliplatin entrapped in the aqueous core. MBP-426 has demonstrated anticancer preclinical activity and recently entered clinical trials.[25] The ongoing Phase 2 trial is an open-label study to evaluate the efficacy of MBP-426 at a dose of 170 mg/m^2 in combination therapy in patients with second line metastatic gastric, gastro-esophageal junction or esophageal adenocarcinoma. MBP-Y003, MBP-Y004, and MBP-Y005 are preclinical candidates based on the

Tf-liposome platform containing methotrexate, docetaxel, and gemcitabine as the actives for the treatment of cancers and rheumatoid arthritis (RA). Similarly to MBP-426, SGT53 (SynerGene Therapeutics, Inc.) is a nanomedicine composed of a wild type p53 gene (plasmid DNA) entrapped into a liposome that is targeted to tumor cells by means of an anti-transferrin receptor single-chain antibody fragment (TfRscFv), attached to the outside of the liposome. Numerous human tumors experience loss or mutation of wild type p53 (wtp53), which plays a vital role in cell cycle control and is a critical component in cell death (apoptosis) and the regulation of angiogenesis. The loss of such critical tumor suppressor activity is believed to be responsible for p53 gene's involvement in a broad array of human tumors and resistance to chemo- and radiotherapy. Preclinical studies have indicated that SGT53 could resensitize tumors to these therapies. The current Phase 1b clinical study is designed to evaluate the safety of SGT53 in combination with docetaxel, determine the recommended Phase 2 doses of both agents, and evaluate the combination effect of SGT53 and docetaxel on tumor size or progression. SynerGene Therapeutics is a spin-off from Georgetown University, and the development of SGT53 seems to be predominantly fueled by U.S. SBIR/STTR grants.[26]

The liposomal formulation of irinotecan-HCl and floxuridine (CPX-1, Celator Pharmaceuticals) extends the platform to true combination therapy, where two anticancer drugs are combined in 1:1 ratio within one liposome. A Phase 2 study in patients with advanced colorectal cancer who have already received chemotherapy that included the drugs oxaliplatin or irinotecan has been completed. All patients received CPX-1 at a dose of 210 units/m^2 over 90 min every two weeks. In addition, the liposomal combination of cytarabine and daunorubicin-HCl in 5:1 ratio (CPX-351) has shown synergistic efficacy against acute myeloid leukemia (AML) in a completed Phase 2 study in newly diagnosed elderly AML patients. Currently, CPX-351 is being evaluated in a Phase 3 trial, comparing the efficacy of CPX-351 to the drugs in (7 + 3) ratio as first line therapy in elderly patients (60–75 years) with high risk (secondary) AML. The primary efficacy endpoint will be overall survival (OS).[27,28]

The delivery of siRNAs using liposomes or "lipid nanoparticles" (LNP), where a bilayer membrane encloses an aqueous core carrying the siRNA, is a recent application of this platform. Three companies have products in Phase 1 and 2 trials, Alnylam Pharmaceuticals (ALN-VSP02, ALN-TTR02, ALN-TTRsc, ALN-PCS), Tekmira Pharmaceuticals (TKM-PLK1, TKM-Ebola), and Silence Therapeutics (Atu027).[29–31] The Phase 1 results of ALN-VSP02, a LNP formulation of two siRNAs targeting vascular endothelial growth factor (VEGF) and kinesin spindle protein (KSP), two genes that are critical for the growth and development of cancer cells, showed that the formulation was generally safe and well-tolerated up to a dose of 1.0 mg/kg, given every two weeks. The study involved 41 patients with advanced solid tumors and liver involvement who had failed to respond to or showed tumor progression after standard treatment.[32] ALN-TTR, a systemically delivered RNAi therapeutic, targets the transthyretin (TTR) gene to treat TTR-mediated amyloidosis (ATTR). ATTR is caused by mutations in the TTR gene, which is expressed predominantly in the liver, causing abnormal amyloid proteins to accumulate and damage body organs and tissue such as the peripheral nerves and heart. The safety, tolerability, PKs and pharmacodynamics of multiple doses of ALN-TTR are being evaluated in an ongoing Phase 2 trial. ALN-TTRsc is a *s.c.* administered RNAi therapeutic comprising a siRNA conjugated to a GalNAc ligand that enables receptor-mediated delivery to the liver. The treatment is targeting familial amyloidotic cardiomyopathy (FAC), which affects at least 40,000 people worldwide. ALN-TTRsc is the first *s.c.* delivered RNAi therapeutic to enter the clinical development stage. The Phase 1 trial is being conducted in the U.K. as a randomized, double-blinded, placebo-controlled, single- and multidose, dose-escalation study, enrolling up to 40 healthy volunteers. ALN-PCS is a systemically delivered RNAi therapeutic targeting the gene proprotein convertase subtilisin/kexin type 9 (PCSK9), a genetically validated target involved in the metabolism of LDL cholesterol ("bad" cholesterol). ALN-PCS has the potential to lower tissue and circulating PCSK9 levels, resulting in higher LDL receptor levels in the liver, and subsequently lower LDL cholesterol levels.

Tekmira has two product candidates in dose-escalating Phase 1 trials. TKM-PLK1 targets polo-like kinase 1, a protein involved in tumor cell proliferation and a validated oncology target. Inhibition of PLK1 expression prevents the tumor cell from completing cell division, resulting in cell cycle arrest and death of the cancer cell. To date, TKM-PLK1 has been administered to 21 patients at doses ranging from 0.15 mg/kg to 0.90 mg/kg on a once-weekly protocol, with each cycle consisting of three doses followed by a rest week. TKM-Ebola aims at the treatment of Ebola virus infections which cause periodic outbreaks of hemorrhagic fever in human populations with mortality rates reaching 90%. Currently, there are no approved treatments for Ebola or other hemorrhagic fever viruses. The Phase 1 clinical trial is a placebo-controlled, single-blinded, single-ascending dose study in healthy human volunteers. The objective of the Phase I trial is to assess the safety and tolerability of TKM-Ebola and evaluate the PKs and systemic exposure following both a single-ascending dose and multiple-ascending doses of TKM-Ebola.

ATU027 for the treatment of solid tumors is Silence Therapeutics most advanced internal drug candidate, targeting PKN3, a protein kinase C-related molecule thought to be an effector mediating malignant cell growth. Atu027 has shown antitumor activity and blockage of cancer cell dissemination, suggesting a profound anti-metastatic effect of the drug. In its open-label, single center, dose finding Phase 1 study Atu027 is given as single treatment followed by repeated treatment (*i.e.* 8 treatments within 4 weeks) to patients with advanced solid cancer. Besides siRNA-only treatment, one very recent study proposes the combination of siRNA (Atu027) with gemcitabine for the treatment of advanced or metastatic pancreatic cancer. The delivery of highly water-soluble siRNA molecules is another extension of the liposome platform, creating new nanomedicines with high expectations for future treatment.

In addition to the RNAi technology, an equally promising but lesser known approach is DNA interference (DNAi). DNAi represents a novel approach to targeting specific disease genes, employing single strands of DNA targeted to treat non-transcribed regions of genomes involved in complex genetic diseases. DNAi constructs consist of single-stranded 24-mer DNA oligonucleotides targeted to the 5 -promoter region of specific oncology genes. Liposomal PNT2258 (ProNAi Therapeutics, Inc.), composed of a 24-mer DNA encapsulated in a Novosom Smarticle® liposome, targets the Bcl-2 gene which has been implicated in a number of cancers through decrease of apoptosis in cancer cells. Hypothesis suggests that simultaneous overexpression of Bcl-2 and the myc proto-oncogene may produce aggressive B-cell malignancies including lymphomas.[33–35] The projected therapeutic effects of DNAi are markedly similar to those of RNAi, which knocks down messenger RNA (mRNA) transcripts, thereby combating the progress of disease or halting it entirely. However, disease genes are continuously transcribing mRNA molecules, which therefore will have to be continuously destroyed by the RNAi therapies. DNAi takes a simpler and theoretically more effective approach because DNAi therapies need only target one to two copies of a disease gene in each cell to cut off their negative effects. The Phase 2 study of PNT2258 is a multicenter, nonrandomized study to characterize antitumor activity and collect safety data on patients with relapsed or refractory non-Hodgkin's lymphoma.

The application of liposomes in vaccine delivery is yet another attempt to develop a successful nanomedicine. Stimuvax® or liposomal BLP25 (L-BLP25) is an investigational therapeutic vaccine developed to provide immunity against cancer cells overexpressing Mucin 1 (MUC-1), a glycoprotein antigen. The synthetic MUC-1 peptide vaccine, consisting of a 25-amino acid sequence of MUC-1, is expected to stimulate the patient's own T cell-mediated immune response and destroy MUC-1 producing cancer cells. The drug is encapsulated into liposomes to enhance the delivery toward the cancer antigen. Stimuvax was originally developed by Cancer Research U.K. The development was further continued by Oncothyreon, formerly known as Biomira. Merck KGaA is developing the drug globally under a license agreement with Oncothyreon. Three Phase 3 trials have employed or are employing Stimuvax, including START for patients suffering from advanced non-small cell lung cancer (NSCLC), STRIDE for treatment against breast cancer, and INSPIRE against NSCLC in Asian countries. The global START trial was designed to evaluate the efficacy and safety of Stimuvax for treatment against unresectable, locally advanced stage IIIA or IIIB NSCLC. It was a randomized, double-blinded, placebo-controlled study that did not meet its primary endpoint, demonstrating a statistically significant improvement in OS. The Phase 3 trial INSPIRE involves Asian NSCLC patients and is similar in design to START. A total of 420 patients are expected to participate in the INSPIRE trial, which is also randomized, double-blinded, and placebo-controlled, aiming at patient survival. The STRIDE Phase 3 trial in patients with non-removable locally advanced, metastatic or recurrent breast cancer, scheduled to enroll more than 900 patients in about 30 countries has been terminated.[36] Several Phase 2 trials on multiple myeloma, and rectal and prostate cancers as well as combination treatment with monoclonal antibody (bevacizumab) are ongoing. The use of vaccines in cancer treatment is a more recent application for nanomedicines with for now uncertain outcome.

The development of nanomedicines is often carried by small companies, who might change ownership or perish, and therefore, the fate of some candidates in early clinical trials is uncertain not just for medical reasons. For example, OSI-211, the liposomal formulation of lurtotecan was developed by OSI Pharmaceuticals, which was bought by Astellas Pharma Inc., in 2010. Astellas sponsored two Phase 2 trials with OSI-211 on recurrent small cell lung cancer (SCLC) and, in combination with topotecan, in patients with relapsed epithelial ovarian cancer. Both studies were completed with last clinical updates from October 2011. No further information about the development of this formulation is readily available; however, a Phase 2 study on metastatic or locally recurrent head and neck cancer was completed with last update from July 2012, so there still seems to be life in liposomal lurtotecan. Two studies on liposomal SN-38 (LE-SN38), the active metabolite of irinotecan, were sponsored by the Cancer and Leukemia Group of the National Cancer Institute (NCI). However, last updates were provided back in midst 2011. The same situation exists for SPI-77, a stealth-liposomal cisplatin formulation sponsored by New York University School of Medicine in cooperation with NCI.

On the positive side, an example of a nanomedicine in development that successfully transferred between companies is the liposomal formulation of irinotecan, MM-398 (Merrimack Pharmaceuticals, Inc.), also known as PEP02 (PharmaEngine, Inc.). PharmaEngine licensed rights to PEP02 in Europe and Asia from Hermes Biosciences, Inc. Hermes was acquired by Merrimack in 2009. In 2011, PharmaEngine granted back to Merrimack the rights to develop, manufacture, and commercialize MM-398 in Europe and Asia with the exception of Taiwan in addition to North America. MM-398 recently achieved its primary efficacy endpoints in Phase 2 clinical trials in pancreatic and gastric cancer. In this single-arm, Phase 2 clinical trial of MM-398 as a monotherapy in 40 metastatic pancreatic cancer patients who had previously failed treatment with gemcitabine, patients treated with MM-398 achieved median OS of 22.4 weeks. Additionally, 20% of the patients in this Phase 2 trial survived for more than one year.[37] Merrimack is currently recruiting participants for a randomized, open label Phase 3 study of MM-398 with or without 5-fluorouracil and leucovorin, versus 5-fluorouracil and leucovorin in patients with metastatic pancreatic cancer who have failed prior gemcitabine-based therapy.

At the time of this review, eight liposome nanomedicines are on the market, six clinical candidates are in Phase 3, eight are in Phase 2, and 13 candidates are in Phase 1 trials (Table 1). It is interesting to note that so far no polymersome-based formulation seems to have translated into the clinic despite the proposed advantages over conventional liposomes. One reason might be the unknown fate of high molecular weight block copolymers in the body which would have to be verified, while the fate of conventional lipids is more established.

2.3. *Polymeric micelles*

Micelles and polymeric micelles are formed by surfactants or amphiphilic block copolymers. Both molecules consist of a hydrophobic and a hydrophilic moiety, which self-assemble in water above a system- and temperature-specific critical micelle concentration (CMC) in a thermodynamically driven process. Lower CMC and therefore higher stability, and larger size (20–100 nm) favor polymeric micelles over surfactant micelles as delivery platform for hydrophobic drugs, encapsulated into the hydrophobic micelle core. Polymeric micelles are large enough to avoid first pass kidney elimination via glomerular filtration but are small enough to capitalize on the EPR effect. Furthermore, the small size of drug-loaded polymeric micelles provides an alternative route

Table 1. Liposomal nanomedicines on the market and under clinical evaluation.[a]

Product	Formulation	Application	Company	Status
AmBisome®	Liposomal amphotericin B	fungal infections	Astellas Pharma/ Gilead Sciences	Marketed
Doxil® Caelyx®	PEGylated liposomal doxorubicin	Kaposi's sarcoma; metastatic breast & ovarian cancers	Janssen Pharmaceuticals	Marketed
Doxisome®, Lipo-Dox®	PEGylated liposomal doxorubicin	Kaposi's sarcoma; metastatic breast & ovarian cancers	Taiwan Liposome Company, Ltd.	Marketed
DaunoXome®	Liposomal daunorubicin	Kaposi's sarcoma	Galen Ltd.	Marketed
DepoCyt®	Liposomal cytarabine	Lymphomatous meningitis	Sigma-Tau Pharmaceuticals	Marketed
DepoDur®	Liposomal morphine	Pain treatment	Pacira Pharmaceuticals	Marketed
Myocet®	Liposomal doxorubicin	Metastatic breast cancer	Cephalon Inc./ Sopherion	Marketed
Visudyne®	Liposomal verteporfin	AMD	Novartis AG/ QLT Inc.	Marketed
ARIKACE®	Inhaled liposomal amikacin	*Pseudomonas* lung infections in cystic fibrosis	Insmed Inc.	Phase 3 (EU, CAN)
L-CsA	Liposomal cyclosporine A	Bronchiolitis obliterans syndrome	Pari Pharma GmbH	Phase 3 NCT01439958*
MM-398 (PEP02)	Liposomal irinotecan	Metastatic pancreatic cancer	Merrimack Pharma, Inc.	Phase 3 NCT01494506*
Marqibo®	Liposomal vincristine sulfate	Acute lymphoblastic leukemia (ALL)	Talon Therapeutics	Phase 3 NCT01439347*
Stimuvax (L-BLP25)	Liposomal vaccine	NSCLC (START)	Merck KGaA	Phase 3 NCT00409188*
	liposomal vaccine	NSCLC in Asians (INSPIRE)	Merck KGaA	Phase 3 NCT01015443*
CPX-351	liposomal cytarabine-daunorubicin combination	acute myeloid leukemia (AML)	Celator Pharmaceuticals	Phase 3 NCT01696084*

(*Continued*)

Table 1. (*Continued*)

Product	Formulation	Application	Company	Status
LEP-ETU	Liposomal paclitaxel	Metastatic pancreatic cancer	Insys Therapeutics	Phase 2 NCT01190982*
LE-DT	Liposomal docetaxel	Metastatic breast cancer	Insys Therapeutics	Phase 2 NCT01186731*
EndoTAG®-1	EndoTAG-1 plus PTX	HER2-negative breast cancer	Medigene AG	Phase 2 NCT01537536*
CPX-1	Liposomal irinotecan-floxuridine combination	Advanced colorectal carcinoma	Celator Pharmaceuticals	Phase 2 NCT00361842*
OSI-211	Liposomal lurtotecan	Metastatic or recurrent head and neck cancer	Astellas Pharma Inc.	Phase 2 NCT00022594*
MBP-426	Transferrin-targeted NP w. oxaliplatin	Gastric, gastro-esophageal or esophageal adenocarcinoma	Mebiopharm Co., Ltd.	Phase 2 NCT00964080*
ALN-TTR02	Liposomal RNAi	Transthyretin (TTR)-mediated amyloidosis	Alnylam Pharmaceuticals	Phase 2 NCT01617967*
PNT2258	Liposomal DNAi	Non-Hodgkin's lymphoma	ProNAi Therapeutics	Phase 2 NCT01733238*
IHL-305	Liposomal irinotecan	Advanced solid tumors	Yakult Honsha Co., Ltd.	Phase 1 NCT00364143*
NL CPT-11	Nanoliposomal CPT-11 (irinotecan)	Recurrent high-grade gliomas	University of California, San Francisco	Phase 1 NCT00734682*
TLI (Brakiva)	Liposomal topotecan	Advanced solid tumors	Talon Therapeutics	Phase 1 NCT00765973*
VLI (Alocrest)	Liposomal vinorelbine	Advanced solid tumors	Talon Therapeutics	Phase 1 NCT00364676*
ATI-1123	Liposomal docetaxel	Advanced solid tumors	Azaya Therapeutics	Phase 1 NCT01041235*
ATI-0918	Liposomal doxorubicin	Ovarian cancer	Azaya Therapeutics	Phase 1 NCT01715168*
SGT53	TfRscFv - liposome w. wild type p53	Neoplasm	SynerGene Therapeutics, Inc.	Phase 1 NCT00470613*

(*Continued*)

Table 1. (*Continued*)

Product	Formulation	Application	Company	Status
ALN-VSP02	Liposomal RNAi	Solid tumors with liver involvement	Alnylam Pharmaceuticals	Phase 1 NCT01158079*
ALN-TTRsc	Liposomal RNAi	Subcutaneous TTR-mediated amyloidosis	Alnylam Pharmaceuticals	Phase 1 (UK)
ALN-PCS	Liposomal RNAi	Elevated LDL cholesterol	Alnylam Pharmaceuticals	Phase 1 NCT01437059*
TKM-PLK1	Liposomal RNAi	Advanced solid tumors	Tekmira Pharmaceuticals	Phase 1 NCT01262235*
TKM-Ebola	Liposomal RNAi	Hemorrhagic fever	Tekmira Pharmaceuticals Corp.	Phase 1 NCT01518881*
Atu027	Liposomal RNAi	Advanced solid tumors	Silence Therapeutics	Phase 1 NCT00938574*

Notes: *ClinicalTrials.gov identifier.

[a]Liposomal formulations whose trials have been completed or are listed as inactive for more than one year with no apparent follow-up have been excluded from Table 1. AMD — age-related macular degeneration; TfRscFv — anti-transferrin receptor single-chain antibody fragment.

of cell internalization via the endosomal pathway, which is speculated to circumvent drug efflux mechanisms associated with multidrug resistance (MDR). PEG chains are the most frequently employed structural motif for the hydrophilic component of the block copolymers due to their high aqueous solubility and nontoxicity.[38–40]

2.4. *Polymeric micelle -based nanomedicines*

To date there are only few nanomedicines based on polymeric micelles on the market.[41] Genexol-PM® is a cremophor-free polymeric micelle formulation using PEGylated poly(D,L-lactide) copolymer as matrix and paclitaxel as the active ingredient. Genexol-PM, developed by Samyang Pharmaceuticals, has been approved in Korea (2007) and other Asian countries for the treatment of breast cancer and NSCLC. In the U.S., Genexol-PM has completed a Phase 2 study in patients with advanced urothelial cancer previously treated with gemcitabine and platinum, and is currently studied in a Phase 3 trial to evaluate the efficacy and safety of Genexol-PM compared to Genexol®, the conventional paclitaxel formulation with Cremorphor EL®, in patients with recurrent or metastatic breast cancer.[42,43] Besides in South Korea, the development of polymeric micelles as nanomedicine platform is a major effort in Japan, fueled by the work of the Kataoka research group at the University of Tokyo. Several polymeric micelles based on methoxy-PEG (mPEG)ylated hydrophobic poly(amino acids) carrying anticancer drugs in their hydrophobic core have been developed. NK105 is a paclitaxel formulation using modified mPEGylated poly(aspartic acid).[44] Originally developed by NanoCarrier Co., Ltd., to treat stomach and breast cancers, NK105 has been licensed to Nippon Kayaku Co., Ltd. and has recently opened a multinational Phase 3 clinical study comparing NK105 versus paclitaxel in patients with metastatic or recurrent breast cancer. Another polymeric micelle formulation under development by Nippon Kayaku is NK012. In this formulation, the irinotecan metabolite SN-38 is covalently bound to a PEG–poly(glutamic acid) block copolymer via an ester bond.[45] NK012 has completed a Phase 2 trial in the U.S. with indications in SCLC, while a second Phase 2 study on triple negative breast cancer is ongoing. The third polymeric micelle formulation from Nippon Kayaku is NK911, in which doxorubicin is covalently bound to mPEG–poly(aspartic acid) block copolymer. NK911 had been evaluated in a Phase 1 trial in Japan; however, the latest publicly available updates for NK911 are several years old.[46]

Two polymeric micelle formulations employing platinum drugs developed by NanoCarrier are NC-6004 and NC-4016. NC-6004 (Nanoplatin) is composed of cisplatin and mPEG–poly(glutamic acid), listed as "currently recruiting" for a Phase 1/2 study in combination with gemcitabine in patients with locally advanced or metastatic pancreatic cancer in Asian countries (NCT00910741). However, the last verification of the study by NanoCarrier dates back to May 2009.[47] NC-4016 is a polymeric micelle containing oxaliplatin (DACH-platin), which had been licensed in 2007 to Debiopharm for worldwide use excluding Japan. This license agreement has been terminated in 2010, and NanoCarrier has decided to start a clinical study at The University of Texas MD Anderson Cancer Center. NanoCarrier is in the process of preparing for the IND application as of early 2013. NC-6301, in which docetaxel is covalently bound via an ester to a mPEG–poly(aspartic acid) block copolymer, is in preclinical evaluation by NanoCarrier.[48] It should be noted that Shin-Etsu Chemical has decided to take a majority stake in NanoCarrier; the impact on Nanocarrier's nanomedicine program needs to be seen. The evaluation of poly(amino acid) micelle nanocarriers in preclinical and clinical studies has recently been reviewed.[49]

In an extension of the polymeric micelle platform, SP1049C, a proprietary *i.v.* composition of BioMod™ polymers and doxorubicin as ingredient, was developed by Supratek Pharma Inc. BioMod polymers, including poly(ethylene oxide)–poly(propylene oxide)–poly(ethylene oxide) (PEO–PPO–PEO) triblock copolymers, have the ability to selectively disrupt mitochondrial functionality in chemoresistant cells. They penetrate cancer cells, reach mitochondrial membranes, compromise their function by depleting ATP, and release reactive oxygen species (ROS) and cytochrome C. The depletion of ATP inactivates the cell's detoxification mechanisms by inhibiting ABC transporters such as P-gp and MRP. The combination of BioMod polymers with an anticancer drug, therefore, results in a new drug that targets chemoresistant cancers. Clinical Phase 2 studies have shown that SP1049C has high activity as a single agent with response rate of 41% (by RESIST) in non-resectable stage IVb patients with adenocarcinoma of the esophagus and median survival of 10 months, comparable to the most efficient three-drug combinations used today. In 2008, a Phase 3 clinical study design has been reviewed and agreed to with the FDA under a Special Protocol Assessment (SPA); however neither SP1049C or Supratek Pharma are currently listed on ClinicalTrials.gov.[50–52]

Current nanomedicines based on polymeric micelles on the market and under active clinical evaluation are listed in Table 2.

Table 2. Nanomedicines based on polymeric micelles on the market and under clinical evaluation.[a]

Product	Formulation	Application	Company	Status
Genexol-PM®	PEG–poly(D,L-lactide) with paclitaxel	Breast cancer and NSCLC	Samyang Pharmaceuticals	Marketed (S. Korea, other Asian countries)
	PEG–poly(D,L-lactide) with paclitaxel	Recurrent or meta-static breast cancer	Samyang Pharmaceuticals	Phase 3 NCT00876486*
NK105	mPEG–poly (aspartic acid) with paclitaxel	Recurrent or meta-static breast cancer	Nippon Kayaku Co., Ltd.	Phase 3 NCT01644890*
NK012	PEG–poly (glutamic acid)–SN38 conjugate	SCLC and triple negative breast cancer	Nippon Kayaku Co., Ltd.	Phase 2 NCT00951613* Phase 2 NCT00951054*

Notes: *ClinicalTrials.gov identifier.

[a]Polymeric micelle formulations whose trials have been completed or are listed as inactive for more than one year with no apparent follow-up have been excluded from Table 2. NSCLC — non-small cell lung cancer; SCLC — small cell lung cancer.

2.5. *Polymeric nanoparticles and prodrugs*

The third class of nanocarriers discussed here consists of polymeric nanoparticles, which are mainly prepared through the use of emulsions or precipitation processes. Their size depends on the preparation method, ranging from 10 to up to several 100 nm. The constituent polymers of nanoparticles are not necessarily amphiphilic as those in micelles, and therefore nanoparticle formation is not caused by self-assembly but requires active processing. Drug molecules are either physically entrapped or chemically conjugated to the hydrophobic core of these nanoparticles. The solid nature of polymeric nanoparticles confers high stability and allows controlled drug release, either through diffusion or through degradation of the polymeric matrix. In addition, stimuli-responsive load-and-release modalities can be incorporated into these solid nanocarriers, activated by an external signal or changes in the local microenvironment within the body.[53–55] Polymeric prodrugs in clinical trials, *i.e.* polymers with drug molecules chemically attached to them, will also be discussed under this platform even when it remains uncertain whether these prodrugs actually assemble to form defined nanoparticles.

2.6. *Polymeric nanoparticles and prodrugs-based nanomedicines*

Although polymeric nanoparticles are a focus of research since the 1970s at least, essentially only one nanoparticle-based nanomedicine has advanced to the market. The approval of drugs for human use by the FDA through the Center for Drug Evaluation and Research (CDER) is a time-consuming and expensive process, and approval rates are low.[56,57] Abraxane® is a paclitaxel formulation based on the nanoparticle albumin bound (nab) technology platform developed by Abraxis BioScience, a wholly-owned subsidiary of Celgene Corporation. The albumin-paclitaxel formulation consists of nanoparticles 130 nm in diameter, and is used as a colloidal suspension of lyophilized drug and HSA diluted in saline solution. Abraxane was approved by the FDA in 2005 and the European Medicines Agency (EMA) in 2008 for the treatment of breast cancer after failure of combination chemotherapy for metastatic disease or relapse within six months of adjuvant chemotherapy. Multiple clinical trials are listed on the NIH clinical trials website involving Abraxane, including Phase 3 trials in first-line NSCLC, metastatic pancreatic and melanoma cancers, and Phase 2 trials in first-line metastatic breast cancer and bladder and ovarian cancers.[58,59] For example, the Metastatic Pancreatic Adenocarcinoma Clinical Trial (MPACT) study involved 861 participants and was set up to evaluate the effectiveness of treating pancreatic patients with

Abraxane and chemotherapy compared to chemotherapy alone. Even though pancreatic cancer is relatively uncommon, making up only 2.1% of all cancer cases, it is one of the most difficult cancers to treat, killing around 38,000 people in the U.S. every year. Results from the MPACT study revealed that 35% people on the combination of Abraxane and chemotherapy were alive at the end of the first year compared to only 22% who just underwent chemotherapy. This translates into a 59% increase in one-year survival for the patients on Abraxane plus chemotherapy.

The nanoparticle formulation Livatag® (doxorubicin transdrug™), developed by BioAlliance Pharma S.A., consists of doxorubicin-loaded poly(isohexylcyanoacrylate) nanoparticles for delivery to chemo-resistant cells such as primary liver cancer cells via the hepatic intra-arterial route. Livatag was granted orphan drug status in the U.S. and Europe.[60] BioAlliance has begun a multicenter, randomized Phase 3 clinical trial (ReLive) in the U.S. and Europe, designed to estimate the efficacy of Livatag on OS in nearly 400 patients suffering from advanced HCC after failure or intolerance to sorafenib. Patients eligible for the ReLive study are receiving either doxorubicin transdrug at 20 or 30 mg/m^2 or the best supportive care (BSC).

The polymeric nanopharmaceutical CRLX101 (formerly IT-101) developed by Cerulean Pharma Inc. is composed of a β-cyclodextrin-PEG alternating copolymer with camptothecin (CPT) drugs conjugated to the polymeric backbone. This polymer–drug conjugate self-assembles to form nanoparticles in the 20–50 nm range.[61,62] CRLX101 has been evaluated in a Phase 2 trial with patients suffering from advanced NSCLC who have failed one or two previous regimens of chemotherapy. The study did not meet its primary efficacy endpoint, OS benefit. Consistent with earlier clinical trials, CRLX101 showed signals of activity, including tumor reductions that meet RECIST criteria, and a favorable safety profile. The study enrolled 157 patients in Russia and the Ukraine, randomized in a two-to-one ratio to two treatment arms, CRLX101 plus best supportive care compared to BSC alone. The trial's primary endpoints were median OS and safety/tolerability. Study participants were also assessed for PFS, objective tumor response, and PK parameters.[63] Currently, CRLX101 is studied in a 2-stage Phase 2 trial in recurrent ovarian, tubal and peritoneal cancer, a Phase 2 study treating recurrent SCLC patients with CRLX101 or topotecan hydrochloride, a Phase 1b study in combination with bevacizumab in the treatment of patients with advanced renal cell carcinoma, and pilot studies treating patients with advanced or metastatic stomach, gastroesophageal, or esophageal cancer that cannot be removed by surgery. Originally developed by Insert Therapeutics, Inc., which then merged with Calando Pharmaceuticals, Inc., CRLX101 was licensed to Cerulean, who further develops this cyclodextrin-PEG (CDP) technology.

The siRNA-delivering polymeric nanoparticle CALAA-01, developed by Calando Pharmaceuticals, Inc, is comprised of a related β-cyclodextrin-based copolymer. Specifically, CALAA-01 is composed of four components: (1) a linear polymer in which positively charged groups alternate with β-cyclodextrin; and (2) siRNA, whose negatively charged backbone complexes with the positively charged polymer segments. Several polymer–siRNA complexes self-assemble into a nanoparticle of less than 100 nm in diameter that fully protects the siRNA from degradation in serum. The resulting nanoparticles are then decorated with (3) adamantane covalently bound to PEG, and (4) adamantane covalently bound to a targeting ligand, *i.e.* Tf, taking advantage of the stable inclusion complex between adamantane and β-cyclodextrin. The siRNA employed in CALAA-01 inhibits tumor growth via RNA interference to reduce expression of the M2 subunit of ribonucleotide reductase (R2). CALAA-01 has completed a Phase 1 trial to determine its safety, toxicity, and MTD when administered intravenously to patients with relapsed or refractory cancer, characterize its PK, provide preliminary evidence of efficacy by evaluating tumor response, recommend the dose for future clinical studies, and evaluate immune response by measuring antibody and cytokine levels, and complement activation.[64,65] To date, CALAA-01 is the only non-lipid siRNA formulation that has translated into the clinic, while siRNA nanomedicines investigated by Alnylam, Tekmira, and Silence are lipid-based, and therefore preferably liver bound. In addition, CALAA-01 is one of two polymeric nanoparticle formulations (PNP) in the clinic that employ ligand (Tf)-receptor targeting instead of EPR-driven targeting.

The second ligand-targeted PNP in the clinic is BIND-014, developed by Bind Therapeutics, Inc. The polymeric matrix consists of PEG–poly(D,L-lactic acid) (PEG5k–PLA16k) mixed with PEG5k–PLA16k-ACUPA in a 97.5% to 2.5% ratio. The ACUPA moiety (S,S-2-[3-[5-amino-1-carboxypentyl]-ureido]-pentanedioic acid) is a prostate-specific membrane antigen (PSMA) substrate analog inhibitor. The active ingredient in this formulation is physically entrapped docetaxel. BIND-014 is completing a Phase 1 trial to assess its dose-limiting toxicities and determine the maximum tolerated dose (MTD).[66] Two Phase 2 studies are listed but not yet recruiting, one study to evaluate the efficacy and safety of BIND-014 in patients with metastatic castration-resistant prostate cancer (mCRPC), the other study to compare the efficacy and safety of BIND-014 in patients with advanced NSCLC.

A docetaxel–polymeric nanoparticle formulation (docetaxel-PNP) is under development by Samyang Pharmaceuticals. The nanoparticles are comprised of a mixture of monovalent metal salts of poly(lactic acid), amphiphilic diblock copolymers, and the drug. In water this mixture forms micelles, which are stabilized by addition of divalent metal ions to form nanoparticles. This formulation has entered a Phase 1 trial in Korea in 2010 with expected trial end around December 2011. A Phase 1 study in the U.S. is of unknown status as of April 2010. Samyang has presented progress with docetaxel–PNP at the Bio Korea 2012 Conference.[67]

Medusa® (pGluVE, or PGA A1) is a nanoparticle hydrogel platform developed by Flamel Technologies SA. The nanoparticles form by self-assembly of hydrophilic, biodegradable polyglutamate polymer chains grafted with hydrophobic Vitamin E in aqueous medium, based on hydrophobic interactions instead of chemical crosslinking. After *s.c.* injection, a depot is formed at the injection site and the drug is slowly released over 1 to 14 days in humans. Medusa hydrogel has been proven to be safe and biodegradable, and Flamel has submitted a drug master file (DMF) for Medusa to the FDA in 2011 (assigned number 024634). These hydrogel nanoparticles can be used to solubilize and stabilize small drug molecules as well as biologics such as peptide and protein drugs. Three Medusa-based products have translated to the clinic, IFN-α XL, a long acting interferon alpha 2b for the treatment of hepatitis C is in Phase 2. FT-105, a once-daily formulation of non-modified recombinant human insulin for the treatment of type 1 and 2 diabetes, and IL-2 XL, a once-weekly human interleukin-2 (IL-2) formulation for the treatment of renal cell carcinoma, are both in Phase 1 trials.[68,69]

The PRINT® (Particle Replication In Non-wetting Templates) technology, developed in the DeSimone lab and licensed to Liquidia Technologies, Inc., is a controlled engineering approach to nanoparticles of well-defined shape, size, composition, and surface functionality. Master templates, fabricated using lithographic techniques, are covered with liquid fluoropolymers which are then cross-linked and peeled off to generate a precise mold of the template having micro- or nanoscale cavities. The cavities are filled with solutions containing polymers and drugs or biological actives. Once the liquid in the cavities is converted to solid particles, the array of particles is transferred into a solution and isolated. The process allows evaluation of the effects of particle size, size ratio, flexibility, and PEG surface density on PK and biodistribution.[70–72] Liquidia is utilizing the PRINT technology for the delivery of vaccines. A completed Phase 1/2 trial was designed to evaluate the safety, tolerability, and immune response of LIQ001 mixed with a commercially available seasonal influenza vaccine (Fluzone) in two populations of subjects; healthy adult subjects 18 to 49 years of age and healthy elderly subjects 65 years of age or older.

The status of active PNP is summarized in Table 3.

Several polymeric prodrug formulations have advanced to the market. Adagen® (pegademase bovine) is a modified enzyme used for enzyme replacement therapy for the treatment of severe combined immunodeficiency disease (SCID) associated with a deficiency of adenosine deaminase (ADA). Developed by Enzon Pharmaceuticals, Inc. (now licensed to Sigma-Tau Pharmaceuticals), Adagen received FDA approval in 1990. Adagen is a polymeric prodrug made by conjugation between mPEG and ADA from bovine intestine.[73] In 2006, the FDA granted approval to Oncaspar® (pegaspargase) from Enzon (licensed to Sigma-Tau) for the first-line treatment of patients with ALL as a component of a multi-agent chemotherapy regimen. Oncaspar was previously approved in 1994 for the treatment of patients with ALL who were hypersensitive to native forms of L-asparaginase. Pegaspargase is the conjugate between mPEG and native L-asparaginase. PEGylation increases the drug hydrodynamic radius, prolongs plasma retention time, decreases proteolysis, decreases renal excretion, and shields antigenic determinants from immune detection without obstructing the substrate-interaction site.[74] Currently more than 80 clinical trials are listed on the NIH clinical trials

Table 3. Nanomedicines based on polymeric nanoparticles on the market and under clinical evaluation.[a]

Product	Formulation	Application	Company	Status
Abraxane®	Albumin-bound paclitaxel	Metastatic breast cancer	Celgene Corp.	Marketed
Livatag®	Doxorubicin nanoparticle	Hepatocellular carcinoma	BioAlliance Pharma S.A.	Phase 3 NCT01655693*
CRLX101	β-CD–PEG–CPT nanopharmaceutical	Advanced NSCLC	Cerulean Pharma Inc.	Phase 2 NCT01380769*
IFN-α XL	Polyglu-Vit E interferon-α	Hepatitis C	Flamel Technologies	Phase 2 NCT01010646*
CALAA-01	Transferrin-β-CD siRNA nanoparticle	Solid tumors	Calando Pharma., Inc.	Phase 1 NCT00689065*
BIND-014	PEG–PLA/ PEG–PLA–ACUPA DTX nanoparticle	Advanced or metastatic cancer	Bind Therapeutics, Inc.	Phase 1 NCT01300533*
LIQ001	PRINT nanoparticles	Influenza vaccine	Liquidia Technologies	Phase 1/2 NCT01224262*

Notes: *ClinicalTrials.gov identifier.

[a]PNP whose trials have been completed or are listed as inactive for more than one year with no apparent follow-up have been excluded from Table 3. NSCLC — non-small cell lung cancer; CPT — camptothecin; DTX — docetaxel.

website involving Oncaspar to extend its indication and study combinations with other treatments. Another example of this PEGylation approach is Neulasta® (pegfilgrastim) from Amgen Inc. Pegfilgrastim is the PEGylated form of filgrastim, a recombinant human granulocyte colony-stimulating factor (GCSF) analog utilized to stimulate the level of white blood cells to fight infections in patients undergoing chemotherapy. The drug is comprised of PEG (20 kDa) conjugated to the N-terminus of filgrastim.[75] The fourth marketed product in this group of PEGylated biomolecules is Somavert® (pegvisomant) from Pfizer Inc. Pegvisomant is a man-made protein similar to human growth hormone. It binds to the same receptor in the body as growth hormone, therefore blocking its effects. Somavert is used to treat acromegaly (a growth disorder caused by too much growth hormone) and is usually given after other medications, surgery, and radiation have been tried without success.[76]

Following this PEGylation approach but extending it to multi-arm PEG, PEGylated SN-38 (EZN-2208), the active metabolite of irinotecan (CPT-11), was developed by Enzon Pharmaceuticals. EZN-2208 has shown increased solubility, higher exposure, and longer apparent half-life of SN-38, as well as more profound DNA damage and inhibition of angiogenesis than irinotecan. EZN-2208 treatment in a Phase 2 metastatic breast cancer clinical trial resulted in prolonged exposure of tumors to SN-38 via preferential accumulation of EZN-2208 in the tumor and prolonged release of SN-38 in the blood.[77,78] A second Phase 2 trial with EZN-2208 alone and in combination with cetuximab for metastatic colorectal carcinoma is ongoing but not recruiting. Several Phase 1 studies are listed as completed or active, not recruiting, with the latest update dating to July 2012. In December 2012, Enzon announced that the company had retained a financial advisor to review the possible sale or disposition of one or more corporate assets, or a sale of the company. In addition, Enzon announced plans to suspend clinical development with the goal to conserve capital and maximize value returned to shareholders.

Etirinotecan pegol (NKTR-102) from Nektar Therapeutics follows a similar approach as EZN-2208 by conjugating four irinotecan molecules to a branched PEG polymer. Etirinotecan pegol is in Phase 3 clinical development for patients with metastatic breast cancer and Phase 2 clinical development for patients with solid tumor malignancies, including ovarian and colorectal cancers. Unlike first generation topoisomerase 1 (TOPO1) inhibitors that exhibit a high initial peak concentration in the blood and short half-life, etirinotecan pegol provides a more continuous concentration of active drug with reduced initial peak. The large etirinotecan pegol molecule (approximately 23 kDa) accumulates in tumor tissue due to the EPR effect. Initially inactive, the drug is slowly released over time through enzymatic cleavage of the linkers within the molecule, continuously freeing active drug.[79] Nektar has four other preclinical or clinical candidates based on the company's polymer conjugation technology listed in the pipeline; however, the docetaxel-carrying prodrug NKTR-105 is no longer listed although it still appears as a clinical candidate in some review articles.

A nanomedicine based on a drug conjugated to the hydrophilic but non-PEG polymer poly(glutamic acid) is OPAXIO™, developed by Cell Therapeutics, Inc. (CTI). OPAXIO (paclitaxel poliglumex, CT-2103; formerly known as XYOTAX) was designed to deliver paclitaxel preferentially to tumor tissue via the EPR effect. Linking paclitaxel to a biodegradable amino acid carrier reduces uptake into healthy tissues, therefore sparing these tissues the toxic side effects of chemotherapy. However, once in the tumor tissue OPAXIO is taken up by tumor cells through endocytosis. The biodegradable polyglutamate backbone is slowly metabolized by lysosomal enzymes, principally cathepsin B, releasing the active drug, paclitaxel.[80] The water-soluble OPAXIO can be administered without solvents or other premedications, and remains stable in the bloodstream for several days after administration. OPAXIO is being studied in a Phase 3 trial as maintenance therapy in ovarian cancer. In January 2013, an independent Data Safety Monitoring Board (DSMB) recommended continuation of the Phase 3 clinical trial of OPAXIO. This trial is the largest maintenance study in this setting, having enrolled approximately 1,000 of the planned 1,100 patients. Enrollment is expected to be completed in 2013. Several Phase 1 and 2 clinical trials for indications including advanced NSCLC, advanced head and neck cancer, metastatic breast cancer, and glioblastoma multiforme (GBM), a malignant brain cancer are listed for paclitaxel poliglumex. For GBM, OPAXIO has been granted orphan drug designation by the U.S. FDA. A related poliglumex prodrug with camptothecin (CT-2106) had advanced to Phase 2 clinical trial for patients with advanced ovarian cancer. The study has been completed and was last verified in February 2010. In addition, CT-2106 is no longer listed in CTIs pipeline.

Another conjugate between a drug and a hydrophilic, non-PEG polymer is XMT-1001. This nanomedicine from Mersana Therapeutics is produced by conjugating CPT molecules to a biodegradable polyacetal, poly(1-hydroxymethylethylene hydroxymethylformal) (Fleximer®). Drug and Fleximer backbone are connected via succinimidoglycinate or other linker molecules, selected based on the desired drug release profile.[81,82] XMT-1001 is being evaluated in a Phase 1b clinical trial in lung cancers, following successful completion of a Phase 1 trial of the safety, tolerability, and PKs of *i.v.* XMT-1001 in patients with advanced solid tumors (SCLC, NSCLC).

N-(2-hydroxypropyl)methacrylamide (HPMA) copolymer–drug conjugates were considered a promising approach to new nanomedicines. For example, HPMA-doxorubicin (PK1, FCE28068), HPMA-doxorubicin-galactosamine (PK2, FCE28069), HPMA-paclitaxel (PNU166945), and HPMA-CPT (MAG-CPT, PNU166148) from Pharmacia and HPMA-DACH platinate (the active moiety of oxaliplatin (ProLindac™, AP5346) from Access Pharmaceuticals, Inc. have all been evaluated in Phase 1 and/or Phase 2 studies. However, none of these polymeric prodrugs have entered the market.[83,84]

The last polymeric prodrug listed here, Renagel® (sevelamer hydrochloride) from Genzyme Corporation, a Sanofi company, is a borderline case. Sevelamer hydrochloride is a copolymer composed of epichlorohydrin and aminopropen. Renagel is being used to treat hyperphosphatemia in patients with chronic kidney disease (CKD). The polymer becomes a prodrug by binding dietary phosphates through hydrogen bonds and ionic interactions, thus preventing phosphate absorption. Contrary to the prodrugs discussed above, the clinical action of Renagel is removal and not delivery of an active substance.[85]

The status of active polymeric prodrugs is summarized in Table 4.

2.7. *Emulsions*

Emulsions have a long tradition in medical, biological, and food applications, mainly to mix essentially immiscible compounds. Water-in-oil (w/o) emulsions are being used to disperse water droplets carrying a hydrophilic compound in an oil phase, while oil-in-water (o/w) emulsions serve the purpose of dissolving lipophilic compounds in water. The development of these emulsions is driven by the desire to solve mixing challenges by producing stable mixtures. These emulsions are not focused on the development of nanostructures with specific delivery properties, and therefore, they are not nanomedicines in the sense of this review. However, some emulsions including water-in-oil-in-water (w/o/w) double emulsions and self-emulsifying drug delivery systems (SEDDS), are common approaches for the preparation of nanoparticles.[86–89]

2.8. *Emulsions-based nanomedicines*

Two amphotericin B formulations are on the market that fall into the emulsion category. Amphotec®, distributed by Three Rivers Pharmaceuticals, LLC, consists of a 1:1 molar ratio complex of amphotericin B and cholesteryl sulfate. It is provided as a sterile, lyophilized powder that forms a colloidal dispersion of

Table 4. Nanomedicines based on polymeric prodrugs on the market and under clinical evaluation.[a]

Product	Formulation	Application	Company	Status
Adagen®	PEG-adenosine deaminase	Enzyme replacement therapy for SCID	Sigma-Tau Pharmaceuticals	Marketed
Oncaspar®	PEG-asparaginase	Acute lymphoblastic leukemia (ALL)	Sigma-Tau Pharmaceuticals	Marketed
Neulasta®	Pegfilgrastim	White blood cell stimulation	Amgen Inc.	Marketed
Somavert®	Pegvisomant	Acromegaly	Pfizer Inc.	Marketed
Renagel®	sevelamer hydrochloride	Hyperphosphatemia	Genzyme Corp.	Marketed
Etirinotecan pegol (NKTR-102)	PEGylated irinotecan	Metastatic breast cancer	Nektar Therapeutics	Phase 3 NCT01492101*
OPAXIO™	PTX poliglumex	Multiple cancers	Cell Therapeutics	Phase 3 NCT00108745*
XMT-1001	Polyacetal-camptothecin	SCLC and NSCLC	Mersana Therapeutics	Phase 1 NCT00455052*

Notes: *ClinicalTrials.gov identifier.
[a]Polymeric prodrug formulations whose trials have been completed or are listed as inactive for more than one year with no apparent follow-up have been excluded from Table 4. NSCLC — non-small cell lung cancer; SCLC — small cell lung cancer.

microscopic, disc-shaped particles upon reconstitution in water. Amphotec received FDA approval in 1996 for the treatment of invasive aspergillosis in patients where renal impairment or unacceptable toxicity precludes the use of amphotericin B deoxycholate in effective doses, and in patients with invasive aspergillosis where prior amphotericin B deoxycholate therapy has failed.[90] The other marketed formulation is Abelcet®, marketed by Sigma-Tau Pharmaceuticals, Inc. Abelcet consists of an amphotericin B complex with two phospholipids in a 1:1 drug-to-lipid molar ratio, *i.e.* l-α-dimyristoylphosphatidylcholine (DMPC) and l-α-dimyristoylphosphatidylglycerol (DMPG) in 7:3 molar ratio. The formulation, provided as a sterile suspension for *i.v.* infusion, received FDA approval in 1995 for the treatment of invasive fungal infections in patients who are refractory to or intolerant of conventional amphotericin B therapy.[91]

Diprivan®, developed by AstraZeneca LP, received FDA approval in 2001 as an *i.v.*, sedative-hypnotic agent for use in induction and maintenance of anesthesia or sedation. Diprivan is an o/w emulsion containing propofol as the active agent and soybean oil, glycerol, egg lecithin, and disodium edetate as the oil phase, as well as sodium hydroxide for pH adjustment to 7.0–8.5.[92]

Neoral®, a formulation of cyclosporin as the active agent and corn oil mono-di-triglycerides, polyoxyl 40 hydrogenated castor oil, DL-α-tocopherol, gelatin, and glycerol as the oil phase forms a microemulsion in aqueous environment. Marketed by Novartis Pharmaceuticals, Neoral is an immunosuppressant which received FDA approval in 1995 for the prevention of organ rejection in kidney, liver, and heart transplants. Neoral is also used to treat psoriasis and RA in certain patients.[93] Restasis® is an ophthalmic emulsion of cyclosporin, indicated to increase tear production in patients whose tear production is suppressed due to ocular inflammation associated with keratoconjunctivitis sicca. Besides cyclosporine, the emulsion is composed of glycerin, castor oil, polysorbate 80, carbomer copolymer type A, purified water, and sodium hydroxide to adjust the pH to 6.5–8.0. Marketed by Allergan, Inc., Restasis received FDA approval in 2002.[94] Vekacia® (Cyclokat; Nova22007) is a cationic emulsion of cyclosporin A, based on Novagali Pharma's Novasorb® technology. Vekacia has completed a Phase 3 trial in patients with moderate to severe dry eye syndrome. The positively charged emulsion electrostatically adheres to the negatively charged epithelial layer of the eye, thereby improving ocular absorption. In addition to dry eye syndrome, Vekacia has received orphan drug designation by the EMA (2006) and FDA (2007) for the treatment of vernal keratoconjunctivitis (VKC).[95] Another marketed ophthalmic emulsion is Durezol® (difluprednate), a sterile, topical anti-inflammatory corticosteroid for ophthalmic use. The oil phase is composed of a mixture of castor oil, glycerin, and polysorbate 80, with boric acid, sodium acetate, and sodium EDTA as additional components, and sodium hydroxide for pH adjustment to 5.2–5.8. Durezol, developed by Sirion Therapeutics, Inc. and FDA approved in 2008 for the treatment of eye redness, swelling, and pain caused by eye surgery and inflammation, is currently marketed by Alcon, a Novartis company.[96]

Cleviprex® (clevidipine) is a dihydropyridine calcium channel blocker which was approved in 2008 by the FDA for the reduction of blood pressure when oral therapy is not feasible or not desirable. Cleviprex is formulated as a lipid o/w emulsion based on soybean oil and glycerin, purified egg yolk phospholipids, and sodium hydroxide to adjust the pH to 6.0–8.0. The *i.v.* emulsion is marketed by The Medicines Company as a ready-to-use formulation.[97]

The EMA recommended in 2009 that Novartis' Focetria® and GlaxoSmithKline's Pandemrix® be granted marketing authorization as vaccines against influenza A (H1N1) ("swine flu") for treatment of adults and children from 6 months up as well as pregnant women. Focetria (Fluvirin® in the U.S.) comes as a suspension for injection in prefilled syringes, while Pandemrix is an emulsion for injection consisting of two vials that must be mixed to form a multidose container usable for 24 h.[98] The NanoStat™ platform developed by NanoBio Corporation is being applied to develop mucosal vaccines and dermatology products with three indications, herpes labialis (cold sores), acne, and onychomycosis. The NanoStat technology employs high-energy, o/w emulsions that are manufactured at droplet sizes of 150–400 nm, stabilized by surfactants. When applied to the skin, the nanoemulsion droplets rapidly penetrate through pores and hair shafts to the site of an infection, and physically disrupt the outer membrane of pathogenic organisms by fusing with and killing lipid-containing organisms. These treatments are selectively toxic to microbes while non-irritating to skin and mucous membranes. The lead product, NB-001, has entered clinical trials and is currently being investigated in a recruiting Phase 3 study for the treatment of recurrent herpes labialis. NB-002, NB-003, and NB-004 are in Phase 1 and 2 trials for the treatment of nail fungus, acne, and molluscum contagiosum. On the vaccine side, NB-1008 has completed a Phase 1 study testing the safety and immune response of this intranasal vaccine against influenza. The vaccine is composed of a licensed vaccine that is normally given as an injection, Fluzone, and the adjuvant W805EC. Other vaccine indications include respiratory syncytial virus (RSV), hepatitis B, smallpox, and anthrax.[99]

ANX-514 is a docetaxel emulsion formulation prepared without the use of polysorbate 80 and ethanol, both present in the marketed docetaxel product Taxotere® and both linked to the occurrence of hypersensitivity. This surfactant-free injectable emulsion, developed by Mast Therapeutics, Inc. (formerly Adventrx Pharmaceuticals, Inc.), has been compared to Taxotere in a Phase 1 study in patients with advanced cancer. The study has been completed in 2009, and Mast Therapeutics is currently looking for partners or investors to continue program development.[100] With a similar goal in mind, Sun Pharma Advanced Research Company Ltd, (SPARC) has developed a self-dispersing nanoparticle platform for paclitaxel (Paclitaxel Injection Concentrate for Nanodispersion (PICN) and Docetaxel Injection Concentrate for Nanodispersion (DICN)). PICN is trying to address the challenges of Taxol (hypersensitivity reactions, neutropenia, and peripheral neuropathies) as well as Abraxane (potential risk of immunogenicity and viral infection from albumin). According to SPARC, the propriety nanodispersion PICN is achieving a higher MTD than Abraxane as well as higher drug concentration in tumor tissue. PICN is being evaluated in a Phase 2/3 trial in India and two recruiting Phase 1 studies in the U.S., one study evaluating the MTD in a 30 min infusion once a week for three weeks, followed by one week of rest cycle, the other study evaluating PICN in combination with carboplatin. The docetaxel formulation DICN has completed a Phase 1 study in India in patients with solid tumors.[101] Another Phase 1 candidate is the emulsion of DM-CHOC-PEN (4-demethyl-cholesteryloxy-carbonyl-penclomedine, a carbonate derivative of 4-demethyl-penclomedine (DM-PEN), studied in patients with advanced cancer. DM-PEN is the antitumor active, non-neurotoxic metabolite of penclomedine (PEN). PEN was active in clinical trials but possessed significant neurotoxicity, and the trial was cancelled. DM-CHOC-PEN was formulated as a soybean oil, lecithin, glycerin, water emulsion that could cross the blood–brain barrier without displaying toxicity. The Phase 1 study, sponsored by DEKK-TEC, Inc. and the NCI through SBIR-STTR funding, is currently recruiting participants.[102]

One interesting example of an o/w emulsion with diagnostic as well as delivery potential is based on the nanodroplet technology developed by Kereos, Inc. Individual nanodroplets consist of a perfluorocarbon core and a lipid monolayer shell, which both stabilizes the droplet and provides anchoring sites for targeting ligands and therapeutic payloads. The droplets have an average size of 250–350 nm and are designed to be both safe and effective using excipients that have been previously studied in humans. The current ligand being incorporated in the emulsion targets the $\alpha\nu\beta3$ integrin, a well-known marker for angiogenesis associated with cancer, cardiovascular, and autoimmune disease. The preclinical diagnostic KER-0001 product uses both the $\alpha\nu\beta3$ integrin biomarker and fluorine-19 imaging to detect the presence of angiogenesis. In addition, therapeutic products can also be imaged by magnetic resonance imaging (MRI) to confirm delivery of the therapeutic payload to the target, and assess clinical response. Drug payloads include small drugs such as paclitaxel (KER-1001) as well as fragile peptide therapeutics.[103]

As with the other delivery platforms, some emulsions have not fulfilled original expectations. Tenifatecan (SN2310; tocopherol succinate-SN38), for example, was an o/w emulsion of tocopherol covalently linked via a succinate linker to SN-38, the active metabolite of irinotecan. Hydrolysis of the succinate linker *in vivo* released SN-38, which then selectively stabilized TOPO1–DNA covalent complexes. This stabilization inhibited religation of TOPO1-mediated single-stranded DNA breaks and induced lethal double-stranded DNA breaks, thus triggering apoptosis.[104] A Phase 1 open-label study of SN2310 injectable emulsion in patients with advanced solid malignancies was completed with last update in 2009 by OncoGenex Pharmaceuticals, Inc. However, no further progress has been reported and tenifatecan is no longer listed in OncoGenex pipeline, which is now focusing on heat shock protein 27 (HSP27) inhibition. Rolofylline (KW-3902) injectable emulsion, another example of an unsuccessful formulation, was comprised of the experimental diuretic rolofylline acting as a selective adenosine A1 receptor antagonist and soybean oil, egg yolk lecithin, oleic acid, and glycerin as the oil phase, mixed with water and sodium hydroxide for pH adjustment. The emulsion was developed by NovaCardia, Inc., which was purchased by Merck & Co. in 2007. Rolofylline was studied in several clinical trials, culminating in three Phase 3 trials including PROTECT-1 and PROTECT-2 trials. Development of rolofylline was terminated in 2009, after results of the large PROTECT trials showed the drug to be no better than placebo control for patients with acute heart failure. The drug did neither prevent kidney damage nor had significant effect on overall treatment success. Rolofylline was also associated with a higher incidence of seizures and stroke.[105,106] CNF1010 from Biogen Idec, a nanoemulsion formulations of 17-(allylamino)-17-demethoxygeldanamycin (17-AAG) was evaluated in a Phase 1, dose-escalating study in patients with ZAP-70 positive B-cell chronic lymphocytic leukemia (CLL). However, the study was terminated due to the toxicity profile of the drug.[107]

The status of active emulsion formulations is summarized in Table 5.

2.9. *Metal ion-containing nanoparticles*

In this section, metal ion-containing nanoparticles that have entered the market or are in clinical translation will be reviewed. The application of these nanoparticles is in imaging and diagnostic rather than drug delivery. This group includes chelates of paramagnetic metals with a large number of unpaired electrons such as transition element manganese(II) (Mn^{2+}) and lanthanide gadolinium(III) (Gd^{3+}). Gd(III)-chelates are positive agents which provide bright images in MRI; however, they suffer from non-specific biodistribution and fast elimination.[108] Moreover, Gd(III)-based agents may be associated with a toxic reaction known as nephrogenic systemic fibrosis (NSF) in patients with severe kidney problems. Release of Gd(III) ions from their chelate complexes is suspected to be responsible for NSF, a hypothesis supported by experiments showing that four different Gd(III) complexes all became lethally toxic when precisely the same quantity of Gd(III) ions was released.[109] The second class of contrast agents is based on superparamagnetic iron oxide (SPIO) nanoparticles. SPIO nanoparticles are negative contrast agents which provide dark images under the magnetic field in MRI. After administration, they are selectively taken up by the MPS, and therefore, are tissue or organ-specific contrast agents. SPIOs are composed of either maghemite (γ-Fe_2O_3) or magnetite (Fe_3O_4) phases that can be prepared by a variety of methods, most commonly by a co-precipitation reaction of a mixture of ferrous and ferric salts in aqueous media in the presence of stabilizers such as hydrophilic polymers. SPIOs ability to shorten T2 relaxation times results

Table 5. Nanomedicines based on emulsions on the market and under clinical evaluation.[a]

Product	Formulation	Application	Company	Status
Amphotec®	Amphotericin B: cholesteryl sulfate 1:1 dispersion	Invasive aspergillosis	Three Rivers Pharmaceuticals	Marketed
Abelcet®	Amphotericin B: (DMPC+DMPG) 1:1 suspension	Invasive fungal infections	Sigma-Tau Pharma, Inc.	Marketed
Diprivan®	Propofol o/w emulsion	Anesthesia	AstraZeneca LP	Marketed
Neoral®	Cyclosporine microemulsion	Immunosuppressant	Novartis	Marketed
Restasis®	Cyclosporine emulsion	Eye treatment	Allergan, Inc.	Marketed
Durezol®	Difluprednate emulsion	Eye treatment	Alcon, a Novartis company	Marketed
Cleviprex®	Clevidipine o/w emulsion	High blood pressure	The Medicines Company	Marketed
Focetria® Fluvirin®	Influenza vaccines (H1N1)	Influenza virus	Novartis	Marketed
Pandemrix®	Influenza vaccines (H1N1)	Influenza virus	GlaxoSmithKline	Marketed
Vekacia®	Cyclosporine cationic emulsion	Dry eye syndrome	Novagali Pharma	Phase 3 NCT00814515*
NB-001	Nanoemulsion	Herpes labialis	NanoBio Corp.	Phase 3 NCT01695187*
NB-1008	Vaccine nanoemulsion	Influenza	NanoBio Corp.	Phase 1 NCT01354379*
ANX-514	DTX nanoemulsion	Solid tumors	Mast Therapeutics	Phase 1 NCT00664170*
PICN	PTX nanodispersion	Solid tumors	SPARC	Phase 1 NCT01305512*
DM-CHOC-PEN	Penclomedine derivative emulsion	Solid tumors	DEKK-TEC, Inc.	Phase 1 NCT01048008*

Notes: *ClinicalTrials.gov identifier.

[a]Emulsion formulations whose trials have been completed or are listed as inactive for more than one year with no apparent follow-up have been excluded from Table 5. DTX — docetaxel; PTX — paclitaxel.

from their large magnetization, causing large enhancements of signal voids (darkening of images). Early clinical targets of SPIOs were liver diseases because of their selective uptake by Kupffer cells, which are specialized liver cells that destroy bacteria, foreign proteins, and worn-out blood cells.[110,111] The third class of imaging agents are semiconductor QDs. QDs are light-emitting nanoparticles that have emerged as a new class of fluorescent labels for biology and medicine. Compared with traditional fluorescent probes, QDs have advantages such as size-dependent light emission, narrow and symmetric emission spectra, and broad absorption spectra that enable the simultaneous excitation of multiple fluorescence colors. These semiconductor nanocrystals can be tuned by the particle size across a broad spectrum of UV, visible, near infrared (IR), and even mid-IR wavelengths. The crystalline nature of QDs also gives rise to large absorption cross-sections, which causes QDs to be 10–100 times brighter than organic dyes and fluorescent proteins. In addition, these particles are dramatically more resistant to photodegradation compared to traditional probes. However, QDs are macromolecules that are an order of magnitude larger than organic dyes, which can limit their use in cellular imaging applications requiring small labels.[112]

2.10. *Metal ion-containing nanoparticles for imaging and diagnostic*

There are at least nine Gd(III) complexes for MRI applications on the market. The first *i.v.* contrast agent available for clinical use was Magnevist® (gadopentetate dimeglumine), the N-methylglucamine salt of the complex between Gd^{3+} and diethylenetriamine pentaacetic acid ($DTPA^{5-}$). In this complex, Gd(III) has nine coordination sites, three occupied by nitrogen atoms and five by oxygen atoms from the carboxylate groups, while the ninth coordination site is occupied by a labile water molecule that rapidly exchanges with water molecules in the immediate vicinity of the GD–DTPA complex. The exchange rate of these water molecules alters the MRI properties of the complex and enables contrast enhancement to be achieved.[113] Magnevist is marketed by Bayer Schering Pharma AG and received its first FDA approval in 1988 for use with MRI in adults and pediatric patients (2 years of age and older) to visualize lesions with abnormal vascularity in the brain (intracranial lesions), spine and associated tissues. One closely related contrast agent is Omniscan® (gadodiamide; Gd-DTPA–BMA) from GE Healthcare, the Gd(III) complex of diethylenetriamine pentaacetic acid–bismethylamide (DTPA–BMA), in which two carboxylic acid groups of DTPA have been transformed to methylamides. Omniscan received FDA approval in 1993 for *i.v.* use in MRI to visualize lesions with abnormal vascularity (or those thought to cause abnormalities in the blood–brain barrier) in the brain (intracranial lesions), spine, and associated tissues. A second closely related imaging agent is Optimark™ (gadoversetamide; Gd-DTPA–BMEA) from Mallinckrodt LLC, FDA approved in 1999 for use with MRI in patients with abnormal blood–brain barrier or abnormal vascularity of the brain, spine and associated tissues as well as patients with abnormal vascularity in the liver. Optimark is the Gd(III) complex with diethylenetriamine pentaacetic acid–bismethoxyethylamide (DTPA–BMEA), in which two carboxylic acid groups of DTPA have been transformed to methoxyethylamides. A third DTPA-related imaging agent, Eovist® (gadoxetate disodium; Gd-EOB–DTPA; in Europe marketed since 2004 under Primovist®) from Bayer HealthCare Pharmaceuticals Inc., is a Gd(III) complex with ethoxybenzyl–DTPA (EOB–DTPA). Eovist received FDA approval in 2008 for *i.v.* use in T_1-weighted MRI of the liver to detect and characterize lesions in adults with known or suspected focal liver disease. A fourth derivative of DTPA, in which one carboxylic acid function (COOH) has been replaced with a phenylmethylether moiety, is MultiHance® (gadobenate dimeglumine; Gd-BOPTA) from Bracco Diagnostics Inc. MultiHance was first approved in the U.K. in 1997 and received FDA approval in 2004 for *i.v.* use in MRI of the central nervous system (CNS) in adults and children over 2 years of age to visualize lesions with abnormal blood–brain barrier or abnormal vascularity of the brain, spine, and associated tissues.[114] The fifth and latest marketed DTPA derivative is Ablavar® (gadofosveset trisodium; Vasovist®; MS-325) from Lantheus Medical. Ablavar is a diphenylcyclohexyl-phosphonooxymethyl-derivative of DTPA. Originally developed as MS-325 and marketed as Vasovist by EPIX Pharmaceuticals, Inc., the contrast agent was approved by the European Commission in 2005 for magnetic resonance angiography (MRA) in adult patients with suspected or known vascular disease of the abdomen and/or extremities, followed by FDA approval in 2008 for marketing as novel blood pool MRA agent. In 2009, Vasovist was purchased by Lantheus and rebranded as Ablavar for the U.S., Canadian, and Australian markets. Ablavar binds reversibly to HSA, which provides higher relaxivity and extended intravascular enhancement, the latter enabling high-resolution steady-state imaging.[115,116] Binding to albumin is one approach to achieve the blood circulation time necessary for blood pool agent properties, another approach utilizes large molecular weights of the imaging agent to reduce rapid clearance. Two examples are Vistarem® (gadomelitol; P792; molecular weight 6.5 kDa) from Guerbet Group (France), and dendrimer-based Gadomer-17 (Gd-DTPA-17; MW ~35 kDa) from Bayer Schering Pharma AG.[117–119] So far no high molecular weight contrast agent has entered the market; the synthetic complexity and resulting high cost of goods (COG) being one reason for this situation.

Gadovist® (gadobutrol; Gd-DO3A-butrol), marketed by Bayer Schering Pharma AG (and as Gadavist® by Bayer HealthCare Pharmaceuticals), has been approved in the European Union (EU) since 1998 and in the U.S. since 2011 for MRI of the brain, spine, vessels, and abnormalities in liver and kidneys. Gadovist is a nonionic Gd(III) complex consisting of a macrocyclic tetraazacyclododecane compound instead of the DTPA linear molecule that forms a cycle only after complexation with Gd(III). Four ring nitrogen atoms and four oxygen atoms from three acetyl and one trihydroxybutyl side groups occupy eight coordination sites of Gd(III), with water occupying the ninth position, giving the DO3A-butrol ring higher complex stability in order to reduce Gd(III) leakage from the contrast agent and minimize the risk of NSF.[120] A closely related nonionic, macrocyclic structure is employed in ProHance® (gadoteridol; Gd-DO3A-HP), consisting of three acetyl and one hydroxypropyl side group. ProHance, marketed by Bracco Diagnostics Inc., received FDA approval in 1992 for use in MRI in adults and children over 2 years of age to visualize lesions with abnormal vascularity in the brain (intracranial lesions), spine and associated tissues. ProHance claims the highest thermodynamic stability constant and highest conditional stability of all contrast agents in its class.[121]

Most MRI contrast agents consist of nonionic complexes, resulting in low osmolality and viscosity at higher concentrations and therefore allowing faster bolus injections. However, some contrast agents are ionic in nature, *e.g.* Magnevist and MultiHance (dimeglumine salts) as well as Eovist (disodium salt). The only ionic and macrocyclic MRI agent on the market is Dotarem® (gadoterate meglumine; Gd-DTOA), manufactured by Guerbet. Dotarem received worldwide approval in 1989, and was FDA approved in 2013 for use in MRI of the brain, spine and associated tissues in patients ages 2 years and older.[122] Artirem®, marketed in some European countries by Guerbet, has the same qualitative formulation as Dotarem but is administered by intra-articular injection directly into the joint at a much lower concentration than Dotarem. A comparative study of six marketed MRI contrast agents as well as a post-marketing surveillance study showed that macrocyclic Gd(III) complexes provided a better safety profile with regard to Gd(III) release and NSF prevention.[123,124]

Motexafin gadolinium (MGd; gadolinium texaphyrin; Xcytrin®) from Pharmacyclics Inc. is a Gd(III)-containing macrocyclic contrast agent with anticancer activity. The aromatic macrocycle has a strong affinity for electrons, and therefore, is easily reduced. In the presence of oxygen, MGd accepts electrons from various cellular reducing metabolites and forms superoxide and other ROS by redox cycling. MGd increases the cytotoxicity of ionizing radiation and various chemotherapy agents and may be directly cytotoxic to tumor cells under certain conditions. MGd selectively localizes in tumors, perhaps due to their metabolic perturbations. In 2007, the FDA issued a non-approvable letter for the company's new drug application (NDA) for the treatment of NSCLC patients with brain metastasis. However, MGd is still being evaluated for clinical activity, for example in a NCI-sponsored Phase 2 trial giving motexafin gadolinium together with radiation therapy for the treatment of young patients with pontine glioma.[125,126]

EP-2104R, a clinical MRI candidate comprised of an 11 amino acid peptide derivatized with two Gd-DOTA-like moieties at both the C- and N-terminus of the peptide, was designed by EPIX Pharmaceuticals to detect blood clots by binding to the protein fibrin, present in all thrombi. Finding blood clots is of medical significance in the evaluation and diagnosis of patients with possible stroke, transient ischemic attack, chest pain, heart attack, irregular heartbeat, deep vein thrombosis and pulmonary embolism. EP-2104R completed a Phase 2a clinical trial in 2006, demonstrating that the contrast agent was well-tolerated and able to detect or enhance clots on MRI images in all six of the body systems studied (pulmonary, deep vein, carotid artery, cardiac atria, left ventricle, and thoracic aorta). Bayer Schering Pharma AG was granted an option to license and develop EP-2104R; however, the company decided in 2006 against exercising the option and the intellectual properties of EP-2104R were auctioned in 2010.[127,128]

The only marketed manganese(II)-based MRI contrast agent, Teslascan® (mangafodipir trisodium; Mn-DPDP) from GE Healthcare, was FDA-approved in 1997 for MRI use in the detection, localization, characterization, and evaluation of lesions of the liver. In this complex, Mn(II) has six coordination sites with the complex-forming agent fodipir (dipyridoxyl diphosphate, DPDP), two sites with nitrogen atoms and four sites with oxygen atoms. However, GE Healthcare withdrew Teslascan from the market due to low demand in the U.S. in 2004 and in the EU in 2012.[129,130]

Two SPIO contrast agents have received clinical approval for liver-specific imaging. Feridex® (ferumoxides; AMI-25; Endorem®) was composed of nanosized (120–180 nm), dextran-coated non-stoichiometric magnetite crystals with the average chemical formula $FeO_{1.44}$. Feridex had been approved in 1996 by the FDA as the first organ-specific MRI contrast agent. Developed by AMAG Pharma (formerly Advanced Magnetics), it had been marketed as Feridex in the U.S. by Bayer HealthCare Pharmaceuticals (formerly Berlex

Laboratories) and in Europe as Endorem by Guerbet S.A. To reduce the incidence of side effects such as hypotension, Feridex was administered as drip infusion over about 30 min. In healthy liver, these SPIO particles are sequestered by phagocytic Kupffer cells but are not retained in lesions lacking Kupffer cells. About 8 min after *i.v.* injection, iron oxide particle uptake by the MPS cells in liver and spleen was approximately 80% and 6–10%, respectively. However, in 2008 AMAG Pharma decided to discontinue Feridex' commercialization.[131,132] The second SPIO contrast agent, Resovist® (ferucarbotran; SHU 555A; Cliavist™), developed by Schering AG and marketed by Bayer Schering Pharma AG, was composed of carboxy-dextran-coated SPIO with a hydrodynamic diameter ranging between 45 and 60 nm. Unlike Feridex, Resovist could be safely injected in a rapid bolus fashion, and the incidence of cardiovascular adverse events was low. Approved in 2001 for the European market, the production of Resovist was abandoned in 2009 because of its competition with Primovist, the other liver imaging agent from Bayer Schering.[132,133]

Combidex® (ferumoxtran-10; AMI-227; Sinerem™ in Europe) had been designed by AMAG Pharmaceuticals for lymph node metastasis evaluation. Combidex consisted of dextran-coated SPIO particles (20–40 nm) that were phagocytosed by macrophages and accumulated over 24 to 36 h in the lymphatic system. Post-contrast imaging was usually obtained 24 h after administration of the contrast agent. However, results of a multicenter study involving 296 patients with prostate cancer with intermediate to high risk for nodal metastases revealed that 24.1% false positive responses were measured, leading to unnecessary surgical intervention. As a consequence, the clinical development for Combidex was stopped in 2007 by AMAG Pharmaceuticals and its European partner, Guerbet S.A. However, in April 2013 AMAG Pharmaceuticals sold the rights to Combidex to Dutch Radboud University Nijmegen Medical Center (RUNMC) for further development.[134–136] Clariscan™ (PEG-fero; Feruglose; NC100150), developed by Nycomed Imaging (now part of GE Healthcare), consisted of ultrasmall SPIO (USPIO) particles (4–7 nm) stabilized with a carbohydrate PEG coat. After suspension in isotonic glucose solution, the final diameter of Clariscan particles was about 20 nm. The imaging agent has been tested for MRA application and for characterization of tumor microvasculature. Clariscan particles are eventually taken up by the MPS and distributed mainly to liver and spleen. Ultimately, however, the development of Clariscan was discontinued in 2006 due to safety concerns.[132,137]

The USPIO Feraheme® (ferumoxytol) from AMAG Pharmaceuticals Inc. is an *i.v.* iron replacement product for the treatment of anemia in CKD. It was approved by the FDA in 2009. Feraheme consists of non-stoichiometric magnetite nanoparticles (17–31 nm), coated with poly (glucose sorbitol carboxymethylether).[138,139] In addition to the anemia treatment, the safety and efficiency of ferumoxytol as MRI and MRA agent in CNS inflammation compared to gadolinium is being investigated in a Phase 2 trial (ClinicalTrials.gov NCT00659776). The expectation is that USPIO nanoparticles might be a safer, alternative MR contrast agent for patients at risk of NSF.[140] NanoTherm® SPIO nanoparticles with average size around 15 nm and stabilized through an aminosilane coating are being developed by MagForce AG for the treatment of solid tumors. The NanoTherm ferrofluid is directly injected into the tumors, followed by treatment with a fluctuating magnetic field. The tumors are either destroyed through thermal ablation or sensitized for concomitant chemo- or radiotherapy (hyperthermia). The first patient with recurrent glioblastoma was treated in Germany in late 2011.[141] Another early stage contrast agent is Supravist™ (SHU 555C) developed by Bayer Schering Pharma for blood pool imaging. Supravist consists of dextran-coated iron oxide nanoparticles with sizes below 20 nm for *i.v.* bolus injection.[110] The contrast agent is currently being studied in a recruiting Phase 1 trial in 50 patients at five French centers for details of the inflammatory process and medical imaging in patients with an inflammatory disease of the CNS, sponsored by Rennes University Hospital and Bayer.

Two oral MRI contrast agents have been developed. GastroMARK™ (ferumoxsil; AMI-121; Lumirem® outside of the U.S.) from AMAG Pharmaceuticals is an oral contrast agent used for delineating the bowel in MRI. Ferumoxsil consists of non-stoichiometric magnetite nanoparticles coated with a silicone polymer. It was FDA approved in 1996; however, AMAG Pharmaceuticals decided to discontinue GastroMARK with any existing supplies in the marketplace expiring in February 2013.[142] The second oral contrast agent was Abdoscan® (ferristene; OMP), developed by Nycomed (now GE Healthcare). Abdoscan consisted of large (≥300 nm) particles, coated with sulfonated styrene–divinylbenzene copolymer. The contrast agent was approved in Europe in 1993 but taken off the market in 2000 with all sales stopped by the end of 2002.[143]

Semiconductor QDs are in the very early stage of clinical translation compared to Gd(III) and iron oxide imaging agents. Many companies are producing QDs on different scales and for many applications, including Life Technologies Corp., eBioscience Inc., Ocean Nanotech LLC., QD Vision Inc., Nanoco Technologies Ltd., Nanosys Inc., Selah Technologies Inc., Samsung Electronics Co. Ltd., Altair Nanotechnologies Inc., Evident Technologies Inc., InVisage Technologies Inc., LG Display Co. Ltd.,

Microvision Inc., Microoled, Nano Axis LLC, NN-labs Inc., Nexxus Lighting Inc., QD Laser Inc., Quantum Material Corp., Sigma-Aldrich Co. LLC, Solexant Corp., and Voxtel Inc. However, the vast majority of biological applications is restricted to *in vitro*. The delay in clinical translation is mainly caused by safety concerns of this new material.[144] A recent pilot study on non-human primates demonstrated that rhesus macaques injected with phospholipid micelle-encapsulated CdSe/CdS/ZnS QDs did not exhibit evidence of toxicity. Blood and biochemical markers remained within normal ranges following treatment, and histology of major organs after 90 days showed no abnormalities, indicating that acute toxicity of these QDs *in vivo* could be minimal. However, chemical analysis revealed that most of the initial dose of cadmium remained in the liver, spleen and kidneys after 90 days, pointing to slow breakdown and clearance of QDs. Longer-term studies will thus be required to determine the ultimate fate of these heavy metals and the impact of their persistence in primates.[145]

There is one early clinical trial involving Cornell Dots (C-dots; (C•spec® particles), named after the university where this material was developed. C-dots consist of silica nanoparticles coated with PEG to provide stealth properties that incorporate organic dyes. Hybrid Silica Technologies, Inc. conducted a Phase 0 safety trial in cooperation with Memorial Sloan-Kettering Cancer Center in New York City in 2011, including five melanoma patients as a springboard to testing C-dot nanoparticles in a larger trial designed to find sentinel lymph nodes in patients.[146,147]

The status of imaging agents is summarized in Table 6.

2.11. *Dendrimer nanoparticles in medical applications*

Dendritic polymers or dendrimers provide an alternative route to create well-defined nanostructures suitable for drug delivery and imaging applications. Dendrimers are core-shell nanostructures with precise architecture and low polydispersity, which are synthesized in a layer-by-layer fashion (expressed in "generations") around a core unit, resulting in high-level of control over size, branching points and surface functionality. The ability to tailor dendrimer properties to therapeutic needs makes them ideal carriers for small molecule drugs and biomolecules. The high number of surface groups provides a scaffold for the attachment of imaging moieties. The three main properties of dendrimers are (i) nanoscale container properties (*i.e.* encapsulation), (ii) nanoscaffolding properties (*i.e.* surface adsorption or attachment), and (iii) biocompatibility. Dendrimers are being considered as additives in several

Table 6. Imaging agents based on metal ion-containing nanoparticles on the market and under clinical evaluation.[a]

Product	Formulation	Application	Company	Status
Magnevist®	Gd-DTPA	MRI (brain, blood vessels, CNS)	Bayer Schering Pharma	Marketed
Omniscan®	Gd-DTPA-BMA	MRI (brain, spine, and associated tissues)	GE Healthcare	Marketed
Optimark™	Gd-DTPA-BMEA	MRI (brain, spine, associated tissues, and liver)	Mallinckrodt LLC	Marketed
Eovist®	Gd-EOB-DTPA	MRI (liver lesions)	Bayer HealthCare Pharmaceuticals	Marketed
MultiHance®	Gd-BOPTA	MRI (brain, spine, and associated tissues)	Bracco Diagnostics	Marketed
Gadovist®	Gd-DO3A-butrol	MRI (brain, spine, associated tissues, and liver)	Bayer Schering Pharma	Marketed
ProHance®	Gd-DO3A-HP	MRI (lesions in brain, spine, and associated tissues)	Bracco Diagnostics	Marketed
Dotarem® (Artirem®)	Gd-DTOA	MRI (brain, spine, associated tissues)	Guerbet Group	Marketed
Ablavar® (Vasovist®)	Gadofosveset trisodium	MRI (blood pool agent)	Lantheus Medical	Marketed
Motexafin gadolinium	MGd	MRI (pontine glioma)	Pharmacyclics Inc.	Phase 2 NCT00387790*
Feraheme®	Iron oxides	Iron therapy in CKD	AMAG Pharmaceuticals	Marketed
GastroMARK™	Iron oxides	MRI (bowel)	AMAG Pharmaceuticals	Marketed until Feb. 2013
Supravist™	Iron oxides	MRI (CNS)	Bayer Schering Pharma	Phase 1 NCT01567553*
NanoTherm®	Iron oxides	Solid tumors	MagForce AG	Phase 1 Germany
C-dots (C•spec®)	Silica-organic dye hybrids	Imaging (lymph nodes)	Hybrid Silica Technologies	Phase 0 Memorial Sloan-Kettering

Notes: *ClinicalTrials.gov identifier.

[a]Imaging agents whose trials have been completed or are listed as inactive for more than one year with no apparent follow-up have been excluded from Table 6. CKD — chronic kidney disease; CNS — central nervous system.

routes of administration, including *i.v*, oral, transdermal, and ocular.[148] However, despite enormous academic research activities and promising results over the last 30 years there is little progress into the clinic for dendrimers. Most applications so far are *in vitro*, including Stratus®CS from Dade Behring (now Siemens; quantitative cardiac assays for fast, cost-effective evaluation of patients presenting with suspected myocardial ischemia); Alert Ticket from the U.S. Army Research Laboratory (diagnostics); NanoJuice® Transfection Reagent Kit from EMD Millipore (efficient *in vitro* transfection of difficult-to-transfect cell lines); and SuperFect® Transfection Reagent from Qiagen® (transient and stable *in vitro* transfection of a broad range of cell lines).

Currently there is only one *in vivo* application of dendrimers, *i.e.* VivaGel® (SPL7013) from Starpharma Pty Ltd. VivaGel is an antimicrobial agent which is being applied to a range of sexual health products in development. The active ingredient of VivaGel, the proprietary sulfonated dendrimer SPL7013, has shown to inhibit infection with viruses, including HIV and genital herpes virus, and the growth of bacteria in bacterial vaginosis (BV). Vivagel has completed two Phase 3 clinical studies in the treatment of BV.[149] Another class of dendrimers aiming at clinical translation are phosphorus dendrimers, supplied by Biodendrimers International.

3. Conclusions and Outlook

There have been many reviews written lately about the translation of nanomedicines into the clinic, so why yet another review? Many (most?) of these reviews seem to center around the same group of products and product candidates, including those that have not seen progression in years and are either dead or dormant, while others, even marketed products, have not been included in those listings. Then, there is one very commendable recent review that included all potential nanomedicines and devices for statistical analyses but also did not differentiate between active and dead/dormant candidates.[150] This present review has tried to pressure-test marketed and potential products in order to get a reliable overview of the preclinical to clinical development of nanomedicines. However, this field is vibrant and changing; for example ThemoDox seemed a strong candidate until it failed in Phase 3 late in 2012, and GastroMark, a marketed product, is being phased out in early 2013.

Where are we in the development of nanomedicines? For some, nanotechnology in medicine is the next revolution. There is considerable hype for often obvious (funding) reasons: liposomes (with the exception of "giant vesicles" ≥ 1 μm), micelles, and dendrimers are by definition particles on the nanoscale; however, there is an increasing number of publications dubbing them nanoliposomes, nanomicelles, and nanodendrimers. This hype is unbecoming to the field and reviewers and editors should push back. Lately, there are other voices that are concerned about "Cancer nanomedicines: So many papers and so few drugs!" and "Nanomedicine: is the wave cresting?"[151,152] As often, the reality is probably somewhat in the middle: more than 30 nanomedicines are on the market and at least 50 nanomedicines are at different stages in clinical translation as apparent from the listings in this review. Therefore, the application of nanotechnology to medicine has clearly advanced beyond academic research and is benefitting (cancer) patients. However, there is still a long way ahead. The NCI Alliance for Nanotechnology in Cancer, approved in July 2004, has done an excellent job supporting the development of nanomedicines by providing US$145 million for Phase I (2005–2010) and additional approximately US$30 million per year for Phase II (2010–2015) of the program. It is important, however, that funding agencies — and more importantly the groups approving their budget — understand that supporting the fight against cancer is still imminently important even in times of budget restraints, although the original, very ambitious goal of eliminating death and suffering from cancer by the year 2015 will not be achieved.

What are the next steps carrying nanomedicines forward? First, the debate over ligand-driven versus EPR-driven targeting (often misleading dubbed "passive" targeting although EPR uptake is driven — and can be actively influenced — by particle size, shape, and surface composition) needs to be addressed. Currently all marketed products and most product candidates rely on EPR-driven targeting. Adding ligands to the surface of nanomedicines is expected to improve cell uptake at the target site and thus the efficacy of the treatment. However, these ligands need to be carefully selected and engineered onto the particle surface because random surface modification can unfavorably affect systemic circulation time, biodistribution, and cellular uptake, as well as product polydispersity and therefore the regulatory pathway. In addition, adding targeting ligands, especially more complex ligands such as monoclonal antibodies and aptamers, will complicate the production process and increase COG. The second area for improvement involves the development of theranostic therapies in which the nanomedicines carry multiple functions, *e.g.* deliver a therapeutic agent and a device that allows imaging of the nanomedicine at the same time to personalize the therapy or carry a diagnostic device to measure the success of the treatment. Research attempts in this area are underway; however, as for targeting, the development of theranostics by random modification of an already existing nanocarrier has little change for success because of regulatory implications. Fortunately, chemistry and engineering provide methods and processes to produce theranostics in a controlled and reproducible manner. Some examples have already surfaced but there is still plenty of room for improvement.[10]

References

1. S. Svenson and R. Prud'homme (Eds.). *Multifunctional Nanoparticles for Drug Delivery Applications: Nanostructure Science and Technology Series.* Springer, New York, Dordrecht, Heidelberg, London, 2012.
2. A. Z. Wang, R. Langer and O. C. Farokhzad, Nanoparticle delivery of cancer drugs. *Annu. Rev. Med.* **63**, 185–198 (2012).
3. D. J. Bharali and S. A. Mousa, Emerging nanomedicines for early cancer detection and improved treatment: Current perspective and future promise. *Pharmacol. Ther.* **128**, 324–335 (2010).
4. J. Fang, H. Nakamura and H. Maeda, The EPR effect: Unique features of tumor blood vessels for drug delivery, factors involved, and limitations and augmentation of the effect. *Adv. Drug Deliv. Rev.* **63**, 136–151 (2011).
5. R. Duncan and R. Gaspar, Nanomedicine(s) under the microscope. *Mol. Pharm.* **8**, 2104–2141 (2011).
6. Y. Li, J. Wang, M. G. Wientjes and J. L. S. Au, Delivery of nanomedicines to extracellular and intracellular compartments of a solid tumor. *Adv. Drug Deliv. Rev.* **64**, 29–39 (2012).
7. F. Gu, L. Zhang, B. A. Teply, N. Mann, A. Wang, A. F. Radovic-Moreno, R. Langer and O. C. Farokhzad, Precise engineering of targeted nanoparticles by using self-assembled biointegrated block copolymers. *Proc. Natl. Acad. Sci. USA* **105**, 2586–2591 (2008).
8. I. K. Kwon, S. C. Lee, B. Han and K. Park, Analysis on the current status of targeted drug delivery to tumors. *J. Control. Release.* **164**, 108–114 (2012).
9. S. Svenson, Clinical translation of nanomedicines. *Curr. Opin. Solid State Mater. Sci.* **16**, 287–294 (2012).
10. S. Svenson, Theranostics-Are we there yet? *Mol. Pharm.* **10**, 848–856 (2013).
11. V. P. Torchilin, Recent advances with liposomes as pharmaceutical carriers. *Nat. Rev. Drug Discov.* **4**, 145–160 (2005).
12. W. T. Al-Jamal and K. Kostarelos, Liposomes: From a clinically established drug delivery system to a nanoparticle platform for theranostic nanomedicine. *Acc. Chem. Res.* **44**, 1094–1104 (2011).
13. F.-X. Contreras, L. Sánchez-Magraner, A. Alonso and F. M. Goñi, Transbilayer (flip-flop) lipid motion and lipid scrambling in membranes. *FEBS Lett.* **584**, 1779–1786 (2010).
14. F. Meng, Z. Zhong and J. Feijen, Stimuli-responsive polymersomes for programmed drug delivery. *Biomacromolecules.* **10**, 196–209 (2009).
15. H. C. Shum, Y.-J. Zhao, S.-H. Kim and D. A. Weitz, Multicompartment polymersomes from double emulsions. *Angew. Chem. Int. Edn. Engl.* **50**, 1648–1651 (2011).
16. A. D. Bangham, M. M. Standish and J. C. Watkins, Diffusion of univalent ions across the lamellae of swollen phospholipids. *J. Mol. Biol.* **13**, 238–252 (1965).
17. Y. Barenholz, Doxil® — The first FDA-approved nanodrug: Lessons learned. *J. Control. Release.* **160**, 117–134 (2012).

18. Available at: http://www.azayatherapeutics.com/about-us/products (accessed on 16 March 2013).
19. Available at: http://celsion.com/docs/technology_overview (accessed on 16 March 2013).
20. Available at: http://investor.celsion.com/releasedetail.cfm?ReleaseID=737033 (accessed on 16 March 2013).
21. Available at: http://www.medigene.com/sites/default/files/downloads/20130219_company_presentation_feb2013_1.pdf (accessed on 18 March 2013).
22. Available at: http://www.talontx.com/index.php (accessed on 18 March 2013).
23. P. Minic, S. Fustik, E. Solyom, H. Mazurek, Y. Antipkin, A. Feketeova, A. Senatorova, E. Csiszer, V. Kostromina, B. Takac, R. Ujhelyi, J. Govan, A. Slee and R. Gupta, A multi-cycle open-label study of nebulized liposomal amikacin (Arikace®) in the treatment of cystic fibrosis patients with chronic *Pseudomonas aeruginosa* lung infection. 34th European Cystic Fibrosis Society (ECFS) Conference, Poster #217 (2011).
24. R. W. Niven, Toward managing chronic rejection after lung transplant: the fate and effects of inhaled cyclosporine in a complex environment. *Adv. Drug Deliv. Rev.* **63**, 88–109 (2011).
25. K. K. Sankhala, A. C. Mita, R. Adinin, L. Wood, M. Beeram, S. Bullock, N. Yamagata, K. Matsuno, T. Fujisawa and A. Phan, A Phase 1 pharmacokinetic (PK) study of MBP-426, a novel liposome encapsulated oxaliplatin. *J. Clin. Oncol.* **27**(15S), 2535 (2009).
26. Available at: http://www.sbir.gov/sbirsearch/detail/322806 (accessed on 16 March 2013).
27. G. Batist, K. A. Gelmon, K. N. Chi, W. H. Miller, S. K. L. Chia, L. D. Mayer, C. E. Swenson, A. S. Janoff and A. C. Louie, Safety, pharmacokinetics, and efficacy of CPX-1 liposome injection in patients with advanced solid tumors. *Clin. Cancer Res.* **15**, 692–700 (2009).
28. W. F. Bayne, L. D. Mayer and C. E. Swenson, Pharmacokinetics of CPX-351 (cytarabine/daunorubicin HCl) liposome injection in the mouse. *J. Pharm. Sci.* **98**, 2540–2548 (2009).
29. D. Haussecker, The business of RNAi therapeutics in 2012. *Mol. Ther. Nucleic Acids* **2**, e8 (2012).
30. X. Yuan, S. Naguib and Z. Wu, Recent advances of siRNA delivery by nanoparticles. *Expert Opin. Drug Deliv.* **8**, 521–536 (2011).
31. Available at: http://www.alnylam.com/Programs-and-Pipeline/index.php, http://www.tekmirapharm.com/Programs/Pipeline.asp and http://silence-therapeutics.com/#!/pipeline/ (accessed on 18 March 2013).
32. J. Tabernero, G. I. Shapiro, P. M. LoRusso *et al.*, First-in-man trial of a RNA interference therapeutic targeting VEGF and KSP in cancer patients with liver involvement. *Cancer Discov.* doi:10.1158/2159-8290.CD-12-0429 (2013).
33. Available at: http://www.iirusa.com/upload/wysiwyg/2009-P-Div/P1412/ProNAiSlides_DDP_FINAL.pdf (accessed on 18 March 2013).
34. Y. Otake, S. Soundararajan, T. K. Sengupta, E. A. Kio, J. P. Smith, M. Pineda-Roman, R. K. Stuart, E. K. Spicer and D. J. Fernandes, Overexpression of nucleolin in chronic lymphocytic leukemia cells induces stabilization of bcl2 mRNA. *Blood.* **109**, 3069–3075 (2007).
35. D. L. Vaux, S. Cory and J. M. Adams, Bcl-2 gene promotes haemopoietic cell survival and cooperates with c-myc to immortalize pre-B cells. *Nature* **335**(6189), 440–442 (1988).

36. Available at: http://www.drugdevelopment-technology.com/projects/stimuvax/ (accessed on 20 March 2013).
37. Available at http://www.merrimackpharma.com/pipeline/mm-398/ (accessed on 20 March 2013).
38. N. Nishiyama and K. Kataoka, Current state, achievements, and future prospects of polymeric micelles as nanocarriers for drug and gene delivery. *Pharmacol. Ther.* **112**, 630–648 (2006).
39. A. B. Ebrahim-Attia, Z. Y. Ong, J. L. Hedrick, P. P. Lee, P. L. R. Ee, P. T. Hammond and Y. Y. Yang, Mixed micelles self-assembled from block copolymers for drug delivery. *Curr. Opin. Colloid Interface Sci.* **16**, 182–194 (2011).
40. C. M. Hu and L. Zhang, Therapeutic nanoparticles to combat cancer drug resistance. *Curr. Drug Metab.* **10**, 836–841 (2009).
41. J. Gong, M. Chen, Y. Zheng, S. Wang and Y. Wang, Polymeric micelles drug delivery system in oncology. *J. Control. Release* **159**, 312–323 (2012).
42. Available at: http://clinicaltrials.gov/ct2/show/NCT00876486?term=Genexol&rank=1, (accessed on 16 February 2013).
43. K. S. Lee, H. C. Chung, S. A. Im, Y. H. Park, C. S. Kim, S. B. Kim, S. Y. Rha, M. Y. Lee and J. Ro, Multicenter Phase 2 trial of Genexol-PM, a cremophor-free, polymeric micelle formulation of paclitaxel, in patients with metastatic breast cancer. *Breast Cancer Res. Treat.* **108**, 241–250 (2008).
44. T. Hamaguchi, Y. Matsumura, M. Suzuki, K. Shimizu, R. Goda, I. Nakamura, I. Nakatomi, M. Yokoyama, K. Kataoka and T. Kakizoe, NK105, a paclitaxel-incorporating micellar nanoparticle formulation, can extend *in vivo* antitumor activity and reduce the neurotoxicity of paclitaxel. *Br. J. Cancer.* **92**, 1240–1246 (2005).
45. Y. Matsumura, Preclinical and clinical studies of NK012, a SN-38-incorporating polymeric micelle, which is designed based on EPR effect. *Adv. Drug Deliv. Rev.* **63**, 184–192 (2011).
46. Y. Matsumura, T. Hamaguchi, T. Ura, K. Muro, Y. Yamada, Y. Shimada, K. Shirao, T. Okusaka, H. Ueno, M. Ikeda and N. Watanabe, Phase 1 clinical trial and pharmacokinetic evaluation of NK911, a micelle-encapsulated doxorubicin. *Br. J. Cancer.* **91**, 1775–1781 (2004).
47. H. Uchino, Y. Matsumura, T. Negishi, F. Koizumi, T. Hayashi, T. Honda, N. Nishiyama, K. Kataoka, S. Naito and T. Kakizoe, Cisplatin-incorporating polymeric micelles (NC-6004) can reduce nephrotoxicity and neurotoxicity of cisplatin in rats. *Br. J. Cancer.* **93**, 678–687 (2005).
48. M. Harada, C. Iwata, H. Saito, K. Ishii, T. Hayashi, M. Yashiro, K. Hirakawa, K. Miyazono, Y. Kato and M. R. Kano, NC-6301, a polymeric micelle optimized for effective release of docetaxel, is potent but less toxic than native docetaxel *in vivo*. *Int. J. Nanomedicine.* 7, 2713–2727 (2012).
49. Y. Matsumura, Poly (amino acid) micelle nanocarriers in preclinical and clinical studies. *Adv. Drug Deliv. Rev.* **60**, 899–914 (2008).
50. T. Minko, E. V. Batrakova, S. Li, Y. Li, R. I. Pakunlu, V. Y. Alakhov and A. V. Kabanov, Pluronic block copolymers alter apoptotic signal transduction of doxorubicin in drug-resistant cancer cells. *J. Control. Release.* **105**, 269–278 (2005).
51. J. W. Valle, A. Armstrong, C. Newman, V. Alakhov, G. Pietrzynski, J. Brewer, S. Campbell, P. Corrie, E. K. Rowinsky and M. Ranson, A phase 2 study of SP1049C, doxorubicin in P-glycoprotein-targeting pluronics, in patients with advanced

adenocarcinoma of the esophagus and gastroesophageal junction. *Invest. New Drugs.* **29**, 1029–1037 (2011).
52. Available at: http://www.supratek.com/pipeline/products#SP1049C (accessed on 20 March 2013).
53. M. Beija, R. Salvayre, N. Lauth-de Viguerie and J. D. Marty, Colloidal systems for drug delivery: From design to therapy. *Trends Biotechnol.* **30**, 485–496 (2012).
54. E. Fleige, M. A. Quadir and R. Haag, Stimuli-responsive polymeric nanocarriers for the controlled transport of active compounds: Concepts and applications. *Adv. Drug Deliv. Rev.* **64**, 866–884 (2012).
55. S. Acharya and S. K. Sahoo, PLGA nanoparticles containing various anticancer agents and tumor delivery by EPR effect. *Adv. Drug Deliv. Rev.* **63**, 170–183 (2011).
56. P. Couvreur, Nanoparticles in drug delivery: Past, present, and future. *Adv. Drug Deliv. Rev.* **65**, 21–23 (2013).
57. A. C. Eifler and C. S. Thaxton, Nanoparticle therapeutics: FDA approval, clinical trials, regulatory pathways, and case study. *Methods Mol. Biol.* **726**, 325–338 (2011).
58. E. Miele, G. P. Spinelli, E. Miele, F. Tomao and S. Tomao, Albumin-bound formulation of paclitaxel (Abraxane® ABI-007) in the treatment of breast cancer. *Int. J. Nanomedicine* **4**, 99–105 (2009).
59. V. Roy, B. R. LaPLant, G. G. Gross, C. L. Bane and F. M. Palmieri, Phase II trial of weekly nab (nanoparticle albumin-bound)-paclitaxel (nab-paclitaxel) (Abraxane®) in combination with gemcitabine in patients with metastatic breast cancer (N0531). *Ann. Oncol.* **20**, 449–453 (2009).
60. L. H. Reddy and P. Couvreur, Nanotechnology for therapy and imaging of liver diseases. *J. Hepatol.* **55**, 1461–1466 (2011).
61. S. Svenson, M. Wolfgang, J. Hwang, J. Ryan and S. Eliasof, Preclinical to clinical development of the novel camptothecin nanopharmaceutical CRLX101. *J. Control. Release* **153**, 49–55 (2011).
62. G. J. Weiss, J. Chao, J. D. Neidhart, R. K. Ramanathan, D. Bassett, J. A. Neidhart, *et al.*, First-in-human phase 1/2a trial of CRLX101, a cyclodextrin-containing polymer-camptothecin nanopharmaceutical in patients with advanced solid tumor malignancies. *Invest New Drugs* .doi10.1007/s10637-012-9921-8 (2013).
63. Available at: http://ceruleanrx.com/wp-content/uploads/2013/03/Top-line-Phase-2b-NSCLC-Results-032213-FINAL.pdf (accessed on 27 March 2013).
64. M. E. Davis, J. E. Zuckerman, C. H. Choi, D. Seligson, A. Tolcher, C. A. Alabi, Y. Yen, J. D. Heidel and A. Ribas, Evidence of RNAi in humans from systemically administered siRNA via targeted nanoparticles. *Nature.* **464**, 1067–1070 (2010).
65. A. Ribas, L. Kalinoski, J. D. Heidel, J. Peterkin, D. B. Seligson, J. E. Zuckerman, C. Choi, Y. Yen, M. E. Davis and A. W. Tolcher, Systemic delivery of siRNA via targeted nanoparticles in patients with cancer: Results from a first-in-class Phase I clinical trial. *J. Clin. Oncol.* **28**(15S), 3022 (2010).
66. J. Hrkach, D. Von Hoff, M. M. Ali, *et al.*, Preclinical development and clinical translation of a PSMA-targeted docetaxel nanoparticle with a differentiated pharmacological profile. *Sci. Transl. Med.* **4**, 128ra39 (2012).
67. Available at: http://biokorea2012.org/2012/korea/businessforum/partnering_print.php?t=0 entry #114, (accessed on 26 March 2013).
68. Available at: http://www.flamel.com/technology-platforms/medusa/ (accessed on 30 March 2013).

69. Y.-P. Chan, R. Meyrueix, R. Kravtzoff, F. Nicolas and K. Lundstrom, Review on Medusa®: A polymer-based sustained release technology for protein and peptide drugs. *Expert Opin. Drug Deliv.* **4**, 441–451 (2007).
70. F. Kersey, T. J. Merkel, J. Perry, *et al.*, The effect of aspect ratio and deformability on nanoparticle extravasation through nanopores. *Langmuir.* **28**, 8773–8781 (2012).
71. T. J. Merkel, K. Chen, S. W. Jones, *et al.*, The effect of particle size on the biodistribution of low-modulus hydrogel PRINT particles. *J. Control. Release.* **162**, 37–44 (2012).
72. J. L. Perry, K. G. Reuter, M. P. Kai, *et al.*, PEGylated PRINT nanoparticles: The impact of PEG density on protein binding, macrophage association, biodistribution, and pharmacokinetics. *Nano Lett.* **12**, 5304–5310 (2012).
73. H. B. Gaspar, E. Bjorkegren, K. Parsley, *et al.*, Successful reconstitution of immunity in ADA-SCID by stem cell gene therapy following cessation of PEG-ADA and use of mild preconditioning. *Mol. Ther.* **14**, 505–513 (2006).
74. P. A. Dinndorf, J. Gootenberg, M. H. Cohen, P. Keegan and R. Pazdur, Pegaspargase (Oncaspar®) for the first-line treatment of children with acute lymphoblastic leukemia (ALL). *Oncologist.* **12**, 991–998 (2007).
75. D. Weycker, J. Malin, R. Barron, *et al.*, Comparative effectiveness of filgrastim, pegfilgrastim, and sargramostim as prophylaxis against hospitalization for neutropenic complications in patients with cancer receiving chemotherapy. *Am. J. Clin. Oncol.* **35**, 267–274 (2012).
76. A. Colao, R. Pivonello, R. S. Auriemma, *et al.*, Efficacy of 12-months treatment with the GH receptor antagonist pegvisomant in patients with acromegaly resistant to long-term, high-dose somatostatin analog treatment: Effect on IGF-I levels, tumor mass, hypertension and glucose tolerance. *Eur. J. Endocrinol.* **154**, 467–477 (2006).
77. P. Sapra, H. Zhao, M. Mehlig, J. Malaby, P. Kraft, C. Longley, L. M. Greenberger and I. D. Horak, Novel delivery of SN38 markedly inhibits tumor growth in xenografts, including a camptothecin-11 refractory model. *Clin. Cancer Res.* **14**, 1888–1896 (2008).
78. R. Kurzrock, S. Goel, J. Wheler, *et al.*, Safety, pharmacokinetics, and activity of EZN-2208, a novel conjugate of polyethylene glycol and SN38, in patients with advanced malignancies. *Cancer* **118**, 6144–6151 (2012).
79. M.-C. Vanderbeeken, P. G. Aftimos and A. Awada, Topoisomerase inhibitors in metastatic breast cancer: Overview of current practice and future development. *Curr. Breast Cancer Rep.* **5**, 31–41 (2013).
80. S. D. Chipman, F. B. Oldham, G. Pezzoni and J. W. Singer, Biological and clinical characterization of paclitaxel poliglumex (PPX, CT-2103), a macromolecular polymer-drug conjugate. *Int. J. Nanomedicine* **1**, 375–383 (2006).
81. A. V. Yurkovetskiy, A. Hiller, S. Syed, M. Yin, X. M. Lu, A. J. Fischman and M. I. Papisov, Synthesis of a macromolecular camptothecin conjugate with dual phase drug release. *Mol. Pharm.* **1**, 375–382 (2004).
82. E. A. Sausville, L. Garbo, G. J. Weiss, D. Shkolny, A. V. Yurkovetskiy, C. Bethune, R. K. Ramanathan and R. J. Fram, Phase 1 study of XMT-1001, a novel water soluble camptothecin conjugate, given as an intravenous infusion once every three weeks to patients with advanced solid tumors. Proc. AACR-NCI-EORTC Int. Conference *Mol. Targets Cancer Ther.* Abstract #B52 (2009).
83. R. Duncan, M. J. Vicent, F. Greco and R. I. Nicholson, Polymer-drug conjugates: Towards a novel approach for the treatment of endrocine-related cancer. *Endocr. Relat. Cancer* **12**, S189–S199 (2005).

84. R. Duncan, The dawning era of polymer therapeutics. *Nat. Rev. Drug Discov.* **2**, 347–360 (2003).
85. P. Evenepoel, R. Selgas, F. Caputo, *et al.*, Efficacy and safety of sevelamer hydrochloride and calcium acetate in patients on peritoneal dialysis. *Nephrol Dial Transplant.* **24**, 278–285 (2009).
86. E. Cohen-Sela, M. Chorny, N. Koroukhov, *et al.*, A new double emulsion solvent diffusion technique for encapsulating hydrophilic molecules in PLGA nanoparticles. *J. Control. Release* **133**, 90–95 (2009).
87. N. Anton and T. F. Vandamme, Nano-emulsions and micro-emulsions: Clarifications of the critical differences. *Pharm. Res.* **28**, 978–985 (2011).
88. M. J. Lawrence and G. D. Rees, Microemulsion-based media as novel drug delivery systems. *Adv. Drug Deliv. Rev.* **64**(Supplement), 175–193 (2012).
89. C. E. Mora-Huertas, H. Fessi and A. Elaissari, Polymer-based nanocapsules for drug delivery. *Int. J. Pharm.* **385**, 113–142 (2010).
90. Available at: http://www.druglib.com/druginfo/amphotec/ (accessed on 30 March 2013).
91. Available at: http://dailymed.nlm.nih.gov/dailymed/lookup.cfm?setid=5587db37-f21a-4a39-a319-e1077032ced9 (accessed on 30 March 2013).
92. Available at: http://www.rxlist.com/diprivan-drug.htm (accessed on 30 March 2013).
93. Available at: http://www.rxlist.com/neoral-drug.htm (accessed on 30 March 2013).
94. Available at: http://www.rxlist.com/restasis-drug.htm (accessed on 30 March 2013).
95. Available at: http://www.novagali.com/en/eye-therapy/allergy/ (accessed on 31 March 2013).
96. Available at: http://www.rxlist.com/durezol-drug.htm (accessed on 31 March 2013).
97. Available at: http://www.cleviprex.com/ (accessed on 31 March 2013).
98. Available at: http://justthevax.blogspot.com/2009/09/eu-approves-gsk-pandemrix-and-novartis.html (accessed on 31 March 2013).
99. Available at: http://www.nanobio.com/Platform-Technology/NanoStat-Platform.html (accessed on 31 March 2013).
100. Available at: http://www.masttherapeutics.com/technology/oncology/anx-514/ (accessed on 31 March 2013).
101. Available at: http://www.sunpharma.in/nanoparticle.htm (accessed on 1 April 2013).
102. Available at: http://www.sbir.gov/sbirsearch/detail/144259_(accessed on 1 April 2013).
103. D. Pan, J. W. Myerson, L. Hu, *et al.*, Programmable nanoparticle functionali-zation for *in vivo* targeting. *FASEB J.* **27**, 255–264 (2013).
104. Available at: http://www.cancer.gov/drugdictionary?cdrid=518315 (accessed on 1 April 2013).
105. T. Hosokawa, M. Yamauchi, Y. Yamamoto, K. Iwata, Y. Kato and E. Hayakawa. Formulation development of a filter-sterilizable lipid emulsion for lipophilic KW-3902, a newly synthesized adenosine A1-receptor antagonist. *Chem. Pharm. Bull (Tokyo)* **50**, 87–91 (2002).
106. J. R. Teerlink, V. J. Iraqui, J. P. Mohr, *et al.,* The safety of an adenosine A1-receptor antagonist, rolofylline, in patients with acute heart failure and renal impairment: Findings from PROTECT. *Drug Saf.* **35**, 233–244 (2012).
107. M. W. Saif, C. Ehrlichman, T. Dragovich, *et al.*, Open-label, dose-escalation, safety, pharmacokinetic study of intravenously administered CNF1010 (17-(allylamino)-17-demethoxygeldanamycin [17-AAG]) in patients with solid tumors. *Cancer Chemother. Pharmacol.* doi10.1007/s00280-013-2134-9 (2013).

108. Y. Liu and N. Zhang, Gadolinium loaded nanoparticles in theranostic magnetic resonance imaging. *Biomaterials.* **33**, 5363–5375 (2012).
109. N. Nainani and M. Panesar, Nephrogenic systemic fibrosis. *Am. J. Nephrol.* **29**, 1–9 (2009).
110. J. E. Rosen, L. Chan, D.-B. Shieh and F. X. Gu, Iron oxide nanoparticles for targeted cancer imaging and diagnostics. *Nanomedicine*: NBM **8**, 275–290 (2012).
111. S. Laurent, J.-L. Bridot, L. Vander Elst and R. N. Muller, Magnetic iron oxide nanoparticles for biomedical applications. *Future Med. Chem.* **2**, 427–449 (2010).
112. A. M. Smith, M. M. Wen and S. Nie, Imaging dynamic cellular events with quantum dots: The bright future. *Biochem.* (*Lond.*) **32**, 12–17 (2010).
113. P. Caravan, J. J. Ellison, Y. J. McMurry and R. B. Lauffer, Gadolinium(III) chelates as MRI contrast agents: Structure, dynamics, and applications. *Chem. Rev.* **99**, 2293–2342 (1999).
114. Z. Seidl, J. Vymazal, M. Mechl, *et al.,* Does higher gadolinium concentration play a role in the morphologic assessment of brain tumors? Results of a multicenter intraindividual crossover comparison of gadobutrol versus gadobenate dimeglimine (the MERIT study). *AJNR Am. J. Neuroradiol.* **33**, 1050–1058 (2012).
115. V. M. Runge, Conventional contrast agents (gadolinium chelates). *Proc. Intl. Soc. Mag. Reson. Med.* 19 (2011).
116. Available at: http://www.diagnosticimaging.com/lantheus-rebrands-mr-agent-vasovist (accessed on 6 April 2013).
117. T. Hompland, C. Ellingsen and E. K. Rofstad, Preclinical evaluation of Gd-DTPA and gadomelitol as contrast agents in DCE-MRI of cervical carcinoma interstitial fluid pressure. *BMC Cancer.* **12**, 544 (2012).
118. J. Tang, Y. Sheng, H. Hu and Y. Shen, Macromolecular MRI contrast agents: Structures, properties, and applications. *Prog. Polym. Sci.* **38**, 462–502 (2013).
119. Z. Zhou and Z.-R. Lu, Gadolinium-based contrast agents for magnetic resonance cancer imaging. WIREs. *Nanomed. Nanobiotechnol.* **5**, 1–18 (2013).
120. Available at: http://www.ncbi.nlm.nih.gov/books/NBK23589/pdf/Gadobutrol.pdf (accessed on 6 April 2013).
121. Available at: http://www.ph.braccosolutions.com/SitePages/Macrocyclic%20Gadolinium%20 Contrast%20Agent-ProHance.aspx_(accessed on 6 April 2013).
122. Available at: http://www.drugs.com/newdrugs/fda-approves-dotarem-new-magnetic-resonance-imaging-agent-3728.html (accessed on 6 April 2013).
123. S. Laurent, L. Vander Elst and R. N. Muller, Comparative study of the physicochemical properties of six clinical low molecular weight gadolinium contrast agents. *Contrast Media Mol. Imaging.* **1**, 128–137 (2006).
124. C. U. Herborn, E. Honold, M. Woplf *et al.*, Clinical safety and diagnostic value of the gadolinium chelate gadoterate meglumine (Gd-DOTA). *Invest Radiol.* **42**, 58–62 (2007).
125. D. Magda and R. A. Miller, Motexafin gadolinium: A novel redox active drug for cancer therapy. *Semin. Cancer Biol.* **16**, 466–476 (2006).
126. S. R. Thomas and D. Khuntia, Motexafin gadolinium: A promising radiation sensitizer in brain metastasis. *Expert Opin. Drug Discov.* **6**, 195–203 (2011).
127. K. Overoye-Chan, S. Koerner, R. J. Looby, *et al.*, EP-2104R: A fibrin-specific gadolinium-based MRI contrast agent for detection of thrombus. *J. Am. Chem. Soc.* **130**, 6025–6039 (2008).
128. Available at: http://www.wikinvest.com/stock/EPIX_Pharmaceuticals_(EPIX)/Ep-2104r (accessed on 7 April 2013).

129. R. T. Gupta, C. M. Brady, J. Lotz, D. T. Boll and E. M, Merkle, Dynamic MR imaging of the biliary system using hepatocyte-specific contrast agents. *AJR. Am. J. Roentgenol.* **195**, 405–413 (2010).
130. Available at: http://www.ema.europa.eu/docs/en_GB/document_library/Public_statement/2012/08/WC500130495.pdf (accessed on 7 April 2013).
131. Available at: http://www.amagpharma.com/products/feridex_iv.php (accessed on 9 April 2013).
132. Y.-X. J. Wang, Superparamagnetic iron oxide based MRI contrast agents: Current status of clinical application. *Quant. Imaging Med. Surg.* **1**, 35–40 (2011).
133. Available at: http://www.medscape.com/viewarticle/733360_3 (accessed on 9 April 2013).
134. R. A. Heesakkers, G. J. Jager, A. M. Hövels, *et al.*, Prostate cancer: Detection of lymph node metastases outside the routine surgical area with ferumoxtran-10-enhanced MR imaging. *Radiology* **251**, 408–414 (2009).
135. Available at: http://www.nanotech-now.com/news.cgi?story_id=26939 (accessed on 9 April 2013).
136. Available at: http://www.sfgate.com/business/prweb/article/Combidex-Technology-Sale-Opens-Door-to-a-4417098.php (accessed on 9 April 2013).
137. Available at: http://www.trademarks411.com/marks/75486789-clariscan (accessed on 9 April 2013).
138. M. Lu, M. H. Cohen, D. Rieves and R. Pazdur, FDA report: Ferumoxytol for intravenous iron therapy in adult patients with chronic kidney disease. *Am. J. Hematol.* **85**, 315–319 (2010).
139. A. Singh, T. Patel, J. Hertel, *et al.*, Safety of ferumoxytol in patients with anemia and CKD. *Am. J. Kidney Dis.* **52**, 907–915 (2008).
140. E. A. Neuwelt, B. E. Hamilton, C. G. Varallyay, *et al.*, Ultrasmall super-paramagnetic iron oxides (USPIOs): A future alternative magnetic resonance (MR) contrast agent for patients at risk for nephrogenic systemic fibrosis (NSF)? *Kidney Int.* **75**, 465–474 (2009).
141. K. Maier-Hauff, F. Ulrich, D. Nestler, *et al.*, Efficacy and safety of intratumoral thermotherapy using magnetic iron oxide nanoparticles combined with external beam radiotherapy on patients with recurrent glioblastoma multiforme. *J. Neurooncol.* **103**, 317–324 (2011).
142. Available at: http://www.fda.gov/drugs/drugsafety/drugshortages/ucm050794.htm (accessed on 10 April 2013).
143. Available at: http://www.mr-tip.com/serv1.php?type=db1&dbs=Abdoscan%AE (accessed on 10 April 2013).
144. M. J. D. Clift and V. Stone, Quantum dots: Insight and perspective of their biological interaction and how this relates to their relevance for clinical use. *Theranostics* **2**, 668–680 (2012).
145. L. Ye, K.-T. Yong, L. Liu, *et al.*, A pilot study in non-human primates shows no adverse response to intravenous injection of quantum dots. *Nat. Nanotechnol.* **7**, 453–458 (2012).
146. J. Choi, A. Burns, R. Williams, *et al.*, Core-shell silica nanoparticles as fluorescent labels for nanomedicine. *J. Biomed. Opt.* **12**, 064007(1–11) (2007).
147. R. Friedman, Nano dot technology enters clinical trials. *JNCI. J. Natl. Cancer Inst.* **103**, 1428–1429 (2011).

148. S. Svenson, Dendrimers as versatile platform in drug delivery applications. *Eur. J. Pharm. Biopharm.* **71**, 445–462 (2009).
149. Available at: http://www.starpharma.com/vivagel/vivagel_clinical_trials, (accessed on 11 April 2013).
150. M. L. Etheridge, S. A. Campbell, A. G. Erdman, *et al.*, The big picture on nanomedicine: The state of investigational and approved nanomedicine products. *Nanomedicine NBM* **9**, 1–14 (2013).
151. V. J. Venditto and F. C. Szoka Jr., Cancer nanomedicines: So many papers and so few drugs! *Adv. Drug Deliv. Rev.* **65**, 80–88 (2013).
152. R. Juliano, Nanomedicine: Is the wave cresting? *Nature Rev. Drug Discov.* **12**, 171–172 (2013).

B:
Novel Concepts in Nanomedicine

Chapter B1

Organelle-Targeted Nanocarriers

*Alexander Koshkaryev**

Department of Pharmaceutical Sciences
Northeastern University, 360 Huntington Avenue,
Boston, MA 02115
**alexander.koshkaryev@gmail.com*

Abstract

Site-specific localization of drugs and bioactive compounds at the subcellular level can strongly improve their therapeutic impact. Nanosized carriers designed specifically for organelle-targeted delivery of the encapsulated cargo represent a new class of delivery systems, which have been intensively studied over the last decade. The chapter highlights the current state in this growing area of organelle-targeted nanocarriers.

1. Introduction

The modern paradigm for targeted drug delivery distinguishes among three different levels of targeting. The first level deals with the positioning of the drug in specific pathogenic areas such a tumor, an infection or infarct. At the second level, a drug must be introduced within the cell. Finally, the last level involves specific targeting on a subcellular level. Therefore, besides other essential properties such as efficient drug loading, stability and longevity in the circulation and biocompatibility, an "ideal" delivery system should be able to: (i) selectively accumulate in the tissue of interest; (ii) effectively internalize in targeted cells, and (iii) localize and release its cargo in a specific cell compartment.

Organelle-targeted nanocarriers that can deliver their pharmaceutical loads to a subcellular site of action represent a relatively new trend in targeted drug delivery. The specific targeting at a subcellular level is especially important for many anticancer drugs that target mitochondria and nuclei, for lysosomal enzymes that should be delivered to lysosomes and for DNA that must be localized in the nucleus or mitochondrion.

Subcellular targeting has been discussed lately in a number of reviews.[1–3] Here, recent developments in organelle-targeted drug delivery will be highlighted using liposome-based nanocarriers as an example of an efficient and flexible delivery platform. Among other pharmaceutical nanocarriers, liposomes possibly represent the best established drug delivery system. Liposomes, phospholipid vesicles with sizes varying from 50 to 1000 nm, can be loaded with a wide variety of drugs.[4] Liposomes have particular advantages over other drug delivery systems, including ease of control of composition and size, a relatively simple preparation and scale up, good drug loading efficiency, high stability and the ability to be specifically targeted with small quantities of targeting molecules. Altogether, these qualities make liposomes a valuable platform for design of organelle-targeted delivery systems. Indeed, it was not a coincidence that a major part of the latest attempts to create such a system was used liposome-based formulations.[3]

A clear understanding of the endocytic mechanisms, intracellular trafficking machinery and internal properties of targeting organelles is a key requirement for the successful development of organelle-targeted nanocarriers. Mammalian cells possess a number of highly regulated endocytotic pathways, which are able to internalize macromolecules from the surrounding environment and traffic them to specific subcellular locations.[5] For example, clathrin-mediated endocytosis (CME), the best-defined endocytotic pathway,[6] can internalize a wide variety of ligand/receptor complexes by formation of clathrin-coated vesicles, which subsequently deliver their cargo to an endosomal sorting machine followed either by return to the plasma membrane or trafficking to lysosomes. Thus, by modification of the nanocarrier's surface with a ligand targeting the receptor-dependent CME, delivery of the nanocarrier-encapsulated drug to lysosomes can be achieved. The same nanocarriers, but consisting of fusogenic lipids, will release their cargo from the endosome-lysosomal compartment, thus providing the cytoplasmic delivery of the encapsulated drug.[7] Other examples of the use of cellular endocytotic machinery for the organelle-targeting include: (i) caveolae-mediated endocytosis (CvME), known for its ability to be directed to endoplasmic reticulum (ER) and Golgi apparatus (GA),[8] and (ii) macropinocytosis, a non-specific pathway for internalization of fluids and membrane.[9] A good overview of cellular mechanisms involved in the internalization and intracellular trafficking of pharmaceutical nanocarriers can be found in Ref. 10.

In this chapter, the latest approaches and design schemes for nanocarriers targeting different subcellular compartments will be reviewed.

2. Cytoplasm-targeted Nanocarriers

Cytosolic deposition can be critical for siRNA and for other therapeutic agents whose site of action is localized in the cytoplasm. Moreover, an effective cytosol-targeted delivery system may prevent cellular efflux of anticancer drugs generated by the action of multidrug resistance proteins (MRP).[11]

The intracellular transport of most therapeutical compounds is restricted by the hydrophobicity of the cell membrane. Unless an active receptor-dependent endocytotic pathway is involved, only a few biologically active agents with a certain size, charge, and polarity can cross the plasma membrane. Furthermore, the cellular internalization of most nanocarriers occurs preferentially via endocytotic pathways followed by their eventual accumulation in the endosomal-lysosomal compartment,[10] which restricts the nanocarrier's deposition into the cytoplasm. A nanocarrier optimized for cytoplasmic targeting has to provide an effective way to cross both cellular and endosomal membranes.

Many strategies have been developed recently to achieve the endosomal escape of internalized nanocarriers. These include "fusogenic" and "pH-sensitive" systems, nanocarriers modified with cell-penetrating peptides (CPPs) and various physical approaches.

2.1. *Nanocarriers modified with "fusogenic" peptides*

The endosomal escape of nanocarriers can be provided by exploitation of the "fusogenic" properties of virus-like peptides.[7] "Fusogenic" peptides mimic viral coat proteins which have evolved to breach the endosome/lysosomal membrane, thus initiating release of the viral particle to the cytoplasm.[12]

Numerous synthetic analogues of the viral peptides, such as diNF7, INF6, DAT and GALA, have been designed and applied to improve the endosomal escape of nanocarriers.[13,14] Upon nanocarrier entry into the endosomal compartment, the fusogenic peptides undergo a structural transition triggered by the endosomal acidic interior. Due to this conformational change, the fusion peptide is inserted into the endosomal membrane and destabilizes it resulting in cytosolic release of the nanocarrier load.[15]

It was shown recently that the endosomal escape of the liposomal nanocarrier can be provided by modification of their surface with a phage fusion coat protein that shares some common structural features with the fusogenic components of viruses and synthetic amphipathic peptides.[16] Such phage-modified liposomes loaded with doxorubicin demonstrated a significantly enhanced association with target cells and a remarkably increased toxicity against MCF-7 breast cancer cells.[17]

Synthetic amphipathic peptide GALA, composed of 30 amino acid residues with a repeated Glu-Ala-Leu-Ala unit, has shown a pH-responsive membrane destabilizing activity and improved cytoplasmic delivery of the GALA-modified small unilamellar phospholipids vesicles.[13] PEGylated GALA-modified liposomes labeled additionally with transferrin (Tf) have been shown to release their contents into the cytosol via endosomal membrane fusion.[18] Such a combination of Tf-labeling and fusogenic activity of GALA strongly (100-fold) increased transfection activity of liposome-packed DNA.[19] It was shown lately that a combination of cationic liposomes (Lipofectamine 2000) with a fusogenic GALA provided increased cytosolic delivery of proteins and nanoparticles.[20]

2.2. *pH-sensitive systems for cytoplasmic delivery*

Liposome-based nanocarriers containing pH-sensitive polyanionic polymers and lipids, such as dioleoylphosphatidylethanolamine (DOPE) and cholesteryl hemisuccinate (CHEMS), represent another common approach for promotion of efficient cytosolic delivery.[21] As reviewed earlier, following endocytosis, pH-sensitive liposomes have a tendency to fuse with the endosomal membrane at acidic pH, releasing their drug content into the cytoplasm.[22] PEGylated pH-sensitive liposome-based nanocarriers, although demonstrating a decreased pH-sensitivity, still effectively deliver their contents into the cytoplasm.[23] New pH-sensitive liposomal additives have been recently described including oleyl alcohol[24] and pH-sensitive morpholine lipids (mono-stearoyl derivatives of morpholine).[25] Lately, pH-sensitive liposomes based on DOPE and stabilized by Pseudomonas aeruginosa dirhamnolipids (diRL) have been shown to be efficient vehicles for cytoplasmic delivery.[26]

A variety of poly(glycidol) derivatives with a PEG-like linear backbone structure and carboxyl groups on the side chains have been used to obtain pH-sensitive liposomes.[27] Such liposomes exhibit considerable destabilization at low pH and deliver their contents into the cytosol by membrane fusion with the endosomal membranes.[28] Liposomes modified with 3-methylglutarylated poly(glycidol) are an efficient carrier of ovalbumin for induction of cellular immunity *in vitro* and *in vivo*.[29]

Ionizable copolymers of N-isopropylacrylamide (NIPAM) have been used to provide pH-sensitivity to the long-circulating liposomal formulation composed of ePC, cholesterol (Chol) and DSPE–PEG.[30,31] The NIPAM-modified liposomes were additionally labeled with anti-CD33 monoclonal antibody to target leukemic cells.[32] Such pH-sensitive immunoliposomes loaded with cytosine arabinoside (ara-C) demonstrated an efficient internalization by CD33+ leukemic cell lines and the highest cytotoxicity against HL60 cells due to the endosomal escape of ara-C.[32]

A new type of micelle-like nanoparticles (MNP) based on covalently conjugated lipids and polyethylenimine (PEI) for the improved cytosolic delivery of siRNA and DNA has been developed recently.[33–35] PEI, a cationic polymer widely used for delivery of nucleic acids, is known for its ability to breach the endosomal membrane due to a "proton sponge" effect.[36] Long-circulating PEG-grafted MNP based on different phospholipid–PEI complexes showed an enhanced intracellular uptake of the encapsulated siRNA following the effective silencing of the targeted genes.[33,34]

Lately, a combination of liposomal pH-sensitivity and tumor specific targeting has become a promising avenue for drug delivery. For example, long circulating liposomes containing DOPE and CHEMS (6:4 molar ratio) and additionally modified with epidermal growth factor receptor (EGFR) antibody have shown an efficient and targeted delivery of gemcitabine to non-small cell lung cancer using A549 cells in a BALB/c-nu/nu mouse tumor model.[37] Folate and Tf-targeted pH-sensitive liposomes that exploit the overexpression of folate and Tf receptors on the surface of cancer cells have been developed recently to provide tumor targeting as well as efficient cytosolic delivery of anticancer drugs and DNA.[38–41]

2.3. *CPP-labeled liposomes for cytoplasmic delivery*

Another approach for delivery of drugs directly to cytoplasm is based on the use of CPPs, such as HIV-1 trans-activating transcriptional activator (TAT) and others short cationic peptides.[3,42] An increased intracellular delivery of TAT-modified liposomes was shown with various cell lines, such as murine Lewis lung carcinoma (LLC), human breast tumor BT-20 and rat cardiac H9C2 myocytes.[43,44] The CPP-modification provides very high rates of intracellular internalization of liposomes proportionally to the number of peptides attached to the liposomal surface.[45]

The recent data suggest that more than one mechanism for CPP-mediated intracellular delivery.[46] For example, individual CPPs or CPP-conjugated small molecules penetrate cells via electrostatic interactions and hydrogen bonding and do not seem to depend on energy,[47] while CPP-mediated intracellular delivery of large particles was proven to proceed via the energy-dependent macropinocytosis with subsequent escape from endosomes into the cell cytoplasm.[48] The internalization of TAT-modified DNA-loaded liposomes has been shown to occur via a Chol-dependent clathrin-mediated pathway.[49] It has been found recently that CPP-mediated escape from the endosome is based on its fusion and subsequent permeabilization of the endosomal membrane.[50]

The high efficiency of CPP-mediated internalization as well as the ability of CPPs to provide endosomal escape makes them a promising candidate for the cytoplasmic delivery of many therapeutic agents. TAT–modified liposomes prepared with the addition of a small quantity of a cationic lipid (DOTAP) and incubated with DNA formed stable non-covalent complexes.[43] Such TATp–liposome–DNA complexes, when incubated with mouse fibroblasts (NIH 3T3) or cardiac myocytes (H9C2), showed substantially higher *in vitro* transfection with lower cytotoxicity than Lipofectin®.

Cationic liposomes modified with $S4_{13}$–PV CPP, a chimeric peptide combined with a cell penetrating sequence of the Dermaseptin S4 peptide and also containing the nuclear localization signal present in the Simian Virus 40 (SV40), have been designed for the cytosolic delivery of splice-switching oligonucleotides targeted toward the mutated splicing site of human β-globin pre-mRNA in the HeLa pLuc/705 splice correction model.[51]

2.4. *Physical approaches for cytoplasmic delivery*

Other strategies promoting endosomal escape followed by cytosolic delivery of the nanocarrier load utilize physical approaches such as photochemical internalization,[52] thermo-sensitivity[53] or ultra-sound exposure.[54]

The photochemical internalization (PCI) method is based on the specific accumulation of a photosensitizer with an amphiphilic nature in the endosomal membrane. Upon illumination, highly reactive singlet oxygen species damage the endosomal membranes causing its permeabilization. Recently, the PCI technique has been successfully introduced for the cytosolic delivery of saporin-loaded liposomes to OVCAR-3.[52]

Thermosensitive immunoliposomes targeted to human epidermal growth factor receptor 2 (Her-2) have been developed lately to provide an effective cytoplasmic delivery to breast cancer cells.[53,55] After internalization, thermosensitive liposomes constructed from dipalmitoylphosphatidylcholine (DPPC) can be induced to release their contents into endosomal compartments at 42°C. Such thermosensitive liposomes were loaded with the pore-forming protein listeriolysin O (LLO) that forms pores in the endosomal membrane allowing a drug to flow out of the endosomal compartment into the cytoplasm.[53]

An interesting approach utilizing ultrasound (US) treatment to provide the cytoplasmic delivery of siRNA has been developed recently.[54] PEG-liposomes loaded with an US-imaging gas effectively delivered liposomal siRNA directly into the cytoplasm, aster short US treatment, and produced effective gene silencing. The mechanism of US-enhanced cytoplasmic delivery depends on the inertial cavitation of the bubbles produced by the pressure oscillations of US. The inertial cavitation is believed to cause transient disruptions in the cell membranes, enabling the transport of extracellular molecules into the cells.[54]

3. Lysosome-targeted Nanocarriers

Lysosomes represent one of the important intracellular targets for delivery of therapeutical enzymes and anticancer agents due to their involvement in pathogenesis of so-called lysosomal storage diseases (LSDs)[56,57] and their key role in the lysosomal cell death pathway (LCDP).[58,59]

Lysosomes are ubiquitous acidic organelles accountable for the recycling and degradation of the majority of cellular components.[60] Lysosomes contain at least 50–60 soluble acidic enzymes (phosphatases, glycosidases, sulphatases, lipases, nucleases and proteases) responsible for degradation of a wide variety of cellular macromolecules including proteins, nucleic acids, oligosaccharides and complex lipids.[61,62]

The involvement of lysosomes in the pathogenesis of LSDs, a set of more than 40 genetic disorders associated with the deficiency of certain lysosomal hydrolyses, makes these organelles a desirable intracellular target for the delivery of therapeutical enzymes.[63] Currently, enzyme replacement therapy (ERT) based on the administration of exogenous enzymes[64,65] is the only approach for the treatment of LSDs. However, this procedure remains limited and highly expensive because of inefficient lysosomal delivery and poor stability of therapeutic enzymes in the circulation. The idea of using nanocarriers such as liposomes for delivery of therapeutical enzymes was introduced long ago[66] and opened new opportunities for ERT. Still, liposome-immobilized enzymes are not in general clinical use for ERT, and there remains a clear need to sharply increase the efficiency of the delivery of the liposomal enzymes to lysosomes, *i.e.* to achieve lysosomal targeting.

Along with LSDs, lysosomes are known to be involved in cellular apoptosis associated with the alternative LCDP.[67,68] Until recently the role of lysosomes in the cells physiology was thought to be restricted mostly to digestive functions. However, over the last decade, the discovery of lysosomal involvement in the cellular death machinery made them an attractive intracellular target for anticancer therapy.

In 1998, Roberg & Ollinger discovered that the oxidative treatment of rat heart myocytes with quinone naphthazarin leads to release of lysosomal enzyme cathepsin D to the cytosol followed by cell apoptosis.[69] It was established lately that the moderate permeabilization of lysosomal membranes results in apoptosis-like cell death, while massive disruption of the lysosomal membrane was associated with necrosis-like cell death.[70–72] Lysosomal membrane permeabilization (LMP) followed by the release of lysosomal enzymes into the cytoplasm can be triggered by several factors including oxidative stress,[69,73] lysosomotropic compounds with detergent activity[72,74] and activation of death receptors of the tumor necrosis factor (TNF) receptor family.[75,76] The relocation of lysosomal enzymes, especially cathepsins, into the cytosol may result in caspase-dependent or -independent activation of apoptosis with or without involvement of the mitochondrion.[77,78]

Hence, lysosomes present a very promising therapeutic target for induction of the LCDP in cancer cells. Cancer therapy may strongly benefit from lysosome-targeted nanocarriers with encapsulated lysosome-destabilizing agents. A variety of methods for improvement of the lysosomal accumulation of pharmaceutical nanocarriers has been developed lately and will be covered in this chapter.

3.1. *Lysosomotropic liposomes modified with octadecyl rhodamine B (Rh) for delivery of therapeutical enzymes*

One of the recent approaches to achieve the lysosomal targeting is based on modification of the liposomal membrane with a lysosomotropic ligand such as rhodamine B (Rh).[79] A large variety of lysosomotropic molecules have been identified the specifically target and accumulate in lysosomes.[80] Among them, Rh is of particular interest. Rh relates to a class of compounds with a high capacity to specifically accumulate in intracellular compartments with an acidic pH through a so-called "ion-trapping" mechanism (see Fig. 1).[81] Because of its lysosomotropic properties, Rh is routinely used for the visualization of lysosomes in live cells.[82,83]

In our proof-of-concept study,[79] we hypothesized that liposomes modified with Rh will significantly improve the accumulation of the liposomal

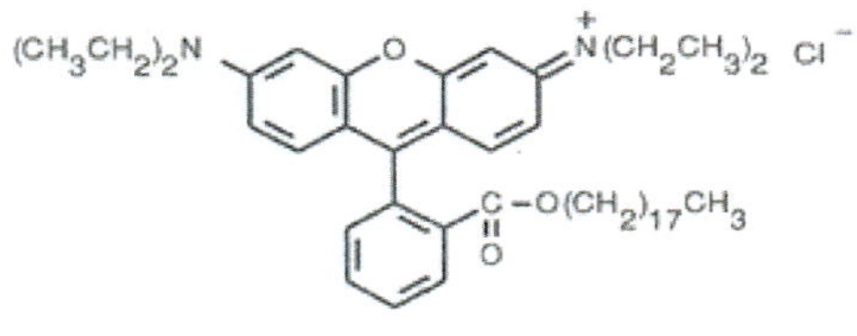

(a)

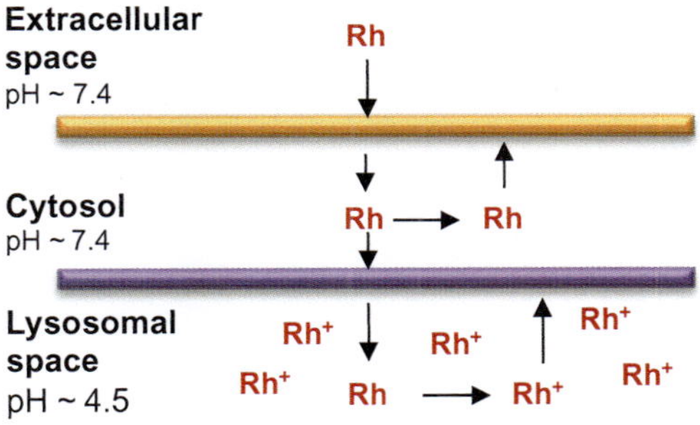

(b)

Fig. 1. Structure (a) and lysosomotropic properties (b) of Rh. At neutral pH, Rh is a membrane-permeable unionized base, while at the low lysosomal pH it forms an ionized state which is unable diffuse through the lipid membrane, therefore becomes trapped within the lysosomal compartment.

content in lysosomes due to the lysosomotropic properties of Rh. With this in mind, we constructed lysosomotropic liposomes (LLs) consisting of egg phosphatidylcholine (PC) and Chol (7:3 molar ratio) and modified with octadecyl-Rh (Fig. 2). Octadecyl-Rh is a commercially available hydrophobized derivative of Rh which has been used widely for monitoring membrane fusion[84–87] and for studying lysosomal metabolism.[88–90]

LLs and plain non-modified liposomes (PLs) were loaded with a model compound, FITC-dextran (FD), and their accumulation in lysosomes was investigated by confocal microscopy, flow cytometry and subcellular fractionation in HeLa cells.[79] We found that despite the same level of FITC-dextran delivered into the liposome-treated cells (as shown by FACS analysis), the confocal microscopy showed that after 4 h of the treatment, LLs were localized preferentially in lysosomes, unlike PLs that had a significantly decreased level of their co-localization with the lysosomal markers.[79]

By comparing the FITC fluorescence of intact lysosome-enriched fractions isolated from liposome-treated cells by subcellular fractionation, and at least 2-fold increased lysosomal delivery of FITC-dextran was clearly demonstrated in cells treated with LLs compared to the cells treated with PLs.[79]

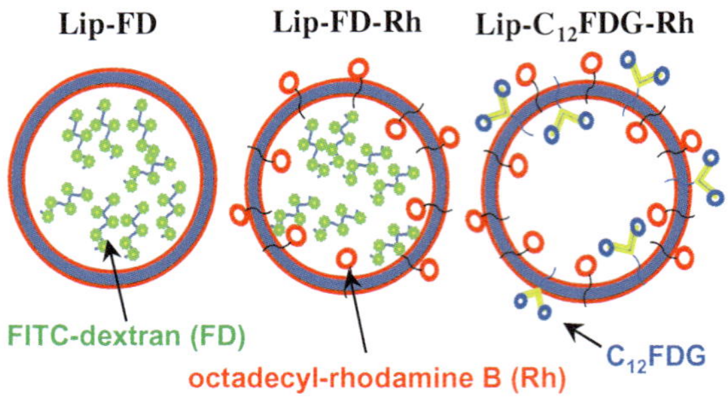

Fig. 2. Schematic structure of plain and LLs modifies with octadecyl-Rh.

Since methods such as cell staining for microscopic analysis and subcellular fractionation of cell lysates are quite disruptive, it was important to confirm the ability of LLs to target lysosomes in intact, live cells. We employed a new method that allows tracking the lysosomal accumulation of the internalized liposomes in intact, live cells by flow cytometry using liposomes loaded with a 12-carbon lipophilic variant of fluorescein di-β-D-galactopyranoside (C_{12}FDG)[91], a substrate for intralysosomal enzyme β-galactosidase (Fig. 2). We predicted that upon undergoing endocytosis (Fig. 3(a)), C_{12}FDG-loaded liposomes would deliver their contents into lysosomes, and the intralysosomal β-galactosidase hydrolyzes the non-fluorescent C_{12}FDG into the fluorescent C_{12}FITC, which would be retained inside lysosomes due to its lipophilic moiety. By a standard flow cytometry procedure, the lysosomal targeting can be thus measured by following the fluorescence intensity of live intact cells.[79]

The treatment of cells with different concentrations of C_{12}FDG-loaded plain liposomes (20 and 200 μg/ml) led to a dose-dependent increase in their FITC fluorescence relative to the control (untreated) cells (Fig. 3(b)). These data suggest that endocytosed liposomes actually deliver C_{12}FDG into lysosomes in amounts sufficient for detection by the flow cytometry. The cells treated with Rh-modified liposomes demonstrated significantly increased C_{12}FITC fluorescence compared to the cells treated with the plain liposomes (Fig. 3(b)). These findings on intact cells show clearly that the modification of the liposomal membrane with lysosomotropic Rh increases the lysosomal targeting compared to non-modified plain liposomes.

The intracellular trafficking and final destination of internalized nanocarriers can vary significantly depending on cell type and physicochemical properties of the nanocarriers.[10] For example, HUVEC and COS-7 cells have been shown to internalize liposomes via both, CvME and CME.[92] In contrast to CME, the caveolae-dependent mechanism is known to be finalized in

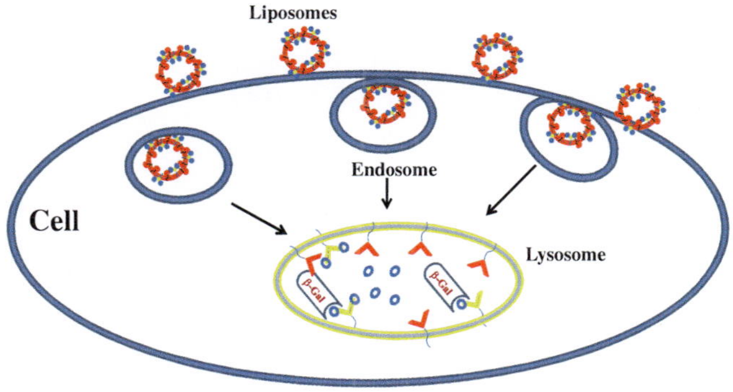

(a)

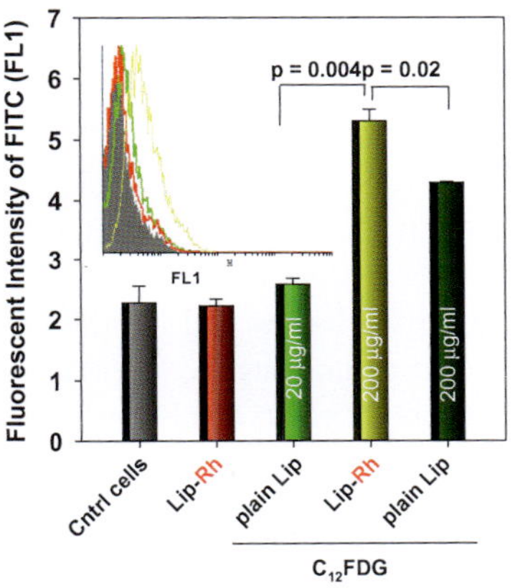

(b)

Fig. 3. Detection of lysosomal targeting using LLs loaded with $C_{12}FDG$ by flow cytometric analysis. (a) A scheme of lysosomal targeting by $C_{12}FDG$-loaded LLs. Intralysosomal β-galactosidase (β-Gal) hydrolyzes non-fluorescent $C_{12}FDG$ delivered by Rh-modified liposomes into the fluorescent $C_{12}FITC$, which accumulates inside lysosomes because of its lipophilic moiety. (b) Flow cytometry analysis of lysosomal targeting by LLs loaded with $C_{12}FDG$. HeLa cells were incubated with PLs (20 and 200 μg/ml) or LLs (200 μg/ml). After 4 h incubation with liposomes, the cells were washed with and additionally incubated for 20 h with liposome-free DMEM. The fluorescence intensity of FITC (channel FL1) was determined by the flow cytometry.[79]

caveosomes, intracellular vesicles with a neutral pH compartment that can be transported directly to smooth ER.[93] Since lysosomes are highly dynamic organelles that can fuse with different cellular compartments,[94] we assumed that the modification of liposomes with lysosomotropic ligands such as Rh could improve their lysosomal targeting due to a more effective fusion of the endocytosed liposomes with lysosomes and the avoidance of the trafficking pathway from endosomes to the GA or ER.

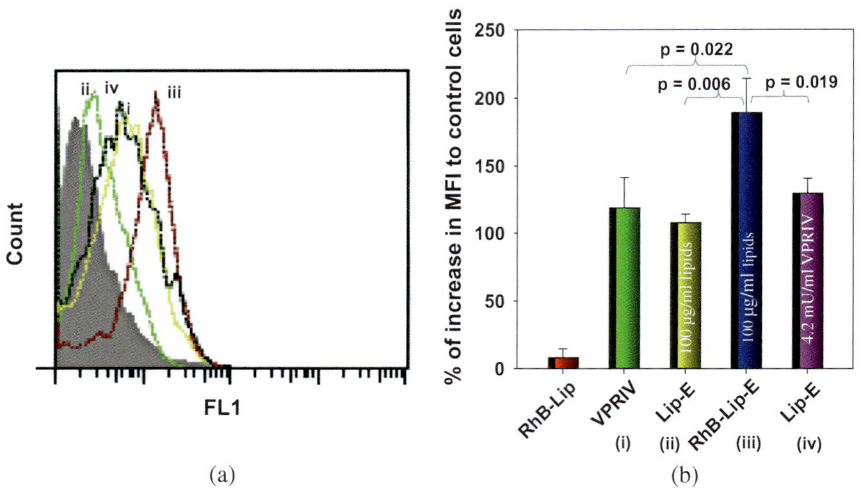

Fig. 4. Lysosomal delivery of VPRIV® by LLs. Gaucher's fibroblasts were treated with VPRIV–free RhB-modified Lip (RhB-Lip), VPRIV–loaded plain (Lip-E) or RhB-modified Lip (RhB-Lip-E), or with free VPRIV for 4 + 20 h. The level of the intralysosomal glucocerebrosidase was measured by flow cytometry (channel FL1) of the cells incubated with PFB-FDGlu. (a) Typical histogram analysis: (i) control (untreated) cells, (ii) Lip-E [100 μg/ml lipids], (iii) Lip-E [4.2 mU/ml], (iv) free VPRIV [4.2 mU/ml], (v) RhB-Lip-E [100 μg/ml lipids]. (b) A percentage increase in mean fluorescence intensity (MFI) normalized to control (untreated) cells.

This proof-of-concept study carried out in HeLa cells,[79] though conclusive, does not entirely apply to conditions of LSDs. Studies performed with LSD cell models were necessary to prove that LLs have the potential for the therapy of LSD. With this in mind, we demonstrated that Rh-modified LLs provided an increased delivery of the therapeutic enzyme velaglucerase alpha (VPRIV) into lysosomes of Gaucher's fibroblasts (Fig. 4) and monocyte-derived macrophages (MDMs) with inhibited activity of lysosomal β-glucocerebrosidase (GlcCerase), as a representative model of Gaucher's macrophages.[95]

In these studies,[79,95] we used non-PEGylated "naked" liposomes, since such liposomes are readily taken up by the liver upon *i.v.* administration, which is among the primary targets for the therapy of LSD.[66] However, taking in consideration that one of the main problems with the traditional ERT is the inability of the administrated enzymes to target bone marrow and brain tissues affected by the disease,[63] we developed PEGylated liposomes with the lysosomotropic agent, Rh, coupled to the distal tip of PEG chains (Fig. 5) for targeting organs other than the liver.[96]

The increased lysosomal targeting of PEGylated LLs relative to PLs was confirmed by confocal microscopy.[96] PEGylated LLs also provided an

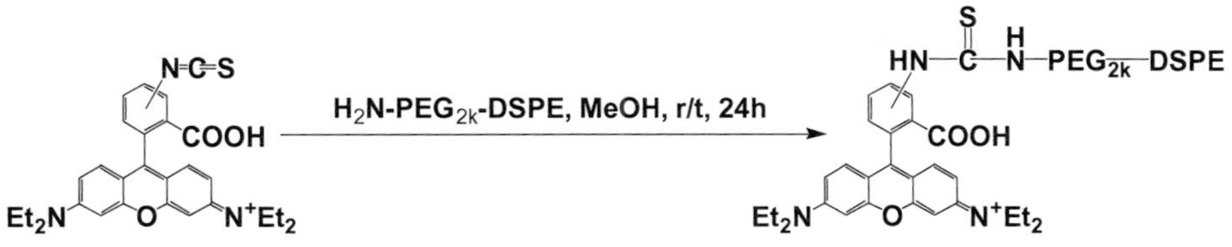

Fig. 5. Synthesis of 6-(3-(DSPE-PEG2k)-thioureido) rhodamine B.

enhancement of lysosomal delivery of their FITC-dextran load (up to 100%) compared to plain liposomes as was shown by the subcellular fractionation data.[96] To further confirm and compare the ability of PEGylated LLs to specifically target lysosomes, we evaluated their lysosomal targeting by flow cytometric analysis of live cells using PEGylated LLs loaded with C_{12}FDG. The lysosomal targeting was evaluated on a number of cultured tumor and normal cell lines including HeLa, B16(F10) melanoma, LLC, NIH/3T3 fibroblasts, and H9C2 myoblasts. For all tested cell lines, the cells incubated with C_{12}FDG-loaded LLs and modified with PEG–Rh conjugates demonstrated significantly increased lysosomal accumulation (up to 155%) compared to the cells treated with plain liposomes.[96]

Summarizing this part, the use of lysosome-targeted nanocarriers may significantly improve delivery of therapeutic enzymes to lysosomes and provide an enhanced therapeutic outcome for the treatment of LSDs. Still, there is an urgent need to evaluate the pharmacokinetic behavior and therapeutical efficiency of the LLs for delivery of therapeutic enzymes in animal models representing LSDs.

3.2. *Transferrin-modified ceramide-loaded liposomes for targeting lysosomal cell death pathway*

In another study,[97] we used a completely different approach for the lysosomal targeting by utilization of transferrin receptor (TfR)-mediated uptake. Tf is a serum glycoprotein responsible for the entry of iron into most cells via the TfR-dependent endocytotic pathway.[98] TfR-mediated uptake presents a very tempting way to provide cancer-targeted drug delivery due to the overexpression of TfR in cancer cells and its extremely high rates of intracellular internalization.[99,100] Previously, Tf-modified liposome (TL)-based nanocarriers have been used for delivery of a variety of anticancer drugs in different tumor models *in vivo*.[101–103]

The only likely limitation for lysosome targeting via the TfR-mediated uptake would involve the constitutively recycling of TfR to the cell membrane after its internalization and release of iron in sorting endosomes.[98] However, as shown previously, a significant portion of TL is trapped in endosome-lysosomal compartments without recycling with TfR.[18] Therefore, we considered that Tf-modification of liposomal nanocarriers could be an effective method for both cancer and lysosome-targeted delivery of anticancer compounds that initiate the LCDP.

As a promising candidate for triggering LMP, we chose ceramide, a sphingolipid signaling molecule, known to be hydrolyzed by the lysosomal enzyme acid ceramidase into sphingosine.[18] Intralysosomal accumulation of sphingosines can initiate LMP by a detergent-like fashion resulting in the cell apoptosis.[70,104–106] With this in mind, we constructed PEGylated TL loaded with short (C6)- or long (C16) N-acyl chain ceramides (Fig. 6).[97]

We hypothesized that modulation of the intralysosomal ceramide level by ceramide-loaded TL would initiate LMP followed by release of the lysosomal cathepsins into the cytoplasm and apoptosis of the cell (Fig. 7).

FACS analysis and fluorescent microscopy of liposome-treated HeLa cells showed that, in contrast to non-modified liposomes, TL were internalized at an increased rate specifically via the TfR-mediated endocytotic pathway with a preferential accumulation in the endosome-lysosomal compartment of the treated cells (Fig. 8(a)).

Ceramide-loaded Tf-liposomes revealed an increased cytotoxic and pro-apoptotic effect toward cancer cells *in vitro* compared to the non-modified liposomes.[97] Our results also indicate that intralysosomal accumulation of the ceramides delivered by TL initiated the permeabilization of the lysosomal membrane (Fig. 8(b)) leading to the release of the lysosomal cathepsins into the cytoplasm with further initiation of cancer cell apoptosis. A strong antitumor and pro-apoptotic (Fig. 8(c)) effect of ceramide-loaded Tf-liposomes was demonstrated also in an A2780 ovarian carcinoma xenograft mouse model.[97]

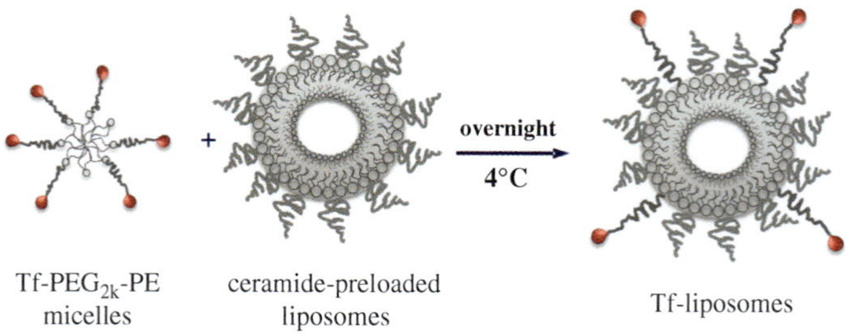

Fig. 6. Modification of the liposomal surface with Tf–PEG–PE conjugates using a post-insertion.

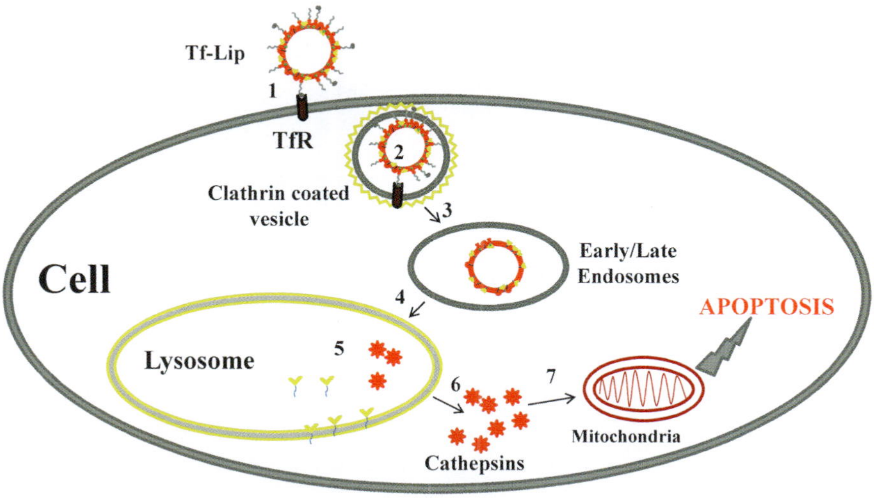

Fig. 7. Targeting of the lysosome cell death pathway by ceramide-loaded Tf-liposomes in cancer cells. (1) Interaction of Tf-liposomes with TfR receptor; (2) Internalization of Tf-liposomes with clathrin-coated vesicles; (3) Maturation of clathrin-coated vesicles into early endosomes (EEs); (4) EEs mature into late endosomes that fuse with lysosome, and thus deliver liposomal ceramides to the lysosomal compartment; (5) Conversion of ceramides into sphingosine by a lysosomal ceramidase; (6) Lysosomal accumulation of sphingosine leads to the lysosome-membrane permeabilization due to its detergent-like properties. (7) Induced LMP releases cathepsins into the cytoplasm with activation of the mitochondrial death pathway leading to cellular apoptosis.

In summary, Tf-modification of the ceramide-loaded liposomes can trigger the lysosome-dependent cell death pathway in cancer cells with the simultaneously targeting at the cellular and the subcellular levels.

3.3. *Other strategies for lysosomal-targeted nanocarriers*

TfR-dependent uptake is not the only possible endocytotic pathway for targeting the lysosomal compartment. Moreover, since the CME is known to be compromised in some LSD cell types,[107] a delivery system that can provide the lysosomal targeting via an alternative endocytotic pathway is highly relevant especially for non-phagocytic cells such as endothelium and muscle tissue.[108]

Intercellular adhesion molecule-1 (ICAM-1), a transmembrane glycoprotein constantly expressed in endothelial cells, is a very promising target molecule that has been recently used for endothelial targeting of different nanocarriers.[109,110] Although ICAM-1 is not an endocytotic receptor per se, nanocarriers coated with multimeric anti-ICAM-1 antibodies show very high

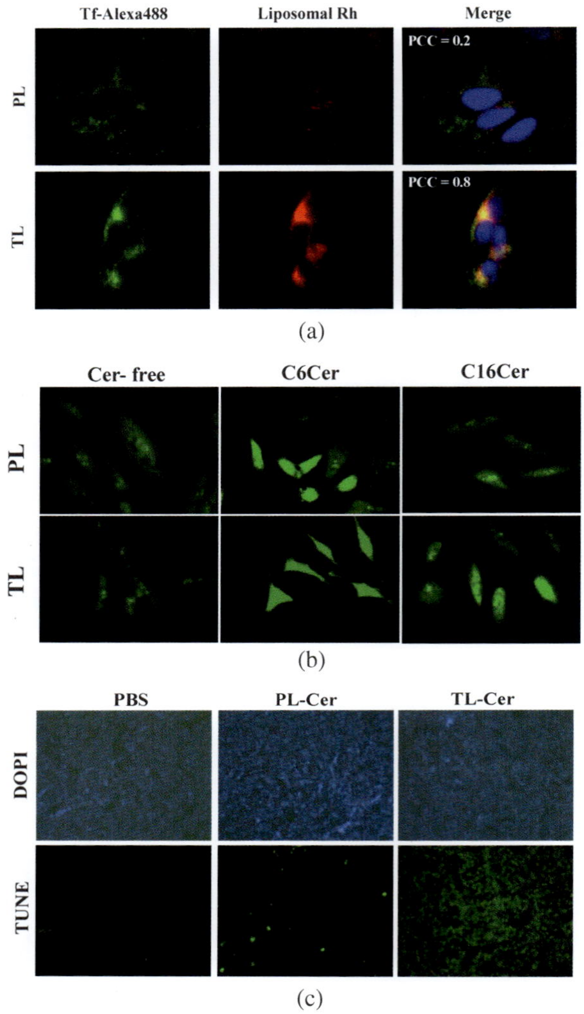

Fig. 8. Ceramide-loaded Tf-liposomes internalized specifically via the TfR-endocytotic pathway followed by LMP *in vitro* and apoptosis in cancer cells *in vivo*. (a) HeLa cells treated with fluorescently labeled PL or TL for 4 h were stained with Tf-Alexa488, as an endosomal marker. Co-localization (yellow) between liposomal (red) and endosomal (green) markers was evaluated by fluorescent microscopy and Pearson's correlation coefficient (PCC) was calculated by Image J. (b) HeLa cells were treated with PL or TL (loaded or non-loaded with C6-ceramide/C16-ceramide) for 24 h, and incubated with a lysosomotropic fluorophore, acridine orange (AO). The incubation of cells with AO leads to its accumulation in the intact lysosomes with the formation of an oligomeric protonated AO (AOH+) that has strong red fluorescence, while the translocation of the lysosomal contents to the cytosol upon the permeabilization of the lysosomal membrane results in formation of the monomeric deprotonated form of AO with an intense green fluorescence. LMP was estimated by an increase in AO-fluorescence in the green channel by fluorescence microscopy. (c) Mice bearing an A2780 tumor were treated with PL-ceramide, TL-ceramide or PBS. The 8-μm frozen sections were extracted, subjected to TUNEL assay and examined by fluorescence microscopy. The upper line shows the sections stained with DAPI and the lower line indicates the TUNEL staining.

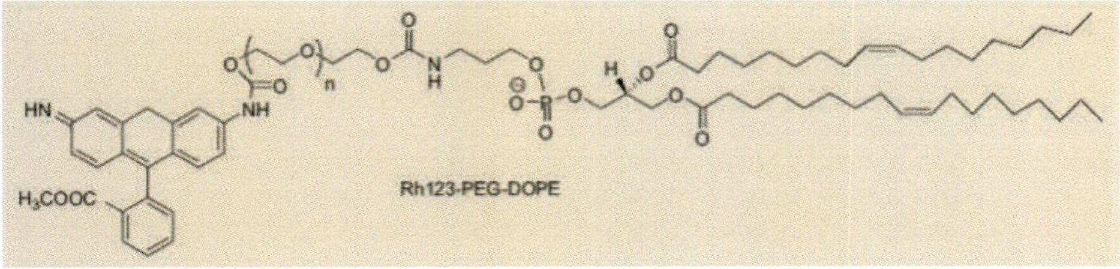

Fig. 9. Structure of mitochondriotropic Rh-123 conjugated to a PEG-PE polymer.

rates of uptake in endothelial cells via an unique clathrin- or caveolar-independent pathway.[111,112] ICAM-1-mediated endocytosis has an ability to internalize submicron-sized particles with an increased lysosomal trafficking and accumulation.[111,113]

Anti-ICAM-1 nanocarriers were an effective lysosome-targeted delivery system for different therapeutical enzymes including acid α-glucosidase,[114] alpha-galactosidase,[115] and acid sphingomyelinase.[108]

4. Mitochondria-targeted Nanocarriers

Mitochondria represent another important destination for targeted delivery of many drugs and DNA. Mitochondrial dysfunction contributes to a variety of human disorders, ranging from neurodegenerative diseases, obesity, diabetes, ischemia-reperfusion injury and cancer.[116] Mitochondria also play a key role in the complex apoptosis mechanism and thus a primary target for anticancer therapy.[117] Many clinically approved anticancer drugs, including paclitaxel, directly act on mitochondria to trigger apoptosis.[118,119] To improve the therapeutic effect of such mitochondria-aimed drugs, ways to specifically deliver those drugs to mitochondria should be sought. Different mitochondrial targeting strategies were discussed recently in a number of reviews.[3,120,121]

One of the key approaches for the use of mitochondria-targeted nanocarriers utilizes the high mitochondrial membrane potential. There exists a variety of amphiphile compounds with delocalized cationic charge which are able to accumulate specifically in the mitochondria in response to its negative membrane potential.[122] Rhodamine 123 (Rh-123), a fluorescent stain for mitochondria in living cells, is the best known representative of this group.[123] Other well-known mitochondriotropic compounds include triphenylphosphonium (TPP)[124] and dequalinium (DQA).[122]

One of the first attempts to create a mitochondriotropic nanocarrier based on DQA complexes (DQAsomes) was reported by Weissig and colleagues.[125–128] DQAsomes, cationic lipid-based vesicles which form stable complexes with plasmid DNA (pDNA), were shown to release the pDNA specifically upon their interaction with the mitochondrial surface.[127] Apparently, DQAsomes escape from endosomes, approach the mitochondria and release the encapsulated pDNA as a result of their interaction with the mitochondrial membrane.[128] Additionally, to the mitochondria-specific delivery of pDNA, DQAsomes have been shown to facilitate the pro-apoptotic activity of paclitaxel.[129]

A mitochondria leader peptide (MLP) has been used in an attempt to target PEI–DNA complexes to mitochondria.[130] The MLP originates from the mitochondria matrix enzyme ornithine transcarbamylas, known to rapidly cross through the inner mitochondrial membrane without any energy requirement.[131] MLP covalently conjugated to DNA complex delivered the DNA to mitochondria via a protein import pathway.[132] PEI–DNA complexes modified with MLP localized at the mitochondrial sites in a cell-free assay and in living cells.[130]

Modification of the liposomal membrane with stearyl triphenylphosphonium (STPP) represents another potent strategy to improve the selective accumulation of nanocarriers in mitochondria.[133] Such mitochondriotropic liposomes loaded with ceramide, to permeablize the mitochondrial membrane via formation of ceramide channels in the mitochondrial membrane,[134–137] elicited a strong apoptotic response *in vivo* in 4T1 tumor-bearing mice.[138] In order to overcome non-specific cytotoxicity of mitochondriotropic STPP-liposomes, a new mitochondriotropic conjugate with a TPP group attached to the terminal end of polyethyleneglycol–phosphatidylethanolamine (PEG–PE) was designed.[139] Liposomes modified with TPP–PEG–PE showed excellent mitochondrial targeting properties with sufficiently decreased non-specific toxicity. These TPP-PEG-PE-modified liposomes strongly facilitated the anticancer efficacy of the liposome-encapsulated paclitaxel in cell culture *in vitro* and *in vivo* in a 4T1 carcinoma xenograft mouse model.[139]

Recently, a novel mitochondria-targeted liposome-based drug delivery system with the liposomal surface modified by (Rh-123)–PEG–PE polymers (Fig. 8) was been introduced.[140] Confocal microscopy as well as the isolation of mitochondrial fractions showed a high degree of accumulation of the mitochondria-targeted liposomes in the mitochondria.[140] To demonstrate that specific delivery of a drug to mitochondria can significantly enhance the drug's action, the mitotic inhibitor, paclitaxel was used. Paclitaxel-loaded Rh-123-liposomes produced significantly higher cytotoxicity toward cancer cells *in vitro* than free paclitaxel or paclitaxel-loaded non-modified liposomes.

It was clearly that shown that Rh-123-modified mitochondriotropic liposomes can target mitochondria efficiently and facilitate the delivery of a therapeutic payload to mitochondria.[140]

A highly sophisticated liposome-based nanocarrier system (MITO-Porter) for delivery of drugs and DNA to mitochondria via a membrane fusion mechanism was recently introduced.[121,141–144] In this approach, octaarginine (R8) moieties were attached to the liposomal surface to provide liposomal internalization into the cell via macropinocytosis. As shown previously, the high density of R8 at the liposomal surface allows the liposomal transfer from macropinosomes to the cytosol with the liposomal structure kept intact.[145] The authors hypothesized that once inside the cell followed by escape from macropinosomes via a membrane fusion, the liposomes will bind to the mitochondria via electrostatic interaction, which can promote fusion between the liposome and mitochondria.[141] The liposomal compositions useful for their ability to fuse with the mitochondria membrane have been evaluated with isolated rat liver mitochondria. Liposomes based on DOPE, sphingomyelin and phosphatidic acid provide the highest fusion activity with mitochondria.[141] The efficient delivery of liposome-encapsulated GFP to mitochondria by the MITO-Porter system was demonstrated by confocal microscopy and by FRET analysis using spectral imaging fluorescence microscopy in live HeLa cells.[141] Additionally, the ability of a MITO-Porter system loaded with propidium iodite to target the intra-mitochondrial DNA was observed by confocal microscopy in live cells.[144] The effective mitochondrial delivery of the bioactive cargo, DNase I protein, by MITO-Porter system was also confirmed.[143]

Thus, a number of mitochondriotropic nanocarriers has been successfully developed for targeted delivery of drugs and DNA to mitochondria and have a potential for clinical applications.

5. Nucleus-targeted Nanocarriers

Efficient nuclear targeting is a crucial step for successful gene delivery. Nuclear translocation of pDNA is relatively inefficient because of the small size of nuclear pores (approximately 25 nm) that are a natural barrier for large molecules. However, DNA can be imported into the nucleus from the cytoplasm using appropriate target signals such as histones, transcription factors, viral proteins and others bearing Nuclear Localization Signal (NLS) peptides that have the ability to be transported into the nucleus by a receptor-mediated process.[146] Different approaches utilizing this strategy for nucleus-targeted delivery of DNA have been described including the association of NLS peptides with lipoplexes[147] or its conjugation with polycations.[148] Lipoplexes, complexes between cationic lipids such as DOTAP/DOPE (Lipofectin®) and DNA, facilitate efficiently DNA transfection because of its slight net positive charge providing enhanced interaction of the complex with negatively charged cells.[149] Despite the endocytosis-mediated uptake of a lipoplex, DNA does not end in the lysosomes but is released in the cytoplasm due to the destabilization of the endosomal membrane provoked by fusogenic DOPE.[150]

The transfection efficiency of the cationic liposomes can be significantly improved by association with NLS peptides such as S4(13)-PV, a combination of a cell penetrating sequence Dermaseptin S4 peptide along with the nuclear localization signal present in the SV40 large T antigen.[151] Cationic liposomes associated with S4(13)–PV/DNA complexes were significantly more efficient in mediating transfection than the corresponding S4(13)-PV/DNA or cationic liposome/DNA complexes.[151]

The Multifunctional Envelope-type Nano Device (MEND), in which DNA is primarily packed with polycations and additionally covered with a fusogenic lipid envelope, has been developed recently to provide a nonviral gene delivery system.[152,153] Such nanocarriers modified with the membrane-permeable peptide R8 provide transfection activity in dividing cells (NIH3T3 and HeLa) as efficiently as an adenovirus without any cytotoxic effect.[154] The R8-modification enhances cellular uptake of the MEND delivery system via macropinocytosis, avoiding the lysosomal degradation, while the envelope consisting of the fusogenic DOPE/CHEMS provides the endosomal escape and thus improves the nuclear delivery of the encapsulated DNA.[70,155]

Despite the good results in dividing cells, the R8-MEND system failed to show an effective transfection in non-dividing cells such as bone marrow-derived dendritic cells (BMDC).[156] A double-membranous nonviral gene delivery system modified with SV-40 T antigen-derived NLS has been designed to promote the nuclear-targeting of the nanocarrier-encapsulated DNA in non-dividing cells.[156] The NLS-modified double membrane MEND (DMEND) was constructed from pH-sensitive fusogenic lipids (DOPE/CHEMS) and stearylated SV-40 T antigen, which was incorporated in the outer leaflet of the liposomal membrane. Condensed DNA particles were incorporated in the liposomal core by their fusion with the DMEND. Such nuclear-targeted NLS-DMEND nanocarrier strongly enhanced the transfection activity in non-dividing dendritic cells.[156] The authors suggested that NLS-DMEND directly entered to the cytoplasm due to fusion with the plasma membrane following by a nuclear transfer of the liposome-encapsulated pDNA provided by the NLS peptides.[156]

The development of NLS-liposome for nuclear targeting has included localization of NLS on the liposomal PEG spacer. Thus, NLS-PEG-liposomes were shown to have a higher affinity binding with the isolated nucleus relative to NLS-liposomes.[157] The R8-MEND system, additionally modified with the pH-sensitive fusogenic peptide GALA, was successfully applied for liver gene delivery *in vivo*.[158]

Using the idea of R8-modification for nuclear-targeted drug delivery, we have recently investigated the ability of PEGylated fusogenic liposomes modified with R8–PEG–PE conjugates to improve the nuclear delivery of an anticancer drug, bleomycin.[159] Bleomycin, a glycopeptide antibiotic widely used in chemotherapy against lymphomas, squamous-cell carcinomas and germ-cell tumors,[160] is a good example of an anticancer drug with poor cellular uptake. Because of its relatively large size (1.5 kDa) and high hydrophilicity, bleomycin cannot diffuse efficiently across the cell membrane, which results in its low intracellular concentration.[161] Moreover, since bleomycin acts by oxidative DNA damage, it should also be targeted to nuclei.

With this in mind, we applied R8-modified PEGylated fusogenic liposomes in an attempt to provide an effective intracellular and nuclear delivery of liposome-encapsulated bleomycin. The surface of bleomycin-loaded liposomes composed of DOPE/CHEMS/DOPE–PEG_{1k} (80/10/7.5 mol%) was modified with R8-DOPE–PEG_{2k} polymers (2.5 mol.%) and their physicochemical properties, cellular uptake, intracellular trafficking and antitumor efficiency were compared with R8-modified non-fusogenic phosphatidylcholine/cholesterol (PC)-based liposomes as well as with non-modified "plain" PC- and DOPE-liposomes.[159] We hypothesized that the CPP R8 would facilitate the liposomal internalization, while the fusogenic DOPE/CHEMS lipids would initiate the endosomal escape of the liposomal bleomycin into the cytosol followed by its eventual transport to the cell nuclei.

As shown by FACS analysis, the R8-modification of both fusogenic DOPE-based and non-fusogenic PC-based liposomes dramatically facilitated their cellular internalization compared to non-modified liposomes (up to 50-fold). Interestingly, in contrast to R8-modified fusogenic compositions, bleomycin in non-fusogenic PC-liposomes did not have any cytotoxic effect toward cancer cells *in vitro*. Therefore, the combination of R8-modification and fusogenic lipids was essential to provide the bleomycin escape from the endosomal compartment. This fact was proved by an experiment, in which we marked the lipid membrane and inner hydrophilic compartments of R8–DOPE liposomes with a lipophilic PE–RhB and hydrophilic FITC-dextran respectively (Fig. 10). Co-localization of both liposomal markers was traced by confocal microscopy in the presence or absence of chloroquine, which is

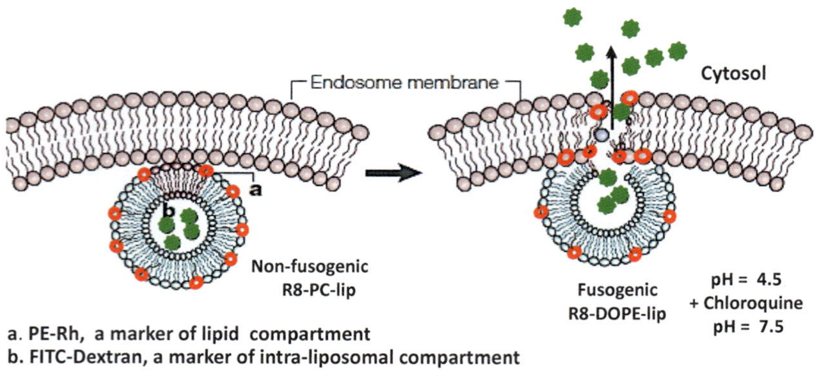

(a)

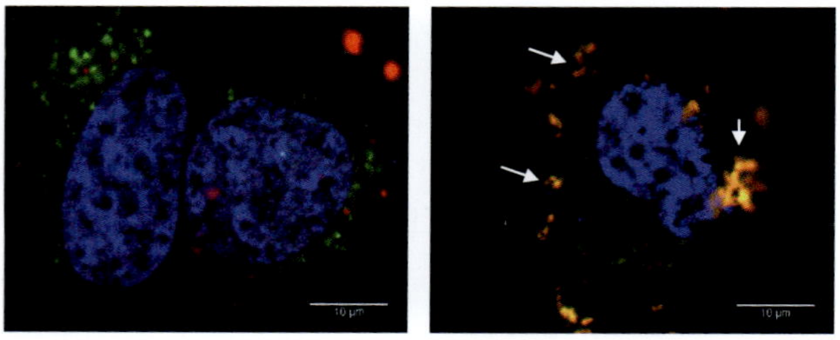

(b) (c)

Fig. 10. Endosomal release of the liposomal loads to the cytoplasm provided by the fusogenic lipids. (a) Schematic presentation for the pH-dependent disruption of the endosomal membrane by fusogenic lipids leading to release of the liposomal cargo to the cytoplasm. HeLa cells were treated with FITC-Dextran-loaded R8–DOPE liposomes in a medium without (b) or in the presence of chloroquine (c) for 4 h. The cells were washed and incubated in liposome/chloroquine-free medium for additional 20 hrs. Co-localization (yellow) of liposomal PE-Rh (red) and FITC-Dextran$_{4kD}$ (green) was evaluated by confocal microscopy. The nuclei were visualized by staining with Hoechst 33342 (blue). The treatment of the cells with R8–DOPE-liposomes in the presence of chloroquine prevented the release of FITC-Dextran from endosomes resulting in its higher co-localization (yellow) with the lipophilic marker PE-Rh.

known to prevent endosomal acidification.[162] It was expected that disruption of the endosomal membrane by the fusogenic lipids would provoke the relocation of FITC-dextran to the cytosol and result in its reduced co-localization with the lipophilic marker PE–RhB, while co-treatment with chloroquine inhibits the pH-dependent membrane fusion. Indeed, we observed that the presence of chloroquine prevented the endosomal escape of FITC-dextran proving that the release of the liposomal loads from R8–DOPE liposomes into the cytoplasm was regulated by the pH-sensitive fusion mechanism (Fig. 10).

Co-localization of fluorescently labeled R8–DOPE liposomes with macropinosomal and lysosomal markers showed that R8–DOPE-liposomes were internalized via macropinocytosis and did not end up in the lysosomal compartment in contrast to non-fusogenic R8–PC liposomes.[159] Moreover, confocal microscopy showed that after internalization R8–DOPE-liposomes formed enlarged structures which were associated with the cell nuclei (Fig. 11). We assumed that the combination of positively charged R8 and fusogenic lipids provoked the fusion/association events of R8–DOPE-containing macropinosomes with the nuclear membrane followed by endosomal escape of liposome-encapsulated bleomycin in the pre-nuclear area to facilitate the eventual nuclear localization of the drug. This conclusion was also supported by strong antitumor and pro-apoptotic effect of bleomycin in R8–DOPE liposomes *in vivo* in a 4T1 xenograft mouse model.[159]

A numerous nuclear targeted liposome-based delivery systems have been developed lately for nuclear transport of DNA and anticancer drugs. However, there remains an urgent need to sharply increase the efficacy of such nuclear targeted nanocarriers, especially for DNA delivery to provide higher transfection activity.

6. Other Organelle-targeted Nanocarriers

Along with others cellular organelles, the ER, GA and peroxisomes are important parts of the cell machinery responsible for the biosynthesis of proteins, lipids, secretory and signaling pathways.[163–165] A wide range of diseases are associated with the dysfunction of these organelles, which also makes them an important target for drug delivery.[163] Unfortunately, very few attempts, if any, have been made so far to create nanocarriers targeted specifically to ER, GA or peroxisomes.

CvME represents an endocytotic pathway by which the targeted transport of drugs and drug-encapsulated nanocarriers to the ER and/or GA can be directed. Along with CME, CvME is one of the major endocytotic pathways,

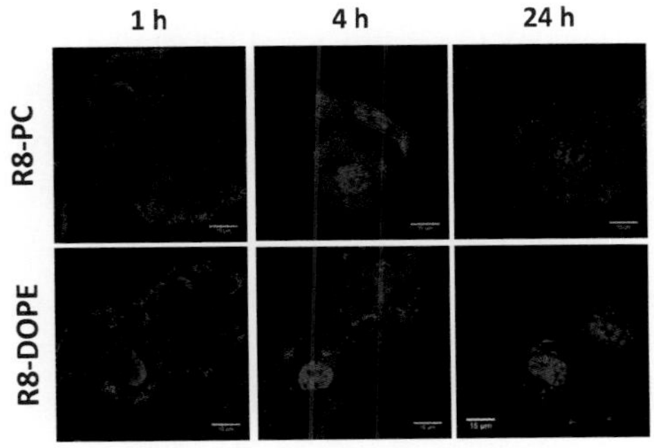

Fig. 11. Intracellular localization of R8-modified liposomes by confocal microscopy.

which is especially abundant in endothelium, smooth muscle cells and fibroblasts.[166] Caveolae are 50–100 nm in diameter uncoated flask-shaped invaginations, which are formed in cholesterol and sphingolipid-rich microdomains of the plasma membrane that can be specified by the presence of a cholesterol-binding protein, caveolin.[5]

During endocytosis, caveolae have been shown to form caveolin-positive endosomes (caveosomes) characterized by a neutral pH and the absence of early and recycling endosomal markers.[167,168] Caveosomes were first described as intermediate endosomal vesicles involved in trafficking of SV40 from the cell membrane to the ER, avoiding the endosome-lysosomal compartment.[93] Caveosomes have also been shown to be responsible for the retrograde transport of sphingolipids and GPI-linked proteins from the cell surface to the GA.[169] Caveosomes contain the molecular machinery that provides their docking and fusion with other organelles[170] including transport routes to ER/GA.[171] Thus, CvME is employed by many bacteria toxins to reach the ER and GA in the host cells.[165] Along with pathogenic toxins, other membrane ligands are known to be internalized via CvME; these includes folic acid,[172] serum albumin,[173] autocrine motility factor (AMF)[174] and others.[8] Thus, the use of nanocarriers able to exploit the CvME can be very useful for the targeted drug delivery to the ER and GA.

An interesting attempt to create an ER-targeted delivery system was recently presented by Pollock and colleagues.[175] The authors developed liposomes capable of specific targeting of the ER and ER-associated membranes. The ER-targeted liposomes were constructed of phosphatidylethanolamine (PE), phosphatidylcholine (PC), phosphatidylinositol (PI) and phosphatidylserine (PS) at molar ratio of 1.5:1.5:1:1 (PE:PC:PI:PS), which closely resembles the composition of ER membrane. The ER-liposomes were internalized by scavenger and low-density lipoprotein receptors, then transported via CvME to the ER as an intact structure to release their encapsulated cargo, followed by fusion with the ER membrane at ~20 min post-incubation.[175] Interestingly, PEGylated ER-liposomes (3 mol.% of PEG–PE) were also capable of trafficking to the ER membrane. It was proposed that the use of ER-targeted liposomes would be beneficial for the antiviral activity of iminosugars, which are known to inhibit the ER resident enzyme α-glucosidases responsible for glycoprotein folding. Indeed, the authors demonstrated that ER-liposomes strongly enhanced the antiviral activity of both water-soluble and lipophilic analogues of iminosugars, as was shown with HIV-1 infected primary peripheral blood mononuclear cells.[175]

In other research, Mustata *et al.*[176] proposed that DOPE/CHEMS-based fusogenic liposomes would be able to deliver their cargo to the

ER via a microtubule-based non-caveosome endocytotic pathway avoiding cytoplasmic exposure. Using live microscopy, they showed that this cholesterol-dependent endocytotic pathway provided the ER targeting of DOPE-based fusogenic liposomes in HeLa and HUH7 cancer cells, and non-transformed MDBK cells. Although these results conflicts with the previous findings,[159,177] which showed that DOPE/CHEMS-based liposomes fuse with the endosomal membrane at acidic pH followed by release their loads to cytoplasm, it implies that intracellular trafficking of fusogenic liposomes can be more complicated than initially assumed.

Another indication that liposome-based systems are capable of ER targeting can be found in the work of Un and coauthors,[178] who investigated uptake and intracellular trafficking of conventional liposomes consisting of phospholipids (dioleoylphosphatidylcholine (DOPC)) and cholesterol (Chol). They found that DOPC/Chol-based liposomes were internalized via CME followed by DOPC localization in the ER and GA, whereas Chol was localized only in the ER.[178]

This data suggest that, by regulation of the lipid composition, the delivery of liposomal nanocarriers (as whole structures or their separate components) to ER/GA can be achieved. This provides a promising platform for development of ER/GA-targeted drug delivery systems.

PEG–PE-based micelles are another example of a lipid-based delivery system for targeting the ER. It was demonstrated recently that a PEG–PE copolymer can be transferred from micelles to the cell membrane and then transported to the ER/GA compartment via a nonspecific endocytotic pathway.[179] More interestingly, the authors found that over-accumulation of PEG–PE polymers in the ER can initiate ER stress, which leads to activation of pro-apoptotic unfolded protein response (UPR) signaling in cancer, but not normal cells. These interesting data may open new opportunities for amplifying toxicity of anticancer drugs encapsulated in ER-targeted lipid-based nanocarriers by disturbing ER lipid homeostasis.[180]

Interestingly, in an attempt to find out how polyplex complexes containing histone tail peptides (H3) and PEI are trafficked to the cell nuclei, it was discovered that H3/PEI–pDNA polyplexes utilize CvME and retrograde trafficking through the GA to ER pathways preceded the nuclear delivery.[181] These findings not only provide new evidence for the role of Golgi-ER pathways in nonviral gene delivery, but also indicate the capability of PEI-based systems to target ER and GA.

An ability to accumulate, at least partially, in the ER/GA compartment was also demonstrated for polylactic co-glycolic acid (PLGA)-based nanoparticles.[182] The intracellular trafficking of PLGA nanoparticles was investigated by immunofluorescence techniques and confocal microscopy in three different epithelial cell lines (***HBE***, Caco-2 and renal OK). PLGA-nanoparticles were capable of escape from the endosome-lysosomal compartment and trafficked in some instances to the GA and ER.[182]

Unfortunately, targeting of nanocarriers to peroxisomes is still an unresolved problem. Peroxisomes (microbodies) are ubiquitous multifunctional organelles responsible for oxidation of specific fatty acids, biosynthesis of ether phospholipids and metabolism of reactive oxygen species.[164] A number of severe human disorders are associated with abnormalities in biogenesis or trafficking of certain peroxisomal enzymes.[183] Therefore, a therapeutical treatment of peroxisomal disorders could strongly benefit from a delivery system specifically targeting peroxisomes. Peroxisomes use a unique translocation mechanism for intraperoxisomal delivery of fully folded, oligomeric, and co-factor-bound proteins by soluble shuttling receptors that recognize a peroxisomal targeting signal (PTS).[184] Theoretically, peroxisomal targeting can be achieved by modification of nanocarriers with a PTS containing serine-lysine-leucine motifs (SKL). At least, this approach was successfully proven for intraperoxisomal delivery of SKL-tagged therapeutical enzymes including catalase[185] and pyruvate aminotransferase.[186]

7. Conclusions

Development of organelle-specific nanocarriers is only in its early stages and remains a considerable challenge for targeted drug delivery. Nevertheless, the latest attempts to create organelle-targeted delivery systems demonstrate that by regulating composition, size, charge and presence of targeting ligands on the nanocarrier's surface, the desirable site-specific delivery of drugs to cytoplasm, lysosomes, mitochondria and nucleus can be achieved. Still, many questions especially related to mechanisms of intracellular trafficking of organelle-targeted nanocarriers remain to be answered. Regrettably, many findings on organelle-specific nanocarriers lack a sufficient explanation of the mechanism of targeting, frequently relating only on results of microscopic experiments and the therapeutical improvement of the nanocarrier-encapsulated drugs. Further progress in this area depends on: (i) better understanding of involvement of cellular endocytotic pathways for the internalization of organelle-specific nanocarriers, (ii) exploration of intracellular trafficking mechanisms involved in the transport of nanocarriers to a desired subcellular compartment and (iii) introduction of new non-invasive analytical methods to measure organelle targeting in living cells.

Some of the other questions that have to be addressed include the following. Is it possible to combine in one nanocarrier the ability to target simultaneously the tissue of interest, improve cellular uptake and provide delivery to specific subcellular compartment? Does an organelle-targeted nanocarrier retain its structure intact through its internalization/trafficking? How does a nanocarrier that is internalized by an endocytotic pathway affect the traffic/fusion of the endosomal vesicles to the targeting organelles? If the nanocarrier's integrity is compromised during the intracellular trafficking, what is the impact of released drug on the observing therapeutical effect? Since most of organelle-targeted systems are characterized by increased cellular uptake, how to differentiate the impact of their improved internalization and organelle localization on the therapeutical effect of the nanocarrier-encapsulated drug.

Despite all the challenges and unresolved tasks, organelle-targeted pharmaceutical nanocarriers should make a very significant impact in the future design of nanomedicines.

References

1. G. G. D'Souza and V. Weissig, Subcellular targeting: a new frontier for drug-loaded pharmaceutical nanocarriers and the concept of the magic bullet, *Expert Opin. Drug Deliv.* **6**(11), 1135–1148 (2009). Epub 2009/08/28.
2. M. Mossalam, A. S. Dixon and C. S. Lim, Controlling subcellular delivery to optimize therapeutic effect, *Ther. Deliv.* **1**(1), 169–193 (2010). Epub 2010/11/30.
3. V. P. Torchilin, Recent approaches to intracellular delivery of drugs and DNA and organelle targeting, *Annu Rev Biomed Eng.* **8**, 343–375 (2006). Epub 2006/07/13.
4. V. P. Torchilin, Recent advances with liposomes as pharmaceutical carriers, *Nat. Rev. Drug Discov.* **4**(2), 145–160 (2005). Epub 2005/02/03.
5. G. J. Doherty and H. T. McMahon, Mechanisms of endocytosis, *Annu. Rev. Biochem.* **78**, 857–902 (2009). Epub 2009/03/26.
6. S. A. Mousavi, L. Malerod, T. Berg and R. Kjeken, Clathrin-dependent endocytosis, *Biochem. J.* **377**(Pt 1), 1–16 (2004). Epub 2003/09/25.
7. M. M. Fretz, E. Mastrobattista, G. A. Koning, W. Jiskoot and G. Storm, Strategies for cytosolic delivery of liposomal macromolecules, *Int J Pharm.* **298**(2), 305–309 (2005). Epub 2005/06/09.
8. A. L. Kiss and E. Botos, Endocytosis via caveolae: alternative pathway with distinct cellular compartments to avoid lysosomal degradation? *J. Cell Mol. Med.* **13**(7), 1228–1237 (2009). Epub 2009/04/23.
9. J. Mercer and A. Helenius, Virus entry by macropinocytosis, *Nat Cell Biol.* **11**(5), 510–520 (2009). Epub 2009/05/01.
10. H. Hillaireau and P. Couvreur, Nanocarriers' entry into the cell: relevance to drug delivery, *Cell Mol. Life Sci.* **66**(17), 2873–2896 (2009). Epub 2009/06/06.
11. K. G. Chen and B. I. Sikic, Molecular pathways: regulation and therapeutic implications of multidrug resistance, *Clin Cancer Res.* **18**(7), 1863–1869 (2012). Epub 2012/02/22.

12. J. M. White, Viral and cellular membrane fusion proteins, *Annual Rev. Physiol.* **52**, 675–697 (1990). Epub 1990/01/01.
13. W. Li, F. Nicol and F. C. Szoka Jr., GALA: a designed synthetic pH-responsive amphipathic peptide with applications in drug and gene delivery, *Adv. Drug Deliv. Rev.* **56**(7), 967–985 (2004). Epub 2004/04/07.
14. M. E. Martin and K. G. Rice, Peptide-guided gene delivery, *The AAPS J.* **9**(1), E18–29 (2007). Epub 2007/04/06.
15. N. K. Subbarao, R. A. Parente, F. C. Szoka Jr., L. Nadasdi and K. Pongracz, pH-dependent bilayer destabilization by an amphipathic peptide, *Biochemistry* **26**(11), 2964–2972 (1987). Epub 1987/06/02.
16. T. Wang, S. Yang, V. A. Petrenko and V. P. Torchilin, Cytoplasmic delivery of liposomes into MCF-7 breast cancer cells mediated by cell-specific phage fusion coat protein, *Mol Pharm.* **7**(4), 1149–1158 (2010). Epub 2010/05/05.
17. T. Wang, G. G. D'Souza, D. Bedi, O. A. Fagbohun, L. P. Potturi, B. Papahadjopoulos-Sternberg, *et al.*, Enhanced binding and killing of target tumor cells by drug-loaded liposomes modified with tumor-specific phage fusion coat protein, *Nanomedicine (Lond).* **5**(4), 563–574 (2010). Epub 2010/06/10.
18. T. Kakudo, S. Chaki, S. Futaki, I. Nakase, K. Akaji, T. Kawakami, *et al.*, Transferrin-modified liposomes equipped with a pH-sensitive fusogenic peptide: an artificial viral-like delivery system, *Biochemistry* **43**(19), 5618–5628 (2004). Epub 2004/05/12.
19. K. Sasaki, K. Kogure, S. Chaki, Y. Nakamura, R. Moriguchi, H. Hamada, *et al.*, An artificial virus-like nano carrier system: enhanced endosomal escape of nanoparticles via synergistic action of pH-sensitive fusogenic peptide derivatives, *Anal. Bioanal. Chem.* **391**(8), 2717–2727 (2008). Epub 2008/03/21.
20. S. Kobayashi, I. Nakase, N. Kawabata, H. H. Yu, S. Pujals, M. Imanishi, *et al.*, Cytosolic targeting of macromolecules using a pH-dependent fusogenic peptide in combination with cationic liposomes, *Bioconjug. Chem.* **20**(5), 953–959 (2009). Epub 2009/04/25.
21. H. Karanth and R. S. Murthy, pH-sensitive liposomes — principle and application in cancer therapy, *J. Pharm. Pharmacol.* **59**(4), 469–483 (2007). Epub 2007/04/14.
22. D. C. Drummond, M. Zignani and J. Leroux, Current status of pH-sensitive liposomes in drug delivery, *Prog. Lipid Res.* **39**(5), 409–460 (2000). Epub 2000/11/18.
23. S. Simoes, J. N. Moreira, C. Fonseca, N. Duzgunes and M. C. de Lima, On the formulation of pH-sensitive liposomes with long circulation times, *Adv. Drug Deliv. Rev.* **56**(7), 947–965 (2004). Epub 2004/04/07.
24. J. J. Sudimack, W. Guo, W. Tjarks and R. J. Lee, A novel pH-sensitive liposome formulation containing oleyl alcohol, *Biochim. Biophys. Acta* **1564**(1), 31–37 (2002). Epub 2002/07/09.
25. A. Asokan and M. J. Cho, Cytosolic delivery of macromolecules. II. Mechanistic studies with pH-sensitive morpholine lipids, *Biochim. Biophys. Acta* **1611**(1–2), 151–160 (2003). Epub 2003/03/28.
26. M. Sanchez, F. J. Aranda, J. A. Teruel and A. Ortiz, New pH-sensitive liposomes containing phosphatidylethanolamine and a bacterial dirhamnolipid, *Chem. Phys. Lipids.* **164**(1), 16–23 (2011). Epub 2010/10/12.
27. N. Sakaguchi, C. Kojima, A. Harada and K. Kono, Preparation of pH-sensitive poly(glycidol) derivatives with varying hydrophobicities: their ability to sensitize stable liposomes to pH, *Bioconjug. Chem.* **19**(5), 1040–1048 (2008). Epub 2008/04/19.

28. E. Yuba, A. Harada, Y. Sakanishi and K. Kono, Carboxylated hyperbranched poly(glycidol s for preparation of pH-sensitive liposomes, *J. Control. Release* **149**(1), 72–80 (2011). Epub 2010/03/09.
29. E. Yuba, C. Kojima, A. Harada, Tana, S. Watarai and K. Kono, pH-Sensitive fusogenic polymer-modified liposomes as a carrier of antigenic proteins for activation of cellular immunity, *Biomaterials* **31**(5), 943–951 (2010). Epub 2009/10/24.
30. N. Bertrand, P. Simard and J. C. Leroux, Serum-stable, long-circulating, pH-sensitive PEGylated liposomes, *Methods Mol. Biol.* **605**, 545–558 (2010). Epub 2010/01/15.
31. E. Roux, C. Passirani, S. Scheffold, J. P. Benoit and J. C. Leroux, Serum-stable and long-circulating, PEGylated, pH-sensitive liposomes, *J. Control. Release* **94**(2–3), 447–451 (2004). Epub 2004/02/24.
32. P. Simard and J. C. Leroux, pH-sensitive immunoliposomes specific to the CD33 cell surface antigen ofleukemic cells, *Int. J. Pharm.* **381**(2), 86–96 (2009). Epub 2009/05/19.
33. G. Navarro, R. R. Sawant, S. Biswas, S. Essex, C. Tros de Ilarduya and V. P. Torchilin, P-glycoprotein silencing with siRNA delivered by DOPE-modified PEI overcomes doxorubicin resistance in breast cancer cells, *Nanomedicine (Lond).* 7(1), 65–78 (2012). Epub 2011/12/24.
34. G. Navarro, R. R. Sawant, S. Essex, C. Tros de Ilarduya and V. P. Torchilin, Phospholipid-polyethylenimine conjugate-based micelle-like nanoparticles for siRNA delivery, *Drug Deliv. Transl. Res.* **1**(1), 25–33 (2011). Epub 2011/02/01.
35. R. R. Sawant, S. K. Sriraman, G. Navarro, S. Biswas, R. A. Dalvi and V. P. Torchilin, Polyethyleneimine-lipid conjugate-based pH-sensitive micellar carrier for gene delivery, *Biomaterials* **33**(15), 3942–3951 (2012). Epub 2012/03/01.
36. O. Boussif, F. Lezoualc'h, M. A. Zanta, M. D. Mergny, D. Scherman, B. Demeneix, *et al.*, A versatile vector for gene and oligonucleotide transfer into cells in culture and in vivo: polyethylenimine, *Proceedings of the National Academy of Sciences of the United States of America* **92**(16), 7297–301 (1995). Epub 1995/08/01.
37. I. Y. Kim, Y. S. Kang, D. S. Lee, H. J. Park, E. K Choi, Y. K Oh, *et al.*, Antitumor activity of EGFR targeted pH-sensitive immunoliposomes encapsulating gemcitabine in A549 xenograft nude mice, *J. Control. Release* **140**(1), 55–60 (2009). Epub 2009/07/21.
38. Z. Cheng, A. K. Chen, H. Y. Lee and A. Tsourkas, Examination of folate-targeted liposomes with encapsulated poly(2-propylacrylic acid) as a pH-responsive nanoplatform for cytosolic drug delivery, *Small* 6(13), 1398–1401 (2010). Epub 2010/06/22.
39. S. Neves, H. Faneca, S. Bertin, K. Konopka, N. Duzgunes, V. Pierrefite-Carle, *et al.*, Transferrin lipoplex-mediated suicide gene therapy of oral squamous cell carcinoma in an immunocompetent murine model and mechanisms involved in the antitumoral response, *Cancer Gene. Ther.* **16**(1), 91–101 (2009). Epub 2008/08/12.
40. J. A. Reddy and P. S. Low, Enhanced folate receptor mediated gene therapy using a novel pH-sensitive lipid formulation, *J. Control. Release* **64**(1–3), 27–37 (2000). Epub 2000/01/21.
41. G. Shi, W. Guo, S. M. Stephenson and R. J. Lee, Efficient intracellular drug and gene delivery using folate receptor-targeted pH-sensitive liposomes composed of cationic/anionic lipid combinations, *J. Control. Release* **80**(1–3), 309–319 (2002). Epub 2002/04/12.
42. V. P. Torchilin, Tatp-mediated intracellular delivery of pharmaceutical nanocarriers, *Biochem. Soc. Trans.* **35**(Pt 4), 816–820 (2007). Epub 2007/07/20.

43. V. P. Torchilin, T. S. Levchenko, R. Rammohan, N. Volodina, B. Papahadjopoulos-Sternberg and G. G. D'Souza, Cell transfection in vitro and in vivo with nontoxic TAT peptide-liposome-DNA complexes, *Proceedings of the National Academy of Sciences of the United States of America* **100**(4), 1972–7 (2003). Epub 2003/02/07.
44. V. P. Torchilin, R. Rammohan, V. Weissig and T. S. Levchenko, TAT peptide on the surface of liposomes affords their efficient intracellular delivery even at low temperature and in the presence of metabolic inhibitors, *Proceedings of the National Academy of Sciences of the United States of America* **98**(15), 8786–91 (2001). Epub 2001/07/05.
45. Y. L. Tseng, J. J. Liu and R. L. Hong, Translocation of liposomes into cancer cells by cell-penetrating peptides penetratin and tat: a kinetic and efficacy study, *Mol Pharmacol.* **62**(4), 864–872 (2002). Epub 2002/09/19.
46. R. Sawant and V. Torchilin, Intracellular delivery of nanoparticles with CPPs, *Methods Mol. Biol.* **683**, 431–451 (2011). Epub 2010/11/06.
47. J. B. Rothbard, T. C. Jessop, R. S. Lewis, B. A. Murray and P. A. Wender, Role of membrane potential and hydrogen bonding in the mechanism of translocation of guanidinium-rich peptides into cells, *J. Am. Chem. Soc.* **126**(31), 9506–9507 (2004). Epub 2004/08/05.
48. I. M. Kaplan, J. S. Wadia and S. F. Dowdy, Cationic TAT peptide transduction domain enters cells by macropinocytosis, *J. Control. Release* **102**(1), 247–53 (2005). Epub 2005/01/18.
49. R. E. Vandenbroucke, S. C. De Smedt, J. Demeester and N. N. Sanders, Cellular entry pathway and gene transfer capacity of TAT-modified lipoplexes, *Biochim. Biophys. Acta.* **1768**(3), 571–579 (2007). Epub 2006/12/26.
50. S. T. Yang, E. Zaitseva, L. V. Chernomordik and K. Melikov, Cell-penetrating peptide induces leaky fusion of liposomes containing late endosome-specific anionic lipid, *Biophys. J.* **99**(8), 2525–2533 (2010).
51. S. Trabulo, S. Resina, S. Simoes, B. Lebleu and M. C. Pedroso de Lima, A non-covalent strategy combining cationic lipids and CPPs to enhance the delivery of splice correcting oligonucleotides, *J. Control. Release* **145**(2), 149–158 (2010). Epub 2010/04/07.
52. M. M. Fretz, A. Hogset, G. A. Koning, W. Jiskoot and G. Storm, Cytosolic delivery of liposomally targeted proteins induced by photochemical internalization, *Pharm. Res.* **24**(11), 2040–2047 (2007). Epub 2007/06/02.
53. M. Kullberg, J. L. Owens and K. Mann, Listeriolysin O enhances cytoplasmic delivery by Her-2 targeting liposomes, *J. Drug Target.* **18**(4), 313–320 (2010). Epub 2010/03/06.
54. Y. Negishi, Y. Endo, T. Fukuyama, R. Suzuki, T. Takizawa, D. Omata, *et al.*, Delivery of siRNA into the cytoplasm by liposomal bubbles and ultrasound, *J. Control. Release* **132**(2), 124–130 (2008). Epub 2008/09/23.
55. M. Kullberg, K. Mann and J. L. Owens, A two-component drug delivery system using Her-2-targeting thermosensitive liposomes, *J. Drug Target.* **17**(2), 98–107 (2009). Epub 2008/12/18.
56. A. H. Futerman and G. van Meer, The cell biology of lysosomal storage disorders, *Nat. Rev. Mol. Cell Biol.* **5**(7), 554–65 (2004). Epub 2004/07/03.
57. M. L. Schultz, L. Tecedor, M. Chang and B. L. Davidson, Clarifying lysosomal storage diseases, *Trends Neurosci.* **34**(8), 401–410 (2011). Epub 2011/07/05.
58. A. C. Johansson, H. Appelqvist, C. Nilsson, K. Kagedal, K. Roberg and K. Ollinger, Regulation of apoptosis-associated lysosomal membrane permeabilization, *Apoptosis: An*

International Journal on Programmed Cell Death **15**(5), 527–540 (2010). Epub 2010/01/16.

59. T. Kirkegaard and M. Jaattela, Lysosomal involvement in cell death and cancer, *Biochim. Biophys. Acta.* **1793**(4), 746–54 (2009). Epub 2008/10/25.
60. C. de Duve, Lysosomes revisited, *Eur. J. Biochem.* **137**(3), 391–7 (1983). Epub 1983/12/15.
61. C. Tardy, P. Codogno, H. Autefage, T. Levade and N. Andrieu-Abadie, Lysosomes and lysosomal proteins in cancer cell death (new players of an old struggle), *Biochim. Biophys. Acta.* **1765**(2), 101–125 (2006). Epub 2006/01/18.
62. B. Winchester, Lysosomal metabolism of glycoproteins, *Glycobiology* **15**(6), 1R–15R (2005). Epub 2005/01/14.
63. G. A. Grabowski and R. J. Hopkin, Enzyme therapy for lysosomal storage disease: principles, practice, and prospects, *Annu. Rev. Genomics Hum. Genet.* **4**, 403–436 (2003). Epub 2003/10/07.
64. R. H. Lachmann, Enzyme replacement therapy for lysosomal storage diseases, *Curr. Opin. Pediatr.* **23**(6), 588–593 (2011). Epub 2011/09/29.
65. M. Rohrbach and J. T. Clarke, Treatment of lysosomal storage disorders: progress with enzyme replacement therapy, *Drugs* **67**(18), 2697–2716 (2007). Epub 2007/12/08.
66. G. Gregoriadis, Liposomes in the therapy of lysosomal storage diseases, *Nature* **275**(5682), 695–6 (1978). Epub 1978/10/26.
67. M. H. Cesen, K. Pegan, A. Spes and B. Turk, Lysosomal pathways to cell death and their therapeutic applications, *Exp. Cell Res.* **318**(11), 1245–1251 (2012). Epub 2012/04/03.
68. G. Kroemer and M. Jaattela, Lysosomes and autophagy in cell death control, *Nat. Rev. Cancer* **5**(11), 886–897 (2005). Epub 2005/10/22.
69. K. Roberg and K. Ollinger, Oxidative stress causes relocation of the lysosomal enzyme cathepsin D with ensuing apoptosis in neonatal rat cardiomyocytes, *Am. J. Pathol.* **152**(5), 1151–1156 (1998). Epub 1998/05/20.
70. K. Kagedal, M. Zhao, I. Svensson and U. T. Brunk, Sphingosine-induced apoptosis is dependent on lysosomal proteases, *Biochem. J.* **359**(Pt 2), 335–343 (2001). Epub 2001/10/05.
71. X. M. Yuan, W. Li, H. Dalen, J. Lotem, R. Kama, L. Sachs, *et al.*, Lysosomal destabilization in p53-induced apoptosis, *Proceedings of the National Academy of Sciences of the United States of America* **99**(9), 6286–6291 (2002). Epub 2002/04/18.
72. P. Boya, K. Andreau, D. Poncet, N. Zamzami, J. L. Perfettini, D. Metivier, *et al.*, Lysosomal membrane permeabilization induces cell death in a mitochondrion-dependent fashion, *J. Exp. Med.* **197**(10), 1323–1334 (2003). Epub 2003/05/21.
73. E. Agostinelli, G. Tempera, L. Dalla Vedova, M. Condello and G. Arancia, MDL 72527 and spermine oxidation products induce a lysosomotropic effect and mitochondrial alterations in tumour cells, *Biochem. Soc. Trans.* **35**(Pt 2), 343–348 (2007). Epub 2007/03/21.
74. S. Nakashima, Y. Hiraku, S. Tada-Oikawa, T. Hishita, E. C. Gabazza, S. Tamaki, *et al.*, Vacuolar H+-ATPase inhibitor induces apoptosis via lysosomal dysfunction in the human gastric cancer cell line MKN-1, *J. Biochem.* **134**(3), 359–364 (2003). Epub 2003/10/17.
75. N. W. Werneburg, M. E. Guicciardi, S. F. Bronk and G. J. Gores, Tumor necrosis factor-alpha-associated lysosomal permeabilization is cathepsin B dependent, *Am. J. Physiol. Gastrointest. Liver Physiol.* **283**(4), G947–956. Epub 2002/09/12.

76. N. W. Werneburg, M. E. Guicciardi, S. F. Bronk, S. H. Kaufmann and G. J. Gores, Tumor necrosis factor-related apoptosis-inducing ligand activates a lysosomal pathway of apoptosis that is regulated by Bcl-2 proteins, *J. Biol. Chem.* **282**(39), 28960–28970 (2007). Epub 2007/08/10.
77. P. Boya and G. Kroemer, Lysosomal membrane permeabilization in cell death, *Oncogene.* **27**(50), 6434–6451 (2008). Epub 2008/10/29.
78. U. Repnik, V. Stoka, V. Turk and B. Turk, Lysosomes and lysosomal cathepsins in cell death, *Biochim. Biophys. Acta* **1824**(1), 22–33 (2012). Epub 2011/09/15.
79. A. Koshkaryev, R. Thekkedath, C. Pagano, I. Meerovich and V. P. Torchilin, Targeting of lysosomes by liposomes modified with octadecyl-rhodamine B, *J. Drug. Target.* **19**(8), 606–614 (2011). Epub 2011/02/01.
80. R. W. Horobin and J. A. Kiernan, (Eds.), *Conn's Biological Stains.* 10th ed. Oxford, UK: BIOS Scientific Publishers (2002).
81. S. D. Goldman, R. S. Funk, R. A. Rajewski and J. P. Krise, Mechanisms of amine accumulation in, and egress from, lysosomes, *Bioanalysis* **1**(8), 1445–1459 (2009). Epub 2010/11/19.
82. R. W. Horobin, In *Xanthenes: Conn's Biological Stains,* R. W. Horobin, J. A. Kiernan, (Eds.), 10th ed. Oxford, UK: BIOS Scientific Publishers (2002), p. 237.
83. C. Minier and M. N. Moore, Rhodamine B accumulation and MXR protein expression in mussel blood cells: effects of exposure to vincristine, *Mar. Ecol. Prog. Ser.* **142**, 165–173 (1996).
84. D. Hoekstra, T. de Boer, K. Klappe and J. Wilschut, Fluorescence method for measuring the kinetics of fusion between biological membranes, *Biochemistry* **23**(24), 5675–5681 (1984). Epub 1984/11/20.
85. D. Hoekstra and K. Klappe, Use of a fluorescence assay to monitor the kinetics of fusion between erythrocyte ghosts, as induced by Sendai virus, *Biosci Rep.* **6**(11), 953–60 (1986).
86. D. Pozzi, A. Lisi, I. De Ros, L. Ferroni, A. Giuliani, G. Ravagnan, *et al.*, Use of octadecylrhodamine fluorescence dequenching to study vesicular stomatitis virus fusion with human aged red blood cells, *Photochem. Photobiol.* **57**(3), 426–430 (1993).
87. R. Daugelavicius, V. Cvirkaite, A. Gaidelyte, E. Bakiene, R. Gabrenaite-Verkhovskaya and D. H. Bamford, Penetration of enveloped double-stranded RNA bacteriophages phi13 and phi6 into Pseudomonas syringae cells, *J. Virol.* **79**(8), 5017–5026 (2005).
88. M. Ohashi, I. Miwako, K. Nakamura, A. Yamamoto, M. Murata, S. Ohnishi, *et al.*, An arrested late endosome-lysosome intermediate aggregate observed in a Chinese hamster ovary cell mutant isolated by novel three-step screening, *J. Cell. Sci.* **112**(Pt 8), 1125–1138 (1999). Epub 1999/03/23.
89. T. Kuwana, B. M. Mullock and J. P. Luzio, Identification of a lysosomal protein causing lipid transfer, using a fluorescence assay designed to monitor membrane fusion between rat liver endosomes and lysosomes, *Biochem. J.* **308**(Pt 3), 937–946 (1995). Epub 1995/06/15.
90. M. Ohashi, M. Murata and S. Ohnishi, A novel fluorescence method to monitor the lysosomal disintegration of low density lipoprotein, *Eur. J. Cell Biol.* **59**(1), 116–126 (1992). Epub 1992/10/01.
91. B. Rotman, J. A. Zderic and M. Edelstein, Fluorogenic substrates for beta-D-galactosidases and phosphatases derived from flurescein (3,6-dihydroxyfluoran) and its

monomethylether, *Proceedings of the National Academy of Sciences of the United States of America* **50**, 1–6 (1963). Epub 1963/07/01.

92. U. S. Huth, R. Schubert and R. Peschka-Suss, Investigating the uptake and intracellular fate of pH-sensitive liposomes by flow cytometry and spectral bio-imaging, *J. Control. Release* **110**(3), 490–504 (2006). Epub 2006/01/03.
93. L. Pelkmans, J. Kartenbeck and A. Helenius, Caveolar endocytosis of simian virus 40 reveals a new two-step vesicular-transport pathway to the ER, *Nat. Cell. Biol.* **3**(5), 473–483 (2001). Epub 2001/05/02.
94. J. P. Luzio, P. R. Pryor and N. A. Bright, Lysosomes: fusion and function, *Nat. Rev. Mol. Cell Biol.* **8**(8), 622–632 (2007). Epub 2007/07/20.
95. R. Thekkedath, A. Koshkaryev and V. P. Torchilin, Lysosome-targeted octadecyl-rhodamine B-liposomes enhance lysosomal accumulation of glucocerebrosidase in Gaucher's cells *in vitro*, *Nanomedicine (Lond).* (2012). Epub 2012/12/04.
96. I. Meerovich, A. Koshkaryev, R. Thekkedath and V. P. Torchilin, Screening and optimization of ligand conjugates for lysosomal targeting, *Bioconjug. Chem.* **22**(11), 2271–2282 (2011). Epub 2011/09/15.
97. A. Koshkaryev, A. Piroyan and V. P. Torchilin, Increased apoptosis in cancer cells in vitro and in vivo by ceramides in transferrin-modified liposomes, *Cancer Biol. Ther.* **13**(1), 50–60 (2012). Epub 2012/02/18.
98. K. M. Mayle, A. M. Le and D. T. Kamei, The intracellular trafficking pathway of transferrin, *Biochim. Biophys. Acta.* **1820**(3), 264–281 (2012). Epub 2011/10/05.
99. J. W. Larrick and P. Cresswell, Modulation of cell surface iron transferrin receptors by cellular density and state of activation, *J. Supramol. Struct.* **11**(4), 579–586 (1979). Epub 1979/01/01.
100. J. Renau-Piqueras, F. Miragall and J. Cervera, Endocytosis of cationized ferritin in human peripheral blood by resting T-lymphocytes, *Cell Tissue Res.* **240**(3), 743–746 (1985). Epub 1985/01/01.
101. X. Li, L. Ding, Y. Xu, Y. Wang and Q. Ping, Targeted delivery of doxorubicin using stealth liposomes modified with transferrin, *Int. J. Pharm.* **373**(1–2), 116–123 (2009). Epub 2009/05/12.
102. T. Kobayashi, T. Ishida, Y. Okada, S. Ise, H. Harashima and H. Kiwada, Effect of transferrin receptor-targeted liposomal doxorubicin in P-glycoprotein-mediated drug resistant tumor cells, *Int. J. Pharm.* **329**(1–2), 94–102 (2007). Epub 2006/09/26.
103. M. Hong, S. Zhu, Y. Jiang, G. Tang and Y. Pei, Efficient tumor targeting of hydroxycamptothecin loaded PEGylated niosomes modified with transferrin, *J. Control. Release.* **133**(2), 96–102 (2009). Epub 2008/10/09.
104. A. C. Johansson, H. Appelqvist, C. Nilsson, K. Kagedal, K. Roberg and K. Ollinger, Regulation of apoptosis-associated lysosomal membrane permeabilization, *Apoptosis* (2010). Epub 2010/01/16.
105. S. Schutze, T. Machleidt, D. Adam, R. Schwandner, K. Wiegmann, M. L. Kruse, *et al.*, Inhibition of receptor internalization by monodansylcadaverine selectively blocks p55 tumor necrosis factor receptor death domain signaling, *J. Biol. Chem.* **274**(15), 10203–10212 (1999). Epub 1999/04/03.
106. C. Ullio, J. Casas, U. T. Brunk, G. Sala, G. Fabrias, R. Ghidoni, *et al.*, Sphingosine mediates TNFalpha-induced lysosomal membrane permeabilization and ensuing programmed cell death in hepatoma cells, *J. Lipid Res.* **53**(6), 1134–1143 (2012). Epub 2012/03/29.

107. R. Dhami and E. H. Schuchman, Mannose 6-phosphate receptor-mediated uptake is defective in acid sphingomyelinase-deficient macrophages: implications for Niemann-Pick disease enzyme replacement therapy, *J. Biol. Chem.* **279**(2), 1526–1532 (2004). Epub 2003/10/15.
108. J. Papademetriou, C. Garnacho, D. Serrano, T. Bhowmick, E. H. Schuchman and S. Muro, Comparative binding, endocytosis, and biodistribution of antibodies and antibody-coated carriers for targeted delivery of lysosomal enzymes to ICAM-1 versus transferrin receptor, *J. Inherit. Metab. Dis.* (2012). Epub 2012/09/13.
109. E. Koren and V. P. Torchilin, Drug carriers for vascular drug delivery, *IUBMB Life* **63**(8), 586–595 (2011). Epub 2011/07/19.
110. S. Muro, M. Koval and V. Muzykantov, Endothelial endocytic pathways: gates for vascular drug delivery, *Curr. Vasc. Pharmacol.* **2**(3), 281–299 (2004). Epub 2004/08/24.
111. S. Muro, C. Gajewski, M. Koval and V. R. Muzykantov, ICAM-1 recycling in endothelial cells: a novel pathway for sustained intracellular delivery and prolonged effects of drugs, *Blood* **105**(2), 650–658 (2005). Epub 2004/09/16.
112. S. Muro, R. Wiewrodt, A. Thomas, L. Koniaris, S. M. Albelda, V. R. Muzykantov *et al.*, A novel endocytic pathway induced by clustering endothelial ICAM-1 or PECAM-1, *J. Cell Sci.* **116**(Pt 8), 1599–1609 (2003). Epub 2003/03/18.
113. S. Muro, C. Garnacho, J. A. Champion, J. Leferovich, C. Gajewski, E. H. Schuchman *et al.*, Control of endothelial targeting and intracellular delivery of therapeutic enzymes by modulating the size and shape of ICAM-1-targeted carriers, *Mol. Ther.: The Journal of the American Society of Gene Therapy* **16**(8), 1450–1458 (2008). Epub 2008/06/19.
114. J. Hsu, L. Northrup, T. Bhowmick and S. Muro, Enhanced delivery of alpha-glucosidase for Pompe disease by ICAM-1-targeted nanocarriers: comparative performance of a strategy for three distinct lysosomal storage disorders, *Nanomedicine: Nanotechnology, Biology, and Medicine* **8**(5), 731–739 (2012). Epub 2011/09/13.
115. J. Hsu, D. Serrano, T. Bhowmick, K. Kumar, Y. Shen, Y. C. Kuo, *et al.*, Enhanced endothelial delivery and biochemical effects of alpha-galactosidase by ICAM-1-targeted nanocarriers for Fabry disease, *J. Control. Release* **149**(3), 323–331 (2011). Epub 2010/11/05.
116. D. C. Wallace, Mitochondrial diseases in man and mouse, *Science* **283**(5407), 1482–1488. Epub 1999/03/05.
117. L. A. Pradelli, M. Beneteau and J. E. Ricci, Mitochondrial control of caspase-dependent and -independent cell death, *Cell Mol. Life Sci.* **67**(10), 1589–1597 (2010). Epub 2010/02/13.
118. P. Costantini, E. Jacotot, D. Decaudin and G. Kroemer, Mitochondrion as a novel target of anticancer chemotherapy, *J. Natl. Cancer Inst.* **92**(13), 1042–1053 (2000). Epub 2000/07/06.
119. D. Pathania, M. Millard and N. Neamati, Opportunities in discovery and delivery of anticancer drugs targeting mitochondria and cancer cell metabolism, *Adv. Drug. Deliv. Rev.* **61**(14), 1250–1275 (2009). Epub 2009/09/01.
120. V. Weissig, Targeted drug delivery to mammalian mitochondria in living cells, *Expert Opin. Drug Deliv.* **2**(1), 89–102 (2005). Epub 2005/11/22.
121. Y. Yamada and H. Harashima, Mitochondrial drug delivery systems for macromolecule and their therapeutic application to mitochondrial diseases, *Adv. Drug Deliv. Rev.* **60**(13–14), 1439–1462 (2008). Epub 2008/07/29.

122. V. Weissig and V. P. Torchilin, Cationic bolasomes with delocalized charge centers as mitochondria-specific DNA delivery systems, *Adv. Drug Deliv. Rev.* **49**(1–2), 127–149 (2001). Epub 2001/05/30.
123. L. B. Chen, I. C. Summerhayes, L. V. Johnson, M. L. Walsh, S. D. Bernal and T. J. Lampidis, Probing mitochondria in living cells with rhodamine 123, *Cold Spring Harb Symp. Quant. Biol.* **46**, 141–155 (1982).
124. D. Rideout, A. Bustamante and J. Patel, Mechanism of inhibition of FaDu hypopharyngeal carcinoma cell growth by tetraphenylphosphonium chloride, *Int. J. Cancer* **57**(2), 247–253 (1994). Epub 1994/04/15.
125. V. Weissig, J. Lasch, G. Erdos, H. W. Meyer, T. C. Rowe and J. Hughes, DQAsomes: a novel potential drug and gene delivery system made from Dequalinium, *Pharmaceutical Research* **15**(2), 334–337 (1998).
126. V. Weissig, C. Lizano and V. P. Torchilin, Selective DNA release from DQAsome/DNA complexes at mitochondria-like membranes, *Drug Deliv.* 7(1), 1–5 (2000).
127. V. Weissig, G. G. D'Souza and V. P. Torchilin, DQAsome/DNA complexes release DNA upon contact with isolated mouse liver mitochondria, *J. Control. Release* **75**(3), 401–408 (2001).
128. G. G. D'Souza, R. Rammohan, S. M. Cheng, V. P. Torchilin and V. Weissig, DQAsome-mediated delivery of plasmid DNA toward mitochondria in living cells, *J. Control. Release* **92**(1–2), 189–197 (2003). Epub 2003/09/23.
129. G. G. D'Souza, S. M. Cheng, S. V. Boddapati, R. W. Horobin and V. Weissig, Nanocarrier-assisted sub-cellular targeting to the site of mitochondria improves the pro-apoptotic activity of paclitaxel, *J. Drug Target.* **16**(7), 578–585 (2008). Epub 2008/08/08.
130. M. Lee, J. S. Choi, M. J. Choi, Y. K. Pak, B. D. Rhee and K. S. Ko, DNA delivery to the mitochondria sites using mitochondrial leader peptide conjugated polyethylenimine, *J. Drug Target.* **15**(2), 115–122 (2007). Epub 2007/03/17.
131. Y. K. Pak and H. Weiner, Import of chemically synthesized signal peptides into rat liver mitochondria, *The Journal of Biological Chemistry* **265**(24), 14298–14307 (1990). Epub 1990/08/25.
132. P. Seibel, J. Trappe, G. Villani, T. Klopstock, S. Papa and H. Reichmann, Transfection of mitochondria: strategy towards a gene therapy of mitochondrial DNA diseases, *Nucleic Acids Research* **23**(1), 10–17 (1995). Epub 1995/01/11.
133. S. V. Boddapati, P. Tongcharoensirikul, R. N. Hanson, G. G. D'Souza, V. P. Torchilin and V. Weissig, Mitochondriotropic liposomes, *J. Liposome Res.* **15**(1–2), 49–58 (2005). Epub 2005/10/01.
134. M. N. Perera, V. Ganesan, L. J. Siskind, Z. M. Szulc, J. Bielawski, A. Bielawska, *et al.*, Ceramide channels: influence of molecular structure on channel formation in membranes, *Biochim. Biophys. Acta* **1818**(5), 1291–1301 (2012). Epub 2012/03/01.
135. L. J. Siskind and M. Colombini, The lipids C2- and C16-ceramide form large stable channels. Implications for apoptosis, *The Journal of Biological Chemistry* **275**(49), 38640–38644 (2000). Epub 2000/10/12.
136. L. J. Siskind, R. N. Kolesnick and M. Colombini, Ceramide channels increase the permeability of the mitochondrial outer membrane to small proteins, *The Journal of Biological Chemistry* **277**(30), 26796–26803 (2002). Epub 2002/05/15.
137. L. J. Siskind, R. N. Kolesnick and M. Colombini, Ceramide forms channels in mitochondrial outer membranes at physiologically relevant concentrations, *Mitochondrion* **6**(3), 118–125 (2006). Epub 2006/05/23.

138. S. V. Boddapati, G. G. D'Souza, S. Erdogan, V. P. Torchilin and V. Weissig, Organelle-targeted nanocarriers: specific delivery of liposomal ceramide to mitochondria enhances its cytotoxicity in vitro and in vivo, *Nano Lett.* **8**(8), 2559–2563 (2008). Epub 2008/07/10.
139. S. Biswas, N. S. Dodwadkar, P. P. Deshpande and V. P. Torchilin, Liposomes loaded with paclitaxel and modified with novel triphenylphosphonium-PEG-PE conjugate possess low toxicity, target mitochondria and demonstrate enhanced antitumor effects in vitro and in vivo, *J. Control. Release.* **159**(3), 393–402 (2012). Epub 2012/01/31.
140. S. Biswas, N. S. Dodwadkar, R. R. Sawant, A. Koshkaryev and V. P. Torchilin, Surface modification of liposomes with rhodamine-123-conjugated polymer results in enhanced mitochondrial targeting, *J. Drug Target* (2011). Epub 2011/02/26.
141. Y. Yamada, H. Akita, H. Kamiya, K. Kogure, T. Yamamoto, Y. Shinohara, *et al.*, MITO-Porter: A liposome-based carrier system for delivery of macromolecules into mitochondria via membrane fusion, *Biochim. Biophys. Acta.* **1778**(2), 423–432 (2008). Epub 2007/12/07.
142. Y. Yamada, R. Furukawa, Y. Yasuzaki and H. Harashima, Dual function MITO-Porter, a nano carrier integrating both efficient cytoplasmic delivery and mitochondrial macromolecule delivery, *Mol. Ther.: The Journal of the American Society of Gene Therapy* **19**(8), 1449–1456 (2011). Epub 2011/06/23.
143. Y. Yamada and H. Harashima, Delivery of bioactive molecules to the mitochondrial genome using a membrane-fusing, liposome-based carrier, DF-MITO-Porter, *Biomaterials* **33**(5), 1589–1595 (2012). Epub 2011/11/23.
144. Y. Yasuzaki, Y. Yamada and H. Harashima, Mitochondrial matrix delivery using MITO-Porter, a liposome-based carrier that specifies fusion with mitochondrial membranes, *Biochemical and Biophysical Research Communications* **397**(2), 181–186 (2010). Epub 2010/06/29.
145. I. A. Khalil, K. Kogure, S. Futaki and H. Harashima, High density of octaarginine stimulates macropinocytosis leading to efficient intracellular trafficking for gene expression, *The Journal of Biological Chemistry* **281**(6), 3544–3551 (2006). Epub 2005/12/06.
146. P. Csermely, T. Schnaider and I. Szanto, Signalling and transport through the nuclear membrane, *Biochim. Biophys. Acta* **1241**(3), 425–451 (1995). Epub 1995/12/20.
147. A. I. Aronsohn and J. A. Hughes, Nuclear localization signal peptides enhance cationic liposome-mediated gene therapy, *J. Drug Target* **5**(3), 163–169 (1998).
148. C. K. Chan and D. A. Jans, Enhancement of polylysine-mediated transferrinfection by nuclear localization sequences: polylysine does not function as a nuclear localization sequence, *Hum. Gene. Ther.* **10**(10), 1695–1702 (1999). Epub 1999/07/31.
149. F. Sakurai, R. Inoue, Y. Nishino, A. Okuda, O. Matsumoto, T. Taga, *et al.*, Effect of DNA/liposome mixing ratio on the physicochemical characteristics, cellular uptake and intracellular trafficking of plasmid DNA/cationic liposome complexes and subsequent gene expression, *J. Control. Release* **66**(2–3), 255–269 (2000). Epub 2000/04/01.
150. I. Koltover, T. Salditt, J. O. Radler and C. R. Safinya, An inverted hexagonal phase of cationic liposome-DNA complexes related to DNA release and delivery, *Science* **281**(5373), 78–81 (1998). Epub 1998/07/04.
151. S. Trabulo, M. Mano, H. Faneca, A. L. Cardoso, S. Duarte, A. Henriques, *et al.*, S4(13)-PV cell penetrating peptide and cationic liposomes act synergistically to mediate intracellular delivery of plasmid DNA, *J. Gene. Med.* **10**(11), 1210–1222 (2008). Epub 2008/08/30.

152. K. Kogure, H. Akita, Y. Yamada and H. Harashima Multifunctional envelope-type nano device (MEND) as a non-viral gene delivery system, *Adv. Drug Deliv. Rev.* **60**(4–5), 559–571 (2008). Epub 2007/11/27.
153. K. Kogure, R. Moriguchi, K. Sasaki, M. Ueno, S. Futaki and H. Harashima, Development of a non-viral multifunctional envelope-type nano device by a novel lipid film hydration method, *J. Control. Release* **98**(2), 317–323 (2004). Epub 2004/07/21.
154. I. A. Khalil, K. Kogure, S. Futaki, S. Hama, H. Akita, M. Ueno, *et al.*, Octaarginine-modified multifunctional envelope-type nanoparticles for gene delivery, *Gene. Ther.* **14**(8), 682–689 (2007). Epub 2007/02/03.
155. I. A. Khalil, K. Kogure, S. Futaki and H. Harashima, Octaarginine-modified liposomes: enhanced cellular uptake and controlled intracellular trafficking, *Int. J. Pharm.* **354** (1–2), 39–48 (2008). Epub 2008/02/05.
156. T. Nakamura, R. Moriguchi, K. Kogure, A. Minoura, T. Masuda, H. Akita, *et al.*, Delivery of condensed DNA by liposomal non-viral gene delivery system into nucleus of dendritic cells, *Biol. Pharm. Bull.* **29**(6), 1290–1293 (2006). Epub 2006/06/07.
157. D. Kurihara, H. Akita, A. Kudo, T. Masuda, S. Futaki and H. Harashima, Effect of polyethyleneglycol spacer on the binding properties of nuclear localization signal-modified liposomes to isolated nucleus, *Biol. Pharm. Bull.* **32**(7), 1303–1306 (2009). Epub 2009/07/03.
158. I. A. Khalil, Y. Hayashi, R. Mizuno and H. Harashima, Octaarginine- and pH sensitive fusogenic peptide-modified nanoparticles for liver gene delivery, *J. Control. Release* **156**(3), 374–380 (2011). Epub 2011/08/26.
159. A. Koshkaryev, A. Piroyan and V. P. Torchilin, Bleomycin in octaarginine-modified fusogenic liposomes results in improved tumor growth inhibition, *Cancer Lett.* (2012). Epub 2012/06/30.
160. J. Chen and J. Stubbe, Bleomycins: towards better therapeutics, *Nat. Rev. Cancer* **5**(2), 102–112 (2005). Epub 2005/02/03.
161. S. N. Roy and S. B. Horwitz, Characterization of the association of radiolabeled bleomycin A2 with HeLa cells, *Cancer Res.* **44**(4), 1541–1546 (1984). Epub 1984/04/01.
162. R. M. Steinman, I. S. Mellman, W. A. Muller and Z. A. Cohn, Endocytosis and the recycling of plasma membrane, *J. Cell. Biol.* **96**(1), 1–27 (1983). Epub 1983/01/01.
163. M. Aridor, Visiting the ER: the endoplasmic reticulum as a target for therapeutics in traffic related diseases, *Adv. Drug Deliv. Rev.* **59**(8), 759–781 (2007). Epub 2007/08/08.
164. M. Schrader and H. D. Fahimi, The peroxisome: still a mysterious organelle, *Histochem. Cell Biol.* **129**(4), 421–440 (2008). Epub 2008/02/16.
165. M. T. Tarrago-Trani and B. Storrie, Alternate routes for drug delivery to the cell interior: pathways to the Golgi apparatus and endoplasmic reticulum, *Adv. Drug Deliv. Rev.* **59**(8), 782–797 (2007). Epub 2007/08/03.
166. S. D. Conner and S. L. Schmid, Regulated portals of entry into the cell, *Nature* **422**(6927), 37–44 (2003). Epub 2003/03/07.
167. G. Bathori, L. Cervenak and I. Karadi, Caveolae — an alternative endocytotic pathway for targeted drug delivery, *Crit. Rev. Ther. Drug Carrier Syst.* **21**(2), 67–95 (2004). Epub 2004/06/19.
168. B. Nichols, Caveosomes and endocytosis of lipid rafts, *J. Cell. Sci.* **116**(Pt 23), 4707–4714 (2003). Epub 2003/11/06.
169. V. Puri, R. Watanabe, R. D. Singh, M. Dominguez, J. C. Brown, C. L. Wheatley, *et al.*, Clathrin-dependent and -independent internalization of plasma membrane sphingolipids

initiates two Golgi targeting pathways, *J. Cell. Biol.* **154**(3), 535–547 (2001). Epub 2001/08/02.

170. J. E. Schnitzer, J. Liu and P. Oh, Endothelial caveolae have the molecular transport machinery for vesicle budding, docking, and fusion including VAMP, NSF, SNAP, annexins, and GTPases, *The Journal of Biological Chemistry* **270**(24), 14399–14404 (1995). Epub 1995/06/16.
171. B. Razani, S. E. Woodman and M. P. Lisanti, Caveolae: from cell biology to animal physiology, *Pharmacol Rev.* **54**(3), 431–467 (2002). Epub 2002/09/12.
172. K. G. Rothberg, Y. S. Ying, J. F. Kolhouse, B. A. Kamen and R. G. Anderson, The glycophospholipid-linked folate receptor internalizes folate without entering the clathrin-coated pit endocytic pathway, *J. Cell Biol.* **110**(3), 637–649 (1990). Epub 1990/03/01.
173. J. E. Schnitzer, P. Oh, E. Pinney and J. Allard, Filipin-sensitive caveolae-mediated transport in endothelium: reduced transcytosis, scavenger endocytosis, and capillary permeability of select macromolecules, *J. Cell Biol.* **127**(5), 1217–1232 (1994). Epub 1994/12/01.
174. N. Benlimame, P. U. Le and I. R. Nabi, Localization of autocrine motility factor receptor to caveolae and clathrin-independent internalization of its ligand to smooth endoplasmic reticulum, *Mol. Biol.Cell.* **9**(7), 1773–1786 (1998). Epub 1998/07/11.
175. S. Pollock, R. Antrobus, L. Newton, B. Kampa, J. Rossa, S. Latham, *et al.*, Uptake and trafficking of liposomes to the endoplasmic reticulum, *FASEB J.* **24**(6), 1866–1878 (2010). Epub 2010/01/26.
176. R. C. Mustata, A. Grigorescu and S. M. Petrescu, Encapsulated cargo internalized by fusogenic liposomes partially overlaps the endoplasmic reticulum, *J. Cell. Mol. Med.* **13**(9B), 3110–3121 (2009). Epub 2009/05/15.
177. J. Connor, M. B. Yatvin and L. Huang, pH-sensitive liposomes: acid-induced liposome fusion, *Proceedings of the National Academy of Sciences of the United States of America.* **81**(6), 1715–1718 (1984). Epub 1984/03/01.
178. K. Un, K. Sakai-Kato, Y. Oshima, T. Kawanishi and H. Okuda, Intracellular trafficking mechanism, from intracellular uptake to extracellular efflux, for phospholipid/cholesterol liposomes, *Biomaterials* **33**(32), 8131–8141 (2012). Epub 2012/08/04.
179. J. Wang, Y. Wang and W. Liang, Delivery of drugs to cell membranes by encapsulation in PEG-PE micelles, *J. Control. Release* **160**(3), 637–651 (2012). Epub 2012/03/13.
180. J. Wang, X. Fang and W. Liang, Pegylated phospholipid micelles induce endoplasmic reticulum-dependent apoptosis of cancer cells but not normal cells, *ACS Nano* **6**(6), 5018–5030 (2012). Epub 2012/05/15.
181. M. J. Reilly, J. D. Larsen and M. O. Sullivan, Polyplexes traffic through caveolae to the Golgi and endoplasmic reticulum en route to the nucleus, *Mol. Pharm.* **9**(5), 1280–1290 (2012). Epub 2012/03/17.
182. M. S. Cartiera, K. M. Johnson, V. Rajendran, M. J. Caplan and W. M. Saltzman, The uptake and intracellular fate of PLGA nanoparticles in epithelial cells, *Biomaterials* **30**(14), 2790–2798 (2009). Epub 2009/02/24.
183. S. Nagotu, V. C. Kalel, R. Erdmann and H. W. Platta, Molecular basis of peroxisomal biogenesis disorders caused by defects in peroxisomal matrix protein import, *Biochim. Biophys. Acta* **1822**(9), 1326–1336 (2012). Epub 2012/05/24.

184. S. Leon, J. M. Goodman and S. Subramani, Uniqueness of the mechanism of protein import into the peroxisome matrix: transport of folded, co-factor-bound and oligomeric proteins by shuttling receptors, *Biochim. Biophys. Acta.* **1763**(12), 1552–1564 (2006). Epub 2006/10/03.
185. J. I. Koepke, K. A. Nakrieko, C. S. Wood, K. K. Boucher, L. J. Terlecky, P. A. Walton, *et al.*, Restoration of peroxisomal catalase import in a model of human cellular aging, *Traffic* **8**(11), 1590–1600 (2007). Epub 2007/09/08.
186. T. Mizuno, K. Ito, C. Uchida, M. Kitagawa, A. Ichiyama, S. Miura, *et al.*, Analyses in transfected cells and in vitro of a putative peroxisomal targeting signal of rat liver serine: pyruvate aminotransferase, *Histochem. Cell Biol.* **118**(4), 321–328 (2002). Epub 2002/10/12.

Chapter B2

Bacteriophage-Targeted Nanomedicine and Molecular Imaging

Tao Wang and Radhika Narayanaswamy*

Center for Pharmaceutical Biotechnology and Nanomedicine, Northeastern University, Boston, MA 02115, USA
**t.wang@neu.edu*

Abstract

Phage display is a powerful technique to generate random peptide libraries containing millions or even billions of different recombinant phage particles for affinity selection of targeted ligands based on their ability to specifically interact with cell surface receptors of interest. While phage display facilitates progress in molecular targeting, the emerging phage nanotechnology uses the genetically- and chemically-modifiable bacteriophage as a robust building block to provide a scaffold for the creation of novel biological materials with diverse nanostructures for nanomedicine fabrication and molecule imaging application. This chapter reviews the development of the phage-derived nanomaterials and their application in molecular imaging and targeted delivery of various cargos, including drugs, genes and vaccines. Strategies designed to overcome biological obstacles and to develop a viable phage theraphy are also discussed.

1. Phage Display

Early in 1985, George P. Smith pioneered the field of phage display by the original cloning of foreign DNA into filamentous M13-derived bacteriophage gene 3 to express a fusion protein at the N-terminal domain of the minor coat protein pIII. The recombinant phage retained phage infectivity and could be enriched by affinity purification on immobilized antibodies, specific for the peptide insert.[1] In 1990, three research groups, including Devlin *et al.*, Smith and Dower *et al.*, independently reported the first generation of phage-display peptide libraries for the screening and identification of targeting ligands.[2–4] Since then, phage display has become a robust technique for the creation of random peptide phage libraries which contain millions or even billions of different recombinant phage particles, each with the transgenic expression of a unique peptide produced by the insertion of a synthetic DNA sequence encoding a peptide into the phage genome.[5] These combinatorial phage peptide libraries have been used for affinity screening of ligands for peptide receptors, epitope discovery, gene delivery, selection of enzyme substrates and screening of cloned antibody repertoires.[6–8]

The most-widely used bacteriophage display system derives from the filamentous phages (Table 1). A filamentous bacteriophage Ff (fd, f1 and M13) is a long, thin virus, which consists of a single-stranded circular DNA packed within a cylindrical shell with a body composed of 2700 copies of major coat protein pVIII, five copies of the minor coat proteins pIII along with the five copies of pVI capping one end of a phage particle and the other end. The major pVIII coat proteins dominate the phage capsid with the 2,700 copies covering the viral DNA, forming a symmetrical repeating pattern and accounting for 87% of total viral mass and 98% of the whole capsid protein content.[5] The minor coat proteins pIII and pVI function in bacterial cell binding and in the termination of phage particle assembly during morphogenesis, while the minor coat proteins pVII and pIX are involved in the initiation and maintenance of phage assembly in the host bacteria.[9] Filamentous phage-display platforms predominantly use type 3 or type 8 display vectors to express guest peptides fused on coat proteins pIII and pVIII,[10] despite the fact that the other coat proteins, including pVI,[11–14] pVII,[15] and pIX[16] are also able to display foreign peptides (Table 1). The type 3 or type 8 display vectors are created by insertion of a gene encoding a foreign polypeptide into the open reading frame of gene 3 or gene 8 of the filamentous genome, expressing guest peptides fused to every copy of the structural phage coat protein displayed on the surface of virions.[17,18] The size of fusion polypeptides varies from a short peptide to that of antibody fragments.[19]

The pVIII-fusion phage particles which are able to display a guest peptide on every pVIII subunit have been entitled "landscape phage" to emphasize the dramatic change in surface architecture caused by arraying thousands of copies of the guest peptide in a dense, repeating pattern around the tubular capsid (Fig. 1).[5] The foreign peptide-decorated phage body serves as an interacting scaffold to constrain the peptide in a particular conformation, creating

Table 1. Summary of phage display platforms.

Phage display system	Displayed peptide	Application	References
Coliphage M 13 PIII	Random 15 peptide residues	Phage-display peptide library	(2)
Filamentous phage PIII	Hexapeptide epitope	Epitope library	(3)
Filamentous phage M13 PIII	Antitetanus toxoid Fab fragments	Screening of antibody repertoires	(44)
Filamentous phage PIII	ScFv fragments	Antibody discovery	(45)
Filamentous phage M 13	Human Interleukin III	Generation of neutralizing Mabs	(46)
Filamentous phage fd PIII	B chain of ricin toxin	Screening of galactose-binding mutants	(47)
Filamentous phage PIII	Human plasminogen activator inhibitor 1	Structure and function analysis of protein	(48)
Filamentous phage M13 PIII	Protease inhibitor variants	Selection of enzyme inhibitors	(49)
Filamentous phage fd PIII	Alkaline phosphatase	Enzyme engineering	(50)
Filamentous phage fd PVIII	Functional T-cell receptors (TCR)	Structure and function analysis of TCR	(11)
Filamentous phage fd PVIII	Circumsporozite protein	Vaccine development	(51)
Filamentous phage f1PVIII	Oligopeptides	Selection of antibody ligands	(52)
Filamentous phage PIX	Human ScFv	Antibody discovery	(16)
Filamentous phage M13 PVII	ScFv fragments	Generation of antibody libraries	(15)
Filamentous phage M13 PVI	cDNA-encoded products from breast cancer T47D or MCF-7 cells	Profiling immune responses in breast cancer patients	(12)
Filamentous phage M13 PVI	cDNA-encoded products from the parasite Ancylostoma caninum	Selection of cDNAs encoding novel serine protease inhibitors	(13)
T4 Bacteriophage Hoc and Soc	PorA peptide from neisseria meningitidis	Vaccine development	(22)
T7 bacteriophage	RNA-binding protein	Selection of RNA-binding protein	(23)
P4 bacteriophage Psu	10-amino acid peptide sequence from the human $p62^{c\text{-}myc}$ protein	Peptide display system	(24)

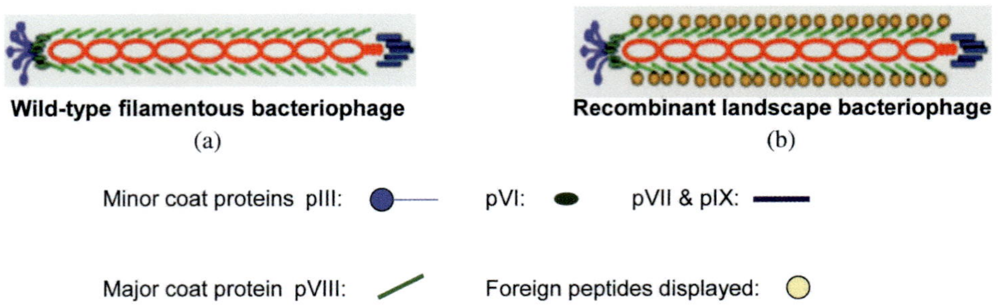

Fig. 1. Structure of filamentous bacteriophages. (a) Wild-type phage; (b) Landscape phage.

a defined organic surface structure ("landscape") that varies from one phage clone to another. Landscape phages with multivalent-displayed foreign peptides endow the recombinant phages with a unique surface architecture and additional properties but with no influence on the general structure of the virions. As a result, landscape phage particles retain the ability to infect *E. coli* and to form phage progeny.[5] A collection of billions of pVIII phage clones with different surface structures and biophysical properties builds up the so-called landscape phage peptide libraries.[5,20]

A major advantage of a filamentous phage as a display vector lies in its genome with a single-stranded DNA amenable to genetic manipulation to create various peptide libraries.[5] But on some occasions, filamentous phage-produced recombinant protein may undergo structural misfolding and lose functional integrity.[21] This is due to the capsid assembly of the non-lytic filamentous phage occurring in the cytoplasmic membrane followed by their secretion from infected bacteria without cell lysis. To overcome the limitation of filamentous phage-based vectors, lytic phage species, including T4,[22] T7,[23] P4[21,24] and bacteriophage lambda (Table 1),[25] have also been utilized as alternative cloning vectors to produce hybrid capsid proteins with structural integrity due to the newly-biosynthesized proteins released via a cell lysis process.[9]

Phage clones produced by host *E. coli* have peptide phenotypes coupled with their genotypes. This unique characteristic allows application of a phage display library for affinity selection of the targeted ligands specific for the receptors on cells. The screening of a random phage-display peptide library is carried out via either *in vitro* affinity selection against purified proteins, antibodies or cultured cells, or *in vivo* biopanning (Fig. 2). In the scheme of an *in vitro* selection, a phage library is incubated with immobilized targets of interest followed by the removal of the non-bound phages. Binding clones are amplified by infection of the host bacteria. Enrichment of the binding

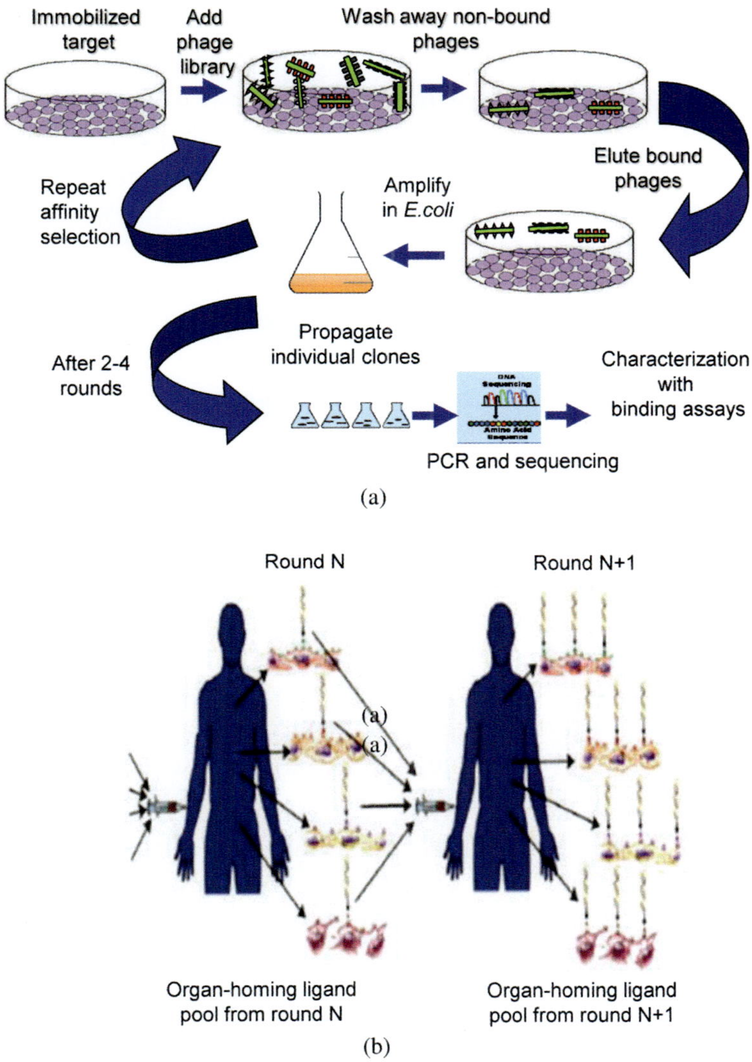

Fig. 2. Schematic illustrations of affinity selection of phage-display libraries for targeted ligands. (a) *In vitro* affinity selection; (b) *In vivo* biopanning. Cited and modified from Refs. 5 and 9.

peptide sequences is carried out by three to five rounds of such repeated *in vitro* panning. The amino acid sequence of the polypeptide fused to the selected binding clone is then deduced based on its DNA sequencing. *In vivo* panning first involves the injection of the whole phage library into an experimental animal[26–30] or human.[19,31] Phage clones with tissue-or organ-specific binding are subsequently captured by dissection of the tissue and organs followed by propagation in *E. coli*. This *in vivo* panning is repeated for several rounds allowing enrichment of targeted clones. Since the selection of phage clones using *in vivo* panning is based on their ability to bind and/or internalize to their targets within a complete organism, the procedure enables rapid selection of human high affinity antibodies and peptides against specific targets such as tumor associated antigens,[19] angiogenic vessels, [27,29, 32,33] muscles,[33] whole organs[27] and the immune system.[34,35] An especial advantage of *in vivo* phage display selection procedures over *in vitro* screening protocols lies in the selection of phages with desirable pharmacokinetic properties and favorable target tissue accumulation.[36] Using such well-established biopanning protocols for affinity-selection from multibillion phage clone libraries based on their abilities to interact very specifically with cell surface receptors of interest,[37] a number of phage-borne peptides specific for a broad array of targeted receptors have been identified without preexisting notions about the nature of the targeting.[37,38]

2. Phage nanobiotechnology

With the enthusiasm for and promise of the development in nanotechnology coupling with the evolution of phage display technique, the bacteriophage has been proposed as a new type of biomaterial for advanced nanobiotechnology. The concept of "phage nanobiotechnology" reflects a paradigm shift in the development of phage display technology. Conventional phage display paradigm predominantly focuses on the use of phage display for the identification of new binding molecular entities by screening libraries composed of different kinds of binding structures.[39] In the scenario of "phage nanobiotechnology," phages serve as more than genetic vehicles for peptide exposure and growth-supporters. Instead they operate as nanoparticles or nanotubes for medical application and other biotechnologies. Phage particles have diverse shapes with well-defined geometry and uniformity in a nanometer range that is suitable for nanofabrication. Thanks to the elucidation of atomic structures of many phage species along with the progress in molecular biology, phages can be precisely engineered according to predetermined geometry and functions with genetic fusion of different functionalities at a nanoscale level.[40] Bacteriophage-derived nanomaterial possesses unique attributes with structure and function encrypted in genomic DNA of bacteriophage, which can be intentionally modified or even rewritten using routine genetic engineering techniques to construct phages with unique surface architectures and emergent properties at will. Within the framework of "phage nanobiotechnology," phage nanoparticles act as versatile building blocks that provide scaffolds for creation of new nanomaterials.[5,41–43]

This chapter reviews the development of the phage-derived nanomaterials and their application in targeted drug delivery for medical diagnosis and molecular imaging. The strategies designed to overcome biological obstacles and to make phage therapy viable are also described.

3. Bacteriophage-Targeting Nanomedicines

3.1. *The use of bacteriophage for targeted drug delivery*

3.1.1. *Whole bacteriophage particles for targeted drug delivery involving the use of chemical conjugation*

Itai Benhar's group has developed a new drug delivery platform involving the use of filamentous bacteriophage with genetic manipulation for displaying of targeting ligands and chemical conjugation for the loading of drugs.[18] This targeted drug delivery system is composed of three critical components, including (1) a filamentous phage to serve as a drug carrier for high loading capacity; (2) a drug linker to provide chemical conjugation or/and control drug release; (3) a phage-displayed targeting ligand to confer host-targeting selectivity (Figs. 3(a) and 3(b)). This phage-based drug system has been tested for targeted delivery of anticancer and antibacterial nanomedicines.

Exploration of the 2,700 copies of the major coat protein pVIII of a filamentous phage for drug conjugation has led to a variety of different enabling techniques utilizing different chemistries that allow a large payload of a variety of drug candidates with different physicochemical properties and for different medical indications.[53,54] In the case of a hydrophilic drug, hygromycin coupled with a phage with an estimated loading capacity of upto 10^4 drug molecules per phage[53] has been prepared using the simple 1-Ethyl-3-[3-dimethylaminopropyl] carbodiimide (EDC) chemistry to form an amide bond between the exposed carboxyl side chains on the phage coat protein pVIII and a free primary amine on the drug. Amphiphilic doxorubicin was also linked to the phage surface using similar chemistry.[54] In order to conjugate a hydrophobic drug, such as hemolytic antibiotic chloramphenicol, to a biological phage entity in aqueous solution, an initial chemical transformation of chloramphenicol into a hydrophilic prodrug is necessary to allow for its conjugation with the PVIII major coat protein of filamentous phage.[53] The resultant chloramphenicol-phage conjugates had 3,000 drug molecules per phage, showing pathogenic bacteria-targeting and bacterial growth inhibition. An alternative protocol using branched, hydrophilic linkers like neomycin containing aminoglycoside for the solvation of hydrophobic drugs, was designed to further improve the effectiveness of this phage-based system in

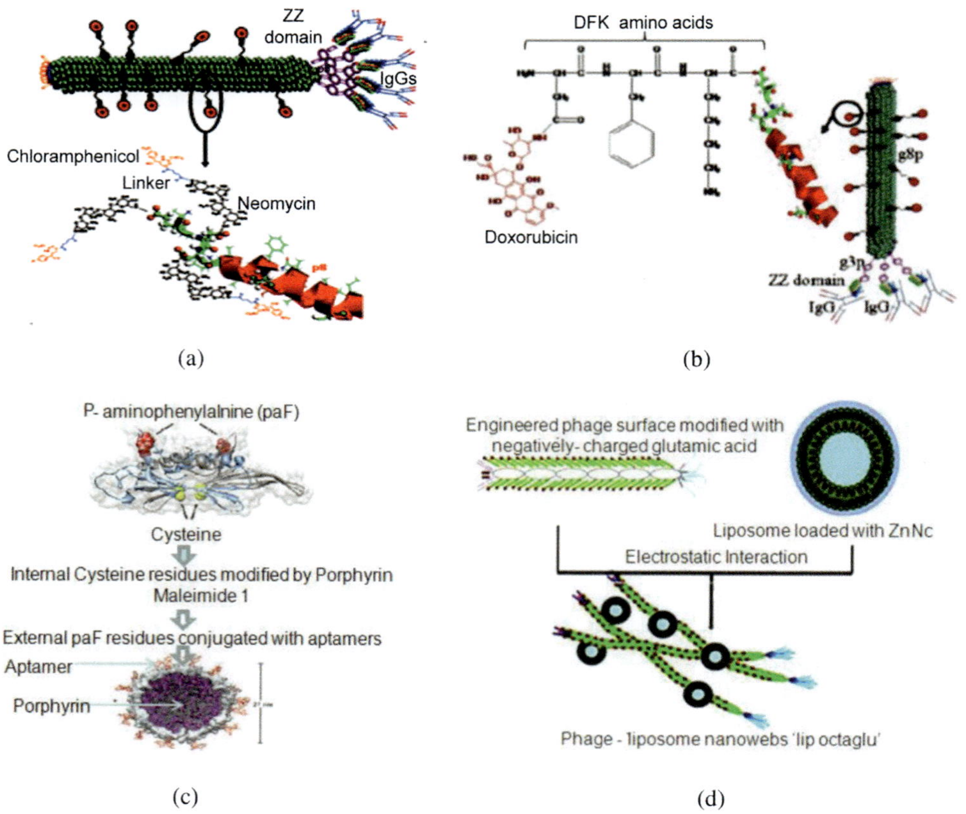

Fig. 3. The whole bacteriophage-mediated targeted drug delivery. (a) Chemical conjugation of a filamentous phage with hydrophobic drug — chloramphenicol via a hydrophilic neomycin linker and indirect immobilization of targeting antibodies (IgG) to phage surface via an IgG-binding ZZ domain genetically-displayed in the tip of PIII minor coat protein; (b) Chemical conjugation of a filamentous phage with amphiphilic drug — doxorubicin via a tripeptide (DFK) linker and indirect immobilization of targeting antibodies (IgG) to phage surface via an IgG-binding ZZ domain genetically-displayed in the tip of PIII minor coat protein. DFK is able to mediate control release of doxorubicin due to its susceptibility to cleavage of lysosomal protease; (c) MS2 phage-based targeted delivery systems with drug conjugated to internal surface of MS2 capsid and cell-specific DNA aptamers coupled to outside surface. (d) Self-assembly of filamentous phages with drug-loaded liposomes driven by electrostatic interaction between engineered filamentous M13 phages displaying anionic peptides and cationic liposomes loading with drug cargoes for photodynamic cancer therapy. Cited from Refs. 53, 54, 56 and 57.

the delivery of water-insoluble drugs (Fig. 3(a)). In an alternative scheme, one of the neomycin-containing six primary amines was linked to a chloramphenicol molecule, while the other amines were left for further conjugation to carboxyl residues of the phage coat proteins by EDC chemistry. This protocol led to beneficial improvement on drug loading capacity with over

40,000 chloramphenicol per phage. The high-drug loading was sufficient to produce complete growth inhibition of bacteria with no impact on the phage integrity.[53,55]

Phage-display platform technology can engineer phage nanoparticles to target various diseases including cancer and infection. The targeted ligands can be either short peptides directly displayed on the coat protein of filamentous phage or antibodies (IgG) indirectly immobilized to the surface via an IgG-binding ZZ domain displayed in the tip of the PIII minor coat protein (Figs. 2(a) and 2(b)). Targeting of hygromycin-conjugated phage nanoparticles via specific antibodies against receptors overexpressed on cancer cell membranes, such as ErbB2 and EGFR, resulted in intracellular delivery and release of drugs, that triggered the growth inhibition of the target cells *in vitro* with a potentiation factor of >1,000 over the free drug.[54] Antibody-targeted phage nanoparticles carrying a large payload of chloramphenicol linked with the aminoglycoside neomycin, induced complete growth inhibition of pathogens *Staphylococcus aureus*, *Streptococcus pyogenes* and *E. coli* with an 200,000-fold increase in chloramphenicol potency when compared to the free drug.[53,55]

Controllable release mechanism has also been incorporated into the design of phage-drug conjugates. By exploring the susceptibility of DFK peptide to cleavage by lysosomal protease-cathepsin-B, an engineered phage (designated fUSE5-ZZ-(g8p) DFK phage) was generated with phage major coat protein pVIII-displayed DFK tripeptide. Doxorubicin was coupled with fUSE5-ZZ-(g8p)-DFK phage via DFK as a linker (Fig. 3(b)). After *in vitro* treatment of the doxorubicin-DFK-phage conjugates with cathepsin-B, ~3,500 doxorubicin molecules were released from each DFK phage, suggesting a complete release of conjugated doxorubicin and potential for maximal theoretical effect.[54] Taking together, the genetically and chemically dual-modified filamentous phage nanoparticles improved the drug potency and showed promise as a multifunctional drug delivery system capable of enhancing drug solubility, enabling target specificity and providing controlled drug release.

In addition to filamentous bacteriophage, chemical modification of a MS2 phage offers another platform for the design of multivalent, targeted drug delivery systems. The bacteriophage MS2 is a spherical virus with a size of 27 nm. MS2 capsid is made of 180 identical protein monomers forming the internal and external surfaces of the hollow protein shell, which has surface-accessible amino acids with reactive moieties (lysine, glutamic acid and cysteine residues), allowing either chemical conjugation or genetic manipulation to display peptides, aptamers, vitamins, glycoproteins, etc.[58–62] MS2 VLPs have a relatively large interior volume able to accommodate diverse cargos.[58,60,63] Thus, MS2 bacteriophage has attracted a considerable interest in targeted drug delivery and molecule/cell imaging.[58] Recently, Francis's group modified the outside surface of the MS2 phage capsid with Jurkat cell receptor-specific DNA aptamers for targeting Jurkat leukemia T-cells and the interior surface of the capsid with porphyrins for photodynamic therapy (Fig. 3(c)). They showed that the bifunctionalized MS2 capsids achieved specific targeting and selective killing of Jurkat cells compared to co-cultured, non-targeted cells. This multivalent system is expected to deliver a variety of cargos to different cellular targets.[56]

3.1.2. *Whole bacteriophage particles for targeted delivery of cytotoxic drugs by complexation with drug carriers*

Complexation of bacteriophage with liposomal drug carriers represents a new design in the use of bacteriophage-based targeted drug delivery. The self-assembly of phages with drug-loaded liposomes is driven by electrostatic interaction between engineered filamentous M13 phages displaying anionic peptides and cationic liposomes loaded with drug cargos. This system has been applied for the delivery of photosensitizer ZnNC in photodynamic cancer therapy, Fig. 3(d).[57,64] Phage-liposome complex encapsulated with ZnNC formed a web-like nanostructure and showed 20% more tumor cell killing than the treatment with liposome-loading ZnNC alone. Further engineering of both cancer-targeting peptides with phage minor coat protein pIII and liposome-binding peptides with phage major coat protein pVIII is expected to achieve ligand-mediated active targeting of drugs to pathogenic sites, such as tumors or inflammations.

3.1.3. *Phage-derived peptides for targeted drug delivery via chemical conjugation with therapeutic agents*

Phage-displayed peptide libraries have been emerging as a powerful resource for the discovery of tumor-specific peptides.[27,28,65] The early use of phage-derived peptides for targeted drug delivery involved the chemical conjugation of targeting peptide with therapeutics. The first report in this regard involved the conjugation of doxorubicin with a tumor blood vessel-homing phage peptide (containing either an α-integrin-binding Arg-Gly-Asp motif or an Asn-Gly-Arg motif) to improve *in vivo* tumor vasculature targeting and the therapeutic index of doxorubicin.[26] Additionally, a prostate-homing phage peptide (designated SMSIARL, identified by *in vivo* biopanning), was coupled with a proapoptotic peptide which has pharmacological effect on mitochondrial membranes. Testing these drug-phage peptide conjugates in human tumor xenografts in mice demonstrated selective damage of the prostate tissue and sparing of non-targeted organs.[66]

3.1.4. *Phage-displayed peptides for targeted drug delivery via chemical conjugation with pharmaceutical nanocarriers*

Despite proof-of-concept of phage peptide-mediated targete drug delivery, the effectiveness of the drug-peptide conjugates is limited due to insufficient drug-carrying capacity with usually less than 10 drug molecules per targeting moiety.[67] The drug-payload can be enhanced by the conjugation of the phage-derived peptides with drug-encapsulated nanoparticles. Phage-derived homing peptides have been successfully used as targeting moieties to deliver pharmaceutical nanocarriers, such as liposomes, to tumors. Tripeptide RGD, selected from phage-displayed libraries against endothelial and melanoma cells, has been incorporated into liposomes after being coupled with a lipid anchor.[68] Doxorubicin-loaded, RGD-targeted liposomes demonstrated targeting to integrin-expressing cells *in vitro* and improved therapeutic efficacy *in vivo* in a C26 colon carcinoma mouse model.[68] Tumor vasculature-targeting peptide PC5–52 discovered using an *in vivo* phage display, was modified with a hydrophobic residue and used to target the doxorubicin-loaded liposomes to tumors in mice prepared with human lung and oral cancer xenografts and increased mouse survival rate.[69]

3.1.5. *Landscape phage fusion protein-based approach*

While direct and fast screening for tumor-targeted peptides via the phage display technology provides a promising approach to promote the targeting of various pharmaceutical agents,[9] traditional phage display approaches for targeted drug delivery assume the necessity of a chemical synthesis of identified tumor-selective peptides and their conjugation with therapeutic agents or drug-loaded pharmaceutical nanocarriers (such as liposomes and micelles). Chemical modification of targeting peptides with drug molecules or a hydrophobic anchor could complicate the preparation process, and be less efficient when scaled up as a standardized pharmaceutical preparation. Most importantly, chemical conjugation potentially alters the properties of the targeting ligands[70] and anticancer propensity.[71]

To simplify the procedure and exclude chemical modifications, our group has developed an innovative approach which relies on the direct use of the landscape phage fusion protein, an intermediate in the phage display process

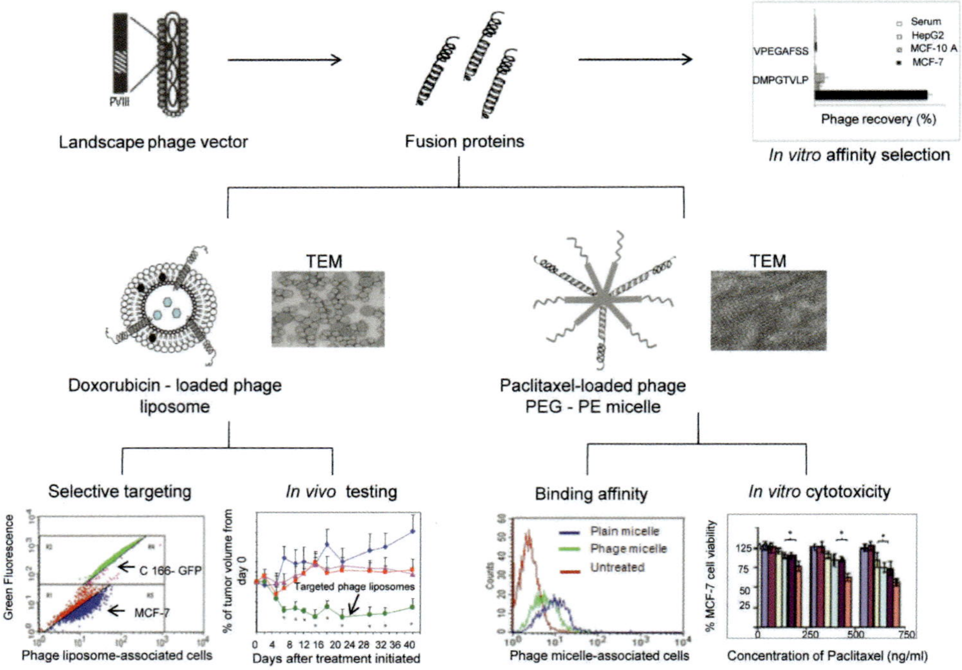

Fig. 4. The use of breast cancer MCF-7 cell- specific landscape phage fusion protein for targeted delivery of liposomal doxorubicin and micellar paclitaxel with enhanced antitumor activity *in vitro* and *in vivo*. Cited from Refs. 71, 77, 78 and 79.

with host-specific peptides fused to the N-terminus of the phage pVIII coat protein, as a targeting moiety for the delivery of drug-loaded liposomes[71–77] or micelles (Fig. 4).[78,79]

3.1.5.1. The use of landscape phage fusion protein for targeted delivery of liposomal drugs

The rationale behind this approach is based on the intrinsically membraneophilic property of the phage pVIII major coat protein, which allows for its spontaneous insertion into bacterial membranes during phage infection[80–82] as well as into the lipid bilayer of liposomes.[83] A thermodynamically favorable interaction of the hydrophobic domain of phage coat protein with the lipid bilayer drives the transmembrane insertion of phage coat protein pVIII.[70] Since the landscape phage ("fusion phage") is constructed via in-frame insertion of the gene encoding for the targeting peptide into the gene of the phage pVIII major coat protein, the membranous nature of the pVIII major coat protein is retained even when a short foreign targeting peptide is fused to its N-terminus.[5] A fusion protein carrying the tumor-cell binding peptide inherits the major structural features of the "wild-type" major coat protein VIII. It has a positively charged C-terminus (amino acids 45–55) which can navigate the protein through the bilayer membrane, a highly hydrophobic "membraneophilic" segment (amino acids 27–40) that allows the protein to accommodate readily in the membrane. The hydrophilic N-terminus (amino acids 1–26) displays the cancer cell-binding sequence on the outer leaflet of the membrane bilayer.[71]

We have used the phage display platform to screen different phage-derived fusion proteins specific for different targets, including MCF-7 breast cancer cells,[71] PC3 prostate cancer cells[75] and streptavidin,[76] and have incorporated different landscape phage fusion proteins into the liposomal membranes in such a fashion that the cancer-specific hybrid phage pVIII coat protein has incorporated into the liposome membrane via its C-terminal hydrophobic segment, while the targeted peptide fused to its N-terminal was exposed on the surface of doxorubicin-loaded liposomes and served as a targeting moiety (Fig. 4). The formed nanoparticles maintained their liposomal morphology and had a substantial drug retention after phage protein incorporation. The resultant phage-liposome formulations, including both MCF-7 breast cancer cell specific-and PC3 prostate cancer cell-specific phage liposomes, showed a high degree of binding affinity and specificity toward the targeted cells and a significantly higher cytotoxicity toward target cells than non-targeted formulations, but this effect was absent with non-target cells.[71,75] Notably, both subcutaneous and orthotopic MCF-7 xenografted mice treated with MCF-7-targeted phage-Doxil had greater tumor shrinkage than those treated with non-targeted formulations.[77] The enhanced anticancer activity was verified by enhanced apoptosis, reduced proliferation and extensive necrosis as a result of improved tumor delivery of doxorubicin.[77] Treatments demonstrated no serious toxicity to mice.[77] These results suggest that MCF-7-targeted phage-Doxil promises to be an active and tolerable chemotherapy for breast cancer therapy.[77]

3.1.5.2. The use of landscape phage fusion protein for targeted delivery of hydrophobic drug with polymeric micelles as a nanocarrier

To meet the need for targeted delivery of hydrophobic drugs, our group has combined a polymeric micelle-based drug delivery system with a cancer cell-specific phage peptide as a targeting ligand (Fig. 4)[78,79]. Polymeric micelles represent an efficient system for the delivery of a broad variety of hydrophobic drugs.[85] Loading such drugs into the hydrophobic micelle core dramatically enhances drug solubility and improves bioavailability. The amphiphilic nature of phage fusion coat protein allows its stable incorporation into the core of polymeric micelles. Using paclitaxel as a model drug, the self-assembled nanopreparation composed of MCF-7-specific landscape phage fusion protein and PEG-PE micelles demonstrated improved cancer cell selectivity and cytotoxicity *in vitro*.[78] Animal studies showed that the targeted phage-micelles triggered a dramatic tumor reduction and extensive necrosis as a result of improved tumor delivery of paclitaxel. The enhanced anticancer effect was also accompanied by enhanced apoptosis and reduced proliferation *in vitro* and *in vivo* (unpublished data). Absence of hepatotoxicity and pathological changes in tissue sections of vital organs, together with the maintenance of overall health of mice following treatment, further support its translational potential as an effective and safe chemotherapy for improved breast cancer treatment (unpublished data).

3.2. *Bacteriophage for targeted delivery of nucleic acids*

3.2.1. *Phage vectors targeting antibiotic-resistant bacteria*

Recently, phages have been explored to increase bacterial susceptibility to antibiotics. The rationale behind this effort is that phages can serve as a vehicle for the delivery of genes required to reverse the pathogens' drug resistance, thereby restoring their sensitivity to antibiotics. One such strategy is based on targeting several non-essential gene networks of the gram-negative bacterium *E. coli* using filamentous phage M13 as a gene vector.[86] The versatility of this phage platform allows for the design of different vectors able to target various gene networks including those regulating non-SOS gene networks, such as SoxR and CsrA, or those modulating sensitivity to antibiotics, such as OmpF. Engineered M13 phage that overexpressed *lexA3 for* targeting the SOS DNA repair system enhanced the killing of antibiotic-resistant bacteria and decreased antibiotic resistance (Fig. 5(a)). The *lexA3* modified phages rendered the bacteria several folds more sensitive to three major classes of bactericidal antibiotics, including ampicillin, a β-lactam antibiotic; ofloxacin, a quinolone antibiotic and gentamicin gentamicin, an aminoglycoside antibiotic. In addition, phage-mediated gene therapy can serve as an antibiotic adjuvant. Its combination with conventional antibiotics increased the survival of mice infected with pathogenic *E. coli*, demonstrating its antibiotic-enhancing effect.[86]

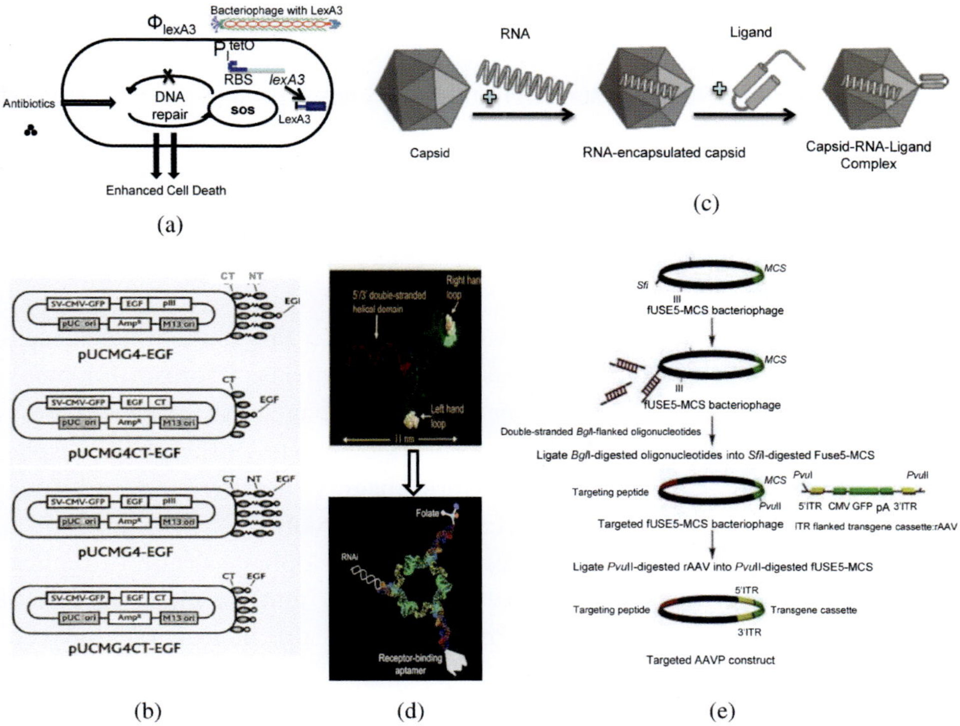

Fig. 5. Selected phage-based gene delivery vectors. (a) Phage vectors targeting antibiotic-resistant bacteria; (b–e) Phage vectors targeting mammalian cells by the use of phagemid vectors (b), MS2-based VLPs (c) and Packaging RNA (pRNA) (d) as bio-inspired gene delivery vectors and (e) Hybrid AAV/Phage vector (AAVP) (Cited from Refs. 86, 89, 101 and 102.)

Instead of direct administration to patients, targeting environmental pathogens is another tactic in the use of phage-mediated gene delivery to reverse bacteria resistance to antibiotics.[87] In this scenario, the specially-constructed phages with cloning of a drug-sensitizing DNA cassette that renders antibiotic-sensitivity to bacteria, were sprayed on the hospital surfaces followed by their lysogenization with the pathogens.[88] The engineered phages gradually reversed the occurrence of drug-resistant pathogens.

Reversal of bacterial resistance to streptomycin or nalidixic acid has been used to support this principle. The engineering of phages carrying sufficient copies of *rpsL* restored the antibiotic sensitivity of streptomycin-resistant bacteria. Additionally, nalidixic acid-resistant bacteria lysogenized with phages encoding *gyrA* showed nalidixic acid sensitivity. Extensive treatment with such specifically-constructed lysogenizing phages on environmental pathogens would enrich antibiotic-treatable pathogens. This enriched sensitive population may further inhibit the incidence of new antibiotic-resistant pathogens by occupying their ecological niche. The use of various phage-constructs carrying versatile sensitizing genes renews the promise of treating antibiotic-resistant infections with antibiotics.[88]

3.2.3. *Targeted gene delivery to mammalian cells*

3.2.3.1. The use of phagemid vectors

A wild-type bacteriophage on its own is unable to target mammalian cells due to its nature as prokaryotic viruses without intrinsic tropism for mammalian cell. However, re-engineering of bacteriophage with surface expressing ligands recognizing mammalian receptors by means of the phage-display technique enables the prokaryotic bacteriophage viruses to bind and internalize into mammalian cells for infection of eukaryotic cells. A phagemid is a plasmid containing DNA sequences and a bacteriophage origin of replication. The use of phagemid vectors has made the genetic manipulation of bacteriophage similar to plasmid vectors used in production of viral vectors.[89,90] Phagemids co-infected with helper phages are packaged into phage particles and allowed to replicate in *E. coli* as a double stranded plasmid. This co-infection with helper phage allows the production of single-stranded phagemid DNA.[91]

In 1998, Larocca's group first investigated the use of phagemid vectors in targeting bacteriophage to mammalian cells for gene delivery.[92] The first prototype was designed to target mammalian cells expressing basic fibroblast growth factor (FGF2) receptor using biotin-modified phages indirectly linked to FGF2 via the interaction between biotin and streptavidin.[92] In later studies, the group endowed the mammalian cell tropism to an M13 phage display vector by inserting a CMV regulated reporter gene (GFP) and gene encoding targeting FGF2 or epidermal growth factor EGF (Fig. 5(b)).[93,94] Those modified phage particles led to genetic transduction of target cells as indicated by GFP expression following their binding to and internalization into targeted cells via receptor-mediated endocytosis. The genetic transduction showed ligand-, dose- and time-dependence and ligand's cognate receptor-specificity.[95] Poul *et al.*'s group also verified specific gene delivery to cells expressing HER2 receptor using a engineered phage displaying a single-chain antibody against HER2.[96] Further optimization of phage vector design and culture conditions for transfection improved the transduction level of phagemid vectors in certain cell lines from an initial 1%–4% to 45% of transfected cells.[95]

3.2.3.2. MS2 bacteriophage capsid as a bioinspired gene delivery vector

MS2 is a RNA bacteriophage with an icosahedral shape and diameter of 27–34 nm. Self-assembly and packaging of viral RNA genome is driven by specific interaction of MS2 coat protein with a 19-nucleotide stem-loop structure-the translational repressor/assembly initiation signal (TR).[97] This self-assembling mechanism of MS2 bacteriophage has been used to construct RNA phage virus-like particles (VLPs) for loading and delivery of various cargoes including antisense oligonucleotides,[63] RNA,[98] microRNA[99] or other therapeutic molecules (Fig. 5(c)).[61,101]

Several strategies have been used to assemble MS2-based VLPs. The synthesized ODNs against p120 messenger RNA were initially modified by covalent coupling with the 19-nucleotide stem-loop TR of MS2 phage. Loading of the synthesized TR-ODNs into MS2 phages is achieved by direct soaking of ODN-TR into capsid shells of the MS2 phages free of wild-type RNA, leading to diffusion of the TR-ODNs into the empty capsids through the pores on the MS2 particle, accompanied by sequence-specific interactions between the TR sequence and every coat protein dimer within the capsid shell. ODNs encapsulated within MS2 capsid make them more stable in cell culture than their free countparts. To confer the tropism to mammalian cells, the synthesized VLP encapsulated with ODNs was surface-modified with transferrin (Fig. 5(c)). The anti-p120 ODNs delivered by the transferrin-targeted MS2-based VLPs were internalized into targeted cells via a receptor-mediated endocytotic pathway and increased cell-specific tumor toxicity.[62] In another study, antisense RNA targeting the 50-untranslated region (UTR) and internal ribosome entry site (IRES) of the hepatitis C virus (HCV) were packed into MS2 VLPs using a two-plasmid coexpression system composed of one plasmid (termed pET- 28(b) encoding the maturase and coat protein of MS2, and another plasmid (pACYCDuet-1) encoding antisense RNA coupled with the modified stem-loop sequence of MS2.[103,104] The synthesized MS2 VLPs were further surface-modified with the HIV-1 Tat_{47-57} cell penetration peptide and showed enhanced intracellular delivery with efficient inhibition on HCV translation.[98] MS2 VLPs have also been used to deliver Micro-RNA. Human pre-miR146a RNA was packaged into MS2 VLP–Tat conjugates using a single-plasmid coexpression system. This construct induced Micro-RNA expression *in vitro* and *in vivo*.[99]

The utility of MS2 VLPs for the delivery of RNA-based nucleic acids offers several advantages. Thanks to the well-understood mechanism regarding the self-assembly and packaging of RNA phage MS2, construction of MS2 VLPs is easy and effective. Packaging of RNAs within bacteriophage

MS2 capsids protects RNA cargoes from nuclease-mediated degradation and other undesirable interactions with host molecules. MS2 VLPs are also desirable as a drug carrier in that they are biocompatible, biodegradable and stable at wide a variety of conditions. MS2 VLPs can be readily scaled-up for production.[62]

3.2.3.3. Packaging RNA (pRNA) as a bioinspired gene delivery vector

Inspired by the essential role of a bacteriophage phi29 encoded 117nucleotide RNA (termed packaging RNA or "pRNA") in the DNA packaging motor of phi29,[105] phi29 pRNA has been explored as a gene vector to escort nucleic acid molecules, including therapeutic siRNA,[101] ribozyme[106] and receptor-binding RNA aptamer.[101]

The structure of bacteriophage phi29 pRNA packaging motor is composed of a central domain and a double-stranded helical domain at its 5'/3'-paired ends (Fig. 5(d)). The 5'/3' helical domain can be manipulated by the addition or deletion of nucleotides at the 5'-end preceding nucleotide #23 and at the 3'-end following nucleotide #97 without influencing the correct folding of the intermolecular interaction domain. Taking advantage of this unique characteristic, chimeric pRNAs can be engineered to serve as a gene delivery vehicle via the replacement of the pRNA helical region with foreign gene sequences.[101,107,108] Additionally, the pRNA central domain contains two interlocking left-hand and right-hand loops able to form stable dimers, trimers or hexamers via hand-in-hand interactions (Fig. 5(d)). These structural and molecular features make phi29 pRNA useful for bottom-up assembly of new bionanomaterials and for construction of multifunctional vehicles for the delivery of siRNA and other therapeutic molecules for advanced medical application.[107]

The feasibility of using of phi29 pRNA in targeting gene delivery has been extensively investigated by Guo *et al.*'s group.[101,107] They constructed various chimeric pRNA nanoparticles harboring different nucleic acids, including pRNA/siRNA(GFP), pRNA/siRNA(luciferase), pRNA/siRNA(BAD), and pRNA/siRNA(survivin) as well as ribozyme(survivin).[101] These chimeric pRNA/siRNA complexes demonstrated cleavability by Dicer and the ability to release functional double-strand siRNAs from phi29 pRNA carriers. Functional assays further showed that pRNA/siRNA (GFP) effectively inhibited GFP gene expression in a dose-dependent manner. Both pRNA/siRNA (survivin) and pRNA/ribozyme (survivin) induced cancer cell death by down-regulation of the anti-apoptosis factor, survivin.[109]

To endow RNA nanoparticles with targeted specificity, this group further constructed chimeric pRNA nanoparticles carrying targeting ligands (including pRNA/CD4-binding aptamer and pRNA/folate.[110] Combining pRNA vehicles carrying targeting ligands with those harboring siRNA or other therapeutic molecules via the interaction of engineered right and left RNA loops, generated either chimeric pRNAs dimers[109] or chimeric pRNAs trimer[101] (Fig. 5(d)). Among them include a dimer containing both pRNA/siRNA (survivin) and pRNA/aptamer(CD4), a dimer composed of pRNA/siRNA(firefly) and pRNA/folate. These chimeric pRNA constructions demonstrated specific cell targeting, effective intracellular delivery and cytoplasmic release of siRNA, which led to gene silencing. Animal trials further justified the effectiveness of chimeric siRNA vectors on tumor growth inhibition.[109]

A chimeric pRNA constructed for the delivery of hepatitis B virus-targeted ribosome, demonstrated that pRNA nanoparticles escorted the hammerhead ribozyme which enhanced HBV inhibition. The pRNA nanocarriers increased stability against enzymatic degradation, retained the structural integrity and enhanced activity of the ribozyme.[106]

A most recent article reported the feasibility of folate-specific pRNA bacterial phage nanoparticles as a vehicle for the delivery of microRNAs (AmiRs) targeting the 3'UTR of the genome of coxsackievirus B3 (CVB3) which cause myocarditis in young infants and adults. This system efficiently inhibited CVB3 replication with little effect on interferon (IFN) induction, for demonstrating the potential for the pRNA-ligand mediated delivery strategy for specific targeting with AmiRs to treat viral infection.[111]

3.2.3.4. The use of hybrid AAV/phage (AAVP)

In general, phage-based vectors have been considered poor gene delivery vehicles despite their potential advantages, presumably due to the undesirable post-targeting fate of the single-stranded DNA of the phage genome.[102] In an attempt to improve phages as gene vectors targeting mammalian cells, a ligand-directed vector as a hybrid between AAV and phage (termed AAVP) has been developed.[112] The idea of creation of the AAVP vector has been derived from favorable combination of the biological characteristic of eukaryotic viruses with prokaryotic ones. Eukaryotic viruses show superior transgene delivery and transduction but suffer from inherent tropism for mammalian cells.[113] Prokaryotic phages with no tropism for mammalian cells,[114] have limited efficiency in the transduction of mammalian cells.[94,96,115] Furthermore, the technical feasibility in generation of novel chimeras combining an AAV with an M13-derived filamentous phage fd-tet[116] displaying short peptides[117] is based on DNA genome's compatibility between recombinant adeno-associated virus (AAV) and single-stranded phage. By the insertion of an eukaryotic gene cassette from AAV into an intergenomic region of an fd-tet phage clone displaying the double-cyclic peptide CDCRGDCFC, which binds to α-integrins, a cell-surface receptor overexpressed in both tumor and endothelial cells,[26,118] targeted chimeric AAV/phage viruses (termed RGD-4C AAVP) were constructed[112] (Fig. 5(e)). The RGD-4C AAVP was bound and internalized into mammalian cells expressing α-integrins, and led to transduction of mammalian cells *in vitro*. *In vivo* systemic administration of the RGD-4C AAVP vector showed specific targeting of tumor vasculature in human Kaposi sarcoma (KS1767 cell) subcutaneous mouse xenografts.[112] The improved mammalian transduction efficiency induced by the targeted AAVP over ordinary phage-based vectors was a result of improved transgenic delivery through maintenance of the entire mammalian transgene cassette, better persistence of episomal DNA and formation of concatamers of the transgene cassette.[102]

3.3. *Bacteriophage-mediated vaccine delivery*

3.3.1. *Bacteriophage-based subunit vaccines*

Carriers are believed to be essential to boost the efficacy of peptide-based vaccination. Filamentous bacteriophage are prone to be uptake by the reticuloendothelium system (RES)[119] and are processed efficiently by action of the MHC class I and II pathways, even in non-professional antigen presenting cells.[120] These remarkable features may explain the unexpected ability of phage virions displaying foreign T-cell epitopes to prime strong T-helper-dependent CTL responses. These findings have important implications in the development of peptide-based subunit vaccines using the filamentous bacteriophage as a scaffold.

Early in 1988, McCutchan TF *et al.*, first proved the concept of application of recombinant phage in the delivery of foreign peptide antigens.[121] Since then, genetically or chemically engineered phages have been developed as useful vectors for the delivery of antigenic peptides for a variety of disease indications including herpes simplex virus infection,[122,123] HIV,[124,125] Alzheimer's disease,[126] melanoma,[127,128] and other cancers (Table 2)[129–131]. Using well-established phage display platform techniques, genetic display of immunogenic peptides on the surface of phage can be achieved by cloning of oligonucletides encoding fusion peptides in a phage vectors followed by

Table 2. Selective examples of recombinant phages as carriers for immunogenic peptides.

Display Platform	Immunogenic peptides	Targeted disease	Test system	Injection route	Results	References
Phage fd-PVIII	Single epitopes of the glycoprotein G of HSV-2	HSV-2*	Balb/c mice	s.c.***	Protective immunity against HSV-2	(138)
Phage fd-PVIII	Conserved antigenic determinants of HIV-1 reverse transcriptase (RTase)	HIV-1**	—		Humoral & cellular response	(120)
Phage fd-PVIII	EFRH residues within α -amyloid peptide	Alzheimer	Transgenic mouse	i.p.***	Reduced myloid plaques	(139)
Phage M13-PVIII	B16-F10 mouse melanoma specific peptide	Melanoma	C57BL/6 bearing s.c B16-F10	p.t.***	Tumor growth Inhibition	(34)
Phage fd-PVIII	Candida albicans heat shock protein 90 (SE-CA-HSP90)	Candidiasis	C57BL/6J mice		Protective humoral & cellular responses	(140)
Phage M13-PVIII	Peptides mimicking epitopes on Rabies Virus glycoprotein	Rabies	BALB/c mice	s.c.	Humoral response and IgG production	(141)
T7 Bacteriophage –10 B	Peptide mimicking M2 protein on Influenza A virus	Influenza A	BALB/c mice	s.c.	Triggers balanced Ig1 and IgG2a response and T cell response	(142)
T7 Bacteriophage –10 B	HER-2 derived immunodominant H-2Kd restricted CTL epitope	Breast cancer	BALB/c mice	s.c.	Humoral & cellular responses	(143)
Hybrid plasmid-phage-PVIII	Candida albicans Aspartyl proteinase epitope VKYTS (sap)	Candidiasis	—	i.v.***	Humoral response	(144)

*Herpes Simplex Virus Type 2; ** Human Immunodeficiency Virus Type 1; ***s.c.; subcutaneous, i.p.; intraperitoneal, i.v.; intravenous, p.t.; peritumoral transcriptional expression on the surface of phage capsid. The production of a transcriptional peptide epitope fused to phage virions is easy to produce and cost-effective.

Unlike the conventional subunit-based vaccine carriers which often need adjuvant for enhanced immune response, filamentous bacteriophage as a vector is able to enhance immunogenicity of peptide antigen in the presence or absence of adjuvant.[132] This inherent immunostimulating ability may be due to particulate nature of phages[51,133] and capacity of phage-mediated polyvalent delivery to generate a more efficient immune response. A phage-mediated subunit vaccine elicits a much more limited anti-phage antibody response when compared to the commonly-used protein carriers. This is due to the restricted distribution of B-cell epitopes within the outer 10–12 amino acid residues of phage pVIII coat protein and outer domain of 5 copies of PIII coat protein.[134]

3.3.2. *Bacteriophage-mediated DNA/RNA vaccination*

Since the early 1990s, DNA-based genetic vaccines have emerged as an attractive approach for immunization.[135–137] A DNA vaccine is composed of a plasmid DNA encoding the antigen of interest under the control of a mammalian promoter (i.e. CMV intron A) and is produced in the bacteria. DNA vaccine stimulates protective response against pathogens by the inoculation of DNA encoding protective antigens into target hosts for antigen expression followed by processing and presentation by antigen presenting cells and stimulation of both humoral and cellular immune responses. This unique capability of DNA vaccination to introduce antigen to the host immune system allowing stimulation of strong Th1 type CD4+ T cells and CD8+ cytotoxic T cells makes it distinct from recombination protein-based vaccines. DNA vaccination overcomes problems in the production of recombinant protein vaccines, including high-cost protein purification and antigenic protein vulnerable to structural destabilization. Genetic vaccines are also safer than traditional attenuated or inactivated vaccines. No anti-DNA antibodies have been elicited even after multiple dosing of DNA vaccines.[145]

Despite its cost-effectiveness and ease-of-production when compared to recombinant protein or subunit vaccines, both preclinical and clinical efficacy of naked DNA immunization is far from ideal.[146] Bacteriophage-mediated DNA vaccination offers an alternative platform for an improved immune response with DNA vaccines. In this approach, DNA vaccination cassettes are first cloned into standard bacteriophage λ vectors which are then propagated in an *E.coli* host. Immune response is stimulated by the injection of these whole phage particles to targeted host. Furthermore, phage-based DNA vaccine can be designed to preferentially target phage particle to particular immune cell types with a phage-display cell-specific protein ligand on the surface of virion, including galactose residues recognizing hepatic receptors in the liver[147,148] or phage-display peptides targeting dendritic cells or Langerhans cells.[149]

This enhanced immune response by bacteriophage-mediated DNA vaccine could be attributable to the several advantages offered by the bacteriophage vectors. First of all, the package of DNA encoding protective antigens within bacteriophage stabilizes and protects DNA from nuclease degradation in the *in vivo* setting. Being particulate, bacteriophage are taken up by antigen presenting cells (APCs), cleared from the circulation and targeted to the spleen and liver Kuppfer cells. Thus, a phage serves as an extremely simple and effective carrier for targeting DNA vaccines to APCs. The standard bacteriophage λ vectors with a large cloning capacity from 15 kb to 20 kb in size allow the use of a single phage for co-delivery of multiple vaccines, including multiple genes of parasites with complicated life cycles and for several disease serotypes.[150,151]

Possible host genomic integration of DNA following administration of DNA vaccine is, however, a major limitation in the use of DNA vaccines. Since mRNA vaccines are risk-free with host genome integration[152,153] and in the generation of autoimmune disease or anti-DNA antibodies[154] it is considered as a safer substitution to DNA vaccines. The naked RNA immunization has been tested for different indications including cancer, infectious disease and allergy. However, naked RNA vaccines fail to show sufficient potency in eliciting an immune response presumably attributable to its intrinsic instability and poor *in vivo* delivery.[155] To overcome the problem, bacteriophage-based RNA vaccine delivery is emerging as a promising platform with the advantages including their genetic tractability, a high degree of stability, high production capacity, inexpensive production and inherent biological safety in mammalian cells.[156,157] Investigation of MS2 VLPs in the development of an anti-HIV vaccine[158] by packaging heterologous antigen-encoding HIV-1 gag mRNA produced in S. cerevisiae has shown that the MS2 VLP-mediated mRNA delivery effectively induced antigen-specific humoral immune responses in BALB/c mice.

4. Phage-mediated Molecular Imaging

With an in-depth understanding and the discovery of disease-related biological markers, coupling of a variety of molecular imaging modalities with biomarkers has emerged as a useful tool in disease diagnosis and in clinical response detection. Among the variety of probes for molecular and cell imaging, small peptides offer advantages over macromolecules, including a desirable pharmacokinetic profile and favorable tissue distribution patterns, enhanced permeability and a relatively lower toxicity and immunogenicity. Peptides are more amenable to chemical modification.[159,160]

Although a great deal of disease-specific peptides have been identified by the screening of phage display peptide libraries, only a few phage-derived peptides tagged with either radiochelators or fluorescent dyes, have been developed into non-invasive imaging agents.[161] This limited success may be attributed to the different binding characteristics between peptides displayed on the phage surface and chemically synthesized peptides.[65] The phage surface provides the microenvironment, allowing a selected fusion peptide to adopt an ideal conformation for targeting specificity. Chemical modification of a short peptide with a radiochelator or fluorescent tracer may alter its binding affinity to its target.

However, phages as biological nanoparticles have unique attributes to serve as *in vivo* imaging agents. Phage is both genetically-modifiable to express multiple copies of disease-targeting peptides and amenable to covalent conjugation or physical complexation with numerous tags such as biotin and fluorophores. The rod-like structure of a filamentous phage also provides a scaffold to assemble multiple imaging nanoparticles, such as gold nanorods[162] and iron oxides,[163,164] and allows for multivalent interaction with one-dimensional optic materials such as SWNTs along the length of the filamentous phage, resulting in multivalent assembly and multimodality imaging.[165] Notably, the structure of filamentous phage can spatially segregate multiple functionalities and so avoid interference between targeted ligands and imaging probes.[163] In addition, a filamentous phage which is able to deliver imaging nanoparticles to receptors with a higher payload, allows amplification of imaging signal with improved detection sensitivity.[163,165] This is especially useful for imaging those hard-to-detect areas with a low density of target receptors.[165]

The phage-based imaging platform can also adapt other signal amplification strategies, including bio-orthogonal chemistry,[166] pretargeting,[167] biomimetic amplification[168] and communicating nanoparticles[169] to further enhance detection sensitivity. Therefore, the integration of recombinant phage nanoparticles with different medical imaging modalities should allow the development of multivalent, multifunctional, biological nanoparticles for *in vivo* targeting and medical imaging.

4.1. *The use of phage in in vivo SPECT radioimaging agents*

Bacteriophage labeled with radiochelators (termed radiolabeled phages) have been developed for imaging infection. This is based on the inherent affinity and specificity of phages for bacteria.[170] 99mTc- radiolabeled filamentous phage showed over four-fold increase in *in vitro* binding specificity towards host bacterial when compared to non-host, as well as the greater accumulation in the infected thigh of mice than in an uninfected thigh,[170] indicating the usefulness of radiolabeled phages serving as SPECT agents in distinguishing infection from inflammation and for the identification of infection type.[171]

However, the use of 99 mTc radiolabel filamentous phages for SPECT imaging suffered from poor resolution due to undesirable *in vivo* biodistribution of radiolabeled phages in non-targeted organs, including liver, gut and spleen[119] and rapid clearance of phage by the RES within 48 h after intravenous (*i.v.*) injection into a mouse.[119]

To minimize the off-targeted accumulation of radiolabeled phages, a two-step, pre-targeting strategy has been developed with the specific aim to increase the signal-to-noise ratio of SPECT imaging.[172] In this strategy, engineered phages displaying tumor-homing peptides and biotin were first injected, followed by the second injection at 4 h later with radiolabeled streptavidin. The time interval of 4 h was to allow the clearance of the majority of unbound phages before injection of the radiolabeled streptavidin, therefore reducing background noise.[172] To provide proof of the concept, engineered phages, displaying α-melanocyte stimulating hormone (α-MSH) specific for the melanocortin-1 receptor overexpressed on malignant melanoma,[172] were utilized for *in vivo* small animal SPECT imaging for melanoma. Although the two-step pretargeting approach achieved improved murine melanomas targeting, a significant level of radioactivity was observed within non-targeted tissues at 24 h post-injection of the ^{111}In-radiolabeled SA. This problem could be attributed to a suboptimal biodistribution and retention of^{111}In-radiolabeled SA in C57 mice bearing B16-F1 mouse melanoma tumors.[173–175]

In a continuing effort to improve the tumor uptake and to promote the clearance of the radiolabels, a three-step pretargeting protocol was designed by the replacement of ^{111}In-radiolabeled SA with^{111}In-labeled DOTA-biotin. The use of the small molecular weight of biotin was thought to favor faster clearance of excess radiolabel as well as less accumulation in non-target tissues. In the three-step pretargeting system, the biotinylated G1 phage was first injected followed by the second injection with avidin 4 h later and a third injection with ^{111}In-labeled DOTA-biotin. Phage-based SPECT/CT imaging with the three-step pretargeting method clearly revealed more efficient visualization of the pretargeted tumor accompanied by an increase in tumor uptake and retention of ^{111}In-labeled biotin in the PC-3 tumors bearing SCID mouse model.[176] These data provided proof of the principle of the utility of bifunctional, multivalent phage as a biological nanoparticle to facilitate the three-step pretargeted radio-imaging of prostate cancer.[176]

4.2. *The use of phage in in vitro and in vivo for optical imaging*

Coupling of disease-targeting phage-derived peptides to a fluorophore emitter offers method for non-invasive molecular imaging for early cancer detection and efficient monitoring of the pharmacodynamic response to anti-tumor therapy. Recently, a phage-displayed 9-mer peptide (designated as GX1GX1) specific for tumor vasculature endothelium, was identified by *in vivo* biopanning and chemically conjugated with a near-infrared (NIR)-emitting fluorescent dye Cy5.5 for small animal optical imaging of tumor angiogenesis, that showed high sensitivity and tumor specific targeting.[160] Phage peptide-based optical imaging probes have also been used to assess the therapeutical responses. A 6-mer peptide (CQRPPR) designated as ApoPep-1 has been identified to selectively target apoptotic cells in tumor tissue and in cell culture by binding to the histone H1 exposed on the surface of apoptotic cells. The significant increase in the accumulation of ApoPep-1-Cy7.5 conjugates in the camptothecin-treated tumors demonstrated the usefulness of ApoPep-1 as an imaging probe for monitoring of tumor apoptosis in response to the chemotherapy.[177] Another molecular imaging probes used to monitor the response to bevacizumab treatment, involved a linear phage peptide sequence (LLADTTHHRPWT) screened by *in vivo* phage display. This 12-mer phage peptide is specific for tumor vascular beds effectively treated with bevacizumab. Its conjugation with NIR fluorescent dyes demonstrated binding selectivity towards tumors treated with bevacizumab rather than the untreated ones.[36]

M13 filamentous bacteriophage has also been explored to assemble the fluorescent tracers for optical imaging of tumors. The rod-like shape of M13 allows multivalent interaction with one-dimensional fluorescent single-walled carbon nanotubes (SWNTs) along the length of the phage, resulting in stable complexes. Bifunctional M13 bacteriophage-SWNTs complexes achieved the second NIR window fluorescence imaging of tumors with specific targetability for the first time and detected deep tissues even at a low dosage (2 μg/mL) and up to 2.5 depth tissue like phantoms, showing the potential for early, non-invasive diagnosis and clinical procedures, such as intraoperative surgery.[178,179]

4.3. *In vivo magnetic resonance imaging*

M13 filamentous bacteriophage also provides a nanoscale scaffold to integrate molecular targeting and MRI imaging for detection of cancers[163] and other pathogenic sites such as atherosclerotic plaques.[180] Recently, Muller's group proposed the direct use of the engineered phage clones as a vehicle for the delivery of MRI contrast agents and synthesized the so-called "magnetophage" by covalent conjugation of a MRI contrast agent such as ultrasmall particles of iron oxide (USPIO) to the capsid of the selected phage E3 specific for the apoptotic biomarker phosphatidylserines (PS).[164] More recently, Kelly's group design a self-assembling magnetophage for the detection of prostate tumor by taking advantages of an engineered M13 filamentous bacteriophage serving as a vector integrating both a cancer-specific peptide for tumor molecule targeting and an anionic peptide motif (triglutamate) as a template for assembly of iron oxide nanoparticles based on electrostatic interaction.[163] These magnetophages retained target selectivity and showed enhanced MRI imaging contrast when compared to several clinically used MRI contrast agents, together with their stability in biological fluid, this makes them promising for *in vivo* MRI imaging.[163]

5. Design of Viable Phage-Based Nanomedicines Bypassing the Biological Obstacles

Despite the fact that phage therapy has been in medical practice for many decades in the former Soviet Union and Poland, clinical application of this therapy has mainly yielded conflicting and unpredictable results.[181] Phage therapy is still on the way to finding a position in mainstream medicine. Major concerns in the use of phage therapy include (1) phages neutralization by the host-defense system; (2) a narrow host range; (3) bacterial resistance to the phage; (4) toxin release;[39,182–185] (5) lack of sufficient pharmacokinetic and efficacy studies in human and a non-human animals. [182,186]

5.1. *Escape from inactivation by the host-defense system*

Given the immunogenicity of phages, they are rapidly-cleared by the RES upon *i.v.* injection.[119] One attempt to prolong *in vivo* viability of phages is based on a series of *in vivo* passages of phage coupled with clonal selection and purification. This approach has yielded a long-circulating bacteriophage with more than 15% longer persistence in the circulatory system of host when compared to the wild-type phage.[185,187] Genetic modification of phages provides another powerful way to confer anti-host-defense properties. By expression of complement-antagonizing peptide, interleukins, cytokines, autocrines and inhibitors of various cellular activating or inhibitory factors (*e.g.* inhibitors of macrophage activating factor), the engineered phages acquire ability to escape from clearance by the host-defense system.[188] In addition, manipulation of phages by the substitution of an amino acid within the phage capsid also produced a long-circulating bacteriophage with a 1000-fold longer persistence in the mouse circulatory system.[189,190]

Another effective technique to prolong the circulating time of phages is based on chemical conjugation of polymers to the phage surface. Surface modification may modulate the pharmacokinetics, biodistribution and the immunogenicity of phage. Indeed, PEGylated bacteriophages showed enhanced resistance against inactivation by the host defense system.[191] A list of synthetic and natural polymers could serve toward this end, includes polyethylene glycol, polypropylene glycol, polypropylene oxide, polyvinylpyrrolidone, polyvinyl alcohol, polybutanediol, polysaccharides, hyaluronic acid, collagen, albumin, dextran, carboxymethylcellulose and poly-D, Lamino acids.

5.2. *Expansion of host-targeting range*

Phages selectively target particular bacteria. The host specificity of phages produces less deleterious effects on the normal microbial balance of patients when compared to wide-spectrum antibiotics. On the other hand, this host specificity requires accurate indentification of targeted bacteria and selecting specific phages in order to make phage therapy effective. Therefore, successful phage therapy largely depends on qualified diagnostic laboratory personnel capable of accurate identification of pathogen and selection of effective phages for therapy. Phage cocktails are also a commonly used clinical practice to expand host range, since the mixed phages are able to collectively fight against a wide spectrum of bacteria.[192–194]

5.3. *Minimization of toxin release*

Toxin release is a major side-effect linked to the use of lytic-phage therapy. Lysates of gram-negative bacteria produce large amounts of target cell wall debris which may induce circulatory shock in patients.[195,196] Interaction of phages with target cells also releases endotoxin. Lysogeny or transduction could possibly transfer undesired, virulent genes into hosts. Compared to the use of lytic-phage, non-lytic phage produces minimal endotoxin release and is less toxic because it retains the structural integrity of bacterial cells during invasion.

Many efforts have been directed towards the construction of modified phages to minimize endotoxin release. Modified phages harboring non-lytic proteins and expressing a gene encoding the restriction endonuclease BglII has led to efficient killing of the host bacteria while reducing endotoxin release. Phages unable to be replicative have been engineered to reduce the risk of leaving lysogenic particles in patients after treatment.[184]

5.4. *Avoidance of phage resistance*

The incidence of bacterial resistance to phages (termed phage resistance) has significantly lower frequency (10^{-7} to 10^{-8} per cell) compared to that of antibiotic resistance (frequency of mutation for one specific gene is 10^{-5} per cell).[197] However, development of phage-resistant bacterial mutants could still be a potential concern. Hagens *et al.*, has reported that bacterial resistance to engineered M13mp18 prototype phage is due to the loss of the F-plasmid required for infection.[183] Phage cocktails targeting a set of distinct bacterial receptors could offer a solution that prevents the incidence of phage resistance since multiple phages invade and infect bacteria using different mechanisms. Using phage cocktails with multiple antibiotics also could enhance bacterial killing and reduce resistance to both phage and antibiotics. Thus, formulation of the "right" cocktail is essential for phage therapy to maximize its efficacy with minimal phage resistance.[197]

6. Conclusion

Well-established phage-display technology and advances in bioconjugation chemistry have enabled manipulation and modification of bacteriophage with unique structural and functional properties. The modifiable and controllable bacteriophage serves as a robust building block and scaffold in the design and development of biological materials and generate novel nanostructures for nanomedicine fabrication and molecule imaging application. Elucidation of the mechanism operating in bacteriophage assembly has led to emergence of vectors or biomemetics for smart drug and gene delivery. Despite progress, successful development of the bacteriophage nanotechnology requires clear regulatory guidelines and potential patentability to tempt pharmaceutical investment.[198]

References

1. G. P. Smith, Filamentous fusion phage: novel expression vectors that display cloned antigens on the virion surface. *Science* **228**, 1315–1317 (1985).
2. J. J. Devlin, L. C. Panganiban and P. E. Devlin, Random peptide libraries: a source of specific protein binding molecules. *Science* **249**, 404–406 (1990).
3. J. K. Scott and G. P. Smith, Searching for peptide ligands with an epitope library. *Science* **249**, 386–390 (1990).
4. S. E. Cwirla, E. A. Peters, R. W. Barrett and W. J. Dower, Peptides on phage: a vast library of peptides for identifying ligands. *Proc. Natl. Acad. Sci. USA* **87**, 6378–6382 (1990).
5. V. Petrenko, Evolution of phage display: from bioactive peptides to bioselective nanomaterials. *Exp. Opin. Drug. Deliv.* **5**, 825–836 (2008).
6. I. Benhar, Biotechnological applications of phage and cell display. *Biotechnol. Adv.* **19**, 1–33 (2001).
7. L. S. Jespers, A. Roberts, S. M. Mahler, G. Winter and H. R. Hoogenboom, Guiding the selection of human antibodies from phage display repertoires to a single epitope of an antigen. *Biotechnology (NY)* **12**, 899–903 (1994).
8. H. R. Hoogenboom, A. P. de Bruine, S. E. Hufton, R. M. Hoet, J. W. Arends and R. C. Roovers, Antibody phage display technology and its applications. *Immunotechnology* **4**, 1–20 (1998).
9. A. Sergeeva, M. G. Kolonin, J. J. Molldrem, R. Pasqualini and W. Arap, Display technologies: application for the discovery of drug and gene delivery agents. *Adv. Drug Deliv. Rev.* **58**, 1622–1654 (2006).
10. J. W. Kehoe and B. K. Kay, Filamentous phage display in the new millennium. *Chem. Rev.* **105**, 4056–4072 (2005).
11. J. A. Weidanz, K. F. Card, A. Edwards, E. Perlstein and H. C. Wong, Display of functional alphabeta single-chain T-cell receptor molecules on the surface of bacteriophage. *J. Immunol. Methods* **221**, 59–76 (1998).
12. M. Sioud and M. H. Hansen, Profiling the immune response in patients with breast cancer by phage-displayed cDNA libraries. *Eur. J. Immunol.* **31**, 716–725 (2001).
13. L. S. Jespers, J. H. Messens, A. De Keyser, D. Eeckhout, I. Van den Brande, Y. G. Gansemans, M. J. Lauwereys, G. P. Vlasuk and P. E. Stanssens, Surface expression and ligand-based selection of cDNAs fused to filamentous phage gene VI. *Biotechnology (NY)* **13**, 378–382 (1995).
14. S. E. Hufton, P. T. Moerkerk, E. V. Meulemans, A. de Bruine, J. W. Arends and H. R. Hoogenboom, Phage display of cDNA repertoires: the pVI display system and its applications for the selection of immunogenic ligands. *J. Immunol. Methods* **231**, 39–51 (1999).
15. P. Kwasnikowski, P. Kristensen and W. T. Markiewicz, Multivalent display system on filamentous bacteriophage pVII minor coat protein. *J. Immunol. Methods* **307**, 135–143 (2005).
16. C. Gao, S. Mao, G. Kaufmann, P. Wirsching, R. A. Lerner and K. D. Janda, A method for the generation of combinatorial antibody libraries using pIX phage display. *Proc. Natl. Acad. Sci. USA* **99**, 12612–12616 (2002).
17. M. B. Irving, O. Pan and J. K. Scott, Random-peptide libraries and antigen-fragment libraries for epitope mapping and the development of vaccines and diagnostics. *Curr. Opin. Chem. Biol.* **5**, 314–324 (2001).

18. I. Yacobyand and I. Benhar, Targeted filamentous bacteriophages as therapeutic agents. *Exp. Opin. Drug Deliv.* **5**, 321–329 (2008).
19. D. N. Krag, G. S. Shukla, G. P. Shen, S. Pero, T. Ashikaga, S. Fuller, D. L. Weaver, S. Burdette-Radoux and C. Thomas, Selection of tumor-binding ligands in cancer patients with phage display libraries. *Cancer Res.* **66**, 7724–7733 (2006).
20. G. A. Kuzmicheva, P. K. Jayanna, I. B. Sorokulova and V. A. Petrenko, Diversity and censoring of landscape phage libraries. *Protein Eng. Des. Sel.* **22**, 9–18 (2009).
21. L. Castagnoli, A. Zucconi, M. Quondam, M. Rossi, P. Vaccaro, S. Panni, S. Paoluzi, E. Santonico, L. Dente and G. Cesareni, Alternative bacteriophage display systems. *Comb. Chem. High Throughput Screen* **4**, 121–133 (2001).
22. J. Jiang, L. Abu-Shilbayeh and V.B. Rao, Display of a PorA peptide from Neisseria meningitidis on the bacteriophage T4 capsid surface. *Infect Immun.* **65**, 4770–4777 (1997).
23. S. Danner and J. G. Belasco, T7 phage display: a novel genetic selection system for cloning RNA-binding proteins from cDNA libraries. *Proc Natl Acad Sci U S A.* **98**, 12954–12959 (2001).
24. B. H. Lindqvist and S. Naderi, Peptide presentation by bacteriophage P4. *FEMS Microbiol. Rev.* **17**, 33–39 (1995).
25. E. Santi, S. Capone, C. Mennuni, A. Lahm, A. Tramontano, A. Luzzago and A. Nicosia, Bacteriophage lambda display of complex cDNA libraries: a new approach to functional genomics. *J Mol Biol.* **296**, 497–508 (2000).
26. W. Arap, R. Pasqualini and E. Ruoslahti, Cancer treatment by targeted drug delivery to tumor vasculature in a mouse model. *Science* **279**, 377–380 (1998).
27. R. Pasqualini and E. Ruoslahti, Organ targeting in vivo using phage display peptide libraries. *Nature* **380**, 364–366 (1996).
28. R. Pasqualini, E. Koivunen and E. Ruoslahti, Alpha v integrins as receptors for tumor targeting by circulating ligands. *Nat. Biotechnol.* **15**, 542–546 (1997).
29. D. Rajotte, W. Arap, M. Hagedorn, E. Koivunen, R. Pasqualini and E. Ruoslahti. Molecular heterogeneity of the vascular endothelium revealed by in vivo phage display. *J. Clin. Invest.* **102**, 430–437 (1998).
30. E. Ruoslahti, Targeting tumor vasculature with homing peptides from phage display. *Semin. Cancer Biol.* **10**, 435–442 (2000).
31. W. Arap, M. G. Kolonin, M. Trepel, J. Lahdenranta, M. Cardo-Vila, R. J. Giordano, P. J. Mintz, P. U. Ardelt, V. J. Yao, C. I. Vidal, L. Chen, A. Flamm, H. Valtanen, L. M. Weavind, M. E. Hicks, R. E. Pollock, G. H. Botz, C. D. Bucana, E. Koivunen, D. Cahill, P. Troncoso, K. A. Baggerly, R. D. Pentz, K. A. Do, C. J. Logothetis and R. Pasqualini, Steps toward mapping the human vasculature by phage display. *Nat. Med.* **8**, 121–127 (2002).
32. M. Essler and E. Ruoslahti, Molecular specialization of breast vasculature: a breast-homing phage-displayed peptide binds to aminopeptidase P in breast vasculature. *Proc. Natl. Acad. Sci. USA* **99**, 2252–2257 (2002).
33. T. I. Samoylova and B. F. Smith, Elucidation of muscle-binding peptides by phage display screening. *Muscle Nerve* **22**, 460–466 (1999).
34. F. Eriksson, W. D. Culp, R. Massey, L. Egevad, D. Garland, M. A. Persson and P. Pisa, Tumor specific phage particles promote tumor regression in a mouse melanoma model. *Cancer Immunol. Immunother.* **56**, 677–687 (2007).
35. P. Laakkonen, K. Porkka, J. A. Hoffman and E. Ruoslahti, A tumor-homing peptide with a targeting specificity related to lymphatic vessels. *Nat. Med.* **8**, 751–755 (2002).

36. Q. Cao, S. Liu, G. Niu, K. Chen, Y. Yan, Z. Liu and X. Chen, Phage display peptide probes for imaging early response to bevacizumab treatment. *Amino Acids* **41**, 1103–1112 (2011).
37. L. R. Krumpe and T. Mori, The use of phage-displayed peptide libraries to develop tumor-targeting drugs. *Int. J. Pept. Res. Ther.* **12**, 79–91 (2006).
38. O. H. Aina, R. Liu, J. L. Sutcliffe, J. Marik, C. X. Pan and K. S. Lam, From combinatorial chemistry to cancer-targeting peptides. *Mol Pharm.* **4**, 631–651 (2007).
39. A. R. M. Bradbury and J. D. Marks, Phage antibody libraries. In *Phage Display a Practical Approach*, T. Clackson and H. B. Lowman (Eds.). Oxford University Press (2004), pp. 253–259.
40. M. Manchester and P. Singh, Virus-based nanoparticles (VNPs): platform technologies for diagnostic imaging. *Adv. Drug Deliv. Rev.* **58**, 1505–1522 (2006).
41. S. Brumfield, D. Willits, L. Tang, J. E. Johnson, T. Douglas and M. Young, Heterologous expression of the modified coat protein of Cowpea chlorotic mottle bromovirus results in the assembly of protein cages with altered architectures and function. *J. Gen. Virol.* **85**, 1049–1053 (2004).
42. M. T. Klem, D. Willits, M. Young and T. Douglas, 2-D array formation of genetically engineered viral cages on au surfaces and imaging by atomic force microscopy. *J. Am. Chem. Soc.* **125**, 10806–10807 (2003).
43. P. A. Suci, D. L. Berglund, L. Liepold, S. Brumfield, B. Pitts, W. Davison, L. Oltrogge, K. O. Hoyt, S. Codd, P. S. Stewart, M. Young and T. Douglas, High-density targeting of a viral multifunctional nanoplatform to a pathogenic, biofilm-forming bacterium. *Chem. Biol.* **14**, 387–398 (2007).
44. C. F. Barbas, 3rd, A. S. Kang, R.A. Lerner and S.J. Benkovic, Assembly of combinatorial antibody libraries on phage surfaces: the gene III site. *Proc Natl Acad Sci U S A.* **88**, 7978–7982 (1991).
45. A. Nissim, H. R. Hoogenboom, I. M. Tomlinson, G. Flynn, C. Midgley, D. Lane and G. Winter, Antibody fragments from a 'single pot' phage display library as immunochemical reagents. *Embo J.* **13**, 692–698 (1994).
46. H. Gram, U. Strittmatter, M. Lorenz, D. Gluck and G. Zenke, Phage display as a rapid gene expression system: production of bioactive cytokine-phage and generation of neutralizing monoclonal antibodies. *J Immunol Methods.* **161**, 169–176 (1993).
47. C. Swimmer, S. M. Lehar, J. McCafferty, D. J. Chiswell, W. A. Blattler and B. C. Guild, Phage display of ricin B chain and its single binding domains: system for screening galactose-binding mutants. *Proc Natl Acad Sci U S A.* **89**, 3756–3760 (1992).
48. H. Pannekoek, M. van Meijer, R. R. Schleef, D. J. Loskutoff and C. F. Barbas, 3rd, Functional display of human plasminogen-activator inhibitor 1 (PAI-1) on phages: novel perspectives for structure-function analysis by error-prone DNA synthesis. *Gene.* **128**, 135–140 (1993).
49. B. L. Roberts, W. Markland, A. C. Ley, R. B. Kent, D. W. White, S. K. Guterman and R. C. Ladner, Directed evolution of a protein: selection of potent neutrophil elastase inhibitors displayed on M13 fusion phage. *Proc Natl Acad Sci U S A.* **89**, 2429–2433 (1992).
50. J. McCafferty, R. H. Jackson and D. J. Chiswell, Phage-enzymes: expression and affinity chromatography of functional alkaline phosphatase on the surface of bacteriophage. *Protein Eng.* **4**, 955–961 (1991).

51. J. Greenwood, A. E. Willis and R. N. Perham, Multiple display of foreign peptides on a filamentous bacteriophage. Peptides from Plasmodium falciparum circumsporozoite protein as antigens. *J Mol Biol.* **220**, 821–827 (1991).
52. F. Felici, L. Castagnoli, A. Musacchio, R. Jappelli and G. Cesareni, Selection of antibody ligands from a large library of oligopeptides expressed on a multivalent exposition vector. *J Mol Biol.* **222**, 301–310 (1991).
53. I. Yacoby, H. Bar and I. Benhar, Targeted drug-carrying bacteriophages as antibacterial nanomedicines. *Antimicrob. Agents Chemother.* **51**, 2156–2163 (2007).
54. H. Bar, I. Yacoby and I. Benhar, Killing cancer cells by targeted drug-carrying phage nanomedicines. *BMC Biotechnol.* **8**, 37 (2008).
55. I. Yacoby, M. Shamis, H. Bar, D. Shabat and I. Benhar, Targeting antibacterial agents by using drug-carrying filamentous bacteriophages. *Antimicrob. Agents Chemother.* **50**, 2087–2097 (2006).
56. N. Stephanopoulos, G. J. Tong, S. C. Hsiao and M. B. Francis, Dual-surface modified virus capsids for targeted delivery of photodynamic agents to cancer cells. *ACS Nano* **4**, 6014–6020 (2010).
57. S. Kalarical Janardhanan, S. Narayan, G. Abbineni, A. Hayhurst and C. Mao, Architectonics of phage-liposome nanowebs as optimized photosensitizer vehicles for photodynamic cancer therapy. *Mol. Cancer Ther.* **9**, 2524–2535 (2010).
58. J. M. Hooker, A. Datta, M. Botta, K. N. Raymond and M. B. Francis, Magnetic resonance contrast agents from viral capsid shells: a comparison of exterior and interior cargo strategies. *Nano Lett.* **7**, 2207–2210 (2007).
59. E. W. Kovacs, J. M. Hooker, D. W. Romanini, P. G. Holder, K. E. Berry and M. B. Francis, Dual-surface-modified bacteriophage MS2 as an ideal scaffold for a viral capsid-based drug delivery system. *Bioconjug. Chem.* **18**, 1140–1147 (2007).
60. G. J. Tong, S. C. Hsiao, Z. M. Carrico and M. B. Francis, Viral capsid DNA aptamer conjugates as multivalent cell-targeting vehicles. *J. Am. Chem. Soc.* **131**, 11174–11178 (2009).
61. M. Wu, W. L. Brown and P. G. Stockley, Cell-specific delivery of bacteriophage-encapsidated ricin A chain. *Bioconjug. Chem.* **6**, 587–595 (1995).
62. C. E. Ashley, E. C. Carnes, G. K. Phillips, P. N. Durfee, M. D. Buley, C. A. Lino, D. P. Padilla, B. Phillips, M. B. Carter, C. L. Willman, C. J. Brinker, C. Caldeira Jdo, B. Chackerian, W. Wharton and D. S. Peabody, Cell-specific delivery of diverse cargos by bacteriophage MS2 virus-like particles. *ACS Nano* **5**, 5729–5745 (2011).
63. M. Wu, T. Sherwin, W. L. Brown and P. G. Stockley, Delivery of antisense oligonucleotides to leukemia cells by RNA bacteriophage capsids. *Nanomedicine* **1**, 67–76 (2005).
64. D. Wohrle, M. Shopova, S. Muller, A. D. Milev, V. N. Mantareva and K. K. Krastev, Liposome-delivered Zn(II)-2,3-naphthalocyanines as potential sensitizers for PDT: synthesis, photochemical, pharmacokinetic and phototherapeutic studies. *J. Photochem. Photobiol. B* **21**, 155–165 (1993).
65. L. A. Landon and S. L. Deutscher, Combinatorial discovery of tumor targeting peptides using phage display. *J. Cell. Biochem.* **90**, 509–517 (2003).
66. W. Arap, W. Haedicke, M. Bernasconi, R. Kain, D. Rajotte, S. Krajewski, H. M. Ellerby, D. E. Bredesen, R. Pasqualini and E. Ruoslahti, Targeting the prostate for destruction through a vascular address. *Proc. Natl. Acad. Sci. USA* **99**, 1527–1531 (2002).
67. R. V. Chari, Targeted delivery of chemotherapeutics: tumor-activated prodrug therapy. *Adv. Drug Deliv. Rev.* **31**, 89–104 (1998).

68. P. Holig, M. Bach, T. Volkel, T. Nahde, S. Hoffmann, R. Muller and R. E. Kontermann, Novel RGD lipopeptides for the targeting of liposomes to integrin-expressing endothelial and melanoma cells. *Protein Eng. Des. Sel.* **17**, 433–441 (2004).
69. T. Y. Lee, C. T. Lin, S. Y. Kuo, D. K. Chang and H. C. Wu, Peptide-mediated targeting to tumor blood vessels of lung cancer for drug delivery. *Cancer Res.* **67**, 10958–10965 (2007).
70. D. F. Emerich and C. G. Thanos, Multifunctional peptide-based nanosystems for improving delivery and molecular imaging. *Curr. Opin. Mol. Ther.* **10**, 132–139 (2008).
71. T. Wang, G. G. D'Souza, D. Bedi, O. A. Fagbohun, L. P. Potturi, B. Papahadjopoulos-Sternberg, V. A. Petrenko and V. P. Torchilin, Enhanced binding and killing of target tumor cells by drug-loaded liposomes modified with tumor-specific phage fusion coat protein. *Nanomedicine* **5**, 563–574 (2010).
72. T. Wang, N. Kulkarni, D. Bedi, G. G. D'Souza, B. Papahadjopoulos-Sternberg, V. A. Petrenko and V. P. Torchilin. *In vitro* optimization of liposomal nanocarriers prepared from breast tumor cell specific phage fusion protein. *J. Drug Target.* **19**, 597–605 (2011).
73. T. Wang, N. Kulkarni, G. G. D'Souza, V. A. Petrenko and V. P. Torchilin, On the mechanism of targeting of phage fusion protein-modified nanocarriers: only the binding peptide sequence matters. *Mol. Pharm.* **8**, 1720–1728 (2011).
74. T. Wang, S. Yang, V. A. Petrenko and V. P. Torchilin, Cytoplasmic delivery of liposomes into MCF-7 breast cancer cells mediated by cell-specific phage fusion coat protein. *Mol. Pharm.* 7, 1149–1158 (2010).
75. P. K. Jayanna, D. Bedi, J. W. Gillespie, P. DeInnocentes, T. Wang, V. P. Torchilin, R. C. Bird and V. A. Petrenko, Landscape phage fusion protein-mediated targeting of nanomedicines enhances their prostate tumor cell association and cytotoxic efficiency. *Nanomedicine* **6**, 538–546 (2010).
76. P. K. Jayanna, V. P. Torchilin and V. A. Petrenko, Liposomes targeted by fusion phage proteins. *Nanomedicine* **5**, 83–89 (2009).
77. T. Wang, W. C. Hartner, J. W. Gillespie, K. P. Praveen, S. Yang, L. A. Mei, V. A. Petrenko and V. P. Torchilin, Enhanced Tumor Delivery and Antitumor Activity in Vivo of Liposomal Doxorubicin Modified with MCF-7-Specific Phage Fusion Protein. *Nanomedicine* **10**, 421–430 (2014).
78. T. Wang, V. A. Petrenko and V. P. Torchilin, Paclitaxel-loaded polymeric micelles modified with MCF-7 cell-specific phage protein: enhanced binding to target cancer cells and increased cytotoxicity. *Mol. Pharm.* 7, 1007–1014 (2010).
79. T. Wang, V. A. Petrenko and V. P. Torchilin, Optimization of Landscape Phage Fusion Protein-Modified Polymeric PEG-PE Micelles for Improved Breast Cancer Cell Targeting. *J Nanomedic Nanotechnol.* **S4**: 008 (2012).
80. M. K. Waldor *et al.* Phages: Their Role in Bacterial Pathogenesis and Biotechnology. Washington, DC: ASM Press (2005).
81. J. K. Broome-Smith, S. Gnaneshan, L. A. Hunt, F. Mehraein-Ghomi, L. Hashemzadeh-Bonehi, M. Tadayyon and E.S. Hennessey, Cleavable signal peptides are rarely found in bacterial cytoplasmic membrane proteins (review). *Mol Membr Biol.* **11**, 3–8 (1994).
82. A. Kuhn, Major coat proteins of bacteriophage Pf3 and M13 as model systems for Sec-independent protein transport. *FEMS Microbiol Rev.* **17**, 185–190 (1995).
83. B. L. Geller and W. Wickner, M13 procoat inserts into liposomes in the absence of other membrane proteins. *J. Biol. Chem.* **260**, 13281–13285 (1985).

84. M. Soekarjo, M. Eisenhawer, A. Kuhn and H. Vogel, Thermodynamics of the membrane insertion process of the M13 procoat protein, a lipid bilayer traversing protein containing a leader sequence. *Biochemistry* **35**, 1232–1241 (1996).
85. V. P. Torchilin, Micellar nanocarriers: pharmaceutical perspectives. *Pharm. Res.* **24**, 1–16 (2007).
86. T. K. Lu and J. J. Collins, Engineered bacteriophage targeting gene networks as adjuvants for antibiotic therapy. *Proc. Natl. Acad. Sci. USA* **106**, 4629–4634 (2009).
87. R. Edgar, N. Friedman, S. Molshanski-Mor and U. Qimron, Reversing bacterial resistance to antibiotics by phage-mediated delivery of dominant sensitive genes. *Appl. Environ. Microbiol.* **78**, 744–751 (2012).
88. J. Lederberg. Streptomycin resistance: a genetically recessive mutation. *J. Bacteriol.* **61**, 549–550 (1951).
89. D. Larocca, K. Jensen-Pergakes, M. A. Burg and A. Baird, Receptor-targeted gene delivery using multivalent phagemid particles. *Mol. Ther.* **3**, 476–484 (2001).
90. Y. Seow and M. J. Wood, Biological gene delivery vehicles: beyond viral vectors. *Mol. Ther.* **17**, 767–777 (2009).
91. S. S. Sidhu, Engineering M13 for phage display. *Biomol. Eng.* **18**, 57–63 (2001).
92. D. Larocca, A. Witte, W. Johnson, G. F. Pierce and A. Baird, Targeting bacteriophage to mammalian cell surface receptors for gene delivery. *Hum. Gene Ther.* **9**, 2393–2399 (1998).
93. P. D. Kassner, M. A. Burg, A. Baird and D. Larocca, Genetic selection of phage engineered for receptor-mediated gene transfer to mammalian cells. *Biochem. Biophys. Res. Commun.* **264**, 921–928 (1999).
94. D. Larocca, P. D. Kassner, A. Witte, R. C. Ladner, G. F. Pierce and A. Baird, Gene transfer to mammalian cells using genetically targeted filamentous bacteriophage. *Faseb J.* **13**, 727–734 (1999).
95. D. Larocca, M. A. Burg, K. Jensen-Pergakes, E. P. Ravey, A. M. Gonzalez and A. Baird, Evolving phage vectors for cell targeted gene delivery. *Curr. Pharm. Biotechnol.* **3**, 45–57 (2002).
96. M. A. Poul and J. D. Marks, Targeted gene delivery to mammalian cells by filamentous bacteriophage. *J. Mol. Biol.* **288**, 203–211 (1999).
97. P. G. Stockley, N. J. Stonehouse and K. Valegard, Molecular mechanism of RNA phage morphogenesis. *Int. J. Biochem.* **26**, 1249–1260 (1994).
98. B. Wei, Y. Wei, K. Zhang, J. Wang, R. Xu, S. Zhan, G. Lin, W. Wang, M. Liu, L. Wang, R. Zhang and J. Li, Development of an antisense RNA delivery system using conjugates of the MS2 bacteriophage capsids and HIV-1 TAT cell-penetrating peptide. *Biomed. Pharmacother.* **63**, 313–318 (2009).
99. Y. Pan, Y. Zhang, T. Jia, K. Zhang, J. Li and L. Wang, Development of a microRNA delivery system based on bacteriophage MS2 virus-like particles. *Febs J.* **279**, 1198–1208 (2012).
100. W. L. Brown, R. A. Mastico, M. Wu, K. G. Heal, C. J. Adams, J. B. Murray, J. C. Simpson, J. M. Lord, A. W. Taylor-Robinson and P. G. Stockley, RNA bacteriophage capsid-mediated drug delivery and epitope presentation. *Intervirology* **45**, 371–380 (2002).
101. A. Khaled, S. Guo, F. Li and P. Guo, Controllable self-assembly of nanoparticles for specific delivery of multiple therapeutic molecules to cancer cells using RNA nanotechnology. *Nano Lett.* **5**, 1797–1808 (2005).

102. A. Hajitou, R. Rangel, M. Trepel, S. Soghomonyan, J. G. Gelovani, M. M. Alauddin, R. Pasqualini and W. Arap, Design and construction of targeted AAVP vectors for mammalian cell transduction. *Nat. Protoc.* **2**, 523–531 (2007).
103. G. G. Pickett and D. S. Peabody, Encapsidation of heterologous RNAs by bacteriophage MS2 coat protein. *Nucleic Acids Res.* **21**, 4621–4626 (1993).
104. Y. Wei, C. Yang, B. Wei, J. Huang, L. Wang, S. Meng, R. Zhang and J. Li. RNase-resistant virus-like particles containing long chimeric RNA sequences produced by two-plasmid coexpression system. *J. Clin. Microbiol.* **46**, 1734–1740 (2008).
105. P. X. Guo, S. Erickson and D. Anderson, A small viral RNA is required for *in vitro* packaging of bacteriophage phi 29 DNA. *Science* **236**, 690–694 (1987).
106. S. Hoeprich, Q. Zhou, S. Guo, D. Shu, G. Qi, Y. Wang and P. Guo, Bacterial virus phi29 pRNA as a hammerhead ribozyme escort to destroy hepatitis B virus. *Gene Ther.* **10**, 1258–1267 (2003).
107. P. Guo, RNA nanotechnology: engineering, assembly and applications in detection, gene delivery and therapy. *J. Nanosci. Nanotechnol.* **5**, 1964–1982 (2005).
108. H. Liu, S. Guo, R. Roll, J. Li, Z. Diao, N. Shao, M. R. Riley, A. M. Cole, J. P. Robinson, N. M. Snead, G. Shen and P. Guo, Phi29 pRNA vector for efficient escort of hammerhead ribozyme targeting survivin in multiple cancer cells. *Cancer Biol. Ther.* **6**, 697–704 (2007).
109. S. Guo, F. Huang and P. Guo, Construction of folate-conjugated pRNA of bacteriophage phi29 DNA packaging motor for delivery of chimeric siRNA to nasopharyngeal carcinoma cells. *Gene Ther.* **13**, 814–820 (2006).
110. S. Guo, N. Tschammer, S. Mohammed and P. Guo, Specific delivery of therapeutic RNAs to cancer cells via the dimerization mechanism of phi29 motor pRNA. *Hum. Gene Ther.* **16**, 1097–1109 (2005).
111. X. Ye, Z. Liu, M. G. Hemida and D. Yang, Targeted delivery of mutant tolerant anti-coxsackievirus artificial microRNAs using folate conjugated bacteriophage Phi29 pRNA. *PLoS One* **6**, e21215 (2011).
112. A. Hajitou, M. Trepel, C. E. Lilley, S. Soghomonyan, M. M. Alauddin, F. C. Marini, 3rd, B. H. Restel, M. G. Ozawa, C. A. Moya, R. Rangel, Y. Sun, K. Zaoui, M. Schmidt, C. von Kalle, M. D. Weitzman, J. G. Gelovani, R. Pasqualini and W. Arap, A hybrid vector for ligand-directed tumor targeting and molecular imaging. *Cell* **125**, 385–398 (2006).
113. N. A. Kootstra and I. M. Verma, Gene therapy with viral vectors. *Annu. Rev. Pharmacol. Toxicol.* **43**, 413–439 (2003).
114. P. A. Barrow and J. S. Soothill, bacteriophage therapy and prophylaxis: rediscovery and renewed assessment of potential. *Trends Microbiol.* **5**, 268–271 (1997).
115. S. Piersanti, G. Cherubini, Y. Martina, B. Salone, D. Avitabile, F. Grosso, E. Cundari, G. Di Zenzo and I. Saggio, Mammalian cell transduction and internalization properties of lambda phages displaying the full-length adenoviral penton base or its central domain. *J. Mol. Med. (Berl)* **82**, 467–476 (2004).
116. A. N. Zacher, 3rd, C. A. Stock, J. W. Golden, 2nd and G. P. Smith, A new filamentous phage cloning vector: fd-tet. *Gene* **9**, 127–140 (1980).
117. G. P. Smith and J. K. Scott, Libraries of peptides and proteins displayed on filamentous phage. *Meth. Enzymol.* **217**, 228–257 (1993).
118. J. D. Hood, M. Bednarski, R. Frausto, S. Guccione, R. A. Reisfeld, R. Xiang and D. A. Cheresh, Tumor regression by targeted gene delivery to the neovasculature. *Science* **296**, 2404–2407 (2002).

119. J. Zou, M. T. Dickerson, N. K. Owen, L. A. Landon and S. L. Deutscher, Biodistribution of filamentous phage peptide libraries in mice. *Mol. Biol. Rep.* **31**, 121–129 (2004).
120. M. Gaubin, C. Fanutti, Z. Mishal, A. Durrbach, P. De Berardinis, R. Sartorius, G. Del Pozzo, J. Guardiola, R. N. Perham and D. Piatier-Tonneau, Processing of filamentous bacteriophage virions in antigen-presenting cells targets both HLA class I and class II peptide loading compartments. *DNA Cell Biol.* **22**, 11–18 (2003).
121. V. F. de la Cruz, A. A. Lal and T. F. McCutchan, Immunogenicity and epitope mapping of foreign sequences via genetically engineered filamentous phage. *J. Biol. Chem.* **263**, 4318–4322 (1988).
122. N. Bourne, L. R. Stanberry, D. I. Bernstein and D. Lew, DNA immunization against experimental genital herpes simplex virus infection. *J. Infect. Dis.* **173**, 800–807 (1996).
123. P. H. Nass, K. L. Elkins and J. P. Weir, Protective immunity against herpes simplex virus generated by DNA vaccination compared to natural infection. *Vaccine* **19**, 1538–1546 (2001).
124. S. Calarota, G. Bratt, S. Nordlund, J. Hinkula, A. C. Leandersson, E. Sandstrom and B. Wahren, Cellular cytotoxic response induced by DNA vaccination in HIV-1-infected patients. *Lancet* **351**, 1320–1325 (1998).
125. V. Kulkarni, M. Rosati, A. Valentin, B. Ganneru, A. K. Singh, J. Yan, M. Rolland, C. Alicea, R. K. Beach, G. M. Zhang, S. Le Gall, K. E. Broderick, N. Y. Sardesai, D. Heckerman, B. Mothe, C. Brander, D. B. Weiner, J. I. Mullins, G. N. Pavlakis and B. K. Felber, HIV-1 p24(gag) derived conserved element DNA vaccine increases the breadth of immune response in mice. *PLoS One* **8**, e60245 (2013).
126. Y. Okura, A. Miyakoshi, K. Kohyama, I. K. Park, M. Staufenbiel and Y. Matsumoto. Nonviral Abeta DNA vaccine therapy against Alzheimer's disease: long-term effects and safety. *Proc. Natl. Acad. Sci. USA* **103**, 9619–9624 (2006).
127. R. Xiang, H. N. Lode, T. H. Chao, J. M. Ruehlmann, C. S. Dolman, F. Rodriguez, J. L. Whitton, W. W. Overwijk, N. P. Restifo and R. A. Reisfeld, An autologous oral DNA vaccine protects against murine melanoma. *Proc. Natl. Acad. Sci. USA* **97**, 5492–5497 (2000).
128. S. T. Tagawa, P. Lee, J. Snively, W. Boswell, S. Ounpraseuth, S. Lee, B. Hickingbottom, J. Smith, D. Johnson and J. S. Weber, Phase I study of intranodal delivery of a plasmid DNA vaccine for patients with Stage IV melanoma. *Cancer* **98**, 144–154 (2003).
129. C. F. Hung, W. F. Cheng, K. F. Hsu, C. Y. Chai, L. He, M. Ling and T. C. Wu, Cancer immunotherapy using a DNA vaccine encoding the translocation domain of a bacterial toxin linked to a tumor antigen. *Cancer Res.* **61**, 3698–3703 (2001).
130. F. K. Stevenson, C. H. Ottensmeier, P. Johnson, D. Zhu, S. L. Buchan, K. J. McCann, J. S. Roddick, A. T. King, F. McNicholl, N. Savelyeva and J. Rice, DNA vaccines to attack cancer. *Proc. Natl. Acad. Sci. USA* **101**(Suppl 2), 14646–14652 (2004).
131. J. Rice, C. H. Ottensmeier and F. K. Stevenson, DNA vaccines: precision tools for activating effective immunity against cancer. *Nat. Rev. Cancer* **8**, 108–120 (2008).
132. R. N. Perham, T. D. Terry, A. E. Willis, J. Greenwood, F. di Marzo Veronese and E. Appella, Engineering a peptide epitope display system on filamentous bacteriophage. *FEMS Microbiol. Rev.* **17**, 25–31 (1995).
133. C. Demangel, S. Rouyre, P. M. Alzari, F. Nato, S. Longacre, P. Lafaye and J. C. Mazie, Phage-displayed mimotopes elicit monoclonal antibodies specific for a malaria vaccine candidate. *Biol Chem.* **379**, 65–70 (1998).

134. S. Kneissel, I. Queitsch, G. Petersen, O. Behrsing, B. Micheel and S. Dubel, Epitope structures recognised by antibodies against the major coat protein (g8p) of filamentous bacteriophage fd (Inoviridae). *J. Mol. Biol.* **288**, 21–28 (1999).
135. F. R. Vogel and N. Sarver, Nucleic acid vaccines. *Clin. Microbiol. Rev.* **8**, 406–410 (1995).
136. D. C. Tang, M. DeVit and S. A. Johnston, Genetic immunization is a simple method for eliciting an immune response. *Nature* **356**, 152–154 (1992).
137. J. B. Ulmer, J. J. Donnelly, S. E. Parker, G. H. Rhodes, P. L. Felgner, V. J. Dwarki, S. H. Gromkowski, R. R. Deck, C. M. DeWitt, A. Friedman *et al.*, Heterologous protection against influenza by injection of DNA encoding a viral protein. *Science* **259**, 1745–1749 (1993).
138. A.M. Grabowska, R. Jennings, P. Laing, M. Darsley, C.L. Jameson, L. Swift, and W.L. Irving, Immunisation with phage displaying peptides representing single epitopes of the glycoprotein G can give rise to partial protective immunity to HSV-2. *Virology.* **269**, 47–53 (2000).
139. D. Frenkel, I. Dewachter, F. Van Leuven, and B. Solomon, Reduction of beta-amyloid plaques in brain of transgenic mouse model of Alzheimer's disease by EFRH-phage immunization. *Vaccine.* **21**, 1060–1065 (2003).
140. G. Wang, M. Sun, J. Fang, Q. Yang, H. Tong, and L. Wang, Protective immune responses against systemic candidiasis mediated by phage-displayed specific epitope of Candida albicans heat shock protein 90 in C57BL/6J mice. *Vaccine.* **24**, 6065–6073 (2006).
141. M. Houimeland and K. Dellagi, Peptide mimotopes of rabies virus glycoprotein with immunogenic activity. *Vaccine.* **27**, 4648–4655 (2009).
142. H. Hashemi, S. Pouyanfard, M. Bandehpour, Z. Noroozbabaei, B. Kazemi, X. Saelens and T. Mokhtari-Azad, Immunization with M2e-displaying T7 bacteriophage nanoparticles protects against influenza A virus challenge. *PLoS One.* 7:e45765 (2012).
143. S. Pouyanfard, T. Bamdad, H. Hashemi, M. Bandehpour and B. Kazemi, Induction of protective anti-CTL epitope responses against HER-2-positive breast cancer based on multivalent T7 phage nanoparticles. *PLoS One.* 7:e49539 (2012).
144. Q. Yang, Q.P. Su, G.Y. Wang, D.Z. Wen, Y.H. Zhang, H.Z. Bao and L. Wang, Production of hybrid phage displaying secreted aspartyl proteinase epitope of Candida albicans and its application for the diagnosis of disseminated candidiasis. *Mycoses.* **50**, 165–171 (2007).
145. H. A. Smith, Regulation and review of DNA vaccine products. *Dev. Biol. (Basel)* **104**, 57–62 (2000).
146. E. Manickan, K. L. Karem and B. T. Rouse, DNA vaccines — A modern gimmick or a boon to vaccinology? *Crit. Rev. Immunol.* **17**, 139–154 (1997).
147. T. J. Molenaar, I. Michon, S. A. de Haas, T. J. van Berkel, J. Kuiper and E. A. Biessen. Uptake and processing of modified bacteriophage M13 in mice: implications for phage display. *Virology* **293**,182–191 (2002).
148. T. J. Curiel, C. Morris, M. Brumlik, S. J. Landry, K. Finstad, A. Nelson, V. Joshi, C. Hawkins, X. Alarez, A. Lackner and M. Mohamadzadeh, Peptides identified through phage display direct immunogenic antigen to dendritic cells. *J. Immunol.* **172**, 7425–7431 (2004).
149. M. J. McGuire, K. F. Sykes, K. N. Samli, L. Timares, M. A. Barry, K. Stemke-Hale, F. Tagliaferri, M. Logan, K. Jansa, A. Takashima, K. C. Brown and S. A. Johnston, A library-selected, Langerhans cell-targeting peptide enhances an immune response. *DNA Cell Biol.* **23**, 742–752 (2004).

150. J. R. Clark and J. B. March, bacteriophage-mediated nucleic acid immunisation. *FEMS Immunol. Med. Microbiol.* **40**, 21–26 (2004).
151. C. D. Jepson and J. B. March, bacteriophage lambda is a highly stable DNA vaccine delivery vehicle. *Vaccine* **22**, 2413–2419 (2004).
152. R. M. Conry, A. F. LoBuglio, M. Wright, L. Sumerel, M. J. Pike, F. Johanning, R. Benjamin, D. Lu and D. T. Curiel, Characterization of a messenger RNA polynucleotide vaccine vector. *Cancer Res.* **55**, 1397–1400 (1995).
153. J. Probst, B. Weide, B. Scheel, B. J. Pichler, I. Hoerr, H. G. Rammensee and S. Pascolo. Spontaneous cellular uptake of exogenous messenger RNA *in vivo* is nucleic acid-specific, saturable and ion dependent. *Gene Ther.* **14**, 1175–1180 (2007).
154. A. Bringmann, S. A. Held, A. Heine and P. Brossart, RNA vaccines in cancer treatment. *J. Biomed. Biotechnol.* **2010**, 623687 (2010).
155. W. F. Cheng, C. H. Hung, C. Y. Chai, K. F. Hsu, L. He, M. Ling and T. C. Wu, Enhancement of sindbis virus self-replicating RNA vaccine potency by linkage of herpes simplex virus type 1 VP22 protein to antigen. *J. Virol.* **75**, 2368–2376 (2001).
156. A. Ghaemi, H. Soleimanjahi, P. Gill, Z. Hassan, S. R. Jahromi and F. Roohvand, Recombinant lambda-phage nanobioparticles for tumor therapy in mice models. *Genet. Vaccines Ther.* **8**, 3 (2010).
157. H. A. Lankes, C. N. Zanghi, K. Santos, C. Capella, C. M. Duke and S. Dewhurst. *In vivo* gene delivery and expression by bacteriophage lambda vectors. *J. Appl. Microbiol.* **102**, 1337–1349 (2007).
158. S. Cellini, C. Fortini, E. Gallerani, F. Destro, E. B. Cofano, A. Caputo and R. Gavioli, Identification of new HIV-1 Gag-specific cytotoxic T lymphocyte responses in BALB/c mice. *Virol. J.* **5**, 81 (2008).
159. K. Chen and X. Chen, Design and development of molecular imaging probes. *Curr. Top Med. Chem.* **10**, 1227–1236 (2010).
160. K. Chen, L. P. Yap, R. Park, X. Hui, K. Wu, D. Fan, X. Chen and P. S. Conti, A Cy5.5-labeled phage-displayed peptide probe for near-infrared fluorescence imaging of tumor vasculature in living mice. *Amino Acids* **42**, 1329–1337 (2012).
161. S. L. Deutscher, Phage display in molecular imaging and diagnosis of cancer. *Chem. Rev.* **110**, 3196–3211 (2010).
162. A. C. Chen, C. McNeilly, A. P. Liu, C. J. Flaim, L. Cuttle, M. Kendall, R. M. Kimble, H. Shimizu and J. R. McMillan, Second harmonic generation and multiphoton microscopic detection of collagen without the need for species specific antibodies. *Burns* **37**, 1001–1009 (2011).
163. D. Ghosh, Y. Lee, S. Thomas, A. G. Kohli, D. S. Yun, A. M. Belcher and K. A. Kelly, M13-templated magnetic nanoparticles for targeted *in vivo* imaging of prostate cancer. *Nat. Nanotechnol.* **7**, 677–682 (2012).
164. J. Segers, C. Laumonier, C. Burtea, S. Laurent, L. V. Elst and R. N. Muller, From phage display to magnetophage, a new tool for magnetic resonance molecular imaging. *Bioconjug. Chem.* **18**, 1251–1258 (2007).
165. H. Yi, D. Ghosh, M. H. Ham, J. Qi, P. W. Barone, M. S. Strano and A. M. Belcher, M13 phage-functionalized single-walled carbon nanotubes as nanoprobes for second near-infrared window fluorescence imaging of targeted tumors. *Nano Lett.* **12**, 1176–1183 (2012).

166. J.B. Haun, N.K. Devaraj, S.A. Hilderbrand, H. Lee, and R. Weissleder. Bioorthogonal chemistry amplifies nanoparticle binding and enhances the sensitivity of cell detection. *Nat Nanotechnol.* **5**, 660–665 (2010).
167. N. K. Devaraj, R. Weissleder and S. A. Hilderbrand, Tetrazine-based cycloadditions: application to pretargeted live cell imaging. *Bioconjug. Chem.* **19**, 2297–2299 (2008).
168. D. Simberg, T. Duza, J. H. Park, M. Essler, J. Pilch, L. Zhang, A. M. Derfus, M. Yang, R. M. Hoffman, S. Bhatia, M. J. Sailor and E. Ruoslahti. Biomimetic amplification of nanoparticle homing to tumors. *Proc. Natl. Acad. Sci. USA* **104**, 932–936 (2007).
169. G. von Maltzahn, J. H. Park, K. Y. Lin, N. Singh, C. Schwoppe, R. Mesters, W. E. Berdel, E. Ruoslahti, M. J. Sailor and S. N. Bhatia, Nanoparticles that communicate *in vivo* to amplify tumour targeting. *Nat. Mater.* **10**, 545–552 (2011).
170. M. Rusckowski, S. Gupta, G. Liu, S. Dou and D. J. Hnatowich, Investigations of a (99m)Tc-labeled bacteriophage as a potential infection-specific imaging agent. *J. Nucl. Med.* **45**, 1201–1208 (2004).
171. M. Rusckowski, S. Gupta, G. Liu, S. Dou and D. J. Hnatowich, Investigation of four (99m)Tc-labeled bacteriophages for infection-specific imaging. *Nucl. Med. Biol.* **35**, 433–440 (2008).
172. J. R. Newton, Y. Miao, S. L. Deutscher and T. P. Quinn, Melanoma imaging with pretargeted bivalent bacteriophage. *J. Nucl. Med.* **48**, 429–436 (2007).
173. B. Schechter, R. Silberman, R. Arnon and M. Wilchek, Tissue distribution of avidin and streptavidin injected to mice. Effect of avidin carbohydrate, streptavidin truncation and exogenous biotin. *Eur. J. Biochem.* **189**, 327–331 (1990).
174. D. S. Wilbur, D. K. Hamlin, K. R. Buhler, P. M. Pathare, R. L. Vessella, P. S. Stayton and R. To, Streptavidin in antibody pretargeting. 2. Evaluation Of methods for decreasing localization of streptavidin to kidney while retaining its tumor binding capacity. *Bioconjug. Chem.* **9**, 322–330 (1998).
175. D. S. Wilbur, D. K. Hamlin, J. Sanderson and Y. Lin, Streptavidin in antibody pretargeting. 4. Site-directed mutation provides evidence that both arginine and lysine residues are involved in kidney localization. *Bioconjug. Chem.* **15**, 1454–1463 (2004).
176. J. R. Newton-Northup, S. D. Figueroa, T. P. Quinn and S. L. Deutscher, Bifunctional phage-based pretargeted imaging of human prostate carcinoma. *Nucl. Med. Biol.* **36**, 789–800 (2009).
177. K. Wang, S. Purushotham, J. Y. Lee, M. H. Na, H. Park, S. J. Oh, R. W. Park, J. Y. Park, E. Lee, B. C. Cho, M. N. Song, M. C. Baek, W. Kwak, J. Yoo, A. S. Hoffman, Y. K. Oh, I. S. Kim and B. H. Lee, *In vivo* imaging of tumor apoptosis using histone H1-targeting peptide. *J. Control. Release* **148**, 283–291 (2010).
178. S. Kim, Y. T. Lim, E. G. Soltesz, A. M. De Grand, J. Lee, A. Nakayama, J. A. Parker, T. Mihaljevic, R. G. Laurence, D. M. Dor, L. H. Cohn, M. G. Bawendi and J. V. Frangioni, Near-infrared fluorescent type II quantum dots for sentinel lymph node mapping. *Nat. Biotechnol.* **22**, 93–97 (2004).
179. M. A. Whitney, J. L. Crisp, L. T. Nguyen, B. Friedman, L. A. Gross, P. Steinbach, R. Y. Tsien and Q. T. Nguyen, Fluorescent peptides highlight peripheral nerves during surgery in mice. *Nat. Biotechnol.* **29**, 352–356 (2011).
180. C. Burtea, S. Laurent, E. Lancelot, S. Ballet, O. Murariu, O. Rousseaux, M. Port, L. Vander Elst, C. Corot and R. N. Muller, Peptidic targeting of phosphatidylserine for the MRI detection of apoptosis in atherosclerotic plaques. *Mol. Pharm.* **6**, 1903–1919 (2009).
181. J. N. Housby and N. H. Mann, Phage therapy. *Drug Discov. Today* **14**, 536–540 (2009).

182. C. R. Merril, D. Scholl and S. L. Adhya, The prospect for bacteriophage therapy in Western medicine. *Nat. Rev. Drug Discov.* **2**, 489–497 (2003).
183. S. Hagens and U. Blasi, Genetically modified filamentous phage as bactericidal agents: a pilot study. *Lett. Appl. Microbiol.* **37**, 318–323 (2003).
184. S. Hagens, A. Habel, U. von Ahsen, A. von Gabain and U. Blasi, Therapy of experimental pseudomonas infections with a nonreplicating genetically modified phage. *Antimicrob. Agents Chemother.* **48**, 3817–3822 (2004).
185. C. R. Merril, B. Biswas, R. Carlton, N. C. Jensen, G. J. Creed, S. Zullo and S. Adhya, Long-circulating bacteriophage as antibacterial agents. *Proc. Natl. Acad. Sci. USA* **93**, 3188–3192 (1996).
186. H. Brussow, Phage therapy: the Escherichia coli experience. *Microbiology* **151**, 2133–2140 (2005).
187. R. Capparelli, I. Ventimiglia, S. Roperto, D. Fenizia and D. Iannelli, Selection of an Escherichia coli O157:H7 bacteriophage for persistence in the circulatory system of mice infected experimentally. *Clin. Microbiol. Infect.* **12**, 248–253 (2006).
188. Z. Moradpour and A. Ghasemian, Modified phages: novel antimicrobial agents to combat infectious diseases. *Biotechnol. Adv.* **29**, 732–738 (2011).
189. C. L. Vitiello, C. R. Merril and S. Adhya, An amino acid substitution in a capsid protein enhances phage survival in mouse circulatory system more than a 1000-fold. *Virus Res.* **114**, 101–103 (2005).
190. R. M. C. Sankar, L. Adhya and C. R. Merril, Antibacterial therapy with bacteriophage genotypically modified to delay inactivation by the host defense system. **WO1997039118** A1, 1997.
191. K. P. Kim, J. D. Cha, E. H. Jang, J. Klumpp, S. Hagens, W. D. Hardt, K. Y. Lee and M. J. Loessner, PEGylation of bacteriophages increases blood circulation time and reduces T-helper type 1 immune response. *Microb. Biotechnol.* **1**, 247–257 (2008).
192. E. Denou, A. Bruttin, C. Barretto, C. Ngom-Bru, H. Brussow and S. Zuber, T4 phages against Escherichia coli diarrhea: potential and problems. *Virology* **388**, 21–30 (2009).
193. M. Merabishvili, J. P. Pirnay, G. Verbeken, N. Chanishvili, M. Tediashvili, N. Lashkhi, T. Glonti, V. Krylov, J. Mast, L. Van Parys, R. Lavigne, G. Volckaert, W. Mattheus, G. Verween, P. De Corte, T. Rose, S. Jennes, M. Zizi, D. De Vos and M. Vaneechoutte, Quality-controlled small-scale production of a well-defined bacteriophage cocktail for use in human clinical trials. *PLoS One* **4**, e4944 (2009).
194. L. D. Goodridge, Designing phage therapeutics. *Curr. Pharm. Biotechnol.* **11**, 15–27 (2010).
195. J. Boratynski, D. Syper, B. Weber-Dabrowska, M. Lusiak-Szelachowska, G. Pozniak and A. Gorski, Preparation of endotoxin-free bacteriophages. *Cell. Mol. Biol. Lett.* **9**, 253–259 (2004).
196. T. Matsuda, T. A. Freeman D. W. Hilbert, M. Duff, M. Fuortes, P. P. Stapleton and J. M. Daly, Lysis-deficient bacteriophage therapy decreases endotoxin and inflammatory mediator release and improves survival in a murine peritonitis model. *Surgery* **137**, 639–646 (2005).
197. C. R. Merril, D. Scholl and S. Adhya (Eds.), *Phage Therapy.* New York: Oxford University Press (2006).
198. L. Gravitz, Turning a new phage. *Nat. Med.* **18**, 1318–1320 (2012).

Chapter B3

Stimuli-Sensitive Nanostructured Systems for Biomedical Applications

*Barbara Blanco-Fernandez, Angel Concheiro and Carmen Alvarez-Lorenzo**

Department of Pharmacy and Pharmaceutical Technology,
University of Santiago de Compostela,
15782 Santiago de Compostela, Spain
**carmen.alvarez.lorenzo@usc.es*

Abstract

Efficient and safe drug delivery to the site of action is still a challenging task. Carriers that combine single components according to a given pattern have been shown advantageous over continuous structures. Nanostructured systems try to mimic the cost-effective way that Nature uses to prepare responsive functional structures that regulate in the body the transport and recognition of endogenous and exogenous substances. This chapter focuses on self-assembly of lipids and polymers as well as other bottom-up strategies (*e.g.* molecular imprinting) that render stimuli-responsive nanostructured particles with a variety of morphologies and capabilities. Applications as self-regulable and externally-tunable drug delivery systems are revisited with a particular attention on the performance under *in vivo* conditions.

1. Introduction

First generation of controlled release medicines was conceived to slowly discharge the drug in the administration via or site, in such a way that the blood concentration could be kept into the therapeutic range for prolonged time. The aim was to improve the adherence to the treatment, making the dosing intervals broader, and to enhance the efficacy/safety ratio, avoiding toxic concentration peaks.[1] These medicines were designed to be administered through diverse routes (*e.g.* oral, transdermal, parenteral) and to release the drug according to a mechanism that should not be altered by the physiological state of the patient. The dosage form should ensure drug access to the systemic blood in a predictable amount and rate, for subsequent distribution in the body (Fig. 1). These environmental-independent preprogrammed-release formulations were followed by site-specific oral release systems conceived for making the drug to enter into contact only with defined absorption areas. Localized release is attained by introducing components in the formulation that change their features along the gastrointestinal tract; namely, the solubility, swelling or chemical/enzymatic degradation of one or more components are dependent on the physicochemical conditions of the different regions of the gastrointestinal tract. Therefore, the release is intended to occur when the formulation "senses" changes in certain variables, such as pH, ionic strength, or enzyme activity. A further generation is that of the nanocarriers that can be absorbed as a whole or directly placed in the blood stream in order that they can control the access and the release of the

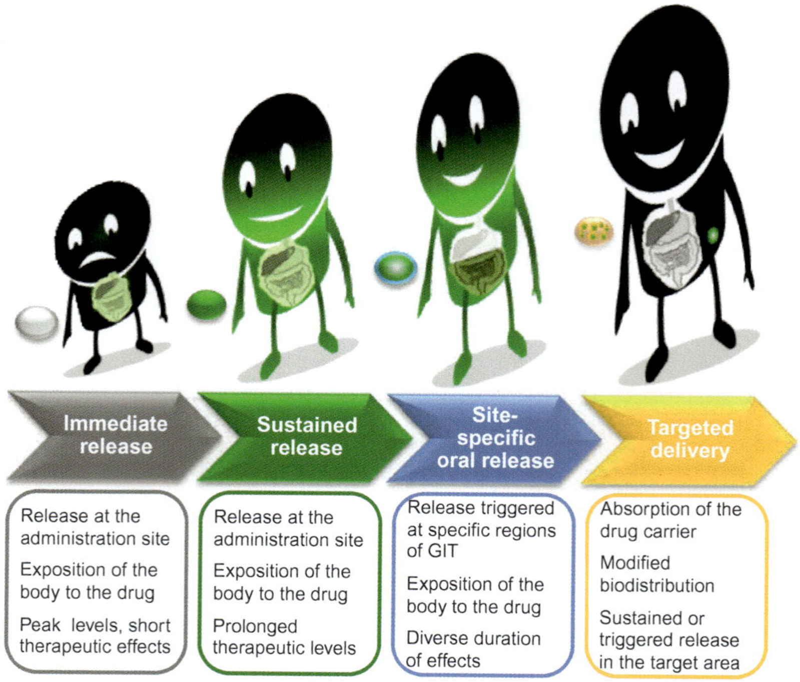

Fig. 1. Main evolution steps from immediate release dosage forms to targeted nanocarriers.

drug in a target site.[2] Thus, far from pursuing an unspecified distribution, the nanocarriers are intended to travel toward definite cells affected by a certain pathology and to deliver the drug to their surroundings or to particular cellular structures (Fig. 1). The purpose is to make accessible the drug only to the tissue where it has to exert the therapeutic effect, minimizing the administered dose, the access of the drug to healthy tissues, and then the untoward collateral effects. In addition to these benefits for the patients, it should be also noticed that this approach may be mandatory for novel active substances that have poor biopharmaceutical features.[3] The targeting can be accomplished by means of passive mechanisms derived of the morphological characteristics of nanosized drug carriers which prompt the capture by the mononuclear phagocyte system or, if stealth, the accumulation in enhanced permeability and retention (EPR) exhibiting tissues. Active mechanisms of targeting rely on the decoration of the nanocarrier with specific bio-adressins or the incorporation of magnetic components that can be externally driven to the tissue of interest. Once in the target region, the drug should be released, which can occur in a sustained, triggered or self-regulated way.[4]

Targeting requires that the drug is not prematurely leaked from the nanocarrier during the trip through the body until the place for delivery. Thus, the permeability of the nanocarrier to the drug should be minimal in the blood stream and the tissue matrix. In the absence of other mechanisms of release, the slow leakiness of the drug from the nanocarriers accumulated in the target area may be not enough for reaching therapeutic levels. This may lead to not only the failure of the treatment, but also the apparition of drug resistance phenomena, which are particularly problematic in the case of tumors and infections. Therefore, slow sustained release is not desirable for most situations. Faster release is possible if the nanocarrier totally or partially destabilizes when arrives to the target region. To do that, the nanocarrier should incorporate components that, when recognize certain variables ("stimuli"), are able to trigger the release.[5] The release could be completed in a single discharge or may fit to pulsate sequences, depending on the therapeutic demands.

Ideally, the nanocarrier should be sensitive to physiopathological changes of one or more internal variables such as pH, temperature or molecular markers. Namely, it should work as closed-loop or self-regulated system that senses the levels of a biological variable and responds switching drug release on/off or automatically adjusting the release rate.[6] Unfortunately for this purpose, in some pathological processes the magnitude of the changes is too small or the variable to be sensed is too unspecific to become a distinguishable stimulus for the nanocarrier. To overcome this limitation, the delivery system can be endowed with responsiveness to an external stimulus, such as light, electric field or ultrasound, that is applied to the target area when the nanocarrier has already accumulated in it. Externally-triggered nanocarriers work in open circuit, switching the release on/off as a function of the intensity and the time of duration of the external stimulus. These latter systems have the advantage of providing on-demand regulation of the release process, which can be adapted to the progress of the illness and the patient conditions.

Overall, the design of stimuli-responsive nanocarriers requires the incorporation of sensitive components at adequate proportion and arrangement in order to ensure their performance as predictable, reproducible, proportional to the intensity of the signal and, ideally, reversible sensors and actuators.[5,7] The sensitive component should bear functional groups that sharply modify their features in the presence of the stimulus of interest and enable the transduction into changes in the material properties.[8,9] Pioneering synthesis of responsive systems in late 1970s focused on cross-linked polymers and biomacromolecules that undergo phase transitions under conditions compatible with the physiological ones, resulting in different states of conformation and association that are macroscopically transmitted to the material as reversible volume transitions between swollen and collapsed states.[10–12] Since then, several other stimuli-responsive systems with different levels of complexity have been developed.[2,13–15] In the last years, combination of stimuli-responsiveness and nanotechnology is opening unprecedented opportunities for improving drug delivery.[16,17]

2. Why Nanostructured Drug Carriers?

From a structural point of view, monolithic matrices of homo components are being replaced by nanostructured assemblies of hetero components that can mimic better the physiological processes of transport and delivery of substances in the body.[18] Natural structures are mostly the result of the hierarchical assembly of small pieces of materials of diverse composition. Namely, evolution-screened combinations of few amino acids, sugars, lipids and inorganic ions provide a wide range of architectures that can perform singular functions. This construction approach, known as "bottom-up", is highly versatile and involves minimum consumption of materials and energy.[19] Interestingly, there are not homogeneous monotonic structures in Nature. By contrast, various materials are arranged according to a strict patterning from the nanoscale to the macroscale, encoded in the genes of the living species. Each compartment plays a specific role in the overall performance of the macrostructure. As an example, reactions inside cells would be not possible if there were not separated compartments with distinctive environmental conditions in their cytoplasm. As a second relevant feature, natural structures commonly exhibit reversible assembly/disassembly which endows them with sophisticated ability to feed-back regulate their functions and in some case to self-healing when damaged. For instance, natural macromolecules, such as antibodies, leucine, collagen, or viral capsides, switch their activity on/off through assembly/disassembly processes. Drug delivery field is trying to incorporate this knowledge about natural materials to the design of advanced nanocarriers bioinspired on their structure and/or functionality.[20] Regarding the structure, carriers in the nano-size to microsize range and with biomimetic shape and surface properties are particularly appealing to attain the targeting to the site where the drug has to be delivered. Regarding the final step of a correct delivery, nanostructured carriers that can regulate the moment and the rate of drug release as a function of the levels of a variable that serves as an index of the pathological process are particularly interesting. Separated domains of components of different features facilitate faster and reversible responses to the stimuli. As recently envisioned by Motornov *et al.*,[8] ideal multifunctional nanocarriers could "resemble a living cell in its complexity where many particle's compartments interact, exchange chemicals, receive energy, perform mechanical work, alter the chemical and physical properties, all of these in response to environmental stimuli" (Fig. 2). Although still far from such a degree of sophistication, important advances already accomplished in this field have notably improved the performance of drug carriers, playing now a relevant role in the therapeutic process.[21]

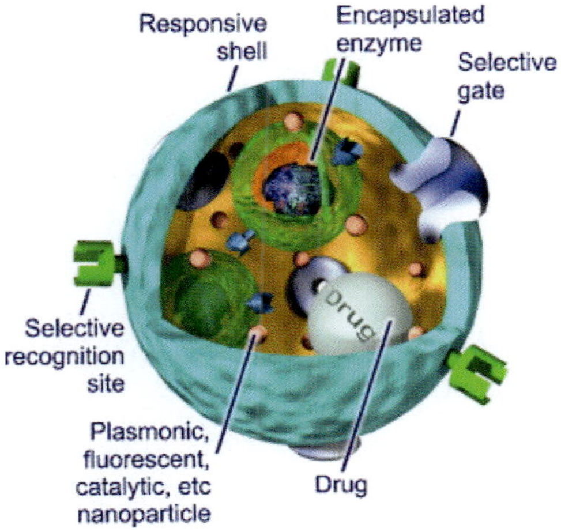

Fig. 2. Futuristic picture of an intelligent multifunctional particle for efficient regulation of drug release. Reproduced from Ref. 8 with permission from Elsevier.

Incorporation of diagnostic elements is also feasible, which has led to the inception of the theragnosis concept.[22] Next sections will focus on nanostructured lipid- or polymer-based systems that can regulate drug release as a function of the stimuli they perceive. Recent examples of their functionality under *in vivo* conditions will be analyzed.

3. Stimuli-Responsive Nanostructured Carriers

Components of advanced drug nanocarriers do not remain inactive in the physiological environment, but they are expected to recognize changes in certain variables and to deliver the drug accordingly. Therefore, they should perform as sensors and actuators.[11,23] To do that, stimuli-responsive components should be correctly incorporated and assembled.

3.1. *Self-assembled lipids: liposomes*

Vesicles with concentric bilayer palisades of lipids can be formed by simple addition of an aqueous liquid at the right temperature to a dry film of amphiphilic lipids.[24] If those lipids do change their conformation as a function of a stimulus or if a responsive component is added to the lipids, stimuli-responsive liposomes can be obtained.

Liposomes responsive to changes in pH have been specifically designed to trigger drug release in response to the acidic environment of cellular endosomes (pH 5.5–6.0) and pathological tissues such as tumors and inflamed or infected areas (pH 6.5–7.0). Responsiveness to the endosomal pH of healthy cells enables cytoplasmic-specific delivery of polar materials, including proteins, DNA and antisense oligonucleotides.[25–27] pH-Sensitive liposomes can be prepared combining a neutral cone-shaped lipid, *e.g.* dioleoylphosphatidyl-ethanolamine (DOPE), and a weakly acidic amphiphile, such as cholesteryl hemisuccinate (CHEMS), with other stabilizing lipids and stealth components (*e.g.* polyethylene glycol(PEG)).[28,29] At pH 7.4, these lipids have negatively charged groups, but become neutral in the acidic environment of the endosome, leading to destabilization, fusion with endosomal membrane, and content release.[30,31] Phosphatidyl ethanolamine (PE) presents a minimally hydrated and small headgroup which occupies a lower volume than the hydrocarbon chains, exhibiting a cone shape (as opposed to the cylinder shape of bilayer stabilizing phospholipids), thus hampering the formation of a lamellar phase. The cone shape conformation favors strong intermolecular interactions between the amine and the phosphate moieties of the polar headgroups, and thus stable liposomes are formed at physiological pH. Acidification triggers protonation of the carboxylic acid groups of the amphiphiles and leads to liposomal destabilization as PE molecules revert into their inverted hexagonal phase. Anionic liposomes can be also prepared with pH-sensitive lipids and synthetic fusogenic peptides/proteins either encapsulated or incorporated in the lipid bilayer, or with the association of pH-sensitive polymers and lipids.[32,33] Endosomal delivery requires the use of liposomes small enough (<100 nm) to penetrate through fine capillaries and be uptaken by cells via endocytosis/phagocytosis. In the case of tumor treatments, endosomal- and tumor-responsiveness can be combined. Exploiting the ability of liposomes to passively accumulate in tumor tissues via EPR effect, the drug can be partially released extracellularly when the liposome destabilizes due to the acidic environment of the tumor tissue. Nevertheless, the intracellular drug delivery is an advantageous way of overcoming the multidrug resistance (MDR) caused by the efflux pumps, one of the major reasons of cancer treatment failure.

Compared to long circulating classical liposomes (soybean phosphatidylch oline:cholesterol:PEG-750-DSPE 47:47:6 molar ratio), pH-sensitive stealth liposomes (DOPE:CHEMS:cholesterol:PEG-750-DSPE 43:21:30:6 molar ratio) have demonstrated *in vitro* improved delivery of peptides to the cytoplasm and internalization into the nucleus of tumorigenic and non-tumorigenic breast cancer cells.[34] Nevertheless, a problem that liposomes have to face up to is the fact that subsequent administration of stealth vesicles leads to a progressively lower accumulation of drug in the tumor site due to an accelerated blood clearance (ABC) phenomenon. PEGylated nanostructures induce the production of anti-PEG IgM, which can bind to the surface of subsequently injected stealth carriers, leading to substantial complement activation and increased accumulation in liver and spleen.[35] This problem could be overcome with the use of PEG derivatives that detach from the liposomes at the tumor site.[36,37] Liposomes simultaneously bearing on their surface cell-penetrating peptides (*e.g.* TAT peptide) and protective PEG chains can accumulate in tumor tissues, then lose the PEG coating due to acid pH-induced hydrolysis, and finally penetrate inside cells via the now-exposed TAT peptide moieties. These liposomes rendered promising results for chemotherapy in cell cultures as well as in ischemic cardiac tissues using the Langendorff perfused rat heart model, and in mice bearing tumors.[38] More selective accumulation of drugs and even of radioactive species in cancer cells can be achieved when the acid-labile PEGylated pH-sensitive liposomes also wear cancer-specific monoclonal antibodies (mAb), estrone or folate, which are "hidden" by the long PEG chains during blood circulation, and then exposed to tumor cells at acid pH.[39,40] Hyaluronic acid has been shown useful as substitute of PEG for providing labile stealth coatings, since hyaluronidases at the tumor extracellular matrix can rapidly degrade the coating and lead to the exposition of cell-penetrating peptides for efficient cell internalization and *in vivo* tumor regression.[41] The pH-sensitive liposomes can also deliver imaging probes, such as ^{99m}Tc or gado-teridol for tumor diagnosis and monitoring of treatment effectiveness.[42,43]

Recently, mitochondrial-targeting has been achieved with the use of zwitterionic oligopeptide liposomes (HHG2C_{18}-L) that exhibit multistage pH response to the mildly acidic tumor microenvironment and to the acidic intracellular endolysosomal compartment. These liposomes consist of phosphatidylcholine (SPC), cholesterol, and a synthetic smart lipid (1,5-dioctadecyl-l-glutamyl 2-histidyl-hexahydro benzoic acid, HHG2C_{18}) that combines two amino acid groups (glutamic acid and histidine) and one pH-cleavable group (hexahydrobenzoic amide). At pH 7.4, the surface of the liposome is strongly negative, while at tumor tissue it becomes positive which favors cellular accumulation. Once in the endosome, the imidazole group of histidine in HHG2C_{18} facilitates proton influx (proton sponge effect) causing the burst of the compartment, while hexahydrobenzoic amide is hydrolyzed to provide permanent positive charge to the liposome. Thus, the liposome does not revert to the negatively charged surface in the cytoplasm at pH 7.2–8.0. The positively charged liposome can electrostatically interact with the mitochondria (Fig. 3). The performance of this type of vesicles has been confirmed *in vivo* using murine renal carcinoma cell xenograft models.[44]

Fliposomes represent a novel type of pH-sensitive liposomes that undergo conformational changes upon protonation, disrupting the liposome membrane and causing instantaneous release of the cargo.[45] Amphiphiles based on trans-2-morpholinocyclohexanol adopt a conformational equilibrium with the morpholine and hydroxyl groups in axial positions. When the amine nitrogen becomes protonized at acid pH, morpholine and hydroxyl groups strongly interact through hydrogen bonds and adopt equatorial positions (Fig. 4). That conformational change is transmitted along the molecule, increasing the spatial separation of the lipid tails and perturbing the structure of the bilayer. The pH sensitivity and consequently the triggering of the release can be finely tuned by modifying the composition of the amphiphile bearing the amino groups.[46]

Liposomes sensitive to temperature have been also successfully tested for tumor therapy. Since the EPR effect provides only limited access for large liposomes, application of an external source of heating could be used to cause a local hyperthermia and make the blood vessels leakier, and also to trigger drug release from temperature-sensitive liposomes, so the drug can readily diffuse throughout the neoplasm.[47] As an example, Thermodox® formulation is based on liposomes formed by dipalmitoylphosphatidylcholine (DPPC) and

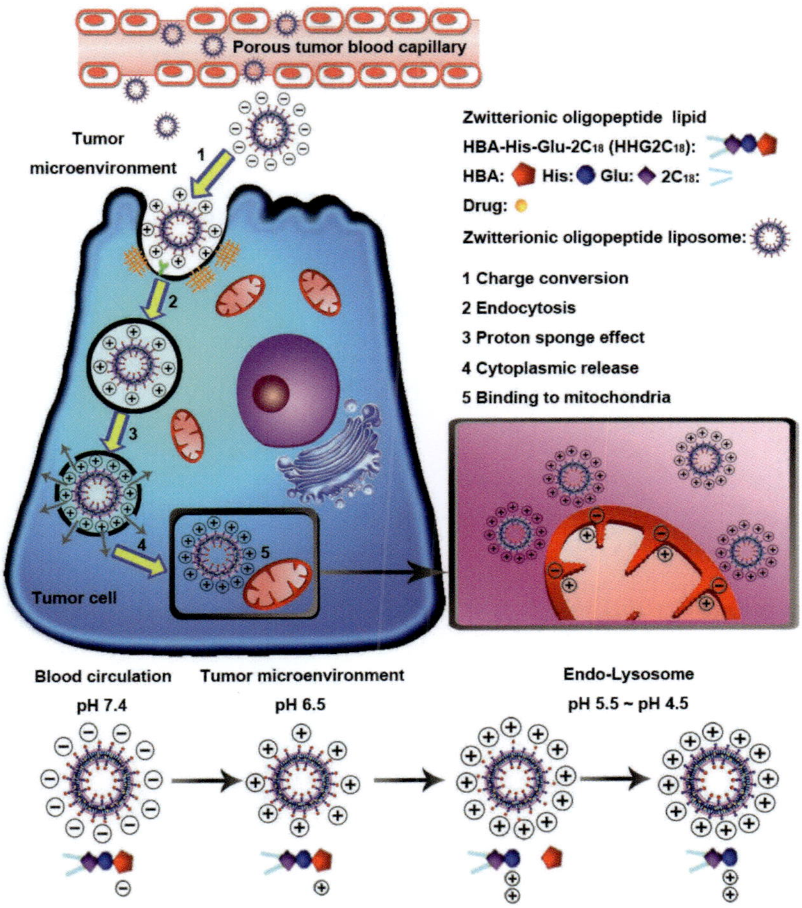

Fig. 3. Schematic design of the smart liposomes ($HHG2C_{18}$ -L) with multistage pH response to the tumor extracellular matrix and intracellular compartments for mitochondrial-targeted anticancer drug delivery. Reproduced from Ref. 44 with permission from John Wiley & Sons.

monostearoylphosphatidylcholine (MSPC), as temperature sensitive components, and distearoylphosphatidylethanolamine–PEG_{2000} (DSPE–PEG_{2000}) as stealth and stabilizer component, which together allow a rapid permeabilization of the liposomes at its phase transition temperature (~41°C). Experiments in animal models have confirmed the capability of these temperature-sensitive liposomes to totally abolish tumor regrowth, providing tumor drug levels up to 30 times higher than those achievable when free drug is administered, and 3–5 times greater than those provided using non-temperature sensitive liposomes.[48,49] This formulation is currently being tested in various human clinical trials.[50,51]

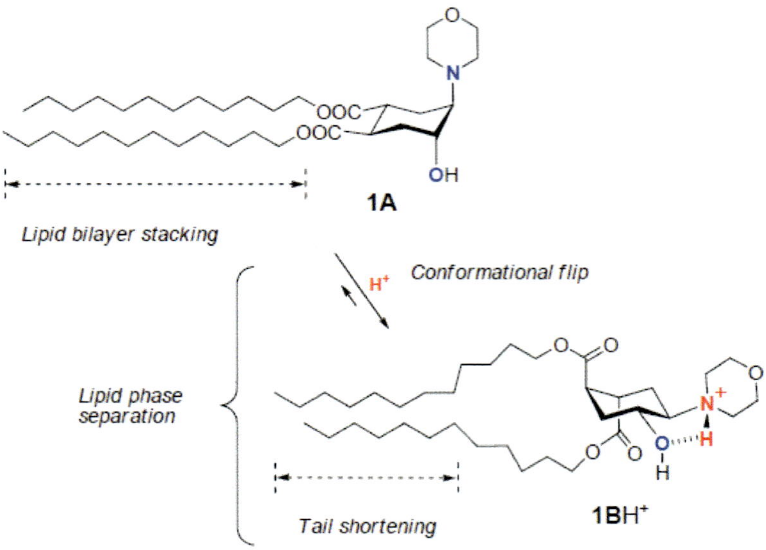

Fig. 4. The protonation-induced conformational shortening of the flipid tails and the change of charge, effective size, and shape of the polar head cause a quick perturbation of the lipid bilayer, lipid phase separation, and fast content leakage from the fliposomes. Reproduced from Ref. 45 with permission of MDPI.

Destabilization of the liposome structure by other external sources of energy, such as ultrasound or light, also attracts interest owing to the spatial and temporal control over the drug release that can be exerted, with minimal exposition of healthy tissues. Echogenic (*i.e.* sensitive to ultrasonic stimulation) liposomes have been prepared co-encapsulating the drug with air or gas bubbles,[52] although DSPE itself can undergo perturbations when exposed to ultrasound that concomitantly enhances drug release rate.[53] Furthermore, there is already a large number of publications on light-responsive liposomes, which can respond to this variable according to photochemical or photophysical mechanisms, namely incorporing sensitive lipids or metallic particles, respectively (Fig. 5).[54,55] The versatility of the responsive mechanisms facilitates the obtaining of liposomes that can respond to UV, visible and near-infrared (NIR) light and, thus, enables the triggering of drug release at different depths in the body. Comprehensive reviews on the performance of light-activated release from liposomes have been recently published.[55–57]

Multicompartment structures of small liposomes in a large liposome, named vesosome, open the possibility of encapsulating different drugs in separate compartments also made of different lipids.[58,59] The external bilayer protects the inner liposomes and also contributes to the regulation of drug release as a function of several variables.[60] Another class of multicompartmentalized assemblies

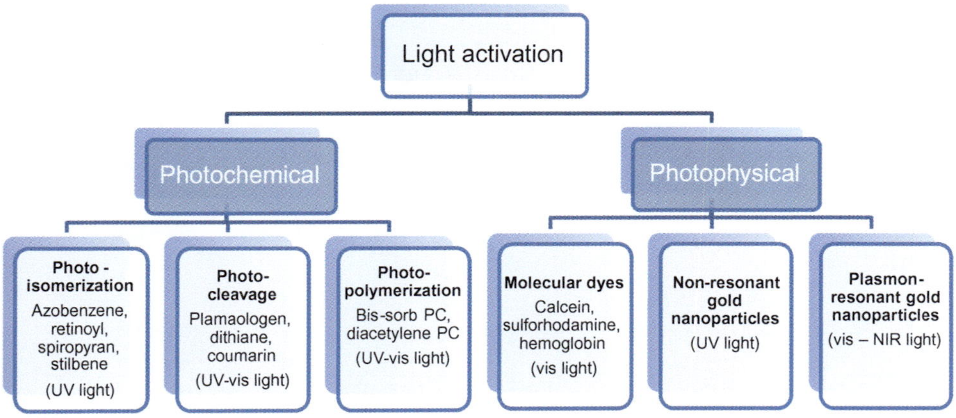

Fig. 5. Main mechanisms of drug release from light-responsive liposomes. Photochemical activation involves lipid-like molecules that destabilize the lipid membrane due to light-induced isomerization, cleavage or polymerization of its components. Photophysical activation refers to photothermal conversion of absorbed light in thermal and/or mechanical processes in the lipid membrane and the surrounding medium.

is that termed capsosomes, which combine the features of liposomes and polymer capsules. They are prepared via sequential deposition of polymers and liposomes onto particle templates according to a layer-by-layer (LbL) approach (Fig. 6). The layers are maintained together by means of cholesterol units grafted to the polymer components. Once an adequate number of layers has been deposited, the template is dissolved and colloidal capsosomes with tunable compartments and internal arrangement are obtained.[61] Since a variety of polymers can be functionalized with cholesterol units, it is feasible to design stimuli-responsive capsosomes.[60] For example, coencapsulation of glutathione reductase, an enzyme that reduces glutathione disulfide (GSSG) into glutathione (GSH, an oxidant), triggers the rupture of a disulfide-linked poly(methacrylic acid)–KP9 oligopeptide when a certain GSSG concentration is reached in the surroundings of the capsosome and the temperature is above the melting temperature of the lipid bilayer (Fig. 6). This platform combines both enzymatic reactions and controlled release of therapeutic agents in a single unit.[62]

3.2. *Self-assembled polymers: micelles, polymersomes, and polypseudorotaxanes*

Amphiphilic copolymers spontaneously aggregate in water forming nanometric structures with a hydrophobic core surrounded by a hydrophilic shell (polymeric micelles) or forming vesicles similar to liposomes but with alternating

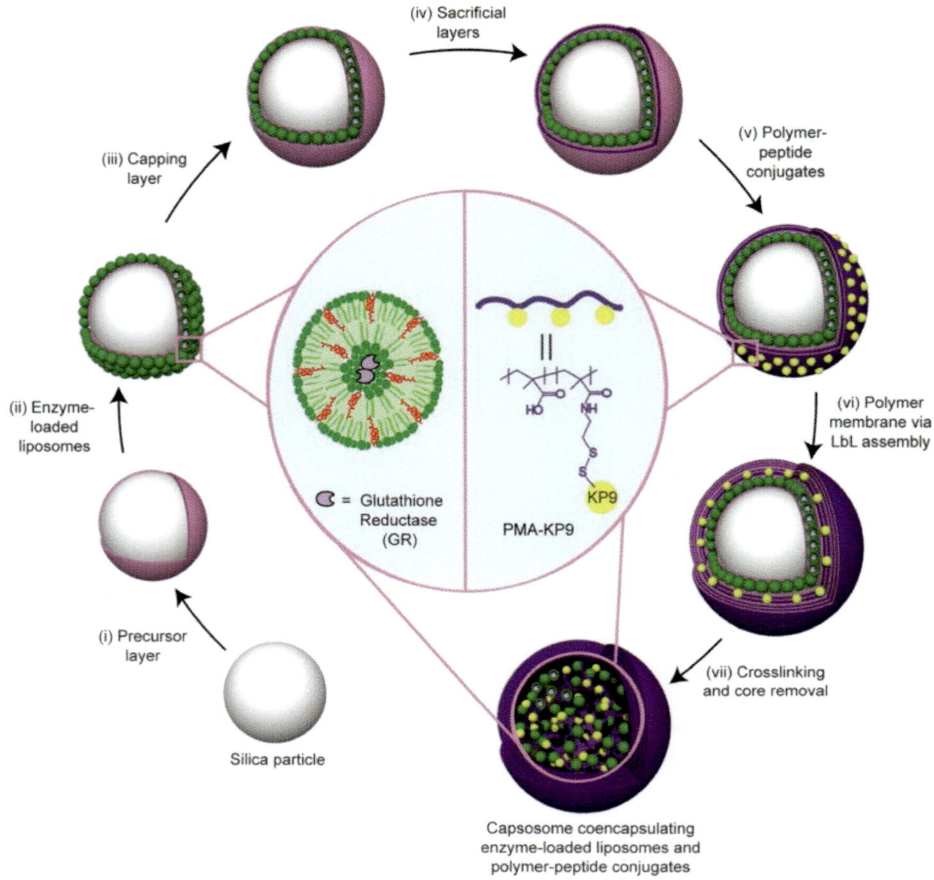

Fig. 6. Schematic illustration of the assembly of capsosomes coencapsulating glutathione reductase-loaded liposomes and PMA-KP9 polymer–peptide conjugates into polymer carrier capsules. Reprinted with permission from Ref. 62. Copyright (2011) American Chemical Society.

layers of water and copolymer organized as a palisade (polymersomes).[63] In addition to the typical spherical and cylindrical structures, the morphology of the micelles and polymersomes can be tuned to obtain novel nanostructures, such as "onions", "knitting ball", "golf ball" or "virus"-like particles.[64]

Polymersomes and polymeric micelles can host nonpolar drugs in the hydrophobic domains and relatively polar substances in the hydrophilic ones.[65] These polymeric nanostructures are thermodynamically and kinetically more stable (*i.e.* lower critical micellar and disassembly rate) than conventional micelles formed by low molecular weight surfactants, and usually exhibit better stealth features than liposomes as the hydrophilic shell efficiently prevents opsonization.[66] As mentioned for liposomes, stimuli-responsive micelles and polymersomes can be designed to not release the drug until a change in the physiological conditions or an external stimulus alters the hydrophilicity or the conformation of the unimers.[67–70] The number of micelles or polymersomes that disintegrate or destabilize and, consequently, the drug release profile depends on the intensity of the stimulus. As soon as the stimulus stops, the colloidal structures can be reformed and the release is switched off.

Most pH-sensitive polymeric assemblies are based on amphiphilic polymers with ionizable groups that are neutral at pH 7.4, favoring the self-aggregation, but become ionized at the tumor tissue or in the endocytic vesicles, increasing the polymer solubility and thus triggering the disassembly.[71,72] That is the case of amphiphilic copolymers bearing amino groups in one of the blocks, such as methyl ether poly(ethylene glycol)-b–poly(β-amino ester), (MPEG–PAE), micelles that can protect camptothecin from hydrolysis of the lactone group and release it more rapidly at pH 6.4 than at pH 7.4.[73] Intravenously injected MPEG–PAE micelles in human breast tumor-bearing mice showed an enhanced accumulation in the tumor compared to non-responsive micelles. These micelles are also advantageous for selective delivery of diagnosis agents to the acidic environment of cerebral ischemic areas.[74] Similar effects can be obtained with micelles and polymersomes formed by polymers bearing acid-labile groups that modify their hydrophilicity.[75,76]

Glutathione concentration can be exploited as a stimulus to trigger drug release inside cells, since this reducing agent is present in intracellular compartments at 2–3 orders higher level (2–10 mM) than in the extracellular fluids (2–20 μM).[77] Disulfide groups used to join different blocks or to cross-linking the polymers in the micelles/polymersomes can be reduced in the presence of glutathione inside micelles, triggering drug release in the cytoplasm (Fig. 7).[78,79] The efficiency of this approach can be enhanced by combining polymers that respond to the low pH and the reductive conditions of the cancer cells, and by decoration of the polymer assemblies surface with targeting elements useful for overcoming the *in vivo* barriers.[80,81]

Poly(N-isopropyl acrylamide) (PNIPAAm), has been widely used as the temperature-responsive block of the copolymers forming micelles/polymersomes.[82] The typical LCST of PNIPAAm at 31°C can be raised above 37°C by copolymerization with hydrophilic polymers. For example, the LCST of PLA-b-(PNIPAAm-co-DMAAm) is close to 40°C. Below the LCST, micelles are formed having PNIPAAm at the shell and slowly release the encapsulated drug. By contrast, above 40°C, the micelles destabilize and the drug is rapidly released.[83] Micellization/demicellization cycles due to changes in temperature and pH have also been reported for poly(2-diethylaminoethyl-methyl

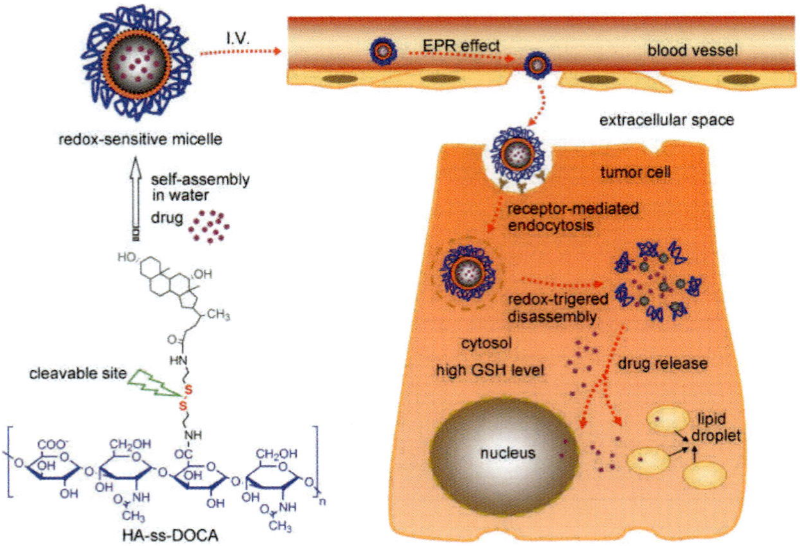

Fig. 7. Illustration of the self-assembly, accumulation at tumor tissue and intracellular trafficking pathway of redox-sensitive micelles based on hyaluronic acid–deoxycholic acid (HA-ss–DOCA) conjugates containing cystamin as bioreducible linkages. The intracellular trafficking pathway includes steps of receptor-meditated cellular internalization, endo/lysosomal escape, reduction triggered micelle disassembly, and drug release. Reproduced from Ref. 79 with permission from Elsevier.

methacrylate)–poly(ethylene oxide)–poly(propylene oxide)–poly(ethylene oxide)–poly(2-diethylaminoethyl-methyl methacrylate) ($PDEAEM_{25}$–PEO_{100}–PPO_{65}–PEO_{100}–$PDEAEM_{25}$),[84] and copolymers of NIPAAm, N,N-dimethylacrylamide and N-acryloylvaline.[85] Similarly, polymersomes consisting of poly(*ε*-caprolactone) and PNIPAAm have shown reversible disintegration in response to tiny changes of temperature.[86]

Polymeric micelles with a dual response to temperature and light have been prepared from copolymers having NIPAAm and azobenzene moieties.[87] Azobenzene groups undergo reversible trans to cis isomerization, changing from hydrophobic to hydrophilic when irradiated at certain wavelength. Under UV light, the hydrophilicity of the copolymer rises and the micelles disintegrate.[56,88] Micelles responsive to NIR have been obtained by grafting naphthoquinone to dextran (Fig. 8). Anticancer drugs can be effectively encapsulated in the micelles and, after internalization, NIR laser can trigger a fast release.[89] Multiple stimuli-responsive micelles that respond to light, temperature and pH have been prepared with pyrene-functionalized poly(dimethylaminoethyl methacrylate),[90] while micelles sensitive to light,

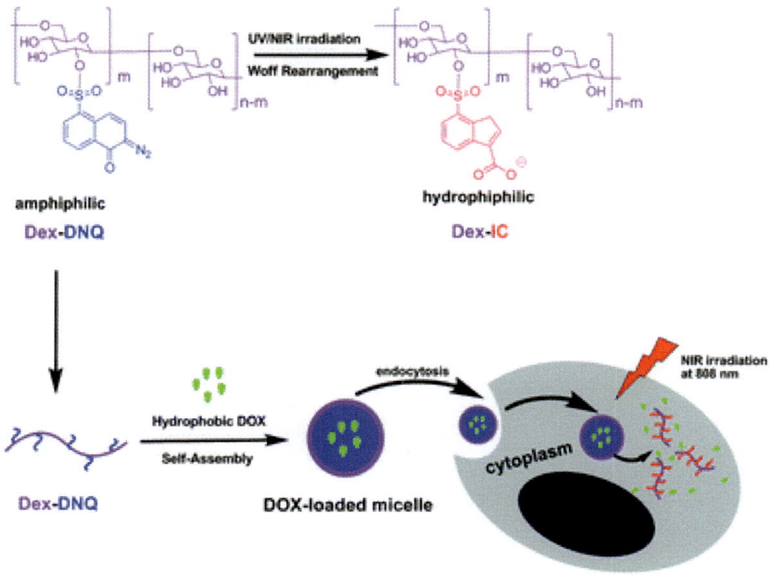

Fig. 8. Schematic illustration of self-assembly and Wolff rearrangement of naphthoquinone-grafted dextran (Dex-DNQ) amphiphilic copolymer. The Wolff rearrangement of the naphthoquinone molecules under NIR irradiation dissociates the micelle structure and enhances the intracellular release of doxorubicin. Reproduced from Ref. 89 with permission of the Royal Society of Chemistry.

temperature, ionic metals and pH can be obtained using poly[N-isopropylacrylamide-b-sodium 2-(acrylamido)-2-methylpropane sulfonate] tagged with a spiropyran dimer.[91]

Responsiveness to ultrasound has been analyzed in detail as a way to externally trigger drug release in profound tissues. Bubble cavitation and local increase in temperature can be regulated through the control of the frequency, power density, pulse length and inter-pulse intervals of ultrasound, triggering one pulse or multiple pulses release from micelles.[92–94] This approach has been successfully tested for treatment of tumors in animal models: the antitumor formulation is systemically administered and, when most micelles have reached the target tissue, ultrasound is applied to disassemble the micelles. The *in vivo* antitumor effectiveness of this approach is also promoted by the cell membrane perturbation caused by ultrasound (sonoporation) which enhances the intracellular uptake of micelles, drugs and genes.[95,96] It has been found that mice transplanted with spontaneous breast adenocarcinoma tumors and that were treated with micellar doxorubicin combined with ultrasound had nearly nine times more doxorubicin in their tumors than mice receiving free drug, and more than three times the amount in mice receiving the micellar drug formulation without sonication. Doxorubicin concentration in other tissues was smaller when the micellar formulation was used.[97]

Recently, multicompartment polymeric micelles and polymersomes with shells that mimic the lipidic structure and protein channels of cell membranes have been explored as a way to achieve reversible stimuli-triggered drug release. Transmembrane protein channels or synthetic replicas prepared with copolymers possessing a common hydrophobic core but different responsive hydrophilic blocks can be inserted in the external shells in order to regulate drug permeability as a function of a variety of stimuli.[98–100] Polymersomes-in-polymersomes can be versatile obtained by combining polymers with different degradation rates and stimulus-responsiveness and that can selectively incorporate active substances/diagnosis agents in the inner and outer compartments.[101–104]

The self-assembly process of responsive amphiphilic polymers can be finely tuned by means of cyclodextrins that spontaneously thread through the polymer chain forming necklace-like complexes with the hydrophilic or the hydrophobic blocks.[105] These polypseudorotaxanes are dynamic, namely the cyclodextrins are free to join and to be released from the stimuli-sensitive polymer chains as a function of the enviormental conditions, which subsequently determine intramolecular and intermolecular interactions of the block copolymers. In the case of poly(ethylene oxide)–poly(propylene oxide) (PEO–PPO), block copolymers, α-cyclodextrins (α-CD) and γ-cyclodextrins (γ-CD) can thread onto one or two PEO chains, respectively, while β-cyclodextrins (β-CD) preferentially interact with PPO blocks.[106,107] At appropriate proportions, complexes with PEO may lead to hydrogel formation, while interactions with PPO results in an increase in the hydrophilicity of the copolymer that makes micellar formation more difficult.[108] Many other responsive copolymers have been shown suitable for forming responsive supramolecular structures with cyclodextrins. For example, triblock copolymers of methoxy-poly(ethylene glycol)-b–poly(ε-caprolactone)-b–poly[2-(dimethylamino)ethyl methacrylate] (MPEG–PCL–PDMAEMA) were prepared to condense pDNA. The MPEG corona imparted stability to the pDNA polyplexes and also served as an anchoring segment to α-CD, forming supramolecular polypseudorotaxane hydrogels able to sustain release of pDNA up to 6 days (Fig. 9).[109]

Interestingly polypseudorotaxanes capping mesoporous silica can function as nanogates that open/close as a function of certain stimuli. PEG-functionalized silica particles via an acid-labile benzoic-imine linker have been

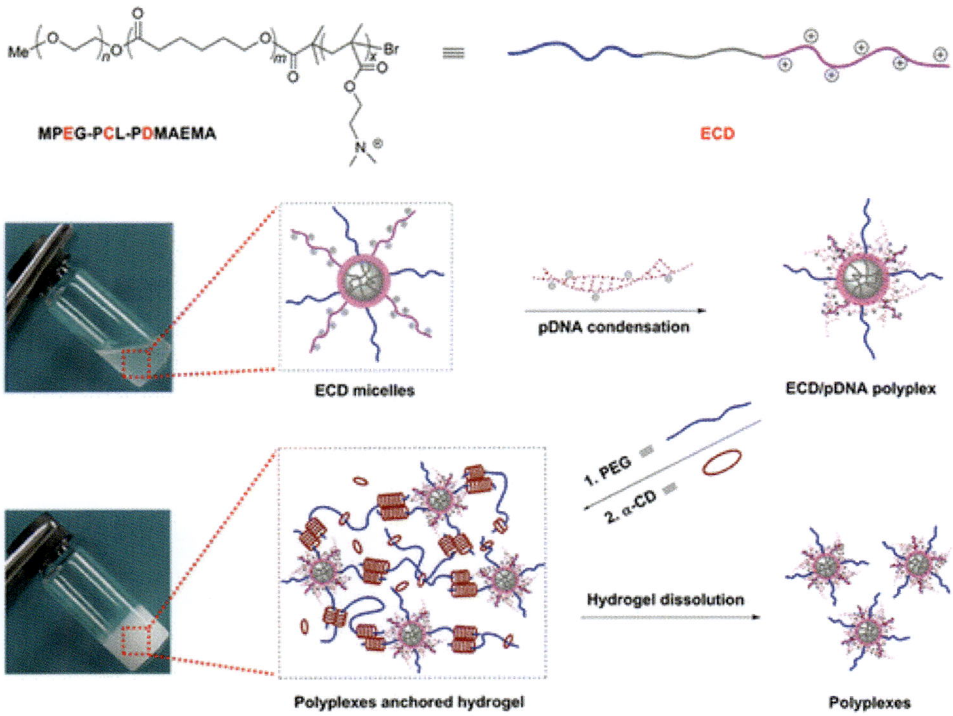

Fig. 9. Amphiphilic MPEG-PCL-PDMAEMA copolymer (EDC) forms micelles in aqueous environment with PCL core and MPEG/PDMAEMA corona. PDMAEMA at the corona is responsible for pDNA binding, while MPEG serves both as stabilizing moiety for the resultant ECD/pDNA polyplex and anchoring segment for α-CD. Sustained release of pDNA polyplexes is achieved via hydrogel dissolution over time. Reprinted with permission from Ref. 109. Copyright (2012) American Chemical Society.

shown suitable to load doxorubicin. After addition of α-CD, the release of the drug is blocked at pH 7.4. By contrast, once the pH decreases to 6.8 (tumor environment), the polypseudorotaxane partially detaches and generates positive amino groups on the carrier surface, which facilitate the cellular uptake. Subsequently, in the more acidic endosomes (pH 5), the hydrolysis of the benzoic–imine bonds is completed and rapid release of the drug occurs into the cytoplasm.[110,111] Responsiveness based on changes in the ionization degree of polyethyleneimine (PEI) has been exploited for a similar purpose. At pH 7.4, α-CD and γ-CD can thread onto one or two PEI chains, respectively, while at acidic pH PEI becomes protonized and the cyclodextrins are released (Fig. 10). Therefore, threading and de-threading switch the release off and on.[112]

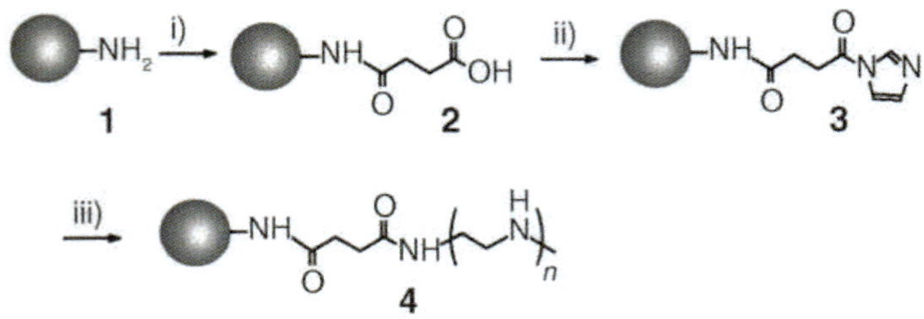

(a)

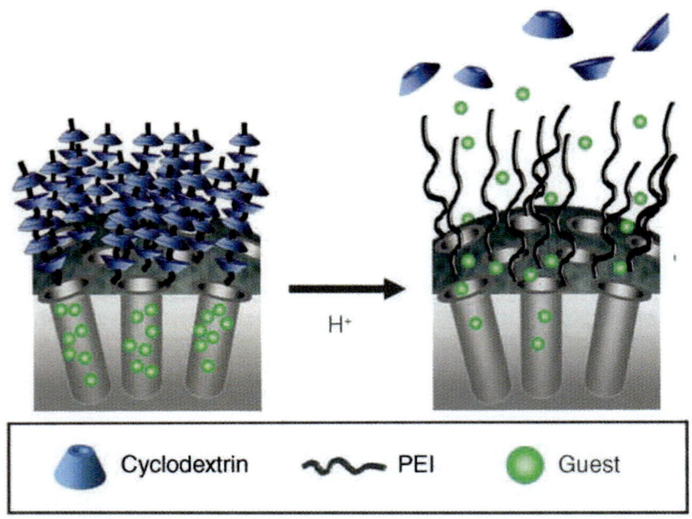

(b)

Fig. 10. (a) Functionalization of mesoporous silica particles with i) succinic anhydride, triethylamine, ii) 1,1′-carbonyldiimidazole; and iii) PEI-NH_2; and (b) schematic of pH-responsive release of guest molecules from the mesoporous silica particles. Reproduced from Ref. 112 with permission from John Wiley & Sons.

3.3. *Dendrimers*

Dendrimers are uniformely ramified polymers with a central core from which repeating branches are built up step-by-step.[113] The composition and the number of branches, named generations, determine the final size of the polymer. The inner regions (building branches) are suitable for hosting drugs, while the various terminal groups (surface branches) can be functionalized with stimuli-responsive components.[114,115] Poly(amidoamine) (PAMAM) dendrimers can serve as carriers of acidic drugs, increasing drug solubility, stability and circulation time and enabling controlled release and even drug targeting.[116,117] Stimuli-responsive dendrimers can be obtained by means of functionalization of the core or the terminal groups. Functionalization with temperature-responsive chains, such as PNIPAAm, oligo(ethylene glycol) or phenylalanine, renders dendritic delivery systems that exhibit temperature-triggered indomethacin release.[118] Amphiphilic dendrimers can even associate in larger structures such as micelles or vesicles that exhibit temperature-dependent morphology. The temperature-responsiveness of collagen triple helix structure can be mimicked by grafting of covalent knots of collagen model peptides with the structure $(\text{Pro-Pro-Gly})_5$ or $(\text{Pro-Hyp-Gly})_n$ at the surface of the dendrimer (Fig. 11). The collagen-like functionalized dendrimers are suitable for encapsulating drugs and to release them faster at high temperature.[119,120]

Dendrimers can be endowed with pH responsiveness by incorporating labile groups that hydrolyze at certain pH and lead to the disintegration of the nanostructure,[121] or by introduction of protonizable amino groups at the external surface of the dendrimers, which can enable the release of the entrapped drug within a narrow pH region.[122,123] Moreover, hyperbranched polymers can undergo pH-triggered supramolecular self-assembly rendering hierarchical structures with novel possibilities for regulating drug release.[124]

Functionalization with azobenzene groups at any region of dendrimer architecture renders light-responsive systems. The process requires starting from azobenzenes with a single reactive group if they are intended to be used as terminal groups, two identical reactive groups of azobenzene is placed at the core, or two different reactive groups of azobenzene is used as branch in order to allow the growing of the structure.[125] The location of the azobenzene group(s) determines the effect of the light irradiation. Azobenzene dendrimers can self-assembly as multilayer (onion-like) vesicles which can

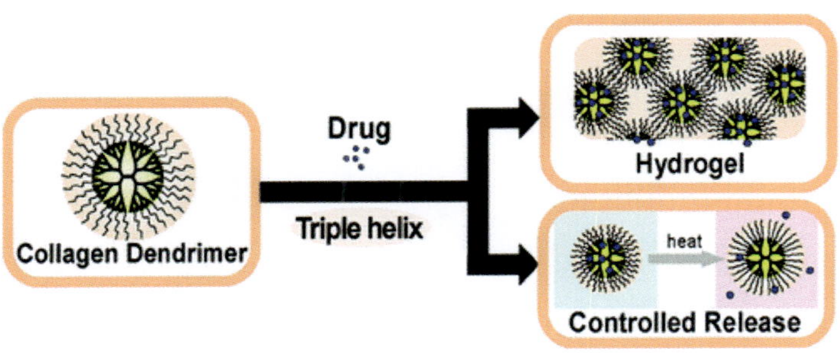

Fig. 11. Collagen model peptide-attached dendrimers can form gels at low temperature, while trigger the release at high temperature. The peptides clustered at the surface of the dendrimer form a thermally reversible collagen-like triple helix. Reprinted with permission from Ref. 119. Copyright (2009) American Chemical Society.

host drugs between the layers or inside each layer. The azobenzene moieties behave as valves with off–on switchability. Upon irradiation, the azobenzene groups in cis conformation increase the permeability and the drug can diffuse out (Fig. 12). Such structures have been shown useful for regulating the release of hydrophilic (calcein) or hydrophobic (Nile red) molecules.[126] Moreover, PEGylated dendrimers encapsulating gold nanoparticles may be suitable for photothermal therapy,[127] while polymers bearing photodegradable units can lead to self-immolative dendrimers that degrade under light irradiation leading to release of the encapsulated drug molecules.[128] Self-immolative dendrimers have been also designed to selectively degrade under the action of single or multienzymes for site-specific release.[129] A comprehensive review on this topic has been recently published.[130]

3.4. *Layer-by-layer assembly of pre-formed polymers*

Layer-by-layer assembly of components that can establish electrostatic interactions, hydrogen bonding or specific recognition bonds is a relatively simple way to obtain stimuli-responsive films and hollow capsules.[131,132] There is a large variety of components that can be deposited, from polymers or self-assembled structures to metallic particles, with the advantage that the successive coats do not negatively affect to the sensitiveness, but in some case can even result in cooperative responses to the stimulus.[133] For example, incorporating magnetic nanoparticles (MNP) into layers deposited on temperature-responsive microgels it is possible to obtain heat-triggered drug release under a mild magnetic field.[134] Moreover, the LbL technique may enable the codelivery of multiple drugs with specific release profiles, being the release of each

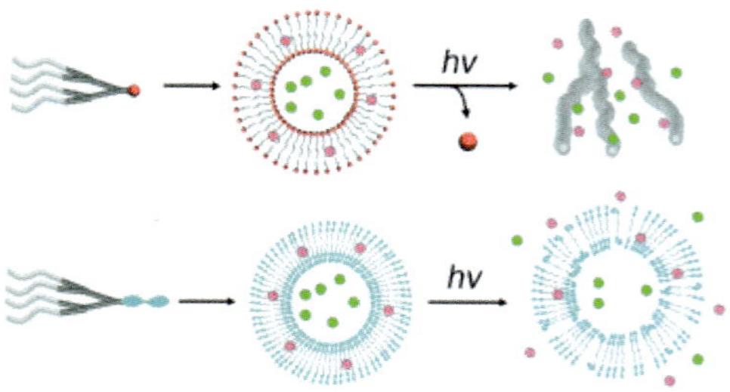

Fig. 12. Light-triggered release from vesicles formed by dendritic building blocks that can undergo a structural transformation (loosing photocleavable units) on application of a external photostimulus (upper draw), or from vesicles of supramolecular aggregates of an amide dendron with a photoisomerizable azobenzene unit that exhibit controlled release of encapsulated molecules upon photoirradiation (lower draw). Reproduced from Ref. 126 with permission from John Wiley & Sons.

drug or of different doses of the same drug triggered by a different stimulus.[135] Temperature-responsive multilayered films have been obtained combining poly(methacrylic acid) and temperature-responsive micelles of poly(N-vinylpyrrolidone)-b-poly(N-isopropylacrylamide) (PVPON-b-PNIPAM),[136] or an elastin-like polymer and chitosan.[137] Thin films that retain the drug at the pH of the stomach, but release it at the weakly acid-neutral pH of the gut were obtained by alternate deposition of insulin and poly(vinyl sulfate), poly(acrylic acid), or dextran sulfate.[138]

Using organic, inorganic or metallic particles as cores, multilayer deposition of complementary materials render core-shell structures that can be used as such, with the core and the shell responsive to different stimuli, or transformed in hollow capsules by dissolving the core, being the shell of the capsule responsible for the stimuli-triggered release.[139] The core could be even the particles of the drug itself, *e.g.* hydrophobic anticancer molecules, which would ensure high encapsulation efficiency.[140,141] LbL assemblies also enable the design of capsules with complex compartmentalized architectures (Fig. 13), very promising for time-controlled or stimuli-triggered release of different encapsulated compounds.[142–144]

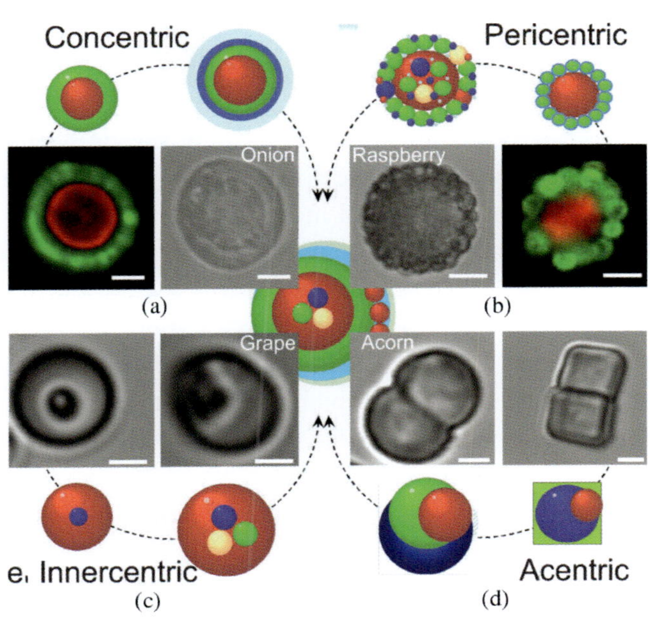

Fig. 13. Multicompartment microcapsules: (a) concentric, (b) pericentric, (c) innercentric, and (d) acentric. The structure in the middle incorporates all four approaches. Reproduced from Ref. 143 with permission of John Wiley & Sons.

3.5. *Particles from pre-formed polymers*

Coacervation/precipitation of polymers in solution is a traditional way to prepare particle-like structures.[8,145] In the absence of permanent cross-linking points, polymer particles resolubilize when enter into contact with the physiological fluids or when the conditions that induced the phase separation are reversed. If the coacervate particles are photo- or chemically-crosslinked, stable smart nano/microgels can be obtained.[146] For example, dually pH- and temperature-responsive particles have been prepared by copolymerization of acrylic acid and NIPAAm monomers in the presence of chitosan.[147] Replacing NIPAAm for Fe_3O_4 cores and cross-linking chitosan rendered pH- and magnetic-responsive particles.[148] Complex coacervation has been also applied to obtain core-shell or micelle-like particles.[149]

Water-in-oil heterogeneous gelation and aqueous homogeneous gelation are other two gentle ways of obtaining micro/nanogels from natural polysaccharides.[150] Polysaccharides and other saccharide derivatives (such as cyclodextrins) possess a high content in functional groups, including hydroxyl, amino, and carboxylic acid groups, suitable for being cross-linked under mild conditions.[151] Cyclodextrins are useful for regulating the nanostructure of coalescent polymers,[152] and to create novel nanostructures and microstructures for drug delivery.[153–156] Combination of the reactivity of the polysaccharides with the self-assembly features of block copolymers opens the possibility of obtaining a variety of nanogels with responsive coatings, cross-linked shells or cross-linked cores, sensitive to diverse stimuli.[157,158]

Trying to avoid toxic solvents and to simplify the preparation of nano/microgels, two novel approaches have appeared:

(i) Micromolding which use solid molds with a hole pattern that isolates aliquots of an aqueous precursor solution, thereby producing nonspherical microgels (Fig. 14).[159,160] A modification of this technique, named particle replication in non-wetting templates (PRINT) methodology, enables the preparation of nanogels from organic solutions of monomers using non-wetting perfluoropolyether molds, which eliminate the formation of residual interconnecting films between molded objects.[161] The precursors are UV cured to render cylindrical nanogels, which are harvested by physical means.[162]

(ii) Rolling and cross-linking of precursor drops on superhydrophobic surfaces,[163] which relies on the hierarchical structure of the surface of certain plant leaves and insect wings to render nearly spherical water drops.[164] Biomimetic superhydrophobic surfaces are used to deposit (with a pipette or a nozzle) aqueous dispersions of monomers/polymers containing the drug, which are then cured/cross-linked under mild conditions at the solid-air

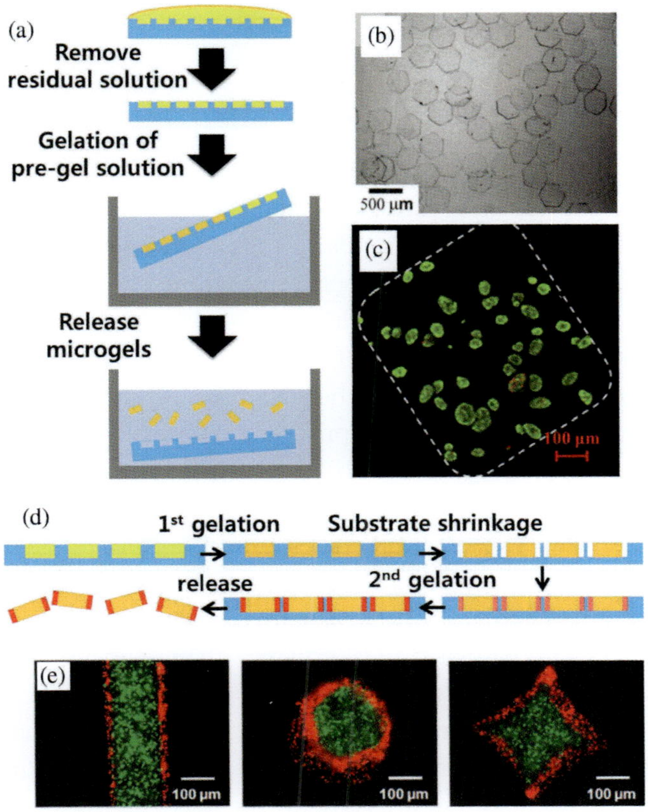

Fig. 14. Micromolding-assisted fabrication of microgels. (a) Alginate gel precursors isolated in the mold were cross-linked upon dipping the pattern into an aqueous solution of $CaCl_2$. Reproduced from Ref. 159 with permission from John Wiley and Sons. (b) Optical microscopy image of hexagon-shaped alginate microgels. Reproduced from Ref. 160 with permission from John Wiley and Sons. (c) Confocal microscopy image of an alginate microgel containing insulin-secreting cells (INS-1) produced using a polypropylene mold. Reproduced from Ref. 167 with permission from Elsevier. (d) Scheme of a procedure used to fabricate multicomponent microgels. A thermoresponsive template was used for the sequential molding of the agarose gel. (e) Fluorescence microscopy images of sequentially molded microgels with strip, cylindrical, and cubic shapes. Reprinted with permission from Ref. 168. Copyright (2011) American Chemical Society.

interface. The versatility of this approach has been already demonstrated by preparing semi-interpenetrating PNIPAAm/dextran microgels that encapsulate proteins and can control the release for several hours.[165] Deposit of additional drops onto preformed microgels/nanogels leads to multilayered systems for simultaneous modulation of the release of various therapeutic substances.[166]

Both methods enable strict control over the particle size, shape, composition, and surface functionality and permits 100% encapsulation yield of delicate cargos.[166–168]

So far, most stimuli-responsive systems rely upon intrinsically sensitive components, namely polymers that undergo reversible phase transitions when activated by the stimulus. An alternative approach of increasing interest is the utilization of dynamic covalent bonds (DCBs), *i.e.* covalent chemical bonds that can be formed and broken under equilibrium control.[169] The obtained structures are much more robust than those obtained using non-covalent bonds and more versatile, since the DCBs can be reformed changing from one partner to another and thus modify the structure, properties and functions of the polymeric particles. If the DCBs appropriately occur inside a single polymer chain or between adjacent polymer chains, nano/microgels can form/cleave/reform in response to changes in physiological parameters, such as pH or redox conditions, as reviewed in detail by Jackson and Fulton.[17] Some examples of DCBs suitable to act as cross-links and to render stimuli-responsive particles are depicted in Fig. 15. For example, dually pH- and redox-responsive nanogels were prepared by cross-linking of polymer chains through both imine and disulfide bonds. If only one stimulus (low pH or presence of reducing agent) act on the nanogels, they do not disgregate and a minor increase in drug release rate is observed because there is still a sufficient cross-linking density. By contrast, when both stimuli are present, the particles disassembly.[170]

3.6. *Polymeric structures from monomers and molecular imprinting*

Polymerization of monomers in a homogeneous phase or in a microscale or nanoscale heterogeneous environment enables a fine control of the morphology and structure of cross-linked networks in the nanosize to macrosize range.[171,172] Regarding nanostructuring, this bottom-up approach enables the application of the molecular imprinting technology for a precise arrangement of the spatial distribution of the monomers, which can enhance the affinity for the drug and a more precise control of the effect of the stimuli on drug release.[173] The molecular imprinting pursues the creation of microdomains (artificial receptors) that are sterically and chemically complementary to the target molecule (*e.g.* the drug or a certain biomarker), recognizing it with high selectivity. This technology involves the use of the target molecule as a template during polymerization in order to induce an adequate arrangement of the monomers, particularly of those suitable to interact with it that are called functional monomers (Fig. 16).[174] The functional monomer/template association can be achieved through: (a) covalent bonds (pre-organized

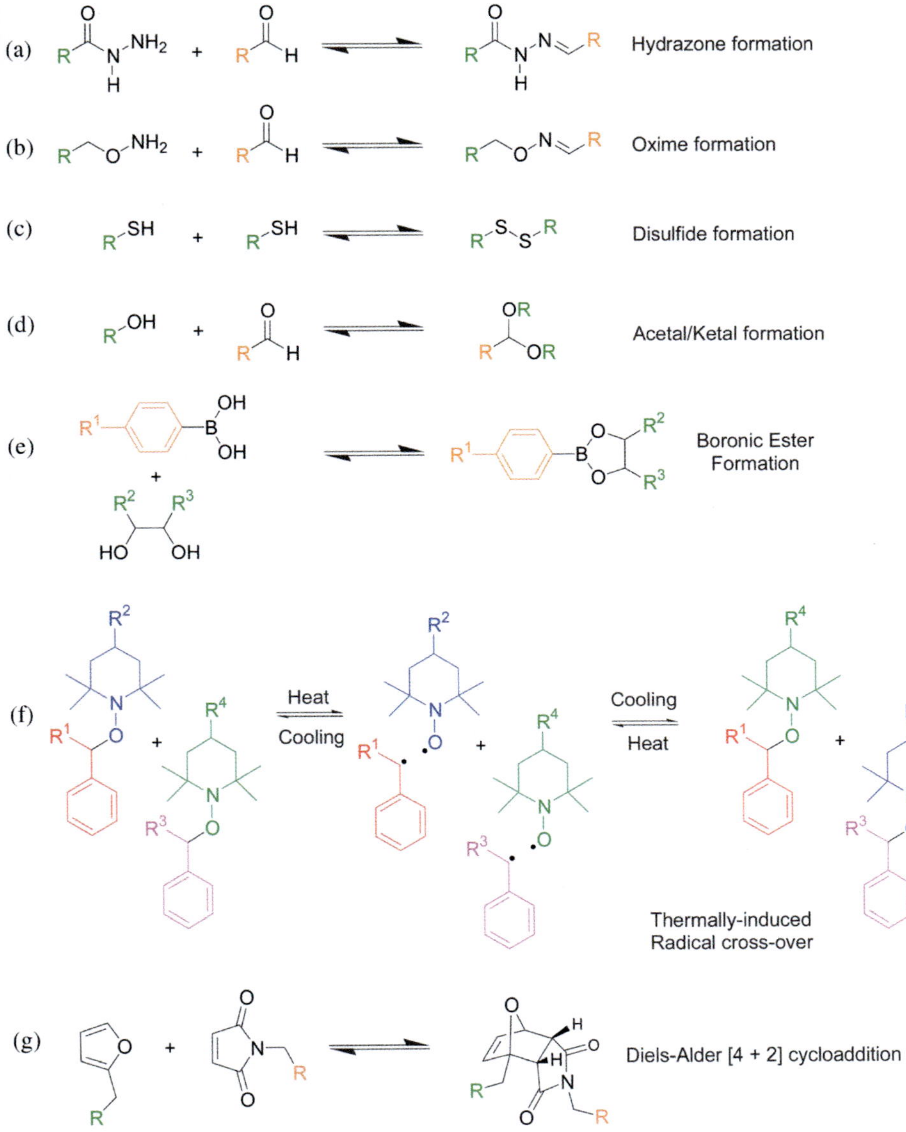

Fig. 15. Examples of DCBs commonly used in to endow polymeric nanoparticles with the virtues of stimuli-responsiveness. (a) Hydrazone formation, (b) oxime formation, (c) disulfide formation, (d) acetal/ketal formation, (e) boronic ester formation, (f) thermally induced radical cross-over reactions, and (g) Diels–Alder [4 + 2] cycloadditions. Reproduced from Ref. 17 with permission of the Royal Society of Chemistry.

approach) or (b) non-covalent interactions, such as hydrogen bonds and ionic, hydrophobic or charge transfer interactions (self-assembly approach). Comprehensive reviews of the molecular imprinting technology have been published elsewhere.[175,176] Conventional imprinted networks are highly rigid

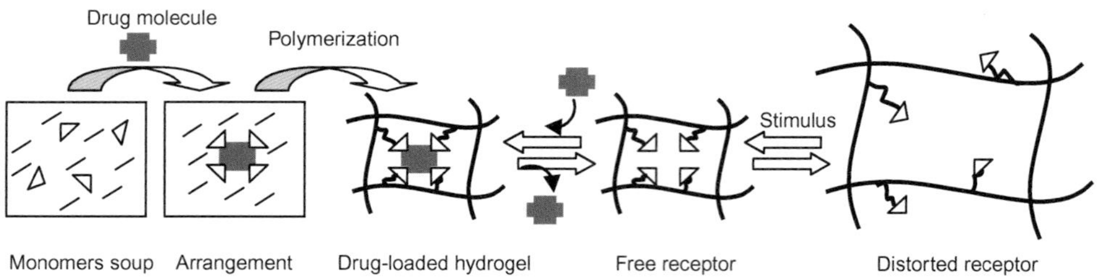

Fig. 16. Diagram of the synthesis of an imprinted hydrogel and the washing out/release and reloading processes. The effect of stimuli on the conformation of the drug-receptors is also depicted.

plastics with physically stable imprinted cavities. Adaptation of the molecular imprinting principles to loosely cross-linked networks is more challenging since the cavities structure has to be thermodynamically memorized in order to maintain the recognition ability after several swelling/collapse cycles.[177]

One of the most traditional methods to prepare polymer particles is the suspension polymerization in aqueous or highly polar solvents.[178] Typically, the monomers are dissolved in an organic solvent and then mixed with a larger volume of water containing a suitable stabilizer. Polymerization occurs in the small organic droplets formed, which determine the size of the resultant polymer particles. Omeprazole-imprinted nanoparticles-on-microspheres were prepared using methacryloyl quinine and methacryloyl quinidine as functional monomers and ethylene glycol dimethacrylate as a cross-linker. The latex obtained was incorporated into a pH-responsive network of poly(hydroxyethyl methacrylate) and polycaprolactone-triol enabling the triggering of drug release at pH above 6.8. The imprinted domains played a relevant role in the control of drug release by selectively regulating the diffusion of each enantiomer (Fig. 17).[179]

Precipitation polymerization is a suitable approach to synthesize uniform microspheres from diluted solutions of monomers that lead to hydrophobic polymers.[180] Monomers and cross-linker with different reactivity ratios may cause heterogeneities in the microgel structure, which can be tuned to modulate stimuli-responsiveness and eventually microgel degradation.[181] Precipitation polymerization synthesis of temperature- and pH-responsive microgels based on PNIPAAm and poly(N-vinylcaprolactam) and the possibilities of tuning size distribution, surface charge and microstructure have been reviewed in detail elsewhere.[182] To obtain smaller nanogels, surfactants are added to the reaction medium for colloidal stability.[183] Temperature-responsive nanohydrogels prepared with poly(N-isopropylacrylamide-co-acrylamide) have been

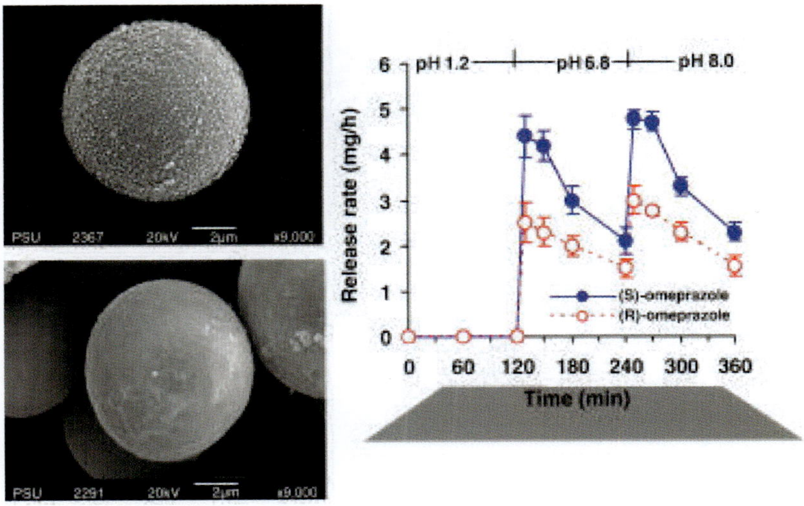

Fig. 17. Molecularly imprinted nanoparticle-on-microsphere cinchona-polymer can be utilized in the formulation of enantioselective-controlled delivery of omeprazole in response to changes in pH. Reproduced from Ref. 179 with permission from Elsevier.

shown able to load 5-fluorouracil or docetaxel and to release them at constant rate at 43°C with improved antitumor efficacy in Kunming mice-bearing S180 sarcoma.[184] Imprinted particles prepared via precipitation polymerization have also a great potential in the controlled drug delivery field. Sulfasalazine-imprinted particles can sustain drug release more efficiently than the corresponding non-imprinted particles, both at pH 1 and pH 6.8.[185] Theophylline-imprinted systems with modulated loading capacities and release rates were obtained using different proportions of methacrylic acid (MAA) and methylmethacrylate (MMA) functional monomers.[186]

Light-responsive imprinted microparticles have been prepared by means of precipitation polymerization of a methacrylate azo functional monomer under dark conditions. The template 2,4-dichlorophenoxyacetic acid only interacts with the functional monomer in the trans conformation. Thus, the affinity and the number of specific binding points notably decrease upon UV light irradiation, whereas they can be recovered after thermal or visible light-induced back-isomerization, enabling repeatable uptake/release of the target molecules under UV light on/off cycles.[187] Recently, magnetic orcinol-imprinted poly(ethylene-co-vinyl alcohol) (EVAL), composite particles have been developed for selective recognition of resveratrol from an extract of *Polygonum cuspidatum* (Fig. 18). Orcinol was used as an analog template of resveratrol due to its chemical similarity but greater commercial accessibility.

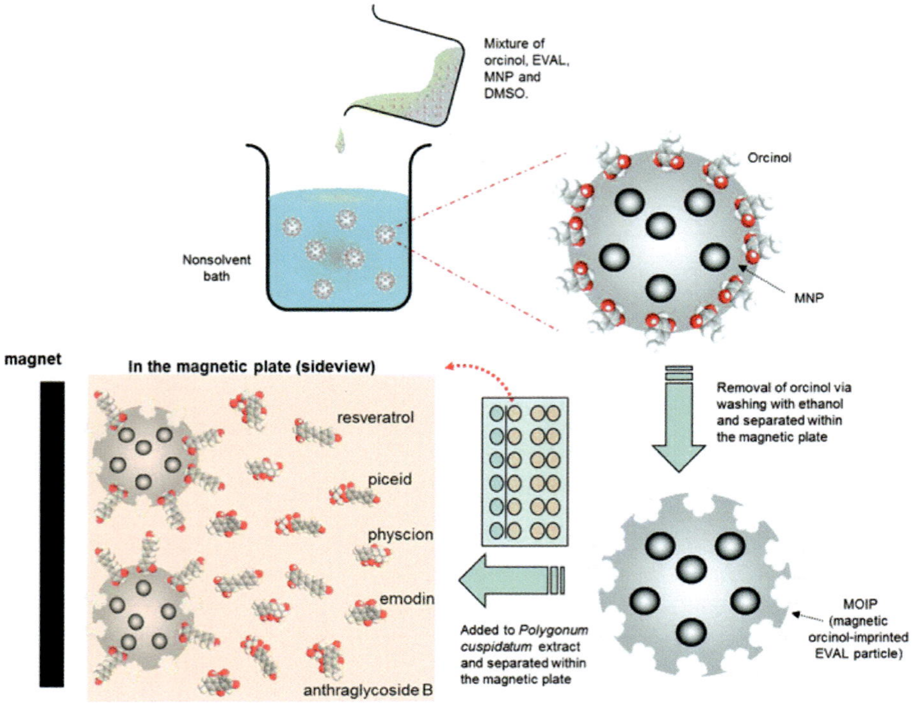

Fig. 18. Steps of the synthesis of magnetic orcinol-imprinted EVAL, nanoparticles: (1) preparation of MNP; (2) mixing of MNP with EVAL solution (EVAL/DMSO = 1 wt%) and 0.1 mg mL^{-1} of orcinol; (3) dispersion of EVAL/MNP system in deionized water/isopropanol = 2/3 at 5°C; (4) recovery of the imprinted magnetic particles and washing for removal of the template. The imprinted particles could selectively recognize resveratrol from an extract of *Polygonum cuspidatum*. Reproduced from Ref. 188 with permission from the Royal Society of Chemistry.

The resveratrol-loaded magnetic particles exhibited a remarkably high cytotoxic effect against cancer cells.[188]

Emulsion polymerization has been also applied to obtain responsive nanosized particles by polymerization of monomers with surface-active features and that distribute at the interface of two liquid phases. The nanoparticles resultant from the emulsion polymerization can be further modified to obtain core-shell particles by dispersion in a solvent containing monomers, cross-linker and initiator.[189] Polystyrene cores with pH- and temperature-responsive shells made of poly[2-(dimethylamino)ethyl methacrylate]-b-poly[methyl methacrylate] (PDMA-b-PMMA) rendered stimuli-sensitive colloidosomes.[190] "Snowman"-like particles with PMMA core and an ultrathin shell of polyaniline have been prepared to combine the electrical responsiveness of polyaniline with non-spherical cores which make the response to the electrical field faster.[191] Nanogels of poly(N-isopropylacrylamide-co-acrylic acid) bearing doxorubicin conjugated through a pH-labile bond exhibited dual temperature/pH-dependent cellular uptake and cytotoxicity. The nanogels shrank when temperature raised from 37 to 43°C favoring cellular internalization, while the pH-responsiveness enabled a fine control of drug release rate.[192] Poly(acrylonitrile-co-4-vinylpyridine) core-shell nanoparticles responsive to pH and magnetic field have been shown able to uptake naproxen and to sustain its release for several hours at pH 7.4.[193] Magnetite nanoparticles have also been used as cores to be coated with a silica layer (by means of sol-gel method) and then with a shell of imprinted polymer for selective recognition of active substances.[194,195]

Continuous covalently cross-linked hydrogel layers can be prepared by free radical polymerization of monomers or applying phase separation, electrospinning, foaming, particle leaching, emulsion freeze-drying, or sintering of preformed polymers.[196,197] The literature about stimuli-responsive hydrogels is already enormous.[2,198,199] Below, we focus on those systems that require a sophisticated degree of nanostructuring for exhibiting multi-responsive switchable and repeatable performance under *in vivo* mimicking conditions,[2,8] or that exhibit domains (imprinted receptors) able to regulate drug release by means of an affinity-driven mechanism.[171,200,201]

Dually pH- and temperature-responsive hydrogels that combine NIPAAm and ionizable components can adjust the release of thrombolytic agents such as heparin or streptokinase, to the small changes of pH and temperature that accompany the formation of the thrombi.[202] They can be also prepared to coat the vaginal tissue and to release an anti-HIV microbicide in response to semen-induced pH change.[203] In addition to these indirect markers, hydrogels can be designed to recognize a direct biomarker of the illness. Molecular recognition can be attained by conjugation of carbohydrates,[204] DNA,[205] enzymes or substrates,[206] antibodies,[207] or other proteins.[208] Competitive binding of the biomarker to the recognition elements may trigger a phase volume transition or the disintegration of the network, as comprehensively reviewed by Miyata.[209,210] Intense response to the biomarker can be attained by combining the molecular recognition with the sensitiveness to other stimuli. Similarly, artificial receptors (imprinted domains) can be created in stimuli-responsive continuous networks as described above.[211–215] For example, applying the imprinting technology, hydrogels of acrylamide, 4-[(4-methacryloyloxy)phenylazo]benzenesulfonic acid and N,N'-hexylene bismethacrylamide were shown able to modulate paracetamol uptake and release through light-responsive changes in the conformation of the imprinted cavities.[216] The affinity of the azobenzene chromophores for paracetamol was strong when the hydrogels were stored at dark or irradiated at 440 nm, enabling a high loading. Oppositely, irradiation at 353 nm caused the photoisomerization of the azobenzene groups losing the affinity for paracetamol and then the release was triggered (Fig. 19). Subsequent irradiation at 440 nm reverted the isomerization and the imprinted receptors recovered their initial conformation, being able to reload the previously released paracetamol. These paracetamol-imprinted hydrogels exhibited a notable selectivity for the template drug, compared to other structural analogs.[216]

In addition to the stimuli-responsive systems, a field where imprinted hydrogels are attracting increasing interest is that of the medicated soft contact lenses (SCL).[200] Although, the main use of commercially-available SCL is for correcting ametropia problems, they may also uptake some drugs and

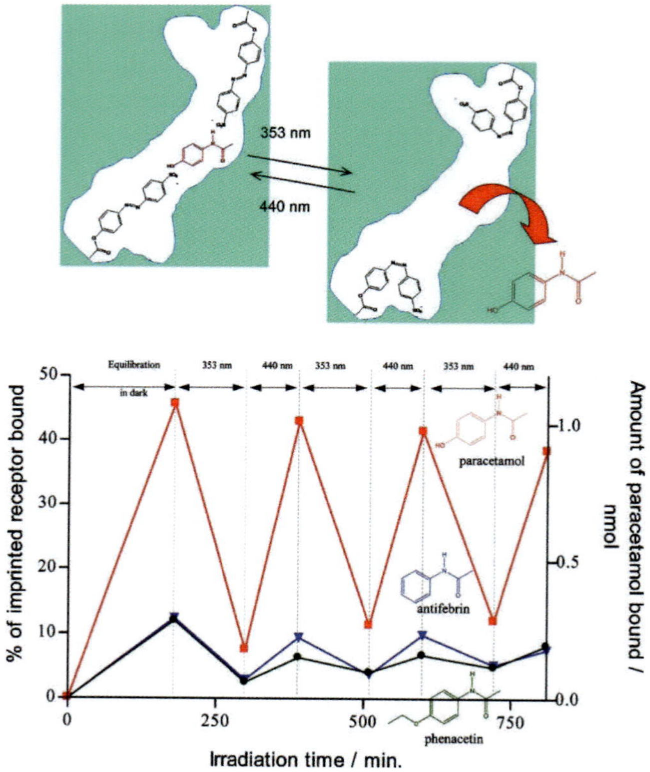

Fig. 19. Photoregulated release and uptake of paracetamol, antifebrin or phenacetin by paracetamol-imprinted 4-[(4-methacryloyloxy)phenylazo]benzenesulfonic acid-containing polyacrylamide hydrogel in HEPES buffer at pH 7.16. Nonspecific binding of substrate to non-imprinted control hydrogels has been subtracted from the binding data. Reprinted with permission from Ref. 216. Copyright (2008) American Chemical Society.

release them into the postlens lachrymal fluid, minimizing the clearance and the sorption through the conjunctiva. Molecular imprinting technology offers the possibility of enhancing the capability of SCL to load drugs and to control the release, fulfilling the requirements of both chronic and acute ocular diseases.[213,217] Since the list of feasible functional monomers and cross-linkers approved for SCL is quite short, the aim of the imprinting is to reorder the components in that way that the interactions between the template and the functional monomers can be maximized and consequently high affinity imprinted domains can be created in the loosely cross-linked network. The success of this approach has been demonstrated for antiglaucoma drugs (timolol, acetazolamide and ethoxzolamide),[218–222] antimicrobial agents (norfloxacin),[223] antihistaminics (ketotifen),[224,225] anti-inflammatories (prednisolone, diclofenac),[226,227] and some comfort ingredients.[228] Two recent preliminary *in vivo* experiments confirmed the prolonged drug ocular levels that can be attained with the imprinted SCL compared to conventional SCL and eyedrops.[217,229] For example, timolol levels in rabbits' lachrymal fluid after the insertion of imprinted lenses (34 μg dose) were sustained for 90 min, while non-imprinted lenses (20 μg dose) and instilled eye-drop solutions of 0.068% (total dose 34 μg) or of 0.25% (commercial solution, total dose 125 μg) provided lower drug levels and for shorter time (Fig. 20).[229] Furthermore, the area under the timolol concentration-time curve (AUC) was 3.3-fold and 8.7-fold greater for imprinted contact lenses than

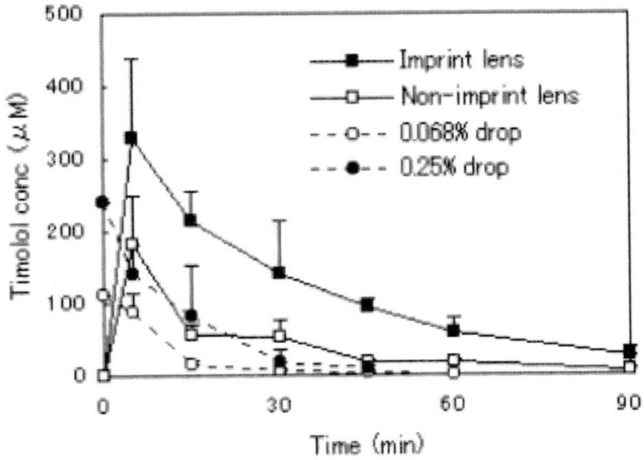

Fig. 20. Timolol tear fluid concentration-time profiles after application of presoaked imprinted and non-imprinted contact lenses and instillation of eyedrops. The doses were 34 μg for imprinted contact lens, 21 μg for non-imprinted contact lens, 34 μg for 0.068% timolol eyedrop, and 125 μg for 0.25% timolol eyedrop. Each point represents the mean ± S.D. (n=3–5). Reproduced from Ref. 229 with permission from Elsevier.

non-imprinted ones and eyedrops, respectively. These results indicate that imprinted SCLs reduce the precorneal elimination of drug compared to the eyedrops and, in consequence, much smaller amount of drug is needed to achieve the desired therapeutic levels.

4. Conclusions and Outlook

A better knowledge of the barriers that therapeutic agents have to overcome during their transit through the body up to the site of action, as well as of the mechanisms that feedback regulate the delivery of physiological substances, has prompted the search of more efficient drug delivery systems. Advanced carriers are becoming to be a reality thanks to the availability of a large variety of materials that are sensitive to physiological and external stimuli, and the feasibility of finely tuning their assembly from the very nanoscale. Nanostructuring has been shown an efficient way to prepare compartmentalized delivery systems, with a variety of degrees of sophistication, using a minimum of resources and costs, as Nature does to create functional structures. Stepwise-triggering of drug release as a function of the intensity of a certain stimulus or of the simultaneous action of different stimuli, and receptor-mimicking affinity controlled uptake and release are two examples of the unprecedented possibilities for regulating the spatiotemporal profile of the drug concentration in the body. Moreover, nanostructured systems can be grafted to or created on the surface of a variety of implantable medical devices, as recently reviewed,[230] enabling the design of smart drug/medical device combination products with improved host/microbial responses and local treatment of hardly accessible pathologies. The next few years will witness a large body of research on the performance of nanostructured systems under *in vivo* conditions and in clinical studies, which should serve to gather useful information for optimizing the design of even more efficient delivery systems. Moreover, a variety of guidance from regulatory agencies may emerge to help companies on the manufacturing, testing and submission of stimuli-responsive nanostructured systems.[231] Advances in this field require multidisciplinary teams able to use current and novel tools for materializing stimuli-reversible systems with enough reproducibility under physiological and pathological conditions.

Acknowledgments

This work was supported by MICINN (SAF2011-22771), FEDER and Xunta de Galicia (10CSA203013PR), Spain. B. Blanco-Fernandez is grateful to the Ministerio de Educación, Cultura y Deporte of Spain for a FPU grant.

References

1. A. S. Hoffman, *J. Control. Release* **132**, 153 (2008).
2. C. Alvarez-Lorenzo and A. Concheiro, *Mini-Rev. Med. Chem.* **8**, 1065 (2008).
3. M. L. Billingsley, *Pharmacology* **82**, 239 (2008).
4. K. Y. Lee and S. H. Yuk, *Prog. Polym. Sci.* **32**, 669 (2007).
5. D. Schmaljohann, *Adv. Drug Del. Rev.* **58**, 1655 (2006).
6. J. Kost and R. Langer, *Adv. Drug Del. Rev.* **46**, 125 (2001).
7. M. Yoshida and J. Lahann, *ACS Nano* **2**, 1101 (2008).
8. M. Motornov, Y. Roiter, I. Tokarev and S. Minko, *Prog. Polym. Sci.* **35**, 174 (2010).
9. G. Pasparakis and M. Vamvakaki, *Polym. Chem.* **2**, 1234 (2011).
10. M. Shibayama and T. Tanaka, in *Responsive Gels: Volume Transitions I*, K. Dusek (Ed.), Berlin: Springer (1993), p. 1.
11. F. Liu and M. W. Urban, *Prog. Polym. Sci.* **35**, 3 (2010).
12. C. Alvarez-Lorenzo and A. Concheiro, In *Smart Materials for Drug Delivery*, C. Alvarez-Lorenzo and A. Concheiro, (Eds.), London: Royal Society of Chemistry (2013), 1 p. 1.
13. C. Alexander and K. M. Shakesheff, *Adv. Mater.* **18**, 3321 (2006).
14. S. Grund, M. Bauer and D. Fischer, *Adv. Eng. Mater.* **13**, B61 (2011).
15. C. Alvarez-Lorenzo, E. Bucio, G. Burillo and A. Concheiro, *Expert Opin. Drug Deliv.* 7, 173 (2010).
16. A. L. Yarin, Mathematical modelling of natural phenomena **3**, 1 (2008).
17. A. W. Jackson and D. A. Fulton, *Polym. Chem.* **4**, 31 (2013).
18. C. Alvarez-Lorenzo, A. M. Puga and A. Concheiro, In *Biomimetic Approaches for Biomaterials Development*, J. F. Mano (Ed.), New York: Wiley-VCH (2012), p. 417.
19. S. A. Kumar and M. I. Khan, *J. Nanosci. Nanotechnol.* **10**, 4124 (2010).
20. C. Alvarez-Lorenzo and A. Concheiro, *Curr. Opin. Biotechnol.* In press (2013).
21. S. Patel, A. A. Bhirde, J. F. Rusling, X. Chen, J. S. Gutkind and V. Patel, *Pharmaceutics* **3**, 34 (2011).
22. H. Cabral, N. Nishiyama and K. Kataoka, *Acc. Chem. Res.* **44**, 999 (2011).
23. M. Yoshida and J. Lahann, *ACS Nano* **2**, 1101 (2008).
24. A. D. Bangham, M. M. Standish and J. C. Watkins, *J. Mol. Biol.* **13**, 238 (1965).
25. K. Kono, M. Iwamoto, R. Nishikawa, H. Yanagie and T. Takagishi, *J. Control. Release* **68**, 225 (2000).
26. E. Roux, C. Passirani, S. Scheffold, J. P. Benoit and J. C. Leroux, *J. Control. Release* **94**, 447 (2004).
27. E. Ducat, J. Deprez, A. Gillet, A. Noel, B. Evrard, O. Peulen and G. Piel, *Int. J. Pharm.* **420**, 319 (2011).
28. M. S. Webb, J. J. Wheeler, M. B. Bally and L. D. Mayer, *Biochim. Biophys. Acta.* **1238**, 147 (1995).
29. Z. Vanic, S. Barnert, R. Suss and R. Schubert, *J. Liposome Res.* **22**, 148 (2012).
30. P. Lutwyche, C. Cordeiro, D. J. Wiseman, M. St-Louis, M. Uh, M. J. Hope, M. S. Webb and B. B. Finlay, *Antimicrob. Agents Chemother.* **42**, 2511 (1998).
31. E. Fattal, P. Couvreur and C. Dubernet, *Adv. Drug Deliv. Rev.* **56**, 931 (2004).
32. S. Simoes, J. N. Moreira, C. Fonseca, N. Duzgunes and M. C. de Lima, *Adv. Drug. Deliv. Rev.* **56**, 947 (2004).

33. E. A. Leite, C. dos Santos Giuberti, A. J. Wainstein, A. P. Wainstein, L. G. Coelho, A. M. Lana, P. R. Savassi-Rocha and M. C. De Oliveira, *Life Sci.* **84**, 641 (2009).
34. E. Ducat, J. Deprez, A. Gillet, A. Noel, B. Evrard, O. Peulen and G. Piel, *Int. J. Pharm.* **420**, 319 (2011).
35. T. Ishida and H. Kiwada, *Int. J. Pharm.* **354**, 56 (2008).
36. D. Chen, W. Liu, Y. Shen, H. Mu, Y. Zhang, R. Liang, A. Wang, K. Sun and F. Fu, *Int. J. Nanomedicine* **6**, 2053 (2011).
37. J. Shin, P. Shum, J. Grey, S. Fujiwara, G. S. Malhotra, A. González-Bonet, S. H. Hyun, E. Moase, T. M. Allen and D. H. Thompson, *Mol. Pharm.* **9**, 3266 (2012).
38. A. Kale and V. P. Torchilin, *J. Liposome Res.* **17**, 197 (2007).
39. E. Koren, A. Apte, A. Jani and V. P. Torchilin, *J. Control. Release* **160**, 264 (2012).
40. S. R. Paliwal, R. Paliwal, H. C. Pal, A. K. Saxena, P. R. Sharma, P. N. Gupta, G. P. Agrawal, and S. P. Vyas, *Mol. Pharm.* **9**, 176 (2012).
41. T. Jiang, Z. Zhang, Y. Zhang, H. Lv, J. Zhou, C. Li, L. Hou and Q. Zhang, *Biomaterials* **33**, 9246 (2012).
42. A. L. de Barros, L. G. Mota, D. C. Soares, M. M. Coelho, M. C. Oliveira and V. N. Cardoso, *Bioorg. Med. Chem. Lett.* **21**, 7373 (2011).
43. E. Torres, F. Mainini, R. Napolitano, F. Fedeli, R. Cavalli, S. Aime and E. Terreno, *J. Control Release* **154**, 196 (2011).
44. R. Mo, Q. Sun, J. Xue, N. Li, W. Li, C. Zhang and Q. Ping, *Adv. Mater.* **24**, 3659 (2012).
45. N. M. Samoshina, X. Liu, B. Brazdova, A. H. Franz, V. V. Samoshin and X. Guo, *Pharmaceutics* **3**, 379 (2011).
46. X. Liu, Y. Zheng, N. M. Samoshina, A. H. Franz, X. Guo and V. V. Samoshin, *J. Liposome Res.* **22**, 319 (2012).
47. D. Needham and M. W. Dewhirst, in *Smart Materials for Drug Delivery*, C. Alvarez-Lorenzo and A. Concheiro, (Eds.), London: Royal society of chemistry (2013), 1, p. 33.
48. D. Needham and M. W. Dewhirst, *Adv. Drug Deliv. Rev.* **53**, 285 (2001).
49. P. S. Yarmolenko, Y. Zhao, C. Landon, I. Spasojevic, F. Yuan, D. Needham, B. L. Viglianti and M. W. Dewhirst, *Int. J. Hyperthermia* **26**, 485 (2010).
50. C.D. Landon, J. Y. Park, D. Needham and M. W. Dewhirst, *Open Nanomedicine Journal.* **3**, 38 (2010).
51. Celsion Corporation, A Phase III, Randomized, Double-Blinded, Dummy-Controlled Study of the Efficacy and Safety of ThermoDox® (Thermally Sensitive Liposomal Doxorubicin) in Combination With Radiofrequency Ablation (RFA) Compared to RFA-Alone in the Treatment of Non-Resectable Hepatocellular Carcinoma (NCT00617981). 2008–2012.
52. S. L. Huang, *Adv. Drug Deliv. Rev.* **60**, 1167 (2008).
53. E. Hagtvet, T. J. Evjen, D. R. Olsen, S. L. Fossheim and E. A. Nilssen, *J. Drug Target.* **19**, 701 (2011).
54. P. Shum, J. M. Kim and D. H. Thompson, *Adv. Drug Deliv. Rev.* **53**, 273 (2001).
55. S. J. Leung and M. Romanowski, *Theranostics* **2**, 1020 (2012).
56. C. Alvarez-Lorenzo, L. Bromberg and A. Concheiro, *Photochem. Photobiol.* **85**, 848 (2009).
57. J. Y. Wang, Q. F. Wu, J. P. Li, Q. S. Ren, Y. L. Wang and X. M. Liu, *Mini Rev. Med. Chem.* **10**, 172 (2010).

58. E. T. Kisat, B. Coldren, C. A. Evans, C. Boyer and J. A. Zasadzinski, *Curr. Med. Chem.* **11**, 199 (2004).
59. B. Wong, C. Boyer, C. Steinbeck, D. Peters, J. Schmidt, R. van Zanten, B. Chmelka and J. A. Zasazinski, *Adv. Mater.* **23**, 2320 (2011).
60. R. Chandrawati and F. Caruso, *Langmuir* **28**, 13798 (2012).
61. L. Hosta-Rigau, B. Städler, Y. Yan, E. C. Nice, J. K. Hearh, F. Albericio and F. Caruso, *Adv. Funct. Mater.* **20**, 59 (2010).
62. R. Chandrawati, P. D. Odermatt, S. F. Chong, A. D. Price, B. Stadler and F. Caruso, *Nano Lett.* **11**, 4958 (2011).
63. K. Letchford and H. Burt, *Eur. J. Pharm. Biopharm.* **65**, 259 (2007).
64. M. Pinna, S. Hiltl, X. Guo, A. Böker and A. V. Zvelindovsky, *ACS Nano* **4**, 2845 (2010).
65. K. Miyata, R. J. Christie and K. Kataoka, *React. Funct. Polym.* **71**, 227 (2011).
66. S. R. Croy and G. S. Kwon, *Curr. Pharm. Design.* **12**, 4669 (2006).
67. C. J. F. Rijcken, O. Soga, W. E. Hennink and C. F. van Nostrum, *J. Control. Release* **120**, 131 (2007).
68. N. Rapoport, *Prog. Polym. Sci.* **32**, 962 (2007).
69. F. Meng, Z. Zhong and J. Feijen, *Biomacromolecules* **10**, 197 (2009).
70. M. Massignani, H. Lomas and G. Battaglia, in *Modern Techniques for Nano- and Microreactors/-reactions*, F. Caruso, (Ed.), Berlin: Springer (2010), p. 115.
71. Y. Bae, N. Nishiyama, S. Fukushima, H. Koyama, M. Yasuhiro and K. Kataoka, *Bioconjug. Chem.* **16**, 122 (2005).
72. H. Wu, L. Zhu and V. P. Torchilin, *Biomaterials* **34**, 1213 (2013).
73. K. H. Min, J. H. Kim, S. M. Bae, H. Shin, M. S. Kim, S. Park, *et al., J. Control. Release* **144**, 259 (2010).
74. G.H. Gao, J. W. Lee, M. K. Nguyen, G. H. Im, J. Yang, H. Heo, *et al., J. Control. Release* **155**, 11 (2011).
75. F. Ahmed, R. I. Pakunlu, A. Brannan, F. Bates, T. Minko and D. E. Discher, *J. Control. Release* **116**, 150 (2006).
76. G. H. Gao, Y. Li and D. S. Lee, *J. Control. Release* (2013) in press, **169**, 180 (2013).
77. F. Q. Schafer and G. R. Buettner, *Free Radic. Biol. Med.* **30**, 1191 (2001).
78. K. M. Park, D. W. Lee, B. Sarkar, H. Jung, J. Kim, Y. H. Ko, K. E. Lee, H. Jeon and K. Kim, *Small* **6**, 1430 (2010).
79. J. Li, M. Huo, J. Wang, J. Zhou, J. M. Mohammad, Y. Zhang, Q. Zhu, A. Y. Waddad and Q. Zhang, *Biomaterials* **33**, 2310 (2012).
80. C. Wei, J. Guo and C. Wang, *Macromol. Rapid. Commun.* **32**, 451 (2011).
81. H. De Oliveria, J. Thevenot and S. Lecommandoux, *Nanomed. Nanobiotechnol.* **4**, 525 (2012).
82. C. Alexander, *Expert Opin. Drug Deliver.* **3**, 573 (2006).
83. Y. Zhang, M. Juang, J. Zhao, X. Ren, D. Chen and G. Zhang, *Adv. Funct. Mater.* **15**, 695 (2005).
84. M. D. Determan, J. P. Cox and S. K. Mallapragada, *J. Biomed. Mater. Res.* **81A**, 326 (2007).
85. B. S. Lokitz, A. W. York, J. E. Stempka, N. D. Treat, Y. Li, W. L. Jarrett and C. L. McCormick, *Macromolecules* **40**, 6473 (2007).

86. Y. Zhang, M. Juang, J. Zhao, X. Ren, D. Chen and G. Zhang, *Adv. Funct. Mater.* **15**, 695 (2005).
87. K. Sugiyama and K. Sono, *J. Appl. Polym. Sci.* **81**, 3056 (2001).
88. X. Tong, G. Wang, A. Soldera and Y. Zhao, *J. Phys. Chem. B* **109**, 20281 (2005).
89. L. Gong-Yan, C. Chao-Jian, L. Dan-Dan, S. S. Wang and J. Ji, *J. Mater. Chem.* **22**, 16865 (2012).
90. J. Dong, Y. Wang, J. Zhang, X. Zhan, S. Zhu, H. Yang and G. Wang, *Soft Matter* **9**, 370 (2013).
91. S. Guragain, B. P. Bastakoti, M. Ito, S. Yusa and K. Nakashima, *Soft Matter* **8**, 9628 (2012).
92. G.A. Husseini, G. D. Myrup, W. G. Pitt, D. A. Christensen and N. Y. Rapoport, *J. Control. Release* **69**, 43 (2000).
93. N. Rapoport, In *Smart Nanoparticles in Nanomedicine*, R. Arshady and K. Kono, (Eds.), London: Kentus Books (2006), p. 305.
94. B. J. Staples, W. G. Pitt, B. L. Roeder, G. A. Husseini, D. Rajeev and G. B. Schaalje, *J. Pharm. Sci.* **99**, 3122 (2010).
95. P. Kamev and N. Rapoport, *Am. J. Phys.* **829**, 543 (2006).
96. N. Rapoport, In *Nanotechnology for cancer therapy*, M. Amighi, (Ed.), Boca Raton Florida: CRC Press (2006), p. 417.
97. H. Hasanzadeh, M. Mokhtari-Dizaji, S. Z. Bathaie and Z. M. Hassan, *Ultrason. Sonochem.* **18**, 1165 (2011).
98. P. Broz, S. Driamov, J. Ziegler, N. Ben-Haim, S. Marsch, W. Meier and P. Hunziker, *Nano Lett.* **6**, 2349 (2006).
99. H. C. Chiu, Y. W. Lin, Y. F. Huang, C. K. Chuang and C. S. Chern, *Angew. Chem. Int. Ed.* **47**, 1875 (2008).
100. R. Ma and L. Shi, *Macromol. Biosci.* **10**, 1397 (2010).
101. Z. Fu, M. A. Ochsner, H. P. M. de Hoog, N. Tomczak and M. Nallani, *Chem. Commun.* **47**, 2862 (2011).
102. M. Marguet, L. Edembe and S. Lecommandoux, *Angew. Chem. Int. Ed.* **51**, 1173 (2012).
103. H. Lomas, A. P. R. Johnston, G. K. Such, Z. Zhu, K. Liang, M. P. van Koeverden, S. Alongkornchotikul and F. Caruso, *Small* 7, 2109 (2011).
104. S. S. Kelkar and T. M. Reineke, *Bioconjug. Chem.* **22**, 1879 (2011).
105. J. J. Li, F. Zhao and J. Li, *Appl. Microbiol. Biotechnol.* **90**, 427 (2011).
106. L. Nogueiras-Nieto, C. Alvarez-Lorenzo, M. I. Sandez-Macho, A. Concheiro and F. J. Otero-Espinar, *J. Phys. Chem.* B **113**, 2773 (2009).
107. T. Higashi, A. Tajima, K. Motoyama and H. Arima, *J. Pharm. Sci.* **101**, 2891 (2012).
108. S. Simoes, F. Veiga, J. J. Torres-Labandeira, A. C. F. Ribeiro, M. I. Sandez-Macho, A. Concheiro and C. Alvarez-Lorenzo, *Eur. J. Pharm. Sci.* **80**, 103 (2012).
109. Z. Li, H. Yin, Z. Zhang, K. L. Liu and J. Li, *Biomacromolecules* **13**, 3162 (2012).
110. Y. Gao, R. Ma, Y. An and L. Shi, *J. Control. Release* **152**, e81 (2011).
111. G. Yaohua, Y. Cuihong, L. Xue, M. Rujiang, K. Deling, S. Linqi, *Macromol. Biosci.* **12**, 251 (2012).
112. C. Park, K. Oh, S. C. Lee and C. Kim, *Angew. Chem. Int. Ed.* **46**, 1455 (2007).
113. D. Astruc, E. Boisselier and C. Ornelas, *Chem. Rev.* **110**, 1857 (2010).
114. C. C. Lee, J. A. Mackay, J. M. J. Frechet and F. C. Szoka, *Nat. Biotechnol.* **23**, 1517 (2005).

115. C. Kojima, *Expert Opin. Drug Deliv.* 7, 307 (2010).
116. R. K. Tekade, T. Dutta, V. Gajbhiye and N. K. Jain, *J. Microencapsul.* **26**, 287 (2009).
117. S. Svenson, *Eur. J. Pharm. Biopharm.* **71**, 445 (2009).
118. Y. Zhao, X. Fan, D. Liu and Z. Wang, *Int. J. Pharm.* **409**, 229 (2011).
119. C. Kojima, S. Tsumura, A. Harada and K. Kono, *J. Am. Chem. Soc.* **131**, 6052 (2009).
120. C. Kojima and T. Suehiro, *Chem. Lett.* **40**, 1249 (2011).
121. S. Xu, Y. Luo and R. Haag, *Macromol. Biosci.* 7, 968 (2007).
122. H. Hui, F. Xiao-dong and C. Zhong-lin, *Polymer* **46**, 9514 (2005).
123. R. K. Tekade, T. Dutta, V. Gajbhiye and N. K. Jain, *J. Microencapsul.* **26**, 287 (2009).
124. H. Jin, Y. Zhou, W. Huang and D. Yan, *Langmuir* **26**, 14512 (2010).
125. R. Deloncle and A. M. Caminade, *J. Photochem. Photobiol. C* **11**, 25 (2010).
126. C. Park, J. Lim, M. Yun and C. Kim, *Angew. Chem. Int. Ed.* **47**, 2959 (2008).
127. Y. Umeda, C. Kojima, A. Harada, H. Horinaka and K. Kono, *Bioconjug. Chem.* **21**, 1559 (2010).
128. R. J. Amir, N. Pessah, M. Shamis and D. Shabat, *Angew. Chem. Int. Ed.* **42**, 4494 (2003).
129. M. Shamis, H. N. Lode and D. Shabat, *J. Am. Chem. Soc.* **126**, 1726 (2004).
130. R. E. Wang, F. Costanza, Y. Niu, H. Wu, Y. Hu, W. Hang, Y. Sun and J. Cai, *J. Control. Release* **159**, 154 (2012).
131. S. A. Sukhisvili, *Curr. Opin. Colloid Interface Sci.* **10**, 37 (2005).
132. Y. Wang, L. Hosta-Rigau, H. Lomas and F. Caruso, *Phys. Chem. Chem. Phys.* **13**, 4782 (2011).
133. S. Pavlukhina and S. Sukhishvili, *Adv. Drug Deliv. Rev.* **63**, 822 (2011).
134. J.E. Wong, A. K. Gaharwar, D. Müller-Schulte, D. Bahadur and W. Richtering, *J. Colloid Interface Sci.* **324**, 47 (2008).
135. A. N. Zelikin, *ACS Nano* **4**, 2494 (2010).
136. Z. Zhu and S. A. Sukhishvili, *ACS Nano* **3**, 3595 (2009).
137. J. S. Barbosa, R. R. Costa, A. M. Testera, M. Alonso, J. C. Rodríguez-Cabello and J. F. Mano, *Nanoscale Res. Lett.* **4**, 1247 (2009).
138. K. Yoshida, K. Sato and J. Anzai, *J. Mater. Chem.* **20**, 1546 (2010).
139. C. Basset, C. Harder, C. Vidaud and C. Déjugnat, *Biomacromolecules* **11**, 806 (2010).
140. Z. Zheng, X. Zhang, D. Carbo, C. Clark, C. A. Nathan and Y. Lvov, *Langmuir* **26**, 7679 (2010).
141. K. Ariga, Y. M. Lvov, K. Kawakami, Q. Ji and J. P. Hill, *Adv. Drug Deliv. Rev.* **63**, 762 (2011).
142. Q. Ji, S. Acharya, J. P. Hill, A. Vinu, S. B. Yoon, J. S. Yu, K. Sakamoto and K. Ariga, *Adv. Funct. Mater.* **19**, 1792 (2009).
143. M. Delcea, A. Yashchenok, K. Videnova, O. Kreft, H. Möhwald and A. G. Skirtach, *Macromol. Biosci.* **10**, 465 (2010).
144. B. M. Wohl and J. F. Engbersen, *J. Control. Release* **158**, 2 (2012).
145. T. Maeda, Y. Akasaki, K. Yamamoto and T. Aoyagi, *Langmuir* **25**, 9510 (2009).
146. R. Barreiro-Iglesias, R. Coronilla, A. Concheiro and C. Alvarez-Lorenzo, *Eur. J. Pharm. Sci.* **24**, 77 (2005).
147. C. Y. Chuang, T. M. Don and W. Y. Chiu, *J. Polym. Sci., Part A: Polym. Chem.* **47**, 2798 (2009).
148. Y. Wu, J. Guo, W. L. Yang, C. C. Wang and S. K. Fu, *Polymer* **47**, 5287 (2006).

149. I. K. Voets, A. Keizer and M. A. C. Stuart, *Adv. Colloid Interface Sci.* **147–148**, 300 (2009).
150. J. K. Oh, D. I. Lee and J. M. Park, *Prog. Polym. Sci.* **34**, 1261 (2009).
151. M. Antoniette and K. Landfester, *Prog. Polym. Sci.* **27**, 689 (2002).
152. A. E. Tonelli, *J. Incl. Phenom. Macrocyclic Chem.* **60**, 197 (2008).
153. J. A. Johnson, N. J. Turro, J. T. Koberstein and J. E. Mark, *Prog. Polym. Sci.* **35**, 332 (2010).
154. Y. Marui, T. Kida and M. Akashi, *Chem. Mater.* **22**, 282 (2010).
155. M. D. Moya-Ortega, C. Alvarez-Lorenzo, A. Concheiro and T. Loftsson, *Int. J. Pharm.* **428**, 152 (2012).
156. M. D. Moya-Ortega, T. F. G. Alves, C. Alvarez-Lorenzo, A. Concheiro, E. Stefánsson, M. Thorsteinsdóttir and T. Loftsson, *Int. J. Pharm.* In press, **441**, 507 (2013).
157. W. Zhang, K. Gilstrap, L. Wu, R. Bahadur, M. A. Moss, Q. Wang, X. Lu and X. He, *ACS Nano* **4**, 6747 (2010).
158. R. T. Chacko, J. Ventura, J. Zhuang and S. Thayumanavan, *Adv. Drug Del. Rev.* **64**, 836 (2012).
159. T. S. Shim, S. H. Kim and S. M. Yang, *Part. Part. Syst. Charact.* **30**, 9 (2013).
160. C. Qiu, M. Chen, H. Yan and H. K. Wu, *Adv. Mater.* **19**, 1603 (2007).
161. J. P. Rolland, B. W. Maynor, L. E. Euliss, A. E. Exner, G. M. Denison and J. M. DeSimone, *J. Am. Chem. Soc.* **127**, 10096 (2005).
162. S. E. Gratton, P. D. Pohlhaus, J. Lee, J. Guo, M. J. Cho and J. M. Desimone, *J. Control. Release* **121**, 10 (2007).
163. W. Song, A. C. Lima and J. F. Mano, *Soft Matter* **6**, 5868 (2010).
164. Z. Guo, W. Liu and B. L. Su, *J. Colloid Interface Sci.* **353**, 335 (2011).
165. A. C. Lima, W. Song, B. Blanco-Fernandez, C. Alvarez-Lorenzo and J. F. Mano, *Pharm. Res.* **28**, 1294 (2010).
166. A. C. Lima, C. A. Custodio, C. Alvarez-Lorenzo and J. F. Mano, *Small* **9**, 2487 (2013).
167. T. T. Dang, Q. B. Xu, K. M. Bratlie, E. S. O'Sullivan, X. Y. Chen, R. Langer and D. G. Anderson, *Biomaterials* **30**, 6896 (2009).
168. H. Tekin, T. Tsinman, J. G. Sanchez, B. J. Jones, G. Camci-Unal, J. W. Nichol, R. Langer and A. Khademhosseini, *J. Am. Chem. Soc.* **133**, 12944 (2011).
169. S. J. Rowan, S. J. Cantrill, G. R. Cousins, J. K. Sanders and J. F. Stoddart, *Angew. Chem. Int. Ed. Engl.* **41**, 898 (2002).
170. A. W. Jackson and D. A. Fulton, *Macromolecules* **45**, 2699 (2012).
171. C. Alvarez-Lorenzo and A. Concheiro, In *Smart Nano- and Microparticles*, R. Arshady and K. Kono, (Eds.), London:Kentus Books (2006), p. 279.
172. N. Sanson and J. Rieger, *Polym. Chem.* **1**, 965 (2010).
173. C. Alvarez-Lorenzo and A. Concheiro, *J. Chromatogr. B* **804**, 231 (2004).
174. A. G. Mayes and M. J. Whitcombe, *Adv. Drug Del. Rev.* **57**, 1742 (2005).
175. P. A. G. Cormack and A. Z. Elorza, *J. Chromatogr. B* **804**, 173 (2004).
176. M. J. Whitcombe, I. Chianella, L. Larcombe, S. A. Piletsky, J. Noble, R. Porter and A. Horgan, *Chem. Soc. Rev.* **40**, 1547 (2011).
177. K. Ito, J. Chuang, C. Alvarez-Lorenzo, T. Watanabe, N. Ando and A.Yu. Grosberg, *Prog. Polym. Sci.* **28**, 1489 (2004).
178. D. Horak, F. Lednicky, V. Rehak and F. Svec, *Polym. Sci.* **49**, 2041 (1993).

179. R. Suedee, C. Jantarat, W. Lindner, H. Viernstein, S. Songkro and T. Srichana, *J. Control. Release* **142**, 122 (2010).
180. L. Ye, R. Weiss and K. Mosbach, *Macromolecules* **33**, 8239 (2000).
181. M. H. Smith, E. S. Herman and L. A. Lyon, *J. Phys. Chem. B* **115**, 3761 (2011).
182. A. Pich and W. Richtering, in *Chemical Design of Responsive Microgels*, A. Pich and W. Richtering, (Eds.), Berlin:Springer (2010), p. 1.
183. J. P. K. Tan, M. B. H. Tan and M. K. C. Tam, *Local Reg. Anesth.* **3**, 93 (2010).
184. J. Zhang, Z. Y. Qian and Y. Q. Gu, *Nanotechnology* **20**, No. 325102 (2009).
185. F. Puoci, F. Iemma, R. Muzzalupo, U. G. Spizzirri, S. Trombino, R. Cassano and N. Picci, *Macromol. Biosci.* **4**, 22 (2004).
186. G. Ciardelli, B. Cioni, C. Cristallini, N. Barbani, D. Silvestri and P. Giusti, *Biosens. Bioelectron.* **20**, 1083 (2004).
187. L. Fang, S. Chen, Y. Zhang and H. Zhang, *J. Mater. Chem.* **21**, 2320 (2011).
188. M. H. Lee, J. L. Thomas, H. Y. Wang, C. C. Chang, C. C. Lin and H. Y. Lin, *J. Mater. Chem.* **22**, 24644 (2012).
189. W. Schartl, *Nanoscale* **2**, 829 (2010).
190. Q. Yuan, O. J. Cayre, S. Fujii, S. P. Armes, R. A. Williams and S. Biggs, *Langmuir* **26**, 18408 (2010).
191. Y. D. Liu, F. F. Fang and H. J. Choi, *Langmuir* **26**, 12849 (2010).
192. W. Xionga, W. Wang, Y. Wang, Y. Zhao, H. Chena, H. Xu and X. Yang, *Colloids Surf., B* **84**, 447 (2011).
193. N. Sahiner and P. Ilgin, *Polymer* **5**, 3156 (2010).
194. L. Chang, S. Chen and X. Li, *Appl. Surf. Sci.* **258**, 6660 (2012).
195. X. Kong, R. Gao, X. He, L. Chen and Y. Zhang, *J. Chromatogr. A* **1245**, 8 (2012).
196. R. P. Shaikh, V. Pillay, Y. E. Choonara, L. C. du Toit, V. M. K. Ndesendo, P. Bawa and S. Cooppan, *AAPS Pharm. Sci. Tech.* **11**, 441 (2010).
197. C. Huang, S. J. Soenen, J. Rejman, B. Lucas, K. Braeckmans, J. Demeester and S. C. De Smedt, *Chem. Soc. Rev.* **40**, 2417 (2011).
198. C. Alexander, *Nat. Mat.* **7**, 767 (2008).
199. M. E. Caldorera-Moore, W. B. Liechty and N. A. Peppas, *Acc. Chem. Res.* **44**, 1061 (2011).
200. C. Alvarez-Lorenzo, F. Yañez and A. Concheiro, *J. Drug Del. Sci. Tech.* **20**, 237 (2010).
201. D. R. Kryscio and N. A. Peppas, *Acta Biomater.* **8**, 461 (2012).
202. S. K. Vakkalanka, C. S. Brazel and N. A. Peppas, *J. Biomat. Sci-Polym. E* **8**, 119 (1996).
203. K. M. Gupta, S. R. Barnes, R. A. Tangaro, M. C. Roberts, D. H. Owen, D. F. Katz and P. F. Kiser, *J. Pharm. Sci.* **96**, 670 (2007).
204. D. S. W. Benoit, S. D. Collins and K. S. Anseth, *Adv. Funct. Mater.* **17**, 2085 (2007).
205. Z. Li, Y. Zhang, P. Fullhart and C. A. Mirkin, *Nano Lett.* **4**, 1055 (2004).
206. R. V. Ulijn, *J. Mater. Chem.* **16**, 2217 (2006).
207. T. Miyata, M. Jige, T. Nakaminami and T. Uragami, *P. Natl. Acad. Sci. U.S.A.* **103**, 1190 (2006).
208. A. S. Hoffman, *Clin. Chem.* **46**, 1478 (2000).
209. T. Miyata, N. Asami and T. Uragami, *J. Polym. Sci. B* **47**, 2144 (2009).
210. T. Miyata, in *Smart Materials for Drug Delivery*, C. Alvarez-Lorenzo and A. Concheiro, (Eds.), London: Royal society of chemistry (2013), 2, p. 261.
211. X. Y. Liu, Y. Guan, X. B. Ding, Y. X. Peng, X. P. Long, X. C. Wang, *et al.*, *Macromol. Biosci.* **4**, 680 (2004).

212. G. Demirel, G. Ozcetin, E. Turan and T. Caykara, *Macromol. Biosci.* **5**, 1032 (2005).
213. C. Alvarez-Lorenzo and A. Concheiro, In *Biotechnology Annual Review* vol. 12, M.R. El-Gewely, (Ed.), Amsterdam: Elsevier (2006), p. 225.
214. L. Ye and K. Mosbach, *Chem. Mater.* **20**, 859 (2008).
215. T. V. Burova, N. V. Grinberg, E. V. Kalinina, R. V. Ivanov, V. I. Lozinsky, C. Alvarez-Lorenzo, *et al., Macromol. Chem. Phys.* **212**, 72 (2011).
216. C. Gong, K. L. Wong and M. H. W. Lan, *Chem. Mater.* **20**, 1353 (2008).
217. A. Tieppo, C. J. White, A. C. Paine, M. L. Voyles, M. K. McBride and M. E. Byrne, *J. Control. Release* **157**, 391 (2012).
218. C. Alvarez-Lorenzo, H. Hiratani, J. L. Gómez-Amoza, R. Martínez-Pacheco, C. Souto and A. Concheiro, *J. Pharm. Sci.* **91**, 2182 (2002).
219. H. Hiratani and C. Alvarez-Lorenzo, *Biomaterials* **25**, 1105 (2003).
220. F. Yañez, A. Chauhan, A. Concheiro and C. Alvarez-Lorenzo, *J. Appl. Polym. Sci.* **122**, 1333 (2011).
221. A. Ribeiro, F. Veiga, D. Santos, J. J. Torres-Labandeira, A. Concheiro and C. Alvarez-Lorenzo, *Biomacromolecules* **12**, 701 (2011).
222. A. Ribeiro, F. Veiga, D. Santos, J. J. Torres-Labandeira, A. Concheiro and C. Alvarez-Lorenzo, *J. Membr. Sci.* **383**, 60 (2011).
223. C. Alvarez-Lorenzo, F. Yañez, R. Barreiro-Iglesias and A. Concheiro, *J. Control. Release* **113**, 236 (2006).
224. S. Venkatesh, S. P. Sizemore and M. E. Byrne, *Biomaterials* **28**, 717 (2007).
225. M. Ali, S. Horikawa, S. Venkatesh, J. Saha, S. Pass, J. W. Hong and M. E. Byrne, *J. Control. Release* **124**, 154 (2007).
226. B. Malaekeh-Nikouei, F. A. Ghaeni, V. S. Motamedshariaty and S. A. Mohajeri, *J. Appl. Polym. Sci.* **126**, 387 (2012).
227. A. Tieppo, K. M. Pate and M. E. Byrne, *Eur. J. Pharm. Biopharm.* **81**, 170 (2012).
228. M. Ali and M. E. Byrne, *Pharm. Res.* **26**, 714 (2009).
229. H. Hiratani, A. Fujiwara, Y. Tamiya, Y. Mizutani and C. Alvarez-Lorenzo, *Biomaterials* **26**, 1293 (2005).
230. C. Alvarez-Lorenzo and A. Concheiro, in *Smart Materials for Drug Delivery*, C. Alvarez-Lorenzo and A. Concheiro, (Eds.), London: Royal society of chemistry (2013), 2, p. 313.
231. K. E. Sapsford, K. Lauritsen and K. M. Tyner, *Ther. Deliv.* **3**, 1383 (2012).

Chapter B4

Biomedical Nanopreparations with Controlled Geometry

Catherine A. Fromen[*,§], *Stuart S. Dunn*[†,¶] *and Joseph M. DeSimone*[‡,||]

[*]Department of Chemical Engineering, North Carolina State University Engineering Building I, 911 Partners Way, Raleigh, NC 27695
[†]Department of Radiology and Biomedical Research Imaging Center, University of North Carolina at Chapel Hill
[‡]Department of Chemistry, University of North Carolina at Chapel Hill Department of Chemical Engineering, North Carolina State University 257 Caudill, South Road, CB# 3290, Chapel Hill, NC 27599
[§]cafromen@ncsu.edu
[¶]ssdunn@email.unc.edu
[||]desimone@unc.edu

Abstract

Discovering the role that particle geometry plays in nanoparticle diagnostic and therapeutic applications is an exciting new thrust in nanomedicine. A plethora of fabrication techniques, both top-down and bottom-up, are emerging to control the physical properties of particles. Enabled by these creative fabrication approaches, studies on the fate of non-spherical particles are beginning to illuminate the role of particle geometry *in vivo*, highlighting the promising applications and benefits for these particles. Much work remains, but initial findings in this area highlight the critical importance particle geometry, composition, flexibility, and surface chemistry have on biodistribution, circulation, and cellular uptake.

1. Significance of Particle Geometry in Drug Delivery

The search for improved efficiency and capabilities in particulate therapeutics, diagnostics, and imaging is driving investigations of nanoparticle physical properties far beyond spherical carriers.[1–7] Historically, particle fabrication techniques have involved methodologies that yield spherical carriers. As a result, most studies on carrier fate have been performed with spherical particles. Spherical particles represent a simplistic geometric system and using such a system may overlook key parameters; most systems in nature are not so simple. Complex geometries and anisotropic systems exist almost everywhere we look; imagine how a lock and key would function without precise definition of the complex geometry of each counterpart. Physical properties are important. In so many cases in Nature, physical geometry is a critical indicator of a particular function or advantage. The seed of a maple tree has a unique geometry which allows it to rotate, create lift as it falls, and spread the seeds over a wider area.[8] The self-cleaning hydrophobic surface of a lotus leaf is attributed to the topography and architecture of the surface.[9] Furthermore, when considering activity on a cellular level, physical properties play another fundamental role. For example, mammalian cells all have unique physical properties that complement their function. Red blood cells have an extremely low elastic modulus (~25 kPa) which allows them to physically deform with up to 100% strain and fit through fenestrations in the vasculature.[10] Modulus is also important in disease indications; cancer cells, especially metastatic ones, are an order of magnitude more deformable than normal cells.[11] Other normal cell types have varying degrees of physical protrusions or branches: (1) neurons are long thin cells that carry and transmit electrical signals, (2) photoreceptor cells in the back of the eye have a protrusion called the outer segment which specializes in absorbing light, (3) dendritic cells have protrusions enabling antigen capture and presentation, and (4) epithelial cells are typically flat sheets that efficiently form tight junctions with their neighbors. Even bacteria are found in a huge variety of different shapes and sizes, from rods and spheres to helices and spikes. In the case of bacteria, the organism's specific shape is tailored for survival in various environmental conditions, which often imparts a selective advantage.[12] To revisit the biological analog of the lock and key, the physical conformation of proteins and enzymes is absolutely critical to each molecule's function and activity.[13] Finally, looking into this incredible diversity of physical properties, the importance of shape in our world should become increasingly apparent.

Considering the tremendous role that physical properties play on the function and biological responses inherently present in nature, it is intuitive that geometry can impact carrier efficacy in drug delivery or diagnostic applications. Of these geometric parameters, size is very well known to be a vital parameter, dictating flow properties, degradation, clearance, and uptake in tissues and cells. Ideal carrier sizes have been identified and depend on the target and administration route. For example, conventional thinking indicates that *i.v.* administration of nanoparticles is ideal for particles of ~100 nm to target solid tumors, as these particles can avoid kidney filtration, minimize recognition by the reticuloendothelial system (RES), and do not cause occlusions in the vasculature. Delivery through a different administration route requires different considerations, *e.g.* pulmonary administration is optimal for particles of 1–5 μm, with particles closer to 1 μm depositing in the alveoli spaces and particles closer to 5 μm depositing in the upper airways.[14] The drug delivery field has a general understanding of guidelines for optimizing particle size, tailored for a specific administration route, and recognizes the significance of this particular physical parameter.

Much less is currently known about particle shape for these same administration routes and applications, despite many indications of established geometrical significance.[1–7] Some of this importance can be seen if we consider a few fundamental processes involved in particulate delivery and their corresponding mathematical representations: (1) particle dynamics in flow, *i.e.* Navier-Stokes Equation, (2) particle diffusion or dissolution, *i.e.* Fick's Law, and (3) particle adhesion, *i.e.* Young-Dupré Equation. All three mathematical descriptions are critically dependent on the system geometry, although difficulty in obtaining analytical solutions for complex geometries often results in assumptions of geometrical simplifications.[15,16] Indeed, many times these problems can only be approached numerically. Supported fundamentally by these mathematical indications, intuition plays a large influence on our increasing consideration of physical properties in drug delivery. For example, one would expect a maple seed and a high aspect ratio rod to flow, diffuse, and adhere differently than an equivalent volume sphere. Yet, despite these and other influences of geometry in the many processes crucial to successful delivery, the investigation of non-spherical particles of well-controlled physical properties in therapeutic and diagnostic carrier deliver is only beginning.

Performing investigations into the control of physical properties of drug and diagnostic carriers, namely shape and flexibility, has required the development of novel approaches. Numerous particle fabrication techniques to control particle size, shape, and flexibility are emerging in order to specifically probe the influence that these properties play in various biological environments. Additional information is being gathered through computational approaches to further investigate individual particle dynamics. In tandem, these studies are illustrating that physical properties are indeed critical and versatile design parameters. Fine tuning carrier geometry to improve and tailor the efficacy of nanomedicines presents an exciting opportunity to rationally design and develop particulate therapeutics and diagnostics.

The focus of this chapter is to inform the reader of recent studies which investigate the role of shape and flexibility on drug and diagnostic particle delivery, as well as provide an overview of relevant fabrication techniques for creating non-spherical particles. There are numerous additional physicochemical properties that influence bioactivity, such as surface topography, surface chemistry, porosity, and particle composition, which will be briefly discussed as well. For additional information on these relevant physiochemical properties, we direct the reader to other chapters. In this chapter, we generally discuss carriers with at least one dimension that is less than one micron, but will also include discussions of relevant microparticle and nanoparticle systems particles of novel geometries.

2. Particle Fabrication Techniques for Controlling Particle Geometry

Increasing interest in the role of particle geometry in drug and diagnostic delivery requires the use of creative approaches to fabricating non-spherical particles. In the past few years, numerous approaches have been adopted. Here, we introduce techniques relevant to biomedical applications.

2.1. *Bottom-up techniques*

Bottom-up fabrication techniques begin with the smallest single base unit, molecular or particulate, which are arranged to create an ordered structure. A wide range of materials can be fabricated through bottom-up approaches, from structured macromolecules to inorganic particulates. The ultimate physical geometries achievable are also diverse. Bottom-up fabrication techniques are often cost effective batch processes and significant advances in recent years have made these approaches appealing for nanoparticle fabrication.

Many conventional relevant bottom-up nanoparticle fabrication approaches have been discussed in detail already: liposomes (Volume 1 – Chapter 1), micelles (Volume 1 – Chapter 3), inorganic nanoparticles (Volume 1 – Chapters 10), metal nanoparticles (Volume 3 – Chapter 2), gold nanoparticles (Volume 1 – Chapters 14, and Volume 3 – Chapter 9), silver nanoparticles (Volume 1 – Chapter 15), silica nanoparticles (Volume 1 – Chapter 11, and Volume 3 – Chapter 8) and carbon nanotubes (Volume 1 – Chapter 12).

2.1.1. Well-controlled chemical syntheses

As discussed in previous chapters, the use of inorganic nanoparticles has been an area of great interest. By controlling the specific reaction conditions, synthesis of inorganic particles can achieve unique geometries.[17,18] Interestingly, the geometry of these nanoparticles often has significant influence on the optical characteristics.[19]

Gold nanoparticles have attracted particular interest for medical applications owing to their tunable optical properties, low toxicity *in vivo*, and ease of synthesis and surface functionalization. Gold nanoparticles have been fabricated by various methods, including solution chemistry and chemical vapor deposition (CVD), and are commercially available.[20] They are readily functionalized by forming gold–thiol bonds, employed for the application of stabilizing poly(ethyleneglycol) (PEG) coatings, antibodies, or other biomolecules.[21–23] Through asymmetrical functionalization approaches, the assembly of gold nanoparticles can further be controlled to yield structures of single chains or "nanonecklaces."[24,25] Unique geometries of the particles themselves have also been achieved. Transmission electron micrographs (TEMs) of gold nanorods and chains of gold nanoparticles in Figs. 1A and 1B illustrate their tunable aspect ratio and self-assembly. Gold nanorods have been widely studied; control of the rod aspect ratio can not only tune optical properties, but results in optical absorbance efficiencies and scattering coefficients greater than the equivalent volume sphere making them ideal for imaging applications.[21,26–29] Additionally, when exposed to specific wavelengths tuned to particle dimensions, heat can be generated, which can be controlled to selectively melt and reshape the particle into new dimensions or to induce hyperthermia and ablate tumor vasculature or pathogens.[30–32]

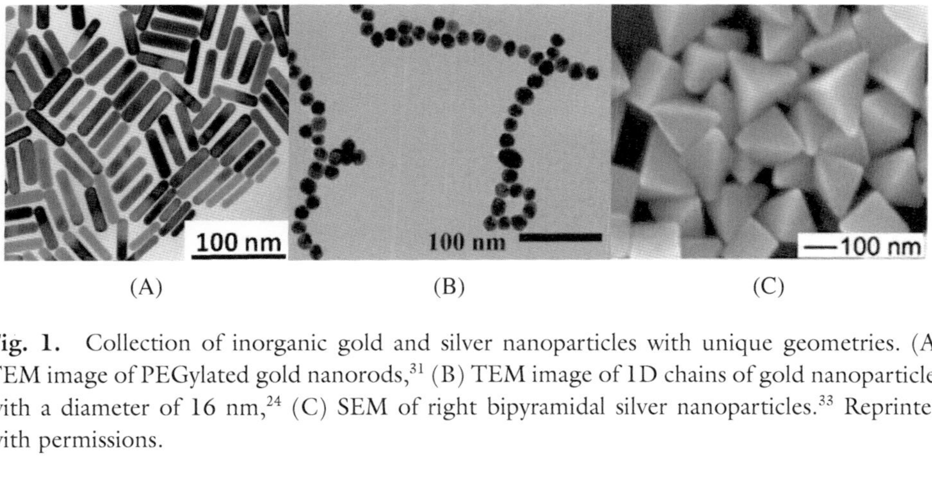

(A) (B) (C)

Fig. 1. Collection of inorganic gold and silver nanoparticles with unique geometries. (A) TEM image of PEGylated gold nanorods,[31] (B) TEM image of 1D chains of gold nanoparticles with a diameter of 16 nm,[24] (C) SEM of right bipyramidal silver nanoparticles.[33] Reprinted with permissions.

Outside of gold nanorods, researchers have generated gold nanoparticles of numerous geometries including nanostars, plates, cubes, urchins and various polyhedrons, through appropriate control of various reaction conditions.[20,34–37] For example, Odom and colleagues fabricated gold nanostars by reducing Au(III) chlorate with a varied amount of 4-(2-hydroxyethyl)-1-piperazineethane-sulfonic acid (HEPES).[35]

Synthetic approaches have been applied to control the geometries of other inorganic nanoparticle types. Similar to gold, silver nanoparticles have been synthesized in various geometries, again with unique geometries offering tunable optical properties.[38] An SEM of right bipyramidal silver nanoparticles in Fig. 1C highlights their tunable shape. Silver is nontoxic to humans in low dosages and has application as a potent antibacterial agent. Studies into the effect that shape exerts on the interaction of silver nanoparticles with different bacteria are ongoing, with particle fabrication achieved by controlling silver reaction conditions in silver reductions.[39] Additional inorganic nanoparticles with controlled geometries include magnetic particles with iron oxide cores fabricated by Park and coworkers. These high aspect ratio, worm-like particles were formed through the reaction of iron II and III salts with dextran, followed by labeling with a FITC-tagged peptide for imaging and targeting.[40,41] Finally, the geometry and aspect ratio of carbon nanotubes (CNT) also has an impact on relevant optical and thermal characteristics, toxicity, and *in vivo* fate. Fabrication of CNTs with controlled aspect ratios has been achieved with CVD and high-pressure carbon monoxide disproportionation (HiPCO) syntheses.[17,42–44]

2.1.2. Self-assembly fabrication

Self-assembly approaches are particularly interesting for biomedical applications and have been the workhorse for core knowledge of drug delivery to date. Typically, non-covalent interactions of low molecular weight lipids are employed to drive the creation of liposomes and micelles, resulting from surface tension minimization and the formation of spherical particles. A colorful variation of particle geometries can be achieved by varying the quality and quantity of the various components used in these particle formations. Bottom-up self-assembly of amphiphilic species has enabled the synthesis of non-spherical particulates. Unique liposome-based structures include cochleates, tubules, and ribbons.[45,46] Assembly of ampliphilic block copolymers can result in the formation of polymeric vesicles, or "polymersomes," with increased thermodynamic stability over liposomes and micelles. These polymersomes have the capacity to form unique geometries outside of spherical vesicles. Tuning the block lengths, hydrophobic and hydrophilic components, solvents, and additives controls the vessel type and resultant structure.[50–57] A variety of particle geometries have been engineered, including spheres, rods, tubes, cylinders, disks, hollow hoops, bowl-shaped, and helices.[55,57–65] Using self-assembling amphiphilic lipid-like diblock copolymers, Disher and coworkers fabricated filamentous micelles with varied aspect ratios, called "filomicelles."[49] These filomicelles were fabricated with equal block lengths of a hydrophilic PEG chain and a hydrophobic chain of either poly(ethylethylene) or poly(caprolactone), which self-assembled in water to form cylindrical micelles. Block lengths and preparation conditions controlled resultant particle dimensions, which were further controlled by fragmenting the micelles during nanoporous filtration steps, impacting the average length. Hydrophobic cargos, including dye molecules and a chemotherapeutic, paclitaxel, were successfully loaded into the hydrophobic core region, making these particles traceable drug delivery vehicles.[49,66–70] Other groups have explored efforts to graft chelating agents into copolymer networks to radiolabel self-assembled nanoparticles for positron emission tomography (PET) imaging modalities, among others.[71] Multiblock polymersome systems have also been explored, accessing novel geometries through kinetic manipulations of charged blocks.[48,72] A series of spherical to elongated micelles were fabricated using poly(acrylic acid) with poly(styrene) (PS) and poly(methyl methacrylate) by Wooley and coworkers who prepared short and long cylinders with different inner diameters. Further stability of these copolymer vesicles was achieved through shell cross-linking via polymerization of pendant chains in the outer hydrophilic region.[73] This

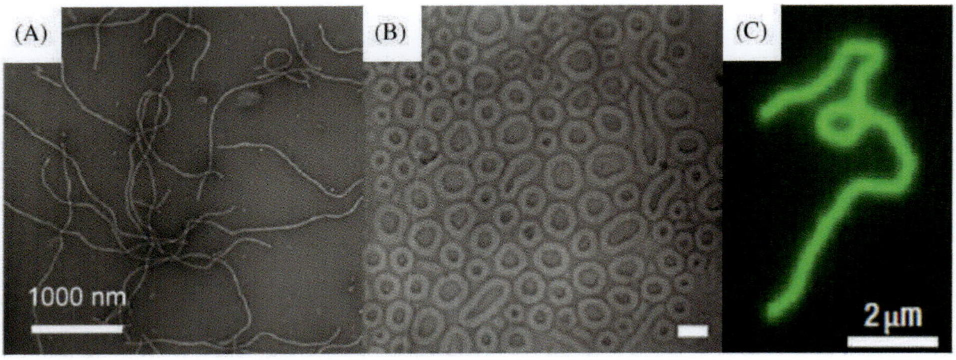

Fig. 2. Collection of self-assembled particles with unique geometries. (A) TEM image of amphiphilic cylindrical micelles,[47] (B) TEM image of toroidal triblock copolymer assembly (scale bar 100 nm),[48] (C) Fluorescence microscopy image of a single filomicelle.[49] Reprinted with permissions.

additional physical stability has been extended from spherical particles to cylindrical vessels, di- and tri-block strategies, and peptide shells.[47,74–76]

The fabrication of mesoporous silica particles has been achieved using bottom-up self-assembly techniques for the creation of particles with controlled interior and exterior geometries. Porous silica has found increasing application for drug delivery due to high surface area and controllable pore characteristics.[77–80] Controlled pore geometries are achieved through the self-assembly of surfactants into micelle templates, around which sol-gel chemistries convert soluble silica precursors into porous particles.[80] Various pore morphologies can be obtained through the selection of the template and tuning the length of the hydrophobic region on the template surfactant, which allows for rational design based on self-assembly knowledge.[79,81,82] Pores can further be functionalized with various organics and metals within the pore structure. Grafting and co-condensation strategies can selectively functionalize both interior and exterior surfaces, often through organo-substituted trialkoxysilanes.[79,83–86] Pores have been loaded with various molecules for imaging and drug delivery.[78,80] Geometries of these pores can range from cylindrical hexagonally packed mesopores to twisted pore morphologies, which are thought to slow diffusion of the encapsulated payload.[78,87–90] Overall macroscopic particle geometry can further be controlled through tuning hydrolysis rates, achieved through selection of reaction pH, co-solvents, and the silica precursor.[79,87] Shaped mesoporous silica nanoparticles have been achieved by numerous groups through variation of reaction conditions.[87,91–95] For example, Meng *et al.* fabricated mesoporous silica nanoparticles with similar pore sizes and morphology but varied particle aspect ratio by tuning

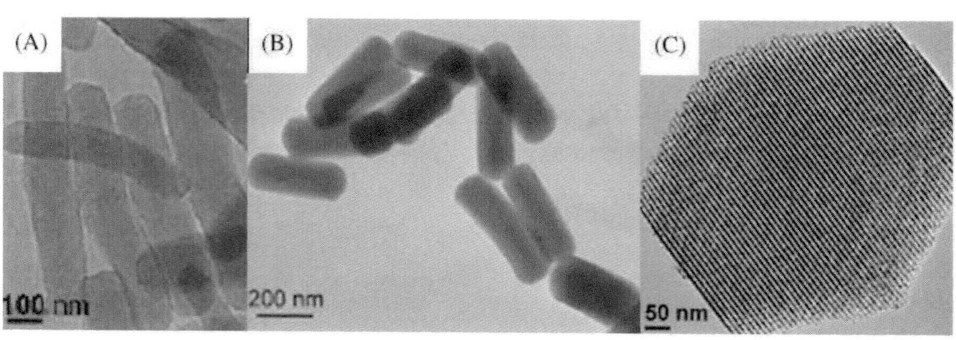

Fig. 3. Collection of mesoporous silicon particles fabricated with bottom-up approaches. (A) TEM image of mesoporous silica nanoparticles with an aspect ratio of ca. 5,[92] (B) TEM image of long rod-shaped mesoporous silica nanoparticles,[97] (C) TEM image of hexagonal plate with 10 nm hexagonal pores.[79] Reprinted with permissions.

the ratio of perfluorooctanoic acid (PFOA) to cetyltrimethylammonium bromide (CTAB) in the template.[96]

2.1.3. Stretching fabrication techniques

Particle stretching is a technique developed by the Mitragotri Lab. In this process, hydrophobic spheres formed independently are dispersed into a film of poly(vinyl alcohol) (PVOH).[98] Films are then stretched and treated with either solvent or heat to liquefy the spheres into the surrounding voids created by stretching. This process can also be performed in reverse, liquefying the spheres prior to stretching the embedded liquid sample. Varying the initial sphere size controls the total particle volume and shaped particles have been fabricated from spheres with radii ranging from 30 nm to 5 μm.[98–101] Multiple stretching steps in tandem can result in a wide variety of shapes and surface textures, such as oblate and prolate ellipsoids, elliptical disks, with exemplar UFOs and needles shown in Fig. 4.[100,102] This film stretching approach has been pursued for the fabrication of hydrophobic thermoplastic particles, including poly(lactic-co-glycolic acid) (PLGA) and PS, which have been prepared to incorporate biologics like small interfering RNA (siRNA), and integrate shape-shifting properties in response to a stimulus.[103–105]

2.2. *Top-down techniques*

Top-down fabrication techniques are the reverse of bottom-up fabrication; processing steps are taken to selectively remove portions of the starting material to design the end product. This type of approach is widely employed through silicon lithographic patterning in the microelectronics industry and

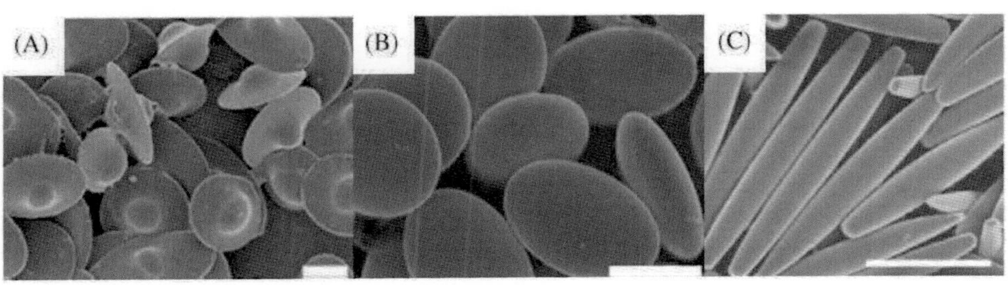

Fig. 4. Collection of particles fabricated through stretching. SEM images of (A) UFO (scale bar 5 μm),[98] (B) oblate ellipsoid (scale bar 5 μm),[98] and (C) needle (scale bar 2 μm) shaped PS particles.[102] Reprinted with permissions.

has recently found application in particle fabrication techniques. Bottom-up fabrication often yields particles of polydisperse size and shape distributions. Inherent to top-down fabrication is an element of controlled design, enabling an additional level of direct control over the precise details of the product.

Photolithography is the most cited example of top-down fabrication and many of the methods described in this section follow from fabrication methods designed for the microelectronics industry. Briefly, photolithography is an optical approach that creates well-defined patterns onto a surface. Geometric patterns, encoded on a mask, are transferred to a light sensitive photoresist coating. Depending on the nature of the photoresist, exposed (positive resist) or unexposed (negative resist) regions are selectively removed to impose the desired pattern. In the case of the microelectronics industry, this process takes places on a silicon wafer. Critical to many methods discussed here, the pattern feature size is limited by state of the art photolithography techniques and controlled by the system optics and energy sources. Increasing market size and the ever increasing need for faster, more powerful microelectronic devices has driven advances in photolithography and facilitated a steady decrease in obtainable feature sizes since the first transistor in 1947.[106,107] Indeed, at the time of this publication, current transistor sizes on the market are Intel's 22 nm tri-gate technology.[108] Further resolution can be achieved through E-beam, ion beam and dip-pen lithography.

The microelectronics industry harnesses the patterning of silicon and deposition of metals, many of which are not ideal for medical applications. Numerous technologies have been inspired by and adapted from traditional photolithography approaches to fabricate particles of relevant materials. These top-down approaches allow for independent control of particle size, shape, and mechanical properties.

2.2.1. Template based fabrication techniques

Soft-lithography techniques evolved from the need to pattern materials outside of traditional lithographic-friendly materials. As the name implies, these techniques employ "soft" materials, often elastomers, and encompass fabrication techniques of molding or imprinting for the creation of microstructures or nanostructures. While the materials and specific approaches vary between researchers, most soft lithographic techniques follow the same general process. A master template is made through traditional photolithography onto a hard surface. An inverse to the original pattern is then made using a low surface energy and low modulus elastomer resin, which is able to conform well to the master. The resin is solidified through a thermal or photochemical curing and then removed from the master, resulting in an exact inverse replica of the original master pattern. This inverse replica can then be used for molding or imprinting new materials, or the formation of a complex device based on the pattern, depending on the application and type of mold material used. Soft-lithography offers unique opportunities to pattern sensitive materials for a huge variety of applications extending beyond nanomedicine and particle fabrication.

Early examples in the literature of elastomer-based soft lithography have their origins from Whitesides and colleagues, who pioneered initial work concerning polydimethyl siloxane (PDMS) in the early to mid-1990s.[113–118] Stemming from this work, many examples of PDMS-based techniques have been employed to pattern surfaces and importantly, fabricate particles. Patterns have been transferred through stamping, imprinting, or molding using the features embedded in the PDMS template. Generally, imprinting leaves the molded material on the original substrate following patterning, while molding results in filled template cavities which maintain integrity once the backing substrate has been removed. PDMS is widely used for particle fabrication, being simple, robust, and inexpensive, and has been successfully applied in patterning of a huge range of materials.[119–125] However, common issues with traditional PDMS-based soft lithography approaches include the occurrence of a residual interconnecting flash layer, which results in an embossed film rather than discrete particles, and chemical permeation of PDMS, leading to swelling and loss of geometric fidelity. This has led to research in developing materials other than PDMS for stamping and molding applications and development of additional processing steps to remove flash layers.

Successful alternatives to PDMS imprinting have been developed. Techniques by Buyukserin *et al.* called bilayer nanoimprint lithography (B-NIL) can create polymeric (model SU-8) particles by replicating features of a silane-modified silicon mold etched through inductively coupled plasma. A sacrificial layer is allowed to partially fill the cavities during the imprinting process and can be dissolved to free imprinted particles.[110] Additional work from Tao *et al.* co-opts B-NIL techniques with photolithography to lithographically define high aspect ratio particles with average feature sizes ranged from 75 nm to 1.1 μm.[109] Step-flash imprint lithography (S-FIL) represents another alternative to PDMS imprinting; a patterned quartz template (coated to prevent adhesion) is used to imprint macromers against a sacrificial (PVOH) layer coated on a silicon wafer. Embossed films are treated with an oxygen plasma etch process to remove the interconnecting flash layer prior to dissolving the sacrificial layer and harvesting the particles.[111] This method can imprint sub-45 nm structures with sub-20 nm features and has great flexibility of particle composition, including enzymatically sensitive particles with biological cargos.[111,126] Pentagonal particles with 400 nm dimensions prepared from PEG diacrylate are illustrated in Fig. 5C. Alternative approaches to PDMS molding have also been developed. The hydrogel template method developed by Acharya *et al.* uses a gelatin-based mold to create particles. These molds are filled with PLGA solution and, following the evaporation of the chloroform, the gelatin molds dissolve to free the particles. Features ranging from 10 μm to 200 nm have been molded using this technique, with examples shown in Fig. 5D.[112]

Another variation on traditional PDMS-based soft-lithography molding techniques is called Particle Replication In Non-wetting Templates (PRINT). Developed in the DeSimone Group, PRINT is a micro molding and nanomolding platform that utilizes molds from fluorinated elastomers.[127–131] The photocurable, perfluoropolyether (PFPE) resin has low surface energy, which enables excellent conformation to the master, clean surface release upon photocure, and improved master-to-template fidelity when compared to PDMS replicas. Combined with the ever advancing field of traditional lithography, these properties endow PRINT molds with superior control over particle geometry. Unlike PDMS, PRINT templates are non-wetting and non-swelling to organic and inorganic materials, which facilitates the molding of a variety of materials as isolated, "scum free" particles and elimination of flash layer. Particles can be readily harvested with a soluble sacrificial harvesting layer. In addition to geometric flexibility, PRINT also affords independent control of particle composition.

PRINT is a versatile and gentle fabrication method, and has been shown to fabricate particles of small molecules, protein biologics, siRNA, and hydrophilic or acid-labile polymers containing therapeutic or diagnostic

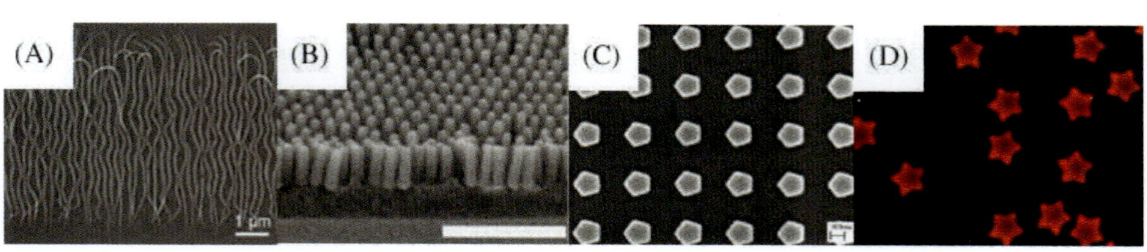

Fig. 5. Images of particles with controlled geometry prepared through top-down template soft-lithography approaches. (A) SEM of free SU-8-based nano-worms showing uniformity and some degree of flexibility,[109] (B) SEM of B-NIL nanopillars from a bilayer substrate of SU-8 on PMMA (scale bar 1 μm),[110] (C) 400 nm pentagonal PEG diacrylate particles by S-FIL (scale bar = 200 nm),[111] (D) 50 μm PLGA particles fabricated through hydrogel templating.[112] Reprinted with permissions.

cargos.[128,134–140] The PRINT process is readily scalable via a roll-to-roll system, by curing molds with a poly(ethylene terephthalate) (PET) backing film. Using PRINT, particles of a variety of geometries can be fabricated, with feature sizes ranging from 20 nm to 10 μm. These shapes range in diversity from rods of varying aspect ratio, to shapes inspired through biomimicry including trapezoids, 80 × 320 nm rice-shaped particles, red blood cell mimicking microparticles, filamentous particles, pollen, toroids, rods, and cylinders, to name a few.[127,133,141–144] Examples of the cylindrical, filamentous, pollen-mimic, and triblock high aspect ratio particles achieved by PRINT can be seen in Fig. 6. The variety of shapes achievable can be further adapted through a process of stretching intermediate master replicas.[131]

2.2.2. Microfluidic particle fabrication

Another technique facilitated by photolithography and the microelectronics industry is that of microfluidic devices. Channels and reservoirs on the micron-scale are created in elasomeric materials through pattern transfer from a hard substrate containing arrays of nanoscale or microscale features patterned through traditional lithographic means. When the patterned surface is tightly adhered to a second flat substrate, specific features in the elastomer remain unobstructed and a microfluidic device is formed. Using such a device, fluids entering the channels can be precisely manipulated through a potentially complicated pathway. Due to the size scale, fluid flow in the channels is laminar, resulting in unique opportunities for precision manipulation of flow streams. Additionally, microfluidic devices are generally low cost, low energy, and potentially high throughput systems. Microfluidic systems show great potential as point-of-care diagnostic applications, especially in

Fig. 6. Images of PRINT particles with controlled geometry and regio-chemistry. (A) SEM of 200 nm cylindrical particles,[132] (B) SEM of 80 × 5000 nm PEG-based hydrogel (scale bar 5 μm),[133] (C) Fluorescence microscopy of 20 × 20 × 240 μm amphiphilic triblock particles (scale bar 200 μm),[134] and (D) SEM of IgG/lactose 10 μm pollen PRINT aerosols.[135] Reprinted with permissions.

developing countries, and recent advances have been reviewed elsewhere.[145–147] Unsurprisingly, microfluidic devices have recently found application specifically for particle fabrication.

To fabricate particles in microfluidic devices, emulsion techniques are adapted to the confined laminar flows of the device. Generally, this approach results in the formation of monodisperse spherical droplets through the minimization of surface energy flows and manipulation of two immiscible fluid phases.[148–150] Whitesides and colleagues utilized this approach to create "microfluidic flow-focusing devices" (MFFDs), in order to additionally control for particle geometry. In these devices, two immiscible liquids are flowed together through a microfluidic channel, in a ratio such that droplets of one material are surrounded in the second continuous material. These droplets can be passed into varying chambers to confine the droplet and instill a designated geometry. While in these specific chambers, particles are solidified either photochemically or thermally. Using MFFDs, monodispersed particles composed of metals, microgels, and polymer carriers were created continuously.[151] Additional particle generating techniques and emulsion approaches have been adapted to take place specifically inside microfluidic devices to enable the synthesis of monodisperse, non-spherical particles. Shum *et al.* discuss relevant strategies for use in microfluidics, such as asymmetric polymer solidification, arrested coalescence, use of single or double emulsions, polymerization under microfluidic flow, and clustering of inner droplets.[152]

Photolithography techniques have been applied in tandem with microfluidic devices to also fabricate particles. This is generally achieved by controlled polymerization using lithographic masks within the continuous flow process of microfluidic devices. Doyle and coworkers created a process called continuous flow lithography (CFL), which utilizes microfluidic devices composed solely of PDMS, in which they flow acrylate oligomers through a pulsed UV source and transparency mask.[153] The oligomer stream polymerizes selectively in the light source, creating polymer particles based on the features of the mask and the height of the channel. Oxygen is able to diffuse through the PDMS, creating an unpolymerized oligomer stream at the channel walls and preventing the formation of any occlusions. This process is appealing in its ability create highly reproducible particles through a continuous, high throughput, readily scalable fabrication method. Triangle-shaped and curved PEG diacrylate particles can be noted in Fig. 7. Further variations on CFL have evolved to further control the light exposure in the channel, including stop-flow lithography (SFL) and stop-flow interference lithography (SFIL) to fabricate 2D and 3D particles. Both techniques are a variation on CFL, creating a stationary flow prior to exposure to allow for higher

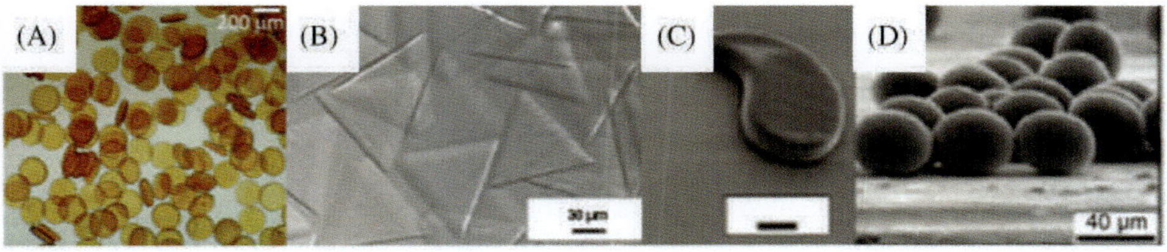

Fig. 7. Examples of particles fabricated using microfluidics (A) Optical image of superparamagnetic STL particles,[156] (B) Differential interference contrast image of PEG diacrylate triangles (scale bar = 30 μm),[153] (C) SEM of curved PEG diacrylate particles (scale bar = 10 μm),[153] and (D) Optical microscopy image of ellipsoids.[151] Reprinted with permissions.

resolution and smaller particle feature sizes.[154,155] Additionally, SFIL uses a PDMS phase mask and collimated light to impose 3D patterns of light into the microfluidic channel onto reactive monomer species; in combination with a patterned transparency mask, this controlled light interference imparts exposure in three dimensions, precisely creating shaped particles following exposure.[155] With these variations, Doyle and colleagues have successfully fabricated particles with defined features ranging from 1–1000 μm of a variety of compositions and loadings, and imply intention to create sub-μm 3D particles.[156]

2.2.3. Additional lithography inspired techniques

In addition to bottom-up fabrication approaches discussed previously, solid and mesoporous silica particles have also been successfully fabricated through lithographic means. Lithography approaches result in superior uniformity of overall particle size and shape, as well as defined pore features. To create these silicon or silicon dioxide particles, a few variations on traditional lithography have been employed.[157] One approach involves patterning a sacrificial silicon nitride later, through which an electrochemical etching process creates defined pores. A high current density through the wafer forms a release layer around the particles, which can be removed following washing and sonication steps.[158] More recent work proceeded in reverse; pores are defined on the silicon wafer before lithographic steps to create and release particles.[159] Like porous silica particles fabricated by bottom-up approaches, these particles are also readily functionalized and loaded with cargos.[160–164]

Mason and colleagues have also adapted traditional silicon lithography to create shape-specific polymeric particles by employing a soluble sacrificial layer. A silicon wafer is coated with a water soluble lift-off layer, prior to coating with light-sensitive monomers. The coated wafer is exposed to light through a patterned mask and high throughput stepper. The unexposed resist and sacrificial layer can then be easily removed to release the "LithoParticles."[165] This method has also been applied on patterned, reusable solid templates, resulting in techniques of Well- and Pillar-Deposition Particle Templating (P-DePT and W-DePT)[168,169]

The final top-down particle fabrication technique for the formation of well controlled geometries discussed here was developed by Lee *et al.* and is known as PEEL: a multistep approach which uses a series of phase-shifting photolithography (PSP), etching, e-beam deposition, and liftoff to fabricate particles with pyramidal geometries. Tipless gold pyramids can be seen in Fig. 8A. The lithography-based approach enables selective functionalization of the various pyramidal faces and is compatible with a variety of metals for particle fabrication.[166]

3. Controlled Particle Geometries for Prolonged Circulation Times and Increased Bioavailability

In the systemic administration of microparticles and nanoparticles, the shape, size, modulus, and surface chemistry of the vector influence biophysical behaviors. Tuning the properties of particles can enable prolonged circulation times, through changes in clearance profiles and tissue biodistribution. While it is still early to draw definitive conclusions for optimized physical characteristics, due in part to the rich variation of particle compositions and formulations studied, promising trends are emerging. Various formulations have successfully modulated physical properties toward increasing circulation times and accessing complex tissue vasculature.

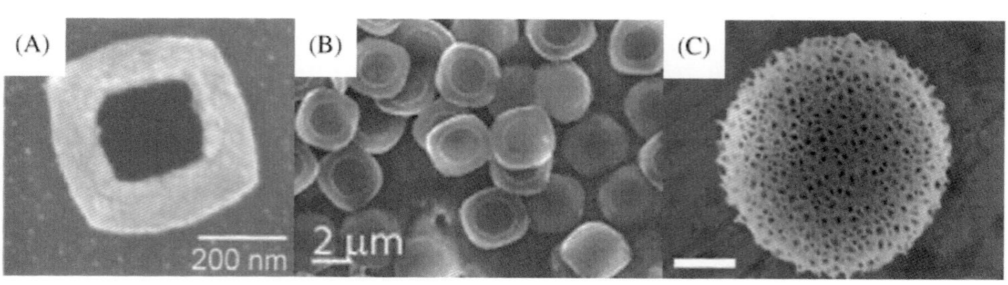

Fig. 8. Examples of particles fabricated using top-down methods: (A) SEM of tipless gold pyramids fabricated using PEEL,[166] Silica (B) discoidal[161] and (C) cylindrical[167] (scale bar 1 μm) mesoporous particles fabricated through lithographic approaches. Reprinted with permissions.

3.1. *Biodistribution studies of non-spherical particles*

Investigation of the fate of non-spherical particles delivered intravenously has proved invaluable in elucidating the crucial role of particle geometry. The following studies have explored biodistribution of particles with varied size, shape, and composition.

Given the potentially overwhelming amount of variables when exploring physical geometries of particle carriers, the first explorations to establish the importance of geometry have involved studies on the effect of aspect ratio on particle biodistribution. A series of mesoporous silica nanoparticles with different aspect ratios (1, 1.5, 2, 5) were synthesized by Huang *et al.* through a co-condensation of reagents for investigation of biodistribution to various organs. PEGylated and bare silica nanoparticles were intravenously administered to mice; over 80% of the inject doses of particles ended up in the liver, spleen, and lung. High aspect ratio silica nanoparticles distributed in the spleen, while shorter particles became trapped in the liver. PEGylation of particles resulted decreased clearance by the RES.[92] Other studies on high aspect ratio particles include flexible filamentous micelles, prepared with different contour lengths to investigate the biophysical behavior *in vivo*. The circulation half-life of filamentous micelles increased with increasing initial particle length and persisted in rodents up to one week after *i.v.* injection, which amounts to half-lives 10 times higher than spherical counterparts. Enhanced circulation of high aspect ratio filomicelles was attributed to evasion of the mononuclear phagocytic system by minimizing interactions with phagocytic cells. Furthermore, filamentous micelles were seen to preferentially accumulate in tumor tissue, while spherical particles entered both normal tissue and tumor tissue (Fig. 9).[49] Similar preferential tumor accumulation was observed with high aspect ratio superparamagnetic iron oxide nanoparticles. Variables on these particles included different peptide ligands, surface coverage of ligand, and particle shapes to optimize tumor targeting. The ability of particles to target tumors *in vivo* was enhanced for high aspect ratio nanoworm particles, compared to spherical counterparts. Additional tumor enhancement was observed for ligands of smaller size and closer to neutral charge, as well as longer PEG linkers (5 kDa versus shorter hydrocarbon chains).[40] Overall, these cases illustrate promising results toward increased circulation through the use of high aspect ratio particles.

Despite the differences observed with the previously discussed high aspect ratio particles, similar studies with gold nanoparticles resulted in less pronounced differences. Gold nanorods fabricated with diameters from 10 to 50 nm and aspect ratios of 1.7 and 5 were not found to have a significant

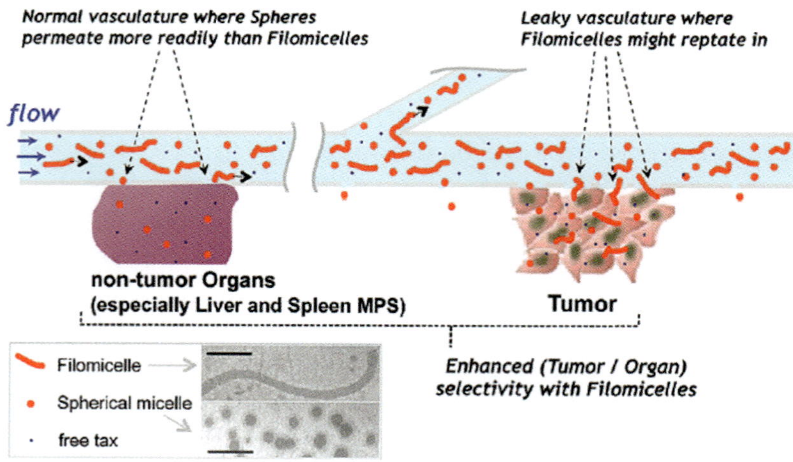

Fig. 9. Sketch of drug delivery by spherical micelles and filomicelles. Filomicelles are more likely to accumulate preferentially in the tumor over other nontumor organs due to the tumor-featured EPR effect. Inset: cryo-TEM of filomicelle and spherical controls (scale bar 100 nm). Reprinted with permission.[49]

difference on biodistribution. Of more importance were surface characteristics, including PEG density, and ζ-potential, which were observed to contribute to differences in bioaccumulation *in vivo*.[30] Another study comparing high aspect ratio single walled CNTs to quantum dots across three different murine tumor models was found to be highly dependent on tumor model. In this work, Smith *et al.* found that quantum dots were able to extravasate through colon adenocarcinomas but not brain glioblastomas, while the reverse was true for CNTs; nanotubes accumulated in brain glioblastomas but only minimally through colon adenocarcinomas. The differences in particle accumulation across the tumor models in this study was attributed interendothelial pores, *e.g.* smaller pores in colon adenocarcinomas allowed for more nanotubes to pass than quantum dots via Brownian diffusive motion, while quantum dots were not able to cross as effectively, due to steric restrictions.[43] Other biodistribution studies performed with CNTs by Villa *et al.* found rapid clearance from blood and significant renal clearance, despite a particle hydrodynamic radius of 230 nm.[170] Clearly, passage of particles into and out of pores can dramatically affect biodistribution and circulation times. This exploration of particle aspect ratio illustrates the complexity faced in predicting circulation and biodistribution; we are only beginning to understand the many variables at play, some of which are difficult to isolate.

Outside of aspect ratio, other parameters of particle geometries have been shown to influence biodistribution; here we highlight studies of shaped mesoporous silicon particles. Godin *et al.* prepared three distinct geometries of porous silicon particles and compared them to spheres with different diameters in breast tumor-bearing mice.[159] The height of the discoidal porous particles was held constant at 400 nm, while the diameter was varied from 600 nm to 1000 nm to 1700 nm. Importantly, tumor accumulation of discoidal porous particles (diameter 600 nm) was found to be five times greater than the spherical control. Particle accumulation in the tumor, liver, and kidney decreased with increasing particle size for discoidal particles, while remaining roughly constant at a low value for spherical particles of varying diameter in tumor tissue. Conversely, for lung and heart, the percentage of the injected dose increased with discoidal particle aspect ratio.[159] A previous study investigated spherical and non-spherical (discoidal, cylindrical, and quasi-hemispherical) silica particles prepared through microfabrication in a similar biodistrubution study in breast tumor-bearing mice. Discoidal particles accumulated in the lung of mice roughly eight times more than cylindrical particles. In the liver, the opposite effect was noted, where cylindrical particles accumulated approximately five times more than discoidal particles in tissue, while spherical and hemispherical particle accumulated intermediately. These data suggest that discoidal particles actively avoided uptake by the RES.[171]

Building on the knowledge from these biodistribution studies, specific tissues can further be targeted using nanoparticles that adopt shapes and surface chemistries to: (1) evade clearance and filtration by the immune system, and (2) promote passage through endothelial linings on blood vessels to specific microenvironments. Provided that particles are larger than ca. 10 nm in diameter to avoid elimination by kidneys, the physical properties of nanoparticles can affect passive targeting to organs after systemic administration. Specific tissues have been preferentially targeted in numerous examples, often through the control of physical characteristics. For example, Devarajan *et al.* reported the critical role of particle shape in enabling passive targeting toward splenotropic drug delivery. Irregular shaped polymer lipid nanoparticles composed of poly(methylvinylether-co-maleic anhydride) and various lipids evaded uptake by macrophages and Kupffer cells, while exhibiting high localization in the spleen relative to spherical counterparts.[65] The shape of particles has also been demonstrated as the major determinant for accumulation in different types of tumors.[43]

Nanoparticulate delivery has benefitted from an ability to exploit local tissue characteristics for passive accumulation, *e.g.* leaky tumor vasculature. Qualities of this particular microenvironment have been used to promote passive diffusion or convection of nanoparticles into tumor parenchyma for enhanced delivery of therapeutics. Subsequent extravasation and accumulation of nanoparticles inside the tumor constitutes the enhanced permeability and retention (EPR) effect, which is commonly harnessed in cancer therapies.[172] Passive targeting through the EPR effect generally requires increased circulation times; decoration of particle surfaces with PEG aids this effort by bestowing a hydrophilic stealth layer to avoid opsonization and clearance by phagocytic cells.[143, 173] As an alternative to PEG, other hydrophilic polymers like PVOH and zwitterions coated onto particle surfaces have been noted to provide similar stealth-like qualities in evading the immune system.[85,86,174] Particle geometry has also been shown to impact accumulation through the EPR effect. Tumor uptake of magnetic nanoparticles was observed to be higher when using nanoworms compared to spheres in human carcinoma and fibrosarcoma tumor-bearing mice.[40,41] Similar increased tumor accumulation has been observed for gold nanorods and filomicelles.[49,175] In a study by Arnida *et al.*, the geometry and surface properties of gold nanoparticles were examined in vital organs of orthotopic ovarian tumor-bearing mice following systemic administration. Spherical gold nanoparticles accumulated notably in the liver and spleen, while nanorods distributed in higher proportions to the other major organs. Gold nanorods demonstrated reduced uptake in liver, increased blood circulation time, and higher accumulation in tumors relative to spherical nanoparticles.[175] Adriani *et al.* utilized an *in vitro* flow chamber system to represent the physical properties of diseased microvasculature, observing preferential accumulation for disk-like particles as compared to rod-shaped ones. Intermediate-sized disk particles adhered to vasculature to a greater degree than smaller and larger disks, illustrating a balance between interfacial interactions and hydrodynamic forces in maximizing adhesion to endothelium.[167] Elliptical disk-shaped particles were further shown by Muro *et al.* to be superior to spheres when specifically targeting pulmonary endothelial cell internalization, having longer circulation half-lives.[101] Passively targeting tumors can depend on the degree of vascularization and angiogenesis, as well as the type of tumor.[176] Furthermore, due to the high interstitial fluid pressure, the size of long-circulating nanovectors can dictate the degree of retention in tumors; larger particles can be retained, while smaller particles have the opportunity to diffuse out of the tumor microenvironment.[177]

3.2. *The role of particle modulus*

As mentioned earlier, it is well-known that the deformability of cells allows them to remain in circulation and navigate the vasculature.[10,11] Unsurprisingly, particle flexibility is also a critical physical parameter for designing circulation profiles. Results from previously discussed biodistribution studies, such as the study by Geng *et al.* on flexible filomicelles, are most likely also dramatically influenced by the inherent particle modulus.[49]

The circulation time of nanoparticles and microparticles can be explored by purposefully tuning the surface and mechanical properties and drawing inspiration from nature. Chambers *et al.* significantly enhanced the circulation time of PS nanoparticles by binding particles with red blood cells, as compared to unbound particles.[178] This advantage was further harnessed by Merkel *et al.* in the preparation of red blood cell mimicking hydrogel microparticles.[179] While an earlier study investigating *i.v.* injection of mice with low molecular weight PEG hydrogel particles showed high uptake in liver and spleen and rapid clearance from blood (similar to a non-PEGylated liposome),[132] varying the modulus of hydrogel particles dramatically changed circulation profiles. Increased deformability resulted in extended circulation, while increased particle stiffness resulted in increased accumulation in the lung due to filtration.[179] The importance of size and deformability in dictating the biodistribution and circulation was further demonstrated for deformable hydrogel microparticles of different diameters *in vivo*. Avoidance and escape of filtration was noted for all microparticles except the largest particles (8.9 μm). The highest persistence in circulation was observed for deformable particles that resembled red blood cells in size, hinting that the natural clearance mechanisms of the body are designed to allow for passage of specific geometric and physicochemical properties of foreign objects.[144] As an alternative to modulating the mechanical and geometrical properties of filamentous particles, the surface properties and corresponding *in vivo* behavior of 80 × 320 nm rod-shaped hydrogel nanoparticles were investigated by tuning the surface PEG density. PEGylation was seen to notably increase the circulation profile of nanoparticles in mice relative to non-PEGylated particles, determined by intravital microscopy.[143]

To further explore carrier propensity to avoid clearance by filtration, Kersey *et al.* studied the extravasation of nanoparticles and microparticles through nanopores using hydrogel nanoparticles with a constant diameter of 80 nm and with lengths varying from 180 to 5000 nm. These particles were prepared with different crosslinking densities, corresponding to varying degrees of deformability. Mechanical and physical properties of nanoparticles were related to their ability to extravasate using a membrane filtration technique (0.2 μm pores). Greater than 77% of particles were recovered for all particle sizes and cross-linker contents, except the highest aspect ratio particles (80 × 5000 nm, which also had a significantly high persistence length) at the highest cross-linker content (96 wt%). This suggests the importance of both shape and stiffness in relation to passive targeting *in vivo*; a threshold existed for clearance for stiff, high aspect ratio particles in the likelihood of filtration, while smaller aspect ratio particles were generally capable of superior extravasation.[133] This is further supported through computational work from Fox *et al.*, who investigated the penetration of polymers with different molecular conformations through pores and found that the shape and flexibility influence entry and passage through pores. In this study, penetration occurred most easily with rigid rods and loose linear coils.[180]

4. Active Targeting implications for Particles of Controlled Geometries

The ability to target specific tissues and cells after *i.v.* administration of particles is governed notably by the surface chemistry, geometry, and deformability of the particulate delivery vector. As discussed, the circulation profiles of particles can be dictated by the physicochemical characteristics of the vector, which in turn impacts uptake into tissues and cellular uptake kinetics. Tissues may be targeted passively by exploiting features of the local microenvironment, such as the leaky vasculature in tumor blood vessels. Additionally, endothelial cells in the vasculature or diseased cells in tumor tissue may be actively targeted by nanoparticles through ligand–receptor interactions by decorating the surface of particles with targeting moieties, *e.g.* peptides, small molecules, and antibodies. Receptors overexpressed on diseased cells may be targeted with complementary ligands functionalized onto nanoparticle surfaces to facilitate receptor-mediated endocytosis. Small molecules, peptides, antibodies, and aptamers may serve as ligands on particles to target the sigma receptor,[181] integrins,[182,183] transferrin receptor,[184] folic acid receptor,[185] and prostate specific membrane antigen[186] on different types of cancer cells. The role of geometry and biophysical features of particles on targeting receptors and corresponding internalization kinetics is important, although may vary depending on cell line and receptor type.

Computational analyses of the receptor-mediated endocytosis of non-spherical particles have been conducted to evaluate critical factors influencing internalization kinetics.[187,188] Through one such model, the internalization of rod-like particles with elliptical cross sections was controlled by tuning the volume and aspect ratio. Particle features were identified that correspond to events of cell wrapping, partial wrapping (frustrated endocytosis), and wrapping by the cell membrane, followed by internalization given appropriate time. Particles with high aspect ratios that lay parallel or orthogonal to cells were not readily internalized; however, ligand density played a significant role in receptor-mediated endocytosis and internalization may occur regardless of particle orientation to the cell membrane, provided a sufficiently high ligand density is present.[187] Relevant *in vitro* studies on ligand density for ellipsoidal and spherical particles were conducted toward improving targeted delivery efficiency of nanovectors.[189] Increased adhesion to endothelial cells was observed for ellipsoidal polymersome particles with high densities of PEG and RGD ligands (oligionucleotides containing an Arg-Gly-Asp sequence) when compared to spherical counterparts.[189]

Endothelial cells have been active targeting with non-spherical carriers *in vivo.* PS spheres of various diameters and elliptical disks ($0.1 \times 1 \times 3$ μm) were coated with an antibody targeted to intercellular adhesion molecule 1 (ICAM-1), which is a transmembrane glycoprotein upregulated in many pathologies.[101] Endothelial cells endocytosed spheres with diameters of 5 and 0.1 μm at relatively equal rates; however, elliptical particles were internalized at a significantly lower rate even though the disks were smaller than 5 μm spheres in their largest dimension. The decreased rate of endocytosis for anti-ICAM disks compared to spheres was found not to be influenced by a difference in endocytic pathways, as seen by a similar influence on internalization by pharmacological inhibitors. Delayed alignment of disks in the orientation of actin bundles was noted relative to spheres. Particle physical geometry has influenced active uptake for other particle types as well. Shuvaev *et al.* targeted endothelial cells with filamentous micelles through the use of an antibody to target platelet-endothelial adhesion cell molecule-1 (PECAM). Specific binding of anti-PECAM filomicelles to endothelial cells was noted *in vitro*, where control (Immunoglobulin G) antibody filomicelles exhibited markedly low binding. Prolonged circulation following *i.v.* injection in mice was noted for Immunoglobulin G-coated filomicelles relative to spherical analogues, while still exhibiting high avidity when coated with particular antibodies to target endothelial cells in vascular beds.[190]

Active targeting of nanoparticles, specifically to tumor cells, has clear therapeutic value and has been attempted with non-spherical particles. Indeed, non-spherical particles have an increased surface area per particle volume, as compared to spherical particles, offering opportunities for improved loading of targeting ligand. For example, Scheinberg and colleagues have shown that the high aspect ratio and periodic structure of CNTs affords unique conjugation strategies for tumor targeting antibodies and peptide antigens.[191–193] Other groups have successfully fabricated non-spherical particles with tumor targeting ligands. A series of block copolymers composed of PS, poly(acrylic acid), and poly(alkyl acrylate)s were self-assembled by Zhang *et al.* to yield short and long cylindrical particles as well as spheres, which were then functionalized with peptide ligands targeting ovarian cells. A greater amount of smaller spherical particles were internalized relative to larger cylindrical counterparts, while higher aspect ratio cylinders were internalized more than short cylinders.[47] Wang *et al.* studied the density of ligands on cylindrical nanoparticle surfaces for targeting efficiency on human tumor cells lines. Hydrogel nanoparticles conjugated with human transferrin and anti-transferrin receptor antibodies were internalized by tumor cells and efficiency increased with ligand density. When transferrin ligand-conjugated nanoparticles were incubated with a human lymphoma cell line, apoptosis occurred and revealed the complex role of multivalency in tumor targeting. Transferrin was thus identified as a potential cancer therapeutic for lymphomas which does not require the delivery of cargo drugs.[194] Ligands grafted on to particle surfaces can enhance the rate and selectivity of internalization by cells; yet, numerous factors still come into play in crossing the cell membrane.

Tuning the geometry, surface properties, and ligand functionality of particles enables a combination of passive and active targeting toward improving the efficiency of tissue- and cell-specific delivery of therapeutics. Simply tuning one particle feature may not make an impact on improving drug delivery efficiency *in vivo*. For example, conjugation of peptide ligands to gold nanorods did not significantly enhance accumulation or uptake into tumor tissue and cells.[195] Success of active targeting is a rather complicated issue, critically dependent on a number of particle variables, as illustrated in Fig. 10.

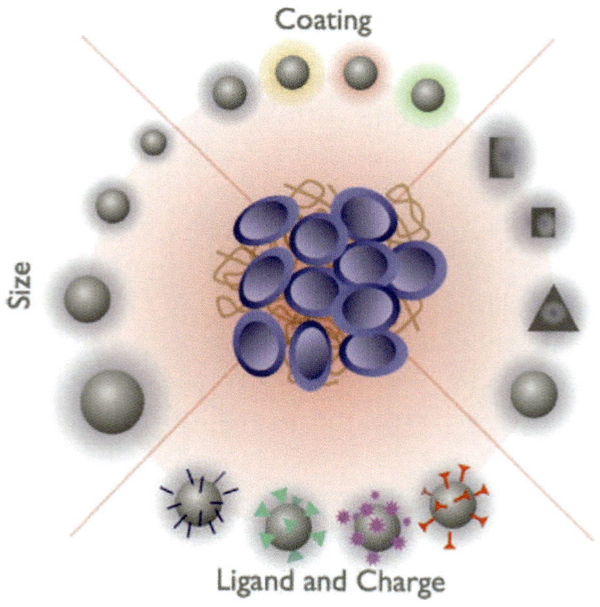

Fig. 10. A diagram of the physical properties of nanoparticles which influence tumor accumulation. Reproduced with permission.[192]

Nonetheless, functionalization of particles with ligands onto non-spherical particles can offer effective targeting of endothelium and promote cellular internalization.

5. Particle Geometries Influence Cellular Uptake

A promising opportunity for particles with engineered geometries involves manipulating physical dimensions to tailor interactions with cells. A tremendous amount of information is conferred to cells and their surroundings during the event of nanoparticle internalization. The various phagocytic pathways by which this event occurs will trigger a different cascade of responses to the internalized cargo, leading to assorted immunological responses.[196] Increasing research into understanding this area has shown the role that particle geometry plays in the kinetics of cellular uptake.

Instead of pursuing ligand–receptor interactions for actively targeting cells, Mitragotri and coworkers screened the shape of particles to promote or avoid phagocytosis by macrophages. For example, the rate of phagocytosis was studied *in vitro* using PS-based oblate and prolate ellipsoids, elliptical disks, UFOs, and worm-like particles. Ellipsoid particles were found to attach more efficiently to macrophages than spheres and the internalization was found to depend largely on local particle shape from the perspective of the phagocyte. Elliptical disks that attached to macrophages along the major axis were internalized significantly faster than when particles attached along the flat side (minor axis). Macrophage actin cups, indicators of initiation of internalization, were absent for internalization along the particle's minor axis. The tendency of macrophages to exhibit spreading or phagocytosis on particles was further elucidated using particles of different shapes and was found to be largely dependent on the local particle shape at the initial point of contact with the cell. Complete inhibition of phagocytosis was shown by modulating particle shape, namely using worm-like particles.[99,100] Interestingly, phagocytosis was found to be optimized for rod-shaped particles on the order of 2–3 microns, which corresponds to typical dimensions of bacteria.[197]

The translocation of different particle shapes across lipid bilayers has been modeled by Yang *et al.*, who compared ellipsoidal and cylindrical particles. The interactions between particles and lipid bilayers were found to depend on shape anisotropy and initial particle orientation. The contact area between the particle and lipid bilayer, as well as the local curvature of the particle at the initial contact point, largely influenced the penetration of particles across the bilayer.[198] Follow-up studies utilized dissipative particle dynamics to further probe lipid bilayer wrapping and internalization of different nanoparticles. The initial orientation and shape of particles both played roles in the internalization process. Rotation was seen to influence the wrapping process, which adjusted the point of contact between the lipid bilayer and particle. Ellipsoidal particles that were initially oriented perpendicular to cell membranes rotated during the wrapping process, which can be attributed to differences in ligand-receptor probabilities.[199]

In addition to particle shape, orientation, and consideration of ligand density, the aspect ratio of particles has been found to affect the degree of internalization by cells. For example, Raffa *et al.* evaluated the internalization of multiwall CNTs with different aspect ratios (diameter ~20 nm, lengths 0.3–2 μm) in hippocampal cells *in vitro*. High aspect ratio particles were significantly less internalized than the shorter particles, both following an energy-independent pathway.[42] Similar results were observed by Aaron *et al.* in a study on the aspect ratio of CdSe quantum dots (~10 nm) interacting with immune cells. Quantum dots of highest aspect ratio diffused significantly slower than lower aspect ratio particles in the plasma membrane of rat basophilic leukemia cells. Internalization of quantum dots by immune cells proceeded more slowly and immobilization inside the membrane of immune cells increased with aspect ratio of particles. Furthermore, co-localization of particles with lysosomes increased with particle size, which contrasts the shape-dependent behavior noted for quantum dots associated with the membrane.[200] Contrasting these two studies, work by Huang *et al.* using mesoporous silicon nanoparticles (~200 nm) with different aspect ratios showed that higher aspect ratio particles had a greater impact on cellular function; the largest aspect ratio particles decreased adhesion to cells, internalized more rapidly, decreased the expression of adhesion molecules, and increased cellular apoptosis.[91] These results highlight the complexity of the role of physical properties on biological systems; over three distinct size ranges, the particle aspect ratio appears to have markedly different effects.

Endocytosis mechanisms and cellular internalization profiles of PRINT nanoparticles were investigated *in vitro* on HeLa cells. Hydrogel nanoparticles and microparticles were fabricated with different sizes, shapes, and aspect ratios. Small, high aspect ratio nanoparticles were internalized faster than cylindrical and cubic particles while proceeding through multiple routes of endocytosis (Fig. 11). After similar incubation times with cells, rod-like particles traveled further into cells compared to shorter and larger cylindrical particles. Smaller dimensions of rod-like particles with the same aspect ratio did not enhance the rate or extent of internalization. However, the ζ-potential of particles was found to notably affect the extent of particle internalization where negating the surface charge resulted in diminished particle uptake.[127] Doshi

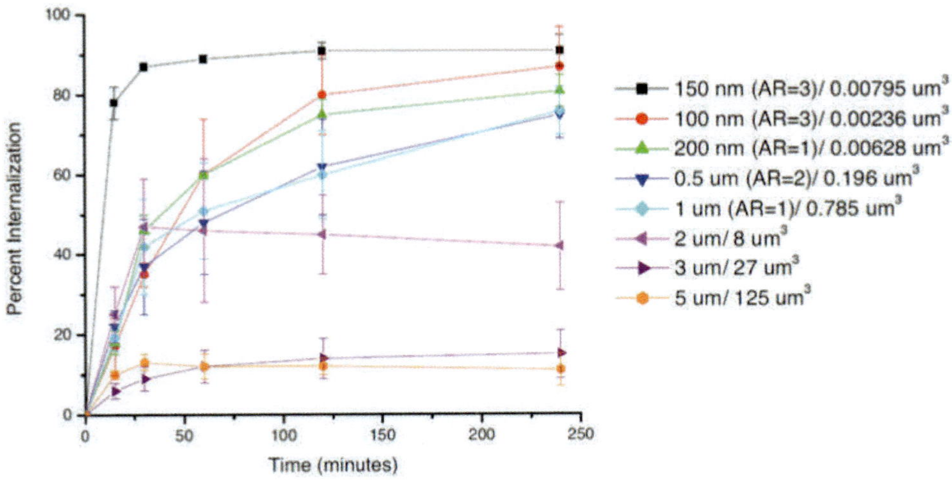

Fig. 11. Internalization profile of PRINT particles with HeLa cells over a 4 h incubation period at 37°C. Legend depicts the particle diameter per particle volume. Reproduced with permission.[127]

et al. studied negatively and positively charged PS-based particles with different shapes and investigated their behavior toward membrane permeation. Needle-shaped particles elicited contractions on endothelial cells in a time-dependent manner relative to elliptical disks and spherical particles, which did not demonstrate notable loss of cell surface area over time. Large-sized needle-shaped particles induced the largest decrease in cell surface area compared to smaller and spherical particles. Rapid cell contraction and return to original cell surface area was observed for positively charged needles, while negatively charged particles caused a slower decrease in cell surface area without recovery of contractions within a two hour timeframe. Endothelial cells were seen to only release notable amounts of lactate dehydrogenase, an indicator of cellular membrane disruption, when incubated with needle-shaped particles compared to spheres without sacrificing the proliferation potential of cells.[102]

The internalization can also be affected by dynamic particle properties. Yoo *et al.* prepared shape-shifting elliptical disks using PLGA of different viscosities for conversion into spheres in response to temperature, pH, or chemicals with tunable dynamics. Phagocytosis of elliptical disks by macrophages did not occur for static particles; however, phagocytosis was made possible by using shape-shifting particles. Time-lapse micrographs of this phenomenon are shown in Fig. 12.[104,105] Dynamic modulation of particle shape enables even further opportunities for controlled and directed interaction and internalization by cells.

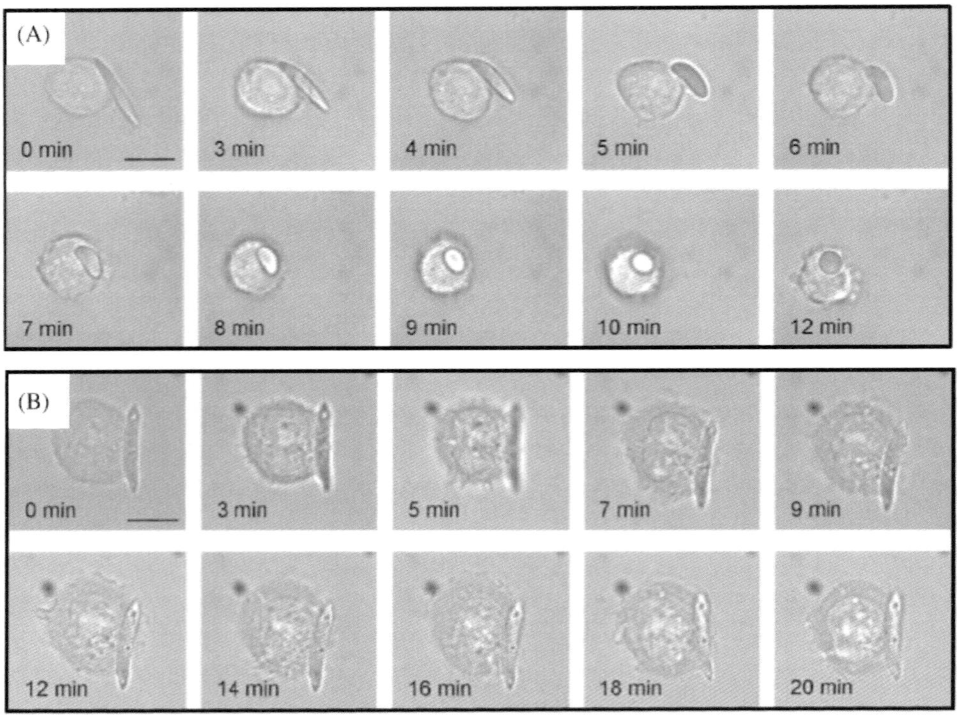

Fig. 12. Time-lapse video microscopy clips of shape-dependent phagocytosis by macrophage. (Scale bars: 10 μm.) (A) A shape-switching PLGA-ester ED (mixture of two PLGAs, AR = 5) was initially attached on a macrophage and not phagocytosed. The macrophage then quickly internalized the particle once shape switched to near-sphere shape. (B) Macrophage spread on a PLGA-acid ED (AR = 5, which do not switch shape) but could not complete phagocytosis. All particles were opsonized with mouse IgG before the experiments. Reproduced with permissions.[105]

Various mechanisms of internalization have been observed for nanoparticle formulations. Actin-dependent pinocytosis of worm-like filamentous micelles was noted through inhibition studies. When incubated with lung epithelial cells, fragmentation of filomicelles was observed, enabling complete internalization after two hours of incubation.[49] Internalization of hemispherical silicon microparticles was found to proceed through micropinocytic and phagocytic routes, while larger particles were only internalized through phagocytosis.[162]

Direct insertion into the cellular membrane has also been hypothesized for certain nanoparticle formulations. Computational analysis of CNT insertion into cellular membranes has been conducted using different techniques, treating individual CNTs and bundles thereof. When CNTs were placed inside the innermost hydrophobic layer of the cellular membrane in horizontal conformations insertion was promoted and modified the structural stability of the bundles.[201]

After effective internalization of particles by cells, passive targeting index drug delivery may require specific subcellular targeting, which has been explored for different particle types. Mesoporous silicon microparticles of different sizes and dimensions were found to migrate to the perinuclear region of cells through a microtubule-driven mechanism; the rate of migration was not influenced by physicochemical properties of particles. When loaded with iron oxide nanoparticles, mesoporous silicon particles were further found to traffic to distinct intracellular locations. Residence in phagosomes was noted for silicon particles, while the iron oxide nanoparticles separately sorted into multivesicular bodies.[164] Differentiation of subcellular targeting was similarly investigated by Xu *et al.* for two hydroxide nanoparticle types: hexagonal sheets (width = 50–150 nm and thickness = 10–20 nm) and rods (width = 30–60 nm, and length = 100–200 nm). Both particle morphologies demonstrated rapid internalization across three mammalian cell types through clathrin-mediated endocytosis and achieved endosomal escape, most likely through the ability of particles to buffer the acidification of the endosome. Afterward, positively-charged nanorods were found to effectively translocate to the nucleus of mammalian cells, while nanosheets remained in the cytoplasm, most likely due to the cylinder-like nuclear pore morphology. A microtubule network directed toward the nucleus was found to facilitate nuclear translocation of rod-shaped particles.[202] Targeting subcellular compartments by tailoring particle geometry should open the window to increasing drug indices for specific cargos, *e.g.* plasmids which require delivery to the nucleus, siRNA which must be delivered to the cytoplasm.

6. Drug Delivery Efficiency of Controlled Geometries

As the physicochemical properties of particulate delivery vectors have been shown to influence biodistribution, passive targeting, cell uptake, and intracellular trafficking, the corresponding efficiency in drug delivery can also be influenced by particle characteristics.

Discher and coworkers have studied how the carrier and shape of particles influenced accumulation of paclitaxel, a chemotherapeutic, in tumors and the resulting effect on efficacy.[49,70] Worm-like filomicelles achieved greater loading of paclitaxel compared to spherical counterparts, while both micellar carriers demonstrated less toxicity to lung cancer cells than the conventional formulation using Cremophor® EL solubilizer.[66] Compared to free paclitaxel and the clinical formulation, a decrease in the half maximal inhibitory concentration required to induce cell death in cancer cells was noted *in vitro* for filomicelles. Filamentous particles enabled greater accumulation of paclitaxel in tumors than spherical micelles and also induced higher levels of apoptosis after a single *i.v.* injection in lung tumor-bearing mice. At the same dose of paclitaxel administered *in vivo*, the length of filomicelles was found to influence therapeutic efficacy; longer filomicelles induced apoptosis to a greater degree and suppressed tumor growth more than shorter filomicelles.[49,66,70]

Similarly, Chaudhuri *et al.* studied the effect of shape of drug-conjugated CNTs and fullerenols on antitumor efficacy *in vivo.* CNTs (ca. 300 nm in length) and fullerenols (sub-100 nm aggregates) conjugated with doxorubicin both inhibited growth of tumors effectively; however, CNTs induced angiogenesis, while dose-dependent inhibition of angiogenesis was seen for fullerenol conjugates. Nevertheless, antitumor activity was observed for both delivery vectors, suggesting that accumulation of doxorubicin dominated angiogenic effects elicited by CNTs. Pro-angiogenic effects of CNTs were accompanied by upregulation and reorganization of integrin receptors on endothelial cells; fullerenols did not elicit similar biological effects. These results indicate that the clustering of integrins can promote activation of pathways implicated in cell survival and angiogenesis.[203]

Further understanding of biological pathways and communication resulting from physical properties can enable increased therapeutic effects. Harnessing some of these pathways, von Maltzahn *et al.* demonstrated amplified tumor passive targeting of gold nanorods to recruit inflammatory cells and provide over 40 times higher accumulation of a chemotherapeutic. Near-infrared absorbing gold nanorods were administered as input signaling modules to target tumors and activate the coagulation cascade. These were then able to broadcast the tumor location and promote reception of nanoworms and liposomes carrying therapeutics. Tumor size reduction and sustained inhibition of tumor growth was most effectively achieved for communicating nanoparticles consisting of nanorods combined with doxorubicin-loaded liposomes which were tethered with a peptide active targeting ligand.[22]

In addition to delivery of chemotherapeutics to induce self-destruction of malignant cells, cancer and diseases may be tackled by delivery of gene regulators, such as siRNA which silences expression of specific genes. Here, physical properties of particulate carriers also play crucial roles.[204,205] For example, the length (aspect ratio) and shape of polymer nanoparticles studied by Kohlar *et al.* impacted transfection efficiency of siRNA in endothelial cells *in vitro.* Needle-shaped PLGA particles provided significantly greater gene knockdown of green fluorescent protein relative to spheres. Transfection efficiency was greater for higher aspect ratio particles compared to spherical or low aspect ratio counterparts. The sharp edges of needles may pierce the lipid bilayer of the cell membrane for greater permeabilization and penetration of particles into the cytoplasm.[206]

Instead of particles only influencing biologics, states or environments in the body may be harnessed to influence particle behavior. Korin *et al.* Passive targeting obstructions in blood vessels using nanoparticle-aggregated microparticles that respond to shear stress. Under high fluid shear stress, microparticle aggregates broke up into nanoparticles which remained localized; these fluid conditions mimicked vascular narrowing, which might arise from the formation of plaques or clots. Particles coated with tissue plasminogen activator, which has affinity for fibrin, yielded rapid clot dissolution and increased survival using a pulmonary embolism model. Significant delay in thrombosis was not observed for control samples, which included dispersed nanoparticles, fused microparticles, uncoated particles, and free activator, suggesting the importance of shear-activation for effectively busting clots.[207] A second example of particle physical properties being further tailored to a specific environment is work by Uskokovic *et al.*, who explored oral transepithelial drug delivery devices. Using silica microparticles of different shapes and coatings, the oral bioavailability of drugs were increased through the physical localization of particles to epithelial cells.[208] The percentage of particles retained on cell monolayers was greater for asymmetric, planar particles, as compared to spherical microparticles. Particles were further coated with nanowires to increase adhesion to epithelial surfaces and subsequent permeability of epithelium. Planar-shaped, nanowire-coated microparticles coated were then loaded with insulin, and accordingly provided the greatest permeation of insulin across the cell layer. These planar nanowire-coated microparticles were explored *in vivo* and improved the bioavailability of insulin orally administered to rabbits. Planar particles provided higher blood insulin concentration after 1 h while spherical particles provided the highest concentration after 2 h. Higher concentration of insulin in blood was attributed to the geometry of the planar particles, allowing more facile passage through the digestive track and superior adhesion characteristics.

7. Imaging Implications for Particles of Controlled Geometries

To evaluate the biodistribution and behavior of drug delivery vehicles *in vitro* and *in vivo*, different imaging and quantitation techniques have been pursued using radioisotopes, fluorescent tags, and inorganic materials. Corresponding imaging techniques include magnetic resonance imaging, PET, single photon emission computed tomography, and near-infrared fluorescence imaging.[209] Each imaging technique provides unique opportunities and possibilities in enhancing spatial resolution, sensitivity, time constraints, and ease of analysis. In many cases, the physicochemical properties of particles can be designed to enhance imaging. For example, dextran-coated magnetic nanoworms provided greater T2 relaxation for magnetic resonance imaging compared to spherical counterparts. Exploring the physical properties of these particles had a two-fold advantage; these iron oxide worms also enhanced attachment to tumor cells and increased accumulation in tumors after *i.v.* injection in mice relative to nanospheres.[40]

Rod-shaped gold nanoparticles are a classic example of how physical properties can enhance imaging opportunities. Gold nanorods have two surface plasmon resonances modes, one along each axis. Typical nanorods with an aspect ratio of 4 have their strongest extinction peak at near-infrared wavelengths, which can be tuned by tailoring the shape of the nanorod.[21] This optical shift as a function of aspect ratio can be seen in Fig. 13. Owing to this tunability, gold nanorods have been imaged in biological systems through dark field microscopy, Raman spectroscopy, two-photon enhanced luminescence, and optoacoustic tomography.[21,29] As with the previous example of iron oxide nanoworms, the physical properties of gold nanoparticles offer two-fold advantages to bioapplications. The shape of gold nanoparticles administered to tumor-bearing mice has influenced the circulation profile, biodistribution, and macrophage uptake. Compared to nanospheres, gold nanorods were internalized less by macrophages, circulated longer, and

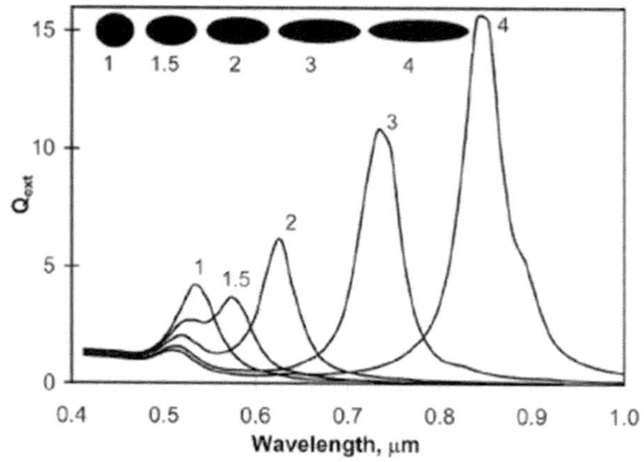

Fig. 13. Calculated optical extinctions of gold nanorods of fixed volume and varied aspect ratios. Reproduced with permission.[210]

accumulated significantly less in the liver and notably more in tumors.[175] The ability to tune optical properties of gold nanorods through physical properties offers unique opportunities for enhancing both diagnostic applications, *e.g.* molecular imaging, and therapeutic ones, *e.g.* photothermal therapy.[29]

Additional approaches to imaging non-spherical particles are many. Fluorescent tags may be installed in particles for real-time intravital microscopy and tissue analysis. Blood vessels and intratumoral space may be imaged after administering particles for direct visualization of particle transport and accumulation. After collecting organs from animals, analysis of tissue fluorescence may be conducted to determine the percentage of the injected dose in each organ.[43,49,144] Whole mouse imaging may be performed using bioluminescence to visualize specific markers or gene expression in gene therapy model experiments.[211] Polymers, lipids, cargos, and nucleic acids may be radiolabeled before incorporation into nanoparticle to allow tracing and quantification of administered delivery vectors and therapies across different organs.[212,213] Inductive coupled plasma atomic emission spectroscopy has provided analysis of silicon content in organs harvested from sacrificed mice administered with silicon particles.[159]

As interest in modulating physical properties of nanocarriers continues to grow, the ability to track and determine the fate of various particle types *in vivo* will continue to be an important evaluation. Imaging and quantification of the biodistribution of different types of particles with various shapes and sizes will direct efforts toward tissue- and cell-specific delivery of therapies and disgnostics.

8. Roadmaps for Dual Targeting of Tissues and Cells

Different diseases and disorders present unique requirements for effective treatment and these biological requirements present definite design criteria for particle therapies. Given the many barriers to successful delivery isolated in previous sections, establishing these design criteria globally for a particular disease is critical to increased efficacy of particulate therapeutic approaches. In the following, we briefly outline current physical design parameters, including distinct geometries, mechanical properties, porosities, and surface chemistries, to enhance the accumulation of particles in specific tissues and cells of therapeutic interest.

8.1. *Kidney*

Delivery to the kidney represents a unique opportunity for nanoparticle carriers. The kidneys perform several life-sustaining roles, such as the removal of waste from the blood, and loss of kidney function in acute kidney injury poses serious threats to health. Removal of particles from circulation across fenestrations can be seen for the kidney through glomerular filtration of small particles (up to 5.5 nm diameter for quantum dots[214] and free molecules (molecular weight up to 50 kDa).[215] However, distinct particles sizes favor bioaccumulation in the support tissue (mesangium) of the kidney. Stolnik *et al.* prepared a series of seven different gold nanoparticles with hydrodynamic diameters (in PBS) ranging from ca. 25 to 164 nm and PEG coatings ranging in molecular weight from approximately 4 to 20 kDa to optimize accumulation in kidney mesangium for application in kidney diseases. Although blood circulation half-life increased with decreasing particle size, uptake of particles into mesangial cells within renal corpsules was maximal for ca. 78 nm particles with a PEG molecular weight of 5 kDa. Uptake of particles into kidney mesangium decreased with increasing particle sizes, showing significant lack of localization for particles at ca. 130 nm in diameter with a PEG molecular weight of 10 kDa.[205]

8.2. *Spleen and liver*

Many nanoparticle therapeutics have physical properties that naturally allow them to accumulate in the spleen or liver. Both organs are critical to clearing toxins and maintaining normal balance in bodily fluids; cancer and bacterial or viral infections present substantial risk to healthy bodily function. Particles larger than 200 nm with rigid stiffness localize notably in the spleen following mechanical filtration, where subsequent phagocytosis will take place.[179,216] In addition to notable accumulation in the spleen, 200 nm cylindrical PEG-based PRINT particles accumulated to a large extent in the liver, which may be attributed to the fact that the sinusoidal walls of the liver and spleen contain discontinuous endothelium that can trap foreign particulates.[127] Notable cells in the liver include Kupffer cells, those lining the walls of sinusoids, and hepatocytes, which are of the main tissue and involved in critical biological processes. Moderately sized microparticles are cleared by Kupffer cells through phagocytosis while larger micron-sized particles may be physically trapped in capillary beds.[217,218] Furthermore, Kupffer cells have been shown to internalize 500 nm multilamellar liposomes more favorably than 80 nm unilamellar liposomes.[219] Smaller-sized particles are favored for hepatic uptake and active targeting receptors on hepatocytes. For example, PS beads of ca. 50 and 140 nm were functionalized with methoxy- or galactose-terminated PEG, yielding slightly negative ζ-potentials, for targeting hepatocytes.[220] An upper size limit was seen by Rensen *et al.* for asialoglycoprotein receptor-mediated endocytosis of lipid-based nanoparticles by hepatocytes at ca. 70 nm. Accumulation of PS-based particles in liver hepatocytes of mice was highest for 50 nm particles functionalized with galactose; larger 140 nm galactose-conjugated particles were mainly found in Kupffer cells.[221] Aknic *et al.* found a similar size dependence using non-targeted lipidoid nanoparticles of different sizes for hepatic gene silencing. Gene silencing efficiency increased with decreasing particle size, reaching ca. 90% accumulation in liver tissue 1 h after systemic administration for 50 nm lipidoids.[212] Active targeting the asialoglycoprotein receptors, which are expressed notably on hepatocytes, has been performed using galactose,[222] lactose,[223] and N-acetylgalactosamine[224] ligands. In addition, proteins and peptides such as lactoferrin[225] and a peptide sequence of T7 phage protein p17[226] have been shown to promote targeting of hepatocytes *in vitro* and *in vivo*. Aside from ligands, the shape of particles can influence preferential targeting of the liver and spleen. Irregularly-shaped polymer lipid nanostructures preferentially targeted the spleen while spherical analogs mainly localized in the liver after *i.v.* administration to dogs.[65]

8.3. *Brain*

In metastatic cancer, nanoparticles for cancer therapeutics may be designed to reach commonly afflicted organs including brain, lung, liver, and lymph nodes. Regarded as the most difficult organ to target with systemically administered nanoparticles, the brain is highly protected by a tight layer of endothelial cells, known as the blood–brain barrier (BBB).[227] Generally, only select species are permitted entry to the cerebrospinal fluid, such as neutrally charged small molecules; yet, certain disease states can modify the permeability of the BBB for transport of therapeutics.[228–231] Access to the brain by particle therapeutics possesses an upper size limit of 15 nm. Although the efficiency of uptake by the brain decreases exponentially with particle size, particles up to 100 nm have been shown in certain cases to still reach the brain.[229] Decorating particle surfaces with lipophilic moieties,[232,233] ligand-functionalized polysaccharides,[234] or antibodies,[235] and other approaches, such as reducing surface charge,[236] and binding apolipoprotein E,[231] can enhance BBB transport or uptake into brain parenchyma. Finally, novel work by van Kasteren *et al.* illustrated the possibility of targeting circulating cells, such as leukocytes, that eventually accumulate at BBB disease sites as an alternative approach to access the brain.[230]

8.4. *Lung*

To access the lungs with particles, *i.v.* and inhalation routes of administration may be pursued for accumulation in target locations. Deposition in the respiratory tract via inhalation is dictated by the mass median aerodynamic diameter (MMAD), which depends on particle size, shape and density. Optimal particle size ranges from 1–5 μm for lung deposition; to obtain this target, dry powder formulations are typically designed by mixing these smaller particles with much larger bulking excipients, such as lactose.[14,237] Recent advances in particle technology have enable unique opportunities for engineering particles and increasing inhalation efficiencies. Pioneered through work by Edwards *et al.*, porous particles of large geometric but small aerodynamic diameter have been shown to deposit efficiently in the lower airways, and are capable of avoiding macrophage uptake and enhancing systemic delivery of biologics, such as insulin.[238] Here, porosity was used to modulate the particles' MMAD through designed physical particle properties. Further modulation of MMAD can come through changes in particle shape, such as work by Garcia *et al.*, who formulated dry powders composed of small molecule and protein therapeutics with shapes such as pollen, torus, and cylinder via PRINT technology.[135] In addition to the size and porosity of particles influencing delivery efficiency to lungs, the surface properties of particles may be harnessed to promote accumulation in particular tissues, including charge and chemistry. For example, positively charged particles delivered both intratracheally and intravenously have been shown to localize in lung tissue more effectively and with longer residences than neutral or negatively charged particles.[239,240] The cationic nature of particles may promote enhanced accumulation and retention in lungs due to enhanced interactions with erythrocytes and serum proteins.[241,242]

8.5. *Lymph*

Contrary to the lung, which can favor uptake of positively charged particles, cells in the lymph nodes internalize negatively charged particles at faster rates than cationic or neutral particles.[243,244] Lymph nodes play a critical role in development of innate immunity and are too often the first site of cancer metastasis. Lymphatic vessels vary in size from 10–60 μm for initial lymph vessels, which are blind-ended structures lined with endothelial cells, while larger lymphatic vessels can be up to 2 mm in diameter.[245] Correspondingly, the geometry of particles can notably affect delivery efficiency to the lymphatic system and mimicking features of pathogens for immune modulation. Particles in the 10–200 nm size range may effectively enter initial lymphatic vessels by diffusing through endothelial junctions.[246–248] For transport of large particles (greater than 500 nm) into lymphatic vessels, uptake by carrier cells, *e.g.* dendritic cells, which may actively facilitate passage into the lymphatic system.[249] From *i.v.* administration, the lymph can be accessed through the active targeting of leukocyctes, often through particle surface decoration with polysaccharides such as dextran,[250,251] peptides, antibodies, and proteins.[250,252] The geometric pattern of antigens or ligands on the surface of particles can also play a crucial role in the efficacy of vaccines, as exemplified by the pathogens recognized by the immune system.[253] For example, specific densities of peptides and small molecules spaced apart by 5–10 nm, the same value for the spacing of relevant receptors, were found to be most effective in cell activation and eliciting an antibody response.[254–256]

9. Outlook for Particles with Controlled Geometry

9.1. *Demands for particle fabrication platforms*

Increasing emphasis will be placed on physical characteristics of therapeutic carriers, requiring particle fabrication platforms to rise to the challenge. Ideally, *a priori* control over the physicochemical properties of particles, including the potential for modifying surface properties, dictating drug loading and release kinetics, and tailoring shape will be realized for each particle fabrication platform. In reality, the physicochemical characteristics of particulate vectors are often interdependent and require fine-tuning to achieve particular features and behaviors, especially for self-assembly approaches. Properties of particles, such as the uniformity in a sample population, geometry (shape and size), three-dimensional distribution of chemical composition, flow dynamics, and responsiveness all may contribute to establishing tailored particulate vectors for *in vivo* applications. Thus, a challenge exists for particle platform approaches to navigate these many demands.

Fabrication approaches through bottom-up or top-down methods that enable both geometric control and chemical composition should further open the door to realizing advantages of engineered drug release profiles and controlled biophysical dynamics. As an example, harnessing face-specific functionalization with ligands should allow for anchoring to cells to unload nanovectors, *e.g.* in vascular delivery of therapeutics.[161] Further demands may arise in inclusion of multicomponent therapies. Delivery of multiple cargos may be critical in enhancing drug indices and effectiveness; therefore, a single particulate vector may need to accommodate multiple, possibly immiscible, drugs. Unique approaches to fabrication may be required to meet combination drug delivery, *e.g.* by tuning the chemical composition. A recent example from Park *et al.* enabled such a combinational delivery using nanoscale liposomal polymeric gels to deliver both hydrophobic molecular inhibitors and hydrophilic protein cytokines. Lyophilized liposomes were loaded with polymer and solubilized cargos for polymerization in "nanoreactors" to provide encapsulation of cargos with specific release rates; tumor immunotherapy was then demonstrated through the activation of innate and adaptive immune responses requiring combination therapy.[257] Adaption of liposomal polymeric gels to additional unique liposomal and micellar structures with different compositions may increase the breadth of accessible particle geometries, surface chemistries, mechanical properties, and drug loadings. Such control of physical and chemical particle characteristics is believed to dramatically impact the scope of nanopreparations in potential applications.

In addition to providing control over the physicochemical and compositional properties of particles, critical factors for a particular platform must be considered. Important aspects of platforms also include ease of fabrication, scalability, biocompatibility, good manufacturing practices (GMP), and FDA approval. Facile approaches to the fabrication of particles and considerations of these restrictions may reduce downstream costs and effort in realizing clinical tests and commercial viability. Biodegradability of materials into endogenous small molecules would be ideal for maximal biocompatibility; however, breakdown of polymers into species that are below the molecular weight cutoff for renal excretion (ca. 30 kDa) should be suitable in meeting safety guidelines. Materials should not elicit adverse cell responses and must be nontoxic to healthy tissues and cells. Additionally, monodispersity, minimization of sample heterogeneity, stability of particles under fabrication, storage, and delivery conditions for each application must be considered to ensure uniform biological responses. Implementation of a GMP process and use of FDA-approved materials should facilitate progression of a platform and delivery system in proceeding through scale-up and clinical studies. Further hurdles to new fabrication approaches are then faced in clearing regulatory guidelines; approval of new drugs takes an average of 12 years and over $350 million starting from the laboratory to over-the-counter.[258] Only 0.1% of the compounds developed for laboratory testing make it to human trials. The success rate for approval is not particularly high due to the rigorous evaluation process, which addresses multiple aspects of drug processing conditions, side effects, and the design of clinical trials.[259] Adherence to guidelines for GMP is necessary to ensure quality and minimize risks involved with products.[260] New particle platforms must measure up to these demands to achieve clinical relevance and success.

Aside from FDA approval and following GMP guidelines, diagnosis of a particular disease represents the true first step in identifying a therapeutic approach which potentially implements a particle platform. Integrating diagnostics with particle platforms directly or through secondary means, *e.g.* based on device readout, can allow for the selection of particle properties tailored to diagnose or treat a disease. Indeed, the theranostic capabilities offered through nanoparticle delivery represent a huge potential step forward toward personalized medicine.[261–265] After diagnosing a disease, effective therapies may be administered; however, research still remains to establish complete guidelines for the most suitable and effective drugs and delivery vector combinations. Embedded in these guidelines will be a fundamental understanding of the role of physical particle properties.

Overall, new particle fabrication techniques which attempt to address physical and compositional control in particle formation face a series of additional requirements on the road to clinical translation. Despite these seemingly overwhelming obstacles to implementing new particle platforms, the benefits of controlling physical characteristics are increasingly compelling and are expected to continue to drive research in this area.

9.2. *Prospects of particle fabrication techniques*

A number of particle fabrication techniques highlighted in this chapter are readily scalable and already commercially viable. For example, Liquidia Technologies has translated the micromolding and nanomolding PRINT technology to a GMP compliant, roll-to-roll scale-up process.[266] Additional high throughput, quality-by-design processes are also in development. Bhaskar *et al.* have fabricated compositionally anisotropic microparticle disks and rods with control over their size and shape using a microcutting process involving electrohydrodynamic co-spinning and cryosectioning different polymers. Exemplar multicompartmental microcylinders include bicompartmental and tricompartmental, pie-shaped, and rosette-like distributions of different phases. The flow rate and concentration of PLGA solutions enabled the tuning of the shape and size of biocompartmental microparticles, produced in high yield followed by filtration to obtain near-perfect particles.[267] Further development of high throughput, GMP compliant manufacture of particles with controlled geometries will facilitate entry into clinical trials and increase the likelihood of therapeutic impact.

While scalability and translation to GMP techniques will further advance the importance of controlled physical properties, novel particle fabrication techniques will continue to inspire new opportunities. One such opportunity may arise from the self-assembly of nucleic acids. Self-assembly of nucleic acids (DNA and RNA) using DNA origami, tiles or bricks has been implemented to form specific 2D and 3D structures.[268,269] Pairing of complementary segments results in highly specific binding, which have been precisely engineered to produce well-defined structures. Further synthetic approaches can be performed to functionalize DNA, making these molecules suitable for diagnostic applications, drug delivery, and biological probes. Recent advances have shown capacity to create 3D self-assembled DNA structures of approximately 20 nm with highly programmable shapes.[268] Such inherently biocompatible systems may prove therapeutically impactful. Coming from a completely different field, an additional example of particle fabrication opportunities comes from advances in polymer chemistry. Yang *et al.* recently obtained high aspect ratio polymer particles with different morphologies and sizes via a mini-emulsion technique using main chain liquid crystalline polymers. The non-spherical shape is an equilibrium feature of the particle due to shape memory of the material when in its microdomain form.[270] Finally, Hancock *et al.* successfully fabricated shape specific hydrogel particles on patterned surfaces with hydrophobic and hydrophilic regions, affording 3D geometric definition.[271] Highlighted by these three diverse examples, research in this area of geometrically programmable particle formation is ongoing in many fields and contributions are expected to arise from the interface of materials science, process engineering, synthetic chemistry, and pharmaceutical sciences.

9.3. *Future avenues to impact biological activity*

Through this chapter, we have highlighted examples in the literature where known physicochemical properties of particles have been shown to influence interactions with cells and behavior *in vivo*. Size, shape, surface properties, composition, and flexibility of particles have been investigated for their effect on cell uptake and targeting, intracellular localization, drug delivery efficiency, and biocompatibility, in addition to *in vivo* behavior, such as circulation times, biodistribution, and accumulation in target tissues and cells. While much ground work has been laid out from these studies, there remains plenty of opportunities to further examine the role of particle properties on biological behaviors. It will be critically important to provide a specific framework to (1) identify key particle parameters and (2) control particle properties needed to effectively treat the requirements of a particular disease diagnosis. In addition to tuning particle properties, specialized systems that respond to specific features of a diseased microenvironment may enhance efficacy. Interesting systems involving communicating particles, shear-responsive systems, and shape-shifting particles have been highlighted in this review as examples of unique mechanisms to elicit distinct biological responses.

Development of nanoparticles for cancer therapeutics will continue to be explored. Approaches to cancer therapy will further involve harnessing and altering the biomechanics of the tumor microenvironment. Given the rigidity of tumor tissues and changes in the stiffness of the extracellular matrix that holds cells together in cancer states, the physical features of tumor environments and their progression differ from those of healthy tissue.[272] High fluid pressure, increased interstitial flow, and growth-induced solid stresses represent additional physical changes noted in tumor environments.[273] To probe and adjust to their microenvironment, cells remain mechanically active; that is, receptors may not only be influenced by biochemical cues, but rather also fluid shear or strain.[274] Resulting conformational changes in receptors and proteins may elicit intracellular signaling cascades that contribute to physiological changes. Attempting to reengineer the tumor microenvironment into normal tissue by implementing materials that mimic the embryonic microenvironment exemplifies an interesting potential route to cancer therapy.[275] Biomaterials with controlled geometric properties may be administered and targeted to tumor tissue wherein self-assembly into three-dimensionally defined scaffolds may occur to reverse the progression of cancer.

There are many additional opportunities to impact particulate efficacy through designed physical properties, as yet unexplored. Aggregated particles with pH- and redox-responsive properties could be an extension of shear-responsive microaggregates in the delivery of therapeutics to diseased tissues. The acidic environment[276] and redox potential[277] of tumor microenvironments may be further harnessed in the design of drug delivery systems for enhanced efficacy. Shape-shifting particles could release drug after exposure to a trigger or a given duration of time to enhance biological responses. Systems that reside in afflicted tissue and release drug at a controlled rate in response to local concentrations of specific biomarkers could enable effective and prolonged activity of a therapy after a single administration. Additional endeavors may entail pursuing convenient and effective modes of administration for enhanced accumulation and efficacy. For instance, oral routes of administration may provide convenience for patient self-administration, while additional therapies may benefit through inhalation or intramuscular injection routes of administration. These represent additional, and relatively underexplored, opportunities for particulate delivery and provide convenient alternatives to *i.v.* systemic administration.

10. Concluding Remarks

In this chapter, we outline some of the most prevalent methodologies for fabricated particles with controlled physical properties and highlight individual studies where these properties have impacted biological processes.

Well-designed particles with controlled compositional and physical properties will continue to be developed and the opportunities still present in this field are vast. Increasingly, these systems will be developed with tailored properties and unique features designed to exploit specific diseased states. We anticipate that enhanced efficacy resulting from these systems will contribute to the advancement of therapies and be actively involved toward improving health and the quality of life.

Acknowledgments

This work was supported by: the Carolina Center for Cancer Nanotechnology Excellence (NIH-U54-CA119343 and NIH-U54-CA151652), National Institutes of Health Director's Pioneer Award (1DP1OD006432) and R01 (R01EB009565), University of North Carolina Cancer Research Fund, the Chancellor's Eminent Professorship of Chemistry at the University of North Carolina at Chapel Hill, and Liquidia Technologies. We would like to thank JF, HJM, KEY, KAP, and CF for valuable comments and edits.

References

1. J. A. Champion, Y. K. Katare and S. Mitragotri, Journal of controlled release: Official journal of the Controlled Release Society **121** (2007).
2. S. Venkataraman, J. L. Hedrick, Z. Y. Ong, C. Yang, P. L. Ee, P. T. Hammond and Y. Y. Yang, *Advanced Drug Delivery Reviews* **63** (2011).
3. N. Daum, C. Tscheka, A. Neumeyer and M. Schneider, *WIREs Nanomedicine and Nanobiotechnology* **4** (2011).
4. P. Decuzzi, R. Pasqualini, W. Arap and M. Ferrari, *Pharmaceutical Research* **26** (2009).
5. L. Tao, W. Hu, Y. Liu, G. Huang, B. D. Sumer and J. Gao, *Exp. Biol. Med.* **236** (2011).
6. S. Mitragotri and J. Lahann, *Nat. Mater.* **8** (2009).
7. E. A. Simone, T. D. Dziubla and V. R. Muzykantov, *Expert Opin. Drug Deliv.* **5** (2008).
8. D. Lentink, W. B. Dickson, J. L. van Leeuwen and M. H. Dickinson, *Science* **324** (2009).
9. P. Roach, N. J. Shirtcliffe and M. I. Newton, *Soft Matter* **4** (2008).
10. M. Diez-Silva, M. Dao, J. Han, C.-T. Lim and S. Suresh, *MRS Bulletin* **35.**
11. S. Suresh, *Acta Biomaterialia* **3** (2007).
12. K. D. Young, *Microbiol. Mol. Biol. Rev.: MMBR* **70** (2006).
13. R. H. Garrett and C. M. Grisham, *Biochemistry,* Belmont, CA: Thomson Brooks/Cole (2005).
14. J. S. Patton and P. R. Byron, *Nature Reviews Drug Discovery* **6** (2007).
15. R. B. Bird, W. E. Stewart and E. N. Lightfoot, *Transport Phenomena*, New York: John Wiley & Sons, Inc. (2007).
16. J. Siepmann, R. A. Siegel and F. Siepmann, In *Fundamentals and Applications of Controlled Release Drug Delivery*, J. Siepmann, R. A. Siegel and M. J. Rathbone (Eds.), Springer (2012), p. 127–152.

17. M. E. Pearce, J. B. Melanko and A. K. Salem, *Pharm. Res.* **24** (2007).
18. R. R. Arvizo, S. Bhattacharyya, R. A. Kudgus, K. Giri, R. Bhattacharya and P. Mukherjee, *Chemical Society Reviews* **41** (2012).
19. S. K. Ghosh and T. Pal, *Chem Rev.* **107** (2007).
20. J. A. Dahl, B. L. S. Maddux and J. E. Hutchinson, *Chem Rev.* **107** (2007).
21. D. Pissuwan, S. M. Valenzuela and M. B. Cortie, *Biotechnol Genet Eng.* **25** (2008).
22. G. von Maltzahn, J.-H. Park, K. Y. Lin, N. Singh, C. Schwoppe, R. Mesters, W. E. Berfel, E. Ruoslahti, M. J. Sailor and S. Bhatia, *Nature Materials* **10** (2011).
23. B. D. Chithrani and W. C. W. Chan, *Nano Letters*, 7 (2007).
24. R. Sardar and J. S. Shumaker-Parry, *Nano Letters* **8** (2008).
25. Q. Dai, J. G. Worden, Trullinger and Q. Huo, *J. Am. Chem. Soc.* **127** (2005).
26. X. Huang, I. H. El-Sayed, W. Qian and M. A. El-Sayed, *J. Am. Chem. Soc.* **128** (2006).
27. B. D. Chithrani, A. A. Ghazani and W. C. W. Chan, *Nano Letters* **6** (2006).
28. T. K. Sau and C. J. Murphy, *Langmuir* **20** (2004).
29. X. Huang, S. Neretina and M. A. El-Sayed, *Adv. Mater.* 21 (2009).
30. Y. Akiyama, T. Mori, Y. Katayama and T. Niidome, *Nanoscale Res. Lett.* 7 (2012).
31. A. J. Gormley, K. Greish, A. Ray, R. Robinson, J. A. Gustafson and H. Ghandehari, *Int. J. Pharm.* **415** (2011).
32. C. Wang and J. Irudayaraj, *Small* **6** (2010).
33. C. M. Cobley, S. E. Skrabalak, D. J. Campbell and Y. Xia, *Plasmonics* **4** (2009).
34. M. R. Langille, M. L. Personick, J. Zhang and C. A. Mirkin, *J. Am. Chem. Soc.* **134** (2012).
35. D. H. M. Dam, J. H. Lee, P. N. Sisco, D. T. Co, M. Zhang, M. R. Wasielewski and T. W. Odom, *ACS Nano* **6** (2012).
36. J. Li, J. Wu, X. Zhang, Y. Liu, D. Zhou, H. Sun, H. Zhang and B. Yang, *J. Phys. Chem. C* **115** (2011).
37. T. K. Sau and C. J. Murphy, *J. Am. Chem. Soc.* **126** (2004).
38. J. J. Mock, M. Barbic, D. R. Smith, D. A. Schultz and S. Schultz, *J. Chem. Phys.* **116** (2002).
39. S. Pal, Y. K. Tak and J. M. Song, *Appl. Environ. Microbiol.* **73** (2007).
40. J. H. Park, G. von Maltzahn, L. Zhang, M. P. Schwartz, E. Ruoslahti, S. N. Bhatia and M. J. Sailor, *Adv. Mater.* **20** (2008).
41. J.-H. Park, G. Von Maltzahn, X. Zhang, A. M. Derfus, D. Simbery, T. J. Harris, E. Ruoslahti, S. Bhattacharyya and M. J. Sailor, *Small* **5** (2009).
42. V. Raffa, G. Ciofani, S. Nitodas, T. Karachalios, D. D'Alessandro, M. Masini and A. Cuschieri, *Carbon* **46** (2008).
43. B. R. Smith, P. Kempen, D. Bouley, A. Xu, Z. Liu, N. Melosh, H. Dai, R. Sinclair and S. S. Gambhir, *Nano Lett.* **12** (2012).
44. S. Beg, M. Rizwan, A. M. Sheikh, M. S. Hasnain, K. Anwer and K. Kohli, *J. Pharm. Pharmacol.* **63** (2011).
45. M. A. Hayes, M. D. Pysher and K. Chen, *J. Nanosci. Nanotechnol.* 7 (2007).
46. A. S. Janoff, L. T. Boni, M. C. Popescu, S. R. Minchey, P. R. Cullis, T. D. Madden, T. Taraschi, S. M. Gruner, E. Shyamsunder, M. W. Tate, R. Mendelsohn and D. Bonner, *Proceedings of the National Academy of Sciences of the United States of America* **85** (1988).
47. K. Zhang, H. Fang, Z. Chen, J.-S. A. Taylor and K. L. Wooley, *Bioconjugate Chem.* **19** (2008).

48. D. J. Pochan, Z. Chen, H. Cui, K. Hales, K. Qi and K. L. Wooley, *Science* **306** (2004).
49. Y. Geng, P. Dalhaimer, S. Cai, R. Tsai, M. Tewari, T. Minko and D. E. Discher, *Nat. Nanotechnol.* **2** (2007).
50. I. W. Hamley, *Soft Matter* **1** (2005).
51. B. M. Discher, Y.-Y. Won, D. S. Ege, J. C.-M. Lee, F. S. Bates, D. E. Discher and D. A. Hammer, *Science* **284** (1999).
52. D. E. Discher and A. Eisenberg, *Science* **297** (2002).
53. M. Antoniette and S. Forster, *Adv. Mater.* 15 (2003).
54. H. Cui, Z. Chen, S. Zhong, K. L. Wooley and D. J. Pochan, *Science* **317** (2007).
55. S. Jain and F. S. Bates, *Science* **300** (2003).
56. P. L. Soo and A. Eisenberg, *J. Polym. Sci. Pol. Phys.* **42** (2004).
57. T. P. Lodge, M. A. Hillmyer, Z. Zhou and Y. Talmon, *Macromolecules* **37** (2004).
58. S. Jain and F. S. Bates, *Macromolecules* **2004** (2004).
59. K. Yu, L. Zhang and A. Eisenberg, *Langmuir* **12** (1996).
60. J. Raez, I. Manners and M. Winnik, *J. Am. Chem. Soc.* **124** (2002).
61. Z. Li, Z. Chen, H. Cui, K. Hales, K. Qi, K. L. Wooley and D. J. Pochan, *Langmuir* **21** (2005).
62. L. Zhang, C. Bartels, Y. Yu, H. Shen and A. Eisenberg, *Phys. Rev. Lett.* **79** (1997).
63. I. C. Riegel and A. Eisenberg, *Langmuir* **18** (2002).
64. J. J. Cornelissen, M. Fischer, N. A. J. M. Sommerdijk and R. J. M. Nolte, *Science* **280** (1998).
65. P. V. Devarajan, A. B. Jindal, R. R. Patil, F. Mulla, R. V. Gaikwad and A. Samad, *J. Pharm. Sci.* **99** (2010).
66. S. Cai, K. Vijayan, D. Cheng, E. M. Lima and D. E. Discher, *Pharmaceutical Research* **24** (2007).
67. P. Dalhaimer, F. S. Bates and D. E. Discher, *Macromolucules* **36** (2003).
68. P. Dalhaimer, H. Bermudez and D. E. Discher, *J. Polym. Sci. Pol. Lett.* **42** (2003).
69. E. A. Simone, T. D. Dziubla, D. E. Discher and V. R. Muzykantov, *Biomacromolecules* **10** (2009).
70. D. A. Christian, S. Cai, O. B. Garbuzenko, T. Harada, A. L. Zajac, T. Minko and D. E. Discher, *Mol. Pharm.* **6** (2009).
71. G. Sun, J. Xu, A. Hagooly, R. Rossin, Z. Li, D. A. Moore, C. J. Hawker, M. J. Welch and K. L. Wooley, *Adv. Mater.* **19** (2007).
72. H. Cui, Z. Chen, K. L. Wooley and D. J. Pochan, *Macromolecules* **39** (2006).
73. K. B. Thurmond II, T. Kowalewski and K. L. Wooley, *J. Am. Chem. Soc.* **118** (1996).
74. E. S. Read and S. P. Armes, *Chem. Commun.* (2007).
75. M. L. Becker, E. E. Remsen, D. Pan and K. L. Wooley, *Bioconjugate Chem.* **15** (2004).
76. T. F. Yang, C. N. Chen, M. C. Chen, C. H. Lai, H. F. Liang and H. W. Sung, *Biomaterials* **28** (2007).
77. M. Vallet-Regi, A. Ramila, R. P. del Real and J. Perez-Pariente, *Chem. Mater.* **13** (2001).
78. I. I. Slowing, B. G. Trewyn, S. Giri and V. S. Y. Lin, *Adv. Funct. Mater.* **17** (2007).
79. I. I. Slowing, J. L. Vivero-Escoto, B. G. Trewyn and V. S. Y. Lin, *J. Mater. Chem.* **20** (2010).
80. F. Tang, L. Li and D. Chen, *Adv. Mater.* **24** (2012).
81. Y. Wan and D. Zhao, *Chem. Rev.* **107** (2007).
82. C. T. Kresge, M. E. Leonowicz, W. J. Roth, J. C. Vartuli and J. S. Beck, *Nature* **359** (1992).

83. R. Mout, D. F. Moyano, S. Rana and V. M. Rotello, *Chemical Society Reviews* **41** (2012).
84. F. Hoffmann, M. Cornelius, J. Morell and M. Froba, *Angew Chem. Int. Ed. Engl.* **45** (2006).
85. Z. G. Estephan, J. A. Jaber and J. B. Schlenoff, *Langmuir* **26** (2010).
86. Z. G. Estephan, P. S. Schlenoff and J. B. Schlenoff, *Langmuir* **27** (2011).
87. S. Huh, J. W. Wiench, J.-C. Yoo, M. Pruski and V. S. Y. Lin, *Chem. Mater.* **15** (2003).
88. S. Che, Z. Liu, T. Ohsuna, K. Sakamoto, O. Terasaki and T. Tatsumi, *Nature* **429** (2004).
89. B. G. Trewyn, C. M. Whitman and V. S. Y. Lin, (2004).
90. S. Yang, L. Zhao, C. Yu, X. Zhou, J. Tang, P. Yuan, D. Chen and D. Zhao, *J. Am. Chem. Soc.* **128** (2006).
91. X. Huang, X. Teng, D. Chen, F. Tang and J. He, *Biomaterials* **31** (2010).
92. X. Huang, L. Li, T. Liu, N. Hao, H. Liu, D. Chen and F. Tang, *ACS Nano* **5** (2011).
93. T. Yu, D. Hubbard, A. Ray and H. Ghandehari, Journal of Controlled Release: official journal of the Controlled Release Society **163** (2012).
94. T. Yu, A. Malugin and H. Ghandehari, *ACS Nano* **5** (2011).
95. B. G. Trewyn, J. A. Nieweg, Y. Zhao and V. S. Y. Lin, *Chem. Eng. J.* **137** (2008).
96. H. Meng, S. Yang, Z. Li, T. Xia, J. Chen, Z. Ji, H. Zhang, X. Wang, S. Lin, C. Huang, Z. H. Zhou, J. I. Zink and A. Nel, *ACS Nano* **5** (2011).
97. X. Huang, X. Peng, Y. Wang, Y. Wang, D. M. Shin, M. A. El-Sayed and S. Nie, *ACS Nano* **4** (2010).
98. J. A. Champion, Y. K. Katare and S. Mitragotri, Proceedings of the National Academy of Sciences of the United States of America **104** (2007).
99. J. A. Champion and S. Mitragotri, *Pharmaceutical Research* **26** (2009).
100. J. A. Champion and S. Mitragotri, Proceedings of the National Academy of Sciences of the United States of America **103** (2006).
101. S. Muro, C. Garnacho, J. A. Champion, J. Leferovich, C. Gajewski, E. H. Schuchman, S. Mitragotri and V. R. Muzykantov, *Mol. Ther. The Journal of the American Society of Gene Therapy* **16** (2008).
102. N. Doshi and S. Mitragotri, *J. R. Soc. Interface. The Royal Society* 7 (Suppl 4) (2010).
103. P. Kolhar and S. Mitragotri, *Adv. Funct. Mater.* **22** (2012).
104. J. W. Yoo, N. Doshi and S. Mitragotri, *Adv. Drug Deliv. Rev.* **63** (2011).
105. J.-W. Yoo, S. Mitragotri and D. A. Tirrell, Proceedings of the National Academy of Sciences of the United States of America **107** (2010).
106. G. E. Moore, *Electronics* **38** (1965).
107. T. Ito and S. Okazaki, *Nature* **406** (2000).
108. Intel(R) 22nm Technoogy. Available at: http://www.intel.com/content/www/us/en/silicon-innovations/intel-22nm-technology.html (Accessed on 20 October 2012),
109. L. Tao, X. M. Zhao, J. M. Gao and W. Hu, *Nanotechnology* **21** (2010).
110. F. Buyukserin, M. Aryal, J. Gao and W. Hu, *Small* **5** (2009).
111. L. C. Glangchai, M. Caldorera-Moore, L. Shi and K. Roy, *J. Controlled Release: Official Journal of the Controlled Release Society* **125** (2008).
112. G. Acharya, C. S. Shin, M. McDermott, H. Mishra, H. Park, I. C. Kwon and K. Park, *J. Controlled Release: Official Journal of the Controlled Release Society* **141** (2010).
113. P. C. Hidber, W. Helbig, E. Kim and G. M. Whitesides, *Langmuir* **12** (1996).
114. E. Kim, Y. Xia and G. M. Whitesides, *Nature* **376** (1995).

115. E. Kim, Y. Xia, X.-M. Zhao and G. M. Whitesides, *Adv. Mater.* **9** (1997).
116. A. Kumar and G. M. Whitesides, *Appl. Phys. Lett.* **63** (1993).
117. Y. Xia and G. S. Whitehead, *Angew. Chem. Int. Ed. Engl.* **37** (1998).
118. Y. Xia and G. S. Whitehead, *Annu. Rev. Mater. Sci.* **28** (1998).
119. Y. Lu and S. C. Chen, *Adv. Drug Deliv. Rev.* **56** (2004).
120. V. N. Truskett and M. P. Watts, *Trends in Biotechnology* **24** (2006).
121. D. C. Pregibon and P. S. Doyle, *Anal. Chem.* **81** (2009).
122. C. L. Lewis, C.-H. Choi, Y. Lin, C.-S. Lee and H. Yi, *Anal. Chem.* **82** (2010).
123. G. P. Cox, K. L. Marshall, J. C. Lambropoulos, M. Leitch, C. Fromen and S. D. Jacobs, *J Appl. Phys.* **106** (2009).
124. C. M. S. Torres, S. Zankovych, J. Seekamp, A. P. Kam, C. C. Cedeno, T. Hoffmann, J. Ahopelto, F. Reuther, K. Pfeiffer, G. Bleidiessel, G. Gruetzner, M. V. Maximov and B. Heidari, *Mater. Sci. Eng. C* **23** (2003).
125. L. J. Guo, *Adv. Mater.* **19** (2007).
126. N. V. Le, W. J. Dauksher, K. A. Gehoski, K. J. Nordquist, E. Ainley and P. Mangat, *Microelectronic Eng.* **83** (2006).
127. S. E. Gratton, P. A. Ropp, P. D. Pohlhaus, J. C. Luft, V. J. Madden, M. E. Napier and J. M. DeSimone, *Proceedings of the National Academy of Sciences of the United States of America* **105** (2008).
128. J. Y. Kelly and J. M. DeSimone, *J. Am. Chem. Soc.* **130** (2008).
129. T. J. Merkel, K. P. Herlihy, J. Nunes, R. M. Orgel, J. P. Rolland and J. M. DeSimone, *Langmuir* **26** (2010).
130. J. P. Rolland, B. W. Maynor, L. E. Euliss, A. E. Exner, G. M. Denison and J. M. DeSimone, *J. Am. Chem. Soc.* **127** (2005).
131. Y. Wang, T. J. Merkel, K. Chen, C. A. Fromen, D. E. Betts and J. M. DeSimone, *Langmuir* **27** (2011).
132. S. E. Gratton, P. D. Pohlhaus, J. Lee, J. Guo, M. J. Cho and J. M. Desimone, *J. Controlled Release: Official Journal of the Controlled Release Society* **121** (2007).
133. F. R. Kersey, T. J. Merkel, J. L. Perry, M. E. Napier and J. M. DeSimone, *Langmuir* **28** (2012).
134. J. Y. Wang, Y. Wang, S. S. Sheiko, D. E. Betts and J. M. DeSimone, *J. Am. Chem. Soc.* **134** (2012).
135. A. Garcia, P. Mack, S. Williams, C. Fromen, T. Shen, J. Tully, J. Pillai, P. Kuehl, M. Napier, J. M. DesImone and B. W. Maynor, *J. Drug Deliv.* **2012** (2012).
136. E. M. Enlow, J. C. Luft, M. E. Napier and J. M. DeSimone, *Nano Lett.* **11** (2011).
137. K. P. Herlihy, J. Nunes and J. M. DesImone, *Langmuir* **24** (2008).
138. J. Nunes, K. P. Herlihy, L. Mair, R. Superfine and J. M. DeSimone, *Nano Lett.* **10** (2010).
139. J. Xu, J. Wang, J. C. Luft, S. Tian, G. Owens, Jr., A. A. Pandya, P. Berglund, P. Pohlhaus, B. W. Maynor, J. Smith, B. Hubby, M. E. Napier and J. M. DeSimone, *J. Am. Chem. Soc.* **134** (2012).
140. H. Zhang, J. K. Nunes, S. E. A. Gratton, K. P. Herlihy, P. D. Pohlhaus and J. M. DeSimone, *New J. Phys.* **11** (2009).
141. K. Chen, T. J. Merkel, A. Pandya, M. E. Napier, J. C. Luft, W. Daniel, S. Sheiko and J. M. DeSimone, *Biomacromolecules* **13** (2012).
142. M. S. Hassan and R. W. Lau, *AAPS Pharm. Sci. Tech.* **10** (2009).

143. J. L. Perry, K. G. Reuter, M. P. Kai, K. P. Herlihy, S. W. Jones, J. C. Luft, M. Napier, J. E. Bear and J. M. Desimone, *Nano Lett.* **12** (2012).
144. T. J. Merkel, K. Chen, S. W. Jones, A. A. Pandya, S. Tian, M. E. Napier, W. E. Zamboni and J. M. DeDimone, *J. Controlled Release: Official Journal of the Controlled Release Society* **162** (2012).
145. S. K. Sia and G. M. Whitesides, *Electrophoresis* **24** (2003).
146. X. Mao and T. J. Huang, *Lab Chip*, **12** (2012).
147. Diagnostics for All. www.dfa.org (01/06/2013),
148. T. Thorsen, R. W. Roberts, F. H. Arnold and S. R. Quake, *Phys. Rev. Lett.* **86** (2001).
149. S. L. Anna, N. Bontoux and H. A. Stone, *Appl. Phys. Lett.* **82** (2003).
150. A. S. Utada, E. Lorenceau, D. R. Link, P. D. Kaplan, H. A. Stone and D. A. Weitz, *Science* **308** (2005).
151. S. Xu, Z. Nie, M. Seo, P. Lewis, E. Kumacheva, H. A. Stone, P. Garstecki, D. B. Weibel, I. Gitlin and G. M. Whitesides, *Angew. Chem. Int. Ed. Engl.* **44** (2005).
152. H. C. Shum, A. R. Abate, D. Lee, A. R. Studart, B. Wang, C. H. Chen, J. Thiele, R. K. Shah, A. Krummel and D. A. Weitz, *Macromolecular Rapid Communications* **31** (2010).
153. D. Dendukuri, D. C. Pregibon, J. Collins, T. A. Hatton and P. S. Doyle, *Nature Materials* **5** (2006).
154. D. Dendukuri, S. S. Gu, D. C. Pregibon, T. A. Hatton and P. S. Doyle, *Lab. Chip* **7** (2007).
155. J. H. Jang, D. Dendukuri, T. A. Hatton, E. L. Thomas and P. S. Doyle, *Angew. Chem. Int. Ed. Engl.* **46** (2007).
156. S. K. Suh, K. Yuet, D. K. Hwang, K. W. Bong, P. S. Doyle and T. A. Hatton, *J. Am. Chem. Soc.* **134** (2012).
157. M. H. Cohen, K. Melnik, A. A. Boiarski, M. Ferrari and F. J. Martin, *Biomed. Microdevices* **5** (2003).
158. C. Chiappini, E. Tasciotti, J. R. Fakhoury, D. Fine, L. Pullan, Y. C. Wang, L. Fu, X. Liu and M. Ferrari, *Chemphyschem: A European Journal of Chemical Physics and Physical Chemistry* **11** (2010).
159. B. Godin, C. Chiappini, S. Srinivasan, J. F. Alexander, K. Yokoi, M. Ferrari, P. Decuzzi and X. Liu, *Adv. Funct. Mater.* **22** (2012).
160. R. E. Serda, S. Ferrati, B. Godin, E. Tasciotti, X. Liu and M. Ferrari, *Nanoscale* **1** (2009).
161. R. E. Serda, B. Godin, E. Blanco, C. Chiappini and M. Ferrari, *Biochimica et biophysica Acta* **1810** (2011).
162. R. E. Serda, J. Gu, R. C. Bhavane, X. Liu, C. Chiappini, P. Decuzzi and M. Ferrari, *Biomaterials* **30** (2009).
163. E. Tasciotti, X. Liu, R. Bhavane, K. Plant, A. D. Leonard, B. K. Price, M. M. Cheng, P. Decuzzi, J. M. Tour, F. Robertson and M. Ferrari, *Nature Nanotechnol.* **3** (2008).
164. S. Ferrati, A. Mack, C. Chiappini, X. Liu, A. J. Bean, M. Ferrari and R. E. Serda, *Nanoscale* **2** (2010).
165. C. J. Hernandez and T. G. Mason, *J. Phys. Chem. C* **111** (2007).
166. J. Lee, W. Hasan, C. L. Stender and T. W. Odom, *Acc. Chem. Res.* **41** (2008).
167. G. Adriani, M. D. de Tullio, M. Ferrari, F. Hussain, G. Pascazio, X. Liu and P. Decuzzi, *Biomaterials* **33** (2012).

168. C. J. Hernandez, K. Zhao and T. G. Mason, *Soft Matter* **5**, 1–11 (2007).
169. C. J. Hernandez, K. Zhao and T. G. Mason, *Soft Matter* **5**, 13–31 (2007).
170. C. H. Villa, M. R. McDevitt, F. E. Escorcia, D. A. Rey, M. Bergkvist, C. A. Batt and D. A. Scheinberg, *Nano Lett.* **8** (2008).
171. P. Decuzzi, B. Godin, T. Tanaka, S. Y. Lee, C. Chiappini, X. Liu and M. Ferrari, *J. Controlled Release: Official Journal of the Controlled Release Society* **141** (2010).
172. M. E. Davis, Z. G. Chen and D. M. Shin, *Nature Reviews Drug Discovery* 7 (2008).
173. H. Otsuka, Y. Nagasaki and K. Kataoka, *Adv. Drug Deliv. Rev.* **55** (2003).
174. H. Takeuchi, H. Kojima, H. Yamamoto and Y. Kawashima, J. Controlled Release: Official Journal of the Controlled Release Society **68** (2000).
175. Arnida, M. M. Janat-Amsbury, A. Ray, C. M. Peterson and H. Ghandehari, European Journal of Pharmaceutics and Biopharmaceutics: Official journal of Arbeitsgemeinschaft fur Pharmazeutische Verfahrenstechnik e.V, 77 (2011).
176. Y. H. Bae, J. Controlled Release: Official Journal of the Controlled Release Society **133** (2009).
177. K. F. Pirollo and E. H. Chang, *Trends Biotechnol.* **26** (2008).
178. E. Chambers and S. Mitragotri, *Exp. Biol. Med.* **232** (2007).
179. T. J. Merkel, S. W. Jones, K. P. Herlihy, F. R. Kersey, A. R. Shields, M. Napier, J. C. Luft, H. Wu, W. C. Zamboni, A. Z. Wang, J. E. Bear and J. M. Desimone, *Proceedings of the National Academy of Sciences of the United States of America* **108** (2010).
180. M. E. Fox, F. Szoka and J. M. Frechet, *Acc.Chem. Res.* **42** (2009).
181. Y. Chen, J. Sen, S. R. Bathula, Q. Yang, R. Fittipaldi and L. Huang, *Mol. Pharm.* **6** (2009).
182. K. Chen, J. Xie, H. Xu, D. Behera, M. H. Michalski, S. Biswal, A. Wang and X. Chen, *Biomaterials* **30** (2009).
183. M. Stevenson, A. B. Hale, S. J. Hale, N. K. Green, G. Black, K. D. Fisher, K. Ulbrich, A. Fabra and L. W. Seymour, *Cancer Gene Therapy* **14** (2007).
184. M. E. Davis, J. E. Zuckerman, C. H. Choi, D. Seligson, A. Tolcher, C. A. Alabi, Y. Yen, J. D. Heidel and A. Ribas, *Nature* **464** (2010).
185. W. A. Henne, D. D. Doorneweerd, A. R. Hilgenbrink, S. A. Kularatne and P. S. Low, *Bioorgan. Med. Chem. Lett.* **16** (2006).
186. E. Levy-Nissenbaum, A. F. Radovic-Moreno, A. Z. Wang, R. Langer and O. C. Farokhzad, *Trends Biotechnol.* **26** (2008).
187. P. Decuzzi and M. Ferrari, *Biophysical Journal* **94** (2008).
188. W. Shi, J. Wang, X. Fan and H. Gao, *Phys. Rev. E* **78** (2008).
189. M.-H. Lai, J. H. Jeong, R. J. DeVolder, C. Brockman, C. Schroeder and H. Kong, *Adv. Funct. Mater.* **22** (2012).
190. V. V. Shuvaev, M. A. Ilies, E. Simone, S. Zaitsev, Y. Kim, S. Cai, A. Mahmud, T. Dziubla, S. Muro, D. E. Discher and V. R. Muzykantov, *ACS Nano* **5** (2011).
191. A. Ruggiero, C. H. Villa, J. P. Holland, S. R. Sprinkle, C. May, J. S. Lewis, D. A. Scheinberg and M. R. McDevitt, *Int. J. Nanomed.* **5** (2010).
192. J. Grimm and D. A. Scheinberg, *Seminars in Radiation Oncology* **21** (2011).
193. C. H. Villa, T. Dao, I. Ahearn, N. Fehrenbacher, E. Casey, D. A. Rey, T. Korontsvit, V. Zakhaleva, C. A. Batt, M. R. Phillips and D. A. Scheinberg, *ACS Nano* **5** (2011).
194. J. Wang, S. Tian, R. A. Petros, M. E. Napier and J. M. DeSimone, *J. Am. Chem. Soc.* **132** (2010).

195. X. Huang, X. Peng, Y. Wang, D. M. Shin, M. A. El-Sayed and S. Nie, *ACS Nano* **4** (2010).
196. D. M. Underhill and H. S. Goodridge, *Nature Reviews Immunology* **12** (2012).
197. N. Doshi and S. Mitragotri, *PloS One* **5** (2010).
198. K. Yang and Y.-q. Ma, *Nature Nanotech.* **5** (2010).
199. K. Yang and Y.-q. Ma, *Aust. J Chem.* **64** (2011).
200. J. S. Aaron, A. C. Greene, P. G. Kotula, G. D. Bachand and J. A. Timlin, *Small* 7 (2011).
201. S. Hofinger, M. Melle-Franco, T. Gallo, A. Cantelli, M. Calvaresi, J. A. Gomes and F. Zerbetto, *Biomaterials* **32** (2011).
202. Z. P. Xu, M. Niebert, K. Porazik, T. L. Walker, H. M. Cooper, A. P. Middelberg, P. P. Gray, P. F. Bartlett and G. Q. Lu, *J. Controlled Release: Official Journal of the Controlled Release Society* **130** (2008).
203. P. Chaudhuri, R. Harfouche, S. Soni, D. M. Hentschel and S. Sengupta, *ACS Nano* **4** (2010).
204. S. M. Elbashir, J. Harborth, W. Lendeckel, A. Yalcin, K. Weber and T. Tuschl, *Nature* **411** (2001).
205. A. Fire, S. Xu, M. K. Montgomery, S. A. Kostas, S. E. Driver and C. C. Mello, *Nature* **391** (1998).
206. P. Kolhar, N. Doshi and S. Mitragotri, *Small* 7 (2011).
207. N. Korin, M. Kanapathipillai, B. D. Matthews, M. Crescente, A. Brill, T. Mammoto, K. Ghosh, S. Jurek, S. A. Bencherif, D. Bhatta, A. U. Coskun, C. L. Feldman, D. D. Wagner and D. E. Ingber, *Science* **337** (2012).
208. V. Uskokovic, K. Lee, L. P. Peng, K. E. Fischer and T. A. Desai, *ACS Nano* **6** (2012).
209. D. A. Canelas, K. P. Herlihy and J. M. Desimone, *Advanced Reviews Wiley*, **1** (2009).
210. X. Xu and M. B. Cortie, *Adv. Funct. Mater.* **16** (2006).
211. T. J. Harris, J. J. Green, P. W. Fung, R. Langer, D. G. Anderson and S. N. Bhatia, *Biomaterials* **31** (2010).
212. A. Akinc, M. Goldberg, J. Qin, J. R. Dorkin, C. Gamba-Vitalo, M. Maier, K. N. Jayaprakash, M. Jayaraman, K. G. Rajeev, M. Manoharan, V. Koteliansky, I. Rohl, E. S. Leshchiner, R. Langer and D. G. Anderson, *Mol. Ther.: The Journal of the American Society of Gene Therapy* **17** (2009).
213. Y. Minakuchi, F. Takeshita, N. Kosaka, H. Sasaki, Y. Yamamoto, M. Kouno, K. Honma, S. Nagahara, K. Hanai, A. Sano, T. Kato, M. Terada and T. Ochiya, *Nucleic Acids Research* **32** (2004).
214. H. S. Choi, W. Liu, P. Misra, E. Tanaka, J. P. Zimmer, B. Itty Ipe, M. G. Bawendi and J. V. Frangioni, *Nat. Biotechnol.* **25** (2007).
215. J. Rappaport, B. Hanss, J. B. Kopp, T. D. Copeland, L. A. Bruggeman, T. M. Coffman and P. E. Klotman, *Kidney International* **47** (1995).
216. S. Stolnik, L. Illum and S. S. Davis, *Adv. Drug Deliv. Rev.* **16** (1995).
217. L. Illum, S. S. Davis, C. G. Wilson, N. W. Thomas, M. Frier and J. G. Hardy, *Int. J. Pharm.*, **12** (1982).
218. Y. Tabata and Y. Ikada, *Adv. Polym. Sci.* **94** (1990).
219. Y. E. Rahman, E. A. Cerny, K. R. Patel, E. H. Lau and B. J. Wright, *Life Sciences* **31** (1982).
220. S. R. Popielarski, S. Hu-Lieskovan, S. W. French, T. J. Triche and M. E. Davis, *Bioconjugate Chem.* **16** (2005).

221. P. C. Rensen, L. A. Sliedregt, M. Ferns, E. Kieviet, S. M. van Rossenberg, S. H. van Leeuwen, T. J. van Berkel and E. A. Biessen, *J. Biol. Chem.*, **276** (2001).
222. C.-H. Lai, C.-Y. Lin, H.-T. Wu, H.-S. Chan, Y.-J. Chuang, C.-T. Chen and C.-C. Lin, *Adv. Funct. Mater.* **20** (2010).
223. M. Oishi, Y. Nagasaki, K. Itaka, N. Nishiyama and K. Kataoka, *J. Am. Chem. Soc.* **127** (2005).
224. D. B. Rozema, D. L. Lewis, D. H. Wakefield, S. C. Wong, J. J. Klein, P. L. Roesch, S. L. Bertin, T. W. Reppen, Q. Chu, A. V. Blokhin, J. E. Hagstrom and J. A. Wolff, *Proceedings of the National Academy of Sciences of the United States of America* **104** (2007).
225. A. H. Weeke-Klimp, M. Bartsch, H. W. Morselt, I. Van Veen-Hof, D. K. Meijer, G. L. Scherphof and J. A. Kamps, *J. Drug Target.* **15** (2007).
226. J. J. Ludtke, A. V. Sokoloff, S. Wong, G. Zhang, D. K. Strickland and J. A. Wolff, *Drug Deliv.* **16** (2009).
227. A. F. Eichler, E. Chung, D. P. Kodack, J. S. Loeffler, D. Fukumura and R. K. Jain, *Nature Review. Clinical Oncology* **8** (2011).
228. B. T. Hawkins and T. P. Davis, *Pharm. Reviews* **57** (2005).
229. P. Calvo, B. Gouritin, H. Chacun, D. Desmaele, J. D'Angelo, J.-P. Noel, D. Georgin, E. Fattal, J. P. Andreux and P. Couvreur, *Pharmaceutical Research* **18** (2001).
230. S. I. van Kasteren, S. J. Campbell, S. Serres, D. C. Anthony, N. R. Sibson and B. G. Davis, *Proceedings of the National Academy of Sciences of the United States of America* **106** (2009).
231. J. Kreuter, D. Shamenkov, V. Petrov, P. Ramge, K. Cychutek, C. Koch-Brandt and R. Alyautdin, *Journal of Drug Targeting* **10** (2002).
232. I. P. Kaur, R. Bhandari, S. Bhandari and V. Kakkar, *J. Controlled Release: Official Journal of the Controlled Release Society* **127** (2008).
233. D. J. Begley, *Acta. Paediatr.* **92** (2007).
234. O. Veiseh, C. Sun, C. Fang, N. Bhattarai, J. Gunn, F. Kievit, K. Du, B. Pullar, D. Lee, R. G. Ellenbogen, J. Olson and M. Zhang, *Cancer Research* **69** (2009).
235. J. Huwyler, D. Wu and Pardridge, Proceedings of the National Academy of Sciences of the United States of America **93** (1996).
236. P. R. Lockman, J. M. Koziara, R. J. Mumper and D. D. Allen, *J.Drug Target.* **12** (2004).
237. C. Bosquillon, *J. Pharm. Sci.* **99** (2010).
238. D. A. Edwards, *Science* **276** (1997).
239. K. J. Polach, M. Matar, J. Rice, G. Slobodkin, J. Sparks, R. Congo, A. Rea-Ramsey, D. McClure, E. Brunhoeber, M. Krampert, A. Schuster, K. Jahn-Hofmann, M. John, H. P. Vornlocher, J. G. Fewell, K. Anwer and A. Geick, *Mol. Ther.: The Journal of the American Society of Gene Therapy* **20** (2012).
240. Z. Ma, Z. Mi, A. Wilson, S. Alber, P. D. Robbins, S. Watkins, B. Pitt and S. Li, *Gene Therapy* **9** (2002).
241. J. H. Senior, K. R. Trimble and R. Maskiewicz, *Biochimica et Biophysica Acta* **1070** (1991).
242. F. Sakurai, T. Nishioka, F. Yamashita, Y. Takakura and M. Hashida, *European Journal of Pharmaceutics and Biopharmaceutics: official journal of Arbeitsgemeinschaft fur Pharmazeutische Verfahrenstechnik* e.V, **52** (2001).

243. M. J. Hsu and R. L. Juliano, *Biochimica et Biophysica Acta* **720** (1982).
244. A. Raz, C. Bucana, W. E. Fogler, G. Poste and I. Fidler, *Cancer Research* **41** (1981).
245. M. F. Bachmann and G. T. Jennings, *Nature Reviews Immunology* **10** (2010).
246. C. Oussoren, J. Zuidema, D. J. A. Crommelin and G. Storm, *Biochim. et Biophys. Acta* **1328** (1997).
247. S. T. Reddy, A. Rehor, H. G. Schmoekel, J. A. Hubbell and M. A. Swartz,. *Controlled Release: Official Journal of the Controlled Release Society* **112** (2006).
248. M. A. Swartz, *Adv. Drug Deliv. Rev.* **50** (2001).
249. H. Pflicke and M. Sixt, *The Journal of Experimental Medicine* **206** (2009).
250. M. G. Harisinghani, J. Barentsz, P. F. Hahn, W. M. Deserno, S. Tabatabaei, C. H. Van de Kaa, J. De La Rosette and R. Weissleder, *New. Engl. J. Med.* **348** (2003).
251. C. Tassa, S. Y. Shaw and R. Weissleder, *Acc. Chem. Res.* **44** (2011).
252. M. C. Finkelstein, S. H. Kuhn, H. Schieren, G. Weissmann and S. Hoffstein, *Biochimica et Biophysica Acta* **671** (1981).
253. H. Cheng, C. J. Kastrup, R. Ramanathan, D. J. Siegwart, M. Ma, S. R. Bogatyrev, Q. Xu, K. A. Whitehead, R. Langer and D. G. Anderson, *ACS Nano* **4** (2010).
254. H. M. Dintzis, R. Z. Dintzis and B. Vogelstein, Proceedings of the National Academy of Sciences of the United States of America **73** (1976).
255. M. F. Bachmann and R. M. Zinkernagel, *Ann. Rev. Immunol* **15** (1997).
256. A. Jegerlehner, T. Storni, G. Lipowsky, M. Schmid, P. Pumpens and M. F. Bachmann, *Eur. J. Immunol* **32** (2002).
257. J. Park, Wrzesinski, E. Stern, M. Look, J. Criscione, R. Ragheb, S. M. Jay, S. L. Demento, A. Agawu, P. L. Limon, A. F. Ferrandino, D. Gonzalez, A. Habermann, R. A. Flavell and T. M. Fahmy, *Nature Materials* (2012).
258. New Drug Approval Process. Available at: http://www.drugs.com/fda-approval-process.html (Accessed on 15 December 2012).
259. The Fda's Drug Review Process: Ensuring Drugs Are Safe and Effective. http://www.fda.gov/drugs/resourcesforyou/consumers/ucm143534.htm.
260. Good Manufacturing Practice (GMP) Resources. http://www.ispe.org/gmp-resources (Accessed on 12/04/2012).
261. T. Lammers, S. Aime, W. E. Hennink, G. Storm and F. Kiessling, *Acc. Chem. Res.* **44** (2011).
262. M. K. Teli, S. Mutalik and G. K. Rajanikant, *Current Pharm. Design* **16** (2010).
263. S. D. Caruthers, S. A. Wickline and G. M. Lanza, *Curr. Opin. Biotechnol.* **18** (2007).
264. G. S. Ginsburg and H. F. Willard, *Translational Research: The Journal of Laboratory and Clinical Medicine* **154** (2009).
265. J. Xie, S. Lee and X. Chen, *Adv. Drug Deliv. Rev.* **62** (2010).
266. Liquidia Technologies. Available at: http://www.liquidia.com/ (Accessed on 16 January 2013).
267. S. Bhaskar, J. Hitt, S. W. Chang and J. Lahann, *Angew Chem. Int. Ed. Engl.* **48** (2009).
268. Y. Ke, L. L. Ong, W. M. Shih and P. Yin, *Science* **338** (2012).
269. S. M. Douglas, I. Bachelet and G. M. Church, *Science* **335** (2012).
270. Z. Yang, W. T. Huck, S. M. Clarke, A. R. Tajbakhsh and E. M. Terentjev, *Nature Materials* **4** (2005).
271. M. J. Hancock, F. Yanagawa, Y. H. Jang, J. He, N. N. Kachouie, H. Kaji and A. Khademhosseini, *Small* **8** (2012).

272. S. Huang and D. E. Ingber, *Cancer Cell* **8** (2005).
273. A. C. Shieh, Annals of Biomedical Engineering **39** (2011).
274. D. Discher, C. Dong, J. J. Fredberg, F. Guilak, D. Ingber, P. Janmey, R. D. Kamm, G. W. Schmid-Schonbein and S. Weinbaum, *Annals of Biomedical Engineering* **37** (2009).
275. D. E. Ingber, *Seminars in Cancer Biology* **18** (2008).
276. I. F. Tannock and D. Rotin, *Cancer Research* **49** (1989).
277. G. Ilangovan, H. Li, J. L. Zweier and P. Kuppusamy, *Mol. Cell Biochem.* **234–235** (2002).

Chapter B5

Biomimetic Systems in Nanomedicine

*Ana Maria Carmona-Ribeiro**

Biocolloids Lab, Departamento de Bioquímica,
Instituto de Química, Universidade de São Paulo
P.O. Box 26077, 05513-970 São Paulo, Brazil
**amcr@usp.br*

Abstract

Mimicking Nature has been a major source of inspiration for producing novel pharmaceutical formulations with improved therapeutic index and reduced toxicity. In Nature, the self-assembly of biomolecules driven by intermolecular interactions leads to highly functional, nanosized biological machines. Novel directions for drug and vaccine delivery to cure or prevent disease somehow have to copy Mother Nature wisdom. In this chapter, smart nanocarriers produced from the self-assembly of biocompatible polymers, lipids or hybrid nanoparticles (NPs) are presented and some of their applications in antimicrobial chemotherapy, drug and vaccine delivery discussed.

1. Self-Assembly and Intermolecular Interactions

Weak intermolecular interactions are important for the rational design of novel biomimetic materials. Main weak interactions are the van der Waals interaction (ion-dipole, dipole-dipole, dipole–induced dipole interactions, induced-dipole induced-dipole), the electrostatic interaction, the hydrogen bonds, the hydrophobic interactions and the steric repulsion.[1,2] The weak but cooperative hydrogen bonding between the two strands of double-stranded DNA is a wise approach used by Nature to keep the strands together without hampering replication and transcription.[3] The repulsion between nonpolar groups and water drives the hydrophobic effect essential for protein folding or formation of supramolecular lipid assemblies such as the bilayer membranes.[4] Whereas the energy needed to break a covalent carbon–carbon bond is 348.6 kJ/mol, only 4–29 kJ/mol are required to break a H-bond, depending on the atoms involved.[1]

The energy involved in the cohesive van der Waals intermolecular interaction between molecules in a liquid fairly corresponds to the energy required to increase the separation distance between the molecules in the liquid and change the physical state from liquid to gas.[1] The boiling point reflects this cohesive interaction and other weak interactions too. For example, the nonpolar ethane, CH_3–CH_3 has a boiling point of –89°C whereas methanol, CH_3–OH, which has a dipole moment, boils at 64°C. Upon increasing the molecular size, the importance of the dispersion forces increases. For example, ethane and n-hexane boil at –89 and 69°C, respectively.

For charge-charge or electrostatic (coulombic) interactions, the free energy of interaction (E) decays with the separation distance (d) as $1/d$. For charge-dipole interactions, E is proportional to $1/d^2$ (fixed dipole) and to $1/d^4$ (freely rotating dipole). For dipole–dipole interactions, E is proportional to $1/d^3$ (two fixed dipoles) or to $1/d^6$ (two freely rotating dipoles). For nonpolar–charge interactions, $E \alpha\ 1/d^4$. For nondipolar–dipole or nonpolar–nonpolar interactions, $E\ \alpha\ 1/d^6$. The covalent interaction is complicated and short ranged in its quantum mechanical nature and the hydrogen bond is also complicated, short-ranged, with E roughly proportional to $1/d^2$.

Water is very effective to reduce the electrostatic interaction between charged atoms or molecules whereas intervening media such as dodecane and other hydrocarbons act as insulators. The interaction between water and a small ion corresponds to the interaction ion–dipole. The ion causes a strong aligning effect on the surrounding dipolar water molecules and becomes "hydrated" forming the "hydrated ions". The first shell of water molecules around a strongly solvated ion is usually referred to as the primary hydration shell. This shell of oriented dipoles causes a second shell of less strongly oriented dipoles and the second causes a third and so on, so that a solvation zone is formed around the ion that is different from the bulk. When solvation zones of two solvated molecules or ions or surfaces overlap, solvation or structural forces arise that are repulsive, short-ranged and known as **hydration forces** if the intervening medium is water.

All atoms and molecules are polarizable. For a nonpolar molecule, the polarizability arises from the displacement of its negatively charged electron cloud relative to its positively charged nucleus under the effect of an external electric field. For a polar molecule, the total polarizability includes an additional term due to its permanent dipole moment. For neutral atoms and molecules, transient polarizabilities cause the induced-dipole induced-dipole interactions also known as **dispersion, London or electrodynamic interactions.**[5,6] The dispersion forces are always present playing a significant role in adhesion, surface tension, physical adsorption, etc. They are the most important contribution to the total van der Waals force. For the interaction between two methane molecules, which are completely neutral and do not bear any permanent dipole moment, the theoretical dispersion energy contribution to the total van der Waals interaction is 100%.[2] In contrast, for two water molecules that do bear permanent dipole moments, the dispersion energy contribution to the total is 24% with the major contribution being due to the orientation force. Water, however, is not the general case. In general, dispersion forces constitute the dominant term in the equation that describes the van der Waals interaction force between molecules.

Dipolar forces are generally small. The attractive or repulsive coulombic interaction has a longer range than most attractive van der Waals forces because the interaction energy varies with $1/d$ whereas interactions between freely rotating dipoles or nonpolar atoms or molecules decay as a function of $1/d^6$. This will be of importance for determining the total interaction energy also for macroscopic particles or surfaces.

Together with electrostatic, steric-polymer, and short-ranged **solvation forces**, the van der Waals forces are among the most important interaction forces acting between macroscopic particles and surfaces in liquids. The net interaction energy becomes proportional to the size of the particles, so that the energy can be much larger than kT even at separations of 100 nm or more.[7]

In liquids, unlike air or vacuum, other forces are usually present: long-ranged electric double layer forces, solvation and steric forces. Similar particles suspended in water or in any liquid of high dielectric constant are usually charged and this prevents their coalescence (coagulation). Because the van der Waals interaction between similar particles is always attractive, they would aggregate if the interaction force in between them were only of the van der Waals type. For charged particles or surfaces, the origin of the charges can be either dissociation of ionizable surface groups *e.g.* $-COOH \rightarrow -COO^- + H^+$, or specific binding of ions from solution onto a previously uncharged surface. In the case of an ionic crystalline solid, like silver iodide in water, its charge can be controlled by the activity (concentration) of Ag+ or I^- ions in solution. These ions are potential determining ions because their activity controls sign and magnitude of the electrostatic potential on the particle surface. Other examples of potential determining ions are H+ and OH− for surfaces of metal oxides (*e.g.* silica particles), polymer latex (*e.g.* carboxylate polystyrene microspheres) or charged bilayer vesicles (*e.g.* dihexadecyl phosphate synthetic vesicles).

On a charged particle or surface, the final surface charge is balanced by an equal but oppositely charged region of counterions, some of which are bound, usually transiently, to the surface within the Stern or Helmholtz layer, while others form an atmosphere of ions in rapid thermal motion close to the surface, the so-called diffuse electric double layer. There is accumulation of counter ions (oppositely charged with respect to the surface) at the surface and the depletion of co-ions for a planar negatively charged surface in the presence of a water solution of a monovalent salt. The counterion number density along the distance from the surfaces can be obtained from the Boltzmann distribution of ions i at a given distance x and from the Poisson equation that is a differential equation relating the net excess charge density at x with the electrostatic potential at x. The solution of the Poisson-Boltzmann equation gives the potential Ψ and thereby the counterion number density ρ at any distance from the surface.

The model for the diffuse double layer initially proposed by Stern[8] and refined by Grahame[9] assumes that all adsorbed counterions are confined to the Stern layer which defines the plane where the diffuse double layer begins: the so-called outer Helmholtz plane. The Stern potential at the outer Helmholtz plane is lower than the surface potential at the physical surface where surface charges are located. A third surface potential definition which is also useful is the zeta potential defined as the potential at the plane of shear for a moving charged particle. The surface potential at the shear plane is a good approximation of the Stern potential.[10]

During interactions of two similar colloidal particles or charged planar surfaces, their electrical double layers overlap so that a repulsive **electrostatic force** develops. The analytical procedures for finding the profile of the electrostatic interaction energy V are rather lengthy and complicated.[10–12] The problem can be solved numerically by using the computer facilities nowadays available. For example, we calculated numerically the electrostatic interaction energy as a function of the separation distance in between charged spherical vesicles or bilayer-covered polystyrene microspheres interacting in water solutions of monovalent electrolyte.[13–15] The calculation of total interaction energies between interacting flat surfaces or particles taking this energy as the sum of the repulsive double layer energy and the attractive van der Waals energy, is the basis of the DLVO theory.[11,16] This theory is successful for many interacting systems but fails for systems where other interaction forces are present such as the solvation forces resulting in structural repulsion at short distances (<1 nm). Also the hydrophobic effect may cause additional attractive forces for which formal treatments are not available.[17] In a single interacting system, several different interaction forces may be contributing to the total interaction force directly measured using a surface force apparatus[18,19] or the osmotic stress technique by Parsegian, Rand and co-workers.[20] This is the case for charged phospholipid bilayers interacting across water solutions for which the osmotic stress technique resulted in measurements in fair agreement with the DLVO theory for $D > 1$ nm, and extrarepulsive forces due to hydration of the phospholipids polar heads for $D < 1$ nm.[21] Another good example of the multiplicity of forces acting in a single system is the case of two dihexadecyl phosphate monolayers adsorbed on hydrophobized mica, interacting across water solutions at different pH and monovalent salt concentrations.[22] At low pH values, an attractive force of unknown origin appears in addition to DLVO forces. At high pH values, adsorption of hydrated sodium ions is the more probable explanation for the repulsive force appearing at short ranges. Direct measurements of the **hydration forces** using the surface force apparatus can define decay lengths, range and force-distance profiles which are be important to ascribe to these forces a real dependence on solvent structure.[17] For uncharged amphiphilic structures such as bilayers of lecithins, phosphatidylethanolamines, dioctadecyldimethylammonium bromide (DODAB), sodium dihexadecylphosphate (DHP) and the oxyethylene surfactants, direct measurements of interaction forces were performed either using the surface force apparatus for layers adsorbed on mica[22,23] or using the osmotic stress/X-ray diffraction method developed by Rand, Parsegian and co-workers.[21,24,25] The principles behind direct surface force measurements performed using these techniques are as follows. Whereas the surface force apparatus measures interactions between amphiphiles adsorbed or deposited on two bare or hydrophobised mica sheets, the Rand–Parsegian method measures interactions between free lipid multibilayers. The first method measures forces from the elastic deflexion of a spring, which is depicted from interferometric patterns formed by the light crossing a sequence of thin, optically transparent films. The second method measures forces from the osmotic pressure applied externally to the interlamellar spacings after this pressure equilibrates the interbilayer repulsive forces. Distances are determined from X-ray diffraction patterns and repeating lamellar distances at each imposed osmotic stress. Both methods yield similar results at least for interaction forces between lipidic layers. Uncharged bilayers composed of lecithins, phosphatidylethanolamines, sugar head-group glycolipids, and the oxyethylene surfactants exhibit limited swelling in water with some short-range hydration repulsion arising from their hydrophilic head-groups. Also the force-distance profiles are expected to include an attractive van der Waals component. Furthermore, the fluid nature of these structures causes local fluctuations of thickness and head-group area[26] so that this thermal motion of relatively lengthy head-groups could generate a kind of "steric" force, which would exhibit a force-distance profile[24,26] similar in shape to the one obtained for polymeric polyethylene oxides.[27,28] The difference between the profiles regards solely the range: polymers, which are macromolecules of very high molecular weight, display their force curve up to 80–120 nm for molecular weights 160,000. For phospholipids the range goes only up to 2–3 nm. Shapes for the curves obtained with different neutral phospholipids are very similar.[2] For more fluid bilayers in the liquid-crystalline state the range for the repulsive steric force is slightly larger than the range displayed by more rigid bilayers in the gel state. Whether the "hydration" repulsion measured for amphiphiles such as lipids is due to thermal motion of polar heads or to solvent (water) structuring around the polar heads or to both is still an open question.

The **hydrophobic effect** is entropic in nature and comes from the immiscibility of inert substances with water.[4] Experiments for silica and mica surfaces rendered hydrophobic by chemical methylation or plasma etching revealed that the force is long ranged, decaying exponentially for the first 10 nm and then more gradually.[29–31] For uncharged phospholipid bilayers on mica, the van der Waals component to describes the interbilayer interaction as long as the polar, hydrophilic headgroups shield the hydrocarbon chains from water.[23] By stressing the bilayers to expose more of their hydrocarbonic moieties to water the attraction acquires characteristics of the hydrophobic interaction: longer ranges and larger attractive forces.[32] Under osmotic or mechanical stress, lipid bilayers may expose hydrophobic regions to water and fuse due to the hydrophobic interaction. Some cationic amphiphile bilayer vesicles composed of dioctadecyldimethylammonium halides may undergo fusion simply by increasing monovalent salt concentration in their medium.[33–35] Figure 1 illustrates the repulsive interaction between two similar particles properly stabilized by a hydrophilic polymer.

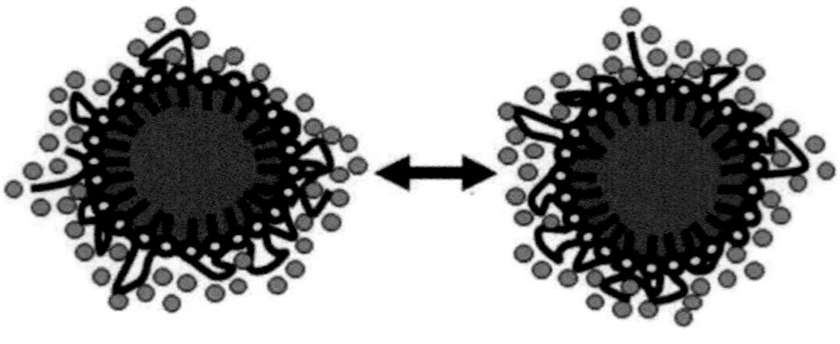

Fig. 1. A schematic illustration of the hydration repulsion between two nanoparticles (NPs) promoting their colloidal stabilization in water solution.

2. Nanomedicines

Molecular self-assembly is ubiquitous in living organisms and central for the formation of a wide variety of complex biological structures. This spontaneous association of molecules under equilibrium conditions leads to stable, structurally well-defined aggregates joined by noncovalent bonds.[36] The cell membranes are a good example of complex entities based on the self-assembly of a variety of lipids, proteins and other biomolecules. Nucleosomes, the repeating units of chromatin, exhibit 10 nm diameter beadlike particles allowing DNA fibers to wound around them. The beads are composed of a core of basic proteins, the histones. Thus, a chromatin fiber is considered to be a flexibly jointed chain of nucleosomes, rather like beads on a string. Biologically produced organic–inorganic composites such as bone, teeth and sea shells are formed through template-assisted self-assembly where self-assembled lipids, or proteins, or both form the scaffolding for the deposition of inorganic material.[37] A common feature for the self-assembly in living organisms is their **hybrid** nature. Self assembly gathers organic and inorganic, soft and hard, polymers and lipids, lipids and proteins, polysaccharides and proteins, etc. to produce the functional bionanomachines of the living organisms.

Despite being in its infancy, the science of nanobiotechnology already produced some man-made supramolecular assemblies with therapeutic applications and small size (up to 100 nm mean diameter), which can be named **nanomedicines.**

Nanobiotechnology has the potential to uncover the structure-function relationship of biosystems at the nanoscale level providing effective tools to treat diseases at a molecular scale and to rationalize delivery of drugs and genes to their targets following systemic administration.[38] In pharmaceutical product development, vehicles include the **polymer-based NPs**, the **lipid-based NPs** (liposomes, nanoemulsions, solid lipid nanoparticles (SLN), bilayer fragments (BF) or disks), and the **hybrid NPs or biomimetic particles**. In the recent years, much research has gone into the characterisation of novel carriers and their biological effects and potential applications. These include molecular self-assembly, biological effects of naked NPs and nanosafety, drug encapsulation and nanotherapeutics, and novel NPs for use in microscopy, imaging and diagnostics.[39] Successful therapeutic supramolecular assemblies are able to reach specific targets and overcome physiological barriers such as endothelial pores and heterogeneous blood supply and tissue architecture that usually occur at the sites of disease.[40]

Carriers have to be tailored for targeting the site of drug action and a multidisciplinary approach is required to achieve this major task. Liposomes,

SLN, polymeric "micelles", organic, inorganic, metallic, biological or biomimetic particles have been used in drug delivery to cells as well as organelle targets. They exploit biological pathways to achieve payload delivery to cellular and intracellular targets and can sometimes overcome the blood–brain barrier (BBB).[41–43] Therefore, they are potentially useful in the treatment of cancers and diseases of the central nervous system, such as neurodegenerative diseases, glioblastoma multiform and neurovascular disorders.

Colloidal nanocarriers, their formulations and applications were comprehensively reviewed recently.[44] Major carriers were classified among the following carrier types: the polymeric particles,[45–50] magnetic particles,[51–53] metal and inorganic particles,[54–57] quantum dots,[58–60] polymer "micelles",[61–64] SLN,[65–72] phospholipid-based micelles[73–75] and liposomes.[76–79] In addition to these carriers, this chapter includes the **hybrid nanostructures** called **biomimetic particles**, which combine **particles and lipids**[80–90] or **synthetic polymeric particles and polysaccharides**[91–93] or **particles, lipids and other biomolecules**[94] such as antigens,[95] receptors[96] or DNA.[97]

3. Nanomedicines Based on Biocompatible Polymers

The NP have unique functional features such as small and controllable particle size, high stability, lower toxicity, tuneable hydrophilic-hydrophobic balance and the possibility of including surface features for target specific localization, steric stabilization, ligand attachment, etc.

Polymeric nanoparticles (PNP) constitute a versatile drug delivery system,[98] which can potentially overcome physiological barriers, and carry the drug to specific cells or intracellular compartments by passive or ligand-mediated targeting approaches.[99] Certain polymers allow sustained drug levels for longer periods of time. Several biocompatible and biodegradable PNP made of biopolymers (natural polymers) such as proteins or polysaccharides have been tried for drug delivery and controlled drug release. More recently, also synthetic biocompatible polymers have been focused as polycationic NPs for encapsulation and controlled release of amphotericin B (AB),[100] encapsulation of curcumin[101] or doxorubicin (DOX) for human cancer therapy.[102] Stable particles of insoluble drugs such as AB with very high drug content can be easily and reproducibly prepared through the application of the layer-by-layer (LbL) technology based on alternate adsorption of oppositely charged polyelectrolytes (PEs) on the surface of drug particles (crystals).[100,103] Such polymeric coatings avoid drug NP aggregation and create a firm polymeric shell on their surface. Drug release rate from such nanocarriers can be easily controlled by assembling multilayer shells with variable shell density and thickness. Methods of preparation of PNP fall into two major classes: one deals with the polymerization of monomers (*e.g.* emulsion and dispersion polymerization), whereas the other essentially involves dispersion of preformed polymers (*e.g.* salting out, emulsification-diffusion, and nanoprecipitation).[104] A high entrapment efficiency in PNPs can be achieved by incorporation of drug during their preparation rather than adsorption on preformed NPs.[105] Drug release takes place through their simultaneous biodegradation, followed by desorption, diffusion, or erosion.[45] The PNP exhibit different structures such as water-containing nanocapsules, oil-containing nanocapsules, nanospheres, PEGylated stealth NPs, PEGylated and targeted NPs, polysaccharide-coated non-stealth NPs.[106] Most pharmaceutical agents have primary targets within cells. Therefore, effective formulations of these agents should reach sites of action within the cell. Selective subcellular delivery is likely to have greater therapeutic benefits.[107] Cytosolic delivery, for instance, is desirable for drugs that undergo extensive exportation from the cell via efflux transporters such as multidrug resistance proteins and P-glycoproteins. These efflux mechanisms continuously reduce therapeutic intracellular drug concentrations.

Multidrug resistance (MDR), which is related to cancer chemotherapy, tumor stem cells, and tumor metastasis, is a huge obstacle for the effective cancer therapy. The MDR cells usually overexpress a P-glycoprotein (P-gp) which increases the efflux of anticancer drugs. Carriers loaded with drugs have been functionalized with the antibody against P-gp (anti-P-gp) and successfully killed cancer cells.

Biodegradable PNP have been considered excellent vehicles for site directed *in-vivo* delivery of drugs and vaccines.[108] Figure 2 presents the chemical structures of major biodegradable polymers which include polysaccharides, proteins or even synthetic biocompatible polyesters often used for *in-vivo* delivery of vaccines, genes, drugs and other biomolecules in the body.

A common feature of most natural polymers is their hydrophilicity so that the specific incorporation of hydrophobic drugs may not be easily achieved in the biodegradable hydrophilic NPs made of these polymers. Tamoxifen or taxol (TAM) were incorporated in NPs of poly-ε-caprolactone or poly-alkyl-cyanoacrylates[48,109,110] since these polymers bear more hydrophobic moieties in their chemical structure. Other polymers used to load anticancer drugs followed the same trend such as poly-isohexylcyanoacrylate NPs loaded with DOX.[111] On the other hand, for the incorporation of protein C antigens, genes or oligonucleotides, the hydrophilic polymeric matrixes have been very successful.[112–116]

Major specific applications of biocompatible NPs have been in tumor targeting using monoclonal antibodies or folate moieties covalently bound to

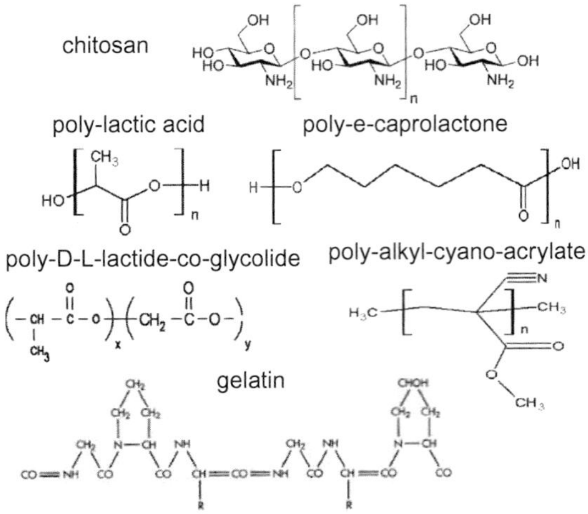

Fig. 2. Chemical structures of some biocompatible polymers.

the NP,[117,118] oral drug delivery exploring the protective effect of NPs against degradation by digestive enzymes and their systemic dissemination,[116,119,120] vaccine and gene delivery due to NPs rapid uptake by cells followed by NPs escape from degradation in the endosomes and cargo release to the cytoplasm[121,122] and drug delivery to the brain due to the ability of certain NPs to cross the BBB.[123–126] Drugs that are normally unable to cross the BBB following *i.v.* injection can be transported across this barrier by binding to poly(butyl cyanoacrylate) NPs coated with polysorbate 80.[41,42] Only drug-loaded NPs coated with polysorbate 80 and/or with apolipoprotein B or E were effective.[42,125] Polysorbate 80-coated NPs apparently adsorbed these apolipoproteins from the blood after injection and thus mimicked lipoprotein particles that could be taken up by the brain capillary endothelial cells via receptor-mediated endocytosis. Bound drugs then may be further transported into the brain by diffusion following release within the endothelial cells or, alternatively, by transcytosis.[41,42]

The complexation of macroions by oppositely charged PEs is a fundamental process in biological systems and many technical applications and can be used to produce therapeutic supramolecular assemblies carrying the therapeutic agent. Polysaccharide NPs formed by complexation of cationic chitosan and anionic alginate have been employed to carry therapeutic peptides and proteins.[127] The complexes are biocompatible, biodegradable, exhibit low toxicity, and adhere to mucosae and other biological structures. The NPs intended for use in the transmucosal delivery of macromolecules were prepared by the ionic gelation of chitosan hydrochloride with pentasodium tripolyphosphate and concomitant complexation with sodium alginate and clearly enhanced the systemic absorption of insulin after nasal administration to rabbits.[127] Chitosan alone has also being useful for mucosal and transmucosal delivery of biomolecules.[128]

Wang and Singh comprehensively reviewed the literature on the selection of optimal adjuvants for vaccines.[129] Because of the limited adjuvant effect of aluminum salts which induce only a Th2 response,[130] other nonaluminum adjuvants (often not approved by the FDA in the U.S.) have been searched and tried extensively over the last 20 years.[131,132] Among these, polymeric particles made of biocompatible polymers such as polylactic acid (PLA), poly(lactide-co-glycolide) acid (PLGA),[133] chitosan,[134] poly(γ-glutamic acid) (γ-PGA),[135] and starch.[136,137] PLGA–CTAB (cetyl trimethylammonium bromide) particles of 0.3 μm were more effective than larger particles (1 and 30 μm) in the elicitation of transgene-specific serum IgG responses.[138] The greater effect with smaller particles could be related to two factors — higher adsorption capacity of smaller particles and greater efficiency of particle uptake by dendritic cells (DCs). It has been shown that the antigens adsorbed on aluminum particles of 3 μm are internalized more effectively than those of 17 μm.[139] Charged NPs could present antigens effectively through electrostatic interactions with antigens. Wang and Singh interpreted the lack of approval of novel and more efficient adjuvants to the basic dogma in vaccinology: a single adjuvant should be used unless it is not effective, that is, keep it simple and stupid.[129] In the forthcoming years, this dogma will have to give way to the hybrid biomimetic adjuvants which are already revealing their superior performance as discussed in the last section of this chapter. For vaccines, targeting means delivering the antigen to the lymph nodes, antigen-presenting cells (APC) or B-cells, and within cells lysosome or cytoplasm.[140] Excellent preclinical results have been achieved with passive targeting, using cationic particles, which interact strongly with negative cell surfaces.[75,95,141,142] The composition of these cationic particles was based on PLGA cationized microparticles with sizes around 1–7 μm of mean diameter[141–143] or on polystyrene sulfate NPs covered by a cationic bilayer[80] which can be obtained over a range of nanometric particle sizes by emulsion polymerization (0.010–1 μm of mean diameter) followed by cationic bilayer deposition and ideally represent an optimal and homodisperse particulate dispersion to evaluate systematically the effect of particle size on the immune response.[82,83]

4. Nanomedicines Based on Lipids

In 1983, the **liposomes** upbringing to the scientific world that occurred in 1965 was enthusiastically documented in the book entitled "Liposomes Letters" edited by Alec Bangham, the liposomes father.[144] Liposomes or vesicles composed of natural lipids are biocompatible, biodegradable, and nonimmunogenic.[145] These properties should make phospholipid liposomes suitable candidates for drug delivery. A low *in vivo* stability, however, was a major challenge solved with the introduction of the sterically stabilized liposomes formed by covalent attachment of poly(ethyleneglycol) (PEG) polymer chains to the phospholipids polar heads.[146] Also liposomes have found applications as immunoadjuvants in vaccines,[147,148] enhancers or carriers in medical diagnostics[149–151] and transfection vectors in genetic engineering.[152–154] In the cosmetics industry, liposomes represent an improvement offering an aqueous and biocompatible matrix to hydrophobic substances in flagrant contrast to gels and creams that employ skin-irritating oils, alcohols and tensioactives.[155–157] Methods to prepare and characterise bilayer vesicles or liposomes were comprehensively presented in the book edited by Roger New in 1989.[158] In general, amphiphilic molecules form assemblies of the bilayer type if they have an approximately cylindrical molecular shape yielding a geometric parameter between 0.5 and 1.0.[159] Not only natural phospholipids form bilayers. Structural and functional aspects of biological membranes have also been reproduced in liposomes and vesicles prepared from a variety of synthetic amphiphiles.[160–162] Vesicles formation is not restricted to lipids but also is the preferential supramolecular assembly for synthetic amphiphiles such as dialkyldimethylammonium bromide or chloride,[163,164] DHP[165] and many others molecules.[166–168] The chemical stability of liposomes is low because acid/base catalyzed hydrolysis may pinch off one or both hydrocarbon chains from the backbone of the lipid[169,170] or oxidation may form cyclic peroxides at adjacent double bonds of the hydrocarbon chains resulting ultimately in the breakage of chains via lipoperoxidation.[171] Hydrolysis rate of soybean lecithin in liposomes is a function of pH and temperature being at highest at extreme pH values where acid-base catalysis is enhanced and/or at the highest temperatures tested.[172] Oxidation may be prevented by using saturated lipids or by adding antioxidants such as vitamin E or butylated hydroxytoluene.[173] Synthetic amphiphiles that form bilayers certainly can be made chemically more stable than natural lipids. The **synthetic amphiphile vesicles** however, in contrast to natural lipids, may show reduced colloidal stability under physiological conditions (150 mM monovalent salt and pH 7.4). Their stability in the biological milieu has been poorly investigated. Furthermore, cytotoxicity for some synthetic amphiphiles such as the cationic lipids in general and the DODAB lipid in particular is high. This isolated property is finding useful applications in the design of liposomal antimicrobials where the liposomes themselves are the active antimicrobial agents.[174–178] In spite of its toxicity, the capability of DODAB bilayers to induce retarded hypersensibility led to its recognition as an efficient immunoadjuvant able to stimulate cellular immune responses.[94,179–183] Large DODAB vesicles have been explored as adjuvants for the development of DNA[184] or protein vaccines.[185–187]

Lipid BF or nanodisks are lipidic nanostructures with 5–90 nm mean hydrodynamic diameter. Some of their uses as drug carriers, coatings for drug granules or antigen presenters are illustrated in Fig. 3.[188] BF preparation, characterization and biomedical applications have been often reviewed in the literature over the last 5 years.[189–194]

Applications for cationic BF include the incorporation and delivery of drugs,[195–197] incorporation of antigens,[95] oligonucleotides,[198,199] peptides,[200] polysaccharides,[201–202] coverage of drug particles,[196,203] and activity against pathogenic microorganisms.[174–178,204] Charged BF can be stabilized in aqueous solution as nanostructures both electrostatically by using a dispersion medium prepared at a low ionic strength (usually at 1 mM NaCl) and/or sterically by interaction with hydrophilic polymers[100,205] or PEGylated lipids.[206–208]

The colloidal stability of BF in aqueous dispersion has been achieved in compositions as BF/proteins, BF/micelle-forming compounds, BF/hydrophilic polymers or BF/polyethylene-glycol lipids or, simply, by using charged BF at low ionic strength. For example, high-density lipoproteins (HDL) and

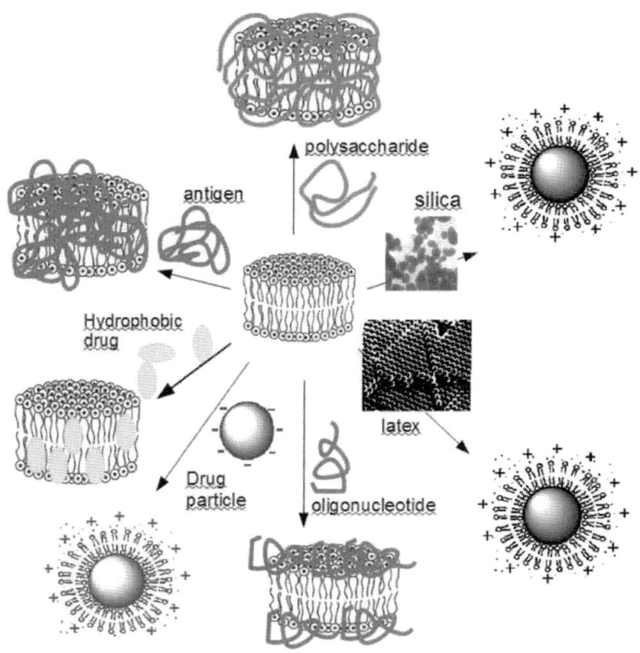

Fig. 3. The multiple uses for some cationic nanosized BF prepared from DODAB.

human apolipoprotein A-I (apoA-I,243 amino acids) have been intensively studied over the past 40 years because HDL is anti-atherogenic and apoA-I is the major protein in HDL.[209–213] The incubation of dimyristoyllecithin (DML) or dipalmitoyllecithin (DPL) multilamellar liposomes with high density lipoproteins over a range of lipid to protein molar ratios produced discoidal complexes in equilibrium with free lipid or free protein as revealed by negative staining electron microscopy and differential scanning calorimetry.[214] Stacked bilayer disks were also observed for HDL lipoproteins.[215] A 30-residue synthetic peptide mimicked the serum lipoproteins and was incorporated in DMPC bilayers at neutral pH undergoing changes in spectroscopic properties that revealed that the peptide acquired a alpha helical structure in the presence of DMPC.[216] The discoidal nanostructure of the dimyristoylphosphatidylcholine (DMPC)–GALA 19/1 (molar ratio) complex was isolated by gel-permeation chromatography and characterized revealing the approximate dimensions of 4.4 nm edge thickness and a 17 to 35 nm diameter.[216] The apoA-1 lipid-binding domain was used as a template for the synthesis of amphipathic helical proteins termed membrane scaffold proteins, employed to self-assemble monodisperse discoidal particles called Nanodiscs.[217,218] The membrane scaffold proteins surrounded the lipid bilayer fragment in a belt-like fashion forming bilayer disks of discrete size and composition. The incorporation of membrane proteins in the BF allowed the determination of single-molecule height by atomic force microscopy (AFM).[219,220] The self-assembly process starts with a mixture of the phospholipids and scaffold proteins in the presence of a detergent. Upon removal of the detergent by dialysis, 10 nm diameter particles were formed.[218] Cytochrome P450 and its reductase became functional upon orientation in BF.[219,220] Upon interaction with dipalmitoylphosphatidylcholine (DPPC) dispersions, the protein SP-B, which associates with pulmonary surfactant, was shown to have little effect on the acyl chains of DPPC by means of magnetic resonance spectroscopy.[221] By means of circular dichroism spectroscopy, the usual alpha-helical content for this protein was shown to remain unaffected by the packing state of the DPPC acyl chains, suggesting the protein interaction with the lipid head groups and/or with the edges of the bilayer disks.[221] The lysis of L-alpha-DML vesicles by cetyltrimethylammonium bromide (CTAB), a micelle-forming amphiphile, also resulted in the formation of BF which were prone to aggregate into large structures depending on the ionic strength of the intervening medium.[222] Aggregation was slow at low ionic strength and practically absent at ionic strength values <0.03, due to stabilizing effect of the electrostatic repulsion between the BF.[222] The phase transition temperatures of pure phospholipid vesicles and a mixture 1:1 of phospholipid and CTAB were similar and equal to 24 and 28°C, respectively, suggesting the bilayer character of the phospholipid-CTAB dispersions.[222] Structural defects in lipid bilayer vesicles would be fault-dislocations in the bilayer structure that would appear when small BF reassemble to form the small bilayer vesicles during the sonication procedure.[223] Cryogenic transmission electron microscopy (cryo-TEM) images of lysolipid-containing thermosensitive liposomes (LTSL) revealed that open liposomes and bilayer disks appeared when liposomes were cycled through the gel to liquid-crystalline phase transition.[224] The amount of bilayer disks depended on the presence of PEGylated lipid and lysolipid in the bilayer with the micelle-forming lysolipid stabilizing the rim of bilayer openings and membrane disks that form when liposomes are cycled through the mean phase transition temperature.[224] The LTSL released their contents instantly when heated to temperatures close to their gel to liquid-crystalline phase transition temperature due to formation of open liposomes, bilayer disks, and pore-like defects.[225] Surfactant stabilized bilayer disks were observed in many instances as products of the interaction between lipid bilayers and neutral (nonionic or zwitterionic) surfactants such as polyethylene glycol n-dodecyl monoethers,[226,227] Triton X-100,[228] octylglucoside,[227,229] dodecylbetaine,[230] and N-dodecyl-N,N-dimethylamine N-oxide (DDAO).[231] In general, a surfactant-mediated increase in membrane permeability occurs below the critical micelle concentration. Above a certain critical surfactant/lipid molar ratio, size and morphology of the lipid vesicles change. The vesicle growth observed at surfactant concentrations just below those inducing bilayer solubilization is explained by fusion of surfactant stabilized bilayer disks.[226] The vesicle-micelle transition of phosphatidylcholine (PC) bilayers in the presence of Triton X-100 studied by static and dynamic light-scattering, transmission electron microscopy (TEM) and permeability changes revealed three steps.[228] In a first step, a sharp initial increase in the surfactant partition coefficient (K) in the bilayer takes place until a maximum value corresponding to bilayer saturation is attained for 50% permeability changes. In a second step, further surfactant additions diminish K until 100% increase in permeability. In a third step, additional amounts of surfactant increase K until bilayer solubilization. The TEM pictures for 50–64 mol% micelle-forming surfactant revealed tubular structures together with open BF.[228] A similar behavior was also described for the structural phase transitions involved in the interaction of phospholipid bilayers with octyl glucoside (OG).[227] The initial steps of bilayer saturation (50–70% permeability) were attained for a lower free surfactant (Sw) than that for its critical micellar concentration (CMC). When Sw reached the glucoside CMC, solubilization started to occur. The TEM pictures for 100% permeability (72 mol/100 mol) still showed unilamellar vesicles.[227] Johnsson and Edwards[227] studied the interactions between sterically stabilized PC liposomes, (EPC/PEG-lipid) liposomes and the nonionic surfactants OG and octa(ethylene glycol) n-dodecyl monoether (C12E8) by means of cryo-TEM. Poly(ethylene glycol)-lipids (PEG-lipids) affected the OG-mediated solubilization of the liposomes. Long-lived small bilayer disks were observed at OG concentrations saturating the bilayers.[227] Mixtures of the amphoteric surfactant DDAO and 1,2-dioleoylphosphatidyl choline (DOPC) in chloroform/methanol were evaporated, dried and hydrated in excess deuterated water.[231] After extrusion through polycarbonate membranes (pores with 50 nm diameter), they were studied by means of small-angle neutron scattering. DDAO destabilizes the bilayer in unilamellar liposomes and solubilizes it into mixed micelles whose shape changes with the DDAO:DOPC molar ratio (r). Bilayers or/and BF have been observed up to $r = 1.5$, cylindric micelles for $2.5 < r < 3.5$, and globular micelles for $r > 4$. In bilayers or/and BF, DDAO modulates the thickness of the bilayer.[231] These bicellar systems[232] have recently been shown to be specially useful for delivery of hydrophilic or hydrophobic drugs to the skin since they retard the drug delivery, a property potentially interesting for the cosmetic industry.[233]

For bile salts, vesicle-micelle transitions are important phenomena during bile formation and intestinal lipid processing. Van Erpecum and coworkers studied detergent-induced vesicle to micelle transitions in PC and sphingomyelin (SM) vesicles by cryo-TEM looking at cholesterol effects on the detergent-induced phase transitions.[234] Equimolar (4 mM) taurocholate added to SM-PC vesicles induces first the formation of open vesicles, then fusion of BF into 200 nm diameter large open structures coexisting with some multilamellar or fused vesicles and thread-like micelles and, finally, transformation into long thread-like micelles. Cholesterol in the SM/PC bilayer changes initial vesicular shape from spherical into ellipsoid and increases detergent resistance.[234]

The phospholipase A2 (PLA2) hydrolysis of phospholipid vesicles was studied by cryo-TEM and high-performance liquid chromatography (HPLC).[235] The hydrolysis product (free fatty acid and 1-lyso-phospholipid) combined with non degraded phospholipids yielded perforated vesicles, BF, and micelles.[235]

It was often observed that the bicelles promote orientation of the bilayer normals parallel to the magnetic field so that studies of isotopically labeled peptide and protein structure in a lipid environment were possible from nuclear magnetic resonance (NMR).[236,237] ^{13}C- and ^{15}N-labeled peptides were synthesized with and without an N-myristoyl anchor.[237] Both myristoylated and non-myristoylated peptides are alfa-helical in a lipid environment and N-myristoylation does not greatly influence the structure of the peptides. The association of the peptides with bilayer disks is through a combination of edge and surface interactions. The presence of the N-linked myristoyl chain is not essential for assembling the peptides with the bilayer disks: highly conserved N-terminal hydrophobic residues play the same role.[237] Also for protein structure studies by NMR, long chain phospholipids such as 1,2-distearoyl-sn-glycero-3-phosphocholine (DSPC) and short chain dihexanoylphosphatidylcholine (DHPC) were assembled into bilayer disks since the short chain phospholipid DHPC breaks up the extended bilayers into bilayered micelles or bicelles that are highly hydrated.[238] Mixtures of DMPC and DHPC which form disks in water were reinvestigated by phosphorus NMR to obtain lipid distribution in the nanostructures.[239] Structural transformations of DMPC/DHPC mixtures inferred from broad band ^{31}P-NMR spectroscopy between 5 and 60°C were described: above a critical temperature, Tv, perforated vesicles progressively replaced alignable structures. There is an increase in DHPC miscibility with the bilayer domain above the DMPC transition temperature producing a "mixed bicelle."[239] The spontaneous alignment in a magnetic field of these bilayers is an attractive property that has been fascinating several laboratories investigating solubilization and reconstitution of membrane and membrane proteins.

In the lung, a complex mixture of lipids, surfactants and proteins synthesized is secreted by pneumocytes cells which reduces the surface tension at the air-liquid alveoli interface.[240,241] During the process of collapse of these alveoli films, individual disks appeared floating in the air just above the interface with a broad range of sizes.[241] Upon compression and expansion, lipids flow continuously from the monolayer into the disks, and vice versa.[241]

In summary, the bilayer disks and fragments were often observed as intermediates in the lipid bilayer dissolution process caused by surfactants and in this case they have been named bicelles since they are bilayered micelles.

Sterically stabilized liposomes prepared from phospholipids with covalently attached PEG lipids are commonly used in the preparation of long circulating liposomes. In contrast to conventional liposomes, the polymer cloud surrounding the liposome provides a steric barrier and thereby effectively stabilizes the liposomes against opsonization *in vivo*. PEG lipids affected the structure of small unilamellar liposomes inducing the formation of open bilayer disks at concentrations well below those where mixed micelles begin to form.[242] The effect of five poly(oxyethylene)–poly(oxypropylene)–poly(oxyethylene) (PEO–PPO–PEO) Pluronic copolymers on PC liposomes of varying composition was also evaluated.[243] Low amounts of Pluronics with long PEO blocks (0.2 mol%) induces the formation of bilayer disks, whereas those with short PEO blocks reduce the liposome size.[243] Cholesterol in the liposomal preparations reduces both the incorporation of copolymers in the lipid bilayer and the morphological changes. Regarding the liposome permeability, there is leakage of carboxyfluorescein caused by all investigated copolymers.[243] The effect of increasing the concentration of C12E8 on DODAB and dioctadecyldimethylammonium chloride (DODAC vesicles seen by dynamic light scattering and cryo-TEM is vesicle swelling due to incorporation of the surfactant into the vesicle bilayer followed by the coexistence of micelles and extended BF.[244] At high drug-to-lipid ratios for encapsulating the drug phenantridine within unilamellar sterically stabilized liposomes of DSPC:Cho:DSPE–PEG 55:40:5 (mol%) using transmembrane pH gradients, drug precipitation in the liposomal aqueous core and formation of small BF occurred due to precipitate growth inside the liposomes.[245] Flat bilayer disks with long-term stability were prepared from DSPC, cholesterol, and 1,2-distearoyl-sn-glycero-3-phosphoethanolamine-N-[methoxy(polyethylene glycol)-5000] [DSPE–PEG(5000)]. By varying the content of the pegylated lipid, disks exhibit sizes ranging from 15 up to 60 nm.[246] The potential for drug partition studies of these sterically stabilized bilayer disks was explored and compared to unilamellar and multilamellar liposomes in a series of drug partition studies.[246] Liposomes or disks were immobilized in chromatographic gel beads and the determination of partitioning for a large set of drugs lead to the conclusion that the sterically stabilized disks can also be loaded with a variety of sparingly soluble drugs.[246]

The DHP and DODAB or DODAC are synthetic lipids which form large vesicles[247,248] or open BF or disks[249,250] depending on the preparation method. Because these bilayers are charged and in the rigid gel state, the BF they form upon sonication are colloidally stable and do not aggregate or fuse at low ionic strength (less than 5 mM monovalent salt). DHP dispersions prepared by sonication with tip (DHP BF) were compared to DHP dispersions obtained by injection in water solution of a DHP chloroformic solution above the bilayer phase transition and above the chloroform boiling point (DHP large vesicles or DHP LV).[249] DHP LV presented more fluid bilayers than DHP BF.[249] From fluorescent probes in the DHP bilayer and visualization by TEM, increasing sonication time or decreasing pH, increased the number of BF in the DHP dispersion.[249] This was the first time that DHP BF obtained by sonication with tip were visualized and images correctly interpreted as corresponding to BF. The mean phase transition temperature (T_m) for the more tightly packed bilayers of DHP BF is higher than the one for large DHP vesicles.[249] Similar findings were reported upon comparison of the phase transition for bilayer disks and vesicles composed of the neutral lipids

DMPC or DPPC.[217] For DODAC, light-scattering experiments pointed out that most of the DODAC molecules are organized in small open structures: flat disk-like bilayers formed by vesicle disruption during the sonication process.[250] Upon annealing or after addition of $Fe(CN)_6^{4-}$, fragments gather to form a population of closed vesicles with a narrow size distribution.[250] Cryo-TEM pictures confirmed the existence of BF for dispersions of DODAC or DHP synthetic lipids obtained by sonication.[251,252] Other revealing property of DODAB or DHP sonicated dispersions is their lack of osmotic response (shrinkage or swelling) upon application of a osmotic gradient, in contrast to the osmometer behavior described for DODAB or DHP LV.[247,248]

Dimyristoylphosphatidylglycerol (DMPG) hydration and storage at ca. 4°C results in a disintegration of bilayers into fragments of different sizes and bicelles with a diameter of ca. 50 nm with a characteristic phase transition at 40°C.[253] Other melting transitions were also obtained and the occurrence of a partial chain interdigitation at the rim of the bicelles was assumed at about 1.4°C above the main transition temperature for the non-interdigitated bilayer phase.[253]

The solubilization of AB by DODAB or DHP BF was followed by means of dynamic light scattering and optical spectroscopy.[195] The very low solubility of AB in water permits the determination of size distributions for AB aggregates in water and in the presence of DODAB or DHP nanosized BF.[195] Large drug aggregates simply disappear upon incubation with both BF types. Light absorption spectra for AB in a poor solvent (water), in a good organic solvent (dimethylsulfoxide:methanol, 1:1), and in different lipid dispersions also show that solubilization strictly depends on the presence of BF.[195] AB is poorly soluble in dispersions formed of closed vesicles of DODAB, DHP, PC, or asolecithin.[195] Its solubilization in BF can be understood from the AB chemical structure and an increased hydrophobicity at the borders of BF. These hydrophobic borders interact with the polyenic moiety of AB leaving the hydroxylated moiety free to face the surrounding water. A light-scattering experiment with miconazole (MCZ) also demonstrated the disappearance of MCZ drug granules in the presence of DODAB BF.[196] Reappearance of drug particles only takes place after attaining the drug loading capacity of the BF, which is ca. 1:10 drug:DODAB molar ratio when the drug is MCZ.[196] Half-dissociation of imidazole moiety in the MCZ molecule occurs at pH 6.5 so that neutral drug molecules can adsorb hydrophobically at the rim of cationic bilayer fragment around this pH value. The interaction between MCZ and DODAB BF was also evaluated from determination of BF loading capacity and effects of drug-to-lipid molar proportion (MP) on particle size, zeta potential and gel-to-liquid-crystalline phase transition (T_m).[254] DODAB BF solubilized MCZ over a range of MP. BF loading capacity was 0.5 mM MCZ at 5 mM DODAB. Above this limit, the drug aggregated in the dispersion. At pH 6.3, BF zeta-potentials decreased with MP, suggesting insertion of deprotonated drug into the bilayer. MCZ optical spectra in BF are similar to those in best organic solvent, confirming drug solubilization. At MP lower than 1:10, BF T_m remained unchanged, suggesting drug capture at BF hydrophobic edges. At MP above 1:10, T_m decreases, showing MCZ insertion into the DODAB bilayer.[254] However, the drug is expelled from the bilayer core upon lowering temperature. Minimal fungicidal concentrations against *C. albicans* are synergically reduced by 10 times for drug/BF.[254]

For AB in DODAB BF, experiments for evaluation of the therapeutic activity and nephrotoxicity were also performed. The formulation is therapeutically effective *in vivo* despite its low drug-to-lipid molar ratio[197] and exhibits very low nephrotoxicity due to the low AB dose.[255] Furthermore, the low cost of such synthetic lipid used in the formulation would justify further testing of this formulation in the clinic.

DODAB antimicrobial activity and its differential cytotoxicity are well described.[174–178,256–259] In addition, both DODAB and DHP BF were employed to stabilize hydrophobic drug particles in water so that formulations at high drug to lipid molar ratios could also be obtained.[196,203] In these formulations, the electrostatic attraction between the charged drug particle and the oppositely charged BF essentially drives the coverage with a lipid layer. In certain cases, the DODAB lipid and the drug being carried act synergistically against the microorganism. This is the case for MCZ and DODAB[196] or rifampicin and DODAB.[204] The DOD/AMB formulation (AB in the DODAB BF) was compared to Fungizone regarding its *in vivo* activity from survival and tissue burden experiments against systemic candidiasis in a mouse model.[197] DOD/AMB and the traditional deoxycholate/AMB formulation called Fungizone (DOC/AMB) have identical effect promoting about 100% mice survival when given by the same route at the same relatively low dose of 0.4 mg/kg/day intraperitoneally for 10 days.[197] From spleen and kidneys tissue burden experiments, similar efficacy of both preparations in reducing tissue colony forming units is obtained. DOD/AMB is as effective as DOC/AMB for treating systemic candidiasis in a mouse model.[197] The problem of nephrotoxicity for the novel DOD/AMB was addressed both *in vitro* against cultured kidney epithelial cells in culture and *in vivo* in a mice model.[255] For the *in vivo* experiments, Swiss Webster female mice were injected intraperitoneally for 10 consecutive days with 0.4 mg/kg/day AMB in the form of traditional bile salt desoxycholate (DOC)/AMB or DOD/AMB.[255] At day 11 after injection, DODAB and DOD/AMB cause loss of body weight and increase in spleen weight, which are not observed for DOC/AMB, although the changes are reversible and weights return to control values at day 180. 10 days after injection, biochemical parameters for hepatic and renal function remain unaltered.[255] The nephrotoxicity of the formulation is lower than that of Fungizone (DOC/AMB), which is the traditional AMB formulation using deoxycholate. Dose-dependent cytotoxicity of DOD/AMB is lower than that exhibited by DOC/AMB. At day 180, renal cortex histopathology reveals leukocytic infiltration and moderate hydropic degeneration of the renal tubules in the DODAB and DOD/AMB groups, in contrast to more severe lesions observed for the DOC/AMB group such as tubular cystic degeneration and glomerular injury, which are absent for the former groups. In conclusion, the DOD/AMB formulation exhibits differential cytotoxicity and low nephrotoxicity, though there are still important aspects of general toxicity that require evaluation with full-scale toxicity protocols.[255]

The stabilizing effect of bilayer coverage on oppositely charged hydrophobic drug particles was also described for cisplatin encapsulation.[260–262] The aqua-species of cisplatin bears a positive charge so that its aggregates obtained by freeze-thawing were covered by anionic phospholipid bilayers. Bean-shaped nanoprecipitates of cisplatin coated by a lipid bilayer were visualized by TEM.[262] The nanocapsules represent a novel lipid formulation of cisplatin characterized by a very high cisplatin-to-lipid ratio and a much higher cytotoxicity against tumor cells *in vitro* as compared to the free drug. The formation of the nanocapsules critically depended on the presence of negatively charged phospholipids and positively charged aqua-species of cisplatin.[260,261] The effect of PEG on the stability of the cisplatin nanocapsules was also studied by incorporating PEG conjugated to phosphatidylethanolamine (DSPE–PEG2000).[262] The release of contents of cisplatin nanocapsules depends on the temperature, the surrounding medium, and the lipid composition of the bilayer coat. Sterically stabilized cisplatin nanocapsules containing 6 mol% DSPE–PEG served as the starting formulation for *in vivo* studies addressing the antitumor efficacy of cisplatin nanocapsules in tumor-bearing mice; there is a requirement of anionic phospholipid for successful nanoencapsulation of the cationic aqua-cisplatin.[262]

There are several lipidic AB formulations which circumvent the problem of nephrotoxicity but high cost remains an issue with all of them.[263–266] For example, amphotericin B colloidal dispersion (ABCD) is a colloidal dispersion of a stable complex of AB with cholesteryl sulphate in a 1:1 proportion, forming uniform disk-shaped particles.[264,265] ABCD is associated with less nephrotoxicity than conventional AB deoxycholate. ABCD is active in the treatment of invasive *Candida* spp. and *Aspergillus* spp. infections in immunocompromised hosts, and approved by the U.S. FDA for the treatment of invasive aspergillosis in patients where renal impairment of unacceptable toxicity precludes the use of AB deoxycholate in effective doses, and in patients with invasive aspergillosis where prior AB deoxycholate therapy has failed.[264,265] The BF composed of the much less expensive synthetic lipids DODAB or DHP have the potential to accommodate a variety of hydrophobic and amphiphatic drugs just by assembly at different regions of the BF or disks. This means that BF formulations might reveal in the near future a broad spectrum of low cost applications to carry and deliver a variety of drugs.

In summary, charged synthetic BF can be of practical use for delivery of drugs both at low and high drug to lipid molar ratios (R). At low R, they offer a large area of hydrophobic nanosurfaces well-dispersed and electrostatically stabilized in water, and solubilize the drug at their rim. At high R, they cover the insoluble drug particles with a bilayer and stabilize these drug particles in a colloidal sense. Coverage of hydrophobic model particles such as latex with lipids is not an entirely novel subject from a physicochemical point of view.[80] It is to be anticipated that early[80–82] and more recent fundamental studies on the interaction between lipid bilayers and hydrophobic particles or surfaces[13,80–84,267,268] will finally prove useful for encapsulating hydrophobic drug particles within lipidic layers.

5. Nanomedicines Based on Hybrid Biomimetic Systems

Over the last 10 years, much work has been devoted to the construction of hybrid nanostructures such as the **biomimetic lipid-covered particles**. Polymeric,[85,95,97,267–281] mineral[86–90,96,282–299] or biological particles such as bacteria,[174–176,179,191–194] fungi,[100,177] mammalian cells[257] or viruses[300,301] have been coated by lipids. The combination of particles or flat surfaces with lipids in different assemblies produced interesting particulates and devices with an outer lipid coating.[13,80–84,302] The so-called Biovector system cross-linked maltodextrin modified with quaternary ammonium moieties in order to produce cationic polysaccharide cores covered with lipid layers with sizes similar to those for viruses which could be used as carriers for proteins and/or peptides for vaccine applications, especially via nasal spray.[303] Other hybrid NPs combined preformed polysaccharides or hydrophilic polymers with synthetic polymeric particles during particle synthesis.[91–93] These were obtained from traditional synthesis of polymeric particles by emulsion polymerization in the presence of natural polysaccharides such as carboxymethylcellulose[92] or chitosan[93] and were excellent for physical adsorption of enzymes.[91]

Particles covered by lipids are also finding many interesting applications in drug and vaccine delivery since they profit of all basic knowledge acquired over the last decades both for liposomes and NPs. The control of physical properties such as size and size distribution, surface composition, charge and homogeneity has been achieved for synthesis of polymeric latexes so that deposition of a bilayer onto these particulates with a narrow size distribution may yield a robust control over size-dependent responses *in vivo*. For example, the immune response changes with the size of the foreigner particles and is tunable in accordance with particle size since the organisms need to respond differently to viruses, to bacteria or to multicellular parasites. On the top-right side of Fig. 4, the narrow size distribution of some polystyrene sulfate nanospheres can be appreciated. These NPs keep this property after coverage with a lipid bilayer.[80,82,87] The importance of the hybrid lipid-particle

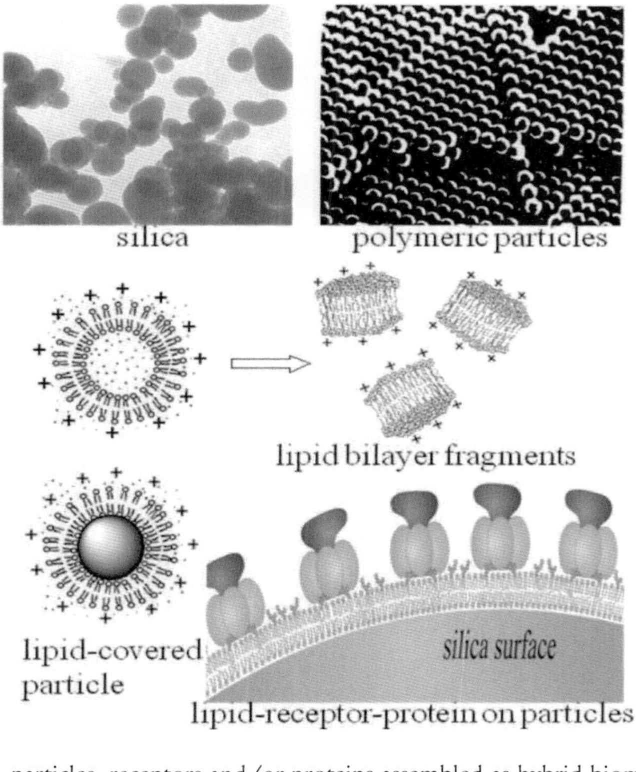

Fig. 4. Lipids, particles, receptors and/or proteins assembled as hybrid biomimetic systems. The proof-of-concept for lipid-receptor-protein reconstitution on silica particles was established by evidencing coverage of silica with one PC bilayer from adsorption isotherms, incorporation of fluorophore-labeled GM1 receptor and, at last, cholera toxin binding to the system.[96]

approach has been realized over the last decade with an exponential growth of research in this area. The supported bilayers on polymeric or mineral particles can display functional cellular receptors in optimized conformational and non-aggregative states.[96,267,268,304] These receptors preferentially bind to and neutralize virulence factors of both bacteria and viruses, thereby promising a new approach to antipathogenic therapy.[304] For instance, the monosialoganglioside GM1 is a glycolipid which was successfully incorporated in supported lipid-bilayers based on latex[267,268] or on silica[96] and used to bind the cholera toxin secreted by *Vibrio cholera* suggesting that these NPs can act as scavengers of dangerous toxins *in vivo*.

Biomimetic NPs can be constructed by the use of interaction forces to drive lipid deposition from bilayers onto spherical solid cores. The NPs such as polystyrene nanospheres,[80,168,81] silica particles[86–88,302] or hydrophobic drug aggregates[196,256] have been coated with lipid bilayers, yielding functional hybrid assemblies.[88,84,89,96,196,203,256] These have found many applications in biomolecular recognition,[96] drug delivery[84,89,96,196,203] and vaccine design.[85,95,294] The principal forces driving bilayer deposition onto hydrophobic or hydrophilic NPs from bilayer vesicles or BF are the electrostatic attraction, the van der Waals attraction and/or the hydrophobic effect.[80,81,86,87,168,302] Medium composition, pH and the ratio between the surface areas of the particles (A_p) and bilayers (A_b) also play a crucial role in achieving the optimal bilayer deposition on the particles and the resultant colloidal stability.[84,88,89,96,196,203,256] For example, if the amount of added lipid is not sufficient to cover all particles with a bilayer, poor colloidal stability may result in aggregation of the primary particles and a broad range of particle sizes. Another important factor is the physical state of the bilayer. In general, bilayer vesicles in the rigid gel state do not get disrupted upon adhesion onto solid particles, so that bilayer coverage does not take place.[90] This difficulty has been circumvented by using BF, which are prone to adsorb onto the solid particles. Charged BFs are easily obtained from vesicles by ultrasonic disruption and have been used for the production of a variety of lipid-based biomimetic particles.

The important physical properties of biomimetic particles are their size distribution, polydispersity index and zeta-potential.[305] These properties typically allow one to ascertain whether bilayer coverage indeed took place and should be determined for both bare and biomimetic NPs. If the lipid bilayers and the particles are oppositely charged, poor colloidal stability occurs at a critical lipid concentration where the sizes are maximum and the zeta-potential is zero.[80,85,88,196,203,294] Above this critical concentration, the particles exhibit zeta-potentials that are similar to that of the charged bilayer and recover the colloidal stability.[60,85,196,294] Aggregates of hydrophobic drugs dispersed in water have also been treated as particles and coated with lipids at high drug-to-lipid molar ratios.[100,196,203] The formation of aggregates of hydrophobic drugs in aqueous solution can be modulated by oppositely charged BF. Under certain experimental conditions (which must be established for each particular system), it is possible to achieve coverage of the drug aggregate with a single lipid bilayer and, at the same time, colloidal stability. Thus, one of the most important applications of this coating procedure is the production of lipid-stabilised NPs of hydrophobic drug. Other important application is the reconstitution of different receptors in the supported bilayer. For example, pathogenic bacteria, viruses and other microorganisms typically entry the host cells by adhering to cell-surface receptors with high-affinity ligands. Host cell receptors that are attractive targets for pathogens typically include glycosylated variants of lipids, proteoglycans, sialic-acid-bearing gangliosides and glycoproteins. The virulence factor of *Vibrio cholerae* is the cholera toxin which binds to the monosialoganglioside GM1, a host glycolipid in the cell membrane.[306] The GP120 protein of the HIV virus targets the galactosyl ceramide glycolipids in the host cell membranes.[307] Therefore, the reconstitution of these receptors in supported bilayer membranes may represent a powerful approach to scavenge pathogenic toxins preventing the host cells invasion that takes place during infection. Furthermore, many orphan receptors may be reconstituted and studied in these NPs.

Solid lipid nanoparticles (SLN) are essentially a dispersed oil phase stabilized in a water phase by an emulsifying layer of surfactant or polymer and very similar to oil-in-water microemulsions.[308–310] The term lipid is used here in a broader sense and includes triglycerides, partial glycerides, fatty acids, steroids, and waxes.[310] However, it is required that the lipid matrix maintains the solid state at room temperature.[311] They have been stabilized with surfactants such as poloxamer 188, polysorbate 80, lecithin, polyglycerol methylglucose distearate, sodium cocoamphoacetate or saccharose fatty acid esters and used to carry drugs, peptides or proteins. The dispersions are prepared by stirring 10% of melted solid lipid such as stearic acid, 15% of surfactant and up to 10% of cosurfactant in hot water. Thereafter, the warm microemulsion is stirred in excess cold water before removal of the water excess by means of ultrafiltration or liophylization.[312–315] Advantages of SLN are their biocompatible compounds, the rapid and efficient scaling-up process, the avoidance of organic solvents and the possibility of obtaining concentrated lipid suspensions. Disadvantages are the limited drug loading capacity of conventional SLN because of the formation of a perfect lipid crystal matrix and the presence of other colloidal structures in the nanoemulsions such as micelles, liposomes, mixed micelles and drug nanocrystals.[311]

Not only lipids have been used to produce NPs. Viruses, ferritin, and other protein cages for ages have self-organized in the nanometer range with biologically relevant functionalities. These bio-inspired systems form monodispersed units that are highly amendable through genetic and chemical modifications. The uses of these **protein-based nanosystems** beyond their native functionalities, and their application as building materials for nanomedicine with particular emphasis on adaptations of protein-based nanostructures for gene delivery, bio-imaging, drug encapsulation, and vaccine development have been reviewed.[316,317] Protein-based nanocarriers meet requirements such as low cytotoxicity, abundant renewable sources, high drug binding capacity and significant uptake into the targeted cells representing promising candidates for efficient drug and gene delivery. Moreover, the unique protein structure offers the possibility of site-specific drug conjugation and targeting using various ligands modifying the surface of protein nanocarriers. Albumin-based NPs have often been used as potential controlled release drug delivery systems.[318] Albumin is indeed a versatile carrier and, has been shown to be nontoxic, non-immunogenic, biocompatible and biodegradable representing an ideal material to fabricate NPs for drug delivery. Albumin NPs have gained considerable attention owing to their high binding capacity of various drugs and being well-tolerated without any serious side-effects.[318] Other examples are recombinant virus-like particles (VLP) derived from human papillomaviruses as multivalent vaccine carriers to immunize patients against genital infections, and vaccinia viruses to treat patients with established cancers.[317] In fact, **biological particles** ranging from pathogens, like viruses and bacteria, to mammalian cells, like red blood cells, macrophages, DCs and stem cells, are highly optimized for their specific functions *in vivo* and possess features that are often desired in drug delivery carriers.[319] Some of these features have indeed been explored to obtain modified biological particles to be used as carriers. The **recombinant bacteria** are bacteria genetically modified by expression systems that encode antigens and biologically active proteins.[320] These are bacteria without their cytoplasmic components. The genetic materials can be removed from the bacterial cells and the resealed ghosts can be used to incorporate drugs.[321] Engineering strategies for viruses involve the replacement of viral genes by the genes of interest to be carried by the viral vector for tumor targeting.[322] Viral capsids derived from viruses can be self-assembled to form VLP which can be loaded with peptides[323] or used for vaccination because their antigenicity is comparable to the parent virus. Human papilloma virus (HPV)-like particles induced protective immune responses against HPV without any side effects so that the vaccines were approved by the FDA and are in the market.[324,325]

The **virosomes** are assemblies of two influenza envelope glycoproteins, hemagglutinin (HA) and neuraminidase (NA) which can be reconstituted in lipid bilayers. These glycoproteins are responsible for targeting, receptor-mediated endocytosis and endosomal escape after endocytosis.[326,327] Virosomes have excellent adjuvant properties and the ability to carry various drugs, including antigens and nucleic acids. In the market, several vaccines based on virosomes are available providing protection against the hepatitis A virus (Epaxal; Crucell) and the influenza viruses (Invivac (Solvay-Influenza) and Inflexal V (Crucell)).[328] Virosomes reconstituted from viral lipids and proteins have been also used to encapsulate small interfering RNA **(siRNA)**.[329] They have also gained attention as potential drug and gene delivery carriers, owing to their ease of production and modification as well as their low toxicity.[330] Their *in vivo* applications for drug delivery, however, are limited by the potential risk of immunogenicity.

Based on the concept that an intracellular and porous NP may act as a drug depot within the cell, **mesoporous silica** has been used. Proteins have been entrapped and protected in carboxylethyl- or aminopropyl-functionalized mesoporous silica with rigid, uniform open-pore geometry.[331] The combination of high immobilization efficiency, protein loading, and stability was attributed to the large and uniform pore structure, and to the optimum environment introduced by the functional groups.[331] Protein molecules were sequestered in or excluded from the porous material, depending on electrostatic interaction with the charged functional groups. In order to protect the contents inside the mesoporous silica additional PE layers have been built on the particles.[332] Multilayered PE [poly (diallyldimethylammonium chloride), PDDA/poly (sodium 4-styrenesulfonate), PSS], or PE/NP [PDDA/silica nanoparticles, SiNP] shells were deposited onto the enzyme-loaded spheres.[332] The activity of the encapsulated enzyme was retained, even after exposure to proteases.[332] By using C12-trimethylammonium bromide versus C16-trimethylammonium bromide in the self-assembly of mesoporous NPs the distribution of pore sizes a parameter that determines the release kinetics of the drug payload — was tuned from a center at 1.8 to 2.5 nm.[333] As a further extension of this work, ibuprofen was introduced into the pores of these particles and placed in a simulated body fluid to determine their potential viability as drug delivery systems. These mesoporous silica materials could be used to deliver relatively large doses of drug in a controlled manner.[334]

Viruses efficiently smuggle DNA into the nucleus incorporating their own DNA in the genome of the host cell. This is precisely what is required in **nanosystems for gene therapy.** The question is whether especially designed carriers could compete with viruses. In fact, only a very modest liposome-mediated gene expression has been obtained via a mechanism that is not completely known and difficult to control. Liposome vectors formulated with cationic lipids and a helper lipid readily form complexes with DNA known as lipoplexes. *In vitro* cationic lipids are being widely used for cell transfection. However, several drawbacks arise in clinical trials for gene therapy.[335] There is aggregation and inactivation by serum which limit the stability and activity of the cationic carriers. By condensing DNA with polymer or changing the surface charge with anionic molecules, protecting the surface with PEG and increasing the target specificity by attaching ligand, efficiency and stability of the vector were improved.[336] However, there is still the problem of an inflammatory toxicity coming from certain motifs in the plasmid DNA that should be reduced.[336] Nonviral vectors including NPs and liposomes can be engineered for reduced immunogenicity, infection or mutagenesis, and can transport large nucleic acid payloads.[337–340] These vectors typically employ cationic liposomes or cationic polymers such as chitosan, polylysine or polyethylenimine (PEI) complexed with anionic nucleic acids.[337,341] Among the cationic polymers, PEI has demonstrated the highest gene delivery efficiency *in vitro* because of its large cationic charge that facilitates a high degree of DNA complexation and an effective ability to escape the cellular endosome upon internalization via the "proton sponge" effect.[342] PEI displays, however, inherent toxicity and lack of cell specificity for selective transfection, a problem partially circumvented by covalent modification of the cationic polymer with PEG.[337,341] PEI was conjugated to antibodies,[343] peptides such as arginine-glycine-aspartic acid (RGD) for integrin receptor targeting of endothelial cells of tumor metastases[344] or small organic molecules such as folic acid for folate receptor targeting in colon adenocarcinoma and oral epidermoid cancer cells[345] or the peptide chlorotoxin (CTX)[346] which binds with specificity and affinity for the vast majority of brain tumors,[347] prostate cancer, sarcoma, and intestinal cancer.[346] CTX peptide has been providing excellent targeting in previous *in vitro* and *in vivo* NP studies[348,349] and has recently been coupled to a branched PEI polymer functionalized with grafted PEG and the DNA loading optimized providing an interesting example of hybrid NPs highly effective against cancer cells in culture.[350] This approach could potentially lead to further use of this carrier system *in vitro* and *in vivo* for bench-side and bedside applications as a selective, stealth, nonviral targeting agent. Similarly, cationic lipids in form of liposomes or biomimetic particles have been evolving to be the basis of systems with targeting and steric stabilization as additional functionalities improving biocompatibility and specificity.

New technologies particularly RNA interference (RNAi)-mediated by **siRNA delivery nanosystems** have become more and more interesting. They provide effective therapeutic entities to silence pathogenic gene products associated with disease, including cancer, viral infections and autoimmune disorders. The Nobel Prize in Physiology or Medicine has been awarded to Andrew Fire and Craig Mello for their discovery of RNAi, *i.e.* the suppression of gene activity by double-stranded RNA.[351] The siRNA molecules, notably the antisense strand, recognize and inhibit the corresponding mRNA, thereby silencing the appropriate gene. RNAi works at a posttranscriptional level by targeting mRNA as a mean for inhibiting the synthesis of the encoded protein.[352] Salient features of RNAi are: (1) the ability of a few molecules of double-stranded RNA to direct destruction of a much larger amount of the corresponding mRNA, suggesting a catalytic mechanism; (2) the transmission of RNAi across generations; (3) the power of RNAi to bring genetics to any organism whose genome sequence is known; (4) the near universality of RNAi among eukaryotes. The discovery that gene expression could be controlled by base-pairing of small RNAs with messenger RNAs containing complementary sequence, a process known as RNAi, has markedly advanced our understanding of eukaryotic gene regulation and function.[353] The modulation of oncogenes expression by siRNA is set to revolutionize the treatment of cancer. Remarkably, despite being just one decade from its discovery, the phenomenon is already being used therapeutically in human clinical trials.[353] Cationic carriers will certainly be useful in novel RNA-interference-based therapeutics.

To assess the stability of siRNAs under conditions relevant to clinical use with particular emphasis on topical delivery considerations, a study of three different unmodified siRNAs was performed.[354] The results indicated that neither repeated freeze/thaw cycles, extended incubations (over 1 year at 21°C), nor shorter incubations at high temperatures (up to 95°C) have any effect on siRNA integrity as measured by nondenaturing polyacrylamide gel electrophoresis and functional activity assays. Degradation was also not observed following exposure to hair or skin at 37°C. However, incubation in fetal bovine or human sera at 37°C led to degradation and loss of activity. Therefore, siRNA in the bloodstream is likely inactivated, thereby limiting systemic exposure. The design of lipid-based formulations of siRNA for systemic use aims at protecting the RNA from degradation in the circulation.[355] Cationic lipids have been essential components of such formulations because they are required to achieve efficient loading of anionic polymers and they facilitate intracellular delivery of polynucleotides such as siRNA by destabilizing biological membranes favoring uptake by endocytosis.

Cationic PNP have also been designed aiming at siRNA protection.[356] In this last approach, NPs were prepared by emulsion polymerization of isobutylcyanoacrylate and isohexylcyanoacrylate with chitosan. Thereafter, loading of the NPs with siRNA was achieved by adsorption. The biological activity of the siRNA-loaded NPs was assessed on mice bearing a papillary thyroid carcinoma after intratumoral and *i.v.* administration. The adsorbed antisense siRNA *in vivo* lead to a strong antitumoral activity so that tumor growth was almost stopped after *i.v.* injection of the antisense siRNA-loaded NPs, while in all control experiments, the tumor size was increased by at least 10 times. *In vivo* studies have also confirmed the relevance of using chitosan-coated poly(isobutylcyanoacrylate) NPs as a delivery system to inhibit the expression of ret/PTC1 oncogene and treat papillary thyroid carcinoma.[357] The siRNA approach was also used for specific targeting and down-regulation of fusion oncogenes in chronic myeloid leukaemia and in a bone cancer, the Ewing sarcoma.[358]

Osteosarcoma is the most common malignant primary bone tumor. Its growth is strongly supported by bone resorption. Antibodies or recombinant proteins have inhibited the cytokine receptor activator of nuclear factor-κB ligand. Recently, its expression was inhibited using genetic approaches based on siRNA.[359] The delivery of siRNA to its cellular target and demonstration of its efficiency *in vivo* relied on two new osteosarcoma models expressing firefly luciferase. These luciferase-expressing osteosarcomas showed conserved osteolytic and osteogenic activities in mice and were detectable by *in vivo* bioluminescence imaging. Bioluminescence analysis revealed extensive cell death in response to ifosfamide treatment. Additionaly, a protocol for *in vivo* administration of siRNA combined with cationic liposome was established.

Hybrid biomimetic systems have been applied in cancer chemotherapy in several instances. Enhancement of the efficacy of chemotherapeutic drugs could be achieved by the suppression of antiapoptotic cellular defense.[360] Combinations using drug and a suppressor of antiapoptotic cellular defense represented successful strategies.[361] A synthetic peptide corresponding to the minimal sequence of BCL-2 homology 3 (BH3) domain, the BH3 peptide, was delivered into cells by fusion with a fusogenic peptide sequence.[361] The cytotoxicity of DOX, Ant-BH3 and Ant-BH3/DOX, mitochondrial transmembrane potential, expression of genes encoding pro- and antiapoptotic members of BCL-2 protein family and caspases, caspases activity and apoptosis induction were assessed in human ovarian carcinoma cells. The combination DOX/Ant-BH3 enhanced apoptosis induced by DOX and prevented the antiapoptotic cellular drug resistance. Other studies also addressed the suppression of MDR and/or anti-apoptotic cellular defense.[361–363] The isolated use of synthetic analogs of the BH3 domain of the pro-apoptotic members of the BCL2 protein family, including BAK, BAX and BAD,[360,364–368] antisense oligonucleotides and siRNA targeted to BCL2, MDR1 and MRP mRNA,[369–371] c-Jun NH2-terminal kinase,[372] ribozyme traditional drugs,[373] several drug groups from a traditional mitomycin C[374] and the exotic plant stress hormones family of jasmonates[375] did not elicit a high anticancer effect. The inhibition of only one contributor to cellular resistance is not sufficient for overcoming all mechanisms of cancer cell resistance to chemotherapy. Co-delivery in one liposomal drug delivery system of DOX simultaneously with siRNA targeted to MDR1/MRP1 and BCL2 mRNA enhanced cell-death induction by simultaneously increasing intracellular DOX concentration and suppressing cellular anti-apoptotic defense.[376] The carrier consisted of cationic liposomes (carrier, 100–140 nm), DOX (anticancer drug) and siRNA targeted to molecules representing pump and nonpump cellular resistance so that the complex was approximately 500 nm in mean diameter and presented a zeta-potential of +4 mV.[376] Optimal efficacy of the treatment required delivery of all cell-death inducer(s) *and* suppressor(s) of both types of cellular-drug resistance inside the cancer cell.[376,377]

A novel conjugate was synthesized between a phospholipid (dioleoylphosphatidylethanolamine) and PEI for siRNA delivery and silencing of P-gp in human breast cancer cells to overcome DOX resistance.[378] The conjugate enhanced the transfection efficacy of low-molecular-weight PEI, which was otherwise totally ineffective. In addition, the PEG/lipid coating of the new complexes gave rise to small micelle-like NPs with improved biocompatibility properties. Both coated and uncoated formulations delivered P-gp-specific siRNA to MDR cells. The combination of DOX and P-gp silencing led to a two-fold increase of DOX uptake and a significant improvement of the therapeutic effect of DOX in resistant cells.[378]

The fate of NPs or liposomes *in vivo* has been determined in real time by means of **quantum dots.**[379,380] In spite of their questionable direct use due to their potential long-term toxicity,[381] they are able to provide monitoring of intracellular and systemic nanocarrier distribution, degradation, drug release, and clearance. They are commercially available, small (a few nanometers in size) and can be incorporated within any drug delivery vehicle with minimal effect on overall characteristics offering excellent optical properties for real-time monitoring of transport and drug release at both cellular and systemic levels.[379]

More biocompatible contrast agents such as **magnetic NPs** with sizes similar to those of quantum dots would be more suitable for direct use in drug delivery. Hydrophobized Fe_3O_4 NPs were synthesized and incorporated into a liposomal membrane using hydrophobic interactions.[382] Hybrid liposomes encapsulating pyranine, a water-soluble fluorescent dye, were prepared from mixtures of phospholipids, the hydrophobized Fe_3O_4 NPs, and the copolymer.[382] While the hybrid liposomes released negligible amounts of pyranine under static conditions, the release of pyranine was drastically enhanced by alternating magnetic field irradiation due to the transition of the thermosensitive segment of the copolymer, which is caused by the release of localized heat from the Fe_3O_4 NPs under magnetic stimuli, rather than the rupture of the capsules. The release rate of the hybrid capsules could be controlled by varying the amount of Fe_3O_4 NPs embedded into the liposomes.[382] Other approaches coated the Fe_3O_4 NPs with rotavirus capside protein to obtain a theranostic (therapeutic plus diagnostic) platform for cellular imaging and drug delivery.[383]

Inorganic hybrid NPs such as the quantum dots and the magnetic NPs are potentially useful in biomedicine. However, in order to avoid long-term toxicity, they need to be cleared from the body within a reasonable timescale.[384] Quantum dots functionalized with high-affinity small-molecule ligands that target tumors are cleared by the kidneys if their hydrodynamic diameter is less than 5.5 nm, which sets an upper limit of 5–10 ligands per quantum dot for renal clearance.[384] Other alternatives for medical imaging are the fluorescent organic dyes (*e.g.* the green fluorescent protein)[385] or some inorganic NPs doped with lanthanides. These doped NPs have narrow emission bandwidths, high photochemical stability, long fluorescence lifetime (up to several milliseconds), different colours available by varying the luminescent center used, *e.g.* Tb, Eu, Dy and just a few atomic% of rare-earth element in order to reduce cytotoxicity.[386–389] Stable luminescent colloidal systems based on doped **biomimetic nanocrystalline apatite** analogous to bone mineral, $Ca_{10}-x(PO_4)_6-x(HPO_4)x(OH)_2-x$ ($0 \leq x \leq 2$) have recently been described.[390] The preparation of such doped colloidal apatite NPs involves the use of ionic salts, some of them at high concentrations, and requires a second step of purification by dialysis for removal of the excess.[391] Hydroxyapatite, $Ca_{10}(PO_4)_6(OH)_2$, is one of the most stable forms of the calcium phosphates and the major inorganic component of bone and teeth in mammals. Biomimetic apatite nanocrystals can be applied in bone implantation and, in addition, function as a local targeted delivery system for anticancer and antimetastatic drugs with controlled release properties. For example, the application of such a material in the chemotherapeutic treatments of osteosarcoma inhibited tumor growth and caused low levels of systemic toxicity. As drug carriers, calcium phosphate NPs have some advantageous properties.[392] They are dissolved at low pH (around 4), *e.g.* in lysosomes after the cellular intake[393] or in the environment of solid tumors, thereby releasing incorporated drugs or biomolecules. Their size can easily be controlled by stabilizing agents, such as polymers or nucleic acids. The NPs can fluoresce by the incorporation of lanthanide ions.[394,395] They were shown to be able to carry insulin,[396] cisplatin,[397] or ceramide.[398] *In vivo* delivery of a suicide gene with calcium phosphate NPs was used to treat nasopharyngeal carcinoma.[399] The evolving applications of biomimetic synthetic calcium phosphates were recently reviewed by Roveli and Iafisco.[400]

In spite of the potential for release of toxic ions such as cadmium (Cd)[381,401] and generation of reactive oxygen species,[402] a few studies have shown that properly coated quantum dots do not exhibit acute toxicity in rodents[403] and rhesus monkey.[404] Nevertheless, long-term toxicity studies should be performed.[405] Similarly to quantum dots, magnetic NPs can be made of a variety of magnetic materials with a wide range of magnetic properties but, many of these materials such as cobalt and chromium are highly toxic and a protective coating should be used to prevent their leakage.[406] Minerals such as magnetite and maghemite, which are iron oxide materials were considered relatively safe and used as contrast agents in the clinic for a while. Coating magnetic NPs has been achieved from biopolymers such as carbohydrates (*e.g.* dextran[407] or starch[408]), proteins (*e.g.* albumin[409]) or synthetic polymers such as PEG[410] or polyvinyl alcohol[411] or lipids.[412] Synthetic polymers have better mechanical strength than many natural polymers. Nevertheless, some coatings formed from synthetic polymers are still porous so that corrosion of the magnetic core may occur. The high mechanical strength of silica was alternatively used to coat the magnetic particles but the coating was not homogeneous since silica spherical particles surrounded the iron oxide nanoparticles (IONPs).[413] Magnetic NPs coated with carbon could be loaded with anticancer drugs in the carbon moiety yielding a synergistic enhancement of cell death in pancreatic cancer cells by combined radio frequency thermolysis and drug delivery.[414] Gold-coated magnetic NPs were also prepared facilitating functionalization by thiol linkers that bind covalently to gold and may bear a functional group (*e.g.* amine) at the other end of the molecule.[415]

The IONPs such as Feridex®[417] and Resovist® were approved for use in the clinic[418] but their long-term toxicity led to withdrawing from use in humans.[419] Degradation of iron oxide increased free radical production in cells and caused cell death.[420,421] Gold coating of magnetic NPs has often circumvented the toxicity of the bare particles imparting a multifunctional platform to the gold-IONPs.[422,423] For ultrasensitive magnetic resonance imaging (MRI), the magnetic NPs are excellent despite their toxicity. Recently, biomimetic IONPs were prepared by coating them with

PEG-phospholipid so that they could be used to label pancreatic islets to be transplanted *in vivo* in diabetes animal models enabling highly sensitive non-invasive assessment after cell transplantation.[424] Pancreatic islets were isolated from animals, purified and labeled by incubation with the NPs and spontaneously endocytosed the coated NPs.[424]

Gold NPs are commercially available over a range of sizes (2–100 nm mean diameter), possess unique optical properties absorbing light and releasing heat at certain frequencies of the incident light,[425,426] and are chemically stable and biocompatible.[427] In nanomedicine, this golden property of gold, absorbing light and releasing heat, has been exploited for applications such as photothermal ablation of tumors and thermosensitive drug delivery.[426] The range of wavelengths for irradiation of the gold NPs is over the near infrared (NIR) region, namely, 750–1400 nm.[428] Over this range, the laser radiation is capable of deep tissue penetration due to the high transmission of water and hemoglobin without any damage to the cells.[428] For other gold nanostructures such as nanoshells, with a hollow architecture, the light absorption increases in intensity as compared to the gold NPs[429] and depends on shell thickness and functionalization.[425,430] Light absorption decreases with increasing shell thickness and so does the production of heat.[425,430]

Gold coating on IONPs previously coated with PEI was achieved by firstly attaching gold seeds to the PEI-NPs surface followed by subsequent reduction of gold onto the surface.[431,427] This method allowed a better control over shell thickness [431,427] and avoided the decrease of saturation magnetization that takes place in the magnetic core when gold directly coats the magnetic NP.[432] The unique physical properties of the magnetic NPs and the gold nanoshell were shown to be preserved using a polymer "cushion" layer between the core and the shell.[422] These improved core-shell NPs have clinical potential as multimodal platforms for a range of biomedical applications such as image guided cellular hyperthermia or thermoresponsive drug delivery. The use of NIR light and gold nanoshells as two benign moieties yielded localized and irreversible photothermal ablation of tumor tissue both *in vitro* and *in vivo* since combining these two therapies produced localized cell death confined to the laser/nanoshell treatment area.[433] Excellent reviews are available from the literature regarding the preparation, characterization and biomedical applications of **gold NPs, nanorods, shells and cages**.[434–438]

Carbon-based nanomaterials (NM), as the nanotubes, nanohorns and nanodiamonds are currently being explored toward different biomedical applications. Similarly to other evolving nanomaterials, their effects and fates in the biological milieu regarding absorption, distribution, metabolism, excretion, and toxicity have to be investigated. Since they lack a specific detection signal, isotopic labeling, with its high sensitivity and specificity, is the first choice to quantify carbon NMs *in vivo*.[439] Examples of isotopes used are ^{13}C, ^{14}C, ^{125}I, ^{131}I, ^{3}H, ^{64}Cu, ^{111}In, ^{86}Y and others. After NMs enter the body, blood will make them travel in the body and be retained in the filtering organs such as lungs, liver, and spleen. Long-term retention was obtained in liver and spleen and a gradual clearance via mucus in the lungs.[439]

Nanodiamonds are obtained by detonation when an explosive mixture is detonated in a closed chamber. Diamond particles with a diameter of ca. 5 nm are formed at the front of detonation wave in time of several microseconds. Their preparation, characterization, functionalization and biomedical applications have been thoroughly described and updated recently.[440] Assays of cell viability such as mitochondrial function (MTT) and luminescent ATP production showed that nanodiamonds were not toxic to a variety of cell types.[441] Furthermore, nanodiamonds did not produce significant reactive oxygen species and cells can grow on nanodiamond-coated substrates without morphological changes compared to controls.[441] Colloidal suspensions of individual diamond particles with diameters of 4–5 nm ("single-digit" nanodiamonds) are available and researchers realized that fluorescent nanodiamonds are a nontoxic alternative to semiconductor quantum dots for biomedical imaging.[442,443] Irradiation of synthetic diamond nanocrystallites with He ions produces luminescent nanodiamonds which have been used in a commercial product.[442] Their chemical reactivity at the surface was used for attaching drugs and biomolecules.[444–446] DNA can be electrostatically attached to nanodiamonds by first covering negatively charged carboxylated nanodiamonds with positively charged PEI800 molecules.[447] A similar electrostatic binding strategy has been used to attach siRNA and DOX to nanodiamonds.[446] The nanodiamond–drug complexes were endocytosed by the cells and the drug passively diffused from the complexes into the cytoplasm since the complexes are able to remain inside the cell and deliver a steady, lethal dose of the drug to the tumor. The photographs of breast cancer tumors resistant to chemotherapy revealed their substantial reduction in size after treatment with the nanodiamonds–DOX complexes.[448,449]

Other carbon nanomaterials as the **fullerenes**, **carbon nanohorns**, and **carbon nanotubes** (CNTs), have been used as carriers in drug delivery[450] due to their facile functionalization,[451–453] high carrier capacity,[454,455] high chemical stability,[456,457] and feasibility of incorporating both hydrophilic and hydrophobic substances.[458,459] However, important safety issues have been raised in the literature regarding their high toxicity. Certain types of CNTs might induce mesothelioma-like lesions in mice, in a manner similar to that observed for mesothelioma induced by asbestos.[460,461,462–464] Following peritoneal injection, long and multiwalled CNTs produce inflammation and fibrosis in the peritoneal cavity similarly to asbestos fibers whereas short tangled ones did not cause any significant inflammation.[461,465] Diameter and rigidity of multiwalled CNTs are critical factors in mesothelial injury and carcinogenesis.[465] The efficacy and safety of CNTs for Nanomedicine was recently reviewed.[466] CNTs do have the potential to deliver drugs directly to targeted cells and tissues in particular as an anticancer platform in oncology.[467] Confocal microscopy showed that the single-walled CNTs have the ability to be internalized into the cells whereas the multiwalled CNTs are excluded from the interior.[468] The long single-walled CNTs are localized only in the cytoplasm but the short single-walled CNTs are transported into the nucleus.[468] The subcellular fate of the CNT would also be size-dependent.[468]

About 4000 papers can be retrieved from the literature over the last 10 years when a search on **dendrimers** is performed. Among these about 400 papers are review articles. Several excellent reviews on dendrimers in nanomedicine are available.[469–473] Dendrimers are routinely synthesized with structural control and can be tuned in accordance with the desired application. Their size depends on the number of generations involved in the polymerization and usually ranges from 1 up to 10 nm of mean diameter. They have been considered nanoscaffolds able to accommodate drugs, biomolecules and NPs such as gold or iron oxide. The most characterized dendrimers are of the polyamidoamine (PAMAM) type and have been used for targeted, diagnostic MRI coupled to near-IR contrast agents, and/or for controlled delivery of cancer therapies. The DOX, cisplatin and DNA have been delivered. Combinations of PAMAM dendrimers functionalized by hyaluronic acid (HA) effectively delivered DOX and MVP targeted small-interfering RNA (MVP-siRNA) to down regulate expression of the MPV protein involved in breast cancer drug resistance.[474] Surface acetylation of the very toxic polypropylenimine dendrimers reduced their cytotoxicity and simultaneously improved their DOX and sodium methotrexate loading capacity.[475] A pH-sensitive dual-targeting drug carrier (G4-DOX–PEG–Tf–TAM) was synthesized with transferrin (Tf)-conjugated on the exterior and TAM in the interior of the fourth generation PAMAM dendrimers for enhancing the BBB transportation and improving the drug accumulation in brain glioma. The carrier was internalized into C6 glioma cells upon crossing the BBB model by the coactions of TfR-mediated endocytosis and the inhibition effect of TAM to the drug efflux transporters. The *in vitro* accumulation of DOX in the avascular C6 glioma spheroids effectively reduced the tumor volume.[476] DOX-conjugated PEGylated polylysin dendrimers showed similar tumoricidal activity but lower systemic toxicity when compared to PEGylated liposome and solution formulations in mouse and rat tumor models. Dendrimer-based DOX formulations may provide advantage over solution and liposomal formulations of DOX via a reduction in systemic toxicity.[477]

Irrespective of their size, anionic and neutral dendrimers did not induce leukocyte procoagulant activity (PCA) *in vitro* whereas cationic ones induced PCA in a size- and charge-dependent manner. The mechanism of PCA induction was similar to that of DOX. Cationic dendrimers were also found to exacerbate endotoxin-induced PCA.[478] Recently, poly(L-glutamic acid)-based star-block copolymers were proposed as pH-responsive release systems with reduced toxicity.[479] Hybrid dendrimers containing both DOX and superparamagnetic IONPs were also synthesized.[480] DOX was conjugated to the dendrimer segments of amino-stabilized IONPs using hydrazine as the linker via hydrazone bonds, which is acid cleavable and can be used as an ideal pH-responsive drug release system. The drug release profiles of DOX–PAMAM dendrimer conjugates were studied at pH 5.0 and 7.4. The results showed that the hydrolytic release profile can be obtained only at the condition of lysosomal pH (pH = 5.0), and IONPs participated in carrying DOX to the tumor by the Enhanced Permeability and Retention (EPR) effect. PEG–oligocholic acid telodendrimer micelles were used for the targeted delivery of DOX to B-cell lymphoma.[481] These telodendrimers can efficiently encapsulate hydrophobic drugs, self-assemble to form stable micelles in aqueous condition, were found to have higher DOX loading capacity, superior stability in physiological condition, and more sustained release profile. DOX-loaded micelles were able to prolong the blood retention time, preferentially accumulated in B-cell lymphomas via the EPR effect.

The efficacy of photodynamic therapy *in vivo* was improved by encapsulating phthalocyanine in dendrimers.[482] In another study, a delivery system based on a dendrimer and a single-strand DNA-A9 prostate-specific membrane antigen (PSMA) RNA aptamer hybrid provided a promising possibility of chemoimmunotherapy against prostate cancer both *in vivo* and *in vitro*.[483]

A folate-functionalized degradable amphiphilic dendrimer-like star polymer (FA-DLSP) with a well-defined poly(L-lactide) (PLLA) star polymer core and six hydrophilic polyester dendrons based on 2, 2-bis (hydroxymethyl) propionic acid was synthesized forming unimolecular micelles in aqueous solution with ca. 15 nm of mean diameter.[484] These FA-DLSP micelles were loaded with DOX and demonstrated a sustained release of DOX due to the hydrophobic interaction between the polymer core and the drug molecules. The cellular binding of the FA-DLSP hybrid against human KB cells with overexpressed folate-receptors was about twice that of the neat DLSP (without FA). The *in vitro* cellular cytotoxicity indicated that the

FA-DLSP micelles (without DOX) had good biocompatibility with KB cells, whereas DOX-loaded micelles exhibited a similar degree of cytotoxicity against KB cells as that of free DOX. These results clearly showed that the FA-DLSP unimolecular micelles could be a promising nanosize anticancer drug carrier with excellent targeting property. In another approach, RGD-modified PEG–PAMAM–DOX conjugates were produced and tested *in vitro* and *in vivo* against glioma.[485] This RGD attachment was expected to target the formulation to integrin receptors which are often overexpressed in tumors. In *in vivo*, the conjugates showed significantly prolonged half-life and higher accumulation in brain tumor than in normal brain tissue. Some toxicity issues regarding the dendrimers have recently been raised. Cationic G7 PAMAM dendrimers disrupt key platelet functions causing alterations in their morphology.[486] This was assumed as the mechanism behind the disseminated coagulopathy and lethality observed after *i.v.* administration of amine-terminated PAMAM dendrimers in mice.[487] In another set of experiments, the hemolytic effect of non-disguised dendrimers on blood cells was evaluated.[488] Hemotoxicity of dendrimers was found to be concentration-, generation-, and time-dependent. Various changes in the RBCs' shape in response to interactions with dendrimers have been observed, from echinocytic transformations through cell aggregation to cluster formation, depending on the dendrimer's type and concentration; anionic dendrimers were found to be less hemotoxic than cationic ones.[488] The genotoxicity and cytotoxicity of poly(propylene imine) (PPI) was recently determined for three types of 4th generation PPI dendrimers.[489] Peripheral blood mononuclear cells (PBMCs) were treated with uncoated (PPI-g4) dendrimers, and dendrimers in which approximately 40% or 90% of peripheral amino groups were coated with maltotriose (PPI-g4-open shell or PPI-g4-dense shell) at concentration of 0.05, 0.5, 5 mg/ml. After 1 h of cell incubation at 37°C, the MTT and comet assays were performed. PPI dendrimers demonstrated surface-modification-degree dependent toxicity, although genotoxicity of PPI-g4 and PPI-g4-OS measured by the comet assay was concentration-dependent up to 0.5 mg/ml and at 5 mg/ml the amount of DNA that left comet's head decreased. This suggested a strong interaction between dendrimers and DNA and also the beneficial effect of coating by maltotriose to reduce their genotoxicity.[489]

The innate immune system recognizes a set of molecular structures produced by microorganisms (pathogen-associated molecular patterns, or PAMPs) through the pattern recognition receptors (PRRs).[490] There are several classes of PRRs, each of which performs distinct functions, such as activation of the complement cascade and phagocytosis. Among these classes is the one of the Toll-like receptors (TLRs) which have the ability to recognize pathogens, or pathogen-derived products, and initiate signaling cascades causing activation of innate host defenses[491] such as the recruitment of neutrophils and activation of macrophages, enhancing the phagocytosis and killing of microbes.[492] In addition to being the primary sensors for invading pathogens, it has now been established that TLRs play a fundamental role in the activation of adaptive immune responses.[493] Activation of cells of the adaptive immune system is controlled, at least in part, by the induction of co-stimulatory molecules and production of various cytokines and chemokines by the cells of the innate immune system, primarily DCs.[494] Although the initial detection and initiation of antimicrobial responses are dependent on the ability of TLRs to sense PAMPs, there is constant interplay between cells of the innate and adaptive immune system through interaction between both cell surface molecules and soluble mediators, such as cytokines and chemokines. Normally, these interactions lead to protective immune responses against invading pathogens. However, inappropriate activation of TLR pathways may lead, under certain conditions, to the initiation of autoimmune responses.[495] A high density of positive charges close to cell surface is likely to be recognized as a signal of danger by cells or contribute to trigger cascades that are classically activated by endogenous cationic compounds. For example, cationic lipids were shown to activate several cellular pathways like pro-apoptotic and pro-inflammatory cascades.[496] A very interesting issue named **Nanotoxicity** just appeared in Volume 64 of the journal *Adv. Drug Deliv. Rev.*, dealing with important *in vivo* responses that may take place when nanomedicines are administered *in vivo*.[497]

6. Conclusions and Outlook

This article has focused on assemblies and NPs with 1–100 nm which have the potential to be used as nanomedicines. The aim of improving the therapeutic index of any novel formulation involves two important issues: the therapeutic effect itself and the side effect(s) often undesirable. The emerging fields of Nanobiotechnology and Nanomedicine are presently shaken by data on Nanotoxicity. However, the right balance will possibly be achieved from the joint effort of a multidisciplinary community as the one that presently handles the task of developing novel biomimetic systems in Nanomedicine.

Acknowledgments

This work was supported by the Conselho Nacional de Desenvolvimento Científico e Tecnológico (CNPq) and by the Fundação de Âmparo à Pesquisa do Estado de São Paulo.

References

1. M. V. Volkenshtein, *Biofísica*, Moscou, Editorial Mir (1985).
2. J. N. Israelachvili, *Intermolecular and surface forces.* 2 edn. San Diego: Academic Press (1992).
3. L. Stryer, *Biochemistry*, 4th edn. New York, Freeman (1995).
4. C. Tanford, *The Hydrophobic Effect*, New York: Wiley (1980).
5. F. London, *Trans. Faraday Soc.* **33**, 8 (1937).
6. J. Mahanty and B.W. Ninham, *Dispersion forces.* New York: Academic Press (1976).
7. H. C. Hamaker, *Physica* IV, **10**, 1058 (1937).
8. O. Stern, *Z. Elektrochem.* **30**, 508 (1924).
9. D. C. Grahame, *Chem. Rev.* **41**, 441 (1947).
10. R. J. Hunter, *Foundations of Colloid Science*, vol. 1. Oxford: Clarendon Press (1989).
11. E. J. W. Verwey and J. T. Overbeek, *Theory of the stability of lyophobic colloids.* Amsterdam: Elsevier (1948).
12. O. F. Devereux and P. L. De Bruyn, *Interaction of plane-parallel double layers.* Cambridge, Mass: MIT Press (1963).
13. E. M. A. Pereira, D. F. S. Petri and A. M. Carmona-Ribeiro, *J. Phys. Chem. B* **106**, 8762 (2002).
14. A. M. Carmona-Ribeiro, *J. Phys. Chem.* **93**, 2630 (1989).
15. A. M. Carmona-Ribeiro, *J. Phys. Chem.* **96**, 9555 (1992).
16. H. Reerink and J.T. Overbeek, *Discuss. Faraday Soc.* **18**, 74 (1954).
17. J. N. Israelachvili, *Chem. Scr.* **25**, 7 (1985).
18. D. Tabor and R. Winterton, *Proc R Soc Lond A Math Phys Sci. Ser. A312*, 1511, 435 (1969).
19. J. N. Israelachvili and G. E. Adams, *J Chem Soc. Faraday Trans.* 1, **74**, 975 (1978).
20. R. P. Rand and V. A. Parsegian, *Biochim. Biophys. Acta* **988**, 351 (1989).
21. D. M. LeNeveu, R. P. Rand and V. A. Parsegian, *Nature* **259** (5544), 601 (1976).
22. P. M. Claesson, A. M. Carmona-Ribeiro and K. Kurihara, *J. Phys. Chem.* **93**, 917 (1989).
23. J. Marra and J. N. Israelachvili, *Biochemistry* **24**(17), 4608 (1985).
24. A. C. Cowley, N. L. Fuller, R. P. Rand and V. A. Parsegian, *Biochemistry* **17**(15), 3163 (1978).
25. R. P. Rand, *Annu. Rev. Biophys. Bioeng.* **10**, 277 (1981).
26. L. J. Lis, M. McAlister, N. Fuller, R. P. Rand and V. A. Parsegian, *Biophys. J.* **37**, 657 (1982).
27. J. Klein and P. F. Luckham, *Macromolecules* **17**, 1041 (1984).
28. J. N. Israelachvili, R. K. Tandon and L. R. White, *Nature* **277**, 120 (1979).
29. Y. I. Rabinovich and B. V. Derjaguin, *Colloids Surf.* **30**, 243 (1988).

30. J. L. Parker, D. L. Cho and P. M. Claesson, *J. Phys. Chem.* **93**, 6121 (1989).
31. H. K. Christenson, J. Fang, B. W. Ninham and J. L. Parker, *J. Phys. Chem.* **94**, 8004 (1990).
32. C. A. Helm, J. N. Israelachvili and P. M. McGuiggan, *Science* **246**(4932), 919 (1989).
33. A. M. Carmona-Ribeiro, L. S. Yoshida and H. Chaimovich, *J. Phys. Chem.* **89**, 2928 (1985).
34. A. M. Carmona-Ribeiro and H. Chaimovich, *Biophys. J.* **50**, 621 (1986).
35. A. M. Carmona-Ribeiro, *J. Phys. Chem.* **97**, 11843 (1993).
36. G. M. Whitesides, J. P. Mathias and C. T. Seto, *Science* **254**, 1312 (1991).
37. I. A. Aksay, M. Trau, S. Manne, I. Honma, N. Yao, L. Zhou, P. Fenter, P. M. Eisenberger and S. M. Gruner, *Science* **273**, 892 (1996).
38. S. Nie, Y. Xing, G. J. Kim and J. W. Simons, *Annu. Rev. Biomed. Eng.* **9**, 257 (2007).
39. M. Soloviev, *J. Nanobiotechnology* **5**, 11 (2007).
40. R. K. Jain, *Adv. Drug Deliv. Rev.* **46**(1–3), 149 (2001).
41. J. Kreuter, *Adv. Drug Deliv. Rev.* **47**(1), 65 (2001).
42. J. Kreuter, D. Shamenkov, V. Petrov, P. Ramge, K. Cychutek, C. Koch-Brandt and R. Alyautdin, *J. Drug Target* **10**(4), 317 (2002).
43. U. Schroeder, P. Sommerfeld, S. Ulrich and B.A. Sabel, *J. Pharm. Sci.* **87**(11), 1305 (1998).
44. B. Mishra, B. B. Patel and S. Tiwari, *Nanomedicine* **6**, 9 (2009).
45. K. S. Soppimath, T. M. Aminabhavi, A. R. Kulkarni and W. E. Rudzinski, *J. Control Release* **70**, 1–20, (2001).
46. Y. I. Jeong, S. J. Seo, I. K. Park, H. C. Lee, I. C. Kang, T. Akaike and C. S. Cho, *Int. J. Pharm.* **296**(1–2), 151 (2005).
47. T. Musumeci, C. A. Ventura, I. Giannone, B. Ruozi, L. Montenegro, R. Pignatello and G. Puglisi, *Int. J. Pharm.* **325**(1–2), 172 (2006).
48. D. B. Shenoy and M. M. Amiji, *Int. J. Pharm.* **293**(1–2), 261 (2005).
49. J. Singh, S. Pandit, V. W. Bramwell and H. O. Alpar, *Methods* **38**(2), 96 (2006).
50. C. Damge, P. Maincent and N. Ubrich, *J. Control Release* **117**(2), 163 (2007).
51. M. Arruebo, R. Fernández-Pacheco, M. R. Ibarra and J. Santamaría, *Nano Today* **2**(3), 22 (2007).
52. A. Ito, M. Shinkai, H. Honda and T. Kobayashi, *J. Biosci. Bioeng.* **100**, 1–11 (2005).
53. T. K. Jain, J. Richey, M. Strand, D. L. Leslie-Pelecky, C. A. Flask and V. Labhasetwar, *Biomaterials* **29**(29), 4012 (2008).
54. P. K. Jain, I. H. El-Sayed and M. A. El-Sayed, *Nano Today* **2**(1), 18 (2007).
55. D. A. Giljohann, D. S. Seferos, P. C. Patel, J. E. Millstone, N. L. Rosi and C. A. Mirkin, *Nano Lett.* 7(12), 3818 (2007).
56. I. I. Slowing, J. L. Vivero-Escoto, C. W. Wu and V. S. Lin, *Adv. Drug Deliv. Rev.* **60**, 1278 (2008).
57. P. Ghosh, G. Han, M. De, C. K. Kim and V. M. Rotello, *Adv. Drug Deliv. Rev.* **60**, 1307 (2008).
58. D. Li, G. Li, W. Guo, P. Li, E. Wang and J. Wang, *Biomaterials* **29**, 2776 (2008).
59. A. M. Smith, H. Duan, A. M. Mohs and S. Nie, *Adv. Drug Deliv. Rev.* **60**, 1226 (2008).
60. M. V. Yezhelyev, X. Gao, Y. Xing, A. Al-Hajj, S. Nie and R. M. O'Regan, *Lancet Oncol*, 7, 657 (2006).
61. L. Bromberg, *J. Control. Release*, **128**, 99 (2008).

62. Y. Bae, T. A. Diezi, A. Zhao and G. S. Kwon, *J. Control. Release* **122**, 324 (2007).
63. W. Y. Seow, J. M. Xue and Y. Y. Yang, *Biomaterials* **28**, 1730 (2007).
64. N. Rapoport, *Prog. Polym. Sci.* **32**, 962 (2007).
65. J. Liu, W. Hu, H. Chen, Q. Ni, H. Xu and X. Yang, *Int. J. Pharm.* **328**, 191 (2007).
66. N. Zhang, Q. Ping, G. Huang, W. Xu, Y. Cheng and X. Han, *Int. J. Pharm.* **327**, 153(2006).
67. J. Liu, T. Gong, H. Fu, C. Wang, X. Wang, Q. Chen, Q. Zhang, Q. He and Z. Zhang, *Int. J. Pharm.* **356**, 333 (2008).
68. C. Puglia, P. Blasi, L. Rizza, A. Schoubben, F. Bonina, C. Rossi and M. Ricci, *Int. J. Pharm.* **357**, 295 (2008).
69. K. A. Shah, A. A. Date, M. D. Joshi and V. B. Patravale, *Int. J. Pharm.* **345**, 163 (2007).
70. I. P. Kaur, R. Bhandari, S. Bhandari and V. Kakkar, *J. Control. Release* **127**, 97 (2008).
71. P. Blasi, S. Giovagnoli, A. Schoubben, M. Ricci and C. Rossi, *Adv. Drug Deliv. Rev.* **59**, 4540 (2007).
72. R. Pandey, S. Sharma and G. K. Khuller, *Tuberculosis* **85**, 415 (2005).
73. O. M. Koo, I. Rubinstein and H. Onyuksel, *Nanomedicine* **1**, 77 (2005).
74. A. Papagiannaros, T. Levchenko, W. Hartner, D. Mongayt and V. Torchilin, *Nanomedicine* **5**, 216 (2009).
75. H. Cesur, I. Rubinstein, A. Pai and H. Onyuksel, *Nanomedicine* **5**, 178 (2009).
76. K. Khatri, A. K. Goyal, P. N. Gupta, N. Mishra, A. Mehta and S. P. Vyas, *Vaccine* **26**, 2225 (2008).
77. K. A. Edwards, F. Duan, A. J. Baeumner and J. C. March, *Anal. Biochem.* **380**, 59 (2008).
78. D. C. Drummond, O. Meyer, K. Hong, D. B. Kirpotin and D. Papahadjopoulos, *Pharmacol. Rev.* **51**, 691 (1999).
79. L. Zhang, H. Gao, L. Chen, B. Wu, Y. Zheng, R. Liao, Y. Jiang and F. He, *Cancer Lett.* **269**, 26 (2008).
80. A. M. Carmona-Ribeiro and B. R. Midmore, *Langmuir* **8**, 801 (1992).
81. A. M. Carmona-Ribeiro and T. M. Herrington, *J. Colloid Interface Sci.* **156**, 19 (1993).
82. L. R. Tsuruta, M. M. Lessa and A. M. Carmona-Ribeiro, *J. Colloid Interface Sci.* **175**, 470 (1995).
83. A. M. Carmona-Ribeiro and M. M. Lessa, *Colloids Surf. A Physicochem. Eng. Aspects* **153**, 355 (1999).
84. E. M. A. Pereira, D. B. Vieira and A. M. Carmona-Ribeiro, *J. Phys. Chem. B* **108**, 11490 (2004).
85. N. Lincopan, N. M. Espíndola, A. J. Vaz and A. M. Carmona-Ribeiro, *Int. J. Pharm.* **340**, 216 (2007).
86. R. Rapuano and A. M. Carmona-Ribeiro, *J. Colloid Interface Sci.* **193**, 104 (1997).
87. R. Rapuano and A. M. Carmona-Ribeiro, *J. Colloid Interface Sci.* **226**, 299 (2000).
88. S. P. Moura and A. M. Carmona-Ribeiro, *Langmuir* **19**, 6664 (2003).
89. S. P. Moura and A. M. Carmona-Ribeiro, *Langmuir* **21**, 10160 (2005).
90. S. P. Moura and A. M. Carmona-Ribeiro, *J. Colloid Interface Sci.* **313**, 519 (2007).
91. D. F. S. Petri and A. M. Carmona-Ribeiro, In *Biomimetic Particles: Polymeric Nanostructures and Their Applications*, H. S. Nalwa (Ed.), Stevenson Ranch: American Scientific Publishers (2007) 485–531.
92. L. B. Castro, K. V. Soares, A. F. Naves, A. M. Carmona-Ribeiro and D. F. S. Petri, *Ind. Eng. Chem. Res.* **43**, 7774 (2004).

93. L. B. Castro, F. F. Silva, A. M. Carmona-Ribeiro, M. Kappl and D. F. S. Petri, *J. Phys. Chem. B* **111**, 8520 (2007).
94. A. M. Carmona-Ribeiro, *Int. J. Nanomedicine* **5**, 249 (2010).
95. N. Lincopan, N. M. Espíndola, A. J. Vaz, M. H. Costa, E. Faquim-Mauro and A. M. Carmona-Ribeiro, *Vaccine* **27**, 5760 (2009).
96. S. P. Moura and A. M. Carmona-Ribeiro, *Cell Biochem. Biophys.* **44**, 446 (2006).
97. H. Rosa, D. F. S. Petri and A. M. Carmona-Ribeiro, *J. Phys. Chem. B* **112**, 16422 (2008).
98. L. Zhang, J. M. Chan, F. X. Gu, J. W. Rhee, A. Z. Wang, A. F. Radovic-Moreno, R. Langer and O. C. Farokhzad, *ACS Nano* **2**, 1696 (2008).
99. V. Labhasetwar, *Curr. Opin. Biotechnol.* **16**, 674 (2005).
100. D. B. Vieira and A. M. Carmona-Ribeiro, *J. Nanobiotechnology* **6**, 1, 6 (2008).
101. S. Bisht, G. Feldmann, S. Soni, R. Ravi, C. Karikar and A. Maitra, *J. Nanobiotechnology.* 5, 1, 3 (2007).
102. R. Chouhan and A. K. Bajpai, *J. Nanobiotechnology* 7, 1, 5 (2009).
103. A. Agarwal, Y. Lvov, R. Sawant and V. Torchilin, *J. Control. Release* **128**, 255 (2008).
104. S. Galindo-Rodriguez, E. Allemann, H. Fessi and E. Doelker, *Pharm. Res.* **21**, 1428 (2004).
105. M. J. Alonso, C. Losa, P. Calvo and J. L. Vila-Jato, *Int. J. Pharm.* **68**, 69 (1991).
106. C. Vauthier and P. Couvreur, *J. Biomed. Nanotechnol.* **3**, 223 (2007).
107. A. H. Faraji and P. Wipf, *Bioorg. Med. Chem.* **17**, 2950 (2009).
108. A. Mahapatro and D. K. Singh, *J. Nanobiotechnology* 9, 1, 55 (2011).
109. S. Y. Kim and Y. M. Lee, *Biomaterials* **22**, 1697 (2001).
110. P. Couvreur, B. Kante, V. Lenaerts, V. Scailteur, M. Roland and P. Speiser, *J Pharm Sci.* **69**(2):199–202 (1980).
111. C. Verdun, F. Brasseur, H. Vranckx, P. Couvreur and M. Roland, *Cancer Chemother. Pharmacol.* **26**, 13 (1990).
112. V. P. Torchilin and V. S. Trubetskoy, *Adv. Drug Deliv. Rev.* **16**, 141 (1995).
113. P. Calvo, C. Remuñan-López, J. L.Vila-Jato and M. J. Alonso, *Pharm. Res.* **14**, 1431 (1997).
114. Q. Gan and T. Wang, *Colloids Surf. B: Biointerfaces* **59**, 24 (2007).
115. J. C. Zillies, K. Zwiorek, F. Hoffmann, A. Vollmar, T. J. Anchordoquy, G. Winter and C. Coester, *Eur. J. Pharm. Biopharm.* **70**, 514 (2008).
116. P. G. Jenkins, K. A. Howard, N. W. Blackball, N. W. Thomas, S. S. Davis and D. T. O'Hagan, *J. Control. Release* **29**, 339 (1994).
117. L. Nobs, F. Buchegger, R. Gurny, E. Allémann, *Bioconjug. Chem.* **17**, 139 (2006).
118. M. O. Oyewumi, R. A. Yokel, M. Jay, T. Coakley and R. J. Mumper, *J. Control. Release* **95**, 613 (2004).
119. C. Damgé, C. Michel, M. Aprahamian, P. Couvreur and J. P. Devissaguet, *J. Control. Release* **13**, 233 (1990).
120. J. H. Eldridge, C. J. Hammond, J. A. Meulbroek, J. K. Staas, R. M. Gilley and T. R. Tice, *J. Control. Release* **11**, 205 (1990).
121. J. Panyam, W-Z Zhou, S. Prabha, S. K. Sahoo and V. Labhasetwar, *FASEB J.* **16**, 1217 (2002).
122. M. L. Hedley, J. Curley and R. Urban, *Nat. Med.* **4**, 365 (1998).
123. C. Chakraborty, B. Sarkar, C. Hsu, Z. Wen, C. Lin and P. Shieh, *J. Neurooncol.* **93**, 285, (2009).
124. Y. Chen, G. Dalwadi and H. A. E. Benson, *Curr. Drug Deliv.* **1**, 361 (2004).

125. W. Sun, C. Xie, H. Wang and Y. Hu, *Biomaterials* **25**, 3065 (2004).
126. P. R. Lockman, M. O. Oyewumi, J. M. Koziara, K. E. Roder, R. J. Mumper and D. D. Allen, *J. Control. Release* **93**, 271 (2003).
127. F. M. Goycoolea, G. Lollo, C. Remuñán-López, F. Quaglia and M. J. Alonso, *Biomacromolecules* **10**, 1736 (2009).
128. M. C. Bonferoni, G. Sandri, S. Rossi, F. Ferrari and C. Caramella, *Expert. Opin. Drug Deliv.* **6**, 923 (2009).
129. W. Wang and M. Singh, In Selection of optimal adjuvants and product factors that affect vaccine immunogenicity, Development of Vaccines: From Discovery to Clinical Testing, M. Singh and I. S. Srivastava (Eds.), 1st ed. John Wiley & Sons (2011).
130. D. T. O'Hagan and R. Rappuoli, *Pharm. Res.* **21**, 1519 (2004).
131. J. H. Wilson-Welder, M. P. Torres, M. J. Kipper, S. K. Mallapragada, M. J. Wannemuehler and B. Narasimhan, *J. Pharm. Sci.* **98**, 1278 (2009).
132. A. Sokolovska, S. L. Hem and H. Hogen Esch, *Vaccine* **25**, 4575 (2007).
133. T. Vesikari, M. Pellegrini, A. Karvonen, N. Groth, A. Borkowski, D. T. O'Hagan and A. Podda, *Pediatr. Infect. Dis. J.* **28**, 563 (2009).
134. P. G. Seferian and M. L. Martinez, *Vaccine* **19**, 661 (2000).
135. X. Wang, T. Uto, T. Akagi, M. Akashi and M. Baba, *J. Med. Virol.* **80**, 11 (2008).
136. L. Stertman, L. Strindelius and I. Sjoholm, *Vaccine* **22**, 2863 (2004).
137. L. Stertman, E Lundgren and I. Sjoholm, *Vaccine* **24**, 3661 (2006).
138. S. A. Moschos, V. W. Bramwell, S. Somavarapu and H. O. Alpar, *Vaccine* **23**, 1923 (2005).
139. G. L. Morefield, D. Jiang, I. Z. Romero-Mendez, R. L. Geahlen, H. Hogenesch and S. L. Hem, *Vaccine* **23**, 1502 (2005).
140. J. P. Amorij, G. F. Kersten, V. Saluja, W. F. Tonnis, W. L. Hinrichs, B. Slütter, S. M. Bal, J. A. Bouwstra, A. Huckriede and W. Jiskoot, *J. Control. Release* **161**, 363 (2012).
141. M. Singh, M. Briones, G. Ott and D. T. O'Hagan, *PNAS* **97**, 811 (2000).
142. V. Saini, V. Jain, M. S. Sudheesh, K. S. Jaganathan, P. K. Murthy, D. V. Kohli, *Int. J. Pharm.* **408**, 50 (2011).
143. D. J. Kirby, I. Rosenkrands, E. M. Agger, P. Andersen, A. G. Coombes and Y. Perrie, *J. Drug Target* **16**, 282 (2008).
144. A. D. Bangham, ed., *Liposome Letters.* London: Academic Press (1983).
145. D. D. Lasic: *Liposomes: from physics to applications.* Amsterdam: Elsevier (1993).
146. D. D. Lasic, *Angew. Chem. Int. Ed. Engl.* **33**, 1685 (1994).
147. A. Katz, C. A. Kraaijeveld and H. Snippe, In *The Theory and Practical Application of Adjuvants*, Chichester: John Wiley & Sons (1994), 37–50.
148. R. Glück, In *Vaccine Design: The Subunit and Adjuvant Approach*, M.F. Powell and M.J. Newman (Eds.), New York: Plenum Press (1995), 325–345.
149. V. S. Trubetskoy and V. P. Torchilin, *J. Liposome Res.* **4**, 961 (1994).
150. V. S. Trubetskoy, J. A. Cannillo, A. Milshtein, G. L. Wolf and V. P. Torchilin, *Magn. Reson. Imaging* **13**, 31 (1995).
151. V. S. Trubetskoy, G. S. Gazelle, G. L. Wolf and V. P. Torchilin, *J. Drug Target.* **4**, 381 (1997).
152. B. Sternberg, K. Hong, W. Zheng and D. Papahadjopoulos, *Biochim. Biophys. Acta* **1375**, 23 (1998).
153. T. Blessing, J. S. Remy and J. P. Behr, *J. Am. Chem. Soc.* **120**, 8519 (1998).

154. M. Chattergoon, J. Boyer and D. B. Weiner, *FASEB J.* **11**, 753 (1997).
155. M. G. Ganesan, N. D. Weiner, G. L. Flynn and N. Ho, *Int. J. Pharm.* **20**, 139 (1984).
156. G. Gregoriadis (Ed.): *Liposome Technology*, 2nd ed. Boca Raton: CRC Press (1993).
157. D. D. Lasic and D. Needham, *Chem. Rev.* **95**, 2601 (1995).
158. R. New (Ed.), *Liposomes: A Practical Approach.* New York: IRL Press, Oxford University Press (1990).
159. J. N. Israelachvili, D. J. Mitchell and B. W. Ninham, *Biochim. Biophys. Acta* **470**(2), 185 (1977).
160. J. H. Fendler, *Acc. Chem. Res.* **13**, 7 (1980).
161. J. H. Fuhrhop and J. Mathieu, *Angew. Chem. Int. Ed. Engl.* **23**, 100 (1984).
162. T. Kunitake and Y. Okahata, *J. Am. Chem. Soc.* **99**, 3860 (1977).
163. T. Kunitake, Y. Okahata, K. Tamaki, F. Kumamaru and M. Takayanagi, *Chem. Lett.* **6**, 387 (1977).
164. T. Kunitake and Y. Okahata, *Bull. Chem. Soc. Jpn.* **51**, 1877 (1978).
165. R. A. Mortara, F. H. Quina and H. Chaimovich, *Biochem. Biophys. Res. Commun.* **81**, 1080 (1978).
166. S. L. Regen, B. Czech and A. Singh, *J. Am. Chem. Soc.* **102**, 6638 (1980).
167. J. H. Fuhrhop and D. Fritsch, *Acc. Chem. Res.* **19**, 130 (1986).
168. A. M. Carmona-Ribeiro, *Chem. Soc. Rev.* **21**, 209 (1992).
169. H. Eibl and A. Blume, *Biochim. Biophys. Acta* **553**, 476 (1979).
170. A. Blume and H. Eibl, *Biochim. Biophys. Acta* **558**, 13 (1979).
171. S. N. Chatterjee and S. Agarwal, *Free Radic. Biol. Med.* **4**, 51 (1988).
172. M. Grit and D. J. A. Crommelin, *Chem. Phys. Lipids* **62**, 113 (1992).
173. F. J. Martin, (Ed.), *Specialised Drug Delivery Systems,* New York: Marcel Dekker (1990).
174. G. N. Tapias, S. M. Sicchierolli, E. M. Mamizuka and A. M. Carmona-Ribeiro, *Langmuir* **10**, 3461 (1994).
175. S. M. Sicchierolli, E. M. Mamizuka and A. M. Carmona-Ribeiro, *Langmuir* **11**, 2991 (1995).
176. L. M. S. Martins, E. M. Mamizuka and A. M. Carmona-Ribeiro, *Langmuir* **13**, 5583 (1997).
177. M. T. N. Campanhã, E. M. Mamizuka and A. M. Carmona-Ribeiro, *J. Phys. Chem. B* **105**, 8230 (2001).
178. M. T. N. Campanhã, E. M. Mamizuka and A. M. Carmona-Ribeiro, *J. Lipid Res.* **40**, 1495 (1999).
179. L. A. Hilgers, H. Snippe, M. Jansze and J. M. Willers, *Cell Immunol.* **92**, 203 (1985).
180. M. D. Prager, M. C. Kanar, J. L. Farmer and J. Vanderzee, *Cancer Lett.* **27**, 225 (1985).
181. M. O. Dailey and R. L. Hunter, *J. Immunol.* **112**, 1526 (1974).
182. R. H. Smith and B. Ziola, *Immunology* **58**, 245 (1986).
183. L. R. Tsuruta, W. Quintilio, M. H. B. Costa and A. M. Carmona-Ribeiro, *J. Lipid Res.* **38**, 2003 (1997).
184. H. Cai, X. Tian, X. D. Hu, S. X. Li, D. H. Yu and Y. X. Zhu, *Vaccine* **23**, 3887 (2005).
185. J. Davidsen, I. Rosenkrands, D. Christensen, A. Vangala, D. Kirby, Y. Perrie, E. M. Agger and P. Andersen, *Biochim. Biophys. Acta* **1718**, 22 (2005).
186. M. Henriksen-Lacey, V. W. Bramwell, D. Christensen, E. M. Agger, P. Andersen and Y. Perrie, *J. Control. Release* **142**, 180 (2010).
187. M. Henriksen-Lacey, A. Devitt and Y. Perrie, *J. Control. Release* **154**, 131 (2011).

188. A. M. Carmona-Ribeiro, *Methods Mol. Biol.* **906**, 283 (2012).
189. A. M. Carmona-Ribeiro, *Curr. Med. Chem.* **13**, 1359 (2006).
190. A. M. Carmona-Ribeiro, *J. Liposome Res.* **17**, 165 (2007).
191. E. M. Mamizuka and A. M. Carmona-Ribeiro, In Cationic liposomes as antimicrobial agents, Communicating Current Research and Educational Topics and Trends in Applied Microbiology, A.M. Vilas (Ed.), Formatex, Badajoz, vol. 2, 636 (2007).
192. A. M. Carmona-Ribeiro, In *Lipid-based Biomimetics in Drug and Vaccine Delivery, Biomimetics: Learning from Nature*, A. Mukherjee (Ed.), InTech (2010). Available from:http://www.intechopen.com/books/biomimetics-learning-from-nature/lipid-based-biomimetics-in-drug-and-vaccine-delivery
193. A. M. Carmona-Ribeiro, L. Barbassa and L. D. de Melo, *Antimicrobial Biomimetics*, In *Biomimetic Based Applications*, A. George (Ed.), InTech (2011). Available from: http://www.intechopen.com/books/biomimetic-based-applications/antimicrobial-biomimetics
194. A. M. Carmona-Ribeiro, In Lipid-based Biomimetics: General Perspectives in Drug and Vaccine Delivery, Lipid Nanocarriers in Cancer Diagnosis and Therapy, E. B. Souto (Ed.), iSmithers (2011).
195. D. B. Vieira and A. M. Carmona-Ribeiro, *J. Colloid Interface Sci.* **244**, 427 (2001).
196. L.F. Pacheco and A.M. Carmona-Ribeiro, *J. Colloid Interface Sci.* **258**, 146 (2003).
197. N. Lincopan, E. M. Mamizuka and A. M. Carmona-Ribeiro, *J. Antimicrob. Chemother.* **52**, 412 (2003).
198. J. H. K. Rozenfeld, T. R. Oliveira, M. T. Lamy and A. M. Carmona-Ribeiro, *Biochim. Biophys. Acta* **1808**, 649 (2011).
199. J. H. K. Rozenfeld, S. R. Silva, P. A. Ranéia, E. Faquim-Mauro and A. M. Carmona-Ribeiro, *J. Control. Release* **160**, 367 (2012).
200. C. A. Carvalho, C. Olivares-Ortega, M. A. Soto-Arriaza and A. M. Carmona-Ribeiro, *Biochim. Biophys. Acta* **1818**, 3064 (2012).
201. F. M. Correia, D. F. S. Petri and A. M. Carmona-Ribeiro, *Langmuir* **20**, 9535 (2004).
202. F. P. Araujo, D. F. S. Petri and A. M. Carmona-Ribeiro, *Langmuir* **21**, 9495 (2005).
203. N. Lincopan and A. M. Carmona-Ribeiro, *J. Antimicrob. Chemother.* **58**(1), 66 (2006).
204. L. Barbassa, E. M. Mamizuka and A. M. Carmona-Ribeiro, *BMC Biotechnol.* 15, 11, 40 (2011).
205. L. D. Melo, E. M. Mamizuka and A. M. Carmona-Ribeiro, *Langmuir* **26**, 12300 (2010).
206. M. Johnsson and K. Edwards, *Biophys. J.* **85**, 3839 (2003).
207. E. Johansson, C. Engvall, M. Arfvidsson, P. Lundahl and K. Edwards, *Biophys. Chem.* **113**, 183 (2005).
208. E. Johansson, A. Lundquist, S. Zuo and K. Edwards, *Biochim. Biophys. Acta* **1768**, 1518 (2007).
209. T. M. Forte and M. R. McCall, *Curr. Opin. Lipidol.* **5**, 354 (1994).
210. J. Y. Letizia and M. C. Phillips, *Biochemistry* **30**, 866 (1991).
211. F. R. Poulain, L. Allen, M. C. Williams, R. L. Hamilton and S. Hawgood, *Am. J. Physiol.* **262**, L730 (1992).
212. G. Walldius and I. Jungner, *J. Intern. Med.* **255**, 188 (2004).
213. H. L. Zhu and D. Atkinson, *Biochemistry* **43**, 13156 (2004).

214. A. R. Tall, D. M. Small, R. J. Deckelbaum and G. G. Shipley, *J. Biol. Chem.* **252**, 4701 (1977).
215. J. B. Ragland, P. D. Bertram and S. M. Sabesin, *Biochem. Biophys. Res. Commun.* **80**, 81 (1978).
216. N. Subbarao, C. Fielding, R. Hamiltona and F. C. Szoka, Jr., *Proteins. Proteins Struct. Funct. Genet.* **3**, 187 (1988).
217. A. W. Shaw, M. A. McLean and S. G. Sligar, *FEBS Lett.* **556**, 260 (2004).
218. T. H. Bayburt, Y. V. Grinkova and S. G. Sligar, *Nano Lett.* **2**, 853 (2002).
219. T. H. Bayburt and S. G. Sligar, *Proc. Natl. Acad. Sci. USA* **99**, 6725 (2002).
220. T. H. Bayburt, J. W. Carlson and S. G. Sligar, *Langmuir* **16**, 5993 (2000).
221. M. R. Morrow, J. Pérez-Gil, G. Simatos, C. Boland, J. Stewart, D. Absolom, V. Sarin, and K. M. W. Keough, *Biochemistry* **32**(16), 4397–4402 (1993).
222. H. De Smedt, H. Olbrechts and R. Borghgraef, *Arch. Internat. Physiol. Biochim.* **84**, 340 (1976).
223. R. Lawaczeck, M. Kainosho and S. I. Chan, *Biochim. Biophys. Acta* **443**, 313 (1976).
224. L. M. Ickenstein, M. C. Arfvidsson, D. Needham, L. D. Mayer and K. Edwards, *Biochim. Biophys. Acta* **1614**, 135 (2003).
225. M. C. Sandstroem, L. M. Ickenstein, L. D. Mayer and K. Edwards, *J. Control. Release* **107**, 131 (2005).
226. K. Edwards and M. Almgren, *Langmuir* **8**, 824 (1992).
227. M. Johnsson and K. Edwards, *Langmuir* **16**, 8632 (2000).
228. A. de la Maza and J. L. Parra, *Biochem. J.* **303**, 907 (1994).
229. A. de la Maza and J. L. Parra, *Eur. J. Biochem.* **226**, 1029 (1994).
230. A. de la Maza, L. Coderch, O. López and J. L. Parra, *Micron* **29**, 175 (1998).
231. D. Uhrikova, N. Kucerka, A. Islamov, V. Gordeliy and P. Balgavy, *Gen. Physiol. Biophys.* **20**, 183 (2001).
232. M. Almgren, *Biochim. Biophys. Acta* **1508**, 146 (2000).
233. L. Rubio, G. Rodríguez, L. Barbosa-Barros, C. Alonso, M. Cócera, A. de la Maza and J. L. Parra, O. López, *Colloids Surf. B: Biointerfaces* **92**, 322 (2012).
234. A. Moschetta, P. M. Frederik, P. Portincasa, G. P. VanBerge-Henegouwen and K. J. Van Erpecum, *J. Lipid Res.* **43**, 1046 (2002).
235. T. H. Callisen and Y. Talmon, *Biochemistry* **37**, 10987 (1998).
236. J. A. Losonczi and J. H. Prestegard, *Biochemistry* **37**, 706 (1998).
237. J. A. Losonczi, F. Tian and J. H. Prestegard, *Biochemistry* **39**, 3804 (2000).
238. E. K. Tiburu, D. M. Moton and G. A. Lorigan, *Biochim. Biophys. Acta* **1512**, 206 (2001).
239. M. N. Triba, D. E. Warschawski and P. F. Devaux, *Biophys. J.* 88, 1887 (2005).
240. J. A. Clements, *Am. Rev. Respir. Dis.* **115**, 67 (1977).
241. W. R. Schief, M. Antia, B. M. Discher, S. B. Hall and V. Vogel, *Biophys. J.* **84**, 3792 (2003).
242. K. Edwards, M. Johnsson, G. Karlsson and M. Silvander, *Biophys. J.* **73**, 258 (1997).
243. M. Johnsson, M. Silvander, G. Karlsson and K. Edwards, *Langmuir* **15**, 6314 (1999).
244. P. C. A. Barreleiro, G. Olofsson, W. Brown, K. Edwards, N. M. Bonassi and E. Feitosa, *Langmuir* **18**, 1024 (2002).
245. M. Johnsson, N. Bergstrand and K. Edwards, *J. Liposome Res.* **9**, 53 (1999).
246. E. Johansson, C. Engvall, M. Arfvidsson, P. Lundahl and K. Edwards, *Biophys. Chem.* **113**, 183 (2005).

247. A. M. Carmona-Ribeiro and H. Chaimovich, *Biochim. Biophys. Acta* **733**, 172 (1983).
248. A. M. Carmona-Ribeiro, L. S. Yoshida, A. Sesso and H. Chaimovich, *J. Colloid Interface Sci.* **100**, 433 (1984).
249. A. M. Carmona-Ribeiro, C. E. Castuma, A. Sesso and S. Schreier, *J. Phys. Chem.* **95**, 5361 (1991).
250. R. B. Pansu, B. Arrio, J. Roncin and J. Faure, *J. Phys. Chem.* **94**, 796 (1990).
251. M. Andersson, L. Hammarstroem and K. Edwards, *J. Phys. Chem.* **99**, 14531 (1995).
252. L. Hammarstroem, I. Velikian, G. Karlsson and K. Edwards, *Langmuir* **11**, 408 (1995).
253. H. W. Meyer, W. Richter, W. Rettig and M. Stumpf, *Colloids Surf. A: Physicochem. Eng. Aspects* **183**, 495 (2001).
254. D. B. Vieira, L. F. Pacheco and A. M. Carmona-Ribeiro, *J. Colloid Interface Sci.* **293**, 240 (2006).
255. N. Lincopan, E. M. Mamizuka and A. M. Carmona-Ribeiro, *J. Antimicrob. Chemother.* **55**, 727 (2005).
256. A. M. Carmona-Ribeiro, *Curr. Med. Chem.* **10**, 2425 (2003).
257. A. M. Carmona-Ribeiro, F. Ortis, R. I. Schumacher and M. C. S. Armelin, *Langmuir* **13**, 2215 (1997).
258. A. M. Carmona-Ribeiro, *An. Acad. Bras. Cienc.* **72**, 39 (2000).
259. A. M. Carmona-Ribeiro, D. B. Vieira and N. Lincopan, Cationic surfactants and lipids as anti-infective agents. *Anti-Infective Agents Med. Chem.* **5**, 33 (2006).
260. K. N. J. Burger, R. W. H. M. Staffhorst, H. C. De Vijlder, M. J. Velinova, P. H. Bomans, P. M. Frederik and B. De Kruijff, *Nat. Med.* **8**, 81 (2002).
261. V. Chupin, A. I. P. M. De Kroon and B. De Kruijff, *J. Am. Chem. Soc.* **126**, 816 (2004).
262. M. J. Velinova, W. H. M. Rutger, W. J. M. Staffhorst, A. S. D. Mulder, B. A. J. Jansen, B. De Kruijff and A. I. P. M. De Kroon, *Biochim. Biophys. Acta* **1663**, 135 (2004).
263. G. Lopez-Berestein, R. L. Hopfer, R. Mehta, K. Mehta, E. M. Hersh and R. L. Juliano, *Antimicrob. Agents Chemother.* **25**, 366 (1984).
264. R. Patel, Expert Opin. *Pharmacother.* **1**, 475 (2000).
265. L.S.S. Guo, *Adv. Drug Deliv. Rev.* **47**, 149 (2001).
266. M. H. Miceli and P. Chandrasekar, *Infection and Drug Resistance* **5**, 9 (2012).
267. S. M. Sicchierolli and A. M. Carmona-Ribeiro, *Colloids Surf. B: Biointerfaces* **5**, 57 (1995).
268. S. M. Sicchierolli and A. M. Carmona-Ribeiro, *J. Phys. Chem.* **100**, 16771, (1996).
269. A. L. Troutier, T. Delair, C. Pichot and C. Ladaviere, *Langmuir* **21**, 4, 1305 (2005).
270. N. Lincopan, H. Rosa and A. M. Carmona-Ribeiro, *Macromol. Symp.* **245**, 485 (2006).
271. A. M. Carmona-Ribeiro, *J. Liposome Res.* **17**, 165 (2007).
272. A. L. Troutier and C. Ladavière, *Adv. Colloid Interface Sci.* **133**, 1 (2007).
273. J. Thevenot, A. L. Troutier, J. L. Putaux, T. Delair and C. Ladavière, *J. Phys. Chem. B* **112**, 13812 (2008).
274. A. Bershteyn, J. Chaparro, R. Yau, M. Kim, E. Reinherz, L. Ferreira-Moita and D. J. Irvine, *Soft Matter* **4**(9), 1787 (2008).
275. E. M. A. Pereira, P. M. Kosaka, D. B. Vieira, H. Rosa, Y. Kawano, D. F. S. Petri and A. M. Carmona-Ribeiro, *J. Phys. Chem. B* **112**(31), 9301 (2008).
276. S. Zuzzi, C. Cametti and G. Onori, *Langmuir* **24**, 6044 (2008).
277. A. L. Troutier-Thuilliez, J. Thevenot, T. Delair and C. Ladavière, *Soft Matter* **5**, 4739 (2009).

278. A. K. Jha and A. Bose, *Langmuir* **25**, 123 (2008).
279. L. C. Salay and A. M. Carmona-Ribeiro, *J. Phys. Chem. B* **102**, 4011 (1998).
280. L. C. Salay and A. M. Carmona-Ribeiro, *J. Adhes. Sci. Technol.* **13**, 1165 (1999).
281. N. Sahai, *J. Colloid Interface Sci.* **252**, 309 (2002).
282. K. Katagiri, M. Hashizume, J. Kikuchi, Y. Taketani and M. Murakami, *Colloids Surf. B Biointerfaces* **38**, 149 (2004).
283. D. J. Lestage, M. Yu and M. W. Urban, *Biomacromol.* **6**, 1561 (2005).
284. S. Mornet, O. Lambert, E. Duguet and A. Brisson, *Nano Lett.* **5**, 281 (2005).
285. E. M. A. Pereira, D. Petri and A. M. Carmona-Ribeiro, *J. Phys. Chem. B* **110**, 10070 (2006).
286. S. K. Parida, S. Dash, S. Patel and B. K. Mishra, *Adv. Colloid Interface Sci.* **121**, 77 (2006).
287. W. T. Al-Jamal and K. Kostarelos, *Nanomedicine* **2**, 85 (2007).
288. S. Bégu, A. A. Pouëssel, D. A. Lerner, C. Tourné-Péteilh and J. M. Devoisselle, *J. Control. Release* **118**, 1 (2007).
289. M. D. Senarath-Yapa, S. Phimphivong, J. W. Coym, M. J. Wirth, C. A. Aspinwall and S. S. Saavedra, *Langmuir* **23**, 12624 (2007).
290. T. A. Oleson and N. Sahai, *Langmuir* **24**, 4865 (2008).
291. I. B. Valentim, M. H. Martins and I. Joekes, *Colloids Surf. B Biointerfaces* **66**, 96 (2008).
292. H. Nabika, B. Takimoto and K. Murakoshi, *Anal. Bioanal. Chem.* **391**, 2497 (2008).
293. J. Xu, M. J. Stevens, T. A. Oleson, J. A. Last and N. Sahai, *J. Phys. Chem. C* **113**, 2187 (2009).
294. N. Lincopan, M. R. Santana, E. Faquim-Mauro, M. H. da Costa and A. M. Carmona-Ribeiro, *BMC Biotechnol.* 9, 1, 5 (2009).
295. S. Ahmed and S. L. Wunder, *Langmuir* **25**, 3682 (2009).
296. G. Nordlund, R. Loenneborg and P. Brzezinski, *Langmuir* **25**, 4601 (2009).
297. S. K. Messerschmidt, A. Musyanovych, M. Altvater, P. Scheurich, K. Pfizenmaier, K. Landfester and R. E. Kontermann, *J. Control. Release* **137**, 69 (2009).
298. G. Nordlund, J. B. Sing Ng, L. Bergström and P. Brzezinski, *ACS Nano* **3**, 2639 (2009).
299. C. Kataoka-Hamai and Y. Miyahara, *Analyst* **135**, 189 (2010).
300. R. Singh, B. Tian and K. Kostarelos, *FASEB J.* **22**, 3389 (2008).
301. R. Singh, K. T. Al-Jamal, L. Lacerda and K. Kostarelos, *ACS Nano* **2**, 1040 (2008).
302. A. M. Carmona-Ribeiro, *Chem. Soc. Rev.* **30**, 241 (2001).
303. P. von Hoegen, *Adv. Drug Deliv. Rev.* **51**, 113 (2001).
304. D. A. Bricarello, M. A. Patel and A. N. Parikh, *Trends Biotechnol.* **30**, 323 (2012).
305. E. Grabowski and I. Morrison, In *Particle size distribution from analysis of quasi-elastic light scattering data.* *Measurements of Suspended Particles by Quasi-Elastic Light Scattering,* B. Dahneke, (Ed.), New York: Wiley-Interscience (1983) 199–236.
306. B. D. Spangler, *Microbiol. Mol. Biol. Rev.* **56**, 622 (1992).
307. S. Bhat, S. L. Spitalnik, F. Gonzalez-Scarano and D. H. Silberberg, *PNAS* **88**, 7131 (1991).
308. R. H. Müller, K. Mäder and S. Gohla, *Eur. J. Pharm. Biopharm.* **50**, 161 (2000).
309. S. Martins, B. Sarmento, D. C. Ferreira and E. B. Souto, *Int. J. Nanomed.* **2**, 595 (2007).
310. P. Severino, T. Andreani, A. S. Macedo, J. F. Fangueiro, M. H. A. Santana, A. M. Silva and E. B. Souto, *J. Drug Delivery* (2012).

311. S. A. Wissing, O. Kayser and R. H. Müller, *Adv. Drug Deliv. Rev.* **56**, 1257 (2004).
312. W. Mehnert and K. Mäder, *Adv. Drug Deliv. Rev.* **47**, 165 (2001).
313. A. Fundaro, R. Cavalli, A. Bargoni, D. Vighetto, G. P. Zara and M. R. Gasco, *Pharmacol. Res.* **42**, 337 (2000).
314. M. A. Hossain, S. Maesaki, M. S. Razzaque, K. Tomono, T. Taguchi and S. Kohno, *J. Antimicrob. Chemother.* **46**, 263 (2000).
315. M. Igartua, P. Saulnier, B. Heurtault, B. Pech, J. E. Proust, J. L. Pedraz and J. P. Benoit, *Int. J. Pharm.* **233**, 149 (2002).
316. L. A. Lee and Q. Wang, *Nanomedicine* **2**, 137 (2006).
317. A. O. Elzoghby, W. M. Samy and N. A. Elgindy, *J. Control. Release* **161**, 38 (2012).
318. A. O. Elzoghby, W. M. Samy and N. A. Elgindy, *J. Control. Release* **157**, 168 (2012).
319. J.-W. Yoo, D. J. Irvine, D. E. Discher and S. Mitragotri, *Nat. Rev. Drug Discov.* **10**, 521 (2011)
320. I. Gentschev, G. Dietrich, S. Spreng, A. Kolb-Mäurer, V. Brinkmann, L. Grode, J. Hess, S. H. E. Kaufmann and W. Goebel, *Vaccine* **19**, 2621 (2001).
321. T. Ebensen, S. Paukner, C. Link, P. Kudela, C. de Domenico, W. Lubitz and C. A. Guzmán, *J. Immunol.* **172**, 6858 (2004).
322. M. Everts, V. Saini, J. L. Leddon, R. J. Kok, M. Stoff-Khalili, M. A. Preuss, C. L. Millican, G. Perkins, J. M. Brown, H. Bagaria, D. E. Nikles, D. T. Johnson, V. P. Zharov and D. T. Curiel, *Nano Lett.* **6(4)**, 587–591 (2006).
323. U. Schmidt, C. Gunther, R. Rudolph and G. Bohm, *FASEB J.* **15**, 1646 (2001).
324. L. L. Villa, R. L. R. Costa, C. A. Petta, R. P. Andrade, K. A. Ault, A. R. Giuliano, C. M. Wheeler, L. A. Koutsky, C. Malm, M. Lehtinen, F. E. Skjeldestad, S.-E. Olsson, M. Steinwall , D. R. Brown, R. J. Kurman, B. M. Ronnett, M. H. Stoler, A. Ferenczy, D. M. Harper, G. M. Tamms, J. Yu, L. Lupinacci, R. Railkar, F. J. Taddeo, K. U. Jansen, M. T. Esser, H. L. Sings, A. J. Saah and E. Barr, *Lancet Oncol.* **6**, 271 (2005).
325. S. M. Garland, M. Hernandez-Avila, C. M. Wheeler, G. Perez, D. M. Harper, S. Leodolter, G. W. Tang, D. G. Ferris, M. Steben, J. Bryan, F. J. Taddeo, R. Railkar, M. T. Esser, H. L. Sings, M. Nelson, J. Boslego, C. Sattler, E. Barr and L. A. Koutsky, *N. Engl. J. Med.* **356**(19), 1928–1943 (2007).
326. M. Lakadamyali, M. J. Rust and X. Zhuang, *Microbes Infect.* **6**, 929 (2004).
327. L. J. Earp, S. E. Delos, H. E. Park and J. M. White, *Curr. Top. Microbiol. Immunol.* **285**, 25 (2005).
328. G. Leroux-Roels, *Vaccine* **28**, 25 (2010).
329. J. De Jonge, M. Holtrop, J. Wilschut and A. Huckriede, *Gene Ther.* **13**, 400 (2006).
330. N. Wegmann, R. Schwaninger, A. Wetterwald, C. Wingenfeld, B. Rothen-Rutishauser and C. D. Gimmi, *Cancer Res.* **62**, 437 (2002).
331. C. Lei, Y. Shin, J. Liu and E. J. Ackerman, *J. Am. Chem. Soc.* **124**, 11242 (2002).
332. Y. Wang and F. Caruso, *Chem. Mater.* **17**, 953 (2005).
333. M. Vallet-Regí and F. Balas, *Open Biomed. Eng. J.* **2**, 1 (2008).
334. N. E. Papen-Botterhuis, Q. Q. Sun, P. Magusin, R. A. Santen and N. Sommerdijk, *Chemistry* **12**, 1448 (2006).
335. D. H. Palmer, L. S. Young and V. Mautner, *Trends Biotechnol.* **24**, 76 (2006).
336. T. Valère, *J. Gene Med.* **1**, 151 (1999).
337. D. W. Pack, A. S. Hoffman, S. Pun and P. S. Stayton, *Nat. Rev. Drug Discov.* **4**, 581 (2005).
338. J. R. Ohlfest, A. B. Freese and D. A. Largaespada, *Curr. Gene Ther.* **5**, 629 (2005).

339. D. J. Glover, H. J. Lipps and D. A. Jans, *Nat. Rev. Genet.* **6**, 299 (2005).
340. M. Ogris and E. Wagner, *Drug Discov. Today* 7, 479 (2002).
341. R. Duncan, *Nat. Rev. Cancer* **6**, 688 (2006).
342. W. T. Godbey, K. K. Wu and A. G. Mikos, *J. Control. Release* **60**, 149 (1999).
343. S. H. Chiu, N. T. Ueno and R. J. Lee, *J. Control. Release* **97**, 357 (2004).
344. K. Kunath, T. Merdan, O. Hegener, H. Haberlein and T. Kissel, *J. Gene Med.* **5**, 588 (2003).
345. J. M. Benns, R. I. Mahato and S. W. Kim, *J. Control. Release* **79**, 255 (2002).
346. M. Veiseh, P. Gabikian, S. B. Bahrami, O. Veiseh, M. Zhang, R. C. Hackman, A. C. Ravanpay, M. R. Stroud, Y. Kusuma, S. J. Hansen, D. Kwok, N. M. Munoz, R. W. Sze, W. M. Grady, N. M. Greenberg and R. G. Elle, *Cancer Res.* **67**, 6882 (2007).
347. S. A. Lyons, J. O'Neal and H. Sontheimer, *Glia* **39**, 162 (2002).
348. C. Sun, O. Veiseh, J. Gunn, C. Fang, S. Hansen, D. Lee, R. Sze, R.G. Ellenbogen, J. Olson and M. Zhang, *Small* **4**, 372 (2008).
349. O. Veiseh, C. Sun, J. Gunn, N. Kohler, P. Gabikian, D. Lee, N. Bhattarai, R. Ellenbogen, R. Sze, A. Hallahan, J. Olson and M. Zhang, *Nano Lett.* **5**, 1003 (2005).
350. O. Veiseh, F. M. Kievit, J. W. Gunn, B. D. Ratner and M. Zhang, *Biomaterials* **30**, 649 (2009).
351. A. Fire, S. Xu, M. K. Montgomery, S. A. Kostas, S. E. Driver and C. C. Mello, *Nature* **391**, 806 (1998).
352. S. Khaliq, S. A. Khaliq, M. Zahur, B. Ijaz, S. Jahan, M. Ansar, S. Riazuddin and S. Hassan, *Biotechnol. Adv.* **28**, 27 (2009).
353. D. Castanotto and J. J. Rossi, *Nature* **457**, 426 (2009).
354. R. P. Hickerson, A. V. Vlassov, Q. Wang, D. Leake, H. Ilves, E. Gonzalez-Gonzalez, C. H. Contag, B. H. Johnston and R. L. Kaspar, *Oligonucleotides* **18**, 345 (2008).
355. S. C. Semple, A. Akinc, J. Chen, A. P. Sandhu, *et al.*, *Nat. Biotechnol.* **28**, 172 (2010).
356. H. de Martimprey, J. R. Bertrand, C. Malvy, P. Couvreur and C. Vauthier, *Pharm. Res.* **27**, 498 (2010).
357. H. de Martimprey, J. R. Bertrand, A. Fusco, M. Santoro, P. Couvreur, C. Vauthier and C. Malvy, *Nucleic Acids Res.* **36**, e2 (2008).
358. M. Scherr, K. Battmer, B. Schultheis, A. Ganser and M. Eder, *Gene Ther.* **12**, 12 (2005).
359. J. Rousseau, V. Escriou, P. Perrot, G. Picarda, C. Charrier, D. Scherman, D. Heymann, F. Rédini and V. Trichet, *Cancer Gene Ther.* **17**, 387 (2010).
360. T. Minko, S. S. Dharap and A. T. Fabbricatore, *Cancer Detect. Prev.* **27**, 193 (2003).
361. A. Aszalos and D. D. Ross, *Anticancer Res.* **18**(4C), 2937 (1998).
362. T. Minko, S. S. Dharap, R. I. Pakunlu and Y. Wang, *Curr. Drug Targets* **5**, 389 (2004).
363. K. Takara, T. Sakaeda and K. Okumura, *Curr. Pharm. Des.* **12**, 273 (2006).
364. S. S. Dharap, P. Chandna, Y. Wang, J. J. Khandare, B. Qiu, S. Stein and T. Minko, *J. Pharmacol. Exp. Ther.* **316**, 992 (2006).
365. P. Chandna, M. Saad, Y. Wang, E. Ber, J. Khandare, A. A. Vetcher, V. A. Soldatenkov and T. Minko, *Mol. Pharm.* **4**, 668 (2007).
366. S. S. Dharap, B. Qiu, G. C. Williams, P. Sinko, S. Stein and T. Minko, *J. Control. Release* **91**, 61 (2003).
367. E. P. Holinger, T. Chittenden and R. J. Lutz, *J. Biol. Chem.* **274**, 13298 (1999).
368. R. J. Lutz, *Biochem. Soc. Trans.* **28**, 51 (2000).

369. V. Jekerle, M. U. Kassack, R. M. Reilly, M. Wiese and M. Piquette-Miller, *J. Pharm. Pharm. Sci.* **8**, 516 (2005).
370. D. E. de Menezes, Y. Hu and L. D. Mayer, *J. Exp. Ther. Oncol.* **3**, 72 (2003).
371. H. Wu, W. N. Hait and J. M. Yang, *Cancer Res.* **63**, 1515 (2003).
372. J. Zhou, M. Liu, R. Aneja, R. Chandra, H. Lage and H. C. Joshi, *Cancer Res.* **66**, 445 (2006).
373. H. Hatanaka, Y. Abe, M. Naruke, S. Asai, H. Miyachi, T. Kawakami, J. Nagata, J. I. Kamochi, H. Kijima, H. Yamazaki, K. J. Scanlon, Y. Ueyama and M. Nakamura, *Anticancer Res.* **21**, 879 (2001).
374. M. A. Ihnat, A. M. Nervi, S. P. Anthony, R. C. Kaltreider, A. J. Warren, C. A. Pesce, S. A. Davis, J. P. Lariviere and J. W. Hamilton, *Oncol. Res.* **11**, 303 (1999).
375. E. Flescher, *Cancer Lett.* **245**, 1 (2007).
376. M. Saad, O. B. Garbuzenko and T. Minko, *Nanomedicine* **3**, 761 (2008).
377. T. Minko, R. I. Pakunlu, Y. Wang, J. J. Khandare and M. Saad, *Anticancer Agents Med. Chem.* **6**, 537 (2006).
378. G. Navarro, R. R. Sawant, S. Biswas, S. Essex, C. T. de Ilarduya and V. P. Torchilin, *Nanomedicine (Lond).* 7, 65 (2012).
379. C. E. Probst, P. Zrazhevskiy, V. Bagalkot and X. Gao, *Adv. Drug Deliv. Rev.* DOI: 10.1016/j.addr.2012.09.036.
380. W. T. Al-Jamal, K. T. Al-Jamal, B. Tian, L. Lacerda, P. H. Bomans, P. M. Frederik and K. Kostarelos, *ACS Nano* **2**, 408 (2008).
381. A. M. Derfus, W. C. W. Chan and S. N. Bhatia, *Nano Lett.* **4**, 11 (2004).
382. K. Katagiri, Y. Imai, K. Koumoto, T. Kaiden, K. Kono and S. Aoshima, *Small* 7, 1683 (2011)
383. W. Chen, Y. Cao, M. Liu, Q. Zhao, J. Huang, H. Zhang, Z. Deng, J. Dai, D. F. Williams and Z. Zhang, *Biomaterials* **33**, 7895 (2012).
384. H. S. Choi, W. Liu, F. Liu, K. Nasr, P. Misra, M. G. Bawendi and J. V. Frangioni, *Nat. Nanotechnol.* **5**, 42 (2010).
385. A. Miyawaki, *Dev. Cell* **4**, 295 (2003).
386. H. Meyssamy, K. Riwotzki, A. Kornowski, S. Naused and M. Haase, *Adv. Mater.* **11**, 840 (1999).
387. K. Riwotzki, H. Meyssamy, A. Kornowski and M. Haase, *J. Phys. Chem.* **104**, 2824 (2000).
388. K. Riwotzki, H. Meyssamy, H. Schnablegger, A. Kornowski and M. Haase, *Angew. Chem.* **40**, 573 ((2001).
389. P. Schuetz and F. Caruso, *Chem. Mater.* **14**, 4509 (2002).
390. A. Al-kattan, P. Dufour, J. Dexpert-Ghys and C. Drouet, *J. Phys. Chem. C* **114**, 2918 (2010).
391. A. Al-kattan, P. Dufour, J. Dexpert-Ghys and C. Drouet, *Colloids Surf. B: Biointerfaces* **82**, 378 (2011).
392. M. Epple, K. Ganesan, R. Heumann, J. Klesing, A. Kovtun, S. Neumann and V. Sokolova, *J. Mater. Chem.* **20**, 18 (2010).
393. A. E. Ewence, M. Bootman, H. L. Roderick, J. N. Skepper, G. McCarthy, M. Epple, M. Neumann, C. M. Shanahan and D. Proudfoot, *Circ. Res.* **103**, e28 (2008).
394. A. Doat, F. Pelle, N. Gardant and A. Lebugle, *J. Solid State Chem.* **177**, 1179 (2004).
395. S. Padilla Mondejar, A. Kovtuna and M. Epple, *J. Mater. Chem.* **17**, 4153 (2007).

396. R. Ramachandran, W. Paul and C. P. Sharma, *J. Biomed. Mater. Res. B Appl. Biomater.* **88**, 41 (2009).
397. X. Cheng and L. Kuhn, *Int. J. Nanomedicine* **2**, 667 (2007).
398. M. Kester, Y. Heakal, T. Fox, A. Sharma, G. P. Robertson, T. T. Morgan, E. I. Altinoğlu, A. Tabaković, M. R. Parette, S. M. Rouse, V. Ruiz-Velasco and J. H. Adair, *Nano Lett.* **8**, 4116 (2008).
399. T. Liu, A. Tang, G. Zhang, A. Sharma, G. P. Robertson, T. T. Morgan, E. I. Altinoğlu, A. Tabaković, M. R. Parette, S. M. Rouse, V. Ruiz-Velasco and J. H. Adair, *Cancer Biother. Radiopharm.* **20**, 141 (2005).
400. N. Roveri and M. Iafisco, *Nanotechnology, Science and Applications* **3**, 107 (2010).
401. Y. Su, Y. He, H. Lu, L. Sai, Q. Li, W. Li, L. Wang, P. Shen, Q. Huang and C. Fan, *Biomaterials* **30**, 19 (2009).
402. S. J. Cho, D. Maysinger, M. Jain, B. Röder, S. Hackbarth and F. M. Winnik, *Langmuir* **23**, 1974 (2007).
403. T. S. Hauck, R. E. Anderson, H. C. Fischer, S. Newbigging and W. C. W. Chan, *Small* **6**, 138 (2010).
404. L.Ye, K.-T.Yong, L. Liu, I. Roy, R. Hu, J. Zhu, H. Cai, W-C. Law, J. Liu, K. Wang, J. Liu, Y. Liu, Y. Hu, X. Zhang, M. T. Swihart and P. N. Prasad, *Nat. Nanotechnol.* **7**, 453 (2012).
405. P. Zrazhevskiy, M. Sena and X.Gao, *Chem. Soc. Rev.* **39**, 4326 (2010).
406. S. C. McBain, H. H. Yiu and J. Dobson, *Int. J. Nanomed.* **3**, 169 (2008).
407. X.-B. Yuan, H. Li, X.-X. Zhu and H.-G. Woo, *J. Chem. Technol. Biotechnol.* **81**, 746 (2006).
408. C. Alexiou, W. Arnold, R. J. Klein, F. G. Parak, P. Hulin, C. Bergemann, W. Erhardt, S. Wagenpfeil and A. S. Lübbe, *Cancer Res.* **60**, 6641 (2000).
409. M. Roser, D. Fischer and T. Kissel, *Eur. J. Pharm. Biopharm.* **46**, 255 (1998).
410. N. Kohler, C. Sun, A. Fichtenholtz, J. Gunn, C. Fang and M. Zhang, *Small* **2**, 785 (2006).
411. K. Schulze, A. Koch, A. Petri-Fink, B. Steitz, S. Kamau, M. Hottiger, M. Hilbe, L. Vaughan, M. Hofmann, H. Hofmann and B. von Rechenberg, *J. Nanosci. Nanotechnol.* **6**, 2829 (2006).
412. T. Matsunaga and H. Takeyama *Supramol Sci.* **5**, 391 (1998).
413. H. H. P. Yiu and P. A. Wright, *J. Mater. Chem.* **15**, 3690 (2005).
414. M. Hu, J. Y. Chen, Z. Y. Li, L. Au, G. V. Hartland, X. Li, M. Marquez and Y. Xia, *Chem. Soc. Rev.* **35**, 1084 (2006).
415. M. Mandal, S. Kundu, S. K. Ghosh, S. Panigrahi, T. K. Sau, S. M. Yusuf and T. Pal, *J. Colloid Interface Sci.* **286**, 187 (2005).
416. Y. Xu, A. Karmakar, W. E. Heberlein, T. Mustafa, A. R. Birisa and A. S. Biris, *Adv. Healthc. Mater.* **1**, 493 (2012).
417. Y. Q. Ke, C. C. Hu, Z. J. Yang, H. W. Zhang, H. M. Ji, L. Y. Zhou, Y. Q. Cai, L. S. Qin and R. X. Xu, *J. Neurosci. Meth.* **179**, 45 (2009).
418. P. Reimer and T. Balzer, *Eur. Radiol.* **13**, 1266 (2003).
419. B. Fadeel and A. E. Garcia-Bennett, *Adv. Drug Deliv. Rev.* **62**, 362 (2010).
420. V. I. Shuayev, T. R. Pisanic and S. Jun, *Adv. Drug Deliv. Rev.* **61**, 467 (2009).
421. G. Minotti and S. D. Aust, *J. Biol. Chem.* **262**, 1098 (1987).
422. C. Hoskins, Y. Min, M. Gueorguieva, C. McDougall, A. Volovick, P. Prentice, Z. Wang, A. Melzer, A. Cuschieri and L. Wang, *J. Nanobiotechnology* **10**, 27 (2012).

423. R. G. Chaudhuri and S. Paria, *Chem. Rev.* **112**, 2373 (2012).
424. N. Lee, H. Kim, S. H. Choi, M. Park, D. Kim, H.-C. Kim, Y. Choi, S. Lin, B. H. Kim, H. S. Jung, H. Kim, K. S. Park, W. K. Moon and T. Hyeon, *PNAS* **108**, 2662 (2011).
425. P. Zijlstra and M. Orrit, *Rep. Prog. Phys.* **74**, 106401 (2011).
426. D. Pissuwan, S. M. Valenzueka and M. B. Cortie, *Trends Biotechnol.* **24**, 62 (2006).
427. I. Y. Goon, L. M. H. Lai, M. Lim, P. Munroe, J. J. Gooding and R. Amal, *Chem. Mater.* **21**, 673 (2009).
428. C. Huang, J. Jiang, C. Muangphat, X. Sun and Y. Hao, *Nanoscale Res. Lett.* **6**, 1 (2011).
429. M. S. Yavuz, Y. Cheng, J. Chen, C. M. Cobley, Q. Zhang, M. Rycenga, J. Xie, C. Kim, K. H. Song, A. G. Schwartz, L. V. Wang and Y. Xia, *Nat. Materials* **8**, 935 (2009).
430. L. R. Hirsch, A. M. Gobin, A. M. Lowery, F. Tam, R. A. Drezek, N. J. Halas and L. West, *Ann. Biomed. Eng.* **34**, 15 (2006).
431. Q. Zhang, J. Ge, J. Goebl, Y. Hu, Y. Sun and Y. Yin, *Adv. Mater.* **22**, 1905 (2010).
432. E. D. Smolensky, M. C. Neary, Y. Zhou, T. S. Berquo and V. C. Pierre, *Chem. Commun.* **47**, 2149 (2011).
433. L. R. Hirsch, R. J. Stafford, J. A. Bankson, S. R. Sershen, B. Rivera, R. E. Price, J. D. Hazle, N. J. Halas and J. L. West, *PNAS* **100**, 13549 (2003).
434. P. K. Jain, X. H. Huang, I. H. El-Sayed and M. A. El-Sayed, *Acc. Chem. Res.* **41**, 1578 (2008).
435. P. Ghosh, G. Han, M. De, C. K. Kim and V. M. Rotello, *Adv. Drug Deliv. Rev.* **60**, 1307 (2008).
436. P. C. Chen, S. C. Mwakwari and A. K. Oyelere, *Nanotechnology, Science and Applications* **1**, 45 (2008).
437. R. Lévy, U. Shaheen, Y. Cesbron and V. Sée, *Nano Rev.* 1, (2010).
438. A. M. Alkilany, S. E. Lohse and C. J. Murphy, *Acc. Chem. Res.* (2012), in press.
439. H. Wang, S. T. Yang, A. Cao and Y. Liu, *Acc. Chem. Res.* (2012), in press.
440. N. M. Vadym, O. Shenderova, D. Ho and Y. Gogotsi, *Nat. Nanotechnol.* 7, 11 (2012).
441. A. M. Schrand, H. Huang, C. Carlson, J. J. Schlager, E. Osawa, S. M. Hussain and L. Dai, *J. Phys. Chem. B* **111**, 2 (2007).
442. Y. R. Chang, H. Y. Lee , K. Chen , C. C. Chang , D. S. Tsai , C. C. Fu , T. S. Lim , Y. K. Tzeng , C. Y. Fang, C. C. Han, H. C. Chang and W. Fann, *Nat. Nanotechnol.* **3**, 284 (2008).
443. V. N. Mochalin and Y. Gogotsi, *J. Am. Chem. Soc.* **131**, 4594 (2009).
444. R. A. Shimkunas, E. Robinson, R. Lam, S. Lu, X. Xu, X.-Q. Zhang, H. Huang, E. Osawa and D. Ho, *Biomaterials* **30**, 5720 (2009).
445. K. V. Purtov, A. I. Petunin, A. E. Burov, A. P. Puzyr and V. S. Bondar, *Nanoscale Res. Lett.* **5**, 631 (2010).
446. A. Alhaddad, M.-P. Adam, J. Botsoa, G. Dantelle, S. Perruchas, T. Gacoin, C. Mansuy, S. Lavielle, C. Malvy, F. Treussart and J.-R. Bertrand, *Small* 7, 3087 (2011).
447. X.-Q. Zhang, M. Chen, R. Lam, X. Xu, E. Osawa and D. Ho, *ACS Nano* **3**, 2609 (2009).
448. T. J. Merkel and J. M. DeSimone, *Sci. Transl. Med.* **3**, 73ps8 (2011).
449. E. K. Chow, X.-Q. Zhang, M. Chen, R. Lam, E. Robinson, H. Huang, D. Schaffer, E. Osawa, A. Goga and D. Ho, *Sci. Transl. Med.* **3**, 73ra21 (2011).
450. A. K. Tiwari, V. Gajbhiye, R. Sharma and N. K. Jain, Drug Deliv. **17**, 605 (2010).

451. E. Martinez-Loran, E. Alvarez-Zauco, V. A. Basiuk, E. V. Basiuk and M. Bizarro, *J. Nanosci. Nanotechnol.* **11**, 5569 (2011).
452. L. Cia and P. M. Ajayan, *J. Nanosci. Nanotechnol.* **10**, 3854 (2010).
453. A. Velamakanni, C. W. Magnuson, K. J. Ganesh, Y. Zhu, J. An, P. J. Ferreira and R. S. Ruoff, *ACS Nano* **4**, 540 (2010).
454. K. Ajima, M. Yudasaka, T. Murakami, A. Maigne, A. K. Shibaaa and S. Iijima, *Mol. Pharm.* **2**, 475 (2005).
455. C. Klumpp, K. Kostarelos, M. Prato and A. Bianco, *Biochim. Biophys. Acta* **1758**, 404 (2006).
456. J. G. Rodriguez-Zavala and R. A. Guirado-Lopez, *J. Phys. Chem. A* **110**, 9459 (2006).
457. E. Heister, C. Lamprecht, V. Neves, C. Tilmaciu, L. Datas, E. Flahaut, B. Soula, P. Hinterdorfer, H. M. Coley, S. R. Silva and J. McFadden, *ACS Nano* **4**, 2615 (2010).
458. D. Iohara, F. Hirayama, K. Higashi, K. Yamamoto and K. Uekama, *Mol. Pharm.* **8**, 1276 (2011).
459. D. R. Weiss, T. M. Raschke and M. Levitt, *J. Phys. Chem. B* **112**, 2981 (2008).
460. A. Takagi, A. Hirose, T. Nishimura, N. Fukumori, A. Ogata, N. Ohashi, S. Kitajima and J. Kanno, *J. Toxicol. Sci.* **33**, 105 (2008).
461. C. A. Poland, R. Duffin, I. Kinloch, A. Maynard, W. A. Wallace, A. Seaton, V. Stone, S. Brown, W. Macneea and K. Donaldson, *Nat. Nanotechnol.* **3**, 423 (2008).
462. A. Magrez, S. Kasas, V. Salicio, N. Pasquier, J. W. Seo, M. Celio, S. Catsicas, B. Schwaller and L. Forro, *Nano Lett.* **6**, 1121 (2006).
463. N. R. Jacobsen, G. Pojana, P. White, P. Moller, C. A. Cohn, K. S. Korsholm, U. Vogel, A. Marcomini, S. Loft and H. Wallin, *Environ. Mol. Mutagen.* **49**, 476 (2008).
464. A. Di Sotto, M. Chiaretti, G. A. Carru, S. Belluccia and G. Mazzanti, *Toxicol. Lett.* **184**, 192 (2009).
465. H. Nagai, Y. Okazaki, S. H. Chew, N. Misawa, Y. Yamashita, S. Akatsuka, T. Ishihara, K. Yamashita, Y. Yoshikawa, H. Yasui, L. Jiang, H. Ohara, T. Takahashi, G. Ichihara, K. Kostarelos, Y. Miyata, H. Shinohara and S. Toyokuni, *PNAS* **108**, E1330 (2011).
466. T. Yamashita, K. Yamashita, H. Nabeshi, T. Yoshikawa, Y. Yoshioka, S. Tsunoda and Y. Tsutsumi, *Materials* **5**, 350 (2012).
467. S. Y. Madani, N. Naderi, O. Dissanayake, A. Tan and A. M. Seifalian, *Int. J. Nanomedicine* **6**, 2963 (2011).
468. B. Kang, S. Chang, Y. Dai, D. Yu and D. Chen, *Small* **6**, 2362 (2010).
469. S. Svenson and D. A. Tomalia, *Adv. Drug Deliv. Rev.* **57**, 2106 (2005).
470. D. A. Tomalia, L. A. Reyna and S. Svenson, *Biochem. Soc. Trans.* **35**, 61 (2007).
471. D. J. Bharali, M. Khalil, M. Gurbuz, T. M. Simone and S. A. Mousa, *Int. J. Nanomedicine* **4**, 1 (2009).
472. M. Shen and X. Shi, *Nanoscale* **2**, 1596 (2010).
473. M. A. Quadir and R. Haag, *J. Control. Release* **161**, 484 (2012).
474. M. Han, Q. Lv, X. J. Tang, Y. L. Hu, D. H. Xu, F. Z. Li, W. Q. Liang and J. Q. Gao, *J. Control. Release* **163**, 136 (2012).
475. F. Wang, X. Cai, Y. Su, J. Hu, Q. Wu, H. Zhang, J. Xiao and Y. Cheng, *Acta Biomater.* **8**, 4304 (2012).
476. Y. Li, H. He, X. Jia, W. L. Lu, J. Lou and Y. Wei, *Biomaterials* **33**, 3899 (2012).
477. L. M. Kaminskas, V. M. McLeod, B. D. Kelly, C. Cullinane, G. Sberna, M. Williamson, B. J. Boyd, D. J. Owen and C. J. H. Porter, *Mol. Pharm.* **9**, 422 (2012).

478. M. A. Dobrovolskaia, A. K. Patri, T. M. Potter, J. C. Rodriguez, J. B. Hall and S. E. McNeil, *Nanomedicine (Lond).* 7, 245 (2012).
479. Y. Yan, L. Liao and D. Liu, *J. Control. Release* **152**, Supp 1:e60-1 (2011).
480. Y. Chang, X. Meng, Y. Zhao, K. Li, B. Zhao, M. Zhu, Y. Li, X Chen and J. Wang, *J. Colloid Interface Sci.* **363**, 403 (2011).
481. K. Xiao, J. Luo, Y. Li, J. S. Lee, G. Fung and K. S. Lam, *J. Control. Release* **155**, 272 (2011).
482. H. L. Lu, W. J. Syu, N. Nishiyama, K. Kataoka and P. S. Lai, *J. Control. Release* **155**, 458 (2011).
483. I. H. Lee, S. An, M. K. Yu, H. K. Kwon, S. H. Im and S. Jon, *J. Control. Release* **155**, 435 (2011).
484. W. Cao, J. Zhou, A. Mann, Y. Wang and L. Zhu, *Biomacromol.* **12**, 2697 (2011).
485. L. Zhang, S. Zhu, L. Qian, Y. Pei, Y. Qiu and Y. Jiang, *Eur. J. Pharm. Biopharm.* **79**, 232 (2011).
486. C. F. Jones, R. A. Campbell, Z. Franks, C. C. Gibson, G. Thiagarajan, A. Vieira-de-Abreu, S. Sukavaneshvar, S. F. Mohammad, D. Y. Li, H. Ghandehari, A. S. Weyrich, B. D. Brooks and D. W. Grainger, *Mol. Pharm.* **9**, 1599 (2012).
487. C. F. Jones, R. A. Campbell, A. E. Brooks, S. Assemi, S. Tadjiki, G. Thiagarajan, C. Mulcock, A. S. Weyrich, B. D. Brooks, H. Ghandehari and D.W. Grainger, *ACS Nano* **6**, 9900 (2012).
488. B. Ziemba, G. Matuszko, M. Bryszewska and B. Klajnert, *Cell Mol. Biol. Lett.* **17**, 21 (2012).
489. B. Ziemba, G. Matuszko, D. Appelhans, B. Voit, M. Bryszewska and B. Klajnert, *Biopolymers* **97**, 642 (2012).
490. C. A. Janeway Jr., Cold Spring Harb. Symp. Quant. Biol. **54**, 1 (1989).
491. R. Medzhitov, *Nat. Rev. Immunol.* **1**, 135 (2001).
492. K. Takeda, T. Kaisho and S. Akira, *Annu. Rev. Immunol.* **21**, 335 (2003).
493. C. Pasare and R. Medzhitov, *Science* **299**, 1033 (2003).
494. J. Banchereau and R. M. Steinman, *Nature* **392**, 245 (1998).
495. C. P. and R. Medzhitov, *Curr. Opin. Immunol.* **15**, 677 (2003).
496. C. Lonez, M. Vandenbranden and J. M. Ruysschaert, *Adv. Drug Deliv. Rev.* **64**, 1749 (2012).
497. Available at: http://www.sciencedirect.com/science/journal/0169409X/64/15

Chapter B6

Nanoreactors for Biomedical Applications

*Fabian Itel, Ionel Adrian Dinu, Pascal Tanner, Ozana Fischer and Cornelia G. Palivan**

Department of Physical Chemistry, University of Basel
Klingelbergstrasse 80, 4056-Basel, Switzerland
**cornelia.palivan@unibas.ch*

Abstract

Nanoreactors based on an encapsulation of active compounds in nanometer-size carriers, where they can act *in situ*, have recently been developed as a new alternative to conventional systems in various domains, such as catalysis or therapeutics. Here, we present carriers used to build nanoreactors, either as compartments (vesicles, capsules, PICsomes) or as templates (dendrimers, micelles) together with the active compounds (enzymes, proteins, mimics, or combinations of them) that provide functionality. Starting with a variety of building blocks (carriers and active compounds) and procedures to generate nanoreactors, we put forth the requirements that must be fulfilled to develop nanoreactors for use in medicine. Finally, we provide examples of nanoreactors already introduced for the detection and treatment of pathologic conditions, even though they are in the early stage of research.

1. Introduction

Efforts to understand the expression of nature's intelligence and to mimic its structures and functions in the design of new materials and active assemblies have yielded powerful methods in use today in various domains including chemistry, electronics, materials science, and medicine. A key challenge in the life- and nanosciences is fusing biomolecules, such as enzymes, with synthetic materials, for example block copolymers, in order to create new, complex bio-synthetic devices. Combining the specificity and efficiency of biological molecules with robustness and the possibility of tailoring synthetic materials allows the design of efficient systems in terms of activity, sensitivity, and rapid reply.[1,2]

An elegant way to protect sensitive compounds, for example enzymes, proteins, or nucleic acids, that have emerged in recent years, consists of shielding them in nanometer range supramolecular assemblies. Bio-adapted, lipid-based supramolecular assemblies, such as liposomes, are already on the market as carriers of active compounds. However, increased, intrinsic mechanical instability and low circulation time have been only partially solved through combination with polymers, for example polyethylene glycol (PEG).[3] Of particular interest are polymer supramolecular structures in the form of particles, micelles, vesicles and films, because they can undergo chemical modification of their properties to support combination with biological molecules, while preserving characteristics of synthetic materials such as stability and mechanical robustness.[4,5] Unlike liposomes, polymer supramolecular assemblies feature morphologies and properties that may be controlled by selecting from among highly diverse block copolymers and choosing different block lengths, providing the molecular architecture for the entire copolymer, thereby making these assemblies more versatile for a wider range of applications.[6,7] In addition, it is possible to tune properties such as charge, toxicity or stimuli-responsiveness.[8]

Here, we will present nanoreactors as an elegant combination of nanometer size, supramolecular assemblies that protect active compounds and support their *in situ* activity.[9,10] Nanoreactors are ideal candidates to counteract drawbacks related to conventional drug carriers, where the active compound may act only after release from the carrier, e.g. a non-controlled release, or is released into a biological compartment other than that desired, due to degradation of the carrier.[11] For nanoreactors, the key point is to provide fully active compounds inside the carrier over efficacious periods of time, and under different, biologically relevant conditions. While sensitivity and overall activity are essential when nanoreactors are used *in vitro*, for *in vivo* applications (therapeutics or theragnostics) more complex requirements are imposed in terms of carrier properties, nanoreactor functionality, and biodistribution.

Even though nanoreactors are at their early stage of research and, until now, have been tested only *in vitro*, we present examples that have already been reported and point out their potential in the medical domain. Based on their simplicity, robustness, and functionality, nanoreactors are expected to lead to significant innovations that will provide solutions adapted to individual conditions in terms of detecting pathology and treating diseases with increased efficacy as dosages decrease, and a combination of these.

2. Nanoreactor Concept

The nanoreactor concept emerged from the natural principle of compartmentalization.[12] In nature, biological processes are well separated in compartments such as cell organelles that are responsible for producing, transporting, storing and disposing of biological molecules. The compartmentalization strategy allows the protection of sensitive biomolecules, exchanging signals for communication and enabling complex biological cascade reactions in order to regulate a diversity of simultaneous processes. Examples of function-specific cell compartments are the endoplasmatic reticulum, the Golgi apparatus, mitochondria, peroxisomes, the nucleus, etc. Researchers introduced the concept of nanoreactors in order to understand such complex biological processes in more detail and to implement new/improved functions in cells for medical applications.

Nanoreactors comprise nanometer-size reaction compartments/spaces for chemical and biological reactions (Fig. 1). Toxic substances can be detoxified[13,14] and monitored,[11] specific drugs can be produced locally,[15] and nanoreactors can be used for enzyme replacement therapy.[16] Since the first presentation of a man-made nanoreactor, their rapid development and importance in medical applications have been summarized in many reviews.[9,17–19] Reactions can be localized to different areas, depending on the nanoreactor architecture used, such as within the building block matrix, in the cavity of the supramolecular assemblies, or at the interface between a compartment and the external environment. Nanoreactors contain active compounds (for example enzymes, proteins, and enzyme mimics) that are permanently

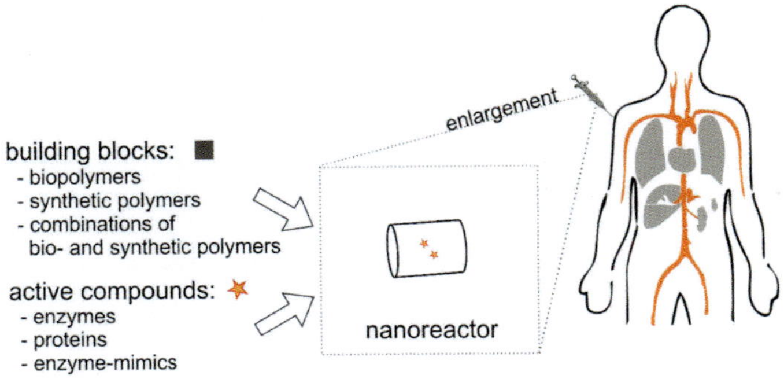

Fig. 1. Schematic overview of production of nanoreactors for biomedical applications using the suitable compounds chosen from the building blocks and the active compounds.

encapsulated or entrapped and have the dual role of protecting contents from proteolytic attack and serving as a reaction space for *in situ* activity.[20] The building blocks of a nanoreactor can either be biopolymers, synthetic polymers or combinations of bio- and synthetic polymers forming different nanoreactor architectures. Depending on the architecture, biologically active molecules can be encapsulated in the polymer matrix of dendrimers, in the core of micelles (based on lipids or polymers), inside the aqueous cavity of vesicular structures (liposomes, polymersomes, capsules, and PICsomes) or in their membranes. When active molecules are present in an aqueous cavity, for example of vesicles, membrane permeability is an important issue, due to the need for transfer of substrates/products across the membrane. To allow substrates and products to pass the membrane, strategies to permeabilize the membrane were developed, either by using substrate-permeable membranes[14,21] or membrane proteins that serve as "gates".[20,22,23] In this respect, synthetic membranes permit the incorporation of membrane proteins, similar to native lipid membranes, even though the dimension of a block copolymer membrane (>10 nm) exceeds the molecular dimension of a membrane protein.[20,24] The site-specific action of a nanoreactor is of interest in medical applications because it could lead to increased efficacy. Such action can be realized through a targeting approach that requires the functionalization of the nanoreactor's surface with small peptides, proteins, antibodies or chelating agents in order to generate molecular recognition points.[25–27] By using the principle of molecular recognition,[11–27] it was shown that surface-functionalized nanoreactors exposing recognition points on their surface can be directed to act at specific locations.[28–30] Molecular engineers are also interested in producing stimuli-responsive nanoreactors to generate activity "on demand".[31] Different stimuli can be applied: (i) physical (light, temperature, mechanical, magnetic, etc), (ii) chemical (solvent, pH, ionic strength) and biological (receptor, enzyme). Nanoreactors can be designed in such a way that either the nanoreactor architecture is rendered permeable by a stimulus, so that the entrapped, active compound reacts with substrate, or the stimulus triggers the function of the active molecule inside the nanoreactor, for example shown by Rose Bengal-encapsulated light-responsive polymersome nanoreactors.[31]

2.1. *Requirements for nanoreactors in biomedical applications*

What considerations are there in the design of nanoreactors that are compatible with medical applications? For simplification, medical applications can be divided into two classes: In *in vivo* applications, where a nanoreactor acts inside a cell, and *ex vivo* applications, where nanoreactors are intended as medical biosensors. For *ex vivo* applications, the demands made on a nanoreactor system are mainly restricted to enhanced system performance achieved by providing optimal conditions with respect to pH, ionic strength, temperature, etc. for the entrapped, active species. For *in vivo* applications, great demands are made to fulfil health safety standards by the US Food and Drug Administration (FDA) and the European Medicines Agency (EMA).

Therefore, a complex scenario of requirements has to be fulfilled for nanoreactor engineering: (i) appropriate nanoreactor size for cell-uptake, (ii) *in situ* biological activity of encapsulated active molecules,[2,32] (iii) stable, permeable matrix for small molecules,[33] (iv) preserved integrity and activity of nanoreactors inside cells, and (v) biocompatibility and biodegradability of the components. The size and morphology of a nanoreactor plays an important role in cellular interactions: Effective cell up-take was observed for nanoreactors with a hydrodynamic radius of around 100 nm.[13,30] In order to preserve *in situ* biological activity of the active species, synthesis strategies for the nanoreactor building blocks and also nanoreactor preparation methods need to be addressed to avoid having the solvents that lower the enzymatic activity.[2] Additionally, the *in situ* activity of the active compound depends on the chemical constitution and properties of the nanoreactor system (charge, flexibility, and density of the nanoreactor building blocks). Furthermore, nanoreactors need to be stable and not release the active compounds. Addressing these requirements, nanoreactor permeability needs to be adjusted depending on the different applications. Stealth-like nanoreactors have prolonged circulation times, positively influencing the efficacy and duration of the active species entrapped in the nanoreactor. Building a nanoreactor system with the ability to interact efficiently on the skin or within the body without having severe side effects or even being toxic requires a highly biocompatible and biodegradable nanoreactor system.[13,34,35] This implies that the nanoreactor building blocks should be non-toxic and biodegradable, and not accumulate in the body.

2.2. *Nanoreactor building blocks*

Nanoreactors can be produced for medical applications using different building blocks such as biopolymers, synthetic polymers, or combinations of bio- and synthetic polymers. We distinguish four different types of natural biopolymers in the field of nanoreactors: lipids, polysaccharides, polypeptides, and polynucleotides. The use of lipids in carrier systems originates historically mainly because of their origin in cell membranes. In this respect, optimal biocompatibility and biodegradability in human bodies, as well as fast, relatively cheap production of lipids is guaranteed.[36] In addition, the self-assembly behavior of lipids is well-known and has been studied over decades, while lipid supramolecular assemblies are also FDA and EMA approved. In addition, lipid membrane stability can be enhanced with cholesterol.[19] Polysaccharides such as chitosan or cellulose are used for nano-assemblies in biotechnical and biomedical applications.[37]

The class of synthetic polymers can be divided into homopolymers, polyelectrolytes, and block copolymers.[9,17,18,38] Homopolymers are built from identical repeating units connected by covalent bonds, either in linear or branched form. In contrast, branched homopolymers belong to the class of dendrimers, which are directly used as nanoreactor architecture. Another type of dendrimer is the Janus dendrimer, which possesses two dendrimeric wedges and two different surface-functionalities,[39] therefore possessing a special, self-assembling property in aqueous solution. Polyelectrolytes are composed of repeating polymer units with charged functional groups. Amphiphilic block copolymers, which are synthetic lipid-mimics, are composed of two or more different polymers covalently linked and with different physical and chemical properties, usually different hydrophilicity. PEG-functionalized lipids are combinations of bio- and synthetic polymers.

2.2.1. Supramolecular assemblies: Architecture

Various supramolecular assemblies have been proposed to serve as compartments/architectures to encapsulate/entrap active compounds and to generate nanoreactors (Fig. 2).

(a) Lipid-based assemblies

Supramolecular assemblies built of natural or synthetic lipids, e.g. phosphatidylethanolamine (PE), dioleoylphosphatidylcholine (DOPC) or phosphatidylglycerol (EPG), self-assemble into micelles, or lipid vesicles (liposomes) in

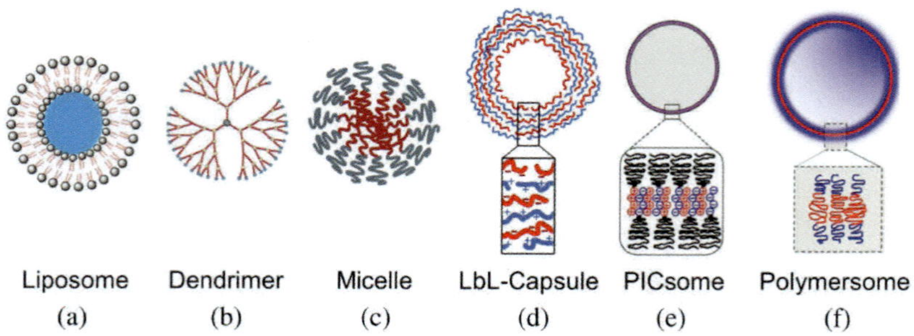

Fig. 2. Supramolecular structures that can serve as nanoreactor architectures in biomedical applications.

aqueous solution (Fig. 2a). Micelles are formed by arranging the polar head groups in the aqueous phase, with the hydrophobic tails in the interior due to the amphipathic nature of the lipids. The driving forces behind this self-assembly are hydrophobic interactions and increased entropy. The size of a micelle is in the range of around 20 nm, and the shape (spherical, rod-like, discoid, *etc.*) depends on the type of lipid used. Inverse micelles were built under "water in oil" conditions, where the non-polar tails face the aqueous environment and the polar head groups build the micellar core. Distinct from micelles, liposomes contain a lipid bilayer, where the polar head groups face the aqueous environment and the hydrophobic tails face each other inside the layer. Depending on the method of preparation, lipid vesicles can be multi-, oligo-, or unilamellar, containing many, few, or one bilayer shell(s), respectively, and under osmotically balanced conditions the vesicles are spherical in shape. The interior of the liposome is an aqueous core, the chemical composition of which corresponds to the chemical composition of the aqueous solution in which the liposomes are prepared. The diameter of the lipid vesicles may vary between about 20 nm and a few hundred micrometers.[19] A limitation of liposomes is their instability in blood plasma, due to their interaction with high and low density lipoproteins. This interaction results in the premature release of the encapsulated drug into the plasma.[40] An effective strategy to increase the circulation time of liposomes is to functionalize lipids with a hydrophilic polymer polyethylene glycol (PEG).[41] Using the biocompatibility of lipids and combining them with highly stable polymers has led to the formation of hybrid PEG-liposomes for biomedical applications. These PEG chains form a protective coat on the liposome surface and decrease recognition by opsonins and uptake by the reticulo-endothelial system (RES),[42] allowing long-circulation time of the "stealth liposomes" (Liposome Technology, Inc.).[40] Although PEG remains the standard polymer for the steric stabilization of liposomes, other polymers also possess stealth-like properties, such as polyvinyl pyrrolidone,[43] polyvinyl alcohol,[44] polyacrylamide[45] and polymethyl- and polyethyl oxazoline.[46] These polymers appear to be attractive alternatives in the design of extended-circulation liposomes.

(b) Polymer-based assemblies

Different morphologies result from the self-assembly process of the above-described building blocks (polyelectrolytes, homopolymers or block copolymers) such as dendrimers, micelles, capsules, PICsomes, and polymersomes.

Dendrimers: Dendrimers are branched, predominantly spherical polymers (Fig. 2b) based on monomers with two- or more functional groups (AB_n) that branch outward, typically starting from one multifunctional core molecule (usually featuring more starting points for branches than the monomer, but the core can be identical to the monomer). The active compounds are usually inserted in the dendrimer architecture via hydrophobic interactions or by exploiting steric hindrance in their core.[47,48] Janus dendrimers were designed by combining two chemically different dendrons, with further reactions leading to a well-defined, hyperbranched, multifunctional dendrimer suitable for medical application.[49] Due to their amphiphilicity, Janus dendrimers are able to self-assemble into various structures such as vesicles, worms, micelles, *etc.*[50]

Micelles: Polymeric micelles (Fig. 2c) self-assemble from amphiphilic block copolymers or polyelectroyltes in aqueous solution above the critical micelle concentration (cmc). Different micelle shapes are possible, such as spherical-, ellipsoidal-, or inverted-micelles, depending on the molecular chemistry (chemical nature of building blocks, molecular weight, hydrophilic-to-hydrophobic ratio), interaction between the surfactants or polymer chains, and the ambient conditions, such as concentration, ionic strength, solvent, temperature, preparation method, *etc.* The use of amphiphilic block copolymer micelles results in several advantages compared to lipid micelles, such as improved stability, lower cmc, slower rate of dissociation, which allows theoretically higher accumulation, and longer retention of cargo.[51]

Capsules: Capsules are polymeric shells (Fig. 2d) generated by the deposition of synthetic polymer layers by the layer-by-layer (LbL) deposition technique[52] on a colloidal particle made of hydrophilic microgels,[53] melamine formaldehyde,[54] polystyrene,[55] carbonate cores[56] and gold- and silica particles,[57,58] which is removed at the end of the deposition process.[59,60] By using oppositely charged polyelectrolytes, the self-assembly of multiple layer of polyelectrolytes can be achieved.[61] Capsules in the size range of a few nanometers to micrometers are feasible, allowing advantageous size adaption for nanoreactor applications.[58,62] The ability to change wall permeability is the most interesting property of capsules.[9] This allows activating a nanoreactor by simply changing physical (light, ultrasound, temperature, *etc.*), chemical (solvent, pH, *etc.*) or biological (receptors, enzymes) conditions.

PICsomes: Polyion complex vesicles (PICsomes) are hollow spheres made of a semi-permeable membrane based on arranged, oppositely charged block copolymers (Fig. 2e).[63] Oppositely charged block copolymer segments are used with matching chain lengths in order to compensate counter charge stoichiometrically and to minimize phase mixing of the PIC middle layer with the outer and inner shell layers of the hydrophobic segment.[63] In particular in medical applications, the precise tunability of the size of PICsomes has great advantage for extended blood circulation times.[64]

Polymersomes: Analogous to liposomes, polymersomes are vesicles possessing a polymeric membrane (Fig. 2f). Vesicle formation is based on the principle of self-assembly of amphiphilic block co-polymers, which is driven by hydrophobic interactions and is influenced by different physical parameters such as cmc, molecular weight (M_w), block length, solubility and glass transition temperature (T_g). The membrane thickness and permeability of the self-assemblies is directly related to the molecular weight.[65–67] For example poly(styrene)-poly(acrylic acid) (PS-*b*-PAA) is able to self-assemble in aqueous solution into polymersomes and encapsulate biomolecules.[68,69] The

Table 1. Polymer-based nanoreactors with different morphologies.

Substrate Polymer-based	Catalyst	Architecture	Compatibility	References
PBE-PEO/PBE-PEG	laccase	Dendrimers	Biocompatible	47
PAMAM	laccase	Dendrimers	Biocompatible	48
PEG-*b*-PAA	lysozyme, trypsin	Micelles	Biocompatible	72–74
PMOXA-*b*-PBOXA	HRP, lipase OF, lipase P	Micelles	Biocompatible	75
PS-*b*-PIAT	CalB, lipase, GOx, HRP, chloroperoxidase	Vesicles	Biocompatible	10, 30, 76–78
Alg-g-PEG/α-CD	L -asparagine	Vesicles	Biocompatible Biodegradable	79
PMOXA-*b*-PDMS-*b*-PMOXA	HRP, trypsin, nucleoside hydrolase, thymidine phosphorylase, SOD, lactoperoxidase	Vesicles	Biocompatible	13, 14, 26, 32, 80–85
PEG-*b*-PSBA	CalB	Vesicles	Biocompatible	86
PS-*b*-PAA	trypsin	Vesicles	Biocompatible	68
PSS/PAH	urease	Capsules	Biocompatible	87
Dextran sulfate/ protamine	peroxidase,GOx	Capsules	Biocompatible Biodegradable	88

underlying principle of the self-assembly of vesicles is highlighted in more detail by Antonietti and Förster.[70] For nanoreactor applications, the use of amphiphilic block copolymer showed to be a very promising approach for further medical uses. This can be explained by the fact that polymer vesicle building blocks show remarkable stability and can be selected to be highly biocompatible and biodegradable, allowing non-toxic, long term medical applications. The combination of polypeptides with polymers was used to produce polypeptide-polymer block copolymers that were able to self-assemble into pH-responsive "schizophrenic" polymersomes.[71]

2.2.2 Active compounds supporting nanoreactor functionality

Nanoreactor functionality derives from active compounds, which are placed in the nanoreactor compartment/space, where chemical reactions can occur, for example entrapped in the polymer matrix of dendrimers, in the core of a micelle (based on lipids or polymers), in capsules, in PICsomes or in the aqueous cavity or membrane of a vesicle (liposome or polymersome). Such active compounds can be (i) enzymes, (ii) enzyme-mimics, (iii) proteins or (iv) photoactive dyes.

Enzymes are known as the most specific and efficient catalysts, containing either ion metal groups or organic functional groups. They are able to speed up metabolic reactions to 10^{11} (diffusion limited). Thereby, different mechanisms of catalysis have been elucidated, such as transition-state binding, entropic control of reactions, acid–base catalysis, *etc.*

Enzyme-mimics are regarded as artificial enzymes that mimic their natural counterparts with improved functionalities[35] or adopt general features of enzymes.[89] In this respect, as an example, due to a short plasma half-life, cell impermeability, and the antigenicity of SOD, SOD has been mimicked by an SOD mimic, which showed promising properties when encapsulated in polymer vesicles.[35]

Photoactive dyes are widely used in therapeutics, for example in cancer therapy, due to their ability to generate reactive oxygen species. Encapsulation in nanoreactors has been shown to lead to reduced toxicity and increased efficacy.

3. Synthesis and Characterization Methods of Nanoreactors

3.1. *Synthesis of supramolecular assemblies*

We present a brief overview of the most commonly used methods to synthesize supramolecular assemblies that form nanoreactors. Some components of nanoreactors are commercially available, for example various lipids and polymers, but many must be synthesized, purified and characterized before nanoreactor formation. Methods of chemical synthesis have advanced rapidly and have been optimized in recent decades. Among these methods, controlled radical-, anionic-, and cationic polymerizations have been intensively explored and extensively used to synthesize polymeric building blocks. However, before a discussion on the synthetic methods, several techniques used to characterize these components will be mentioned first.

Usually NMR–, IR, UV-Vis spectroscopy and mass spectrometry are the basic techniques used to investigate the structure and composition of lipids and polymers. The average molecular weights and polydispersity indexes of polymers are also measured by chromatographic methods (SEC, GPC, HPLC).

Using UV-Vis spectroscopy, turbidity measurements, as a function of water content, are also often used to observe the behavior of molecules and macromolecules in solution. This method can demonstrate self-assembly of amphiphilic molecules and macromolecules in a selective solvent, and can be used to follow morphological transitions of aggregates in response to changes in medium properties.[90–93] A stepwise increase in turbidity corresponds to formation of self-assembled particles of various sizes, with larger structures leading to increased light scattering. Also, turbidity changes are related to the different morphologies of aggregates, which change from one to another upon further addition of water.[90] Nevertheless, scattering methods and microscopy techniques are mainly preferred when investigating the formation of different nanoarchitectures by the self-assembly of amphiphiles.

3.1.1 Lipids and their conjugates

Molecules used in the preparation of lipid nanoreactors are derived from, or are based on, the structure of biological membrane lipids.[19,94,95] Two hydrocarbon chains are usually esterified to form a glycerol backbone (glycerolipids, or plasmalogens in the case of an α-β unsaturated ether), or to constitute the hydrophobic ceramide moiety (sphingolipids). This hydrophobic part is linked to a hydrophilic headgroup containing either a phosphate (phospholipids) or certain carbohydrate units (glycolipids).

There exists a wide variety of synthetic and naturally occurring phospholipids, but the most commonly used phosholipids are: egg phosphatidylcholine, synthetic dipalmitoyl–DL- α -phosphatidylcholine, phosphatidylglycerol, phosphatidylserine, phosphatidylethanolamine, sphingomyelin, phosphatidylinositol, and ovolecithin. Saturated acyl chains vary in length from 10 carbons (lauryl) to 16 (stearoyl), and the longer 18-carbon chains typically contain one (oleoyl), two (linoleyl) or three (linolenyl) unsaturated cis-double bonds.[94] Usually, a zwitterionic or non-ionic lipid is used for the preparation of micelles or vesicles, but the net surface charge on the supramolecular nanocompartment can be modified by incorporating positively or negatively charged lipids. The presence of negative or positive charges on the surface leads to a greater overall volume for aqueous entrapment and reduces the likelihood of aggregation after preparation of the liposomes.[94]

The use of saturated phospholipids in the formulation of lipid-based compartments cannot fully overcome their binding with serum components. Moreover, these types of assemblies feature the disadvantage of low aqueous entrapment volume, and the use of charged liposomes, especially those cationic, can sometimes be toxic. An effective strategy to increase the circulation time of lipid-based assemblies is to attach a polymer to the lipids.[41] Among the different polymers investigated, polyethylene glycol (PEG) has been widely used as a polymeric steric stabilizer. Surface modification of liposomes with PEG can be achieved in several ways: physical adsorbtion of the polymer onto the surface of the lipid-based compartments; incorporation of a PEG-lipid conjugate during self-assembly; or covalent attachment to the surface.[3,19,96] Several other polymers have been used successfully as steric protectors, such as poly(vinyl pyrrolidone), poly(2-methyl-2-oxazoline) or poly(2-ethyl-2-oxazoline),[43,46,97] poly[*N*-(2-hydroxypropyl) methacrylamide],[98] L-amino-acid-based biodegradable polymer,[99] and polyvinyl alcohol.[100] These polymers are expected to be biocompatible, soluble, hydrophilic, and to exhibit highly flexible main chains.[3,101]

3.1.2 Polymers

Dendrimers: Dendrimers are technologically advanced synthetic compounds with a monodisperse-, highly branched structure and well-defined, three-dimensional architecture.[102–104] Features of dendrimers, including multivalency, well-defined, large globular structures, mono- or low polydispersities, densities of radial branches, and distribution of chemical moieties located inside the molecular labyrinth or at the periphery of the molecules, endow them with many properties that are significantly different from those of corresponding linear polymers.[105–110]

Mainly, two major strategies have evolved for dendrimer synthesis. The first involves a divergent process of the growth of a dendron that originates from a core site. This approach involves a two-step, iterative reaction sequence that produces concentric shells of branches (generations) around a central initiator core. This core-shell architecture grows linearly in diameter as a function of added generations, while the surface groups amplify exponentially at each generation.[107,111] For example, the synthesis of an ammonia core dendrimer requires a two-step process: exhaustive Michael addition to a suitable amine initiator core with methyl acrylate, followed by extensive amidation of the resulting esters with a large excess of ethylenediamine.[104] Unfortunately, multiple reactions are simultaneously required in this method, causing incomplete growth steps and side reactions, and leading to imperfect samples. Also, it is extremely difficult to isolate pure dendrimers from side products of similar structures.

The second strategy follows a convergent growth process.[112] The reaction begins from the surface of the future dendrimer and proceeds to the core, leading to the formation of a single, reactive dendron. To obtain a dendrimer structure, several dendrons react with a multifunctional core. This method allows intermediate purification and results in a final product of high purity, and thus to a nearly perfect dendrimer structure.

Variation of initiator cores, repeating units, and surface groups allow for the synthesis of a variety of dendrimers. Examples include poly(propyleneimine), polyethers, polyesters, polyamides, poly(alkanes), poly(siloxanes) and polypeptides.[113–115]

Most dendrimers possess a single type of terminal group, because such a group leads to easier synthesis. However, it appears that having two (or more) types of terminal groups is also highly desirable to combine several properties in a single molecule. This can be achieved by grafting two terminal groups at defined area on the surface. This can be accomplished by patches of one type of function interspersed among others but, in general, the dendrimeric structure can be composed of two halves, possibly having different sizes and numbers of terminal groups.[39,50,110,116] The simplest method of synthesis of these compounds, also called Janus dendrimers, consists of reacting two dendrons with complementary functions as the core. Another possibility is to react a first dendron with a multifunctional core, followed by grafting the second dendron to the remaining function(s) of the core.[39] When synthesized with judiciously tailored hydrophilic and hydrophobic elements, Janus dendrimers can function as powerful structure-directing amphiphiles, with greater versatility than simple lipids, surfactants, or block copolymers.

Amphiphilic block copolymers: Polymers are attractive compounds for nanoreactor applications because they exist in a wide variety of compositions, molecular weights, dispersities, functionalities, and shapes. They can be designed using a large number of monomers and their properties can be well-defined by using the most appropriate polymerization reaction. Many synthesis techniques have been used to prepare homo- and block copolymers with well-defined composition, molecular weight, and structure, such as: (a) living anionic and cationic polymerization and (b) controlled radical polymerization,[117] which includes three main types of methods: (i) atom transfer radical polymerization (ATRP), (ii) reversible addition-fragmentation chain transfer polymerization (RAFT) and (iii) nitroxide mediated radical polymerization (NMP), (c) "click" chemistry.[118–126]

Amphiphilic block copolymers are macromolecules consisting of at least two blocks that tend to aggregate in solvents selective for one of the constituent blocks.[18] When a hydrophilic- (A) block is connected to a hydrophobic (B) block, diblock copolymers are formed, and by combining two hydrophilic and one hydrophobic block (ABA, or ABC if one hydrophilic block has a different chemical structure), or two hydrophobic blocks and one hydrophilic (BAB), triblock copolymers are generated.[127–129] The polymers, as constituents of block copolymers, can be synthetic, natural, or a combination thereof.[123,130] The hydrophobic block is typically based on polystyrene (PS), polybutadiene (PB) or polydimethylsiloxane (PDMS), whereas the hydrophilic block is usually poly(acrylic acid) (PAA), polyethyleneglycol (PEG), polymethyloxazoline (PMOXA) or poly(N-hydroxypropyl)methacrylamide (PHPMA).[123,126,131–133]

As an example of the above, by initiating the anionic polymerization of ethylene oxide with a functionalized initiator, diblock copolymers bearing a targeting moiety such as H_2N-PEG-OH,[134] HOOC-PEG-OH,[135] aldehyde-PEG-OH,[136]and monosaccharide-PEG-OH, [137] were successfully prepared. α-methoxy-ω-amino-PEG has been reported as a macroinitiator in the preparation of polyester block copolymers of D,L-lactide, glycolide, ε-caprolactone, and δ-valerolactone.[138–140] After reaction with 2-bromoisobutyryl bromide, such α-methoxy-ω-amino-PEG served as a macroinitiator for controlled copolymerization of ethyl acrylate (EA), ethyl methacrylate (EMA), methyl methacrylate (MMA), n-butyl acrylate (nBA), tert-butyl methacrylate (tBMA), DMAEMA, and 2-(N,N-diethylamino)ethyl methacrylate (DEAEMA).[141–143]

Poly(2-oxazoline)s are used as blocks in different applications as a result of their biocompatibility and stealth behavior.[144] Recent studies on the biodistribution and excretion of well-defined, radiolabeled poly(2-methyl-2-oxazoline) (PMOXA) and poly(2-ethyl-2-oxazoline)(PEtOz) in mice demonstrated no accumulation in tissue and rapid clearance from the bloodstream.[145] PEtOz has been connected to biodegradable poly(L-lactide) (PLA) and poly(ε-caprolactone) (PCL) to form amphiphilic block copolymers.[146,147]

Polypeptides and conjugates: Polypeptides, due to their special secondary structures, biocompatibility and different pendant functional groups, such as the pH-responsive groups of carboxylic acid on poly(L-glutamic acid) (PGA) and primary amine on poly(L-lysine) (PLL), are used as building units for supramolecular structures, either as homopolymers or as hydrophobic/hydrophilic blocks in block copolymers, namely polypeptide diblock copolymers and polypeptide-polymer hybrids or conjugates.[71,148–152] There are different routes for the synthesis of polypeptides and peptide-polymer conjugates. They involve the use of living radical-, anionic-, cationic-, or ring-opening polymerizations, as well as coupling reactions,[149,151,152] and can be classified into four main approaches: (i) coupling peptide segments to preformed polymers using one or multiple reactive sites, (ii) starting from a polypeptide used as a macroinitiator, (iii) constructing peptidic blocks starting from a preformed synthetic polymer, and (iv) polymerization of short, oligopeptidic segments used as (macro)monomers.[151]

The most economical process for synthesis of polypeptide chains is the ring-opening polymerization of α-amino acid-N-carboxyanhydrides (NCAs).[71,148] This method involves very simple reagents, and polypeptides or polypeptide block copolymers can be prepared with good yield and without detectable racemization at the chiral centers.[148] For example, poly(L-glutamic acid)-*b*-poly(L-lysine) (PGA-*b*-PLL) was synthesized through this method. At low pH, the poly(L-glutamic acid) constitutes the hydrophobic block and, by increasing the pH, it becomes hydrophilic due to the transition to the charged coil conformation. In contrast, deprotonated poly(L-lysine) changes from hydrophilic to hydrophobic. The stimuli-responsive behavior of polypeptides depending on pH and ionic strength is used to produce stimuli-responsive nanocompartments of controlled size and shape, and these are biocompatible.[71]

3.2 *Preparation of nanoreactors*

The preparation method to generate nanoreactors is a complex process and has to fulfill many requirements that depend on the specificity of the building blocks, the active compounds, the architecture, and also on the medical application. The safety issues increase considerably when nanoreactors are intended for use in the human body. The preparation of nanoreactors usually depends on two main points: (i) the formation of the architecture using the main building blocks and (ii) combining the active compound with the architecture. The key point is to keep the active compound in its functional form, because this is specific to each biomolecule and can depend strongly on the preparation method, mainly on pH, ionic strength, temperature, traces of organic solvents, *etc.*[9] Therefore, the preparation method to form the architecture has to be adapted to the active compound. Also, the building blocks are always treated with organic solvents, which can influence the activity of biological compounds dramatically. Organic solvent use should be decreased to a minimum where possible, or solvent-free preparation methods should be used.[153]

Lipid-based nanoreactors: There are different methods for the preparation of lipid nanoreactors (Fig. 3), including the film rehydration technique starting from a dried lipid film, dialysis of mixed detergent-lipid micelles, lipids in water-immiscible solvent emulsions, lipids in water-miscible solvents, and injection techniques. All of these methods produce vesicles (multilamellar and unilamellar) and micelles containing the desired active component when it is included in the aqueous phase. Preparation procedures have been reviewed in detail in various publications.[19,154] Solvent-free methods (film rehydration and dialysis) are advantageous for use with biological compounds and should be tested first. A homogenization step is needed to form unilamellar and monodispersed vesicle solutions by extrusion through polycarbonate filters with defined pore sizes, or by sonication, and another step is required for removal of extravesicular material, either by size exclusion,[155] dialysis[154] or by ultracentrifugation that pellets the vesicles, with subsequent re-dispersion in pure aqueous solution.[156]

PEG-liposomes are obtained by adding a small percentage (5%) of PEGylated lipids to the lipid mixture and can be handled in the same way as pure lipid vesicles.

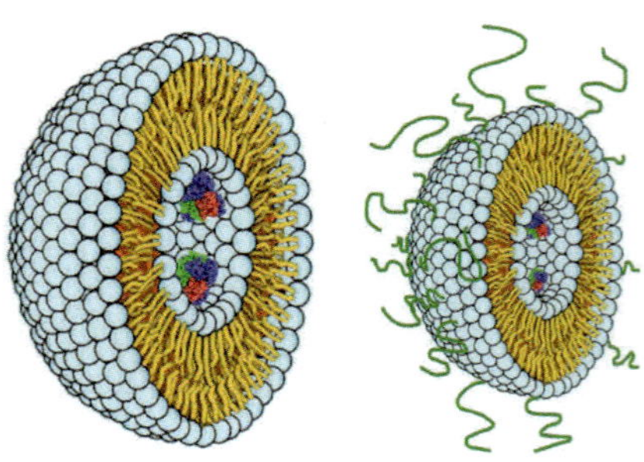

Fig. 3. Cross-section of a liposome-based nanoreactor with encapsulated enzymes. (a) Standard liposome and (b) PEG surface modified liposome (stealth-liposome).

Dendrimer-based nanoreactors: The preparation of dendritic nanoreactors (Fig. 4) relies mainly on the synthesis, because the active compound conjugates to the dendrimers by chemical reaction: either the surface of an enzyme is modified with dendrimers,[47] or a catalytically active center is chemically placed in the core of surrounded dendrimers.[157]

Janus dendrimers, as an amphiphilic moiety, self-assemble in aqueous solution to dendrimersomes. They can also incorporate pore-forming proteins, co-assemble with structure-determining lipids and block copolymers, and offer a molecular periphery suitable for further chemical functionalization.[39,50]

Polymeric micelle-based nanoreactors: Amphiphilic polymers self-assemble in a manner similar to lipids above their cmc, to form core-shell micelles in aqueous solutions (Fig. 5). In contrast to lipid micelles, polymeric micelles are generally more stable, with a considerably lower cmc, and a slower dissociation rate. Micelle-formation strongly depends on hydrophobic to hydrophilic block ratio.[158] Polymeric micelles contain a hydrophobic core and, thus, only hydrophobic active compounds can be encapsulated.[9]

Capsule-based nanoreactors: Capsules (Fig. 6) are formed by the layer-by-layer deposition technique. A colloidal particle of nanometer-to-micrometer size is used as a template for the controlled, consecutive deposition of

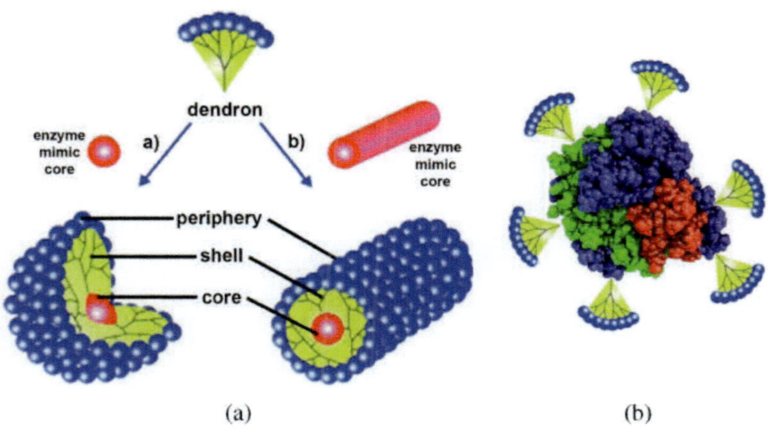

Fig. 4. (Color online) Dendrimer-based nanoreactors. (a) Multiple dendrons with terminal groups (blue) and inner blocks (yellow) attached to a catalytically active enzyme-mimic core (red), resulting in (a) spherical or (b) cylindrical assemblies (Adapted with permission from Ref. 157. Copyright 2001 WILEY-VCH). (b) Dendritic nanoreactor with a biological active enzyme surface functionalized with single dendrons to increase the stability of the enzyme in solution.

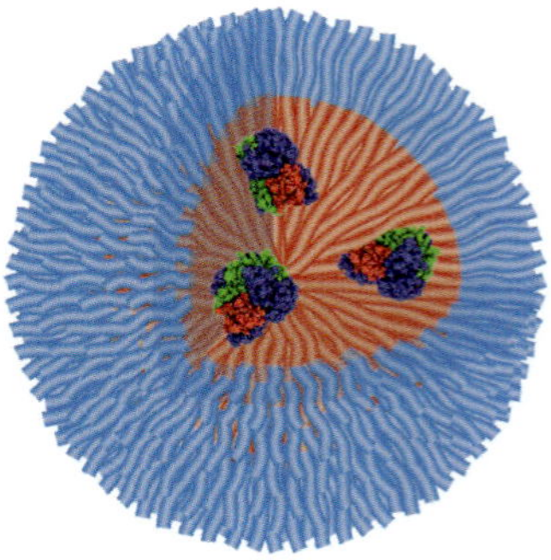

Fig. 5. (Color online) Cross-section through a polymeric micelle-based nanoreactor with enzymes entrapped in the inner hydrophobic core (red) and protected by the hydrophilic outer shell (blue).

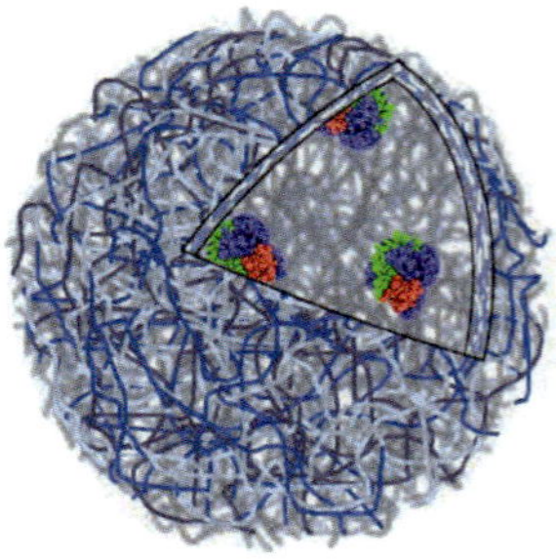

Fig. 6. Capsule-based nanoreactor with encapsulated enzymes. The capsule membrane is made by the layer-by-layer deposition technique and, depending on the number of layers, permeability to small molecules can be controlled.

polymeric layers.[159] The layers consist of oppositely charged polymers and each layer can encapsulate different substances.[88] The layer formation is time consuming, and each adsorption cycle requires three centrifugation- and resuspension steps in water for cleaning. Control of the particle- and polyion concentration, pH and ionic strength ensures uniform coating and avoids aggregate formation. After the deposition process, the core is removed by thermal or chemical processes, leaving a hollow, stable sphere. The core-removal process has to be adapted to the active-entity entrapped, because the conditions are relatively harsh (heat, very acidic/basic). Enzyme entrapment can be realized either during capsule preparation or after the removal of the core. In the latter case, capsule wall permeability needs to be changed by altering the physical parameters of the capsule solution, such as pH, ionic strength, solvent and temperature.

PICsome-based nanoreactors: Oppositely charged block copolymers conjugated to a PEG segment self-assemble in water to form polyion complex (PIC) membranes (Fig. 7). The membrane is composed of the stable PIC lamellae, with matched chain lengths exactly compensating the counter charge in order to minimize the ability to mix with the aqueous environment or the hydrophilic outer segments.[63] It is important, that the block ionomers have equal ratios of negative and positive charges in order to form a neutral, hydrophobic PIC membrane. After the formation of the PIC membrane in water containing the active compound, the particles are subjected to sonication cycles, resulting in hollow particles with sizes up to 10 μm in diameter, depending on the PEG weight fraction.[160] In addition, PICmicelles are obtained with high PEG fractions (10–40%), while PICsomes are formed with PEG fractions between 5–15%.[160] PICsomes are quick to produce and there is no organic solvent involved in the production procedure.[161]

Polymersome-based nanoreactors: Polymersomes (Fig. 8) made of amphiphilic block copolymers are usually prepared in a manner similar to liposomes. The techniques also include the film rehydration method, dialysis, and solvent displacement techniques.[153] In the film rehydration method (film swelling), the amphiphile is dissolved in an organic solvent and evaporated in a glass flask to form a uniform, smooth film on the surface.[162] Addition of aqueous solution containing the desired active entity to be encapsulated will result in film swelling and encapsulation of the entities. For vesicle formation, additional agitation forces (stirring, vortexing, and sonication) are applied.

Solid rehydration, similar to film rehydration, of bulk polymer particles is also possible, performed by extended agitation (usually stirring) of the polymer powder in aqueous solution, until the polymer particles are completely hydrated.[153]

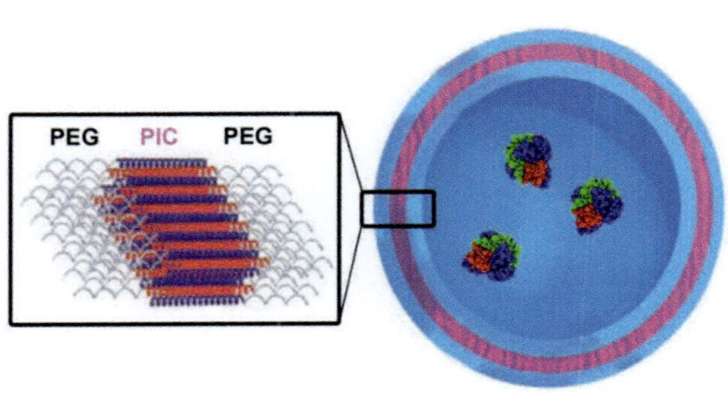

Fig. 7. PICsome-based nanoreactor made of oppositely charged polymers forming the polyion complex (PIC) membrane, with a layer of PEG on each side (Adapted with permission from Ref. 161. Copyright 2007 WILEY-VCH).

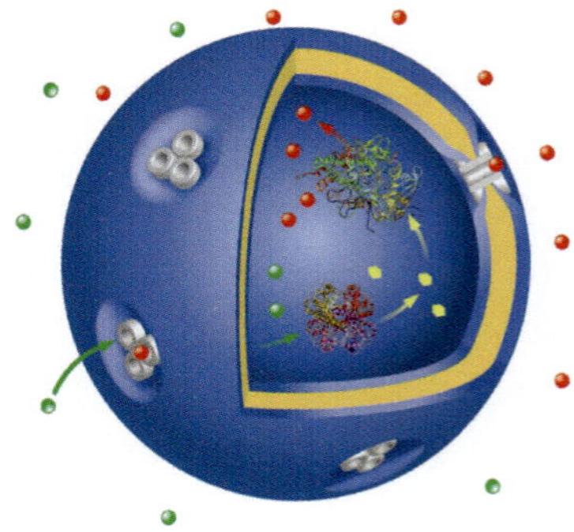

Fig. 8. Schematic representation of a polymersome-based nanoreactor containing enzymes. The membrane is permeabilized with membrane proteins for the diffusion of substrates and products, while retaining the bigger enzymes inside the nanoreactor (Reprinted with permission from Ref. 13. Copyright 2011 WILEY-VCH).

The detergent dialysis method can also be used with block copolymers to obtain polymersomes.[163] The polymer is dissolved in a detergent aqueous solution and then dialyzed against the same buffer in dialyzing tubings with a pore-size cut-off between the molecular weight of the detergent and the polymer. Because detergent molecules are much smaller than polymer molecules, the detergent will slowly diffuse out and this reduces the detergent concentration until the detergent can no longer fully dissolve the polymer molecules, resulting in the formation of self-assembled structures of the polymers.[153,163]

In the solvent-displacement technique, the amphiphile is dissolved in an organic solvent at a high concentration and then added dropwise to an aqueous buffer under vigorous stirring.[153] For chemical reactions to occur inside the compartments, substrates must diffuse through the membranes of the polymersomes. However, the shells of most polymersomes turn out to be impermeable to small molecules.[164] On one hand, the impermeability of polymersomes is a drawback to efficient catalysis when the reservoir of available substrate is depleted for these nanoreactors [165] but, on the other hand, it can also be regarded as an advantage of such systems, because the membrane protects the encapsulated guest, e.g., from degrading agents such as proteases.[78,161] The permeability of the membrane can be increased with the help of plasticizers,[166] peptides[86] or with membrane proteins[165,167]. Membrane proteins can tune the permeability of the membrane in a selective manner when specific transporters are inserted[22,23,82,168] or non-selectively when unspecific channel proteins[20,29,85] are used.

Despite the thicker membranes of polymersomes of more than 10 nm compared to lipid membranes (4 nm), amphiphilic block copolymers closely resemble biological membranes and therefore are able to embed biological membrane proteins in the hydrophobic part.[24] The production of a membrane protein stock solution involves overexpression by a suitable bacterial strain (e.g. E. coli), extraction from bacterial cells by detergent solubilization, and purification of the desired membrane protein in its functional form.[169]

Most channel proteins used in nanoreactors are β-barrel proteins,[170] providing better stability and handling compared to α-helical membrane proteins. For example, the outer membrane protein F (OmpF), possessing a β-barrel structure, allows passive passage of molecules smaller than 600 Da. It can be isolated in reasonable quantities from E. coli cultures (tens of mg/L culture),[171] is extremely robust[172] (stable up to 70°C in 2% sodium dodecyl sulphate) and can be dissolved, for example, in pure ethanol for nanoreactor preparation. The integration of protein channels into block copolymer membranes was initially realized by Meier *et al.* [24] by integrating OmpF in PMOXA-*b*-PDMS-*b*-PMOXA copolymer membranes.[83]

In contrast to general porins, channel proteins that specifically transport ions (e.g. FhuA) or nucleosides and nucleotides (Tsx) are available for nanoreactor permeabilization. By genetic engineering, FhuA channel protein was modified to increase the channel pore size from around 10 Å to 45 Å, allowing passage of much larger molecules.[85]

Protein insertion in polymeric membranes is achieved interdependent with the stability of the membrane protein used. For β-barrel proteins, the protein can be mixed together with an ethanol-solution containing the amphiphilic polymers.[26] Upon solvent evaporation, the protein will partially integrate into the polymer layer. For less stable membrane proteins (α-helical), the incorporation procedure has to be performed by detergent-destabilization of preformed vesicles and the subsequent removal of the detergent with biobeads, by the dialysis method[163] or by the film rehydration method.[22,173]

3.3 *Characterization of nanoreactors*

Various methods are used to characterize nanoreactors both in terms of their building blocks (lipids, peptides, and polymers) and of overall architecture, properties (stability, flexibility, size, charge, *etc.*) and activity (Fig. 9). Besides the investigation of the supramolecular self-assembly of amphiphiles in aqueous solution, UV-Vis spectroscopy is used to study the activity of

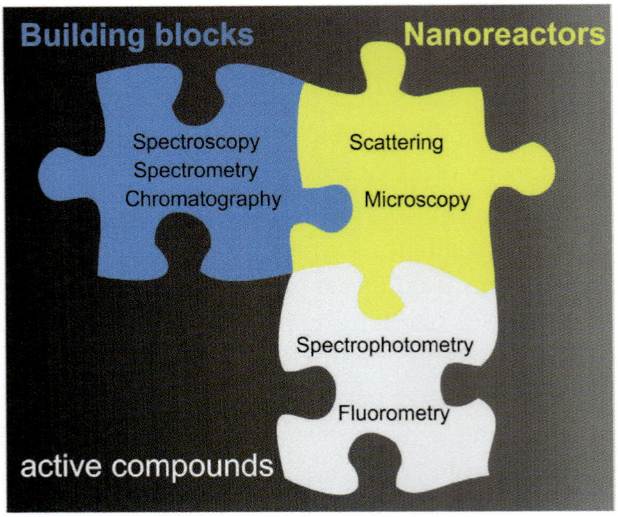

Fig. 9. (Color online) Overview of methods needed for characterization of nanoreactors. The methods are divided in characterization of the building blocks (blue), characterization of the nanoreactors (yellow) and the characterization of the active compounds (white).

encapsulated enzymes or hemoglobin in nanocompartments,[30,47,81,174–177] or the formation of metal nanoparticles inside of nanoreactors.[178]

Due to their availability and simplicity, static and dynamic light scattering (SLS and DLS) measurements have commonly been used to determine particle size and size distribution[153,179–181] (polydispersity index), and also the critical aggregation concentration,[20,182] particle morphology[183] and change of size or shape with variation of pH, temperature or detergent addition.[184–186] The zeta potential of nanoparticles can be determined through DLS by measuring their electrophoretic mobility in a capillary cell.[187]

From light scattering measurements, structural parameters such as weight averaged molar mass, radius of gyration and hydrodynamic radius are available for many particles with various shapes and different composition.[180,183,188] Also, information about particle–particle and particle–solvent interactions can be obtained from SLS.[183]

Various studies on supramolecular assemblies in solution that used light scattering are reported in the literature.[22,127,182,184,189–192] For example, an investigation of vesicle formation by the self-assembly of triblock copolymers based on PMOXA–*b*–PDMS–*b*–PMOXA or PDMAEMA-*b*-PBMA-*b*-PDMAEMA has been done by light scattering in order to obtain information on size polydispersity of vesicles and vesicular morphology.[182,191] Studies on the self-assembly of block copolymers possessing a peptide sequence as a hydrophilic block have revealed that the hydrodynamic radii of particles depended strongly on the conformation of the poly(L-glutamate) segment.[189] The size of nanocapsules formed of PEG-*b*-PLA block copolymer and used for encapsulation of polyhemoglobin-tyrosinase was determined using DLS.[193] It was shown that the mean diameter can be controlled by the stirring speed of the reaction mixture or by the size of the PLA block in the copolymer.

Information on the structural properties of nanoparticles has also been obtained by small-angle X-ray scattering (SAXS). This scattering method probes structures in the nanometer range (length scales between approximately one and several hundreds of nm) and have been used to investigate the phase behavior of block copolymers in water across a concentration gradient. A larger range of sizes (especially at the low end) can be determined by using this method, due to the shorter wavelength of X-rays (0.1–0.2 nm) compared to light (500–700 nm). SAXS offers information similar to that obtained with light scattering, i.e. on the radius of gyration, weight average molar mass, and particle–particle and particle–solvent interactions,[183,194–196] but it has a higher resolution with respect to the shape features of particles. In the case of aggregated samples, this means that it is possible to determine, by SAXS, the mean diameter of the agglomerates, and also the mean diameter of the primary particles within agglomerates, which is not possible with LS. Also, there is only slight interference from multiple scattering in dispersions with high solid concentrations compared to LS, with which only diluted suspensions can be measured in order to prevent multiple scattering of the light after scattering off of the particle.

SAXS was used to confirm the formation of nanoarchitectures,[197] or even to study the binding of proteins to polymer nanocompartments.[198] Moreover, using this technique, a multilamellar structure has been indicated for the assemblies formed by self-directed reconstitution of proteorhodopsin with quaternized amphiphilic block copolymers, for example P4VP-*b*-PBD-*b*-P4VP.[199] Microcapsules created by the layer-by-layer (LbL) polyelectrolyte adsorption technique have been subcompartmentalized by embedding cubic mesophase lipid nanoparticles (cubosomes) in the capsule shell wall.[200] Small angle X-ray scattering experiments proved that cubosomes remained structurally intact when embedded within the polyelectrolyte wall of capsules.

The small-angle neutron scattering (SANS) technique allows investigation of the morphology and thermodynamics of polymer blends and copolymers.[195,196,201,202] SANS measurements have also been performed to elucidate membrane distortions caused by inorganic nanoparticles embedded in the membrane of PEG-ylated liposomes.[203] Microscopy is another important method extensively used for characterization of self-assembled structures in a broad range of length scale (from few nanometers to several micrometers) but, in comparison with scattering techniques, information on size, morphology and homogeneity are obtained by direct visualization. The simplest possibility to visualize the assemblies under "physiological" conditions is offered by optical microscopy, but the limited resolution (around 200 nm) of these devices represents a big disadvantage compared with electron microscopy. Therefore, only large assemblies such as giant lipid or polymer vesicles can be observed by optical microscopy.[204,205]

For assemblies or aggregates that exhibit low absorption of light in the visible part of the spectrum, such as vesicles, enhanced contrast is necessary in order to visualize structural details. Phase contrast microscopy and differential interference contrast microscopy represent excellent methods to characterize unstained samples, especially giant lipid or polymer vesicles, and to obtain information on their structures.[206–209]

Electron microscopy techniques, such as transmission electron microscopy (TEM), scanning electron microscopy (SEM), cryo-TEM and cryo-SEM, are used as vizualization methods to study self-assembled templates and nanoreactors,[30,81,174,178,191,210–212] or to investigate the influence of water content on vesicle formation.[92,208,213] Formation of vesicles, thickness and structure of a vesicle membrane, and their phase behavior were observed by cryo-TEM, cryo-SEM,[22,25,214] and the freeze-fracture technique.[215]

For the investigation of nanocompartment shape and size, atomic force microscopy (AFM) is a very useful tool.[191,216,217]

Another microscopy technique that can be used to investigate the size and shape of nanocompartments is fluorescence microscopy. In order for a sample to be suitable for fluorescence microscopy, it should be labeled with fluorophores or with a fluorescent protein. This technique has been used to visualize polymer assemblies, evaluate dye encapsulation efficiency and to investigate fluorescent dye attachment to polymer molecules.[14,25,26,84,178,211,212,218,219] Confocal fluorescence microscopy offers improved sensitivity for investigation of nano-assemblies.[4,10,30,68,161,178,211,220] This method provides true three-dimensional (3D) optical resolution for analyzed samples using laser illumination.

Fluorescence correlation spectroscopy (FCS) is an experimental technique to investigate the dynamic properties of fluorescent molecules such as dyes, macromolecules and nanoparticles.[221] Monitoring fluorescence intensity fluctuations in a small observation volume — a confocal volume — and performing correlation analysis of the fluctuations can yield information on the fluorescence brightness, diffusion coefficients and times, concentrations, *etc.*

For example, the size and number of encapsulated fluorescent dyes or proteins in nanocontainers have been determined.[14,25,221] Also, a large variety of studies concerning encapsulation/entrapment/binding of fluorescent molecules in/on polymer assemblies has been done in order to determine the sensitivity and stability of nanoreactors.[14,27,81,191,212]

4. Applications of Nanoreactors in Biomedicine

In today's medicine, there are numerous challenges that require new strategies for treatment: (i) side effects of drugs, (ii) uncontrolled drug distribution in the body and (iii) induced toxic side effects arising from imaging agents (contrast agents for MRI imaging, fluorescent dyes, radioactive labels). Nanoreactors have emerged as valuable candidates in diagnostic, therapeutic and theragnostic applications.[11] Nanoreactors are intended to solve or reduce these problems, because of their ability to protect, transport, and provide an enclosed reaction space for *in situ* activity of bio compounds.[9] In addition, nanoreactors can be modified for targeting approaches to deliver them specifically to desired sites *in vivo*.[28,222,223] Targeting molecular moietes include antibodies, folate, transferrin, peptides (cell penetrating peptides) and molecules specifically overexpressed on a specific site (tumor cells).[28,30,224] Although nanoreactors are still at the early research stage (models, or *in vitro* and *in vivo* studies), essential points have to be elucidated before reaching the clinical trial stage; biodistribution, bioaccumulation and degradation — all of these points remain unknown.

4.1. *Nanoreactors for diagnostic applications*

Nanoreactors for purely diagnostic applications are still rare, because, typically, detection simply requires a contrast agent or fluorescent molecule that can be detected. In this respect, the use of nanoreactors to detect pathological conditions is based on rapid, selective reply to a stimulus from a specific signal/molecule/reaction that triggers an active compound in the nanoreactor. For example, an enzymatic reaction *in situ*, which generates fluorescent molecules, paramagnetic nanoparticles or quantum dots as products, can be simply used as a very specific biosensor for detection methods.

In situ production of a fluorescent compound was demonstrated with a nanoreactor based on PMOXA-*b*-PDMS-*b*-PMOXA block-copolymers functionalized with OmpF pores for passive substrate diffusion. (84) An enzyme, acid phosphatase, was encapsulated and produced the insoluble fluorescent product (ELF 97) from a soluble, non-fluorescent substrate that diffused through the porous nanoreactor membrane. Due to the enzyme's activity at low pH values between 4 and 6.5, the fluorescent signal can only be detected when the nanoreactors are taken into biological compartments with a low pH, such as in endosomes.

Enzyme-micelle supramolecular structures, with improved catalytic activities compared to free enzymes, are of increased interest for the detection of low concentrations of chemicals in diagnostic applications. A step forward in designing capsule-based nanoreactors is multicompartment capsules; each capsule compartment (inner and outer) contains a different type of enzyme that works in tandem with the other to produce a fluorescently monitored cascade reaction (Fig. 10).[225]

The enzymes studied in LbL capsules are mainly model enzymes, but the fact that they all produce fluorescent molecules makes the LbL capsule system

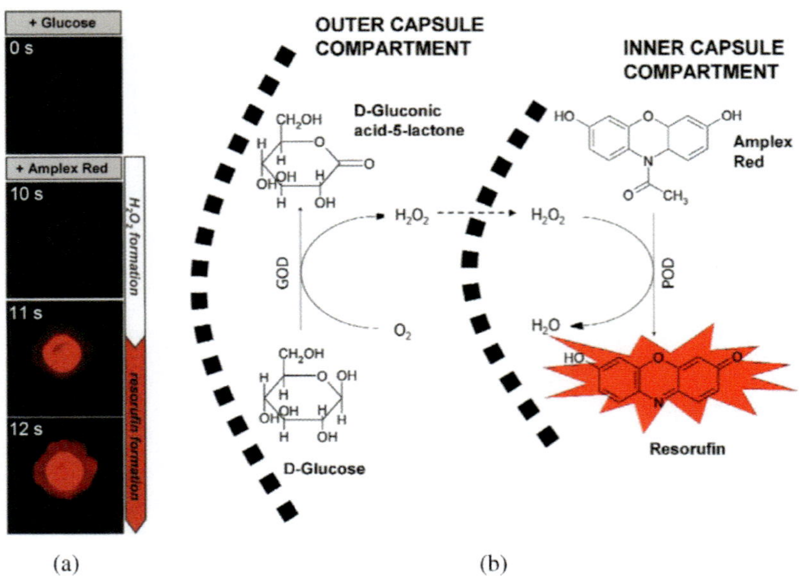

Fig. 10. (Color online) Coupled enzymatic reaction using glucose oxidase (GOx) and peroxidase (POD) inside layer-by-layer (LbL) capsules for detection of glucose. (a) CLSM imaging of resorufin formation *in situ*. (b) Illustration of a coupled, enzymatic reaction using GOx and POD in a multicompartment capsule. All substrates and products of GOx and POD can freely diffuse through the polyelectrolyte shells, whereas GOx and POD are retained in separated capsule compartments. Oxidation of glucose by GOx in the outer capsule compartment is accompanied by the generation of H_2O_2. After diffusing through the inner compartment shell, H_2O_2 is coupled to the conversion of Amplex Red to fluorescent resorufin by POD. (Reprinted with permission from Ref. 225. Copyright 2007 WILEY-VCH).

of great use in detecting various compounds found at low concentrations, which would otherwise be difficult to identify with other methods. For example, glucose oxidase is widely used to detect glucose in body fluids, and increasing its stability by encapsulation would represent a step forward in developing stable and sensitive biosensors. It is important to mention here that there are no *in vivo* studies on LbL capsules.

4.2 *Lipid-based nanoreactors for therapeutic applications*

Liposomal drug carriers are already on the market, while many others are in clinical trials.[226] Because such carriers are FDA and EMA approved, lipid-based nanoreactors also have potential in medical applications, including as biosensors, enzyme-replacement therapy, and prodrug therapy. Considerable advances have been made with liposomal-based nanoreactors up to the end of the last century, and have been reviewed in detail.[19] Due to the relatively low stability of liposomes and their short systemic circulation time, recent advances in more stable systems based on lipid-polymer conjugates/hybrids have led to a decrease in research with purely lipidic vesicles. Research on stealth-liposomes, with the addition of PEG chains to increase stability, have led to improvements in these nanoreactor systems. In this respect, we will present latest liposomal nanoreactors made for medical applications.

Prodrug-activating enzymes are encapsulated in liposomes for enzyme prodrug therapy in cancer treatment.[227] Two strategies are currently under development: antibody-directed enzyme prodrug therapy (ADEPT) and gene-directed enzyme prodrug therapy (GDEPT). The principle of GDEPT relies on the delivery of a gene with the information to produce a prodrug-activating enzyme in a target cell, and the prodrug can be delivered to the whole body (systemically).[228] A toxin is then produced locally in the tissue to which the gene is delivered, and systemic toxicity is thereby significantly reduced. Instead of delivering an enzyme-encoding gene, prodrug-activating enzymes can be encapsulated in nanoreactors and targeted to the cells to be treated. Toxin will be produced as they dock to target cells. Because enzymes cannot be administered intravenously due to immunogenicity, the advantage of encapsulating these enzymes in nanocontainers is obvious, preventing immune responses in humans and thus leading to long circulation times.

An example shown to serve as an alternative to GDEPT made use of nucleoside hydrolase from *Trypanosoma vivax*,[227] which catalyses the hydrolysis of the non-toxic prodrug 6-methyl-purine riboside and releases the toxic purine base 6-methylpurine, known as an anti-tumor drug.[229] To obtain prodrug-permeable liposomes, the bacterial outer membrane protein OmpF was incorporated in the lipid membranes of nanoreactors. A simple analogue, p-nitrophenylriboside, was used to determine the efficiency of the enzymatic activity of the nanoreactor.

The nitric oxide synthase (NOS) enzyme is involved in nitric-oxide (NO) deficiency, which is known to cause hypertension, impotence, hyperglycemia, arteriosclerosis, *etc.*[230] By encapsulating NOS in liposomes, the resulting nanoreactors were stable for at least 15 days and retained an enzyme activity of around 75% as compared to free enzyme. Work is currently ongoing to show the efficiency of these NOS-nanoreactors for targeted delivery to NO-deficient model cells.[230]

Artificial red blood cells represent a large field of research in the use of oxygen therapeutics, and lipid vesicles with encapsulated hemoglobin (HbV) are currently being developed for clinical trials.[231,232] There are various problems associated with such lipid-based blood substitutes for applications in humans: stability of the liposomes, blood circulation time, blood compatibility, and degradation in the reticuloendothelial system. However, animal tests showed good efficacy of HbV as transfusion alternatives.[233] Co-encapsulation of reductants (glutathione, homocysteine) or catalase enhanced oxygen uptake efficiency through the suppression of methemoglobin (metHb) formation.[234] Because artificial red blood cells based on liposomes represent such a large field of research, the reader is referred to the latest reviews on this topic.[232,235–237]

PEG-liposomes were introduced as alternatives to pure lipid vesicles in order to enhance blood-circulation time.[40] The antioxidant enzyme superoxide dismutase (SOD) has been used for the treatment of rheumatoid arthritis, because it catalytically degrades reactive oxygen species (ROS) involved in the pathogenesis of several diseases, such as Parkinson's disease, Alzheimer's disease, and rheumatoid arthritis, *etc.*[238] SOD-encapsulated in PEGylated liposomes were tested on arthritic rats,[239] but only the release and uptake of SOD to the inflamed site was studied. Enzymosomes, i.e. liposomes that contain SOD and a hydrophobic derivative of SOD (called Ac-SOD), have also been proposed as nanoreactors.[240] Adding Ac-SOD to liposomes results in vesicle-surface-presenting SOD enzymes. Conventional liposomes and PEGylated liposomes were compared with respect to their pharmacokinetic- and accumulation properties at arthritic inflammation sites in rats after a single dose injection. The PEGylated liposomes showed prolonged circulation times compared to the conventional liposomes, thus demonstrating better accumulation efficiency for SOD at inflammation sites. The therapeutic anti-inflammatory effect was superior with PEG-liposomes, showing the effect irrespective of dose or enzyme form. In addition, the anti-inflammatory activity was faster with Ac-SOD, because 50% of the enzyme activity was presented at the liposomal surface, thereby showing that the release of enzymes is not necessary. Repeated dose administration might improve the therapeutic effect compared to the single administration.[240]

4.3 *Polymer-based nanoreactors for therapeutic applications*

4.3.1 Dendrimer-based nanoreactors

Recently, the concept of catalysis as performed by functionalized dendrimers that contain small molecules has been extended to enzymes. Laccase was the enzyme of choice and was surface-modified with polymeric dendrimers intended to be used as nanoreactors. The dendrimers were formed of AB or ABA copolymers based on polybenzylether and polyethyleneoxide. Compared to the free enzyme, the laccase-dendrimer supramolecular structure showed improved activity (six-fold) with phenolic compounds. The improved activity of the enzyme with more hydrophobic compounds was explained by improved binding of substrates favored by the dendrimer enzyme assemblies (Fig. 11).[47] The improved catalytic properties of lac-

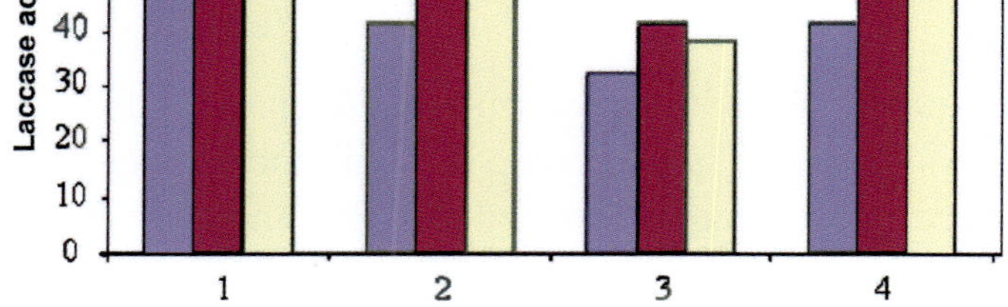

Fig. 11. Effects of organic solvents on free (native) and modified laccase activity: 1: buffer; 2: ethanol; 3: acetone; 4: acetonitrile. Laccase was surface modified with different dendritic polymers containing polybenzyl-ether third generation dendron (G3) linked to PEG polymer (11 kDa or 13 kDa). (Reprinted with permission from Ref. 47. Copyright 2008 American Chemical Society).

case-dendrimer assemblies open up new possibilities for obtaining pharmaceutical molecules via biotransformations.

Laccase-based dendrimer nanoreactors were used to detect catechin, an anticancer agent.[48] For this purpose, in a first step, gold nanoparticles were encapsulated in dendrimers made of polyamidoamine (PAMAM) functionalized with diamine-tertiophene (PDATT). The second step consisted of covalently conjugating laccase to Au-dendrimers to improve sensitivity, stability, and direct electron transfer. The laccase-Au-dendrimer complex was shown to detect submicromolar catechin concentrations and was in good agreement with standard methods for measuring catechin in human urine;[48] this can be considered only the first step in the use of electrochemical measurements for catechin detection in complex fluids.

4.3.2 Micelle-based nanoreactors

In recent decades, micelles have been studied mainly for drug delivery purposes and only recently as nanoreactors. These applications are well suited to micelles, because their chemical and physical properties can be easily tailored and their cores, or their surfaces, can be functionalized with chemical or biological molecules.[18,241,242]

A comprehensive review describing micelles containing active entities has already been published;[243] however, the catalytic reactions described are not suitable for medical applications.

An interesting example of polymeric micelles based on PMOXA-*b*-PBOXA block copolymers has been reported for the entrapment of horseradish peroxidase and two types of lipases (OF and P).[75] Upon entrapment in the micellar structure, the lipases increased their catalytic activity in organic solvents, due to increased substrate concentration in the vicinity of the enzymes, or to a conformational change in the enzyme in the presence of the polymer.[75]

The structure of micelles makes them suitable only for entrapment of hydrophobic compounds. To overcome this limitation, reversed micelles have been introduced; in the case of reversed micelles, the hydrophilic part of the block copolymer is segregated in the micelle core, while the hydrophobic part is found in the outer part of the micelle. Laccase and α-chemotrypsin were encapsulated in reverse micelles made of modified poly(ethyleneimine); while the activity and stability of α-chemotrypsin was improved by encapsulation, in the case of laccase the situation was the opposite.[244] There are not many examples of enzymes encapsulated in reverse micelles, and the disadvantage of the concept is the use of organic solvents, which makes it difficult to be applied to *in vitro* medical applications.

4.3.3 Capsule-based nanoreactors

LbL capsules have been studied in the formation of nanoreactors for more than two decades.[245] The first enzyme entrapped by LbL PSS/PAH capsules was catalase, as a model enzyme. The multilayered capsule concept was extended to other enzymes such as α-chemotrypsin,[246,247] urease,[87,139] HRP,[88,248] laccase,[249] and glucose oxidase.[250]

The enzyme HRP was encapsulated or entrapped between polymeric layers.[88,251] The encapsulated enzyme retained 57% of initial activity; after seven days of storage the encapsulated HRP featured a loss of 4% activity, whereas the value for the free enzyme was 17%. This work has demonstrated that the capsules helped stabilize the enzyme for days of storage.[88]

The use of capsules in therapeutics might not be a first choice, as their size range, in micrometer scale, is considered too large for delivery purposes.[9] The permeability of the capsule wall is an important factor when dealing with nanoreactors — in the case of capsules, the permeability of the membrane is controlled by the number of layers that form the membrane; permeability decreases with an increasing number of layers. A limitation of permeability control resulting from too few layers is the risk of leakage of the encapsulated molecules. For example, in the case of HRP encapsulated in LbL capsules, enzyme leakage decreased when the number of layers increased from four to six.[88]

4.3.4 PICsome-based nanoreactors

PEG-*b*-poly(aspartic acid) block copolymers form polymeric capsules (called PICsomes) in aqueous solution and are able to encapsulate fluorescent, water-soluble molecules[63,64] and proteins.[161] Myoglobin was encapsulated in PICsomes and the system was proven to be a successful, reversible oxygen carrier; up to three cycles of alternating O_2/argon bubbling demonstrated the feasibility of the system (Fig. 12).[161] Oxygen and other small molecules permeated the PICsomes membrane. The permeability of the vesicles increased with time; thus, reducing permeability by cross-linking might be required. When PICsomes were cross-linked, circulation times in blood were comparable to the those of polymersomes.[64]

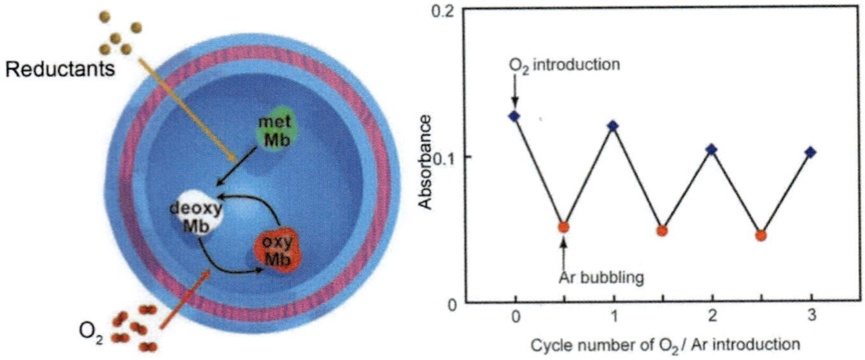

(a) (b)

Fig. 12. Reversible myoglobin (Mb) oxygenation inside the PICsome. (a) Self-assembled from a pair of oppositely charged block ionomers. (b) Change in absorbance of the Mb-PICsome at 434 nm by the alternating introduction of O_2/Ar gas to the solution. (Reprinted with permission from Ref.(161). Copyright 2007 WILEY-VCH).

4.3.5 Polymersome-based nanoreactors

Since their first description in 2000,[252] polymersomes have been intensively studied for the encapsulation of various cargos (small chemical entities, proteins, DNA), and for membrane functionalization.[26,223] The permeability of polymersome-based nanoreactors is essential in their development; there are examples of nanoreactors made of polymers permeable to small molecules (e.g. PS-*b*-PIAT, PEG-*b*-PLA, PEG-*b*-PCL). Other examples consist of nanoreactors made of PMOXA-*b*-PDMS-*b*-PMOXA block copolymers that are selectively permeable to certain ions such as superoxide anions or hydrophobic molecules (e.g. BZiPAR).[253]

There are numerous examples of proteins that have been encapsulated in polymersomes for protection, while acting *in vitro*. Hemoglobin and myoglobin were entrapped in polymersomes made of PEO-*b*-PEE, PEO-*b*-PBD or PMOXA-*b*-PDMS-*b*-PMOXA.[254–257] Oxygen carriers based on polymersomes (PEO-*b*-PEE, PEO-*b*-PDB) were stable for five days in blood plasma at room temperature; this represents a first step in designing successful oxygen replacement therapy.[254] The oxygen binding of polymersomes was significantly decreased compared to free hemoglobin, and this is explained by the low permeability of the PBD-*b*-PEO membrane.[255]

Nanoreactors based on hemoglobin encapsulated in PMOXA-*b*-PDMS-*b*-PMOXA vesicles have been designed for dual-function: as artificial blood and to detoxify peroxinitrites. The permeability of PMOXA-*b*-PDMS-*b*-PMOXA membranes has been enhanced by OmpF insertion; this resulted in functional nanoreactors in which oxygen binding properties were preserved compared to free hemoglobin.[257]

Nanoreactors that fight reactive oxygen species (ROS) were obtained by encapsulation of superoxide dismutase in PMOXA-*b*-PDMS-*b*-PMOXA vesicles.[13,14] Superoxide dismutase is responsible for the dismutation of superoxide anions. The activity of SOD-containing nanoreactors was optimized using PMOXA-*b*-PDMS-*b*-PMOXA block copolymers with different block lengths, thereby tuning the permeability of the nanoreactor membrane.[81] Nanoreactors based on SOD are anticipated for use in antioxidant therapy to prevent aging and age-related pathological conditions where ROS play an important role (e.g. cancer, cardiovascular disease, immune-system decline, brain dysfunction, cataracts,[258–260] and age-related macular degeneration).[261]

Catalase is another therapeutic enzyme involved in the antioxidant mechanism; it decomposes toxic oxygen peroxide to oxygen and water. Catalase was encapsulated in PEG-*b*-PLA polymer vesicles to increase clearance and avoid proteolysis for use in antioxidant therapy.[262]

On the other hand, the production of oxidative species in a high amount in a localized manner represents a strategy known as photodynamic therapy (PDT).[263,264] Nanoreactors containing a photosensitizer conjugated to BSA were used to produce ROS and caused cell death when illuminated with a particular wavelength of light.[31] Therapy based on nanoreactors provides localized production of ROS without the release of the photosensitizer itself, a great advantage compared with PDT therapies already in clinical phases or on the market.[265,266]

Other therapeutic strategies include the use of nanoreactors in treatments targeted at abnormal proteins and enzymes such as enzyme replacement therapy (ERT) and prodrug therapy.

Nanoreactors with trypsin encapsulated in polymersomes made of PS-*b*-PAA or PMOXA-*b*-PDMS-*b*-PMOXA block copolymers have been proposed for enzyme prodrug therapy.[32,68,267] Other nanoreactors encapsulating nucleoside hydrolase and reconstituting channel proteins have proven to be suitable candidates for prodrug therapy, as demonstrated in the case of the prodrug (2-fluoroadenosine), which was converted to a cytotoxic molecule 2-fluoroadenine.[82] Recently, penicillin acylase-containing PMOXA-*b*-PDMS-*b*-PMOXA nanoreactors were designed for antibiotic production. Nanoreactors showed their antimicrobial properties by inhibiting the growth of E. coli cells.[15] Nanoreactors can produce antibiotics in a controlled manner, tuned by the amount of substrate added, which itself has no antibiotic effect; using this approach in therapy will increase the efficacy of antibiotics that can be produced locally by the nanoreactors, thereby reducing overall side-effects.

Nanoreactors based on PMOXA-*b*-PDMS-*b*-PMOXA triblock copolymer vesicles and encapsulating thymidine phosphorylase have been designed for enzyme replacement therapy for mitochondrial neurogastrointestinal encephalomyopathy (MNGIE), a genetic disease that affects the digestive and the nervous systems in particular.[9,80] In patients affected by the disease, the activity of the enzyme is reduced and the thymidine and deoxyuridine are elevated dramatically in plasma, leading to toxic effects.[268] Thymidine phosphorylase catalyzes the phosphorolysis of thymidine to thymine and deoxyribose 1-phosphate inside nanoreactors. These nanoreactors were stable in blood serum at 37°C, and non-toxic in hepatocytes and in macrophages.[80]

Nanoreactors based on polymeric vesicles and encapsulated enzymes have been studied with regard to other model enzymes such as lipase B,[76] glucose oxidase,[130,269] HRP,[85,130] chloroperoxidase,[10] acid phosphatase.[26,29] The enzymes accept substrates that become fluorescent after the catalysis, therefore offering a good basis for their use in biosensing.

A nanoreactor that acts against oxidative stress and simultaneously monitors such activity by producing a fluorescent product is one step towards developing tools for dual purpose: therapy combined with diagnosis.[13] Two different enzymes encapsulated in nanoreactors created a cascade reaction by which radical oxygen species were detoxified by one enzyme to produce water and oxygen (Fig. 13). A co-substrate entered the nanoreactor through the channel protein OmpF incorporated in the polymer membrane and, inside, it was converted by the second enzyme into a fluorescent product, allowing real time detection of the nanoreactor` activity.[13]

A further step was accomplished through the bio-inspired approach of mimicking cell organelles with nanoreactors taken up by cells.[2,30,32,35] The aim of this approach is to develop "cell implants" in order to repair cell dysfunction, which would have a dramatic impact on medicine in terms of correcting disorders and in the design of artificial cells. Complex requirements, such as appropriate nanoreactor size for cell-uptake, maintenance of *in situ* activity of entrapped/encapsulated enzymes, stable nanoreactor architecture and enzyme activity preserved in cells, must be fulfilled in artificial organelle design. Until now, no example of an organelle claimed to be artificial has shown activity that mimics a specific type of natural organelle, which would be a milestone in the development of *in vivo* replacements for cell organelles and a new frontier in medical therapy.

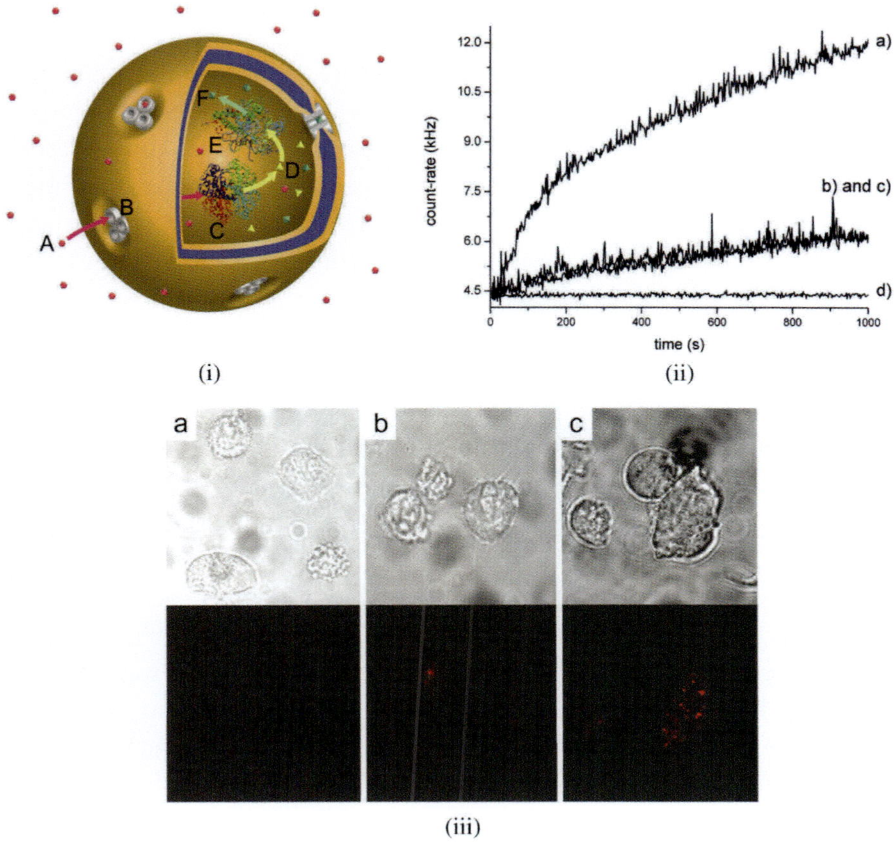

Fig. 13. (i) Enzymatic cascade reactions shown inside a polymeric nanocontainer to detect and combat oxidative stress. The combination of enzymes acts inside of polymeric nanocontainers in a cascade reaction to detoxify reactive oxygen species. Substrates, products, and reactive oxygen species penetrate the polymeric membrane, due to the insertion of protein channels; (ii) (a) Cascade reaction of SOD/LPO-coencapsulated nanocontainers with OmpF in the polymer membrane; (b) SOD/LPO-coencapsulated nanocontainers without OmpF in the polymer membrane; (c) LPO-containing vesicles with OmpF in the polymer membrane; and (d) empty polymer vesicles in the presence of amplex red. (iii) Cell-uptake and activity of SOD/LPO-coencapsulated nanocontainers as artificial organelles: (a) Transmission and confocal fluorescence image of THP-1 cells preincubated with SOD/LPO-coencapsulated nanocontainers without OmpF, after treatment with amplex red and paraquat; (b) THP-1 cells preincubated with SOD/LPO-coencapsulated nanocontainers with reconstituted OmpF, after treatment with amplex red; (c) THP-1 cells preincubated with SOD/LPO-coencapsulated nanocontainers with functionally reconstituted OmpF, after treatment with amplex red and paraquat. The confocal fluorescence images of THP-1 cells were background corrected for autooxidation of amplex red. (Reprinted with permission from Ref. 13. Copyright 2011 WILEY-VCH).

5 Conclusion

We have shown that the concept of nanoreactors, if successfully applied, represents a novel strategy in the medical domain, and will eliminate drawbacks of conventional drug delivery systems or the compounds usually used in diagnostics. In this respect, supramolecular assemblies, which constitute the compartments/templates of nanoreactors, play a dual role: to protect and to simultaneously allow active compounds (enzymes, proteins, mimics) to act *in situ*. Protection from proteolytic attack is essential in the case of protein replacement therapy, while confinement inside compartments serves to increase the local concentration of active compounds and avoids side effects related to the inherent toxicity of active compounds used in cancer therapy when released in other regions of the body. Supramolecular assemblies, especially those based on polymers, allow functionalization of their outer surface to support targeting approaches or to exhibit special properties (size, shape, charge) that favor a specific administration pathway, or localization in certain regions. A straightforward way to use nanoreactors is to develop biosensors for diagnostics, which will benefit from biomolecule specificity, sensitivity and rapidity while featuring the robustness and stability of the supramolecular assemblies. The few examples already reported indicate these advantages, and we expect, in the near future, the design of other types of nanoreactors to enlarge their use in detection methods.

The main application of nanoreactors is oriented toward providing solutions in therapeutics and theragnostics by introducing multifunctionality (cascade reactions inside one supramolecular assembly), and reply "on demand" (based on stimuli-responsiveness properties). Very recently, the first examples of nanoreactors behaving as simple artificial organelles have been introduced, even if, until now, none have mimicked a natural organelle. Artificial organelles will serve as "cell implants" in cellular dysfunctions and, in the future, as building blocks for the design of artificial cells.

Keeping in mind all of these advantages and novel approaches that we have included here via specific examples, it is important to note that nanoreactors are only in their early stage of research, and substantial efforts are still necessary to bring them into clinical trials or to market. In this respect, there are various steps still to be studied in detail: up-take, toxicity, trafficking, stability and *in situ* activity in cells, biodistribution, elimination and long terms effects *in vivo*.

Acknowledgments

Financial support was provided by the Swiss National Science Foundation, and the National Center of Competence in Nanoscale Science and this is gratefully acknowledged. Authors thank Prof. W. Meier from Basel University, for useful discussions and Mr. M. Inglin for reading the manuscript.

References

1. S. M. Borisov and O. S. Wolfbeis, Optical biosensors. *Chem Rev.* **108**(2), 423–461 (2008).
2. P. Tanner, S. Egli, V. Balasubramanian, O. Onaca, C. G. Palivan and W. Meier, Can polymeric vesicles that confine enzymatic reactions act as simplified organelles? *FEBS Lett.* **585**(11), 1699–1706 (2011).
3. M. L. Immordino, F. Dosio and L. Cattel Stealth liposomes: review of the basic science, rationale, and clinical applications, existing and potential. *Int. J. Nanomed.* **1**(3), 297–315 (2006).
4. Z. Fu, M. A. Ochsner, H-P. M. de Hoog, N. Tomczak and M. Nallani, Multicompartmentalized polymersomes for selective encapsulation of biomacromolecules. *Chem. Commun.* **47**(10), 2862–2864 (2011).
5. P. Baumann, P. Tanner, O. Onaca and C. G. Palivan, Bio-Decorated Polymer Membranes: A New Approach in Diagnostics and Therapeutics. *Polymers* **3**(1), 173–192 (2011).
6. T. D. Dziubla, A. Karim and V. R. Muzykantov, Polymer nanocarriers protecting active enzyme cargo against proteolysis. *J. Control Release* **102**(2), 427–439 (2005).
7. J. Wu and A. Eisenberg, Proton Diffusion across Membranes of Vesicles of Poly(styrene-b-acrylic Acid) Diblock Copolymers. *J. Am. Chem. Soc.***128**(9), 2880–2884 (2006).
8. C. G. Palivan, C. Vebert, F. Axthelm and W. Meier, Responsive Self-Assembled Nanostructures. *Nanotechnology, Biology and Medicine* (2007).
9. C. G. Palivan, O. Fischer-Onaca, M. Delcea, F. Itel and W. Meier, Protein-polymer nanoreactors for medical applications. *Chem Soc Rev.* **41**(7), 2800–2823 (2012).
10. H. M. de Hoog, M. Nallani, J. J. Cornelissen, A. E. Rowan, R. J. Nolte and I. W. Arends, Biocatalytic oxidation by chloroperoxidase from Caldariomyces fumago in polymersome nanoreactors. *Org. Biomol. Chem.* 7(22), 4604–4610 (2009).
11. P. Tanner, P. Baumann, R. Enea, O. Onaca, C. Palivan and W. Meier, Polymeric Vesicles: From Drug Carriers to Nanoreactors and Artificial Organelles. *Acc. Chem. Res.* **44**(10), 1039–1049 (2011).
12. Walter H. Consequences of phase separation in cytoplasm. *Int. Rev. Cytol.* **192**, 331–343 (2000).
13. P. Tanner, O. Onaca, V. Balasubramanian, W. Meier and C. G. Palivan, Enzymatic Cascade Reactions inside Polymeric Nanocontainers: A Means to Combat Oxidative Stress. *Chem. Eur. J.* **17**, 4552–4560 (2011).
14. F. Axthelm, O. Casse, Koppenol WH, Nauser T, Meier W, Palivan CG. Antioxidant Nanoreactor Based on Superoxide Dismutase Encapsulated in Superoxide-Permeable Vesicles. *J. Phys. Chem. B* **112**(28), 8211–8217 (2008).
15. K. Langowska, C. G. Palivan and W. Meier, Polymer nanoreactors shown to produce and release antibiotics locally. *Chem. Commun.* **49**(2), 128–130 (2013).
16. C. De Vocht, A. Ranquin, J. Van Ginderachter, T. Vanhaecke, V. Rogiers, P. Van Gelder *et al.*, Polymeric nanoreactors for enzyme replacement therapy of MNGIE. *J.Control. Release* **148**(1), e19–e20 (2010).

17. M. Karlsson, M. Davidson, R. Karlsson, A. Karlsson, J. Bergenholtz, Z. Konkoli *et al.*, Biomimetic Nanoscale Reactorsand Networks. *Annu. Rev. Phys. Chem.* **55**(1), 613–649 (2004).
18. D. M.Vriezema, M. C.Aragones, J. A. A. W. Elemans, J. J. L. M.Cornelissen, A. E. Rowan and R. J. M. Nolte, Self-Assembled Nanoreactors. *Chem. Rev.* **105**(4), 1445–1489 (2005).
19. P. Walde and S. Ichikawa, Enzymes inside lipid vesicles: preparation, reactivity and applications. *Biomol. Eng.* **18**(4), 143–177 (2001).
20. C. Nardin, S. Thoeni, J. Widmer, M. Winterhalter and W. Meier, Nanoreactors based on (polymerized) ABA-triblock copolymer vesicles. *Chem. Commun.* **15**, 1433–1434 (2000).
21. F. Ahmed and D. E. Discher, Self-porating polymersomes of PEG-PLA and PEG-PCL: hydrolysis-triggered controlled release vesicles. *J. Control. Release* **96**(1), 37–53 (2004).
22. M. Kumar, M. Grzelakowski, J. Zilles, M. Clark and W. Meier, Highly permeable polymeric membranes based on the incorporation of the functional water channel protein Aquaporin Z. *Proc. Natl. Acad. Sci.* **104**(52), 20719–20724 (2007).
23. A. Graff, M. Sauer, P. Van Gelder and W. Meier, Virus-assisted loading of polymer nanocontainer. *Proc. Natl. Acad. Sci.* **99**(8), 5064–5068 (2002).
24. W. Meier, C. Nardin and M. Winterhalter, Reconstitution of channel proteins in (polymerized) ABA triblock copolymer membranes. *Angew. Chem. Int. Ed.* **39**(24), 4599–4602 (2000).
25. S. Egli, M. G. Nussbaumer, V. Balasubramanian, M. Chami, N. Bruns, C. Palivan *et al.*, Biocompatible Functionalization of Polymersome Surfaces: A New Approach to Surface Immobilization and Cell Targeting Using Polymersomes. *J. Am. Chem. Soc.* **133**(12), 4476–4483 (2011).
26. M. Grzelakowski, O. Onaca, P. Rigler, M. Kumar and W. Meier, Immobilized Protein-Polymer Nanoreactors. *Small* **5**(22), 2545–2548 (2009).
27. P. Tanner, M. Ezhevskaya, R. Nehring, S. Van Doorslaer, W. Meier and C. Palivan, Specific His6-tag attachment to metal-functionalized polymersomes relies on molecular recognition. *J. Phys. Chem. B* **116**(33), 10113–10124 (2012).
28. V. P. Torchilin, Recent advances with liposomes as pharmaceutical carriers. *Nat. Rev. Drug Discov.* **4**(2), 145–160 (2005).
29. P. Broz, S. M. Benito, C. Saw, P. Burger, H. Heider, M. Pfisterer *et al.*, Cell targeting by a generic receptor-targeted polymer nanocontainer platform. *J. Control. Release* **102**(2), 475–488 (2005).
30. S. F. M. van Dongen, W. P. R. Verdurmen, R. J. R. W. Peters, R. J. M. Nolte, R. Brock and J. C. M. van Hest, Cellular Integration of an Enzyme-Loaded Polymersome Nanoreactor. *Angew Chem. Int. Ed.* **49**(40), 7213–7216 (2010).
31. P. Baumann, V. Balasubramanian, O. Onaca-Fischer, A.Sienkiewicz and C. G. Palivan, Light-responsive polymer nanoreactors: a source of reactive oxygen species on demand. *Nanoscale* **5**(1), 217–224 (2013).
32. N. Ben-Haim, P. Broz, S. Marsch, W. Meier and P. Hunziker, Cell-Specific Integration of Artificial Organelles Based on Functionalized Polymer Vesicles. *Nano Lett.* **8**(5), 1368–1373 (2008).
33. R. Roodbeen and J. C. M. van Hest. Synthetic cells and organelles: compartmentalization strategies. *BioEssays* **31**(12), 1299–1308 (2009).

34. V. Balasubramanian, O. Onaca, R. Enea, D. W. Hughes and C. G. Palivan, Protein delivery: From conventional drug delivery carriers to polymeric nanoreactors. *Exp. Opin. Drug Del.* 7(1), 63–78 (2010).
35. V. Balasubramanian, O. Onaca, M. Ezhevskaya, S. Van Doorslaer, B. Sivasankaran and C. G. Palivan, A surprising system: polymeric nanoreactors containing a mimic with dual-enzyme activity. *Soft Matter* 7(12), 5595–5603 (2011).
36. P-Y. Bolinger, D. Stamou and H.Vogel, An integrated self-assembled nanofluidic system for controlled biological chemistries. *Angew Chem. Int. Ed.* **47**(30), 5544–5549 (2008).
37. A. N. Singh, S. Singh, N. Suthar and V. K. Dubey, Glutaraldehyde-Activated Chitosan Matrix for Immobilization of a Novel Cysteine Protease, Procerain B. *J. Agr Food. Chem.* **59**(11), 6256–6262 (2011).
38. A. Graff, M. Winterhalter and W. Meier, Nanoreactors from Polymer-Stabilized Liposomes. *Langmuir* **17**(3), 919–923.
39. A-M. Caminade, R. Laurent, B. Delavaux-Nicot and J.-P. Majoral, "Janus" dendrimers: syntheses and properties. *New J. Chem.* **36**(2), 217–226 (2012).
40. TM. Allen, C. Hansen, F. Martin, C. Redemann and A. Yau-Young, Liposomes containing synthetic lipid derivatives of poly(ethylene glycol) show prolonged circulation half-lives *in vivo*. *Biochimica et Biophysica Acta (BBA) Biomembranes* **1066**(1), 29–36 (1991).
41. A. L. Klibanov, K. Maruyama, V. P. Torchilin and L. Huang, Amphipathic polyethyleneglycols effectively prolong the circulation time of liposomes. *FEBS Lett.* **268**(1), 235–237 (1990).
42. G. Blume and G. Cevc, Molecular mechanism of the lipid vesicle longevity *in vivo*. *Biochimica et Biophysica Acta (BBA) Biomembranes* **1146**(2), 157–168 (1993).
43. V. P. Torchilin, T. S. Levchenko, K. R. Whiteman, A. A. Yaroslavov, A. M. Tsatsakis, A. K.Rizos, *et al.* Amphiphilic poly-N-vinylpyrrolidones:: synthesis, properties and liposome surface modification. *Biomaterials* **22**(22), 3035–3044 (2001).
44. X. Mu and Z. Zhong, Preparation and properties of poly(vinyl alcohol)-stabilized liposomes. *International Journal of Pharmaceutics* **318**(1–2), 55–61 (2006).
45. V. P.Torchilin, M. I.Shtilman, V. S.Trubetskoy, K.Whiteman and A. M.Milstein, Amphiphilic vinyl polymers effectively prolong liposome circulation time *in vivo*. *Biochimica et Biophysica Acta (BBA) Biomembranes* **1195**(1), 181–184 (1994).
46. M. C. Woodle, C. M. Engbers and S. Zalipsky, New amphipatic polymer-lipid conjugates forming long-circulating reticuloendothelial system-evading liposomes. *Bioconjugate Chem.* **5**(6), 493–496 (1994).
47. I. Gitsov, J. Hamzik, J. Ryan, A. Simonyan, J. P. Nakas, S. Omori, *et al.* Enzymatic Nanoreactors for Environmentally Benign Biotransformations. 1. Formation and Catalytic Activity of Supramolecular Complexes of Laccase and Linear-Dendritic Block Copolymers. *Biomacromolecules* **9**, 804–811 (2008).
48. M. A. Rahman, H-B. Noh and Y-B. Shim, Direct Electrochemistry of Laccase Immobilized on Au Nanoparticles Encapsulated-Dendrimer Bonded Conducting Polymer: Application for a Catechin Sensor. *Anal. Chem.* **80**, 8020–8027 (2008).
49. Ct. Ornelas, R. Pennell, L. F. Liebes and M. Weck, Construction of a Well-Defined Multifunctional Dendrimer for Theranostics. *Org. Lett.* **13**(5), 976–979 (2011).
50. V. Percec, D. A. Wilson, P. Leowanawat, C. J. Wilson, A. D. Hughes, M. S. Kaucher, *et al.* Self-Assembly of Janus Dendrimers into Uniform Dendrimersomes and Other Complex Architectures. *Science* **328**(5981), 1009–1014 (2010).

51. K. Kataoka, G. S. Kwon, M. Yokoyama, T. Okano and Y. Sakurai, Block copolymer micelles as vehicles for drug delivery. *J. Control. Release* **24**, 119–132 (1993).
52. F. Caruso, H. Lichtenfeld, E. Donath and H. Moehwald, Investigation of Electrostatic Interactions in Polyelectrolyte Multilayer Films: Binding of Anionic Fluorescent Probes to Layers Assembled onto Colloids. *Macromolecules.* **32**, 2317–2328 (1999).
53. B. G. De Geest, N. N. Sanders, G. B. Sukhorukov, J. Demeester and S. S. C. De, Release mechanisms for polyelectrolyte capsules. *Chem. Soc. Rev.* **36**, 636–649 (2007).
54. G. B. Sukhorukov, E. Donath, S. Davis, H. Lichtenfeld, F. Caruso, Popov VI, *et al.* Stepwise polyelectrolyte assembly on particle surfaces: a novel approach to colloid design. *Polym. Adv. Technol.* **9**, 759–767 (1998).
55. C. Dejugnat and G. B. Sukhorukov, pH-Responsive Properties of Hollow Polyelectrolyte Microcapsules Templated on Various Cores. *Langmuir.* **20**, 7265–7269 (2004).
56. A. A. Antipov, D. Shchukin, Y. Fedutik, A. I. Petrov, G. B. Sukhorukov and H. Mohwald, Carbonate microparticles for hollow polyelectrolyte capsules fabrication. *Colloids Surf. A* **224**, 175–183 (2003).
57. Y. Itoh, M. Matsusaki, T. Kida and M. Akashi, Preparation of biodegradable hollow nanocapsules by silica template method. *Chem. Lett.* **33**, 1552–1553 (2004).
58. G. Schneider and G. Decher, From Functional Core/Shell Nanoparticles Prepared via Layer-by-Layer Deposition to Empty Nanospheres. *Nano Lett.* **4**, 1833–1839 (2004).
59. G. Decher, Fuzzy Nanoassemblies: Toward Layered Polymeric Multicomposites. *Science* **277**(5330), 1232–1237 (1997).
60. G. Decher and J. D. Hong, Buildup of ultrathin multilayer films by a self-assembly process. 1. Consecutive adsorption of anionic and cationic bipolar amphiphiles on charged surfaces. *Makromol. Chem. Macromol. Symp.* **46**, 321–327 (1991).
61. M. Delcea, H. Moehwald and A. G. Skirtach, Stimuli-responsive LbL capsules and nanoshells for drug delivery. *Adv. Drug Deliv. Rev.* **63**, 730–747 (2011).
62. D. B. Shenoy, A. A. Antipov, G. B. Sukhorukov and H. Mohwald, Layer-by-Layer Engineering of Biocompatible, Decomposable Core-Shell Structures. *Biomacromolecules.* **4**, 265–272 (2003).
63. A. Koide, A. Kishimura, K. Osada, W. D. Jang, Y. Yamasaki and K. Kataoka, Semipermeable polymer vesicle (PICsome) self-assembled in aqueous medium from a pair of oppositely charged block copolymers: Physiologically stable micro-/nanocontainers of water-soluble macromolecules. *J. Am. Chem. Soc.* **128**(18), 5988–5989 (2006).
64. Y. Anraku, A. Kishimura, M. Oba, Y. Yamasaki and K. Kataoka, Spontaneous Formation of Nanosized Unilamellar Polyion Complex Vesicles with Tunable Size and Properties. *J. Am. Chem. Soc.* **132**(5), 1631–1636 (2010).
65. D. E. Discher and A. Eisenberg, Polymer vesicles. *Science* **297**(5583), 967–973 (2002).
66. A. Blanazs, M. Massignani, G. Battaglia, S. P. Armes and A. J. Ryan,Synthesis of pH-responsive amphiphilic ABC block copolymers: The generation of asymmetric vesicle membranes and their cell uptake kinetics. *Abstr Pap Am. Chem Soc.* **236**, 393 (2008).
67. K. Kita-Tokarczyk, J. Grumelard, T. Haefele and W. Meier, Block copolymer vesicles-using concepts from polymer chemistry to mimic biomembranes. *Polymer.* **46**(11), 3540–3563 (2005).
68. Q. Chen, H. Schoenherr and G. J. Vancso, Block-copolymer vesicles as nanoreactors for enzymatic reactions. *Small* **5**(12), 1436–1445 (2009).

69. D. Hvasanov, J. Wiedenmann, F. Braet and P. Thordarson, Induced polymersome formation from a diblock PS-b-PAA polymer via encapsulation of positively charged proteins and peptides. *Chem. Commun.* **47**, 6314–6316 (2011).
70. M. Antonietti and S. Foerster, Vesicles and liposomes: A self-assembly principle beyond lipids. *Adv. Mater.* **15**(16), 1323–1333 (2003).
71. F. Chécot, J. Rodríguez-Hernández, Y. Gnanou and S. Lecommandoux, pH-responsive micelles and vesicles nanocapsules based on polypeptide diblock copolymers.*Biomol. Eng.* **24**(1), 81–85 (2007).
72. A. Harada and K. Kataoka, Pronounced activity of enzymes through the incorporation into the core of polyion complex micelles made from charged block copolymers. *J. Control. Release* **72**(1–3), 85–91 (2001).
73. A. Harada and K. Kataoka, Switching by pulse electric field of the elevated enzymatic reaction in the core of polylon complex micelles. *J. Am. Chem. Soc.* **125**(50), 15306–15307 (2003).
74. M. Jaturanpinyo, A. Harada, X. F. Yuan and K. Kataoka, Preparation of bionanoreactor based on core-shell structured polyion complex micelles entrapping trypsin in the core cross-linked with glutaraldehyde. *Bioconjugate Chem.* **15**(2), 344–348 (2004).
75. K. Naka, Y. Kubo, A. Ohki and S. Maeda, Aggregate of Amphiphilic Block-Copolymer Derived from Poly[(N-Acylimino)Ethylene]s and Guest-Binding Properties for 8-Anilino-1-Naphthalenesulfonic Acid, Pyrene, and Enzymes. *Polym. J.* **26**(3), 243–249 (1994).
76. D. M. Vriezema, J. Hoogboom, K. Velonia, K. Takazawa, P. C. M. Christianen, J. C. Maan, *et al.* Vesicles and Polymerized Vesicles from Thiophene-Containing Rod–Coil Block Copolymers. *Angew Chem.* **115**(7), 796–800 (2003).
77. M. Nallani, H-P. M. de Hoog, J. J. L. M. Cornelissen, A. R. A. Palmans, J. C. M. van Hest and R. J. M. Nolte, Polymersome nanoreactors for enzymatic ring-opening polymerization. *Biomacromolecules* **8**(12), 3723–3728 (2007).
78. S. M. Kuiper, M. Nallani, D. M. Vriezema, J. J. L. M. Cornelissen, J. C. M. van Hest, R. J. M. Nolte *et al.*, Enzymes containing porous polymersomes as nano reaction vessels for cascade reactions. *Org. Biomol. Chem.* **6**(23), 4315–4318 (2008).
79. W. Ha, X.-W. Meng, Q. Li, M-M. Fan, S-L. Peng, L-S.Ding, *et al.* Self-assembly hollow nanosphere for enzyme encapsulation. *Soft Matter* **6**(7), 1405–1408 (2010).
80. C. De Vocht, A. Ranquin, R. Willaert, J. A. Van Ginderachter, T. Vanhaecke, V. Rogiers, *et al.* Assessment of stability, toxicity and immunogenicity of new polymeric nanoreactors for use in enzyme replacement therapy of MNGIE. *J. Control. Release* **137**(3), 246–254 (2009).
81. O. Onaca, D. W. Hughes, V. Balasubramanian, M. Grzelakowski, W. Meier and C. G. Palivan, SOD antioxidant nanoreactors: influence of block copolymer composition on the nanoreactor efficiency. *Macromol. Biosci.* **10**(5), 531–538 (2010).
82. A. Ranquin, W. Versees, W. Meier, J. Steyaert and P. Van Gelder, Therapeutic nanoreactors: Combining chemistry and biology in a novel triblock copolymer drug delivery system. *Nano Lett.* **5**(11), 2220–2224 (2005).
83. C. Nardin, J. Widmer, M. Winterhalter and W. Meier, Amphiphilic block copolymer nanocontainers as bioreactors. *Eur. Phys. J. E* **4**(4), 403–410 (2001).
84. P. Broz, S. Driamov, J. Ziegler, N. Ben-Haim, S. Marsch, W. Meier *et al.*, Toward intelligent nanosize bioreactors: A pH-switchable, channel-equipped, functional polymer nanocontainer. *Nano Lett.* **6**(10), 2349–2353.

85. M. Nallani, S. Benito, O. Onaca, A. Graff, M. Lindemann, M. Winterhalter *et al.*, A nanocompartment system (Synthosome) designed for biotechnological applications. *J. Biotechnol.* **123**(1), 50–59.
86. K. T. Kim, J. J. L. M. Cornelissen, R. J. M. Nolte and J. C. M. van Hest, A Polymersome Nanoreactor with Controllable Permeability Induced by Stimuli-Responsive Block Copolymers. *Adv. Mater.* **21**(27), 2787–2791 (2009).
87. Y. Lvov, A. Antipov, A. Mamedov, H. Möhwald and G. B. Sukhorukov, Urease Encapsulation in Nanoorganized Microshells. *Nano Lett.* **1**, 125–128 (2001).
88. N. G. Balabushevich, O. P. Tiourina, D. V. Volodkin, N. I. Larionova and G. B. Sukhorukov, Loading the multilayer dextran sulfate/protamine microsized capsules with peroxidase. *Biomacromolecules* **4**(5), 1191–1197 (2003).
89. B. Geibel, M. Merschky, C. Rether and C. Schmuck, (Eds.), *Artificial enzyme mimics* 2012*:* John Wiley & Sons Ltd.
90. A. Choucair and A. Eisenberg, Control of amphiphilic block copolymer morphologies using solution conditions. *Eur. Phys. J. E.* **10**(1), 37–44 (2003).
91. A. Choucair, C. Lavigueur and A. Eisenberg, Polystyrene-b-poly(acrylic acid) Vesicle Size Control Using Solution Properties and Hydrophilic Block Length. *Langmui r***20**(10), 3894–3900 (2004).
92. P. Lim Soo and A. Eisenberg, Preparation of block copolymer vesicles in solution. *J. Polym. Sci. A2.* **42**(6), 923–938 (2004).
93. J. del Barrio, L. Oriol, C. Sánchez, J. L. Serrano, A. Di Cicco, P. Keller, *et al.* Self-Assembly of Linear–Dendritic Diblock Copolymers: From Nanofibers to Polymersomes. *J. Am. Chem. Soc.* **132**(11), 3762–3769 (2010).
94. A. Ulrich, Biophysical Aspects of Using Liposomes as Delivery Vehicles. *Biosci. Rep.* **22**(2), 129–150 (2002).
95. A. Samad, Y. Sultana and M. Aqil, Liposomal drug delivery systems: an update review. *Curr. Drug Deliv.* **4**(4), 297–305 (2007).
96. C. Allen, N. Dos Santos, R. Gallagher, G. N. Chiu, Y. Shu, W. M. Li, *et al.* Controlling the physical behavior and biological performance of liposome formulations through use of surface grafted poly(ethylene glycol). *Biosci Rep.* **22**(2), 225–250 (2002).
97. V. P. Torchilin, V. S. Trubetskoy, K. R. Whiteman, P. Caliceti, P. Ferruti and F. M. Veronese, New synthetic amphiphilic polymers for steric protection of liposomes *in vivo. J. Pharm. Sci.* **84**(9), 1049–1053 (1995).
98. K. R. Whiteman, V. Subr, K. Ulbrich and V. P. Torchilin, Poly(Hpma)-coated liposomes demonstrate prolonged circulation in mice. *J. Lipos. Res.* **11**(2–3), 153–164 (2001).
99. J. M. Metselaar, P. Bruin, L. W. de Boer, T. de Vringer, C. Snel, C. Oussoren *et al.*, A novel family of L-amino acid-based biodegradable polymer-lipid conjugates for the development of long-circulating liposomes with effective drug-targeting capacity. *Bioconjugate Chem.* **14**(6), 1156–1164 (2003).
100. H. Takeuchi, H. Kojima, H. Yamamoto and Y. Kawashima, Evaluation of circulation profiles of liposomes coated with hydrophilic polymers having different molecular weights in rats. *J. Control. Release.* **75**(1–2), 83–91 (2001).
101. V. P. Torchilin, Multifunctional nanocarriers. *Adv. Drug Deliv. Rev.* **64**, Supplement(0) 302–315 (2012).
102. E. R. Gillies and J. M. Frechet, Designing macromolecules for therapeutic applications: polyester dendrimer-poly(ethylene oxide) "bow-tie" hybrids with tunable molecular weight and architecture. *J. Am. Chem. Soc.* **124**(47), 14137–14146 (2002).

103. D. A. Tomalia, Birth of a new macromolecular architecture: dendrimers as quantized building blocks for nanoscale synthetic polymer chemistry. *Prog. Polym. Sci.* **30**(3–4), 294–324 (2005).

104. G. R. Newkome and C. D. Shreiner, Poly(amidoamine), polypropylenimine, and related dendrimers and dendrons possessing different 1→2 branching motifs: An overview of the divergent procedures. *Polymer.* **49**(1), 1–173 (2008).

105. R. Esfand and D. A. Tomalia, Poly(amidoamine) (PAMAM) dendrimers: from biomimicry to drug delivery and biomedical applications. *Drug Discov. Today.* **6**(8), 427–436 (2001).

106. C. C. Lee, J. A. MacKay, J. M. J. Frechet and F. C. Szoka, Designing dendrimers for biological applications. *Nat. Biotechnol.* **23**(12), 1517–1526 (2005).

107. S. Svenson and D. A. Tomalia, Dendrimers in biomedical applications — reflections on the field. *Adv. Drug Deliv. Rev.* **57**(15), 2106–2129.

108. B. M. Rosen, C. J. Wilson, D. A. Wilson, M. Peterca, M. R. Imam and V. Percec, Dendron-Mediated Self-Assembly, Disassembly, and Self-Organization of Complex Systems. *Chem. Rev.* **109**(11), 6275–6540 (2009).

109. M. A. Mintzer and M. W. Grinstaff, Biomedical applications of dendrimers: a tutorial. *Chem. Soc. Rev.* **40**(1), 173–190 (2011).

110. J. Pan, M. Wen, D. Yin, B. Jiang, D. He and L. Guo, Design and synthesis of novel amphiphilic Janus dendrimers for bone-targeted drug delivery. *Tetrahedron* **68**(14), 2943–2949.

111. D. A. Tomalia, Starburstr dendrimers — Nanoscopic supermolecules according to dendritic rules and principles. *Macromol. Sy.* **101**(1), 243–255 (1996).

112. C. J. Hawker and J. M. J. Frechet, Preparation of polymers with controlled molecular architecture. A new convergent approach to dendritic macromolecules. *J. Am. Chem. Soc.* **112**(21), 7638–7647 (1990).

113. H. Yang and W. J. Kao, Dendrimers for pharmaceutical and biomedical applications. *J. Biomat. Sci. Polym. E.* **17**(1–2), 3–19 (2006).

114. F. Vögtle, G. Richardt and N. Werner, Synthetic Methods for Dendritic Molecules. Dendrimer Chemistry: Wiley-VCH Verlag GmbH & Co. KGaA; 2009. pp. 25–48.

115. J-L. Reymond and T. Darbre, Peptide and glycopeptide dendrimer apple trees as enzyme models and for biomedical applications. *Org. Biomol Chem.* **10**(8), 1483–1492 (2012).

116. M. Peterca, V. Percec, P. Leowanawat and A. Bertin, Predicting the Size and Properties of Dendrimersomes from the Lamellar Structure of Their Amphiphilic Janus Dendrimers. *J. Am. Chem. Soc.* **133**(50), 20507–20520 (2011).

117. K. Matyjaszewski, Macromolecular engineering by controlled/living ionic and radical polymerizations. *Macromol. Sy.* **174**(1), 51–68 (2001).

118. S. Aoshima and S. Kanaoka, A Renaissance in Living Cationic Polymerization. *Chem. Rev.* **109**(11), 5245–5287 (2009).

119. A. E. Smith, X. Xu and C. L. McCormick, Stimuli-responsive amphiphilic (co)polymers via RAFT polymerization. *Prog. Polym. Sci.* **35**(1–2), 45–93 (2010).

120. N. Hadjichristidis, H. Iatrou, M. Pitsikalis and G. Sakellariou, Synthesis of Well-Defined Polypeptide-Based Materials via the Ring-Opening Polymerization of α-Amino Acid N-Carboxyanhydrides. *Chem. Rev.* **109**(11), 5528–5578 (2009).

121. W. A. Braunecker and K. Matyjaszewski, Controlled/living radical polymerization: Features, developments, and perspectives. *Prog. Polym. Sci.* **32**(1), 93–146 (2007).

122. O. Dechy-Cabaret, B. Martin-Vaca and D. Bourissou, Controlled Ring-Opening Polymerization of Lactide and Glycolide. *Chem. Rev.* **104**(12), 6147–6176 (2004).
123. M. van Dijk, D. T. S. Rijkers, R. M. J. Liskamp, C. F. van Nostrum and W. E. Hennink, Synthesis and Applications of Biomedical and Pharmaceutical Polymers via Click Chemistry Methodologies. *Bioconjugate Chem.* **20**(11), 2001–2016 (2009).
124. A. B. Lowe and C. L. McCormick, Reversible addition–fragmentation chain transfer (RAFT) radical polymerization and the synthesis of water-soluble (co)polymers under homogeneous conditions in organic and aqueous media. *Prog. Polym. Sci.* **32**(3), 283–351 (2007).
125. M. A. Harvison and A. B. Lowe, Combining RAFT Radical Polymerization and Click/Highly Efficient Coupling Chemistries: A Powerful Strategy for the Preparation of Novel Materials. *Macromol. Rapid Commun.* **32**(11), 779–800 (2011).
126. M. J. Isaacman, K. A. Barron and L. S. Theogarajan, Clickable amphiphilic triblock copolymers. *J. Polym. Sci. Pol. Chem.* **50**(12), 2319–2329 (2012).
127. T. Ruysschaert, A. F. P. Sonnen, T. Haefele, W. Meier, M. Winterhalter and D. Fournier, Hybrid Nanocapsules: Interactions of ABA Block Copolymers with Liposomes. *J. Am. Chem. Soc.* **127**(17), 6242–6247 (2005).
128. J. Grumelard, A. Taubert and W. Meier, Soft nanotubes from amphiphilic ABA triblock macromonomers. *Chem. Commun.* **13**, 1462–1463 (2004).
129. Y. Matter, R. Enea, O. Casse, C. C. Lee, J. Baryza and W. Meier, Amphiphilic PEG-b-PMCL-b-PDMAEMA Triblock Copolymers: From Synthesis to Physico-Chemistry of Self-Assembled Structures. *Macromol. Chem. Phys.* **212**(9),937–949 (2011).
130. S. F. M. van Dongen, M. Nallani, J. L. L. M. Cornelissen, R. J. M. Nolte and J. C. M. van Hest, A Three-Enzyme Cascade Reaction through Positional Assembly of Enzymes in a Polymersome Nanoreactor. *Chem. Eur. J.* **15**(5), 1107–1114 (2009).
131. A. Taubert, A. Napoli and W. Meier, Self-assembly of reactive amphiphilic block copolymers as mimetics for biological membranes. *Curr. Opin. Chem. Biol.* **8**(6), 598–603 (2004).
132. D. Lensen, D. M. Vriezema and J. C. M. van Hest, Polymeric Microcapsules for Synthetic Applications. *Macromol. Biosci.* **8**(11), 991–1005 (2008).
133. C. Wei, K. Wu, J. Li, W. Ma, J. Guo, J. Hu, *et al.* Folate-Conjugated Poly(N-(2-hydroxypropyl) methacrylamide)-block-Poly(benzyl methacrylate): Synthesis, Self-Assembly and Drug Release. *Macromol. Chem. Phys.* **213**(5), 557–565 (2012).
134. J. K. Tessmar, A. G. Mikos and A. Göpferich, Amine-Reactive Biodegradable Diblock Copolymers. *Biomacromolecules* **3**(1), 194–200 (2001).
135. S. Zhang, J. Qing, C. Xiong and Y. Peng, Synthesis of end-functionalized AB copolymers. II. Synthesis and characterization of carboxyl-terminated poly(ethylene glycol)–poly(amino acid) block copolymers. *J. Polym. Sci. Pol. Chem.* **42**(14), 3527–3536 (2004).
136. Y. Nagasaki, T. Kutsuna, M. Iijima, M. Kato, K. Kataoka, S. Kitano, *et al.* Formyl-Ended Heterobifunctional Poly(ethylene oxide): Synthesis of Poly(ethylene oxide) with a Formyl Group at One End and a Hydroxyl Group at the Other End. *Bioconjugate Chem.* **6**(2), 231–233 (1995).
137. T. Nakamura, Y. Nagasaki and K. Kataoka, Synthesis of Heterobifunctional Poly(ethylene glycol) with a Reducing Monosaccharide Residue at One End. *Bioconjugate Chem.* **9**(2), 300–303 (1998).

138. H. M. Burt, X. Zhang, P. Toleikis, L. Embree and W. L. Hunter, Development of copolymers of poly(d,l-lactide) and methoxypolyethylene glycol as micellar carriers of paclitaxel. *Colloids Surf. B* **16**(1–4), 161–171 (1999).
139. W-J. Lin, L.-W. Juang and C-C. Lin, Stability and Release Performance of a Series of Pegylated Copolymeric Micelles. *Pharm. Res.* **20**(4), 668–673 (2003).
140. G. Gaucher, M-H. Dufresne, V. P. Sant, N. Kang, D. Maysinger and J-C. Leroux, Block copolymer micelles: preparation, characterization and application in drug delivery. *J. Control. Release.* **109**(1–3), 169–188 (2005).
141. M. Ranger, M-C. Jones, M-A. Yessine and J.-C. Leroux, From well-defined diblock copolymers prepared by a versatile atom transfer radical polymerization method to supramolecular assemblies. *J. Polym. Sci. Pol. Chem.* **39**(22), 3861–3874 (2001).
142. M. H. Dufresne and J. C. Leroux, Study of the micellization behavior of different order amino block copolymers with heparin. *Pharm. Res.* **21**(1), 160–169 (2004).
143. V. P. Sant, D. Smith and J.-C. Leroux, Novel pH-sensitive supramolecular assemblies for oral delivery of poorly water soluble drugs: preparation and characterization. *J. Control. Release* **97**(2), 301–312 (2004).
144. R. Hoogenboom, Poly(2-oxazoline)s: A Polymer Class with Numerous Potential Applications. *Angew Chem. Int. Ed.* **48**(43), 7978–7994 (2009).
145. F. C. Gaertner, R. Luxenhofer, B. Blechert, R. Jordan and M. Essler, Synthesis, biodistribution and excretion of radiolabeled poly(2-alkyl-2-oxazoline)s. *J. Control. Release* **119**(3), 291–300 (2007).
146. S. C. Lee, Y. Chang, J.-S. Yoon, C. Kim, I. C. Kwon, Y-H. Kim, *et al.* Synthesis and Micellar Characterization of Amphiphilic Diblock Copolymers Based on Poly(2-ethyl-2-oxazoline) and Aliphatic Polyesters1. *Macromolecules* **32**(6), 1847–1852 (1999).
147. C-H. Wang and G.-H. Hsiue, New Amphiphilic Poly(2-ethyl-2-oxazoline)/ Poly(l-lactide) Triblock Copolymers. *Biomacromolecules* **4**(6), 1487–1490 (2003).
148. T. Deming, In *Polypeptide and Polypeptide: Hybrid Copolymer Synthesis via NCA Polymerization. Peptide Hybrid Polymers*: H-A.Klok and H.Schlaad, (Eds.), Berlin Heidelberg: Springer (2006), pp. 1–18.
149. J. Nicolas, G. Mantovani and D. M. Haddleton, Living Radical Polymerization as a Tool for the Synthesis of Polymer-Protein/Peptide Bioconjugates. *Macromol. Rapid Comm.* **28**(10), 1083–1111 (2007).
150. M. A. Gauthier and H. A. Klok, Peptide/protein-polymer conjugates: synthetic strategies and design concepts. *Chem.Commun.* **21**(23), 2591–2611 (2008).
151. J.-F.Lutz and H. G.Börner, Modern trends in polymer bioconjugates design. *Prog. Polym. Sci.* **33**(1), 1–39 (2008).
152. B. Le Droumaguet and K. Velonia, Click Chemistry: A Powerful Tool to Create Polymer-Based Macromolecular Chimeras. *Macromol. Rapid Comm.* **29**(12–13), 1073–1089 (2008).
153. K. Kita-Tokarczyk, J. Grumelard, T. Haefele and W. Meier, Block copolymer vesicles — using concepts from polymer chemistry to mimic biomembranes. *Polymer* **46**(11), 3540–3563 (2005).
154. M. R. Mozafari, Liposomes: an overview of manufacturing techniques. *Cell Mol. Biol. Lett.* **10**(4), 711–719 (2005).
155. C. Grabielle-Madelmont, S. Lesieur and M. Ollivon, Characterization of loaded liposomes by size exclusion chromatography. *J. Biochem. Biophys. Methods* **56**(1–3), 189–217 (2003).

156. New RRC, Editor. Liposomes: A Practical Approach. 2003.
157. S. Hecht and J. M. J. Fréchet, Dendritic Encapsulation of Function: Applying Nature's Site Isolation Principle from Biomimetics to Materials Science. *Angew Chem. Int. Ed.* **40**(1), 74–91 (2001).
158. J. N. Israelachvili, *20 — Soft and Biological Structures. Intermolecular and Surface Forces.* 3rd edn. San Diego: Academic Press (2011), pp. 535–576.
159. Caruso F. In *Hollow Inorganic Capsules via Colloid-Templated Layer-by-Layer Electrostatic Assembly: Colloid Chemistry II.* M.Antonietti (Ed.) Berlin Heidelberg: Springer (2003). pp. 145–168.
160. W-F. Dong, A. Kishimura, Y. Anraku, S. Chuanoi and K. Kataoka, Monodispersed Polymeric Nanocapsules: Spontaneous Evolution and Morphology Transition from Reducible Hetero-PEG PICmicelles by Controlled Degradation. *J. Am. Chem. Soc.* **131**(11), 3804–3805 (2009).
161. A. Kishimura, A. Koide, K. Osada, Y. Yamasaki and K. Kataoka, Encapsulation of Myoglobin in PEGylated Polyion Complex Vesicles Made from a Pair of Oppositely Charged Block Ionomers: A Physiologically Available Oxygen Carrier. *Angew Chem. Int. Ed.* **46**(32), 6085–6088 (2007).
162. E. Evans and D. Needham, Physical properties of surfactant bilayer membranes: thermal transitions, elasticity, rigidity, cohesion and colloidal interactions. *J. Phys. Chem.* **91**(16), 4219–4228 (1987).
163. M. Kumar, J. E. O. Habel, Y.-X. Shen, W. P. Meier and T. Walz, High-Density Reconstitution of Functional Water Channels into Vesicular and Planar Block Copolymer Membranes. *J. Am. Chem. Soc.* **134**(45), 18631–18637 (2012).
164. D. E. Discher and F. Ahmed, Polymersomes. *Annu. Rev. Biochem.* **8**, 323–841 (2006).
165. K. Renggli, P. Baumann, K. Langowska, O. Onaca, N. Bruns and W. Meier, Selective and Responsive Nanoreactors. *Adv. Funct. Mater.* **21**(7), 1241–1259 (2011).
166. A. Choucair, P. Lim Soo and A. Eisenberg, Active Loading and Tunable Release of Doxorubicin from Block Copolymer Vesicles. *Langmuir* **21**(20), 9308–9313 (2005).
167. O. Onaca, P. Sarkar, D. Roccatano, T. Friedrich, B. Hauer, M. Grzelakowski *et al.*, Functionalized nanocompartments (synthosomes) with a reduction-triggered release system. *Angew Chem. Int. Ed.* **47**, 7029–7031 (2008).
168. A. Graff, C. Fraysse-Ailhas, C. G. Palivan, M. Grzelakowski, T. Friedrich, C. Vebert *et al.*, Amphiphilic copolymer membranes promote NADH:ubiquinone oxidoreductase activity: Towards an electron-transfer nanodevice. *Macromol. Chem. Physic.* **211**(2), 229–238 (2010).
169. S. M. Smith, In *Strategies for the Purification of Membrane Proteins: Protein chromatography, methods and protocols,* D.Walls and S. T. Loughran (eds.) (2011), pp. 485–496.
170. R. Koebnik, K. P.Locher and P.Van Gelder, Structure and function of bacterial outer membrane proteins: barrels in a nutshell. *Mol. Microbiol.* **37**(2), 239–253 (2000).
171. A. Prilipov, P. S. Phale, P. Van Gelder, J. P. Rosenbusch and R. Koebnik, Coupling site-directed mutagenesis with high-level expression: large scale production of mutant porins from E. coli. *FEMS Microbiol. Lett.* **163**(1), 65–72 (1998).
172. P. S. Phale, A. Philippsen, T. Kiefhaber, R. Koebnik, V. P. Phale, T. Schirmer, *et al.* Stability of Trimeric OmpF Porin: The Contributions of the Latching Loop L2. *Biochemistry* **37**(45), 15663–15670 (1998).

173. R. Stoenescu, A. Graff and W. Meier, Asymmetric ABC-Triblock Copolymer Membranes Induce a Directed Insertion of Membrane Proteins. *Macromol. Biosci.* **4**(10), 930–935 (2004).
174. N. G. Balabushevich, G. B. Sukhorukov and N. I. Larionova, Polyelectrolyte Multilayer Microspheres as Carriers for Bienzyme System: Preparation and Characterization. *Macromol. Rapid Comm.* **26**(14), 1168–1172 (2005).
175. D. M. Vriezema, P. M. L. Garcia, N. S. Oltra, N. S. Natzakis, S. M. Kuiper, R. J. M. Nolte *et al.*, Positional assembly of enzymes in polymersome nanoreactors for cascade reactions. *Angew Chem.* **46**(39), 7378–7382 (2007).
176. W. Qi, X. Yan, L. Duan, Y. Cui, Y. Yang and J. Li, Glucose-Sensitive Microcapsules from Glutaraldehyde Cross-Linked Hemoglobin and Glucose Oxidase. *Biomacromolecules* **10**(5), 1212–1216 (2009).
177. S. Rameez, U. Banerjee, J. Fontes, A. Roth and A. F. Palmer, Reactivity of Polymersome Encapsulated Hemoglobin with Physiologically Important Gaseous Ligands: Oxygen, Carbon Monoxide, and Nitric Oxide. *Macromolecules* **45**(5), 2385–2389 (2012).
178. H. Sami, A. K. Maparu, A. Kumar and S. Sivakumar, Generic Delivery of Payload of Nanoparticles Intracellularly via Hybrid Polymer Capsules for Bioimaging Applications. *PLoS ONE* 7(5), e36195 (2012).
179. N. Robinson, Molecular Size and Shape. *J. Pharm. Pharmacol.* **12**(1), 193–218 (1960).
180. R. M. Murphy, Static and dynamic light scattering of biological macromolecules: what can we learn? *Curr. Opin Biotechnol.* **8**(1), 25–30 (1997).
181. V. Malinova, S. Belegrinou, D. Bruyn Ouboter and W. Meier, In *Biomimetic Block Copolymer Membranes*: *Organic Electronics*, T. Grasser, G. Meller and L. Li (eds.)Berlin Heidelberg: Springer(2010), pp. 213–258.
182. C. Nardin, T. Hirt, J. Leukel and W. Meier, Polymerized ABA Triblock Copolymer Vesicles. *Langmuir* **16**(3), 1035–1041 (2000).
183. W. Burchard, Static and dynamic light scattering from branched polymers and biopolymers. *Adv. Polym. Sci.* **48**, 1–124 (1983).
184. M. Sauer, T. Haefele, A. Graff, C. Nardin and W. Meier, Ion-carrier controlled precipitation of calcium phosphate in giant ABA triblock copolymer vesicles. *Chem. Commun.* 2452–2453 (2001).
185. K. Bryskhe, J. Jansson, D. Topgaard, K. Schillén and U. Olsson, Spontaneous Vesicle Formation in a Block Copolymer System. *J. Phys. Chem. B* **108**(28), 9710–9719 (2004).
186. V. Pata, F. Ahmed, D. E. Discher and N. Dan, Membrane Solubilization by Detergent: Resistance Conferred by Thickness. *Langmuir* **20**(10), 3888–3893 (2004).
187. J. S. Lee, M. Ankone, E. Pieters, R. M. Schiffelers, W. E. Hennink and J. Feijen, Circulation kinetics and biodistribution of dual-labeled polymersomes with modulated surface charge in tumor-bearing mice: Comparison with stealth liposomes. *J. Control. Release* **155**(2), 282–288 (2011).
188. J. Holoubek, Some applications of light scattering in materials science. *J. Quant. Spectrosc. R. A.* **106**(1–3), 104–121 (2007).
189. H. Kukula, H. Schlaad, M. Antonietti and S. Foerster, The Formation of Polymer Vesicles or \"Peptosomes\" by Polybutadiene-block-poly(L-glutamate)s in Dilute Aqueous Solution. *J. Am. Chem. Soc.* **124**(8), 1658–1663 (2002).
190. F. Checot, S. Lecommandoux, H. A. Klok and Y. Gnanou, From supramolecular polymersomes to stimuli-responsive nano-capsules based on poly(diene-b-peptide) diblock copolymers. *Eur. Phys. J. E* **10**(1), 25–35 (2003).

191. E. Rakhmatullina, T. Braun, M. Chami, V. Malinova and W. Meier, Self-organization behavior of methacrylate-based amphiphilic di- and triblock copolymers. *Langmuir* **23**(24), 12371–12379 (2007).
192. J. Gaitzsch, D. Appelhans, L. Wang, G. Battaglia and B. Voit, Synthetic Bio-nanoreactor: Mechanical and Chemical Control of Polymersome Membrane Permeability. *Angew Chem. Int. Ed.* **51**(18), 4448–4451 (2012).
193. Y. Wang and T. M. S. Chang, Nanobiotechnological Nanocapsules Containing Polyhemoglobin-Tyrosinase: Effects on Murine B16F10 Melanoma Cell Proliferation and Attachment. *J. Skin Cancer.* **2012**, 9 (2012).
194. B. Chu and B. S. Hsiao, Small-Angle X-ray Scattering of Polymers. *Chem. Rev.* **101**(6), 1727–1762 (2001).
195. M. Ballauff, SAXS and SANS studies of polymer colloids. *Curr. Opin. Colloid Interface Sci.* **6**(2), 132–139 (2001).
196. I. W. Hamley and V. Castelletto, Small-angle scattering of block copolymers: in the melt, solution and crystal states. *Prog. Polym. Sci.* **29**(9), 909–948 (2004).
197. H. Schlaad, L. You, R. Sigel, B. Smarsly, M. Heydenreich, A. Mantion *et al.*, Glycopolymer vesicles with an asymmetric membrane. *Chem. Commun.* **12**,1478–1480 (2009).
198. R. Nehring, C. G. Palivan, S. Moreno-Flores, A. Mantion, P. Tanner, J. L. Toca-Herrera *et al.*, Protein decorated membranes by specific molecular interactions. *Soft Matter.* **6**(12), 2815–2824 (2010).
199. D. Hua, L. Kuang and H. Liang, Self-Directed Reconstitution of Proteorhodopsin with Amphiphilic Block Copolymers Induces the Formation of Hierarchically Ordered Proteopolymer Membrane Arrays. *J. Am. Chem. Soc.* **133**, 2354–2357 (2011).
200. C. D. Driever, X. Mulet, A. P. R. Johnston, L. J. Waddington, H. Thissen, F. Caruso, *et al.* Converging layer-by-layer polyelectrolyte microcapsule and cubic lyotropic liquid crystalline nanoparticle approaches for molecular encapsulation. *Soft Matter.* **7**(9), 4257–4266 (2011).
201. K. Mortensen, Structural properties of self-assembled polymeric micelles. *Curr. Opin. Colloid Interface Sci.* **3**(1), 12–19 (1998).
202. I. W. Hamley, Nanostructure fabrication using block copolymers. *Nanotechnology* **14**(10), R39 (2003).
203. E. Amstad, J. Kohlbrecher, E. Müller, T. Schweizer, M. Textor and E. Reimhult, Triggered Release from Liposomes through Magnetic Actuation of Iron Oxide Nanoparticle Containing Membranes. *Nano Lett.* **11**(4), 1664–1670 (2011).
204. P. Samyn, Jr. Rühe, and M. Biesalski, Polymerizable Biomimetic Vesicles with Controlled Local Presentation of Adhesive Functional DOPA Groups. *Langmuir.* **26**(11), 8573–8581 (2010).
205. J. Nam, P. A. Beales and T. K. Vanderlick, Giant Phospholipid/Block Copolymer Hybrid Vesicles: Mixing Behavior and Domain Formation. *Langmuir* **27**(1), 1–6 (2010).
206. P. Girard, J. Pécréaux, G. Lenoir, P. Falson, J.-L. Rigaud and P. Bassereau, A New Method for the Reconstitution of Membrane Proteins into Giant Unilamellar Vesicles. *Biophys. J.* **87**(1), 419–429 (2004).
207. J. R. Howse, R. A. L. Jones, G. Battaglia, R. E. Ducker, G. J. Leggett and A. J. Ryan, Templated formation of giant polymer vesicles with controlled size distributions. *Nat. Mater.* **8**, 507–511 (2009).
208. H. R. Marsden, L. Gabrielli and A. Kros, Rapid preparation of polymersomes by a water addition/solvent evaporation method. *Polym. Chem.* **1**(9), 1512–1518 (2010).

209. A. Carlsen, N. Glaser, J-Fo. Le Meins and SB. Lecommandoux, Block Copolymer Vesicle Permeability Measured by Osmotic Swelling and Shrinking. *Langmuir* **27**(8), 4884–4890 (2011).
210. R. Stoenescu and W. Meier, Vesicles with asymmetric membranes from amphiphilic ABC triblock copolymers. *Chem. Commun.* **24**, 3016–3017 (2002).
211. A. Wittemann, T. Azzam and A. Eisenberg, Biocompatible Polymer Vesicles from Biamphiphilic Triblock Copolymers and Their Interaction with Bovine Serum Albumin. *Langmuir* **23**, 2224–2230 (2007).
212. S. Egli, B. Fischer, S. Hartmann, P. Hunziker, W. Meier and P. Rigler, Towards Targeted Drug Delivery by Covalent Ligand-Modified Polymeric Nanocontainers. *Macromol. Sy.* **296**(1), 278–285 (2010).
213. H. Shen and A. Eisenberg, Block Length Dependence of Morphological Phase Diagrams of the Ternary System of PS-b-PAA/Dioxane/H2O. *Macromolecules* **33**(7), 2561–2572 (2000).
214. G. Kickelbick, J. Bauer, N. Hüsing, M. Andersson and A. Palmqvist, Spontaneous Vesicle Formation of Short-Chain Amphiphilic Polysiloxane-b-Poly(ethylene oxide) Block Copolymers. *Langmuir.* **19**(8), 3198–3201 (2003).
215. U. B. Sleytr and A. W. Robards, Freeze-fracturing: a review of methods and results. *J. Microsc.* **111**(1), 77–100 (1977).
216. C. LoPresti, M. Massignani, C. Fernyhough, A. Blanazs, A. J. Ryan, J. Madsen *et al.*, Controlling Polymersome Surface Topology at the Nanoscale by Membrane Confined Polymer/Polymer Phase Separation. *ACS Nano.* **5**(3), 1775–1784 (2011).
217. G. Battaglia, C. LoPresti, M. Massignani, N. J. Warren, J. Madsen, S. Forster, *et al.* Wet Nanoscale Imaging and Testing of Polymersomes. *Small* 7(14), 2010–2015 (2011).
218. F. Ahmed, R. I. Pakunlu, A. Brannan, F. Bates, T. Minko and D. E. Discher, Biodegradable polymersomes loaded with both paclitaxel and doxorubicin permeate and shrink tumors, inducing apoptosis in proportion to accumulated drug. *J. Control. Release.* **116**(2), 150–158 (2006).
219. T. Rosenkranz, A. Katranidis, D. Atta, I. Gregor, J. Enderlein, M. Grzelakowski *et al.*, Observing Proteins as Single Molecules Encapsulated in Surface-Tethered Polymeric Nanocontainers. *Chem. Bio. Chem.* **10**(4), 702–709 (2009).
220. N. A. Christian, M. C. Milone, S. S. Ranka, G. Li, P. R. Frail, K. P. Davis *et al.*, Tat-Functionalized Near-Infrared Emissive Polymersomes for Dendritic Cell Labeling. *Bioconjugate Chem.* **18**(1), 31–40 (2006).
221. P. Rigler and W. Meier, Encapsulation of fluorescent molecules by functionalized polymeric nanocontainers: investigation by confocal fluorescence imaging and fluorescence correlation spectroscopy. *J. Am. Chem. Soc.* **128**(1), 367–373 (2006).
222. J. W. Park, D. B. Kirpotin, K. Hong, R. Shalaby, Y. Shao, U. B. Nielsen, *et al.* Tumor targeting using anti-her2 immunoliposomes. *J. Control. Release.* **74**(1–3), 95–113 (2001).
223. S. Egli, H. Schlaad, N. Bruns and W. Meier, Functionalization of Block Copolymer Vesicle Surfaces. *Polymers.* **3**(1), 252–280 (2011).
224. Z. Cheng, A. Al Zaki, J. Z. Hui, V. R. Muzykantov and A. Tsourkas, Multifunctional Nanoparticles: Cost Versus Benefit of Adding Targeting and Imaging Capabilities. *Science* **338**(6109), 903–910 (2012).

225. O. Kreft, M. Prevot, H. Möhwald and G. B. Sukhorukov, Shell-in-Shell Microcapsules: A Novel Tool for Integrated, Spatially Confined Enzymatic Reactions. *Angew Chem. Int. Ed.* **46**(29), 5605–5608 (2007).
226. M. Slingerland, H-J. Guchelaar and H. Gelderblom, Liposomal drug formulations in cancer therapy: 15 years along the road. *Drug Discov. Today.* **17**(3–4),160–166 (2012).
227. G. Huysmans, A. Ranquin, L. Wyns, J. Steyaert and P. Van Gelder, Encapsulation of therapeutic nucleoside hydrolase in functionalised nanocapsules. *J. Control. Release.* **102**(1), 171–179 (2005).
228. G. W. Both, Gene-directed enzyme prodrug therapy for cancer: a glimpse into the future? *Discov. Med.* **8**(42), 97–103 (2009).
229. W. B. Parker, P. W. Allan, S. C. Shaddix, L. M. Rose, H. F. Speegle, G. Y. Gillespie *et al.*, Metabolism and Metabolic Actions of 6-Methylpurine and 2-Fluoroadenine in Human Cells. *Biochem. Pharmacol.* **55**(10), 1673–1681 (1998).
230. R. H. Perera, P. M. Peiris, S. F. Peteu and M. Bayachou, Nitric Oxide Synthase Encapsulation in Liposomes: a Potential Delivery Platform to (Nitric Oxide)-Deficient Targets. *Electroanalysis.* **24**(1), 37–41 (2012).
231. E. Tsuchida, H. Sakai, H. Horinouchi and K. Kobayashi, Hemoglobin-vesicles as a transfusion alternative. *Artif Cells Blood Substit. Biotechnol.* **34**(6), 581–588 (2006).
232. T. M. S. Chang, From artificial red blood cells, oxygen carriers, and oxygen therapeutics to artificial cells, nanomedicine, and beyond. Artificial Cells, Blood Substitutes and *Biotechnology* **40**(3), 197–199 (2012).
233. H. Sakai, K. Sou, H. Horinouchi, K. Kobayashi and E. Tsuchida, Haemoglobin-vesicles as artificial oxygen carriers: present situation and future visions. *J. Intern. Med.* **263**(1), 4–15 (2008).
234. T. Li, X. Jing and Y. Huang, Polymer/Hemoglobin Assemblies: Biodegradable Oxygen Carriers for Artificial Red Blood Cells. *Macromol. Biosci.* **11**(7), 865–875 (2011).
235. K. Taguchi, T. Maruyama and M. Otagiri, Pharmacokinetic properties of hemoglobin vesicles as a substitute for red blood cells. *Drug Metab. Rev.***43**(3), 362–373 (2011).
236. E. Tsuchida, K. Sou, A. Nakagawa, H. Sakai, T.Komatsu and K. Kobayashi, Artificial Oxygen Carriers, Hemoglobin Vesicles and Albumin-Hemes, Based on Bioconjugate Chemistry. *Bioconjugate Chem.* **20**(8), 1419–1440 (2009).
237. H. Sakai, K. Sou, H. Horinouchi, K. Kobayashi and E. Tsuchida, Review of Hemoglobin-Vesicles as Artificial Oxygen Carriers. *Artif Organs.* **33**(2), 139–145 (2009).
238. R. Noor, S. Mittal and J. Iqbal, Superoxide dismutase--applications and relevance to human diseases. *Med Sci Monit. Int. Med. J. Exp. Clinical Res.* **8**(9), RA210–RA215 (2002).
239. M. Luisa Corvo, J. C. S. Jorge, R. van't Hof, M. E. M. Cruz, D. J. A. Crommelin and G. Storm, Superoxide dismutase entrapped in long-circulating liposomes: formulation design and therapeutic activity in rat adjuvant arthritis. *Biochim. Biophys. Acta–Biomembranes.* **1564**(1), 227–236 (2002).
240. M. M. Gaspar, O. C. Boerman, P. Laverman, M. L. Corvo, G. Storm and M. E. M. Cruz Enzymosomes with surface-exposed superoxide dismutase: *In vivo* behaviour and therapeutic activity in a model of adjuvant arthritis. *J. Control. Release* **117**(2), 186–195 (2007).
241. K. Kataoka, A. Harada and Y. Nagasaki, Block copolymer micelles for drug delivery: design, characterization and biological significance. *Adv. Drug Deliv. Rev.* **47**(1), 113–131 (2001).

242. K. Miyata, R. J. Christie and K. Kataoka, Polymeric micelles for nano-scale drug delivery. *React Funct Polym.* **71**(3), 227–234 (2011).
243. A. Harada and K. Kataoka, Supramolecular assemblies of block copolymers in aqueous media as nanocontainers relevant to biological applications. *Prog. Polym. Sci.* **31**(11), 949–982 (2006).
244. Y. L. Khmelnitsky, A. K. Gladilin, V. L. Roubailo, K. Martinek and A. V. Levashov, Reversed micelles of polymeric surfactants in nonpolar organic solvents. A new microheterogeneous medium for enzymatic reactions. *Eur. J. Biochem.* **206**(3), 737–745 (1992).
245. F. Caruso, D. Trau, H. Mohwald and R. Renneberg, Enzyme encapsulation in layer-by-layer engineered polymer multilayer capsules. *Langmuir.* **16**(4), 1485–1488 (2000).
246. O. P. Tiourina, A. A. Antipov, G. B. Sukhorukov, N. I. Larionova, Y. Lvov and H. Möhwald, Entrapment of α-Chymotrypsin into Hollow Polyelectrolyte Microcapsules. *Macromol. Biosci.* **1**(5), 209–214 (2001).
247. A. I. Petrov, D. V. Volodkin and G. B. Sukhorukov, Protein-calcium carbonate coprecipitation: a tool for protein encapsulation. *Biotechnol. Prog.* **21**(3), 918–925 (2005).
248. R. Ghan, T. Shutava, A. Patel, V. T. John and Y. Lvov, Enzyme-catalyzed polymerization of phenols within polyelectrolyte microcapsules. *Macromolecules.* **37**(12), 4519–4524 (2004).
249. G. Roman-Gusetu, K. C. Waldron and D. Rochefort, Development of an enzymatic microreactor based on microencapsulated laccase with off-line capillary electrophoresis for measurement of oxidation reactions. *J. Chromatogr. A* **1216**(47), 8270–8276 (2009).
250. C. Schuler and F. Caruso, Preparation of enzyme multilayers on colloids for biocatalysis. *Macromol. Rapid Commun.* **21**(11), 750–753 (2000).
251. C. Schüler and F. Caruso, Decomposable Hollow Biopolymer-Based Capsules. *Biomacromolecules.* **2**, 921–926 (2001).
252. B. M. Discher, Y. Y.Won, D. S. Ege, J. C. M. Lee, F. S.Bates, D. E.Discher *et al.*, Polymersomes: Tough vesicles made from diblock copolymers. *Science* **284**(5417), 1143–1146 (1999).
253. P. Broz, N. Ben-Haim, M. Grzelakowski, S. Marsch, W. Meier and P. Hunziker, Inhibition of Macrophage Phagocytotic Activity by a Receptor-targeted Polymer Vesicle-based Drug Delivery Formulation of Pravastatin. *J. Cardiovasc. Pharmacol.* **51**, 246–252 (2008).
254. J. C. M. Lee, H. Bermudez, B. M. Discher, M. A. Sheehan, Y. Y. Won, F. S. Bates, *et al.* Preparation, stability, and *in vitro* performance of vesicles made with diblock copolymers. *Biotechnol Bioeng.* **73**(2), 135–145 (2001).
255. D. R. Arifin and A. F. Palmer, Polymersome encapsulated hemoglobin: a novel type of oxygen carrier. *Biomacromolecules.* **6**(4), 2172–2181 (2005).
256. S. Rameez, H. Alosta and A. F. Palmer, Biocompatible and biodegradable polymersome encapsulated hemoglobin: a potential oxygen carrier. *Bioconjugate Chem.* **19**(5), 1025–1032 (2008).
257. D. Dobrunz, A. C. Toma, P. Tanner, T. Pfohl and C. G. Palivan, Polymer nanoreactors with dual functionality: simultaneous detoxification of peroxynitrite and oxygen transport. *Langmuir.* **28**(45), 15889–15899 (2012).
258. J. E. Slemmer, J. J. Shacka, M. I. Sweeney and J. T. Weber, Antioxidants and free radical scavengers for the treatment of stroke, traumatic brain injury and aging. *Curr. Med. Chem.* **15**(4), 404–414 (2008).
259. W. Droege, Oxidative stress in HIV infection. Oxid Stress, *Dis Cancer* 885–895 (2006).

260. A. Lomri, Role of reactive oxygen species and superoxide dismutase in cartilage aging and pathology. *Future Rheumatol.* **3**(4), 381–392 (2008).
261. J. M. Seddon, W. C. Willett, F. E. Speizer and S. E. Hankinson, A prospective study of cigarette smoking and age-related macular degeneration in women. JAMA-*J. Am. Med. Assoc.* **276**(14), 1141–1146.
262. E. A. Simone, T. D. Dziubla, F. Colon-Gonzalez, D. E. Discher and V. R. Muzykantov, Effect of polymer amphiphilicity on loading of a therapeutic enzyme onto protective fulamentous and spherical polymer nanocarrlers. *Biomacromolecules* **8**(12), 3914–3921 (2007).
263. D. Bechet, P. Couleaud, C. Frochot, M-L. Viriot, F. Guillemin and M. Barberi-Heyob, Nanoparticles as vehicles for delivery of photodynamic therapy agents. *Trends Biotechnol.* **26**(11), 612–621 (2008).
264. S. A. Sibani, P. A. McCarron, A. D. Woolfson and R. F. Donnelly, Photosensitiser delivery for photodynamic therapy. Part 2: systemic carrier platforms. *Expert Opin. Drug Del.* **5**(11), 1241–1254 (2008).
265. R. F. Donnelly, P. A. McCarron and D. Woolfson, Drug delivery systems for photodynamic therapy. *Recent Pat. Drug Deliv. Formul.* **3**(1), 1–7 (2009).
266. W. M. Sharman, C. M. Allen and J. E. van Lier, Photodynamic therapeutics: basic principles and clinical applications. *Drug Discov. Today.* **4**(11), 507–517 (1999).
267. W. Chen, F. H. Meng, R. Cheng and Z. Y. Zhong, pH-Sensitive degradable polymersomes for triggered release of anticancer drugs: A comparative study with micelles. *J. Control. Release.* **142**(1), 40–46 (2010).
268. Hirano M, Martí R, Spinazzola A, Nishino I, Nishigaki Y. Thymidine Phosphorylase Deficiency Causes MNGIE: An Autosomal Recessive Mitochondrial Disorder. *Nucleosides NucleotidesNucleic Acids.* **23**(8–9), 1217–1225 (2004).
269. A. Napoli, M. Valentini, N. Tirelli, M. Muller and J. A. Hubbell, Oxidation-responsive polymeric vesicles. *Nat Mater.* **3**(3), 183–189 (2004).

Chapter B7

"Click" Reactions: An Emerging Tool for Biology

*Annie Castonguay and Dusica Maysinger**

Department of Pharmacology and Therapeutics,
McGill University 3655 Promenade Sir-William-Osler,
Montreal, Quebec H3G 1Y6, Canada
**dusica.maysinger@mcgill.ca*

Abstract

The so-called "click" reactions are a collection of efficient tools designed to facilitate the synthesis and functionalization of various molecules including biologically relevant ones, that meets several specific *green chemistry* criteria, such as atom economy and selectivity. During the last decade, particular attention has been paid toward a few of those reactions, notably the copper-catalyzed alkyne azide cycloaddition (CuAAC) and the Diels–Alder cycloaddition (DA), the two "click" reactions discussed in this chapter. Relevant examples of the use of these convenient reactions to prepare molecules of biological interest are presented, with a focus on their use in biology, e.g. for imagery and therapy.

1. Introduction

Since the beginning of the 21st century, extensive work has been done in the area of *green chemistry*, in particular to achieve the production of agents and to design processes that share the common attribute of consuming economically justifiable amount of energy while retaining an environmental friendly character. "Green" reactions are generally atom efficient (meaning that all of the material used in the process is found in the products), are fast (sometimes benefiting from the use of catalysts), are selective, and can occur under a variety of conditions, *e.g.* in water or even without solvent, etc. Some of these properties are characteristic for the "click chemistry" reactions.[1] The two most common reactions among the several "click" reactions that have attracted considerable attention during the last decade are the copper-catalyzed alkyne azide coupling (CuAAC) and the Diels–Alder cycloaddition (DA). These reactions, which are described in Secs. 1.1 and 1.2 below, are currently exploited for numerous applications, notably for drug discovery[2] and for the functionalization of macromolecules,[3–5] carbon nanotubes,[6] nanoparticles,[7–9] glass[10] and biomaterials.[11,12] All types of DA reactions will be discussed in this chapter, although only irreversible DA reactions are usually considered as part of the "click" chemistry toolbox.

Since a variety of molecules with multiple functionalities are present in the human body (amines, carboxylic acids, thiols, etc), it is a challenge for chemists to engineer molecules/reactions which could potentially be used to accomplish certain well-defined tasks, especially without losing their efficiency and selectivity. Specific "click" reactions have the advantage to be bioorthogonal, and are increasingly investigated and utilized for various applications in biology. In this chapter, we show how CuAAC and DA reactions are exploited to synthesize molecules of biological interest, but the chapter focuses on examples that illustrate their utilization for cell and tissue imaging as well as for drug delivery. In the first and most extensive section, relevant examples from the literature demonstrate how "click" reactions can be used to link "bioshuttle" molecules to imaging agents (fluorescent dyes or magnetic resonance agents) in order to track components or processes of interest (viruses, proteins, cancer cells, DNA synthesis, drugs, etc). In the second section, examples of "bioshuttle" molecules linked to therapeutic agents such as antioxidants or anticancer drugs (Fig. 1) are presented. A brief description of these two selected reactions is provided below.

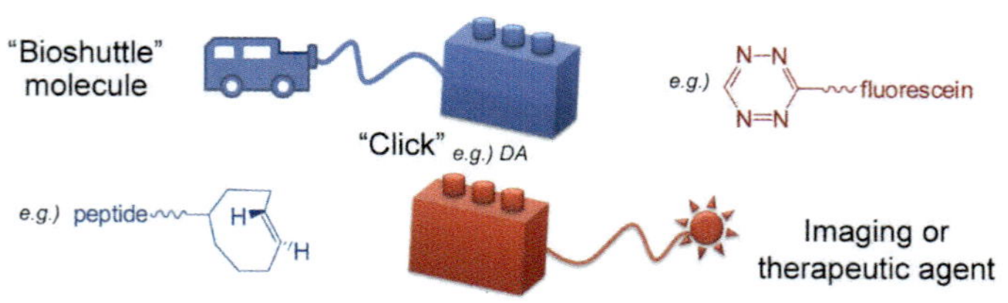

Fig. 1. Illustration of how "click" chemistry can be used in biology.

1.1. *The CuAAC reaction*

The CuAAC reaction is by far the most popular of the "click" reactions, and is now used in various fields of research, such as macromolecules functionalization, synthesis of pharmaceutical agents, etc. Despite appearances, the 1,3-dipolar cycloaddition between alkynes and azides is not a newly discovered reaction. The reaction was noted for the first time more than a century ago, by Micheal (1893), and studied in the middle of the last century by Huisgen.[13] The reaction is very exothermic ($\Delta H°$ = from −50 to −65 kcal/mol), but relatively slow, even at the elevated temperatures, due to its high activation energy barrier (for example, E_a = 25 kcal/mol for methyl azide and propyne).[14] It is only in 2001 that the alkyne azide coupling reaction received considerable attention, when Sharpless introduced the concept of "click" chemistry, with a Cu(I)-catalyzed version of the reaction (for which he received a Noble Prize in 2002). This catalyzed version allows high reaction rates at relatively low-temperatures and regioselectivity of the reaction products (Fig. 2a).[15] CuAAC reactions can be performed at ambient temperature, in water-containing medium, and the purification of the final product is usually straightforward. An obvious limitation of the exploitation of this reaction in biology is the recognized cytotoxicity of Cu. Nevertheless, the CuAAC reaction has been widely applied in biology to *in vitro*, but also to *in vivo* studies. An alternative procedure which avoids Cu has been extensively exploited in biology; however, this "Cu-free version", known to require the use of strained alkynes, is a much slower process than the Cu-catalyzed reaction. The details of Cu-free reactions for biological applications are not elaborated in this chapter but the relevant information is available elsewhere.[16,17]

1.2. *The DA cycloaddition reaction*

The DA cycloaddition was discovered in 1928 and described by Diels and Alder (for which they received the Nobel Prize in 1950). The DA reaction (which can be performed in water) occurs between a conjugated diene and an alkene (Fig. 2b) *via* a concerted mechanism (meaning that bonds are formed and broken simultaneously). The alkene, called the dienophile, usually bears electron withdrawing substituents and acts as an electophile. In contrast, the diene component usually bears electron donating substituents and acts as a nucleophile.

Interestingly, a DA cycloaddition can also occur when the alkene bears electron donating substituents and that the diene bears electron withdrawing substituents; this inverse set of electronic conditions is known as "inverse

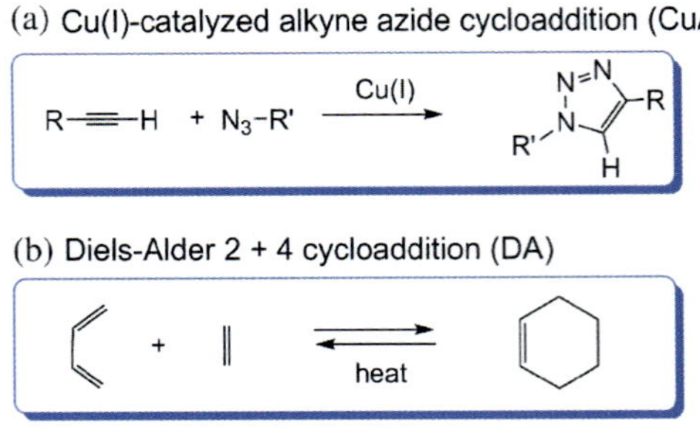

Fig. 2. Illustration of two types of reaction discussed in this chapter: (a) CuAAC; (b) DA.

electron demand". An example of such a DA reaction is the cycloaddition of alkenes and tetrazines, which is so far the most widely (and almost the only) exploited DA reaction for biology applications. It is a DA reaction of choice to link "bioshuttle" molecules to imaging agents (Fig. 1), due to its irreversible nature; the reaction releases a dinitrogen molecule and so, prevents the reverse process to take place. Irreversible DA cycloadditions are considered as part of the "click" chemistry toolbox.

Nevertheless, in most cases, the DA cycloaddition is atom efficient and generates no by-product.[18] An interesting feature of the DA reaction is that, some conditions can drive its equilibrium toward the reactants and favor a reverse process called a "retro DA" (rDA) reaction. Physical means such as applied heat or irradiation can favor the rDA reaction, which makes it an appealing reaction for other biological applications, such as drug delivery. Surprisingly, this reaction has not been extensively exploited for that purpose.

2. Applications of "Click" Chemistry in Biology

The study of biomolecules in their original environment is a great challenge for chemists and biologists, given the complexity of living systems and the presence of large numbers of chemical functionalities.[19] Especially desirable is the development of bioorthogonal reactions which are rapid and selective under physiological conditions. Over the past few years, new technologies for the selective modification of biological species in living systems have emerged. The Nobel Prize for chemistry awarded to Shimomura, Chalfie and Tsien in 2008 "*for the discovery and development of the green fluorescent protein*" illustrates the importance of studying intracellular processes by employing novel approaches to follow them visually. "Click" chemistry is playing an increasingly important role in the development of new techniques[20] for tracking biomolecules, as well as for killing cancer cells or rescuing cells from oxidative stress. Some of the most impressive examples of the utilization of click chemistry are presented below.

2.1. *"Click" chemistry to track cells, cellular components and therapeutics*

2.1.1. Viruses

The bioorthogonality of the CuAAC "click" reaction was first demonstrated by Finn *et al.*[21] They showed the efficiency of the "click" reaction in the attachment of fluorescent tags to the surface of the Cowpea mosaic virus (CPMV), a particle isolated from plants. The CPMV is characterized by an icosahedrally-arrayed coat protein structure, sometimes compared to the one of a dendrimer (*vide infra*), which can undergo reactions at specific sites to provide particles with tailored reactivities.[22] Alkyne or azide functionalities were pre-linked onto the coat protein surface of CPMV, via cystein and lysine residues, and further reacted with fluorescein-labeled azide or alkyne moieties (Fig. 3). $CuSO_4$ was used (Cu(II)), along with Cu wire (comproportionation reaction) or tris(carboxyethyl)phosphine to generate the Cu(I) reaction catalyst. The use of the ligand tris(benzyltriazolyl)methyl amine (TBTA), also prepared by using the "click" reaction, was not only found to greatly enhance the CuAAC reaction rates, but also to prevent the virus from decomposing. The virus was found to be unstable when triazole rings were formed at its surface, in the presence of Cu(II).

The introduction of the TBTA ligand was an important finding and has been widely used later on for CuAAC reactions in the presence of nucleic acids, proteins, E. Coli bacteria, and mammalian cells (*vide infra*). Modified versions of the TBTA ligands such as TTTA, tris(tert-butyltriazolylmethyl) amine, or THPTA, tris(hydroxypropyltrizaolyl)methyl amine, are also very potent ligands for the CuAAC, in part due to their increased water solubility. TBTA has also been immobilized on a resin,[23] and its complex with Cu(I) acts as an efficient air-stable catalyst for the CuAAC and prevents the contamination of products by Cu salts. The numerous interesting results obtained for the CuAAC reaction using the TBTA-type ligand have promoted research toward the development of improved ligands and have led to the discovery of sulfonated bathophenanthroline (illustrated in Fig. 4). This ligand is now part of the fastest CuAAC water-soluble catalytic system known, under dilute aqueous conditions.

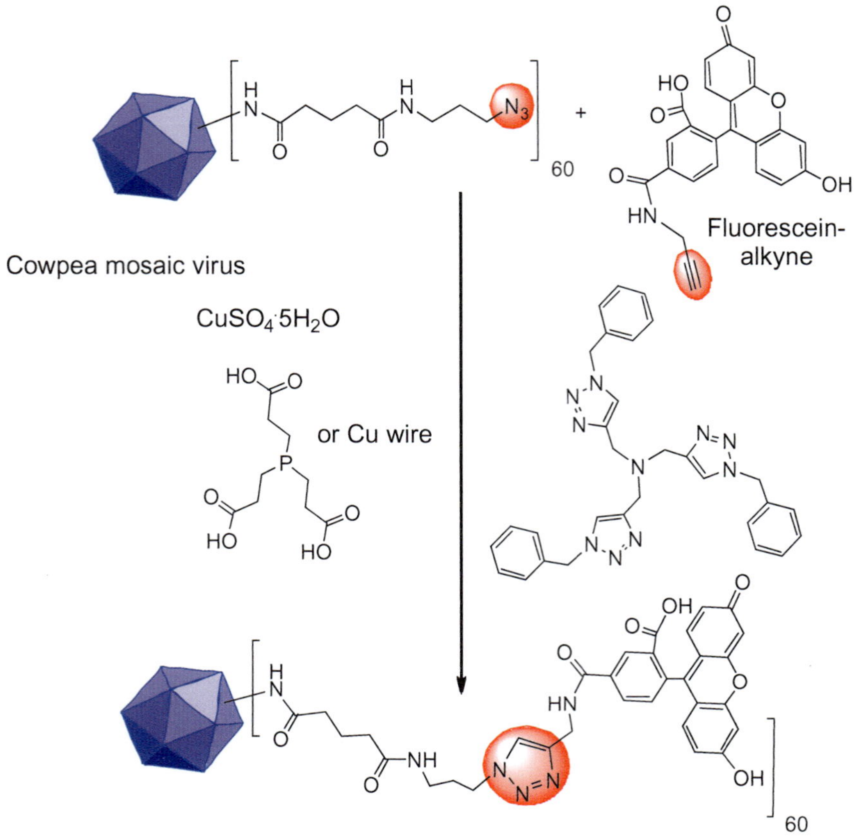

Fig. 3. Virus labeling with fluorescein, illustrating the bioorthogonality of the CuAAC reaction.

Following the work of Finn *et al.*, the functionalization of viruses using "click" chemistry has been applied to other viruses as well, such as the rod-like tobacco mosaic virus (TMV), by Wang *et al.* (2008).[24]

In order to tag biological components in live cells using a fluorescent dye, an ideal bioorthogonal reaction needs to be fast, compatible with the biological environment and able to provide a strong signal-to-background ratio, since washing out the unreacted probe is not an acceptable option. A strong signal-to-background ratio of fluorescence intensity (between the signal intensity before and after the tag reaction with the target) provides the detection of a specifically attached imaging agent and reduces the probability of non-specific labeling of other molecules. Some non-fluorescent azides can be used as fluorogenic activatable "turn on" probes, using "click chemistry", to tag various biological components (DNA, cancer cells, etc., *vide infra*),[25]

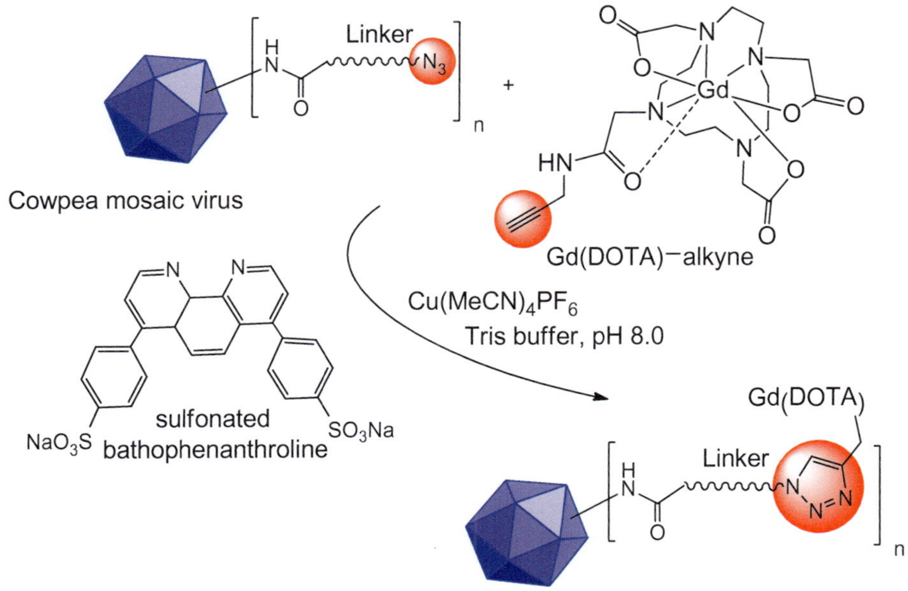

Fig. 4. MRI application using CuAAC "click" chemistry.

including CPMV and TMV viruses.[26] For instance, azidoanthracenes fluorescence can be greatly enhanced upon the formation of the "click" triazole linkage.

The "click" strategy has also been applied to image viruses by magnetic resonance imaging (MRI). MRI is a technique which uses harmless radiation to produce high contrast images for medical diagnostics, to differentiate between normal and pathological tissues. The aim of the research in this area is to improve contrast agents displaying enhanced T1 relaxivities. Finn *et al.* have prepared a gadolinium (Gd^{3+}) paramagnetic complex (Gd(DOTA), DOTA: 1,4,7,10-tetraazacyclododecanetetraacetic acid) bearing a pendant alkyne, and linked it via CuAAC "click" chemistry to azide groups at the surface of CPMV (Fig. 4).[27] Gd(DOTA) is a widely used contrast agent and the T1 relaxivity of the new Gd(DOTA)-virus system was found to be enhanced by 2 to 3 times per Gd (12.8 $mM^{-1}s^{-1}$ at 64 MHz), relative to Magnevist (DTPA-Gd, DTPA: diethylenetriaminepentaacetic acid), the first clinically approved MRI contrast agent (relaxivity of 5.2 $mM^{-1}s^{-1}$ at 64 MHz). The sulfonated bathophenanthroline system was used to link the imaging agent via "click" chemistry, in order to achieve fast reaction rates with small amounts of the coupling partners.[27] The authors successfully demonstrated the possible encapsulation of Gd^{3+} ions inside the viral structure, thereby obtaining a superior contrast imaging tool.

Recently, Douglas *et al.* (2012) reported the use of CuAAC for the introduction of dendron-like molecules at the surface of P22 bacteriophages, each consisting of multiple Gd-DTPA complex units.[28] The modified P22 capsid nanocomposite displayed very high relaxivity, a six-fold increase compared to the well-known Gd-DTPA complex alone. This finding is very promising for the development of improved contrast agents.

2.1.2. Proteins

As proteins are involved in most of biological functions, efforts were made to characterize, quantify and compare protein expressions, their interactions with other cellular components and their functions. To investigate the animal, plant and human proteome, a number of techniques are required, some of which involve click chemistry, particularly when the function of the protein needs to be determined *in vivo*. Several examples of the use of CuAAC "click" chemistry to analyze biological samples are provided in the following section.[29]

Activity-Based Protein Profiling (ABPP) is an important proteomic chemical method to visualize changes in the activity of many proteins in cells and tissues. Importantly, the technique led to the elaboration of probes for different enzymes such as serine hydrolases, cystein proteases, oxidoreductases and kinases. Tracking enzyme activities using ABPP also contributed to important findings about cancer pathogenesis, obesity, as well as malaria, and is used for the elaboration of new enzyme inhibitors. ABPP consists in directing a reactive group toward enzyme active sites and as a second step, in sending a reporter tag for detection or enrichment (using a fluorophore or biotin). Unfortunately, the traditional ABPP method requires homogenization of cells and tissues prior to their treatment with probes, which can alter the protein functions and consequently, lead to errors in the interpretation of the obtained results. In addition, the use of large reporter tags can inhibit cellular uptake and influence probe distribution within the cell. CuAAC "click" chemistry provides the way to access live cells *in vitro* and in animals without these limitations. For instance, small proteomic probes can now be easily "clicked" and enzymatic activities can be routinely determined *in vivo*. Moreover, one probe can bear more than one reporter tag, enhancing the efficiency of the measurements. For example, detection and enrichment can be achieved using one single probe, with the use of a fluorophore (detection) as well as biotin (enrichment). The first demonstration of the use of "click" chemistry in ABPP was done by Cravatt *et al.* (Fig. 5).[30] In that study, the authors compared the traditional one-step ABPP technique to the

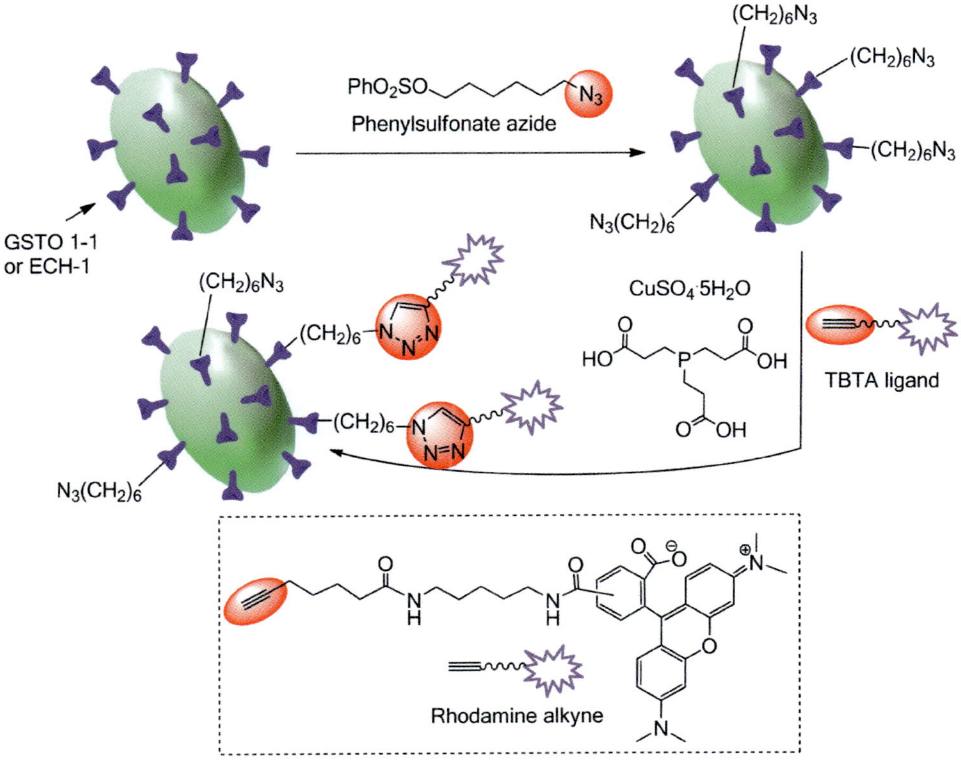

Fig. 5. "Click" chemistry ABPP probes assay.

new two-step "click" chemistry-ABPP. Both methods were designed to target glutathione-S-transferases omega 1-1 (GSTO 1-1) overexpressed at the surface of COS7 cells using a phenylsulfonate ester moiety. The traditional method simply involved the use of a phenylsulfonate ester rhodamine probe, whereas the "click" method was based on the use of a phenylsulfonate azide, designed to eventually react with a rhodamine alkyne tag. The CuAAC "click" reaction was performed using Cu(II), the reducing agent tris(carboxyethyl)phosphine and the TBTA ligand introduced earlier by Finn *et al.* (*vide supra*). The possibility of using the "click" ABPP technique in live cells was demonstrated in COS7 cells.[30] COS7 cells were treated with the "click" reagents in two steps, and proteins were isolated and separated by SDS-PAGE. Fluorescence scanning signals for GSTO 1-1 were found to be more intense than when the cells were reacted with phenylsulfonate ester rhodamine. This result suggested that using a smaller tag increases the reaction rate between GTSO 1-1 and phenylsulfonate. These promising results from the *in vitro* studies led to the subsequent studies *in vivo* (in mice) with phenylsulfonate azide.[30] Since Cu-catalyzed reactions cannot take place in living animals, the labeling has to be done in tissues' post mortem, an hour after the injection, before heart tissues are reacted with rhodamine alkyne. SDS-PAGE analysis showed that another phenylsulfonate target enzyme, enoyl CoA hydratase-1 (ECH-1), was intensely labeled in mice that underwent a phenylsulfonate azide treatment.[a] At the time, a high background signal was noted compared to the traditional method using phenylsulfonate rhodamine, which was attributed, although to a small extent, to the modification of the proteins by the rhodamine alkyne tag. The problem was easily solved by interchanging the "click" partners, using instead phenylsulfonate alkyne and rhodamine azide. Although it is unclear why this new strategy led to a lower background signal, it was later on also successfully adopted by others.

2.1.3. DNA synthesis and functionalization

Disregulation of a cell cycle is one of the hallmarks of cancer cells and acquired immunodeficiency syndrome (AIDS). In both of these pathologies there is a failure of the normal cell cycle regulation and cell turnover. To reveal these abnormalities, DNA is tracked *in vivo* by exposing the cells to nucleotide analogs bearing tags. The tagged molecules get inserted into DNA during the S-phase of the cell cycle, thereby allowing for the detection of the labeled DNA. There are several ways of labeling DNA: one of the most common is to employ tritiated thymidine molecules and to detect labeled DNA by autoradiography.[31] This technique is sensitive, but not convenient and time consuming. An alternative method is based on immunofluorescence detection of halogen-based thymidine analogs and use of antibodies.[32] This immunofluorescence technique has been used extensively over the past decades. However, in this case the access to the nuclear DNA by the antibodies follows the fixation and denaturation of DNA. These steps in the process limit the wide use of tracking proliferating cells and their DNA in living cells.

The CuAAC "click" reaction was proposed as a new approach to track cell division.[33–35] The technique consists of incorporating an analog to thymidine, 5-ethynyl-2'-deoxyuridine (EdU), into DNA during the S-phase. The analog bears an alkyne moiety which can react with a fluorescently-labeled azide in the presence of a copper catalyst, linking the fluorescent dye to the DNA,

[a] Enzymes other than glutathione-S-transferases (GTSO 1-1), such as aldehyde dehydrogenases (ALDH-1) or enoyl CoA hydratases (ECH-1), are also known to react with phenylsulfonate probes.

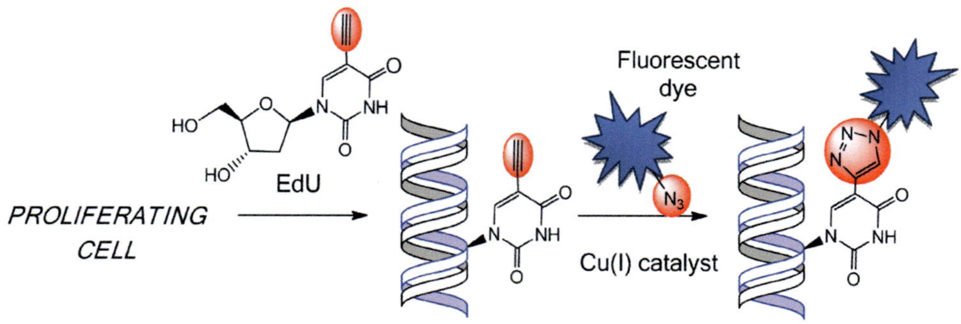

Fig. 6. Tagging DNA synthesis using CuAAC "click" chemistry.

forming a triazole ring (Fig. 6). This new approach has many advantages over the initially utilized methods. Notably, it allows the detection of EdU in intact double-stranded DNA after its insertion. The availability of small size and permeable fluorescent probes allows the labelling of EdU even in the thick tissue sections. The method using a cell permeable azide such as tetramethylrhodamine azide was validated in different cell types in cultures and in the tissues (*e.g.* intestine and brain) in whole animals.[35]

The "click" EdU staining reaction can be performed using the combination of $CuSO_4$ and ascorbic acid, and is usually found to be extremely rapid, which can be explained by the Cu(I) chelating effect of the neighboring newly formed triazole rings.[36] Interestingly, when an azide-functionalized EdU was used, such as 5-azido-3-deoxyuridine (AdU), and that staining was done with an Alexa568-fluorophore alkyne after its incorporation into cellular DNA, the background was found to be much higher than in the case of EdU. This was previously noted by others when "click" reactions were attempted on azide-labeled cells.[37]

The use of fluorogenic probes, such as 3-azido-7-hydroxycoumarin, were found helpful to decrease background fluorescence (*vide supra*).[25] Thus, by careful choice of the probe, removal of the unreacted azide can be prevented, and direct quantification/visualization of the "click" reaction product is then made possible. Indeed, the use of the CuAAC reaction to track DNA synthesis in living systems is limited by the toxicity of copper and methods using the "Cu-free" version of the reaction have been investigated.[38]

The EdU and CuAAC "click" chemistry technique has also been applied to "high-content screening" (HCS), defined as "multiplexed functional screening based on imaging multiple targets in the physiological context of intact cells by extraction of multicolor fluorescence information".[39] The technique has proven to be convenient for siRNA, miRNA and small-molecule library screening.[40]

2.1.4. Cancer cells

At first glance, the DA cycloaddition does not seem to be a reaction of choice for biological molecules because this type of reaction is, in most cases, reversible under certain conditions and one would anticipate poor labeling of the target molecules. Interestingly, some DA reactions are not reversible. For example, the "inverse electron demand" DA cycloaddition of an alkene and a tetrazine, proceeds irreversibly due to the subsequent loss of a nitrogen molecule, forming a dihydropyridazine. For instance, the alkene-tetrazine "click" DA reaction was used for live cell and imaging.[41] Another example of a DA reaction is its use for multistep labeling of monoclonal antibodies to increase the ratio of the target-to-background signal. Hilderbrand *et al.* has demonstrated the efficiency of the technique *in vitro* by targeting Her2/neu receptors in live human breast cancer cells (SKBR3) with a norbornene-modified antibody (trastuzumab), co-labeled with tetramethylrhodamine. SKBR3 cancer cells overexpress Her2/neu receptors for that specific antibody. Tetrazine conjugated to a near-infrared (NIR) fluorophore (VT680) successfully labeled the norbornene-modified antibody in the presence of serum. To demonstrate the specificity of the tetrazine imaging agent for norbornene-modified antibody in living cells, cells were incubated with a control antibody which contained only rhodamine but no norbornene. No NIR labeling was noted after exposure to the tetrazine-fluorophore.

Although these results were promising for eventual *in vivo* applications, the rate of the process was found to be comparable to the previously reported rate for the optimized cycloaddition of azides and cyclooctynes (1.6 $M^{-1}s^{-1}$ at 20°C). The required concentrations for these reactions were in the micromolar range. The rate, however, can be increased when more strained dienophiles (*e.g. trans*-cyclooctene, Fig. 7) were used. An increase in the rate constant up to 6000 $M^{-1}s^{-1}$ at 37°C,[42] was noted in this case. This multifold increase is significantly larger than when less strained dienophiles are used, as well as larger than the rate constant values reported for other bioorthogonal "click" reactions used to label live cells covalently. Tetrazine did not degrade in water at low nanomolar concentrations, and allowed real time probe imaging *in vivo*. Hilderbrand *et al.* have also used strained dienophiles for live-cell imaging to visualize epidermal growth factor receptor (EGFR). EGFR is expressed at the surface of A549 lung cancer cells. The receptor was first exposed to the anti-EGFR monoclonal antibody cetuximab-trans cyclooctene conjugate followed by tetrazine linked to a fluorophore (VT680). The tetrazine cycloaddition to trans-cyclooctene-labeled cells is fast and can be amplified by increasing the loading of the dienophile on the antibody.

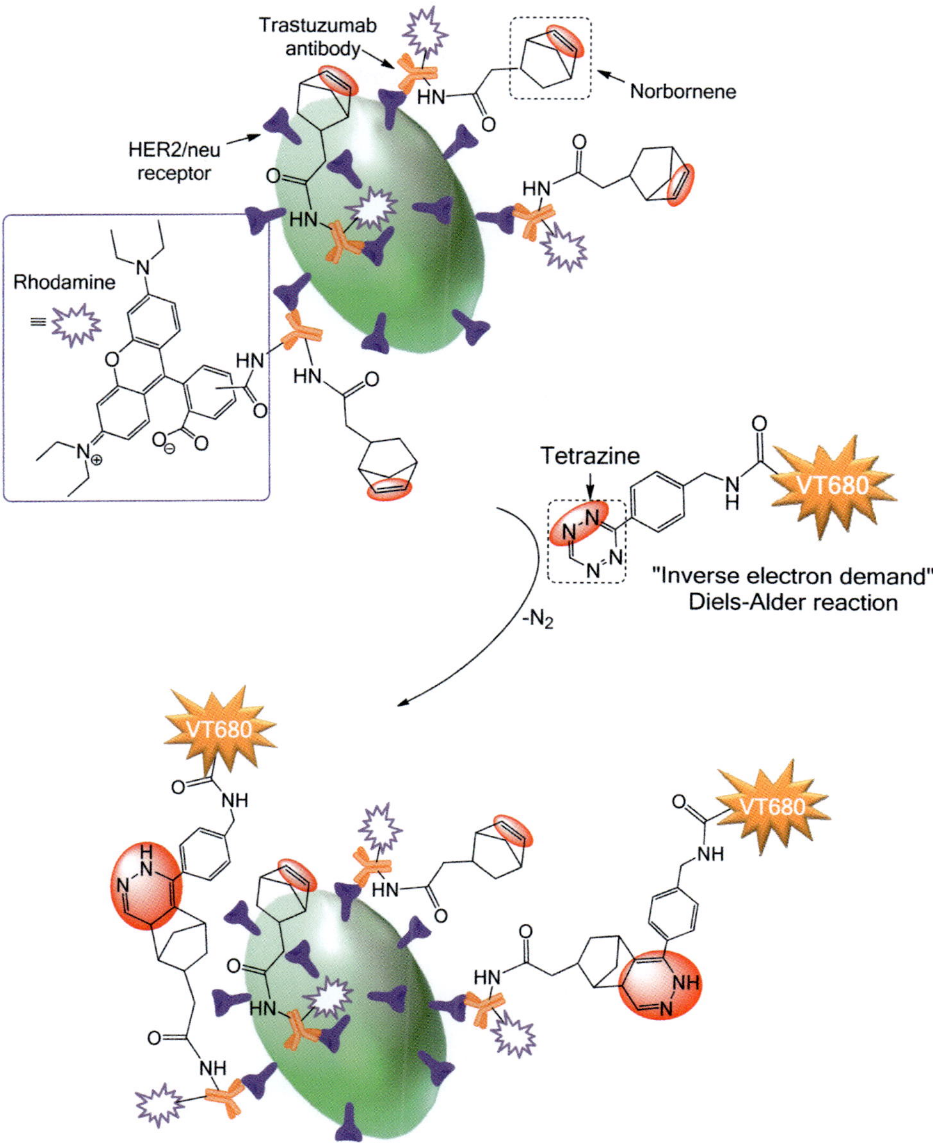

Fig. 7. Pretargeting of SKBR3 cells with norbornene and tetramethylrhodamine co-labeled trastuzumab, followed by tagging the lice cells with tetrazine-VT680 via an inverse electron demand DA cycloaddition.

2.1.5. Therapeutics

DA reactions employing fluorescent molecules have been exploited to track drugs, drug internalization processes and drug interactions with cellular components. An advantage of using tetrazine as a "click" reagent for biology is that it often leads to the creation of fluorogenic probes (*vide supra*). Upon linkage to tetrazine, the emission wavelength of some fluorophores can be absorbed, thereby quenching its fluorescence. When a fluorescently "turned-off" tetrazine-fluorophore adduct undergoes an inverse-electron-demand [4+2] DA cycloaddition with a cycloalkene dienophile, the probe fluorescence can be "turned back on". For example, the probes resulting from the reaction of 3-(4-benzylamino)-1,2,4,5-tetrazine with the succinimidyl esters of boron-dipyrromethene (BODIPY) fluorescent dyes exhibit a strongly reduced fluorescence, when compared to the BODIPY esters alone.[43] Upon DA "click" reaction of the probes with the strained dienophiles trans-cyclooctenol or norbornene, the fluorescence was "switched back on" (Fig. 8). Green emitting dyes are usually more suitable than red emitting ones; the optimal absorption maxima of the tetrazine chromophores are within the 500–525 nm range.

The novel "turn-on" probes were used for intracellular live-cell imaging of microtubuli-binding *trans*-cyclooctene modified paclitaxel (taxol) in PtK2 kangaroo rat kidney. Taxol is a drug known to bind to microtubules, and PtK2 kangaroo rat kidney cells are commonly used for microtubule studies because of their flattened morphology.[44] Thus, *trans*-cyclooctene taxol

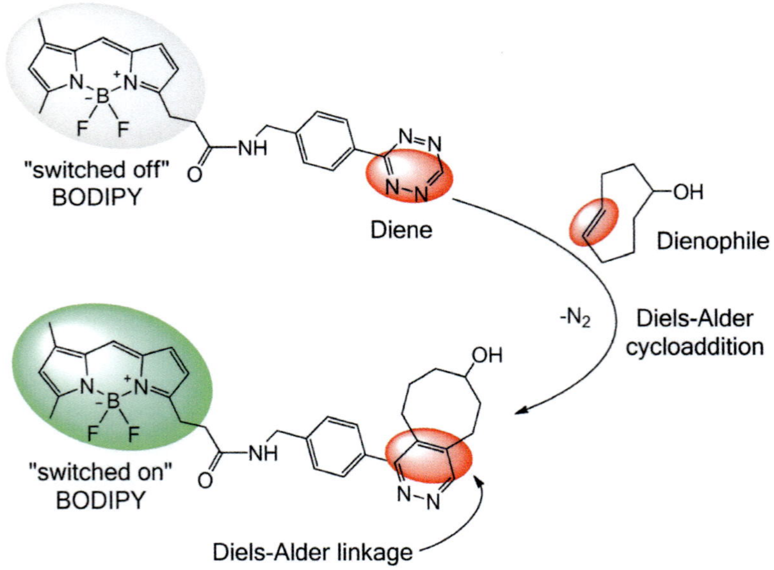

Fig. 8. Development of a fluorogenic probe which reacts with strained dienophiles via an irreversible inverse electron demand DA "click" reaction, and which was applied for intracellular live-cell imaging of microtubuli-binding *trans*-cyclooctene modified paclitaxel.

molecules can induce a formation of tubule bundles which can be visualized by fluorescence microscopy after staining with tetrazine fluorophore probes. These probes covalently link the microtubule-bound trans-cyclooctene and taxol molecules via DA "click" reactions. An example of the use of these reactions for studying cellular internalization of fluorescent probes was reported for tetrazine-BODIPY-FL.[43] This result was in contrast to the lack of internalization found when the membrane impermeable sulfonated tetrazine–VT680 probe was used, a probe found to be inefficient at staining the microtubule-bound trans-cyclooctene taxol molecules.

Until now, we have only looked at examples where "click" reactions were used to track biological components as the last step of a two-reaction sequence. We should also draw some attention to examples where "click" reactions allow the synthesis of new multifunctional nanocarriers (*i*) which can, in one step, act as a "bioshuttle", as a therapeutic agent and as an imaging agent, and (*ii*) which, without the use of "click" chemistry, would have been very complicated or impossible to prepare. Hyperbranched and monodisperse macromolecules of nanodimensions, commonly referred to as dendrimers,[45] and a class of star-shaped polymers whose arms can be varied in terms of their chemical identity and/or molecular weight, commonly called mikoarm polymers,[46] (*i*) can both act as multifunctional nanocarriers for biology applications, and (*ii*) can be efficiently prepared thanks to "click" chemistry strategies.

For instance, trifunctional dendrimers and miktoarm polymers, in which solubilizing (polyethylene glycol, PEG), therapeutic (lipoic acid, LA) and imaging (BODIPY dye) agents are incorporated, can be prepared with relative ease using a series of CuAAC "click" reactions (using $CuSO_4$/sodium ascorbate to generate the catalyst), Fig. 9.[47] LA (1,2-dithiolane-3-pentanoic acid) is a therapeutic agent introduced into clinics as an antioxidant, chelating agent and transcription factor regulator. LA can facilitate re-establishing redox homeostasis in stressed cells with disregulated intracellular antioxidant actions involving vitamins C and D, glutathione and redox responsive transcription factors.[48] We showed that the trifunctionnal dendrimers/miktoram polymers to which LA units were covalently linked, effectively rescued insulted cells from oxidative stress by increasing the intracellular glutathione concentrations. Glutathione is one of the most abundant endogenous thiols with antioxidant properties. BODIPY served as an imaging moiety to visualize intracellular location of the dendrimer/polymer structures. PEG units provided the necessary hydrophilicity to the new multifunctional scaffolds, which would otherwise be highly lipophilic.

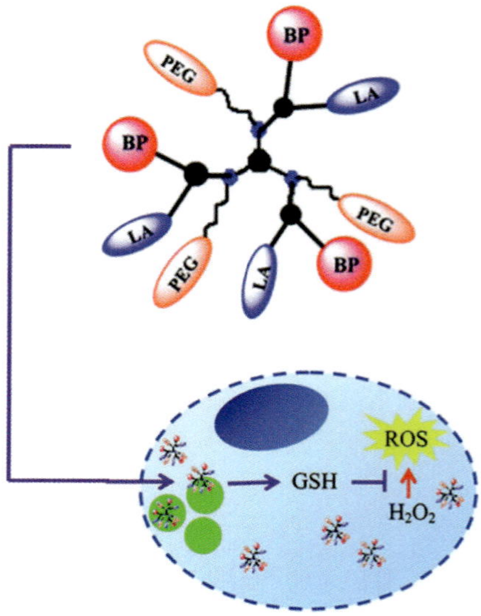

Fig. 9. Trifunctional dendrimer composed of solubilizing (PEG), imaging (BODIPY) and cell rescuing (LA) agents. Reprinted (adapted) from Ref. 47. Copyright (2011) American Chemical Society.

2.2. *"Click" chemistry to rescue or to kill cells*

2.2.1. "Click" chemistry to rescue cells

Among the two "click" reactions (CuAAC and DA cycloadditions) discussed in the previous sections, DA reaction is the only one which is reversible under certain conditions. So far, we have exclusively discussed examples of the irreversible "click" DA cycloaddition between tetrazine and cyclooctene moieties, since the irreversibility is of extreme importance for imagery purposes. As discussed in Sec. 1.2, in most cases, the DA reaction is an equilibrium which can easily be driven to the reactants side, by using physical means such as applied heat or irradiation. The most obvious role that this reversible reaction can play in biology is the release of therapeutics. Surprisingly, not many researchers have used this powerful tool in biology, which could lead to interesting applications. Using reversible DA linkages, drugs can be slowly released in the body at physiological/pathological temperatures, or at much faster rates using irradiation.

In the course of our ongoing studies on drug delivery systems, we have been interested in evaluating the potential of the DA reaction reversibility using a dendrimer framework (Fig. 10).[49] Dendrimers are monodisperse macromolecules of spherical shape (*vide supra*), extensively studied for various

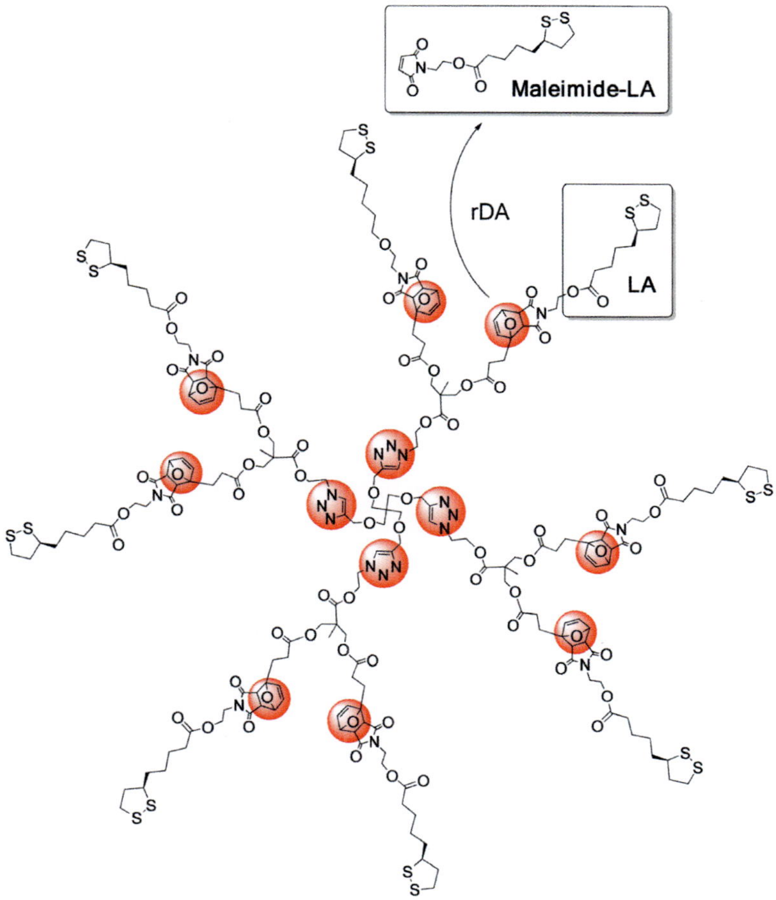

Fig. 10. Dendrimer framework with maleimide-LA release potential from peripherial rDA reactions, prepared via two types of reactions, CuAAC and DA.

applications, and well-studied for their potential in the field of drug delivery. They can act as "bioshuttle" macromolecules and encapsulate drugs inside their internal cavities, which can be eventually released by events such as redox reactions or pH changes when the biological targets have been reached. As discussed previously, dendrimers can also be used as multitple-drug entities, since many drug units can be covalently linked to their periphery. As a proof of concept, we have shown that LA can be efficiently linked at the periphery of dendrimers, and that it can be released from them via a rDA reaction at 37°C–42°C (physiological/pathological range). A maleimide moiety was linked to LA via a Steglish esterification and the new maleimide-LA molecule was then reacted (in excess) with a furan-terminated dendrimer (constructed using the CuAAC reaction). The rDA dendrimer disassembly was studied by ^{1}H NMR and it was shown that in deuterated 3 mM dimethylsulfoxide solutions, about 20%–30% of the maleimide-LA moieties were released after 24 h at physiological/pathological temperatures. Our studies have demonstrated that those dendrimers are not toxic and are able to rescue cells from oxidative stress. The maleimide/furan DA reaction has rarely been used in biology and this study suggests that it could be used for the delivery of other types of drugs, as long as the maleimide-drug released units display a considerable therapeutic effect.

2.2.2. "Click" chemistry to kill cells

In an effort to combat cancer, characterized by an uncontrolled cell growth, different new reactions have been developed and new therapeutics introduced into clinics. One of the requirements for the usefulness of these reactions is their efficiency. Efficient reactions are required for introduction of chemical modifications into existing anticancer drugs in order to enhance their therapeutic effect, to change their hydrophilicity, or their size, or to link them to "bioshuttle" molecules. In the following section, several selected examples of chemotherapeutic agents for which CuAAC or DA reactions were applied are presented. These examples illustrate how drug modifications can enhance more effective killing of rapidly dividing cancer cells.

Among numerous other examples of the use of such reactions for the discovery of new therapeutics, San Feliciano *et al.*[50] reported the preparation of several diacetylated monoterpenyl-1,4-naphthohydroquinones, using the reversible DA reaction between α-myrcene(diene) and 1,4-benzoquinone, 2-acetyl-1,4-benzoquinone or 2-chloro-1,4-benzoquinone (dienophiles). The technique has been used until recently, by the same researchers,[51] to screen for more potent drugs against A-549 lung, HT-29 colon, and MB-231 breast cancer cells. Lesyk *et al.*[52] have used the reversible hetero DA reaction between 4-thioxo-2-thiazolidinones and 5-norbornene-2,3-dicarboxylic acid imido-derivatives to screen for anticancer activity against various cell types. One of the thiopyrano[2,3-d]thiazole-based compounds was found to be particularly active against leukemia cells.

DA reactions were also used to enhance drug potency or to reduce the side effects of certain drugs. A good example in this case is temozolomide (TMZ). TMZ is a DNA-alkylating agent, which is a highly active therapeutic used against a very aggressive type of brain cancer called glioblastoma[53] multiforme. Braun *et al.* [54] showed that inverse electron demand "click" DA reaction

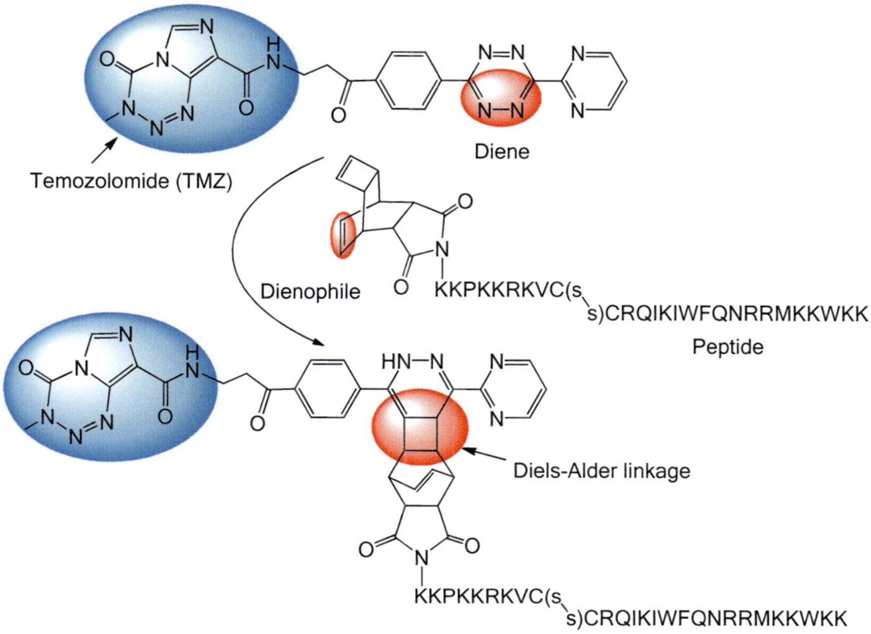

Fig. 11. TMZ-peptide as an improved therapeutics against glioblastoma multiform.

between a tetrazine and a cyclo-octene can be used to irreversibly link TMZ to a transmembrane transport peptide (Fig. 11). The resultant "bioshuttle" was significantly more potent than TMZ against glioblastoma multiforme. It also reduced the occurrence of adverse reactions. A similar "bioshuttle" strategy was applied for the treatment of prostate cancer, also employing TMZ.[55]

3. Conclusions and Outlook

"Click" reactions, established and recently developed ones, constitute a new toolbox for various applications in chemistry, biology, pharmacology and medicine. They are increasingly exploited in biology, sometimes even in combination. "Click" reactions have been successfully used for the synthesis and the functionalization of molecules, in particular macromolecules such as polymers and dendrimers. The copper-catalyzed alkyne azide 1,3-dipolar cycloaddition (CuAAC), and the DA cyclo-additon are two especially efficient reactions that can be carried out in water at both physiological and pathological temperatures, and that are widely used in biology.

In this chapter, we have shown that those reactions can be used to link "bioshuttle" molecules to fluorescent dyes in order to track biomolecules of interest, or to therapeutic agents, such as antioxidants, to rescue cells from oxidative stress, or such as anticancer agents, to promote and/or enhance the killing of cancer cells. More specifically, CuAAC "click" reactions have been used to track viruses, proteins and DNA synthesis, whereas irreversible "click" DA reactions have been mainly exploited to track and kill cancer cells. Due to copper cytotoxicity, further developments of the "copper free" version of the CuAAC reaction are expected to set the stage for more detailed applications in the area.

For drug delivery purposes, the use of reversible DA reactions is very limited. Since reverse DA reactions can be promoted by applied heat or irradiation, they seem to be a useful approach to trigger the release of therapeutics from "bioshuttle" molecules, when a particular target would be reached. In the years to come, we expect further developments of reversible DA systems in the field of drug delivery.

Acknowledgments

This work was supported by NSERC (Natural Sciences and Engineering Research Counsil) and CIHR (Canadian Institutes of Health Research). We would like to thank Dr. Ashok Kakkar (Department of Chemistry, McGill) for helpful discussions.

References

1. J. E. Hein and V. V. Fokin, Copper-catalyzed azide-alkyne cycloaddition (CuAAC) and beyond: new reactivity of copper(i) acetylides. *Chem. Soc. Rev.* **39**(4), 1302–1315 (2010).
2. H. C. Kolb and K. B. Sharpless, The growing impact of click chemistry on drug discovery. *Drug Discov. Today.* **8** (Copyright (C) 2012 American Chemical Society (ACS). All Rights Reserved.) 1128–1137 (2003).
3. A. J. Dirks, J. J. L. M. Cornelissen, D. F. L. van, H. J. C. M. van, R. J. M. Nolte, A. E. Rowan *et al.*, From (bio)molecules to biohybrid materials with the click chemistry approach. *QSAR Comb. Sci.* **26** (Copyright (C) 2012 American Chemical Society (ACS). All Rights Reserved.) 1200–1210 (2007).
4. R. Goto, K. Okura, H. Sakazaki, T. Sugawara, S. Matsuoka and M. Inoue, Synthesis and biological evaluation of triazole analogs of antillatoxin. *Tetrahedron* **67** (Copyright (C) 2012 American Chemical Society (ACS). All Rights Reserved.) 6659–6672 (2011).
5. J.-F. Lutz and H. G. Boerner, Modern trends in polymer bio-conjugates design. *Prog. Polym. Sci.* **33** (Copyright (C) 2012 American Chemical Society (ACS). All Rights Reserved.) 1–39 (2008).
6. S. Rana and J. W. Cho, Functionalization of carbon nanotubes via Cu(I)-catalyzed Huisgen [3 + 2] cycloaddition "click chemistry". *Nanoscale* **2** (Copyright (C) 2012 American Chemical Society (ACS). All Rights Reserved.) 2550–2556 (2010).
7. N. Li and W. H. Binder, Click-chemistry for nanoparticle-modification. *J. Mater. Chem.* **21** (Copyright (C) 2012 American Chemical Society (ACS). All Rights Reserved.) 16717–16734 (2011).

8. M. Shi, J. H. Wosnick, K. Ho, A. Keating and M. S. Shoichet, Immuno-polymeric nanoparticles by Diels-Alder Chemistry. *Angew. Chem. Int. Ed.* **46** (Copyright (C) 2012 American Chemical Society (ACS). All Rights Reserved.) 6126–6131, S/1–S/12 (2007).
9. C. Xu and L. Ye, Clickable molecularly imprinted nanoparticles. *Chem. Commun.* **47** (Copyright (C) 2012 American Chemical Society (ACS). All Rights Reserved.) 6096–6098 (2011).
10. A. Bertin and H. Schlaad, Mild and Versatile (Bio-)Functionalization of Glass Surfaces via Thiol-Ene Photochemistry. *Chem. Mater.* **21** (Copyright (C) 2012 American Chemical Society (ACS). All Rights Reserved.) 5698–5700 (2009).
11. T. Pauloehrl, G. Delaittre, V. Winkler, A. Welle, M. Bruns, H. G. Boerner *et al.*, Adding Spatial Control to Click Chemistry: Phototriggered Diels-Alder Surface (Bio)functionalization at Ambient Temperature. *Angew. Chem. Int. Ed.* **51** (Copyright (C) 2012 American Chemical Society (ACS). All Rights Reserved.) 1071–1074, S/1–S/30 (2012).
12. C. Chu and R. Liu, Application of click chemistry on preparation of separation materials for liquid chromatography. *Chem. Soc. Rev.* **40** (Copyright (C) 2012 American Chemical Society (ACS). All Rights Reserved.) 2177–2188 (2011).
13. R. Huisgen, 1,3-Dipolar Cycloadditions. Past and Future. *Angew. Chem. Int. Ed.* **2**(10), 565–598 (1963).
14. F. Himo, T. Lovell, R. Hilgraf, V. V. Rostovtsev, L. Noodleman, K. B. Sharpless *et al.* Copper(I)-Catalyzed Synthesis of Azoles. DFT Study Predicts Unprecedented Reactivity and Intermediates. *J. Am. Chem. Soc.* **127**(1), 210–216 (2004).
15. H. C. Kolb, M. G. Finn and K. B. Sharpless, Click chemistry: diverse chemical function from a few good reactions. *Angew. Chem. Int. Ed.* **40**(11), 2004–2021 (2001).
16. M. F. Debets, D. C. W. J. van d, F. P. J. T. Rutjes and D. F. L. van, Azide: A Unique Dipole for Metal-Free Bio-orthogonal Ligations. *ChemBioChem.* **11** (Copyright (C) 2012 American Chemical Society (ACS). All Rights Reserved.) 1168–1184 (2010).
17. P. V. Chang, J. A. Prescher, E. M. Sletten, J. M. Baskin, I. A. Miller, N. J. Agard *et al.* Copper-free click chemistry in living animals. *Proc. Natl. Acad. Sci.* **107**(5),1821–1826 (2010).
18. C. J. Li, Organic reactions in aqueous media — with a focus on carbon-carbon bond formation. *Chem. Rev.* **93**(6), 2023–2035 (1993).
19. E. M. Sletten and C. R. Bertozzi, Bioorthogonal chemistry: fishing for selectivity in a sea of functionality. *Angew. Chem. Int. Ed.* **48**(38), 6974–6998 (2009).
20. M. D. Best, Click chemistry and bioorthogonal reactions: unprecedented selectivity in the labeling of biological molecules. *Biochemistry* **48**(28), 6571–6584 (2009).
21. Q. Wang, T. R. Chan, R. Hilgraf, V. V. Fokin, K. B. Sharpless and M. G. Finn, Bioconjugation by copper(I)-catalyzed azide-alkyne [3 + 2] cycloaddition. *J. Am. Chem. Soc.***125**(11), 3192–3193 (2003).
22. Q. Wang, T. Lin, L. Tang, J. E. Johnson and M. G. Finn, Icosahedral Virus Particles as Addressable Nanoscale Building Blocks. *Angew. Chem. Int. Ed.* **41**(3), 459–462 (2002).
23. T. R. Chan and V. V. Fokin, Polymer-supported copper(I) catalysts for the experimentally simplified azide–alkyne cycloaddition. *QSAR & Comb. Sci.* **26**(11&12), 1274–1279 (2007).
24. M. A. Bruckman, G. Kaur, L. A. Lee, F. Xie, J. Sepulveda, R. Breitenkamp *et al.*, Surface modification of tobacco mosaic virus with "Click" chemistry. *ChemBioChem.* **9**(4), 519–523 (2008).

25. C. Le Droumaguet, C. Wang and Q. Wang, Fluorogenic click reaction. *Chem. Soc. Rev.* **39**(4), 1233–1239 (2010).
26. F. Xie, K. Sivakumar, Q. Zeng, M. A. Bruckman, B. Hodges and Q. Wang, A fluorogenic 'click' reaction of azidoanthracene derivatives. *Tetrahedron* **64**(13), 2906–2914 (2008).
27. J. D. E. Prasuhn, R. M. Yeh, A. Obenaus, M. Manchester and M. G. Finn, Viral MRI contrast agents: coordination of Gd by native virions and attachment of Gd complexes by azide-alkyne cycloaddition. *Chem. Commun.* (12), 1269–1271 (2007).
28. S. Qazi, L. O. Liepold, M. J. Abedin, B. Johnson, P. Prevelige, J. A. Frank *et al.*, P22 Viral Capsids as Nanocomposite High-Relaxivity MRI Contrast Agents. *Mol. Pharm.* (2012).
29. C. M. Salisbury and B. F. Cravatt, Click chemistry-led advances in high content functional proteomics. *QSAR Comb. Sci.* **26** (Copyright (C) 2012 American Chemical Society (ACS). All Rights Reserved.) 1229–1238 (2007).
30. A. E. Speers, G. C. Adam, B. F. Cravatt, Activity-based protein profiling *in vivo* using a copper(I)-catalyzed azide-alkyne [3 + 2] cycloaddition. *J. Am. Chem. Soc.* **125**(16), 4686–4687 (2003).
31. J. H. Taylor, P. S. Woods and W. L. Hughes, The organization and duplication of chromosomes as revealed by autoradiographic studies using tritiumlabeled thymidinee. *Proc. Natl. Acad. Sci.* **43**(1), 122–128 (1957).
32. H. G. Gratzner, R. C. Leif, D. J. Ingram and A. Castro, The use of antibody specific for bromodeoxyuridine for the immunofluorescent determination of DNA replication in single cells and chromosomes. *Exp. Cell Res.* **95**(1), 88–94 (1975).
33. J. Gierlich, G. A. Burley, P. M. E. Gramlich, D. M. Hammond and T. Carell, Click chemistry as a reliable method for the high-density postsynthetic functionalization of alkyne-modified DNA. *Org. Lett.* **8**(17), 3639–3642 (2006).
34. B. L. Cavanagh, T. Walker, A. Norazit and A. C. B. Meedeniya, Thymidine analogues for tracking DNA synthesis. *Molecules* **16** (Copyright (C) 2012 American Chemical Society (ACS). All Rights Reserved.) 7980–7993 (2011).
35. A. Salic and T. J. Mitchison, A chemical method for fast and sensitive detection of DNA synthesis *in vivo*. *Proc. Natl. Acad. Sci.* **105**(7), 2415–2420 (2008).
36. V. O. Rodionov, V. V. Fokin and M. G. Finn, Mechanism of the Ligand-Free CuI-Catalyzed Azide–Alkyne Cycloaddition Reaction. *Angew. Che. Int. Ed.* **44**(15), 2210–2215 (2005).
37. A. E. Speers and B. F. Cravatt, Profiling Enzyme Activities *In Vivo* Using Click Chemistry Methods. *Chem. Biol.* **11**(4), 535–546 (2004).
38. J. M. Baskin, J. A. Prescher, S. T. Laughlin, N. J. Agard, P. V. Chang, I. A. Miller *et al.*, Copper-free click chemistry for dynamic *in vivo* imaging. *Proc. Natl. Acad. Sci.* **104**(43), 16793–16797 (2007).
39. K. A. Giuliano, R. L. DeBiasio, R. T. Dunlay, A. Gough, J. M. Volosky, J. Zock *et al.*, High-content screening: a new approach to Easing Key Bottlenecks in the Drug Discovery Process. *J. Biomol. Screen.* **2**(4), 249–259 (1997).
40. M. J. Chen, D. Z. Qu, W. L. Chi, W. Wang, X. S. Ren, S. J. Cong *et al.*, 5-Ethynyl-2′-deoxyuridine as a molecular probe of cell proliferation for high-content siRNA screening assay by "click" chemistry. *Sci. China Chem.* **54** (Copyright (C) 2012 American Chemical Society (ACS). All Rights Reserved.) 1702–1710 (2011).
41. N. K. Devaraj, R. Weissleder and S. A. Hilderbrand, Tetrazine-based cycloadditions: application to pretargeted live cell imaging. *Bioconjug. Chem.* **19** (Copyright (C) 2012 American Chemical Society (ACS). All Rights Reserved.) 2297–2299 (2008).

42. N. K. Devaraj, R. Upadhyay, J. B. Haun, S. A. Hilderbrand and R. Weissleder, Fast and sensitive pretargeted labeling of cancer cells through a tetrazine/trans-cyclooctene cycloaddition. *Angew. Chem. Int. Ed.* **48**(38), 7013–7016 (2009).
43. N. K. Devaraj, S. Hilderbrand, R. Upadhyay, R. Mazitschek and R. Weissleder, Bioorthogonal Turn-On Probes for Imaging Small Molecules inside Living Cells. *Angew. Chem. Int. Ed.* **49**(16), 2869–2872 (2010).
44. A. A. Souto, A. U. Acuña, J. M. Andreu, I. Barasoain, M. Abal and F. Amat-Guerri, New fluorescent water-soluble taxol derivatives. *Angew. Chem. Int. Ed.* **34**(23&24), 2710–2712 (1996).
45. G. Franc and A. Kakkar, Dendrimer design using CuI-catalyzed alkyne-azide "click-chemistry". *Chem. Commun.* (42), 5267–5276 (2008).
46. K. Khanna, S. Varshney and A. Kakkar, Designing miktoarm polymers using a combination of "Click" reactions in sequence with ring-opening polymerization. *Macromolecules* **43**(13), 5688–5698 (2010).
47. A. Sharma, K. Neibert, R. Sharma, R. Hourani, D. Maysinger and A. Kakkar, Facile construction of multifunctional nanocarriers using sequential click chemistry for applications in biology. *Macromolecules* **44**(3), 521–529 (2011).
48. L. Packer and E. Cadenas, Lipoic acid: energy metabolism and redox regulation of transcription and cell signaling. *J. Clin. Biochem. Nutr.* **48**(1), 26–32 (2010).
49. A. Castonguay, E. Wilson, N. Al-Hajaj, L. Petitjean, J. Paoletti, D. Maysinger *et al.*, Thermosensitive dendrimer formulation for drug delivery at physiologically relevant temperatures. *Chem. Commun.* **47**(44), 12146–12148 (2011).
50. A. Molinari, A. Oliva, P. Reinoso, J. M. del Corral, M. A. Castro, M. Gordaliza *et al.*, Cytotoxic-antineoplastic activity of hydroquinone derivatives. *Eur. J. Med. Chem.* **37**(2), 177–182 (2002).
51. A. Molinari, A. Oliva, C. Ojeda, J. M. M. del Corral, M. A. Castro, C. Cuevas *et al.*, Cytotoxic-Antineoplastic Derivatives of Prenyl-1,2-naphthohydroquinone. *Arch. Pharm.* **341**(5), 301–306 (2008).
52. D. Atamanyuk, B. Zimenkovsky and R. Lesyk, Synthesis and anticancer activity of novel thiopyrano[2,3-d]thiazole-based compounds containing norbornane moiety. *J. Sulfur. Chem.* **29** (Copyright (C) 2012 American Chemical Society (ACS). All Rights Reserved.) 151–162 (2008).
53. P. Y. Wen and S. Kesari, Malignant gliomas in adults. *New Engl. J. Med.* **359**(5), 492–507 (2008).
54. K. Braun, M. Wiessler, V. Ehemann, R. Pipkorn, H. Spring, J. Debus *et al.*, Treatment of glioblastoma multiforme cells with temozolomide-BioShuttle ligated by the inverse Diels-Alder ligation chemistry. *Drug Des. Dev. Ther.* **2** (Copyright (C) 2012 American Chemical Society (ACS). All Rights Reserved.) 289–301 (2008).
55. R. Pipkorn, W. Waldeck, B. Didinger, M. Koch, G. Mueller, M. Wiessler *et al.*, Inverse-electron-demand Diels-Alder reaction as a highly efficient chemoselective ligation procedure: synthesis and function of a BioShuttle for temozolomide transport into prostate cancer cells. *J. Pept. Sci.* **15**(3), 235–241 (2009).

Index

Vol. 3

Frontiers in Nanobiomedical Research

HANDBOOK OF NANOBIOMEDICAL RESEARCH

Fundamentals, Applications and Recent Developments

❶ Materials for Nanomedicine

Frontiers in Nanobiomedical Research

ISSN: 2251-3965

Series Editors: Martin L. Yarmush *(Harvard Medical School, USA)*
Donglu Shi *(University of Cincinnati, USA)*

Published

Vol. 1: Handbook of Immunological Properties of Engineered Nanomaterials
edited by Marina A. Dobrovolskaia and Scott E. McNeil (SAIC-Frederick, Inc., USA)

Vol. 2: Tissue Regeneration: Where Nano-Structure Meets Biology
edited by Qing Liu (3D Biotek, USA and Tongji University, China) and Hongjun Wang (Stevens Institute of Technology, USA)

Vol. 3: Handbook of Nanobiomedical Research: Fundamentals, Applications and Recent Developments (In 4 Volumes)
edited by Vladimir Torchilin (Northeastern University, USA)

Forthcoming titles

Cancer Therapeutics and Imaging: Molecular and Cellular Engineering and Nanobiomedicine
edited by Kaushal Rege (Arizona State University, USA)

Nano Vaccines
edited by Balaji Narasimhan (Iowa State University, USA)

Nano Pharmaceuticals
edited by Rajesh N. Dave (New Jersey Institute of Technology, USA)

Thermal Aspects in Nanobiomedical Systems and Devices
by Dong Cai (Boston College, USA)

Nano Mechanochemistry in Biology
edited by Jeffrey Ruberti (Northeastern University, USA)

Nanomaterial Probes of Biological Processes and Systems
edited by David Mast (University of Cincinnati, USA)

Handbook of Biomaterials
edited by Donglu Shi (University of Cincinnati, USA) and Xuejun Wen (Clemson University, USA)

HANDBOOK OF NANOBIOMEDICAL RESEARCH

Fundamentals, Applications and Recent Developments

1 Materials for Nanomedicine

editor

Vladimir Torchilin

Northeastern University, USA

Vol. 3

Frontiers in Nanobiomedical Research

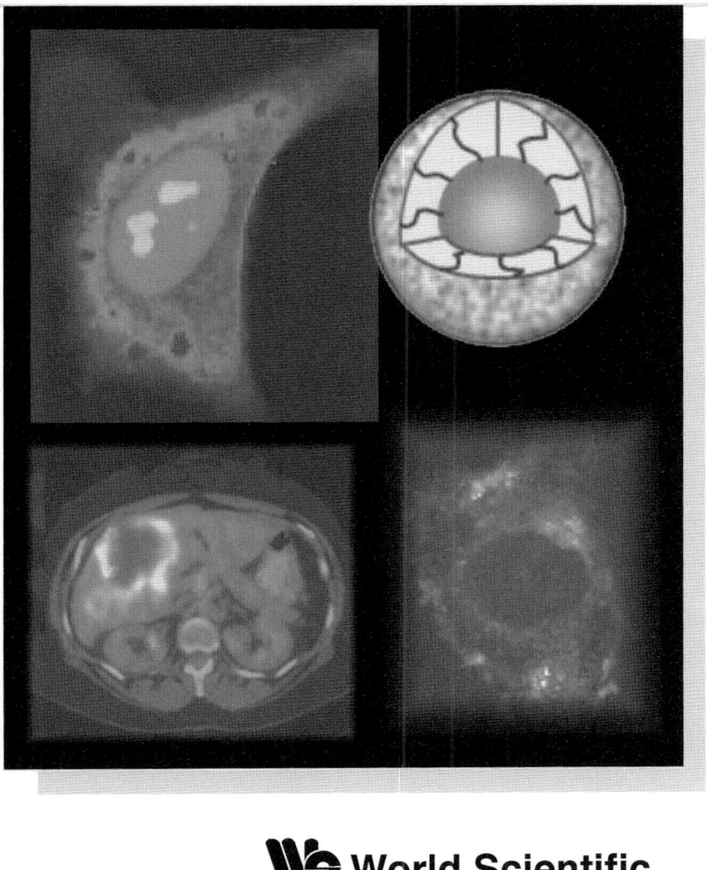

World Scientific

NEW JERSEY • LONDON • SINGAPORE • BEIJING • SHANGHAI • HONG KONG • TAIPEI • CHENNAI

Published by

World Scientific Publishing Co. Pte. Ltd.
5 Toh Tuck Link, Singapore 596224
USA office: 27 Warren Street, Suite 401-402, Hackensack, NJ 07601
UK office: 57 Shelton Street, Covent Garden, London WC2H 9HE

Library of Congress Cataloging-in-Publication Data
Handbook of nanobiomedical research : fundamentals, applications, and recent developments / editor, Vladimir Torchilin.
p. ; cm. -- (Frontiers in nanobiomedical research ; vol. 3)
Includes bibliographical references and index.
ISBN 978-9814520645 (set : alk. paper) -- ISBN 978-9814520676 (volume 1 : alk. paper) --
ISBN 978-9814520683 (volume 2 : alk. paper) -- ISBN 978-9814520690 (volume 3 : alk. paper) --
ISBN 978-9814520706 (volume 4 : alk. paper)
I. Torchilin, V. P., editor. II. Series: Frontiers in nanobiomedical research ; v. 3. 2251-3965
[DNLM: 1. Nanomedicine. 2. Nanostructures. 3. Nanotechnology. QT 36.5]
R857.N34
610.28--dc23

2014024090

British Library Cataloguing-in-Publication Data
A catalogue record for this book is available from the British Library.

Typeset by Stallion Press
Email: enquiries@stallionpress.com

Printed in Singapore

Contents

Chapter 1

Liposomal Nanomedicines

Amr S. Abu Lila[*,†], *Tatsuhiro Ishida*[*] *and Theresa M. Allen*[‡,§]

[*]*Department of Pharmacokinetics and Biopharmaceutics, Subdivision of Biopharmaceutical Sciences, Institute of Health Biosciences, The University of Tokushima; 1-78-1, Sho-machi, Tokushima 770-8505, Japan*
[†]*Department of Pharmaceutics and Industrial Pharmacy, Faculty of Pharmacy, Zagazig University, Zagazig, Egypt*
[‡]*Department of Pharmacology, University of Alberta, Edmonton, AB T6G 2H7, Canada*
[§]*tallen@ualberta.ca*

1. Introduction

Since Bangham's original description of phospholipid bilayer vesicles in 1965,[1] liposomes have emerged as archetypal nanoscale drug carriers, and they have received much attention as transporters of pharmacological agents.[2–7] Liposomes, sometimes called lipidic nanoparticles (LNP), are phospholipid bilayer vesicles that self-assemble when naturally occurring, or synthetic, phospholipids (PLs) are hydrated with excess water or aqueous salt solutions. Hydration can occur from dried preparations or in the presence of organic solvent such as ethanol, and various therapeutic molecules can be included in the hydration solution.

Some examples of the PLs commonly used in liposome production include phosphatidylcholine (PC), in particular hydrogenated soybean phosphatidylcholine (HSPC), sphingomyelin (SM) and phosphatidylglycerol (PG), and also the lipid cholesterol (Chol) is often included in the preparations. Liposomes can range in diameter from 0.025 μm to greater than

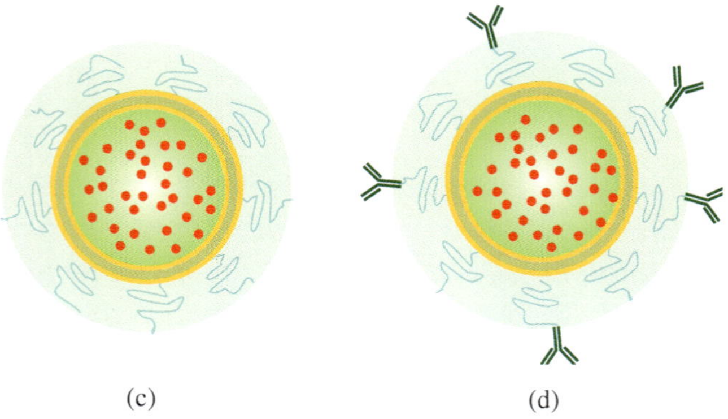

Fig. 1. Structures of different liposomal preparations, (a) classical liposome encapsulating lipid soluble drugs, (b) classical liposome encapsulating aqueous soluble drugs, (c) sterically stabilized liposomes and (d) ligand-targeted liposome containing an aqueous soluble drug.

20 μm, and are composed of a single bilayer surrounding an aqueous core, or of multiple bilayers, called lamellae, which are separated by aqueous compartments. Because of their amphiphilic nature, liposomes can accommodate a variety of drugs with different physicochemical characteristics Figs. 1(a)–1(b). Hydrophilic drugs — polar and ionic compounds — can be encapsulated within the internal aqueous compartments,[8] while lipophilic drugs, are usually associated with the fatty acyl chains of the lipid bilayers[9] Moreover, drugs may partition between the enclosed aqueous volume and the phospholipid bilayer membranes according to the solubility and ionization characteristics of the drug, which is greatly affected by the pH of the surrounding medium.[10–12]

The net surface charge of a liposome can be varied by incorporation of lipids with negative or positive charges. For example, inclusion of a long-chain amine in the bilayer will give positively charged vesicles, and inclusion of diacetyl phosphate will give negatively charged vesicles. Positively charged liposomes have been used experimentally as carriers for anionic genetic materials.[13–15]

2. Inherent Problems Relating to Liposomal Formulations

2.1. *Non-optimal drug release rates*

Drug release rates can have implications for the therapeutic effects of all types of drug delivery systems, including liposomes. It is well-known that drugs entrapped in liposomes, or other types of particles, are not bioavailable; they only become bioavailable when they are released. Therefore, optimization of the release rate of entrapped drugs from the liposomes is crucial for ensuring the delivery of minimal therapeutic concentrations of bioavailable drugs within the target tissue, at appropriate rates, for sufficient periods of time, to achieve an acceptable therapeutic outcome.[16–18]

Soon after the development of liposomes as drug delivery systems, several problems associated with the *in vivo* use of the 1st generation "classical" liposomes were identified. The challenges include: the difficulty in retaining some types of molecules, in particular hydrophobic substances, in association with the liposomes, and an inappropriate rate of drug release from liposomes. Drug release from liposomes was shown to be mediated by factors such as interactions with serum proteins,[19–21] the drug loading method, and the physicochemical properties of the lipid membrane and the entrapped drug.[22] Drugs with extremely low octanol/buffer partition coefficients exhibited prolonged liposomal retention, whereas molecules with log p ranging from −0.3 to 1.7 were, in contrast, released rapidly.[23] Alteration of the membrane composition of liposomes was found to affect the release rate of the encapsulated drug. Switching from a fluid phase phospholipid bilayer to a solid phase bilayer,[24] and incorporation of cholesterol[19,25,26] were shown to reduce the leakage of drugs from liposomes.

With the development of active "remote" loading procedures for encapsulation of weak bases and weak acids, and by the careful choice of drugs with physical characteristics that made them amenable to retention in liposomes, control over the release rate of entrapped drugs could be achieved. The retention properties of drugs in liposomes are dependent on the drug properties and the bilayer properties. For example, weak bases such as doxorubicin (DXR), which are hydrophobic at physiological pH, can be retained for long periods of time when present as sulfate or citrate precipitates inside liposomes that have an interior/exterior pH gradient.[27] Other drugs that lack an ionizable group, including paclitaxel or ciprofloxacin, cannot be remotely loaded and are released rapidly from liposomes.[28–31] Alteration of the drug-to-lipid ratio was found to affect the drug release rate, in a drug-dependent manner. Mayer *et al.*[32] have reported that increasing the drug-to-lipid ratio resulted in *decreased* drug retention of DXR. On the other hand, increasing drug-to-lipid ratio resulted in *increased* retention of vincristine and irinotecan within liposomes.[33,34]

2.2. *Rapid clearance of liposomes*

Another problem of the *in vivo* use of "classical" liposomes was their rapid elimination from the blood through uptake, primarily in the liver and spleen, by the cells of the mononuclear phagocyte system (MPS).[35-37] Except for the treatment of diseases where there was an MPS involvement, e.g., leishmanaisis,[38] the rapid uptake of liposomes by the MPS substantially limited their application as drug carriers to other tissues and/or cells,[39–41] and this increased uptake could result in toxicity to the MPS organs.[42–44]

Various strategies have been developed in order to avoid MPS uptake. Initially, large pre-doses of classical liposomes containing no drug ("empty" liposomes) were administered to saturate the uptake capacity of MPS cells and thereby increase the circulation half-life of subsequently injected "classical" liposomes.[45–47] Later, modification of the liposomal surface with natural polysaccharides such as mannan, amylopectin and dextran[48] was found to slightly prolong the circulation characteristics of "classical" liposomes. Coating the liposomal surface with inert, biocompatible substances, such as ganglioside GM1, phosphatidylinositol or lipid-conjugated polyethylene glycol (PEG)[49–51] has also been reported to efficiently prolong the blood circulation times of liposomes. These substances work by forming a protective, hydrophilic barrier at the liposome surface (termed steric stabilization) that slows down the binding to the liposome surface of various opsonins such as immunoglobulin G (IgG) and complement-related components that are responsible for the rapid uptake of liposomes by the cells of MPS in the liver and spleen.[52,53]

Nevertheless, although PEG remains the gold standard for the steric stabilization of liposomes, many attempts have focused on the use of other polymers to prepare long-circulating liposomes. Polymers such as polyglycerols,[54] poly[N-(2-hydroxypropyl)meth-acrylamide)],[55] poly-N-vinylpyrrolidones,[56] polyvinyl alcohol[57] or L-amino-acid-based biodegradable polymer-lipid conjugates[58] have been identified as promising alternatives to PEG in preparing long-circulating liposomes, but till date, none of the polymers has been proven superior to PEG.

2.3. *Inefficient intracellular delivery of drugs*

Another problem for the use of liposomes as drug delivery systems is their difficulty in delivering the encapsulated drugs across cellular membranes to intracellular sites of drug action, such as mitochondria or the nucleus. Liposomes are large hydrophilic structures that are too large to cross membranes, and they do not fuse with cell membranes in normal circumstances. Their uptake into cells is normally mediated via phagocytosis and/or endocytosis.[59] In recent years, ligands bound to the liposome surface have been used to trigger receptor-mediated internalization of liposomes and their contents by target cells in desired tissues and organs.[60–63] However, following internalization of liposomes via the endocytotic pathway, the liposomes and their associated drugs can be degraded by lysosomal enzymes. For drugs that escape this fate, efficient release of liposomal drugs from endosomal compartments constitutes a major hurdle for liposomal contents to enter the cytoplasm (e.g., chemotherapeutic agents and RNAi molecules), which is necessary for them to diffuse to their intracellular sites of action.[64] Therefore, many strategies have been proposed to enhance the endosomal escape of liposomes following endocytosis. The pH-sensitive peptides or cationic lipids can enhance the cytoplasmic delivery of liposomal cargo by disrupting the endosome/lysosome membrane.[56,65–70] Incorporation of fusogenic lipids into liposomes, or the use of membrane active peptides (also called cell penetrating peptides), can also increase cytoplasmic delivery, and bypass the endocytotic pathway, by causing fusion of liposomes with the plasma membrane.[71,72] However, these non-endocytotic pathways are less prominent compared to those using endocytosis.

3. Liposome Pharmacokinetics

3.1. *Clearance of liposomes*

As described in Sec. 2, encapsulation of anticancer drugs in a liposomal delivery system results in substantial changes in the pharmacokinetics and biodistribution of the drug.[73] Until it is released from the liposome, the drug adopts the pharmacokinetics of the liposomes.[74,75] Many reports have described the *in vivo* behaviors of liposomal drugs after *i.v.* administration. These reports demonstrate that the pharmacokinetics; absorption, biodistribution and clearance, of liposome-associated drugs can differ significantly from those of the free form of the drug. The PK alterations for a liposomal drug, compared to the free drug, depend to a large extent on the carrier's physiochemical properties including lipid composition, particle size, surface charge, surface modifications, the administrated dose, as well as the route of administration.[8,49,76,77]

Generally, the clearance of liposomes and their associated drugs from the blood circulation is via the cells of MPS, comprised principally of Kupffer cells in the liver and fixed macrophages in the spleen, although uptake of the liposome by bone marrow and lymph node also occurs.[41,78] After *i.v.* administration, opsonins such as plasma proteins adsorb onto the surface of liposomes, triggering recognition and liposome uptake by the cells of MPS through receptors such as the complement C3b receptor, the Fc receptor, CD47 and others.[79–81] The interaction of liposome with plasma proteins is determined mainly by physicochemical properties of the liposomes.[40,53]

Of critical importance to any therapeutic outcome is that the liposomes have a circulation half-life long enough to allow them to reach their target. Manipulation of liposomal physicochemical properties (see Sec. 3.2) was found to efficiently protect liposomal delivery systems for rapid clearance from circulation following *i.v.* administration. Although, short circulating characteristics may seem to be a major hurdle against the application of liposomes as therapeutic delivery systems, rapid uptake and retention by target tissues, combined with rapid clearance from the blood circulation, would be a desirable characteristic for liposomes carrying imaging agents, in order to minimize the background signal of normal tissues and to produce increased contrast.[82–84]

3.2. *Long-circulating PEGylated liposomes*

An important advance in prolonging circulation half-lives of liposomes, by decreasing their uptake into MPS cells of the liver and spleen, was achieved by the application of "Stealth" technologies, which involved the grafting of a hydrophilic polymer such as PEG onto the liposome surface (Fig. 1(c)).[51,61,85,86] This approach not only increased the residence time of liposomes in the circulation but also changed the pharmacokinetics from dose-dependent (saturable) to the relatively dose-independent (log-linear) within the clinical dose range.[41,87]

The "Stealth" effect imparted by the inclusion of PEG–lipid conjugates in liposomes is believed to be due to the formation of a hydration zone, at the liposome surface. The hydrophilic surface has an anti-opsonic effect by diminishing liposome–protein interactions, resulting in prolonged liposome circulation times.[85,88] However, the concentration of PEG lipid incorporated into liposomes was found to be critical for the prolonged circulation properties of the liposomes. The inclusion of more than 15 mol% of PEG_{2000}-lipid was found to have a detergent-like effect on the liposomes that increased their clearance from the circulation.[89] For this reason, approximately 5–10 mol% of PEG_{2000}-distearoylphosphatidylethanolamine (PEG_{2000}-DSPE) is normally included in the preparation of PEGylated liposomes. This amount has been found to be sufficient to delay the cell recognition of MPS, resulting in prolonged liposomal circulation time. This amount will change as the molecular weight of the PEG increases or decreases, and with the nature of the lipid anchor.

3.3. *Accelerated blood clearance phenomenon of PEGylated liposomes*

It has been reported that, in some circumstances, an *i.v.* injection of PEGylated liposomes causes a second dose of PEGylated liposomes, (injected a few days later) to circulate for a shorter time and to accumulate extensively in the liver. This is known as the "accelerated blood clearance (ABC) phenomenon" (Fig. 2).[90–92]

The ABC phenomenon is triggered mainly by anti-PEG IgM production by splenic B cells in response to an initial dose of PEG liposomes.[93,94] This anti-PEG IgM selectively binds to PEG when a second dose of PEG liposomes is administered, and activates the complement system. As a consequence, the liposomes are taken up by the Kupffer cells.[95,96]

The ABC phenomenon, which involves anti-PEG IgM production, may present a challenge to the use of some PEGylated nanoparticle-based therapeutic regimens that involve repeat administration of PEG-coated nanoparticles. However, many studies have demonstrated that the encapsulation of cytotoxic agents such as anticancer drugs in PEG liposomes significantly abrogates the induction of the ABC phenomenon. Abu Lila *et al.* have recently revealed that the *i.v.* injection of empty PEG-coated cationic

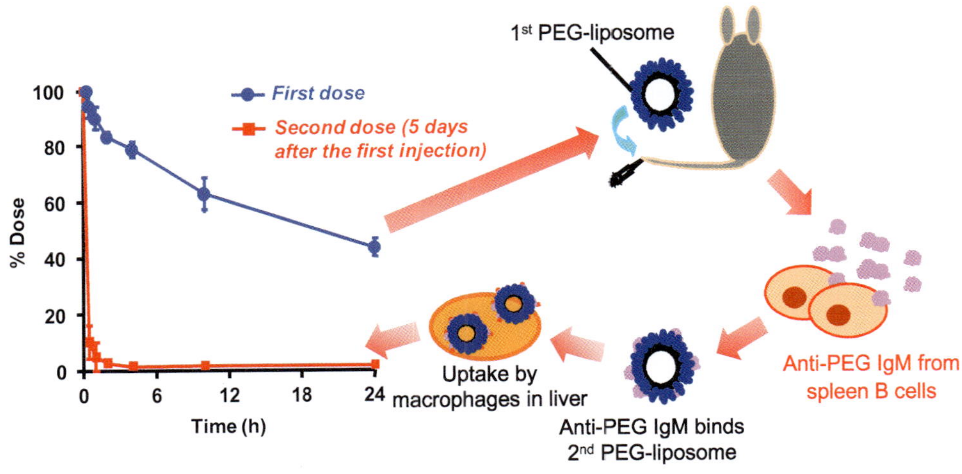

Fig. 2. Representation of the sequence of events leading from anti-PEG IgM induction to accelerated clearance of PEGylated liposomes.

liposomes (PCL) into mice induces anti-PEG IgM production, and there was a strong inverse relationship between the dose of initially injected PEG-coated liposomes and the extent to which the ABC phenomenon was induced.[97] Interestingly, the encapsulation of oxaliplatin (l-OHP) in PCL substantially suppressed the anti-PEG IgM production even at a very low lipid dose (0.25 μmol PLs/kg). The authors assumed that l-OHP released from liposomes accumulating in the spleen might impair the production of PEG-specific IgM as a consequence of inhibition of B cell proliferation and/or killing of proliferating B cells, and thus would prevent the induction of the enhanced clearance of a second dose of PEG-coated liposomes. Similar results have been obtained upon the encapsulation of cytotoxic agents such as DXR[98] and mitoxantrone[99] within PEGylated liposomes.

4. Targeting Strategies for Liposomes

Passive and "active" (i.e., ligand-mediated) targeting of drug delivery systems has been tried for both low and high molecular drugs as a means to increase the selectivity and the therapeutic effect of the drugs.

4.1. *Passive targeting of liposomes*

Most solid tumors, as well as regions of infection and inflammation, possess unique pathophysiological characteristics compared to normal tissue, such as extensive recruitment of new blood vessels (angiogenesis), leaky vascular architecture, and impaired lymphatic drainage.[100,101] The increased permeability of tumor vasculature has been exploited extensively for the passive accumulation of small drug-loaded nanocarriers (~50–200 nm in diameter) in solid tumors[102,103] via a phenomenon known as the "enhanced permeability and retention (EPR) effect".[104,105] The EPR effect can result in selective tumor targeting of anticancer drugs associated with the nanocarrier system in those tumors that are well vascularized with "leaky" blood vessels (Fig. 3). Drug accumulation in tumor cells resulting from the EPR effect can be increased up to several-fold when using nanoparticles as delivery systems, compared with that observed with the free drug.[106] However, drug accumulation is considerably less in poorly vascularized solid tumors, and for even well vascularized tumors it has been recognized that nanocarriers result in a focal distribution of drugs within the tumor, and rarely do the drugs spread uniformly throughout the tumor tissue following passive accumulation via EPR effect, due to extensive pathophysiological heterogeneity of solid

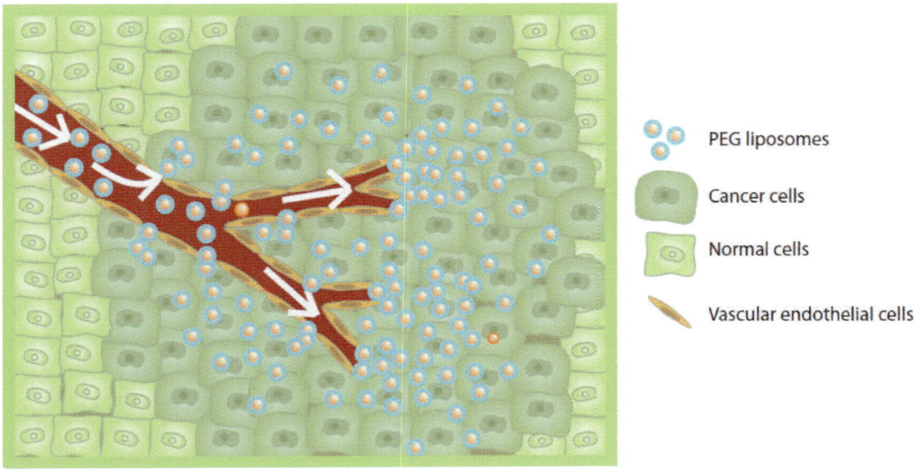

Fig. 3. Cartoon depicting passive targeting of PEGylated liposomes to tumors via extravasation through gaps in the tumour vasculature endothelium, resulting in localization of the liposomes within the tumor interstitial space.

tumor.[107,108] Such temporal and spatial heterogeneity in the tumor microenvironment, and the resulting non-homogenous intratumor distribution of drug, may compromise the therapeutic effect, but till date few studies have quantitated this.

Several approaches to reorganize the abnormal tumor vasculature or tumor microenvironment have been attempted in order to enhance the accumulation of nanocarriers in solid tumors.[109–111] Kano and colleagues reported that a TGF-β type 1 receptor (TβR-1) inhibitor caused increased accumulation of nanocarriers (e.g., DXR-containing PEGylated liposomes) in tumors, accompanied by a pronounced tumor response in a murine solid tumor model.[110] It appeared that a low dose of a TβR-1 inhibitor induced an alteration in the tumor-associated neovasculature in favor of increased permeability to nanocarriers, with less pericyte coverage and fewer endothelial cells. In addition, Ten Hagen and co-workers have reported a similar observation with tumor necrosis factor α (TNF-α) in both rat and murine models.[112] It appeared that TNF-α increases the leakiness of the vasculature by increasing the gaps between adjacent endothelial cells. Recently, Doi *et al.* showed that weekly metronomic oral dosing of S-1 enhanced tumor accumulation of I-OHP-containing PEGylated liposomes via alteration of the tumor microenvironment, resulting in increased therapeutic efficacy in Colon 26 murine colorectal tumor models.[113]

4.2. *Ligand-mediated targeting*

Although a number of passively targeted liposomal products have received approved clinical approval, the search continues for ways to improve the selectivity and therapeutic effect of nanocarriers. There has been considerable activity in the area of ligand-mediated targeting. This is sometimes called "active" targeting, but this is a misnomer since ligand-targeted particles still reach tumors to the same extent and by the same route as passively targeted carriers. In ligand-mediated targeting, selective binding and internalization of the carrier by the target tissue can be substantially enhanced by surface conjugation with specific targeting moieties.[114] A variety of molecules can be used as targeting ligands, including: monoclonal antibodies or antibody fragments such as antigen-binding fragments (Fab') or single chain variable fragments (scFv), lectins, proteins, peptides, lipoproteins, hormones, charged molecules, mono-, oligo- and polysaccharides and some low molecular-weight ligands, such as folate, and nucleic acid-based aptamers. Some of these are discussed in Sec. 4.2.1.

4.2.1. **Antibody-mediated liposome targeting** (i.e., **immunoliposomes**)

Various immunoliposomes, with antibodies or antibody fragments conjugated either to the surface of liposome, or, more effectively, to the distal end of PEG at the liposome surface, have used to deliver liposomal drug to a variety of targets (Fig. 1(d)). However, the majority of research in this area has been focused on improving uptake of drugs by cancer cells. Relatively few antibody molecules per PEGylated liposome (20–40 $\mu g/\mu mol$ phospholipid, roughly equivalent to 20–40 molecules of mAb per liposome) are required for the selective delivery of encapsulated drugs to target cells *in vivo*.[73] In addition, the selection of an antibody against an internalizing receptor or antigen on the target cell, is required to obtain increased therapeutic efficacy of immunoliposomal drugs relative to passively targeted carriers.[115]

Targeting ligands used in the formulation of immunoliposomes may be whole antibodies or antibody fragments (e.g., Fab' or scFv fragments).[116] An advantage to using whole monoclonal antibodies, compared to antibody fragments, is their higher binding avidity due to the presence of two or more binding domains on the same molecule, and their greater storage stability and ease of preparation.[117] However, the multivalent display of monovalent fragments on the surface of immunoliposomes can restore binding avidity.[17]

The presence of an Fc domain in whole antibody molecules can result in increased Fc-mediated clearance of immunoliposomes from circulation, and increases in immunogenic responses.[118] Taking all of these factors into account, smaller antibody fragments (Fab' or scFv), which lack the Fc domain, are preferentially used to reduce immunogenicity of the immunoliposomes, as well as to increase their circulation times. An interesting example is the development of anti-HER2 liposomes targeting HER2-overexpressing tumors. Kirpotin *et al.* have demonstrated that DXR-containing anti-HER2 Fab' immunoliposomes show superior *in vivo* antitumor activity compared to DXR-containing non-targeted liposomes in female NCR homozygous nude mice bearing MCF-7 breast cancer cells highly overexpressing HER2, although the intratumor accumulation levels of immunoliposomes, versus nontargeted liposomes, were similar.[119] They attributed the enhanced *in vivo* antitumor efficacy of DXR-containing immunoliposomes to the rapid internalization of the immunoliposomes, with subsequent specific intracellular drug delivery in the target cells. On the other hand, nontargeted liposomes were localized in the extracellular areas of tumor stroma and tissue macrophages.

Targeting to tumor vascular using immunoliposomes was also found to be effective in suppressing tumor growth as well. The major targeted receptors/antigens in these studies have been endoglins (CD105),[116] vascular endothelial growth factor receptor 2,[120] intercellular adhesion molecule 1 (ICAM-1),[121] E-selectin,[122] membrane type-1 matrix metalloproteinase (MT1-MMP),[123,124] vascular cell adhesion molecule 1 (VCAM-1)[125] and aminopeptidase N.[126]

The coupling of multiple targeting molecules on the surface of individual liposomes can increase the binding avidity for the target antigens.[18,127] Mack *et al.* have recently developed dual-targeted immunoliposomes using genetically engineered scFv molecules recognizing two different tumor-associated antigens, epidermal growth factor receptor (EGFR) and carcinoembryonic antigen (CEA).[127] The prepared dual-targeted immunoliposomes exhibited specific binding to cells expressing either EGFR or CEA, or both antigens *in vitro*. In addition, they proposed that these dual-targeted immunoliposomes might be suitable for delivering therapeutic payloads to a panel of different tumor cells expressing EGFR and/or CEA *in vivo*.

4.2.2. Liposomal targeting with biological ligands

4.2.2.1. Folate-mediated liposome targeting

The overexpression of folate receptors (FR) in a wide variety of human tumors, including ovarian, colorectal, breast, lung, renal and brain metastases, makes FR a viable candidate for achieving active targeting using liposomes.[128] Binding of folate to the FR mediates internalization of the liposomes and their contents into cells.[129] Many studies have demonstrated the delivery of liposomal drugs into cells using FR endocytosis, and this route of delivery is also claimed to bypass the multi-drug resistance pump.[128] DXR- or daunorubicin-containing liposomes targeted via folate demonstrated enhanced cytotoxicity for FR-positive tumor cell lines *in vitro*, compared to nontargeted liposomes containing these drugs.[130,131] Folate-targeted liposomes have also been utilized as delivery vehicles for the boron neutron capture therapy[132] and for targeting tumors with haptens for tumor immunotherapy.[133] Within the field of gene therapy, folate-targeted liposomes have been used for gene targeting to tumor cells[134] and for targeting tumors with antisense oligonucleotides.[135]

4.2.2.2. Transferrin-mediated liposome targeting

Receptor-mediated endocytosis of the complex of transferrin receptor (TfR) and transferrin (Tf)-bound iron is a major route of cellular iron uptake. Because TfR are overexpressed on the surface of many tumor cells, this pathway has been exploited for the targeted delivery of Tf-conjugated liposomes to tumor cells overexpressing TfR.[136] Ishida *et al.* have coupled antibodies against the TfR, as well as Tf itself, to the distal end of PEG in PEGylated liposomes to enhance their targetability to solid tumors.[137] Tf-coupled DXR-containing liposomes demonstrated increased binding and toxicity against C6 glioma cells *in vitro*.[138] Also, Tf-mediated liposome delivery was successfully used to enhance the intracellular delivery of cisplatin into gastric cancer cells in nude mice with peritoneal dissemination of human gastric cancer cells.[139] Suzuki *et al.* have confirmed the tumor-selective delivery of l-OHP in a C-26 colon cancer murine model, using Tf-conjugated PEGylated liposomes.[140] This resulted in enhanced extravasation of liposomes into tumors, and increased tumor suppression, compared with l-OHP-containing PEGylated liposomes, l-OHP-containing classical liposomes or free l-OHP.

4.2.2.3. Miscellaneous ligand-targeted liposomes

The search for new ligands that improve the selective targeting of liposomes to target tissues continues. Many studies have exploited the overexpression of specific markers on the surface of tumor cells, such as vitamin and growth factor receptors, for achieving selective targeting of liposomes to tumor tissue.[141] In addition, vasoactive intestinal peptide (VIP) has been used to target radionuclide-containing PEGylated liposomes to tumor cells with surface VIP receptors; this has resulted in enhanced growth inhibition of breast cancer in rats.[142] Further, the ability of galactosylated liposomes to accumulate specifically in liver parenchymal cells led to their application in both drug and gene delivery to the liver for the treatment of liver cancer or metastases.[143,144]

Despite the potential promise of biological ligands for selective delivery of liposomes to cancer cells, receptors to folate, Tf and EGFR, etc., are not only expressed on cancer cells, but also on fast-growing normal cells such as fibroblasts, endothelial and epithelial cells, and the free molecules are normally present in the circulation. Therefore, competition for binding and nonspecific targeting with subsequent increased toxicity for healthy cells could occur.[145,146]

4.2.3. Liposomal targeting using peptides

It is generally recognized that immunoliposomes have not lived up to their clinical potential; their clinical use has been compromised by their immunogenicity, and by their toxicity to liver and bone marrow from nonspecific uptake.[147,148] These limitations might be circumvented by using peptide ligands, which are smaller, less immunogenic, and easier to produce and manipulate. Peptide ligands have moderate affinity to antigens, which may give them an advantage over high affinity ligands which can impair the penetration of ligand-targeted liposomes into tumor tissue via the binding site barrier.[149,150] Moreover, biopanning procedures using phage display has yielded a number of peptide motifs that show preferential binding to specific antigens expressed at the surface of tumor cells.[151] The coupling of peptides such as NGR to PEGylated liposomes has been shown to enhance the accumulation of peptide-modified liposomes into tumor tissue.[152]

The tumor tissue and/or tumor vascular targetability of liposomes coupled with peptides has been investigated using several experimental animal models. These peptides include RGD motifs binding to $\alpha v\beta 3$ integrins,[153,154] NGR motifs binding to aminopeptidase N,[155] CREKA binding to fibrinogen or fibrin,[156] or GPLPLR binding to (MT1-MMP).[157]

Interestingly, a combination of modified liposomes and endosome-disruptive peptide was found to improve the cytosolic delivery of liposomal drugs to target cells. Mastrobattista *et al.* have revealed that liposomes targeted to the internalizing EGFR on the surface of ovarian carcinoma cells (OVCAR- 3) containing encapsulated diphtheria toxin A (DTA) did not show any cytotoxicity toward OVCAR-3 cells *in vitro*.[158] Cytotoxicity was only observed when the pH-dependent fusogenic peptide diINF-7 was co-encapsulated in the modified liposomes. This suggests that diINF-7 peptides entrapped inside liposomes significantly enhanced the cytosolic delivery of liposomal macromolecules by pH-dependent destabilization of endosomal membranes after cellular uptake of the liposomes.

Recently, Oku's group has identified two novel peptides homing to angiogenic blood vessels, namely, Ala–Pro–Arg–Pro–Gly (APRPG) and Gly–Asn–Gly–Arg–Gly (GNGRG). They showed that dual-targeting liposomes modified with these two peptides exerted a substantial avidity for tumor angiogenic vessels. DXR encapsulated within these dual-targeted liposomes strongly suppressed tumor growth in C-26 NL-17 colon carcinoma-bearing mice, compared to APRPG-modified or GNGRG-modified single-targeted liposomes.[62] Such dual-targeting strategy using liposomal delivery system has considered a promising approach for a number of targeting therapies because most of target organs express multiple address molecules.

4.2.4. Liposomal targeting using electrostatic interactions

Nearly two decades have passed since the introduction of cationic liposomes in gene delivery.[13,14,159,160] Cationic lipoplexes have several advantages over viruses as gene transfer vectors *in vitro*. Unlike viral vectors, they can be used to transfer DNA of essentially unlimited size. In addition, they are technically simple and quick to formulate, have low immunogenicity, and are readily available commercially.[161] There are, however, some drawbacks with lipid vectors, especially for *in vivo* use, including rapid removal from circulation, lower efficiency than viral vectors in gene transfer, nonspecific binding with cells, and the very transient nature of the gene expression.[162–164] Such drawbacks prompted a re-evaluation of their use as gene vectors.

Recently, there has been renewed interest in cationic liposomes, mainly due to their inherent ability to bind to tumor vasculature. This has provided an opportunity for the development of many anti-angiogenic and/or anticancer formulations. Kunstfeld and colleagues demonstrated that paclitaxel, encapsulated in cationic liposomes, strongly suppressed tumor angiogenesis and inhibited orthotopic melanoma growth in severe combined immunodeficient (SCID) mice.[165] By contrast, free paclitaxel was unable to suppress angiogenesis and tumor growth. Strieth and coworkers demonstrated that *i.v.* administration of paclitaxel, encapsulated in cationic liposomes, significantly retarded tumor growth, as compared with treatment using free paclitaxel, in the dorsal skinfold model.[166] They showed that the potent antitumor activity of paclitaxel-containing cationic liposomes resulted from a decrease in tumor vessel density and a reduction in the microcirculatory perfusion index in these animals. Schmitt-Sody *et al.* also showed that, in a dorsal skinfold model, vascular targeting of paclitaxel could be achieved by encapsulating the drug in cationic liposomes.[167] In subcutaneous A-Mel-3 tumors, the growth was significantly suppressed, and the appearance of regional lymph node metastases was significantly delayed, by treatment with paclitaxel-containing cationic liposomes in comparison with free paclitaxel. The use of DXR-loaded cationic liposomes to selectively target activated tumor endothelium increased the survival rate of tumor-bearing mice compared with either free DXR or DXR-containing PEGylated neutral liposomes.[168] Recently, Abu Lila *et al.* reported that PEGylated cationic liposomes containing l-OHP preferentially bound to angiogenic vessels, and strongly suppressed tumor angiogenesis in a dorsal air sac (DAS) mouse model.[169] Neither free l-OHP nor l-OHP encapsulated in PEGylated neutral liposomes showed such a strong suppressive effect. The strong anti-angiogenic effect of l-OHP-containing PEGylated cationic liposomes was attributed to the selective delivery of the drug to angiogenic vessels in the tumor, and its subsequent uptake by tumor endothelial cells.[169]

4.2.5. Liposomal targeting with aptamers

Aptamers, which are single-stranded nucleic acids with well-defined three dimensional structures, are starting to show potential for the development of both targeted diagnostics and therapeutics.[170] They can be selected *in vitro* from a large random DNA or RNA library to recognize different cells using a technique called cell-SELEX (systematic evolution of ligands by exponential enrichment). The selected aptamers can bind a number of molecules with high affinity and specificity.[171,172] Currently, they are thought to rival antibodies in a number of ways, mainly due to their high selectivity and the convenience of their generation and modification.[173,174] Compared with ~150 KD antibodies, aptamers exhibit lower immunogenicity, and their smaller size enables better tissue penetration.[175,176] In addition, aptamers are easily modified by incorporating 2'-amino, 2'-fluoro, or 2'-O-alkyl nucleotides in their backbone. This chemical modification of aptamers should improve their serum stability.[177] Furthermore, aptamers can be easily synthesized and chemically modified for molecular conjugation. These excellent properties, therefore, make aptamers attractive for the preparation of multifunctional target-specific nanoparticles. However, their large surface charge may lead to problems *in vivo*, as discussed below. Aptamers selective for the prostate-specific membrane antigen (PSMA) have been widely used for the targeted delivery of quantum dots (QDs),[178] gold nanoparticles[179] and polymeric nanoparticles[180] to prostate cancer cells *in vitro*.

Farokhzad *et al.* have reported that bioconjugates composed of docetaxel-encapsulated aptamer-targeted nanoparticles (Dtxl-NP-Apt), targeted against the PSMA protein expressed on the surface of LNCaP prostate epithelial cells, showed enhanced *in vitro* cellular toxicity as compared with nontargeted nanoparticles that lack the PSMA aptamers.[181] In addition, after a single intratumoral injection of Dtxl-NP-Apt bioconjugates, superior tumor growth inhibitory effects were observed in LNCaP xenograft nude mice, accompanied by prolonged survival times, compared with mice treated with nontargeted nanoparticles that lack the PSMA aptamers. Recently, Kang *et al.*[182] developed aptamer-conjugated multifunctional target-specific liposomes. They selected the scg8 aptamer, which has high binding affinity towards leukemia CEM-CCRF cells. They demonstrated that aptamer-targeted liposomes can specifically bind with target cells and release encapsulated drug during a 30 min incubation time.

Although direct coupling of aptamers to nanocarriers containing chemotherapeutic agents has improved the intracellular delivery of nanocarriers *in vitro,* or *in vivo* following intratumoral injection ,[178,180–182] the highly charged nature of aptamers results in their rapid clearance from circulation following their *i.v.* administration.[183]

5. New Generations of Liposomes

5.1. *pH-sensitive liposomes*

Conventional PEGylated immunoliposomes that are targeted to specific cell surface internalizing receptors using ligands like peptides or antibodies are mainly taken up by cells via receptor-mediated endocytosis. They progress through the lysosmal apparatus of cells, ending up in endosomes and then lysosomes, and finally release their payload from the lysosomes, assuming the payload has not been degraded.[184–186] This routing of drug-containing liposomes through the cell will influence the delivery of peptides, nucleic acid based drugs or cytotoxic agents that are membrane impermeable and/or unstable in the acidic enzyme environment present in lysosomes. There has been a need for a strategy to increase the rate at which drugs delivered to endosomes or lysosomes are released into the cytoplasm and subsequently proceeds to their intracellular site of action. One strategy that has been tried involves the use of pH-sensitive liposomes that can destabilize endosomes and lysosomes, under the mildly acidic conditions present in these vesicles, and promote drug release into the cytoplasm.[187,188] Many studies have reported the utilization of pH-sensitive liposomes to deliver cytotoxic agents or nucleic acid based drugs to intracellular target site. Connor *et al.* have demonstrated that encapsulating Ara-C in a pH-sensitive liposome was efficient in enhancing the cytoplasmic delivery of Ara-C in tumor cells, while protecting the drug from degradation in the lysosomes.[189] Fattal *et al.* also demonstrated that antisense oligonucleotide can be efficiently delivered into cell cytoplasm by anionic pH-sensitive liposomes that are stable in the blood, but that undergo a phase transition at the acidic pH of endosomes.[190]

Recently, the approach of using pH-sensitive liposomes together with ligand-mediated targeting to specific cell surface receptors such as folate, Tf, CD33 or EGFR, has been exploited for increasing the intracellular delivery of liposomal payloads.[191–194]

Although pH-sensitive liposomes appear effective for the *in vitro* cytoplasmic delivery of chemotherapeutic agents, proteins, antigens, antisense oligonucleotides and plasmid DNAs[189] in cell culture, significant obstacles, relating to their rapid clearance from blood circulation, hinder their effective *in vivo* application.[195]

5.2. *Thermosensitive liposomes*

Another approach is the use of thermosensitive liposomes. The formulation of such liposomes is based on pioneering work of Yatvin and Weinstein in the late 1970s,[196] who demonstrated that liposomes composing of PLs that undergo a gel-to-liquid crystalline phase transition at temperatures of around 44°C, could efficiently release a liposome-encapsulated water-soluble payload. From this basic formulation, thermosensitive liposomes have been developed to the point where they are undergoing clinical trials. Generally, for the design of thermosensitive liposomes, it is necessary to choose PLs such as DSPC (1,2-distearoyl-sn-glycero-3-phospho-choline) and DPPC (1,2-dihexadecanoyl-sn-glycero-3-phosphocholine), which undergo a solid-to-liquid phase change at around 42–44°C, to trigger drug release from acidic cell compartments.[197] In addition, to increase the permeability of the membrane during transition from the gel to a liquid, additional lipid compounds such as lysolipids and oligoglycerols could be added to optimize this change of state.[198,199] Furthermore, hydrophilic polymers such as PEG-conjugated lipids or mono-sialoganglioside (G_{M1}) could be incorporated to extend the circulation lifetimes of thermosensitive liposomes via steric stabilization.[200–202]

Currently, thermosensitive liposome formulations are under investigation in the clinic for the delivery of cytotoxic agents to solid tumors.[203,204] ThermoDox®, a proprietary heat-activated liposomal encapsulation of DXR, is an example of thermosensitive liposomes that has progressed to Phase III clinical trials for the treatment of hepatocellular carcinoma,[205] but this study did not meet its primary endpoint. ThermoDox®, used in combination with radiofrequency ablation, releases entrapped DXR from thermosensitive liposomes at temperatures of 39.5–42°C. It is also in clinical trials for recurrent chest wall breast cancer (http://celsion.com/pages/pipeline).

Nevertheless, since thermosensitive liposomes require the addition of an external stimulus, in this case heat, in a spatially well-defined region, surrounding and including the tumor, the delivery of thermosensitive liposomal cargo may be limited to primary tumors or larger metastatic lesions or diseases where the tissue-sparing properties of local treatment provide clinical benefit.[206]

5.3. *Magnetic liposomes*

Another targeting strategy that has been used for targeting liposomes is the use of lipid-coated magnetic nanoparticles (magnato-liposomes). This strategy utilizes magnetic field to mediate the local accumulation of liposomes loaded with a drug and a ferromagnetic material. Several recent studies have shown antitumor efficacy of magnetic liposomes in animal models.[207,208] Nobuto *et al.* have demonstrated that magnetic liposomes containing DXR efficiently deliver DXR to osteosarcomas in hamsters, with a four-fold increase in intratumor drug concentration when the tumor-implanted limb was exposed to a magnetic field for 60 min.[209] More recently, Bealle *et al.* have described the preparation of ultramagnetic liposomes (UMLs) encapsulating iron oxide nanoparticles.[210] They demonstrated that these magnetic liposomes are rapidly and efficiently internalized by cultured tumor cells and, when administered to mice, they can be vectorized to tumors by an external magnet. Based on their results, they suggested that these vesicles are not only promising vectors for tumor targeting, but also as tools to mimic and monitor tumor growth pressure *in vivo*.

Nonetheless, similar to thermosensitive liposomes, magnetic liposomes require the local application of a magnetic field in the vicinity of tumor to direct the liposomes towards the tumor tissue. Therefore, deliver of drug to metastases will be unlikely, raising the chance of disease recurrence.

5.4. *Virosomes*

Virosomes have a long history for use as delivery systems for peptides, nucleic acid-based drugs, and small molecular drugs such as anticancer agents. Structurally, virosomes are spherical unilamellar vesicles in which the vesicle surface is modified with fusogenic viral envelope glycoproteins such as influenza virus hemagglutinin and neuraminidase.[211,212] Viral envelope glycoproteins have receptor-binding and membrane-fusion properties that enable the cellular delivery of encapsulated molecules.[213] Unlike liposomal delivery systems, virosomes protect their payload from proteolytic degradation and low pH within endosomes/lysosomes, allowing their contents to remain intact until they reach the cell cytoplasm.[212,214] Similar to liposomes, virosomes can be optimized for enhanced encapsulation of the drug by modifying the content or type of membrane lipids used. For example, charged PLs might be incorporated into the viral membrane to improve its ability to function as a carrier for genetic materials such as antisense-oligonucleotides. In addition, to enhance virosome targetability, various ligands such as peptides and monoclonal antibodies can be incorporated.

The feasibility of targeted delivery of nucleic acid-based drugs and cytotoxic drugs by means of virosomes has been demonstrated in a number of studies.[215,216] Daemen *et al.* have used virosomes as gene delivery systems and shown that incorporated nucleic acid-based drugs are efficiently delivered into the host cell cytoplasm upon fusion of the virosome with endosomes or the plasma membrane.[217] Felnerova *et al.* have also demonstrated that virosomes, with a cross-linked monoclonal antibody on their surface, were able to mediate the targeted delivery of the anticancer drug, DXR to human cancer cells.[218]

Recently, the ability of virosomes to deliver proteins to cells has been exploited for the development of new vaccines.[219,220] The delivery of protein antigens to the immune system, especially dendritic cells (DCs), by virosomes was found to be effective in stimulating the host immune cells. An interesting example is the delivery of influenza vaccine using virosomes containing the spike protein of influenza virus.[221] It elicited high titers of influenza-specific antibodies, while being well tolerated.[222]

5.5. *Multi-functional liposomes*

Multi-functional liposomes encompass liposomes that have been developed to achieve more than one useful property, e.g., increased circulation time to allow their efficient accumulation at the target site, selective targeting and enhanced uptake via surface-attached ligands, *in vivo* imaging via an encapsulated image contrast agent, and/or drug release triggered via stimulation under certain physiological conditions such as temperature or pH.[7,223,224] Multifunctional liposomes designed to combine several of these properties in a single system have been explored as a means to circumvent many of the

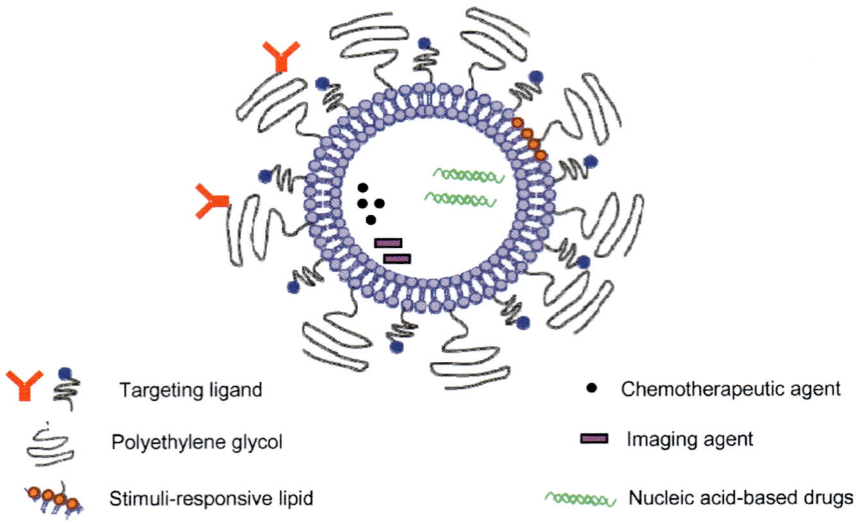

Fig. 4. Schematic picture of a multifunctional liposome.

physiological barriers associated with the delivery of the therapeutic payload or image contrast agents to target sites (Fig. 4). They can facilitate the online monitoring of tumor location, tumor targeting properties, intratumor localization and drug release kinetics during treatment, and tumor response.[225]

Recently, several reports have appeared on the use of multifunctional liposomes in anticancer therapy.[226–228] Kale *et al.* have demonstrated that, in mice bearing Lewis lung cancer cell (LLC), liposomal nanocarriers modified with a cell-penetrating peptide (TATp) and a pH-sensitive PEG-shield have the ability to provide both improved systemic circulation times and site-specific localization of TATp-containing liposomes.[226] In order for the TATp peptide to mediate the delivery of the liposome-encapsulated drug into the cell cytoplasm, the TATp-containing PEGylated liposomes need to lose their PEG coating. The liposomes accumulate in solid tumors via the EPR effect, and in the slightly acid interstitial space of solid tumors (or ischemic tissues) acid hydrolysis cleaves the PEG linker, exposing the TATp peptide and mediating cytoplasmic delivery. This strategy could overcome the "PEG-dilemma", i.e., inhibition of cellular uptake and endosomal escape by pegylation of the surface of nanocarriers, but still provide the very important contribution of PEG to increasing circulation half-lives. More recently, Zhu *et al.* have developed novel "smart" multifunctional liposomes that respond to up-regulated matrix metalloprotease 2 (MMP2) in the tumor microenvironment and improve cancer cell-specific delivery of encapsulated drugs.[228] In this research, liposomal surfaces were modified with two different PEG–lipid conjugates: a long chain PEG, linked with MMP2 via a cleavable octapeptide linker and a short chain PEG, linked with TATp. In addition, the tumor cell-specific anti-nucleosome monoclonal antibody (mAb 2C5) was attached to the liposomal surface. Hence, several functions were built into these liposomes including: passive tumor targeting by the EPR effect, due to the longer circulation times imparted by the long PEG chains; steric shielding of TATp moieties while in circulation with the long-chain PEG; specific binding of the particle to the tumor via a surface-attached monoclonal antibody (mAb 2C5); release of the PEG at the tumor via the cleavable peptide substrate of MMP2 exposing the cell-penetrating TATp; and finally the enhanced cellular internalization by TATp-mediated endocytosis.

However, the expense of production, and the demands of reproducible manufacturing and quality control increases with each added functionality. Therefore, the added advantages must to be sufficient to justify the higher costs and the greater demands of manufacturing a complex system.

6. Selective Examples of Current Applications of Liposomes

6.1. *Selective examples of liposomes in anticancer drug delivery*

Traditional chemotherapeutic anticancer drugs kill rapidly growing healthy normal cells in the body cancer cells, as well as diseased cells. This leads to side effects on tissues such as blood and hair cells, and on cells lining the intestine,[229] and imposes practical limits on the drug dose and dosing frequency. Furthermore, a number of physiological barriers have to be overcome to deliver anticancer drugs to solid tumors.[230,231] These include the unique properties of tumor tissue itself, such as irregular vasculature, variable permeability of blood vessels, and high interstitial fluid pressure.[100,232] In addition, in certain solid tumors, overexpression of the *P*-glycoprotein gene, the MRP gene and others, cause drug resistance to emerge against most, if not all, anticancer drugs, e.g., DXR and cisplatin.[233] Some evidence is emerging that the use of lipidic and polymer nanoparticles can help to overcome drug resistance, and also can increase the therapeutic index of some anticancer drugs by increasing the concentration of the anticancer drugs in the tumor tissues, while, at the same time, decreasing drug concentrations in healthy normal tissues such as heart.[146,234] An oft-quoted example is the clinical use of Doxil®, a PEGylated liposomal formulation of DXR, a potent anticancer drug active against a wide range of human tumors including lymphomas, leukemia and solid tumors. Treatment of patients with the conventional (free) drug produces severe side effects including bone marrow depression, alopecia, oral ulceration and irreversible cardiotoxicity.[235–237] The substitution of Doxil® for the free drug results in higher therapeutic effects in some types of tumors, e.g., Kaposi's sarcoma and refractory ovarian cancer, because liposome encapsulation changes the pharmacokinetics and biodistribution of the drug, leading to enhanced drug accumulation in solid tumors and decreased accumulation in normal healthy tissues, and to reductions in peak drug concentrations in the heart.[238] After the approval of Doxil® in 1995, a number of cytotoxic agent-containing liposomes have emerged for clinical use. The most recently FDA-approved liposomal formulation for a cytotoxic agent is Marqibo®, a liposomal formulation of vincristine for the treatment of acute lymphoblastic leukemia at second or greater relapse. In addition, several more liposomal formulations of anticancer agents are currently in various stages of clinical trials (Table 1).

Table 1. Selected list of clinically approved liposomal formulations of anticancer agents, and some formulations that have proceeded to clinical trials.

Product	Drug	Indications	References
		FDA-approved liposomal formulations	
Doxil/Caelyx	doxorubicin	Kaposi's sarcoma Ovarian cancer Breast cancer	239, 240, 241
DaunoXome	daunorubicin	Kaposi's sarcoma	242
DepoCyt	Cytarabine	Lymphomatous meningitis	243
Myocet	doxorubicin	Breast cancer in combination with cyclophosphamide	244
Lipo-Dox	doxorubicin	Kaposi's sarcoma Ovarian cancer Breast cancer	http://www.tlcbio.com/en/product_lipo_dox.html
Marquibo	vincristine	Acute lymphoblastic leukemia	245 246
		Liposomal formulations that have progressed to clinical investigation	
SPI-077	cisplatin	Solid tumor (*Phase II*)	247 248
CPX-351	Cytarabine: daunor ubicine	Acute myeloid leukemia (*Phase II*)	249

(*Continued*)

Table 1. *(Continued)*

Product	Drug	Indications	References
CPX-1	Irinotecan HCl:floxuridine	Colorectal cancer (*Phase II*)	250 http://www.celatorpharma.com/new/products.html
MM-398	CPT-11	Gastric and pancreatic cancer/ Colon cancer (*Phase I*)	251
MM-302	ErbB2/ErbB3-targeted doxorubicin	ErbB2-positive breast cancer (*Phase I*)	252
MBP-436	Transferrin-targeted oxaliplatin	Gastric cancer Gastro-esophageal junction (*Phase II*)	140
Brakiva	Topotecan	Relapsed solid tumors (*Phase I*)	http://www.talontx.com/pipeline.php?divid=brakiva
Alocrest	Vinorelbine	Newly diagnosed or relapsed solid tumors (*Phase I*)	http://www.talontx.com/pipeline.php?divid=alocrest
Lipoplatin	Cisplatin	Non-small cell lung cancer (*Phase III*)	253, 254
ThermoDox	Thermosensitive doxorubicin	Primary hepatocellular carcinoma (*Phase III*)	205, 255
Endo-Tag-1	Cationic liposomal paclitaxel	Pancreatic cancer (*Phase II*)	256

A novel dual targeting approach, involves using a liposomal delivery system that targets two separate cell populations within tumor tissue[115] (Fig. 5). Proof-of-principle for the concept involved the combined administration of two populations of liposomal anticancer drugs, one that targets the tumor cells themselves, and one that targets the tumor vasculature. For the targeting of tumor cells, the researchers used an anti-GD2 monoclonal antibody against disialoganglioside GD2 that is widely expressed on cancer cells of neuronal origin, including neuroblastoma cells; expression is at low levels in cerebellum and peripheral nerves.[257] For targeting tumor vasculature, DXR-containing liposome was modified with NGR peptides that target the angiogenic endothelial cell marker aminopeptidase N (CD13).[258] In an orthotopic neuroblastoma xenograft model, the combined formulations GD2-immunoliposomal DXR (tumor-targeted) and NGR-liposomal DXR

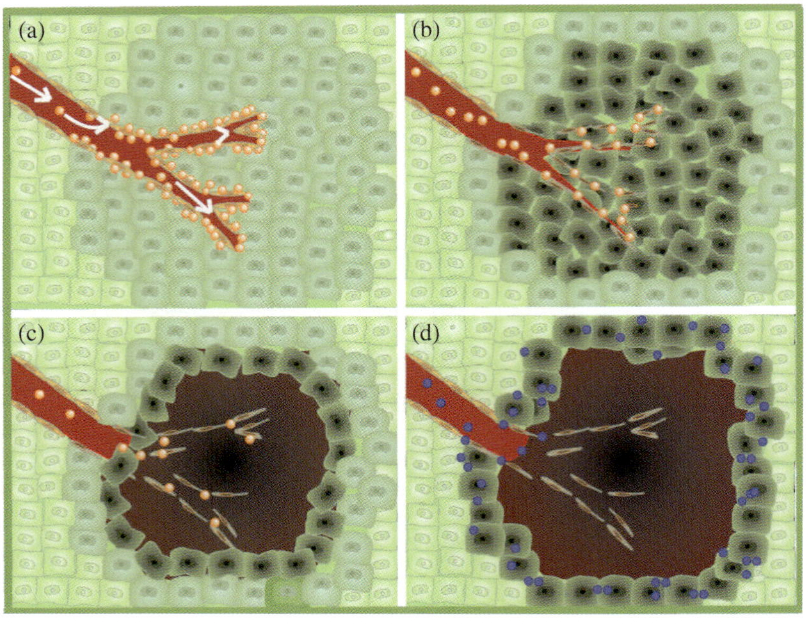

Fig. 5. Cartoon depicting the dual targeting approach, in which populations of tumor cells and tumor endothelial cells are targeted with PEG-coated liposomes bearing ligands specific to each cell population. (a) Targeting to vascular endothelial cells, e.g., with NGR-targeted liposomes or cationic liposomes. (b) Destruction of tumor vasculature, starving the tumor cells of oxygen and nutrients. (c) Tumor necrosis in the interior of the tumor, but the outer rim tumor cells survive. (d) Targeting of outer rim cells, e.g., with anti-GD2 or anti-HER2 liposomes leading to potential cure. The temporal sequencing in this example shows the anti-tumor cell targeting as occurring after the vascular targeting, but both treatments could be given together, or the order could be reversed (Ref. 259).

(vascular-targeted) showed superior antitumor efficiency over either tumor- or vascular-targeted liposomal formulations. This dual targeting strategy, which in this case involved two different cell populations, could be applied in a number of other situations, e.g., two different ligands at the liposome surface, targeted against two different receptors on the same cell surface, or on different cell surfaces in heterogeneous tumor cell populations.

Recently, Hare *et al.*[259] have used a dual-targeted, two drug approach against human epidermal growth factor receptor 2 (HER2)-positive breast cancer (Fig. 5). Two ligand-modified liposomal drugs were used; liposomal DXR, targeted to tumor cells via αHER2 Fab fragments, and liposomal vincristine, targeted to CD13 (aminopeptidase N) on tumor vasculature cells via NGR peptides, to selectively deliver two different chemotherapeutics to tumor cells and tumor vasculature endothelial cells in an orthotopic mouse model of HER2-positive breast cancer. Their results confirmed that combined treatment with tumor vasculature-targeted plus tumor cell-targeted therapies was therapeutically superior to either single agent. In addition, the dose of simultaneously administered combination therapy was halved without compromising therapeutic effect.

Abu Lila *et al.* confirmed that a PEGylated cationic liposome is a promising carrier for the delivery of l-OHP to tumor endothelium, resulting in a potent anti-angiogenic effect in a DAS model.[169] Subsequently, they demonstrated that l-OHP-containing PEGylated cationic liposomes have a superior antitumor activity over either free l-OHP or l-OHP-containing PEGylated neutral liposomes in a murine tumor-xenograft model. The superior antitumor activity of l-OHP encapsulated in PEGylated cationic liposomes was confirmed to be due to the dual-targeting activity of PEGylated cationic liposomes against both tumor endothelial cells and tumor cells.[260,261]

6.2. *Liposomes in antimicrobial therapy*

Classical liposomes are preferentially taken up by the cells of MPS, mainly by the macrophages of the liver and spleen; this makes them ideal carriers for the delivery antimicrobial drugs to the cells of MPS, which are considered to be the main reservoirs of parasites.[262,263] Many studies have utilized classical liposomes for the efficient delivery of anti-infective agents such as antifungals and antibiotics to their desired sites of action.[264–268]

Liposomal formulations of the antifungal drug, amphotericin B, have been shown to be effective in treating the systemic fungal infections, while reducing the severe renal toxicities.[269] These formulations have been instrumental in clearing infections from many patients who could no longer tolerate the dose intensity of the free drug necessary to control their infection.[269–271] AmBisome®, a liposomal formulation of amphotericin B has been a top-selling product for the treatment of severe systemic fungal infections in immunocompromised patients, and has recently come off patent.[272]

The unique properties possessed by mycobacteria, such as the complex lipid-rich cell envelope that protects the organism from both the host response and antimycobacterial therapy, and their localization deep in intracellular sites, hinder the effective treatment of mycobacteria.[273] To be effective, drugs must penetrate the host macrophage, travel to the intracellular locale, and preferably have an increased therapeutic index to allow prolonged therapy.[274] Many studies have demonstrated that liposomes could function as efficient carriers for the delivery of antibiotics into infected intracellular sites, while reducing the associated side effects. Incorporation of rifampin in liposomes was found to enhance the antimicrobial activity of rifampin against *Mycobacterium avium* infections compared to free rifampin. In addition, a further increase in activity was achieved upon grafting the macrophage activator tetrapeptide tuftsin to the surface of drug-containing liposomes, which significantly enhanced the uptake of liposomes by macrophages, and lowered the bacterial load in infected animals, compared with the free drug.[275]

Currently, many antibiotic-containing liposomal formulations are under clinical investigation. Arikace™, a liposomal amikacin formulation, is an example of antibiotic-containing-liposomal formulation that showed promising results in treating pulmonary *P. aeruginosa* infections in cystic fibrosis (CF) patients. In a Phase II clinical study, daily administration of Arikace for 28 consecutive days significantly decreased the numbers of *P. aeruginosa* colonies in sputum and produced significant improvement in lung function, compared to placebo. This potent antimicrobial efficacy was attributed to the ability of liposomes in delivering high concentrations of amikacin into the bacterial microenvironment.[276] Liposomal ciprofloxacin is another example of antibiotic-containing liposomes that underwent clinical investigation. A phase II trial revealed that daily administration of liposomal ciprofloxacin, via a nebulizer, for 2 weeks in CF and for 28 days in the non-CF bronchiectasis patients resulted in a 1.5 log reduction in bacterial colony forming unit (CFU), compared to free ciprofloxacin. In addition, liposomal ciprofloxacin was well tolerated and no serious side effects were noted.[277]

Macrophages have mannosyl–fucosyl receptors that can be targeted via mannosylated liposomes. Mannosylated liposomes encapsulating the antileishmanial drug, andrographolide, were found to be effective in treating experimental leishmaniasis in the hamster model.[278] The targeted liposomes showed a potent efficacy in reducing the parasitic burden in the spleen, as well as in reducing hepatic and renal toxicities. Similarly, Frezard *et al.* have demonstrated efficacy of freeze-dried liposomes encapsulating the antimonial drug, meglumine antimoniate, in reducing the liver parasite burdens in hamsters experimentally infected with *Leishmania chagasi*, compared with animals treated with free meglumine antimoniate.[265] DXR encapsulated in immunoliposomes, prepared by grafting the Fab'$_2$ of an antibody against an anti-51-kDa *Leishmania* species-specific protein onto the liposomal surface, resulted in complete eradication of the spleen parasite burden in mice, compared to 45% and 84% parasite suppression with similar doses of free and liposomal DXR, respectively.[279]

6.3. *Gene delivery*

The delivery of drugs based on nucleic acids across cellular membrane has been a difficult challenge. Nucleic acid-based materials exhibit a short half-life because of their rapid degradation by circulating and intracellular nucleases.[280] In addition, rapid systemic clearance, low selectivity for the desired tissue and poor cellular uptake limit their clinical applications. Many vectors, either viral or non-viral, have been developed to achieve intracellular delivery of nucleic acid-based drugs such as DNA and RNA.

Liposomes, in particular cationic liposomes, can function as non-viral vectors, and they have proven especially efficient for the transfection of nucleic acid-based materials into cells in culture. They have several advantages over viruses as gene transfer vectors. Unlike viral vectors, they can carry DNA of essentially unlimited size. In addition, they are technically simple and quick to formulate, have low immunogenicity, and are readily available commercially.[161] Cationic liposomes, by virtue of their surface positive charge, form complexes with the negatively charged nucleic acid-based materials through electrostatic interactions. Such complexes (also called lipoplexes) are thought to enter the cell through fusion with the plasma membrane, and they can also promote release of nucleic acids from endosomal membranes following inernalizaion.[281] The ratio of anionic charge (nucleic acids) to cationic charge (cationic lipids) determines the net zeta potential of the complexes, and hence their biodistribution, e.g., positively charged complexes showed higher lung and liver accumulation compared to negatively charged, or neutral complexes.[282] In general, cationic lipoplexes with a positive surface zeta potential are rapidly removed from circulation *in vivo*, and, unlike their success in *in vitro* applications, have not proven very successful *in vivo*. A number of researchers have used a variety of techniques to develop cationic lipid-containing liposomes that efficiently entrap nucleic acids in their interior, but which have a net neutral or anionic surface charge. These include CCL,[283] stable nucleic acid lipid particle (SNALP),[284] LNP[285] and liposome-polycation-hyaluronic acid particles (LPD).[286]

The cell selectivity of the surface neutral liposome–nucleic acid complex can be improved by the application of cell-specific targeting technology to cationic liposomes.[287] Many reports have emphasized the efficacy of disialoganglioside (GD2)-targeted long-circulating CCLs entrapping c-myb antisense oligodeoxynucleotides in inhibiting the tumor growth and metastases in murine models of melanoma and neuroblastoma.[152,288,289] CCLs contain a central core of a cationic phospholipids bound to oligonucleotides and an outer shell of neutral lipids, PEG and anti-GD2 as a targeting ligand.

Liposomes are also used for the targeting of RNA interference (RNAi) molecules to specific tissues.[284,290,291] Arnold *et al.* have demonstrated that siRNA lipoplexes selectively inhibited β1-adrenoreceptor expression and markedly reduced blood pressure for up to 12 days in spontaneously hypertensive rats.[294] Zimmermann *et al.* have developed cationic liposomes termed SNALPs that have been stabilized by PEGylation for improved pharmacokinetics.[284] They showed that the SNALPs were efficient in delivering siRNA systemically to silence the hepatic apoB gene in mice and non-human primates. Two applications of LNP siRNA systems have reached the clinic. In one case the target is PCSK9, a gene responsible for modulation of low density lipoprotein (LDL) levels in the circulation. This LNP siRNA system has found to be effective in inducing rapid and significant reduction in LDL levels without exerting toxic side effects (http://www.alnylam.com/Programs-and-Pipeline/Alnylam-5×15/Hypercholestoralemia.php). Another example is LNP siRNA system targeting transthyretin (TTR), which is currently under clinical investigation for the treatment of TTR-induced amyloidosis (http://www.alnylam.com/Programs-and-Pipeline/Alnylam-5×15/TTR-Amyloidosis.php). Recently, a new technique for the formation of LNP siRNA systems was developed through the application of microfluidic mixing.[292] These new systems are presently under evaluation and optimization.

6.4. *Co-delivery of chemotherapeutic agents and siRNA*

Recently, a novel approach has been proposed for co-delivering both anticancer drugs and siRNA to tumor tissue using the same liposomal delivery system.[293–297] Saad *et al.* studied the possibility of treating multidrug-resistant (MDR) cancer by combining an anticancer drug with suppressors of cellular resistance, within one multifunctional nanocarrier-based delivery system.[293] For this purpose, they synthesized cationic liposomes encapsulating DXR, and complexed with siRNA targeted to MRP1 and BCL2 mRNA (suppressors of pump and non-pump cellular-resistance, respectively). They demonstrated that a liposomal delivery system could provide effective co-delivery of both DXR and siRNA, inducing suppression of cellular resistance in MDR LLC and cell death *in vitro*.

The Huang laboratory[295] have demonstrated efficient delivery of c-Myc siRNA to the CD13-expressing cell line HT-1080, by using a PEGylated cationic liposome complexed with polycation-DNA, and decorated with the NGR (aspargine–glycine–arginine) moiety (LPD–PEG–NGR) targeted to CD13, which was expressed on tumor cells, as well as tumor vascular endothelium. Upon the co-delivery of c-Myc siRNA and DXR, enhanced intratumoral accumulation of siRNA and DXR and potent antitumor efficacy were observed in HT-1080 tumor-bearing mice.

Kang *et al.* have developed a cationic liposome containing a novel cationic lipid, N′, N″-dioleylglutamide (DG), for the systemic co-delivery of PD0325901, an inhibitor of mitogen-activated protein kinase (MEK) that participates in cancer cell proliferation and resistance to apoptosis, and a therapeutic siRNA against the myeloid cell leukemia sequence 1 (MCl-1) gene (siMCl-1) that is overexpressed in tumor cells resistant to chemotherapeutic agents. Co-delivery of siMCl-1 and the MEK inhibitor, PD0325901, using cationic liposomes in mice bearing KB (human epithelial carcinoma) cells suppressed the tumor volume by 79% when compared to the control group.[298]

Yu *et al.* reported the utility of 1,2-dioleoyl-sn-glycero-3-ethylphosphocholine (EDOPC)-based cationic lipid nanoparticles (cSLN) for the co-delivery of the cytotoxic agent, paclitaxel, and siRNA against MCl-1 gene to KB cancer cells both *in vitro* and *in vivo*. They showed that, in mice bearing a KB tumor, treatment with cSLN liposomes carrying both paclitaxel and siMCl-1 induced a superior tumor growth inhibitory effect, compared to either control group or mice treated with free paclitaxel and cSLN liposomes carrying siMCl-1.[297]

6.5. *Liposomal vaccines*

Almost four decades ago, Gregoriadis and Allison first reported the use of liposomes as immunological adjuvants.[299,300] Since that time, liposomes have been established as robust systems for the induction of humoral and cell-mediated immunity for a broad spectrum of infectious diseases and cancers.[301–303] Currently, several vaccines containing liposome-based adjuvants have been approved for human use, or have reached late stage clinical evaluation. Virosomes, composed of reconstituted viral membranes supplemented with PC, are the main components of licensed influenza and hepatitis A vaccines.[304–306] Liposomes with encapsulated protein or peptide antigen are phagocytosed by macrophages and eventually accumulate in lysosomes. After processing in the lysosomes, degraded peptides become accessible to the major histocompatibility complex class II (MHCII) on the macrophage surface, resulting in the stimulation of specific T-helper cells and specific B cells[307] and the subsequent secretion of antibodies specific for the peptides (and the proteins they were derived from). Structural modification of liposomes, such as the formulation of pH-sensitive liposomes, provides liposomes with additional advantages over traditional adjuvants by faciliatating the escape of the peptide antigen from endosomes into the cytoplasm,[308] allowing the association of antigen with the MHC-I complex, which induces a cytotoxic T-lymphocyte (CTL) response. Such stimulatory effects were not observed with traditional adjuvants such as Freund's adjuvant.[263]

Interestingly, lipid constituents of liposomal vesicles can be tailored to achieve particular immunogenic responses. Certain cationic lipids originally synthesized as transfection reagents have been discovered to have specific adjuvant activity. Lipid–protamine–DNA complexes were observed to induce robust IFN-γ and TNF-α responses in serum, and induce antitumor immunity in murine models of pulmonary metastasis and fibrosarcoma.[309,310] In addition to the immuno-stimulatory effect of CpG motifs of DNA, it was observed that DOTAP, a cationic lipid, exerted an enantiospecific adjuvant effect in DCs through the induction of reactive oxygen species and downstream activation of ERK and p38 pathways.[311,312] In recent studies, liposomes have been successfully used for the delivery of peptide vaccines and CTL epitopes to human DCs to improve the immune response to antigenic peptides.[313,314] Activation of DCs has been proposed as the mechanism of action of many viral and cancer liposomal-based vaccines under development.[315]

6.6. *Liposomes in diagnostic imaging*

A relatively new and promising application of liposomal delivery system is their use for molecular and diagnostic imaging. Recently, many efforts have been made to provide imaging functions to nanocarriers by incorporating various imaging probes such as QDs,[316,317] magnetic iron oxide nanoparticles,[318–320] and radionuclides such as ^{18}F[321,322] for efficient fluorescence imaging, magnetic resonance imaging (MRI) and positron emission tomography (PET) imaging, respectively. Liposomes are considered to be attractive delivery vehicles in the imaging area from the viewpoint of biodegradability, size control, and high encapsulation efficacy for contents. In addition, the multifunctional character of liposomes has significant advantages because it allows for the inclusion of a variety of imaging agents, and the inclusion of specific ligands for efficient targeting to the desired tissue.[323]

Recent reports[324,325] have explored the possibility of using liposomes as carriers of contrast agents for CT and MRI imaging. These studies emphasize the ability of liposomes containing contrast agents to circulate in the bloodstream for prolonged times, compared to free contrast agents. CT and MRI signal enhancement mediated by liposomal formulations were evaluated. Danila *et al.* have developed immunoliposomes capable of encapsulating iodinated contrast agents (iohexol) for use in CT imaging of atheromatous plaques in the activated human coronary artery endothelial cells (HCAEC).[323] The surface of the liposomes was modified by attaching antibodies against ICAM-1, which is expressed preferentially on the surface of inflamed endothelial cells during the early stages of atherosclerosis.[326] They showed that the targeted immunoliposomes bound to the surface of activated HCAEC with high specificity, and that the iodinated immunoliposomes showed promise for use in CT imaging of vulnerable atherosclerotic plaques, relative to free contrast agents.

Progress is being made for the use of multifunctional liposomes for imaging and treatment. Liposomal encapsulation of apomorphine circumvented its inherent instability and its short half-life (about 41 min) and resulted in a 2.4-fold increase in its accumulation in the brain.[327] Subsequently, Wen *et al.* have developed a new theranostic multimodal nanocarrier, in which an imaging agent and drug are co-assembled, allowing simultaneous diagnosis and therapy.[328] This multimodal carrier is composed of liposomes, QDs, and an anti-Parkinson's drug (apomorphine), and has the ability to achieve passive targeting, imaging and drug delivery with one formulation. Liposomal encapsulation of QDs resulted not only in reduced uptake of QDs by the heart and liver, but enhanced the targeted delivery of QDs to brain of nude mice, as indicated by the greater fluorescence intensity observed in mouse brains treated with liposomes, compared to free QDs. Hence, multifunctional liposomes integrating QDs and apomorphine may be an effective tool for the simultaneous diagnosis and treatment of brain disorders.

Mikhaylov *et al.* showed that magnetic nanoparticle clusters encapsulated within a liposome (ferri-liposomes) can, under the influence of an external magnet, target both the tumour and its microenvironment.[329] They used drug-loaded ferri-liposomes to non-invasively monitor the targeting of the tumour microenvironment *in vivo*. The ferri-liposomes were shown to deliver a cathepsin protease inhibitor to a mammary tumour and its microenvironment in a mouse, which substantially reduced the size of the tumour compared with systemic delivery of the free drug.

Kono *et al.* have prepared multi-functional liposomes by incorporating a thermo-sensitive polymer (poly(2-ethoxy(ethoxyethyl) vinyl ether, having a low critical solution temperature of around 40°C, and an imaging lipid (polyamidoamine G3 dendron-based lipids having Gd^{3+} chelate residues) into PEGylated liposomes.[224] The high resolution MRI images from these liposomes could be used to monitor their *in vivo* accumulation. When the liposomes loaded with DXR, they were stable and retained DXR at or below physiological temperatures, but rapidly released DXR at temperature at or above 40°C. After *i.v.* administration into mice bearing C-26 colon tumors, following the T1-weighted MRI signal intensity in tumors showed the accumulation of the liposomes in the tumors increased with time and reached a constant level at 8 h after administration. Such multi-functional liposomes having temperature-triggered drug release combined with imaging functions have been proposed to enable personalized chemotherapy that is optimized to individual patients.

7. Conclusions

In the last quarter century, liposomes have been extensively used to increase the therapeutic efficacy of encapsulated drugs, while minimizing toxicities against healthy normal tissues. The commercial success of the anticancer drug formulation Doxil®, and the antifungal drug formulation Ambisome®, are notable, as is the recent success of liposomes for use as non-viral vectors for intracellular delivery of nucleic acid-based drugs to the liver. Extra-hepatic delivery of nucleic acid-based drugs has been more challenging. However, new generations of liposomes, and new applications continue to be developed. Among them, multifunctional liposomes have garnered a lot of attention for "therasnostic" applications that combine diagnosis and therapy within the same formulation. To date, although multifunctional liposomes have achieved proof-of-principle in preclinical studies, the utilization of multifunctional liposomes in the clinical setting may be limited by complexities in the development, manufacturing, quality control and cost of these formulations, and in the difficulty in controlling the switching on and off of different functionalities at the right time and place *in vivo.* Alternatively, combination of a liposomal formulation having a single functionality (e.g., Doxil® or Marquibo®) with free drugs for combination chemotherapy is already showing success. Liposomes containing two drugs at optimized ratios and release rates in the same liposomes are also showing clinical success, e.g., CPX-351. Combinations of different liposomes, each having a single functionality (e.g., combining two different optimized liposomal formulations of two different cancer drugs, or a cancer drug formulation with a separate imaging formulation), may be simpler to achieve and as effective as multifunctional liposomes, but this remains to be seen. Liposomal nanomedicines and, more frequently, nanomedicines containing both lipids and polymers in the same formulation, are still growth areas, and a number of promising candidates are working their way through the pipelines of biotechnology and pharmaceutical companies. Increasingly nanomedicines are using the principles of combination chemotherapy to deliver synergistic combinations of drugs to diseased cells. We can look forward to seeing more of these on the market in the near future.

References

1. D. Bangham, M. M. Standish and J. C. Watkins, Diffusion of univalent ions across the lamellae of swollen phospholipids, *J. Mol. Biol.* **13**(1), 238–252 (1965).
2. G. Gregoriadis, The carrier potential of liposomes in biology and medicine (first of two parts), *N. Engl. J. Med.* **295**(13), 704–710 (1976a).
3. G. Gregoriadis, The carrier potential of liposomes in biology and medicine (second of two parts), *N. Engl. J. Med.* **295**(14), 765–770 (1976b).
4. Chonn and P. R. Cullis, Recent advances in liposomal drug-delivery systems, *Curr. Opin. Biotechnol.* **6**(6), 698–708 (1995).
5. N. Maurer, D. B. Fenske and P. R. Cullis, Developments in liposomal drug delivery systems, *Expert Opin. Biol. Ther.* **1**(6), 923–947 (2001).
6. T. M. Allen and P. R. Cullis, Drug delivery systems: entering the mainstream, *Science* **303**(5665), 1818–1822 (2004).
7. V. P. Torchilin, Multifunctional nanocarriers, *Adv. Drug Deliv. Rev.* **58**(14), 1532–1555 (2006).
8. L. D. Mayer, L. C. Tai, D. S. Ko, D. Masin, R. S. Ginsberg, P. R. Cullis and M. B. Bally, Influence of vesicle size, lipid composition, and drug-to-lipid ratio on the biological activity of liposomal doxorubicin in mice, *Cancer Res.* **49**(21), 5922–5930 (1989).
9. D. B. Fenske and P. R. Cullis, Entrapment of small molecules and nucleic acid-based drugs in liposomes, *Methods Enzymol.* **391**, 7–40 (2005).
10. L. D. Mayer, M. B. Bally and P. R. Cullis, Uptake of adriamycin into large unilamellar vesicles in response to a pH gradient, *Biochim. Biophys. Acta.* **857**(1), 123–126 (1986).
11. T. D. Madden, P. R. Harrigan, L. C. Tai, M. B. Bally, L. D. Mayer, T. E. Redelmeier, H. C. Loughrey, C. P. Tilcock, L. W. Reinish and P. R. Cullis, The accumulation of drugs within large unilamellar vesicles exhibiting a proton gradient: a survey, *Chem. Phys. Lipids* **53**(1), 37–46 (1990).
12. T. Nii and F. Ishii, Encapsulation efficiency of water-soluble and insoluble drugs in liposomes prepared by the microencapsulation vesicle method, *Int. J. Pharm.* **298**(1), 198–205 (2005).
13. P. L. Felgner, T. R. Gadek, M. Holm, R. Roman, H. W. Chan, M. Wenz, J. P. Northrop, G. M. Ringold and M. Danielsen, Lipofection: a highly efficient, lipid-mediated DNA-transfection procedure, *Proc. Natl. Acad. Sci. USA* **84**(21), 7413–7417 (1987).
14. P. L. Felgner and G. M. Ringold, Cationic liposome-mediated transfection, *Nature* **337**(6205), 387–388 (1989).
15. N. Duzgunes, C. T. De Ilarduya, S. Simoes, R. I. Zhdanov, K. Konopka and M. C. Pedroso de Lima, Cationic liposomes for gene delivery: novel cationic lipids and enhancement by proteins and peptides, *Curr. Med. Chem.* **10**(14), 1213–1220 (2003).
16. N. L. Boman, L. D. Mayer and P. R. Cullis, Optimization of the retention properties of vincristine in liposomal systems, *Biochim. Biophys. Acta* **1152**(2), 253–258 (1993).
17. G. J. Charrois and T. M. Allen, Drug release rate influences the pharmacokinetics, biodistribution, therapeutic activity, and toxicity of pegylated liposomal doxorubicin formulations in murine breast cancer, *Biochim. Biophys. Acta* **1663**(1–2), 167–177 (2004).
18. K. M. Laginha, S. Verwoert, G. J. Charrois and T. M. Allen, Determination of doxorubicin levels in whole tumor and tumor nuclei in murine breast cancer tumors, *Clin. Cancer Res.* **11**(19 Pt 1), 6944–6949 (2005).

19. G. Scherphof, F. Roerdink, M. Waite and J. Parks, Disintegration of phosphatidylcholine liposomes in plasma as a result of interaction with high-density lipoproteins, *Biochim. Biophys. Acta* **542**(2), 296–307 (1978).
20. T. M. Allen and L. G. Cleland, Serum-induced leakage of liposome contents, *Biochim. Biophys. Acta* **597**(2), 418–426 (1980).
21. J. Senior and G. Gregoriadis, Is half-life of circulating liposomes determined by changes in their permeability? *FEBS Lett.* **145**(1), 109–114 (1982).
22. D. Stamp and R. L. Juliano, Factors affecting the encapsulation of drugs within liposomes, *Can J. Physiol. Pharmacol.* **57**(5), 535–539 (1979).
23. F. Defrise-Quertain, P. Chatelain, J. M. Ruysschaert and M. Delmelle, Spin label partitioning in lipid vesicles. A model study for drug encapsulation, *Biochim. Biophys. Acta* **628**(1), 57–68 (1980).
24. G. Storm, F. H. Roerdink, P. A. Steerenberg, W. H. de Jong and D. J. Crommelin, Influence of lipid composition on the antitumor activity exerted by doxorubicin-containing liposomes in a rat solid tumor model, *Cancer Res.* **47**(13), 3366–3672 (1987).
25. P. R. Cullis, Lateral diffusion rates of phosphatidylcholine in vesicle membranes: effects of cholesterol and hydrocarbon phase transitions, *FEBS Lett.* **70**(1), 223–228 (1976).
26. T. J. McIntosh, The effect of cholesterol on the structure of phosphatidylcholine bilayers, *Biochim. Biophys. Acta* **513**(1), 43–58 (1978).
27. E. Bolotin, R. Cohen, L. Bar, N. Emanuel, S. Ninio, Y. Barenholz, and D. Lasic, Ammonium Sulfate gradients for efficient and stable remote loading of amphipathic weak bases into liposomes and ligandoliposomes, *J. Liposome Res.* 4(1), 455–479 (1994).
28. M. H. Bartoli, M. Boitard, H. Fessi, H. Beriel, J. P. Devissaguet, F. Picot and F. Puisieux, In vitro and in vivo antitumoral activity of free, and encapsulated taxol, *J. Microencapsul.* 7(2), 191–197 (1990).
29. G. Haran, R. Cohen, L. K. Bar, and Y. Barenholz, Transmembrane ammonium sulfate gradients in liposomes produce efficient and stable entrapment of amphipathic weak bases, *Biochim. Biophys. Acta* **1151**(2), 201–215 (1993).
30. J. Conley, H. Yang, T. Wilson, K. Blasetti, V. Di Ninno, G. Schnell and J. P. Wong, Aerosol delivery of liposome-encapsulated ciprofloxacin: aerosol characterization and efficacy against Francisella tularensis infection in mice, *Antimicrob. Agents Chemother.* **41**(6), 1288–1292 (1997).
31. Cabanes, K. E. Briggs, P. C. Gokhale, J. A. Treat and A. Rahman, Comparative in vivo studies with paclitaxel and liposome-encapsulated paclitaxel, *Int. J. Oncol.* **12**(5), 1035–1040 (1998).
32. R. J. Mayer, Acute leukemias in adults: an overview of recent strategies, *J. Cancer Res. Clin. Oncol.* **116**(1), 94–96 (1990).
33. V. Zhigaltsev, N. Maurer, Q. F. Akhong, R. Leone, E. Leng, J. Wang, S. C. Semple and P. R. Cullis, Liposome-encapsulated vincristine, vinblastine and vinorelbine: a comparative study of drug loading and retention, *J. Control. Release* **104**(1), 103–111 (2005).
34. M. J. Johnston, S. C. Semple, S. K. Klimuk, K. Edwards, M. L. Eisenhardt, E. C. Leng, G. Karlsson, D. Yanko and P. R. Cullis, Therapeutically optimized rates of drug release can be achieved by varying the drug-to-lipid ratio in liposomal vincristine formulations, *Biochim. Biophys. Acta* **1758**(1), 55–64 (2006).
35. G. Gregoriadis and D. E. Neerunjun, Control of the rate of hepatic uptake and catabolism of liposome-entrapped proteins injected into rats. Possible therapeutic applications, *Eur. J. Biochem.* **47**(1), 179–185 (1974).

36. R. L. Juliano and D. Stamp, The effect of particle size and charge on the clearance rates of liposomes and liposome encapsulated drugs, *Biochem. Biophys. Res. Commun.* **63**(3), 651–658 (1975).
37. H. K. Kimelberg, T. F. Tracy Jr., S. M. Biddlecome and R. S. Bourke, The effect of entrapment in liposomes on the in vivo distribution of [3H]methotrexate in a primate, *Cancer Res.* **36**(8), 2949–2957 (1976).
38. C. R. Alving, E. A. Steck, W. L. Chapman Jr., V. B. Waits, L. D. Hendricks, G. M. Swartz Jr. and W. L. Hanson, Therapy of leishmaniasis: superior efficacies of liposome-encapsulated drugs, *Proc. Natl. Acad. Sci. USA* **75**(6), 2959–2963 (1978).
39. G. Poste and R. Kirsh, Liposome-encapsulated macrophage activation agents and active non-specific immunotherapy of neoplastic disease, *Prog. Clin. Biol. Res.* **102** pt A, 309–319 (1982).
40. H. Senior, Fate and behavior of liposomes in vivo: a review of controlling factors, *Crit. Rev. Ther. Drug Carrier Syst.* **3**(2), 123–193 (1987).
41. T. M. Allen, G. A. Austin, A. Chonn, L. Lin and K. C. Lee, Uptake of liposomes by cultured mouse bone marrow macrophages: influence of liposome composition and size, *Biochim. Biophys. Acta* **1061**(1), 5664 (1991).
42. H. Ellens, E. Mayhew and Y. M. Rustum, Reversible depression of the reticuloendothelial system by liposomes, *Biochim. Biophys. Acta* **714**(3), 479–485 (1982).
43. M. J. Hsu and R. L. Juliano, Interactions of liposomes with the reticuloendothelial system. II: Nonspecific and receptor-mediated uptake of liposomes by mouse peritoneal macrophages, *Biochim. Biophys. Acta* **720**(4), 411–419 (1982).
44. T. M. Allen, L. Murray, S. MacKeigan and M. Shah, Chronic liposome administration in mice: effects on reticuloendothelial function and tissue distribution, *J. Pharmacol Exp. Ther.* **229**(1), 267–275 (1984).
45. R. M. Abra, M. E. Bosworth and C. A. Hunt, Liposome disposition in vivo: effects of pre-dosing with lipsomes, *Res. Commun. Chem. Pathol. Pharmacol.* **29**(2), 349–360 (1980).
46. Y. J. Kao and R. L. Juliano, Interactions of liposomes with the reticuloendothelial system. Effects of reticuloendothelial blockade on the clearance of large unilamellar vesicles, *Biochim. Biophys. Acta* **677**(3–4), 453–461 (1981).
47. R. T. Proffitt, L. E. Williams, C. A. Presant, G. W. Tin, J. A. Uliana, R. C. Gamble and J. D. Baldeschwieler, Liposomal blockade of the reticuloendothelial system: improved tumor imaging with small unilamellar vesicles, *Science* **220**(4596), 502–505 (1983).
48. V. Sihorkar and S. P. Vyas, Potential of polysaccharide anchored liposomes in drug delivery, targeting and immunization, *J. Pharm. Pharm. Sci.* **4**(2), 138–158 (2001).
49. T. M. Allen and A. Chonn, Large unilamellar liposomes with low uptake into the reticuloendothelial system, *FEBS Lett.* **223**(1), 4246 (1987).
50. A. Gabizon and D. Papahadjopoulos, Liposome formulations with prolonged circulation time in blood and enhanced uptake by tumors, *Proc. Natl. Acad. Sci. USA* **85**(18), 6949–6953 (1988).
51. L. Klibanov, K. Maruyama, V. P. Torchilin and L. Huang, Amphipathic polyethyleneglycols effectively prolong the circulation time of liposomes, *FEBS Lett.* **268**(1), 235–237 (1990).
52. G. Blume and G. Cevc, Molecular mechanism of the lipid vesicle longevity in vivo, *Biochim. Biophys. Acta* **1146**(2), 157–168 (1993).
53. T. Ishida, H. Harashima and H. Kiwada, Liposome clearance, *Biosci. Rep.* **22**(2), 197–224 (2002a).

54. Maruyama, S. Okuizumi, O. Ishida, H. Yamauchi, H. Kikuchi and M. Iwatsuru, Phosphatidyl polyglycerol prolongs liposome circulation in vivo, *Int. J. Pharm.* **111**(1), 103–107 (1994).
55. R. Whiteman, V. Subr, K. Ulbrich and V. P. Torchilin, Poly(Hpma)-coated liposomes demonstrate prolonged circulation in mice, *J. Liposome Res.* **11**(2–3), 153–164 (2001).
56. V. P. Torchilin, T. S. Levchenko, K. R. Whiteman, A. A. Yaroslavov, A. M. Tsatsakis, A. K. Rizos, E. V. Michailova and M. I. Shtilman, Amphiphilic poly-N-vinylpyrrolidones: synthesis, properties and liposome surface modification, *Biomaterials* **22**(22), 3035–3044 (2001a).
57. H. Takeuchi, H. Kojima, H. Yamamoto and Y. Kawashima, Evaluation of circulation profiles of liposomes coated with hydrophilic polymers having different molecular weights in rats, *J. Control. Release* **75**(1–2), 83–91 (2001).
58. J. M. Metselaar, P. Bruin, L. W. de Boer, T. de Vringer, C. Snel, C. Oussoren, M. H. Wauben, D. J. Crommelin, G. Storm and W. E. Hennink, A novel family of L-amino acid-based biodegradable polymer-lipid conjugates for the development of long-circulating liposomes with effective drug-targeting capacity, *Bioconjug. Chem.* **14**(6), 1156–1164 (2003).
59. T. Soldati and M. Schliwa, Powering membrane traffic in endocytosis and recycling, *Nat. Rev. Mol. Cell Biol.* 7(12), 897–908 (2006).
60. T. D. Heath, R. T. Fraley and D. Papahdjopoulos, Antibody targeting of liposomes: cell specificity obtained by conjugation of F(ab')2 to vesicle surface, *Science* **210**(4469), 539–541 (1980).
61. T. M. Allen, Ligand-targeted therapeutics in anticancer therapy, *Nat. Rev. Cancer* **2**(10), 750–763 (2002).
62. Y. Murase, T. Asai, Y. Katanasaka, T. Sugiyama, K. Shimizu, N. Maeda and N. Oku, A novel DDS strategy, "dual-targeting", and its application for antineovascular therapy, *Cancer Lett.* **287**(2), 165–171 (2010).
63. A. Koshkaryev, R. Sawant, M. Deshpande and V. Torchilin, Immunoconjugates and long circulating systems: Origins, current state of the art and future directions, *Adv. Drug Deliv. Rev.* **65**(1), 24–35 (2013).
64. J. Li, Y. Liu, J. Qian, L. Wu, J. Kemp, M. Nii, H. Tomkinson, Y. Zuo, M. Ranson and M. Usami, Single- and multiple-dose pharmacokinetics, safety and tolerability of zibotentan (ZD4054) in Chinese men with advanced solid tumors, *Cancer Chemother. Pharmacol.* **70**(1), 57–63 (2012).
65. G. Weissmann, C. Cohen and S. Hoffstein, Introduction of missing enzymes into the cytoplasm of cultured mammalian cells by means of fusion-prone liposomes, *Trans. Assoc. Am. Physicians.* **89**, 171–183 (1976).
66. M. Ozawa and A. Asano, The preparation of cell fusion-inducing proteoliposomes from purified glycoproteins of HVJ (Sendai virus) and chemically defined lipids, *J. Biol. Chem.* **256**(12), 5954–5956 (1981).
67. R. A. Parente, S. Nir and F. C. Szoka Jr., pH-dependent fusion of phosphatidylcholine small vesicles. Induction by a synthetic amphipathic peptide, *J. Biol. Chem.* **263**(10), 4724–4730 (1988).
68. A. L. Bailey, M. A. Monck and P. R. Cullis, pH-induced destabilization of lipid bilayers by a lipopeptide derived from influenza hemagglutinin, *Biochim. Biophys. Acta* **1324**(2), 232–244 (1997).

69. W. Li, F. Nicol and F. C. Szoka Jr., GALA: a designed synthetic pH-responsive amphipathic peptide with applications in drug and gene delivery, *Adv. Drug Deliv. Rev.* **56**(7), 967–985 (2004).
70. M. Jayaraman, S. M. Ansell, B. L. Mui, Y. K. Tam, J. Chen, X. Du, D. Butler, L. Eltepu, S. Matsuda, J. K. Narayanannair, K. G. Rajeev, I. M. Hafez, A. Akinc, M. A. Maier, M. A. Tracy, P. R. Cullis, T. D. Madden, M. Manoharan and M. J. Hope, Maximizing the potency of siRNA lipid nanoparticles for hepatic gene silencing in vivo, *Angew Chem. Int. Ed. Engl.* **51**(34), 8529–8533 (2012).
71. R. E. Pagano and J. N. Weinstein, Interactions of liposomes with mammalian cells, *Annu. Rev. Biophys. Bioeng.* **1978**, 7, 435–468 (2012).
72. V. P. Torchilin, R. Rammohan, V. Weissig and T. S. Levchenko, TAT peptide on the surface of liposomes affords their efficient intracellular delivery even at low temperature and in the presence of metabolic inhibitors, *Proc. Natl. Acad. Sci. USA* **98**(15), 8786–8791 (2001b).
73. P. Sapra and T. M. Allen, Ligand-targeted liposomal anticancer drugs, *Prog. Lipid Res.* **42**(5), 439–462 (2003).
74. R. L. Juliano and D. Stamp, Pharmacokinetics of liposome-encapsulated antitumor drugs. Studies with vinblastine, actinomycin D, cytosine arabinoside, and daunomycin, *Biochem. Pharmacol.* **27**(1), 21–27 (1978).
75. T. M. Allen, W. W. Cheng, J. I. Hare and K. M. Laginha, Pharmacokinetics and pharmacodynamics of lipidic nano-particles in cancer, *Anticancer Agents Med. Chem.* **6**(6), 513–523 (2006).
76. M. Barza, M. Stuart and F. Szoka Jr., Effect of size and lipid composition on the pharmacokinetics of intravitreal liposomes, *Invest Ophthalmol. Vis. Sci.* **28**(5), 893–900 (1987).
77. S. Qian, C. Li and Z. Zuo, Pharmacokinetics and disposition of various drug loaded liposomes, *Curr. Drug Metab.* **13**(4), 372–395 (2012).
78. M. C. Woodle and D. D. Lasic, Sterically stabilized liposomes, *Biochim. Biophys. Acta.* **1113**(2), 171–199 (1992).
79. Y. Takakura, M. Nishikawa, F. Yamashita and M. Hashida, Influence of physicochemical properties on pharmacokinetics of non-viral vectors for gene delivery, *J. Drug Target.* **10**(2), 99–104 (2002).
80. M. L. Immordino, F. Dosio and L. Cattel, Stealth liposomes: review of the basic science, rationale, and clinical applications, existing and potential, *Int. J. Nanomedicine* **1**(3), 297–315 (2006).
81. P. L. Rodriguez, T. Harada, D. A. Christian, D. A. Pantano, R. K. Tsai and D. E. Discher, Minimal "Self" peptides that inhibit phagocytic clearance and enhance delivery of nanoparticles, *Science* **339**(6122), 971–975 (2013).
82. V. J. Caride, Liposomes as carriers of imaging agents, *Crit. Rev. Ther. Drug. Carrier Syst.* **1**(2), 121–153 (1985).
83. M. Willis and E. Forssen, Ligand-targeted liposomes, *Adv. Drug Deliv. Rev.* **29**(3), 249–271 (1998).
84. N. Mitchell, T. L. Kalber, M. S. Cooper, K. Sunassee, S. L. Chalker, K. P. Shaw, K. L. Ordidge, A. Badar, S. M. Janes, P. J. Blower, M. F. Lythgoe, H. C. Hailes and A. B. Tabor, Incorporation of paramagnetic, fluorescent and PET/SPECT contrast agents into liposomes for multimodal imaging, *Biomaterials* **34**(4), 1179–1192 (2013).

85. J. Senior, C. Delgado, D. Fisher, C. Tilcock and G. Gregoriadis, Influence of surface hydrophilicity of liposomes on their interaction with plasma protein and clearance from the circulation: studies with poly(ethylene glycol)-coated vesicles, *Biochim. Biophys. Acta* **1062**(1), 77–82 (1991).
86. V. P. Torchilin, Polymer-coated long-circulating microparticulate pharmaceuticals, *J. Microencapsul.* **15**(1), 1–19 (1998).
87. D. Papahadjopoulos, T. M. Allen, A. Gabizon, E. Mayhew, K. Matthay, S. K. Huang, K. D. Lee, M. C. Woodle, D. D. Lasic and C. Redemann *et al.*, Sterically stabilized liposomes: improvements in pharmacokinetics and antitumor therapeutic efficacy, *Proc. Natl. Acad. Sci. USA* **88**(24), 11460–11464 (1991).
88. D. D. Lasic, F. J. Martin, A. Gabizon, S. K. Huang D. Papahadjopoulos, Sterically stabilized liposomes: a hypothesis on the molecular origin of the extended circulation times, *Biochim. Biophys. Acta* **1070**(1), 187–192 (1991).
89. T. S. Levchenko, R. Rammohan, A. N. Lukyanov, K. R. Whiteman and V. P. Torchilin, Liposome clearance in mice: the effect of a separate and combined presence of surface charge and polymer coating, *Int. J. Pharm.* **240**(1–2), 95–102 (2002).
90. E. T. Dams, P. Laverman, W. J. Oyen, G. Storm, G. L. Scherphof, J. W. van Der Meer, F. H. Corstens and O. C. Boerman, Accelerated blood clearance and altered biodistribution of repeated injections of sterically stabilized liposomes, *J. Pharmacol. Exp. Ther.* **292**(3), 1071–1079 (2000).
91. P. Laverman, M. G. Carstens, O. C. Boerman, E. T. Dams, W. J. Oyen, N. van Rooijen, F. H. Corstens and G. Storm, Factors affecting the accelerated blood clearance of polyethylene glycol-liposomes upon repeated injection, *J. Pharmacol. Exp. Ther.* **298**(2), 607–612 (2001).
92. T. Ishida, R. Maeda, M. Ichihara, Y. Mukai, Y. Motoki, Y. Manabe, K. Irimura and H. Kiwada, The accelerated clearance on repeated injection of pegylated liposomes in rats: laboratory and histopathological study, *Cell. Mol. Biol. Lett.* 7(2), 286 (2002b).
93. T. Ishida, M. Ichihara, X. Wang, K. Yamamoto, J. Kimura, E. Majima and H. Kiwada, Injection of PEGylated liposomes in rats elicits PEG-specific IgM, which is responsible for rapid elimination of a second dose of PEGylated liposomes, *J. Control. Release* **112**(1), 15–25 (2006c).
94. H. Koide, T. Asai, K. Hatanaka, S. Akai, T. Ishii, E. Kenjo, T. Ishida, H. Kiwada, H. Tsukada and N. Oku, T cell-independent B cell response is responsible for ABC phenomenon induced by repeated injection of PEGylated liposomes, *Int. J. Pharm.* **392**(1–2), 218–223 (2010).
95. T. Ishida, M. Ichihara, X. Wang and H. Kiwada, Spleen plays an important role in the induction of accelerated blood clearance of PEGylated liposomes, *J. Control. Release* **115**(3), 243–250 (2006b).
96. X. Wang, T. Ishida and H. Kiwada, Anti-PEG IgM elicited by injection of liposomes is involved in the enhanced blood clearance of a subsequent dose of PEGylated liposomes, *J. Control. Release* **119**(2), 236–244 (2007).
97. A. S. Abu Lila, N. E. Eldin, M. Ichihara, T. Ishida and H. Kiwada, Multiple administration of PEG-coated liposomal oxaliplatin enhances its therapeutic efficacy: a possible mechanism and the potential for clinical application, *Int. J. Pharm.* **438**(1–2), 176–183 (2012).
98. T. Ishida, K. Atobe, X. Wang and H. Kiwada, Accelerated blood clearance of PEGylated liposomes upon repeated injections: effect of doxorubicin-encapsulation and high-dose first injection, *J. Control. Release* **115**(3), 251–258 (2006a).

99. J. Cui, C. Li, C. Wang, Y. Li, L. Zhang, and H. Yang, Repeated injection of pegylated liposomal antitumour drugs induces the disappearance of the rapid distribution phase, *J. Pharm. Pharmacol.* **60**(12), 1651–1657 (2008).
100. R. K. Jain, Physiological barriers to delivery of monoclonal antibodies and other macromolecules in tumors, *Cancer Res.* **50**(3 Suppl), 814s–819s (1990).
101. H. Hashizume, P. Baluk, S. Morikawa, J. W. McLean, G. Thurston, S. Roberge, R. K. Jain and D. M. McDonald, Openings between defective endothelial cells explain tumor vessel leakiness, *Am. J. Pathol.* **156**(4), 1363–1380 (2000).
102. D. D. Lasic and D. Papahadjopoulos, Liposomes revisited, *Science* **267**(5202), 1275–1276 (1995).
103. K. Maruyama, O. Ishida, T. Takizawa and K. Moribe, Possibility of active targeting to tumor tissues with liposomes, *Adv. Drug. Deliv. Rev.* **40**(1–2), 89–102 (1999).
104. Y. Matsumura and H. Maeda, A new concept for macromolecular therapeutics in cancer chemotherapy: mechanism of tumoritropic accumulation of proteins and the antitumor agent smancs, *Cancer Res.* **46**(12 Pt 1), 6387–6392 (1986).
105. H. Maeda, G. Y. Bharate and J. Daruwalla, Polymeric drugs for efficient tumor-targeted drug delivery based on EPR-effect, *Eur. J. Pharm. Biopharm.* **71**(3), 409–419 (2009).
106. N. Desai, V. Trieu, Z. Yao, L. Louie, S. Ci, A. Yang, C. Tao, T. De, B. Beals, D. Dykes, P. Noker, R. Yao, E. Labao, M. Hawkins and P. Soon-Shiong, Increased antitumor activity, intratumor paclitaxel concentrations, and endothelial cell transport of cremophor-free, albumin-bound paclitaxel, ABI-007, compared with cremophor-based paclitaxel, *Clin. Cancer Res.* **12**(4), 1317–1324 (2006).
107. S. Maki, T. Konno and H. Maeda, Image enhancement in computerized tomography for sensitive diagnosis of liver cancer and semiquantitation of tumor selective drug targeting with oily contrast medium, *Cancer* **56**(4), 751–757 (1985).
108. A. Nagamitsu, K. Greish and H. Maeda, Elevating blood pressure as a strategy to increase tumor-targeted delivery of macromolecular drug SMANCS: cases of advanced solid tumors, *Jpn. J. Clin. Oncol.* **39**(11), 756–766 (2009).
109. W. L. Monsky, D. Fukumura, T. Gohongi, M. Ancukiewcz, H. A. Weich, V. P. Torchilin, F. Yuan and R. K. Jain, Augmentation of transvascular transport of macromolecules and nanoparticles in tumors using vascular endothelial growth factor, *Cancer Res.* **59**(16), 4129–4135 (1999).
110. M. R. Kano, Y. Bae, C. Iwata, Y. Morishita, M. Yashiro, M. Oka, T. Fujii, A. Komuro, K. Kiyono, M. Kaminishi, K. Hirakawa, Y. Ouchi, N. Nishiyama, K. Kataoka and K. Miyazono, Improvement of cancer-targeting therapy, using nanocarriers for intractable solid tumors by inhibition of TGF-beta signaling, *Proc. Natl. Acad. Sci. USA* **104**(9), 3460–3465 (2007).
111. A. L. Seynhaeve, S. Hoving, D. Schipper, C. E. Vermeulen, G. de Wiel-Ambagtsheer, S. T. van Tiel, A. M. Eggermont and T. L. Ten Hagen, Tumor necrosis factor alpha mediates homogeneous distribution of liposomes in murine melanoma that contributes to a better tumor response, *Cancer Res.* **67**(19), 9455–9462 (2007).
112. T. L. Ten Hagen, A. H. Van Der Veen, P. T. Nooijen, S. T. Van Tiel, A. L. Seynhaeve and A. M. Eggermont, Low-dose tumor necrosis factor-alpha augments antitumor activity of stealth liposomal doxorubicin (DOXIL) in soft tissue sarcoma-bearing rats, *Int. J. Cancer* **87**(6), 829–837 (2000).
113. Y. Doi, T. Okada, H. Matsumoto, M. Ichihara, T. Ishida and H. Kiwada, Combination therapy of metronomic S-1 dosing with oxaliplatin-containing polyethylene glycol-coated

liposome improves antitumor activity in a murine colorectal tumor model, *Cancer Sci.* **101**(11), 2470–2475 (2010).
114. P. Sapra, P. Tyagi and T. M. Allen, Ligand-targeted liposomes for cancer treatment, *Curr. Drug. Deliv.* **2**(4), 369–381 (2005).
115. P. Sapra and T. M. Allen, Internalizing antibodies are necessary for improved therapeutic efficacy of antibody-targeted liposomal drugs, *Cancer Res.* **62**(24), 7190–7194 (2002).
116. T. Volkel, P. Holig, T. Merdan, R. Muller and R. E. Kontermann, Targeting of immunoliposomes to endothelial cells using a single-chain Fv fragment directed against human endoglin (CD105), *Biochim. Biophys. Acta* **1663**(1–2), 158–166 (2004).
117. W. W. Cheng and T. M. Allen, Targeted delivery of anti-CD19 liposomal doxorubicin in B-cell lymphoma: a comparison of whole monoclonal antibody, Fab' fragments and single chain Fv, *J. Control. Release* **126**(1), 50–58 (2008).
118. M. Voinea, I. Manduteanu, E. Dragomir, M. Capraru and M. Simionescu, Immunoliposomes directed toward VCAM-1 interact specifically with activated endothelial cells--a potential tool for specific drug delivery, *Pharm. Res.* **22**(11), 1906–1917 (2005).
119. D. B. Kirpotin, D. C. Drummond, Y. Shao, M. R. Shalaby, K. Hong, U. B. Nielsen, J. D. Marks, C. C. Benz and J. W. Park, Antibody targeting of long-circulating lipidic nanoparticles does not increase tumor localization but does increase internalization in animal models, *Cancer Res.* **66**(13), 6732–6740 (2006).
120. P. G. Bloemen, P. A. Henricks, L. van Bloois, M. C. van den Tweel, A. C. Bloem, F. P. Nijkamp, D. J. Crommelin and G. Storm, Adhesion molecules: a new target for immunoliposome-mediated drug delivery, *FEBS Lett.* **357**(2), 140–144 (1995).
121. D. D. Spragg, D. R. Alford, R. Greferath, C. E. Larsen, K. D. Lee, G. C. Gurtner, M. I. Cybulsky, P. F. Tosi, C. Nicolau and M. A. Gimbrone Jr., Immunotargeting of liposomes to activated vascular endothelial cells: a strategy for site-selective delivery in the cardiovascular system, *Proc. Natl. Acad. Sci. USA* **94**(16), 8795–8800 (1997).
122. S. Kessner, A. Krause, U. Rothe and G. Bendas, Investigation of the cellular uptake of E-Selectin-targeted immunoliposomes by activated human endothelial cells, *Biochim. Biophys. Acta* **1514**(2), 177–190 (2001).
123. K. Atobe, T. Ishida, E. Ishida, K. Hashimoto, H. Kobayashi, J. Yasuda, T. Aoki, K. Obata, H. Kikuchi, H. Akita, T. Asai, H. Harashima, N. Oku and H. Kiwada, In vitro efficacy of a sterically stabilized immunoliposomes targeted to membrane type 1 matrix metalloproteinase (MT1-MMP), *Biol. Pharm. Bull* **30**(5), 972–978 (2007).
124. H. Hatakeyama, H. Akita, E. Ishida, K. Hashimoto, H. Kobayashi, T. Aoki, J. Yasuda, K. Obata, H. Kikuchi, T. Ishida, H. Kiwada and H. Harashima, Tumor targeting of doxorubicin by anti-MT1-MMP antibody-modified PEG liposomes, *Int. J. Pharm.* **342**(1–2), 194–200 (2007).
125. I. Ahmad, M. Longenecker, J. Samuel and T. M. Allen, Antibody-targeted delivery of doxorubicin entrapped in sterically stabilized liposomes can eradicate lung cancer in mice, *Cancer Res.* **53**(7), 1484–1488 (1993).
126. F. Pastorino, D. Di Paolo, F. Piccardi, B. Nico, D. Ribatti, A. Daga, G. Baio, C.E. Neumaier, C. Brignole, M. Loi, D. Marimpietri, G. Pagnan, M. Cilli, E. A. Lepekhin, S. V. Garde, R. Longhi, A. Corti, T. M. Allen, J. J. Wu and M. Ponzoni, Enhanced antitumor efficacy of clinical-grade vasculature-targeted liposomal doxorubicin, *Clin. Cancer Res.* **14**(22), 7320–7329 (2008).

127. K. Mack, R. Rüger, S. Fellermeier, O. Seifert and R. E. Kontermann, Dual Targeting of Tumor Cells with Bispecific Single-Chain Fv-Immunoliposomes, *mAbs.* **1**(2), 199–214 (2012).
128. Y. Lu and P. S. Low, Folate targeting of haptens to cancer cell surfaces mediates immunotherapy of syngeneic murine tumors, *Cancer Immunol. Immunother.* **51**(3), 153–162 (2002a).
129. X. Q. Pan, H. Wang and R. J. Lee, Antitumor activity of folate receptor-targeted liposomal doxorubicin in a KB oral carcinoma murine xenograft model, *Pharm Res.* **20**(3), 417–422 (2003).
130. R. J. Lee and P. S. Low, Folate-mediated tumor cell targeting of liposome-entrapped doxorubicin in vitro, *Biochim. Biophys. Acta* **1233**(2), 134–144 (1995).
131. S. Ni, S. M. Stephenson and R. J. Lee, Folate receptor targeted delivery of liposomal daunorubicin into tumor cells, *Anticancer Res.* **22**(4), 2131–2135 (2002).
132. S. M. Stephenson, W. Yang, P. J. Stevens, W. Tjarks, R. F. Barth and R. J. Lee, Folate receptor-targeted liposomes as possible delivery vehicles for boron neutron capture therapy, *Anticancer Res.* **23**(4), 3341–3345 (2003).
133. Y. Lu and P. S. Low, Folate-mediated delivery of macromolecular anticancer therapeutic agents, *Adv. Drug Deliv. Rev.* **54**(5), 675–693 (2002b).
134. J. A. Reddy, C. Abburi, H. Hofland, S. J. Howard, I. Vlahov, P. Wils and C. P. Leamon, Folate-targeted, cationic liposome-mediated gene transfer into disseminated peritoneal tumors, *Gene Ther.* **9**(22), 1542–1550 (2002).
135. C. P. Leamon, S. R. Cooper and G. E. Hardee, Folate-liposome-mediated antisense oligodeoxynucleotide targeting to cancer cells: evaluation in vitro and in vivo, *Bioconjug. Chem.* **14**(4), 738–747 (2003).
136. H. Hatakeyama, H. Akita, K. Maruyama, T. Suhara and H. Harashima, Factors governing the in vivo tissue uptake of transferrin-coupled polyethylene glycol liposomes in vivo, *Int. J. Pharm.* **281**(1–2), 25–33 (2004).
137. O. Ishida, K. Maruyama, H. Tanahashi, M. Iwatsuru, K. Sasaki, M. Eriguchi and H. Yanagie, Liposomes bearing polyethyleneglycol-coupled transferrin with intracellular targeting property to the solid tumors in vivo, *Pharm. Res.* **18**(7), 1042–1048 (2001).
138. D. A. Eavarone, X. Yu and R. V. Bellamkonda, Targeted drug delivery to C6 glioma by transferrin-coupled liposomes, *J. Biomed. Mater. Res.* **51**(1), 10–14 (2000).
139. H. Iinuma, K. Maruyama, K. Okinaga, K. Sasaki, T. Sekine, O. Ishida, N. Ogiwara, K. Johkura and Y. Yonemura, Intracellular targeting therapy of cisplatin-encapsulated transferrin-polyethylene glycol liposome on peritoneal dissemination of gastric cancer, *Int. J. Cancer* **99**(1), 130–137 (2002).
140. R. Suzuki, T. Takizawa, Y. Kuwata, M. Mutoh, N. Ishiguro, N. Utoguchi, A. Shinohara, M. Eriguchi, H. Yanagie and K. Maruyama, Effective anti-tumor activity of oxaliplatin encapsulated in transferrin-PEG-liposome, *Int. J. Pharm.* **346**(1–2), 143–150 (2008).
141. D. C. Drummond, M. Zignani and J. Leroux, Current status of pH-sensitive liposomes in drug delivery, *Prog. Lipid Res.* **39**(5), 409–460 (2000b).
142. S. Dagar, A. Krishnadas, I. Rubinstein, M. J. Blend and H. Onyuksel, VIP grafted sterically stabilized liposomes for targeted imaging of breast cancer: in vivo studies, *J. Control. Release* **91**(1–2), 123–133 (2003).
143. M. Hashida, M. Nishikawa, F. Yamashita and Y. Takakura, Cell-specific delivery of genes with glycosylated carriers, *Adv. Drug Deliv. Rev.* **52**(3), 187–196 (2001).

144. I. Matsuda, H. Konno, T. Tanaka and S. Nakamura, Antimetastatic effect of hepatotropic liposomal adriamycin on human metastatic liver tumors, *Surg. Today.* **31**(5), 414–420 (2001).
145. P. Ekblom, I. Thesleff, V. P. Lehto and I. Virtanen, Distribution of the transferrin receptor in normal human fibroblasts and fibrosarcoma cells, *Int. J. Cancer* **31**(1), 111–117 (1983).
146. D. Peer, J. M. Karp, S. Hong, O. C. Farokhzad, R. Margalit and R. Langer, Nanocarriers as an emerging platform for cancer therapy, *Nat. Nanotechnol.* **2**(12), 751–760 (2007).
147. K. Maruyama, S. J. Kennel and L. Huang, Lipid composition is important for highly efficient target binding and retention of immunoliposomes, *Proc. Natl. Acad. Sci. USA* **87**(15), 5744–5748 (1990).
148. N. C. Phillips and A. Emili, Immunogenicity of immunoliposomes, *Immunol. Lett.* **30**(3), 291–296 (1991).
149. J. N. Weinstein and W. van Osdol, Early intervention in cancer using monoclonal antibodies and other biological ligands: micropharmacology and the "binding site barrier", *Cancer Res.* **52**(9 Suppl), 2747s–2751s (1992).
150. G. P. Adams, R. Schier, A. M. McCall, H. H. Simmons, E. M. Horak, R. K. Alpaugh, J. D. Marks and L. M. Weiner, High affinity restricts the localization and tumor penetration of single-chain fv antibody molecules, *Cancer Res.* **61**(12), 4750–4755 (2001).
151. A. J. Zurita, A. J. W. Arap and R. Pasqualini, Mapping tumor vascular diversity by screening phage display libraries, *J. Control. Release* **91**(1–2), 183–186 (2003).
152. F. Pastorino, C. Brignole, D. Marimpietri, M. Cilli, C. Gambini, D. Ribatti, R. Longhi, T. M. Allen, A. Corti and M. Ponzoni, Vascular damage and anti-angiogenic effects of tumor vessel-targeted liposomal chemotherapy, *Cancer Res.* **63**(21), 7400–7409 (2003a).
153. R. M. Schiffelers, G. A. Koning, T. L. ten Hagen, M. H. Fens, A. J. Schraa, A. P. Janssen, R. J. Kok, G. Molema and G. Storm, Anti-tumor efficacy of tumor vasculature-targeted liposomal doxorubicin, *J. Control. Release.* **91**(1–2), 115–122 (2003).
154. G. A. Koning, M. M. Fretz, U. Woroniecka, G. Storm and G. C. Krijger, Targeting liposomes to tumor endothelial cells for neutron capture therapy, *Appl. Radiat. Isot.* **61**(5), 963–967 (2004).
155. F. Pastorino, C. Brignole, D. Di Paolo, B. Nico, A. Pezzolo, D. Marimpietri, G. Pagnan, F. Piccardi, M. Cilli, R. Longhi, D. Ribatti, A. Corti, T. M. Allen and M. Ponzoni, Targeting liposomal chemotherapy via both tumor cell-specific and tumor vasculature-specific ligands potentiates therapeutic efficacy, *Cancer Res.* **66**(20), 10073–10082 (2006).
156. D. Simberg, T. Duza, J. H. Park, M. Essler, J. Pilch, L. Zhang, A. M. Derfus, M. Yang, R. M. Hoffman, S. Bhatia, M. J. Sailor and E. Ruoslahti, Biomimetic amplification of nanoparticle homing to tumors, *Proc. Natl. Acad. Sci. USA.* **104**(3), 932–936 (2007).
157. M. Kondo, T. Asai, Y. Katanasaka, Y. Sadzuka, H. Tsukada, K. Ogino, T. Taki, K. Baba and N. Oku, Anti-neovascular therapy by liposomal drug targeted to membrane type-1 matrix metalloproteinase, *Int. J. Cancer* **108**(2), 301–306 (2004).
158. E. Mastrobattista, G. A. Koning, L. van Bloois, A. C. Filipe, W. Jiskoot and G. Storm, Functional characterization of an endosome-disruptive peptide and its application in cytosolic delivery of immunoliposome-entrapped proteins, *J. Biol. Chem.* **277**(30), 27135–27143 (2002).

159. X. Gao and L. Huang, Cationic liposome-mediated gene transfer, *Gene Ther.* **2**(10), 710–722 (1995).
160. T. Serikawa, N. Suzuki, H. Kikuchi, K. Tanaka and T. Kitagawa, A new cationic liposome for efficient gene delivery with serum into cultured human cells: a quantitative analysis using two independent fluorescent probes, *Biochim. Biophys. Acta* **1467**(2), 419–430 (2000).
161. G. D. Schmidt-Wolf and I. G. Schmidt-Wolf, Non-viral and hybrid vectors in human gene therapy: an update, *Trends Mol. Med.* **9**(2), 67–72 (2003).
162. M. Bessodes, C. Dubertret, G. Jaslin and D. Scherman, Synthesis and biological properties of new glycosidic cationic lipids for DNA delivery, *Bioorg. Med. Chem. Lett.* **10**(12), 1393–1395 (2000).
163. D. Niculescu-Duvaz, J. Heyes and C. J. Springer, Structure-activity relationship in cationic lipid mediated gene transfection, *Curr. Med. Chem.* **10**(14), 1233–1261 (2003).
164. M. Sainlos, M. Hauchecorne, N. Oudrhiri, S. Zertal-Zidani, A. Aissaoui, J. P. Vigneron, J. M. Lehn and P. Lehn, Kanamycin A-derived cationic lipids as vectors for gene transfection, *Chembiochem.* **6**(6), 1023–1033 (2005).
165. R. Kunstfeld, G. Wickenhauser, U. Michaelis, M. Teifel, W. Umek, K. Naujoks, K. Wolff and P. Petzelbauer, Paclitaxel encapsulated in cationic liposomes diminishes tumor angiogenesis and melanoma growth in a "humanized" SCID mouse model, *J. Invest. Dermatol.* **120**(3), 476–482 (2003).
166. S. Strieth, M. E. Eichhorn, B. Sauer, B. Schulze, M. Teifel, U. Michaelis and M. Dellian, Neovascular targeting chemotherapy: encapsulation of paclitaxel in cationic liposomes impairs functional tumor microvasculature, *Int. J. Cancer.* **110**(1), 117–124 (2004).
167. M. Schmitt-Sody, S. Strieth, S. Krasnici, B. Sauer, B. Schulze, M. Teifel, U. Michaelis, K. Naujoks and M. Dellian, Neovascular targeting therapy: paclitaxel encapsulated in cationic liposomes improves antitumoral efficacy, *Clin. Cancer Res.* **9**(6), 2335–2341 (2003).
168. J. Wu, A. Lee, Y. Lu and R. J. Lee, Vascular targeting of doxorubicin using cationic liposomes, *Int. J. Pharm.* **337**(1–2), 329–335 (2007).
169. A. Abu-Lila, T. Suzuki, Y. Doi, T. Ishida and H. Kiwada, Oxaliplatin targeting to angiogenic vessels by PEGylated cationic liposomes suppresses the angiogenesis in a dorsal air sac mouse model, *J. Control. Release* **134**(1), 1825 (2009).
170. A. D. Keefe, S. Pai and A. Ellington, Aptamers as therapeutics, *Nat. Rev. Drug Discov.* **9**(7), 537–550 (2010).
171. A. D. Ellington and J. W. Szostak, In vitro selection of RNA molecules that bind specific ligands, *Nature* **346**(6287), 818–822 (1990).
172. C. Tuerk and L. Gold, Systematic evolution of ligands by exponential enrichment: RNA ligands to bacteriophage T4 DNA polymerase, *Science* **249**(4968), 505–510 (1990).
173. S. D. Jayasena, Aptamers: an emerging class of molecules that rival antibodies in diagnostics, *Clin. Chem.* **45**(9), 1628–1650 (1999).
174. K. Maehashi, T. Katsura, K. Kerman, Y. Takamura, K. Matsumoto and E. Tamiya, Label-free protein biosensor based on aptamer-modified carbon nanotube field-effect transistors, *Anal Chem.* **79**(2), 782–787 (2007).
175. O. C. Farokhzad, J. Cheng, B. A. Teply, I. Sherifi, S. Jon, P. W. Kantoff, J. P. Richie and R. Langer, Targeted nanoparticle-aptamer bioconjugates for cancer chemotherapy in vivo, *Proc. Natl. Acad. Sci. USA* **103**(16), 6315–6320 (2006a).

176. E. Levy-Nissenbaum, A. F. Radovic-Moreno, A. Z. Wang, R. Langer and O. C. Farokhzad, Nanotechnology and aptamers: applications in drug delivery, *Trends Biotechnol.* **26**(8), 442–449 (2008).
177. A. Potti, C. P. Rusconi, B. A. Sullenger and T. L. Ortel, Regulatable aptamers in medicine: focus on antithrombotic strategies, *Expert Opin. Biol. Ther.* **4**(10), 1641–1647 (2004).
178. V. Bagalkot, L. Zhang, E. Levy-Nissenbaum, S. Jon, P. W. Kantoff, R. Langer and O. C. Farokhzad, Quantum dot-aptamer conjugates for synchronous cancer imaging, therapy, and sensing of drug delivery based on bi-fluorescence resonance energy transfer, *Nano Lett.* 7(10), 3065–3070 (2007).
179. D. Kim, Y. Y. Jeong and S. Jon, A drug-loaded aptamer-gold nanoparticle bioconjugate for combined CT imaging and therapy of prostate cancer, *ACS Nano.* **4**(7), 3689–3696 (2010).
180. N. Kolishetti, S. Dhar, P. M. Valencia, L. Q. Lin, R. Karnik, S. J. Lippard, R. Langer and O. C. Farokhzad, Engineering of self-assembled nanoparticle platform for precisely controlled combination drug therapy, *Proc. Natl. Acad. Sci. USA* **107**(42), 17939–17944 (2010).
181. O. C. Farokhzad, J. M. Karp and R. Langer, Nanoparticle-aptamer bioconjugates for cancer targeting, *Expert Opin. Drug Deliv.* **3**(3), 311–324 (2006b).
182. H. Kang, M. B. O'Donoghue, H. Liu and W. Tan, A liposome-based nanostructure for aptamer directed delivery, *Chem. Commun. (Camb).* **46**(2), 249–251 (2010).
183. J. M. Healy, S. D. Lewis, M. Kurz, R. M. Boomer, K. M. Thompson, C. Wilson and T.G. McCauley, Pharmacokinetics and biodistribution of novel aptamer compositions, *Pharm. Res.* **21**(12), 2234–2246 (2004).
184. R. M. Straubinger, K. Hong, D. S. Friend and D. Papahadjopoulos, Endocytosis of liposomes and intracellular fate of encapsulated molecules: encounter with a low pH compartment after internalization in coated vesicles, *Cell.* **32**(4), 1069–1079 (1983).
185. D. L. Daleke, K. Hong and D. Papahadjopoulos, Endocytosis of liposomes by macrophages: binding, acidification and leakage of liposomes monitored by a new fluorescence assay, *Biochim. Biophys. Acta.* **1024**(2), 352–366 (1990).
186. R. M. Straubinger, D. Papahadjopoulos and K. L. Hong, Endocytosis and intracellular fate of liposomes using pyranine as a probe, *Biochemistry* **29**(20), 4929–4939 (1990).
187. A. A. Kale and V. P. Torchilin, "Smart" drug carriers: PEGylated TATp-modified pH-sensitive liposomes, *J Liposome Res.* **17**(3–4), 197–203 (2007).
188. S. Ganta, H. Devalapally, A. Shahiwala and M. Amiji, A review of stimuli-responsive nanocarriers for drug and gene delivery, *J. Control. Release* **126**(3), 187–204 (2008).
189. J. Connor and L. Huang, pH-sensitive immunoliposomes as an efficient and target-specific carrier for antitumor drugs, *Cancer Res.* **46**(7), 3431–3435 (1986).
190. E. Fattal, P. Couvreur and C. Dubernet, "Smart" delivery of antisense oligonucleotides by anionic pH-sensitive liposomes, *Adv. Drug Deliv. Rev.* **56**(7), 931–946 (2004).
191. G. Shi, W. Guo, S. M. Stephenson and R. J. Lee, Efficient intracellular drug and gene delivery using folate receptor-targeted pH-sensitive liposomes composed of cationic/anionic lipid combinations, *J. Control. Release* **80**(1–3), 309–319 (2002).
192. C. Fonseca, J. N. Moreira, C. J. Ciudad, M. C. Pedroso de Lima and S. Simoes, Targeting of sterically stabilised pH-sensitive liposomes to human T-leukaemia cells, *Eur. J. Pharm. Biopharm.* **59**(2), 359–366 (2005).

193. M. J. Kim, H. J. Lee, I. A. Lee, I. Y. Kim, S. K. Lim, H. A. Cho and J. S. Kim, Preparation of pH-sensitive, long-circulating and EGFR-targeted immunoliposomes, *Arch. Pharm. Res.* **31**(4), 539–546 (2008).
194. P. Simard and J. C. Leroux, pH-sensitive immunoliposomes specific to the CD33 cell surface antigen of leukemic cells, *Int. J. Pharm.* **381**(2), 86–96 (2009).
195. J. X. Zhang, S. Zalipsky, N. Mullah, M. Pechar and T. M. Allen, Pharmaco attributes of dioleoylphosphatidylethanolamine/cholesterylhemisuccinate liposomes containing different types of cleavable lipopolymers, *Pharmacol. Res.* **49**(2), 185–198 (2004).
196. M. B. Yatvin, J. N. Weinstein, W. H. Dennis and R. Blumenthal, Design of liposomes for enhanced local release of drugs by hyperthermia, *Science* **202**(4374), 1290–1293 (1978).
197. G. N. Chiu, S. A. Abraham, L. M. Ickenstein, R. Ng, G. Karlsson, K. Edwards, E. K. Wasan and M. B. Bally, Encapsulation of doxorubicin into thermosensitive liposomes via complexation with the transition metal manganese, *J. Control. Release* **104**(2), 271–288 (2005).
198. D. Needham, G. Anyarambhatla, G. Kong and M. W. Dewhirst, A new temperature-sensitive liposome for use with mild hyperthermia: characterization and testing in a human tumor xenograft model, *Cancer Res.* **60**(5), 1197–1201 (2000).
199. L. H. Lindner, M. E. Eichhorn, H. Eibl N. Teichert, M. Schmitt-Sody, R. D. Issels and M. Dellian, Novel temperature-sensitive liposomes with prolonged circulation time, *Clin. Cancer Res.* **10**(6), 2168–2178 (2004).
200. K. Maruyama, S. Unezaki, N. Takahashi and M. Iwatsuru, Enhanced delivery of doxorubicin to tumor by long-circulating thermosensitive liposomes and local hyperthermia, *Biochim. Biophys. Acta.* **1149**(2), 209–216 (1993).
201. S. Unezaki, K. Maruyama, N. Takahashi, M. Koyama, T. Yuda, A. Suginaka and M. Iwatsuru, Enhanced delivery and antitumor activity of doxorubicin using long-circulating thermosensitive liposomes containing amphipathic polyethylene glycol in combination with local hyperthermia, *Pharm. Res.* **11**(8), 1180–1185 (1994).
202. N. Dos Santos, L. D. Mayer, S. A. Abraham, R. C. Gallagher, K. A. Cox, P. G. Tardi and M. B. Bally, Improved retention of idarubicin after intravenous injection obtained for cholesterol-free liposomes, *Biochim. Biophys. Acta.* **1561**(2), 188–201 (2002).
203. D. Needham and M. W. Dewhirst, The development and testing of a new temperature-sensitive drug delivery system for the treatment of solid tumors, *Adv. Drug Deliv. Rev.* **53**(3), 285–305 (2001).
204. G. A. Koning, A. M. Eggermont, L. H. Lindner and T. L. ten Hagen, Hyperthermia and thermosensitive liposomes for improved delivery of chemotherapeutic drugs to solid tumors, *Pharm. Res.* **27**(8), 1750–1754 (2010).
205. R. T. Poon and N. Borys, Lyso-thermosensitive liposomal doxorubicin: an adjuvant to increase the cure rate of radiofrequency ablation in liver cancer, *Future Oncol.* 7(8), 937–945 (2011).
206. D. C. Drummond, K. Hong, J. W. Park, C. C. Benz and D. B. Kirpotin, Liposome targeting to tumors using vitamin and growth factor receptors, *Vitam. Horm.* **60**, 285–332 (2000a).
207. T. Kubo, T. Sugita, S. Shimose, Y. Nitta, Y. Ikuta and T. Murakami, Targeted systemic chemotherapy using magnetic liposomes with incorporated adriamycin for osteosarcoma in hamsters, *Int. J. Oncol.* **18**(1), 121–125 (2001).

208. T. Kikumori, T. Kobayashi, M. Sawaki and T. Imai, Anti-cancer effect of hyperthermia on breast cancer by magnetite nanoparticle-loaded anti-HER2 immunoliposomes, *Breast Cancer Res. Treat.* **113**(3), 435–441 (2009).
209. H. Nobuto, T. Sugita, T. Kubo, S. Shimose, Y. Yasunaga, T. Murakami and M. Ochi, Evaluation of systemic chemotherapy with magnetic liposomal doxorubicin and a dipole external electromagnet, *Int. J. Cancer.* **109**(4), 627–635 (2004).
210. G. Bealle, R. Di Corato, J. Kolosnjaj-Tabi, V. Dupuis, O. Clement, F. Gazeau, C. Wilhelm and C. Menager, Ultra magnetic liposomes for MR imaging, targeting, and hyperthermia, *Langmuir* **28**(32), 11834–11842 (2012).
211. J. D. Almeida, D. C. Edwards, C. M. Brand and T. D. Heath, Formation of virosomes from influenza subunits and liposomes, *Lancet.* **2**(7941), 899–901 (1975).
212. Y. Kaneda, Virosomes: evolution of the liposome as a targeted drug delivery system, *Adv. Drug Deliv. Rev.* **43**(2–3), 197–205 (2000).
213. A. Huckriede, L. Bungener, T. Daemen and J. Wilschut, Influenza virosomes in vaccine development, *Methods Enzymol.* **373**, 74–91 (2003a).
214. A. Huckriede, L. Bungener, T. Stegmann, T. Daemen, J. Medema, A. M. Palache and J. Wilschut, The virosome concept for influenza vaccines, *Vaccine* **23** Suppl 1, S26–S38 (2005).
215. D. P. Sarkar, K. Ramani and S. K. Tyagi, Targeted gene delivery by virosomes, *Methods Mol. Biol.* **199**, 163–173 (2002).
216. M. G. Cusi, C. Terrosi, G. G. Savellini, G. Di Genova, R. Zurbriggen and P. Correale, Efficient delivery of DNA to dendritic cells mediated by influenza virosomes, *Vaccine* **22**(5–6), 735–739 (2004).
217. T. Daemen, A. de Mare, L. Bungener, J. de Jonge, A. Huckriede and J. Wilschut, Virosomes for antigen and DNA delivery, *Adv. Drug Deliv. Rev.* **57**(3), 451–463 (2005).
218. D. Felnerova, J. F. Viret, R. Gluck and C. Moser, Liposomes and virosomes as delivery systems for antigens, nucleic acids and drugs, *Curr. Opin. Biotechnol.* **15**(6), 518–529 (2004).
219. L. Bungener, A. Huckriede, J. Wilschut and T. Daemen, Delivery of protein antigens to the immune system by fusion-active virosomes: a comparison with liposomes and ISCOMs, *Biosci. Rep.* **22**(2), 323–338 (2002a).
220. L. Bungener, J. Idema, W. ter Veer, A. Huckriede, T. Daemen and J. Wilschut, Virosomes in vaccine development: induction of cytotoxic T lymphocyte activity with virosome-encapsulated protein antigens, *J. Liposome Res.* **12**(1–2), 155–163 (2002b).
221. A. Huckriede, L. Bungener, W. ter Veer, M. Holtrop, T. Daemen, A. M. Palache and J. Wilschut, Influenza virosomes: combining optimal presentation of hemagglutinin with immunopotentiating activity, *Vaccine* **21**(9–10), 925–931 (2003b).
222. C. Herzog, I. C. Metcalfe and U. B. Schaad, Virosome influenza vaccine in children, *Vaccine* **20** Suppl 5, B24–B28 (2002).
223. L. Jabr-Milane, L. van Vlerken, H. Devalapally, D. Shenoy, S. Komareddy, M. Bhavsar and M. Amiji, Multi-functional nanocarriers for targeted delivery of drugs and genes, *J. Control. Release* **130**(2), 121–128 (2008).
224. K. Kono, S. Nakashima, D. Kokuryo, I. Aoki, H. Shimomoto, S. Aoshima, K. Maruyama, E. Yuba, C. Kojima, A. Harada and Y. Ishizaka, Multi-functional liposomes having temperature-triggered release and magnetic resonance imaging for tumor-specific chemotherapy, *Biomaterials* **32**(5), 1387–1395 (2011).

225. G. A. Koning and G. C. Krijger, Targeted multifunctional lipid-based nanocarriers for image-guided drug delivery, *Anticancer Agents Med. Chem.* 7(4), 425–440 (2007).
226. A. A. Kale and V. P. Torchilin, Environment-responsive multifunctional liposomes, *Methods Mol. Biol.* **605**, 213–242 (2010).
227. B. Goldenbogen, N. Brodersen, A. Gramatica, M. Loew, J. Liebscher, A. Herrmann, H. Egger, B. Budde and A. Arbuzova, Reduction-sensitive liposomes from a multifunctional lipid conjugate and natural phospholipids: reduction and release kinetics and cellular uptake, *Langmuir* **27**(17), 10820–10829 (2011).
228. L. Zhu, P. Kate and V. P. Torchilin, Matrix metalloprotease 2-responsive multifunctional liposomal nanocarrier for enhanced tumor targeting, *ACS Nano.* **6**(4), 3491–3498 (2012).
229. A. J. Primeau, A. Rendon, D. Hedley, L. Lilge and I. F. Tannock, The distribution of the anticancer drug Doxorubicin in relation to blood vessels in solid tumors, *Clin. Cancer Res.* **11**(24 Pt 1), 8782–8788 (2005).
230. R. K. Jain, The next frontier of molecular medicine: delivery of therapeutics, *Nat. Med.* **4**(6), 655–657 (1998).
231. A. Pluen, Y. Boucher, S. Ramanujan, T. D. McKee, T. Gohongi, E. di Tomaso, E. B. Brown, Y. Izumi, R. B. Campbell, D. A. Berk and R. K. Jain, Role of tumor-host interactions in interstitial diffusion of macromolecules: cranial vs. subcutaneous tumors, *Proc. Natl. Acad. Sci. USA* **98**(8), 4628–4633 (2001).
232. P. A. Netti, L. T. Baxter, Y. Boucher, R. Skalak and R. K. Jain, Time-dependent behavior of interstitial fluid pressure in solid tumors: implications for drug delivery, *Cancer Res.* **55**(22), 5451–5458 (1995).
233. T. Tsuruo and A. Tomida, Multidrug resistance, *Anticancer Drugs* **6**(2), 213–218 (1995).
234. T. Lammers, W. E. Hennink and G. Storm, Tumour-targeted nanomedicines: principles and practice, *Br. J. Cancer* **99**(3), 392–397 (2008).
235. F. Olson, E. Mayhew, D. Maslow, Y. Rustum and F. Szoka, Characterization, toxicity and therapeutic efficacy of adriamycin encapsulated in liposomes, *Eur J. Cancer Clin. Oncol.* **18**(2), 167–176 (1982).
236. E. H. Herman, A. Rahman, V. J. Ferrans, J. A. Vick and P. S. Schein, Prevention of chronic doxorubicin cardiotoxicity in beagles by liposomal encapsulation, *Cancer Res.* **43**(11), 5427–5432 (1983).
237. Q. G. van Hoesel, P. A. Steerenberg, D. J. Crommelin, A. van Dijk, W. van Oort, S. Klein, J. M. Douze, D. J. de Wildt and F. C. Hillen, Reduced cardiotoxicity and nephrotoxicity with preservation of antitumor activity of doxorubicin entrapped in stable liposomes in the LOU/M Wsl rat, *Cancer Res.* **44**(9), 3698–3705 (1984).
238. B. Uziely, S. Jeffers, R. Isacson, K. Kutsch, D. Wei-Tsao, Z. Yehoshua, E. Libson, F. M. Muggia and A. Gabizon, Liposomal doxorubicin: antitumor activity and unique toxicities during two complementary phase I studies, *J. Clin. Oncol.* **13**(7), 1777–1785 (1995).
239. N. D. James, R. J. Coker, D. Tomlinson, J. R. Harris, M. Gompels, A. J. Pinching and J. S. Stewart, Liposomal doxorubicin (Doxil): an effective new treatment for Kaposi's sarcoma in AIDS, *Clin. Oncol. (R Coll Radiol)* **6**(5), 294–296 (1994).
240. J. Blade, P. Sonneveld, J. F. San Miguel, H. J. Sutherland, R. Hajek, A. Nagler, A. Spencer, T. Robak, K. C. Lantz, S. H. Zhuang, J. L. Harousseau and R. Z. Orlowski, Efficacy and safety of pegylated liposomal Doxorubicin in combination with bortezomib

for multiple myeloma: effects of adverse prognostic factors on outcome, *Clin. Lymphoma Myeloma Leuk.* **11**(1), 44–49 (2011).
241. Y. Barenholz, Doxil(R)--the first FDA-approved nano-drug: lessons learned, *J. Control. Release* **160**(2), 117–134 (2012).
242. C. E. Petre and D. P. Dittmer, Liposomal daunorubicin as treatment for Kaposi's sarcoma, *Int. J. Nanomedicine* **2**, 277–288 (2007).
243. M. J. Glantz, K. A. Jaeckle, M. C. Chamberlain, S. Phuphanich, L. Recht, L. J. Swinnen, B. Maria, S. LaFollette, G. B. Schumann, B. F. Cole and S. B. Howell, A randomized controlled trial comparing intrathecal sustained-release cytarabine (DepoCyt) to intrathecal methotrexate in patients with neoplastic meningitis from solid tumors, *Clin. Cancer Res.* **5**(11), 3394–3402 (1999).
244. G. Batist, G. Ramakrishnan, C. S. Rao, A. Chandrasekharan, J. Gutheil, T. Guthrie, P. Shah, A. Khojasteh, M. K. Nair, K. Hoelzer, K. Tkaczuk, Y. C. Park and L. W. Lee, Reduced cardiotoxicity and preserved antitumor efficacy of liposome-encapsulated doxorubicin and cyclophosphamide compared with conventional doxorubicin and cyclophosphamide in a randomized, multicenter trial of metastatic breast cancer, *J. Clin. Oncol.* **19**(5), 1444–1454 (2001).
245. A. H. Sarris, F. Hagemeister, J. Romaguera, M. A. Rodriguez, P. McLaughlin, A. M. Tsimberidou, L. J. Medeiros, B. Samuels, O. Pate, M. Oholendt, H. Kantarjian, C. Burge and F. Cabanillas, Liposomal vincristine in relapsed non-Hodgkin's lymphomas: early results of an ongoing phase II trial, *Ann. Oncol.* **11**(1), 69–72 (2000).
246. M. A. Rodriguez, R. Pytlik, T. Kozak, M. Chhanabhai, R. Gascoyne, B. Lu, S. R. Deitcher and J. N. Winter, Vincristine sulfate liposomes injection (Marqibo) in heavily pretreated patients with refractory aggressive non-Hodgkin lymphoma: report of the pivotal phase 2 study, *Cancer* **115**(15), 3475–3482 (2009).
247. M. S. Newman, G. T. Colbern, P. K. Working, C. Engbers and M. A. Amantea, Comparative pharmacokinetics, tissue distribution, and therapeutic effectiveness of cisplatin encapsulated in long-circulating, pegylated liposomes (SPI-077) in tumor-bearing mice.,*Cancer Chemother. Pharmacol.* **43**(1), 1–7 (1999).
248. N. Seetharamu, E. Kim, H. Hochster, F. Martin and F. Muggia, Phase II study of liposomal cisplatin (SPI-77) in platinum-sensitive recurrences of ovarian cancer, *Anticancer Res.* **30**(2), 541–545 (2010).
249. K. Riviere, H. M. Kieler-Ferguson, K. Jerger and F. C. Szoka Jr., Anti-tumor activity of liposome encapsulated fluoroorotic acid as a single agent and in combination with liposome irinotecan, *J. Control. Release* **153**(3), 288–296 (2011).
250. A. Dicko, L. D. Mayer and P. G. Tardi, Use of nanoscale delivery systems to maintain synergistic drug ratios in vivo, *Expert Opin. Drug Deliv.* 7(12), 1329–1341 (2010).
251. D. C. Drummond, C. O. Noble, Z. Guo, K. Hong, J. W. Park and D. B. Kirpotin, Development of a highly active nanoliposomal irinotecan using a novel intraliposomal stabilization strategy, *Cancer Res.* **66**(6), 3271–3277 (2006).
252. C. F. McDonagh, A. Huhalov, B. D. Harms, S. Adams, V. Paragas, S. Oyama, B. Zhang, L. Luus, R. Overland, S. Nguyen, J. Gu, N. Kohli, M. Wallace, M. J. Feldhaus, A. J. Kudla, B. Schoeberl and U. B. Nielsen, Antitumor activity of a novel bispecific antibody that targets the ErbB2/ErbB3 oncogenic unit and inhibits heregulin-induced activation of ErbB3, *Mol. Cancer Ther.* **11**(3), 582–593 (2012).

253. M. Fantini, L. Gianni, C. Santelmo, F. Drudi, C. Castellani, A. Affatato, M. Nicolini and A. Ravaioli, Lipoplatin treatment in lung and breast cancer, *Chemother Res. Pract.* **2011,** 125–192 (2011).
254. G. P. Stathopoulos, D. Antoniou, J. Dimitroulis, J. Stathopoulos, K. Marosis and P. Michalopoulou, Comparison of liposomal cisplatin versus cisplatin in non-squamous cell non-small-cell lung cancer, *Cancer Chemother. Pharmacol.* **68**(4), 945–950 (2011).
255. R. Staruch, R. Chopra and K. Hynynen, Localised drug release using MRI-controlled focused ultrasound hyperthermia, *Int. J. Hyperthermia.* **27**(2), 156–171 (2011).
256. U. Fasol, A. Frost, M. Buchert, J. Arends, U. Fiedler, D. Scharr, J. Scheuenpflug and K. Mross, Vascular and pharmacokinetic effects of EndoTAG-1 in patients with advanced cancer and liver metastasis, *Ann. Oncol.* **23**(4), 1030–1036 (2012).
257. G. Schulz, D. A. Cheresh, N. M. Varki, A. Yu, L. K. Staffileno and R. A. Reisfeld, Detection of ganglioside GD2 in tumor tissues and sera of neuroblastoma patients, *Cancer Res.* **44**(12 Pt 1), 5914–5920 (1984).
258. F. Pastorino, C. Brignole, D. Marimpietri, G. Pagnan, A. Morando, D. Ribatti, S. C. Semple, C. Gambini, T. M. Allen and M. Ponzoni, Targeted liposomal c-myc antisense oligodeoxynucleotides induce apoptosis and inhibit tumor growth and metastases in human melanoma models, *Clin. Cancer Res.* **9**(12), 4595–4605 (2003b).
259. J. I. Hare, E. H. Moase and T. M. Allen, Targeting combinations of liposomal drugs to both tumor vasculature cells and tumor cells for the treatment of HER2-positive breast cancer, *J. Drug Target.* **1**(1), 87–96 (2013).
260. A. S. Abu Lila, S. Kizuki, Y. Doi, T. Suzuki, T. Ishida and H. Kiwada, Oxaliplatin encapsulated in PEG-coated cationic liposomes induces significant tumor growth suppression via a dual-targeting approach in a murine solid tumor model, *J. Control. Release* **137**(1), 814 (2009).
261. A. S. Abu Lila, Y. Doi, K. Nakamura, T. Ishida and H. Kiwada, Sequential administration with oxaliplatin-containing PEG-coated cationic liposomes promotes a significant delivery of subsequent dose into murine solid tumor, *J. Control. Release* **142**(2), 167–173 (2010).
262. M. Owais and C. M. Gupta, Targeted drug delivery to macrophages in parasitic infections, *Curr. Drug Deliv.* **2**(4), 311–318 (2005).
263. V. P. Torchilin, Recent advances with liposomes as pharmaceutical carriers, *Nat. Rev. Drug Discov.* **4**(2), 145–160 (2005).
264. G. Bodey, B. Bueltmann, W. Duguid, D. Gibbs, H. Hanak, M. Hotchi, G. Mall, P. Martino, F. Meunier and S. Milliken *et al.*, Fungal infections in cancer patients: an international autopsy survey, *Eur. J. Clin. Microbiol. Infect. Dis.* **11**(2), 99–109 (1992).
265. F. Frezard, M. S. Michalick, C. F. Soares and C. Demicheli, Novel methods for the encapsulation of meglumine antimoniate into liposomes, *Braz. J. Med. Biol. Res.* **33**(7), 841–846 (2000).
266. M. M. Gaspar, S. Neves, F. Portaels, J. Pedrosa, M. T. Silva and M. E. Cruz, Therapeutic efficacy of liposomal rifabutin in a Mycobacterium avium model of infection, *Antimicrob. Agents Chemother.* **44**(9), 2424–2430 (2000).
267. B. Dupont, Overview of the lipid formulations of amphotericin B, *J. Antimicrob. Chemother.* **49** (Suppl 1), 31–36 (2002).

268. P. Goyal, K. Goyal, S. G. Vijaya Kumar, A. Singh, O. P. Katare and D. N. Mishra, Liposomal drug delivery systems–clinical applications, *Acta Pharm.* **55**(1), 1–25 (2005).
269. J. Tollemar, K. Hockerstedt, B. G. Ericzon, H. Jalanko and O. Ringden, Liposomal amphotericin B prevents invasive fungal infections in liver transplant recipients. A randomized, placebo-controlled study, *Transplantation* **59**(1), 45–50 (1995).
270. V. Heinemann, B. Kahny, A. Debus, K. Wachholz and U. Jehn, Pharmacokinetics of liposomal amphotericin B (AmBisome) versus other lipid-based formulations, *Bone Marrow Transplant.* **14** Suppl 5, S8–S9 (1994).
271. R. Russo, L. C. Nigro, S. Minniti, A. Montineri, L. Gradoni, L. Caldeira and R. N. Davidson, Visceral leishmaniasis in HIV infected patients: treatment with high dose liposomal amphotericin B (AmBisome), *J. Infect.* **32**(2), 133–137 (1996).
272. J. Adler-Moore and R. T. Proffitt, AmBisome: liposomal formulation, structure, mechanism of action and pre-clinical experience, *J. Antimicrob. Chemother.* **49** (Suppl 1), 2130 (2002).
273. W. W. Barrow, Treatment of mycobacterial infections, *Rev. Sci. Tech.* **20**(1), 55–70 (2001).
274. G. Gregoriadis, Overview of liposomes, *J. Antimicrob. Chemother.* **28** (Suppl B), 39–48 (1991).
275. A. Agarwal, H. Kandpal, H. P. Gupta, N. B. Singh and C. M. Gupta, Tuftsin-bearing liposomes as rifampin vehicles in treatment of tuberculosis in mice, *Antimicrob. Agents Chemother.* **38**(3), 588–593 (1994).
276. O. O. Okusanya, S. M. Bhavnani, J. Hammel, P. Minic, L. J. Dupont, A. Forrest, G. J. Mulder, C. Mackinson, P. G. Ambrose and R. Gupta, Pharmacokinetic and pharmacodynamic evaluation of liposomal amikacin for inhalation in cystic fibrosis patients with chronic pseudomonal infection, *Antimicrob. Agents Chemother.* **53**(9), 3847–3854 (2009).
277. Z. Drulis-Kawa, and A. Dorotkiewicz-Jach, Liposomes as delivery systems for antibiotics, *Int. J. Pharm.* **387**(1–2), 187–198 (2010).
278. J. Sinha, S. Mukhopadhyay, N. Das and M. K. Basu, Targeting of liposomal andrographolide to L. donovani-infected macrophages in vivo, *Drug Deliv.* 7(4), 209–213 (2000).
279. S. Mukherjee, L. Das, L. Kole, S. Karmakar, N. Datta and P. K. Das, Targeting of parasite-specific immunoliposome-encapsulated doxorubicin in the treatment of experimental visceral leishmaniasis, *J. Infect. Dis.* **189**(6), 1024–1034 (2004).
280. S. H. Chen and G. Zhaori, Potential clinical applications of siRNA technique: benefits and limitations, *Eur. J. Clin. Invest.* **41**(2), 221–232 (2011).
281. M. T. da Cruz, S. Simoes, P. P. Pires, S. Nir and M. C. de Lima, Kinetic analysis of the initial steps involved in lipoplex--cell interactions: effect of various factors that influence transfection activity, *Biochim. Biophys. Acta* **1510**(1–2), 136–151 (2001).
282. H. Ishiwata, N. Suzuki, S. Ando, H. Kikuchi and T. Kitagawa, Characteristics and biodistribution of cationic liposomes and their DNA complexes, *J. Control. Release* **69**(1), 139–148 (2000).
283. D. D. Stuart, G. Y. Kao and T. M. Allen, A novel, long-circulating, and functional liposomal formulation of antisense oligodeoxynucleotides targeted against MDR1, *Cancer Gene Ther.* 7(3), 466–475 (2000).
284. T. S. Zimmermann, A. C. Lee, A. Akinc, B. Bramlage, D. Bumcrot, M. N. Fedoruk, J. Harborth, J. A. Heyes, L. B. Jeffs, M. John, A. D. Judge, K. Lam, K. McClintock, L. V.

Nechev, L. R. Palmer, T. Racie, I. Rohl, S. Seiffert, S. Shanmugam, V. Sood, J. Soutschek, I. Toudjarska, A. J. Wheat, E. Yaworski, W. Zedalis, V. Koteliansky, M. Manoharan, H. P. Vornlocher and I. MacLachlan, RNAi-mediated gene silencing in non-human primates, *Nature* **441**(7089), 111–114 (2006).

285. G. Basha, T. I. Novobrantseva, N. Rosin, Y. Y. Tam, I. M. Hafez, M. K. Wong, T. Sugo, V. M. Ruda, J. Qin, B. Klebanov, M. Ciufolini, A. Akinc, Y. K. Tam, M. J. Hope and P. R. Cullis, Influence of cationic lipid composition on gene silencing properties of lipid nanoparticle formulations of siRNA in antigen-presenting cells, *Mol. Ther.* **19**(12), 2186–2200 (2011).
286. Y. Chen, X. Zhu, X. Zhang, B. Liu and L. Huang, Nanoparticles modified with tumor-targeting scFv deliver siRNA and miRNA for cancer therapy, *Mol. Ther.* **18**(9), 1650–1656 (2010c).
287. S. Ramishetti and L. Huang, Intelligent design of multifunctional lipid-coated nanoparticle platforms for cancer therapy, *Ther. Deliv.* **3**(12), 1429–1445 (2012).
288. G. Pagnan, D. D. Stuart, F. Pastorino, L. Raffaghello, P. G. Montaldo, T. M. Allen, B. Calabretta and M. Ponzoni, Delivery of c-myb antisense oligodeoxynucleotides to human neuroblastoma cells via disialoganglioside GD(2)-targeted immunoliposomes: antitumor effects, *J. Natl. Cancer Inst.* **92**(3), 253–261 (2000).
289. C. Brignole, G. Pagnan, D. Marimpietri, E. Cosimo, T. M. Allen, M. Ponzoni and F. Pastorino, Targeted delivery system for antisense oligonucleotides: a novel experimental strategy for neuroblastoma treatment *Cancer Lett.* **197**(1–2), 231–235 (2003).
290. U. N. Verma, R. M. Surabhi, A. Schmaltieg, C. Becerra and R. B. Gaynor, Small interfering RNAs directed against beta-catenin inhibit the in vitro and in vivo growth of colon cancer cells, *Clin. Cancer Res.* **9**(4), 1291–1300 (2003).
291. A. S. Arnold, Y. L. Tang, K. Qian, L. Shen, V. Valencia, M. I. Phillips and Y. C. Zhang, Specific beta1-adrenergic receptor silencing with small interfering RNA lowers high blood pressure and improves cardiac function in myocardial ischemia, *J. Hypertens.* **25**(1), 197–205 (2007).
292. A. K. Leung, I. M. Hafez, S. Baoukina, N. M. Belliveau, I. V. Zhigaltsev, E. Afshinmanesh, D. P. Tieleman, C. L. Hansen, M. J. Hope and P. R. Cullis, Lipid Nanoparticles Containing siRNA Synthesized by Microfluidic Mixing Exhibit an Electron-Dense Nanostructured Core, *J. Phys. Chem. C Nanomater Interfaces* **116**(34), 18440–18450 (2012).
293. M. Saad, O. B. Garbuzenko and T. Minko, Co-delivery of siRNA and an anticancer drug for treatment of multidrug-resistant cancer, *Nanomedicine (Lond)* **3**(6), 761–776 (2008).
294. Y. Chen, S. R. Bathula, J. Li and L. Huang, Multifunctional nanoparticles delivering small interfering RNA and doxorubicin overcome drug resistance in cancer, *J. Biol. Chem.* **285**(29), 22639–22650 (2010a).
295. Y. Chen, J. J. Wu and L. Huang, Nanoparticles targeted with NGR motif deliver c-myc siRNA and doxorubicin for anticancer therapy, *Mol. Ther.* **18**(4), 828–834 (2010b).
296. R. S. Chang, M. S. Suh, S. Kim, G. Shim, S. Lee, S. S. Han, K. E. Lee, H. Jeon, H. G. Choi, Y. Choi, C. W. Kim and Y. K. Oh, Cationic drug-derived nanoparticles for multifunctional delivery of anticancer siRNA, *Biomaterials* **32**(36), 9785–9795 (2011).
297. Y. H. Yu, E. Kim, D. E. Park, G. Shim, S. Lee, Y. B. Kim, C. W. Kim and Y. K. Oh, Cationic solid lipid nanoparticles for co-delivery of paclitaxel and siRNA, *Eur. J. Pharm. Biopharm.* **80**(2), 268–273 (2012).

298. S. H. Kang, H. J. Cho, G. Shim, S. Lee, S. H. Kim, H. G. Choi, C. W. Kim and Y. K. Oh, Cationic liposomal co-delivery of small interfering RNA and a MEK inhibitor for enhanced anticancer efficacy, *Pharm. Res.* **28**(12), 3069–3078 (2011).
299. A. G. Allison and G. Gregoriadis, Liposomes as immunological adjuvants, *Nature* **252**(5480), 252 (1974).
300. G. Gregoriadis and A. C. Allison, Entrapment of proteins in liposomes prevents allergic reactions in pre-immunised mice, *FEBS Lett.* **45**(1), 71–74 (1974).
301. G. Gregoriadis, Liposomes as immunoadjuvants and vaccine carriers: antigen entrapment, *Immunomethods* **4**(3), 210–216 (1994).
302. C. R. Alving and M. Rao, Lipid A and liposomes containing lipid A as antigens and adjuvants, *Vaccine* **26**(24), 3036–3045 (2008).
303. M. Henriksen-Lacey, K. S. Korsholm, P. Andersen, Y. Perrie and D. Christensen, Liposomal vaccine delivery systems, *Expert Opin. Drug Deliv.* **8**(4), 505–519 (2011).
304. V. Usonis, V. Bakasenas, R. Valentelis, G. Katiliene, D. Vidzeniene and C. Herzog, Antibody titres after primary and booster vaccination of infants and young children with a virosomal hepatitis A vaccine (Epaxal), *Vaccine* **21**(31), 4588–4592 (2003).
305. P. A. Bovier, Epaxal: a virosomal vaccine to prevent hepatitis A infection, *Expert Rev. Vaccines* **7**(8), 1141–1150 (2008).
306. C. Herzog, K. Hartmann, V. Kunzi, O. Kursteiner, R. Mischler, H. Lazar and R. Gluck, Eleven years of Inflexal V-a virosomal adjuvanted influenza vaccine, *Vaccine* **27**(33), 4381–4387 (2009).
307. G. Raposo, D. Tenza, S. Mecheri, R. Peronet, C. Bonnerot and C. Desaymard, Accumulation of major histocompatibility complex class II molecules in mast cell secretory granules and their release upon degranulation, *Mol. Biol. Cell.* **8**(12), 2631–2645 (1997).
308. J. Y. Legendre and F. C. Szoka Jr., Delivery of plasmid DNA into mammalian cell lines using pH-sensitive liposomes: comparison with cationic liposomes, *Pharm. Res.* **9**(10), 1235–1242 (1992).
309. S. Li, S. P. Wu, M. Whitmore, E. J. Loeffert, L. Wang, S. C. Watkins, B. R. Pitt and L. Huang, Effect of immune response on gene transfer to the lung via systemic administration of cationic lipidic vectors, *Am. J. Physiol.* **276**(5 Pt 1), L796–L804 (1999).
310. M. Whitmore, S. Li and L. Huang, LPD lipopolyplex initiates a potent cytokine response and inhibits tumor growth, *Gene Ther.* **6**(11), 1867–1875 (1999).
311. W. Yan, W. Chen and L. Huang, Reactive oxygen species play a central role in the activity of cationic liposome based cancer vaccine, *J. Control. Release* **130**(1), 22–28 (2008).
312. E. A. Vasievich, W. Chen and L. Huang, Enantiospecific adjuvant activity of cationic lipid DOTAP in cancer vaccine, *Cancer Immunol. Immunother.* **60**(5), 629–638 (2011).
313. G. Chikh and M. P. Schutze-Redelmeier, Liposomal delivery of CTL epitopes to dendritic cells, *Biosci. Rep.* **22**(2), 339–353 (2002).
314. M. J. Copland, M. A. Baird, T. Rades, J. L. McKenzie, B. Becker, F. Reck, P. C. Tyler and N. M. Davies, Liposomal delivery of antigen to human dendritic cells, *Vaccine* **21**(9–10), 883–890 (2003).
315. B. Ludewig, F. Barchiesi, M. Pericin, R. M. Zinkernagel, H. Hengartner and R. A. Schwendener, In vivo antigen loading and activation of dendritic cells via a liposomal peptide vaccine mediates protective antiviral and anti-tumour immunity, *Vaccine* **19**(1), 23–32 (2000).

316. C. Walther, K. Meyer, R. Rennert and I. Neundorf, Quantum dot-carrier peptide conjugates suitable for imaging and delivery applications, *Bioconjug. Chem.* **19**(12), 2346–2356 (2008).
317. K. C. Weng C. O. Noble, B. Papahadjopoulos-Sternberg, F. F. Chen, D. C. Drummond, D. B. Kirpotin, D. Wang, Y. K. Hom, B. Hann and J. W. Park, Targeted tumor cell internalization and imaging of multifunctional quantum dot-conjugated immunoliposomes in vitro and in vivo, *Nano Lett.* **8**(9), 2851–2857 (2008).
318. M. Liong, J. Lu, M. Kovochich, T. Xia, S. G. Ruehm, A. E. Nel, F. Tamanoi and J. I. Zink, Multifunctional inorganic nanoparticles for imaging, targeting, and drug delivery, *ACS Nano.* **2**(5), 889–896 (2008).
319. C. Sun, J. S. Lee and M. Zhang, Magnetic nanoparticles in MR imaging and drug delivery, *Adv. Drug Deliv. Rev.* **60**(11), 1252–1265 (2008).
320. O. Veiseh, J. W. Gunn and M. Zhang, Design and fabrication of magnetic nanoparticles for targeted drug delivery and imaging, *Adv. Drug Deliv. Rev.* **62**(3), 284–304 (2010).
321. T. Urakami, S. Akai, Y. Katayama, N. Harada, H. Tsukada and N. Oku, Novel amphiphilic probes for [18F]-radiolabeling preformed liposomes and determination of liposomal trafficking by positron emission tomography, *J. Med. Chem.* **50**(26), 6454–6457 (2007).
322. M. M. Herth, M. Barz, D. Moderegger, M. Allmeroth, M. Jahn, O. Thews, R. Zentel and F. Rosch, Radioactive labeling of defined HPMA-based polymeric structures using [18F]FETos for in vivo imaging by positron emission tomography, *Biomacromolecules* **10**(7), 1697–1703 (2009).
323. D. Danila, R. Partha, D. B. Elrod, M. Lackey, S. W. Casscells and J. L. Conyers, Antibody-labeled liposomes for CT imaging of atherosclerotic plaques: in vitro investigation of an anti-ICAM antibody-labeled liposome containing iohexol for molecular imaging of atherosclerotic plaques via computed tomography, *Tex. Heart Inst. J.* **36**(5), 393–403 (2009).
324. J. Zheng, G. Perkins, A. Kirilova, C. Allen and D. A. Jaffray, Multimodal contrast agent for combined computed tomography and magnetic resonance imaging applications, *Invest. Radiol.* **41**(3), 339–348 (2006).
325. J. Zheng, J. Liu, M. Dunne, D. A. Jaffray and C. Allen, In vivo performance of a liposomal vascular contrast agent for CT and MR-based image guidance applications. *Pharm. Res.* **24**(6), 1193–1201 (2007).
326. A. J. Hamilton, S. L. Huang, D. Warnick, M. Rabbat, B. Kane, A. Nagaraj, M. Klegerman and D. D. McPherson, Intravascular ultrasound molecular imaging of atheroma components in vivo, *J. Am. Coll. Cardiol.* **43**(3), 453–460 (2004).
327. T. L. Hwang, Y. K. Lin, C. H. Chi, T. H. Huang and J. Y. Fang, Development and evaluation of perfluorocarbon nanobubbles for apomorphine delivery, *J. Pharm. Sci.* **98**(10), 3735–3747 (2009).
328. C. J. Wen, L. W. Zhang, S. A. Al-Suwayeh, T. C. Yen and J. Y. Fang, Theranostic liposomes loaded with quantum dots and apomorphine for brain targeting and bioimaging, *Int. J. Nanomedicine* 7, 1599–1611 (2012).
329. G. Mikhaylov, U. Mikac, A. A. Magaeva, V. I. Itin, E. P. Naiden, I. Psakhye, L. Babes, T. Reinheckel, C. Peters, R. Zeiser, M. Bogyo, V. Turk, S. G. Psakhye, B. Turk and O. Vasiljeva, Ferri-liposomes as an MRI-visible drug-delivery system for targeting tumours and their microenvironment, *Nat. Nanotechnol.* **6**(9), 594–602 (2011).

Chapter 2

Solid Lipid Nanoparticles for Biomedical Applications

Karsten Mäder

Institute of Pharmacy, Martin-Luther-University Halle-Wittenberg
Wolfgang-Langenbeck-Strasse 4, D-01620 Halle(Saale), Germany
maeder@pharmazie.uni-halle.de

Abstract

Solid lipid nanoparticles (SLNs) and the related nanostructured lipid carriers (NLCs) compete with other nanoscaled drug delivery systems (DDSs). This chapter will discuss, how nanoparticles are manufactured, characterized and how they perform *in vitro* and *in vivo*.

Although the manufacturing process is often quite simple, SLN and NLC are complex and often heterogeneous systems. Temperature, storage, dilution and other factors might impact the physical state of the lipid matrix (e.g. supercooled melts, liquid crystalline, solid in different modifications), the size, viscosity and the drug location and release pattern heavily. Unfortunately, many publications do not reflect these issues and oversimplify SLN as drug carriers. This chapter discusses the general concept of SLN and pays attention to suitable materials and manufacturing processes. Necessary and appropriate analytical characterization techniques are conferred. The performance of SLN and NLC with respect to different applications routes and the resulting delivering performance and biofate will be discussed (oral, parenteral, dermal and ocular). The current state and the future of the clinical use are summarized at the end of the chapter.

1. Introduction

Drug delivery science aims the improvement and control of the drug transport and release to the desired target tissue. A reproducible performance of the drug delivery system (DDS) with the control of location and the kinetics drug release is required to obtain a safe and reliable optimization of drug therapy. Many diverse systems with different sizes, materials and principles of drug release are used clinically or are in the preclinical phase. The size of DDS might range from few nanometers (e.g. micelles) over micrometers (microparticles) to the mm–cm scale (implants, pellets, tablets). During the last years, special focus was spent on nanoscaled systems, which in former years were classified as "colloidal drug carriers". In the field of submicron sized DDS, solid lipid nanoparticles (SLNs) compete with liposomes, polymeric nanoparticles, nanocapsules, "nanostructured lipid carriers" (NLCs), nanodispersed liquid crystals (e.g. cubosomes), micelles, lipid nano-emulsions, drug nanosuspensions, carbon nanotubes, inorganic nanoparticles, amorphous systems, polymer conjugates and other structures and materials.

Pioneering work on the SLN field has been done by the groups of Speiser[1] in Switzerland, Gasco[2] in Italy, Domb[3,4] in Israel and the groups of Westesen[5,6] and Müller[7] in Germany. The first papers and patents date back to the mid-80s and early 1990s. In the following years, more and more groups joined the field and the number of publications increased almost exponentially. The importance of SLN is also reflected by the fact that one SLN review is within the top-ten papers of the most cited papers in the journal of Advanced Drug Delivery Reviews.[8]

The original concept of SLN was derived from the advantages of lipid nano-emulsions. Submicron emulsions (nano-emulsions) are in clinical use for decades for parenteral nutrition and also as drug carriers (e.g. Etomidate-Lipuro®, Diazepam-Lipuro®, Diprivan®). They are manufactured by high pressure homogenization (HPH), a process which is scalable, cheap and fast. Nano-emulsions are typically composed of liquid triglycerides and contain lecithin as stabilizer. However, they are not suitable for controlled release due to liquid nature of the droplets. The general SLN concept was to keep the advantages of nano-emulsions (scalable and fast manufacturing process and the use of safe and low cost materials) and to introduce the feature of controlled release by the use of lipids, which are solid at room and body temperature. SLNs have been developed as "frozen nano-emulsions" to aim controlled drug delivery and to increase the chemical stability of incorporated drugs.

It was claimed that SLN combine the advantages and avoid the disadvantages of other colloidal carriers. The proposed features of SLN include:

- Possibility of controlled drug release and drug targeting
- Increased drug stability
- High drug payload
- Incorporation of lipophilic and hydrophilic drugs feasible
- No biotoxicity of the carrier
- Avoidance of organic solvents
- No problems with respect to large scale production and sterilization

The proposed advantages have been copied and claimed in many publications. The question arises; do we need to investigate or to develop other carrier systems if SLN provide all advantages? Should we stop research and clinical evaluation of liposomes, nanoparticles, nanocapsules and other carriers? What is the status of SLN after more than 20 years of research? What are the reasons to propose follow up generations of SLN and NLC? There is no simple answer to these questions. One of the main points to consider is that SLN in reality are much more complex systems than many people assume. This statement will be discussed in the subchapter "SLN as complex nanodispersions". As a preparation for this discussion, aspects of the material selection, the manufacturing process, the necessity and challenges of SLN characterization, the application routes and the biofate of the drug and the carrier will be discussed.

2. Excipients for the Production of SLN

The selection of the matrix lipid is of crucial importance for the characteristics and the performance of the final SLN dispersion. The term "lipid" includes quite different materials. The most important aspects include:

- Melting point/melting range
- Degree of crystalinity
- Presence of polymorphs
- Biodegradability
- Degree of polarity/hydrophobicity; interactions with water
- Heterogeneity of chemical composition
- Presence of impurities, degradation products
- Mobility of aliphatic chains in aqueous environment at 37°C

In general, the use of lipids with a defined chain length of the aliphatic chain (e.g. Tristearine, Tripalmitine, Cetylpalmitate) and a high purity leads to rigid materials with a narrow melting point. In contrast, the use of heterogenous fatty acids (e.g. mixtures of palmitic acid, stearic acid, oleic acid) and different substitution degrees (e.g. mono-, di- and triglycerides) leads to less perfect crystals, softer materials and a melting range instead of a melting point. Not only the ratio of the fatty acids, but also the pattern of substitution is important. For example, cacao butter contains oleic acid mainly in the second position of the glycerol backbone. Transesterification of the fatty acids leads to a lipid mixture with a more random substitution pattern (e.g. oleic acid is localized also on the first and third position. This material containing the same fatty acids and the same fatty acid/glycerol ratio is much softer and melts over a broad range instead of a small temperature range.

The majority of the used materials are mixtures of several compounds which differ in their physicochemical properties (e.g. the melting) point, which often leads to segregation within the bottle or vial. Therefore, careful homogenization of the material is advised. For many of the used materials monographs in the pharmacopeia do exist. However, the specifications are not sufficient to guarantee a comparable performance as SLN matrix. Examples of commonly used excipients for the lipid matrix are given in Table 1.

Because crystalline lipids show by their nature as crystals poor drug incorporation capacities, liquid lipids have been added. These compositions have been named as "NLC". NLC somehow are hybrid systems between nano-emulsions and SLNs. The goal was — compared to pure SLN — to achieve higher drug loadings (by drug dissolution in the oil) and — compared to nano-emulsions — to decrease the release rate.

In addition to the lipid matrix, surface active substances (emulsifiers) have to be used in most cases to stabilize the water-lipid interface. The selection is made with respect to the administration route. The degree of free selection decreases from oral over dermal to parenteral administration. Anionic, amphoteric, cationic and non-ionic emulsifiers of different molecular weight have been used to stabilize the SLN dispersions. Many publications and patents describe solid lipid nanodispersions which are stabilized by Poloxamers, Polysorbates, PEG-hydroxystearate, lecithin and bile acids. In many cases the combination of emulsifiers (e.g. a fast low molecular surfactant and slower polymeric surfactant) was more efficient to achieve and to preserve small particle sizes compared to the use of a single surfactant.

Table 1. Examples of commonly used excipients for the manufacture of SLN.

Lipid class	Substance	Melting point/range	Literature
Synthetic Monoacid Triglycerides	Tricaprin	~33*	9,10
	Trilaurin (Dynasan 112)	~45**	9–13
	Trimyristin (Dynasan 114)	55–58**	9,11,13–19
	Tripalmitin (Dynasan 116)	61–65	9,11,12,20–26
	Tristearin (Dynasan 118)	70–74	9,17–19,27–31
Hard fats and glyceride mixtures	Witepsol W35	33–35*	34
	Witepsol H35	33–35*	35,36
	Witepsol H42	41–43	36
	Witepsol E85	42–44**	35–37
	Softisan 100	~35*	37,38
	Softisan 142	~43**	39
	Softisan 154	55–60	40–43
	Glyceryl behenate (*Compritol 888 ATO1)	~70	41–45
	Glyceryl palmitostearate (**Precirol ATO 52)	52–55	46–48
Glyceride-monoesters	Glyceridemonostearate (Imwitor 900)	~61	49,50
Waxes	Cetylpalmitate	43–53	50–52
	Emulsifying wax (cetylstearylalcolhol and polysorbate 60 4:1 w/w)	50–55°C	52–54
	Carnauba wax	78–85	54–56
Fatty acids	Stearic acid	70	12,56,57
	Behenic acid	74–78	58
Others	Cholesterol	148–150	59,60
	Brij 72 (polyoxyl(2)stearyl ether)	43	61,62

* Compritol 888 ATO is a mixture of approximately 15% mono-, 50% di- and 35% triglycerides of behenic acid.

** Precirol ATO 5 is a mixture of mono-, di- and triglycerides of C16 and C18 fatty acids.

3. Methods for SLN Manufacturing and Sterilization

Several methods and combinations of them have been proposed for the manufacture of SLN. In general, two different approaches are possible to achieve the desired small particle size of the solid lipid dispersion. The first family of approaches (top–down processes) involves the downsizing of the dispersion by means of commonly used dispersion techniques (e.g. High Pressure Homogenization, Ultrasound, high shear rotor-stator techniques). Because the desired particle size is in the nanometer range, rather high shear forces have to be used. In most cases, the dispersion steps are made with molten lipids at higher temperatures because the downsizing of solid particles is much harder compared to the liquid state. A typical flow chart is shown in Fig. 1.

The second family of approaches is based on bottom up techniques. In most cases, they require not high shear forces. The SLN dispersion is formed from a solution. For example, the lipid is solubilized in a hydrophilic organic solvent, in a microemulsion or supercritical carbon dioxide. By changing the environment (e.g. by adding the aqueous phase, decrease of the temperature or pressure change) the solubilization capacity is decreased and the SLNs are formed. These methods require no or only moderate shear stress. However, this advantage is often counterbalanced by several disadvantages (e.g. lower SLN concentrations which can be achieved, requirement of purification steps, increased processing times).

The most important methods for SLN manufacturing will be discussed in more detail below.

High Shear Homogenization (HPH, High Speed Rotor Stator)

HPH is clearly the most frequently applied technique to produce SLN. The flow chart of the general process is shown in Fig. 1. In most cases, the emulsification and homogenization is conducted at temperatures which are

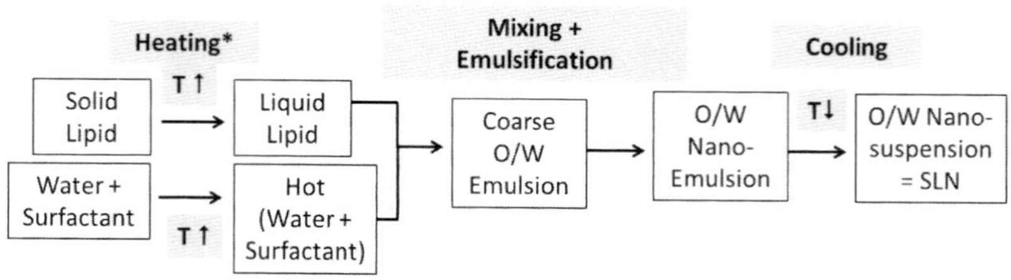

Fig. 1. Typical flow chart for top–down approaches for the production of SLN. The downsizing step is achieved by high energy dispersion techniques, in most cases by HPH.* The melting and heating step are in most cases necessary to achieve the desired submicron size.

5–10°C above the melting temperature of the lipid phase ("Hot Homogenization" or "Melt Homogenization"). The temperature is selected to facilitate an effective emulsification and to avoid (or minimize) thermal degradation. It should also kept in mind that the hydrophilicity–lipophilicity balance (HLB) of many surfactants is temperature dependent (especially PEG-based structures), which might impact the resulting particle size heavily. In most cases, a direct emulsification by HPH of the lipid water mixture is not possible. At least one preparatory emulsification step by "classical" rotor-stator machines is necessary. Typical particle sizes of the "coarse" emulsion obtained in this step are in the lower micrometer range with often bi- or multimodal size distributions. The quality of coarse emulsion greatly affects the results of the following HPH. Therefore, droplet sizes in the upper micrometer range should be avoided. The HPH leads to a further decrease of the particle sizes. In most cases, several cycles (3–5) with moderate pressures (300–1000 bar) are used. Commonly, the pressure is increased between the first and the last cycles. Higher homogenization pressures are not desired for reasons of time and energy, and might even lead to increased particle sizes. Different homogenizers are commercially available from small to large scale. Also different geometries of the homogenization chambers do exist. HPH is an established technique which has been used for decades for the production of nano-emulsions for parenteral nutrition. The mean emulsification principle is disruption due to very high shear forces. In addition, cavitation forces might contribute. A strong control of the inlet and outlet temperature is necessary to achieve reproducible results, because during HPH, a significant increase of product temperature occurs. In many cases, continuous homogenization in several cycles is performed with an external tempering of the sample. Care has to be taken, that all parts of the sample have the same number of passages through the homogenizer. Otherwise, multimodal particle size will be obtained. The standard HPH process involves emulsification of the molten lipid emulsion at higher temperatures. Therefore, a nano-emulsion and not a nanosuspension are obtained after the emulsification process (Fig. 1). Upon cooling to room temperature (RT), the liquid lipid is expected to solidify and therefore, the nano-emulsion (lipid in the molten, mobile state) transforms into a solid lipid nanosuspension. As it will be discussed later, the crystallization process might be complex and long lasting. It might lead to different polymorphic states of the lipid with impact on drug location, release rate, stability and viscosity. In addition, crystallization might not take place despite the storage temperature is significant below the melting temperature of the lipid.[6,10,19,63] Under these circumstances, a nanodispersion of a supercooled lipid melt is obtained with a liquid state of the lipid. These systems are not SLN dispersions, but rather nano-emulsions. Storage at lower temperatures[6,10,19] or isostatic high pressure treatment[63] forces the transformation from the liquid to the crystalline state.

Potential disadvantages of the melt homogenization technique include: (1) Temperature induced drug degradation, (2) Relocation of the drug into the aqueous phase during homogenization and (3) the complexity of the crystallization step. As an alternative the "Cold Homogenization" process at temperatures below the melting range of the lipids has been developed already in the early years of the SLN.[59,60] Cold homogenization has been used scarcely since this time, because the particle sizes are usually larger and often polydisperse suspensions of micro- and nanoparticles are obtained. Although the homogenization is processed at low temperatures (often between 4°C and RT), a preparatory melting step of the drug and the bulk lipid is necessary in order to produce drug loaded nanoparticles. Therefore, thermal stress is not avoided totally and also the crystallization of the drug loaded lipid bulk might be a tedious and complex process. It must be kept in mind that the rapid cooling of the lipid melt in the first step favors metastable lipid modifications (with higher drug loading capacity) which might transform with time in more stable polymorphs (with expulsion of the incorporated drug). Slow cooling, on the other hand, might lead to stable modifications, but to micro-segregation of the drug from the lipid.

The solidified bulky drug-lipid mixture is downsized at lower temperatures to the micrometer range by means of ball milling or other means. The lipid microparticles are then dispersed in an aqueous surfactant solution and subjected to HPH at low temperature. Temperatures slightly above 0°C are preferred because lower temperatures favor the brittleness of the lipid and the disruption of the particles. A stringent control of the outlet temperature (and cooling) is often required to compensate the HPH induced temperature increase. Compared to the melt homogenization, higher pressures are often required.

In the context of SLN formulations, rotor-stator based emulsification (e.g. by an Ultra-Turrax) is mainly applied as a preceding step prior to HPH. However, it has been shown that it is possible to produce SLN by means of rotor-stator emulsification only.[64,65] However, in general better results have been obtained by HPH and therefore, rotor-stator emulsification plays only a minor role as a nanosizing process. It is, however, very commonly used and has significant importance as a preparatory emulsification step for HPH.

Ultrasound Homogenization

Ultrasound sonication is a further dispersing method which can be used as an alternative or in combination to HPH and high shear mixing to achieve particle size distributions in the nanometer range.[66] Compared to SLN preparations made by HPH, ultrasound manufactured SLN often show larger particle sizes and/or multimodal size distributions.

Precipitation from microemulsions

The preparation of SLN from microemulsions was described already in the early years of the SLN by Gasco and coworkers.[2] It is a bottom up approach, which avoids the use of high shear energy.

The general flow chart of the processing is shown in Fig. 2.

There are quite different definitions and opinions about the structure and dynamics of microemulsion within the scientific community. An interesting and balanced review article has been published by Moulik and Paul.[67] Gasco and other scientists describe microemulsions as two-phase systems composed of an inner and outer phase (e.g. o/w-microemulsions). Other people regard microemulsions as "critical solutions". Microemulsions as precursor for SLN are typically made by stirring an optically transparent mixture at 65–70°C which is typically composed of a low melting long chain fatty acid (e.g. stearic acid), emulsifier(s) (e.g. Tween 20, Tween 60, lecithin, bile acids), co-emulsifiers (e.g. butanol, ethanol) and water. Please note that often a high amount of amphiphilic substances are required for the process. If the amount of non-polar lipids (e.g. long chain triglycerides) is too high, no microemulsion will be formed. The hot microemulsion is dispersed in cold water (2–3°C) under stirring. This leads to dilution of the microemulsion. Amphiphilic surfactants and co-surfactants leave (at least partially) the lipid and translocate into the water. This translocation of the amphiphilic molecules and the dramatic decrease of the temperature cause a solidification of the lipid.

Typical volume ratios of the hot microemulsion to cold water are in the range of 1:25 to 1:50. This dilution is necessary, which is one of the major

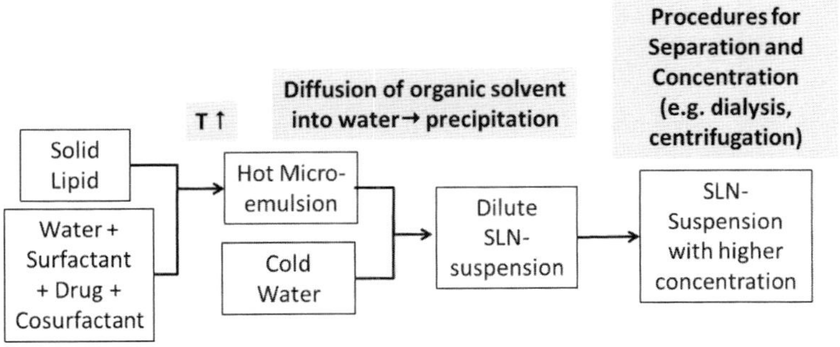

Fig. 2. Typical flow chart for the SLN production via the cold dilution of hot microemulsions.

drawbacks of this method. Lower dilution ratios end up in larger particles. The required dilution ratio is critically determined by the composition of the microemulsion. Because this translocation process is energetically favorable, no energy is required to achieve submicron particle sizes.

After the dilution step, further processing takes place to concentrate the SLN. Commonly used methods include centrifugation, dialysis and freeze drying. The composition of the SLN will change by the concentration processes, because the amphiphilic substances (surfactants) will redistribute. This aspect has not been investigated in detail so far. Therefore, there is a lack of knowledge about the real composition of the final sample, which is of course a crucial parameter to understand the quality of the SLN. In addition, many (and sometimes all) components used for the manufacture of SLN via hot microemulsions are of amphiphilic nature and known to form several liquid crystalline phases alone or in combination. A dispersed lamellar phase cannot be regarded as a SLN, because of the major differences between a classical solid crystals and a liquid crystal. Drug loading capacity will be increased for liquid crystalline phases and also the interaction with the biological environment will be different.

Therefore, serious efforts should be made to characterize carefully the final composition and the physicochemical state.

Solvent Diffusion

A further possibility for the production of SLN is the solvent diffusion method. In several aspects it is similar to the SLN production via a microemulsion dilution process. Both approaches represent bottom up techniques. In both procedures, the solubilized lipid is precipitated via a dilution step. The solvent diffusion technique is well-known for the manufacture of polymeric nanoparticles[68,69] and nanocapsules.[70]

The flow chart of the solvent diffusion method is shown in Fig. 3.

The organic solvent must be water-miscible to permit spontaneous emulsification. Because the organic solvent molecules spontaneously translocate

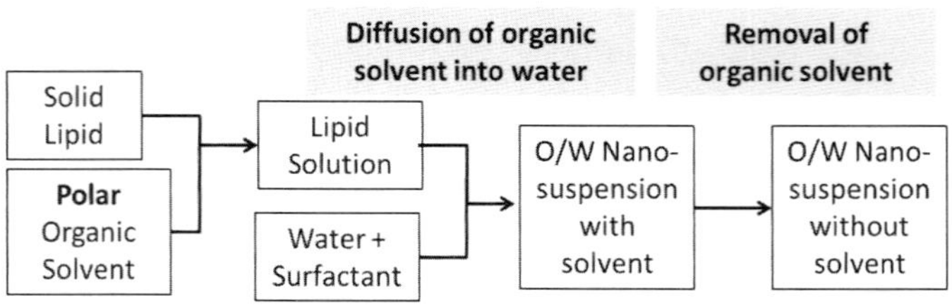

Fig. 3. Typical flow chart for the SLN production via the solvent diffusion method.

into the water phase, the solubilization capacity for the lipid is lost and the lipid precipitates. Less polar organic solvents will phase-separate. This requirement decreases the selection of the solvent and the lipid. Many non-polar lipids do not dissolve sufficiently in polar organic solvents. Therefore, long chain triglycerides are less or not suitable for this procedure. Monoglycerides like glycerolmonostearate and fatty acids are commonly used. In addition to the limited number of excipients with the required polarity/solubility profile, the necessary dilution and removal steps are further intrinsic disadvantages of the production method.

Examples of SLN produced via solvent diffusion are listed in Refs. 71–76. Typical solvents include acetone, ethanol, ethylacetate, methanol, isopropanol and glycerol. A detailed study about the requirements and optimization was published by the group of Müller-Goyman.[35]

Supercritical fluids

Supercritical fluids (e.g. carbon dioxide or water) have attracted increased attention in the pharmaceutical field. Carbon dioxide (CO_2) plays the most important role due to it physical properties, low costs and good availability. Supercritical fluid technology has been applied in several fields of pharmaceutics, including drug micro- and nanoparticles, plant extracts, polymeric micro- and nanoparticles. Over the time, supercritical fluid technology has been diversified to many different approaches, which include for example the rapid expansion of supercritical solutions (RESS), static supercritical fluid (SSF) process, particles from gas saturated solutions (PGSS), gas antisolvent process (GAS), supercritical antisolvent process (SAS), precipitation from compressed antisolvent (PCA), aerosol solvent extraction system (ASES), solution enhanced dispersion by supercritical fluids (SEDS), supercritical antisolvent process with enhanced mass transfer (SAS-EM), depressurization of an expanded liquid organic solution (DELOS), supercritical assisted atomization (SAA), hydrothermal synthesis under supercritical conditions via flow reactor (HTSSF), supercritical fluid extraction emulsions (SFEE).[72] Detailed studies show the impact of the CO_2 pressure and temperature on the melting behavior of lipid excipients.[76–79] Because CO_2 acts as a solvent for the lipid, a depression of the melting point is observed initially by increasing the CO_2 pressure. Typical values are in the range between 10°C and 20°C. However, at a certain point, higher pressures lead to a small increase of the melting point of the lipid (e.g. 1–3°C). The solubility's of lipids supercritical CO_2 are quite high and increase with the applied pressure. For example, more than 35% solubility of Precirol© were observed at a pressure of 20 MPa.[78,79] The solubility of Gelucire© was around 50% and of Compritol© around 30%.[75] Using different subtypes of supercritical CO_2-technologies, the successful production of SLN via supercritical CO_2 has been reported by several groups.[76–80] Particle sizes are often comparable, or even better than HPH techniques. Using a high pressure pre-homogenization step and a SFEE technology, very small particle sizes (20–80 nm) have been obtained.[77]

Solvent evaporation

Solvent evaporation of homogenized nano-emulsions represents another option for the manufacturing of SLN. The general procedure is shown in Fig. 4.

The main difference to the melt homogenization technique (using HPH or ultrasound) is the presence of a non-polar, water immiscible organic solvent. The organic solvent solubilizes the lipid and the drug. Therefore, melting is not necessary and higher temperatures can be avoided. It should, however, kept in mind, that HPH will lead to increased temperatures. The main disadvantages of the method are (1) the requirement for efficient removal of the organic solvent and (2) typical low lipid concentration. Besides physicochemical aspects, toxicological aspects are important, because often halogenated solvents are used (e.g. $CHCl_3$). The final particle size is determined by mainly two parameters: (i) the efficiency of the homogenization step and (2) the lipid/solvent ratio. Low lipid/solvent ratios favor smaller particle sizes, because more material will be removed in the evaporation step. However, it increases the costs of the whole process as evaporation requires more time and efforts.

The solvent evaporation method has been first described in the early 1990s by Sjöström and Bergenståhl.[81,82] In the following years, other groups published on this topic[83–90] and also multiple w/o/w emulsion techniques were developed to incorporate hydrophilic drugs.[90] In a special case, solvent evaporation has also been reported for SLN manufacture in waterfree oil/oil

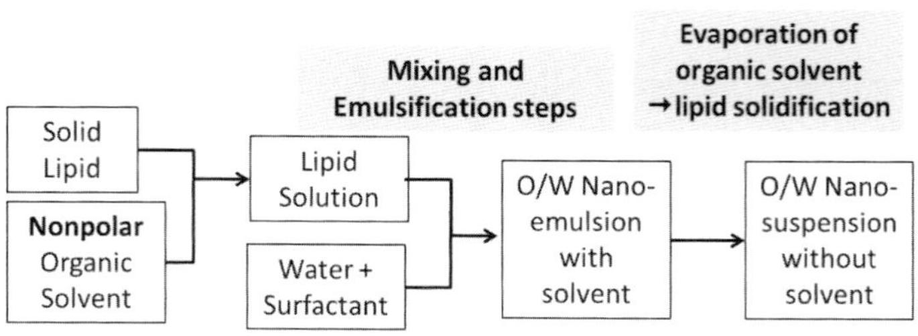

Fig. 4. Typical flow chart for the SLN production via the solvent evaporation method.

emulsion. SLN were obtained by (1) dissolving the lipid drug complex and polar lipids in hot ethanol, (2) dispersing the ethanolic solution in span containing paraffin by ultrasound and (3) evaporating ethanol from an homogenized ethanol/paraffin emulsion.[91]

4. The Necessity and Challenge of SLN Characterization

Several times nanodispersed systems have been claimed as "SLN", although the dispersed phase was clearly in the liquid state. Examples include Vitamin-E-TPGS/Vitamin-E-acetate 1:1 mixtures[92] and Gelucire 44/14 based formulations.[93,94] Although published in high ranked journals, Refs. 93 and 94 have misleading statements of the chemical and physical properties of the used materials (Gelucire is NOT Glycerolbehanate). Instead, Gelucire 44/14 is a mixture of amphiphilic lipids composed of fatty acid esters with PEG and glycerol. The number 44 indicates the melting point in the dry state, the 14 is the HLB value. Gelucire does, as many other lipid excipients do, show a complex phase behavior as a function of water content and temperature. The nice paper by Svensson *et al.* clearly shows that in Gelucire 44/14 water mixtures no solid phase exists at body temperature and high water concentrations.[95] Therefore, the description as "SLN" is totally misleading. Unfortunately, many publications lack the appropriate characterization and as a result, many "SLN" systems are in reality not solid nanodispersions, but liquid nano-emulsions, mixed micelles or nanodispersed liquid crystalline phases. In addition, very often the superposition of several phases has to be considered. Surfactants are necessary to stabilize the lipid–water interface. However, they will also form alternative structures such as micelles, mixed micelles or vesicles. These structures are in many cases considerable smaller than the SLN. Because in the lower nanometer range the light scattering intensity increases by a factor of 10^6 times for particles with a 10-fold diameter, small structures are very difficult to detect in the presence of larger ones. For example, if we assume a diameter of 200 nm for the SLNs and 40 nm for liposomes, the SLN will scatter light roughly 15,000 times more intense.

The choice of the appropriate matrix lipid is of crucial importance for the performance of the delivery system. Generally the matrix lipid should have a melting temperature high enough to assure the solid state of the lipid matrix in the colloidal state. Furthermore, the high degree of supercooling of the melt in the nanodispersed, colloidal state has to be taken into account when the nanoparticles are prepared from a molten lipid phase or heated above the melting temperature of the matrix afterwards, e.g. upon autoclaving. The supercooled state might also occur for lipid nanodispersions produced via solvent evaporation derived processes. The increased supercooling effect in micro- and nanodispersions compared to the bulk material is a common phenomenon,[96] which is also observed for non-lipidic materials.[97,98] The high incidence of supercooling effects has several reasons:

(1) A decreased probability of the presence of embryonic nuclei. A certain number of embryonic nuclei are necessary to start the crystallization process. The crystallization nuclei can be derived from the lipid itself (homogeneous nucleation) or from foreign material (heterogeneous nucleation caused by small impurities, interface to the vessel etc.). For bulk materials, heterogeneous nucleation is often faster than homogeneous nucleation. The probability of the presence of embryonic nuclei is much higher in the bulk compared to the dispersed phase.[97]

(2) Lack of crystallization propagation. After the start of the crystallization process, crystallization can spread within the continuous bulk phase. However, a direct propagation of the crystallization is not possible or less likely (and emulsifier/interface dependent) between nanodispersed lipids.

(3) In addition to the reasons (1) and (2), the physicochemical properties, including the solubility, surface tension and the melting point, change as a function of the particle size at the nanoscale.

As discussed above, several publications claimed "SLN" with excipients which never will solidify at room or body temperature as aqueous dispersion. The majority of the publications use lipids which are solid (in bulk!) at temperatures below 40°C. However, many authors claim "SLN"-dispersions by taking crystallization for granted. Westesen and Bunjes were the first to look in this complex issue in more detail and to show that this aspect has been overlooked by other groups in previous publications. They published a paper in 1995 as a wake-up call.[10] This paper was followed by several other publications of the same and other groups which showed that supercooling is a common phenomenon for lipid nanodispersions.[19,63,97–103] The literature clearly shows, that:

- depression of the melting and the crystallization temperature occurs[100,101,105]
- crystallization occurs over several polymorphic states[28,31,97,100,101,105–107]
- crystallization is enforced by low temperatures[101,106,19] or isostatic high pressure[63]
- is also strongly emulsifier dependent, high melting surfactants solidify at the lipid interface and induce crystallization of the lipid phase.[97,100,105]

Many of the lipids used for SLN belong to the class of long chain triglycerides. They are solid at RT in the bulk. They show a complex phase behavior and can crystallize in three basic polymorphic forms: alpha (α), beta-prime (β') and beta (β). The polymorphic transformations are monotropic and take place from the less stable α-polymorph over the metastable β'-status to the thermodynamic stable β-form.[108] The kinetics of the polymorphic transitions is strongly dependent on the particular sample composition and the storage conditions. Frequently (e.g. for monacid triglycerides, the transformation from α to β' is slow and from β' to β fast and as a result, a transformation from α to β is observed in practice due to the low steady state concentration of the β' polymorph. The β'-modification is commonly observed in more complex triglycerides (with mixed chain lengths of the fatty acid), triglyceride mixtures and triglycerides containing larger fractions of partial glycerides (e.g. glyceryl behenate and many hard fats). For complex triglyceride mixtures like hard fats, an intermediate crystal form (β_i) between the β'- and β-modification has also been published.[6,100]

The solidification of the lipid (from liquid to solid) and the polymorphic transitions (e.g. from α- over β'- to β-modification) take place after production and during storage. They are connected with changes of the mobility of the lipid molecules, the density and the change of the melting temperatures and enthalpies. Most importantly, these changes will decrease the drug accommodation capacity of the lipid matrix. Therefore, drug expulsion induced by crystallization and/or polymorphic transitions are very common phenomena.[100,101,106–108] Sometimes, polymorphic transitions in SLN dispersions are also connected with dramatic changes of the macroviscosity.[6,28,97,100,105,109,110] The change from a low viscous aqueous dispersion to a solid, stiff gel can occur in the vial. Light, temperature, shaking (transport) and the surface characteristics of the vial are some factors which impact this process. In addition, to the storage, also the application itself might cause the transformation. It has been observed, that during injection through a syringe needle SLN dispersions might be transformed to a gel due to shear stress. In case of an intravenous (*i.v.*) application, it could cause a lethal embolic event. The reasons for the impressive change of the macroviscosity have been elucidated. Early studies by Freitas and Müller prove the link between gelation and polymorphic transitions.[110,111] More recent studies confirm these findings in and show that polymorphic transitions from the α-phase to the β'- or β-phase are connected with the transformation from a spherical shape to a highly anisotropic (platelet or needle) shape.[97,112,113,114] It has also been demonstrated, that during the transformation isotropic shape $\rightarrow$ anisotropic shape the lipid/water interface is increasing and that this new interface is covered by surfactant molecules from the aqueous phase (formerly present as monomers or micelles).[113] The gelation indicates the formation of a 3D-interconnected solid lipid phase. At lower lipid concentrations, the nano-dispersed state of the lipid and therefore the low viscous state might be preserved even though highly anisotropic platelets are formed.[115] It has also been shown that higher surfactant concentrations might retard or prevent polymorphic transition induced gel formation.[113]

It should be kept in mind, that the crystal classification into α, β' and β is a rough classification system and does not alone sufficiently describe the quality of the lipid nanocrystals. The regularity of the lipid crystal is important. Monoacid triglycerides or Cetylpalmitate form more dense and perfect crystal structures than glycerides with mixed fatty acids or mixtures of mono-, di- and triglycerides. Drug incorporation will be much lower in more perfect crystal structures. Another important aspect is the number of crystal layers. Several publications show, that sometimes only a very small number of lipid layers is formed. Very thin platelets have been reported by Jores in the case of Glyceryl behenate SLN.[115] The thickness of approximately 15 nm translates into only three layers! Also very thin platelets (10 nm to 5 nm) were observed by the Bunjes and Westesen group for tripalmitin SLN.[116,117] It has also been shown in the same publications that SLN with a low number of discrete lipid layers (1 to 5) melt below SLN with a higher number of lipid layers.

Long chain fatty acids are also commonly used for the preparation of SLN. Especially stearic acid and palmitic acid are popular. They also show a complex polymorphic behavior and can exist in different modifications (A, B, C, D).[118–122] In aqueous environments, also liquid crystalline phases can exist.[123,124] The polymorphic transitions are strongly dependent from surfactants (e.g. Span). In addition, it is known that long chain fatty acids interact with lecithin (which is often used in SLN preparations as surfactant).[125–127] This interaction is pH and ion-dependent. The protonation of the carboxylic group changes the phase behavior. It is also known, that the pKa of a fatty acid is strongly dependent from the environment and it might range from 4.5 to 11. Fatty acids are more polar than tri- or diglycerides. For example, they dissolve in mixed micelles during the digestion of food. Due to the higher polarity, interactions of fatty acids with surfactants, drugs and the physiological environment are much more likely compared to triglycerides. It has also been reported that stearic acids form eutectic mixtures with other compounds, e.g. iboprofen[128] and stearyl alcohol.[129] Therefore, we are faced with a very complex phase behavior of fatty acids in SLN dispersion. The fatty acids will distribute in different locations (e.g. different polymorphs of the crystalline state, several liquid crystal states and in the (mixed) micellar phase. The main parameters are the temperature, the pH, the presence of ions (e.g. formation of poorly soluble calcium salts), the presence of amphiphilic compounds. Unfortunately, this complexity is not considered in most of the SLN publications.

A similar statement can be made on monoglycerides, which also might form crystalline and liquid crystalline phases.[130–133] In the lamellar α-phase the lipid chains are molten and water can be incorporated between the polar groups of the lipid lamellae. This status will be reached for glycerol monopalmitate and glycerol monostearate at temperatures >60°C, which are typically used for HPH. Therefore, not an emulsion, but a dispersed lamellar α-phase is formed. The cooling should lead to the crystallization of the lipid chains, however, water is still present between the polar groups of the monoglycerides. This metastable form is called the α-gel phase. Although the lipid chains are in the crystal state, quite large amounts of water are still present within the monoglyceride crystal. With time, the α-gel phase can expel the water and transform into the thermodynamic stable β-crystal form, which contains almost no water within the crystal. Both transformations (α-lamellar→α-gel→β-gel) can be highly retarded. In addition, a small amount of negative charges (e.g. by free fatty acid impurities) can increase the stiffness and stability of the α-gel phase.[133]

In conclusion, lipid nanodispersions which are described as SLN in the literature are in general composed of lipid components (as matrix). These matrix lipids range from hydrophobic to more amphiphilic materials, depending on their general structure and the environment (e.g. temperature, pH, ions). Even hydrophobic lipids (triglycerides) contribute to the complexity due to the problems of supercooled melts and the polymorphic behavior. In the case of more polar lipids, e.g. monoglycerides and fatty acids even more complex behavior arises from their increased hydrophilicity. These substances are capable by themselves to form several liquid crystalline phases (mobility in 1 or 2 directions) instead of crystalline phases (rigid in all three dimensions). In addition, it is proven and well-known, that they interact with surfactants including lecithin, bile acids, Tween, Poloxamer and other commonly used stabilizers. Unfortunately, many publications do not pay sufficient attention to this topic.

The following table gives an overview about the general properties of the excipients and the phases which can be expected.

	Excipients	Interaction with water	Typical structures dispersed in water (1) alone or (2) due to high interactions with other excipients
Lipidic Matrix	Triglycerides and long chain waxes	Small	(1) Dispersed solid crystals with different polymorphs (α,β',β) (1) Liquid supercooled melts
	Diglycerides	Small	(1) Dispersed solid crystals with different polymorphs (α,β',β) (1) Liquid supercooled melts
	Mono-glycerides	High	(1) Dispersed solid gel state with different polymorphs (1) Dispersed liquid crystals (1) Liquid supercooled melts (2) Mixed micelles with fatty acids, bile acids, hydrophilic surfactants (2) Lamellar phase with lecithin
	Fatty acids	High ($pH \geq pKa$) Low ($pH < pKa$)	(1) Dispersed solid gel state with different polymorphs (1) Several liquid crystalline phases (2) Lamellar phase with lecithin (2) Mixed micelles (e.g. with bile acids, and monoglycerides, hydrophilic surfactants, lecithin)
Emulsi-fier(s)	Lecithin	High	(1) Liposomes (lamellar phase) (2) Mixed micelles (with bile acids, fatty acids)
	Bile acids	High	(1) Monomer (1) Different bile acid aggregates "micelles" (no liquid crystals) (2) mixed micelles (with fatty acids, monoglycerides, lecithin)
	Micelle formers (HLB14–18) (e.g. Poloxamer, Gelucire, Soluplus, Tween, TPGS)	High	(1) Monomer (1) Micelles (2) Lamellar phase with lecithin (2) Mixed micelles (with bile acids, and monoglycerides, lecithin)

As a consequence of the physicochemical properties of the SLN ingredients, the coexistence of several nanoscaled structures is a reality. In many publications, the amount of surfactants equals or does even exceed the amount of the lipid phase. Possible coexisting structures are schematically shown in the Figure 5.

These structures might serve as alternative location for incorporated drugs. The structures formed are sensitive to time and storage conditions (e.g. crystallization and polymorphic transitions, water expulsion from gel states) and to dilution (e.g. dilution will shift equilibria, mixed micelles might transform in liposomes). Dilution does occur after administration, but also several analytical techniques (e.g. light scattering techniques) require in many cases dilution. Therefore, possible dilution induced changes and artifacts have to be kept in mind.

Because lipid nanodispersions are so complex, time and environment sensitive systems, reliable analytical characterization is a must. Particle sizing is a mandatory, but not sufficient parameter to characterize lipid nanodispersions.

In particle sizing, micelles and small liposomes are easily overlooked due to their small scattering intensities. A 150 nm lipid particle scatters light 10^6 times more intense than a 15 nm micelle. In addition, many particle sizing methods (e.g. dynamic light scattering) assume spheres, but SLN might occur also as highly anisotropic platelets or even needles. The use of several sizing techniques is advisable. For example, it has been shown on samples

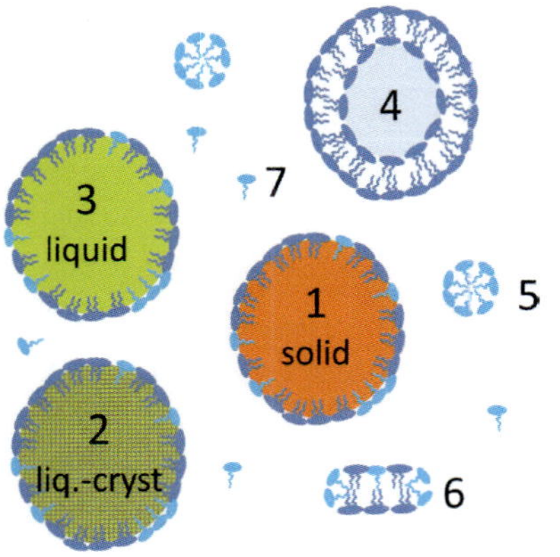

Fig. 5. Nanoscaled structures commonly present in SLN dispersions. (1) SLN lipid nanoparticle with mixed emulsifier layer, (2) Liquid crystalline nanodispersed phase with mixed emulsifier layer, (3) Liquid supercooled melt nanodroplet with mixed emulsifier layer (4) Liposome, (5) Micelle, (6) Mixed Micelle, (7) molecular dissolved surfactant molecule.

derived from the same SLN production batch, but stored at different temperatures, that supercooled melt nanodispersions (liquid lipid spherical nanodroplets, stored at RT) gave similar sizes as their crystallized (induced by lower storage temperatures) nanoplatelet counterparts in dynamic light scattering experiments, but they behaved dramatically different in field-flow-fractionation experiments.[134] Electron microscopy based methods are very powerful tools to gain information about the existing structures. But also these methods often fight with problems: tedious and complex sample preparation, poor contrasts, artifacts, poor statistics (for particle sizes), non-representive pictures. Simple light microscopy should also be used for SLN investigations although nothing should be seen for real lipid nanodispersions. Light microscopy is however, able to detect microparticles. A polarization filter should also be used to check for the existence of larger areas with liquid crystalline (lamellar or hexagonal) structures.

The physical state of the lipid is of course of crucial importance. Differential Scanning Calorometry (DSC) is widely used. The interpretation of the experimental results is however, not every time straightforward. Very often broad signals with low intensities are observed. In addition, the change of temperature might cause artifacts. Many emulsifiers show less stabilization and different phases at higher temperatures. Results of own ESR and fluorescent measurements show that Poloxamer, a widely used surfactant to stabilize SLN forms lipophilic nanodomains, which are not present at RT. Also the melting of the lipids (often in the range of 50–70°C) leads to different situation compared to RT or body temperature. A lipophilic drug (for example with a higher point >70°C) might dissolve during the DSC measurement in the new molten lipid phase, but not in the crystalline phase. Therefore, the absence of a melting peak for a drug does not — against many claims in publications — prove molecular dispersion or drug incorporation at RT or body temperature. X-ray diffraction can give further information about crystalline and liquid-crystalline structures. Due to the small number of crystal layers and the imperfectness, the signals might be rather broad and measurements difficult and time consuming. Synchrotron irradiation has much shorter measurement times, but its access is restricted. Simple proton NMR provides valuable information, because solid lipids are not visible in standard 1H-NMR. In contrast, supercooled melts give clear NMR signals and they can therefore be identified easily.[19,63,135] The use of fluorescent and ESR probes as model drugs permits the characterization of the microenvironment. Both methods show that lipid crystallization leads to expulsion of the probes from the lipid to a more polar environment.[109,135]

5. SLN Application Routes and Biomedical Applications

Almost all application routes have been explored and numerous papers have been published. Controlled release properties and targeting features have been claimed. However, many authors did work with insufficient characterized samples. The physicochemical state of the lipid and the coexistence of alternative structures (mixed micelles, liposomes, liquid crystalline phases) have often been neglected. In the following sentences, the main application routes will be explored with the respect of the requirements and the fate of the lipid nanodispersions.

The peroral route is certainly in the focus of SLN applications. SLN should enhance the bioavailability of poorly soluble drugs and decrease the food dependency of the oral uptake. In animal studies, mainly aqueous SLN dispersions have been given by gavage. A final product, however, would preferably be administrated as tablet or capsule. Liquid filled capsules are difficult to realize because aqueous dispersions will dissolve the capsule shell (gelatin, HPMC) and oily SLN dispersions will extract the drug from the SLN into the oily liquid phase. PEG-dispersion might be an option, but the solubility of the drug in PEG must be low and PEG interactions with the capsule shell remain as a problem. Therefore, tablets, pellets or sachets are preferred. The solidification can be achieved by adsorption on colloidal silica dioxide Mg–Al silicates, by wet granulation (using the SLN dispersion as aqueous granulation fluid), spray drying[136,137] or lyophilization.[138–140]

Preferably, water soluble well-established excipients should be used to solidify SLN. Otherwise, incomplete release might be observed as it has been described for polymeric nanoparticles from tablets.[141] After oral ingestion, the SLN are exposed to the GI tract and the content. The pH might vary between acidic (pH 1) and neutral values, the ionic strength and ionic composition might vary. Acidic pH values and multivalent ions (Ca^{2+}, Mg^{2+}) might induce particle aggregation. Dilution might change the phase behavior by decreasing the local concentration of water soluble components in the lipid mixtures (hydrophilic amphiphiles, e.g. Tween, Poloxamer). Interactions with the food components (and their degradation products), the gastric mucosa and secreted GI-fluids might change the matrix composition and the surface properties. Especially monoglycerides and fatty acids, which are sometimes the main components of "SLN"-dispersions, will interact with bile fluids. Bile acids and lecithin are responsible for the solubilization of food fat degradation products, which are fatty acids and monoglycerides. They will also perform the job on non-food served fatty acids and monoglycerides. Enzymatic digestion of the stabilizers and the lipid matrix might already start in the stomach and continue in the small intestine. Food derived triglycerides are almost completely digested to two fatty acids and the 2-monoglycerides, which are solubilized by the bile fluids and absorbed close to 100%. Food lipids are digested mainly by the gastric and pancreatic lipases. The gastric lipase digests lipids incompletely (to around 10%–25%) because it has a negative feedback mechanism and is inhibited by free fatty acids. In addition other enzymes (e.g. carboxyl ester hydrolase) might contribute or even play the major role for the digestion of lipid excipients.[142] *In vitro* tests show that SLN dispersions are, at least partially, digested.[20,143] For mixed lipids, a complete digestion is very likely. Highly crystalline and high melting glycerides or waxes might be only partially digested or non-digested, because the enzymatic digestion of perfect crystals is much more difficult. Also the absorption of very long chain fatty acids and monoglycerides (behenic acid, glycerol monobehenate) might be incomplete. The impact of dilution and digestion has been investigated heavily in academia and industry at the field of self-emulsifying systems. The results clearly show the importance and complexity. The scientists working at the SLN field are encouraged to apply the existing methods and to explore the impact of dilution in physiological relevant media (e.g. in SGF, FASSIF, FESSIF) and digestion (pancreatin assay) on the properties of the SLN.[20,143]

A second major field, and probably the most successful one so far, has been the dermal application. SLN have reached the cosmetic market.[144] SLN can be either incorporated into other semisolid dermal systems or the formulated in a semisolid consistency. SLN might act as carrier system for drugs (e.g. steroids[145,146] or retinoids[147,148]) or act as sunblockers.[149–151] Also occlusive properties of SLN have been described.[152] As described also for other nanoparticles, an enhanced localization in the hair follicles might be observed.[153]

The *i.v.* route is also of great interest. Intravenous application requires sterility, good tolerability. SLN formulations should be isotonic, non-hemolytic and within the physiological pH-range. Only a very limited number of surfactants are used for parenteral applications (e.g. lecithin, poloxamer 188, Tween 20 and 80, Solutol). Dangers of *i.v.* applications are embolic events due to (1) injection induced gelation caused by shear enforced polymorphic transition and/or (2) blockage of blood vessels due to the accumulation and aggregation of SLN. At this point, the commonly observed high anisotropy of SLN should be recalled. A PCS size of 200 nm might be measured from thin SLN platelets with a layer thickness of only 15 nm but a much larger dimension of the platelet area. This statement is even stronger for needle shaped dispersions. The fate of the SLN is complex. Dilution occurs, which might change the structure of the surface and the presence of mixed micelles. Adsorption of proteins will take place. In many cases, rapid recognition (within few minutes) by the RES and the uptake by the liver, spleen and lung will be seen. The SLN uptake by macrophages will lead to the SLN localization in the phagolysosomes, which is a rather harsh environment due to an acidic pH and the high activity of degrading enzymes. Pegylation, which is also commonly used for proteins and polymeric nanoparticles, might decrease the RES uptake and increase the circulation time. The impact of lipolysis on the SLN fate after *i.v.* application has not been published so far. The knowledge concerning interactions with lipoproteins and other substances is still very limited.

Many paper describe the *i.v.* application of SLN, some of them claim advantages (e.g. increased stability, controlled release and increased delivery to the brain). Frequently, positive results required the complex formation of the drug with a long-alkyl chain containing counterions (hexadecylphosphate for doxorubicin[154] and dimethyl-distearyl-ammoniumbromide for chlorambucil.[155] Without the counterion, a rapid degradation was seen for chlorambucil. Unfortunately, many papers lack a real characterization of the sample which was finally administered. For example, Ref. 154 produced SLN by the hot microemulsion technique. The hot mixture did consist of 0.2 mM soy lecithin (purity 95%), 0.7 mM stearic acid, sodium taurocholate 0.68 mM, the 1:2 doxorubicin/dimethyl-distearyl-ammoniumbromide (DDAB) complex (and in the stealth SLN formulation also PEG-stearate). The sample preparation involved mixing, heating, cooling and ultrafiltration. The authors describe the particle size, the drug content (about 5%) and the drug loss during preparation (about 1%). However, no information is given about the final content of the excipients in the applied sample. Because the interactions between bile acids, lecithin and stearic acid are quite complex, it is impossible to know which phases (liquid crystalline, solid polymorphs, mixed micelles) have been formed. So far it is impossible to draw valid conclusions from the data which have been obtained. In this study, the SLN have been claimed for longer drug release times. However, this conclusion can be questioned because a drug complex was given. Further references (SLN with Doxorubicin or Doxorubicin/DDAB complexin nano-emulsion or Doxorubicin/DDAB complex as nanosuspension) would be required. Despite the quite large number of existing papers on SLN pharmacokinetics, the detailed knowledge is still limited mainly due to insufficient characterization of the administrated samples.

Another emerging application route is the pulmonal administration, which has been explored for example for local cancer treatment, hypertension treatment and insulin delivery.[77,156–160]

6. Controlled Release — Wishful Thinking or Reality?

SLN have been proposed initially as "frozen nano-emulsions". The solid lipid should provide a controlled release. After more than 20 years of research the question arises whether or not this goal has been achieved. Many publications claim controlled release properties. Other scientists show immediate release from SLN dispersions. These conflicting data should be discussed with respect to the general laws of dissolution. In 1897, the Noyes–Whitney was published which described the dissolution kinetics:

$$\frac{dm}{dt} = \frac{DA\left(c_{\text{sat}} - c\right)}{h}.$$

The amount of the mass (m) dissolved at a certain time (t) equals the product of the diffusion coefficients (D), the area (A), divided by the thickness of the dissolving layer (h) and multiplied by the difference between the saturation solubility (c_{sat}) and the actual concentration in the solvent (c). Therefore, poorly soluble drugs — which have intrinsically a low value of (c_{sat}) — will dissolve slowly. Many studies did use release media where the drugs have low solubility's. Therefore, a "controlled release" rather reflects the Noyes–Whitney law and not the performance of the lipid matrix. It is therefore recommended to work under sink conditions ($c < 10\%\ c_{\text{sat}}$) by either (1) constant flow conditions (open dissolution) or (2) by offering the dissolved drug an acceptor compartment, where it can accumulate. Care has to be taken if dialysis is used as the penetration through the membrane might be slow.

A second consideration should be given to the diffusion kinetics. It is well-known, that polymeric or lipid microparticles can control the release by diffusion. The basic of diffusion is the Einstein equation, which tells us that the square of the average diffusion length equals twice the product of the diffusion coefficient and the time t.

$$\overline{x^2} = 2Dt.$$

Therefore, doubling the distance translates in a factor of four in time. Downsizing from 10 microns to 100 nm means a factor of 100 in distance but 10,000 in time. A rough estimation, based on typical diffusion coefficients, reveals that diffusion at the nanoscale is rather fast even if the diffusion coefficient is small. Even if the diffusion coefficient is very low (e.g. $2*10^{-12}$ cm^2s^{-1}), the molecule will diffuse 20 nm in 2 s and 200 nm in 200 s. For 2 microns already a time of 20,000 s (roughly 5.5 h) and for a distance of 20 microns about 23 days are needed. These numbers illustrate that diffusion control is unlikely to be achieved on nanoscaled objects. In reality, the diffusion coefficients are expected to be higher and the diffusion lengths shorter than 200 nm. In addition, due to the anisotropic shapes, many molecules are located on the interface and not in the lipid layers.

A third consideration is linked to the difference between drug incorporation and drug association. It is well-known in the physical chemistry that crystals have a low capacity to incorporate foreign molecules. Therefore, in many cases segregation occurs, as shown in Figure 6.

Phase segregation has been observed by electron microscopy.[100,134,161] Therefore, SLN could be seen as a drug nanosuspension which is lipid associated.

The three considerations discussed above are supported by experimental data, which show clearly that crystallization leads to a relocation into more polar environments.[109,135] It is also supported by *in vitro* release data.[162]

In summary, the solid matrix is not expected to control the release rate from lipid nanodispersions. The release is mainly triggered by solubility and partition.

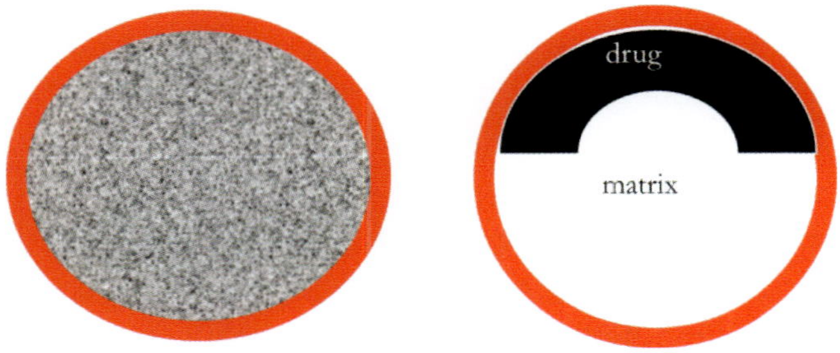

Fig. 6. (Color online) Schematic localization of the drug inside lipid nanoparticles. Left: The drug (black color) is homogeneously distributed within the lipid matrix (white); Right: The drug and the lipid form segregated phases, they are only associated. In both cases, the particle interface is covered by a surfactant layer (red color).

7. Concluding Remarks — SLN as Complex Nanodispersions

In conclusion, SLN (and also NLC) are complex nanoscaled lipid dispersions. Many claims of the existing literature should be taken with care, because many papers suffer from one or several of the following drawbacks:

— Insufficient description of sample composition (e.g. change of the lipid/emulsifier composition due to separation steps by ultrafiltration are not followed).
— Insufficient characterization with respect to the lipid phases which have been formed (supercooled melts, different crystalline and liquid crystalline phases).
— Although high amounts of emulsifiers are used, the coexistence of alternative structures is neglected (mixed micelles, liposomes).

Comparative studies should include well-characterized samples of liposomes, mixed micelles, nano-emulsions and drug nanoparticles. Studies which have been done in this direction do not support the general advantages of SLN and NLC.

The following statement, made in 1997 by Westesen, Bunjes and Koch is still valid[100]:

"Drug carrier systems based on lipid nanosuspensions prepared by melt emulsification present a number of severe stability problems such as a high gelation tendency, considerable particle growth and drug expulsion. Destabilization of the emulsified lipidic carriers is related to recrystallization of the lipids. In addition, the drug payload of crystalline nanosuspension particles is generally low."

There is certainly no single answer to all requirements and every DDS has its own advantages and disadvantages. However, the broad versatility of all different colloidal carrier systems provides the formulator with the possibility to select the most suitable system for the special drug and therapeutic question of interest. The accurate knowledge of the physicochemical properties of the nanocarriers is, however, a prerequisite for a proper selection. SLN and NLC are important developments in the field of nanoscaled DDS, but they are not generally superior to alternative systems (liposomes, mixed micelles, polymeric nanoparticles, nanocapsules, nano-emulsions).

References

1. P. Speiser, EP 0167825 A3, Lipid nano pellets as drug carriers for oral administration (1987).
2. M. R. Gasco and S. Morel, **45**, 1127–1128 (1990).

3. A. J. Domb, United States patent 5,188,837. Lipospheres for controlled delivery of substances. (1993).
4. A. J. Domb, *Int. J. Pharm.* (1995).
5. K. Westesen and B. Siekmann, *PCT Int. Appl.* (1994), 78 pp.
6. K. Westesen, B. Siekmann and M. H. J. Koch, *Int. J. Pharm.* **93**, 189–199 (1993).
7. C. Schwarz, W. Mehnert, J. S. Lucks and R. H. Müller, *J. Control. Release* **30**, 83–96 (1994).
8. W. Mehnert and K. Mäder, *Adv. Drug Deliv. Rev.* **64**(suppl.) 83–101 (2012).
9. S.-E. Jin and C.-K. Kim, *J. Pharm. Pharm. Sci.* **15**, 467–482 (2012).
10. H. Bunjes and K. Westesen, *Int. J. Pharm.* **115**, 129–131 (1995).
11. C. W. How, A. Rasedee and R. Abbasalipourkabir, *IEEE Trans. Nanobiosci.* **12**, 72–78 (2013).
12. C. Chen, T. Fan, Y. Jin, Z. Zhou, Y. Yang, X. Zhu, Z.-R. Zhang, Q. Zhang and Y. Huang, *Nanomedicine* **8**, 1085–1100 (2013).
13. P. Wasutrasawat, H. Al-Obaidi, S. Gaisford, M. J. Lawrence and W. Warisnoicharoen, *Eur. J. Pharm. Biopharm.* (2013).
14. H. Heiati, N. C. Phillips and R. Tawashi, *Pharm. Res.* **13**, 1406–1410 (1996).
15. P. Boonme, E. B. Souto, N. Wuttisantikul, T. Jongjit, and W. Pichayakorn, *Eur. J. Lipid Sci. Technol.* **115**, 820–824 (2013).
16. P. Severino, S. C. Pinho, E. B. Souto and M. H. A. Santana, *J. Therm. Anal. Calorim.* **108**, 101–108 (2012).
17. S. Joseph and H. Bunjes, *J. Pharm. Sci.* **101**, 2479–2489 (2012).
18. S. Joseph and H. Bunjes, *Int. J. Pharm.* **446**, 59–62 (2013).
19. A. Noack, G. Hause and K. Mäder, *Int. J. Pharm.* **423**, 440–451 (2012).
20. A. Noack, J. Oidtmann, J. Kutza, and K. Mäder, *J. Nanoparticle Res.* 14 (2012).
21. M. Burra, R. Jukanti, K. Y. Janga, S. Sunkavalli, A. Velpula, S. Ampati and K. N. Jayaveera, *Adv. Powder Technol.* **24**, 393–402 (2013).
22. S. P. Bhattacharyya and I. Bhattacharyya, *Int. J. Pharm. Pharm. Sci.* **4**, 103–109 (2012).
23. R. Lopes, C. V. Eleutério, L. M. D. Gonalves, M. E. M. Cruz and A. J. Almeida, *Eur. J. Pharm. Sci.* **45**, 442–450 (2012).
24. C. Qi, Y. Chen, J.-H. Huang, Q.-Z. Jin and X.-G. Wang, *J. Sci. Food Agric.* **92**, 787–793 (2012).
25. P. Fadda, M. Monduzzi, F. Caboi, S. Piras and P. Lazzari, *Int. J. Pharm.* **446**, 166–175 (2013).
26. M. R. Patel and M. F. San Martin-Gonzalez, *J. Food Sci.* **77**, N8–N13 (2012).
27. N. Kishore, U. M. Dhanalekshmi, M. D. Raja, S. Bhavani and P. N. Reddy, *J. Dispers. Sci. Technol.* **33**, 96–102 (2012).
28. C. Qian, E. A. Decker, H. Xiao and D. J. McClements, *J. Am. Oil Chem. Soc.* **89**, 17–28 (2012).
29. Y.-P. Fang, P.-C. Wu, C.-C. Tzeng, Y.-L. Chen, H.-L. Lin and Y.-H. Tsai, *Curr. Nanosci.* **8**, 266–273 (2012).
30. R. Nair, C. K. Ashok Kumar, K. S. Arun Kumar, K. Vishnu Priya, M. Thusahra Bindu and Y. Prasannaraju, *J. Pharm. Sci. Res.* **4**, 1797–1802 (2012).
31. H. Salminen, T. Helgason, B. Kristinsson, K. Kristbergsson and J. Weiss, *Food Chem.* **141**, 2934–2943 (2013).
32. E. Esposito, M. Fantin, M. Marti, M. Drechsler, L. Paccamiccio, P. Mariani, E. Sivieri, F. Lain, E. Menegatti, M. Morari and R. Cortesi, *Pharm. Res.* **25**, 1521–1530 (2008).

33. Y. Yang, A. Corona and M. A. Henson, *J. Colloid Interface Sci.* **374**, 297–307 (2012).
34. P. Ahlin, J. Kristl and J. Smid-Korbar, *Acta Pharm.* **48**, 257–267 (1998).
35. M. A. Schubert and C. C. Müller-Goymann, *Eur. J. Pharm. Biopharm.* **55**, 125–131 (2003).
36. K. Westesen, M. H. Koch and H. Bunjes, *J. Control. Release* **48**, 223–236 (1997).
37. J. F. Fangueiro, E. Gonzalez-Mira, P. Martins-Lopes, M. A. Egea, M. L. Garcia, S. B. Souto and E. B. Souto, *Pharm. Dev. Technol.* **18**, 545–549 (2013).
38. E. B. Souto, S. B. Souto, J. F. Fangueiro, M. A. Egea, E. Gonzalez-Mira, M. L. Garcia and P. Martins-Lopes, *Pharm. Dev. Technol.* 1–5 (2011).
39. S. I. Abdelwahab, B. Y. Sheikh, M. M. E. Taha, C. W. How, R. Abdullah, U. Yagoub, R. El-Sunousi and E. E. M. Eid, *Int. J. Nanomedicine* **8**, 2163–2172 (2013).
40. M. A. Schubert and C. C. Müller-Goymann, *Eur. J. Pharm. Biopharm.* **61**, 77–86 (2005).
41. S. U. Upadhyay, J. K. Patel, V. A. Patel and A. K. Saluja, *J. Pharm. Bioallied Sci.* **4**, 112–113 (2012).
42. C. Laura, S. Milena, B. Giovanna, B. M. Cristina, S. Giuseppina and B. Giampiero, *J. Therm. Anal. Calorim.* **111**, 2149–2155 (2013).
43. E. H. Gokce, E. Korkmaz, E. Dellera, G. Sandri, M. Cristina Bonferoni and O. Ozer, *Int. J. Nanomed.* **7**, 1841–1850 (2012).
44. K. Hippalgaonkar, G. R. Adelli, M. A. Repka and S. Majumdar, *J. Ocul. Pharmacol. Ther.* **29**, 216–228 (2013).
45. B. Rohit and K. I. Pal, *Curr. Nanosci.* **9**, 211–220 (2013).
46. K. Vivek, H. Reddy and R. S. R. Murthy, *AAPS Pharm Sci Tech.* **8**, E83 (2007).
47. F. Cerreto, P. Paolicelli, S. Cesa, H. M. A. Amara, F. D. D'Auria, G. Simonetti and M. A. Casadei, *AAPS Pharm Sci Tech.* **14**, 847–853 (2013).
48. A. Estella-Hermoso De Mendoza, M. A. Campanero, H. Lana, J. A. Villa-Pulgarin, J. De La Iglesia-Vicente, F. Mollinedo and M. J. Blanco-Prieto, *Nanomedicine.* **7**, 679–690 (2012).
49. P. Prombutara, Y. Kulwatthanasal, N. Supaka, I. Sramala and S. Chareonpornwattana, *Food Control* **24**, 184–190 (2012).
50. E. Zimmermann, R. H. Müller and K. Mäder, *Int. J. Pharm.* **196**, 211–213 (2000).
51. A. Lippacher, R. H. Müller and K. Mäder, *Eur. J. Pharm. Biopharm.* **58**, 561–567 (2004).
52. P. Blasi, A. Schoubben, G. V. Romano, S. Giovagnoli, A. Di Michele, and M. Ricci, *Colloids Surf. B Biointerfac.* **110**, 130–137 (2013).
53. J. M. Koziara, J. J. Oh, W. S. Akers, S. P. Ferraris, and R. J. Mumper, *Pharm. Res.* **22**, 1821–1828 (2005).
54. J. M. Koziara, P. R. Lockman, D. D. Allen and R. J. Mumper, *Pharm. Res.* **20**, 1772–1778 (2003).
55. S. Kheradmandnia, E. Vasheghani-Farahani, M. Nosrati and F. Atyabi, *Nanomedicine* **6**, 753–759 (2010).
56. A. L. R. De Souza, T. Andreani, F. M. Nunes, D. L. Cassimiro, A. E. De Almeida, C. A. Ribeiro, V. H. V. Sarmento, M. P. D. Gremião, A. M. Silva and E. B. Souto, *J. Therm. Anal. Calorim.* **108**, 353–360 (2012).
57. M. Ghadiri, S. Fatemi, A. Vatanara, D. Doroud, A. R. Najafabadi, M. Darabi and A. A. Rahimi, *Int. J. Pharm.* **424**, 128–137 (2012).

58. J.-S. Baek, S.-C. Shin and C.-W. Cho, *J. Pharm. Investig.* **42**, 279–283 (2012).
59. A. zur Mühlen, C. Schwarz and W. Mehnert, *Eur. J. Pharm. Biopharm.* **45**, 149–155 (1998).
60. A. zur Mühlen, E. zur Mühlen, H. Niehus and W. Mehnert, *Pharm. Res.* **13**, 1411–1416 (1996).
61. P. R. Lockman, J. Koziara, K. E. Roder, J. Paulson, T. J. Abbruscato, R. J. Mumper and D. D. Allen, *Pharm. Res.* **20**, 705–713 (2003).
62. M. O. Oyewumi and R. J. Mumper, *Drug Dev. Ind. Pharm.* **28**, 317–328 (2002).
63. C. Blümer and K. Mäder, *Pharm. Res.* **22**, 1708–1715 (2005).
64. P. Ahlin, J. Kristl, M. Sentjurc, J. Strancar and S. Pecar, *Int. J. Pharm.* **196**, 241–244 (2000).
65. P. Ahlin, J. Kristl, S. Pecar, J. Strancar, M. Sentjurc, *J. Pharm. Sci.* **92**, 58–66 (2003).
66. H. Jesús, B. Juana, L.-R. J. Manuel, M. De Yuso Maria Del Valle and R.-C. Enrique, *Med. Chem. (Los. Angeles).* **8**, 541–548 (2012).
67. B. K. Paul and S. P. Moulik, *Adv. Colloid Interface Sci.* **78**, 99–195 (1998).
68. H. Murakami, M. Kobayashi, H. Takeuchi and Y. Kawashima, *Int. J. Pharm.* **187**, 143–152 (1999).
69. H.-Y. Kwon, S.-W. Choi, W.-S. Kim and J.-H. Kim, *Colloids Surfaces A Physicochem. Eng. Asp.* **201**, 283–289 (2002).
70. A. Rübe, G. Hause, K. Mäder and J. Kohlbrecher, *J. Control. Release* **107**, 244–252 (2005).
71. M. Gupta and S. P. Vyas, *Chem. Phys. Lipids* **165**, 454–461 (2012).
72. K. Byrappa, S. Ohara and T. Adschiri, *Adv. Drug Deliv. Rev.* **60**, 299–327 (2008).
73. V. Trivedi, R. Bhomia, J. C. Mitchell, N. J. Coleman, D. Douroumis and M. J. Snowden, *J. Chem. Eng. Data* **58**, 1861–1866 (2013).
74. M. Calderone, E. Rodier, J.-J. Letourneau and J. Fages, *J. Supercrit. Fluids* **42**, 189–199 (2007).
75. A. R. S. de Sousa, M. Calderone, E. Rodier, J. Fages and C. M. M. Duarte, *J. Supercrit. Fluids* **39**, 13–19 (2006).
76. T. P. P. Freire, A. S. Pedro, R. Fialho, E. C. Albuquerque, A. Bertucco and G. M. N. Costa, *J. Chem. Thermodyn.* (2013).
77. P. Chattopadhyay, B. Y. Shekunov, D. Yim, D. Cipolla, B. Boyd and S. Farr, *Adv. Drug Deliv. Rev.* **59**, 444–453 (2007).
78. S. Salmaso, N. Elvassore, A. Bertucco and P. Caliceti, *J. Pharm. Sci.* **98**, 640–650 (2009).
79. J. W. Park, J. M. Yun, E. S. Lee, Y. S. Youn, K. S. Kim, Y. T. Oh and K. T. Oh, *Arch. Pharm. Res.* 1–8 (2013).
80. C. Y. Acevedo-Morantes, M. T. Acevedo-Morantes, D. Suleiman-Rosado and J. E. Ramírez-Vick, *Drug Deliv.* **20**, 338–348 (2013).
81. B. Sjöström, K. Westesen and B. Bergenståhl, *Int. J. Pharm.* **94**, 89–101 (1993).
82. B. Sjöström and B. Bergenståhl, *Int. J. Pharm.* **84**, 107–116 (1992).
83. M. Nabi-Meibodi, B. Navidi, N. Navidi, A. Vatanara, M. Reza Rouini and V. Ramezani, *J. Drug Deliv. Sci. Technol.* **23**, 225–230 (2013).
84. M. Wang, L. Qin, K. Li, R. Zhu, W. Wang and S. Wang, *AAPS PharmSciTech.* **13**, 1348–1354 (2013).
85. P. M. Dandagi, P. D. Patel, A. P. Gadad and A. K. Aravapalli, *Asian J. Pharm.* **6**, 116–123 (2012).

86. J. Varshosaz, F. Hassanzadeh, H. Sadeghi and M. Shakery, *Curr. Nanosci.* **8**, 249–258 (2012).
87. C. Tao, H. Cheng, K. Zhou, Q. Luo, L. Guo and W. Chen, *Asian J. Pharm.* **6**, 275–281 (2012).
88. E. Troncoso, J. M. Aguilera and D. J. McClements, *J. Colloid Interface Sci.* **382**, 110–116 (2012).
89. D. Sutaria, B. K. Grandhi, A. Thakkar, J. Wang and S. Prabhu, *Int. J. Oncol.* **41**, 2260–2268 (2012).
90. M. García-Fuentes, D. Torres and M. Alonso, *Colloids Surfaces B Biointerfaces* **27**, 159–168 (2003).
91. H. Yuan, S.-P. Jiang, Y.-Z. Du, J. Miao, X.-G. Zhang and F.-Q. Hu, *Colloids Surfaces B Biointerfaces* **70**, 248–253 (2009).
92. A. Hanafy, H. Spahn-Langguth, G. Vergnault, P. Grenier, M. Tubic Grozdanis, T. Lenhardt and P. Langguth, *Adv. Drug Deliv. Rev.* **59**, 419–426 (2007).
93. C. Charcosset, A. El-Harati and H. Fessi, *J. Control. Release* **108**, 112–120 (2005).
94. A. A. El-Harati, C. Charcosset and H. Fessi, *Pharm. Dev. Technol.* **11**, 153–157 (2006).
95. A. Svensson, C. Neves and B. Cabane, *Int. J. Pharm.* **281**, 107–118 (2004).
96. M. Talati, P. K. Jha and S. K. Gupta, *Mater. Sci. Forum* **570**, 132–137 (2008).
97. D. J. McClements, *Adv. Colloid Interface Sci.* **174**, 1–30 (2012).
98. F. Broto and D. Clausse, *J. Phys. C Solid State Phys.* (1976).
99. H. Bunjes and K. Westesen, in ... *emulsions in drug targeting and* ..., (1998).
100. K. Westesen, H. Bunjes and M. H. Koch, *J. Control. Release* **48**, 223–236 (1997).
101. H. Bunjes, K. Westesen and M. H. J. Koch, *Int. J. Pharm.* **129**, 159–173 (1996).
102. R. Shukat, C. Bourgaux and P. Relkin, *J. Therm. Anal. Calorim.* **108**, 153–161 (2012).
103. J. Kuntsche, K. Westesen, M. Drechsler, M. H. J. Koch and H. Bunjes, *Pharm. Res.* **21**, 1834–1843 (2004).
104. M. Trotta, F. Debernardi and O. Caputo, *Int. J. Pharm.* **257**, 153–160 (2003).
105. H. Bunjes and M. H. J. Koch, *J. Control. Release* **107**, 229–243 (2005).
106. K. Westesen and H. Bunjes, *Int. J. Pharm.* **115**, 129–131 (1995).
107. C. Qian, E. A. Decker, H. Xiao and D. J. McClements, *Food Res. Int.* **52**, 342–349 (2013).
108. J. Weiss, E. A. Decker, D. J. McClements, K. Kristbergsson, T. Helgason and T. Awad, *Food Biophys.* **3**, 146–154 (2008).
109. K. Jores, A. Haberland, S. Wartewig, K. Mäder and W. Mehnert, *Pharm. Res.* **22**, 1887–1897 (2005).
110. C. Freitas and R. H. Müller, *Eur. J. Pharm. Biopharm.* **47**, 125–132 (1999).
111. C. Freitas and R. H. Müller, *Int. J. Pharm.* **168**, 221–229 (1998).
112. H. Bunjes, F. Steiniger and W. Richter, *Langmuir* **23**, 4005–4011 (2007).
113. T. Helgason, T. S. Awad, K. Kristbergsson, D. J. McClements and J. Weiss, *J. Colloid Interface Sci.* **334**, 75–81 (2009).
114. H. Salminen, T. Helgason, B. Kristinsson, K. Kristbergsson and J. Weiss, *Food Chem.* **141**, 2934–2943 (2013).
115. K. Jores, W. Mehnert, M. Drechsler, H. Bunjes, C. Johann and K. Mäder, *J. Control. Release* **95**, 217–227 (2004).
116. T. Unruh, H. Bunjes, K. Westesen and M. H. J. Koch, *J. Phys. Chem. B* **103**, 10373–10377 (1999).

117. H. Bunjes, M. H. J. Koch and K. Westesen, *Langmuir* **16**, 5234–5241 (2000).
118. W. Beckmann, *J. Cryst. Growth* **79**, 797–800 (1986).
119. W. Beckmann and R. Boistelle, *J. Cryst. Growth* **72**, 621–630 (1985).
120. D. Aquilano, R. Cavalli and M. R. Gasco, *Thermochim. Acta* **230**, 29–37 (1993).
121. N. Garti, E. Wellner and S. Sarig, *J. Cryst. Growth* **57**, 577–584 (1982).
122. N. Garti and K. Sato, *J. Am. Oil Chem. Soc.* **63**, 236–239 (1986).
123. M. Heppenstall-Butler and M. F. Butler, *Langmuir* **19**, 10061–10072 (2003).
124. V. Luzzati, H. Mustacchi and A. Skoulios, *Discuss. Faraday Soc.* **25**, 43 (1958).
125. A. B. Kohn and S. E. Schullery, *Chem. Phys. Lipids* **37**, 143–153 (1985).
126. F. V. Mercado, B. Maggio and N. Wilke, *Chem. Phys. Lipids* **164**, 386–392 (2011).
127. S. Mabrey and J. M. Sturtevant, *Biochim. Biophys. Acta — Lipids Lipid Metab.* **486**, 444–450 (1977).
128. S. Lerdkanchanaporn, D. Dollimore and S. J. Evans, *Thermochim. Acta* **367**, 1–8 (2001).
129. F. G. Gandolfo, A. Bot and E. Flöter, *Thermochim. Acta* **404**, 9–17 (2003).
130. A. Zetzl, M. Ollivon and A. Marangoni, *Cryst. Growth Des.* **9**, 3928–3933 (2009).
131. E. S. Lutton, *J. Am. Oil Chem. Soc.* **42**, 1068–1070 (1965).
132. J. Ubbink, R. Mezzenga, L. Sagalowicz, M. E. Leser, H. J. Watzke and M. Michel, *Trends Food Sci. Technol.* **17**, 204–214 (2006).
133. I. Heertje, E. Roijers and H. A. C. Hendrickx, *LWT — Food Sci. Technol.* **31**, 387–396 (1998).
134. K. Jores, W. Mehnert, M. Drechsler, H. Bunjes, C. Johann and K. Mäder, *J. Control. Release* **95**, 217–227 (2004).
135. K. Jores, W. Mehnert and K. Mäder, *Pharm. Res.* **20**, 1274–1283 (2003).
136. C. Freitas and R. H. Müller, *Eur. J. Pharm. Biopharm.* **46**, 145–151 (1998).
137. M. C. L. Marchiori, A. F. Ourique, C. D. B. Da Silva, R. P. Raffin, A. R. Pohlmann, S. S. Guterres and R. C. R. Beck, *J. Nanosci. Nanotechnol.* **12**, 2059–2067 (2012).
138. C. Schwarz and W. Mehnert, *Int. J. Pharm.* **157**, 171–179 (1997).
139. S. Doktorovova, R. Shegokar, L. Fernandes, P. Martins-Lopes, A. M. Silva, R. H. Müller and E. B. Souto, *Pharm. Dev. Technol.* **7450**, 1–8 (2013).
140. E. Zimmermann, R. H. Müller and K. Mäder, *Int. J. Pharm.* **196**, 211–213 (2000).
141. C. Schmidt and R. Bodmeier, *J. Control. Release* **57**, 115–125 (1999).
142. S. Fernandez, J.-D. Rodier, N. Ritter, B. Mahler, F. Demarne, F. Carrière and V. Jannin, *Biochim. Biophys. Acta — Mol. Cell Biol. Lipids* **1781**, 367–375 (2008).
143. A. M. Nik, S. Langmaid and A. J. Wright, *Food Funct.* **3**, 234–245 (2012).
144. J. Pardeike, A. Hommoss and R. H. Müller, *Int. J. Pharm.* **366**, 170–184 (2009).
145. C. S. Maia, W. Mehnert and M. Schäfer-Korting, *Int. J. Pharm.* **196**, 165–167 (2000).
146. M. Schäfer-Korting, W. Mehnert and H.-C. Korting, *Adv. Drug Deliv. Rev.* **59**, 427–443 (2007).
147. G. Shiva, M. Somaye and J. Mahmoud Reza, *Acta Pharm.* **62**, 547–562 (2012).
148. H. S. Jeon, J. E. Seo, M. S. Kim, M. H. Kang, D. H. Oh, S. O. Jeon, S. H. Jeong, Y. W. Choi and S. Lee, *Int. J. Pharm.* (2013).
149. S. A. Wissing and R. H. Muller, *J Control Release* **81**, 225–233 (2002).
150. S. A. Wissing and R. H. Müller, *Cosmetic applications for solid lipid nanoparticles (SLN).* **254** (2003).
151. S. A. Wissing and R. H. Müller, *Int. J. Cosmet. Sci.* **23**, 233–243 (2001).

152. S. Wissing and R. Müller, *Int. J. Pharm.* **242**, 377–379 (2002).
153. H. K. Chan, T. Armer, T. W. Prow, J. E. Grice, L. L. Lin, R. Faye, M. Butler, W. Becker, E. M. T. Wurm, C. Yoong, T. A. Robertson, H. P. Soyer and M. S. Roberts, *Adv. Drug Deliv. Rev.* **63**, 470–491 (2011).
154. A. Fundarò, R. Cavalli, A. Bargoni, D. Vighetto, G. P. Zara and M. R. Gasco, *Pharmacol. Res.* **42**, 337–343 (2000).
155. P. Sharma, S. Ganta, W. A. Denny and S. Garg, *Int. J. Pharm.* **367**, 187–194 (2009).
156. M. Videira, A. J. Almeida and T. Fabra, *Nanomedicine Nanotechnology Biol. Med.* **8**, 1208–1215 (2012).
157. X. Yang, Y. Liu, C. Liu and N. Zhang, *J. Biomed. Nanotechnol.* **8**, 834–842 (2012).
158. S. Weber, A. Zimmer and J. Pardeike, *Eur. J. Pharm. Biopharm.* (2013).
159. M. A. Videira, A. C. Santos and M. F. Botelho, *Curr. Radiopharm.* **5**, 158–165 (2012).
160. M. Paranjpe, V. Neuhaus, J. H. Finke, C. Richter, T. Gothsch, A. Kwade, S. Büttgenbach, A. Braun and C. C. Müller-Goymann, *Inhal. Toxicol.* **25**, 536–543 (2013).
161. H. Bunjes, M. Drechsler, M. H. J. Koch and K. Westesen, *Pharm. Res.* **18**, 287–293 (2001).
162. K. M. Rosenblatt and H. Bunjes, *Mol. Pharm.* **6**, 105–120 (2008).

Chapter 3

Micellar Nanopreparations for Medicine

Rupa Sawant and Aditi Jhaveri*

Center for Pharmaceutical Biotechnology and Nanomedicine
Northeastern University, Boston, MA 02115, USA
**ru.sawant@neu.edu*

Abstract

Polymeric micelles are association core–shell colloidal carriers formed in aqueous media from self-assembling amphiphilic block-co-polymers, which have been studied and developed as drug carriers for the targeting of solid tumors. In this chapter, the polymeric micelles used for passive and active tumor targeting, for stimuli-sensitivity and multifunctionality are summarized. In the final section, polymeric micelle formulations in clinical trials are discussed.

1. Introduction

The lack of selectivity of anticancer drugs toward cancer cells is still a major problem for chemotherapy in cancer patients. Many conventional chemotherapy regimens have been discontinued because of the significant adverse effects due to the non-selective targeting. Currently, the poor water solubility of many potential new drug candidates is rapidly becoming the principal obstacle for their clinical application. Polymeric micelles have lately emerged as an important new type of drug carrier. The study of polymeric micelle drug carriers started in the 1980s.[1] These systems were recognized as among the most potent drug carrier types in the 1990s, and in the 2000s, several micellar drug formulations entered clinical trials.[2–4]

Polymeric micelles are composed of an inner core and an outer shell. The outer shell is responsible for their *in vivo* pharmacokinetic behavior, while the inner core is responsible for the drug loading capacity. The size of polymeric micelles allows extravasation from tumor blood vessels, with minimal extravasation through normal vessel walls and renal excretion. This effect is due to the pathophysiological characteristics of solid tumor tissues, including hypervascularity, incomplete vascular architecture and absence of effective lymphatic drainage that leads to the enhanced permeability and retention (EPR) effect,[5,6] which is the basis for the passive targeting of polymeric micelles. Polymeric micelles can also be actively targeted to a tumor by attachment of specific targeting ligands. Certain stimulus-sensitive functions can also be incorporated in polymeric micelles to respond to the stimuli that are characteristic of pathological sites.

In this chapter, polymeric micellar medicines for passive and active targeting of chemotherapeutic drugs and the advances in development of stimuli-sensitive and multifunctional micelles are reviewed. Polymeric micellar formulations now in clinical trials are also discussed.

2. Polymeric Micelles

In the early 1980s, Ringsdorf and coworkers reported the use of polymeric micelles as drug delivery vehicles.[1] Polymeric micelles are association core–shell colloidal carriers with sizes ranging from 10 to 100 nm formed in aqueous media from self-assembling amphiphilic block-co-polymers that consist of hydrophilic and hydrophobic monomer units.[7–9] The core is formed by hydrophobic blocks stabilized by the corona, or shell, of hydrophilic polymeric chains. The hydrophobic core can incorporate drugs, protein or DNA. The core can also be made up of a water-soluble polymer chemically conjugated to a water-insoluble drug and thus be rendered hydrophobic[10] and by complexation of two oppositely charged polyions, called polyion complex (PIC) micelles.[11] The hydrophilic corona allows the carrier to be unrecognized by the reticuloendothelial system (RES), provides longevity in blood and the possibility for further micelle derivatization, such as attachment of targeting moieties with surface reactive groups.[9] The most commonly used polymers for polymeric micelles preparation are amphiphilic diblock copolymers. However, triblock copolymers and graft copolymers are also reported.[9] To avoid micelle or unimer accumulation in the body, micelles should be composed either of biodegradable components, or the unimers should be of a size to allow renal excretion (molecular mass < 40 kDa). Polyethylene glycol (PEG), which is a water-soluble, biocompatible and highly hydrated flexible polymer with low toxicity, is the most commonly used corona-forming polymer.[9] Other hydrophilic polymers reported as shell-forming polymers are poly(ethylene oxide), poly(acryloylmorpholine), poly(trimethylene carbonate) and poly(vinylpyrrolidone).[9,12,13] Core-forming blocks typically consist of poly(propylene oxide) (PPO), poly(ε -caprolactone) (PCL), poly(D,L-lactic acid) (PDLLA) and poly(L-aspartic acid).[9]

The stability of micelles depends on their critical micellar concentration (CMC). Generally, the higher the molecular weight and more hydrophobic a micelle is, the lower the CMC.[9] The CMC of block co-polymer micelles is in the 10^{-6} to 10^{-7} M range, which is about 1000 times lower than that of low-molecular-weight surfactants,[14] and thus provides the stability of micellar formulations upon extreme dilutions after intravenous administration. Micelles will solubilize poorly soluble non-polar pharmaceuticals, mainly within the micelle core. Polar molecules are readily adsorbed on the micelle surface. Substances with intermediate polarity distribute along the amphiphilic surfactant molecules in intermediate positions.[9,15] Solubilization of drug in micelles offers many advantages, including increased water solubility of sparingly soluble drugs, improved bioavailability, reduced toxicity and other adverse effects, enhanced permeability across physiological barriers and better biodistribution.[16–20]

Due to their small size, polymeric micelles have the ability to passively target their payload to specific tissues via the enhanced permeability and retention (EPR) effect.[5,6] The drug-delivery potential of polymeric micelles can be further enhanced by attachment of targeting ligands or incorporation of stimulus sensitive moieties for triggered release. Various ligands such as antibodies, sugar moieties, transferrin, folate residues and peptides have been attached to polymeric micelles for active targeting.[21–26] A variety of internal and external stimuli have been investigated in attempts to destabilize drug-loaded polymeric micelles, including temperature, pH, enzymatic reactions, redox processes, light, ultrasound and their combinations.[27,28]

Sometimes, short, but extremely hydrophobic residues provided by two long-chain fatty acyl groups, can be used as hydrophobic core-forming groups. These groups contribute to the increased hydrophobic interactions between the polymeric chains in the micelle core, and thus provide additional advantages for particle stability when compared to conventional amphiphilic micelles. Diacyllipid-PEG conjugates such as PEG-phosphatidylethanolamine (PE) represent a characteristic amphiphilic polymer with a bulky PEG portion and a short, hydrophobic diacyllipid part that can form micelles in an aqueous environment.[29,30] PEG-PE micelles are structured in such a way that the outer PEG corona, known to be highly water soluble and highly hydrated, serves as

Table 1. Particle diameter and CMC values of PEG-PE micelles.

Micelle-forming Conjugate	Particle size (nm)	CMC (M)
PEG_{750}-DSPE	7–15	1×10^{-5}
PEG_{2000}-DSPE	7–20	1×10^{-5}
PEG_{5000}-DSPE	10–40	6×10^{-6}
PEG_{2000}-DOPE	7–20	9×10^{-6}
PEG_{5000}-DOPE	10–35	7×10^{-6}

Note: DSPE: 1,2-distearoyl-*sn*-glycero-3-phosphoethanolamine; DOPE: 1, 2-dioleoyl-*sn*-glycero-3-phosphoethanolamine.

an efficient steric protector in biological media. On the other hand, the phospholipid residues, which represent the micelle core, are extremely hydrophobic and can solubilize various hydrophobic compounds, including drugs and contrast agents in their core with a loading efficiency that varies from 1.5% to 50% by weight depending on the hydrophobicity of the agent.[31]

All versions of PEG-PE conjugates form micelles with a spherical shape, uniform size distribution (7–40 nm) and very low CMC values (in a high nanomolar to low micromolar range) because of the strong hydrophobic interactions between the double acyl chains of the phospholipid residues (Table 1).

Ideally, the polymeric micellar system should function as a solubilizer and demonstrate good stability and drug retention *in vivo* in the presence of blood components. Micelles with very poor stability following *i.v.* injection will not alter the pharmacokinetic profile of the drug itself and thus function mainly as solubilizers.[32] In this regard, it has been observed that PEG_{2000}-PE and PEG_{5000}-PE micelles retain their size characteristics after 48 h incubation in blood plasma.[33] This suggests good stability of PEG-PE micelles in biological fluids upon parenteral administration.

Various attempts have been made to further increase the solubilization efficiency of micelles by forming mixed micelles with the addition of another surfactant or hydrophobic material. One should expect that these mixed micelles will allow better solubilization of poorly soluble drugs because of some loosening of the micelle core (e.g. with egg phosphatidyl choline) [34] or an increased volume of the lipophilic core of mixed micelles (e.g with D-α-tocopheryl polyethylene glycol 1000 succinate/vitamin E/Pluronic).[24,35–37]

Extensive studies have also been performed with Poly(ethylene oxide)-block-poly(L-amino acid) micelles.[38]

Luxenhofer *et al.* reported poly(2-oxazoline) (Pox) based amphiphilic diblock (poly(2-ethyl-2-oxazoline)-b-poly(2-butyl-2-oxazoline), Mn = 11.5 kDa) and a series of triblock copolymers (poly(2-methyl-2-oxazoline)- poly(2-butyl-2-oxazoline)- poly(2-methyl-2-oxazoline), Mn = 8.5 – 10.4 kDa) for the solubilization of paclitaxel.[39] The size of the paclitaxel-loaded Pox micelles was in the range of 20–30 nm and an extraordinary high drug-loading capacity of up to 45% was found (a nearly 1:1 drug:polymer ratio). The unloaded Pox amphiphiles exhibited very low cytotoxicity with human breast carcinoma cells (MCF7/ADR) and immunogenicity as determined by the activation of the C3a complement fraction was significantly lower as compared to Cremophor EL or Taxol® (commercial formulation of paclitaxel with Cremophor EL). In *in vivo* study using tumor-bearing mice, the Pox-paclitaxel formulation showed superior growth inhibition than Taxol®. Pox micelles were also reported as a promising high capacity delivery platform for multidrug cancer chemotherapy.[40] It was shown that a unprecedented high total loading capacities of up to 50 wt % drug per final formulation was achieved with a variety of binary and ternary drugs combinations of paclitaxel, docetaxel (DTX), 17-allylamino-17-demethoxygeldanamycin (17-AAG), etoposide (ETO) and bortezomib (BTZ). Drug-ratio–dependent synergistic cytotoxicity of micellar ETO/17-AAG was observed in MCF-7 cancer cells and of micellar BTZ/17-AAG in MCF-7, PC3, MDA-MB-231 and HepG2 cells.

Currently, the use of polymeric micelles is being explored in various phases of clinical trials. The formulation, NK911, which consists of PEG–poly(L-aspartic acid) micelles containing doxorubicin (DOX) showed impressive preclinical antitumor efficacy.[3] Another micellar formulation in clinical trials, Genexol-PM, consisting of paclitaxel encapsulated within PEG–poly(D,L lactide) micelles which was much more tolerable than the formulation used clinically that contained Cremephor® EL, an excipient shown to lead to hypersensitivity reactions.[41]

3. Micelle Formation and Drug Loading

Polymeric micelles are formed by a self-assembly process. Self-association of these amphiphilic polymers is driven by the decrease in the free energy of the system due to removal of hydrophobic fragments from the aqueous environment and the re-establishment of a hydrogen-bond network in water. Formation of van der Waals bonds between the hydrophobic blocks in the core of the formed micelles results in an entropy gain.[42,43] At low concentration in aqueous medium, the amphiphiles exist as individual monomers, but as their concentration increases, they begin to aggregate. The concentration of a monomeric amphiphile at which micelles begin to appear is called the critical micellar concentration (CMC). The lower the CMC of an amphiphile

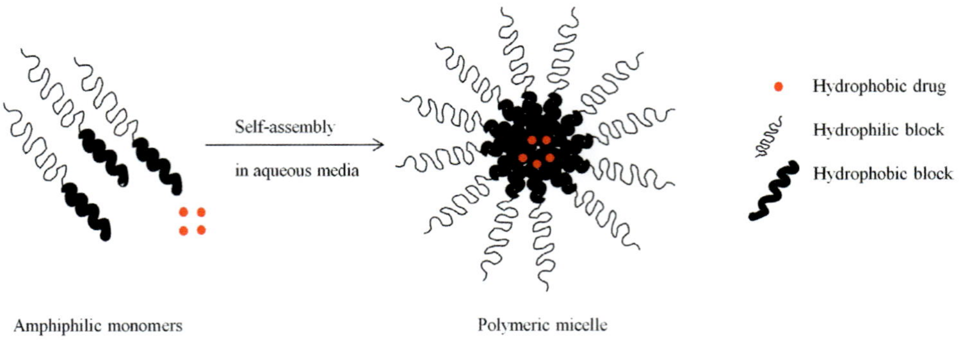

Fig. 1. Formation of polymeric micelles in aqueous media by self-assembly of amphiphilic monomers. Hydrophobic blocks form the core of micelles and hydrophilic blocks form the shell. Hydrophobic drugs are encapsulated within the micelle core.

polymer, the more stable micelles are at a low net concentration of the amphiphile in a medium.[44] Figure 1 shows the formation of polymeric micelles from amphiphilic monomers.

Drug loading can be done by either chemical conjugation or physical entrapment.[9] The typical methods used for loading of poorly soluble drugs include dialysis,[45] solid dispersion,[46] oil-in-water emulsion solvent evaporation,[47,48] direct dissolution,[49] complexation,[50] chemical conjugation[51] and solvent evaporation.[52] It is also possible to calculate drug loading efficiency using the Flory–Huggins interaction parameter, where the compatibility between the drug and hydrophobic block of the polymer can be assessed.[53]

Sometimes, the polymeric micelles are further stabilized by crosslinking either the core [54,55] or shell of the polymeric micelles.[56] This crosslinking stabilizes the micellar morphology, and the release of the drug can be controlled.

4. Drug Delivery Using Polymeric Micelles

4.1. *Passive targeting*

Some of the characteristic features of tumors that allow large molecules and nanocarriers such as polymeric micelles to accumulate by the EPR effect form the basis for passive targeting.[57] Tumors often display increased permeability of their blood vessels.[58–61] The enhanced permeation of solid tumors is often in response to the mediators of angiogenesis such as vascular endothelial growth factor (VEGF),[62] and elevated levels of vascular permeability factors such as bradykinin,[63,64] nitric oxide,[65] peroxynitrite[66] and matrix metalloproteinases (MMPs).[67] The new blood vessels formed in tumors have large gaps in their endothelium and are often relatively leaky. This allows increased accumulation of nanocarriers (depending on size, shape, charge) in tumors. Further, the nanocarriers are retained within the tumor area due to the absence of functioning lymphatics.[61] Although the threshold vesicle size of ~400 nm was reported for extravasation into tumors,[68] some studies have shown more effective extravasation with < 200 nm particles.[58,69] The EPR effect was first reported by Matsumura and Maeda in 1986[70] and later was validated and described in greater detail by Maeda *et al.*[6,17,71–73]

However, it should be noted that the permeability of vessels in tumors varies with tumor type, anatomical location of the tumor and tumor progression. Therefore, delivery of micellar-loaded drug depends not only on the properties of carrier itself but also on the physiological features specific for the tumor.

Polymeric micelles smaller than 10 nm are easily cleared by kidneys[74] and large micelles (50–100 nm) are removed by liver and spleen.[75] It should be kept in mind that high accumulation of micelles in tumor tissue due to the EPR effect cannot guarantee accumulation of encapsulated drug at cellular targets. This is often due to premature drug release from micelles before reaching the desired tumor target or insufficient intracellular uptake.[76] To overcome this challenge, proper design of micelles to avoid premature drug release in blood, promoting cellular internalization and/or triggered drug release at the tumor site is envisioned. Micellar stability and premature drug release can be overcome by chemical conjugation of drug to the micelles core and by crosslinking.[77–79] Attachment of tumor-targeting ligands to the micellar surface and/or incorporation of stimulus-sensitive functions in the micellar structure can be used to enhance cellular internalization and increase drug release at a site.[21,28]

Our group has shown increased accumulation of PEG-PE micelles in solid tumors due to the EPR effect.[33] The circulation half-lives of *i.v.* PEG-PE micelles were from 1.2 to 2.0 h in mice.[33] Thus, the PEG-PE micelles clearly demonstrate longevity. The peak tumor accumulation time of PEG-PE micelles prepared from all versions of PEG-PE conjugates observed was at about 5 h post-injection. The increase in the size of PEG blocks increased the micelle circulation time. This could be due to better steric protection against opsonin penetration to micelle core. However, micelles prepared from PEG-PE conjugates with shorter versions of PEG, might be more efficient carriers of poorly soluble drugs because of their greater hydrophobic-to-hydrophilic phase ratio and their more efficient drug loading on a weight-to-weight basis.

Since transport across the tumor vasculature depends somewhat on the particular type of tumor,[58] the use of micelles as drug carriers can be particularly useful for tumors, whose vasculature has a low cutoff size (below 200 nm). Thus, 15–20 nm PEG-PE micelles effectively delivered a model protein drug to a solid tumor with a very low cut off size (Lewis lung carcinoma) in mice,[80] while even small (100 nm) long-circulating liposomes did not show an increased accumulation of their liposome encapsulated drug.[81]

We have recently reported palmitoyl ascorbate (PA)-loaded PEG-PE micelles, which selectively target and kill cancer cells.[82] Human pharmacokinetics data indicate that *i.v.* ascorbic acid (ascorbate) in pharmacologic concentrations could have an unexpected role in cancer treatment. However, the *in vivo* application of ascorbate is currently limited by the need for very high plasma concentrations to achieve therapeutic levels in tumors. The limited chemical stability and strict physiologic regulation of the native ascorbate demands for the high doses of ascorbate necessary for its anticancer activity. We hypothesized and proved that loading of micellar nanocarriers with hydrophobized derivatives of ascorbate such as PA is ideally suited for *in vivo* applications and allows for selective accumulation in solid tumors by virtue of the EPR effect.[82] It was also shown that PA-loaded PEG-PE micelles specifically bound to cancer cells and delivered substantial anticancer cell toxicity. The mechanism of cell death was due to generation of reactive oxygen species (ROS). A significant difference in the IC50 value was obtained when compared between a BT20, MCF-7, RAG, RENCA, A2780, 4T1 (cancer-cell lines) and NIH3T3 (a non-cancer cell line). The greater sensitivity of cancer cells to ascorbate could be due to a higher basal status of intracellular ROS[83–85] and/or a low antioxidant status of various cancer cell lines.[86–88]

It was reported that free ascorbic acid suppressed tumor growth after daily *i.p.* administration in TLT-tumor bearing mice at a dose of 1g/Kg[89] and 4g/Kg in nude mice bearing either Ovcar5, Pan02, and 9L tumors.[90] We showed that the PA micelles with a much lower dose of 15 mg/kg were effective in reducing tumor growth rate in 4T1 tumor-bearing mice and thus represented a viable alternative to free ascorbate for anticancer therapy. Table 2 lists examples of micelles based on passive targeting.

4.2. *Active targeting*

Active targeting aims to increase the micellar accumulation at the target site or intracellular delivery to target cells. The micellar surface can be modified with tumor-targeting ligands that bind to tumor-associated antigens overexpressed on tumor cells compared to normal cells. Ligands that have been tried for active targeting of polymeric micelles include sugar moieties such as galactose,[101,102]

Table 2. Some recent examples of micelles based on passive targeting.

Polymers used in micelle formation	Therapeutic agent	Results	Ref.
Poly (ethylene oxide)-poly [(R)-3-hydroxybutyrate]-poly(ethylene oxide)	Doxorubicin	Improved cytotoxicity in monolayer cells, effective penetration of micellar DOX in 3D spheroids, tumor regression *in vivo*.	91
PEG2000-PE	Paclitaxel and Elacridar	Reversal of multi-drug resistance and improved sensitivity to paclitaxel *in vitro*.	92
	Doxorubicin	Improved biodistribution, *in vitro* cytotoxicity, and antitumor metastasis *in vivo*.	93
	Quercetin	Enhanced peroral anticancer activity of nanomicellar drug.	94
	Thiostrepton	*In vivo* suppression of oncogenic FOXM1, reduced tumor growth in mice.	95
Styrene–maleic acid copolymer	Pirarubicin	80% tumor volume reduction in a mouse liver metastasis model with improved survival.	96
PEG-co-poly(D,L-lactic acid)	β-Lapachone	Enhanced antitumor efficacy in treating subcutaneous and orthotopic lung tumors overexpressing NQO1.	97
PEG–oligocholic acid	Doxorubicin	Micelles more efficacious than free DOX or Doxil® in lymphoma bearing mice, reduced DOX-associated cardiotoxicity.	98
Stearate grafted dextran	Doxorubicin	Excellent cellular uptake, prolonged release *in vitro*, reversal of drug resistance of MCF-7/Adr cells, activity comparable to commercial DOX *in vivo* with reduced toxicity.	99
Polycaprolactone grafted chondroitin sulfate	Doxorubicin	Significantly improved *in vivo* activity of micellar DOX over free DOX in lung tumor bearing mice.	100

epidermal growth factor,[103] folate,[104] peptides[26] and antibodies.[21,24] Figure 2 shows some modifications of drug loaded micelles for targeting.

Various *in vitro* studies have shown that targeted micelles selectively accumulate at target cells via receptor-mediated endocytosis and can also

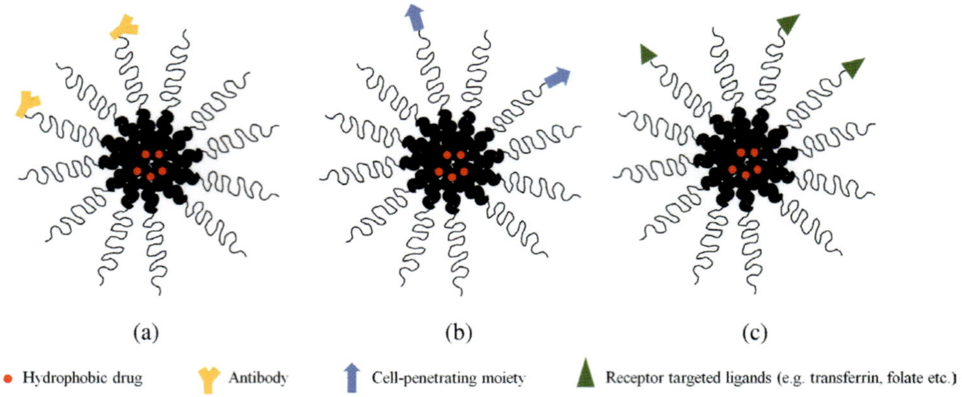

Fig. 2. Targeted micelles loaded with drug (a) Antibody-targeted micelle; (b) micelles bearing cell-penetrating moieties for intracellular targeting; (c) micelles targeted with ligands to specific receptors on the cell surface.

overcome multidrug resistance and allow delivery to specific subcellular organelles.

Asialogycoproteins (ASGP), which are carbohydrate receptors on hepatocytes and mannose receptors on Kupffer and liver endothelial cells can be targeted using sugar-modified micelles. Cho *et al.* reported higher anticancer drug delivery to ASGP-expressing HePG2 cells using galactose-modified micelles.[102]

Folate receptor is overexpressed in a variety of cancer cells. Folate-modified polymeric micelles have been shown to enhance the delivery and efficacy of anticancer drug-loaded polymeric micelles both *in vitro* and *in vivo*.[104–106]

RGD sequence-containing peptides can target integrins overexpressed on the angiogenic endothelial cells of the tumor vasculature. Thus, RGD-modified micelles have been shown to increase the delivery of anticancer drugs to angiogenic tumor endothelial cells and metastatic tumor cells that overexpress $\alpha v\beta 3$ integrins.[26,107,108]

Zeng *et al.* reported an EGF-conjugated PEG-b-poly(δ-valerolactone) micelle system that targets the EGF receptors overexpressed by breast cancer cells. EGF is a ligand with nuclear translocation properties. These targeted micelles localized in the nucleus and perinuclear region of MDA-MB-468 breast cancer cells.[109]

Vasoactive intestinal peptide-grafted sterically stabilized phospholipid mixed nanomicelles (VIP-SSMM; size, approximately 15 nm), was reported as a novel nanosized actively targeted drug delivery platform for breast cancer.[110,111]

Among all specific ligands, antibodies have provided the broadest opportunities in terms of diversity of targets and specificity of interaction. To ensure the absence of any steric hindrances for antibody targeting, antibody-targeted micelles or immunomicelles can be prepared by chemical attachment of antibodies or antibody fragments to the activated water-exposed free terminus of a hydrophilic block of the micelle-forming polymer.[9,112] It is important to note that modification with antibodies does not increase micelle size, since diffusion and accumulation of carriers in tumors are both dependent on the cutoff size of the tumor vasculature, which in turn varies among tumors.[58]

For attachment of mAb to the PEG-PE–based micelles, a simple one-step procedure has been adapted that utilizes PEG-PE with the free PEG terminus activated with a p-nitrophenylcarboxyl group, that remains stable at pH values below 6 and efficiently interacts with amino groups of various ligands (antibodies and their fragments) at pH values above 7.5, to yield a stable carbamate bond.[80,113]

PEG-PE micelles modified with monoclonal antinuclear antibodies with a nucleosome-restricted specificity (monoclonal antibody 2C5 (mAb 2C5)) enhanced accumulation in tumors.[21,23,24,34,114,115] Paclitaxel loaded 2C5-modified mixed immunomicelles prepared from PEG_{2000}-PE and egg phosphatidylcholine (egg PC) provided the most efficient killing of cancer cells compared to the free drug or paclitaxel-loaded plain micelles.[34] In another study, dramatic improvement in *in vitro* cancer cell killing by mAb 2C5-modified mixed micelles composed of PEG-PE and Vitamin E for solubilization of paclitaxel or camptothecin compared to antibody-free micelles or free drug mixed immunomicelles was reported in various cell lines.[24,34,116] The 2C5-targeted immunomicelles showed up to 30% enhanced accumulation over free paclitaxel, or paclitaxel in plain micelles at 30 min and at 2 h post-injection in LLC tumors in mice.[21]

DOX-loaded polymeric immunomicelles made from poly (D,L-lactic-co-glycolic acid)-PEG (DOX-PLGA-PEG) and targeted with the bivalent fragment, HAb18 $F(ab')_2$, against hepatocellular carcinoma (HCC) have been reported.[117] The therapeutic action of these micelles improve owing to the specific molecular targeting and a higher uptake by the HCC cell lines HepG2 and Huh7 *in vitro*. The targeted micelles also suppressed *in vivo* tumor growth significantly (63.9%) when compared to free DOX (39.8%) or DOX-PLGA-PEG (50.2%) in HepG2 xenograft-bearing nude mice.[117] Table 3 shows recent examples of targeted polymeric micelles.

Table 3. Examples of actively targeted polymeric micelles.

Polymers used	Therapeutic agent	Targeting ligand	Target	Results	Ref.
Poly (D,L-lactic-co-glycolic acid)-PEG	DOX	HAb18 $F(ab')_2$	Hepatocellular carcinoma	Significant tumor growth suppression *in vivo* over non-targeted micelles and free DOX, higher uptake by HCC cell lines *in vitro*.	117
PEG2000-PE	DOX	2C5	Cancer cell surface associated nucleosomes	Superior accumulation in cancer cell spheroids compared to free DOX and non-targeted DOX micelles, higher toxicity in Bcl-2 overexpressing ADR-resistant spheroids.	118
	Paclitaxel	CB86	Peripheral benzodiazepine receptor (PBR)	Enhanced toxicity *in vitro* relative to non-targeted drug-loaded micelles due to synergistic effect of CB86 with drug.	119
d-α-tocopheryl polyethylene glycol succinate 2000	Docetaxel	Folic acid	Folate receptor	Increased cellular uptake of targeted micelles, improved cytotoxicity *in vitro* compared with Taxotere® or non-targeted $TPGS_{2k}$ micelles.	120
poly(ε-caprolactone)-*b*-poly(ethylene oxide)	DOX	cyclic RGDfK	$\alpha_v\beta_3$ integrin	Significant interaction of targeted micelles with bladder cancer cells in contrast to normal cells, greater cell uptake relative to non-targeted micelles *in vitro*.	121

(*Continued*)

Table 3. (*Continued*)

Polymers used	Therapeutic agent	Targeting ligand	Target	Results	Ref.
N-succinyl-N- octyl chitosan	10-hydroxy camptothecin	Folate	Folate receptor	Significantly enhanced cell uptake compared to free drug or non-targeted micelles, higher toxicity against FR positive hepatoma cells, greater antitumor efficacy in mice.	122
Pluronic P123 (PEO_{35}–PPO_{39}-PEO_{35})	Paclitaxel	anti-HIF-1α Ab	HIF-1α	Targeted micelles specifically bound to stomach cancer cells, resulted in selective killing of cells overexpressing HIF-1α.	123
PEG–Chitosan–Stearic acid	Doxorubicin	RGD	$\alpha_v\beta_3$ and $\alpha_v\beta_5$ integrins	RGD-targeted micelles selectively increased DOX concentration in integrin-overexpressing cells BEL-7402, significant cytotoxicity relative to non-targeted micelles.	124

4.3 *Stimulus sensitive micelles for triggered release*

Applying a certain internal or external stimulus can enhance drug release from micelles at the targeted area (Fig. 3). Several stimuli for triggered release have been reported, including pH, temperature, ultrasound and light.[125] Some examples of polymeric micelles which respond to various stimuli are summarized in Table 4.

4.3.1. *pH-sensitive micelles*

Differences in pH between healthy tissue and tumor tissue, as well as the environment in endosomal and lysosomal compartments, can be exploited as an internal stimulus for triggered drug release. Tumors have slightly acidic intratumoral pH (around 6.5) but close to blood capillaries the pH is still around 7.2. The acidic pH is as a result of hypoxia and cell death within tumors.[126,127] The acidic pH occurs deep within tumors, and so it is difficult to diffuse through for large particles.[128,129] pH is lowered further inside cells, especially in the endosomes, to about 5.5.[130] Thus, for tumor targeting, micelles with a phase transition at a pH of approximately 7.0 are useful and for subcellular targeting, micelles that are responsive to pH values ranging from 5.0 to 6.0 (endosomal-lysosomal pH) are beneficial.

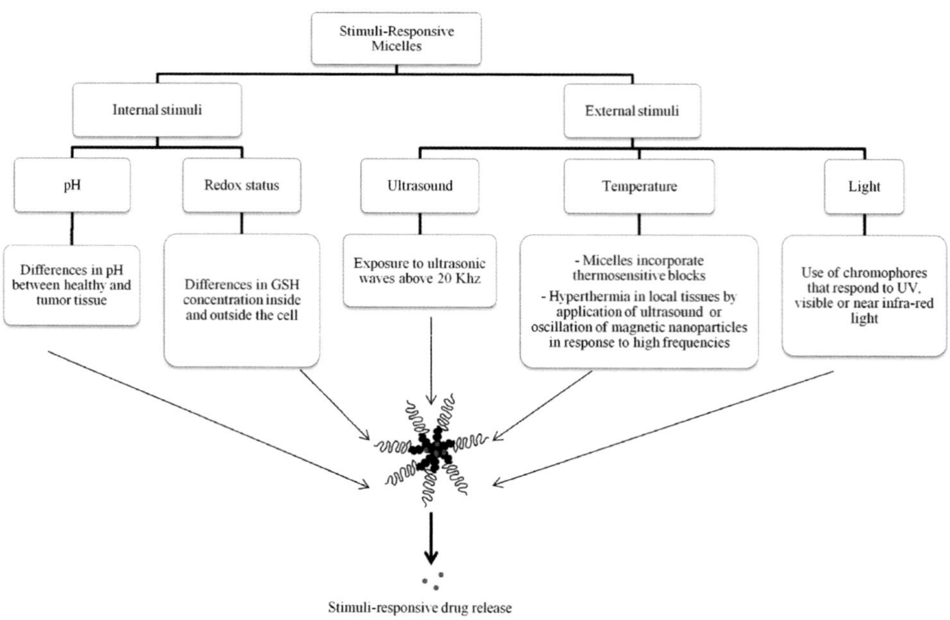

Fig. 3. An outline of stimuli-responsive polymeric micelles with various mechanisms used to trigger drug release.

Table 4. Examples of stimuli-sensitive micelles.

pH- sensitive micelles			
Micelle components	Mechanism	Results	Ref.
Methyl ether poly (ethylene glycol)-poly (β-amino ester) block copolymer	Sharp pH-dependent micellization/demicelli-zation at the acidic tumor pH 6.4.	Efficient release of drug/dye in acidic pH, 11-fold higher therapeutic efficacy in mice with pH-sensitive micelles compared to free drug or drug-loaded, non-pH sensitive micelles.	140
Poly(ethylene glycol)-poly(aspartate hydrazide) block copolymer	Hydrazide groups on side chains used as pH-sensitive binding linkers for drugs with ketone groups.	Successful combination chemotherapy with DOX and 17-hydroxyethylamino-17-demethoxygeldanamycin in pH-sensitive micelles compared to combination in DMSO formulations.	148
PEG-oxime tethered polycaprolactone-PEG (PEG-OPCL-PEG)	pH-sensitivity imparted by oxime bonds in the polymer backbone.	DOX release from micelles accelerated at acidic pH of 5.0 compared to physiological pH (7.4), efficient internalization and high anticancer efficacy of micelles.	152
Stearic acid grafted chitosan modified with PEG via cis-aconityl linkage	pH-sensitivity due to the cis-aconityl anhydride linkage between PEG and polymer backbone.	Cellular uptake of micelles enhanced under acidic conditions due to cleavage of PEG, greater accumulation and antitumor activity of pH-sensitive micelles versus pH-insensitive counterparts.	267
Poly(ethylene oxide)-*hyperbranched*-polyglycerol –DOX	DOX conjugated to PEO-*hb*-PG via an acid-labile hydrazone linkage which cleaved under acidic conditions to release DOX.	Increased cytotoxicity and efficacy of pH-sensitive DOX micelles in HeLa cells.	268

(*Continued*)

Table 4. (*Continued*)

Thermosensitive micelles

Micelle components	Mechanism	Results	Ref.
Poly(*N*-isopropylacrylamide-*co*-*N*,*N*-dimethylacrylamide)-*b*-poly(**d**,**l**-lactide)	Intracellular uptake of thermoresponsive micelles promoted by heating target area above micellar LCST, 39.4°C.	Uptake above micellar LCST caused by enhanced interactions between micelles and cell membranes through dehydration of P(IPAAm-DMAAm)chains.	269
Poly(ethylene glycol) (PEG), 1,3-bis(carboxyphenoxy) propane (CPP) and sebacic acid (SA)	Irreversible sol-nanogel transition at body temperature (above LCST), resulting in deformation of micelle structure and drug release.	DOX-loaded micelles effectively inhibited growth of tumor cells *in vitro*.	270
Y-shaped amphiphilic copolymers with two hydrophobic poly(solketal acrylate) branches and one MPEG2000 block	Possible shrinkage of the of micellar aggregates with increasing temperature generated contraction stress for squeezing out dye molecules from hydrophobic compartment.	Decreasing micelle diameter with increase in temperature, significantly enhanced release of nile red from micelles at 37°C versus that at 25°C under different pH conditions.	271
β-cyclodextrin-modified poly(ε-caprolactone)-poly(ethylene glycol)-poly(ε-caprolactone) co-polymer (PCEC-β-CD)	Micelles undergo temperature dependent sol-gel transition which is a free flowing solution at low temperatures and non-flowing gel at physiological temperature.	Indomethacin loaded PCEC-β-CD micelles exhibited controlled release *in vitro*, and significantly improved anti-inflammatory effect of indomethacin in rat arthritis models.	272

(*Continued*)

Table 4. (*Continued*)

Ultrasound-sensitive micelles

Micelle components	Ultrasound Frequency	Results	Ref.
Pluronic 105	Low frequency, 70 kHz	Release of DOX from folate targeted micelles increased as power intensity of ultrasound increased, maximum drug release (14%) at 5.4 W/cm^2, drug release triggered by cavitation.	273
Methoxy poly(ethylene glycol)-*b*-poly (D,L-lactide)	High frequency, 4 MHz	Two-fold increase in intracellular PTX , decreased drug efflux and greater cytotoxicity for ultrasound treated drug-sensitive as well as drug-resistant cell lines compared with no ultrasound treatment.	274
Pluronic105-DSPE-PEG2000	Low frequency, 20 kHz	Low frequency, high power ultrasound released 7–10% DOX from mixed micelles, potential of forming cytotoxic DNA adducts at tumor sites with co-administered formaldehyde releasing drugs.	275

(*Continued*)

Table 4. (*Continued*)

Redox-sensitive micelles

Micelle components	Redox-sensitive link	Results	Ref.
poly(ethylene oxide)-*b*-poly(N-methacryloyl-N′-(t-butyloxycarbonyl) cystamine) (PEO-b-PMABC)	Intermediate disulfide linkages in the core of micelles (formed by PMABC units).	Micelles readily cleaved into free copolymers under reducing environment. Higher anticancer efficacy and faster DOX release from redox-sensitive micelles versus non-sensitive controls.	217
Hexa-arm star-shaped poly(ε-caprolactone)-SS- poly(ethyl glycol) (6sPCL-SS-PEG).	Intermediate disulfide linkages between the core and shell result in shedding of PEG shell under reductive conditions.	High stability and rapid destabilization of micelles in reducing conditions, GSH-dependent cytotoxicity of DOX-loaded formulations *in vitro.*	219
PEG-SS-poly(ε -benzyloxycarbonyl-L-lysine) (mPEG-SS-PzLL)	Micellar rearrangement due to reductive cleavage of disulfide linked PEG causes rapid release of encapsulated DOX.	DOX release from micelles accelerated under increasing GSH concentrations.	276
Monomethoxy-poly(ethylene glycol)-S-S-hexadecyl (mPEG-S-S-C16)	Disulfide bridge linkage between mPEG and C16 imparts redox sesntivity in reducing environment.	Faster DOX release, significantly enhanced cytotoxicity and efficient internalization into HeLa cells along with DOX release in the cytoplasm and entry into nuclei for the redox-sensitive micelles compared to the non-redox-sensitive counterparts.	277

(*Continued*)

Table 4. (*Continued*)

Light-responsive micelles			
Micelle type	Chromophore/photosensitizer	Response to light	Ref.
Poly(N-isopropylacrylamide)-azobenzene	Azobenzene moieties inserted in the main chain.	Micelle core formed by azobenzene undergoes reversible *cis-trans* isomerism due to polarity change.	213
Dialkoxycyanostilbene polymethacrylate-b-poly(ethylene oxide) (PDACS-b-PEO)	Cyano stilbene units as pendant groups on the hydrophobic polymethacrylate moiety.	UV irradiation converts linear trans-stilbene to bent cis form and this change of polarity reduces the hydrophobicity of the polymer causing micellar disruption to release encapsulated curcumin.	252
Poly(ethylene oxide)-b-poly(4,5-dimethoxy-2-nitrobenzyl methacrylate) with $NaYF_4$.TmYb upconverting nanoparticles (UCNPs) encapsulated in micelles.	Nitrobenzyl groups on the micelle-core forming block undergo photocleavage.	Exposure to 980nm (NIR) light caused UCNPs to emit photons in the UV region, which are absorbed by o-nitrobenzyl groups, activating photocleavage and dissociation of micelles to release co-loaded nile red.	278
Poly(2-ethyl-2-oxazoline)-b-poly(D,L-lactide) (PEOz-b-PLA)	Photosensitizer: meta-tetra (hydroxyphenyl) chlorin (m-THPC).	*In vitro* light exposure: 60s (0.96 J/cm2), *in vivo*: single dose of 10 J/cm^2 (652nm light source), light spot diameter:1cm. m-THPC released under acidic conditions due to pH sensitivity of PEOz-b-PLA micelles. Phototoxicity *in vitro* and *in vivo*, with significantly reduced skin photosensitivity due to micelle encapsulation of photosensitizer.	279

In general, two strategies have been used to prepare pH-sensitive micelles. The first strategy involves use of pH-sensitive polymers bearing "titratable" groups such as amines, which display reversible protonation–deprotonation. The second is the use of acid-cleavable bonds between the chemotherapeutic agents and the micelle-forming copolymers.

The pH-triggered destabilization of micelles can be established by using a pH-sensitive polybase,[131] using pH-sensitive linkages (hydrazine linkage),[132] or by introducing titrable groups (methacrylic acid)[133] into the copolymer at the hydrophilic end while modifying the micelles for pH-sensitive targeting.

Poly(L-histidine) with a pKa of around 6.5 is the most commonly used pH-sensitive component in pH-triggered micelle-based systems since this polymer contains an imidazole ring that provides pH-dependent amphoteric properties.[105,131,134–136]

The protonation of the imidazole group at lower pH causes a reduction in the hydrophobicity of poly(L-histidine) leading to an increase in the CMC at pH below 7.2, which results in micelle dissociation. Thus, poly(L-histidine) is ionized at the lower pH of the tumors or acidic intracellular vesicles such as endosomes and lysosomes. In addition, poly(L-histidine) has an endosomal membrane disruption capability induced by its "proton sponge" mechanism and its interaction with negatively-charged membrane phospholipids.[137]

Bae *et al.* synthesized a novel diblock copolymer poly(L-histidine) -*b*-PEG for the preparation of pH-sensitive micelles. However, these micelles were less stable at pH 7.4, resulting in a shorter *in vivo* circulation time.[131] Later they synthesized a mixed poly(L-lactide) (PLLA)/PEG block copolymer with poly(L-histidine)-*b*-PEG to regulate the pH.[138] The results showed that, depending on the amount of PLLA/PEG, the triggered release of adriamycin occurred in the pH range of 7.2–6.6. Furthermore, to improve drug retention in multidrug-resistant (MDR) tumor cells, folate[103] or biotin[105] was selected as a specific ligand. Later, tumor-targeting DOX-loaded folate-pH sensitive mixed micelles were reported, which not only recognized the cancer cells but also released the drug at tumor sites, inhibiting cancer growth and even metastasis.[139]

Similarly, other pH-sensitive micelles using poly(β-amino ester)[140] and poly-sulfonamide have also been successfully developed.[141,142]

Wu *et al.* recently reported a mixed micellar system used to achieve a rapid intracellular drug release in response to the acidity in endosomes that was based on three grafted copolymers, including PEG2000-DSPE, mAb 2C5-modified DSPE-PEG3400 and PEG-coupled poly(L-histidine) (PHIS-PEG2000).[143] Mixed micelles loaded with paclitaxel showed about 75–95% drug release at a pH of around 5.5 within 2 h. The MTT assay showed

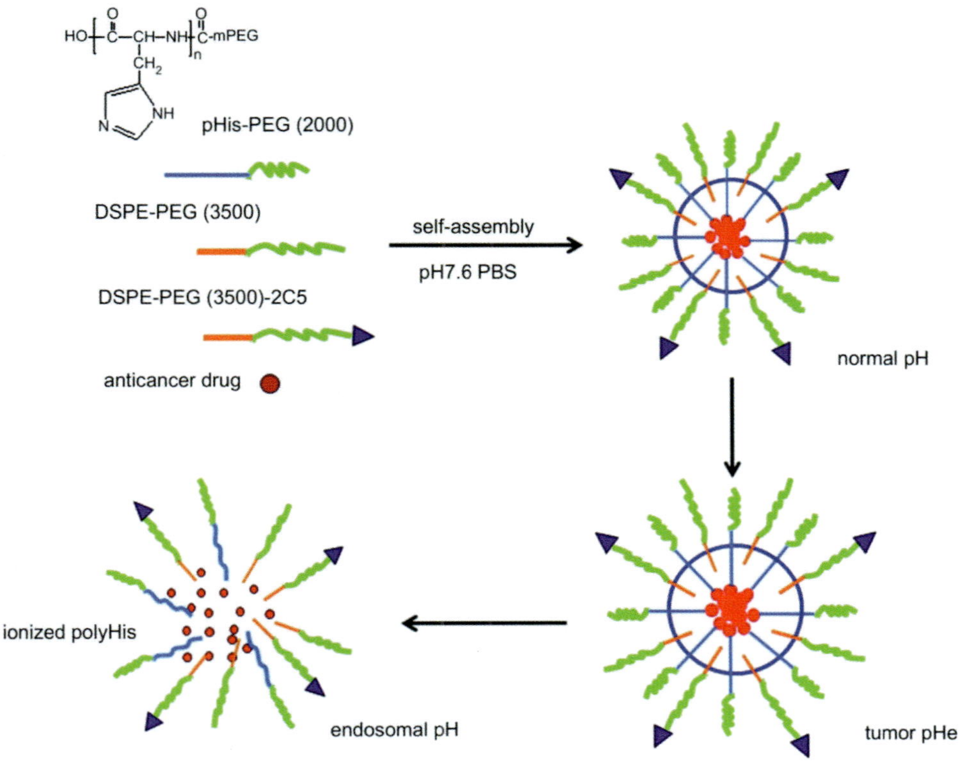

Fig. 4. Acid-triggered drug release from pH-sensitive mixed micelles. (Reprinted from Ref. 143, with permission from Elsevier.)

paclitaxel-loaded mixed micelles had higher cytotoxicity at pH 5.8 than that at pH 7.4. Further modification of the mixed micelles with anti-cancer nucleosome-specific monoclonal antibody 2C5 significantly increased their cellular uptake efficiency and cytotoxicity (Fig. 4).

Another strategy involved pH-dependent cleavage of the chemotherapeutic agent from micellar nanoconjugates obtained by linking a pH-sensitive moiety between the drug and copolymer. Several acid-sensitive bonds have been investigated for the development of polymer-drug conjugates such as Schiff's base,[144,145] hydrazone,[146–149] acetal,[150] ortho ester[151] and oxime.[152] In addition to direct conjugation of drugs, these linkages were inserted into the main chain, side chain or at the terminal of the core-forming blocks, resulting in triggered drug release through structural changes of the micelle-forming copolymers.

Introduction of titratable groups is another approach to disrupt micelles in an acidic environment. Leroux *et al.* introduced the hydrophilic titratable monomer methacrylic acid at the hydrophilic end of poly(N-isopropylacrylamine)-b-poly(O-octadecyl acrylate) copolymer which showed a pH phase transition and structural changes in the micelle core at a pH of approximately 5.8.[153]

Polymeric micelles with stably crosslinked ionic cores (*cl*-micelles) composed of poly(ethylene oxide)-*b*-poly(methacrylic acid) (PEO-*b*-PMA) diblock copolymers and with cores of crosslinked PMA network surrounded by a shell of PEO chains were reported.[154,155] They can swell in aqueous environments and display pH and ionic-strength responsive, hydrogel-like behavior characteristic of ionic nanogels.[156] Various anti-cancer drugs such as DOX or cisplatin were successfully encapsulated into the *cl*-micelle ionic cores with high loading efficiency.[154,157,158] However, these *cl*-micelles are not degradable and hence, there is a concern of their long-term accumulation in the body.

4.3.2. *Thermosensitive micelles*

Thermosensitive micelles are prepared from a thermosensitive block of block copolymer-forming micelles, which have a distinct lower critical solution temperature (LCST) or cloud point (CP). Below the LCST, a thermosensitive polymer is water soluble, but above this temperature the polymer becomes insoluble.

The most commonly studied polymers for the preparation of thermosensitive micelles are poly(N-isopropylacrylamide) (PNIPAAm) and Pluronic (PEG-*b*- PPO-*b*- PEG, triblock copolymers of polypropylene oxide [PPO] middle blocks flanked by polyethylene glycol [PEG] blocks).[159,160] Pluronics, due to the thermosensitive PPO block, form micelles above the LCST of this block.

PNIPAAm has a reversible and sharp phase transition at around 32°C. At this temperature, it switches from a water-soluble hydrophilic polymer to a hydrophobic insoluble polymer.[161] Block polymers of pNIPAAm with hydrophobic or hydrophilic blocks can form micelles with pNIPAAm as the micellar shell or core, respectively. Thus, the copolymers that include the LCST as a hydrophobic block, such as PEG–PNIPAAm copolymers, form micelles and can incorporate lipophilic drugs only above the LCST of PNIPAAm. In this case, local cooling of the tumor tissue would trigger the drug release rather than heating, which is a difficult problem. Conversely, when the pNIPAAm block is combined with a permanently hydrophobic block, it forms the micellar corona at temperatures below the LCST. However, pNIPAAm has its LCST below body temperature and the micelles with pNIPAAm as a shell are expected to rapidly aggregate upon injection. Thus, it is not suitable for *in vivo* applications.

Therefore, NIPAAm was copolymerized with hydrophilic monomers to yield polymers with an LCST above 37°C. p(NIPAAm *-co-* acrylamide)*-b-* (D,L, lactide) with an LCST of 41°C was used for the design of thermosensitive polymeric micelles for the triggered release of docetaxel by hyperthermia resulting in enhanced *in vitro* and *in vivo* effects.[152] For example, co-polymerization of NIPAAm with *N* -hydroxymethylacrylamide and subsequent coupling to poly(methyl methacrylate) resulted in poly(NIPAAm*-co-*HMAAm)*-b-* pMMA) with an LCST of 42.8°C that formed micelles with a size of 250 nm. Micelles of this polymer were loaded with an anticancer drug (methotrexate) and showed enhanced release when heated to 43°C.[163]

The major drawback of pNIPAAm block copolymers for the development of polymeric micelles for pharmaceutical applications is that they are not degradable and local hypothermia or hyperthermia is necessary for micellar destabilization and subsequent release of the encapsulated drug, which is not always feasible in clinical settings. Thus, copolymers of NIPAAm with biodegradable monomers have been synthesized. In these systems, side-chain hydrolysis leads to a phase transition of the thermosensitive block from the collapsed to the dissolved state, resulting in micelle destabilization and concomitant release of the payload.[164]

Soga *et al.* synthesized a copolymer of poly(N-(2-hydroxypropyl) methacrylamide lactate) and poly(ethylene glycol) (pHPMAmDL-b-PEG).[165,166] The poly(HPMAm-lactate) showed a LCST in aqueous solution and the CP can be well controlled between 10 and 65° C by the length of the lactate side group (e.g. monolactate or dilactate) and copolymer composition.[167] AB block copolymers of poly(HPMAm-dilactate) and PEG (pHPMAmDL-b -PEG) form polymeric micelles which gradually dissolve due to hydrolysis of the lactic acid side groups.[168] When loaded with paclitaxel, these micelles showed a release of about 70% of the incorporated paclitaxel during 20 h at 37 °C and at pH 7.4. The antitumor efficacy of this micelle formulation in B16F10 tumor-bearing mice after *i.v.* administration was comparable to the Cremophor EL formulation of paclitaxel but without the toxic side- effects associated with Cremophor EL. However, no significant accumulation of micelles in tumor tissue was observed after 24 h. This could be due to premature micelle destabilization.

Hennink *et al.* synthesized and extensively studied biodegradable thermosensitive PEG-b-poly[N-(2-hydroxypropyl) methacrylamide-lactate] (mPEG-b-p(HPMAm-Lacn)) diblock copolymers which consist of a permanently hydrophilic (PEG) and a thermosensitive (p(HPMAm-Lacn)) block.[168–170] This block copolymer self-assembles into micellar structures in aqueous solution when heated above the critical micelle temperature (CMT).[168] These micelles demonstrated tunable biodegradability due to hydrolysis of the lactic acid side chains under physiological conditions, resulting in micelle disintegration. However, polymeric micelles of mPEG-b-p(HPMAm-Lacn) were shown to be rapidly removed from the circulation *in vivo*, probably due to premature micelle dissociation.[54]

An increased micellar stability can be established by either physical or chemical crosslinking of the shell, the intermediate layer or the core of the micelles.[54,171] Thus, in this way, the circulation time of the micelles can be enhanced. Also, crosslinking enables controlled release of the entrapped drug.[172] Rijcken *et al.* prepared core-crosslinked thermosensitive micelles based on N-(2-hydroxyethyl) methacrylamide (HEMAm) block copolymers which were derivatized with various monodispersed oligolactates (HEMAm-Lacn). Part of the hydroxyl end groups of the lactic acid groups were derivatized with methacrylic acid to allow crosslinking of the micellar core by radical polymerization.[54] These core-crosslinked micellar system showed excellent physical stability and a superior circulation profile compared to non-crosslinked micelles. More than 50% of the micelles remained in the circulation 6 h post-injection, and an increased tumor accumulation was observed.

In another study, a DOX derivative was covalently conjugated to the micellar core through a pH-sensitive hydrazine linker that is relatively stable at pH 7.4, but readily cleaves at pH 5 (in the endo/lysosomes of cells). An increased cytotoxicity, compared with free DOX, was observed in two different cell lines, as well as an increased antitumor activity and prolonged survival of B16 melanoma-bearing mice.[55] Very recently, the same system was actively targeted via conjugation of an anti-EGF receptor nanobody on the micellar surface and increased cellular binding and uptake was observed by EGF receptor-overexpressing cells, compared to micelles without the nanobody.[173] This opens a route to actively targeted thermosensitive polymeric micelles, which is expected to result in even more enhanced tumor accumulation and thus, increased efficacy, owing to the targeting moiety coupled on their surface.

A different approach to the development of thermosensitive polymeric micelles consists in polymerization of temperature-responsive LCST hydrogels inside micelle cores.[174] Plurogels were developed with an interpenetrating network of poly(propylene oxide) blocks of Pluronic P-105 and poly(N-alkylacrylamide) in Pluronic micelle cores. These Plurogels can be loaded with substantial amounts of drug at room temperature. At the transition temperature, a sharp reversible volume decrease occurs, which "locks" micelle cores and substantially decreases micelle degradation rate even at extreme dilutions. Unfortunately, these systems have proven to be toxic due to the dramatic dehydration of live cells.[175] In another approach, thermoresponsive Pluronic micelles were produced by crosslinking shells with gold nanoparticles; the micelles exhibited reversible swelling/shrinking behavior during temperature cycling between 15°C and 37°C.[176]

4.3.3. *Ultrasound sensitive micelles*

Ultrasound refers to the transmission of pressure waves at frequencies above human hearing, or above 20 kHz.[177] The usefulness of ultrasound derives from its non-invasiveness, deep penetration into tissues and the ability to control and focus ultrasonic waves at tissues or tumor sites.[178,179] It also enhances drug absorption and membrane permeability of drugs.[180,181] The mechanisms for the biological effects of ultrasound are related to the generation of thermal energy (hyperthermia), perturbation of cell membranes due to formation of oscillating or cavitating bubbles and enhanced permeability of blood capillaries.[180,182]

Ultrasound-responsive drug delivery systems can accumulate in the desired areas, which can be made leaky by locally applied ultrasound to liberate incorporated drugs or genes.[28] Micellar drug delivery using ultrasound was initially performed with Pluronic-based micelle systems and has been studied extensively.[183,184] Munshi *et al.* reported that the IC_{50} of DOX in HL-60 (human leukemia) cells could be lowered significantly from 2.35 to 0.19 μg/ml when it was encapsulated in Pluronic® P105 micelles and used in combination with ultrasound [183]. The same group reported that DOX led to increased DNA damage and cell death in HL-60 cells when ultrasound was applied, whereas no significant DNA damage occurred upon exposure of cells to DOX for up to 9 h in the presence of P105 (10 wt%) in the absence of ultrasound.[185] In other studies involving DOX-loaded Pluronic micelles, Marin *et al.* showed that exposure to ultrasound not only increased the intracellular uptake of Pluronic micelles but also enhanced their trafficking to the nucleus.[181] In addition, they also studied the mechanism of ultrasound enhanced intracellular uptake of drugs from polymeric micelles. Higher amounts of free drug were found in the cells than those retained in the micelles, when exposed to ultrasound, which pointed to the role of ultrasound in micelle perturbation and release of drug from the micelle core. Ultrasound also offered the advantage of enabling the same degree of drug uptake at a much lower concentration of drug in the incubation medium.[181,186]

A number of factors affect the ultrasound-triggered release of drugs from polymeric micelles. These include time of application, frequency of the ultrasound, application of pulsed versus continuous wave ultrasound, power density, drug lipophilicity as well as the concentration of drug and polymer.[186,187] Low-frequency ultrasound is more effective at drug release from micelles than high-frequency ultrasound.[188] Although low-frequency (20–100 kHz) ultrasound can penetrate deeper into tissue than high-frequency ultrasound (1–3 MHz), it is not amenable to sharp focusing.[188,189] For *in vivo* applications, low-frequency ultrasound is more appropriate for larger and deep-seated tumors, albeit it may damage healthy tissues owing to a strong cavitation effect, which may limit its use in the clinic. High-frequency ultrasound, on the other hand, may be useful for small, superficial tumors.[188,189]

One of the key considerations when injecting micelle-encapsulated drugs *in vivo* is their stability. Micelles diluted manyfold in the body disintegrate into unimers, thereby releasing the encapsulated drug prematurely.[182] To prevent this, micelles can be stabilized with highly hydrophobic cores or solid-like cores. PEG- diacyl phoshospholipids like $mPEG_{2000}$-DSPE and crosslinking agents like N, N- diethylacrylamide (NNDEA) have been used to stabilize Pluronic P105 micelles.[175,182,190] Plurogels are an example of Pluronic P105 micelles stabilized by an interpenetrating network of the thermoresponsive polymer NNDEA within the hydrophobic core. Poly(NNDEA) has a LCST of 28°C, below which it exists in an expanded hydrophilic state that enables drug loading, while above the LCST, the interpenetrating network collapses into a tight hydrophobic core in which the drug is entrapped.[190] These stabilized Pluronic micelles released DOX in response to 70 kHz ultrasound like their non-stabilized counterparts.[175,191,192] Nelson *et al.* showed in a rat tumor model using the DHD/K12/TRb colon carcinoma cell line, that application of 20 and 70 kHz low-frequency ultrasound reduced the tumor size significantly compared to non-insonated controls. Rats were treated with stabilized micelles or free DOX. Free DOX also showed significant tumor reduction as a result of enhancement of passive diffusion using ultrasound.[191] Gao *et al.* used mPEG2000-DSPE combined in a 1:1 ratio with Pluronic P105 to decrease the CMC and stabilize micelles. *In vivo,* stabilized and non-stabilized micelles loaded with DOX, when exposed to low-frequency ultrasound, resulted in a significant decrease in tumor volumes compared to mice not treated with ultrasound.[182,193] Biodistribution of stabilized and non-stabilized Pluronic P105 micelles was studied in a mouse model of ovarian cancer using fluorescently labeled compounds. A 30s ultrasonic irradiation with 1 or 3 MHz ultrasound resulted in an enhanced accumulation of the labeled micelles at the tumor sites.[193] The use of high-frequency ultrasound (1 MHz) *in vivo* allows better focusing on the tumor site and causes less cavitation and hence less damage to healthy tissues.[194]

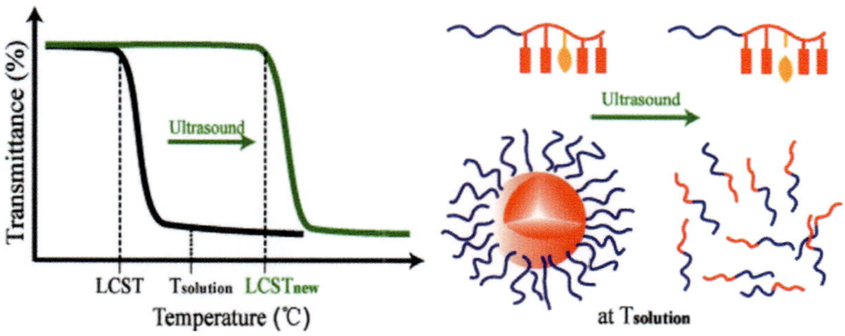

Fig. 5. Schematic illustration of the amplification mechanism for ultrasound-disrupted block copolymer micelles based on an ultrasound-induced increase in the lower critical solution temperature (LCST) of the hydrophobic block. THPMA groups undergo hydrolysis upon application of ultrasound and convert hydrophobic THPMA comonomer units into hydrophilic methacrylic acid, resulting in disruption of micelles. (From Ref. 196, Copyright © 2012 American Chemical Society.)

Although the majority of the research has focused on Pluronic micelles, non-Pluronic micelles such as poly(ethylene glycol)-co-poly(β-benzyl-L-aspartate) (PEG-b-PBLA)[182] and poly(lactic acid)-*b*-poly(ethylene glycol) (PLA-b-PEG) copolymer micelles[195] have also been studied for ultrasound-mediated drug release. Recently, Xuan *et al.* suggested the use of a diblock copolymer made up of a water-soluble poly(ethylene oxide) PEO block and a block of poly(2-(2-methoxyethoxy)ethyl methacrylate) ($PMEO_2MA$) that is hydrophobic at temperatures above its LCST.[196] A small amount of HIFU-labile 2-tetrahydropyranyl methacrylate (THPMA) co-monomer units was incorporated into $PMEO_2MA$ to form the micelle core above its LCST. Under the effect of HIFU (1.1 MHz), the THPMA groups hydrolyzed, and caused the conversion of hydrophobic THPMA co-monomer units into hydrophilic methacrylic acid, and the block co-polymer micelles disassembled without changing the solution temperature (Fig. 5).[196]

Ultrasound is also used to enhance the efficacy of nucleic acid delivery either with or without the use of microbubbles (gas-filled bubbles used as ultrasound contrast agents).[179,197,198] Chen *et al.* used three Pluronics F127, L61 and P85 for gene delivery *in vitro* and showed that when used alone, they were unable to mediate gene transfer, but in the presence of ultrasound (1MHz) for 20 s at a power density of 1 W/cm^2, the pluronic micelles significantly enhanced gene transfer efficacy.[198] F127 and P85 increased the transfection rates up to 30% under the influence of ultrasound at concentrations ranging between 0.005% and 0.1%. The use of polymeric micelles combining ultrasonic tumor imaging and chemotherapy has also been reported.[199,200] Rapoport *et al.* developed multifunctional formulations, which consisted of mixtures of DOX-loaded polymeric micelles (PEG-PLLA) and nanoemulsion droplets formed by perfluoropentane (PFP) at room temperature. The droplets were stabilized by the same biodegradable block copolymer which was used to form micelles. At physiological temperatures, the nanodroplets converted into stable nano/microbubbles, with the drug partitioned between micelle cores and microbubble surfaces. Upon sonication, the drug transferred from the microbubble surface to the interior of cells due to the collapse of microbubbles via inertial cavitation.[199] *In vivo*, this effect led to a significant tumor regression related to enhancement of drug uptake by tumor cells.[199,201] Moreover, the micelle/microbubble formulation provided a strong and long-lasting contrast under the effect of ultrasound for imaging. Liu *et al.* developed a novel remotely triggered system with multimodal imaging which was capable of magnetic resonance (MR) imaging and ultrasound imaging, encapsulated camptothecin, and also showed ultrasound-triggered release.[200] This platform consisted of superparamagnetic iron oxide (SPIO) nanoparticles self-assembled with an amphiphilic chitosan derivative, carboxymethyl hexanoyl chitosan (CHC), to form CHC/SPIO micelles, which were then loaded with camptothecin, an anticancer agent. The micelles were further conjugated with albumin-based microbubbles (MBs) to form superparamagnetic micelle-decorated MBs (CHC/SPIO-decorated MBs). *In vitro,* this system showed significant concentration-dependent ultrasound imaging contrast, MR T2 imaging contrast and ultrasound-induced MR T2 signal change. The release of camptothecin encapsulated in the micelles could be triggered by therapeutic, but not diagnostic, ultrasound imaging. The timing and location of drug release could be determined by MR or diagnostic ultrasound images.[200] Other studies also have combined HIFU with MRI and have shown that ultrasound-induced hyperthermia can be controlled using MRI thermometry to make this technique suitable for triggered drug release from both ultrasound- and thermosensitive micelles.[202]

4.3.4. *Redox-sensitive micelles*

There exists a significant difference in the redox potential (100–1000 fold) between the intracellular and the extracellular compartments, which results in a reducing environment within the cell and an oxidizing environment in the extracellular space.[203,204] This change in redox potential is attributed to the difference in concentration of glutathione (GSH) in the cytoplasm (1–11 mM) from that in the plasma (about 10 μM).[205,206] In addition, cytosolic GSH concentration in tumor cells is elevated and is several times higher than in normal cells.[207] These differences in redox potential can be exploited for the design of thiol/redox-sensitive nanocarriers for delivery of drugs and genes.[204] A key step in many biological processes, the interconversion of thiols and disulfides, has been utilized synthetically for various bioconjugation protocols.[203,208] The cargo (drug/DNA) can be encapsulated in the nanocarrier which is held together by disulfide (S–S) bonds under normal conditions. Once disulfide bonds are reduced to thiols due to the high GSH within cells, the carrier is destabilized and can release the entrapped cargo.[28,209]

In the 1980s Saji *et al.* initially demonstrated the concept of redox-responsive delivery systems [210,211] using ferrocene-based, single-aliphatic chain surfactants, the CMC of which could be altered by oxidation of neutral ferrocene to cationic ferrocenium. These micelles could be oxidized by electrochemical or chemical methods, causing an encapsulated hydrophobic azo dye to be precipitated from solution. The process was reversible, and resolubilization of dye occurred upon reduction of the solution.[211] Various strategies have been devised for developing redox-sensitive micelles, including incorporation of reduction-responsive links at the junction of hydrophilic and hydrophobic blocks (a cleavable backbone), utilizing reduction-cleavable side chains and employing the crosslinking approach in the micelle core, shell or at the core–shell interface.[205,212]

Li *et al.* developed a novel disulfide-containing prodrug of camptothecin (CPT) covalently linked to a polymer (CPT-SS-PEG-SS-CPT) which self-assembled into nanomicelles that showed a redox-sensitive drug release mechanism under tumor-relevant reductive conditions.[213] PEG provided the hydrophilic component and CPT formed the hydrophobic portion in these nanomicelles. A disulfide linker and ester bond were incorporated between PEG and CPT which, when cleaved, generated native CPT due to ester degradation. The pharmacological efficacy of CPT released from the prodrug micelles was confirmed with cytotoxicity assays on HepG2 (human hepatoma) cancer cells in the presence of 10 mM GSH.[213] In another study, targeted, redox-sensitive micelles based on hyaluronic acid–deoxycholic acid (HA-ss-DOCA) conjugates were developed for the intracellular delivery of paclitaxel.[214] These micelles exhibited rapid drug release and an increased uptake in breast cancer cells via HA-receptor-mediated endocytosis under reducing conditions *in vitro* and also possessed higher tumor-targeting capacity relative to redox insensitive controls *in vivo* in MDA-MB-231 tumor-bearing mice.[214] A number of other groups have also employed novel polymers which incorporate reduction responsive links either in the backbone or side chains for redox-responsive release of anticancer agents from micelles (Fig. 6(A)).[215–220]

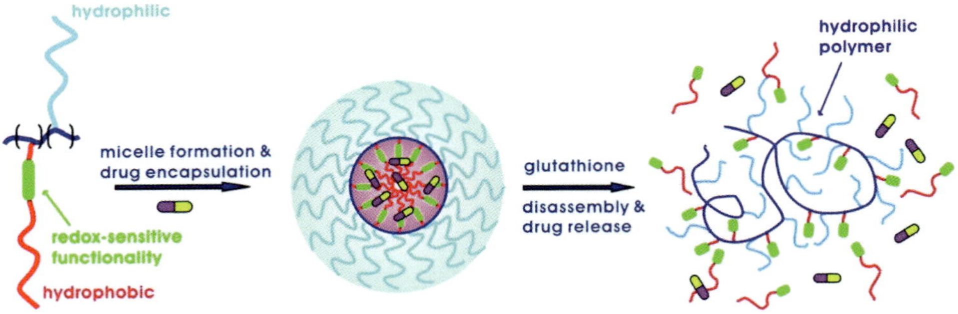

(a)

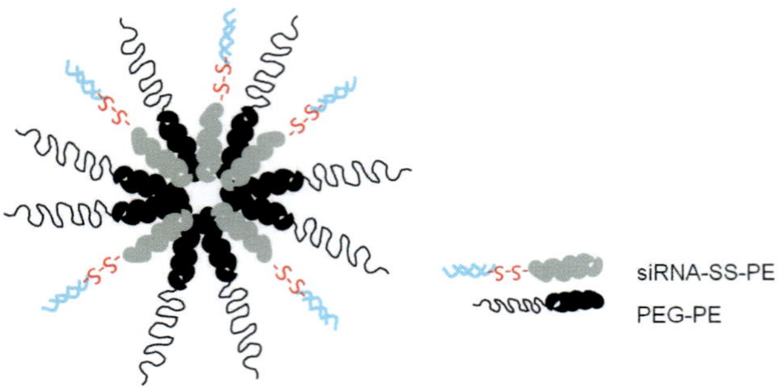

(b)

Fig. 6. Redox-sensitive micelles for GSH triggered release of drugs and siRNA. (a) Schematic representation of drug release via disassembly of amphiphilic polymer micelle triggered by GSH (From Ref. 216, Copyright © 2010 American Chemical Society) (b) Structure of siRNA-SS-PE/PEG-PE mixed micelles. Under reductive conditions, disulfide bonds are cleaved to release siRNA . (Modified from Ref. 228.)

The practical applicability of micelles *in vivo* is often hampered due to dissociation of micelles under shear forces, dilution in body fluids as well as interaction with blood components when administered intravenously.[171,205,221] To overcome stability issues and enhance the temporal control of micelles, crosslinkable groups can be used for functionalization of block copolymers.[221] Redox-sensitive micelles utilizing the crosslinking approach have been investigated extensively for delivery of small molecules and nucleic acids using either core-crosslinked, shell-crosslinked or micelles crosslinked at the core–shell interface.[222-225] Li *et al.* reported a reversible disulfide crosslinked micellar system (DCM) for targeted paclitaxel delivery using cysteine containing telodendrimers (a linear dendritic polymer composed of PEG and dendritic clusters of cholic acids). This crosslinked micellar system proved to be more efficacious in SKOV-3 (ovarian cancer) xenograft-bearing mice than the non-crosslinked counterparts as well as Taxol at similar doses. The paclitaxel-DCMs showed high drug-loading capacity, improved stability, prolonged circulation, preferential tumor targeting and controlled release, which could be subsequently enhanced in a reducing environment.[222] Core–shell–corona micelles with redox-responsive, shell-specific crosslinks were developed for delivery of docetaxel (DTX) for cancer therapy.[225] These consisted of poly(ethylene glycol)-b-poly(L-lysine)-b-poly(L-phenylalanine)(PEG-PLys-PPhe). Shell crosslinking was carried out by reaction of disulfide-containing crosslinkers with Lys moieties in the middle shells. These micelles exhibited enhanced efficacy in MDA-MB231 tumor-bearing mice compared with free DTX and non-crosslinked micelles.[225]

One of the key features for any successful gene delivery system is to ensure that the cargo remains stable in the circulation and is released only when it reaches the cell after overcoming cellular barriers.[226,227] Redox-sensitive micelles present a promising platform for intracellular delivery of genes and siRNA. Musacchio *et al.* reported an effective, non-toxic, redox-sensitive micellar system for stabilization and delivery of siRNA.[228] GFP-siRNA was reversibly modified with phosphothioethanol (PE) via a reducible disulfide bond (siRNA-SS-PE) and incorporated in PEG-PE micelles. In the resultant siRNA-SS-PE/PEG-PE mixed micelles, siRNA was protected against degradation by nucleases for at least 24 h and could be released from the micelles in free form in the presence of intracellularly relevant GSH concentrations. When tested in GFP-C166 endothelial cells, the mixed micelles downregulated GFP production 50-fold more effectively compared to free siRNA and also lacked the cytotoxicity often associated with delivery systems based on electrostatic association of siRNA with cationic carriers (Fig. 6(B)).[228] Kataoka and colleagues have investigated redox-sensitive micelles for delivery of plasmid DNA, antisense oligonucleotides and siRNA.[224,229–231] Antisense oligonucleotide (asODN) was conjugated to PEG with a disulfide linkage (PEG-SS-asODN) and assembled into polyion complex micelles (PICs) through complexation with branched poly(ethyleneimine) B-PEI . *In vitro*, this system exhibited an efficient antisense effect against luciferase gene expression in HuH-7 cells. B-PEI allowed successful transport of micelles from endosome to cytoplasm because of its buffering effect and permitted release of active asODN molecules in the cell interior due to cleavage of the disulfide linkage under the reductive conditions in the cytoplasm.[231] Disulfide-crosslinked PIC micelles for effective siRNA delivery were made by assembly of iminothiolane-modified poly(ethylene glycol)-block-poly(L-lysine) (PEG-b-(PLL-IM)) and siRNA at optimum mixing ratios. When compared to controls, they achieved a 100-fold higher transfection efficacy *in vitro* relative to non-crosslinked PICs prepared from PEG-b-poly(L-lysine).[224] Other block copolymers like poly(ethylene glycol)-block-poly{N-[N-(2-aminoethyl)-2-aminoethyl]aspartamide} have been investigated as gene carriers by forming PEG-detachable, disulfide-crosslinked micelles (PEG-SS-P[Asp(DET)]) with plasmid DNA for redox-responsive delivery.[230,232] Many other groups have also developed redox-responsive polyelectrolyte micellar systems for the delivery of siRNA.[233–235]

4.3.5. *Light-sensitive micelles*

Light can be used as an external stimulus to spatially and temporally control and trigger drug release by disruption or local destabilization of micelles in the body.[202,206,236] Ultraviolet (UV), visible or near-infrared (NIR) light can be used to trigger the drug release from micelles.[164] UV light is absorbed by the skin and does not penetrate deeply into the body, whereas NIR light has deeper penetration through tissues (up to 10 cm) and water, and has minimal effects on healthy cells which make it a feasible option for biomedical applications.[164,202,237] Light or photoresponsive processes can be controlled by adjusting the wavelength and intensity of the light. [236]

Light-responsive micelles can be designed based on the type and location of the chromophore or photoresponsive group within the micelles. These groups could be introduced either in the micellar core, shell or at the core–shell interface. However, in most cases, the chromophore is incorporated within or conjugated to the hydrophobic micellar core (as side chain or at the block junction).[236,237] When light-responsive micelles are illuminated, their dissociation occurs owing to a shift in the hydrophilic/hydrophobic balance as a result of photoreaction of the chromophore. Two types of photoreactions lead to these effects. In the first type, reaction of the chromophore can reversibly change the polarity of the hydrophobic polymer, so that it is unable to maintain the micellar association and undergoes destabilization. However, the polarity can be regained and micelles reform when the reaction stops.[202,237] In the second type, the photoreaction may lead to a structural change by irreversible cleavage of the chromophore from the hydrophobic block, converting it to a hydrophilic polymer and causing micelle destabilization.[164,237] In addition to shifting the hydrophobic–hydrophilic balance, a design strategy that is the most widely studied, light-sensitive block copolymer micelles can be grouped into three additional types.[238] In the first approach, micellar disruption occurs upon photoinduced cleavage at the junction of the hydrophilic and hydrophobic blocks. In the second type, photocleavable units are repeatedly placed along the main chain of the hydrophobic block to ensure photoinduced degradation of the micelles, while the last approach is used in cases where stability of micelles is desired. This involves a reversible photocrosslinking reaction for optical stabilization and destabilization of micelles. Here, micelles undergo photoinduced crosslinking of associated polymer chains, which provides structural integrity, while photoinduced de-stabilization can occur under a different wavelength due to polymer chain de-crosslinking.[238]

Zhao and colleagues pioneered the development of light-sensitive micelles by synthesizing a novel amphiphilic diblock copolymer with poly(ethylene oxide) PEO as hydrophilic block, linked to a hydrophobic poly(methacrylate) (PMA) block bearing a pyrene moiety in the side group (PPy) via an ester linkage.[239] A hydrophobic dye, Nile Red was incorporated into the PEO-*b*-PPy micelles. When an aqueous micellar solution was illuminated using UV light, photosolvolysis of the pyrenylmethyl esters caused cleavage of the ester bond, converting the hydrophobic PPy to hydrophilic poly(methacrylic acid) (PMA) resulting in irreversible disruption of micelles and the release of Nile Red (Fig. 7).[239] The same group also employed photocleavable moieties like 2-nitrobenzyl esters for investigating photocontrolled release from micelles [240] and esters of [7-(diethylamino)coumarin-4-yl]methyl (DEACM) to design polymer micelles that could be controlled with NIR light.[241]

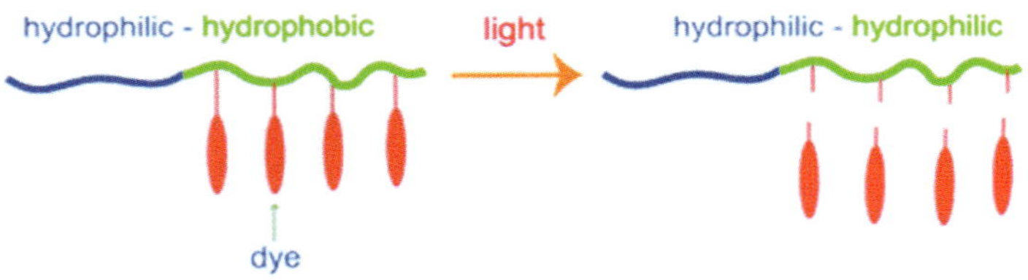

(a)

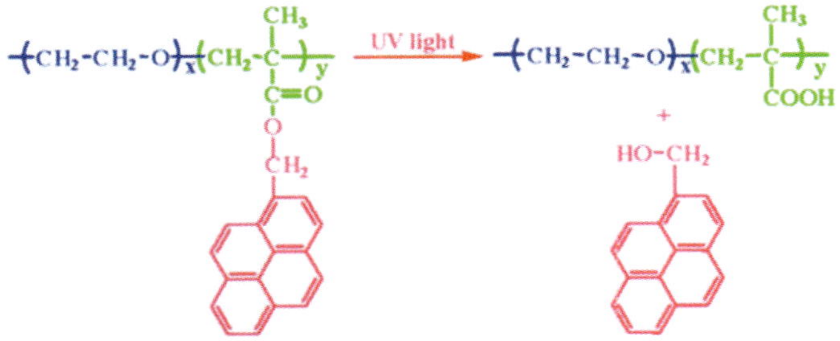

(b)

Fig. 7. (a) Schematic illustration of light-induced detachment of dye pendant groups resulting in the hydrophobic-to-hydrophilic switch. (b) Chemical structure of the pyrene-containing amphiphilic diblock copolymer and its photosolvolysis under UV light irradiation. (From Ref. 239, Copyright © 2005, American Chemical Society.)

Jin *et al.* reported a biocompatible, photoresponsive micellar system for delivery of 5-FU under physiological conditions using coumarin-functionalized block co-polymers.[242] The block copolymers poly(ethylene oxide)-b-poly(n-butylmethacrylate-co-4-methyl-[7-(methacryloyl)oxyethyloxy] coumarin)) (PEO-*b*-P(BMA-*co*-CMA) were synthesized using atom transfer radical polymerization (ATRP) and micelle-drug conjugates were formed by covalent bonding of 5-FU to coumarin using UV irradiation above 310 nm. They exhibited controlled release of 5-FU under UV light (254 nm) and also showed excellent biocompatibility in *in vitro* cytotoxicity assays.[242]

Light-induced disruption or formation of micelles can be designed so as to be reversible. Azobenzenes and their derivatives are the most widely studied reversible photoresponsive groups.[204,236] Azobenzenes can undergo *trans-cis* photoisomerization of their nitrogen-nitrogen double bond (N=N). UV light irradiation converts the apolar *trans* isomer to polar *cis* isomer, where as visible light can trigger the reverse isomerization.[237] Reversible light-responsive micelles with azobenzene-containing amphiphilic copolymers have been investigated by a number of groups.[213,243–246] NIR light-sensitive polymeric micelles formed from dextran-graft-(2-diazo-1,2-naphthoquinone) (Dex-DNQ) amphiphilic copolymers were developed for intracellular delivery of DOX. DNQ, a hydrophobic molecule, was the phototrigger group, which upon UV/NIR irradiation undergoes Wolff rearrangement to a hydrophilic 3-indene carboxylic acid (3-IC) molecule. *In vitro* studies showed a significant growth inhibition of HepG2 cancer cells as a consequence of higher intracellular DOX release from micelles under NIR irradiation.[214] In a novel approach, Sun *et al.* combined sugar-based targeting and two-photon sensitivity in a degradable, dendritic micellar nanocarrier for DOX. A third-generation propargyl focal point poly(amido amine) dendron (D3) was used to form D3-PCL-DNQ, and the propargyl focal point was conjugated with sugar-targeting groups (glucose, Glu; lactose, Lac) using click chemistry. Glu/Lac-D3-PCL-DNQ amphiphiles formed micelles with a PCL-DNQ core and hydrophilic D3 corona linked by free sugar-targeting groups. The sugar-targeted micelles could bind to lectins concanavalin A and ricinus communis agglutinin and NIR triggered release of DOX occurred in a controlled manner.[247] In another study, NIR light was used to trigger release of rifampicin and paclitaxel from biocompatible polypeptide block copolymer micelles prepared from poly(ethylene oxide) and poly (L-glutamic acid) bearing a number of 6-bromo-7-hydroxycoumarin-4-ylmethyl groups. The micelles were disrupted due to high two-photon absorption and NIR-induced removal of coumarin groups from the polypeptide block, causing a shift of hydrophobic–hydrophilic balance towards destabilization of micelles.[248] Spiropyran derivatives also exhibit a photoresponsive behavior, and undergo a reversible isomerization between colorless spiropyran (SP) and colored zwitterionic merocyanine (ME). SP incorporation into micelles allows better light responsiveness due to a larger difference in the polarity between hydrophobic SP units and hydrophilic, zwitterionic ME units.[204] Polymeric micelles using SP-containing blocks have also been reported.[249,250]

In addition to the photochromes discussed (azobenzene, DNQ and spiropyran), other groups such as dithienylethene,[251] stilbene[252] and cinnamoyl derivatives[253] have also been employed, and most show reversible photoisomerization upon UV and visible light absorption. Recently, Yan *et al.* reported the design and synthesis of a visible-light responsive polymer system using dialkoxyanthracene (DN) (a photocleavable linker) containing diblock copolymers and a green light absorbing sensitizer, eosin. They self-assemble into core–shell micelles, and DN provides them with weak light energy-induced disassembly.[254]

A light stimulus, in addition to being used for triggering drug release from micelles, can also be used to activate drugs (photosensitizers). Photodynamic therapy (PDT) combines the action of a photosensitizer and a specific light source for the treatment of cancer.[255] When exposed to light, the photosensitizer is activated from the ground state to an excited state. When it returns to the ground state, it releases energy, which is transferred to oxygen to generate ROS like singlet oxygen and free radicals that cause cellular toxicity. The mechanisms by which PDT can destroy tumors include (i) direct tumor cell killing[256]; (ii) ROS-induced damage to the vascular system and deprivation of nutrients and oxygen to tumors[257,258] and (iii) immunosuppressive effects on the tumor.[259,260] Some of the most commonly used photosensitizing agents include porphyrins, chlorins and phthalocyanines.[261] Many photosensitizers exhibit poor solubility in water, which limits their application. A number of studies have employed polymeric micelles, both targeted and non-targeted, to effectively solubilize these agents and enhance their photodynamic efficacy, simultaneously minimizing side effects.[23,114,262–266]

4.4. *Multifunctional micelles*

As discussed in the previous sections, polymeric micelles possess several favorable characteristics, including good stability and biocompatibility *in vitro* and *in vivo*, the ability to solubilize poorly soluble drugs and contrast agents, the ability to be complexed with nucleic acids,

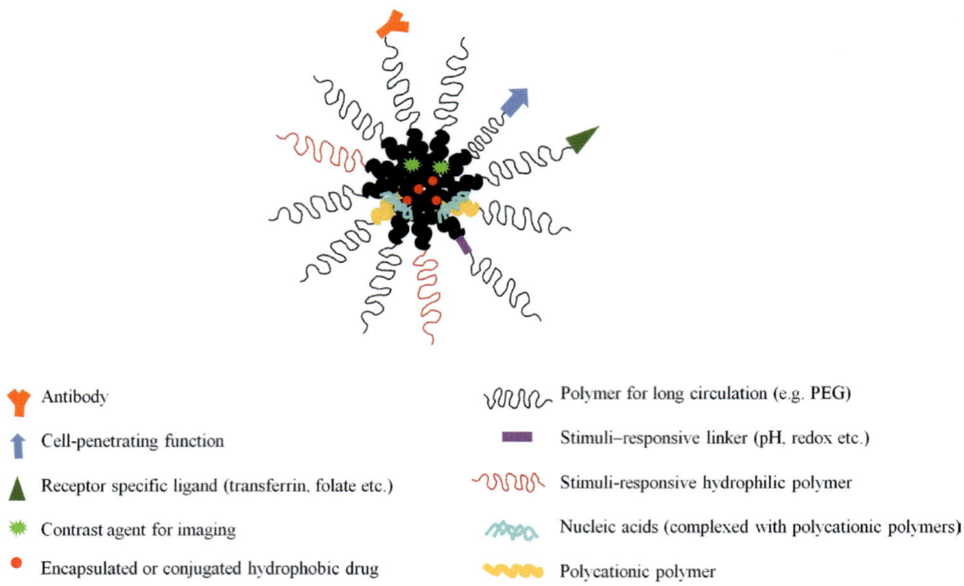

Fig. 8. Schematic of multifunctional micellar nanocarriers. Polymeric micelles can be constructed by combining two or more of the functions listed above to create multifunctional micelles.

passive targeting via an EPR effect or active targeting into the required pathological zone via surface-attached ligands and responsiveness to local physiological stimuli like pH, temperature or redox state at the pathological site.[43,280]

There is increasing interest in engineering polymeric micelles, which combine two or more of the above functions to produce multifunctional micelles (Fig. 8). To provide the desired combination of useful properties, it is essential that various functionalities are introduced in the micelles in a coordinated manner.[280]

Polymeric micelles have been investigated for the delivery of nucleic acids as well as drugs as already discussed. The key challenges in gene therapy with non-viral vectors include their poor transfection efficiency, instability *in vivo*, and in some cases, toxicity, which limit their clinical translation.[226] Polymeric micelles engineered with multiple integrated functions are potentially useful for gene delivery.[281] Of the many nanocarriers available for gene delivery, polyion complex micelles (PICs) formed by electrostatic interaction between charged block copoloymers and oppositely charged macromolecules [282,283] have been investigated extensively.[281,284,285] Moreover, efforts have been made to incorporate multiple functions like stimuli sensitivity (pH or redox), site-specific targeting via attachment of ligands and enhanced stability via crosslinking in these PIC micelles.[145,224,286]

We developed a multifunctional pH-responsive micellar system for the delivery of plasmid DNA conjugated with low-molecular-weight polyethyleneimine (PEI-1.8 kDa) and modified by dioleoyl phosphatidylethanolamine (PEI-PE/DNA).[286] To form micelles, the PEI-PE–pDNA complexes were mixed with low-pH–degradable PEG-Hydrazone (Hz)-PE (PEG-Hz-PE), which provided a tunable transfection efficiency and improved biocompatibility. At an optimized N/P ratio of 16, the PEI-PE–pDNA complexes had a particle size of 225 nm, a zeta potential of 31 mv and effectively protected the pDNA from the action of DNase I. The PEI-PE–pEGFP complexes showed increased transfection efficiency in B16 F10 cells compared with PEI-25 (high MW PEI, 25 kDa) or unmodified PEI-1.8 vectors, and additionally exhibited no toxicity compared to PEI-25. To study the cellular uptake of the complexes, YOYO-1 labeled pEGFP was used, which showed an enhanced uptake of the plasmid with PEI-PE compared to PEI-1.8 and PEI-25. The biocompatibility of PEI-PE/DNA complexes improved by mixing them with PEG-PE, which resulted in particles with a neutral surface charge, resistance to salt-induced aggregation and good transfection even in the presence of serum in the media. In the presence of low-pH–degradable PEG-Hz-PE, the complexes enhanced transfection efficiency at pH 5.4 than at pH 7.4 due to the deshielding of PEI after removal of the PEG. This established the suitability of pH-sensitive polymers for targeting the relatively acidified tumor environment.[286] Other groups have also reported multifunctional PIC-based micelles for gene delivery.[287,288]

In another example, Xiong *et al.* reported a novel, traceable, multifunctional polymeric micellar system constructed from PEO-*b*-PCL block copolymers which integrated multiple functions, including an ability to co-deliver siRNA and DOX, passive and active cancer targeting, cell membrane translocation and pH-triggered drug release. The micelle core stably complexed mdr1-siRNA and conjugated DOX (through a pH-sensitive Hz bond) and fluorescent imaging probes (for tracking micelles *in vitro* and *in vivo*). The virus-like shell had two ligands attached, one for active targeting (via $\alpha v\beta$ 3 specific ligand RGD4C) and another for efficient intracellular uptake (TAT-peptide) (Fig. 9). The system effectively targeted P-gp-mediated DOX resistance *in vitro* and $\alpha v\beta 3$-positive tumors *in vivo*.[289] Multifunctional polymeric micelles for siRNA delivery have been developed by a number of other groups as well.[290,291]

Additionally, multifunctional polymeric micelles have been developed for simultaneous imaging and gene delivery applications, or as “theranostics”.[292,293]

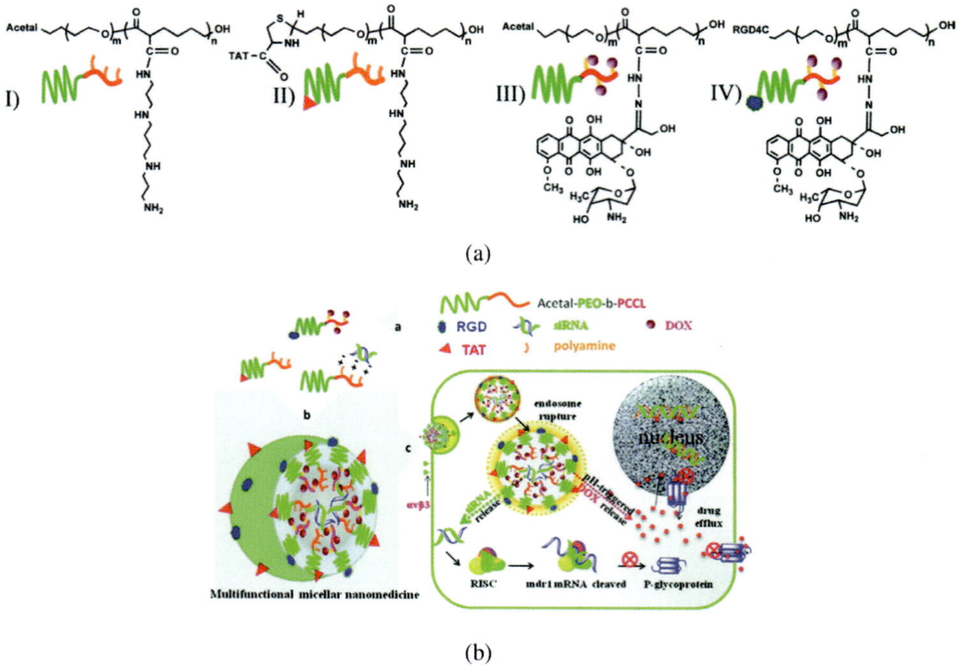

Fig. 9. Example of a multifunctional polymeric micellar system.

(A) Schematic illustration of acetal- and TAT-PEO-b-P(CL-g-SP) (I and II) and acetal- and RGD4C-PEO-b-P(CL-Hyd-DOX) (III and IV). (B) Rational design of a multifunctional micellar nanomedicine for targeted co-delivery of siRNA and DOX to overcome multidrug resistance. (a) Chemical structure of functionalized copolymers with ligands at the end of the PEO block and conjugated moieties on the PCL block. (b) Assembly of multifunctional micelles with DOX and siRNA in the micellar core and RGD and/or TAT on the micellar shell. (c) Schematic diagram showing the proposed model for the intracellular processing of targeted micelles in MDR cancer cells after receptor-mediated endocytosis, leading to cytoplasmic siRNA delivery followed by P-gp down-regulation on the cellular and nuclear membrane and endosomal DOX release, followed by DOX nuclear accumulation. (From Ref. 289, Copyright © 2011, American Chemical Society.)

Dual-purpose chitosan and PEI coated magnetic micelles (CP-mag-micelles) self-assembled from block copolymers of poly(D,L-lactide) (PLA) and mPEG that simultaneously deliver DNA and enable MR imaging were developed. Superparamagnetic iron oxide nanoparticles (SPIONs) were loaded into the core of these micelles. A single injection of CP-mag-micelles carrying reporter plasmids *in vivo* expressed genes for at least one week.[293] Howell *et al.* used oleic-acid–coated, MnO-loaded, lipid-based theranostic micelles (MR contrast) composed of PEG-2000-PE, DC-cholesterol and DOPE for simultaneous drug (DOX) and gene (pDNA) delivery to lungs.[292]

Table 5. Examples of multifunctional polymeric micelles

Polymers	Therapeutic agents	Targeting agents	Imaging function	Trigger	Ref.
Polyethyleneimine–polyethylene glycol (PEI-PEG)	DOX, DNA	TAT peptide , NGR peptide (targeting CD13 on tumor vasulature) (TAT shielded by NGR-Sulfamerazine-PEG)	—	pH	288
Poly(oligo(ethylene glycol) monomethyl ether methacrylate-*b*-Poly(N-isopropylacrylamide)-*co*-o-nitrobenzyl acrylate-*co*-Gd (POEGMA-*b*-P(NIPAM-*co*-NBA-*co*-Gd)	DOX	—	Gadolinium (for MRI)	Light (UV)	294
poly(ethylene glycol)-*block*-poly(l-lysine) (PEG-*b*-PLL) containing lysine amines modified with 2-iminothiolane (2IT)	siRNA	cRGD (targets integrin receptors)	—	Redox Disulfide crosslinking in the micelle core	295
poly(*N*-isopropylacrylamide-*co* acrylamide)-*block*-poly(ε-caprolactone) (P(NIPAAm-*co*-AAm)-*b*-PCL)	DOX	integrin β4 monoclonal antibody (recognizes A9 antigen)	Superparamag-netic iron oxide Fe_3O_4 nanoparticles (SPIONs) for MRI.	Thermo-responsive micelles LCST: 43°C	296
Pluronic P123/F127 mixed micelles	PTX	Folic acid (targets folate receptors)	—	—	297
PEG2000-Phosphatidylethanolamine	PTX, cyclosporine A	Antinucleosome mAb 2C5	—	—	298

(*Continued*)

Table 5. (*Continued*)

Polymers	Therapeutic agents	Targeting agents	Imaging function	Trigger	Ref.
polyethylene glycol–poly(N-(N′,N′ diisopropylaminoethyl) aspartamide)–cholic acid (PEG-PAsp(DIP)-CA)	PTX	Folic acid	Quantum dots	pH	299
Poly(styrene–acrylic acid–ethylene glycol) (PS-PAA-PEG), with calcium phosphate mineralization on PAA	Cisplatin	—	Nile red dye	pH	300
DSPE-PEG, biotin-DSPE-PEG, Lissamine rhodamine (Phospholipid-encapsulated SIPP cores)	PTX	J591 (mAb against PSMA)	Superparamag -netic iron platinum nanoparticles (SIPP) for MRI.		301
Octreotide-polyethylene glycol-monostearate and N-octyl-N-succinyl-O-carboxymethyl chitosan mixed micelles (OPMS-OSCC)	DOX	Octreotide (targets somatostatin receptors, SSTRs)	—	pH	302
Folate–chitosan (FA-CS)	Pyrrolidinedithi-ocarbamate (PDTC) (DOX)	Folic acid	—	pH	303

5. Micelles in the Clinic

Although polymeric micelles are being investigated extensively in laboratory settings, few have reached the clinic to date.[304] A list of polymeric micelle formulations of anticancer drugs currently under clinical investigation is shown in Table 6. Paclitaxel and DOX have been the most common anti-cancer agents under investigation both in the pre-clinical and clinical phases so far. Some of these polymeric micelles are discussed briefly in this section.

5.1. *Genexol®-PM*

Currently under Phase III and IV clinical trials, Genexol®-PM contains paclitaxel physically encapsulated in monomethoxy-PEG-b-poly($_{D,L}$-lactide) (mPEG-PDLLA) using the solid dispersion technique. The final formulation has a drug loading of about 16.7%. [41] The response rate for Genexol®-PM (administered over 3 h every 3 weeks at 300 mg/m^2) was higher than that of Abraxane® and Taxol® in Phase II and III clinical trials.[2] Due to its superior efficacy and lower adverse reactions, Genexol® is the first polymeric micelle

Table 6. Polymeric-micelle–based anticancer agents in clinical trials.

Formulation	Polymer	Drug	Size (nm)	Clinical phase	Indication	Refs.
Genexol®-PM	mPEG-PDLLA	Paclitaxel	20–50	IV	Breast cancer	41
NK 105	PEG-P(Asp)	Paclitaxel	85	III	Advanced stomach cancer	308, 309
SP1049C	Pluronic L61 and F127	DOX	22–27	III	Adenocarcinoma of esophagus, gastroesophageal junction and stomach	306, 307
NK012	PEG-P(Glu)-SN38	SN-38	20	II	Breast cancer	4, 310
NC-6004	PEG-P(Glu) cisplatin	Cisplatin	30	I/II	Solid tumors	313
NC-4016	PEG-P(Glu) (DACHPt)	DACHPt	40	I	Colorectal cancer	316, 317
NK911	PEG-P(Asp)-DOX	DOX	40	II	Solid tumors	3, 318

Note: PDLLA- poly($_{D,L}$ lactide),P(Asp)- poly(aspartic acid), P(Glu)- poly (γ-benzyl-L glutamate) or poly(glutamate), DACHPt -dichloro-(1,2-diaminocyclohexane) platinum (II).

formulation to be commercially available for the treatment of NSCLC, ovarian cancer, breast cancer and gastric cancer.[304]

5.2. *SP1049C*

SP1049C is a Pluronic®-based polymeric micelle formulation of DOX.[305] It is prepared by reconstituting DOX, to obtain a final concentration of 2 mg/mL; with a 0.9% sodium chloride solution containing non-ionic Pluronic block copolymers L61 (0.25% w/v) and F127 (2% w/v).[306] The DOX loading in this micellar formulation was about 8.2%. In a Phase I study, the maximum tolerated dose (MTD) of SP1049C was determined to be 70 mg/m^2, about the same as systemic DOX, but it showed a significantly higher antitumor activity and a favorable safety profile compared to DOX.[306] A Phase II clinical trial concluded that SP1049C monotherapy was effective in adenocarcinoma of the esophagus.[307] SP1049C is currently under evaluation in Phase III clinical trials and was designated an orphan drug by the FDA in 2008.[304]

5.3. *NK105*

NK105, currently in Phase III trials, is a polymeric micellar formulation self-assembled from PEG-poly(aspartic acid) (PEG-P(Asp)) modified with 4-phenyl-1-butanol with PTX physically entrapped in the hydrophobic block.[3] It shows PTX loading of about 23 %w/w. In a dose escalation study (10–180 mg/m^2), NK105 had reduced toxicity and adverse reactions compared to Taxol® in patients with pancreatic, bile duct, gastric and colonic cancers.[308] The MTD for NK105 was 150 mg/m^2 and it showed 32-fold higher AUC and 72-fold lower clearance compared with that of Genexol®-PM at 300 mg/m^2, indicating a higher blood stability.[41] In Phase II clinical studies, NK105 showed modest activity and tolerability, with an overall response rate of 25% and did not compromise the activity of paclitaxel in patients previously treated for advanced or recurrent gastric cancer.[309]

5.4. *NK012*

NK012 represents a polymeric micellar formulation of a block co-polymer of PEG and polyglutamate (PGlu) in which 7-ethyl-10-hydroxy-camptothecin (SN-38), a camptothecin analog, is covalently conjugated to the PGlu segment.[310] The drug loading of NK012 was about 20% (w/w). In Phase I studies, NK012 was infused *i.v.* over 30 min every 3 weeks. The MTD and recommended dose (RD) were 37 mg/m^2 and 28 mg/m^2, respectively. Antitumor activity of NK012 was significant and the dose-limiting toxicity was mild (neutropenia, diarrhea).[4,308]

5.5. *NC-6004*

Cisplatin (cis-dichlorodiammineplatinum (II) CDDP), is a widely used anticancer drug and results in severe adverse effects like nephron- and neurotoxicity as well as drug resistance.[311,312] To reduce side effects and improve its efficacy, a micellar formulation NC-6004 composed of PEG and poly(γ-benzyl L-glutamate)/CDDP complex was developed.[313] A high drug-loading capacity of about 39% was reported.[314] It was well tolerated after *i.v.* administration (1 h every 3 weeks with a dose range of 10–120 mg/m^2) in a Phase I study in patients suffering from colorectal carcinoma, upper gastrointestinal cancers, NSCLC, melanoma and other tumor types. The MTD was 120 mg/m^2 and RD for Phase II studies was 90 mg/m^2. Phase I and Phase II studies are ongoing for advanced or metastatic pancreatic cancer in combination with gemcitabine.[315]

6. Conclusion

Overall, the use of polymeric micellar drug carriers has been recognized as a promising strategy for clinical translation of potential but undeliverable water insoluble drugs. These systems offer advantages such as drug solubilization, tumor targeting and decreased systemic side-effects. With advancements in molecular biology and oncology, polymeric micellar systems have been extended to gene delivery and multifunctional systems. Thus, polymeric micellar medicine has become increasingly important for cancer therapy.

References

1. L. Gros, H. Ringsdorf and H. Schupp, *Angew. Chem. Int. Ed.* 305 (1981).
2. W. J. Gradishar, S. Tjulandin, N. Davidson, H. Shaw, N. Desai, P. Bhar *et al.*, *J. Clin. Oncol.* 7794 (2005).
3. Y. Matsumura, T. Hamaguchi, T. Ura, K. Muro, Y. Yamada, Y. Shimada *et al.*, *Br. J. Cancer* 1775 (2004).
4. Y. Matsumura and K. Kataoka, *Cancer Sci.* 572 (2009).
5. H. Maeda, J. Wu, T. Sawa, Y. Matsumura and K. Hori, *J. Control. Release* 271 (2000).
6. H. Maeda, *Adv. Enzyme Regul.* 189 (2001).
7. R. R. Sawant and V. P. Torchilin, *Mol. Membr. Biol.* 232 (2010).
8. V. P. Torchilin, *Cell. Mol. Life Sci.* 2549 (2004).
9. V. P. Torchilin, *J. Control. Release* 137 (2001).

10. M. Yokoyama, M. Miyauchi, N. Yamada, T. Okano, Y. Sakurai, K. Kataoka *et al.*, *Cancer Res.* 1693 (1990).
11. Y. Kakizawa and K. Kataoka, *Adv. Drug. Deliv. Rev.* 203 (2002).
12. Z. Zhang, D. W. Grijpma and J. Feijen, *J. Control. Release* 57 (2006).
13. V. P. Torchilin, V. S. Trubetskoy, K. R. Whiteman, P. Caliceti, P. Ferruti and F. M. Veronese, *J. Pharm. Sci.* 1049 (1995).
14. G. S. Kwon, M. Naito, K. Kataoka, M. Yokoyama, Y. Sakurai and T. Okano, *Colloids Surf. B Biointerfaces* 429 (1994).
15. R. R. Sawant and V. P. Torchilin, *Methods Mol. Biol.* 131 (2010).
16. A. V. Kabanov, V. P. Chekhonin, V. Alakhov, E. V. Batrakova, A. S. Lebedev, N. S. Melik-Nubarov *et al.*, *FEBS Lett.* 343 (1989).
17. H. Maeda, G. Y. Bharate and J. Daruwalla, *Eur. J. Pharm. Biopharm.* 409 (2009).
18. Y. Matsumura, M. Yokoyama, K. Kataoka, T. Okano, Y. Sakurai, T. Kawaguchi *et al.*, *Jpn. J. Cancer Res.* 122 (1999).
19. S. Y. Lin and Y. Kawashima, *J. Parenter. Sci. Technol.* 83 (1987).
20. G. S. Kwon, M. Yokoyama, T. Okano, Y. Sakurai and K. Kataoka, *Pharm. Res.* 970 (1993).
21. V. P. Torchilin, A. N. Lukyanov, Z. Gao and B. Papahadjopoulos-Sternberg, *Proc. Natl. Acad. Sci. U. S. A.* 6039 (2003).
22. S. Vinogradov, E. Batrakova, S. Li and A. Kabanov, *Bioconjug. Chem.* 851 (1999).
23. I. Skidan, P. Dholakia and V. Torchilin, *J. Drug Target* 486 (2008).
24. R. R. Sawant, R. M. Sawant and V. P. Torchilin, *Eur. J. Pharm. Biopharm.* 51 (2008).
25. D. Kim, Z. G. Gao, E. S. Lee and Y. H. Bae, *Mol. Pharm.* 1353 (2009).
26. N. Nasongkla, X. Shuai, H. Ai, B. D. Weinberg, J. Pink, D. A. Boothman *et al.*, *Angew. Chem. Int. Ed. Engl.* 6323 (2004).
27. D. Sutton, N. Nasongkla, E. Blanco and J. Gao, *Pharm. Res.* 1029 (2007).
28. V. Torchilin, *Eur. J. Pharm. Biopharm.* 431 (2009).
29. V. P. Torchilin, V. G. Omelyanenko, M. I. Papisov, A. A. Bogdanov, Jr., V. S. Trubetskoy, J. N. Herron *et al.*, *Biochim. Biophys. Acta* 11 (1994).
30. K. Edwards, M. Johnsson, G. Karlsson and M. Silvander, *Biophys. J.* 258 (1997).
31. A. N. Lukyanov and V. P. Torchilin, *Adv. Drug Deliv. Rev.* 1273 (2004).
32. S. C. Kim, D. W. Kim, Y. H. Shim, J. S. Bang, H. S. Oh, S. Wan Kim *et al.*, *J. Control. Release* 191 (2001).
33. A. N. Lukyanov, Z. Gao, L. Mazzola and V. P. Torchilin, *Pharm. Res.* 1424 (2002).
34. Z. Gao, A. N. Lukyanov, A. R. Chakilam and V. P. Torchilin, *J. Drug Target.* 87 (2003).
35. H. Arimura, Y. Ohya and T. Ouchi, *Biomacromolecules* 720 (2005).
36. R. D. Dabholkar, R. M. Sawant, D. A. Mongayt, P. V. Devarajan and V. P. Torchilin, *Int. J. Pharm.* 148 (2006).
37. L. Li and Y. B. Tan, *J. Colloid Interface Sci.* 326 (2008).
38. A. Lavasanifar, J. Samuel and G. S. Kwon, *Adv. Drug Deliv. Rev.* 169 (2002).
39. R. Luxenhofer, A. Schulz, C. Roques, S. Li, T. K. Bronich, E. V. Batrakova *et al.*, *Biomaterials* 4972 (2010).
40. Y. Han, Z. He, A. Schulz, T. K. Bronich, R. Jordan, R. Luxenhofer *et al.*, *Mol. Pharm.* (2012).
41. T. Y. Kim, D. W. Kim, J. Y. Chung, S. G. Shin, S. C. Kim, D. S. Heo *et al.*, *Clin. Cancer Res.* 3708 (2004).

42. M. Jones and J. Leroux, *Eur. J. Pharm. Biopharm.* 101 (1999).
43. V. P. Torchilin, *Biochem. Soc. Trans.* 816 (2007).
44. Z. Gao and A. Eisenberg, *Macromolecules* 7353 (1993).
45. K. Van Butsele, P. Sibret, C. A. Fustin, J. F. Gohy, C. Passirani, J. P. Benoit *et al.*, *J. Colloid Interface Sci.* 235 (2009).
46. J. Taillefer, M. C. Jones, N. Brasseur, J. E. van Lier and J. C. Leroux, *J. Pharm. Sci.* 52 (2000).
47. Y. B. Patil, U. S. Toti, A. Khdair, L. Ma and J. Panyam, *Biomaterials* 859 (2009).
48. H. Chu, N. Liu, X. Wang, Z. Jiao and Z. Chen, *Int. J. Pharm.* 190 (2009).
49. V. Alakhov, E. Moskaleva, E. V. Batrakova and A. V. Kabanov, *Bioconjug. Chem.* 209 (1996).
50. N. Nishiyama and K. Kataoka, *J. Control. Release* 83 (2001).
51. Y. Li and G. S. Kwon, *Pharm. Res.* 607 (2000).
52. A. Lavasanifar, J. Samuel and G. S. Kwon, *J. Control. Release* 155 (2001).
53. S. Patel, A. Lavasanifar and P. Choi, *Biomacromolecules* 3014 (2008).
54. C. J. Rijcken, C. J. Snel, R. M. Schiffelers, C. F. van Nostrum and W. E. Hennink, *Biomaterials* 5581 (2007).
55. M. Talelli, M. Iman, A. K. Varkouhi, C. J. Rijcken, R. M. Schiffelers, T. Etrych *et al.*, *Biomaterials* 7797 (2010).
56. Y. Kim, M. H. Pourgholami, D. L. Morris and M. H. Stenzel, *Biomacromolecules* 814 (2012).
57. V. Torchilin, *Adv. Drug. Deliv. Rev.* 131 (2011).
58. S. K. Hobbs, W. L. Monsky, F. Yuan, W. G. Roberts, L. Griffith, V. P. Torchilin *et al.*, *Proc. Natl. Acad. Sci. U. S. A*, 4607 (1998).
59. R. K. Jain, *Annu. Rev. Biomed. Eng.* 241 (1999).
60. R. K. Jain, *J. Natl. Cancer Inst.* 570 (1989).
61. F. Yuan, M. Leunig, S. K. Huang, D. A. Berk, D. Papahadjopoulos and R. K. Jain, *Cancer Res.* 3352 (1994).
62. D. R. Senger, S. J. Galli, A. M. Dvorak, C. A. Perruzzi, V. S. Harvey and H. F. Dvorak, *Science* 983 (1983).
63. H. Maeda, Y. Matsumura and H. Kato, *J. Biol. Chem.* 16051 (1988).
64. Y. Matsumura, M. Kimura, T. Yamamoto and H. Maeda, *Jpn. J. Cancer Res.* 1327 (1988).
65. H. Maeda, Y. Noguchi, K. Sato and T. Akaike, *Jpn. J. Cancer Res.* 331 (1994).
66. J. Wu, T. Akaike and H. Maeda, *Cancer Res.* 159 (1998).
67. H. R. Reichman, C. L. Farrell and R. F. Del Maestro, *J. Neurosurg.* 233 (1986).
68. F. Yuan, M. Dellian, D. Fukumura, M. Leunig, D. A. Berk, V. P. Torchilin *et al.*, *Cancer Res.* 3752 (1995).
69. V. P. Torchilin, *Nat. Rev. Drug Discov.* 145 (2005).
70. Y. Matsumura and H. Maeda, *Cancer Res.* 6387 (1986).
71. H. Maeda, *Adv. Drug. Deliv. Rev.* 169 (2001).
72. J. Fang, T. Sawa and H. Maeda, *Adv. Exp. Med. Biol.* 29 (2003).
73. H. Maeda, *Bioconjug. Chem.* 797 (2010).
74. A. V. Kabanov, E. V. Batrakova and V. Y. Alakhov, *J. Control. Release* 189 (2002).
75. R. M. Schiffelers, I. A. Bakker-Woudenberg, S. V. Snijders and G. Storm, *Biochim. Biophys. Acta* 329 (1999).

76. H. M. Aliabadi and A. Lavasanifar, *Expert Opin. Drug Deliv.* 139 (2006).
77. A. Lavasanifar, J. Samuel and G. S. Kwon, *J. Control. Release* 165 (2002).
78. X. Shuai, T. Merdan, A. K. Schaper, F. Xi and T. Kissel, *Bioconjug. Chem.* 441 (2004).
79. X. Yuan, A. Harada, Y. Yamasaki and K. Kataoka, *Langmuir* 2668 (2005).
80. V. P. Torchilin, R. Rammohan, V. Weissig and T. S. Levchenko, *Proc. Natl. Acad. Sci. U. S. A.* 8786 (2001).
81. M. J. Parr, D. Masin, P. R. Cullis and M. B. Bally, *J. Pharmacol. Exp. Ther.* 1319 (1997).
82. R. R. Sawant, O. Vaze, G. G. D'Souza, K. Rockwell and V. P. Torchilin, *Pharm. Res.* 301 (2011).
83. O. Vafa, M. Wade, S. Kern, M. Beeche, T. K. Pandita, G. M. Hampton *et al.*, *Mol. Cell* 1031 (2002).
84. M. Sattler, S. Verma, G. Shrikhande, C. H. Byrne, Y. B. Pride, T. Winkler *et al.*, *J. Biol. Chem.* 24273 (2000).
85. A. Laurent, C. Nicco, C. Chereau, C. Goulvestre, J. Alexandre, A. Alves *et al.*, *Cancer Res.* 948 (2005).
86. T. D. Oberley and L. W. Oberley, *Histol. Histopathol.* 525 (1997).
87. J. Yang, E. W. Lam, H. M. Hammad, T. D. Oberley and L. W. Oberley, *J. Oral Pathol. Med.* 71 (2002).
88. J. Verrax, J. Cadrobbi, C. Marques, H. Taper, Y. Habraken, J. Piette *et al.*, *Apoptosis* 223 (2004).
89. J. Verrax and P. B. Calderon, *Free Radic. Biol. Med.* 32 (2009).
90. Q. Chen, M. G. Espey, A. Y. Sun, C. Pooput, K. L. Kirk, M. C. Krishna *et al.*, *Proc. Natl. Acad. Sci. U. S. A.* 11105 (2008).
91. T. H. Kim, C. W. Mount, W. R. Gombotz and S. H. Pun, *Biomaterials* 7386 (2010).
92. C. Sarisozen, I. Vural, T. Levchenko, A. A. Hincal and V. P. Torchilin, *Drug Deliv.* 363 (2012).
93. X. Wei, Y. Wang, W. Zeng, F. Huang, L. Qin, C. Zhang *et al.*, *Pharm. Res.* 1977 (2012).
94. B. J. Tan, Y. Liu, K. L. Chang, B. K. Lim and G. N. Chiu, *Int. J. Nanomed.* 651 (2012).
95. M. Wang and A. L. Gartel, *Mol. Cancer Ther.* 2287 (2011).
96. J. Daruwalla, M. Nikfarjam, K. Greish, C. Malcontenti-Wilson, V. Muralidharan, C. Christophi *et al.*, *Cancer Sci.* 1866 (2010).
97. E. Blanco, E. A. Bey, C. Khemtong, S. G. Yang, J. Setti-Guthi, H. Chen *et al.*, *Cancer Res.* 3896 (2010).
98. K. Xiao, J. Luo, Y. Li, J. S. Lee, G. Fung and K. S. Lam, *J. Control. Release* 272 (2011).
99. Y. Z. Du, Q. Weng, H. Yuan and F. Q. Hu, *ACS Nano* 6894 (2010).
100. Y. J. Lin, Y. S. Liu, H. H. Yeh, T. L. Cheng and L. F. Wang, *Int. J. Nanomed.* 4169 (2012).
101. Y. I. Jeong, S. J. Seo, I. K. Park, H. C. Lee, I. C. Kang, T. Akaike *et al.*, *Int. J. Pharm.* 151 (2005).
102. C. S. Cho, A. Kobayashi, R. Takei, T. Ishihara, A. Maruyama and T. Akaike, *Biomaterials* 45 (2001).
103. G. Mohajer, E. S. Lee and Y. H. Bae, *Pharm. Res.* 1618 (2007).
104. J. You, X. Li, F. de Cui, Y. Z. Du, H. Yuan and F. Q. Hu, *Nanotechnology* 045102 (2008).
105. E. K. Park, S. Y. Kim, S. B. Lee and Y. M. Lee, *J. Control. Release* 158 (2005).
106. H. S. Yoo and T. G. Park, *J. Control. Release* 247 (2004).
107. X. B. Xiong, A. Mahmud, H. Uludag and A. Lavasanifar, *Biomacromolecules* 874 (2007).

108. X. B. Xiong, A. Mahmud, H. Uludag and A. Lavasanifar, *Pharm. Res.* 2555 (2008).
109. F. Zeng, H. Lee and C. Allen, *Bioconjug. Chem.* 399 (2006).
110. I. Rubinstein, I. Soos and H. Onyuksel, *Chem. Biol. Interact.* 190 (2008).
111. H. Onyuksel, P. S. Mohanty and I. Rubinstein, *Int. J. Pharm.* 157 (2009).
112. V. P. Torchilin, *Adv. Drug. Deliv. Rev.* 235 (2002).
113. V. P. Torchilin, T. S. Levchenko, A. N. Lukyanov, B. A. Khaw, A. L. Klibanov, R. Rammohan *et al.*, *Biochim. Biophys. Acta* 397 (2001).
114. A. Roby, S. Erdogan and V. P. Torchilin, *Eur. J. Pharm. Biopharm.* 235 (2006).
115. T. A. Elbayoumi, S. Pabba, A. Roby and V. P. Torchilin, *J. Liposome Res.* 1 (2007).
116. L. Mu, A. Chrastina, T. Levchenko and V. P. Torchilin, *J. Biomed. Nanotech.* 190 (2005).
117. C. Jin, N. Qian, W. Zhao, W. Yang, L. Bai, H. Wu *et al.*, *Biomacromolecules* 2422 (2010).
118. F. Perche, N. R. Patel and V. P. Torchilin, *J. Control. Release* 95 (2012).
119. T. Musacchio, V. Laquintana, A. Latrofa, G. Trapani and V. P. Torchilin, *Mol. Pharm.* 468 (2009).
120. Y. Mi, Y. Liu and S. S. Feng, *Biomaterials* 4058 (2011).
121. D. Zhou, G. Zhang and Z. Gan, *J. Control. Release* (2013).
122. H. Zhu, J. Cao, S. Cui, Z. Qian and Y. Gu, *J. Pharm. Sci.* 1318 (2013).
123. H. Song, R. He, K. Wang, J. Ruan, C. Bao, N. Li *et al.*, *Biomaterials* 2302 (2010).
124. L. L. Cai, P. Liu, X. Li, X. Huang, Y. Q. Ye, F. Y. Chen *et al.*, *Int. J. Nanomed.* 3499 (2011).
125. K. Na, V. T. Sethuraman and Y. H. Bae, *Anticancer Agents Med. Chem.* 525 (2006).
126. J. L. Wike-Hooley, J. Haveman and H. S. Reinhold, *Radiother. Oncol.* 343 (1984).
127. F. Kallinowski, K. H. Schlenger, S. Runkel, M. Kloes, M. Stohrer, P. Okunieff *et al.*, *Cancer Res.* 3759 (1989).
128. R. K. Jain and T. Stylianopoulos, *Nat. Rev. Clin. Oncol.* 653 (2010).
129. G. Helmlinger, F. Yuan, M. Dellian and R. K. Jain, *Nat. Med.* 177 (1997).
130. L. E. Gerweck and K. Seetharaman, *Cancer Res.* 1194 (1996).
131. E. S. Lee, H. J. Shin, K. Na and Y. H. Bae, *J. Control. Release* 363 (2003).
132. J. Sudimack and R. J. Lee, *Adv. Drug. Deliv. Rev.* 147 (2000).
133. E. R. Gillies and J. M. Frechut, *Pure Appl. Chem.* 1295 (2004).
134. E. S. Lee, Z. Gao and Y. H. Bae, *J. Control. Release* 164 (2008).
135. H. Yin and Y. H. Bae, *Eur. J. Pharm. Biopharm.* 223 (2009).
136. H. Yin, E. S. Lee, D. Kim, K. H. Lee, K. T. Oh and Y. H. Bae, *J. Control. Release* 130 (2008).
137. D. W. Pack, D. Putnam and R. Langer, *Biotechnol. Bioeng.* 217 (2000).
138. E. S. Lee, K. Na and Y. H. Bae, *J. Control. Release* 103 (2003).
139. Z. G. Gao, L. Tian, J. Hu, I. S. Park and Y. H. Bae, *J. Control. Release* 84 (2011).
140. K. H. Min, J. H. Kim, S. M. Bae, H. Shin, M. S. Kim, S. Park *et al.*, *J. Control. Release* 259 (2010).
141. V. A. Sethuraman, M. C. Lee and Y. H. Bae, *Pharm. Res.* 657 (2008).
142. V. A. Sethuraman and Y. H. Bae, *J. Control. Release* 216 (2007).
143. H. Wu, L. Zhu and V. P. Torchilin, *Biomaterials* 1213 (2013).
144. S. Xu, Y. Luo and R. Haag, *Macromol. Biosci.* 968 (2007).
145. C. Ding, J. Gu, X. Qu and Z. Yang, *Bioconjug. Chem.* 1163 (2009).
146. Y. Bae, S. Fukushima, A. Harada and K. Kataoka, *Angew. Chem. Int. Ed. Engl.* 4640 (2003).

147. Y. Bae, N. Nishiyama, S. Fukushima, H. Koyama, M. Yasuhiro and K. Kataoka, *Bioconjug. Chem.* 122 (2005).
148. Y. Bae, A. W. Alani, N. C. Rockich, T. S. Lai and G. S. Kwon, *Pharm. Res.* 2421 (2010).
149. Y. Bae, N. Nishiyama and K. Kataoka, *Bioconjug. Chem.* 1131 (2007).
150. V. Knorr, V. Russ, L. Allmendinger, M. Ogris and E. Wagner, *Bioconjug. Chem.* 1625 (2008).
151. J. Heller and J. Barr, *Biomacromolecules* 1625 (2004).
152. Y. Jin, L. Song, Y. Su, L. Zhu, Y. Pang, F. Qiu *et al.*, *Biomacromolecules* 3460 (2011).
153. J. Leroux, E. Roux, D. Le Garrec, K. Hong and D. C. Drummond, *J. Control. Release* 71 (2001).
154. S. Bontha, A. V. Kabanov and T. K. Bronich, *J. Control. Release* 163 (2006).
155. T. K. Bronich, P. A. Keifer, L. S. Shlyakhtenko and A. V. Kabanov, *J. Am. Chem. Soc.* 8236 (2005).
156. A. V. Kabanov and S. V. Vinogradov, *Angew. Chem. Int. Ed. Engl.* 5418 (2009).
157. J. O. Kim, A. V. Kabanov and T. K. Bronich, *J. Control. Release* 197 (2009).
158. H. S. Oberoi, F. C. Laquer, L. A. Marky, A. V. Kabanov and T. K. Bronich, *J. Control. Release* 64 (2011).
159. A. Chilkoti, M. R. Dreher, D. E. Meyer and D. Raucher, *Adv. Drug. Deliv. Rev.* 613 (2002).
160. E. V. Batrakova and A. V. Kabanov, *J. Control. Release* 98 (2008).
161. R. Pelton, *Adv. Colloid Interface Sci.* 1 (2000).
162. M. Yang, Y. Ding, L. Zhang, X. Qian, X. Jiang and B. Liu, *J. Biomed. Mater. Res. A* 847 (2007).
163. C. Cheng, H. Wei, X. Z. Zhang, S. X. Cheng and R. X. Zhuo, *J. Biomed. Mater. Res. A* 814 (2009).
164. C. J. Rijcken, O. Soga, W. E. Hennink and C. F. van Nostrum, *J. Control. Release* 131 (2007).
165. O. Soga, C. F. van Nostrum and W. E. Hennink, *J. Control. Release* 383 (2005).
166. O. Soga, C. F. van Nostrum, M. Fens, C. J. Rijcken, R. M. Schiffelers, G. Storm *et al.*, *J. Control. Release* 341 (2005).
167. O. Soga, C. F. van Nostrum and W. E. Hennink, *Biomacromolecules* 818 (2004).
168. O. Soga, C. F. van Nostrum, A. Ramzi, T. Visser, F. Soulimani, P. M. Frederik *et al.* *Langmuir* 9388 (2004).
169. M. Talelli, C. J. Rijcken, C. F. van Nostrum, G. Storm and W. E. Hennink, *Adv. Drug Deliv. Rev.* 231 (2010).
170. C. J. Rijcken, T. F. Veldhuis, A. Ramzi, J. D. Meeldijk, C. F. van Nostrum and W. E. Hennink, *Biomacromolecules* 2343 (2005).
171. R. K. O'Reilly, C. J. Hawker and K. L. Wooley, *Chem. Soc. Rev.* 1068 (2006).
172. F. Q. Hu, G. F. Ren, H. Yuan, Y. Z. Du and S. Zeng, *Colloids Surf B Biointerf.* 97 (2006).
173. M. Talelli, C. J. Rijcken, S. Oliveira, R. van der Meel, P. M. van Bergen En Henegouwen, T. Lammers *et al.*, *J. Control. Release* 183 (2011).
174. O. Soga, C. F. van Nostrum, M. Fens, C. J. Rijcken, R. M. Schiffelers, G. Storm *et al.*, *J. Control. Release* 341 (2005).
175. G. A. Husseini, D. A. Christensen, N. Y. Rapoport and W. G. Pitt, *J. Control. Release* 303 (2002).

176. K. H. Bae, S. H. Choi, S. Y. Park, Y. Lee and T. G. Park, *Langmuir* 6380 (2006).
177. W. G. Pitt, G. A. Husseini and B. J. Staples, *Expert. Opin. Drug Deliv.* 37 (2004).
178. G. A. Husseini and W. G. Pitt, *J. Pharm. Sci.* 795 (2009).
179. G. A. Husseini and W. G. Pitt, *Adv. Drug Deliv. Rev.* 1137 (2008).
180. G. A. Husseini, C. M. Runyan and W. G. Pitt, *BMC Cancer* 20 (2002).
181. A. Marin, M. Muniruzzaman and N. Rapoport, *J. Control. Release* 69 (2001).
182. Z. G. Gao, H. D. Fain and N. Rapoport, *J. Control. Release* 203 (2005).
183. N. Munshi, N. Rapoport and W. G. Pitt, *Cancer Lett.* 13 (1997).
184. N. Rapoport, *Colloids Surf. B Biointerf.* 93 (1999).
185. G. A. Husseini, R. I. El-Fayoumi, K. L. O'Neill, N. Y. Rapoport and W. G. Pitt, *Cancer Lett.* 211 (2000).
186. A. Marin, M. Muniruzzaman and N. Rapoport, *J. Control. Release* 239 (2001).
187. G. A. Husseini, G. D. Myrup, W. G. Pitt, D. A. Christensen and N. Y. Rapoport, *J. Control. Release* 43 (2000).
188. A. Marin, H. Sun, G. A. Husseini, W. G. Pitt, D. A. Christensen and N. Y. Rapoport, *J. Control. Release* 39 (2002).
189. N. Rapoport, *Int. J. Hypertherm.* 374 (2012).
190. J. D. Pruitt, G. Husseini, N. Rapoport and W. G. Pitt, *Macromolecules* 9306 (2000).
191. J. L. Nelson, B. L. Roeder, J. C. Carmen, F. Roloff and W. G. Pitt, *Cancer Res.* 7280 (2002).
192. J. D. Pruitt and W. G. Pitt, *Drug Deliv.* 253 (2002).
193. Z. Gao, H. D. Fain and N. Rapoport, *Mol. Pharm.* 317 (2004).
194. N. Y. Rapoport, D. A. Christensen, H. D. Fain, L. Barrows and Z. Gao, *Ultrasonics* 943 (2004).
195. H. Zhang, H. Xia, J. Wang and Y. Li, *J. Control. Release* 31 (2009).
196. J. Xuan, O. Boissiere, Y. Zhao, B. Yan, L. Tremblay, S. Lacelle *et al.*, *Langmuir* 16463 (2012).
197. J. M. Warram, A. G. Sorace, R. Saini, A. V. Borovjagin, K. Hoyt and K. R. Zinn, *Cancer Gene Ther.* 545 (2012).
198. Y. C. Chen, H. D. Liang, Q. P. Zhang, M. J. Blomley and Q. L. Lu, *Ultrasound Med. Biol.* 131 (2006).
199. N. Rapoport, Z. Gao and A. Kennedy, *J. Natl. Cancer Inst.* 1095 (2007).
200. T. Y. Liu, M. Y. Wu, M. H. Lin and F. Y. Yang, *Acta Biomater.* 5453 (2013).
201. N. Rapoport, *Prog. Polym. Sci.* 962 (2007).
202. C. Oerlemans, W. Bult, M. Bos, G. Storm, J. F. Nijsen and W. E. Hennink, *Pharm. Res.* 2569 (2010).
203. G. Saito, J. A. Swanson and K. D. Lee, *Adv. Drug Deliv. Rev.* 199 (2003).
204. D. Roy, J. N. Cambre and B. S. Sumerlin, *Prog. Polym. Sci.* 278 (2010).
205. H. Wei, R.-X. Zhuo and X.-Z. Zhang, *Prog. Polym. Sci.* (2012).
206. E. Fleige, M. A. Quadir and R. Haag, *Adv. Drug Deliv. Rev.* 866 (2012).
207. A. Russo, W. DeGraff, N. Friedman and J. B. Mitchell, *Cancer Res.* 2845 (1986).
208. P. C. Jocelyn, *Meth. Enzymol.* 246 (1987).
209. S. Ganta, H. Devalapally, A. Shahiwala and M. Amiji, *J. Control. Release* 187 (2008).
210. T. Saji, K. Hoshino and S. Aoyagui, *J. Am. Chem. Soc.* 6865 (1985).
211. T. Saji, K. Hoshino and S. Aoyagui, *J. Chem. Soc., Chem. Commun.* 865 (1985).
212. Y. Li, K. Xiao, J. Luo, W. Xiao, J. S. Lee, A. M. Gonik *et al.*, *Biomaterials* 6633 (2011).

213. O. Boissiere, D. Han, L. Tremblay and Y. Zhao, *Soft Matter* 9410 (2011).
214. G.-Y. Liu, C.-J. Chen, D.-D. Li, S.-S. Wang and J. Ji, *J. Mater. Chem.* 16865 (2012).
215. J. Liu, Y. Pang, W. Huang, Z. Zhu, X. Zhu, Y. Zhou *et al.*, *Biomacromolecules* 2407 (2011).
216. J. H. Ryu, R. Roy, J. Ventura and S. Thayumanavan, *Langmuir* 7086 (2010).
217. P. Sun, D. Zhou and Z. Gan, *J. Control. Release* e85 (2011).
218. H. Sun, B. Guo, R. Cheng, F. Meng, H. Liu and Z. Zhong, *Biomaterials* 6358 (2009).
219. T.-B. Ren, Y. Feng, Z.-H. Zhang, L. Li and Y.-Y. Li, *Soft Matter* 2329 (2011).
220. L. Li, X. Jiang and R. Zhuo, *J. Polym. Sci. A Polym. Chem.* 5989 (2009).
221. A. Rosler, G. W. Vandermeulen and H. A. Klok, *Adv. Drug Deliv. Rev.* 95 (2001).
222. X. Q. Li, H. Y. Wen, H. Q. Dong, W. M. Xue, G. M. Pauletti, X. J. Cai *et al.*, *Chem. Commun.* 8647 (2011).
223. S. Cajot, N. Lautram, C. Passirani and C. Jerome, *J. Control. Release* 30 (2011).
224. S. Matsumoto, R. J. Christie, N. Nishiyama, K. Miyata, A. Ishii, M. Oba *et al.*, *Biomacromolecules* 119 (2009).
225. A. N. Koo, K. H. Min, H. J. Lee, S. U. Lee, K. Kim, I. C. Kwon *et al.*, *Biomaterials* 1489 (2012).
226. D. W. Pack, A. S. Hoffman, S. Pun and P. S. Stayton, *Nat. Rev. Drug Discov.* 581 (2005).
227. D. Putnam, *Nat. Mater.* 439 (2006).
228. T. Musacchio, O. Vaze, G. D'Souza and V. P. Torchilin, *Bioconjug. Chem.* 1530 (2010).
229. Y. Kakizawa, A. Harada and K. Kataoka, *Biomacromolecules* 491 (2001).
230. S. Takae, K. Miyata, M. Oba, T. Ishii, N. Nishiyama, K. Itaka *et al.*, *J. Am. Chem. Soc.* 6001 (2008).
231. M. Oishi, T. Hayama, Y. Akiyama, S. Takae, A. Harada, Y. Yamasaki *et al.*, *Biomacromolecules* 2449 (2005).
232. M. Kumagai, S. Shimoda, R. Wakabayashi, Y. Kunisawa, T. Ishii, K. Osada *et al.*, *J. Control. Release* 542 (2012).
233. S. H. Lee, S. H. Kim and T. G. Park, *Biochem. Biophys. Res. Commun.* 511 (2007).
234. S. H. Kim, J. H. Jeong, S. H. Lee, S. W. Kim and T. G. Park, *Bioconjug. Chem.* 2156 (2008).
235. A. M. Al-Abd, S. H. Lee, S. H. Kim, J. H. Cha, T. G. Park, S. J. Lee *et al.*, *J. Control. Release* 130 (2009).
236. J. F. Gohy and Y. Zhao, *Chem. Soc. Rev.* (2013).
237. Y. Zhao, *Chem. Rec.* 286 (2007).
238. Y. Zhao, *Macromolecules* 3647 (2012).
239. J. Jiang, X. Tong and Y. Zhao, *J. Am. Chem. Soc.* 8290 (2005).
240. J. Jiang, X. Tong, D. Morris and Y. Zhao, *Macromolecules* 4633 (2006).
241. J. Babin, M. Pelletier, M. Lepage, J. F. Allard, D. Morris and Y. Zhao, *Angew. Chem. Int. Ed. Engl.* 3329 (2009).
242. Q. Jin, F. Mitschang and S. Agarwal, *Biomacromolecules* 3684 (2011).
243. G. Wang, X. Tong and Y. Zhao, *Macromolecules* 8911 (2004).
244. X. Tong, G. Wang, A. Soldera and Y. Zhao, *J. Phys. Chem. B* 20281 (2005).
245. X. Tang, L. Gao, X. Fan, X. Liang and Q. Zhou, *Macromol. Chem. Phys.* 1556 (2009).
246. F. D. Jochum and P. Theato, *Chem. Commun.* 6717 (2010).
247. L. Sun, Y. Yang, C. M. Dong and Y. Wei, *Small* 401 (2011).
248. S. Kumar, J.-F. Allard, D. Morris, Y. L. Dory, M. Lepage and Y. Zhao, *J. Mater. Chem.* 7252 (2012).

249. H. I. Lee, W. Wu, J. K. Oh, L. Mueller, G. Sherwood, L. Peteanu *et al.*, *Angew. Chem. Int. Ed. Engl.* 2453 (2007).
250. C.-J. Chen, Q. Jin, G.-Y. Liu, D.-D. Li, J.-L. Wang and J. Ji, *Polymer* 3695 (2012).
251. Z. Chen, Y. He, Y. Wang and X. Wang, *Macromol. Rapid Commun.* 977 (2011).
252. S. Menon, R. Thekkayil, S. Varghese and S. Das, *J. Polym. Sci. A Polym. Chem.* 5063 (2011).
253. A. Laschewsky and E. D. Rekaï, *Macromol. Rapid Commun.* 937 (2000).
254. Q. Yan, J. Hu, R. Zhou, Y. Ju, Y. Yin and J. Yuan, *Chem. Commun.* 1913 (2012).
255. D. E. Dolmans, D. Fukumura and R. K. Jain, *Nat. Rev. Cancer* 380 (2003).
256. B. W. Henderson, S. M. Waldow, T. S. Mang, W. R. Potter, P. B. Malone and T. J. Dougherty, *Cancer Res.* 572 (1985).
257. Q. Chen, H. Chen and F. W. Hetzel, *Photochem. Photobiol.* 128 (1996).
258. B. Krammer, *Anticancer Res.* 4271 (2001).
259. S. O. Gollnick, X. Liu, B. Owczarczak, D. A. Musser and B. W. Henderson, *Cancer Res.* 3904 (1997).
260. F. H. van Duijnhoven, R. I. Aalbers, J. P. Rovers, O. T. Terpstra and P. J. Kuppen, *Immunobiology* 105 (2003).
261. Z. Huang, *Technol. Cancer Res. Treat.* 283 (2005).
262. H. C. Tsai, C. H. Tsai, S. Y. Lin, C. R. Jhang, Y. S. Chiang and G. H. Hsiue, *Biomaterials* 1827 (2012).
263. W. J. Syu, H. P. Yu, C. Y. Hsu, Y. C. Rajan, Y. H. Hsu, Y. C. Chang *et al.*, *Small* 2060 (2012).
264. A. M. Master, Y. Qi, N. L. Oleinick and A. S. Gupta, *Nanomedicine* 655 (2012).
265. A. Roby, S. Erdogan and V. P. Torchilin, *Cancer Biol. Ther.* 1136 (2007).
266. N. Nishiyama, Y. Nakagishi, Y. Morimoto, P. S. Lai, K. Miyazaki, K. Urano *et al.*, *J. Control. Release* 245 (2009).
267. F. Q. Hu, Y. Y. Zhang, J. You, H. Yuan and Y. Z. Du, *Mol. Pharm.* 2469 (2012).
268. S. Lee, K. Saito, H. R. Lee, M. J. Lee, Y. Shibasaki, Y. Oishi *et al.*, *Biomacromolecules* 1190 (2012).
269. J. Akimoto, M. Nakayama, K. Sakai and T. Okano, *Mol. Pharm.* 926 (2010).
270. A. Zhao, S. Zhou, Q. Zhou and T. Chen, *Pharm. Res.* 1627 (2010).
271. J. Yang, D. Zhang, S. Jiang and J. Nie, *J. Colloid Interface Sci.* 405 (2010).
272. X. Wei, X. Lv, Q. Zhao and L. Qiu, *Acta Biomater.* (2013).
273. G. A. Husseini, D. Velluto, L. Kherbeck, W. G. Pitt, J. A. Hubbell and D. A. Christensen, *Colloids Surf. B Biointerf.* 153 (2013).
274. R. R. Sawant, O. S. Vaze, T. Wang, G. G. D'Souza, K. Rockwell, K. Gada *et al. Pharm. Res.* 375 (2012).
275. M. Ugarenko, C. K. Chan, A. Nudelman, A. Rephaeli, S. M. Cutts and D. R. Phillips, *Oncol. Res.* 283 (2009).
276. H. Y. Wen, H. Q. Dong, W. J. Xie, Y. Y. Li, K. Wang, G. M. Pauletti *et al. Chem. Commun.* (*Camb.*) 3550 (2011).
277. C. Cui, Y. N. Xue, M. Wu, Y. Zhang, P. Yu, L. Liu *et al.*, *Biomaterials* 3858 (2013).
278. B. Yan, J. C. Boyer, N. R. Branda and Y. Zhao, *J. Am. Chem. Soc.* 19714 (2011).
279. M. J. Shieh, C. L. Peng, W. L. Chiang, C. H. Wang, C. Y. Hsu, S. J. Wang *et al.*, *Mol. Pharm.* 1244 (2010).
280. V. P. Torchilin, *Adv. Drug Deliv. Rev.* 1532 (2006).
281. N. Nishiyama and K. Kataoka, *Pharmacol. Ther.* 630 (2006).

282. A. Harada and K. Kataoka, *Macromolecules* 5294 (1995).
283. A. Harada and K. Kataoka, *Science* 65 (1999).
284. M. Harada-Shiba, K. Yamauchi, A. Harada, I. Takamisawa, K. Shimokado and K. Kataoka, *Gene Ther.* 407 (2002).
285. Y. T. Ko, A. Kale, W. C. Hartner, B. Papahadjopoulos-Sternberg and V. P. Torchilin, *J. Control. Release* 132 (2009).
286. R. R. Sawant, S. K. Sriraman, G. Navarro, S. Biswas, R. A. Dalvi and V. P. Torchilin, *Biomaterials* 3942 (2012).
287. M. Oishi, K. Kataoka and Y. Nagasaki, *Bioconjug. Chem.* 677 (2006).
288. C. Liu, F. Liu, L. Feng, M. Li, J. Zhang and N. Zhang, *Biomaterials* 2547 (2013).
289. X. B. Xiong and A. Lavasanifar, *ACS Nano* 5202 (2011).
290. M. C. Palanca-Wessels, A. J. Convertine, R. Cutler-Strom, G. C. Booth, F. Lee, G. Y. Berguig *et al.*, *Mol. Ther.* 1529 (2011).
291. A. E. Felber, B. Castagner, M. Elsabahy, G. F. Deleavey, M. J. Damha and J. C. Leroux, *J. Control. Release* 159 (2011).
292. M. Howell, J. Mallela, C. Wang, S. Ravi, S. Dixit, U. Garapati *et al.*, *J. Control. Release* (2013).
293. C. Wang, S. Ravi, G. V. Martinez, V. Chinnasamy, P. Raulji, M. Howell *et al.*, *J. Control. Release* 82 (2012).
294. Y. Li, Y. Qian, T. Liu, G. Zhang and S. Liu, *Biomacromolecules* 3877 (2012).
295. R. J. Christie, Y. Matsumoto, K. Miyata, T. Nomoto, S. Fukushima, K. Osada *et al.*, *ACS Nano* 5174 (2012).
296. D. H. Kim, E. A. Vitol, J. Liu, S. Balasubramanian, D. J. Gosztola, E. E. Cohen, *et al.*, *Langmuir* (2013).
297. W. Zhang, Y. Shi, Y. Chen, J. Ye, X. Sha and X. Fang, *Biomaterials* 2894 (2011).
298. C. Sarisozen, I. Vural, T. Levchenko, A. A. Hincal and V. P. Torchilin, D*rug Deliv.* 169 (2012).
299. W. Wang, D. Cheng, F. Gong, X. Miao and X. Shuai, *Adv. Mater.* 115 (2012).
300. B. P. Bastakoti, K. C. Wu, M. Inoue, S. I. Yusa, K. Nakashima and Y. Yamauchi, *Chemistry* (2013).
301. R. M. Taylor and L. O. Sillerud, *Int. J. Nanomed.* 4341 (2012).
302. J. Niu, Z. Su, Y. Xiao, A. Huang, H. Li, X. Bao *et al. Eur. J. Pharm. Sci.* 216 (2012).
303. L. Fan, F. Li, H. Zhang, Y. Wang, C. Cheng, X. Li *et al. Biomaterials* 5634 (2010).
304. Y. Lu and K. Park, *Int. J. Pharm.* (2012).
305. V. Alakhov, E. Klinski, S. Li, G. Pietrzynski, A. Venne, E. Batrakova *et al.*, *Colloids Surf. B Biointerf.* 113 (1999).
306. S. Danson, D. Ferry, V. Alakhov, J. Margison, D. Kerr, D. Jowle *et al.*, *Br. J. Cancer* 2085 (2004).
307. J. W. Valle, A. Armstrong, C. Newman, V. Alakhov, G. Pietrzynski, J. Brewer *et al. Invest. New Drugs* 1029 (2011).
308. T. Hamaguchi, K. Kato, H. Yasui, C. Morizane, M. Ikeda, H. Ueno *et al.*, *Br. J. Cancer* 170 (2007).
309. K. Kato, K. Chin, T. Yoshikawa, K. Yamaguchi, Y. Tsuji, T. Esaki *et al.*, *Invest. New Drugs* 1621 (2012).
310. F. Koizumi, M. Kitagawa, T. Negishi, T. Onda, S. Matsumoto, T. Hamaguchi, *et al. Cancer Res.* 10048 (2006).

311. N. Nishiyama, S. Okazaki, H. Cabral, M. Miyamoto, Y. Kato, Y. Sugiyama *et al. Cancer Res.* 8977 (2003).
312. J. T. Hartmann and H. P. Lipp, *Expert Opin. Pharmacother.* 889 (2003).
313. R. Plummer, R. H. Wilson, H. Calvert, A. V. Boddy, M. Griffin, J. Sludden *et al.*, *Br. J. Cancer* 593 (2011).
314. H. Uchino, Y. Matsumura, T. Negishi, F. Koizumi, T. Hayashi, T. Honda *et al.*, *Br. J. Cancer* 678 (2005).
315. J. Gong, M. Chen, Y. Zheng, S. Wang and Y. Wang, *J. Control. Release* 312 (2012).
316. H. Cabral, N. Nishiyama, S. Okazaki, H. Koyama and K. Kataoka, *J. Control. Release* 223 (2005).
317. H. Cabral, N. Nishiyama and K. Kataoka, *J. Control. Release* 146 (2007).
318. T. Nakanishi, S. Fukushima, K. Okamoto, M. Suzuki, Y. Matsumura, M. Yokoyama *et al.*, *J. Control. Release* 295 (2001).

Chapter 4

Nanoemulsions in Medicine

*William B. Tucker and Sandro Mecozzi**

School of Pharmacy and Department of Chemistry,
University of Wisconsin, Madison, WI 53705, USA
**smecozzi@wisc.edu*

Abstract

Herein, the use of nanoemulsions (NEs) — described as metastable emulsions consisting of sub-500-nm sized droplets — in medicine is discussed. The ambiguity associated with defining NEs is elaborated upon, and the nature of NEs is contrasted with that of microemulsions (MEs). The various high- and low-energy methods, by which NE are prepared, are discussed, and stability issues are elaborated upon. Finally, specific examples concerning the use of NEs in medicine are given, including: drug delivery vehicles, mucosal vaccine vectors, topical antimicrobials and MRI/ ultrasonography contrast agents.

1. Introduction. What is a Nanoemulsion?

Nanoemulsions (NEs) are oil-in-water (O/W) or water-in-oil (W/O) emulsions that usually consist of sub-500-nm mean droplet size (Fig. 1). Various terms are used in the literature — mini-emulsions,[1,2] ultrafine emulsions,[3] submicron emulsions,[4,5] and translucent emulsions[6,7] — which all correspond to the same system; however, a point should be made to differentiate NEs from microemulsions (MEs). The NE and ME are terms that are often used interchangeably in the literature but from a physicochemical standpoint that are quite different systems.[8] A NE is a kinetically-stable colloidal emulsion, which can be prepared by a variety of means (*vide infra*).[9] In contrast, a ME

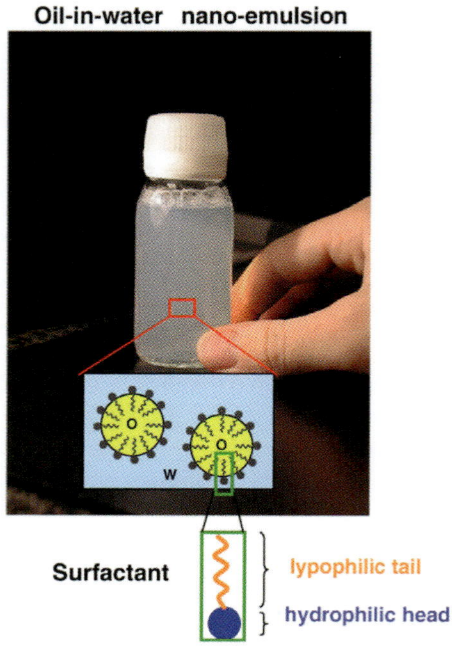

Fig. 1. Visual appearance of a typical O/W NE, with diagram of droplet structure.[17]

corresponds to a system under thermodynamic equilibrium, between oil, water and surfactant(s), which spontaneously emulsifies.[10] This chapter will focus solely on NEs, as mentioned in the Abstract and Introduction. It is noteworthy that while the difference between NEs and MEs is generally agreed upon, no coherent size-cutoff has been ascribed to NEs.

Nanoparticles are generally considered to be between 0.1–100 nm,[11] as such this range has been used to describe NEs.[12] Other size ranges given in the literature for NEs includes 20–300 nm,[9] 50–500 nm,[13] and 50–1000 nm.[14] Here NE are defined as those emulsions possessing sub-500 nm diameters. Given the ambiguity present in the literature, the sub-500 nm diameter cutoff has been purposefully chosen. The reason for this choice comes from USP <729>,[15] and subsequent monograph,[16] which specifies that any lipid-based emulsion to be used *in vivo* needs to have an average diameter of 500 nm or less. Since this chapter is dedicated to a discussion of NEs in medicine, the 500 nm diameter cutoff seems appropriate.

Since NEs are metastable, non-equilibrium systems, energy must be added to the system to achieve emulsification of the oil, water and surfactant components. Two different methodologies exist to achieve emulsification: high-energy and low-energy. High-energy emulsification methods are the traditional method for NE preparation.[18] Typically, high-shear stirring,[19] high-pressure homogenization,[20] ultrasonic emulsification,[21] and/or micro-fluidization[22,23] are used. Each method uses mechanical force to disrupt the dispersed and continuous phases causing nanodroplets formation of the dispersed phase. The high interfacial energy between the nanodroplets and continuous phase is stabilized by adsorption of surfactant into the dispersed-phase, stabilizing the NE. If surfactant adsorption is slow, droplets will coalesce and mean droplet diameters will be larger.[24]

Since the 1960s,[25] low-energy emulsification methodologies have been developed. Instead of using inputs of mechanical energy to form nanodroplets, low-energy techniques utilize the change in chemical potential upon changes in temperature[26,27] or composition[28,29] to affect a spontaneous (self-) emulsification. The energy change associated with the change in chemical potential (upon variance in temperature or composition) is leveraged to form nanodroplets. Both methods have their advantages and drawbacks, which will be discussed in the following sections.

Given the small average size of NE droplets, Brownian motion is able to overcome both gravity and buoyancy thus limiting sedimentation and creaming, respectively.[30] Furthermore, the use of appropriate surfactants — ionic or non-ionic — limits flocculation due to electrostatic and/or steric repulsion.[31] Therefore, Ostwald ripening — the diffusion of dispersed phase molecules from smaller (high Laplace pressure droplets) to larger (low Laplace pressure droplets) — is the sole method for reduction of the free energy of the system and, as such, the sole mechanism for NE coarsening.[30] Various methods to inhibit or slow Ostwald ripening have been developed, and thus extend NE lifetimes, including the use of insoluble additives,[32,33] evaporation of a volatile, oil phase,[34] and co-surfactants.[35,36] These methodologies will also be discussed in this chapter.

Nanoemulsions have been found to be useful in a wide range of medical applications from drug delivery vehicles[17] to mucosal vaccines,[37] to topical antimicrobials,[38] to contrast agents for ultrasonography and MRI.[39] In terms of drug-delivery applications, NEs have been used in parental, ocular, oral, dermal, and nasal deliveries.[17] The attractiveness of NEs is related, in part, to the variety of formulations in which NEs can be prepared (from water-like fluids to gels)[40] and their small, narrow size distribution.

2. Preparation of Nanoemulsions

The NEs are non-equilibrium colloidal emulsions; consequently, preparation involves the input of energy to emulsify the water, oil, and surfactant components.[41] To this end, high-energy and low-energy methods are available. High-energy methodologies utilize mechanical force — high-shear stirring, high-pressure homogenization, ultrasonic emulsification, or microfluidization — to disrupt the oil and water phases to form nano-sized droplets (Fig. 2). The particle size and distribution depends on the instrument and operating conditions, e.g. pressure, temperature and homogenization cycles.[42,43] High-energy emulsification allows for a greater control of particle size and a larger variance in the composition, but may not be suitable for use with temperature-sensitive drugs.

Low-energy emulsification methods, which have only been more recently developed,[44] can also be used to prepare NEs. The principals which form the basis for low-energy emulsification methods were elucidated in the late 1960's.[25,45] These processes take advantage of the changes in phase behavior and interfacial properties of various surfactants with changes in temperature, component concentration, and solvent polarity.

Techniques that are considered low-energy includes the phase inversion temperature (PIT) method,[26,27] the phase inversion composition (PIC) method[28,29] (Fig. 3) and the solvent displacement method.[46,47] The low-energy methods are attractive because of their reduced energy requirements, potentially enhancing their scalability. Furthermore, low-energy emulsification is less intensive of a process and as such can be used to emulsify temperature sensitive drugs and biologics. The NEs prepared by low-energy methods can, however, suffer from long-term instability inherent in the preparation methodology.[12]

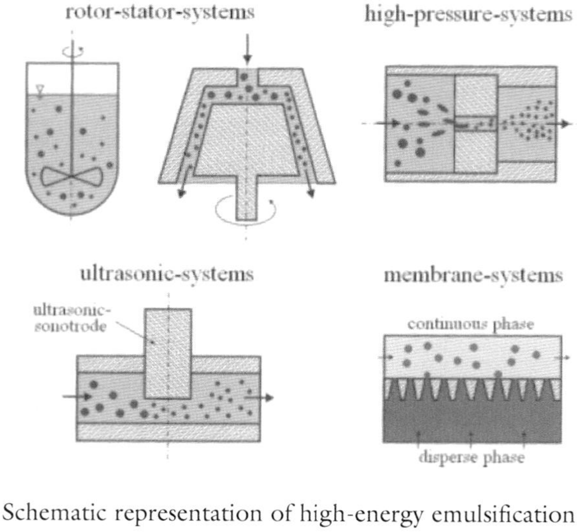

Fig. 2. Schematic representation of high-energy emulsification methods.[20]

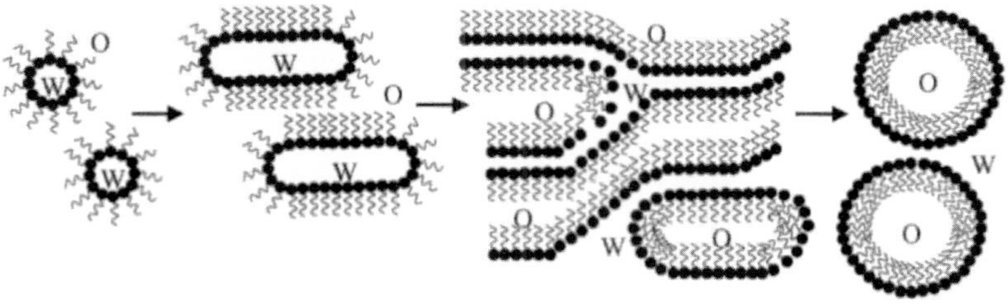

Fig. 3. Phase inversion composition method (W = water phase, O = oil phase).[6] Moving left to right corresponds to the addition of water to a mixture of oil and surfactant. The system goes from inverse micelles (left) to a bicontinuous, lamellar structure (middle) before collapsing to a submicron O/W emulsion (right).

2.1. *High-energy emulsification techniques*

2.1.1. *High-shear stirring*

High-energy mixers and rotor-stators were some of the first systems used for the preparations of NEs.[19] The shearing forces imparted by the rotor-stators continually disrupt the dispersed and continuous phases, and submicron droplets are generated. Increasing the mixing intensity can decrease droplet size, but achieving sizes smaller than 200–300 nm can be difficult.[48] As such, high-shear stirring is typically used to generate a coarse emulsion from the starting components, which are then refined into a NE using high-pressure homogenization, ultrasonication or microfluidization.[49]

2.1.2. *High-pressure homogenization*

High-pressure homogenization is the preferred method for the emulsification of low- and medium-viscosity systems.[50] As mentioned in Section 2.1.1, high-shear mixing is first used to generate a coarse emulsion, which is then subjected to high-pressure homogenization to reduce the mean droplet size (as small as 1 nm). High-pressure homogenization is the preferred method of high-energy emulsification as it allows for a greater variance in composition of NEs. Furthermore, controlling the homogenization pressure and the number of homogenization cycles can modulate the size and polydispersity of the emulsion. High-pressure homogenization yields small droplet sizes through a mixture of hydraulic shear, intense turbulence and cavitation.[20]

2.1.3. *Ultrasonic emulsification*

Ultrasonic emulsification forms nanometer-sized droplets by sonication due to cavitation and the local release of energy.[21] Up to a certain limit, increases in ultrasonic emulsification can lead to decreases in droplet size.[51] Further increases lead to insignificant changes in droplet size. Furthermore, high-energy sonication is undesirable due to the formation of hydrogen and hydroxyl radicals from the decomposition of water. The formation of these radical species can lead to the decomposition of surfactant molecules on the surface of cavitation bubbles.[52,53] Furthermore, ultrasonic emulsification is limited to the vicinity of the waveguide radiator. For large volumes, mechanical stirring is required to emulsify the mixture. It has been shown that O/W NEs generated by sonication typically have much lower mean droplet diameters than those prepared by high-shear stirring.[54]

2.1.4. *Microfluidization*

Microfluidization is a patented mixing technology that uses a microfluidizer to formulate the NE.[22] The emulsification occurs through collision of the flow streams of two immiscible fluids moving in microchannels under high pressure (500–20,000 psi).[23] The two phases (water and oil) are combined together in an inline homogenization process. Dispersion of the oil phase into the continuous aqueous phase occurs under the action of shear deformation, impact stress, and cavitation. Microfluidizers have been shown to form fairly monodisperse droplets under moderate flow.[55,56] Microfluidization, however, faces several disadvantages compared to other high-energy emulsification methods: highly specialized apparatuses are required, high rates of flow are not well tolerated within the microchannels, and only O/W emulsions have been successfully prepared.[12]

2.2. *Low-energy emulsification techniques*

2.2.1. *The phase inversion temperature and phase inversion composition methods*

Low-energy methods of emulsification are based on the phase inversion in emulsion upon a change in the composition or temperature. For both methods, emulsion formation is achieved by first forming a bicontinuous, or lamellar assembly, at a given temperature or composition (Fig. 2).[6,25,45] The emulsion then forms upon changing the composition (changing the electrolyte concentration, changing the pH, adding more water, or adding more surfactant) or changing the temperature, such that the bicontinuous structure collapses to an O/W emulsion. This method works well for labile drugs or biologics, but can suffer from rapid Ostwald ripening due to the need for the oil and water phases to be fairly miscible.[28] Furthermore, both the temperature and composition methods have stringent requirements in terms of surfactant types (typically non-ionic) and temperature ranges over which the methods are effective, and relative concentrations of surfactant, electrolyte and oil phase.[29]

2.2.2. *Solvent displacement method*

The solvent displacement method for spontaneous fabrication of NEs has been adopted from the nanoprecipitation method used for polymeric nanoparticles.[46] In this method, the oil phase is dissolved in a water-miscible solvent — ethanol, acetone, or ethyl methyl ketone — the organic phase is then poured into an aqueous phase containing surfactant to yield spontaneous NEs – this process is also known as self-emulsification.[47] The organic solvent is then removed by vacuum evaporation.

Spontaneous emulsification has also been reported for dilute solutions of oil, in a water miscible solvent, upon mixing with large volumes of water.[57] This is known as the "ouzo effect" because of the well-known cloudiness that results from diluting anise aperitifs, e.g. ouzo, with water.[57–59] Here, the water-insoluble anethole immediately forms nanodroplets without the presence of surfactant.[57] This same effect has been observed for divinyl benzene, *N,N*-dimethylaniline, fluorotoluene, fluorostyrene and benzyl alcohol dissolved in DMSO, ethanol, and/or acetonitrile.[62]

The solvent displacement method can yield NEs at room temperature with only simple stirring which makes this method appealing for small scale, batch emulsion formulation. Disadvantages of NEs formulated by this method requires large amount of solvent to achieve small droplet sizes and the presence of an organic solvent requires the input of energy for its removal.[63] This may be trivial in a laboratory setting, but limit the scalability of such methodology for industrial emulsion preparation.

2.3. *Ostwald ripening and emulsion stability*

Since NEs possess particles small enough such that creaming and sedimentation are overcome by Brownian motion, Ostwald ripening is generally the only route by which NE destabilize. Ostwald ripening arises from polydispersity among the emulsion droplets, with high Laplace pressure in smaller droplets and lower Laplace pressure in larger droplets. The dispersed phase will diffuse from smaller to larger droplets (higher to lower Laplace pressure) thus fomenting particle size growth and eventual phase separation.[30]

Several methods have been developed to impede or inhibit Ostwald ripening and thus enhance the lifetimes of NEs, a point of key importance in developing clinically useful NEs. These methods are all based on the fact that when dispersed-phase molecules have extremely low solubility in the continuous phase, Ostwald ripening is suppressed.[62]

The first method, the two-component dispersed phase method (trapped species method) utilizes a second, oil-phase additive, which is less soluble in water than the first dispersed-phase liquid. The additive is then trapped within the droplet and any transport of the first dispersed-phase liquid creates an osmotic pressure that counterbalances the Laplace pressure, inhibiting net transport out of the droplets thus stabilizing the droplet.[63]

The second method, the evaporative ripening method, has been developed to further reduce the size of NEs. Aside from the desired dispersed phase, a second volatile oil phase is added, which is then removed by heating and addition, of continuous phase. This reduces the overall size of droplets down to micellar dimensions and leads to osmotic pressure great enough within the micelle-sized droplets that Ostwald ripening is slowed, if not stopped.[12] This method thus functions not only to slow Ostwald ripening, but also to greatly diminish droplet size.

3. Nanoemulsions in Medicine

The NEs have found a wide range of promising applications in the pharmaceutical and medical fields, including: drug delivery, mucosal vaccines, topical antiseptics and contrast agents for ultrasonography and MRI. A short list of commercially available NEs is given in Table 1. The interest in NEs arises from their unique characteristics, as described in the previous section. First, since they can be prepared from a variety of components by a variety of methods, preparation and development can be both facile and rapid. Second, NEs' small size and potential long-term stability (either inherent to the system or achieved through direct formulation modification) make NEs attractive for medical formulations where long-shelf lives are required: FDA regulations demand at least 6–12 months, depending on storage conditions.[64] Furthermore, their small size gives NEs greater surface area and higher free energy than traditional macroemulsions, which enhances their drug delivery ability. Also, the small size of most NEs qualifies them for *in vivo* use, as determined by USP<729>.[15]

Table 1. Examples of commercially available NEs.[11,14,65]

Brand	Emulsified agent	Indication	Manufacturer
Diprivan	Propofol	Anesthetic	Astra Zeneca
Limethason	Dexamethasone	Steroid	Mitsubishi Pharmaceutical, Japan
Liple	Palmitate alprostadil	Vasodilator, platelet inhibitor	Mitsubishi Pharmaceutical, Japan
Ropion	Flurbiprofen axetil	NSAID	Kaken Pharmaceuticals, Japan
Sandimmune	Cyclosporin A	Immunosuppressant	Novartis
Restasis	Cyclosporin A	Ophthalmic Anti-inflammatory	Allergan
Norvir	Ritonavir	Protease Inhibitor	AbbVie
Diazepam-Lipuro	Diazepam	Sedative	Braun Melsungen
Vitalipid	Vitamins A, D, E, K	Parenteral Nutrition	Fresenius Kabi, Europe
EMLA Cream	Lidocaine, prilocaine	Topical anesthetic	Astra, Sweden
Oxycyte	Fluorocarbons	Blood supplement, O_2 carrier	Oxygen Biotherapeutics

Finally, it should be noted that while NEs have been proposed for use in a variety of fields, there has been a relative dearth of publications presenting clinically useful formulations.[18] The main difficulty facing the development of NEs is particle size growth and eventual phase separation due to Ostwald ripening. The focus in clinical formulations, therefore, has slowly shifted from bulk-prepared, long-shelf-life NEs, to NEs, which can be quickly prepared prior to use.[17,18] As such, low-energy emulsification methods have proven attractive because of the rapidity and easiness with which NEs can be prepared by such methods — typically right before an emulsion's use. The recent and promising developments in NE applications in the medical and pharmaceutical fields, e.g. drug delivery, mucosal vaccines, topical antimicrobials and contrast agents for imaging are presented in the following Sections.

3.1. *Drug delivery*

The NE have been studied and used as drug-delivery vehicles for a variety of routes of administration. There are several advantages to using NE formulations for the delivery of lipophilic pharmaceuticals. First, the dispersed, oil phase can solubilize the hydrophobic drug, thus dissolving a larger quantity than would be possible in an aqueous solution. Second, the encapsulated drugs are protected from hydrolysis and enzymatic degradation from both tissues and bodily fluids due to their sequestration within the oil phase.[65,66] Together, dispersion of the lipophilic species, within a NE formulation, and protection of *in vivo* degradation enhances the bioavailability of dispersed therapeutics.[67] In terms of orally delivered NEs, enhanced bioavailability has also been observed but it is not yet clearly understood; several different potential mechanisms have been proposed in the literature.[68–72] Furthermore, the use of NEs limits the exposure of gut-wall tissue[73] or epithelial cells[3] — depending on the route of administration — to the dispersed therapeutics, thus reducing irritation.

The broad variance in NE formulations and delivery methods also allows for the targeted delivery of therapeutics. One means of affecting targeted delivery includes changing the dispersed oil phase. It has been found that the use of soybean oil in gadolinium-containing NEs lead to higher blood and tumor gadolinium concentrations than other oil phases in melanoma-bearing hamsters.[74] The high gadolinium concentrations were then used to suppress tumor growth using neutron-capture therapy. Alternatively, controlled release, generally using ultrasonography, of drugs from stable nanodroplets has been shown effective in selectively releasing camptothecin, a topoisomerase I inhibitor, once the NE droplets had been internalized within tumor cells.[75] Lastly, site specific delivery can be achieved by varying the means of administration, examples of which include the high drug transport efficiency of risperidone (antipsychotic) NEs by intranasal delivery in rats[76] and the selective deep skin tissue delivery of paclitaxel (PCL) in rat models through the use of a dermal NE.[77] Applications of NEs in the field of drug delivery are presented in the Sec. 3.1.1.

3.1.1. *Parenteral drug delivery*

Oil-in-water nanoemulsions (O/W NEs) are attractive in parenteral delivery for both nutrition (Table 1) and the delivery of poorly water-soluble drugs. As discussed, this is a result of the enhanced solubilization of lipophilic drugs in O/W NEs and because of the enhanced penetration rate of medicines using NEs as compared to other delivery vehicles — macroemulsions, micelles, etc.[78–80] Finally, NEs have been shown to sustain controlled release of drugs over a long period, which allows for dosages and frequencies of drug administration can be reduced.[11]

One of the most widely studied applications for parenteral NEs is that of anticancer formulations,[8,11,14,18,77] including: photodynamic therapy agents,[81,82] neutron capture agents,[74,83] and traditional chemotherapy drugs.[84,85] Aside from enhanced bioavailability and sequestration from potential *in vivo* degradation, drug loaded NEs have demonstrated prolonged release for drugs with a short half-life *in vivo*.[86] As an example, Cheliensisin A (GC-51) has been emulsified in a lipid emulsion by high-shear stirring followed by high pressure homogenization.[87] Compared to Cheliensisin A (delivered as a 1:1:5 (v/v/v) cremophor/ethanol/PBS solution), the GC-51 NE demonstrated enhanced solubility. Furthermore, the enhanced tumor inhibition both *in vitro* and *in vivo* in BALB/c mice-bearing lung carcinoma (Fig. 4) was demonstrated, compared to GC-51 solution. Finally, the lyophilized NE both improved the stability of GC-51 *in vivo* and demonstrated prolonged sustained release.

The NEs have been shown to improve therapeutic efficacy of poorly water-soluble drugs in anticancer therapies. However, the development of multidrug resistance severely limits the ability of all agents — both chemical and biological — to be therapeutically effective.[88] To overcome this problem, the co-administration of multiple chemotherapeutic agents has been shown to be very effective *in vitro*.[89,90] The NEs have been investigated as potential vehicles for delivering a PCL and curcumin.[91] The PCL and curcumin were dispersed in an O/W emulsion with flaxseed oil and 1, 2-distearoyl-*sn*-glycero-3-phosphoethanolamine-*N*-[methoxy(polyethylenegylcol)-2000]

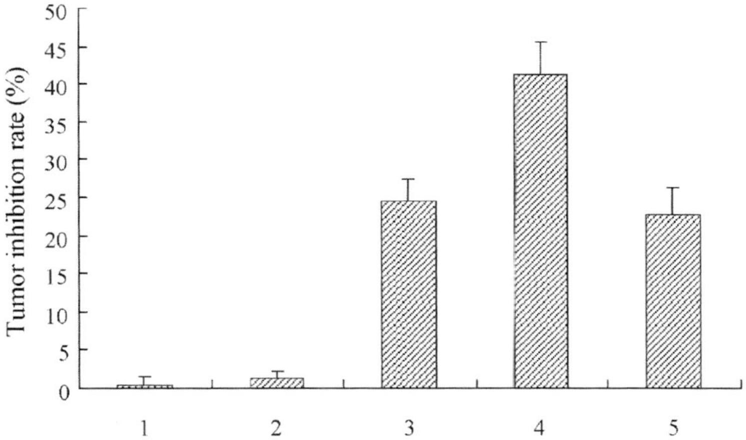

Fig. 4. Antitumor effect of GC-51 modalities administered into mice-bearing lung carcinoma by *i.v.* administration ($n = 8$); (1) blank submicron emulsion; (2–4) GC-51 submicron emulsion 5, 15, and 30 mg/kg; and (5) GC-51 drug solution in 1:1:5 (v/v/v) cremophor/ethanol/PBS 30 mg/kg.[87]

($DSPE-PEG_{2000}$) by microfluidization to give particles < 200 nm.[91] It was shown, *in vitro*, that the co-NE-administered PCL and curcumin showed significant enhancement in cytotoxicity in drug-resistant $SKOV3_{TR}$ ovarian adenocarcinoma cells.[91]

Poorly water-soluble drugs dispersed in O/W NEs have been found to increase bioavailability *in vivo*. As discussed, this is a direct result of enhanced solubilization of and protection of the drug within the NE. Concomitantly, dispersion within a NE can help lower the cytotoxicity against normal cells.[92] Doxorubicin (DOX), as either a solution in DMSO or as the commercial formulation, was shown to be significantly more cytotoxic than NE DOX.[92] This reduction in cytotoxicity against normal cells could help increase DOX's — and other drug's — clinical usefulness, which is currently limited by its systemic cytotoxicity.[93]

While there are many examples of parenteral delivery of poorly, water-soluble drugs, NEs have also been used for the delivery of DNA, oligonucleotides, and proteins.[94–97] The use of cationic, polyamide-based surfactants yields nanodroplets that are positively charged. Simple mixing of DNA, oligonucleotides or RNA yields an electrostatic complex, which has been shown to be stable and serum-resistant.[98] Lipid-based DNA vectors are the most promising non-viral vectors, and *in vivo* administration of cationionic/DNA conjugates has shown higher transfection and lower toxicity than liposomes.[99] PEGylation has been shown to not only further enhance *in vivo* stability, but also reduce clumping of DNA/nanodroplets.[100]

The *i.v.* delivery of anesthetics is of interest because it offers several unique pharmacodynamic improvements, including faster induction.[101,102] Direct injection of neat anesthetic, however, has lead to severe complications, if not death, while emulsified anesthetics have shown no such toxicity.[103] A NE of propofol is commercially available and is widely used to induce anesthesia.[11,14]

While a lipid-based NE has been shown to stably emulsify propofol, this use of traditional lipid and hydrocarbon surfactants fails to stably emulsify clinically useful concentrations of fluorinated, volatile anesthetics such as sevoflurane. The maximum concentration (v/v) of sevoflurane emulsifiable in intralipid is only 3.5%.[104] An issue is the mismatch between the lipophilicity of the traditional NE system and the fluorophilicity of sevoflurane. The synthesis and use of a linear perfluorocarbon surfactant, was able to stably emulsify 25% (v/v) sevoflurane.[105] This is a result of the greater affinity between the fluorous anesthetic and the fluorous hydrophobic region of the surfactants. To overcome the fast Ostwald ripening rate, perfluorooctyl bromide (PFOB) was added as a less water-soluble dispersed phase, which increased the stability to several months.[105] Further, rational modification of the surfactant lead to a dibranched polymer, which, with 10% PFOB, showed no Ostwald ripening over the course of more than 12 months.[106] The sevoflurane emulsion was shown to be effective at inducing anesthesia — from which recovery was fast and rapid — in bolus dosing in rats.[105]

3.1.2. *Ocular drug delivery*

Oil-in-water (O/W) NEs are attractive vehicles for ocular drug delivery due to their ability to sustain drug efficacy over time,[107] increase penetration into deep ocular tissue,[107] and they reduce vision-blurring effects.[43] The extent of greater, ocular penetration has been demonstrated using ^{14}C-indomethacin loaded NE, which showed a permeation coefficient 3.65[108]–3.8[109] times greater than the commercial eye drop formulation. Studies have also clearly demonstrated that both neutral and cationic NEs were tolerated without any toxic or inflammatory response in rabbit eye.[110,111]

A commercially available O/W NE has been developed, Restasis (Table 1). Restasis is used to deliver cyclosporine A in a 0.05% solution and is marketed as an anti-inflammatory for the treatment of keratoconjunctivitis sicca (dry eye syndrome).[112] Restasis has also demonstrated dramatic diminution in the recovery time for post-operative LASIK patients to regain uncorrected visual acuity.[113]

3.1.3. *Oral drug delivery*

The NE drug delivery systems offer many advantages in the oral delivery of poorly permeable and poorly water-soluble drugs. Primarily, O/W NEs dramatically increase the bioavailability of lipophilic drugs delivered orally. The enhanced bioavailability is not yet clearly understood, however, several different potential mechanisms have been proposed in the literature.[68–72] Further increases in the drug activity of NE formulations has been demonstrated when the size of droplets is reduced below 100 nm.[114] Finally, NEs reduce intersubject and intrasubject variability for some drugs, including increased reproducibility of plasma concentrations.[115–118]

Aside from increasing the bioavailability of drugs, oral-delivery NEs protect drugs from *in vivo* metabolism. Drugs dispersed in NEs are shielded from hydrolysis and enzymatic degradation.[119,120] Together with the higher availability, this reduced *in vivo* breakdown allows for lower dosing compared to conventional oral administration. The antimalarial primaquine NE was shown to be as efficacious as the traditional oral formulation, but at 25% lower dosing. The lower dosing helps to minimize systemic toxicity of the drug.[121]

The use of self-nanoemulsifying drug delivery systems (SNEDDs) are water-free mixtures delivered orally in capsular form, which upon dissolution in the gastrointestinal tract spontaneously emulsifies (solvent-displacement method emulsification).[122,123] The use of SNEDDs has shown increased permeability of drugs. Examples include atenolol, danazol and metoprolol, which shows 1.4- to 3.2-fold permeability in the Caco-2 monolayer model of the small intestine.[124] This type of methodology has also been used in the oral delivery of protein drugs. The oral absorption of a β-lactamase, a model protein, SNEDD formulation showed significant increase compared to the free solution.[125–127]

The Paclitaxel is an antineoplastic agent with high therapeutic efficacy against breast and ovarian cancers.[128] Although PCL is clinically active, its delivery is therapeutically limited by poor solubility and permeability.[129] The oral bioavailability of Taxol is low 6.5–10.62% (Fig. 5),[77,130] but formulated as a NE, the bioavailability increases 6-fold and approaches that of *i.v.* administered Taxol.[77] Combined with the ability to localize PCL delivery by dermal application, the NE formulation of PCL offers unique advantages over traditional formulations.

Acquired immunodeficiency syndrome (AIDS) is a debilitating disease caused by the HIV virus. There has been significant improvement over the past 25 years with highly active antiretroviral therapy (HAART) improving

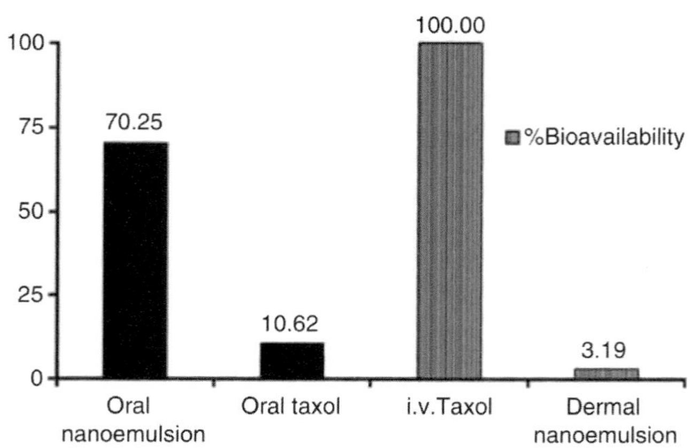

Fig. 5. The absolute bioavailability of PCL administered from various formulation via different routes of administration. The numbers indicate the absolute bioavailability of formulations. The absolute bioavailability was calculated from the ratio of $AUC_{\alpha-\infty}$ of each formulation with that obtained upon *i.v.* administration of same dose of PCL.[77]

average survival from 1 to 10 years.[131] However, concentration of viral particles in reservoir sites including the central nervous system (CNS) — accompanied with the low bioavailability and poor permeability of many HAART compounds — impedes effective management of HIV.[132] The NE formulation of Saquinavir, a protease inhibitor, in a flax-seed oil based NE has been shown to improve the bioavailability from 42.2%, for the orally administered suspension, to 86.69% (Fig. 6).[133] Additionally, the NE formulation enhances the concentration of Saquinavir in the brain, demonstrating the higher systemic bioavailability and permeability of the NE formulation.[133]

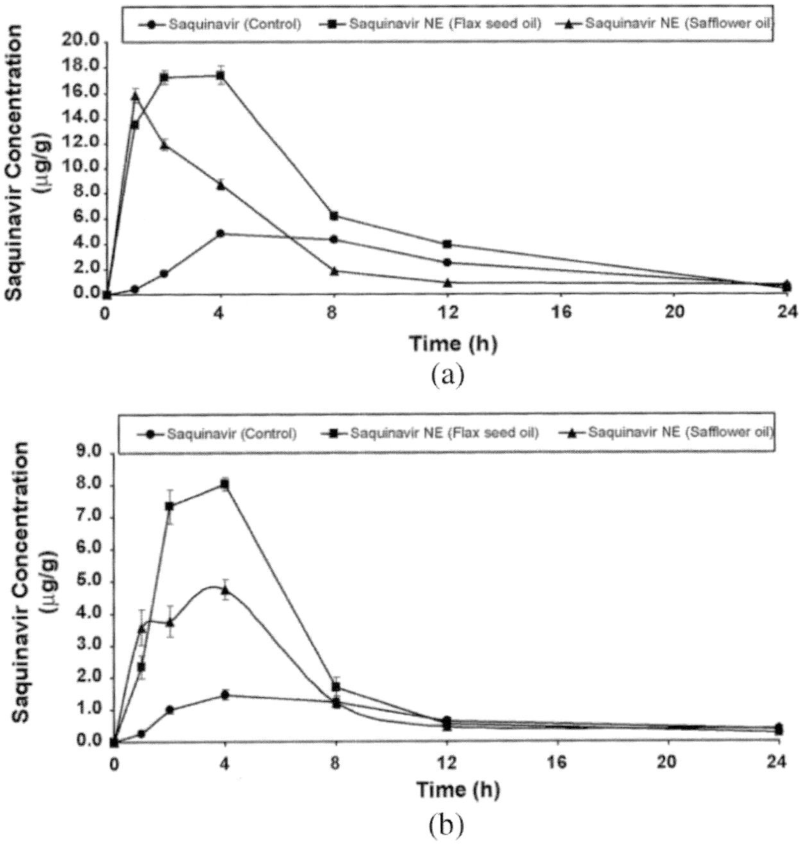

Fig. 6. Plasma and brain Saquinavir concentration versus time profiles following oral administration of the drug in aqueous suspension or NEs formulations to BALB/c mice. The plots represent plasma concentrations versus time following oral administration (a) and brain concentrations versus time following oral administration (b). The BALB/c mice were dosed with 0.50 mL of the control aqueous suspension or NE formulations by oral gavage. The administered dose contained 1 μCi radioactivity as titrated [^{3}H]-Sasquinavir.[133]

3.1.4. *Transdermal drug delivery*

Transdermal drug delivery is challenging because of the poor skin penetration of many drugs, which results in poor efficacy. Nanoemulsion formulations have been shown to improve transdermal drug permeation.[4] The improved permeability of transdermal NEs is attributable to the high surface area and small droplet size,[134] low viscosity and low surface tension.[135] While improving permeability, the low viscosity of NE limits their potential applicability. Some systems have been developed that include hydrogel thickeners,[136,137] which offer improved prospects for the development of commercial NEs for transdermal delivery. Aside from the higher permeability, NE have demonstrated lower irritation, without diminution of drug-loading capacity, than liposomes, classical emulsions or gels.[40,136,138]

Two studies have investigated the potential application of topical, NSAID NEs. Aceclofenac[139] and aspirin[140] have both been prepared as O/W NEs. The development of NSAID NEs for dermal applications stems from the side effects of chronic oral administration (including gastrointestinal irritation).[139,140] The aceclofenac NE showed improved anti-inflammatory behavior than a conventional gel, which has been ascribed to the greater permeability.[139] A topical, aspirin NE formulation demonstrated anti-inflammatory efficacy in mice and offers an alternative to oral delivery of aspirin for anti-inflammatory needs. Together, these two emulsions offer the potential to apply NSAIDs dermally to the effective area, reducing oral administration and thus reducing the side effects associated with chronic use.[140]

Transdermal NE formulations have also been developed for anticancer therapies. The development of a NE with dispersed Foscan — a classical photodynamic agent — showed improved penetration of Foscan throughout the skin of a pig-ear *in vitro* model, including significant concentration of drug in deep tissue layers.[81] The NE of highly lipophilic dacarbazine (an antineoplastic therapeutic), when applied topically, significantly reduced the tumor size in human-melanoma-xenografted mice.[141] Similarly, a tamoxifen (an estrogen antagonist) NE inhibited cell proliferation and increased apoptosis in the HP-20 breast cancer cell line, compared to a tamoxifen suspension.[142]

3.1.5. *Nasal drug delivery*

Intranasal drug delivery has emerged as a reliable route for the administration of drugs. Nasal mucosa are a therapeutically viable channel for systemic drug administration,[143] and delivery to the nasal cavity is attractive for its reduced enzymatic activity, high availability of immunoactive sites and its epithelial cells' moderate permeability.[144] Finally, the direct connection between the nose and the brain offers a unique route for delivery of therapeutics to the brain.[145] The direct olfactory-to-brain route of administration has been probed in mice, where it was shown that nasally administered morphine on polysaccharide Biovector™ particles increased the analgesic effect without increasing the blood concentration of morphine. Delivery of morphine to the brain is believed to follow uptake by the olfactory bulb.[146]

Risperidone is an atypical antipsychotic typically used in the treatment of schizophrenia. When administered as a chitosan-containing NE, risperidone was shown to increase from 62% nasal mucosa to brain delivery — for a risperidone suspension — to 78% in rat models.[76] The mechanism of nose-to-brain uptake was not discussed, but assumed to follow that mechanism described for morphine uptake.[145]

3.2. *Mucosal vaccines*

While intranasal drug delivery has been found to be an appealing route for delivery of lipophilic drugs to the brain, the same cannot be said for traditional vaccinations. Vaccination through introduction of antigens to the nasal mucosa would offer the potential to enhance protection against respiratory illness; however, intranasal vaccination has failed to produce a strong immune response when using killed influenza virus.[147] Currently, a nasal spray has been clinically approved, which uses partially attenuated virus; nevertheless, complications can easily arise.[147] Approved influenza antigen NE adjuvant has provoked a stronger immune response via intranasal vaccination; in a first-in-human study the NE vaccine was found to be safe, well-tolerated and elicited both a strong mucosal and systemic immune response.[147] The success in using the influenza antigen NE is due to the NE acting as both carrier and adjuvant, i.e. a non-biological compound that enhances the immune system's response during immunization and enhances immunity.

Nasal immunization with a recombinant HIV gp120 NE adjuvant has been shown to produce neutralizing antibodies to primary HIV type 1 Isolates in a guinea pig model.[148] Similar to the influenza vaccine, both systemic and mucosal antibody responses were observed; the HIV/NE vaccine, however, also showed significant, cross-reactive neutralizing immunity. These results are promising — the best HIV vaccine would show high titers of antibodies that could neutralize many different HIV isolates — and are considered the basis for future HIV vaccine development.[148]

The currently available anthrax vaccine requires multiple injections, and yearly booster shots, for efficacy and has multiple side effects due to its alum adjuvant. These factors ultimately limit the utility of such a vaccine, which lead to the development of a NE adjuvant, recombinant protective antigen anthrax vaccine, which has been shown effective in mice and guinea pigs.[37] Similarly, hepatitis B remains a global health concern, yet large scale vaccinations are limited by a rigorous treatment regimen.[149] A new recombinant hepatitis B surface antigen NE adjuvant vaccine has been studied in preclinical tests in mice, rats and guinea pigs.[150] The new vaccine shows safe and effective nasal vaccination and could help boost the administration of the vaccine, especially in the field since the formulation is stable at elevated temperatures.[150]

3.3. *Topical antimicrobials*

Nanoemulsions have been shown to possess antimicrobial activity, both broad-based[11,151] and against Ebola virus specifically.[152] For surface cleaning, O/W and W/O NEs (both diluted with water before use) stabilized by long-chain alcohols, have been developed. Antimicrobial ability is imparted by fusion of surfactant molecules with lipid membranes. Surfactant-membrane fusion can be enhanced by the use of cationic lipids, which facilitate binding to anionic pathogen membranes.[11,151] Fusion of surfactant to the lipid membrane then causes cell lysis and death. BCTP (O/W NE of soybean oil with tri-*n*-butyl phosphate emulsified with Triton X-100) has been shown to be bactericidal (against Gram-positive bacteria),[153] sporicidal,[154] and antiviral (against enveloped viruses).[155] Furthermore, it has been shown efficacious in inhibiting the biofilm of four bacteria (Fig. 7).[151]

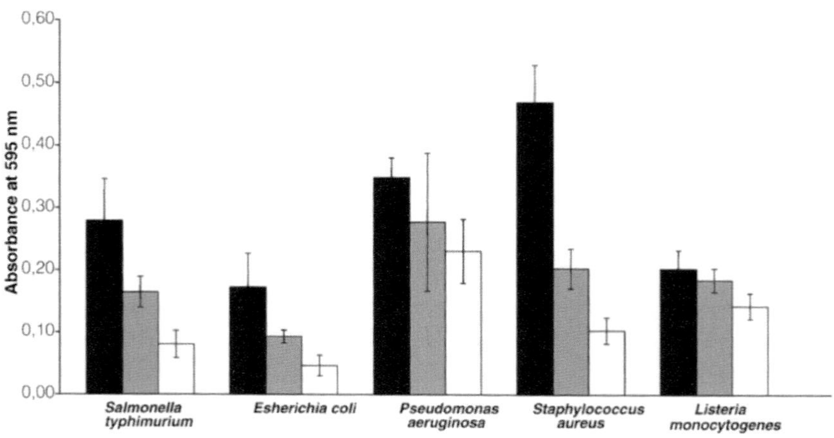

Fig. 7. Effect of the BCTP NE on biofilms at 0 min (black bars), 5 min (grey bars), and 30 min (white bars).[151]

A W/O emulsion ATB (comprised of Triton X-100, cetylpyridinium chloride, tetra-*n*-butyl phosphate, soybean oil, and water) has been shown to be virucidal against Ebola (subtype Zaire, strain Zaire).[152] Ebola is both extremely contagious and particularly hard to inactivate.[156–158] ATB was tested against Ebola virus (used as a 1:10 or 1:100 solution in water) and demonstrated virucidal activity within 20 min in both growth medium and in suspensions of infected tissue.[152]

3.4. *MRI and ultrasonography contrast agents*

The ability to detect a specific analyte or follow its progress *in vivo* is of key importance in biomedical research.[159] Furthermore the ability to clearly — and non-invasively – image or visualize internal tissues and organs is vital to modern medicine.[152] Liquid perfluorocarbon NEs have found applications as both MRI and ultrasonography contrast agents.[39,159,160]

3.4.1. *MRI contrast agents*

Due to its non-invasive nature, contrast among soft tissues at high spatial resolution and lack of penetration limitations, MRI is a widely used tool in clinical diagnosis.[160] ^{19}F-MRI offers several advantages in MRI diagnostic and tracing experiments. First, the ability to directly detect ^{19}F-nulcei dramatically increases the sensitivity compared to indirectly detected super paramagnetic iron oxide particles.[161,162] Second, ^{19}F has high natural abundance (100%), and ^{19}F has a resonant frequency, NMR sensitivity, and signal-to-noise ratio 94%, 83% and 89%, respectively, that of ^{1}H. Together with the lack of background ^{19}F-MRI signals, this makes ^{19}F-MRI uniquely sensitive and an attractive tool for diagnostics.[163]

Perfluorocarbon NEs have found several unique uses in providing MRI contrast agents. Tracking cells *in vivo* is a challenging and yet crucial challenge in utilizing cell therpeutics.[164,165] Typical procedures that have been developed utilize *ex vivo* labeled cells, which are inserted into the organism and then followed *in vivo* using ^{19}F-MRI.[164,165] These techniques have shown great potential, but rely on the direct labeling and then successful insertion of labeled cells into an organism. Recent work has developed a directly-injected perfluorocarbon NEs, which has been shown to be taken up by monocytes/macrophages.[156] The monocytes/macrophages are then enriched in fluorine content, perfluoro-15-crown-5 has been used with no ill effect to the cells, and can then be followed *in vivo*.[166] These *in situ* labeled monocytes/macrophages have been used to provide positive signals

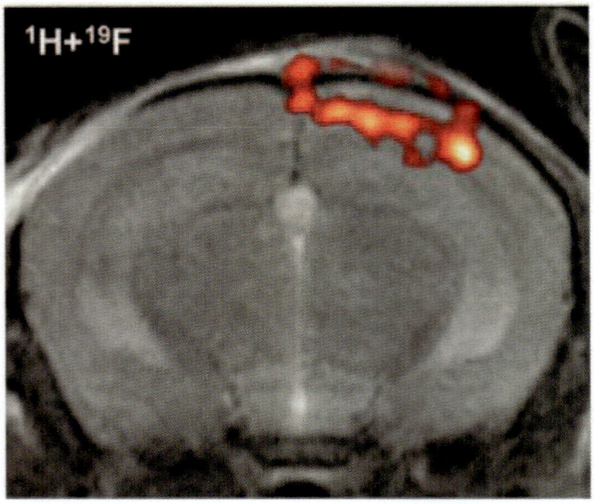

Fig. 8. Merged ^{1}H and ^{19}F (red) images of C57BL/6 male mouse brain 7 days after induction of focal cerebral ischemia by photothrombosis.[166]

at sites of inflammation in animal models (Fig. 8).[166] Finally, further emulsion development has lead to a ^{129}Xe loaded perfluorocarbon NE, which has also demonstrated detection at sub-picomolar concentrations, thus further enhancing the role that perfluorocarbon emulsions may come to play in *in vivo* diagnostics.[159]

3.4.2. *Ultrasound contrast agents*

Similar to MRI, ultrasonography is non-invasive, but unlike MRI it can suffer from poor resolution of soft tissues (similar echogenicity). Moreover, there exists a trade-off between high penetration (low power) and high resolution (high power).[39] A compensating benefit of ultrasound is its low cost.[39] Contrast agent bubbles increase the echogenicity of the tissue, in which they exist, but must be stable, inert and small enough to pass through the circulatory system. As such, perfluorocarbon NEs are particularly attractive because of their small, stable, inert nature.[39] Also the liquid perfluorocarbon nanodroplets serve as efficient sound-wave reflectors, i.e. have excellent echogenicity.[165]

Because of their high echogenicity, perfluorocarbon NEs offer the ability for assessing ventricular function and wall-motion abnormalities, which is otherwise difficult to image.[168] In addition, labeled-surfactants have been used to develop a targeting perfluorocarbon NE contrast agent.[169] In low doses, high echogenicity is only observed upon accumulation of nanodroplets at the target site. This offers the ability to target and identify a variety of cells that express the corresponding receptor. Finally, ultrasonography has been shown increase drug release from NE droplets,[75] and specific NE formulation can offer theranostic (therapeutic and diagnostic) behavior.[85] Ultrasound theranostic NEs offers the possibility of both increasing diagnostic imaging capabilities, using perfluoro-15-crown-5 ether NEs, and concomitantly increasing drug efficacy, *vis-à-vis* drug release.[170]

4. Conclusion

The NEs considered in this chapter are a special class of dispersed, liquid systems. The nanoemulsions are kinetically stable systems that possess an average particle size less than or equal to 500 nm, a cutoff that is not with contention in the field but which is based on the instructions of USP<729>. Given their small size, Brownian motion is able to prevent sedimentation and creaming, while steric and/or ionic repulsion prevents flocculation. As such, Ostwald ripening becomes the main mechanism by which NEs coarsen; however, several methodologies exist for slowing or stopping this process.

Given NEs' thermodynamic instability, energy must be input to induce emulsion formation. Two possible methodologies exists: high-energy methods, which utilize mechanical force to disrupt the dispersed and continuous phase and drive emulsion formation; and low-energy methods, also called self-emulsification methods, whereby changes in temperature or composition drive the formation of NEs by leveraging the energy change associated with the change in chemical potential. High-energy methods are limited in terms of the types of drugs that can be emulsified (thermolabile or biological material can be destroyed in the process), while low-energy methods are dependent on the physicochemical properties of the system for effective formulation. Low-energy methods continue to receive increased attention due to their ease of preparation, which allows for rapid formulation before use.

The small size, ease of formulation and potentially high stability has brought intense study into their potential medical applications. The NEs have been studied as drug delivery vehicles for their ability to deliver large doses of lipophilic therapeutics with enhanced bioavailability. The O/W and W/O NEs have found roles in many potential routes of drug delivery from dermal applications to intranasal and parenteral. The NEs have demonstrated the ability to act as vector and adjuvant in intranasal NEs and as antimicrobials against a host of pathogens. Finally, NE have been used as contrast agents in both MRI and ultrasonography. Altogether, NEs represent an expanding field with significant potential for a variety of medical and medical-related fields.

References

1. P. P. Costantinides, M. V Chaubal and R. Shorr. *Adv. Drug Deliver. Rev.* **60**, 757–767 (2008).
2. C. Solans, P. Izquerdo, J. Nolla, N. Azemar and M. Garcia-Celma. *Curr. Opin. Colloid Interface Sci.* **10,** 102–110 (2005).
3. G. Guglielmini. *Clin. Dermatol.* **26**, 341–346 (2008).
4. S. Benita. *Biomed. Pharmacother.* **53**, 193–206 (1999).
5. S. Klang and S. Benita, In *Submicron Emulsions in Drug Targeting and Delivery*, S. Benita (Ed). Amsterdam: Harwood Academic (1998), pp. 199–152.
6. P. Fernandez, V. André, J. Rieger and A. Kühnle. *Colloids Surf., A***251**, 53–58 (2004).
7. T. F. Tadros, P. Izquierdo, J. Esquena and C. Solans. *Adv. Colloid Interface Sci.* **108–109**, 303–318 (2004).
8. N. Anton and T. F. Vandamme. *Pharm. Res.* **28**, 978–985 (2011).
9. N. Anton and T. Vandamme. *Int. J. Pharm.* **377**, 142–147 (2009).
10. Nanotechnology 101: what it is and how it works? National Nanotechnology Initiative 2000. Available from: **http://www.nano.gov/nanotech-101/what.**
11. P. Shah, D. Bhalodia and P. Shelat. *Syst. Rev. Pharm.* **1**, 24–32 (2010).
12. M. M. Fryd and T. G. Mason. *Annu. Rev. Phys. Chem.* **63**, 493–518 (2012).
13. P. Bhatt and S. Madhav. *Int. J. Pharm. Sci. Res.* **2**, 2482–2489 (2011).
14. C. Lovelyn and A. A. Attma. *J. Biomater. Bionanotech.* **2**, 626–639 (2011).
15. Chapter <729>: *United States Pharmacopeia and National Formulary* (USP 30-NF 25). Rockville, MD: United States Pharmacopeial Convention; 2007.
16. Lipid Injectable Emulsion Monograph. *Pharm. Forum.* **32**, 350–353 (2006).
17. J. M. Gutiérrez, C. González, A. Maestro, I. Solè, C. M. Pey and J. Nolla. *Curr. Opin. Colloid Interface Sci.* **13**, 245–251 (2008).
18. M. Y. Koroleva and E. V. Yurtov. *Russ. Chem. Rev.* **81**, 21–43 (2012).
19. J. Bałdyga, W. Orciuch, L. Makowski, M. Malski-Brodzicki and K. Malik. *Chem. Eng. Process* **46**, 851–861 (2007).
20. S. Schultz, G. Wagner, K. Urban and J. Ulrich. *Chem. Eng. Technol.* **27**, 361–368 (2004).
21. T. S. H. Leong, T. J. Wooster, S. E. Kentish and M. Ashokkumar. *Ultrason. Sonochem.* **16**, 721–727 (2009).
22. S. Okushima, T. Nisisako, T. Torii and T. Higuchi. *Langmuir* **20**, 9905–9908 (2004).
23. V. Barbier M. Tatoulian, H. Li, F. Arefi-Khonsari, A. Ajdari and P. Tabeling. *Langmuir* **22**, 5230–5232 (2006).
24. S. Brösel and H. Schubert. *Chem. Eng. Process* **38,** 533–540 (1999).
25. K. Shinoda and H. Saito. *J. Colloid Interface Sci.* **30**, 258–263 (1969).
26. P. Izquierdo, J. Esquena, T. Tadros, J. C. Dederen, J. Feng, M. J. Garcia-Celma, N. Azemar and C. Solans. *Langmuir* **20**, 6594–6598 (2004).
27. P. Izquierdo, J. Esquena, T. Tadros, J. C. Dederen, M. J. Garcia, N. Azemar and C. Solans. *Langmuir* **18**, 26–30 (2002).
28. I. Solè, C. M. Pey, A. Maestro, C. González, M. Porras, C. Solans and J. M. Gutiérrez. *J. Colloid Interface Sci.* **344**, 417–423 (2010).
29. S. Sajjadi. *Langmuir* **22**, 5597–5603 (2006).
30. P. Taylor. *Adv. Colloid Interface Sci.* **75**, 107–163 (1998).
31. T. F. Tadros, Applied Surfactants: Principles and Applications. Weinheim: Wiley-VCH (2005).

32. W. I. Higuchi and J. Misra. *J. Pharm. Sci.* **51**, 459–466 (1962).
33. A. S. Kabal'nov, A.V. Pertzov and E. D. Shchukin. *Colloids Surf.* **24**, 19–32 (1987).
34. M. M. Fryd and T. G. Mason. *J. Phys. Chem. Lett.* **1**, 3349–3353 (2010).
35. I. Solè, C. Solans, A. Maestro, C. González and J. M. Gutiérrez. *J. Colloid Interface Sci.* **376**, 133–139 (2012).
36. S. M Bertilla, J.-L. Thomas, P. Marie and M. P. Krafft. *Langmuir* **20**, 3920–3924 (2004).
37. A. U. Bielinska, K. W. Janczak, J. J. Landers, P. Makidon, L. E. Sower, J. W. Peterson and J. R. Baker Jr. *Infect. Immun.* **75**, 4020–4029 (2007).
38. T. Hamouda, A. Myc, B. Donovan, A. Y. Shih, J. D. Reuter and J. R. Baker Jr. *Microbiol. Res.* **156**, 1–7 (2001).
39. R. Díaz-López, N. Tsapis and E. Fattal. *Pharm. Res.* **27**, 1–16 (2009).
40. O. Sonneville-Aubrun, J.-T. Simonnet and F. L'Alloret. *Adv. Colloid Interface Sci.* **108–109**, 145–149 (2004).
41. K. Meleson, S. Graves and T. G. Mason. *Soft Mater.* **2**, 109–123 (2004).
42. S. Gramdorf, S. Hermann, A. Hentschel, K. Schrader, R. H. Müller and M. Kraume. *Colloids Surf., A***331**, 108–113 (2008).
43. M. B. Schulz and R. Daniels. *Eur. J. Pharm. Biopharm.* **49**, 231–236 (2000).
44. N. Anton, J. P. Benoit and P. Saulnier. *J. Control. Release* **128**, 185–199 (2008).
45. K. Shinoda and H. Saito. *J. Colloid Interface Sci.* **26**, 70–74 (1968).
46. C. Chaix, E. Pacard, A. Elaïssari, J. F. Hilaire and C. Pichot. *Colloids Surf., B***29**, 39–52 (2003).
47. R. G. Kelmann, G. Kuminek, H. F. Teixeira and L. S. Koester. *Int. J. Pharm.* **342**, 231–239 (2007).
48. H. P. Grace. *Chem. Eng. Commun.* **14**, 225–277 (1982).
49. Y. Yuan, Y. Gao, L. Mao and J. Zhao. *Food Chem.* **107**, 1300–1306 (2008).
50. C. Washington and S. S. Davis. *Int. J. Pharm.* **44**, 169–176 (1988).
51. B. Abismaïl, J. P. Canselier, A. M. Wilhelm, H. Delmas and C. Gourdon. *Ultrason. Sonochem.* **7**, 187–192 (2000).
52. A. E. Alegria, Y. Lion, T. Kondo and P. Riesz. *J. Phys. Chem.* **93**, 4908–4913 (1989).
53. K. Makino, M. M. Mossoba and P. Riesz. *J. Am. Chem. Soc.* **104**, 3537–3539 (1982).
54. Y. Imai, N. Ujiie, A. Nakamura, M. Koshinuma and K. Tajima. *Colloids Surf., A***276**, 134–142 (2006).
55. S. Takeuchi, P. Garsteck, D. B. Weibel and G. M. Whitesides. *Adv. Mater.* **17**, 1067–1072 (2005).
56. P. B. Umbanhowar, V. Prasad and D. A. Weitz. *Langmuir* **16**, 347–351 (2000).
57. F. Ganachaud and J. L. Katz. *Chem Phys Chem.* **6**, 209–216 (2005).
58. I. Grillo. *Colloids Surf., A***225**, 153–160 (2003).
59. N. L. Sitnikova, R. Sprik, G. Wegdam and E. Eiser. *Langmuir* **21**, 7083–7089 (2005).
60. S. A. Vitale and J. L. Katz. *Langmuir* **19**, 4105–4110 (2003).
61. N. S. Santo-Magalhães, A. Pontes, V. M. W. Pereira and N. P. Caetano. *Int. J. Pharm.* **208**, 71–80 (2000).
62. R. Buscall, S. S. Davis and D. S. Potts. *Colloid. Polym. Sci.* **257**, 636–644 (1979).
63. Y. Meliana, N. A. Cala C. T. Lin and C. S. Cern. *J. Dispersion Sci. Technol.* **31**, 1568–1573 (2010).
64. Guidance for Industry, 2003. Q1A(R2) Stability Testing of New Drug Substances and Products, United States Department of Health and Human Services, Food and Drug

Administration. Accessed 6/20/2013, Revision 2. Available at: < http://www.fda.gov/downloads/regulatoryinformation/guidances/ucm128204.pdf>
65. S. Tamilvanan. *Prog. Lipid Res.* **43**, 489–533 (2004).
66. A. T. Serajuddin, P. C. Sheen, D. Mufson, D. F. Bernstein and M. A. Augustine. *J. Pharm. Sci.* 77, 414–417 (1988).
67. D. L. Burcham, M. B. Maurin, E. A. Hausner and S. M. Huang. *Biopharm. Drug Dispos.* **18**, 737–742 (1997).
68. W. W. Yang and I. Q. Wang. *Eur. J. Pharm. Biopharm.* **63**, 288–294 (2006).
69. A. A. Date and M. S. Nagarsenker. *Int. J. Pharm.* **329**, 166–172 (2007).
70. E. I. Taha, S. Al-Saidan, A. M. Samy and M. A. Khan. *Int. J. Pharm.* **285**, 109–119 (2004).
71. V. B. Patravale, A. A. Date and A. A. Kale. *Pharm. Technol.* **29**, 44–48 (2003).
72. S. M. Khoo, A. J. Humberstone, C. J. Porter, G. A. Edward and W. N. Charman. *Int. J. Pharm.* **160**, 11–19 (1998).
73. R. N. Gursoy and S. Benita. *Biomed. Pharmacother.* **8**, 173–182 (2006).
74. H. Ichikawa, T. Watanabe, H. Tokumitsu and Y. Fukumori. *Curr. Drug. Deliv.* **4**, 131–140 (2007).
75. J.-Y. Fang, C.-F. Hung, S.-C. Hua and T.-L. Hwang. *Ultrasonics.* 39–46 (2009).
76. M. Kumar, A. Misra, A. K. Babbar, A. K. Mishra, P. Mishra and K. Pathak. *Int. J. Pharm.* **358**, 285–291 (2008).
77. S. Khandavilli and R. Panchagnula. *J. Invest Dermatol.* **127**, 154–162 (2007).
78. J.-B. Tagne, S. Kakumanu, D. Ortiz, T. Shea and R. J. Nicolosi. *Mol. Pharm.* **5**, 280–286 (2008).
79. T. Kotyla, F. Kuo, V. Moolchandani, T. Wilson and R. Nicolosi. *Int. J. Pharm.* **347**, 144–148 (2008).
80. J. M. Janjic, M. Srinivas, D. K. K. Kadayakkara and T. E. Ahrens. *J. Am. Chem. Soc.* **130**, 2832–2841 (2008).
81. F. L. Primo, L. Michieleto, M. Rodrigues, P. Macaroff, P. C. Morais, Z. G. M. Lacava, M. V. Bentley and A. C. Tudesco. *J. Magn. Magn. Mater.* **311**, 354–357 (2007).
82. T. Maisch, F. Santarelli, S. Schreml, P. Babilas and R.-M. Szeimies. *Exp. Dermatol.* **19**, 302–305 (2010).
83. T. Watanabe, H. Ichikawa and Y. Fukumori. *Eur. J. Pharm. Biopharm.* **54**, 119–124 (2002).
84. M. Dias, J. Carvalho, D. Rodrigues, S. Graziani and R. Maranhao. *Cancer Chemoth. Pharm.* **59**, 105–111 (2007).
85. M. H. Oh, J. S. Kim, J. Y. Lee, T. G. Park and Y. S. Nam. *RSC Adv.* [Just Accepted]. DOI: 10.1039/C3RA40883C (2013).
86. J. J. Wang, C. F. Hung, C. H. Yeh and J. Y. Fang. *J. Drug Target.* **16**, 294–301 (2008).
87. D. Zhao, T. Gong, Y. Fu, Y. Nie, L.-L. He, J. Liu and Z.-R. Zhang. *Int. J. Pharm.* **357**, 139–147 (2008).
88. G. Szakacs, J. K. Paterson, J. A. Ludwig, C. Booth-Genthe and M. M. Gottesman. *Nat. Rev. Drug. Discovery* **5**, 219–234 (2006).
89. H. Devalapally, Z. Duan, M. V. Seiden and M. M. Amiji. *Int. J. Cancer* **121**, 1830–1838 (2007).
90. L. S. Jabr-Milane, L. E. van Vlerken, S. Yadav and M. M. Amiji. *Cancer Treat. Rev.* **34**, 592–602 (2008).

91. S. Ganta and A. Mansoor. *Mol. Pharm.* **6**, 928–939 (2009).
92. M. H. Alkhatib and H. M. AlBishi. *J. Nanopart. Res.* **15**, 1489–1504 (2013).
93. K. Laginha, S. Verwoert, G. Charrois and T. Allen. *Clin. Cancer Res.* **11**, 6944–6949 (2005).
94. S. Tamilvanan, S. Schmidt, R. H. Müller and S. Benita. *Eur. J. Pharm. Biopharm.* **59**, 1–7 (2005).
95. H. Teixeira, C. Dubernet, V. Rosilio, A. Laigle, J. R. Deverre, D. Scherman, S. Benita and P. Couvreur. *J. Control. Release.* **70**, 243–255 (2001).
96. A. R. Gascon and J. L. Pedraz. *Expert Opin. Ther. Pat.* **18**, 515–524 (2008).
97. W. M. Pardridge. *Bioconjugate Chem.* **19**, 1327–1338 (2008).
98. S. W. Yi, T. K. Yune, T. W. Kim, H. Chung, Y. W. Choi, I. C. Kwon, E. B. Lee and S. Y. Jeong. *Pharm. Res.* **17**, 314–320 (2000).
99. S. M. Kwon, H. Y. Nam, T. Nam, P. Kyeongsoon, L. Seulki, K. Kim, I. C. Kwon, J. Kim, D. Kang, J. H. Park and S. Y. Jeong. *J. Control. Release.* **128**, 89–97 (2008).
100. L. M. Verissimo, L. F. Lima, L. C. Egito, A. G. de Oliveira, and E. S. do Egito. *J. Drug Target.* **18**, 333–342 (2010).
101. B. Biber, G. Johannesson, O. Lennander, J. Martner, H. Sonander and O. Werner. *Acta Anesthesiol. Scand.* **28**, 385–389 (1984).
102. R. P. Eger and B. A. Macleod. *Can. J. Anaesth.* **42**, 173–176 (1995).
103. J. B. Musser, J. L. Fontana and P. D. Mongan. *Anesth. Analg.* **88**, 671–675 (1999).
104. J. Zhou, N. Luo, X. Liang and J. Liu. *Anesth. Analg.* **102**, 129–134 (2006).
105. J. P. Fast, M. G. Perkins, R. A. Pearce and S. Mecozzi. *Anesthesiology* **109**, 651–656 (2008).
106. M. C. Parlato, J.-P. Jee, M. Teshite and S. Mecozzi. *J. Org. Chem.* **76**, 6584–6591 (2011).
107. H. O. Ammar, H. A. Salama, M. Ghorab and A. A. Mahmoud. *AAPS Pharm Sci Tech.* **10**, 808–819 (2009).
108. P. Calvo, J. L. Vila-Jato and M. J. Alonso. *J. Pharm. Sci.* **85**, 530–536 (1996).
109. S. Muchtar, M. Abdulrazik, J. Frucht-Perry and S. Benita. *J. Control. Release* **44**, 55–64 (1997).
110. T. Hagigit, T. Nassar, F. Behar-Cohen, G. Lambert and S. Benita. *E. J. Pharm. Biopharm.* **70**, 248–259 (2008).
111. S.H. Klang, S. Siganos, S. Benita and J. Frucht-Perry. *J. Control. Release.* **57**, 19–27 (1999).
112. S. C. Pflugfelder. *Int. Ophthalmol. Clin.* **46**, 101–103 (2006).
113. R. Ursea, T. L. Purcell, B. U. Tan, A. Nalgirkar, M. E. Lovaton, M. R. Ehrenhaus and D. J. Schanzlin. *J. Refract. Surg.* **24**, 473–476 (2008).
114. X. Wang, Y. Jiang, Y. W. Wang, M. T. Huang and C. T. Ho. *Food Chem.* **108**, 419–424 (2008).
115. K. Kawakami, T. Yoshikawa, T. Hayashi, Y. Nishihara and K. Masuda. *J. Control. Release* **81**, 65–74 (2002).
116. K. Kawakami, T. Yoshikawa, Y. Moroto, E. Kanaoka, K. Takahashi, Y. Nishihara and K. Masuda. *J. Control. Release* **81**, 75–82 (2002).
117. T. R. Kommuru, B. Gurley, M. A. Khan and I. K. Reddy. *Int. J. Pharm.* **212**, 233–246 (2001).
118. M. J. Lawrence and G. D. Rees. *Adv. Drug Deliv. Rev.* **45**, 89–121 (2000).

119. S. Crauste-Manciet, D. Brossard, M. O. Decroix, R. Farinotti and J. C. Chaumeil. *Int. J. Pharm.* **165**, 97–106 (1998).
120. G. Nicolaso, S. Crauste-Manciet, R. Farinotti and D. Brossard. *Int. J. Pharm.* **262**, 165–171 (2003).
121. K. S. Singh and S. K. Vingkar. *Int. J. Pharm.* **347**, 136–143 (2008).
122. S. Nazzal and M. A. Khan. *AAPS PharmSciTech.* **3**, 23–31 (2002).
123. R. P. Dixit and M. S. Nagarsenker. *Eur. J. Pharm. Sci.* **35**, 183–192 (2008).
124. C. Brüsewitz, A. Schendler, A. Funke, T. Wagner and R. Lipp. *Int. J. Pharm.* **329**, 173–181 (2007).
125. S. V. R. Rao, P. Agarwal and J. Shao. *Int. J. Pharm.* **362**, 10–15 (2008).
126. S. V. R. Rao and J. Shao. *Int. J. Pharm.* **362**, 2–9 (2008).
127. S. V. R. Rao, K. Yajurvedi and J. Shao. *Int. J. Pharm.* **362**, 10–15 (2008).
128. R. M. Straubinger, In *Taxol: Science and Applications*, M. Suffness (ed). New York: CRC Press (1995), pp. 237–258.
129. N. A. Kasim, M. Whitehouse, C. Ramachandran, M. Barnejo, H. Lennernas, A. S. Hussain, H. E. Junginger, S. A. Stavchansky, K. K. Midha, V. P. Shah and G. L. Amidon. *Mol. Pharm.* **1**, 85–96 (2003).
130. S. Peltier, J.-M. Oger, F. Lagarce, W. Couet and J.-P. Benoît. *Pharmaceut. Res.* **23**, 1243–1250 (2006).
131. C. Frezzini, J. C. Leao and S. J. Porter. *J. Oral. Pathol. Med.* **34**, 513–531 (2005).
132. J. Kulkovsky and S. Bray. *Curr. HIV Res.* **4**, 199–208 (2006).
133. T. K. Vyas, A. Shahiwala and M. M. Amiji. *Int. J. Pharm.* **347**, 93–101 (2008).
134. D. I. Friedman, S. Schwarz and M. Weisspapir. *J. Pharm. Sci.* **84**, 324–329 (1995).
135. K. Bouchemal, S. Briançok, E. Perrier and H. Fessi. *Int. J. Pharm.* **280**, 173–181 (2004).
136. D. Mou, H. Chen, D. Du, C. Mao, J. Wan, H. Xu and X. Yang. *Int. J. Pharm.* **353**, 270–276 (2008).
137. E. Yilmaz and H. H. Borchert. *Eur. J. Pharm. Biopharm.* **60**, 91–98 (2005).
138. S. Hatziantoniou, G. Deli, Y. Nikas, C. Demetzos and G. T. Papioannou. *Micron.* **38**, 819–823 (2007).
139. F. Shakeel, S. Baboota, A. Ahuja, J. Ali, M. Aqil and S. Shafiq. *AAPS Pharm Sci Tech.* **8**, 191–199 (2007).
140. B. Subramanian, F. Kuo and E. Ada. *Int. Immunopharm.* **8**, 1533–1539 (2008).
141. J. B. Tagne, S. Kakumanu and J. Nicolosi. *Mol. Pharm.* **5**, 1055–1063 (2008).
142. J. B. Tagne, S. Kakumanu, D. Ortiz, T. Shea and R. J. Nicolosi. *Mol. Pharm.* **5**, 280–286 (2008).
143. A. Pires, A. Fortuna, G. Alves and A. Falcao. *J. Pharm. Pharm. Sci.* **12**, 388–311 (2009).
144. M. I. Ugwoke, U. Agu, N. Verbeke and R. Kinget. *Adv. Drug Deliver. Rev.* **57**, 1640–1665 (2005).
145. A. Mistry S. Stolnik and L. Illum. *Int. J. pharm.* **379**, 146–157 (2009).
146. D. Betbeder, S. Sperandio, J.-P. Latapie, J. de Nadai, A. Etienne, J.-M. Zajac and B. Frances. *Pharmaceut. Res.* **17**, 743–748 (2000).
147. L. R. Stanberry, J. K. Simon, C. Johnson, P. L. Robinson, J. Morry, M. R. Flack, S. Gracon, A. Myc, T. Hamouda and J. R. Baker Jr. *Vaccine* **30**, 307–316 (2012).
148. A. U. Bielinska, K. W. Janczak, J. J. Landers, D. M. Markovitz, D. C. Montefiori and J. R. Baker Jr. *AIDS Res. Hum. Retrov.* **24**, 271–281 (2008).

149. A. Floreani, V. Baldo, M. Cristofoletti, G. Ranzulli, A. Valeri, C. Zanetti and R. Trivello. *Vaccine* **22**, 608–611 (2004).
150. P. E. Makidon, A. U. Bielinska, S. S. Nigavekar, K. W. Janczak, J. Knowlton, A. J. Scott, N. Mank, Z. Cao, S. Rathinavelu, M. R. Beer, J. E. Wilkinson, L. P. Blanco, J. J. Landers and J. R. Baker Jr. *PLOS One* **3**, e2954–e2955 (2008).
151. P. C. Teixeira, G. M Leite, R. J. Domingues, J. Silva, P. A. Gibbs and J. P. Ferreira. *Int. J. Food Microbiol.* **118**, 15–19 (2007).
152. A. A. Chepurnov, L. F. Bakulina, A. A. Dadaeva, E. N. Ustinova, T. S. Chepurnova and J. R. Kaber Jr. *Acta Trop.* **87**, 315–320 (2003).
153. T. Hamouda, and J. R. Baker Jr. *J. Appl. Microbiol.* **89**, 397–403 (2000).
154. T. Hamouda, M. M. Hayes, Z. Cao, R. Tonda, K. Johnson, D. C. Wright, J. Brisker and J.R. Baker Jr. *J. Infect. Dis.* **180**, 1939–1949 (1999).
155. B. W. Donovan, J. D. Reuter, Z. Cao, A. Myc, K. J. Jonson, and J. R. Baker Jr. *Antiviral Chem. Chemother.* **11**, 41–49 (2001).
156. M. Bray, K. Davis, T. Geisbert, C. Schmaljohn and J. Huggins. *J. Infect. Dis.* **179**, 248–258 (1999).
157. A. A. Chepurnov, Y. P. Chuev, O. V. P'iankov and I. V. Efimova. *Vopr. Virusol.* **40**, 74–76 (1995).
158. G. Van der Croen and L. H. Elliott. *Ann. Soc. Belg. Med. Trop.* **62**, 49–54 (1982).
159. T. K. Stevens, R. M. Ramirez and A. Pines. *J. Am. Chem. Soc.* [Just Accepted] doi: 10.1021/ja402885q (2013).
160. J. M. Janjic and E. T. Ahrens. *Wiley Interdiscip. Rev. Nanomed. Nanobiotechnol.* **1**, 492–501 (2009).
161. E. T. Ahrens, R. Flores, H. Xu and P. A. Morel. *Nat. Biotechnol.* **23**, 983–987 (2005).
162. K. C. Partlow, J. Chen, J. A. Brant, A. M. Neubauer, T. E. Meyerrose, M. H. Creer, J. A. Nolta, S. D. Caruthers, G. M. Lanza and S. A. Wickline. *FASEB J.* **21**, 1647–1654 (2007).
163. M. Cametti, B. Crousse, P. Metrangolo, R. Milani and G. Resnati. *Chem. Soc. Rev* **41**, 31–42 (2012).
164. E. T. Ahrens, R. Flores, H. Xu and P. A. Morel. *Nat. Biotechnol.* **23**, 983–987 (2005).
165. F. Bonetto, M. Srinivas, A. Heerschap, R. Mailliard, E. T. Ahrens, C. G. Figdor and I. J. M. de Vries. *Int. J. Cancer.* **129**, 365–373 (2011).
166. U. Flögel, Z. Ding, H. Hardung, S. Jander, G. Reichmann, C. Jacoby, R. Schubert and J. Schrader. *Circulation* **118**, 140–148 (2008).
167. J. R. Lindner. *Nat. Rev. Drug Discov.* **3**, 527–533 (2004).
168. S. A. Wickline, M. Hughes, F. C. Ngo, C. S. Hall, J. N. Marsh, P. A. Brown, J. S. Allen, M. D. McLean, M. J. Scott, R. W. Fuhrhop and G. M. Lanza. *Acad. Radiol.* **9**, S290–S293 (2002).
169. G. M. Lanza, K. D. Wallace, M. J. Scott, W. P. Cacheris, D. R. Abendschein, D. H. Christy, A. M. Sharkey, J. G. Miller, P. J. Gaffney and S. A. Wickline. *Circulation* **94**, 3334–3340 (1996).
170. N. Rapoport, K.-H. Nam, R. Gupta, Z. Gao, P. Mohan, A. Payne, N. Todd, X. Liu, T. Kim, J. Shea, C. Scaife, D. L. Parker, E. K. Jeong and A. M. Kennedy. *J. Control. Release* **153**, 4–15 (2011).

Chapter 5

Drug Nanocrystals and Nanosuspensions in Medicine

Leena Peltonen, Jouni Hirvonen and Timo Laaksonen*

Division of Pharmaceutical Technology
University of Helsinki, P.O. Box 56, FI-00014 Helsinki, Finland
**leena.peltonen@helsinki.fi*

Abstract

Drug nanocrystals have been widely studied during the last two decades for improving dissolution properties of poorly soluble drug materials. Nanocrystals are nanosized particles, where the core is formed from pure (crystalline) drug, which is covered with a stabilizer layer. Typically, drug nanocrystal sizes vary from 100 nm to 400 nm. The most benefit of drug nanocrystals as compared to other nanotechnological approaches in medicine is their simple structure, as was shown by their rapid entry to the market. The first commercial product was introduced less than 10 years after research began in the field. The first nanocrystal products on the market were formulated as tablets, where dried nanocrystals were utilized in a powder form. Later on, nanocrystals have also been studied and formulated for ocular, parenteral, dermal, pulmonary and buccal drug delivery routes. Nanocrystals can be produced either by top-down or bottom-up techniques, and the mostly studied production methods are wet pearl milling, high-pressure homogenization and antisolvent precipitation. The biggest problem with nanocrystals is their inherent instability (aggregation tendency), and the selection of an efficient stabilizer is very critical for successful nanocrystallization.

1. Introduction

Pharmaceutical nanocrystals are solid nanosized drug particles that are covered by a stabilizing agent layer. The most important benefit achieved with nanocrystals is improved and/or faster dissolution and, hence, enhanced bioavailability. Nanocrystals have been studied since the beginning of 1990s, and the first patent related to nanocrystals was by Gary Liversidge.[1] The main benefit of nanocrystals compared to other drug nanosystems is their considerable simplicity, e.g. structure without too many excipients. This is also the main reason for the early entrance of commercial products: the first product, Rapamune by Wyeth, was already accepted in 2000 by the US FDA.

In the beginning, the commercial products were product improvements to existing drugs, but later, new chemical entities have also been under research.[2] The first nanocrystal products on the market were formulated in tablet form, but nanocrystals have also been studied for ocular,[3] parenteral,[4] dermal,[5] pulmonary[6] and buccal[7] drug delivery. Both dry nanocrystals as well as nanocrystals in suspension have been utilized, and even peptides such as insulin have been formulated as nanocrystals.[4,8]

Nanocrystals can be produced either by bottom-up or top-down methods, and for example, the stability of the nanosupension or final particle size can vary, depending on the production method.[9–11] As with other nanosized systems, long-term stability is the biggest problem with nanocrystals.[12] The most common stability problem is aggregation; changes in the degree of crystallinity or polymorphic form may also be faced during processing or storage.[13] Chemical stability is often good, although the high drug solubility (>1 mg/ml) and low drug concentration may decrease it. Analytics is also demanding due to the small particle size.[14]

After all the intense research during the past two decades, drug nanocrystals are today a real option for pharmaceutical industry for solving formulation problems with poorly soluble drug materials. In this chapter, drug nanocrystals and their nanosuspensions are introduced, and their advantages/disadvantages, production and analysis are discussed. Finally, examples for different drug delivery routes as well as for nanocrystal-based drug formulations are given.

2. Benefits of Using Drug Nanocrystals in Drug Delivery

The benefits with using drug nanocrystals have often been classified into three categories[2]:

(i) Enhanced dissolution rate
(ii) Increased solubility
(iii) Increased adhesion to surfaces

The results have shown that the combined effect of (i)–(iii) have indeed considerably increased the *in vivo* performance of various drugs.[15] Better dissolution can lead to faster absorption, shorter T_{max} and increased C_{max}. This may even cause toxic blood levels or unexpectedly high plasma peak levels, if the higher bioavailability is not taken into account when calculating the doses. Nevertheless, it has been unclear to what extent each aspect helps in drug delivery. Below, each point will be discussed in detail.

An increase in the dissolution rate for nanocrystals is the most direct effect of size decrease. As particle size decreases, the surface area available for drug release per mass of drug increases inversely proportional to the radius. This relation is easily obtained from the consideration of spherical surface area versus volume ($A/V = 3/r$). Thus, a reduction in particle size from 30 μm to 300 nm would increase the dissolution rate by 100-fold. This large difference is the main mechanism of action in increasing oral bioavailability, since it means that more drug has time to dissolve during the time-limited transit through the gastro-intestinal tract and that the increased concentration gradient improves the permeability and, hence, bioavailability. In fact, the dissolution rate is so fast that it causes problems for conventional dissolution testing of the nanocrystalline formulations.

The hypothesis of solubility increase has been based on the Ostwald–Freundlich theory, which describes the solubility increase for very small particles (see Eq. (1)). The theory was first derived for liquid droplets in a gas phase, but has been considered valid also for solid particles in liquids. As can be seen, it predicts exponential relation between solubility and particle size.

$$c_{\mathrm{NP}} = c_b \exp\left(\frac{2V_m\gamma}{RTr}\right) \tag{1}$$

where c_{NP} is the solubility of nanoparticles with a radius r, c_b is the solubility of bulk material, V_m is the molar volume of the drug, γ is the interfacial tension between the solvent and the drug material (for realistic values for this parameter, Lindfors *et al.*[16] gives a range between 5 and 50 mN/m), R is the gas constant, and T is temperature. For a recent review of the theory, see for example Kaptay's work on the exact derivation of the theory and its corrected forms.[17]

Recently, several authors have pointed out the discrepancy between the exact solution of Eq. (1) and reported drug nanocrystal solubility values in the literature. The theory predicts only a modest solubility increase of 5–20% for particles above *ca.* 100 nm in size,[18–20] whereas much larger experimental values have sometimes been reported. And indeed with realistic parameters, Eq. (1) does not predict significant solubility increase until the particle size is well below 100 nm (shown in Fig. 1), which seems to be the current practical limit for top-down production of drug nanocrystals and very

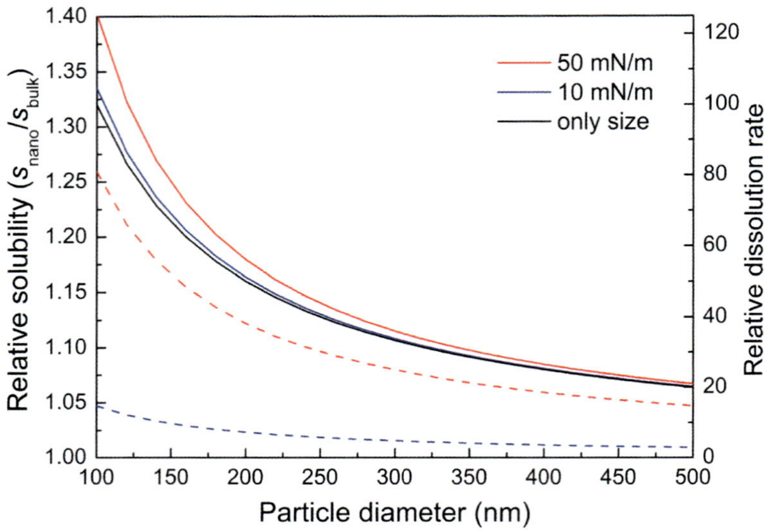

Fig. 1. Relative solubility of drug nanocrystals calculated with Eq. (1) for a model compound with a molecular weight of 400 g/mol and density of 1.4 g/cm^3 and with the surface tension term either as 10 mN/m (dashed blue line) or 50 mN/m (dashed red line). An estimated increase in the dissolution rate is also shown. This was estimated as the solubility multiplied with the increase in surface area compared to a "bulk" particle with a diameter of 10 μm. As a comparison, a dissolution rate increase without considering the solubility increase is also shown (black).

close to what can be achieved with bottom-up methods as well. Typically drug nanocrystals are between 100 and 400 nm in size.

The reasons for the observed higher solubility values can arise from several factors. One is the incomplete removal of very small nanoparticulate matter from the extracted samples by simple centrifugation and filtration processes. This can be avoided by careful ultracentrifugation, but even this approach suffers from inaccuracies due to the adsorption of the stabilizing agents onto the separated particles.[20] Another reason for the apparent high solubility values reported might be that the particles have been partly transformed into amorphous form. This can be problematic, as one of the advantages of nanocrystals over amorphous drug forms is the longer and predictable storage stability due to the crystallinity. A third cause for the inaccurate determination of solubility arises from the stabilizers used. An excess of surface active species, which are the main category of stabilizers used, can easily increase the solubility of poorly soluble drugs through improved wetting and micellization.

It is unlikely that the drug nanocrystals could markedly improve drug delivery by increased saturation solubility alone. But increased concentration gradient could still be obtained based on the faster dissolution rate. In the gastro-intestinal tract, a saturated or supersaturated drug solution will most likely not be reached, and thus the fast dissolution from drug nanocrystals can result in much higher concentrations in the gastro-intestinal tract than conventional formulations, as higher dissolved amounts will lead to higher drug concentrations. This in turn can increase the total permeation rate of the drug through the gastro-intestinal wall, which is dependent on the concentration difference across the intestinal wall, not on the solubility of the drug as such. Furthermore, as Fig. 1 shows, the dissolution rate is tremendously increased and will benefit the overall drug delivery.

The third proposed benefit of the nanocrystals, increased cell adhesion of the drug nanocrystals in the absorption epithelium, is also due to the increased drug surface area available for interactions in a nanosuspension. It is unclear, however, how much this would really help in the oral delivery of drugs, since the particles are not expected to be able to cross the gut wall barrier, and will most likely dissolve quickly enough for this process to have a low significance. Thus, it must be concluded that the main benefit of drug nanocrystals in terms of drug release and delivery is due to the increased surface area per mass.

Based on the reasoning stated above, the best candidates for nanocrystals are BCS (Biopharmaceutics Classification System) class II drugs, where the poor bioavailability is caused by the low solubility. With these drugs, higher concentration due to the faster dissolution improves the permeability. Faster dissolution is also important for materials with narrow absorption window, when higher absorption rate is related to the improved bioavailability. The faster dissolution rate of nanocrystals has also been shown to decrease the deviation in bioavailability between the fed and fasted state, and hence lipophilic drugs, which require food for solubilization, may benefit from nanocrystallization.

3. Production Techniques

Drug nanocrystals can be produced either by top-down or bottom-up methods. In top-down methods the starting point is bulk material and the particle size is decreased, for example, by milling or high-pressure homogenization (HPH). In bottom-up methods, the particles are formed from predissolved molecules through various methods, most commonly precipitation. Also combination of different techniques, such as first pre-milling followed by HPH, can be utilized (Fig. 2).[21,22]

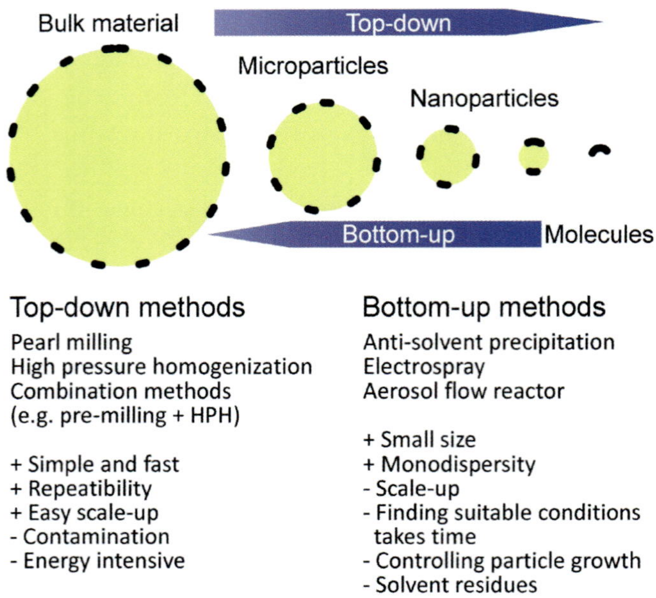

Fig. 2. Schematic presentation of different techniques for nanocrystal production and their advantages and disadvantages.

After the nanocrystal production, drug nanocrystals are either in suspension or as dry powder. For further formulation, drying of suspension is mostly performed by freeze drying[23] or spray drying techniques.[24,25] Other techniques, for example, coating of sugar beads with nanocrystals in fluid bed equipment[26] or inkjet-type printing of nanocrystal suspensions have been utilized for the production of personalized medicines.[27]

3.1. *Bottom-up methods*

The bottom-up techniques are antisolvent precipitation[9,28–30] or liquid atomization based techniques.[31,32] In antisolvent precipitation, the drug material and stabilizer is first dissolved in a solvent. When an antisolvent is added, the drug precipitates as its solubility is rapidly decreased. If an efficient stabilizer is found, appropriate (usually fast) mixing is achieved, the drug concentration is optimal and the right temperature is found, very small particles with narrow size deviations can be formed (Fig. 3). Often, the poor solubility of drug limits the utilization of this technique as some new compounds are difficult to dissolve in any solvent, especially since it has to be miscible with the antisolvent. Also, solvent removal after the process may be tedious, and possible solvent residues can pose problems. Finally, the control over the particle size

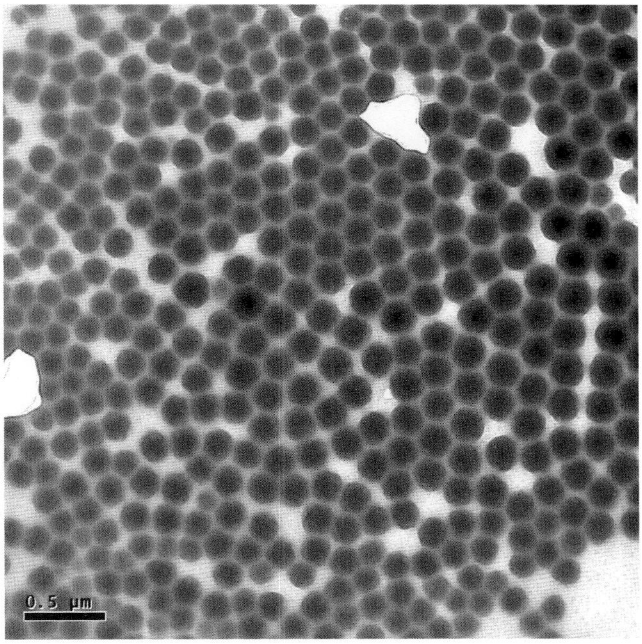

Fig. 3. TEM figure of hydrophobin-stabilized itraconazole nanocrystals produced by antisolvent precipitation method with a similar approach as used in Valo *et al.* in 2010.[29] Scale bar is 500 nm.

is often difficult to achieve; if the precipitation is performed under a high-energy zone, like in a microfluidizer, this can help in stopping the particle growth.[33] Precipitation techniques can also produce amorphous particles, which may be a benefit for the case of solubility but is problematic for the case of stability.

Liquid atomization-based techniques for producing nanocrystals are spray drying, electrospraying and aerosol flow reactor methods.[31,32,34,35] Spray drying is mostly used to formulate microparticles, but nanoparticle production is also possible.[32] In spray drying, the particles are formed by forcing the fluid through a nozzle, which produces a mist that is dried to fine powder. In aerosol flow reactor, the aerosol is formed by a collision type air jet atomizer,[31] and typically very small particles can be produced.

In electrospray atomization, a high voltage is applied to a microcapillary nozzle, which causes the solution interface at the tip to change shape due to the electrostatic charge.[34,35] The electrostatic forces are increased as the potential gets higher, and the effect of surface tension on the shape of the interface is reduced. After the equalization of the two forces, further charge will cause spraying of the liquid into small droplets, and solid nanoparticles are formed after the evaporation of the solvent.

After liquid atomization, the product is often a dry powder, although, for example, electrospraying can be performed in the liquid phase.[36] The product is often in an amorphous form, but after the process, the drug nanoparticles can be crystallized, for example, by annealing, in order to improve the stability.[34] So far, liquid atomization techniques are only rarely utilized in the production of drug nanocrystals.

3.2. *Top-down methods*

The most used top-down methods are pearl milling and HPH. With these techniques, particle sizes around 100 nm can be easily achieved, but smaller particles are more difficult to produce. The product is often more polydisperse compared to the bottom-up techniques, but homogeneity of the product can be increased with longer processing times, e.g. longer milling or increased number of homogenization cycles.

In milling, the solid drug is dispersed in an aqueous stabilizer solution, and the milling is performed with the aid of milling pearls.[37,38] Depending on the process parameters, e.g. the amount and size of milling pearls, milling speed, milling time, temperature, the amounts and properties of the drug and stabilizer, different sized materials can be achieved. Earlier low milling speeds (80–90 rpm) with long milling times (1–5 days) were typically used.[39] However, the problem with contamination by the milling pearl fragments is increased with longer time. Nowadays mostly high milling speeds (1000–4000 rpm) combined with very short milling times (even some minutes) are favored.[40,41] Milling can be performed with very small amounts of drug (e.g. only 50 mg) depending on the milling vessel used, and scaling-up of milling is possible.[41] Milling is a very reproducible process and so far most of the marketed products have been produced by different kinds of milling techniques. Drug concentration in a milling vessel can be as high as 400 mg/ml.[38]

During milling, material fragmentation and particle growth occur simultaneously. Particle growth can be minimized with a proper stabilizer, but with certain process parameters (milling pearl size, rotation speed, number of milling pearls, etc.) a constant minimum particle size is achieved after some time period.[42–44] If the milling is continued after this time point, the particle size and polydispersity of the sample can even be increased.[45,46] The finest particles are produced with the smallest pearls. The total collision energy is determined by the pearl weight and number of collisions,[47] but the number of collisions seems to be the most important, hence, the smaller the pearls, the smaller is the final product particle size.[41] Pearls are typically from 0.3 mm to 1 mm. Even smaller pearls have been used, but the separation of pearls from the suspension is very demanding.[10]

The most important drawback with milling is the contamination risk.[48] During the milling process, high shear forces are induced, which also increases the temperature. This can cause wearing of the milling vessel or milling pearls. To avoid this, mainly hard ceramics such as zirconium oxide are used for the vessel and pearl material, although also coated pearls/vessel can be used.[38,40,44,48] Ductile drug materials are also problematic because they tend to deform rather than to fragment,[49] as well as abrasive materials.[50] Variations of existing techniques, like more efficient milling set ups, have been developed to overcome the problems faced with the challenging raw materials.[51]

HPH can be performed by piston gap (in water or in non-aqueous medium) or microfluidizer techniques.[52,53] In the piston gap technique, the drug is dispersed in a stabilizer solution, and this dispersion is forced by a piston through a small gap with high speed. Typically the pressure is 1500–2000 bar, and up to 20 cycles are used. Dynamic pressure inside the narrow gap is increased by the high flow speed, and the static pressure at the same point is lowered below the vapor pressure of water according to Bernoulli's equation. The water starts to boil, but the boiling ends immediately after the homogenization gap, when the suspension enters the normal pressure area. This cavitation induces shock waves, which cause particle size reduction together with the high shear forces, particle collisions and turbulent flow. If the HPH by piston gap method is performed in oils or in polyethylene glycols, cavitation is not reached, and the particle size decrease is due to the particle collisions, shear forces and turbulent flow.

Another HPH technique is a microfluidizer, which is based on jet stream principle. In this technique, two fluid flows with high velocity (e.g. 1000 m/s) are colliding frontally under high pressure (e.g. 2000 bar), and as a result, nanocrystals are formed by particle collisions, shear forces and cavitation forces. Compared to the piston gap method, considerably higher number of homogenization cycles is required, typically 50–100.[22,52,53]

In HPH techniques, the particle size is affected by the homogenization pressure, number of homogenization cycles and drug properties. During the homogenization step, drug crystals break down from weaker crystal imperfections, and the force needed to break the crystals is higher with smaller particle size. When the force caused by homogenization pressure is equal to the attraction forces in the crystal, the crystal is not broken down any further, and the minimum particle size at the given homogenization pressure is reached.[10] The stabilizer does not affect the particle size during the processing, but it stabilizes the particles after the particle formation. The process is considerably energy intensive and contamination due to the wearing of the process equipment is possible. HPH techniques are very effective in lab-scale, but so far only one product on the market is based on this technique.

3.3. *Combination techniques*

New techniques are based on a combination of different methods. For example, a pre-precipitated sample can be further homogenized by HPH (cavi-precipitation technique)[10,24] or a pre-milled material can be still homogenized.[53] Combination techniques can be used to shorten the production time, for example, by reducing the number of passes through the homogenizer or to produce very small nanocrystals (smaller than 100 nm), which can be difficult with one technique only.

3.4. *Comparison of different production techniques*

Besides the process parameters, the process method also affects the properties of the end product. For example, when milling, HPH and cavi-precipitation (combination technique with first the precipitation step and after that homogenization) were compared in the production of quercetin nanocrystals.[10,11] Nanocrystallization was successful with all the techniques, but the smallest particles were achieved with milling and the lowest polydispersity index with HPH. These two techniques also showed the best stability. Stability problems related to cavi-precipitation processed nanocrystals were assumed to be caused by the solvent residues from the precipitation step: the solvent residues competed with the surfactant for hydration, causing partial surfactant dehydration and lowered stability of the nanocrystals produced with this technique.

For ophthalmic drug delivery, hydrocortisone nanosuspensions were produced by both microfluidic nanoprecipitation and milling methods.[9] The particle size of the nanosuspensions was around 300 nm with both the approaches. After milling, the particles were crystalline, while after precipitation, predominantly amorphous particles were formed. The AUC values in rabbits were comparable with both the nanosuspensions, and almost doubled compared to the corresponding value with a hydrocortisone solution. The stability of the milled sample was good after a two-month period, while the particle size of the precipitated sample increased to 440 nm during the two-month stability testing period.

4. Stabilization of Drug Nanocrystals with Stabilizers

The stabilization of nanocrystals is important not only for the production of the drug nanocrystals but also for their long-term storage.[54] The size reduction creates higher-energy surfaces, which will lead to aggregation and Ostwald ripening unless properly protected. A suitable excipient able to bind to the surface and disperse the particles in water is, therefore, normally needed as a stabilizer. The smaller the particle size, the more important is the role of the efficient stabilizer. Very rarely nanosuspensions can be formed without any stabilizer. However, sometimes, for example, the zeta potential of the material itself is so high that self-stabilization is possible; an example is 2-devinyl-2(1-hexyloxyethyl)pyropheophorbide nanocrystals, where the zeta potential value is −40 mV, which is high enough for efficient stabilization.[55]

Usually amphiphilic molecules are used as stabilizers, since most of the poorly water soluble drugs are very hydrophobic and the hydrophilic part of the molecule will then enhance the wetting of the particles. The overall stability can then be described in terms of the classical DLVO-theory,[56,57] with the stability being either achieved through electrostatic forces or steric barriers. Many simple and pharmaceutically accepted polymers, such as cellulose derivatives (HPMC, HPC), PVP and poloxamers[40,58,59] or simple surfactants such as Tween, TPGS and SDS, have been used successfully in stabilizing a wide variety of different drugs.[59] Also lecithin, amino acid copolymers and tyloxapol have been used as stabilizers (Table 1).[60–62]

Often the most suitable stabilizer is found by an empirical process, because the factors determining the efficacy and success of nanocrystal stabilization are not totally clear.[59,67,68] Efficient surface adsorption is critical for the stabilization. For example, in a study conducted on ibuprofen nanocrystals, interaction surface adsorption studies made by AFM confirmed a successful stabilization of the nanocrystals with HPMC and HPC.[54] On the other hand, the surface adsorption of PVP and poloxamer were inadequate and consequently, they were not good stabilizers for ibuprofen crystals.

The problem with small surfactant molecules is that they may solubilize the drug inside micelles, which disturbs the nanocrystals formation. With amphiphilic polymers, the hydrophobic chains might have good adsorption tendency, and hence, stabilize the nanocrystal structures.[41] If polymers are used to stabilize the nanocrystals by forming a steric barrier, the efficient molecular weight is approximately 5000–25,000 g/mol.

Mostly the stabilizer is attached to the drug particle surfaces via physisorption,[69,70] and the driving force for surface adsorption is the hydrophobic/hydrophilic interactions, and hence, the hydrophobicity of the

Table 1. Examples of drugs and stabilizers studied with different nanocrystallization techniques.

Drug	Stabilizer	Particle size/nm	Production method	Ref
itraconazole, fenofibrate, griseofulvin, ibuprofen, azodicarbonamide, sulfamethoxazole	HPMC, Tween 80, SDS, sodium alginate	>200	milling	63
itraconazole	chitosan, N-trimethyl polyethyleneimine, chitosan, poloxamer 188	270–550	HPH	64
ibuprofen	HPMC, PVP, SDS	270–480	milling	25
quarcetin	Tween 80	250–790	HPH, milling, cavi-precipitation	10
phenytoin, nifedipine, pranlukast	PVP, PVA, HPC, HPMC, Tween 80, SDS	290–350	milling	41
itraconazole	Tween 80, Cremophor EL, Poloxamer 338	>130	milling	23
hydrocortisone	Tween 80, HPMC, PVP	300	precipitation, milling	9
hydrocortisone, prednisolone, dexamethasone	Pluronic F68	210–930	HPH	3
phenylephrine	mannitol	500–830	milling, HPH	7
itraconazole	Poloxamers 188 and 407, HPMC, sodium deoxycholate	450–540	microprecipitation-HPH	24
nitrendipine	chitosan	175	precipitation, HPH	65
Amoitone B	Poloxamer 188, HPMC, lecithin	260	microfluidization HPH	53
insulin	Myglyol 812	3700	HPH	8
zinc-insulin	Poloxamer 188, sodium deoxycholate	<150	milling	4
folic acid	Tween 20	<5000	HPH	27
piposulfan, camptothecin, paclitaxel, etoposide	Tween 80, Span 80, Pluronics F108 and F127,	210–280	milling	66
beclomethasone dipropionate	tyloxapol	164	milling	60
RMKP22	lecithin	510–580	HPH	61
naproxen	amino acid copolymeers	200–300	milling	62

stabilizer (the formed drug particle surfaces are hydrophobic) seems to be one of the most important properties for efficient stabilization. Other factors affecting the absorption tendency are drug–stabilizer interactions between the functional groups, such as phenols, amines, hydroxyls, ethers and carboxylic acids. Appropriate viscosity of the nanosuspension is important not only during the process but also with the end product; molecular weight of the stabilizer affects both the viscosity and the steric layer thickness.

If the stabilization is based purely on electrostatic stabilization, zeta potential values lower than −30 mV or higher than +30 mV are needed. Surface charge is related to environmental factors, and, for example, addition of charged materials or pH changes can destroy the stabilization effect.[64] Typical electrostatic stabilizers are ionic surfactants and polymers, and hence, depending on the stabilizer molecular weight, often steric stabilization can also be involved. Non-ionic surfactants and polymers are often utilized as steric stabilizers. Steric stabilization is based on the physical barrier formed by the long molecular chains on particle surfaces. These barriers prevent close contact between particles. Electrostatic stabilization is efficient in an aqueous medium, but stability problems may be faced during drying, when the ionized state is altered. Steric stabilization is more sensitive to temperature changes.

As already described, nanocrystals can be produced with different techniques and, accordingly, the role of the stabilizer is also related to the production technique. In milling, the stabilizer is needed during the process to protect the newly formed surfaces against aggregation, while in HPH, the nanosized particles are formed independently and the stabilizer is only needed to stabilize the end product. In precipitation, the stabilizer controls the particle formation and stops the particle growth to a specific size. However, in all processes, the stabilizer absorption on the drug particle surfaces is very important.

Often only one stabilizer is used, but sometimes, a mixture of two may be beneficial. In this case, two stabilization methods can be combined, e.g. electrostatic and steric. Especially, if a non-ionic stabilizer is combined with an ionic one, the self-repulsion between the ionic surfactants is lowered, which leads to closer packing of the stabilizer molecules.[71] For example, PVP and SDS have been used in combination to stabilize phenytoin nanocrystals produced by a milling technique,[41] and Poloxamer 188 with sodium deoxycholate for the stabilization of itraconazole nanoparticles.[24] Two stabilizers with the same stabilization mechanism can also be combined, such as Poloxamer 188 and Poloxamer 407 for itraconazole nanocrystals.[24]

The amount of stabilizer is important to optimize. With a too small amount of stabilizer, nanocrystals may not be formed, and too high an amount may cause enhanced solubility, and hence, increased Ostwald ripening and increased particle size during storage. Some stabilizers may also form sticky layers and enhance aggregation, if they are used in excess. The required stabilizer amount varies for each combination of the drug and stabilizer, but usually between 5% and 50% of the drug amount is needed.[44] With surfactants, the critical micelle concentration (CMC) can also have an influence. Below the CMC, the individual stabilizer molecules have higher affinity to the particle surfaces and the stabilization effect is good, but above the CMC, monomers start to form micelles and the affinity to particle surfaces can be lowered.[70] Temperature can also affect the affinity, for example, via changes in CMC, because CMC-value is temperature related.

Besides stability, the stabilizers can also impart some other functions to the nanocrystals. For instance, the stabilizer can be labeled in order to localize the particles,[29] or it can be used to link the crystals to an external matrix for better stabilization and handling,[72,73] as was recently shown with an amphiphilic hydrophobin protein. A higher amount of steric stabilizer may function as a cryoprotectant in a later freeze drying process[23]; the steric layer surrounding the particles is denser with a higher steric stabilizer concentration, which separates the particles efficiently during the drying. Sometimes, the stabilizer layer on the top of the crystal surfaces can even enhance the chemical stability of the drug. Paclitaxel nanocrystals stabilized by Pluronic F68 have been shown to stand four years storage due to the degraded drug monolayer that is formed on the nanocrystals surface immediately when they are in contact with oxygen and water.[74] This tight layer will prevent further degradation of the drug molecules in the crystal, very much like in the case of an oxide layer formed in many metals.

Normally all the stabilizer is added before the actual nanocrystallization process takes place, but periodic addition during the process is also possible.[63] The periodic addition has been shown to produce smaller particles with a narrow particle size deviation. When a polymer is used as a stabilizer, adding it all at once may increase the viscosity of the system, which especially in the case of milling reduces the size reducing effect of the pearls, and may increase the processing time and energy consumption.[62] During milling, the particle size is decreased, which usually increases the viscosity of the suspension, and hence, increases the energy demand. With high viscosity, the oscillation velocity and kinetic energy are lowered, which leads to a decreased efficiency in particle comminution. Accordingly, when the viscosity is increased, the higher energy demand may increase breakage rate, but further increase in viscosity causing a decrease in oscillation velocity and kinetic energy may reduce the breakage rate. The viscosity of the formed nanosuspension may affect the suspension stability and also the bioavailability; especially in ocular drug delivery, the suspension viscosity has been correlated to bioavailability, with higher viscosity increasing the duration of the drug action.[3]

5. Characterisation of Nanocrystals

Repeatability of the formulation process and the stability of nanocrystals in the formulation demand thorough characterization of the crystals/particles. Proper analytical techniques and even *in situ* process analysis have been highlighted by drug authorities (e.g. FDA) in the form of PAT (process analytical technology) approach.[14] Critical parameters, including the particle size and shape, surface charge and crystal morphology, play significant roles in the formulation behavior and in the drug efficacy and safety profiles in *in vivo* animal models and in patients. The small size of the nanocrystals imposes extra demands for the characterization techniques that are applicable. Suitable techniques include standard pharmaceutical methods such as microscopic techniques, thermal analysis, zeta potential measurements and dissolution testing, but also supplementary techniques such as surface tension/pressure measurements and spectroscopic methods should be taken into account.

5.1. *Mean particle size and particle size distribution*

The particle size of nanosuspensions and nanoparticles can be measured with different techniques depending on the particle size and particle size deviation. The most common techniques are light scattering[75] and electron microscopy techniques such as transmission electron microscopy (TEM), scanning electron microscopy (SEM) and environmental scanning electron microscopy (ESEM).[76] Also, Coulter Counter,[77] sedimentation,[78] atomic force microscopy (AFM)[79] and capillary hydrodynamic fractionation[80] have been used for the particle size measurements. With electron microscopy techniques, the morphology and shape of the particles can be measured simultaneously. However, coating of the sample may be necessary, which may change the properties of the sample.[81] High vacuum and the electron beam may also destroy fragile samples and cause water loss, which changes the microstructures. To avoid these problems, freeze fracturing or cryo-TEM can be utilized,[82] but this complicates the measurements and interpretations. Compared to more automated techniques, the EM analyses are often time consuming and the interpretation of the results needs specific expertise.

Light scattering methods are fast, have a large dynamic range and are precise and sensitive, making them more suitable, e.g. for process control purposes compared to the electron microscopy techniques above. Large particle size deviation may impose a true problem to light scattering techniques,[83] as the larger particles are dominating the results and their interpretation. If the samples are polydisperse, the measurement requests a pre-handling step, e.g. by size-fractioning, which may cause misinterpretations in the results. Scattering intensity depends on the scattering angle, the absorption and the size of the particles as well as on the refractive indices of both the particles and the dispersion medium.[84] In addition to the fact that the last terms are difficult to determine, also dynamic processes such as Brownian motion affect the results with time. With respect to the particle form, special algorithms have been formulated to transform the light scattering results to particle size information. These algorithms often assume a spherical particle shape, sufficient dilution and a large difference between the refractive indices of the two phases. Therefore, the light scattering results should be confirmed by other techniques such as SEM.[83]

5.2. *Morphology of nanocrystals*

Depending on the particle size and the scale of the morphology measurements, electron microscopy techniques and AFM are recommended, although the latter is clearly a more demanding technique.[85] Besides the surface morphology, the cross-section and internal structure of the particles may also be determined by SEM/TEM: with TEM, the surrounding stabilizer layer on drug crystal surfaces may even be detectable.[40] During the drying step of nanocrystals, nanostructured microparticles can be formed,[86] where the drug material may reside on the particle surface and/or inside the larger encapsulating particles. Vandelli *et al.*[87] used SEM, DSC and XRD techniques to show the morphology of needle-like crystals of nimesulide in a microparticle formulation. The surfaces of the particles were smooth and free of nimesulide crystals, confirming the fact that the drug was loaded inside the particles. Peltonen *et al.*[88] showed, with cromolyn sodium, the opposite behavior: drug crystals could be seen outside the poly (l-lactic acid) nanoparticles indicating that the drug was not entirely entrapped inside the biodegradable PLA-nanoparticles.

Surface area of the particles may be measured by gas adsorption based on the Brunauer, Emmett and Teller (BET) technique.[89] A monolayer of liquid nitrogen is adsorbed onto the surface of the particles, and the amount of nitrogen that has been released when that monolayer is vaporized is determined, and the surface area of the sample can be calculated based on this quantity. Also mean pore sizes, pore size distributions and total pore volumes can be obtained with this technique.

5.3. *Particle–particle interactions*

By studying the behavior of surface films formed by the colloidal nanoparticles/ nanocrystals on liquid–air interface, it is possible to achieve information concerning, e.g. the stability and aggregation tendency of the particles.[90] The surface pressure at a given area has been found out to be the highest with uncoated particles, and the surface pressure was dependent on the coating material.[91] As the surface pressure is a function of interfacial forces, the interparticulate forces vary strongly with the particle composition and surface properties, which is an important factor in the evaluation of the colloidal stability of different particle formulations. Interactions of nanoparticles with a lipid layer can also be studied with Langmuir through.[91] An increase in the surface pressure shows that the nanoparticles have penetrated into the lipid monolayer, and the penetration rate was increased by the increased hydrophobicity of the particle surface. Particle–particle interaction forces and changes in these forces as a function of ionic composition or pH have been determined with surface pressure/area isotherms.[92] Furthermore, the Langmuir films of particles formed on the liquid surface can be transformed to a solid substrate, e.g. silica plate, by Langmuir–Blodgett (LB) technique.[93] The LB films formed may be analyzed for example by SEM, and information about the packing of the particles is obtained. This enables information gathering, whether the particles are packed tightly along each other or if there exists channels between the particles under investigation; this information can be related to the aggregation tendency of the system.

5.4. *Chemical and physical composition of particles*

Surface analysis of nanoparticles may be performed by X-ray photoelectron spectroscopy (XPS, which is also called electron spectroscopy for chemical analysis, ESCA),[94] and the chemical composition of the particle surfaces can be determined. The technique is based on the irradiation of photons, which causes emission of electrons from the material and ionization of the electrons at the core level.[95] The electrons are emitted at atom characteristic energies, and the atoms from which they are emitted can be recognized.

Differential scanning calorimetry (DSC) is an excellent tool to analyze thermal behavior of crystalline and amorphous materials: for example, glass transition temperature, melting point, solvents/hydrates or polymorphic changes are detected with the DSC.[89] DSC gives accurate and precise quantitative data of the energy that is adsorbed (endothermic) or released (exothermic) by the sample during the measurement. In a study by Hyvönen *et al.*,[96] the glass transition temperature of PLA polymer was increased by the presence of beclomethasone dipropionate, and the increase in the glass transition temperature was due to the amorphous drug, according to the DSC measurements. Wang *et al.*[97] constructed and evaluated puerarin (a drug for coronary heart disease) nanocrystals *in vitro* and *in vivo*. The nanocrystals prepared were characterized using DLS, AFM, TEM, SEM and DSC. The results revealed narrow size-distributed nanocrystals composed of crystallized spherical particles with a mean particle size of 425 nm, a poly-dispersity index of 0.13 ± 0.07 and a negative charge around −30 mV. The DSC analyses revealed an important fact that the crystallinity of puerarin was preserved during the HPH and freeze-drying processes. Furthermore, puerarin dissolution velocity and saturation solubility were enhanced by the nanocrystals, leading to an improved absorption *in vitro* and *in vivo*.

A confirmatory and complementary technique to DSC is variable temperature X-ray diffraction (VT-XRD) measurements, which enables the analysis of crystalline/amorphous forms and degrees of crystallinity.[98] During nanoprecipitation experiments of PLA-nanoparticles, the degree of crystallinity was dependent on the selected solvents. Also the physical state of the drug material may change during the particle formation, as exemplified by XRPD studies of the anhydrate and monohydrate forms of beclomethasone during the formation of the PLA nanoparticles.[96] Bates *et al.*[99] measured the X-ray amorphous powder diffraction patterns of microcrystalline cellulose, indomethacin and piroxicam. The analysis of the data was carried out using a combination of direct methods, such as pair distribution functions (PDF), and indirect material modeling techniques, including Rietveld, total scattering and amorphous packing. The X-ray amorphous powder patterns indicated the presence of amorphous, glassy or disordered nanocrystalline material in the samples, and Rietveld modeling of the microcrystalline cellulose (Avicel PH102) indicated that it was predominantly disordered crystalline cellulose Form I with some amorphous contribution. The average crystallite size of the disordered nanocrystalline cellulose was determined to be 10.9 nm.

Infrared (IR and FT-IR) spectra can be utilized to recognize specific functionalities of chemicals and to determine the interactions between different materials. Moon *et al.*[100] confirmed the size-controlled synthesis and optical properties of monodisperse colloidal magnesium oxide nanocrystals by this technique. The resulting nanocrystals were single crystalline, and they exhibited size-dependent and efficient blue luminescence. Gupta *et al.*[101] designed a dual-drug delivery system based on *in situ* gel-forming nanosuspension of forskolin to enhance antiglaucoma efficacy. Forskolin, poloxamer 407 and Noveon® AA-1 polycarbophil were analyzed using FTIR spectrophotometer for characteristic absorption bands indicative of their interaction. All the characteristic peaks in the individual forskolin and poloxamer 407 were present in forskolin nanocrystals, confirming the fact that all the groups were present in the prepared nanocrystals. All the characteristic peaks of forskolin were also present in forskolin combination with poloxamer and Noveon® AA-1 polycarbophil, and there was no change in the structure of forskolin molecule, indicating a proof that forskolin was compatible with both the poloxamer 407 and Noveon® AA-1 polycarbophil. Overall, the FT-IR technique was an important tool to show that the pH and thermo reversible polymeric in situ gel-forming nanosuspension was suitable for controlled drug release, exhibiting a potential for improved glaucoma therapy.

5.5. *Surface charge and electrophoretic mobility*

Zeta potential determinations are based on the electrophoretic mobility measurements of particles, and dynamic light scattering has been utilized for the electromobility measurements.[102,103] For example the Smoluchowski approximation of the Henry equation for electrophoretic mobility that is often used to calculate zeta potential values from the light scattering data is based on the assumption of rigid spherical colloids.[104] This means that the particle radius is considerably larger compared to the Debye screening length (double layer thickness). In pharmaceutical applications the Smoluchowski approximation is the most used model[102,103] as it applies to particles in aqueous media in moderate salt concentrations. The determination of the zeta potential of a nanosuspension is essential as it gives an idea about the physical stability of the nanosuspension. The zeta potential of a nanosuspension is governed by both the stabilizer and the drug itself. One has to remember that it is of outmost importance for reliable and comparative zeta potential measurements and calculations to take into account the nature of the material that is characterized.[105] Besides the particle characteristics, one needs to carefully control the environment into which the particles are suspended. For example if the viscosity, salt concentration or the nature of the media is changed, the values of zeta potential will change.[106] Also, the total ion concentration has an impact on the results. For example the zeta potential of poly(lactic acid) nanoparticles changed 30 mV when the NaCl concentration was altered from 0.02 to 0.1 mol/l, and it changed 50 mV when the pH was altered from 3 to 6.[107]

5.6. *Drug dissolution and release*

The dissolution rate of nanoparticles can be considered to be in the order of 100 times faster than that of the bulk material. This means that a dissolution test of a poorly soluble drug, where a complete dissolution could normally take a few hours, can be complete in a few minutes, and conventional dissolution methods are not sensitive enough to follow drug dissolution from nanocrystals. Anhalt *et al.*[18] recently considered this problem and pointed out the problems of the traditional sampling and separation methods. Simple filtrations cannot remove the smallest fraction of nanoparticles from the extracted samples, and furthermore, most separation techniques such as filtration and centrifugation take some time to perform, while the dissolution will still proceed in the extracted sample volume until the solids have been removed. This causes severe temporal inaccuracies in the dissolution data. Methods utilizing, e.g. filtration bags, suffer from slow diffusion through the membrane, causing lag times that are even longer than the actual time to complete dissolution and cannot be used for accurate dissolution testing of nanocrystals either. Therefore, *in situ* dissolution methods seem to be the most promising and accurate way to measure drug dissolution from nanocrystals. Limitations apply to these methods as well. UV methods are sensitive to scattering and/or absorption of light by the nanocrystals and cannot be used for all samples.[108] On the other hand, light-scattering has proved very useful for the accurate determination of solubility from nanocrystals.[18,20,109] This currently seems like the most sensitive and reliable method for dissolution testing of drug nanocrystals. Electrochemical *in situ* approaches[110–112] are limited to charged species, while the most difficult poorly soluble compounds have neutral charge. Other methods proposed include calorimetric measurements[113] or following the turbidity of the nanosuspension.[114]

Drug release from nanocrystals is assumed to be diffusion limited, as the particles are mostly composed of the drug itself and no controlling polymer is used. In drug formulations, however, the nanocrystals can be adsorbed onto a carrier particle surface, entrapped or dissolved in a polymer matrix or surrounded by a polymer film. These can then further be formulated into tablets or capsules. In these cases, the release kinetics from drug nanocrystals is further complicated by external factors such as the type and structure of the polymer matrix. According to Kreuter,[115] drug release may take place by (i) desorption of the surface-bound drug, (ii) diffusion through the (nano) particle matrix, (iii) diffusion through the polymer wall of (nano)capsules, (iv) erosion of the particulate matrix or (v) combined diffusion-erosion processes. According to Siepmann *et al.*,[116] there are many factors that may influence the drug release rate from the particles, e.g. drug dissolution, drug diffusion with non-constant diffusivities and moving boundary conditions, polymer degradation/erosion, time-dependent system porosities and three-dimensional geometry of the devices. In most controlled-release systems, the nanocrystals would act as a drug reservoir which would keep the free drug concentration high inside the formulation. The release properties and mechanisms would then be mainly based on the diffusion coefficient of the drug and (bio)degradation rate of the polymer. In general, release tests using nanoparticles or microparticles are more demanding than with conventional pharmaceutical dosage forms (tablets or capsules). Effective and rapid separation of the dissolved/released drug from the colloidal carrier is a prerequisite for a reliable determination of drug release profiles.[14]

6. Delivery Routes for the Nanocrystals

Administration of nanosuspensions and nanocrystals has been studied via different drug delivery routes.[117–119] The parenteral route (*i.v., i.m., s.c.*) of administration provides a quick onset of action (especially *i.v.*), and reduced dosage of the drug. It is the preferred route for drugs undergoing first-pass metabolism and those that are not absorbed or degraded in the gastrointestinal tract. In blood circulation, the pharmacokinetic profile and tissue distribution of nanocrystal suspensions are similar to that of drug solutions, if the drug is soluble in blood.[120] But with insoluble drugs, nanosuspensions behave like other particulate systems. As described earlier, Wang *et al.*[97] formulated and evaluated the pharmacokinetics parameters of puerarin nanocrystals for intravenous delivery. Puerarin is used to treat coronary heart disease, but has poor water solubility and causes adverse drug reactions by cosolvents. Drug delivery and absorption results in beagle dogs revealed that the puerarin nanocrystals exhibited a significantly reduced C_{max} and clearance, and a significantly greater MRT (mean residence time), clearance and elimination half-life compared to the puerarin solution, indicating higher safety of the nanocrystals.

Aqueous nanosuspensions can be nebulized using mechanical or ultrasonic nebulizers for lung (pulmonary) delivery. Müller and Jacobs[121] delivered budenoside drug nanoparticles with a nebulizer and the pharmacokinetic parameters showed comparable AUC, and higher C_{max} and lower T_{max} — values as that of the control dosage form. Because of their small size, it is likely that at least one drug particle was contained in each aerosol droplet, leading to a more uniform distribution of the drug in the lungs. Nanocrystals also increased the adhesiveness and prolonged the pulmonary residence time.

As described above, ocular delivery of an antiglaucoma drug forskolin as a nanosuspension has been utilized to provide a more rapid release and can be further formulated for sustained release of the drug through, e.g. lyophilization. According to Araújo *et al.*,[122] the poor bioavailability of drugs from ocular dosage forms is mainly due to precorneal loss factors (e.g. tear dynamics, non-productive absorption, transient residence time in the cul-de-sac and relative impermeability of the corneal epithelial membrane). Therefore, there is a true need for effective topical formulations capable of promoting drug penetration and maintaining therapeutic levels with a reasonable frequency of application and, at the same time, reducing the unwanted side effects. The delivery of a drug via a nanotechnology-based product fulfils these objectives: enhancement of drug permeation and controlled release.

Oral drug delivery route is the most popular and convenient. Nanoparticles have been orally administered either in dried form or in the form of liquid nanosuspensions. In a recent review article, Gao *et al.*[123] have listed the benefits of nanocrystalline and nanoparticulate dosage forms in oral drug delivery: enhanced oral bioavailability, reduced fasted/fed variation, potentially improved transcellular uptake of the nanoparticles (at least improved mucoadhesion of the nanoparticles) and the improved safety profiles of the nanocrystal formulations.

The mucoadhesive and gastroretentive properties of coated nanoparticles can be modified, e.g. with hydrophobin proteins.[124] Upon entry of the nanoparticles from the stomach into the intestines, it was observed that the particles may experience a delay; the hydrophobin-coated nanoparticles were found to be retained in the stomach up to three hours after administration, whereas the uncoated nanoparticles were released in two hours in the same conditions. Valo *et al.*[72] used the hydrophobins to stabilize oral itraconazole nanoformulations. Nanocrystals with size ~100 nm were produced by anti-solvent precipitation technique and hydrophobins and hydrophobin — double cellulose binding domain fusion proteins were used as the stabilizers. Nanocrystals were further stabilized by immobilizing them into nanofibrillar cellulose matrices. *In vivo* studies were performed with rats, and Sporanox® solution (amorphous itraconazole with cyclodextrin) was used as a positive control. All the itraconazole nanocrystal suspensions showed comparable plasma concentration-time profiles with Sporanox® and *ca.* 15 fold increase in drug absorption compared to a simple microsuspension. The study demonstrated the possibility of functional coating of nanocrystals.

Nanoparticles are thought to have potential as novel intravascular probes for both diagnostic (e.g. imaging) and therapeutic purposes (e.g. drug delivery). Critical issues for successful nanoparticle delivery include the ability to target specific tissues and cell types and escape from the biological particulate filter known as the reticuloendothelial system.[125] So far, no targeted nanocrystal formulations have been developed, but this is highly probable in the future. Targeting may be based on the surface properties of the nanoformulation and changing of the stabilizer can easily alter the *in vivo* behavior.[119] The mononuclear phagocytic system will uptake the drug to allow regional-specific delivery. This can be used for targeting antimycobacterial, fungal or leishmanial drugs to the macrophages, as often the infectious pathogens are persisting intracellularly. All in all, intravenous administration may result in advantages, such as administering poorly soluble drugs without using high concentrations of toxic co-solvents, improving the therapeutic effect of the drug available as conventional oral formulations and targeting the drug to macrophages and the pathogenic microorganisms residing in the macrophages.

7. Nanocrystalline Products

Current drug discovery technologies provide a large number of drug candidates with high *in vitro* efficiency, but their clinical applications have been restricted due to poor water solubility, which has become one of the leading challenges for the formulation of these compounds. Nanocrystals provide a potential way to overcome this issue. Several successful products are already on the market and in the research pipelines (Table 2).[117,126] Other benefits of these nanocrystalline products include large increase in surface area due to the nano-size leading to a rapid *in vivo* dissolution, fast absorption and increased bioavailability; bioavailability of a drug (in the case of fenofibrate) that is independent of a simultaneous meals; and enhanced bioavailability via a decreased dose and reduced viscosity (megestrol) that leads to a better patient compliance.[117,126,127]

The first products on the market containing nanocrystals were prepared by pearl mill technology developed by Elan nanosystems. Rapamune® (2000) was the first marketed product containing sirolimus nanocrystals (Table 2), with the main advantages of more convenient coated tablet dosage form and a higher bioavailability compared to the corresponding solution formulation. Sirolimus is an effective macrocyclic immunosuppressive drug. Emend® was the second nanocrystal product introduced in 2003. It is a capsule containing pellets of aprepitant, a drug used for the treatment of emesis. Aprepitant has a narrow absorption window in the upper gastrointestinal tract only, and nanosizing of the drug improves the absorption of this drug. Tricor® was launched in 2004. The active drug compound is fenofibrate that reduces

Table 2. Nanocrystalline products.[117,126]

Tradename®	Drug	Company	Preparation method	Formulation
Rapamune	rapamycin	Wyeth	milling	tablet
Emend	aprepitant	Merck	milling	capsule
Tricor	fenofibrate	Abbott	milling	tablet
Megace ES	megestrol	Par Pharmaceutical Companies	milling	oral nanosuspension
Avinza	morphine sulfate	King Pharmaceutical	milling	capsule
Focalin XR	dexmethylphenidate HCl	Novartis	milling	capsule
Zanaflex Capsules	tizanidine HCl	Acorda	milling	capsule
Triglide	fenofibrate	Sciele Pharma Inc	microfluidizer HPH	tablet
Ritalin LA	methylphenidate HCl	Accorda	milling	ER capsule
Azopt	brinzolamide	Alcon Laboratories Inc.	milling	eye drops
Invega Sustenna	paliperidone palmitate	Janssen Pharmaceuticals	milling	ER injectable suspension
Xeplion	paliperidone palmitate	Janssen-Cilag	milling	injection

cholesterol levels in patients at risk of cardiovascular disease. Triglide® is another nanocrystalline fenofibrate product produced, instead of ball milling, by a microfluidizer technology. Megace ES® (ES stands for Enhanced Stability) is prescribed for the treatment of appetite loss, severe malnutrition or significant weight loss in AIDS patients. Avinza® is a strong opioid pain medicine that is used to treat moderate to severe around-the-clock pain. Focalin® XR is a once-daily capsule to help patients 6 years and older with attention deficit hyperactivity disorder (ADHD). Ritalin® LA contains psychostimulant nanocrystalline methylphenidate approved for the treatment of ADHD, postural orthostatic tachycardia syndrome and narcolepsy. Zanaflex® capsules are delivering a short-acting drug for the management of spasticity. Azopt® eye drops contain nanocrystalline brinzolamide with tyloxapol as the stabilizer. The product is used to treat glaucoma. Invega® Sustenna® and Xeplion® are used for the treatment of schizophrenia.

Nanocrystallization techniques are considerably simple, cost-effective and fast, and their scale-up is relatively easy. This is very different compared to other nanotechnological approaches for medical applications. This is one reason that many drug products based on top-down nanocrystallization processes are already approved worldwide by regulatory agencies in, e.g. USA, Canada, EU and Japan.

8. Conclusions and Outlook

Drug nanocrystals have been widely studied for improving dissolution properties of poorly soluble drug materials since the beginning of the 1990s. Nanocrystals are formed from 100% drug which is covered a stabilizer layer, and due to the simple manufacturing process, the first product entered the market already in 2000. Although oil or microemulsion filled capsules are still the first option for the formulation of poorly soluble drug materials in pharmaceutical industry, nanocrystal formulations are a very competitive next option.

References

1. G. G. Liversidge, K. C. Cundy, J. F. Bishop and D. A. Czekai, USP patent 5145684 (1992).
2. R. H. Müller, S. Gohla and C. M. Keck, *Eur. J. Pharm. Biopharm.* **78**, 1 (2011).
3. M. A. Kassem, A. A. Abdel Rahman, M. M. Ghorab, M. B. Ahmed and R. M. Khalil, *Int. J. Pharm.* **340**, 126 (2007).
4. E. Merisko-Liversidge, S. L. McGurk and G. G. Liversidge, *Pharm. Res.* **21**, 1545 (2004).
5. K C. Bhol and P. J. Schechter, *Br. J. Dermatol.* **152**, 1235 (2005).
6. J. Zhang, L. Wu, H.-K. Chan and W. Watanabe, *Adv. Drug Deliv. Rev.* **63**, 441 (2011).
7. S. Rao, Y. Song, F. Peddie and A. M. Evans, *Int. J. Nanomed.* **6**, 1245 (2011).
8. A. Maschke, N. Calí, B. Appel, J. Kiermaier, T. Blunk and A. Göpferich, *Pharm. Res.* **23**, 2220 (2006).
9. H. S. M. Ali, P. York, A. M. A. Ali and N. Blagden, *J. Control. Release* **149**, 175 (2011).
10. M. Kakran, R. Shegokar, N. G. Sahoo, L. Al Shaal, L. Li and R. H. Müller, *Eur. J. Pharm. Biopharm.* **80**, 113 (2012).
11. M. Kakran, R. Shegokar, N. G. Sahoo, S. Gohla, L. Li and R. H. Müller, *J. Pharm. Pharmacol.* **64**, 1394 (2012).
12. L. Wu, J. Zhang and W. Watanabe, *Adv. Drug Deliv. Rev.* **63**, 456 (2011).
13. P. Sharma, W. A. Denny and S. Garg, *Int. J. Pharm.* **380**, 40 (2009).
14. L. Peltonen and J. Hirvonen, *Curr. Nanosci.* **4**, 101 (2008).
15. L. Gao, G. Liu, J. Ma, X. Wang, L. Shou and X. Li, *J. Control. Release* **160**, 418 (2012).
16. L. Lindfors, S. Forssen, J. Westergren and U. Olsson, *J. Colloid Interface Sci.* **325**, 404 (2008).
17. G. Kaptay, *Int. J. Pharm.* **430**, 253 (2012).

18. K. Anhalt, S. Geissler, M. Harms, M. Weigandt, and G. Fricker, *Pharm. Res.* **29**, 2887 (2012).
19. F. Kesisoglou and Y. Wu, *AAPS J.* **10**, 516 (2008).
20. B. van Eerdenbrugh, J. Vermant, J. A. Martens, L. Froyen, J. van Humbeeck, G. van den Monter and P. Augustijns, *Mol. Pharm.* 7, 1858 (2010).
21. P. Sharma, Z. D. Zujovic, G. A. Bowmaker, W. A. Denny and S. Garg, *Int. J. Pharm.* 408, 138 (2011).
22. G. Liu, D. Zhang, Y. Jiao, D. Zheng, Y. Liu, C. Duan, L. Jia, Q. Zhang and H. Lou, *Int. J. Pharm.* **422**, 516 (2012).
23. J. Beirowski, S. Inghelbrecht, A. Arien and H. Gieseler, *J. Pharm. Sci.* **100**, 1958 (2011).
24. M. V. Chaubal and C. Popescu, *Pharm. Res.* **25**, 2302 (2008).
25. S. Plakkot, M. de Matas, P. York, M. Saunders and B. Sulaiman, *Int. J. Pharm.* **415**, 307 (2011).
26. P. Kayaert, M. Anné and G. van den Mooter, *J. Pharm. Pharmacol.* **63**, 1446 (2011).
27. J. Pardeike, D. M. Strohmeier, N. Schrödl, C. Voura, M. Gruber, J. G. Khinast and A. Zimmer, *Int. J. Pharm.* **420**, 93 (2011).
28. M. List and H. Sucker, GBPatent 2200048 (1988).
29. H. K. Valo, P. H. Laaksonen, L. J. Peltonen, M. B. Linder, J. T. Hirvonen and T. J. Laaksonen, *ACS Nano* **4**, 1750 (2010).
30. S. M. D'Addio and R. K. Prud'homme, *Adv. Drug Deliv. Rev.* **63**, 417 (2011).
31. H. Eerikäinen, W. Watanabe, E. I. Kauppinen and P. P. Ahonen, *Eur. J. Pharm. Biopharm.* **55**, 357 (2003).
32. S. H. Lee, D. Heng, W. K. Ng, H.-K. Chan and R. B. H. Tan, *Int. J. Pharm.* **403**, 192 (2011).
33. R. H. Müller and J. Moschwitzer, Patent PCT/EP2006/009930 (2006).
34. M. Wang, G. C. Rutledge, A. S. Myerson and B. L. Trout, *J. Pharm. Sci.* **101**, 1178 (2012).
35. L. Peltonen, H. Valo, R. Kolakovic, T. Laaksonen and J. Hirvonen, *Expert Opin. Drug Deliv.* 7, 705 (2010).
36. H. Valo, L. Peltonen, S. Vehviläinen, M. Karjalainen, R. Kostiainen, T. Laaksonen and J. Hirvonen, *Small* **5**, 1791 (2009).
37. G. G. Liversidge and K. C. Cundy, *Int. J. Pharm.* **125**, 91 (1995).
38. E. Merisko-Liversidge, G. G. Liversidge and E. R. Cooper, *Eur. J. Pharm. Sci.* **18**, 113 (2003).
39. G. C. Na, H. J. Stevens Jr. B. O. Yuan and N. Rajagopalan, *Pharm. Res.* **16**, 569 (1999).
40. P. Liu, X. Y. Rong, J. Laru, B. van Veen, J. Kiesvaara, J. Hirvonen, T. Laaksonen and L. Peltonen, *Int. J. Pharm.* **411**, 215 (2011).
41. T. Niwa, S. Miura and K. Danjo, *Int. J. Pharm.* **405**, 218 (2011).
42. C. Knieke, M. Sommer and W. Peukert, *Powder Technol.* **195**, 25 (2009).
43. C. Knieke, C. Steiborn, S. Romeis, W. Peukert, S. Breitung-Faes and A. Kwade, *Chem. Eng. Technol.* **33**, 1401 (2010).
44. E. Merisko-Liversidge and G. G. Liversidge, *Adv. Drug Deliv. Rev.* **63**, 427 (2011).
45. K. Brodka-Pfeiffer, P. Langguth, P. Graβ and H. Häusler, *Eur. J. Pharm. Biopharm.* **56**, 393 (2003).

46. J. Lee, J.-Y. Choi and C. H. Park, *Int. J. Pharm.* **355**, 328 (2008).
47. T. Takatsuka, T. Endo, Y. Jianguo, K. Yuminoki and N. Hashimoro, *Chem. Pharm. Bull.* **57**, 1061 (2009).
48. M. Juhnke, D. Märtin and E. John, *Eur. J. Pharm. Biopharm.* **81**, 214 (2012).
49. I. Larsson and H. G. Kristensen, *Powder Technol.* **107**, 175 (2000).
50. R. H. Müller, C. Jacobs and O. Kayser, *Adv. Drug Deliv. Rev.* **47**, 3 (2001).
51. B. Sulaiman, Patent no WO/2007/020407 (2007).
52. C. M. Keck and R. H. Müller, *Eur. J. Pharm. Biopharm.* **62**, 3 (2006).
53. L. Hao, X. Wang, D. Zhang, Q. Xu, S. Song, F. Wang, C. Li, H. Guo, Y. Liu, D. Zheng and Q. Zhang, *Int. J. Pharm.* **433**, 157 (2012).
54. S. Verma, B. D. Huey and D. J. Burgess, *Langmuir* **25**, 12481 (2009).
55. K. Baba, H. E. Pudavar, I. Roy, T. Y. Ohulchansky, Y. Chan, R. K. Pandery and P. N. Prasad, *Mol. Pharm.* **4**, 289 (2007).
56. B. Derjaguin and L. Landau, *Prog. Surf. Sci.* **50**, 633 (1941).
57. E. J. W. Verwey and J. T. G. Overbeek, *Trans. Faraday Soc.* **42B**, 117 (1946).
58. J.-Y. Choi, J. Y. Yoo, H.-S. Kwak, B. Uk Nam and J. Lee, *Curr. Appl. Phys.* **5**, 472 (2005).
59. B. van Eerdenbrugh, J. Vermant, J. A. Martens, L. Froyen, J. van Humbeeck, P. Augustijns, G. and van Den Mooter, *J. Pharm. Sci.* **98**, 2091 (2009).
60. K. D. Ostrander, H. W. Bosch and D. M. Bondanza, *Eur. J. Pharm. Biopharm.* **48**, 207 (1999).
61. K. Peters, R. H. Müller and D. Q. M. Craig, *Int. J. Pharm.* **184**, 53 (1999).
62. J. Lee, S.-J. Lee, J.-Y. Choi, J. Y. Yoo and C.-H. Ahn, *Eur. J. Pharm. Sci.* **24**, 441 (2005).
63. A. Bhakay, M. Merwade, E. Bilgili and R. N. Dave, *Drug Dev. Ind. Pharm.* **37**, 963 (2011).
64. W. Sun, W. Tian, Y. Zhang, J. He, S. Mao and L. Fang, *Nanomedicine: NBM* **8**, 460 (2012).
65. P. Quan, K. Shi, H. Piao, H. Piao, N. Liang, D. Xia and F. Cui, *Int. J. Pharm.* **430**, 366 (2012).
66. E. Merisko-Liversidge, P. Sarpotdar, J. Bruno, S. Hajj, L. Wei, N. Peltier, J. Rake, J. M. Shaw, S. Pugh, L. Polin, J. Jones, T. Corbett, E. Cooper and G. G. Liversidge, *Pharm. Res.* **13**, 272 (1996).
67. B. van Eerdenbrugh, G. van den Mooter and P. Augustijns, *Int. J. Pharm.* **364**, 64, (2008).
68. L. Peltonen and J. Hirvonen, *J. Pharm. Pharmacol.* **62**, 1569 (2010).
69. B. E. Rabinow, *Nat. Rev. Drug Discov.* **3**, 785 (2004).
70. J. Deng, L. Huang and F. Liu, *Int. J. Pharm.* **390**, 242 (2010).
71. M. J. Rosen, in *Surfactants and Interfacial Phenomena* (John Wiley & Sons, New Jersey, 2004), p. 337.
72. H. Valo, M. Kovalainen, P. Laaksonen, M. Häkkinen, S. Auriola, L. Peltonen, M. Linder, K. Järvinen, J. Hirvonen and T. Laaksonen, *J. Control. Release* **156**, 390 (2011).
73. K. Fuhrmann, J. D. Schulz, M. A. Gauthier and J.-C. Leroux, *ACS Nano* **6**, 1667 (2012).
74. R. H. Müller and C. M. Keck, *J. Biotechnol.* **113** (2004) 151–170.

75. C. Donini, D. N. Robinson, P. Colombo, F. Giordano and N. A. Peppas, *Int. J. Pharm.* **245**, 83 (2002).
76. W. Abdelwahed, G. Degobert and H. Fessi, *Int. J. Pharm.* **309**, 178 (2006).
77. V. Guérin, S. Geiger, D. Robic, A. Didierlaurent, C. André, G. Couarraze and R. Bourbouze, *STP Pharma Sci.* **9**, 351 (1999).
78. V. Vogel, K. Langer, S. Balthasar, P. Schuck, W. Machtle, W. Haase, J. A. van den Broek, C. Tziatzios and D. Schubert, *Prog. Colloid Polymer Sci.* **119**, 31 (2002).
79. P. Shahgaldian, J. Gualbert, K. Aïssa and A. W. Coleman, *Eur. J. Pharm. Biopharm.* **55**, 181 (2003).
80. A. J. P. van Zyl, D. de Wet-Roos, R. D. Sanderson, B. Klumperman, *Eur. Polym. J.* **40**, 2717 (2004).
81. T. Ito, L. Sun, M. A. Bevan and R. M. Crooks, *Langmuir*, **20**, 6940 (2004).
82. C. C. Mueller-Goymann, in *Encyclopedia of Pharmaceutical Technology*, Ed. J. Swarbrick and J. S. Boylan (Marcel Dekker, New York, 2001), p. 117.
83. B. A. Yegin and A. Lamprecht, *Int. J. Pharm.* **320**, 165 (2006).
84. C. F. Bohren and D. R. Huffman, *Absorption and Scattering of Light by Small Particles*, (John Wiley & Sons, New York, 1983).
85. D. Quintanar-Guerrero, E. Allemann, E. Doelker and H. Fessi, *Pharm. Res.* **15**, 1056 (1998).
86. T. Laaksonen, P. Liu, A. Rahikkala, L. Peltonen, E. I. Kauppinen, J. Hirvonen, K. Järvinen and J. Raula, *Pharm. Res.* **28**, 2403 (2011).
87. M. Vandelli, B. Ruozi and F. Forni, *S.T.P. Pharma Sci.* **9**, 567 (1999).
88. L. Peltonen, P. Koistinen and J. Hirvonen, *S.T.P Pharma Sci.* **13**, 299 (2003).
89. A. G. Hausberger and P. P. DeLuca, *J. Pharm. Biomed. Anal.* **13**, 747 (1995).
90. O. Gómez-Guzmán and J. Ruiz-Garcia, *J. Colloid Interface Sci.* **291**, 1 (2005).
91. L. Mu and P. H. Seow, *Colloids Surf. B: Biointerfaces* **47**, 90 (2006).
92. P. Massé and S. Ravaine, *Colloids Surf. A: Physicochem. Eng. Aspect* **270**, 148 (2005).
93. A. Deák, B. Bancsi, A. L. Tóth, A. L. Kovács and Z. Hórvölgyi Z, *Colloids Surf. A: Physicochem. Eng. Aspect*, **278**, 10 (2006).
94. K. S. Soppimath, T. M. Aminabhavi, A. R. Kulkarni and W. E. Rudzinski, *J. Control. Release* **70**, 1 (2001).
95. H. Virieux, M. Le Troedec, A. Cros-Gagneux, W. S. Ojo, F. Delpech, C. Nayral, H. Martinez and B. Chaudret, *J. Am. Chem. Soc.* **134**, 19701 (2012).
96. S. Hyvönen, L. Peltonen, M. Karjalainen and J. Hirvonen, *Int. J. Pharm.* **295**, 269 (2005).
97. Y. Wang, Y. Ma, Y. Ma, Y. Du, Z. Liu, D. Zhang and Q. Zhang, *J. Nanosci. Nanotechnol.* **12**, 6176 (2012).
98. L. Peltonen, P. Koistinen, M. Karjalainen, A. Häkkinen and J. Hirvonen, *AAPS PharmSciTech* **3**, 32 (2002).
99. S. Bates, G. Zografi, D. Engers, K. Morris, K. Crowley and A. Newman, *Pharm. Res.* **23**, 2333 (2006).
100. H. R. Moon, J. J. Urban and D. J. Milliron, *Angew. Chem. Int. Ed.* **48**, 6278 (2009).
101. S. Gupta, M. K. Samanta and A. M. Raichur, *AAPS PharmSciTech* **11**, 322 (2010).
102. Y. F. Fan, Y. N. Wang, Y. G. Fan and J. B. Ma, *Int. J. Pharm.* **324**, 158 (2006).
103. W. L. J. Hinrichs, F. A. Manceñido, N. N. Sanders, K. Braeckmans, S. C. De Smedt, J. Demeester and H. W. Frijlink, *Int. J. Pharm.* **311**, 237 (2006).

104. R. W. O'Brien and L. R. White, *J. Chem. Soc. Faraday Trans. 2*, **74**, 1607 (1978).
105. V. B. Patravale, A. A. Date and R. M. Kulkarni, *J. Pharm. Pharmacol.* **56**, 827 (2004).
106. Y. S. Kang, D. K. Lee, C. S. Lee and P. Stroeve, *J. Phys. Chem. B* **106**, 9341 (2002).
107. S. Hirsjärvi, L. Peltonen and J. Hirvonen, *Colloids Surf. B: Biointerfaces* **49**, 93 (2006).
108. B. van Eerdenbrugh, D. E. Alonzo and L. S. Taylor, *Pharm. Res.* **28**, 1643 (2011).
109. L. Lindfors, S. Forssen, P. Skantze, U. Skantze, A. Zackrisson and U. Olsson, *Langmuir* **22**, 911 (2006).
110. N. Charalampopoulos, K. Avgoustakis and C. G. Kontoyannis, *Anal. Chim. Acta* **491**, 57 (2003).
111. L. Mora, K. Y. Chumbimuni-Torres, C. Clawson, L. Hernandez, L. Zhang and J. Wang, *J. Control. Release* **140**, 69 (2009).
112. K. M. Rosenblatt, D. Douroumis and H. Bunjes, *J. Pharm. Sci.* **96**, 1564 (2007).
113. P. Kayaert, B. Li, I. Jimidar, P. Rombaut, F. Ahssini and G. van den Mooter, *Eur. J. Pharm. Biopharm.* **76**, 507 (2010).
114. M. T. Crisp, C. J. Tucker, T. L. Rogers, R. O. III Williams and K. P. Johnston, *J. Control. Release* **117**, 351 (2007).
115. J. Kreuter, in *Encyclopedia of Pharmaceutical Technology, Vol. 10*, Ed. J. Swarbrick (Marcel Dekker Inc., New York, 1994), p. 165.
116. J. Siepmann, N. Faisant and J.-P. Benoit, *Pharm. Res.* **19**, 1885 (2002).
117. J.-U. A. H. Junghanns and R. H. Müller, *Int. J. Nanomed.* **3**, 295 (2008).
118. N. Arunkumar, M. Deecaraman and C. Rani, *Asian J. Pharm.* **3**, 168 (2009).
119. P. Lakhsmi and G. A. Kumar, *Int. J. Pharm. Sci.* **2**(Suppl. 4), 35 (2010).
120. S. Pace, G. W. Pace, I. Parikh and A. Misra, *Pharm. Technol.* **23**, 116 (1999).
121. R. H. Müller and C. Jacobs, *Pharm. Res.* **19**, 189 (2002).
122. J. Araújo, E. Gonzalez., M. A. Egea, M. L. Garcia and E. B. Souto, *Nanomedicine* **5**, 394 (2009).
123. L. Gao, G. Liu, J. Ma, X. Wang, L. Zhou, X. Li and F. Wang, *Pharm. Res.* DOI 10.1007/s11095-012-0889-z (2012).
124. M. P. Sarparanta, L. M. Bimbo, E. M. Mäkilä, J. J. Salonen, P. H. Laaksonen, K. A. M. Helariutta, M. B. Linder, J. T. Hirvonen, T. J. Laaksonen, H. A. Santos and A. J. Airaksinen, *Biomaterials* **33**, 3353 (2012).
125. M. E Åkerman, W. C. W. Chan, P. Laakkonen and S. N. E. Bhatia, *PNAS* **99**, 12617 (2002).
126. S. Bansal, M. Bansal and R. Kumria, *Int. J. Res. Pharm. Biomed. Sci.* **3**, 406 (2012).
127. J. Chingunpituk, *Walailak J. Sci. Tech.* **4**, 139 (2007).

Chapter 6

Polymeric Nanosystems for Integrated Image-Guided Cancer Therapy

*Amit Singh, Arun K. Iyer and Mansoor M. Amiji**

Department of Pharmaceutical Sciences
School of Pharmacy, Bouvé College of Health
Sciences Northeastern University
Boston, MA 02115, USA
**m.amiji@neu.edu*

Abstract

Despite significant advances in molecular understanding of diseases, such as cancer and inflammation, there are still challenges in furthering the knowledge into effective treatment strategies that can benefit patients at the bedside. Cancer mortality rates, for example, will eclipse heart disease and other major killers in both the US and several other countries around the world in the near future. The failure of significantly affecting clinical outcomes in cancer can in part be attributed to the inability to diagnose the disease early, especially before anatomical abnormalities prevail and the tumor cells disseminate to other regions of the body. Additionally, at the advanced stage of the disease, a number of cancers become refractory to drug treatment due to development of acquired multi-drug resistant phenotype, making their complete eradication highly challenging. Therefore, there is an urgent unmet need for simultaneous early detection and treatment of cancers. In this chapter, we discuss some of the recent advances in the imaging technologies and the development of polymeric image-guided nanosystems or the so called *"theranostic nanosystems"* with illustrative examples from contemporary literature. The collective advancement in the genetic understanding of the disease, coupled with the technological breakthroughs in imaging/diagnostic instrumentation and novel synthetic routes for designing multifunctional image-guided nanotechnology-based systems portend to hold promising potentials for combating this deadly disease.

1. Biomedical Challenges and Need for Image-Guided Delivery

1.1. *Genomic instability and multidrug resistance in cancers*

The diagnosis of many cancers stills remains elusive and poses a major barrier in its effective treatment. Hanahan and Weinberg, in a recent review, discuss the "*hallmark*" characteristics of cancers and the dysregulated signaling pathways involved in the progression and development of the disease.[1] Cancer cells, unlike normal cells, are highly invasive and possess a unique ability to divide and proliferate very rapidly, resisting cell death or apoptosis. Tumor cells can infiltrate and invade surrounding normal tissues or metastasize to multiple distant sites via the extensive lymphatic networks and/or blood vessels (called circulating tumor cells), forming highly refractory secondary tumors.[1–2] This process of tumor metastasis usually occurs at an advanced stage of the disease. At this point, it becomes highly challenging to treat the disease or resort to other options such as surgical resection. Even in the case where surgery is viable, there can be a high risk of tumor relapse after resection due to resistant microscopic subpopulations of virulent tumor cells that are speculated to possess "stem-cell–like" characteristics.[3] Moreover, most tumors do not respond to a broad spectrum of structurally and functionally variant anticancer drugs due to prolonged exposure to drug treatment, a process called acquired tumor multi-drug resistance (MDR).[4] MDR is known to play a pivotal role in self-renewal and formation of new types of tumors.[5] In addition, the tumor cells posses the highly sophisticated ability to "reprogram" its cellular machinery to adapt to highly challenging microenvironment, such as very low levels of oxygen or hypoxic conditions[6] as well as abnormal low pH environments.[7] These micro-environmental selection pressures that leads to the acquired MDR is also considered to be a major player in drug treatment failures.[8] Overall, the genomic instability in cancers and acquired MDR leads to a cascade of multiple mutational events, which in turn leads to successive generations of heterogeneous tumor types,[9] making the diagnosis and management of this malignant disease highly complex.

1.2. *Systemic delivery challenges for small molecule drugs and imaging agents*

Conventional forms of chemotherapy involve the use of small molecules that in general are low-molecular-weight anticancer compounds. Such agents inherently suffer from short plasma half-lives and their effective concentration in the blood drops very quickly on systemic administration, leading to sub-optimal drug utilization[10]. Also, anticancer agents, in general, face systemic delivery challenges because they lack selectivity and tumor-targeting ability. When injected into the bloodstream via the intravenous (*i.v.*) route, they indiscriminately diffuse across blood vessels of normal and tumor tissues and cells, causing severe dose limiting toxicities.[11] In this regard, although the recent advances made in target-specific alternative strategies, such as the ones based on RNA interference (RNAi) mechanism, hold greater promise, they also suffer from similar drawbacks as the small molecule drug counterparts. Furthermore, naked siRNAs, DNAs, proteins and peptides are prone to rapid degradation by enzymes and nucleases present in the blood and are quickly eliminated via the renal excretion. Similarly, fluorescently tagged or radionuclide-conjugated drugs and imaging agents that can be used for diagnostic imaging or for, early detection as well as for therapeutic interventions also face similar shortfalls that make their systemic delivery a great challenge.

1.3. *Delivery strategies for cancer therapy: Nanotechnology for passive and active targeting*

As discussed in Sec. 1.2, conventional cytotoxic systemic therapeutic agents fail to distinguish between cancer and normal tissues. The non-specific distribution of active compounds and imaging agents alike suffer from off-target effects. Fortunately, the tumor microenvironment has several unique characteristics that differentiates it from normal tissues and thus could be exploited for targeted drug delivery. For instance, it is possible to design macromolecular/nanoparticle-based delivery systems that can specifically target tumor tissues.[12] In this regard, drugs encapsulated in nanoparticles can also reduce systemic toxicity by protecting the cytotoxic drug from non-target organs and protect the labile payload (such as anticancer drug, peptides, RNA and DNA) from degradation during circulation after intravenous injection. Nanoparticles made of biodegradable matrices could also help accomplish spatial and temporal control on drug release, along with targeted drug delivery.[13] Drugs and imaging agents encapsulated in nanosystems can also be endowed with long plasma circulation half-life, thereby reducing the dose administration to patients and improving the quality of life. In general, nanoparticles-based mechanisms that are majorly contributing to tumor drug delivery are classified into two main categories, namely passive and active targeting.[14] The schematic representation of the targeting mechanisms is illustrated in Fig. 1.

The passive tumor targeting of macromolecular drugs and nanoparticles relies on a phenomenon called the "*enhanced permeability and retention*" (EPR) effect, first observed by Matsumura and Maeda, caused by the selective accumulation of a polymer-conjugated drug, SMANCS in solid tumors.[15] This

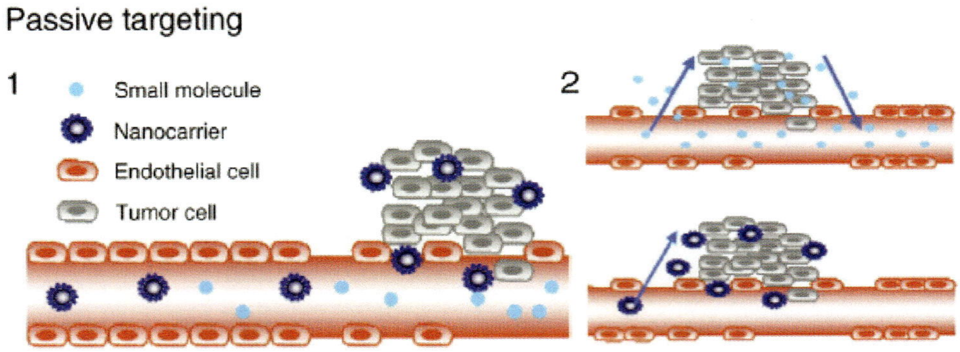

(a)

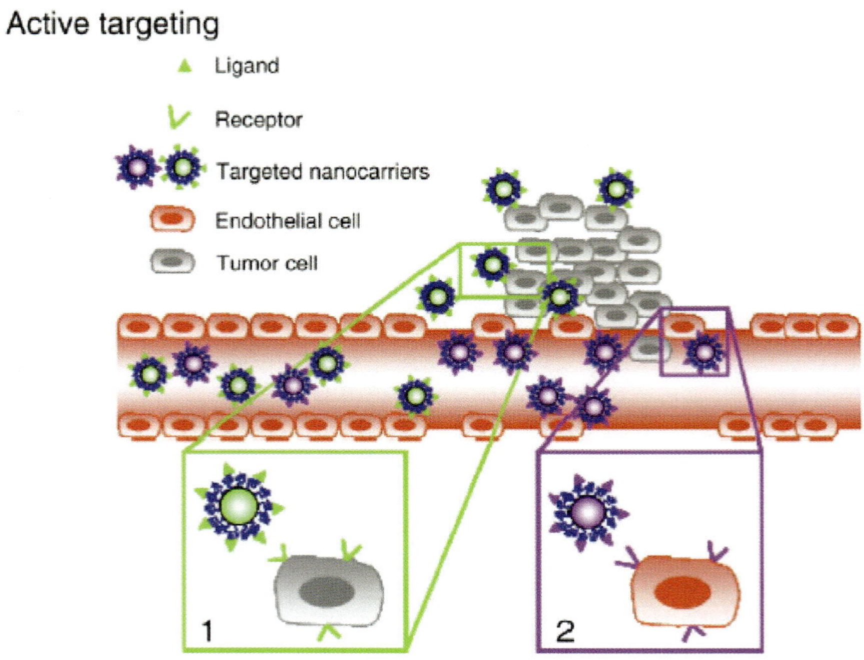

(b)

Fig. 1. Targeting strategies for macromolecules and nanosystems. a. Passive tumor targeting based on the EPR–effect (1) nanosystems extravasate into the tumor interstitium through the leaky vasculature, (2) effect of cut-off size for retention in the tumor tissue. Small molecules and imaging/contrast agents diffuse freely in and out the tumor blood vessels due to their small size and thus the effective drug concentration in the tumor drops rapidly. However, large-sized nanosystems cannot diffuse back into blood stream, leading to enhanced accumulation. Also the defective lymphatic drainage system assists in prolonged retention of nanosystems, promoting better therapeutic outcome, the EPR effect. b. Active targeting — targeting moiety present on the nanosystem bind to surface receptors expressed by tumor cells. Reprinted with permission from Ref. 12.

phenomenon was found to be a result of leaky tumor microvasculature and defective lymphatic drainage that drives the macromolecular drug into the tumor interstitium and cells and enables their prolonged retention (Fig. 1). More strikingly, SMANCS-Lipiodol® (the first approved polymer-conjugated anticancer drug for human use in Japan) when injected into patients with hepatocellular carcinoma via the tumor-feeding artery resulted in >2000-fold drug retention in the tumor tissue than in blood plasma, and the drug remained in the tumor for several weeks.[16] The EPR mechanism was later observed almost universally in almost all human solid tumors, with the exception of necrotic and hypovascular tumors (such as prostate or pancreatic cancer).[17] The EPR-effect is now considered a gold standard for tumor drug delivery, including polymeric nanoparticles, micelles and liposomal systems depicting this behavior.[12] In general, nanosystems with a size of ~20–300 nm and having a molecular weight >40 kDa (above the renal excretion threshold) have been shown to extravagate and accumulate in the tumor tissues selectively.[12] However, the cut-off size of abnormal vasculature may vary from tumor to tumor, and the size of the nanosystems has to be manipulated in the right range for achieving optimal passive targeting.[18] The EPR phenomenon can be enhanced several folds if the nanosystems are designed to evade opsonization by the immune surveillance system or the mononuclear phagocytic system (MPS) after *i.v.* administration. In this regard, conjugating or grafting of poly(ethylene glycol) (PEG) chains on the surface of the nanosystems provides stealth characteristics and allows for MPS escape, thereby rendering long plasma circulation half-life.[19]

Interestingly, accumulation of nanosystems based on the EPR-effect can be remarkably improved when tumor homing ligands or targeting peptides are conjugated to the delivery system for more selective tumor targeting. Such nanosystems can anchor or bind to overexpressed receptors or antigens present on the tumor cell surfaces for more specific uptake or internalization, subsequent to their tumor accumulation by the EPR-effect.[20] Such systems are termed "actively targeted" systems (Fig. 1). One such example is a folic-acid–grafted nanoparticle that was designed to target tumor cells that overexpress folate receptors. The folate-decorated nanoparticles can internalize by the folate-receptor–mediated endocytosis after EPR-based "passive targeting" to tumor tissues.[21] In another example, arginine–glycine–aspartic acid (RGD) tripeptide sequence containing peptides was conjugated on the surface of nanoparticle to target integrin receptors ($\alpha_5\beta_5$ or $\alpha_5\beta_3$) which are overexpressed on angiogenic vascular endothelial cells and tumor cells. Active and passive targeting could also be used for screening of cancers as well as for non-invasive diagnosis of treatment response. For instance, in one study, we engineered a optically labeled multifunctional particle decorated with RGD peptide that specifically targeted the colon tumors and enhanced the fluorescence imaging contrast for colon cancer screening.[22] Similarly researchers from other groups have used aptamers that selectively bind to prostate specific membrane antigen for prostate tumor targeting.[23] In one early example, the response to drug therapy was performed using SMANCS-Lipiodol® which serves both as a contrast agent and therapeutic drug for image-guided tumor monitoring and measuring tumor regression based on X-ray CT.[24]

1.4. *Multifunctional image guided nanosystems*

Nanomedical technologies are emerging at a rapid pace and are poised to play an important role in management of diseases such as cancer. The combination of diagnostic and therapeutic interventions performed using the same nanoconstruct coined *"theranostic nanosystems"* is the culmination of the efforts toward the development of targeted image-guided therapies that can lead to favorable diagnostic and therapeutic clinical outcomes. The inclusion of image guidance as an integral component of drug and gene delivery system has major implication in disease diagnosis and treatment. For instance, in the conventional settings, when a therapeutic drug or gene is systemically administered to cancer patients, there is neither any means of (*in situ*) tracking or imaging the *in vivo* sites of its localization, nor the ability to measure delivery efficiency of drugs/genes. Also, the bioavailability, dose response of drug/gene and efficacy has to be estimated based on separate sets of individual experiments. For example, in the clinical setting, imaging the regression of tumors in patients after drug treatment by radiographic examination will in itself be a separate intervention. In this regard, co-administering the image guiding molecules as part of nanodelivery system can help accomplish several multiple goals in a single step. Some such combinations included non-invasive, real-time assessment of drug delivery efficiency and targeting, *in vivo* fate of drugs/genes and sites of localization/accumulation, modes of excretion as well as imaging/monitoring the efficacy of drug treatment. Such multifunctional agents are already in development and have the potential to revolutionize the area of image-guided drug/gene delivery.[25]

Theranostic nanomedicines, such as the one shown in Fig. 2, can be designed to simultaneously deliver a large cargo of drugs and imaging/contrast agents to the diseased site. These nanosystems can inherently target tumor tissues utilizing the pathophysiological abnormalities of the leaky tumor vasculature, or the EPR effect, as described earlier.[26] More importantly these multifunctional nanosystems can be decorated with targeting ligands

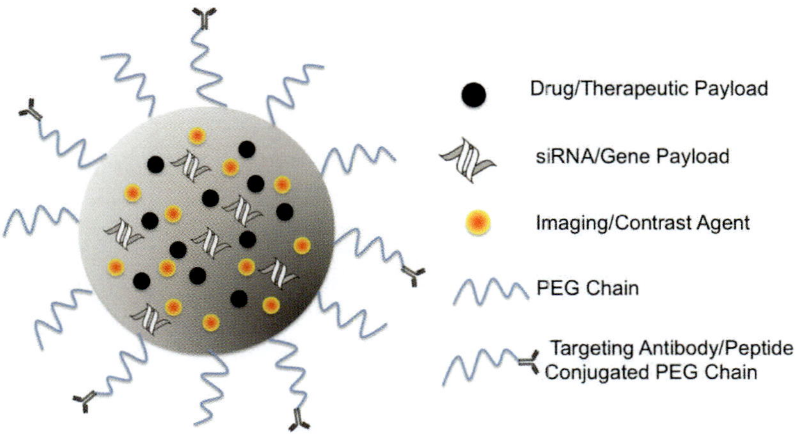

Fig. 2. A schematic representation of multifunctional image-guided targeted nanosystem.

and tumor homing antibodies/peptides for more specific "active" tumor targeting. Such precision-guided multifunctional nanosystems portend to revolutionize the field of cancer therapy. On a similar context, the development of "custom-configured" nanosystems that can be engineered with a cocktail of drugs and imaging agents, having specific tumor homing ligands for targeting particular cancer types, as well as made of particular chemical composition and size, based on tumor heterogeneity, stage and location, can be formulated based on individual needs of patients and the type of disease. Such a multifunctional nanosystem approach offers new possibilities toward the development of personalized medicine, the conceptual reform in medicine to treat each patient based on individual profiling. In this regard, some of the most sought-after imaging modalities centered around the development of multimodal nanosystems, including optical imaging, magnetic resonance imaging, nuclear imaging and computed tomography, that are discussed below.[27]

2. Clinical Imaging Modalities

The use of molecular and functional imaging modalities to identify and characterize diseases such as cancers has great clinical utility.[28] For instance, the ability to image and diagnose the early progression of the disease using novel diagnostic tools and imaging agents can be one of the most important steps for achieving a cure for several diseases.[29] Also non-invasive methods of whole-body imaging using novel imaging probes and targeted nanosystems can not only be regarded as a clinical utility for early detection but are also highly beneficial for management of metastatic diseases, including efficacy validation of drug treatment and a rational development of new molecular target based therapeutics.[30] In this regard, the imaging systems has already been an integral part of disease screening, diagnosis and staging of many cancers.[31] Among the imaging modalities, the ones using magnetic resonance (MR) has been highly popular among clinicians for tumor detection mainly because of its success due to high spatial resolution as well as high depth penetration, with the added benefit of high soft tissue contrast.[32] More importantly, MRI-based imaging modalities is completely devoid of the use of harmful ionizing radiation which can significantly benefit patient screening.[33] With regard to the contrast agents development for MR imaging, gadolinium (Gd^{3+})-based chelates outweighs several others agents and have remained the most popular thus far.[34] In the clinical and preclinical settings, colloidal iron oxides nanoparticles are actively explored as magnetic resonance imaging (MRI) contrast agents.[35] The magnetic iron oxide nanoparticles coated with a biocompatible polymeric shell offer several additional advantages that are attributed to the magnetic responsiveness and MRI visibility apart from the tumor targeting based on the EPR-effect.[26] In principle, development of biocompatible iron oxide nanosystems is promising because of their demonstrated utility in crossing the blood–brain barrier (BBB), for targeting gliomas.[35b,36]

X-ray computed tomography (CT) has also been extensively utilized for screening, detection and measuring the progression of diseases such as lung cancers in the clinics.[37] Iodinated contrast agents such as Lipiodol® have been used to enhance the contrast for CT imaging, and one such example is the intra-arterial (*i.a.*) administration of the patients with the first polymer conjugated drug, SMANCS dissolved in Lipiodol®.[24] SMANCS-Lipiodol® served both as a diagnostic tool and an anticancer drug for therapeutic treatment of hepatocellular carcinoma. The lipid contrast agent Lipiodol® helped detect the tumor accumulation of SMANCS and also facilitated imaging/monitoring the regression of tumors based on X-ray CT.[24] In the recent past, CT has been combined with nuclear modality imaging such as positron emission tomography (PET-CT) and single photon emission tomography (SPECT-CT) to arrive at fused hybrid imaging systems.[38] In this regard, nuclear modality imaging with high detection sensitivities have been highly sought after for staging and follow-up of solid tumors in patients in the clinics.[39] In addition, the availability of radionuclides and relatively fast conjugation chemistries makes them a preferred candidate for imaging applications.[40] Nuclear modality imaging has also been used to image/monitor biological processes such as metabolic activity in organs and tissues. For instance, a fluorinated PET isotope (^{18}F-FDG) with a half-life of ~110 min and high-energy positron emission (0.6335 MeV) is routinely used in the clinic to detect tumors that in general has high metabolic activity.[41] However, one significant drawback of PET imaging is the difficulty in the production of positron emitting radionuclides. PET isotopes, in general, require a dedicated and costly cyclotron. PET isotopes also have relatively short half-lives with limited scope.[42] Recently, ^{68}Ga isotope has now being actively pursued as alternative to ^{18}F for PET imaging mainly because the production of ^{68}Ga does not require an on-site cyclotron. Also, ^{68}Ga has high positron emission of 1.899 MeV that can be ideal for PET imaging applications.[42,43]

As described, another alternative to PET, is the use of SPECT imaging. Although PET has higher detection efficiency compared to SPECT, recent developments in multi-aperture pinhole technology significantly enhances the detection efficiency making SPECT a more useful technique.[44] For SPECT imaging applications, ^{99m}Technetium (^{99m}Tc) has been the most versatile radioisotope that emits 140 KeV γ-rays with ~6-h half-life.[45] The time frame of ^{99m}Tc decay is ideal for the clinical settings because it keeps the radiation exposure to patients at the minimum while still producing sufficiently high dose for imaging.[45] ^{99m}Tc is still used as the most common isotope for disease diagnosis, in the clinics. Another gamma-emitting isotope ^{111}In is popularly used for SPECT imaging because of its favorable half-life of ~3.8 days. In the preclinical settings, ^{111}In proves to be more advantageous, for applications such as pharmacokinetics and tissue biodistribution of radiolabeled drugs and nanosystems in small animal models, using a micro-SPECT/CT.[46]

Recently, there has been tremendous interest and growth in cross-functional imaging technology development that portend to revolutionize the field of radiology and biomedical nanotechnology. In one such example, instrumentation that can provide structural information such as CT or MRI is routinely combined with radionuclide imaging techniques such as PET/SPECT or optical imaging to enhance the rich features they bring forward.[47] Some such examples include the clinical hybrid/fused PET/CT and SPECT/CT[48] and preclinical small animal micro-PET/SPECT-CT imaging systems, which have the ability to co-register both functional and anatomical features previously lacking in individual SPECT or PET instrumentation.[49] Other combinations of functional and structural imaging modalities such as PET/MRI and optical/MRI are also under development for both preclinical and clinical settings.[50]

Among the non-radioactive imaging tools, optical imaging using fluorescently labeled probes and nanoconstructs are increasingly becoming attractive for preclinical *in vitro* screening as well for clinical assessments,[51] mainly due to the availability of vast choice of imaging probes, ease of synthesis/

fabrication as well as accessibility and low cost of instrumentation. Indeed, the application of fluorescence imaging for detecting cancers is not new and was demonstrated as early as 1948 by Moore *et al.* using sodium fluorescein as the fluorophore.[52] In recent times, the two most commonly utilized optical imaging techniques are based on 2D-fluorescence reflectance imaging (FRI) and 3D-fluorescence molecular tomography (FMT) imaging.[53] In an *in vivo* animal study, Vinegoni *et al.* showed that FMT could clearly provide three-dimensional tomographic maps of fluorescence signal throughout the entire body of mice using biocompatible nanoparticles.[54] Such nanosystems, when translated to clinical setting, can be ideally suited for whole body imaging. The three-dimensional tomography using FMT is in a way similar to CT scans but without the use of ionizing radiations,[55] which makes them quite attractive from the safety perspective.

In the current scenario of molecular-marker–based cancer screening/identification and detection, fluorescence imaging is proving to be an indispensable tool for rapid assessment. Furthermore, the commercial availability of a wide range of near-infrared (NIR) optical probes, the relative ease of photolabeling and development of reliable chemical conjugation chemistries, coupled with low cost and relative easy access to optical imaging stations, makes them a preferred choice for molecular imaging.[51b] However, fluorescence imaging also have their own problems and pitfalls, such as low sensitivity due to low penetration of fluorescent light in tissues, rapid bleaching or quenching of the fluorophore and auto-fluorescence (or false positive background signal) from soft tissues. Despite these limitations, optical imaging probes are making progress with several product pipelines under preclinical development.[56]

3. Polymeric Nanosystems for Image-Guided Delivery

Several different hybrid nanomaterials such as gold and silica nanoparticles, quantum dots, carbon nanotubes and iron-oxide–based systems have been developed as imaging agents as well as materials for drug delivery systems. In this respect, the nanosystems constructed using polymeric backbone or as the base building block has unique properties with several advantages. For example, the relative ease of formulating polymer-based nanosystems, which possess abundant surface functional groups for chemical conjugation/modification, makes them an ideal candidate for such applications. Also, advances in high-throughput screening and combinatorial methods of monomer and polymer synthesis have resulted in fast-paced rational development of novel biocompatible and biodegradable materials, well suited for clinical translation. Some of the recent developments in this area are discussed.

The development of image-guided nanoparticulate delivery systems aimed at resolving both diagnostic and therapeutic requirements need thorough optimization. The image guidance component, such as radioactive ligand or fluorescence probe, can be conjugated, complexed or encapsulated in the carrier polymer along with the therapeutic agent. In this regard, several biocompatible polymers already exist that are ideally suited for such application.[57] Also, many of the synthetic and biopolymers have abundant and diverse functional groups that could be utilized for building amphiphilic block copolymers systems. Such systems can assist encapsulation of diverse payloads via self-assembly,[57,58] affording polymeric nanosystems and micellar nanoparticles. Additionally, the functional domains on the backbone of the polymer could also be used for conjugation of multiple different payloads such as drugs and imaging agents as reported elsewhere.[59] The added benefit of using polymeric nanosystems is particularly favorable for conjugating or decorating the nanoparticle surface with targeting ligands such as tumor homing peptides and antibodies for active targeting as described earlier. Such systems can be utilized for multimodal targeted-imaging.[60] In a related application, the multifunctional nanosystems also provides the ability to combine dual or multiple-labeled probes onto the therapeutic molecules and polymers, providing an opportunity for simultaneous dual probe or multimodality imaging. Such nanoparticles are also of great value for evaluating the stability and *in vivo* fate of the delivery systems or for tracking the localization of carrier polymers and labeled drugs using single or multiple instrumentation techniques.

3.1. *Combinatorial polymer synthesis for drug and gene delivery*

The strategy for designing novel polymeric materials for biomedical application using combinatorial synthesis evolved from its highly successful paradigm in drug discovery research. Some such examples are listed below. Vogel *et al.* showed that combinatorial synthesis could be applied for designing polymers libraries using polycondensation reactions.[61] A sebacic anhydride (SA) and 1,6-bis(*p*-carboxyphenoxy)hexane (CPH) was used to synthesize the polyanhydride copolymer library that showed promising potentials as drug carriers for controlled delivery.[61] Anderson *et al.* at the Massachusetts Institute of Technology developed the first combinatorial library of degradable acrylate-terminated poly(*b*-amino ester)s using photo-crosslinking methodology.

These polymeric materials were reported to be useful for a wide array of drug delivery applications.[62]

Although numerous researchers have demonstrated the prospects of combinatorial polymer synthesis for drug delivery application,[61,63] Dr. Langer's group was the first to introduce parallel synthesis for arriving at degradable polymer library for gene transfection.[64] Anderson *et al.* from the same group utilize a semi-automated high-throughput screening method for the synthesis of a large library of polymers for gene delivery applications (Fig. 3).[65] The library of polymers consisted of biodegradable poly(β-amino ester) backbone having low cytotoxicity. These derivatives could be efficiently synthesized by the conjugate addition reaction of a primary amines or bis(secondary amine) with diacrylates. The methodology also introduced synthesis of the polymers as well as their testing in the same "reaction-well" without the need for purification, thus allowing for large library screens with minimum steps involved. In fact, a large library of over 2,500 polymers could be synthesized and tested concomitantly for gene transfection at an astounding rate of about 1,000 per day.[66]

Researchers from Langer's laboratory further expanded the combinatorial library of diacrylates and amines with more diversity in molecular weights.[67] The data from this report showed that poly(β-amino esters) could be useful for cardiovascular disease and cancers. Furthermore, additional work in Dr. Langer's Laboratory yielded several derivatives that showed highly improved transfection ability both *in vitro* and *in vivo* settings.[67] In this regard, the technological breakthrough in polymer library synthesis as well as screening not only reduces human capital but also affords a high degree of success in arriving at the right "hits" that can be carried forward for further evaluation. The methods developed in Dr. Langer's laboratory can thus be more broadly applied toward the identification of materials for nucleic acid, protein and drug delivery in general.

In another example of library synthesis, Thomas *et al.* developed a polymer library of 144 biodegradable macrostructures of polyethyleneimine with oligo-acrylate esters.[68] From the library screenings, it was found that some of the least toxic polymers showed promising hits for high transfection both in the *in vitro* and *in vivo* settings holding, promising future potentials. In another recent report, Barua *et al.* used ring-opening polymerization (ROP) between epoxide groups of diglycidyl ethers and the (poly)amines for the parallel synthesis and screening of a polymer library for gene delivery.[69] From the study, lead candidate polymers were identified that exhibited significantly higher transfection efficacies and lower cytotoxicity than polyethyleneimine. These library screens show promise for furthering the gene delivery applications utilizing combinatorial designed screening methods.

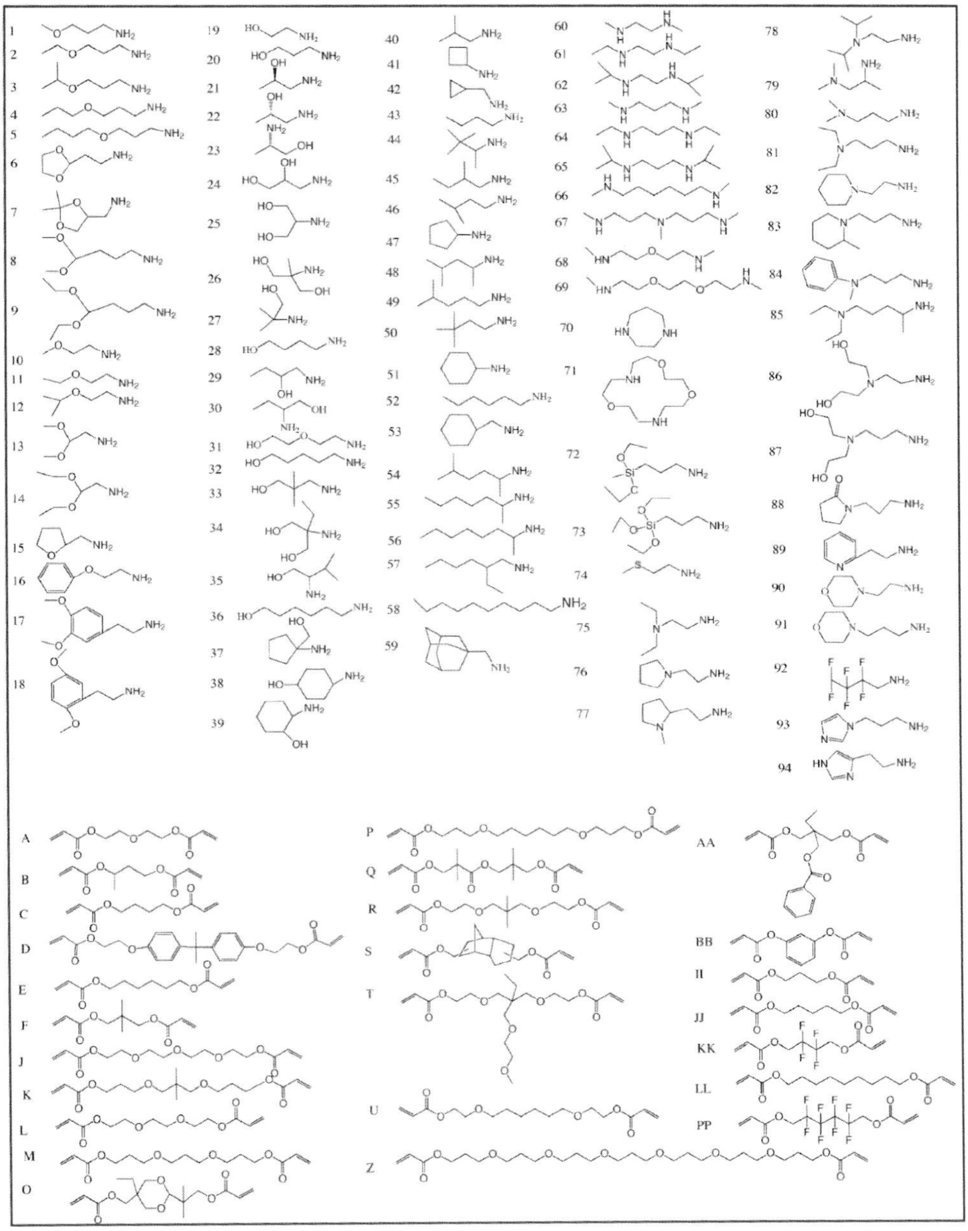

Fig. 3. Library synthesis poly(β-amino esters). Amino (numbers) and diacrylate (letters) monomers used for the library synthesis of diverse poly((β-amino esters) for gene delivery applications. Reprinted with permissions from Ref. 65, While VCH.

In yet another novel combinatorial approach, researchers synthesized a library of cationic N-substituted glycine oligomers with varying charge densities hydrophobicity, side-chain make up and chain lengths to identify candidate polymers that could facilitate gene transfection.[70] Indeed, the library screening resulted in identification of an oligomer that showed transfection efficiencies that could match or rival commercial transfection agents. However, the most interesting feature was that the derivatives were not only serum-insensitive but also protease-resistant, heralding an era of combinatorial polymeric library screening for new class of materials for gene delivery applications.

3.2. *Combinatorial designed nanosystems for image guided drug delivery*

As discussed in Sec. 3.1, traditional polymer synthesis approaches, in general, have poor time and resource utilization. For the quick and efficient development of diverse, multifunctional, image-guided delivery systems, one cannot rely on traditional methods. In this respect, embracing combinatorial approaches and high-throughput polymer synthesis can be highly beneficial, as discussed above.[71] Apart from the utility of high-throughput screening in arriving at diverse polymers with favorable properties, such techniques have recently given rise to rational design of polymer-based building blocks, especially in understanding the structure–property relationship between the components that form the nanosystems, such as the carrier polymer building blocks and the encapsulated drug/gene payload. In one such endeavor in our laboratory, we are developing a novel "*mix and match*" type combinatorial library screening for encapsulation and delivery of diverse drug and gene payload.[71b] By the choice of biodegradable/biocompatible carrier polymers and their functional variants synthesized in our laboratory, we are trying to study the effect of variables such as polymer backbone, its modification with various functional blocks, molecular weight, surface charge, hydrophilicity and hydrophobicity on their ability to stably encapsulate drug and gene payload.[71b] In this regard, as described earlier, the polymeric nanosystems decorated with targeting ligands (biomarkers, antibodies, peptides), and tagged with imaging probes (MRI, optical, radioactive) can provide quick and reliable assessment on assessing the right ratios of the building blocks for efficient and stable encapsulation of drugs and genes forming self-assembled nanosystems.[72] Such novel approaches will therefore expedite the rational design and development of multifunctional nanosystems for clinical use. In this context, an important criterion in selection of materials (such as the ones used in our testing) is the inherent toxicity/bio-compatibility of some of the components used in building the multi-component delivery systems. The combinatorial "mix and match" type screening can thus be useful in developing and identifying nanosystems that, in general, need to be devoid of immunogenicity/toxicity or complement activation *in vivo*. Combinatorial nanosystem screens can afford additional information on the right choice/selection of materials such as incorporation of counter ion-containing moieties to reduce surface charge or including PEG-modified blocks to prevent immune recognition.[73] We intend to shed light on the some such variables in the development of such multifunctional image-guided nanosystems currently being evaluated in our laboratory.

4. Illustrative Examples — Spherical Polymeric Nanoparticles

Polymers have been the most extensively explored delivery vehicles as conjugates or in nanoparticulate formulations for small molecules as well as macromolecules such as proteins, peptides and genes. Spherical polymeric nanoparticles are commonly used for delivery and imaging applications, though other-shaped nanoparticles have also been explored and shape-dependent delivery characteristics have been studied.[74] Polymeric nanoparticles, often being submicron in size, offer several advantages such as ease of drug/macromolecule solubilization, well understood surface chemistry for functionalization, improved stability, half-life and bioavailability of payload and increased and sustained tissue targeting, as discussed earlier. Incorporation of a suitable imaging moiety along with the therapeutic agent realizes a multipronged platform for imaging, diagnostic and therapy of dreaded diseases like cancer. Development of a targeted system is another key aspect that facilitates rapid uptake and localization of these particles in a specific organ of interest. The biodegradable poly(D, L-lactic-co-glycolic acid) (PLGA) particles, for example, loaded with magnetite (Fe_3O_4) particles for magnetically guided delivery and MRI contrast agent, doxorubicin (DOX) as chemo-therapeutic agent as well as semiconductor CdSe/ZnS quantum dots as optical imaging component were synthesized by oil-in-water emulsion process followed by solvent evaporations.[75] The targeting of these particles could be achieved by conjugating folate derivitized PEG chain to PLGA particles to target KB cancer cells *in vitro* that overexpress folate receptor on their surface. MRI and confocal microscopy could confirm the uptake of these multifunctional polymeric systems in the cells by folate receptor mediated endocytosis.

Amphiphilic block copolymers have similarly been used to design multifunctional systems with targeting and imaging modalities. $MnFe_2O_4$ (or

Fe_3O_4) were co-encapsulated as MRI contrast agents and DOX as therapeutic drug in amphipbilic block copolymer to obtain 69.7±8.7 nm (72.8±9.3 nm) size multifunctional magneto-polymeric nanohybrids (MMPN).[76] These nanoparticles were successfully employed for targeting human epidermal growth factor receptor 2 (HER2), a molecular marker for the treatment of metastatic breast cancer. Anti-HER antibody modified (HER-MMPNs) targeted nanoparticles show 320% and 57% better MRI contrast in NIH3T3 cells (high expression of HER2) and MDA-MB-231 (lower expression of HER2) respectively, compared to an irrelevant antibody modified nanoparticles (IRR-MMPNs). *In vivo* MRI and tumor growth inhibition studies reveal better contrast and therapeutic efficacy of HER-MMPNs due to receptor target-mediated endocytosis compared to IRR-MMPNs that rely on EPR effect for tumor localization.

Multifunctional polymeric nanoparticles have been also been successfully used to deliver their imaging and therapeutic payload across BBB for treatment of brain tumors. An acrylamide-based polymeric nanoparticle (~40 nm) was encapsulated with imaging agent (iron oxide magnetic nanoparticles as MRI contrast agent), a photosensitizer agent (Photo-frin) for photodynamic therapy and a targeting agent (f3-peptide).[77] f3-peptide (a vascular homing peptide) is a 31–amino-acid sequence from the NH_2-terminal of the human high-mobility group protein 2,[78] which aids particles uptake through nucleolin, a shuttle protein expressed on some cell membrane surface that traffics to the nucleus.[79] Photo-frin, a mixture of porphyrin oligomers, is used as a photosensitive agent that, upon light activation, generates short-lived singlet oxygen from molecular oxygen that has cell apoptotic and necrotic activity.[80] These particles were further appended with Alexa Fluor 594 to study intracellular uptake analysis and with PEG to increase *in vivo* circulation time. Initial *in vitro* analysis in MDA-435 breast cancer cells demonstrated nearly 95% cell death by 10 μM/L dose concentration of photo-frin–loaded targeted nanoparticles incubated for 4 h followed by exposure to 1500 mW power of laser (630 nm) irradiation for 5 min. MRI measurement with *i.v.* dose (7 mg/kg) of targeted nanoparticles in glioma (9L) tumor-bearing mice confirmed contrast enhancement by a factor of 2 and increase in half-life to 123 min compared to the non-targeted system ($t_{1/2}$ = 39 min). The efficacy study of the targeted nanoparticles showed marked necrosis in the tumor area and improved survival rate, with complete eradication of tumor in 3 out of 5 test mice.

Application of polymeric nanoparticles as a delivery system does not limit to small drug molecules but has been found to be equally efficient in image-guided gene transport *in vitro* and *in vivo*. Yao *et al.* recently reported the synthesis of folate appended PEI_{600}-cyclodextrin (Fol-PEI_{600}-CyD) polyplexes of 120 nm size as excellent image guided plasmid delivery vehicle.[81] Grafting β-cyclodextrin compensated the toxicity of PEI,[82] and the cell targeting could be achieved by conjugation of folic acid to the purified PEI_{600}-CyD derivative. The targeted conjugate loaded with plasmid encoding enhanced green fluorescent protein (pEGFP) or firefly luciferase gene (pLuc) was transfected in human glioblastoma cells (U87 and U138), mouse melanoma B16, human colorectal Lovo cells as well as human lung cancer A549 cells. Presence of EGFP (or luciferase) enables fluorescence (or chemiluminiscence)-based imaging to ascertain particle uptake and payload release in the cells. pLuc loaded polyplexes transfected in B16 cells demonstrated 5, 17 and 33-fold increased luciferase activity at N/P ratio of 10, 20 and 30, respectively, by targeted system compared to non-targeted. Cytotoxicity analysis of Fol-PEI_{600}-CyD, PEI_{600}-CyD and Fol-PEI_{25k}-CyD in U87 and B16 cells further revealed that the targeted and non-targeted polyplexes are considerably less toxic (LD50 ~1 mM) compared to Fol-PEI_{25k}-CyD. The subsequent *in vivo* analysis of pLuc-loaded polyplexes injected intratumorally in B16 xenograft tumor-bearing C57BL/6 mice showed a 13-fold better transfection compared to the plasmid alone and the transfection efficiency was comparable to adenovirus-mediated transfection (Fig. 4). Immunohistochemical analysis of the tumor sections did not show any toxic effect of the particles *in vivo* confirming the biocompatibility of the material.

5. Illustrative Examples — Self-Assembled Polymeric Nanosystems

Self-assembling polymeric materials are extremely popular candidates in designing multifunctional delivery systems, such as the ones being developed in our laboratory.[72,83] These materials generally are composed of two or more polymers with distinct hydrophobic and hydrophilic domains, which aids in the self-assembly to form particles with hydrophobic core that can be used as reservoir for poorly water soluble drug and a hydrophilic shell that can be easily tailored to improve bio-stability and impart targeting functionality. Such systems increase the drug half-life in biological *milieu*, prevent biofouling of the nanoparticles, resulting in increased circulation time and facilitating a sustained release of the drug.[84] Besides, the ease of synthesis and scalability of these nanostructures make them a prefered choice of material for drug and gene delivery applications. There is a multitude of literature on self-assembled polymeric micelle-based delivery systems that are already under various phases of clinical trials (Table 1). This section will provide some case examples of attempts made toward development of self-assembled image-guided polymeric drug delivery systems.

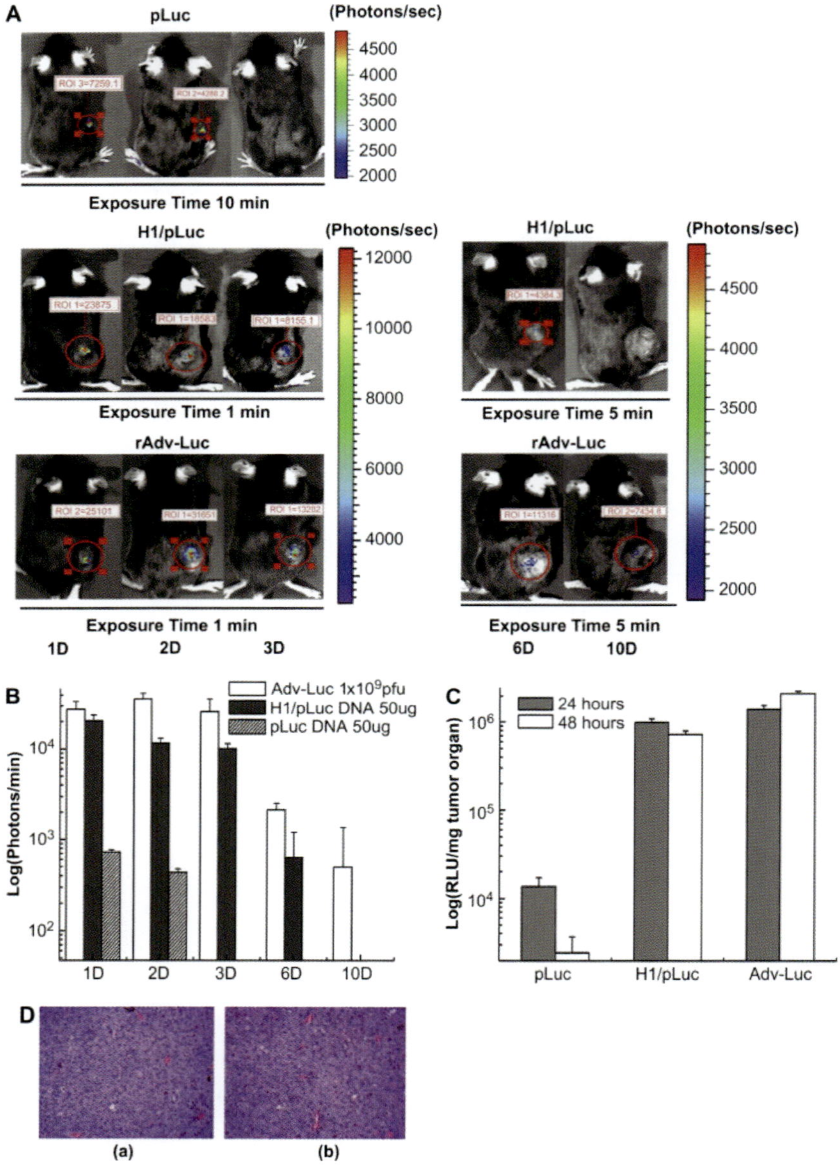

Fig. 4. The transfection efficiency and toxicity of H1 in vivo. (a) Representative images of C57BL/6 mice with B16 xenografts exhibiting in vivo bioluminescence on days (D) 1, 2, 3, 6 and 10 following intratumoral injection with plasmid DNA expressing luciferase (pLuc; 50 mg), H1/pLuc (50 mg pLuc, N/P ratio 20:1) and adenovirus-Luc (rAdv-Luc; 1 _109 pLuc), respectively. The total photon flux (photons/second) in a region of interest (ROI) was shown in the red boxes. (b) The total flux of ROI over time (photons/minute) was used as a parameter to compare the transfection efficiency of pLuc injection, H1-mediated pLuc transfection, and rAdv-Luc transduction on various time points. (c) Determination of relative luciferase unit (RLU) of tumor tissue homogenates at 24 h after transfection. The level of luciferase

Table 1. Polymeric self-assembled nanodelivery systems in clinical trials.

Product	Polymer component	Drug Component	Clinical Phase	Patients/Cancer	Ref
NK105	PEG-pAsp	Paclitaxel	I	Pancreatic (11), bile (5), gastric (2) and colonic (1) cancer]	85
			II	56 patients with gastric cancer and tolerance to at least 1 front line drug except taxane	86
NK911	PEG-pAsp	Doxorubicin	I	23 patients with solid and refractory tumor	87
NK012	PEG-pGlu(S38)	SN-38	I	colorectal (12), pancreatic (4), small cell lung cancer (3), esophageal (3), lung carcinoid (1) and non–small-cell lung cancer (1)	88
SP1049C	Pluronic F127 and L61	Doxorubicin	I	[colorectal (5), esophageal (5), soft tissue sarcoma (4), lung (3), mesothelioma (2) and others (7)]	89
			II	21 patients with advanced stage adenocarcinoma of esophagus or gastroesophagus junction.	90
NC-6004	PEG-pGlu-cisplatin	Cisplatin	I	17 patients with solid tumors	91
Genexol-PM	PEG-p(D,L -lactide)	Paclitaxel	IV	Breast Cancer	92
			II	Pancreatic Cancer	93
			II	Non–small-cell lung cancer (combination with Carboplatin)	92a

Fig. 4. (*Continued*) expression induced by H1/pLuc transfection was similar to that mediated by rAdv- Luc transduction. (D) HE staining did not detect any toxic response in tumor tissue injected with (a) H1/pLuc (50 mg; N/P ratio 20:1) and (b) polymer-free 5% glucose solution. Reprinted with permission from Ref. 81, Elsevier.

Nasongkla *et al.* designed a PEG-PLA–based self-assembling system loaded with super-paramagnetic iron oxide nanoparticles (SPION) as the MRI contrast agent, DOX as chemotherapeutic agent and cRGD peptide as targeting ligand against $\alpha_v\beta_3$ integrins receptors on tumor endothelial cells.[94] Their *in vitro* studies in SLK endothelial cells (human Kaposi's sarcoma cell line) over-expressing $\alpha_v\beta_3$ integrins receptors show enhanced uptake with a marked decrease in the MRI contrast by targeted nanoparticles compared to non-targeted system. Genexol-PM™ is one such commercialized formulation of paclitaxel encapsulated in biodegradable block copolymer, monomethoxy PEG-block-poly(D, L-lactide) (PEG-PLA) micelles made by solid dispersion technique.[95] Several successful Phase-I and Phase-II clinical trials with Genexol-PM™ formulation have been undertaken for non–small-cell lung cancer,[92a] solid tumors[96] and metastatic breast cancer.[92b]

Yang *et al.* designed a smart and versatile folate-targeted polymeric system using heterofunctional amphiphilic triblock copolymers loaded with hydrophilic SPION as the MRI contrast agent inside the core and DOX as therapeutic drug conjugated to the polymer backbone through a pH-responsive bond.[97] They used R(R = folic acid or methoxy group)-PEG_{5000}-poly(glutamate hydrozone doxorubicin)-PEG_{2000}-acrylate (i.e. R-PEG_{114}-P(Glu-Hyd-DOX)-PEG_{46} acrylate) as self-assembling triblock polymer to form vesicles such that the longer PEG chain segregated outward, exposing targeting ligand while shorter PEG chain could be cross-linked by acrylate groups to provide *in vivo* stability and robustness. This targeted multifunctional nanoparticle system showed an improved MRI contrast *in vitro* compared to commercially used Feridex while the flow cytometry and confocal laser scanning microscopic (CLSM) imaging demonstrated higher nanoparticle uptake, resulting in increased cell cytotoxicity. The same group used R (R = folic acid or methoxy group)-PEG_{5000}-PLA-PEG_{2000}-acrylate (i.e. R-PEG_{114}-PLA-PEG_{46} acrylate) to load hydrophilic SPION in the core and DOX in the hydrophobic pockets of worm-like vesicles.[98] They further demonstrated enhanced receptor mediated uptake of the targeted vesicles in HeLa cells (human cervical tumor cell line) compared to non-targeted vesicles, resulting in improved MRI contrast and DOX mediated cell cytotoxicity.

Formation of self-assembled delivery systems does not rely on use of single component polymeric moieties but can also be realized by use of mixed derivatives. One such demonstration exploited the use of poly(2-hydroxyletheyl methacrylate-histidine)-g-PLA and diblock copolymer PEG–PLA for encapsulating DOX as antitumor drug.[99] Folic-acid–derived PEG-PLA facilitated the targeting in folate-receptor expressing cells while Cy5.5-derived PEG-PLA was used as fluorescence-assisted imaging modality in the nanoparticles. *In vitro* assessment of drug-induced cell toxicity in HeLa cells confirmed an IC_{50} of 0.92, 9.49 and 1.60 μg/mL for free DOX, non-targeted micelles and folate-targeted micelles, respectively, confirming that folate-mediated micelle uptake shows high potency compared to non-targeted system. Targeted-mixed micelles also showed an enhanced tumor accumulation (Fig. 5), and efficacy studies revealed 80% inhibition in tumor growth compared to the control mice while non-targeted and free DOX showed only 60% reduction of the tumor burden. Mixed micelle approach thus elucidated tremendous potential as nanocarrier in chemotherapy and diagnosis.

The success of self-assembling polymeric nanoparticles in the clinical setting could be affirmed by NK105, a micellar formulation of paclitaxel encapsulated in hydrophobic core of modified polyaspartate, while the shell was composed of PEG giving the stealth effect.[100] Phase-I clinical trials in 19 patients with tumors showed a maximum tolerated dose (MTD) of 180 mg m^{-2}, and the resulting dose limiting toxicity (DLT) of neutropenia was found to be grade 4, resulting for more than 5 days.[85] The maximum concentration (C_{max}) and area under concentration (AUC) was dose-dependent and plasma AUC of NK105 at 150 mg m^{-2} was 15-fold higher than the conventional paclitaxel formulation. Based on the Phase-I studies, the recommended dose for Phase-II trials were 150 mg m^{-2} every 3 weeks. A total of 56 patients with measurable level of gastric cancer and resistance to at least one first line of therapy except taxane were enrolled for the study. 150 mg m^{-2} of NK105 was administered for 30 min by *i.v.* infusion every 3 weeks without any antiallergic infusion. Two complete recoveries and twelve partial responses were recorded, with an overall response rate of 25%.[86] NK911, a micellar system consisting of block copolymer PEG and poly(aspartic acid) (PEG-PAA) conjugated with DOX as anticancer drug was similarly studied. Preclinical evaluations affirmed NK911 half-life to be 2.8 h while the plasma and intratumor AUC was found to be 15–30 and 2–3 fold higher, respectively, compared to free DOX.[87] Phase-I clinical trials among 23 patients revealed a MTD and DLT of neutropenia to be 67 mg m^{-2}. Among the 23 participating patients with solid tumors, one patient with metastatic pancreatic tumor showed partial response to the formulations. Based on Phase I clinical data, a dose of 50 mg m^{-2} every 3 weeks was recommended for Phase II studies.

6. Illustrative Examples — Hybrid Systems

Polymers and lipids remain the most popular material of choice for designing nanoparticles for drug delivery applications. It was therefore natural to integrate polymer–lipid hybrid materials to compliment their

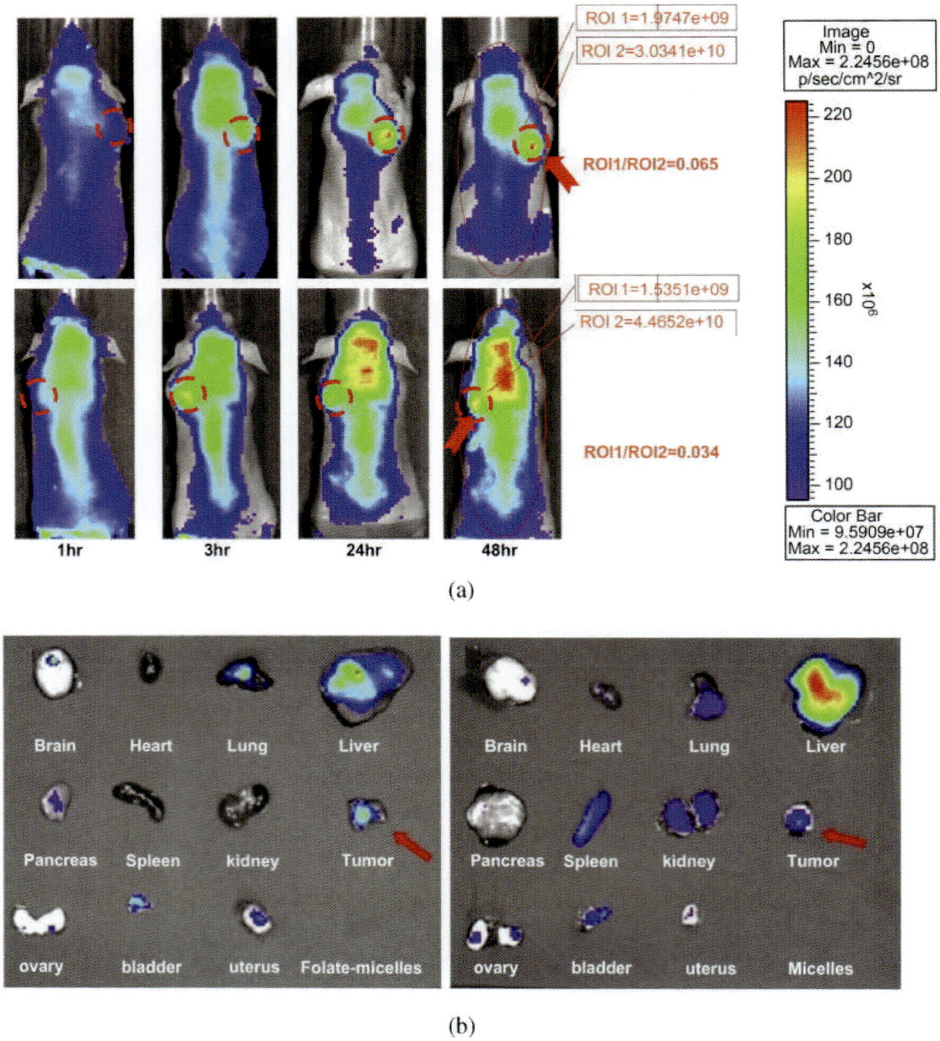

Fig. 5. (a) *In vivo* optical fluorescence imaging of Hela tumor xenografted nude mice administrated with micelles and Folate-micelles. Red circle specifies the location of tumor. (b) Photoimaging of several organs and tumor, which were taken after sacrifice (at 48 h post-injection) of the nude mice (injected with 5 mg/kg of Cy5.5 basis micelles). The fluorescence images were acquired with an exposure time 1 s and used the Cy5.5 filter channel. The color indicates the relative fluorenscence. Reprinted with permission from Ref. 99, Elsevier.

strengths in formulating nanosystems. Zhang *et al.*, for example, used polymer–lipid–polymer conjugation for the formation of a nanodelivery system, where the hydrophobic polymer core ensures encapsulation of poorly water-soluble drug, hydrophilic polymeric outer shell with antibiofouling effect to enhance nanoparticle stability and circulation, while the lipid monolayer at the interface acts as molecular shield to enhance drug retention within the polymeric core.[101] Such self-assembling systems have several advantages such as ease of preparation and scale-up, high drug encapsulation efficiency, controlled and sustained release and surface modification possibilities for targeting ligand functionalization. In a similar effort, DSPE-PEG_{2000} ((1,2-distearoyl-sn-glycero-3-phosphoethanolamine-N-carboxy PEG_{2000}) hybrid was utilized to make 83-nm particles with PLGA core followed by functionalization with anticarcinoembryonic antigen (anti-CEA) half antibody tagged to Alexa Fluor 488 dye. PLGA was simultaneously tagged with Alexa Fluor 647 prior to use, thus enabling incorporation of two-fluorophore to visualize nanoparticle uptake. Paclitaxel was encapsulated as anticancer therapeutic agent and thus formed multifunctional nanosystem was assayed in human pancreatic cells BxPC-3 (CEA^{+}) and XPA-3 (CEA^{-}). *In vitro* analysis by fluorescence microscopy reveals concurrent receptor-mediated delivery of PLGA core and DSPE-PEG_{2000} shell into the BxPC-3 cells by targeted system, while XPA-3 cells showed negligible nanoparticle uptake. Cytotoxicity studies in BxPC-3 cells further confirmed higher nanoparticle uptake and subsequent cell death by targeted system (IC_{50} ~251 nM) compared to non-targeted system (IC_{50} ~526 nM).

Alternatively, Park *et al.* employed a more straightforward approach by using phospholipid–PEG conjugate as self-assembling micellar hybrid nanoparticles (MHNs) for encapsulation of SPION as MRI contrast agent, quantum dots (QDs) as imaging agent and DOX as therapeutic component.[102] They could tailor the loading of SPIONs and QDs in the micelles by varying their relative ratio during the nanoparticle synthesis. Specific targeting of MDA-MB-435 human breast cancer cell line could be achieved by functionalizing the nanoparticle surface with f3 peptide. *In vitro* analysis demonstrated dramatic increase in NIR fluorescence and MRI contrast enhancement of the targeted nanoparticles compared to the non-targeted system that did not show any significance fluorescence and MRI contrast. Nanoparticles loaded with SPION and QD (emission ~800 nm) administered in MDA-MB-435 sub-cutaneous tumor-bearing mice showed enhanced circulation time (~3h). MRI and optical imaging confirmed an enhanced accumulation of the nanoparticles in the tumor up to 20 h post injection (Fig. 6). Such nanoparticles are promising drug delivery systems with multimodal imaging capabilities.

7. Conclusions and Future Out Look

Application of polymeric hybrid systems does not limit to image-guided drug delivery for cancer therapeutics alone but for targeting other diseased conditions as well. Peters *et al.* have devised a DSPE-PEG micelle-based multifunctional

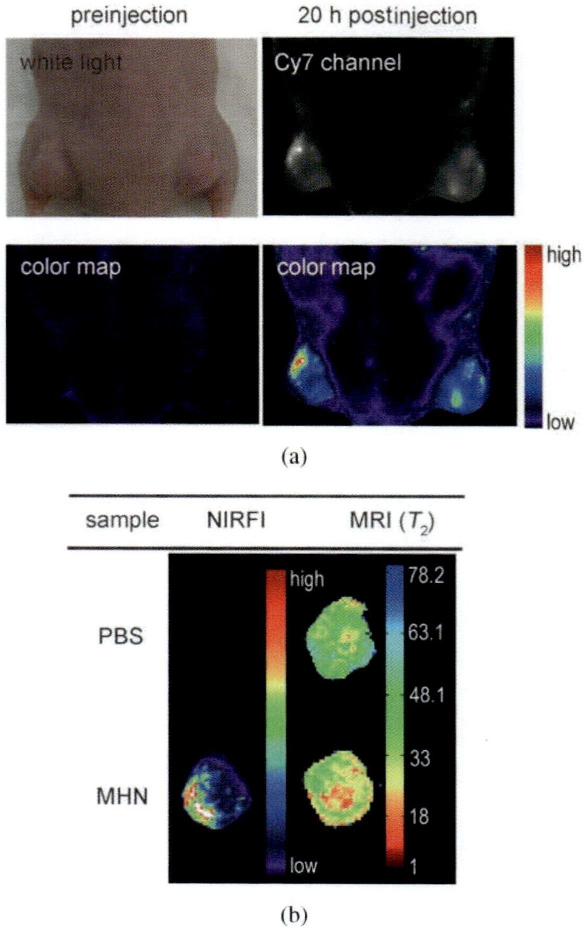

Fig. 6. (a) NIR fluorescence images showing the passive accumulation of MHNs containing QDs (emission at 800 nm, MHN(800)) in a mouse with MDA-MB-435 tumors. The mouse was imaged pre-injection and 20 h pos-tinjection (injection dose: 10 mg kg^{-1}). (b) Image table describing the results of multimodal imaging (by MRI and NIR fluorescence) of the tumor harvested from the mouse in (a). PBS: a control in which a mouse with a tumor was injected with phosphate buffered saline; NIRFI: near-infrared fluorescence image; MRI(T_2): T_2 values from T_2-weighted MRI.

delivery vehicle with fluorophore for optical imaging, surface functionalized 5-aminoacid clot-binding peptide (CREKA) for targeting and hirolog as an anticoagulant for imaging and treatment of atherosclerosis.[103] *Ex-vivo* imaging of the isolated aortic tree from the atherosclerosis-induced *ApoE*-null mice post-injection showed a marked increase in the plaque staining and thus fluorescence intensity from the targeted nanoparticle (I ~209,000 ± 59,000) compared to the non-targeted particles (I ~ 5,100 ± 3300). Hirulog loaded CREKA targeted micelles showed 3 h circulation with high accumulation in atherosclerotic aorta and increased antithrombin activity in the excised aorta tree (1.8 μg/mg of tissue) compared to non-targeted micelles (1.2 μg/mg of tissue).

A plethora of literature has been published, outlining the advantages associated with polymeric-particles–based drug delivery systems compared to conventional form of free drug administration, both in the pre-clinical and clinical settings. Polymeric systems significantly improve the drug half-life *in vivo*, prevent biofouling of the delivery vehicle, facilitate controlled and sustained drug release, thereby improving therapeutic efficacy with reduced side effects. More importantly, polymeric nanosystems can be engineered to selectively accumulate in target organs and tissues and internalize in cells of interest by utilizing the passive and active targeting strategies. Nearly 20 polymeric nanoparticle formulations are currently in pre-clinical and clinical trials for anticancer treatment, and recently, Genexol-PM received FDA approval for the treatment of breast cancer. However, all these delivery vehicles have largely relied on passive accumulation based on the EPR effect, which is effective but not efficient. Of concern in this regard is the emergence of drug resistance in cancer that has led to failure of most of the first-line drugs and resurgence of the disease after initial response to therapy. The condition is further compounded by the fact that refractory tumor acquires resistance not only to the drug used for therapeutic regime but also to several other drugs, a condition popularly known as multi-drug resistance (MDR). In order to tackle MDR in cancer, the delivery strategies has to rely on a more robust, multipronged therapeutic approach, such as inclusion of targeting ligand for efficient delivery, imaging probes to confirm the successful delivery to the target site and future monitoring of disease progression and use of combination therapy with two or more therapeutic modalities. Recent research endeavors have been channeled in design of such multifunctional delivery systems and the advancement in the understanding of material property, functionalization strategies, imaging capabilities and assay methods has immensely benefited the cause. Polymeric delivery system has come to age, with several reports of successful multifunctional targeted vehicles that have exploited clinically relevant MRI, PET/SPECT-CT, FRI, FMT, optical coherence tomography, etc., as imaging modality.

Screening one polymer at a time is a very time-consuming, labor-intensive and cost-ineffective method for the development of drug delivery systems. A combinatorial approach to synthesis of polymers with diverse properties, therefore, is another avenue that has been aggressively explored. Such libraries of customized polymeric materials provide a flexible platform for high-throughput screening to design disease target, drug/imaging agent specific delivery system. The rationale for material selection depends on optimized payload encapsulation and subsequent release profile, physical properties of the formed delivery system as well as ease of synthesis, functionalization and scale-up. Combinatorial design provides an insight into the relationship trend between material and payload for designing an efficient delivery system and promises a diverse platform for the rapid selection of material for the development of a multifunctional drug delivery system.

Despite the increasing effort in developing novel and versatile methods for improved multifunctional polymeric delivery vehicles, material-related toxicity remains a major concern. Long-term safety is pertinent to any polymeric material used for drug delivery application. Besides, development of multifunctional systems involves complex chemistries and a variety of individual material components that could have long-term adverse effects. Literature on multifunctional drug delivery systems is in abundance, but there is a dearth of studies on the safety aspect of these materials. Care should also be taken in the clinical translation of preclinically tested materials since most studies rely on safety data from animal models, which could be toxic to human due to inter- and intra-species physiological differences.

Mass production of material from lab-scale to manufacturing level with consistent properties and high yield is yet an important and challenging aspect for the development of multifunctional nanosystems. Scale-up of such formulations with negligible batch-to-batch variation is often a very complex process due to involvement of several individual components with complicated synthesis steps. Poor understanding of polymer properties and involvement of closely interacting different components also contribute to issues with bulk manufacturing of the material for clinical applications. The US Food and Drug Administration has laid current good laboratory and manufacturing practice (cGLP/cGMP) guidelines to ensure reproducibility of manufacturing products for clinical trials. Bulk production of multifunctional delivery vehicles with stringent quality control is germane to meet these guidelines and maintain a smooth translation of research from the lab to the clinical level. Much work is needed to address these critical issues before multifunctional therapeutics could become a reality; however, clinical translation of even a handful of such agents would herald a new era in the rational development of image-guided polymeric nanosystems for personalized therapies.

References

1. D. Hanahan and R. A. Weinberg, Hallmarks of cancer: The next generation, *Cell* **144**(5), 646–674 (2011).
2. D. Hanahan and R. A. Weinberg, The hallmarks of cancer, *Cell* **100**(1), 57–70 (2000).

3. P. B. Gupta, C. L. Chaffer and R. A. Weinberg, Cancer stem cells: Mirage or reality? *Nat. Med.* **15**(9), 1010–1012 (2009).
4. (a) A. Dutour, D. Leclers, J. Monteil, F. Paraf, J. L. Charissoux, R. Rousseau, and M. Rigaud, Non-invasive imaging correlates with histological and molecular characteristics of an osteosarcoma model: Application for early detection and follow-up of MDR phenotype, *Anticancer Res.* **27**(6B), 4171–4178 (2007); (b) J. L. Biedler, Genetic aspects of multidrug resistance, *Cancer* **70**(6 Suppl), 1799–809 (1992); (c) L. S. Jabr-Milane, L. E. van Vlerken, S. Yadav and M. M. Amiji, Multi-functional nanocarriers to overcome tumor drug resistance, *Cancer Treat. Rev.* **34**(7), 592–602 (2008).
5. M. Al-Hajj and M. F. Clarke, Self-renewal and solid tumor stem cells, *Oncogene* **23**(43), 7274–7282 (2004).
6. (a) K. Asosingh, H. De Raeve, M. de Ridder, G. A. Storme, A. Willems, I. Van Riet, B. Van Camp and K. Vanderkerken, Role of the hypoxic bone marrow microenvironment in 5T2MM murine myeloma tumor progression, *Haematologica* **90**(6), 810–817 (2005); (b) C. V. Dang, B. C. Lewis, C. Dolde, G. Dang and H. Shim, Oncogenes in tumor metabolism, tumorigenesis, and apoptosis, *J. Bioenerg. Biomembr.* **29**(4), 345–354 (1997); (c) P. Vaupel and A. Mayer, Hypoxia in cancer: Significance and impact on clinical outcome, *Cancer Metastasis Rev.* **26**(2), 225–239 (2007); (d) L. Milane, Z. Duan and M. Amiji, Role of hypoxia and glycolysis in the development of multi-drug resistance in human tumor cells and the establishment of an orthotopic multi-drug resistant tumor model in nude mice using hypoxic pre-conditioning, *Cancer Cell. Int.* **11**, 3 (2011).
7. (a) R. Martinez-Zaguilan, E. A. Seftor, R. E. Seftor, Y. W. Chu, R. J. Gillies and M. J. Hendrix, Acidic pH enhances the invasive behavior of human melanoma cells, *Clin. Exp. Metastasis* **14**(2), 176–186 (1996); (b) R. A. Gatenby and E. T. Gawlinski, The glycolytic phenotype in carcinogenesis and tumor invasion: Insights through mathematical models, *Cancer Res.* **63**(14), 3847–3854 (2003).
8. V. S. Donnenberg and A. D. Donnenberg, Multiple drug resistance in cancer revisited: The cancer stem cell hypothesis, *J. Clin. Pharmacol.* **45**(8), 872–877 (2005).
9. (a) D. P. Cahill, K. W. Kinzler, B. Vogelstein and C. Lengauer, Genetic instability and Darwinian selection in tumours, *Trends Cell Biol.* **9**(12), M57–M60 (1999); (b) L. A. Loeb, K. R. Loeb and J. P. Anderson, Multiple mutations and cancer, *Proc. Natl. Acad. Sci. USA* **100**(3), 776–781 (2003).
10. T. M. Allen and P. R. Cullis, Drug delivery systems: Entering the mainstream, *Science* **303**(5665), 1818–1822 (2004).
11. (a) E. K. Rowinsky, V. Chaudhry, A. A. Forastiere, S. E. Sartorius, D. S. Ettinger, L. B. Grochow, B. G. Lubejko, D. R. Cornblath and R. C. Donehower, Phase I and pharmacologic study of paclitaxel and cisplatin with granulocyte colony-stimulating factor: Neuromuscular toxicity is dose-limiting, *J. Clin. Oncol.* **11**(10), 2010–2020 (1993); (b) L. J. Steinherz, P. G. Steinherz, C. T. Tan, G. A. Heller and M. L. Murphy, Cardiac toxicity 4 to 20 years after completing anthracycline therapy, *JAMA* **266**(12), 1672–1677 (1991).
12. F. Danhier, O. Feron and V. Préat, To exploit the tumor microenvironment: Passive and active tumor targeting of nanocarriers for anti-cancer drug delivery, *J. Control Release* **148**(2), 135–146 (2010).
13. P. Couvreur and C. Vauthier, Nanotechnology: Intelligent design to treat complex disease, *Pharm. Res.* **23**(7), 1417–1450 (2006).

14. S. Ganta, H. Devalapally, A. Shahiwala and M. Amiji, A review of stimuli-responsive nanocarriers for drug and gene delivery, *J. Control Release* **126**(3), 187–204 (2008).
15. Y. Matsumura and H. Maeda, A new concept for macromolecular therapeutics in cancer chemotherapy: Mechanism of tumoritropic accumulation of proteins and the antitumor agent smancs, *Cancer Res.* **46**(12 Pt 1), 6387–6392 (1986).
16. K. Iwai, H. Maeda, H. and T. Konno, Use of oily contrast medium for selective drug targeting to tumor: Enhanced therapeutic effect and X-ray image, *Cancer Res.* **44**(5), 2115–2121 (1984).
17. (a) H. Maeda, T. Sawa and T. Konno, Mechanism of tumor-targeted delivery of macromolecular drugs, including the EPR effect in solid tumor and clinical overview of the prototype polymeric drug SMANCS, *J. Control. Release* **2001, 74** (1–3), 47–61; (b) R. K. Jain, Transport of molecules in the tumor interstitium: A review, *Cancer Res.* **47**(12), 3039–3051 (1987).
18. V. P. Torchilin, Drug targeting, *Eur. J. Pharm. Sci.* **11**(Suppl 2), S81–S91 (2000).
19. (a) A. A. Gabizon, Selective tumor localization and improved therapeutic index of anthracyclines encapsulated in long-circulating liposomes, *Cancer Res.* **52**(4), 891–896 (1992); (b) S. S. Davis, Biomedical applications of nanotechnology--implications for drug targeting and gene therapy, *Trends Biotechnol.* **15**(6), 217–224 (1997).
20. F. Marcucci and F. Lefoulon, Active targeting with particulate drug carriers in tumor therapy: Fundamentals and recent progress, *Drug Discov. Today* **9**(5), 219–228 (2004).
21. B. Rihova, Receptor-mediated targeted drug or toxin delivery, *Adv. Drug Deliv. Rev.* **29**(3), 273–289 (1998).
22. N. Lue, S. Ganta, D. X. Hammer, M. Mujat, A. E. Stevens, L. Harrison, R. D. Ferguson, D. Rosen, M. Amiji and N. Iftimia, Preliminary evaluation of a nanotechnology-based approach for the more effective diagnosis of colon cancers, *Nanomed. (Lond)* **5**(9), 1467–1479 (2010).
23. O. C. Farokhzad, J. M. Karp and R. Langer, Nanoparticle-aptamer bioconjugates for cancer targeting, *Expert Opin. Drug Deliv.* **3**(3), 311–324 (2006).
24. H. Maeda, SMANCS and polymer-conjugated macromolecular drugs: Advantages in cancer chemotherapy, *Adv. Drug Deliv. Rev.* **46**(1–3), 169–185 (2001).
25. (a) G. A. Koning and G. C. Krijger, Targeted multifunctional lipid-based nanocarriers for image-guided drug delivery, *Anticancer Agents Med. Chem.* 7(4), 425–440 (2007); (b) T. Lammers, V. Subr, P. Peschke, R. Kuhnlein, W. E. Hennink, K. Ulbrich, F. Kiessling, M. Heilmann, J. Debus, P. E. Huber and G. Storm, Image-guided and passively tumour-targeted polymeric nanomedicines for radiochemotherapy, *Br. J. Cancer* **99**(6), 900–910 (2008); (c) T. Lammers, F. Kiessling, W. E. Hennink and G. Storm, Nanotheranostics and image-guided drug delivery: Current concepts and future directions, *Mol. Pharm.* 7(6), 1899–1912 (2010).
26. A. K. Iyer, G. Khaled, J. Fang and H. Maeda, Exploiting the enhanced permeability and retention effect for tumor targeting, *Drug Discov. Today* **11**(17–18), 812–818 (2006).
27. S. M. Janib, A. S. Moses and J. A. MacKay, Imaging and drug delivery using theranostic nanoparticles, *Adv. Drug Deliv. Rev.* **62**(11), 1052–1063 (2010).
28. K. Glunde, A. P. Pathak and Z. M. Bhujwalla, Molecular-functional imaging of cancer: to image and imagine, *Trends Mol. Med.* **13**(7), 287–297 (2007).
29. F. R. Hirsch, W. A. Franklin, A. F. Gazdar, A. F. and P. A. Bunn, Jr. Early detection of lung cancer: Clinical perspectives of recent advances in biology and radiology, *Clin. Cancer Res.* 7(1), 5–22 (2001).

30. D. A. Mankoff, F. O'Sullivan, W. E. Barlow and K. A. Krohn, Molecular imaging research in the outcomes era: measuring outcomes for individualized cancer therapy, *Acad. Radiol.* **14**(4), 398–405 (2007).
31. F. A. Jaffer and R. Weissleder, Molecular imaging in the clinical arena, *JAMA* **293**(7), 855–862 (2005).
32. (a) R. Weissleder and M. J. Pittet, Imaging in the era of molecular oncology, *Nature* **452** (7187), 580–589 (2008); (b) C. Balu-Maestro, C. Chapellier, A. Bleuse, I. Chanalet, C. Chauvel and R. Largillier, Imaging in evaluation of response to neoadjuvant breast cancer treatment benefits of MRI, *Breast Cancer Res. Treat.* **72**(2), 145–152 (2002).
33. R. C. Semelka, D. M. Armao, J. Elias Jr. and W. Huda, Imaging strategies to reduce the risk of radiation in CT studies, including selective substitution with MRI, *J. Magn. Reson. Imaging* **25**(5), 900–909 (2007).
34. (a) S. Aime, C. Cabella, S. Colombatto, S. Geninatti Crich, E. Gianolio and F. Maggioni, Insights into the use of paramagnetic Gd(III) complexes in MR-molecular imaging investigations, *J. Magn. Reson. Imaging* **16**(4), 394–406 (2002); (b) P. Caravan, J. J. Ellison, T. J. McMurry and R. B. Lauffer, Gadolinium(III) chelates as MRI contrast agents: Structure, dynamics, and applications, *Chem. Rev.* **99**(9), 2293–2352 (1999).
35. (a) J. S. Weinstein, C. G. Varallyay, E. Dosa, S. Gahramanov, B. Hamilton, W. D. Rooney, L. L. Muldoon and E. A. Neuwelt, Superparamagnetic iron oxide nanoparticles: diagnostic magnetic resonance imaging and potential therapeutic applications in neurooncology and central nervous system inflammatory pathologies, a review, *J. Cereb. Blood Flow Metab.* **30**(1), 15–35 (2010); (b) B. Chertok, B. A. Moffat, A. E. David, F. Yu, C. Bergemann, B. D. Ross and V. C. Yang, Iron oxide nanoparticles as a drug delivery vehicle for MRI monitored magnetic targeting of brain tumors, *Biomaterials* **29**(4), 487–496 (2008).
36. O. Veiseh, C. Sun, J. Gunn, N. Kohler, P. Gabikian, D. Lee, N. Bhattarai, R. Ellenbogen, R. Sze, A. Hallahan, J. Olson and M. Zhang, Optical and MRI multifunctional nanoprobe for targeting gliomas, *Nano Lett.* **5**(6), 1003–1008 (2005).
37. (a) M. Kaneko, K. Eguchi, H. Ohmatsu, R. Kakinuma, T. Naruke, K. Suemasu and N. Moriyama, Peripheral lung cancer: screening and detection with low-dose spiral CT versus radiography, *Radiology*, *201* (3), 798–802 1996; (b) S. Sone, S. Takashima, F. Li, Z. Yang, T. Honda, Y. Maruyama, M. Hasegawa, T. Yamanda, K. Kubo, K. Hanamura and K. Asakura, Mass screening for lung cancer with mobile spiral computed tomography scanner, *Lancet* **351**(9111), 1242–1245 (1998).
38. A. Bockisch, L. S. Freudenberg, D. Schmidt and T. Kuwert, Hybrid imaging by SPECT/CT and PET/CT: Proven outcomes in cancer imaging, *Seminars Nucl. Med.* **39**(4), 276–289 (2009).
39. (a) M. Bangerter, J. Kotzerke, M. Griesshammer, K. Elsner, S. N. Reske and L. Bergmann, Positron emission tomography with 18-fluorodeoxyglucose in the staging and follow-up of lymphoma in the chest, *Acta. Oncol.* **38**(6), 799–804 (1999); (b) Z. Keidar, O. Israel and Y. Krausz, SPECT/CT in tumor imaging: Technical aspects and clinical applications, *Semin. Nucl. Med.* **33**(3), 205–218 (2003).
40. (a) F. Berger and S. Gambhir, Recent advances in imaging endogenous or transferred gene expression utilizing radionuclide technologies in living subjects: applications to breast cancer, *Breast Cancer Res.* **3**(1), 28–35 (2001); (b) M. Phelps, PET: The merging of biology and imaging into molecular imaging, *J. Nucl. Med.* **41**(4), 661–681 (2000).

41. (a) D. Pan, S. Gambhir, T. Toyokuni, M. Iyer, N. Acharya, M. Phelps and J. Barrio, Rapid synthesis of a 5 -fluorinated oligodeoxy-nucleotide: A model antisense probe for use in imaging with positron emission tomography (PET), *Bioorg. Med. Chem. Lett.* **8**(11), 1317–1320 (1998); (b) R. L. Wahl, L. E. Quint, R. L. Greenough, C. R. Meyer, R. I. White and M. B. Orringer, Staging of mediastinal non-small cell lung cancer with FDG PET, CT, and fusion images: preliminary prospective evaluation, *Radiology* **191**(2), 371–377 (1994); (c) H. A. Nabi and J. M. Zubeldia, Clinical applications of (18)F-FDG in oncology, *J. Nucl. Med. Technol.* **30**(1), 3–9; quiz 10–1 (2002).
42. M. Mitterhauser, S. Toegel, W. Wadsak, R. Lanzenberger, L. Mien, C. Kuntner, T. Wanek, H. Eidherr, D. Ettlinger and H. Viernstein, Pre vivo, ex vivo and in vivo evaluations of [68Ga]-EDTMP, *Nucl. Med. Biol.* **34**(4), 391–397 (2007).
43. H. Maecke, M. Hofmann and U. Haberkorn, 68Ga-labeled peptides in tumor imaging, *J. Nucl. Med.* **46**(1 suppl), 172S–178S (2005).
44. T. Funk, P. Despres, W. C. Barber, K. S. Shah and B. H. Hasegawa, A multipinhole small animal SPECT system with submillimeter spatial resolution, *Med. Phys.* **33**(5), 1259–1268 (2006).
45. S. Banerjee, A. Pillai, M. Raghavan and N. Ramamoorthy, Evolution of Tc-99m in diagnostic radiopharmaceuticals, *Semin. Nucl. Med.* **31**, 260–277 (2001).
46. (a) M. Hamoudeh, M. A. Kamleh, R. Diab and H. Fessi, Radionuclides delivery systems for nuclear imaging and radiotherapy of cancer, *Adv. Drug Deliv. Rev.* **60**(12), 1329–1346 (2008); (b) T. H. Chow, Y. Y. Lin, J. J. Hwang, H. E. Wang, Y. L. Tseng, V. F. Pang, R. S. Liu, W. J. Lin, C. S. Yang and G. Ting, Therapeutic efficacy evaluation of ^{111}In-labeled PEGylated liposomal vinorelbine in murine colon carcinoma with multi-modalities of molecular imaging, *J. Nucl. Med.* **50**(12), 2073–2081 (2009); (c) G. Ting, C. H. Chang and H. E. Wang, Cancer nanotargeted radiopharmaceuticals for tumor imaging and therapy, *Anticancer Res.* **29**(10), 4107–4118 (2009); (d) A. K. Iyer, Y. Su, J. Feng, X. Lan, X. Zhu, Y. Liu, D. Gao, Y. Seo, H. F. Vanbrocklin, V. Courtney Broaddus, B. Liu and J. He, The effect of internalizing human single chain antibody fragment on liposome targeting to epithelioid and sarcomatoid mesothelioma, *Biomaterials* **32**(10), 2605–2613 (2011).
47. E. F. Jones, J. He, H. F. VanBrocklin, B. L. Franc and Y. Seo, Nanoprobes for medical diagnosis: Current status of nanotechnology in molecular imaging, *Curr. Nanosci.* **4**(1), 17–29 (2008).
48. G. Mariani, L. Bruselli, T. Kuwert, E. E. Kim, A. Flotats, O. Israel, M. Dondi and N. Watanabe, A review on the clinical uses of SPECT/CT, *Eur. J. Nucl. Med. Mol. Imaging* **37**(10), 1959–1985 (2010).
49. A. Chatziioannou, Instrumentation for molecular imaging in preclinical research: Micro-PET and Micro-SPECT, *Proc. Am. Thorac. Soc.* **2**, 533–536 (2005).
50. M. S. Judenhofer, H. F. Wehrl, D. F. Newport, C. Catana, S. B. Siegel, M. Becker, A. Thielscher, M. Kneilling, M. P. Lichy, M. Eichner, K. Klingel, G. Reischl, S. Widmaier, M. Rocken, R. E. Nutt, H. J. Machulla, K. Uludag, S. R. Cherry, C. D. Claussen and B. J. Pichler, Simultaneous PET-MRI: A new approach for functional and morphological imaging, *Nat. Med.* **14**(4), 459–465 (2008).
51. (a) E. L. Kaijzel, G. van der Pluijm and C. W. Lowik, Whole-body optical imaging in animal models to assess cancer development and progression, *Clin. Cancer Res.* **13**(12), 3490–3497 (2007); (b) G. Choy, P. Choyke and S. K. Libutti, Current advances in

molecular imaging: Noninvasive in vivo bioluminescent and fluorescent optical imaging in cancer research, *Mol. Imaging* **2**(4), 303–312 (2003); (c) K. Sokolov, M. Follen, J. Aaron, I. Pavlova, A. Malpica, R. Lotan and R. Richards-Kortum, Real-time vital optical imaging of precancer using anti-epidermal growth factor receptor antibodies conjugated to gold nanoparticles, *Cancer Res.* **63**(9), 1999–2004 (2003).

52. R. L. Moore, Treatment of an infant with paroxysmal auricular tachycardia, *Pediatrics* **2**(3), 266–271 (1948).
53. C. Bremer, V. Ntziachristos, B. Weitkamp, G. Theilmeier, W. Heindel and R. Weissleder, Optical imaging of spontaneous breast tumors using protease sensing 'smart' optical probes, *Invest. Radiol.* **40**(6), 321–327 (2005).
54. C. Vinegoni, D. Razansky, S. A. Hilderbrand, F. Shao, V. Ntziachristos and R. Weissleder, Transillumination fluorescence imaging in mice using biocompatible upconverting nanoparticles, *Opt. Lett.* **34**(17), 2566–2568 (2009).
55. E. E. Graves, R. Weissleder and V. Ntziachristos, Fluorescence molecular imaging of small animal tumor models, *Curr. Mol. Med.* **4**(4), 419–430 (2004).
56. S. A. Hilderbrand and R. Weissleder, Near-infrared fluorescence: application to in vivo molecular imaging, *Curr. Opin. Chem. Biol.* **14**(1), 71–79 (2010).
57. A. K. Iyer, K. Greish, J. Fang, R. Murakami and H. Maeda, High-loading nanosized micelles of copoly(styrene-maleic acid)-zinc protoporphyrin for targeted delivery of a potent heme oxygenase inhibitor, *Biomaterials* **28**(10), 1871–1881 (2007).
58. F. Wang, T. K. Bronich, A. V. Kabanov, R. D. Rauh and J. Roovers, Synthesis and evaluation of a star amphiphilic block copolymer from poly(epsilon-caprolactone) and poly(ethylene glycol) as a potential drug delivery carrier, *Bioconjug. Chem.* **16**(2), 397–405 (2005).
59. E. R. Gillies and J. M. Frechet, Dendrimers and dendritic polymers in drug delivery, *Drug Discov. Today* **10**(1), 35–43 (2005).
60. (a) D. Peer, J. M. Karp, S. Hong, O. C. Farokhzad, R. Margalit and R. Langer, Nanocarriers as an emerging platform for cancer therapy, *Nat. Nanotechnol.* **2**(12), 751–760 (2007); (b) J. Kim, Y. Piao and T. Hyeon, Multifunctional nanostructured materials for multimodal imaging, and simultaneous imaging and therapy, *Chem. Soc. Rev.* **38**(2), 372–390 (2009).
61. B. M. Vogel, J. T. Cabral, N. Eidelman, B. Narasimhan and S. K. Mallapragada, Parallel synthesis and high throughput dissolution testing of biodegradable polyanhydride copolymers, *J. Comb. Chem.* **7**(6), 921–928 (2005).
62. D. Anderson, C. Tweedie, N. Hossain, S. Navarro, D. Brey, K. Van Vliet, R. Langer and J. Burdick, A combinatorial library of photocrosslinkable and degradable materials, *Adv. Mater.* **18**(19), 2614–2618 (2006).
63. S. Brocchini, K. James, V. Tangpasuthadol and J. Kohn, Structure-property correlations in a combinatorial library of degradable biomaterials, *J. Biomed. Mater. Res.* **42**(1), 66–75 (1998).
64. D. Lynn, D. Anderson, D. Putnam and R. Langer, Accelerated discovery of synthetic transfection vectors: parallel synthesis and screening of a degradable polymer library, *J. Am. Chem. Soc.* **123**(33), 8155–8156 (2001).
65. D. Anderson, D. Lynn and R. Langer, Semi automated synthesis and screening of a large library of degradable cationic polymers for gene delivery, *Angewandte Chemie* **115**(27), 3261–3266 (2003).

66. A. Akinc, A. Zumbuehl, M. Goldberg, E. S. Leshchiner, V. Busini, N. Hossain, S. A. Bacallado, D. N. Nguyen, J. Fuller, R. Alvarez, A. Borodovsky, T. Borland, R. Constien, A. de Fougerolles, J. R. Dorkin, K. Narayanannair Jayaprakash, M. Jayaraman, M. John, V. Koteliansky, M. Manoharan, L. Nechev, J. Qin, T. Racie, D. Raitcheva, K. G. Rajeev, D. W. Sah, J. Soutschek, I. Toudjarska, H. P. Vornlocher, T. S. Zimmermann, R. Langer and D. G. Anderson, A combinatorial library of lipid-like materials for delivery of RNAi therapeutics, *Nat. Biotechnol.* **26**(5), 561–569 (2008).
67. J. Green, G. Zugates, N. Tedford, Y. Huang, L. Griffith, D. Lauffenburger, J. Sawicki, R. Langer and D. Anderson, Combinatorial modification of degradable polymers enables transfection of human cells comparable to adenovirus, *Adv. Mater.* **19**(19), 2836–2842 (2007).
68. M. Thomas, J. Lu, C. Zhang, J. Chen and A. Klibanov, Identification of novel superior polycationic vectors for gene delivery by high-throughput synthesis and screening of a combinatorial library, *Pharm. Res.* **24**(8), 1564–1571 (2007).
69. S. Barua, A. Joshi, A. Banerjee, D. Matthews, S. T. Sharfstein, S. M. Cramer, R. S. Kane and K. Rege, Parallel synthesis and screening of polymers for nonviral gene delivery, *Mol. Pharm.* **6**(1), 86–97 (1998).
70. J. Murphy, T. Uno, J. Hamer, F. Cohen, V. Dwarki and R. Zuckermann, A combinatorial approach to the discovery of efficient cationic peptoid reagents for gene delivery, *Proc. Natl. Acad. Sci. USA* **95**(4), 1517 (1998).
71. (a) M. Goldberg, K. Mahon and D. Anderson, Combinatorial and rational approaches to polymer synthesis for medicine, *Adv. Drug Deliv. Rev.* **60**(9), 971–978 (2008); (b) S. C. Abeylath, S. Ganta, A. K. Iyer and M. Amiji, Combinatorial-designed multifunctional polymeric nanosystems for tumor-targeted therapeutic delivery, *Acc. Chem. Res.* (2011); (c) J. J. Green, R. Langer and D. G. Anderson, A combinatorial polymer library approach yields insight into nonviral gene delivery, *Acc. Chem. Res.* **41**(6), 749–759 (2008).
72. M. Susa, A. K. Iyer, K. Ryu, E. Choy, F. J. Hornicek, H. Mankin, L. Milane, M. M. Amiji and Z. Duan, Inhibition of ABCB1 (MDR1) expression by an siRNA nanoparticulate delivery system to overcome drug resistance in osteosarcoma, *PLoS One* **5**(5), e10764 (2010).
73. S. H. Kim, J. H. Jeong, S. H. Lee, S. W. Kim and T. G. Park, Local and systemic delivery of VEGF siRNA using polyelectrolyte complex micelles for effective treatment of cancer, *J. Control. Release* **129**(2), 107–116 (2008).
74. J. A. Champion, Y. K. Katare and S. Mitragotri, Particle shape: A new design parameter for micro- and nanoscale drug delivery carriers, *J. Control Release* **121**(1–2), 3–9 (2007).
75. J. Kim, J. E. Lee, S. H. Lee, J. H. Yu, J. H. Lee, T. G. Park and T. Hyeon, Designed fabrication of a multifunctional polymer nanomedical platform for simultaneous cancer-targeted imaging and magnetically guided drug delivery, *Adv. Mater.* **20**(3), 478–483 (2008).
76. J. Yang, C. H. Lee, H. J. Ko, J. S. Suh, H. G. Yoon, K. Lee, Y. M. Huh and S. Haam, Multifunctional magneto-polymeric nanohybrids for targeted detection and synergistic therapeutic effects on breast cancer, *Angew. Chem. Int. Ed. Engl.* **46**(46), 8836–8839 (2007).
77. G. R. Reddy, M. S. Bhojani, P. McConville, J. Moody, B. A. Moffat, D. E. Hall, G. Kim, Y. E. Koo, M. J. Woolliscroft, J. V. Sugai, T. D. Johnson, M. A. Philbert,

R. Kopelman, A. Rehemtulla and B. D. Ross, Vascular targeted nanoparticles for imaging and treatment of brain tumors, *Clin. Cancer Res.* **12**(22), 6677–6686 (2006).

78. K. Porkka, P. Laakkonen, J. A. Hoffman, M. Bernasconi and E. Ruoslahti, A fragment of the HMGN2 protein homes to the nuclei of tumor cells and tumor endothelial cells in vivo, *Proc. Natl. Acad. Sci. USA* **99**(11), 7444–7449 (2002).
79. S. Christian, J. Pilch, M. E. Akerman, K. Porkka, P. Laakkonen and E. Ruoslahti, Nucleolin expressed at the cell surface is a marker of endothelial cells in angiogenic blood vessels, *J. Cell Biol.* **163**(4), 871–878 (2003).
80. C. Abels, Targeting of the vascular system of solid tumours by photodynamic therapy (PDT), *Photochem. Photobiol. Sci.* **3**(8), 765–771 (2004).
81. H. Yao, S. S. Ng, W. O. Tucker, Y. K. Tsang, K. Man, X. M. Wang, B. K. Chow, H. F. Kung, G. P. Tang and M. C. Lin, The gene transfection efficiency of a folate-PEI600-cyclodextrin nanopolymer, *Biomaterials* **30**(29), 5793–5803 (2009).
82. S. H. Pun, N. C. Bellocq, A. Liu, G. Jensen, T. Machemer, E. Quijano, T. Schluep, S. Wen, H. Engler, J. Heidel and M. E. Davis, Cyclodextrin-modified polyethylenimine polymers for gene delivery, *Bioconjug. Chem.* **15**(4), 831–840 (2004).
83. (a) E. Kobayashi, A. K. Iyer, F. J. Hornicek, M. M. Amiji and Z. Duan, Lipid-functionalized dextran nanosystems to overcome multidrug resistance in cancer: A pilot study, *Clin. Orthopaed. Rel. Res.* **2012**; (b) M. Susa, A. K. Iyer, K. Ryu, F. J. Hornicek, H. Mankin, M. M. Amiji and Z. Duan, Doxorubicin loaded polymeric nanoparticulate delivery system to overcome drug resistance in osteosarcoma, *BMC Cancer* **9**, 399 (2009); (c) S. C. Abeylath, S. Ganta, A. K. Iyer and M. Amiji, Combinatorial-designed multifunctional polymeric nano-systems for tumor-targeted therapeutic delivery, *Acc. Chem. Res.* **44**(10), 1009–1017 (2011); (d) S. C. Abeylath and M. M. Amiji, 'Click' synthesis of dextran macrostructures for combinatorial-designed self-assembled nanoparticles encapsulating diverse anticancer therapeutics, *Bioorganic Med. Chem.* **19**(21), 6167–6173 (2011).
84. (a) R. Tong and J. Cheng, Anticancer polymeric nanomedicines, *J. Macromol. Sci. Part C: Polymer Rev.* **47**(3), 345–381 (2007); (b) V. Torchilin, Micellar nanocarriers: Pharmaceutical perspectives, *Pharm. Res.* **24**(1), 1–16 (2007).
85. T. Hamaguchi, K. Kato, H. Yasui, C. Morizane, M. Ikeda, H. Ueno, K. Muro, Y. Yamada, T. Okusaka, K. Shirao, Y. Shimada, H. Nakahama and Y. Matsumura, A phase I and pharmacokinetic study of NK105, a paclitaxel-incorporating micellar nanoparticle formulation, *Br. J. Cancer* **97**(2), 170–176 (2007).
86. K. Kato, K. Chin, T. Yoshikawa, K. Yamaguchi, Y. Tsuji, T. Esaki, K. Sakai, M. Kimura, T. Hamaguchi, Y. Shimada, Y. Matsumura and R. Ikeda, Phase II study of NK105, a paclitaxel-incorporating micellar nanoparticle, for previously treated advanced or recurrent gastric cancer, *Invest. New Drugs* **30**(4), 1621–1627.
87. Y. Matsumura, T. Hamaguchi, T. Ura, K. Muro, Y. Yamada, Y. Shimada, K. Shirao, T. Okusaka, H. Ueno, M. Ikeda and N. Watanabe, Phase I clinical trial and pharmacokinetic evaluation of NK911, a micelle-encapsulated doxorubicin, *Br. J. Cancer* **91**(10), 1775–1781 (2004).
88. T. Hamaguchi, T. Doi, T. Eguchi-Nakajima, K. Kato, Y. Yamada, Y. Shimada, N. Fuse, A. Ohtsu, S. Matsumoto, M. Takanashi and Y. Matsumura, Phase I Study of NK012, a Novel SN-38-incorporating micellar nanoparticle, in adult patients with solid tumors, *Clin. Cancer Res.* 16(20), 5058–5066 (2010).

89. S. Danson, D. Ferry, V. Alakhov, J. Margison, D. Kerr, D. Jowle, M. Brampton, G. Halbert and M. Ranson, Phase I dose escalation and pharmacokinetic study of pluronic polymer-bound doxorubicin (SP1049C) in patients with advanced cancer, *Br. J. Cancer* **90**(11), 2085–2091 (2004).
90. J. W. Valle, A. Armstrong, C. Newman, V. Alakhov, G. Pietrzynski, J. Brewer, S. Campbell, P. Corrie, E. K. Rowinsky and M. Ranson, A phase 2 study of SP1049C, doxorubicin in P-glycoprotein-targeting pluronics, in patients with advanced adenocarcinoma of the esophagus and gastroesophageal junction, *Invest. New Drugs* **29**(5), 1029–1037 (2011).
91. R. Plummer, R. H. Wilson, H. Calvert, A. V. Boddy, M. Griffin, J. Sludden, M. J. Tilby, M. Eatock, D. G. Pearson, C. J. Ottley, Y. Matsumura, K. Kataoka and T. Nishiya, A Phase I clinical study of cisplatin-incorporated polymeric micelles (NC-6004) in patients with solid tumours, *Br. J. Cancer* **104**(4), 593–598 (2011).
92. (a) D. W. Kim, S. Y. Kim, H. K. Kim, S. W. Kim, S. W. Shin, J. S. Kim, K. Park, M. Y. Lee and D. S. Heo, Multicenter phase II trial of Genexol-PM, a novel Cremophor-free, polymeric micelle formulation of paclitaxel, with cisplatin in patients with advanced non-small-cell lung cancer, *Ann. Oncol.* **18**(12), 2009–2014 (2007); (b) K. S. Lee, H. C. Chung, S. A. Im, Y. H. Park, C. S. Kim, S. B. Kim, S. Y. Rha, M. Y. Lee and J. Ro, Multicenter phase II trial of Genexol-PM, a Cremophor-free, polymeric micelle formulation of paclitaxel, in patients with metastatic breast cancer, *Breast Cancer Res. Treat.* **108**(2), 2412–50 (2008).
93. M. W. Saif, N. A. Podoltsev, M. S. Rubin, J. A. Figueroa, M. Y. Lee, J. Kwon, E. Rowen, J. Yu and R. O. Kerr, Phase II clinical trial of paclitaxel loaded polymeric micelle in patients with advanced pancreatic cancer, *Cancer Inv.* **28**(2), 186–194 (2009).
94. N. Nasongkla, E. Bey, J. Ren, H. Ai, C. Khemtong, J. S. Guthi, S. F. Chin, A. D. Sherry, D. A. Boothman and J. Gao, Multifunctional polymeric micelles as cancer-targeted, MRI-ultrasensitive drug delivery systems, *Nano Lett.* **6**(11), 2427–2430 (2006).
95. S. C. Kim, D. W. Kim, Y. H. Shim, J. S. Bang, H. S. Oh, S. Wan Kim and M. H. Seo, In vivo evaluation of polymeric micellar paclitaxel formulation: toxicity and efficacy, *J. Control Release* 72(1–3), 191–202 (2001).
96. W. T. Lim, E. H. Tan, C. K. Toh, S. W. Hee, S. S. Leong, P. C. Ang, N. S. Wong and B. Chowbay, Phase I pharmacokinetic study of a weekly liposomal paclitaxel formulation (Genexol-PM) in patients with solid tumors, *Ann. Oncol.* **21**(2), 382–328 (2010).
97. X. Yang, J. J. Grailer, I. J. Rowland, A. Javadi, S. A. Hurley, V. Z. Matson, D. A. Steeber and S. Gong, Multifunctional stable and pH-responsive polymer vesicles formed by heterofunctional triblock copolymer for targeted anticancer drug delivery and ultrasensitive MR imaging, *ACS Nano* **4**(11), 6805–6817 (2010).
98. X. Yang, J. J. Grailer, I. J. Rowland, A. Javadi, S. A. Hurley, D. A. Steeber and S. Gong, Multifunctional SPIO/DOX-loaded wormlike polymer vesicles for cancer therapy and MR imaging, *Biomaterials* **31**(34), 9065–9073 (2010).
99. H. C. Tsai, W. H. Chang, C. L. Lo, C. H. Tsai, C. H. Chang, T. W. Ou, T. C. Yen and G. H. Hsiue, Graft and diblock copolymer multifunctional micelles for cancer chemotherapy and imaging, *Biomaterials* **31**(8), 2293–2301 (2010).
100. T. Hamaguchi, Y. Matsumura, M. Suzuki, K. Shimizu, R. Goda, I. Nakamura, I. Nakatomi, M. Yokoyama, K. Kataoka and T. Kakizoe, NK105, a paclitaxel-incorporating micellar nanoparticle formulation, can extend in vivo antitumour activity and reduce the neurotoxicity of paclitaxel, *Br. J. Cancer* **92**(7), 1240–1246 (2005).

101. L. Zhang, J. M. Chan, F. X. Gu, J. W. Rhee, A. Z. Wang, A. F. Radovic-Moreno, F. Alexis, R. Langer and O. C. Farokhzad, Self-assembled lipid-polymer hybrid nanoparticles: A robust drug delivery platform, *ACS Nano* **2**(8), 1696–1702 (2008).
102. J. H. Park, G. von Maltzahn, E. Ruoslahti, S. N. Bhatia and M. J. Sailor, Micellar hybrid nanoparticles for simultaneous magnetofluorescent imaging and drug delivery, *Angew. Chem. Int. Ed. Engl.* **47**(38), 7284–7288 (2008).
103. D. Peters, M. Kastantin, V. R. Kotamraju, P. P. Karmali, K. Gujraty, M. Tirrell and E. Ruoslahti, Targeting atherosclerosis by using modular, multifunctional micelles, *Proc. Natl. Acad. Sci. USA* **106**(24), 9815–9819 (2009).

Chapter 7

Polysaccharide-Based Nanocarriers for Drug Delivery

Carmen Teijeiro, Adam McGlone, Noemi Csaba, Marcos Garcia-Fuentes and María J. Alonso

CIMUS Research Center, Faculty of Pharmacy University of Santiago de Compostela (USC) Santiago de Compostela Spain

1. Introduction

Polysaccharides are complex carbohydrate molecules typically composed of mono- or disaccharide units linked by glycosidic bonds, which allows their structural organization into either linear or highly branched homo- or hetero-polysaccharides. These naturally occurring biopolymers represent a large family of structurally and functionally different molecules; many of them play critical roles in plant, animal and human physiology.

Due to their high diversity and abundance in nature, polysaccharides have been applied for several hundred years as traditional pharmaceutical excipients (e.g. stabilizers, lubricants, coagulants etc.). In addition, over the past decades polysaccharides have also been implemented in biomedical research as biomaterials for advanced drug delivery applications and tissue engineering.

Within this context, polysaccharides of interest are typically biodegradable, biocompatible and preferentially water soluble. In addition, they also display additional characteristics such as specific mechanical-rheological properties and/or ease of chemical transformations, which can further expand the possibility of their application as biomaterials in diverse drug delivery strategies. Most relevant examples of natural polysaccharides exhibiting these features are reflected in Table 1.

Table 1. Relevant examples of polysaccharides as biomaterials and their main characteristics.

Name	Origin	Composition	Structure	Charge
Alginate	Algae	Mannuronate Guluronate	Linear	Negative
Dextran	Plant	Glucose	Branched	Neutral
Hyaluronic acid	Animal/microbial	Glucuronic acid N-acetyl-glucosamine	Linear	Negative
Chitosan	Animal — semisynthetic	Glucosamine N-acetyl glucosamine	Linear	Positive

As reflected by the steadily increasing number of scientific reports (Fig. 1), over the past 10 years these biomaterials have been extensively studied for biomedical applications in many different forms. These include the formulation of films, hydrogels, beads, or micro- and nanostructured carriers, alone or in combination with other polymeric or lipidic materials.[1–3] Within the specific field of nanomedicine and nanocarrier mediated drug delivery, chitosan (CS) and hyaluronic acid (HA) are two polysaccharides that have emerged as the most prominent candidates. At first glance, CS and HA may seem very different: CS is produced by deacetylation of chitin, a polysaccharide obtained from crustacean shells, and is polycationic in nature. On the other hand, HA is a polyanionic biomacromolecule widely distributed in the human body as part of the extracellular matrix.

They both, however, share a few common features that explain their extraordinary success in biomedical research. Both materials are hydrophilic, structurally flexible affording possible interaction with other molecules through various types of non-covalent bonds. Another important characteristic is their high gelling capacity, which further expands the possibilities of their formulation as biomaterials for drug delivery. In addition, CS and HA are highly functional biomacromolecules due to their capacity to interact with the biological environment. These properties, together with their biocompatibility and biodegradability, have rendered these two polysaccharides very attractive for a number of different drug delivery strategies such as transmucosal drug delivery, gene therapy and anticancer therapy, among others. Within this chapter, we present an updated review of their biological and toxicological profile, followed by the current biomedical applications and future potential of CS and HA as functional materials in nanomedicine.

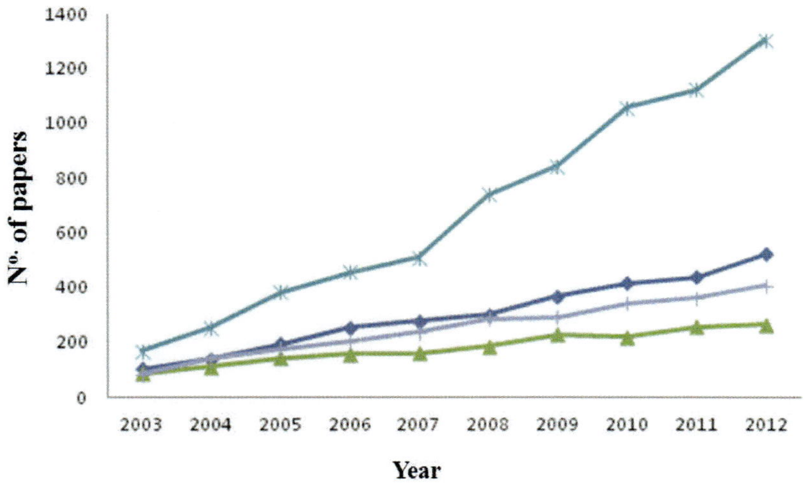

Fig. 1. Evolution of the number of scientific reports on the application of polysaccharides for drug delivery and tissue engineering; legend: (∗) CS, (+) HA, (♦) alginate, (▲) dextran.

2. Biopharmaceutical and Toxicological Profile

2.1. *Chitosan*

CS, derived from deacytalation of chitin, is a copolymer of β-(−14)-linked D-glucosamine and N-acetyl-D-glucosamine linked through glycosidic bonds. It has basic amine groups with a pK_a around 6.5, which results in various pH-dependent behaviors regarding solubility and polymer conformation. In addition to the protonation of its amines, the amount of acetylated monomers in CS and their distribution throughout the polymer backbone can also affect its solubility and properties in aqueous media. CS can partake in both electrostatic and hydrogen bonding interactions with other entities and molecules, and this is an advantage for the formation of CS nanocarriers.[4,5]

Not only are these the molecular features of interest in the formation of CS-based supramolecular assemblies, they are also critical features by which CS interacts with biological membranes. For example, CS is able to open epithelial tight junctions, an important action for their drug absorption promoting properties. Recent studies in the *in vitro* Caco-2 cell model have shown that this opening of cell junctions is related to CS electrostatic interaction with integrin-αV-$\beta 3$, which in turn induces surface clustering of integrin and reorganization of *F*(actin).[6,7] However, this could be just one of the transport enhancing mechanisms for CS. In fact, TEER based studies suggest that there may also exist a degree of transcellular transport for molecules in the presence of CS.[8]

In addition to its permeation properties, CS is considered as a mucoadhesive polymer. This is due to its ability to interact with the negatively charged mucin strands via electrostatic and hydrogen bonding interactions. Most delivery routes, including ocular, nasal and oral are protected by mucus, and thus the interest in building drug delivery systems around CS becomes obvious.[3]

This interest in CS as a pharmaceutical excipient and dietary supplement has led to a number of toxicological studies. CS toxicity in various cell models has been characterized by IC_{50} values between 0.2 and 2 $mg.mL^{-1}$, depending on the specific cell type, but also on CS deacetylation and molecular weight.[3,9,10] *In vivo* toxicity values have been determined for several administration routes: toxic levels upon sub-cutaneous administration into rabbits and dogs were in the 5–50 $mg.kg^{-1}.day^{-1}$ range; oral toxicity in rats was estimated at 1–15 $g.kg^{-1}.day^{-1}$.[11–13] This *in vivo* data seems to translate to humans where oral doses of 6.75 $g.day^{-1}$ have been administered without adverse effects.

Various enzymes present in human mucosas and other physiological fluids participate in CS biodegradation: lysozyme, di-N-acetylchitobiase, chitiotriosidase and N-acetyl-β-D-glucosaminidase.[9,10,14,15] While CS is not currently listed as an excipient in any FDA approved formulations, it is included in the FDA GRAS list. CS is also used as a dietary supplement to reduce fat adsorption and has been used in wound dressings,[16] indicating that a good understanding of its topical and oral tolerance exists.

The above toxicological data were obtained from CS in solution or in bulk form, which although indicative, cannot be reliably extrapolated to colloidal forms of CS. However, some studies have shown similar cytotoxicity in cell cultures with CS nanocarriers as with CS solutions.[17,18] These results suggest that the safety profile of CS in colloidal form may not be radically different from that of CS solutions.

The described properties have led to intensive study of CS as a biomaterial in a wide range of applications. In this chapter, however, we focus on CS drug nanocarriers, their preparation and their promise in the transmucosal delivery of biological macromolecules such as DNA, antigens and peptides.

2.2. *Hyaluronic acid*

HA is an endogenous biomolecule found in all tissues and body fluids of vertebrates. It is a structural component of the extracellular matrix, sinovial fluids, and soft tissue, which connects and support structures and organs of the body. As in the case of CS, HA is composed by D-glucuronic acid and N-acetylglucosamine linked through alternating $\beta(1 \rightarrow 4)$ and $\beta(1 \rightarrow 3)$ glycosidic bonds (Fig. 2). Native HA has a molecular weight (Mw) of

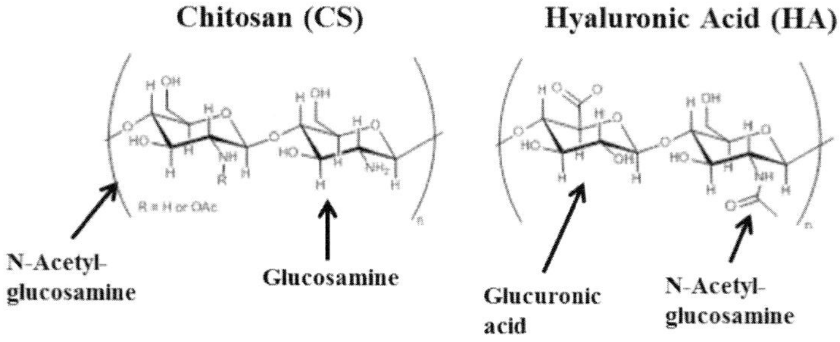

Fig. 2. Chemical structure of CS and HA.

10000–100000 kDa, however, it can be obtained from commercial sources in a large range of Mws. Within this broad range, HA of Mw up to 10 kDa is normally named as oligomeric HA[19] and HA of Mw up to 45 kDa[20,21] is generally considered as LMw–HA.

As shown in Fig. 2, the bulky hydroxyl, carboxylate and amide groups are fixed in an equatorial position due to steric repulsion in HA. When placed in physiological solutions HA folds into a random coil structure, the resulting conformation of HA in solution produces hydrophobic patches consisting of the axial hydrogens in the polysaccharide backbone.[22] These regions are also considered to play a part in the mucoadhesivity of HA.[23]

In addition to the described interactions, HA is involved in several biological functions mediated by its interaction with specific proteins. For example, its role as a structural tissue component, HA binds to proteoglycans, thus forming complexes with lubricant and viscoelastic properties.[20,24] On the other hand, HA interacts with a number of protein cell receptors, this interaction being critical for HA degradation and elimination from the body.

The best known HA receptor, CD44, is a transmembrane glycoprotein found in lymphocytes, epithelial and endothelial cells, ocular tissue (conjunctival cells), among others. This protein is also overexpressed in many cancer cells participating in angiogenesis, tumor invasion and metastasis.[25,26] Specific HA-CD44 binding is involved in several cellular functions such as cell differentiation, proliferation and migration and contributes as well to HA degradation.[26–28]).

The binding affinity of HA to CD44 increases with the HA Mw where 6–10 oligosaccharide units are considered as the minimum amount for effective receptor binding. Paradoxically, the higher affinity of HMw-HA toward the CD44 receptor does not necessarily lead to its enhanced cellular internalization. Due to the cluster nature of HA-CD44 binding, the increased number of linkages between receptor and ligand may lead to the saturation of the receptor incapacitating it for further binding and internalization of more HA.[29] CD44 mediated HA degradation is controlled by two types of hyaluronidases: Hyal-1 and Hyal-2. The degradation process starts extracellularly with Hyal-2 that is only active in presence of CD44. Once HA is bound to CD44, the degradation is predominantly mediated by Hyal-1.[26]

One mechanism for the elimination of HA from blood is via the HA Receptor for Endocytosis (HARE).[26,30] This receptor is located mostly in sinusoidal endothelial cells of liver and spleen and its affinity for HA is Mw-dependent. For example, Jadin *et al.*[31] have shown that upon intravenous (*i.v.*) administration, high Mw HA (1.2 MDa) was accumulated in liver and kidney up to a much greater extent than low Mw HA (<160 kDa). Taking this into account, some authors have indicated the possibility to saturate the HARE receptor with a specific compound (e.g. chondroitin sulfate with high HARE affinity and no CD44 affinity) in order to reduce the elimination rate of HA. This strategy might be of interest for the use of HA as a targeting polymer carrier.[32,33]

Another receptor of interest that participates in the tissue elimination of HA is the Lymphatic Vessel Endothelial Receptor 1 (LYVE-1), mainly located in lymph vessel endothelium. The interaction with this receptor is responsible for the continuous transit of HA from tissues into the lymphatic system, which is the main route of degradation of tissue-derived HA. In fact, the capacity of the lymphatic system to extract HA is known to be up to 90%.[34–37] The binding mechanism of HA-LYVE-1 is similar to that of HA-CD44 (both share similar domains).[36] Although this HA receptor has not yet been exploited for the targeting of HA-based nanocarriers, targeting the lymphatic system with HA may offer therapeutic opportunities in several pathologies such as inflammation, autoimmunity and also for metastatic cancer dissemination via lymph nodes, since LYVE-1 has recently been related to tumor lymphangiogenesis.

TSG-6 is another HA receptor, found in synovial fluids and also detected in the serum of patients with inflammatory or autoimmune diseases. Its up-regulation is related with inflammation processes like arthritis and tissue remodeling,[38,39] in which TSG-6 and hyaluronan binding suppresses cartilage degeneration and normalize synovial fluids.[20] A schematic overview of the receptors involved HA-binding and their main physiological role is presented in Fig. 3.

HA, being both non-toxic and biodegradable, is considered to be a biocompatible material with a wide number of clinical applications and commercial products for topical, ophthalmic, nasal, pulmonary and also parenteral administration. Toxicological studies were performed in order to evaluate the toxicity, immunogenic response, genotoxicity, neurotoxicity, carcinogenicity

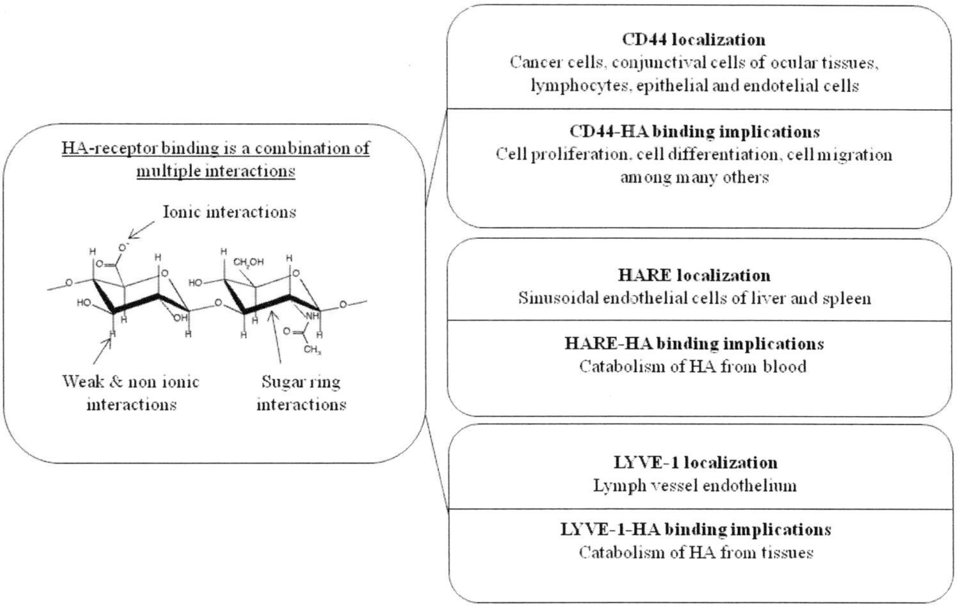

Fig. 3. Schematic overview of the receptors involved in HA-binding and their main physiological role.

or effects on reproductive and organogenesis in a wide range of animal species and with different exposure routes (oral, peritoneal, inhalation, topical, intramuscular etc.). In all cases high doses of HA did not induce any immunogenic or adverse toxicological effects either in the short or long term.[40] Although it is important to mention that recent studies have suggested the potential immunogenic properties of LMw-HA.[41,42]

Because of its interesting physicochemical and biological properties, both HA and CS are regarded as very interesting materials for a wide variety of biomedical applications. Both are currently being used in wound healing and skin care in different forms. Additionally, HA is being employed in intra-articular injections for osteoarthritis treatment, and as eye-drops for dry-eye treatment, whereas CS is being evaluated for this use in clinical trials. In this chapter, however, we focus on CS- and HA-based drug nanocarriers and their potential for transmucosal drug delivery, anticancer-drug targeting and gene therapy.

3. Types of CS and HA-based Nanocarriers

Over the last five years, the reactivity of HA and CS has inspired the synthesis of polymer-drug conjugates,[33,43–47] amphiphilic copolymers[21,48–64] that have been used to form micelles, and phospholipid linked HA to form hyaluronized liposomes.[65–71] For the purpose of simplicity within this chapter we will focus on the description of polymer-based nanostructures, namely micelles, liposomes, nanoparticles and nanocapsules. Schematic representations of these structures are shown in Figure 4. Some specific references covering these nanostructures and their applications are presented in Tables 2 and 3. Table 4 outlines and summarises the properties and biomedical applications of both CS and HA.

Table 2. Overview of HA-nanocarriers compositions and their main applications.

Description	Application	Refs.
HA micelles		
HA-cholanic acid copolymer	Antitumoral drugs/imaging to CD44 tumor cells	51, 60 and 62
HA-cholanic acid–PEG copolymer	Antitumoral drugs/imaging to CD44 tumor cells	50, 52, 56 and 57
HA-cholanic acid–PEG-aminopropylmethacrylamide copolymer	Antitumoral drugs/imaging to CD44 tumor cells	48
HA-cholanic acid–PEG and without PEG copolymer mineralized with Ca^{2+} y PO_4^{3-}	Antitumoral drugs/imaging to CD44 tumor cells	49 and 55
HA-ceramide copolymer and Pluronic® P85	Antitumoral drugs/imaging to CD44 tumor cells	53 and 58
HA-polybenzyl glutamate copolymer	Antitumoral drugs/imaging to CD44 tumor cells	59
HA-spermine-amines copolymer	Gene therapy to CD44 tumor cells	21 and 54
HA nanoparticles		
HA, CS or CS–PEG & TPP	Gene therapy through ocular mucosal barriers to CD44 cells	76–78 and 106
HA & CS	Gene therapy to CD44 tumor cells	61
HA, trimethyl CS & TPP with and without PEG	Antigen through nasal mucosal barrier	107

(*Continued*)

(*Continued*)

HA, adipic acid dihydrazide and TweenR 80	Peptide through oral mucosal barrier	108
HA and adipic acid dihydrazide	Visco supplementation of synovial fluid for osteoarthritis treatment	109
HA & CS-ammonium and CS-thiol derivatives	Drug delivery through oral mucosal barrier	110
HA:CS or CS derivative	Gene therapy through oral mucosal barriers	111
HA liposomes		
HA covalently linked to ethanolamine derived phospholipid	Antitumoral drugs to CD44 tumor cells	71
HA covalently linked to ethanolamine derived phospholipid	Antitumoral drugs to CD44 tumor cells	69
HA covalently linked to ethanolamine derived phospholipid	Gene therapy to CD44 tumor cells	65
HA covalently linked to stearylamine	Drug delivery to HARE endothelial liver cells	66
HA covalently linked to ethanolamine derived phospholipid	Antitumoral drugs to CD44 tumor cells	67
HA covalently linked to ethanolamine derived phospholipid	Gene therapy to CD44 tumor cells	68
HA electrostatically coated onto cationic liposomes	Imaging agents to CD44 tumor cells	112
HA electrostatically coated onto cationic liposomes	Antitumoral drugs to CD44 tumor cells	113
HA nanocapsules		
HA, soybean core and surfactants	Antitumoral drugs to CD44 tumor cells	104
HA, Miglyol core and surfactants	Antitumoral drugs to CD44 tumor cells	100
HA & hydroxiethyl starch, aqueous core and surfactants	Wound care	114
HA, polycaprolactone core with surfactants	Drug delivery through ocular mucosal barriers	115

Table 3. Overview of CS-nanocarriers compositions and their main applications.

Description	Application	Ref.
CS Micelles		
Doxorubicin conjugated and hydrophobized glycol CS	Antitumoral	47
Cholanic acid modified glycol CS	Antitumoral	64
Targeted (transferrine) self-assembled GCPQ chitosan micelles	Antitumoral	116
Self-assembled GCPQ CS micelles	Oligonucleotide	75
Self-assembled GCPQ CS micelles	Oral peptide	73
CS nanoparticles		
TMC based nanoparticles	Nasal antigen	117–123
CS based nanoparticles	Nasal antigen	124–127
PEGylated CS/TPP	Nasal oligonucleotide	128
bfFP-CS/pDNA, dendritic cell targeting	Nasal oligonucleotide	129
Mannosylated CS/pDNA	Nasal oligonucleotide	130
CS/TPP	Nasal peptide	131
CS/TPP/cyclodextrin	Nasal peptide	92
CS/TPP/alginate	Nasal peptide	83
Thiolated CS, binary polyelectrolyte complex	Nasal peptide	132
CS-PVA-PLA interpolymer complex	Nasal peptide	133
TMC-CS, binary polyelectrolyte complex	Nasal peptide	134
CS/TPP nanogel particles	Nasal peptide	131
CS-pDNA	Ocular oligonucleotide	135 and 136
TMC/pDNA	Oral oligonucleotide	137
Oleoyl-TMC/pDNA/TPP	Oral oligonucleotide	138
Thiolated-TMC/pDNA	Oral oligonucleotides	139
CS-insulin binary polyelectrolyte complex	Oral peptide	140
CS-insulin dispersed in oil or surfactant.	Oral peptide	141 and 142
CS-alginate Ca^{2+} crosslinked nanoparticles	Oral peptide	80 and 143
CS-albumin coated alginate-dextran nanoparticles	Oral peptide	144–147
Binary polyelectrolyte complex with CS targeting peptide conjugate	Oral peptide	148
CS-poly-γ-glutamate nanoparticles	Oral peptide	85, 86, 89, 149

(*Continued*)

(*Continued*)

Description	Application	Ref.
CS Nanocapsules		
CS coated emulsion	Nasal antigen	127
Miglyol-lecithin-CS coated nanocarriers	Nasal peptide	150
CS/poly-L-lysine coated caprolactone nanocapsules	Ocular	99
Alginate-CS coated W/O/W nanoemulsion	Oral peptide	105
Miglyol-lecithin-CS coated nanocarriers.	Oral peptide	151–153

Table 4. Outline of CS & HA properties and their main biomedical applications.

Property	CS	HA	Applications
Positive charge	√	—	• Ionic interactions with biological surfaces • Condensation of oligonucleotides
Viscoelasticity	√	√	• Dermal applications • Ocular applications • Intra-articular applications
Penetration enhancement	√	—	• Mucosal drug delivery
Mucoadhesion	√	√	• Mucosal drug delivery
Targeting capacity	—	√	• Anticancer drug delivery to CD44 cancer cells • Drug delivery to HARE epithelial liver cells

3.1. *HA and CS-based micelles*

Hydrophobic moieties, e.g. cholanic acid or ceramide, can be conjugated to HA's carboxylic acid groups through simple reactions to yield amphiphillic copolymers. The resulting polymers can then self-assemble into micelles in physiological conditions.[21,48–62] Similarly, CS has been chemically modified forming, for example, quaternary ammonium palmitoyl glycochitosan (GCPQ), which induces the formation of micelles upon dispersion in water.[72,73] The hydrophobic part of copolymer is oriented inwards, forming a core suited for the encapsulation of hydrophobic drugs.[48–53,55–60,62,72] In the case of CS, the positive charge has also led to the efficient entrapment of negatively charged macromolecules, i.e. peptides and nucleic acid-based molecules.[74,75] An illustration of this type of structures is presented in Fig. 4.

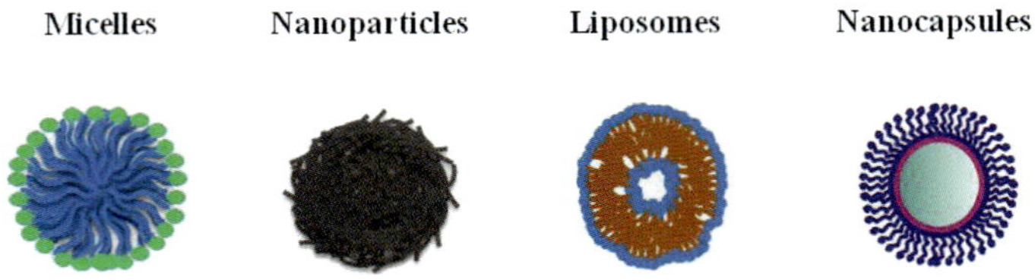

Fig. 4. Schematic representation of main CS and HA-based nanocarriers in the literature.

3.2. *HA and CS-based nanoparticles*

Electrostatically formed hybrid HA–CS nanoparticles can be achieved while avoiding the use of organic solvents by simply ionic gelation using a crosslinking agent like tripolyphosphate or adipic acid dihydrazide. This method takes advantage of the gelling properties of polysaccharides and permits the association of delicate macromolecules without the use of organic solvents or high energy sources.[76–79] In the case of CS, other polymers besides hyaluronate, i.e. alginate and poly-γ-glutamate have been used in the formation of these nanoparticles.[61,80–89] Such techniques also permit the incorporation of additional excipients to modify the properties of CS nanoparticles, for example, cyclodextrins.[90–94]

Given that solid matrix nanoparticles are formed and stabilized mainly through ion pair interactions it is important that the pH is controlled during preparation, as alterations in pH will change the degree and state of ionization of the polymers and macromolecules.

Overall, these different nanoparticle types have been designed with the aim of taking advantage of their bioadhesive properties to improve the interactions and contact time with mucosal barriers. A schematic illustration of this type of nanocarrier is shown in Fig. 4.

3.3. *HA and CS-coated liposomes*

The reactivity of HA also allows its conjugation with chemically activated phospholipids with molecules like ethanolamine or stearylamine to form HA liposomes.[41,65–67,69,71] This so-called hyaluronization can be achieved prior or post liposome formation. In addition to this strategy, liposomes can be readily coated either with CS, or HA or their derivatives based upon the ion-pair formation between the lipid's groups and the polysaccharide's groups.[95–98] Therefore, the formation of the HA or CS-coated liposomes can be easily achieved upon incubation of liposomes in an aqueous solution of the polysaccharide. The resulting system enables the association of hydrophilic molecules (in the aqueous core) or hydrophobic drugs (within the lipid bilayer). An illustration of this type of structures is presented in Fig. 4.

3.4. *HA and CS nanocapsules*

These are vesicular systems with an oily core stabilized by surfactants and a surrounding polymer shell. The formation of the HA or CS polymer shell is mediated by hydrophobic and ionic interactions of the polysaccharides with the surfactants and can be easily achieved using simple emulsification methods, such as the solvent displacement technique.[99,100] Thus, the electrostatic interaction of CS with the oily core has been mediated by negatively charged phospholipids,[101–103] whereas that of HA has been achieved by the incorporation of a cationic surfactant such as cetyl trimethylammoium bromide or bezalkonium chloride.[100,104] The oily core provides high capacity for the encapsulation of hydrophobic cancer drugs while the polymer coating barrier confers capacity to control drug release and optionally allows the association of hydrophilic substances (e.g. nucleic acids). Potential also exists for the formation of nanocapsules via solvent-free methods as is the case with a described self-double-emulsifying system with an alginate-CS coating.[105] An illustration of this type of structures is presented in Fig. 4.

4. Biomedical Applications of HA and CS-based Nanocarriers

CS and HA-based nanocarriers have been proposed for a wide range of biopharmaceutical applications. Because they share some interesting biological properties, notably their bioadhesiveness, they have been widely studied regarding their ability to overcome cellular and epithelial barriers. However, while CS-based nanocarriers have been mainly used for overcoming mucosal barriers, those based on HA have mainly been conceived for targeting to specific cell populations, i.e. cancer cells. On the other hand, due to its positive charge, CS has also attracted a great deal of attention in the gene delivery field. In this section, we will summarize the advances made toward those specific aims.

4.1. *Transmucosal drug/antigen delivery*

The bioadhesive properties of HA have been studied in several cases in combination with CS. The aim has been to form nanocarriers with enhanced residence time and contact[77,154–157] with different mucosal surfaces, achieving higher doses of the associated bioactive compound crossing the mucosal barrier.

4.1.1. *Ocular drug delivery*

CS-based nanocarriers have also been explored as a way to increase the residence time and the penetration of drug either into the corneal or the conjunctival epithelium. The first evidence of this potential was shown with CS-nanocapsules, which exhibited the ability to enter the corneal epithelium and favor the retention of drugs, i.e. indomethacin and cyclosporin A.[99,158] A comparative study of fluorescently marked CS and PEG-CS coated nanocapsules demonstrated that while unmodified CS enhanced the retention in the mucosal layer, the PEGylated derivative accelerated the transport of the nanocarrier through the epithelium.[159]

Similarly, CS nanoparticles and CS-lipid hybrid nanostructures were also found to interact with ocular mucosa[160] showing good ocular tolerance and cellular uptake.[161–163] Other specific examples are indicated in Sec. 4.3.

Despite the extensive use of HA solutions in the form of artificial tears and the known presence of CD44 receptors in the conjunctival epithelial cells, there has been only minor activity of the use of HA-based nanocarriers for ocular drug delivery. One of the few examples of HA nanocarriers in ocular drug delivery are HA coated polycaprolactone nanocapsules reported by Barbault–Foucher.[115,164] Ocular administration of cyclosporine A loaded nanocapsules yielded a significant increase in the conjunctival and corneal concentration of cyclosporine A when compared to a castor oil solution. Furthermore, cyclosporine A was present for up to 24 h post-administration in the case of the nanocapsules. This corresponded to a lower cyclosporine A concentration in tear fluid for the nanocapsules than the castor oil control. Both observations show that the ability of the nanocarrier to adhere to the ocular tissues is imperative in the resulting efficacy of the formulations.

Of particular interest is the combination of both polysaccharides, HA and CS, in the form of nanoparticles. *In vivo* studies performed with fluorescently labeled HA/CS nanoparticles in rabbit ocular mucosa showed the ability of these nanocarriers to efficiently interact with and penetrate the corneal and conjunctival cells.[76–78] These nanoparticles have been widely tested as non-viral gene delivery systems in ocular tissues improving transfection efficiency by facilitating interaction with conjunctival CD44 receptors for the treatment of several diseases such as dry eye, immune-mediated diseases and ocular allergy.

4.1.2. *Nasal drug/antigen delivery*

The nasal route is of particular interest in the development of needle-free vaccines. Also, the increasing prevalence of peptide drugs in research pipelines and marketed products is behind an ever increasing interest in obtaining efficacious transmucosal delivery systems.

While HA-based nanocarriers have barely been studied for nasal drug delivery, the previously described properties of CS have led to its extensive study in the nasal delivery of drugs and antigens. Early work with insulin-loaded CS nanoparticles demonstrated the potential of CS nanocarriers in nasal insulin delivery.[131] This base system led to the development of hybrid nanoparticles consisting of CS and caboxymethyl or sulfobutyl β-CDs,[165] CS and alginate[83] or CS and polylactic acid,[133] which significantly enhanced insulin response upon nasal administration. Other examples of successful nasal administration of insulin, in terms of pharmacological response, include gold nanoparticles coated with CS.[166]

On the other hand, nanoparticles formed with chemically modified CS have proven successful with both TMC and thiolated CS derivatives.[132,134] For example, in a study it was found that thiolated CS nanoparticles produced an improvement in hypoglyceamia compared to the non-thiolated analog in healthy rats.[132] Similarly, the degree of quaternization of TMC was observed to positively affect the absorption of insulin associated to TMC based nanoparticles in rats.[134]

Further to the solid matrix nanoparticle systems, salmon calcitonin (sCT) loaded CS nanocapsules produced a hypocalcaemic effect after nasal administration, where the CS coating resulted in an enhanced effect when compared to the control nanoemulsion.[150]

Beyond the enhanced absorption of therapeutic peptides, the potential of CS-based nanocarriers as antigen delivery systems is well established. Following the initial studies demonstrating significant transport of tetanus toxoid across the nasal mucosa when associated to CS nanoparticles,[126,167] a range of CS-based nanocarriers have emerged. For example, CS nanocapsules containing the antigen associated to the CS shell have been reported as useful carriers for the nasal delivery of the model antigen, ovalbumin[127] and also the recombinant hepatitis B surface antigen (rHBsAg).[168–169]

Trimethyl CS based nanocarriers have also yielded promising results regarding nasal antigen delivery. Ovalbumin conjugated to TMC nanocarriers induced a marked increase in the IgG response when compared to a TMC ovalbumin solution.[170–172] This positive *in vivo* performance of TMC nanocarries has been corroborated with other antigens, namely, influenza and tetanus toxoid.[117,122,123,173] Nevertheless, a comparative study of CS and TMC nanoparticles loaded with rHBsAg showed that both were similarly efficacious after either intranasal or intramuscular administration.[118] Therefore, the potential advantage of TMC nanoparticles versus CS nanoparticles for antigen delivery remains to be elucidated.

Other immuno stimulating molecules have also been co-associated to CS nanocarriers, where further improvements in immune response have been observed. This was the case when CpG ODN was co-encapsulated with ovalbumin in CS nanoparticles. Interestingly, this combination led to an improved cellular immune response as confirmed by the increased ovalbumin specific T-cell response in the spleen.[121] More recently, the same group evaluated a battery of immunostimulants, among which LPS and MDP were found to be the most efficacious in combination with TMC nanocarriers.[120]

Finally, HA–CS nanoparticles have been studied for nasal vaccination, however this preliminary study in mice did not demonstrate an enhanced immune response when compared to the plain antigen.[107]

4.1.3. *Oral administration*

The results on the potential of nanocarriers for nasal administration of peptides can serve as a stepping stone toward developing successful nanocarriers for oral peptide delivery. However, this delivery route does pose some additional complications such as the high dilution of the nanocarriers in the harsh gastro-intestinal environment. Therefore, it is imperative that oral delivery platforms offer protection against proteolysis in addition to facilitating mucosal transport.

The main focus of attention regarding oral peptide delivery using nanocarriers has been diabetes treatments, chiefly with insulin, although GLP-1 analogs are also of interest. This has been driven by the current and projected socioeconomic impact of diabetes. Further to showing promising PK/PD results in animal studies CS nanocarriers have also yielded a bank of information regarding the behavior of colloidal nanocarriers administered orally.

Insulin-loaded CS-alginate nanoparticles showed a clear hypoglycemic effect upon oral administration to rats. A feature of interest of these nanocarriers was that insulin release was found to be pH-dependent: i.e. inhibited at gastric pH and sustained at intestinal pH.[80,81,143] The same authors have more recently presented a second-generation of nanoparticles with a dextran-alginate core coated with CS, and with an additional albumin coating to protect the nanoparticle from enzymatic degradation.[144–147] *In vitro* and *ex vivo* models were used to elucidate the mechanism of absorption of the nanocarriers.[147] Overall, the results indicated that the epithelial permeation of Caco-2, Caco-2/HT29MTX cell models and excised rat intestinal tissue increased in the presence of nanoparticles. Fluorescent labeling of alginate and insulin in *ex vivo* experiments indicated that both alginate and insulin were co-localized in enterocytes.

Another approach that has been explored relies on the formation of CS-insulin or CS-sodium lauryl sulphate insulin nanocomplexes.[140–142,174] A clear hypoglycemic effect was observed upon administration of these formulations, this effect being more pronounced in the case of CS-Insulin nanocomplexes dispersed in an oil.[141–142] In both architectures, either the oily dispersion or the presence of sodium lauryl sulphate would afford protection to insulin from pepsin degradation.

Poly-γ-glutamate has been used to form interpolymer crosslinked CS nanoparticles for oral peptide delivery. When administered to rats in enteric coated capsules these prototypes showed *in vivo* data among the best reported in terms of oral insulin bioavailabilities (15% to 20%).[85,86,175] Similar results were obtained for the GLP-1 analog Exendin-4 with a bioavailability value of 14%.[176] In this study it was found that the presence of Fe^{3+} increased greatly the encapsulation efficiency of the peptide due to the coordination between Fe^{3+}, CS and Exendin-4. Further to the high bioavailability, additional SPECT studies provided important information regarding the mechanism of action of CS nanocarriers. Namely, the results indicated that while insulin was absorbed into the villi, CS remained in the mucosa.

Chemical derivatization of CS offers an interesting pathway to control the behavior of nanocarriers *in vivo*. The effect of incorporating targeting ligands onto trimethylated CS (TMC)-insulin nanoparticles was recently tested.[148] Confocal microscopy of Caco-2/HT29MTX monolayers after incubation with the fluorescently labeled nanoparticles showed an increased cellular uptake of the nanoparticles targeted to the goblet cells when compared to the ligand free TMC-insulin nanoparticles. Indeed, this translated to a moderate improvement in relative bioavailability from 3.69% to 5.66% due to the conjugated ligand.

Besides offering the potential for improvements in bioavailibity chemical modification of CS can also significantly alter the uptake pathway of the polymer. One of the strongest examples is the GCPQ CS polymer that merits special mention here[73] GCPQ is an amidoylalkyl glycochitosan derivative whose amphiphatic character induces self-assembly into micelles. Oral administration of Leu-enkephalin-loaded GCPQ micelles resulted not only in an increased absorption of the peptide but also in its transport across the blood–brain barrier.

In addition to the various solid matrix nanocarriers described previously, CS coated lipid core nanocarriers have also been studied for oral peptide delivery. CS and PEG-CS nanocapsules have been studied for oral sCT delivery. Interestingly, these nanocapsules gave rise to an enhanced and extended hypocalcaemic effect of at least 24 h (duration of experiment).[17,151–153] In a subsequent study, CS was PEGylated via amidation of the primary amine (1% and 0.5% degrees of PEGylation) and its effect on the pharmacological response to sCT-loaded CS nanocapsules was evaluated.[152] The results showed that a high PEGylation degree affected negatively the efficacy of the nanocarriers. In all the examples reported in this series of studies, CS was found to be fundamental to the eventual pharmacological activity of the formulations. A more recent approach has been to produce nanocapsules with a mixed CS-alginate polymer shell. This composition has been used to form W/O/W nanocapsules for oral insulin delivery, and resulted in a hypoglycaemic nadir of ~30% at 7 h post administration.[105]

Regarding oral administration with HA-based nanocarriers; the work of Zambito[110] uses the combination of CS derivatives and HA to form mucoadhesive nanoparticles for oral drug delivery. CS derivatives consist of quaternary and primary ammonium pendant moieties available for covalent attachment of thiol-bearing compounds. Quaternary ammonium has the ability to open epitelial tight junctions whilst thiols can form disulphide bonds with epithelial mucus increasing the mucoadhesion capacity of the system. Nanoparticles were found to be non-toxic in endothelial progenitor cells. Mucoadhesion studies with fluorescently labeled nanoparticles using rat intestinal mucosa, showed a high absorption of nanoparticles and furthermore that this absorption was concentration independent. Moreover, the thiolated nanoparticles proved to be more mucoadhesive than their non-thiolated analogs demonstrating the potential advantages of thiol containing nanoparticles for transmucosal delivery.

Again hybrid HA/CS nanoparticles have also been tested to deliver siRNA by the oral route.[111] Trimethyl CS or thiolated TMC derivatives were used in order to improve stability of CS against disolution at acidic pH. Stability studies in gastrointestinal fluids showed that the non-HA containing nanoparticles' size increased to a much larger extent than the HA/CS nanoparticles. Whilst naked siRNA was rapidly degraded in endogenous nucleases supplemented medium, HA/CS nanoparticles were able to partially protect siRNA from nuclease degradation. The degradation of a fraction of the siRNA loaded into the nanoparticles can be attributed to polysaccharide degradation; more specifically to the presence of chitosanase in colonic mucosa. Permeation studies with fluorescent labeled nanoparticles using rat ileum showed significant enhancement of siRNA transport, this was confirmed by biodistribution studies in which nanoparticle concentration in the intestine decreased with time while increasing in plasma, liver and spleen, demonstrating that the nanocarrier is capable of promoting adsorption and subsequent entry into systemic circulation and the reticuloendothelial system.

Finally, Han *et al.* proposed HA nanoparticles for insulin delivery. These nanoparticles protected insulin from enzymatic degradation by gastrointenstinal enzymes (97% free insulin degradation versus 31% degradation inside nanoparticles). Transepithelial transport capacity studied by the Caco-2 monolayer cell model and in rat small intestine showed higher efficiency in transport compared with insulin solution. A reduction in blood glucose was observed after oral administration of these nanoparticles to diabetic rats helped by the nanocarrier's aforementioned ability to enhance epithelial permeation and protect insulin from degradation.[108]

The systems discussed above show that a wide variety of CS based nanocarriers are effective in facilitating the transmucosal delivery of proteins and peptides, where CS plays a fundamental role in their action. Besides showing good PK/PD data in animal models, which in some cases is on the top end of the reported performances for oral protein delivery, a better understanding of the behavior of CS in the absorption of the nanocarriers is beginning to emerge. In the case of nanoparticulate CS, the current evidence suggests that the polymer may not translocate into the serosal side of the villi. Instead, it tends to accumulate in the mucosa adjacent to the epithelia. This behavior, from a toxicological standpoint, is preferred given that internalization of the nanocarrier or its components may imply a more complicated safety evaluation.

4.2. *Anticancer drug delivery*

Opposite to what was described for transmucosal drug delivery, CS-based nanocarriers have received little attention for anticancer drug delivery; Refs. 177 and 178 from Alonso's group being some of the limited examples available. The main goal in the delivery of anticancer drugs is to target them to the cancer cells, this enhancing their efficacy and reducing their toxicity. For this specific purpose, in general, the delivery carrier needs to be administered intravenously and the positive charge of CS represents a hurdle from the point of view of the biological stability of the nanocarrier. To overcome this limitation some authors have proposed the use of chemically modified CS such as hydrophobic derivatives,[63,64] or PEGylation.[178] Doxorubicin has also been conjugated to glycol CS in nanoparticles.[47]

Both specificity and cellular uptake can be improved by incorporating targeting ligands such as folate or transferrine. The previously described GCPQ glycol CS derivative has been further modified to include targeting ligands.[116]

Here, the authors studied the efficacy of both transferrine and glucosamine, where the uptake and cytotoxicity of docetaxel loaded vesicles were improved by conjugating the transferrine ligand. Folate-CS has also been employed to take advantage of folate receptor mediated endocytosis in cancer cells.[63,179] Conjugation of OX26 to CS-PEG based nanoparticles facilitated brain delivery of an antitumoral[180] and dendritic cell targeting for cancer immunotherapy.[181]

In contrast, the specific ability of HA to target CD44 overexpressed in many tumors and its shielding effect make HA-based nanocarriers particularly attractive for developing more effective cancer therapies.[31,62,182,183] As a consequence of this capacity, a number of different HA based nanocarriers have been developed to deliver antitumoral drugs, gene therapy or imaging contrast agents to many types of cancer like breast, lung, colon, carcinoma, melanoma, liver, among others.

A critical factor to consider for the design of nanocarriers with optimal CD44 affinity is the selection of HA Mw. As detailed in Sec. 1, the affinity of HA for CD44 increases with Mw of HA. In accordance with this, studies undertaken with HA-conjugated liposomes of varying HA Mw showed that conjugation with higher Mw HA resulted in stronger CD44 binding.[41] This observation is also supported by studies with similar liposomes also linked to HA, where increased uptake was achieved with higher Mw in a CD44 cancer cell line.[67] However, recent studies suggest that the cluster nature of HA CD44 binding results in a limit to the number of linkages that can be formed before the CD44 receptor is saturated and incapacitated preventing further internalization.[29] This is in agreement with Wojcicki *et al.*[65] where the transfection efficiency of DNA loaded HA-liposomes with two different HA concentrations (10% and 15%) were studied. Here, the authors found that the lower concentration of HA gave higher levels of transfection.

Besides CD44's targeting potential, the hydrophilicity of HA can diminish carrier opsonization by steric repulsion increasing nanocarrier circulation time after *i.v.*,[69,70,184] in turn promoting nanocarrier accumulation in tumors through passive mechanisms.[182,183]

Micelles composed of amphiphilic copolymers of HA and hydrophobic molecules are the most studied systems for tumor targeting. Choi *et al.* carried out extensive studies with micelles of HA conjugated with the bile salt, 5β-cholanic acid. HA-cholanic acid micelles were loaded with antitumoral drugs, e.g. irinotecan, chlorine e6, doxorubicin, paclitaxel, camptothecin and image contrasts agents and were used for the targeting of carcinoma, breast, colon or liver cancer.[51,60,62] Tumor cells overexpressing CD44 could internalize these nanocarriers more readily than normal fibroblasts. Moreover, drug loaded HA-micelles exhibited higher cytotoxicity than the free drug, thus suggesting the role of the nanocarrier for increasing the drug internalization. Finally, following *i.v.* administration of fluorescently labeled HA-cholanic acid micelles in tumor bearing mice it was observed that micellar HA remained in the blood circulation for a much longer time than HA in solution. This prolonged circulation time produced a preferential accumulation of the micelles in tumoral tissue and consequently a significant increase in antitumoral efficacy, measured in terms or reduction of tumoral volume. Although a significant amount of nanocarriers were also detected in the liver. Similar results have also been obtained with other HA nanocarriers such as polybenzyl glutamate conjugated HA micelles.[59]

The hepatic accumulation of HA micelles was attributed to their uptake of the Mononuclear Phagocytic System (MPS) and also to their specific interaction with the HARE receptors. Based on these results, HA micelles were proposed for hepatic cancer imaging.[185] Nevertheless, hepatic accumulation may represent a major drawback for cancer therapies directed to organs. In order to reduce the hepatic uptake of HA-cholanic acid micelles, PEGylation of the HA backbone[50,52,56,57] has been studied. *In vitro* experiments showed a reduced internalization of PEGylated micelles, both in tumor cells and normal fibroblasts, most likely due to a shielding effect of PEG.

In vivo studies with breast and colon tumor bearing mice demonstrated a reduced hepatic accumulation and increased circulation time of the PEGylated micelles in comparison with the non-PEGylated analog. This difference in nanocarrier biodistribution and pharmacokinetics was associated to an improvement in the accumulation of the PEGylated nanocarrier in tumoral tissue. Although PEGylation may be of interest for inhibiting hepatic acumulation it may also reduce binding to CD44, thus reducing in part the benefits of HA based nanocarriers.

Additional strategies aimed at improving the behavior of HA micelles have been oriented to improve their stability and controlled release properties. One strategy relied on coating micelles with calcium and phosphate salts, termed by the authors as mineralization.[49,55] These minerals acted as a stabilizing barrier which inhibited drug release at pH 7.4. However, at the more acidic pH environment of tumor tissues, the dissolution of this mineral barrier facilitated the release of the encapsulated drug. Biodistribution studies with fluorescent labeled micelles showed significant enhancement of the tumor/liver accumulation ratio compared with non-mineralized micelles. Overall, the combination of these factors allows drug-loaded mineralized HA micelles to reduce tumor volume more efficiently than non-mineralized micelles or the free drug.

A further strategy to improve the stability and release properties of HA micelles has been the crosslinking of carboxylates of HA with aminopropylmethacrylamide.[48] The result was a slower drug release rate, without altering the affinity for the CD44 receptors. The *in vitro* studies in Hyal-1 supplemented medium showed that drug release can be achieved intracellularly (80% of drug released in these conditions). Fluorescent labeled crosslinked HA micelles also showed increased accumulation in tumor tissues and improved antitumoral efficacy.

Other authors have developed oligomeric HA micelles, PEGylated or not, conjugated to ceramide,[53,58] a cell membrane sphingolipid involved in apoptosis of cancer cells.[186] These micelles also contain Pluronic® 85, a surfactant with abilities to overcome multidrug resistance, a major cause in the lack of efficacy in many chemotherapeutic treatments. The resulting micelles were found to be stable to dilution and exhibited controlled release properties, partially attributed to Pluronic® 85. *In vitro* studies in tumoral cell lines confirmed that oligomeric HA-ceramide micelles presented a high CD44 affinity. Again, the control drug release and enhanced structural stability of these HA micelles resulted in a markedly preferential accumulation in tumor and increased efficacy compared with free drug for both PEGylated and non-PEGylated system.

HA-coated liposomes have also been investigated for targeting the CD44 receptor in cancer cells. In some instances, hyaluronized liposomes loaded with anticancer drugs, e.g. doxorubicin and mytomicin C, presented an increased cytotoxicity in cancer cell lines when compared with non-hyaluronized liposomes or free drug.[69,71] *In vivo* these drug-loaded liposomes exhibited a significant reduction in toxicity — assessed by loss of body weight — a prolongation of their circulation time and an increase in their tumor uptake as compared to the free drug. This favorable biodistribution pattern of HA liposomes resulted in a two-fold decrease in the tumor volume as compared to the free drug.

HA nanocapsules represent a new type of nanocarrier consisting of an oily core conveniently stabilized and surrounded by a HA shell, which are particularly attractive for the encapsulation of hydrophobic antitumoral drugs. The first report presented by Orzayun-Ampuero *et al.* showed the efficient internalization and increase of cytotoxicity of docetaxel-loaded HA nanocapsules as compared to the free drug.[100] A similar type of nanocarrier consisting of a solid lipid core and soybean oil[104] was used for the delivery of paclitaxel. Upon systemic administration to tumor bearing mice, paclitaxel-loaded HA nanocapsules exhibited longer circulation time, reduced toxicity (in terms of body weight loss), and enhanced efficacy as compared to the commercial paclitaxel formulation (Taxol®).

4.3. *Gene delivery*

Both CS and HA have been used as carriers in gene therapy. However, while CS is positively charged and, thus, able to condense oligonucleotide-based molecules, HA has to be associated to other positively charged molecules in order to be useful for gene therapy approaches. In some instances both, CS and HA have been included together in order to entrap oligonulceotides, such as DNA and siRNA.[76–78,106] These combinations are particularly interesting where CD44 targeting or mucosal delivery are involved; such as ocular delivery or cancer cell targeting. Entrapment of oligonucleotides inside solid matrix nanoparticles can protect them from nuclease degradation, further, the specific interaction with CD44 increase cellular uptake resulting in higher transfection efficiency.

HA has also been combined with other positively charged molecules for the targeting of gene therapy approaches to cancer cells. For example, Shen *et al.* developed siRNA-loaded HA micelles in which low Mw HA was conjugated with hydrophobic alkylamines and spermines.[21,54] Here the hydrophobic and cationic spermine formed a hydrophobic core within which the oligonucleotide was condensed, due to the polycationic nature of spermine, thus protecting the genetic material from enzymatic degradation. This resulted in improved transfection and gene silencing in cancer cells when compared to other established transfection agents like lipofectamine and polyethyleneimine.

In another study, Wojcicki *et al.*[65] developed HA liposomes with HA covalently linked to ethanolamine derived phospholipid in which plasmid DNA is electrostatically complexed to the lipid chains of the liposome. In this case the HA shell was found to protect DNA from nuclease degradation and the transfection studies in a cancer cell line showed a significant expression of the encoded protein (green fluorescent protein, GFP). Similarly, HA liposomes loaded with anti-telomerasa siRNA targeted to CD44 lung cancer cells resulted in an improved protection of siRNA from degradation and an higher transfection efficiency *in vitro*.[68]

In the case of CS, its great ability to complex genetic material (e.g. pDNA or siRNA) has led not only to a great amount of literature but also to the commercial development of a CS-based transfection agent. Early work carried out by Leong's research group [187–189] pioneered the concept of CS-mediated gene delivery and the attention to this concept has increased over the years.[3,190] Several studies have shown that the good *in vitro* observations can be translated into excellent *in vivo* transfection efficiencies.[136,191–193] The high performance of these systems has been linked to CS's capacity to enter cells and deliver the associated genetic material in a proper form intracellularly. CS is able to promote endosomal escape, aided by its buffering capacity, i.e. CS acting as a proton sponge.[189,194] A recent study suggests that CS's buffering capacity may be closer to that of polyethlyeneimine (a well established proton sponge) than originally thought.[195]

As with protein delivery, a wide variety of CS based vectors have been studied. The majority are based on direct complexation between CS and the nucleic acid,[187] although other sophisticated systems incorporating lipids or polymers, e.g. HA, have also been reported.[78,128,196–199]

CS nanocarriers have shown high transfection levels not only in HEK293 cell lines, [137,200,201] but also in HeLa, CaLu-3 and MDCK cell lines, known to be more challenging for transfection.[78,92,137,191,202–207] Supporting this high efficacy, siRNA-loaded CS nanocarriers have also been reported to reduce gene expression to similar levels to those obtained with lipofectaminee (~50%).[196,208]

Further to determining the transfection capacity of CS nanocarriers the current body of work gives some indications regarding the design features required for effective transfection. First, a reduced particle size and cationic surface charge are essential for transfection.[209] Various studies have highlighted the role played by CS Mw, where shorter polymer chains have proven to be more effective due to the easier decomplexation and release of pDNA after internalization.[61,203,210–213] Although CS chain length, particle size and surface charge appear to have clear correlations with transfection, the role of deacetylation degree cannot be correlated with the same clarity. Whereas, *in vitro* studies showed better results with highly deacetylated CS,[214,215]), better *in vivo* results (intramuscular administration) were found with a low degree of deacetylation.[215]

The colloidal stability of CS nanocarriers is of importance in obtaining effective transfection vectors. One of the more salient properties is the pH dependence of surface charge where the pH of physiological fluids reduces the degree of protonation, thus neutralizing surface charge and affecting stability. Although, this destabilization can be offset by the coating of CS nanocarriers with serum proteins,[216] some other strategies have been investigated. These include chemical modifications of CS such as quarternization of the amine groups, ensuring that the surface charge remains unaffected by pH,[201,202,213,217] PEG-CS conjugated polymers have also been reported.[128]

A further alternative has been the preparation branched CS derivatives with higher water solubility.[206,218,219] Both transfection efficiency and cell uptake were found to be improved by these modifications. Grafting of polymers, such as cell penetrating peptides (CPP) or PEG, onto a CS polymer backbone yielded some interesting observations regarding cellular uptake. For example, CPP such as TAT[220–222] and poly-arginines[223–229] have been employed to improve cellular uptake. It was also observed that poly-arginine grafted CS was internalized via a caveolae mechanism, while CS nanoparticles were taken up via a clathrin-mediated pathway.[227,228] Similarly, Reitan *et al.* studied the mechanisms behind the different transfection efficiencies observed for glycosylated oligomeric CS and oligomeric CS.[230] Here glycosylated CS internalized DNA to a higher degree and over a longer period than oligomeric CS.

Various systems have been reported with the aim of further improving endosomal escape and cytosol-specific release characteristic of CS based systems. Thiolation of CS is one such strategy where S-S bridges are reduced in the intracellular environment thus inducing site-specific release in the cytosol.[231,232] CS has also been modified with polyethyleneimine, imidazole and histidines to further improve its endosomolytic activity.[206,233,234] Both temperature dependence and biodegradation have also been exploited as mechanisms for improving intracellular release of gene medicines.[198,207]

An alternative to the aforementioned chemical modifications is the formation of gene-loaded CS nanoparticles prepared via ionotropic gelation (e.g. with TPP). As indicated above, these nanoparticles can be further optimized using additional polymers, e.g. HA. Various studies have demonstrated that this approach results in an improved transfection when compared to classical CS-oligonucleotides complexes.[191,196,235,236] Folate has also been postulated as a receptor mediated approach to improve cellular uptake, although this has not been studied to the same extent as HA.[192,237,238]

As indicated in Sec. 4.1, the potential of CS as a gene delivery vector is well established and studied with various covalent and non-covalent modifications being employed to improve stability, uptake and targeting.[128–130,135,136,138,139,213] This has led to the emergence of various *in vivo* studies supporting the potential of CS based gene delivery vectors in this field.[181,192,193,238–241]

5. Comparative Analysis of the Potential of CS and HA-based Nanocarriers

Based on the reviewed works, it becomes clear that the potential of CS and HA for nanotechnology-based drug delivery is there. The field of application of these technologies has been, however, driven by differences in the regulatory and safety profile of both materials. CS is approved as a dietary fiber and as hemostatic bandages, but, from our knowledge, no regulatory study regarding human parenteral administration or implantation has been performed. On the other hand, HA is FDA-approved since 2003 for injection as tissue filler during cataract removal and there are currently other clinical applications, such as intra-articular injection and topical ocular application, as well as a myriad of clinical studies involving its administration by different modalities.

CS is well-known for its capacity to interact strongly with mucus, and recent studies have also shown specific cell binding through integrins. This specific cell interaction seems to be critical for CS permeation enhancing properties,[6,7] and we expect further research to clarify whether it may also drive CS nanoparticle endocytosis. On this basis, it is not surprising that CS nanocarriers have become some of the most successful transmucosal drug delivery systems, with proven efficacy for nasal and oral delivery. Nasal delivery has been applied mostly to vaccination since this route is more suitable to biopharmaceuticals active at low doses, and not requiring chronic administration. Antigens administered intranasaly in CS nanocarriers elicit a strong mucosal immune response, but also systemic immunity levels well above the protective level.[118,126] Viral challenge studies have proven the capacity of CS nanocarrier-based vaccines to prevent several viral infections.[122]

Another area of extensive research for CS nanocarriers has been oral protein delivery. Beginning in the late 1990s, a few groups including ours reported effective oral delivery of therapeutic peptides and proteins upon integration in a CS-based nanocarrier.[131,242] Further developments in this line have optimized these basic prototypes, reporting some of the best bioavailability results for oral insulin formulations.[86] As indicated, these positive results are explained by the capacity of CS nanoparticles to penetrate the mucus and to bioadhere to the apical membrane of enterocytes.[243] However, how CS nanocarriers behave in the body is not yet fully understood. For example, there is still limited information regarding the fate of CS nanoparticles upon oral administration.[73]

On the other hand, beyond the current application to the ocular mucosa, HA degradation in the presence of the gut's microflora questions the potential of HA for oral delivery.[244] However, as a major component of the extracellular matrix, HA has shown capacity to bind to a few cell receptors with a broad spectrum of activities indicating potential for other medical uses.[26–28,30,34–37] This specific binding of HA has stimulated research regarding its potential for cell or tissue targeting, and to improve cellular uptake by receptor-mediated endocytosis. It has also opened the possibility to use nanocarriers in combined therapies where HA receptor activation would act as co-signaling supporting specific cellular pathways. For instance, CD44 activation can result in regenerative effects and also stimulate cell migration.[245] HA activation of TSG-6 could have an effect on immunomodulation.[38,39] Based on this profile, and considering its excellent biocompatibility, HA-based nanomedicines have found the greatest interest for the design of new oncological therapies, however, we could anticipate increased interest in HA nanomedicines for immune diseases and regenerative medicine.

Future breakthroughs in HA-based nanomedicines will require a better understanding of HA interactions and processing in the body. Also an improvement of HA targeting abilities could arise from chemically controlling the binding of HA to several potential receptor targets. This would avoid competition between receptors of interest and possible off-targets. For instance, suppressing HARE binding while preserving CD44 affinity could improve current HA nanocarrier delivery to most cancer tumors, since the active targeted carrier will not have extensive affinity for the liver. Apart from the areas currently under research, we speculate that targeting to cancer stem cells and lymphatic targeting could be two emerging areas of application of HA in cancer therapy.

Despite the different profile of CS and HA, there are two common areas where both nanocarriers could have a bright future, these are gene therapy and ocular drug delivery. The use of CS nanocarriers for gene therapy is largely documented and there is now a considerable body of knowledge and a marketed transfection agent supporting this application. The success of CS is based in the reported ability of CS to enhance its penetration through cellular and epithelial barriers, its cationic nature is capable of compacting gene medicines, and its buffering properties are important for endosomal escape.[189,195] A potential weakness of CS for gene therapy is its general cationic character that would drive opsonization upon parenteral administration. Due to this limitation, some of the most successful applications of CS in gene delivery are related to mucosal administration.[128,135,138,139] Alternatively, several groups have reported the PEGylation of CS as a method to improve the behavior upon parenteral administration of these nanocarriers.[128,206]

Although much less explored than CS, HA can also be used to generate nanocarriers that are taken up by specific cell receptor endocytosis. HA offers no major limitations for parenteral administration, but has not shown the same performance upon transmucosal administration as CS. The main limitation of HA for gene therapy relies on its polyanionic nature, making it incapable of condensing oligonucleotides. To address this, researchers are trying to synthesize HA derivatives bearing cationic groups,[21,54,65,68] or to engineer nanocarriers combining HA with other cationic condensing agents.[21,54,65,68]

Ocular drug delivery is often seen as a niche application, but its industrial interest has grown steadily in the last decade; many drugs would benefit from improved retention on the ocular surface, and enhanced corneal and conjunctival penetration. This is an application where both CS and HA have already shown promising data. CS nanocapsules, with their intrinsic capacity to overcome mucosal barriers have shown enhanced levels of indomethatin and cyclosporin A upon ocular instillation.[99,158] HA nanoparticles have also shown high penetration, in this case of gene medicines, a result that can be related to the widespread expression of CD44 in corneal tissue.[76–78]

In conclusion, HA- and CS-based nanocarriers have shown outstanding results for several applications, justifying the intense interest and study in the development of CS nanocarriers. In the future, a general challenge for both nanocarrier platforms would be translating the promising results, mostly obtained in rodents, to large mammals and ultimately to humans. Interestingly, despite the difficulties for analyzing the clinical development of CS and HA in a "nano" format, we have identified a number of nanocarrier prototypes undergoing early clinical trials. It is therefore expected that, if successful, the clinical development of these technologies will open interesting avenues for the exploitation of polysaccharide-based nanomedicines.

References

1. Y. Wang and P. G. Wang, Polysaccharide-based systems in drug and gene delivery. *Adv. Drug Deliv. Rev.* **65**(9), 1121–1122 (2013).
2. F. Khan and S. R. Ahmad, Polysaccharides and their derivatives for versatile tissue engineering application. *Macromol. Biosci.* **13**(4), 395–421 (2013).
3. M. Garcia-Fuentes and M. J. Alonso, Chitosan-based drug nanocarriers: where do we stand? *J. Control. Release* **161**(2), 496–504 (2012).
4. G. F. Payne and S. R. Raghavan, Chitosan: a soft interconnect for hierarchical assembly of nano-scale components. *Soft Matter* **3**(5), 521–527 (2007).
5. H.-J. Schneider and R. M. Strongin, Supramolecular interactions in chemomechanical polymers. *Acc. Chem. Res.* **42**(10), 1489–1500 (2009).
6. L.-W. Hsu, P.-L. Lee, C.-T. Chen, F.-L. Mi, J.-H. Juang, S.-M. Hwang *et al.*, Elucidating the signaling mechanism of an epithelial tight-junction opening induced by chitosan. *Biomaterials* **33**(26), 6254–6263 (2012).
7. L.-W. Hsu, Y.-C. Ho, E.-Y. Chuang, C.-T. Chen, J.-H. Juang, F.-Y. Su *et al.*, Effects of pH on molecular mechanisms of chitosan-integrin interactions and resulting tight-junction disruptions. *Biomaterials* **34**(3), 784–793 (2013).
8. R. Rosenthal, D. Guenzel, C. Finger, S. M. Krug, J. F. Richter, J.-D. Schulzke *et al.*, The effect of chitosan on transcellular and paracellular mechanisms in the intestinal epithelial barrier. *Biomaterials* **33**(9), 2791–2800 (2012).

9. C. Gorzelanny, B. Poeppelmann, K. Pappelbaum, B. M. Moerschbacher and S. W. Schneider, Human macrophage activation triggered by chitotriosidase-mediated chitin and chitosan degradation. *Biomaterials* **31**(33), 8556–8563 (2010).
10. T. Kean and M. Thanou, Biodegradation, biodistribution and toxicity of chitosan. *Adv. Drug Deliv. Rev.* **62**(1), 3–11 (2010).
11. S. Minami, M. OhOka, Y. Okamoto, K. Miyatake, A. Matsuhashi, Y. Shigemasa *et al.*, Chitosan-inducing hemorrhagic pneumonia in dogs. *Carbohydr. Polym.* **29**(3), 241–246 (1996).
12. B. CarrenoGomez and R. Duncan, Evaluation of the biological properties of soluble chitosan and chitosan microspheres. *Int. J. Pharm.* **148**(2), 231–240 (1997).
13. P. Baldrick, The safety of chitosan as a pharmaceutical excipient. *Regul. Toxicol. Pharmacol.* **56**(3), 290–299 (2010).
14. R. A. A. Muzzarelli, Human enzymatic activities related to the therapeutic administration of chitin derivatives. *Cell. Mol. Life Sci.* **53**(2), 131–140 (1997).
15. K. M. Varum, M. M. Myhr, R. J. N. Hjerde and O. Smidsrod, In vitro degradation rates of partially N-acetylated chitosans in human serum. *Carbohydr. Res.* **299**(1&2), 99–101 (1997).
16. J. S. Boateng, K. H. Matthews, H. N. E. Stevens and G. M. Eccleston, Wound healing dressings and drug delivery systems: a review. *J. Pharm. Sci.* **97**(8), 2892–923 (2008).
17. C. Prego, M. Garcia, D. Torres and M. J. Alonso, Transmucosal macromolecular drug delivery. *J. Control. Release* **101**(1–3), 151–162 (2005).
18. M. Garcia-Fuentes, C. Prego, D. Torres and M. J. Alonso, A comparative study of the potential of solid triglyceride nanostructures coated with chitosan or poly(ethylene glycol) as carriers for oral calcitonin delivery. *Eur. J. Pharm. Sci.* **25**(1), 133–143 (2005).
19. C. Yang, M. Cao, H. Liu, Y. He, J. Xu, Y. Du *et al.*, The high and low molecular weight forms of hyaluronan have distinct effects on CD44 clustering. *J. Biol. Chem.* **287**(51), 43094–43107 (2012).
20. K. S. Girish and K. Kemparaju, The magic glue hyaluronan and its eraser hyaluronidase: a biological overview. *Life Sci.* **80**(21), 1921–1943 (2007).
21. Y. Shen, Q. Li, J. Tu and J. Zhu, Synthesis and characterization of low molecular weight hyaluronic acid-based cationic micelles for efficient siRNA delivery. *Carbohydr. Polym.* **77**(1), 95–104 (2009).
22. J. E. Scott, Supramolecular organization of extracellular matrix glycosaminoglycans, in vitro and in the tissues. *FASEB J.* **6**(9), 2639–2645 (1992).
23. K. Pritchard, A. B. Lansley, G. P. Martin, M. Helliwell, C. Marriott and L. M. Benedetti, Evaluation of the bioadhesive properties of hyaluronan derivatives: detachment weight and mucociliary transport rate studies. *Int. J. Pharm.* **129**(1), 137–145 (1996).
24. T. C. Laurent, U. B. G. Laurent and J. R. E. Fraser, Functions of hyaluronan. *Ann. Rheum. Dis.* **54**(5), 429–432 (1995).
25. J. Lesley, R. Hyman and P. W. Kincade, CD44 and its interaction with extracellular matrix. In *Advances in Immunology*, J. D. Frank (Ed.), Academic Press (1993), pp. 271–335.
26. H. Harada and M. Takahashi, CD44-dependent intracellular and extracellular catabolism of hyaluronic acid by hyaluronidase-1 and-2. *J. Biol. Chem.* **282**(8), 5597–2607 (2007).

27. C. B. Knudson, Hyaluronan and CD44: strategic players for cell-matrix interactions during chondrogenesis and matrix assembly. *Birth Defects Res. C Embryo Today* **69**(2), 174–196 (2003).
28. H. Li, L. Guo, J. Li, N. Liu, R. Qi and J. Liu, Expression of hyaluronan receptors CD44 and RHAMM in stomach cancers: relevance with tumor progression. *Int. J. Oncol.* **17**(5), 927–959 (2000).
29. A. Almalik, S. Karimi, S. Ouasti, R. Donno, C. Wandrey, P. J. Day *et al.*, Hyaluronic acid (HA) presentation as a tool to modulate and control the receptor-mediated uptake of HA-coated nanoparticles. *Biomaterials* **34**(21), 5369–5380 (2013).
30. B. Zhou, J. A. Weigel, L. Fauss and P. H.Weigel, Identification of the hyaluronan receptor for endocytosis (HARE). *J. Biol. Chem.* **275**(48), 37733–37741 (2000).
31. L. Jadin, L. H. Bookbinder and G. I. Frost, A comprehensive model of hyaluronan turnover in the mouse. *Matrix Biol.* **31**(2), 81–89 (2012).
32. E. N. Harris, S. V. Kyosseva, J. A. Weigel and P. H. Weigel, Expression, processing, and glycosaminoglycan binding activity of the recombinant human 315-kDa hyaluronic acid receptor for endocytosis (HARE). *J. Biol. Chem.* **282**(5), 2785–2797 (2007).
33. Y. Luo, M. R. Ziebell and G. D. Prestwich, A hyaluronic acid–taxol antitumor bioconjugate targeted to cancer cells. *Biomacromolecules* **1**(2), 208–218 (2000).
34. J. R. Fraser, W. G. Kimpton, T. C. Laurent, R. N. Cahill and N. Vakakis, Uptake and degradation of hyaluronan in lymphatic tissue. *Biochem. J.* **256**, 153–158 (1988).
35. J. R. E. Fraser, T. C. Laurent and U. B. G. Laurent, Hyaluronan: its nature, distribution, functions and turnover. *J. Intern. Med.* **242**(1), 27–33 (1997).
36. D. G. Jackson, R. Prevo, S. Clasper and S. Banerji, LYVE-1, the lymphatic system and tumor lymphangiogenesis. *Trends Immunol.* **22**(6), 317–321 (2001).
37. D. G. Jackson, The lymphatics revisited: new perspectives from the hyaluronan receptor LYVE-1. *Trends Cardiovasc. Med.* **13**(1), 1–7 (2003).
38. D. Kohda, C. J. Morton, A. A. Parkar, H. Hatanaka, F. M. Inagaki, I. D. Campbell *et al.*, Solution structure of the link module: a hyaluronan-binding domain involved in extracellular matrix stability and cell migration. *Cell* **86**(5), 767–775 (1996).
39. C. M. Milner and A. J. Day, TSG-6: a multifunctional protein associated with inflammation. *J. Cell Sci.* **116**(10), 1863–1873 (2003).
40. L. C. Becker, W. F. Bergfeld, D. V. Belsito, C. D. Klaassen, J. G. Marks, R. C. Shank *et al.*, Final report of the safety assessment of hyaluronic acid, potassium hyaluronate, and sodium hyaluronate. *Int. J. Toxicol.* **28**(4 Suppl.), 5–67 (2009).
41. S. Mizrahy, S. R. Raz, M. Hasgaard, H. Liu, N. Soffer-Tsur, K. Cohen *et al.*, Hyaluronan-coated nanoparticles: the influence of the molecular weight on CD44-hyaluronan interactions and on the immune response. *J. Control. Release* **156**(2), 231–238 (2011).
42. C. Ke, D. Wang, Y. Sun, D. Qiao, H. Ye and X. Zeng, Immunostimulatory and antiangiogenic activities of low molecular weight hyaluronic acid. *Food Chem. Toxicol.* **58**, 401–407 (2013).
43. Y. Luo and G. D. Prestwich, Synthesis and selective cytotoxicity of a hyaluronic acid-antitumor bioconjugate. *Bioconjug. Chem.* **10**(5), 755–763 (1999).
44. S. C. Gosh, E. Auzenne, M. Khodadadian, B. Rivera, D. Farquhar, R. E. Price, M. Ravoori, V. Kundra, R. S. Freedman and J. Klostergaard, Hyaluronic acid-paclitaxel: antitumor efficacy against CD44(+) human ovarian carcinoma xenografts. *Neoplasia* **9**(6), 479–486 (2007).

45. A. Rosato, A. Banzato, G. De Luca, D. Renier, F. Bettella, C. Pagano *et al.*, HYTAD1-p20: a new paclitaxel-hyaluronic acid hydrosoluble bioconjugate for treatment of superficial bladder cancer. *Urol. Oncol.* **24**(3), 207–215 (2006).
46. H. Lee, H. Mok, S. Lee, Y.-K. Oh and T. G. Park, Target-specific intracellular delivery of siRNA using degradable hyaluronic acid nanogels. *J. Control. Release* **119**(2), 245–252.
47. J. Hyung Park, S. Kwon, M. Lee, H. Chung, J.-H. Kim, Y.-S. Kim *et al.*, Self-assembled nanoparticles based on glycol chitosan bearing hydrophobic moieties as carriers for doxorubicin: in vivo biodistribution and anti-tumor activity. *Biomaterials* **27**(1), 119–126 (2006).
48. H. Y. Yoon, H. Koo, K. Y. Choi, I. Chan Kwon, K. Choi, J. H. Park *et al.*, Photo-crosslinked hyaluronic acid nanoparticles with improved stability for in vivo tumor-targeted drug delivery. *Biomaterials* **34**(21), 5273–5280 (2013).
49. H. S. Han, J. Lee, H. R. Kim, S. Y. Chae, M. Kim, G. Saravanakumar *et al.*, Robust PEGylated hyaluronic acid nanoparticles as the carrier of doxorubicin: mineralization and its effect on tumor targetability in vivo. *J. Control. Release* **168**(2), 105–114 (2013).
50. H. Y. Yoon, H. Koo, K. Y. Choi, S. J. Lee, K. Kim, I. C. Kwon *et al.*, Tumor-targeting hyaluronic acid nanoparticles for photodynamic imaging and therapy. *Biomaterials* **33**(15), 3980–3989 (2012).
51. D.-E. Lee, A. Y. Kim, H. Y. Yoon, K. Y. Choi, I. C. Kwon, S. Y. Jeong *et al.*, Amphiphilic hyaluronic acid-based nanoparticles for tumor-specific optical/MR dual imaging. *J. Mater. Chem.* **22**(21), 10444–10447 (2012).
52. K. Y. Choi, E. J. Jeon, H. Y. Yoon, B. S. Lee, J. H. Na, K. H. Min *et al.*, Theranostic nanoparticles based on PEGylated hyaluronic acid for the diagnosis, therapy and monitoring of colon cancer. *Biomaterials* **33**(26), 6186–6193 (2012).
53. H.-J. Cho, I.-S. Yoon, H. Y. Yoon, H. Koo, Y.-J. JinJ, S.-H. Ko *et al.*, Polyethylene glycol-conjugated hyaluronic acid-ceramide self-assembled nanoparticles for targeted delivery of doxorubicin. *Biomaterials* **33**(4), 1190–1200 (2012).
54. Y. Shen, B. Wang, Y. Lu, A. Ouahab, Q. Li and J. Tu, A novel tumor-targeted delivery system with hydrophobized hyaluronic acid–spermine conjugates (HHSCs) for efficient receptor-mediated siRNA delivery. *Int. J. Pharm.* **414**(1&2), 233–243 (2011).
55. S.-Y. Han, H. S. Han, S. C. Lee, Y. M. Kang, I.-S. Kim, and J. H. Park, Mineralized hyaluronic acid nanoparticles as a robust drug carrier. *J. Mater. Chem.* **21**(22), 7996–8001 (2011).
56. K. Y. Choi, H. Y. Yoon, J.-H. Kim, S. M. Bae, R.-W. Park, Y. M. Kang *et al.*, Smart nanocarrier based on PEGylated hyaluronic acid for cancer therapy. *ACS Nano* **5**(11), 8591–8599 (2011).
57. K. Y. Choi, K. H. Min, H. Y. Yoon, K. Kim, J. H. Park, I. C. Kwon *et al.*, PEGylation of hyaluronic acid nanoparticles improves tumor targetability in vivo. *Biomaterials* **32**(7), 1880–1889 (2011).
58. H.-J. Cho, H. Y. Yoon, H. Koo, S.-H. Ko, J.-S. ShimS, J.-H. Lee *et al.*, Self-assembled nanoparticles based on hyaluronic acid-ceramide (HA-CE) and pluronic® for tumor-targeted delivery of docetaxel. *Biomaterials* **32**(29), 7181–7190 (2011).
59. K. K. Upadhyay, A. N. Bhatt, E. Castro, A. K. Mishra, K. Chuttani, B. S. Dwarakanath *et al.*, In vitro and in vivo evaluation of docetaxel loaded biodegradable polymersomes. *Macromol. Biosci.* **10**(5), 503–512 (2010).

60. K. Y. Choi, H. Chung, K. H. Min, H. Y. Yoon, K. Kim, J. H. Park *et al.*, Self-assembled hyaluronic acid nanoparticles for active tumor targeting. *Biomaterials* **31**(1), 106–114 (2010).
61. N. Duceppe and M. Tabrizian, Factors influencing the transfection efficiency of ultra low molecular weight chitosan/hyaluronic acid nanoparticles. *Biomaterials* **30**(13), 2625–2631 (2009).
62. K. Y. Choi, K. H. Min, J. H. Na, K. Choi, K. Kim, J. H. Park *et al.*, Self-assembled hyaluronic acid nanoparticles as a potential drug carrier for cancer therapy: synthesis, characterization, and in vivo biodistribution. *J. Mater. Chem.* **19**(24), 4102–4107.
63. H. Wang, P. Zhao, X. Liang, X. Gong, T. Song, R. Niu *et al.*, Folate-PEG coated cationic modified chitosan — Cholesterol liposomes for tumor-targeted drug delivery. *Biomaterials* **31**(14), 4129–4138 (2010).
64. K. H. Min, K. Park, Y.-S. Kim, S. M. Bae, S. Lee, H. G. Jo *et al.*, Hydrophobically modified glycol chitosan nanoparticles-encapsulated camptothecin enhance the drug stability and tumor targeting in cancer therapy. *J. Control. Release* **127**(3), 208–218 (2008).
65. A. Dufaÿ Wojcicki, H. Hillaireau, T. L. Nascimento, S. Arpicco, M. Taverna, S. Ribes *et al.*, Hyaluronic acid-bearing lipoplexes: physico-chemical characterization and in vitro targeting of the CD44 receptor. *J. Control. Release* **162**(3), 545–552 (2012).
66. N. Toriyabe, Y. Hayashi, M. Hyodo and H. Harashima, Synthesis and evaluation of stearylated hyaluronic acid for the active delivery of liposomes to liver endothelial cells. *Biol. Pharm. Bull.* **34**(7), 1084–1089 (2011).
67. H. S. S. Qhattal and X. Liu, Characterization of CD44-mediated cancer cell uptake and intracellular distribution of hyaluronan-grafted liposomes. *Mol. Pharm.* **8**(4), 1233–1246 (2011).
68. S. Taetz, A. Bochot, C. Surace, S. Arpicco, J. M. Renoir, U. F. Schaefer *et al.*, Hyaluronic acid-modified DOTAP/DOPE liposomes for the targeted delivery of anti-telomerase siRNA to CD44-expressing lung cancer cells. *Oligonucleotides* **19**(2), 103–116.
69. D. Peer and R. Margalit, Loading mitomycin C inside long circulating hyaluronan targeted nano-liposomes increases its antitumor activity in three mice tumor models. *Int. J. Cancer* **108**(5), 780–789 (2004).
70. D. Peer and R. Margalit, Tumor-targeted hyaluronan nanoliposomes increase the antitumor activity of liposomal doxorubicin in syngeneic and human xenograft mouse tumor models. *Neoplasia* **6**(4), 343–353 (2004).
71. R. E. Eliaz and F. C. Szoka, Liposome-encapsulated doxorubicin targeted to CD44: a strategy to kill CD44-overexpressing tumor cells. *Cancer Res.* **61**(6), 2592–2601 (2001).
72. A. Siew, H. Le, M. Thiovolet, P. Gellert, A. Schatzlein and I. Uchegbu, Enhanced oral absorption of hydrophobic and hydrophilic drugs using quaternary ammonium palmitoyl glycol chitosan nanoparticles. *Mol. Pharm.* **9**(1), 14–28 (2012).
73. A. Lalatsa, N. L. Garrett, T. Ferrarelli, J. Moger, A. G. Schaetzlein and I. F. Uchegbu, Delivery of peptides to the blood and brain after oral uptake of quaternary ammonium palmitoyl glycol chitosan nanoparticles. *Mol. Pharm.* **9**(6), 1764–1774 (2012).
74. C. J. Thompson, L. Tetley, I. F. Uchegbu and W. P. Cheng, The complexation between novel comb shaped amphiphilic polyallylamine and insulin — Towards oral insulin delivery. *Int. J. Pharm.* **376**(1&2), 46–55 (2009).

75. I. F. Uchegbu, L. Sadiq, A. Pardakhty, M. El-Hammadi, A. I. Gray, L. Tetley *et al.*, Gene transfer with three amphiphilic glycol chitosans — The degree of polymerisation is the main controller of transfection efficiency. *J. Drug Target.* **12**(8), 527–539 (2004).
76. M. de la Fuente, B. Seijo and M. J. Alonso, Design of novel polysaccharidic nanostructures for gene delivery. *Nanotechnology* **19**, 075105 (2008).
77. M. de la Fuente, B. Seijo and M. J. Alonso, Bioadhesive hyaluronan–chitosan nanoparticles can transport genes across the ocular mucosa and transfect ocular tissue. *Gene Ther.* **15**(9), 668–676 (2008).
78. M. de la Fuente, B. Seijo and M. J. Alonso, Novel hyaluronic acid-chitosan nanoparticles for ocular gene therapy. *Invest. Ophthalmol. Vis. Sci.* **49**(5), 2016–2024 (2008).
79. P. Calvo, C. Remuñan-López, J. Vila-Jato and M. Alonso, Chitosan and chitosan/ethylene oxide-propylene oxide block copolymer nanoparticles as novel carriers for proteins and vaccines. *Pharm. Res.* **14**(10), 1431–1436 (1997).
80. B. Sarmento, A. Ribeiro, F. Veiga, P. Sampaio, R. Neufeld and D. Ferreira, Alginate/chitosan nanoparticles are effective for oral insulin delivery. *Pharm. Res.* **24**(12), 2198–2206 (2007).
81. B. Sarmento, A. Ribeiro, F. Veiga, D. Ferreira and R. Neufeld, Oral bioavailability of insulin contained in polysaccharide nanoparticles. *Biomacromolecules* **8**(10), 3054–3060 (2007).
82. F.-L. Mi, Y.-Y. Wu, Y.-H. Lin, K. Sonaje, Y.-C. Ho, C.-T. Chen *et al.*, Oral delivery of peptide drugs using nanoparticles self-assembled by poly(gamma-glutamic acid) and a chitosan derivative functionalized by trimethylation. *Bioconjug. Chem.* **19**(6), 1248–1255 (2008).
83. F. M. Goycoolea, G. Lollo, C. Remunan-Lopez, F. Quaglia and M. J. Alonso, Chitosan-alginate blended nanoparticles as carriers for the transmucosal delivery of macromolecules. *Biomacromolecules* **10**(7), 1736–1743 (2009).
84. M. Ravina, E. Cubillo, D. Olmeda, R. Novoa-Carballal, E. Fernandez-Megia, R. Riguera *et al.*, Hyaluronic acid/chitosan-g-poly(ethylene glycol) nanoparticles for gene therapy: an application for pDNA and siRNA delivery. *Pharm. Res.* **27**(12), 2544–2555 (2010).
85. K. Sonaje, K.-J. Lin, S.-P. Wey, C.-K. Lin, T.-H. Yeh, H.-N. Nguyen *et al.*, Biodistribution, pharmacodynamics and pharmacokinetics of insulin analogs in a rat model: oral delivery using pH-Responsive nanoparticles vs. subcutaneous injection. *Biomaterials* **31**(26), 6849–6858 (2010).
86. K. Sonaje, Y.-J. Chen, H.-L. Chen, S.-P. Wey, J.-H. Juang, H.-N. Nguyen *et al.*, Enteric-coated capsules filled with freeze-dried chitosan/poly(gamma-glutamic acid) nanoparticles for oral insulin delivery. *Biomaterials* **31**(12), 3384–3394 (2010).
87. K. Sonaje, K.-J. Lin, J.-J. Wang, F.-L. Mi, C.-T. Chen, J.-H. Juang *et al.*, Self-assembled pH-sensitive nanoparticles: a platform for oral delivery of protein drugs. *Adv. Funct. Mater.* **20**(21), 3695–3700 (2010).
88. H.-J. Moon, J.-S. Lee, M. R. Talactac, M. Y. E. Chowdhury, J.-H. Kim, M.-E. Park *et al.*, Mucosal immunization with recombinant influenza hemagglutinin protein and poly gamma-glutamate/chitosan nanoparticles induces protection against highly pathogenic influenza A virus. *Vet. Microb.* **160**(3–4), 277–289 (2012).
89. E.-Y. Chuang, G. T. H. Nguyen, F.-Y. Su, K.-J. Lin, C.-T. Chen, F.-L. Mi *et al.*, Combination therapy via oral co-administration of insulin- and exendin-4-loaded

nanoparticles to treat type 2 diabetic rats undergoing OGTT. *Biomaterials* **34**(32), 7994–8001 (2013).
90. F. Maestrelli, M. Garcia-Fuentes, P. Mura and M. J. Alonso, A new drug nanocarrier consisting of chitosan and hydoxypropyleyclodextrin. *Eur. J. Pharm. Biopharm.* **63**(2), 79–86 (2006).
91. A. Trapani, M. Garcia-Fuentes and M. J. Alonso, Novel drug nanocarriers combining hydrophilic cyclodextrins and chitosan. *Nanotechnology* **19**(18), (2008).
92. D. Teijeiro-Osorio, C. Remunan-Lopez and M. Jose Alonso, Chitosan/cyclodextrin nanoparticles can efficiently transfect the airway epithelium in vitro. *Eur. J. Pharm. Biopharm.* **71**(2), 257–263 (2009).
93. A. Trapani, A. Lopedota, M. Franco, N. Cioffi, E. Ieva, M. Garcia-Fuentes *et al.*, A comparative study of chitosan and chitosan/cyclodextrin nanoparticles as potential carriers for the oral delivery of small peptides. *Eur. J. Pharm. Biopharm.* **75**(1), 26–32 (2010).
94. M. Nahaei, H. Valizadeh, B. Baradaran, M. R. Nahaei, D. Asgari, S. Hallaj-Nezhadi *et al.*, Preparation and characterization of chitosan/beta-cyclodextrin nanoparticles containing plasmid DNA encoding interleukin-12. *Drug Res.* **63**(1), 7–12 (2013).
95. M. Werle and H. Takeuchi, Chitosan-aprotinin coated liposomes for oral peptide delivery: development, characterisation and in vivo evaluation. *Int. J. Pharm.* **370** (1&2), 26–32 (2009).
96. H. Takeuchi, Y. Matsui, H. Yamamoto and Y. Kawashima, Mucoadhesive properties of carbopol or chitosan-coated liposomes and their effectiveness in the oral administration of calcitonin to rats. *J. Control. Release* **86**(2&3), 235–242 (2003).
97. W. Liu, J. Liu, W. Liu, T. Li and C. Liu, Improved physical and *in vitro* digestion stability of a polyelectrolyte delivery system based on layer-by-layer self-assembly alginate-chitosan-coated nanoliposomes. *J. Agric. Food Chem.* **61**(17), 4133–4144 (2013).
98. J. Kowapradit, A. Apirakaramwong, T. Ngawhirunpat, T. Rojanarata, W. Sajomsang and P. Opanasopit, Methylated N-(4-N,N-dimethylaminobenzyl) chitosan coated liposomes for oral protein drug delivery. *Eur. J. Pharm. Sci.* **47**(2), 359–366 (2012).
99. P. Calvo, J. L. VilaJato and M. J. Alonso, Evaluation of cationic polymer-coated nanocapsules as ocular drug carriers. *Int. J. Pharm.* **153**(1), 41–50 (1997).
100. F. A. Oyarzun-Ampuero, G. R. Rivera-Rodríguez, M. J. Alonso and D. Torres, Hyaluronan nanocapsules as a new vehicle for intracellular drug delivery. *Eur. J. Pharm. Sci.* **49**(4), 483–490 (2013).
101. P. Calvo, C. RemunanLopez, J. L. VilaJato and M. J. Alonso, Development of positively charged colloidal drug carriers: chitosan coated polyester nanocapsules and submicron-emulsions. *Colloid Polym. Sci.* **275**(1), 46–53 (1997).
102. M. J. Santander-Ortega, J. M. Peula-Garcia, F. M. Goycoolea and J. L. Ortega-Vinuesa, Chitosan nanocapsules: effect of chitosan molecular weight and acetylation degree on electrokinetic behaviour and colloidal stability. *Colloids Surf. B Biointerfaces* **82**(2), 571–580 (2011).
103. F. M. Goycoolea, A. Valle-Gallego, R. Stefani, B. Menchicchi, L. David, C. Rochas *et al.*, Chitosan-based nanocapsules: physical characterization, stability in biological media and capsaicin encapsulation. *Colloid Polym. Sci.* **290**(14), 1423–1434 (2012).
104. X.-Y. Yang, Y.-X. Li, M. Li, L. Zhang, L.-X. Feng and N. Zhang, Hyaluronic acid-coated nanostructured lipid carriers for targeting paclitaxel to cancer. *Cancer Lett.* **334**(2), 338–345 (2013).

105. X. Li, J. Qi, Y. Xie, X. Zhang, S. Hu, Y. Xu *et al.*, Nanoemulsions coated with alginate/chitosan as oral insulin delivery systems: preparation, characterization, and hypoglycemic effect in rats. *Int. J. Nanomed.* **8**, 23–32 (2013).
106. M. Raviña, E. Cubillo, D. Olmeda, R. Novoa-Carballal, E. Fernandez-Megia, R. Riguera *et al.*, Hyaluronic acid/chitosan-g-poly(ethylene glycol) nanoparticles for gene therapy: an application for pDNA and siRNA delivery. *Pharm. Res.* **27**(12), 2544–2555 (2010).
107. R. J. Verheul, B. Slütter, S. M. Bal, J. A. Bouwstra, W. Jiskoot and W. E. Hennink, Covalently stabilized trimethyl chitosan-hyaluronic acid nanoparticles for nasal and intradermal vaccination. *J. Control. Release* **156**(1), 46–52 (2011).
108. L. Han, Y. Zhao, L. Yin, R. Li, Y. Liang, H. Huang *et al.*, Insulin-loaded pH-sensitive hyaluronic acid nanoparticles enhance transcellular delivery. *AAPS PharmSciTech.* **13**(3), 836–845 (2012).
109. A. Fakhari, Q. Phan, S. V. Thakkar, C. R. Middaugh and C. Berkland, Hyaluronic acid nanoparticles titrate the viscoelastic properties of viscosupplements. *Langmuir* **29**(17), 5123–5131 (2013).
110. Y. Zambito, F. Felice, A. Fabiano, R. Di Stefano and G. Di Colo, Mucoadhesive nanoparticles made of thiolated quaternary chitosan crosslinked with hyaluronan. *Carbohydr. Polym.* **92**(1), 33–39 (2013).
111. J. Zhang, C. He, C. Tang and C. Yin, Ternary polymeric nanoparticles for oral siRNA delivery. *Pharm. Res.* **30**(5), 1228–1239 (2013).
112. G. Esposito, S. Geninatti Crich and S. Aime, Efficient cellular labeling by CD44 receptor-mediated uptake of cationic liposomes functionalized with hyaluronic acid and loaded with MRI contrast agents. *ChemMedChem* **3**(12), 1858–1862 (2008).
113. T. Jiang, Z. Zhang, Y. Zhang, H. Lv, J. Zhou, C. Li *et al.*, Dual-functional liposomes based on pH-responsive cell-penetrating peptide and hyaluronic acid for tumor-targeted anticancer drug delivery. *Biomaterials* **33**(36), 9246–9258 (2012).
114. G. Baier, A. Cavallaro, K. Vasilev, V. Mailänder, A. Musyanovych and K. Landfester, Enzyme responsive hyaluronic acid nanocapsules containing polyhexanide and their exposure to bacteria to prevent infection. *Biomacromolecules* **14**(4), 1103–1112 (2013).
115. S. Barbault-Foucher, R. Gref, P. Russo, J. Guechot and A. Bochot, Design of poly-ε-caprolactone nanospheres coated with bioadhesive hyaluronic acid for ocular delivery. *J. Control. Release* **83**(3), 365–375 (2002).
116. C. Dufes, J. M. Muller, W. Couet, J. C. Olivier, I. F. Uchegbu and A. G. Schatzlein, Anticancer drug delivery with transferrin targeted polymeric chitosan vesicles. *Pharm. Res.* **21**(1), 101–107 (2004).
117. R. J. Verheul, N. Hagenaars, T. van Es, E. V. B. van Gaal, P. H. de Jong, S. Bruijns *et al.*, A step-by-step approach to study the influence of N-acetylation on the adjuvanticity of N,N,N-trimethyl chitosan (TMC) in an intranasal nanoparticulate influenza virus vaccine. Eur. *J. Pharm. Sci.* **45**(4), 467–474 (2012).
118. M. Tafaghodi, V. Saluja, G. F. A. Kersten, H. Kraan, B. Sluetter, J.-P. Amorij *et al.*, Hepatitis B surface antigen nanoparticles coated with chitosan and trimethyl chitosan: impact of formulation on physicochemical and immunological characteristics. *Vaccine* **30**(36), 5341–5348 (2012).
119. R. Subbiah, P. Ramalingam, S. Ramasundaram, D. Y. Kim, K. Park, M. K. Ramasamy *et al.*, N,N,N-Trimethyl chitosan nanoparticles for controlled intranasal delivery of HBV surface antigen. *Carbohydr. Polym.* **89**(4), 1289–1297 (2012).

120. S. M. Bal, B. Slutter, R. Verheul, J. A. Bouwstra and W. Jiskoot, Adjuvanted, antigen loaded N-trimethyl chitosan nanoparticles for nasal and intradermal vaccination: adjuvant- and site-dependent immunogenicity in mice. *Eur. J. Pharm. Sci.* **45**(4), 475–481 (2012).
121. B. Slutter and W. Jiskoot, Dual role of CpG as immune modulator and physical crosslinker in ovalbumin loaded N-trimethyl chitosan (TMC) nanoparticles for nasal vaccination. *J. Control. Release* **148**(1), 117–121 (2010).
122. N. Hagenaars, R. J. Verheul, I. Mooren, P. H. de Jong, E. Mastrobattista, H. L. Glansbeek *et al.*, Relationship between structure and adjuvanticity of N,N,N-trimethyl chitosan (TMC) structural variants in a nasal influenza vaccine. *J. Control. Release* **140**(2), 126–133 (2009).
123. M. Amidi, S. G. Romeijn, J. C. Verhoef, H. E. Junginger, L. Bungener, A. Huckriede *et al.*, N-Trimethyl chitosan (TMC) nanoparticles loaded with influenza subunit antigen for intranasal vaccination: biological properties and immunogenicity in a mouse model. *Vaccine* **25**(1), 144–153 (2007).
124. S. Sharma, H. A. E. Benson, T. K. S. Mukkur, P. Rigby and Y. Chen, Preliminary studies on the development of IgA-loaded chitosan-dextran sulphate nanoparticles as a potential nasal delivery system for protein antigens. *J. Microencapsul.* **30**(3), 283–294 (2013).
125. Y. Wu, S. Wu, L. Hou, W. Wei, M. Zhou, Z. Su *et al.*, Novel thermal-sensitive hydrogel enhances both humoral and cell-mediated immune responses by intranasal vaccine delivery. *Eur. J. Pharm. Biopharm.* **81**(3), 486–497 (2012).
126. A. Vila, A. Sanchez, K. Janes, I. Behrens, T. Kissel, J. L.V. Jato *et al.*, Low molecular weight chitosan nanoparticles as new carriers for nasal vaccine delivery in mice. *Eur. J. Pharm. Biopharm.* **57**(1), 123–131 (2004).
127. T. Nagamoto, Y. Hattori, K. Takayama, Y. Maitani, Novel chitosan particles and chitosan-coated emulsions inducing immune response via intranasal vaccine delivery. *Pharm. Res.* **21**(4), 671–674 (2004).
128. N. Csaba, M. Koping-Hoggard, E. Fernandez-Megia, R. Novoa-Carballal, R. Riguera and M. Jose Alonso, Ionically crosslinked chitosan nanoparticles as gene delivery systems: effect of PEGylation degree on in vitro and in vivo gene transfer. *J. Biomed. Nanotechnol.* **5**(2), 162–171 (2009).
129. D. Raghuwanshi, V. Mishra, D. Das, K. Kaur and M. R. Suresh, Dendritic cell targeted chitosan nanoparticles for nasal DNA Immunization against SARS CoV nucleocapsid protein. *Mol. Pharm.* **9**(4), 946–956 (2012).
130. W. Yao, Y. Peng, M. Du, J. Luo and L. Zong, Preventative vaccine-loaded mannosylated chitosan nanoparticles intended for nasal mucosal delivery enhance immune responses and potent tumor immunity. *Mol. Pharm.* **10**(8), 2904–2914 (2013).
131. R. Fernandez-Urrusuno, P. Calvo, C. Remunan-Lopez, J. L. Vila-Jato and M. J. Alonso, Enhancement of nasal absorption of insulin using chitosan nanoparticles. *Pharm. Res.* **16**(10), 1576–1581 (1999).
132. X. Wang, C. Zheng, Z. Wu, D. Teng, X. Zhang, Z. Wang *et al.*, Chitosan-NAC nanoparticles as a vehicle for nasal absorption enhancement of insulin. *J. Biomed. Mater. Res. B Appl. Biomater.* **88B**(1), 150–161 (2009).
133. M. Simon, M. Wittmar, T. Kissel and T. Linn, Insulin containing nanocomplexes formed by self-assembly from biodegradable amine-modified poly(vinyl alcohol)-graft-poly(L-lactide): bioavailability and nasal tolerability in rats. *Pharm. Res.* **22**(11), 1879–1886 (2005).

134. L. H. du Plessis, A. F. Kotze and H. E. Junginger, Nasal and rectal delivery of insulin with chitosan and N-trimethyl chitosan chloride. *Drug Deliv.* **17**(6), 399–407 (2010).
135. M. Jiang, L. Gan, C. Zhu, Y. Dong, J. Liu and Y. Gan, Cationic core-shell liponanoparticles for ocular gene delivery. *Biomaterials* **33**(30), 7621–7630 (2012).
136. E. A. Klausner, Z. Zhang, R. L. Chapman, R. F. Multack and M. V. Volin, Ultrapure chitosan oligomers as carriers for corneal gene transfer. *Biomaterials* **31**(7), 1814–1820 (2010).
137. D. Lee, W. Zhang, S. A. Shirley, X. Kong, G. R. Hellermann, R. F. Lockey *et al.*, Thiolated chitosan/DNA nanocomplexes exhibit enhanced and sustained gene delivery. *Pharm. Res.* **24**(1), 157–167 (2007).
138. L. Yin, Z. Song, Q. Qu, K. H. Kim, N. Zheng, C. Yao *et al.*, Supramolecular self-assembled nanoparticles mediate oral delivery of therapeutic TNF-alpha siRNA against systemic inflammation. *Angew. Chem. Int. Ed. Engl.* **52**(22), 5757–5761 (2013).
139. C. He, L. Yin, C. Tang and C. Yin, Multifunctional polymeric nanoparticles for oral delivery of TNF-alpha siRNA to macrophages. *Biomaterials* **34**(11), 2843–2854 (2013).
140. P. Mukhopadhyay, K. Sarkar, M. Chakraborty, S. Bhattacharya, R. Mishra and P. P. Kundu, Oral insulin delivery by self-assembled chitosan nanoparticles: in vitro and in vivo studies in diabetic animal model. *Mater. Sci. Eng. C Mater. Biol. Appl.* **33**(1), 376–382 (2013).
141. A. Elsayed, M. Al-Remawi, N. Qinna, A. Farouk, K. A. Al-Sou'od and A. A. Badwan, Chitosan-sodium lauryl sulfate nanoparticles as a carrier system for the in vivo delivery of oral insulin. *AAPS PharmSciTech.* **12**(3), 958–964 (2011).
142. A. Elsayed, M. Al Remawi, N. Qinna, A. Farouk and A. Badwan, Formulation and characterization of an oily-based system for oral delivery of insulin. *Eur. J. Pharm. Biopharm.* **73**(2), 269–279 (2009).
143. B. Sarmento, D. C. Ferreira, L. Jorgensen and M. van de Weert, Probing insulin's secondary structure after entrapment into alginate/chitosan nanoparticles. *Eur. J. Pharm. Biopharm.* **65**(1), 10–17 (2007).
144. C. B. Woitiski, R. J. Neufeld, A. J. Ribeiro and F. Veiga, Colloidal carrier integrating biomaterials for oral insulin delivery: influence of component formulation on physicochemical and biological parameters. *Acta Biomater.* **5**(7), 2475–2484 (2009).
145. C. B. Woitiski, F. Veiga, A. Ribeiro and R. Neufeld, Design for optimization of nanoparticles integrating biomaterials for orally dosed insulin. *Eur. J. Pharm. Biopharm.* **73**(1), 25–33 (2009).
146. C. B. Woitiski, R. J. Neufeld, A. F. Soares, I. V. Figueiredo, F. J. Veiga and R. A. Carvalho, Evaluation of hepatic glucose metabolism via gluconeogenesis and glycogenolysis after oral administration of insulin nanoparticles. *Drug Dev. Ind. Pharm.* **38**(12), 1441–1450 (2012).
147. C. B. Woitiski, B. Sarmento, R. A. Carvalho, R. J. Neufeld and F. Veiga, Facilitated nanoscale delivery of insulin across intestinal membrane models. *Int. J. Pharm.* **412**(1&2), 123–131 (2011).
148. Y. Jin, Y. Song, X. Zhu, D. Zhou, C. Chen, Z. Zhang *et al.*, Goblet cell-targeting nanoparticles for oral insulin delivery and the influence of mucus on insulin transport. *Biomaterials* **33**(5), 1573–1582 (2012).
149. K. Sonaje, Y.-H. Lin, J.-H. Juang, S.-P. Wey, C.-T. Chen and H.-W. Sung, In vivo evaluation of safety and efficacy of self-assembled nanoparticles for oral insulin delivery. *Biomaterials* **30**(12), 2329–2339 (2009).

150. C. Prego, D. Torres and M. J. Alonso, Chitosan nanocapsules: a new carrier for nasal peptide delivery. *J. Drug Deliv. Sci. Technol.* **16**(5), 331–337 (2006).
151. C. Prego, D. Torres and M. J. Alonso, Chitosan nanocapsules as carriers for oral peptide delivery: effect of chitosan molecular weight and type of salt on the in vitro behaviour and in vivo effectiveness. *J. Nanosci. Nanotechnol.* **6**(9&10), 2921–2928 (2006).
152. C. Prego, D. Torres, E. Fernandez-Megia, R. Novoa-Carballal, E. Quinoa and M. J. Alonso, Chitosan-PEG nanocapsules as new carriers for oral peptide delivery — Effect of chitosan pegylation degree. *J. Control. Release* **111**(3), 299–308 (2006).
153. C. Prego, M. Fabre, D. Torres and M. J. Alonso, Efficacy and mechanism of action of chitosan nanocapsules for oral peptide delivery. *Pharm. Res.* **23**(3), 549–556 (2006).
154. V. Grabovac, D. Guggi and A. Bernkop-Schnürch, Comparison of the mucoadhesive properties of various polymers. *Adv. Drug Deliv. Rev.* **57**(11), 1713–1723 (2005).
155. D. Snyman, J. Hamman and A. Kotze, Evaluation of the mucoadhesive properties of N-trimethyl chitosan chloride. *Drug Dev. Ind. Pharm.* **29**(1), 61–69 (2003).
156. J. L. Greaves and C. G. Wilson, Treatment of diseases of the eye with mucoadhesive delivery systems. *Adv. Drug Deliv. Rev.* **11**(3), 349–383 (1993).
157. D. Snyman, J. H. Hamman and A. F. Kotze, Evaluation of the mucoadhesive properties of N-trimethyl chitosan chloride. *Drug Dev. Ind. Pharm.* **29**(1), 61–69 (2003).
158. A. M. De Campos, A. Sanchez and M. J. Alonso, Chitosan nanoparticles: a new vehicle for the improvement of the delivery of drugs to the ocular surface. Application to cyclosporin A. *Int. J. Pharm.* **224**(1&2), 159–168 (2001).
159. A. M. De Campos, A. Sanchez, R. Gref, P. Calvo and M. J. Alonso, The effect of a PEG versus a chitosan coating on the interaction of drug colloidal carriers with the ocular mucosa. *Eur. J. Pharm. Sci.* **20**(1), 73–81 (2003).
160. A. M. de Campos, Y. Diebold, E. L. S. Carvalho, A. Sanchez and M. J. Alonso, Chitosan nanoparticles as new ocular drug delivery systems: in vitro stability, in vivo fate, and cellular toxicity. *Pharm. Res.* **21**(5), 803–810 (2004).
161. Y. Diebold, M. Jarrin, V. Saez, E. L. S. Carvalho, M. Orea, M. Calonge *et al.*, Ocular drug delivery by liposome-chitosan nanoparticle complexes (LCS-NP). *Biomaterials* **28**(8), 1553–1564 (2007).
162. E. L. S. Carvalho, A. Grenha, C. Remunan-Lopez, M. Jose Alonso and B. Seijo, Mucosal delivery of liposome-chitosan nanoparticle complexes. In *Methods in Enzymology Liposomes*, N. Duzgunes (Ed.), Pt G2009. pp. 289–312.
163. L. Battaglia, I. D'Addino, E. Peira, M. Trotta and M. Gallarate, Solid lipid nanoparticles prepared by coacervation method as vehicles for ocular cyclosporine. *J. Drug Deliv. Sci. Technol.* **22**(2), 125–130 (2012).
164. İ. Yenice, M. C. Mocan, E. Palaska, A. Bochot, E. Bilensoy, İ. Vural *et al.*, Hyaluronic acid coated poly-ε-caprolactone nanospheres deliver high concentrations of cyclosporine A into the cornea. *Exp. Eye Res.* **87**(3), 162–167 (2008).
165. D. Teijeiro-Osorio, C. Remunan-Lopez and M. Jose Alonso, New generation of hybrid poly/oligosaccharide nanoparticles as carriers for the nasal delivery of macromolecules. *Biomacromolecules* **10**(2), 243–249 (2009).
166. D. R. Bhumkar, H. M. Joshi, M. Sastry and V. B. Pokharkar, Chitosan reduced gold nanoparticles as novel carriers for transmucosal delivery of insulin. *Pharm. Res.* **24**(8), 1415–1426 (2007).
167. A. Vila, A. Sanchez, M. Tobio, P. Calvo and M. J. Alonso, Design of biodegradable particles for protein delivery. *J. Control. Release* **78**(1–3), 15–24 (2002).

168. C. Prego, P. Paolicelli, B. Diaz, S. Vicente, A. Sanchez, A. Gonzalez-Fernandez *et al.*, Chitosan-based nanoparticles for improving immunization against hepatitis B infection. *Vaccine* **28**(14), 2607–2614 (2010).
169. S. Vicente, M. Peleteiro, B. Díaz-Freitas, A. González-Fernández Á. Sanchez and M. J. Alonso, Co-delivery of viral proteins and TLR7 agonist from polysaccharide nanocapsules: a needle-free vaccination strategy. *J. Control. Release* (2013), in press.
170. B. Slutter, L. Plapied, V. Fievez, M. Alonso Sande, A. des Rieux, Y.-J. Schneider *et al.*, Mechanistic study of the adjuvant effect of biodegradable nanoparticles in mucosal vaccination. *J. Control. Release* **138**(2), 113–121 (2009).
171. B. Slutter, P. C. Soema, Z. Ding, R. Verheul, W. Hennink and W. Jiskoot, Conjugation of ovalbumin to trimethyl chitosan improves immunogenicity of the antigen. *J. Control. Release* **143**(2), 207–214 (2010).
172. B. Slutter, S. Bal, C. Keijzer, R. Mallants, N. Hagenaars, I. Que *et al.*, Nasal vaccination with N-trimethyl chitosan and PLGA based nanoparticles: nanoparticle characteristics determine quality and strength of the antibody response in mice against the encapsulated antigen. *Vaccine* **28**(38), 6282–6291 (2010).
173. B. Sayin, S. Somavarapu, X. W. Li, D. Sesardic, S. Senel and O. H. Alpar, TMC-MCC (N-trimethyl chitosan-mono-N-carboxymethyl chitosan) nanocomplexes for mucosal delivery of vaccines. *Eur. J. Pharm. Sci.* **38**(4), 362–369 (2009).
174. S. M. Assaf, N. D. Al-Jbour, A. F. Eftaiha , A. M. Elsayed, M. M. Al-Remawi, N. A. Qinna *et al.*, Factors involved in formulation of oily delivery system for proteins based on PEG-8 caprylic/capric glycerides and polyglyceryl-6 dioleate in a mixture of oleic acid with chitosan. *J. Dispers. Sci. Technol.* **32**(5), 623–633 (2011).
175. K. Sonaje, K.-J. Lin, M. T. Tseng, S.-P. Wey, F.-Y. Su, E.-Y. Chuang *et al.*, Effects of chitosan-nanoparticle-mediated tight junction opening on the oral absorption of endotoxins. *Biomaterials* **32**(33), 8712–8721 (2011).
176. H.-N. Nguyen, S.-P. Wey, J.-H. Juang, K. Sonaje, Y.-C. Ho, E.-Y. Chuang *et al.*, The glucose-lowering potential of exendin-4 orally delivered via a pH-sensitive nanoparticle vehicle and effects on subsequent insulin secretion in vivo. *Biomaterials* **32**(10), 2673–2682 (2011).
177. K. A. Janes, M. P. Fresneau, A. Marazuela, A. Fabra and M. J. Alonso, Chitosan nanoparticles as delivery systems for doxorubicin. *J. Control. Release* **73**(2&3), 255–267 (2001).
178. M. V. Lozano, D. Torrecilla, D. Torres, A. Vidal, F. Dominguez and M. J. Alonso, Highly efficient system to deliver taxanes into tumor cells: docetaxel-loaded chitosan oligomer colloidal carriers. *Biomacromolecules* **9**(8), 2186–2193 (2008).
179. A. Rata-Aguilar, P. Sanchez-Moreno, A. B. Jodar-Reyes, A. Martin-Rodriguez, H. Boulaiz, J. A. Marchal-Corrales *et al.*, Colloidal stability and "in vitro" antitumor targeting ability of lipid nanocapsules coated by folate-chitosan conjugates. *J. Bioact. Compat. Polym.* **27**(4), 388–404 (2012).
180. Y. Aktas, M. Yemisci, K. Andrieux, R. N. Gursoy, M. J. Alonso, E. Fernandez-Megia *et al.*, Development and brain delivery of chitosan-PEG nanoparticles functionalized with the monoclonal antibody OX26. *Bioconjug. Chem.* **16**(6), 1503–1511 (2005).
181. Q.-Q. Zhao, Y.-L. Hu, Y. Zhou, N. Li, M. Han, G.-P. Tang *et al.*, Gene-carried hepatoma targeting complex induced high gene transfection efficiency with low toxicity and significant antitumor activity. *Int. J. Nanomed.* 7, 3191–3202 (2012).
182. A. K. Iyer, G. Khaled, J. Fang and H. Maeda, Exploiting the enhanced permeability and retention effect for tumor targeting. *Drug Discov. Today* **11**(17&18), 812–818 (2006).

183. V. Torchilin, Tumor delivery of macromolecular drugs based on the EPR effect. *Adv. Drug Deliv. Rev.* **63**(3), 131–135 (2011).
184. P. Dan, M. K. Jeffrey, H. Seungpyo, C. F. Omid, M. Rimona and L. Robert, Nanocarriers as an emerging platform for cancer therapy. *Nat. Nanotechnol.* **2**(12), 751–760 (2007).
185. H. S. Min, S. Son, T. W. Lee, H. Koo, H. Y. Yoon, J. H. Na *et al.*, Liver-specific and echogenic hyaluronic acid nanoparticles facilitating liver cancer discrimination. *Adv. Funct. Mater.* (2013).
186. S. Saddoughi, P. Song and B. Ogretmen, Roles of bioactive sphingolipids in cancer biology and therapeutics. In *Lipids in Health and Disease*, P. Quinn and X. Wang (Eds.), Netherlands: Springer (2008), pp. 413–440.
187. K. W. Leong, H. Q. Mao, V. L. Truong-Le, K. Roy, S. M. Walsh and J. T. August, DNA-polycation nanospheres as non-viral gene delivery vehicles. *J. Control. Release* **53**(1–3), 183–193 (1998).
188. K. Roy, H. Mao, S. K. Huang and K. W. Leong, Oral gene delivery with chitosan-DNA nanoparticles generates immunologic protection in a murine model of peanut allergy. *Nat. Med.* **5**(4), 387–391 (1999).
189. H. Q. Mao, K. Roy, V. L. Troung-Le, K. A. Janes, K. Y. Lin, Y. Wang *et al.*, Chitosan-DNA nanoparticles as gene carriers: synthesis, characterization and transfection efficiency. *J. Control. Release* **70**(3), 399–421 (2001).
190. S. K. Shukla, A. K. Mishra, O. A. Arotiba and B. B. Mamba, Chitosan-based nanomaterials: a state-of-the-art review. *Int. J. Biol. Macromol.* **59**, 46–58 (2013).
191. N. Csaba, M. Koeping-Hoeggard and M. Jose Alonso, Ionically crosslinked chitosan/tripolyphosphate nanoparticles for oligonucleotide and plasmid DNA delivery. *Int. J. Pharm.* **382**(1&2), 205–214 (2009).
192. H.-L. Jiang, C.-X. Xu, Y.-K. Kim, R. Arote, D. Jere, H.-T. Lim *et al.*, The suppression of lung tumorigenesis by aerosol-delivered folate-chitosan-graft-polyethylenimine/Akt1 shRNA complexes through the Akt signaling pathway. *Biomaterials* **30**(29), 5844–5852 (2009).
193. A. D. Steg, A. A. Katre, B. Goodman, H.-D. Han, A. M. Nick, R. L. Stone *et al.*, Targeting the notch ligand jagged1 in both tumor cells and stroma in ovarian cancer. *Clin. Cancer Res.* **17**(17), 5674–5685 (2011).
194. T. Ishii, Y. Okahata and T. Sato, Mechanism of cell transfection with plasmid/chitosan complexes. *Biochim. Biophys. Acta* **1514**(1), 51–64 (2001).
195. I. Richard, M. Thibault, G. De Crescenzo, M. D. Buschmann and M. Lavertu, Ionization behavior of chitosan and chitosan-DNA polyplexes indicate that chitosan has a similar capability to induce a proton-sponge effect as PEI. *Biomacromolecules* **14**(6), 1732–1740 (2013).
196. H. Katas and H. O. Alpar, Development and characterisation of chitosan nanoparticles for siRNA delivery. *J. Control. Release* **115**(2), 216–225 (2006).
197. J. Haas, M. Kumar, G. Borchard, U. Bakowsky and C. M. Lehr, Preparation and characterization of chitosan and trimethyl-chitosanmodified poly-(epsilon-caprolactone) nanoparticles as DNA carriers. *AAPS PharmSciTech.* **6**(1), (2005).
198. X. Yuan, B. A. Shah, N. K. Kotadia, J. Li, H. Gu and Z. Wu, The development and mechanism studies of cationic chitosan-modified biodegradable PLGA nanoparticles for efficient siRNA drug delivery. *Pharm. Res.* **27**(7), 1285–1295 (2010).

199. M. de la Fuente, B. Seijo and M. J. Alonso, Bioadhesive hyaluronan-chitosan nanoparticles can transport genes across the ocular mucosa and transfect ocular tissue. *Gene Ther.* **15**(9), 668–676 (2008).
200. S. P. Strand, M. M. Issa, B. E. Christensen, K. M. Varum and P. Artursson, Tailoring of chitosans for gene delivery: novel self-branched glycosylated chitosan oligomers with improved functional properties. *Biomacromolecules* **9**(11), 3268–3276 (2008).
201. Y. Gao, Z. Zhang, L. Chen, W. Gu and Y. Li, Synthesis of 6-N,N,N-trimethyltriazole chitosan via "click chemistry" and evaluation for gene delivery. *Biomacromolecules* **10**(8), 2175–2182 (2009).
202. O. Germershaus, S. Mao, J. Sitterberg, U. Bakowsky and T. Kissel, Gene delivery using chitosan, trimethyl chitosan or polyethylenglycol-graft-trimethyl chitosan block copolymers: establishment of structure-activity relationships in vitro. *J. Control. Release* **125**(2), 145–154 (2008).
203. K. Turan and K. Nagata, Chitosan-DNA nanoparticles: the effect of cell type and hydrolysis of chitosan on in vitro DNA transfection. *Pharm. Dev. Technol.* **11**(4), 503–512 (2006).
204. A. Fattahi, K. Sadrjavadi, M. A. Golozar, J. Varshosaz, M.-H. Fathi and H. Mirmohammad-Sadeghi, Preparation and characterization of oligochitosan-tragacanth nanoparticles as a novel gene carrier. *Carbohydr. Polym.* **97**(2), 277–283 (2013).
205. B. Wang, S. Zhang, S. Cui, B. Yang, Y. Zhao, H. Chen *et al.*, Chitosan enhanced gene delivery of cationic liposome via non-covalent conjugation. *Biotechnol. Lett.* **34**(1), 19–28 (2012).
206. S. K. Tripathi, R. Goyal, P. Kumar and K. C. Gupta, Linear polyethylenimine-graft-chitosan copolymers as efficient DNA/siRNA delivery vectors in vitro and in vivo. *Nanomed. Nanotechnol. Biol. Med.* **8**(3), 337–345 (2012).
207. H. Bao, Y. Ping, Y. Pan, L. Li, J. Li and L. H. Gan, Thermo-responsive transfection of DNA complexes with well-defined chitosan terpolymers. *Soft Matter* **8**(8), 2518–2526 (2012).
208. T. Rojanarata, P. Opanasopit, S. Techaarpornkul, T. Ngawhirunpat and U. Ruktanonchai, Chitosan-thiamine pyrophosphate as a novel carrier for siRNA delivery. *Pharm. Res.* **25**(12), 2807-2814 (2008).
209. V. M. Gaspar, F. Sousa, J. A. Queiroz and I. J. Correia, Formulation of chitosan-TPP-pDNA nanocapsules for gene therapy applications. *Nanotechnology* **22**(1), (2011).
210. S. P. Strand, S. Lelu, N. K. Reitan, C. de Lange Davies, P. Artursson and K. M. Varum, Molecular design of chitosan gene delivery systems with an optimized balance between polyplex stability and polyplex unpacking. *Biomaterials* **31**(5), 975–987 (2010).
211. M. Thibault, S. Nimesh, M. Lavertu and M. D. Buschmann, Intracellular trafficking and decondensation kinetics of chitosan-pDNA polyplexes. *Mol. Ther.* **18**(10), 1787–1795 (2010).
212. X. Yang, X. Yuan, D. Cai, S. Wang and L. Zong, Low molecular weight chitosan in DNA vaccine delivery via mucosa. *Int. J. Pharm.* **375**(1&2), 123–132 (2009).
213. F. Zheng, X.-W. Shi, G.-F. Yang, L.-L. Gong, H.-Y. Yuan, Y.-J. Cui *et al.*, Chitosan nanoparticle as gene therapy vector via gastrointestinal mucosa administration: results of an in vitro and in vivo study. *Life Sci.* **80**(4), 388–396 (2007).
214. M. Huang, C. W. Fong, E. Khor and L. Y. Lim, Transfection efficiency of chitosan vectors: effect of polymer molecular weight and degree of deacetylation. *J. Control. Release* **106**(3), 391–406 (2005).

215. T. Kiang, H. Wen, H. W. Lim and K. W. Leong, The effect of the degree of chitosan deacetylation on the efficiency of gene transfection. *Biomaterials* **25**(22), 5293–5301 (2004).
216. S. Nimesh, M. M. Thibault, M. Lavertu and M. D. Buschmann, Enhanced gene delivery mediated by low molecular weight chitosan/DNA complexes: effect of pH and serum. *Mol. Biotechnol.* **46**(2), 182-196 (2010).
217. A. K. Varkouhi, R. M. Schiffelers, M. J. van Steenbergen, T. Lammers, W. E. Hennink and G. Storm, Photochemical internalization (PCI)-mediated enhancement of gene silencing efficiency of polymethacrylates and N,N,N-trimethylated chitosan (TMC) based siRNA polyplexes. *J. Control. Release* **148**(1), E98-E99 (2010).
218. J. Malmo, K. M. Varum and S. P. Strand, Effect of chitosan Chain architecture on gene delivery: comparison of self-branched and linear chitosans. *Biomacromolecules* **12**(3), 721-729 (2011).
219. J. Deng, Y. Zhou, B. Xu, K. Mai, Y. Deng and L.-M. Zhang, Dendronized chitosan derivative as a biocompatible gene delivery carrier. *Biomacromolecules* **12**(3), 642–649 (2011).
220. H. Katas, N. N. S. N. Dzulkefli and S. Sahudin, Synthesis of a new potential conjugated TAT-peptide-chitosan nanoparticles carrier via disulphide linkage. *J. Nanomater.* (2012).
221. M. Malhotra, C. Tomaro-Duchesneau, S. Saha, I. Kahouli and S. Prakash, Development and characterization of chitosan-PEG-TAT nanoparticles for the intracellular delivery of siRNA. *Int. J. Nanomed.* **8** (2013).
222. M. Malhotra, C. Tomaro-Duchesneau and S. Prakash, Synthesis of TAT peptide-tagged PEGylated chitosan nanoparticles for siRNA delivery targeting neurodegenerative diseases. *Biomaterials* **34**(4), 1270–1280 (2013).
223. V. M. Gaspar, J. G. Marques, F. Sousa, R. O. Louro, J. A. Queiroz and I. J. Correia, Biofunctionalized nanoparticles with pH-responsive and cell penetrating blocks for gene delivery. *Nanotechnology* **24**(27), (2013).
224. J. Chang, X. Xu, H. Li, Y. Jian, G. Wang, B. He *et al.*, Components simulation of viral envelope via amino acid modified chitosans for efficient nucleic acid delivery: in vitro and in vivo study. *Adv. Funct. Mater.* **23**(21), 2691–2699 (2013).
225. S. Park, E. J. Jeong, J. Lee, T. Rhim, S. K. Lee and K. Y. Lee, Preparation and characterization of nonaarginine-modified chitosan nanoparticles for siRNA delivery. *Carbohydr. Polym.* **92**(1), 57–62 (2013).
226. S. Plianwong, P. Opanasopit, T. Ngawhirunpat and T. Rojanarata, Chitosan combined with poly-l-arginine as efficient, safe, and serum-insensitive vehicle with RNase protection ability for siRNA delivery. *Biomed. Res. Int.* (2013).
227. H. Zhang, D. Zhu, L. Song, L. Liu, X. Dong, Z. Liu *et al.*, Arginine conjugation affects the endocytic pathways of chitosan/DNA nanoparticles. *J. Biomed. Mater. Res. A* **98A**(2), 296–302 (2011).
228. S. M. NohM, M. O. Park, G. Shim, S. E. Han, H. Y. Lee, J. H. Huh *et al.*, Pegylated poly-L-arginine derivatives of chitosan for effective delivery of siRNA. *J. Control. Release* **145**(2), 159–164 (2010).
229. H. Zheng, C. Tang and C. Yin, The effect of crosslinking agents on the transfection efficiency, cellular and intracellular processing of DNA/polymer nanocomplexes. *Biomaterials* **34**(13), 3479–3488 (2013).
230. N. K. Reitan, B. Sporsheim, A. Bjorkoy, S. Strand and C. de Lange Davies, Quantitative 3-D colocalization analysis as a tool to study the intracellular trafficking and dissociation of pDNA-chitosan polyplexes. *J. Biomed. Opt.* **17**(2), (2012).

231. A. K. Varkouhi, R. J. Verheul, R. M. Schiffelers, T. Lammers, G. Storm and W. E. Hennink, Gene silencing activity of siRNA polyplexes based on thiolated N,N,N-trimethylated chitosan. *Bioconjug. Chem.* **21**(12), 2339–2346 (2010).
232. X. Zhao, Z. Li, H. Pan, W. Liu, M. Lv, F. Leung *et al.*, Enhanced gene delivery by chitosan-disulfide-conjugated LMW-PEI for facilitating osteogenic differentiation. *Acta Biomater.* **9**(5), 6694–6703 (2013).
233. C. Moreira, H. Oliveira, L. R. Pires, S. Simoes, M. A. Barbosa and A. P. Pego, Improving chitosan-mediated gene transfer by the introduction of intracellular buffering moieties into the chitosan backbone. *Acta Biomater.* **5**(8), 2995-3006 (2009).
234. K.-L. Chang, Y. Higuchi, S. Kawakami, F. Yamashita and M. Hashida, Efficient gene transfection by histidine-modified chitosan through enhancement of endosomal escape. *Bioconjug. Chem.* **21**(6), 1087–1095 (2010).
235. Z.-X. Liao, Y.-C. Ho, H.-L. Chen, S.-F. Peng, C.-W. Hsiao and H.-W. Sung, Enhancement of efficiencies of the cellular uptake and gene silencing of chitosan/siRNA complexes via the inclusion of a negatively charged poly(gamma-glutamic acid). *Biomaterials* **31**(33), 8780–8788 (2010).
236. S.-F. Peng, M.-J. Yang, C.-J. Su, H.-L. Chen, P.-W. Lee, M.-C. Wei *et al.*, Effects of incorporation of poly(gamma-glutamic acid) in chitosan/DNA complex nanoparticles on cellular uptake and transfection efficiency. *Biomaterials*, **30**(9), 1797–1808 (2009).
237. J. C. Fernandes, X. Qiu, F. M. Winnik, M. Benderdour, X. Zhang, K. Dai *et al.*, Low molecular weight chitosan conjugated with folate for siRNA delivery in vitro: optimization studies. *Int. J. Nanomed.* **7**, 5833–5845 (2012).
238. Y.-K. Kim, A. Minai-Tehrani, J.-H. Lee, C.-S. Cho, M.-H. Cho and H.-L. Jiang, Therapeutic efficiency of folated poly(ethylene glycol)-chitosan-graft-polyethylenimine-Pdcd4 complexes in H-ras12V mice with liver cancer. *Int. J. Nanomed.* **8**, 1489–1498 (2013).
239. Z. Liu, D. Lv, S. Liu, J. Gong, D. Wang, M. Xiong *et al.*, Alginic acid-coated chitosan nanoparticles loaded with legumain DNA vaccine: effect against breast cancer in mice. *Plos One* **8**(4), (2013).
240. X. Liang, X. Li, J. Chang, Y. Duan and Z. Li, Properties and evaluation of quaternized chitosan/lipid cation polymeric liposomes for cancer-targeted gene delivery. *Langmuir* **29**(27), 8683–8693 (2013).
241. H. D. Han, L. S. Mangala, J. W. Lee, M. M. K. Shahzad, H. S. Kim, D. Shen *et al.*, Targeted gene silencing using RGD-labeled chitosan nanoparticles. *Clin. Cancer Res.* **16**(15), 3910–3922 (2010).
242. P. Calvo, C. RemunanLopez, J. L. VilaJato and M. J. Alonso, Chitosan and chitosan ethylene oxide propylene oxide block copolymer nanoparticles as novel carriers for proteins and vaccines. *Pharm. Res.* **14**(10), 1431–1436 (1997).
243. K. Sonaje, E.-Y. Chuang, K.-J. Lin, T.-C. Yen, F.-Y. Su, M. T. Tseng *et al.*, Opening of epithelial tight junctions and enhancement of paracellular permeation by chitosan: microscopic, ultrastructural, and computed-tomographic observations. *Mol. Pharm.* **9**(5), 12710–1279 (2012).
244. L. V. Hooper, T. Midtvedt and J. I. Gordon, How host-microbial interactions shape the nutrient environment of the mammalian intestine. *Ann. Rev. Nutr.* **22**, 283–307 (2002).
245. A. J. Meinel, K. E. Kubow, E. Klotzsch, M. Garcia-Fuentes, M. L. Smith, V. Vogel *et al.*, Optimization strategies for electrospun silk fibroin tissue engineering scaffolds. *Biomaterials* **30**(17), 3058–3067 (2009).

Chapter 8

Dendrimers for Biomedical Applications

*Lisa M. Kaminskas, Victoria M. McLeod, Seth A. Jones, Ben J. Boyd and Christopher J. H. Porter**

Drug Delivery Disposition Dynamics,
Monash Institute of Pharmaceutical Sciences
381 Royal Parade, Parkville, Victoria, Australia, 3052
**chris.porter@monash.edu*

Abstract

Dendrimers are globular polymers with well-defined molecular structures and polyfunctional surfaces. The ability to functionalize the surface of dendrimers with a range of chemical species and to tailor physicochemical properties has resulted in the exploration of dendrimers for a wide range of biomedical applications. The clinical application of dendrimers is still in the early stages of development, although data generated thus far have suggested that they have the potential to advance the fields of drug delivery, gene therapy, imaging and wound healing. This chapter provides an overview of the current state-of-the-art of research into the biomedical application of dendrimers.

1. Introduction to Dendrimers

The development of biocompatible nanomedicines with defined molecular structure and narrow polydispersity is a challenge that continues to engage the polymer and pharmaceutical research communities. To this end, dendritic polymers (dendrimers) have attracted increasing recent interest as potential nanomedicines, imaging agents and scaffolds for a wide range of biomedical applications. Dendrimers are constructed from a central core unit, containing at least two functional groups, that provides the attachment point for further layers or "generations" (G) of monomers (Fig. 1). Provided the conjugated monomers contain a single point for attachment to the core, and multiple functional groups to allow conjugation of a further two or more monomers to the scaffold, a globular polymeric structure can be constructed with good control over the final size and surface properties. Tomalia and coworkers described the first family of dendritic polymers in the early 1980s. These were constructed with N-(2-aminoethyl)acrylamide monomers and an ammonia or ethylenediamine "initiator" core[1] and ultimately gave rise to the first commercial dendrimers (PAMAM or Starburst dendrimers, Fig. 2) that were constructed based on a poly(amidoamine) scaffold. This method of polymeric construction gives rise to protein-like polymers that contain a defined number of generations and surface functional groups and excellent monodispersity. Indeed, dendrimers have been referred to as "artificial proteins" due to similarities in size (approximately 1–20 nm in diameter) and shape, with some common proteins. By convention, a dendrimer structure is built up on both sides of a core to give rise to a regular, spherical structure. In contrast, construction of the dendritic structure on only one side of the core gives rise to a more "tree-like" structure or "dendron" (Fig. 1). As a general rule, the attainment of ideal monodispersity is limited to dendrimers up to approximately six generations in size, since beyond G6, the addition of subsequent layers or generations is partially hindered by steric constraints at the surface.

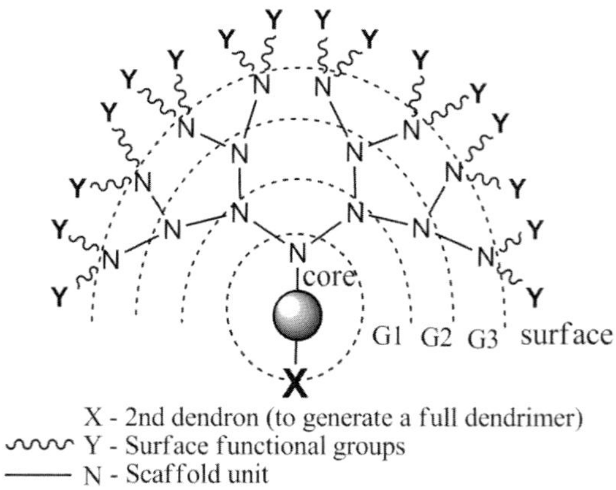

Fig. 1. The basic structure of a dendrimer, where the scaffold can be built up in only one direction from the core (to construct a half dendron structure) or on both sides of the core to create a full dendrimer. Each additional layer of monomers (or scaffold units, N) is one additional generation (G). The number of surface groups (Y) is dictated by the number of functional groups present on the monomers, generation, and dendrimer structure (dendron or full dendrimer). Figure is modified with permission from Ref. 2.

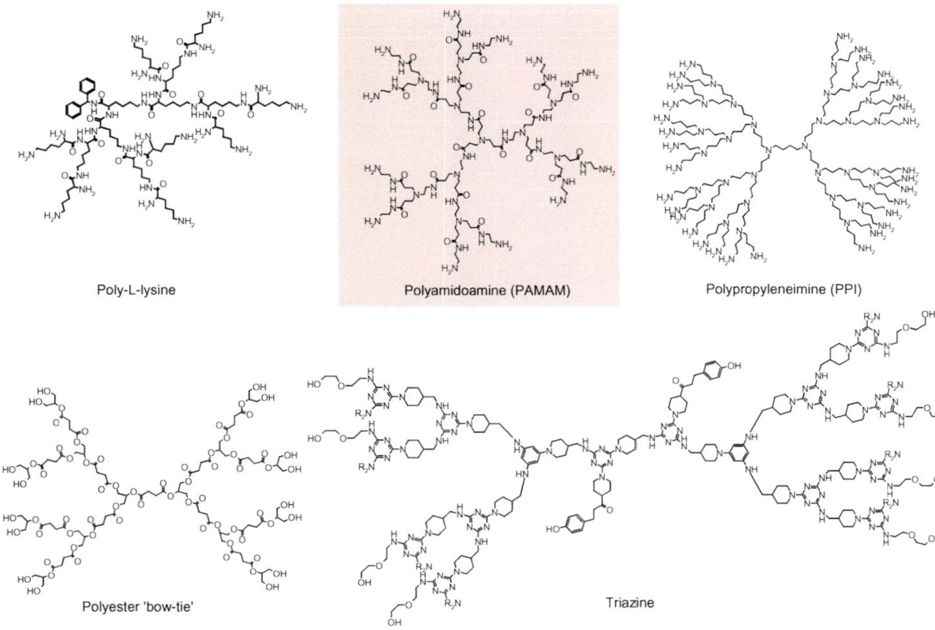

Fig. 2. Example of some of the most common dendrimer structures. Early dendrimers constructed were based on the PAMAM scaffold shown in red.

Nevertheless, dendrimers containing >200 reactive surface sites may be produced, allowing for facile conjugation of drugs, targeting groups, imaging agents or antifouling functionalities to modify the chemical or *in vivo* properties of the macromolecule.

Historically, dendrimers have been constructed in one of two ways: i.e. via convergent or divergent synthesis. The convergent approach was developed more recently by Hawker and Frechet[3] in 1990 and involves the construction of the dendrimer scaffold by first building individual dendron segments containing many generations that are then conjugated to the core to generate a full dendrimer. The divergent synthetic techniques that were used to construct the first PAMAM dendrimers, were developed by Tomalia[1] and involve the step-wise addition of monomers to a central core, effectively building one generation at a time. The final structure generally contains a hydrophobic scaffold (that may be used to non-covalently encapsulate small molecules) and a polyfunctional surface that may be modified via the covalent conjugation of additional functionalities. From a commercial viability standpoint, the complexity of the original dendrimer synthetic processes had the potential to lead to high costs of goods. This has led to the more recent development of alternate (and simpler) synthetic methods. For example, methods employing click chemistry based on Cu(I) catalysis of alkyne-triazole reactions[4] and what has been described as the "lego" approach, that employs highly branched monomers attached to polyfunctionalized cores.[5]

PAMAM dendrimers are the most commonly explored dendrimers in the literature. This in part stems from their commercial availability and a long history of exploration for a range of biomedical applications. However, a wide range of alternate dendritic structures have been synthesized, since chemical modifications can be made to the core, the scaffold and the surface leading to the generation of dendritic structures with widely varying biological and chemical properties (Fig. 2). Whilst PAMAM and poly (propylene imine) (PPI) dendrimers represent excellent and relatively biologically inert scaffolds for biomedical applications, they can also have an intrinsic effect on biological processes, and PAMAM dendrimers, for example, can inhibit the activity of acetylcholinesterase. Concerns about the potential toxicity associated with prolonged *in vivo* exposure of nanosized drug carriers and accumulation in certain susceptible organ systems has also driven recent interest in the development of biodegradable dendrimers that are constructed from biogenic esters or amino acids. As such, the range of dendrimers that have been explored for potential biomedical applications continues to increase.

This chapter describes some of the common biological applications for which dendrimers are being explored. To this point, interest has focused largely on potential utility as carriers and targeting vectors for drugs and genetic material, and as improved magnetic resonance imaging agents. Dendrimers have also been designed to provide anti-infective properties against virus transmission and bacterial growth and have been suggested to help purge the body of disease-causing proteins such as PrP^{Sr}, which causes Prion disease, and β-amyloid, which can precipitate Alzheimer's disease. There is also a significant research focus on using dendrimers in tissue engineering; as protein mimetics and scaffolds for wound repair and tissue growth. Finally, dendrimers have a wide range of other potential biomedical applications, including as agents to remove metal toxins from the body, as bioenzymes, biosensors, biomedical coatings for artificial joints and artificial red blood cells.

2. Dendrimer-Based Companies

Given the diverse applications and interest in dendrimer-based nanotechnology, a small number of companies have emerged that specialize in the manufacture of dendrimers for research purposes, or that have been established to conduct research and development into dendrimer-based products. The first of these was Dendritech, which was started in 1992 by the co-founder of dendrimer synthesis, Donald Tomalia. Dendritech remains active and is focused on the production and supply of PAMAM dendrimers. Subsequently, two further dendrimer-based companies were established in the US and Australia. Dendritic Nanotechnologies Inc started around 2001 in the US and developed the Priostar™ dendrimer portfolio and produced and sold Starburst™ dendrimers. Dendritic Nanotechnologies Inc became a full subsidiary of Starpharma Pty. Ltd in 2006.

Starpharma Pty. Ltd. was established in 1996 in Melbourne, Australia, and is a research and development company with a focus on the commercialization of dendrimers. To date, Starpharma has been the only biotech company to have advanced a dendrimer-based product into FDA approved clinical trials. Starpharma have a microbicidal gel "VivaGel," currently in late-stage trials, for bacterial vaginosis that has applicability in both treatment and prevention. The gel is also in the clinic as a vaginal microbicide to offer safer sex by preventing transmission of genital herpes (HSV-2), HIV and other sexually transmitted infections (STIs) including human papillomavirus (HPV), a virus implicated in the epidemiology of cervical cancer. Starpharma has also developed a growing pipeline of broader applications of their dendrimer technology, notably in drug delivery and agrochemicals. Within the drug delivery arena, Starpharma has applied dendrimer-based technologies in its own development programs, in collaboration with academic groups including ourselves (e.g. in the improved delivery of doxorubicin, methotrexate and docetaxel), and in partnership with pharmaceutical companies, including Eli Lilly, GSK's Stiefel, Astra Zeneca and a number of other undisclosed partners. The technology has applicability for many different diseases and drug classes.

More recently, companies such as The Dendrimer Company and Colcom have begun commercializing dendritic scaffolds for research purposes. The Dendrimer Company manufacture dendrimers based on a 2,2-bis(methylol) propanoic acid scaffold, whilst Colcom are commercial suppliers for poly L lysine-based dendrimers. Most recently, Donald Tomalia has founded NanoSynthons, a company dedicated to the manufacture of precise nanoscale materials.

3. Dendrimers as Drug Delivery Vectors

3.1. *Overview of dendrimer pharmacokinetics and biodistribution*

Understanding the relationship between the physicochemical properties of drug carriers and their *in vivo* behavior is central to the drug delivery system design. For dendrimers, *in vivo* disposition is, in large part, a function of dendrimer size and surface properties. This relationship has been reviewed in detail recently,[6,7] but is summarized briefly below. In general, dendrimers with unmodified cationic surfaces (such as amine-terminated constructs) display very rapid plasma clearance after intravenous administration, with elimination half-lives of only a few minutes. This appears to reflect rapid and extensive binding to anionic sites on the vascular endothelium and likely results in biodistribution behavior that reflects the density of blood vessels within any given organ, rather than any particular site-selective dispositional properties of the dendrimer. This behavior is also a function of generational size, where higher-generation dendrimers, with more surface cationic groups, display more rapid and extensive tissue binding. As a result, cationic dendrimers are generally not excreted via the urine, and where biodegradation is slow, dendrimers are retained in the body for prolonged periods of time. In contrast, cationic dendrimers constructed with natural amino acids or esters that are rapidly degraded *in vivo*, show markedly differing profiles. For dendrimers constructed from L-lysine, for instance, liberation of L-lysine monomers leads to reincorporation into resynthetic pathways and appearance of macromolecular materials containing lysine fragments (presumably proteins) in the plasma.[8]

The high cell- and tissue-binding affinity of cationic dendrimers provides opportunities for alternate applications, and cationic dendrimers have been widely studied as a means of enhancing transdermal and oral delivery of associated drugs (discussed in a later section). The electrostatic interaction of the cationic surface with anionic sites on the cell membrane may also promote internalization of the dendrimer, usually via clathrin-mediated endocytosis.[9] This is the basis for the commercial success of SuperFect, an amine-terminated PAMAM dendrimer that is used as an *in vitro* gene transfection reagent.[10] Conversely, the high tissue binding of cationic dendrimers and potential to permeabilize cell membranes has raised questions regarding the *in vivo* toxicity of these materials. In general, therefore, the rapid clearance and questionable toxicity of intravenously administered unmodified cationic dendrimers limit *in vivo* application.

Dendrimers with anionic surface functionality appear to display more prolonged blood residence times than similar generation/scaffold cationic dendrimers.[8,11] Blood residence times, however, are a function of both overall anionic charge and size, and smaller (<10 kDa) dendrimers with lower numbers of surface charged groups are rapidly eliminated from the blood via the urine with elimination half-lives of less than 1 h (reviewed in Refs. 6 and 7). In contrast, larger dendrimers, with larger numbers of charged surface groups, appear to bind plasma proteins more readily (opsonisation) and are retained in blood for longer periods (with terminal half lives of ~ 1 h), possibly by virtue of the increased size of the opsonized construct reducing renal filtration.[11] The polyanionic surface may also hinder glomerular filtration when compared to uncharged dendrimers of similar molecular weight due to electrostatic repulsion by sialic acid contained in the glomerular basement membrane.[12] The anionic charge present in the glomerulus prevents the unwanted passage of endogenous plasma proteins, such as albumin,[12] that are required to maintain fluid homeostasis. The rate of biodegradation of anion-terminated polyaminoacid dendrimers, such as those constructed from L-lysine, is also slower than for equivalent cation terminated systems, presumably as a result of modification of the outer amino acid layer masking cleavage points within the dendrimer scaffold. The reduced cellular affinity of anionic dendrimers dictates that these systems are relatively non-toxic. Indeed, anion-capped poly-L lysine dendrimers are currently in clinical trials for a number of applications as antiinfectives (see section "Dendrimers as Drugs").

For uncharged dendrimers, plasma circulation times and excretion are a function of molecular weight, renal clearance and patterns of recognition with cells of the mononuclear phagocyte system (MPS). Thus, small-molecular-weight dendrimers (<20 kDa) display relatively short plasma circulation times and are excreted rapidly via the urine. In contrast, higher-molecular-weight dendrimers avoid renal filtration, display more prolonged circulation times and can have terminal plasma half-lives of several days if recognition by the MPS is suppressed. This is discussed in more detail below.

3.2. *Long-circulating, sustained-release drug delivery systems*

For many drug molecules, low water solubility and rapid clearance present a challenge to effective development. PEGylated dendrimer–drug conjugates provide one means to both improve drug solubility and to prolong drug circulating times. Similar to the well-described ability of PEGylation to mask interaction of proteins, liposomes and nanoparticles with opsonic factors and phagocytic cells, and to prolong *in vivo* half-life, PEGylation can enhance the *in vivo* exposure of dendrimers. In this case, PEGylation provides two major advantages; first: PEGylation increases molecular weight and therefore reduces renal clearance, and second, the PEG chains serve to inhibit recognition by the MPS. We have previously described a relationship between the molecular weight of the PEGylated complex and plasma half-life and describe a molecular weight cut-off of approximately 20 kDa as the minimum size to significantly reduce renal clearance.[6,13]

Drugs may be associated with PEGylated dendrimers by covalently conjugating to the surface via stable or labile chemical linkers, or by non-covalently encapsulating drug within the dendrimer structure. These approaches have been applied with varying degrees of success to improving the solubility and circulation times of chemotherapeutic, anti-inflammatory, antimicrobicidal and antihypertensive drugs.[14] Either approach may be useful for prolonging the plasma exposure of small molecule drugs, and each has its own advantages and disadvantages. These have been reviewed recently.[15] In brief, covalent drug attachment enables better control over the rate of drug liberation but can provide a challenge in developing approaches to liberate unmodified drug at sufficient rates to provide therapeutic concentrations at the disease site. For example, enzyme-mediated cleavage of peptide-based linkers may result in the release of drug conjugated to a peptide "stub" i.e. the cleavage product of the linker, and the peptide-modified drug may display reduced *in vivo* activity when compared to the unmodified parent drug. In contrast, non-covalent entrapment provides a facile means of drug loading of dendrimer-based delivery systems, but commonly results in burst release profiles and poor control over the rate of drug release, particularly in biological fluids. For non-covalent entrapment, drug release profiles are dictated by dendrimer size, the nature of the scaffold and surface and the physicochemical properties of the drug to be entrapped. For instance, drug entrapment may occur via H-bonding or hydrophobic interactions within the scaffold, or via electrostatic interactions with charged interfacial species. Drug entrapment within PEG layers has also been reported. Variations in solvent properties (such as pH or salt content), the physicochemical properties of the drug (especially pKa) and the chemistries of the dendrimer scaffold and surface all influence the efficiency by which drugs are entrapped and subsequently released (reviewed in Ref. 15).

Much of the attention focused on long-circulating, dendrimer-based drug delivery systems has been directed toward the formulation of chemotherapeutic and non-steroidal anti-inflammatory drugs (NSAIDs). In the case of anticancer therapy, prolonging the plasma exposure of chemotherapeutics can lead to more sustained drug delivery to cancer cells (and therefore improved cancer killing), however, long-circulating nanomedicine approaches, including dendrimer-based delivery systems, have been more commonly assembled to take advantage of passive tumor-targeting properties. These are discussed in more detail in the following section. In contrast, for NSAIDs, long-circulating dendrimer-based "depot" formulations have been employed in an attempt to overcome solubility limitations and to prolong the duration of analgesia and inflammation suppression. For example, flurbiprofen release from a G4

PAMAM dendrimer was sustained for more than 8 h *in vitro* and after *i.v.* administration of the drug–dendrimer complex, the terminal half-life of the drug was increased three-fold when compared to *i.v.* administration of flurbiprofen alone.[16] This led to a more powerful pharmacodynamic response (decreased paw oedema) for a longer period of time when compared to *i.v.* administration of the drug alone. In a more recent study, entrapment of ketoprofen in a PAMAM dendrimer moderately prolonged plasma exposure of the drug after oral administration when compared to ketoprofen alone.[17] The resulting antinociceptive response in a mouse acetic acid writhing model was also significantly prolonged when compared to the oral administration of ketoprofen alone, with 50% or greater antinociceptive activity lasting for ~ 8 h for the dendrimer–drug complex versus 3 h for the drug alone. However, whether the latter effect represented sustained absorption of free ketoprofen or absorption of the dendrimer complex followed by release of ketoprofen was not identified. In general, although a number of studies have described the complexation and *in vitro* drug release kinetics of dendrimer–drug complexes, examples of *in vivo* pharmacodynamic studies are far less prevalent.

3.3. *Site specific delivery to tumors*

The *i.v.* chemotherapy using simple drug solution formulations is often effective in killing cancer cells, however, the lack of pharmacological or dispositional specificity dictates that non-target toxicity typically limits effective therapy. For this reason, drug delivery approaches that reduce circulating concentrations of free drug, and maximize (or target) drug exposure to cancer cells, are highly sought.

After parenteral administration, long-circulating dendrimers preferentially accumulate in solid tumors, and the extent of tumor uptake is, to some extent, correlated with the elimination half-life of the dendrimer (reviewed in Ref. 6). The passive targeting properties of dendrimers (and other long-circulating drug delivery systems) result from architectural defects in tumor blood vessels that increase vascular permeability and promote extravasation of macromolecules and nanomaterials. The increase in vascular permeability and tumor uptake of macromolecules occurs coincidently with a reduction in lymphatic drainage from the tumor interstitium, due to reduced lymphatic functionality. This phenomenon is termed the enhanced permeation and retention (EPR) effect and was first described by Maeda.[18]

Dendrimer-based drug delivery systems also take advantage of differences between tumor and normal tissue microenvironments, such as reduced pH or tumor-specific overexpression of enzymes, in order to stimulate site-selective drug liberation. In combination with the EPR effect, these properties have been widely exploited to develop dendrimer-based chemotherapeutic nanomedicines that more selectively deliver cytotoxic drugs to tumors, but that limit exposure to non-target sites and sites of toxicity (reviewed in Ref. 6). Dendrimer-based delivery systems also have the potential to enhance chemotherapy by overcoming drug resistance mechanisms. This is achieved by promoting drug uptake into the tumor by mechanisms (endocytosis) that limit drug access to efflux mechanisms.[19] The relatively small size of dendrimers, in contrast to larger nanoparticles and colloidal systems (approximately 10 nm diameter for the dendrimer compared to ~100 nm diameter or greater for liposomes) may also enable better penetration of the drug into a tumor mass, away from the site of extravasation, a process that is disfavored for larger systems.[20]

While the EPR effect is most commonly explored as a means to improve the delivery of nanomedicines to solid tumors, it is not unique to cancer. Macromolecules also display enhanced organ uptake in disease states that have pathologies involving rapid blood vessel growth or increased vascular permeability such as sites of inflammation[21] and infarction.[22,23]

By associating chemotherapeutic drugs with long-circulating dendrimer carriers, concentrations of free drug in the blood are reduced/minimized, and drugs are distributed more specifically to solid tumors. For example, previous work from our laboratories have shown that *i.v.* administration of a PEGylated G5 polylysine dendrimer decorated on 50% of the surface with doxorubicin via an acid-labile hydrazone linker leads to approximately 10-fold lower concentrations of free drug in the blood, and three-fold increases in tumor uptake when compared to *i.v.* administration of free doxorubicin[24] (Fig. 3). In this case the antitumor efficacy of the dendrimer (using a Walker 256 rat breast carcinoma xenograft) was similar to that of *i.v.* doxorubicin, however, systemic toxicity was reduced. In contrast, others have reported significantly enhanced antitumor efficacy following the administration of drug-loaded PEGylated dendrimers when compared to equivalent doses of the unconjugated drug.[19,25] The efficacy of tumor-targeted drug–dendrimer complexes is commonly thought to reflect how effectively and rapidly drug is liberated from the dendrimer carrier once within the tumour environment.[15] For instance, covalent association of a drug with a dendrimer via a chemical linker that releases drug slowly in the acidic environment of a solid tumor may have reduced antitumor activity when compared to a linker that rapidly liberates drug at low pH or in the presence of enzymes that are overexpressed in the tumor microenvironment, such as matrix metalloproteinases.[26] However, the ideal design of drug–dendrimer conjugation platforms remains

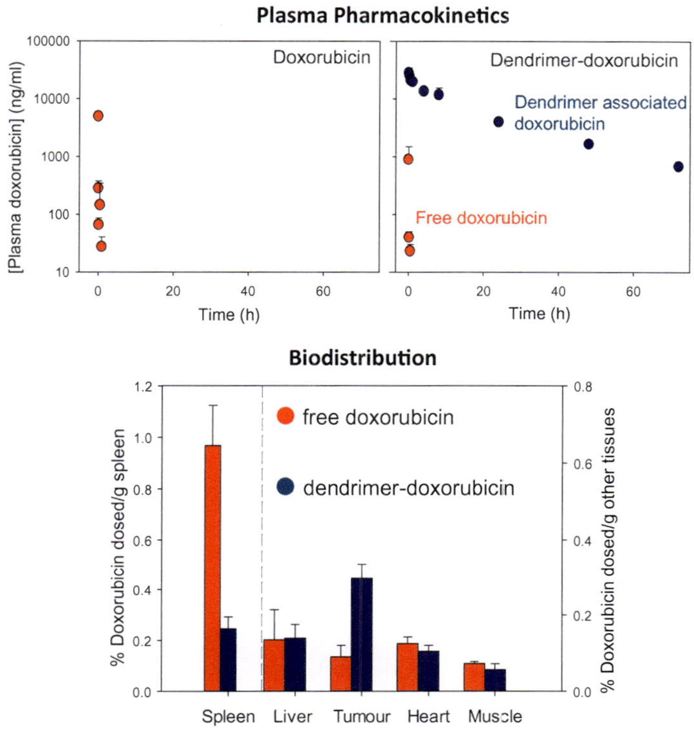

Fig. 3. Plasma pharmacokinetics and biodistribution of a solution formulation of doxorubicin and a PEGylated G5 polylysine dendrimer conjugated with doxorubicin via an acid-labile hydrazone linker. Figure is adapted and modified with permission from Ref. 24.

contentious and may, for example, vary depending on whether drug release occurs pre-endocytosis (where slower release may allow continued exposure at the target site) or post-endocytosis, where retention at the site is unlikely to be a limitation and, therefore, more rapid release is likely to be beneficial.

Passive tumor targeting therefore provides a simple means of improving the tumor-specific disposition of cytotoxic drugs. The specificity of passive targeting, however, is limited, and the requirement for a long-circulating system can ultimately lead to uptake into organs of the MPS such as the liver and spleen. To enhance targeting efficiency, nanomaterials, including dendrimers, can also be conjugated with ligands or antibodies that target receptors overexpressed on the surface of cancer cells (common examples include receptors for folate, epidermal growth factor (EGF) and leutinizing hormone releasing hormone). This approach has the potential for rapid and site specific "active" targeting to cancer cells and may reduce the requirement for long-circulating properties. In turn, this may be beneficial for dendrimers that are associated with chemotherapeutic drugs via non-covalent encapsulation and that display burst release properties. However, when combined with long-circulating carriers that also display passive tumor targeting, the tumor uptake of dendrimer-based drug carriers can be significantly enhanced. For example, a folate-targeted G5 PAMAM dendrimer administered to mice-bearing KB tumors displayed three-fold higher tumor uptake when compared to a non-targeted construct that relied on passive uptake alone.[27] Other examples of actively targets systems are reviewed in Ref. 28.

Actively targeted systems are also likely to be more avidly internalized via receptor-mediated endocytocis, potentially enabling higher intracellular concentrations of drug to be achieved. Drug linkers commonly employed to enable drug liberation from dendrimers within lyso-endosomes include cathepsin-cleavable peptides or acid labile linkers such as hydrazones and esters. The advantage of using acid-labile drug linkers in combination with cancer-targeting ligands is that the pH within lyso-endosome (pH 5–6) is lower than that in the tumour interstitium (pH 5.7–7.8) and may therefore promote enhanced drug liberation.

Although actively targeted dendrimer-based drug delivery systems provide a means to enhance delivery to tumor cells, disadvantages are also evident. Notably, targeted delivery systems also have the potential to deliver the drug to non-target organs by virtue of the elevated expression of the same receptors elsewhere in the body. For instance, while cancer cells often overexpress folate receptors to maintain their increased metabolic activity, folate receptors are also overexpressed in the lung and kidneys, and the use of folate as a targeting agent runs the risk of promoting parallel uptake into these organs. Antibody-based targeting, where antibody specificity is greater than that of an overexpressed enzyme, may provide for additional selectively and for antibodies with anticancer activity (for instance a humanized monoclonal antibody to Her-2) antibodies have the potential to actively deliver drug to a solid tumor and also to inhibit tumor growth and metastasis. However, antibodies are large, limiting the number of targeting groups that might be conjugated to a single dendrimer, and synthetic conjugation is complex and highly hindered. Under these circumstances, antibody fragments may provide the benefits of specific targeting with the advantages of a smaller, more synthetically tractable species. Finally, although active targeting provides a high affinity means of capture of dendrimer-based drug delivery systems by target cells, the efficiency of this process may limit further diffusion of the delivery system into the tumor mass. The relative benefits of active versus passive targeting approaches remain contentious.

3.4. *Methotrexate and folate conjugated dendrimers as macrophage-targeting systems for the treatment of rheumatoid arthritis*

Thomas *et al.* have reported a PAMAM dendrimer, surface functionalized with a mixture of folic acid (FA) and methotrexate (MTX) that displays activity in the treatment of inflammatory arthritis.[29] The G5-PAMAM dendrimer (G5-PAMAM-FA-MTX) was found to bind to, and be internalized by, folate receptor β-expressing macrophages. Rats with collagen-induced arthritis were treated with G5-PAMAM-FA-MTX, and the dendrimer was effective in reducing inflammation parameters such as bone resorption, cartilage damage and ankle damage. In this rat model, G5-PAMAM-FA-MTX was found to have a greater therapeutic index than MTX alone.

3.5. *Delivery applications in the CNS*

3.5.1. *Dendrimer-mediated transport across the blood–brain barrier*

The blood–brain barrier (BBB) remains a major obstacle to the effective pharmacological treatment of many neurological disorders, including glioma, neuroinflammation, neurodegenerative diseases and viral infections. The BBB acts to physically impede the diffusion of drugs and foreign substances from the systemic circulation into the brain. It is made up of neurovascular units that consist of endothelial cells separated by tight intercellular junctions and underpinned by a basement membrane, that are bordered by pericytes and astrocyte end feet (Fig. 4). This physical barrier is further strengthened by the presence of multidrug resistance proteins (e.g. the efflux pump P-glycoprotein (P-gp)) that are highly expressed on the BBB endothelium and act as a biochemical barrier to transcytosis. The combination of physical and biochemical barriers to delivery at the BBB results in poor delivery of many otherwise effective drugs to the brain.

Dendrimer-based drug delivery systems provide a means to overcome diffusion and efflux-related barriers to BBB transport by stimulating alternate routes of drug transport (Fig. 5). For example, dendrimer transport across the BBB has been reported to occur via clathrin and caveolin mediated endocytosis[30] and via specific receptor-mediated endocytosis pathways driven by brain-targeting ligands conjugated to the dendrimer surface. The (im)permeability of the BBB is reflected in high transendothelial electrical resistance (TEER), approximately three-fold higher than in other tissues. Cationic dendrimers, however, have been shown, at least *in vitro*, to penetrate epithelial

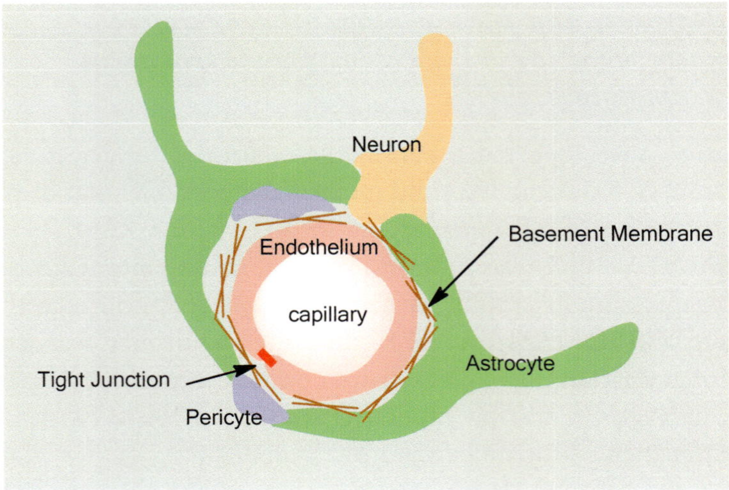

Fig. 4. Schematic representation of the blood–brain barrier showing a cross section through a brain capillary. The neurovascular unit consist of an endothelial cell enclosed by tight junctions with an underlying basement membrane surrounded primarily by pericytes and astrocyte end feet.

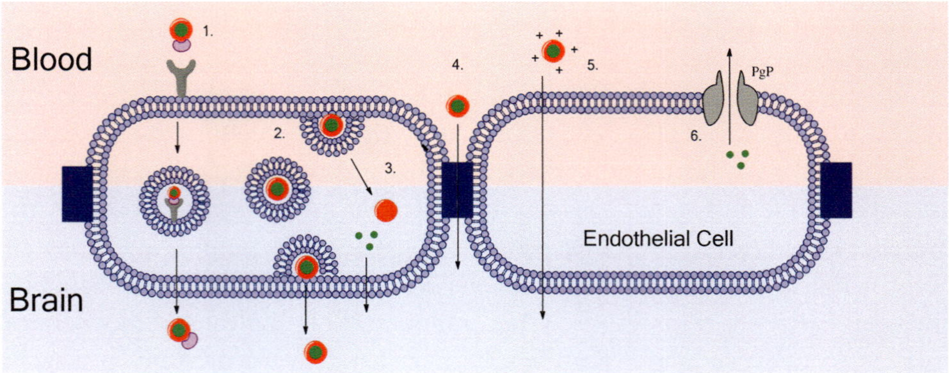

Fig. 5. Potential pathways of dendrimers across the blood–brain barrier. 1. Receptor-mediated endocytosis/transcytosis. 2. Membrane transcytosis. 3. Intracellular degradation/lysosome uptake/drug release. 4. Paracellular uptake/disruption of tight junctions. 5. Electrostatic interactions. 6. Bypass drug efflux pumps.

monolayers following initial electrostatic binding to the cell membrane and transient reduction in transepithelial electrical resistance values, indicative of an opening of tight junctions.[31] While this has not been explored specifically at the BBB, similar uptake mechanisms might be expected to occur with cationic dendrimers as seen with other cationic nanoparticles.[32] There is also evidence suggesting that cationic surface charge can also involve uptake at the BBB via transcellular pathways. Nonetheless, whilst cationic dendrimers are capable of opening tight junctions *in vitro*, their applicability *in vivo* may be limited by rapid clearance and toxicity concerns.

A key issue in evaluating the potential for the transport of drugs and drug delivery constructs across the BBB is the potential for pathological changes to the BBB to alter permeability. One successful demonstration of improved therapeutic outcomes for a brain-targeted dendrimer is provided by the utilization of PAMAM-based dendrimers to improve the delivery of the anti-inflammatory drug N-acetylcysteine (NAC) to the brains of rabbits with experimentally induced cerebral palsy. In these studies by Kannan and colleagues,[33] a NAC-conjugated, hydroxyl-terminated G4 PAMAM dendrimer selectively targeted inflammatory activated microglia and astrocytes in diseased brains but did not do so in healthy controls. Furthermore, when administered systemically, the dendrimer–drug conjugate resulted in significantly improved motor function in affected rabbits when compared to data obtained after administration of up to 10-fold higher doses of free NAC. Importantly, the dendrimer–drug complex showed no evidence of toxicity at the therapeutically relevant doses employed. This study highlights the potential for dendrimers to deliver neuroprotective agents to the brain and also suggests that dendrimer-based constructs may have benefits in the treatment of brain-related inflammatory disorders, such as inflammation resulting from injury or neurological disorders.

3.5.2. *Enhanced drug delivery to brain tumours*

The properties of the vasculature surrounding brain tumors can vary depending on the grade/stage of the tumor and can also affect the targeting mechanism for successful delivery across the BBB. Current dendrimer-based treatment strategies against glioma (an aggressive form of brain cancer with generally poor prognosis) are focused on employing dual targeting mechanisms for both drug and gene delivery applications.[34,35] In low-grade glioma, the BBB remains largely intact, and active targeting mechanisms are required to penetrate the BBB and to specifically target glial cells. In high-grade gliomas, however, BBB properties are often attenuated, allowing dendrimer uptake by passive targeting mechanisms, such as the EPR effect. PEGylation of dendrimers has been used as an approach to achieve accumulation into gliomas through EPR. It also provides a neutral surface charge for dendrimers such as PAMAMs, and the lower polarity of the PEG group favors endocytic uptake into cells rather than electrostatic interaction. Brain targeting ligands such as transferrin, lactoferrin and angiopep-2, which are highly expressed on glioma cells, are commonly used to help target glioma cells. P-gp efflux inhibitors may also assist in accumulation in the brain.

3.5.3. *Solubilization of disease-causing proteins in the brain*

Protein aggregates or amyloids are present in the brain in many neurological diseases as a result of protein misfolding. Alterations in the secondary structure of these proteins occurs most notably through increased folding into β-pleated sheet conformations, stimulating aggregation to form large insoluble fibrils that are resistant to proteolysis. The accumulation of these insoluble proteins causes inflammation and neurotoxicity that is associated with progressive neurological dysfunction. Since dendrimers have the capacity to solubilize hydrophobic drugs and to bind to proteins, a number of investigators have sought to explore their ability to "trap" disease-causing proteins. Two proteins have been explored in some detail; β-amyloid (Aβ) (the primary component of plaques found in Alzheimer's disease), and the infectious "scrapie" isoform of prion protein (PrPSc) that causes prion disease.

Supattapone and colleagues[36,37] demonstrated that PAMAM and PPI dendrimers were able to clear pre-existing PrPSc from neurons in cell culture by modifying the protein aggregates into protease-sensitive forms while maintaining cell viability. The properties of the dendrimers that were found to be important in mediating effective modification of the aggregates were dendrimer size or generation, where more effective results were achieved with G4 or larger dendrimers, and where the surface density of primary amine groups on PAMAM and PPI dendrimers only stimulated protein degradation at low pH (<4) suggesting that protonation of surface amines was important for activity. Similarly, cationic phosphorous containing dendrimers (CPD) were shown to stimulate degradation of prion aggregates, although in this case, activity was evident at neutral pH, most likely due to more effective retention of surface charge at pHs higher than that required for PAMAM and PPI dendrimers. Klajnert and colleagues[38–40] have shown similar findings for Aβ aggregates *in vitro* with protein aggregates being reduced by CPDs, PAMAM and PPI dendrimers. It appears, therefore, that slight changes in dendrimer structure or composition can have a significant impact on the susceptibility of different peptide formations to disaggregate.

Much of the current work on β-amyloid has focused on preventing the formation of early fibrils by "trapping" the protein prior to aggregation. Similar to other fibril-disrupting agents, dendrimers have been shown to stimulate the de-aggregation of pre-formed fibrils in a concentration-dependent manner.[41] Specifically, at higher concentrations, dendrimers were shown to stimulate break up and prevent reformation of amyloid fibrils *in vitro*, whereas at low concentrations, dendrimers appeared to accelerate fibril formation by providing a larger amount of broken fibril ends which serve as precursors to further fibrillogenesis. To this point, however, there have been no reports of the ability of dendrimers to remove soluble β-amyloid or fibril fragments from the brain, thereby preventing incorporation into fibrils and plaques. The utility of dendrimers in providing a means to enhance the treatment of Alzheimer's therefore remains unknown. For all CNS applications, dendrimer transport across the BBB and access to the brain *in vivo* will continue to provide a challenge to utility.

3.6. *Drug delivery vectors for dosing via non-intravenous routes (transdermal, oral, lungs)*

Although dendrimers have the ability to improve the pharmacokinetics and site selective disposition of drugs after *i.v.* administration, the *i.v.* route requires administration by trained medical professionals. There is, therefore, considerable interest in delivering dendrimer-based therapeutics to the systemic circulation via less invasive routes (such as subcutaneous [*s.c.*], intramuscular [*i.m.*], oral, transdermal or pulmonary administration) that can be self-administered. In some cases, non-intravenous delivery routes also provide a means for targeting to a specific site of action (e.g., the lungs or lymphatics).

3.6.1. *Interstitial (s.c., i.m.) delivery*

Much of the work conducted on interstitially injected dendrimers has focused on their capability to resolve draining lymphatics and sentinel lymph nodes in an attempt to identify lymph-resident metastases via magnetic resonance lymphangiography (MRL). This work is described in more detail in a later section of this chapter. However, some of the work conducted in this area has described the pharmacokinetic behavior of dendrimers after *s.c.* administration, and their potential to act as systemic drug delivery vectors via this less invasive route.

The earliest studies in this area were conducted by Kobayashi and coworkers and described the absorption patterns of PAMAM dendrimers after interstitial administration into mice.[42–45] The work was mainly focused on the use of PAMAM dendrimers containing contrast agents (e.g. Gd(III)) to resolve the lymphatics draining an injection site. The authors showed that small G2 dendrimers (3 nm in diameter) rapidly spread from the injection site, leading to poor resolution of the draining lymphatic ducts and sentinel nodes. These dendrimers were also cleared rapidly via urinary excretion. Larger G6 dendrimers (approximately 9 nm in diameter) showed limited spreading from the injection site, but did appear to drain into the local lymphatics and as such gave excellent resolution of draining lymph ducts and sentinel lymph nodes. Much larger (G8, ~12 nm diameter) dendrimers were retained at the injection site, and disposition into lymph ducts and sentinel lymph nodes was considerably slower and gave only weak lymphatic signals. In addition, Kobayashi showed that more hydrophobic dendrimers (such as PPI dendrimers containing a diaminobutane core) displayed better lymphatic resolution properties by MRI than equivalently sized hydrophilic PAMAM dendrimers.[45] Given the propensity for MPS uptake of hydrophobic dendrimers, it is possible that improved lymphatic resolution resulted from increased retention of the PPI dendrimers by macrophages in sentinel lymph nodes. This work provided a preliminary indication of the likely drainage patters of dendrimers from interstitial injection sites, and suggested that dendrimer size and hydrophilicity dictate (i) the tendency for dendrimers to spread from the injection site and (ii) the extent of lymphatic absorption. From a drug delivery perspective, however, this early data gave no indication of the bioavailability of interstitially injected dendrimers and did not attempt to quantify predisposition towards lymphatic or vascular absorption.

More recently, we have attempted to delineate these issues in greater detail using a thoracic lymph duct cannulated rat model.[46] In these studies, PEGylated G4 polylysine dendrimers were completely absorbed from subcutaneous injection sites, however, the rate and route of absorption was dependent on molecular weight (which for a series of G4-based dendrimer cores was dictated by the degree of PEGylation) (Fig. 6). Thus, small (10 kDa) PEGylated polylysine dendrimers were rapidly absorbed from the injection site via the blood, whereas a similarly sized 10 kDa anionic arylsulphonated dendrimer, displayed poor absorption from the s.c. injection site and was only ~25% bioavailable. As the molecular weight increased, the propensity for slower absorption and preferential absorption into the lymph increased, such that, for a 68 kDa dendrimer, approximately equal quantities of dendrimer were absorbed into the lymph and the blood. The data suggest that the interstitial delivery route provides an effective means by which dendrimer-associated drugs may be delivered to the systemic circulation, that PEGylation provides a means to enhance drainage from s.c. injection sites and that plasma exposure and maximal plasma concentrations can be controlled in part by changing the molecular weight of the dendrimer carrier. There is also the potential to more specifically target drugs to the lymphatics via the s.c.

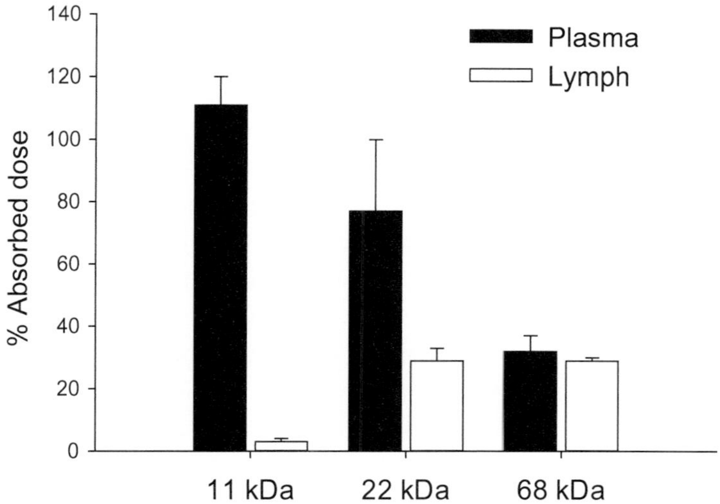

Fig. 6. Proportion of a subcutaneous dose of fully PEGylated polylysine dendrimers (11–68 kDa) administered to rats that was absorbed via the blood (reported value is fraction absorbed extrapolated to infinity) and via the lymph (reported value represents % absorbed dose over 30 h only). Rats were injected with 5 mg/kg dendrimer into the inner left ankle and lymph was collected from an indwelling cannula implanted in the thoracic lymph duct. Data are extrapolated from Ref. 46. Data represents mean ± s.d.

delivery of larger uncharged dendrimer carriers. This may have some benefit in optimizing the delivery of drugs toward lymphatic sites of disease progression, such as sites of lymphatic cancer metastasis or microbial infection.

3.6.2. *Oral delivery*

Oral administration is the benchmark for most drug delivery applications, as it enables facile and non-invasive systemic drug delivery and is well tolerated by patients. Many drugs, however, display poor oral bioavailability and in many cases, this reflects poor water solubility or low intestinal permeability. For this reason, there has been considerable interest in the use of dendrimers as a means to improve oral bioavailability. Accurate delineation of the mechanisms by which individual dendrimer–drug complexes improve the oral bioavailability of associated drugs is difficult, however, since most investigations monitor only the permeation of the drug and not the dendrimer. Indeed, to this point, *in vivo* evidence of significant dendrimer absorption is scant. Nevertheless, several investigations suggest the potential for dendrimers to promote oral bioavailability and these studies are described below.

Most of the work in this area has focused on examining the permeability of dendrimers (or dendrimer associated drug) across *in vitro* models of the human intestine, most commonly, the passage of labeled dendrimers across a monolayer of immortalized Caco-2 cells. Caco-2 cells are derived from a colon cell line but differentiate to form a polarized cell layer that displays a brush border on the apical surface and intercellular tight junctions and is similar to that of an intestinal absorptive cell. One limitation of the Caco-2 model, particularly in the context of absorption enhancement, is the lack of a protective mucous layer, as is evident *in vivo*. Nonetheless, using this model, several studies have shown that charged dendrimers promote drug permeability across Caco-2 cells, typically in the order cationic>anionic>uncharged dendrimers.[47] This has been suggested to reflect the capability of the charged dendrimers to reduce the integrity of the epithelial barrier by loosening tight junctions (cationic dendrimers) or by promoting cellular endocytosis following electrostatic binding to the cell membrane (cationic and anionic dendrimers). An elegant *in vitro* study with PAMAM dendrimers also recently reported that dendrimer endocytosis may be followed by the opening of epithelial tight junctions to promote a further enhancement in monolayer permeability via a positive feedback mechanism.[48] Dendrimer size (generation) has also been reported to have an impact on monolayer permeability, although conflicting evidence is apparent.[49,50] In general, increasing the size of cationic dendrimers appears to reduce Caco-2 monolayer permeability, whereas for anionic dendrimers, increasing size increases permeability (up to G4), and this has been suggested to reflect more efficient disruption of gap junctional proteins.

As described above, *in vivo* evidence of dendrimer absorption after oral administration is limited, although a number of studies have described increased bioavailability for drugs after administration of a drug–dendrimer complex. For example, *in vivo* studies have examined the oral bioavailability of silybin, after complexation with an amine-terminated G2 PAMAM dendrimer, and doxorubicin after complexation with an amine-terminated G3 PAMAM dendrimer in rats.[51,52] In these studies, complexation of silybin with the PAMAM dendrimer led to a 2-fold increase in oral bioavailability when compared to the administration of drug alone. However, the most significant improvement in oral bioavailability for a drug administered orally as a dendrimer formulation, was described with doxorubicin. Complexation of doxorubicin with a G3 PAMAM dendrimer resulted in a 300-fold increase in oral bioavailability in rats when compared to the oral administration of the free drug.[52] The mechanism by which this occurred is unclear, but may be due to either an increase in the solubility of the drug in the intestinal environment or the transport of the dendrimer–drug complex across the intestinal barrier in a manner that effectively masks doxorubicin interaction with intestinal P-gp efflux transporters. In summary, therefore, dendrimers show potential for improving the absorption of poorly bioavailable drugs; however, considerably more *in vivo* data is needed to provide evidence to support significant absorption of dendrimer (as opposed to co-administered drug) and to evaluate whether this technology can be translated into human use.

3.6.3. *Transdermal delivery*

In some instances, transdermal drug delivery may provide a means of enhancing systemic delivery of drugs, usually where oral bioavailability is limited by first pass metabolism. Transdermal delivery may also provide delivery advantages including avoidance of gastric disturbance or irritation and localized delivery to, for example, topical sites of infection or inflammation. However, percutaneous absorption is typically limited to low molecular weight (<300 Da), moderately lipophilic (log P 1–3) drugs with low melting points and low doses that are chemically stable on and in the skin. The formulation of drugs into nanosized drug carriers, provides some opportunity to improve chemical stability, may promote skin penetration and may also provide for continuous absorption when compared to direct application of the drug to the skin. Similar to the studies described above for oral administration, however, whether dendrimers themselves are absorbed across the skin, or whether they simply facilitate drug absorption is less clear. The increased barrier properties of the skin when compared to the intestine suggest that absorption of intact dendrimer in significant quantities is unlikely.

Many of the studies reported to this point have explored the potential for dendrimers to improve the percutaneous absorption of dendrimer encapsulated drugs and have almost exclusively examined drug absorption from PAMAM dendrimers through excised portions of porcine and rat skin. The data suggest that the ability of dendrimers to enhance transdermal delivery decreases in the order cationic>uncharged>anionic dendrimers.[53,54] Cationic dendrimers have been proposed to enhance percutaneous absorption due to their ability to adhere to the lipid bilayers in the skin. A more recent study, however, that directly examined the percutaneous absorption of FITC-labeled dendrimers, suggested that the percutaneous absorption of cationic dendrimers was, in fact, reduced when compared to uncharged and anionic dendrimers, that were absorbed via extracellular pathways. This was a result of cationic dendrimers displaying stronger cellular retention, rather than membrane permeation.[55] As expected, smaller dendrimers appear to show better permeability properties when compared to larger dendrimers that have a reduced capability to migrate between cells. The *in vitro* studies described above have also been supported by several *in vivo* studies that have examined the drug penetrating capabilities of PAMAM dendrimers containing 8-methoxypsoralene (as used in photochemotherapy) and the NSAIDs ketoprofen or diflunisal.[56,57] In the latter study, the transdermal delivery of the NSAIDs in a PAMAM dendrimer formulation led to prolonged analgesia in a mouse acetic acid induced writhing model when compared to application of the free drug. Bioavailability of the NSAIDs was also approximately 2.5-fold higher after skin application of the PAMAM-NSAID complex when compared to the free drug solution.

3.6.4. *Pulmonary delivery*

In addition to exploring the potential of dendrimers as drug delivery vectors via the oral and transdermal routes, there has also been some very recent work looking at the potential of dendrimer-based delivery systems to deliver drugs via inhalation. In this study, cationic PAMAM dendrimers were explored as a means to enhance the absorption of insulin and calcitonin as model peptide and protein drugs after pulmonary instillation to rats.[58] The absorption enhancing effects were in the order of G3>G2>G1>G0 dendrimers, and the dendrimers appeared to promote absorption by increasing the apparent positive charge of the proteins. The exact mechanism by which this increased lung permeability was not identified, but the PAMAM dendrimers were well tolerated, suggesting that the enhanced permeability was not accompanied by significant toxicity.

4. Dendrimers as Drugs

Multivalency is useful as a strategy to increase the affinity of compounds towards cellular targets that contain multiple binding sites.[59] The macromolecular structure of dendrimers makes them ideal as scaffolds for development of multivalent therapeutics.[60,61] Consequently, this concept has been exploited for the development of intrinsically biologically active dendrimers with antiviral, anti-bacterial, anti-cancer or anti-inflammatory properties.

4.1. *Polyanionic dendrimers as inhibitors of viral replication*

Polyanionic dendrimers have emerged as a particularly promising class of dendrimer therapeutics. A number have reported antiviral activity, and most appear to function by acting as viral entry inhibitors. Polyanionic dendrimers based on polylysine and PAMAM scaffolds have been reported to show antiviral activity against both HIV and Herpex Simplex Virus (HSV).[62–64] These dendrimers are decorated with carboxylic or sulfonate moieties and function predominantly as fusion inhibitors, disrupting the binding of viral surface proteins to host cell receptors. Dendrimers capped with sodium disulfonate containing moieties inhibit both HIV-1 viral entry, and HIV-1 reverse transcriptase (HIV-1 RT) activity.[62,63] Interestingly, polylysine disulfonate dendrimers did not inhibit HIV-1 RT activity within HIV infected cells; possibly due to poor cellular permeability and access.[60,61]

Tyssen *et al.* have described a structure–activity relationship for a series of polyanionic polylysine dendrimers as antivirals.[62] This study varied the generation size and surface group functionality for a series of constructs and correlated structure with activity against HIV-1 and HSV-2. Dendrimers with a generation size greater than two, and with higher anionic surface charge were found to exhibit the broadest potent antiviral activity. The most promising of these dendrimers, VivaGel®, developed by Starpharma Pty Ltd, has been found to be effective in the prevention of genital herpes infection (HSV-2), HIV and other sexually transmitted infections (STIs), including human papillomavirus (HPV). VivaGel® has also shown activity in the treatment of bacterial vaginosis, and is in clinical development as a topical microbicide for the prevention of bacterial vaginosis recurrence.[61,65] Other applications of dendrimers as antibacterials are described below.

4.2. *Polycationic dendrimers as antibacterials*

Several examples of polycationic dendrimers have been explored as potential antibacterials.[66] An early example was a poly(propyl imine) dendrimer, surface functionalized with dimethyldodecylammonium chloride moieties, that displayed antibacterial activity against a bioluminescent gram-negative *E. coli* strain TV 1048.[67] An EC_{50} (the concentration which reduces bioluminescence by 50%) of 12 μg/mL for the dendrimer was reported in comparison to 2000 μg/mL for the (non-dendrimer associated) small molecule analog, n-dodecyl-trimethylammonium chloride. The data reinforce the potential for multivalency to amplify monovalent activity. Later examples of polycationic–dendrimer–based antibacterials include an unmodified amine terminated G5-PAMAM dendrimer that showed activity against *Pseudomonas aeruginosa* (PA) and *Staphylococcus aureus* (SA).[68]

The mechanism by which polycationic dendrimers inhibit bacterial growth is believed to be by physical disruption of the cell membrane rather than interaction with specific proteins.[66] As such, polycationic dendrimers are also often toxic toward eukaryotic cells and are in general more toxic than the equivalent polyanionic species.[60,66,69] Consequently, attempts have been made to reduce toxicity and partial PEGylation of amine functionalities in G3-G5 PAMAM dendrimers has been shown to reduce cytotoxicity. In this case, however, the reduction in toxicity was also accompanied by reduced potency against PA and SA.[70]

In an attempt to reduce cytotoxicity while maintaining potency, Xue *et al.* have described a smaller (G2) amine terminated PAMAM dendrimer.[71] The G2 dendrimer displayed antibacterial activity against 15 of 17 bacterial strains investigated. *In vitro* cell viability assays were performed against the human gastric epithelial cell line GES-1, and G2 and PAMAM-NH_2 dendrimers displayed only limited toxicity. Interestingly, higher-generation dendrimers reduced cell viability more significantly. *In vivo* antibacterial activity was also investigated and G2 PAMAM-NH_2 increased the survival of *Escherichia coli* (EC) infected mice in a dose dependent manner. Mice treated with 10 mg/kg before and after infection with EC saw a survival rate of 100% after 24 h compared with 0% for control animals.[71]

4.3. *Peptide-functionalised dendrimers as inhibitors of metastasis*

The functionalization of dendrimers with peptides known to inhibit tumor growth and metastasis has been explored as a means of exploiting multivalency. For example, Nomizu *et al.* prepared polylysine dendrimers functionalized with Ac-YIGSR (a peptide derived from the laminin B1 chain) to give a series of dendrimers with intrinsic antitumor activity.[72] The dendrimer design was inspired by the use of polylysine dendrimers as scaffolds for the presentation of peptide antigens (multiple antigenic peptide systems (MAPs)) in order to promote the production of antibodies to the proteins from which the peptide was derived.[73,74] Thus, Nomizu prepared a series of polylysine dendrimers functionalized with 16 ($(\text{Ac-YIGSR})_{16}K_8K_4K_2KG$ — designated Y16), 8 ($(\text{Ac-YIGSR})_8K_4K_2KG$ — designated Y8), and 4 ($(\text{Ac-YIGSR})_4K_2KG$ — designated Y4) peptides. Lung colony counts in mice co-injected with B16-F10 melanoma cells and Y16 dendrimer were reduced by 97% from control. In comparison, co-injection with the same dose of single peptide (Y1) reduced colony counts by only 50%. Significant *s.c.* tumour growth inhibition was observed in mice administered Y16, while no inhibition was seen in mice treated with Y1.[72] Y16 has also been shown to inhibit angiogenesis and tumor growth of HT1080 human fibrosarcoma cells *s.c.* injected into mice, and again this inhibition was greater than that of the monomeric Y1.[75]

The acetylated pentapeptide Ac-PHSCN is a ligand for $\alpha5\beta1$ integrin that has been shown to inhibit cancer cell invasion and metastasis. In a similar manner to the data above with Ac-YIGSR, multimeric polylysine dendrimers functionalized with eight Ac-PHSCN peptides ($(\text{Ac-PHSCNGGK})_8K_4K_2K$-$\beta$-Ala — designated P8) were more than 100 times more potent than monomeric Ac-PHSCN in reducing lung cancer colony formation in mice injected with DU 145 or PC-3 prostate cancer cells.[76] Further experiments revealed that P8 also reduced metastasis and lung colony formation in mice injected with SUM-149 PT or MDA-MB-231 breast cancer cells.[76]

5. Imaging Agents and Cancer Diagnostics

Conventional magnetic resonance and CT contrast agents commonly employ gadolinium (III) and iodine ions that are rapidly cleared from the body after i.v. or s.c. administration. This dictates the need to rapidly image the patient in order to capture well-resolved images of blood vessels before the contrast agents begin to "bleed" out of the vessels, leading to poorer tissue resolution. For this reason, the potential to improve the resolution of tissue structures and to enhance the relaxivity of contrast agents has been explored by coupling conventional contrast agents (such as Gd(III), Mn(III) and In(111)) to macromolecules, including dendrimers, that are better retained within blood vessels and can complex multiple ions. The first reports of dendrimer-based MRI contrast agents appeared in the literature in the mid-1990s. The earliest dendrimer-based contrast agents were based on various sized PAMAM dendrimers complexed with Gd(III) or Dy (III) via DOTA, DTPA or Do3A chelating ligands.[77–80] This initial work showed that, by complexing multiple contrast ions onto dendrimers, molecular relaxivities were increased in a generation-dependent manner (with increased generation translating into increased relaxivity) and proton relaxation enhancements in imaging efficiency. One study also reported improvements in the ratio or relaxivity per Gd(III) upon conjugation to a dendrimer when compared to the Gd(III)-chelate alone and Gd(III) chelated albumin or linear polymers.[80] Several studies also explored the potential to improve tumor retention or cancer cell targeting of contrast agents by conjugating cancer-targeting ligands such as EGF and folic acid to the dendrimer.[81,82]

Following the success of these early studies, Kobayashi and coworkers sought to further characterize how the physicochemical properties of Gd(III)-decorated dendrimers affected *in vivo* biodistribution profiles.[83–85] The goal of this work was to characterize how dendrimers could best be utilized to resolve particular organs or structures. This work demonstrated that small

PAMAM dendrimers could usefully resolve the bladder and kidneys but were cleared rapidly from the blood, limiting their utility as vascular contrast agents. When compared to the use of conventional small molecule Gd-DTPA agents, however, G2-G4 PAMAM and diaminobutane dendrimers led to a different resolution pattern in the kidneys.[86] Specifically, the dendrimers were better retained by the renal blood vessels and urinary tracts and showed limited diffusion into the extravascular kidney tissue. In addition, following filtration at the glomerulus, the dendrimer-based agents strongly resolved the proximal tubules, enabling the more efficient diagnosis of renal tubule dysfunction and renal failure when compared to Gd-DTPA.

As the size (generation) of the dendrimers was increased, they displayed more prolonged retention in the blood, improved vascular resolution and limited clearance via the urine. This latter point had the advantage of enabling the resolution of organ structures near the bladder and kidneys. For very large dendrimers (G8), however, blood clearance was increased and the dendrimers were more avidly targeted toward the liver.[83] Similarly, more hydrophobic dendrimers based on a diaminobutane core were also targeted toward the liver and resolved the liver parenchyma more effectively than the more hydrophilic dendrimers (Fig. 7).[87] As a result, a G5 diaminobutane dendrimer was able to resolve liver metastases in mice as small as 0.3 mm in diameter, by specifically highlighting the liver surrounding the metastasis (Fig. 7). Alternatively, the data suggest that mid-sized and relatively hydrophilic G6 dendrimers have the greatest potential as blood pool contrast reagents for MRI.

Later work from the same group sought to characterize the potential for dendrimers to resolve lymphatic vessels with the aim of better identifying lymphatic cancer metastases.[42,43,45] As described in a previous section, G6 PAMAM dendrimers displayed the best magnetic resonance imaging (MRI) properties by allowing good drainage from the injection site and effective resolution of draining lymphatic vessels and sentinel lymph nodes. This approach can be used as a diagnostic tool for lymphatic dysfunction and cancer staging and also as a means to monitor node locations prior to surgical resection of sentinel lymph nodes draining a primary cancer. Several investigators have also demonstrated that this technology can be applied to the development of fluorescently labeled dendrimers that similarly target draining lymph nodes after an intracutaneous or subcutaneous injection[88–90] and again provide a powerful probe for visualizing the lymph nodes that drain a primary tumor during the surgical resection of sentinel lymph nodes. This is in contrast to MRI which occurs prior to surgery. This has the potential to improve cancer survival rates by enabling the improved detection of target lymph nodes during routine lymph node resection.

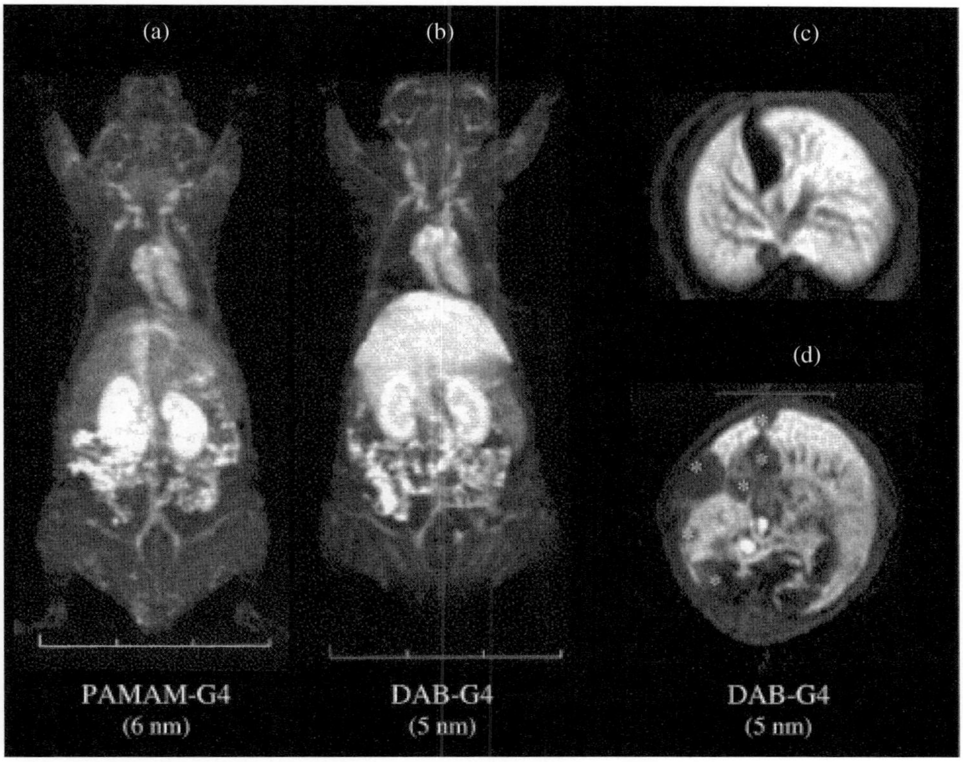

Fig. 7. Whole body MRI (MIP) showing the pharmacokinetic properties of IV administered 0.03 mmol Gd/kg PAMAM dendrimer (a), or G4 diaminobutane core dendrimer (b) in mice 1 min post injection. Panels (c) and (d) show 2D images of normal and tumour-burdened liver 9 min post injection. *Represents micronodules of tumour metastases. Reproduced with permission.[87]

A similar approach can be applied to the visualization of cancers during surgical removal (Fig. 8). The surgical removal of solid tumors can be problematic, since the surgeon must rely on their ability/instinct to visualize the edges of a tumor and to remove the entire cancer without disturbing the surrounding tissues. Decoration of the surface of dendrimers with fluorescent or luminescent probes and cancer targeting ligands as a means of visualizing tumors during surgery has therefore been evaluated. This approach was expected to enable the labeling of cancer cells within a tumor mass and subsequent visualization of the tumor margins during surgery.[89,91–93] This technology has been successfully applied to the surgical removal of tumors in pre-clinical cancer models, but has yet to find routine application in the clinic. Part of the difficulty is in identifying the optimal targeting ligand, a problem shared with the use of targeted dendrimer-based constructs for drug delivery.

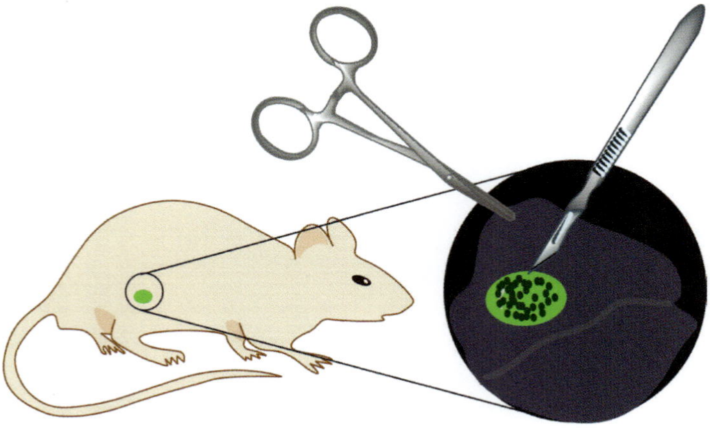

Fig. 8. Diagramatic representation of the principle behind tumor-targeted fluorescent/luminescent dendrimer administration during the surgical removal of solid tumours. The fluorescent/luminescent probe labels cancer cells contained within a tumor enabling good visual distinction between cancerous and normal tissues, enabling less damage to normal tissues.

For instance, whilst ligands such as folic acid, EGF, avidin, RGD peptide, various hormones and antibodies directed toward overexpressed cell surface ligands have been explored, there is no one ligand that targets all cancers. The heterogeneity among tumors dictates that, for this approach to be most useful, preliminary tumor biopsies would need to be taken to identify key proteins that are overexpressed on that tumor in order to tailor a decision as to which cancer-targeting probe to use during surgery.

6. Gene Delivery Vectors

A number of pathological conditions result from differences in protein expression. Whilst the increased production of disease-causing proteins may be managed using small molecule inhibitors, complications persist, such as a lack of specificity, toxicity and inappropriate pharmacokinetics. For disease states reflecting protein deletion or reduction, therapies largely surround administration of replacement protein, however, this approach is limited by poor protein absorption and stability and therefore the need for parenteral administration. There is therefore considerable interest in the delivery of genes (DNA) to facilitate protein replacement or silencing technologies such as small interfering (si)RNAs to block the translation of messenger (m)RNA into an overexpressed target protein. The appeal of siRNA in contrast to small molecule protein inhibitors stems from the fact that siRNA is relatively simple to manufacture in cell-free systems and can evoke highly specific responses *in vivo*. A classic example of the desire to downregulate the translation of mRNA into a particular target protein is in the treatment of multiple drug resistant and invasive cancers. Cancer cells commonly upregulate the expression of proteins that enable the cancer to survive in the presence of cytotoxic drugs and to promote invasion into the surrounding tissue. For example, P-gp is often upregulated in cancer cells, leading to expulsion of internalized chemotherapeutic drugs that are P-gp substrates (such as doxorubicin and vinblastine) and reduced cancer killing. The *in vivo* or cellular delivery of genetic material, however, is restricted by inefficient transport of DNA and siRNA across the cell membrane (due to their anionic charge and high molecular weight), low *in vivo* stability of nucleic acid complexes and immunogenicity. There is therefore considerable interest in the development of delivery systems that promote efficient delivery of DNA or siRNA into target cells, either *ex vivo* or *in vivo*.

The most effective means of introducing genetic material into target cells is via the use of viral gene delivery vectors,[94,95] however, concerns regarding toxicity have led to increasing interest in the use of non-viral delivery systems including liposomes, linear polymers and dendrimers. DNA and siRNA are polyanionic and therefore form electrostatic complexes with polycationic particles and polymers. Amine-terminated PAMAM and PPI dendrimers have been demonstrated to efficiently form complexes with siRNA and DNA (termed dendriplexes).[96] Larger, more highly cationic dendrimers typically complex higher loads of nucleic acid, and also retain polycationic charge better than smaller dendrimers that can become charge-neutral upon complexation with large quantities of nucleic acid.[97] In contrast, however, one study has proposed that smaller G2 PAMAM dendrimers may more effectively bind to DNA due to higher structural fluidity of the smaller generation dendrimer.[97] Given that the cell membrane contains an overall negative charge, cationic dendrimers promote initial cell binding and endocytosis-mediated internalization of the dendriplex. As such, larger, amine-terminated dendrimers are typically more efficient in delivering nucleic acids into cells when compared to smaller dendrimers. The polycationic nature of large amine-terminated dendrimers also promotes eventual escape from late endosomes, by taking advantage of the decreasing pH of the endosomal compartment.[98] Specifically, polycationic dendrimers become protonated as the pH decreases, leading to an influx of Cl- ions, endosome swelling and the eventual rupture of the endosomal membrane. In this way the endosomal content is released into the cytoplasm where siRNA evokes its effects (Fig. 9). For DNA-loaded systems, polycationic dendrimers have been suggested to enhance trafficking toward and accumulation in the nucleus.[98] These properties have led to the

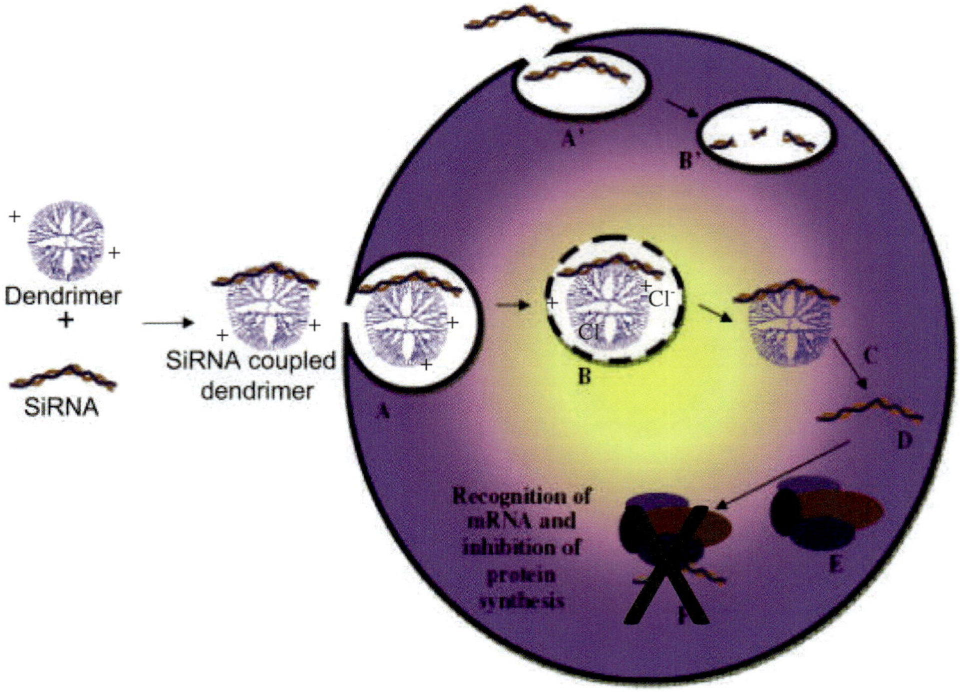

Fig. 9. Schematic representation of dendrimer-mediated delivery of siRNA to the cytosol. Dendrimer-coupled siRNA initially binds to the cell membrane via electrostatic attraction, followed by endocytosis (A). The reduction in endosomal pH for maturing endosomes leads to the protonation of amine residues on the dendrimer surface and the influx of Cl^- ions. The resulting influx of water leads to endosomal swelling and rupture, leading to the liberation of the dendrimer–siRNA complex (B). Following the liberation of siRNA from the dendrimer (C, D), siRNA binds to the target mRNA (E) and blocks RNA translation (F). In contrast, if native siRNA is internalized by the cell (A′) there is no mechanism for efficient endosomal escape (B') and cytosolic exposure. Figure is reproduced and modified with permission.[100]

commercial development of the amine-terminated PAMAM-based *in vitro* cell transfection reagents Superfect and Polyfect (Qiagen). These agents are reportedly able to facilitate cell transfection even in the presence of serum, unlike many liposome-based transfection systems.

After *in vivo* delivery, dendriplexes based on high-generation (G9) PAMAM dendrimers are reportedly targeted largely toward the lungs, whereas for smaller G3 dendriplexes, the main target is the spleen.[99] As expected, the organs of the MPS appear to be significant targets for the disposition of cationic dendriplexes.

This may have advantage in the treatment of organ-specific diseases (such as cystic fibrosis which largely affects the lungs), but limits the use of these systems for diseases which affect more diverse organs.

Dendrosomes have also been explored as gene-delivery vectors. Dendrosomes are dendriplexes that are encased within a lipid shell to improve biocompatibility and reduce haemolytic toxicity. By formulating the dendriplexes into the dendrosomes, the DNA is protected from anionic plasma proteins, which can lead to premature liberation of DNA from the dendrimer complex and degradation. In many cases, dendrosomes show enhanced transfection efficiency when compared to the dendriplex alone.[101,102] The reason for this is unknown, but it appears that by encasing the dendrimer in a lipid membrane, the cellular uptake of DNA is increased. Conversely, PEGylation has been explored as an approach to prolonging the circulation time of dendrimer–gene complexes, and to improve the flexibility and biocompatibility of the system.[103,104] The more hydrophilic PEGylated systems, however, typically require some form of cellular targeting approach to promote cell binding and endocytosis. This is most effectively achieved using cell-specific targeting ligands that allow the transfection of a defined cell type. This may be achieved in cancer cell lines that express high levels of receptors that promote cell survival, invasion and accelerated growth. For instance, ligands to cell surface receptors including EGF, $\alpha v\beta 3$ integrin and leutinizing hormone releasing hormone have all been employed to target dendrimer-based nucleic acid delivery vectors to cancer cells *in vitro*.[105–107]

Unfortunately, there is limited *in vivo* evidence of utility for the efficient delivery of siRNA or DNA to target cells and eventual alterations in protein expression in these *in vivo* targets. Many studies have described the *in vivo* transfection of siRNA into cancer cells within a solid tumor or the reduced translation of target mRNA into protein, but these typically involved the intratumoral injection of dendrimer–siRNA complexes, rather than the targeted delivery of siRNA following intravenous injection of the particles. One study, however, has described a G3 PAMAM dendrimer conjugated to lactose-bearing α-cyclodextrin to promote *in vivo* hepatocyte targeting of an siRNA to human amyloidogenic transthyretin, (the causative protein for transthyretin-related familial amyloidotic polyneuropathy).[108] Hepatocytes were specifically targeted due to the predominant formation of this protein by the liver. The study showed a significant decrease in mRNA expression for the protein in the liver 48 h after a 9 mg/kg equivalent *i.v.* dose of siRNA. This demonstrates the potential to achieve cellular targeting and transfection *in vivo* with dendrimer-based delivery vectors; however, functional assessment is also needed to demonstrate that *in vivo* transfection translates into a therapeutic response.

One explanation for the lack of examples of *in vivo* utility for dendrimer based transfection agents is the lack of data describing *in vivo* biodistribution for dendriplexes and dendrosomes. The translational potential of dendrimer-based gene delivery vectors is critically dependent on efficient delivery of DNA or siRNA to target cells, efficient internalization of the nucleic acid complex and liberation from the endosome. The competing requirements for long-circulating delivery systems (where design criteria seek to avoid cellular interaction) and the need to promote cellular uptake and endosomal escape in the target tissue, dictate that the challenges associated with *in vivo* gene delivery remain significant.

7. Tissue Engineering

7.1. *Background and role of collagen in tissue engineering*

Collagen is the most abundant protein in the human body, accounting for approximately 30% of total protein content.[109] It is a major component of connective tissues, around which all organ systems are formed and structured, and plays a significant role in wound healing. Collagen therefore has a wide range of biomedical applications, including the regeneration of cartilage lost as a result of injury, ageing, rheumatoid arthritis, etc.; in bone grafts; as a scaffold for tissue regeneration and wound repair and in cosmetic surgery. The use of collagen for biomedical applications, however, is in part limited by (i) its weak mechanical properties and rapid biodegradation (which often dictate the need for crosslinking); (ii) the potential for cytotoxicity induced by unreacted crosslinking reagents; (iii) the formation of weak crosslinks that have poor structural integrity and mechanical strength when using less cytotoxic crosslinking reagents; (iv) changes to the structural configuration of collagen during purification and sterilization for *in vivo* use and (v) the potential to induce hypersensitivity reactions and the transmission of diseases via the use of collagen obtained from cows and human cadavers.[110,111] Alternatives to the use of collagen, and commonly used collagen crosslinking agents, are therefore being sought. Dendrimers have properties that enable deployment as both protein mimetics and cross-linking reagents, and for this reason, interest in the use of dendrimers in tissue engineering and wound repair has increased. Current advances in the application of dendrimers to tissue engineering and wound repair are summarized briefly below.

7.2. *Dendrimers as collagen mimetics*

The first dendrimers to be explored as collagen mimetics were based on the general Gly-X (repeat)-Y (repeat) trimer sequence found in collagen, where

X and Y are proline and hydroxyproline.[112] Collagen mimetics composed of Gly-Pro-Nleu and Gly-Nleu-Pro sequences were attached either directly, or separated by a *β*-alanine spacer to a N-(benzyloxycarbonyl)-tris (carboxyethoxymethyl) aminomethane (TRIS[OH]3) core (Fig. 10). Both dendrimers assembled into a triple helix structure similar to collagen; however, dendrimers based on Gly-Neu-Pro showed better thermal stability than dendrimers based on Gly-Pro-Nleu. Later work by the same authors resulted in the development of the first collagen mimetic dendrimer based on a G1 PAMAM scaffold and incorporating the Gly-Pro-Nleu peptide.[113] The authors suggested that since PAMAM dendrimers have the potential to form complexes with Cu^{+2} and Ni^{+2}, collagen mimetic structures based on PAMAM scaffolds may provide advantage, given the role of Cu^{+2} and other ions in the crosslinking of fibril collagens and in biological processes, such as wound healing and bone mineralization. This first G1 structure, however, had a melt transition temperature of 25°C, limiting *in vivo* utility.

More recent work has utilized a G2 PAMAM scaffold to construct more intricate and biologically active dendrimer-based collagen mimetics that display tissue transglutaminase-mediated crosslinking and cellular adhesion properties.[114] In this later study, the dendrimer design was based on the realization that the hexapeptide sequence Gly-Phe-Hyp-Gly-Glu-Arg (GFOGER) in collagen is a major cell binding domain. Enzyme-mediated crosslinking of collagen plays an important part in stabilizing extracellular matrix proteins *in vivo* and in modulating cellular activity. The authors therefore built GFOGER peptides into the dendrimer structure to promote cellular adhesion and

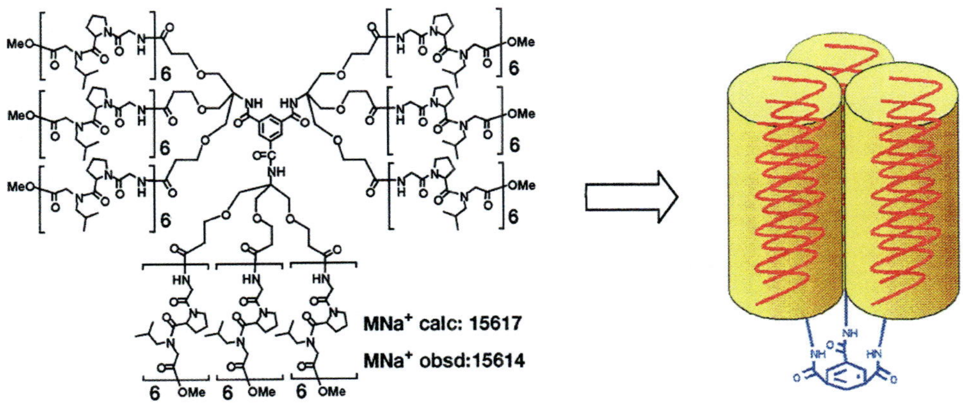

Fig. 10. Dendrimer-based collagen mimetic composed of TMA[TRIS[$(\text{Gly-Pro-Nleu})_6$-$\text{OMe}]_3]_3$ synthesized by Kinberger *et al* (2002) and a schematic representation of the triple helix structure formed around the trimesic acid core. Image reproduced with permission.[112]

APQQEA and EDGFFKI peptides to promote dendrimer crosslinking. The dendrimers formed again assembled into triple helix structures that are characteristic of collagen. The dendrimer complex was crosslinked through incubation with tissue glutaminase, was not toxic against L929 mouse fibroblasts up to 50 μg/ml, and Hep3G cells bound to the dendrimer matrix with similar affinity to pure collagen. Upon incubation with the crosslinked dendrimer scaffold or collagen, Hep3G cells also developed actin cytoskeletal structures and focal adhesion contacts on these surfaces suggesting good matrix binding. The data suggest that biologically active collagen mimetics based on dendrimers can be developed by incorporating peptide functions that mimic the natural activity of collagen *in vivo*.

7.3. *Dendrimers as collagen co-crosslinking reagents*

Despite recent advances in the development of dendrimer-based collagen mimetics, biomedical application of the technology remains in its infancy. A more immediate approach to provide practical application is to improve the safety, biological stability and mechanical strength of collagen-based crosslinked scaffolds, by developing more effective and non-toxic crosslinking approaches. Currently used crosslinking agents include [3-dimethyl aminopropyl] carbodiimide (EDC), glutaraldehyde and riboflavin in combination with UVA to induce the collagen crosslinking. The later crosslinking approach is being used to treat keratoconus, a degenerative condition characterized by a thinning of the cornea. Glutaraldehyde generates chemically and thermally stable protein crosslinks, but can lead to cytotoxicity.[115] In contrast, EDC is more biocompatible but leads to the generation of weaker and more biodegradable crosslinks. The mechanical properties of EDC crosslinked collagens is limited by the relatively low availability of primary amines (3% of collagen) and acidic groups (12% of collagen) (EDC bridges amine functionalities and acidic groups).[116] The polyfunctional nature of dendrimers can significantly enhance the mechanical properties of collagen crosslinks created with glutaraldehyde or EDC, by providing an additional source of primary amines to enable crosslinking with acid groups on collagen. For example, Duan and coworkers[116] demonstrated that crosslinking collagen with EDC in the presence of G2–3 PPI-octaamine dendrimers, resulted in an increase in the mechanical strength of the gels formed when compared to those constructed using glutaraldehyde or EDC crosslinked collagen alone. This appeared to reflect an increase in crosslinking efficacy, as demonstrated by decreases in absorbed water content in the gels and higher denaturation temperatures. The introduction of the G2–3 dendrimers into the collagen and EDC reaction mixtures at a ratio of 10 parts collagen to 1 part dendrimer improved the biostability of the gel when compared to EDC crosslinked gels alone, and provided comparable stability to glutaraldehyde crosslinked gels. The G2 dendrimer–EDC scaffolds also displayed lower cytotoxicity than glutaraldehyde crosslinked collagen gels and promoted the infiltration of corneal epithelial cells.[117] Later work by the same investigators incorporated heparin into the G2 PPI-collagen-EDC gels to provide stability and maintain bioactivity of fibroblast growth factor (FGF) incorporated within the gels. FGF was incorporated to promote fibroblast infiltration and wound healing.[118] Heparin is known to bind to FGF via specific heparin binding sites and stabilizes FGF in its active form. The incorporation of heparin had no impact on gel strength or stability but did stabilize FGF incorporation. *In vitro*, approximately 40% of absorbed FGF was released from the gel over 2 weeks, suggesting the potential to lead to more effective wound healing beneath the collagen matrix. G2 PAMAM dendrimers have been similarly reported to improve the crosslinking efficiency of collagen in the presence of EDC and to promote improved fibroblast proliferation when compared to EDC or glutaraldehyde crosslinked collagen gels alone.[119] However, the addition of the dendrimer into collagen-glutaraldehyde gels did not result in significant improvements to biostability or crosslinking efficiency of collagen- glutaraldehyde gels alone. These studies suggest that incorporating amine-terminated G2 dendrimers into collagen gels crosslinked with EDC improves crosslinking efficiency and results in mechanical and biostability properties that are at least as good as glutaraldehyde crosslinked collagen gels. In this way, dendrimer–collagen–EDC gels may be used as alternatives to collagen-glutaraldehyde gels, since they display similarly good mechanical and stability properties, but have reduced cytotoxicity and better capacity for cellular infiltration and proliferation.

7.4. *Joint repair in osteoarthritis*

Osteoarthritis results from the erosion of cartilage lining the joints and leads to an increase in friction between the bones of the joint on movement, ultimately resulting in pain. The ultimate exposure of bare bone surfaces eventually leads to severe pain, immobility and the need for joint replacement, arthroplasty or fusion surgery. Current medical treatments are focused on the use of anti-inflammatory drugs and there is a growing interest in supplementary remedies, including oral chondroitin and glucosamine sulfate. Dendrimers have also been explored as both drugs and drug delivery systems for arthritis and these are discussed in previous sections.

Cartilage may be regrown or replaced in arthritis affected joints via the injection of cells (specifically chondrocytes, the matrix producing cells within cartilage), artificial tissue scaffolds and growth factors into the joint to promote *in vivo* synthesis and deposition of collagen, or via the insertion of cartilage grown *in vitro* into the defect site. These treatments have proven to be effective at replacing cartilage and in returning mobility to the affected joints[120] but are limited to the treatment of relatively small defects. Furthermore, the identification of appropriate tissue scaffold materials is challenging since scaffolds need to be sufficiently biodegradable to allow progression of cell growth and collagen/glycosaminoglycan deposition through the scaffold while maintaining sufficient mechanical strength to withstand the significant pressures placed on the joint. Scaffolds must also display limited swelling properties since this can lead to the gel dislodging from the defect site. Hydrogels have been explored as favorable scaffolds due to their high saturable water content and there is therefore considerable scope to improve the design of hydrogel-based scaffolds that enable the more efficient regrowth of cartilage *in vivo*.

Dendrimer-based hydrogels have potential to provide better frameworks than those based on linear functionalized polymers, with higher branching allowing for a greater crosslink density at lower polymer concentrations as well as increased biodegradation sites. Dendrimer-based hydrogels have been explored as support scaffolds for chondrocytes for the treatment of osteochondral defects that are characterized by damage to cartilage and underlying bone resulting from lesioning or injury. Hydrogels (containing chondrocytes) were constructed based on dendrons comprising a PEG core conjugated to succinate and glycerol via an ester linker, and including a methacrylate functionality incorporated into the dendron surface to promote photo-inducible crosslinking.[121] Hydrogels composed of 7.5% to 15% w/w dendron showed limited swelling over 30 days in PBS (properties that are advantageous for cartilage repair since swelling may promote dislodgement) with gel-entrapped chondrocytes promoting the *in vitro* deposition of proteoglycan and collagen throughout the hydrogel scaffold. However, these first-generation dendron-based scaffolds degraded too rapidly for effective use *in vivo*.

Second-generation hydrogels were therefore constructed that contained both ester and carbamate linkages to slow the rate of biodegradation (Fig. 11).[122] The hydrogels formed were able to withstand increasingly high compressive forces when increasing w/v proportions of dendron were included (within the range of 5–20% w/v dendron), suggesting that gels

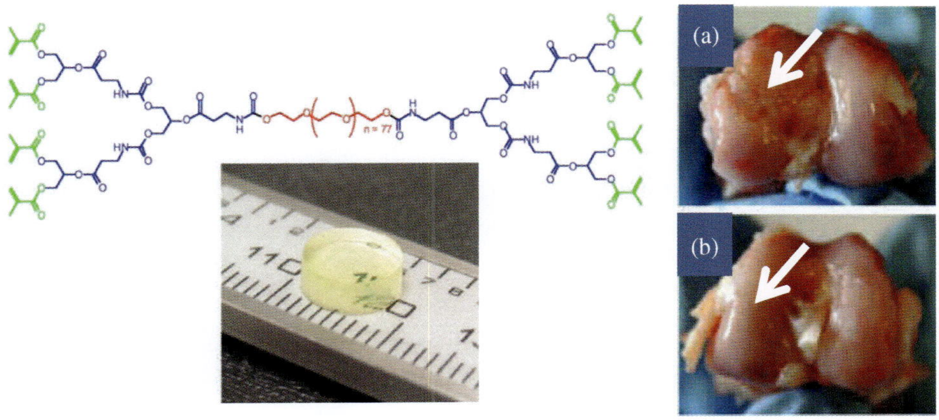

Fig. 11. An ABA triblock dendrimer composed of a PEG core (red), G1 dendron wedges composed of poly(glycerol beta-alanine) (blue) and crosslinkable methacrylate groups (green) and the resulting hydrogel formed following photopolymerization. Images on the right show pictures of the femurs of rabbits taken 24 weeks after induction of full-thickness osteochondral defects that remained untreated (a) or filled with a hydrogel composed of 10% w/v poly(glycerol beta-alanine) (b). Images are reproduced with permission.[122]

composed of high weight percentages of dendron would be able to withstand the high pressures present within joints. Hydrogels composed of 10–20% w/v dendron copolymers with poly(glycerol succinic acid) and poly(glycerol beta-alanine) showed high degrees of swelling (~12–27% increase in swelling) following incubation in PBS over 30 days (with the attendant risk of dislodgement). In contrast, hydrogels with lower dendron loads (5%) and gels composed only of poly(glycerol succinic acid) or poly(glycerol beta-alanine) showed no gel swelling over 30 days. In both cases, the gels were able to completely fill simulated osteochondral defects in an *ex vivo* rabbit model and remained intact in the defect site on repeated mechanical loading (loading representative of the body weight of a 3 kg rabbit). Pilot *in vivo* studies in rabbits with experimentally induced full-thickness osteochondral defects in both medial femoral condyles showed that, by filling the defect with 10% w/v poly(glycerol beta-alanine) dendron gels, the knees healed considerably more effectively than unfilled defects. Specifically, the treated joints showed increased levels of collagen and glycosaminoglycan deposition and smoother cartilage surfaces (Fig. 11, A/B). Ongoing work is focused on the identification of biocompatible dendrimer-based hydrogels with limited swelling properties *in vivo* and the ability to tolerate the high compressive forces experienced in joints, particularly in the knees and pelvis.

8. Corneal Tissue Engineering and Wound Repair

Several groups have described the use of dendrimers to deliver drugs to[123] and across the cornea.[124–128] In addition, dendrimers are starting to be employed in corneal tissue engineering and repair. The principal use of dendrimers in this application is, similar to the applications described above, as collagen crosslinking agents and seeks to improve cellular adhesion and proliferation of corneal tissues.

For example, G2 PPI dendrimers modified by the attachment of a peptide sequence (YIGSR), found in the cell adhesive protein laminin, have been used as a means to incorporate growth promoting peptides into collagen gels through their action as collagen crosslinking agents, resulting in improved scaffolds for corneal tissue engineering. In these studies, introduction of the dendrimer cross linker promoted the incorporation of the peptide into the collagen scaffold, enabling improved cell growth and adhesion on the collagen gels.[129,130] Princz *et al.*[131] describes similar studies to crosslink collagen gels using second-generation PPI dendrimers, but in this case, employed dendrimers functionalized with heparin. Heparin was used to promote loading of EGF into the gel. Uptake of EGF was substantially increased in the heparin-functionalized dendrimer gels, and the subsequent release of the growth factor from the gels was shown to stabilize corneal cell proliferation.

Another application of dendrimers as collagen crosslinking agents is in corneal wound repair. Givens *et al.*[132] have described the functionalization of a first-generation PAMAM dendrimer with a diazopyruvoyl group enabling light activated crosslinking that resulted in greater effectiveness than an equivalent non-dendronized bis-PEG based crosslinking reagent. Corneal tissue crosslinking was greatest at low hydration levels, and nearly twice as strong as the non-dendrimer equivalent.

Dendrimers have also been investigated as a means of modifying the surfaces of engineered tissue implants to promote cellular interaction via the presentation of poly-hydroxy groups at the surface.[133] In model systems assembled via association of hydroxy- and methoxy-terminated dendrons to a gold surface using PEG spacers (to provide a template for dendrimer adhesion), dendrons with free hydroxyl groups led to cell adhesion and proliferation of corneal epithelial cells, whereas proliferation was greatly suppressed when the hydroxyl groups were capped with the methoxy functionality in both the presence and absence of serum in growth media. The mechanism by which this occurred was not characterized in this study but it was clear that the polyhydroxyl surface promoted cellular adhesion.

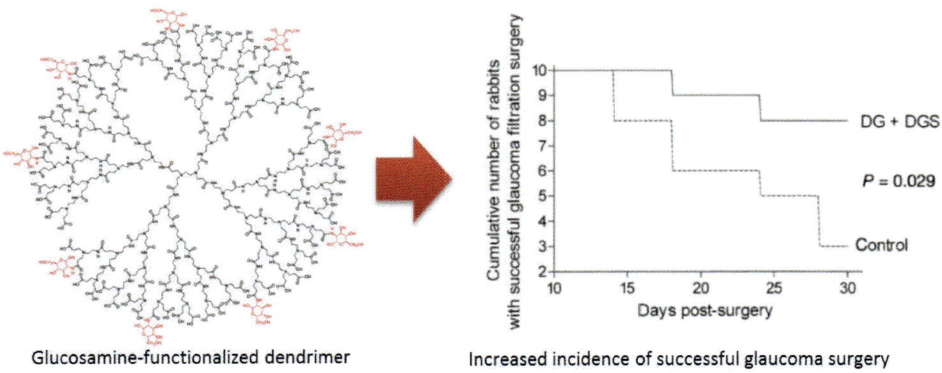

Fig. 12. Representative structure of protonated G3.5 PAMAM glucosamine functionalized dendrimer and the cumulative graph of the number of rabbits with successful glaucoma filtration surgery after combined administration of dendrimer-glucosamine (DG) and dendrimer-glucosamine 6 phosphate (DGS) or untreated rabbits (control). Adapted with permission.[134]

In one of the first applications of dendrimers as a therapeutic treatment for ocular wound healing, Shaunak *et al.*[134] showed that combined administration of two anionic dendrimers functionalized with glucosamine or glucosamine sulfate were able to significantly reduce scar tissue formation after glaucoma surgery in rabbits. The immunomodulatory and antiangiogenic properties of these respective dendrimer functionalities acted synergistically when administered as subconjunctival and intraperitoneal injections to rabbits following experimental glaucoma filtration surgery, a model representative of clinical surgical treatment in glaucoma patients (Fig. 12).[134] Anionic PAMAM dendrimers appeared to provide a toxicity advantage over cationic dendrimers.

9. Conclusions and Outlook

The ability to create a diverse range of dendrimer structures with variations in size, shape, surface functionality, biological activity and physicochemical properties has led to their exploration for a large number of applications, spanning biomedical, industrial and agricultural uses. Dendrimers are similar in size, and in some cases *in vivo* metabolic properties, to proteins, and as such have significant potential as drug delivery systems and protein mimetics. In addition, some dendrimers have intrinsic biological activity, and for example, the anionic dendrimer, Vivagel®, is in clinical development as a topical microbicide and as a means of preventing the spread of STIs.

The development of dendrimers for biomedical use is perhaps most advanced for applications that seek to improve the pharmacokinetic behavior of small molecule drugs or to enhance the efficiency of *in vivo* imaging. For example, a significant body of work has shown proof of principle that the development of dendrimer-based nanochemotherapeutics can, in some cases, provide therapeutic and/or toxicity benefits in the treatment of cancer when compared to other nanomedicine approaches. Dendrimers may also be useful for prolonging the plasma exposure of rapidly cleared small molecule drugs after parenteral administration.

In contrast, the utility of dendrimers as drug carriers after oral and transdermal administration, or as a means of enhancing drug exposure to the brain, is less clear and several questions remain. Notably, whether dendrimer–drug complexes are able to traverse *in vivo* epithelial barriers (or endothelial barriers such as the BBB) intact, or whether dendrimer co-administration simply enhances membrane permeability allowing improved cellular access of associated drug (rather than the complex). Delineation of these possibilities is further complicated by the lack of *in vitro* models that recapitulate the complex barrier functionality of many *in vivo* epithelial barriers, especially with respect to mucous production and the generation of representative sink conditions.

Dendrimers show potential as improved *in vivo* imaging vectors and have produced well-resolved images of tissue structures that persist for longer periods than those obtained with conventional contrast agents. They may also be developed to aid in the removal of solid tumors during surgery, achieved by administering fluorescent dyes that targeting and define the edges of a solid tumor, thereby enabling surgical resection with minimal damage to the surrounding normal tissue.

Dendrimers have been commercialized as *in vitro* reagents and show considerable potential as transfection reagents for promoting the cellular internalization of siRNA and DNA. Superfect® and Polyfect® are commercially available *in vitro* gene transfection agents, and considerable efforts continue in an attempt to translate *in vitro* utility into *in vivo* application. However, highly charged (typically cationic) dendrimers that display good *in vitro* gene transfection also suffer from widespread cell and membrane affinity *in vivo*, limiting site selective gene transfer and enhancing the propensity for *in vivo* toxicity. More work is therefore needed to enable cell-specific targeting and internalization of siRNA or DNA, to address the instability of the dendrimer–siRNA/DNA complex and to avoid *in vivo* toxicity.

Finally, dendrimers show potential as novel crosslinking reagents and matrixes for wound healing and tissue regeneration. In some cases, clinical utility is apparent, whereas in others, for instance, the use of dendrimer scaffolds for the repair of osteochondral defects, more evidence is needed to demonstrate that these systems are compatible with the joint pressures experienced in humans.

In summary, dendrimers provide a highly flexible, multivalent means of ligand presentation and provide a scaffold to assist in drug and gene delivery across a range of applications. As a biomedical technology, however, dendrimers remain relatively young and more data, and more experience with the synthesis of clinical grade material, is required before the potential clinical advantages are more broadly realized.

Acknowledgments

L.M. Kaminskas is supported by an NHMRC Career Development fellowship. The authors are funded by an Australian Research Council Linkage grant and by Starpharma Pty. Ltd. The authors would like to thank David Owen and Paul Barrett for proofreading sections in this chapter.

References

1. D. A. Tomalia, H. Baker, J. Dewald, M. Hall, G. Kallos, S. Martin, J. Roeck, J. Ryder and P. Smith, A new class of polymers — Starburst-dendritic macromolecules, *Polym. J.* **17,** 117–132 (1985).
2. L. M. Kaminskas and C. J. H. Porter Targeting the lymphatics using dendritic polymers (dendrimers). *Adv. Drug Deliv. Rev.* **63,** 890–900 (2011).
3. C. J. Hawker and J. M. J. Frechet, Preparation of polymers with controlled molecular architecture — A new convergent approach to dendritic macromolecules, *J. Am. Chem. Soc.* **112,** 7638–7347 (1990).
4. P. Wu, A. K. Feldman, A. K. Nugent, C. J. Hawker, A. Scheel, B. Voit, J. Pyun, M. J. Frechet, K. B. Sharpless and V. V. Fokin, Efficiency and fidelity in a click-chemistry route to triazole dendrimers by the copper(I)-catalyzed ligation of azides and alkynes, *Angew Chem. Int. Edit.* **43,** 3928–3932 (2004).
5. V. Maraval, J. Pyzowski, A. M.Caminade and J. P. Majoral, "Lego" chemistry for the straightforward synthesis of dendrimers, *J. Organic Chem.* **68,** 6043–6046 (2003).
6. L. M. Kaminskas, B. J. Boyd and C. J. Porter, Dendrimer pharmacokinetics: The effect of size, structure and surface characteristics on ADME properties, *Nanomedicine (Lond)* **6,** 1063–1084 (2011).
7. W. Wijagkanalan, S. Kawakami and M. Hashida, Designing dendrimers for drug delivery and imaging: Pharmacokinetic considerations, *Pharm. Res.* **28,** 1500–1519 (2011).
8. B. J. Boyd, L. M. Kaminskas, P. Karellas, G. Krippner, R. Lessene and C. J. Porter, Cationic poly-L-lysine dendrimers: Pharmacokinetics, biodistribution, and evidence for metabolism and bioresorption after intravenous administration to rats, *Mol. Pharm.* **3,** 614–627 (2006).

9. K. M. Kitchens, A. B. Foraker, R. B. Kolhatkar, P. W. Swaan and H. Ghandehari, Endocytosis and interaction of poly (amidoamine) dendrimers with Caco-2 cells, *Pharm. Res.* **24,** 2138–2145 (2007).
10. S. Ohashi, T. Kubo, T. Ikeda, Y. Arai, K. Takahashi, Y. Hirasawa, M. Takigawa, E. Satoh, J. Imanishi and O. Mazda, Cationic polymer-mediated genetic transduction into cultured human chondrosarcoma-derived HCS-2/8 cells, *J. Orthop. Sci.* **6,** 75–81 (2001).
11. L. M. Kaminskas, B. J. Boyd, P. Karellas, S. A. Henderson, M. P. Giannis, G. Y. Krippner, and C. J. Porter, Impact of surface derivatization of poly-L-lysine dendrimers with anionic arylsulfonate or succinate groups on intravenous pharmacokinetics and disposition, *Mol. Pharm.* **4,** 949–961 (2007).
12. J. A. Bertolatus and D. Klinzman, Macromolecular sieving by glomerular basement membrane *in vitro*: Effect of polycation or biochemical modifications, *Microvasc. Res.* **41,** 311–327 (1991).
13. L. M. Kaminskas, B. J. Boyd, P. Karellas, G. Y. Krippner, R. Lessene, B. Kelly, C. J. Porter, The impact of molecular weight and PEG chain length on the systemic pharmacokinetics of PEGylated poly l-lysine dendrimers, *Mol. Pharm.* **5,** 449–463 (2008).
14. N. K. Jain and U. Gupta, Application of dendrimer-drug complexation in the enhancement of drug solubility and bioavailability, *Expert Opin. Drug Met.* **4,** 1035–1052 (2008).
15. L. M. Kaminskas, V. M. McLeod, C. J. Porter and B. J. Boyd, Association of chemotherapeutic drugs with dendrimer nanocarriers: An assessment of the merits of covalent conjugation compared to noncovalent encapsulation, *Mol. Pharm.* **9,** 355–373 (2012).
16. A. Asthana, A. S. Chauhan, P. V. Diwan and N. K. Jain, Poly(amidoamine) (PAMAM) dendritic nanostructures for controlled site-specific delivery of acidic anti-inflammatory active ingredient, *AAPS PharmSciTech* **6,** E536–E542 (2005).
17. M. Na, C. Yiyun, X. Tongwen, D. Yang, W. Xiaomin, L. Zhenwei, C. Zhichao, H. Guanyi, S. Yunyu and W. Longping, Dendrimers as potential drug carriers. Part II. Prolonged delivery of ketoprofen by *in vitro* and *in vivo* studies, *Eur. J. Med. Chem.* **41,** 670–674 (2006).
18. Maeda, H. SMANCS and polymer-conjugated macromolecular drugs: Advantages in cancer chemotherapy, *Adv. Drug Deliv. Rev.* **46,** 169–185 (2001).
19. C. Lee, E. Gillies, M. Fox, S. Guillaudeu, L. Frechet, E. Dy, F. Szoka, A single dose of doxorubicin-functionalized bow-tie dendrimer cures mice bearing C-26 colon carcinomas. *Proc Natl Acad Sci USA* **103**, 16679–16754 (2006).
20. Y. Tsukioka, Y. Matsumura, T. Hamaguchi, H. Koike, F. Moriyasu and T.Kakizoe Pharmaceutical and biomedical differences between micellar doxorubicin (NK911) and liposomal doxorubicin (Doxil), *Jpn J. Cancer Res.* **93,** 1145–1153 (2002).
21. E. A. Azzopardi, E. L. Ferguson and D. W. Thomas The enhanced permeability retention effect: a new paradigm for drug targeting in infection, *J. Antimicrob. Chemother.* **68,** 257–274 (2013).
22. A. N. Lukyanov, W. C. Hartner and V. P. Torchilin, Increased accumulation of PEG-PE micelles in the area of experimental myocardial infarction in rabbits, *J. Control Release* **94**, 187–193 (2004).
23. R. Topakian, T. R. Barrick, F. A. Howe and H. S. Markus. Blood-brain barrier permeability is increased in normal-appearing white matter in patients with lacunar stroke and leucoaraiosis, *J. Neurol. Neurosurg. Psych.* **81**, 192–197 (2010).

24. L. M. Kaminskas, V. M. McLeod, B. D. Kelly, G. Sberna, B. J. Boyd, M. Williamson, D. J. Owen and C. J. Porter, A comparison of changes to doxorubicin pharmacokinetics, antitumor activity, and toxicity mediated by PEGylated dendrimer and PEGylated liposome drug delivery systems, *Nanomedicine* **8,** 103–111 (2011).
25. M. E. Fox, S. Guillaudeu, J. M. J. Frechet, K. Jerger, N. Macaraeg, F. C. Szoka, Synthesis and *in vivo* antitumor efficacy of PEGylated poly(L-lysine) dendrimer-camptothecin conjugates *Mol. Pharm.* **6,** 1562–1572 (2009).
26. C. C. Lee, A. T. Cramer, F. C. Szoka and J. M. J. Frechet, An intramolecular cyclization reaction is responsible for the *in vivo* inefficacy and apparent pH insensitive hydrolysis kinetics of hydrazone carboxylate derivatives of doxorubicin, *Bioconjug. Chem.* **17,** 1364–1368 (2006).
27. J. F. Kukowska-Latallo, K. A. Candido, Z. Cao, S. S. Nigavekar, I. J. Majoros, T. P. Thomas, L. P. Balogh, M. K. Khan and J. R. Baker, Jr. Nanoparticle targeting of anticancer drug improves therapeutic response in animal model of human epithelial cancer, *Cancer Res.* **65**, 5317–5324 (2005).
28. A. Agarwal, S. Saraf, A. Asthana, U. Gupta, V. Gajbhiye and N. K. Jain, Ligand based dendritic systems for tumor targeting, *Int. J. Pharm.* **350**, 3–13 (2008).
29. T. P. Thomas, S. N. Goonewardena, I. J. Majoros, A. Kotlyar, Z. Cao, P. R. Leroueil, J. R. Baker Jr. Folate-targeted nanoparticles show efficacy in the treatment of inflammatory arthritis, *Arthritis Rheum.* **63**, 2671–2680 (2011).
30. R. S. Dhanikula, T. Hammady and P. Hildgen, On the mechanism and dynamics of uptake and permeation of polyether-copolyester dendrimers across an *in vitro* blood-brain barrier model, *J. Pharm. Sci.* **98,** 3748–3760 (2009).
31. R. Jevprasesphant, J. Penny, D. Attwood, N. B. McKeown, A. D'Emanuele, Engineering of dendrimer surfaces to enhance transepithelial transport and reduce cytotoxicity, *Pharm. Res.* **20,** 1543–1550 (2003).
32. A. Agarwal, N. Lariya, G. Saraogi, N. Dubey, H. Agrawal, G. P. Agrawal, Nanoparticles as novel carrier for brain delivery: A review, *Curr. Pharm. Design.* **15**, 917–925 (2009).
33. S. Kannan, H. Dai, R. S. Navath, B. Balakrishnan, A. Jyoti, J. Janisse, R. Romero, R. M. Kannan, Dendrimer-based postnatal therapy for neuroinflammation and cerebral palsy in a rabbit model, *Sci. Transl. Med.* **4**, 130–146 (2012).
34. S. X. Huang, J. F. Li, L. Han, S. H. Liu, H. J. Ma, R. Q. Huang and C. Jiang, Dual targeting effect of Angiopep-2-modified, DNA-loaded nanoparticles for glioma, *Biomaterials* **32,** 6832–6838 (2011).
35. Y. Li, H. He, X. R. Jia, W. L. Lu, J. N. Lou and Y. Wei, A dual-targeting nanocarrier based on poly(amidoamine) dendrimers conjugated with transferrin and tamoxifen for treating brain gliomas, *Biomaterials* **33**, 3899–908 (2012).
36. S. Supattapone, H. Wille, L. Uyechi, J. Safar, P. Tremblay, F. C. Szoka, F. E. Cohen, S. B. Prusiner and M. R. Scott, Branched polyamines cure prion-infected neuroblastoma cells, *J. Virol.* **75**, 3453–3461 (2001).
37. S. Supattapone, J. R. Piro and J. R. Rees Complex polyamines: Unique prion disaggregating compounds, *CNS Neurol. Disord. Drug Targets* **8**, 323–328 (2009).
38. B. Klajnert, J. Cladera and M. Bryszewska, Molecular interactions of dendrimers with amyloid peptides: pH dependence, *Biomacromolecules* **7**, 2186–2191 (2006).
39. B. Klajnert, M. Cangiotti, S. Calici, J. P. Majoral, A. M. Caminade, J. Cladera, M. Bryszewska, M. F. Ottaviani, EPR study of the interactions between dendrimers

and peptides involved in Alzheimer's and prion diseases, *Macromol. Biosci.* 7, 1065–1074 (2007).

40. M. Fischer, D. Appelhans, S. Schwarz, B. Klajnert, M. Bryszewska, B. Voit and M. Rogers Influence of surface functionality of poly(propylene imine) dendrimers on protease resistance and propagation of the scrapie prion protein, *Biomacromolecules* **11,** 1314–1325 (2010).
41. B. Klajnert, T. Wasiak, M. Ionov, M. Fernandez-Villamarin, A. Sousa-Herves, J. Correa, R. Riguera and E. Fernandez-Megia, Dendrimers reduce toxicity of A beta 1–28 peptide during aggregation and accelerate fibril formation, *Nanomed. Nanotechnol.* **8**, 1372–1378 (2012).
42. H. Kobayashi, S. Kawamoto, R. A. Star, T. A. Waldmann, Y. Tagaya and M. W. Brechbiel, Micro-magnetic resonance lymphangiography in mice using a novel dendrimer-based magnetic resonance imaging contrast agent, *Cancer Res.* **63**, 271–276 (2003).
43. H. Kobayashi, S. Kawamoto, M. Bernardo, M. W. Brechbiel, M. V. Knopp and P. L. Choyke, Delivery of gadolinium-labeled nanoparticles to the sentinel lymph node: comparison of the sentinel node visualization and estimations of intra-nodal gadolinium concentration by the magnetic resonance imaging, *J. Control. Release* **111,** 343–351 (2006).
44. H. Kobayashi, S. Kawamoto, T. Saga, N. Sato, T. Ishimori, J. Konishi, K. Ono, K. Togashi and M. W. Brechbiel, Avidin-dendrimer-(1B4M-Gd)(254): A tumor-targeting therapeutic agent for gadolinium neutron capture therapy of intraperitoneal disseminated tumor which can be monitored by MRI, *Bioconjug. Chem.* **12,** 587–593 (2001).
45. H. Kobayashi, S. Kawamoto, P. L. Choyke, N. Sato, M. V. Knopp, R. A. Star, T. A. Waldmann, Y. Tagaya and M. W. Brechbiel, Comparison of dendrimer-based macromolecular contrast agents for dynamic micro-magnetic resonance lymphangiography *Magn. Reson. Med.* **50**, 758–766 (2003).
46. L. M. Kaminskas, J. Kota, V. M. McLeod, B. D. Kelly, P. Karellas and C. J. Porter, PEGylation of polylysine dendrimers improves absorption and lymphatic targeting following SC administration in rats, *J. Control. Release* **140**, 108–116 (2009).
47. S. Sadekar and H. Ghandehari, Transepithelial transport and toxicity of PAMAM dendrimers: Implications for oral drug delivery, *Adv. Drug Deliv. Rev.* **64,** 571–588 (2012).
48. D. S. Goldberg, H. Ghandehari and P. W. Swaan Cellular entry of G3.5 Poly (amido amine) dendrimers by clathrin- and dynamin-dependent endocytosis promotes tight junctional opening in intestinal epithelia, *Pharm. Res.* **8**, 1547–1557 (2010).
49. M. El-Sayed, M. Ginski, C. Rhodes and H. Ghandehari, Transepithelial transport of poly(amidoamine) dendrimers across Caco-2 cell monolayers, *J. Control. Release* **81**, 355–365 (2002).
50. K. M. Kitchens, R. B. Kolhatkar, P. W. Swaan, N. D. Eddington and H. Ghandehari, Transport of poly(amidoamine) dendrimers across Caco-2 cell monolayers: Influence of size, charge and fluorescent labeling, *Pharm. Res.* **23**, 2818–2826 (2006).
51. X. Huang, Z. Wu, W. Gao, Q. Chen and B. Yu Polyamidoamine dendrimers as potential drug carriers for enhanced aqueous solubility and oral bioavailability of silybin, *Drug Dev. Ind. Pharm.* **37:** 419–27 (2011).
52. W. Ke, Y. Zhao, R. Huang, C. Jiang and Y. Pei, Enhanced oral bioavailability of doxorubicin in a dendrimer drug delivery system, *J. Pharm. Sci.* **97,** 2208–2216 (2008).

53. A. S. Chauhan, S. Sridevi, K. B. Chalasani, A. K. Jain, S. K. Jain, N. K. Jain and P. V. Diwan, Dendrimer-mediated transdermal delivery: enhanced bioavailability of indomethacin, *J. Control. Release* **90,** 335–343 (2003).
54. V. V. Venuganti and O. P. Perumal, Poly(amidoamine) dendrimers as skin penetration enhancers: Influence of charge, generation, and concentration, *J. Pharm. Sci.* **98**, 2345–2356 (2009).
55. V. V. Venuganti, P. Sahdev, M. Hildreth, X. Guan and O. Perumal, Structure-skin permeability relationship of dendrimers, *Pharm. Res.* **28**, 2246–2260 (2011).
56. K. Borowska, S. Wolowiec, A. Rubaj, K. Glowniak, E. Sieniawska and S. Radej, Effect of polyamidoamine dendrimer G3 and G4 on skin permeation of 8-methoxypsoralene–*in vivo* study, *Int. J. Pharm.* **426**, 280–283 (2012).
57. Y. Cheng, N. Man, T. Xu, R. Fu, X. Wang, L. Wen, Transdermal delivery of nonsteroidal anti-inflammatory drugs mediated by polyamidoamine (PAMAM) dendrimers, *J. Pharm. Sci.* **96**, 595–602 (2007).
58. Z. Dong, K. A. Hamid, Y. Gao, Y. Lin, H. Katsumi, T. Sakane and A. Yamamoto, Polyamidoamine dendrimers can improve the pulmonary absorption of insulin and calcitonin in rats, *J. Pharm. Sci.* **100**, 1866–1878 (2011).
59. M. Mammen, S. K. Choi and G. M. Whitesides, Polyvalent interactions in biological systems: Implications for design and use of multivalent ligands and inhibitors, *Angew. Chem. Int. Ed.* **37**, 2755–2794 (1998).
60. M. A. Mintzer, E. L. Dane, G. A. O'Toole and M. W. Grinstaff, Exploiting dendrimer multivalency to combat emerging and re-emerging infectious diseases, *Mol. Pharm.* **9**, 342–354 (2012).
61. T. D. McCarthy, P. Karellas, S. A. Henderson, M. Giannis, D. F. O'Keefe, G. Heery, J. R. A. Paull, B. R. Matthews and G. Holan, Dendrimers as drugs: Discovery and preclinical and clinical development of dendrimer-based microbicides for HIV and STI prevention, *Mol. Pharm.* **2,** 312–318 (2005).
62. D. Tyssen, S. A. Henderson, A. Johnson, J. Sterjovski, K. Moore, J. La, M. Zanin, S. Sonza, P. Karellas, M. P. Giannis, G. Krippner, S. Wesselingh, T. McCarthy, P. R. Gorry, P. A. Ramsland, R. Cone, J. R. A. Paull, G. R. Lewis and G. Tachedjian, Structure activity relationship of dendrimer microbicides with dual action antiviral activity, *PLoS One* **5**, e12309 (2010).
63. M. Witvrouw, V. Fikkert, W. Pluymers, B. Matthews, K. Mardel, D. Schols, J. Raff, Z. Debyser, E. De Clercq, G. Holan and C. Pannecouque, Polyanionic (i.e., polysulfonate) dendrimers can inhibit the replication of human immunodeficiency virus by interfering with both virus adsorption and later steps (Reverse transcriptase/integrase) in the virus replicative cycle, *Mol. Pharm.* **58**, 1100–1108 (2000).
64. N. Bourne, L. R. Stanberry, E. R. Kern, G. Holan, B. Matthews and D. I. Bernstein, Dendrimers, a new class of candidate topical microbicides with activity against herpes simplex virus infection, *Antimicrob. Agents Chemother.* **44**, 2471–2474 (2000).
65. S. Telwatte, K. Moore, A. Johnson, D. Tyssen, J. Sterjovski, M. Aldunate, P. R. Gorry, P. A. Ramsland, G. R. Lewis, J. R. A. Paull, S. Sonza and G. Tachedjian, Virucidal activity of the dendrimer microbicide SPL7013 against HIV-1, *Antivir. Res.* **90**, 195–199 (2011).
66. A. Castonguay, E. Ladd, T. G. M. van de Ven and A. Kakkar, Dendrimers as bactericides, *New J. Chem.* **36**, 199–204 (2012).

67. C. Z. Chen, N. C. B. Tan and S. L. Cooper, Incorporation of dimethyldodecylammonium chloride functionalities onto poly(propylene imine) dendrimers significantly enhances their antibacterial properties, *Chem Commun.* 1585–6 (1999).
68. M. K. Calabretta, A. Kumar, A. M. McDermott and C. Z. Cai, Antibacterial activities of poly(amidoamine) dendrimers terminated with amino and poly(ethylene glycol) groups, *Biomacromolecules* **8**, 1807–1811 (2007).
69. N. Malik, R. Wiwattanapatapee, R. Klopsch, K. Lorenz, H. Frey, J. W. Weener, E. W. Meijer, W. Paulus and R. Duncan, Dendrimers: Relationship between structure and biocompatibility *in vitro*, and preliminary studies on the biodistribution of I-125-labelled polyamidoamine dendrimers *in vivo*, *J. Control. Release* **65**, 133–148 (2000).
70. A. I. Lopez, R. Y. Reins, A. M. McDermott, B. W. Trautner and C. Z. Cai, Antibacterial activity and cytotoxicity of PEGylated poly(amidoamine) dendrimers, *Mol. Biosyst.* **5**, 1148–1156 (2009).
71. X. Xue, X. Chen, X. Mao, Z. Hou Z, Zhou Y, Bai H, Meng J, Da F, Sang G, Wang Y, Luo X. Amino-terminated generation 2 poly(amidoamine) Dendrimer as a potential broad-spectrum, nonresistance-inducing antibacterial agent, *AAPS J.* **15**, 132–142 (2013).
72. M. Nomizu, K. Yamamura, H. K. Kleinman and Y. Yamada, multimeric forms of Tyr-Ile-Gly-Ser-Arg (YIGSR) peptide enhance the inhibition of tumor-growth and metastasis, *Cancer Res.* **53**, 3459–3461 (1993).
73. J. P. Tam, Synthetic Peptide vaccine design — synthesis and properties of a high-density multiple antigenic peptide system, *Proc. Natl. Acad. Sci. USA* **85**, 5409–5413 (1988).
74. P. Veprek and J. Jezek, Peptide and glycopeptide dendrimers. Part II, *J. Pept. Sci.* **5**, 203–220 (1999).
75. Y. Iwamoto, M. Nomizu, Y. Yamada, Y. Ito, K. Tanaka and Y. Sugioka, Inhibition of angiogenesis, tumour growth and experimental metastasis of human fibrosarcoma cells HT1080 by a multimeric form of the laminin sequence Tyr-Ile-Gly-Ser-Arg (YIGSR), *Brit. J. Cancer* **73**, 589–595 (1996).
76. H. R. Yao, D. M. Veine, K. S. Fay, E. D. Staszewski, Z. Z. Zeng and D. L. Livant, The PHSCN dendrimer as a more potent inhibitor of human breast cancer cell invasion, extravasation, and lung colony formation, *Breast Cancer Res. Tr.* **125**, 363–375 (2011).
77. J. W. Bulte, C. Wu, M. W. Brechbiel, R. A. Brooks, J. Vymazal, M. Holla and J. A. Frank, Dysprosium-DOTA-PAMAM dendrimers as macromolecular T2 contrast agents. Preparation and relaxometry, *Invest. Radiol.* **33**, 841–845 (1998).
78. K. E. Kellar, P. M. Henrichs, R. Hollister, S. H. Koenig, J. Eck, and D. Wei, High relaxivity linear Gd(DTPA)-polymer conjugates: The role of hydrophobic interactions, *Magn. Reson. Med.* **38**, 712–716 (1997).
79. M. W. Bourne, L. Margerun, N. Hylton, B. Campion, J. J. Lai, N. Derugin and C. B. Higgins, Evaluation of the effects of intravascular MR contrast media (gadolinium dendrimer) on 3D time of flight magnetic resonance angiography of the body, *J. Magn. Reson. Imaging* **6**, 305–310 (1996).
80. E. C. Wiener, M. W. Brechbiel, H. Brothers, R. L. Magin, O. A. Gansow, D. A. Tomalia and P. C. Lauterbur, Dendrimer-based metal chelates: A new class of magnetic resonance imaging contrast agents, *Magn. Reson. Med.* **31**, 1–8 (1994).
81. E. C. Wiener, S. Konda, A. Shadron, M. Brechbiel and O. Gansow, Targeting dendrimer-chelates to tumors and tumor cells expressing the high-affinity folate receptor, *Invest. Radiol.* **32**, 748–754 (1997).

82. G. Wu, W. Yang, R. F. Barth, S. Kawabata, M. Swindall, A. K. Bandyopadhyaya, W. Tjarks, B. Khorsandi, T. E. Blue, A. K. Ferketich, M. Yang, G. A. Christoforidis, T. J. Sferra, P. J. Binns, K. J. Riley, M. J. Ciesielski and R. A. Fenstermaker, Molecular targeting and treatment of an epidermal growth factor receptor-positive glioma using boronated cetuximab, *Clin. Cancer Res.* **13**, 1260–1268 (2007).
83. H. Kobayashi, S. Kawamoto, T. Saga, N. Sato, A. Hiraga, J. Konishi, K. Togashi and M. W. Brechbiel, Micro-MR angiography of normal and intratumoral vessels in mice using dedicated intravascular MR contrast agents with high generation of polyamidoamine dendrimer core: Reference to pharmacokinetic properties of dendrimer-based MR contrast agents, *J. Magn. Reson. Imaging* **14**, 705–713 (2001).
84. H. Kobayashi, N. Sato, A. Hiraga, T. Saga, Y. Nakamoto, H. Ueda, J. Konishi, K. Togashi and M. W. Brechbiel 3D-micro-MR angiography of mice using macromolecular MR contrast agents with polyamidoamine dendrimer core with reference to their pharmacokinetic properties, *Magn. Reson. Med.* **45**, 454–460 (2001).
85. H. Kobayashi, C. C. Wu, M. K. Kim, C. H. Paik, J. A. Carrasquillo and M. W. Brechbiel, Evaluation of the *in vivo* biodistribution of indium-111 and yttrium-88 labeled dendrimer-1B4M-DTPA and its conjugation with anti-Tac monoclonal antibody, *Bioconjug. Chem.* **10**, 103–111 (1999).
86. H. Kobayashi and M. W. Brechbiel, Nano-sized MRI contrast agents with dendrimer cores, *Adv. Drug Deliv. Rev.* **57**, 2271–2286 (2005).
87. H. Kobayashi, S. Kawamoto, T. Saga, N. Sato, A. Hiraga, T. Ishimori, Y. Akita, M. H. Mamede, J. Konishi and K. Togashi, Novel liver macromolecular MR contrast agent with a polypropylenimine diaminobutyl dendrimer core: Comparison to the vascular MR contrast agent with the polyamidoamine dendrimer core, *Magn. Reson. Med.* **46**, 795–802 (2001).
88. V. S. Talanov, C. A. Regino, H. Kobayashi, M. Bernardo, P. L. Choyke and M. W. Brechbiel, Dendrimer-based nanoprobe for dual modality magnetic resonance and fluorescence imaging, *Nano Lett.* **6**, 1459–1463 (2006).
89. Y. Koyama, V. S. Talanov, M. Bernardo, Y. Hama, C. A. Regino, M. W. Brechbiel, P. L. Choyke and H. Kobayashi, A dendrimer-based nanosized contrast agent dual-labeled for magnetic resonance and optical fluorescence imaging to localize the sentinel lymph node in mice, *J Magn Reson Imaging* **25:** 866–71 (2007).
90. Mounzer R, Shakarin P, Papademetris X, Constable T, Ruddle NH, Fahmy TM. Dynamic imaging of lymphatic vessles and lymph nodes using a bimodal nanoparticulate contrast agent, *Lymphat. Res. Biol.* **5**, 151–158 (2007).
91. H. Yan, L. Wang, J. Wang, X. Weng, H. Lei, X. Wang, L. Jiang, J. Zhu, W. Lu, X. Wei and C. Li, Two-order targeted brain tumor imaging by using an optical/paramagnetic nanoprobe across the blood brain barrier, *ACS Nano* **6**, 410–420 (2012).
92. J. O. McIntyre, B. Fingleton, K. S. Wells, D. W. Piston, C. C. Lynch, S. Gautam and L. M. Matrisian, Development of a novel fluorogenic proteolytic beacon for *in vivo* detection and imaging of tumour-associated matrix metalloproteinase-7 activity, *Biochem. J.* **377**, 617–628 (2004).
93. M. A. Alcala, S. Y. Kwan, C. M. Shade, M. Lang, H. Uh, M. Wang, S. G. Weber, D. L. Bartlett, S. Petoud and Y. J. Lee, Luminescence targeting and imaging using a nanoscale generation 3 dendrimer in an *in vivo* colorectal metastatic rat model, *Nanomedicine* 7, 249–258 (2011).

94. M. A .Kay, J. C. Glorioso and L. Naldini. Viral vectors for gene therapy: The art of turning infectious agents into vehicles of therapeutics, *Nat. Med.* 7, 33–40 (2001).
95. X. Zhang and W. T. Godbey, Viral vectors for gene delivery in tissue engineering, *Adv. Drug Deliv. Rev.* **58**, 515–534 (2006).
96. J. Haensler and F. C. Szoka FC, Polyamidoamine cascade polymers mediate efficient transfection of cells in culture, *Bioconjug. Chem.* **4**, 372–379 (1993).
97. V. A. Kabanov, V. G. Sergeyev, O. A. Pyshkina, A. A. Zinchenko, A. B. Zezin, J. G. H. Joosten, J. Brackman and K. Yoshikawa, Interpolyelectrolyte complexes formed by DNA and astramol poly(propylene imine) dendrimers, *Macromolecules* **33**, 9587–9593 (2000).
98. W. T. Godbey, K. K. Wu and A. G. Mikos, Tracking the intracellular path of poly(ethylenimine)/DNA complexes for gene delivery, *Proc. Natl. Acad. Sci. USA* **96**, 5177–5181 (1999).
99. F. Kihara, H. Arima, T. Tsutsumi, F. Hirayama and K. Uekama, *In vitro* and *in vivo* gene transfer by an optimized alpha-cyclodextrin conjugate with polyamidoamine dendrimer, *Bioconjug. Chem.* **14**, 342–350 (2003).
100. P. Kesharwani, V. Gajbhiye and N. K. Jain, A review of nanocarriers for the delivery of small interfering RNA, *Biomaterials* **33**, 7138–7150 (2012).
101. T. Dutta, M. Garg M and N. K. Jain, Poly(propyleneimine) dendrimer and dendrosome mediated genetic immunization against hepatitis B, *Vaccine* **26**, 3389–3394 (2008).
102. T. Dutta, H. B. Aghase, P. Vijayarajkumar, M. Joshi and N. K. Jain, Dendrosome-based gene delivery, *J. Exp. Nanosci.* **1**, 235–248 (2006).
103. M. L. Patil, M. Zhang, S. Betigeri, O. Taratula, H. He and T. Minko, Surface-modified and internally cationic polyamidoamine dendrimers for efficient siRNA delivery, *Bioconjug. Chem.* **19**, 1396–1403 (2008).
104. S. Biswas, P. P. Deshpande, G. Navarro, N. S. Dodwadkar and V. P. Torchilin, Lipid modified triblock PAMAM-based nanocarriers for siRNA drug co-delivery, *Biomaterials* **34**, 1289–1301 (2013).
105. C. L. Waite and C. M. Roth, PAMAM-RGD conjugates enhance siRNA delivery through a multicellular spheroid model of malignant glioma, *Bioconjug. Chem.* **20**, 1908–1916 (2009).
106. H. Yu, Y. Nie, C. Dohmen, Y. Li and E. Wagner, Epidermal growth factor-PEG functionalized PAMAM-pentaethylenehexamine dendron for targeted gene delivery produced by click chemistry, *Biomacromolecules* **12**, 2039–2047 (2011).
107. M. L. Patil, M. Zhang, O. Taratula, O. B. Garbuzenko, H. He and T. Minko, Internally cationic polyamidoamine PAMAM-OH dendrimers for siRNA delivery: effect of the degree of quaternization and cancer targeting, *Biomacromolecules* **10**, 258–266 (2009).
108. Y. Hayashi, Y. Mori, S. Yamashita, K. Motoyama, T. Higashi, H. Jono, Y. Ando and H. Arima, Potential use of lactosylated dendrimer (G3)/alpha-cyclodextrin conjugates as hepatocyte-specific siRNA carriers for the treatment of familial amyloidotic polyneuropathy, *Mol. Pharm.* **9**, 1645–1653 (2012).
109. C. L. Jenkins and R. T Raines, Insights on the conformational stability of collagen, *Nat. Prod. Rep.* **19**, 49–59 (2002).
110. L. R. Ellingsworth, F. DeLustro, J. E. Brennan, S. Sawamura and J. McPherson, The human immune response to reconstituted bovine collagen, *J. Immunol.* **136**, 877–882 (1986).

111. M. A. Mintzer and M. W. Grinstaff, Biomedical applications of dendrimers: A tutorial, *Chem. Soc. Rev.* **40**, 173–190 (2011).
112. G. A. Kinberger, W. Cai and M. Goodman, Collagen mimetic dendrimers, *J. Am. Chem. Soc.* **124**, 15162–15163 (2002).
113. G. A. Kinberger, J. P. Taulane and M. Goodman, The design, synthesis, and characterization of a PAMAM-based triple helical collagen mimetic dendrimer, *Tetrahedron* **62**, 5280–5286 (2006).
114. S. T. Khew, Q. J. Yang and Y. W. Tong, Enzymatically crosslinked collagen-mimetic dendrimers that promote integrin-targeted cell adhesion, *Biomaterials* **29**, 3034–3045 (2008).
115. J. E. Gough, C. A. Scotchford and S. Downes, Cytotoxicity of glutaraldehyde crosslinked collagen/poly(vinyl alcohol) films is by the mechanism of apoptosis, *J. Biomed. Mater. Res.* **61**, 121–130 (2002).
116. X. Duan and H. Sheardown, Crosslinking of collagen with dendrimers, *J. Biomed. Mater. Res. A* **75**, 510–518 (2005).
117. X. Duan and H. Sheardown, Dendrimer crosslinked collagen as a corneal tissue engineering scaffold: mechanical properties and corneal epithelial cell interactions, *Biomaterials* **27**, 4608–4617 (2006).
118. M. A. Princz and H. Sheardown, Heparin-modified dendrimer cross-linked collagen matrices for the delivery of basic fibroblast growth factor (FGF-2), *J. Biomater. Sci. Polym. Ed.* **19**, 1201–1218 (2008).
119. S. Zhong and L. Y. Yung, Enhanced biological stability of collagen with incorporation of PAMAM dendrimer, *J. Biomed. Mater. Res. A* **91**, 114–122 (2009).
120. P. Buma, T. van Tienen and R. Veth, The collagen meniscus implant, *Expert Rev. Med. Devices* **4**, 507–516 (2007).
121. S. H. M. Sontjens, D. L. Nettles, M. A. Carnahan, L. A. Setton and M. W. Grinstaff, Biodendrimer-based hydrogel scaffolds for cartilage tissue repair, *Biomacromolecules* **7**, 310–316 (2006).
122. L. Degoricija, P. N. Bansal, S. H. M. Sontjens, N. S. Joshi, M. Takahashi, B. Snyder and M. W. Grinstaff, Hydrogels for osteochondral repair based on photocrosslinkable carbamate dendrimers, *Biomacromolecules* **9**, 2863–2872 (2008).
123. T. F. Vandamme and L. Brobeck, Poly(amidoamine) dendrimers as ophthalmic vehicles for ocular delivery of pilocarpine nitrate and tropicamide, *J. Control. Release* **102**, 23–38 (2005).
124. W. J. Yao, K. X. Sun, Y. Liu, N. Liang, H. J. Mu, C. Yao, R. C. Liang and A. P. Wang, Effect of poly(amidoamine) dendrimers on corneal penetration of puerarin, *Biol. Pharm. Bull.* **33**, 1371–1377 (2010).
125. W. Yao, K. Sun, H. Mu, N. Liang, Y. Liu, C. Yao, R. Liang and A. Wang, Preparation and characterization of puerarin-dendrimer complexes as an ocular drug delivery system, *Drug Dev. Ind. Pharm.* **36**, 1027–1035 (2010).
126. C. Durairaj, R. S. Kadam, J. W. Chandler, S. L. Hutcherson and U. B. Kompella, Nanosized dendritic polyguanidilyated translocators for enhanced solubility, permeability, and delivery of gatifloxacin, *Invest. Ophthalmol. Vis. Sci.* **51**, 5804–5816 (2010).
127. C. A. Holden, P. Tyagi, A. Thakur, R. Kadam, G. Jadhav, U. B. Kompella and H. Yang, Polyamidoamine dendrimer hydrogel for enhanced delivery of antiglaucoma drugs, *Nanomedicine* **8**, 776–783 (2012).

128. H. Yang, P. Tyagi, R. S. Kadam, C. A. Holden and U. B. Kompella, Hybrid dendrimer hydrogel/PLGA nanoparticle platform sustains drug delivery for one week and antiglaucoma effects for four days following one-time topical administration, *ACS Nano* **6**, 7595–7606 (2012).
129. X. Duan and H. Sheardown, Incorporation of cell-adhesion peptides into collagen scaffolds promotes corneal epithelial stratification, *J. Biomater. Sci. Polym. Ed.* **18**, 701–711 (2007).
130. X. Duan, C. McLaughlin, M. Griffith and H. Sheardown, Biofunctionalization of collagen for improved biological response: scaffolds for corneal tissue engineering, *Biomaterials* **28**, 78–88 (2007).
131. M. A. Princz and H. Sheardown, Heparin-modified dendrimer crosslinked collagen matrices for the delivery of heparin-binding epidermal growth factor, *J. Biomed. Mater. Res. A* **100**, 1929–1937 (2012).
132. R. S. Givens, A. L. Yousef, S. Yang and G. T. Timberlake, Collagen cross linking agents: Design and development of a multifunctional cross linker, *Photochem. Photobiol.* **84**, 185–92 (2008).
133. S. R. Benhabbour, H. Sheardown and A. Adronov, Cell adhesion and proliferation on hydrophilic dendritically modified surfaces, *Biomaterials* 29, 4177–4186 (2008).
134. S. Shaunak, S. Thomas, E. Gianasi, A. Godwin, E. Jones, I. Teo, K. Mireskandari, P. Luthert, R. Duncan, S. Patterson, P. Khaw and S. Brocchini Polyvalent dendrimer glucosamine conjugates prevent scar tissue formation, *Nat. Biotechnol.* **22,** 977–984 (2004).

Chapter 9

Layer-by-Layer Nanopreparations for Medicine — Smart Polyelectrolyte Multilayer Capsules and Coatings

Rawil F. Fakhrullin, Gleb B. Sukhorukov[†] and Yuri M. Lvov[‡,§]*

**Kazan Federal University, Kazan, Republic of Tatarstan, Russian Federation*
[†]Queen Mary University of London, London, UK
[‡]Louisiana Tech University, Ruston, USA
[§]ylvov@coes.latech.edu

1. Introduction

A variety of techniques have been developed to address specific encapsulation and release requirements for biomedical applications. It comprises liposomes, polymer-based delivery systems with adhered active molecules, block-copolymer forming vesicles for encapsulation, organic micro- and nanoparticles enabling release upon degradation and other core/shell systems.[1–3] Active substances were encapsulated into confined volumes and drug release kinetic was found to depend upon external conditions. Such delivery systems and bio-microreactors should possess multiple functionalities, such as combination of easy processing, high storage capacity and protection of encapsulated substances, and delivery with targeting and controlled release should be tailored in one tiny vehicle. This allows for more efficient drug transport by guiding the carriers to designated types of cells or tissues. Such carriers release could also be activated by external triggers. There are two most important functions of the micro/nano vehicle: the capsule's ability to carry and controlled release the drug and its targeting to the thick organ. Needs for a long-term blood circulation and high drug-capsule concentrations necessary for injection are often serious obstacles for drug nanoparticulation. Magnetic liposomes or magnetic carriers with immobilized receptors have been used. Liposomes could also be engineered to be responsive to surrounding. The polymers can bear the drug and release it once they reach an area with certain pH. We describe here layer-by-layer (LbL) architecting of nano-organized polyelectrolyte shells drug cores; these microshells can be equipped with elements to allow slow controlled release and targeting through attached antibodies.

2. Polyelectrolyte LbL Assembled Capsules

Originally proposed in the early 1990s in works by Decher, Möhwald, Lvov, Rubner, Hammond and other researchers,[4–11] LbL assembly utilized a variety of polyelectrolytes, including many biocompatible ones, i.e. proteins and DNA (Fig. 1). Typically, one starts LbL assembly with polyanion layer deposited on the positively charged surface; it is followed by the adsorption of an oppositely charged polycation reversing surface charge, then again polyanion, and so on. In each cycle, a stable polycation/polyanion complex is formed, resulting in the appearance of a very thin (5–50 nm depending on the number of cycles) but stable polymeric multilayer. By varying the charge density on each polymer or the number of coating cycles, multilayers with a different surface charge and different polymeric composition can be prepared. Polycations can be represented by poly(ethylene imine) (PEI), poly(allylamine) (PAH), polylysine, chitosan, gelatin B, amino-dextran, protamine sulfate;

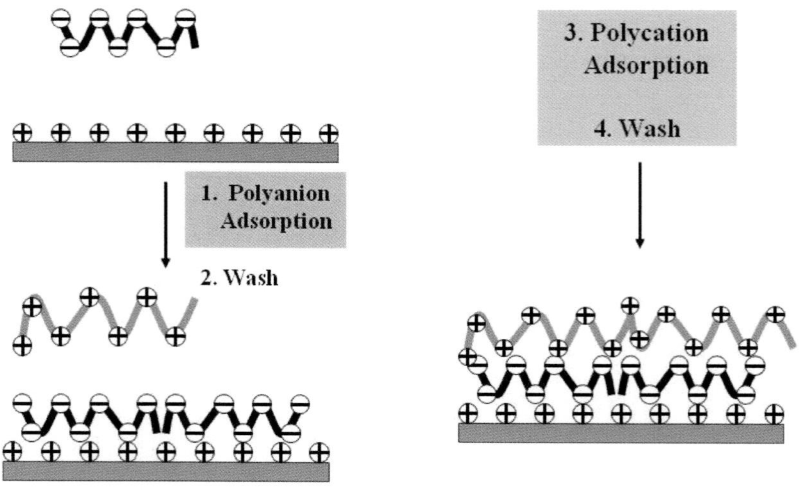

Fig. 1. Electrostatic charge driven LbL assembly of charged polymers.

polyanions — by poly(styrene sulfonate) (PSS), poly(acrylic acid), dextran sulfate, carboxymethyl cellulose, sodium alginate, hyaluronic acid, gelatin A, chondroitin and heparin. At present, LbL assembly is widely used to produce functional biocompatible coating and ultrathin free-standing films.

First, a drug delivery platform can be devised by a planar LbL multilayered coating on the surface of drug molecules for which drug release is triggered through destruction of the film structure by specific stimuli, such as pH variation, temperature variation and illumination by light or natural biodegradation.[9] Such LbL multilayers may be based not only on electrostatic interlayer bonding but also on hydrogen bonding and may incorporate prefabricated tiny containers (or drug particles) for hierarchical nanosystems. Planar LbL multilayers may contain 1–5 wt. % drugs and have a release over a span of 1–100 h. These medical applications are currently limited by designing of drug loaded polyelectrolyte coating on bone implants or patches for skin applications.

In 1997–1998, an approach to fabricate nanoengineered capsules was introduced by the group at the Max Planck Institute of Colloids and Interfaces.[7,8] The method is based on the LbL adsorption of oppositely charged macromolecules, mainly polyelectrolytes on colloidal particles. If the core gets dissolved, the product of core decomposition expels out, and the polyelectrolyte multilayer constructs remain and form the empty shell. These multilayers can be made from a variety of layer constituents, such as synthetic and natural polyelectrolytes, nanoparticles, lipids and multivalent dyes, and hence the build-up is extremely modular with regards to tailoring the composition of the capsule wall. Shell thickness is determined by the layer number and defined in the range of a few nanometers. Each of the involved components brings functionality to the multilayer capsules. The hollow capsules can be re-filled with substances of interest (proteins, drugs). The variable size of the capsules can be fully controlled from 50 nm to tens of microns. This provides a way to control drug release from such particles by design of the shell architecture at the nanometer level. The use of a polymer containing reactive groups (such as amino or carboxylic groups) to form the last "outer" surface layer will allow for the attachment of specific ligands, or reporter groups, and other moieties of interest to drug nanoparticles.

We focus on the application of LbL for coating micron and submicron objects such as decomposable core particles and cells. By moving from microscopic flat substrates toward micron-sized curved surfaces, we aim to explore the possibilities of tailoring different functionalities by charge-driven adsorption within multilayers, alternating their mechanical properties and impregnating inorganic and organic substances both inside the capsule volume and

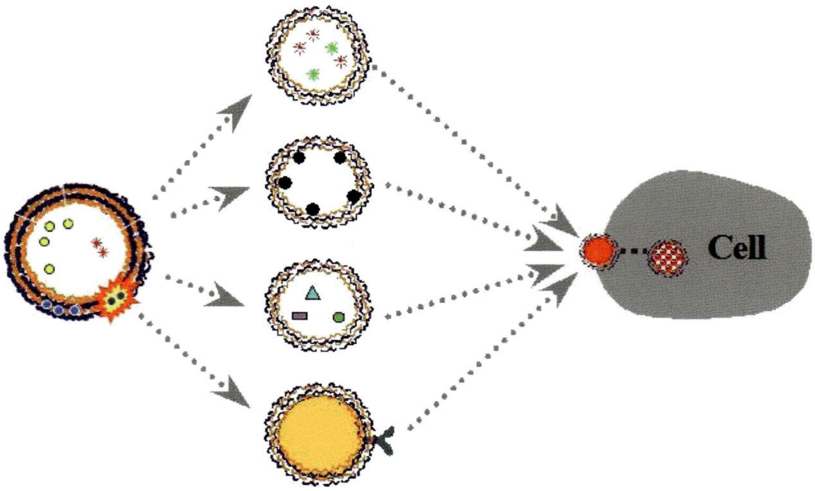

Fig. 2. Multifunctional capsules made of LbL assembly: various functions tailored in one capsules (left side) and loaded with sensing elements (for identification), magnetic nanoparticles (to drive the capsule macroscopically into a specific region), enzymes (working as a protection tool to catalyze specific reactions inside the cell) or drugs (to be delivered at a given site). The outer wall can be functionalized with receptors or antibodies for targeting on a molecular level.

in the polyelectrolyte shell, and control release of encapsulated material,[12,13] providing continuous scientific and industrial interest in employing capsules as microcontainers and microreactors as well as using them as sensing elements residing in living cells (Fig. 2).[14]

The synthesis of inorganic particles within the multilayer makes capsules rigid. Additional functionality to capsules could be incorporated, for instance, by inclusion magnetic particles or hydroxyapatite nanoparticle, and are reviewed in Refs. 15 and 16. Composite capsules of inorganic and organic components could combine the advantages of both: storage, susceptibility to external forces for inorganic and responsiveness for polymeric constituents. Comparing our capsules with traditional and new alternatives (capsules or other types of vesicles of size 0.1–10 μm, for instance, liposomes and polysomes[17,18]) the most characteristic outstanding features of the proposed multilayer capsules are the ability to tailor different properties in one structure and to create novel multifunctional materials.

2.1. *Encapsulation and release of biologically active molecules*

Capsules made out of polymers with functional groups are able to undergo conformational changes, and hence the permeability of such capsules is varied.

For instance, polyelectrolyte capsules containing at least one weak polyelectrolyte are stable at the defined pH range there the capsules have electrostatic balanced charges. The pH change results in imbalance of charges inside a polyelectrolyte shell followed by its swelling and increase in permeability.[19,20] Permeability of shells can be reverted to the original state, returning pH to the normal values. Thus it is possible to realize a reversible "switch" of capsule shells permeability between "open" and "closed" states via pH change.[4,19] The influence of pH value on permeability of capsules of mostly used $(PAH/PSS)_4$ was investigated in Ref. 19, and resulted in loss of encapsulated materials at high pH, where amino-groups become non-charged. Basically, polyelectrolytes with chargeable groups, i.e. any weak polyelectrolyte, are responsive to pH. Capsules made of polyelectrolytes exhibit response once the pH of the media has changed. Swelling of the capsules containing polymers with amino groups at alkaline pH was observed.[19] The amino group becomes uncharged, and the entire charge of the capsule becomes negative and, as a consequence, the capsule swells and, depending on shell thickness and charge imbalance, dissolves with time. If capsules are made of two weak polyelectrolytes such as PAH and PAA (polyacrylic acid), the responsiveness occurs at "both ends" of the pH region reflecting charging carboxylic and amino-groups. Any charge disbalance breaks the structure of a shell and were open for high-molecular compounds, for example, dextran.[20] Permeability change caused by pH is reversible (Fig. 3), thus it making it possible to let macromolecules inside capsules and enclose them afterward by further pH change.

The permeability of polyelectrolyte shells can be changed by addition of organic solvent to water suspension of microcapsules.[21] The decrease in dielectric permittivity of microenvironment leads to the increase of electrostatic interaction between polyionic parts. The permeability of polyelectrolyte shells increases because of compression and curtailing polymeric chains. The microcapsule shell permeability can also be varied by the change of an ionic force of their media surrounded. It was shown that the size of the capsules did not vary at the low ionic force. However, permeability of their shells increased because of partial break of ionic bonds between the oppositely charged polyions. Electrostatic interactions between oppositely charged polyions at high salt concentrations weakened enough to cause reorganization of the shell structure. As a result, compression of polyionic layers took place.[22] It was shown[23,24] that morphological changes of polyelectrolyte shell at high ionic force led to its loosening.

In general, tuning permeability for a wide class of substances can be reached by designed layer thickness.[13] However, the capsule wall can be subjected to densification after their preparation has been completed. As

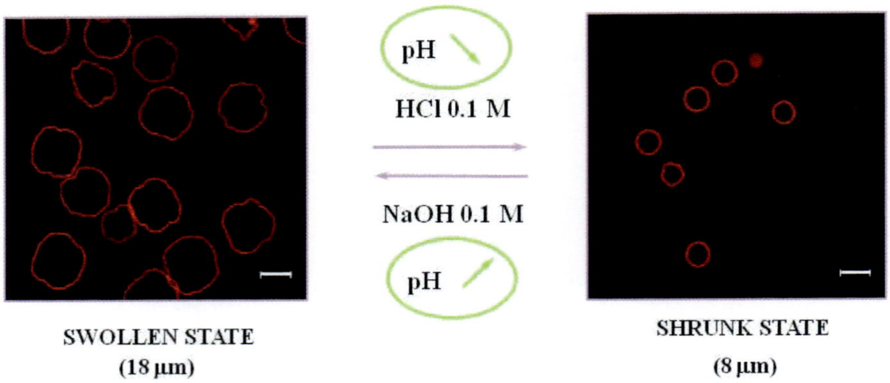

Fig. 3. pH-sensitive behavior of hollow capsules (Scale bar is 10 μm). Adapted from Ref. 19.

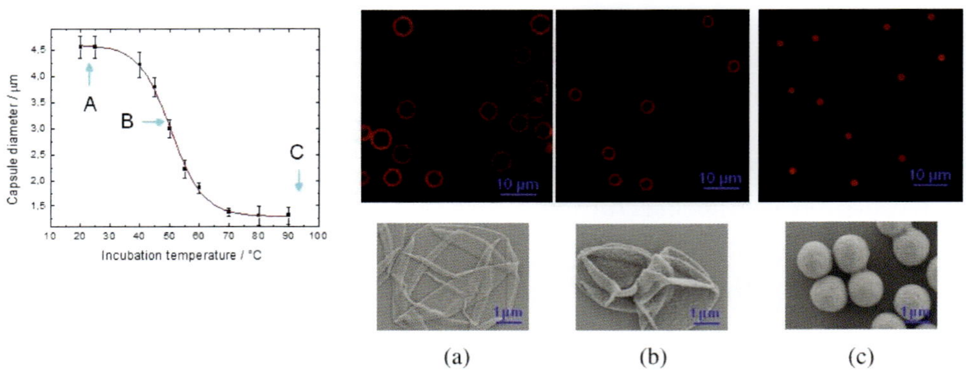

Fig. 4. Temperature induced capsule shrinking; top: confocal microscopy, bottom line is SEM. Images A, B and C are from correspondent temperature 1 h treatment: left. Adapted from Ref. 25.

shown in Ref. 25, higher temperature leads to shrinking of capsules, with compensated charges, which results in increasing wall thickness and drastically reduced permeability (Fig. 4). Keeping the microcapsules at raised temperature can lead to the change in their size, which can be used in entrapping of biologically important molecules.[26] Such a temperature-induced change of capsule's shape is irreversible.

Alternatively, photosensitive polymers used as layer constituents could make capsules sensitive to light. Azobenzene group containing polymers, such as sodium poly(carboxyl-hydroxyphenyl azobenzene sulfonamide ethanediyl (PAZO), are used to cause changes in spatial orientation of polymers in the shell, induced mobility of the chain and, as a result, the capsules shrink and macromolecules can be entrapped.[27] Photoinduced crosslinking of polymers in the multilayer results in shell tightening and less permeability.[28]

2.2. *Routes of encapsulations*

The opportunity of inclusion of various biomedical substances inside LbL microcapsules is the most challenging. All possible routes of encapsulation can be classified into two basic groups: (i) capsulation during the formation of polyelectrolyte microcapsule shells and (ii) capsulation after core dissolution, which is mainly done by alternation of permeability by temperature, light, pH, ionic strength or solvent as described above. As far as biomacromolecules are concerned, the general guideline is that if the biomacromolecules are stable in alternated condition for opening and closing capsules, this can be applicable. However, it would never reach anywhere close to 100% encapsulation efficiency.

The alternative method of the first approach is coating crystals or units of capsulated substances by a multilayer film. However, such an application is limited by the quality of the substances which are capable to form regular conglomerates of a certain size and form, which are precipitates or poorly soluble in the conditions of polyelectrolyte coating deposition.

The most prominent method to encapsulate proteins and DNA is co-precipitation with porous calcium carbonate particles which are formed with about 50% internal volumes in voids capable to host proteins.[29] Followed by LbL assembly on surface of $CaCO_3$ particles containing substances of interest, instances of dissolution of core, resulting in escape of calcium ions, and retaining proteins in the capsules are very less in terms of protein structure preservation and could result to up to 80% of encapsulation efficiency. Thus, a so-called strategy, "the ship in a bottle", uses variation of microcapsule shell permeability to fill a ready-made capsule with macromolecules incorporated directly inside it either as crystals or precipitates or co-precipitated with inorganic particles (Fig. 5).

Encapsulation of substances with smaller molecular weight is more complicated and could be done by adsorption onto a polymeric matrix inside a capsule[30] or by direct precipitation inside a shell. In the first case, strong interactions exclude free release of capsulated molecules. This is especially important for "drug delivery" application of capsules. The most interesting way of capsulation is filling already created capsules by means of shell permeability change varying microenvironment properties, i.e. ionic force, pH value, structure of solvent.[21,22] However, the described encapsulation procedures are not so simple. Certain substances need to be added to capsule

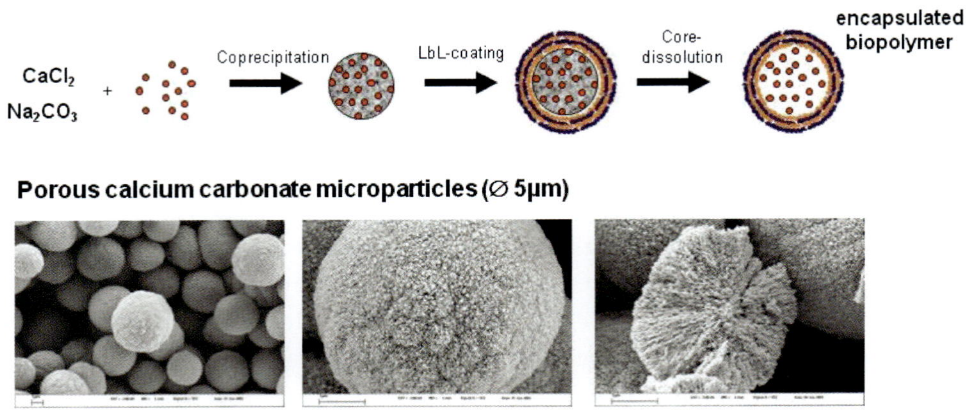

Fig. 5. Encapsulation of macromolecules after incorporation into calcium carbonate microparticles ("co-precipitation method"). Image adapted from Ref. 29.

suspensions in all methods, which can cause deformation of the shells because of an osmotic shock. The applicability of these methods is limited by the nature of encapsulated substances and/or structures of capsules. Such structural changes lead to the reduction of permeability. The capsules prepared from a broad variety of polyelectrolytes were investigated for a more general view on the thermally induced rearrangements [25].

2.3. *Stimuli and externally triggered release*

Inorganic nanoparticles incorporated in polyelectrolyte shells enable remote activated release by ultrasound, infrared or microwave radiation. Such remote-activated release systems exhibit response to remote physical influence (such as light or ultrasound). LbL assembly provides a unique opportunity to combine different properties in single entity of micro-packaged delivery system to make it susceptible to remote physical influence for navigation and activation. The concept is that inorganic nanoparticles embedded in polyelectrolyte films are susceptible to external physical impact, and their response is damage or even rupture of the films. Light absorption leads to local heat and hence disruption of polyelectrolyte shells. The susceptibility of metals or semiconductors nanoparticles for vis/near-infrared irradiation can be tuned in a range of a few nanometers, which makes possible the activation of different capsules and the release of different encapsulated materials optically by switching the wavelengths of laser irradiation. Use of ultrasound is also envisaged because composite organic/inorganic materials undergo significant stress due to the density gradient across the films, while being exposed to

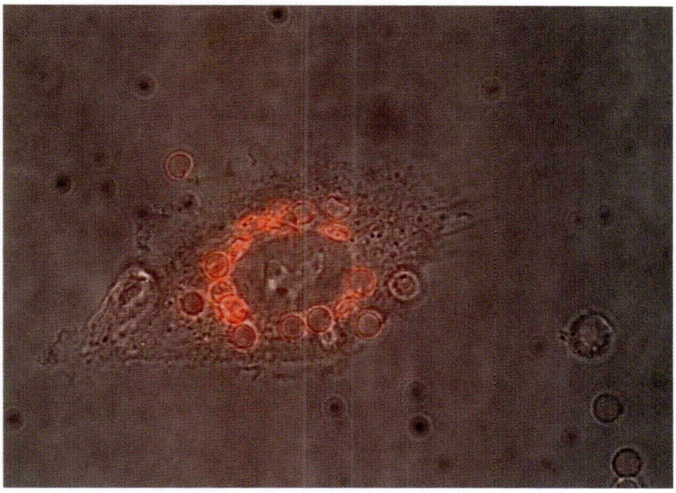

Fig. 6. LbL microcapsules uptake by human breast cancer cells (note the spherical red microcapsules releasing encapsulated cargo inside the cell).

sonication. Inorganic nanoparticles embedded within a polyelectrolyte multilayer lead to breakage of the entire multilayer capsule wall and release of encapsulated materials (Fig. 6).[31,32]

As reviewed in Ref. 31, some data reported in very recent years clearly show perspectives for magnetic addressing and navigation of filled capsules as well as electromagnetic stimulation for induced release. In a cell culture, we have seen magnetic localization and light stimulation of capsules while they are internalized by cells (Fig. 6). In the case of cell study, the capsules can be accessed well. Of more concern is the applicability of remotely activated capsules for animal study and foreseen clinical trials. So far, we are still in lack of *in vivo* data on release from remotely addressable capsules. The penetration depth of near-infrared light is limited with few millimeters and can be used just for under-skin treatment. These capsules can serve as depot systems residing under-skin in "stand-by" regime awaiting time to be activated by a remote trigger.[32] As per controlling capsules located deeper in body, optical fiber waveguides might improve the situation and allow to access areas with delivered capsules. So, there is still a motivation for improvement of light-addressable capsules. As per other remote addressing approaches, the magnetic navigation of capsules is rather optimistic due to the presence of similar *in vivo* data on the other magnetically modified delivery systems. However, ultrasound activated release is still to be further elaborated before it can be brought in agenda for *in vivo* experiments.

3. Combination of Drug Interpolyelectrolyte Complexation with LbL Shell

The nano-organized LbL shell formed around drug microparticles was considered earlier as an adjustable diffusion barrier, and no specific interaction with microcores was taken into account. In another approach, interpolyelectrolyte complexation of drugs with polyelectrolytes to form spontaneously microparticles was developed[33,34] but no LbL assembly was previously used. In the current work on sustained release from pre-formed insulin microparticles, we combined these two approaches. For LbL shells on soft insulin cores, we combined these two conceptions: interpolyelectrolyte complexation and formation of organized multilayer shell for prolonged release. Formation of 10- to 50-nm-thick LbL multilayers on the soft permeable insulin cores was not sufficient to provide a tight physical barrier to slow down insulin diffusion into aqueous solution. We established conditions for complexation of the soft insulin core with the first polyelectrolyte layer, which played a critical role for sustained release.

First, insulin PROMAXX® particles of 2 μm diameter were coated with the following polyanions: polydiallyldimethyl ammonium chloride (PDDA, model polycation), protamine sulfate (ProtS), poly-l-arginine (PLA) and poly-l-lysine (PLL). The zeta-potential of the preformed insulin microspheres at pH 7.0 showed a negative net surface charge. The resulting charge reversal of zeta potential values upon coating with all polyions under investigation confirmed successful deposition of the coating compounds (Fig. 7(a)). One can see that, in this case, fluorescently labeled polycation penetrated in depth of the microcore, so that these polyelectrolyte "roots" are binding insulin and practically blocking its release. After a complex of insulin in the core conjugated with the first layer of polycation is formed, the particle was stabilized, and the addition of the following polyanion layer was performed to ease this conjugation. Balancing strong interaction of the first polymer layer with protein microcore, and easing this interaction by depositing the second oppositely charge polyelectrolyte (polycation/polyanion for negative insulin core at pH 7.0, or polyanion/polycation for positive insulin core at pH 5.8) allowed to regulate complexation of the encapsulated proteins and have sustained insulin release. Finer tuning of the release profile was reached by using third and fourth polyelectrolyte layer addition to the complex. The general idea of the controllable complex formation is as follows: adsorption of the first polyelectrolyte layer gives strong complex and very slow release, addition of the second oppositely charged polyelectrolyte layer results in relaxed complex and quicker release, the third oppositely charged layer again

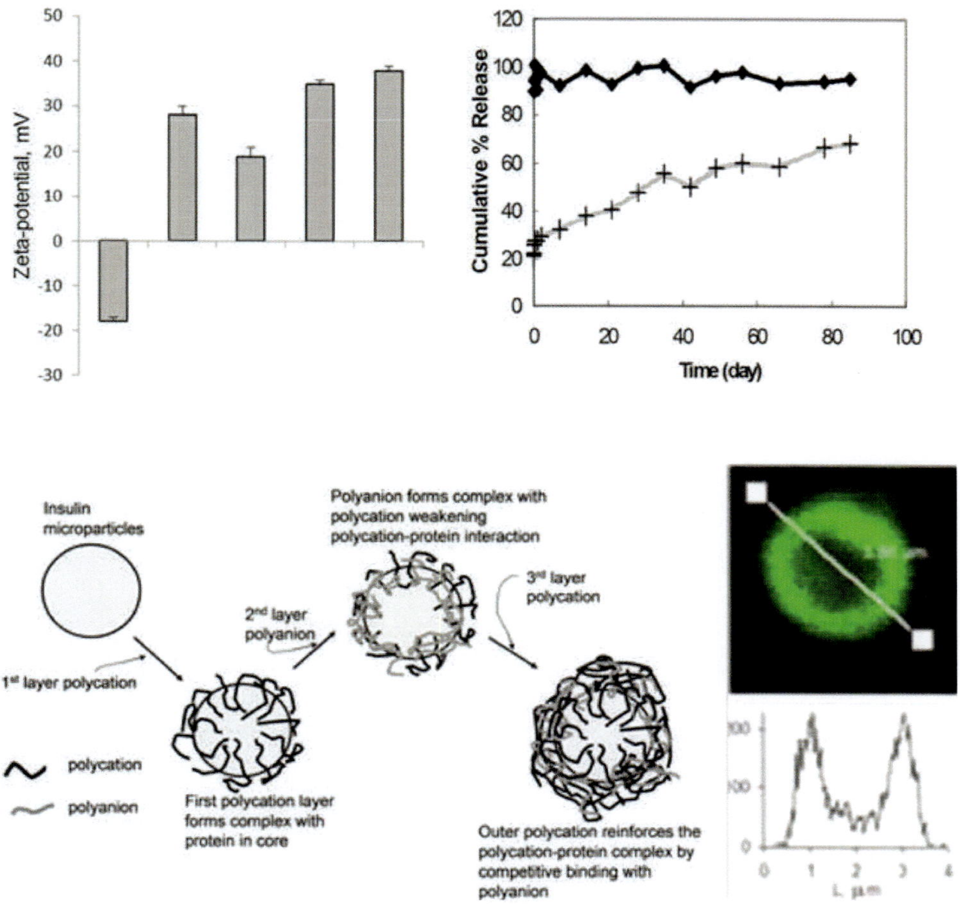

Fig. 7. (a) Surface charge modification of uncoated PROMAXX® insulin microspheres (A) through deposition of one monolayer of various polycationic compounds: B) PDDA, C) ProtS, D) PLA, and E) PLL. (b) *In vitro* release profiles of insulin from the particles; black: uncoated, and grey: coated, with protamine sulfate/alginic acid shell; (b) a scheme of the assembly of polyelectrolyte layers on insulin particles.

gives slower release, and so on. Therefore, step-wise application of polyion complex formation resulted in fine tuning of insulin release from microparticle (Figs. 7 (b)–(d)).[35] The observed serum glucose depression in rat injection model was significantly extended from 3 h to at least 8 h upon surface modification of the microspheres with LbL shell of protamine/alginic acid. The pharmaco-dynamic data was in good agreement with the insulin pharmacokinetic data obtained from the corresponding animal. The *in vivo* release time of protamine-modified formulation was 2.60 ± 0.55 h as compare with 0.55 ± 0.41 h for the uncoated insulin PROMAXX® formulation.[35]

Just growing thickness of the capsule walls within the 5–6 bilayers (30–50 nm) did not result in slowing of release from insulin microparticles. With this, we developed the concept formulated in a Max Planck Institute,[7,8,14–16] where LbL multilayer was considered as a tight diffusion barrier with adjustable thickness in the range of 10–15 bilayers, to the new approach where the main role in the controlled release from protein micro-aggregates is given to adjustable interpolyelectrolyte complex formation by alternate polycation/polyanion coating of a few layers.

4. Using Preformed Gelatin Nanocores

Gelatin-based 200-nm nanoparticles consisting of a soft gel-like interior which strongly bind poorly soluble compounds of the polyphenol family were developed. These nanocores were coated with LbL shells (e.g. poly-L-glutamic acid/poly-L-lysine, dextran sulfate/protamine sulfate, carboxymethyl cellulose/gelatin) (Fig. 8). These LbL nanocapsules encased a wide range of polyphenolic compounds with demonstrated anticancer potential. To overcome low solubility of theaflavin and curcumin in water, they were loaded from 25% acetone or 25% ethanol: 12.5% acetone in water, while EGCG and tannic acid were adsorbed from aqueous solutions. Adsorption of the polyphenols into gelatin nanoparticles equalizes their extremely different solubility and makes the features of polyphenol loading and release to be a function of general rules of polyphenol–protein interaction. Adsorption of polyphenols with higher molecular weights and a larger number of phenolic –OH groups was found to be higher. For theaflavin, which has the highest molecular weight, it reached 70% of the mass of nanoparticulated material on

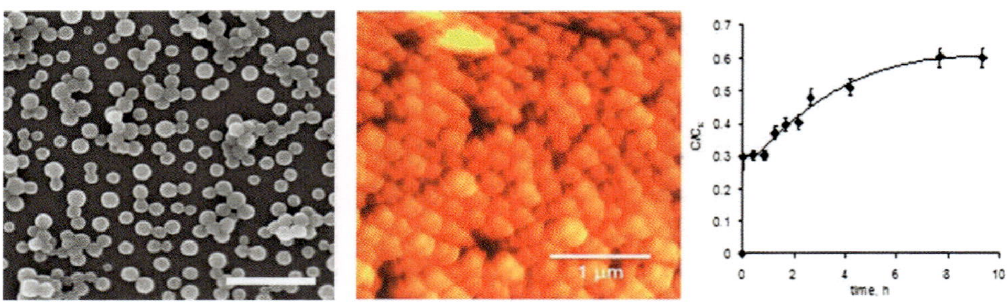

(a) (b) (c)

Fig. 8. SEM (a) and AFM (b) images of 200 nm $(PGA/PLL)_2$-coated gelatin nanoparticles loaded with polyphenol; PGA: poly-L-glutamic acid, PLL: poly-L-lysine. (c) EGCG release from 250-nm gelatin A nanoparticles coated with PSS/PAH shells. pH of release 6.0, time of loading 48 h.

a dry weight basis. The degree of tannic acid and (-)-epigallacathechin gallate (EGCG) encapsulation was lower (25–30%). All encapsulated polyphenols retained their antioxidant properties, as was proved with the ABTS cation-radical assay, and biological activity.[36]

5. Low Soluble Drug Cores Coated with PEGylated Polyelectrolytes

Many organic and inorganic compounds with good application potential are limited in admixing/dispersing in liquids because of their poor (lower than <0.005 mg/mL) or moderate solubility in aqueous and non-aqueous media, thus excluding needed composite formulations. For liquid hydrophobic materials, such as oils, there is a well-established procedure of emulsification, which allows for the stable dispersion of microdroplets in water. We successfully converted solid materials into stable aqueous nanocolloids with a particle size of less than 200 nm through simultaneous powerful ultrasonication and LbL polyelectrolyte coating (sonication-assisted LbL encapsulation). This was a development of LbL microencapsulation pioneered by Sukhorukov, Donath, Caruso, Möhwald and other researchers.[7,8,19–25] Most of these works exploit the formation of nano/engineered polyelectrolyte shells on pre-formed microtemplates with much larger diameters of 2 to 5 μm.

Simultaneous ultrasonication and electrostatic LbL polyelectrolyte coating converts many poorly soluble inorganic and organic solid materials into stable, aqueous, 100–200-nm colloids with up to 80 wt. % loading coated with 2–3 polyelectrolyte bilayer shells of 10–15 nm thickness. This treatment provides good colloidal stability in buffers and presents a platform for building sophisticated capsule wall architecture for advanced properties. We described the recent progress in preparation of LbL coated nanoparticles of anticancer drugs, proteins, dye pigments, anticorrosion agents, salts, oxides, etc. The top-down approach based on rupturing material powder with ultrasonication and the sequential adsorption of oppositely charged polyelectrolytes, often combined with a non-washing LbL procedure, is efficient in the preparation of large quantities of *ca.* 200-nm diameter, polyelectrolyte-coated colloids. The bottom-up approach allows for a variety of conditions for nanoparticulation and involves drug nucleation from an organic solvent by its gradual worsening, with the addition of an aqueous polyelectrolyte assisted by ultrasonication. Drug nanocolloids of 120-nm diameter were prepared with pre-PEGylated polycations alternated with heparin. We also demonstrated poorly soluble compound adsorption into preformed soft protein nanocores and the possibility of LbL nano-encapsulation from solutions with an excess of uncharged pharmaceutical polymers traditionally used for colloid stabilization.[37–42]

Bottom-up techniques, or nanoprecipitation methods, start from real solutions of substances of interest in an appropriate organic solvent such as acetone, tetrahydrofuran, DMSO, ethanol, N-methyl-2-pyrrolidone, etc. Seed development is initiated by adding this highly concentrated solution into a miscible antisolvent (often water) in the presence of a stabilizer, and then slowly increasing the amount of antisolvent in the mixture. The latter can be achieved by means of a secondary effect of powerful sonication (elevated temperature of the reaction media), which enhances evaporation of an organic solvent, increases the percentage of water in the media that leads to slow oversaturation, multiple seeds, and, finally, nanocrystal formation. Growth of such seeded nanocrystals is limited by the adsorption of amphiphilic molecules and polyelectrolytes, resulting in 100–200-nm cores. In another approach, a sonication method of preparation of aqueous 50-nm beta-lactoglobulin particles based on thermal denaturation of the protein was proposed.

Stabilizers versus LbL polyelectrolyte shell. When particle dimensions are at the nanoscale range, the surface area of the nanoparticles increases, thus increasing the Gibbs free energy of the system. The system prefers to reduce it by agglomeration, especially in concentrated suspensions. It results in a particles size increase and poorer colloidal stability. To minimize surface energy and prevent re-aggregation, all the aforementioned techniques require the presence of surfactants, both ionic (e.g. sodium lauryl sulfate and sodium docusate) and non-ionic (polysorbates, polyethylene glycols, polyvinyl pyrrolidone).[41] Among ionic polymeric stabilizers, several belong to the class of polyelectrolytes, such as carboxymethyl cellulose, chitosan and its quaternized derivatives, carrageenans, amine derivatives of dextran, etc. For better performance, surfactants are applied in a combination that provides both electrostatic and steric repulsions between nanoparticles. Typical formulations contain up to 5–10 wt % of stabilizers in supernatant that cannot be removed or exchanged without losing colloidal stability.

LbL polyelectrolyte-stabilized nanoparticles are different. After adsorption of the first, usually polycation, layer, one has a colloidal dispersion of materials coated with a layer, which provides a high surface ξ-potential, usually of +35 mV. Deposition of the second anionic polyelectrolyte ties chains of the first polyelectrolyte together, preventing their detachment from the surface while bringing nanoparticle ξ-potential to a negative value (−40 mV). Such multilayer polyelectrolyte shells provide colloidal stability for months.

A novel nanoform of the poorly soluble anticancer drug paclitaxel, suitable for intravenous administration, has been developed using sonication-assisted LbL assembly. The nanoparticles contain up to 70 wt.% of the drug and consist of 150–180 nm solid crystal paclitaxel cores coated with 20- to 30-nm-thick shells of alternative layers of PEGylated poly-L-lysine and heparin. Dispersions of paclitaxel nanoparticles are stable in different biological solvents at concentrations up to 5 mg/ml. Intravenous administration of the LbL-coated nanocrystalline paclitaxel to mice via tail injection provided the preliminary confirmation of their safety and tolerability.

Paclitaxel, a diterpenoid derived from the needles and bark of the Pacific Yew tree (Taxus brevifolia), is a clinically proven anticancer drug for treating a broad range of cancers including breast, lung, and ovarian cancer.[36] The poor solubility of the drug in aqueous media hinders its application for tumor treatment. Some of the successfully commercialized products of paclitaxel are: a self-emulsifying combination of dehydrated ethanol and Cremophor EL (TAXOL, Bristol-Myers Squibb, USA), polymeric micelles of copolymer of polyethylene glycol (PEG) and poly(L,D-lactide) (Genexol-PM, Samyang Co., Korea) and albumin-bound nanoparticles (Abraxane, American Bioscience, USA), which brought strategies to promote drug bioavailability and facilitate controlled drug release. Other paclitaxel delivery systems include a wide range of lipid- and polymer-based nanoparticles, but most of them are still in the development stages.

To be considered as a promising candidate for *in vivo* tests and to compete with already marketed intravenously administered paclitaxel, newly developing particulated formulations must comply with several criteria as well as with the FDA regulations. The formulation should be aqua-based with a particle size less than 200 nm, and should retain colloidal stability in phosphate buffered saline (PBS) and biological fluids (e.g. blood serum). The content of paclitaxel in a nanoparticle should exceed 30 wt. %, while drug concentration in the injection should be above 2 mg/mL.

Our architectural approach includes the preparation of hydrophobic paclitaxel nanocores in the presence of amphiphilic and polymeric surfactants by solvent replacement method followed by deposition of thin-layered shells of polyelectrolytes using the LbL assembly technique. LbL shell plays a dual role in the nanoparticle stabilization. The inner layers and amphiphilic surfactants minimize free surface energy of nanoparticles and prevent their coalescence and paclitaxel recrystallization, whilst the outermost layers enhance colloid stability due to high hydrophilicity of the polyelectrolyte shell. Incorporation of polyethylene glycol (PEG) moieties in LbL shell by using PEG-modified polyelectrolytes is aimed to increase colloidal stability of the dispersions and attain antifouling properties.

To prepare nanocores of paclitaxel, bulk paclitaxel crystals were dissolved in acetone at 80 mg/mL, and 60 μL of the concentrated solution was added to 2.8 mL of PBS buffer containing 2.85 mg/mL polyvinylpyrrolidone, 1.2 mg/mL AOT and 0.35 mg/mL Polysorbate 80 at 10°C under constant sonication in ultrasound bath. The mixture was sonicated for 20 min to allow the formation of paclitaxel cores that are less than 150 nm diameter.

The drug nanocores were coated using a washless LbL procedure: polyelectrolytes in 10–20 μL aliquots of a concentrated 60 mg/mL solution was added to ~2.8 mL of 1.8 mg/mL paclitaxel dispersion under constant sonication until the surface of the nanoparticles was recharged; the assembly was followed with electrical surface ξ-potential (in water) and particle size (in PBS) measurements. For the measurements, 10–25 μL aliquots of the dispersion in supernatant were taken and re-dispersed in 2 mL of the appropriated solvent. No intermediate washing of nanoparticles was done until the shell was complete; usually, three to four polycation/polyanion pairs were deposited. The coated particles of *ca.* 200-nm diameters were separated from the supernatant by centrifugation, re-dispersed in PBS buffer, and kept as stable colloids of 3–5 mg/mL paclitaxel concentration. Paclitaxel concentration in the LbL-colloids was estimated after extraction of the drug with ethanol, using UV-vis spectroscopy (227 nm). The content of paclitaxel on a dry weight basis was evaluated in lyophilized LbL-coated dispersions.

The *in vitro* release studies for the paclitaxel nanocolloids were done at sink conditions at 37°C in PBS buffer containing 0.2 mg/mL Polysorbate 80, which was added to increase drug solubility limit. The LbL-coated paclitaxel nanoparticles were added in the release medium; aliquots were withdrawn after certain time intervals.

To assure colloidal stability of LbL-coated particles in PBS buffer, the cores were prepared in the solution of the same ionic strength. The size of paclitaxel cores is controlled by the concentrations of PVP, AOT and Polysorbate 80, which were optimized in our procedure to obtain cores with diameters lower than 200 nm with a narrow size distribution. The nanoparticles were stable when the amphiphilic surfactants/paclitaxel ratio was 0.5–0.7 g/g. Even though the exact function of PVP in the mixture is unclear, concentrations of the polymer as high as 2–3 mg/mL, along with AOT and Polysorbate 80, is required. By increasing the surfactant/paclitaxel ratio, the initial core size can be decreased to 120 nm, but long-term stability of LbL-coated particles with such small cores is lower.

The surfactants in the initial paclitaxel dispersion are loosely bound to the surface and were partially replaced by polyelectrolytes adsorbed. The estimated experimentally amount of amphiphilic surfactants remaining in a LbL-coated nanoparticles sample do not exceed *ca.* 0.2 g/g of paclitaxel. The polar side of adsorbed surfactants serves as anchors for polyelectrolyte shell.

Polyelectrolytes were added step-wise in small aliquots to a dispersion of the paclitaxel nanocores in the presence of the surfactants (Fig. 9). It has been found that ~0.3 mg of the polycation per 1 mg of the drug colloid is required to reverse ξ-potential to the maximal positive value, of 30 ± 5 mV. The procedure was applied to concentrated nanoparticle dispersions (2 mg/mL for paclitaxel) and avoided multiple washing steps used in a classical centrifugation-based LbL assembly. Thus, the loss of the small colloidal particles from supernatant upon wash is minimized. The alternation of surface ξ-potential upon addition of proper amounts of PLB and Hep confirms the formation of polycation/polyanion shell on the surface of nanocores (Fig. 10). The hydrodynamic diameter of the colloid is 180 nm and slightly increases with addition of each polyelectrolyte (estimated as 4–5 nm for the bilayer in hydrated shell).

The alternate adsorption of PEG ylated PLL and Hep on planar surfaces was studied by quartz crystal microbalance (QCM-R); the data was processed using Sauerbrey equation (rigid film approximation). The thickness of a Hep/PLB bilayer adsorbed from 0.1 mg/mL aqueous polyelectrolytes with intermediate washing was 0.64 ± 0.5 and 0.90 ± 0.5 nm for PLB16-5 and PLB16-20, which was in the same range as that for non-PEGylated PLL

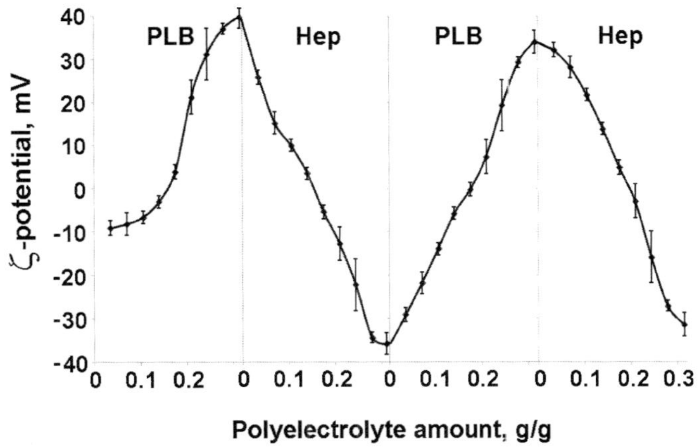

Fig. 9. Dependence of the ξ-potential value of 200-nm diameter paclitaxel nanoparticles on the amount of polycation PLB/polyanion (Hep) sequentially added to a dispersion of the drug.

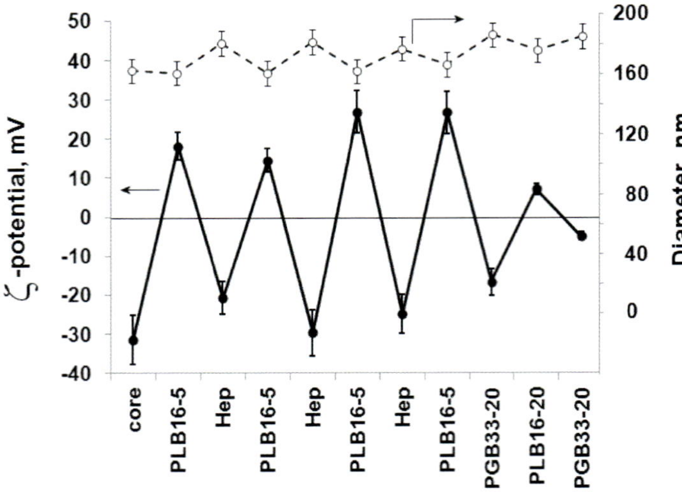

Fig. 10. Changes of *x*-potential and hydrodynamic diameter of paclitaxel nanoparticles upon adsorption of PLB16-5 and Hep on their surface.

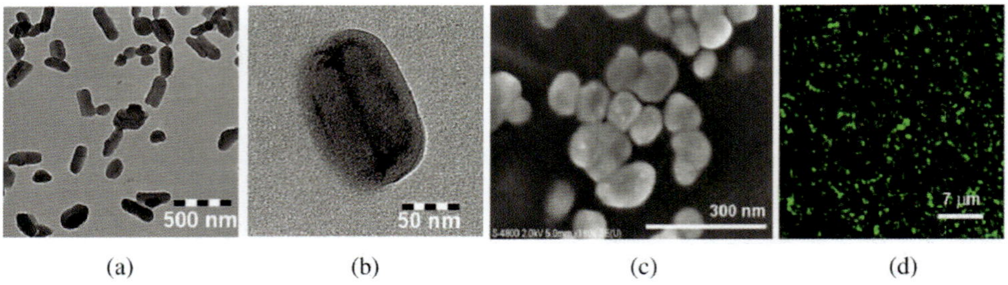

Fig. 11. (a) and (b): TEM images of paclitaxel nanoparticles coated with $(PLB16\text{-}5/Hep)_3/$ PgP shells (stained with $(NH_4)_2Mo_2O_7$); (c) SEM image of nanoparticles with $(PLB16\text{-}5/Hep)_{3.5}$ shell (coated with 0.5 nm gold); (d) CLSM images of paclitaxel nanoparticles with $(PLB32\text{-}20/BSA\text{-}FITC)_{2.5}$ shell.

(0.66 ± 0.5 nm). These values are in good agreement with polylysine/heparin bilayer thicknesses reported for LbL films.

Paclitaxel cores in the nanoparticles prepared in the presence of the surfactants and coated with PEG-modified polyelectrolyte shells have elongated shape with a length of *ca.* 250 × 60 nm (Figs. 11(a)–(b)). On a TEM close-up image of a nanoparticle, one can see crystal faces and borders inside of a halo of less dense material, which can be attributed to polyelectrolyte shell of 10–20 nm thickness. A top-view SEM image confirms presence of 100–250-nm paclitaxel particles. Laser confocal image (Fig. 11(d)) shows well-dispersed nanoparticles with minimal aggregation. One sees their slightly enlarged particle diameters due to the image convolution with laser wavelength function (operating below confocal microscope resolution limit), as shown in Fig. 11. The drug loading in 200-nm LbL-coated nanoparticles was 60–70 wt. % for a $(PLB16\text{-}5/Hep)_{3.5}$ coating and 30–50 wt. % for a $(PLB16\text{-}20/BSA)_{2.5}$ coating. A minor fraction (less than 2%) of smaller particles with lower paclitaxel contents was detected in supernatant with light scattering sample sizing diagram.

The molecular weight of polyelectrolytes tested as components of the LbL assembly was less than 65 kDa to avoid bridging of neighboring particles. Different combinations of PEG-modified and unmodified polyelectrolytes were tested. PEGylated and non-PEGylated polyelectrolytes were combined in the shell. It was found that the shell composition providing stable colloids cannot contain more than one bilayer in which both polycation and polyanions are not PEGylated. The shell providing the most stable in PBS buffer paclitaxel nanoformulation consisted of PLL-PEG block-copolymers of different length (PLB) as a positive component and heparin or albumin (BSA) as a negative component. Polyethylene glycol is a highly hydrophilic molecule, and its surface layer significantly improves colloidal stability of microparticles. The best colloidal stability of concentrated paclitaxel dispersions in PBS was obtained while using PLB copolymer at every bilayer combined with non-PEGylated heparin with the shell architecture of $(PLB/Hep)_{3.5}$, rather than using PEGylated polyelectrolyte only in the capsule outermost layer. However, long-time stability of nanoparticles with inner shells made of highly PEGylated polyelectrolytes, for example, PLB16-20/PGA32-20 or PgP, was lower.

In vitro Stability and Drug Release. The paclitaxel nanoparticles coated with PEG-modified polyelectrolyte shells were stable in PBS buffer up to 200 h at concentration of 3 mg/mL. To evaluate the nanoparticles' aggregation in blood upon intravenous injection, colloidal stability of LbL-coated nanoparticles in fetal bovine serum was tested. In serum, paclitaxel nanoparticles coated with $(PLB16\text{-}5)/Hep)_{3.5}$ shell or having one more outermost polyanion layer to convert particle charge to negative (heparin, dextran sulfate, albumin, gelatin B, or PGA33-20) retained the particle diameter below 250 nm for at least 24 h. Paclitaxel posess an extremly low solubility in aquous solutions (5.6 mg/L) and without a solubility enhancer, its concentration is practically undetectable. For all tested paclitaxel colloids, a fast initial dissolution (65–70%) of the drug was observed 15 min after addition of the nanoparticles to a 0.2 mg/mL polysorbate-80–PBS release medium. Then the rest of paclitaxel dissolved very slowly; a complete dissolution of the nanoparticles was observed within 65–70 h. No distinguishable influence on the release rate was found for shells consisting of less than 3.5 PLB16-5/Hep bilayers. In another experiment with paclitaxel nanoparticles, LbL-coated with 1, 4, 8 and 12 bilayers of PLL and heparin, a moderate decrease in the drug release rate with number of layers in the shell was detected.

In vivo experiments: The experiments on administration of the LbL-coated paclitaxel nanoparticles in mice via the tail vein have provided a preliminary confirmation of their safety and tolerability at concentrations up to 40 mg/kg, which is two times higher when compared with the commercial formulation of paclitaxel Taxol with MTD of 20 mg/kg. Experimental animals showed no apparent signs of preparation toxicity for up to 24 h.

Paclitaxel possesses an extremely low solubility in aqueous solutions (5.6 mg/L) and without a solubility enhancer, their released concentrations are very difficult to monitor. For all tested paclitaxel colloids, a fast initial dissolution (more than 20%) of the drug was observed within 15 min after addition of the nanoparticles to a 0.2 mg/mL polysorbate 80 release medium. The rest of the drug was released very slowly; a complete dissolution of the nanoparticles was observed within 65–70 h (Fig. 12). No influence on release rate was found for shells thinner than 3.5 PLB16-5/Hep bilayers. With larger number of polyelectrolyte layers (4, 8 and 12 bilayers of PLL and heparin), a slight decrease in the drug release rate with increasing number of layers in the shell was detected.

Therefore, we elaborated stable nanocolloids of paclitaxel encapsulated in 200-nm LbL shells of PEGylated polylysine alternated with heparin. This core–shell nanoformulation contains up to 5 mg/mL of paclitaxel and is

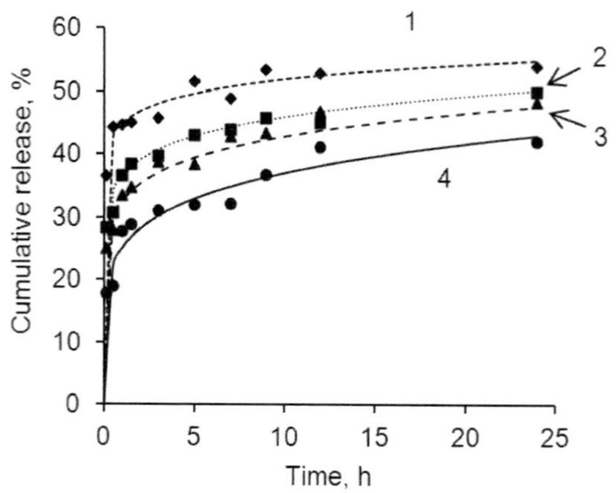

Fig. 12. Paclitaxel release from 300-nm nanocolloids prepared using top-down, sonication-assisted LbL method and coated with a (PLL/Hep)$_n$ shell. Number of bilayers in shell, *n*: 1–0.5, 2–4, 3–8, 4–12; 0.2% polysorbate 80 in water at 37°C.

stable in PBS buffer and blood serum. It has shown good tolerability and therapeutic effect for mice injection. The combination of high drug-loading capacity and colloidal stability in salt containing buffers suggests that nanocores of paclitaxel coated with amphiphiles surrounded with multilayer shell of PEG-modified polysaccharides or proteins are promising vehicles for sustained delivery of poorly water-soluble drugs.

6. LbL Modification of Biological Cells

Biological cells are considered as a fascinating object to be coupled with the LbL technique. Biological cells constitute the main "building block" in all multicellular organisms (starting form the most primitive ones and finishing with humans). In addition, a great number of microbial unicellular species exist. Here, we will briefly review the main features of biological cells which are utilized during the LbL modification of cells. Essentially, all cells can be regarded as composite microscopic particles built from cytoplasm separated from the environment with the cellular membrane. Unicellular organisms (bacteria, fungi, microalgae, protozoa) typically have cell dimensions from 1–20 microns (with some exceptions), and normally possess a thick cell wall encircling the membrane and protecting the cells from the harsh environments. Most plant cells also possess the cell walls, whereas animal tissues are built from cell wall lacking cells, which are protected by more complex outer coatings, such as exoskeletons, cuticle or skin. Unicellular species are the independent organisms acting on their own, while the very complex regulation pathways orchestrate the cells in multicellular organisms. However, in the first approximation, all cells behave as solid colloid microparticles, therefore they can be easily subjected to LbL modification. The major advantages of using living cells as LbL templates:

(i) The cells have a great number of different shapes and sizes, including very complex ones (i.e. astrocyte red blood cells, anisotropic bacterial cells).
(ii) The cells, unlike many colloid microparticles, typically have a very narrow size distribution.
(iii) The cells can be cultivated in large numbers using relatively cheap growth infrastructure.
(iv) As living entities, cells can be easily destroyed using various treatments (i.e. calcination, chemical or enzymatic treatment).
(v) Most of the cells, due to the cellular membrane and cell wall chemical composition, are negatively charged, which allows for the effective polyelectrolyte deposition.

These features inspired the use of biological cells as templates for the fabrication of multilayered LbL microcapsules having shapes and sizes similar to those of the biological cells. The main reason to do this was the potential use of convectional LbL microcapsules as novel drug carriers. One may anticipate that microcapsules, which resemble the cells at least to a certain extent, will have better biocompatibility, rheological and structural properties, which will be beneficial in injectable drugs encapsulation.

6.1. *Cells-templated hollow microcapsules*

The first human cells to be encapsulated with polyelectrolyte pairs were red blood cells (erythrocytes).[43] Typically, red blood cells are discoid cells with diameters around 6 microns. These cells can be isolated from the whole blood samples in large amounts, then fixed and coated via the sequential deposition of polyelectrolytes from aqueous solutions. The LbL technique applied to red blood cell requires the preliminary treatment of cells with glutaraldehyde to avoid the strong interactions of the cellular membrane with the deposited polyelectrolyte molecules, which can lead to membrane rupture with the subsequent breakdown of the template structure during the coating procedure. Chemical fixation protects the originally fragile cellular membranes of human cells (note that the human cells lack the cell wall). After the deposition of polymer multilayers, the cells were completely destroyed using chemical decomposition with aqueous NaOCl of cellular proteins, the cell-mimicking microcapsules were obtained.[44] The chemical decomposition results in removal of proteins and lipids from the cytoplasm, which consequently leads to a certain increase of the osmotic pressure inside the cell-templated capsule. As a result, the increase of the final volume of a microcapsule is observed. The microcapsules were thoroughly investigated using the appropriate microscopy techniques (i.e. SEM, TEM, AFM and confocal microscopy) as well as zeta-potential monitoring (Fig. 13). The resulting microcapsules were exact copies of the biological templates, exhibiting cell-like elastic properties and experimentally controlled permeability. The cell-mimicking microcapsules preserve their integrity and resembled the original shapes of erythrocytes. Interestingly, even the elaborate spiky shapes of echinocytes (red blood cells subclass) were completely reconstituted by the LbL-deposited polymer coating. The functional properties of cell-mimicking capsules depend on the polymer multilayers, which can be finely tuned to allow the controllable loading and subsequent release of a cargo (i.e. drugs, MRI contrast agents, enzymes, antibodies).

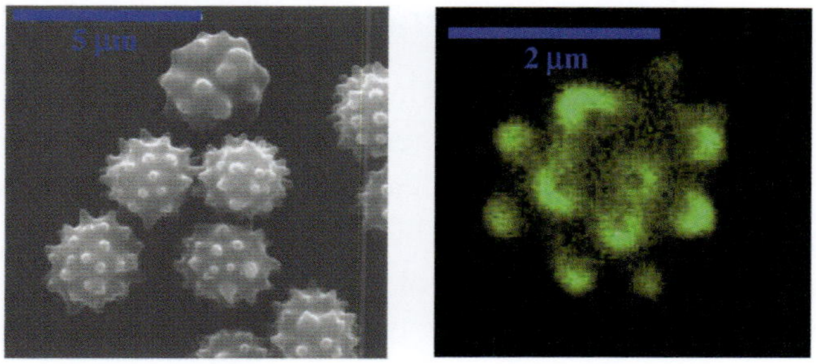

(a) (b)

Fig. 13. (a) SEM and (b) confocal images of echinocytes-templated microcapsules.

The effective encapsulation of fluorescence-labeled DNA and human serum albumin as the model substances in hollow microcapsules templated on human erythrocytes has been successfully demonstrated.[45] The erythrocyte-templated microcapsules are believed to be prospective reaction vessels for multiparallel DNA amplification by polymerase chain reaction (PCR). Obviously, the potential areas of application of cell-templated LbL polyelectrolyte microcapsules are not limited to PCR vessels and can be further extended to perform chemical reactions on the microliter scale.

The most important features of LbL-encapsulated cells are: (i) control of transport of molecules into and from cells; (ii) control of cellular adhesive properties; and (iii) control of cells division and germination.

The advantage of biological cells for templating hollow microcapsules is that the microcapsules prepared using the cells as sacrificial cores have sizes and shapes exactly as the cells, which means that they can easily travel inside small capillaries. Moreover, the cell-templated microcapsules lack any potentially toxic component which cannot be completely removed if non-biological templates are used. Finally, red blood cells are available from whole blood, which means, theoretically, the blood of the patient who later will be subjected to drug-carrying LbL-built microcapsules administration can be used, thus almost completely reducing the risks of contamination or immune reaction. So far, there were no reports on clinical trials of LbL microcapsules template on cells, however, this approach is a promising means of effective drug delivery. Apart from using human cells, microbial cells (i.e. *E. coli* bacteria) can also be subjected to LbL deposition of polyelectrolyte multilayers to prepare hollow microcapsules after the decomposition of cellular templates. In this case, the decomposition procedure should be sufficient to remove the cell walls of bacterial cells.

6.2. *Living cell surface modification*

Fabrication of artificial LbL shells on living microbial cells leads to alteration of cell functional properties (such as membrane transport, nutrition, germination times and adhesive properties). Bacteria and fungi (i.e. yeast) are extensively used in biotechnology and in biomedicine, therefore, it is crucial for some applications to precisely control the immobilization of the cells, the properties of cellular surfaces and other functional properties of cells. Although there are several reports demonstrating the use of microbial cells as sacrificial cores, we are mostly concerned with LbL deposition of functional polymers as a means of cell surface engineering, in other words, using polyelectrolyte multilayers as functional modifiers of cellular properties. So far, microbial cells were encapsulated both in pure polymer multilayers and in LbL shells doped with functional nanomaterials (such as nanoparticles, nanorods and nanotubes). Here the added layer(s) of polyelectrolytes serve as a semipermeable artificial membrane covering the cell walls and controlling the cellular interactions with environment and with other cells. As long as we need to alter or enhance the functionalities of living cells, the serious limitation on the LbL technique is applied in this case — the deposition of polyelectrolytes should not jeopardize the viability of the coated cells. Fortunately, most of the reported systems so far demonstrate high viability levels in functionalized cells.[46]

Diaspro and co-workers were the first to report the LbL encapsulation of living yeast cells (*Saccharomyces cerevisiae*) using the alternating layers of PAH and PSS.[47,48] They employed a similar procedure for the sequential deposition of polyelectrolytes as was used to fabricate cell-based microcapsules. The deposition of 4–6 bilayers of synthetic polyelectrolytes was performed via the sequential incubation of living cells in polymer aqueous buffered solutions. Since the cells possess negative surface charge, typically, the uncoated cells are first introduced into a polycation solution (i.e. PAH), which is then repeated using a polyanion (i.e. PSS), and the procedure can be repeated until the required thickness of the nanocoating is achieved. Noteworthy, the composition of the polyelectrolytes can be chosen according to the required architecture, for example, the outer polymers might be selected in order to provide the LbL-functionalized cells with the desired adhesive properties, as required for the immobilization of cells onto polyelectrolyte pre-coated surfaces.[48] After each deposition step, the cells should be separated by centrifugation and washed several times with water to remove the excessive unbound polymer molecules. The conventional LbL-coating of cells in large volumes allows for scalable functionalization of cells, however, each polymer deposition and washing step is followed by the uncontrolled adsorption of cells onto the centrifuge tubes, which reduces the overall number of the coated cells at the end of the coating procedure. To avoid this, the exact amount of polyelectrolytes can be used to coat the cells without the subsequent excess polymer removal (so called "washless LbL method"). In addition, deposition/washing cycles are time-consuming, which stimulates the search for alternative deposition methods, such as microfluidic-device–based cell labeling with polyelectrolytes, where the cells pass through the laminar flow of polyelectrolytes in a microfluidic chip and are effectively labeled by polymers without any washing/centrifugation steps.[49]

Apart from yeast, many other types of biological cells were coated in a similar way with polyion nanoshells. Currently, researchers are mostly focused on biotechnologically relevant species, in order to utilize the LbL-functionalized cells in practical applications. A good example is the use of nanocoating to control the spore germination times,[50] where the LbL coating allows controlling the ability of nanocoated bacterial spores to grow. A similar approach allows to study and to control the nutritional pathways in LbL-coated microorganisms.[51]

LbL functionalization of living cells can be monitored using the same experimental techniques as applied to micro- and nanoparticle LbL encapsulation. Zeta-potential monitoring has been utilized to effectively demonstrate the cells' zeta-potential reversion due to the deposition of oppositely charged polymers (Fig. 14(a)). A very simple approach allows for the effective visualization of LbL-coated cells using fluorescence or confocal microscopy. Fluorescently labeled polyelectrolytes introduced as layers in the LbL shells can be used to demonstrate the nanoshells on the cell walls. Figures 14 (b)–(d) demonstrates microalgae and bacteria cells coated with 4 PAH/PSS bilayers, with the FITC-labeled PAH constituting the outermost layer. As one may see in the figure, the LbL-coated cells are uniformly encircled by polyelectrolytes, modifying the cell walls.

Besides using electrostatic interactions between polyions, another modification of LbL-nanocoating of microbial cells was recently proposed. This approach is based on hydrogen bonding of uncharged polymers, yielding in essentially the same multilayered structures built from the constituent polymer monolayers.[52] The hydrogen-bonding LbL assembly was first proposed in order to avoid the interaction of cells with potentially toxic polycations (such as PAH), although it was shown several times that the toxicity of PAH is negligible.[48] From the practical point of view, the hydrogen-bonded LbL on cells is achieved via the same deposition/washing approach. Yeast cells were LbL-coated using consequently deposited poly(N-vinyl pyrrolidone)

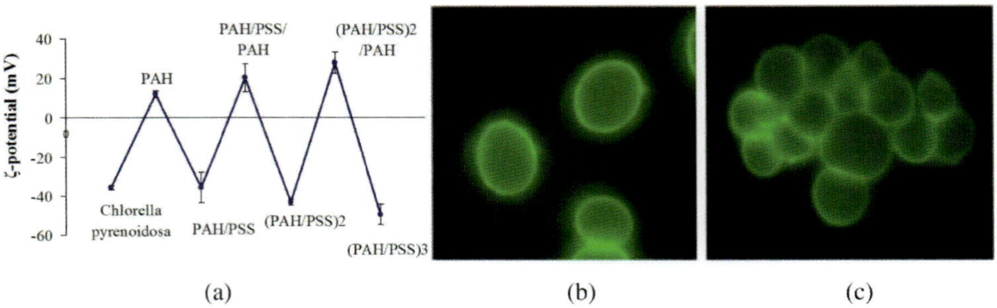

Fig. 14. (a) zeta-potential monitoring during *Chlorella* cells modification with polyelectrolytes. (b) and (c) Fluorescence microscopy images of yeast and *Chlorella* cells coated with $(PAH/PSS)_4$ LbL shells.

(PVPON) and tannic acid (TA) layers. In order to perform the adhesion of the following uncharged TA/PVPON multilayers onto the negatively charged cells, a first positively charged polyelectrolyte (polyethyleneimine (PEI)) add-layer was deposited onto the microbial cell walls. This technique was further expanded to functionalize more fragile mammal (including human) cells using the LbL approach.[53]

In addition to controlling the surface permeability of coated cells, LbL assembly of polyelectrolytes can be utilized to support the fabrication of inorganic (i.e. calcium phosphate or calcium carbonate) hard shells on cells. For instance, the preliminary LbL-modification of yeast cells with polyelectrolyte multilayers (i.e. $(PAH/PSS)_4$) was used to fabricate mineralized shells on cells. It is supposed that the precursor LbL-deposited polyelectrolyte shell supported the electrostatically driven concentration of Ca^{2+} ions within the multilayers, which, in turn, stimulated the formation of 1-μm-thick calcium phosphate structures in the presence of the calcification solution.[54] A similar polyelectrolyte-facilitated approach was recently employed in the deposition of silica shells on individual yeast cells, where the (PDDA/PSS) multilayers were deposited onto yeast prior to fabrication of 50–nm-thick uniform silica mineral shells.[55] Such applications of LbL-coated cells led to the concept of artificial spores.[56]

Viruses. We encapsulated the 78-nm diameter bacteriophage T7 with a PAH/PSS/PAH/5-nm silica shell and, using transmission electron microscopy (TEM), demonstrated an even coating of bacteriophage capsid and its short tail with polyelectrolyte/silica shell. The encapsulation was performed with the non-washing LbL assembly technique, and the goal was to demonstrate the possibility of virus encapsulation, which follows with efforts to encapsulate spherical viruses for drug delivery (e.g. 200-nm diameter modified herpes). Similar approaches are under way to LbL encapsulate adenovirus to decrease its immune respond.

6.3. *LbL-facilitated tissue engineering*

LbL-coated cells can be controllably arranged in larger three-dimensional (3D) free-standing multicellular clusters, mimicking the primitive multicellular microorganisms. Biomimetic free-standing yeast biofilms were fabricated using LbL-assembled polyelectrolyte nanofilms.[57] These biofilm-like multicellular structures were fabricated using calcium carbonate supportive layers coated with polyelectrolytes. The future directions of this research will be focused on increasing the functionalities of artificial biofilms, i.e. introduction of different cell compositions in a controllable way, doping with functional nanoparticles and fabrication of an elaborate internal structure mimicking the natural vessels existing in plant and animal tissues.

Apart from fabrication of artificial microbial biofilms, the LbL approach was effectively utilized in mammal tissue engineering. Isolated LbL-functionalized mammal cells were controllably assembled in tissue-mimicking structures consisting of up to four layers of mouse fibroblast cells.[58] To perform a more biocompatible functionalization, instead of using synthetic polyelectrolytes, biogenic polymers fibronectin and gelatine were used. These polyelectrolytes were employed to fabricate the extracellular matrix-mimicking structures, which facilitated the attachment of fibronectin-coated fibroblasts. This resulted in the fabrication of the artificial tissue built from the four consecutive layers of viable cells (around 25 microns). Furthermore, this technique was extended to heterogeneous populations of human cells. Human endothelium cells and myositis supported by LbL-assembled polyelectrolytes were arranged into xenogeneic multicellular clusters, resembling the elaborate structure of human tissues.[58]

LbL-coated cells were also employed in the fabrication of spherical, spindle-like and cubic 3D multicellular living clusters (*cellosomes*).[59] Polyelectrolyte-coated charged yeast cells were controllably assembled over oppositely charged microscopic templates (such as air microscopic bubbles or calcium carbonate microcrystals), which resulted in the fabrication of a monolayer of living cells having the shapes of the original templates.[60] The produced 3D microscopic clusters were very stable and remembered the spherical, spindle-like or rhombohedral shapes of the sacrificial templates. Remarkably, the yeast cells assembled in cellosomes were viable for several weeks.[61] Microscopic images of cellosomes, to a certain extent, resemble several colonial microorganisms, i.e. *Volvox* microalgae (Fig. 15).

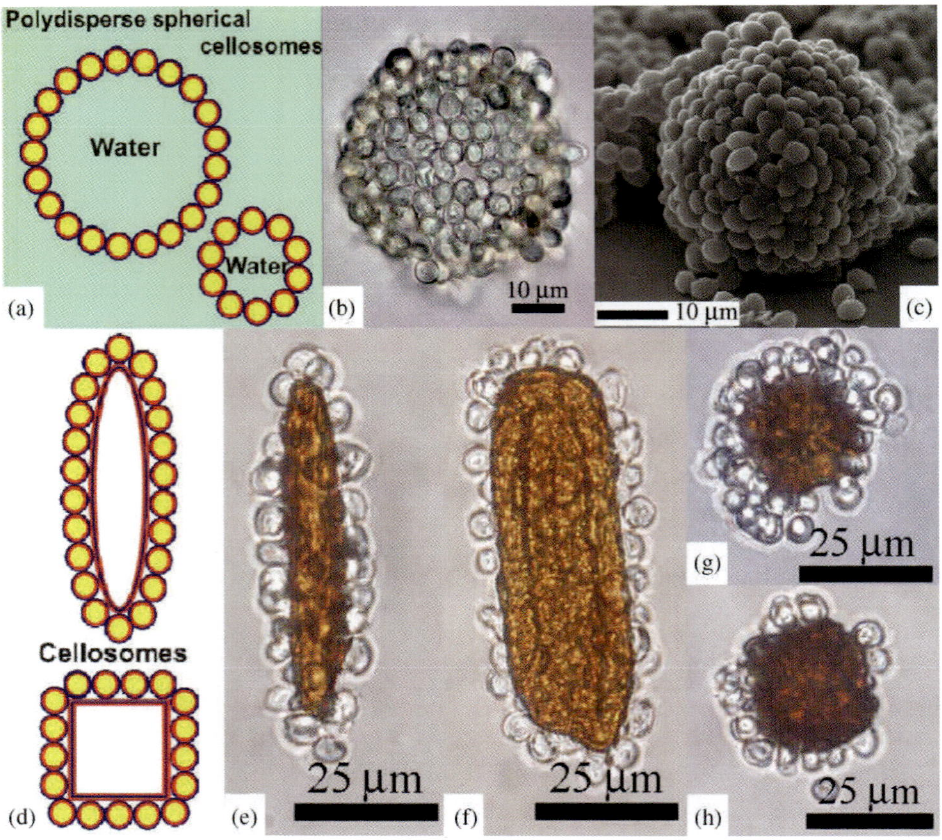

Fig. 15. Schematics ((a) and (d)) and optical and SEM images of spherical ((b) and (c)), spindle-like (e), planar (f) and rhombohedral ((g) and (h)) cellosomes fabricated using LbL-coated yeast cells. The images are reproduced from Refs. 60 and 61, with permission from the Royal Society of Chemistry.

6.4. *LbL-coating of cells with nanomaterials*

Another important extension of the LbL functionalization of living cells is the immobilization of functional nanoparticles on the surfaces of microorganisms. Here, the polyelectrolyte multilayers serve as glue, which immobilizes the nanoparticles to the coated cells. As a result, the detachment of nanoparticles from the cells is avoided; in addition, the use of polymer shells serves as a semipermeable membrane which prevents the internalization of the deposited nanomaterials into the cytoplasm. A sketch in Fig. 16 demonstrates the sequential LbL assembly of polyelectrolytes followed by the deposition of nanoparticles and nanotubes onto cells. Here, the negatively charged gold nanoparticles and multiwall carbon nanotubes are embedded between PAH

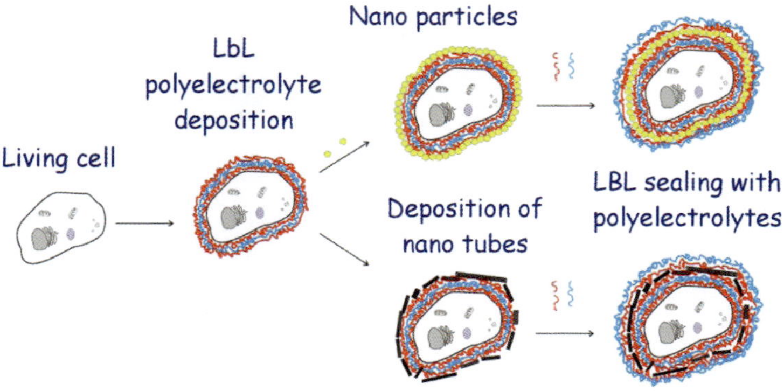

Fig. 16. A scheme demonstrating the LbL functionalization of living cells with oppositely charged polycations (red) and polyanions (blue) multilayered films doped with carbon nanotubes and gold nanoparticles. Image reprinted from Ref. 46, by permission of the Royal Society of Chemistry.

layers; however, positively charged nanoparticles can be assembled between oppositely charged polymers in the same way. The zeta-potential on nanoparticles should be considered prior to deposition, since the most effective immobilization occurs if the nanomaterials are embedded between oppositely charged polyelectrolytes.

Typically, after the deposition of nanoparticles onto the precursor layers, one needs to coat the cells with several other polyelectrolyte bilayers to prevent the disassembly of nanomaterials from the cell walls due to mechanical or chemical impact. The method so far has been employed to deposit of multiwalled carbon nanotubes,[62] magnetic iron oxide nanoparticles and nanorods,[63,64] graphene oxide sheets[65] and halloysite nanotubes[66] onto the living cells. Human platelets were also encapsulated within the PDDA/PSS multilayers functionalized with silica nanoparticles and immunoglobulin G macromolecules.[67]

High-resolution microscopy techniques (such as AFM, SEM and TEM) were employed to demonstrate the incorporation of nanomaterials into polyelectrolyte shells over the living cells. Figure 17 demonstrates the even distribution of nanomaterials-doped LbL shells on yeast cells.

Immobilization of functional nanomaterials onto living cells is an emerging area of modern research where chemistry meets biology. Nanocoated cells exhibit functional properties which differ considerably from their intact counterparts. However, LbL-coating is a non-penetrating technique, which does not require genetic manipulations. Importantly, LbL-coated cells are able to divide[66,67]; therefore, the functional properties provided by nanocoating are not retained by the offspring cells after several generations, which minimize

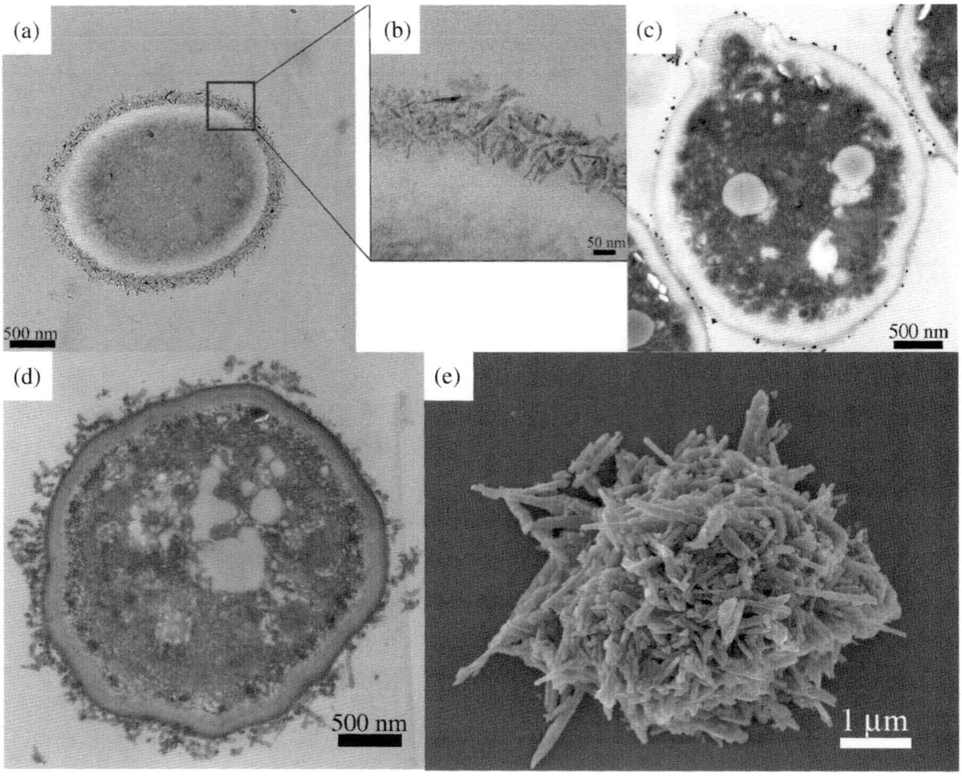

Fig. 17. Arrangement of functional nanomaterials on yeast cells: (a) TEM image of yeast cell coated with PAH/PSS multilayers doped with magnetic nanorods (The image is reproduced from Ref. 63, with permission from the Royal Society of Chemistry); (b) TEM image of yeast cell coated with PAH/PSS multilayers doped with 20-nm gold nanoparticles; (c) TEM image of yeast cell coated with PAH/PSS multilayers doped with multiwall carbon nanotubes; (d) SEM image of yeast cell coated with PAH/PSS multilayers doped with halloysite nanotubes.

the environmental impact and allows for a better control over the populations. Currently, nanocoated cells are used as models in investigation of cellular properties[55] or are employed in biosensors and bio-analytical and microfluidic devices.[49,63,64] In addition, it is speculated that LbL-modified cells may find potential applications in magnetically facilitated cell delivery and tissue engineering.

7. Conclusions

LbL nanoassembly is a simple method allowing the design of multilayer architectures with nanometer precision. The technique is based on the alternating adsorption of oppositely charged components, such as synthetic and natural polyelectrolytes, proteins and nanoparticles, which saturate and recharge the available surface. The multilayer film, therefore, comprises a defined number of layers spanning only a few nanometers in thickness that are bonded through electrostatic forces (positive/negative), hydrogen bonding or other dispersion forces. Until recently, the majority of these studies were devoted to the LbL assembly using predominantly sodium poly (styrene sulfonate)-PSS, poly(acrylic acid)-PAA, poly(allylamine) hydrochloride-PAH and poly(ethyleneimine)-PEI. Lately, researchers have begun exploring natural polyelectrolytes and proteins to improve the biocompatibility of LbL coatings such as alginic acid, hyaluronic acid, chondroitin sulfate, heparin, dextran sulfate, carboxymethyl cellulose, polyaminoacids and DNA in alternation with cationic colloids, including chitosan, dextran amine, polylysine, collagens, protamine sulfate, gelatin A and others.

The formation of LbL multilayers on tiny templates and the use of LbL encapsulation were first introduced in 1998 at the Max Planck Institute, Germany. LbL encapsulation is a variation of planar LbL coating; it is applied to small drug particles that are 2–10 μm in diameter or to sacrificial cores (polymeric, $MnCO_3$, $CaCO_3$ or silica) to produce empty shells. There are two methods used in LbL encapsulation: (1) direct methods of encapsulation of poorly soluble sub-micrometer or -nanometer diameter drug-containing particles and (2) the more common method of production, in which LbL capsules are formed on sacrificial cores, which are subsequently removed, preceding loading of the empty LbL shells by variation of permeability for drugs or proteins through pH-dependent, temperature or light opening and closing capsule wall. A number of drugs have been encapsulated by LbL coating, including ibuprofen, furosemide, nifedipine, naproxen, biotin, vitamin K3, insulin, dexamethasone, tamoxifen, paclitaxel, camptothecin, curcumin, DNA fragments and peptides. Coatings for vaccination applications are also beginning to emerge. An architectural LbL assembly makes it possible not only to design polyelectrolyte capsules to encase drugs but also to add functionalities to such capsules through antibody attachment for targeted delivery or through imaging (fluorescent dyes, NMR agents), facilitating the tracking of drug delivery.

There are two LbL strategies. One is developing microcapsules with a multifunctional shell that is 30–50 nm thick containing 5–10 wt. % drugs and with the ability to control the release of those drugs over time (from a few hours up to 100 hours). Such LbL capsules have diameter of a few micrometers, being too large for intravenous injection. However, if these capsules have very thin flexible walls, then they can squeeze through narrow capillary walls. Another application is LbL coating of poorly soluble drug nanoparticles (often anticancer drugs) for their solubilization. Then, the LbL coating provides a strong particle charge (zeta potential of *ca.* –40 mV), producing stable nanocolloids with drug concentration 60–70%. Such LbL capsules have 100–200-nm diameter and may be injectable. Higher-ionic strength buffers (such as PBS) drastically decreases electrostatic repulsion of LbL shells inducing aggregation. Capsule surface PEGylation is considered a necessary step in LbL encapsulation to avoid aggregation, premature uptake and elimination of the capsules by macrophages. In the top-down approach, nanocolloids were prepared by rupturing material powder using ultrasonication and simultaneous sequential adsorption of oppositely charged polyelectrolytes (often with a non-washing procedure). In the bottom-up approach, drugs were dissolved in an organic solvent (ethanol or acetone), and drug nucleation was initiated by gradually worsening the solution with the addition of an aqueous polyelectrolyte assisted by ultrasonication. The release rates from such nanocapsules were controlled by assembling multilayer shells with variable thicknesses and are in the range of 10–20 h. Small real nanosize of the particles, stability in bio-buffers at high concentration (5 mg/mL) and the use of PEGylated biodegradable natural polyelectrolytes in the shell allowed for the development of new drug nanoformulations which have a potential use in medical injections. This process has the potential to facilitate the production of polyelectrolyte capsules appropriate for injectable drug nanoformulations with controlled release and targeting properties.

We also described LbL encapsulation of biological cells and viruses with nano-organized shells containing polyelectrolytes, nanoparticles and proteins arranged in a way to provide cell with additional functionality. We foresee in this approach not only drug delivery possibility, like for example, in LbL encapsulated adenoviruses to suppress immune-respond but, even more challenging perspective to design cyborg cell with supernatural abilities due to equipment with nanotools. LbL surface modification can make living cells "healthier": to enhance their nutrition paths, to better protect them against phagocytosis/digestion, to provide them with the ability to surpass a harmful media (e.g., acidic stomach on the way to the higher pH digestion tract), to protect them against ultraviolet (UV) radiation with a layer of polyphenols (i.e. cell sun-screen) and to supply cells with additional instrumentation for their functionality (i.e. magnetic function and mechanical robustness). Biocells may be "dressed" with a functional "cloth," preserving these newly acquired properties for one to two generations.

Acknowledgments

This work was partially supported by 1R01CA134951 grant from the National Cancer Institute (NCI). The content is solely the responsibility of the authors and does not necessarily represent the official views of the NCI or National Institute of Health. Funding from RFBR grants (12-04-33290 Mol_ved_a and 12-03-93939G8) is also appreciated. The authors thank S. Konnova, R. Minullina and R. Kareev (Kazan Federal University) for their help in the manuscript preparation.

References

1. R. Langer and D. A. Tirrell, Designing materials for biology and medicine, *Nature* **428**, 487–492 (2004).
2. V. P. Torchilin, Recent advances with liposomes as pharmaceutical carriers, *Nat. Rev. Drug Discov.* **4**(2), 145–160 (2005).
3. C. J. Hawker and K. L. Wooley, The convergence of synthetic organic and polymer chemistries, *Science* **309**, 1200–1205 (2005).
4. G. Decher, Fuzzy nanoassemblies: Toward layered polymeric multicomposites. *Science* **227**, 1232–1237 (1997).
5. Y. Lvov, G. Decher and H. Möhwald, Assembly, structural characterization and thermal behavior of layer-by-layer deposited ultrathin films of polyvinylsulfonate and polyallylamine. *Langmuir* **9**, 481–486 (1993).
6. Y. Lvov, K. Ariga, I. Ichinose and T. Kunitake, Assembly of multicomponent protein films by means of electrostatic layer-by-layer adsorption. *J. Am. Chem. Soc.* **117**, 6117–6122 (1995).
7. G. B. Sukhorukov, E. Donath, S. Davis, H. Lichtenfeld, F. Caruso, V. I. Popov and H. Mohwald, Stepwise polyelectrolyte assembly on particle surfaces: A novel approach to colloid design. *Polym. Adv. Technol.* **9**, 759–767 (1998).
8. E. Donath, G. B. Sukhorukov, F. Caruso, S. A. Davis and H. Mohwald, Novel hollow polymer shells by colloid-templated assembly of polyelectrolytes. *Angew. Chem. Int. Ed.* **37**(16), 2202–2205 (1998).
9. N. Shah, J. Hong, M. Hyder and P. Hammond, Osteophilic multilayer coatings for accelerated bone tissue growth, *Adv. Mater.* **24**, 1445–1450 (2012).
10. Y. Lvov, G. Decher and G. Sukhorukov, Assembly of thin films by means of successive deposition of alternate layers of DNA and poly(allylamine). *Macromolecules* **26**, 5396–5399 (1993).
11. M. Ferreira, J. H. Cheung and M. F. Rubner, Molecular self-assembly of conjugated polyions: A new process for fabricating multilayer thin film heterostructures. *Thin Solid Films* **244**, 806–809 (1994).
12. A. Antipov and G. B. Sukhorukov, Polyelectrolyte multilayer capsules as vehicles with tunable permeability. *Adv. Colloid Interface Sci.* **111**, 49–61 (2004).
13. H. Ai, S. A. Jones, M. M. de Villiers and Y. M. Lvov, Nano-encapsulation of furosemide microcrystals for controlled drug release. *J. Control. Release* **86**(1), 59–68 (2003).

14. O. Kreft, A. Javier Muñoz, G. B. Sukhorukov and W. J. Parak, Polymer microcapsules as mobile local pH-sensors. *J. Mater. Chem.* **42**, 4471–4476 (2007).
15. D. G. Shchukin, I. L. Radtchenko and G. B. Sukhorukov, Micron-scale hollow polyelectrolyte capsules with nanosized magnetic Fe_3O_4 inside. *Mater. Lett.* **57**, 1743–1747 (2003).
16. D. G. Shchukin and G. B. Sukhorukov, Nanoparticle synthesis in engineered organic nanoscale reactors. *Adv. Mater.* **16**(8), 671–682 (2004).
17. W. Meier, Polymer nanocapsules. *Chem. Soc. Rev.* **29**, 295–303 (2000).
18. S. Forster and M. Konrad, From self-organizing polymers to nano- and biomaterials. *J. Mat. Chem.* **13**, 2671–2688 (2003).
19. C. Déjugnat and G. B. Sukhorukov, pH-responsive properties of hollow polyelectrolyte microcapsules templated on various cores. *Langmuir* **20**(17), 7265–7269 (2004).
20. T. Mauser, C. Dejugnat and G. B. Sukhorukov, Balance of hydrophobic and electrostatic forces in the pH response of weak polyelectrolyte capsules. *J. Phys. Chem. B.* **110**(41), 20246–20253 (2006).
21. A. A. Antipov, G. B. Sukhorukov, S. Leporatti, I. L. Radtchenko, E. Donath and H. Möhwald, Polyelectrolyte multilayer capsule permeability control. *Colloids Surf. A.* **198**, 535–541 (2001).
22. Y. Lvov, A. A. Antipov, A. Mamedov, H. Möhwald and G. B. Sukhorukov, Urease encapsulation in nanoorganized microshells. *Nano Lett.* **1**(3), 125–128 (2001).
23. R. Georgieva, R. Dimova, G. Sukhorukov, G. Ibarz and H. Möhwald, Influence of different salts on micro-sized polyelectrolyte hollow capsules, *J. Mater. Chem.* **15**, 4301–4310 (2005).
24. C. Y. Gao, H. Möhwald and J. C. C. Shen, Enhanced biomacromolecule encapsulation by swelling and shrinking procedures. *Chem. Phys. Chem.* **5**, 116–120 (2004).
25. K. Köhler, D. G. Shchukin, H. Möhwald and G. B. Sukhorukov, Thermal behavior of polyelectrolyte multilayer microcapsules. 1. The effect of odd and even layer number, *J. Phys. Chem. B.* **109**(39), 18250–18259 (2005).
26. J. Heuvingh, M. Zappa and A. Fery, Salt softening of polyelectrolyte multilayer capsules *Langmuir* **21**(7), 3165–3171 (2005).
27. M. Bédard, A. G. Skirtach and G. B. Sukhorukov, Optically driven encapsulation using novel polymeric hollow shells containing an azobenzene polymer. *Macromol. Rapid. Commun.* **28**(15), 1517–1521 (2007).
28. Q. Yi, D. Wen and G. B. Sukhorukov, UV-cross-linkable multilayer microcapsules made of weak polyelectrolytes. *Langmuir* **28**(29), 10822–10829 (2012).
29. A. I. Petrov, D. V. Volodkin and G. B. Sukhorukov, Protein-calcium carbonate coprecipitation: A tool for protein encapsulation. *Biotechnol. Progr.* **21**(3), 918–925 (2005).
30. L. Dähne, S. Leporatti, E. Donath and H. Möhwald, Fabrication of micro reaction cages with tailored properties. *J. Am. Chem. Soc.* **123**(23), 5431–5436 (2001).
31. M. N. Antipina and G. B. Sukhorukov, Remote control over guidance and release properties of composite polyelectrolyte based capsules. *Adv. Drug Deliv. Rev.* **63**(9), 716–729 (2011).
32. G. B. Sukhorukov, A. L. Rogach, B. Zebli, T. Liedl, A. G. Skirtach, K. Köhler, A. A. Antipov, N. Gaponik, A. S. Susha, M. Winterhalter and W. J. Parak, Nanoengineered polymer capsules: Tools for detection, controlled delivery, and site-specific manipulation. *Small* **1**(2), 194–200 (2005).

33. J. Xia and P. L. Dubin, "Protein-Polyelectrolyte Complexes Macromolecular Complexes in Chemistry and Biology." Ch. 15 In P. L. Dubin, J. Bock, R. M. Davis, D. Schulz and C. Thies (eds.) Springer-Verlag: Berlin 1994, pp. 236–271.
34. V. A. Kabanov and A. V. Kabanov, Interpolyelectrolyte and block ionomer complexes for gene delivery: Physico-chemical aspects. *Adv. Drug Delivery Rev.* **30**, 49–60 (1998).
35. J. Rashba-Step, R. Darvari, Q. Lin, J. Kelly, T. Shutava, Y. Lvov and T. Scott, *33rd Annual Controlled Release Society Meeting*, Vienna, Austria, July 26, pp. 97–98 .
36. T. Shutava, S. Balkundi, P. Vangala, P. O'Neal, J. Steffan, R. Bigelow, J. Cardelli and Y. Lvov, Layer-by-layer coated gelatin nanoparticles as a vehicle for natural polyphenols. *ACS Nano* **3**(7), 1877–1885 (2009).
37. V. Vergara, A. Tinelli, V. Lorusso, M. Maffia, X. Zhang, Y. Lvov and S. Leporatti Lapatinib/Paclitaxel-loaded nanocolloids by sonicated-assisted layer-by-layer assembly for overcoming multidrug resistance in ovarian cancer. *Nanomedicine* **8**, 891–899 (2012).
38. M. de Villiers and Y. Lvov, Layer-by-layer self-assembled nanoshells for drug delivery *Adv. Drug Delivery Rev.* **63**(9), 699–700 (2011).
39. P. Pattekari, Z. Zheng, X. Zhang, T. Levchenko, V. Torchilin and Y. Lvov, Top-down and bottom-up approaches in production aqueous nanocolloids of paclitaxel, *Phys. Chem. Phys.* **13**(19), 9014–9019 (2011).
40. Y. Lvov, P. Pattekari, X. Zhang and V. Torchilin, Converting poorly soluble materials into stable aqueous nanocolloids. *Langmuir* **27**, 1212–1217 (2011).
41. A. Agarwal, Y. Lvov, R. Sawant and V. Torchilin, Stable nanocolloids of poorly soluble drugs with high drug content prepared using sonicated layer-by-layer technology. *J. Control. Release* **128**, 255–260 (2008).
42. T. Shutava, K. Arapov, V. Torchilin and Y. Lvov, Architectural design of drug nanocapsules with PEGylated polyelectrolytes. *Soft Matter* **8**(36), 9418–9427 (2012).
43. E. Donath, S. Moya, B. Neu, G. B. Sukhorukov, R. Georgieva, A. Voigt, H. Baumler, H. Kiesewetter and H. Mohwald, Hollow polymer shells from biological templates: Fabrication and potential applications. *Chem. Eur. J.* **8**(23), 5481–5485 (2002).
44. B. Neu, A. Voigt, R. Mitlohner, S. Leporatti, C. Y. Gao, E. Donath, H. Kiesewetter, H. Mohwald, H. J. Meiselman and H. Baumler, Biological cells as templates for hollow microcapsules. *J. Microencapsul.* **18**(3), 385–395 (2001).
45. O. Kreft, R. Georgieva, H. Baümler, M. Steup, B. Müller-Röber, G. B. Sukhorukov, H. Möhwald, Red blood cell templated polyelectrolyte capsules: A novel vehicle for the stable encapsulation of DNA and proteins. *Macromol. Rapid Comm.* **27**(6), 435–440 (2006).
46. R. F. Fakhrullin, A. I. Zamaleeva, R. T. Minullina, S. A. Konnova and V. Paunov, Cyborg cells: Functionalisation of living cells with polymers and nanomaterials. *Chem. Soc. Rev.* **41**, 4189–4206 (2012).
47. A. Diaspro, D. Silvano, S. Krol, O. Cavalleri and A. Gliozzi, Single living cell encapsulation in nano-organized polyelectrolyte shells, *Langmuir* **18**(13), 5047–5050 (2002).
48. S. Krol, M. Nolte, A. Diaspro, D. Mazza, R. Magrassi, A. Gliozzi and A. Fery, Encapsulated living cells on microstructured surfaces. *Langmuir* **21**, 705–770 (2005).
49. M. Tarn, R. F. Fakhrullin, V. N. Paunov and N. Pamme. Microfluidic device for the rapid coating of magnetic cells with polyelectrolytes. *Mater. Lett.* **95**, 182–185 (2013).
50. S. S. Balkundi, N. G. Veerabadran, D. M. Eby, G. R. Johnson and Y. M. Lvov, Encapsulation of bacterial spores in nanoorganized polyelectrolyte shells. *Langmuir* **25**(24), 14011–14016 (2009).

51. B. Franz, S. S. Balkundi, C. Dahl, Y. M. Lvov and A. Prange, Layer-by-layer nano-encapsulation of microbes: Controlled cell surface modification and investigation of substrate uptake by bacteria. *Macromol. Biosci.* **10**, 164–172 (2010).
52. V. Kozlovskaya, S. Harbaugh, I. Drachuk, O. Shchepelina, N. Kelley-Loughnane, M. Stone and V. V. Tsukruk, Hydrogen-bonded LbL shells for living cell surface engineering. *Soft Matter* **7**, 2364–2372 (2011).
53. J. T. Wilson, W. Cui, V. Kozlovskaya, E. Kharlampieva, D. Pan, Z. Qu, V. R. Krishnamurthy, J. Mets, V. Kumar, J. Wen, Y. Song, V. V. Tsukruk and E. L. Chaikof, Cell surface engineering with polyelectrolyte multilayer thin films. *J. Am. Chem. Soc.* **133**, 7054–7064 (2011).
54. B. Wang, P. Liu, W. Jiang, H. Pan, X. Xu and R. Tang, Yeast cells with an artificial mineral shell: protection and modification of living cells by biomimetic mineralization. *Angew. Chem. Int. Ed.* **47**(19), 3560–3564 (2008).
55. S. H. Yang, K. -B. Lee, B. Kong, J. -H. Kim, H. -S. Kim and I. S. Choi, Biomimetic encapsulation of individual cells with silica. *Angew. Chem. Int. Ed.* **48**, 9160–9163 (2009).
56. S. H. Yang, D. Hong, J. Lee, E. H. Ko and I. S. Choi, Artificial spores: Cytocompatible encapsulation of individual living cells within thin, tough artificial shells. *Small* **9**, 178–186 (2013).
57. M. Matsusaki, K. Kadowaki, Y. Nakahara and M. Akashi, Fabrication of cellular multilayers with nanometer-sized extracellular matrix films, *Angew. Chem. Int. Ed.* **46**, 4689–4692 (2007).
58. S. A. Konnova, M. Kahraman, A. I. Zamaleeva, M. Culha, V. N. Paunov and R. F. Fakhrullin, Functional artificial free-standing yeast biofilms. *Colloid Surf. B.* **88**, 656–663 (2011).
59. R. F. Fakhrullin and V. N. Paunov, Fabrication of living cellosomes of rod-like and rhombohedral morphologies based on magnetically responsive templates. *Chem. Comm.* 2511–2513 (2009).
60. M. -L. Brandy, O. J. Cayre, R. F. Fakhrullin, O. D. Velev and V. N. Paunov, Directed assembly of yeast cells into living celloidosomes by microbubble templating. *Soft Matter* **6**, 3494–3498 (2010).
61. R. F. Fakhrullin, M. -L. Brandy, O. J. Cayre, O. D. Velev and V. N. Paunov, Live celloidosome structures based on the assembly of individual cells by colloid interactions *Phys. Chem. Chem. Phys.* **12**, 11912–11922 (2010).
62. A. I. Zamaleeva, I. R. Sharipova, A. V. Porfireva, G. A. Evtugyn and R. F. Fakhrullin, Polyelectrolyte-mediated assembly of multiwalled carbon nanotubes on living yeast cells *Langmuir* **26**, 2671–2679 (2010).
63. R. F. Fakhrullin, J. Garcia-Alonso and V. N. Paunov, A direct technique for preparation of magnetically functionalised living yeast cells. *Soft Matter* **6**, 391–397 (2010).
64. D. Zhang, R. F. Fakhrullin, M. Özmen, H. Wang, J. Wang, V. N. Paunov, G. Li and W. E. Huang, Functionalisation of whole cell bacterial reporters with magnetic nanoparticles. *Microb. Biotech.* **4**, 89–97 (2011).
65. S. H. Yang, T. Lee, E. Seo, E. H. Ko, I. S. Choi and B. -S. Kim, Interfacing living yeast cells with graphene oxide nanosheaths. *Macromol. Biosci.* **12**, 61–66 (2012).

66. S. A. Konnova, I. R. Sharipova, T. A. Demina, Y. N. Osin, D. R. Yarullina, O. N. Ilinskaya, Y. M. Lvov and R. F. Fakhrullin, Biomimetic cell-mediated three-dimensional assembly of halloysite nanotubes. *Chem. Commun.* DOI: 10.1039/C2CC38254G (2013).
67. H. Ai, M. Fang, S. A. Jones and Y. M. Lvov, Electrostatic layer-by-layer nanoassembly on biological microtemplates: Platelets. *Biomacromolecules* **3**, 560–562 (2002).

Chapter 10

Inorganic Nanopreparations for Nanomedicine

James Ramos and Kaushal Rege*[†,‡]

**Biomedical Engineering, School of Biological and Health Systems Engineering Arizona State University*
501 E. Tyler Mall, ECG 303, Tempe, Arizona 85287-6106, USA
[†]Chemical Engineering, School for Engineering of Matter, Transport, and Energy
Arizona State University
501 E. Tyler Mall, ECG 303, Tempe, Arizona 85287-6106, USA
[‡]Kaushal.Rege@asu.edu

Abstract

Rapid developments in nanomedicine can potentially transform disease detection and treatment. Inorganic nanopreparations or nanoparticles have been increasingly investigated for several applications in nanomedicine. These include multimodal bioimaging techniques that can be used for disease detection and tracking, and delivery applications for disease therapy. Advances of nanopreparations, including carbon nanotubes, quantum dots, plasmonic/noble metal nanoparticles (gold, silver, copper, platinum) and magnetic nanoparticles, will be discussed in this chapter.

1. Introduction

Nanomedicine is a rapidly expanding field at the intersection of several fields, including biology, chemistry, engineering, electronics and biomaterials.[1] Inorganic nanomedicine is the practice of applying inorganic nanotechnology-based approaches for the detection/diagnosis and treatment of diseases.

When reduced to the nanoscale, many metallic materials exhibit unique properties when compared to the bulk state. These unique physical, optical and/or chemical properties have resulted in the use of nanoparticles for a wide range of applications in biomedicine. Nanoparticles exist in sub-micron sizes, smaller than biological organelles, with dimensions similar to biological macromolecules. Additionally, as a result of their high surface area-to-volume ratios and often facile surface functionalization/modification it is possible to load increased cargo payloads amounts on nanoparticles for delivery to diseased tissues on interest. Nanoparticles have been used for targeted delivery of chemotherapeutic agents, plasmid DNA and molecules capable of RNA interference (RNAi). In addition to their delivery capabilities, nanoparticles have also been employed for multimodal bioimaging techniques and disease detection. The engineering of multifunctional nanosystems capable of both delivery and imaging has also been explored.[2]

Organic nanoparticles have been extensively investigated in drug and gene delivery. For example, polymersomes, liposomes,[3–6] polymeric constructs for protein and macromolecule controlled release,[7–9] long-circulating polymeric nanoparticles and polymeric micelles, are currently being investigated in various stages of both preclinical and clinical development[10] and have also found their way to the clinic. However, due to their unique physiochemical properties, nanoparticles formulated with inorganic building blocks have received increased attention in the recent past. Specifically, optical, magnetic and other physical properties, in concert with biological inertness, stability and ease of functionalization, make inorganic nanoparticles attractive alternatives to organic nanoparticles.[11] Inorganic nanoparticles or materials can consist of metals, metal oxides, metal alloys, core/shell structures, metal phosphides, metal chalocogenides, metal oxide ceramics, non-oxide ceramics, silicates, semiconductor nanocrystals and many other materials/combination of materials.[12] Inorganic nanoparticles are currently being widely investigated for drug and biomarker discovery, molecular diagnostics, drug delivery and therapeutic treatments. In this chapter, we will highlight inorganic nanoparticles that are promising for applications in medicine (Table 1).

2. Carbon Nanotubes

Carbon nanotubes (CNTs) are cylindrical graphene sheets that exist as single-walled CNTS (SWCNTs), which are 1–3 nm in diameter and 5–30 nm in length, and multi-walled CNTs (MWCNTs), which are 10–150 nm in diameter and 200 nm to a few microns in length. CNTs have many unique physical and chemical properties which include, high aspect ratios, high electrical and thermal conductivities, mechanical strength and ultra-light weight.[13] As a result of their unique optical properties, they have been used as photoluminescent, Raman, and photoacoustic (PA) contrast agents for bioimaging *in vitro* and *in vivo*.[14] Due to their strong optical absorption in the near-infrared (NIR) wavelength region and high surface areas, they have potential applications in radiofrequency (RF) and NIR thermal therapy and as delivery vehicles.[15]

2.1. *Bioimaging*

Photoacoustic tomography (PAT) is a non-invasive, hybrid, non-ionizing imaging technique that utilizes a short pulse laser to illuminate biological tissue. Light absorption by the area of interest results in thermoelastic expansion, which generates PA waves.[16] SWNT-assisted PA signals have been found to be up to two-fold stronger than blood *in vivo*, following excitation in the NIR range.[17] Recently, CNTs were utilized as PA molecular imaging agents for *in vivo* brain xenograft models.[18] Additionally, SWNTs conjugated with indocyanine green (ICG) dye, an FDA-approved NIR clinical imaging agent[19] has been found to enhance PA sensitivity significantly compared to SWNTs alone. Using SWNT-ICG, Koo *et al.*[16] demonstrated *in vivo* visualization of sentinel lymph nodes (SLNs) and bladders in healthy rats.

2.2. *Therapeutic and delivery agents*

Due to their unique optical and physicochemical properties, CNTs have been investigated for thermal ablation of cancer cells. SWCNTs have been found to release heat following exposure to a focused external RF field.[20] Hepatic VX2 tumors in rabbits were injected with Kentera functionalized SWCNTs and exposed to an RF field. Following histopathology sectioning, it was found that the tumor tissue underwent complete thermal necrosis with a surrounding 2–5 mm zone of thermal injury.[21] In addition to applications in RF-thermal therapy, both MWCNTs and SWCNTs have been found to strongly absorb energy from continuous NIR irradiation and efficiently convert the energy to heat.[22] In one study, MWCNTs were used to treat mice kidney cancer models and were found to enable complete ablation of tumors and over 3.5-month durable remission in ~80% of treated mice.[23] Complete PC3 xenograft tumor ablation was observed in mouse models of human cancer following NIR irradiation of DNA-encapsulated MWCNTs.[24] Iron-containing MWCNTs were used as a bifunctional imaging and thermal ablation material.[25] MWCNTs entrapping iron were injected into mice bearing murine renal carcinoma tumor and tracked using magnetic resonance imaging (MRI). Significant tumor regression was observed following NIR laser irradiation.

CNTs have also been investigated as effective carriers for therapeutic drugs. Polyethylene glycol (PEG) chains on SWCNTs were conjugated with paclitaxel (PTX) and injected subcutaneously in BALB/c mice containing breast tumor xenografts. Tumor growth was inhibited by almost 60%, which was significantly higher than Taxil.[26] In a different study, integrin-targeted SWCNTs were found to have long blood circulation time as well as tumor accumulation with minimal reticulo-endothelial system (RES) uptake.[27] In a similar study, doxorubicin-loaded PEGylated SWCNTs were injected into the tail vein of Raji lymphoma tumor xenograft containing mice. Although reduction in tumor volume was observed, complete elimination was not achieved.[28] RGD-conjugated PEGylated SWCNTs were found to have tumor-specific uptake upon targeting integrin overexpression in tumor neo-vasculature.[29] EGF-conjugated SWCNTs loaded with the anticancer agent, cisplatin, were taken up in head and neck squamous cancer cell tumors in mouse models. Tumor growth over 10 days was seen to decrease following treatment (Fig. 1).[30]

CNTs have also been used for the delivery of nucleic acids. For example, SWCNTs were non-covalently complexed with hypoxia-inducible factor 1 alpha (HIF-1α) transcription factor siRNA. HIF-1α is overexpressed under hypoxic conditions and regulates transcription of several antiapoptotic genes. HIF-1α activity in the tumor was greatly inhibited following delivery of SWCNT- HIF-1α to mice-bearing pancreatic tumors.[31] In a study by Podesta *et al.*, amine-functionalized MWCNTs were used for intratumorally administered delivery of siRNA in human lung carcinoma models *in vivo*. Tumor growth was arrested and survival was prolonged in the tumor-bearing mice following delivery.[32]

2.3. *Toxicity*

Despite their potential therapeutic applications, there remain major concerns for the use of CNTs due to their biodistribution, toxicity and pharmacokinetics profiles. These concerns are further aggravated due to the high rate of accumulation of CNTs in the kidney and liver.[28,30] Furthermore, intraperitoneal injection indicated that the carcinogenic activity of CNTs may be comparable to that of asbestos.[33,34] In a study by Fraczek *et al.*, SWCNTs appeared to be a more favorable therapeutic agents as they demonstrated more effective excretion and transport from the implantation site to

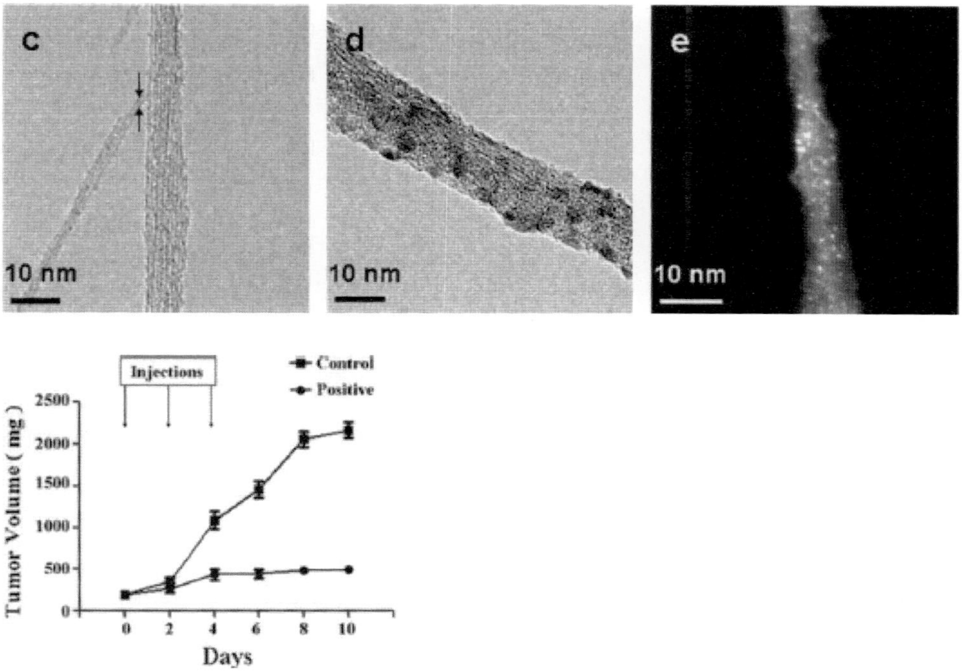

Fig. 1. (Top right) Oxidized SWCNT bundles. Arrows designate a single SWCNT. (Top middle) SWCNT bundle bioconjugated with Qdot and EGF. (Top left) STEM image of SWCNT bundle loaded with cis-platin shown as bright spots. (Bottom) Tumor progression over time of HN12HNSCC tumor growth by SWCNT-cisplatin-EGF bioconjugated following intravaneous tail injection (error bars represent S.E.M, $n = 3$). Modified and reprinted with permission from Ref. 30.

lymph nodes in skeletal rat muscle, compared to MWCNTs, which formed large aggregates and accumulated in the muscle tissue.[35] Aggregation and a resulting inflammatory response of MWCNTs was also observed on the peritoneal side of the diaphragm in C57B1/6 mice females following intraperitoneal injection.[33] It is therefore important to address issues concerning toxicity, stability, dosage, inflammation and circulation times before CNTs can be translated to the clinic.

3. Quantum Dots

Quantum dots (QDs) are nanoscale (<10 nm) inorganic semiconductor nanocrystal fluorophores composed of atoms of elements from II–IV groups (i.e. Cd, Zn, Se, Te) and have emerged as a versatile tool for bioimaging.[36–40] QDs have emerged as attractive bioimaging alternatives to conventional organic dyes due to their increased brightness and photostability, broad excitation wavelength range and narrow symmetric emission spectrums.[40–42] QD absorption spectra and emission wavelengths are tunable from the ultraviolet to NIR region by confining electrons in variable-sized dots.[40] Quantum yields of these nanoparticles can be enhanced by the addition of a high band-gap semiconductor material (i.e. zinc sulfide or ZnS) around the semiconductor core (i.e. CdSe).[43] QD cores generally exist in sizes that can be rapidly cleared by the kidney, with those at ~2 nm in diameter having fluorescence emission in the blue region, and those at ~5 nm in diameter having fluorescence emittion in the red region.[44–46] However, due to addition of the high-bad gap layer, use of ligands for enhancing dispersion in aqueous media, and conjugation of biomolecules, increases their sizes to the 10–100 nm range. Several functionalization chemistries have been explored in order to facilitate the dispersion of QDs in aqueous media.[47,48]

3.1. *Bioimaging*

Using histological sections of different organs, Akerman *et al.*[49] found that QD–peptide conjugates targeted to tumor vasculature or tumor lymphatic vessels accumulated in the tumor xenograft after intravascular injection. Mice bearing JHU-31 prostate cancer tumors were intravenously injected with poly-*N*-isopropylacrimide (PNIPAM) conjugated QDs via the tail vein. Mice were sacrificed three hours following injection, and tumor tissues were recovered and cryo-sectioned. Following imaging, it was found that the QD peptide conjugates localized at the periphery of the tumor tissue sections, whereas unmodified QDs were not found in tumor tissue.[50] QDs conjugated with the Prostate Specific Membrane Antigen (PSMA) were used to target prostate tumor xenograft models in mice. Following tail-vein injection, PSMA-QDs showed intense signal in tumor sites, while unmodified QDs showed no to weak signals in tumors.[39] QDs conjugated to EGFR and the human epidermal growth factor receptor 2 (HER2) accumulated in MIA-PaCa-2 pancreatic and human breast cancer cells in mice, respectively.[51,52]

Recent developments have resulted in the synthesis of NIR-emitting QDs in order to address concerns with light penetration and deep tissue auto fluorescence. QD705 (QDs with emission peak at 705) conjugated to argine-glycine-aspartic acid (RGD) were used to target tumor vasculature overexpressing integrin receptors following tail vein injection in mice bearing human glioblastoma xenografts. (Fig. 2).[53] QD800 were conjugated with anti EGFR in order to target tumor sites overexpression EGFR in colorectal cancer cell xenograft bearing mice.[54]

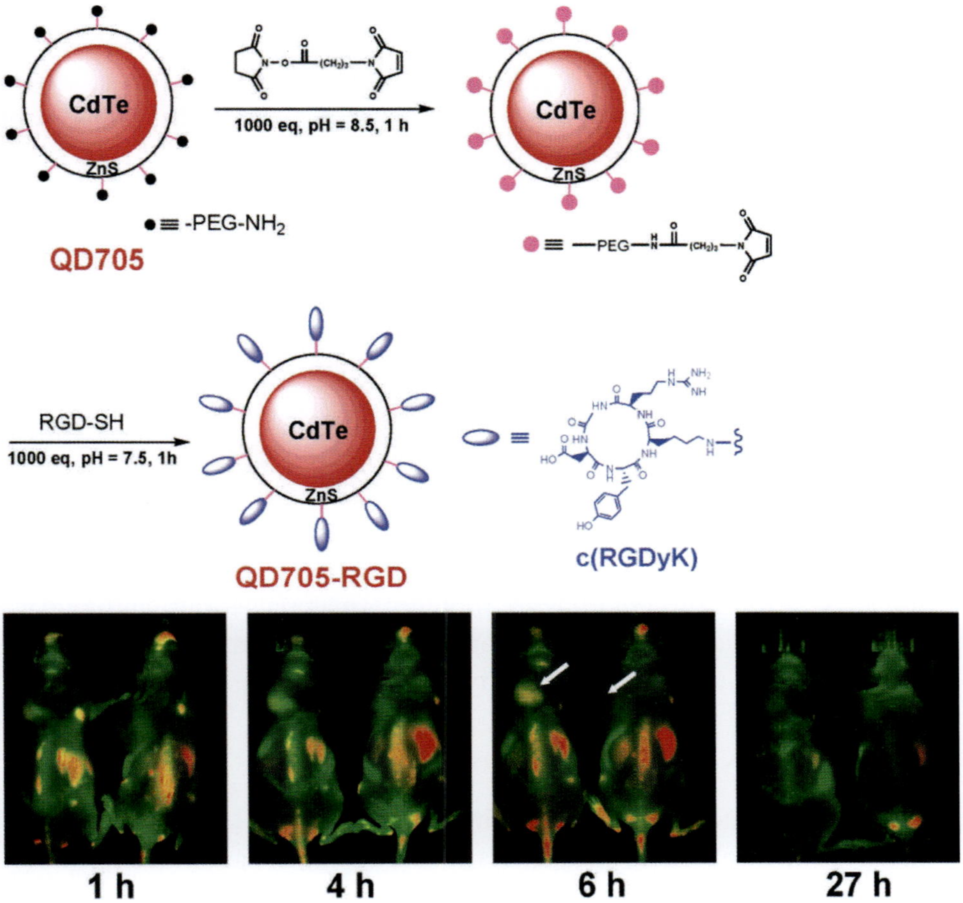

Fig. 2. (Top) Schematic representation of Synthesis of QD705-RGD, PEG denotes poly(ethylene glycol) (MW2000). (Bottom) *In vivo* NIR fluorescence imaging of mice bearing U87MG tumor in their left shoulder (white arrow) injected with 200 pmol of QD705-RGD (left mouse) and QD705 (right mouse). Mice autofluorescence is pseudo-colored green and the unmixed QD signal is pseudo-colored red. Uptake in the liver, bone marrow and lymph nodes was also observed. Adapted and reprinted with permission from Ref. 53.

NIR emitting CdZTe QDs and polyethylenimine (pEI)-DNA complexes ("quantoplexes") were intravenously injected in mice for their gene carrying ability. They were found to accumulate in the lungs, liver, and spleen five minutes after injection.[55] Following tail-vein injection in mice, QDs conjugated with hyaluronic acid and pEI for siRNA delivery were found to accumulate in liver and B16F1 tumor tissue. Inhibition of tumor growth was observed following delivery of antivascular endothelial growth factor (VEGF) targeted siRNA.[56]

3.2. *Limitations*

In addition to the previously mentioned issue of low penetration of visible light, low resolution in deep tissues and biodistribution profiles, toxicity of the CdSe cores of QDs are major concerns[40, 46]. These issues, in combination with metabolism and excretion characteristics have resulted in challenges for employing QDs as clinical imaging agents. Although non-cadmium (InAs, InP, ZnSe or $CuInS_2/ZnS$) containing QDs with NIR fluorescence properties have been synthesized, concerns related to their stability and use of hazardous precursors still persist.[57–59] Additionally, modifications to QDs, such as with hydrazinonicotinaminde (HYNIC) have been able to prolong QD blood circulation and reduce hepatosplenic uptake.[60] Following injection of phospholipid micelle-encapsulated CdSe/CdS/ZnS QDs, rhesus macaques exhibited no signs of acute toxicity.[61] Biochemical markers in the blood remained within normal ranges following 90 days after injection. However, longer-term studies are still necessary as degradation and excretion of QDs in the formulation may be slow. Although QDs are promising materials for use in bioimaging and image-guided surgery,[62,63] concerns related to their photostability, biocompatibility and toxicity need to be addressed before clinical use.

4. Plasmonic/Noble Metal Nanoparticles

Nanoscale noble metal particles exhibit great potential in biomedical applications due to their plasmonic properties. These properties are a result of localized surface plasmons resonance (LSPR) which are collective oscillations of their surface conduction electrons in response to electromagnetic irradiation.[64] Emerging understanding on the fundamental physics of plasmonic nanoparticles,[65] most commonly gold and silver, has led to investigation into their applications in catalysis,[66] energy harvesting[67] and medicine.[68]

4.1. *Gold nanoparticles*

Plasmonic gold nanoparticles exhibit unique optical and photothermal properties. Surface plasmon resonance of gold nanoparticles allows them to strongly absorb and scatter incident light significantly higher when compared to molecular dyes.[69] Thus, they have emerged as promising contrast agents for biological imaging. Some classes of gold nanoparticles are also able to convert resonant energy to heat. Electron–phonon and phonon–phonon interactions can convert incident NIR light to heat.[70,71] Several classes of gold nanoparticles with the ability to absorb NIR light have been synthesized using chemical[71–80] and electrochemical[81–84] methods. These nanoparticles have been widely investigated for uses in bioimaging, nucleic acid and drug delivery and photothermal ablation.

4.1.1. Bioimaging and disease detection/diagnostics

Non- and minimally invasive bioimaging techniques are invaluable tools for clinical diagnostics of diseased tissue. Currently, bioimaging techniques, from the molecular to the tissue scales, are employed for detection of early-stage cancer, guiding drug and biological therapies and surgery. Gold nanoparticles are attractive as contrast agents since they can be engineered to overcome limitations of chemical and photostability, detection limits and compatibility across various imaging modalities, typically associated with conventional contrast agents.[85] The use of imaging modalities based on NIR lasers has been increasingly investigated since NIR light can penetrate tissue at greater depths compared to visible light, in addition to exhibiting minimal tissue scattering and reduced phototoxicity. Thus, gold nanoparticles, tuned to absorb NIR light, have been widely explored for use in multiple imaging modalities.[86–88]

Optical coherence tomography (OCT) is an NIR-based interferometer imaging modality that uses optical scattering media to provide cross-sectional subsurface images. Using NIR irradiation for maximal light penetration into tissue, images are formed based on differences between the absorption to scattering profiles of the medium, with resolution in the micrometer range at penetration depths of 2–3 mm.[89] Using mouse xenograft colon tumor models, Gobin *et al.*[90] injected poly(ethylene glycol) (PEG)-conjugated gold nanoshells (PEGylated gold nanoshells) via tail vein. Following 20 h after the injection, OCT images were acquired, and accumulation of the gold nanoshells in the tumor resulted in enhanced brightness (Fig. 3, top). Similarly, in mouse xenograft epidermoid tumor models, Kah *et al.*[91] demonstrated that antibody-functionalized PEGylated gold nanoshells increased that fractional concentration of injected gold nanoshells in tumor regions compared to untargeted nanoshells. A reduction in time from 6 h to 2 h was also observed for delivery of nanoshells to the tumor site. Although OCT has deeper tissue penetration compared to that observed with confocal microscopy, there are still limitations for the application of gold nanoparticle OCT in deep tissue imaging.[11]

Two-photon luminescence (TPL) imaging uses femto- to pico-second pulsed, high-intensity NIR Ti:S lasers in order to deliver a high photon flux to specimens of interest.[92] Due to their surface plasmon resonance property,

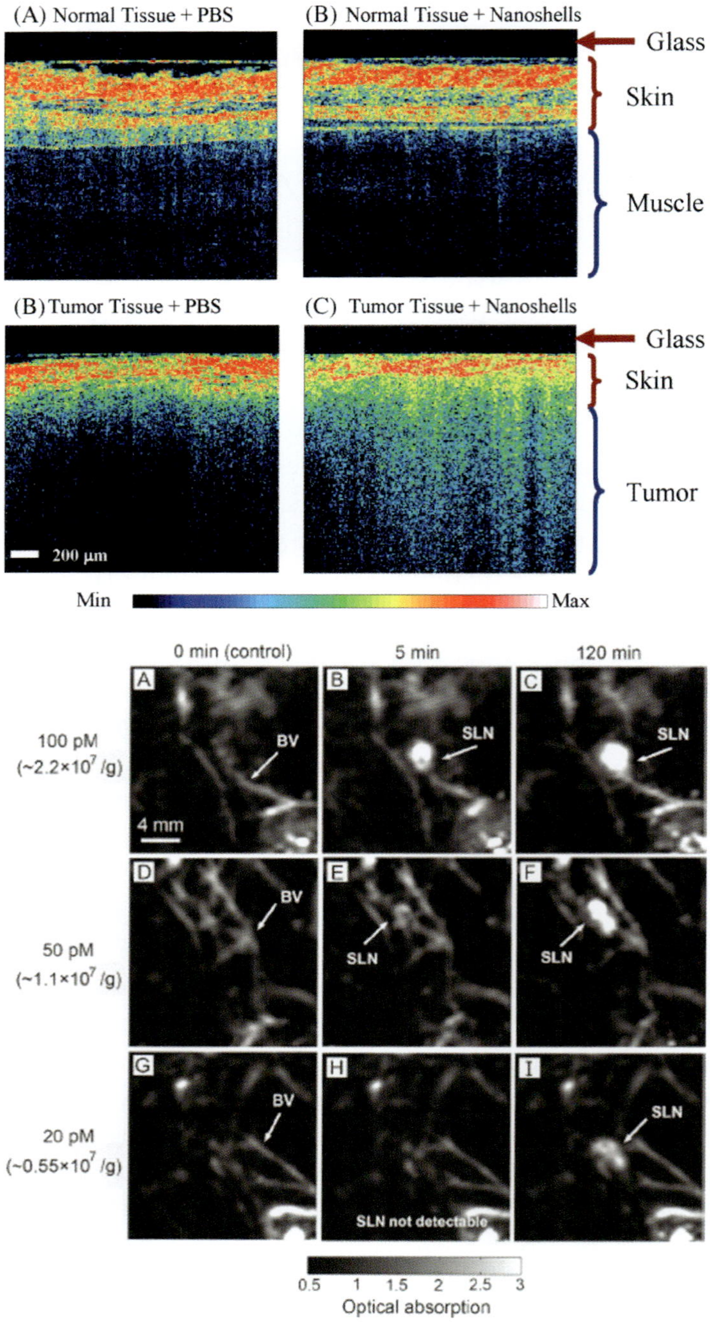

Fig. 3. (Top) Representative OCT images of normal skin and muscle tissue of mice systemically injected with (a) PBS or (b) nanoshells. Representative OCT images of tumors in mice systemically injected with (c) PBS or (d) nanoshells/ Adapted and reprinted with permission from Ref. 90. (Bottom) PA images of axillary reagion or rats following 0, 5 and 120 min after AuNCs injection at different concentrations, Adapted and reprinted with permission from Ref. 203.

noble metal nanoparticles have emerged as viable contrast agents for TPL. TPL is a serial process in which photons are sequentially absorbed and electrons from the d-band are excited to unoccupied states of the sp-conduction band. Luminescent emission results from the relaxation of the electron–hole pair. Gold nanorods (GNRs) have emerged as particularly appealing for TPL due to facile synthesis, tunable longitudinal plasmon absorbance in the NIR region and reduced plasmon dampening.[86] In a study performed by Wang *et al.*,[88] picomolar concentrations of sulfide-treated gold nanorods were injected into the tail vein of mice and visualized flowing through the mouse earlobe vessels in real-time using TPL imaging.

Gold nanoparticles have proven to be useful contrast agents for PAT due to their unique optical absorption properties.[93,94] In a study performed by the Wang group,[95] PEGylated gold nanoshells were utilized as PAT contrast agents for imaging rat brain vasculature and demonstrated up to 63% enhancement in optical absorption of brain vasculature following three sequential nanoshell treatments. Additionally, 80% enhancement in cerebral cortex absorption was found following systemic injection of PEGylated gold nanocages.[96] In a rat model study performed by the Xia group, PAT was used to quantitatively evaluate gold nanocage transport in the lymphatic system and uptake in the lymph nodes. This also enabled mapping of sentinel lymph nodes up to 3.3 cm below skin surface. Gold nanocages were transported to the second and third axillary lymph nodes, which can have potential implications for use in metastatic lymph node mapping (Fig. 3, bottom). Bioconjugated gold nanocages were employed for active targeting to subcutaneous inoculated melanomas in mouse models, resulting in 300% higher signal enhancement compared to PEG-gold nanocages which demonstrate passive targeting due to the EPR effect.[97]

4.1.2. **Drug and nucleic acid delivery**

The emergence of novel nanomaterials, including gold nanoparticles, has led to new options for delivery of therapeutic agents.[98–104] Due to various controlled delivery mechanisms such as pulsed-laser–induced gold–thiol bond cleavage,[105, 106] acidic-environment–induced release,[107] diffusion release,[108] photothermal modulation and surface coating shrinkage,[109] gold nanoparticles have been extensively investigated as drug and gene delivery vehicles.

Cetuximab (C225) is an FDA-approved anti-EGFR antibody for treatment of EGFR-positive colorectal cancer and is currently in different phases of clinical trials for other malignancies.[110,111] Gold nanoparticles using Cetuximab as a targeting agent for delivery of low doses of gemcitabine showed significant tumor growth inhibition in an orthotopic model of pancreatic cancer.[112] AuriTax is a multivalent nanosystem composed of tumor necrosis factor (TNF), paclitaxel, and PEG assembled on 25-nm gold nanoparticles (GNPs). Similarly, Aurimune-T is assembled on a 25-nm GNP with only TNF and PEG. Aurimune-T demonstrated a 10-fold increase in reducing tumor burden compared to TNF alone in animal studies.[113,114]

Gold nanoparticles are promising vectors for non-viral transgene delivery primarily due to the availability of facile surface-modification methods.[115] Surface functionalization of gold nanoparticles allows for tuning of hydrophobicity and charge, which can enhance transgene expression and decrease cytotoxicity.[102] Additionally, their high surface area to volume ratios facilitate high payload-carrying capabilities. Gold nanoparticles have been extensively investigated for *in vitro* transgene[116–119] delivery and expression and siRNA[116,120,121]/shRNA[122] delivery for gene silencing.

Folate receptor-targeted hollow gold nanospheres were used for NIR-induced delivery of siRNA against NF-κB p65 subunits both *in vitro* and *in vivo*.[105] NIR irradiation induced a shape transformation in the gold nanoparticles, resulting in siRNA release into the cytosol. This *photothermal transfection* resulted in downregulation of NF-κB only in tumors irradiated with NIR light. This downregulation of NF-κB was exploited to sensitize tumors to the chemotherapeutic drug, irinotecan, resulting in growth inhibition. In a study done by Bonoiu *et al.*,[123] gold nanorods with electrostatically bound siRNA were used to silence approximately 70% of the housekeeping gene, glyceraldehyde 3-phosphate dehydrogenase (GAPDH) in the CA1 hippocampal region of the rat brain at >10 days post-injection. The Prasad group[124] demonstrated that downregulation of the antiapoptotic sphingosine kinase (Sphk1) gene in head and neck squamous cell carcinoma in subcutaneous tumors grown in mice resulted in radiosensitization and over 50% tumor regression compared to control treatments. In these cases, the downregulation was achieved using gold nanorod-siRNA "nanoplexes." Additionally, tumor regression was possible with radiation doses much lower than those commonly used in clinical settings. Ryou *et al.*[125] reported studies in which mice with subcutaneous xenograft human colon carcinoma tumor models were subcutaneously injected with shRNA-conjugated gold nanoparticles for silencing the Mcl-1L gene. A decrease in tumor size of approximately 36%

was observed when compared to control treatments following direct administration into the tumor every 3 days for 30 days, indicating that shRNA treatment reduced tumor growth.

4.1.3. **Thermo-chemotherapy**

Thermal or hyperthermic therapy is an attractive alternate or co-approach for the ablation of diseased (e.g. cancer) tissue. Due to their unique optical properties, several classes of gold nanoparticles have been engineered for use in NIR and radio-frequency (RF)-based hyperthermic therapy.[128–135] External tunable control of NIR irradiation can be employed to achieve hyperthermic temperatures (>43°C) at localized sites, resulting in significant advantages over whole-body hyperthermia. As a result, gold nanoparticles (GNPs) have received increased attention for localized administration of hyperthermia for ablation of cancer cells[126] and bacteria.[127]

Pancreatic carcinoma growth in nude balb/c mice xenografts was inhibited for up to 7 weeks following treatment with intraperitoneal-injected gold nanospheres in concert with radio-frequency field exposure.[128] Gold nanoshells consisting of a dielectric core surrounded by an ultra-thin gold shell demonstrate photothermal responses that are tunable in the NIR region.[129] Nanoshells are often coated with PEG and, in cancer therapy investigations, can preferentially localize to tumor sites due to the enhanced permeability and retention (EPR) effect.[130] In female non-obese diabetic mice, Hirsch *et al.*[131] interstitially injected PEGylated gold nanoshells, and following NIR laser irradiation demonstrated irreversible thermal destruction of xenograft canine transmissible venereal tumors. O'Neal *et al.*[132] injected PEGylated gold nanoshells intravenously into female albino mice bearing murine colon carcinoma tumors. The mice were tumor free for at least a month following NIR laser irradiation. In a related study, PC3-tumor bearing mice were intravenously injected with PEGylated gold nanoshells and demonstrated 93% tumor regression three weeks after NIR laser irradiation. Currently, the most advanced application using gold nanoshells for photothermal therapy is currently in early clinical trials for ablation of recurrent head and neck cancer.[133]

Cylindrical gold nanorods possess a transverse absorption at 520 nm as well as a longitudinal absorption that is tunable in the NIR region. PEGylated gold nanorods injected intratumorally in nude (nu/nu) mice with subcutaneous squamous cell carcinoma xenographs demonstrated over 96% decrease in average tumor growth following NIR laser irradiation; more than

74% decrease in average tumor growth was observed following intravenous injection.[134] In murine colon cancer models, intravenous injection of PEGylated gold nanorods and NIR irradiation resulted in survival of approximately 44% of the initial mice population 60 days after treatment.[135]

Hyperthermic treatment is limited by cellular thermotolerance, whereby cells are able to maintain viability due to an array of cell survival responses.[136,137] Combinatorial/synergistic treatments for cancer therapy can be developed by engineering nanoparticles that simultaneously deliver chemotherapeutic drugs in concert with hyperthermic treatment for enhanced ablation of cancer cells. In addition, hyperthermia has been shown to increase tumor tissue perfusion, which allows for easier and deeper absorption of chemotherapeutic drugs in tumor tissue, resulting in greater efficacies between the two treatments. Hyperthermia also lowers tumor resistance to various chemotherapeutic drugs, including doxorubicin, cisplatin, bleomycin, nitrosoureas and cyclophosphamide.[138–140] As such, potential for higher efficacies of cancer cell ablation can be realized by combining the drug delivery and photothermal therapy properties on gold nanoparticles via a photo-chemotherapy approach. Multiple studies have employed different gold nanoparticle platforms for dual delivery of cancer cell-targeted chemotherapeutics and NIR light-based induction of hyperthermia. For example, the antineoplastic drug doxorubicin, has been loaded into hollow gold nanospheres,[141,142] hollow gold nanoshells,[143] and poly(ethylene glycol)–poly(lactic-co-glycolic acid)–Au half-shell nanoparticles[144,145] for simultaneous delivery of the drug and NIR photothermal therapy. Additionally, gold nanoshells on silica nanorattles[146] have been investigated for delivery of docetaxel along with simultaneous NIR photothermal therapy. Elastin-like polypeptide (ELP)-gold nanorod nanocomposites have also been investigated for dual delivery of the anticancer drug, 17-AAG simultaneously with NIR photothermal therapy.[147]

Doxorubicin (DOX)-loaded thermally sensitive micelles/liposomes were found to accumulate greater in xenograft MDA-MB-435 human tumors in mice following GNR-mediated hyperthermia. Additionally, DOX molecule release was triggered via hyperthermia, and it was found that the combination of the two treatments resulted in higher tumor growth suppression compared to hyperthermia alone and hyperthermia plus free DOX controls.[148] In murine fibrosarcoma-bearing mice, tail-vein injection was used to administer PEG-grafted gold nanospheres covalently linked with tumor necrosis factor alpha (TNF-α). After hyperthermic treatment (by exposure or tumor and surrounding tissue to water bath), it was found that tumor recession was increased when compared to each treatment alone. More recently,

PEG-grafted gold nanospheres have been used for delivery of recombinant Tumor Necrosis Factor-alpha (rhTNF a.k.a CYT-6091) in Phase-I clinical trials; Phase-II clinical trials are currently in progress.[149]

Although promising, one major limitation of administering hyperthermic treatment is the depth of penetration of NIR light. Currently, penetration depths in soft tissue are limited to approximately 2 cm in the most advanced NIR laser systems.[150] This results in applications limited to topical or near-surface conditions. However, alternative techniques have been investigated in order increase potential treatment depths. This includes the use of minimally invasive interventional techniques, where fibers inserted directly into tumors/ tissues are used for administration of laser-based hyperthermic treatment. This approach has been applied to hyperthermic treatment for primary and metastatic liver cancer[161–163] as well as in lung metastases.[151]

4.2. *Silver nanoparticles*

Though not as widely studied as gold nanoparticles, applications of silver nanoparticles (AgNPs) in biomedicine have also been explored.[152] AgNPs have been found to be plasmonically superior to gold, however, their uses have been limited due to their potential toxicity.[153,154] However, recent investigations have involved the encapsulation of AgNPs within nanothin silica coatings in order to increase their biocompatibility for increased use.[155,156]

The antibacterial activity of nanoscale silver has resulted in investigations in hospital treatment against microbial infections, burns, diabetic skin ulcers and medical devices. The antimicrobial spectrum of AgNPs is broader than that of most common antibiotics.[157] This activity is most widely attributed to release of positively charged silver ions that then bind with negatively charged proteins, RNA, and/or DNA which can limit cell replication.[158] Encapsulating silver(1) carbene complexes in nanofibers has shown promising results for sustained and effective silver ion delivery over long periods of time. This provides bactericidal activity and have primarily been applied for wound-or burn-dressings.[159] Nano-Ag sepiolite, which is a soda-lime glass powder containing monodispersed AgNPs with a low melting point, possesses high antibacterial activity as well as antifungal activity.[160] No report of resistance effect has been reported against AgNPs in the clinic to date.[157] Additionally, AgNPs have been found to be effective virucidal agents against cell-free and cell-associated viruses. For example, AgNPs have been found to exert anti-HIV activity at early stages of the viral replication. Additionally, they inhibit the post-entry stages of the HIV-1 life cycle.[161]

Further studies into the cytotoxicity, antibacterial mechanisms, bio-distribution, organ accumulation, degradation and adverse effects due to AgNP treatment are still needed. It was been widely accepted that AgNPs in high concentrations can cause apoptosis in mammalian cells.[162] This toxicity is attributed to AgNPs facilitating generation of intracellular reactive oxygen species (ROS) that result in oxidative damage to proteins and DNA.[163–165] As characterization and investigation into the application of AgNPs continues to progress, it is anticipated that they will find significant application in biomedicine.

4.3. *Platinum nanoparticles*

Although platinum complexes have been widely investigated for treatment of several diseases, the use of platinum nanoparticles for biomedical applications is a fairly new emerging field.[152] Currently, platinum nanoparticles have been mostly used as a composite of other nanoparticles on core/shell formulations.[1] For example, Yolk-shell nanoparticle consisting of a FePt core and a CoS_2 shell were found to be more efficient at killing HeLa cells compared to cis-platin alone.[166]

4.4. *Copper nanoparticles*

The use of alloy or shell/core multi-metallic nanoparticles is an emerging field of study. In one such study, Au_3Cu_1 nanoshells were synthesized and found to have potential as blood pool agents in magnetic resonance angiography as they were found to enhance contrast of blood vessels *in vivo* (Fig. 4).[167] CuS nanoparticles have also been used as photothermal agents, since they demonstrate SPR peaks in the NIR region.[168,169] Following intratumoral or intravenous injection of PEG-CuS nanoparticles and laser treatment, a significant necrotic response was observed in nude mice bearing subcutaneous U87 glioma xenograft tumors.[169]

5. Magnetic Nanoparticles

Magnetic nanoparticles (MNPs) are investigated for several applications in biomedicine, including contrast agents in bioimaging using MRI, RF-based thermal therapy and drug delivery vehicles.[183–185] In presence of a magnetic field, iron oxide nanoparticles acquire a large magnetic moment, making them superparamagnetic. Only small amounts of super paramagnetic

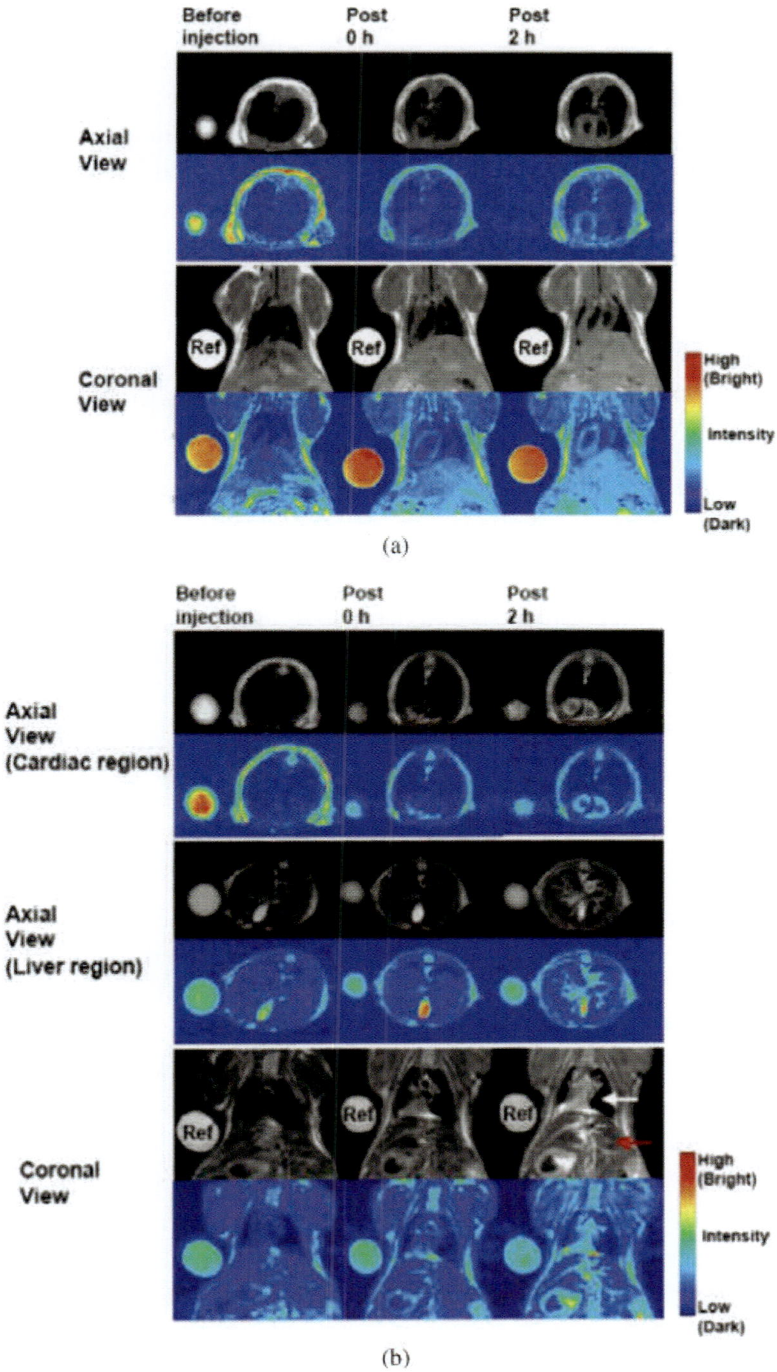

Fig. 4. *In vivo* progress MRI events and color map of Au_3Cu_1 nanocapsules injected in male BALB/c mice at indicated temporal points of (a) T_1-weighted (b) T_2-weighted images. Adapted and reprinted with permission from Ref. 167.

iron-oxide nanoparticle (SPIONs) are needed due to increased signal change/contrast, which is a result of their large magnetic moments. SPIONs can range in size from a few to several hundred nanometers in diameter and are most commonly synthesized via alkaline co-precipitation of $Fe(OH)_2$ and Fe(OH)3 suspension.[170,171] Additionally, their biocompatibility, stability and circulation half-life can be improved by functionalization with dextran,[172] polysaccharides,[173] PEG and polyethylene oxide[174] coatings. This, in concert with their inherent biocompatibility, and biodegradability, makes them attractive for *in vivo* applications. When degraded, the iron ions get incorporated into hemoglobin in erythrocytes and are degraded along normal iron recycling pathways,[175] although iron overloading can lead to toxicity.

5.1. *MRI Imaging*

SPIONs have been employed as contrast agents primarily due to their large magnetic moments. MNPs have primarily been used for diagnostic applications, including detection of diseased tissue as well as pathogenic bioaggregates.[12] For example, tumors as small as 2–3 mm in size have been detectable using RES macrophage-mediated uptake of SPIONs.[176] Additionally, SPIONs have been used to diagnose lymph node metastases due to clearance by macrophages of the lymphatic system and accumulation in the lymph nodes.[177–179] In clinical investigations, ultra-small SPIO (USPIO), with diameters less than 40 nm, have been used as contrast agents to delineate brain tumors on magnetic resonance mages.[179]

Several surface-modified SPIONs have also been used for detection and diagnosis in solid tumor investigations.[180,181] Targeted SPIONs can be employed for accurate diagnosis and detection of various disease states. For example, SPIONs modified with antibodies have been used to image retinal carcinoma,[182] as well as breast cancer.[11] In addition to using antibodies, SPIONs modified with targeting peptides such as chlorotoxin have been found to accumulate in glioma tumors that over express MMP-2.[181] SPIONs conjugate with folic acid have also been employed to target different human tumor cells.[183,184] In addition, the RGD peptide has been commonly used as a targeting ligand for integrin and has also been successfully used to modify SPIONs for active targeting to neoplasia.[185,186] The use of iron oxide MNPs as MRI signal enhancers has received approval by the US Food and Drug Administration (FDA) for the detection of liver metastasis.[187]

PEGylated iron-oxide-core–gold-shell nanoparticles were investigated in an effort to increase circulation time with an eye toward increasing tumor uptake by the EPR effect and reducing RES recognition. These nanoparticles

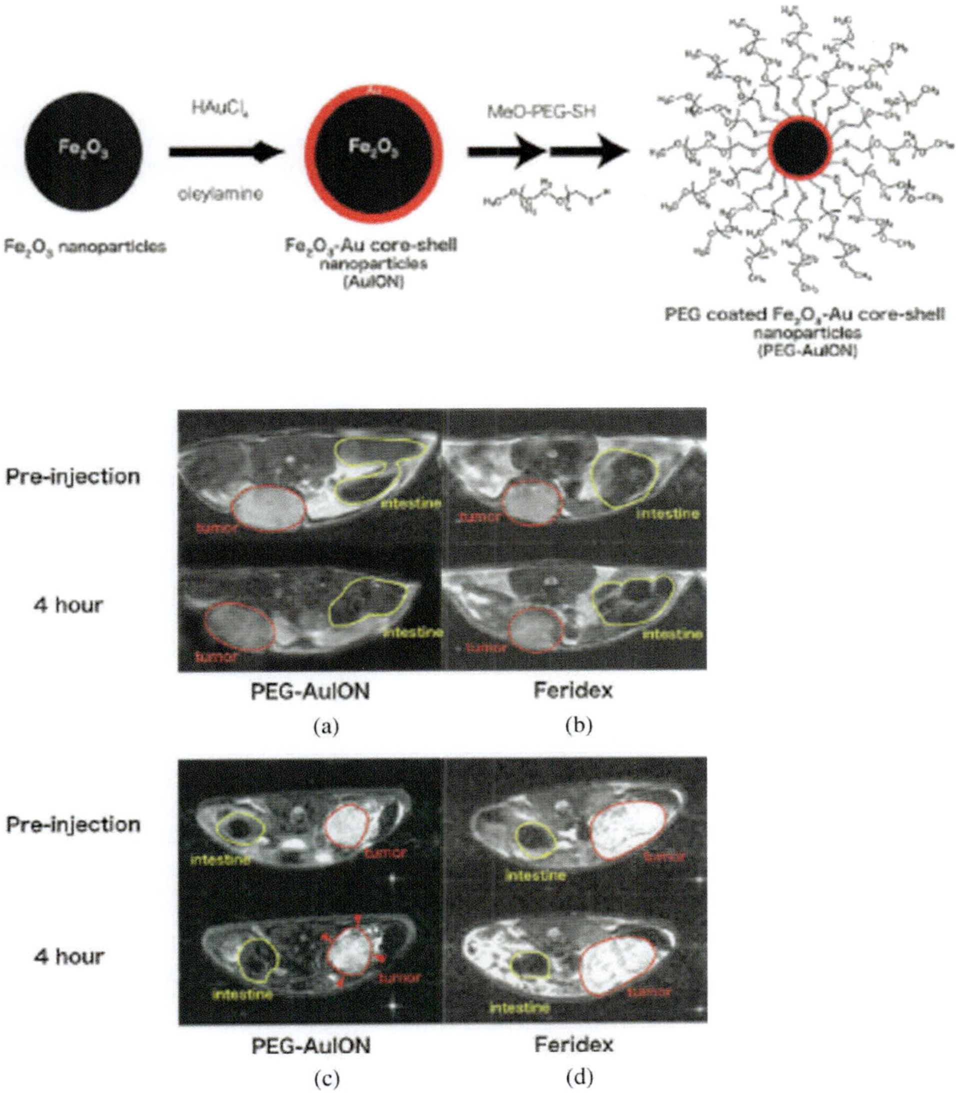

Fig. 5. (Top) Schematic of Au shell formation onto Fe_2O_3 nanoparticles and coating of MeO-PEG-SH. (Bottom) *In vivo* MR images of tumors. (a) and (b): T_2-weighted images of subcutaneous C26 murine carcinoma at 4 h intravenous injection of (a) Peg-AuIONs and (b) Feridex. (c) and (d) T_2-weighted images of orthoscopic MiaPaCa-2 human pancreatic cancer model at 4 h intravenous injection of (c) PEG-AuIONs and (d) Feridex. Adapted and reprinted from Ref. 204.

were found to be promising MRI contrast agents for imaging pancreatic cancer (Fig. 5).[188] Similarly, SPIONs have been modified with a silica coat that is then functionalized with amino groups. These, in turn, facilitate the formation of gold nanocrystal seeds, resulting in the formation of SPIO core-silica-Au shell nanoparticle (SPIO@silica-AuNPs),[189] which were used to treat nude mice implanted with A431 epidermoid carcinoma tumors. Due to NIR light absorption by the Au core, these nanoparticles were able to mediate an efficient photothermal effect and the SPIO core facilitated MR imaging of these particles thereby demonstrating their combing therapeutic and imaging or "theranostic" potential.[190]

5.2. *Thermal ablation*

In addition to MRI applications, SPIONs are capable of converting electromagnetic energy into heat, when subjected to an alternating magnetic field. This can result in magneto-thermal ablation of diseased tissue. Successful tumor ablation has been achieved by using only 5–10 mg of SPIONs/cm^3 of tissue.[191,192] This property, in concert with their MRI applications, makes SPIONs attractive theranostic agents, particularly in image-guided therapy applications.[193] The use of SPIONs as thermal agents is beginning to show promise in clinical studies.[194]

5.3. *Drug and gene delivery*

In addition to their dual imaging + thermal therapy applications, SPIONs have also been employed for dual imaging + drug delivery. Thermally cross-linked-SPIONs (TCL-SPIONs) consisting of a surface PEG layer and the anticancer drug, doxorubicin, incorporated into the polymeric shell, were used for the detection and treatment of Lewis lung carcinoma. They were also found to be capable of delivering a sufficient amount of the drug to tumor tissues, which was released faster under mildly acidic conditions commonly found in the tumor microenvironment as compared to neutral pH conditions of the vasculature.[195] In a study by Perlstein *et al.*,[196] TRAIL was conjugated to MNPs and used to label U251 cell-derived glioma xenografts in rats. TRAIL-MNPs localized in tumors and induced a significant decrease in tumor volume, increase in glioma cell apoptosis, and increase in animal survival. In another study, PEGylated Fe_3O_4/Au core/shell nanoparticles (PGMNPs) were used for external magnetic field (EMF)-guided delivery of the anticancer drug doxorubicin (DOX-PGMPs). Tumor volume significantly decreased and an increase in life span was observed [197] following intravenous injection of DOX-PGMNPs to H22 hepatoma-bearing mice under EMF-guided delivery.

MNPs have also been employed as delivery vectors for nucleic acids. This is often accomplished with a technique termed magnetofection, wherein nucleic acid delivery by MNPs is guided by an EMF.[198] The Yang group[199] synthesized polyethylenimine (PEI)-modified magnetic nanoparticles (GPEI) and demonstrated their potential as drug/gene delivery vectors to brain tumor lesions by intra-carotid administration. Magnetic entrapment was used in order to accumulate GPEI nanoparticles at brain tumor lesions. The Shuai group[200] synthesized a PEG-grafted PEI SION (PEG-g-PEI-SPION) as a nanocarrier for siRNA delivery for downregulation of CD44v6 expression in SGC-7901 gastric carcinoma cells. PEG-g-PEI-SPION was effective in downregulating the target *in vitro* and is a highly effective contrast agent for MRI *in vivo*. In an investigation by the Tada group,[201] oleic-acid–coated magnetic nanocrystal core with a cationic lipid shell, termed LipoMag, was used to deliver siRNA against the epidermal growth factor receptor to mice bearing gastric tumors; a therapeutic benefit was achieved with no evident adverse effects. Kumar *et al.*[202] employed a magnetic nanoparticle labeled with the NIR dye Cy5, conjugated to the EPPT peptide, which targets tumor-specific antigen uMUC-1. This construct was used to deliver siRNA targeted against the antiapoptotic *birc5* gene which encodes for the pro-survival protein, survivin. Image-guided delivery of the siRNAwsa employed for delivery to subcutaneous breast tumors in mice, resulting in a significant therapeutic effect.

6. Conclusions and Outlook

The field of nanomedicine has benefited from nanotechnology and the various classes of inorganic nanoparticles that have been investigated for applications in nanomedicine. Inorganic nanoparticles for biomedical applications have the potential to transform healthcare in the future. These classes of materials have applications directly relevant to biomedicine, due to their unique optical and physiochemical properties. In addition, their applications in therapeutic regimes including drug and gene delivery and thermal conversion, make them attractive theranostic agents.

Nevertheless, despite their great potential, several issues concerning inorganic nanoparticles will need to be addressed. For example, long-term effects related to tissue damage, toxicity, carcinogenicity, immunogenicity and inflammation following administration of inorganic nanoparticles are still not fully understood. However, comprehensive studies on these promising materials will begin to address some of these concerns and investigate their clinical translational potential.

Table 1. Applications and concerns/drawback of several inorganic nanoparticles.

	Classes	Applications	Concerns/Drawbacks
Carbon Nanotubes	Single Walled	Bioimaging	Toxicity
	Multi-Walled	RF/NIR Thermal Therapy	Stability
		Drug/Gene Delivery	
Quantum Dots	NIR emitting	Bioimaging	Imaging resolutions and penetration depth
			Toxicity
			Stability
			Clearance characteristics
Plasmonic/ Noble Metals	Gold	Bioimaging (Au,Cu)	Limited NIR light penetration depth (Au-based imaging and photothermal therapy)
	Silver	RF/NIR Thermal Therapy (Au, Cu)	Toxicity (Ag)
	Platinum	Drug/Gene Delivery (Au, Pt)	
	Copper	Anti-bacterial/fungal/ viral (Ag)	
Magnetic	SPIONs	Magnetic Resonance Imaging	Rapid clearance b phagocytes
	Plasmonic Shelled	Magneto/NIR Thermal Therapy	Synthesis

References

1. S. Bhattacharyya, R. A. Kudgus, R. Bhattacharya and P. Mukherjee, Inorganic nanoparticles in cancer therapy, *Pharm. Res.* **28**, 237–259 (2011).
2. S. R. Kamboj and B. S. Sekhon, Inorganic nanomedicine-Part 2, *Nanomed. Nanotechnol. Biol. Med.* **6**, 612–618 (2010).
3. T. Levchenko, W. Hartner, D. Verma, E. Bernstein and V. Torchilin, ATP-loaded liposomes for targeted treatment in models of myocardial ischemia, *Meth. Mol .Biol.* **605**, 361–375 (2010).
4. T. Wang, G. G. M. D'Souza, D. Bedi, O. A. Fagbohun, L. P. Polturi, B. Papahadjopoulos-Stember, V. A. Petrenko and V. P. Torchilin, Enhanced binding and killing of target tumor cells by drug-loaded liposomes modified with tumor-secific phage fusion coat protein, *Nanomedicine* **5**(4), 563–574 (2010).

5. A. D. Bangham, M. M. Standish, J. C. Watkins, Diffusion of univalent ions across lamellae of swollen phospholipids, *J. Mol. Biol.* **13**(1), 238–252 (1965).
6. V. P. Torchilin, Recent advances with liposomes as pharmaceutical carriers, *Nat. Rev. Drug. Discov.* **4**(2), 145–160 (2005).
7. M. Susa, A. K. Iyer, K. Ryu, F. J. Hornicek, H. Mankin, M. M. Amiji and Z. F. Duan, Doxorubicin loaded polymeric nanoparticuate delivery systems to overcome drug resistance in osteosarcoma, *BMC Cancer* **9**. DOI 10.1186/1471-2407-9-399 (2009).
8. L. E. van Vlerken, Z. F. Duan, S. R. Little, M. V. Seiden and M. M. Amiji, Augmentation of therapeutic efficacy in drug-resistant tumor models using cermide coadministration in temporal-controlled polymer-blend nanoparticle delivery systems, *AAPS J.* **12**(2), 171–180 (2010).
9. R. Langer and J. Folkman, Polymers for sustained-release of proteins and other macromolecules, *Nature* **263**(5580), 797–800 (1976).
10. R. F Service, Nanotechnology nanoparticle Trojan Horses gallop from the lab into the clinic, *Science* **330**(6002), 314–315 (2010).
11. H.-C. Huang, S. Barua, G. Sharma, S. K. Dey and K. Rege, Inorganic nanoparticles for cancer imaging and therapy, *J. Control. Release* **155**, 344–357 (2011).
12. B. S. Sekhon and S. R. Kamboj, Inorganic nanomedicine-Part 1, *Nanomed.: Nanotechnol. Biol. Med.* **6**, 516–522 (2010).
13. S. R. Ji, C. Liu, B. Zhang, F. Yang, J. Xu, J. Long, C. Jin, D. L. Fu, Q. X. Ni and X. J. Yu, Carbon nanotubes in cancer diagnosis and therapy, *Biochim. Biophys. Acta* **1**, 29–35 (2010).
14. Z. Liu, S. Tabakman, K. Welsher and H. Dai, Carbon nanotubes in biology and medicine: *In vitro* and *in vivo* detection, imaging and drug delivery, *Nano Res.* **2**, 85–120 (2009).
15. M. Terrones, N. Grobert, J. Olivares, J. P. Zhang, H. Terrones, K. Kordatos, W. K. Hsu, J. P. Hare, P. D. Townsend, K. Prassides, A. K. Cheetham, H. W. Kroto and D. R. M. Walton. Controlled production of aligned-nanotube bundles, *Nature* **388**(6637), 52–55 (1997).
16. J. Koo, M. Jeon, Y. Oh, H. W. Kang, J. Kim, C. Kim and J. Oh, *In vivo* non-ionizing photoacoustic mapping of sentinel lymph nodes and bladders with ICG-enhanced carbon nanotubes, *Phy. Med. Biol.* **57**, 7853–7862 (2012).
17. M. Pramanik, K. H. Song, M. Swierczewska, D. Green, B. Sitharaman and L. Wang, *In vivo* carbon nanotube-enhanced non-invasive photoacoustic mapping of the entinel lymph node, *Phys. Med. Biol.* **54**, 3291–3301 (2009).
18. A. de la Zerda, Z. Liu, S. Bodapati, R. Teed, S. Vaithilingam, B. T. Khuri-Yakub, X. Chen, H. Dai and S. S. Gambhir, Ultrahigh sensitivity carbon nanotube agents for photoacoustic molecular imaging in living mice, *Nano Lett.* **10**, 2168–2172 (2010).
19. X. Zheng, D. Xing, F. Zhou, B. Wu and W. R. Chen, Indocyanine green-containing nanostructure as near infrared dual-functional targeting probed for optical imaging and photothermal therapy, *Mol. Pharm.* **8**, 447–456 (2011).
20. J.-G. Yu, F.-P. Jiao, X.-Q. Chen, X.-Y Jiang, Z.-G. Peng, D.-M. Zeng and D.-S. Huang, Irradiation-mediated carbon nanotubes' use in cancer therapy, *J. Cancer Res. Therap.* **8**(3), 348–354 (2013).
21. C. J. Gannon, P. Cherukuri, B. I. Yahobson, L. Cognet, J. S. Kanzius, C. Kittrell, R. B. Weisman, M. Pasquali, H. K. Schmidt, R. E. Smalley and S. A. Curley, Carbon-nanotube enhanced thermal destruction of cancer cells in noninvasive radiofrequency field, *Cancer* **110**, 2654–2665 (2007).

22. V. V. Chaban, T. I. Savchenko, S. M. Kovalenko and O. V. Prezhdo, Heat-driven release of a drug molecule from carbon nanotubes: A molecular dynamics study, *J. Phys. Chem. B* **114**, 13481–13486 (2010).
23. A. Burke, X. Ding, R. Singh, R. A. Kraft, N. Levi-Polyachenko, M. N. Rylander, C. Szot, C. Buchanan, J. Whitney, J. Fisher, H. C. Hatcher, R. D'Agostino, N. D. Kock, P. M. Ajayan, D. L. Carrol, S. Akman, F. M. Torti and S. V. Torti, Long-term survival following treatment of kidney tumors with multiwalled carbon nanotubes and near-infrared radiation, *Proc. Natl. Acad. Sci. USA* **106,** 12897–12902 (2009).
24. S. Ghosh, S. Dutta, E. Gomes, D. Carroll, R. D'Agostino and J. Olson, Increased heating efficiency and selective thermal ablation of malignant tissue with DNA-encased multiwalled carbon nanotubes, *ACS Nano.* **3**, 2667–2673 (2009).
25. X. Ding, R. Singh, A. Burke, H. Hatcher, J. Olson, R. Kraft, M. Schmid, D. Carroll, J. D. Bourland, S. Akman, F. M. Torti and S. V. Torti, Development of iron-containing multiwalled carbon nanotubes for MR-guided laser-induced thermotherapy, *Nanomed. (Lond)* **6**(8), 1341–1352 (2011).
26. Z. Liu, K. Chen, C. Davis, S. Sherlock, Q. Z. Cao, X. Y. Chen and H. J. Drug delivery with carbon nanotubes for *in vivo* cancer treatment, *Cancer Res.* **68**(16), 6652–6660 (2008).
27. Z. Liu, W. B. Cai, L. N. He, N. Nakayama, K. Chen, X. M. Sun, X. Y. Chen and H. J. Dai, *In vivo* biodistribution and highly efficient tumor targeting of carbon nanotubes in mice, *Nat. Nanotechnol.* **2**(1), 47–52 (2007).
28. Z. Liu, A. C. Fan, K. Rakhra, S. Sherlock, A. Goodwin, X. Y. Chen, Q. W. Yang, D. W. Felsher and H. J. Dai, Supramolecular stacking of doxorubicin on carbon nanotubes for *in vivo* cancer therapy, *Angew. Chem. Int. Ed.* **48**(41), 7668–7672 (2009).
29. L. Z. Xiang, Y. Yuan, D. Xing, Z. M. Ou, S. H. Yang and F. F. Zhou, Photoacoustic molecular imaging with antibody-functionalized single-walled carbon nanotubes for early diagnosis of tumor, *J. Biomed. Opt.* **14**(2), Doi: 10.1117/1.3078809 (2009).
30. A. A. Bhirde, V. Patel, J. Gavard, G. F. Zhang, A. A. Sousa, A. Masedunskas, R. Leapman, R. Weigert, J. S. Gutkind and J. F. Rusling, Targeted killing of cancer cells *in vivo* and *in vitro* with EGF-directed carbon nanotube-based drug delivery, *ACS Nano* **3**(2), 307–316 (2009).
31. G. B. P. Cherukuri, J. Kingston, L. Cognet, R. Lemos, T. K. Leeuw, L. Gumniner-Russo, R. B. Weisman and G. Powis, *In vivo* therapeutic silencing of hypoxia-inducible factor 1 alpha (HIF-1 alpha) using single walled carbon nanotubes noncovalently coated with siRNA, *Nano Res.* **2**(4), 279–291 (2009).
32. J. E. Podesta, K. T. Al-Jamal, M. A. Herrero, B. W. Tian, H. Ali-Boucetta, V. Hegde, A. Bianco, M. Prato and K. Kostarelos, Antitumor activity and prolonged survival by carbon-nanotube-mediated therapeutic siRNA silencing in human lung xenograft model, *Small* **5**(10), 1176–1185 (2009).
33. C. A. Poland, R. Duffin, I. Kinloch, A. Maynard, W. A. H. Wallace, A. Seaton, V. Stone, S. Brown, W. MacNee and K. Donaldson, Carbon nanotubes introduced into the abdominal cavity of mice show asbestos-like pathogenicity in pilot study, *Nat. Nanotechnol.* **3**(7), 423–428 (2008).
34. A. Takagi, A. Hirose, T. Nishimura, N. Fukumori, A. Ogata, N. Ohashi, S. Kitajima and J. Kanno, Induction of mesothelioma in p53 +/_ mouse by intraperitoneal application of multi-wall carbon nanotube, *J. Toxicol. Sci.* **33**(1), 105–116 (2008).

35. A. Fraczek, E. Menaszek, C. Paluszkiewicz and M. Blazewicz, Competitive *in vivo* biocompatibility study of single- and multi-wall carbon nanotubes, *Acta Biomater.* **4**(6), 1593–1602 (2008).
36. M. V. Yezhelyev, L. Qi, R. M. O'Regan, S. Nie and X. Gao, Proton-sponge coated quantum dots for siRNA delivery and intracellular imaging, *J. Am. Chem. Soc.* **130**(28), 9006–9012 (2008).
37. L. Qi, X. Gao, A. Senpa, S. Ghosh, S. A. Wickline, L. V. Wang and G. M. Lanza, Quantum dot-amphipol nanocomplex for intracellular delivery and real-time imaging of siRNA, *ACS Nano.* 1403–1410 (2008).
38. A. M. Derfus, A. A. Chen, D.-H. Min, E. Ruoslahti and S. N. Bhatia, Targeted quantum dot conjugates for siRNA delivery, *Bioconjug. Chem.* **18**(5), 1391–1396 (2007).
39. X. Gao, L. Yang, J. A. Petros, F. F. Marshall, J. W. Simons and S. Nie, *In vivo* molecular and cellular imaging with quantum dots, *Curr. Opin. Biotechnol.* **16**(1), 63–72 (2005).
40. A. P. Alivisatos, Semiconductor clusters, nanocrystals, and quantum dots, *Science* **271**, 933–937 (1996).
41. K. T. Yong, J. Qian, I. Roy, H. H. Lee, E. J. Bergey, K. M. Tramposch, S. He, M. T. Swihart, A. Maitra and P. N. Prasad, Quantum rod bioconjugates as targeted probes for confocal and two-photon fluorescence imaging of cancer cells, *Nano Lett.* 7(3), 761–765 (2007).
42. B. Dubertret, P. Skourides, D. J. Norris, V. Noireaux, A. H. Brivanlou and A. Libchaber, *In vivo* imaging of quantum dots encapsulated in phospholipid micelles, *Science* **298**(5599), 1759–1762 (2002).
43. J. Aldana, Y. A. Wang and X. Peng, Photochemical instability of CdSe nanocrystals coated by hydrophilic thiols, *J. Am. Chem. Soc.* **123**(36), 8844–8850 (2001).
44. C. B. Murray, D. J. Norris and M. G. Bawendi, Synthesis and characterization of nearly monodisperse CDE (E=S, SE, TE) semiconductor nanocrystallites, *J. Am. Chem. Soc.* **115**(19), 8706–8715 (1993).
45. M. Bruchez, M. Moronne, P. Gin, S. Weiss and A. P. Alivisatos, Semiconductor nanocrystals as fluorescent biological labels, *Science* **281**(5385), 2013–2016 (1998).
46. H. S. Choi, W. Liu, P. Misra, E. Tanaka, J. P. Zimmer, B. I. Ipe, M. G. Bawendi and J. V. Fragioni, Renal clearance of quantum dots, *Nat. Biotechnol.* **25**(10), 1165–1170 (2007).
47. W. C. Chan and S. Nie Quanum dot bioconjugates for ultrasensitive nonisotopic detection, *Science* **281**(5385), 2016–2018 (1998).
48. H. Mattoussi, J. M. Mauro, E. R. Goldman, G. P. Anderson, V. C. Sundar, F. V. Mikulec and M. G. Bawendi, Self-assembly of CdSe-ZnS quantum dot bioconjugates using an engineered recombinant protein, *J. Am. Chem. Soc.* **122**(49), 12142–12150 (2000).
49. M. E. Akerman, W. C. W. Chan, P. Laakkonen, S. N. Bhatia and E. Ruoslahti, Nanocrystal targeting *in vivo*, *Proc. Natl. Acad. Sci.* **99**(20), (2002).
50. A. Nair, J. H. Shen, P. Thevenot, L. Zou, T. Cai, Z. B. Hu and L. P. Tang, Enhanced intratumoral uptake of quantum dots concealed within hydrogel nanoparticles, *Nanotechnology* **19**(48) (2008).
51. H. Tada, H. Higuchi, T. M. Wanatabe and N. Ohuchi, *In vivo* real-time tracking of single quantum dots conjugated with monoclonal anti-HER2 antibody in tumors of mice, *Cancer Res.* **67**(3), 1138–1144 (2007).

52. L. L. Yang, H. Mao, Y. A. Wang, Z. H. Cao, X. H. Peng, X. X. Wang, H. W. Duan, C. C. Ni, Q. G. Yuan, G. Adams, M. Q. Smith, W. C. Wood, X. H. Gao and S. M. Nie, Single chain epidermal growth factor receptor antibody conjugated nanoparticles for *in vivo* tumor targeting and imaging, *Small* **5**(2), 235–243 (2009).
53. W. B. Cai, D. W. Shin, K. Chen, O. Gheysems, Q. Z. Cao, S. X. Wang, S. S. Gambhir and X. Y. Chen, Peptide-labeled near-infrared quantum dots for imaging tumor vasculator in living subjects, *Nano Lett.* **6**(4), 669–676 (2006).
54. P. Diagaradjane, J. M. Orenstein-Cardona, N. E. Colon-Casasnovas, A. Deorukhkar, S. Shentu, N. Kuno, D. L. Schwartz, J. G. Gelovania and S. Krishnan, Imaging epidermal growth factor receptor expression *in vivo*: Pharmacokinetic and biodistribution characterization of a bioconjucated quantum dot nanoparticle, *Clin. Cancer. Res.* **14**(3), 731–741 (2008).
55. A. Zintchenko, A. S. Susha, M. Concia, J. Feldman, E. Wagner, A. L. Rogach and M. Ogris, Drug nanocarrier labeled with near-infrared-emitting quantum dots (quantoplexes): Imaging fast dynamics of distribution in living animals, *Mol. Ther.* **17**(11), 1849–1856 (2009).
56. G. Jiang, K. Park, J. Kim, K. S. Kin and S. K. Hahn, Target specific intracellular delivery od siRNA/PEI-HA complex by receptor mediated endocytosis, *Mol. Pharm.* **6**(3), 727–737 (2009).
57. J. H. Gao, K. Chen, R. G. Xie, J. Xie, S. Lee, Z. Cheng, X. G. Peng, and X. Y. Chen, Ultrasmall near-infrared non-cadmium quantum dots for *in vivo* tumor imaging, *Small* **6**(2), 256–261 (2010).
58. Xie, RG, Chen, L, Chen, XY, Peng, XG. InAs/InP/ZnSe core/shell/shell quantum dots as near-infrared emitters: Bright, narrow-band, non-cadmium containg and biocompatible, *Nano Res.* **1**(6), 457–464 (2008).
59. L. Li, T. J. Daou, I. Texier, T. K. C. Tran, Q. L. Nguyen and P. Reiss, Highly luminescent CuIS2/ZnS core/shell nanocrystals: cadmium-free quantum dots for *in vivo* imaging, *Chem. Mater.* **21**(12), 2422–2429 (2009).
60. K.-H. Jung, J. W. Park, J.-Y. Paik, E. J. Lee, Y. S. Choe and K.-H. Lee, Hydrazinonicotinamide prolongs quantum dot circulation and reduces reticuloendothelial system clearance by suppressing opsonization and phaocyte engulfment, *Nanotechnology* **23**(495102), (2012).
61. L. Ye, K.-T. Yong, L. Liu, I. Roy, R. Hu, J. Zhu, H. Cai, W.-C. Lay, J. Liu, K. Wang, J. Liu, Y. Liu, Y. Hu, X. Zhang, M. T. Swihart and P. N. Prasad, A pilot study in non-human primates shows no adverse response to intravenous injection of quantum dots *Nat. Nanotechnol.* 7, 453–458 (2012).
62. H. Daneshvar, J. Nelms, O. Muhammad, H. Jackson, J. Tkach, W. Davros, T. Peterson, M. A. Vogelbaum, M. P. Bruchez and S. A. Toms, Imaging characteristics of zinc shell, cadmium telluride core quantum dots, *Nanomedicine (Lond).* **3**(1), 21–29 (2008).
63. Y. Charles and Y. C. Cao, Nanomaterials for biomedical applications, *Nanomed.* **3**, 467–469 (2008).
64. K. A. Willets and R. P. Van Duyne, Localized surface plasmon resonance spectroscopy, *Annu. Rev. Phys. Chem.* **58**(26), 267–297 (2007).
65. A. B. Dahlin, T. Sannomiya, R. Zahn, G. A. Sotiriou and J. Voros, Electrochemical crystallization of plasmonic nanostructures, *Nano. Lett.* **11**, 1337–1343 (2011).

66. N. Liu, M. I. Tang, M. Hentshel, H. Giessen and A. P. Alivisatos, Nanoantenna-enhanced gas sensing in a single tailored nanofocus, *Nat. Mater.* **10**, 631–636 (2011).
67. P. Nagal, N. C. Lindquist, S. H. and Oh, D. J. Norris, Ultra-smooth patterned metals for plasmonics and metamaterials, *Science* **325**, 594–597 (2009).
68. G. A. Sotiriou, Biomedical applications of multifunctional plasmonic nanoparticles, *WIREs Nanomed. Nanobiotechnol.* **5**, 19–30 (2013).
69. S. Link and M. A. El-Sayed, Spectral properties and relaxation dynamics of surface plasmon electronic oscillations in gold and silver nanodots and nanorods, *J. Phys. Chem. B* **103**(40), 8410–8426 (1999).
70. P. Jain, X. Haung, I. H. El-Sayed and M. A. El Sayed, Review of some interesting surface plasmon resonance-enhanced properties of noble metal nanoparticles and their applications to biosystems, *Plasmonics* **2**, 107–118 (2007).
71. S. J. Oldenburg, J. B. Jackson, S. L. Westcott and N. J. Halas, Infrared extinction properties of gold nanoshells, *Appl. Phys. Lett.* **75**(19), 2897–2899 (1999).
72. N. R. Jana, L. Gearheart and C. J. Murphy, Seed-mediated growth approach for shape-controlled synthesis of spheroidal and rod-like gold nanoparticles using a surfactant template, *Adv. Mater.* **13**(18), 1389–1393 (2001).
73. N. R. Jana, L. Gearheart and C. J. Murphy, Wet chemical synthesis of high aspect ratio cylindrical gold nanorods, *J. Phys. Chem. B* **105**(19), 4065–4067 (2001).
74. N. R. Jana, L. Gearheart, S. O. Obare and C. J. Murphy, Anisotropic chemical reactivity of gold spheroids and nanorods, *Langmuir* **18**(3), 922–927 (2002).
75. S. J. Oldenburg, R. D. Averitt, S. L. Westcott and N. J. Halas, Nanoengineering of optical resonances, *Chem. Phys. Lett.* **288**(2–4), 243–247 (1998).
76. S. J. Oldenburg, S. L. Westcott, R. D. Averitt and N. J Halas, Surface enhanced Raman scattering in the near infrared using metal nanoshell substrates, *J. Chem. Phys.* **111**(10), 4729–4735 (1999).
77. Y. G. Sun, B. T. Mayers and Y. N. Xia, Template-engaged replacement reaction: A one-step approach to the large-scale synthesis of metal nanostructures with hollow interiors, *Nano Lett.* **2**(5), 481–485 (2002).
78. J. Y. Chen, J. M. McLellan, A. Siekkinen, Y. J. Xiong, Z. Y. Li and Y. N. Xia, Facile synthesis of gold-silver nanocages with controllable pores on the surface, *J. Amer. Chem. Soc.* **128**(46), 14776–14777 (2006).
79. S. E. Skrabalak, L. Au, X. D. Li and Y. Xia, Facile synthesis of Ag nanocubes and Au nanocages, *Nat. Protocols* **2**(9), 2182–2190 (2007).
80. B. Nikoobakht and M. A. El-Sayed, Preparation and growth mechanism of gold nanorods (NRs) using seed-mediated growth method, *Chem. Mater.* **15**(10), 1957–1962 (2003).
81. C. R. Martin, Nanomaterials — a membrane-based synthetic approach, *Science* **266**(5193), 1961–1966 (1994).
82. J. Burdick, E. Alonas, H. C. Huang, K. Rege and J. Wang, High-throughput templated multisegment synthesis of gold nanowires and nanorods, *Nanotechnol.* **20**(6) (2009).
83. Y. Y. Yu, S. S. Chang, C. L. Lee and C. R. C. Wang, Gold nanorods: Electrochemical synthesis and optical properties, *J. Phys. Chem. B* **101**(34), 6661–6664 (1997).
84. S. S. Chang, C. W. Shih, C. D. Chen, W. C. Lai and C. R. C. Wang, The shape transition of gold nanorods, *Langmuir* **15**(3), 701–709 (1999).

85. M. A. Hahn, A. K. Singh, P. Sharma, S. C. Brown and B. M. Moudgil, Nanoparticles as contrast agents for *in-vivo* bioimaging: Current status and future perspectives, *Anal. Bioanal. Chem.* **399**, 3–27 (2011).
86. N. J. Durr, T. Larson, D. K. Smith, B. A. Korgel, K. Sokolov and A. Ben-Yakar, A. Two-photon luminescence imaging of cancer cells using molecularly targeted gold nanorods, *Nano Lett.* 7(4), 941–945 (2007).
87. J.-L. Li and M. Gu, Surface plasmonic gold nanorods for enhanced two-photon microscopic imaging and apoptosis induction of cancer cells, *Biomaterials* **31**, 9492–9498 (2010).
88. H. F. Wang, T. B. Huff, D. A. Zweifel, W. He, P. S Low, A. Wei and J. X. Cheng, *In vitro* and *in vivo* two-photon luminescence imaging of single gold nanorods, *Proc. Natl Acad. Sci. USA* **102**(44), 15752–15756 (2005).
89. A. F. Fercher, W. Drexler, C. K. Hitzenberger and T. Lasser, Optical coherence tomography — principles and applications, *Report Prog. Phys.* **66**(2), 239–303 (2003).
90. A. M. Gobin, M. H. Lee, N. J. Halas, W. D. James, R. A. Drezek and J. L. West, Near-infrared resonant nanoshells for combined optical imaging and photothermal cancer therapy, *Nano Lett.* 7(7), 1929–1934 (2007).
91. J. C. Y. Kah, M. Olivo, T. H. Chow, K. S. Song, K. Z. Y. Koh, S. Mhaisalkar and C. J. R. Sheppard, Control of optical contrast using gold nanoshells for optical coherence tomography imaging of mouse xenograft tumor model *in vivo*, *J. Biomed. Opt.* **14**(5) (2009).
92. S. W. Perry, R. M. Burke and E. B. Brown, Two-photon and second harmonic microscopy in clinical and translational cancer research, *Ann. Biomed. Eng.* **40**(2), 277–291 (2012).
93. X. Yang, E. W. Stein, S. Ashkenazi and L. V. Wang, Nanoparticles for photoacoustic imaging, *Wiley Interdiscip. Rev. Nanomed. Nanobiotechnol.* **1**, 360–368 (2009).
94. S. Mallidi, T. Larson, J. Tam, P. P. Joshi, A. Karpiouk, K. Sokolov and S. Emelianov, Multiwavelength photoacoustic imaging and plasmon resonance coupling of gold nanoparticles for selective detection of cancer, *Nano Lett.* **9**(8), 2825–2831 (2009).
95. Y. W. Wang, X. Y. Xie, X. D. Wang, G. Ku, K. L. Gill, D. P. O'Neal, G. Stoica and L. V. Wang, Photoacoustic tomography of a nanoshell contrast agent in the *in vivo* rat brain, *Nano Lett.* **4**(9), 1689–1692 (2004).
96. X. M. Yang, S. E. Skrabalak, Z. Y. Li, Y. N. Xia and L. H. V. Wang, Photoacoustic tomography of a rat cerebral cortex *in vivo* with au nanocages as an optical contrast agent, *Nano Lett.* 7(12), 3798–3802 (2007).
97. C. Kim, E. C. Cho, J. Chen, K. H. Song, L. Au, C. Favazza, Q. Zhang, C. M., Cbloey, F. Gao, Y. Xia and L. V. Wang, *In Vivo* molecular photoacoustic tomography of melanomas targeted by bioconjugated gold nanocages, *ACS Nano.* **4**(8), 4559–4564 (2010).
98. L. Jabr-Milane, L. van-Vlerken, H. Devalapally, D. Shenoy, S. Komareddy, M. Bhavsar and M. Amiji, Multi-functional nanocarriers for targeted delivery of drugs and genes, *J. Control. Release* **130**(2), 121–128 (2008).
99. J. Heath and M. Davis, Nanotechnology and cancer, *Annu. Rev. Med.* **59**, 251–265 (2008).
100. K. Cho, X. Wang, S. Nie, Z. Chen and D. Shin, Therapeutic nanoparticles for drug delivery in cancer, *Clin. Cancer Res.* **14**(5), 1310–1316 (2008).

101. A. Smith, H. Duan, A. Mohs and S. Nie, Bioconjugated quantum dots for *in vivo* molecular and cellular imaging, *Adv. Drug Deliv. Rev.* **60**(11), 1226–1240 (2008).
102. P. Ghosh, G. Han, M. De, C. Kyu Kim and V. M. Rotello, Gold nanoparticles in delivery applications, *Adv. Drug Deliv. Rev.* **60**, 1307–1315 (2008).
103. P. Chinnaiyan, S. Varambally, S. A. Tomlins, S. Ray, S. M. Huang, A. M. Chinnaiyan and P. M. Harari, Enhancing the antitumor activity of ErbB blockade with histone deacetylase (HDAC) inhibition, *Int. J. Cancer* **118**(4), 1041–1050 (2006).
104. H. C. Huang, S. Barua, G. Sharma, S. K. Dey and K. Rege, Inorganic nanoparticles for cancer imaging and therapy, *J. Control. Release* **155**(3), 344–357 (2011).
105. W. Lu, G. D. Zhang, R. Zhang, L. G. Flores, Q. Huang, J. G. Gelovani, C. Li, Tumor site-specific silencing of NF-kappa B p65 by targeted hollow gold nanosphere-mediated photothermal transfection, *Cancer Res.* **70**(8), 3177–3188 (2010).
106. J. You, G. D. Zhang and C. Li, Exceptionally high payload of doxorubicin in hollow gold nanospheres for near-infrared light-triggered drug release, *ACS Nano.* **4**(2), 1033–1041 (2010).
107. W. T. Wu, T. Zhou, A. Berliner, P. Banerjee and S. Q. Zhou, Smart core-shell hybrid nanogels with ag nanoparticle core for cancer cell imaging and gel shell for pH-regulated drug delivery, *Chem. Mater.* **22**(6), 1966–1976 (2010).
108. Y. Cheng, A. C. Samia, J. Li, M. E. Kenney, A. Resnick and C. Burda, Delivery and efficacy of a cancer drug as a function of the bond to the gold nanoparticle surface, *Langmuir* **26**(4), 2248–2255 (2010).
109. S. R. Sershen, S. L. Westcott, N. J. Halas and J. L. West, Temperature-sensitive polymer-nanoshell composites for photothermally modulated drug delivery, *J. Biomed. Mater. Res.* **51**(3), 293–298 (2000).
110. N. E. Hynes and H. A. Lane, ERBB receptors and cancer: The complexity of target inhibitors, *Nat. Rev. Cancer* **5**(341–354), (2005).
111. R. S. Herbst, E. S. Kim and P. M. Harari, IMC-C225, an anti-epidermal growth factor receptor monoclonal antibody, for treatment of head and neck cancer, *Expert Opin. Biol. Ther.* **1**, 719–732 (2001).
112. C. R. R. Patra, E. Wang, A. Katarya, J. S. Lau and S. Dutta, Targeted delivery of gemcitabine to pancreatic adenocarcinoma using cetuximab as a targeting agent, *Cancer Res.* **68**, 1970–1978 (2008).
113. G. F. Paciotti, L. Myer, D. Weinreich, D. Goia, N. Pavel, R. E. McLaughlin and L. Tamarkin, Colloidal gold: A novel nanoparticle vector for tumor directed drug delivery, *Drug Deliv.* **11**(3), 169–183 (2004).
114. C. R. Patra, R. D. M. Bhattacharya and P. Mukherjee, Applications of gold nanoparticles for targeted therapy in cancer, *J. Biomed. Nanotechnol.* **4**, 99–132 (2008).
115. X. Huang, S. Neretina and M. El-Sayed, Gold nanorods: From synthesis and properties to biological and biomedical applications, *Adv. Biomater.* **27**, 1–31 (2009).
116. A. C. Bonoiu, S. D. Mahajan, H. Ding, I. Roy, K.-T. Yong, R. Kumar, R. Hu, E. J. Bergey, S. A. Schwartz and P. N. Prasad, Nanotechnology approach for drug addiction therapy: Gene silencing using delivery of gold nanorod-siRNA nanoplex in dopaminergic neurons, *Proc. Natl. Acad. Sci. USA* **106**(14), 5546–5550 (2009).
117. K. K. Sandhu, C. M. McIntosh, J. M. Simard, S. W. Smith and V. M. Rotello, Gold nanoparticle-mediated transfection of mammalian cells, *Bioconj. Chem.* **13**(1), 3–6 (2002).

118. M. Thomas and A. M. Klibanov, Conjugation to gold nanoparticles enhances polyethylenimine's transfer of plasmid DNA into mammalian cells, *Proc. Natl. Acad. Sci. USA* **100**(16), 9138–9143 (2003).
119. M. M. Ow Sullivan, J. J. Green and T. M. Przybycien, Development of a novel gene delivery scaffold utilizing colloidal gold-polyethylenimine conjugates for DNA condensation, *Gene. Ther.* **10**(22), 1882–1890 (2003).
120. S. R. A. Mahajan, J. Reynolds, B. Nair, D. Sykes, A. Bonoiu, I. Roy, K. Yong, W. Law, E., Bergey, P. Prasad and S. Schwartz, Suppression of MMP-9 expression in brain microvascular endothelial cell (BMVEC) using gold nanorod (GNR)-siRNA nanoplex, *Immunol. Invest.* **41**(4), 337–355 (2012).
121. W. Zhang, J. Meng, Y. Ji, X. Li, H. Kong, X. Wu and H. Xu, Inhibiting metastasis of breast cancer cells *in vitro* using gold nanorod-siRNA delivery system, *Nanoscale* **3**(9), 3923–3932 (2011).
122. S. Ryou, S. Kim, H. Jang, J. Kim, J. Yeom, M. Eom, J. Bae, M. Han and K. Lee, Delivery of shRNA using gold nanoparticle-DNA oligonucleotide conjugated as a universal carrier, *Biochem. Biophys. Res. Commun.* **398**(3), 542–546 (2010).
123. A. Bonoiu, E. Bergey, H. Ding, R. Hu, R. Kumar, P. Prasad, S. Mahajan, K. Picchione, A. Bhattacharjee and T. Ignatowski, Gold nanorod-siRNA induces efficient *in vivo* gene silencing in the rat hippocampus, *Nanomedicine* **6**(4), 617–630 (2011).
124. R. Masood, I. Roy, S. Zu, C. Hochstim, K. Yong, W. Law, H. Ding, U. Sinha and P. Prasad, Gold nanorod-sphingosine kinase siRNA nanocomplexes: A novel therapeutic tool for potent radiosensitization of head and neck cancer, *Integr. Biol. (Camb)* **4**(2), 132–141 (2012).
125. S. Ryou, M. Park, J. Kim, C. Jeon, Y. CH, S. Han, S. Kim, Y. Lee, S. Kim, M. Han, J. Bae and K. Lee, Inhibition of xenograft tumor growth in mice by goldnanoparticle-assisted delivery of short hairpin RNAs against Mcl-1L. **2**, 89–94 (2011).
126. G. Cao, *Nanostructures & Nanomaterials: Synthesis, Properties & Applications.* 1st ed. London: Imperial College Press (2004).
127. M. B. Cortie, The weird world of nanoscale gold, *Gold Bull.* **37**, 12–19 (2004).
128. E. S. Glazer, C. H. Zhu, K. L. Massey, C. S. Thompson, W. D. Kaluarachchi, A. N. Hamir and S. A. Curley, Noninvasive radiofrequency field destruction of pancreatic adenocarcinoma xenografts treated with targeted gold nanoparticles, *Clin. Cancer Res.* **16**(23), 5712–5721 (2010).
129. R. D. Averitt, S. L. Westcott and N. J. Halas, Linear optical properties of gold nanoshells, *J. Opt. Soc. Am. B-Opt. Phys.* **16**(10), 1824–1832 (1999).
130. H. Maeda, J. Wu, T. Sawa, Y. Matsumura and K. Hori, Tumor vascular permeability and the EPR effect in macromolecular therapeutics: A review. (2000).
131. L. R. Hirsch, R. J. Stafford, J. A. Bankson, S. R. Sershen, B. Rivera, R. E. Price, J. D. Hazle, N. J. Halas and J. L. West, Nanoshell-mediated near-infrared thermal therapy of tumors under magnetic resonance guidance, *Proc. Natl Acad. Sci. USA* **100**(23), 13549–13554 (2003).
132. D. P. O'Neal, L. R. Hirsch, N. J. Halas, J. D. Payne and J. L. West, Photo-thermal tumor ablation in mice using near infrared-absorbing nanoparticles, *Cancer Lett.* **209**(2), 171–176 (2004).
133. Pilot Study of AuroLase(tm) Therapy in Refractory and/or Recurrent Tumors of the Head and Neck. October (2012).

134. E. B. Dickerson, E. C. Dreaden, X. H. Huang, I. H. El-Sayed, H. H. Chu, S. Pushpanketh, J. F. McDonald and M. A. El-Sayed, Gold nanorod assisted near-infrared plasmonic photothermal therapy (PPTT) of squamous cell carcinoma in mice, *Cancer Lett.* **269**(1), 57–66 (2008).
135. G. P. Goodrich, L. L. Bao, K. Gill-Sharp, K. L. Sang, J. Wang and J. D. Payne, Photothermal therapy in a murine colon cancer model using near-infrared absorbing gold nanorods, *J. Biomed. Opt.* **15**(1).
136. P. Cherukuri, E. S. Glazer and S. A. Curley, Targeted hyperthermia using metal nanoparticles, *Adv. Drug Deliv. Rev.* **62**(3), 339–345 (2010).
137. B. Hildebrandt, P. Wust, O. Ahlers, A. Dieing, G. Sreenivasa, T. Kerner, R. Felix and H. Riess, The cellular and molecular basis of hyperthermia, *Crit. Rev. Oncol./Hematol.* **43**, 33–56 (2002).
138. K. S. Martirosyan, Thermosensitive magnetic nanoparticles for self-controlled hyperthermia cancer treatment, *J. Nanomed. Nanotechnol.* **3**, 6 (2012).
139. H. Kampinga, Cell biological effects of hyperthermia alone or combined with radiation or drugs: A short introductionto newcomers in the field, *Int. J. Hypertherm.* **22**(3), 191–196 (2006).
140. B. Dayanc, S. Beachy, J. Ostberg and E. Repasky, Dissecting the role of hyperthermia in natural killer cell mediated anti-tumor responses, *Int. J. Hypertherm.* **24**(1), 41–56 (2008).
141. J. You, R. Zhang, C. Xong, M. Zhong, M. Melancon, S. Gupta, A. M. Nick, A. K. Sood and C. Li, Effective photothermal chemotherapy using doxorubicin-loaded gold nanospheres that target EphB4 receptors in tumors, *Cancer Res.* **72**(18), 4777–4786 (2012).
142. J. You, R. Zhang, G. Zhang, M. Zhong, Y. Liu, C. S. Van Pelt, D. Liang, W. Wei, A. K. Sood and C. Li, Photothermal-chemotherapy with doxorubicin -loaded hollow gold nanospheres: A platform for near-infrared light-triggered drug release, *J. Control. Release* **158**, 319–328 (2012).
143. C. Wu, C. Yu and M. Chu, A gold nanoshell with a silica inner chell synthesized using liposome templates for doxorubicin loading and near-infrared photothermal therapy, *Int. J. Nanomed.* **6**, 807–813 (2011).
144. S.-M. Lee, H. Park, J.-W. Choi, Y. N. Park, C.-O. Yun, K.-H. Yoo, Multifunctional nanoparticles for targeted chemophotothermal treatment of cancer cells, *Angew. Chem. Int. Ed.* **50**, 7581–7586 (2011).
145. H. Park, J. Yang, J. Lee, S. Haam, I.-H. Choi and K.-H. Yoo, Multifunctional nanoparticles for combined doxorubicin and photothermal treatments, *ACS Nano* **3**(10), 2919–2926 (2009).
146. H. Liu, D. Chen, L. Li, T. Liu, L. Tan, X. Wu and F. Tang, Multifunctional gold nanoshells on silica nanorattles: A platform for combination of photothermal therapy and chemotherapy with low systemis toxicity, *Angew. Chem. Int. Ed.* **50**, 891–895 (2011).
147. H.-C. Huang, Y. Yang, A. Nanda, P. Koria and K. Rege, Synergistic administration of photothermal therapy and chemotherapy to cancer cells using polypeptide-based degradable plasmonic matrices, *Nanomedicine (Lond).* **6**(3), 459–473 (2011).
148. J. H. Park, G. von Maltzahn, L. L. Ong, A. Centrone, T. A. Hatton, E. Ruoslahti, S. N. Bhatia and M. J. Sailor, Cooperative nanoparticles for tumor detection and photothermally triggered drug delivery, *Adv. Mater.* **22**(8), 880 (2010).

149. S. K. Libutti, G. F. Paciotti, A. A. Byrnes, H. R. Alexander, S. A Gannon, M. Walker, G. D. Seidel, N. Yuldasheva and L. Tamarkin, Phase I and pharmacokinetic studies of CYT-6091, a novel PEGylated colloidal gold-rhTNF nanomedicine, *Clin. Cancer Res.* **16**(24), 6139–6149 (2010).
150. Y. Gu, W. Chen, M. Xia, S. Jeong and H. Liu, Effect of photothermal therapy on breast tumor vascular contents: noninvasive monitoring by near-infrared spectroscopy, *Photochem. Photobiol.* **81**(4), 1002–1009 (2005).
151. C. Rosenberg, R. Puls, K. Hegenscheid, J. Kuehn, T. Bollman, A. Westerholt, C. Weigel and N. Hosten, Laser ablation of metastatic lesions of the lung: long-term outcome, *AJR Am. J. Roentgenol.* **192**(3), 785–792 (2009).
152. R. R. Arvizo, S. Bhattacharyya, R. A. Kudgus, K. Giri, R. Bhattacharya and P. Mukherjee, Intrinsic therapeutic applications of noble metal nanoparticles: Past, present and future, *Chem. Soc. Rev.* **41**, 2943–2970 (2012).
153. S. Singh, V. D'Britto, A. A. Prabhune, C. V. Ramana, A. Dhawan and B. L. V. Prasad, Cytotoxic and genotoxic assessment of glycolipid-reduced and -capped gold and silver nanoparticles, *New J. Chem.* **34**, 294–301 (2010).
154. G. A. Sotiriou and S. E. Pratsinis, Engineering nanosilver as an antibacterial, biosensor and bioimaging material, *Opin. Chem. Eng.* **1**, 3–10 (2011).
155. G. A. Sotiriou, T. Sannomiya, A. Teleki, F. Krumeich, J. Voros and S. E. Pratsinis, Non-toxic dry-coated nanosilver for plasmonic biosensors, *Adv. Funct. Mater.* **20**, 4250–4257 (2010).
156. G. A. Sotiriou, A. M. Hirt, P. Y. Lozach, A. Teleki, F. Krumeich and S. E. Pratsinis, Hybrid, silica-coated, Janus-like plasmonic-magnetic nanoparticles, *Chem. Mater.* **23,** 1985–1992 (2011).
157. C. You, C. Han, Z. Wang, Y. Zheng, Q. Li, X. Hu and H. Sun, The progress of silver nanoparticles in the antibacterial mechanism, clinical application and cytotoxicity, *Mol. Biol. Rep.* **39**, 9193–9201 (2012).
158. S. Silver, L. T. Phung and G. Silver, Silver as biocides in burn and wound dressing and bacterial resistance to silver compounds, *J. Ind. Microbiol. Biotechnol.* **33**, 627–634 (2006).
159. R. R. Butorac, S. S. Al-Deyab and A. H. Cowley, Antimicrobial properties of some bis(iminoacenaphthene) (BIAN)-supported N-heterocyclic carbene complexes of silver and gold, *Molecules* **16**, 2285–2292 (2011).
160. L. Esteban-Tejada, F. Malpartida, A. Esteban-Cubillo, C. Pecharroman and J. S. Moya, The antibacterial and antifungal activity of a soda-lime glass containing silver nanoparticles, *Nanotechnology* **20**, (2009).
161. A. Panacek, M. Kolar, R. Vacerova, R. Prucek, J. Soukupova, V. Krystof, P. Hamal, R. Zboril and L. Kvitek, Antifungal activity of silver nanoparticles against, *Candida. Biomaterials* **30**, 6333–6340 (2009).
162. Wei, L, J. Tang, Z. Zhang, Y.Chen, G. Zhou and T. Xi, Investigation of the cytotoxicity mechanism of silver nanoparticles *in vitro*, *Biomed. Mater.* **5**, 044103 (2010).
163. B. A. Vanwinkle, K. L. D. Bently, J. M. Malecki, K. K. Gunter, I. M. Evans, A. Elder, J. N. Finkelstein, G. Oberdorster and T. E. Gunter, Nanoparticle (NP) uptake by type I alveolar epithelial cells and their oxidant stress response, *Nanotoxicology* **3**, 307–318 (2009).

164. Y. H. Hsin, C. F. Chen, S. Huang, T. S. Shih, P. S. Lai and P. J. Chueh, The apoptotic effect of nanosilver is mediated by a ROS- and JNK-dependent mechanism involving the mitochondrial pathway in NIH3T3 cells, *Toxicol. Lett.* (179), 130–139 (2008).
165. J. Boonstra and J. A. Post, Molecular events associated with reactive oxygen species and cell cycle progression in mammalian cells, *Gene.* **337**, 1–13 (2004).
166. J. Gao, G. Liang, B. Zhang, Y. Kuang, X. Zhang and B. Xu, FePt yolk-shell nanocrystals as a potent agent to kill HeLa cells, *J. Am. Chem. Soc.* **129**, 3403–3411 (2007).
167. C.-H. Su, H.-S. Sheu, C.-Y. Lin, C.-C. Huang, Y.-W. Lo, Y.-C. Pu, J.-C. Wneg, D.-B. Shieh, J.-H. Chen, and C.-S. Yeh, Nanoshell magnetic resonance imaging contrast agents, *J. Am. Chem. Soc.* **129**(7), 2139–2146 (2007).
168. M. Zhou, R. Zhang, M. Huang, W. Lu, S. Song, M. P. Melancon, M. Tian, D. Liang and C. Li, A chelator-free multifunctional [64Cu]CuS nanoparticle platform for simultaneous micro-PET/CT imaging and photothermal ablation therapy, *J. Am. Chem. Soc.* **132,** 15351–15358 (2010).
169. Y. B. Li, W. Lu, M. Huang, W. Chen and C. Li, CuS nanoparticles for photothermal ablation of tumor cells, *Nanomedicine* **5,** 1161–1171 (2010).
170. J. Lee, T. Isobe and M. Senna, Preparation of ultrafine Fe_3O_4 particles by precipitation in the presence of PVA at high pH, *J. Colloid Interface Sci.* **177**(2), 490–494 (1996).
171. A. K. Gupta and M. Gupta, Synthesis and surface engineering of iron oxide nanoparticles for biomedical applications, *Biomaterials* **26**(18), 3995–4021 (2005).
172. P. Reimer and T. Balzer, Ferucarbotran (Resovist): A new clinically approved RES-specific contrast agent for contrast-enhanced MRI of the liver: Properties, clinical development, and applications, *Eur. Radiol.* **13**(6), 1266–1276 (2003).
173. C. Gruttner and J. Teller, New types of silica-fortified magnetic nanoparticles as tools for molecular biology applications, *J. Magn. Magn. Mater.* **194**, 8–15 (1999).
174. C. Berry and A. S. C. Curtis, Functionalization of magnetic nanoparticles for applications in biomedicine, *J. Phys. D.: Appl. Phys.* **36**, R198–R206 (2003).
175. R. Weisslender, D. D. Stark, B. L. Engelstad, B. R. Bacon, C. C., Compton, D. L. White, P. Jacobs and J. Lewis, Superparamagnetic iron oxide: Pharmacokinetics and toxicity, *AJR* **152**, 167–173 (1989).
176. M. G. Marisinghani and R. Weissleder, Sensitive, noninvasive detection of lymph node metastases, *PLoS Med.* **1**(3), e66 (2004).
177. O. Clement, R. Guimaraes, E. de Kerviler and G. Frija, Magnetic resonance lymphography. Enhancement patterns using superparamagnetic nanoparticles, *Invest. Radiol.* **2**(Suppl. 2), S226–S228 (1994).
178. M. F. Bellin, L. Lebleu and J. B. Meric, Evaluation of retroperitoneal and pelvic lymph node metastases with MRI and MR lymphangiography, *Abdom. Imaging* **28**(2), 155–163 (2003).
179. M. G. Harisinghani, J. Barentsz, P. F. Hahn, W. M. Deserno, S. Tabatabaei, C. H. van de Kaa, J. de la Rosette and R. Weissleder, Noninvasive detection of clinically occult lymph-node metastases in prostate cancer, *N. Engl. J. Med.* **348**(25), 2491–2499 (2003).
180. P. Reimer and B. Tomback, Hepatic MRI with SPIO: Detection and characterization of focal liver lesions, *Eur. Radiol.* **8**(7), 1198–1204 (1998).

181. C. Sun, O. Veiseh, J. Gunn, C. Fang, S. Hansen, D. Lee, R. Sze, R. G. Ellenbogen, J. Olson and M. Zhang, *In vivo* MRI detection of gliomas by chlorotoxin-conjugated superparamagnetic nanoprobes, *Small* **4**(3), 372–379 (2008).
182. A. Toma, E. Otsuji, Y. Kuriu, K. Okamoto, D. Ichikawa, A. Hagiwara, H. Ito, T. Nishimura and H. Yamagishi, Monoclonal antibody A7-superparamagnetic iron oxide as contrast agent of MR imaging of rectal carcinoma, *Br. J. Cancer* **93**(1), 131–136 (2005).
183. S. D. Weitman, R. H. Lark, L. R. Coney, D. W. Fort, V. Frasca, V. R. Zurawski jr. and B. A. Kamen, Distribution of the folate receptor GP38 in normal and malignant cell lines and tissues, *Cancer Res.* **52**(12), 3396–3401 (1992).
184. J. F. Ross, P. K. Chaudhuri and M. Ratnam, Differential regulation of folate receptor isoforms in normal and malignant tissues *in vivo* and in established lines. Physiologic and clinical implications, *Cancer* **73**(9), 2432–2443 (1994).
185. X. Montet, K. Montet-Abou, F. Reynolds, R. Weissleder and L. Josephson, Nanopartilce imaging of integrins on tumor cells, *Neoplasia* **8**(3), 214–222 (2006).
186. C. Zhang, M. Jugold, E. C. Woenne, T. Lammers, B. Morgenstern, M. M. Mueller, H. Zentgraf, M. Bock, M. Eisenhut, W. Semmler and F. Kiessling, Specific targeting of tumor angiogenesis by RGD-conjugated ultrasmall superparamagnetic iron oxide particles using a clinical 1.5-T magnetic resonance scanner, *Cancer Res.* **67**(4), 1555–1562 (2007).
187. J. Riegler, A. Liew, S. O. Hynes, D. Ortega, T. O'Brien, R. M. Day, T. Richards, F. Shariff, Q. A. Pankhurst and M. F. Lythgoe, Superparamagnetic iron oxide nanoparticle targeting of MSCs in vascular injury, *Biomaterials* **34**, 1987–1994 (2013).
188. M. Kumagai, T. K. Sarma, H. Cabral, S. Kaida, M. Sekino, N. Herlambang, K. Osada, M. R. Kano, N. Nishiyama and K. Kataoka, Enhanced *in vivo* magnetic resonance imaging of tumors by PEGylated ironoxide-gold core-shell nanoparticles with prolonged blood circulation properties, *Macromol. Rapid Commun.* **31**, 1521–1528 (2010).
189. X. Ji, R. P. Shao, A. M. Elliott, R. J. Stafford, E. Esparza-Coss, J. A. Bankson, G. Liang, Z. P. Luo, K. Park, J. T. Markert and C. Li, Bifunctional gold nanoshells with a superparamagnetic iron oxide-silica core suitable for both MR imaging and photothermal therapy, *J. Phys. Chem. C* **111**, 6245–6251 (2007).
190. M. P. Melancon, W. Lu, Z. Meng, M. Zhou, G. Liang, A. M. Elliott, J. D. Hazle, J. N. Myers, C. Li and R. J. Stafford, Targeted multi-functional gold-based nanoshells for MR-guided laser ablation of head and neck cancer, *Biomaterials* **32**, 7600–7608 (2011).
191. K. Overgaard and J. Overgaard, Investigations on the possibility of a thermic tumour therapy. I. Short-wave treatment of a transplanted isologous mouse mammary carcinoma, *Eur. J. Cancer* **8**(1), 65–78 (1972).
192. Q. A. Pankhurst, J. Connolly, S. K. Jones and J. Dobson, Applications of magnetic nanoparticles in biomedicine, *J. Phys. D: Appl. Phys.* **36**, R167–R181 (2003).
193. E. Duguet, S. Vasseur, S. Mornet and J. M. Devoisselle, Magnetic nanoaprticles and their applications in medicine, *Nanomedicine (Lond).* **1**(2), 157–168 (2006).
194. M. Johannsen, U. Gneveckow, K. Taymoorian, B. Thiesen, N. Waldofner, R. Scholz, K. Jung, A. Jordan, P. Wust and S. A. Loening, Morbidity and quality of life during termotherapy using magnetic nanoparticles in locally recurrent prostate cancer: Results of a prospective phase I trial, *Int. J. Hypertherm.* **23**(3), 315–323 (2007).

195. M. Mikhaylova, Y. S. Jo, D. K. Kim, N. Bobrysheva, Y. Anderson, T. Eriksson, M. Osmolowsky, V. Semenov and M. Muhammed, M. The effect of biocompatible coating layers on magnetic properties of superparamagnetic iron oxide nanoparticles, *Hyperfine Interact* **156–157**(4), 257–263 (2004).
196. B. Perlstein, S. Finniss, C. Miller, H. Okhrimenko, G. Kazimirsky, S. Cazacu, H. K. Lee, N. Lemke, S. Brodite, F. Umansky, R. Sandra, M. Rosemblum, T. Mikklesen, S. Margel and C. Brodie, TRAIL conjugated to nanoparticles exhibits increased anti-tumor activities in glioma cells and glioma stem cells *in vitro* and *in vivo*, *Nero-Oncol.* **15**(1), 29–40 (2013).
197. X. Chao, Z. Zhang, L. Guo, J. Zhu, M. Peng, A. J. M. Vermorken, W. J. M. Van de Ven, C. Chen and Y. Cui, A novel magnetic nanoparticle drug carrier for enhanced cancer chemotherapy, *PLos One* 7(10), e40388 (2012).
198. C. Plank, O. Zelphati and O. Mykhaylyk, O. Magnetically enhanced nucleic acid delivery. Ten years of magnetofection-Progress and prospects, *Adv. Drug. Deliv. Rev.* **63,** 1300–1331 (2011).
199. B. Chertok, A. E. David and V. C. Yang, Polyethyleneimine-modified iron oxide nanoparticles for brain tumor drugdelivery using magnetic targeting and intracarotid administration, *Biomaterials* **31**(24), 6317–6324 (2010).
200. Y. Chen, G. Lian, C. Liao, W. Wang, L. Zeng, C. Qian, K. Huang and X. Shuai, Characterization of polyethylene glycol-grafted polyethylenimine and superparamagnetic iron oxide nanoparticles (PEG-g-PEI-SPION) as an MRI-visible vector for siRNA delivery in gastric cancer in tiro and *in vivo*, *J. Gastroenterol.* DOI 10.1007/s00535-012-0713-x (2012).
201. Y. Namiki, T. Namiki, H. Yoshida, Y. Ishii, A. Tsubota, S. Koido, K. Nariai, M. Mitsunaga, S. Yanagisawa, H. Kashiwagi, Y. Mabashi, Y. Yumoto, S. Hoshina, K. Fujise and N. Tada, A novel magnetic crystal-lipid nanostructure for magnetically guided *in vivo* gene delivery, *Nat. Nanotechnol.* **4**, 598–606 (2009).
202. M. Kumar, M. Yigit, G. Dai, A. Moore and Z. Medarove, Image-guided breast tumor therapy using small interfering RNA nanodrug, *Cancer Res.* **70**, 7553–7561 (2010).
203. X. Cai, W. Li, C.-H. Kim, Y. Yuan, L. Wang and Y. Xia, *In vivo* quantitative evaluation of the transport kinetics of gold nanocages in a lymphatic system by noninvasive photoacoustic tomography, *ACS Nano.* **5**(12), 9658–9667 (2011).

Chapter 11

Silica-Based Nanoparticles for Biomedical Imaging and Drug Delivery Applications

Stephanie A. Kramer[†] *and Wenbin Lin*[‡,*]

[†]*Department of Chemistry, University of North Carolina at Chapel Hill, CB #3290, Chapel Hill, NC 27599*
[‡]*Department of Chemistry, University of Chicago, 929 E 57th St, Chicago, IL 60637*
**wenbinlin@uchicago.edu*

Abstract

Silica-based nanoparticles have well-defined and tunable structures and surface chemistry, making them suitable for nanomedicine applications. They can be designed to achieve long blood circulation times and functionalized to target specific sites in the body. In this chapter, the latest progress in designing silica-based nanoparticles for biomedical imaging and drug delivery is summarized. The synthesis of both solid silica and mesoporous silica nanoparticles is described, along with different approaches for functionalization. The use of silica-based nanoparticles in optical, magnetic resonance and multimodal imaging and as drug delivery agents is discussed. Additionally, this chapter highlights recent advances in the use of silica-based nanoparticles as theranostic agents.

1. Introduction

Nanoparticles, due to their unique chemical and physical properties, have received an increasing amount of attention for imaging and therapeutic applications over the past 20 years.[1–3] In particular, silica-based nanoparticles are a promising class of materials due to their well-defined and tunable structures (i.e. size, morphology, and porosity) and surface chemistry. To impart diagnostic and therapeutic capabilities, surface functionalization of silica nanoparticles can be achieved using well-established siloxane chemistry.[4] Additionally, silica nanoparticles are inexpensive, easily synthesized, relatively chemically inert, biocompatible and water dispersible. This chapter will focus on the development of silica-based nanoparticles for biomedical imaging and drug delivery applications.

2. Synthesis of Silica-Based Nanoparticles

There are two major types of silica-based nanoparticles, solid silica nanoparticles (SiNPs) and mesoporous silica nanoparticles (MSNs). Both possess many desirable properties. SiNPs are biocompatible and photochemically stable, and they have favorable colloidal properties. The surface of SiNPs can be modified with various functional groups, such as antibodies, aptamers, and polymers.[5] Functionalized SiNPs have extensively been utilized as optical imaging agents and drug delivery vehicles. To synthesize SiNPs, a sol-gel process is utilized to produce the nanoparticles of the desired size. The sol-gel process involves hydrolysis of a silicon alkoxide precursor followed by polycondensation of the monomers to yield silica nanoparticles. Either the Stöber method or the reverse microemulsion method can be used to synthesize SiNPs.[6,7] The Stöber method requires a shorter reaction time than the reverse microemulsion method and typically results in larger, non-uniform particles. In contrast, the reverse microemulsion method can readily yield monodisperse silica particles that are smaller than 100 nm.

In contrast to SiNPs, MSNs have many unique properties, such as high surface areas, tunable pore sizes, and large pore volumes. For MSNs, the imaging or therapeutic agent can either be loaded into the pores, directly incorporated into the silica matrix, or grafted onto the outer surface of the particle. MSNs are typically synthesized by a surfactant-templated approach.[8,9]

2.1. *Solid silica nanoparticles*

In 1968, Stöber *et al.* reported a method to synthesize monodisperse solid silica particles ranging in size from 50 nm to 2 μm.[6] Ever since Stöber published this system, many literature reports have surfaced evolving from his work. This synthesis involves the controlled hydrolysis and condensation of an alkyl silicate, such as tetraethyl orthosilicate (TEOS), under alkaline conditions in alcoholic solutions (Fig. 1). The reaction conditions can be tuned to control the size of the particles.[10] For instance, Wang and coworkers

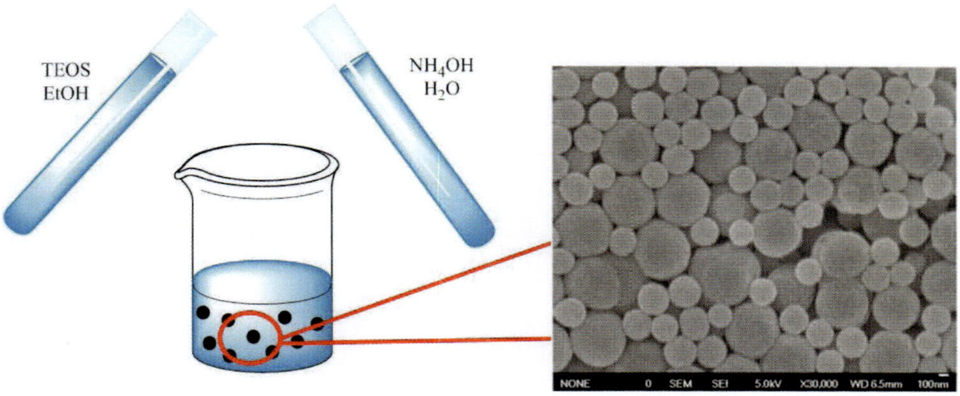

Fig. 1. The Stöber method, in which the hydrolysis and condensation of TEOS is facilitated by base in ethanol/water. TEM micrographs show silica particles with an average diameter of 403 nm.[10]

reported that by varying the TEOS concentration, silica nanoparticles ranging from 30 to 1000 nm can be synthesized. The negatively charged silica particles remain suspended in solution due to electrostatic repulsion.

Another method used to synthesize monodisperse SiNPs is the reverse phase, or water-in-oil, microemulsion method (Fig. 2).[7] This method involves small amounts of water, a large volume of organic solvent (oil) and a surfactant. Nanometer-sized water droplets form in the predominant oil phase and are stabilized by the surfactant. The water droplets serve as "nanoreactors," controlling the kinetics of particle nucleation and growth, for the synthesis of the SiNPs. The size of the water droplets affects the size of the silica particles. To carefully control the size and polydispersity of the nanoparticles, the water to surfactant ratio (W) can be adjusted. In general, the higher the value of W, the larger the silica particles will be.[11]

To functionalize SiNPs, polar hydrophilic functional molecules can be added to the microemulsion to incorporate the organic species within the silica matrix via electrostatic attraction of the functional molecules to the negatively charged silica. Bagwe *et al.* demonstrated this approach by entrapping luminescent dyes, such as $Ru(bpy)_3^{2+}$, within silica nanoparticles.[12] One disadvantage of this method is that the entrapped dye molecules can leach out of the nanoparticle under physiological conditions. An alternative approach to functionalize SiNPs is to covalently incorporate the imaging or therapeutic agents into the nanoparticles, ensuring protection from the environment.[5] The agent must contain trialkoxysilyl functional groups to allow for chemical incorporation within the silica matrix via silanol linkages. Using this approach, stable hybrid silica nanoparticles with homogeneous functionalization can be synthesized. Lastly, surface functionalization can be achieved

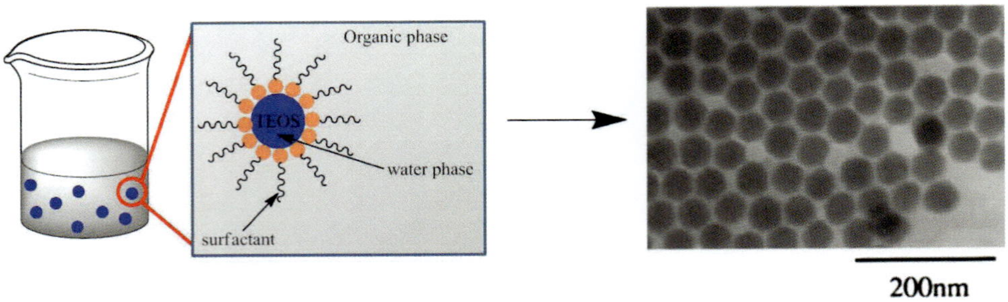

Fig. 2. The reverse microemulsion method, in which TEOS is hydrolyzed at the micelle interface and then enters the water droplet to form a silica nanoparticle within the micelle.[11]

by post-synthetic modification. The surface silanols can react with trialkoxysilane molecules to attach the desired functionalities to the surface. This approach is particularly useful for modifying the particle surface with selected agents that are not stable during the particle synthesis.

The synthetic versatility of silica nanoparticles allows for the incorporation of imaging and therapeutic agents in multistep sequences. After a silica nanoparticle is modified with different functional groups, it can be grafted with inert polymers or biological moieties (dextran, antibodies, etc.) using standard covalent bioconjugation strategies. For instance, an amine-modified particle can be treated with various carboxylate-containing molecules to form a stable amide bond. In addition to providing reaction sites for conjugation, the additional functionality can also change the colloidal stability of the silica particles in solution. For example, phosphonate groups on the surface improve the long term stability of the particle by increasing the repulsive forces between the nanoparticles in solution.[13] Additionally, silica nanoparticles with core–shell structures allow multiple functionalities to be incorporated within a single platform.[14,15]

2.2. *Mesoporous silica nanoparticles*

The first reported mesoporous material was MCM-41, which is composed of hexagonally ordered cylindrical mesopores. It is a micrometer-sized particle without a well-defined morphology.[16] Since then, improved procedures have been developed to control the size and shape of MCM-41 particles.[8,17] Different morphologies — including spheres, rods and twisted columns — and diameters ranging from 60 to 1,000 nm have been produced.

The synthesis of MCM-41 involves hydrolysis and condensation of silica precursors around ordered arrays of amphiphilic surfactants (Fig. 3).[18] In aqueous solution, the surfactant molecules self-assemble into micelles and

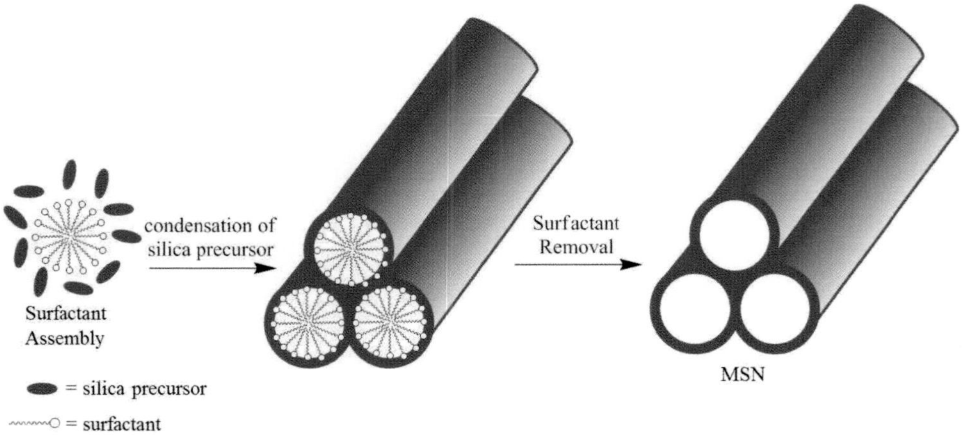

Fig. 3. Synthesis of MCM-41 type MSN particles, in which (1) the surfactant assembles into hexagonal arrays of micelles, (2) the silica precursors hydrolyze and condense at the micelle interface, and (3) the surfactant is extracted to produce a mesoporous material.

then into liquid crystal mesophases. The silica precursors concentrate around the mesophases through electrostatic interactions, and the nanoparticles are formed by base-catalyzed sol-gel condensation around the mesophase structures. The surfactant is subsequently removed by calcination or solvent extraction to produce the mesoporous silica.

MSNs can be functionalized by either covalent bonding or electrostatic interactions. In the case of covalent bonding, functional groups can be attached to MSNs using post-synthetic grafting or co-condensation. The functional groups can be introduced on external, internal or both surfaces of the MSNs. The most popular method to covalently functionalize MSNs is post-synthetic grafting. After synthesis of MSNs, the particle surfaces can be modified with organosilanes. Using this approach, the addition of functional groups can be achieved without altering the morphology or pore structure of the silica particles. However, most materials that are modified using this method contain an inhomogeneous coverage of organic functional groups. Using this approach, MSNs were functionalized with a targeting ligand for drug delivery.[19]

The other approach used to covalently functionalize MSNs is the co-condensation method. Victor Lin and coworkers reported a one-pot method to co-condense various organic functional groups into the pores of MSNs.[18,20] The desired trialkoxysilane is condensed into the pores of the nanoparticles during the particle synthesis, resulting in homogeneous incorporation of the functional group throughout the material. For example, Gd(III) chelates were co-condensed into MSNs to produce MRI contrast agents with high loadings of Gd(III).[21,22] By adjusting the reagent concentration, size and hydrophobicity or hydrophilicity of the co-condensing reagent, the degree of functionalization and the particle size can be controlled. Unfortunately, not all types of MSNs can be made using this approach; the triethoxysilane can destabilize the surfactant templates during the synthesis, resulting in large, rod-like particles (~1,000 nm).[20] The combination of both functionalization approaches has resulted in the synthesis of multifunctional platforms with a wide variety of applications in biomedicine.[17,23]

Electrostatic interactions can also be used to incorporate functional groups onto MSNs. This method makes use of the negative charges from free silanol groups on the particle surface. For instance, bare MSNs can be electrostatically coated with a cationic polymer, such as polyethyleneimine.[24] Surface coating with polymers or other functional groups can be used to retain cargo loaded within the silica pores.

3. Silica Nanoparticles for Biomedical Imaging

A number of imaging approaches, such as computed tomography (CT), magnetic resonance (MR), positron emission tomography (PET), single photon emission tomography (SPECT) and optical imaging, have provided non-invasive techniques to image and diagnose various diseases. However, each of these techniques still has issues. For instance, the ionizing radiation from PET/SPECT imaging agents and CT scanning can be harmful to patients. Nanoparticles provide a way to overcome some of the issues of conventional contrast agents. By incorporating a contrast agent into a nanomaterial, sufficiently long blood circulation time can be achieved. The particle shields the contrast agent from the environment, increasing its chemical stability under a range of physiological conditions. Additionally, the nanoparticle can be designed to have high selectivity for the target of interest (i.e. cell, antigen or tissue). Ideally, these nanoparticle-based contrast agents should be suitable for long-term quantitative imaging at low doses and should be safely cleared from the body after imaging is complete.

Silica nanoparticles can be loaded or functionalized with contrast agents to allow for bioimaging. Optical and MR imaging are the two most common imaging modalities that silica-based probes are used for. Optical imaging allows for low cost, rapid screening, while MR imaging provides high resolution images and the ability to obtain both anatomical and physiological information.

Optical imaging is a common research technique that utilizes photons emitted from bioluminescent or fluorescent probes. Advantages of optical imaging include low cost of instrumentation, high sensitivity, and avoidance of radiopharmaceuticals. Unfortunately, this imaging modality suffers from poor resolution. As radiation is scattered and absorbed by the body, the resolution is limited to 1–2 mm. Additionally, the use of fluorescent imaging results in significant noise due to tissue autofluorescence and light absorption by biological molecules. To reduce these issues, NIR fluorophores can be used to increase the penetration depth, reduce autofluorescence, and reduce tissue scattering.

Magnetic resonance imaging is based on the precession of water protons in an applied magnetic field. Radiofrequency pulses are applied, and then the nuclei return to their original state. This relaxation process can be used to produce an image. This modality is advantageous due to high spatial resolution (sub-millimeter scale) and unlimited tissue penetration. However, MRI has an overall low sensitivity in detecting molecular probes (10^{-3}–10^{-5} M), but this can be overcome with the use of contrast agents.[25] Most of the available MRI contrast agents are paramagnetic complexes that contain Mn(II), Mn(III) or Gd(III) chelates, with the most common being Gd-based agents.[26] The Gd-based agents lack sensitivity, requiring high concentrations to be used. When high doses of contrast agents are administered, the accumulation and toxicity become a concern. By incorporating the Gd-based agent into a nanoparticle, a high payload of contrast agent can be delivered, which increases the effective relaxivity per nanoparticle. The increased size results in reduced tumbling rates, which increases the relaxivity on a per metal basis.

3.1. *Solid silica nanoparticles for imaging applications*

Solid silica nanoparticles are an ideal platform for the development of contrast agent-doped particles, such as those used in MR and optical imaging applications. Dye-doped SiNPs were first synthesized by Blaaderen using the Stöber method to conjugate fluorescein isothiocyanate (FITC) with (3-aminopropyl)trimethoxysilane.[27] However, these particles were polydisperse and large (> 400 nm in diameter). In 2004, Tan *et al.* reported the synthesis of FITC-doped silica nanoparticles, which were 70 nm in diameter, using the reverse microemulsion method.[13] The SiNPs were conjugated with a cellular penetrating peptide (TAT) and used to label human lung cancer cells *in vitro*. To demonstrate the bioimaging capability of this platform *in vivo*, rat brain blood vessels were selectively labeled via intra-arterial administration. By using the TAT peptide, the authors were able to deliver diagnostic agents to the brain without compromising the blood–brain barrier (BBB).

Luminescent metal complexes, such as $Ru(bpy)_3^{2+}$, have also been incorporated into SiNPs for biomedical imaging. Not only is $Ru(bpy)_3^{2+}$ more stable than organic fluorophores such as fluorescein, it has a large Stokes shift, which significantly reduces the background signal from biological systems. In 2001, Tan and coworkers synthesized monodisperse, spherical $Ru(bpy)_3^{2+}$-doped SiNPs that were 63 nm in diameter using the reverse microemulsion method.[28] The doped SiNPs demonstrated several advantages over small-molecule dyes; they demonstrated enhanced signals, increased photostability, homogenous fluorophore distribution, and protection from the external environment. The SiNPs were conjugated with an antibody specific to leukemia cells, allowing for efficient, easy identification of leukemia cells *in vitro*. The reaction conditions were further optimized, by changing reagent types and concentrations, to study the effects on the fluorescence spectra, particle size, and size distribution of the SiNPs.[12] The *in vivo* biodistribution and urinary excretion of this platform was investigated.[29]

Weisner and coworkers further built upon the use of SiNPs in imaging applications by developing highly fluorescent core-shell SiNPs referred to as Cornell dots (C dots).[30] These 20–30 nm particles contain a fluorophore-rich core and a silica-based shell. The particles were ~20 times brighter than the constituent dye. The toxicity and biodistribution of C dots was carried out, and no apparent toxicity in mice was observed.[31] Surface modification of C dots with polymers or targeting ligands increases the particle's success in biological applications. For example, C dots that were electrostatically coated with polyethyleneimine showed enhanced intracellular delivery compared to unmodified versions.[32] HeLa cells internalized twice the amount of coated particles relative to the uncoated C dots.

Near-infrared (NIR) dyes were also incorporated into C dots. The synthesis of C dots containing Cy5 ($\lambda_{em} > 650$ nm) was recently reported.[33] These nanoparticles had hydrodynamic diameters of 3.3 and 6.0 nm and improved photophysical characteristics compared to the parent dye. Biodistribution studies in mice indicated that these small particles were excreted through the kidney; however, significant accumulation in the liver was observed. Surface modification with poly(ethylene glycol) (PEG) showed increased renal excretion and decreased liver uptake (Fig. 4). The 6.0 nm particles had a longer blood circulation time, longer urination half-life, and greater organ retention. The particles were completely cleared after 48 hours. The next generations of C dots (7 nm in diameter) were recently synthesized to afford nanoparticles with greater tumor selectivity, less toxicity, enhanced targeting, and favorable clearance profiles.[34] The particles were functionalized with RGD peptide, a targeting agent for the $\alpha_v\beta_3$ integrin overexpressed in a variety of malignant

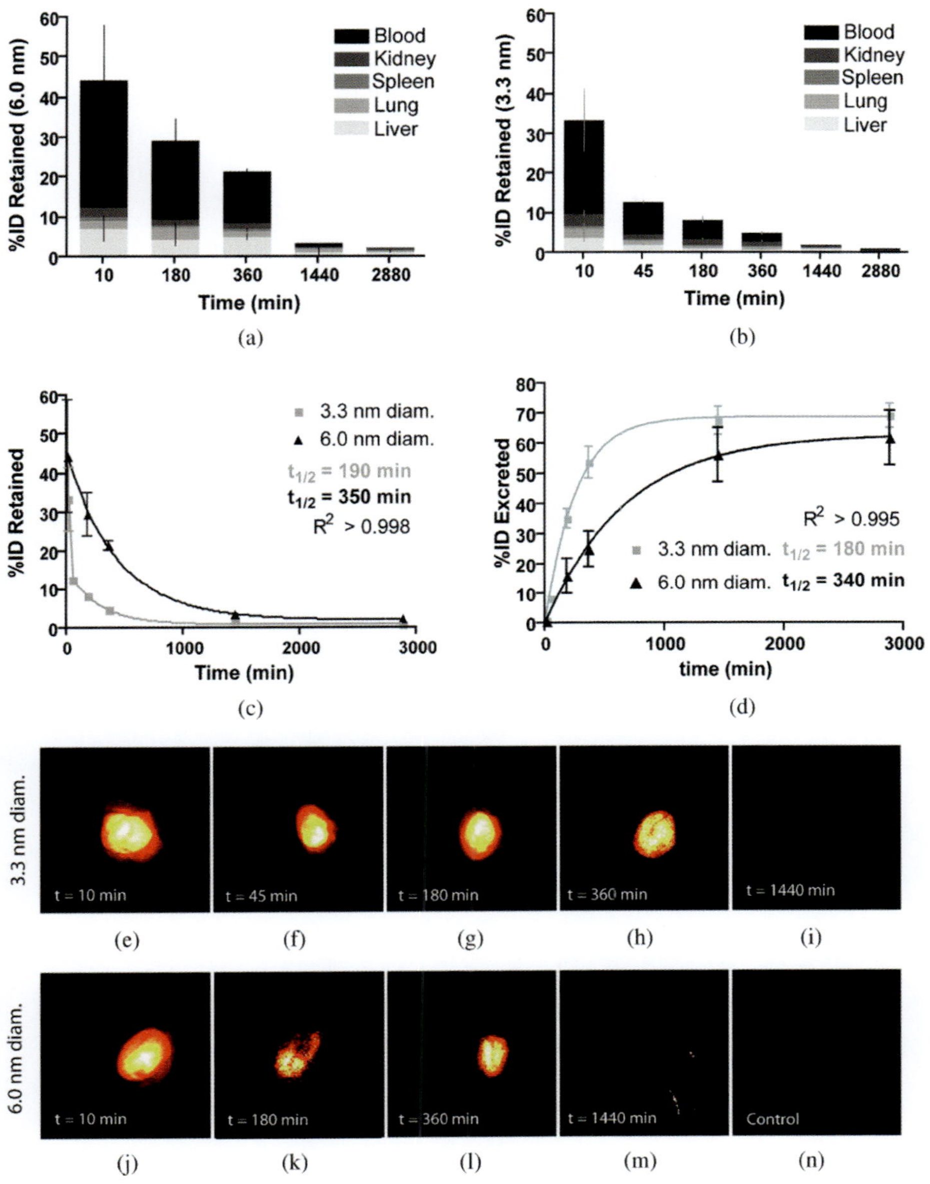

Fig. 4. Biodistribution analysis of C dots in mice (top) and fluorescence pseudo-color imaging of a mouse bladder *ex vivo* (bottom). Percent of initial particle dose (%ID) retained for 6.0 nm (a) and 3.3 nm (b) C dots. (c) Retained particle concentration for 3.3 nm (light gray) and 6.0 nm (black) C dots. (d) Estimated particle excretion for 3.3 nm (light gray) and 6.0 nm (black) C dots. (e)–(i) Pseudo-color images of Cy5 fluorescence in intact mouse bladders showing the accumulation of 3.3 nm diameter dots over the first 6 h post-injection (e–h), followed by the negligible particle fluorescence seen at 24 h post-injection (i). (j–m) Pseudo-color images of Cy5 fluorescence in intact mouse bladders showing the accumulation of 6.0 nm diameter dots over the first 6 h post-injection (j–l) and at the 24 h end point (m). Reprinted from Ref. 33, with permission from the American Chemical Society.

epithelial cell lines, and labeled with the radio-nuclide ^{124}I. As shown in Fig. 5, the modified C dots demonstrated enhanced tumor-selective accumulation in $\alpha_v\beta_3$ integrin-expressing melanoma xenografts in mice. Additionally, this platform showed high sensitivity, real-time detection and imaging of lymphatic drainage patterns and differential tumor burden in a large animal model (miniswine). The FDA has recently allowed this multimodal optical-PET imaging system to be tested as a cancer-specific optical probe in a first-in-human trial.[34]

Another approach to develop silica-based optical imaging contrast agents was reported by Prasad and coworkers.[35] Multifunctional organically modified silica (ORMOSIL) nanoparticles that were 20–25 nm in diameter were synthesized using a modified Stöber method. The particles contained

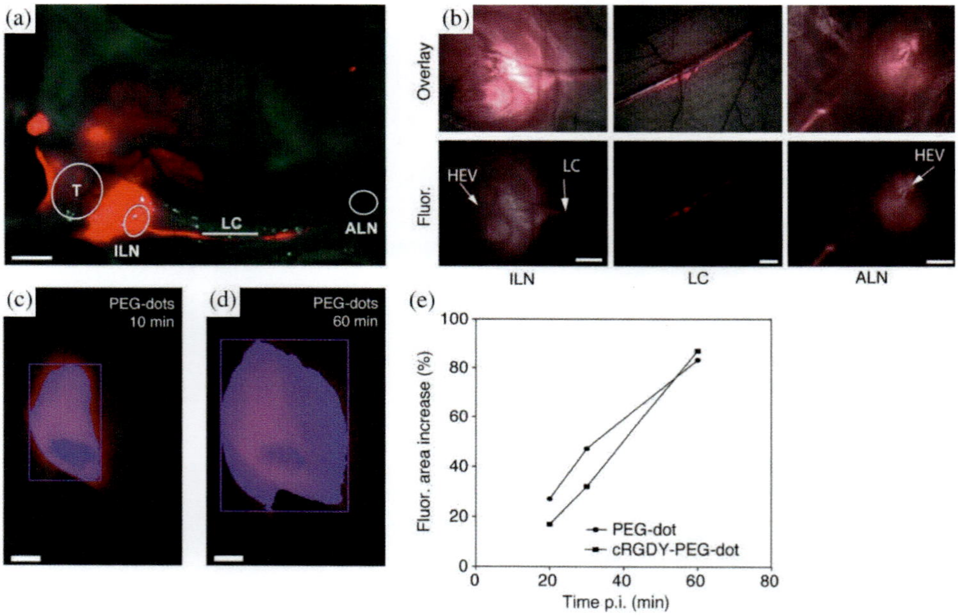

Fig. 5. Nodal mapping using multiscale NIR optical fluorescence imaging. (a) Whole-body fluorescence imaging of the tumor site (T), draining inguinal (ILN) and axillary (ALN) nodes, and communicating lymphatics channels (LCs) 1-hour post-injection in a surgically exposed living animal. (b) Corresponding coregistered white-light and high-resolution fluorescence images (top row) and fluorescence images only (bottom row), revealing nodal infrastructure of local and distant nodes, including high endothelial venules (HEVs). (c) Whole-body fluorescence image of the tumor site 10 minutes after subdermal PEG-dot injection. (d) Delayed whole-body fluorescence image of the tumor site 1 hour after PEG-dot injection. (e) Percentage increase in the area of fluorescence relative to that measured at 10-minutes post-injection for targeted and nontargeted probes. Scale bars: 1.0 cm (a); 500 microns (b); 3 mm (c) and (d). Reprinted from Ref. 34, with permission from the American Chemical Society.

the NIR fluorophore DY776 and a radio-nuclide ^{124}I for use in optical and PET imaging *in vivo*. The biodistribution of non-targeted particles was study in non-tumored mice using optical and PET imaging. The results from these studies indicated a greater accumulation of nanoparticles in the liver, spleen and stomach than in the kidneys, heart, and lungs. A clearance study was carried out over a 15-day period, and it showed hepatobiliary excretion of the nanoparticles. No apparent toxicity in the mice was observed.

Solid silica nanoparticles have also been used as a platform in developing improved contrast for magnetic resonance (MR) imaging. Nanoparticle agents provide several advantages over small molecule agents, including reduced tumbling rates and increased relaxivity. In addition, each nanoparticle is able to deliver a high payload of paramagnetic materials. Several groups have developed multimodal imaging probes that combine both optical and MR contrast enhancing ability, allowing for two modes of imaging within a single platform.[15, 36–41]

The Lin group prepared 37-nm SiNPs containing an embedded $[Ru(bpy)_3]Cl_2$ luminophore and a Gd-chelate shell for use in optical and MR imaging.[36] The surface-immobilized Gd chelates effectively relaxed water molecules, and the resulting particles possessed relaxivities four orders of magnitude higher than the constituent Gd complex, presumably due to enhanced per Gd relaxivity as a result of reduced tumbling rates. Interestingly, when the Gd chelates were immobilized on the surface of the SiNPs in a multilayer fashion, reduced relaxivities on a per Gd basis were observed. This was likely due to reduced water accessibility of the Gd chelates. Murine monocyte cells were successfully labeled with the hybrid SiNPs and T_1- and T_2-weighted signal enhancement of the cell pellet was observed (Fig. 6).

Using murine monocyte cells, the efficiency of these SiNPs as a multimodal *in vitro* imaging agent was tested.[39] The monocytes were preferentially labeled with the hybrid SiNPs, as indicated by flow cytometry and phantom MR image studies (Fig. 7). Using a collagen-induced arthritis (CIA) mouse model, optical and MR imaging demonstrated that this platform can be used to specifically target inflammation by selective trafficking of labeled monocyte cells. Arthritic mouse paws were labeled with the SiNPs, and the level of luminescence correlated to the degree of disease activity, suggesting that this platform could be utilized as a sensitive, selective method to monitor arthritis in patients. MR imaging showed a 33–35% reduction in the T_2 signal of mice treated with the nanoparticle agent; however, no T_1 enhancement was observed (Fig. 8). In addition, tissue histopathological analysis of arthritic

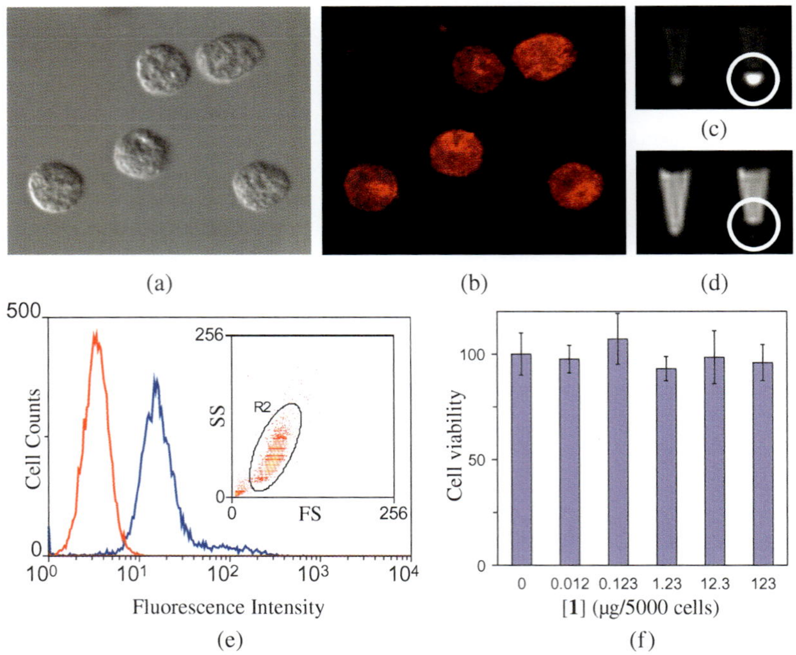

Fig. 6. Microscopic images of labeled monocyte cells: (a) optical; (b) laser scanning confocal fluorescence. (c) and (d) MR images of unlabeled (left) and labeled (right) monocyte cells: (c) T_1-weighted and (d) T_2-weighted. (e) Flow cytometric results for the unlabeled (red) and labeled (blue) monocyte cells indicating greater than 98% labeling efficiency (the inset shows the purity of the labeled cells; SS = side scatter, FS = forward scatter). (f) MTS cell viability assay of the monocyte cells incubated with different amounts of nanoparticle. Reprinted from Ref. 36, with permission from John Wiley & Sons Inc.

paws was conducted using light and fluorescent microscopy, confirming that the SiNPs were located within the synovial tissue where the disease was prevalent.

Additionally, the Lin group has developed a polyelectrolyte layer-by-layer (LbL) self-assembly approach to evade the reduced relaxivities found in the SiNPs with multilayer coatings.[38] Highly anionic SiNPs, containing a $Ru(bpy)_3^{2+}$ core and a Gd-DTTA (DTTA = diethylenetriamine tetraacetate) monolayer, were coated with positively charged Gd chelate oligomers via electrostatic interactions. The resulting positively charged particle was further treated with negatively charged polystyrene sulfonate to provide a net negatively charged layer. These steps were repeated to produce the LbL multilayer architecture (Fig. 9). The formation of LbL SiNPs was confirmed using TEM and fluorescence studies. MR phantom studies showed that nanoparticles

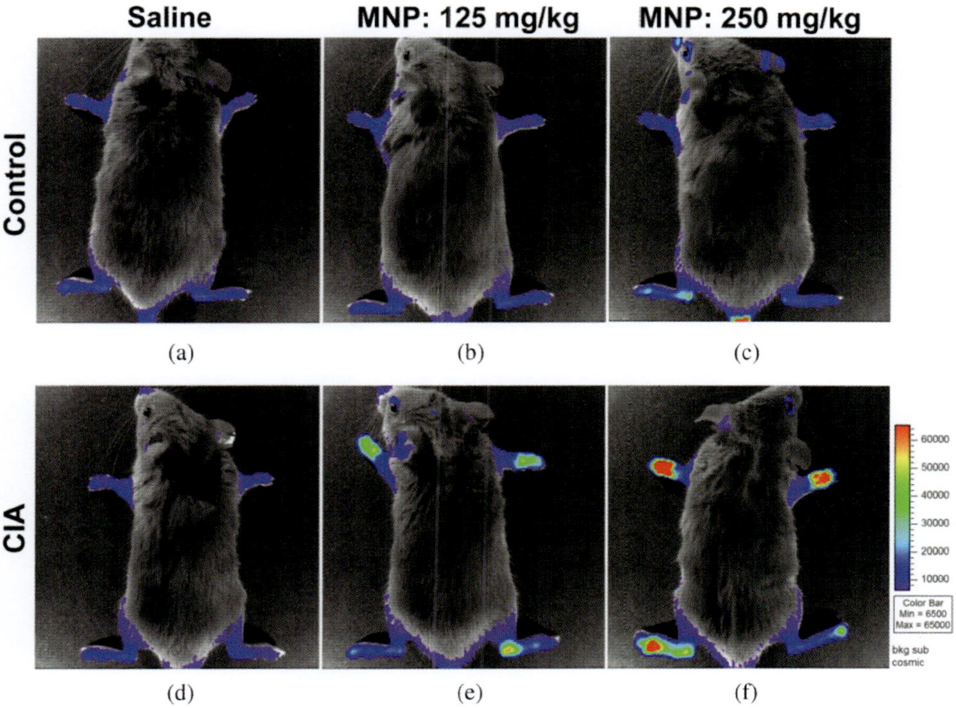

Fig. 7. (Top) Control animals (without arthritis) that were intravenously injected with two separate doses of (a) saline, (b) 125 mg SiNPs per kg, or (c) 250 mg SiNPs per kg 12 h before imaging. (Bottom) CIA animals with arthritis that were intravenously injected with (d) saline, (e) 125 mg SiNPs per kg, or (f) 250 mg SiNPs per kg 12 h before imaging. Reprinted from Ref. 39, with permission from *Clinical and Experimental Rheumatology*.

with more layers of the positively-charged Gd chelate (Gd-DOTA, DOTA = cyclen-1,4,7,10-tetraacetate) had higher relaxivities on a per particle basis. The relaxivities on a per Gd basis remained constant, regardless of the number of layers deposited. Thus, the LbL assembly method increased the Gd payload without affecting the MR relaxivity on a per Gd basis, most likely due to the flexible and disordered nature of the polyelectrolyte LbL assembly which allows the surrounding water molecules access to the Gd(III) centers. In order to provide targeting capabilities to this system, the K_7RGD targeting peptide was absorbed onto the negatively charged particles. The RGD peptide is able to bind strongly to the integrin cell surface receptors that are upregulated in angiogenic cancer cells. *In vitro* optical and MR imaging studies demonstrated efficient targeting by the RGD-terminated nanoparticles to HT-29 human colon adenocarcinoma cells (Fig. 10).

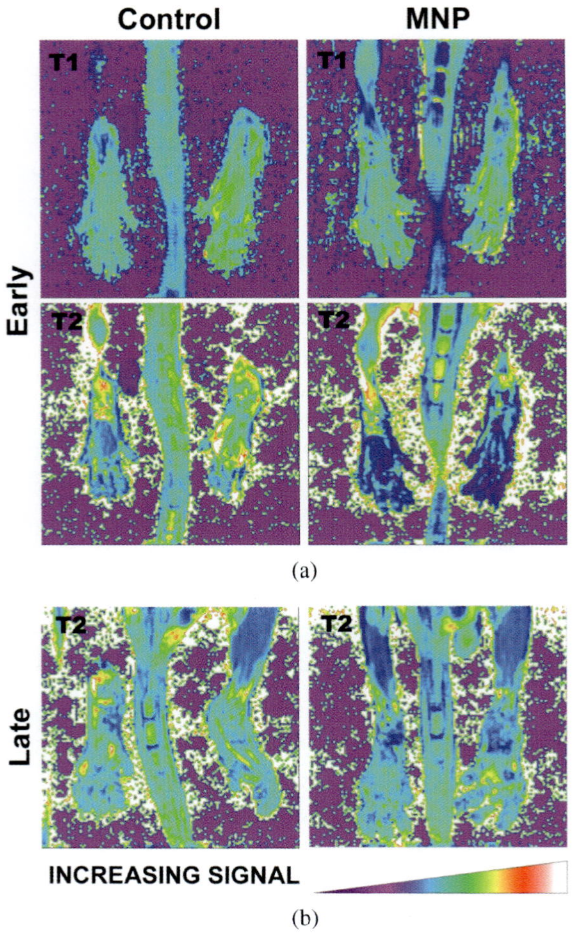

Fig. 8. (a) Representative T_1 and T_2 relaxation map of the hind limbs of a mouse with early stage CIA before and after SiNP administration. (b) T_2 relaxation maps of the hind limbs of a mouse with later stage CIA before and after receiving a SiNP-based contrast agent. Reprinted from Ref. 39, with permission from *Clinical and Experimental Rheumatology.*

Furthermore, Gd chelates have been incorporated onto quantum dots, iron oxide nanoparticles and gold nanoparticles.[42,43] For these systems, the particle core provided the optical and T_2-weighted contrast, and the Gd-chelate shell contributed to T_1-weighted enhancements. The biodistribution profile and blood circulation of these nanoparticles were studied. Silica-based luminescent spore detectors with lanthanide chelates on the surface were also developed.[44] In addition, nanosized metal-organic frameworks (NMOFs) with silica shells have been explored for use in optical, MR imaging, and biosensing applications.[45–47]

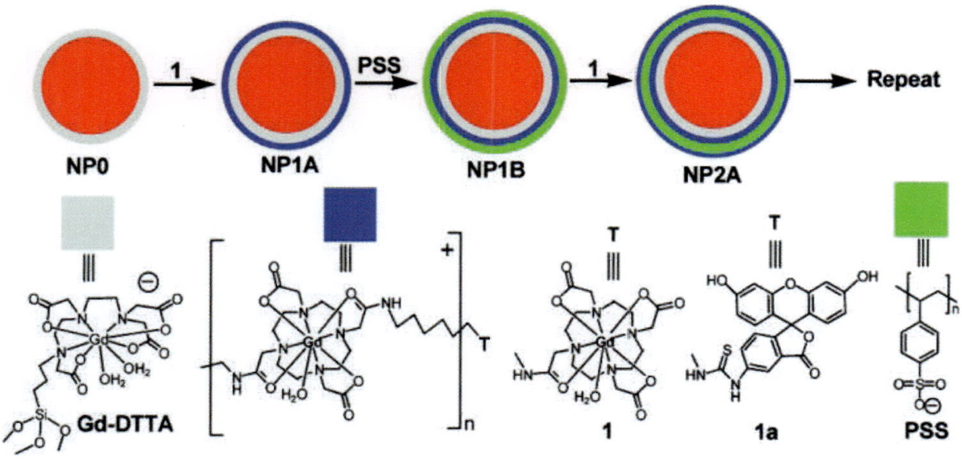

Fig. 9. Schematic representation of the LbL self-assembly strategy for a dual optical and MR imaging multimodal contrast agent. Reprinted from Ref. 38, with permission from the American Chemical Society.

3.2. *Mesoporous silica nanoparticles for imaging applications*

Mesoporous silica nanoparticles incorporating dye molecules into their framework have also been used as a tool to monitor cellular internalization and cell tracking.[23] Amino-functionalized MSNs containing fluorescein isothiocyanate (FITC) were surface-modified with polyethyleneimine *via* surface polymerization.[48] Then, folic acid was conjugated to the particles to specifically target cancer cells. *In vitro* confocal microscopy experiments using HeLa cells demonstrated that the targeted MSNs were internalized at a higher amount than the non-targeted particles.

Although optical imaging is a low-cost, sensitive technique to rapidly image organisms, it suffers from attenuation of proton propagation in living tissue and poor signal to noise ratio due to tissue autofluorescence. To overcome these issues, Lo and coworkers developed NIR MSN-based probes.[49] Electrostatic interactions between indocyanine green (ICG) and the MSNs were used to entrap the dye molecules inside the particle. Indocyanine green was selected as the NIR contrast agent due to its FDA approval for use in clinical applications. The high dispersion of ICG throughout the high surface areas of the MSNs reduced particle aggregation and decreased self-quenching. Additionally, the ICG molecules are protected from environmental degradation when they are confined inside the pores of the MSNs. The *in vivo* biodistribution in a rat revealed that the particles immediately accumulated in the liver, followed by the heart, lungs, spleen, and kidneys

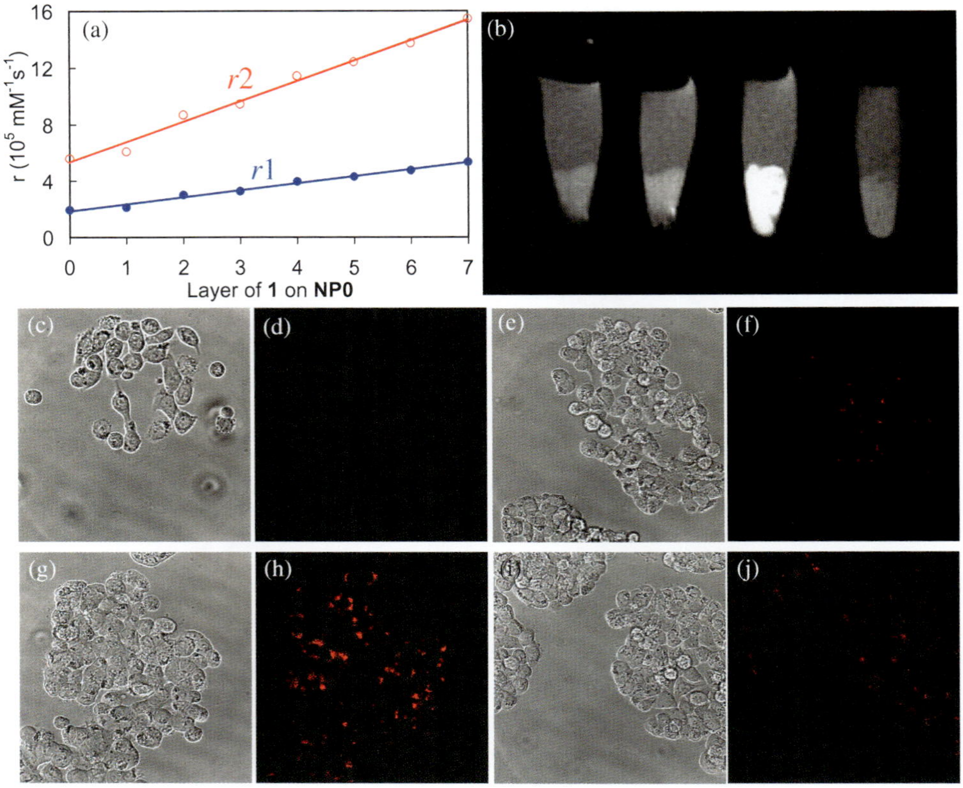

Fig. 10. (a) Dependence of per particle *r*1 and *r*2 values on the number of deposited Gd–DOTA oligomer layers. (b) T_1-weighted MR images of HT-29 cells that have been incubated with various nanoparticles. From left to right: control cells without any nanoparticle, cells with LbL particles, cells with LbL particles that have been noncovalently functionalized with K_7RGD, and cells with LbL particles that have been noncovalently functionalized with K_7GRD. (c)–(j) Phase contrast optical (c, e, g and i) and LSCFM microscopic images (d, f, h and j) of HT-29 cells that have been incubated with various nanoparticles: control cells without any nanoparticle (c) and (d), cells with LbL particles (e) and (f), cells with LbL particles that have been noncovalently functionalized with K_7RGD (g) and (h), and cells with LbL particles that have been noncovalently functionalized with K_7GRD (i) and (j). Reprinted from Ref. 38, with permission from the American Chemical Society.

(Fig. 11). The Lo group recently studied the effect of surface charge on the biodistribution of ICG-MSNs, concluding that both the biodistribution and rate of excretion can be controlled by tailoring the surface charge.[50]

Nanoparticle MR contrast agents based on MSNs have also been developed to take advantage of the high internal surface area.[51,52] Lin and coworkers reported the synthesis of an MR contrast agent based on MSN particles with an MCM-41 type pore structure.[51] The high surface areas of MSNs

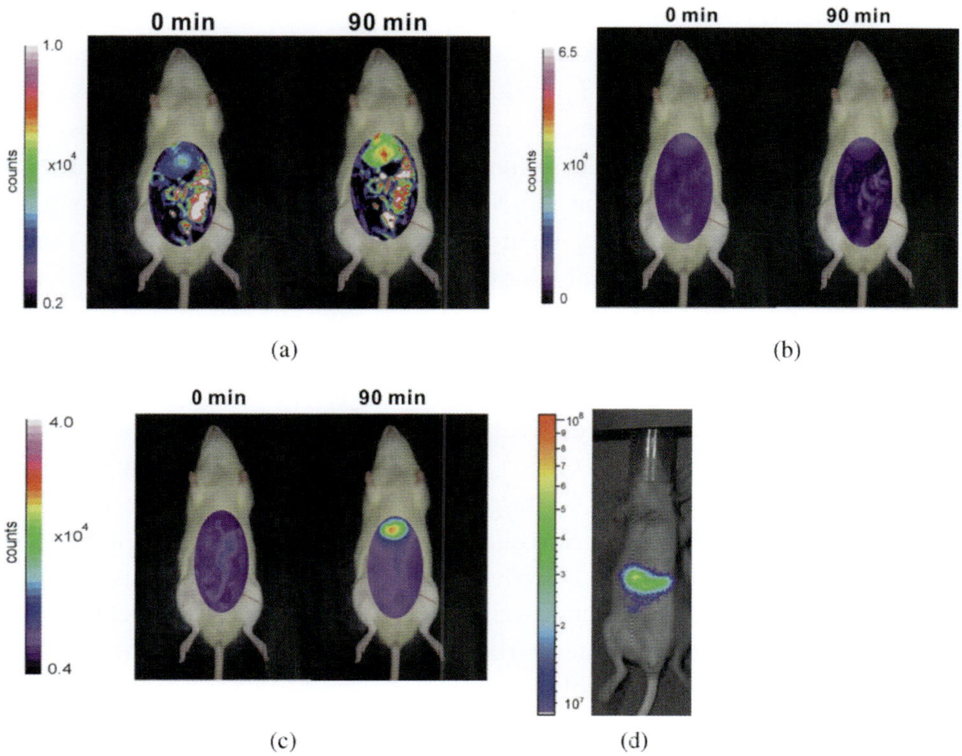

Fig. 11. Biodistribution of FITC-MSNs in an anesthetized rat before and after (90 min) intravenous injection. The experimental conditions were set at 492 nm excitation and 518 nm emission and (a) a longer shutter time (60 s) for visible imaging and (b) a 30 s shutter time. (c) Biodistribution of ICG-MSNs in an anesthetized rat before and after (90 min) intravenous injection. The ICG-MSNs showed less interference from autofluorescence in a shorter shutter period (30 s). (d) ICG-MSNs in nude mice after intravenous injection for 3 h. Reprinted from Ref. 49 with permission from John Wiley & Sons Inc.

provided a high payload of Gd chelates with enhanced water accessibility, resulting in an efficient MR contrast agent. The MSN nanoparticles were synthesized using a surfactant-templated, base-catalyzed condensation of TEOS. The particles were then coated with a monolayer of Gd-chelates. Exceptionally high relaxivities on both per Gd and per particle basis were observed. On a per mM Gd basis, the Gd-MSN exhibited an r_1 of 28.8 $mM^{-1}s^{-1}$ and an r_2 of 65.6 $mM^{-1}s^{-1}$ at 3 T. Each particle contains ~36,900 Gd ions; therefore, the r_1 and r_2 values on a per mM of particle basis were 1.06 $\times 10^6$ $mM^{-1}s^{-1}$ and 2.42 $\times$ 10^6 $mM^{-1}s^{-1}$, respectively. The *in vivo* contrast enhancing ability of these particles showed significant T_1-weighted signal enhancement in the aorta of a mouse 15 min after intravenous injection of a dose of Gd-MSN corresponding to 2.1 μmol Gd per kg of body weight

(Figs. 12(a) and (b)). This dose is approximately 100-fold lower than what is currently used for commercially available Gd-chelate contrast agents. At a higher dose of 31 μmol/kg, T_2-weighted signal enhancement was observed in the liver 1 hour post-administration (Figs. 12(c) and (d)). Signal loss was observed in the liver several days after administration of the Gd-MSN material. The slow excretion of the Gd-MSN nanoprobe after the MR imaging study and the consequent long-term tissue accumulation of Gd^{3+} ions could potentially lead to toxicological issues.

To overcome the potential toxicity hazards, the Lin group synthesized a multifunctional MSN-based MRI contrast agent that was quickly excreted by the renal pathway after imaging.[22] The MSNs contained Gd(III)-chelates that were covalently linked via a redox-responsive linker. The nanoprobe was further functionalized with PEG to improve its biocompatibility and an anisamide ligand to improve its target specificity. The anisamide ligand is known to target epithelial cancer cells that overexpress sigma receptors. The *in vitro* targeting ability and effectiveness as a contrast agent was demonstrated using human colon adenocarcinoma and pancreatic cancer cells. Rapid renal Gd clearance *in vivo*, due to the thiol-activated release of the Gd(III) chelates from the MSNs, was observed.

The ability to monitor cell trafficking and biodistribution *in vivo* is a necessity for successful stem-cell therapy development. Currently, MR imaging is an ideal noninvasive technique that gives excellent spatial resolution. However, cells must be magnetically labeled by endocytic internalization prior to transplantation. Therefore, it is desirable to develop more efficient,

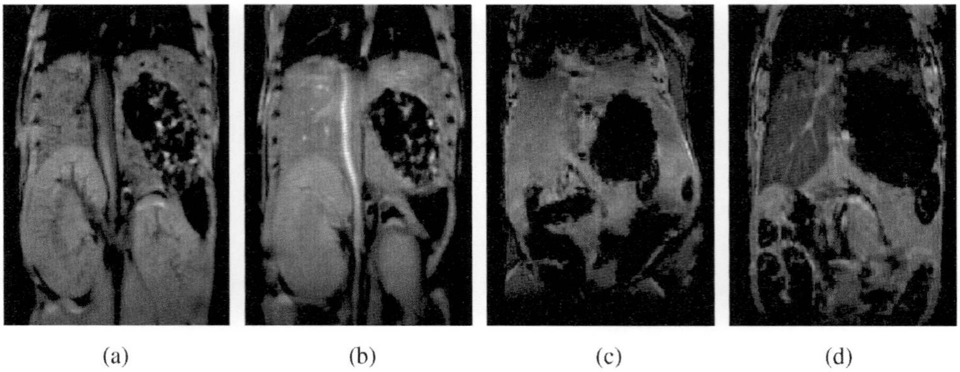

Fig. 12. (a) Pre-contrast and (b) post-contrast (2.1 μmol/kg dose) T_1-weighted mouse MR image showing aorta signal enhancement. (c) Pre-contrast and (d) post-contrast (31 μmol/kg dose) mouse MR images showing liver signal loss due to T_2-weighted enhancement. Reprinted from Ref. 51, with permission from the American Chemical Society.

noninvasive cellular-internalizing platforms. Huang and coworkers reported the synthesis of multimodal MSNs for tracking stem cells.[52, 53] The dual-mode platform combined a fluorescent dye FITC and an MR contrast agent. Both T_1- and T_2-weighted MR contrast agents were incorporated into the material. The first system was composed of silica-coated core-shell superparamagnetic iron oxide nanoparticles co-condensed with FITC-incorporated mesoporous silica (Mag-Dye@MSNs).[53] The small particles of iron oxide behave as negative contrast agents. The second system was composed of MSNs with a Gd-chelate grafted on the surface to afford Gd-Dye@MSNs.[52] Both systems were efficiently internalized into human mesenchymal stem cells (hMSCs) without affecting cell viability, growth, or differentiation. MRI was used *in vivo* to detect the labeled cells after long-term growth or differentiation, providing evidence of the biocompatibility and durability of both the Mag-Dye and Gd-Dye@MSNs (Fig. 13). The dual-modality Mag-Dye@MSNs were used to follow the biodistribution of MSNs *in vivo* after eye vein injection in a mouse model.[53] The Mag-Dye@MSNs were found to accumulate in organs, mainly the liver and spleen, belonging to the reticuloendothelial system (RES) and were detectable for 3 months. A long-term MR tracking study in the liver showed that these MSNs are highly resistant to decomposition and not easily excreted from the body.

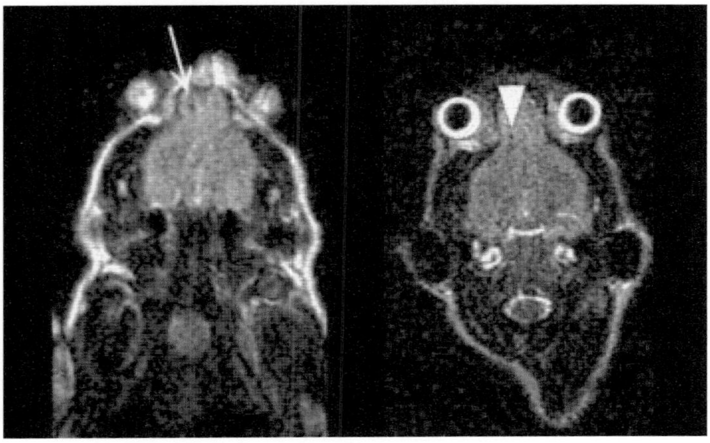

(a) (b)

Fig. 13. Clinical 1.5 T MR images of nude mouse (a) 8 h and (b) 9 days after implantation of Mag-Dye@MSN-labeled hMSCs at the frontal cortex. (a) Under clinical 1.5 T MR scanning, the hMSCs revealed a dark dot at the frontal cortex (arrow). (b) Repeated MR scanning was carried out 9 days after hMSC implantation. The stem cells could still be visualized as a black dot at the frontal cortex. No migration of these cells is found (arrow head). Reprinted from Ref. 53, with permission from John Wiley & Sons Inc.

4. Silica Nanoparticles for Drug Delivery

Silica nanoparticles have been used in drug delivery and therapy applications. By incorporating a drug into a nanoparticle, it is shielded from the physiological environment. Ideally, no drug release before reaching the target site, yet fast drug release in the selected site, is desirable in a nanoparticle system. These systems should increase the blood circulation time of the drug, yet be safely cleared from the body. Nanoparticle silica-based drug delivery agents are a promising platform to meet the mentioned requirements.

Nanoparticle delivery agents can passively target a tumor region due to the enhanced permeability and retention (EPR) effect.[54] Due to the enhanced permeability of tumor vasculature, nanoparticles accumulate at the tumor site. However, after particles reach the tumor region, they may not be internalized by cancer cells. To impart cell-specific interactions, silica nanoparticles can be designed to actively target a specific cell receptor. By attaching ligands that have a high affinity to cell-specific receptors, nanoparticle binding and cell internalization is promoted.

4.1. *Solid silica nanoparticles for drug delivery applications*

In terms of passive targeting, Jiang and coworkers synthesized hollow chitosan-silica nanospheres that exhibit controllable drug release in physiological and intracellular environments.[55] Silica was deposited onto the surface of biocompatible hollow chitosan-poly(acrylic acid) polyelectrolyte nanospheres. The polyelectrolyte inner core was designed to be pH-sensitive, so that the loaded doxorubicin was released once the particles reached the more acidic tumor site. *In vitro* cytotoxicity assays showed that the doxorubicin-loaded nanospheres were more effective than free doxorubicin, presumably due to more efficient drug uptake.

Controlling the release of the therapeutic agent from silica nanoparticles can sometimes prove challenging. Ortega and coworkers reported the synthesis of nanoparticles composed of an organic liposomal phase containing the active agent, doxorubicin, and surrounded by a polysilsesquioxane (PSQ) shell.[56] The organosilica shell aids in stabilizing the internal liposomal phase and protecting the drug molecules. *In vitro* experiments using human glioma cells have demonstrated that the nanoparticles were able to release doxorubicin and kill the cells.

Because PSQ materials exhibit similar biocompatibility to pure silica nanoparticles, additional systems have been studied for use in drug delivery applications. The Lin group has demonstrated that a much higher drug loading can be achieved using PSQ materials compared to silica nanoparticles with drugs grafted only on the surface.[57] A Pt(IV) prodrug containing siloxy functional groups was synthesized and condensed into nanoparticles with a diameter of ~80 nm. The active agent loading was confirmed to be 47% by ICP-MS. Release experiments showed that the Pt(IV) prodrug was released in the presence of a reducing agent, such as cysteine. Thus, the platin cargo will presumably be released when the nanoparticle reaches the highly reducing tumor region (Fig. 14(a)). The surface of the PSQ particles was functionalized with both a targeting ligand (either cRGD or anisamide) and PEG (Fig. 14(b)). The PSQ particles demonstrated superior toxicity to oxaliplatin against several human colon and pancreatic cancer cell lines *in vitro*, with the targeting ligands further enhancing the cytotoxicity of the particles. The PSQ particles were evaluated *in vivo* against a pancreatic cancer xenograft mouse model; the anisamide- and RGD-targeted PSQ particles showed superior efficacy to oxaliplatin in inhibiting tumor growth.

4.2. *Mesoporous silica nanoparticles for drug delivery applications*

Mesoporous silica nanoparticles have emerged as a highly robust and tunable delivery platform for the controlled release of therapeutic agents. To control the drug release kinetics, the cargo must be strongly adsorbed, or the pores need to be sealed after loading. One of the first systems loaded ibuprofen into MCM-41 materials, and the release of the drug was studied in simulated body fluid.[58] This study showed that large quantities of drug molecules could be loaded into MSNs and delivered over a period of time.

To avoid premature release, it is desirable to develop drug delivery systems that can be activated by external stimuli to release their loaded cargo. Victor Lin and coworkers have developed several MSN-based delivery systems containing stimuli-responsive pore caps (such as CdS, Au, and superparamagnetic Fe_3O_4 nanoparticles) to allow for controlled release of entrapped molecules.[59–61] One system developed in the Victor Lin group used an *o*-nitroester group that was electrostatically linked to gold nanoparticles to cap MSNs (Fig. 15). This system was used to deliver paclitaxel to human liver and fibroblast cells.[61] After exposure to UV radiation, the *o*-nitroester linker was cleaved, uncapping the MSN channels and releasing the loaded paclitaxel. Significant decreases in cell viability were observed after UV radiation. Using a similar strategy, other drug molecules and neurotransmitters, such as vancomycin and adenosine triphosphate (ATP), were delivered to cells.[59,60]

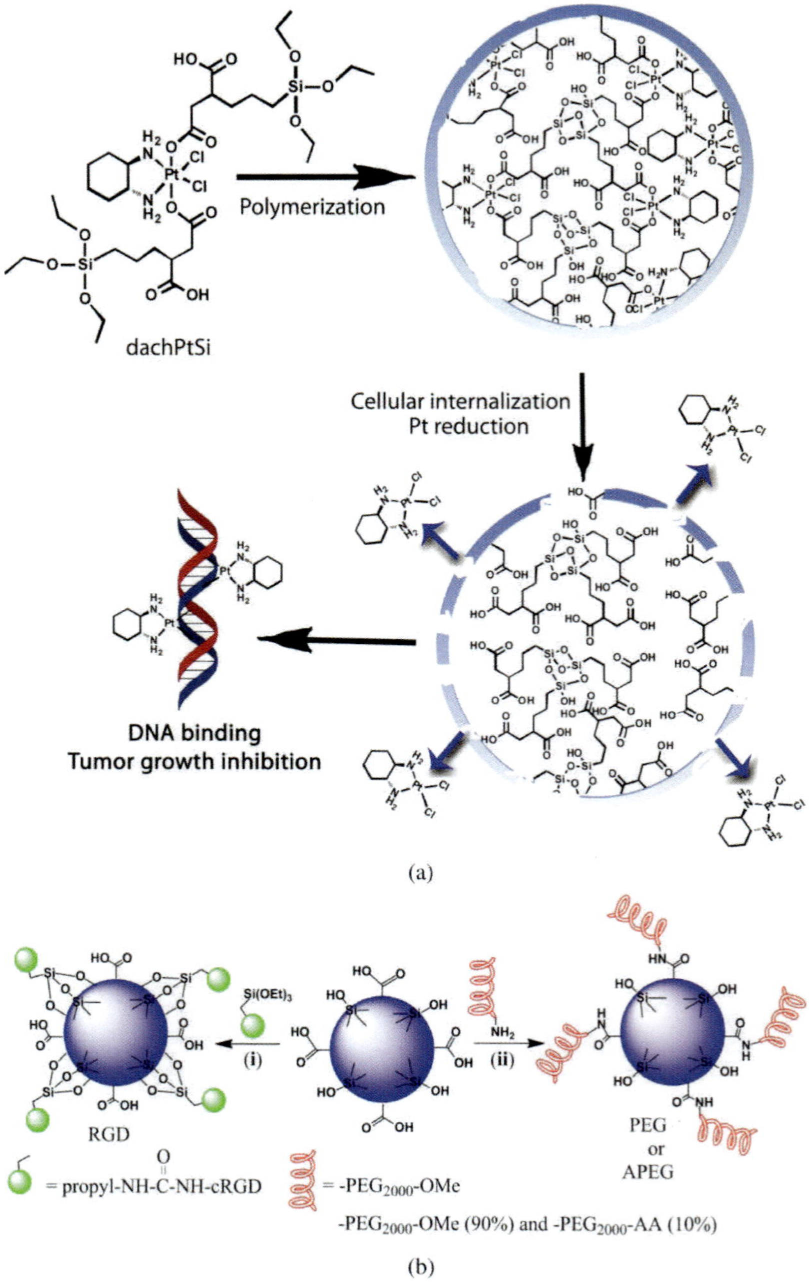

Fig. 14. (a) Generalized scheme showing the formation of PSQ particles from the Pt(IV) precursor. Upon cellular internalization and reaction with endogenous biomolecules, the Pt(IV) complexes in the particle will be reduced, thereby releasing the active Pt(II) agent. (b) Surface functionalization of PSQ particles by 1) the siloxane linkage formed between silanol groups and silyl-derived molecules, and 2) the amide linkage formed between carboxylic acid groups and amine-containing molecules. Reprinted from Ref. 57, with permission from the John Wiley & Sons Inc.

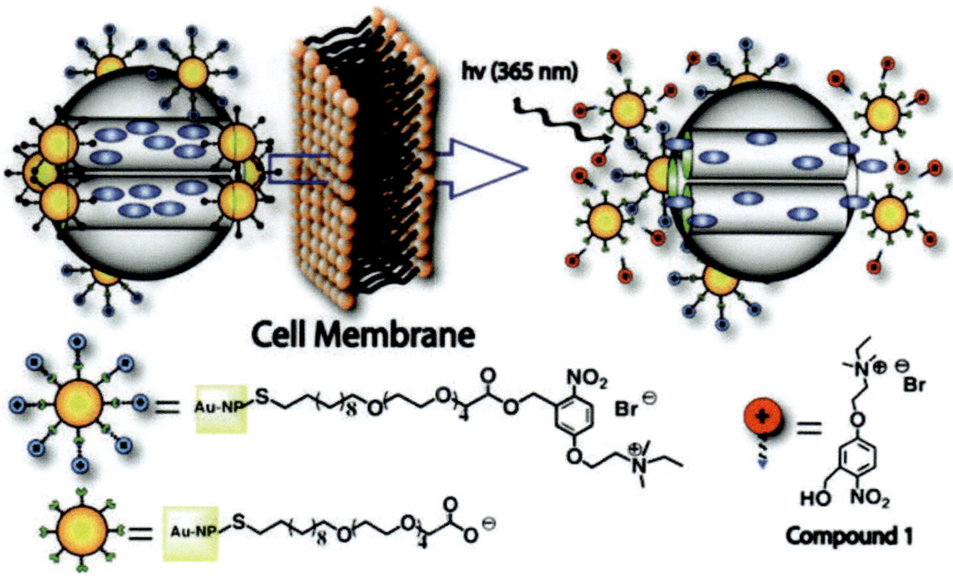

Fig. 15. Schematic representation of the Au nanoparticle-capped MSN-based drug delivery system. The controlled release mechanism is based on UV radiation. Reprinted from Ref. 61, with permission from the American Chemical Society.

MSNs have also been used to sequentially co-deliver two different therapeutic agents. The synthesis of boronic-acid functionalized MSNs for glucose-responsive controlled release of both insulin and cyclic adenosine monophosphate (cAMP) was reported.[62] Gluconic acid-modified insulin was used to cap the MSNs, entrapping cAMP molecules inside the pores. Release from the MSNs was triggered by the introduction of saccharides, which form more stable conjugates with the boronic acids present on the nanoparticles. In the presence of glucose, 80% of the entrapped cAMP was released within 20 hours. However, in the absence of glucose, less than 10% was released in the same time frame. The cytotoxicity of the drug loaded particles was examined; the materials exhibited low cytotoxicity and efficient delivery of cAMP to pancreatic cells in a dose-dependent manner.

Additionally, cross-linked polymers have been utilized to control molecule release from MSNs.[63] Feng *et al.* attached polymers to the surface of MCM-41 particles to form a redox-responsive nanogate (Fig. 16). The ability to release cargo in a controlled manner was demonstrated by loading rhodamine B inside the mesopores before cross-linking the polymers. Another example where nanogates were used to control release from MSNs was reported by Zink and coworkers.[64] The authors used light-activated,

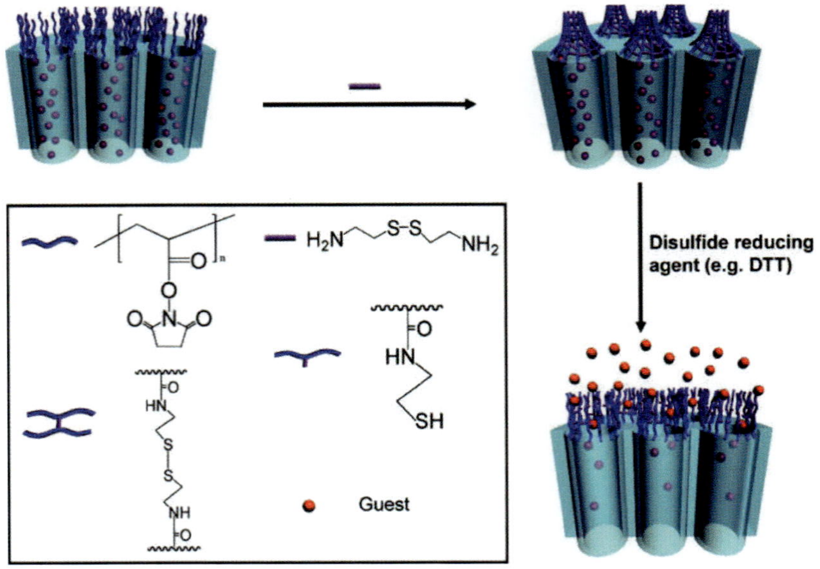

Fig. 16. Scheme showing the redox responsive nanogated, polymer-capped MSN. Reprinted from Ref. 63 with permission from the American Chemical Society.

nanoimpeller-controlled MSNs that undergo azobenzene-driven release once the particles are internalized by cancer cells. By tethering azobenzene molecules to the interior of the mesopores, azobenzene photoisomerization can be utilized to release loaded molecules from the material. When the azobenzene molecules are irradiated at a specific wavelength, photoisomerization is activated. The tethered azobenzene molecules then wag, making a "nanoimpeller." To demonstrate this, camptothecin was loaded into the MSN pores. After incubating cancer cells with the drug-loaded MSN, cell death was only observed after irradiation at 413 nm. Recently, a related triggered release system based on a modified MSN that contains a *trans*-azobenzene amphiphilic copolymer structure on the surface was reported.[65]

In addition to controlled release of therapeutic agents, targeted delivery of the drugs is desirable. By targeting only cancer cells, the toxic side effects can be reduced and the drug efficacy can be enhanced. Passive targeting of drug-loaded MSNs can be achieved due to the EPR effect, as demonstrated by Zink and coworkers.[66] Active targeting can be achieved by attaching a targeting ligand, such as folic acid, onto the surface of the MSN. Zink and coworkers developed MSNs that were functionalized with folic acid and then loaded with camptothecin.[67] The efficacy of the MSNs on two different human pancreatic cancer xenografts mouse models was reported. The targeted nanoparticles demonstrated enhanced tumor inhibition compared to the nontargeted versions.

Most nanoparticles are designed to target the cancer cells; however, Shi and coworkers recently reported a method to target nuclei, killing cancer cells more directly and efficiently.[68] In this work, doxorubicin was loaded into the nanoparticles, and TAT-peptide was conjugated to the surface of MSNs. The TAT-peptide facilitates nuclear internalization and the subsequent release of the loaded doxorubicin within the nucleoplasm. *In vitro* cytotoxicity assays using HeLa cells demonstrated a significant enhancement in the anticancer activity of doxorubicin when the drug was loaded into the targeted particles.

5. Silica Nanoparticles as Theranostic Agents

Multifunctional nanoparticles for simultaneous imaging and therapy, also known as "theranostic" agents, have attracted a remarkable level of research in recent years. The term theranostic is defined by the combination of diagnostic and therapeutic agents in a single system.[69] Theranostic nanoparticles have the potential to simultaneously monitor and treat a disease of interest. Imaging agents have already been incorporated into nanomaterials, so they can easily be further functionalized to include therapeutic functionalities. Both diagnostic and therapeutic agents require sufficient accumulation in the diseased area; thus, silica nanoparticles are a promising candidate for theranostic applications due to their ability to load a variety of imaging and therapeutic functionalities.

5.1. *Solid silica nanoparticles for theranostic applications*

In the past decade, SiNPs have been used in photodynamic therapy (PDT) applications. When a photosensitizer is activated by light, a reactive oxygen species (ROS), such as singlet oxygen ($^1O_2$), is formed.[70] The ROS destroys targeted cells, resulting in an observed therapeutic effect. The role of the photosensitizer is not only to generate singlet oxygen but to evaluate its internalization and therapeutic effect through fluorescence. Most photosensitizers function as fluorophores, often emitting in the NIR region. Thus, nanoparticles loaded with photosensitizers can be used as theranostic agents for PDT.

Prasad and coworkers developed a fluorescence resonance energy transfer (FRET) system composed of ORMOSIL nanoparticles that incorporate both photosensitizers and organic dyes.[71] The ORMOSIL particles served as a platform to deliver photosensitizers with poor water solubility. The photosensitizing anticancer drug 2-devinyl-2-(1-hexyloxyethyl) pyropheophorbide (HPPH) was entrapped within the particles. Because the entrapped drug is more fluorescent than the free drug, the loaded ORMOSIL materials can be used in fluorescence bioimaging studies. The authors demonstrated efficient singlet oxygen generation after the particles were irradiated with light. *In vivo*, cancer cells were shown to uptake the drug-doped nanoparticles, and after irradiation, significant damage to the cancer cells was observed.

Prasad and coworkers went on to then co-encapsulate HPPH and a two-photon absorbing dye, 9,10-bis-[4′-(4″-aminostyryl)styryl]anthracene (BDSA), into silica nanoparticles. Two-photon absorption-induced excitation of photosensitizers enables greater light penetration depths to be achieved. BDSA was able to efficiently up-convert the NIR light and transfer the energy to the co-encapsulated HPPH to activate the PDT function.

Hai *et al.* synthesized bifunctional silica nanoparticles doped with methylene blue for use in optical imaging and PDT.[72] The SiNPs effectively shielded the dye from the environment, preventing leakage of methylene blue from the particles. Irradiation to the doped nanoparticles resulted in singlet oxygen generation and photodynamic damage to HeLa cells. The therapeutic efficacy was also demonstrated *in vivo* using fluorescence imaging on a mouse model. After intratumoral injection of the methylene blue-loaded particles, optical images showed that the tumor region was clearly defined (Fig. 17).

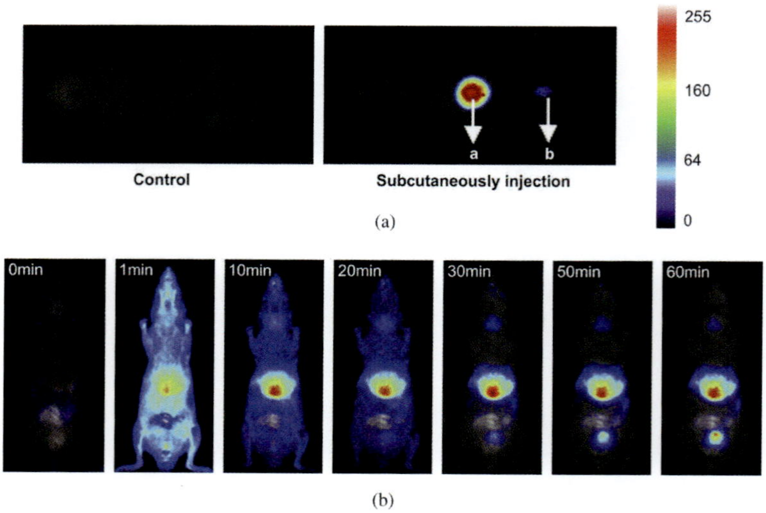

Fig. 17. (a) Images of the two subcutaneous injections of 100 microliters MB-encapsulated SiNPs with concentration of 44 mg/mL (a) and 4.4 mg/mL (b). The acquisition was performed 2 min after injection. (b) Real-time *in vivo* abdomen imaging of intravenously injected with 200 mL of MB-encapsulated SiNPs (44 mg/mL) at different time points, post injection. Reprinted from Ref. 72 with permission from Elsevier Ltd.

Silica-coated NMOFs have also been developed for theranostic applications. Iron(III) carboxylate NMOFs with the MIL-101 structure were post-synthetically modified to introduce both therapeutic and imaging moieties.[73] An optical contrast agent, a BODIPY dye, and an ethoxysuccinato-cisplatin anticancer prodrug were incorporated into the framework.

5.2. *Mesoporous silica nanoparticles for theranostic applications*

Mesoporous silica nanoparticles have a unique topology with three different components — silica matrix, internal, and external surfaces — available for functionalization. Additionally, the pores can be loaded with drugs or imaging agents. The ability to incorporate several different functionalities onto MSNs is advantageous in the development of theranostic nanoparticles. For example, Lo and coworkers prepared an MSN that incorporated three different functional groups: a NIR dye (ATTO 647N) into the silica matrix, a photosensitizer (*meso*-tetratolylporphyrin-Pd, PdTPP) within the channels, and RGD-PEG on the exterior surface. *In vitro* analysis of this theranostic platform demonstrated efficient targeting and uptake by U87-MG glioblastoma cells, which overexpress the $\alpha_v\beta_3$ integrin. Additionally, the PDT efficacy was tested using bioavailability assays under irradiation and non-irradiation conditions. The results showed a significant difference in the post-irradiation cytotoxic response of cells treated with the RGD-PdTPP-MSNs.

Recently, Brinker and coworkers have reported theranostic MSNs covered with lipid bilayers (protocells) (Fig. 18).[74] The protocells are loaded with combinations of therapeutic agents (drugs, siRNA, toxins) and diagnostic agents (quantum dots). After agent loading, the mesoporous particles are fused with liposomes to seal the cargo within the pores. Moreover, the protocell can be modified with targeting peptides to selectively bind to cancer cells or with fusogenic peptides to promote endosomal escape.

Moon and coworkers have developed magnetic, fluorescent core-shell nanoparticles consisting of a single iron oxide nanocrystal and a dye-doped mesoporous silica shell. The iron oxide center provides T_2-weighted MR contrast enhancement, while the mesoporous silica serves as a drug carrier. Fluorescent dyes were covalently attached to the silica shell and the anticancer drug doxorubicin was loaded into the pores. For sizes under 100 nm, the particle size can be carefully controlled. The cellular uptake of these MSNs in MCF-7 breast cancer cells was demonstrated using fluorescence and confocal microscopy. Also, *in vivo* assays on a murine xenograft demonstrated a strong T_2 signal reduction at the tumor site up to 24 h post-injection. To evaluate

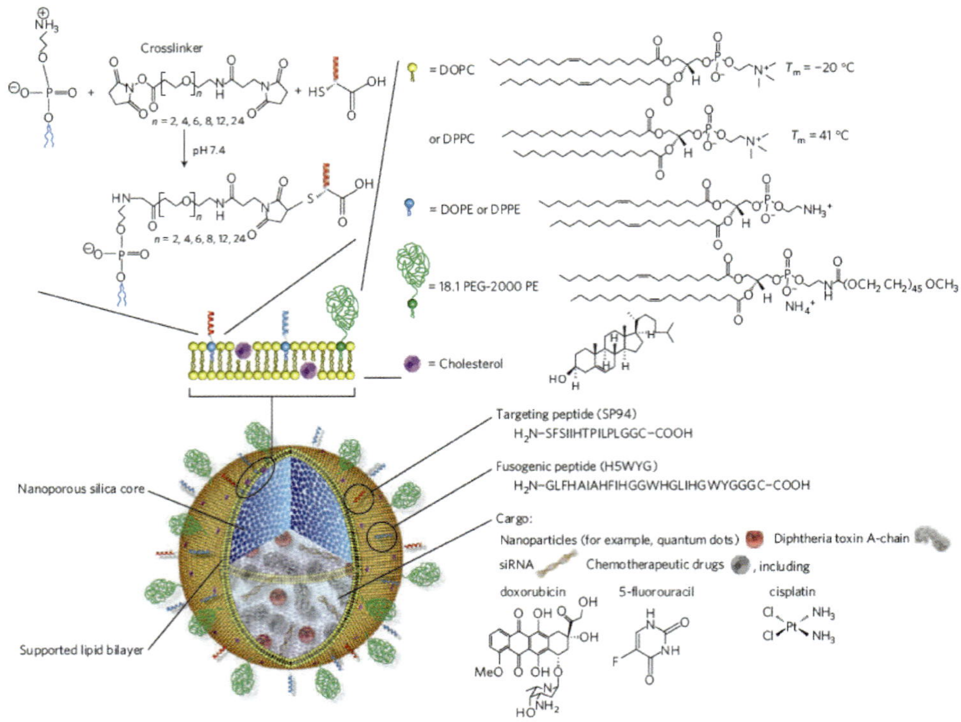

Fig. 18. Schematic illustration of the nanoporous particle-supported lipid bilayer, depicting the disparate types of therapeutic and diagnostic agents that can be loaded within the nanoporous silica core, as well as the ligands that can be displayed on the surface of the single-lipid bilayer. Reprinted from Ref. 74, with permission from Macmillan Publishers Ltd.

the therapeutic index of the MSNs, SK-BR-3 cells were dosed with the doxorubicin-loaded particles. The results showed a dose-dependent cytotoxicity profile. Recently, the authors reported the synthesis of dye-doped iron oxide-capped MSNs for multimodal imaging and drug delivery applications.[75] The fluorophore (either FITC or TRITC) was loaded into the channels of the MSN, and the iron oxide nanocrystals were integrated onto the silica surface. Similar to their previous study, the nanoparticle was functionalized with PEG. When the iron oxide nanoparticles were attached to the silica surface, a remarkable T_2-weighted MR contrast enhancement. Doxorubicin was loaded into the particles, and the *in vivo* passive targeting and tumor accumulation of the MSNs was confirmed using T_2 MR and fluorescence imaging. At 3 h post-injection, the T_2-weighted MR signal decreased at the tumor site, demonstrating passive targeting of iron oxide-MSNs by the EPR effect. Furthermore, the antitumor effect was evaluated using terminal deoxynucleotidyl transferase-mediated nick end labeling (TUNEL) assay (Fig. 19). After

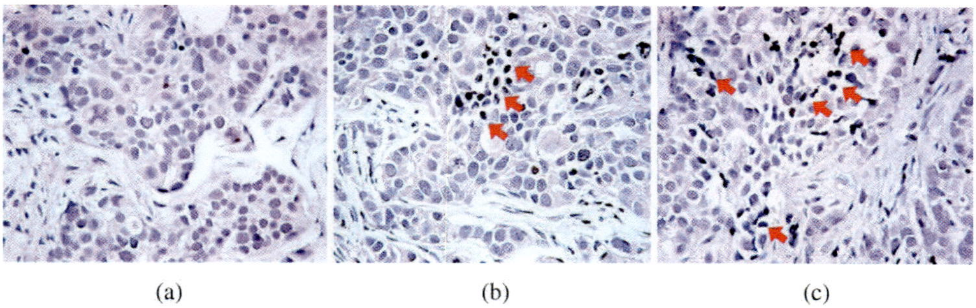

Fig. 19. TUNEL assay for apoptotic cell death. Tumor section from mouse that was given intravenous injection of (a) unloaded iron oxide-MSN as control, (b) doxorubicin-loaded iron oxide-MSN (DOX 2 mg/kg), and (c) doxorubicin-loaded iron oxide-MSN (DOX 4 mg/kg). Arrows indicate examples of TUNEL-positive (brown color) cells with apoptotic morphology. Reprinted from Ref. 75, with permission from the American Chemical Society.

treatment with doxorubicin-loaded particles, TUNEL-positive tumor cell nuclei with apoptotic morphology were detected in tumor tissues of mice, while apoptotic cells were not detectable in the case of unloaded MSNs.

Mesoporous silica nanoparticles that contain an optical imaging agent (FITC) and the anticancer drug camptothecin have also been reported.[66] The authors demonstrate that MSNs are well tolerated *in vivo*, as shown by serological, hematological, and histopathological examinations of blood samples and mouse tissues. *In vivo* imaging, fluorescence microscopy imaging, and inductively coupled plasma mass spectroscopy confirmed that MSNs preferentially accumulate in the tumor region. Analysis of renal and hepatobiliary routes of excretion showed that MSNs are cleared through both pathways. Lastly, the therapeutic efficiency of the camptothecin-loaded MSNs *in vivo* was evaluated using an MCF-7 xenograft model; significant tumor growth inhibition was observed after administration of the theranostic particles at the conclusion of the experiment.

6. Conclusions and Outlook

Silica-based nanoparticles, both solid and porous, are promising candidates for biomedical imaging and drug delivery applications. The ability of silica nanoparticles to be functionalized with desirable moieties allows for the design of multimodal platforms. Additionally, integrating different functionalities into the material can alter the physical and chemical properties of the particles. The advantages of silica-based platforms over small molecule agents have been extensively demonstrated *in vitro* and *in vivo*. Further work demonstrating *in vivo* efficacy and lack of toxicity is still necessary to advance these materials into clinical applications.

Acknowledgments

We thank the National Cancer Institute for support (U01-CA151455 and U54-CA151562).

References

1. K. Cho, X. Wang, S. Nie, Z. Chen and D. M. Shin, Therapeutic nanoparticles for drug delivery in cancer, *Clin. Cancer Res.* **14**(5), 1310–1316 (2008).
2. D. Peer, J. M. Karp, S. Hong, O. C. Farokhzad, R. Margalit and R. Langer, Nanocarriers as an emerging platform for cancer therapy, *Nat Nano* **2**(12), 751–760 (2007).
3. S. Laurent, D. Forge, M. Port, A. Roch, C. Robic, L. Vander Elst and R. N. Muller, Magnetic iron oxide nanoparticles: Synthesis, stabilization, vectorization, physicochemical characterizations, and biological applications, *Chem. Rev.* (*Washington, DC*) **108**(6), 2064–2110 (2008).
4. K. M. L. Taylor-Pashow, J. Della Rocca, R. C. Huxford and W. Lin, Hybrid nanomaterials for biomedical applications, *Chem. Commun.* **46**(32), 5832–5839 (2010).
5. L. Wang, W. Zhao and W. Tan, Bioconjugated silica nanoparticles: Development and applications, *Nano Res.* **1**(2), 99–115 (2008).
6. W. Stober, A. Fink and E. Bohn, Controlled growth of monodisperse silica spheres in micron size range, *J. Colloid Interf. Sci.* **26**(1), 62 (1968).
7. F. J. Arriagada and K. Osseo-Asare, Phase and dispersion stability effects in the synthesis of silica nanoparticles in a non-ionic reverse microemulsion, *Colloids Surf.* **69** (2–3), 105–115 (1992).
8. I. I. Slowing, J. L. Vivero-Escoto, B. G. Trewyn and V. S. Y. Lin, Mesoporous silica nanoparticles: Structural design and applications, *J. Mater. Chem.* **20**(37), 7924–7937 (2010).
9. Y. Wan and D. Zhao, On the controllable soft-templating approach to mesoporous silicates, *Chem. Rev.* (*Washington, DC,*) **107**(7), 2821–2860 (2007).
10. X.-D. Wang, Z.-X. Shen, T. Sang, X.-B. Cheng, M.-F. Li, L. -Y. Chen and Z.-S. Wang, Preparation of spherical silica particles by Stöber process with high concentration of tetra-ethyl-orthosilicate, *J. Colloid Interf. Sci.* **341**(1), 23–29 (2010).
11. F. J. Arriagada and K. Osseo-Asare, Synthesis of nanosize silica in a nonionic water-in-oil microemulsion: Effects of the water/surfactant molar ratio and ammonia concentration, *J. Colloid Interf. Sci.* **211**(2), 210–220 (1999).
12. R. P. Bagwe, C. Yang, L. R. Hilliard and W. Tan, Optimization of dye-doped silica nanoparticles prepared using a reverse microemulsion method, *Langmuir* **20**(19), 8336–8342 (2004).
13. S. Santra, H. Yang, D. Dutta, J. T. Stanley, P. H. Holloway, W. Tan, B. M. Moudgil and R. A. Mericle, TAT conjugated, FITC doped silica nanoparticles for bioimaging applications, *Chem. Commun.* (*Cambridge, U. K.*) (24), 2810–2811 (2004).
14. D. Ma, J. Guan, F. Normandin, S. Denommee, G. Enright, T. Veres, and B. Simard, Multifunctional nano-architecture for biomedical applications, *Chem. Mater.* **18**(7), 1920–1927 (2006).

15. J.-L. Bridot, A.-C. Faure, S. Laurent, C. Riviere, C. Billotey, B. Hiba, M. Janier, V. Josserand, J.-L. Coll, L. Vander Elst, R. Muller, S. Roux, P. Perriat and O. Tillement, Hybrid gadolinium oxide nanoparticles: Multimodal contrast agents for in vivo imaging, *J. Am. Chem. Soc.* **129**(16), 5076–5084 (2007).
16. J. S. Beck, J. C. Vartuli, W. J. Roth, M. E. Leonowicz, C. T. Kresge, K. D. Schmitt, C. T. W. Chu, D. H. Olson and E. W. Sheppard, A new family of mesoporous molecular sieves prepared with liquid crystal templates, *J. Amer. Chem. Soc.* **114**(27), 10834–10843 (1992).
17. S.-H. Wu, Y. Hung and C.-Y. Mou, Mesoporous silica nanoparticles as nanocarriers, *Chem. Commun.* **47**(36), 9972–9985 (2011).
18. B. G. Trewyn, I. I. Slowing, S. Giri, H.-T. Chen and V. S. Y. Lin, Synthesis and functionalization of a mesoporous silica nanoparticle based on the sol-gel process and applications in controlled release, *Acc. Chem. Res.* **40**(9), 846–853 (2007).
19. M. Liong, J. Lu, M. Kovochich, T. Xia, S. G. Ruehm, A. E. Nel, F. Tamanoi and J. I. Zink, Multifunctional inorganic nanoparticles for imaging, targeting, and drug delivery, *ACS Nano* **2**(5), 889–896 (2008).
20. S. Huh, J. W. Wiench, J.-C. Yoo, M. Pruski and V. S. Y. Lin, Organic functionalization and morphology control of mesoporous silicas via a co-condensation synthesis method, *Chem. Mater.* **15**(22), 4247–4256 (2003).
21. K. M. L. Taylor-Pashow, J. Della Rocca and W. Lin, Mesoporous silica nanoparticles with co-condensed gadolinium chelates for multimodal imaging, *Nanomaterials* **2**, 1–14 (2012).
22. J.-L. Vivero-Escoto, J. L.; Taylor-Pashow, K. M. L.; Huxford, R. C.; Della Rocca, J.; Okoruwa, C.; An, H.; Lin, W.; Lin, W., Multifunctional mesoporous silica nanospheres with cleavable Gd(III) chelates as MRI contrast agents: Synthesis, characterization, target-specificity, and renal clearance, *Small* 7(24), 3519–3528 (2011).
23. J. L. Vivero-Escoto, I. I. Slowing, B. G. Trewyn and V. S. Y. Lin, Mesoporous silica nanoparticles for intracellular controlled drug delivery, *Small* **6**(18), 1952–1967 (2010).
24. T. Xia, M. Kovochich, M. Liong, H. Meng, S. Kabehie, S. George, J. I. Zink and A. E. Nel, Polyethyleneimine coating enhances the cellular uptake of mesoporous silica nanoparticles and allows safe delivery of siRNA and DNA constructs, *ACS Nano* **3**(10), 3273–3286 (2009).
25. T. F. Massoud and S. S. Gambhir, Molecular imaging in living subjects: Seeing fundamental biological processes in a new light, *Genes Dev.* **17**(5), 545–580 (2003).
26. H. B. Na, I. C. Song and T. Hyeon, Inorganic nanoparticles for MRI contrast agents, *Adv. Mater.* (*Weinheim, Ger.*) **21**(21), 2133–2148 (2009).
27. A. Van Blaaderen and A. Vrij, Synthesis and characterization of colloidal dispersions of fluorescent, monodisperse silica spheres, *Langmuir* **8**(12), 2921–2931 (1992).
28. S. Santra, P. Zhang, K. Wang, R. Tapec and W. Tan, Conjugation of biomolecules with luminophore-doped silica nanoparticles for photostable biomarkers, *Anal. Chem.* **73**(20), 4988–4993 (2001).
29. X. He, H. Nie, K. Wang, W. Tan, X. Wu and P. Zhang, *In vivo* study of biodistribution and urinary excretion of surface-modified silica nanoparticles, *Anal. Chem.* (*Washington, DC*) **80**(24), 9597–9603 (2008).
30. H. Ow, D. R. Larson, M. Srivastava, B. A. Baird, W. W. Webb and U. Wiesner, Bright and stable core-shell fluorescent silica nanoparticles, *Nano Lett.* **5**(1), 113–117 (2005).

31. J. Choi, A. A. Burns, R. M. Williams, Z. Zhou, A. Flesken-Nikitin, W. R. Zipfel, U. Wiesner and A. Y. Nikitin, Core-shell silica nanoparticles as fluorescent labels for nanomedicine, *J. Biomed. Opt.* **12**(6), 064007/1–064007/11 (2007).
32. J. E. Fuller, G. T. Zugates, L. S. Ferreira, H. S. Ow, N. N. Nguyen, U. B. Wiesner and R. S. Langer, Intracellular delivery of core-shell fluorescent silica nanoparticles, *Biomaterials* **29**(10), 1526–1532 (2008).
33. A. A. Burns, J. Vider, H. Ow, E. Herz, O. Penate-Medina, M. Baumgart, S. M. Larson, U. Wiesner and M. Bradbury, Fluorescent silica nanoparticles with efficient urinary excretion for nanomedicine, *Nano Lett.* **9**(1), 442–448 (2009).
34. M. Benezra, O. Penate-Medina, P. B. Zanzonico, D. Schaer, H. Ow, A. Burns, E. DeStanchina, V. Longo, E. Herz, S. Iyer, J. Wolchok, S. M. Larson, U. Wiesner and M. S. Bradbury, Multimodal silica nanoparticles are effective cancer-targeted probes in a model of human melanoma, *J. Clin. Invest.* **121**(7), 2768–2780 (2011).
35. R. Kumar, I. Roy, T. Y. Ohulchanskyy, L. N. Goswami, A. C. Bonoiu, E. J. Bergey, K. M. Tramposch, A. Maitra and P. N. Prasad, Covalently dye-linked, surface-controlled, and bioconjugated organically modified silica nanoparticles as targeted probes for optical imaging, *ACS Nano* **2**(3), 449–456 (2008).
36. W. J. Rieter, J. S. Kim, K. M. L. Taylor, H. An, W. Lin, T. Tarrant and W. Lin, Hybrid silica nanoparticles for multimodal imaging, *Angew. Chem., Int. Ed.* **46**(20), 3680–3682 (2007).
37. D. Gerion, J. Herberg, R. Bok, E. Gjersing, E. Ramon, R. Maxwell, J. Kurhanewicz, T. F. Budinger, J. W. Gray, M. A. Shuman and F. F. Chen, Paramagnetic silica-coated nanocrystals as an advanced MRI contrast agent, *J. Phys. Chem. C* **111**(34), 12542–12551 (2007).
38. J. S. Kim, W. J. Rieter, K. M. L. Taylor, H. An, W. Lin and W. Lin, Self-assembled hybrid nanoparticles for cancer-specific multimodal imaging, *J. Am. Chem. Soc.* **129**(29), 8962–8963 (2007).
39. J. S. Kim, H. An, W. J. Rieter, D. Esserman, K. M. L. Taylor-Pashow, R. B. Sartor, W. Lin and T. K. Tarrant, Multimodal optical and Gd-based nanoparticles for imaging in inflammatory arthritis, *Clin. Exp. Rheumatol.* **27**(4), 580–586 (2009).
40. C. Louis, R Bazzi, C. A. Marquette, J.-L. Bridot, S. Roux, G. Ledoux, B. Mercier, L. Blum, P. Perriat and O. Tillement, Nanosized hybrid particles with double luminescence for biological labeling, *Chem. Mater.* **17**(7), 1673–1682 (2005).
41. P. Voisin, E. J. Ribot, S. Miraux, A.-K. Bouzier-Sore, J.-F. Lahitte, V. Bouchaud, S. Mornet, E. Thiaudiere, J.-M. Franconi, L. Raison, C. Labrugere, M.-H. Delville, Use of lanthanide-grafted inorganic nanoparticles as effective contrast agents for cellular uptake imaging, *Bioconjugate Chem.* **18**(4), 1053–1063 (2007).
42. C.-C. Huang, C.-Y. Tsai, H.-S. Sheu, K.-Y. Chuang, C.-H. Su, U. S. Jeng, F.-Y. Cheng, C.-H. Su, H.-Y. Lei and C.-S. Yeh, Enhancing transversal relaxation for magnetite nanoparticles in MR imaging using Gd3+-chelated mesoporous silica shells, *ACS Nano* **5**(5), 3905–3916 (2011).
43. M. M. van Schooneveld, E. Vucic, R. Koole, Y. Zhou, J. Stocks, D. P. Cormode, C. Y. Tang, R. E. Gordon, K. Nicolay, A. Meijerink, Z. A. Fayad and W. J. M. Mulder, Improved biocompatibility and pharmacokinetics of silica nanoparticles by means of a lipid coating: A multimodality investigation, *Nano Lett.* **8**(8), 2517–2525 (2008).

44. K. M. L. Taylor and W. Lin, Hybrid silica nanoparticles for luminescent spore detection, *J. Mater. Chem.* **19**(35), 6418–6422 (2009).
45. D. Liu, R. C. Huxford and W. Lin, Phosphorescent nanoscale coordination polymers as contrast agents for optical imaging, *Angew. Chem., Int. Ed.* **50**(16), 3696–3700; S3696/1–S3696/14 (2011).
46. W. J. Rieter, K. M. L. Taylor and W. Lin, W., Surface modification and functionalization of nanoscale metal-organic frameworks for controlled release and luminescence sensing, *J. Am. Chem. Soc.* **129**(32), 9852–9853 (2007).
47. K. M. L. Taylor, W. J. Rieter and W. Lin, Manganese-based nanoscale metal-organic frameworks for magnetic resonance imaging, *J. Am. Chem. Soc.* **130**(44), 14358–14359 (2008).
48. J. M. Rosenholm, E. Peuhu, J. E. Eriksson, C. Sahlgren and M. Linden, Targeted Intracellular delivery of hydrophobic agents using mesoporous hybrid silica nanoparticles as carrier systems, *Nano Lett.* **9**(9), 3308–3311 (2009).
49. C.-H. Lee, S.-H. Cheng, Y.-J. Wang, Y.-C. Chen, N.-T. Chen, J. Souris, C.-T. Chen, C.-Y. Mou, C.-S. Yang and L.-W. Lo, Near-infrared mesoporous silica nanoparticles for optical imaging: Characterization and *in vivo* biodistribution, *Adv. Funct. Mater.* **19**(2), 215–222 (2009).
50. J. S. Souris, C.-H. Lee, S.-H. Cheng, C.-T. Chen, C.-S. Yang, J.-a. A. Ho, C.-Y. Mou and L.-W. Lo, Surface charge-mediated rapid hepatobiliary excretion of mesoporous silica nanoparticles, *Biomaterials* **31**(21), 5564–5574 (2010).
51. K. M. L. Taylor, J. S. Kim, W. J. Rieter, H. An, W. Lin and W. Lin, Mesoporous silica nanospheres as highly efficient MRI contrast agents, *J. Amer. Chem. Soc.* **130**(7), 2154–2155 (2008).
52. J.-K. Hsiao, C.-P. Tsai, T.-H. Chung, Y. Hung, M. Yao, H.-M. Liu, C.-Y. Mou, C.-S. Yang, Y.-C. Chen and D.-M. Huang, Mesoporous silica nanoparticles as a delivery system of gadolinium for effective human stem cell tracking, *Small* **4**(9), 1445–1452 (2008).
53. H.-M. Liu, S.-H. Wu, C.-W. Lu, M. Yao, J.-K. Hsiao, Y. Hung, Y.-S. Lin, C.-Y. Mou, C.-S. Yang, D.-M. Huang and Y.-C. Chen, Mesoporous silica nanoparticles improve magnetic labeling efficiency in human stem cells, *Small* **4**(5), 619–626 (2008).
54. H. Maeda, J. Wu, T. Sawa, Y. Matsumura and K. Hori, Tumor vascular permeability and the EPR effect in macromolecular therapeutics: A review, *J. Control. Release* **65**(1–2), 271–284 (2000).
55. E. Yan, Y. Ding, C. Chen, R. Li, Y. Hu and X. Jiang, Polymer/silica hybrid hollow nanospheres with pH-sensitive drug release in physiological and intracellular environments, *Chem. Commun.* (*Cambridge*) (19), 2718–2720 (2009).
56. A. Corma, U. Díaz, M. Arrica, E. Fernández and Í. Ortega, Organic–Inorganic nanospheres with responsive molecular gates for drug storage and release, *Ang. Chem. Int. Ed.* **48**(34), 6247–6250 (2009).
57. J. Della Rocca, R. C. Huxford, E. Comstock-Duggan and W. Lin, Polysilsesquioxane nanoparticles for targeted platin-based cancer chemotherapy by triggered release, *Ang. Chem. Int. Ed.* **50**(44), 10330–10334 (2011).
58. M. Vallet-Regi, A. Ramila, R. P. del Real and J. Perez-Pariente, A new property of MCM-41: Drug delivery system, *Chem. Mater.* **13**(2), 308–311 (2001).
59. C.-Y. Lai, B. G. Trewyn, D. M. Jeftinija, K. Jeftinija, S. Xu, S. Jeftinija, V. S. Y. Lin, A mesoporous silica nanosphere-based carrier system with chemically removable CdS

nanoparticle caps for stimuli-responsive controlled release of neurotransmitters and drug molecules, *J. Amer. Chem. Soc.* **125**(15), 4451–4459 (2003).
60. S. Giri, B. G. Trewyn, M. P. Stellmaker and V. S. Y. Lin, Stimuli-responsive controlled-release delivery system based on mesoporous silica nanorods capped with magnetic nanoparticles, *Angew. Chem., Int. Ed.* **44**(32), 5038–5044 (2005).
61. J. L. Vivero-Escoto, I. I. Slowing, C.-W. Wu and V. S. Y. Lin, Photoinduced intracellular controlled release drug delivery in human cells by gold-capped mesoporous silica nanosphere, *J. Am. Chem. Soc.* **131**(10), 3462–3463 (2009).
62. Y. Zhao, B. G. Trewyn, I. I. Slowing and V. S. Y. Lin, Mesoporous silica nanoparticle-based double drug delivery system for glucose-responsive controlled release of insulin and cyclic AMP, *J. Amer. Chem. Soc.* **131**(24), 8398–8400 (2009).
63. R. Liu, X. Zhao, T. Wu and P. Feng, Tunable redox-responsive hybrid nanogated ensembles, *J. Amer. Chem. Soc.* **130**(44), 14418–14419 (2008).
64. J. Lu, E. Choi, F. Tamanoi and J. I. Zink, Light-activated nanoimpeller-controlled drug release in cancer cells, *Small* **4**(4), 421–426 (2008).
65. X. Mei, S. Yang, D. Chen, N. Li, H. Li, Q. Xu, J. Ge and J. Lu, Light-triggered reversible assemblies of azobenzene-containing amphiphilic copolymer with β-cyclodextrin-modified hollow mesoporous silica nanoparticles for controlled drug release, *Chem. Commun.* (*Cambridge*) **48**(80), 10010–10012 (2012).
66. J. Lu, M. Liong, Z. Li, J. I. Zink and F. Tamanoi, Biocompatibility, biodistribution, and drug-delivery efficiency of mesoporous silica nanoparticles for cancer therapy in animals, *Small* **6**(16), 1794–1805 (2010).
67. J. Lu, Z. Li, J. I. Zink and F. Tamanoi, *In vivo* tumor suppression efficacy of mesoporous silica nanoparticles-based drug-delivery system: enhanced efficacy by folate modification, *Nanomed. Nanotechnol. Biol. Med.* **8**(2), 212–220 (2012).
68. L. Pan, Q. He, J. Liu, Y. Chen, M. Ma, L. Zhang and J. Shi, Nuclear-targeted drug delivery of TAT peptide-conjugated monodisperse mesoporous silica nanoparticles, *J. Amer. Chem. Soc.* **134**(13), 5722–5725 (2012).
69. S. M. Janib, A. S. Moses and J. A. MacKay, Imaging and drug delivery using theranostic nanoparticles, *Adv. Drug Deliv. Rev.* **62**(11), 1052–1063 (2010).
70. K. Berg, P. K. Selbo, A. Weyergang, A. Dietze, L. Prasmickaite, A. Bonsted, B. O. Engesaeter, E. Angell-Petersen, T. Warloe, N. Frandsen and A. Hogset, Porphyrin-related photosensitizers for cancer imaging and therapeutic applications, *J. Microsc.* (*Oxford*) **218**(2), 133–147 (2005).
71. I. Roy, T. Y. Ohulchanskyy, H. E. Pudavar, E. J. Bergey, A. R. Oseroff, J. Morgan, T. J. Dougherty and P. N. Prasad, Ceramic-based nanoparticles entrapping water-insoluble photosensitizing anticancer drugs: A novel drug-carrier system for photodynamic therapy, *J. Am. Chem. Soc.* **125**(26), 7860–7865 (2003).
72. X. He, X. Wu, K. Wang, B. Shi and L. Hai, Methylene blue-encapsulated phosphonate-terminated silica nanoparticles for simultaneous in vivo imaging and photodynamic therapy, *Biomaterials* **30**(29), 5601–5609 (2009).
73. K. M. L. Taylor-Pashow, J. Della Rocca, Z. G. Xie, S. Tran and W. B. Lin, Postsynthetic modifications of iron-carboxylate nanoscale metal-organic frameworks for imaging and drug delivery, *J. Amer. Chem.l Soc.* **131**(40), 14261–14263 (2009).
74. C. E. Ashley, E. C. Carnes, G. K. Phillips, D. Padilla, P. N. Durfee, P. A. Brown, T. N. Hanna, J. Liu, B. Phillips, M. B. Carter, N. J. Carroll, X. Jiang, D. R. Dunphy,|

C. L. Willman, D. N. Petsev, D. G. Evans, A. N. Parikh, B. Chackerian, W. Wharton, D. S. Peabody and C. J. Brinker, The targeted delivery of multicomponent cargos to cancer cells by nanoporous particle-supported lipid bilayers, *Nat. Mater.* **10**(5), 389–397 (2011).

75. J. E. Lee, N. Lee, H. Kim, J. Kim, S. H. Choi, J. H. Kim, T. Kim, I. C. Song, S. P. Park, W. K. Moon and T. Hyeon, Uniform Mesoporous dye-doped silica nanoparticles decorated with multiple magnetite nanocrystals for simultaneous enhanced magnetic resonance imaging, fluorescence imaging, and drug delivery, *J. Am. Chem. Soc.* **132**(2), 552–557 (2010).

Chapter 12

Carbon Nanotubes in Biomedical Applications

Krunal K. Mehta[*,†], *Elena E. Paskaleva*[*,†],
Jonathan S. Dordick[*,†,‡,§,¶,**] *and Ravi S. Kane*[*,†,||]

Howard P. Isermann Department of Chemical and Biological Engineering, Rensselaer Polytechnic Institute, Troy, NY, USA
[†]*Center for Biotechnology and Interdisciplinary Studies, Rensselaer Polytechnic Institute, Troy, NY, USA*
[‡]*Department of Materials Science and Engineering, Rensselaer Polytechnic Institute, Troy, NY, USA*
[§]*Department of Biology, Rensselaer Polytechnic Institute, Troy, NY, USA*
[¶]*Department of Biomedical Engineering, Rensselaer Polytechnic Institute, Troy, NY, USA*
[||]*kaner@rpi.edu*
***dordick@rpi.edu*

Abstract

Carbon nanotubes (CNTs) have attracted considerable interest in recent years on account of their unique properties. CNTs have an exceptionally high surface area-to-volume ratio, which makes them a convenient platform for the surface presentation of biological molecules, ranging from antibodies to enzymes. They have a higher tensile strength than steel fibers with equivalent dimensions and good biocompatibility, which has made them a focus of research in regenerative medicine. This chapter serves as a review of the latest advances related to biomedical applications of CNTs. Specifically, we focused on the newest materials and their applications that were developed in the past decade, such as targeted drug delivery for cancer therapy, vaccine delivery and the delivery of nucleic acids for gene therapy. CNTs have been incorporated into alternative anticancer strategies such as thermal ablation and radiotherapy. CNTs are also a promising scaffolding material for biomaterial composites and diagnostic nanosensors. Herein, we also discuss our studies of protein stabilization on nanomaterials and the development of nanocomposites incorporating enzyme–nanotube conjugates that have antimicrobial properties.

1. Introduction

In recent years, there has been considerable progress in the synthesis of nanomaterials of many different shapes and sizes with unique mechanical, electrical, thermal, optical, physicochemical and structural properties.[1] In particular, following the discovery of carbon-based nanomaterials such as CNTs, carbon-based nanotechnology has rapidly become a platform technology for a variety of uses, including biomedical applications.[2] CNTs are ordered, hollow nanostructures consisting of carbon atoms bonded to each other via *sp*2 bonds. An individual CNT has the structure of a graphene sheet rolled into a cylinder. CNTs are broadly categorized as either single-walled carbon nanotubes (SWNTs) or multi-walled carbon nanotubes (MWNTs), the latter consisting of multiple concentric tubes.[2] SWNTs usually have diameters ranging from 0.5 to 1.5 nm and lengths ranging from 100 nm to several centimeters.[3] MWNTs have larger diameters (can be more than 100 nm) than SWNTs due to multilayer structures.[4] CNTs can be metallic or semiconducting, depending on their diameter and helicity. The ability to introduce defects into the hexagonal network of a single CNT, thereby altering its helicity and electronic properties, may help to create nanodevices with applications in nano-electronics and sensing. In addition, their low coefficient of thermal expansion and high capacity to absorb heat make CNT-based composites very attractive for use as high-performance thermal management materials in heat sinks and energy storage materials.[5] The exceptional mechanical properties of CNTs may be useful for the design of light-weight and rugged composites and fibers.[6] Finally, the ability to effectively interface various biomolecules with CNTs[2,7,8] and the discovery that biomolecule function on nanomaterials differs from that on macroscopic supports has resulted in numerous applications of biofunctionalized CNTs.

This chapter serves as a review of some of the promising recent advances related to biomedical applications of CNTs. Specifically, we discuss applications of CNTs in the delivery of biomolecules, the development of scaffolds for tissue engineering, sensing, the design of antifouling nanocomposites and the design of antimicrobial materials.

2. CNTs in Targeted Drug Delivery Systems for Cancer Therapy

Despite significant advances over the past few years, cancer therapy strategies still suffer from severe drawbacks. One of the main limitations is the lack of selectivity, which results in severe side effects and limits patient compliance. Multi-drug resistance (MDR) is another major problem and is often ascribed to the activity of P-glycoprotein (P-gp), an efflux pump that can recognize the internalized drug and transport it out of the cell, thereby preventing it from exerting its cytotoxic action.[9]

Nanomaterials can help to overcome such limitations by targeting cancer cells either in non-specific, passive ways or via specific ligand–receptor interactions. The passive targeting is mediated by the so-called enhanced permeability and retention (EPR) effect (Fig. 1).[9] In particular, functionalized CNTs have shown great promise as novel delivery systems based on their ability to cross biological barriers. Although different mechanisms of internalization (endocytosis and needle-like penetration) have been proposed, it is generally recognized that CNTs are able to enter cells independent of the cell type and the functional groups on the CNT surface.[9–11] Cancer tissue is usually characterized by leaky and defective blood vessels with large fenestrations, due to extensive endothelial cell disorganization.[9] The consequence is the enhanced extravasation of macromolecules in the tumor tissue, in contrast to low-molecular-weight molecules that undergo rapid renal clearance.[9] Moreover, slower venous return and poor lymphatic drainage can lead to the retention of accumulated macromolecules in the tumor (Fig. 1).[9] Consequently, nanoscale drug delivery systems are likely to be more effective than the free

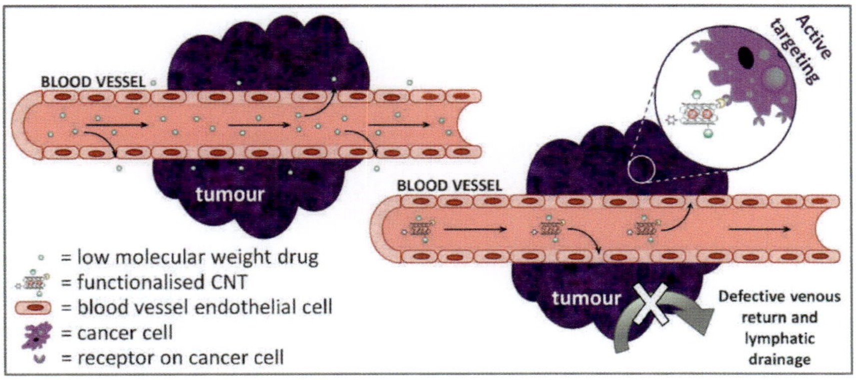

Fig. 1. A schematic representation of the enhanced permeability and retention (EPR) effect (reproduced with permission from Fabbro *et al.*[9] Copyright (2012). The Royal Society of Chemistry).

drug alone (Fig. 1).[9] In addition, the delivery of a drug through a carrier could involve different metabolic and cellular pathways, thus offering a way to elude MDR.

The high surface area of CNTs provides multiple attachment sites for biomolecules, allowing for polyvalent derivatization.[12–15] As a result, the effectiveness of CNTs can be further enhanced by attaching targeting ligands to the CNTs such as antibodies or other targeting agents that can recognize specific tumor markers.[9] Moreover, multiple *in vitro* and *in vivo* studies have shown that many types of chemically functionalized CNTs are biocompatible with the biological milieu, highlighting how the *in vivo* behavior of this material could be modulated by the degree and type of functionalization, both critical aspects that need to be accurately controlled.[9,16–19] We describe below some of the most significant results for CNT-based drug delivery for cancer treatment.[9]

2.1. *Delivery of chemotherapeutics*

2.1.1. **CNT–doxorubicin complexes**

In some of the earlier investigations in this field, doxorubicin (Dox) was conjugated with polyethylene glycol (PEG)-functionalized SWNTs or MWNTs coated with biocompatible amphiphilic diblock copolymer such as poly(ethylene glycol-b-propylene sulfide).[20–22] When tested *in vitro* against human breast cancer cells (MCF7), increased cell death was observed with the MWNT–Dox complex compared to that for treatment with Dox alone.[21] The *in vivo* pharmacokinetics and biodistribution of SWNT-DOX (doxorubicin conjugated with branched polyethylene glycol (PEG)-functionalized-SWNTs) was investigated by injecting SWNT–DOX *via* the tail vein into SCID (Severe Combined Immune Deficient) mice bearing Raji lymphoma xenografts. A greater inhibition of tumor growth was seen with SWNT-DOX compared to the free drug, probably due to the higher tumor uptake and prolonged circulation half-life of the complex.[23]

Targeted Dox delivery was further tested using a cyclic arginine–glycine–aspartic acid (RGD) peptide, which acts as a recognition motif for integrin $\alpha_v\beta_3$ receptors that are overexpressed in a wide range of solid tumors.[9,24] SWNTs were non-covalently functionalized with phospholipid–PEG (PL-PEG) followed by conjugation with thiolated RGD *via* a sulfo-SMCC linker.[24] These conjugates were labeled with ^{64}Cu by the formation of a stable complex of ^{64}Cu with DOTA, which was conjugated to SWNT-PEG *via* a sulfo-NHS linker.[24] The biodistribution of radiolabeled conjugates in mice was investigated by *in vivo* positron emission tomography (PET), *ex vivo* biodistribution and Raman spectroscopy.[24] When mice bearing subcutaneous integrin $\alpha_v\beta_3$-positive U87MG tumors were intravenously injected with these conjugates, they showed tumor uptake of ~10–15% injected dose (ID) *per* gram of weight of mouse, a significant increase compared to the derivative without RGD.[24] The tumor/muscle uptake ratio was > 15. The high tumor accumulation was attributed to the long blood circulation time, the high tumor-binding affinity of RGD-functionalized SWNT–PEG and the multivalency effect (i.e. multiple RGDs along a SWNT binding to multiple integrin $\alpha_v\beta_3$ simultaneously).[24]

Dox and mitoxantrone have been shown to attach to the surface of CNTs by π-π interactions. In addition, both drugs contain pH-sensitive amine groups, which makes their binding to nanotubes pH-dependent (enhanced at high pH and reduced at low pH[20]). This pH-dependence may enable drug release in acidic endosomes and in the microenvironment of tumors.[9,25] A triple-functionalized SWNT derivative was prepared, bearing Dox, fluorescein and a targeting antibody that recognizes the carcinoembryonic antigen (CEA), a tumor marker for a variety of adenocarcinomas. This construct was designed to target CEA-overexpressing cancer cells, to be internalized by endocytosis and to release Dox at lysosomal acidic pH. Fluorescein was introduced to study the internalization of the SWNT construct by WiDr human colon cancer cells, and it was observed that the control fluorescein-labeled BSA did not undergo internalization to the same high extent as seen for the SWNT construct, indicating the ability of CNTs to deliver desired cargo.[9,25,26]

With the two-fold aim of obtaining a targeted drug delivery system while overcoming MDR, SWNTs were functionalized via amidation with an antibody (Ab) directed toward P-gp, one of the main players in MDR.[9,27,28] Subsequently the Ab–SWNT construct was non-covalently loaded with Dox via π-π stacking.[9,29] This construct was tested *in vitro* on human leukemia cells that overexpressed P-gp and were therefore resistant to Dox.[9,29] Cells were exposed to near-infrared (NIR) radiation to trigger doxorubicin release.[9,29] Given the ability of CNTs to absorb NIR radiation and the fact that Dox desorption from SWNTs is an endothermic process, exposure of Ab-SWNTs to NIR radiation was expected to improve the release of Dox from Ab-SWNTs.[29] Indeed, exposure to NIR radiation resulted in a tripling of the amount of doxorubicin released in 24 h, although partial release was observed without NIR radiation as well.[9,29] The Ab–SWNT construct appeared bound to the cell membrane, while Dox was localized inside cells in much higher quantities than for treatments with the free drug. The SWNT–Dox construct also inhibited cell growth more efficiently than an equivalent concentration of the free drug.[9,29] Such non-covalent drug-delivery is a promising strategy for delivering drugs to different types of cells that combines minimal loss of drug activities with the controllable release of drugs from CNTs by NIR radiation.[29]

2.1.2. CNT–methotrexate constructs

Methotrexate (MTX), another clinically used anticancer drug, also suffers from low bioavailability and toxicity.[9] MWNTs were covalently functionalized simultaneously with methotrexate and fluorescein. Although rapid internalization of the constructs was observed in human Jurkat T lymphocytes, the efficacy of the CNT–MTX construct was lower than that of the free drug.[12] A more recent study used MTX that was bound to MWNTs through a peptide linker recognized by intracellular proteases. Administration of these conjugates resulted in a significant decrease in the observed viability of the MCF-7 breast cancer cell line, compared to that seen with free (unbound) MTX. The use of the cleavable peptide linker may have facilitated the release of the drug from the conjugate after its internalization.[30]

2.1.3. CNT–taxane constructs

Paclitaxel (PTX), a member of the taxane class of antineoplastic drugs, was covalently attached to non-covalently PEGylated SWNTs and was intravenously injected into xenograft tumor-bearing mice (a PTX resistant 4T1 murine breast cancer mice model).[31] The introduction of SWNT–PTX resulted in higher apoptosis and lower cellular proliferation. The inhibition of the tumor growth with SWNT–PTX constructs was two to three times higher than that for PTX alone.[31] Analysis of the pharmacokinetics of SWNT–PTX conjugates showed that the conjugates had a longer half-life in blood circulation, compared to the drug alone. The increased half-life resulted in a much higher concentration of PTX in the reticuloendothelial system organs (liver/spleen) and the intestine 2 h after injection, compared to that for free PTX, possibly raising concerns of toxicity towards these organs.[9] However, the release of drug from the conjugate, due to ester cleavage by carboxylesterases, not only made the drug immediately available but also resulted in rapid excretion of the free drug, thereby lowering its toxicity.[9] It is important to emphasize the lack of toxicity in mice from the PEGylated SWNTs themselves, which has been demonstrated by the same authors in another study.[32] The PTX levels in tumors were 10 times higher for the SWNT–PTX derivative than for the free PTX, 2 h post-injection, and the tumor-to normal organ/

tissue PTX uptake ratio was higher, indicating the better selectivity of the CNT–drug conjugate.[9]

Another study with the same class of therapeutic compounds emphasizing the selectivity of CNT–drug conjugates is described below. Fluorescently labeled polymers linked to targeting agents such as vitamin B_{12}, folate or biotin were used to examine the relative uptake of these polymers by various types of tumor cell lines both *in vitro* and *in vivo*.[33] These studies have shown that while most tumor types over-express receptors for folate, or vitamin B12, in either case, there is also a greatly increased expression of a yet-to-be identified biotin receptor.[33] Chen *et al.* prepared bifunctional biotin–SWNT–taxoid conjugates by first attaching biotin to oxidized SWNTs and then covalently attaching taxoid molecules to these biotin–SWNT conjugates via a linker containing a disulfide bond.[34] This construct was intended to be a prodrug that would bind to cancer cells which overexpress the receptor for biotin on their surface and be internalized by receptor-mediated endocytosis.[9,34] Once inside the cell, the prodrug was designed to liberate its cargo upon reduction of the disulfide bond by endogenous thiols such as glutathione (GSH), whose concentrations are typically more than one order of magnitude higher in tumor tissues than in blood plasma.[9] As a result, the system should act specifically on cancer cells, leading the taxoid to carry out its mitosis inhibition action only in the desired sites.[9] The biotin conjugate, as foreseen, was internalized by receptor-mediated endocytosis, the uptake being far higher than in cells not expressing the biotin receptor.[9] The release of the drug by GSH and its binding to the microtubule network, where it exerts its cytotoxic action, was efficient, with an IC_{50} value smaller than the one for the drug alone, probably due to an increase in intracellular delivery.[9,34]

2.1.4. **CNT–platinum constructs**

There are several studies on the use of Pt-based targeted prodrug systems for cancer treatment.[35–37] One such study reported a route for preparing Pt (IV) complex containing a cell-receptor–targeting moiety and also a functional group that could be conjugated to a nanoscale delivery system. The Pt(IV) compound is a derivative of cisplatin (*cis*-dichloro diammineplatinum(II)) or *c*-$[Pt(NH3)_2Cl_2]$, to which two axial ligands, L1 and L2 (Fig. 2) are introduced, one of which can be attached to folic acid for cellular targeting while the other axial site offers a position for conjugation to a nanocarrier for efficient delivery.[38] Dhar *et al.* conjugated folic acid to Pt (IV) complex to target human cells highly overexpressing the folate receptor.[39] This conjugate was then tethered through an amide bond to the terminal amines of

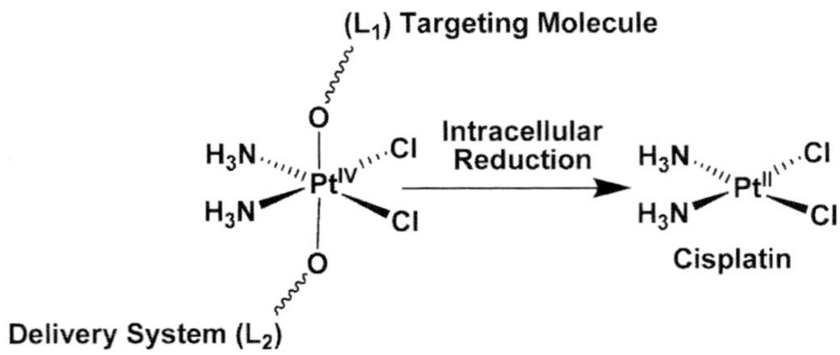

Fig. 2. A schematic representation of Pt (IV) compound and its conversion to Cisplatin upon internalization into cells. L_1 is ($O_2CCH_2CH_2CONH$–PEG–FA), L_2 is ($O_2CCH_2CH_2CO_2H$) (reproduced with permission from Dhar *et al.*[38] Copyright (2008) American Chemical Society).

phospholipid–PEG chains wrapped around SWNTs to use SWNTs as the delivery system.[38]

The Pt(IV) can be reduced inside endosomes, resulting in the formation of the active Pt(II) compound (cisplatin). The authors demonstrated that the internalization of this construct takes place by folate receptor-mediated endocytosis.[38] The system was active in killing cancer cells (human choriocarcinoma and human nasopharyngeal carcinoma) with IC_{50} values more than eight times lower than that for cisplatin alone. This compound, therefore, acts as a prodrug, generating the cytotoxic derivative (cisplatin) once internalized, and thus selectively killing cells that over-express the folate receptor.[38] Along these lines, Bhirde *et al.* described the targeted killing of cancer cells using a cisplatin-delivery system based on drug–SWNT covalent conjugates wherein epidermal growth factor (EGF) was used to target CNTs to the receptor (EGFR), which is over-expressed in cancer cells.[40] *In vitro* imaging studies with head and neck squamous carcinoma cells (HNSCC) overexpressing EGFR using quantum dot luminescence and confocal microscopy showed that SWNT–Qdot–EGF conjugates internalized rapidly into the cancer cells.[40] Limited uptake occurred for control cells without EGFR, and uptake was blocked by siRNA knockdown of EGFR in cancer cells, revealing the importance of EGF–EGFR binding.[40] *In vivo* imaging revealed that SWNT–Qdot–EGF conjugates injected into live mice were selectively taken up by HNSCC tumors, but SWNT–Qdot controls with no EGF were cleared from the tumor region in less than 20 min.[40] Most significantly, regression of tumor growth was rapid in mice treated with targeted SWNT–cisplatin–EGF relative to non-targeted SWNT–cisplatin that did not feature EGF–EGFR binding. This result demonstrated that SWNTs can effectively and selectively deliver cisplatin to EGFR over-expressing tumors.[40] Similar studies with the antineoplastic drug, 10-hydroxycamptothecine (HCPT), covalently attached to MWNTs showed that tumor growth inhibition was much more efficient with the conjugates than with HCPT alone, without any severe toxicity.[41]

2.2. *Vaccine delivery*

The basic concept for utilizing CNTs in vaccine delivery is to link the antigen to CNTs while retaining the antigen's native conformation, thereby inducing an antibody response with the correct specificity. Pantarotto *et al.* covalently linked a neutralizing B cell epitope from the foot-and-mouth disease virus (FMDV) to mono- and bis-derivatized CNTs.[42] Peptide–CNTs elicited strong antipeptide–antibody responses in mice with no detectable cross-reactivity to the CNTs. However, only the mono-derivatized CNT conjugate induced high levels of virus-neutralizing antibodies (Fig. 3).[42]

The ability of CNTs to present biologically important epitopes in an appropriate conformation both *in vitro* and *in vivo* opens up the possibility for their use in the delivery of the immunogenic component of the vaccine.[43,44] Current immunotherapies employ tumor cell vaccines (TCVs) made of inactivated cancer cells or dendritic cells presenting tumor antigens to trigger the immune response of the patient against the tumor itself.[45] While these treatments have proven to be effective, there is growing interest in improving the efficacy of immunotherapy.[46] To that end, a number of researchers have

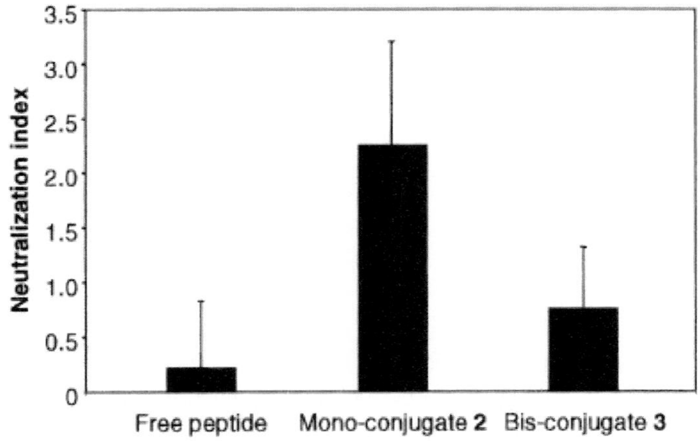

Fig. 3. Neutralization indices of serum samples of immunized mice (reproduced with permission from Pantarotto *et al.*[42] Copyright (2003). Elsevier Science Ltd).

investigated the antitumor immunogenicity of tumor lysate proteins.[47–50] Tumor lysate proteins contain a mixture of various proteins that are readily obtainable, unlike tumor-specific antigens that are rare and difficult to obtain. Moreover, tumor lysate proteins are a very effective adjuvant in antitumor immunotherapy.[45–50]

With the aim of improving the efficacy of TCV, Meng *et al.* covalently coupled tumor lysate proteins to oxidized MWNTs via an amide bond and subcutaneously injected the conjugates into H22 hepatoma-bearing mice treated with TCV. Various treatments were performed, and the treatment groups included the group that received TCV only (TCV), the group that received TCV and H22p (TCV+H22p), the group that received TCV and CNTs (TCV+CNT), and the group that received TCV and CNTs conjugated to H22p (TCV+CNT-H22p).[45] The cure rate for (TCV+CNT-H22p) was significantly increased with respect to animals treated only with TCVs or with TCVs in conjunction of tumor lysate proteins (TCV+H22p), although a partial effect was exerted by the CNTs alone (TCV+CNT) (Fig. 4).[45] To assess whether the immunity was tumor-specific, the animals that survived after treatment with tumor lysate proteins bearing MWNTs were challenged again with a subcutaneous injection of tumor cells. In the case of H22 cells, the animals rejected the tumor, while they did not reject cancerous cells from another mouse breast cancer, demonstrating the specificity of the immune response.

Similarly, Villa *et al.* studied the ability of SWNTs to act as antigen-presenting carriers, in order to improve the response to weakly immunogenic

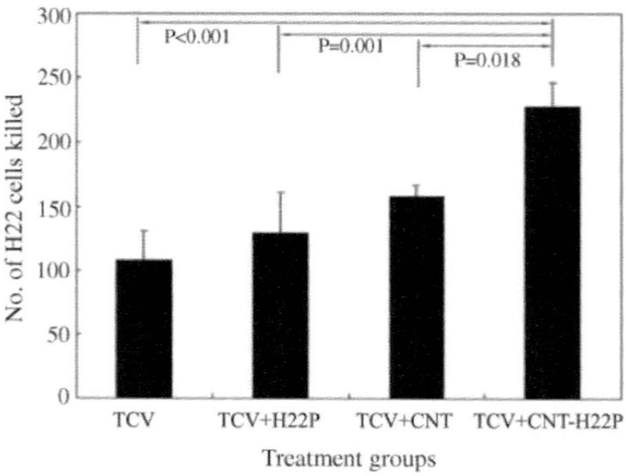

Fig. 4. Neutralization indices of serum samples of immunized mice (reproduced with permission from Meng *et al.*[45] Copyright (2008). Wiley-VCH Verlag GmbH & Co. KGaA, Weinheim).

peptides.[44] The Wilm's tumor protein (WT1), which has attracted attention as a vaccine target for many types of human leukemia and cancers, was covalently conjugated to SWNTs. It was found that SWNT–WT1 conjugates were rapidly taken up *in vitro* by antigen-presenting cells in a dose-dependent manner. Interestingly, *in vivo* immunization of BALB/c mice using SWNT–WT1 and an adjuvant caused a specific humoral immune response, which was not seen against the peptide alone or against the peptide mixed with the adjuvant, proving the potential of the SWNT to enhance the immunogenicity of peptides.

2.3. *Delivery of nucleic acids*

The interaction of a wide range of biomolecules, including nucleic acids with CNTs, has been extensively studied for their potential applications in treating many diseases, such as cancer, HIV and hepatitis[51–54] that have a genetic component.[52,55,56] CNTs have been used in experimental settings for gene delivery to correct genetic disorders as well as for the delivery of silencing RNA to cancerous tissue. The use of antisense oligonucleotides and small interfering RNAs (siRNAs) represent promising approaches to inhibit protein expression and block cellular pathways. Cancer therapy is one possible application, wherein the targets are oncogenes or genes involved in angiogenesis or chemotherapy resistance. Cationic SWNTs have been used to complex siRNA that had been designed to silence the expression of telomerase reverse transcriptase. These conjugates inhibited cell growth *in vitro*, in both murine and human cell lines, and *in vivo*, after intra-tumoral injection in xenografted mice.[57] Dai and co-workers demonstrated SWNTs as non-viral molecular transporters for the delivery of siRNA into human T cells and primary cells. The delivery ability and RNA interference efficiency of CNTs exceeded those of several non-viral transfection agents, including various liposome formulations. It was suggested that CNTs could be used as generic molecular transporters for various types of biologically important cells, from cancer cells to T cells and primary cells, with superior silencing effects over conventional liposome-based non-viral agents.[24,58]

Targeted delivery of nucleic acids using CNTs as a carrier and employing the use of a targeting unit like folic acid has also been demonstrated, wherein an increased uptake for the derivative with the folic acid was observed compared to the one without the targeting unit. Dendrimer–CNT constructs have also been exploited for gene delivery. The antisurvivin oligonucleotide was complexed to the polyamidoamine (PAMAM) dendrimer covalently attached to CNTs and used to transfect MCF-7 cells. The conjugate was distributed mainly in the cytoplasm, endosomes and lysosomes and was able to release the antisense oligonucleotide, inducing apoptosis. The PAMAM dendron was also directly grown on a MWNT surface and used as an anchoring point for siRNA, after introducing a trimethylammonium unit on each branch termination. Different dendron generations were prepared, and an improved efficacy in delivering a fluorescent oligonucleotide was observed with an increase in the degree of dendritic branching. The gene silencing capability of the system was also confirmed.[59,60]

2.4. *Alternative anticancer strategies: thermal ablation and radiotherapy*

Thermal ablation, which makes use of the intrinsic optical properties of various nanomaterials such as gold nanoshells, nanoparticles, nanorods and CNTs to develop hyperthermia, is another interesting strategy for treating cancer.[61–67] The optical absorbance of CNTs is very high in the NIR region, 700–1100 nm, where biological systems are transparent.[68,69] As a result, CNTs require lower laser power and shorter radiation times to effect cancer cell destruction compared to other NIR-absorbing nanomaterials studied.[66,70] Dai and co-workers demonstrated that oligonucleotides transported inside living cells by CNTs can be translocated into the cell nucleus upon endosomal rupture triggered by NIR laser pulses.[70] They also showed that continuous NIR radiation can cause cell death because of excessive heating of SWNTs *in vitro*.[70] Furthermore, selective cancer cell destruction was demonstrated by the functionalization of SWNTs with a folate moiety, which facilitates selective internalization of SWNTs inside cells labeled with folate receptor tumor markers.[70] Irradiation with NIR light triggered cell death without harming the receptor-free cells.[70] Thus, the transporting capabilities of CNTs, combined with suitable functionalization chemistry, and their intrinsic optical properties can lead to development of new classes of nanomaterials for drug delivery and cancer therapy.[70]

Another promising anticancer strategy is radiotherapy. The main purpose of attaching radioisotopes to CNTs is for imaging studies to determine their biodistribution following radioisotope traces. However, recent focus has shifted toward the use of radioisotope–nanomaterial conjugates for therapeutic application. A SWNT derivative bearing an ^{111}In chelate used in biodistribution studies was then further functionalized with a tumor-specific monoclonal antibody, and delivered *in vivo* to a murine model of disseminated human lymphoma, showing selective tumor targeting.[71] Therefore, the replacement of indium with an alpha-particle–emitting radionuclide such as actinium-225 would render these constructs ideal for application in radiotherapy. For instance, Ruggiero *et al.* used antibody E4G10, that recognizes a monomeric form of cadherin expressed on angiogenic endothelial cells, to target the tumor neovasculature for radiotherapy.[72] They prepared doubly functionalized SWNTs by binding both the specific antibody and the alpha-particle–emitting ^{225}Ac to CNTs and tested the conjugate *in vivo*, via intravenous injection in xenografted tumor mice. The conjugates reduced tumor growth and improved mice survival compared with controls bearing a different antibody.[72] This approach is promising, as in principle, the targeting of tumor vasculature renders these constructs suitable for different kinds of tumors.

3. CNTs as Improved Biomaterials

As the field of tissue engineering advances, new tools for better monitoring and evaluating of engineered tissues along with new biomaterials to direct tissue growth are needed.[73] In the past decade, major advances in stem cell biology have transformed the world of tissue engineering and regenerative medicine.[4] Carbon-based nanomaterials may also make significant contributions to these fields (Table 1).[4] CNTs may enable improved tracking of cells, sensing of microenvironments, delivering of transfection agents and biocompatible scaffolding for tissue- and organ-engineering.[73] Using CNTs functionalized with fluorophores, heavy elements or radiotracer contrast agents for optical, magnetic resonance and radiotracer modalities would provide better means of evaluating tissue formation.[73] In addition, monitoring and altering intracellular and intercellular processes would be useful for the design of engineered tissues.[73] CNTs can also be incorporated into tissue scaffolds, providing the two-fold advantage of structural reinforcement and utilizing their properties such as electrical conductivity to aid and direct the cell growth.[73]

Many studies have shown that mechanical properties are significantly improved when CNTs are dispersed in conventional polymer solutions; some of the key results are summarized here. MWNTs blended with chitosan showed significant improvements in mechanical properties (increasing tensile modulus from 1.08 GPa to 2.15 GPa and from 37.7 MPa to 74.3 MPa for tensile strength) compared with those of chitosan alone.[74] Highly crosslinked, three-dimensional (3D) hydrogel microenvironments have high stiffness, which might limit cellular proliferation, migration and morphogenesis. To create 3D hydrogels with controlled mechanical properties, Shin *et al.* coated CNTs with a thin layer of photo-crosslinkable gelatin methacrylate (GelMA), which is known to allow for cell encapsulation and proliferation.[75] The addition of CNTs significantly improved the mechanical properties of GelMA

Table 1. Extracellular matrix (ECM) complexities and use of CNTs to create biomaterials that mimic the native ECM.[4]

Organ example	Natural tissue structure	Engineered CNT scaffold
Heart	The ECM of muscle tissue, like the heart, facilitates cardiomyocytes to couple mechanically to each other and to form elongated and aligned cell bundles that create an anisotropic syncytium	Nanogrooved surfaces and aligned CNTs embedded into a matrix are suitable for cardiac tissue engineering applications because they force cardiomyocytes to align.
Bone	Bone is a nanocomposite material that consists primarily of a collagen-rich organic matrix and inorganic hydroxyapatite nanocrystallites which serve as a chelating agent for ECM mineralization by osteoblasts.	Integrating CNTs and hydroxyapatite to create a stiff and porous scaffold.
Liver	Cells composing epithelial tissue are polarized and contact three types of surfaces for efficient mass transfer.	Scattered and non-aligned CNTs modified with surface molecules embedded in a matrix can promote cell adhesion and tissue polarity.

Source: Reproduced with permission from Webster *et al.*[4] Copyright (2012) Elsevier Ltd.

hydrogels without decreasing their porosity. Apart from polymer and hydrogel enhancement, CNTs can also be used to reinforce ceramic matrices.[76] Hence, CNTs can be used to enhance the mechanical properties of hydrogels, polymers and ceramics to create better scaffolds to mimic native tissues. Yana *et al.* explored the possibility of using polydopamine-coated CNTs (CNT–Pdop) for the deposition of apatite and evaluated their properties as a new improved biomaterial compared to pristine CNTs.[77] The improvement in the bone-forming activity of the CNT–Pdop was confirmed by soaking both CNTs and CNT–Pdop in simulated body fluid (SBF) for 24 h and comparing apatite formation in each case using X-ray diffraction (XRD), Fourier transform infrared spectroscopy (FTIR), high-resolution transmission electron microscopy (HR-TEM) and field emission scanning electron microscopy (FE-SEM). The cytocompatibility of the CNTs and CNTs–Pdop was also assessed using the 3-(4, 5)-dimethylthiahiazo (-z-y1)-3,5-di-phenytetrazoliumromide (MTT) assay.[77] The results of these tests demonstrate that CNT–Pdop yields better bone-forming activity and cytocompatibility than pristine CNTs (Fig. 5). The improvement of bioactive behavior was attributed to the structural advantages of CNTs as a framework material and the presence of

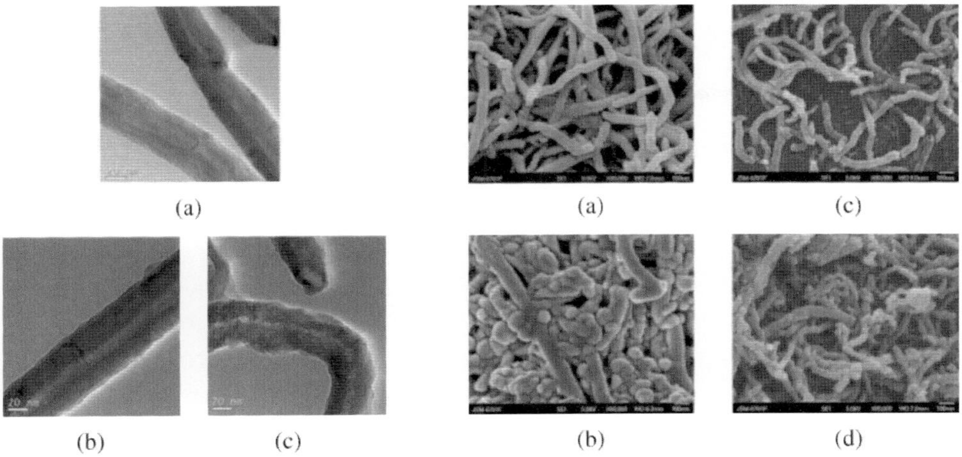

Fig. 5. **(left panel)** TEM images of the CNTs (a), CNT–Pdop before (b) and after (c) being soaked in SBF **(right panel).** SEM images of the CNTs (a, b) and CNT–Pdop (c, d) before and after being soaked in SBF (reproduced with permission from Yan *et al.*[77] Copyright (2011) Elsevier B.V).

abundant catecholamine moieties in Pdop that not only act as molecular anchors between CNTs and Pdop but also facilitate the binding of Ca^{2+} ions by chelation.[78] The polydopamine-coated CNTs may find potential applications in bone tissue engineering and other biomedical areas.[77]

In another line of investigation, composite materials comprising a collagen matrix with embedded CNTs were prepared by mixing solubilized Type-I collagen with solutions of SWNTs functionalized with carboxyl groups at concentrations of 0, 0.2, 0.4, 0.8 and 2.0 wt. %. Living smooth muscle cells were incorporated at the time of collagen gelation to produce cell-seeded collagen–CNT composite matrices. While constructs containing 0–0.8 wt % CNT compacted at the same rate as pure collagen controls, those containing 2.0 wt % CNT exhibited delayed gel compaction. Such collagen–CNT composite matrices may have utility as scaffolds in tissue engineering, or as components of biosensors or other medical devices.[79]

Vertically aligned carbon nanotubes (VACNTs) also hold promise for tissue regeneration.[80] Marsi *et al.* introduced a new biomimetic mineralization routine employing superhydrophilic VACNT films as highly stable template materials.[80] Biomineralization was facilitated by soaking VACNTs in a synthetic body fluid known to favor the calcification process.[80] The resulting biomineralized VACNT nanocomposites showed excellent bioactivity and human osteoblast adhesion.[80] Owing to the excellent material properties of biomineralized superhydrophilic VACNT nanocomposites and straightforward processing via biomimetic mineralization using SBF, this approach offers a new opportunity for carbon-based mineral nanocomposites for bone tissue regeneration.[80]

4. CNTs as Biosensors

A biosensor is a self-contained device that provides analytical information using a specific biological recognition element (a primary transducer) which is in contact with a non-biological part — an electrochemical transduction element — that acts as a signal conversion unit.[81] CNTs are convenient secondary transducers to detect physical changes and transform them into a measurable output, because their electronic properties are altered in a specific and measurable way by the analyte.[82–85] CNTs are on the same size scale as the molecules they detect, which opens an opportunity for the detection of single-molecule events.[86] The biochemical compatibility of CNTs with biological systems offers the opportunity to detect the presence and concentration of analytes through either non-covalent adsorption on the carbon surface through hydrophobic interactions[87] or by covalent attachment to the functional groups produced through chemical reactions.[88]

4.1. *Detection of small physiologically relevant molecules*

CNT-based sensors have been designed to monitor clinically relevant analytes such as glucose,[89–91] cholesterol[92,93] and ethanol.[94] For example, the superior electronic properties of CNTs were harnessed in devising a non-invasive glucose sensor that detects the analyte iontophoretically, or through the skin, owing to the better transfer of electrons between the biological recognition element (glucose oxidase) and the transducer.[95] In another recent development, a CNT-based sensor was used to detect blood cholesterol with considerably higher sensitivity than previously reported.[96] The use of CNTs in the detection of biological analytes has led to the development of an integrated sensing platform for the simultaneous detection of glucose, glycosylated hemoglobin, cholesterol and triglycerides in a single blood sample[93] — a potential gateway to monitoring a wide spectrum of clinical parameters with minimal inconvenience to the patient.

4.2. *Detection of biological macromolecules (nucleic acids and proteins)*

CNT-based sensors have been developed to selectively detect DNA[97] and map the genetic fingerprint of pancreatic cancer.[98] An electrochemical sandwich-type, gene-sensing electrode has been recently designed, where SWNTs could significantly amplify the electrochemical signal, enabling target DNA to be detected selectively and sensitively.[99]

Given that many biomedical analyses rely on the accurate and sensitive detection of specific proteins, a rapidly growing field for application of the properties of the CNT[100] has recently emerged. The developments include an ultra-sensitive electrochemical sensor for the detection of cancer-related proteins present in nanogram to femtogram quantities in a mixed sample.[101] An important application of this type of biosensor is in the quantification of prostate-specific antigen.[102] Another study has focused on detection of immunoglobulin concentration in blood plasma, to assist in the diagnosis of immunological disorders.[103]

4.3. *Pathogen detection*

Rapid and sensitive detection of bacterial and viral pathogens is paramount for the prevention of epidemics and in parasitology. The last several years have seen a number of advances in the field of CNT-based pathogen detection. A disposable multiplex sensor was proposed for the simultaneous detection of several pathogenic bacteria in food samples.[104–106] A method for the colorimetric detection of *Staphylococcus aureus* in homogenous samples has been proposed.[107] A system for the detection of viruses has been proposed, that detects a model virus (T7 bacteriophage) within 5 min.[108] There have been reports of detection devices to monitor algae[109] and pathogenic fungi.[110]

4.4. *Olfactory sensors*

The olfactory sensors, often called "artificial noses," employ both the superior sensitivity of CNT-based sensors and the ability to create nanotube forests of primary transducers.[111,112] The mammalian olfactory system can detect many more chemical stimuli than the number of receptors it possesses, because it utilizes cross-reactive receptors that can generate a unique response to each stimulus.[113] Artificial noses, therefore, are being developed based on cross-reactive sensor arrays of CNTs.[114,115]

4.5. *Single-molecule detection*

CNT-based biosensors capable of resolving the adsorption and desorption of single molecules and measuring the local concentration of an analyte have been developed.[116] CNTs fluoresce in the NIR range (820–1600 nm), in wavelengths that are minimally absorbed by biological tissues. This property of the material has made it possible to incorporate it into sensors to directly measure analytes, such as reactive oxygen species (ROS) implicated in apoptotic cascades indicative of angiogenesis.[117]

5. CNT-Based Nanocomposite Films

5.1. *Structure and function of proteins on CNTs*

In one of the earliest reports on protein–nanotube interactions, Karajanagi *et al.*[118] studied the structure and function of proteins adsorbed onto SWNTs, using enzymatic activity as a highly sensitive probe of protein function.[8,118] Two structurally and functionally distinct enzymes — soybean peroxidase (SBP) and α-chymotrypsin (CT) were adsorbed onto SWNTs, to assess the influence of the hydrophobic, nanoscale environment on protein structure and function.[8] Activity measurements for the SWNT–SBP conjugates indicated that SBP retained ~30% of its native activity when loaded onto nanotubes at 50% of maximal surface coverage, while α-chymotrypsin retained no greater than 1% of its native activity.[8,118] The results of these studies suggested that attachment onto SWNTs affects the structure of the two enzymes in a fundamentally different manner.[8,118] Transmission mode Fourier transform infrared (FT-IR) spectroscopy and atomic force microscopy (AFM) were used to characterize the changes in secondary structure of the proteins and overall structural perturbations, respectively.[8]

In the case of SWNT–SBP conjugates, distinct globular structures were observed as shown in Fig. 6(a); line scans suggested that the height of these structures ranged from 4 to 6 nm, which matched very well with the molecular dimensions of SBP (6.1 × 3.5 × 4.0 nm).[8] On the other hand, the AFM image of SWNT–CT conjugates showed a layer of CT on the SWNT surface, as shown in Fig. 6(b). The height of the layer varied between 2 and 20 nm, which is far different from the molecular size of CT (5.1 × 4.0 × 4.0 nm), suggesting that significant unfolding and aggregation of CT occurred on the nanotubes.[118] Collectively, these results suggest that retention of protein structure and function were related, and that the extent of retention of protein structure and activity on CNTs was protein-specific.[8]

5.2. *Enhanced protein stability on nanomaterials*

The curvature of a nanoscale support can influence not just protein structure and activity but also protein stability. Asuri *et al.*[119,120] reported the ability of SWNTs to stabilize proteins under harsh conditions (e.g. at elevated

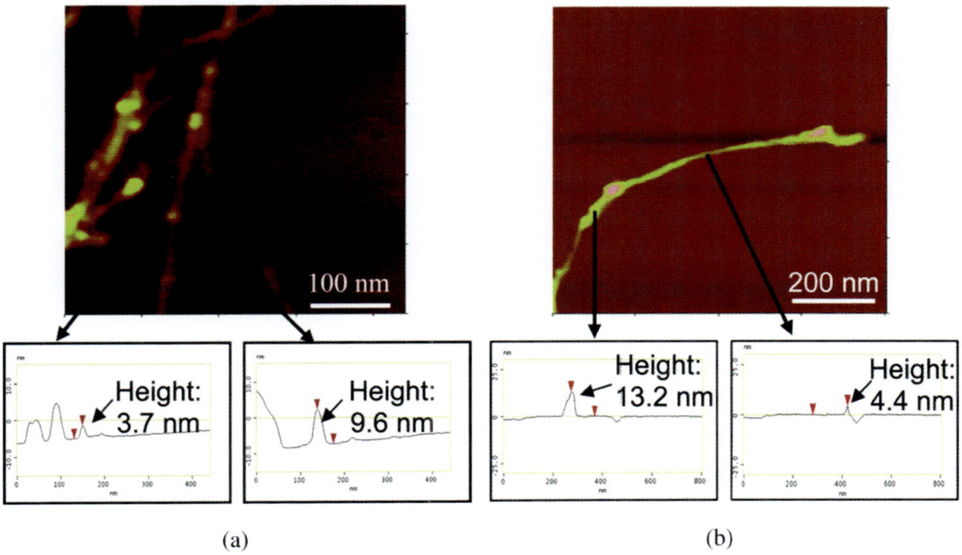

Fig. 6. AFM images of SBP and CT adsorbed onto SWNTs at half-maximal surface coverage. The globular structures seen on SWNTs in image (a) represent SBP molecules. Image (b) shows a non-uniform layer of CT on SWNT (reproduced with permission from Karajanagi *et al.*[118] Copyright (2004) American Chemical Society).

temperatures or in solutions containing organic solvents) to a greater extent than conventional flat supports.[8] For instance, the half-life of SBP adsorbed onto SWNTs at 95°C or in nearly neat methanol was approximately two-fold longer than that of SBP adsorbed onto flat supports under similar conditions (data shown in Fig. 7).[120]

These results suggest that lateral interactions between adjacent adsorbed proteins contribute to protein deactivation in harsh environments and that these unfavorable lateral protein–protein interactions are suppressed on highly curved supports such as SWNTs relative to flat surfaces (as shown in Fig. 8).[120] These results also suggest that the enhancements in protein stability should not be unique to SWNTs; similar enhancements in SBP stability were observed on other curved supports, including silica and gold nanoparticles and fullerene aggregates.[119]

5.3. *Polymer–nanotube–enzyme composites for antifouling applications*

As discussed above, depending on the enzyme, nanotube–enzyme conjugates were found to be active and stable. In this section, we show that CNTs serve

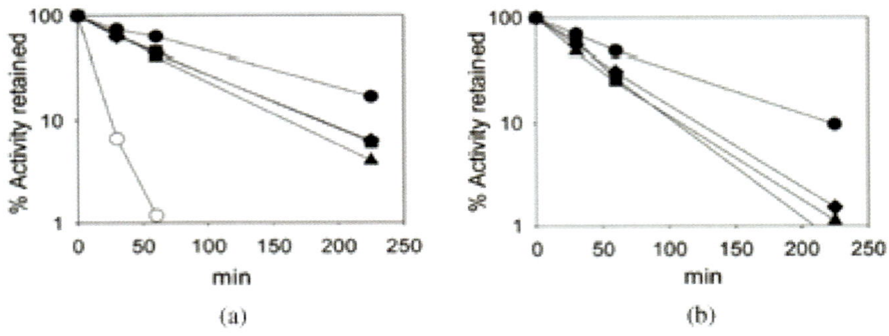

Fig. 7. Time-dependent deactivation of native SBP (open circles) and SBP on various supports — SWNTs (filled circles), highly ordered pyrolytic graphite or HOPG (diamonds), self-assembled monolayer (SAM) of undecanethiolate on gold (squares) and graphite flakes (triangles) (a) at 95°C and (b) in 100% methanol. The activities were normalized relative to the initial activity (activity at $t = 0$ min). Figure (b) does not contain % activity data for native SBP as the enzyme shows no activity in 100% methanol (reproduced with permission from Asuri *et al.*[120] Copyright (2006) American Chemical Society).

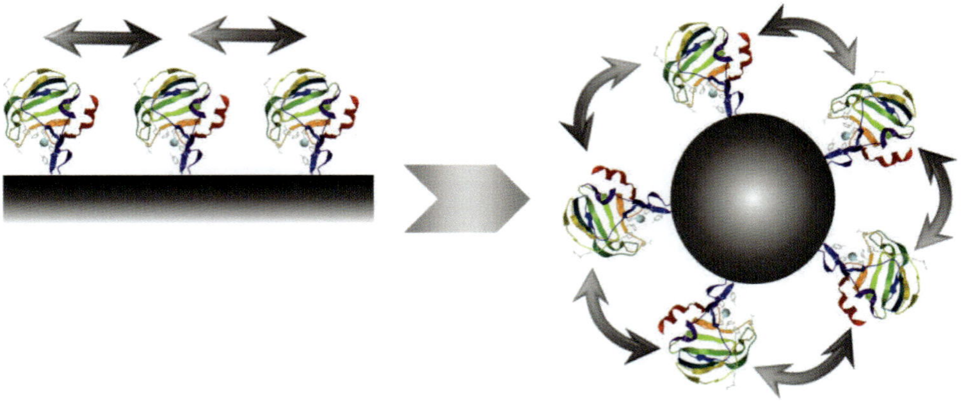

Fig. 8. Schematic depicting the reduced lateral interactions between enzyme molecules on a cylindrical support (such as a CNT) compared to a flat support. (Figure inspired by figure in Asuri *et al.*[120] Copyright (2006) American Chemical Society.)

as ideal supports for incorporating enzymes into polymeric matrices due to their large surface area that leads to large functional enzyme loadings and their high aspect ratio that aids in the retention of enzyme–nanotube conjugates in the matrix.

We have previously described the generation of highly stable and active polymer–nanotube–enzyme composites that prevent the fouling of surfaces by proteins (Fig. 6).[121] These nanocomposite films contained

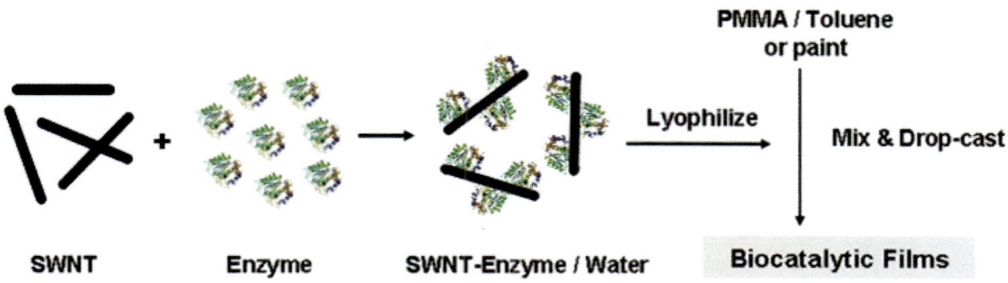

Fig. 9. Preparation of polymer–nanotube–enzyme-based biocatalytic films with antifouling properties (schematic representation) (reproduced with permission from Pangule *et al.*[8] Copyright (2009), Springer Science + Business Media, LLC).

proteases that carried out proteolytic degradation of adsorbed proteins thereby preventing non-specific protein adsorption.[121] Such antifouling nanocomposite films may ultimately find use in preventing the fouling of a variety of surfaces, ranging from membranes for downstream bioprocessing and water purification to medical instruments and implantable materials.[121]

5.4. *Nanocomposites incorporating nanotube–enzyme conjugates for antimicrobial applications*

We have designed nanocomposites incorporating conjugates of CNTs with cell-wall–degrading enzymes that have been shown to have potent bactericidal activity, even against antibiotic-resistant pathogens. To specifically target antibiotic resistant pathogens such as methicillin-resistant *Staphylococcus aureus* (MRSA), which is one of the primary causes of hospital-acquired infections and deaths, we incorporated conjugates of CNTs with lysostaphin, a cell-wall–degrading enzyme, into films to impart bactericidal properties against *S. aureus*.[123] We fabricated and characterized nanocomposites containing different conjugate formulations and enzyme loadings. These enzyme-based composites were highly efficient in killing MRSA (>99% within 2 h, tested against 4 different MRSA strains), as shown in Fig. 10. Additionally, these films were reusable and stable under dry storage conditions for a month (Fig. 11). Such enzyme-based film formulations may be used to prevent the growth of pathogenic and antibiotic-resistant microorganisms on various common surfaces in hospital settings. Polymer and paint films containing such antimicrobial conjugates, in particular, could be advantageous to prevent the risk of infection and biofouling.[122]

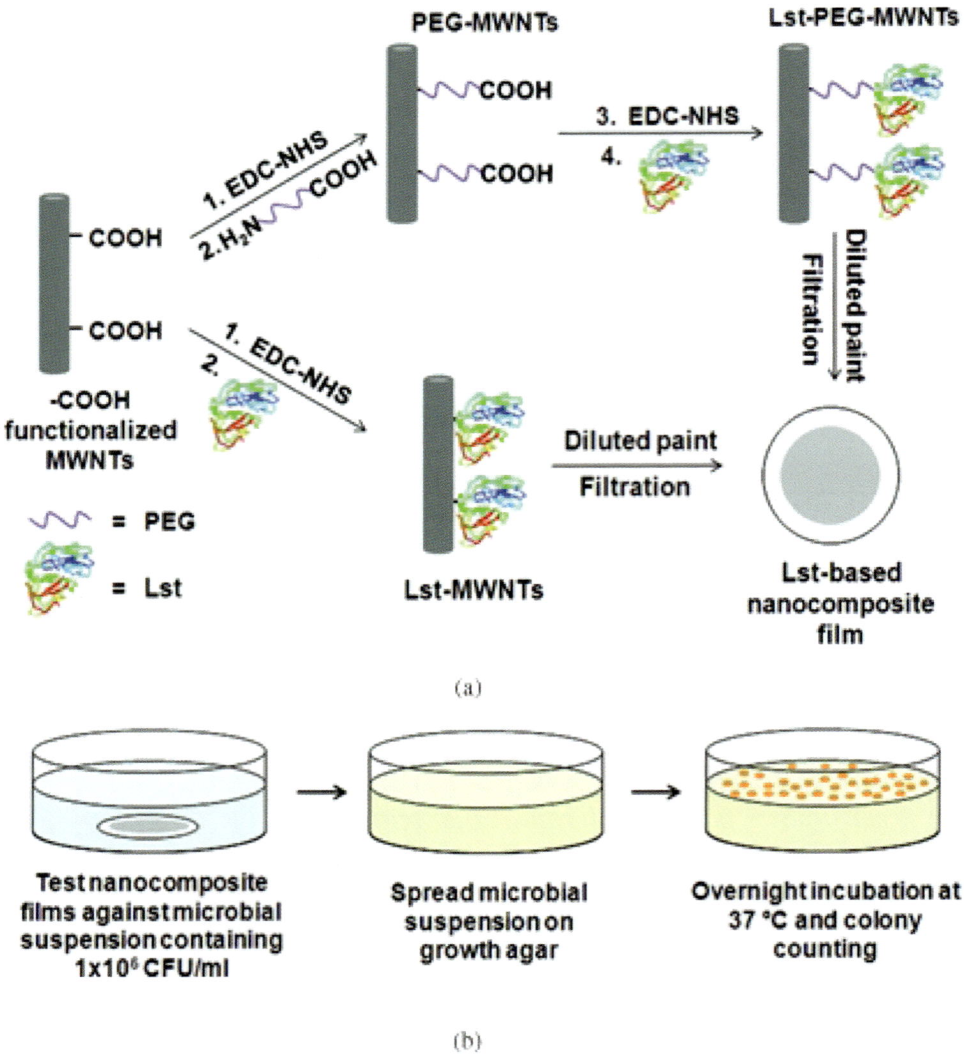

Fig. 10. Antimicrobial nanocomposite films containing Lst–nanotube conjugates (not drawn to scale). (a) Preparation of Lst–nanotube conjugates via two (no PEG linker) and four (with PEG linker) step EDC-NHS coupling reactions. Nanocomposite films were prepared by filtering a mixture of diluted paint and conjugates onto a polycarbonate membrane. (b) Cell-based assay to determine bactericidal activity of Lst-containing nanocomposite films (reproduced with permission from Pangule *et al.*[122] Copyright (2010) American Chemical Society).

5.5. *Nanocomposites incoroporating nanotube–enzyme conjugates for sporicidal applications*

We have extended the approach discussed above to the incorporation of Perhydrolase S54V (AcT) into nanocomposite films that may have sporicidal

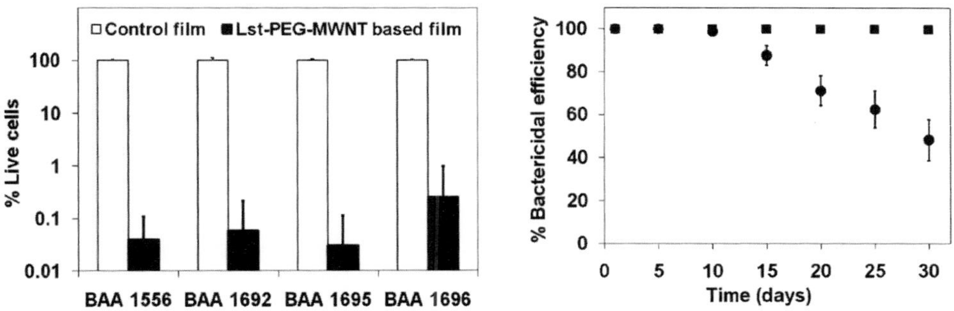

Fig. 11. **(Left figure)** Percent viable cells obtained after exposing the microbial suspension containing 106 CFU/mL MRSA to control and Lst-containing films (4% w/w Lst in the form of Lst-PEG-MWNT) for 2 h. The results are normalized with respect to the colony counts obtained for control samples. **(Right figure)** Operational (•) and storage stability (▪) of films containing 4% w/w Lst in the form of Lst–PEG–MWNT. The films were stored in dry conditions and at room temperature in between the two use cycles (reproduced with permission from Pangule *et al.*[122] Copyright (2010) American Chemical Society).

applications. AcT effectively catalyzes the perhydrolysis of propylene glycol diacetate (PGD) to generate peracetic acid (PAA). PAA is a potent oxidant used for sanitization and disinfection, with broad effectiveness against bacteria, yeasts, fungi and spores. When AcT was immobilized onto MWNTs directly, the conjugates retained only about 7% of the native AcT activity. This poor retention of activity could be due to the large block-like structure and extensive hydrophobicity of AcT, which could lead to substantial non-specific hydrophobic interactions between the AcT surface and the non-functionalized hydrophobic regions of the MWNTs.[124] To improve the retention of enzyme activity for the conjugates, a bifunctional amino-dPEG12-acid (dPEG, 4.7 nm in contour length) spacer was first covalently attached to the acid-treated MWNTs and subsequently AcT was attached to the free end of the spacer.[124] The dPEG spacer was effective at increasing the specific activity of the resulting AcT–dPEG–MWNT conjugates to ~24% of that of free AcT.[124] Other PEG spacers could be also employed; however, similar effects in terms of blocking non-specific protein binding should be observed.[124] Longer spacers (MW > 2000), however, should be avoided, since they could potentially create porous composites and thus alter their mechanical properties.[124]

The AcT–dPEG–MWNT conjugates were then incorporated into polymer and latex-based paints to develop active and stable composites.[124] At a conjugate loading of 0.16% (w/v), the composite generated 11 mM PAA in 20 min and was capable of killing more than 99% spores initially charged at

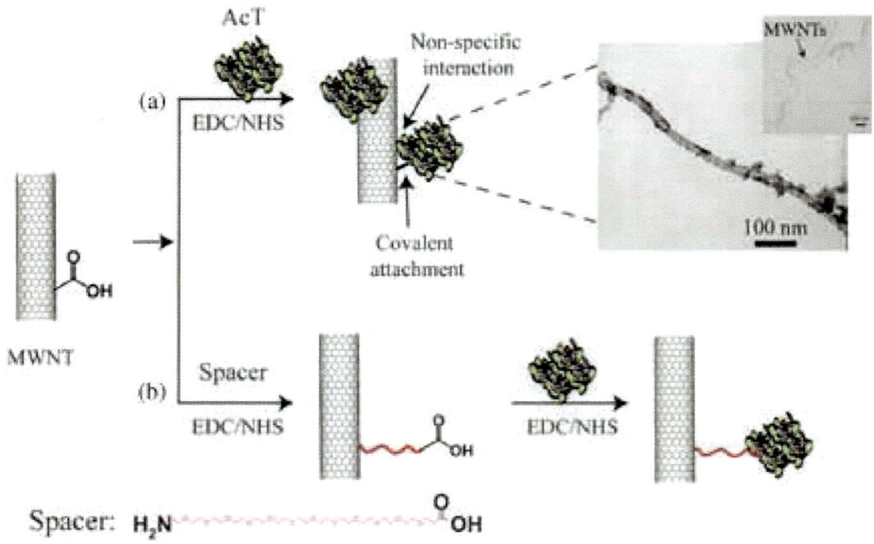

Fig. 12. Covalent attachment of AcT onto MWNTs. (a) Direct attachment of AcT onto MWNTs. In addition to covalent binding, non-specific hydrophobic interaction also occurs due to the large size and hydrophobic nature of AcT. Inset: TEM image of AcT–MWNT conjugates. (b) Attachment of AcT onto MWNTs using dPEG as spacer (reproduced with permission from Dinu *et al.*[124] Copyright (2009) WILEY-VCH Verlag GmbH & Co. KGaA, Weinheim).

10^6 colony-forming units (CFU) per milliliter.[124] AcT activity in the PMMA and PVAc composites obtained by direct evaporation was quite low, <10% of that of the AcT–dPEG– MWNT conjugates in aqueous solution (Fig. 13).[124] However, these polymer films had a thickness of ~200 μm, which could have severely limited the diffusion of the two substrates PGD and H_2O_2 to the immobilized AcT.[124] To overcome mass transfer limitations in the polymeric composites, we dramatically reduced the film thickness from 200 μm to ~2 μm by spin-coating the mixed solution of AcT–dPEG–MWNT conjugate and polymer onto a cover glass.[124] The corresponding activity of the AcT–dPEG–MWNT polymer films increased to ~40% and more than 90% of the conjugate activity for PMMA and PVAc, respectively (Fig. 13).[124]

The relatively hydrophilic nature of PVAc film may contribute to its higher activity when compared to the more hydrophobic and dense PMMA film.[124] The more porous latex paint composite was not nearly as diffusionally limited as the polymers with ~40% of the conjugate activity even at a thickness of 400 μm.[124] The MWNT-based formulations enabled high retention of the active and stable enzyme in the polymeric film or in paints with no leaching from the composite.[124] This, in turn, enabled the active enzyme to be retained

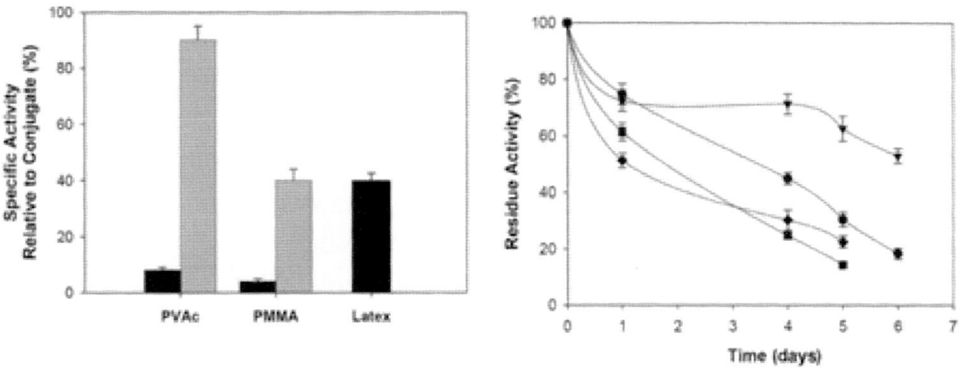

Fig. 13. (a) Specific activity of the composites relative to that of AcT– dPEG–MWNT conjugates. Black bars indicated thick films obtained by direct evaporation (200 mm for polymeric composites and 400 mm for paint composite) and grey bars indicated thin films obtained by spin-coating (~2 mm for polymeric composites). (b) Stability of the paint composite tested under different conditions: (•) dry state at room temperature; (▾) immersed in water at room temperature; (▪) dry state at 50°C; (♦) immersed in water at 50°C (reproduced with permission from Dinu *et al.*[124] Copyright (2009) WILEY-VCH Verlag GmbH & Co. KGaA, Weinheim).

in the composite and is expected to reduce or eliminate potential adverse health effects that could result from release of MWNTs into the environment.[124] The stability and reusability of the composites was shown to be sufficient for use as a microbicidal and sporicidal decontaminating surface.[124] Finally, the capability of generating sufficiently high amount of potent PAA makes these composites useful as coating materials for disinfection.[124]

We further developed a strategy to preserve the activity and operational stability of a large multi-subunit enzyme immobilized onto CNTs and incorporated into latex paint. The strategy was based on the hypothesis that the operational and thermal stability of a large multi-subunit enzyme such as AcT may be increased on a surface if greater rigidity of the enzyme were induced to avoid the formation of non-specific protein–protein associations.[125] Along these lines, we used homo-, bi- or poly-functional aldehydes, such as glutaraldehyde or aldehyde dextran, respectively, to crosslink the multimeric enzyme, AcT, a homo-octamer and then subsequently immobilized the crosslinked AcT onto SWNTs.[126] The activity of AcT crosslinked with aldehyde dextran and covalently attached to SWNTs (AcT-dex-SWNTs) was ~40% of that of native AcT and more than two-fold higher than that of enzyme immobilized directly, i.e. without crosslinking. This relatively high retention of AcT activity was consistent with the nearly complete retention of the enzyme's secondary structure upon attachment to the nanoscale support.[126,127] It was also shown that the intramolecularly crosslinked AcT attached to SWNTs showed dramatically improved enzyme operational stability, thermostability, and enzyme activity.[127] Further incorporation of the conjugates into paint led to composites that showed complete decontamination of 10^6 CFU/mL of *Bacillus cereus* spores in 60 min. We have also shown in another study that these AcT-based paint composites showed a 6-log reduction in the viability of spores of *Bacillus cereus* and *Bacillus anthracis-ΔSterne* within 60 min[126,127] and >4-log reduction in the titer of influenza virus (X-31) within 10 min (initially challenged with 10^7 PFU/mL).[127] The operational stability of paint composites was also studied at different relative humidities and temperatures to simulate real-life operation.[127]

In other work, we have immobilized laccase and chloroperoxidase (CPO) separately onto MWNTs and subsequently mixed each of these conjugates with a commercial eco-friendly paint to generate biocatalytic coatings. Films incorporating laccase–nanotube conjugates showed >99% bactericidal activity against *Escherichia coli* and *Staphylococcus aureus* (both challenged with 10^6 CFU/mL) within 30 min and > 98% sporicidal activity against *B. cereus* and *B. anthracis-ΔSterne* (initially challenged with 10^4 CFU/mL) within 120 min. Films containing CPO-nanotube conjugates also showed >99% antimicrobial activity within 30 min against *E. coli* and *S. aureus* (both challenged with 10^6 CFU/mL).[128] Moreover, these enzyme–nanotube-containing formulations provide an eco-friendly route to generate biocidal compounds, which can prevent the growth of a broad spectrum of bacterial pathogens, including spores. These enzyme-containing paints may be envisioned to be applied as self-decontaminating coatings onto a wide range of surfaces, such as hospital infrastructure, medical devices and equipment, food processing and packaging, etc.; in all cases, effective killing of a variety of infectious organisms is critical.

5.6. *CNTs in antimicrobial photodynamic applications*

The localized generation of heat resulting from the NIR irradiation of nanomaterials such as nanoshells and CNTs has been used to effect cell death[66,70] as described earlier. It is also known that the irradiation of carbon-based nanomaterials such as fullerenes and CNTs can result in the generation of reactive oxygen species (ROS).[15,129–133] The efficiency of ROS generation can be further increased by using nanotubes as a scaffold for the attachment of photoactive moieties such as porphyrin.[134] The mechanism of action of the resulting conjugates would be somewhat analogous to that of some of the enzyme conjugates discussed in the previous section — they would generate reagents that have antimicrobial activity. Moreover, the resulting facile way to incorporate porphyrins into thin coatings that are compatible with a larger variety of surfaces may lead to a more effective and widespread use of porphyrins in hospitals, the food industry, and community settings.[134] Along these lines, Banerjee *et al.* have synthesized MWNT-PPIX (Protoporphyrin IX) conjugates and investigated their antimicrobial activity.[134] The conjugates showed potent bactericidal activity against *S. aureus* cells in solution upon irradiation with visible light (Fig. 14(a)).[134] Large-area nanocomposite films composed of MWNT-PPIX conjugates were fabricated and also showed effective bactericidal activity (Fig. 14(b)).[134]

Banerjee *et al.* further extended the approach discussed above for antiviral applications. Specifically, PPIX, was attached to acid-functionalized MWNTs and the nanotube–porphyrin conjugates (NT-P) in the presence of visible light significantly reduced the ability of Influenza A virus to infect mammalian cells.[135] NT-P may be used effectively against influenza virus, with little or no chance of them developing resistance to the treatment. CNTs were used as scaffolds because of their ease of recovery from a solution through filtration. As a result, NT-P can be easily recovered through filtration, which offers a facile strategy to reuse the active porphyrin moiety to its fullest extent.

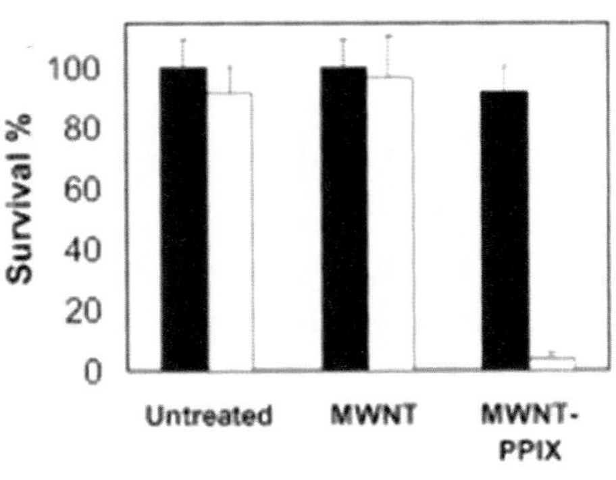

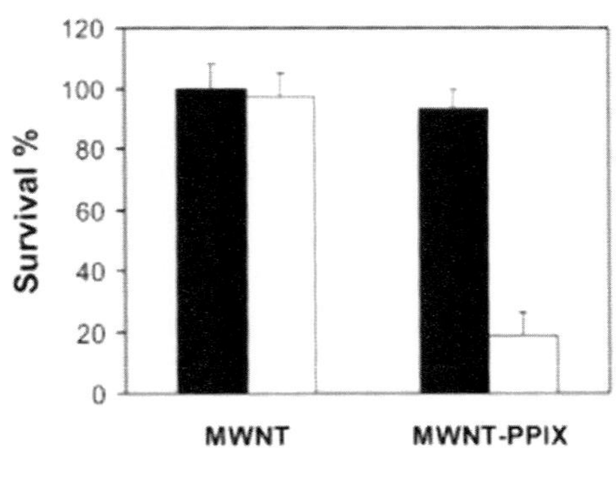

Fig. 14. Survival of *S. aureus* cells exposed to MWNT-PPIX and acid-functionalized MWNTs in saline buffer: (a) Survival of *S. aureus* cells when left untreated, or when treated with either acid-functionalized MWNTs, or MWNT-PPIX in dark (indicated by black bars), or in the presence of light (indicated by white bars) for 15 min. (b) Survival of *S. aureus* treated with films containing acid-functionalized MWNT or MWNT-PPIX in the dark (black bars) and on irradiation with light for 1 h (white bars) (reproduced with permission from Banerjee *et al.*[134] Copyright (2010) American Chemical Society).

Thus NT-P conjugates represent a new approach for preparing *ex vivo* reusable antiviral agents.

6. Biocompatibility and Biodegradation of CNTs

The growing interest in CNTs in biomedical applications has led to a large number of studies that address their biocompatibility, ecological properties and toxicity. In particular, the toxicity of CNTs has been a subject of numerous studies, and the topic has been reviewed recently.[136–141] Given their proposed use as scaffolds for tissue engineering[142–144] and as implantable biosensors,[145–147] there has been an extensive effort in increasing the biocompatibility of CNTs. Modifications to improve the biocompatibility and to reduce toxicity includes covalent[148,149] and non-covalent[150,151] functionalization. Recent reports have demonstrated that PEG-functionalization renders CNT non-toxic[32,152] in biological systems.

While CNTs are inert nanomaterials, natural enzymatic biodegradation under physiological conditions have been shown using horse radish peroxidase.[153] Recently, it has been reported that a myeloperoxidase found in human neutrophils (hMPO) was able to degrade CNTs to the extent that the characteristic fibrillar structure of the nanotubes was lost and the CNTs disappeared after 24 h, as determined by Raman spectroscopy and scanning electron microscopy. This nature-inspired approach holds a promise for enhancing the rate of clearance and reducing the pulmonary toxicity.[154]

7. Conclusion

In this chapter, we reviewed the existing biomedical applications based on CNTs, as well as the emerging technologies and the future trends in using this exciting and versatile nanomaterial in biology and medicine. Owing to their unique chemical and mechanical properties, CNTs have been successfully incorporated in delivery, scaffolding and sensing applications. Bioactive molecules have been delivered by attaching them through a variety of chemistries to functionalized nanotubes. CNT-based scaffolds are an expanding branch of tissue engineering, with nanotubes being used both for directed tissue growth and for structural reinforcement. The electronic properties of CNTs, combined with their nanoscale size, allow their use in disposable or implantable biosensors. A growing trend in the biosensor development is that of CNT-based sensing devices that can detect single-molecule events. Biocompatibility research, aimed to reduce any adverse effects of CNTs *in vivo*, is going on in parallel. Finally, we have described our work to characterize biomolecule–nanotube interactions and to develop novel nanocomposites that take advantage of the properties of CNTs (stable and inert material) and enzymes (catalytically active, high specificity) for the decontamination of microbial pathogens. The studies described in this review highlight the great promise of CNTs in numerous biomedical applications.

References

1. C. Burda *et al.*, Chemistry and properties of nanocrystals of different shapes, *Chem. Rev.* **105**(4), 1025–1102 (2005).
2. S. Polizu *et al.*, Applications of carbon nanotubes-based biomaterials in biomedical nanotechnology, *J. Nanosci. Nanotechnol.* **6**(7), 1883–1904 (2006).
3. X. Wang *et al.*, Fabrication of ultralong and electrically uniform single-walled carbon nanotubes on clean substrates, *Nano Lett.* **9**(9), 3137–3141 (2009).
4. D. A. Stout and T. J. Webster, Carbon nanotubes for stem cell control, *Mater. Today* **15**(7–8), 312–318 (2012).
5. K. Chu *et al.*, Thermal properties of carbon nanotube-copper composites for thermal management applications, *Nanoscale Res. Lett.* **5**(5), 868–874 (2010).
6. M. S. Dresselhaus *et al.*, Electronic, thermal and mechanical properties of carbon nanotubes, *Philos. Transact. A Math Phys. Eng. Sci.* **362**(1823), 2065–2098 (2004).
7. N. Saifuddin, A. Z. Raziah and A. R. Junizah, Carbon nanotubes: A review on structure and their interaction with proteins, *J. Chem.* **2013**, 18 (2012).
8. R. C. Pangule, S. S. Bale, A. D. Shah, A. Joshi, P. Asuri, J. S. Dordick and R. S. Kane, Biomolecule-nanomaterial interactions: Effect on biomolecule structure, function, and stability. In Biological Interactions on Materials Surfaces, pp. 97–114, Springer US.
9. C. Fabbro *et al.*, Targeting carbon nanotubes against cancer, *Chem. Commun.* (*Camb.*) **48**(33), 3911–3926 (2012).
10. N. W. Kam, Z. Liu and H. Dai, Carbon nanotubes as intracellular transporters for proteins and DNA: an investigation of the uptake mechanism and pathway, *Angew. Chem. Int. Ed. Engl.* **45**(4), 577–581 (2006).
11. K. Kostarelos *et al.*, Cellular uptake of functionalized carbon nanotubes is independent of functional group and cell type, *Nat. Nanotechnol.* **2**(2), 108–113 (2007).
12. G. Pastorin *et al.*, Double functionalization of carbon nanotubes for multimodal drug delivery, *Chem. Commun.* (*Camb.*) (11), 1182–1184 (2006).
13. M. Prato, K. Kostarelos and A. Bianco, Functionalized carbon nanotubes in drug design and discovery, *Acc. Chem. Res.* **41**(1), 60–68 (2008).
14. D. Vance *et al.*, The design of polyvalent scaffolds for targeted delivery, *Adv. Drug Deliv. Rev.* **61**(11), 931–939 (2009).
15. A. Joshi *et al.*, Nanotube-assisted protein deactivation, *Nat. Nanotechnol.* **3**(1), 41–45 (2008).
16. L. Lacerda *et al.*, Tissue histology and physiology following intravenous administration of different types of functionalized multiwalled carbon nanotubes, *Nanomedicine* (*Lond.*) **3**(2), 149–161 (2008).
17. R. Singh *et al.*, Tissue biodistribution and blood clearance rates of intravenously administered carbon nanotube radiotracers, *Proc Natl. Acad. Sci. USA* **103**(9), 3357–3362 (2006).

18. H. Dumortier *et al.*, Functionalized carbon nanotubes are non-cytotoxic and preserve the functionality of primary immune cells, *Nano Lett.* **6**(7), 1522–1528 (2006).
19. A. Bianco, K. Kostarelos and M. Prato, Making carbon nanotubes biocompatible and biodegradable, *Chem. Commun.* (*Camb*). **47**(37), 10182–10188 (2011).
20. Z. Liu *et al.*, Supramolecular chemistry on water-soluble carbon nanotubes for drug loading and delivery, *ACS Nano* **1**(1), 50–56 (2007).
21. H. Ali-Boucetta *et al.*, Multiwalled carbon nanotube-doxorubicin supramolecular complexes for cancer therapeutics, *Chem. Commun.* (*Camb*) (4), 459–461 (2008).
22. A. Di Crescenzo *et al.*, Biocompatible dispersions of carbon nanotubes: a potential tool for intracellular transport of anticancer drugs, *Nanoscale* **3**(3), 925–928 (2011).
23. Z. Liu *et al.*, Supramolecular stacking of doxorubicin on carbon nanotubes for *in vivo* cancer therapy, *Angew. Chem. Int. Ed. Engl.* **48**(41), 7668–7672 (2009).
24. Z. Liu *et al.*, *In vivo* biodistribution and highly efficient tumour targeting of carbon nanotubes in mice, *Nat. Nanotechnol.* **2**(1), 47–52 (2007).
25. E. Heister, V. N. Constanze Lamprecht, S. Ravi, P. Silva, H. M. Coley and J. McFadden, Drug loading, dispersion stability, and therapeutic efficacy in targeted drug delivery with carbon nanotubes, *Carbon* **50**, 622–632 (2012).
26. E. Heister, V. Neves, C. Tilmaciu, K. Lipert, V. S. Beltran, H. M. Coley, S. R. P. Silva and J. McFadden, Triple functionalisation of single-walled carbon nanotubes with doxorubicin, a monoclonal antibody, and a fluorescent marker for targeted cancer therapy, *Carbon* **47**(9), 2152–2160 (2009).
27. X. Li *et al.*, Recent advances in P-glycoprotein-mediated multidrug resistance reversal mechanisms, *Meth. Find Exp. Clin. Pharmacol.* **29**(9), 607–617 (2007).
28. R. Clarke, F. Leonessa and B. Trock, Multidrug resistance/P-glycoprotein and breast cancer: Review and meta-analysis, *Semin. Oncol.* **32**(6 Suppl 7), S9–S15 (2005).
29. R. Li *et al.*, P-glycoprotein antibody functionalized carbon nanotube overcomes the multidrug resistance of human leukemia cells, *ACS Nano* **4**(3), 1399–1408 (2005).
30. C. Samori *et al.*, Enhanced anticancer activity of multi-walled carbon nanotube-methotrexate conjugates using cleavable linkers, *Chem. Commun.* (*Camb.*) **46**(9), 1494–1496 (2009).
31. Z. Liu *et al.*, Drug delivery with carbon nanotubes for *in vivo* cancer treatment, *Cancer Res.* **68**(16), 6652–6660 (2008).
32. Z. Liu *et al.*, Circulation and long-term fate of functionalized, biocompatible single-walled carbon nanotubes in mice probed by Raman spectroscopy, *Proc. Natl. Acad. Sci. USA* **105**(5), 1410–1415 (2008).
33. G. Russell-Jones *et al.*, Vitamin-mediated targeting as a potential mechanism to increase drug uptake by tumours, *J. Inorg. Biochem.* **98**(10), 1625–1633 (2004).
34. J. Chen *et al.*, Functionalized single-walled carbon nanotubes as rationally designed vehicles for tumor-targeted drug delivery, *J. Am. Chem. Soc.* **130**(49), 16778–166785 (2008).
35. K. R. Barnes, A. Kutikov and S. J. Lippard, Synthesis, characterization, and cytotoxicity of a series of estrogen-tethered platinum(IV) complexes, *Chem. Biol.* **11**(4), 557–564 (2004).
36. K. R. Barnes and S. J. Lippard, Cisplatin and related anticancer drugs: recent advances and insights, *Met. Ions. Biol. Syst.* **42**, 143–177 (2004).

37. M. Galanski, M. A. Jakupec and B. K. Keppler, Update of the preclinical situation of anticancer platinum complexes: Novel design strategies and innovative analytical approaches, *Curr. Med. Chem.* **12**(18), 2075–2094 (2005).
38. S. Dhar *et al.*, Targeted single-wall carbon nanotube-mediated Pt(IV) prodrug delivery using folate as a homing device, *J. Am. Chem. Soc.* **130**(34), 11467–11476 (2008).
39. S. Wang and P. S. Low, Folate-mediated targeting of antineoplastic drugs, imaging agents, and nucleic acids to cancer cells, *J. Control Release* **53**(1–3), 39–48 (1998).
40. A. A. Bhirde *et al.*, Targeted killing of cancer cells *in vivo* and *in vitro* with EGF-directed carbon nanotube-based drug delivery, *ACS Nano* **3**(2), 307–316 (2009).
41. W. Wu *et al.*, Covalently combining carbon nanotubes with anticancer agent: preparation and antitumor activity, *ACS Nano* **3**(9), 2740–2750 (2009).
42. D. Pantarotto *et al.*, Immunization with peptide-functionalized carbon nanotubes enhances virus-specific neutralizing antibody responses, *Chem. Biol.* **10**(10), 961–966 (2003).
43. P. D. Marcato and N. Duran, New aspects of nanopharmaceutical delivery systems, *J. Nanosci. Nanotechnol.* **8**(5), 2216–2229 (2008).
44. C. H. Villa *et al.*, Single-walled carbon nanotubes deliver peptide antigen into dendritic cells and enhance IgG responses to tumor-associated antigens, *ACS Nano* **5**(7), 5300–5311 (2011).
45. J. Meng *et al.*, Carbon nanotubes conjugated to tumor lysate protein enhance the efficacy of an antitumor immunotherapy, *Small* **4**(9), 1364–1370 (2008).
46. F. Levy and S. Colombetti, Promises and limitations of murine models in the development of anticancer T-cell vaccines, *Int. Rev. Immunol.* **25**(5–6), 269–295 (2006).
47. H. Iwaki *et al.*, Vaccinia virus-infected C-C36 colon tumor cell lysates stimulate cellular responses *in vitro* and protect syngeneic Balb/c mice from tumor cell challenge, *J. Surg. Oncol.* **40**(2), 90–96 (1989).
48. M. Sivanandham *et al.*, Therapeutic effect of a vaccinia colon oncolysate prepared with interleukin-2-gene encoded vaccinia virus studied in a syngeneic CC-36 murine colon hepatic metastasis model, *Cancer Immunol. Immunother.* **38**(4), 259–264 (1994).
49. V. K. Sondak and J. A. Sosman, Results of clinical trials with an allogenic melanoma tumor cell lysate vaccine: Melacine, *Semin. Cancer Biol.* **13**(6), 409–415 (2003).
50. J. A. Sosman and V. K. Sondak, Melacine: an allogeneic melanoma tumor cell lysate vaccine, *Expert Rev. Vaccine* **2**(3), 353–368 (2003).
51. M. F. Tang *et al.*, Recent progress in nanotechnology for cancer therapy, *Chin. J. Cancer* **29**(9), 775–780 (2010).
52. A. Nunes *et al.*, Hybrid polymer-grafted multiwalled carbon nanotubes for *in vitro* gene delivery, *Small* **6**(20), 2281–2291 (2010).
53. Y. Chen, G. Cheng and R. I. Mahato, RNAi for treating hepatitis B viral infection, *Pharm. Res.* **25**(1), 72–86 (2008).
54. P. Kumar *et al.*, T cell-specific siRNA delivery suppresses HIV-1 infection in humanized mice, *Cell* **134**(4), 577–586 (2008).
55. S. Beg *et al.*, Advancement in carbon nanotubes: basics, biomedical applications and toxicity, *J. Pharm. Pharmacol.* **63**(2), 141–163 (2011).
56. W. Cheung *et al.*, DNA and carbon nanotubes as medicine, *Adv. Drug Deliv. Rev.* **62**(6), 633–649 (2010).

57. Z. Zhang *et al.*, Delivery of telomerase reverse transcriptase small interfering RNA in complex with positively charged single-walled carbon nanotubes suppresses tumor growth, *Clin. Cancer Res.* **12**(16), 4933–4939 (2006).
58. W. Zhang, Z. Zhang and Y. Zhang, The application of carbon nanotubes in target drug delivery systems for cancer therapies, *Nanoscale Res. Lett.* **6**, 555 (2011).
59. M. A. Herrero *et al.*, Synthesis and characterization of a carbon nanotube-dendron series for efficient siRNA delivery, *J. Am. Chem. Soc.* **131**(28), 9843–9848 (2009).
60. D. Cui *et al.*, Dendrimer-modified gold nanorods as efficient controlled gene delivery system under near-infrared light irradiation, *J. Control. Release* **152**(Suppl 1), e137–e139 (2011).
61. C. Iancu *et al.*, Enhanced laser thermal ablation for the in vitro treatment of liver cancer by specific delivery of multiwalled carbon nanotubes functionalized with human serum albumin, *Int. J. Nanomed.* **6**, 129–141 (2011).
62. R. Marches *et al.*, The importance of cellular internalization of antibody-targeted carbon nanotubes in the photothermal ablation of breast cancer cells, *Nanotechnology* **22**(9), 095101 (2011).
63. R. Marches *et al.*, Specific thermal ablation of tumor cells using single-walled carbon nanotubes targeted by covalently-coupled monoclonal antibodies, *Int. J. Cancer* **125**(12), 2970–2977 (2009).
64. P. Chakravarty *et al.*, Thermal ablation of tumor cells with antibody-functionalized single-walled carbon nanotubes, *Proc. Natl. Acad. Sci. USA* **105**(25), 8697–702 (2008).
65. D. P. O'Neal *et al.*, Photo-thermal tumor ablation in mice using near infrared-absorbing nanoparticles, *Cancer Lett.* **209**(2). 171–176 (2004).
66. L. R. Hirsch *et al.*, Nanoshell-mediated near-infrared thermal therapy of tumors under magnetic resonance guidance, *Proc. Natl. Acad. Sci. USA* **100**(23), 13549–13554 (2003).
67. X. Huang *et al.*, Cancer cell imaging and photothermal therapy in the near-infrared region by using gold nanorods, *J. Am. Chem. Soc.* **128**(6), 2115–2120 (2006).
68. K. Konig, Multiphoton microscopy in life sciences, *J. Microsc.* **200** (Pt 2), 83–104 (2000).
69. M. J. O'Connell *et al.*, Band gap fluorescence from individual single-walled carbon nanotubes, *Science*, **297**(5581), 593–596 (2002).
70. N. W. Kam *et al.*, Carbon nanotubes as multifunctional biological transporters and near-infrared agents for selective cancer cell destruction, *Proc. Natl. Acad. Sci. USA* **102**(33), 11600–11605 (2005).
71. M. R. McDevitt *et al.*, Tumor targeting with antibody-functionalized, radiolabeled carbon nanotubes, *J. Nucl. Med.* **48**(7), 1180–1189 (2007).
72. A. Ruggiero *et al.*, Imaging and treating tumor vasculature with targeted radiolabeled carbon nanotubes, *Int. J. Nanomed.* **5**, 783–802 (2010).
73. B. S. Harrison and A. Atala, Carbon nanotube applications for tissue engineering, *Biomaterials*, **28**(2), 344–353 (2007).
74. S. F. Wang *et al.*, Preparation and mechanical properties of chitosan/carbon nanotubes composites, *Biomacromolecules* **6**(6), 3067–3072 (2005).
75. S. R. Shin *et al.*, Carbon nanotube reinforced hybrid microgels as scaffold materials for cell encapsulation, *ACS Nano* **6**(1), 362–372 (2012).
76. K. Balani *et al.*, Plasma-sprayed carbon nanotube reinforced hydroxyapatite coatings and their interaction with human osteoblasts *in vitro, Biomaterials* **28**(4), 618–624 (2007).

77. J. W. Penghua Yan, L. Wang, B. Liu, Z. Lei and S. Yang, The *in vitro* biomineralization and cytocompatibility of polydopamine coated carbon nanotubes, *Appl. Surf. Sci.* **257**, 4849–4855 (2011).
78. S. H. K. Jungki Ryu, H. Lee and C. B. Park, Mussel-inspired polydopamine coating as a universal route to hydroxyapatite crystallization, *Adv. Funct. Mater.* **20**, 2132–2139 (2010).
79. R. A. MacDonald *et al.*, Collagen-carbon nanotube composite materials as scaffolds in tissue engineering, *J. Biomed. Mater. Res. A* **74**(3), 489–496 (2005).
80. T. C. Marsi *et al.*, Biomineralization of superhydrophilic vertically aligned carbon nanotubes, *Langmuir* **28**(9), 4413–4424 (2012).
81. D. R. Thevenot *et al.*, Electrochemical biosensors: recommended definitions and classification, *Biosens. Bioelectron.* **16**(1–2), 121–131 (2001).
82. X. Tang *et al.*, Carbon nanotube DNA sensor and sensing mechanism, *Nano Lett.* **6**(8), 1632–1636 (2006).
83. E. L. Gui *et al.*, DNA sensing by field-effect transistors based on networks of carbon nanotubes, *J. Am. Chem. Soc.* **129**(46), 14427–14432 (2007).
84. R. J. Chen *et al.*, An investigation of the mechanisms of electronic sensing of protein adsorption on carbon nanotube devices, *J. Am. Chem. Soc.* **126**(5), 1563–1568 (2004).
85. H. R. Byon and H. C. Choi, Network single-walled carbon nanotube-field effect transistors (SWNT-FETs) with increased Schottky contact area for highly sensitive biosensor applications, *J. Am. Chem. Soc.* **128**(7), 2188–2189 (2006).
86. J. G. Coroneus *et al.*, Mechanism-guided improvements to the single molecule oxidation of carbon nanotube sidewalls, *ChemPhysChem* **9**(7), 1053–1056 (2008).
87. N. Nuraje *et al.*, Biological bottom-up assembly of antibody nanotubes on patterned antigen arrays, *J. Am. Chem. Soc.* **126**(26), 8088–8089 (2004).
88. X. Q. He *et al.*, Enhanced mechanical properties of single-walled carbon nanotubes due to chemical functionalization, *J. Phys. Condens. Matter* **21**(21), 215301 (2009).
89. D. C. Klonoff, Overview of fluorescence glucose sensing: a technology with a bright future, *J. Diabetes Sci. Technol.* **6**(6), 1242–1250 (2012).
90. A. P. Periasamy, Y. J. Chang and S. M. Chen, Amperometric glucose sensor based on glucose oxidase immobilized on gelatin-multiwalled carbon nanotube modified glassy carbon electrode, *Bioelectrochemistry* **80**(2), 114–120 (2011).
91. S. Fei *et al.*, Electrochemical behavior of L-cysteine and its detection at carbon nanotube electrode modified with platinum, *Anal. Biochem.* **339**(1), 29–35 (2005).
92. A. Wisitsoraat *et al.*, Fast cholesterol detection using flow injection microfluidic device with functionalized carbon nanotubes based electrochemical sensor, *Biosens. Bioelectron.* **26**(4), 1514–1520 (2010).
93. S. Viswanathan *et al.*, Protein-carbon nanotube sensors: single platform integrated micro clinical lab for monitoring blood analytes, *Meth. Enzymol.* **509**, 165–194 (2012).
94. S. Zhen *et al.*, A novel microassay for measuring blood alcohol concentration using a disposable biosensor strip, *Forensic Sci. Int.* **207**(1–3), 177–182 (2011).
95. T. P. Sun *et al.*, Carbon nanotube composites for glucose biosensor incorporated with reverse iontophoresis function for noninvasive glucose monitoring, *Int. J. Nanomed.* **5**, 343–349 (2010).
96. A. Wisitsoraat *et al.*, High sensitivity electrochemical cholesterol sensor utilizing a vertically aligned carbon nanotube electrode with electropolymerized enzyme immobilization, *Sensors (Basel)* **9**(11), 8658–8668 (2009).

97. K. M. Abu-Salah *et al.*, Nanomaterials as analytical tools for genosensors, *Sensors (Basel)* **10**(1), 963–993 (2010).
98. Q. Liu *et al.*, Coupling technique of random amplified polymorphic DNA and nanoelectrochemical sensor for mapping pancreatic cancer genetic fingerprint, *Int. J. Nanomed.* **6**, 2933–2939 (2011).
99. X. Yang *et al.*, DNA electrochemical sensor based on an adduct of single-walled carbon nanotubes and ferrocene, *Biotechnol. Lett.* **29**(11), 1775–1779 (2007).
100. D. Cai *et al.*, A molecular-imprint nanosensor for ultrasensitive detection of proteins, *Nat. Nanotechnol.* **5**(8), 597–601 (2010).
101. B. V. Chikkaveeraiah *et al.*, Electrochemical immunosensors for detection of cancer protein biomarkers, *ACS Nano* **6**(8), 6546–6561 (2012).
102. J. Okuno *et al.*, Label-free immunosensor for prostate-specific antigen based on single-walled carbon nanotube array-modified microelectrodes, *Biosens. Bioelectron.* **22**(9–10), 2377–2381 (2007).
103. S. Khezrian *et al.*, Label-free electrochemical IgE aptasensor based on covalent attachment of aptamer onto multiwalled carbon nanotubes/ionic liquid/chitosan nanocomposite modified electrode, *Biosens. Bioelectron.* **43C**, 218–225 (2012).
104. S. Viswanathan, C. Rani and J. A. Ho, Electrochemical immunosensor for multiplexed detection of food-borne pathogens using nanocrystal bioconjugates and MWCNT screen-printed electrode, *Talanta* **94**, 315–319 (2012).
105. S. Subramanian *et al.*, Rapid, sensitive and label-free detection of Shiga-toxin producing *Escherichia coli* O157 using carbon nanotube biosensors, *Biosens. Bioelectron.* **32**(1), 69–75 (2012).
106. G. A. Zelada-Guillen *et al.*, Real-time potentiometric detection of bacteria in complex samples, *Anal. Chem.* **82**(22), 9254–9260 (2010).
107. X. Ding *et al.*, A novel homogenous detection method based on the self-assembled DNAzyme labeled DNA probes with SWNT conjugates and its application in detecting pathogen, *Biosens. Bioelectron.* **26**(11), 4596–4600 (2011).
108. C. Garcia-Aljaro *et al.*, Carbon nanotubes-based chemiresistive biosensors for detection of microorganisms, *Biosens. Bioelectron.* **26**(4), 1437–1441 (2010).
109. F. N. Ishikawa *et al.*, Rapid and label-free cell detection by metal-cluster-decorated carbon nanotube biosensors, *Biosens. Bioelectron.* **24**(10), 2967–2972 (2009).
110. M. A. Fernandez-Baldo *et al.*, Screen-printed immunosensor modified with carbon nanotubes in a continuous-flow system for the Botrytis cinerea determination in apple tissues, *Talanta* **79**(3), 681–686 (2009).
111. A. Venkatanarayanan *et al.*, High sensitivity carbon nanotube based electrochemiluminescence sensor array, *Biosens. Bioelectron.* **31**(1), 233–239 (2012).
112. R. Malhotra, F. Papadimitrakopoulos and J. F. Rusling, Sequential layer analysis of protein immunosensors based on single wall carbon nanotube forests, *Langmuir* **26**(18), 15050–15056 (2010).
113. S. E. Stitzel, M. J. Aernecke and D. R. Walt, Artificial noses, *Annu. Rev. Biomed. Eng.* **13**, 1–25 (2011).
114. S. Kladsomboon *et al.*, Hybrid optical-electrochemical electronic nose system based on Zn-porphyrin and multi-walled carbon nanotube composite, *J. Nanosci. Nanotechnol.* **12**(7), 5240–5244 (2012).

115. Z. Guo *et al.*, Single-walled carbon nanotubes based quenching of free FAM-aptamer for selective determination of ochratoxin A, *Talanta* **85**(5), 2517–2521 (2011).
116. A. A. Boghossian *et al.*, The chemical dynamics of nanosensors capable of single-molecule detection, *J. Chem. Phys.* **135**(8), 084124 (2011).
117. A. A. Boghossian *et al.*, Near-infrared fluorescent sensors based on single-walled carbon nanotubes for life sciences applications, *ChemSusChem* **4**(7), 848–863 (2011).
118. S. S. Karajanagi *et al.*, Structure and function of enzymes adsorbed onto single-walled carbon nanotubes, *Langmuir* **20**(26), 11594–11599 (2004).
119. P. Asuri *et al.*, Enhanced stability of enzymes adsorbed onto nanoparticles, *J. Nanosci. Nanotechnol.* 7(4–5), 1675–1678 (2007).
120. P. Asuri *et al.*, Increasing protein stability through control of the nanoscale environment, *Langmuir* **22**(13), 5833–5836 (2006).
121. P. Asuri *et al.*, Polymer-nanotube-enzyme composites as active antifouling films, *Small* **3**(1), 50–53 (2007).
122. R. C. Pangule *et al.*, Antistaphylococcal nanocomposite films based on enzyme-nanotube conjugates, *ACS Nano* **4**(7), 3993–4000 (2010).
123. C.A. Schindler and V. T. Schuhardt, Lysostaphin: A new bacteriolytic agent for the Staphylococcus, *Proc. Natl. Acad. Sci. USA*, **51**, 414–421 (1964).
124. G. Z. Cerasela Zoica Dinu, S. S. Bale, G. Anand, P. J. Reeder, K. Sanford, G. Whited, R. S. Kane and J. S. Dordick, Enzyme-based nanoscale composites for use as active decontamination surfaces, *Adv. Funct. Mater.* **20,** 392–398 (2010).
125. M. Fuentes *et al.*, Determination of protein-protein interactions through aldehyde-dextran intermolecular cross-linking, *Proteomics* **4**(9), 2602–2607 (2004).
126. I. V. B. Cerasela Zoica Dinu, S. S. Bale, A. S. Campbell, R. S. Kane and J. S. Dordick, Perhydrolase-nanotube-paint sporicidal composites stabilized by intramolecular crosslinking, *J. Mol. Catal. B: Enzymatic* **75,** 20–26 (2012).
127. N. Grover *et al*, Perhydrolase-nanotube paint composites with sporicidal and antiviral activity, *Appl. Microbiol. Biotechnol*, 1–9 (2012).
128. N. Grover *et al.*, Laccase- and chloroperoxidase-nanotube paint composites with bactericidal and sporicidal activity, *Enzyme Microb Technol.* **50**(6–7), 271–279.
129. E. M. Hotze *et al.*, Mechanisms of photochemistry and reactive oxygen production by fullerene suspensions in water, *Environ. Sci. Technol.* **42**(11), 4175–4180 (2008).
130. Z. Markovic and V. Trajkovic, Biomedical potential of the reactive oxygen species generation and quenching by fullerenes (C60), *Biomaterials* **29**(26), 3561–3573 (2008).
131. K. D. Pickering and M. R. Wiesner, Fullerol-sensitized production of reactive oxygen species in aqueous solution, *Environ. Sci. Technol.* **39**(5), 1359–1365 (2005).
132. Z. Markovic *et al.*, The mechanism of cell-damaging reactive oxygen generation by colloidal fullerenes, *Biomaterials* **28**(36), 5437–5448 (2007).
133. T. Murakami *et al.*, Photodynamic and photothermal effects of semiconducting and metallic-enriched single-walled carbon nanotubes, *J. Am. Chem. Soc.* **134**(43), 17862–17865 (2012).
134. I. Banerjee *et al.*, Photoactivated antimicrobial activity of carbon nanotube-porphyrin conjugates, *Langmuir* **26**(22), 17369–17374 (2010).
135. I. Banerjee *et al.*, Light-activated nanotube-porphyrin conjugates as effective antiviral agents, *Nanotechnology* **23**(10), 105101 (2012).

136. S. A. Love *et al.*, Assessing nanoparticle toxicity, *Annu. Rev. Anal. Chem.* (*Palo Alto Calif*) **5,** 181–205 (2012).
137. J. Ai *et al.*, Nanotoxicology and nanoparticle safety in biomedical designs, *Int. J. Nanomed.* **6,** 1117–1127 (2011).
138. A. Bianco, K. Kostarelos and M. Prato, Making carbon nanotubes biocompatible and biodegradable, *Chem. Commun.* (*Camb.*) **47**(37), 10182–10188 (2011).
139. N. Saito *et al.*, Carbon nanotubes: Biomaterial applications, *Chem. Soc. Rev.* **38**(7), 1897–1903 (2009).
140. S. Sharifi *et al.*, Toxicity of nanomaterials, *Chem. Soc. Rev.* **41**(6), 2323–2343 (2011).
141. M. van der Zande *et al.*, Carbon nanotubes in animal models: a systematic review on toxic potential, *Tissue Eng. Part B Rev.* **17**(1), 57–69 (2010).
142. B. Dorj *et al.*, Robocasting nanocomposite scaffolds of poly(caprolactone)/hydroxyapatite incorporating modified carbon nanotubes for hard tissue reconstruction, *J. Biomed. Mater. Res. A* (2012).
143. S. L. Edwards, J. A. Werkmeister and J. A. Ramshaw, Carbon nanotubes in scaffolds for tissue engineering, *Expert Rev. Med. Dev.* **6**(5), 499–505 (2009).
144. E. Zawadzak *et al.*, Polyurethane foams electrophoretically coated with carbon nanotubes for tissue engineering scaffolds, *Biomed. Mater.* **4**(1), 015008 (2009).
145. D. J. Lipomi *et al.*, Skin-like pressure and strain sensors based on transparent elastic films of carbon nanotubes, *Nat. Nanotechnol.* **6**(12), 788–792 (2011).
146. L. He *et al.*, Electroactive SWNT/PEGDA hybrid hydrogel coating for bio-electrode interface, *Colloids Surf. B Biointerfaces* **87**(2), 273–279 (2011).
147. C. X. Liu and J. W. Choi, An embedded PDMS nanocomposite strain sensor toward biomedical applications, *Conf. Proc. IEEE Eng. Med. Biol. Soc.* **2009,** 6391–6394 (2009).
148. S. Jain *et al.*, Toxicity of multiwalled carbon nanotubes with end defects critically depends on their functionalization density, *Chem. Res. Toxicol.* **24**(11), 2028–2039 (2011).
149. H. F. Cui *et al.*, Interfacing carbon nanotubes with living mammalian cells and cytotoxicity issues, *Chem. Res. Toxicol.* **23**(7), 1131–1147 (2010).
150. M. Roldo *et al.*, N-Octyl-O-sulfate chitosan stabilises single wall carbon nanotubes in aqueous media and bestows biocompatibility, *Nanoscale* **1**(3), 366–373 (2009).
151. Z. Ji *et al.*, Targeted therapy of SMMC-7721 liver cancer in vitro and in vivo with carbon nanotubes based drug delivery system, *J. Colloid Interf. Sci.* **365**(1), 143–149 (2012).
152. A. A. Bhirde *et al.*, Distribution and clearance of PEG-single-walled carbon nanotube cancer drug delivery vehicles in mice, *Nanomedicine* (*Lond*) **5**(10), 1535–1546 (2010).
153. B. L. Allen *et al.*, Biodegradation of single-walled carbon nanotubes through enzymatic catalysis, *Nano Lett.* **8**(11), 3899–3903 (2008).
154. V. E. Kagan *et al.*, Carbon nanotubes degraded by neutrophil myeloperoxidase induce less pulmonary inflammation, *Nat. Nanotechnol.* **5**(5), 354–359 (2010).

Chapter 13

Core-Shell Nanoparticles for Biomedical Applications

Mahmoud Elsabahy[*,†,‡] *and Karen L. Wooley*[*,§]

[*] *Department of Chemistry, Department of Chemical Engineering, Laboratory for Synthetic-Biologic Interactions, Texas A&M University, P.O. Box 30012, 3255 TAMU, College Station, Texas 77842-3012, USA*
[†] *Department of Pharmaceutics, Faculty of Pharmacy, Assiut Clinical Center of Nanomedicine, Al-Rajhy Liver Hospital, Assiut University, Assiut, Egypt*
[‡] *mahmoud.elsabahy@chem.tamu.edu*
[§] *wooley@chem.tamu.edu*

Abstract

Nanomaterials are emerging as a powerful tool to improve therapeutic outcomes of the currently available medications. Among the various types of nanoparticles, core-shell nanodevices of different structures and compositions (organic-(lipid, polymer, *etc.*) or inorganic-based) have proven potential for several biomedical applications. These nanomaterials often have core-shell architectures; for organic-based nanoparticles, the core is usually loaded with therapeutic and/or diagnostic payload, while the stabilizing shell can be decorated with cell recognition moieties and other functionalities. These nanomaterials are designed to evade the biological barriers while maintaining their characteristics and being able to navigate through the body until delivering their cargoes to the target sites. Optimizing nanoparticles that have high therapeutic efficacy and an ability to accumulate in target tissue, but exhibit low off-target toxicity and are able to evade an immune response, is still challenging. The shape, structure, composition and surface chemistry dictate the *in vivo* pharmacokinetics, biodistribution and toxicity of nanoparticles. This chapter discusses some of the common types of core-shell nanoparticles with focus on their clinical applications.

1. Introduction

Nanomedicine can be defined as the design of materials, devices or structures on the nanoscale with therapeutic and/or diagnostic agents being physically loaded or covalently attached.[1] These materials might consist of an undistinguished solid matrix, or distinguished into a core and shell (Fig. 1). Core-shell nanoparticles include several types of materials (*i.e.* polymeric, lipidic, graphitic or metallic) with architecture consisting of a core surrounded by a shell that is usually hydrophilic.[2] Nanoparticles are emerging as nanocarriers for delivery of nucleic acids, proteins and small molecules for treatment of a wide diversity of diseases.[1–7] The structure and type of the core usually dictate the type of cargo to be encapsulated, and the core may serve as diagnostic or therapeutic agent *per se*. The core of nanoparticles greatly influences the

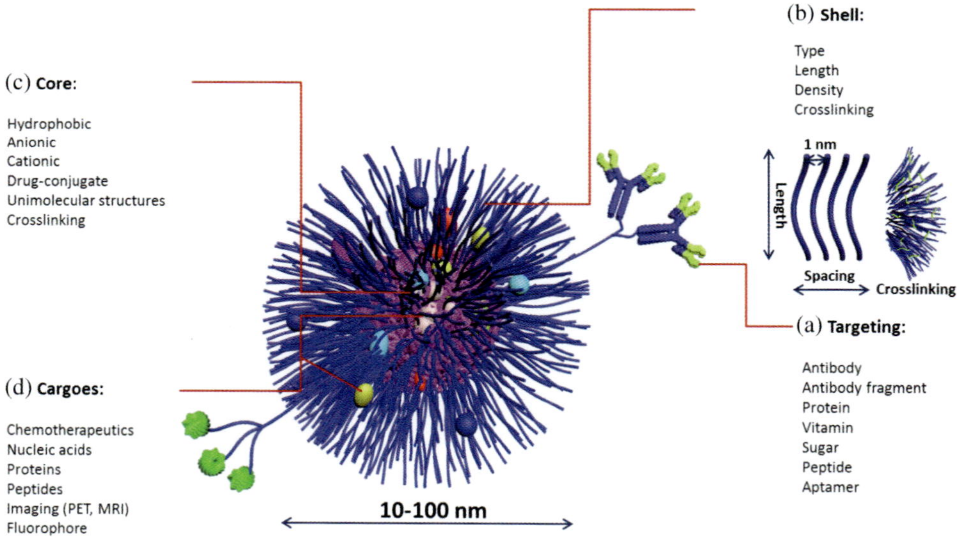

Fig. 1. Composition of multifunctional nanoparticles for biomedical delivery applications: (a) clusters of targeting moieties have been shown to be important for multivalent binding to receptors for enhanced cellular uptake; the use of various ligands (antibody, antibody fragment, peptide) depends on the therapeutic application and disease type, (b) shell: length, spacing and crosslinking of the shell are critical parameters that dictate the blood circulation time and stability of nanoparticles with ~1 nm spacing found to be efficient in preventing protein adsorption, (c) core: nature of the core dictates the type of the drug to be encapsulated. Crosslinking and conjugation of drugs to the core-forming polymer are common strategies for enhancing the stability of nanoparticles and drug-encapsulation efficiency, respectively. (d) drug: a wide range of therapeutics can be used ranging from small molecules to macromolecular cargoes. Reproduced from Ref. 1, with permission from The Royal Society of Chemistry.

toxicity, immunogenicity of the vehicle and the release kinetics of the embedded payload. Chemical modifications of the core are routinely utilized to accommodate with the drug of interest, for instance, to impart cationic character into the core to complex negatively charged nucleic acids. The hydrophilic corona imparts solubility to nanoparticles and is considered essential for steric stabilization and to impart stealth properties upon *in vivo* administration. The hydrophilicity and flexibility of the shell minimize the adsorption of opsonins (*e.g.* plasma proteins) in the blood and thus reduce the clearance by the mononuclear phagocyte system (MPS) and prolong the *in vivo* circulation time. Prolonged circulation time allows the passive targeting of the nanoparticles into tumors or inflamed tissues *via* the enhanced permeability and retention (EPR) effect, due to the vascular leakage and impaired lymphatic drainage at these areas. Importantly, the hydrophilic layer provides a possibility for grafting surface molecules for targeting and/or imaging applications. Stimuli-responsive linkages between the core, shell and/or surface molecules are often utilized to allow the release of the drug in the diseased tissues in response to internal or external stimuli and minimize toxicity to other healthy tissues, thereby enhancing the therapeutic efficacy of nanoparticulates.

Advantages of utilizing nanoparticles for biomedical applications include the protection of drugs against enzymatic or hydrolytic degradation, providing microenvironments suitable for solubilizing or dispersing the encapsulated cargoes, and allow for control over the release of the drug selectively in the targeted organs, tissues, cells or subcellular compartments. Further control over the polymer synthetic chemistries and supramolecular assemblies allowed the construction of multimodal nanoparticles for combinational therapy to deliver more than one cargo to the target site.[8–13] The main focus of the current research is directed toward the design and optimization of nanoparticles that have high efficacy, low toxicity and high biocompatibility for coincident transport and delivery of the loaded drugs, with control over their navigation within the body. In this chapter, we will introduce some of the common types of core-shell nanoparticles with focus on their biomedical and clinical applications. Some examples of nanodevices in the market or in different phases of clinical trials will be reviewed with a focus on their rational design and future prospects.

2. Types of Core-Shell Nanoparticles

Although there are several types of core-shell nanoparticles, some examples of polymeric, lipidic and metallic nanoparticles with core-shell architectures will briefly be discussed. Later, some clinical examples of these nanoparticles, which have been approved by the US Food and Drug Administration (FDA) or still in different phases of clinical trials, will be highlighted.

2.1. *Polymeric micelles*

Self-assembly of amphiphilic copolymer chains in aqueous milieu forms particles with core/shell architecture that have a hydrophobic core suitable for encapsulation of hydrophobic drugs and hydrophilic layer that act as a stabilizing interface between the core and surrounding environment (Fig. 2).[14,15] Hydrophobic associations of the core forming polymers are the main driving

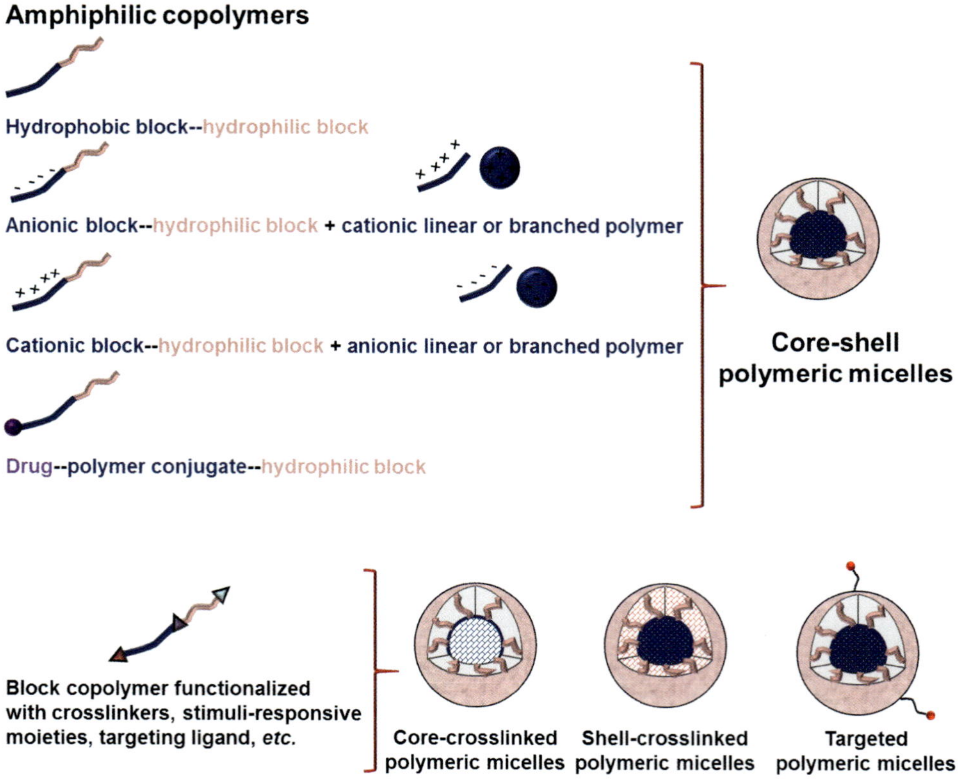

Fig. 2. Self-assembly of amphiphilic copolymers into core-shell polymeric micelles. The hydrophilic block is usually conjugated to hydrophobic, anionic or cationic polymers. For charged blocks, complexation with oppositely-charged polymers or drugs, or formation of coordination complexes is essential for supramolecular assembly into polyion complex micelles and polymer–metal complex micelles. Conjugation of drug to the hydrophobic block also is utilized to drive self-assembly, and/or to enhance the core hydrophobicity. Functionalization of amphiphilic copolymers is usually exploited to construct stimuli-responsive, crosslinked and/or targeted polymeric micelles.

force for the self-assembly process. Upon solubilization of amphiphilic copolymers in aqueous milieu, hydrophobic blocks tend to coalescence together to form a hydrophobic core (to reduce their contact with water molecules) isolated from the aqueous media by the hydrophilic layer. The micellization process is usually balanced and controlled by the repulsive forces brought by shell forming polymers and attractive forces of the hydrophobic blocks. The repulsive forces hinder micelle growth and further aggregation. Neutral hydrophilic corona is preferred over ionic shells to minimize adsorption of opsonins, prolong *in vivo* circulation time and to reduce toxicity. Other forces might also be involved in the formation of micelles. For instance, chemical modifications of the core can drive the formation of other subtypes of polymeric micelles, such as, polyion complex micelles and polymer–metal complex micelles *via* other mechanisms of self-assembly (Fig. 2).[14–16] Polyion complex micelles are formed *via* electrostatic interactions between oppositely charged polymers, where one of the charged polymers has to be attached to the hydrophilic segment of the block copolymer.[7,17–19] Electrostatic interactions form neutralized (*i.e.* hydrophobic) condensates that tend to segregate into the micelle core. Another contributing factor to the formation of polyion complex micelles could be the cooperative electrostatic interactions between polyions which hold the core in place with the hydrophilic polymers covalently conjugated to the core. Ligand substitution reactions between ligands on metallic drugs (*e.g.* cis-dichlorodiamineplatinum (II) (cisplatin)) and charged groups of the copolymer trigger formation of polymer–metal complex micelles through the formation of coordination linkages.[20,21]

Polymeric micelles are dissociated when diluted to a concentration less than their critical micellar concentration, the lowest concentration at which polymeric micelles are formed. Polyion complex micelles and polymer–metal complex micelles are less sensitive to dilution than polymeric micelles because they are formed *via* electrostatic interactions. However, they are subjected to immediate dissociation upon exposure to a medium rich in counter ions (*e.g.* proteins and ions in the blood). Polymeric micelles usually exhibit nanometer size in the range of 10–100 nm with narrow size distribution. The hydrophilic/hydrophobic balance, charge type and density of every block, solvent pH and ionic strength, and interaction with drugs are among the factors dictating the size, morphology, thermodynamic and kinetic stabilities of the formed micelles.

Several polymers have been utilized to form the shell and core of polymeric micelles, as well as for construction of other polymeric nanostructures. Among the various polymers utilized to form the shell, such as, poly(acrylic acid),[22] poly(*N*-vinylpyrrolidone) (PVP),[23,24] poly(carboxybetaine)[25,26] and poly(*N*-isopropylacrylamide),[27,28] poly(ethylene glycol) (PEG) is the most commonly utilized polymer that has been successfully used for biomedical delivery applications. PEG has been approved by the FDA, and is included in several of the products which are already in the market or in different phases of clinical trials. Hydrophobic (*e.g.* poly(ε-caprolactone) (PCL),[10,11] poly(lactic acid)[23,26,29]), cationic (*e.g.* poly(L-lysine),[30,31] polyamidoamine[19,32]) and anionic (*e.g.* poly(aspartic acid),[33,34] antisense oligonucleotide,[32] short-interfering RNA (siRNA)[35]) polymers, have been utilized to form the core of polymeric nanoparticles. In addition, entirely degradable nanoparticles based on polyphosphoester copolymers have recently been developed.[36]

Limitations of polymeric micelles are the rapid dissociation upon dilution. In addition, they do not afford encapsulation of hydrophilic drugs, unless reverse micelles are being utilized.[37,38] Crosslinking of the core and/or shell of polymeric micelles has been utilized to provide greater kinetic stability of polymeric micelles against dissociation and in particular upon *in vivo* administration.[5,31,39,40] Recently, it has been found that crosslinking is associated with lower toxicity.[40] Crosslinking usually maintain the size, polydispersity and drug-loading properties of the non-crosslinked micelles. The crosslinkers can be made degradable and/or carrying additional functionalities, such as, fluorescent or radioactive probes, *etc.*[39,41–44] Nanocages have been also prepared from shell crosslinked micelles after chemical degradation of the core.[45–47]

2.2. *Protein-based and polymeric nanoparticles*

The main difference between polymeric micelles and nanoparticles is that the core of the latter usually contains homopolymer in which the drug is solubilized or dispersed (Fig. 3). The size of nanoparticles is usually larger than polymeric micelles and characterized by a wider distribution. Inclusion of hydrophobic homopolymer is the core might reduce the dissociation of nanoparticles and slow down the drug release. Polymeric nanoparticles can be prepared from the same amphiphilic copolymers constituting the polymeric micelles.[48–51] In this case, the amphiphilic copolymer acts as emulsifier between the oil phase (hydrophobic core matrix) and the external aqueous phase. Gaucher *et al.* have utilized PVP-*b*-poly(D,L-lactide) (PVP-*b*-PDLLA) and PEG-*b*-PDLLA as emulsifiers to prepare polymeric nanoparticles (<250 nm) *via* oil/water emulsion-solvent evaporation method, and loaded the nanoparticles with two anticancer drugs, paclitaxel and etoposide.[49] The organic phase contained the PDLLA homopolymer and the drug (paclitaxel or etoposide), while the block copolymers were dissolved in the aqueous

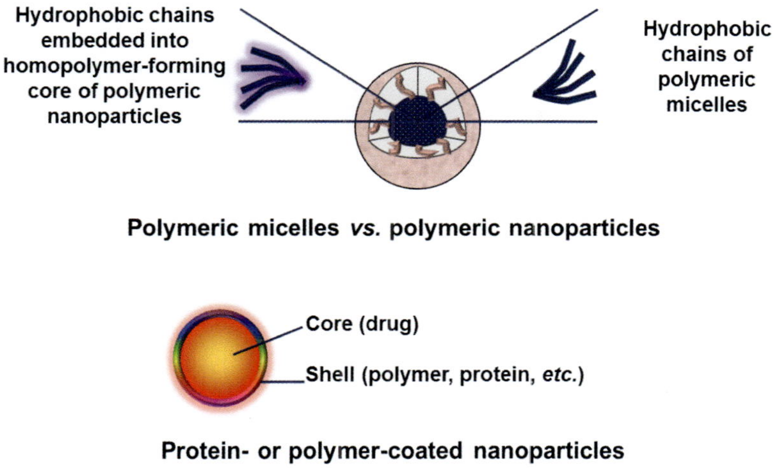

Fig. 3. Polymeric nanoparticles usually have homopolymer in the core, in which drug molecules and hydrophobic blocks are dispersed. On the contrary, polymeric micelle core usually consists only of hydrophobic chains. Other nanoparticles consist of drug nanocrystals (core) coated with a layer of proteins (*e.g.* albumin) or polymers.

medium. After emulsification, the organic solvent was removed by evaporation to yield nanoparticles with PDLLA core and PVP or PEG shell, followed by the removal of the free drug and copolymers. After *i.v.* administration of the PEG-*b*-PDLLA and PVP-*b*-PDLLA nanoparticles in rats demonstrated longer circulation time for the PEG-coated nanoparticles.[48]

Drug nanocrystals coated with polymers are yet another interesting approach for forming nanoparticles that can be either polymeric (coated with polymers) or protein-coated. Fuhrmann *et al.* have recently prepared paclitaxel nanocrystals coated with stabilizer (PEG-*b*-PCL-based copolymer), which were then crosslinked for further stabilization of the nanoparticles.[47] Paclitaxel nanocrystals coated with human albumin non-covalently are currently approved in the market (Abraxane) for treatment of various types of cancer.[52,53]

2.3. *Lipidic nanoparticles*

Liposomes consist of aqueous core and hydrophobic lipid bilayers and they are usually spherical in shape (Fig. 4). Liposomes loaded with both hydrophilic and hydrophobic drugs have previously been prepared for drug delivery applications.[54–56] The lipids of liposomes have been also modified with hydrophilic polymers to prolong the *in vivo* circulation time, and cell recognition

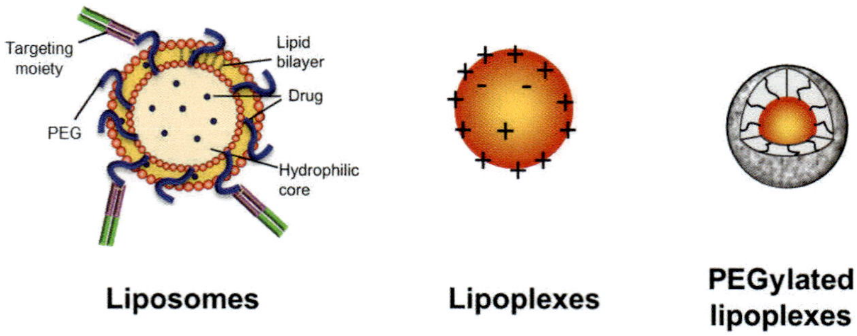

Fig. 4. Liposomes consist of lipid bilayers and aqueous core, which can be utilized to solubilize both hydrophobic and hydrophilic drugs, respectively. PEGylation and functionalization (*e.g.* stimuli-responsive moieties, targeting ligands) of lipids are routinely applied to construct multifunctional liposomes. Lipoplexes consist of cationic lipids, together with other helper lipids, and nucleic acids. PEGylation of lipoplexes yields PEGylated lipoplexes.

moieties have been tethered to the distal end of the hydrophilic polymers for active targeting into diseased tissues.[54,57] Stimuli-responsive lipids have been also incorporated into the constituents of liposomes to enhance the selectivity of drug release.[58] Limitations of liposomes include low stability, batch-to-batch variability, difficulties in sterilization due to relatively larger size and mainly the low drug-loading capacities. Like polymeric assemblies, cationic lipids can be exploited to impart cationic character for complexation of negatively charged drugs, such as nucleic acids.[59–61] In this case, cationic lipids are complexed directly with DNA/RNA, and modification of the lipids (*e.g.* with hydrophilic polymers or targeting ligands) can be performed prior to assembly or after preparation. PEGylated lipoplexes and stable nucleic acid-lipid particles (SNALP) are considered as core-shell type lipidic nanoparticles.[62–68] PEGylated lipoplexes are complexes between nucleic acids and cationic lipids (lipoplexes) coated with PEG. SNALP are considered as PEGylated lipoplexes but with specific composition. SNALP usually consist of siRNA complexed with lipid mixture, cationic lipid, neutral helper lipid, cholesterol and PEG-lipid to provide hydrophilicity to the particles, stability against degradation, and allow for prolonged *in vivo* circulation time.[69]

2.4. *Carbon, metal and metal oxide nanoparticles*

Core-shell nanoparticles of this type usually have core based on metal, metal oxide or carbon nanotubes. The core is usually then coated with hydrophilic layer that can be further modified to incorporate other functionalities. Metal- and metal oxide-based nanoparticles have the added advantage of using them as contrast agents, due to their interesting electrical and optical properties, which can be manipulated by controlling their sizes and morphologies. For instance, magnetic particles can provide direct imaging in addition to the control over the biodistribution of the delivery vehicles by application of magnetic field,[70,71] and can be also used for other clinical applications.[72] The advantages of utilizing these materials have been combined with surface modification chemistries and supramolecular assemblies to yield core-shell nanoparticles as a platform for various biomedical delivery applications. Surface modification of magnetic nanoparticles (*e.g.* paramagnetic iron oxide nanoparticles), single-walled or multi-walled cylindrical graphene sheets, quantum dots, gold nanoparticles and other kinds of nanoparticles have been carried out to improve solubility, stability, selectivity and pharmacokinetics of nanoparticles.[73–79] Coating with hydrophilic layer can enhance their biocompatibility and solubility, and being exploited for further functionalization. Although the focus of this chapter is beyond detailing the extensive list of core-shell nanoparticles, few examples will be given to highlight the role of various modifications in providing nanoparticles suitable for various biomedical applications. Carbon nanotubes have been functionalized with amine or carboxylic acid groups. The functionalized nanotubes have been utilized to carry drugs (*e.g.* cisplatin, doxorubicin and siRNA) and cell recognition moieties (*e.g.* folic acid) for targeted delivery of therapeutics.[76–78] In attempts to control the drug release, carbon nanotubes were loaded with cisplatin and coated with Pluronic surfactant to reduce the passive diffusion of the drug from the carbon nanotubes. Heat generated by radiofrequency application could disrupt the surfactant layer and thus controlling the release of cisplatin.[76] Surface chemistry of carbon nanotubes could also be modified to allow complexation with nucleic acids (*e.g.* siRNA) and targeted delivery of the complexed siRNA.[78] Magnetic iron oxide nanoparticles have been utilized to decorate multiwalled carbon nanotubes that were already functionalized with poly(acrylic acid). The nanoparticle hybrid was loaded with doxorubicin and functionalized with folic acid moieties to achieve active targeting *via* receptor-mediated endocytosis and by the application of magnetic field.[77]

3. Biomimicry

The synthesis and construction of core-shell nanomaterials borrow analogy from the synthesis and functional activity of viruses, proteins and lipoproteins. Both naturally-driven molecules and synthetic nanoparticles are constructed from small molecules to form polymers, which are further assembled into covalently or non-covalently stabilized functional materials. The building units of natural systems are ranging from amino acids, polypeptides, proteins, DNA and RNA nucleotides and polysaccharides. Due to the efficiency of natural molecules in transporting their payloads to the target sites in the body or in transfecting cells with high efficiency, great efforts are continued to be made to build up nanoparticles that mimic their compositions, functionalities and assembly/dissociation patterns.[80–83] Advances in synthetic chemistries and controlled supramolecular assembly have enabled the control over the construction of complex polymeric systems that can assemble into sophisticated architectures. These nanostructures have certain functionalities at specific regions to enable the delivery of their loaded cargoes to a particular target that located either cellularly or intracellularly.

Several groups have designed multifunctional core-shell nanoparticles that resemble the structures and dynamic activities of lipoproteins and viral vectors for delivery of chemotherapeutics and nucleic acids.[84,85] Building principles of polymeric micelles from amphiphilic block copolymers, liposomes from amphiphilic lipids, and further stabilization of these nanoscopic objects through surface functionalization, crosslinking or changing their compositions, borrow analogy from the synthesis of natural viruses and lipoproteins. The dimensions of these synthetic assemblies are similar to the size of viruses and lipoproteins (10–100 nm) (Fig. 5). The detailed similarities between the synthetic chemistries, supramolecular assemblies and functional properties of natural molecules and various nanostructures have recently been reviewed by our group and others.[5,84–86]

4. Design of Core-Shell Nanoparticles: Essential Requirements Versus Smart Design

Nanoparticles can be designed to control size, surface charge, polarity of the nanoenvironment inside nanomaterials, surface chemistry, targeting, release kinetics of the embedded cargoes, *in vivo* circulation time, localization into the target sites, and dissociation under specific conditions (*e.g.* pH) or in response to internal or external stimuli (Table 1). Depending on the number of monomers in the copolymers, relative length of different segments of the copolymers or lipids, relative solubility of each part and the surrounding environment (*e.g.* polarity and ionic strength of the dissolving medium), various morphologies can be attained.[14,41,87–92] These shapes can range from spherical (most common), cylindrical, vesicular, worm shape, *etc.*[93–95] The selection of various criteria depends on the drug to be delivered, therapeutic application and route of administration. The design of smart nanocarriers with smart features to evade the various biological barriers has recently been

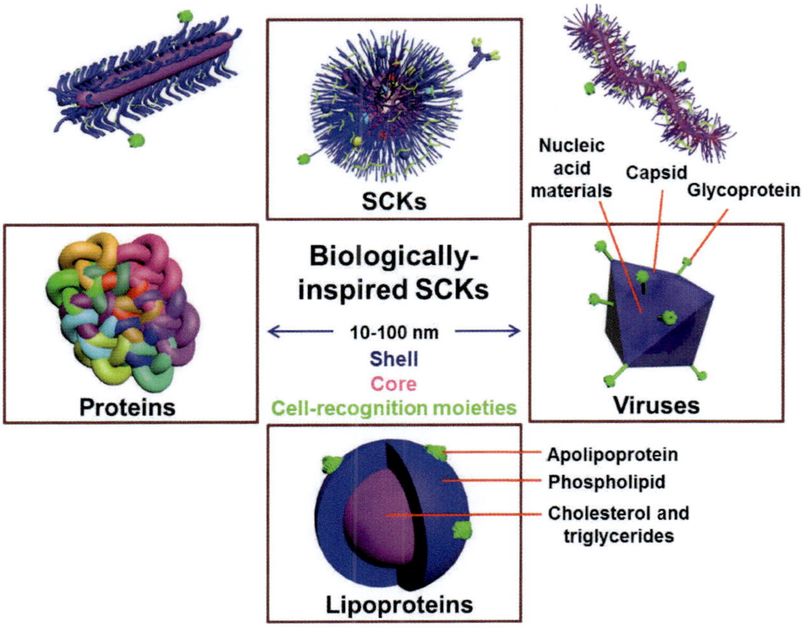

Fig. 5. Biomimicry approaches utilized for the synthesis, supramolecular assembly, stabilization and further functionalization of "smart, biomimic" SCKs, in which the synthetic SCKs are of similar dimensions (10–100 nm) and can be made to present surface features resembling those of proteins, *e.g.* cationic moieties of the histone core proteins of nucleosomes for packaging of nucleic acids, hydrophilic surfaces and hydrophobic cores like lipoproteins for the function of packaging water-insoluble guests, or internal sites for loading of the nucleic acids in a manner analogous to the behavior of viral capsids. SCKs, viruses and lipoproteins share the features of a nanosized (10–100 nm) core (in red)/shell (in blue) morphology with surface-decorating moieties (in green). Reproduced with permission from Ref. 5. Copyright © 2012 Wiley Periodicals, Inc.

reviewed.[1,5,96,97] Few specific examples will be given in this section to highlight the recent development of synthesis and construction of core-shell nanoparticles, with emphasis on the control over their structure and supramolecular assembly.

Wooley group has developed cationic shell crosslinked knedel-like nanoparticles (cSCKs) for delivery of nucleic acids of various structures (antisense oligonucleotide, siRNA, plasmids).[9,86,98–100] The cSCKs have a shell that consists of poly(acrylamidoethylamine) (PAEA) shell, necessary for electrostatic complexation with the negatively charged nucleic acids, and a polystyrene core. The core of these nanoparticles has recently been replaced with a biodegradable core of PDLLA.[29] Modification of these nanoparticles by

Table 1. Summary of the important features of polymeric nanoparticles designed for biomedical delivery applications. Reproduced from Ref. 1, with permission from The Royal Society of Chemistry.

Feature	Description
Stealth	• A dense brush of neutral hydrophilic flexible polymer to prevent adsorption of plasma proteins and avoid recognition by the MPS and prolong the blood circulation time, and thereby allowing the accumulation of the nanoparticles at sites with impaired vasculature *via* the EPR effect • PEG is the most commonly utilized polymer
Smart ingredients	• Cell-recognition moieties enhance the selective cellular uptake • Targeting subcellular compartments (*e.g.* nuclear localization sequence for nuclear delivery) • Skin, mucus (mucolytic) or cell-penetration enhancers • Endosomolytic polymers or lipids to increase the intracellular delivery and avoid lysosomal degradation • Stimuli-responsiveness (temperature, pH, oxidation, reduction, light, magnetic field, ultrasound, *etc.*) allows the controlled-drug release at specific sites or under applied conditions • "Multivalency", "bio-mimicking" or "self-communication" are recently-applied approaches to enhance the delivery efficiency • Any added component to the composition of the nanoparticle impacts their biodistribution, clearance and immunogenicity
Size	• Intermediate size (20–200 nm) have the highest potential for *in vivo* applications
Shape/aspect ratio	• Various morphologies and/or aspect ratios can have different solubilization capacities, blood circulation time, biodistribution, toxicity, cellular uptake and intracellular fate
Surface chemistry	• The surface chemistry of the nanomaterials greatly impact their toxicity, immunogenicity and biodistribution • Excess positive charge results in rapid opsonization and clearance with possible blood vessels and capillary occlusion
Route of administration	• The route of administration influences the stability and delivery efficiency of the nanoparticles • There are different barriers at the various entry routes of the body (*e.g.* pH, enzymes, *etc.*) • Auxiliary devices can enhance the delivery efficiency by localizing the nanoparticles at the target sites
Stability	• Drug leakage, dissociation of nanoparticles and detachment of the surface-decorating moieties are detrimental to the efficacy of the nanoparticles • Stereocomplexation, non-covalent interactions and crosslinking are efficient techniques to enhance the kinetic stability of nanoparticles

replacing some of the primary amines in the shell with tertiary amine and histamine groups has provided nanoparticles of lower toxicity and higher transfection efficiency.[98,100] The higher transfection efficiency is mostly attributed to the increased buffering capacities of nanoparticles that contain two species of low and high p*K*a's which facilitate the endosomal escape and enhance the intracellular bioavailability of the encapsulated nucleic acids. At physiological pH, the high pK_a part (*e.g.* primary amines) preferentially interacts with DNA/RNA *via* electrostatic interactions, which is important for complexation of nucleic acids into compact particles. The moieties of low pK_a (*e.g.* histamine or tertiary amine) remain unionized at the physiological pH, but become protonated in the acidic environment of the endosomes and aid in pumping protons and chloride ions (proton sponge effect), or interact with the endosomal membrane.[18,32,101] Any of these mechanisms will disrupt the endosomes and facilitate the intracellular delivery of DNA/RNA. In addition, incorporation of histamine functionalities into the shell resulted in lower levels of the secreted cytokines, probably due to lowering the charge density on the shell.[100] The latter has been confirmed by the lower siRNA and cellular binding affinity for the histamine-modified nanoparticles. However, great efforts have to be made into tuning the structure of these structures to find the optimal composition. For instance, increasing the histamine content of these nanoparticles to >15% abolished the ability of nanoparticles to complex siRNA and to transfect cells, although was associated with further reduction of cytotoxicities. No cellular uptake or transfection efficiency was observed in cells treated with the cSCKs modified with 50% or 100%-histamine.

Recently, our group has combined the experiences gained from designing spherical and cylindrical nanoparticles for encapsulation and delivery of chemotherapeutics and nucleic acids, into constructing supramolecular assembly that consists of both spherical and cylindrical compartments.[9] The individual nanostructures were crosslinked in the shell region to provide stabilization and enhanced kinetic stability, and were functionalized with various smart features (*e.g.* decoration with cell recognition moieties, built-on sites for radiolabeling or equipped with stimuli-responsive functionalities).[100] The advantages of utilizing two nanoparticles with different shapes (spherical versus cylindrical) and surface charges (anionic versus cationic) were then taken to build multifunctional theranostic nanostructures *via* templating cSCKs onto anionic shell crosslinked anionic cylinders (Fig. 6).[9] These structures had an intermediate zeta-potential value (25 mV) between the cationic spheres (35 mV) and anionic cylinders (30 mV), and their formation was further confirmed by transmission electron microscopy (Fig. 6). Both the hierarchical assemblies and cationic SCKs were able to bind siRNA and deliver it

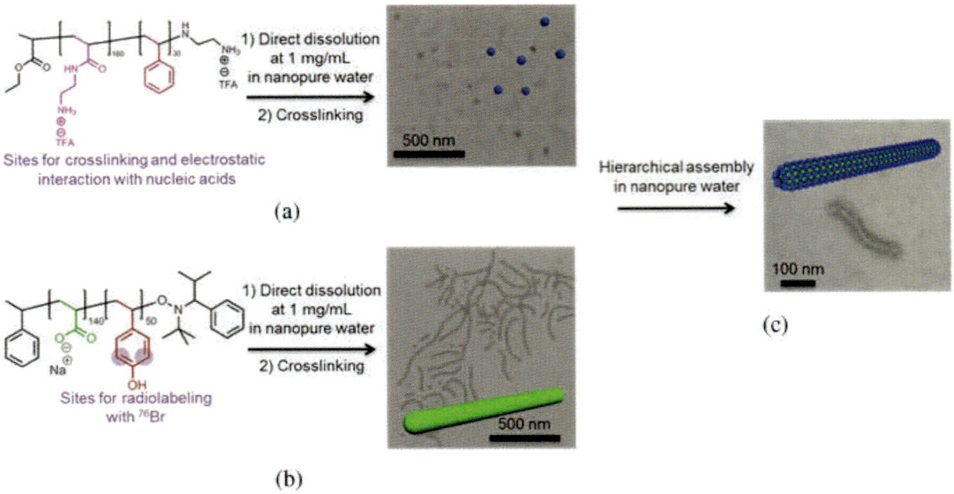

Fig. 6. (a) Preparation of cationic shell crosslinked nanostructures by directly dissolving polymers into water to obtain micelles followed by crosslinking selectively in the shell region to obtain cSCKs; (b) Preparation of anionic nanorods by direct dissolution of polymers followed by crosslinking in the shell region and (c) Hierarchical-assembly of cSCKs on SCRs *via* supramolecular assembly in aqueous medium. Reprinted with permission from Ref. 9. Copyright 2012 American Chemical Society.

intracellularly but not the anionic cylinders due to the lack of positively charged functionalities. The hierarchical assemblies retained the advantage of elongated morphologies which might be associated with prolonged circulation *in vivo*, as reported previously.[93,102] In addition, these structures were loaded with chemotherapeutics and decorated with targeting ligands to further advance their use (work *in progress*). The differences in *in vivo* pharmacokinetics and intracellular delivery mechanisms of these assemblies are currently under investigation.

The complications of constructing efficient nanocarriers arise from integrating multifunctionalities that although improving the performance of nanoparticulates, could adversely influence other pharmacological properties. For instance, decoration of nanoparticles with targeting moieties such as antibodies or proteins, may enhance specificity and cellular uptake of the modified particles, but may also accelerate *in vivo* clearance and/or enhance toxicity/immunogenicity of the carriers.[58,103] Simplicity of preparation is yet another perquisite for clinical translation of nanomaterials from bench to the bedside. Recently, we have developed a rapid and efficient retrosynthetic approach to construct unimodal nanosized core-shell particles based on biodegradable polyphosphoester block copolymers (Fig. 7).[36] The protocol

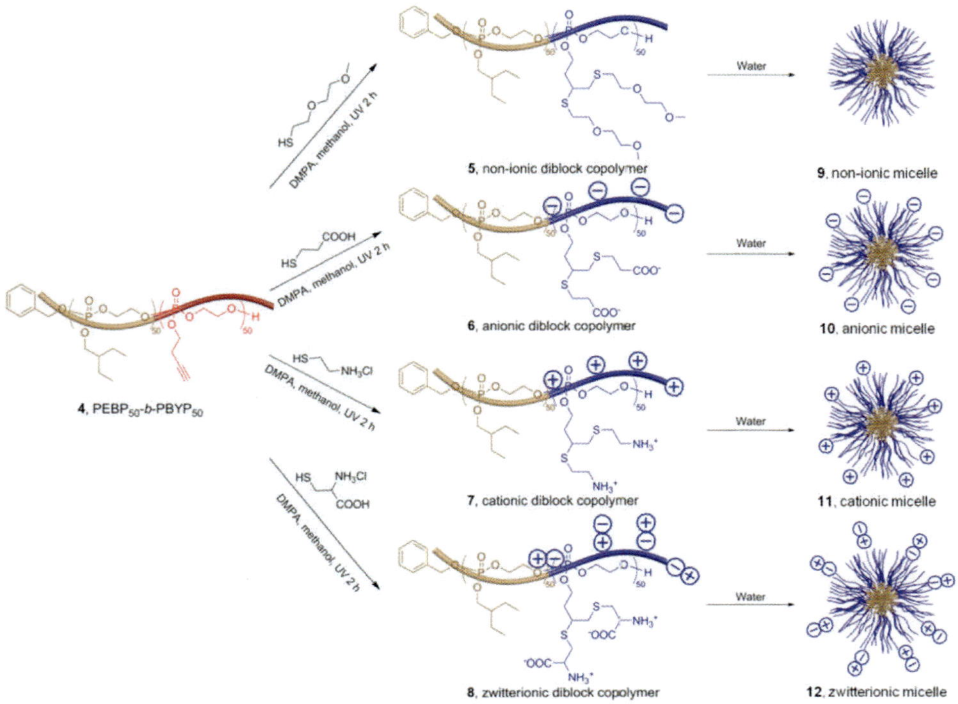

Fig. 7. Schematic representation of the functionalizations of $PEBP_{50}$-*b*-$PBYP_{50}$ with four different charged or non-charged thiols and the self-assembly of four resulting amphiphilic diblock copolymers: non-ionic block copolymer (**5**), anionic block copolymer (**6**), cationic block copolymer (**7**) and zwitterionic block copolymer (**8**) into four micelles, **9–12**, respectively, with different surface charges. Reprinted with permission from Ref. 36. Copyright 2012 American Chemical Society.

could be adapted to yield neutral, anionic, cationic and zwitterionic functional polymeric particles. All nanoparticles demonstrated low cytotoxicity. The cationic ones exhibited a six-fold lower toxicity as compared to Lipofectamine (a commercially available transfection agent), probably due to the biodegradability or the surface characteristics of the nanoparticles. This study was among the first trials to control the surface charge of polymeric nanoparticles as opposed to the facile control of the surface chemistries of inorganic nanoparticles.[104–106]

5. Toxicity

Although interactions between nanomaterials and cells are required for efficient uptake and delivery to the target sites, nanoparticles are capable of inducing cellular toxicity, and to interact with the various components of the immune system and results in immunomodulatory effects. Hence, we are always looking for designing particles with high therapeutic efficacy and low toxicity, *i.e.* to avoid recognition by the immune system and being internalized by the target cells without inducing irreversible toxicity.

Detailed studies on the mechanisms of toxicity and immunomodulatory effects of nanoparticles have recently been reviewed.[107–110] Although there are several mechanisms for induction of nanoparticles-related toxicities and immunogenicities, production of reactive oxygen species and oxidative stress are the most common mechanisms (or at least the most studied) for these toxicities.[110] Interactions of nanoparticles with mitochondria could result in generation of reactive oxygen species, which can also be produced from the nanoparticles *per se*.[110,111] Reactive oxygen species at high levels cause cell damage by peroxidizing various biomolecules (*e.g.* DNA, proteins). Cell damage can results in various adverse reactions, such as, renal, pulmonary, cardiovascular diseases and cancer. Biocompatibility is another important parameter that influences the toxicity of nanoparticles. Interactions with blood components (red blood cells, leukocytes, proteins and platelets) may induce hemolysis, platelet aggregation and blood coagulation.[112,113]

The induction of immunotoxicity both *in vitro* and *in vivo* is commonly tested by measuring the levels of secreted cytokines, along with other immunomodulatory markers.[114–117] The structure, composition, shape and surface chemistries of nanomaterials dictate the type and extent of their interactions with the immune system components and thus the resulting immune response.[114,118,119] Toxicity of nanomaterials is greatly influenced by the route of administration, biodistribution in organs (amount of drugs in a particular organ) and residence time in the organs and cells until clearance or excretion. It is equally important to consider both the nanoparticle-forming material and the drug/payload.[118,120,121] Table 2 summarizes some of the common immunotoxicities induced by nanoparticles and demonstrates dependence of immunomodulatory effects on nanoparticle composition. Shorter contact/residence times with cells and faster cellular uptake have usually been observed for spherical nanoparticles, as compared to cylindrical nanoparticles. The high aspect ratio of cylindrical nanoparticles usually hinders the efficient cellular uptake.[122,123] *In vivo*, it has been reported that cylindrical nanoparticles have lower rate of phagocytosis and tend to adhere to the wall of blood vessels which might lead to damage of macrophages and initiate thrombosis.[124] However, physicochemical properties can alter this rule, e.g., presence of cationic charges or targeting ligands on the surface of nanoparticles can significantly modify the previously mentioned rule.[9,122,123] For instance, although

Table 2. Examples of some of the cytokines that were induced upon the treatment with nanomaterials loaded with nucleic acids or other payloads.

Nanomaterial	Composition	Immunotoxicity	Remarks	Reference
Lipoplexes	*pDNA carrying various transgenes (*e.g.* CFTR) Cationic lipids (*e.g.* DOTAP, GL67) Protamine Cholesterol	1. Increased levels of inflammatory cells and cytokines in a CpG-dependent manner 2. Production of large amounts of proinflammatory cytokines (TNF-α, IL1-β, IL-6, IL-12 and IFN-γ).** 3. Adverse reactions include: 'flu-like' inflammatory syndrome, fever, myalgias and arthralgia. 4. Immunostimulation is associated with reduction of gene expression in a dose-dependent manner. 5. A single CpG can elicit inflammation 6. Methylation of pDNA or the use of CpG-free pDNA reduces the release of cytokines, inflammation and enhance gene expression.		125–129
Cationic liposomes 165 nm 24 mV	DOTAP, cholesterol, protamine and pDNA	***Stimulated the expression of CD80/CD86 on dendritic cells ***Induced the expression of TNF-α	Both DNA and the cationic lipid (DOTAP) are required for full immunostimulation	118
siRNA/lipid complexes	*siRNA targeting various genes (*e.g.* β-galactosidase, luciferase, apolipoprotein B) PEG-lipid Cationic lipids (*e.g.* Oligofectamine) Cholesterol	1. Unmodified siRNA induced induction of cytokines and toxicity in the treated animals in a dose and sequence-dependent manner. 2. Induced the release of several cytokines (IL-1α, IL1-β, IL-6, IL-10, keratinocyte-derived cytokine, TNF-α, IFN-α, IFN-γ and MCP-1).** 3. Chemically modified siRNA contains 2'-*O*-methyl modified nucleosides (<20%) into one of the siRNA strands prevented the immunostimulatory responses induced by siRNA without affecting their gene-silencing activity		119, 121, 130, 131

(*Continued*)

Table 2. (*Continued*)

Nanomaterial	Composition	Immunotoxicity	Remarks	Reference
Targeted anisamide-PEGylated liposomes 114–118 nm 21–28 mV	siRNA-targeting luciferase *Condensing agent: Hyaluronic acid or calf thymus DNA and protamine* DOTAP Cholesterol DSPE-PEG$_{2K}$ DSPE-PEG$_{2K}$-anisamide	1. The use of hyaluronic reduced the immunotoxicity of the nanoparticles, as compared to calf thymus DNA (measured by the induction of IL-6 and IL-12 in the serum of the treated mice). 2. Both formulations had similar characteristics and gene-silencing activity.		132, 133
Targeted anisamide-PEGylated liposomes 150 nm 40 mV	siRNA-targeting luciferase *Condensing agent: Calcium phosphate* DOTAP Cholesterol DSPE-PEG$_{2K}$ DSPE-PEG$_{2K}$-anisamide	1. Resulted in 3–4 folds higher *in vitro* gene-silencing as compared the same nanoparticles formed with DNA/protamine core 2. Similar *in vivo* gene-silencing was observed between the two formulations 3. The calcium phosphate-based nanoparticles had lower *in vivo* immunogenicity as compared to the DNA/protamine-based nanoparticles as indicated by the measurement of the IL-6 and IL-12 cytokines		134
Protamine: 220, 500 or 1200 nm ~ 7 mV	Single-stranded RNA and protamine	Smaller particles elicited antiviral-like response (released IFN-α), whereas the larger particles developed antibacterial-like immune response (released TNF-α)	RNA or protamine alone could not increase the levels of any of the tested cytokines	117

(*Continued*)

Table 2. (*Continued*)

Nanomaterial	Composition	Immunotoxicity	Remarks	Reference
Solid-core particles–antigen conjugates: 20–2000 nm	Polystyrene ovalbumin or other antigens	40–50 nm particles had the highest: a. Immunogenicity b. IFN-γ production c. Uptake and localization into the dendritic cells in the draining lymph nodes	The nanoparticle size influenced the extent of immune response The higher immunogenicity explained by the similarity between this size range of the beads and viral particles	135
Solid-core particles–antigen conjugates: 20, 40, 49, 67, 93, 101 or 123 nm	Polystyrene ovalbumin or other antigens	40 and 49 nm beads induced higher induction of IFN-γ (Type 1 immune response, T_H1) 93, 101 and 123-nm beads induced production of IL-4 (Type 2 immune response, T_H2)	The nanoparticle size dictate the type of the immune response	136

*Composition includes some of the indicated components but varies depending on the study. In addition, these formulations were mainly delivered to the lung of mice or patients *via* aerosols.
**The cytokines profile varies depending on the formulation composition and the animal model.
***Indicators of toll-like receptors (TLRs) signaling pathway.

spherical SCKs were internalized by cells more rapidly than cylindrical nanoparticles, the folate-functionalized cylinders were taken up to a greater extent by cells than the folate-functionalized spheres, probably due to the multiple receptor binding.[122,123] The cellular uptake of Cy3-labeled siRNA complexed with cSCKs and cationic hierarchically assembled nanostructures with overall higher aspect ratio in human ovarian adenocarcinoma cells was comparable, whereas no uptake observed when the siRNA was mixed with the anionic cylinders.[9]

6. Biomedical Applications

Core-shell nanoparticles have been utilized for treatment of several diseases, such as, cancer, viral and bacterial infections, dermatological disorders, although most of applications have focused on cancer therapy. In addition, they have been exploited for imaging, and currently under investigation as theranostic agents (carry therapeutic and diagnostic agents, for concurrent therapy and diagnosis, respectively).[3,9,137–139] This section focuses on some of the clinical examples of core-shell nanoparticles and highlights the important features in their rational design and characteristics.

Genexol, NK-105, *NK*-012, *NK*-911, *NC*-6004 *and NC*-4016: Many polymeric micelles formulations have either been approved or in different phases of clinical trials for the delivery of many potent anticancer drugs (Table 3).[140,141] The common characteristics of these formulations are the unimodal size distribution, nanometer size and core-shell architectures. The average sizes of these micelles (20–85 nm) allow their passage into tumor tissues of leaky vasculature and accumulate in these tissues due to the impaired lymphatic drainage, and thus leading to higher accumulation in these areas, as compared to healthy tissues (EPR effect). The main differences between these polymeric carriers are the structure of the hydrophobic block and the loading method or the self-assembly mechanism. Different block copolymers have been utilized for preparation of these polymeric micelles, such as, PEG-*b*-poly(D,L-lactide) (Genexol, paclitaxel),[142–144] PEG-*b*-polyaspartate, where the polyaspartate is esterified with 4-phenyl-1-butanol to increase the core hydrophobicity (NK-105, paclitaxel),[145] PEG-*b*-poly(aspartic acid)–doxorubicin conjugate (NK-911),[146] PEG-*b*-poly(glutamic acid) (NC-6004, cisplatin[21,147] and NC-4016, oxaliplatin[148–150]) and PEG-*b*-poly(glutamic acid)-conjugated to 7-ethyl-10-hydroxycamptothecin (SN-38) (NK-012, SN-38).[151] NC 4016 is in Phase I, NC-6004 in phase I/II, NK-105, NK-012, NK-911 and Genexol are in Phase II clinical trials. Genexol has been approved since 2007 in Korea. Genexol and NK-105 are examples of physical

Table 3. Clinical applications of polymeric micelles prepared by physical loading, chemical conjugation and ligand substitution reactions.

Clinical example	Type	Loaded-drug	Composition	Clinical applications	Remarks	References
NK-105	Polymeric micelles	Paclitaxel	PEG-*b*-polyaspartate esterified with 4-phenyl-1-butanol	Gastric cancer Phase II (Japan)	Physical loading 85 nm	145
Genexol		Paclitaxel	PEG-*b*-poly(D,L-lactide)	Breast, lung, and ovarian cancer In the market in Korea since 2007, in Phase II in US and Russia	Biodegradable Physical loading 20–50 nm	142–144
NK-012	Polymeric micelles (drug-conjugate)	7-ethyl-10-hydroxyca-mptothecin (SN-38)	PEG-*b*-poly(glutamic acid)–SN-38 conjugate	Breast cancer Phase II (Japan and USA)	Biodegradable 20 nm	151
NK-911		Doxorubicin	PEG-*b*-poly(aspartic acid)–doxorubicin conjugate	Pancreatic and colorectal cancers Phase II (Japan)	Biodegradable Chemical conjugation and physical loading 40 nm	146
NC-4016	Polymeric micelles (polymer–metal complex micelles)	Oxaliplatin	PEG-*b*-poly(glutamic acid)	Bowel cancer Phase I (Japan)	Biodegradable Ligand substitution 40 nm	148–150
NC-6004		Cisplatin	PEG-*b*-poly(glutamic acid)	Advanced solid tumors Phase I/II (UK)	Biodegradable Ligand substitution 28 nm	21, 147

loading of paclitaxel into PEG-*b*-poly(D,L-lactide) and PEG-*b*-polyaspartate esterified with 4-phenyl-1-butanol, respectively. NK-012 and NK911 are polymeric conjugates of PEG-*b*-poly(glutamic acid) and PEG-*b*-poly(aspartic acid) with SN-38 and doxorubicin, respectively, which are then self-assembled into polymeric micelles. In NK-911, doxorubicin was both free and conjugated, where the conjugated drug molecules increase the core hydrophobicity and allow higher loading of the free doxorubicin. NC-4016 and NC-6004 are polymer–metal complex micelles formed between PEG-*b*-poly(glutamic acid) and oxaliplatin and cisplatin, respectively.[16,152]

Polymeric micelles formulations could alter the pharmacokinetics and reduce the toxicity of the free drugs or the side effects associated with the low molecular weight surfactants utilized for their delivery (*e.g.* Cremophor-EL and Tween 80). The lower toxicity associated with polymeric vehicles allowed increase in the maximum tolerated dose thus leading to higher concentration of the drug in the tumors and better tumor responses among patients. For instance, cisplatin loaded into PEG-*b*-poly(glutamic acid) had prolonged circulation time and 20-fold higher accumulation in the tumor tissues than the free drug, and completed tumor regression was observed only in the tumor-bearing mice treated with the cisplatin-loaded micelles.[20] In clinical studies, the NC-6004 was well-tolerated with minimal nephrotoxicity and myelosuppression.[16,153] Similarly, NC-4016 micelle formulation has demonstrated higher tumor accumulation and antitumor activity in tumor-bearing mice than the free oxaliplatin.[16,150]

Abraxane: Abraxane, also known as *nab*-paclitaxel, is protein nanoparticles that consist of albumin-bound paclitaxel with a mean particle size diameter of 130 nm (Table 4). It has been approved by the FDA since 2005 for the treatment of metastatic breast cancer. The albumin coating, in addition to enhancing the paclitaxel solubility, it facilitates transport of paclitaxel into tumors through binding of albumin to SPARC (secreted protein acid and rich in cysteine), an extracellular matrix glycoprotein that is overexpressed and associated with poor prognosis in a variety of cancers.[52,53] When Abraxane and Cremophor-EL-paclitaxel were administered at equivalent doses of paclitaxel, higher intratumoral accumulation of the drug was observed for the albumin nanoparticles.[154] In addition, premedication to reduce the hypersensitivity reactions associated with the low molecular weight surfactants is not required with Abraxane.[155–157]

BIND-014: BIND-014 is the first targeted polymeric nanoparticle to enter clinical trials. BIND-014 is targeted to prostate-specific membrane antigen (PSMA), a cell surface antigen abundantly expressed on the surface of prostate cancer cells and on newly formed blood vessels (Table 5).[4,50,51] These

Table 4. Clinical applications of protein nanoparticles.

Clinical example	Type	Loaded-drug	Composition	Clinical applications	Remarks	References
Abraxane	Protein nanoparticles	Paclitaxel	Albumin-bound paclitaxel	Metastatic breast cancer, head and neck, lung, ovarian, prostate and colon cancer	On the market Physical loading 130 nm	52, 53

Table 5. Clinical applications of polymeric nanoparticles.

Clinical example	Type	Loaded-drug	Composition	Clinical applications	Remarks	References
BIND-014	Polymeric nanoparticles	Docetaxel	Poly(lactic acid) core coated with PEG Surface functionalization with targeting ligands directed to PSMA	Advanced solid tumors and metastatic cancers Phase I (USA)	Biodegradable First active targeted polymeric formulation in clinical trials Physical loading 100 nm	4, 50, 51

nanoparticles have an average size around 100 nm and contain the cytotoxic agent, docetaxel, which is approved in advanced solid tumors, breast, prostate and other types of cancer, embedded into poly(lactic acid) core that is coated with PEG. BIND-014 was shown to deliver up to 10 times more docetaxel to the tumor site in multiple animal models than an equivalent dose of Taxotere, and have shown high tolerability and efficacy.[50] BIND-014 is currently in Phase I clinical trials in patients with advanced solid tumors and metastatic cancer.

Aurimune (*CYT*-6091): Aurimune is based on colloidal gold nanoparticles (27 nm) with PEG and recombinant human tumor necrosis factor alpha (rhTNF) bound onto their surfaces (Table 6).[158] It is currently in Phase I clinical trials for treatment of pancreatic cancer, melanoma, soft tissue sarcoma, ovarian, and breast cancer. Higher doses of rhTNF formulated as CYT-6091 could be given as compared to previous clinical trials with the free drug, and has exceeded the lethal dose of the protein alone. The small size and chemical conjugation of the drug could reduce the premature release of the drug and allowed higher accumulation of the drug in the tumor tissues through the EPR effect. The side effects associated with the rhTNF were reduced. Tissue biopsies of the treated patients had confirmed the accumulation of the nanoparticles in the tumor tissues.

AuroLase: AuroLase nanoparticles, also known as gold nanoshells, are *ca.* 120 nm nanoparticles based on gold shell and silica particle core that are currently in clinical trials for photothermal tumor ablation, for treatment of refractory head and neck cancers (Table 6).[159] After tumor accumulation of the nanoparticles, laser irradiation is usually utilized to heat nanoparticles and thus resulting in tumor ablation. This kind of treatment has the advantages of localized and controlled therapy. In addition, magnetic resonance imaging can be utilized for monitoring the nanoparticles. However, the difficultly of thermal ablation therapy is in reaching deeper tissues.

ALN-VSP: SNALP that consist of lipids and two siRNA (Table 7).[65,160] It is produced by Alnylam and currently in Phase I clinical trials for treatment of liver cancer. The two siRNAs target kinesin spindle protein (KSP) and vascular endothelial growth factor (VEGF), which are involved in cancer proliferation and growth of new blood vessels (angiogenesis), respectively, with the purpose of increasing therapeutic benefits. The SNALP formulations were well-tolerated and have demonstrated antiproliferative activity due to KSP and VEGF genes silencing and led to reduction of tumor growth in the liver, in the preclinical studies. This high therapeutic efficacy was associated with an increase in the survival of the treated animals. In Phase I clinical trials, the

Table 6. Clinical applications of gold nanoparticles.

Clinical example	Type	Loaded-drug	Composition	Clinical applications	Remarks	References
Aurimune (CYT-6091)	Gold nanoparticles	TNF-α	PEGylated gold nanoparticles	Pancreatic cancer, melanoma, soft tissue sarcoma, ovarian, and breast cancer Phase I	Non-degradable Chemical conjugation 27 nm	158
AuroLase	Gold/silica nanoparticles	Gold	Gold shell and silica particle core	Head and neck cancer Pilot study in clinical trials	Non-degradable 120 nm	159

Table 7. Clinical applications of lipid nanoparticles.

Clinical example	Type	Loaded-drug	Composition	Clinical applications	Remarks	References
ALN-VSP	Lipid nanoparticles (SNALP)	Two siRNA (KSP and VEGF)	Lipids and siRNA	Liver cancer Phase I		65,160
Allovectin-7	Lipid nanoparticles (Lipoplexes)	Plasmid	Cationic lipid mixture (DMRIE/ DOPE)	Metastatic melanoma Phase III (USA)	DMRIE is a cationic lipid necessary for complexation of DNA DOPE is a helper lipid to enhance transfection efficiency	161,162
CPX-351	Lipid nanoparticles (Liposomes)	Cytarabine and daunorubicin (5:1 synergistic molar ratio)	Distearoyl phosphatidy lglycerol, distearoyl phosphatidy lcholine and cholesterol	Acute myeloid leukemia Phase II	107 nm	55, 163, 164

formulations were well-tolerated and pharmacologically relevant levels were detected in the tumor biopsies from the treated patients.

Allovectin-7: Allovectin-7 formulations are lipoplexes developed by Vical Inc. They consist of plasmid DNA, cationic lipid, 1, 2-dimyristyloxypropyl-3-dimethylhydroxyethyl ammonium bromide (DMRIE) and dioleoyl-phosphatidylethanolamine (DOPE) (Table 7).[161,162] DMRIE is a cationic lipid necessary to complex the negatively charged plasmid, whereas DOPE is a helper lipid to enhance the transfection efficiency of the nanoparticles. The plasmid encodes for two genes, namely, foreign human HLA-B7 heavy chain and β2-microglobulin. HLA-B7 is highly immunogenic cell surface protein and thus can stimulate the immune system against tumor cells expressing the antigen.

CPX-351: CPX-351 is currently in Phase II clinical trials. It is based on liposomes that contain two anticancer drugs, namely, cytarabine and daunorubicin with size of *ca.* 107 nm (Table 7).[55,163,164] A specific 5:1 molar ratio of cytarabine and daunorubicin has been utilized to provide synergistic effect of combinational therapy against acute myeloid leukemia. The bilayer of CPX-351 consists of distearoylphosphatidylglycerol and distearoylphosphatidylcholine phospholipids and cholesterol. The liposomes had extended blood circulation time *in vivo* and improved efficacy.

Feraheme: Feraheme is based on superparamagnetic iron oxide nanoparticles (17–31 nm) coated with carbohydrate (polyglucose sorbitol carboxymethylether) that have been approved by the FDA (Table 8).[165,166] These nanoparticles have been utilized to treat iron deficiency anemia in adult patients with chronic kidney disease.

Feridex: Feridex superparamagnetic iron oxide nanoparticles (120–180 nm; 80–150 nm) have been approved by FDA as a contrast agent for imaging of liver and spleen cancers (Table 8).[72,167,168] These nanoparticles are firstly opsonized by the plasma proteins and then taken up by the MPS, and thus being phagocytozed by the liver Kupffer cells or in the spleen (organs rich in phagocytic cells). Cancerous tissues (*e.g.* hepatocellular carcinoma) do not have phagocytic activity and thus providing the contrast in magnetic resonance imaging (*i.e.* darkening of healthy tissues as compared to tumor tissues).

7. Core-Shell Nanoparticles for Inhalational Antimicrobial Applications

Although the main attention of this chapter is directed toward the development of nanomaterials for cancer therapy, core-shell nanoparticles have also

Table 8. Clinical applications of iron oxide nanoparticles.

Clinical example	Type	Loaded-drug	Composition	Clinical applications	Remarks	References
Feraheme	Iron oxide nanoparticles	Iron oxide	Polyglucose sorbitol carboxymethylether-coated superparamagnetic iron oxide	Iron replacement therapy for the treatment of iron deficiency anemia FDA approved	17–31 nm	165, 166
Feridex	Iron oxide nanoparticles	Iron oxide	Dextran-coated superparamagnetic iron oxide	Magnetic resonance imaging agent in the liver FDA approved	120–180 nm; 80–150 nm	72, 167, 168

been utilized for delivery of drugs, other than chemotherapeutics, *via* other routes of administration.[1,5] Mucosal routes are among the attractive non-invasive routes of administration, and there are many advantages for inhalation-based and/or catheter-based routes of entry that allows for direct encounter of the nanomaterials with the sites of injury or disease, such as in the treatment of pulmonary or urinary tract diseases, as opposed to *i.v.* introduction into the bloodstream. For instance, one of the evolving eras in the field of nanomedicine is the development of nanotherapeutics as carriers for antimicrobial agents for treatment of pulmonary or urinary tract infections. Advantages of utilizing inhalational or catheter delivery over systemic delivery is the reduced exposure of the healthy tissues to the drug, while concentrating the major fraction of the therapeutics into the diseased tissues in the lung or bladder. In addition, it can improve the patient compliance as it can be nebulized by patients, for instance *via* metered-dose inhaler, on the contrary to the injection requirements and complications. Furthermore, these routes of administration avoid the hepatic first-pass metabolism, have faster onset of action and are considered as a non-invasive routes of administration. It should always be considered, however, that the therapeutics delivered in these manners can reach the blood through the lung or bladder epithelium, requiring rigorous biodistribution studies. Combination of more than one drug or inclusion of diagnostic agents into the composition of nanomaterials is feasible, and can facilitate such studies of efficacy coincidentally with pharmacokinetics. The usefulness of these particular deliveries is evidenced in the treatment of pulmonary infectious diseases, such as tuberculosis, and cystic fibrosis or other lung diseases, such as asthma and lung injury, and in the treatment of recurrent urinary tract infections; such nanomedicines are being actively developed in our laboratory together with several collaborators.

Nanomaterials utilized for inhalational therapy should be equipped with smart features, as discussed previously for core-shell nanoparticles for other biomedical application.[1] The smart features are required to bypass the physiological barriers encountered during their delivery, in particular to circumvent the mucociliary clearance, adhesiveness, steric hindrance and enzymes found in the mucosal surfaces.[169] These barriers hinder the uptake and delivery of the nanoparticulates to their target sites. In addition, they can greatly influence the stability and intactness of nanoparticles. Mucus is a viscoelastic hydrogel that consists of a network of negatively-charged mucin fibers loaded with cells, bacteria, proteins and other macromolecules, with a composition varying according to the physiological region.[169] Coating nanoparticles with PEG partially reduces interactions with mucus and extracellular matrix and decreases the mucosal clearance. In addition to PEG, other moieties to hydrolyze mucin and enhance the mucosal penetration of nanomaterials can also be incorporated into the composition of nanoparticles.[170]

Poly(acrylic acid)-*block*-polystyrene copolymer-based SCKs have been utilized for the co-delivery of silver and silver–carbene complexes and demonstrated high antimicrobial activity against respiratory isolates of the Gram-negative bacteria, *Pseudomonas aeruginosa*.[171] The latter is considered as a cystic fibrosis relevant bacteria. Recently, the same nanoparticles were modified with targeting ligands onto their surfaces (uropathogenic *E. coli* type 1 pilus adhesin FimHA) *via* amidation chemistry to allow higher binding and uptake into the bladder epithelial cells, as a mechanism that is being explored to treat recurrent urinary tract infections.[172] The same concept can be adopted to construct multifunctional inhalational nanoparticles for treatment of lung diseases. Silver nanoparticles also were utilized for the treatment of bronchial inflammation and hyperresponsiveness in mice *via* the inhalational administration.[173] In this study, the nanoparticles were able to reduce the intracellular reactive oxygen species, inflammatory cells and airway hyperresponsiveness (induced by ovalbumin inhalation). Sustained release of silver–carbene complexes over several days have also been demonstrated from L-tyrosine polyphosphate nanoparticles, and were associated with high antimicrobial activity against *Pseudomonas aeruginosa* both *in vitro* and *in vivo* in mice with only two doses.[174] Several other inhalable antitubercular agents were developed based on various polymeric and lipidic nanoparticles and have demonstrated high efficiency.[175]

8. Conclusions

Core-shell nanoparticles constitute the major category of nanomaterials that are currently in the market, in different phases of clinical trials or still under investigation in several research laboratories. The core of these nanoparticles can have different compositions, ranging from mobile, rigid or solid matrix, aqueous, to hollow cage, and can be constructed from various polymeric, lipidic or metallic materials. Chemical modifications of the core were performed to accommodate payloads of various charges, solubilities and molecular weights. The shell is mainly composed of hydrophilic polymers and usually dictates *in vivo* pharmacokinetics and biodistribution of the nanocarriers, and provides active sites for further functionalization with cell-recognition moieties, therapeutic and/or diagnostic agents. Polymeric nanoparticles might have the added advantages of precise control over their structures, compositions and properties (being degradable or stimuli-responsive), in contrast to other metallic nanoparticles. However, the latter impart a robust structure to nanomaterials and can also be utilized as diagnostic or contrast agents. Hence, a combination of both kinds of materials "hybrid nanoparticles" might enhance the therapeutic efficacy of nanoparticles and facilitate their clinical translation. Although there are several examples of core-shell nanoparticles in the market or under clinical trials, further development and characterizations are continuing to allow control over their pharmacological properties and enhance the drug accumulation into the diseased sites *versus* healthy tissues. It is extremely important to consider the entire life-time of nanomaterials from production to monitoring their physicochemical and biological properties, to their *in vivo* and intracellular fate.

9. Acknowledgments

We gratefully acknowledge financial support from the National Heart Lung and Blood Institute of the National Institutes of Health as a Program of Excellence in Nanotechnology (HHSN268201000046C), and the National Institute of Diabetes and Digestive and Kidney Diseases of the National Institutes of Health (R01-DK082546). The Welch Foundation is gratefully acknowledged for support through the W. T. Doherty-Welch Chair in Chemistry, Grant No. A-0001.

References

1. M. Elsabahy and K. L. Wooley, Design of polymeric nanoparticles for biomedical delivery applications, *Chem. Soc. Rev.* **41**, 2545–2561 (2012).
2. R. Ghosh Chaudhuri and S. Paria, Core/shell nanoparticles: classes, properties, synthesis mechanisms, characterization, and applications, *Chem. Rev.* **112**, 2373–2433 (2012).
3. A. M. Nystrom and K. L. Wooley, The importance of chemistry in creating well-defined nanoscopic embedded therapeutics: devices capable of the dual functions of imaging and therapy, *Acc. Chem. Res.* **44**, 969–978 (2011).
4. J. Shi, Xiao, Z. N. Kamaly and O. C. Farokhzad, Self-assembled targeted nanoparticles: evolution of technologies and bench to bedside translation, *Acc. Chem. Res.* **44**, 1123–1134 (2011).
5. M. Elsabahy and K. L. Wooley, Strategies toward well-defined polymer nanoparticles inspired by nature: chemistry versus versatility, *J. Polym. Sci. Part A: Polym Chem* **50**, 1869–1880 (2012).
6. E. C. Dreaden, M. A. Mackey, X. Huang, B. Kang and M. A. El-Sayed, Beating cancer in multiple ways using nanogold, *Chem. Soc. Rev.* **40**, 3391–3404 (2011).
7. K. Miyata, N. Nishiyama and K. Kataoka, Rational design of smart supramolecular assemblies for gene delivery: chemical challenges in the creation of artificial viruses, *Chem. Soc. Rev.* **41**, 2562–2574 (2012).

8. Y. H. Yu, E. Kim, D. E. Park, G. Shim, S. Lee, Y. B. Kim, C. W. Kim and Y. K. Oh, Cationic solid lipid nanoparticles for co-delivery of paclitaxel and siRNA, *Eur. J. Pharm. Biopharm.* **80**, 268–273 (2012).
9. R. Shrestha, M. Elsabahy, H. Luehmann, S. Samarajeewa, S. Florez-Malaver, N. S. Lee, M. J. Welch, Y. Liu and K. L. Wooley, Hierarchically assembled theranostic nanostructures for siRNA delivery and imaging applications, *J. Am. Chem. Soc.* **134**, 17362–17365 (2012).
10. X. B. Xiong and A. Lavasanifar, Traceable multifunctional micellar nanocarriers for cancer-targeted co-delivery of MDR-1 siRNA and doxorubicin, *ACS Nano* **5**, 5202–5213 (2011).
11. T. M. Sun, J. Z. Du, Y. D. Yao, C. Q. Mao, S. Dou, S. Y. Huang, P. Z. Zhang, K. W. Leong, E. W. Song and J. Wang, Simultaneous delivery of siRNA and paclitaxel via a "two-in-one" micelleplex promotes synergistic tumor suppression, *ACS Nano* **5**, 1483–1494 (2011).
12. K. H. Bae, K. Lee, C. Kim and T. G. Park, Surface functionalized hollow manganese oxide nanoparticles for cancer targeted siRNA delivery and magnetic resonance imaging, *Biomaterials* **32**, 176–184 (2011).
13. J. H. Lee, K. Lee, S. H. Moon, Y. Lee, T. G. Park and J. Cheon, All-in-one target-cell-specific magnetic nanoparticles for simultaneous molecular imaging and siRNA delivery, *Angew. Chem. Int. Ed. Engl.* **48**, 4174–4179 (2009).
14. M. Elsabahy, M. H. Dufresne and J. C. Leroux, in Materials for Nanomedicine, V. Torchilin and M. Amiji (Eds.), Hackensack: Pan Stanford Publishing (2010), pp. 169–234.
15. K. Kataoka, A. Harada and Y. Nagasaki, Block copolymer micelles for drug delivery: design, characterization and biological significance, *Adv. Drug Deliv. Rev.* **47**, 113–131 (2001).
16. X. Wang and Z. Guo, Targeting and delivery of platinum-based anticancer drugs, *Chem. Soc. Rev.* **42**, 202–224 (2012).
17. Y. I. Jeong, S. J. Seo, I. K. Park, H. C. Lee, I. C. Kang, T. Akaike and C. S. Cho, Cellular recognition of paclitaxel-loaded polymeric nanoparticles composed of poly(gamma-benzyl L-glutamate) and poly(ethylene glycol) diblock copolymer end-capped with galactose moiety, *Int. J. Pharm.* **296**, 151–161 (2005).
18. M. Elsabahy, N. Wazen, N. Puxan, G. Deleavey, M. Servant, M. J. Damha and J. C. Leroux, Delivery of nucleic acids through the controlled disassembly of multifunctional nanocomplexes, *Adv. Funct. Mater.* **19**, 3862–3867 (2009).
19. A. E. Felber, B. Castagner, M. Elsabahy, G. F. Deleavey, M. J. Damha and J. C. Leroux, siRNA nanocarriers based on methacrylic acid copolymers, *J. Control. Release* **152**, 159–167 (2011).
20. N. Nishiyama, S. Okazaki, H. Cabral, M. Miyamoto, Y. Kato, Y. Sugiyama, K. Nishio, Y. Matsumura and K. Kataoka, Novel cisplatin-incorporated polymeric micelles can eradicate solid tumors in mice, *Cancer Res.* **63**, 8977–8983 (2003).
21. H. Uchino, Y. Matsumura, T. Negishi, F. Koizumi, T. Hayashi, T. Honda, N. Nishiyama, K. Kataoka, S. Naito and T. Kakizoe, Cisplatin-incorporating polymeric micelles (NC-6004) can reduce nephrotoxicity and neurotoxicity of cisplatin in rats, *Br. J. Cancer* **93**, 678–687 (2005).

22. L. Y. Lin, N. S. Lee, J. Zhu, A. M. Nystrom, D. J. Pochan, R. B. Dorshow and K. L. Wooley, Tuning core vs. shell dimensions to adjust the performance of nanoscopic containers for the loading and release of doxorubicin, *J. Control. Release* **152**, 37–48 (2011).
23. D. Le Garrec, S. Gori, L. Luo, D. Lessard, D. C. Smith, M. A. Yessine, M. Ranger and J. C. Leroux, Poly(N-vinylpyrrolidone)-block-poly(D,L-lactide) as a new polymeric solubilizer for hydrophobic anticancer drugs: *in vitro* and *in vivo* evaluation, *J. Control. Release* **99**, 83–101 (2004).
24. L. Luo, M. Ranger, D. G. Lessard, D. L. Garrec, S. Gori, J. C. Leroux, S. Rimmer and D. Smith, Novel amphiphilic diblock copolymer of low molecular weight Poly (N-vinylpyrrolidone)-block-poly(D,L-lactide): synthesis, characterization, and micellization, *Macromolecules* **37**, 4008–4013 (2004).
25. A. J. Keefe and S. Jiang, Poly(zwitterionic)protein conjugates offer increased stability without sacrificing binding affinity or bioactivity, *Nat. Chem.* **4**, 59–63 (2012).
26. A. Li, H. P. Luehmann, G. Sun, S. Samarajeewa, J. Zou, S. Zhang, F. Zhang, M. J. Welch, Y. Liu and K. L.Wooley, Synthesis and *in vivo* pharmacokinetic evaluation of degradable shell cross-linked polymer nanoparticles with poly(carboxybetaine) versus poly(ethylene glycol) surface-grafted coatings, *ACS Nano* **6**, 8970–8982 (2012).
27. M. Nakayama and T. Okano, Polymer terminal group effects on properties of thermoresponsive polymeric micelles with controlled outer-shell chain lengths, *Biomacromolecules* **6**, 2320–2327 (2005).
28. B. Liu, M. Yang, R. Li, Y. Ding, X. Qian, L. Yu and X. Jiang, The antitumor effect of novel docetaxel-loaded thermosensitive micelles, *Eur. J. Pharm. Biopharm.* **69**, 527–534 (2008).
29. S. Samarajeewa, R. Shrestha, Y. Li and K. L. Wooley, Degradability of poly(Lactic Acid)-containing nanoparticles: Enzymatic access through a cross-linked shell barrier, *J. Am. Chem. Soc.* **134**, 1235–1242 (2012).
30. K. Miyata, Y. Kakizawa, N. Nishiyama, A. Harada, Y. Yamasaki, H. Koyama and K. Kataoka, Block catiomer polyplexes with regulated densities of charge and disulfide cross-linking directed to enhance gene expression, *J. Am. Chem. Soc.* **126**, 2355–2361 (2004).
31. K. Miyata, Y. Kakizawa, N. Nishiyama, Y. Yamasaki, T. Watanabe, M. Kohara and K. Kataoka, Freeze-dried formulations for *in vivo* gene delivery of PEGylated polyplex micelles with disulfide crosslinked cores to the liver, *J. Control. Release* **109**, 15–23 (2005).
32. M. Elsabahy, M. Zhang, S. M. Gan, K. C. Waldron and J. C. Leroux, Synthesis and enzymatic stability of PEGylated oligonucleotide duplexes and their self-assemblies with polyamidoamine dendrimers, *Soft. Matter* **4**, 294–302 (2008).
33. M. Y. Nobuhiro Nishiyama, Takao Aoyagi, Teruo Okano, Yasuhisa Sakurai and Kazunori Kataoka, Preparation and characterization of self-assembled polymer-metal complex micelle from cis-Dichlorodiammineplatinum(II) and Poly(ethylene glycol)-Poly (,-aspartic acid) block copolymer in an aqueous medium, *Langmuir* **15**, 377–383 (1999).
34. T. Govender, S. Stolnik, C. Xiong, S. Zhang, L. Illum and S. S. Davis, Drug-polyionic block copolymer interactions for micelle formation: physicochemical characterisation, *J. Control. Release* **75**, 249–258 (2001).

35. S. W. Choi, S. H. Lee, H. Mok and T. G. Park, Multifunctional siRNA delivery system: polyelectrolyte complex micelles of six-arm PEG conjugate of siRNA and cell penetrating peptide with crosslinked fusogenic peptide, *Biotechnol. Prog.* **26**, 57–63 (2010).
36. S. Zhang, J. Zou, F. Zhang, M. Elsabahy, S. E. Felder, J. Zhu, D. J. Pochan and K. L. Wooley, Rapid and Versatile Construction of Diverse and Functional Nanostructures Derived from a Polyphosphoester-Based Biomimetic Block Copolymer System, *J. Am. Chem. Soc.* **134**, 18467–18474 (2012).
37. M. C. Jones, P. Tewari, C. Blei, K. Hales, D. J. Pochan and J. C. Leroux, Self-assembled nanocages for hydrophilic guest molecules, *J. Am. Chem. Soc.* **128**, 14599–14605 (2006).
38. H. Gao, M.-C. Jones, J. Chen, R. E. Prud'homme and J. C. Leroux, Core cross-linked reverse micelles from star-shaped polymers, *Chem. Mater.* **20**, 3063–3067 (2008).
39. J. O. Kim, G. Sahay, A. V. Kabanov and T. K. Bronich, Polymeric micelles with ionic cores containing biodegradable cross-links for delivery of chemotherapeutic agents, *Biomacromolecules* **11**, 919–926 (2010).
40. N. L. Klyachko, D. S. Manickam, A. M. Brynskikh, S. V. Uglanova, S. Li, S. M. Higginbotham, T. K. Bronich, E. V. Batrakova and A. V. Kabanov, Cross-linked antioxidant nanozymes for improved delivery to CNS, *Nanomedicine* **8**, 119–129 (2012).
41. G. Sun, H. Cui, L. Y. Lin, N. S. Lee, C. Yang, W. L. Neumann, J. N. Freskos, J. J. Shieh, R. B. Dorshow and K. L. Wooley, Multicompartment polymer nanostructures with ratiometric dual-emission pH-sensitivity, *J. Am. Chem. Soc.* **133**, 8534–8543 (2011).
42. G. Sun, M. Y. Berezin, J. Fan, H. Lee, J. Ma, K. Zhang, K. L. Wooley and S. Achilefu, Bright fluorescent nanoparticles for developing potential optical imaging contrast agents, *Nanoscale* **2**, 548–558 (2010).
43. C. J. Rijcken, C. J. Snel, R. M. Schiffelers, C. F. van Nostrum and W. E. Hennink, Hydrolysable core-crosslinked thermosensitive polymeric micelles: synthesis, characterisation and *in vivo* studies, *Biomaterials* **28**, 5581–5593 (2007).
44. N. S. Lee, G. Sun, L. Y. Lin, W. L. Neumann, J. N. Freskos, A. Karwa, J. J. Shieh, R. B. Dorshow and K. L. Wooley, Tunable dual-emitting shell-crosslinked nano-objects as single-component ratiometric pH-sensing materials, *J. Mater. Chem.* **21**, 14193–14202 (2011).
45. J. L. Turner, Z. Chen and K. L. Wooley, Regiochemical functionalization of a nanoscale cage-like structure: robust core-shell nanostructures crafted as vessels for selective uptake and release of small and large guests, *J. Control. Release* **109**, 189–202 (2005).
46. A. O. Moughton and R. K. O'Reilly, Noncovalently connected micelles, nanoparticles, and metal-functionalized nanocages using supramolecular self-assembly, *J. Am. Chem. Soc.* **130**, 8714–8725 (2008).
47. K. Fuhrmann, J. D. Schulz, M. A. Gauthier and J. C. Leroux, PEG Nanocages as non-sheddable stabilizers for drug nanocrystals, *ACS Nano* **6**, 1667–1676 (2012).
48. G. Gaucher, K. Asahina, J. Wang and J. C. Leroux, Effect of poly(N-vinyl-pyrrolidone)-block-poly(D,L-lactide) as coating agent on the opsonization, phagocytosis, and pharmacokinetics of biodegradable nanoparticles, *Biomacromolecules* **10**, 408–416 (2009).
49. G. Gaucher, M. Poreba, F. Ravenelle and J. C. Leroux, Poly(N-vinyl-pyrrolidone)-block-poly(D,L-lactide) as polymeric emulsifier for the preparation of biodegradable nanoparticles, *J. Pharm. Sci.* **96**, 1763–1775 (2007).

50. N. Kamaly, Z. Xiao, P. M. Valencia, A. F. Radovic-Moreno and O. C. Farokhzad, Targeted polymeric therapeutic nanoparticles: design, development and clinical translation, *Chem. Soc. Rev.* **41**, 2971–3010 (2012).
51. J. Hrkach, D. Von Hoff, M. Mukkaram Ali, E. Andrianova, J. Auer, T. Campbell, D. De Witt, M. Figa, M. Figueiredo, A. Horhota, S. Low, K. McDonnell, E. Peeke, B. Retnarajan, A. Sabnis, E. Schnipper, J. J. Song, Y. H. Song, J. Summa, D. Tompsett, G. Troiano, T. Van Geen Hoven, J. Wright, P. LoRusso, P. W. Kantoff, N. H. Bander, C. Sweeney, O. C. Farokhzad, R. Langer and S. Zale, Preclinical development and clinical translation of a PSMA-targeted docetaxel nanoparticle with a differentiated pharmacological profile, *Sci. Transl. Med.* **4**, 128ra139 (2012).
52. V. Roy, B. R. LaPlant, G. G. Gross, C. L. Bane and F. M. Palmieri, Phase II trial of weekly nab (nanoparticle albumin-bound)-paclitaxel (nab-paclitaxel) (Abraxane) in combination with gemcitabine in patients with metastatic breast cancer (N0531), *Ann. Oncol.* **20**, 449–453 (2009).
53. M. J. Hawkins, P. Soon-Shiong and N. Desai, Protein nanoparticles as drug carriers in clinical medicine, *Adv. Drug Deliv. Rev.* **60**, 876–885 (2008).
54. C. Mamot, R. Ritschard, A. Wicki, G. Stehle, T. Dieterle, L. Bubendorf, C. Hilker, S. Deuster, R. Herrmann and C. Rochlitz, Tolerability, safety, pharmacokinetics, and efficacy of doxorubicin-loaded anti-EGFR immunoliposomes in advanced solid tumours: a phase 1 dose-escalation study, *Lancet Oncol.* **13**, 1234–1241 (2012).
55. E. J. Feldman, J. E. Kolitz, J. M. Trang, B. D. Liboiron, C. E. Swenson, M. T. Chiarella, L. D. Mayer, A. C. Louie and J. E. Lancet, Pharmacokinetics of CPX-351; a nano-scale liposomal fixed molar ratio formulation of cytarabine: daunorubicin, in patients with advanced leukemia, *Leuk. Res.* **36**, 1283–1289 (2012).
56. D. Bedi, T. Musacchio, O. A. Fagbohun, J. W. Gillespie, P. Deinnocentes, R. C. Bird, L. Bookbinder, V. P. Torchilin and V. A. Petrenko, Delivery of siRNA into breast cancer cells via phage fusion protein-targeted liposomes, *Nanomedicine* **7**, 315–323 (2011).
57. A. Koshkaryev, R. Sawant, M. Deshpande and V. Torchilin, Immunoconjugates and long circulating systems: Origins, current state of the art and future directions, *Adv. Drug Deliv. Rev.* **65**, 24–35 (2013).
58. P. Simard and J. C. Leroux, *In vivo* evaluation of pH-sensitive polymer-based immunoliposomes targeting the CD33 antigen, *Mol. Pharm.* **7**, 1098–1107 (2010).
59. W. De Haes, G. Van Mol, C. Merlin, S. C. De Smedt, G. Vanham and J. Rejman, Internalization of mRNA lipoplexes by dendritic cells, *Mol. Pharm.* **9**, 2942–2949 (2012).
60. H. Zhang, A. Mitin and S. V. Vinogradov, Efficient transfection of blood–brain barrier endothelial cells by lipoplexes and polyplexes in the presence of nuclear targeting NLS-PEG-acridine conjugates, *Bioconjugate Chem.* **20**, 120–128 (2009).
61. C. Tros De Ilarduya, M. Bunuales, C. Qian and N. Duzgunes, Antitumoral activity of transferrin-lipoplexes carrying the IL-12 gene in the treatment of colon cancer, *J. Drug Target* **14**, 527–535 (2006).
62. M. Oishi, Y. Nagasaki, K. Itaka, N. Nishiyama and K. Kataoka, Lactosylated poly(ethylene glycol)-siRNA conjugate through acid-labile beta-thiopropionate linkage to construct pH-sensitive polyion complex micelles achieving enhanced gene silencing in hepatoma cells, *J. Am. Chem. Soc.* **127**, 1624–1625 (2005).
63. M. Oishi, K. Kataoka and Y. Nagasaki, pH-responsive three-layered PEGylated polyplex micelle based on a lactosylated ABC triblock copolymer as a targetable and endosome-disruptive nonviral gene vector, *Bioconjugate Chem.* **17**, 677–688 (2006).

64. S. Fukushima, K. Miyata, N. Nishiyama, N. Kanayama, Y. Yamasaki and K. Kataoka, PEGylated polyplex micelles from triblock catiomers with spatially ordered layering of condensed pDNA and buffering units for enhanced intracellular gene delivery, *J. Am. Chem. Soc.* **127**, 2810–2811 (2005).
65. T. S. Zimmermann, A. C. Lee, A. Akinc, B. Bramlage, D. Bumcrot, M. N. Fedoruk, J. Harborth, J. A. Heyes, L. B. Jeffs, M. John, A. D. Judge, K. Lam, K. McClintock, L. V. Nechev, L. R. Palmer, T. Racie, I. Rohl, S. Seiffert, S. Shanmugam, V. Sood, J. Soutschek, I. Toudjarska, A. J. Wheat, E. Yaworski, W. Zedalis, V. Koteliansky, M. Manoharan, H. P. Vornlocher and I. MacLachlan, RNAi-mediated gene silencing in non-human primates, *Nature* **441**, 111–114 (2006).
66. D. V. Morrissey, J. A. Lockridge, L. Shaw, K. Blanchard, K. Jensen, W. Breen, K. Hartsough, L. Machemer, S. Radka, V. Jadhav, N. Vaish, S. Zinnen, C. Vargeese, K. Bowman, C. S. Shaffer, L. B. Jeffs, A. Judge, I. MacLachlan and B. Polisky, Potent and persistent *in vivo* anti-HBV activity of chemically modified siRNAs, *Nat. Biotechnol.* **23**, 1002–1007 (2005).
67. S. C. Semple, A. Akinc, J. Chen, A. P. Sandhu, B. L. Mui, C. K. Cho, D. W. Sah, D. Stebbing, E. J. Crosley, E. Yaworski, I. M. Hafez, J. R. Dorkin, J. Qin, K. Lam, K. G. Rajeev, K. F. Wong, L. B. Jeffs, L. Nechev, M. L. Eisenhardt, M. Jayaraman, M. Kazem, M. A. Maier, M. Srinivasulu, M. J. Weinstein, Q. Chen, R. Alvarez, S. A. Barros, S. De, S. K. Klimuk, T. Borland, V. Kosovrasti, W. L. Cantley, Y. K. Tam, M. Manoharan, M. A. Ciufolini, M. A. Tracy, A. de Fougerolles, I. MacLachlan, P. R. Cullis, T. D. Madden and M. J. Hope, Rational design of cationic lipids for siRNA delivery, *Nat. Biotechnol.* **28**, 172–176 (2010).
68. A. D. Judge, M. Robbins, I. Tavakoli, J. Levi, L. Hu, A. Fronda, E. Ambegia, K. McClintock and I. MacLachlan, Confirming the RNAi-mediated mechanism of action of siRNA-based cancer therapeutics in mice, *J. Clin. Invest.* **119**, 661–673 (2009).
69. S. A. Barros and J. A. Gollob, Safety profile of RNAi nanomedicines, *Adv. Drug. Deliv. Rev.* **64**, 1730–1737 (2012).
70. P. Pouponneau, J. C. Leroux, G. Soulez, L. Gaboury and S. Martel, Co-encapsulation of magnetic nanoparticles and doxorubicin into biodegradable microcarriers for deep tissue targeting by vascular MRI navigation, *Biomaterials* **32**, 3481–3486 (2011).
71. P. Pouponneau, J. C. Leroux and S. Martel, Magnetic nanoparticles encapsulated into biodegradable microparticles steered with an upgraded magnetic resonance imaging system for tumor chemoembolization, *Biomaterials* **30**, 6327–6332 (2009).
72. E. Amstad, M. Textor and E. Reimhult, Stabilization and functionalization of iron oxide nanoparticles for biomedical applications, *Nanoscale* **3**, 2819–2843 (2011).
73. E. C. Dreaden, A. M. Alkilany, X. Huang, C. J. Murphy and M. A. El-Sayed, The golden age: gold nanoparticles for biomedicine, *Chem. Soc. Rev.* **41**, 2740–2779 (2012).
74. Wahajuddin and S. Arora, Superparamagnetic iron oxide nanoparticles: magnetic nanoplatforms as drug carriers, *Int. J. Nanomedicine* **7**, 3445–3471 (2012).
75. J. Xu, C. Wang, J. Wang, Q. Huo, N. F. Crawford, E. A. Veliz and R. M. Leblanc, Smooth surface roughness of silanized CdSe(ZnS) quantum dots, *J. Colloid. Interface Sci.* **393**, 21–28 (2013).
76. M. Raoof, B. T. Cisneros, A. Guven, S. Phounsavath, S. J. Corr, L. J. Wilson and S. A. Curley, Remotely triggered cisplatin release from carbon nanocapsules by radiofrequency fields, *Biomaterials* **34**, 1862–1869 (2013).

77. Y. J. Lu, K. C. Wei, C. C. Ma, S. Y. Yang and J. P. Chen, Dual targeted delivery of doxorubicin to cancer cells using folate-conjugated magnetic multi-walled carbon nanotubes, *Colloids Surf. B Biointerfaces* **89**, 1–9 (2012).
78. L. Wang, J. Shi, H. Zhang, H. Li, Y. Gao, Z. Wang, H. Wang, L. Li, C. Zhang, C. Chen, Z. Zhang and Y. Zhang, Synergistic anticancer effect of RNAi and photothermal therapy mediated by functionalized single-walled carbon nanotubes, *Biomaterials* **34**, 262–274 (2013).
79. J. Wang, S. Han, D. Ke and R. Wang, Semiconductor Quantum Dots Surface Modification for Potential Cancer Diagnostic and Therapeutic Applications, *J. Nanomaterials*, 1–8 (2012).
80. D. Simberg, T. Duza, J. H. Park, M. Essler, J. Pilch, L. Zhang, A. M. Derfus, M. Yang, R. M. Hoffman, S. Bhatia, M. J. Sailor and E. Ruoslahti, Biomimetic amplification of nanoparticle homing to tumors, *Proc. Natl. Acad. Sci. USA* **104**, 932–936 (2007).
81. C. Wolfrum, S. Shi, K. N. Jayaprakash, M. Jayaraman, G. Wang, R. K. Pandey, K. G. Rajeev, T. Nakayama, K. Charrise, E. M. Ndungo, T. Zimmermann, V. Koteliansky, M. Manoharan and M. Stoffel, Mechanisms and optimization of *in vivo* delivery of lipophilic siRNAs, *Nat. Biotechnol.* **25**, 1149–1157 (2007).
82. N. Murthy, J. Campbell, N. Fausto, A. S. Hoffman and P. S. Stayton, Bioinspired pH-responsive polymers for the intracellular delivery of biomolecular drugs, *Bioconjugate Chem.* **14**, 412–419 (2003).
83. R. Esfand and D. A. Tomalia, Poly(amidoamine) (PAMAM) dendrimers: from biomimicry to drug delivery and biomedical applications, *Drug. Discov. Today* **6**, 427–436 (2001).
84. K. K. Ng, J. F. Lovell and G. Zheng, Lipoprotein-inspired nanoparticles for cancer theranostics, *Acc. Chem. Res.* **44**, 1105–1113 (2011).
85. T. Nakamura, H. Akita, Y. Yamada, H. Hatakeyama and H. Harashima, A multifunctional envelope-type nanodevice for use in nanomedicine: concept and applications, *Acc. Chem. Res.* **45**, 1113–1121 (2012).
86. K. Zhang, H. Fang, G. Shen, J. S. Taylor and K. L. Wooley, Well-defined cationic shell crosslinked nanoparticles for efficient delivery of DNA or peptide nucleic acids, *Proc. Am. Thorac. Soc.* **6**, 450–457 (2009).
87. M. Elsabahy, M. E. Perron, N. Bertrand, G. E. Yu, and J. C. Leroux, Solubilization of Docetaxel in Poly(ethylene oxide)-block-poly(butylene/styrene oxide) Micelles, *Biomacromolecules* **8**, 2250–2257 (2007).
88. H. Cui, Z. Chen, S. Zhong, K. L. Wooley and D. J. Pochan, Block copolymer assembly via kinetic control, *Science* **317**, 647–650 (2007).
89. A. Choucair and A. Eisenberg, Control of amphiphilic block copolymer morphologies using solution conditions, *Eur. Phys. J. E. Soft Matter* **10**, 37–44 (2003).
90. N. S. Lee, L. Y. Lin, W. L. Neumann, J. N. Freskos, A. Karwa, J. J. Shieh, R. B. Dorshow and K. L. Wooley, Influence of nanostructure morphology on host capacity and kinetics of guest release, *Small* **7**, 1998–2003 (2011).
91. Q. Ma, E. E. Remsen Jr., C. G. Clark, T. Kowalewski and K. L. Wooley, Chemically induced supramolecular reorganization of triblock copolymer assemblies: trapping of intermediate states via a shell-crosslinking methodology, *Proc. Natl. Acad. Sci. USA* **99**, 5058–5063 (2002).
92. A. O. Moughton, J. P. Patterson and R. K. O'Reilly, Reversible morphological switching of nanostructures in solution, *Chem. Commun.* **47**, 355–357 (2011).

93. Y. Geng, P. Dalhaimer, S. Cai, R. Tsai, M. Tewari, T. Minko and D. E. Discher, Shape effects of filaments versus spherical particles in flow and drug delivery, *Nat. Nanotechnol.* **2**, 249–255 (2007).
94. P. D. Younghoon Kim, David A. Christian and Dennis E. Discher, Polymeric worm micelles as nano-carriers for drug delivery, *Nanotechnology* **16**, S484–S491 (2005).
95. D. E. Discher and A. Eisenberg, Polymer vesicles, *Science* **297**, 967–973 (2002).
96. R. de la Rica, D. Aili and M. M. Stevens, Enzyme-responsive nanoparticles for drug release and diagnostics, *Adv. Drug Deliv. Rev.* **64**, 967–978 (2012).
97. L. M. Randolph, M. -P. Chien and N. C. Gianneschi, Biological stimuli and biomolecules in the assembly and manipulation of nanoscale polymeric particles, *Chem. Sci.* **3**, 1363–1380 (2012).
98. K. Zhang, H. Fang, Z. Wang, Z. Li, J. S. Taylor and K. L. Wooley, Structure-activity relationships of cationic shell-crosslinked knedel-like nanoparticles: shell composition and transfection efficiency/cytotoxicity, *Biomaterials* **31**, 1805–1813 (2010).
99. K. Zhang, H. Fang, Z. Wang, J. S. Taylor and K. L. Wooley, Cationic shell-crosslinked knedel-like nanoparticles for highly efficient gene and oligonucleotide transfection of mammalian cells, *Biomaterials* **30**, 968–977 (2009).
100. R. Shrestha, M. Elsabahy, S. Florez-Malaver, S. Samarajeewa and K. L. Wooley, Endosomal escape and siRNA delivery with cationic shell crosslinked knedel-like nanoparticles with tunable buffering capacities, *Biomaterials* **33**, 8557–8568 (2012).
101. A. Akinc, M. Thomas, A. M. Klibanov and R. Langer, Exploring polyethylenimine-mediated DNA transfection and the proton sponge hypothesis, *J. Gene Med.* 7, 657–663 (2005).
102. S. Cai, K. Vijayan, D. Cheng, E. M. Lima and D. E. Discher, Micelles of different morphologies–advantages of worm-like filomicelles of PEO-PCL in paclitaxel delivery, *Pharm. Res.* **24**, 2099–2109 (2007).
103. M. Shokeen, E. D. Pressly, A. Hagooly, A. Zheleznyak, N. Ramos, A. L. Fiamengo, M. J. Welch, C. J. Hawker and C. J. Anderson, Evaluation of multivalent, functional polymeric nanoparticles for imaging applications, *ACS Nano* **5**, 738–747 (2011).
104. C. M. Goodman, C. D. McCusker, T. Yilmaz and V. M. Rotello, Toxicity of gold nanoparticles functionalized with cationic and anionic side chains, *Bioconjug. Chem.* **15**, 897–900 (2004).
105. R. R. Arvizo, O. R. Miranda, M. A. Thompson, C. M. Pabelick, R. Bhattacharya, J. D. Robertson, V. M. Rotello, Y. S. Prakash and P. Mukherjee, Effect of nanoparticle surface charge at the plasma membrane and beyond, *Nano Lett.* **10**, 2543–2548 (2010).
106. C. D. Walkey, J. B. Olsen, H. Guo, A. Emili and W. C. Chan, Nanoparticle size and surface chemistry determine serum protein adsorption and macrophage uptake, *J. Am. Chem. Soc.* **134**, 2139–2147 (2012).
107. B. S. Zolnik, A. Gonzalez-Fernandez, N. Sadrieh and M. A. Dobrovolskaia, Nanoparticles and the immune system, *Endocrinology* **151**, 458–465 (2010).
108. S. Hussain, J. A. Vanoirbeek and P. H. Hoet, Interactions of nanomaterials with the immune system, *Wiley Interdiscip. Rev. Nanomed. Nanobiotechnol.* **2**, 169–183 (2012).
109. M. A. Dobrovolskaia, D. R. Germolec and J. L. Weaver, Evaluation of nanoparticle immunotoxicity, *Nat. Nanotechnol.* **4**, 411–414 (2009).
110. S. Sharifi, S. Behzadi, S. Laurent, M. L. Forrest, P. Stroeve and M. Mahmoudi, Toxicity of nanomaterials, *Chem. Soc. Rev.* **41**, 2323–2343 (2012).

111. T. Xia, M. Kovochich, J. Brant, M. Hotze, J. Sempf, T. Oberley, C. Sioutas, J. I. Yeh, M. R. Wiesner and A. E. Nel, Comparison of the abilities of ambient and manufactured nanoparticles to induce cellular toxicity according to an oxidative stress paradigm, *Nano Lett.* **6**, 1794–1807 (2006).
112. M. A. Dobrovolskaia, J. D. Clogston, B. W. Neun, J. B. Hall, A. K. Patri and S. E. McNeil, Method for analysis of nanoparticle hemolytic properties *in vitro*, *Nano Lett.* **8**, 2180–2187 (2008).
113. S. Lin, F. Du, Y. Wang, S. Ji, D. Liang, L. Yu and Z. Li, An acid-labile block copolymer of PDMAEMA and PEG as potential carrier for intelligent gene delivery systems, *Biomacromolecules* **9**, 109–115 (2008).
114. B. C. Heng, X. Zhao, E. C. Tan, N. Khamis, A. Assodani, S. Xiong, C. Ruedl, K. W. Ng and J. S. Loo, Evaluation of the cytotoxic and inflammatory potential of differentially shaped zinc oxide nanoparticles, *Arch. Toxicol.* **85**, 1517–1528 (2011).
115. C. A. Poland, R. Duffin, I. Kinloch, A. Maynard, W. A. Wallace, A. Seaton, V. Stone, S. Brown, W. Macnee and K. Donaldson, Carbon nanotubes introduced into the abdominal cavity of mice show asbestos-like pathogenicity in a pilot study, *Nat. Nanotechnol.* **3**, 423–428 (2008).
116. S. Hussain, S. Boland, A. Baeza-Squiban, R. Hamel, L. C. Thomassen, J. A. Martens, M. A. Billon-Galland, J. Fleury-Feith, F. Moisan, J. C. Pairon and F. Marano, Oxidative stress and proinflammatory effects of carbon black and titanium dioxide nanoparticles: role of particle surface area and internalized amount, *Toxicology* **260**, 142–149 (2009).
117. L. Rettig, S. P. Haen, A. G. Bittermann, L. von Boehmer, A. Curioni, S. D. Kramer, A. Knuth and S. Pascolo, Particle size and activation threshold: a new dimension of danger signaling, *Blood* **115**, 4533–4541 (2010).
118. Z. Cui, S. J. Han, D. P. Vangasseri and L. Huang, Immunostimulation mechanism of LPD nanoparticle as a vaccine carrier, *Mol. Pharm.* **2**, 22–28 (2005).
119. W. Tao, X. Mao, J. P. Davide, B. Ng, M. Cai, P. A. Burke, A. B. Sachs and L. Sepp-Lorenzino, Mechanistically probing lipid-siRNA nanoparticle-associated toxicities identifies Jak inhibitors effective in mitigating multifaceted toxic responses, *Mol. Ther.* **19**, 567–575 (2011).
120. J. Y. Kim, S. Choung, E. J. Lee, Y. J. Kim and Y. C. Choi, Immune activation by siRNA/liposome complexes in mice is sequence- independent: lack of a role for Toll-like receptor 3 signaling, *Mol. Cells* **24**, 247–254 (2007).
121. A. D. Judge, G. Bola, A. C. Lee and I. MacLachlan, Design of noninflammatory synthetic siRNA mediating potent gene silencing *in vivo*, *Mol. Ther.* **13**, 494–505 (2006).
122. K. Zhang, H. Fang, Z. Chen, J. S. Taylor and K. L. Wooley, Shape effects of nanoparticles conjugated with cell-penetrating peptides (HIV Tat PTD) on CHO cell uptake, *Bioconjugate Chem.* **19**, 1880–1887 (2008).
123. K. Zhang, R. Rossin, A. Hagooly, Z. Chen, M. J. Welch and K. L. Wooley, Folate-mediated cell uptake of shell-crosslinked spheres and cylinders, *J. Polym. Sci. Part A: Polym. Chem.* **46**, 7578–7583 (2008).
124. K. H. Park, M. Chhowalla, Z. Iqbal and F. Sesti, Single-walled carbon nanotubes are a new class of ion channel blockers, *J. Biol. Chem.* **278**, 50212–50216 (2003).
125. S. C. Hyde, I. A. Pringle, S. Abdullah, A. E. Lawton, L. A. Davies, A. Varathalingam, G. Nunez-Alonso, A. M. Green, R. P. Bazzani, S. G. Sumner-Jones, M. Chan, H. Li, N. S. Yew, S. H. Cheng, A. C. Boyd, J. C. Davies, U. Griesenbach, D. J. Porteous,

D. N. Sheppard, F. M. Munkonge, E. W. Alton and D. R. Gill, CpG-free plasmids confer reduced inflammation and sustained pulmonary gene expression, *Nat. Biotechnol.* **26**, 549–551 (2008).

126. Y. Tan, S. Li, B. R. Pitt and L. Huang, The inhibitory role of CpG immunostimulatory motifs in cationic lipid vector-mediated transgene expression *in vivo*, *Hum. Gene Ther.* **10**, 2153–2161 (1999).
127. F. E. Ruiz, J. P. Clancy, M. A. Perricone, Z. Bebok, J. S. Hong, S. H. Cheng, D. P. Meeker, K. R. Young, R. A. Schoumacher, M. R. Weatherly, L. Wing, J. E. Morris, L. Sindel, M. Rosenberg, F. W. van Ginkel, J. R. McGhee, D. Kelly, R. K. Lyrene and E. J. Sorscher, A clinical inflammatory syndrome attributable to aerosolized lipid-DNA administration in cystic fibrosis, *Hum. Gene Ther.* **12**, 751–761 (2001).
128. N. S. Yew, K. X. Wang, M. Przybylska, R. G. Bagley, M. Stedman, J. Marshall, R. K. Scheule and S. H. Cheng, Contribution of plasmid DNA to inflammation in the lung after administration of cationic lipid: pDNA complexes, *Hum. Gene Ther.* **10**, 223–234 (1999).
129. B. D. Freimark, H. P. Blezinger, V. J. Florack, J. L. Nordstrom, S. D. Long, D. S. Deshpande, S. Nochumson and K. L. Petrak, Cationic lipids enhance cytokine and cell influx levels in the lung following administration of plasmid: cationic lipid complexes, *J. Immunol.* **160**, 4580–4586 (1998).
130. A. D. Judge, V. Sood, J. R. Shaw, D. Fang, K. McClintock and I. MacLachlan, Sequence-dependent stimulation of the mammalian innate immune response by synthetic siRNA, *Nat. Biotechnol.* **23**, 457–462 (2005).
131. M. T. Abrams, M. L. Koser, J. Seitzer, S. C. Williams, M. A. DiPietro, W. Wang, A. W. Shaw, X. Mao, V. Jadhav, J. P. Davide, P. A. Burke, A. B. Sachs, S. M. Stirdivant and L. Sepp-Lorenzino, Evaluation of efficacy, biodistribution, and inflammation for a potent siRNA nanoparticle: effect of dexamethasone co-treatment, *Mol. Ther.* **18**, 171–180 (2010).
132. S. Chono, S. D. Li, C. C. Conwell and L. Huang, An efficient and low immunostimulatory nanoparticle formulation for systemic siRNA delivery to the tumor, *J. Control. Release* **131**, 64–69 (2008).
133. S. D. Li, S. Chono and L. Huang, Efficient oncogene silencing and metastasis inhibition via systemic delivery of siRNA, *Mol. Ther.* **16**, 942–946 (2008).
134. J. Li, Y. C. Chen, Y. C. Tseng, S. Mozumdar and L. Huang, Biodegradable calcium phosphate nanoparticle with lipid coating for systemic siRNA delivery, *J. Control. Release* **142**, 416–421 (2010).
135. T. Fifis, A. Gamvrellis, B. Crimeen-Irwin, G. A. Pietersz, J. Li, P. L. Mottram, I. F. McKenzie and M. Plebanski, Size-dependent immunogenicity: therapeutic and protective properties of nano-vaccines against tumors, *J. Immunol.* **173**, 3148–3154 (2004).
136. P. L. Mottram, D. Leong, B. Crimeen-Irwin, S. Gloster, S. D. Xiang, J. Meanger, R. Ghildyal, N. Vardaxis and M. Plebanski, Type 1 and 2 immunity following vaccination is influenced by nanoparticle size: formulation of a model vaccine for respiratory syncytial virus, *Mol. Pharm.* **4**, 73–84 (2007).
137. T. Lammers, F. Kiessling, W. E. Hennink and G. Storm, Nanotheranostics and image-guided drug delivery: current concepts and future directions, *Mol. Pharm.* **7**, 1899–1912 (2010).
138. N. Ahmed, H. Fessi and A.Elaissari, Theranostic applications of nanoparticles in cancer, *Drug Discov. Today* **17**, 928–934 (2012).

139. M. P. Melancon, R. J. Stafford and C. Li, Challenges to effective cancer nanotheranostics, *J. Control. Release* **164**, 177–182 (2012).
140. C. Deng, Y. Jiang, R. Cheng, F. Meng and Z. Zhong, Biodegradable polymeric micelles for targeted and controlled anticancer drug delivery: Promises, progress and prospects, *Nano Today* 7, 467–480 (2012).
141. M. Beija, R. Salvayre, N. Lauth-de Viguerie and J. D. Marty, Colloidal systems for drug delivery: from design to therapy, *Trends Biotechnol.* **30**, 485–496 (2012).
142. T. Y. Kim, D. W. Kim, J. Y. Chung, S. G. Shin, S. C. Kim, D. S. Heo, N. K. Kim and Y. J. Bang, Phase I and pharmacokinetic study of Genexol-PM, a cremophor-free, polymeric micelle-formulated paclitaxel, in patients with advanced malignancies, *Clin. Cancer Res.* **10**, 3708–3716 (2004).
143. K. S. Lee, H. C. Chung, S. A. Im, Y. H. Park, C. S. Kim, S. B. Kim, S. Y. Rha, M. Y. Lee and J. Ro, Multicenter phase II trial of Genexol-PM, a Cremophor-free, polymeric micelle formulation of paclitaxel, in patients with metastatic breast cancer, *Breast Cancer Res. Treat* **108**, 241–250 (2008).
144. D. W. Kim, S. Y. Kim, H. K. Kim, S. W. Kim, S. W. Shin, J. S. Kim, K. Park, M. Y. Lee, and D. S. Heo, Multicenter phase II trial of Genexol-PM, a novel Cremophor-free, polymeric micelle formulation of paclitaxel, with cisplatin in patients with advanced non-small-cell lung cancer, *Ann. Oncol.* **18**, 2009–2014 (2007).
145. T. Hamaguchi, Y. Matsumura, M. Suzuki, K. Shimizu, R. Goda, I. Nakamura, I. Nakatomi, M. Yokoyama, K. Kataoka and T. Kakizoe, NK105, a paclitaxel-incorporating micellar nanoparticle formulation, can extend *in vivo* antitumour activity and reduce the neurotoxicity of paclitaxel, *Br. J. Cancer* **92**, 1240–1246 (2005).
146. Y. Matsumura, T. Hamaguchi, T. Ura, K. Muro, Y. Yamada, Y. Shimada, K. Shirao, T. Okusaka, H. Ueno, M. Ikeda and N. Watanabe, Phase I clinical trial and pharmacokinetic evaluation of NK911, a micelle-encapsulated doxorubicin, *Br. J. Cancer* **91**, 1775–1781 (2004).
147. R. Plummer, R. H. Wilson, H. Calvert, A. V. Boddy, M. Griffin, J. Sludden, M. J. Tilby, M. Eatock, D. G. Pearson, C. J. Ottley, Y. Matsumura, K. Kataoka and T. Nishiya, A Phase I clinical study of cisplatin-incorporated polymeric micelles (NC-6004) in patients with solid tumours, *Br. J. Cancer* **104**, 593–598 (2011).
148. M. Rafi, H. Cabral, M. R. Kano, P. Mi, C. Iwata, M. Yashiro, K. Hirakawa, K. Miyazono, N. Nishiyama and K. Kataoka, Polymeric micelles incorporating (1,2-diaminocyclohexane)platinum (II) suppress the growth of orthotopic scirrhous gastric tumors and their lymph node metastasis, *J. Control. Release* **159**, 189–196 (2012).
149. H. Cabral, N. Nishiyama, S. Okazaki, H. Koyama and K. Kataoka, Preparation and biological properties of dichloro(1,2-diaminocyclohexane)platinum(II) (DACHPt)-loaded polymeric micelles, *J. Control. Release* **101**, 223–232 (2005).
150. H. Cabral, N. Nishiyama and K. Kataoka, Optimization of (1,2-diamino-cyclohexane) platinum(II)-loaded polymeric micelles directed to improved tumor targeting and enhanced antitumor activity, *J. Control. Release* **121**, 146–155 (2007).
151. T. Hamaguchi, T. Doi, T. Eguchi-Nakajima, K. Kato, Y. Yamada, Y. Shimada, N. Fuse, A. Ohtsu, S. Matsumoto, M. Takanashi and Y. Matsumura, Phase I study of NK012, a novel SN-38-incorporating micellar nanoparticle, in adult patients with solid tumors, *Clin. Cancer Res.* **16**, 5058–5066 (2010).
152. Y. Matsumura, Polymeric micellar delivery systems in oncology, *Jpn. J. Clin. Oncol.* **38**, 793–802 (2008).

153. Y. Matsumura, Poly (amino acid) micelle nanocarriers in preclinical and clinical studies, *Adv. Drug Deliv. Rev.* **60**, 899–914 (2008).
154. N. Desai, V. Trieu, Z. Yao, L. Louie, S. Ci, A. Yang, C. Tao, T. De, B. Beals, D. Dykes, P. Noker, R. Yao, E. Labao, M. Hawkins and P. Soon-Shiong, Increased antitumor activity, intratumor paclitaxel concentrations, and endothelial cell transport of cremophor-free, albumin-bound paclitaxel, ABI-007, compared with cremophor-based paclitaxel, *Clin. Cancer Res.* **12**, 1317–1324 (2006).
155. I. F. Uchegbu and A. Siew, Nanomedicines and nanodiagnostics come of age, *J. Pharm. Sci.* **102**, 305–310 (2013).
156. W. J. Gradishar, Taxanes for the treatment of metastatic breast cancer, *Breast Cancer (Auckl)* **6**, 159–171 (2012).
157. W. J. Gradishar, S. Tjulandin, N. Davidson, H. Shaw, N. Desai, P. Bhar, M. Hawkins and J. O'Shaughnessy, Phase III trial of nanoparticle albumin-bound paclitaxel compared with polyethylated castor oil-based paclitaxel in women with breast cancer, *J. Clin. Oncol.* **23**, 7794–7803 (2005).
158. S. K. Libutti, G. F. Paciotti, A. A. Byrnes, H. R. Alexander Jr., W. E. Gannon, M. Walker, G. D. Seidel, N. Yuldasheva and L. Tamarkin, Phase I and pharmacokinetic studies of CYT-6091, a novel PEGylated colloidal gold-rhTNF nanomedicine, *Clin. Cancer Res.* **16**, 6139–6149 (2010).
159. A. S. Thakor, J. Jokerst, C. Zavaleta, T. F. Massoud and S. S. Gambhir, Gold nanoparticles: a revival in precious metal administration to patients, *Nano Lett.* **11**, 4029–4036 (2011).
160. L. Huang and Y. Liu, *In vivo* delivery of RNAi with lipid-based nanoparticles, *Annu. Rev. Biomed. Eng.* **13**, 507–530 (2011).
161. R. Chowdhery and R. Gonzalez, Immunologic therapy targeting metastatic melanoma: allovectin-7, *Immunotherapy* **3**, 1721 (2011).
162. A. Y. Bedikian and M. Del Vecchio, Allovectin-7 therapy in metastatic melanoma, *Expert Opin. Biol. Ther.* **8**, 839–844 (2008).
163. W. F. Bayne, L. D. Mayer and C. E. Swenson, Pharmacokinetics of CPX-351 (cytarabine/daunorubicin HCl) liposome injection in the mouse, *J. Pharm. Sci.* **98**, 2540–2548 (2009).
164. A. Dicko, S. Kwak, A. A. Frazier, L. D. Mayer and B. D. Liboiron, Biophysical characterization of a liposomal formulation of cytarabine and daunorubicin, *Int. J. Pharm.* **391**, 248–259 (2010).
165. P. Geisser and S. Burckhardt, The Pharmacokinetics and Pharmacodynamics of Iron Preparations, *Pharmaceutics* **3**, 12–33 (2011).
166. M. Lu, M. H. Cohen, D. Rieves and R. Pazdur, FDA report: Ferumoxytol for intravenous iron therapy in adult patients with chronic kidney disease, *Am. J. Hematol.* **85**, 315–319 (2010).
167. J. E. Rosen, L. Chan, D. B. Shieh and F. X. Gu, Iron oxide nanoparticles for targeted cancer imaging and diagnostics, *Nanomedicine* **8**, 275–290 (2012).
168. A. J. Cole, V. C. Yang and A. E. David, Cancer theranostics: the rise of targeted magnetic nanoparticles, *Trends Biotechnol.* **29**, 323–332 (2011).
169. S. K. Lai, Y. Y. Wang and J. Hanes, Mucus-penetrating nanoparticles for drug and gene delivery to mucosal tissues, *Adv. Drug Deliv. Rev.* **61**, 158–171 (2009).

170. X. Wang, C. Zheng, Z. Wu, D. Teng, X. Zhang, Z. Wang and C. Li, Chitosan-NAC nanoparticles as a vehicle for nasal absorption enhancement of insulin, *J. Biomed. Mater. Res. Part B* **88**, 150–161 (2009).
171. Y. Li, K. Hindi, K. M. Watts, J. B. Taylor, K. Zhang, Z. Li, D. A. Hunstad, C. L. Cannon, W. J. Youngs and K. L. Wooley, Shell crosslinked nanoparticles carrying silver antimicrobials as therapeutics, *Chem. Commun.* **46**, 121–123 (2010).
172. L. Y. Lin, K. M. Tiemann, Y. Li, J. S. Pinkner, J. N. Walker, S. J. Hultgren, D. A. Hunstad and K. L. Wooley, Synthetic polymer nanoparticles conjugated with FimH(A) from E. coli pili to emulate the bacterial mode of epithelial internalization, *J. Am. Chem. Soc.* **134**, 3938–3941 (2012).
173. H. S. Park, K. H. Kim, S. Jang, J. W. Park, H. R. Cha, J. E. Lee, J. O. Kim, S. Y. Kim, C. S. Lee, J. P. Kim and S. S. Jung, Attenuation of allergic airway inflammation and hyperresponsiveness in a murine model of asthma by silver nanoparticles, *Int. J. Nanomedicine* **5**, 505–515 (2010).
174. K. M. Hindi, A. J. Ditto, M. J. Panzner, D. A. Medvetz, D. S. Han, C. E. Hovis, J. K. Hilliard, J. B. Taylor, Y. H. Yun, C. L. Cannon and W. J. Youngs, The antimicrobial efficacy of sustained release silver-carbene complex-loaded L-tyrosine polyphosphate nanoparticles: characterization, *in vitro* and *in vivo* studies, *Biomaterials* **30**, 3771–3779 (2009).
175. R. Pandey and G. K. Khuller, Antitubercular inhaled therapy: opportunities, progress and challenges, *J. Antimicrob. Chemother.* **55**, 430–435 (2005).

Chapter 14

Structure–Activity Relationships for Tumor-Targeting Gold Nanoparticles

Erik C. Dreaden, Ivan H. El-Sayed and Mostafa A. El-Sayed

**Koch Institute for Integrative Cancer Research, Massachusetts Institute of Technology 77 Massachusetts Avenue, Cambridge, Massachusetts, 02139, USA.*
†Otolaryngology-Head and Neck Surgery, Comprehensive Cancer Center, University of California at San Francisco, California, 94143, USA
‡School of Chemistry and Biochemistry, Georgia Institute of Technology 901 Atlantic Drive NW, Atlanta, GA 30332–0400, USA
§edreaden@mit.edu
¶ielsayed@ohns.ucsf.edu
‖melsayed@gatech.edu

Abstract

Gold nanoparticles are rapidly emerging platforms for multifunctional diagnostic and therapeutic nanotechnologies. Tumor-specific cellular delivery of a gold nanoparticle, however, is a complex, multistep process involving systemic circulation, transvascular flux, intratumoral penetration and intracellular delivery. Each of these steps present unique constrains which determine the optimal architecture for gold nanoparticle construct, including at its most fundamental level, size, shape and surface chemistry. Gold nanoparticles, given their highly well-defined chemical synthesis and surface chemistry, represent one of the most versatile and reliable platforms for designing and understanding tumor-targeting nanotechnologies. Here, we review a recent body of work examining the effects of gold nanoparticle size, shape and surface chemistry on circulation kinetics, tumor extravasation, interstitial tumor transport and cellular uptake with the aim of delineating optimal design parameters for each of these processes. Discussions here focus primarily on poly(ethylene glycol)-functionalized gold nanoparticles with an emphasis on practical design considerations for new researchers in the field.

1. Introduction

More than 1.6 million individuals in the US were diagnosed with cancer in 2012, and more than 1,500 died from the disease every day.[1] Our improved understanding of cancer biology, genetics and risk factors for the disease, as well as the development of advanced diagnostic and therapeutic technologies, have greatly reduced the burden of cancer in the US. Nonetheless, cancer remains the leading cause of non–cardiovascular-related death in this country, with roughly 1 in 4 males and 1 in 5 females succumbing to cancer during their lifetime.[2] In addition to the tremendous personal toll on both patients and care-givers, the disease also presents substantial financial burden, and the total cost of new cases diagnosed each year in the US exceed $15.8 bn,[3] roughly equivalent to the gross domestic product of Iceland.

Nanotechnologies hold great promise for the improved detection and treatment of solid tumors.[4–9] These materials, with structural features ranging in size from 1 to 100 nm, can dramatically alter the pharmacokinetic and pharmacodynamic profiles of drug compounds, improving therapeutic index and diminishing unintended side effects. These structures are generally larger than small-molecule drugs and proteins, yet small enough to systemically circulate for long periods and efficiently enter cells. Given their relatively large sizes, these constructs often contain large quantities of diagnostic or therapeutic moieties and display surface ligands in a high-affinity, multivalent fashion. Thus, these platforms can exhibit a high degree of multifunctionality, providing potential for both active (ligand-directed) and passive (size-dependent) tumor targeting, intrinsically or extrinsically triggered activation, controlled and delivery sequence/timing and simultaneous adjuvant imaging or therapeutic modalities (e.g. radiation, photodynamic, photothermal therapy, etc.).

Gold nanoparticles (AuNPs) exhibit a unique combination of chemical, optical and electronic properties that can be leveraged in a wide range of diagnostic and therapeutic strategies for the treatment of solid tumors (Fig. 1).[10–28] Given their well-defined structure and reactivity, colloidal gold nanoparticles are well suited to perform fundamental biological studies with nanomaterials. For example, unlike polymeric and liposomal nanostructures, gold nanoparticle surfaces are both crystalline and highly reproducible, thus presenting a high density of surface sites through which ligands can be displayed in a multivalent fashion, typically, 100–1000 times higher in density than that achievable with liposomal and polymeric nanoparticles, respectively.[27] Due to the absence of Au in the body, pharmacokinetics and disposition can be easily monitored with virtually zero background, typically by standard inductively coupled plasma mass spectrometry (ICP-MS) methods. Because of their high contrast under electron and optical microscopy, these structures can also be

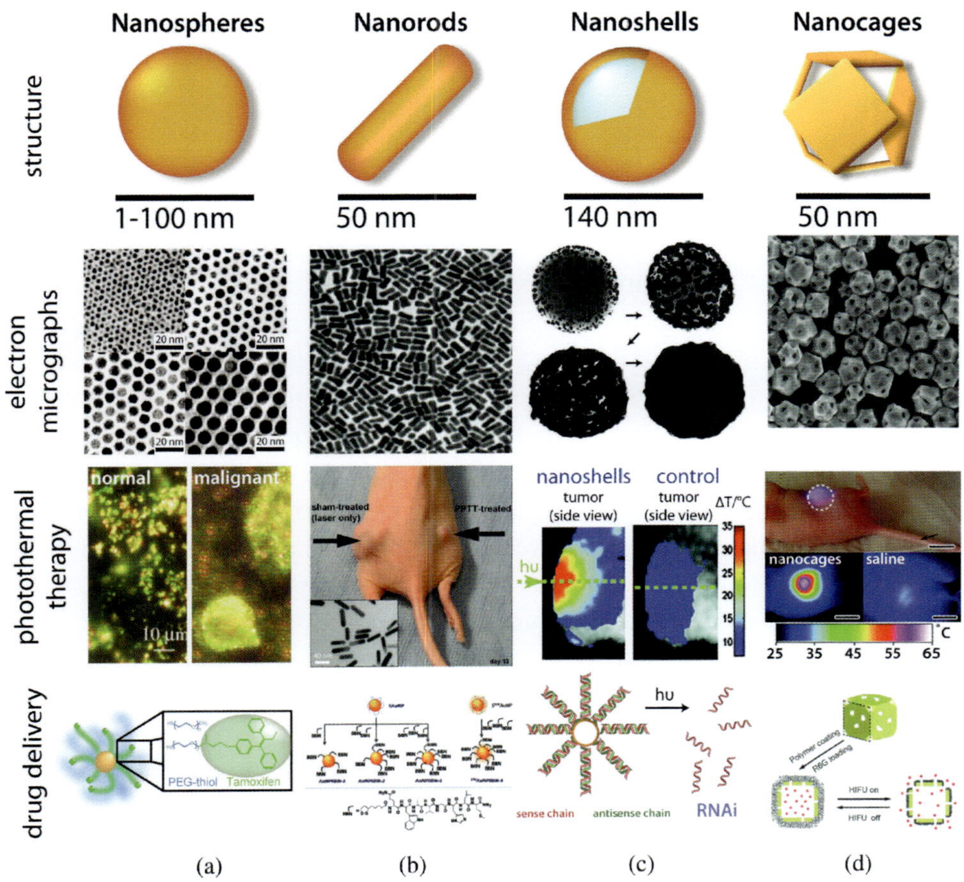

Fig. 1. Applications of colloidal gold nanoparticles in drug delivery and laser photothermal therapy. (a) Gold nanospheres, (b) gold nanorods, (c) gold nanoshells, and (d) gold nanocages. Reproduced with permission from Ref. 28. Copyright 2012 Future Science Ltd. Original adapted with permission from (a) Refs. 38, 47 and 48, (b) Refs. 30, 49 and 50, (c) Refs. 51 and 52 and (d) Refs. 53–55. Copyright (a) 2003, 2009, and 2006 American Chemical Society, (b) 2004 American Chemical Society, 2010 National Academy of Sciences, and 2008 Elsevier Science BV, (c) 1998 Elsevier Science BV and 2003 National Academy of Sciences, and (d) 2006 American Chemical Society, 2011 Royal Society of Chemistry, and 2010 Wiley-VCH Verlag GmbH & Co. KGaA.

used to visualize gold conjugate disposition in humans[29] as well as that of other low-contrast nanomaterials such as polysaccharide siRNA nanocarriersin clinical studies.[29] Because AuNPs are typically small compared with light waves, localization in tissue sections determined by optical microscopy can be easily enhanced by electroless silver deposition,[30, 31] while those for electron microscopy are often sufficient without modification. Using simple and highly robust thiol conjugation chemistry,[32–35] ligands can be coupled to gold nanoparticle surfaces reproducibly and in high density, providing improved ligand solubility in mixed monolayer configurations, and tunable affinity for therapeutic agonists/antagonists or tumor-targeting ligands. Although, sterically hindered dithiolates are often less preferred given their propensity for desorption (thiol oxidation), monothiolate-functionalized Au surface have been shown to be stable for as long as 21 days under physiological conditions.[36] Per unit volume, gold nanoparticles can be more than 11-fold brighter than semiconductor quantum dots[a] and cost roughly six times less than typical monoclonal antibodies per unit mass.[37] For example, in previous work, we have found that receptor-binding affinity and therapeutic potency of endocrine receptor antagonists currently used in clinical therapy for breast and prostate cancer could be enhanced more than 3–4 orders of magnitude simply through gold nanoparticle conjugation.[38–40] Another unique aspect of these structures is their intrinsic multifunctionality. Due to their novel physical and electronic properties,[41–45] gold nanoparticles can enhance diagnostic imaging contrast, provide controlled or sequential drug delivery, improve solubility and tumor targeting and facilitate adjuvant therapeutic modalities, including radiotherapy contrast, photothermal therapy and photodynamic therapy. In another example, we have shown that passively[30] and actively[39] tumor-targeted gold nanoparticles can also achieve dramatic treatment response following non-invasive laser photothermal therapy *in vitro* and *in vivo*, and that active nuclear- and cytoplasmic-targeted subcellular localization of these particles can facilitate rapid cell death.[46]

Modern biomedical investigations of gold nanoparticles in drug delivery began in 2001 with the discovery that gold nanoparticle formulations could greatly improve the biodistribution, tumor-specific accumulation, and therapeutic index of tumor necrosis factor alpha (TNFα).[56, 57] Today, among the more than 40 nanomedicines that have reached the clinic (see Table 1 for a brief history),[9, 58] gold nanotechnologies represent a notably versatile and multifunctional platform: diagnostic applications range from high-stability labels for lateral-flow biomarker assays (First Response®, Carter–Wallace; Duopath®, EMD Chemical Inc.), vectors for transdermalvaccination[59] (Gene Gun®, Pfizer Inc.), high contrast labels for cytometric T cell monitoring (PointCare CD4NOW®, Pointcare Inc.), high efficiency probes for mRNA detection (SmartFlare®, Millipore Inc.), high-sensitivity immunolabels for electron microscopy (Immunogold®, BBI International), and imaging contrast agents for X-ray computed tomography (AuroVist®, Nanoprobes Inc.), photoacoustic

---

[a] Data calculated from the extinction cross section of an 80 nm gold nanosphere in water ($\lambda = 549/549$ nm, ε_{ext}~6.90×10^{10}, QY = 0.41) as compared to a 10–20 nm diameter and Qdot® 800 ($\lambda_{ex} = 350$ nm, ε_{abs}~15.8×10^6, QY = 1).
http://www.invitrogen.com/site/us/en/home/References/Molecular-Probes-The-Handbook/tables/Extinction-Coefficients-of-Qdot-Streptavidin-Conjugates-at-Common-Wavelengths.html retrieved 24 October, 2011.

Table 1. A selected timeline of nanoscale drug development.

Nanoscale drug	Initial dev.	Seminal reference(s)	Pharmaceutical	Trade Name®	Carrier	Drug
micelle	1958, 1983	Stüttgen G, Wust H. Serum concentrations of vitamin A following variations in forms of application, dosage, and solubilizers. *Arzneim Forsch.* **8**(161–162) (1958). Toronto Lung Transplant Group. Unilateral lung transplantation for pulmonary fibrosis. *New Eng I. J. Med.* **314**(18), 1140–1145 (1986).	cyclosporine emulsion	Sandimmune	Cremophor EL (polyoxyethyl ated castor oil)	cyclosporine
polymer-drug nanoconjugate	1954, 1977	Jatzkewitz H. Incorporation of physiologically-active substances into a colloidal blood plasma substitute.I. Incorporation of mescaline peptide into polyvinylpyrrolidone. *Hoppe Seylers Z Physiol. Chem* **297**(3–6), 149-156 (1954). Abuchowski A, McCoy JR, Palczuk NC, van Es T, Davis FF. Effect of covalent attachment of polyethylene glycol on immunogenicity and circulating life of bovine liver catalase. *J. Biol.Chem.* **252**(11), 3 582–6 (1977).	pegademase bovine	Adagen	poly(ethylene glycol) (PEG)	adenosine deaminase (ADA)
liposome	1965	Bangham AD, Stabdish MM, Watkins JC. Diffusion of univalent ions across the lamellae of swollen phospholipids. *J.Mol. Biol.* **13**(1), 238–252 (1965).	Doxorubicin liposomal	Doxil	PEGylatged phospholipid bilayer (artificial vesicle)	doxorubicin

(*Continued*)

Table 1. (*Continued*)

Nanoscale drug	Initial dev.	Seminal reference(s)	Pharmaceutical	Trade Name®	Carrier	Drug
iron oxide	1985(6)	Olsson M BE, Persson BRB, Salford LG, Schröder U. Ferro magnetic particles as contrast agent in T2 NMR imaging. *Magn. Reson. Imaging* **4**(5), 437–440 (1986).	imaging agent	Feridex	iron oxide	n/a
protein-drug nanoconjugate	1975, 2000	Ringsdorf H. Structure and properties of pharmacologically active polymers. *J. Polym. Sci. Pol. Sym.* 51(1), 135–153 (1975). Desai NP, Louie L, Ron N, Magdassi S, Soon-Shiong P. Protein-based nanoparticles for drug delivery of paclitaxel. *Trans. World Biomater. Congr.* 1, **199** (2000).	nab-Paclitaxel	Abraxane	albumin	paclitaxel
polymeric nanocarrier	1976, 1999	Langer RS, Folkman J. Polymers for the sustained release of proteins and other macromolecules. Nature 263(5580), 797–800 (1976). Gonzalez H, Hwang SJ, Davis ME. New class of polymers for the delivery of macromolecular therapeutics. *Bioconjugate Chem.* 10(6), 1068–1074 (1999).	CRLX101, CALAA-01	n/a	poly(cyclodextrin)	camptothecin (CRLX101), siRNA (CALAA-01)

(*Continued*)

Table 1. (*Continued*).

Nanoscale drug	Initial dev.	Seminal reference(s)	Pharmaceutical	Trade Name®	Carrier	Drug
dendrimer	1978	Buhleier E, Wehner W, Vögtle F. "Cascade" and "Nonskid-Chain-like" Syntheses of Molecular Cavity Topologies. *Synthesis* 1978(02), 155–158 (1978). Haensler J, Szoka FC. Polyamidoamine cascade polymers mediate efficient transfection of cells in culture. *Bioconjugate Chem.* **4**(5), 372–379 (1993).	SPL7013	VivaGel	SPL7013 L-lysine dendrimer	intrinsic activity
gold	1954	Root SW, Andrews GA, Kniseley RM, Tyor M P. The distribution and radiation effects of intravenously administered colloidal Au^{198} in man. *Cancer* 7(5), 856–866 (1954). Paciotti GF, Myer LD, Kim TH et al. Colloidal gold: A novel colloidal nanoparticle vector for tumor-directed drug delivery. *Clin. Cancer Res* 7(11), 3673S–3674S (2001). Hirsch LR, Stafford RJ, Bankson JA *et al.* Nanoshell-mediated near-infrared thermal therapy of tumors under magnetic resonance guidance *Proc. Natl. Acad. Sci.* USA. **100**(23), 13549–13554 (2003).	CYT-6091	Aurimune	solid gold nanoparticles	tumor necrosis factor α

(*Continued*)

Table 1. (*Continued*)

Nanoscale drug	Initial dev.	Seminal reference(s)	Pharmaceutical	Trade Name®	Carrier	Drug
silica	1968	Stöber W, Fink A, Bohn E. Controlled growth of monodisperse silica spheres in the micron size range. *J. Colloid Interface Sci.* **26**(1), 62–69 (1968). Yanagisawa T, Shimizu T, Kuroda K, Kato C. The preparation of alkyltriinethylaininonium-kaneinite complexes and their conversion to microporous materials. *Bul.Chem.Soc. Japan* **63**(4), 988–992 (1990).	Variable label	cornell dots (C dots)	cRGDY-mesoporous silica ^{128}PET and Cy5 Fluorophore label	Variable

Details	First FDA approval	Disease	Subsequent clinical use/development
imparts water-solubility to cyclosporine for the prevention of organ transplant rejection; allowed 1st successful lung transplant in 1983	1983	n/a (immunosu ppressive)	Genexol-PM (paclitaxel, metastatic breast cancer, 2006/S. Korea), NK012 (SN-38 topoisomerase I inhibitor, relapsed small cell lung cancer, Phase II complete, Phase I with Pt), NK 105 (paclitaxel, gastric cancer, Phase II compete), NK911 (doxorubicin, solid tumors, Phase II), NC-6004 (cisplatin, pancreatic cancer, Phase II with gemcitabine)
(11–17) 5 kDa mPEG groups per ADA, 100 × improved half life, diminished immune response	1990	severe combined immuno deficiency disease (SCID)	Zinostatin stimalamer (hepatocellular carcinoma, Japan/1993), Oncospar (acute lymphoblastic leukemia; 1994/USA), PEG-Intron (chronic hepatitis C; 2000/EU, 2001/USA), Pegasys (chronic hepatitis C; 2002/USA&EU), Neulasta (febrile neutropenia; 2002/USA & EU), Somavert (acromegaly; 2002/EU, 2003/USA), Pegamotecan (topoisomerase I-inhibitor, gastric or gastrooesophageal adenocarcinoma, Phase II complete), NKTR-102 (topoisomerase I-inhibitor, breast/ovarian cancer, Phase II complete)

(*Continued*)

Table 1. (*Continued*)

Details	First FDA approval	Disease	Subsequent clinical use/development
ca.3-fold increased drug half life, >90% drug retention in circulation, EPR tumor accumulation, granted orphan drug exclusivity Oct 2009-May 2014	1995	Kaposi's sarcoma	DaunoXome (daunorubicin, kaposi's sarcoma, 1996/USA), Ambisome (amphotericin B, visceral leishmaniasis, USA/1997), Visudyne/verteporfin, photodynamic therapy (age-related macular degeneration; 2000/USA & EU), Myocet (doxorubicin, breast/ovarian cancer & Kaposi's sarcoma, 2000/EU, 2001/Canada), Thermodox (doxorubicin, liver/breast cancer, Phase III complete), Marqibo (vincristine, acute lymphoblastic leukemia, Phase III), Onco TCS (vincristine, non-Hodgkin's lymphoma, Phase III complete), ATI-1123 (docetaxel, ovarian cancer, AIDS-related Kaposi's sarcoma & multiple myeloma) Phase II, MCC-465 (doxorubicin, stomach cancers, Phase I complete), MBP-42 (oxaplatin, solid tumors, Phase Completed), SGT-53 (p53 gene plasmid, solid tumors, Phase I complete)
120 to 180 nm diameter, carboxydextran-coated superparamagnetic iron oxide (SPIO) for MRI of focal liver lesions	1996	n/a (Imaging)	Resovist (40–60 nm HD, carboxydextran-SPIO, MRI of focal liver lesions, 2009/EU), Ferumoxtran-10 (20–50 nm HD, dextran-SPIO, liver MRI, limited experimental use in Netherlands)
hydrophobic association with human serum albumin (67 kDa); 33% increased tumor accumulation, higher MTD, reduced neutropenia; 130 nm diameter, –31 mV	2005	breast cancer	ABI-008 (nab-docetaxel, metastatic prostate cancer, Phase I), ABI-009 (nab-rapamycin, solid tumors, preclinical complete), ABI-010 (nab-17AAG, solid tumors, preclinical), ABI-011 (dual microtubule & topoisomerase-1 inhibitor, solid tumors, preclinical)

(*Continued*)

Table 1. (*Continued*)

Details	First FDA approval	Disease	Subsequent clinical use/development
transferrin-targeted, PEG-stabilized, DNA topoisomerase I inhibitor or RNAi/Oligonucleotide Nanoparticle Delivery (RONDEL) system (siRNA for M2 subunit of ribonucleotide reductase, RRM2); CRLX101 ca. 20–50 nm & CALAA-01 ca. 70 nm diameter	phase II CRLX 10 1), Phase I (CALAA 01)	solid tumors	XYOTAX (paclitaxel, non-small cell lung cancer, Phase III complete), CRLX101 (formerly IT-101; camptothecin nanoconjugate, solid tumors, Phase II), FCE28068 (PK1, doxorubicin, breast/colorectal/non-small cell lung cancer, Phase II complete), AP5280 (Pt^{4+}, solid tumors, Phase I complete), AP5346 (Pt^{4+} solid tumors, Phase II), PNU166945 (paclitaxel, solid tumors, Phase I complete), MAG-CPT (camptothecin, solid tumors, Phase I complete), CT-2106 (camptothecin, colorectal/ovarian cancer, Phase I/II discontinued), FCE28069 (PK2, galactosamine, hepatocellular carcinoma, Phase I discontinued)
L-lysine based SPL7013 dendrimer (16.6 kDa) inhibits viral infection (incl HIV and genital herpes) and the growth of bacteria	Phase II (BV)	bacterial vaginosis (BV) & STIs	ATI-001 (G5 PAMAM dendrimer, <5 nm diameter, folate-targeted, methotrexate-loaded; solid tumors, preclinical complete)
systemic administration of 27 nm diameter spherical gold nanoparticles well tolerated; 3-fold increased drug MTD; tumor-accumulation of particles	Phase I complete	solid tumors	AuroLase (ca. 150 nm diameter silica-gold core-shell nanoparticles, laser photothermal ablation therapy of head & neck tumors, human pilot studies recruiting)

(*Continued*)

Table 1. (*Continued*)

Details	First FDA approval	Disease	Subsequent clinical use/development
cRGDY-labeled, Cy5-loaded, PEGylated mesoporous silica nanoparticles; 6.8 nm HD, renal clearance; FDA IND approved; tumor-selective accumulation; imaging of lymphatic drainage, nodal metastases, & tumor burden	n/a	solid tumors	Zink & Stoddart (100 nm diameter mesoporous silica nanoparticles with 2 nm nanopores & a pH-responsive release valve); Ferrari (3.5 nm diameter hemispherical mesoporous silica with 24.2 nm nanopores); both commercialization pending

Abbreviations: DNA (Deoxyribonucleic acid), EPR (enhanced permeability and retention), G5 (generation 5), HD (hydrodynamic diameter), HIV (human immunodeficiency virus), IND (investigational new drug), MRI (magnetic resonance imaging), MTD (maximum tolerated dose), nm (nanometer), PAMAM (poly(amido amine)), PET (positron emission tomography), Pt (Platinum), SPIO (superparamagnetic iron oxide), STIs (sexually transmitted infections)

Source: Reproduced with permission from Dreaden *et al.*[28] Copyright 2012 Future Science Ltd.

tomography, optical coherence tomography and Raman spectroscopy (Spectropen®, Spectropath Inc.). Therapeutic applications range from high stability carriers for poorly soluble drug compounds (Aurimmune®, Cytimmune Inc.), contrast agents for laser photothermal hyperthermia/ablation (Aurolase®, Nanospectra Biosciences) and radiotherapy, as well as highly potent multivalent targeted therapies.[38, 40] In clinical studies, gold nanotechnologies have been applied as contrast agents for laser photothermal ablation of skin melanoma, prostate carcinoma, head and neck tumors (NCT00848042), lung carcinoma (NCT01679470) and atherosclerotic plaques (NCT01436123, NCT01270139), as well as delivery vectors for TNFα (NCT00356980, NCT00436410).

Given their versatility and multifunctionality, gold nanotechnologies are rapidly accelerating in their clinical translation. In this review, we highlight how these novel nanoparticles have shed unprecedented light on optimal design parameters for systemic circulation, tumor extravasation, intratumoral penetration and cellular uptake of these and other tumor-targeting colloidal nanotechnologies. These findings are presented in the context of practical design considerations for new researchers in the field. For a more thorough discussion of the chemistry, photophysics and general biomedical applications of gold nanoparticles, readers are encouraged to consult Refs 23, 25–28, 60 and 61.

2. Physiological Barriers for Tumor-Targeted Gold Nanotechnologies

The tumor microenvironment often differs greatly from that of normal and otherwise non-inflamed bodily tissues. In order to realize efficient delivery at or within the cells of a solid tumor, colloidal gold nanotechnologies must overcome a myriad of physiologic barriers[6, 62–64]: (1) intravenously administered nanoparticles must first circulate from the injection site to various and distant blood vessels surrounding the tumor; (2) when localized about the tumor, these nanoparticles must traverse the blood vessel walls and into the tumor interstitium where (3) they must efficiently permeate throughout the tumor cell mass and its surrounding stromal tissue matrix. Once perfused throughout the tumor microenvironment, (4) nanoparticles must either release their diagnostic/therapeutic payload or efficiently cross the cell membrane, escape endosomal sequestration, and intracellularly release their payload in a controlled fashion.

2.1. *Long-circulating gold nanoparticles*

2.1.1. **Solubility**

Following intravenous administration, colloidal gold nanotechnologies partition into various regions of the body depending on their size, shape, rigidity, charge, lipophilicity, immunogenicity and affinity/avidity for various biomolecules. Perhaps the most obvious and most important benefit of these and other nanoscale delivery platforms is their ability to greatly enhance both the solubility and stability of drugs and diagnostic agents which they encapsulate or append. About 70% of new active pharmaceutical ingredients are reported to fail in drug development due to poor solubility[65] that results in dose-limiting toxicity and poor therapeutic index. Treatment modalities such as small interfering RNA[66] are also susceptible to chemical or enzymatic degradation and are likewise hampered by poor pharmacokinetics following systemic administration. Doxil®, or liposomally encapsulated doxorubicin, represents the archetype for nanoparticle-enhanced pharmacokinetics. Here, PEGylated liposomal encapsulation confers a roughly 5-fold increase in circulation half-life, while methods such as polmyer-, protein-, and gold-nanoparticle-conjugation achieve roughly 3- to 50-fold enhanced circulation half-life (e.g. Adagen®, PEGylated adenosine deaminase;[67,68] Abraxane®, albumin-bound paclitaxel[69–71] and CYT-6091, gold-nanoparticle–bound TNFα,[29] respectively). This increased blood residence time increases the probability that the diagnostic or therapeutic agent will reach its intended site of action, thus presumably resulting in a more favorable treatment outcome.[72] Other examples solubility-enhanced nanoparticles currently in drug development include DepoDur® analgesic (liposomal morphine, SkyePharma/Endo) and Relenza® for the treatment of influenza (zanamivir, Liquidia).

2.1.2. Size, charge and circulation

Ideally, a systemically administered gold nanoparticle resists protein adsorption, elicits no immunogenic responses, is relatively hydrophilic (octanol:water logP roughly ≤ 2),[73] weakly anionic in surface charge (zeta potential ζ roughly -10 to 0 mV) and approximately 10–200 nm in hydrodynamic diameter (HD). During intravenous circulation, nanoparticles are subject to both size-dependent renal (urinary) and splenic clearance, factors which constrain optimal dimensions for intravenously administered tumor-targeting nanoparticles within the 10 to 200 nm hydrodynamic diameter range, as typically measured by dynamic light scattering (DLS). During circulation, the kidney glomular basement membrane (GBM) acts as a size-selective barrier, separating circulating blood from urinary filtrates. Rigid nanoparticles with anionic to neutral surface charge (zeta potential) are efficiently cleared with the urine at hydrodynamic diameters less than roughly 6 nm. Although the upper limit for renal clearance of rigid, cationic nanoparticles is somewhat larger given the net negative charge of the kidney GBM, intravenous pharmacokinetics from these particles are generally suboptimal and kidney sequestration is often observed.[74,75]

Splenic clearance generally dictates the upper size limit for circulating nanoparticles, where the average width of interendothelial cell slits of venous sinuses are approximately 200–250 nm.[76, 77] Using PEGylated gold nanoparticles, Davis and coworkers have examined disposition and pharmacokinetics of intravenously administered PEGylated gold nanoparticles (5–98 nm diameter; 26–167 nm HD; conjugated at 1.5×10^{15} PEG cm^{-2} excess) and found a nearly linear increase in RES organ accumulation (liver and spleen) with hydrodynamic diameter (Fig. 2)[78]; circulation half-life correspondingly decreased semi-exponentially with hydrodynamic diameter. Liver fenestrae also play a significant role in nanoparticle disposition and clearance. Here, particles with a least one dimension < 100 nm HD can pass directly from the sinusoidal lumen to the liver hepatocyte cells. Readers may note, that in humans, this value (107 ± 1.5 nm) is notably similar to that previously reported in New Zealand White (NZW) rabbits (103 ± 1.3 nm), but significantly smaller than in C57BL/6 mice (141 ± 5.4 nm) and Sprague–Dawley rats (161 ± 2.7 nm); thus murine models may overestimate liver accumulation or dose-limiting hepatotoxicity.[79] Although particles >100 nm in all dimensions are susceptible to endocytosis by liver sinusoidal endothelial cells and Kupffer cells and thus, diminished blood residence time and passive targeting capacity, potential liver-specific delivery should be considered when designing gold nanoparticles 10–100 nm in total HD that are

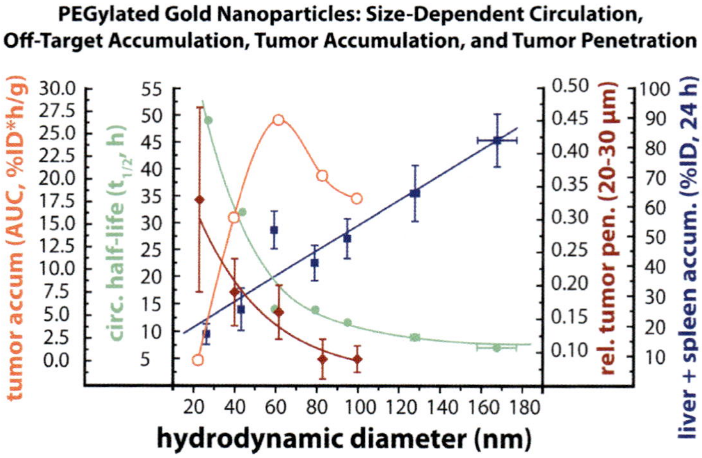

Fig. 2. Relationship between hydrodynamic diameter and circulation half-life (green), off-target accumulation (blue), tumor-specific accumulation (orange) and tumor penetration efficiency (red) following intravenous administration of poly(ethylene glycol)-functionalized gold nanospheres in BALB/c (green, blue) or CD1 (orange, red) mice. Lines are added for clarity. Data adapted with permission from Ref. 31 (green, blue) and Ref. 84 (orange, red). Copyright 2011 National Academy of Sciences and 2009 American Chemical Society.

conjugated with potentially hepatotoxic or compounds or those which may illicit chronic inflammatory responses. Other size-dependent physiological barriers toward the delivery of biomedical gold nanoparticle conjugates include the blood–brain barrier (BBB) (*ca.* 15 nm), brain extracellular space (*ca.* 114 nm dia),[80] mucus (e.g. cervicovaginal[81, 82] *ca.* 340 nm), lung periciliary brush (*ca.* 20–40 nm), the lymphatics (dependent on the route of administration), and the nuclear pore complex[83] (*ca.* 20–40 nm) (see Table 2).

Table 2. Size-dependent physiological barriers relevant to biomedical gold nanoparticles.

Barrier	Size (nm)	Comments	Refs
vessels (normal)	<5–12	larger in tumor tissues	92, 93
kidney	<6–8	IV, neutral/anionic: <6 nm cleared in the urine; IV, cationic: <8 nm cleared in the urine	5, 62, 63, 94
lymphatics	6–34; <80	IT: 6–34 nm, non-cationic; SC: < 80 nm	95–97
blood–brain	<15	IV: <15nm efficiently pass; 15–100 nm also reported	187, 188
nuclear pore	<20–40	midplane Nup54 ·Nup58 ring expandable up to 20–40 nm; 39 nm diameter transport reported	83
lung PCL	<20–40	pulmonary delivery: penetration across periciliary brush to pulmonary epithelium; dextran, non-PEGylated NPs	98
liver	<107	IV: <100 nm delivered to hepatocytes; >100 nm susceptible to endocytosis by liver sinusoidal endothelial cells and Kupffer cells; mice/rats larger (*ca.* 141–161 nm); larger in tumors	79
brain	<114	human and rat brain extracellular space (ECS)	80
spleen	>200–250	IV: diminished blood residence if larger	99,100
vessels (tumor)	<200–1200	varies with tumor type, stage, grade, etc.	6, 101
lung	>300	IV: large particles trapped in lung capillaries	102
mucus	<340	e.g. cervicovaginal mucus; requires dense brush layer of PEG	81

Note: Sizes denote hydrodynamic (Stokes–Einstein) diameters necessary for extravasation, transport, or clearance. IV, intravenous; IT, intratracheal; SC, subcutaneous; periciliary layer (PCL), nanoparticles (NPs), PEG, poly(ethylene glycol).

Highly anisotropic and/or mechanically deformable[85,86] nanoparticles can circumvent size constraints to an extent; however, these effects are often less pronounced in gold nanoparticle constructs whose core sizes are generally larger than 6 nm in order to support optical plasmon resonance modes or efficient drug loading capacity. Generally, deformable nanoparticles >300 nm HD and anisotropic nanoparticles 100–200 nm HD in any dimension are suitable for long circulation.[5] For example, DeSimone and coworkers synthesized red blood cell-mimetic nanoparticles (*ca.* 6 μm diameter, 1.5 μm thick) from 2-hydroxyethyl acrylate (HEA) photo-crosslinked with poly(ethylene glycol) diacrylate (PEGDA). Interestingly, they found that with decreasing crosslink density (decreasing bulk modulus) pharmacokinetics and biodistribution could be dramatically modulated by deformability, increasing circulation and elimination half-life *ca.* 10- and 30-fold, respectively, while also avoiding pulmonary accumulation typically observed at these sizes splenic accumulation *ca.* 7-fold.[87] In addition to deformability, shape also plays an important role in circulation kinetics; for example, high aspect ratio cylindrical micelles (e.g. filomicelles) *ca.* $\leq$8 μm in length have also been found to persist in systemic circulation for as long as one week[88] due in part to hydrodynamic shears that align their longitudinal axes parallel with the blood vessels and decrease interactions with peripheral and/or splenic red pulp phagocytes. Biomedical applications of near-infrared (NIR) absorbing rod-shaped gold nanoparticles is also an area of great research interest.[20, 23, 25–28] To-date the dimensions of their minor axes have been limited to 10–20 nm using traditional seed/surfactant-mediated growth methods,[30,89] a fact which limits their potential for renal excretion and contributes to potential liver inflammation.[90] Newly developed ultrasmall NIR-absorbing gold nanorods (*ca.* 2.5–5 nm in width) may circumvent potential hepatotoxicity[91]; however, these effects have yet to be investigated.

2.1.3. Surface chemistry and RES clearance

During circulation, protein adsorption also plays a critical role in determining the capacity of theranostic gold nanoparticles for tumor site-specific accumulation. Although electrostatic stabilization of colloids is often effective in dilute aqueous solvents, physiological conditions presenting high salt and high serum concentrations often screen or neutralize charge–charge repulsion interactions, respectively, causing irreversible particle aggregation and leading to efficient splenic clearance or pulmonary accumulation[94] (the first capillary bed encountered after parenteral administration). Further, multicomponent serum protein adsorption — the so-called nanoparticle corona[103–105] or adsorbome[106] — can dramatically alter the pharmacokinetic profiles of gold nanoparticles. Surface opsonization, the deposition of proteins that facilitate nanoparticle recognition by the reticuloendothelial system (RES), can facilitate clearance by circulating phagocytes and tissue macrophages in direct contact with the blood (e.g. hepatic Kupffer cells and the marginal zone and red-pulp macrophages in the spleen).[107–111] Opsonins including antibodies and compliment activation products such as C3b and iC3b facilitate interactions with macrophage receptors (e.g. antibody Fc domain receptors) and other cells which mediate rapid nanoparticle clearance. Thus, non-ionic synthetic polymers such as poly(ethylene glycol) (PEG) are often used to provide both steric stabilization and high water solubility. Given the large hydration sphere and high flexibility of their ethylene oxide repeat units, PEGyalted surfaces present a so-called "conformational cloud" which screens both protein adsorption and interactions such as enzymatic degradation.[72] Thus, systemically administered gold nanoparticles are most frequently utilized in combination with thiolated methoxy PEG surface ligands and can be obtained in gram-scale from a variety of suppliers (e.g. Laysan Bio, Quanta BioDesign, Creative PEGWorks, etc.). In these cases, AuNP-PEG grafting density can be determined through a variety of techniques, including Raman spectroscopy, nuclear magnetic resonance and gel electrophoresis[112]; however, methods such as thermogravimetric analysis (TGA) and enzyme-linked immunosorbent assay (ELISA) using IgG toward PEG subunits is often preferred (see Roffler and coworkers[208] for a thorough review of PEG quantification). Using known nanoparticle concentrations (see nanoparticle extinction coefficient data in Table 4) and surrogate chromophore-conjugated or end-functional PEG (e.g. NH_2-PEG-SH), ligand coverages can be estimated based on spectral changes following ligand or chromophore/fluorophore conjugation (e.g. succinimidyl ester addition[89]). Note than in the latter cases, ratios of ligand to particle surface area should remain constant and that changes in monolayer density due to hydrophobic–hydrophobic (chromophore–chromophore) interactions and/or electrostatic repulsion (e.g. RNH_2^+) may occur.

Route of administration also plays an important role in modulating the disposition of biomedical gold nanoparticles. Although oral administration is generally ineffective[62], Rotello *et al.* have recently examined the role administration route on the pharmacokinetics and biodistribution of 9–10 nm HD gold nanoparticles functionalized with tetra(ethyleneoxide)-ligands bearing neutral (–OH), cationic (RNH_3^+), anionic ($RCOO^-$) and zwitterionic ($RN_2^+RRRSO_3^-$) groups (Fig. 3).[113] By both intraperotineal (*i.p.*) and intravenous (*i.v.*) administration, neutral and zwiterionic nanoparticles achieved

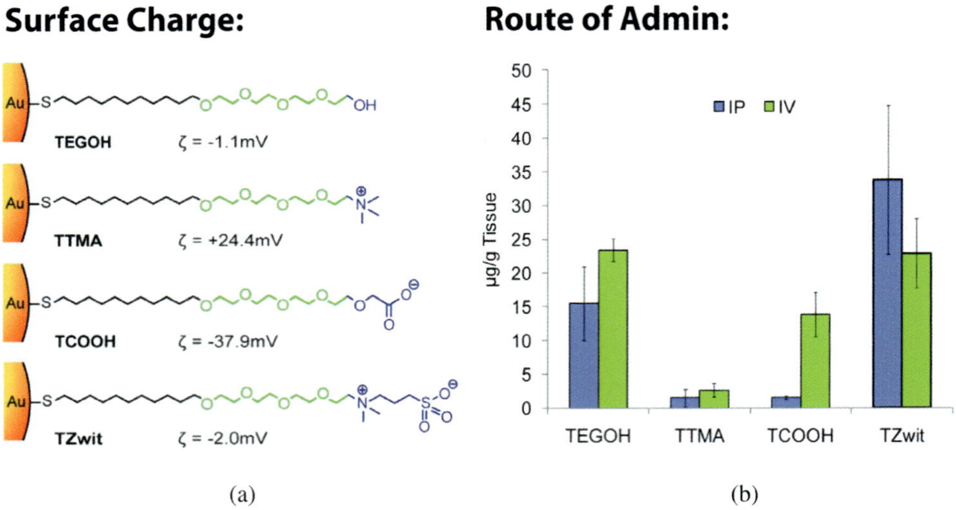

Fig. 3. The effects of surface charge and route of administration on tumor accumulation. (a) Gold nanoparticles functionalized with thiolated tetra(ethylene oxide) ligands bearing neutral, cationic, anionic, and zwitterionic surface charge (top to bottom, respectively; 9–10 nm hydrodynamic diameter). (b) Tumor accumulation 24 h following equivalent dosing of the particles in (a) by intraperotineal (*i.p.*, blue) or intravenous (*i.v.*, green) delivery. Reproduced from Ref. 113 under CCAL open access license.

higher accumulation in subcutaneous tumor xenografts (area under the curve, AUC) and generally longer circulation versus anionic or cationic gold nanoparticles. The *i.p.* tumor delivery was consistently lower than its *i.v.* counterpart, in keeping with passive/EPR targeting of by the particles (78%, 70%, 86% and 14% versus *i.v.* for neutral, zwitterionic, cationic and anionic, respectively). Biodistribution *following i.v.* administration was primarily observed in the liver and to a lesser extent in the spleen, while *i.p.* administration resulted in predominant splenic accumulation, presumably due to lymphatic drainage. Although poor performance by the cationic nanoparticles can be understood in the context of non-specific adsorption with negatively charged laminar surfaces of the blood vessels,[23] poor performance of the highly anionic nanoparticles may be due to lipoproteins such as (e.g. apolipoprotein H) which are known to bind negatively charged surfaces, leading to liver sequestration/clearance.[114] Thus, particles with weakly anionic in surface charge are preferred by both delivery methods. The *i.p.* delivery provides opportunities for tumors that have metastasized or are present within the perotineal cavity (e.g. ovarian carcinoma); however, *i.v.* delivery may be preferred in other cases.

In general, circulation half-life and tumor accumulation of AuNPs following *i.v.* administration scale with both PEG grafting density and molecular weight. For example, Niidome and coworkers examined pharmacokinetics and tissue disposition of PEGylated gold nanorods over a *ca.* 50-fold range of grafting densities (5 kDa PEG-SH) in tumor-bearing mice and observed *ca.* 5-fold decreased relative liver accumulation and *ca.* 80% increased relative splenic accumulation with increasing grafting density. These effects can be attributed to density-dependent changes in end-grafted polymer conformations. PEGylated surfaces exist in two general configurations, so-called "mushroom" and "brush" conformations, which occur at low and high densities, respectively. As the ligand density increases, the distance between ligand attachment points approaches a critical value known as the Flory radius (F; Table 3), the approximate crossover point at which steric crowding decreases the effective interaction volume of each chain, thus impeding surface adsorption and increasing blood residence time. Chan and coworkers have also investigated the effects of PEG molecular weight (2, 5, 10 kDa) on the circulation half-life of intravenously administered AuNPs cores 18–87 nm in diameter.[84] Although the blood half-life of smaller 18-nm AuNP cores was significantly extended with increasing PEG molecular weight (from 4 , 30,

Table 3. Useful physiochemical parameters for biomedical gold nanoparticles and Auf surfaces.

Property	Value	References
Au–S bond strength	40–45 kcal mol^{-1}	32, 118
Au–PPh_3 bond strength	16 kcal mol^{-1}	119
Au–NH_2 bond strength	8 kcal mol^{-1}	120
Au–COO- bond strength	2 kcal mol^{-1}	120
Au surface atom density; Au(111)	1.5×10^{15} atoms cm^{-2}	121
Typical Au surface adsorbate density	10^{-10} to 10^{-9} mol cm^{-2}; 10 to 100 μC cm^{-2}	121
Au Fermi energy	−5.1 eV (v EF^{VAC})	122
Mushroom-to-brush ligand transition radius[a]	$F = \alpha n^{3/5}$	112, 123
Fraction of ligands on nanospheres in contact with a cell surface[b]	d/R	78

[a]Flory radius (F) where n is the number of monomers per polymer chain and a is the length of one monomer in Angstroms (a = 3.5 Å for PEG).

[b]For d = Stokes–Einstein (hydrodynamic) diameter of the ligand, R = hydrodynamic radius of the nanoparticle for $d \ll R$.

and 52 h, for 2, 5, and 10 kDa PEG, respectively), diminishing returns in residence time for molecular weights >5 kDa were observed for larger particles, presumably due to curvature effects on steric screening by brush conformers. Thus, PEG molecular weights ≥10 kDa are recommended particles <20 nm and ≥5 kDa for particles >20 nm in tumor-targeting intravenous delivery applications. In practice, ligand excesses employed during conjugation are reported in experimental methods such that subsequent grafting densities can be compared across multiple data sets. This is often reported as the ratio of initial ligand concentration per unit surface area or surface atom, easily calculated from particle concentrations determined using experimentally derived extinction coefficient data (Table 4), particle dimensions obtained from electron micrographs or interpolated from optical spectra (size-dependent λ_{max}, Table 4), and surface atom density, often approximated as Au(111) for spherical AnNPs (Table 3). Typically, ligand functionalization can be performed by rapid injection of gold nanoparticle concentrates into an aqueous solution of thiols under sonication or stirring at room temperature. Surface monolayer formation can occur over milliseconds to minutes[33]; however, dense packing/reordering of the adsorbate layer (e.g. that necessary for PEG brush conformations) can require several hours.[12,18,115] Xia and coworkers investigated PEGylation of various gold nanoparticle architectures and found that PEG grafting density decreased with polymer molecular weight and that binding affinity of the native capping ligand (see Table 3) determined the subsequent extent of subsequent ligand exchange (i.e. coverage). They found grafting density plateaued after *ca.* 120 min and that PEGylation efficiency decreased in the order of citrate-capped nanoparticles > PVP-capped nanocages ≈ CTAC-capped nanoparticles > CTAB-capped nanorods, where PVP, CTAC and CTAB stand for poly(vinyl pyrrolidone), cetyltrimethylammonium chloride and cetyltrimethylammonium bromide, respectively.[116] In another study, the authors also observed significantly augmented apparent *in vitro* cell uptake from particles prone to sedimentation in serum-containing growth media, highlighting the importance of spectroscopic monitoring (e.g. UV-Vis, DLS) of particle stability over the time course of the experiment.[117]

2.2. *Tumor-targeting gold nanoparticles*

Gold nanoparticles can be synthesized in a variety of sizes and shapes depending on their intended uses. Often, structural architecture is varied to facilitate light-matter interactions at discreet wavelengths, properties useful in both imaging and therapeutic applications; however, size and shape also

Table 4. Extinction coefficient values for biomedical gold nanoparticles.

Property	Value	References
Gold nanorod longitudinal molar extinction coefficient[a,b]	$\varepsilon_{600-850\,nm}=\lambda_{(nm)}* 12.3 \times 10^{6} - 5.02 \times 10^{9}$	Adapted from 130
Hollow gold nanoshell extinction coefficient[b]	$\varepsilon_{(30)808\,nm} = 8.3 \times 10^{9}\ cm^{-1}\ M^{-1}$	131
Gold-silver nanocage extinction coefficient[b]	$\varepsilon_{(32)650\,nm} = 7.82 \times 10^{9}\ cm^{-1}\ M^{-1}$	132
Silica-gold core-shell nanoparticle extinction coefficient[b]	$\varepsilon_{(45)650\,nm} = 1.90 \times 10^{10}\ cm^{-1}\ M^{-1}$	133
Citrate-AuNP molar extinction coefficient[b,c]	$\varepsilon_{(150)810\,nm} = 2.26 \times 10^{11}\ cm^{-1}\ M^{-1}$	133
	$\varepsilon_{(1.8)505-10\,nm} = 6.90 \times 10^{5}\ cm^{-1}\ M^{-1}$	207
	$\varepsilon_{(2.2)505-510\,nm} = 1.25 \times 10^{5}\ cm^{-1}\ M^{-1}$	207
	$\varepsilon_{(3)505-510\,nm} = 3.13 \times 10^{6}\ cm^{-}\ M^{-1}$	207
	$\varepsilon_{(4)505-510\,nm} = 7.35 \times 10^{6}\ cm^{-}\ M^{-1}$	207
	$\varepsilon_{(5)514-520\,nm} = 1.43 \times 10^{7}\ cm^{-}\ M^{-1}$	207
	$\varepsilon_{(10)514-520\,nm} = 1.12 \times 10^{8}\ cm^{-}\ M^{-1}$	207
	$\varepsilon_{(15)528\,nm} = 3.6 \times 10^{8}\ cm^{-1}\ M^{-1}$	134
	$\varepsilon_{(20)514-520\,nm} = 8.80 \times 10^{8}\ cm^{-1}\ M^{-1}$	207
	$\varepsilon_{(25)514-520\,nm} = 1.71 \times 10^{9}\ cm^{-1}\ M^{-1}$	207
	$\varepsilon_{(30)530\,nm} = 3.0 \times 10^{9}\ cm^{-1}\ M^{-1}$	134
	$\varepsilon_{(35)526\,nm} = 4.65 \times 10^{9}\ cm^{-1}\ M^{-1}$	207
	$\varepsilon_{(40)533\,nm} = 6.7 \times 10^{9}\ cm^{-1}\ M^{-1}$	134
	$\varepsilon_{(45)529\,nm} = 9.80 \times 10^{9}\ cm^{-1}\ M^{-1}$	207
	$\varepsilon_{(50)535\,nm} = 1.5 \times 10^{10}\ cm^{-1}\ M^{-1}$	134
	$\varepsilon_{(55)533\,nm} = 1.79 \times 10^{10}\ cm^{-1}\ M^{-1}$	207
	$\varepsilon_{(60)540\,nm} = 2.9 \times 10^{10}\ cm^{-1}\ M^{-1}$	134
	$\varepsilon_{(65)539\,nm} = 2.94 \times 10^{10}\ cm^{-1}\ M^{-1}$	207
	$\varepsilon_{(70)542\,nm} = 3.70 \times 10^{10}\ cm^{-1}\ M^{-1}$	207
	$\varepsilon_{(75)545\,nm} = 4.55 \times 10^{10}\ cm^{-1}\ M^{-1}$	207
	$\varepsilon_{(80)550\,nm} = 6.9 \times 10^{10}\ cm^{-1}\ M^{-1}$	134
	$\varepsilon_{(85)553\,nm} = 6.67 \times 10^{10}\ cm^{-1}\ M^{-1}$	207
	$\varepsilon_{(90)558\,nm} = 7.69 \times 10^{10}\ cm^{-1}\ M^{-1}$	207
	$\varepsilon_{(95)563\,nm} = 9.09 \times 10^{10}\ cm^{-1}\ M^{-1}$	207
	$\varepsilon_{(100)569\,nm} = 1.00 \times 10^{11}\ cm^{-1}\ M^{-1}$	207
	$\varepsilon_{(125)600\,nm} = 2.00 \times 10^{11}\ cm^{-1}\ M^{-1}$	207

[a]Obtained via surfactant (cetyltrimethylammonium bromide, CTAB) mediated colloidal growth.[135,136]

[b]Optical Density (OD) = $\varepsilon_{(size/diameter,nm)\lambda max,nm} \times b \times c$ for b = path length in cm and c = concentration in moles particles per L of aqueous solvent.

[c]Obtained via Turkevich–Frens reduction.[26,137,138]

substantially effect nanoparticle disposition in the body, from circulation and clearance (Sec. 2.1), to tumor extravasation, perfusion, retention and cellular uptake. Although we review the synthesis[26,28] and biomedical applications[23,25–27,39] of novel gold nanostructures elsewhere, we focus our discussion here on the impact of size, shape, charge and surface chemistry on tumor-specific targeting of gold nanoparticles.

2.2.1. **Nanoparticle size and passive tumor extravasation**

Tumor-targeting phenomena are classified as either "passive" or "active," depending on the mechanisms by which they spatially localize at or within tumors. While active targeting utilizes nanoparticle ligands (e.g. small molecules,[38–40,149] peptides/proteins,[46,50,78,150–155] aptamers,[156–160] antibodies[48,131,161–167] and fragments[155,168] thereof) or responsiveness toward the tumor microenvironment (hypoxic pH,[169] elevated temperature, proteolytic degradation[170]), passive targeting relies on the fact the angiogenic neovasculature is highly disordered (i.e. fenestrated) and is thus hyperpermeable to large macromolecules such as nanoscale particles (relative to normal healthy tissues). For example, transvascular pore diameter in normal tissues range from 5 to 12 nm,[92, 93] while those in tumor tissues range from 200 nm to 1.2 μm; Fig. 4);[6,171] Because effective pore diameters and macromolecular transvascular flux rates across the angiogenic endothelium is significantly increased, selective delivery of high-molecular-weight compounds and nanoscale constructs can be achieved. Once localized within the tumor interstitium, high-molecular-weight compounds are retained due to characteristically diminished lyphatic drainage from the tumor site and size-proportional diffusivity, which decreases transport away from the site of extravasation. The latter phenomena, first termed by Maeda in 1986[172] and Jain in 1987[173] as the enhanced permeability and retention (EPR) effect forms the basis for selective delivery of a wide range of diagnostic and therapeutic compounds in use today and involved many seminal leaders in the nanomedicinal field including Torchilin,[101,171,174] Griffith[171] and Fukumura.[101, 174–179] Although "passive" and "EPR" targeting are often used interchangeably, the former describes only extravasation, while the latter describes both interstitial localization and retention.[180,181] Readers may also note that treatment regimens involving vasculature normalization (e.g. vascular endothelial growth factor receptor-2, VEGFR2, blockade) generally improve the delivery of low-molecular-weight compounds and nanoparticles *ca.*12 nm in diameter, while diminishing those of larger nanospheres *ca.* 125 nm.[178]

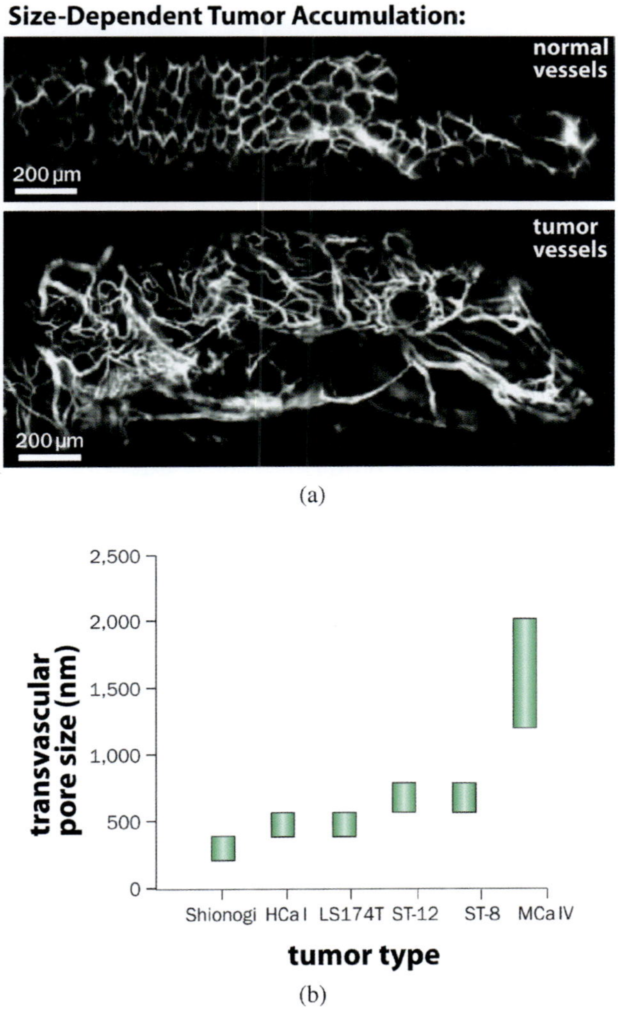

Fig. 4. The physiological basis for passive tumor-targeting and the enhanced permeability and retention (EPR) effect. (a) In contrast to normal blood vessels (top panel) which exhibit well-ordered endothelia and transvascular pore sizes 5–12 nm, the angiogenic neovasculature in tumors (bottom panel) exhibits a disordered endothelial cell lining, abnormally thick or thin basement membrane and detached or absent pericytes, all of which increase transvascular pore diameter to 200–1200 nm, promoting extravasation of larger compounds and nanoparticles at the tumor site. (b) Effective transvascular pore size ranges widely across various tumor types. Intravital fluorescence microscopy in (a) reproduced with permission from Ref. 182. Murine xenograft data in (b) reproduced with permission from Ref. 171. Copyright 2010 Macmillan Publishers Ltd and 1998 National Academy of Sciences.

Although the aforementioned size ranges reflect what one may often observe, transvascular pore cutoff sizes can range widely depending on the tumor type, stage,[174, 183] location[9] as well as its degree of stromal compaction (desmoplasia)[2, 173, 184–186] and interstitial fluid pressure[6,178] (IFP). Delivery to the brain, for example, requires penetration of the tightly compacted endothelial cell layer known as the blood–brain barrier (BBB) which separates blood from cerebrospinal fluid. Because of the tight endothelial junctions in the BBB, only very small molecules including gases (e.g. CO_2 and O_2), metabolic products (e.g. glucose) and hormones, as well as small, neutrally charged and lipophilic molecules can pass. Thus, efficient penetration of the BBB is restricted to hydrodynamic diameters less than 15 nm,[187,188] although lower efficiency penetration by larger particles (15–100 nm) has been reported.[188, 189] Targeting of the lyphatics is another scenario requiring special attention in the design of a gold nanoparticles construct. Intratrachael administration of non-cationic nanoparticles in the 6–34 nm range has been shown to facilitate rapid traffic to the local lymph nodes (<1 h),[95] while others have shown that subcutaneously administered nanospheres <80 nm are likewise transported to the lymphatics.[96, 97] In addition, conjugation with polysaccharides[190–193] (e.g. dextran) and IgG[194] have been shown to promote lymphatic targeting, while PEGylation appears to diminish such accumulation.[195,196] As discussed in Sec. 2.1.2, liver-specific nanoparticle targeting requires HD <100 nm in order to cross liver fenestrae.[9]

Given the above described size constraints, it becomes clear that size of a tumor-targeting gold nanoparticle greatly dictates its subsequent capacity for passive targeting. Nanospheres with comparable levels of PEGylation typically exhibit a semi-linear dependence between hydrodynamic diameter and RES organ accumulation (liver and spleen; *ca.* 17–82% ID for 26–167 nm HD, respectively).[31] Similarly, circulation half-life decays semi-exponentially with hydrodynamic diameter (Fig. 2).[31] For conjugates of a given particle diameter, blood half-life and subcutaneous tumor xenograft AUC (area under the curve) increase with PEG molecular weight, with diminishing returns after 5000 Da for constructs with relatively low radii of curvature (<20 nm).[84] For particles of a given hydrodynamic (Stokes–Einstein) diameter and PEG grafting density, circulation half-life likewise scales proportionately with PEG molecular weight. Perrault *et al.*, for example, find that tumor accumulation of PEGylated gold nanoparticles peaks in *ca.* 40–80 nm HD range.[84]

2.2.3. **Active targeting**

In contrast to passive targeting methods, active targeting utilizes ligands with binding specificity or responsiveness toward tumor cells or their microenvironment. PEGylated gold nanoparticles, for example, typically exhibit between 2% and 7 %ID g^{-1} tumor-specific accumulation with passive/EPR targeting and 6% to 13%ID g^{-1} with active targeting (initial dose, ID).[197] Although active targeting often, but not always,[78,155] facilitates increasingly specific tumor accumulation, this specificity is also accompanied by diminished steric stabilization and thus, decreased blood residence time and capacity for passive/EPR targeting. For example, Huang *et al.* observed 26–48% decreased circulation half-life following conjugation with antibody fragments and peptides exhibiting specificity for the tumor microenvironment.[155] In practice, a range of targeting ligand densities should be investigated to balance tradeoffs in circulation half-life, ligand-specific targeting and cellular uptake.[78]

2.2.4. Shape-dependent tumor-targeting

In addition to size, shape also plays a critical role in modulating systemic transport and transvascular flux across the tumor neovasculature. Generally, nanoparticles with at least one dimension above/below the mean vascular fenestrae size within a given organ can achieve efficient transvascular flux (e.g. kidney ≤ *ca.* 6 nm; liver ≤ *ca.* 100 nm; spleen ≥ 200 nm). High anisotropy, however, can also diminish extravasation due to hydrodynamic shear-induced alignment along the blood vessel (e.g. long-circulating filomicelles *ca.* 20 nm × 8–18 μm)[88]; however, such alignment effects are expected to be minimal in tumors with characteristically sluggish blood flow and low transvascular fluid flux.[176] Fukumura *et al.* investigated tumor-specific transvascular transport of PEGylated semiconductor nanoparticles (nanospheres 35 nm HD; nanorods, 15 × 54 nm HD).[176] Although they observed no differences in blood half-life, the authors found that nanorods penetrated tumor xenografts 4.1 times more rapidly than nanospheres, and that nanorods penetrated to 1.7 times the volume over which the nanospheres distributed (1 h postinjection). The authors observed similar effects across porous membranes and collagen matrix, concluding that reduced steric hindrance from and viscous drag near vessel pore walls contributed to superior permeability by the nanorods. Dai *et al.* have similarly examined tumor extravasation using PEGylated quantum dots (20–25 nm HD) and single-walled carbon nanotubes (SWCNT; 2–3 × 200 nm). While SWNTs diffused 3.5-times more rapidly than quantum dots across 100-nm porous membranes, diffusion of SWNTs and quantum dots through 200-nm pores was approximately the same. Interestingly, the authors found that neither nanoparticle extravasated into SKOV-3 tumors; in LS174T tumors, however, quantum dots extravasated more than SWNTs, while SWNTs extravasated much more than quantum dots in U87MG tumors. Together, these data indicate a high degree of complexity in determining the optimal shape for tumor-targeting gold nanoparticles.[198]

2.3. *Tumor penetration and cell uptake*

Size, shape and surface chemistry also play critical roles in determining tumor penetration, cellular delivery and subsequent diagnostic or therapeutic efficacy of a gold nanoparticle conjugate. Constructs with varying size, shape, charge and lipophilicity will invariably partition over different areas within the complex tumor extracellular matrix and are taken up by cells at different rates. In an early study, Dreher *et al.* investigated intratumoral penetration of fluorescently labeled dextrans with varying molecular weight (*ca.* 3–49 nm HD, 3.3 kDa–2 MDa).[199] Interestingly, they found that polymer conjugates in the 11–15 nm range (40–70 kDa) displayed the highest tumor accumulation, but also the lowest tumor penetration depth, due in part to the fact that increased circulation half-life was observed with increasing molecular weight. They observed maximal tumor penetration (>35 μm from the vessel walls) in the 3–5 nm range (3.3–10 kDa), with 15 μm penetration in the 11–15 nm (40–70 kDa) range, and only 5 μm penetration by the larger polymer conjugates (49 nm, 2 MDa). Chan and coworkers have examined intratumoral penetration of gold nanospheres across a larger size range (22–99 nm HD, Fig. 2), each PEGylated at equivalent grafting densities.[84] In this size range, they found that nanoparticle penetration was inversely correlated with particle size. In all cases, accumulation 10–20 μm from the vessel walls was roughly half that directly adjacent to it, with penetration decreasing in a size-dependent manner at 20–30, 30–40 and 40–50 μm distances. Importantly, the smallest (22 nm HD) gold nanoparticles penetrated significantly deeper than the larger constructs with 15–20% relative accumulation over the 30–50 μm range. In related work, Wong *et al.* attempted to exploit size-dependent tumor penetration by designing gelatin nanoparticle *ca.* 100 nm HD that could be degraded by tumor proteases to release small *ca.* 10 nm HD nanoparticles for efficient tumor penetration.[177, 179] Indeed, intratumoral injection of the multistage constructs yielded significantly higher tumor penetration (*ca.* 300 v. 50 μm) relative to non-degradable control nanoparticles of comparable diameter. Results from the multistage construct agreed well with prior investigations of size-dependent penetration by individual nanoparticles (12–125 nm HD).[200] These studies highlight the delicate balance between prolonged circulation, preferential extravasation and tumor penetration that must be considered when designing a tumor-targeting gold nanoparticle. Although limited data is available regarding the effects of shape on tumor perfusion, readers may note that rod-shaped nanoparticles penetrated 1.7 times greater tumor volume than comparably sized nanospheres in the previously discussed studies by Chauhan *et al.*[176]

Once a gold nanoparticle conjugate has traveled from the injection site to deep within the tumor interstitium, these constructs must enter cells with high efficiency. Here both size and shape play prominent roles, as surface chemistry largely dictates prior pharmacokinetics and disposition. Early work by Aoyama and coworkers investigated size-dependent cellular uptake of virus-like saccharide-terminal micelles and quantum dots and found that cellular uptake in the 5–100 nm HD range peaked at 50 nm HD.[201,202] Freund and coworkers later investigated size-dependent nanoparticle endocytosis using a highly advanced theoretical model based on cell membrane wrapping that accounts for receptor density and mobility, as well as nanoparticle size/shape, ligand binding affinity and a variety of other parameters. By considering the cost of elevated elastic energy associated with increased local curvature of the membrane and reduced configurational entropy associated with receptor immobilization, the authors derived an optimal particle diameter of 54–60 nm (HD) based on the minimum wrapping time required for a receptor-mediated process.[203] Chithrani *et al.* have examined size-dependent cellular uptake of anionic (citrate-capped) gold nanospheres (14–100 nm physical dia) and found peak particle uptake at 50 nm (Fig. 5(a)).[204] Later work by the authors investigating size-dependent, receptor-mediated endocytosis of anti-HER2/neu-conjugated gold nanospheres also yielded a similar size dependence, peaking at 40–50 nm HD.[205] Chithrani *et al.* also found that cellular uptake of citrate-functionalized CTAB-conjugated gold nanorods decreased with increasing aspect ratio (Fig. 5(b)).[204] Nanorods with an aspect ratio of 3 and 5 were taken up by HeLa cells at roughly 2- and 3-fold lower amounts compared with citrate nanospheres equivalent in diameter to their minor axes. In contrast, DeSimone and coworkers have examined shape-dependent cellular uptake of photolithographically patterned polymer particles in the larger size regime and observed that cellular internalization increased with increasing aspect ratio, finding that 450 × 150 nm particles were taken up 4-fold faster than symmetric particles similar in size to the minor axes of the nanorods (200 × 200 nm). Surprisingly, for a fixed aspect ratio, that authors also found that cell uptake increased with increasing size, where 150 × 450 nm particles were internalized to a greater extent than 100 × 300 nm particles.[206] Together, these findings suggest that tumor penetration is often most efficient in the *ca.* 15–30 nm HD size range, with

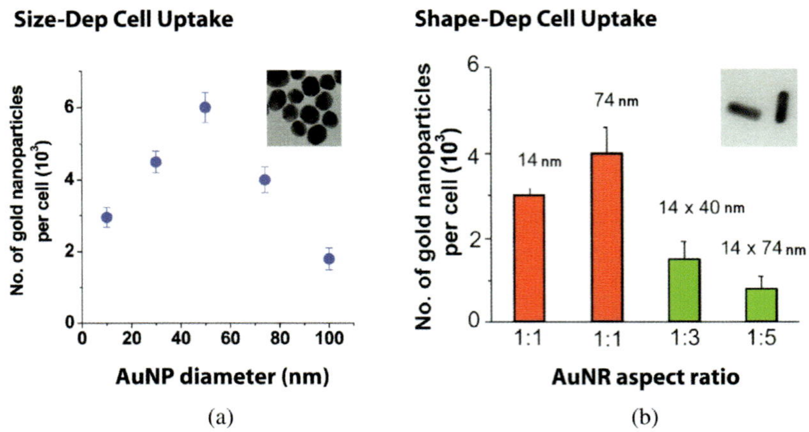

Fig. 5. Size and shape-dependent cellular uptake of gold nanoparticles. (a) Cellular uptake of spherical nanoparticles generally peaks in the 40–60 nm range with (b) nanospheres favored over rod-shaped gold nanoparticles with equivalently sized minor axes. Reproduced with permission from Ref. 204. Copyright 2006 American Chemical Society.

optimal receptor-mediated cellular uptake in the 40–60 nm HD range (although optimal sizes likely depend on specific characteristics of the tumor, mechanism of cellular uptake, etc.). Shape-dependent tumor penetration and cell uptake remains to be fully explored.

2.4. *Practical considerations*

In practice, overnight agitation of the particle conjugates with additional sonication or gentle heating (+10–20°C) is often sufficient to achieve optimal surface monolayer formation. Particles can be purified by equilibrium dialysis or more rapidly by centrifugation or centrifugal diafiltration (e.g. 100 kDa MWCO, Amicon). Based on the so-called "standard model" for gold thiolate coordination (monothiolate bridge-site binding),[34] a minimum ligand excess of *ca.* 2 400, 59 000, and 240 000 per particle would be required for full (111) surface coverage of a 10-, 50- and 100-nm diameter AuNP core, thus efficient functionalization requires experimental ratios exceeding these values. Although thiol ligand exchange may be easily performed (i.e. Murray place exchange[12,18,115]), in mixed thiol monolayer configurations, it is often easiest to prepare thiol solutions at desired stoichiometric ratios followed by subsequent rapid introduction of particles. Given that Au–S bond strengths of common thiolate ligands are typically comparable, subsequent stoichiometry of the surface monolayer can often be assumed preserved. In cases where ligand exchange is desired, bond strengths reported in Table 3 indicate relevant propensity for exchange. Authors may note that quaternary ammonium halide surfactants (e.g. CTAB-conjugated gold nanorods) are generally resistant to high affinity (e.g. Au–S) ligand exchange except at low energy crystalline facets, for example, those which comprise the tips of colloidal gold nanorods.[124] Highly amphiphilic mixed monolayers can also yield unique phase separation patterns that may dramatically alter cellular and biomolecular interactions[125–127]; in these cases, researchers are encouraged to consider such possible contributions in the interpretation of their data. Readers may also note that conjugation protocols often include or omit steps to reduce disulfide moieties that may have formed as a result of thiol oxidation by exposure to ambient air, light or trace metal ions. Although the disulfide bond strength[128] (60 kcal mol^{-1}) is only slightly stronger that of Au–S[32, 118] (40–45 kcal mol^{-1}) omission of disulfide reduction does not often significantly affect subsequent results. While disulfide reduction is encouraged, common biochemical reagents such as glutathione (GSH) and dithiothreitol (DTT) are themselves thiols and can passivate gold surfaces without proper removal by dialysis, column chromatography, etc. Similarly, disulfide reducing tris(2-carboxyethyl)phosphine (TCEP) and other disulfide-reducing, triply-substituted phosphines may also passivate Au surfaces.[129]

Because insufficient stabilizer grafting density or molecular weight can promote particle sedimentation in high serum/salt, it is also recommended that particles be evaluated for changes in particle size/charge, as measured by DLS and laser Doppler electrophoretic measurements, respectively, as well as aggregation state, as measured by UV-Vis–NIR absorption spectroscopy and typically indicated by a spectral red-shift. We find that the conjugation of hydrophobic small molecules (e.g. tamoxifen, logP 6.58; nilutamide, logP 1.72) can be achieved at ≤10% monolayer densities without significant particle destabilization or receptor screening by high molecular weight PEG (e.g. 5 kDa).[38–40] In addition to physiochemical properties of gold nanoparticle conjugates themselves, microbial and/or enzymatic contaminants can also dramatically affect both the accuracy of *in vitro* assays at subsequent *in vivo* pharmacokinetics/efficacy. Although aseptic handling practices and gamma-irradiated or autoclaved equipment can often mitigate potential hazards, sterile filtering of final products is highly recommended, albeit with small losses, and particles that are recommend to be stored as concentrated solutions at 4°C in dark. In situations where ribonuclease (RNase) or deoxyribonuclease (DNAse) is of particular concern, particles (e.g. citrate-capped AuNPs) can be autoclaved, though investigators should be cautioned that other structures (e.g. CTAB-capped gold nanorods) typically aggregate under such high heat conditions. Given that Turkevich–Frens[31, 79, 80] syntheses of gold nanospheres is performed under reflux, one might expect intrinsically low nuclease activity, however, Giljohann and Mirkin[139] recently found that treatment with diethylpyrocarbonate (DEPC), a common commercial additive in nuclease-free water, was required to maintain stability of RNA–gold nanoparticle conjugates (*ca.* 24-mer thiol terminal RNA), presumably due to residual nuclease contamination. Here, the authors treated as-prepared citrate-AuNPs with 0.1% diethylpyrocarbonate DEPC for 12 h with stirring, followed by autoclaving at 121°C for 60 min. (Note that DEPC decomposes to ethanol and carbon dioxide at with high heat, *ca.* 155°C). The authors also employ brief sonication (*ca.* 10 s) and a salt "aging" process, whereby a gradual increase in solvent Debye screening in the presence of a mild surfactant, 0.01% sodium dodecylsulfate (SDS), was used to increase subsequent oligonucleotide coverage density[139–141] (e.g. gradual increase from 0 M to 0.2–1.0 M NaCl over 24 to 40 h). The use of 0.01% SDS is also helpful in stabilizing gold and other nanoparticle conjugates while performing carbodiimide (i.e. EDC) coupling to carboxylate-terminal surface ligands (e.g. SH-PEG-COOH), whose uncharged, reactive succinimidyl ester leads to diminished electrostatic stabilization and/or is increasingly prone to aggregation following centrifugation.[142] Readers may also note that common citrate-AuNPs are typically unstable in 1x concentrated phosphate buffered saline (PBS) and that neutral (pH 6.8–8.2) buffering can often be achieved in 10 mM HEPES (note that HEPES buffers should be stored in dark as UV/light exposure can result in the production of cytoxic peroxides[143, 144]).

Gold nanoparticles are most often stored as concentrated aqueous solutions; however, in commercial settings, lyophilization of nanoconjugates can be highly desirable. Dried powers provide significant packaging, transportation and storage advantages over more cumbersome aqueous solutions, and in practice, both freezing and drying of gold nanoparticle solutions leads to some degree of irreversible aggregation, which is often substantial. While very few reports describe methods for the lyophilization of gold nanoparticles with size, surface chemistry, spectral response or functional activity preserved following reconstitution, these methods have been applied in industrial settings. For example, CYT-6091[29, 145] (TNFα) which is stored as a lyophilized powder utilizes human serum albumin as a blocking agent during conjugation, possibly[146] in addition to mannitol (20 mg/mL) and/or human serum albumin (5 mg/mL) as a cryoprotectant. Others report that re-dispersible gold nanosphere powders conjugated with polyvinylpyrrolidone (PVP; 10–40 kDa) containing and 0.1–0.5% PVP as excipient.[147]

In cases where particle storage or ligand exchange is required under harsh conditions (e.g. high salt/heat) that typically result in aggregation of citrate-functionalized AuNPs, nanoparticle cores can be separately prepared and stored via conjugation of 4,4'-(phenylphosphinidene)bis(benzenesulfonic acid) dipotassium salt hydrate. Because the Au–PR_3 bond is significantly weaker than the Au–S bond strength (16 kcal mol^{-1} versus 40–45 kcal mol^{-1}), phosphine can serve labile (i.e. thiol-displaceable) stabilizer capable of providing solvent stability greater than carboxylate-terminal ligands. Taton recommends incubating Turkevich–Frens (citrate-functionalized) AuNPs at *ca.* $\geq$ 500,000 phosphine:particle excess (see Table 4) for 10 h at RT with stirring.[148] NaCl is added to 2–3 M and particle color shifts from deep red to purple. The particle suspension is centrifuged for 30 min at 500 × g, and the pellet is re-dispersed in 0.5 mM phosphine. Methanol is then added to 33% v/v in order to precipitate the particles, and the suspension is centrifuged again 30 min at 500 × g at RT. The recovered pellet is redispersed in 0.5 mM phosphine. Alternatively, ligand exchange of a thiol-PEGylated stabilizing ligand may be performed; however, this methods typically requires that the nascent ligand be present in large excess, given that the thermodynamic driving force for Au–S bond formation is comparable for both the incoming and outgoing ligand.

3. Conclusions and Outlook

Tumor-specific cellular delivery of a gold nanoparticle is a complex, multistep process involving systemic circulation, vascular extravasation, tumor penetration and intracellular delivery. Each step presents unique constrains on optimal architectures for the nanoparticle construct, including size, shape and surface chemistry. Although a comprehensive and sometimes conflicting account of these parameters are presented here, some general conclusions can be drawn; however, readers may note that these values likely vary depending on the particular situation at hand (e.g. nanoparticle polydispersity, ligand grafting density, serum adsorption, tumor location/stage/vascularization, cell type, receptor expression/density, mechanism of cell uptake, etc.). During circulation, weakly negative surface charge (*ca.* -10 to 0 mV), steric stabilization (high-density brush conformation, $\geq$5 KDa PEG) and small core size (5–50 nm; 25–75 nm HD) promote long circulation and decreased RES accumulation. Intravenous administration appears more effective than intraperotineal methods for tumor-specific delivery outside the peritoneal cavity, with higher aspect ratio promoting longer circulation and diminished clearance for particles with minor axes comparable to the aforementioned spherical dimensions. In terms of extravasation, spherical nanoparticle accumulation appears to peak in the 40–80 nm HD range, while reports of shape-dependent transvascular flux are conflicting. Although surface chemistry is largely dictated by circulation constraints, its effects on transvascular flux have yet to be fully explored. Size-dependent tumor penetration appears highly dependent on angiogenic signaling and generally favors nanoparticle constructs in the 15–30 nm HD range, with limited data suggesting preferential penetration by high-aspect-ratio nanoparticles. Again, surface-chemistry–dependent effects here are less well characterized; however, architectures which can sterically destabilize or deshield cationic charge or lipophilic surfaces in response to the tumor microenvironment (e.g. hypoxic pH, protease activity) are currently areas of promising investigation. Size-dependent cellular uptake, both receptor-mediated and otherwise, generally peak in the 40–60 nm HD range, with limited data suggesting that spherical nanoparticles are favored in this size range and rod-like particles at larger hydrodynamic diameters.

Given the above discussion, it is clear that there is no singularly optimal architecture for a tumor-targeting gold nanoparticle. Areas of further exploration for gold nanotechnologies include long-term clearance as well as genetic and reproductive effects. Shape-dependent extravasation/tumor penetration and modulation of the tumor vasculature/stroma are also important areas for exploration. The simultaneous delivery of nanoparticles with varying size/shape/chemistry and those exerting synergistic responses will likely become increasingly important. Dynamic nanoparticle assemblies whose physiochemical properties can shift in response to the particular stage of their delivery or can participate in cooperative or emergent physiological interactions are expected to contribute to next-generation tumor-targeting nanotechnologies. Going forward, gold nanoparticles, given their unparalleled utility in both designing and understanding cancer nanotechnologies, in addition to their intrinsically unique and novel physiochemical and pharmacodynamics properties, will play an increasingly important role in the field.

Acknowledgments

This work was supported by the US National Institutes of Health (1U01CA1518020-1: MAE, ECD; Kirschstein NRSA 1F32EB017614-01). The authors also wish to thank Lenore Beadsman, Norman Bombardini, and Dr. Curtis Jay for valuable comments on the early drafts.

References

1. R. Siegel, D. Naishadham and A. Jemal, Cancer statistics, 2012, *CA Cancer J. Clin.* **62**, 10 (2012).
2. Surveillance Epidemiology and End Results (SEER) Database (2005–2007). US National Cancer Institute, Bethesda, MD. Available at: http://seer.cancer.gov.
3. N. Beaulieu, D. E. Bloom, L. R. Bloom, R. M. Stein, L. R. Aronson, M. O. Harhay, E. Cafiero, M. Ozolins, J. Ferlay, L. Abruzzese, R. Powell and M. Kenny, Breakaway: The global burden of cancer — challenges and opportunities. (2009).
4. M. E. Davis, Z. Chen and D. M. Shin, Nanoparticle therapeutics: An emerging treatment modality for cancer, *Nat. Rev. Drug Discov.* **7**, 771 (2008).
5. R. A. Petros and J. M. DeSimone, Strategies in the design of nanoparticles for therapeutic applications, *Nat. Rev. Drug Discov.* **9**, 615 (2010).
6. R. K. Jain and T. Stylianopoulos, Delivering nanomedicine to solid tumors, *Nat. Rev. Clin. Oncol.* **7**, 653 (2010).
7. B. Y. S. Kim, J. T. Rutka and W. C. W. Chan, Nanomedicine, *New Engl. J. Med.* **363**, 2434 (2010).
8. J. V. Jokerst and S. S. Gambhir, Molecular imaging with theranostic nanoparticles, *Acc. Chem. Res.* **44**, 1050 (2011).
9. A. Schroeder, D. A. Heller, M. M. Winslow, J. E. Dahlman, G. W. Pratt, R. Langer, T. Jacks and D. G. Anderson, Treating metastatic cancer with nanotechnology, *Nat. Rev. Cancer* **12**, 39 (2012).
10. N. L. Rosi and C. A. Mirkin, Nanostructures in biodiagnostics, *Chem. Rev.* **105**, 1547 (2005).
11. M. Hu, J. Chen, Z.-Y. Li, L. Au, G. V. Hartland, X. Li, M. Marquez and Y. Xia, Gold nanostructures: Engineering their plasmonic properties for biomedical applications, *Chem. Soc. Rev.* **35**, 1084 (2006).
12. A. C. Templeton, M. P. Wuelfing and R. W. Murray, Monolayer protected cluster molecules, *Acc. Chem. Res.* **33**, 27 (2000).
13. M. Grzelczak, J. Perez-Juste, P. Mulvaney and L. M. Liz-Marzan, Shape control in gold nanoparticle synthesis, *Chem. Soc. Rev.* **37**, 1783 (2008).
14. S. Lal, S. E. Clare and N. J. Halas, Nanoshell-enabled photothermal cancer therapy: Impending clinical impact, *Acc. Chem. Res.* **41**, 1842 (2008).
15. C. J. Murphy, A. M. Gole, J. W. Stone, P. N. Sisco, A. M. Alkilany, E. C. Goldsmith and S. C. Baxter, Gold nanoparticles in biology: Beyond toxicity to cellular imaging, *Acc. Chem. Res.* **41**, 1721 (2008).
16. S. E. Skrabalak, J. Chen, Y. Sun, X. Lu, L. Au, C. M. Cobley and Y. Xia, Gold nanocages: Synthesis, properties, and applications, *Acc. Chem. Res.* **41**, 1587 (2008).
17. E. Boisselier and D. Astruc, Gold nanoparticles in nanomedicine: Preparations, imaging, diagnostics, therapies and toxicity, *Chem. Soc. Rev.* **38**, 1759 (2009).
18. R. Sardar, A. M. Funston, P. Mulvaney and R. W. Murray, Gold nanoparticles: Past, present, and future, *Langmuir* **25**, 13840 (2009).
19. J. Zhao, A. O. Pinchuk, J. M. McMahon, S. Li, L. K. Ausman, A. L. Atkinson and G. C. Schatz, Methods for describing the electromagnetic properties of silver and gold nanoparticles, *Acc. Chem. Res.* **41**, 1710 (2008).

20. X. Huang, S. Neretina and M. A. El-Sayed, Gold nanorods: From synthesis and properties to biological and biomedical applications, *Adv. Mater.* **21**, 4880 (2009).
21. D. A. Giljohann, D. S. Seferos, W. L. Daniel, M. D. Massich, P. C. Patel and C. A. Mirkin, Gold nanoparticles for biology and medicine, *Angew. Chem. Int. Ed.* **49**, 3280 (2010).
22. Y. Xia, W. Li, C. M. Cobley, J. Chen, X. Xia, Q. Zhang, M. Yang, E. C. Cho and P. K. Brown, Gold nanocages: From synthesis to theranostic applications, *Acc. Chem. Res.* **44**, 914 (2011).
23. E. C. Dreaden, M. A. Mackey, X. Huang, B. Kang and M. A. El-Sayed, Beating cancer in multiple ways using nanogold, *Chem. Soc. Rev.* **40**, 3391 (2011).
24. A. S. Thakor, J. Jokerst, C. Zavaleta, T. F. Massoud and S. S. Gambhir, Gold nanoparticles: A revival in precious metal administration to patients, *Nano Lett.* **11**, 4029 (2011).
25. E. C. Dreaden, M. A. El-Sayed and I. H. El-Sayed, in *Nanomedicine and Cancer*, V. R. Preedy and R. Srirajaskanthan (Eds.), Enfield, NH: Science Publishers (2011).
26. E. C. Dreaden, A. M. Alkilany, X. Huang, C. J. Murphy and M. A. El-Sayed, The golden age: Gold nanoparticles for biomedicine, *Chem. Soc. Rev.* **41**, 2740 (2012).
27. E. C. Dreaden and M. A. El-Sayed, Detecting and destroying cancer cells in more than one way with noble metals and different confinement properties on the nanoscale, *Acc. Chem. Res.* **45**, 1854 (2012).
28. E. C. Dreaden, L. A. Austin, M. A. Mackey and M. A. El-Sayed, Size matters: Gold nanoparticles in targeted cancer drug delivery, *Therap. Deliv.* **3**, 457 (2012).
29. S. K. Libutti, G. F. Paciotti, A. A. Byrnes, H. R. Alexander, W. E. Gannon, M. Walker, G. D. Seidel, N. Yuldasheva and L. Tamarkin, Phase I and pharmacokinetic studies of CYT-6091, a novel PEGylated colloidal gold-rhTNF nanomedicine, *Clin. Cancer Res.* **16**, 6139 (2010).
30. E. B. Dickerson, E. C. Dreaden, X. Huang, I. H. El-Sayed, H. Chu, S. Pushpanketh, J. F. McDonald and M. A. El-Sayed, Gold nanorod assisted near-infrared plasmonic photothermal therapy (PPTT) of squamous cell carcinoma in mice, *Cancer Lett.* **269**, 57 (2008).
31. C. H. J. Choi, J. E. Zuckerman, P. Webster and M. E. Davis, Targeting kidney mesangium by nanoparticles of defined size, *Proc. Natl. Acad. Sci. U.S.A.* **108**, 6656 (2011).
32. A. Ulman, Formation and structure of self-assembled monolayers, *Chem. Rev.* **96**, 1533 (1996).
33. J. C. Love, L. A. Estroff, J. K. Kriebel, R. G. Nuzzo and G. M. Whitesides, Self-assembled monolayers of thiolates on metals as a form of nanotechnology, *Chem. Rev.* **105**, 1103 (2005).
34. H. Hakkinen, The gold-sulfur interface at the nanoscale, *Nat. Chem.* **4**, 443 (2012).
35. E. Pensa, E. Cortés, G. Corthey, P. Carro, C. Vericat, M. H. Fonticelli, G. Benítez, A. A. Rubert and R. C. Salvarezza, The chemistry of the sulfur–gold interface: In search of a unified model, *Acc. Chem. Res.* **45**, 1183 (2012).
36. N. T. Flynn, T. N. T. Tran, M. J. Cima and R. Langer, Long-term stability of self-assembled monolayers in biological media, *Langmuir* **19**, 10909 (2003).
37. G. Walsh, Biopharmaceuticals: Recent approvals and likely directions, *Trends Biotechnol.* **23**, 553 (2005).

38. E. C. Dreaden, S. C. Mwakwari, Q. H. Sodji, A. K. Oyelere and M. A. El-Sayed, Tamoxifen-Poly(ethylene glycol)-Thiol Gold Nanoparticle Conjugates: Enhanced Potency and Selective Delivery for Breast Cancer Treatment, *Bioconjugate Chem.* **20**, 2247 (2009).
39. A. K. Oyelere, M. A. El-Sayed and E. C. Dreaden, in *12/890, 519 pending*. (2010).
40. E. C. Dreaden, B. E. Gryder, L. A. Austin, B. A. Tene Defo, S. C. Hayden, M. Pi, L. D. Quarles, A. K. Oyelere and M. A. El-Sayed, Antiandrogen Gold Nanoparticles Dual-Target and Overcome Treatment Resistance in Hormone-Insensitive Prostate Cancer Cells, *Bioconjugate Chem.* **23**, 1507 (2012).
41. E. C. Dreaden, R. D. Near, T. Abdallah, M. H. Talaat and M. A. El-Sayed, Multimodal plasmon coupling in low symmetry gold nanoparticle pairs detected in surface-enhanced Raman scattering, *Appl. Phys. Lett.* **98**, 183115 (2011).
42. S. Neretina, E. C. Dreaden, W. Qian, M. A. El-Sayed, R. A. Hughes, J. S. Preston and P. Mascher, The dependence of the plasmon field induced nonradiative electronic relaxation mechanisms on the gold shell thickness in vertically aligned CdTe-Au core-shell nanorods, *Nano Lett.* **9**, 3772 (2009).
43. S. Neretina, W. Qian, E. C. Dreaden, M. A. El-Sayed, R. A. Hughes, J. S. Preston and P. Mascher, Plasmon field effects on the nonradiative relaxation of hot electrons in an electronically quantized system: CdTe-Au Core-shell nanowires, *Nano Lett.* **8**, 2410 (2008).
44. S. Neretina, W. Qian, E. C. Dreaden, M. A. El-Sayed, R. A. Hughes, J. S. Preston and P. Mascher, Exciton lifetime tuning by changing the plasmon field orientation with respect to the exciton transition moment direction: CdTe-Au core-shell nanorods, *Nano Lett.* **9**, 1242 (2009).
45. C.-W. Yen, S. C. Hayden, E. C. Dreaden, P. Szymanski and M. A. El-Sayed, Tailoring plasmonic and electrostatic field effects to maximize solar energy conversion by bacteriorhodopsin, the other natural photosynthetic system, *Nano Lett.* **11**, 3821 (2011).
46. X. Huang, B. Kang, W. Qian, M. A. Mackey, P. C. Chen, A. K. Oyelere, I. H. El-Sayed and M. A. El-Sayed, Comparative study of photothermolysis of cancer cells with nuclear-targeted or cytoplasm-targeted gold nanospheres: Continuous wave or pulsed lasers, *J. Biomed. Opt.* **15**, 058002 (2010).
47. T. Shimizu, T. Teranishi, S. Hasegawa and M. Miyake, Size evolution of alkanethiol-protected gold nanoparticles by heat treatment in the solid state, *J. Phys. Chem. B* **107**, 2719 (2003).
48. X. Huang, I. H. El-Sayed, W. Qian and M. A. El-Sayed, Cancer cell imaging and photothermal therapy in the near-infrared region by using gold nanorods, *J. Am. Chem. Soc.* **128**, 2115 (2006).
49. T. K. Sau and C. J. Murphy, Seeded high yield synthesis of short Au nanorods in aqueous solution, *Langmuir* **20**, 6414 (2004).
50. N. Chanda, V. Kattumuri, R. Shukla, A. Zambre, K. Katti, A. Upendran, R. R. Kulkarni, P. Kan, G. M. Fent, S. W. Casteel, C. J. Smith, E. Boote, J. D. Robertson, C. Cutler, J. R. Lever, K. V. Katti and R. Kannan, Bombesin functionalized gold nanoparticles show in vitro and in vivo cancer receptor specificity, *Proc. Natl. Acad. Sci. U.S.A.* **107**, 8760 (2010).
51. S. J. Oldenburg, R. D. Averitt, S. L. Westcott and N. J. Halas, Nanoengineering of optical resonances, *Chem. Phys. Lett.* **288**, 243 (1998).

52. L. R. Hirsch, R. J. Stafford, J. A. Bankson, S. R. Sershen, B. Rivera, R. E. Price, J. D. Hazle, N. J. Halas and J. L. West, Nanoshell-mediated near-infrared thermal therapy of tumors under magnetic resonance guidance, *Proc. Natl. Acad. Sci. U.S.A.* **100**, 13549 (2003).
53. J. Chen, J. M. McLellan, A. Siekkinen, Y. Xiong, Z.-Y. Li and Y. Xia, Facile Synthesis of Gold/Silver Nanocages with Controllable Pores on the Surface, *J. Am. Chem. Soc.* **128**, 14776 (2006).
54. W. Li, X. Cai, C. Kim, G. Sun, Y. Zhang, R. Deng, M. Yang, J. Chen, S. Achilefu, L. V. Wang and Y. Xia, Gold nanocages covered with thermally-responsive polymers for controlled release by high-intensity focused ultrasound, *Nanoscale* **3**, 1724 (2011).
55. J. Chen, C. Glaus, R. Laforest, Q. Zhang, M. Yang, M. Gidding, M. J. Welch and Y. Xia, Gold nanocages as photothermal transducers for cancer treatment, *Small* **6**, 811 (2010).
56. G. F. Paciotti, L. D. Myer, T. H. Kim, S. Wang, H. R. Alexander, D. Weinreich and L. Tamarkin, Colloidal gold: A novel colloidal nanoparticle vector for tumor-directed drug delivery, *Clin. Cancer Res.* **7**, 3673S (2001).
57. G. F. Paciotti, D. G. I. Kingston and L. Tamarkin, Colloidal gold nanoparticles: A novel nanoparticle platform for developing multifunctional tumor-targeted drug delivery vectors, *Drug Dev. Res.* **67**, 47 (2006).
58. R. Duncan and R. Gaspar, Nanomedicine(s) under the Microscope, *Mol Pharm.* **8**, 2101 (2011).
59. C. Medley, B. Muralidhara, S. Chico, S. Durban, P. Mehelic and C. Demarest, Quantitation of plasmid DNA deposited on gold particles for particle-mediated epidermal delivery using ICP-MS, *Anal. Bioanal. Chem.* **398**, 527 (2010).
60. I. H. El-Sayed, X. Huang, M. Fima, J. Oren Humstoe, R. Kramer and M. El-Sayed, *Effect of Plasmonic Gold Nanoparticles on Benign and Malignant Cellular Autofluorescence: A Novel Probe for Fluorescence Based Detection of Cancer.* (Adenine Press, Schenectady, NY, ETATS-UNIS, 2007), Vol. 6, pp. 10.
61. I. El-Sayed, Nanotechnology in head and neck cancer: The race is on, *Curr. Oncol. Rep.* **12**, 121 (2010).
62. N. Khlebtsov and L. Dykman, Biodistribution and toxicity of engineered gold nanoparticles: A review of in vitro and in vivo studies, *Chem. Soc. Rev.* **40**, 1647 (2011).
63. M. Longmire, P. L. Choyke and H. Kobayashi, Clearance properties of nano-sized particles and molecules as imaging agents: Considerations and caveats, *Nanomedicine* **3**, 703 (2008).
64. S. M. Moghimi, A. C. Hunter, T. L. Andresen, Factors controlling nanoparticle pharmacokinetics: An integrated analysis and perspective, *Ann. Rev. Pharmacol. Toxicol.* **52**, 481 (2012).
65. A. Yarnell, in *Chem. Eng. News.* (2012), Vol. 90, pp. 30–31.
66. M. E. Davis, J. E. Zuckerman, C. H. J. Choi, D. Seligson, A. Tolcher, C. A. Alabi, Y. Yen, J. D. Heidel and A. Ribas, Evidence of RNAi in humans from systemically administered siRNA via targeted nanoparticles, *Nature* **464**, 1067 (2010).
67. S. Davis, A. Abuchowski, Y. K. Park and F. F. Davis, Alteration of the circulating life and antigenic properties of bovine adenosine deaminase in mice by attachment of polyethylene glycol, *Clin. Exp. Immunol.* **46**, 649 (1981).

68. J. M. Harris and R. B. Chess, Effect of pegylation on pharmaceuticals, *Nat. Rev. Drug Discov.* **2**, 214 (2003).
69. E. K. Rowinsky, in *Cancer Chemotherapy and Biotherapy: Principles and Practice*, B. A. Chabner, D. L. Longo (Eds.), Philadelphia, PA: Lippincott Williams & Wilkins (2010).
70. K. N. Sugahara, T. Teesalu, P. P. Karmali, V. R. Kotamraju, L. Agemy, D. R. Greenwald and E. Ruoslahti, Coadministration of a tumor-penetrating peptide enhances the efficacy of cancer drugs, *Science* **328**, 1031 (2010).
71. M. R. Green, G. M. Manikhas, S. Orlov, B. Afanasyev, A. M. Makhson, P. Bhar and M. J. Hawkins, Abraxane®, a novel Cremophor®-free, albumin-bound particle form of paclitaxel for the treatment of advanced non-small-cell lung cancer, *Ann. Oncol.* **17**, 1263 (2006).
72. K. Knop, R. Hoogenboom, D. Fischer and U. S. Schubert, Poly(ethylene glycol) in Drug delivery: Pros and cons as well as potential alternatives, *Angew. Chem. Int. Ed.* **49**, 6288 (2010).
73. R. S. Obach, F. Lombardo and N. J. Waters, Trend analysis of a database of intravenous pharmacokinetic parameters in humans for 670 drug compounds, *Drug Metab. Dispos.* **36**, 1385 (2008).
74. J. C. Roberts, M. K. Bhalgat and R. T. Zera, Preliminary biological evaluation of polyamidoamine (PAMAM) StarburstTM dendrimers, *J. Biomed. Mater. Res.* **30**, 53 (1996).
75. J. E. Zuckerman, C. H. J. Choi, H. Han and M. E. Davis, Polycation-siRNA nanoparticles can disassemble at the kidney glomerular basement membrane, *Proc. Natl. Acad. Sci. U.S.A.* **109**, 3137 (2012).
76. S. Moghimi, C. Porter, I. Muir, L. Illum and S. Davis, Non-phagocytic uptake of intravenously injected microspheres in rat spleen: Influence of particle size and hydrophilic coating, *Biochem. Biophys. Res. Commun.* **177**, 861 (1991).
77. X. Dong and R. J. Mumper, Nanomedicinal strategies to treat multidrug-resistant tumors: Current progress, *Nanomedicine* **5**, 597 (2010).
78. C. H. J. Choi, C. A. Alabi, P. Webster and M. E. Davis, Mechanism of active targeting in solid tumors with transferrin-containing gold nanoparticles, *Proc. Nat. Acad. Sci. U.S.A.* **107**, 1235 (2010).
79. E. Wisse, F. Jacobs, B. Topal, P. Frederik and B. De Geest, The size of endothelial fenestrae in human liver sinusoids: Implications for hepatocyte-directed gene transfer, *Gene Ther.* **15**, 1193 (2008).
80. E. A. Nance, G. F. Woodworth, K. A. Sailor, T.-Y. Shih, Q. Xu, G. Swaminathan, D. Xiang, C. Eberhart and J. Hanes, A dense poly(ethylene glycol) coating improves penetration of large polymeric nanoparticles within brain tissue, *Sci. Transl. Med.* **4**, 149ra119 (2012).
81. S. K. Lai, Y.-Y. Wang, K. Hida, R. Cone and J. Hanes, Nanoparticles reveal that human cervicovaginal mucus is riddled with pores larger than viruses, *Proc. Natl. Acad. Sci. U.S.A.* **107**, 598 (2010).
82. L. M. Ensign, B. C. Tang, Y.-Y. Wang, T. A. Tse, T. Hoen, R. Cone and J. Hanes, Mucus-penetrating nanoparticles for vaginal drug delivery protect against Herpes Simplex virus, *Sci. Transl. Med.* **4**, 138ra79 (2012).

83. R. Sozanne, R. Solmaz, G. Chauhan and I. Blobel, Mel ák: Molecular architecture of the transport channel of the nuclear pore complex, *Cell* **147**, 590 (2011).
84. S. D. Perrault, C. Walkey, T. Jennings, H. C. Fischer and W. C. W. Chan, Mediating tumor targeting efficiency of nanoparticles through design, *Nano Lett.***9**, 1909 (2009).
85. G. R. Hendrickson and L. A. Lyon, Microgel translocation through pores under confinement, *Angew. Chem. Int. Ed.* **49**, 2193 (2010).
86. D. A. Holden, G. R. Hendrickson, W.-J. Lan, L. A. Lyon and H. S. White, Electrical signature of the deformation and dehydration of microgels during translocation through nanopores, *Soft Matter* **7**, 8035 (2011).
87. T. J. Merkel, S. W. Jones, K. P. Herlihy, F. R. Kersey, A. R. Shields, M. Napier, J. C. Luft, H. Wu, W. C. Zamboni, A. Z. Wang, J. E. Bear and J. M. DeSimone, Using mechanobiological mimicry of red blood cells to extend circulation times of hydrogel microparticles, *Proc. Natl. Acad. Sci. U.S.A.* (2011).
88. Y. Geng, P. Dalhaimer, S. Cai, R. Tsai, M. Tewari, T. Minko and D. E. Discher, Shape effects of filaments versus spherical particles in flow and drug delivery, *Nat. Nanotechnol.* **2**, 249 (2007).
89. G. von Maltzahn, J.-H. Park, A. Agrawal, N. K. Bandaru, S. K. Das, M. J. Sailor and S. N. Bhatia, Computationally guided photothermal tumor therapy using long-circulating gold nanorod antennas, *Cancer Res.* **69**, 3892 (2009).
90. W.-S. Cho, M. Cho, J. Jeong, M. Choi, H.-Y. Cho, B. S. Han, S. H. Kim, H. O. Kim, Y. T. Lim, B. H. Chung and J. Jeong, Acute toxicity and pharmacokinetics of 13 nm-sized PEG-coated gold nanoparticles, *Toxicol. Appl. Pharmacol.* **236**, 16 (2009).
91. M. R. K. Ali, B. Snyder and M. A. El-Sayed, Synthesis and optical properties of small Au nanorods using a seedless growth technique, *Langmuir* **28**, 9807 (2012).
92. H. Lum and A. B. Malik, Regulation of vascular endothelial barrier function, *Am. J. Physiol.* **267**, L223 (1994).
93. H. Sarin, Physiologic upper limits of pore size of different blood capillary types and another perspective on the dual pore theory of microvascular permeability, *J. Angigenesis Res.* **2**, 1 (2010).
94. S. M. Moghimi, A. C. Hunter and J. C. Murray, Long-circulating and target-specific nanoparticles: Theory to practice, *Pharmacol. Rev.* **53**, 283 (2001).
95. H. S. Choi, Y. Ashitate, J. H. Lee, S. H. Kim, A. Matsui, N. Insin, M. G. Bawendi, M. Semmler-Behnke, J. V. Frangioni and A. Tsuda, Rapid translocation of nanoparticles from the lung airspaces to the body, *Nat. Biotechnol.* **28**, 1300 (2010).
96. T. M. Allen, C. B. Hansen and L. S. S. Guo, Subcutaneous administration of liposomes — a comparison with the intravenous and intraperitoneal routes of injection, *Biochim. Biophys. Acta* **1150**, 9 (1993).
97. S. T. Reddy, A. J. van der Vlies, E. Simeoni, V. Angeli, G. J. Randolph, C. P. O'Neill, L. K. Lee, M. A. Swartz and J. A. Hubbell, Exploiting lymphatic transport and complement activation in nanoparticle vaccines, *Nat. Biotechnol.* **25**, 1159 (2007).
98. B. Button, L.-H. Cai, C. Ehre, M. Kesimer, D. B. Hill, J. K. Sheehan, R. C. Boucher and M. Rubinstein, A periciliary brush promotes the lung health by separating the mucus layer from airway epithelia, *Science* **337**, 937 (2012).
99. L.-T. Chen and L. Weiss, The role of the sinus wall in the passage of erythrocytes through the spleen, *Blood* **41**, 529 (1973).

100. S. M. Moghimi, C. J. H. Porter, I. S. Muir, L. Illum and S. S. Davis, Non-phagocytic uptake of intravenously injected microspheres in rat spleen: Influence of particle size and hydrophilic coating, *Biochem. Biophys. Res. Commun.* **177**, 861 (1991).
101. F. Yuan, M. Dellian, D. Fukumura, M. Leunig, D. A. Berk, V. P. Torchilin and R. K. Jain, Vascular permeability in a human tumor xenograft: Molecular size dependence and cutoff size, *Cancer Res.* **55**, 3752 (1995).
102. S. Azarmi, W. H. Roa and R. Loebenberg, Targeted delivery of nanoparticles for the treatment of lung diseases, *Adv. Drug Del. Rev.* **60**, 863 (2008).
103. G. W. Doorley and C. K. Payne, Cellular binding of nanoparticles in the presence of serum proteins, *Chem. Comm.* **47**, 466 (2011).
104. C. C. Fleischer and C. K. Payne, Nanoparticle surface charge mediates the cellular receptors used by protein–nanoparticle complexes, *J. Phys. Chem. B* **116**, 8901 (2012).
105. G. W. Doorley and C. K. Payne, Nanoparticles act as protein carriers during cellular internalization, *Chem. Comm.* **48**, 2961 (2012).
106. C. D. Walkey and W. C. W. Chan, Understanding and controlling the interaction of nanomaterials with proteins in a physiological environment, *Chem. Soc. Rev.* **41**, 2780 (2012).
107. M. Lundqvist, J. Stigler, G. Elia, I. Lynch, T. Cedervall and K. Dawson, Nanoparticle size and surface properties determine the protein corona with possible implications for biological impacts, *Proc. Natl. Acad. Sci. U.S.A.* **105**, 14265 (2008).
108. I. Hamad, O. Al-Hanbali, A. Hunter, K. Rutt and T. Andresen, Distinct polymer architecture mediates switching of complement activation pathways at the nanosphere-serum interface: implications for stealth nanoparticle engineering, *ACS Nano* **4**, 6629 (2010).
109. P. Karmali and D. Simberg, Interactions of nanoparticles with plasma proteins: Implications on clearance and toxicity of drug delivery systems, *Expert Opin. Drug Deliv.* **8**, 343 (2011).
110. M. Monopoli, D. Walczyk, A. Campbell, G. Elia and I. Lynch, Physical-chemical aspects of protein corona: Relevance to in vitro and in vivo biological impacts of nanoparticles, *J. Am. Chem. Soc.* **133**, 2525 (2011).
111. S. Moghimi and J. Szebeni, Stealth liposomes and long circulating nanoparticles: Critical issues in pharmacokinetics, opsonization and protein-binding properties, *Prog. Lipid Res.* **42**, 463 (2003).
112. J. V. Jokerst, T. Lobovkina, R. N. Zare and S. S. Gambhir, Nanoparticle PEGylation for imaging and therapy, *Nanomedicine* **6**, 715 (2011).
113. R. R. Arvizo, O. R. Miranda, D. F. Moyano, C. A. Walden, K. Giri, R. Bhattacharya, J. D. Robertson, V. M. Rotello, J. M. Reid and P. Mukherjee, Modulating pharmacokinetics, tumor uptake and biodistribution by engineered nanoparticles, *PLoS ONE* **6**, e24374 (2011).
114. D. Simberg, J.-H. Park, P. P. Karmali, W.-M. Zhang, S. Merkulov, K. McCrae, S. N. Bhatia, M. Sailor and E. Ruoslahti, Differential proteomics analysis of the surface heterogeneity of dextran iron oxide nanoparticles and the implications for their in vivo clearance, *Biomater.* **30**, 3926 (2009).
115. M. J. Hostetler, A. C. Templeton and R. W. Murray, Dynamics of place-exchange reactions on monolayer-protected gold cluster molecules, *Langmuir* **15**, 3782 (1999).

116. X. Xia, M. Yang, Y. Wang, Y. Zheng, Q. Li, J. Chen and Y. Xia, Quantifying the coverage density of poly(ethylene glycol) chains on the surface of gold nanostructures, *ACS Nano* **6**, 512 (2011).
117. E. C. Cho, Q. Zhang and Y. Xia, The effect of sedimentation and diffusion on cellular uptake of gold nanoparticles, *Nat. Nanotechnol.* **6**, 385 (2011).
118. R. G. Nuzzo, B. R. Zegarski and L. H. Dubois, Fundamental studies of the chemisorption of organosulfur compounds on gold(111). Implications for molecular self-assembly on gold surfaces, *J. Am. Chem. Soc.* **109**, 733 (1987).
119. H.-F. Zhang, M. Stender, R. Zhang, C. Wang, J. Li and L.-S. Wang, Toward the solution synthesis of the tetrahedral Au20 cluster, *J. Phys. Chem. B* **108**, 12259 (2004).
120. F. Tarazona-Vasquez and P. B. Balbuena, Complexation of the lowest generation poly(amidoamine)-NH2 dendrimers with metal ions, metal atoms, and Cu(II) hydrates: An ab initio study, *J. Phys. Chem. B* **108**, 15992 (2004).
121. A. J. Bard and L. R. Faulkner, *Electrochemical Methods: Fundamentals and Applications*, 2nd edn. New York: John Wiley & Sons, Inc. (2001).
122. G. Greczynski and W. R. Salaneck, Photoelectron spectroscopy of hybrid interfaces for light emitting diodes: Influence of the substrate work function, *Appl. Phys. Lett.* **79**, 3185 (2001).
123. H. Lee, A. H. de Vries, S.-J. Marrink and R. W. Pastor, A coarse-grained model for polyethylene oxide and polyethylene glycol: Conformation and hydrodynamics, *J. Phys. Chem. B* **113**, 13186 (2009).
124. J. Perez-Juste, I. Pastoriza-Santos, L. M. Liz-Marzan and P. Mulvaney, Gold nanorods: Synthesis, characterization and applications, *Coord. Chem. Rev.* **249**, 1870 (2005).
125. A. M. Jackson, J. W. Myerson and F. Stellacci, Spontaneous assembly of subnanometre-ordered domains in the ligand shell of monolayer-protected nanoparticles, *Nat. Mater.* **3**, 330 (2004).
126. C. Singh, P. K. Ghorai, M. A. Horsch, A. M. Jackson, R. G. Larson, F. Stellacci and S. C. Glotzer, Entropy-mediated patterning of surfactant-coated nanoparticles and surfaces, *Phys. Rev. Lett.* **99**, 226106 (2007).
127. L. Xiang, Y. Miao, K. Hyewon, M. Marta and S. Francesco, Determination of monolayer-protected gold nanoparticle ligand–shell morphology using nmR, *Nat. Comm.* **3**, 1182 (2012).
128. R. Singh and G. M. Whitesides, in *Sulphur-Containing Functional Groups (1993).* (John Wiley & Sons, Inc., 2010), pp. 633–658.
129. C. J. Ackerson, P. D. Jadzinsky, G. J. Jensen and R. D. Kornberg, Rigid, specific, and discrete gold nanoparticle/antibody conjugates, *J. Am. Chem. Soc.* **128**, 2635 (2006).
130. C. J. Orendorff, C. J. Murphy, Quantitation of metal content in the silver-assisted growth of gold nanorods, *J. Phys. Chem. B* **110**, 3990 (2006).
131. M. P. Melancon, W. Lu, Z. Yang, R. Zhang, Z. Cheng, A. M. Elliot, J. Stafford, T. Olson, J. Z. Zhang and C. Li, In vitro and in vivo targeting of hollow gold nanoshells directed at epidermal growth factor receptor for photothermal ablation therapy, *Mol. Cancer Ther.* 7, 1730 (2008).
132. E. C. Cho, C. Kim, F. Zhou, C. M. Cobley, K. H. Song, J. Chen, Z.-Y. Li, L. V. Wang and Y. Xia, Measuring the optical absorption cross sections of Au–Ag nanocages and Au nanorods by photoacoustic imaging, *J. Phys. Chem. C* **113**, 9023 (2009).

133. P. Puvanakrishnan, J. Park, D. Chatterjee, S. Krishnan and J. W. Tunnell, In vivo tumor targeting of gold nanoparticles: Effect of particle type and dosing strategy, *Int. J. Nanomed.* 7, 1251 (2012).
134. J.-S. Lee, S. I. Stoeva and C. A. Mirkin, DNA-induced size-selective separation of mixtures of gold nanoparticles, *J. Am. Chem. Soc.* **128**, 8899 (2006).
135. N. R. Jana, L. Gearheart and C. J. Murphy, Seed-mediated growth approach for shape-controlled synthesis of spheroidal and rod-like gold nanoparticles using a surfactant template, *Adv. Mater.* **13**, 1389 (2001).
136. B. Nikoobakht and M. A. El-Sayed, Preparation and growth mechanism of gold nanorods (NRs) using seed-mediated growth method, *Chem. Mater.* **15**, 1957 (2003).
137. J. Turkevich, P. C. Stevenson and J. Hillier, A Study of the nucleation and growth processes in the synthesis of colloidal gold, *Discuss. Faraday Soc.* 55 (1951).
138. G. Frens, Controlled nucleation for the regulation of the particle size in monodisperse gold suspensions, *Nature* **241**, 20 (1973).
139. D. A. Giljohann, D. S. Seferos, A. E. Prigodich, P. C. Patel and C. A. Mirkin, Gene regulation with polyvalent siRNA-nanoparticle conjugates, *J. Am. Chem. Soc.* **131**, 2072 (2009).
140. S. J. Hurst, A. K. R. Lytton-Jean and C. A. Mirkin, Maximizing DNA loading on a range of gold nanoparticle sizes, *Anal. Chem.* **78**, 8313 (2006).
141. S. I. Stoeva, J.-S. Lee, C. S. Thaxton and C. A. Mirkin, Multiplexed DNA Detection with biobarcoded nanoparticle probes, *Angew. Chem. Int. Ed.* **45**, 3303 (2006).
142. G. T. Hermanson, *Bioconjugate Techniques* (Elsevier, Amsterdam, 2008).
143. G. T. Spierenburg, F. T. J. J. Oerlemans, J. P. R. M. van Laarhoven and C. H. M. M. de Bruyn, Phototoxicity of N-2-Hydroxyethylpiperazine-N'-2-ethanesulfonic acid-buffered culture media for human leukemic cell lines, *Cancer Res.* **44**, 2253 (1984).
144. J. L. Lepe-Zuniga, J. S. Zigler Jr and I. Gery, Toxicity of light-exposed Hepes media, *J. Immunol. Meth.* **103**, 145 (1987).
145. S. K. Libutti, G. F. Paciotti, L. Myer, R. Haynes, W. Gannon, M. Walker, G. Seidel, A. Byrnes, N. Yuldasheva and L. Tamarkin, Results of a completed phase I clinical trial of CYT-6091: A pegylated colloidal gold-TNF nanomedicine, *J. Clin. Oncol.* **27**, 3586 (2009).
146. G. Paccioti and L. Tamarkin. (WO Patent 2,005,072,893, 2005).
147. Available at: http://nanocomposix.com/products/gold/dried. Accessed 29 December 2012.
148. T. A. Taton, *Current Protocols in Nucleic Acid Chemistry.* (John Wiley & Sons, Inc., 2001).
149. E. C. Dreaden, S. C. Mwakwari, L. A. Austin, M. J. Kieffer, A. K. Oyelere and M. A. El-Sayed, Small molecule–gold nanorod conjugates selectively target and induce macrophage cytotoxicity towards breast cancer cells, *Small* **8**, 2819 (2012).
150. A. G. Tkachenko, H. Xie, D. Coleman, W. Glomm, J. Ryan, M. F. Anderson, S. Franzen and D. L. Feldheim, Multifunctional gold nanoparticle-peptide complexes for nuclear targeting, *J. Am. Chem. Soc.* **125**, 4700 (2003).
151. A. K. Oyelere, P. C. Chen, X. H. Huang, I. H. El-Sayed and M. A. El-Sayed, Peptide-conjugated gold nanorods for nuclear targeting, *Bioconjugate Chem.* **18**, 1490 (2007).

152. W. Lu, C. Xiong, G. Zhang, Q. Huang, R. Zhang, J. Z. Zhang and C. Li, Targeted photothermal ablation of murine melanomas with melanocyte-stimulating hormone analog–conjugated hollow gold nanospheres, *Clin. Cancer Res.* **15**, 876 (2009).
153. C. Kim, E. C. Cho, J. Chen, K. H. Song, L. Au, C. Favazza, Q. Zhang, C. M. Cobley, F. Gao, Y. Xia and L. V. Wang, In vivo molecular photoacoustic tomography of melanomas targeted by bioconjugated gold nanocages, *ACS Nano* **4**, 4559 (2010).
154. B. Kang, M. A. Mackey and M. A. El-Sayed, Nuclear targeting of gold nanoparticles in cancer cells induces dna damage, causing cytokinesis arrest and apoptosis, *J. Am. Chem. Soc.* **132**, 1517 (2010).
155. X. Huang, X. Peng, Y. Wang, D. M. Shin, M. A. El-Sayed and S. Nie, A reexamination of active and passive tumor targeting by using rod-shaped gold nanocrystals and covalently conjugated peptide ligands, *ACS Nano* **4**, 5887 (2010).
156. H. Xu, X. Mao, Q. X. Zeng, S. F. Wang, A. N. Kawde and G. D. Liu, Aptamer-functionalized gold nanoparticles as probes in a dry-reagent strip biosensor for protein analysis, *Anal. Chem.* **81**, 669 (2009).
157. V. Pavlov, Y. Xiao, B. Shlyahovsky and I. Willner, Aptamer-functionalized Au nanoparticles for the amplified optical detection of thrombin, *J. Am. Chem. Soc.* **126**, 11768 (2004).
158. J. Liu and Y. Lu, Preparation of aptamer-linked gold nanoparticle purple aggregates for colorimetric sensing of analytes, *Nat. Protoc.* **1**, 246 (2006).
159. S. Dhar, F. X. Gu, R. Langer, O. C. Farokhzad and S. J. Lippard, Targeted delivery of cisplatin to prostate cancer cells by aptamer functionalized Pt(IV) prodrug-PLGA–PEG nanoparticles, *Proc. Natl. Acad. Sci. U.S.A.* **105**, 17356 (2008).
160. S. Dhar, N. Kolishetti, S. J. Lippard and O. C. Farokhzad, Targeted delivery of a cisplatin prodrug for safer and more effective prostate cancer therapy in vivo, *Proc. Natl. Acad. Sci. U.S.A.* (2011).
161. I. H. El-Sayed, X. Huang and M. A. El-Sayed, Surface plasmon resonance scattering and absorption of anti-EGFR antibody conjugated gold nanoparticles in cancer diagnostics: Applications in oral cancer, *Nano Lett.* **5**, 829 (2005).
162. I. H. El-Sayed, X. Huang and M. A. El-Sayed, Selective laser photo-thermal therapy of epithelial carcinoma using anti-EGFR antibody conjugated gold nanoparticles, *Cancer Lett.* **239**, 129 (2006).
163. I. H. El-Sayed, X. Huang and M. A. El-Sayed, Surface plasmon resonance scattering and absorption of anti-EGFR antibody conjugated gold nanoparticles in cancer diagnostics: Applications in oral cancer, *Nano Lett.* **5**, 829 (2005).
164. I. H. El-Sayed, X. Huang and M. A. El-Sayed, Surface plasmon resonance scattering and absorption of anti-EGFR antibody conjugated gold nanoparticles in cancer diagnostics: Applications in oral cancer, *Nano Lett.* **5**, 829 (2005).
165. I. H. El-Sayed, X. Huang and M. A. El-Sayed, Selective laser photo-thermal therapy of epithelial carcinoma using anti-EGFR antibody conjugated gold nanoparticles, *Cancer Lett.* **239**, 129 (2006).
166. I. H. El-Sayed, X. H. Huang and M. A. El-Sayed, Surface plasmon resonance scattering and absorption of anti-EGFR antibody conjugated gold nanoparticles in cancer diagnostics: Applications in oral cancer, *Nano Lett.* **5**, 829 (2005).
167. X. Huang, El-Sayed IH, Qian W and El-Sayed MA, Cancer cells assemble and align gold nanorods conjugated to antibodies to produce highly enhanced, sharp and

polarized surface Raman spectra: A potential cancer diagnostic marker, *Nano Lett.* 7, 1591 (2007).

168. B. Van de Broek, N. Devoogdt, A. D'Hollander, H.-L. Gijs, K. Jans, L. Lagae, S. Muyldermans, G. Maes and G. Borghs, Specific cell targeting with nanobody conjugated branched gold nanoparticles for photothermal therapy, *ACS Nano* **5**, 4319 (2011).
169. Z. Poon, D. Chang, X. Zhao and P. T. Hammond, Layer-by-Layer Nanoparticles with a pH-sheddable layer for in vivo targeting of tumor hypoxia, *ACS Nano* **5**, 4284 (2011).
170. T. J. Harris, G. von Maltzahn, A. M. Derfus, E. Ruoslahti and S. N. Bhatia, Proteolytic actuation of nanoparticle self-assembly, *Angew. Chem. Int. Ed.* **45**, 3161 (2006).
171. S. K. Hobbs, W. L. Monsky, F. Yuan, W. G. Roberts, L. Griffith, V. P. Torchilin and R. K. Jain, Regulation of transport pathways in tumor vessels: Role of tumor type and microenvironment, *Proc. Natl. Acad. Sci. U.S.A.* **95**, 4607 (1998).
172. Y. Matsumura, H. Maeda, A new concept for macromolecular therapeutics in cancer chemotherapy: Mechanism of tumoritropic accumulation of proteins and the antitumor agent Smancs, *Cancer Res.* **46**, 6387 (1986).
173. R. K. Jain, Transport of molecules in the tumor interstitium: A review, *Cancer Res.* **47**, 3039 (1987).
174. W. L. Monsky, D. Fukumura, T. Gohongi, M. Ancukiewcz, H. A. Weich, V. P. Torchilin, F. Yuan and R. K. Jain, Augmentation of transvascular transport of macromolecules and nanoparticles in tumors using vascular endothelial growth factor, *Cancer Res.* **59**, 4129 (1999).
175. Z. Popović, W. Liu, V. P. Chauhan, J. Lee, C. Wong, A. B. Greytak, N. Insin, D. G. Nocera, D. Fukumura, R. K. Jain and M. G. Bawendi, A nanoparticle size series for in-vivo fluorescence imaging, *Angew. Chem. Int. Ed.* **49**, 8649 (2010).
176. V. P. Chauhan, Z. Popović, O. Chen, J. Cui, D. Fukumura, M. G. Bawendi and R. K. Jain, Fluorescent nanorods and nanospheres for real-time in vivo probing of nanoparticle shape-dependent tumor penetration, *Angew. Chem. Int. Ed.* **50**, 11417 (2011).
177. C. Wong, T. Stylianopoulos, J. Cui, J. Martin, V. P. Chauhan, W. Jiang, Z. Popović, R. K. Jain, M. G. Bawendi and D. Fukumura, Multistage nanoparticle delivery system for deep penetration into tumor tissue, *Proc. Natl. Acad. Sci. U.S.A.* **108**, 2426 (2011).
178. V. P. Chauhan, T. Stylianopoulos, J. D. Martin, Z. Popovic, O. Chen, W. S. Kamoun, M. G. Bawendi, D. Fukumura and R. K. Jain, Normalization of tumour blood vessels improves the delivery of nanomedicines in a size-dependent manner, *Nat. Nanotechnol.* **7**, 383 (2012).
179. T. Stylianopoulos, C. Wong, M. G. Bawendi, R. K. Jain and D. Fukumura, in *Methods in Enzymology,* D. Nejat (Ed.), (Academic Press, 2012), Vol. 508, pp. 109–130.
180. A. K. Iyer, G. Khaled, J. Fang and H. Maeda, Exploiting the enhanced permeability and retention effect for tumor targeting, *Drug Discov. Today* **11**, 812 (2006).
181. H. Maeda, Tumor-selective delivery of macromolecular drugs via the EPR Effect: Background and future prospects, *Bioconjugate Chem.* **21**, 797 (2010).
182. P. Kim, E. Chung, H. Yamashita, K. E. Hung, A. Mizoguchi, R. Kucherlapati, D. Fukumura, R. K. Jain and S. H. Yun, In vivo wide-area cellular imaging by side-view endomicroscopy, *Nat. Meth.* **7**, 303 (2010).

183. P. P. Adiseshaiah, J. B. Hall and S. E. McNeil, Nanomaterial standards for efficacy and toxicity assessment, *Wiley Interdiscip. Rev. Nanomed. Nanobiotechnol.* **2**, 99 (2010).
184. K. P. Olive, M. A. Jacobetz, C. J. Davidson, A. Gopinathan, D. McIntyre, D. Honess, B. Madhu, M. A. Goldgraben, M. E. Caldwell, D. Allard, K. K. Frese, G. DeNicola, C. Feig, C. Combs, S. P. Winter, H. Ireland-Zecchini, S. Reichelt, W. J. Howat, A. Chang, M. Dhara, L. Wang, F. Rückert, R. Grützmann, C. Pilarsky, K. Izeradjene, S. R. Hingorani, P. Huang, S. E. Davies, W. Plunkett, M. Egorin, R. H. Hruban, N. Whitebread, K. McGovern, J. Adams, C. Iacobuzio-Donahue, J. Griffiths and D. A. Tuveson, Inhibition of hedgehog signaling enhances delivery of chemotherapy in a mouse model of pancreatic cancer, *Science* **324**, 1457 (2009).
185. C. M. Rudin, G. J. Weiss, A. Chang, S. Gettinger, W. H. Miller, B. Eigl, A. Savage, G. Loewen, R. W. Ross and A. Jimeno, A Phase 1 Study of IPI-926, An inhibitor of the hedgehog pathway, in patients (PTS) with advanced or metastatic solid tumors, *Ann. Oncol.* **21**, 164 (2010).
186. P. M. LoRusso, C. M. Rudin, J. C. Reddy, R. Tibes, G. J. Weiss, M. J. Borad, C. L. Hann, J. R. Brahmer, I. Chang, W. C. Darbonne, R. A. Graham, K. L. Zerivitz, J. A. Low and D. D. Von Hoff, Phase I trial of hedgehog pathway inhibitor vismodegib (GDC-0449) in Patients with refractory, locally advanced or metastatic solid tumors, *Clin. Cancer Res.* **17**, 2502 (2011).
187. W. S. Enochs, G. Harsh, F. Hochberg and R. Weissleder, Improved delineation of human brain tumors on MR images using a long-circulating, superparamagnetic iron oxide agent, *J. Mag. Res. Imaging* **9**, 228 (1999).
188. O. Veiseh, C. Sun, C. Fang, N. Bhattarai, J. Gunn, F. Kievit, K. Du, B. Pullar, D. Lee, R. G. Ellenbogen, J. Olson and M. Zhang, Specific targeting of brain tumors with an optical/magnetic resonance imaging nanoprobe across the blood-brain barrier, *Cancer Res.* **69**, 6200 (2009).
189. P. Calvo, B. Gouritin, H. Chacun, D. Desmaele, J. D'Angelo, J. P. Noel, D. Georgin, E. Fattal, J. P. Andreux and P. Couvreur, Long-circulating PEGylated polycyanoacrylate nanoparticles as new drug carrier for brain delivery, *Pharm. Res.* **18**, 1157 (2001).
190. S. B, M. U, W. A, H. E, M. M, Mannan-binding lectin activates C3 and the alternative complement pathway without involvement of C2, *J. Clin. Investig.* **116**, 1425 (2006).
191. M. G. Harisinghani, M. Saksena, R. W. Ross, S. Tabatabaei, D. Dahl, S. McDougal and R. Weissleder, A pilot study of lymphotrophic nanoparticle-enhanced magnetic resonance imaging technique in early stage testicular cancer: A new method for noninvasive lymph node evaluation, *Urology* **66**, 1066 (2005).
192. M. Nahrendorf, E. Keliher, B. Marinelli, F. Leuschner, C. S. Robbins, R. E. Gerszten, M. J. Pittet, F. K. Swirski and R. Weissleder, Detection of macrophages in aortic aneurysms by nanoparticle positron emission tomography-computed tomography, *Arteriosclerosis Thrombosis Vascular Biol.* **31**, 750 (2011).
193. C. Tassa, S. Y. Shaw, R. Weissleder, Dextran-coated iron oxide nanoparticles: A versatile platform for targeted molecular imaging, molecular diagnostics, and therapy, *Acc. Chem. Res.* **44**, 842 (2011).
194. M. C. Finkelstein, S. H. Kuhn, H. Schieren, G. Weissmann and S. Hoffstein, Liposome uptake by human-leukocytes — enhancement of entry mediated by human-serum and aggregated immunoglobulins, *Biochimica Et Biophysica Acta* **673**, 286 (1981).

195. R. Gref, Y. Minamitake, M. T. Peracchia, V. Trubetskoy, V. Torchilin and R. Langer, Biodegradable long-circulating polymeric nanospheres, *Science* **263**, 1600 (1994).
196. A. A. Gabizon, Stealth liposomes and tumor targeting: One step further in the quest for the magic bullet, *Clin. Cancer Res.* 7, 223 (2001).
197. E. C. Dreaden, Ph.D Thesis, Georgia Insitute of Technology (2012).
198. B. R. Smith, P. Kempen, D. Bouley, A. Xu, Z. Liu, N. Melosh, H. Dai, R. Sinclair and S. S. Gambhir, Shape matters: Intravital microscopy reveals surprising geometrical dependence for nanoparticles in tumor models of extravasation, *Nano Lett.* **12**, 3369 (2012).
199. M. R. Dreher, W. Liu, C. R. Michelich, M. W. Dewhirst, F. Yuan and A. Chilkoti, Tumor vascular permeability, accumulation, and penetration of macromolecular drug carriers, *J. Natl. Cancer Inst.* **98**, 335 (2006).
200. Z. Popović, W. Liu, V. P. Chauhan, J. Lee, C. Wong, A. B. Greytak, N. Insin, D. G. Nocera, D. Fukumura, R. K. Jain and M. G. Bawendi, A nanoparticle size series for in vivo fluorescence imaging, *Angew. Chem. Int. Ed.* **49**, 8649 (2010).
201. T. Nakai, T. Kanamori, S. Sando and Y. Aoyama, Remarkably size-regulated cell invasion by artificial viruses. Saccharide-dependent self-aggregation of glycoviruses and its consequences in glycoviral gene delivery, *J. Am. Chem. Soc.* **125**, 8465 (2003).
202. F. Osaki, T. Kanamori, S. Sando, T. Sera and Y. Aoyama, A quantum dot conjugated sugar ball and its cellular uptake. On the size effects of endocytosis in the subviral region, *J. Am. Chem. Soc.* **126**, 6520 (2004).
203. H. Gao, W. Shi and L. B. Freund, Mechanics of receptor-mediated endocytosis, *Proc. Natl. Acad. Sci. U.S.A.* **102**, 9469 (2005).
204. B. D. Chithrani, A. A. Ghazani and W. C. W. Chan, Determining the size and shape dependence of gold nanoparticle uptake into mammalian cells, *Nano Lett.* **6**, 662 (2006).
205. W. Jiang, B. Y. S. Kim, J. T. Rutka and W. C. W. Chan, Nanoparticle-mediated cellular response is size-dependent, *Nat. Nanotechnol.* **3**, 145 (2008).
206. S. E. A. Gratton, P. A. Ropp, P. D. Pohlhaus, J. C. Luft, V. J. Madden, M. E. Napier and J. M. DeSimone, The effect of particle design on cellular internalization pathways, *Proc. Natl. Acad. Sci. U.S.A.* **105**, 11613 (2008).
207. http://www.nanopartz.com/bare_spherical_gold_nanoparticles.asp retrieved 26 December, 2012.
208. T.-L. Cheng, K.-H. Chuang, B.-M. Chen, S. R. Roffler, Analytical Measurement of PEGylated Molecules. Bioconjugate Chem. 23, 881 (2012).

Chapter 15

Silver Nanoparticles as Novel Antibacterial and Antiviral Agents

Stefania Galdiero,§, Annarita Falanga*, Marco Cantisani*, Avinash Ingle†, Massimiliano Galdiero‡ and Mahendra Rai‡*

**Department of Pharmacy, Via Mezzocannone 16, Università degli Studi di Napoli Federico II 80134, Naples, Italy*
†Department of Biotechnology, SGB Amravati University, Amravati, 444 602, Maharashtra, India
‡Department of Experimental Medicine, II University of Naples, Via De Crecchio 7, 80138, Naples, Italy
§stefania.galdiero@unina.it

Abstract

Advanced and innovative technologies are continuously dedicated to the understanding of the mechanisms of diseases and to the design of new drugs, but infectious diseases continue to be one of the greatest health burdens worldwide. The main drawbacks for conventional antimicrobial agents are the development of multiple drug resistance and adverse side effects. Advances in nanotechnology have produced novel horizons in nanomedicine, enabling the synthesis of nanoparticles that could be assembled into complex architectures. Among these, silver nanoparticles are attracting great interest owing to their potent antibacterial and antiviral activity. This chapter summarizes emerging efforts to tackle the current challenges in treating infectious diseases, particularly using antimicrobial silver nanoparticles. Silver nanoparticles toxicity is also addressed to disclose the concerns against widespread application in the medical field.

1. Introduction

Nanotechnology is the branch of science correlated to physics, chemistry, biology and medicine and dedicated to materials with dimensions in the order of 100 nm or less. It finds application in multiple aspects of research and in everyday life, such as electronics and new material design. However, nanomedicine is emerging as the fastest growing field in medical research in which the functional mechanisms of nanoparticles and, especially, metal-based nanoparticles, are just beginning to be exploited (Fig. 1). Nanotechnology is used to develop nanoparticle-based targeted drug carriers,[1,2] rapid pathogen detection[3,4] and biomolecular sensing,[5] as well as nanoparticle-based cancer therapies.[6,7]

Nanoparticles are defined as clusters of atoms in the range of 1–100 nm, which possess defined chemical, optical and mechanical properties and are extremely promising for their antimicrobial activity, which is believed to be linked to their large surface area to volume ratio, thus opening up a whole new world of science and discovery. Reducing the particle size of materials has demonstrated to be an efficient and reliable tool for improving their biocompatibility; furthermore, nanomaterials can also be modified for better efficiency, thanks to the possibility of decorating their surface with functional groups, to facilitate their applications in bioscience and medicine.

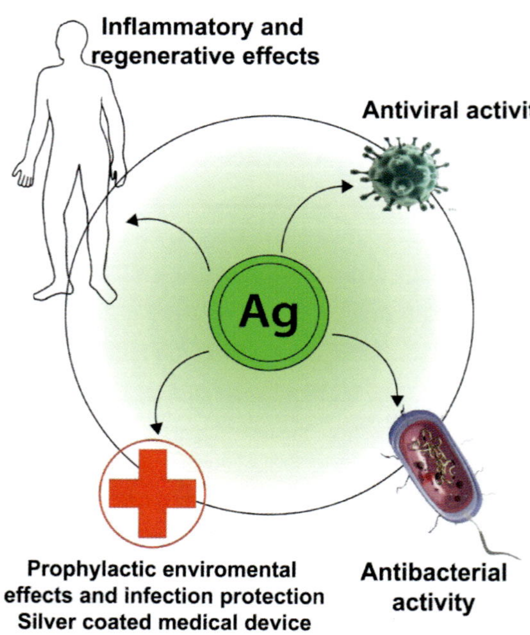

Fig. 1. Silver nanoparticles applications and human health.

Nanoparticles have attracted much attention because of their unique physical (e.g. plasmonic resonance, fluorescent enhancement) and chemical (e.g. catalytic activity enhancement) properties which derive from the high quantity of surface atoms and the high area/volume relation; in fact, as their diameter decreases, the available surface area of the particle itself increases dramatically and, as a consequence, there is an increase over the original properties of their corresponding bulk materials, and this feature make them superior and indispensable in many areas of human activity and excellent candidates for biomedical applications as a variety of biological processes occur at nanometer level. It suffices to consider that cells are typically 10 μm size while proteins are usually 5 nm, which are dimensions comparable with those of the smallest synthetic nanoparticle.

The biological interactions are generally multivalent, and the interplay between microbes and host cells often involves multiple copies of receptors and ligands that bind in a coordinated manner, resulting in enhanced specificities, efficiencies and strengths of such interactions that allow the microbial agent to take possession of the cell under attack. In this context, metal nanoparticles seem to be very versatile in that they can be involved in multiple interactions.

Different types of metal nanomaterials like copper, zinc, titanium, magnesium, gold and silver have been studied for their antimicrobial potential and shown to be antibacterial agents against both Gram-negative and Gram-positive bacteria[8–11]; but silver nanoparticles have proved to be the most promising for their impressive antimicrobial activity against bacteria, viruses and some eukaryotic micro-organisms.

Silver has been widely used for many different purposes since antiquity, such as production of utensils, jewelry, dental alloy, photography, monetary currency, explosives, etc. Silver has a history of successes in medicine going back more than 7,000 years and its properties in preventing many diseases have been exploited, generally without even knowing the scientific basis. It was used in the ancient world to accelerate the healing of wounds and fight infections but first and foremost to preserve drinking water and to prevent food spoilage. Especially meaningful examples are: the introduction of a silver coin in milk for retarding the fermentation process by reducing lactic acid bacteria; the fact that Alexander the Great (356–323 BC) used to drink only from silver vessels; the fact that the Romans used silver nitrate therapeutically and silver was entered in the official Roman book of medicines. Also in the ancient Indian medical system (Ayurveda) silver was considered a therapeutic agent for many diseases. As a matter of fact, before the beginning of antibiotic therapy, silver was used worldwide for its antiseptic activity, specifically for the treatment of open wounds and burns. When penicillin was discovered and antibiotics were subsequently developed, silver was no longer at the heart of medical applications of the broad public since there was a widespread euphoria on the supposed omnipotence of antibiotics used for all sorts of diseases; notwithstanding the decline of its use, prophylaxis against gonococcal ophthalmia neonatorum with silver ions was considered the standard of care in many countries until the end of the 20th century. At the beginning of the 1990s, the problem of certain strains of bacteria generating resistance to antibiotics became so life-threatening that silver has made a remarkable comeback as a supplemental or alternative medicine.[12] In the most recent years, there has been a search for new unconventional antimicrobial agents and an increased use of silver as an efficacious antibacterial and antifungal agent in wound care product[13] and medical devices such as dental work and catheters[14–16]; it has also been used to obtain water disinfecting filters[17]; and in textiles[18] and cosmetics.[19] Silver compounds have also been found non-toxic to humans when used at very small concentrations.

Silver nanoparticles are now becoming the most common used nanoparticles in science, thanks to the fact that their biological activity couples with the knowledge that silver itself is a safe substance to man, and scientists are dedicating great efforts in exploiting silver nanoparticles to prevent the spread of a wide range of diseases.

Silver nanoparticles, thanks to their unique features and small size, are able to interact with viruses, bacteria and fungi and block the functional operations of the microorganism.[20,21] Silver nanoparticles showed strong antiviral activity against several viruses[22] with lower possibility of developing resistance compared to conventional antivirals; they are effective broad-spectrum bactericidal agents, and many researchers reported their activity as a new generation of antimicrobials which are also powerful to address the problem of multidrug resistant bacteria.[23] The multiresistant pathogens are ineffectively managed with current medications and represent a serious problem in public health, with an urgent need to develop new bactericides and virucides.

The present chapter aims at a description of the reported antibacterial and antiviral activities of silver nanoparticles.

2. Nanoparticles Properties

Particle size, shape, zeta potential, concentration and dose are the most important factors influencing the antimicrobial activity of nanoparticles. These properties (Table 1) determine the *in vivo* distribution, biological fate, toxicity and the targeting ability of the nanoparticles.

Table 1. Nanoparticle properties.

Nanoparticle properties	Effect
Size	*In vivo* distribution
Shape	Biological fate
Zeta potential	Toxicity
Concentration	Targeting
Dose	
Surface area/volume	
Aggregation	

Small nanoparticles (<30 nm) appear to be the most capable of penetrating into bacteria cell bodies; in fact, interactions with cell membranes, and subsequent membrane disruption, bacteria flocculation and loss of cell viability, is attributed to nearly all varieties of nanoparticles, but particularly to those with a small diameter and positive zeta potential. The electrostatic interaction of nanoparticles with positive zeta potential and negatively charged bacteria surfaces pushes the particle close to the bacteria and promotes penetration into the membrane. As a matter of fact, zeta potential, along with particle size, is a highly relevant parameter controlling antimicrobial activity, and the smallest nanoparticles with positive charge have always proved to be the most effective antimicrobials.

Compared to bulk material, small nanoparticles possess a large surface-area to volume ratio, providing larger surfaces that may come in contact with the bacterial cells and determine a higher percentage of interaction.[8] The smallest nanoparticle tested in each of the reviewed study generally had the strongest activity. The size dependence of antimicrobial activity was studied by several authors. Panacek *et al.*[24] reported that nanoparticles of 25 nm possess the highest antibacterial activity.

The bactericidal activity is influenced also by the shape as demonstrated by Pal *et al.*,[25] who studied the effect of spherical, rod and triangular nanoparticles against *E. coli*. Dose-dependent inhibition of bacteria activity was observed for all shapes of silver nanostructures, with triangular silver nanoparticles exhibiting the greatest antibacterial activity at low doses. The authors suggested that the process of producing triangular nanoparticles may have resulted in the addition of a positive charge to the particle surface that enhanced electrostatic interactions between particles and cells. This, along with the increased number of active facets on a triangular-shaped particle, was the explanation proposed for the superior antibacterial activity.

Silver nanoparticles have been synthesized through an array of methods and the development of better experimental procedure for the synthesis of nanoparticles of different size, shape and controlled polydispersity is pivotal.[26] Nanoparticles produced through different processes and for different purposes may vary in surface charge and agglomeration state and thus in activity.

Each synthetic method has its own merits and limitations: an impressive number of methods for nanoparticle synthesis have been reported in literature. For an extensive coverage of the chemical and biological synthesis of metal nanoparticles, there are several recent reviews available.[27–32]

3. Silver Nanoparticles and Antibacterial Activity

Antibiotic resistance represents a big challenge for the pharmaceutical and biomedicine research, forcing the scientific community to constantly develop new drugs and drug targets. Development of resistance to antimicrobial agents may involve changes in bacterial cell wall permeability, removal of antimicrobial agents through efflux pumps of membrane, drug action site modification and inactivation of the antimicrobial agent. Thus the multi-resistant pathogens are ineffectively managed with current medications.

Non-traditional antibiotic agents are creating huge interest in overcoming resistance that is developed by several pathogenic microorganisms against most of the commonly used antibiotics.[33] The development of novel applications of nanoparticles in this field makes them an attractive alternative to antibiotics and seems to have a high potential to solve the problem of multidrug resistant bacteria.[23] In fact, microorganisms are unlikely to develop resistance against nanoparticles, in comparison to antibiotics, as nanoparticles tackle multiple targets in the microbe, which means that they would have to develop simultaneously a set of mutations to protect themselves.

There is a broad literature describing silver, as a bulk material, effective against a wide range of pathogens; silver nanoparticles have demonstrated to be extremely appealing and hence represent a new generation of antimicrobials[9,34] and have opened up whole new strategies against a wide array of pathogens. The key to the broad-acting and potent antibacterial activity of silver nanoparticles is the multifaceted mechanism by which they act on microbes. As a naturally antibacterial metal, a silver nanoparticle is likely to have multiple mechanisms of antibacterial activity. In particular, the physical structure of a nanoparticle itself and the modality of interaction with and penetration into bacteria appear to provide unique and underexplored bactericidal mechanisms.

Although the antibacterial effect of silver nanoparticles has been extensively studied, their mechanism of action is still not clearly understood. Several hypotheses have been proposed and the bactericidal effect of silver nanoparticles has been attributed to a number of mechanisms (Fig. 2). The three most common mechanisms proposed are: (i) gradual release of silver ions followed by disruption of ATP production and DNA replication; (ii) direct damage of cell membranes; (iii) generation of reactive oxygen specie.

Cationic silver binds to negatively charged components of proteins and nucleic acids, thereby causing structural changes and deformations in bacterial cell walls, membranes and nucleic acids. In fact, silver ions interact with a number of electron donor functional groups like thiols, phosphates, hydroxyls, imidazoles and indoles. The ability of silver ions to strongly interact[35] with thiol groups in vital enzymes or their interaction with the phosphorus-containing bases of the DNA, may prevent cell division and DNA replication, leading to cell death. Klueh *et al.*[36] also hypothesized that the bactericidal activity is due to silver ions entering cells and intercalating between the purine and pyrimidine bases of DNA, leading to denaturation of the DNA molecule.

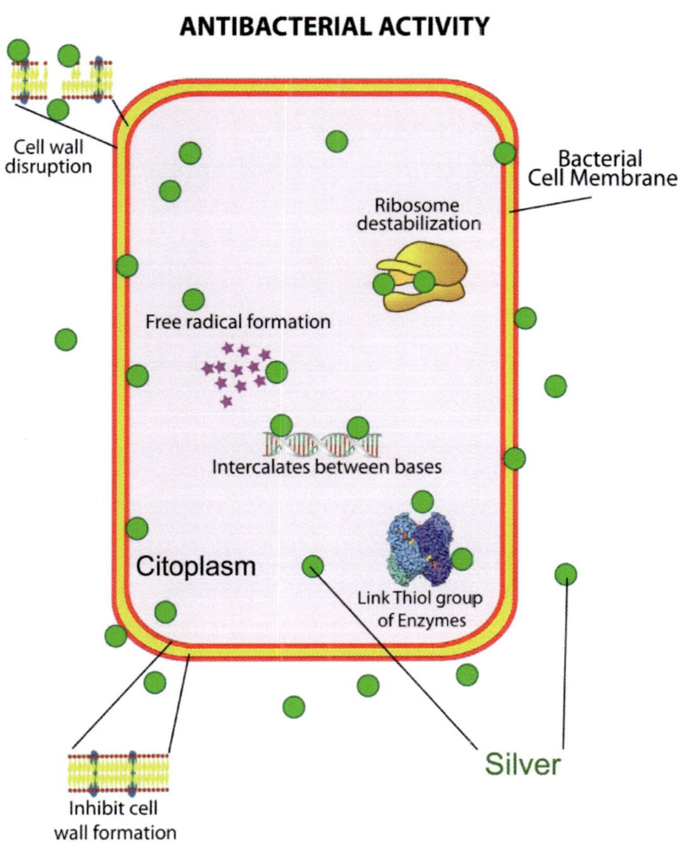

Fig. 2. Mechanism of bactericidal activity of silver nanoparticles.

Shrivastava[37] proposed that silver nanoparticles modulate the phosphotyrosine profile of putative bacterial peptides that can affect cellular signaling, which leads to inhibition of bacterial growth. Klueh *et al.*[36] support the hypothesis of silver interaction with the thiol groups in enzymes. In particular, the formation of stable S-Ag bonds with the thiol groups of enzymes alters their function in the cell membrane, which is fundamental for transmembrane energy generation and ion transport.

Silver nanoparticles attack Gram-negative bacteria by anchoring and penetrating the cell wall and thus causing structural changes in the cell membrane and increasing cell permeability; as a consequence the uncontrolled transport across cytoplasmic membrane causes the cell death.[10,35] The presence of a large number of nanoparticles inside the bacteria suggests that this is an important mechanism. Kim *et al.*[8] also proposed the formation of free radicals which induce membrane damage. Reactive oxygen species (ROS) generation is also a nearly universally described mechanism of nanoparticle antibacterial activity.

It was also demonstrated that nanoparticles have higher antibacterial potency than free silver ions, suggesting that nanoparticles have intrinsic antibacterial properties that do not depend on the elution of silver ions.[38] Thus the responses to ionic silver and nanoparticles are different, but both are essential for a complete understanding of the antibacterial activity of silver nanoparticles. There is much conflict in the literature regarding which action of silver nanoparticles has a major antibacterial effect, but it is likely a combined effect in which each mechanism contributes to provide broad spectrum antibacterial activity. Furthermore, the fact that bacterial resistance to elemental silver is extremely rare[39] emphasizes the presence of multiple bactericidal mechanisms acting in synergy.

Compared to other metals, silver nanoparticles demonstrated higher activity toward broad-spectrum microorganisms, including multidrug-resistant bacteria, while exhibiting lower toxicity to mammalian cells.[40] In particular, the antimicrobial effectiveness of silver nanoparticles was shown for both Gram-positive and Gram-negative bacteria.[10,41]

The antibacterial activity of silver nanoparticles was mainly demonstrated by *in vitro* experiments (Table 2).

Wright *et al.*[42] demonstrated that wound dressing coated with sputtered nanoscale silver reduced infections in burn patients. In 2006, Ip *et al.*[43] studied antibacterial activities of five commercially available silver-coated/impregnated dressings against nine common burn-wound pathogens, namely methicillin-sensitive and -resistant *Staphylococcus aureus* (MRSA), *Enterococcus faecalis, Pseudomonas aeruginosa, Escherichia coli, Enterobacter cloacae,*

Table 2. Selected studies on the antibacterial activity of silver nanoparticles.

Authors	Functionalization	Size (nm)	Organism	Effect	Ref.
Kim J.S. *et al.*	unfunctionalized	13.5	*E. coli*	MIC between 3.3–6.6 nM	8
			S. aureus	MIC>33 nM	
Morones *et al.*	unfunctionalized	21	*E. coli* *V. cholerae* *S. typhi* *P. aeruginosa*	MIC 75 μg/mL	35
Pal *et al.*	unfunctionalized	50	*E. coli*	MIC_{99} 0.1 μg/mL	25
Sondi *et al.*	unfunctionalized	12	*E. coli*	MIC_{70} 10 μg/mL	10
Shrivastava *et al.*	unfunctionalized	10–15	*E. coli* *S. typhi* *S. aureus*	MIC 25 μg/mL MIC 25 μg/mL 100μg/mL partial inhibition	37
Stefan *et al.*	poly(amide-hydroxyurethane)-coated	23	*E. coli* *S. aureus*	MIC 10 μg/mL	49
Kim S.H. *et al.*	unfunctionalized	?	*E. coli* *S. aureus*	MIC 100 μg/mL	54
Zhou *et al.*	citrate	30	*E. coli*	MIC 5–10 μg/mL	56

Proteus vulgaris, Acinetobacter baumannii and a multidrug, efflux-positive *Acinetobacter baumannii,* using a broth culture method. The rapidity and extent of killing of these pathogens under *in vitro* conditions were evaluated. All five silver-impregnated dressings investigated exerted bactericidal activity against all the tested bacteria, but a more potent activity was found against Gram-negative bacteria, including *Enterobacter species, Proteus species* and *E. coli.* Lee *et al.*[44] reported the bactericidal activity of nanosized silver particles in colloidal solution on cotton and polyester wovens against Gram-positive and Gram-negative bacteria. Reduction in the number of bacteria was observed in all specimens and were excellent against *S. aureus* and *K. pneumoniae.*

Sondi and Salopek-Sondi[10] used *E. coli* as a model for Gram-negative bacteria and evaluated the antibacterial activity of silver nanoparticles against it. The results showed a significant inhibition in bacterial growth. Shrivastava *et al.*[37] also reported the potent antibacterial activity of silver nanoparticles against Gram negative bacteria, suggesting that the activity of silver nanoparticles depends on its concentration.[35] They studied the effect of different concentrations of silver nanoparticles against different bacteria, and the results demonstrated that the concentration of silver nanoparticles that prevents bacteria growth is different for each type of bacteria. *P. aeruginosa* and *V. cholera* were found to be more resistant than *E. coli* and *S. typhi.* However, at concentrations above 75 μg/mL, bacterial growth was almost completely abolished. Similar concentrations of silver nanoparticles with different shapes were demonstrated to have different activities by Pal *et al.*[25] They investigated the antibacterial properties of differently shaped silver nanoparticles against the Gram-negative bacterium *E. coli* and reported a higher activity of triangular nanoparticles.

As far as the antibacterial activity of silver nanoparticles is concerned, *E. coli* and *S. aureus* are the most studied bacteria. Kim *et al.*[8] investigated the antimicrobial activity of silver nanoparticles against *E. coli* and *S. aureus* on Mullar Hinton agar plates. In this study, *E. coli* was inhibited at low concentration of silver nanoparticles, whereas the growth inhibitory effects on *S. aureus* were less profound. Similarly, Shahverdi *et al.*[11] reported the activity of silver nanoparticles against the same organisms and in combination with several antibiotics. Duran *et al.*[45] used silver nanoparticles for the preparation of cotton fabrics and evaluated its antibacterial activity against *S. aureus.*

Dose-dependent activity of silver nanoparticles in the size range of 10–15 nm was studied against *E. coli* and multidrug-resistant strains of *S. typhi*,[37] where increased stability and enhanced antibacterial activity was reported. The effect was more pronounced against Gram-negative than Gram-positive bacteria. In an attempt to study the mechanism of action, it was found that bacterial cell lysis could be one of the reasons for antibacterial activity; nevertheless, nanoparticles also modulated the phosphotyrosine profile of putative bacterial peptides, which could thus affect bacterial signal transduction and inhibit the growth of the organisms.[37]

The mycofabricated silver nanoparticles were also tested for antibacterial activity against human pathogenic bacteria like *E. coli, S. typhi, Staphylococcus epidermidis* and *S. aureus* using the well diffusion method. Silver nanoparticles proved to be toxic to each of the above species, and the effect was found to be 1.4–1.9 times stronger than that of pure silver ions. In the study, *S. aureus* showed higher sensitivity compared to the other bacteria.[46] Kumar *et al.*[47] developed paint from metal nanoparticles, which showed potent bactericidal activity against both Gram-positive (*S. aureus*) and Gram-negative bacteria (*E. coli*).

Gade *et al.*[48] synthesized silver nanoparticles from *Opuntia ficus-indica* and evaluated antibacterial activity of silver nanoparticles against *E. coli* and *S. aureus.* They found that the silver nanoparticles, in combination with commercially available antibiotics, demonstrated strong antibacterial activity against these bacteria. Antimicrobial activity of the silver nanoparticles of 19–23 nm size prepared by electrochemical synthesis in polyamide-hydroxyurethane media was investigated by disc diffusion method against *S. aureus* and *E. coli*.[49] They found that silver nanoparticles of size 23 nm at concentrations 5 μg/mL have strong bactericidal effect against *S. aureus*. In another study performed by Raheman *et al.*[50] silver nanoparticles synthesized from endophytic fungus *Pestalotia sp.* showed a remarkable activity against some human pathogenic bacteria. Knetsch *et al.*[51] studied the process of surface coating of medical devices with silver nanoparticles; the methodology was helpful to prevent bacterial adhesion and subsequent biofilm formation on the medical devices. These nanoparticles can directly be coated and/or deposited on the device surface, where silver is slowly released from the surface and kills the bacterial population around it.

Nam[52] prepared a combination of silver nanoparticles with tissue conditioner and tested its antibacterial activity against *S. aureus*, *Streptococcus mutans* and *C. albicans*. It was found that combination of silver nanoparticles with tissue conditioner showed significant bactericidal effect against two bacterial strains (*S. aureus* and *S. mutans*). Namasivayam *et al.*[53] reported the synthesis of silver nanoparticles mainly by using *Candida glabrata* and *Fusarium oxysporum*. The authors evaluated these silver nanoparticles against drug-resistant pathogenic bacteria *S. aureus, E. coli, Klebsiella pneumoniae, Bacillus subtilis, E. faecalis* and *P. aeruginosa*. They observed that the synthesized silver nanoparticles showed significant antibacterial activity against all the pathogenic bacteria. Kim *et al.*[54] studied the activity of silver nanoparticles against *S. aureus* and *E. coli*, where they treated bacterial cells with silver nanoparticles powder, and the growth rates were investigated under varying silver nanoparticles concentrations, incubation times, incubation temperatures and pH. The results showed that both *S. aureus* and *E. coli* were substantially inhibited by silver nanoparticles and the antibacterial activity of silver nanoparticles did not fluctuate with temperature or pH.

Recently, phytosynthesis of silver nanoparticles from the leaf extract of *Murraya koenigii*, an Indian curry leaf tree, was reported. The bactericidal potential of synthesized nanoparticles alone and in combination with commercially available antibiotics (gentamycin, ampicillin, tetracycline and streptomycin) against pathogenic bacteria (*E. coli*, *S. aureus* and *P. aeruginosa*) was studied.[55] They observed that silver nanoparticles, in combination with gentamycin, showed the maximum activity against *E. coli* with an increase of activity of 4.06 times. Tetracycline combination with nanoparticles showed maximum activity against *S. aureus* (2.16 times), indicating that activity of standard antibiotics was significantly enhanced in the presence of silver nanoparticles and that this strategy can be effectively used against antibiotic-resistant pathogens.

Zhou *et al.*[56] investigated effects of gold and silver nanoparticles on *Bacillus Calmette-Guérin* (BCG) and *E. coli.* Different concentrations of nanoparticles were applied in bacterial culture, and the growth of *E. coli* was monitored through colony forming units (CFU). However, the antibacterial effects on BCG were observed by recording fluorescent protein expression levels. The results showed a significant activity against both microorganisms. Shameli *et al.*[57] demonstrated the antibacterial activity of different sized silver nanoparticles in PEG against Gram-positive (*S. aureus*) and Gram-negative bacteria (*S. typhimurium*) by the disk diffusion method using Mueller–Hinton Agar. They reported a significant inhibition in growth of both these pathogens. They also concluded that the antibacterial activities of silver nanoparticles in PEG can be modified by controlling nanoparticle size, because the activity of silver nanoparticles decreases with the increase in the particle size.

Devi and Joshi[58] screened 53 different fungi isolated from soils of different microhabitats of the Eastern Himalayan range for mycosynthesis of silver nanoparticles and also studied their efficacy as antimicrobials alone and in combination with commonly used antibiotics against *S. aureus*, *Streptococcus pyogenes*, *Salmonella enterica* and *Enterococcus faecalis*. The fungi isolates able to synthetize silver nanoparticles were *Aspergillus terreus*, *Paecilomyces lilacinus* and *Fusarium sp.* The mycosynthesized nanoparticles showed potent antibacterial activity and, interestingly, their syngergistic effect with erythromycin, methicillin, chloramphenicol and ciprofloxacin was significantly higher as compared to inhibitions by silver nanoparticles alone.

4. Silver Nanoparticles and Antiviral Activity

Viruses still represent one of the leading causes of disease and death worldwide. Viral diseases were largely untreatable no more than 50 years ago. Nowadays, effective and safe therapies are available, and this has led to a significant improvement of the life for a large numbers of patients.[59,60] New viral diseases, however, continue to emerge, and established viruses have been shown to develop resistance to available therapies, thus rendering the field of antiviral compound discovery an extremely fertile research area.

Emerging and re-emerging viruses are to be considered a continuing threat to human health because of their amazing ability to adapt to their current host, to switch to a new host and to evolve strategies to escape antiviral measures.[61]

Viral infections are governed by an interplay of interactions between the virus and the host cell apparatus. All viruses are strictly dependent upon a living host for their mRNA production and protein synthesis. Thus, all viruses replicate via a broadly similar sequence of events. The virus must first attach and penetrate into the cell. Once into the cell cytoplasm, the genome is liberated from the protective capsid and, either in the nucleus or in the cytoplasm, it is transcribed, and viral mRNA directs protein synthesis, in a generally well-regulated fashion. Finally, the virus undergoes genome replication and, together with viral structural proteins, assembles new virions, which are then released from the cell.

Each of the described phase represents a possible target for inhibition, and silver nanoparticles have been found to possess a strong antiviral potential, as they can inhibit the viral multiplication inside the host cell by preventing the replication or blocking the penetration of virus particles inside the cell cytoplasm.[62–64]

In the future, an ideal antiviral agent will need to exert a broad-spectrum action against viruses of different families to be used as first-aid compound against unforeseen viral epidemics or pandemics. Due to the pressure of emerging infectious diseases caused by new pathogenic viruses and the development of antiviral resistance to classical antiviral drugs, pharmaceutical companies and numerous researchers in search of novel antiviral agents have turned their interest to nanoscale materials, among which silver nanoparticles have received considerable attention due to their particular intrinsic properties.

The mechanism of action of silver nanoparticles as an antiviral and/or virucidal agents has been studied against several enveloped viruses (Table 3); nevertheless, the interaction with viruses is still an unexplored field.

Recently, several studies showed that metal nanoparticles can be effective antiviral agents against viruses belonging to the following families: retroviridae,[22,65–67] hepadnaviridae,[68] paramyxoviridae,[69] herpesviridae,[70,71] poxviridae,[72] orthomyxoviridae[73] and arenaviridae.[74]

Several studies have demonstrated that silver nanoparticles have *in vitro* antiviral activity against human immunodeficiency virus type 1 (HIV-1).[22,65,66] Elechiguerra *et al.*[65] analyzed the effect of different silver nanoparticles whose physicochemical properties were differently modified by the applied surface coating, respectively, foamy carbon, poly(N-vinyl-2-pyrrolidone) (PVP), and bovine serum albumin (BSA). A dose-dependent inhibition of HIV-1 infectivity of the different silver nanoparticles was determined by *in vitro* assays. Furthermore, they observed that the interaction between HIV and silver nanoparticles was clearly influenced by the size of the silver nanoparticle

Table 3. Antiviral silver nanoparticles.

Family	Virus	Nanoparticle characteristics	Size (nm)	Mechanism of action	Ref.
Retroviridae	HIV-1	PVP-coated silver nanoparticles	1–10	Interaction with gp120	22,66,67
Hepadnaviridae	HBV	Silver nanoparticles	10–50	Interaction with double-stranded DNA and/or binding with viral particles	68
Herpesviridae	HSV-1	MES-coated silver and gold nanoparticles	4	Competition for the binding of the virus to the cell	70
	HSV-1 and-2	Mycosynthesized silver nanoparticles	4–46	Interference with early steps of penetration	78
Paramyxoviridae	RSV	PVP-coated silver nanoparticles	69	Interference with viral attachment	69
	HPIV-3	Mycosynthesized silver nanoparticles	4–46	Interference with early steps of penetration	78
Poxviridae	Monkey pox virus	Silver nanoparticles and polysaccharide-coated Silver nanoparticles	10–80	Block of virus-host cell binding and penetration	72
Orthomyxoviridae	Influenza virus	Silver nanoparticles	Not reported	Possible interaction with hemagglutinin	93
		Silver nanoparticles	5–20	Interaction with viral particles	94
Arenaviridae	Tacaribe virus (TCRV)	Silver nanoparticles and polysaccharide-coated Silver nanoparticles	10	Inactivation of virus particles prior to entry	74
Enteroviridae	Coxakie-virus B3	Plant-derived silver nanoparticles	Up to 1000	Not reported	95

since only nanoparticles within the range of 1–10 nm were able to bind to the virus.

A different group[66] also reported about the antiviral activity of silver nanoparticles that had been fabricated using Hepes buffer. They showed that silver nanoparticles exhibited potent cytoprotective and post-infected anti-HIV-1 activities (at 50 mM, a 98% reduction was achieved) in a dose-dependent fashion.

Lara *et al.*[22] further demonstrated that silver nanoparticles inhibited a variety of HIV-1 strains regardless of their tropism, clade and resistance to antiretrovirals. The fact that silver nanoparticles inhibited a number of HIV-1 isolates (including laboratory strains, clinical isolates, macrophage [M]-tropic and T lymphocyte [T]-tropic strains, and resistant strains) suggest that their mode of action does not depend on cell tropism and that silver nanoparticles are broad-spectrum anti-HIV-1 agents. A cell-based fusion assay using Env expressing cells and CD4 expressing cells mixture demonstrated that silver nanoparticles efficiently blocked cell-cell fusion in a dose-dependent manner within the 1.0–2.5 mg/mL dose range.

The concentration of silver nanoparticles at which infectivity was inhibited by 50% (IC_{50}) ranged from 0.44 to 0.91 mg/mL. The observed antiviral effect of silver nanoparticles could be ascribed to the nanoparticles. In fact, silver salts exerting antibacterial effect through silver ions, has an inhibitory effect 12 times lower than the one of silver nanoparticles.

Since the cell-based fusion assay showed that silver nanoparticles inhibit the initial stages of the HIV-1 infection cycle, efficient inhibitory activity of silver nanoparticles against gp120-CD4 interaction was also measured in a competitive gp120-capture ELISA. Evidence was obtained by the combined experimental procedures that silver nanoparticles inhibit HIV-1 infection by blocking the viral entry, particularly the gp120-CD4 interaction.[22]

To assess antiviral mechanisms, many other studies were involved to analyze the interactions between the viral shell and the nanoparticle. Transmission electron microscopy (TEM) was used to show that only nanoparticles smaller than 16 nm were able to interact with HIV-1 cells and that the nanoparticle population is heterogeneous being composed of nanoparticles with different morphologies: decahedra and icosahedra, truncated pyramids, etc. High-angle annular dark field (HAADF) scanning transmission electron microscopy was employed to study the interaction of silver nanoparticles with HIV-1. Interestingly, the size of nanoparticles bound to the virus was exclusively within the range of 1–10 nm. Additionally, HAADF microscopy showed that nanoparticles are not randomly attached to the virus, in fact, all the three species of nanoparticles established regular spatial relationships with the viral envelope. Both the spatial arrangement of nanoparticles and the size dependence of interaction find an explanation based on the structure of the HIV-1 envelope. Glycoprotein spikes formed by trimers consisting of two subunits are embedded into the viral envelope: the gp120 surface glycoprotein subunit, which forms knobs on the exposed surface, and the gp41 subunit, which spans the viral envelope connecting the external gp120 subunit with the internal p17 matrix protein.[65] The main function of these protruding gp120 knobs is to bind to CD4 receptor sites on host cells. Silver nanoparticles appear to be positioned at specific sites, with a regular spatial relation that correlate well with the positions of the gp120 knobs. These data also suggested that, with the sulfur-bearing residues of the gp120 glycoprotein knobs being the most likely sites for interaction, only a limited number of bonds are freely available for nanoparticles. This may result in a limited number of stabilizing sites; therefore, particles larger than 10 nm were not able to bind to the viral envelope. BSA- and PVP-coated nanoparticles were shown to possess a slightly lower inhibition efficacy, probably because the surface of the nanoparticle is directly bound to and encapsulated by the capping agent. In contrast, the silver nanoparticles released from the carbon matrix showed a greater inhibitory effect due to their essentially free surface area.

Silver nanoparticles proved to be virucidal to cell-free and cell-associated HIV-1 as judged by viral infectivity assays. In fact, HIV infectivity was effectively eliminated following short exposure of isolated virus to silver nanoparticles.

These properties make silver nanoparticles a potential broad-spectrum agent not prone to inducing resistance that could be used preventively against a wide variety of circulating HIV-1 strains.

PVP-coated silver nanoparticles, being an interesting virucidal candidate drug, have been further investigated as a potential topical vaginal microbicide to prevent transmission of HIV-1 infection[67] using an *in vitro* human cervical tissue-based organ culture that simulates *in vivo* conditions.[75,76] Results have been very encouraging since a short pre-treatment of the cervical explant with 0.025–0.15 mg/mL of PVP-coated silver nanoparticles prevented the transmission of cell-associated HIV-1 and cell-free HIV-1 isolates. A long-lasting protective effect by the PVP-coated silver nanoparticles in the cervical explants conferring almost total protection against HIV-1 transmission for 48 h was also reported.

A recent application of these experimental evidence has been the development of a silver-nanoparticle–coated polyurethane condom (PUC) in order to investigate its antimicrobial potential, including the inactivation of HIV and herpes simplex virus (HSV) infectiousness.[77] The contact of the silver nanoparticles coated PUC with HIV-1 (both M-tropic and T-tropic strains) and HSV-1/2 was able to efficiently inactivate their infectivity. Furthermore, the growth of bacteria and fungi was also inhibited. These results demonstrated that the application of silver nanoparticles to condoms can directly inactivate infectious microorganisms and can provide a further defense against sexually transmitted microbial diseases.

Besides, silver nanoparticles are able to inhibit post-entry stages of the HIV-1 life cycle, in fact, the antiviral activity was maintained also when the silver nanoparticles were added 12 h after the cell had been infected with HIV.[22] Since silver ions can form complexes with electron donor groups containing sulfur, oxygen or nitrogen that are normally present as thiols or phosphates on amino acids and nucleic acids, they might inhibit post-entry stages of infection by blocking HIV-1 proteins other than gp120, or reducing reverse transcription or proviral transcription rates by directly binding to the RNA or DNA molecules. A further explanation for a post-entry effect could be related to the suppression of TNF-α expression exerted by silver nanoparticles. This cytokine plays a critical role in HIV-1 pathogenesis incrementing HIV-1 transcription, therefore, the inhibition of the TNF-α activated transcription might represent a further target for the anti- HIV activity of silver nanoparticles.

The antiviral activity of silver nanoparticles on the hepatitis B virus (HBV) has been reported.[68] Silver nanoparticles of different dimension ~10 nm, ~50 nm and ~800 nm were synthesized and tested for their ability to inhibit HBV replication in a human hepatoma cell line. Cytotoxicity studies ruled out the possible use of the larger nanoparticles being too toxic in a cell-culture–based assay, while the smaller nanoparticles (10 and 50 nm) only showed a minor toxicity, and were, therefore, used for antiviral studies. The 10-nm particles at a concentration of 5 μM were able to reduce HBV replication of 38%, while at a higher concentration (50 μM) inhibition reached 80%. Interestingly, larger nanoparticles (50 nm) were more efficient in inhibiting HBV infectivity, which was reduced by 53% and 92% at concentrations of 5 μM and 50 μM, respectively. The inhibitory activity could be ascribed to nanosilver since infectivity inhibition of compounds with silver in different oxidation states was obviously less important. Lu *et al.*[68] showed evidence of the high binding affinity of silver nanoparticles (both the 10 and 50 nm sizes) for HBV DNA and extracellular virions, therefore, their inhibitory activity was probably due to a specific interaction between the nanoparticles and the double-stranded DNA of HBV and/or direct binding with viral particles and consequent prevention of virions from entering into host cells.

Baram-Pinto *et al.* described[70] the anti-Herpes simplex virus (HSV) potential of silver nanoparticles coated with mercaptoethanesulfonate (MES).

Their anti-HSV-1 activity depends on their ability to mimic HS and thus compete for the binding of the virus to the cell and is amplified by the presence of the inner core of nanosilver. The high local concentration of binding molecules may explain the improved inhibitory effect. The silver-MES nanoparticles were tested in antiviral assays to analyze the different stages of the viral infection that may be blocked, and taken together, their results indicate that sulfonate-capped silver nanoparticles inhibit HSV-1 infections by blocking the attachment and thereby the entrance of the virus into the cells and/or by preventing the cell-to-cell spread of the virus. Other studies have also shown the ability of uncapped silver nanoparticles to inhibit both HSV-1 and HSV-2.[78] These results highlight the importance of interactions between the virus and nanoparticles or inhibition events happening during the early stages of the infection cycle. HSV is a quite complex virus with several virus-encoded proteins on its lipidic envelope. The entry of HSV into host cells involves five viral glycoproteins: gB, gC, gD, gH and gL. The initial interaction, or binding to cells, is mediated via interactions of gC and/or gB with heparan sulfate (HS).[79] Following binding, HSV entry is achieved through fusion of the viral envelope with a host cell membrane. The core fusion machinery is composed by gB and gH/gL.[80,81] Glycoprotein B may act as the premier fusogen,[82–86] but it seems to need the cooperation of several membranotropic sequences harboured in gH.[87–92] Interactions with these glycoproteins could be used to explain the inhibitory activity of silver nanoparticles.

An interesting approach was undertaken by Sun *et al.*,[69] which used different proteins: poly(N-vinyl-2-pyrrolidone) (PVP), bovine serum albumin (BSA) and a recombinant F protein (RF 412) from Respiratory Syncytial Virus (RSV) as capping agents of silver nanoparticles.

Nanoparticles were mixed with RSV and added to HEp-2 cells in order to evaluate the effectiveness of viral infectivity inhibition by microscopic examination for syncytia formation and by immunofluorescence microscopy. PVP-coated silver nanoparticles showed the highest activity, in fact, they were able to bind to the viral surface with a regular spatial arrangement, suggesting a possible interaction with G proteins that are evenly distributed on the envelope of the RSV virion. Due to their small size and uniformity (4–8 nm), compared to the other (BSA and RF 412) coated nanoparticles (3–38 nm) their binding was more effective.

Mehrbod *et al.*[93] investigated the inhibitory effect of silver nanoparticles against the influenza virus A/H1N1. Inhibition of virus replication in the presence of the compound was assessed by a hemagglutination assay. They observed that nanosilver at concentration up to 1 μg/mL is non-toxic to

MDCK (Madin-Darbey Canin Kidney) cells and a concentration of 0.5 μg/mL is sufficient to reduce infectivity. Nanosilver may interfere with the fusion of the viral membrane, inhibiting viral penetration into the host cell.

Further experiments to investigate the activity of silver nanoparticles against influenza virus were performed by Xiang *et al.*[94] They synthesized silver nanoparticles with sizes varying from 5 to 20 nm and tested them with a number of assays, including hemagglutination inhibition (HAI) tests and embryo inoculation assays and transmission electron microscopy (TEM) analysis combined with a flow cytometry (FCM) assay to determine whether silver nanoparticles could reduce H1N1 influenza A virus-induced apoptosis in MDCK cells. Collectively, they suggested that silver nanoparticles could provide a strong protection against influenza virus infections without the risk of cell toxicity.

Rogers *et al.*[72] described silver nanoparticles able to inhibit poxviruses cell infectivity. Nanoparticles (10–80 nm, with or without polysaccharide coating) at concentrations of 100, 50, 25, and 12.5 μg/mL were evaluated for efficacy against Monkeypox virus using a plaque reduction assay. Silver nanoparticles of approximately 10 nm exhibited a significant dose-dependent reduction on the mean number of plaque-forming units. Concentrations of 100 μg/mL are toxic to Vero cells and hampered the evaluation of the antiviral activity. The mechanism by which this inhibition occurs has not been further investigated. The mechanism of poxviruses entry is still largely unknown. Probably the silver nanoparticles may interfere in the early steps of binding and penetration by blocking virus-host cell binding by physical obstruction, or they may block endocytosis related mechanisms, or they can disrupt host cell biochemical pathways important for virus replication.

Other studies also showed that AgNPs at non-toxic concentrations effectively inhibit arenavirus replication during the early phases of viral replication.[74] The activity of two types of silver nanoparticles (uncoated and polysaccharide coated silver nanoparticles) against Tacaribe virus (TCRV) was investigated. The polysaccharide coating protects the cell from the toxic effects of the silver nanoparticles but it reduces its activity. Uncoated silver nanoparticles seem to interact with TCRV prior to cellular exposure, resulting in a decrease in viral infectivity with 10- and 25-nm nanoparticles. This interaction can either prevent the internalization of the viral particle by interfering with cellular receptor binding or may facilitate the internalization of the silver nanoparticle together with the virus and produce an inhibitory effect on viral replication interfering with the TCRV RNA-dependent RNA polymerase.

In the global effort to reduce generated hazardous waste, the demand of "green" synthesis methods has recently spurred the interest of scientists in the field of biological metal nanoparticle production. The synthesis of silver nanoparticles derived from *Ricinus communis* aqueous extracts has been reported.[95] Silver nanoparticles up to the size of about 1000 nm were tested for their inhibitory effect on Coxakievirus B3 (Enteroviridae), and showed a dose response effect. The biosynthesized nanoparticles in the present study presented a size greater than the virus and lower than the cell size. Thus, the antiviral activity of *Ricinus communis* fruit silver nanoparticles may be due to fusion inhibition between virus and cells.

Biological synthesis of silver nanoparticles (size range 5–20 nm) was obtained using fungal cell filtrate (*Alternaria* sp., *Fusarium oxysporum, Curveleria* sp., *Chaetomium indicum* and *Phoma* sp.) when treated with aqueous silver nitrate.[78] It was shown that production of silver nanoparticles is feasible in different fungi, and the antiviral activity is dependent on the production system used. Some of the fungi strains used in that study resulted in production of silver nanoparticles with considerable toxicity in order to be considered as eventual antivirals; on the contrary some of them, namely *F. oxysporum, Curvularia* sp. and *C. Indicum*, have indeed low toxicity levels and promising antiviral activity.

5. Silver Nanoparticles Toxicity

The antibacterial activity of silver nanoparticles has determined their incorporation into numerous consumer products and the continuous evolution in the field of metal-based nanoparticle applications in our society for drug delivery, medical imaging, diagnostics, therapeutics and engineering technology. Unfortunately, there is a serious lack of information about the impact of metal nanoparticles on human health and environment. In addition to toxic effects on microbes, silver nanoparticles can also induce cytotoxicity in mammalian cells, thus their potential *in vivo* use as antimicrobial agents is limited by their cytotoxycity to eukaryotic cells. The concentrations of nanoparticles needed to kill bacteria cannot exceed concentrations which would substantially compromise the functions of eukaryotic cells. Further research is required to provide the necessary warranties to allow a safe exploitation of the remarkable *in vitro* antimicrobial properties of silver nanoparticles and their transfer to the clinical setting.

Most studies focus on the interaction of silver nanoparticles with single cells, indicating that they have the ability to induce the genes associated with cell cycle progression, DNA damage and mitochondria-associated apoptosis. For metal nanoparticles to be effective as antimicrobial pharmaceuticals, it is imperative to gain a better understanding of their biodistribution/

accumulation in living systems and toxicity, but there are still very few studies on possible adverse effects and toxicity *in vivo*.[96] These studies give the idea of the probable harmful effect up to a certain extent. Therefore, *in vivo* studies are very important to obtain a closer look into the exposure effects.

Several studies have demonstrated the cytotoxic effects of metal nanoparticles[97–100]; in fact, silver nanoparticles were found to be highly cytotoxic to mammalian cells based on the evaluation of mitochondrial function, membrane leakage of lactate dehydrogenase and abnormal cell morphologies. At a cellular level, metal nanoparticles interact with biological molecules within mammalian cells and can interfere with the antioxidant defense mechanism, leading to the generation of ROS. Such species, in excess, can cause damage to biological components through oxidation of lipids, proteins and DNA. Oxidative stress may have a role in the induction or the enhancement of inflammation through upregulation of redox sensitive transcription factors (e.g. NF-κB), activator protein-1 and kinases involved in inflammation.[101–104]

The generation of reactive oxygen species by cells exposed to silver nanoparticles[98] has been showed in human lung fibroblast and human glioblastoma cells, and as a consequence, DNA damage and cell cycle abnormalities have been observed. Accumulation of silver nanoparticles in various organs (lungs, kidneys, brain, liver and testes) has been evidenced in animal studies.[105] Most of the *in vitro* studies show dose dependence, in fact, higher doses of silver induce a stronger cellular toxicity. Nevertheless, it should be considered that *in vitro* concentrations of nanoparticles are often much higher than the ones used in *in vivo* experiments, therefore, such exposures do not represent a reproduction of the conditions expected for *in vivo* exposure. A recent study[106] showed that mice exposed to silver nanoparticles showed minimal pulmonary inflammation or cytotoxicity following sub-acute exposures, but longer exposures with higher lung burdens of nanosilver were not investigated, and therefore, eventual cronic effects may be underscored.

The biological consequences of metal nanoparticles also depend on the same set of parameters (chemical composition, surface structure, solubility, shape and aggregation) which fundamentally determine the antimicrobial activity, indicating that there is a subtle balance among the various parameters and their positive/negative effect. All of these features can modify cellular uptake, protein binding, translocation to the target site and most of the biological interactions, with the possibility of causing tissue injury. Therefore, in terms of safety, the effect of silver nanoparticles on host cells is a major consideration; in fact, though they have antimicrobial activity, they may not be beneficial if they present adverse effects to humans or animals.

One of the most debated issues is whether the toxicity of silver nanoparticles is due to the nanoparticles themselves or the release of silver ions when they are suspended in the suitable solvent. Singh *et al.*[107] suggested toxicity is due to the silver ions which cause the mitochondrial damage, and thereby induce apoptosis and cell death; they also showed that silver nanoparticles dissolve 50 times faster inside the cell cytoplasm as compared to the water. Asharani *et al.*[98] also support the hypothesis that the surface oxidation of silver nanoparticles, after contact with cell culture medium or proteins in the cytoplasm, determine the release of silver ions which disrupt the normal functioning of the cell. On the contrary, Grosse *et al.*[108] showed that nanosilver particles have a greater effect on the colony formation of rat brain endothelial (RBE4) cells compared to silver ions.

The size of metal nanoparticles together with surface area, modifies the physicochemical properties of the metal and can also influence the reactivity of nanoparticles with themselves or with the cellular environment, leading to different modes of cellular uptake and further processing, to adverse biological effects in living cells that would not otherwise be possible with the same material in larger forms. In fact, as the particle size decreases, some metal nanoparticles show increased toxicity, even if the same material is relatively inert in its bulk form (e.g., Ag, Au and Cu). It is also a well-accepted theory that smaller sized nanoparticles cause more harm in a size and dose-dependent manner, as compared to their larger counterparts. This conclusion has been also supported by various studies including those performed on human periodontal fibroblasts.[109] The toxicity of silver nanoparticles also depends on their surface chemistry.[110] A commonly used strategy to reduce toxicity is to use various capping agents to prevent the direct contact of the metal with the cells.

Potential routes of human exposure to metal nanoparticles used as antimicrobial therapeutic compounds include the gastrointestinal tract, skin, lungs and systemic administration. The safest and easiest route can be the topical use of nanoparticles as microbicide for direct microbial inactivation and the dermal route seems the perfect choice. However, in the presence of damaged skin, micrometer-size particles gain access to the dermis and regional lymph nodes. A further concern would be the potential of nanoparticles translocation to the brain via the olfactory nerve as a consequence of the vicinity of the nasal mucosa to the olfactory bulb. Whether silver nanoparticles have any pathological or clinical significance in these tissues is unclear, therefore, more data is needed to properly address the safety concern on the use of silver nanoparticles as pharmaceuticals.

The inhalation of silver nanoparticles in rats for longer time periods can lead to its accumulation in the lung, where it could cause harmful effects. Ji *et al.*[111] evaluated the inhalation effects of silver nanoparticles for 4 weeks and showed the accumulation and transport of silver nanoparticles from lung to the gastrointestinal tract and liver. Additionally the ingestion of colloidal silver was also found to cause liver and kidney damage.[8,112] The silver nanoparticles were reported to hamper the normal development of the fish embryo[113] being able to cross the blood–brain barrier (BBB)[114] as well as the blood–testes barrier,[97] thus causing harmful effects.

Nanoparticle concentrations required to inhibit bacteria activity may have local cytotoxic effects, indicating that concentrations of nanoparticles necessary for the treatment of an infection must be limited to the proximity of the bacteria being treated. Relatively large doses of nanoparticles introduced to the infection site will become less concentrated as they dissipate throughout the body. Ideally, techniques to confine the nanoparticles to the site of infection can be developed.

While the population exposed to silver nanoparticles continues to increase with ever new applications, their toxicity remains a controversial research issue because the exact mechanism of toxicity has not been fully understood. Such data are very important for helping the design of biocompatible silver nanoparticles to facilitate their application in the medical field. Therefore, the biological applications employing silver nanoparticles should be given special attention besides adapting the antimicrobial potential,[115] with a need to define the boundaries beyond which silver nanoparticles would be toxic and within which they would be safe for use.

6. Conclusions and Future Prospects

The potential benefits of nanotechnology in biomedical and industrial applications for human health and environment are now widely accepted and represent the most promising field for generating new applications in medicine. Nanoparticles are the magic tool which have specific physicochemical features different from their bulk materials, and recent developments in nanotechnology has provided a tremendous impetus toward their applications due to the possibility of forging metals into differently shaped nanosizes, which drastically modulate their chemical, physical and optical properties and thus their applications.

Among the different nanoparticles, it has been proved that silver nanoparticles have strong antibacterial and antiviral activities, which have been evidenced by various studies. Many researchers attribute their highly efficient antibacterial effect to their ultrafine size and larger surface area, by which they can easily destroy the membrane, pass through the microbial body and then convert to silver ions in the cytoplasm to damage the intracellular structure as a secondary result. The potential activity of silver nanoparticles can be further increased by understanding the exact site of interaction and mode of action, leading to modifications of the nanoparticle surface for a broader and more effective use.

The exploitation of silver nanoparticles is already established for some medical applications, with many new potential applications currently being investigated at the preclinical stage. A wide range of silver nanoparticles has emerged in consumer products, ranging from medical and healthcare devices to home appliances and cosmetics. Silver is used worldwide for its antimicrobial activity in a huge number of applications (Table 4) such as personal care products, medical devices, wound dressings, washing machines and computer keyboards.[116–118]

One of the most common applications is in the textile industry.[45] Sterile cloths and materials play an important role in hospitals; in fact, to prevent and reduce infections, various antibacterial disinfection techniques have been developed for all types of textiles, and silver nanoparticles have been incorporated in cotton and silk clothes. Currently, these are regularly used in the coatings of biomedical devices, such as silver impregnated catheters. This is a very important application since contaminated or infected catheters are a major source of nosocomial infections and are responsible for more than 40% of sepsis cases in hospitals.

The emergence and spread of antibiotic-resistant pathogens is an alarming concern because it has invalidated major antimicrobial drugs that are

Table 4. Examples of silver nanoparticle applications.

Technology	Applications
Healthcare	Wounds dressings, antiseptics, hospital beds and furniture
Home Products	Baby bottles, food storage containers, food packaging, kitchen plastics, vacuum cleaner, bathroom products, dishwashers, refrigerators, computer keyboards, hand dryers
Cosmetics	Toothpaste, cosmetic deodorants, toothbrushes, tissue paper, epilator, electric shaver, pet shampoos
Disinfectants	Agricultural disinfectants, industrial disinfectants, aquaculture disinfectants, pool disinfectants
Clothing and Fabrics	Apparel, footwear
Construction	Paints

currently used in clinical practice. Despite extensive efforts in research and enormous investment of resources, the rate of drug development has not kept up with the development of resistance, and the increasing rates of bacterial resistance has also invalidated the utility of antibiotics, resulting in mortality due to failure in infection control and high health-care costs. There is thus a need of a cheap broad-active agent that can be used against a variety of pathogens, and silver nanoparticles have been found to be effective against many viruses and bacterial species without causing resistance. Combined use of silver nanoparticles with antibiotics, such as penicillin G, amoxicillin, erythromycin and vancomycin, also resulted in enhanced and synergistic antimicrobial effects against Gram-positive and Gram-negative bacteria (e.g. *E. coli* and *S. aureus*).[9,11,119]

Moreover, the use of silver nanoparticles for cancer therapy is emerging as a novel approach. It is now widely accepted that silver nanoparticles can kill cells in a manner similar to chemotherapeutics, giving them the potential to act as antineoplastic agents.

Thus, there is an enormous interest in exploiting silver nanoparticles in various biomedical applications since their size scale is similar to that of biological molecules (e.g. proteins, DNA) and structures (viruses and bacteria). Notably, silver nanoparticles are the most powerful antimicrobial agents that exhibit a strong toxicity toward a broad range of microorganism and simultaneously a remarkable low human toxicity. Nonetheless, before using silver nanoparticles for therapeutic applications, one should keep in mind their cytotoxic level to avoid human and environmental risks.

Exposure to nanosilver in the body is becoming increasingly widespread, and silver nanoparticles gain increasing access to tissues, cells and biological molecules within the human body. Although metallic silver has been proved to have minimal health risks, once in the nanoscale, it may exhibit significant toxicity to mammalian cells, so more accurate attention should be paved to aspects such as their entry portals into human body, biodistribution, organ accumulation as well as their potential interactions with tissues, cells and molecules and their relevant toxicological implications. Toxicity of silver nanoparticles to eukaryotic cells is a legitimate concern but may be addressed by targeting nanoparticles to a specific site and confining cytotoxic nanoparticle concentrations to the local environment of an infection.

Compared to conventional antibiotics and other drugs, preparation of antimicrobial nanoparticles could be cost-effective and they are stable enough for long-term storage with a prolonged shelf-life. Recently, intracellular or extracellular synthesis of silver nanoparticles using microbial cells or enzymes has also been explored as novel biological and environment-friendly nanoparticle preparation.[120–123] In addition, some nanoparticles can withstand harsh conditions, such as high-temperature sterilization, under which conventional antibiotics are inactivated.

Recent research has revealed exciting new biological properties of silver nanoparticles that could be translated into new therapeutic and pharmacological treatments. The full potential of this technology has yet to be discovered and is barely in its infancy; the mechanisms underlying the impressive biological properties of silver nanoparticles are still not understood and represent a priority for future research.

References

1. A. Falanga, M. T. Vitiello, M. Cantisani, R. Tarallo, D. Guarnieri, E. Mignogna, P. Netti, C. Pedone, M. Galdiero and S. Galdiero, *Nanomedicine* 7, 925 (2011).
2. S. Hallaj-Nezhadi, F. Lotfipour and C. R. Dass, *Pharmazie* **65**, 855 (2010).
3. C. Cao, L. C. Gontard, L. Thuy Tram le, A. Wolff and D. D. Bang, *Small*, **7**, 1701 (2011).
4. G. G. Daaboul, A. Yurt, X. Zhang, G. M. Hwang, B. B. Goldberg and M. S. Unlu, *Nano Lett.* **10**, 4727 (2010).
5. O. R. Miranda, B. Creran and V. M. Rotello, *Curr. Opin. Chem. Biol.* **14**, 728 (2010).
6. L. C. Kennedy, L. R. Bickford, N. A. Lewinski, A. J. Coughlin, Y. Hu, E. S. Day, J. L. West and R. A. Drezek, *Small*, 7, 169 (2011).
7. N. G. Portney and M. Ozkan, *Anal. Bioanal. Chem.* **384**, 620 (2006).
8. J. S. Kim, E. Kuk, K. N. Yu, J. H. Kim, S. J. Park, H. J. Lee, S. H. Kim, Y. K. Park, Y. H. Park, C. Y. Hwang, Y. K. Kim, Y. S. Lee, D. H. Jeong and M. H. Cho, *Nanomedicine*, **3**, 95 (2007).
9. M. Rai, A. Yadav and A. Gade, *Biotechnol. Adv.* **27**, 76 (2009).
10. I. Sondi and B. Salopek-Sondi, *J. Colloid Interf. Sci.* **275**, 177 (2004).
11. A. R. Shahverdi, A. Fakhimi, H. R. Shahverdi and S. Minaian, *Nanomedicine*, **3**, 168 (2007).
12. U. B. Hoyme, *Infect Dis Obstet Gynecol*, **1**, 32 (1993).
13. K. Chaloupka, Y. Malam and A. M. Seifalian, *Trends Biotechnol.* 28, 580 (2010).
14. O. Salata, *J. Nanobiotechnol.* **2**, 3 (2004).
15. J. H. Crabtree, R. J. Burchette, R. A. Siddiqi, I. T. Huen, L. L. Hadnott and A. Fishman, *Perit. Dial. Int.* **23**, 368 (2003).
16. M. D. Khare, S. S. Bukhari, A. Swann, P. Spiers, I. McLaren and J. Myers, *J. Infect.* **54**, 146 (2007).
17. P. Jain and T. Pradeep, *Biotechnol. Bioeng.* **90**, 59 (2005).
18. B. Tang, J. Wang, S. Xu, T. Afrin, W. Xu, L. Sun and W. X., *J. Colloid Interf. Sci.* **356**, 513 (2011).
19. A. B. Lansdown, *Adv. Pharmacol. Sci.* 2010, 910686 (2010).
20. H. H. Lara, E. N. Garza-Trevino, L. Ixtepan-Turrent and D. K. Singh, *J. Nanobiotechnol.* **9**, 30 (2011).
21. A. J. Huh and Y. J. Kwon, *J. Control Release* **156**, 128 (2011).

22. H. H. Lara, N. V. Ayala-Nunez, L. Ixtepan-Turrent and C. Rodriguez-Padilla, *J. Nanobiotechnol.* **8**, 1 (2010).
23. M. K. Rai, S. D. Deshmukh, A. P. Ingle and A. K. Gade, *J. Appl. Microbiol.* **112**, 841 (2012).
24. A. Panacek, L. Kvitek, R. Prucek, M. Kolar, R. Vecerova, N. Pizurova, V. K. Sharma, T. Nevecna and R. Zboril, *J. Phys. Chem. B* **110**, 16248 (2006).
25. S. Pal, Y. K. Tak and J. M. Song, *Appl. Environ. Microbiol.* **73**, 1712 (2007).
26. R. Bhattacharya and P. Mukherjee, *Adv. Drug Deliv. Rev.* **60**, 1289 (2008).
27. K. B. Narayanan and N. Sakthivel, *Adv. Colloid Interf. Sci.* **156**, 1 (2010).
28. A. K. Mittal, Y. Chisti and U. C. Banerjee, *Biotechnol. Adv.* (2013).
29. Z. Huang, X. Jiang, D. Guo and N. Gu, *J. Nanosci. Nanotechnol.* **11**, 9395 (2011).
30. M. A. Faramarzi and A. Sadighi, *Adv. Colloid Interf. Sci.* (2012).
31. M. J. Sweet, A. Chesser and I. Singleton, *Adv. Appl. Microbiol.* **80**, 113 (2012).
32. R. Vaidyanathan, K. Kalishwaralal, S. Gopalram and S. Gurunathan, *Biotechnol Adv*, **27**, 924 (2009).
33. J. T. Seil and T. J. Webster, *Int. J. Nanomed.* **7**, 2767 (2012).
34. C. You, C. Han, X. Wang, Y. Zheng, Q. Li, X. Hu and H. Sun, *Mol. Biol. Rep.* **39**, 9193 (2012).
35. J. R. Morones, J. L. Elechiguerra, A. Camacho, K. Holt, J. B. Kouri, J. T. Ramirez and M. J. Yacaman, *Nanotechnology*, **16**, 2346 (2005).
36. U. Klueh, V. Wagner, S. Kelly, A. Johnson and J. D. Bryers, *J. Biomed. Mater. Res.* **53**, 621 (2000).
37. S. Shrivastava, T. Bera, A. Roy, G. Singh, P. Ramchandrarao and D. Dash, *Nanotechnology* **18**, 225103 (2007).
38. O. Choi and Z. Hu, *Environ. Sci. Technol.* **42**, 4583 (2008).
39. S. Silver, *FEMS Microbiol. Rev.* **27**, 341 (2003).
40. M. Chen, Z. Yang, H. Wu, X. Pan, X. Xie and C. Wu, *Int. J. Nanomed.* **6**, 2873 (2011).
41. D. Wei, W. Sun, W. Qian, Y. Ye and X. Ma, *Carbohydr. Res.* **344**, 2375 (2009).
42. J. B. Wright, K. Lam, A. G. Buret, M. E. Olson and R. E. Burrell, *Wound Repair Regen.* **10**, 141 (2002).
43. M. Ip, S. L. Lui, V. K. Poon, I. Lung and A. Burd, *J. Med. Microbiol.* **55**, 59 (2006).
44. H. J. Lee, S. Y. Yeo and S. H. Jeong, *J. Mater. Sci.* **38**, 2199 (2003).
45. N. Duran, P. D. Marcarto, G. I. H. De Souza, O. L. Alves and E. Esposito, *J. Biomed. Nanotech.* **3**, 203 (2007).
46. A. Ingle, A. Gade, S. Pierrat, C. Sonnichsen and M. Rai, *Curr. Nanosci.* **4**, 141 (2008).
47. A. Kumar, P. K. Vemula, P. M. Ajayan and G. John, *Nat. Mater.* 7, 236 (2008).
48. A. Gade, S. Gaikwad, V. Tiwari, A. Yadav, A. Ingle and M. Rai, *Curr. Nanosci.* **6**, 370 (2010).
49. M. Stefan, L. Hritcu, M. Mihasan, D. Pricop, I. Gostin, R. I. Olariu, S. Dunca and V. Melnig, *J. Mater. Sci. Mater. Med.* **22**, 789 (2011).
50. F. Raheman, S. Deshmukh, A. Ingle, A. Gade and M. Rai, *Nano Biomed. Eng.* **3**, 174 (2011).
51. M. L. W. Knetsch and L. H. Koole, *Polymers*, **3**, 340 (2011).
52. K. Y. Nam, *J. Adv. Prosthodont.* **3**, 20 (2011).
53. S. K. R. Namasivayam and S. Ganesh Avimanyu, *Int. J. Med. Res.* **1**, 130 (2011).
54. S. H. Kim, H. S. Lee, D. S. Ryu, S. J. Choi and D. S. Lee, *Korean J. Microbiol. Biotechnol.* **39**, 77 (2011).

55. S. R. Bonde, D. P. Rathod, A. P. Ingle, R. B. Ade, A. K. Gade and M. K. Rai, *Nanosci. Meth.* **1**, 25 (2012).
56. Y. Zhou, Y. Kong, S. Kundu, J. D. Cirillo and H. Liang, *J. Nanobiotechnol.* **10**, 19 (2012).
57. K. Shameli, M. B. Ahmad, S. D. Jazayeri, P. Shabanzadeh, P. Sangpour, H. Jahangirian and Y. Gharayebi, *Chem. Cent. J.* **6**, 73 (2012).
58. L. S. Devi and S. R. Joshi, *Mycobiol.* **40**, 27 (2012).
59. D. A. Henderson, *Bull. World Health Organ.* **65**, 535 (1987).
60. H. F. Hull, N. A. Ward, B. P. Hull, J. B. Milstien and C. de Quadros, *Lancet* **343**, 1331 (1994).
61. E. Domingo, *Vet. Res.* **41**, 38 (2010).
62. D. S. Dimitrov, *Nat. Rev. Microbiol.* **2**, 109 (2004).
63. M. Vitiello, E. Finamore, A. Falanga, K. Raieta, M. Cantisani, F. Galdiero, C. Pedone, M. Galdiero and S. Galdiero, *Intervirol.* **53**, 381 (2010).
64. T. Melby and M. Westby, *Handb. Exp. Pharmacol.* **177** (2009).
65. J. L. Elechiguerra, J. L. Burt, J. R. Morones, A. Camacho-Bragado, X. Gao, H. H. Lara and M. J. Yacaman, *J. Nanobiotechnol.* **3**, 6 (2005).
66. R. W. Sun, R. Chen, N. P. Chung, C. M. Ho, C. L. Lin and C. M. Che, *Chem. Commun. (Camb)*, **40**, 5059 (2005).
67. H. H. Lara, L. Ixtepan-Turrent, E. N. Garza-Trevino and C. Rodriguez-Padilla, *J. Nanobiotechnol.* **8**, 15 (2010).
68. L. Lu, R. W. Sun, R. Chen, C. K. Hui, C. M. Ho, J. M. Luk, G. K. Lau and C. M. Che, *Antivir. Ther.* **13**, 253 (2008).
69. L. Sun, A. K. Singh, K. Vig, S. Pillai, R. Shreekumar and S. R. Singh, *J. Biomed. Biotech.* 4, 149 (2008).
70. D. Baram-Pinto, S. Shukla, N. Perkas, A. Gedanken and R. Sarid, *Bioconjug. Chem.* **20**, 1497 (2009).
71. D. Baram-Pinto, S. Shukla, A. Gedanken and R. Sarid, *Small*, **6**, 1044 (2010).
72. J. V. Rogers, C. V. Parkinson, Y. W. Choi, J. L. Speshock and S. M. Hussain, *Nanoscale Res. Lett.* **3**, 129 (2008).
73. I. Papp, C. Sieben, K. Ludwig, M. Roskamp, C. Bottcher, S. Schlecht, A. Herrmann and R. Haag, *Small*, **6**, 2900 (2010).
74. J. L. Speshock, R. C. Murdock, L. K. Braydich-Stolle, A. M. Schrand and S. M. Hussain, *J. Nanobiotechnol.* **8**, 19 (2010).
75. K. B. Collins, B. K. Patterson, G. J. Naus, D. V. Landers and P. Gupta, *Nat. Med.* **6**, 475 (2000).
76. A. Zussman, L. Lara, H. H. Lara, Z. Bentwich and G. Borkow, *AIDS*, **17**, 653 (2003).
77. A. Mohammed Fayaz, Z. Ao, M. Girilal, L. Chen, X. Xiao, P. Kalaichelvan and X. Yao, *Int. J. Nanomed.* 7, 5007 (2012).
78. S. Gaikwad, A. Ingle, A. Gade, M. Rai, A. Falanga, N. Incoronato, S. Galdiero and M. Galdiero, *Int. J. Nanomed.* **8**, 1 (2013).
79. P. G. Spear, *Cell Microbiol.* **6**, 401 (2004).
80. E. E. Heldwein and C. Krummenacher, *Cell Mol. Life Sci.* **65**, 1653 (2008).
81. A. Turner, B. Bruun, T. Minson and H. Browne, *J. Virol.* **72**, 873 (1998).
82. E. E. Heldwein, H. Lou, F. C. Bender, G. H. Cohen, R. J. Eisenberg and S. C. Harrison, *Science* **313**, 217 (2006).

83. B. P. Hannah, E. E. Heldwein, F. C. Bender, G. H. Cohen and R. J. Eisenberg, *J. Virol.* **81**, 4858 (2007).
84. S. D. Stampfer, H. Lou, G. H. Cohen, R. J. Eisenberg and E. E. Heldwein, *J. Virol.* **84**, 12924 (2010).
85. A. Falanga, R. Tarallo, G. Vitiello, M. Vitiello, E. Perillo, M. Cantisani, G. D'Errico, M. Galdiero and S. Galdiero, *PLoS One*, 7, e32186 (2012).
86. S. Galdiero, A. Falanga, G. Vitiello, M. Vitiello, C. Pedone, G. D'Errico and M. Galdiero, *Biochim. Biophys. Acta* **1798**, 579 (2010).
87. S. Galdiero, A. Falanga, M. Vitiello, L. Raiola, R. Fattorusso, H. Browne, C. Pedone, C. Isernia and M. Galdiero, *J. Biol. Chem.* **283**, 29993 (2008).
88. S. Galdiero, M. Vitiello, M. D'Isanto, A. Falanga, C. Collins, K. Raieta, C. Pedone, H. Browne and M. Galdiero, *J. Gen. Virol.* **87**, 1085 (2006).
89. S. Galdiero, A. Falanga, M. Vitiello, M. D'Isanto, C. Collins, V. Orrei, H. Browne, C. Pedone and M. Galdiero, *Chembiochem.* **8**, 885 (2007).
90. S. Galdiero, A. Falanga, M. Vitiello, H. Browne, C. Pedone and M. Galdiero, *J. Biol. Chem.* **280**, 28632 (2005).
91. S. Galdiero, M. Vitiello, A. Falanga, M. Cantisani, N. Incoronato and M. Galdiero, *Curr. Drug Metab.* **13**, 93 (2012).
92. S. Galdiero, A. Falanga, M. Vitiello, L. Raiola, L. Russo, C. Pedone, C. Isernia and M. Galdiero, *J. Biol. Chem.* **285**, 17123 (2010).
93. P. Mehrbod, N. Motamed, M. Tabatabaian, S. E. R. E. Amini, M. Shahidi and M. T. Kheiri, *DARU*, **17**, 88 (2009).
94. D. X. Xiang, Q. Chen, L. Pang and C. L. Zheng, *J. Virol. Meth.* **178**, 137 (2011).
95. A. N. B. Salem, R. Zyed, M. A. Lassoued, S. Nidhal, S. Sfar and A. Mahjoub, *Dig. J. Nanomater. Bios* 7, 737 (2012).
96. S. Chernousova and M. Epple, *Angew. Chem. Int. Ed. Engl.* (2012).
97. L. Braydich-Stolle, S. Hussain, J. J. Schlager and M. C. Hofmann, *Toxicol. Sci.* **88**, 412 (2005).
98. P. V. AshaRani, G. Low Kah Mun, M. P. Hande and S. Valiyaveettil, *ACS Nano*, **3**, 279 (2009).
99. S. M. Hussain, K. L. Hess, J. M. Gearhart, K. T. Geiss and J. J. Schlager, *Toxicol. In Vitro* **19**, 975 (2005).
100. K. Kawata, M. Osawa and S. Okabe, *Environ. Sci. Technol.* **43**, 6046 (2009).
101. I. Rahman, S. K. Biswas, L. A. Jimenez, M. Torres and H. J. Forman, *Antioxid. Redox Signal* 7, 42 (2005).
102. I. Rahman, *Biochem. Pharmacol.* **60**, 1041 (2000).
103. S. Lanone and J. Boczkowski, *Curr. Mol. Med.* **6**, 651 (2006).
104. K. L. Aillon, Y. Xie, N. El-Gendy, C. J. Berkland and M. L. Forrest, *Adv. Drug Deliv. Rev.* **61**, 457 (2009).
105. Y. S. Kim, J. S. Kim, H. S. Cho, D. S. Rha, J. M. Kim, J. D. Park, B. S. Choi, R. Lim, H. K. Chang, Y. H. Chung, I. H. Kwon, J. Jeong, B. S. Han and I. J. Yu, *Inhal. Toxicol.* **20**, 575 (2008).
106. L. V. Stebounova, A. Adamcakova-Dodd, J. S. Kim, H. Park, P. T. O'Shaughnessy, V. H. Grassian and P. S. Thorne, *Part Fibre Toxicol.* **8**, 5 (2011).
107. R. P. Singh and P. Ramarao, *Toxicol. Lett.* **213**, 249 (2012).
108. S. Grosse, L. Evje and T. Syversen, *Toxicol. In Vitro* (2012).

109. J. F. Hernandez-Sierra, O. Galicia-Cruz, S. A. Angelica, F. Ruiz, M. Pierdant-Perez and A. J. Pozos-Guillen, *J. Clin. Pediatr. Dent.* **36**, 37 (2011).
110. I. Sur, M. Altunbek, M. Kahraman and M. Culha, *Nanotechnology* **23**, 375102 (2012).
111. J. H. Ji, J. H. Jung, S. S. Kim, J. U. Yoon, J. D. Park, B. S. Choi, Y. H. Chung, I. H. Kwon, J. Jeong, B. S. Han, J. H. Shin, J. H. Sung, K. S. Song and I. J. Yu, *Inhal. Toxicol.* **19**, 857 (2007).
112. Y. S. Kim, M. Y. Song, J. D. Park, K. S. Song, H. R. Ryu, Y. H. Chung, H. K. Chang, J. H. Lee, K. H. Oh, B. J. Kelman, I. K. Hwang and I. J. Yu, *Part Fibre Toxicol.* 7, 20 (2010).
113. P. V. Asharani, Y. Lian Wu, Z. Gong and S. Valiyaveettil, *Nanotechnology* 19, 255102 (2008).
114. H. S. Sharma, S. F. Ali, S. M. Hussain, J. J. Schlager and A. Sharma, *J. Nanosci. Nanotechnol.* **9**, 5055 (2009).
115. M. Ghosh, M J. S. Sinha, A. Chakraborty, S. K. Mallick, M. Bandyopadhyay and A. Mukherjee, *Mutat. Res.* **749**, 60 (2012).
116. S. Silver, T. Phung le and G. Silver, *J. Ind. Microbiol. Biotechnol.* **33**, 627 (2006).
117. V. Edwards-Jones, *Lett. Appl. Microbiol.* **49**, 147 (2009).
118. J. Y. Maillard and P. Hartemann, *Crit. Rev. Microbiol.* (2012).
119. A. M. Fayaz, K. Balaji, M. Girilal, R. Yadav, P. T. Kalaichelvan and R. Venketesan, *Nanomedicine* **6**, 103 (2010).
120. P. Jain and V. Aggarwal, *Int. J. Nano Mat. Sci.* **1**, 108 (2012).
121. A. Nanda and M. Saravanan, *Nanomedicine*, **5**, 452 (2009).
122. D. S. Balaji, S. Basavaraja, R. Deshpande, D. B. Mahesh, B. K. Prabhakar and A. Venkataraman, *Colloids Surf. B Biointerf.* **68**, 88 (2009).
123. M. Saravanan and A. Nanda, *Colloids Surf B Biointerf.* 77, 214 (2010).

Chapter 16

Magnetic Nanoparticles for Drug Delivery

*Rainer Tietze, Harald Unterweger and Christoph Alexiou**

Else Kröner-Fresenius-Stiftung Professorship, ENT-Departement
Section for Experimental Oncology and Nanomedicine
University Hospital Erlangen, Germany
**c.alexiou@web.de*

Abstract

Targeted drug delivery, as a vehicle to transport an efficient agent to a specific site of action, has been developed to improve treatment for various diseases, such as cancer or inflammation, since conventional drug administration causes systemic side effects. One possible way to reduce those side effects is to increase the local efficiency by incorporating magnetic particles into drug delivery vehicles. With the use of an external magnetic field, the drug can be accumulated in the desired tissue or organ. Magnetoliposomes and magnetic nanoparticles are presented in this literature overview as commonly used vehicles, having the potential of being designed to release the drug upon an external stimulus, like temperature. Furthermore, additional targeting strategies are discussed, which can be applied on the carrier for a further increase of the local efficiency, as well as recent researches for magnetic drug delivery systems, in terms of their carrier system, drug and targeting strategies. Due to the rising interest and knowledge in toxicological aspects of applied (magnetic) nanomaterial, this issue is also addressed in brief within this chapter.

1. Introduction

In the past two decades, developments in particulate nanotechnology had a huge impact on many research fields, like electronics, tissue engineering, medicine and especially pharmaceutics. The urgent need for new strategies to defeat diseases like cancer is exemplified in statistics from the Robert Koch

Institute in Germany, which revealed that, in 2008, more than 469,800 new cancer cases were reported and for 2012, further 486,000 diseases are expected.[1] In the European Union, the annual number of new cancer incidences was 2.4 million, and about 1.2 million deaths were reported in 2008.[2] Taking a global view, in the year 2007, 7.9 million deaths (13% of all deaths) have been recorded, and WHO predicts that this number will be increasing.[3] The tendency is on the rise, in spite of all efforts that have been made, because of an aging population. Half of all the incidences that have been diagnosed are in a state of a local tumor extension. About 75% of these patients can be treated with established therapies like resection of the tumor and irradiation of the affected area plus systemic chemotherapy or even treatment with tumor antibodies. To the other half of the patients, no promising surgery can be offered. These locally far-extended tumors are treated using a combined schedule of both systemic chemotherapy and irradiation. Conventional drug administration causes systemic side effects, and also it is difficult to maintain the required concentration window to avoid toxicity from overdose, as well as ineffective treatment from underdose.[4,5]

Therefore, targeted drug delivery, as a vehicle to transport a drug to a specific site of action, emerged from this development to improve the treatment for various diseases such as cancer or inflammation. One possible way to increase the local efficiency is to incorporate magnetic particles into drug delivery vehicles and apply an external magnetic field, which is believed to be harmless to biological systems, to direct the accumulation of the pharmaceutical in the desired tissue.[6] Commonly used magnetic particles for this purpose are magnetite (Fe_3O_4), maghemite (γ-Fe_2O_3) or other ferrites.[7] They are biocompatible and have, depending on their size, superparamagnetic characteristics, which lead to an extensive use of these particles in bioseparation, hyperthermia, magnetic resonance imaging (MRI) and drug targeting.[4,7,8] Possible forms of magnetic drug delivery systems are magnetic nanoparticles, nanocapsules and liposomes.[9] Transportation in nanosized ranges makes slightly soluble or instable agents even available to address tumor cells.[10] A multitude of different antitumor drugs, radiotherapeutic nuclides, genetic material and antibodies as macromolecular agents can use nanosized delivery vehicles for a focused enrichment.[11] Thus, high-potential substances can be applied that have not yet been applicable due to their high systemic toxicity or because of metabolic barriers.

In addition to magnetic targeting, it is possible to design the carrier in a way that it responds to external or internal stimuli to release the drug at a specific site of action and thus increases the therapeutic benefit.[5]

In this chapter, general magnetic drug delivery systems and their properties, as well as the principle of active and passive targeting, are presented. Since most studies published are focused on cancer treatment, the presented aspects are discussed from that point of view. Moreover, recent researches for magnetic drug delivery systems are presented, in terms of their carrier system, drug and targeting strategies.

2. General Magnetic Drug Delivery Systems

A promising application in the fields of nanomedicine is drug targeting by the means of a magnetic field. The basic principle of physically directing a magnetic drug carrier system to a specific organ or tissue, is to use an externally applied magnetic field, which is schematically presented in Fig. 1.

The carrier system is applied by intravasal administration through a catheter to the vessel which supplies blood to the desired organ or tissue.[6] As a consequence of the external magnetic field in the proximity of the targeted tissue, an accumulation of drug-loaded magnetic particles appears. Depending on the drug and its linkage to the carrier, the subsequent desired release of the drug can occur through diffusion or desorption or can be triggered by an additional external stimulus.[5] This leads, on one hand, to a reduced drug concentration in overall body compartments, thus decreasing the systemic side effects, whereas on the other hand, the concentration in the affected tissue is effectively increased, resulting in an enhanced therapeutic effect.[4]

There are various possible systems which can be used for magnetic drug targeting. One of the first clinical trials of magnetic particles took place 1966 under the aegis of Alksne *et al.* by generating a closure of intracranial

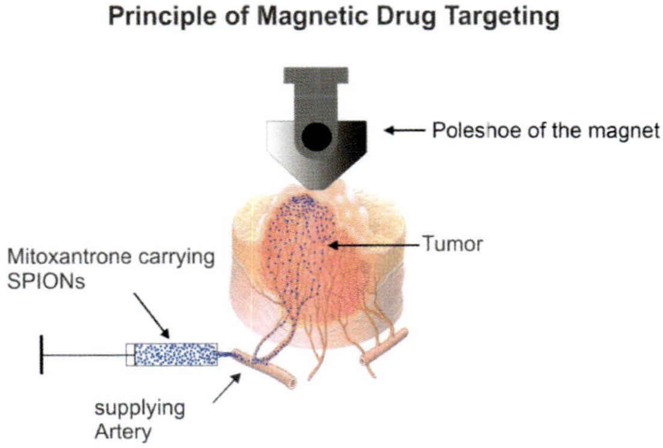

Fig. 1. Schematic presentation of the principle of magnetic drug delivery for tumor targeting.

aneurysms using ferromagnetic iron.[12] By means of magnetic albumin microspheres as the vehicle, magnetic particles and doxorubicin as the cytostatic drug, tumor remissions could be achieved in animal experiments with rats.[13] Carbon-coated iron oxide particles carrying doxorubicin have also been developed for local chemotherapy. These particles (MTC-Dox, Ferex, USA) were applied intravascularly, followed by attraction of an external magnetic field in the scope of a Phase I/II multicenter study on patients with hepatocellular carcinoma.[14] Earlier, Lübbe *et al.* did a first-time human trial with starch-coated magnetic nanoparticles in a Phase I/II study, whereby epirubicin was ionically bound to the particles. The formulation was applied intravenously and its tolerance could be demonstrated.[15] Both magnetic nanoparticles and liposomes are the most prominent candidates for this purpose and will be presented in the following section.

For the concept of magnetic drug targeting (MDT) with intra-arterial administered magnetic nanoparticles, a sufficient particle size (80–150 nm) is important to attract them by the means of an external magnetic field.[16] The advantage of this kind of application is the major local enrichment of particles in the region of interest due to by-passing the metabolic barriers of the mononuclear phagocyte system (MPS). The magnetic force affecting the particles strongly depends on the radius ($\sim r^3$). Besides the particle size, the charge and chemical features are relevant. For *in vivo* applications, the nanoparticles usually are coated for colloidal stability and to prevent segregation into particles and carrier medium.[17,18] A collapse of the colloidal stability would end in the emergence of embolized thrombi. A study on the attraction of particles was performed with a powerful electromagnet with a magnetic field gradient up to 72 T/m at the tip of the pole shoe.[19] Figure 2 exhibits a concrete example of MDT connected with the histological evidence of particle enrichment in tumor tissue. Thereby, biocompatible coated iron oxide nanoparticles electrostatically functionalized with mitoxantrone are administered intra-arterially into tumor-supplying vessels. Hence, a capture by the reticular endothelial system (RES) is avoided as it is observed for intravenous application. For the direction of the particles, an electron magnet with a magnetic field gradient of 72 T/m directly on the tip of the pole shoe, was installed.[19]

Within animal studies, an enrichment of the applied SPIONs in an VX2-squamous cell carcinoma could be demonstrated using histology, MRI and computed microtomography (μCT) after MDT application.[20–22] The same is true for the co-present chemotherapeutic agent.[16,23,24] Further morphological investigations showed no pathological alteration in liver, kidneys, spleen, lung and brain.[25] Applying only 7%–10% of the regular required dose in comparison to conventional chemotherapy complete tumor remission could be very

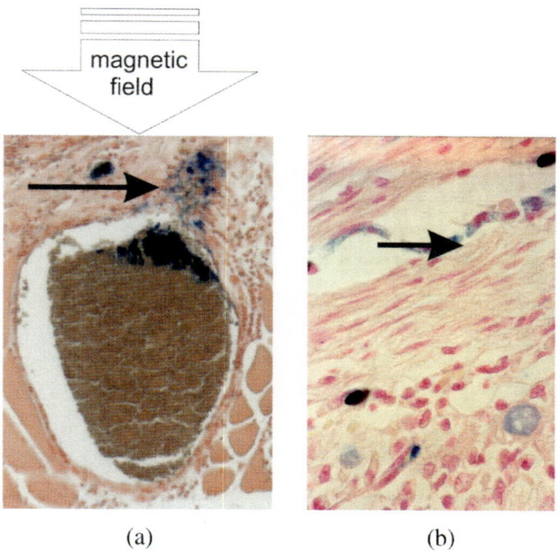

Fig. 2. (a) Particle infiltration into the tumor tissue after MDT. Visible: tumor vessel and extravasations of particles into the tumor tissue (magnification 400-fold, Prussian blue staining) iron oxide appears blue (arrow). (b) Histologic investigation of the tumor tissue after MDT (magnification 1000-fold, Prussian Blue staining) blue stained iron oxide particles are highlighted by an arrow.

convincingly achieved in animal experiments. Angiographic imaging clearly demonstrates that only one MDT application could lead to complete remission of tumor tissue (Fig. 3).[26]

2.1. *Magnetic nanoparticles*

Magnetic iron oxide nanoparticles have already found their way into medical fields. For diagnostic purposes, nanoparticles are used for magnetic cell separation, whereby antibodies featuring a specific affinity are attached to particle surfaces. Hence, cells, DNA or bacteria can be separated magnetically. Moreover, this technique is useful to verify the therapeutic effect after chemotherapeutic treatment.[28] Colloidal magnetic iron oxide particles (ferrofluid) are used already as contrast agents for MRI.[29,30] Commercially available contrast media have been developed, such as Resovist®, Endorem®, Sinerem® and Combidex®. Magnetite and maghemite particles are most commonly used in medicine and they are in general well tolerated.[31] They possess magnetic properties only during magnetic field application. One of the important parameters for *in vivo* application is the size of the particles. Particles at a size of about 100 nm seem to be especially useful. Bigger

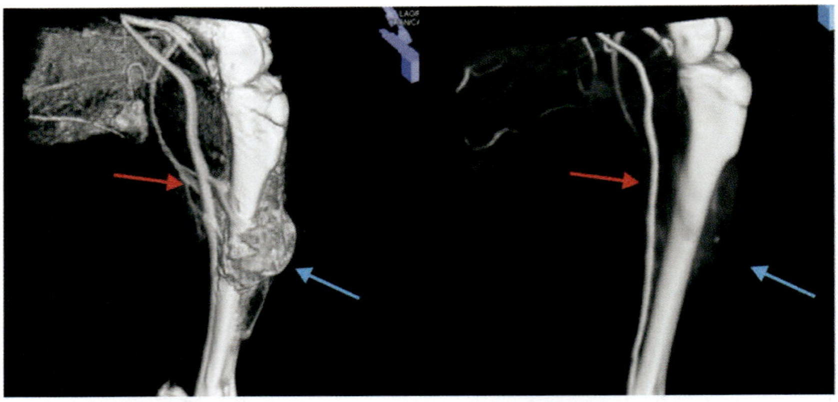

(a) (b)

Fig. 3. Tumor remission after MDT: Angiographic imaging before (a) and after (b): blue arrow = tumor. red arrow = Arteriafemoralis.[27]

particles can easily be caught by MPS cells and enriched in liver and spleen. This is especially true for systemic (intravenous) application.[17] Bigger particles can furthermore lead to vascular obliterations in contrary to smaller ones.[32]

Magnetic nanoparticles are particles containing polymers and drugs along with magnetic particles, for example, magnetite, with diameter between 10 and 500 nm.[7,9] Further characteristics are size and shape distribution, surface charge, magnetite and drug content as well as drug-release behavior. Typical polymers used in this system are poly(D,L-lactic-co-glycolic acid) (PLGA), albumin, ethylcellulose, casein, poly(ethylene glycol) (PEG) and gelatin.[9,33,34] The first step in the production of magnetic nanoparticles is the fabrication of colloidal magnetite, usually by a co-precipitation method first proposed by Khalafalla and Reimers.[35] Possible routes for the succeeding incorporation of these particles into the polymer are interfacial polymerization, enzymatical hydrolyzation, emulsion polymerization or oil-in-water emulsion and subsequent solvent evaporation, as depicted in Fig. 4.[9,34] The drug can be either dispersed inside the polymer particle or attached to its surface, which leads to different release behaviors. In the first case, the drug release mainly depends on diffusion of the drug through the polymer as well as polymer degradation. If the drug is linked to the surface, the breaking of this linkage will be the main release mechanism. To improve the local efficiency and behavior in the vascular circulation, targeting or functional groups can be

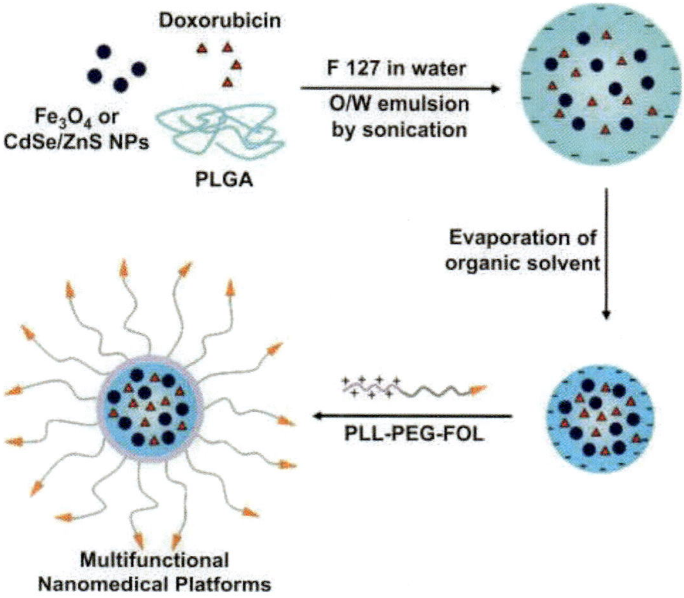

Fig. 4. Synthesis scheme for magnetic nanoparticles with targeting group.[33]

further added to the surface of the particles (the importance of those targeting groups will be discussed in Chapter 3).

2.2. *Magnetic liposomes*

Magnetic liposomes, also referred to magnetoliposomes, have the form of uni- or multilamellar, concentric bilayer vesicles with an aqueous volume entirely closed by a membrane, composed of lipid molecules. The diameter is in the range of 25 nm up to 4 μm. Their main ingredients are phospholipids and cholesterol, components of natural human cell membranes as well as magnetite particles entrapped in their core (see Fig. 5).[6,9,36] This carrier provides a highly flexible system and good chemical functionality due to covalent attachment of ligands to the surface or by incorporation of target lipids. Additionally, both hydrophobic and hydrophilic drugs can be delivered via lipid mediated exchange between the liposomes and the cell membrane.[37]

A huge advantage for magnetic liposomes is the possibility to design various trigger mechanism for the drug release, including pH-, temperature-, or enzyme-triggered release.[5] Especially, temperature-sensitive liposomes have a high potential for cancer treatment. Due to the magnetic particles, they can combine thermally stimulated, reversible permeability for drugs, with the

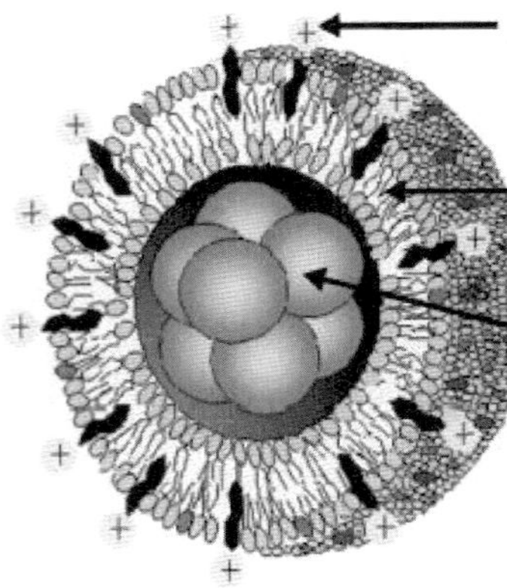

Fig. 5. Magnetic cationic liposome, which can be used for drug targeting or hyperthermia.[38]

possibility of induced hyperthermia, the controlled heating of tissue to promote cell necrosis.[39–41]

3. Strategies in Active and Passive Targeting

Once administered to the body, specific targeting strategies have to be applied to the drug delivery vehicle in order to increase the drug concentration in the relevant tissue. This is especially applicable for most conventional anticancer drugs, as they do not distinguish between healthy and cancerous cells.

In general such strategies can be differentiated into two classes — active and passive targeting — which are described in the following two sections.

3.1. *Passive targeting*

If foreign particles are administered to the body, it is likely that they will be taken up through phagocytosis/endocytosis by macrophages of the reticuloendothelial system (RES).[34] This effect distinctly reduces the efficacy of the drug delivery to cancer cells or inflammation sites. One strategy to reduce this uptake and thus increase circulation time is to attach polymers, e.g. PEG, or macromolecules, e.g. dextran, to the vehicle surface.[34,36] The increased circulation time and the structure of the targeted tissue can now be used for passive targeting, which is demonstrated hereafter for cancerous cells.

Tumor tissue is diffusion-limited, which impacts its metabolism. This state is circumvented by the production of vascular endothelial growth factors (VEGFs) to promote angiogenesis of tumor endothelial cells. The endothelial wall of blood vessels is encapsulated by those cells, which are, in healthy tissue, bound together to prevent leakage of large particles. However, tumor vessels lack the same seal between cells and are considered as leaky for larger particles. Furthermore, tumors do not have a well-defined lymphatic drainage.[36] The combination of those two factors leads to the so-called enhanced permeation and retention (EPR) effect, which means that drug delivery vehicles can enter the interstitium for being entrapped in the tumor.[36,42]

A different possible strategy for passive targeting is the use of congenital properties of the drug delivery vehicle, e.g. surface charge. For example, cationic liposomes (see Fig. 5) can interact electrostatically with negatively charged phospholipid headgroups, preferably expressed on tumor endothelial cells.[42]

3.2. *Active targeting*

Active targeting implies the attachment of specific targeting moieties, ligands or functional groups on the vehicle surface in order to evoke an interaction with the targeted tissue to enhance, in combination with magnetic targeting, the local efficiency. Since most current researches in this field address the treatment of cancer, this section describes active targeting strategies for cancer-related diseases, which, according to Byrne *et al.*, can be divided into angiogenesis-associated and uncontrolled cell proliferation targeting strategies.[42]

3.2.1. Angiogenesis-associated targeting strategies

One characteristic of tumor tissue is described by the promotion of angiogenesis of tumor endothelial cells. The angiogenesis process is believed to be one of the important steps in tumor translation to the metastatic form, capable of spreading in the body. Therefore, angiogenesis-associated targeting strategies are effective in limiting tumor growth.[43] In this context, the main angiogenic targets include vascular endothelial growth factor receptors (VEGFRs), $\alpha_v\beta_3$ integrins, vascular cell adhesion molecules (VCAMs) and matrix metalloproteinases (MMPs).[42]

Vascular endothelial growth factor receptors

VEGFRs are considered to be important inducer of tumor angiogenesis. The most relevant receptor for active targeting is VEGFR-2, which is tyrosine phosphorylated upon ligand binding by VEGF, initiating the signaling cascade for angiogenesis and highly expressed on tumor endothelial cells. Hence, there are two treatment possibilities: first, targeting VEGFR-2 to decrease VEGF-binding or second, targeting VEGF to circumvent ligand binding to VEGFR-2.[44,45]

For the first approach, it is possible to use boronated dendrimers or radioisotope-labeled polymerized liposomes, e.g. anti-VEGFR-2 mAb-NP-90Y.[44] Radiolabeled ^{131}I-anti-VEGF mAbs (Sc-7269) on the other hand, can be utilized in the second approach.[45] Thus, magnetic targeting coupled with VEGF/VEGFR targeting is a successful system for the internalization of drug delivery vehicles by endothelial tumor cells.

$\alpha_v\beta_3$ integrins

Highly expressed on tumor endothelial cells, the $\alpha_v\beta_3$ integrin is important for the calcium-dependent signaling pathway to endothelial cell migration. Moreover, it enacts an endothelial cell receptor for the extracellular matrix (ECM) proteins, harboring the three amino acid sequence arginine–lycine–aspartic acid (RGD). Therefore, RGD sequences harbored on peptides, for instance cyclic RGD pentapeptide and non-peptide mimetic, can be used as an active targeting scheme for $\alpha_v\beta_3$ integrins. Blocking of $\alpha_v\beta_3$ integrin binding results in decreased VEGF signaling and thus reduces angiogenesis.[46]

Vascular cell adhesion molecules

During inflammation and cancer, VCAM-1, an immunoglobulin-like transmembrane glycoprotein, is expressed on the surface of endothelial cells to promote cell-to-cell adhesion. A peculiar feature of this molecule is its virtual absence on normal human vasculature, but it is inducible by angiogenesis. A possibility to target VCAM-1 is given by immunoliposomes conjugated to anti-VCAM-1 mAb (M/K-271).[42,47]

Matrix metalloproteinases

MMPs are a family of structurally related zinc-dependent endopeptidases, playing an essential role for tissue repair, morphogenesis and angiogenesis. They are related to pathological conditions of rheumatoid arthritis, periodontitis, autoimmune blistering of the skin and tumor invasion. Especially MT1-MMP is found to be expressed on endothelial cells and certain types of tumor cells, including lung, gastric, breast and colon cancer. By targeting MT1-MMP, angiogenesis can be reduced as well as the ligand-binding activity of $\alpha_v\beta_3$ integrins. An example for this strategy is conjugating Fab^{222}-1D8 fragments of anti-human MT1-MMP mAb to doxorubicin immunoliposomes.[42,48]

3.2.2. Uncontrolled cell proliferation targeting strategies

Since cell proliferation markers are highly overexpressed on certain tumor cells, they can be used as targets for drug delivery vehicles, functionalized with monoclonal antibodies. To utilize monoclonal antibodies, four basic targeting criteria have to be met: (i) the antigen of choice is overexpressed on tumor cells; (ii) it is a component, playing a role in the progression of the disease; (iii) it is stable upon the tumor surface; and (iv) a large variety of tumors as well as a high percentage of tumor cells express the antigen. Common targets for this strategy are human epidermal, transferrin or folate receptors.[42]

Human epidermal receptors

The epidermal growth factor receptor (EGFR) and human epidermal receptor-2 (HER-2) belong to the family of human epidermal receptors (HER). They are known to trigger growth and proliferation, as a response to the binding of the growth factor ligand. On tumor cell surfaces, HERs are highly upregulated and offer therefore a promising target. With boronated immunoliposomes conjugated with Fab′ fragments of cetuximabm Ab, it is possible to target EFGR, whereas trastuzumab serves as an HER-2 inhibitor.[42,49]

Transferrin receptors

Transferrin is an iron-binding glycoprotein, which helps to transport iron to proliferating cells. Due to angiogenesis and the resulting need of iron, transferrin receptors are overexpressed on tumor cells. Thus, transferrin–polymer–drug conjugates, like transferrin-PEG-admantane (Tf-PEG-AD) have the potential to target malignant tumors.[50]

Folate receptors

The vitamin folic acid is required for essential cell functions, as for example, the synthesis of purines and pyrimidines. Compared to normal tissue, the folate receptor is significantly upregulated on many cancerous cells, in some cases, by two orders of magnitude. As a consequence, it is feasible to use folic acid–liposomal conjugates as a mean for active targeting.[51]

4. Examples of Recent Researches for Magnetic Drug Delivery

Further developments of nanoscaled materials are still in the scope, and this process has been continued for more than three decades. Fullerenes and compounds derived from that, the so-called carbon nanotubes, as well as other condensed carbon clusters, are the most popular species of a class of substances that once have been implicated with "nanotechnology." Those roots date back to the 1970s and early 1980s.[52] Among many other potential applications, using them as tiny carrier to cargo therapeutic substances has become one of the most promising possibilities.[53] Meanwhile, a bulky number of structures in the nanosized scale have been developed to face this challenge. The most important ones are well known as liposomes, micells, encapsulations, emulsions and furthermore, metallic nanoparticles. At first, acting as pure vehicles for carrying drugs effectively over metabolic barriers, later on nano-systems have been developed in order to release agents in a controlled way in the targeted area. The release occurs by alternating the physical and chemical parameters in the area of interest. Transportation and controlled release of therapeutic agents using nano- and micro-carrier systems in the area of interest is a promising approach in current medicine.[11,54] The operation mode thereby can vary considerably. For example it can base on specific attributes of diseased tissues.[55,56] Thus, inflamed tissue is discerned from healthy tissue by lowered pH-value, increased temperature and a higher permeability of vessels.[57] These basic factors can be used for the "targeting" with temperature- and pH-responsive nanoparticles.[58,59] Most of these thermo-responsive systems base on polymer-coated, drug-loaded nanoparticles or capsules.

In this chapter, a selection of recent researches for magnetic drug delivery systems are presented, in terms of their carrier system, drug and targeting strategies. The results are summarized in Table 1.

4.1. *Examples for magnetic liposomes*

This section deals with the description of magnetic liposome-based drug carriers. The first example is given by Babincová *et al.*, who utilized unilamellar magnetoliposomes, consisting of dipalmitoyl-phosphatdylcholine (lipid) and colloidal dispersed iron oxide particles, as a drug delivery vehicle.[39] Using a pH gradient method, the anticancer drug, doxorubicin, was encapsulated inside the liposomes. In an *in vitro* study, the drug release was determined with and without applied alternating magnetic field (AC-field) with a frequency of 3.5 MHz. The group could prove that, without AC-field, the release was minuscule. But an applied AC-field resulted in an increase of the temperature from 37°C to 42°C via Néel relaxation, which lead to the phase transition of the lipid and the following complete release of the drug within minutes. A huge advantage of this system for

Table 1. Summary of discussed research papers for magnetic drug delivery.

Author	Targeted disease	Drug	Targeting group	Drug release mechanism trigger	Carrier
Babincová *et al.*	Cancer	Doxorubicin	n/a	AC-magnetic field controlled heating	Magnetoliposomes
Bothun *et al.*	Cancer	Doxorubicin	Folic acid	AC-magnetic field controlled heating	Magnetoliposomes
Benyettou *et al.*	Cancer	Alendronate	n/a	n/a	Magnetoliposomes
Jain *et al.*	Inflammation	Diclofenac sodium	RGD-sequences	n/a	Magnetoliposomes
Shen *et al.*	Cancer	Adriamycin	n/a	Temperature	Magnetoliposomes
Lee *et al.*	various	Caffein (model)	n/a	Temperature	
Zhang *et al.*	Cancer	Doxorubicin	n/a	Temperature, acidic pH	Magnetic nanoparticles
Dorniani *et al.*	Cancer	Gallic acid	n/a	n/a	Magnetic nanoparticles
Kim *et al.*	Cancer	Doxorubicin	Folic acid	n/a	Magnetic nanoparticles
Arias *et al.*	Cancer	Squalenoyl gemcitabine	n/a	n/a	Magnetite nanoparticles
Stover *et al.*	Cancer	Ceramides	Si-RNA	Temperature	Polymer Nanoparticles
Bakhtiari *et al.*	Various	Model substance	n/a	Photothermal	Magnetic nanoparticles

cancer treatment is the potential use of temperature as a drug release mechanism and as a source for hyperthermia as well.

A similar approach was done by Bothun *et al.*, who also used magnetoliposomes with an AC-field to deliver and release encapsulated doxorubicin.[60] Instead of unilamellar liposomes, they used DOPC1, cationic DPTAP2, cholesterol (Chol), PEG-DMPE3 and maghemite, to fabricate multilamellar magnetoliposomes. With folate-modified PEG2000-DSPE4 as a targeting group, they examined the *in vitro* uptake of this system by human cervical cancer cells (HeLa) and human breast carcinoma cells (ZR-75-1). Since HeLa cells exhibit a high expression of folate receptors (FR), surface binding and cellular uptake of the vehicle could be observed. No uptake could be noticed for lower FR expressing ZR-75-1, proving the importance of active targeting combined with magnetic targeting for an increased local efficiency.

Instead of encapsulating the free drug in the liposome, Benyettou *et al.* decided in their concept to graft their anticancer drug, alendronate, to maghemite through phophonate groups and then encapsulate it to multilamellar liposomes, consisting of DOPC and Chol (9:1).[61] They used an *in vitro* cytotoxicity assay on MDA-MB-231, U87-MG and A431 human cancer cell lines to find out, whether an applied magnetic field, or an encapsulation of the alendronate–maghemite particle by liposomes is necessary for the cellular uptake. The result of this study was that both the magnetic field and the liposomes support the cytotoxicity for all cell lines.

A completely different strategy for a drug delivery vehicle was conducted by Jain *et al.*, by choosing relatively inaccessible inflammatory sites, i.e. the brain, as potential targets.[62] For this purpose, they fabricated negatively charged magnetoliposomes with soya lecithin, Chol, phosphatidyl serine and magnetite, as well as diclofenac sodium, an anti-inflammation agent, as the encapsulated drug. On the surface of the liposomes, RGD sequences were anchored, using carbodiimide as coupling agent. The negative charge and the RGD-sequences lead, *in vivo*, to a selective uptake of the drug delivery vehicle by circulating blood monocytes via receptor-mediated endocytosis. Afterwards, they used a magnetic field to guide the loaded monocytes, known for their capability to penetrate the blood–brain barrier (BBB), to inflammation sites in mice brains. The authors could show that the drug levels for their vehicle was 9.1 times higher compared to the free drug and 6.6 times higher compared to non-magnetic liposomes.

Shen *et al.* developed temperature-sensitive, polymer-conjugated albumin nanospheres being co-polymerized with a poly(N-isopropylacrylamid) moiety.[63] In this case, the release occurs by a dehiscing of the interspaces of the polymer network on the outside of the nanoparticles due to the increased temperature. As a consequence the nanospheres can be attacked and hydrolyzed easily by trypsin.

Poly(N-isopropylacrylamid) is well known for facile alternations of conformation in a temperature level about 32°C in aqueous media and therefore it is often used as co-block polymer for thermo labile surfaces. As Lee *et al.* described in their work, by connecting alternative polymers featuring different conditions for switching their conformation, defined profiles for a release on demand can be arranged.[64] Due to multivalent properties of the surrounding polymer network, temperature- as well as pH-responsive release systems are possible.[59,65] Nevertheless, thermosensitive polymer systems often lack in sharply differentiated thresholds. So the rate of drug release merely differs marginally.

4.2. *Examples for magnetic particles*

In this section drug carrier based on magnetic particles are presented.

The first example for this type is given by Zhang *et al.*, who used a magnetic core–shell nanoparticle carrier.[66] The surface of magnetite was functionalized by chemical bonding with 3-mercaptopropionic acid hydrazide via Fe-S covalent bonds, representing the core. The shell was formed by doxorubicin, which was encapsulated with the thermosensitive polymer dextran-g-poly(NIPAAm-co-DMAAm)5 and attached to the core through an acid-labile hydrazine bond. In an *in vitro* study, they investigated that the collapse of the thermosensitive polymer, and cleavage of the acid-labile hydrazine bond can be stimulated by increasing the temperature and decreasing the pH. As a consequence of the stimuli, an initially rapid drug release occurs.

Protonated chitosan coated on negatively charged magnetite was the drug delivery vehicle for Dorniani *et al.*[7] With gallic acid, a drug against cancer, directly conjugated onto chitosan, the group made a controlled-release study (*in vitro*). They could show that the release of gallic acid increases with decreasing pH. In this context, they fitted the release profile for pH 7.4 with a second-order kinetics model and the profile for pH 4.8 with a parabolic equation.

Kim *et al.* utilized biodegradable PLGA nanoparticles as a dispersion medium for magnetite and doxorubicin.[33] PLL-PEG-FOL6, which was electrostatically attached to the particles, served as a folate-based targeting group. An *in vitro* study revealed a relatively rapid release in the beginning, followed by a sustained release pattern, due to re-protonation of the amino group of doxorubicin and degradation of the PLGA matrix. A further *in vitro* study with KB cancer cells proved that nanoparticles were more readily internalized into the cells through receptor-mediated endocytosis mechanism, compared to the free drug, which was transported to cells by passive diffusion mechanisms.

In a further example, Arias *et al.* employed a magnetite core to synthesize core–shell nanoparticles of squalenoyl gemcitabine (SQdFdC), a squalenoyl prodrug of gemcitabine.[67] A prodrug is a type of drug which is activated by specific conditions in the target-site, otherwise it is inactive.[68] Van der Waals and acid–base interactions lead to the adsorption of SQdFdC onto magnetite nanoparticles. After examining the surface characteristics of the carrier, the group concluded that its surface is comparable to that of pure SQdFdC nanoparticles, which displayed appreciable anticancer activity. Furthermore, the carriers possess the property of being magnetizable. Hence, they are potential candidates for the targeted delivery of gemcitabine to solid tumors.

A different approach is picked up by Stover *et al.*, where ceramides triggering apoptosis are thereby used as effective anti-cancer agents. Nanoparticles with bioactive sphingolipid-like second messenger sequences are incorporated depending on temperature.[69] Ceramides hardly penetrate the cell membrane and are degraded very easily. Linear-dendritic nanoparticles contain a thermoresponsivepoly(N-isopropylacrylamid) moiety that changes the whole polymer conformation above a threshold temperature. Due to the more hydrophobic property, the particles now pass the membrane, become hydrolyzed and release the effective ceramide.

Other concepts (such as that by Bakhtiari *et al.*), which are still away from clinical trials, involve performing drug release by absorbing light. More precisely, the absorption of light causes heat development and leads to bond breaking between the drug and coating substance. As a proof of principle, substances were used to feature fluorescence only in the unbound state. These model compounds display fluorescence after bond breaking to the nanoparticles surface.[70] Above a temperature of 60°C, a Retro-Diels–Alder reaction divides the compound into a terminal maleimid linker and a furane moiety. A Retro-Diels–Adler ring opening could also be achieved enabled by alternating magnetic fields causing heat induction. This kind of defined bond breaking enables an abrupt release of considerably amounts of therapeutic substances and offers new ways of progressed drug release systems. Currently, research is moving toward multifunctional particle systems that can be understood as a platform technique for different concepts.

5. Toxicological Considerations of Magnetic Nanomaterial

5.1. *General toxicological aspects of nanomaterials*

Nanoparticles are commonly categorized into carbon-based nanoparticles, metals, metal oxides and quantum dots. Due to their high surface to volume ratio, nanoparticles exhibit unique properties compared to their larger counterparts. Thus they are very reactive, highly catalytic and may possess dangerous effects. Extensive systematic toxicological analyses of each nanoparticle class have to be performed to provide information about safety and hazardous potential. In addition to environmental exposure and uptake of nanoparticles by inhalation or via the skin, nanoparticles are used in diagnostic and therapeutic approaches. Besides the appearance as single objects, nanoparticles tend to form agglomerates and aggregates, which may influence the biological response.[71] A number of investigations have been conducted concerning toxicological investigations on nanoscaled materials. In general, prior to toxicological experiments, the physical and chemical characteristics of nanoparticles as size, size distribution, morphology, composition and surface area must be accurately defined to ensure reproducibility.

Most of the works about the biological behavior of nanoparticles concern TiO_2-, ZnO-, and Ag-particles; moreover, carbon-based objects (fullerenes, carbon-nanotubes, nano diamonds etc.) are in the scope of toxicological research. These particles own specific reactivities and therefore having specific risks. Titanium dioxide can enrich oxygen species and subsequently could lead to photocatalyzed reactions that have an impact on the surrounding tissue. These effects are strongly dependent from the particle size. TiO nanoparticles have been extensively studied and have served as model particles in various experiments.

Nano silver is commonly used for different applications. In principle, the efficacy is based on the silver ion release from the substrate, which have an antimicrobial effect.

Some fullerenes could cause pro-oxidative reactions and thereby alter the lipid structure of cell membranes. Moreover, well known for their risk potential are multi-walled carbon nanotubes that have morphological similarities to asbestos fibers and finally cause serious inflammation processes. This also applies to fragments of nano-diamonds.[72]

5.2. *Special toxicological aspects of magnetic particles*

Magnetic drug delivery mainly bases on iron oxide and therefor specific considerations have to be taken into account. One is the investigation of applied iron itself and the other is the analysis of iron oxide as a nano-scaled object.

5.2.1. Elemental Iron

Naturally, humans have a Fe content of 45–160 μg/dl in serum. Iron metabolism is mediated by various transport proteins (e.g. ferritin and transferrin). Usually, the daily need for iron can be satisfied by nutritional intake. Therefore, soluble Fe III is reduced to Fe II by cytochrome-b-reductaseand subsequently absorbed by duodenal enterocytes or directly absorbed if Fe II is available. Later on, for passing the blood vessels, Fe II has to be oxidized to Fe III again. In clinics, intravenously applied iron preparations are in a widespread use to treat different kinds of anemia and especially for patients with renal insufficiency.[73] They become more popular, as oral application is not useful in many cases, and *i.v.* application leads to a quick increase of ferritin and saturation of transferrin. Concerning compatibility, high molecular iron complexes show commonly more negative side effects like hypotension, tachycardia, uricaria and dyspnoe than low molecular complexes. Especially, the allergenic potential of Fe-dextran complexes has been proven and makes test injections necessary.[74–76] However, side effects are also observed for labile Fe complexes like Fe-gluconate. The so-induced release of free iron excesses the cargo capacity of transferrin and subsequently promotes pro-oxidative processes. Increasing malon-di-aldehyde release and an overexpression of glutathion peroxidase are parameters that have been recorded for the lipid peroxidation and the presence of free radicals.[77] These effects can be relevant for any intravascular application of iron-containing materials.

5.2.2. Iron as particulate material

Superparamagnetic iron oxide nanoparticles (SPIONs) are in a widespread usage for biomedical applications, even as contrast agents[30,78] and they are generally well tolerated.[31]

In combination with some biocompatible surface coatings, cell damages could be observed but there is a strong dependency on sizing and coating material. However, long-term risks cannot be estimated at the moment. Due to the variety of applied SPIONs, risk profiles are only valid for very specific particle formulations and their discrete parameters.[79] Risk assessment is also impeded as official guidelines concerning rules for the application of SPIONs are not available yet, so they have to meet the overridden requirements for medical devices. In the last years, more efforts have been made to investigate the toxicological aspects of SPIONs.[80]

The main parameters for *in vitro* evaluation that are frequently requested are viability of cells, cytotoxicity, oxidative stress, inflammatory reactions, and genotoxicity. In early studies, varying degrees of cell death, modifications of the vacuoles and damages in fibroblasts have been reported and reconfirmed.[81–83] A number of *in vitro* investigations have been conducted, and a correspondence with the surface coating, influencing size, aggregation and subsequently cellular uptake and biological impact can commonly subsumed.[79] Formation of reactive oxygen species (ROS) is frequently observed, having an impact on cells in spite of distinct defense strategies. Four original sources of oxidative stress can be distinguished: direct generation of ROS from the surface of the nanoparticles, production of ROS via leaching of iron molecules from the surface degradation of SPIONs by enzymatic degradation, altering mitochondrial and other organelle functions as well as induction of cell signaling pathways together with their consequence of activating inflammation, which results in the generation of ROS and reactive nitrogen species.[79] Mechanistically, ROS generation is described as a result of the iron-catalyzed Fenton reaction building highly reactive hydroxyl radicals (see Eq. (1)). Free radicals often arise from redox processes that are connected with single electron transitions. These reactive agents can occur on the surface of iron oxide nanoparticles at distinct particle sizes of 20–30 nm[84] and alternatively, with catalytic amounts of intracellular free iron (especially in its reduced form Fe(II)) can lead to the production of the hydroxyl radical (OH·)).[85] Direct contact of SPIONs to mitochondria could affect the electron flow and also cause cellular damages.[86] Upto a certain extent protein coronas of any intravascular applied particles can be observed. Therefore a lot of subsequent SPION-related toxicity e.g. induction of cell signaling pathways, redox reactions and protein up-regulation could happen.[87–89]

$$H_2O_2 + Fe^{2+} \rightarrow HO \cdot + HO^- + Fe^{3+} \quad (1)$$

In the *in vivo* stage, SPIONs are often characterized for having only scarce toxic effects.[90] Biodistribution studies demonstrate that after *i.v.* administration, the majority of the particles is trapped in liver and spleen.[16] The decomposition of the magnetic structure[91] and the elemental embedding of therefore applied iron into metabolites could be demonstrated.[92] Hence an increasing amount of free iron ions could be monitored having the consequences mentioned in Sec. 5.2.1.

6. Conclusions and Outlook

The main purpose of magnetic drug delivery is the externally controlled accumulation of an effective agent in a specific site of action. Since conventional drug administration causes systemic side effects and it is difficult to maintain the required concentration window, magnetic drug delivery vehicles can increase the local efficiency. For a further increase, active targeting strategies can be put to use. Magnetoliposomes and magnetic nanoparticles commonly serve as vehicles, having the potential of being designed to release the drug upon an external stimulus, like temperature.

Hence, magnetic drug delivery has an enormous potential in the field of therapeutics, in general, and in the treatment of cancer as well as inaccessible inflammation-sites, in particular. Nanomedicine offers a splendid prospect for directed drug application to avoid the negative side effects of conventional tumor therapy mentioned above. This is especially true for eliminating severe side effects of systemic chemotherapy. As other nanoscaled material, SPIONs have to undergo further toxicological considerations to expand the rather scarce knowledge about the hazardous potential.

Acknowledgments

The authors wish to thank Else Kröner-Fresenius Stiftung-Bad Homburg v. d. H., Germany Deutsche Forschungsgemeinschaft (AL552/3-3), Bundesministerium für Bildung und Forschung (FKZ:01EX1012B), Emerging Fields Initiative of the University Erlangen-Nuremberg and Bayerisches Staatsministerium für Umwelt und Verbraucherschutz (74-U8793-2012/7-35).

References

1. Robert Koch-Institute and the Society of Epidemiological Cancer Index in Germany (Eds.), *Cancer in Germany 2007/2008*, Vol. 8 (2012).
2. J. Ferlay, D. M. Parkin and E. Steliarova-Foucher, Estimates of cancer incidence and mortality in Europe in 2008, *Eur. J. Cancer* **46**, 765–781 (2010).
3. *Cancer* (2007). Available at: www.who.int/mediacentre/factsheets/fs297/en/index.html and http://www.who.int/mediacentre/factsheets/fs297/en/index.html.
4. G. Bealle, C. R. Di, J. Kolosnjaj-Tabi, V. Dupuis, O. Clement, F. Gazeau, C. Wilhelm and C. Menager, Ultra magnetic liposomes for MR imaging, targeting, and hyperthermia, *Langmuir* **28**, 11834–11842 (2012).
5. B. P. Timko, T. Dvir and D. S. Kohane, Remotely triggerable drug delivery systems, *Adv. Mater.*, **22**, 4925–4943 (2010).
6. T. P. Vinod, J. H. Chang, J. Kim and S. W. Rhee, Self-assembly and photopolymerization of diacetylene molecules on surface of magnetite nanoparticles, *Bull. Korean Chem. Soc.* **29**, 799–804 (2008).
7. D. Dorniani, H. M. Z. Bin, A. U. Kura, S. Fakurazi, A. H. Shaari and Z. Ahmad, Preparation of Fe3O4 magnetic nanoparticles coated with gallic acid for drug delivery, *Int. J. Nanomed.* 7, 5745–5756 (2012).

8. S. Shi, J.-f. Jia, X.-k. Guo, Y.-p. Zhao, B.-y. Liu, D.-s. Chen, Y.-y. Guo and X.-l. Zhang, Toxicity of iron oxide nanoparticles against osteoblasts, *J. Nanopart. Res.* **14**, 1091/1–1091/11 (2012).
9. A. Aggarwal, P. Chhajer and S. Maheshwari, Magnetic drug delivery in therapeutics, *Int. J. Pharm. Sci. Res.* **3**, 4670–4680 (2012).
10. W. R. Perkins, I. Ahmad, X. G. Li, D. J. Hirsh, G. R. Masters, C. J. Fecko, J. K. Lee, S. Ali, J. Nguyen, J. Schupsky, C. Herbert, A. S. Janoff and E. Mayhew, Novel therapeutic nano-particles (lipocores): trapping poorly water soluble compounds, *Int. J. Pharm.* **200**, 27–39 (2000).
11. Q. A. Pankhurst, N. K. T. Thanh, S. K. Jones and J. Dobson, Progress in applications of magnetic nanoparticles in biomedicine, *J. Phys. D Appl. Phys.* **42** (2009).
12. F. A. Alksne and R. Rand, Magnetically controlled metallic thrombosis of intracranial aneurysms, *Surgery* **60**, 212–218 (1965).
13. M. P. A., K. J. Widder, R. M. Morris, D. P. Howard, G. A. Poore and A. E. Senyei, Selective targeting of magnetic albumin microspheres to the Yoshida sarcoma: ultrastructural evalutation of microsphere distribution, *Eur. J. Cancer. Clin. Oncol.* **19**, 141–147 (1993).
14. M. W. Wilson, R. K. Kerlan, Jr., N. A. Fidelman, A. P. Venook, J. M. LaBerge, J. Koda and R. L. Gordon, Hepatocellular carcinoma: Regional therapy with a magnetic targeted carrier bound to doxorubicin in a dual MR imaging/ conventional angiography suite-initial experience with four patients, *Radiology* **230**, 287–293 (2004).
15. A. S. Lubbe, C. Bergemann, H. Riess, F. Schriever, P. Reichardt, K. Possinger, M. Matthias, B. Dorken, F. Herrmann, R. Gurtler, P. Hohenberger, N. Haas, R. Sohr, B. Sander, A. J. Lemke, D. Ohlendorf, W. Huhnt and D. Huhn, Clinical experiences with magnetic drag targeting: A phase I study with 4'-epidoxorubicin in 14 patients with advanced solid tumors, *Cancer Res.* **56**, 4686–4693 (1996).
16. C. Alexiou, R. Jurgons, R. J. Schmid, C. Bergemann, J. Henke, W. Erhardt, E. Huenges and F. Parak, Magnetic drug targeting — Biodistribution of the magnetic carrier and the chemotherapeutic agent mitoxantrone after locoregional cancer treatment, *J. Drug Target* **11**, 139–149 (2003).
17. G. Storm, S. O. Belliot, T. Daemen and D. D. Lasic, Surface modification of nanoparticles to oppose uptake by the mononuclear phagocyte system, *Adv. Drug Deliv. Rev.* **17**, 31–48 (1995).
18. I. Brigger, J. Morizet, L. Laudani, G. Aubert, M. Appel, V. Velasco, M. J. Terrier-Lacombe, D. Desmaele, J. d'Angelo, P. Couvreur and G. Vassal, Negative preclinical results with stealth((R)) nanospheres-encapsulated Doxorubicin in an orthotopic murine brain tumor model, *J. Control. Release* **100**, 29–40 (2004).
19. C. Alexiou, D. Diehl, P. Henninger, H. Iro, R. Rockelein, W. Schmidt and H. Weber, A high field gradient magnet for magnetic drug targeting, *IEEE Transact. Appl. Superconduct.* **16**, 1527–1530 (2006).
20. C. Alexiou, R. Jurgons, R. Schmid, W. Erhardt, F. Parak, C. Bergemann and H. Iro, Magnetic drug targeting — A new approach in locoregional tumortherapy with chemotherapeutic agents. Experimental animal studies, *HNO* **53**, 618–622 (2005).
21. C. Alexiou, W. Arnold, P. Hulin, R. J. Klein, H. Renz, F. G. Parak, C. Bergemann and A. S. Lubbe, Magnetic mitoxantrone nanoparticle detection by histology, X-ray and MRI after magnetic tumor targeting. *J. Magnet. Magnetic Mater,* **225**, 187–193 (2001).

22. H. Rahn, I. Gomez-Morilla, R. Jurgons, C. Alexiou and S. Odenbach, Microcomputed tomography analysis of ferrofluids used for cancer treatment, *J. Phys. Condensed Matter.* **20**, 204152–204156 (2008).
23. C. Alexiou Ch., P. Hulin, R. J. Klein, Ch. Bergemann and W. Arnold, Magnetic drug targeting: Biodistribution and dependency on magnetic field strength, *J. Magnet. Magnetic Mater.* **252**, 363–366 (2002).
24. R. Tietze, E. Schreiber, S. Lyer and C. Alexiou, Mitoxantrone loaded superparamagnetic nanoparticles for drug targeting: a versatile and sensitive method for quantification of drug enrichment in rabbit tissues using HPLC-UV, *J. Biomed. Biotechnol.* **2010**, 597304 (2010).
25. C. Alexiou, R. J. Schmid, R. Jurgons, M. Kremer, G. Wanner, C. Bergemann, E. Huenges, T. Nawroth, W. Arnold and F. G. Parak, Targeting cancer cells: magnetic nanoparticles as drug carriers, *Eur. Biophys. J. Biophys. Lett.* **35**, 446–450 (2006).
26. S. Lyer, E. Schreiber, R. Tietze, J. Mann, T. Struffert, T. Engelhorn, H. Rahn, K. Gitter, F. Wiekhorst, U. Steinhoff, L. Trahms, S. Odenbach, A. Dorfler, M. Ostermeier, S. Britzen, W. Schmidt, R. Hellinger and C. Alexiou, Nanotechnology and cancer treatment: Magnetic nanoparticles for a new and innovative drug delivery system, *Anticancer Res.* **31**, 1991–1992 (2011).
27. S. Lyer, R. Tietze, R. Jurgons, T. Struffert, T. Engelhorn, E. Schreiber, A. Dorfler and C. Alexiou, Visualisation of tumour regression after local chemotherapy with magnetic nanoparticles — a pilot study, *Anticancer Res.* **30**, 1553–1557 (2010).
28. U. Bilkenroth, H. Taubert, D. Riemann, U. Rebmann, H. Heynemann and A. Meye, Detection and enrichment of disseminated renal carcinoma cells from peripheral blood by immunomagnetic cell separation, *Int. J. Cancer* **92**, 577–582 (2001).
29. M. Taupitz, S. Wagner, B. Hamm, D. Dienemann, R. Lawaczeck and K. J. Wolf, MR Lymphography using iron oxide particles. Detection of lymph node metastases in the VX2 rabbit tumour model, *Acta Radiologica* **34**, 10–15 (1993).
30. M. G. Harisinghani, J. Barentsz, P. F. Hahn, W. M. Deserno, S. Tabatabaei, C. H. van de Kaa, J. de la Rosette and R. Weissleder, Noninvasive detection of clinically occult lymph-node metastases in prostate cancer, *N. Engl. J. Med.* **348**, 2491–2499 (2003).
31. Y. Gao, "Biofunctionalization of magnetic nanoparticles," in *Biofunctionalization of Nanomaterials*. Vol. 1, C. S. Kumar (Ed.), 1st edn. (Weinheim: Wiley-VCH Verlag GmbH & Co. KGaA, 2005) pp. 72–89.
32. R. H. Muller, C. Jacobs and O. Kayser, Nanosuspensions as particulate drug formulations in therapy. Rationale for development and what we can expect for the future, *Adv. Drug Deliv. Rev.* **47**, 3–19 (2001).
33. J. Kim, J. E. Lee, S. H. Lee, J. H. Yu, J. H. Lee, T. G. Park and T. Hyeon, Designed fabrication of a multifunctional polymer nanomedical platform for simultaneous cancer-targeted imaging and magnetically guided drug delivery, *Adv. Mater.* **20**, 478 (2008).
34. S. M. Moghimi, A. C. Hunter and J. C. Murray, Long-circulating and target-specific nanoparticles: Theory to practice, *Pharmacol. Rev.* **53**, 283–318 (2001).
35. S. Khalafalla and G. Reimers, Preparation of dilution-stable aqueous magnetic fluids, *IEEE Transact. Magnetics* **16**, 178–183 (1980).
36. H. K. Sajja, M. P. East, H. Mao, Y. A. Wang, S. Nie and L. Yang, Development of multifunctional nanoparticles for targeted drug delivery and noninvasive imaging of therapeutic effect, *Curr. Drug Discovery Technol.* **6**, 43–51 (2009).

37. M. Gonzales and K. M. Krishnan, Synthesis of magnetoliposomes with monodisperse iron oxide nanocrystal cores for hyperthermia, *J. Magnet. Magnetic Mater.* **293**, 265–270 (2005).
38. A. Ito, M. Shinkai, H. Honda and T. Kobayashi, Medical application of functionalized magnetic nanoparticles, *J. Biosci. Bioeng.* **100**, 1–11 (2005).
39. M. Babincová, P. Čičmanec, V. Altanerová, Č. Altaner and P. Babinec, AC-magnetic field controlled drug release from magnetoliposomes: design of a method for site-specific chemotherapy, *Bioelectrochemistry* **55**, 17–19 (2002).
40. M. Shinkai, M. Yanase, H. Honda, T. Wakabayashi, J. Yoshida and T. Kobayashi, Intracellular hyperthermia for cancer using magnetite cationic liposomes: In vitro study, *Cancer Sci.* **87**, 1179–1183 (1996).
41. M. Yanase, M. Shinkai, H. Honda, T. Wakabayashi, J. Yoshida and T. Kobayashi, Intracellular hyperthermia for cancer using magnetite cationic liposomes: An in vivo study,*Cancer Sci.* **89**, 463–470 (1998).
42. J. D. Byrne, T. Betancourt and L. Brannon-Peppas, Active targeting schemes for nanoparticle systems in cancer therapeutics, *Adv. Drug Deliv. Rev.* **60**, 1615–1626 (2008).
43. J. Folkman, Fighting cancer by attacking its blood supply, *Sci Am.* **275**, 150–154 (1996).
44. M. V. Backer, T. I. Gaynutdinov, V. Patel, A. K. Bandyopadhyaya, B. T. S. Thirumamagal, W. Tjarks, R. F. Barth, K. Claffey and J. M. Backer, Vascular endothelial growth factor selectively targets boronated dendrimers to tumor vasculature, *Mol. Cancer Therap.* **4**, 1423–1429 (2005).
45. J. Chen, H. Wu, D. Y. Han and C. S. Xie, Using anti-VEGF McAb and magnetic nanoparticles as double-targeting vector for the radioimmunotherapy of liver cancer, *Cancer Lett.* **231**, 169–175 (2006).
46. E. Ruoslahti, Specialization of tumour vasculature, *Nature Rev. Cancer* **2**, 83–90 (2002).
47. S. Gosk, T. Moos, C. Gottstein and G. Bendas, VCAM-1 directed immunoliposomes selectively target tumor vasculature *in vivo*, *Biochimica Et Biophysica Acta-Biomembr.* **1778**, 854–863 (2008).
48. H. Hatakeyama, H. Akita, E. Ishida, K. Hashimoto, H. Kobayashi, T. Aoki, J. Yasuda, K. Obata, H. Kikuchi, T. Ishida, H. Kiwada and H. Harashima, Tumor targeting of doxorubicin by anti-MT1-MMP antibody-modified PEG liposomes, *Int. J. Pharm.* **342**, 194–200 (2007).
49. V. Grunwald and M. Hidalgo, Developing inhibitors of the epidermal growth factor receptor for cancer treatment, *J. Natl. Cancer Institute* **95**, 851–867 (2003).
50. N. C. Bellocq, S. H. Pun, G. S. Jensen and M. E. Davis, Transferrin-containing, cyclodextrin polymer-based particles for tumor-targeted gene delivery, *Bioconjug. Chem.* **14**, 1122–1132 (2003).
51. H. Shmeeda, L. Mak, D. Tzemach, P. Astrahan, M. Tarshish and A. Gabizon, Intracellular uptake and intracavitary targeting of folate-conjugated liposomes in a mouse lymphoma model with up-regulated folate receptors, *Mol. Cancer Therap.* **5**, 818–824 (2006).
52. S. Iijima, Helical microtubules of graphitic carbon. *Nature* **354**, 56–58 (1991).
53. K. Kostarelos, A. Bianco and M. Prato, Promises, facts and challenges for carbon nanotubes in imaging and therapeutics, *Nat. Nanotechnol.* **4**, 627–633 (2009).
54. Q. A. Pankhurst, J. Connolly, S. K. Jones and J. Dobson, Applications of magnetic nanoparticles in biomedicine, *J. Phys. D Appl. Phys.* **36**, R167–R181 (2003).

55. W. Arap, W. Haedicke, M. Bernasconi, R. Kain, D. Rajotte, S. Krajewski, H. M. Ellerby, D. E. Bredesen, R. Pasqualini and E. Ruoslahti, Targeting the prostate for destruction through a vascular address, *Proc. Natl. Acad. Sci. U S A* **99**, 1527–1531 (2002).
56. V. P. Torchilin, Drug targeting, *Eur. J. Pharm. Sci.* **11**, S81–S91 (2000).
57. L. E. Gerlowski and R. K. Jain, Microvascular permeability of normal and neoplastic tissues. *Microvasc, Res.* **31**, 288–299 (1986).
58. Z. Shen, W. Wei, Y. J. Zhao, G. H. Ma, T. Dobashi, Y. Maki, Z. G. Su and J. P. Wan, Thermosensitive polymer-conjugated albumin nanospheres as thermal targeting anti-cancer drug carrier, *Eu. J. Pharm. Sci.* **35**, 271–282 (2008).
59. C. Y. Chuang, T. M. Don and W. Y. Chiu, Synthesis of chitosan-based thermo- and pH-responsive porous nanoparticles by temperature-dependent self-assembly method and their application in drug release, *J. Polymer Sci. A Polymer Chem.* **47**, 5126–5136 (2009).
60. G. D. Bothun, A. Lelis, Y. Chen, K. Scully, L. E. Anderson and M. A. Stoner, Multicomponent folate-targeted magnetoliposomes: Design, characterization, and cellular uptake, *Nanomed. Nanotechnol. Biol. Med.* **7**, 797–805 (2011).
61. F. Benyettou, I. Chebbi, L. Motte and O. Seksek, Magnetoliposome for alendronate delivery, *J. Mater. Chem.* **21**, 4813–4820 (2011).
62. S. Jain, V. Mishra, P. Singh, P. K. Dubey, D. K. Saraf and S. P. Vyas, RGD-anchored magnetic liposomes for monocytes/neutrophils-mediated brain targeting, *Int. J. Pharm.* **261**, 43–55 (2003).
63. Z. Y. Shen, G. H. Ma, T. Dobashi, Y. Maki and Z. G. Su, Preparation and characterization of thermo-responsive albumin nanospheres. *Int, J. Pharm.* **346**, 133–142 (2008).
64. C. F. Lee, C. J. Wen, C. L. Lin and W. Y. Chiu, Morphology and temperature responsiveness-swelling relationship of poly(N-isopropylamide-chitosan) copolymers and their application to drug release, *J. Polymer Sci. A Polymer Chem.* **42**, 3029–3037 (2004).
65. C. Y. Chuang, T. M. Don and W. Y. Chiu, Synthesis and properties of chitosan-based thermo- and pH-responsive nanoparticles and application in drug release, *J. Polymer Sci. A Polymer Chem.* **47**, 2798–2810 (2009).
66. J. Zhang and R. D. K. Misra, Magnetic drug-targeting carrier encapsulated with thermosensitive smart polymer: Core-shell nanoparticle carrier and drug release response, *Acta Biomaterialia* **3**, 838–850 (2007).
67. J. L. Arias, L. H. Reddy and P. Couvreur, Magnetoresponsive squalenoyl gemcitabine composite nanoparticles for cancer active targeting. *Langmuir* **24**, 7512–7519 (2008).
68. J. Khandare and T. Minko, Polymer-drug conjugates: Progress in polymeric prodrugs, *Prog. Polym. Sci.* **31**, 359–397 (2006).
69. T. C. Stover, Y. S. Kim, T. L. Lowe and M. Kester, Thermoresponsive and biodegradable linear-dendritic nanoparticles for targeted and sustained release of a pro-apoptotic drug, *Biomaterials* **29**, 359–369 (2008).
70. A. B. S. Bakhtiari, D. Hsiao, G. X. Jin, B. D. Gates and N. R. Branda, An efficient method based on the photothermal effect for the release of molecules from metal nanoparticle surfaces, *Ang. Chem.-Int. Ed.* **48**, 4166–4169 (2009).
71. A. Lankoff, W. J. Sandberg, A. Wegierek-Ciuk, H. Lisowska, M. Refsnes, B. Sartowska, P. E. Schwarze, S. Meczynska-Wielgosz, M. Wojewodzka and M. Kruszewski, The effect of agglomeration state of silver and titanium dioxide nanoparticles on cellular response of HepG2, A549 and THP-1 cells, *Toxicol. Lett.* **208**, 197–213 (2012).

72. H. F. Krug and P. Wick, Nanotoxicology: An interdisciplinary challenge. *Angew. Chem. Int. Ed. Engl.* **50**, 1260–1278 (2011).
73. H. P. Lipp, Peroral und intravenös anwendbare Eisenpräparate, 08/2011 (2011).
74. M. Auerbach and K. Al Talib, Low-molecular weight iron dextran and iron sucrose have similar comparative safety profiles in chronic kidney disease. *Kidney Int.* **73**, 528–530 (2008).
75. G. M. Chertow, P. D. Mason, O. Vaage-Nilsen and J. Ahlmen, Update on adverse drug events associated with parenteral iron. *Nephrol. Dialysis Transplant.* **21**, 378–382 (2006).
76. R. Fletes, J. M. Lazarus, J. Gage and G. M. Chertow, Suspected iron dextran-related adverse drug events in hemodialysis patients. *Amer. J. Kidney Disease* **37**, 743–749 (2001).
77. S. D. Anker, J. C. Colet, G. Filippatos, R. Willenheimer, K. Dickstein, H. Drexler, T. F. Luscher, B. Bart, W. Banasiak, J. Niegowska, B. Kirwan, C. Mori, B. V. Rothe, S. J. Pocock, P. A. Poole-Wilson, P. Ponikowski and F.-H. T. Investigators, Ferric carboxymaltose in patients with heart failure and iron deficiency. *New Engl. J. Med.* **361**, 2436–2448 (2009).
78. J. M. Perez, Iron oxide nanoparticles — Hidden talent, *Nature Nanotechnol.* **2**, 535–536 (2007).
79. M. Mahmoudi, H. Hofmann, B. Rothen-Rutishauser and A. Petri-Fink, Assessing the *in vitro* and *in vivo* toxicity of superparamagnetic iron oxide nanoparticles. *Chem. Rev.* **112**, 2323–2338 (2012).
80. US National Library of Medicine National Institutes of Health, PubMed database. "SPION" and "toxicity" [Online]. Available: http://www.ncbi.nlm.nih.gov/pubmed?term=SPION%2Btoxicity. Accessed on 5 February 2013.
81. A. K. Gupta, C. Berry, M. Gupta and A. Curtis, Receptor-mediated targeting of magnetic nanoparticles using insulin as a surface ligand to prevent endocytosis. *IEEE Trans. Nanobiosci.* **2**, 255–261 (2003).
82. A. K. Gupta and A. S. Curtis, Lactoferrin and ceruloplasmin derivatized superparamagnetic iron oxide nanoparticles for targeting cell surface receptors. *Biomaterials* **25**, 3029–3040 (2004).
83. A. K. Gupta and M. Gupta, Synthesis and surface engineering of iron oxide nanoparticles for biomedical applications. *Biomaterials* **26**, 3995–4021 (2005).
84. M. A. Voinov, J. O. S. Pagan, E. Morrison, T. I. Smirnova and A. I. Smirnov, Surface-mediated production of hydroxyl radicals as a mechanism of iron oxide nanoparticle biotoxicity. *J. Amer. Chem. Soc.* **133**, 35–41 (2011).
85. M. Levy, F. Lagarde, V. A. Maraloiu, M. G. Blanchin, F. Gendron, C. Wilhelm and F. Gazeau, Degradability of superparamagnetic nanoparticles in a model of intracellular environment: follow-up of magnetic, structural and chemical properties. *Nanotechnology* **21** (2010).
86. M. Li, H. S. Kim, L. Tian, M. K. Yu, S. Jon and W. K. Moon, Comparison of two ultrasmall superparamagnetic iron oxides on cytotoxicity and MR imaging of tumors. *Theranostics* **2**, 76–85 (2012).
87. A. Afaf El, A. S. Warsy and S. A. Hassan, On the biological phenomenon of nanoparticle-cellular protein interactions. *Int. J. Pharma Bio Sci.* **3**, 645–657 (2012).

88. M. Mahmoudi, I. Lynch, M. R. Ejtehadi, M. P. Monopoli, F. B. Bombelli and S. Laurent, Protein-nanoparticle interactions: Opportunities and challenges, *Chem. Rev.* (*Washington, DC*), **111**, 5610–5637 (2011).
89. L. Treuel and G. U. Nienhaus, Toward a molecular understanding of nanoparticle-protein interactions. *Biophys, Rev.* **4**, 137–147 (2012).
90. T. K. Jain, M. K. Reddy, M. A. Morales, D. L. Leslie-Pelecky and V. Labhasetwar, Biodistribution, clearance, and biocompatibility of iron oxide magnetic nanoparticles in rats. *Mol. Pharm,* **5**, 316–327 (2008).
91. M. Levy, N. Luciani, D. Alloyeau, D. Elgrabli, V. Deveaux, C. Pechoux, S. Chat, G. Wang, N. Vats, F. Gendron, C. Factor, S. Lotersztajn, A. Luciani, C. Wilhelm and F. Gazeau, Long term *in vivo* biotransformation of iron oxide nanoparticles, *Biomaterials* **32**, 3988–3999 (2011).
92. D. Pouliquen, J. J. Le Jeune, R. Perdrisot, A. Ermias and P. Jallet, Iron oxide nanoparticles for use as an MRI contrast agent: Pharmacokinetics and metabolism, *Magnet. Reson. Imaging* **9**, 275–283 (1991).

Chapter 17

Quantum Dots as a Platform Nanomaterial for Biomedical Applications

Eleonora Petryayeva[*], *Roza Bidshahri*[*,†], *Kate Liu*[*], *Charles A. Haynes*[†], *Igor L. Medintz*[‡,§], *and W. Russ Algar*[*,¶]

[*]*Department of Chemistry, University of British Columbia 2036 Main Mall, Vancouver, BC V6T 1Z1 Canada*
[†]*Chemical and Biological Engineering Department, University of British Columbia 2185 East Mall, Vancouver, BC V6T 1Z1 Canada*
[‡]*Center for Bio/Molecular Science and Engineering, U.S. Naval Research Laboratory Code 6900, 4555 Overlook Avenue SW, Washington, DC 20375 USA*
[§]*igor.medintz@nrl.navy.mil*
[¶]*algar@chem.ubc.ca*

Abstract

Quantum dots (QDs), or colloidal semiconductor nanocrystals, possess a set of optical properties that are cumulatively unique and highly advantageous. These optical properties arise from the nanoscale size of QDs through quantum confinement. This chapter begins by describing the optical properties, synthesis, surface functionalization, and bioconjugation of QDs. Various biomedical applications of QDs are then discussed, including *in vitro* diagnostics and biosensing, with an emphasis on multiplexing, energy transfer, and charge transfer; tissue and cellular imaging applications such as immunohistochemistry, immunocytochemistry, and tracking or sensing ligand-receptor interactions; and finally *in vivo* imaging and research toward drug delivery and photodynamic therapy with QDs. Challenges such as the potential toxicity of QDs and their biodistribution are also discussed, along with a brief outlook on future directions for research and development with QDs.

1. Introduction

Quantum dots (QDs), or colloidal semiconductor nanocrystals, are one of many promising nanomaterials for biomedical applications. Like most other nanoparticles, QDs have the ability to serve as scaffolds or carriers that can access cellular environments; however, QDs also have a set of optical properties that are both highly advantageous and cumulatively unique amongst other nanomaterials. The most striking of these properties is bright luminescence that depends on the size and composition of the QD nanocrystal, and which, while robust, can also be modulated in response to biological stimuli. Most, if not all of the prospective biomedical applications of QDs aim to harness their optical properties for generating image contrast or analytical signals. These properties can be harnessed by tailoring the interfacial properties of the QD to impart biological activity. Interestingly, biomedical applications for QDs were not foreseen upon their initial discovery but accounted for $48M of the $67M market for QD materials in 2010.[1] Since many researchers synthesize their own QDs or obtain materials through collaborations, this amount significantly underestimates the overall interest in QDs for biomedical applications, which has been growing since Bruchez *et al.* and Chan *et al.* first demonstrated cellular labeling with QDs in 1998.[2,3]

This chapter briefly summarizes the current state-of-the-art in engineering and applying QDs for biomedical applications. Topics include the basic chemistry and properties of QDs; applications such as assay development, biosensing, cellular and tissue labeling, *in vivo* imaging, and therapeutic delivery; as well as ongoing challenges such as mitigating the potential for toxicity and controlling biodistribution. While QDs have not yet made the leap from the laboratory bench to the clinic, QDs have made their impact by facilitating the development of multiplexed bioanalyses, by elucidating cellular structure and function, and by demonstrating promise for the diagnosis and treatment of disease. These advances have been the product of concerted interdisciplinary efforts that have tackled challenges such as synthesis, functionalization and characterization; methods for generating optical signals in response to physical and molecular processes; measuring the response of biological systems to QDs; and integrating the above knowledge into the design of "smart" multifunctional nanoparticles. Only an introduction to these concepts can be provided here, and the interested reader is referred to the cited references and current literature for more information on this exciting and diverse field of research.

2. Optical Properties of QDs

QDs are colloidal semiconductor nanocrystals with diameters between 2–10 nm (Fig. 1a) and are most commonly composed of II–VI (e.g. CdSe, CdTe) or III–V (e.g. InP, InAs) materials. The crystalline structure of QDs is analogous to that of the bulk material, except that a very large fraction of its constituent atoms are located at the surface of the crystal. This feature makes QD cores particularly susceptible to degradation and modification of their electronic properties by adsorbates or other interfacial phenomena. To overcome these issues and improve optical properties, cores are typically coated with a thin shell of another wider band-gap semiconductor material (e.g. core/shell: CdSe/ZnS, CdTe/ZnS). It should be noted, however, that even with a high-quality inorganic shell, the optical properties of QDs are still somewhat influenced by surface ligands and other environmental factors such as pH and temperature.[4]

The absorption of light by a QD generates an exciton (electron-hole pair) and subsequent recombination leads to photoluminescence (PL). The renowned brightness of QDs is a result of high quantum yields (0.1–0.9) and

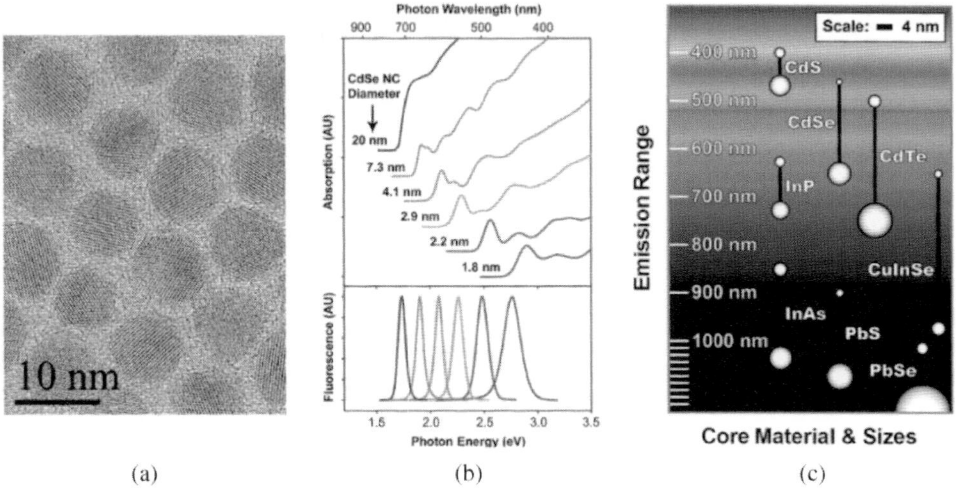

Fig. 1. (a) TEM image of CdSe/ZnS QDs (courtesy Dr. E. Oh, U.S. NRL). (b) Size-dependent absorption and fluorescence spectra of CdSe QDs. The 20 nm particle represents bulk material (no quantum confinement). QD PL spectra (fluorescence) are shown below (and slightly red-shifted from) the corresponding absorption spectra (1.8–7.3 nm, from left to right). Reprinted with permission from Ref. 5. Copyright 2010 American Chemical Society. (c) Approximate PL emission ranges for various QD core sizes and semiconductor materials. Reprinted with permission from Ref. 6. Copyright 2011 American Chemical Society.

broad absorption spectra with extinction coefficients (10^4–10^7 $M^{-1}cm^{-1}$) that progressively increase into the near-UV (Fig. 1b). Narrow, symmetric QD PL is red-shifted relative to the first exciton absorption peak, which, in combination with the broad absorption spectrum, can lead to *effective* Stokes shifts >100 nm (Fig. 1b). The best known characteristic of QDs is their size- and material-tunable PL, which allows tuning of emission wavelengths between 380–2000 nm (Fig. 1c), albeit that the size-tuning range is ~150 nm for a given material composition.[7] The combination of broad absorption spectra, which allows multiple colors of QD to be excited with a common wavelength of incident light, and narrow symmetric PL emission (full-width-at-half-maximum, FWHM, 25–35 nm), which permits resolution of many optical signals within a limited spectral range, is very important for multicolor imaging and multichannel detection. These two unique properties of QDs have facilitated numerous multiplexed bioanalyses that would be very technically challenging with conventional fluorescent dyes and other luminescent materials. These latter materials are typically plagued by broad, overlapping emission spectra, or the need for multiple excitation wavelengths.

In addition, to the above optical properties, QDs have relatively long fluorescence lifetimes (>10 ns), excellent resistance to photobleaching, and very large two-photon absorption cross-sections (10^3–10^4 GM). The latter two properties make QDs particularly suitable for long-term cellular imaging and multiphoton imaging, respectively.

The size-tunable PL properties of QDs are a consequence of quantum confinement. At nanocrystal diameters smaller than the Bohr exciton diameter (e.g. 9.6 nm for CdSe),[8] the density of electronic states decreases to a level intermediate to bulk materials and molecules, transitioning to discrete states at the semiconductor band edges with a corresponding increase in band-gap energy and exciton confinement potential as nanocrystal size decreases.[9–12] Since PL originates from exciton recombination at the band edges, the PL of, for example, a CdSe QD, can be tuned across the visible spectrum from blue to red as size increases from *ca.* 2 to 6 nm. Given that bulk CdSe has a near-infrared band-gap, the effect of quantum confinement is quite evident. Selection of a semiconductor material with a certain band-gap energy ultimately determines wavelength range over which QD PL can be size-tuned. For example, CdTe[13] and InP,[14] with bulk band-gap energies of 1.43 and 1.34 eV, respectively, compose QDs with long-wavelength visible and near-infrared (NIR) emission that is useful for tissue imaging.[15]

In addition to binary QDs, alloyed QDs such as $CdSe_xTe_{1-x}$ and CdS_xSe_{1-x} provide the possibility of tuning band-gap energy and the peak PL emission of QDs without changing the nanocrystal size, but rather by controlling the relative composition of the core material (*i.e.* shift PL emission by varying the mole fraction, *x*, of an elemental component).[16,17]

3. Synthesis of QDs

Following the discovery of colloidal QDs by Brus in 1980s,[18,19] optimized synthetic methods for preparing monodisperse cadmium chalcogenide QDs were developed by the groups of Alivisatos, Bawendi, Guyot-Sionnest, and others during the 1990s.[9,20–25] These advances were critical to pushing QDs to their current frontiers in biomedical research.

The breakthrough method for synthesizing monodisperse CdX QDs (X = S, Se, Te), and the basis for many current methods, involved the rapid injection of dimethyl cadmium (Me_2Cd) and trioctylphosphine chalcogenide (TOPSe, TOPS, TOPTe) into a hot (300°C) coordinating mixture of trioctylphosphine oxide (TOPO) and trioctylphosphine (TOP).[20] QDs with narrow size distribution and good PL quantum yield were produced with this method,[20] and average particle size was controlled through the reaction time. The hot-injection method succeeded in separating the particle nucleation and growth phases, which had been proposed as the necessary condition for making monodisperse colloids.[26,27] Injection of precursors into the hot solvent instantaneously nucleates nanocrystals ("nucleation burst") from the sudden supersaturation of Cd/X monomers. The degree of supersaturation decreases as monomers are consumed, leading to the end of the nucleation process and the beginning of growth stage. Smaller particles grow faster than larger ones, resulting in "focusing" of the size distribution to yield nearly monodisperse QDs.[21]

A second major advance in the synthesis of highly luminescent QDs came with the development of core/shell type nanoparticles.[22,28] Passivation of surface states with an overlayer of structurally similar semiconductor with higher band-gap energy (e.g. CdSe/ZnS, CdTe/ZnS) can eliminate low energy traps for the exciton, offers superior confinement of the exciton, and promotes radiative recombination. The synthesis of a ZnS shell around a CdSe QD is similar to the core synthesis described above, but involves the slower addition of shell precursors, such as bis(trimethylsilyl)sulfide ($(TMS)_2S$) and diethyl zinc (Et_2Zn), at lower temperature, after the core synthesis and purification.

Another important refinement to the synthesis of high-quality QDs was made by Peng's group at the turn of the millennium: the highly toxic and pyrophoric Me_2Cd precursor was replaced by cadmium oxide (CdO).[29] This alternative, and the use of other solid cadmium precursors (e.g. cadmium acetylacetonate), remain very popular.

Use of the TOPO/TOP solvent system for QD synthesis is often associated with variability in growth dynamics and the final particle shape, which has been attributed to impurities in TOPO. Indeed, technical grade TOPO (90%) has been reported to provide better conditions for the growth of QDs than distilled TOPO.[24] Variations of the synthetic method have therefore introduced additional coordinating ligands (e.g. alkylphosphonic acids, stearic acid, hexadecylamine) to improve focusing of the size distribution.[24,30–32] Non-coordinating solvents, such as octadecene (ODE), have also been investigated with the goal of better controlling growth kinetics by carefully varying the concentration of oleic acid or other coordinating ligands in the ODE.[33] As a solvent, ODE also offers the advantage of being more environmentally benign.[33]

Although most high quality nanocrystals are synthesized in organic solvents at high temperature, under anhydrous and anaerobic conditions, there is ongoing interest in synthesizing high-quality QDs in aqueous media. The most success in this endeavour has been with CdTe QDs, which can be synthesized in water with quantum yields up to *ca.* 40% and reasonable monodispersity (FWHM 30–60 nm).[34–36] Many other efforts have been made to improve the safety and simplicity of QD synthesis by using more environmentally-friendly starting materials and milder reaction conditions, ideally without compromising the quality of the QDs obtained. For example, Beri *et al.* synthesized ZnSe QDs from 1,2,3-selenadiazole, zinc acetate, and oleic acid, eliminating the use of toxic TOPO.[37] Siy *et al.* developed a low-temperature organometallic method to synthesize high quality CdSe QDs with narrow size distribution (FWHM 28–35 nm) and good optical properties (quantum yields up to 10%); this method could be advantageous for scale-up production (from milligrams to grams).[38]

4. Quantum Dot Surface Chemistry and Bioconjugation

4.1. *Coatings*

QDs synthesized in organic solvents are coated with hydrophobic stabilizing ligands and cannot be dispersed in aqueous media, which is a necessary condition for bioassays and imaging. The most prominent strategies for aqueous solubilization are (1) encapsulation with amphiphilic block copolymers,[39–43] and (2) ligand exchange with bifunctional molecules.[3,44,45] The polymers used for encapsulation typically incorporate hydrophobic alkyl chains and hydrophilic functional groups such as carboxylic acids (Fig. 2(i)) or poly(ethylene glycol) (PEG) oligomers (Fig. 2(ii)). The hydrophobic chains interdigitate with the alkyl chains of the native ligands from QD synthesis while the

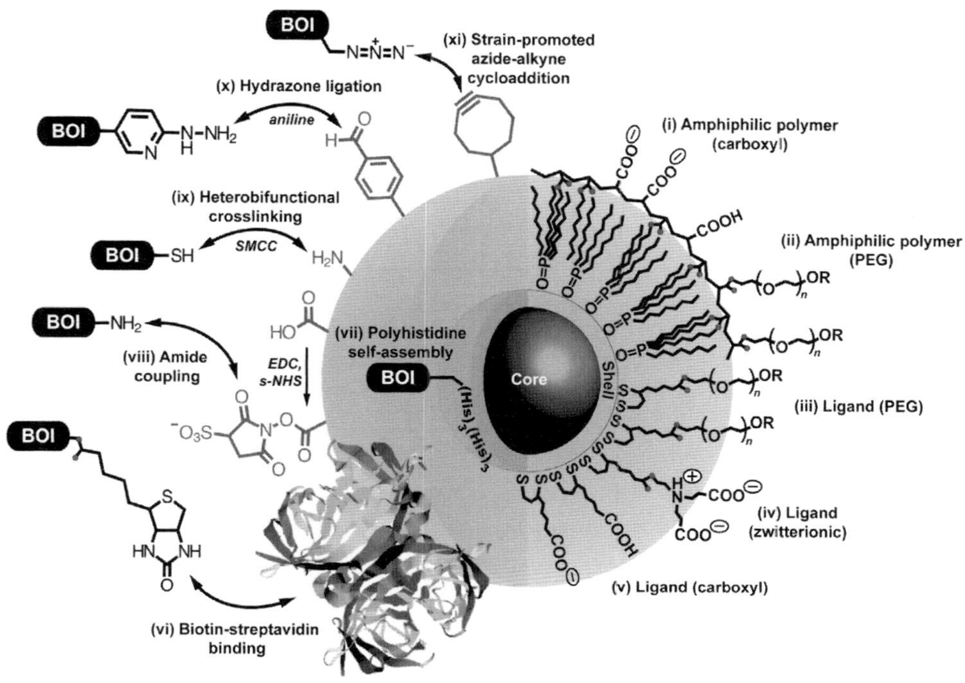

Fig. 2. An illustration of selected chemistries on core-shell QDs. Coatings for aqueous solubility: (i) amphiphilic polymer coating with carboxyl(ate) groups; (ii) amphiphilic polymer coating with PEG oligomers; (iii) dithiol ligand with a distal PEG oligomer; (iv) dithiol ligand with a distal zwitterionic functionality; (v) dithiol ligand with a distal carboxyl(ate) group. Common R groups include carboxyl, amine, and methoxy, although many others can be introduced (*e.g.* see vi, x, xi). Methods for conjugating biomolecules of interest (BOI); (vi) biotin–streptavidin binding; (vii) polyhistidine self-assembly to the inorganic shell of the QD; (viii) amide coupling using EDC/s-NHS activation; (ix) heterobifunctional crosslinking using succinimidyl-4-(*N*-maleimidomethyl) cyclohexane-1-carboxylate (SMCC; structure not shown); (x) aniline-catalyzed hydrazone ligation; and (xi) strain-promoted azide-alkyne cycloaddition. The double arrows are intended to represent conjugation between the functional groups and, in principle, their interchangeability (not reaction mechanisms or reversibility). Not drawn to scale. [Reprinted with permission from Ref. 48. Copyright 2013 Society for Applied Spectroscopy.]

hydrophilic groups are exposed to aqueous solution to mediate solubility and provide functional groups for bioconjugation (*vide infra*). Polymeric coatings largely retain the initial brightness of QDs[42] but typically yield hydrodynamic diameters (HDs) between 20–30 nm, thus hindering their intracellular mobility, reducing their effectiveness in certain assay/biosensing formats (e.g. energy transfer, *vide infra*), and preventing renal clearance *in vivo* (the approximate size threshold for clearance is a HD <5.5 nm).

The ligands used to prepare water-soluble QDs are quite diverse in the nature of their distal functional groups (e.g. carboxylic acids, amine and hydroxyl groups, PEG, biotin, *etc.*), but generally have one or more thiol groups that anchor to the QD (Figs. 2(iii–v)).[46] Thiol ligands bind to the QD by coordinating to the metal atoms at its inorganic interface (*i.e.* Zn^{2+} for most core/ZnS shell QDs, Cd^{2+} for CdX core-only QDs). Exchange of the native QD ligands for thiol ligands is driven by mass-action in presence of an excess of monodentate thiols (e.g. mercaptopropionic acid, cysteine) or bidentate thiols (e.g. dihydrolipoic acid and an extensive library of PEG and other derivatives).[47] The latter provide more stable coordination and improve the long-term colloidal stability of aqueous QDs.

Selection of the hydrophilic functional group(s) used to coat QDs is important for colloidal stability. Carboxyl-terminated ligands are susceptible to aggregation as ionic strength increases and at neutral or acidic pH,[49–51] and tend to be prone to the non-specific adsorption of proteins and other biomolecules. The use of PEGylated coatings largely ameliorates these issues. For example, PEGylated ligands with a tetradentate thiol anchoring moiety have been synthesized and shown to maintain aqueous colloidal stability across a very wide range of pH (1–14) and ionic strength (up to 2 M).[52] Zwitterionic ligands are an alternative to PEG ligands and offer many similar advantages (e.g. colloidal stability over a wide pH range, minimal non-specific adsorption, *etc.*) with the additional benefit of a more compact coating.[10,53,54] Overall, the ideal coating is one that confers aqueous solubility and biocompatibility, provides means for bioconjugation (*vide infra*), protects the QD from the surrounding environment (and *vice versa*), and supports the necessary biological activity/function. Coatings that meet all of these criteria are essential to the development of next generation QD materials for biomedical applications.

4.2. *Bioconjugation strategies*

Common bioconjugation strategies can be broadly classified as either (1) covalent coupling or (2) self-assembly, and have been used to attach enzymes, proteins, antibodies, peptides, oligonucleotides, carbohydrates, and other bioactive materials (e.g. drugs) to QDs. The ideal bioconjugate chemistry preserves the optical properties and colloidal stability of the QD, provides reproducible control over the orientation and number of conjugated biomolecules per QD, and largely retains the native activity of the conjugated biomolecule. Commonly utilized bioconjugate methods address these criteria to varying degrees.

Covalent coupling or crosslinking methods, such as those commonly done with reagents such as carbodiimides, succinimidyl esters, and maleimides, (Figs. 2(viii–ix)) are most effective with mono-reactive biomolecules such as custom synthesized oligonucleotides and peptides. In contrast, proteins often have multiple reactive functional groups (*i.e.* carboxyls and amines) that can induce crosslinking and aggregation between nanoparticles, requiring careful optimization of protocols. Another potential challenge is the limited control over the orientation of the conjugated protein, potentially resulting in a sub-population with, for example, an occluded binding site and reduced biological activity. Similarly, mixed conjugate valence (*i.e.* a large variation in the number of biomolecules per QD across the ensemble) can result in mixed avidity and poor reproducibility.[10]

The limitations of conventional conjugation methods have stimulated interest in developing bioorthogonal and chemoselective ligation chemistries that are suitable for more controlled nanoparticle functionalization.[55] Three prominent examples that have been applied with QDs include the copper-free azide-alkyne cycloaddition (an example of 'click' chemistry) (Fig. 2(xi)),[56,57] hydrazone coupling (Fig. 2(x)),[58] and intein-mediated chemical ligation.[59] Enzyme-catalyzed ligation methods have also been used to functionalize QDs, including those based on biotin ligase, acyl carrier protein, and the HaloTag system. While the above conjugation methods are highly specific and efficient, they may be limited by slow reaction rates, poor scalability (high cost), or the requirement of modifying the native biomolecule of interest to introduce a bioorthogonal functional group (e.g. protein fusions or the incorporation of artificial amino acids).[55]

Self-assembly conjugation methods are typically based on the spontaneous coordination of proteins appended with polyhistidine tags (Fig. 2(vii)) or metallothionein domains to the inorganic surface of a QD (e.g. binding to zinc atoms associated with a ZnS shell). These interactions have nanomolar dissociation constants, provide excellent control over the average conjugate valence on the basis of mixing stoichiometry (no excess reagents or purification), and reproducibly orient the bound protein. Polyhistidine self-assembly has been extended to the conjugation of oligonucleotides and peptides to QDs.[60–62] Similarly, peptides with a terminal cysteine residue or oligonucleotides with a thiolated linker can also self-assemble to the inorganic interface of QDs. While self-assembly methods generally require a ligand coating so as to permit the biomolecule to access the inorganic surface of the QD, polyhistidine self-assembly has also been adapted to polymer coatings *via* modification of the polymer with Ni^{2+}-nitrilotriacetic acid (NTA)[63] or direct coordination to Ni^{2+} ions.[64]

Perhaps the most frequently applied bioconjugate method with QDs has been the use of biotinlyated biomolecules with commercially available streptavidin-modified QDs (Fig. 2(vi)). While this method is facile and robust, the degree of control is limited by the nature of the underlying streptavidin modification, which tends to have a heterogeneous distribution of protein orientation and conjugate valence across the ensemble. The conjugates are also quite large due to the combined polymer coating and streptavidin modification.[6]

4.3. *Bioconjugation to enable active tumor targeting*

A prospective biomedical application of QDs is utilization as contrast agents to light up tumors near the surface of the body (e.g. lymph nodes, skin, breast, testes, colon). Such targeting is based on functionalizing QDs to selectively bind to cancer cells *via* the upregulation of certain surface receptors or antigens relative to healthy cells. To this end, QDs are commonly modified with a cognate peptide or antibody. Peptides are often preferred due to their smaller size and their facility for forming polyvalent assemblies with QDs, which can increase the conjugate affinity for cell surface receptors. The known library of "homing peptides" is extensive; a few prominent examples include F3 (a fragment of HMGN2 protein that targets blood vessels and tumor cells), RGD/cyclic RGD (arginine-glycine-aspartate; targets integrin $\alpha_v\beta_3$ and $\alpha_v\beta_5$ proteins associated with tumor angiogenesis and metastasis), GFE (targets membrane dipeptidase of endothelial cells in lung blood vessels), and Lyp-1 (targets lymphatic vessels and tumors).[65–68] QD–antibody conjugates are equally, if not more, utilized for targeting QDs to cancerous tissue. Two common examples are the use of monoclonal anti-prostate specific membrane antigen (PSMA) to target QDs to (mouse) prostate tumor xenographs,[69] and anti-human epidermal growth factor receptor 2 (HER2) to target QDs to (mouse) breast tumors,[70] amongst many others.

5. Biomedical Applications of QDs

The favorable optical properties of QDs paved the way for their use as fluorescent labels (e.g. immunoassays, optical barcodes, cell and tissue imaging), active reporters (e.g. energy and charge transfer (CT) assays), and vectors for drug delivery and therapy. This section highlights representative examples of the above and frequently emphasizes the multiplexing advantages afforded by QDs.

5.1. *In Vitro diagnostics and biosensing*

5.1.1. *Quantum dots as optical labels for multiplexed detection*

The importance of screening for multiple biomarkers in biomedical assays and diagnostics is widely recognized and, as noted previously, QDs have ideal properties for optical multiplexing. One method of harnessing these properties is to associate different biomarkers of interest with different colors of QDs (*i.e.* N colors for N biomarkers). For example, a four-plex immunoassay for the simultaneous detection of ricin, staphylococcal enterotoxin B, shiga-like toxin I, and cholera toxin has been developed using four different colors of QD–antibody conjugates.[71] The primary advantage of the method was that the mixture of toxins was analyzed in a single microtiter well (with UV excitation at a single wavelength), permitting higher throughput and analysis of smaller sample volumes. Many other examples of similar QD-based multiplexed immunoassays have been reported in the literature for various toxins and biomarkers.

An excellent example of multiplexed flow cytometric analysis has been reported by Roederer's laboratory. This group used QDs to label and immunophenotype antigen-specific T cells from HIV-positive individuals using flow cytometry.[72] A combination of 9 organic fluorophores and 8 colors of QDs was used to achieve a total multiplexing capacity of 17. The QDs and dyes were used as labels on different antibodies, each specific to a different cell surface protein.[72] The study found that T cells exhibit complex phenotypic heterogeneity for a specific pathogen or a single epitope on that pathogen, indicating that T-cell clones may influence cell differentiation. This heterogeneity would have went unobserved without the information-rich 17-plex analysis, highlighting the value of QDs for facilitating multiplexed bioanalysis.

In addition to the direct use of QDs as labels on biological reporter elements (e.g. antibodies), QDs have been combined with microbead technology to facilitate highly multiplexed assays using spectral barcodes. Although optically encoded bead technology based on organic fluorophores is a commercially successful technology, the narrow PL emission of QDs and the ability to excite multiple colors with a single excitation source is highly advantageous for barcode reading technology. Typically, 2–4 colors of QDs are incorporated into microbeads at well-defined ratios to achieve a library of barcodes based on PL color and intensity. The overall multiplexing capacity of such a system is N^m-1, where m is the number of QD colors and N is the intensity level. The QD PL properties allow higher numbers of barcodes to be more easily produced, read, and resolved, and further help to reduce instrument complexity (*i.e.* one laser instead of the two lasers used with

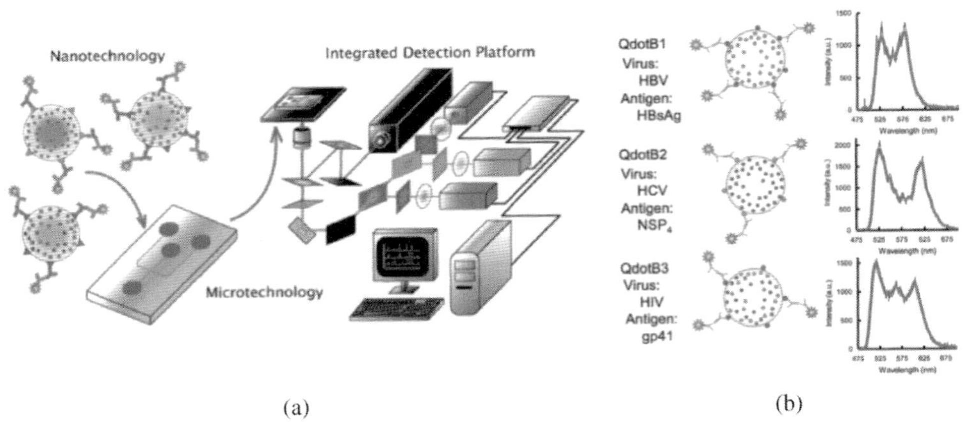

Fig. 3. (a) Schematic diagram illustrating the integration of QD-barcodes and solution-phase sandwich immunoassays with a microfluidic fluorescence detection platform for blood–borne pathogens. (b) Three types of QD barcodes and the corresponding pathogen biomakers: hepatitis B antibody (HBsAg), HCV nonstructural protein 4 (NSP4), and HIV glycoprotein 41 (gp41). The PL spectra for positive-test barcodes are shown at the right; the peak at 520 nm is the reporter signal from Alexa Fluor 488 modified antibodies, and the peaks at 570 nm and 615 nm are from the QD barcodes. Reprinted with permission from Ref. 73. Copyright 2007 American Chemical Society.

dye-based barcodes). QD-barcode assays have been demonstrated for the rapid detection of immunogenic[73] and genetic[74] markers for infectious diseases such as HIV, hepatitis B, hepatitis C, syphilis, and malaria. Up to a nine-plex assay was performed in less than 1 h, either within a microfluidic platform (Fig. 3) or using a flow cytometer. The assays had detection limits in the femtomolar range, which were superior to those of screening methods currently approved by the U.S. Food and Drug Administration. High-throughput 10-plex genotyping of single nucleotide polymorphisms (SNPs) has also been reported with QD barcodes, requiring as little as 0.02 ng of genomic DNA per SNP genotype.[75] In a separate study, Eastman *et al.* analyzed 92 genes using four colors of QDs (525, 545, 565 and 585 nm) and twelve intensity levels.[76] The analysis was sensitive to 1×10^4 gene copies, which was an order of magnitude lower than 20-plex assays based on commercially available fluorophore-encoded bead technology.[76,77]

5.1.2. *Quantum dot-based energy transfer assays*

Beyond assays where QDs are used as labels, there is considerable interest in designing more sophisticated signal transduction formats based on the modulation of QD PL with molecular biorecognition processes. These formats are promising for biosensing applications, and typically utilize Förster resonance energy transfer (FRET), bioluminescence energy transfer (BRET), chemiluminescence energy transfer (CRET), nanosurface energy transfer (NSET), or CT to modulate QD PL.[78] Among these approaches, the most developed and best characterized is FRET. While examples of CT-based sensing have been reported, the physical mechanism(s) of CT is much less understood than FRET.

QDs are ideal FRET donors when paired with organic fluorophores as acceptors: the strong QD absorption in the UV-blue region allows selection of the excitation wavelength to minimize direct excitation of the acceptor dye; the narrow and size-tunable QD PL permits optimization of spectral overlap integrals with minimal crosstalk between donor and acceptor emission channels; and the potentially high quantum yield of a QD helps to maximize Förster distances.[6,78,79] Additionally, the large surface area of each QD supports bioconjugation with an (approximately) centrosymmetric array of multiple copies of acceptor dye-labeled biomolecules, enhancing the energy transfer efficiency in comparison to a single donor–single acceptor FRET pair. To date, QD-FRET methods have been designed for a diverse range of analyses, those for metal ions, small molecules, toxins, protease and nuclease activity, as well as immunoassays and hybridization assays — both in the ensemble and at the single nanoparticle level. In each sensing application, whether *in vitro* or intracellular, transduction is driven by the dissociation or association of biomolecules that alter the donor–acceptor separation, which can be monitored through changes in QD PL quenching, FRET-sensitized emission from the acceptor, or the relative ratio of these two measurements.[78] Some examples of QD-bioconjugate probes that function *via* the above mechanisms include those designed to detect protease and kinase activity, or monitor biologically relevant molecules such as cocaine, dopamine, glucose, and maltose.[80–85] Lowe *et al.* demonstrated the simultaneous detection of protease and kinase activity, which are two important classes of enzymes for disease diagnostics and therapeutic intervention.[86] Multiplexed detection was achieved by combining a QD–FRET-pair (QD_{655}–Alexa Fluor 660) with a QD–NSET-pair (QD_{525}–Au NP), as illustrated in Fig. 4(a). Conceptually, NSET is similar to FRET process, but exhibits a weaker distance dependence (γ^{-4} versus γ^{-6}) and occurs over 1–20 nm (*cf.* <10 nm for FRET). A biotinylated peptide substrate specific to urokinase-type plasminogen activator (uPA), a serine protease was labeled with a 1.4 nm Au NP and, upon assembly on streptavidin-coated QD_{525}, induced QD PL quenching that was inversely proportional to the enzyme activity. Similarly, a polyhistidine-terminated peptide that incorporated a tyrosine residue as a substrate for phosphorylation by HER2

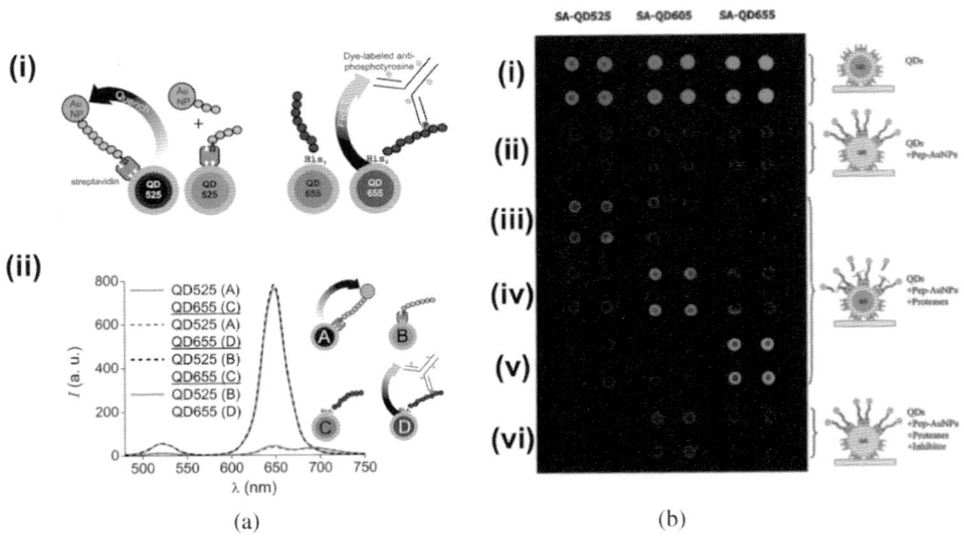

Fig. 4. (a) FRET/NSET-based two-plex detection of protease and kinase activity using two bioorthogonal methods for assembling QD–peptide conjugates (biotin-streptavidin and poly-histidine coordination). (i) Monitoring the activity of uPA and HER2 based on QD PL quenching by Au NPs (NSET) or Alexa Fluor 647 (A647) labeled anti-phosphotyrosine antibody (FRET). (ii) QD PL in the absence and presence of enzymatic activity. Reprinted with permission from Ref. 86. Copyright 2011 American Chemical Society. (b) A solid-phase multiplexed protease assay assembled using streptavidin coated QD_{525}, QD_{605} and QD_{655} with biotinylated Au NP-labeled substrates for MMP-7, caspase-3 and thrombin, respectively. PL images from spots of (i) QD only; (ii) QD–peptide conjugates; QD–peptide conjugates exposed to (iii) MMP-7, (iv) caspase-3, (v) thrombin and (vi) a mixture of corresponding protease and its inhibitor. Reprinted with permission from Ref. 87. Copyright 2008 American Chemical Society.

kinase was assembled to ligand-coated QD_{655}. With HER2 activity, Alexa Fluor 660 labeled anti-phosphotyrosine bound the phosphorylated tyrosine residue and generated a FRET signal. The selection of two orthogonal assembly methods between the peptides and QDs permitted direct ensemble measurements with sensitivity well below the clinical thresholds used for positive/negative breast cancer prognosis. Similarly, Kim *et al.* developed a high-sensitivity multiplexed chip-based QD-NSET assay for detecting the activity of matrix metalloproteinases (MMP-2 and MMP-7), caspase-3, and thrombin (Fig. 4(b)).[87]

Single QD-based sensors offer several advantages over their ensemble counterparts, including higher sensitivity, wider dynamic range, and less sample consumption. For example, to detect cocaine, Zhang and Johnson used a three-way FRET-relay system with QD donor, a fluorescent Cy5 acceptor, and a non-fluorescent Iowa Black (IaB) dye acceptor. An IaB modified oligonucleotide complimentary to the cocaine-binding aptamer, which was conjugated to the QD through a Cy5 labeled oligonucleotide linker, quenched the QD and FRET-sensitized Cy5 PL. Displacement of the IaB-oligonucleotide when the aptamer bound cocaine restored the FRET-sensitized Cy5 PL, which was used as the analytical signal.[82] The detection limit was comparable to other aptameric sensors (limited by their intrinsic affinity) and the method required only a small amount of reagent (25 pM QD_{605}). Subsequently, Zhang and Hu demonstrated multiplexed detection of genetic markers of HIV-1 and HIV-2 in a QD-based sandwich hybridization assay.[88] The combined direct excitation of an Alexa Fluor 488 labeled HIV-1 DNA reporter probe and FRET-sensitized emission from an A647 labeled HIV-2 DNA reporter probe — both excited by a 488 nm laser line — permitted measurement *via* a three-color (QD, Alexa Fluor 488 and 647) fluorescence burst coincidence analysis.

In addition to acting as FRET donors, QDs can also act as FRET acceptors when paired with appropriate donors. The strong and broad absorption of QDs implies that they can be very effective FRET acceptors; however, the challenge is avoiding direct optical excitation in conventional FRET formats. This challenge has been addressed by utilizing QDs as (1) FRET acceptors with lanthanide ion donors (e.g. Tb^{3+}, Eu^{3+}), which have excited state lifetimes (10^{-4}–10^{-3} s) that are orders of magnitude longer than QDs, thereby permitting time-gated measurements following the decay of the QD excited state; (2) BRET acceptors with Renilla luciferase (Rluc) enzyme, which catalyzes the oxidation of coelentrazine with concomitant bioluminescent emission at *ca.* 480 nm, avoiding the need for direct optical excitation; and (3) CRET acceptors following chemiluminogenic reactions such as that between luminol and H_2O_2.

Multiplexed detection of DNA hybridization has been designed around a single color of QD_{620} acting as both a FRET acceptor for a Tb^{3+} complex and a donor for an A647 dye in a time-gated FRET relay.[62] Due to the millisecond excited state lifetime of Tb^{3+}, signal acquisition was performed in two modes with either 0 μs (prompt) or 55 μs (time-gated) delays after pulsed UV-violet optical excitation. In the prompt mode, the energy was directly transferred from the QD to the A647, while in the time-gated mode, energy was transferred from the Tb^{3+} to the QD, then to the A647. This format provided two orthogonal optical signals for the detection of two different DNA sequences, one of which was associated with Tb^{3+} and the other with A647 (Fig. 5). Similar prompt and time-gated measurements were applied to the quantitative two-plex detection of protease activity *in vitro*, including downstream

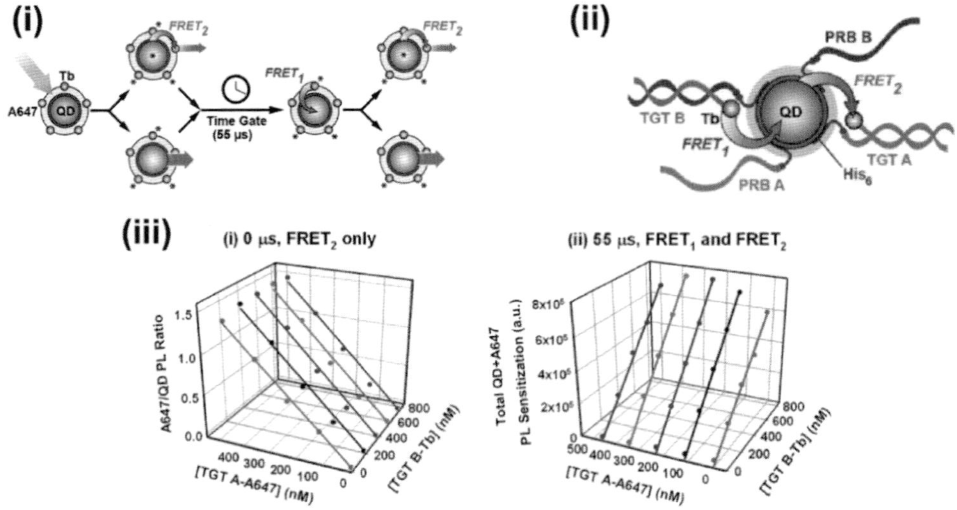

Fig. 5. Time-gated FRET relay for the two-plex detection of DNA hybridization using one QD vector. (i) Schematic of the time-gated FRET relay concept, showing energy transfer from a luminescent Tb^{3+} complex (Tb) to the QD, and from the QD to a fluorescent dye (A647). (ii) QD–probe oligonucleotide conjugates and hybridization to assemble the FRET relay. (iii) Orthogonal calibration curves for detection of the two DNA targets. Reprinted with permission from Ref. 62. Copyright 2012 American Chemical Society.

activation of a zymogen by an upstream protease.[89] The time-gated relay was also useful in that, when the Tb^{3+} was not associated with a biorecognition process, it transferred the optical signal from the QD-A647 FRET pair to the 10^{-4}–10^{-3} s time-domain, which is suitable for time-gated rejection of scattering and autofluorescence background from tissue and other complex biological samples.

The development of QD-BRET methods using Rluc–QD bioconjugates can be largely attributed to Rao's group. "Self-illuminating" *in vivo* imaging (*i.e.* no external excitation source) has been demonstrated in mice injected with luciferase (Luc8) bioconjugates of QDs emitting almost entirely within the optical window for biological tissue (~650–900 nm), thereby overcoming the challenge of tissue autofluorescence with optical excitation.[90] Rao's group has further demonstrated sensitive multiplexed assays for matrix metalloproteinase-7 (MMP-7) and uPA activity in mouse serum and tumor secretions (Fig. 6a).[59] The limits of detection were 1 ng mL^{-1} and 500 ng mL^{-1}, respectively.

QD-CRET-based sensing formats for the detection of metal ions, aptamer–substrate complexes, DNA, thrombin, glucose oxidase, and ATP have been developed by Willner's group.[91–95] For example, the sensitive

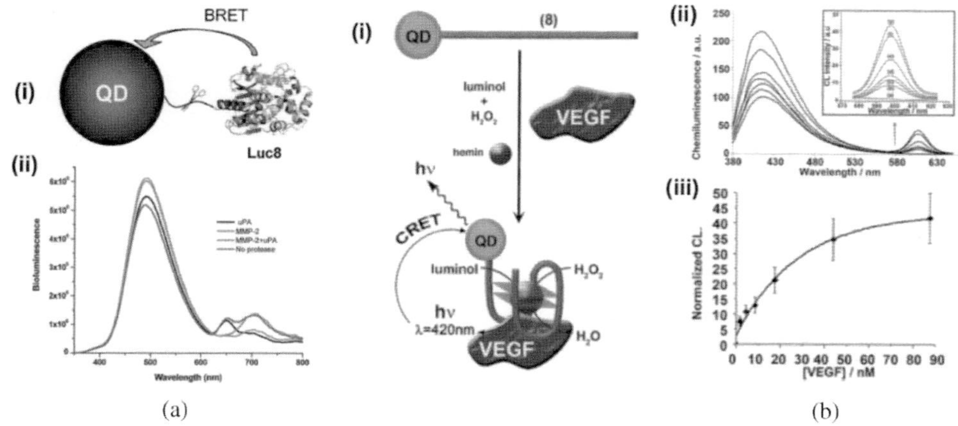

Fig. 6. (a) (i) Schematic of a QD-BRET system comprising a QD conjugated with mutant luciferase protein (Luc8) *via* a peptide substrate linker. (ii) PL spectra for the simultaneous detection of MMP-2 and uPA using QD655 and QD705, respectively. Reprinted with permission from Ref. 59. Copyright 2008 American Chemical Society. (b) (i) Example of a QD-CRET construct for detection of vascular endothelial growth factor (VEGF) *via* luminol oxidation in hemin/G-quadruplex aptamer complexes. (ii) Luminescence spectra and CRET signal (inset) for varying concentrations of VEGF. (iii) Corresponding calibration curve for the CRET signal at 600 nm. Reprinted with permission from Ref. 91. Copyright 2012 American Chemical Society.

detection of VEGF (an important biomarker associated with tumor growth, brain injuries, and neurological disorders such as Parkinson's and Alzheimer's diseases) was demonstrated using a QD-hemin/G-quadruplex aptasensor.[91] QDs were modified with anti-VEGF aptamer containing a G-quadruplex capable of binding hemin, which catalyzed the oxidation of luminol by H_2O_2. The result was chemiluminescence in close proximity to QDs, leading to CRET and a QD PL signal (Fig. 6b). Analogous to optical excitation, CRET-sensitization of QD PL can be readily multiplexed.

5.1.3. *Quantum dot-based charge transfer assays*

Despite a limited understanding of the CT mechanism at present, there is significant potential in developing CT quenching-based sensing applications with QDs. In contrast to FRET-based assays, CT quenching appears to have no strong requirement for spectral overlap, and enhanced sensitivity is expected in some applications due to the putative exponential distance dependence of CT rates (*cf.* inverse-sixth-power for FRET) or the greater sensitivity of CT to the local microenvironment (which differs from that of

FRET).[96,97] A variety of redox-active dyes, metal complexes, and organic molecules have been used as electron acceptors or donors (and PL quenchers) with QDs, including Ru^{2+}-phenanthroline (Ru-phen),[96,98] Ru^{2+}–polypyridine complexes,[99] ferrocene,[100,101] bipyridinium dyes,[102–105] and quinones,[81,106,107] among others.[108,109]

In an example of an *in vitro* sensing application, Medintz *et al.* conjugated Ru-phen-labeled peptide substrates to QDs to measure the proteolytic activity of chymotrypsin and thrombin. The Ru-phen complex, which quenched the QD as a putative hole acceptor when bound in close proximity, was cleaved off the QD with proteolysis, restoring QD PL.[98] This design can be used in parallel across up to 8 colors of QD (peak PL at 510, 537, 555, 565, 581, 590, 610 and 635 nm) with resolution of the CT quenching of each QD (Fig. 7a).[110] Medintz *et al.* have also developed an intracellular pH sensor based on QD PL quenching with electron transfer from a QD to dopamine.[106] QDs were functionalized with peptides labeled at their *N*-terminus with dopamine, which exhibits complex pH-dependent redox chemistry. At low pH, the predominant hydroquinone form of a dopamine is a poor electron acceptor, and quenching of the QD is minimal. As pH increases, the dopamine converts to its quinone form and becomes a good electron acceptor, resulting in efficient CT quenching of QD PL (Fig. 7b). Using pH-insensitive FLX fluorescent microspheres as a reference signal, the QD PL was observed to decrease linearly in the range from pH 6 to 12. Progressive changes in intracellular pH could be tracked upon exchange of slightly acidic cell medium for basic cell medium, in the presence of a cell membrane permeabilizing drug.

5.2. *Tissue and cellular imaging*

5.2.1. *Immunohistochemistry and immunocytochemistry*

Immunohistochemical (IHC) and immunocytochemical (ICC) labeling of biomarkers in fixed cells and tissues are two prominent biomedical applications of QDs. In clinical pathology, the classification of disease progress is determined based on the level of expression and distribution of molecular antigens. The subjective nature of the current qualitative morphological assessments of tissue samples — where conclusions may vary between pathologists — necessitates improved methodologies for objective classification, preferably by quantitation of tissue biomarkers. The parallel detection of multiple markers is important for reliable diagnoses and prognoses, but is often hindered by small sample sizes (e.g. fine needle or core biopsy). The optical multiplexing advantages of QDs are potentially ideal for IHC/ICC analysis of clinical specimens. For example, Fountaine *et al.* demonstrated sequential

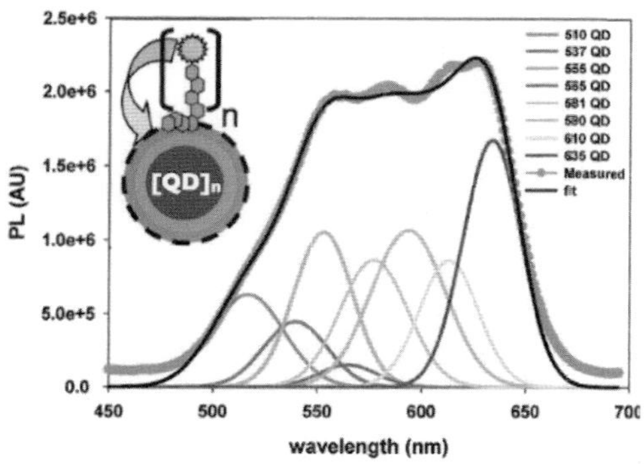

(a)

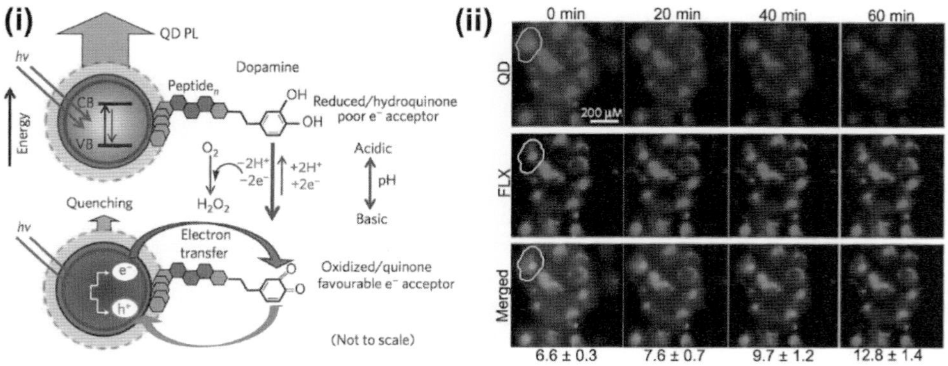

(b)

Fig. 7. (a) PL spectra with deconvoluted contributions from a mixture of eight different colors of QD quenched by Ru-phen to different degrees. The inset illustrates the CT process between the QD and Ru-phen labeled peptide. Reprinted with permission from Ref. 110. Copyright 2009 American Chemical Society. (b) QD-CT intracellular pH sensor. (i) Schematic illustration of the pH nanosensor and transduction mechanism *via* electron transfer between the QD and the dopamine modified peptide. (ii) Intracellular pH sensing over time when acidic media bathing COS-1 cells is exchanged for basic media. Red fluorescent nanospheres (FLX) are used as an internal standard. Adapted with permission from Ref. 106. Copyright © 2010, Rights Managed by Nature Publishing Group.

staining of CD20, IgD, Ki67 (MIB-1 clone), CD3 and CD68 antigens in human lymph nodes using QDs.[111] The staining of the surface, cytoplasmic, and nuclear markers was found to be consistent with their expected expression patterns. Similarly, Xing *et al.* used QDs for five-color molecular profiling of a human prostate cancer tissue biopsy specimen.[112] Four biomarkers, including vimentin, N-cadherin, receptor activator of nuclear factor κB ligand (RANKL), and E-cadherin were labeled with secondary antibody conjugates of QD_{525}, QD_{565}, QD_{605} and QD_{655}, respectively (Fig. 8a). The fifth channel

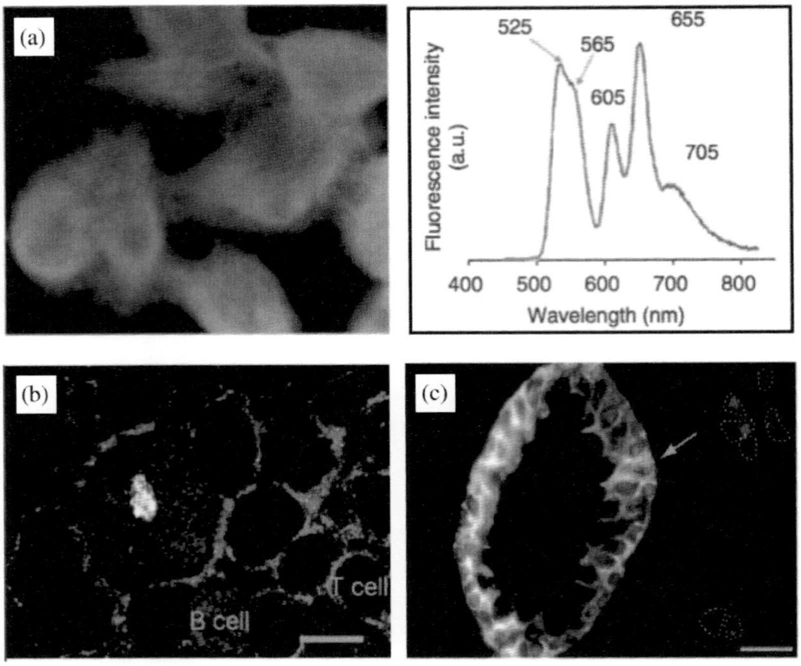

Fig. 8. (a) Five-plex staining of human prostate cancer cells and corresponding wavelength-resolved spectra from cell specimens. The cells were labeled with QD_{525}, QD_{565}, QD_{605}, QD_{655} and QD_{705}–antibody conjugates targeted to vimentin, N-cadherin, RANKL, E-cadherin, and EF-1α (elongation factor)-1, respectively. Adapted with permission from Ref. 112. Copyright © 2007, Rights Managed by Nature Publishing Group. (b) Detection of rare HRS tumor cells in Hodgkin's lymphoma. The image shows HRS malignant cells and infiltrating immune cells on lymph node tissue specimens. The HRS cells exhibited characteristic staining pattern: membrane staining (CD30 positive), golgi staining (CD15 positive), nuclear staining (Pax5 positive). Staining patterns were clearly distinct from infiltrating B cells and T cells. The scale bar is 10 μm. Reprinted with permission from Ref. 114. Copyright 2010 American Chemical Society. (C) Composite spectral image of a single malignant cell (arrow) in the basal layer of a largely benign prostate gland, surrounded by malignant cells. The scale bar is 20 μm. Four different protein biomarkers, labeled with different colors of QD–antibody conjugates: E-cadherin, CK HMW, AMACR and p63. Reprinted with permission from Ref. 115. Copyright 2010 American Chemical Society.

(QD_{705}) was used to stain elongation factor-1α, a housekeeping gene expressed at a constant level, which served as an internal standard. Jennings *et al.* utilized five colors of QD–antibody conjugate (QD_{525}, QD_{565}, QD_{605}, QD_{625} and QD_{650}) in combination with multispectral imaging to label and characterize fixed mouse spleen tissue sections.[113] B cells, T cells, leukocytes, thymocytes and macrophages were labeled with the QDs *via* antibodies targeting CD45 (type C protein tyrosine phosphatase receptor), CD11c (integrin αX), CD31 (platelet endothelial cell adhesion molecule), CD4 (a glycoprotein) and CD11b (integrin αM) antigens, respectively.[113] Liu *et al.* similarly utilized multispectral imaging with multiple colors of QDs to reliably identify low abundance (~1%) Hodgkin's and Reed-Sternberg (HRS) cells, which is essential for differentiating Hodgkin's lymphoma from non-Hodgkin's lymphoma and benign lymphoid hyperplasia.[114] Four protein biomarkers (CD15, CD30, CD45 and Pax5) were simultaneously mapped in human lymph node biopsy specimens (Fig. 8b) using four colors of QD–secondary antibody conjugates (QD_{525}, QD_{565}, QD_{605} and QD_{655}).[114]

Advances in multiplexed QD-IHC have led to a deeper understanding of the complex nature of tumor development. For example, Nie's group quantitatively mapped biomarkers in biopsy specimens revealing molecular and morphological details that were unseen with traditional staining methods.[115] The authors used four protein markers associated with prostate cancer — E-cadhedrin, high-molecular weight cytoketarin (CK HMW), p63 and α-methylacyl CoA racemase (AMACR) — and visualized those markers with four different colors of QD–secondary antibody conjugates (QD_{565}, QD_{605}, QD_{655} and QD_{705}). The staining patterns (Fig. 8c) revealed that progressive changes in benign prostate glands began with a single malignant cell and ultimately lead to a malignant gland.[115]

Such studies elucidating tumor heterogeneity, which exists at the molecular, cellular, and tissue-architecture levels, as well as between individuals, are invaluable for gaining insight into mechanisms of tumor growth. Similarly, Peng *et al.* helped improve our understanding of cancer progression by studying the effect of microenvironment (*i.e.* dynamic tumor–stromal interactions) on patterns of tumor invasion.[116] Fixed tissue samples from gastric cancer and breast cancer patients were stained with QD–antibody conjugates (QD_{525}, QD_{585} and QD_{655}) to target type IV collagen, tumor infiltrating macrophages, tissue destructive matrix metalloproteinase-9 (MMP-9) enzyme, and CD105 on neovessels associated with tumor angiogenesis. One of the most important observations was that, for each of four identified tumor invasion patterns (*i.e.* washing, ameba-like, polarity and linear), the expression level of each of the aforementioned markers differed.

5.2.2. *Cellular sensing and imaging*

A myriad of processes function in concert to maintain homeostasis in mammalian cells. A full understanding of the spatiotemporal organization of these processes requires methods for real-time, quantitative visualization and tracking. QDs can be used to visualize cellular and subcellular structures, monitor ligand-cell surface receptor binding, protein–protein interactions, and membrane dynamics, to name only a few applications.

5.2.2.1. Ligand–receptor interactions

Several studies have used QDs to visualize ligand binding to cognate receptor proteins. QDs functionalized with receptor ligands can, for example, activate membrane receptors and permit tracking of the formed complexes during cellular internalization through endosomal pathways. Jovin's group investigated erbB/HER receptor-mediated signal transduction using QDs conjugated with epidermal growth factor (EGF).[117] QD–EGF conjugates specifically activated ebrB1 receptor tyrosine kinase and underwent rapid clathrin pit-mediated endocytosis at the base of filopodia in CHO and A431 cell lines. In addition, tracking of QD PL from QD–EGF–ebrB1 complexes formed on the filopodia revealed the existence of a new mechanism of retrograde transport to the cell body.[117] Active transport was triggered by the dimerization or oligomerization of EGF-activated receptors, as observed from an overlay of a time series of images of the emission from two colors of QD–EGF conjugates (QD_{525} and QD_{605}).[118] The rate of transport was similar to the rate of filopodial actin flow, indicating that cytoskeletal components actively participated in the transport of membrane proteins. The authors confirmed that receptor complexes were internalized after retrograde transport by monitoring FRET between streptavidin coated QD_{525}–EGF conjugates and membrane-impermeable biotinylated Alexa Fluor 546 dye.

In neuroscience, QDs have been used to study the diffusion dynamics of membrane proteins associated with neuronal cells, including the distribution and surface mobility of neuroreceptors for glycine,[119] γ-aminobutyric acid (GABA),[120] $\alpha 7$ neuronal nicotinic acetylcholine,[121] and 2-amino-3-(3-hydroxy-5-methyl-isoxazol-4-yl)propanoic acid (AMPA),[122] at both synaptic and extrasynaptic sites. Interestingly, evidence obtained by labeling presynaptic vesicles with QDs has provided support for a long supposed but sometimes controversial mechanism of "kiss-and-run" (K&R) vesicle fusion at synapses upon neuronal stimulation. Previous studies that have used fluorescent dyes to label mammalian nerve terminals have suffered from poor signal-to-noise ratios and conclusive evidence of a prevalent of K&R mechanism has been elusive. Zhang *et al.* studied the vesicular secretion of neurotransmitters in an elegant way by tracking individual synaptic vesicles loaded with QD_{605}.[123,124] A small change in QD PL intensity (±15%) between vesicles (pH 5.5) and the extracellular matrix (pH 7.3) was used to distinguish between the full collapse fusion (FCF; complete vesicle integration with membrane) and K&R (transient vesicular fusion with membrane, retrieval, and release) mechanisms. The authors tracked a single vesicle through multiple cycles of K&R and vesicle reuse, concluding that K&R was more prominent under high-activity demand conditions. Importantly, QDs were small enough (diameter *ca.* 15 nm) to fit inside vesicular lumen (*ca.* 24 nm) and large enough to be rejected by putative K&R fusion pores (1–5 nm).

5.2.2.2. Intracellular tracking and sensing

Lowe *et al.* used QD probes to elucidate the selectivity mechanism for translocation of transport receptor/cargo complexes across the nuclear pore complex. QDs were conjugated with multiple copies of the importin-β binding (IBB) domain of snurportin-1, the import receptor for small molecule nucleoriboproteins (Fig. 9a).[125] The import or rejection of the QD cargos in HeLa cells was monitored using time-resolved fluorescence imaging, and the authors were able to derive a multistep model for cargo translocation, which comprised cargo capture, filtering, translocation and release into the nucleus. Wells *et al.* demonstrated that QDs could be used to track the three-dimensional motions of proteins inside live cells and on their surface (Fig. 9b).[126] QD–immunoglobulin E (IgE) conjugates were bound to high-affinity IgE receptors (FcεRI) on rat tumor mast cells. The motion of single QD–IgE–FcεRI complexes were tracked using a specially built imaging platform, revealing the three-dimensional surface topology of the cell membrane and permitting extraction of diffusion rates (0.21 $\mu m^2\ s^{-1}$). The observed rate of vesicular transit during ligand-mediated endocytosis (upon cross-linking QD–IgE–FcεRI complexes with dinitrophenyl-BSA) was 950 nm s^{-1}.[126]

Intracellular sensing with QDs is another prominent area of research. Rather than serving only as a tracer for biological molecules, QD PL responds to bio/physicochemical changes in these applications. As an example, increases in temperature alter electron–lattice interactions in QDs, leading to a red-shift in QD PL as temperature increases.[127] The dependence of the peak position of QD PL on the temperature of the local microenvironment was used to map intracellular heat generation in NIH/3T3 murine fibroblast cells following Ca^{2+} stress and cold shock.[128] Extracellular temperature perturbations induced a transient, inhomogeneous intracellular temperature response.[128] Considering chemical stimuli, Wang *et al.* developed an intracellular QD-based sensor to monitor sodium (Na^+) and chloride (Cl^-) ions at physiological concentrations *via* QD PL quenching.[129,130] QDs appended with crown-ether terminated ligands complexed Na^+ with high specificity and without interference from K^+ or Ca^{2+}.[129]

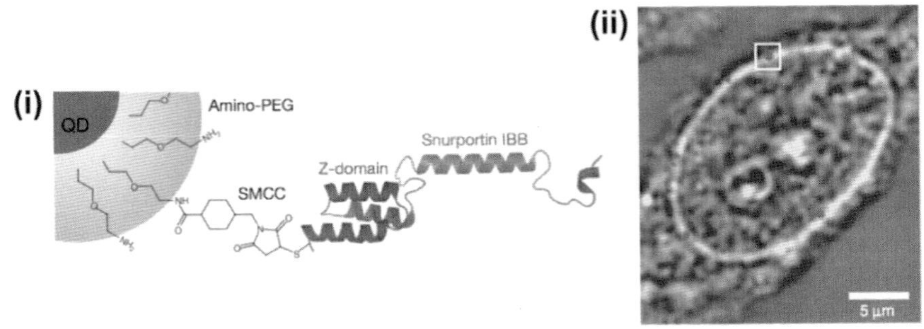

(a)

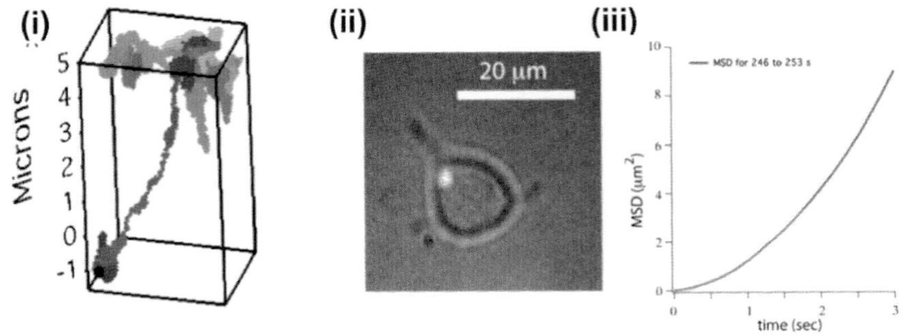

(b)

Fig. 9. (a) QD cargo and transport across the nuclear pore complex. (i) Amino-PEG coated QDs functionalized with snurportin-1 IBB/Z-domain fusion protein. (ii) Bright-field image of a HeLa cell nucleus with PL emission from single QD605 conjugate indicated (box). Figure adapted with permission from Ref. 125. Copyright © 2010, Rights Managed by Nature Publishing Group. (b) (i) 3D tracking of a single patch of QDs labeled with IgE–FcεRI over time. (ii) image of the cell for which the trajectory was generated. (iii) Parabolic fit of the mean-square-displacement (MSD) indicative of directed motion used to determine transport velocity of 950 nm s^{-1}. Reprinted with permission from Ref. 126. Copyright 2010 American Chemical Society.

Dynamic physiological responses of Na^+ ion concentrations were monitored in HEK-293F cells and primary rat cardiac myocytes. Similarly, to monitor intracellular Cl^- concentration in epithelial cells, the crown ether was substituted with an anion receptor, 1-(2-mercapto-ethyl)-3-phenyl-thiourea (MEPTU), that quenched QDs *via* an electron transfer mechanism upon binding chloride ions.[130] Many other examples of sensing based on FRET and CT have been published in the literature, including FRET methods for measuring intracellular pH[131] and intracellular levels of nicotinamide adenine dinucleotide phosphate (NADPH).[132] A few of these examples were discussed earlier in this chapter.

5.3. *In Vivo imaging*

Standard medical imaging systems include computed tomography (CT), magnetic resonance imaging (MRI), positron emission tomography (PET), single-photon emission computed tomography (SPECT), and combined CT-PET imaging. These methods frequently rely on contrast agents (e.g. gadolinium, iron oxide nanoparticles for MRI; iodine for CT; and ^{18}F, ^{11}C, ^{16}O for PET; and ^{99m}Tc for SPECT) to enhance the visibility of internal organs and other bodily structures. Deep tissue anatomy can be visualized with resolution >10 μm. In contrast, fluorescence imaging depths are typically limited to *ca.* 10^{-2}–1 cm; however, spatial resolution is <1 μm and provides detail at the cellular level. The optical properties of QDs — particularly those materials with emission tunable in the NIR — are well suited for certain *in vivo* imaging applications. There is a tissue transparency window between *ca.* 650–950 nm that offers several advantages for optical imaging: (i) reduced light scattering, (ii) minimal absorption by water, hemoglobin, and other tissue chromophores, and (iii) minimal tissue autofluorescence. Recently, Helle *et al.* used NIR (780 nm)-emitting $CuInS_2/ZnS$ QDs encapsulated within PEGylated lipid micelles for sentinel lymph node mapping.[133] PL from the lymph nodes was observed 5 min after injection of 20 pmol of QDs and progressively decreased over 8 h. In a separate study, Pons *et al.* evaluated the accumulation of similar PEGylated lipid micelle-encapsulated $CuInS_2/ZnS$ QDs (peak PL at 800 nm; average diameter 5.5 ± 2 nm) within sentinel lymph nodes (Fig. 10).[134] The minimum injected dose (ID) for measurable QD PL over a 7 day period was 2 pmol, which is 50 times lower than the amount required to induce inflammation. A study by Kim *et al.* demonstrated that NIR (850 nm)-emitting QDs could be used for real-time sentinel lymph node mapping during light assisted surgery in a pig.[135] Importantly, good contrast with a penetration depth of about 1 cm was achieved with low power excitation (5 mW cm^{-2}) and only 400 pmol of QDs. The smaller amounts of QD minimize exposure and the potential for toxicity (*vide infra*).

The very large two-photon absorption cross-section of QDs provides another means of taking advantage of the NIR optical tissue window. Larson *et al.* used injected CdSe/ZnS QDs (encapsulated with an amphiphilic polymer) to image mouse vasculature at a tissue depth of hundreds of micrometers using two-photon excitation PL.[136] In a direct comparison with conventional fluorescein isothiocyanate (FITC)-labeled dextran, the QDs revealed greater vasculature detail per unit depth, even though the excitation source used with FITC was operated at five times the power of that used for the QDs.

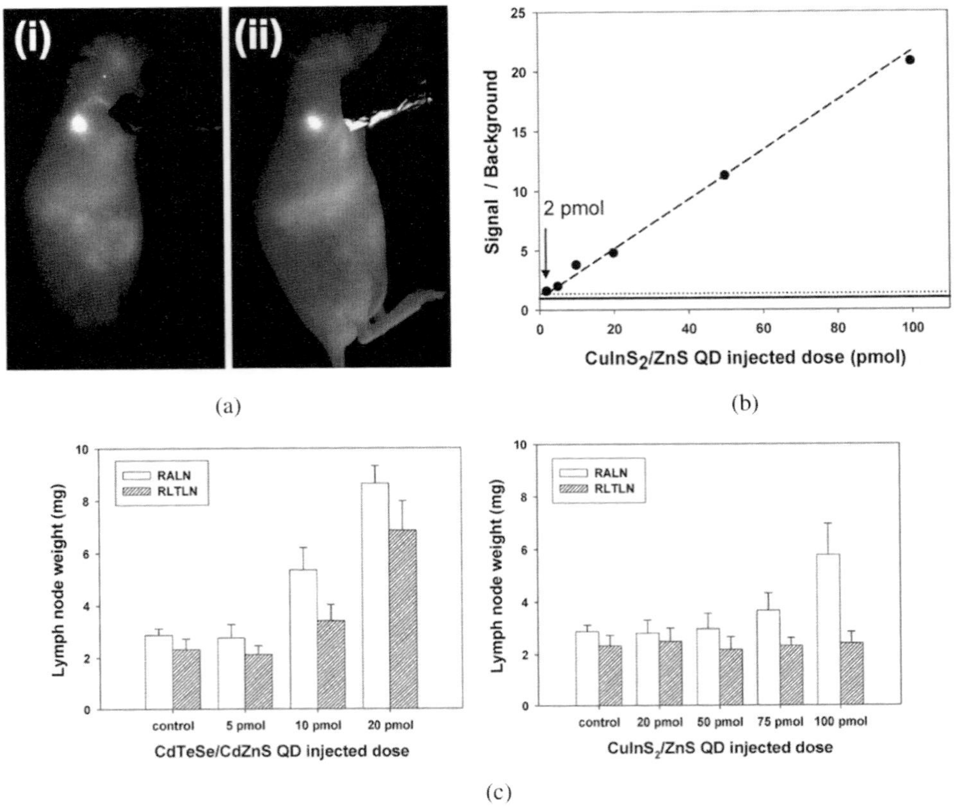

Fig. 10. Sentinel lymph node imaging in mice using $CuInS_2$/ZnS QDs. (a) NIR fluorescence imaging of right axillary lymph node (RALN) after (i) 15 min and (ii) 7 days post-injection of 20 pmol micelle encapsulated QDs. (b) Ratio of RALN NIR fluorescence versus autofluorescence 24 h post-injection. Estimated limit of detection corresponds to 2 pmol QD injected dose (ID). (c) Weight of the RALN and right lateral thoracic lymph node (RLTLN) 7 days post-injection with CdTeSe/CdZnS (left) and $CuInS_2$/ZnS (right) QDs. Signs of inflammation are observed at 10 pmol dose with Cd-based QDs, while same level of inflammation with $CuInS_2$/ZnS is observed at 100 pmol ID. Reprinted with permission from Ref. 134. Copyright 2010 American Chemical Society.

Other *in vivo* applications of QDs have included imaging embryogenesis,[137–139] lymphatic tracking,[133,135,140–142] stem cell tracking,[143,144] and cell tracking in mice upon extravasation in lung tissue.[145] Indeed, embryogenesis was the first example of *in vivo* imaging with QDs, and was demonstrated by Dubertret *et al.* with biocompatible micelle-encapsulated CdSe/ZnS QD_{550} and *Xenopus* (frog) embryos.[138] In a subsequent study, commercially available NIR-emitting QD_{800} were used to visualize and quantify mesoderm migration with single cell resolution in *Xenopus* embryos.[139]

Both of these studies relied on microinjection to deliver the QDs into embryonic cells.

Since intracellular delivery by microinjection (or other active techniques such as electroporation) is limited to embryos and cultured cells, the majority of *in vivo* studies exploit either passive and targeted delivery methods or *in vivo* tracking of QD-labeled cells. For example, the *in vivo* fate of stem cells was evaluated using injection of stem cells labeled with QD–cell penetrating peptide conjugates (e.g. TAT-peptide, polyarginine).[143,144] To evaluate the accumulation of adipose tissue-derived stem cells during transplantation in mice with acute liver failure, Yukawa *et al.* injected 1×10^6 cells labeled with QD_{655} or QD_{800} in saline with or without heparin.[143] While in the absence of heparin, cells preferentially accumulated in lung tissue, the presence of heparin increased the liver accumulation rate by 30%.

The targeted delivery of QDs to tumor cells takes advantage of overexpressed cell surface receptors. Tada *et al.* visualized the blood circulation, extravasation, cell membrane binding, and cellular internalization of QD–anti-HER2 conjugates.[70] Single QD–antibody complexes were tracked in real-time from a capillary vessel to tumor cells following injection of 200 pmol of QD in the tail vein of a mouse. The data provided information about the velocity, directionality, and transport mode of the nanoparticles. Specific delivery of the QD conjugates to the perinuclear region of tumor cells was observed in six stages: (i) vessel circulation, (ii) extravasation, (iii) movement into the extracellular region, (iv) binding to HER2 on the cell membrane, (v) movement from cell membrane to the perinuclear region after endocytosis and (vi) localization in the perinuclear region. Similarly, Ruan *et al.* prepared CdTe QD–HER2 antibody conjugates functionalized with Ribonuclease-A (RNase-A) for simultaneous targeted imaging and therapy of gastric cancer.[146] The *in vivo* distribution of HER2–(RNase-A)–QD conjugates 24 h post-injection in mouse models, estimated from inductively coupled plasma mass spectrometry (ICP-MS) measurements, was 10.2 ppm in gastric tumor tissue and <5 ppm in blood, heart, spleen, kidney, lung and liver tissue. Therapeutic effects from RNase-A, which acts by degrading cellular mRNA to prevent translation and inhibit tumor cell growth by inducing apoptosis, was confirmed by a reduction in the tumor size. The median survival time of the treated group of mice increased to eleven weeks from eight weeks with the control group.

The suitability of QDs for multiplexed *in vivo* imaging was elegantly demonstrated by Kobayashi *et al.* who obtained five-color wavelength-resolved images of lymphatic drainages in the neck and the upper trunck of a mouse by selecting appropriate injection sites (middle phalanges, upper extremities, ears and chin).[141] The lymphatic flow was mapped using carboxyl polymer-coated CdX/ZnS QDs, where X = Se (565, 605, 655 nm) and X = Te (705, 800 nm), each with similar hydrodynamic radius (15–19 nm).

5.4. *Multimodal in vivo imaging*

Composite nanoprobes with a QD component for PL imaging and a second component for other imaging modalities (e.g. MRI, PET and CT) have great potential in biomedical applications. Bakalova *et al.* constructed a dual MRI/optical imaging probe by encapsulating QD–gadolinium (Gd^{3+}) complex conjugates within silica shells.[147] The large pores in the silica shell allowed water molecules to interact with the gadolinium complexes for MRI contrast. The close proximity between the Gd^{3+} complex and QD neither affected the paramagnetic properties of the Gd^{3+} nor the optical properties of the QD. An intravenous injection of the dual-MRI/optical probe (diameter ~15–20 nm) permitted MRI and two-photon imaging of blood vasculature in rats. Recently, Stasiuk *et al.* reported cell-permeable InP/ZnS QDs that were functionalized with Ln(III) chelates as effective multimodal reporters.[148] The QDs were co-functionalized with penicillamine ligands, dithiol modified lanthanide (Gd^{3+}, Tb^{3+} or Eu^{3+}) chelates (up to 80 per QD), and cell-penetrating peptides derived from maurocalcine (a scorpion toxin) in a one-pot reaction. Further, retention of the dual mode QD probe within striatum (brain tissue) over 4 h was found to be superior in comparison to Dotarem, a commercial contrast agent.

In addition dual mode MRI/optical probes, multimodal optical/PET probes have been designed around QDs. For example, Cai *et al.* developed a PET/NIR emitting probe to visualize tumor vasculature in mice.[66] CdTe/ZnS QDs coated with PEG ligands were conjugated with RGD-cell penetrating peptides and 1,4,7,10-tetraazacyclodocecane-*N,N',N',N''*-tetraacetic acid (DOTA)-^{64}Cu chelates. The RGD–QD–DOTA–^{64}Cu conjugates primarily targeted integrin $\alpha_V\beta_3$ on tumor vasculature and uptake of QD-DOTA was minimal. Quantification of organ and tumor uptake based on ^{64}Cu radioactivity and QD PL intensity were in good agreement. The majority of the QD conjugates were taken up by the reticuloendothelial system (RES), accumulating in the liver, spleen and bone marrow. Similarly, Tu *et al.* used manganese doped silicon QDs conjugated with ^{64}Cu-DOTA chelates to evaluate the biodistribution of nanoparticles in mice.[149] The Si_{Mn} QDs were coated with neutral dextran and functionalized with a derivative of DOTA to achieve a final HD of *ca.* 15 nm. Within 2 min post-injection, >95% of the ID of QDs had been cleared from bloodstream. At 2 h, *ca.* 1.5% ID remained in blood, and at 48 h post-injection the highest uptake of QDs was observed in the liver (12% ID of Si_{Mn} QDs), kidney, heart, and intestine (each with *ca.* 4% ID). Comparable uptake has also been observed with dual mode PET/optical labels designed around CdSe/ZnS QDs functionalized with zwitterionic cysteine ligands, RGD peptides, and ^{99m}Tc.[150] However, these QDs conjugates had HDs less than 5.5 nm, permitting efficient renal clearance in mice. At 4 h post-injection, 65% ID had been excreted in urine and, at 24 h post-injection, only 9% ID was retained.

5.5. *Quantum dots as vectors for drug delivery and therapy*

5.5.1. *Targeted drug delivery*

Many current cancer treatments rely on drugs that are highly toxic but have little or no intrinsic specificity for cancer cells. As a consequence of systemic toxicity, cancer patients must often endure numerous unpleasant side effects. Liposomes and polymeric micelles have been explored for targeted drug delivery with some success; however, a vibrant area of research is the development of multifunctional, nanoscale vectors that can concurrently provide targeted imaging contrast and targeted drug delivery. Such vectors are often referred to as theranostic (therapy + diagnostic) probes. Bagalkot *et al.* designed a QD probe for targeted cancer imaging, therapy, and sensing (Fig. 11a).[151] QDs were functionalized with A10 RNA aptamer, which selectively binds PSMA, and aptamer-intercalated doxorubicin (Dox). When all three components were bound together, the Dox quenched the QD PL *via* energy transfer and intercalation in the aptamer quenched the Dox fluorescence. QDs functionalized with A10 RNA aptamer exhibited high specificity of uptake in PSMA-expressing LNCaP cells when compared to PSMA-negative PC3 prostate adenocarcinoma cells, indicating selective targeting and delivery. Intracellular Dox release was monitored by tracking the changes in QD and Dox PL intensities.

Short double-stranded oligoribonucleotides called small-interfering-RNA (siRNA) exhibit therapeutic effects by inhibiting translation of target genes into proteins. While siRNA is a potentially powerful therapeutic due to its high specificity and flexibility in targeting different signaling pathways, a significant challenge is the selective delivery of siRNA to target cells and tracking of siRNA-mediated target-gene silencing. Derfus *et al.* demonstrated a proof-of-concept model for QD-mediated siRNA delivery by conjugating QDs with tumor-homing peptide (F3) and siRNA targeting the enhanced green fluorescent protein (EGFP) gene.[152] EGFP-transfected HeLa cells demonstrated

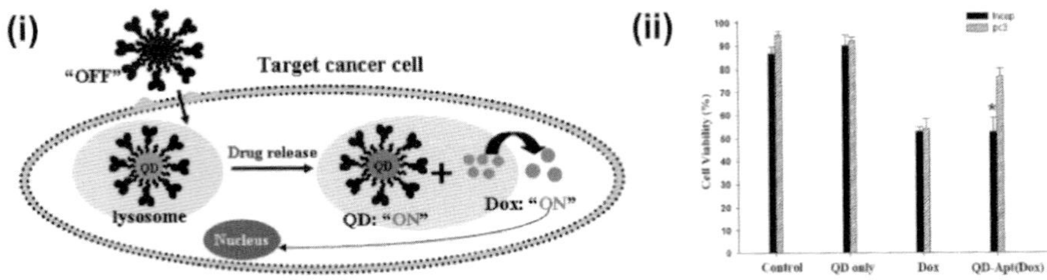

(a)

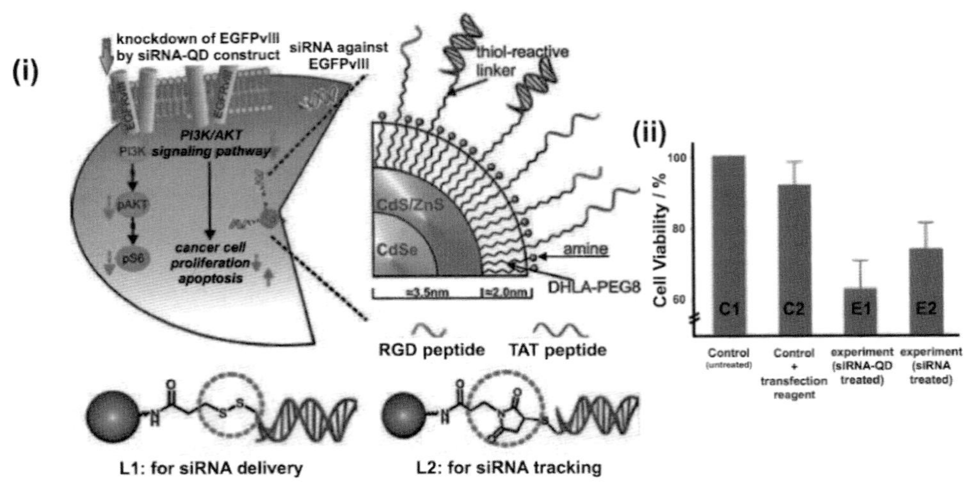

(b)

Fig. 11. (a) Schematic illustration of QD–aptamer conjugates for targeted delivery and sensing of Dox to prostate cancer cells. (i) Upon endocytosis, Dox is released restoring QD and Dox PL ("ON" configuration). (ii) Growth inhibition assay with prostate cancer cell lines: LNCap (PSMA positive) and control PC3 (PSMA negative) upon 3 h exposure and 24 h incubation. Reprinted with permission from Ref. 151. Copyright 2007 American Chemical Society. (b) Multifunctional QD–based siRNA delivery to brain tumor cells (U87) tracked by EGFRvIII gene knockdown for down-regulation of PI3K/Akt signaling pathway. (i) QDs were functionalized with RGD and TAT peptide, as well as two types of siRNA (one type for delivery; one type for tracking). (ii) Growth inhibition assay showing the therapeutic efficacy of delivered siRNA with knockdown of the EGFRvIII gene important for cancer cell proliferation and apoptosis. Reprinted with permission from Ref. 153. Copyright © 2010 WILEY-VCH Verlag GmbH & Co. KGaA, Weinheim.

a 29% reduction in EGFP expression when incubated with QD–F3–siRNA. Jung *et al.* described a multifunctional siRNA–QD construct that inhibited the expression of epidermal growth factor receptor variant III (EGFRvIII) in human U87 glioblastoma cells, including tracking of the subsequently down-regulated PI3K/Akt signaling pathway (Fig. 11b).[153] QDs coated with

PEG ligands were conjugated with two siRNA sequences along with RGD and HIV-TAT derived peptides to enhance transfection into cells. The siRNA sequences were linked to the QD *via* a disulfide linker, which could be cleaved by intracellular enzymes or exchange with glutathione. After 48 h of incubation with siRNA–QD conjugates, cells exhibited a distinct change in morphology, marking the onset of apoptosis. A significant decrease in the number of viable cells was observed in cells transfected with QD–siRNA–RGD–TAT, while QDs alone did not exhibit any cytotoxicity. Decreased expression of EGFRvIII and down-regulation of phospho-Akt and phospho-S6, which are important for cancer cell proliferation and apoptosis, was confirmed with Western immunoblotting.

Given the myriad of nanomaterials with their own unique properties, there is strong interest in developing multifunctional hybrid nanostructures. QDs can be integrated with other types of nanomaterial to assemble nanohybrids for enhanced performance and function. Al-Jamal *et al.* encapsulated PEG-coated QDs into functionalized liposomes and injected the QDs into a solid tumor.[154] The hybrid QD-lipid vesicles showed stronger fluorescence in comparison to QDs alone due to better internalization into tumor cells and penetration through the tumor interstitium. This proof-of-concept study demonstrated the potential for long-term cell imaging in a format that could be easily adapted for theranostic applications (e.g. co-encapsulation of drug within a single vesicle for simultaneous drug delivery and imaging). Hsu *et al.* took the idea of multifunctional nanohybrids a step further by designing multimodal silica nanohybrids suitable for MRI and optical contrast plus drug delivery. Both NIR emitting $CuInS_2/ZnS$ QDs and superparamagnetic Fe_3O_4 magnetite nanocrystals were co-encapsulated within the silica matrix to provide fluorescence emission within the tissue optical window and MRI contrast (spin-spin relaxivity, r_2 = 214 $mM^{-1}s^{-1}$).[155] The average diameter of hybrid nanoparticles was <30 nm. Therapeutic function was investigated by covalently coupling a Pt(IV) anticancer drug to the silica nanohybrids, which demonstrated higher therapeutic efficacy in comparison to free drug when tested with MCF-7 cells.

5.5.2. *Photodynamic therapy*

Photodynamic therapy (PDT) has attracted attention as an alternative, less-invasive treatment for cancer. Sensitizers, activated by light, produce free radicals that rapidly react with oxygen to generate highly cytotoxic reactive oxygen species (ROS) such as singlet oxygen ($^1O_2$). While the possibility of direct generation of singlet oxygen by illuminated QDs has been suggested,[156–159] a more promising strategy for QD-based PDT is the use of QDs as FRET donors for traditional photosensitizers (PS) such as phthalocyanines, Rose Bengal, and various porphyrins.[156,157,159,160] QDs are advantageous in that they can be engineered to (i) achieve optimal spectral overlap with the PS, (ii) act as a carrier for a large number of (typically poorly soluble) PS molecules, (iii) facilitate targeting of PS to tumor cells through co-derivatization with antibodies or other targeting ligands, and (iv) permit deep tissue activation of PS upon two-photon excitation-induced FRET due to their large two-photon absorption cross-section.

In the first demonstration of QD-FRET-based PDT, Samia *et al.* linked hydrophobic CdSe QDs (in organic solvent) to silicon phthalocyanine (Pc4) and stimulated singlet oxygen production through excitation of QDs at 488 nm (5% efficiency).[156] Tsay *et al.* demonstrated production of singlet oxygen with water-soluble CdSe/CdS/ZnS QDs coupled with either Rose Bengal or chlorin e6 *via* a peptide linker.[158] The spectral overlap with the PS was optimized by selection of the peak PL wavelength for the QD (QD_{538} for Rose Bengal; QD_{620} for chlorin e6). Singlet oxygen quantum yields as high as 0.31 were achieved. Importantly, Fowley *et al.* demonstrated that CdSe/ZnS–Rose Bengal conjugates could produce singlet oxygen with two-photon excitation of the QDs (at 800 nm), taking advantage or the NIR optical window in biological tissue.[159] Direct two-photon excitation of the PS would be very inefficient, highlighting the advantage of the QD conjugate. Cytotoxic effects from QD–PS conjugates were observed in tests with HeLa cells; QDs alone did not exhibit appreciable cytotoxicity. Despite these initial successes, much work remains to be done before QD-PDT will be ready for clinical assessment, including optimization in singlet oxygen generation efficiencies, additional cellular studies, and *in vivo* testing with animal models.

6. Current Challenges

6.1. *Toxicity of quantum dots*

With a growing list of cellular and *in vivo* applications for QDs, and ambitions for clinical use, there is ongoing debate over their potential toxicity. Superficial concerns are typically associated with Cd or other heavy metal constituents.[161,162] Various studies have found differing levels of cytotoxicity under different conditions, including the absence of significant cytotoxicity. One response to concerns about potential Cd exposure has been a growing number of synthetic methods and applications for Cd-free QDs (e.g. InP, $CuInS_2$), and these materials have been reported to have minimal cytotoxicity. However, the nanotoxicology is very complex since, in addition to dose and exposure, the *complete* chemical and physical properties of the QD must be considered, including any heterogeneity in those properties across a population. Important considerations include the inorganic core and shell, the organic coating, any biomolecular ligands, and factors such as particle size, shape, and other morphological characteristics (e.g. roughness), surface charge, and undetected impurities or endotoxins in the QD preparation. Drawing unambiguous conclusions from the literature has been hindered by the absence of a standard set of protocols for characterizing the structure and properties of QD samples. This limitation has been exacerbated by the poor continuity and congruence between synthesized QDs, whether batch-to-batch within the same laboratory, or between laboratories and different commercial suppliers. As Hardman succinctly noted, "Not all QDs are alike."[162]

Toxicity assessment studies have been conducted with different cell lines and animal models, which, in addition to the above, have made it challenging to compare data and arrive at a systematic model for predicting toxicity. In studies that observed toxicity, suggested mechanisms have included particle breakdown with leaching of toxic metals[163–165] and generation of ROS.[166,167] In general, it has been found that the presence of relatively benign shell materials such as ZnS can significantly reduce the potential toxicity of CdSe and CdTe QD cores,[163,165] although a few studies have reported the contrary,[168,169] further highlighting the discrepancies that arise between different materials, experimental conditions, and protocols. Another general conclusion has been that organic coatings play a key role in the toxicity and biological fate of QDs. An earlier study evaluated the effect of the surface charge of surfactant-coated QDs on the viability of HepG2 cells.[170] Although, it was noted that negatively charged CdSe QDs were less toxic in comparison to positively charged QDs, it is known that, at sufficient dose and exposure, surfactants alone can induce toxic effects. Furthermore, these ligands are not necessarily representative of capping ligands used in a majority of assays. Importantly, Hoshino *et al.* found that QD cytotoxicity is primarily influenced by the surface ligands rather than the core material in experiments with TOPO-capped CdSe/ZnS QDs.[171] Among the capping ligands they studied — mercaptoundecanoic acid (MUA), cysteamine, and thioglycerol — only a small amount of cell death was observed with negatively charged MUA-QDs. The induction of apoptosis was correlated with DNA damage from free MUA (100–400 μg/mL), whereas no DNA damage was observed with amine, hydroxyl, or mixed coatings on QDs. Importantly, the purity of the QD samples was found to be important in cytotoxicity assays. For instance, crude (membrane filtered)

QDs caused greater DNA damage in comparison to ultrafiltration purified QDs at 1–2 μM concentrations.

The Rotello Laboratory has attempted to develop label-free method for quantifying the uptake of QDs and the intracellular stability of their surface ligands in HeLa cells.[172] Four different sizes (between 2.9–5.9 nm) of QDs with emission at 535, 565, 595 and 630 nm were capped with cationic (to facilitate cellular uptake) mono and bidentate thiol-PEG ligands. A combination of laser desorption-ionization mass spectrometry (LDI-MS) and ICP-MS revealed that ligand stability decreased with increasing particle size, with the total ligand loss reaching 70%. Monothiolate ligands were 40% less stable than bidentate thiol ligand over a 24 h period within cells.

Since cultured cells cannot replicate the complexity of a living organism, several groups have conducted systematic studies of QD toxicity in animal models.[173] Two recent studies, one dosing non-human primates (Rhesus macaques) with CdSe/CdS/ZnS QDs[174] and another dosing Sprague-Dawley rats with CdSe/ZnS QDs,[175] concluded that QDs were non-toxic *in vivo*. Results indicated no acute toxicity, inflammation or alteration in organ function over 80–90 days, consistent with earlier short-term exposure studies in mice.[176–178] Despite the absence of acute toxicity, chemical analysis indicated that most of the initial cadmium dose remained in the RES system (liver, kidneys and spleen). Although these are promising results, much work remains to be done to assess the safety of the variety of QD materials proposed for various biomedical applications, the effect of the method of *in vivo* administration, and the long-term fate of QDs sequestered in the RES.[173]

6.2. *Biodistribution of quantum dots*

Another impediment to the use of QDs in clinical applications is the need for a better understanding of the pharmacokinetics of nanoparticles with different routes of administration, and control over their accumulation in the body over long periods of time. QDs administered *in vivo* preferentially accumulate in the organs associated with the RES, including the liver, kidneys, spleen, lymph nodes, and bone marrow. The mechanisms of biodistribution, renal filtration and urinary elimination have been strongly correlated with QD size and surface properties.[150,179] QDs with a final HD less than 5.5 nm are rapidly excreted in urine, whereas no renal excretion is observed with HDs >15 nm. Zwitterionic or neutral coatings can minimize the adsorption of plasma proteins, which typically leads to a much increased HD for positively or negatively charged QDs.[179] For example, QDs capped with MUA (anionic) increased in HD from 25 nm to 80 nm upon exposure to bovine serum albumin.[180] In contrast, QDs capped with zwitterionic and PEG-based ligand did not show a significant change in HD in the presence of proteins.[179,181] The limitation of PEG-coated QDs is their larger HD compared to most zwitterionic QDs and, therefore, limited renal clearance. Nonetheless, the excellent colloidal stability and biocompatibility of PEGylated QDs has made them useful for many *in vivo* studies.[69] Ballou *et al.* and Al-Jamal *et al.* correlated the circulation half-life of QDs (12–70 min;[177] and 5 h[182]) with the size of the PEG ligands (0.75–5 kDa). Accumulation in liver and spleen was observed 24 h post-injection and, after one month, QDs were primarily observed in lymph nodes, bone marrow, and intestine.[177] It should be noted that elimination rates are important pharmacokinetic parameter, as sufficiently long blood circulation times are essential for efficient drug delivery, tracking, and imaging *in vivo*. However, accumulation rates in RES system are not necessarily directly correlated with elimination rates. For instance, although larger sized nanoparticles accumulate more rapidly in the liver, an increased HD for nanoparticles significantly improves blood circulation lifetime (decreased elimination rate).[179,180]

7. Conclusions and Outlook

The role of QDs in biomedical applications has grown tremendously over the last two decades. Their unique optical and physical properties — the brightness required for sensitive detection, the photostability required for long-term tracking and imaging, and the multiplexed capacity required for unraveling complex biological processes — have led to utility well beyond use as an alternative to fluorescent dyes. Advances in QD materials, surface coatings, and bioconjugation strategies have facilitated tailoring of nanomaterials for specific biomedical applications. The successful use of QDs in these applications requires an understanding of how to rationally control processes at the nanoscale, whether the continuity of synthesized QD materials, surface modifications for colloidal stability in biological media, or bioconjugation for active sensing, bioimaging and drug delivery. This field of research is truly interdisciplinary in its scope.

Among the myriad of QD applications, significant progress has been made in the area of assays and biosensing, particularly in the design of configurations that modulate QD PL in response to biorecognition events. These methods tend to be based on Förster-type energy transfer processes (FRET, BRET, CRET), NSET and CT quenching. QDs have also made great strides in the area of IHC and ICC imaging, and have provided the opportunity to address new challenges and gain insight into the details of various biological processes that might otherwise remain unseen. The same can be said for the use of QDs in live cell imaging and sensing, albeit that a remaining challenge is the reliable delivery of QD constructs to the cellular cytosol with escape from the endolysomal system. Additionally, the size and surface area of QDs has permitted engineering of multifunctional nanoprobes for targeted drug delivery, PDT, multimodal *in vivo* imaging, and theranostic applications. Despite these significant advances, and the entry of QD technology into the optoelectronics market (display and lighting technologies), the commercialization of biomedical technologies based on QDs (for non-research purposes) remains somewhere on the horizon as an attainable, high-reward goal.

9. Acknowledgments

The authors acknowledge support from the University of British Columbia, the Natural Sciences and Engineering Research Council of Canada (NSERC), the Office of Naval Research (ONR), the Naval Research Laboratory (NRL), the NRL Nanosciences Institute (NSI), the Defense Advanced Research Projects Agency (DARPA), and the Defense Threat Reduction Agency Joint Science and Technological Office (DTRA-JSTO) Military Interdepartmental Purchase Requisition (MIPR) #B112582M.

References

1. K. Bourzac, *Nature* 283 (2013).
2. M. Bruchez, M. Moronne, P. Gin, S. Weiss and A. P. Alivisatos, *Science* 2013 (1998).
3. W. C. W. Chan and S. M. Nie, *Science* 2016 (1998).
4. P. A. S. Jorge, M. A. Martins, T. Trindade, J. L. Santos and F. Farahi, *Sensors* 3489 (2007).
5. A. M. Smith and S. Nie, *Acc. Chem. Res.* 190 (2009).
6. W. R. Algar, K. Susumu, J. B. Delehanty and I. L. Medintz, *Anal. Chem.* 8826 (2011).
7. I. L. Medintz, H. T. Uyeda, E. R. Goldman and H. Mattoussi, *Nat. Mater.* 435 (2005).
8. D. J. Norris and M. G. Bawendi, *Phys. Rev. B: Condens. Matter* 16338 (1996).
9. A. P. Alivisatos, *J. Phys. Chem.* 13226 (1996).
10. W. R. Algar, K. Susumu, J. B. Delehanty and I. L. Medintz, *Anal. Chem.* 8826 (2011).
11. B. O. Dabbousi, J. Rodriguez-Viejo, F. V. Mikulec, J. R. Heine, H. Mattoussi, R. Ober, K. F. Jensen and M. G. Bawendi, *J. Phys. Chem. B* 9463 (1997).
12. W. J. Parak, T. Pellegrino and C. Plank, *Nanotechnology* R9 (2005).
13. L. Zou, Z. Y. Gu, N. Zhang, Y. L. Zhang, Z. Fang, W. H. Zhu and X. H. Zhong, *J. Mater. Chem.* 2807 (2008).
14. D. Battaglia and X. Peng, *Nano Lett.* 1027 (2002).
15. Q. Ma and X. Su, *Analyst* 1867 (2010).
16. M. D. Regulacio and M. Y. Han, *Acc. Chem. Res.* 621 (2010).
17. L. A. Swafford, L. A. Weigand, M. J. Bowers, J. R. McBride, J. L. Rapaport, T. L. Watt, S. K. Dixit, L. C. Feldman and S. J. Rosenthal, *J. Am. Chem. Soc.* 12299 (2006).

18. R. Rossetti, S. Nakahara and L. E. Brus, *J. Chem. Phys.* 1086 (1983).
19. A. I. Ekimov, A. L. Efros and A. A. Onushchenko, *Solid State Commun.* 921 (1985).
20. C. B. Murray, D. J. Norris and M. G. Bawendi, *J. Am. Chem. Soc.* 8706 (1993).
21. X. Peng, J. Wickham and A. P. Alivisatos, *J. Am. Chem. Soc.* 5343 (1998).
22. M. A. Hines and P. Guyot-Sionnest, *J. Phys. Chem.* 468 (1996).
23. A. P. Alivisatos, *Science* 933 (1996).
24. X. G. Peng, L. Manna, W. D. Yang, J. Wickham, E. Scher, A. Kadavanich and A. P. Alivisatos, *Nature* 59 (2000).
25. X. Peng, M. C. Schlamp, A. V. Kadavanich and A. P. Alivisatos, *J. Am. Chem. Soc.* 7019 (1997).
26. V. K. LaMer and R. H. Dinegar, *J. Am. Chem. Soc.* 4847 (1950).
27. Z. A. Peng and X. Peng, *J. Am. Chem. Soc.* 3343 (2002).
28. B. O. Dabbousi, J. RodriguezViejo, F. V. Mikulec, J. R. Heine, H. Mattoussi, R. Ober, K. F. Jensen and M. G. Bawendi, *J. Phys. Chem. B* 9463 (1997).
29. Z. A. Peng and X. Peng, *J. Am. Chem. Soc.* 183 (2000).
30. D. V. Talapin, A. L. Rogach, A. Kornowski, M. Haase and H. Weller, *Nano Lett.* 207 (2001).
31. Z. A. Peng and X. G. Peng, *J. Am. Chem. Soc.* 183 (2001).
32. L. Qu, Z. A. Peng and X. Peng, *Nano Lett.* 333 (2001).
33. W. W. Yu and X. Peng, *Angew. Chem. Int. Ed.* 2559 (2002).
34. W. J. Zhang, G. J. Chen, J. Wang, B. C. Ye and X. H. Zhong, *Inorg. Chem.* 9723 (2009).
35. Z. M. Yuan, Q. Ma, A. Y. Zhang, Y. Q. Cao, J. Yang and P. Yang, *J. Mater. Sci.* 3770 (2012).
36. Z. T. Deng, O. Schulz, S. Lin, B. Q. Ding, X. W. Liu, X. X. Wei, R. Ros, H. Yan and Y. Liu, *J. Am. Chem. Soc.* 5592 (2010).
37. R. K. Beri, P. More, B. G. Bharate and P. K. Khanna, *Curr. App. Phys.* 553 (2010).
38. J. T. Siy, E. M. Brauser and M. H. Bartl, *Chem. Commun.* 364 (2011).
39. D. Janczewski, N. Tomczak, M. Y. Han and G. J. Vancso, *Nat. Protoc.* 1546 (2011).
40. L. C. Mattheakis, J. M. Dias, Y. J. Choi, J. Gong, M. P. Bruchez, J. Q. Liu and E. Wang, *Anal. Biochem.* 200 (2004).
41. T. Pellegrino, L. Manna, S. Kudera, T. Liedl, D. Koktysh, A. L. Rogach, S. Keller, J. Radler, G. Natile and W. J. Parak, *Nano Lett.* 703 (2004).
42. W. W. Yu, E. Chang, J. C. Falkner, J. Y. Zhang, A. M. Al-Somali, C. M. Sayes, J. Johns, R. Drezek and V. L. Colvin, *J. Am. Chem. Soc.* 2871 (2007).
43. E. E. Lees, T. L. Nguyen, A. H. A. Clayton, P. Mulvaney and B. W. Muir, *ACS Nano* 1121 (2009).
44. Y. H. Zhang, H. S. Zhang, X. F. Guo and H. Wang, *Microchem. J.* 142 (2008).
45. K. Susumu, H. T. Uyeda, I. L. Medintz, T. Pons, J. B. Delehanty and H. Mattoussi, *J. Am. Chem. Soc.* 13987 (2007).
46. Y. J. Zhang and A. Clapp, *Sensors* 11036 (2011).
47. K. Susumu, B. C. Mei and H. Mattoussi, *Nat. Protoc.* 424 (2009).
48. E. Petryayeva, I. L. Medintz and W. R. Algar, *Appl. Spectrosc.*, DOI: 10.1366/12 06948 (in press) (2013).
49. U. Resch-Genger, M. Grabolle, S. Cavaliere-Jaricot, R. Nitschke and T. Nann, *Nat. Methods* 763 (2008).
50. W. R. Algar and U. J. Krull, *Chem. Phys. Chem.* 561 (2007).
51. J. Aldana, N. Lavelle, Y. Wang and X. Peng, *J. Am. Chem. Soc.* 2496 (2005).

52. M. H. Stewart, K. Susumu, B. C. Mei, I. L. Medintz, J. B. Delehanty, J. B. Blanco-Canosa, P. E. Dawson and H. Mattoussi, *J. Am. Chem. Soc.* 9804.
53. V. V. Breus, C. D. Heyes, K. Tron and G. U. Nienhaus, *ACS Nano*, 2573 (2009).
54. E. Muro, T. Pons, N. Lequeux, A. Fragola, N. Sanson, Z. Lenkei and B. Dubertret, *J. Am. Chem. Soc.* 4556.
55. W. R. Algar, D. E. Prasuhn, M. H. Stewart, T. L. Jennings, J. B. Blanco-Canosa, P. E. Dawson and I. L. Medintz, *Bioconjugate Chem.* 825 (2011).
56. H. S. Han, N. K. Devaraj, J. Lee, S. A. Hilderbrand, R. Weissleder and M. G. Bawendi, *J. Am. Chem. Soc.* 7838 (2010).
57. A. Bernardin, A. Cazet, L. Guyon, P. Delannoy, F. Vinet, D. Bonnaffe and I. Texier, *Bioconjugate Chem.* 583 (2010).
58. D. E. Prasuhn, J. B. Blanco-Canosa, G. J. Vora, J. B. Delehanty, K. Susumu, B. C. Mei, P. E. Dawson and I. L. Medintz, *ACS Nano* 267 (2010).
59. Z. Y. Xia, Y. Xing, M. K. So, A. L. Koh, R. Sinclair and J. H. Rao, *Anal. Chem.* 8649 (2008).
60. I. L. Medintz, L. Berti, T. Pons, A. F. Grimes, D. S. English, A. Alessandrini, P. Facci and H. Mattoussi, *Nano Lett.* 1741 (2007).
61. A. M. Dennis, D. C. Sotto, B. C. Mei, I. L. Medintz, H. Mattoussi and G. Bao, *Bioconjugate Chem.* 1160 (2010).
62. W. R. Algar, D. Wegner, A. L. Huston, J. B. Blanco-Canosa, M. H. Stewart, A. Armstrong, P. E. Dawson, N. Hildebrandt and I. L. Medintz, *J. Am. Chem. Soc.* 1876 (2012).
63. H. Duan, M. Kuang and Y. A. Wang, *Chem. Mater.* 4372 (2010).
64. K. Susumu, I. L. Medintz, J. B. Delehanty, K. Boeneman and H. Mattoussi, *J. Phys. Chem. C* 13526 (2010).
65. M. E. Åkerman, W. C. W. Chan, P. Laakkonen, S. N. Bhatia and E. Ruoslahti, *Proc. Natl. Acad. Sci. USA* 12617 (2002).
66. W. Cai, K. Chen, Z.-B. Li, S. S. Gambhir and X. Chen, *J. Nucl. Med.* 1862 (2007).
67. W. Cai, D.-W. Shin, K. Chen, O. Gheysens, Q. Cao, S. X. Wang, S. S. Gambhir and X. Chen, *Nano Lett.* 669 (2006).
68. J. A. J. Fitzpatrick, S. K. Andreko, L. A. Ernst, A. S. Waggoner, B. Ballou and M. P. Bruchez, *Nano Lett.* 2736 (2009).
69. X. H. Gao, Y. Y. Cui, R. M. Levenson, L. W. K. Chung and S. M. Nie, *Nat. Biotechnol.* 969 (2004).
70. H. Tada, H. Higuchi, T. M. Wanatabe and N. Ohuchi, *Cancer Res.* 1138 (2007).
71. E. R. Goldman, A. R. Clapp, G. P. Anderson, H. T. Uyeda, J. M. Mauro, I. L. Medintz and H. Mattoussi, *Anal. Chem.* 684 (2004).
72. P. K. Chattopadhyay, D. A. Price, T. F. Harper, M. R. Betts, J. Yu, E. Gostick, S. P. Perfetto, P. Goepfert, R. A. Koup, S. C. De Rosa, M. P. Bruchez and M. Roederer, *Nat. Med.* 972 (2006).
73. J. M. Klostranec, Q. Xiang, G. A. Farcas, J. A. Lee, A. Rhee, E. I. Lafferty, S. D. Perrault, K. C. Kain and W. C. W. Chan, *Nano Lett.* 2812 (2007).
74. S. Giri, E. A. Sykes, T. L. Jennings and W. C. W. Chan, *ACS Nano* 1580 (2011).
75. H. Xu, M. Y. Sha, E. Y. Wong, J. Uphoff, Y. Xu, J. A. Treadway, A. Truong, E. O'Brien, S. Asquith, M. Stubbins, N. K. Spurr, E. H. Lai and W. Mahoney, *Nucleic Acids Res.* e43 (2003).

76. P. S. Eastman, W. Ruan, M. Doctolero, R. Nuttall, G. de Feo, J. S. Park, J. S. F. Chu, P. Cooke, J. W. Gray, S. Li and F. F. Chen, *Nano Lett.* 1059 (2006).
77. J. M. Naciff, B. D. Richardson, K. G. Oliver, M. L. Jump, S. M. Torontali, K. D. Juhlin, G. J. Carr, J. R. Paine, J. A. Tiesman and G. P. Daston, *Environ. Health Perspect.* 1164 (2005).
78. W. R. Algar, A. J. Tavares and U. J. Krull, *Anal. Chim. Acta* 1 (2010).
79. I. L. Medintz and H. Mattoussi, *Phys. Chem. Chem. Phys.* 17 (2009).
80. I. L. Medintz, A. R. Clapp, F. M. Brunel, T. Tiefenbrunn, H. T. Uyeda, E. L. Chang, J. R. Deschamps, P. E. Dawson and H. Mattoussi, *Nat. Mater.* 581 (2006).
81. R. Freeman, T. Finder, R. Gill and I. Willner, *Nano Lett.* 2192 (2010).
82. C. Y. Zhang and L. W. Johnson, *Anal. Chem.* 3051 (2009).
83. R. Freeman, L. Bahshi, T. Finder, R. Gill and I. Willner, *Chem. Commun.* 764 (2009).
84. B. Tang, L. Cao, K. Xu, L. Zhuo, J. Ge, Q. Li and L. Yu, *Chem. Eur. J.* 3637 (2008).
85. I. L. Medintz, A. R. Clapp, H. Mattoussi, E. R. Goldman, B. Fisher and J. M. Mauro, *Nat. Mater.* 630 (2003).
86. S. B. Lowe, J. A. G. Dick, B. E. Cohen and M. M. Stevens, *ACS Nano* 851 (2011).
87. Y.-P. Kim, Y.-H. Oh, E. Oh, S. Ko, M.-K. Han and H.-S. Kim, *Anal. Chem.* 4634 (2008).
88. C. Y. Zhang and J. Hu, *Anal. Chem.* 1921 (2010).
89. W. R. Algar, A. P. Malanoski, K. Susumu, M. H. Stewart, N. Hildebrandt and I. L. Medintz, *Anal. Chem.* 10136 (2012).
90. M. K. So, C. J. Xu, A. M. Loening, S. S. Gambhir and J. H. Rao, *Nat. Biotechnol.* 339 (2006).
91. R. Freeman, J. Girsh, A. F. J. Jou, J. A. A. Ho, T. Hug, J. Dernedde and I. Willner, *Anal. Chem.* 6192 (2012).
92. X. Q. Liu, R. Freeman, E. Golub and I. Willner, *ACS Nano* 7648 (2011).
93. A. Niazov, R. Freeman, J. Girsh and I. Willner, *Sensors* 10388 (2011).
94. R. Freeman, X. Q. Liu and I. Winner, *J. Am. Chem. Soc.* 11597 (2011).
95. E. Golub, A. Niazov, R. Freeman, M. Zatsepin and I. Willner, *J. Phys. Chem. C* 13827 (2012).
96. M. D. Swain, J. Octain and D. E. Benson, *Bioconjugate Chem.* 2520 (2008).
97. B. P. Aryal and D. E. Benson, *J. Am. Chem. Soc.* 15986 (2006).
98. I. L. Medintz, T. Pons, S. A. Trammell, A. F. Grimes, D. S. English, J. B. Blanco-Canosa, P. E. Dawson and H. Mattoussi, *J. Am. Chem. Soc.* 16745 (2008).
99. M. Sykora, M. A. Petruska, J. Alstrum-Acevedo, I. Bezel, T. J. Meyer and V. I. Klimov, *J. Am. Chem. Soc.* 9984 (2006).
100. M. Malicki, K. E. Knowles and E. A. Weiss, *Chem. Commun.* (2013).
101. D. Dorokhin, N. Tomczak, A. H. Velders, D. N. Reinhoudt and G. J. Vancso, *J. Phys. Chem. C* 18676 (2009).
102. M. J. Ruedas-Rama and E. A. H. Hall, *Analyst* 1556 (2008).
103. I. Yildiz, M. Tomasulo and F. M. Raymo, *Proc. Natl. Acad. Sci. USA* 11457 (2006).
104. A. J. Morris-Cohen, M. T. Frederick, L. C. Cass and E. A. Weiss, *J. Am. Chem. Soc.* 10146 (2011).
105. F. Scholz, L. Dworak, V. V. Matylitsky and J. Wachtveitl, *Chem. Phys. Chem.* 2255 (2011).

106. I. L. Medintz, M. H. Stewart, S. A. Trammell, K. Susumu, J. B. Delehanty, B. C. Mei, J. S. Melinger, J. B. Blanco-Canosa, P. E. Dawson and H. Mattoussi, *Nat. Mater.* 676 (2010).
107. X. Ji, G. Palui, T. Avellini, H. B. Na, C. Yi, K. L. Knappenberger and H. Mattoussi, *J. Am. Chem. Soc.* 6006 (2012).
108. J. Callan, R. Mulrooney, S. Kamila and B. McCaughan, *J. Fluoresc.* 527 (2008).
109. J. Huang, D. Stockwell, Z. Huang, D. L. Mohler and T. Lian, *J. Am. Chem. Soc.* 5632 (2008).
110. I. L. Medintz, D. Farrell, K. Susumu, S. A. Trammell, J. R. Deschamps, F. M. Brunel, P. E. Dawson and H. Mattoussi, *Anal. Chem.* 4831 (2009).
111. T. J. Fountaine, S. M. Wincovitch, D. H. Geho, S. H. Garfield and S. Pittaluga, *Mod. Pathol*, 1181 (2006).
112. Y. Xing, Q. Chaudry, C. Shen, K. Y. Kong, H. E. Zhau, L. WChung, J. A. Petros, R. M. O'Regan, M. V. Yezhelyev, J. W. Simons, M. D. Wang and S. Nie, *Nat. Protoc.* 1152 (2007).
113. T. L. Jennings, S. G. Becker-Catania, R. C. Triulzi, G. Tao, B. Scott, K. E. Sapsford, S. Spindel, E. Oh, V. Jain, J. B. Delehanty, D. E. Prasuhn, K. Boeneman, W. R. Algar and I. L. Medintz, *ACS Nano* 5579 (2011).
114. J. A. Liu, S. K. Lau, V. A. Varma, B. A. Kairdolf and S. M. Nie, *Anal. Chem.* 6237 (2010).
115. J. Liu, S. K. Lau, V. A. Varma, R. A. Moffitt, M. Caldwell, T. Liu, A. N. Young, J. A. Petros, A. O. Osunkoya, T. Krogstad, B. Leyland-Jones, M. D. Wang and S. M. Nie, *ACS Nano* 2755 (2010).
116. C.-W. Peng, X.-L. Liu, C. Chen, X. Liu, X.-Q. Yang, D.-W. Pang, X.-B. Zhu and Y. Li, *Biomaterials* 2907 (2011).
117. D. S. Lidke, P. Nagy, R. Heintzmann, D. J. Arndt-Jovin, J. N. Post, H. E. Grecco, E. A. Jares-Erijman and T. M. Jovin, *Nat. Biotechnol.* 198 (2004).
118. D. S. Lidke, K. A. Lidke, B. Rieger, T. M. Jovin and D. J. Arndt-Jovin, *J. Cell Biol.* 619 (2005).
119. M. Dahan, S. Levi, C. Luccardini, P. Rostaing, B. Riveau and A. Triller, *Science* 442 (2003).
120. C. Bouzigues, M. Morel, A. Triller and M. Dahan, *Proc. Natl. Acad. Sci. USA* 11251 (2007).
121. T. Buerli, K. Baer, H. Ewers, C. Sidler, C. Fuhrer and J.-M. Fritschy, *PLoS One* e11507 (2010).
122. M. Heine, L. Groc, R. Frischknecht, J. C. Beique, B. Lounis, G. Rumbaugh, R. L. Huganir, L. Cognet and D. Choquet, *Science* 201 (2008).
123. Q. Zhang, Y. Q. Cao and R. W. Tsien, *Proc. Natl. Acad. Sci. USA* 17843 (2007).
124. Q. Zhang, Y. Li and R. W. Tsien, *Science* 1448 (2009).
125. A. R. Lowe, J. J. Siegel, P. Kalab, M. Siu, K. Weis and J. T. Liphardt, *Nature* 600 (2010).
126. N. P. Wells, G. A. Lessard, P. M. Goodwin, M. E. Phipps, P. J. Cutler, D. S. Lidke, B. S. Wilson and J. H. Werner, *Nano Lett.* 4732 (2010).
127. Y. P. Varshni, *Physica* 149 (1967).
128. J.-M. Yang, H. Yang and L. Lin, *ACS Nano* 5067 (2011).
129. Y. Wang, H. Mao and L. B. Wong, *Talanta* 694 (2011).
130. Y. C. Wang, H. Mao and L. B. Wong, *Nanotechnology*, (2010).

131. A. M. Dennis, W. J. Rhee, D. Sotto, S. N. Dublin and G. Bao, *ACS Nano* 2917 (2012).
132. R. Freeman, R. Gill, I. Shweky, M. Kotler, U. Banin and I. Willner, *Angew. Chem. Int. Ed.* 309 (2009).
133. M. Helle, E. Cassette, L. Bezdetnaya, T. Pons, A. Leroux, F. Plénat, F. Guillemin, B. Dubertret and F. Marchal, *PLoS One* e44433 (2012).
134. T. Pons, E. Pic, N. Lequeux, E. Cassette, L. Bezdetnaya, F. Guillemin, F. Marchal and B. Dubertret, *ACS Nano* 2531 (2010).
135. S. Kim, Y. T. Lim, E. G. Soltesz, A. M. De Grand, J. Lee, A. Nakayama, J. A. Parker, T. Mihaljevic, R. G. Laurence, D. M. Dor, L. H. Cohn, M. G. Bawendi and J. V. Frangioni, *Nat. Biotechnol.* 93 (2004).
136. D. R. Larson, W. R. Zipfel, R. M. Williams, S. W. Clark, M. P. Bruchez, F. W. Wise and W. W. Webb, *Science* 1434 (2003).
137. S. Rieger, R. P. Kulkarni, D. Darcy, S. E. Fraser and R. W. Koster, *Dev. Dyn.* 670 (2005).
138. B. Dubertret, P. Skourides, D. J. Norris, V. Noireaux, A. H. Brivanlou and A. Libchaber, *Science* 1759 (2002).
139. P. Stylianou and P. A. Skourides, *Mech. Dev.* 828 (2009).
140. J. V. Frangioni, S.-W. Kim, S. Ohnishi, S. Kim and M. G. Bawendi, *Methods Mol. Biol.* 147 (2007).
141. H. Kobayashi, Y. Hama, Y. Koyama, T. Barrett, C. A. S. Regino, Y. Urano and P. L. Choyke, *Nano Lett.* 1711 (2007).
142. S. H. Bhang, N. Won, T.-J. Lee, H. Jin, J. Nam, J. Park, H. Chung, H.-S. Park, Y.-E. Sung, S. K. Hahn, B.-S. Kim and S. Kim, *ACS Nano* 1389 (2009).
143. H. Yukawa, M. Watanabe, N. Kaji, Y. Okamoto, M. Tokeshi, Y. Miyamoto, H. Noguchi, Y. Baba and S. Hayashi, *Biomaterials* 2177 (2012).
144. Y. Lei, H. Tang, L. Yao, R. Yu, M. Feng and B. Zou, *Bioconjugate Chem.* 421 (2007).
145. E. B. Voura, J. K. Jaiswal, H. Mattoussi and S. M. Simon, *Nat. Med.* 993 (2004).
146. J. Ruan, H. Song, Q. Qian, C. Li, K. Wang, C. Bao and D. Cui, *Biomaterials* 7093 (2012).
147. R. Bakalova, Z. Zhelev, I. Aoki, K. Masamoto, M. Mileva, T. Obata, M. Higuchi, V. Gadjeva and I. Kanno, *Bioconjugate Chem.* 1135 (2008).
148. G. J. Stasiuk, S. Tamang, D. Imbert, C. Poillot, M. Giardiello, C. Tisseyre, E. L. Barbier, P. H. Fries, M. de Waard, P. Reiss and M. Mazzanti, *ACS Nano* 8193 (2011).
149. C. Tu, X. Ma, A. House, S. M. Kauzlarich and A. Y. Louie, *ACS Med. Chem. Lett.* 285 (2011).
150. H. S. Choi, W. H. Liu, F. B. Liu, K. Nasr, P. Misra, M. G. Bawendi and J. V. Frangioni, *Nat. Nanotechnol.* 42 (2010).
151. V. Bagalkot, L. Zhang, E. Levy-Nissenbaum, S. Jon, P. W. Kantoff, R. Langer and O. C. Farokhzad, *Nano Lett.* 3065 (2007).
152. A. M. Derfus, A. A. Chen, D.-H. Min, E. Ruoslahti and S. N. Bhatia, *Bioconjugate Chem.* 1391 (2007).
153. J. Jung, A. Solanki, K. A. Memoli, K.-i. Kamei, H. Kim, M. A. Drahl, L. J. Williams, H.-R. Tseng and K. Lee, *Angew. Chem. Int. Ed.* 103 (2010).
154. W. T. Al-Jamal, K. T. Al-Jamal, P. H. Bomans, P. M. Frederik and K. Kostarelos, *Small* 1406 (2008).
155. J.-C. Hsu, C.-C. Huang, K.-L. Ou, N. Lu, F.-D. Mai, J.-K. Chen and J.-Y. Chang, *J. Mater. Chem.* 19257 (2011).

156. A. C. S. Samia, X. B. Chen and C. Burda, *J. Am. Chem. Soc.* 15736 (2003).
157. J. Ma, J.-Y. Chen, M. Idowu and T. Nyokong, *J. Phys. Chem. B* 4465 (2008).
158. J. M. Tsay, M. Trzoss, L. X. Shi, X. X. Kong, M. Selke, M. E. Jung and S. Weiss, *J. Am. Chem. Soc.* 6865 (2007).
159. C. Fowley, N. Nomikou, A. P. McHale, P. A. McCarron, B. McCaughan and J. F. Callan, *J. Mater. Chem.* 6456 (2012).
160. H. D. Duong and J. I. Rhee, *Chem. Phys. Lett.* 496 (2011).
161. M. Bottrill and M. Green, *Chem. Commun.* 7039 (2011).
162. R. Hardman, *Environ. Health Perspect.* 165 (2006).
163. A. M. Derfus, W. C. W. Chan and S. N. Bhatia, *Nano Lett.* 11 (2003).
164. C. Kirchner, T. Liedl, S. Kudera, T. Pellegrino, A. Muñoz Javier, H. E. Gaub, S. Stölzle, N. Fertig and W. J. Parak, *Nano Lett.* 331 (2004).
165. Y. Su, Y. He, H. Lu, L. Sai, Q. Li, W. Li, L. Wang, P. Shen, Q. Huang and C. Fan, *Biomaterials* 19 (2009).
166. J. Lović, S. J. Cho, F. M. Winnik and D. Maysinger, *Chem. Biol.* 1227 (2005).
167. H. Chibli, L. Carlini, S. Park, N. M. Dimitrijevic and J. L. Nadeau, *Nanoscale* 2552 (2011).
168. N. Chen, Y. He, Y. Su, X. Li, Q. Huang, H. Wang, X. Zhang, R. Tai and C. Fan, *Biomaterials* 1238 (2012).
169. A. Shiohara, A. Hoshino, K. Hanaki, K. Suzuki and K. Yamamoto, *Microbiol. Immunol.* 669 (2004).
170. G. Guo, W. Liu, J. Liang, Z. He, H. Xu and X. Yang, *Mater. Lett.* 1641 (2007).
171. A. Hoshino, K. Fujioka, T. Oku, M. Suga, Y. F. Sasaki, T. Ohta, M. Yasuhara, K. Suzuki, and K. Yamamoto, *Nano Lett.* 2163 (2004).
172. Z. J. Zhu, Y. C. Yeh, R. Tang, B. Yan, J. Tamayo, R. W. Vachet and V. M. Rotello, *Nat. Chem.* 963 (2011).
173. L. Y. T. Chou and W. C. W. Chan, *Nat. Nanotechnol.* 416 (2012).
174. L. Ye, K. T. Yong, L. W. Liu, I. Roy, R. Hu, J. Zhu, H. X. Cai, W. C. Law, J. W. Liu, K. Wang, J. Liu, Y. Q. Liu, Y. Z. Hu, X. H. Zhang, M. T. Swihart and P. N. Prasad, *Nat. Nanotechnol.* 453 (2012).
175. T. S. Hauck, R. E. Anderson, H. C. Fischer, S. Newbigging and W. C. W. Chan, *Small* 138 (2010).
176. E. B. Voura, J. K. Jaiswal, H. Mattoussi and S. M. Simon, *Nat. Med.* 993 (2004).
177. B. Ballou, B. C. Lagerholm, L. A. Ernst, M. P. Bruchez and A. S. Waggoner, *Bioconjugate Chem.* 79 (2003).
178. A. Hoshino, K.-i. Hanaki, K. Suzuki and K. Yamamoto, *Biochem. Biophys. Res. Commun.* 46 (2004).
179. H. S. Choi, W. Liu, P. Misra, E. Tanaka, J. P. Zimmer, B. I. Ipe, M. G. Bawendi and J. V. Frangioni, *Nat. Biotechnol.* 1165 (2007).
180. H. C. Fischer, L. Liu, K. S. Pang and W. C. W. Chan, *Adv. Funct. Mater.* 1299 (2006).
181. H. S. Choi, B. I. Ipe, P. Misra, J. H. Lee, M. G. Bawendi and J. V. Frangioni, *Nano Lett.* 2354 (2009).
182. W. T. Al-Jamal, K. T. Al-Jamal, A. Cakebread, J. M. Halket and K. Kostarelos, *Bioconjugate Chem.* 1696 (2009).

Index

Vol. 3

Frontiers in Nanobiomedical Research

HANDBOOK OF NANOBIOMEDICAL RESEARCH

Fundamentals, Applications and Recent Developments

❸ Applications in Diagnostics

Frontiers in Nanobiomedical Research

ISSN: 2251-3965

Series Editors: Martin L. Yarmush *(Harvard Medical School, USA)*
Donglu Shi *(University of Cincinnati, USA)*

Published

Vol. 1: Handbook of Immunological Properties of Engineered Nanomaterials
edited by Marina A. Dobrovolskaia and Scott E. McNeil (SAIC-Frederick, Inc., USA)

Vol. 2: Tissue Regeneration: Where Nano-Structure Meets Biology
edited by Qing Liu (3D Biotek, USA and Tongji University, China) and Hongjun Wang (Stevens Institute of Technology, USA)

Vol. 3: Handbook of Nanobiomedical Research: Fundamentals, Applications and Recent Developments (In 4 Volumes)
edited by Vladimir Torchilin (Northeastern University, USA)

Forthcoming titles

Cancer Therapeutics and Imaging: Molecular and Cellular Engineering and Nanobiomedicine
edited by Kaushal Rege (*Arizona State University, USA*)

Nano Vaccines
edited by Balaji Narasimhan (Iowa State University, USA)

Nano Pharmaceuticals
edited by Rajesh N. Dave (New Jersey Institute of Technology, USA)

Thermal Aspects in Nanobiomedical Systems and Devices
by Dong Cai (Boston College, USA)

Nano Mechanochemistry in Biology
edited by Jeffrey Ruberti (Northeastern University, USA)

Nanomaterial Probes of Biological Processes and Systems
edited by David Mast (University of Cincinnati, USA)

Handbook of Biomaterials
edited by Donglu Shi (University of Cincinnati, USA) and Xuejun Wen (Clemson University, USA)

HANDBOOK OF NANOBIOMEDICAL RESEARCH

Fundamentals, Applications and Recent Developments

❸ Applications in Diagnostics

editor
Vladimir Torchilin
Northeastern University, USA

Vol. 3
Frontiers in Nanobiomedical Research

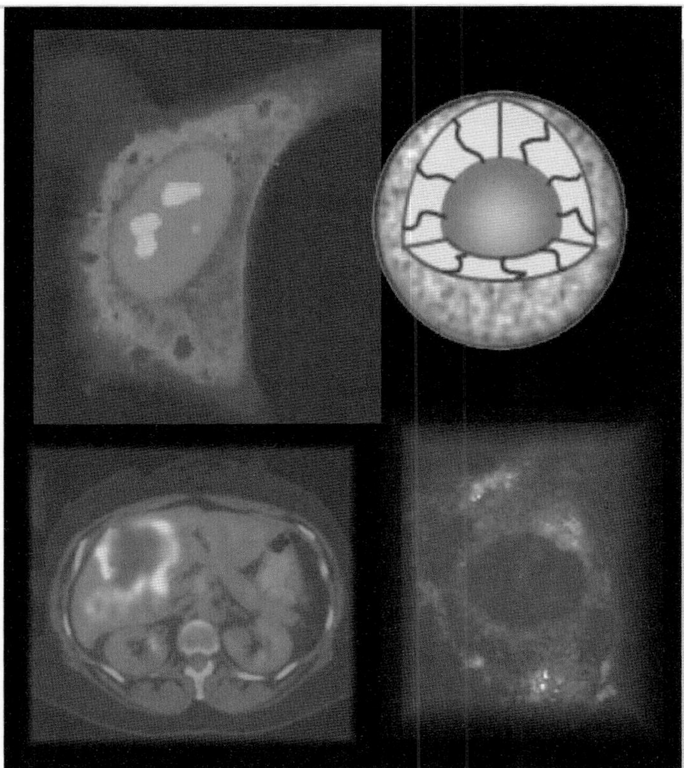

World Scientific

NEW JERSEY • LONDON • SINGAPORE • BEIJING • SHANGHAI • HONG KONG • TAIPEI • CHENNAI

Published by

World Scientific Publishing Co. Pte. Ltd.
5 Toh Tuck Link, Singapore 596224
USA office: 27 Warren Street, Suite 401-402, Hackensack, NJ 07601
UK office: 57 Shelton Street, Covent Garden, London WC2H 9HE

Library of Congress Cataloging-in-Publication Data
Handbook of nanobiomedical research : fundamentals, applications, and recent developments / editor, Vladimir Torchilin.
p. ; cm. -- (Frontiers in nanobiomedical research ; vol. 3)
Includes bibliographical references and index.
ISBN 978-9814520645 (set : alk. paper) -- ISBN 978-9814520676 (volume 1 : alk. paper) --
ISBN 978-9814520683 (volume 2 : alk. paper) -- ISBN 978-9814520690 (volume 3 : alk. paper) --
ISBN 978-9814520706 (volume 4 : alk. paper)
I. Torchilin, V. P., editor. II. Series: Frontiers in nanobiomedical research ; v. 3. 2251-3965
[DNLM: 1. Nanomedicine. 2. Nanostructures. 3. Nanotechnology. QT 36.5]
R857.N34
610.28--dc23

2014024090

British Library Cataloguing-in-Publication Data
A catalogue record for this book is available from the British Library.

Typeset by Stallion Press
Email: enquiries@stallionpress.com

Printed in Singapore

Contents

Chapter 1

Nano Advantages in Diagnostic Imaging

William T. Phillips and Beth Goins

Department of Radiology, University of Texas Health Science Center at San Antonio, 7703 Floyd Curl Drive, San Antonio, Texas 78229–3800
phillips@uthscsa.edu

Abstract

The *in vivo* imaging of nanoparticles can provide important diagnostic information for the medical management of patients. This chapter reviews some of the specific advantages provided by nanoparticles as carriers for contrast agents for diagnostic imaging. These advantages include: (1) the ability of nanoparticles to carry multiple contrast agents enabling imaging with a variety of imaging technologies that each provide specific advantages, (2) the ability of nanoparticles to have sustained retention in the blood pool compared with small molecules, and (3) the ability of nanoparticles to image specific organ systems and pathological processes of the body including inflammation, tumors, lymphatics and bone marrow. Additionally, the imaging of nanoparticles can provide a theranostic advantage facilitating the development of nanoparticle drug carriers and customization of drug delivery in individual patients.

1. Introduction

Nanoparticles have many promising uses for diagnostic imaging. This chapter will review the numerous advantages of nanoparticles for diagnostic imaging as well as their limitations. Many potential uses for nanoparticles in diagnostic imaging have already been described and many more will undoubtedly be developed as new nanoparticle technologies are developed. Nanoparticles have been described as carriers for a wide variety of diagnostic contrast agents including X-ray computed tomography (CT) contrast agents,[1–4] nuclear imaging agents with single photon emission computed tomography (SPECT)[5–8] and positron emission computed tomography (PET),[9–12] optical contrast agents,[13,14] magnetic resonance imaging (MRI)[15] and ultrasound contrast agents.[16,17]

This chapter will review some of the most promising current uses of nanoparticles for diagnostic and theranostic imaging applications. Nanoparticle imaging can provide valuable diagnostic imaging for the localization of numerous pathological processes, such as cancer and infection, as well as vital information regarding the physiologic function of normal organs. In addition, nanoparticles can carry multiple diagnostic agents that can be localized with a variety of imaging techniques at both the macro whole body imaging level as well as at a tissue and cellular micro imaging level. Finally, in a theranostic application, the imaging of nanoparticles can provide guidance for the proper administration and intra-lesional distribution of nanoparticle carried therapies.

2. Specific Advantages of Nanoparticles for Diagnostic Imaging

2.1. *Nanoparticles can carry very high quantities of multiple diagnostic and therapeutic agents*

Nanoparticles are very flexible vehicles for carrying diagnostic agents for *in vivo* diagnosis. A single nanoparticle can carry more than 1 million contrast molecules which can be important for certain diagnostic modalities, such as X-ray CT and MRI that have decreased detection sensitivity in comparison with nuclear SPECT and PET imaging. This advantage is one of the most useful characteristics of nanoparticles especially when compared with other small molecule or protein agents which generally can only bind one or two contrast molecules per each targeting molecule before the targeting molecule loses its targeting effectiveness. The large carrying capacity of nanoparticles allows them to be excellent theranostic agents capable of carrying significant quantities of therapeutic agents as well as diagnostic agents. Nanoparticles carrying multiple diagnostic agents or a therapeutic agent in addition to a diagnostic agent have been called multifunctional or multimodality nanoparticles.[7,17–26]

2.2. *Theranostic nanoparticle imaging can provide radiation dosimetry and chemodosimetry*

The imaging of nanoparticles can be performed to determine the dose of therapy specifically delivered to different regions of a tumor. This type of individualized dosing to a specific location of the body is routinely performed in traditional external beam radiation therapy administered for the treatment of cancer. Nanoparticles can be radiolabeled with therapeutic radionuclides that emit both radioactive beta particles as well as gamma photons. The beta particles deliver damaging radiation that treats the cancer while the gamma photons emitted can be imaged with SPECT nuclear cameras. The SPECT imaging permits determination of the distribution of the nanoparticles in the region of the tumor in three dimensions (3D). This imaging permits accurate calculation of the radiation dose delivered to all regions of the tumor in radiation dosage units (Grays). Because radiation response of tumors is generally well-understood, it is possible to calculate the amount of radiation delivered to the tumor and adjust the dose of radiolabeled nanoparticles in real time in order to deliver the radiation required to kill the tumor cells.

Although these techniques are well-established for radiation therapy, they have not generally been applied to the delivery of chemotherapeutic drugs and there is very little understanding of the regional distribution of these drugs in a tumor. Imaging of the distribution of drug carrying nanoparticles *in vivo* can provide real-time determination of drug delivered to a specific organ or pathologic process. This type of theranostic monitoring has been called chemodosimetry.[27] Several studies have shown a very good correlation of the uptake of chemotherapy carrying radiolabeled nanoparticles in targeted tissues with measured tissue drug levels obtained from chemical analysis of extracted tissue samples.[27,28] In a specific example, doxorubicin carrying liposomes were radiolabeled with the common clinical imaging agent, technetium-99m (^{99m}Tc). Very good correlation was found with image-derived activity and tumor levels of doxorubicin.[27,28] This type of imaging addresses a current deficiency in the understanding of specific tissue level drug delivery, which is not currently available with traditional pharmacokinetic analysis of drug blood levels. This use of diagnostic imaging of nanoparticles can provide treatment planning information for individualized patient dosing of a drug for optimal treatment of a cancer.

2.3. *Nanoparticles can have very long intravenous circulation half-life*

Nanoparticles can have very long circulation half-lives as compared with free molecules. Extended circulation times have been shown for nanoparticles that have their surfaces covered with polyethylene glycol (PEG) as a so-called "stealth" modification which was developed in the 1990s.[29,30] Our group has shown that liposome nanoparticles with the stealth modification can have circulation half-lives of up to 90 h.[31,32] Long circulating nanoparticles that carry imaging contrast agents can be useful for diagnostic imaging of the heart and blood vessels. They can provide localization and assessment of many different pathological processes such as tumors, infection and inflammation that frequently have increased vascularity.[33]

2.4. *Tracer carrying nanoparticles can diagnose and localize focal regions of infection, inflammation and tumor by accumulation and retention in regions of increased vascular permeability*

One of the most striking advantages of nanoparticles is their ability to accumulate and be very well-retained in regions of the body that have increased permeability such as infection, inflammation and tumors. This uptake in the region of increased permeability has been termed an enhanced permeability and retention (EPR) effect.[34]

Microscopic studies have shown that the liposomes become mainly localized in macrophages and, to a lesser extent, endothelial cells in the region of the increased permeability. Although a variety of agents including small molecules will accumulate initially in a region of increased permeability in the body, only nanoparticles appear to be well-retained in these regions. An example is shown in work by the authors in which radiolabeled liposome nanoparticles were injected intravenously into rats having a focal staphylococcal thigh infection.[35] In Fig. 1, it can be seen that radiolabeled large molecule, albumin (molecular weight ~ 67,000) initially accumulated in the focal thigh infection at 4 h; however, by 24 h, the majority of the initially accumulated activity cleared from the thigh infection and little remained at 24 h. Conversely, the radiolabeled liposome nanoparticles continued to accumulate in the focal infection from 4 h to 24 h due to their superior retention at the site of the infection associated increased permeability. The mechanism for increased nanoparticle retention in tumor and inflammation

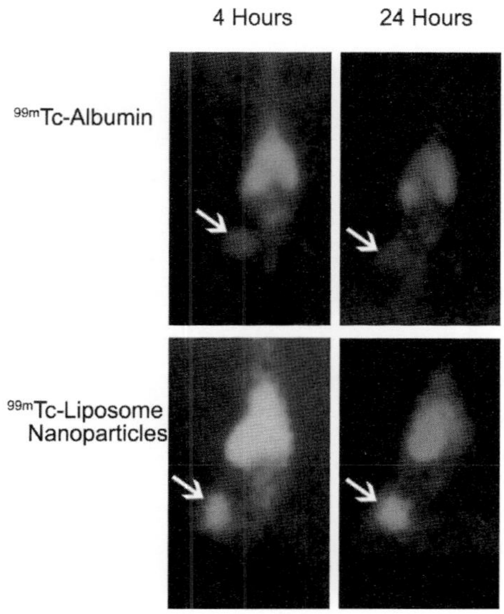

Fig. 1. Gamma camera images of a rat model with an induced thigh bacterial abscess after injection with either ^{99m}Tc-liposome nanoparticles or the large protein molecule, ^{99m}Tc-albumin. The ^{99m}Tc-liposome nanoparticles had significant uptake in the abscess at 4 h which increased by 24 h, demonstrating the excellent retention of the liposome nanoparticles at the site of their uptake. Conversely, the ^{99m}Tc-albumin had some initial uptake in the abscess, but the ^{99m}Tc-albumin did not stay retained in the region of the abscess and there was minimal retention after 24 h. These images demonstrate one of the important essential advantages of nanoparticles for imaging and therapy.

is not completely understood, however it is likely that poor lymphatic drainage of the nanoparticles and nanoparticle phagocytosis by macrophages and neutrophils are important factors in their high retention at sites of infection and inflammation. Similar prolonged retention of radiolabeled liposome nanoparticles has been shown in focal tumors.[36]

The prolonged retention of intravenously administered nanoparticles at sites of focal enhanced permeability is a specific advantage that nanoparticles have for the diagnosis and drug delivery to pathologic processes in the body associated with increased permeability such as infection, inflammation and tumors. Since the vast majority of pathologic processes in the body are associated with some degree of inflammation associated with increased permeability, the ability of nanoparticles to accumulate in these regions and be well-retained offers important advantages for diagnostic imaging as well as drug delivery to these regions.

2.5. *Nanoparticles can mimic the distribution of natural agents of similar size such as viruses*

Nanoparticles can mimic the behavior of natural agents such as viruses, most of which have a similar size to nanoparticles. Viruses have been shown to accumulate in the first draining lymph nodes encountered and this is where an immune response is generated.[37] An example of this type of targeting is the application of nanoparticles for the targeting of lymph nodes. Subcutaneously injected nanoparticles have been shown to be cleared through the lymphatic vessels.[24,38–47] This occurs because nanoparticles are too large to be directly absorbed into blood capillaries and thus are cleared through the lymphatic system. The clearance of nanoparticles through the lymphatic system is thought to mimic the clearance pattern of subcutaneously absorbed viruses. Nanoparticles labeled with contrast agents can therefore serve as agents for the diagnostic imaging of lymph nodes as well as carriers for the delivery of therapeutic agents to lymph nodes.

An important clinical use for imaging agents that trap in lymph nodes is the determination of the location of the sentinel lymph node. The sentinel lymph node is the first lymph node that drains from the lymphatics of a tumor and as such, it is the most likely lymph node to contain metastatic tumor cells that frequently metastasize from the primary tumor via the lymphatics. Since there are many possible lymph nodes in the region of the tumor, specific localization of the sentinel node becomes important in order to surgically remove this lymph node to inspect it microscopically for tumor cells. A potential use of nanoparticles that are injected adjacent to a tumor is as carriers for contrast agents to localize the sentinel node.

As discussed in Sec. 2.1., nanoparticles can carry more than one type of diagnostic agent. In our work, we have labeled liposomes with both a blue dye and radiolabeled imaging agent for localization by imaging in the Radiology Department and by a hand held detection device in the operating room.[41] The hand held imaging device allows the surgeon to precisely determine where to operate on the patient while the visually detectable dye contained within the nanoparticle can allow the surgeon to visually detect the node once the surgery has begun.

Many studies of radiolabeled liposomes have been performed on liposome nanoparticles injected subcutaneously.[48,49] These studies generally show that 30–40% of the injected dose remains in the region of the injection while the remaining portion of the subcutaneously injected liposomes move into the lymphatic vessels. After moving into the lymphatics, only 1–2% of each injected dose of liposome nanoparticles becomes trapped in each draining lymph node in a lymph node chain.[48] Specific methodologies have been investigated to increase liposome retention in the lymph nodes receiving lymphatic drainage. In one approach, it was found that liposomes containing phosphatidylserine (PS) in their liposomal bilayers have significantly higher uptake in lymph nodes (up to 3.6% of the subcutaneously injected dose).[50] Another method in which the liposomes were coated with the antibody, IgG, has been shown to increase lymph node localization of liposomes to 4.5% of the injected dose at 1 h, but this level decreased to 3% by 24 h.[51] Studies indicate that macrophages play a major role in the retention of liposomal nanoparticles within lymph nodes.

The relatively low retention of liposomes in lymph nodes led our group to search for new ways to improve liposome retention in lymph nodes. This research resulted in a new method of increasing lymph node retention of subcutaneously injected liposomes to ~10% of the injected dose.[40] This lymph node targeting method utilizes the high affinity ligands, biotin and avidin. Biotin is a naturally occurring cofactor and avidin is a protein derived from eggs. The four receptor sites on avidin permit the binding of multiple biotin molecules on the surface of the liposomes which causes their aggregation by being cross-linked with avidin. When the biotin–liposomal nanoparticles and avidin are injected subcutaneously at nearby sites, both the liposomes and the avidin drain into the same lymphatic vessels. As the liposomes and avidin encounter each other, the liposomes become aggregated and then heavily trapped in the next lymph node encountered.

Potential applications of this lymph node trapping technology include imaging for the detection of the sentinel lymph nodes. Biotin–liposomes have been manufactured which can be radiolabeled with ^{99m}Tc for imaging and blue dye for visual detection with the potential application of sentinel lymph node detection.[41] This delivery system also has potential application for the delivery of high quantities of drug to lymph nodes for the treatment of diseases such as micrometastatic cancer cells in lymph nodes as well as infections and parasitic disease of lymph nodes. This methodology may also have application for the delivery of antigens to lymph nodes as a vaccine adjuvant approach because lymph nodes are one of the main sites of immune response development to infectious disease. This type of lymph node trapping may be mimicking a natural process that occurs with viruses and multivalent antibody molecules so that when viral particles enter the lymphatic vessels in a high enough density and antibody molecules are present that react with the viral particles, they become aggregated in the lymph nodes. Although this approach has only been investigated for the targeting of liposomal nanoparticles, it should not be viewed as limited to liposome nanoparticles. It is very likely that other nanoparticles with their surfaces modified with ligands, which are co-administered with multivalent ligand binding molecules, will become trapped in lymph nodes. The nanoparticles can be labeled with a variety of contrast agents to permit non-invasive imaging. Imaging studies will be very useful for the development of lymph node targeting methodologies of this type.

2.6. *Nanoparticles can be targeted to specific tissue by placing targeting ligands on their surface to increase accumulation in specific body tissue or pathologic processes*

Nanoparticles can be targeted to specific locations in the body or to specific disease processes by placing targeting ligands on their surface. When these specifically targeted nanoparticles are tracked with an imaging modality, this imaging has the potential to provide important diagnostic information for patient management. An example of this type of targeting is the placement of an antibody on the surface of a nanoparticle for the imaging of cancer. Several studies have used imaging to show the promise of modifying the surface of liposomes with molecules specifically targeted for the treatment of cancer. [52,53]In one study, a monoclonal antibody against nucleosomes was also attached to these liposomes. Tumor-to-muscle ratio for the liposomes with the specific anti-nucleosome antibodies was 13.9 at 24 h versus lower ratios determined for the non-specifically targeted liposomes that were either modified with non-specific polyclonal antibodies (4.3) or liposomes with an unmodified surface (3.0) in rats that had Lewis lung carcinomas implanted on their thighs. This comparison with both types of control liposomes provides convincing evidence of the specificity of this antibody modified nanoparticle for targeting the lung tumor. This same antibody was also shown to have a significantly increased tumor uptake in rats that had been implanted with colon carcinoma cells demonstrating that anti-nucleosome antibodies may enhance the targeting of liposomes to a wide variety of cancer types.

Specifically targeted nanoparticles appear to have significant promise for targeting the process of angiogenesis.[19] This may be because the nanoparticles have better access to ligands appearing on the surface of vessels that are part of the vasculature. One of the most targeted markers of angiogenesis in which nanoparticles have been targeted is the $\alpha_v\beta_3$ integrin. The $\alpha_v\beta_3$ integrin is upregulated on angiogenic tumor vasculature but not on normal tissue vasculature.[19] Nanoparticles with their surfaces modified with the arginine-glycine-aspartic acid (RGD) ligand target the $\alpha_v\beta_3$ integrin. These targeted nanoparticles have shown significantly increased uptake in regions of tumor angiogenesis as compared with non-targeted controls. The $\alpha_v\beta_3$ integrin angiogenesis based tumor targeting has been shown to be successful in animal models for a wide variety of nanoparticles including quantum dots for optical and PET imaging,[54] perfluorocarbon nanoparticles radiolabeled with indium-111 for nuclear imaging,[55] single wall nanotubes imaged with PET[56], superparamagnetic iron oxide nanoparticles for MRI[57] and as shown by our group, gold nanoshells for PET imaging with copper-64 (^{64}Cu).[58] Figure 2 demonstrates PET imaging of control and $\alpha_v\beta_3$ targeted- ^{64}Cu-gold nanoshells in a rat head and neck squamous cell carcinoma xenograft model.

2.7. *Nanoparticles can target specific macrophage populations at very high percentages following intravenous injection such as bone marrow, liver and spleen*

Nanoparticles have particular advantages for the targeting of macrophage populations. This is because macrophages are natural targets for nanoparticle therapies because of their important role in the body as cells that phagocytize naturally occurring nanoparticle sized entities including cellular debris from dying and apoptotic cells, lipids from the blood following meal ingestion as well as infectious agents when the body's natural barriers have been penetrated. The largest repositories of macrophages in the body are located in the

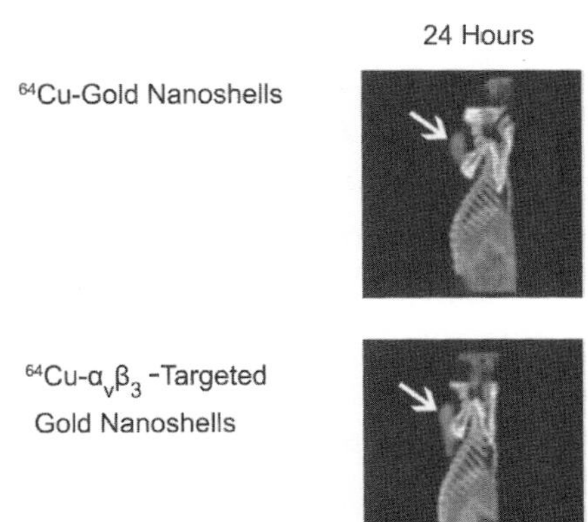

Fig. 2. These co-registered PET/CT images provide visual and quantitative demonstration of the moderately improved tumor targeting of $\alpha_v\beta_3$ targeted- ^{64}Cu-gold nanoshells compared with non-targeted control ^{64}Cu-gold nanoshells. Imaging of targeted ^{64}Cu-gold nanoshells could be used to determine the optimal time for treatment of the tumor with infrared laser therapy to induce heating of the tumor. In our studies, ^{64}Cu-gold nanoshells had the highest concentration in the tumor at 48 h.

spleen, liver, lymph nodes and in the bone marrow, and therefore these important organs are natural targets for macrophages. In addition, recent studies have indicated the important of macrophages in the growth of tumors.[59,60] These recent studies have shown that macrophages may well be an important target for tumor therapy.[61]

The ability of macrophages to deliver therapeutic agents contained in nanoparticles to macrophages in the liver and spleen has been known for a long time since the beginning of *in vivo* studies with nanoparticles. Our group has recently shown that it is also possible to specifically target macrophage populations in the bone marrow.[62] Bone marrow imaging has potential for clinical determination of the location of marrow distribution. Currently, bone marrow imaging is performed with small sulfur colloid particles radiolabeled with ^{99m}Tc, which is done to distinguish normal bone distribution of radiolabeled white blood cells from abnormal uptake of white blood cells due to bone marrow infection, i.e. osteomyelitis.

Specific surface modified nanoparticles can have surprisingly high uptake in the bone marrow. Our group has shown that liposomes with a particular surface modification have very high uptake in bone marrow. [63] The surface modification of the liposome was the addition of a negative charge by adding a non-phospholipid anionic amphiphile, L-glutamic acid, N-(3-carboxy-1-oxopropyl)-1,5-dihexadecyl ester (abbreviated SA for succinic acid) to the liposome composition. This very high uptake can be seen with whole body nuclear imaging of ^{99m}Tc-labeled liposomes in rabbit and primate models at 6 h as shown in Fig. 3. The uptake of liposome nanoparticles modified with the SA-lipid had much more uptake in rabbits as compared with control neutral liposomes. The addition of a small amount of PEG to this liposome surface also appeared to modestly enhance the already high uptake of the liposomes in the marrow which was due to mild prolongation of circulation time and the shifting of liposome uptake from the liver to the bone marrow. Imaging was used to determine the most ideal concentration of PEG to have on the surface of these SA-liposomes for a maximized bone marrow uptake. Based on imaging studies, the uptake in the marrow was fairly prompt with a high percentage of the infused dose taken up by the marrow at 6 h post intravenous (*i.v.*) administration in rabbits and rhesus primates.[63] Interestingly, uptake of bone marrow targeted liposomes in rats and mice was greatly reduced, although they remained greater than control liposome formulations showing important species differences.[62] Although the precise mechanism by which these liposomes accumulate in marrow is unknown, electron microscopy studies as well as fluorescently labeled liposome confocal microscopy studies have

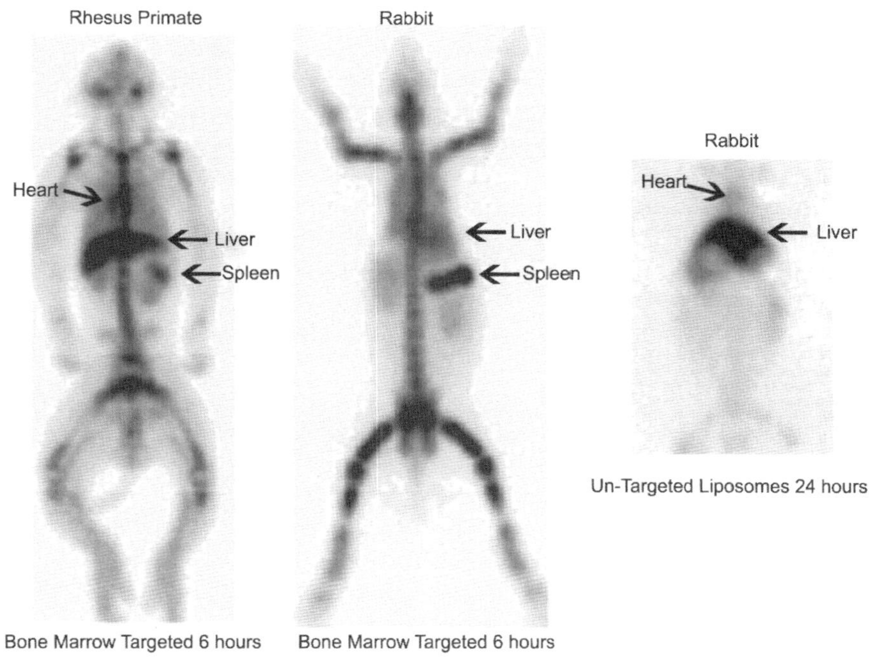

Fig. 3. Gamma camera images depicting high uptake of bone marrow targeted liposomal nanoparticles in the bone marrow of rabbits and primates at 6 h. Note the differences in organ distribution in the rabbit model between bone marrow liposome nanoparticles at 6 h as compared with control neutral liposomal nanoparticles at 24 h.

demonstrated that the liposomes taken up in the marrow are located in bone marrow macrophages. It is believed that the special surface modification of the liposomes results in specific targeting of the scavenger receptors on the surface of the bone marrow macrophages. In addition to their diagnostic imaging uses, the specific targeting of therapeutic agents to bone marrow using these SA-liposomes may be a promising approach for the delivery of drugs to the bone marrow.

3. Conclusion and Future Directions

In the above review, we have covered many of the most promising uses of nanoparticles for diagnostic and theranostic imaging. Undoubtedly, many new imaging applications for nanoparticles will be developed as more nanotechnological innovations evolve. Although nanoparticles are not applicable in all imaging situations, their nano advantages are becoming ever more apparent for many important diagnostic and theranostic imaging uses.

References

1. B. H. Hasegawa, K. H. Wong, K. Iwata, W. C. Barber, A. B. Hwang, A. E. Sakdinawat, M. Ramaswamy, D. C. Price and R. A. Hawkins, *Technol. Cancer Res. Treat.* **449**, (2002).
2. E. Karathanasis, S. Suryanarayanan, S. R. Balusu, K. McNeeley, I. Sechopoulos, A. Karellas, A. V. Annapragada and R. V. Bellamkonda, *Radiology* **398**, (2009).
3. O. Rabin, J. Manuel Perez, J. Grimm, G. Wojtkiewicz and R. Weissleder, *Nat. Mater.* **118**, (2006).
4. J. Zheng, D. Jaffray and C. Allen, *Mol. Pharm.* **571**, (2009).
5. L. C. Chen, C. H. Chang, C. Y. Yu, Y. J. Chang, W. C. Hsu, C. L. Ho, C. H. Yeh, T. Y. Luo, T. W. Lee and G. Ting, *Nucl. Med. Biol.* **415**, (2007).
6. L. Lacerda, *Adv. Mater.* **225,** (2008).
7. H. Lee, B. Hoang, H. Fonge, R. M. Reilly and C. Allen, *Pharm. Res.* 2343 (2010).
8. W. T. Phillips, B. A. Goins and A. Bao, *Wiley Interdiscip. Rev. Nanomed. Nanobiotech.* **69**, (2009).
9. M. Silindir, A. Y. Ozer and S. Erdogan, *Drug Deliv.* **68**, (2012).
10. J. Marik, M. S. Tartis, H. Zhang, J. Y. Fung, A. Kheirolomoom, J. L. Sutcliffe and K. W. Ferrara, *Nucl. Med. Biol.* **165**, (2007).
11. A. L. Peterson, T. Binderup, P. Rasmussen, J. R. Henriksen, D. R. Elema, A. Kjaer and T. L. Andresen, *Biomaterials*, in press (2011).
12. J. W. Seo, H. Zhang, D. L. Kukis, C. F. Meares and K. W. Ferrara, *Bioconjug. Chem.* **2577**, (2008).
13. S. Santra, *Methods Mol. Biol.* **151**, (2010).
14. S. Srinivasan, R. Manchanda, A. Fernandez-Fernandez, T. Lei and A. J. McGoron, *J. Photochem. Photobiol. B, Biol.* **52**, (2013).
15. W. J. Mulder, G. J. Strijkers, G. A. van Tilborg, A. W. Griffioen and K. Nicolay, *NMR Biomed.* **142**, (2006).
16. M. R. Bohmer, A. L. Klibanov, K. Tiemann, C. S. Hall, H. Gruell and O. C. Steinbach, *Eur. J. Radiol.* **242**, (2009).
17. T. Y. Liu, M. Y. Wu, M. H. Lin and F. Y. Yang, *Acta Biomater.* **5453**, (2013).
18. E. Kluza, G. J. Strijkers and K. Nicolay, *Recent Results in Cancer Res. Fortschritte der Krebsforschung. Progres dans les recherches sur le cancer*, **151**, (2013).
19. W. Cai and X. Chen, *J. Nucl. Med.* **113S**, (2008).
20. Z. Chen, M. F. Penet, S. Nimmagadda, C. Li, S. R. Banerjee, P. T. Winnard, Jr., D. Artemov, K. Glunde, M. G. Pomper and Z. M. Bhujwalla, *ACS Nano* **7752**, (2012).
21. J. Y. Hwang, J. Park, B. J. Kang, D. J. Lubow, D. Chu, D. L. Farkas, K. K. Shung and L. K. Medina-Kauwe, *PloS one* **e34463**, (2012).
22. Y. K. Lee, J. M. Jeong, L. Hoigebazar, B. Y. Yang, Y. S. Lee, B. C. Lee, H. Youn, D. S. Lee, J. K. Chung and M. C. Lee, *J. Nucl. Med.* **1462**, (2012).
23. S. Li, B. Goins, L. Zhang and A. Bao, *Bioconjug. Chem.* (2012).
24. R. Madru, P. Kjellman, F. Olsson, K. Wingardh, C. Ingvar, F. Stahlberg, J. Olsrud, J. Latt, S. Fredriksson, L. Knutsson and S. E. Strand, *J. Nucl. Med.* **459**, (2012).
25. N. Mitchell, T. L. Kalber, M. S. Cooper, K. Sunassee, S. L. Chalker, K. P. Shaw, K. L. Ordidge, A. Badar, S. M. Janes, P. J. Blower, M. F. Lythgoe, H. C. Hailes and A. B. Tabor, *Biomaterials* **1179**, (2013).

26. M. Yang, K. Cheng, S. Qi, H. Liu, Y. Jiang, H. Jiang, J. Li, K. Chen, H. Zhang and Z. Cheng, *Biomaterials* **2796**, (2013).
27. M. M. Kleiter, D. Yu, L. A. Mohammadian, N. Niehaus, I. Spasojevic, L. Sanders, B. L. Viglianti, P. S. Yarmolenko, M. Hauck, N. A. Petry, T. Z. Wong, M. W. Dewhirst and D. E. Thrall, *Clin. Cancer Res.* **6800**, (2006).
28. H. W. Head, G. D. Dodd, 3rd, A. Bao, A. Soundararajan, X. Garcia-Rojas, T. J. Prihoda, L. M. McManus, B. A. Goins, C. A. Santoyo and W. T. Phillips, *Radiology* **405**, (2010).
29. T. M. Allen, C. Hansen, F. Martin, C. Redemann and A. Yau-Young, *Biochim. Biophys. Acta.* **29**, (1991).
30. V. P. Torchilin and M. I. Papisov, *J. Liposome Res.* **725**, (1994).
31. W. T. Phillips, R. W. Klipper, V. D. Awasthi, A. S. Rudolph, R. Cliff, V. Kwasiborski and B. A. Goins, *J. Pharmacol. Exp. Ther.* **665**, (1999).
32. V. D. Awasthi, D. Garcia, B. A. Goins and W. T. Phillips, *Int. J. Pharm.* **121**, (2003).
33. O. Boerman, W. Oyen, L. vanBloois, *J Nucl Med*, **489**, (1997).
34. H. Maeda, G. Y. Bharate, J. Daruwalla, *European journal of pharmaceutics and biopharmaceutics : official journal of Arbeitsgemeinschaft fur Pharmazeutische Verfahrenstechnik e.V*, **409**, (2009).
35. B. Goins, R. Klipper, R. Cliff, R. Blumhardt and W. Phillips, *J. Nucl. Med.* **2160**, (1993).
36. B. Goins, R. Klipper, A. S. Rudolph and W. T. Phillips, *J. Nucl. Med.* **1491**, (1994).
37. M. R. Olson, D. S. McDermott and S. M. Varga, *PLoS pathogens*, **e1003054**, (2012).
38. A. Hagiwara, T. Takahashi, K. Sawai, A. Iwamoto, M. Shimotsuma, C. Yoneyama, K. Seiki, M. Itoh, T. Sasabe and M. Lee, *Lymphology* **84**, (1992).
39. W. T. Phillips and B. Goins, *J. Liposome Res.* **71**, (2002).
40. W. T. Phillips, R. Klipper and B. Goins, *J. Pharmacol. Exp. Ther.* **309**, (2000).
41. W. T. Phillips, R. Klipper and B. Goins, *J. Nucl. Med.* **446**, (2001).
42. V. S. Trubetskoy, K. R. Whiteman, V. P. Torchilin and G. L. Wolf, *J. Control Release* **13**, (1998).
43. K. Hiraiwa, M. Ueda, H. Takeuchi, T. Oyama, T. Irino, T. Yoshikawa, A. Kondo and Y. Kitagawa, *J. Surg. Res.* **48**, (2012).
44. L. M. Kaminskas and C. J. Porter, *Adv. Drug Deliv. Rev.* **890**, (2011).
45. N. Kosaka, M. Mitsunaga, P. L. Choyke and H. Kobayashi, *Contrast Med. Mol. Imag.* **96**, (2013).
46. B. E. Ocampo-Garcia, M. Ramirez Fde, G. Ferro-Flores, L. M. De Leon-Rodriguez, C. L. Santos-Cuevas, E. Morales-Avila, C. A. de Murphy, M. Pedraza-Lopez, L. A. Medina and M. A. Camacho-Lopez, *Nucl. Med. Biol.* **1**, (2011).
47. C. Oussoren and G. Storm, *Adv. Drug Deliv. Rev.* **143**, (2001).
48. W. T. Phillips, T. Andrews, H. Liu, R. Klipper, A. J. Landry, R. Blumhardt and B. Goins, *Nucl. Med. Biol.* **435**, (2001).
49. C. Oussoren, M. Velinova, G. Scherphof, J. J. van der Want, N. van Rooijen and G. Storm, *Biochim. Biophys. Acta.* **259**, (1998).
50. C. Oussoren, J. Zuidema, D. J. Crommelin and G. Storm, *Biochim. Biophys. Acta.* **261**, (1997).
51. S. Mangat and H. M. Patel, *Life Sci.* **1917**, (1985).
52. T. A. Elbayoumi and V. P. Torchilin, *Eur. J. Nucl. Med. Mol. Imag.* **1196**, (2006).
53. S. Erdogan, A. Roby and V. P. Torchilin, *Mol. Pharm.* **525**, (2006).

54. W. Cai, K. Chen, Z. B. Li, S. S. Gambhir and X. Chen, *J. Nucl. Med.* **1862**, (2007).
55. G. Hu, M. Lijowski, H. Zhang, K. C. Partlow, S. D. Caruthers, G. Kiefer, G. Gulyas, P. Athey, M. J. Scott, S. A. Wickline and G. M. Lanza, *Int. J. Cancer*, In *Journal international du cancer*, **1951**, (2007).
56. Z. Liu, W. Cai, L. He, N. Nakayama, K. Chen, X. Sun, X. Chen and H. Dai, *Nat. Nanotechnol.* **47**, (2007).
57. C. Zhang, M. Jugold, E. C. Woenne, T. Lammers, B. Morgenstern, M. M. Mueller, H. Zentgraf, M. Bock, M. Eisenhut, W. Semmler and F. Kiessling, *Cancer Res.* **1555**, (2007).
58. H. Xie, P. Diagaradjane, A. A. Deorukhkar, B. Goins, A. Bao, W. T. Phillips, Z. Wang, J. Schwartz and S. Krishnan, *Int. J. Nanomed.* **259**, (2011).
59. M. de Vries, I. Briaire-de Bruijn, M. J. Malessy, S. F. de Bruine, A. G. van der Mey and P. C. Hogendoorn, *Otology & neurotology : official publication of the American Otological Society, Amer. Neurotology Society [and] European Academy of Otology and Neurotology*, **347**, (2013).
60. K. L. Tan, D. W. Scott, F. Hong, B. S. Kahl, R. I. Fisher, N. L. Bartlett, R. H. Advani, R. Buckstein, L. M. Rimsza, J. M. Connors, C. Steidl, L. I. Gordon, S. J. Horning and R. D. Gascoyne, *Blood*, **3280**, (2012).
61. L. A. Lievense, K. Bezemer, J. G. Aerts and J. P. Hegmans, *Lung Cancer* (*Amsterdam, Netherlands*), (2013).
62. K. Sou, B. Goins, B. O. Oyajobi, B. L. Travi and W. T. Phillips, *Exp. Opinion on Drug Deliv.* **317**, (2011).
63. K. Sou, B. Goins, S. Takeoka, E. Tsuchida and W. T. Phillips, *Biomaterials* **2655**, (2007).

Chapter 2

Superparamagnetic Iron Oxides as Imaging Probes of Metastases and Vulnerable Atherosclerotic Plaques

*Mary L. Mazzanti and Alexei A. Bogdanov, Jr.**

The Laboratory of Molecular Imaging Probes, Department of Radiology, University of Massachusetts Medical School, Worcester, MA 01655, USA
**Alexei.Bogdanov@umassmed.edu*

Abstract

Stabilized superparamagnetic iron oxide (SPIO) particles were used in the past as contrast agents (CAs) for magnetic resonance imaging (MRI) in a variety of applications ranging for gastrointestinal imaging to labeling of live cells. SPIOs are unique in that they represent the only example of true nano-sized (between 20 nm and 300 nm) diagnostic agents with a history of extensive clinical trials and clinical use. SPIOs have high magnetic susceptibility and influence predominantly spin–spin or transverse (T2/T2*) relaxation of water protons by dephasing their coordinated precession thereby dramatically shortening proton relaxation times at very low concentrations of iron. This shortening results in an attenuation of the magnetic resonance (MR) signal in the microenvironment surrounding the particles and appears as a darkening of the anatomy in the resultant MR image. Much of the diagnostic use of SPIOs has been based on their selective uptake in normal versus diseased tissue, which in many cases is based on the selective accumulation of SPIOs in monocytes/macrophages and other hematopoietic cells. The development of ultrasmall SPIOs (SPIOs with diameters of 30±10 nm) has widened the diagnostic potential of these agents by increasing their volume of distribution throughout the body, however concerns about their safety have thus far precluded their routine use in clinical settings. Here we focus mainly on the capability of ultrasmall SPIOs to detect disease in two highly relevant areas of medicine — the imaging of lymph nodes for the staging of cancer and the imaging of blood vessels for assessment of vascular inflammation and plaque formation/vulnerability in cardiovascular disease.

1. Overview of SPIOs as MR CAs

The rationale for using various formulations of iron as CAs is clearly evident given that even the small amounts of iron inherent in the human body (mainly in the form of haeme) are sufficient to influence the signal intensity and thus contrast in MR images. This effect can be exploited to produce diagnostic images as in the case of BOLD MRI where paramagnetic deoxygenated hemoglobin (dHb) produces less MR signal in areas with lower brain activity (thus oxygen usage) as compared to oxyhemoglobin in brain areas with higher activity. The core of most iron based MR CAs consists of iron oxides (IOs), which exist in multiple forms (i.e. magnetite, Fe_3O_4 or maghemite, γFe_2O_3). When suspended in solution, nano-sized IOs can exert either ferromagnetic, superparamagnetic or paramagnetic properties depending on their exact chemical structure, particle size, and other factors. Ferromagnetic substances which are reduced to a very small particle size lose their permanent magnetism and become superparamagnetic. Because IOs with a smaller core diameter (below 100 nm) are more biocompatible than larger particles, IOs used as MR CAs are generally superparamagnetic. Superparamagnetic agents have zero magnetization outside external magnetic fields (i.e. the unpaired dipoles of iron atoms flip randomly and do not align in the absence of external magnetic field), but produce a large net magnetization when inside the magnet of an MRI system, due to the alignment of the electron spins of the iron cations of the core. This magnetization gives rise to local field gradients that accelerate the loss of phase coherence of the spins contributing to the MR signal (i.e. resulting in increased spin–spin relaxation). MR pulse sequences (namely gradient recalled echo (GRE)) can detect these local magnetic differences and the resulting image intensity is attenuated in the location of the IO particles. The magnetization produced by SPIOs is larger than that produced by an equal molar amount of paramagnetic gadolinium cations (Gd^{3+}), making SPIOs very effective CAs.

SPIOs have been the focus of extensive pre-clinical and clinical research over the last two decades and a central theme to emerge from these studies is that SPIOs can have substantially different pharmacokinetic and magnetic properties *in vivo* based on their core size, coatings, and overall electrochemical properties.[1–5] The ability to modify these characteristics lends SPIOs to being somewhat customizable and in turn useful for a number of different MR applications. For example, although SPIOs generally are used to shorten T2/T2* relaxation times and produce negative contrast using T2 weighted MR pulse sequences, smaller diameter particles can be used to shorten T1 relaxation time and produce positive contrast on T1-weighted MR images, and as such have been used for vascular imaging. A recent development that should further widen the applications of SPIOs is the use of a newly developed MR pulse sequence for generating higher spatial sensitivity positive contrast with these agents[6] (for detailed review see Ref. 7). The principle of this method is to use selective RF pulses and gradient compensation to excite and refocus the off-resonance water surrounding cells labeled with IOs so that only the fluid and tissue immediately adjacent to the labeled cells are visible in the image. This technique is similar to the way that typical slice encoding is done in that a magnetic gradient (in this case the one produced by the *S*) produces a range of proton spin frequencies, and a specific bandwidth RF pulse is used to excite only spins within a narrow band of that range. The benefits of this strategy are two-fold: (1) because the signal produced by the SPIO is positive (brighter signal), the problem of distinguishing a SPIO-mediated signal void from a normally dark tissue background is avoided, and; (2) because a narrower range of SPIO-produced magnetic susceptibility is sampled (only the disturbed frequencies very near the particles), the spatial localization of the signal is improved. Cunningham and colleagues[6] labeled cultured cells with Feridex and then injected these cells into gel phantoms, excised porcine myocardium, or the hindlimb of a live mouse. Using the newly developed pulse sequence, they were easily able to detect injected cells ($\sim 5 \times 10^5$), (which suggest that the limit of the method is lower), and a significant linear correlation ($r = 0.87$, $P < 0.005$) between the estimated number of cells and the signal was observed. The authors conclude that these results provide hope that this positive contrast method will allow more straight forward quantification of an SPIO-labeled cell volume than has been possible with conventional negative contrast methods.

The small size and flexibility of coatings also makes SPIOs ideal for attachment of ligands used to target these CAs to specific extracellular receptors, thereby affording true "molecular imaging" applications. They may potentially serve as an alternative in patients with altered kidney function who are not candidates for Gd^{3+}-based CA scans.

The diagnostic utility of most SPIOs is based on their selective uptake in monocytes and macrophages of the reticuloendothelial system. One notable exception is the use of Ferumoxsil for gastrointestinal imaging. Ferumoxsil (based on experimental compound AMI-121 and OMP and sold under the brand names GastroMARK™ (AMAG Pharmaceuticals) and Abdoscan™ (Granulat)) is formulated using large IO particles (with hydrodynamic diameters in the range of 0.3–3 μm) that are coated with a non-biodegradable matrix composed of siloxane or polystyrene. It is administered as an oral suspension and used to produce negative contrast of the GI lumen. It is

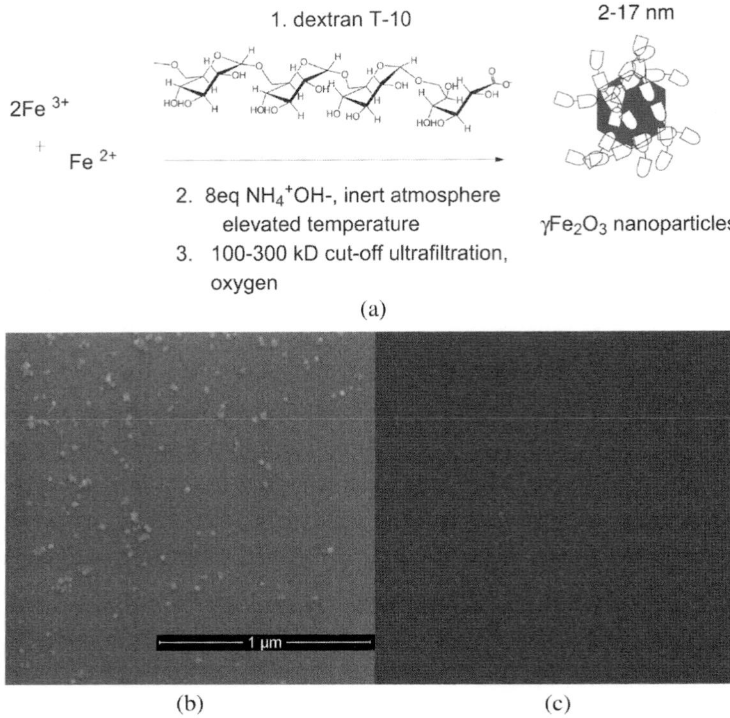

Fig. 1. A schematic illustration of the dextran (T10)-coated ultrasmall SPIO synthesis (a) and scanning electron micrographs depicting the final nanoparticles (b,c). A — the synthesis includes treating the mixture of Fe(II) and Fe(III) salts with a strong base (ammonium hydroxide) in the presence of an excess of a polysaccharide (Dextran T-10) in the absence of oxygen and under heating. During the purification the magnetite, Fe_3O_4 nanoparticle cores oxidize to maghemite, γFe_2O_3.

The resultant dextran-stabilized nanoparticles are purified using 100–300 kD cut-off hollow-fiber ultrafiltration and appear as individual, non-aggregating nanoparticles (b). The nanoparticles have non-metallic (IO) cores resulting in a low level of X-ray backscattering (c).

routinely given before MR cholangiopancreatography (MRCP) so that the signal from fluid in the stomach and duodenum is nulled, thereby allowing an unobstructed view of the pancreaticobiliary tree.

SPIOs which are taken up by cells of the RES are generally divided into two classes based on size; standard SPIOs have IO core diameters of approximately 5 nm and hydrated diameters of 60–150 nm,[1] whereas ultrasmall SPIOs have core diameters of 1–3 nm and hydrated diameters of 20–40 nm (Fig. 1). This classification is not arbitrary since the hydrated particle size, in combination with other characteristics of the IO particle, such as the surface coating, largely determines the volume of distribution in the body.

Ferumoxide AMI-25 (Endorem®, Guerbet and Feridex IV® Berlex Laboratories developed by AMAG Pharmaceuticals) was the first SPIO marketed in the US in the mid 1990s for liver imaging, namely for detection of small focal lesions. This was followed by the introduction of ferucarbotran (SHU555A) (Resovist®, Bayer Healthcare; Schering AG) in Europe in 2001. These SPIO based CAs are encapsulated in a dextran or carboxydextran coating which confers biocompatibility and stability to the particle, making it suitable for intravenous (*i.v.*) administration. Following either bolus (Resovist) or slow (Feridex *i.v.*) *i.v.* injection (0.56 mg iron/0.05ml per kg body weight), these SPIO formulations are carried in the blood stream where they interact with opsonizing proteins which bind to the outer coating and in this way tag the SPIO particles for removal from the body through the reticuloendothelial system (RES) of the liver, spleen, bone marrow, and lymph nodes. Uptake into monocytes and phagocytes in liver and spleen is facilitated by the presence of a discontinuous endothelium lining and fenestrated basal membrane in these organs. Approximately 10 min following injection, roughly 80% of SPIOs are taken up via phagocytosis by liver macrophages (Kupffer cells) in the RES, where they reside for up to 6 h. In this time period, the dextran coating erodes in the cells' lysosomes together with the superficial IO layers due to the effect of low local pH. Eventually dextran is eliminated (primarily in the urine) while the iron is added to the body's normal iron store. The amount of iron in a single dose of SPIO contrast is small (50–200 mg) compared to the total amount of iron in the body (~3500 mg), and posses no safety risk. Ferumoxide is cleared from the body within 24 h of administration. In normal liver and spleen, standard SPIOs will produce a large net magnetic moment and local magnetic susceptibility in the presence of a static magnetic field. This net magnetization accelerates spin–spin relaxation following an MR excitation pulse, and this reduces the MR signal generated by T2*/T2 weighted pulse sequences and renders the liver and spleen dark in the MR image.

In the case of liver pathology (e.g. metastases, primary liver cancer, cysts and various benign tumors, adenomas, and hyperplasia) macrophages are absent from the diseased tissue hence there is lower SPIO uptake in these tissues and they appear brighter on the resulting T2/T2* weighted MR image. SPIO-contrast enhanced MRI has been used successfully to detect liver malignancies and other pathologies with high sensitivity and specificity and with a good safety profile.[8] The sensitivity and specificity of Feridex for liver lesion detection compares favorably with other imaging methods including spiral and multidetector CT,[9–11] and Gd^{3+} enhanced MRI.[12] In one early multicenter trial, Feridex-enhanced T2-weighted images revealed additional lesions not seen on unenhanced images in 27% of cases and additional lesions not seen by conventional (non-spiral) computed tomography (CT) scans in 40% of cases; the additional information would have changed therapy in 59% of cases.[13] Unlike Feridex which must be administered as a slow infusion, Resovist can be bolus injected and can also generate some T1-weighted contrast (positive enhancement) which has been tested as a means to perform vascular perfusion imaging in addition to lesion detection.[14–16]

Ultrasmall SPIOs (AMI-277, ferumoxtran-10; Sinerem®, Combidex®) are chemically similar to larger SPIOs but their small size (20–40 nm) and hydrophilic coating allows them to escape immediate uptake by the liver and spleen and endows them with a longer blood half-life (on the order of 24 h to 36 h). This long blood half-life makes ultrasmall SPIOs available for phagocytic uptake into monocytes, macrophages and olligodendroglial of more distant organs, allowing their use for detection of a range of diseases having an inflammatory component. It is important to note that in applications apart from liver and lymph node imaging, the ultrasmall SPIO-generated attenuation of MR signal (negative contrast) indicates the *presence* of immune cells and thus the occurrence of pathology, whereas in detecting liver lesions and malignant lymph nodes it is the *absence* of the normal RES which causes the lesions and nodes to appear bright in contrast to normal tissue which has accumulated SPIOs and appear dark (Fig. 2). The smallest ultrasmall SPIOs developed thus far are monocrystalline IO nanoparticles (MION-47, and Fergulose (NC100150), Clariscan™ Nycomed Imaging/GE healthcare) which are composed of singular IO crystals coated with dextran, carboxylated dextran, or starch and have a core size of only 5 nm and a diameter in blood of 20 nm.[17]

Ultrasmall SPIOs have been used most widely for imaging lymph nodes in the detection of nodal metastatic spread in cancer patients (discussed further in Sec. 3). Other uses include imaging atherosclerosis[18–20] (Sec. 4), stroke,[21] multiple sclerosis[22] and cancer including head and neck cancers.[23,24] Studies comparing ultrasmall SPIOs and conventional MR CAs have shown that the two CAs can detect different lesions in the case of a single disease within the same individual suggesting the potential utility of ultrasmall SPIOs as an adjunctive CA for some Gd^{3+} enhanced MRI exams. Because they circulate in the vascular space for extended time periods, ultrasmall SPIOs have also been tested as CAs for vascular imaging and MR angiography,[25–27] especially for use in detecting slow moving blood and small vessels not easily imaged with non-contrast methods. For example, ultrasmall SPIOs increased the specificity and sensitivity of detection of small diameter vessel occlusion. The comparison of the data obtained using high-relaxivity SPIO (Feridex,

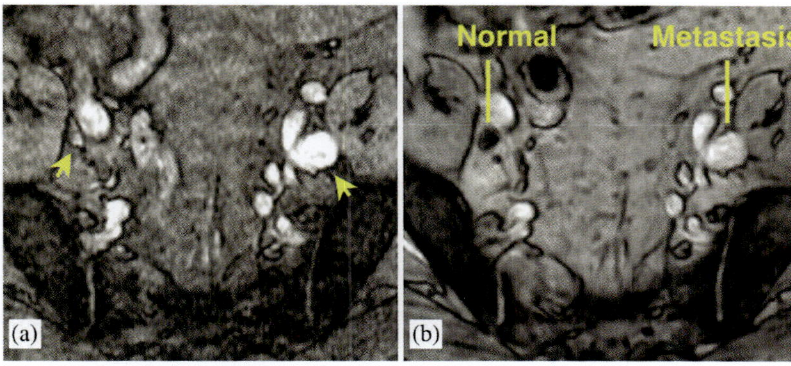

Fig. 2. Lymphadenopathy in the pelvis in prostate cancer patient. The right pelvic lymph nodes are normal and their MR signal intensity decrease 24 h after the injection of Combidex (2.6 mg total iron/kg). The left lymph node is enlarged and remain bright after the injection of IO suggesting that the normal tissue has been replaced by metastatic cancer cells.

AMI-25) particles[28] and ultrasmall SPIOs[29] showed obvious advantages of both types of nanoparticles in terms of small vessel delineation.

Although a variety of ultrasmall SPIO formulations have been tested at length in animals and humans and have demonstrated advantages for contrast-enhanced MRI applications, adverse side effects have also been documented. Patients in several clinical studies reported lower back pain and other symptoms after being administered *i.v.* ultrasmall SPIOs.[30] One death was reported after bolus administration of an undiluted ultrasmall SPIOs and subsequently this method of administration is no longer used. These safety issues coupled with the incidence of a high false positive rate for metastatic lymph node detection[31] raised concerns about the safety and effectiveness of the leading ultrasmall SPIO candidate Combidex®, and resulted in the termination of this compound for approval by the FDA for use in the US. Despite these problems, the utility of ultrasmall SPIO-enhanced MRI for diagnosing disease in humans is still being investigated.[3,32] Modification of these agents, especially with the addition of more biocompatible carriers and coatings, will be needed in order to lower side effects, increase diagnostic reliability, and bring them to the clinical setting. The remainder of this chapter will discuss in greater detail the synthesis of these agents and their potential uses in contrast-enhanced MRI.

2. SPIO Chemical Characteristics and *In Vivo* Kinetics

Because of the obvious relationship between disease progress and immune cell involvement, it was recognized early on that SPIOs which could target immune cells outside of the liver and spleen would be potentially useful for the detection of a wider range of diseases. The small size and chemical makeup (particularly surface characteristics) of ultrasmall SPIOs are responsible for their capability to escape sequestration by the liver and distribute to other areas of the body, making them useful for applications beyond liver imaging.[17,33] Nevertheless, the relationship between ultrasmall SPIO form and function (final distribution and magnetization in tissue, including elimination and overall safety) is complex and it is still not possible to predict with certainty the behavior of newly developed ultrasmall SPIOs based solely on their chemical makeup.[3]

Ultrasmall SPIOs are generally obtained from mixtures of larger IO particles which are synthesized by alkaline co-precipitation of ferrous (Fe^{2+}) and ferric (Fe^{3+}) iron ions in aqueous solution in the presence of a polymer that acts as a capping agent and a stabilizer.[17,33–35] Physical separation allows sizing of the particles and is accomplished by filtering the particles through a series of membranes. This removes impurities (the excess of the capping polymer) and highly aggregated particles and leaves IOs that form multicrystal cores several nm in diameter. Electron microscopy reveals these cores to be hexagonal in shape with a diameter of between 2 nm and 8 nm (mean size is 4.6 nm). Subsequently these ultrasmall cores are coated with a surface layer which acts to stabilize the molecule and also allows for attachment (covalent or non-covalent linking) of cell targeting moieties to the cores. The final hydrodynamic size of the ultrasmall SPIO *in vivo* is less than 50 nm. In general, altering the size and composition of the core determines the overall magnetic properties of the SPIO whereas altering the surface coating will change the distribution and elimination of the particle, especially as it relates to hepatic uptake. Lager ultrasmall SPIOs (Feridex) have relaxivites of $r^2 = 98.3$ and $r^1 = 23.9\ \text{mM}^{-1}\text{s}^{-1}$, respectively, whereas smaller IOs (Combidex/ Sinerem) have relaxivities of $r^2 = 60$ and $r^1 = 10\ \text{mM}^{-1}\text{s}^{-1}$. Because the r^2/r^1 ratio decreases with decreasing particle size, smaller ultrasmall SPIO agents are more advantageous for use as blood pool imaging and for magnetic resonance angiography, where the blood signal appears bright on short TE/TR gradient echo imaging sequences.

The addition of a surface coating is critical for clinical applications because it allows the particles to be suspended in solution for *i.v.* administration, and more importantly, it reduces the surface charge and hydrophobicity of the particles which would otherwise cause them to aggregate. In addition, the surface coating can prevent attachment of proteins (opsonisation) which mark the particles for removal of the ultrasmall SPIOs from the circulation in the liver and spleen. The surface coating is generally made up of biodegradable polymers (dextran, carboxymethylated dextran, carboxydextran or polyethylene glycol, and silica[36,37]) or small organic molecules used to target ultrasmall SPIOs to specific extracellular receptors.[17] Polysaccharide-coated IOs are amenable to chemical stabilization using cross-linked agents, e.g. epichlorohydrin. The epichlorohydrin treated IO-associated dextran carry multiple cross-links which results in a more stable dextran coating. In addition, epoxides can be subjected to ammonolysis with the resultant formation of primary amines. These functionalized particles (amino-CLIO, i.e. cross-linked IOs) found multiple uses for immobilizing molecules required for cellular targeted imaging applications (see below).[38] It should be noted that many other polysaccharides, including carboxylated dextran,[39] oxidized starch[40] and co-polymers of chitosan[41] can be used for stabilizing nanoparticles during their synthesis. Apart from the "single-pot" IO nanoparticle synthesis strategy described above, a separate group of methods include pre-formed particle coatings made by using water soluble dimercaptosuccinic acid[42] or amphiphilic copolymers that enable stabilization and formation of colloidal solutions of IO nanoparticles.[43,44]

Ultrasmall SPIOs of different diameters and with a variety of coatings and surface characteristics have been tested in animals, and some have been tested in humans (e.g. ferumoxtran-10 (dextran)[45,46]; VSOP (citrate)[47]; Feruglose (pegylated starch)[48]; and SHU555C (carboxydextran).[5,49] Ferumoxtran (AMI-227 –diameter 30 nm) (Combidex in the USA, Sinerem in Europe) has an IO core that is surrounded by an 8 mm to 12 mm layer of T-10 dextran. Although dextran coatings are routinely used in pharmaceuticals, the interactions between small IO cores and dextran type coatings are complex and relatively few studies have investigated this interaction at the molecular level.[3] It has been speculated that some of the adverse effects in patients given Combidex may have been related to the surface coating. Immune responses to dextran in other pharmaceuticals has been reported[50,51] and low back pain (one of the more common side effects of Combidex) is a clinical symptom of complement activation. Efforts to develop Ultrasmall SPIO formulations devoid of side effects and with increased solubility, safety, and stability are ongoing.[37,52–57]

Following *i.v.* administration, ultrasmall SPIOs are not immediately taken up by the liver and spleen, leaving them free to circulate and reach deeper tissues. Several hours after administration ultrasmall SPIOs can be found in lymph nodes, kidney, brain, bone marrow, osteoarticular tissue, and vascular plaques,[21–23,58–60,61–63] where they accumulate primarily in resident or infiltrating macrophages.[64–67] Although the exact means of cell uptake has not been established in all of these tissues, it is likely that ultrasmall SPIOs are taken through the capillary wall by vesicular transport/transcytosis through inter-endothelial junctions and, once inside the tissue, are taken up by macrophages via fluid-phase endocytosis,[65] or in some cases possibly by receptor mediated uptake.[68] Alternatively, ultrasmall SPIOs may be taken up by activated blood monocytes while in the bloodstream and delivered to the tissues along with immune cell infiltration.[4,66,68–71] Many lesions, especially tumors, are characterized by leaky vasculature which allow passive passage of particles through the vessel wall, and this probably explains the build-up of ultrasmall SPIO in tumors, especially those in the brain. This pattern of distribution can be used to image tumor-associated microvessels during the early time points after ultrasmall SPIO administration, and macrophage accumulation in tumors at later time points.[14,72,73] However, longer circulation times are not necessarily the only factor determining the overall uptake of particles as experiments in culture have shown that larger ultrasmall SPIO particles as opposed to smaller ultrasmall SPIOs are preferentially taken up by human monocytes via phagocytosis.[74]

The accumulation of ultrasmall SPIOs at sites of inflammation makes possible MRI detection of diseases that are characterized by inflammatory responses including atherosclerosis,[18–20] stroke,[21] multiple sclerosis[22] and cancer.[23,24] Applications for ultrasmall SPIO detection of other pathologies have been tested in animal models, e.g. spinal nerve injury,[75] degenerative disease,[76] kidney disease,[77] transplant rejection,[78] atherosclerosis[18–20] and cancer.[64,79] The prolonged blood half-life of ultrasmall SPIOs also make them useful for perfusion imaging and MR angiography, in which case T1-weighted MR pulse sequences are used and the effect of ultrasmall SPIOs is to decrease the T1 relaxation time and produce positive contrast enhancement.[25–27] When used to detect tumors, ultrasmall SPIO perfusion imaging results in a distinguishable ring-like enhancement around the tumor border of malignant lesions — a sign that is absent from benign lesions.[80]

Monocrystalline IO nanoparticles (MION-47) are one of the smallest ultrasmall SPIO developed to date.[81] They are composed of singular IO crystals coated with dextran and have a core size of only 5 nm and a diameter in blood of 20 nm.[17] We previously performed experiments with these compounds (also known as ultrasmall SPIO MION-46 and MION-47) which were synthesized in the presence of high concentrations of dextran T10, enabling trapping small IO crystals of several nm in diameter in a dense carbohydrate shell.[81] The latter nanoparticles never aggregated in the presence of fresh blood plasma, which is consistent with the extended circulation in the bloodstream. However, gradient density separation showed that after incubations at 37°C, MION-47 contained adhered plasma proteins. Size-exclusion

HPLC of these opsonized as well as non-opsonized preparations of MION-47 showed that as a result of the binding of plasma proteins to the surface of MION-47 there was a 25% increase of effective hydrodynamic diameter, which corresponded to a shift in size from a 0.75 ×.10^3 kD globular protein to a 1.01 ×.10^3 kDa globular protein. The immunoblotting of proteins resolved by sodium dodecyl sulfate electrophoresis of the above opsonized nanopartilcles showed that fibronectin, vitronectin and C3 fragments were associated with their coating.[65] However, no transferrin binding was observed suggesting that these IO were different from larger IO particles that had imperfect coating with dextran and short blood half-lives *in vivo*.[82] MIONs have been studied in animals for applications of MR receptor targeted imaging[17,83,84] perfusion imaging,[14] MR lymphography[64] fMRI[85] and cell labeling and tracking[86,87] but testing in humans has been performed using alternative IO preparations, e.g. citrate-stabilized IO nanoparticles.[88]

3. Ultrasmall SPIOs in the Detection of Metastases

Ultrasmall SPIO-contrast enhanced MRI has made the most meaningful clinical contribution in the detection of metastatic disease, especially with the use of MR lymphography (MRL). Many organ tumors metastasize via the lymph nodes, and detection of such spread greatly alters prognosis and quality of life of cancer patients. The determination of metastatic involvement of lymph nodes is key in cancer staging and therapeutic planning and assessment, especially in relation to considering surgical interventions. Furthermore, effective imaging methods obviate the need for more invasive and higher risk surgical lymph node dissection. Thus CAs that increase diagnostic precision are essential, especially for MRI, since this imaging method is often chosen due to the safety of performing multiple exams during the course of the disease. Although more conventional diagnostic methods (such as gadolinium contrast enhanced MRI, CT, PET, and ultrasonography) have been tested and used for detection of nodal metastasis, these methods generally are based on determinations of lymph node size and shape for a positive diagnosis, even though such measures are not always indicative of lymph node involvement or metastatic spread.[89–93] Although the use of SPIOs recently has been thwarted due to safety concerns, increases in the safety profile of new agents should hasten the return of these agents to the clinic.

The sensitivity of ultrasmall SPIO-based imaging for detection of metastases is based on differential uptake of ultrasmall SPIOs into normal versus malignant lymph nodes. Ultrasmall SPIOs extravasate into the interstitium and accumulate in lymphatic sinuses with uptake by lymph node resident macrophages.[64,94] Normal lymph nodes that harbor lymphocytes and macrophages filter debris, cell waste, and pathogens (if present) from the body. They also frequently serve as immediate homing sites for cancer cells that are carried by lymphatic flow away from the primary tumors. Because of their high macrophage content, normal lymph nodes appear dark on T2/T2* weighted MR images after administration of ultrasmall SPIOs due to the high uptake of the CA in the macrophages. Through mechanisms not yet entirely understood, highly malignant cells will alter their physiology and repertoire of extracellular receptors so that they begin to escape capture by immune cells of the lymphatic system. In many cases, these alterations are associated with epithelial mesenchymal transition in circulating cancer cells.[95] Progressively, cancer cells will overtake and replace the normal tissue of the lymph node (Fig. 2). This stage of disease progression is highly detectable by ultrasmall SPIO-based MR imaging due to the fact that these nodes are largely devoid of macrophages and thus do not accumulate ultrasmall SPIOs, making them appear brighter on T2/T2* MR images. This is in contrast to conventional Gd^{3+}-based agents that delineate nodal involvement by their sequestration into malignant nodes. These agents, when carried through the systemic circulation, can permeate into thin-walled fenestrated lymphatic capillaries and subsequently be transported with lymphatic fluid to the lymph nodes. Because malignancy is often accompanied by insufficient lymphatic drainage, Gd^{3+}-based agents accumulate in sentinel lymph nodes, causing the nodes to appear bright on T1-weighted images (MR lymphangiography).

The first studies showing nodal uptake of ultrasmall SPIOs in animals were reported in the early 1990s,[1,33,96,97] and were soon followed by demonstrations in humans.[98] Since that time, ultrasmall SPIO-based methods for detection of metastatic disease spread have been reported for many types of cancer,[99] including breast,[100–102] prostate,[46,90,103,104] rectal,[105] pelvic,[106] and head and neck cancer[107–109] and guidelines for their use in this capacity have been established.[89,105,106,110–112] These studies have examined use of IOs alone as well as in combination with Gd^{3+}-based CAs. A recent addition to the application of ultrasmall SPIOs to tumor detection and staging has been the use of macrophages which are labeled, *ex vivo*, with IOs and then re-injected into the animal for subsequent delivery to tumor and lymph nodes.[113–115] This method will no doubt be useful in the study of tumor formation and progression as well as for subsequent detection of malignancy. It was recently demonstrated that labeled macrophages are also capable of homing in on tumors such as gliomas in rat brain following *i.v.* injection[116] which suggests that macrophage labeling with IOs could be used for combined treatment and assessment (theranostic) of brain disease, which is often resistant to *i.v.* administered agents.

Although measures of diagnostic efficacy of ultrasmall SPIO-enhanced MRI vary somewhat across studies and between body regions and cancer types, the overall consensus has been that this imaging method performs well and provides additional diagnostic information to that obtained with conventional contrast enhanced MRI, CT and PET, and is capable of detecting malignant nodes as small as 5 mm. The most recent meta-analysis of 34 clinical studies by Wu and colleagues,[117] showed ultrasmall SPIO-enhanced imaging to have a high sensitivity (0.90, 95% CI 0.88–0.91) and specificity 0.96 (95% CI: 0.95–0.97), with a Q value (reflecting diagnostic value) of 0.9195. This performance was statistically higher than that obtained with non-contrast MRI which showed an overall sensitivity, specificity and Q value of 0.39 (95% CI: 0.34–0.43), 0.90 (95% CI: 0.89–0.91), and 0.63, respectively. It was noted however that the advantages gained in diagnostic precision from ultrasmall SPIO-based imaging must be considered in light of the fact that this type of examination involves two separate MRI examinations (non-enhanced and ultrasmall SPIO-enhanced, with the latter performed 24 h after the administration of ultrasmall SPIO) and thus incurs additional time and cost. Nonetheless, because metastatic detection is often critical in establishing a care plan (especially one involving surgery), the extra exam may not be a major concern. A second meta-analysis evaluating the diagnostic accuracy of ultrasmall SPIO-enhanced imaging for identification of axillary metastases in early stage newly diagnosed breast cancer patients, yielded similar findings.[118] Sensitivity and specificity were meta-analyzed using a bivariate random effects approach, utilizing pathological diagnosis via node biopsy as the comparative gold standard. Mean sensitivity was 98% (95% CI: 61e100%) and mean specificity 96% (95% CI: 72e100%), whereas across three studies of gadolinium-enhanced MRI (n 1/4 187), mean sensitivity was 88% (95% CI: 78e94%) and mean specificity 73% (95% CI: 63e81%).

A number of clinical studies have shown even higher success with ultrasmall SPIO-enhanced MRI for specific types of cancers. Remarkable sensitivity — 100% with 80% specificity was achieved in patients with pelvic and urological cancers spread to their lymph nodes as detected by a relatively small dose of AMI-227.[119] Metastatic nodal detection for prostate cancer using this method yielded a sensitivity of 90.5% and a specificity of 97.8%.[46] When combined with high-resolution MRI and 3D mapping, ultrasmall SPIO-enhanced MRI is an effective tool for surgical planning.[46] Positive results have also been reported for ultrasmall SPIO-enhanced detection of metastatic spread in prostate carcinoma patients.[104] In these patients, radical resection of the prostate is a completely inefficient procedure in terms of potential curative effect and thus makes metastatic detection critical for replacing surgery with alternative treatments that can improve quality of life and save treatment costs. In 61 patients positively identified as having lymph-node metastases, 50 were detected by MR lymphography, and of these, 40 patients (80%) had metastases in normal-sized lymph nodes. The negative predictive value of ultrasmall SPIO-enhanced lymphography was determined to be so great that patients with a negative MRL had only a 4% chance of having a malignant positive lymph node, while patients with a positive MRL had a 68% chance of having metastases.[91]

As with all diagnostic imaging modalities, experience of the radiologist in reading ultrasmall SPIO-enhanced images will be key. False negative readings can be caused by nodal fat, which remains hyperintense after ultrasmall SPIO uptake, therefore comparative T1 weighted sequences that identify nodal fat content can increase the accuracy of diagnosis. False positives may be higher in cases where there is avid inflammation of the tissue (and increased macrophage content).[120,121] Another factor that must be taken into account is that the spatial extent of the ultrasmall SPIO mediated MR signal reduction is not rigidly confined to the spatial distribution of the actual particles because the induced susceptibility effect (blooming) extends beyond the particle. This effect can be somewhat controlled by proper choice of imaging parameters (amount of T2 and T2* weighting). An understanding of the expected distribution of ultrasmall SPIOs is helpful in overall interpretation of images collected with different imaging parameters i.e. dispersed particles imaged with short TE and TR (T1 weighted) produce hyperintensity signal whereas higher concentration particles imaged with longer TE and TR (T2 weighted) produce hypointensity signal. In some cases these two effects may overlap and in effect null ultrasmall SPIO induced contrast. Multi-echo gradient echo sequences acquire multiple images at different TEs and thus are useful for ultrasmall SPIO imaging applications. The specificity of the method (but not the sensitivity) may be higher at higher magnetic field strengths,[122] and the size of the IO particle will also affect the route of uptake into macrophages and thus will produce maximal contrast enhancement at different times after administration.[123] Newer ultrasmall SPIO formulations may also prove more effective in disease detection.[124]

4. Applications of Ultrasmall SPIOs in Vascular Imaging: Detection of Inflammatory Unstable Plaques

A second area of human disease diagnosis and treatment evaluation that has a potential of benefiting from the use of IO nanoparticles is cardiovascular disease. Imaging of vascular lesions by ultrasmall SPIOs is feasible because of their selective uptake into phagocytes, and the ability of these CAs to target this aspect of vascular disease is not only somewhat unique, but also highly useful for detecting what is most clinically relevant for preventing future cardiovascular events (i.e. plaque rupture and resultant ischemia due to vascular occlusion) via monitoring the inflammatory burden of specific vascular lesions.

Overwhelming evidence has accumulated over more than a decade suggesting that atherosclerosis is in large part an inflammatory disease.[125–127] The initial sign of vessel lesion, "the fatty streak" is made up of infiltrating granulocytes, macrophages and T-cells. Although the steps in the progression to end stage disease are not the same across all cases, a general pattern has been observed. Early perturbations cause the endothelium to become dysfunctional which further recruits immune cells, smooth muscle cells, platelets, and other factors such as low-density lipoproteins (LDL) to the site. Once LDL is oxidized, it is phagocytosed by the infiltrated macrophages, which ultimately transform into foam cells and form the core of an atherosclerotic plaque. This necrotic core is covered by a fibrous cap, that when disrupted, will lead to a loss of the core contents leading to thrombo-embolism and eventually disrupting distal blood flow resulting in ischemia, (stroke, myocardial infarction). Because inflammation is key in lesion progression and destabilization, it is regarded as a major risk factor for future cardiovascular events,[125,127,128] and the degree of risk increases for lesions that show macrophage infiltration either at the shoulders of the plaque or in the necrotic lipid core.[129]

Macrophage accumulation of ultrasmall SPIOs (which have a diameter similar to LDL) makes these agents ideal for diagnostic imaging of high risk atherosclerotic plaques, especially given that more conventional imaging methods are based on determination of the degree of vessel lumen stenosis which is less relevant in predicting future cardiovascular events.[130] Ferumoxtran-10 mediated MR signal reduction in atherosclerotic plaques was first observed in patients that had been given the CA for the detection of lymph node metastases.[131] Areas of signal loss confined to the arterial wall were clearly detectable in the aortic and pelvic arteries of elderly patients. Concurrent findings in an animal model of atherosclerosis were verified with histopathology and showed the site of IO accumulation to be macrophages.[132] Soon afterward, histology of excised plaques from patients undergoing endarterectomy confirmed the site of ultrasmall SPIO accumulation to be primarily in macrophages (10 of 11 patients).[58] The latter study was significant in that it showed the specificity for ultrasmall SPIO accumulation and higher contrast enhancement in ruptured or rupture-prone carotid atherosclerotic lesions (75%) versus stable lesions (7%); a result that has since been replicated numerous times.[19,133–136] A similar study showed 24 to 36 h post injection to be the optimal time point for imaging carotid lesions using ultrasmall SPIOs,[133] and more recently it has been shown that diagnostically meaningful images can be acquired with both T2*-weighted and T1-weighted MR sequences at 36–48 h post contrast.[136] Although ultrasmall SPIOs accumulate to a greater extent in higher rupture-risk plaques, the technique can also detect stable disease in unsymptomatic patients.[19,134,137] Ongoing research in animal models should produce IO agents that have greater specificity for plaque imaging.[138–140] The use of ultrasmall SPIO-contrast enhanced MRI may allow for greater risk assessment in patients when used in combination with other diagnostic measures such as wall stress analysis,[134] PET[141,142] and MR angiography,[143] especially given that each of these methods has advantages and can detect different risk related pathologies.[144] Furthermore, the technique has proven capable of detecting treatment effects after therapeutic interventions with statins.[62,145,146,147] Another variation of cardiovascular disease, aneurysm formation, also involves infiltration of macrophages and as such may be diagnosable or more precisely characterized with ultrasmall SPIO contrast enhanced MRI.[148–150] Risk stratification of these patients has also been made possible by this approach since the uptake of ultrasmall SPIOs in abdominal aortic aneurysms was correlated with a three-fold higher aneurysm growth rate in these lesions.[148]

Continued work in animal models will likely suggest new strategies for the use of ultrasmall SPIOs in vascular imaging,[151] such as the use of ligand-conjugated IOs for receptor targeted imaging[152–155] and new applications of the method to known pathologies, such as detection of acute inflammation following invasive vascular procedures[156] and stroke.[157] Conversely, ultrasmall SPIO contrast enhanced imaging in animals will likely give us a better understanding of the mechanisms underlying the progression of vascular disease.[158] Of particular interest in this respect is the use of ultrasmall SPIO labeled macrophages. We observed early on that MION when internalized by macrophages *in vitro* could produce the same proton relaxation and MR signal changes as MION in the culture medium but at a concentration that was an order of magnitude lower,[159] suggesting that IO magnetically labeled cells could be a viable option for *in vivo* cell tracking, and for use in highly localized CA delivery. Subsequent studies have been carried out to determine the optimal IO particle size,[160,161] and loading method[162] for cell labeling, as well as optimized MR pulse sequences for this application.[163] Recent studies have shown the utility of the IO cell labeling method for exploring the mechanisms of aortic atheroma progression in a murine model of vascular disease (ApoE2

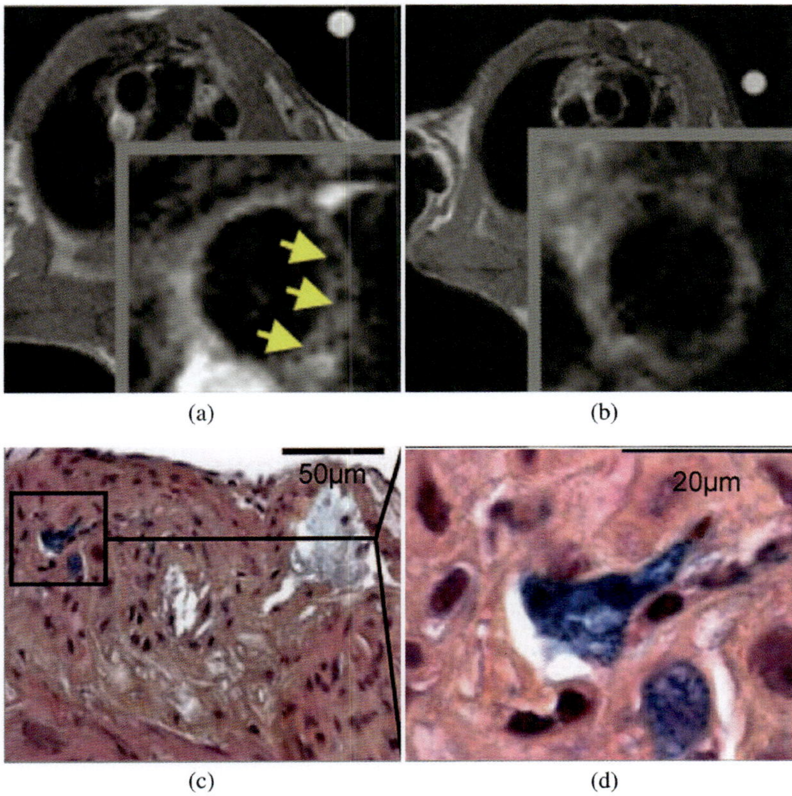

Fig. 3. (Color online) MR imaging at 4.7 T of activated IO-labeled macrophages (shown with arrows) visible as focal dark signal losses in the aortic arch wall plaque of ApoE2 knock-in transgenic (a), and to a lesser degree in the control C57Bl/6 syngenic mouse (b). Histology corroboration performed by using Prussian blue staining show iron labeled cells in the atheroma of ApoE2 mouse (c, d). Adapted from Ref. 164.

knock-in mice)[164] (Fig. 3). Activated macrophages derived from bone marrow cell culture were magnetically labeled with iron nanoparticles via simple incubation following addition of IO to the medium and subsequently injected into mice. MR imaging performed on the three days following administration of cells showed localized signal loss regions in the thickened aortic wall, with a maximal effect at day 2 (34% ± 7.3% $P < 0.001$ compared with baseline), which corresponded to the location of the injected cells. In addition, after recruitment into the plaque, the labeled activated macrophages continued to migrate over the course of days deeper into the adventitia, and spread from the shoulders to the core of the plaque. The authors concluded that the adventitial macrophage localization toward the back of atherosclerotic plaques can be related either to their migration through the lesion from their point of recruitment or migration to highly vascularized adventitia and the perivascular fat, with further inflammation triggers in these regions.

Subsequent studies have also shown advantages in plaque imaging afforded by injection of endothelial cells labeled with SPIO,[165] SPIO-labeled high density lipoprotein (HDL),[166] and the use of the above approach for the study of aneurysm formation[167] inflammation and infection,[168–170] reperfusion injury,[171] graft rejection,[172] and nerve injury.[173] Molecular targeting of plaque related ligands, such as those found on oxidized low-density lipoproteins, has also been achieved by using antibody conjugated ultrasmall SPIOs.[174]

5. Conclusions

IO nanoparticles comprise a class of versatile MRI CAs that can be tailored to the study and diagnosis of a vast array of human diseases. So far a substantial amount of evidence has proven adaptability of SPIO-contrast enhanced MRI for detection of anatomical abnormalities and disease processes. The advantages of SPIO include accessibility, ease and reproducibility of synthesis of nanoparticles, and established efficacy in clinical diagnosis. Remaining challenges include engineering nanoparticles that have very low toxicity and improving MR imaging techniques (pulse sequences) that will result in more reliable MR image interpretation, such as the development of "positive contrast" methods[6] that improve the detection of a low concentration of IO labeled cells *in vivo*. Therefore, there is little doubt that SPIO based MR imaging methods will eventually become part of the routine MR exam in the coming years.

References

1. R. Weissleder, D. D. Stark, B. L. Engelstad, B. R. Bacon, C. C. Compton, D. L. White, P. Jacobs and J. Lewis, Superparamagnetic iron oxide: pharmacokinetics and toxicity. *Am. J. Roentgenol.* **152**(1), 167–173 (1989).
2. L. E. LaConte, N. Nitin, O. Zurkiya, D. Caruntu, C. J. O'Connor, X. Hu and G. Bao, Coating thickness of magnetic iron oxide nanoparticles affects R2 relaxivity. *J. Magn. Reson. Imaging* **26**(6), 1634–1641 (2007).
3. M. Di Marco, C. Sadun, M. Port, I. Guilbert, P. Couvreur and C. Dubernet, Physicochemical characterization of ultrasmall superparamagnetic iron oxide particles (USPIO) for biomedical application as MRI contrast agents. *Int. J. Nanomedicine.* **2**(4), 609–622 (2007).
4. O. Lunov, V. Zablotskii, T. Syrovets, C. Rocker, K. Tron, G. U. Nienhaus and T. Simmet, Modeling receptor-mediated endocytosis of polymer-functionalized iron oxide nanoparticles by human macrophages. *Biomaterials* **32**(2), 547–555 (2011), LID - 510.1016/j.b.

5. S. Saito, M. Tsugeno, D. Koto, Y. Mori, Y. Yoshioka, S. Nohara and K. Murase, Impact of surface coating and particle size on the uptake of small and ultrasmall superparamagnetic iron oxide nanoparticles by macrophages. *Int. J. Nanomedicine* 7, 5415–5421 (2012), LID - 5410.2147/IJN.
6. C. H. Cunningham, T. Arai, P. C. Yang, M. V. McConnell, J. M. Pauly and S. M. Conolly, Positive contrast magnetic resonance imaging of cells labeled with magnetic nanoparticles. *Magn. Reson. Med.* **53**(5), 999–1005 (2005).
7. C. Burtea, S. Laurent, L. Vander Elst and R. N. Muller,Contrast agents: magnetic resonance. *Handb. Exp. Pharmacol.* **185**(Pt 1), 135–165 (2008).
8. S. Saini, R. Sharma, R. L. Baron, D. A. Turner, P. R. Ros, P. F. Hahn, W. C. Small, E. E. Delange, A. E. Stillman, R. R. Edelman, V. M. Runge and E. K. Outwater, Multicentre dose-ranging study on the efficacy of USPIO ferumoxtran-10 for liver MR imaging. *Clin. Radiol.* **55**(9), 690–695 (2000).
9. Y. Tang, Y. Yamashita, A. Arakawa, T. Namimoto, K. Mitsuzaki, Y. Abe, K. Katahira and M. Takahashi, Detection of hepatocellular carcinoma arising in cirrhotic livers: comparison of gadolinium- and ferumoxides-enhanced MR imaging. *Am. J. Roentgenol.* **172**(6), 1547–1554 (1999).
10. P. Reimer, N. Jahnke, M. Fiebich, W. Schima, F. Deckers, C. Marx, N. Holzknecht and S. Saini, Hepatic lesion detection and characterization: value of nonenhanced MR imaging, superparamagnetic iron oxide-enhanced MR imaging, and spiral CT-ROC analysis. *Radiology* **217**(1), 152–158 (2000).
11. S. H. Kim, D. Choi, S. H. Kim, J. H. Lim, W. J. Lee, M. J. Kim, H. K. Lim and S. J. Lee, Ferucarbotran-enhanced MRI versus triple-phase MDCT for the preoperative detection of hepatocellular carcinoma. *AJR Am. J. Roentgenol.* **184**(4), 1069–1076 (2005).
12. D. Pauleit, J. Textor, R. Bachmann, R. Conrad, S. Flacke, G. Layer, B. Kreft and H. Schild, Hepatocellular carcinoma: detection with gadolinium- and ferumoxides-enhanced MR imaging of the liver. *Radiology* **222**(1), 73–80 (2002).
13. P. R. Ros *et al.*, Hepatic MR imaging with ferumoxides: a multicenter clinical trial of the safety and efficacy in the detection of focal hepatic lesions. *Radiology* **196**(2), 481–488 (1995).
14. C. Bremer, M. Mustafa, A. Bogdanov Jr., V. Ntziachristos, A. Petrovsky and R. Weissleder, Steady-state blood volume measurements in experimental tumors with different angiogenic burdens a study in mice. *Radiology* **226**(1), 214–220 (2003).
15. T. Persigehl, A. Wall, J. Kellert, J. Ring, S. Remmele, W. Heindel, H. Dahnke and C. Bremer, Tumor blood volume determination by using susceptibility-corrected Delta R2* multiecho MR. *Radiology* **255**(3), 781–789 (2010), LID - 710.1148/radi.
16. M. A. Fischer, O. F. Donati, C. S. Reiner, R. Hunziker, D. Nanz and A. Boss, Feasibility of semiquantitative liver perfusion assessment by ferucarbotran bolus injection in double-contrast hepatic MRI. *J. Magn. Reson. Imaging* **36**(1), 168–176 (2012), LID - 110.1002/jmr.
17. R. Weissleder, A. S. Lee, B. A. Khaw, T. Shen and T. J. Brady, Antimyosin-labeled monocrystalline iron oxide allows detection of myocardial infarct: MR antibody imaging. *Radiology* **182**(2), 381–385 (1992).
18. C. Corot, K. G. Petry, R. Trivedi, A. Saleh, C. Jonkmanns, J. F. Le Bas, E. Blezer, M. Rausch, B. Brochet, P. Foster-Gareau, D. Baleriaux, S. Gaillard and V. Dousset, Macrophage imaging in central nervous system and in carotid atherosclerotic plaque

using ultrasmall superparamagnetic iron oxide in magnetic resonance imaging. *Invest. Radiol.* **39**(10), 619–625 (2004).

19. T. Tang, S. P. Howarth, S. R. Miller, R. Trivedi, M. J. Graves, J. U. King-Im, Z. Y. Li, A. P. Brown, P. J. Kirkpatrick, M. E. Gaunt and H. J. Gillard, Assessment of inflammatory burden contralateral to the symptomatic carotid stenosis using high-resolution ultrasmall, superparamagnetic iron oxide-enhanced MRI. *Stroke* **37**(9), 2266–2270 (2006).
20. R. A. Trivedi, C. Mallawarachi, J. M. U-King-Im, M. J. Graves, J. Horsley, M. J. Goddard, A. Brown, L. Wang, P. J. Kirkpatrick, J. Brown and J. H. Gillard, Identifying inflamed carotid plaques using *in vivo* USPIO-enhanced MR imaging to label plaque macrophages. *Arterioscler. Thromb. Vasc. Biol.* **26**(7), 1601–1606 (2006).
21. A. Saleh, M. Schroeter, A. Ringelstein, H. P. Hartung, M. Siebler, U. Modder and S. Jander, Iron oxide particle-enhanced MRI suggests variability of brain inflammation at early stages after ischemic stroke. *Stroke* **38**(10), 2733–2737 (2007).
22. V. Dousset, B. Brochet, M. S. Deloire, L. Lagoarde, B. Barroso, J. M. Caille and K. G. Petry, MR imaging of relapsing multiple sclerosis patients using ultra-small-particle iron oxide and compared with gadolinium. *AJNR Am. J. Neuroradiol.* **27**(5), 1000–1005 (2006).
23. Y. Anzai, S. McLachlan, M. Morris, R. Saxton and R. B. Lufkin, Dextran-coated superparamagnetic iron oxide, an MR contrast agent for assessing lymph nodes in the head and neck. *AJNR Am. J. Neuroradiol.* **15**(1), 87–94 (1994).
24. E. A. Neuwelt, P. Varallyay, A. G. Bago, L. L. Muldoon, G. Nesbit and R. Nixon, Imaging of iron oxide nanoparticles by MR and light microscopy in patients with malignant brain tumours. *Neuropathol. Appl. Neurobiol.* **30**(5), 456–471 (2004).
25. H. Frank, R. Weissleder and T. J. Brady, Enhancement of MR angiography with iron oxide: preliminary studies in whole-blood phantom and in animals. *AJR Am. J. Roentgenol.* **162**(1), 209–213 (1994).
26. W. W. Mayo-Smith, S. Saini, G. Slater, J. A. Kaufman, P. Sharma and P. F. Hahn, MR contrast material for vascular enhancement: value of superparamagnetic iron oxide. *AJR Am. J. Roentgenol.* **166**(1), 73–77 (1996).
27. Y. Anzai, M. R. Prince, T. L. Chenevert, J. H. Maki, F. Londy, M. London and S. J. McLachlan, MR angiography with an ultrasmall superparamagnetic iron oxide blood pool agent. *J. Magn. Reson. Imaging* 7(1), 209–214 (1997).
28. S. A. Schmitz, T. Albrecht and K. J. Wolf, MR angiography with superparamagnetic iron oxide: feasibility study. *Radiology* **213**(2), 603–607 (1999).
29. K. H. Ahlstrom, L. O. Johansson, J. B. Rodenburg, A. S. Ragnarsson, P. Akeson and A. Borseth, Pulmonary MR angiography with ultrasmall superparamagnetic iron oxide particles as a blood pool agent and a navigator echo for respiratory gating: pilot study. *Radiology* **211**(3), 865–869 (1999).
30. H. Bernd, E. De Kerviler, S. Gaillard and B. Bonnemain, Safety and tolerability of ultrasmall superparamagnetic iron oxide contrast agent: comprehensive analysis of a clinical development program. *Invest. Radiol.* **44**(6), 336–342 (2009), LID - 310.1097/RLI.
31. R. A. Heesakkers, G. J. Jager, A. M. Hovels, B. de Hoop, B. oH. van den, F. Raat, J. A. Witjes, P. F. Mulders, Ka. C. van der and J. O. Barentsz, Prostate cancer: detection of lymph node metastases outside the routine surgical area with ferumoxtran-10-enhanced MR imaging. *Radiology* **251**(2), 408–414 (2009), LID - 410.1148/rad.

32. T. Tourdias, S. Roggerone, M. Filippi, M. Kanagaki, M. Rovaris, D. H. Miller, K. G. Petry, B. Brochet, J. P. Pruvo, E. W. Radue and V. Dousset, Assessment of disease activity in multiple sclerosis phenotypes with combined gadolinium- and superparamagnetic iron oxide-enhanced MR imaging. *Radiology* **264**(1), 225–233 (2012), LID - 210.1148/rad.
33. R. Weissleder, G. Elizondo, J. Wittenberg, C. A. Rabito, H. H. Bengele and L. Josephson, Ultrasmall superparamagnetic iron oxide: characterization of a new class of contrast agents for MR imaging. *Radiology* **175**(2), 489–493 (1990).
34. S. Benderbous, C. Corot, P. Jacobs and B. Bonnemain, Superparamagnetic agents: physicochemical characteristics and preclinical imaging evaluation. *Acad. Radiol.* **3**(Suppl 2), S292–S294 (1996).
35. A. K. Gupta and M. Gupta. Synthesis and surface engineering of iron oxide nanoparticles for biomedical applications. *Biomaterials* **26**(18), 3995–4021 (2005).
36. Q. Gao, J. Zhang, G. Hong and J. Ni, One-pot reaction to synthesize PEG-coated hollow magnetite nanostructures with excellent magnetic properties. *J. Nanosci. Nanotechnol.* **10**(10), 6400–6406 (2010).
37. A. Kunzmann, B. Andersson, C. Vogt, N. Feliu, F. Ye, S. Gabrielsson, M. S. Toprak, T. Buerki-Thurnherr, S. Laurent, M. Vahter, H. Krug, M. Muhammed, A. Scheynius and B. Fadeel, Efficient internalization of silica-coated iron oxide nanoparticles of different sizes by primary human macrophages and dendritic cells. *Toxicol. Appl. Pharmacol.* **253**(2), 81–93 (2011), LID - 10.1016/j.t.
38. P. Wunderbaldinger, L. Josephson and R. Weissleder, Tat peptide directs enhanced clearance and hepatic permeability of magnetic nanoparticles. *Bioconjug. Chem.* **13**(2), 264–268 (2002).
39. Y. X. Wang, S. M. Hussain and G. P. Krestin, Superparamagnetic iron oxide contrast agents: physicochemical characteristics and applications in MR imaging. *Eur. Radiol.* **11**(11), 2319–2331 (2001).
40. K. E. Kellar, D. K. Fujii, W. H. Gunther, K. Briley-Saebo, A. Bjornerud, M. Spiller and S. H. Koenig, NC100150 Injection, a preparation of optimized iron oxide nanoparticles for positive-contrast MR angiography. *J. Magn. Reson. Imaging* **11**(5), 488–494 (2000).
41. O. Veiseh, C. Sun, C. Fang, N. Bhattarai, J. Gunn, F. Kievit, K. Du, B. Pullar, D. Lee, R. G. Ellenbogen, J. Olson and M. Zhang, Specific targeting of brain tumors with an optical/magnetic resonance imaging nanoprobe across the blood-brain barrier. *Cancer Res.* **69**(15), 6200–6207 (2009).
42. J. H. Lee, Y. M. Huh, Y. W. Jun, J. W. Seo, J. T. Jang, H. T. Song, S. Kim, E. J. Cho, H. G. Yoon, J. S. Suh and J. Cheon, Artificially engineered magnetic nanoparticles for ultra-sensitive molecular imaging. *Nat Med.* **13**(1), 95–99 (2007).
43. L. Yang, H. Mao, Z. Cao, Y. A. Wang, X. Peng, X. Wang, H. K. Sajja, L. Wang, H. Duan, C. Ni, C. A. Staley, W. C. Wood, X. Gao and S. Nie, Molecular imaging of pancreatic cancer in an animal model using targeted multifunctional nanoparticles. *Gastroenterology* **136**(5), 1514–1525 (2009), e1512.
44. L. Yang, X. H. Peng, Y. A. Wang, X. Wang, Z. Cao, C. Ni, P. Karna, X. Zhang, W. C. Wood, X. Gao, S. Nie and H. Mao, Receptor-targeted nanoparticles for *in vivo* imaging of breast cancer. *Clin. Cancer Res.* **15**(14), 4722–4732 (2009).
45. O. Clement and A. Luciani, Imaging the lymphatic system: possibilities and clinical applications. *Eur. Radiol.* **14**(8), 1498–1507 (2004).

46. M. G. Harisinghani, J. Barentsz, P. F. Hahn, W. M. Deserno, S. Tabatabaei, C. H. van de Kaa, J. de la Rosette and R. Weissleder, Noninvasive detection of clinically occult lymph-node metastases in prostate cancer. *N. Engl. J. Med.* **348**(25), 2491–2499 (2003).
47. S. Wagner, J. Schnorr, H. Pilgrimm, B. Hamm and M. Taupitz, Monomer-coated very small superparamagnetic iron oxide particles as contrast medium for magnetic resonance imaging: preclinical *in vivo* characterization. *Invest. Radiol.* **37**(4), 167–177 (2002).
48. A. M. Taylor, J. R. Panting, J. Keegan, P. D. Gatehouse, D. Amin, P. Jhooti, G. Z. Yang, S. McGill, E. D. Burman, J. M. Francis, D. N. Firmin and D. J. Pennell, Safety and preliminary findings with the intravascular contrast agent NC100150 injection for MR coronary angiography. *J. Magn. Reson. Imaging* **9**(2), 220–227 (1999).
49. B. Tombach, P. Reimer, C. Bremer, T. Allkemper, M. Engelhardt, M. Mahler, W. Ebert and W. Heindel, First-pass and equilibrium-MRA of the aortoiliac region with a superparamagnetic iron oxide blood pool MR contrast agent (SH U 555 C): results of a human pilot study. *NMR Biomed.* **17**(7), 500–506 (2004).
50. S. Anastase, D. Letourneur and J. Jozefonvicz, Affinity chromatography of human antidextran antibodies. Isolation of two distinct populations. *J. Chromatogr. B Biomed. Appl.* **686**(2), 141–150 (1996).
51. C. E. Zinderman, L. Landow and R. P. Wise, Anaphylactoid reactions to Dextran 40 and 70: reports to the United States Food and Drug Administration, 1969 to 2004. *J. Vasc. Surg.* **43**(5), 1004–1009 (2006).
52. S. Tong, S. Hou, Z. Zheng, J. Zhou and G. Bao, Coating optimization of superparamagnetic iron oxide nanoparticles for high T2 relaxivity. *Nano Lett.* **10**(11), 4607–4613 (2010), LID - 4610.1021/nl4601.
53. W. Xiao, J. Lin, M. Li, Y. Ma, Y. Chen, C. Zhang, D. Li and H. Gu, Prolonged *in vivo* circulation time by zwitterionic modification of magnetite nanoparticles for blood pool contrast agents. *Contrast Media Mol. Imaging* **7**(3), 320–327 (2012), LID - 310.1002/cmmi.
54. B. H. Kenzaoui, C. C. Bernasconi, H. Hofmann and L. Juillerat-Jeanneret, Evaluation of uptake and transport of ultrasmall superparamagnetic iron oxide nanoparticles by human brain-derived endothelial cells. *Nanomedicine (Lond)* **7**(1), 39–53 (2012), LID - 10.2217/nnm.
55. H. Wei, N. Insin, J. Lee, H. S. Han, J. M. Cordero, W. Liu and M. G. Bawendi, Compact zwitterion-coated iron oxide nanoparticles for biological applications. *Nano Lett.* **12**(1), 22–25 (2012), LID - 10.1021/nl1020.
56. A. Masoudi, H. R. Hosseini, S. M. Reyhani, M. A. Shokrgozar, M. A. Oghabian and R. Ahmadi, Long-term investigation on the phase stability, magnetic behavior, toxicity, and MRI characteristics of superparamagnetic Fe/Fe-oxide core/shell nanoparticles. *Int. J. Pharm.* **439**(1&2), 28–40 (2012), LID - 10.1016/j.i.
57. K. Tsuchiya, N. Nitta, A. Sonoda, A. Nitta-Seko, S. Ohta, M. Takahashi, K. Murata, K. Mukaisho, M. Shiomi, Y. Tabata and S. Nohara, Evaluation of atherosclerotic lesions using dextran- and mannan-dextran-coated USPIO: MRI analysis and pathological findings. *Int. J. Nanomed.* **7**, 2271–2280 (2012), LID - 2210.2147/IJN.
58. M. E. Kooi, V. C. Cappendijk, K. B. Cleutjens, A. G. Kessels, P. J. Kitslaar, M. Borgers, P. M. Frederik, M. J. Daemen and J. M. van Engelshoven, Accumulation of ultrasmall superparamagnetic particles of iron oxide in human atherosclerotic plaques can be detected by *in vivo* magnetic resonance imaging. *Circulation* **107**(19), 2453–2458 (2003).

59. A. M. Lutz, K. Gopfert, W. Jochum, D. Nanz, J. M. Frohlich and D. Weishaupt, USPIO-enhanced MR imaging for visualization of synovial hyperperfusion and detection of synovial macrophages: preliminary results in an experimental model of antigen-induced arthritis. *J. Magn. Reson. Imaging* **24**(3), 657–666 (2006).
60. O. Hauger, N. Grenier, C. Deminere, C. Lasseur, Y. Delmas, P. Merville and C. Combe, USPIO-enhanced MR imaging of macrophage infiltration in native and transplanted kidneys: initial results in humans. *Eur. Radiol.* **17**(11), 2898–2907 (2007).
61. M. M. Vellinga, R. D. Oude Engberink, A. Seewann, P. J. Pouwels, M. P. Wattjes, PO. S. van der, C. Pering, C. H. Polman, H. E. de Vries, J. J. Geurts and F. Barkhof, Pluriformity of inflammation in multiple sclerosis shown by ultra-small iron oxide particle enhancement. *Brain* **131**(Pt 3), 800–807 (2008), LID - 810.1093/brai.
62. T. Y. Tang, S. P. Howarth, S. R. Miller, M. J. Graves, A. J. Patterson, J. M. U-King-Im, Z. Y. Li, S. R. Walsh, A. P. Brown, P. J. Kirkpatrick, E. A. Warburton, P. D. Hayes, K. Varty, J. R. Boyle, M. E. Gaunt, A. Zalewski and J. Gillard, The ATHEROMA (Atorvastatin Therapy: Effects on Reduction of Macrophage Activity) Study. Evaluation using ultrasmall superparamagnetic iron oxide-enhanced magnetic resonance imaging in carotid disease. *J. Am. Coll. Cardiol.* **53**(22), 2039–2050 (2009), LID - 2010.1016/j.j.
63. S. Lefevre, D. Ruimy, F. Jehl, A. Neuville, P. Robert, C. Sordet, M. Ehlinger, J. L. Dietemann and G. Bierry, Septic arthritis: monitoring with USPIO-enhanced macrophage MR imaging. *Radiology* **258**(3), 722–728 (2011), LID - 710.1148/radi.
64. R. Weissleder, J. F. Heautot, B. K. Schaffer, N. Nossiff, M. I. Papisov, A. Bogdanov Jr., and T. J. Brady, MR lymphography: study of a high-efficiency lymphotrophic agent. *Radiology* **191**(1), 225–230 (1994).
65. A. Moore, R. Weissleder and A. Bogdanov Jr., Uptake of dextran-coated monocrystalline iron oxides in tumor cells and macrophages. *J. Magn. Reson. Imaging* 7(6), 1140–1145 (1997).
66. V. Dousset, C. Delalande, L. Ballarino, B. Quesson, D. Seilhan, M. Coussemacq, E. Thiaudiere, B. Brochet, P. Canioni and J. M. Caille, *In vivo* macrophage activity imaging in the central nervous system detected by magnetic resonance. *Magn. Reson. Med.* **41**(2), 329–333 (1999).
67. T. Matsushita, Y. Kusakabe, H. Fujii, K. Murase, Y. Yamazaki and K. Murase, Inflammatory imaging with ultrasmall superparamagnetic iron oxide. *Magn. Reson. Imaging* **29**(2), 173–178 (2011), LID - 110.1016/j.mr.
68. Mu. C. von Zur, D. von Elverfeldt, N. Bassler, I. Neudorfer, B. Steitz, A. Petri-Fink, H. Hofmann, C. Bode and K. Peter, Superparamagnetic iron oxide binding and uptake as imaged by magnetic resonance is mediated by the integrin receptor Mac-1 (CD11b/CD18): implications on imaging of atherosclerotic plaques. *Atherosclerosis.* **193**(1), 102–111 (2007).
69. E. Schulze, J. T. Ferrucci Jr., K. Poss, L. Lapointe, A. Bogdanova and R. Weissleder, Cellular uptake and trafficking of a prototypical magnetic iron oxide label *in vitro. Invest. Radiol.* **30**(10), 604–610 (1995).
70. I. Raynal, P. Prigent, S. Peyramaure, A. Najid, C. Rebuzzi and C. Corot, Macrophage endocytosis of superparamagnetic iron oxide nanoparticles: mechanisms and comparison of ferumoxides and ferumoxtran-10. *Invest. Radiol.* **39**(1), 56–63 (2004).
71. R. D. Oude Engberink, E. L. Blezer, C. D. Dijkstra, Po. S. van der, To. A. van der and H. E. de Vries, Dynamics and fate of USPIO in the central nervous system in

experimental autoimmune encephalomyelitis. *NMR Biomed.* **23**(9), 1087–1096 (2010), LID - 1010.1002/nbm.

72. W. Reichardt, D. Hu-Lowe, D. Torres, R. Weissleder and A. Bogdanov Jr., Imaging of VEGF receptor kinase inhibitor-induced antiangiogenic effects in drug-resistant human adenocarcinoma model. *Neoplasia.* 7(9), 847–853 (2005).
73. A. Moore, E. Marecos, A. Bogdanov Jr. and R. Weissleder, Tumoral distribution of long-circulating dextran-coated iron oxide nanoparticles in a rodent model. *Radiology* **214**(2), 568–574 (2000).
74. S. Metz, G. Bonaterra, M. Rudelius, M. Settles, E. J. Rummeny and H. E. Daldrup-Link, Capacity of human monocytes to phagocytose approved iron oxide MR contrast agents *in vitro*. *Eur. Radiol.* **14**(10), 1851–1858 (2004).
75. G. Stoll, C. Wesemeier, R. Gold, L. Solymosi, K. V. Toyka and M. Bendszus, *In vivo* monitoring of macrophage infiltration in experimental autoimmune neuritis by magnetic resonance imaging. *J. Neuroimmunol.* **149**(1&2), 142–146 (2004).
76. J. Yang, Y. Z. Wadghiri, D. M. Hoang, W. Tsui, Y. Sun, E. Chung, Y. Li, A. Wang, M. de Leon and T. Wisniewski, Detection of amyloid plaques targeted by USPIO-Abeta1–42 in Alzheimer's disease transgenic mice using magnetic resonance microimaging. *Neuroimage* **55**(4), 1600–1609 (2011), LID - 1610.1016/j.ne.
77. S. K. Jo, X. Hu, H. Kobayashi, M. Lizak, T. Miyaji, A. Koretsky and R. A. Star, Detection of inflammation following renal ischemia by magnetic resonance imaging. *Kidney Int.* **64**(1), 43–51 (2003).
78. S. Kanno, P. C. Lee, S. J. Dodd, M. Williams, B. P. Griffith and C. Ho, A novel approach with magnetic resonance imaging used for the detection of lung allograft rejection. *J. Thorac. Cardiovasc. Surg.* **120**(5), 923–934 (2000).
79. Y. X. Wang, S. M. Hussain and G. P. Krestin, Superparamagnetic iron oxide contrast agents: physicochemical characteristics and applications in MR imaging. *Eur. Radiol.* **11**(11), 2319–2331 (2001).
80. P. Reimer and B. Tombach, Hepatic MRI with SPIO: detection and characterization of focal liver lesions. *Eur. Radiol.* **8**(7), 1198–1204 (1998).
81. T. Shen, R. Weissleder, M. Papisov, A. Bogdanov Jr. and T. J. Brady, Monocrystalline iron oxide nanocompounds (MION): physicochemical properties. *Magn. Reson. Med.* **29**(5), 599–604 (1993).
82. A. A. Bogdanov Jr., M. Papisov, R. Weissleder, T. Shen and T. J. Brady, Opsonization of dextran-stabilized iron oxides with plasma proteins. *11th Annual Scientific Meeting of Society of Magnetic Resonance in Medicine SMRM,* Berkley, CA Berlin, Germany, pp. 864 (1992).
83. B. K. Schaffer, C. Linker, M. Papisov, E. Tsai, N. Nossiff, T. Shibata, A. Bogdanov Jr., T. J. Brady and R. Weissleder, MION-ASF: biokinetics of an MR receptor agent. *Magn. Reson. Imaging* **11**(3), 411–417 (1993).
84. H. W. Kang, D. Torres, L. Wald, R. Weissleder and A. A. Bogdanov Jr., Targeted imaging of human endothelial-specific marker in a model of adoptive cell transfer. *Lab. Invest.* **86**(6), 599–609 (2006).
85. I. Y. Zhou, M. M. Cheung, C. Lau, K. C. Chan and E. X. Wu, Balanced steady-state free precession fMRI with intravascular susceptibility contrast agent. *Magn. Reson. Med.* **68**(1), 65–73 (2012), LID - 10.1002/mrm.

86. R. Weissleder, M. Simonova, A. Bogdanova, S. Bredow, W. S. Enochs and A. Bogdanov Jr., MR imaging and scintigraphy of gene expression through melanin induction. *Radiology* **204**(2), 425–429 (1997).
87. TT. Shen, A. Bogdanov, Jr., A. Bogdanova, K. Poss, TJ. Brady, and R. Weissleder. Magnetically labeled secretin retains receptor affinity to pancreas acinar cells. *Bioconjug Chem.* 1996; 7(3):3113–16.
88. M. Taupitz, S. Wagner, J. Schnorr, I. Kravec, H. Pilgrimm, H. Bergmann-Fritsch, and B. Hamm. Phase I clinical evaluation of citrate-coated monocrystalline very small superparamagnetic iron oxide particles as a new contrast medium for magnetic resonance imaging. *Invest Radiol.* 2004;**39**(7):3944–05.
89. M. Torabi, SL. Aquino, and MG. Harisinghani. Current concepts in lymph node imaging. *J Nucl Med.* 2004;**45**(9):15091–518.
90. RW. Ross, AL. Zietman, W. Xie, JJ. Coen, DM. Dahl, WU. Shipley, DS. Kaufman, T. Islam, AR. Guimaraes, R. Weissleder, and M. Harisinghani. Lymphotropic nanoparticle-enhanced magnetic resonance imaging (LNMRI) identifies occult lymph node metastases in prostate cancer patients prior to salvage radiation therapy. *Clin Imaging.* 2009;**33**(4):301–305.
91. WM. Deserno, OA. Debats, T. Rozema, AS. Fortuin, RA. Heesakkers, Y. Hoogeveen, PG. Peer, JO. Barentsz, and van EN. Lin. Comparison of nodal risk formula and MR lymphography for predicting lymph node involvement in prostate cancer. *Int J Radiat Oncol Biol Phys.* 2011;**81**(1):81–5 LID - 10.1016/j.i.
92. JM. Froehlich, M. Triantafyllou, A. Fleischmann, P. Vermathen, GN. Thalmann, and HC. Thoeny. Does quantification of USPIO uptake-related signal loss allow differentiation of benign and malignant normal-sized pelvic lymph nodes? *Contrast Media Mol Imaging.* 2012;7(3):346–355 LID - 310.1002/cmm.
93. M. Triantafyllou, UE. Studer, FD. Birkhauser, A. Fleischmann, LJ. Bains, G. Petralia, A. Christe, JM. Froehlich, and HC. Thoeny. Ultrasmall superparamagnetic particles of iron oxide allow for the detection of metastases in normal sized pelvic lymph nodes of patients with bladder and/or prostate cancer. *Eur J Cancer.* 2013;**49**(3):616–624 LID - 610.1016/j.e.
94. R. Guimaraes, O. Clement, J. Bittoun, F. Carnot and G. Frija, MR lymphography with superparamagnetic iron nanoparticles in rats: pathologic basis for contrast enhancement. *AJR Am. J. Roentgenol.* **162**(1), 201–207 (1994).
95. G. Barriere, A. Riouallon, J. Renaudie, M. Tartary and M. Rigaud, Mesenchymal and stemness circulating tumor cells in early breast cancer diagnosis. *BMC Cancer* **12,** 114 (2012).
96. P. Vassallo, C. Matei, W. D. Heston, S. J. McLachlan, J. A. Koutcher and R. A. Castellino, AMI-227-enhanced MR lymphography: usefulness for differentiating reactive from tumor-bearing lymph nodes. *Radiology* **193**(2), 501–506 (1994).
97. R. Guimaraes, O. Clement, J. Bittoun, F. Carnot and G. Frija, MR lymphography with superparamagnetic iron nanoparticles in rats: pathologic basis for contrast enhancement. *AJR Am. J. Roentgenol.* **162**(1), 201–207 (1994).
98. Y. Anzai, K. E. Blackwell, S. L. Hirschowitz, J. W. Rogers, Y. Sato, W. T. Yuh, V. M. Runge, M. R. Morris, S. J. McLachlan and R. B. Lufkin, Initial clinical experience with dextran-coated superparamagnetic iron oxide for detection of lymph node metastases in patients with head and neck cancer. *Radiology* **192**(3), 709–715 (1994).

99. O. Will, S. Purkayastha, C. Chan, T. Athanasiou, A. W. Darzi, W. Gedroyc and P. Tekkis, Diagnostic precision of nanoparticle-enhanced MRI for lymph-node metastases: a meta-analysis. *Lancet Oncol.* 7(1), 52–60 (2006).
100. S. C. Michel, T. M. Keller, J. M. Frohlich, D. Fink, R. Caduff, B. Seifert, B. Marincek and R. A. Kubik-Huch, Preoperative breast cancer staging: MR imaging of the axilla with ultrasmall superparamagnetic iron oxide enhancement. *Radiology* **225**(2), 527–536 (2002).
101. T. W. Stadnik, H. Everaert, S. Makkat, R. Sacre, J. Lamote and C. Bourgain, Breast imaging. Preoperative breast cancer staging: comparison of USPIO-enhanced MR imaging and 18F-fluorodeoxyglucose (FDC) positron emission tomography (PET) imaging for axillary lymph node staging — initial findings. *Eur. Radiol.* **16**(10), 2153–2160 (2006).
102. M. Memarsadeghi, C. C. Riedl, A. Kaneider, A. Galid, M. Rudas, W. Matzek and T. H. Helbich, Axillary lymph node metastases in patients with breast carcinomas: assessment with nonenhanced versus uspio-enhanced MR imaging. *Radiology* **241**(2), 367–377 (2006).
103. M. G. Harisinghani, M. A. Saksena, P. F. Hahn, B. King, J. Kim, M. T. Torabi and R. Weissleder, Ferumoxtran-10-enhanced MR lymphangiography: does contrast-enhanced imaging alone suffice for accurate lymph node characterization? *AJR Am. J. Roentgenol.* **186**(1), 144–148 (2006).
104. R. A. Heesakkers, A. M. Hovels, G. J. Jager, H. C. van den Bosch, J. A. Witjes, H. P. Raat, J. L. Severens, E. M. Adang, C. H. van der Kaa, J. J. Futterer and J. Barentsz, MRI with a lymph-node-specific contrast agent as an alternative to CT scan and lymph-node dissection in patients with prostate cancer: a prospective multicohort study. *Lancet Oncol.* **9**(9), 850–856 (2008).
105. M. J. Lahaye, S. M. Engelen, A. G. Kessels, A. P. de Bruine, M. F. von Meyenfeldt, J. M. van Engelshoven, C. J. van de Velde, G. L. Beets and R. G. Beets-Tan, USPIO-enhanced MR imaging for nodal staging in patients with primary rectal cancer: predictive criteria. *Radiology* **246**(3), 804–811 (2008).
106. M. A. Saksena, J. Y. Kim and M. G. Harisinghani, Nodal staging in genitourinary cancers. *Abdom. Imaging* **31**(6), 644–651 (2006).
107. M. G. Mack, J. O. Balzer, R. Straub, K. Eichler and T. J. Vogl, Superparamagnetic iron oxide-enhanced MR imaging of head and neck lymph nodes. *Radiology* **222**(1), 239–244 (2002).
108. R. Sigal, T. Vogl, J. Casselman, G. Moulin, F. Veillon, R. Hermans, F. Dubrulle, J. Viala, J. Bosq, M. Mack, M. Depondt, C. Mattelaer, P. Petit, P. Champsaur, S. Riehm, Y. Dadashitazehozi, T. de Jaegere, G. Marchal, D. Chevalier, L. Lemaitre, C. Kubiak, R. Helmberger and P. Halimi, Lymph node metastases from head and neck squamous cell carcinoma: MR imaging with ultrasmall superparamagnetic iron oxide particles (Sinerem MR) — results of a phase-III multicenter clinical trial. *Eur. Radiol.* **12**(5), 1104–1113 (2002).
109. L. Curvo-Semedo, M. Diniz, J. Migueis, M. J. Juliao, P. Martins, A. Pinto and F. Caseiro-Alves, USPIO-enhanced magnetic resonance imaging for nodal staging in patients with head and neck cancer. *J. Magn. Reson. Imaging* **24**(1), 123–131 (2006).
110. R. Weissleder, G. Elizondo, J. Wittenberg, A. S. Lee, L. Josephson and T. J. Brady, Ultrasmall superparamagnetic iron oxide: an intravenous contrast agent for assessing lymph nodes with MR imaging. *Radiology* **175**(2), 494–498 (1990).

111. M. A. Saksena, A. Saokar and M. G. Harisinghani, Lymphotropic nanoparticle enhanced MR imaging (LNMRI) technique for lymph node imaging. *Eur. J. Radiol.* **58**(3), 367–374 (2006).
112. N. Gharehaghaji, M. A. Oghabian, S. Sarkar, S. Amirmohseni and H. Ghanaati, Optimization of pulse sequences in magnetic resonance lymphography of axillary lymph nodes using magnetic nanoparticles. *J. Nanosci. Nanotechnol.* **9**(7), 4448–4452 (2009).
113. Y. Y. Shih, Y. H. Hsu, T. Q. Duong, S. S. Lin, K. P. Chow and C. Chang, Longitudinal study of tumor-associated macrophages during tumor expansion using MRI. *NMR Biomed.* **24**(10), 1353–1360 (2011).
114. H. R. Cho, S. H. Choi, N. Lee, T. Hyeon, H. Kim and W. K. Moon, Macrophages homing to metastatic lymph nodes can be monitored with ultrasensitive ferromagnetic iron-oxide nanocubes and a 1.5T clinical MR scanner. *PLoS One* **7**(1), e29575 (2012).
115. T. Liu, H. Zhou, R. Xia, J. Liao, C. Wu, H. Wang, H. Ai, F. Bi and F. Gao, Tracking tumor cells in lymphatics in a mice xenograft model by magnetic resonance imaging. *Mol. Imaging* **11**(6), 451–460 (2012).
116. S. Valable, E. L. Barbier, M. Bernaudin, S. Roussel, C. Segebarth, E. Petit and C. Remy, *In vivo* MRI tracking of exogenous monocytes/macrophages targeting brain tumors in a rat model of glioma. *Neuroimage* **40**(2), 973–983 (2008).
117. L. Wu, Y. Cao, C. Liao, J. Huang and F. Gao, Diagnostic performance of USPIO-enhanced MRI for lymph-node metastases in different body regions: a meta-analysis. *Eur. J. Radiol.* **80**(2), 582–589 (2011), LID - 510.1016/j.ej.
118. S. E. Harnan, K. L. Cooper, Y. Meng, S. E. Ward, P. Fitzgerald, D. Papaioannou, C. Ingram, E. Lorenz, I. D. Wilkinson and L. Wyld, Magnetic resonance for assessment of axillary lymph node status in early breast cancer: a systematic review and meta-analysis. *Eur. J. Surg. Oncol.* **37**(11), 928–936 (2011), LID - 910.1016/j.e.
119. M. F. Bellin, C. Roy, K. Kinkel, D. Thoumas, S. Zaim, D. Vanel, C. Tuchmann, F. Richard, D. Jacqmin, A. Delcourt, E. Challier, T. Lebret and P. Cluzel, Lymph node metastases: safety and effectiveness of MR imaging with ultrasmall superparamagnetic iron oxide particles — initial clinical experience. *Radiology* **207**(3), 799–808 (1998).
120. H. K. Pannu, K. P. Wang, T. L. Borman and D. A. Bluemke, MR imaging of mediastinal lymph nodes: evaluation using a superparamagnetic contrast agent. *J. Magn. Reson. Imaging* **12**(6), 899–904 (2000).
121. D. Suzuki, M. Yamaguchi, T. Furuta, Y. Okuyama, K. Yoshikawa and H. Fujii, Central high signal in inflammatory swollen lymph nodes on SPIO-enhanced interstitial MR lymphograms: a mimic of lymph node metastasis. *Magn. Reson. Med. Sci.* **11**(1), 61–63 (2012).
122. S. H. Choi, K. H. Kim, W. K. Moon, H. C. Kim, J. H. Cha, J. H. Paik and K. H. Chang, Comparison of lymph node metastases assessment with the use of USPIO-enhanced MR imaging at 1.5 T versus 3.0 T in a rabbit model. *J. Magn. Reson. Imaging* **31**(1), 134–141 (2010), LID - 110.1002/jmr.
123. Y. Mori, M. Umeda, M. Fukunaga, K. Ogasawara and Y. Yoshioka, MR contrast in mouse lymph nodes with subcutaneous administration of iron oxide particles: size dependency. *Magn. Reson. Med. Sci.* **10**(4), 219–227 (2011).
124. S. Kinner, S. Maderwald, J. Albert, N. Parohl, C. Corot, P. Robert, H. A. Baba and J. Barkhausen, Comparison of two different iron oxide-based contrast agents for discrimination of benign and malignant lymph nodes. *Invest. Radiol.* **47**(9), 511–515 (2012), LID - 510.1097/RLI.

125. R. Ross, Atherosclerosis — an inflammatory disease. *N. Engl. J. Med.* **340**(2), 115–126 (1999).
126. B. H. Maskrey, I. L. Megson, P. D. Whitfield and A. G. Rossi, Mechanisms of resolution of inflammation: a focus on cardiovascular disease. *Arterioscler. Thromb. Vasc. Biol.* **31**(5), 1001–1006 (2011).
127. S. Tsimikas and Y. I. Miller, Oxidative modification of lipoproteins: mechanisms, role in inflammation and potential clinical applications in cardiovascular disease. *Curr. Pharm. Des.* **17**(1), 27–37 (2011).
128. P. Libby, Inflammation and cardiovascular disease mechanisms. *Am. J. Clin. Nutr.* **83**(2), 456S–460S (2006).
129. N. Nighoghossian, L. Derex and P. Douek, The vulnerable carotid artery plaque: current imaging methods and new perspectives. *Stroke* **36**(12), 2764–2772 (2005).
130. T. Y. Tang, S. P. Howarth, S. R. Miller, M. J. Graves, J. M. UK-I, Z. Y. Li, S. R. Walsh, A. J. Patterson, P. J. Kirkpatrick, E. A. Warburton, K. Varty, M. E. Gaunt and J. H. Gillard, Correlation of carotid atheromatous plaque inflammation using USPIO-enhanced MR imaging with degree of luminal stenosis. *Stroke* **39**(7), 2144–2147.
131. S. A. Schmitz, M. Taupitz, S. Wagner, K. J. Wolf, D. Beyersdorff and B. Hamm, Magnetic resonance imaging of atherosclerotic plaques using superparamagnetic iron oxide particles. *J. Magn. Reson. Imaging* **14**(4), 355–361 (2001).
132. S. G. Ruehm, C. Corot, P. Vogt, S. Kolb and J. F. Debatin, Magnetic resonance imaging of atherosclerotic plaque with ultrasmall superparamagnetic particles of iron oxide in hyperlipidemic rabbits. *Circulation* **103**(3), 415–422 (2001).
133. R. A. Trivedi, J. M. UK-I, M. J. Graves, J. J. Cross, J. Horsley, M. J. Goddard, J. N. Skepper, G. Quartey, E. Warburton, I. Joubert, L. Wang, P. J. Kirkpatrick, J. Brown and J. H. Gillard, *In vivo* detection of macrophages in human carotid atheroma: temporal dependence of ultrasmall superparamagnetic particles of iron oxide-enhanced MRI. *Stroke* **35**(7), 1631–1635 (2004).
134. S. P. Howarth, T. Y. Tang, M. J. Graves, J. M. UK-I, Z. Y. Li, S. R. Walsh, M. E. Gaunt and J. H. Gillard, Non-invasive MR imaging of inflammation in a patient with both asymptomatic carotid atheroma and an abdominal aortic aneurysm: a case report. *Ann. Surg. Innov. Res.* **1**, 4 (2007).
135. S. P. Howarth, T. Y. Tang, R. Trivedi, R. Weerakkody, J. UK-I, M. E. Gaunt, J. R. Boyle, Z. Y. Li, S. R. Miller, M. J. Graves and J. H. Gillard, Utility of USPIO-enhanced MR imaging to identify inflammation and the fibrous cap: a comparison of symptomatic and asymptomatic individuals. *Eur. J. Radiol.* **70**(3), 555–560 (2009).
136. T. Y. Tang, A. J. Patterson, S. R. Miller, M. J. Graves, S. P. Howarth, J. M. UK-I, Z. Y. Li, U. Sadat, V. E. Young, S. R. Walsh, J. R. Boyle, M. E. Gaunt and J. H. Gillard, Temporal dependence of *in vivo* USPIO-enhanced MRI signal changes in human carotid atheromatous plaques. *Neuroradiology* **51**(7), 457–465 (2009).
137. T. Y. Tang, S. P. Howarth, S. R. Miller, M. J. Graves, J. M. UK-I, Z. Y. Li, S. R. Walsh, P. D. Hayes, K. Varty and J. H. Gillard, Comparison of the inflammatory burden of truly asymptomatic carotid atheroma with atherosclerotic plaques in patients with asymptomatic carotid stenosis undergoing coronary artery bypass grafting: an ultrasmall superparamagnetic iron oxide enhanced magnetic resonance study. *Eur. J. Vasc. Endovasc. Surg.* **35**(4), 392–398 (2008).
138. A. D. Yancy, A. R. Olzinski, T. C. Hu, S. C. Lenhard, K. Aravindhan, S. M. Gruver, P. M. Jacobs, R. N. Willette and B. M. Jucker, Differential uptake of ferumoxtran-10 and

ferumoxytol, ultrasmall superparamagnetic iron oxide contrast agents in rabbit: critical determinants of atherosclerotic plaque labeling. *J. Magn. Reson. Imaging* **21**(4), 432–442 (2005).

139. C. U. Herborn, F. M. Vogt, T. C. Lauenstein, O. Dirsch, C. Corot, P. Robert and S. G. Ruehm, Magnetic resonance imaging of experimental atherosclerotic plaque: comparison of two ultrasmall superparamagnetic particles of iron oxide. *J. Magn. Reson. Imaging* **24**(2), 388–393 (2006).
140. K. Morishige, D. F. Kacher, P. Libby, L. Josephson, P. Ganz, R. Weissleder and M. Aikawa, High-resolution magnetic resonance imaging enhanced with superparamagnetic nanoparticles measures macrophage burden in atherosclerosis. *Circulation* **122**(17), 1707–1715 (2010), LID - 1710.1161/CIR.
141. T. Y. Tang, R. R. Moustafa, S. P. Howarth, S. R. Walsh, J. R. Boyle, Z. Y. Li, J. C. Baron, J. H. Gillard and E. A. Warburton, Combined PET-FDG and USPIO-enhanced MR imaging in patients with symptomatic moderate carotid artery stenosis. *Eur. J. Vasc. Endovasc. Surg.* **36**(1), 53–55 (2008).
142. A. Millon, S. D. Dickson, A. Klink, D. Izquierdo-Garcia, J. Bini, E. Lancelot, S. Ballet, P. Robert, J. Mateo de Castro, C. Corot and Z. A. Fayad, Monitoring plaque inflammation in atherosclerotic rabbits with an iron oxide (P904) and F-FDG using a combined PET/MR scanner. *Atherosclerosis* (2013).
143. S. Metz, A. J. Beer, M. Settles, J. Pelisek, R. M. Botnar, E. J. Rummeny and P. Heider, Characterization of carotid artery plaques with USPIO-enhanced MRI: assessment of inflammation and vascularity as *in vivo* imaging biomarkers for plaque vulnerability. *Int. J. Cardiovasc. Imaging* **27**(6), 901–912 (2011), LID - 910.1007/s1010.
144. D. R. Owen, A. C. Lindsay, R. P. Choudhury and Z. A. Fayad, Imaging of atherosclerosis. *Ann. Rev. Med.* **62,** 25–40 (2011).
145. A. J. Degnan, A. J. Patterson, T. Y. Tang, S. P. Howarth and J. H. Gillard, Evaluation of ultrasmall superparamagnetic iron oxide-enhanced MRI of carotid atherosclerosis to assess risk of cerebrovascular and cardiovascular events: follow-up of the ATHEROMA trial. *Cerebrovasc. Dis.* **34**(2), 169–173 (2012).
146. A. J. Patterson, T. Y. Tang, M. J. Graves, K. H. Muller and J. H. Gillard, *In vivo* carotid plaque MRI using quantitative T2* measurements with ultrasmall superparamagnetic iron oxide particles: a dose-response study to statin therapy. *NMR Biomed.* **24**(1), 89–95 (2011).
147. F. Hyafil, L. Feldman, D. Le Guludec and Z. A. Fayad, Evaluating efficacy of pharmaceutical interventions in atherosclerosis: role of magnetic resonance imaging and positron emission tomography. *Mt. Sinai. J. Med.* **79**(6), 689–704 (2012).
148. J. M. Richards, S. I. Semple, T. J. MacGillivray, C. Gray, J. P. Langrish, M. Williams, M. Dweck, W. Wallace, G. McKillop, R. T. Chalmers, OJ. Garden, and DE. Newby. Abdominal aortic aneurysm growth predicted by uptake of ultrasmall superparamagnetic particles of iron oxide: a pilot study. *Circ. Cardiovasc. Imaging* **4**(3), 274–281 (2011), LID - 210.1161/CIR.
149. M. Truijers, J. J. Futterer, S. Takahashi, R. A. Heesakkers, J. D. Blankensteijn and J. O. Barentsz, *In vivo* imaging of the aneurysm wall with MRI and a macrophage-specific contrast agent. *AJR Am. J. Roentgenol.* **193**(5), W437–W441 (2009), LID - 410.2214/AJR.
150. A. Nchimi, O. Defawe, D. Brisbois, T. K. Broussaud, J. O. Defraigne, P. Magotteaux, B. Massart, J. M. Serfaty, X. Houard, J. B. Michel and N. Sakalihasan, MR imaging of

iron phagocytosis in intraluminal thrombi of abdominal aortic aneurysms in humans. *Radiology* **254**(3), 973–981 (2010).

151. G. Korosoglou, R. G. Weiss, D. A. Kedziorek, P. Walczak, W. D. Gilson, M. Schar, D. E. Sosnovik, D. L. Kraitchman, R. C. Boston, J. W. Bulte, R. Weissleder and M. Stuber, Noninvasive detection of macrophage-rich atherosclerotic plaque in hyperlipidemic rabbits using "positive contrast" magnetic resonance imaging. *J. Am. Coll. Cardiol.* **52**(6), 483–491 (2008).
152. C. Corot, P. Robert, J. M. Idee and M. Port, Recent advances in iron oxide nanocrystal technology for medical imaging. *Adv. Drug Deliv. Rev.* **58**(14), 1471–1504 (2006).
153. C. Burtea, S. Ballet, S. Laurent, O. Rousseaux, A. Dencausse, W. Gonzalez, M. Port, C. Corot, L. Vander Elst and R. N. Muller, Development of a magnetic resonance imaging protocol for the characterization of atherosclerotic plaque by using vascular cell adhesion molecule-1 and apoptosis-targeted ultrasmall superparamagnetic iron oxide derivatives. *Arterioscler. Thromb. Vasc. Biol.* **32**(6), e36–e48 (2012), LID - 10.1161/ATV.
154. M. A. McAteer, K. Mankia, N. Ruparelia, A. Jefferson, H. B. Nugent, L. A. Stork, K. M. Channon, J. E. Schneider and R. P. Choudhury, A leukocyte-mimetic magnetic resonance imaging contrast agent homes rapidly to activated endothelium and tracks with atherosclerotic lesion macrophage content. *Arterioscler. Thromb. Vasc. Biol.* **32**(6), 1427–1435 (2012).
155. S. Wen, D. F. Liu, Z. Liu, S. Harris, Y. Y. Yao, Q. Ding, F. Nie, T. Lu, H. J. Chen, Y. L. An, F. C. Zang and G. J. Teng, OxLDL-targeted iron oxide nanoparticles for *in vivo* MRI detection of perivascular carotid collar induced atherosclerotic lesions in ApoE-deficient mice. *J. Lipid Res.* **53**(5), 829–838 (2012), LID - 810.1194/jlr.
156. Z. Yu, J. Geng, Y. Tan, Q. Wang, J. Zhu and M. Zhang, Non-invasive assessment of acute vascular inflammation after PCI using USPIO enhanced MRI *in vivo*. *Int. J. Cardiol.* **151**(1), 110–112 (2011), LID - 110.1016/j.ij.
157. T. D. Farr, J. U. Seehafer, M. Nelles and M. Hoehn, Challenges towards MR imaging of the peripheral inflammatory response in the subacute and chronic stages of transient focal ischemia. *NMR Biomed.* **24**(1), 35–45 (2011), LID - 10.1002/nbm.
158. M. Sigovan, A. Bessaad, H. Alsaid, E. Lancelot, C. Corot, B. Neyran, N. Provost, Z. Majd, M. Breisse and E. Canet-Soulas, Assessment of age modulated vascular inflammation in ApoE-/- mice by USPIO-enhanced magnetic resonance imaging. *Invest. Radiol.* **45**(11), 702–707 (2010), LID - 710.1097/RLI.
159. R. Weissleder, H. C. Cheng, A. Bogdanova and AQ. Bogdanov Jr., Magnetically labeled cells can be detected by MR imaging. *J. Magn. Reson. Imaging* 7(1), 258–263 (1997).
160. E. M. Shapiro, S. Skrtic and A. P. Koretsky, Sizing it up: cellular MRI using micron-sized iron oxide particles. *Magn. Reson. Med.* **53**(2), 329–338 (2005).
161. J. B. Williams, Q. Ye, T. K. Hitchens, C. L. Kaufman and C. Ho, MRI detection of macrophages labeled using micrometer-sized iron oxide particles. *J. Magn. Reson. Imaging* **25**(6), 1210–1218 (2007).
162. P. Walczak, J. Ruiz-Cabello, D. A. Kedziorek, A. A. Gilad, S. Lin, B. Barnett, L. Qin, H. Levitsky and J. W. Bulte, Magnetoelectroporation: improved labeling of neural stem cells and leukocytes for cellular magnetic resonance imaging using a single FDA-approved agent. *Nanomedicine* **2**(2), 89–94 (2006).
163. N. Mascheri, R. Dharmakumar, Z. Zhang, T. Paunesku, G. Woloschak and D. Li, Fast low-angle positive contrast steady-state free precession imaging of USPIO-labeled

macrophages: theory and *in vitro* experiment. *Magn. Reson. Imaging* **27**(7), 961–969 (2009).

164. A. Bessaad, M. Sigovan, H. Alsaid, G. De Souza, N. Provost, Z. Majd, C. Menager, J. Honnorat, F. Lagarde, N. Nighoghossian, S. Nataf and E. Canet-Soulas, M1-activated macrophages migration, a marker of aortic atheroma progression: a preclinical MRI study in mice. *Invest. Radiol.* **45**(5), 262–269 (2010).
165. Q. Zhou, K. R. Yang, P. Gao, W. L. Chen, D. Y. Yang, M. J. Liang and L. Zhu, An experimental study on MR imaging of atherosclerotic plaque with SPIO marked endothelial cells in a rabbit model. *J. Magn. Reson. Imaging* **34**(6), 1325–1332 (2011).
166. T. Skajaa, D. P. Cormode, P. A. Jarzyna, A. Delshad, C. Blachford, A. Barazza, E. A. Fisher, R. E. Gordon, Z. A. Fayad and W. J. Mulder, The biological properties of iron oxide core high-density lipoprotein in experimental atherosclerosis. *Biomaterials* **32**(1), 206–213 (2011).
167. N. Miyama, M. M. Dua, G. M. Schultz, H. Kosuge, M. Terashima, L. J. Pisani, R. L. Dalman and M. V. McConnell, Bioluminescence and magnetic resonance imaging of macrophage homing to experimental abdominal aortic aneurysms. *Mol. Imaging* **11**(2), 126–134 (2012).
168. J. S. Lee, H. J. Kang, G. Gong, H. D. Jung, K. H. Lim, S. T. Kim and T. H. Lim, MR imaging of *in vivo* recruitment of iron oxide-labeled macrophages in experimentally induced soft-tissue infection in mice. *Radiology* **241**(1), 142–148 (2006).
169. J. S. Lee, J. Y. Sohn, H. D. Jung, S. T. Kim, K. G. Lee and H. J. Kang, Pace of macrophage recruitment during different stages of soft tissue infection: semi-quantitative evaluation by *in vivo* magnetic resonance imaging. *Eur. Radiol.* **18**(10), 2033–2039 (2008).
170. A. Al Faraj, N. Luciani, J. Kolosnjaj-Tabi, E. Mattar, O. Clement, C. Wilhelm and F. Gazeau, Real-time high-resolution magnetic resonance tracking of macrophage subpopulations in a murine inflammation model: a pilot study with a commercially available cryogenic probe. *Contrast Media Mol. Imaging* **8**(2), 193–203 (2013).
171. Q. Y. Cai, H. Lee, E. J. Kim, H. Moon, K. Chang, J. Rho and K. S. Hong, Magnetic resonance imaging of superparamagnetic iron oxide-labeled macrophage infiltrates in acute-phase renal ischemia-reperfusion mouse model. *Nanomedicine* **8**(3), 365–373 (2012).
172. E. Y. Chae, E. J. Song, J. Y. Sohn, S. T. Kim, C. W. Woo, G. Gong, H. J. Kang and J. S. Lee, Allogeneic renal graft rejection in a rat model: *in vivo* MR imaging of the homing trait of macrophages. *Radiology* **256**(3), 847–854 (2010).
173. M. Bendszus and G. Stoll, Caught in the act: *in vivo* mapping of macrophage infiltration in nerve injury by magnetic resonance imaging. *J. Neurosci.* **23**(34), 10892–10896 (2003).
174. K. C. Briley-Saebo, Y. S. Cho, P. X. Shaw, S. K. Ryu, V. Mani, S. Dickson, E. Izadmehr, S. Green, Z. A. Fayad and S. Tsimikas, Targeted iron oxide particles for *in vivo* magnetic resonance detection of atherosclerotic lesions with antibodies directed to oxidation-specific epitopes. *J. Am. Coll. Cardiol.* **57**(3), 337–347 (2011).

Chapter 3

Nanoparticulate Contrast Agents for CT, SPECT and PET Imaging

Suna Erdogan and Mine Silindir

Hacettepe University, Faculty of Pharmacy,
Department of Radiopharmacy, 06100 Sıhhıye-Ankara, Turkey
serdogan@hacettepe.edu.tr

Abstract

The application of nanotechnology in non-invasive diagnosis comprises many fields such as diagnostic radiology (i.e. Computed tomography (CT), and nuclear medicine [i.e. Radionuclidic Imaging, Single Photon Emission Computed Tomography (SPECT) and Positron Emission Tomography (PET)] for imaging, characterization, staging and therapy monitoring. Medical diagnostic imaging requires accumulation of contrast agent in the target site, often achieved most efficiently by steering targeted nanoparticles containing contrast agent into the target. For imaging modalities with low sensitivity, nanoparticles bearing multiple contrast groups provide signal amplification. The materials used to synthesize diagnostic nanoparticles include lipids, polymers, and metals. Signal generators incorporated into nanoparticles include iodine, gold, bismuth, and radionuclides. Recently, multimodal imaging techniques such as SPECT/CT, PET/CT, and SPECT/CT/MRI are becoming popular as imaging modalities depending on the ability to detect functional information and anatomical information about the disease within a single imaging procedure in less time.

1. Introduction

Nanotechnology is improving into a broad application in many different fields, depending on the latest investigations and improvements in the fields of chemistry, physics, and biology.[1] The application of nanotechnology in non-invasive diagnosis comprises many fields such as diagnostic radiology and nuclear medicine for imaging, characterization, staging, and therapy monitoring.

The probes in nanotechnology can be liposomes, lipid and block co-polimer micelles, polymeric nanoparticles, metal nanoparticles, quantum dots, dendrimers and different nano-assemblies. Their use in diagnostic and/or therapeutic field aims to target the diseased area, prevent undesirable side effects, increase the fraction of contrast agent and/or drug delivered in the pathological area, perform *in vitro/in vivo* stability of radiolabeled and/or contrast encapsulated agent and obtain better contrast signal to accurately diagnose the disease. In general, diagnostic drug carriers are expected to be biodegradable, easily formulated, reasonably cheap, small in particle size, high loading capacity, prolonged circulation time and, ideally, specifically or non-specifically accumulation in required diseased sites within the body. By design of nanosized, hydrophilic polymer (such as PEG) coated and active targeted nanocarrier delivery system to molecular targets such as receptors, antigens, hormones related with the disease which can be formed during the disease period, increase or decrease or even disappeared, molecular imaging can also be performed recently.[2,3]

Diagnostic nanocarriers can be targeted to desired tissue and organ by passive and active targeting. But one of the most important properties of any diagnostic nanocarriers is short blood circulation time of these nanocarriers resulting from rapid and efficient recognition and removal from blood by cells of the mononuclear phagocyte system (MPS), particularly those in the liver and spleen. When imaging agent containing nanocarriers applied to blood circulation, plasma proteins so-called opsonins recognize and adhere to surface of these particles, provoking the non-specific elimination (phagocytosis) of all exogenous material in the circulation by MPS-macrophages.[4] Since they are rapidly removed from the circulation, they do not have any chance to reach the action of site. That is why design of long-circulating carriers is currently an important and still growing area of biomedical application.[4–7]

Long circulated nanocarriers can be targeted by passive targeting. Passive targeting refers to the accumulation of drug or drug-carrier system at a particular site due to physicochemical factors such as size, surface charge and surface hydrophobicity or pathophysiological opportunities such as enhanced permeability and retention (EPR) effect and pharmacological factors (Fig. 1).[8,9]

To achieve a better selective targeting, tremendous effort has been directed toward the development of active approaches for drug targeting. Different targeting agents such as antibody,[10–12] aptamers,[13–15] albumin,[16,17]

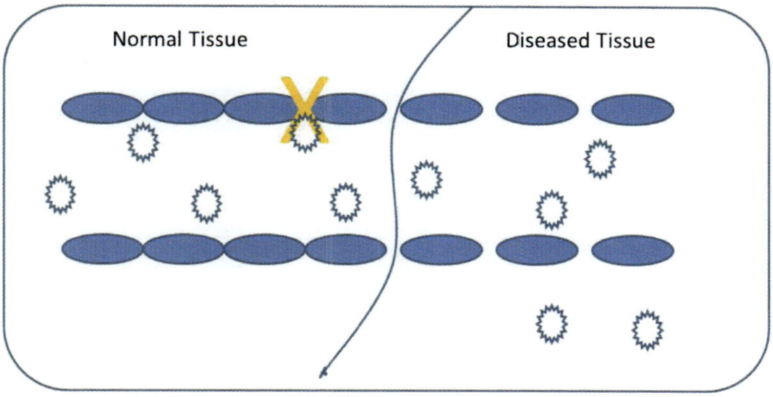

Fig. 1. Long-circulating drug carriers penetrate through the leaky pathological vasculature and result in its high local concentration by EPR effect.

transferrin,[18,19] lectin,[20,21] lipoprotein,[22,23] folic acid,[24–27] and saccharide[28] having selective affinity for recognizing and interacting with a specific cell, tissue or organ in the body have been used for active targeting.[29]

Medical diagnostic imaging modalities can be basically classified under four main sections; Computed tomography (CT), Magnetic Resonance Imaging (MRI), Radionuclidic Imaging [scintigraphy, Single Photon Emission Computed Tomography (SPECT) and Positron Emission Tomography (PET)] and Ultrasonography (US), each associated with strengths and limitations. For example, PET is a highly sensitive radionuclide based imaging technique but spatially limited in resolution, CT is comparatively insensitive methods of imaging, but present high spatial resolution. In this chapter, we will discuss nanoparticle based contrast agent used for CT, SPECT, and PET imaging modalities.

2. CT and Contrast Agents Used for CT

Nowadays, CT is one of the leading radiological technologies in the biomedical imaging field. The discrimination competence of different tissues in CT depends on the fact that X-ray attenuation of different tissues. CT provides high quality imaging of bone tissues due to the contrast difference within the electron dense bones and soft tissues.[30]

Soft living tissues present very close densities, making it difficult to clearly distinguish them. It is, therefore, essential to use contrast media based on heavy elements, like iodine or barium, which allow significant enhancement of the differences in contrast between anatomic compartments.[31]

To obtain better CT contrast, the key point is the selection of CT contrast with high atomic number. Contrast agent with high atomic number enable imaging in lesser doses and with better sensitivity and specificity.

Another significant parameter in the development of CT contrast agent is relatively low sensitivity of CT. The tissue concentration that must be achieved for successful imaging is relatively high (about 10^{-2} M). Therefore, in order to induce sufficient contrast in desired tissue/organs, a larger amount of molecules with high atomic weight is needed, since CT contrast is directly proportional with the total amount of high atomic number molecules in voxel.

Radiocontrast agents are a type of medical contrast medium used to improve the visibility of internal bodily structures in X-ray based imaging techniques such as CT.

Modern intravenous (*i.v.*) contrast agents are typically based on a tri-iodinated benzene ring. This may be bound either in an organic (non-ionic) compound or an ionic compound. Ionic agents were developed first and are still in widespread use. Non-ionic (organic) compounds that dissolve in water but do not dissociate are covalently bind iodine. The most recent class of agents is dimers that consist of a molecule with two benzene rings (again, each with three iodine atoms) that does not dissociate in water (non-ionic).

Significant improvements in their composition during the past few decades have made them safer and better tolerated. Nonetheless, risks associated with contrast agents have not been eliminated, and adverse reactions of varying degree continue to occur. Many of the side effects are due to the hyperosmolar solution. Hence, with fewer particles in solution, non-ionic compounds are designated low or iso-osmolar contrast media and cause fewer side effects. Adding calcium ions and altering the number and distribution of –OH ions are other additional modification to reduce the cardiac and neural toxicity, respectively.

Another limitation of currently used iodine based contrast agent is short blood half-life. They are cleared by glomerular filtration and their circulatory half-life is about 1–2 h, assuming normal renal function.

2. SPECT and PET Imaging and Radiotracers

SPECT which is a radionuclide-based imaging technique is a proper modality to assess diseases on the basis of functional and metabolic information obtained from organs and cells and even in molecular basis.[32]

SPECT images are acquired by photon collection and subsequent mapping of radiotracer distribution, this modality is characterized by radioisotopes emitting a single high-energy gamma photon. The sensitivity and spatial resolution of SPECT are lower than other radionuclide based imaging modalities so-called PET described below.

One of the most significant superiority of radiolabeling drugs and drug delivery systems with SPECT tracers is the cost-effectiveness. Second one is the easier production of SPECT tracers in small radiopharmacy and hospital Nuclear Medicine departments without need to any advanced centers such as cyclotrons required for PET tracers. Some simply prepared SPECT tracers such as commonly used ^{99m}Tc-radiopharmaceuticals can give the opportunity to image, stage and differentiate cancers and monitor therapy easily. The radioisotopes (i.e. ^{133}Xe, ^{99m}Tc and ^{123}I) are more readily available and typically have longer decay times than those used in PET.

PET is a nuclear medical imaging technique that produces a three-dimensional image or picture of functional processes in the body. The system detects pairs of gamma rays emitted indirectly by a positron-emitting radionuclide (tracer), which is introduced into the body on a biologically active molecule.[33,34]

To acquire PET images, the emitted positrons pass through the tissues of the body and interact with the electrons and nuclei by the side of atoms. When one of the positrons combines with an electron, a pair forms and this pair annihilates and converts into two photons with high energy (each having 511 keV). These two photons convey 180° apart. PET device determines both of the coincident photons having 511 keV. Although clinical PET scanners are proper to obtain better images of the patient with a spatial resolution of 5 mm, small animal PET tomographs are available to obtain the image of the desired tissue/organ of the small animal for preclinical investigations with a spatial resolution of 1.5 mm. Although small animal PET tomography has relatively low spatial resolution, it has higher molecular sensitivity among 10^{-11}–10^{-12} mol/L.[35,36]

PET provides quantitative and highly sensitive imaging and also it is able to obtain an accurate, reproducible data for the targeted probe, molecular information such as receptor expression and specificity, absorption, *in vivo* biodistribution, specific targeting, blood or tissue residence times, bioavailability, metabolism, clearance, quantitative data, and radiation dosimetry.[37]

Although generally short half-lived positron emitting radionuclides are routinely used in PET clinics such as ^{11}C, ^{13}N, ^{15}O, and ^{18}F, in research long half lived positron emitting radionuclides like ^{64}Cu, ^{68}Ga, ^{86}Y, and ^{89}Zr are used for monitoring biomolecular targeting vectors and for tracing antibodies, oligopeptides, or oligonucleotides. The concept of radiometallation varies from ^{68}Ga-labeled pentapeptides to ^{89}Zr-labeled mAbs. Apart from these positron emitting radionuclides, ^{64}Cu is a promising one which is especially used for antibody-targeted imaging.[38,39]

SPECT and PET are more sensitive rather than other modalities such as optical imaging, CT and MRI. SPECT and PET have higher sensitivity and ability to detect in picomolar levels and while they can provide quantitative detection, no tissue penetration limit was detected. Nonetheless, the drawback of these radionuclide detecting modalities is their resolution which is not as high as other imaging modalities such as MRI and CT.[40]

3. Nanoparticulate Drug Delivery Systems as Contrast Agents

The contrast agents which are specific for each imaging modality are used to achieve sufficient attenuation/signal intensity in the case of small lesions. To further increase local spatial concentration of a contrast agent for better imaging, nanoparticulate carriers, able to carry contrast moieties for efficient delivery of contrast agents to areas of interest and enhancement of signals from these areas, are used.[41–43] Reporter moiety can be encapsulated into core of carries, incorporate in bilayer, anchored surface of bilayer via chelating agent or attached to the distal end of PEG chain (Fig. 2).

The key component in CT imaging of biological systems is to design suitable contrast agents that have good biocompatibility, longer imaging time, desirable imaging sensitivity and specificity.[44] Due to the drawbacks of clinically used iodine-based CT contrast agents such as short imaging time, renal toxicity at a high concentration, and non-specificity, development of various

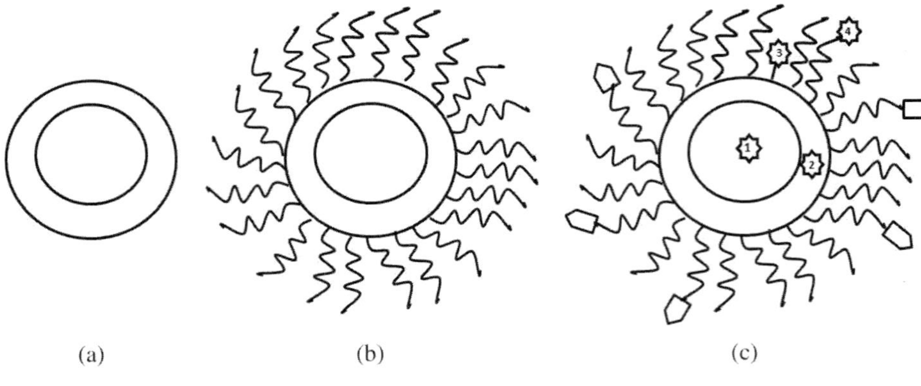

Fig. 2. Unmodified nanosized diagnostic carrier (a), long circulating carrier (b) and antibody conjugated long circulating carrier (c). Reporter moiety can be encapsulated into core of carries (1), incorporate in bilayer (2), anchored surface of bilayer via chelating agent (3) or attached to the distal end of PEG chain (4).

nanoparticle-based contrast agents for CT imaging is emerging and has received considerable attention in recent years. For instance, gold nanoparticles (AuNPs) and bismuth sulfide nanoparticle (Bi_2S_3NPs) have been investigated as contrast agents due to their higher X-ray absorption coefficients than that of iodine based contrast agents.[45–48] On the other hand, commercially available iodine based CT contrast agents encapsulated liposomes,[49–52] polymeric particles,[53,54] micelles[55–57] have also been used to develop long circulated and specifically targeted contrast agents.

Drug targeting mechanism was first defined as "magic bullet" concept by Paul Erhlich in 1891.[58] As widely known, the main goal of drug targeting is to deliver the drug or imaging agent to the desired area at sufficient concentration for a definite period of time. Non-invasive early *in vivo* detection provides to apply early and most appropriate treatment to that disease, and by this way improving life quality.[59]

Multifunctional nanosized drug delivery systems so-called theragnostic drug delivery systems that can simultaneously serve as an imaging agent and a drug carrier could allow diseased tissue to be treated with encapsulated pharmaceutical agents while monitoring the treatment of the disease site through different imaging modalities. Additionally, these nanocarriers can be functionalized with some active targeting ligands such as antibodies, peptides, proteins, lipids, oligonucleotides to target the specific receptor on the desired disease tissue/organ or even cell.[40,60]

3.1. *Liposomes*

Liposomes are one of the most desirable systems in the delivery of drugs and/or diagnostic agents to the intended target tissue/organ depending on their convenient properties. Liposomes are self-assembling vesicles containing lipid bilayers composed of phospholipids including aqueous buffer in hydrophilic core. The idea of liposome production had gained its roots in 1965 in mimicking the behavior of natural membranes by phospholipid vesicles which are the basic structures of the body and they are biodegredable, biocompatible and non-toxic.[61] Subsequently, a huge improvement has been performed in the development of liposomal drug delivery systems.

Liposomes have the ability to deliver both hydrophilic and hydrophobic drugs and/or diagnostic contrast agents to the desired diseased area depending on containing both the aqueous compartment and lypophilic bilayers. Due to the physicochemical properties of drugs and/or contrast agents, they can be encapsulated into the aqueous core, incorporated in lipid bilayer or anchored onto the lipid bilayer. Liposomes have been used as drug delivery systems for many years due to their well-characterized, biocompatible and biodegredable properties. Liposomes have been investigated for diagnostic purpose since 1980s. First, researches were initiated with liposomes in micrometer particle sizes.[62–66]

Longer blood circulation time can be achieved by decreasing particle sizes to nanometer sizes and altering surface properties such as PEG coating of liposomes which cause longer circulation property depending on lesser removal by macrophages of RES organs. This modification is defined as passive targeting of liposomes.[67–70] Active targeting to the diseased area can be performed by attaching target specific ligands such as antibody, antibody fragments, peptides, lipids, hormones on the lipid bilayer of the liposomes.[71–74]

There are several liposomal formulations existing in the market or in different stages of phase trials, for therapy or imaging purposes.[75,76]

The biodistribution of commercially available iodine based-contrast agents can be altered by encapsulating into liposomes. Iopromide containing liposomes were prepared by Erdogan *et al.*[77] and their biodistributon was compared with that of free iopromide. While free iopromide was observed to accumulate within the kidneys, liposome encapsulated iopromide was observed to accumulate in many organs mainly within the liver.

Another advantage of encapsulation into liposomes is to increase the plasma half-life which is very short (in minutes) normally. Liposomally encapsulated contrast media allow for persistence of tissue opacification for several hours.[43]

One of the most important difficulties to prepare the iodine containing liposomes is the encapsulation of commercially available contrast agent into liposomes due to high osmolality of them. The influence of concentration in encapsulated contrast agents, preparation method, the composition of the liposome bilayer and the lipid concentration in the preparation on the encapsulation rate was shown by Schneider *et al.*[78]

It has been reported that encapsulated iodinated contrast agent concentration can be increased by modification of preparation method. For example lipid mixture and other ingredients were dissolved in ethanol (10%, v/v) at a temperature higher than the phase transition of phospholipids used in the formulation. The lipid-ethanol solution was then hydrated with an aqueous solution of iodine based contrast agent at the same temperature.[51,52,79]

Seltzer *et al.*[80] administered centrifugation procedure for both washing and throwing away the free iodine and also for concentrating the iodine amount into iotrolan carrying integration-fusion liposomes. Mukundan *et al.*[79] used tangential flow filtration cassette to increase the encapsulated iodine concentration resulting to an increase in the iodine concentration from

43.8 mgI/mL to 118.6 mgI/mL. Schuhmann–Giampieri used lyophilization procedure to increase both stability of liposomes by resuspension in a proper buffer and iodine concentration encapsulated within the liposome vesicles.[81]

Even though CT images have been improved by encapsulating currently used, water-soluble iodine contrast agent in liposomes, it is still required to administer more iodine to patient since obtaining diagnostically significant imaging requires the iodine concentration on the order of mmol per ml of tissue.[55,82,83] To overcome these limitations different approaches have been tried. Liposomes co-loaded with iopamidol, a water-soluble iodinated compound, and lipiodol, an iodized oil, were prepared with modified reverse-phase evaporation method. Simultaneous loading of iopamidol and lipiodol into liposomes resulted in entrapped iodine concentrations as high as 49.2 iodine mg/ml.[84]

Alternatively, to increase the lipid amount associated to liposome membrane chemically modified phosphatidylcholine lipid (Fig. 3) was used for liposome formations. Diiodophosphatidylcholine containing liposomes are structurally organized such that the iodinated moieties are contained within the vesicle's bilayer, leaving the liposomal interior unoccupied and making it possible to encapsulate the commercially available iodinated contrast agent inside the iodoliposomes' aqueous core to further enhance their imaging contrast.[85] An amount of 120 mg/mL of total iodine concentration was achieved by iodinated liposomes while 30–38 mg iodine per milliliter was obtained using conventional liposomes consisting of L-α-phosphatidylcholine and cholesterol. The X-ray attenuation of the resulting iohexol-encapsulated iodoliposomes might be nearly twice as great as that was observed for conventional iohexol-encapsulated liposomes.[85]

For SPECT and/or PET imaging, radionuclides are anchored to the lipid bilayer[86,87] or encapsulated inside the hydrophilic core of liposomes.[88–94]

The diagnostic ability of various liposomal formulations labeled with different radionuclides has been demonstrated in different diseases such as cancer[95–98] infection and inflammation[99–101] cardiovascular systems.[102,103]

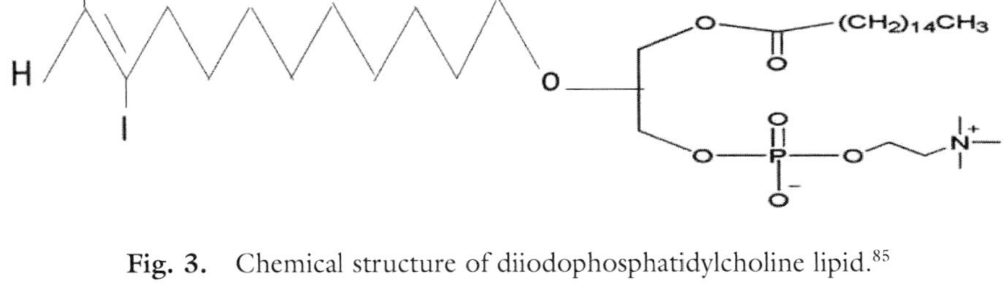

Fig. 3. Chemical structure of diiodophosphatidylcholine lipid.[85]

Radiolabeled liposomes have also potential applications in monitoring of therapy and/or biodistribution of therapeutic agents non-invasively.[104–109]

^{111}In-DTPA-labeled PEGylated liposomes were evaluated for cancer diagnosis in 17 patients with different kinds of locally advanced cancers. The biodistribution of liposomes was monitored by gamma camera. Tumor uptake of liposomes was found to be approximately 0.5%–3.5% of the injected dose at 72 h and the highest uptake was seen in patients with head and neck cancers.[96]

The main point for the application of diagnostic nanocarriers is to increase the signal intensity coming from interested tissue/organs. One possible way of increasing the signal intensity is to increase the injected amount of liposomes loaded with diagnostic moieties but in this case there is always a risk to also increase the side effects. The other way is to increase the amount of diagnostic groups without increasing the applied lipid concentration. For this purposes, Polychelating Amphiphilic Polymers (PAPs) were synthesized. These polymers consist of a hydrophilic main chain with multiple side chelating groups capable of binding to many reporter metal atoms and a hydrophobic terminal group allowing for polymer adsorption onto hydrophobic nanoparticles or incorporation into hydrophobic domains of liposomes or micelles (Fig. 4). Such surface modification of targeted nanocarriers allows to a sharp increase in the number of bound reporter metal atoms per particle and the image signal intensity (Fig. 5).[51,110–115]

Radiolabelling procedures are also used to visualize pharmacokinetics of liposomes as well as the other nanoparticles, *in vivo*. Long half-lived PET-isotopes such as ^{18}F and ^{64}Cu are used for radiolabeling of nanoparticles and visualized by PET imaging modalities.

^{18}F-labeled diglyceride was synthesized by nucleophilic substitution of p-toluenesulfonyl moiety and its potential as a reagent for radiolabeling

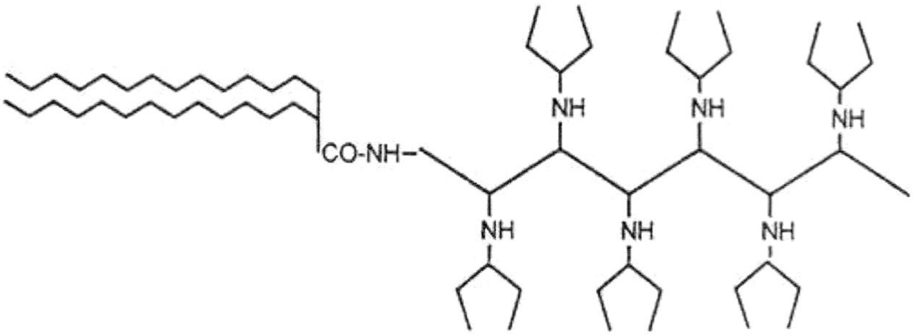

Fig. 4. Chemical structure of PAPs, DTPA-PL-NGPE consisting of hydrophilic DTPA-polylysyl moiety and hydrophobic N-glutaryl phosphatidyl ethanolamine moiety.[111]

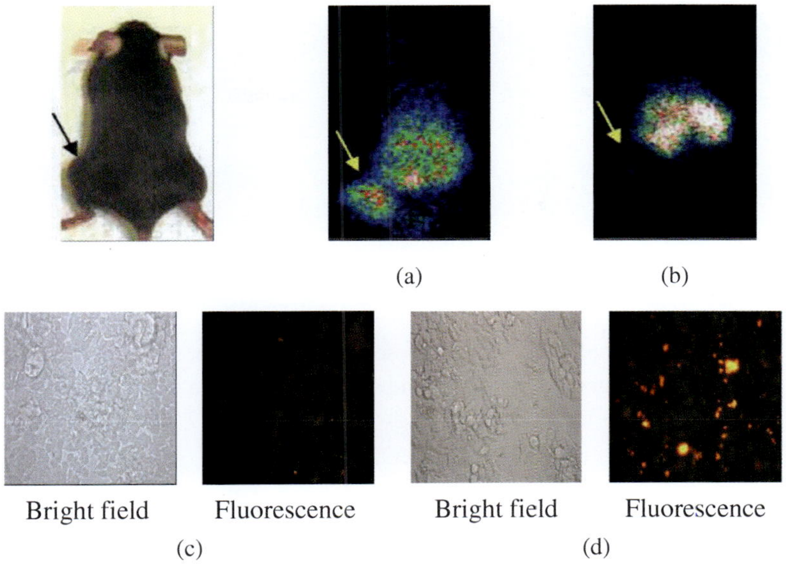

Fig. 5. Whole body imaging of LLC tumor bearing mice at different time points after the injection of ^{111}In-labeled PAP-containing 2C5-modified (a) and unmodified (b) PEGylated liposomes[111] and *in vitro* cell binding of Rh-PE-labeled DTPA-PE (c) and PAP (d) containing PEGylated 2C5 modified liposomes to breast adenocarcinoma (MCF-7) cells.[110]

long-circulating liposomes was investigated. It was found that while free [^{18}F] FDP had highest uptake in the liver, spleen and lungs, liposomal [^{18}F]FDP remained in blood circulation at a constant level for at least 90 min. Successful incorporation of [(^{18})F]FDP into the phospholipid bilayer showed the potential use of this method for radiolabeling a variety of lipid-based drug carriers.[116]

In another study, Octreotate (TATE) was covalently attached to the distal end of DSPE-PEG(2000) on PEGylated liposomes with an encapsulated positron emitter ^{64}Cu that can be utilized for PET imaging. The biodistribution and pharmacokinetics of either TATE modified or non-modified liposomes were evaluated. Results designated that TATE modification results in faster blood clearance, however PEGylated liposomes either TATE modified or non-modified showed higher accumulations on neuroendocrine tumors than free peptide. The highest tumor to muscle ratio was obtained by TATE modified ^{64}Cu-loaded PEGylated liposomes (12.7 ± 1.0) when compared with non-modified ones (8.9 ± 0.9) and ^{64}Cu-DOTA-TATE free peptide (7.2 ± 0.3). Results showed that ^{64}Cu-loaded PEGylated liposomes with TATE conjugated to the surface could be promising new imaging agents for visualizing tumor tissue and especially neuroendocrine tumors using PET (Fig. 6).[95]

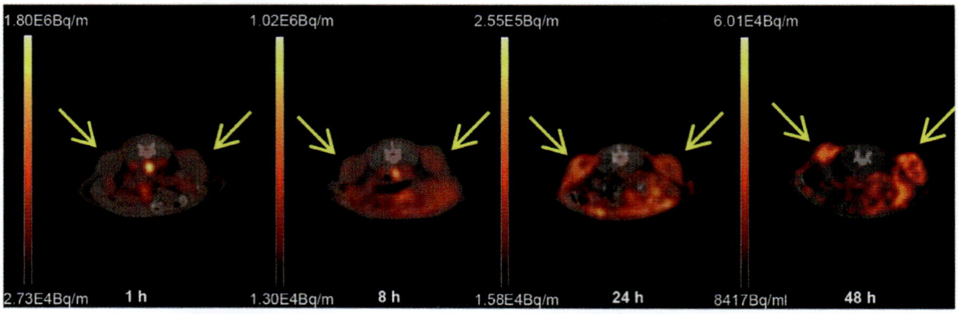

Fig. 6. PET/CT-images of 64Cu-TATE-liposome distribution in a mouse xenograft model. Axial PET/CT fusion images 1 h, 8 h, 24 h and 48 h post-injection of 64Cu-TATE-liposomes into a mouse with tumors (neuroendocrine carcinoma NCI-H727; marked with arrows) on right and left flank.[95]

3.2. *Micelles*

Micelles are self-assembled, mainly spherical shaped, molecular clusters in water. When the concentration of surfactants reaches to a critical concentration so-called critic micelle concentration (CMC), lipophilic hydrocarbon chains of surfactant are oriented toward the interior of micelle, leaving the hydrophilic groups on contact with the water. Micelles used as drug carriers should be stable in blood circulation and not disintegrate upon contact with blood components. Because of this, CMC must be as low as possible. Due to their amphiphilic properties, block or graft copolymers exhibit surfactant behavior and form micelles. One of the most important advantages of these polymers is CMC, is very low, can be 1000-folds lower than one of surfactant micelles.[117,118] That is why di- or tri-block co-polymers are frequently used to prepare micelles served as pharmaceutical carriers. Both di-block and tri-block-copolymers generally contain PEG as hydrophilic blocks, since this polymer is well soluble, highly hydrated and able to serve as an efficient steric protector for various drug carriers.[119] While di-block copolymers consist of hydrophobic blocks conjugated with PEG chains, tri-block copolymers known as Pluronic consist of a central poly(propylene oxide) (PPO) block with terminal poly(ethylene oxide) (PEO) blocks. These micelles contain a hydrophilic corona of PEO and a hydrophobic core of PPO.[120–122]

Micelles are used as diagnostic agents or drug carriers by either covalent incorporation of the contrast agent or drugs within the amphiphilic polymer/lipid structure or solubilization into the hydrophobic core of the micelles.

Polymeric micelles have a size larger than fenestrated capillaries (>10 nm), the reporter moiety structurally incorporated within the particulate, and PEG coated micelles are resistant to phagocytosis providing the ability to stay in the blood circulation long enough to obtain clinically useful images.[55]

Since use of conventional CT contrast agent for blood pool imaging is limited due to rapid extravasation and rapid clearance of contrast agent, polymeric micelles which are coated with PEG especially attractive for blood pool imaging.

Torchilin *et al.*[56] prepared iodinated polymeric micelles used particularly as blood pool contrast agent for micro-CT. Synthesized block-copolymer of methoxy-poly(ethylene glycol) (MPEG) and iodine-substituted poly-L-lysine can be easily micellize in the solution forming stable and heavily iodine-loaded particles (up to 35% of iodine by weight) with a size below 100 nm (Fig. 7).

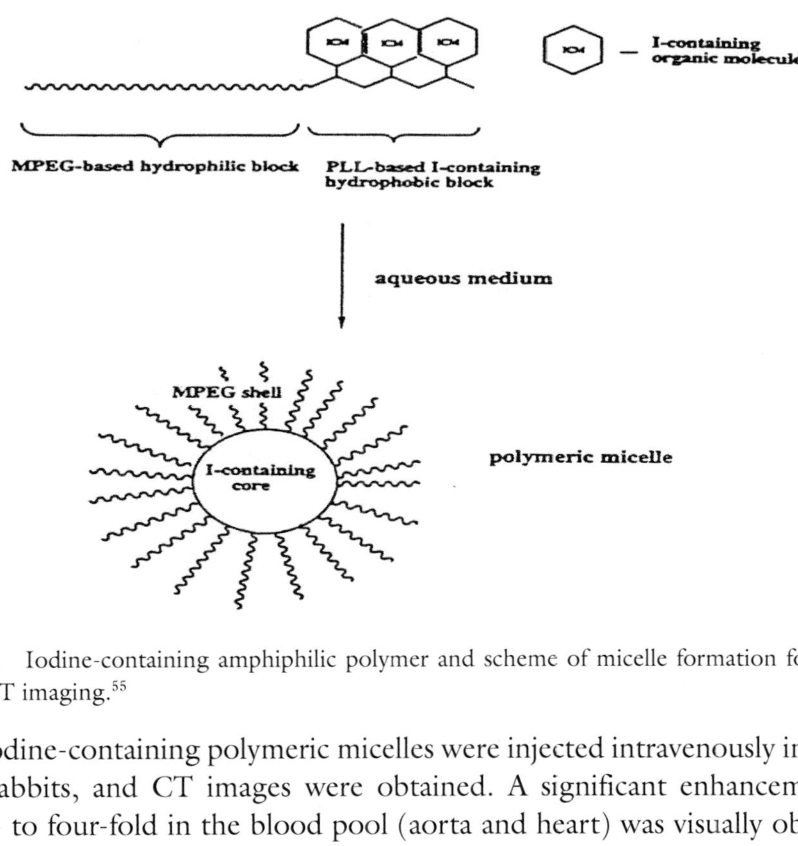

Fig. 7. Iodine-containing amphiphilic polymer and scheme of micelle formation for blood pool CT imaging.[55]

Iodine-containing polymeric micelles were injected intravenously into rats and rabbits, and CT images were obtained. A significant enhancement of three- to four-fold in the blood pool (aorta and heart) was visually observed in rats for at least a period of 2 h following the injection and a similar pattern was also observed in rabbits (Fig. 8).[56]

Active targeting strategies are possible with micelles, by grafting the specific targeting function onto the micelle surface, i.e. peripheral part of the hydrophilic polymer block.[123,124]

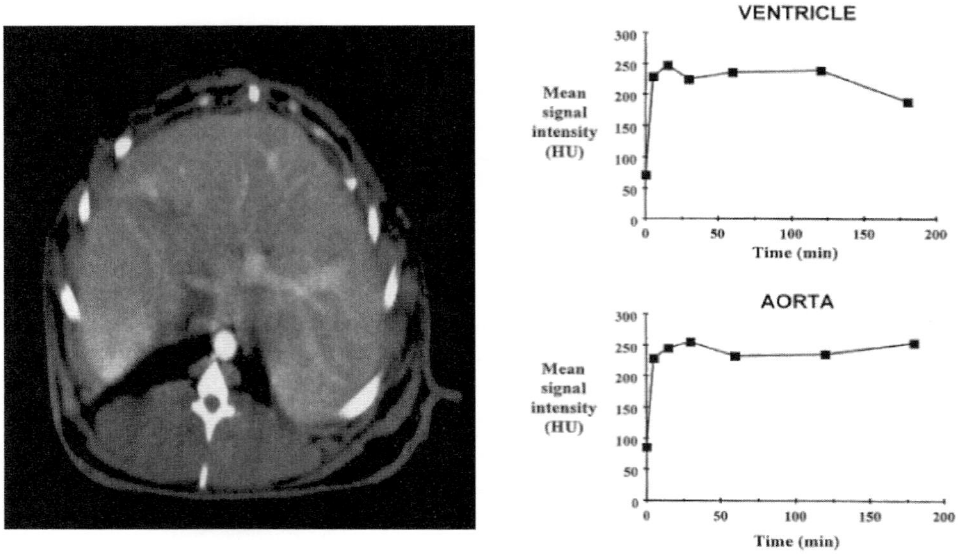

(a) (b)

Fig. 8. CT section of the rabbit liver 2 h after *i.v.* administration of iodo-containing polymeric micelles (250 mg iodine /kg) (a), opacification of heart and aorta at different post-injection times after *i.v.* administration of iodine-containing polymeric micelles (injected into rat tail vein at a dose of 170 mg I/kg).[56]

PEG-lipid micelles is attractive systems for certain diagnostic imaging application such as lymphatic imaging. Because the optimum size of the colloidal particles still remains an unresolved issue for lymphoscintigraphy. While large particles fail to migrate, too small particles move very fast and only a fraction is retained in the sentinel lymph node. Due to small size and surface properties imparted by the PEG corona, the micellar particulates can move with ease from the injection site along the lymphatics to the systemic circulation with the lymph flow. ^{111}In-DTPA-PE incorporated PEG(5 kDa)-PE micelles having 20 nm of size successfully used for the experimental percutaneous lymphography in rabbits by gamma-scintigraphy.[125]

3.3. *Dendrimers*

Dendrimers have been identified as a class of highly branched, monodispersed, synthetic macromolecules with well-defined composition, dimension, and structural architecture. They consist of three major architectural components: core which can encapsulate various chemical species, branches which can solubilize various small guest molecules such as contrast agent and end groups which can potentially be functionalized to create specific interactions with the surrounding environment.[126]

Dendrimers are polymeric complexes composed of a series of polymeric branches linked together by an inner core. They are characterized by the presence of internal cavities and by a large number of reactive end groups.[127]

Dendrimers possess some advantages in diagnosis and therapy. For example;

- controllable size,[128,129]
- high density and a well-defined number of surface functional groups providing multifunctional modification for multiple purposes,[130–132]
- reproducible pharmacokinetics due to monodispersity,[133–135]
- enhanced biological, chemical, or physical activities of the compounds bound on the surface of the dendrimer due to the "synergistic effect" or "multivalency effect",[136,137]
- excellent cellular uptake ability,[138–140]
- ability to administer in different routes such as oral, injectable, transdermal, ocular, nasal, colon, rectal, and pulmonary.[134,141]

Generally used dendrimers, polyamidoamine dendrimers, consist of ammonia core (PAMAMs) and poly(propylene imine) dendrimers consist of butylenediamine core (PPI) are commercially available. While amine terminated PAMAM and PPI called as cationic dendrimers are generally hemolytic and cytotoxic, anionic dendrimers bearing a carboxylate surface are not cytotoxic over a broad concentration range.[141] Biocompatible dendrimers are synthesized by PEGylation, acetylation, and glycosylation methods.[132,143]

Use of dendrimers in diagnostic field is especially important for cancer diagnosis. Since the ultimate goal of cancer diagnosis is to detect the cancer in its earliest stage, the multivalency effect of dendrimers and the amplification property of the surface functionalities provide the chance to detect the cancer in its early stage and more importantly these amplification factors can be further increased by choosing higher generation dendrimers.[132]

However, dendrimers generally preferred as contrast agent for MR,[144–146] optical[147,148] and radionuclidic imaging,[149,150] potential of iodinated dendrimers[151] and gold nanoparticles (AuNPs)[152,153] or gold-silver alloy nanoparticles[154] entrapped dendrimers has also been investigated for CT imaging.

To prepare AuNPs entrapped PEG-attached dendrimers, AuNPs were grown in the PEGylated dendrimer by adding gold ions and ascorbic acid at different ratios. By increasing of AuNPs size with adding gold ions, X-ray attenuation of the AuNPs also increased. With prolonged circulation time,

AuNPs entrapped PEG-attached dendrimers showed similar contrast enhancement with commercial iodine agent.[155]

Yordanov *et al.*[156] prepared water-soluble iodinated dendritic nanoparticle, G-4-(DMAA-IPA)37 consisting of a Starburst PAMAM Generation 4 (G-4) dendrimer that has on its surface thirty-seven (37) 3-N-[(N¢,N¢-dimethylaminoacetyl) amino]-R-ethyl-2,4,6-triiodobenzene propanoic acid (DMAA-IPA) molecules covalently linked by an amide bond. Thus, the total number of iodine atoms per dendrimer unit is 111. DMAA-IPA was specifically designed so that the total positive charge (and the solubility) of the dendrimer were not affected after the attachment of the iodine-containing moieties to its surface.

PEGylated dendrimer-entrapped gold nanoparticles (Au DENPs) were synthesized using PEGylated dendrimers as templates with a size range of 2–4 nm. These PEGylated Au DENPs were found to be stable in different media and also under different temperature and pH conditions. X-ray absorption coefficient measurements designated that the attenuation intensity of PEGylated Au DENPs is much higher than that of the iodine-based contrast agent at the same molar concentration of the radiodense element (Au or iodine). Results showed that with the partial PEGylation of the dendrimer periphery, high loading of Au content within the dendrimers can be achieved to form Au DENPs with good stability, significantly enhanced biocompatibility, and efficient X-ray attenuation property for CT imaging.[48]

In another study, to develop large-molecular CT contrast agent, PEG-cored dendrimer was conjugated with clinically used, non-ionic, small molecular CT contrast agent, iobitridol, via a hydroxyl protection/deprotection strategy. PEG12000-carbamate-Gen4-IOB conjugate, 27% (w%) rich in iodine, demonstrated strong and persistent intravascular enhancement (Fig. 9) with prolonged blood half-life (about 35 min) and also showed high water solubility (> 550 mg/mL at 25°C), large molecular size, high hydrophilicity, and stability.[151]

Folic acid-modified dendrimer entrapped[153] or dendrimer-stabilized[157] gold nanoparticles showed good biocompatibility at tested concentration range and results obtained from *in vitro* and *in vivo* studies suggested that Au DENPs-FA have a great potential to be used as imaging probes for targeted CT imaging of cancer cells.

Folic acid modified acetylated PAMAM (G5) dendrimers (Ac-G5-FA) were also investigated for SPECT imaging. For stabile radiolabelling, Ac-G5-FA conjugated with 2-(p-isothiocyanatobenzyl)-6-methyl-diethylenetria minepentaacetic acid (1B4M DTPA). Biodistribution and *in vivo* SPECT imaging studies showed that uptake of 99mTc-labelled Ac-G5-FA-1B4M

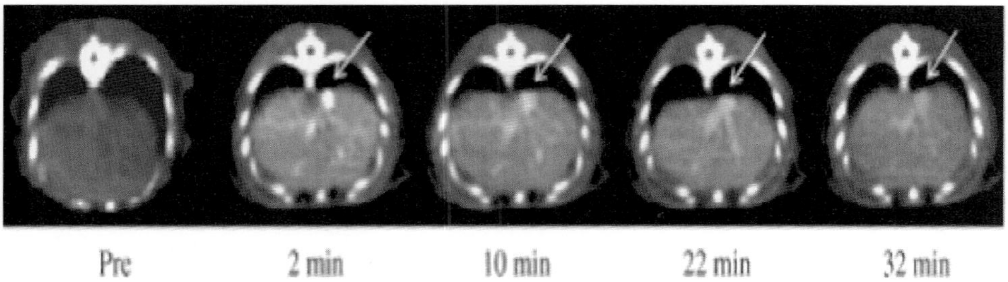

Fig. 9. Serial axial CT images of a normal Sprague–Dawley rat at the level of liver prior to, 2, 10, 22, and 32 min after *i.v.* injection of PEG12000-carbamate-Gen4-IOB as a CT contrast agent at a dose of 450 mg iodine/Kg body weight. Strong and persistent enhancement of the inferior vena cava (see the arrows) and other vessels (hepatic veins) within the liver were observed.[151]

DTPA was increased in tumor while decreased in blood. Because of the presence of folate receptors in the proximal tubules of the kidneys, high amount of radioactivity was measured in kidney.[158] To avoid the high uptake of folic acid modified nanoparticles by folate receptor presence organ such as kidney, avidin was used for tumor targeting.[148] Because, it is well-known that avidin is a quickly internalizing molecule into cancer cells especially b-D-galactose receptors expresses cells such as ovarian and colorectal adenocarcinoma cells as well as normal hepatocytes.[159]

3.4. *Nanoparticles*

Like other delivery systems, nanoparticles can also successfully be used as multifunctional systems in either imaging or therapy.[160] Nanoparticles were formulated in nanosizes by synthesizing in systemic fashion. Like other delivery systems, surface modification provides them to decorate with imaging agents, therapeutic substances and specific targeting ligands.

Metal nanoparticles are promising CT contrast agents due to their unique physical, chemical and biological properties. These unique properties strongly depend on their size, shape, composition, and structure. The atomic number of the recently proposed metal tantalum, gold and bismuth (73, 79 and 83 respectively) is much higher than that of currently used iodine (atomic number of iodine is 53) and these proposed metals can induce stronger X-ray attenuation. Therefore, this higher atomic number provides better resultant CT contrast (proportional to Z^3) and lesser radiation exposure to the patients.[161] Their surface has a strong binding affinity toward thiol, disulfide and amine groups, which allows by simple chemistry surface conjugation with various targeting moieties described above.

Among the metal nanoparticles, gold nanoparticles (AuNPs) in particular are attracting increasing interest. Gold nanoparticles are prepared in different morphologies such as nanocubes,[162] nanorods, [163,164] nanocomposites,[165] nanocrystals,[166,167] nanowires,[168] nanobelts. [169] The reduction of gold salts (HAuCl4) using sodium citrate, is the simplest and most commonly employed method for the synthesis of AuNPs.[170] A large variety of stabilizers such as ligands, surfactants, polymers, dendrimers can be used to stabilize the AuNPs, most robust particles to be stabilized by thiolates using the strong Au–S bond between the soft acid Au and soft thiolate base.[171] Solubility of AuNPS is controlled by the solubilizing properties of terminal group of the thiolate ligands. Water soluble AuNPs typically contain terminal carboxylate groups at their periphery.[172] These molecules provide colloidal stability due to their negative charges, in addition they can also be used as anchor points for the further attachment of biological molecules. Biological molecules can be attached to the particles in several ways. If the biological molecules have a functional group such as thiols or specific peptide sequences which can bind to the gold molecules, they can replace some of the original stabilizer molecules when they are added directly to the particle solution. Alternatively biological molecules can also be attached to the shell of stabilizer molecules around the AuNPs by bioconjugate chemistry. [173]

Like the other targeted drug delivery systems, gold nanoparticles can selectively tag a wide range of medically important targets, for example, cancer cells, atherosclerotic plaque with high affinity and specificity, when they are linked to specific-targeting ligands.

Popovtzer *et al.*[174] prepared antiepidermal growth factor receptor conjugated gold nanoparticles (30 nm) and injected to mice implanted human squamous cell carcinoma head and neck cancer. The results clearly demonstrate that a small tumor, which is currently undetectable through anatomical CT, is enhanced and becomes clearly visible by the molecularly-targeted GNPs.

Lim *et al.*[175] were prepared dextran coated gold nanoparticles for detection of atherosclerosis via CT imaging. First, they synthesized pyrenyl dextran which could simultaneously act as a stabilizer for specific targeting of macrophages and as a reducing agent. Then, dextran coated AuNPs were formulated by a chemical reduction method using pyrenyl dextran and their feasibility as a CT imaging agent for atherosclerosis imaging were evaluated. It was reported that dextran coated AuNPs exhibited excellent colloidal stability and optical properties. They could be effectively internalized into macrophages due to their surface properties along with their biocompability. Enhanced atherosclerosis-specific imaging was achieved by dextran coated AuNPs (Fig. 10).

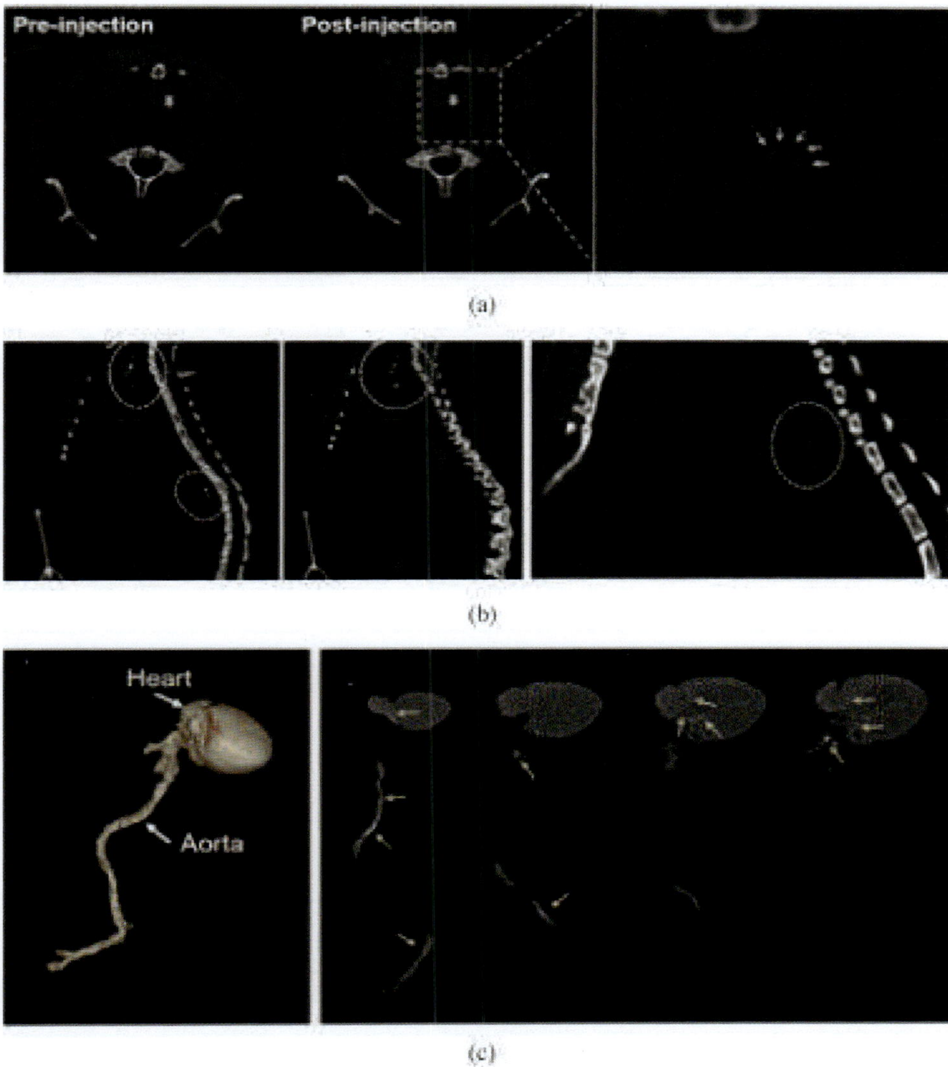

Fig. 10. Micro CT images of a Apo E–/– mouse after injection of the DGNPs: (a) cross-sectional CT images pre-/post-injection of the DGNPs and (b) longitudinal sectional CT images after injection of the DGNPs (c) 3D-micro CT images of extracted tissue (aorta and heart) from a Apo E–/– mouse (left) and *ex vivo* micro CT images after DGNPs treatment (cross-sectional images of right).[175]

Chanda *et al.*[176] prepared bombesin (BBN) functionalized gold nanoparticles (BBN–AuNPs) (115–155 nm) used for tumor imaging by CT. BBN peptides have demonstrated high affinity toward gastrin-releasing peptide (GRP) receptors *in vivo* that are overexpressed in prostate, breast, and small-cell lung carcinoma. They evaluated the cellular interactions and binding affinities of BBN–AuNPs with PC-3 cells, which are human prostate cancer cells. They reported that cell binding affinity of conjugates depends on the degree of BBN peptide coated over AuNPs and gold nanoparticles having more BBN on the surface exhibit higher cell binding affinity. *In vivo* CT imaging studies showed that the selective uptake of this new generation of GRP-receptor-specific AuNP–BBN has demonstrated clinical potential in molecular imaging by CT as the contrast numbers in prostate tumor sites are several fold higher as compared to the pretreatment group.

In another study, Chanda *et al.*[177] prepared cinnamon-coated gold nanoparticles (Cin-AuNPs) to explore the utilization of Cin-AuNPs as CT/optical contrast-enhancement agents for detection of cancer cells. The results indicate that these non-toxic Cin-AuNPs can serve as excellent CT/photoacoustic contrast enhancement agents and may provide a novel approach toward tumor detection through nanopharmaceuticals.

In addition to gold nanoparticles other metal particles such as bismuth sulphide (Bi_2S_3), silver (Ag), tantalum oxide or alloy metal (i.e. Ag–Au) have been investigated to develop CT contrast agent.

Polymer-coated Bi_2S_3 nanoparticles were prepared as an injectable CT imaging agent. This preparation demonstrates excellent stability at high concentrations (0.25 M Bi3+), high X-ray absorption (five-fold better than iodine), very long circulation times (>2 h) *in vivo* and an efficacy/safety profile comparable to or better than iodinated imaging agents. The results showed the utility of these polymer-coated Bi_2S_3 nanoparticles for enhanced *in vivo* imaging of the vasculature, liver, and lymph nodes in mice.[178]

Liu *et al.*[154] used amine-terminated fifth-generation poly(amidoamine) dendrimers as templates or stabilizers to synthesize dendrimer-entrapped or dendrimer-stabilized Au–Ag alloy nanoparticles (NPs) with different gold atom/silver atom/dendrimer molar ratios with the assistance of sodium borohydride reduction chemistry. They showed that dendrimer-entrapped or dendrimer-stabilized Au–Ag alloy nanoparticles should have a promising potential for CT imaging and other biomedical applications with the controllable particle size and optical property, metal composition-dependent X-ray attenuation characteristics, and improved biocompatibility after acetylation.

In another study, Liu *et al.*[179] have been reported a facile size-controlled synthesis of dendrimer-stabilized silver nanoparticles (Ag DSNPs) for X-ray CT imaging applications. They showed that the size of Ag DSNPs can be changed by dendrimer/Ag salt molar ratio. The formed Ag DSNPs are stable in different medium such as water, PBS buffer, fetal bovine serum, and also at different pH conditions (pH 5–8) and temperatures (20–50°C). X-Ray absorption coefficient measurements show that the attenuation of Ag DSNPs is size-dependent, and the Ag DSNPs with a diameter of 16.1 nm display an X-ray attenuation intensity close to that of a clinically used iodine-based contrast agent (Omnipaque) at the same molar concentration of the active element (Ag versus iodine).

Tantalum oxide is another alternative nanoparticle based X-ray contrast agents. They are less expensive and also more safe for clinical use when compared to other metal nanoparticles such as gold and Bi_2S_3, respectively. Hyeon *et al.*[180] prepared uniformly sized tantalum oxide nanoparticles using a micro-emulsion method and modified the surface of particles with PEG and fluorescent dye. The resulting nanoparticles exhibited remarkable performances in the *in vivo* X-ray CT angiography and bimodal image-guided lymph node mapping. The results of *in vivo* toxicity studies of tantalum oxide nanoparticles revealed that the nanoparticles did not affect normal functioning of organs.

Liu *et al.*[181] described novel PEGylated Ytterbium (Yb)-based nanoparticulate (PEG-UCNPs) CT contrast agent demonstrated the excellent biocompatibility. By virtue of a high atomic number and a well-positioned K-edge energy of ytterbium, the Yb-based nanoparticulate CT contrast agent provided higher contrast efficacy than routine iodine-based agents in clinic and other metal-based nanoparticulate CT contrast agents previously reported. Yb-based nanoparticles showed great potential for angiography and cancer-target imaging.

Radiolabeled polymer based nanoparticles have been used for imaging of different disease as well as following biodistribution of carriers by SPECT or PET imaging modalities.

Liver targeted glycyrrhetinic acid (GA) containing chitosan/poly(ethylene glycol)-glycyrrhetinic acid (CTS/PEG-GA) nanoparticles were prepared by ionic chelation process. The biodistribution of nanoparticles was assessed by SPECT and the cellular uptake of doxorubicin HCl-loaded nanoparticles (DOX-loaded nanoparticles) by human hepatic carcinoma cells (QGY-7703 cells) were also evaluated both *in vitro* and *in vivo.* CTS/PEG-GA nanoparticles designated high liver accumulation depending on targeting effectiveness which was seen 2.6 times more accumulation in liver than that of nanoparticles without GA after 3 h. Both *in vitro* QGY-7703 tumor cell culture experiments and *in vivo* tumor bearing mice experiments designated that DOX-loaded CTS/PEG-GA nanoparticles have inhibitory effect on tumor growth.[182]

Polymeric n-butyl-2-cyanoacrylate (PBCA) NPs can be formulated by encapsulating radiolabeled clioquinol (CQ, 5-chloro-7-iodo-8-hydroxyquinoline) to enhance penetration of BBB, to increase amyloid plaque retention and to achieve specific imaging and/or therapy. This specific radio-iodinated CQ-PBCA NPs can be used specifically for either SPECT or PET imaging purposes by radiolabeling with ^{123}I or ^{124}I, respectively.[59]

A novel class of chelator free PEG coated ^{64}Cu-labeled copper sulfide (CuS) nanoparticles were prepared for theranostic application. These [64Cu] CuS nanoparticles are simple to make, possess excellent stability, and allow robust non-invasive micro-PET imaging (Fig. 11). Furthermore, improved tumor uptake of PEG-[64Cu]CuS NPs increased the efficacy of photothermal ablation therapy, decreased the energy dose of the laser, and minimized the potential damage to surrounding normal tissues.[183]

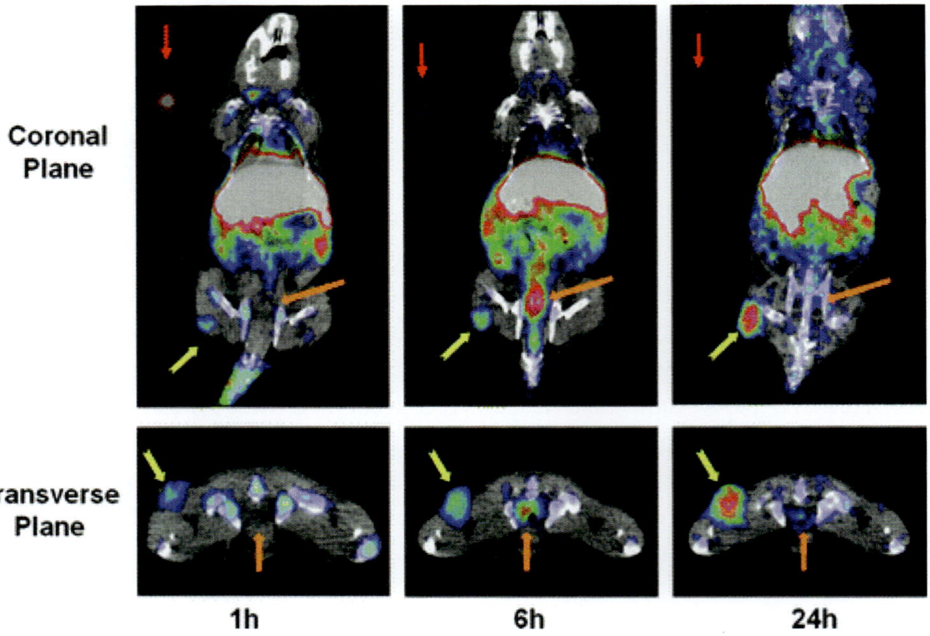

Fig. 11. Micro-PET/CT images of nude mice bearing subcutaneous U87 glioma xenografts acquired at 1, 6, and 24 h after *i.v.* injection of PEG–[64Cu]CuS NPs. Yellow arrow, tumor; orange arrow, bladder; red arrow, standard.[183]

4. Future Perspectives

Contrast agent containing nanoparticles have been used for imaging, characterization, staging, and therapy monitoring as well as tracking of biodistribution of drug molecules and/or drug encapsulated nanocarriers. Especially targeted nanoparticles as contrast agents may permit imaging enhanced signal sensitivity, better spatial resolution and accuracy to give a chance to the physician a better diagnosis.

Current technologies are mostly more non-specific and depend on macroscopic physical and physiological alterations and these are mostly responsible for distinction of pathological tissue than normal tissue rather than detection of the molecular issues responsible for disease such as gene expression. Molecular imaging maintains diagnosis and imaging of pathological irregularities in genomic and proteomic pathways, and it provides earlier diagnosis of diseases and tracing of the effectiveness of treatment and it provides a perspective to fundamental biology. It is a considerable technique for early diagnosis and characterization of diseases such as cancer in order to diagnose, before it develops to a definite extent, while it is in molecular and cellular level. Molecular imaging can be acquired by different imaging modalities used in a variety of disciplines. Molecular imaging is an interdisciplinary field composed of many disciplines such as biology, biochemistry, electronic engineering and biophysics, and medicine and it facilitates imaging of cellular and molecular processes. The enhancement in the development of molecular imaging technology had led to investigate both new imaging instruments and new imaging probs. [3,184]

Hybrid imaging technologies conjugate two different vital imaging modality such as PET, SPECT, and CT which can decrease their drawbacks and increase the effectiveness of their individual advantages. With the access of these multimodal imaging modalities to the routine use in clinics, it induces the need of multifunctional imaging probes that can be specific to both systems and can also target to the desired tissue/organ and by this way better and accurate images can be obtained from that tissue, organ, cell, subcell, or even molecules.[185]

While the technology is getting improved, many researchers are studying on the engineering of the surface and core structure of the formulations and performing better characterization of multifunctional systems by labeling them with an imaging agent. Hence, different imaging modalities are used for their detection, especially hybrid imaging systems can be efficiently utilized by decreasing the deficiencies of each other and acquiring better images.

In addition, some nanoparticles, so-called theranostics, can deliver both contrast agents and therapeutics allowing monitoring of biodistribution and therapeutic activity simultaneously. They are highly advantageous as they offer *in vivo* imaging and tracking of drug efficacy.

References

1. M. G. Pomper and J. G. Gelovani, in *Nanoparticle-Based Molecular Imaging in Living Subjects,* B. R. Smith and S. S. Gambhir (Eds.) New York: Informa Helathcare (2008), p. 261.
2. R. Gref, Y. Minamitake, M. T. Peracchia, V. Trubetskoy, V. P. Torchilin and R. Langer, *Science* **263**, 16001–603 (1994).

3. M. Silindir, S. Erdogan, A. Y. Ozer and S. Maia, *J. Drug Target.* **20**, 401–415 (2012).
4. J. H. Senior, *Crit. Rev. Ther. Drug Carrier Syst.* **3**, 123–193 (1987).
5. D. D. Lasic and F. J. Martin (Eds.), *Stealth Liposomes.* Boca Raton: CRC Press (1995).
6. V. P. Torchilin and V. S. Trubetskoy, *Adv. Drug Deliv. Rev.* **16**, 141–155 (1995).
7. V. P. Torchilin, *J. Lip. Res.* **9**, 99–116 (1996).
8. M. C. Garnett, *Adv. Drug Deliv. Rev.* **53**, 171–216 (2001).
9. H. Maeda, T. Sawa and T. Konno, *J. Control. Release* **74**, 47–61 (2001).
10. L. Wang, W. Su, Z. Liu, M. Zhou, S. Chen, Y. Chen, D. Lu, Y. Liu, Y. Fan, Y. Zheng, Z. Han, D. Kong, J. C. Wu, R. Xiang and Z. Li, *Biomaterials* **33**, 5107–5114 (2012).
11. S. Erdogan, A. Roby, R. Sawant, J. Hurley and V. P. Torchilin, *J. Liposome Res.* **16**, 45–55 (2006).
12. S. Li, B. Goins, B. A. Hrycushko, W. T. Phillips and A. Bao, *Mol. Pharm.* **9**, 2513–2522 (2012).
13. Y. F. Huang, H. T. Chang and W. Tan, *Anal. Chem.* **80**, 567–572 (2008).
14. Y. F. Huang, Y. W. Lin, Z. H. Lin and H. T. Chang, *J. Nanopar. Res.* **11**, 775–783 (2009).
15. B. A. Teply, F. G. Rocha, E. Levy-Nissenbaum, R. Langer and O. C. Farokhzad, *Am. J. Drug Deliv.* **4**, 123–130 (2006).
16. S. F. Colombo, R. Longhi and N. Borgese, *J. Cell Sci.* **122**, 2383–2392 (2009).
17. Y. L. Xie, W. Lu and X. G. Jiang, *Behav. Brain Res.* **173**, 76–84, (2006).
18. J. D. Heidel, Z. Yu, J. Y. C. Liu, S. M. Rele, Y. Liang, R. K. Zeidan, D. J. Kornbrust and M. E. Davis, *Proc. Natl. Acad. Sci. USA* **104**, 5715–5721 (2007).
19. H. Iinuma, K. Maruyama, K. Okinaga, K. Sasaki, T. Sekine, O. Ishida, N. Ogiwara, K. Johkura and Y. Yonemura, *Int. J. Cancer* **99**, 130–137 (2002).
20. I. Paschkunova-Martic, C. Kremser, K. Mistlberger, N. Shcherbakova, H. Dietrich, H. Talasz, Y. Zou, B. Hugl, M. Galanski, E. Sölder, K. Pfaller, I. Höliner, W. Buchberger, B. Keppler and P. Debbage, *Histochem. Cell Biol.* **123**, 283–301 (2005).
21. M. A. Arangoa, G. Ponchel, A. M. Orecchioni, M. J. Renedo, D. Duchêne and J. M. Irache, *Eur. J. Pharm. Sci.* **11**, 333–341 (2000).
22. J. Kreuter, *Curr. Med. Chem. Central Nervous Syst. Agents* **2**, 241–49 (2002).
23. K. Michaelis, M. M. Hoffmann, S. Dreis, E. Herbert, R. N. Alyautdin, M. Michaelis, J. Kreuter and K. Langer, *J. Pharmacol. Exp. Ther.* **317**, 1246–1253 (2006).
24. A. Yamada, Y. Taniguchi, K. Kawano, T. Honda, Y. Hattori and Y. Maitani, *Clin. Cancer Res.* **14**, 8161–8168 (2008).
25. G. Xiang, J. Wu, Y. Lu, Z. Liu and R. J. Lee, *Int. J. Pharm.* **356**, 29–36 (2008).
26. A. Gabizon, A. T. Horowitz, D. Goren, D. Tzemach, H. Shmeeda and S. Zalipsky, *Clin. Cancer Res.* **9**, 6551–6559 (2003).
27. R. Rossin, D. Pan, K. Qi, J. L. Turner, X. Sun, K. L. Wooley and M. J. Welch, *J. Nucl. Med.* **46**, 1210–1218 (2005).
28. J. Sunamoto and K. Iwamoto, *Crit. Rev. Ther. Drug Carrier Syst.* **2**, 117–136 (1986).
29. K. Maruyama, *Nat. Rev. Cancer* **2**, 750–763 (2002).
30. P. L. McLain, in *Computed Tomography of The Temporal Bone and Orbit: Technique of Direct Multiplanar, High-resolution CT and Correlative Cryosectional Anatomy*, F. W. Zonneveld (Ed.) Munich: Urban & Schwarzenberg (1987), p. 125.
31. U. Speck, in *Contrast Media-Overview, Use and Pharmaceutical Aspects*, Heidelberg, New York: Springer-Verlag (1993), p. 27.

32. M. G. Pomper and J. G. Gelovani, in *Molecular SPECT Imaging Instrumentation and Techniques Oncology*, B. M. W. Tsui, Y. Wang and S. P. Mok (Eds.) new York: Informa Healthcare (2008), p. 93.
33. S. Vallabhajosula, in *PET and SPECT Scanners*, E. Schrodinger (Ed.) New York: Springer (2009), p. 59.
34. M. G. Pomper and J. G. Gelovani, in *Molecular PET Instrumentation and Imaging Techniques*, R. Lecomte (Ed.) New York: Informa Healthcare (2008), p. 67.
35. M. Lecchi, L. Ottobrini, C. Martelli, A. Del Sole and G. Lucignani, *Q. J. Nucl. Med. Mol. Imaging* **51**, 111–126 (2007).
36. K. P. Schafers, A. J. Reader, M. Kriens, C. Knoess, O. Schober and M. Schafers, *J. Nucl. Med.* **46**, 455–463 (2005).
37. M. Benezra, O. Penate-Medina, P. B. Zanzonico, D. Schaer, H. Ow, A. Burns, E. DeStanchina, V. Longo, E. Herz, S. Iyer, J. Wolchok, S. M. Larson, U. Wiesner and M. S. Bradbury, *J. Clin. Invest.* **121**, 2768–2780 (2011).
38. S. D. Voss, S. V. Smith, N. DiBartolo, L. J. McIntosh, E. M. Cyr, A. A. Bonab, J. L. Dearling, E. A. Carter, A. J. Fischman, S. T. Treves, S. D. Gillies, A. M. Sargeson, J. S. Huston and A. B. Packard, *Proc. Natl. Acad. Sci. USA* **104**, 17489–17493 (2007).
39. Y. Zhou, K. E. Baidoo and M. W. Brechbiel, *Adv. Drug Del. Rev.* (2012).
40. H. Ding and F. Wu, *Theranostics* **2**, 1040–1053 (2012).
41. V. P. Torchilin (Ed.), *Handbook of Targeted Delivery of Imaging Agents*, Boca Raton, FL: CRC Press (1995), p.153.
42. D. A. Groneberg, M. Giersig, T. Welte and U. Pison, *Curr. Drug Targets* **7**, 643–648 (2006).
43. S. E. Seltzer, *Radiology* **171**, 19–21 (1989).
44. W. Krause, *Adv. Drug Deliv. Rev.* **37**, 159–173 (1999).
45. C. J. Xu, G. A. Tung and S. H. Sun, *Chem. Mater.* **20**, 4167–4169 (2008).
46. Q. Y. Cai, S. H. Kim, K. S. Choi, S. Y. Kim, S. J. Byun, K. W. Kim, S. H. Park, S. K. Juhng and K. H. Yoon, *Invest. Radiol.* **42**, 797–806 (2007).
47. O. Rabin, J. M. Perez, J. Grimm, G. Wojtkiewicz and R. Weissleder, *Nat. Mater.* **5**, 118–122 (2006).
48. C. Peng, L. Zheng, Q. Chen, M. Shen, R. Guo, H. Wang, X. Cao, G. Zhang and X. Shi, *Biomaterials* **33**, 1107–1119 (2012).
49. L. Zhuang, J. Gao, Y. Zeng, F. Yu, B. Zhang, M. Li M, H. Derendorf and C. Liu, *Eur. J. Drug Metab. Pharmacokinet.* **36**, 61–69 (2011).
50. W. Krause, A. Schönborn and K. Rupp, *J. Liposome Res.* **21**, 229–236 (2011).
51. M. Silindir, S. Erdoğan, A. Y. Ozer, A. L. Doğan, M. Tuncel, Ö. Uğur and V. P. Torchilin, *J. Lip. Res.* **23**, 20–27 (2013).
52. J. Zheng, G. Perkins, A. Kirilova, C. Allen and D. A. Jaffray, *Invest. Radiol.* **41**, 339–348 (2006).
53. Y. Liu, K. Ai and L. Lu, *Acc. Chem. Res.* **45**, 1817–1827 (2012).
54. W. H. Kong, W. J. Lee, Z. Y. Cui, K. H. Bae, T. G. Park, J. H. Kim , K. Park K and S. W. Seo, *Biomaterials* **28**, 5555–5561 (2007).
55. V. P. Torchilin, *Adv. Drug Del. Rev.* **54**, 235–252 (2002).
56. V. P. Torchilin, M. Frank-Kamenetsky and G. Wolf, *Acad. Radiol.* **6**, 61–65 (1999).
57. A. de Vries, E. Custers, J. Lub, S. van den Bosch, K. Nicolay and H. Grüll, *Biomaterials* **31**, 6537–6544 (2010).

58. P. Ehrlich P, *Proc. R. Soc. Lond.* **66**, 424–448 (1900).
59. P. V. Kulkarni, C. A. Roney, P. P. Antich, F. J. Bonte, A. V. Raghu and T. M. Aminabhavi, *WIREs Nanomed. Nanobiotechnol.* **2**, 35–47 (2010).
60. A. Mukerjee, A. P. Ranjan and J. K. Vishwanatha, *Curr. Med. Chem.* **19**, 3714–3721 (2012).
61. A. D. Bangham, M. M. Standish and J. C. Watkins, *J. Mol. Biol.* **13**, 238–252 (1965).
62. J. U. Leike, A. Sachse and K. Rupp, *Invest. Radiol.* **36**, 303–308 (2001).
63. T. S. Desser, D. L. Rubin, H. Muller, G. L. McIntire, E. R. Bacon and J. L. Toner, *Acad. Radiol.* **6**, 176–183 (1999).
64. J. Petersein, B. Franke, X Fouillet and B. Hamm, *Invest. Radiol.* **34**, 401–409 (1999).
65. S. E. Seltzer, *Invest Radiol.* **1**, S122–S125 (1988).
66. S. Erdogan, A. Y. Ozer and A. A. Hıncal, *STP Pharma. Sci.* **8**, 127–132 (1998).
67. N. Oku, Y. Tokudome, T. Asaia and H. Tsukada, *Curr. Pharm. Des.* **6**, 1669–1691 (2000).
68. G. Blume and G. Cevc, *Biochim. Biophys. Acta.* **1029**, 91–97 (1990).
69. A. L. Klibanov, K. Maruyama, V. P. Torchilin and L. Huang, *FEBS Lett.* **268**, 235–237 (1990).
70. T. M. Allen, C. Hansen, F. Martin, C. Redemann and A. Yau-Young, *Biochim. Biophys, Acta.* **1066**, 29–36 (1991).
71. V. P. Torchilin, *Nat. Rev. Drug Discov.* **4**, 145–60 (2005).
72. S. M. Moghimi, A. C. Hunter and J. C. Murray, *Pharmacol. Rev.* **53**, 283–318 (2001).
73. P. Sapra, P. Tyagi and T. M. Allen, *Curr. Drug Deliv.* **2**, 369–381 (2005).
74. O. P. Medina, Y. Zhu and K. Kairemo, *Curr. Pharm. Des.* **10**, 2981–2989 (2004).
75. M. Silindir, A. Y. Ozer and S. Erdoğan, *Drug Deliv.* **19**, 68–80 (2012).
76. P. Goyal, K. Goyal, S. G. Kumar, A. Singh, O. P. Katare and D. N. Mishra, *Acta Pharmaceut.* **55**, 1–25 (2005).
77. S. Erdogan, A. Y. Ozer, M. T. Ercan, K. Aydın and A. A. Hıncal, *STP Pharma. Sci.* **8**, 133–7 (1998).
78. T. Schneider, A. Sachse, G. Rossling and M. Brandl, *Int. J. Pharm.* **117**, 1–12 (1995).
79. S. Mukundan, K. B. Ghaghada, C. T. Badea, C. Y. Kao, L. W Hedlund, J. M. Provenzale, G. A. Johnson, E. Chen, R. V. Bellamkonda and A. Annapragada, *Am. J. Roentgenol.* **186**, 100–7 (2006).
80. S. E. Seltzer, M. Blau, L. W. Herman, R. L. Hooshmand, L. A. Herman, D. F. Adams, S. R. Minchey and A. S. Janoff, *Radiology* **194**, 775–781 (1995).
81. G. Schuhmann-Giampieri, J. Like, A. Sachse and W. Krause, *J. Lip. Res.* **4**, 239–253 (1994).
82. G. L. Wolf, in *Handbook of Targeted Delivery of Imaging Agents,* V. P. Torchilin (Ed.) Boca Raton, FL: CRC Press (1995), p. 3.
83. G. L. Wolf, *Adv. Drug Deliv. Rev.* **37**, 1–12 (1999).
84. S. Kweon, H. J. Lee, W. J. Hyung, J. Suh, J. S. Lim and S. J. Lim, *Pharm. Res.* **27**, 1408–1415 (2010).
85. D. B. Elrod, R. Partha, D. Danila, S. W. Casscellsa and J. L. Conyers, *Nanomed. Nanotechnol. Biology Med.* **5**, 42–45 (2009).
86. C. Tilcock, Q. F. Ahkong and D. Fisher, *Nucl. Med. Biol.* **21**, 89–96 (1994).
87. V. S. Trubetskoy, M. D. Frank-Kamenetsky, K. R. Whiteman, A. Bao, B. Goins, R. Klipper, G. Negrete and W. T. Phillips, *J. Nucl. Med.* **44**, 1992–1999 (2003).

88. A. Bao, B. Goins, R. Klipper, G. Negrete, M. Mahindaratne and W. T. Phillips, *J. Pharm. Sci.* **92**, 1893–1904 (2003).
89. A. Bao, B. Goins, R. Klipper, G. Negrete and W. T. Phillips, *J. Pharmacol. Exp. Ther.* **308**, 419–425 (2004).
90. A. Bao, B. Goins, R. Klipper, G. Negrete and W. T. Phillips, *J. Nucl. Med.* **44**, 1992–1999 (2003).
91. W. T. Phillips, A. S. Rudolph, B. Goins, J. H. Timmons, R. Klipper and R. Blumhardt, *Int. J. Rad. Appl. Instrum. B* **19**, 539–547 (1992).
92. M. C. Woodle, *Nucl. Med. Biol.* **20**, 149–155 (1993).
93. H. Essien and K. J. Hwang, *Biochim. Biophys. Acta* **944**, 329–336 (1988).
94. S. Erdogan, A. Y. Ozer, M. T. Ercan and A. A. Hincal, *J. Microencapsul.* **17**, 459–465 (2000).
95. A. L. Petersen, T. Binderup, R. I. Jølck, P. Rasmussen, J. R. Henriksen, A. K Pfeifer, A. Kjær and T. L. Andresen, *J. Control. Release* **160**, 254–263 (2012).
96. K. J. Harrington, S. Mohammadtaghi, P. S. Uster, D. Glass, A. M. Peters, R. G. Vile and J. S. W. Stewart, *Clin. Cancer Res.* 7, 243–254 (2001).
97. H. Belhaj-Tayeb, D. Briane, J. Vergote, S. Kothan, G. Léger, S. E. Bendada, M. Tofighi, F. Tamgac, A. Cao and J. L. Moretti, *Eur. J. Nucl. Med. Mol. Imaging* **30**, 502–509 (2003).
98. S. Dagara, A. Krishnadasa, I. Rubinstein, M. J. Blend and H. Önyüksel, *J. Control. Release* **91**, 123–133 (2003).
99. O. C. Boerman, P. Laverman, W. J. G. Oyen, F. H. M. Corstens and G. Storm, *Prog. Lipid Res.* **39**, 461–475 (2000).
100. V. D. Awasthi, B. Goins, R. Klipper and W. T. Phillips, *Nucl. Med. Biol.* **25**, 155–160 (1998).
101. J. R. Morgan, L. A. Williams and C. B. Howard, *Br. J. Radiol.* **58**, 35–39 (1985).
102. C. Tilcock, *J. Lip. Res.* **4**, 909–936 (1994).
103. D. D. Spragg, D. R. Alford, R. Greferath, C. E. Larsen, K. D. Lee, G. C. Gurtner, M. I. Cybulsky, P. F. Tosi, C. Nicolau and M. A. Gimbrone, *Proc. Natl. Acad. Sci.*, **94**, 8795–800 (1997).
104. L. C. Chen, C. H. Chang, C. Y. Yu, Y. J. Chang, W. C. Hsu, C. L. Ho, C. H. Yeh, T. Y. Luo, T. W. Lee and G. Ting, *Nucl. Med. Biol.* **34**, 415–423 (2007).
105. E. W. Izaguirre, M. Sun, D. C. Drummond, D. B. Kirpotin, T. Funk, S. Thompson, J. Carver, M. E. Hayes, M. Wendland, M. Knudsen and B. H. Hasegawa, *IEEE Nucl. Sci. Symp. Conf. Rec.* (2005).
106. Y. J. Chang, C. H. Chang, T. J. Chang, C. Y. Yu, L. C. Chen, M. L. Jan, T. Y. Luo and T. W. Lee, *Anticancer Res.* **27**, 2217–2225 (2007).
107. U. K. Dar, I. Khan, M. Javed, M. Ali, S. W. Hyder and S. Murad, *Pak. J. Pharm. Sci.* **26**, 333–337 (2013).
108. P. McQuade, L. C. Knight and M. J. Welch, *Bioconjug. Chem.* **15**, 988–996 (2004).
109. D. Wang, M. Sima, R. L. Mosley, J. P. Davda, N. Tietze, S. C. Miller, P. R. Gwilt, P. Kopecková and J. Kopecek, *Mol. Pharm.* **3**, 717–725 (2006).
110. M. Silindir, S. Erdoğan, A. Y. Özer, A. L. Doğan, M. Tuncel, Ö. Uğur and V. P. Torchilin, *Eur.J. Pharm. Sci.* **38**, 129–130 (2009).
111. S. Erdogan, A. Roby and V. P. Torchilin, *Mol. Pharm.* **3**, 525–530 (2006).
112. S. Erdogan, Z. O. Medarova, A. Roby, A. Moore and V. P. Torchilin, *J. Magn. Reson. Imaging* **27**, 574–580 (2008).

113. S. Erdogan and V. P. Torchilin, *Methods Mol. Biol.* **605**, 321–334 (2010).
114. V. V. Weissig, J. Babich and V. P. Torchilin, *Colloids Surf. B Biointerfaces* **18**, 293–299 (2000).
115. V. S. Trubetskoy, G. S. Gazelle, G. L. Wow and V. P. Torchilin, *J. Drug Target.* **4**, 381–388 (1997).
116. J. Marik, M. S. Tartis, H. Zhang, J. Y. Fung, A. Kheirolomoom, J. L. Sutcliffe and K. W. Ferrara, *Nucl. Med. Biol.* **34**, 165–171 (2007).
117. N. Nishiyama and K. Kataoka, *Pharmacol. Ther.* **112**, 630–648 (2006).
118. M. Malmsten and B. Lindman, *Macromolecules* **25**, 5440–5445 (1992).
119. G. Kwon and K. Kataoka, *Adv. Drug Deliv. Rev.* **16**, 295–309 (1995).
120. M. Almgren, W. Brown and S. Hvidt, *Colloid Polym. Sci.* **273**, 2–15 (1995).
121. X. Y. Xiong, K. C. Tam and L. H. Gan, *J. Nanosci. Nanotechnol.* **6**, 2638–2650 (2006).
122. A. V. Kabanov, E. V. Batrakova and V. Y. Alakhov, *J. Control. Release* **82**, 189–212 (2002).
123. A. Roby, S. Erdogan and V. P. Torchilin, *Cancer Biol. Ther.* **6**, 1136–1142 (2007).
124. T. A. Elbayoumi, S. Pabba, A. Roby and V. P. Torchilin, *J. Lip. Res.* **17**, 1–14 (2007).
125. V. S. Trubetskoy, M. D. Frank-Kamenetsky, K. R. Whiteman, G. L. Wolf and V. P Torchilin, *Acad. Radiol.* **3**, 232–238 (1996).
126. A. Bosman, H. Janssen and E. Meijer, *Chem. Rev.* **99**, 1665–1688 (1999).
127. B. Klajnert and M. Bryszewska, *Acta Biochimica Polonica* **48**, 199–208 (2001).
128. M. Liu and J. M. Fréchet, *Pharm. Sci. Technol. Today* **2**, 393–401, (1999).
129. A. M. Caminade and J. P. Majoral, *Acc. Chem. Res.* **37**, 341 (2004).
130. S. Wen, H. Liu, H. Cai, M. Shen and X. Shi, *Adv. Health. Mater.* (2013).
131. A. Saovapakhiran, A. D'Emanuele, D. Attwood and J. Penny, *Bioconjug. Chem.* **20**, 693–701 (2009).
132. Y. Cheng, L. Zhao, Y. Li and T. Xu, *Chem. Soc. Rev.* **40**, 2673–2703 (2011).
133. A. Caminade, R. Laurent and J. Majoral, *Adv. Drug Deliv. Rev.* **57**, 2130–2146 (2005).
134. S. Mignani S. El Kazzouli, M. Bousmina and J. P. Majoral, *Adv. Drug Deliv. Rev.* (2013).
135. Y. Y. Cheng and T. W. Xu, *Eur. J. Med. Chem.* **43**, 2291–2297 (2008).
136. Y. W. Li, Y. Y. Cheng and T. W. Xu, *Curr. Drug Discov. Technol.* **4**, 246 (2007).
137. U. Boas and P. M. H. Heegaard, *Chem. Soc. Rev.* **33**, 43 (2004).
138. H. Y. Nam, K. Nam, M. Lee, S. W. Kim and D. A. Bull, *J. Control. Release* **160**, 592–600 (2012).
139. H. Arima, M. Arizono, T. Higashi, A. Yoshimatsu, H. Ikeda, K. Motoyama, K. Hattori, T. Takeuchi, F. Hirayama and K. Uekama, *Cancer Gene Ther.* **19**, 358–366 (2012).
140. M. T. Morgan, Y. Nakanishi, D. J. Kroll, A. P. Griset, M. A. Carnahan, M. Wathier, N. H. Oberlies, G. Manikumar, M. C. Wani and M. W. Grinstaff, *Cancer Res.* **66**, 11913–11921 (2006).
141. U. Gupta, H. B. Agashe, A. Asthana and N. K. Jain, *Nanomedicine* **2**, 66–73 (2006).
142. N. Malik, R. Wiwattanapatapee, R. Klopsch, K. Lorenz, H. Frey, J. W. Weener, E. W. Meijer, W. Paulus and R. Duncan, *J. Control. Release* **65**, 133–148 (2000).
143. S. G Sampathkumar and K. J Yarema, in *Nanotechnologies for the Life Sciences*, Vol. 6, C. S. S. R. Kumar (Ed.) Hoboken, NJ: John Wiley & Sons (2007), p. 1.
144. K. Sano, T. Temma, T. Azuma, R. Nakai, M. Narazaki, Y. Kuge and H. Saji, *Mol. Imaging Biol.* **13**, 1196–1203 (2011).
145. Z. Huang, R. S. Sengar, A. Nigam, M. C. Abadjian, D. M. Potter, D. B. Grotjahn and E. C. Wiener, *Invest. Radiol.* **45**, 641–654 (2010).

146. A. Bumb, M. W. Brechbiel and P. Choyke, *Acta. Radiol.* **51**, 751–767 (2010).
147. N. Kosaka, M. Bernardo, M. Mitsunaga, P L. Choyke and H. Kobayashi, *Contrast Media Mol. Imaging* 7, 247–253 (2012).
148. H. Kobayashi, M. Ogawa, N. Kosaka, P. L. Choyke and Y. Urano, *Nanomedicine* (*Lond*) **4**, 411–419 (2009).
149. X. Xu, Y. Zhang, X. Wang, X. Guo, X. Zhang, Y. Qi and Y. M. Shen, *Bioorg. Med. Chem.* **19**, 1643–1648 (2011).
150. Y. Zhang, Y. Sun, X. Xu, X. Zhang, H. Zhu, L. Huang, Y. Qi and Y. M. Shen, *J. Med. Chem.* **53**, 3262–3272 (2010).
151. Y. Fu, D. E. Nitecki, D. Maltby, G. H. Simon, K. Berejnoi, H. J. Raatschen, B. M. Yeh, D. M. Shames and R. C. Brasch, *Bioconjug. Chem.* **17**, 1043–1056 (2006).
152. C. Peng, K. Li, X. Cao, T. Xiao, W. Hou, L. Zheng, R. Guo, M. Shen, G. Zhang and X. Shi, *Nanoscale* **4**, 6768–6778 (2012).
153. H. Wang, L. Zheng, C. Peng, M. Shen, X. Shi and G. Zhang, *Biomaterials* **34**, 470–480 (2013).
154. H. Liu, M. Shen, J. Zhao, R. Guo, X. Cao, G. Zhang and X. Shi, *Colloids Surf. B Biointerfaces* **94**, 58–67 (2012).
155. C. Kojima, Y. Umeda, M. Ogawa, A. Harada, Y. Magata and K. Kono, *Nanotechnology* **21**, 245104 (2010).
156. A. T. Yordanov, A. L. Lodder, E. K. Woller, M. J. Cloninger, N. Patronas, D. Milenic and M. W. Brechbiel, *NanoLett.* **2**, 595–599 (2002).
157. H. Liu, Y. Xu, S. Wen, Q. Chen, L. Zheng, M. Shen, J. Zhao, G. Zhang and X. Shi, *Chemistry* (2013).
158. Y. Zhang, Y. Sun, X. Xu, H. Zhu, L. Huang, X. Zhang, Y. Qi and Y. M. Shen, *Bioorg. Med. Chem. Lett.* **20**, 927–931 (2010).
159. M. Mamede, T. Saga, T. Ishimori, T. Higashi, N. Sato, H. Kobayashi, M. W. Brechbiel and J. Konishi, *J. Control. Release* **95**, 133–141 (2004).
160. S. K. Lee and X. Chen, *Mol. Imaging* **8**, 87–100 (2009).
161. M. Shilo, T. Reuveni, M. Motiei and R. Popovtzer, *Nanomedicine* (*Lond.*) 7, 257–269 (2012).
162. J. Gong, F. Zhou, Z. Li and Z. Tang, *Langmuir* **28**, 8959–8964 (2012).
163. J. Choi, J. Yang, J. Park, E. Kim, J. Suh, Y. M. Huh and S. Haam, *Adv. Funct. Mater.* **21**, 1082–1088 (2011).
164. R. Choi, J. Yang, J. Choi, E. K. Lim, E. Kim, J. S. Suh, Y. M. Huh and S. Haam, *Langmuir* **26**, 17520–17527 (2010).
165. J. Lee, J. Yang, H. Ko, S. Oh, J. Kang, J. Son, K. Lee, S. W. Lee, H. G. Yoon, J. S. Suh, Y. M. Huh and S. Haam, *Adv. Funct. Mater.* **18**, 258–264 (2008).
166. E. Hao, R. C. Bailey, G. C. Schatz, J. T. Hupp and S. Li, *Nano Lett.* **4**, 327–330 (2004).
167. L. Manna, D. J. Milliron, A. Meisel, E. C. Scher and A. P. Alivisatos, *Nat. Mater.* **2**, 382–385 (2003).
168. J. He, Y. Wang, Y. Feng, X. Qi, Z. Zeng, Q. Liu, W. S. Teo, C. L. Gan, H. Zhang and H. Chen, *ACS Nano.* 7, 2733–2740 (2013).
169. Y. Su, B. Mayers and Y. Xia, *Nano Lett.* **3**, 675–679 (2003).
170. J. Turkevich, P. C. Stevenson and J. Hillier, *Discuss. Faraday Soc.* **11**, 55–75 (1951).
171. M. Giersig and P. Mulvaney, *Langmuir* **9**, 3408–3413 (1993).
172. E. Boisselier and D. Astruc, *Chem. Soc. Rev.* **38**, 1759–1782 (2009).

173. R. A. Sperling, P. R. Gil, F. Zhang, M. Zanella and W. J. Parak, *Chem. Soc. Rev.*, **37**, 1896–1908 (2008).
174. T. Reuveni, M. Motiei, Z. Romman, A. Popovtzer and R. Popovtzer, *Int. J. Nanomed.* **6**, 2859–2864 (2011).
175. E. K Lim, E. Jang, J. Kim, T. Lee, E. Kim, H. S. Park, J. S. Suh, Y. M. Huh and S. Haam, *J. Mater. Chem.* **22**, 17518–17524 (2012).
176. N. Chanda, V. Kattumuri, R. Shukla, A. Zambre, K. Katti, A. Upendran, R. R. Kulkarni, P. Kan, G. M. Fent, S. W. Casteel, C. J. Smith, E. Boote, J. D. Robertsond, C. Cutlerd, J. R. Lever, K. V. Katti and R. Kannan, *Proc. Natl. Acad. Sci.* **107**, 8760–8765 (2010).
177. N. Chanda, R. Shukla, A. Zambre, S. Mekapothula, R. R. Kulkarni, K. Katti, K. Bhattacharyya, G. M. Fent, S. W. Casteel, E. J. Boote, J. A. Viator, A. Upendran, R. Kannan and K. V. Katti, *Pharm. Res.* **28**, 279–291 (2011).
178. O. Rabin, J. M. Perez, J. Grimm, G. Wojtkiewicz and R. Weissleder, *Nat. Mater.* **5**, 118–122 (2006).
179. H. Liu, H. Wang, R. Guo, X. Cao, J. Zhao, Y. Luo, M. Shen, G. Zhang and X. Shi, *Polym. Chem.* **1**, 1677–1683 (2010).
180. M. H. Oh, N. Lee, H. Kim, S. P. Park, Y. Piao, J. Lee, S. W. Jun, W. K. Moon, S. H. Choi and T. Hyeon, *J. Am. Chem. Soc.* **133**, 5508–5515 (2011).
181. Z. Liu, Z. Li, J. Liu, S. Gu, Q. Yuan, J. Ren and X. Qu, *Biomaterials* **33**, 6748–6757 (2012).
182. Q. Tian, C. N. Zhang, X. H. Wang, W. Wang, W. Huang, R. T. Cha, C. H. Wang, Z. Yuan, M. Liu, H. Y. Wan and H. Tang, *Biomaterials* **31**, 4748–4756 (2010).
183. M. Zhou, R. Zhang, M. Huang, W. Lu, S. Song, M. P. Melancon, M. Tian, D. Liang and C. Li, *J. Am. Chem. Soc.* **132**, 15351–15358 (2010).
184. C. R. M. Glaus, Development and analysis of radiolabeled magnetic nanoparticles for positron emission tomography and magnetic resonance imaging, Georgia Institute of Technology Department of Biomedical Engineering, PhD Thesis (2008).
185. J. Cheon and J. H. Lee. *Acc. Chem. Res.* **41**, 1630–1640 (2008).

Chapter 4

Nanotargeted Radiopharmaceuticals for Nuclear Imaging and Radiotherapy

*Gann Ting**,‡, *Yun-Long Tseng*†, *Chih-Hsien Chang** *and Te-Wei Lee**

**Institute of Nuclear Energy Research, Taoyuan, Taiwan*
†*Taiwan Liposome Company, Ltd, Taipei, Taiwan*
‡*gann.ting@msa.hinet.net*

Abstract

Nanotechnology holds significant promise for improving the development of new and paradigm shift tools for the targeted personalized diagnostics and therapeutics of cancer and other diseases. Current progress in nanotechnology has exploited the possibility of designing tumor-targeted nanocarriers that can deliver radionuclide payloads in a site or molecular selective manner to improve the efficacy and safety of nuclear imaging and radiotherapy. Radionuclides of auger electron-, α-, β-, γ- and positron-emitters have been after-loaded, inserted and surface bioconjugated in nanoparticles (NPs) to improve the efficacy and to reduce the toxicity in preclinical studies. In addition, the combining of disease-specific multifunctional, multimodality and multivalent delivery of nanocarriers will hopefully achieve earlier disease detection and better treatment.

The aim of this chapter is to provide an overview of nanotargeted radionuclides, up-to-date research and development, current status of applications, advantages, problems and future prospects and challenges. Two major strategies of passively and actively nanotargeted radiopharmaceuticals with illustrating examples are briefly reviewed and summarized. Current advances in research of passive targeting co-delivery of radionuclides and chemotherapeutics, and advantages of nanocarriers in multifunctional and multimodality application for nuclear imaging and radiotherapy are discussed. Finally, radiotoxicology/nanotoxicology, clinical research and regulatory issues are also briefly explored. Research on the combining different modes of selective delivery of radionuclides with targeted nanocarriers offers the new possibility of increasing personalized diagnostic efficacy and radiotherapeutic index. However, further efforts and challenges in preclinical and clinical efficacy and toxicity studies are required to translate those advanced new technologies to the clinical applications for the healthcare benefit of the patients.

1. Introduction

Cancer is an upregulated biological process of cell growth with an ability of tumor cells to invade and metastasize. A century ago, Paul Ehrlich hypothesized that a "magic bullet" could be developed selectively to target cancer.[1] Over the past few decades, the progress in molecular biology and the understanding of malignant transformation and tumorigenesis have revealed the recognition of the widespread applicability of the hallmarks of cancer concepts,[2] which will increasingly affect the development of new means to treat human cancer in personalized and translational medicine. There are two major classes of antitumor therapeutic strategies: (i) application of molecularly targeted therapeutics to block major hallmarks of cancer cell pathways, and (ii) employment of novel drug delivery systems through tumor-targeted nanomedicines to improve the pharmacokinetics and bioavailability of vehicle-carried radionuclides or drugs.

Conventional anticancer drugs exhibit a lack of specificity, unfavorable pharmacokinetics and high normal tissue damage or toxicity. In addition, many antineoplastic agents show low stability, poor solubility, low permeability and/or a short half-life, and as a consequence the nanodelivery strategies are critical for improving the efficacy and safety of the targeted drugs. Nanotechnology can bring fundamental changes to the study and understanding of biological processes in health and disease, as well as enable novel personalized diagnostics and therapeutics for treating cancer and many other diseases. Thus, advances made based on the nanotechnology research and development could result in progress of disease treatment and healthcare. Targeted drug delivery systems such as passive and active targeting NPs or nanocarriers, with diameters ranging from 10–200 nm have been developed to improve the biodistribution, pharmacology, therapeutic and toxicity properties of agents used in cancer and disease diagnostics and therapeutics.[3–15]

Metastasis accounts for the vast majority of cancer deaths; nanocarriers have many potential benefits for diagnosing and treating metastatic cancer, including the ability to transport complex molecular cargoes to the major tumor sites of metastasis, such as the lungs, liver and lymph nodes, as well as targeting to specific cell populations within these organs.[12] Many current therapies are not reaching the sites of metastases. Nanocarriers have the potential to combine multiple diagnostic and therapeutic functions into a single platform, and it can be targeted to specific sites or tissues, can reach particular subcellular compartments and can treat the complex problem of metastatic cancer.[12] The status of the development of targeting delivery systems, including targeting strategies, potential applications and the prospects of tumor-targeted nanocarriers have been reviewed and discussed.[3–15] Nanotechnology is increasing the attention of the biomedical community, owing to unique prospects for targeted delivery in imaging, therapy and drug delivery. Nanotechnology is expected to transform current treatment systems by providing more efficient diagnostics and therapeutics to cancer and other disease. Today, nanocarriers are used in detecting cancer at an early stage, delivering anticancer drugs specifically to malignant cells and determining concurrently if these drugs are killing malignant cells.[5–7,9–15]

Many therapeutic nanocarrier liposomes and albumin NPs have been approved by US Food and Drug Administration (FDA) for clinical practices.[7,8,10] Pegylated liposomal doxorubicin represents a new frontier class of nanochemotherapy delivery system that significantly improves the therapeutic index of doxorubicin through improving therapeutic pharmacokinetics.[3] The "Abraxane®" for injectable suspension (paclitaxel protein-bound particles for injectable suspension) (albumin-bound) has also approved by FDA as the first protein-bound NP drug. Because tumor tissues have a greater concentration of blood vessels than healthy tissues, there is a greater opportunity for long-circulating liposomal drugs and albumin-bound drugs to reach tumors versus healthy tissues.[4] As nanocarriers are evaluated for safety and efficacy, nanotechnology will bring significant advances in molecular imaging and specific targeting of tumor therapeutic agents, elevating therapeutic efficacy and finally achieving the goal of early detection and control of cancer. Key factors relevant to the design of practical nanomedicines and the regulatory mechanisms designed to ensure safe and timely realization of healthcare benefits are discussed.[11] Personalized and customized passive and active nanoscale constructs can serve as targeted drug delivery vehicles capable of delivering large doses of radionuclide or chemotherapeutic agents into malignant cells, while sparing normal tissues, greatly reducing the side effects that usually accompany many current cancer therapies.[5–16]

Monoclonal antibody (mAb)-guided internal radiation therapy, or radioimmunotherapy, demonstrated promise in preclinical and clinical anticancer applications.[17–21] The principles and applications of molecular targeting involving radionuclide methods for tumor nuclear imaging and therapy are reviewed and discussed.[22] Two radiolabeled anti-CD20 mAbs ^{90}Y-ibritumomab

(Zevalin®) and ^{131}I-tositumomab (Bexxar®) were approved by US FDA in 2002 and 2003, respectively, for treatment of B-cell non-Hodgkin lymphoma, which indicates the potential benefit of antibody-guided systemic radionuclide-targeted therapy.[18,20,21] However, tumor-targeting studies with radiolabeled mAbs also showed some limitations, such as, inefficient targeting and low accumulation in tumor sites (e.g. <0.1% of injection dose/g (%ID/g) of tumor for human) and irradiation of normal tissues by long circulation of mAbs. Emerging new methods for improving the specific uptake of radionuclides in tumor cells while sparing the normal tissues need to be established. Several advanced strategies for radionuclide-targeted delivery have been extensively studied, including the combination of chemotherapy agents with particle-emitting radionuclides and the development of novel multimodality and multifunctional nanotargeted therapeutics.[23,24] Optimization of treatment protocols has significantly improved the therapeutic efficacy and reduced toxicity in normal tissues. NPs delivering radionuclides for improving the pharmacokinetics and therapeutic efficacy of cancer have been presented elsewhere.[6,23–25] Cancer nanotargeted radiopharmaceuticals for tumor imaging and therapy have been presented and summarized.[26] The goal of this chapter is to review and summarize the update research and development progress, the basic concept of nanotechnology applications in diagnostic and therapeutic radiopharmaceuticals, the future opportunities and challenges of advanced NPs to deliver radionuclides for cancer and other disease, as well as the problems and advantages of nuclear imaging and internal targeted radiotherapeutics translation from preclinical to clinical applications.

2. NPs and Targeted Principle

Major challenges of drug delivery carriers in cancer diagnostics and therapeutics are the low drug bioavailability within tumors and the high toxicities to normal organs.[4–9] Targeted radionuclide therapy is often limited by insufficient delivery of radionuclides to tumor sites using the currently available targeting strategies, such as mAbs and peptides, due to relatively low and heterogeneous expression of receptors on tumor cells, as well as dose-limiting toxicities to normal tissues.[17,19] To maximize the therapeutic index and to minimize the side effects, it is very important to deliver the radionuclides to the right site at the right concentration and at the right time. The rapidly advancing field of cancer nanotechnology has generated several innovative radionuclides and drug delivery systems, such as liposomes,[27–37] iron oxide (IO),[38–43] polymers,[44–46] dendrimers,[47] quantum dots (QD),[48–52] carbon nanotubes,[53,54] gold nanoshells[55] and latex,[56] to improve and enhance targeted transport of cytotoxic drugs or radionuclides to tumor lesions.[23–26,57–59] It is estimated that ~240 nano-enabled products entered pharmaceutical research pipelines in 2006.[60] These nanocarrier systems could provide the delivery platforms needed for improving the delivery of drugs and radionuclides to tumor sites.

NP or nanocarrier delivery systems have also revealed enhanced imaging and therapeutic efficacy by targeted delivery of drugs to the tumor site and by reducing their toxic side effects.[4–10,12,14] Major advantages of nanocarriers are that they can be prepared in sizes <100 nm and it can ensure minimal drug or radionuclide leaking out from the carriers during circulation to enhance passive targeting and to increase the localization of drugs and radionuclides in the tumor through enhanced permeability and retention (EPR) effect to the leaky tumor tissues,[4,6,9] or through the NP surface bioconjugation, while sparing nontargeted tissue, to facilitate intracellular drug or radionuclide delivery and uptake by molecular active targeting.[9,34,58,59,61] Two major mechanisms for radionuclide- or drug-targeted accumulation delivery system of NPs to tumor tissue sites are: (i) site-specific passive tumor targeting and (ii) molecular affinity and site-specific active tumor targeting for tumor diagnostics and therapy as illustrated in Fig. 1. For chemotherapeutics, the challenge is to design safe, approvable and easily scaled-up nanocarrier products that simultaneously meet the requirements of drug "retention" in blood circulation versus "release" in tumor tissue or intracellular, and stealthy in blood circulation versus "sticky" in tumor cell-binding at the right places in order to deliver a cytosolic drug dose lethal to cancer cells with minimized side effects.[14]

There are three generations of nanocarriers for radionuclides delivery[26,28–30,32,35]: (i) the first generation of nanocarriers (passive targeting) which are rapidly trapped in the reticuloendothelial system (RES) organs (e.g. liver and/or spleen); (ii) the second generation of sterically stabilized PEGylated nanocarriers (passive targeting to angiogenesis or inflammatory sites), which can evade the RES of the liver and spleen, enjoys a prolonged circulation in the blood and allows for passive targeting through the EPR effect in leaky tumor tissues; and (iii) the third generation of nanocarriers with a bioconjugated surface modification of the NPs using specific antibodies, peptides, aptamers or ligand-like molecules to actively target specific tumor or tissues through molecular interaction or affinity (active targeting).[6,26,38–40,43,44,47–54,58,59,61–69] The pharmacokinetics and bioavailability of drugs and radionuclides delivered by the third-generation nanocarriers have been much improved. There are three major challenges of applying NPs for delivery of drugs or radionuclides: (i) synthesizing nanocarriers with stealth characteristics which can decrease

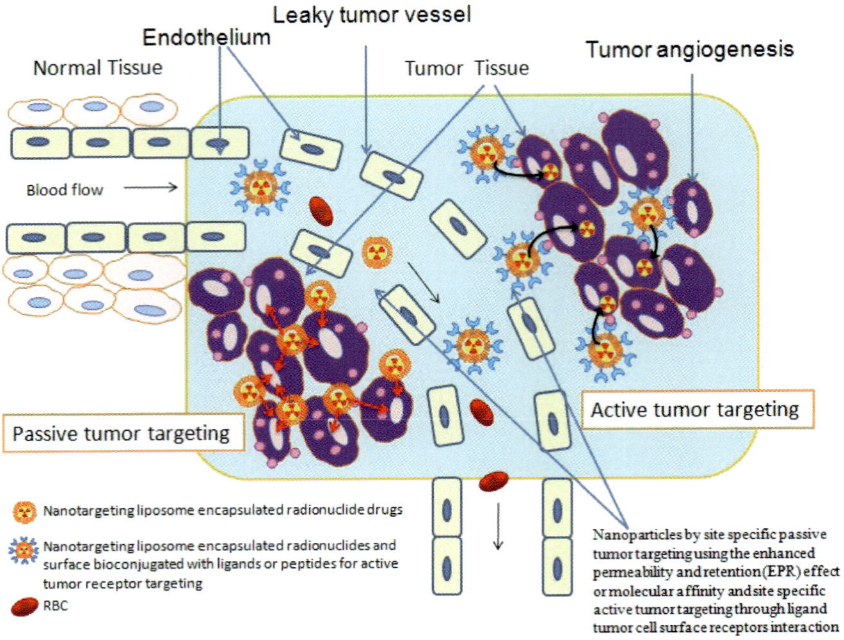

Fig. 1. Schematic illustration showing the possible mechanism or strategies for radionuclide drug accumulation delivery system of NPs by site-specific passive tumor targeting using the EPR effect for leaky tumor angiogenesis vessel, and molecular affinity and site-specific active tumor targeting through ligand–tumor cell surface receptors interaction, internalization and intracellular action for nanotargeted tumor diagnostics and therapy.

uptake of delivered diagnostic and therapeutic agents in the RES and prolong blood circulation; (ii) producing multifunctional NPs with improved *in vivo* targeting capabilities and interaction with disease biomarkers at the molecular level; (iii) developing reliable approaches to test these new materials *in vitro* and *in vivo* for a fast translation from the bench to the bedside.[11,15,24,26,59]

The advantages of nanomedicine over conventional medicines rely essentially on the EPR effect and their improved pharmacokinetics, loading capacity, release rates, lower systemic toxicity and ability to bypass multidrug resistance (MDR) mechanisms that involve cell-surface protein pumps.[70] However, only a few nanotargeted chemotherapeutics based on the EPR principle have reached the clinic applications over the past two decades. Nuclear imaging and therapeutic radiopharmaceuticals have not yet reached clinical use worldwide. Some variables can influence the outcome of EPR strategies, such as, the validation of animal data (animal model, tumor origin and site of injection), accurate characterization and scale-up of the nanoproducts, the chronic toxicity (the immunotoxicity and nonspecific target toxicity), and the combinational theranostics with multifunctional and multimodal properties. The optimal design and consideration of the NPs and EPR-based drugs should be further enhanced in the future personal and translational preclinical and clinical studies.[14,15,70]

3. Nanopreparations for Nuclear Imaging and Ratherapy

Traditionally the most fundamental and simple method is to entrap the radionuclides in the nanocarriers during liposomes or NPs formed by direct hydration or solvent injection in which the radioactivity is added in the buffer. However, this method is inadvisable and has a low labeling efficiency, generally <10%, and batch production variability. There are two more advanced approaches generally used for nanopreparation, labeling or encapsulating radionuclides on liposomes or NPs (Table 1). In the first approach, the

Table 1. Examples of methods of nanopreparation and radiolabeling of nanotargeted liposome radiopharmaceuticals

Radionuclide	Nanopreparation and radiolabeling method	Radiolabeling strategies	References
^{99m}Tc, ^{111}In, ^{67}Ga, ^{64}Cu, ^{186}Re, ^{188}Re, ^{90}Y, ^{225}Ac.	Encapsulation and after-loading	Radionuclide is loaded into the aqueous liposomal core with membrane permeable lipophilic chelators and changed to hydrophilic chelator in the liposomal core	35, 37, 72–76
^{64}Cu, ^{111}In, ^{99m}Tc and, ^{18}F.	Bilayer insertion, surface bioconjugation or chelation	Radiolabeled lipids are incorporated into the liposomal bilayer or radionuclides, thus covalently attaching the chelator on the surface of the NPs	71, 73, 77–83

radionuclides are after-loaded with membrane permeable lipophilic chelators and trapped in the enclosed aqueous liposomal core with a change to a hydrophilic chelator, this approach to labeling NPs has an important advantage in that it does not leave the radiolabel on the surface of the liposomes or NPs where it might interfere with specific targeting molecules or where it would have the opportunity to interact with natural metal-binding molecules in the body. In the design of the radiolabeling nanoliposome preparation, the major factors to be considered are *in vitro* and *in vivo* radiopharmaceutical purity and stability, higher specific activity and convenient and easy preparation.[23,33,35,54,71]

The radionuclide after-loading method has provided higher labeling efficiencies (>90%) and the greatest *in vivo* stability for ^{99m}Tc, ^{111}In and ^{67}Ga radionuclides for nuclear imaging[23,24,27,3–33,35,59,84–86] as well as ^{186}Re and ^{188}Re radionuclides for radiotherapy.[72,87–92] This kind of approach is also useful in preparing polymeric NP, lutetium-177-loaded poly(D,L-lactide-*co*-glycolide) (PLGA) NPs are formulated by simply incubation of pre-formulated blank PLGA 50:50 or 75:25 NPs by incubating with ^{177}Lu- DOTA-Tyr^3-octreotate (DOTATATE), although the percentage of encapsulation was about 77%.[93] In the second approach, the radionuclides are bounded to a lipid chelator, inserted and loaded into the liposomal bilayer or radionuclide chelators are covalently attached to the surface of the liposomes or NPs after forming.[71,73,77–83] The radionuclides ^{64}Cu, ^{18}F and ^{99m}Tc for imaging were attached to the liposome surface using chelators covalently linked to lipid soluble anchors.[73,77–79,81] Chelating compound can be chemically derivatized by the incorporation of a hydrophobic group, which can anchor the chelating moiety onto the nanoparticle surface during or after preparation. Different chelators and different hydrophobic anchors were tried for the preparation of ^{111}In and ^{99m}Tc.[82] In general, external conjugation suffers from the potential for *in vivo* interactions between the labels and other biomolecules as well as surface crowding that could affect interactions with other functional groups on the liposome, and surface modification may also affect the circulation time and biodistribution. Accessing to the most up-to-date information, the lipid-based NPs, which are combined with auto-assembly and easier size control properties with innovative design for isotope loading into preformed nanocarriers and post-insertion of lipid-conjugated ligand on vesicle surface, are useful for passive and active targeting (Fig. 2).

Ideally, NPs or nanocarriers for targeted therapeutics can be prepared to include the desired entity, e.g. an anticancer agent or radionuclides with characteristics of (i) a high loading efficiency, i.e. high percentage of encapsulated entity relative to the total amount of the entity used in the encapsulation

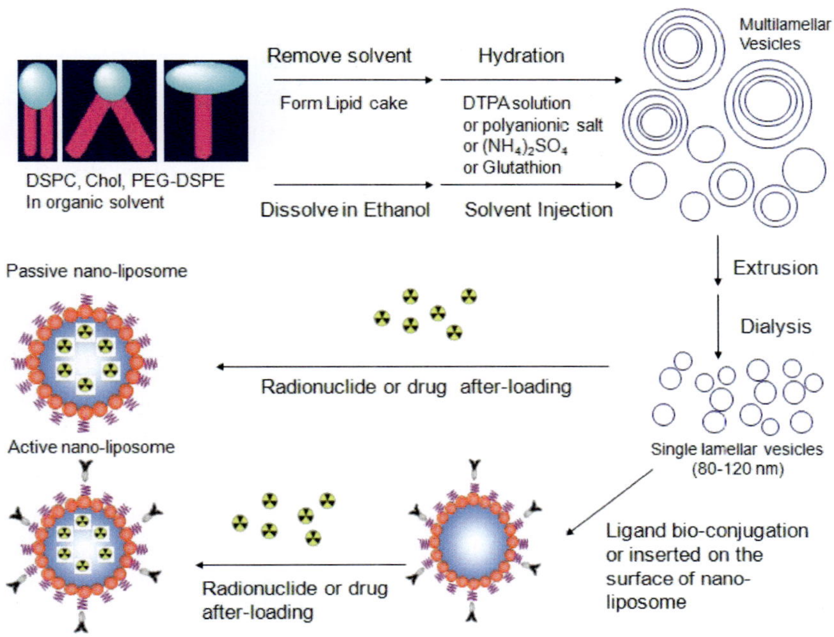

Fig. 2. Schematic illustration of examples showing the nanopreparations of passively and actively targeted liposome radiopharmaceuticals by auto-assembly and after-loading processes, the site and molecular specific target ligand can be bioconjugated or inserted on the bilayer or surface of the nanoliposomes.

process, and (ii) in a stable form, *i.e.* with little release (leakage) of the encapsulated entity on storage or generally before the vehicle reaches the site or the environment where the NP-entrapped entity is expected to exert its intended activity. For radiotherapeutics, ideal radioisotopes are those with an abundance of low γ-penetrating radiations, example: high specific activity β-emitters, α-particle emitters or auger electron emitters.[94–97] This is because when the radiopharmaceuticals reach the disease target, the energy from the radioisotope should be deposited at that site and not irradiate nearby normal tissues. The energy of particles from different radioisotopes and their ranges in tissues will vary, as well as their half-lives, and the most appropriate radioisotope will be different depending on the application, the disease and the accessibility of the disease tissue.[94] Radiopharmaceuticals labeled with low-energy electron emitters, such as ^{111}In, have several key advantages over traditional agents that emit higher-energy particles.[98–100] Unfortunately, the majority of such low-energy electron emitters described in the literature to date have harnessed only a small percentage of the actual cytotoxic potential of auger-emitting radionuclides because of poor drug design. Therefore, there is a need in the art to provide various NPs with passive/active targeting properties that are useful for delivery of a variety of radiotherapeutic, bimodality radiochemotherapeutic, diagnostic or imaging entities. The design of nanopreparation combined with the radionuclide selection involves the optimization of the pharmacokinetics, the biodistribution, the physical half-life and the emission properties of the radionuclide toward maximizing therapeutic effects.[97]

Another advancement in nanoradiopharmaceutics development is that many studies use combined radioisotopes with chemotherapeutic liposomal drugs (i.e. ^{111}In-labeled PEGylated liposomal vinorelbine and ^{188}Re-labeled PEGylated liposomal doxorubicin)[101–104]; this kind of design have the potential to increase the therapeutic index and as the theranostics. Nanopreparations for co-delivery of chemotherapeutics and radiopharmaceuticals are feasible by after-loading approach. The radioisotope can be loaded into anticancer-loaded NPs with reserved loading capacity to achieve a good encapsulated efficiencies (>90%) and the greatest *in vivo* stability[101–104] (Fig. 2). The methods of nanopreparation and labeling strategies for positron-emitting radionuclides of ^{68}Ga, ^{64}Cu, ^{124}I, etc., have been reviewed recently.[105]

One of key elements for successfully translating a nanopreparation from laboratory research into clinical development is scale-up in regard to feasible process, good quality and stability in use and storage. These basic elements hold true for any clinical and commercial development of nanotechnology-based drugs, but even more so for the field of cancer nanomedicine given the challenges, complexities and consequences of optimizing nanoscale technology for good manufacturing practices (GMP) production. Chemistry, manufacturing and controls (CMC) requirements are mandated by current good manufacturing practices (cGMP) and the regulatory agency. These requirements assure among other things batch-to-batch reproducibility and shelf-life stability based on testing a variety of properties (particle size, specific activity and purity, drug release rates, targeting ligand content and activity (if applicable), stability of NPs and drug under storage and in-use conditions).

4. Nanotargeted Radionuclides for Nuclear Imaging and Radiotherapy

The research on tumor-targeted diagnostic and therapeutic radionuclides is one of the potential areas of cancer drug development. Normally, targeted radionuclides consist of two components, a targeting carrier and a trace amount of radionuclide with a specific radiation emitter. The tumor therapeutic efficacy and diagnostic quality are determined by the selectivity or specificity of the targeted delivery systems and the radionuclide radiation characteristics. The selection of potential targeted radionuclides for tumor imaging (Table 2)[22–24,26,31,32,105] and targeted radionuclide for internal radiotherapy (Table 3)[17,19,23–26,98–100,101–104,106,107] involves the physical half-life, decay mode and the emission properties of the radionuclides. Gamma emitters with energy range between 130 and 370 keV can be used for gamma imaging or single photon emission tomography (SPECT).[22–24,31–33] The high-energy positron emitters with annihilation energy at 511 keV can be applied for positron-emission tomography (PET).[23,24,35,105] The major characteristics of nanotargeted nuclear imaging modalities are listed in Table 4.[22–26,57] In functional and molecular imaging, the *in vivo* radionuclide SPECT and PET imaging is the

Table 2. Characteristics of potential radionuclides for nanotargeted nuclear imaging.[22–24,26,31,32,105]

Radionuclide	Production	Emission type	Half-life	$E_{max}(\gamma)$, (keV)	Imaging type
^{131}I	$^{130}I(n, \gamma)^{131}Te$ (β) ^{131}I	γ (81.2%), β	8 days	284, 364, 637	SPECT
^{67}Ga	^{68}Zn (n, p) ^{67}Ga	γ	78.3 h	93, 184, 300, 393	SPECT
^{111}In	^{111}Cd (p, n) ^{111}In	Auger, γ	67.2 h	171, 245	SPECT
^{123}I	^{121}Sn (α, 2n) ^{123}I	Auger, γ	13.2 h	159	SPECT
^{99m}Tc	$^{99}Mo/^{99m}Tc$-generator	γ	6 h	140	SPECT
^{125m}Te	$^{124}Sn(n,\gamma)^{125m}Sn(\beta)^{125m}Te$	I T	58d	28, 36, 109	SPECT
^{11}C	$^{14}N(p,\alpha)^{11}C$	Positron	20.4 min	960	PET
^{18}F	^{18}O (p, n) ^{18}F	Positron	1.83 h	E_β+635	PET
^{64}Cu	$^{64}Ni(p, n)^{64}Cu$	Positron	12.7 h	E_β+653	PET
^{76}Br	$^{76}Se(p, n)^{76}Br$	Positron	16.2 h	E_β+3941	PET
^{124}I	$^{124}Te(p, n)^{124}I$	Positron	100.2h	E_β+2134, 1533	PET
^{68}Ga	$^{68}Ge/^{68}Ga$-generator	Positron	67.7 min	E_β+1900, 822	PET
^{86}Y	$^{86}Sr(p,n)^{86}Y$	Positron	14.7 h	E_β+3141	PET
^{89}Zr	$^{89}Y(p,n)^{89}Zr$	Positron	3.3d	E_β+901	PET

Table 3. Characteristics of potential radionuclides for nanotargeted radiotherapy.[17,19,23–26,101–104,98–100,106,107]

Rd[a]	Production	Emission type	Half-life	E_{max} (MeV)	R_{max} (mean)[b]	Size of tumor cells[c]
^{186}Re	^{185}Re (n, γ) ^{186}Re	β, γ (9.4%)	89.2 h	1.07	5 mm (0.9 mm)	Intermediate clusters
^{188}Re	^{188}W/^{188}Re-generator	β, γ (15.1%)	17 h	2.12	11 mm (2.4 mm)	L clusters
^{198}Au	^{197}Au(n,γ)^{198}Au	β, γ (95.5%)	2.7 d	0.96, 0.28		I clusters
^{177}Lu	^{176}Lu (n, γ)^{177}Lu	β, γ	6.7 d	0.49	1.6 mm (0.67 mm)	S clusters
^{131}I	^{131}Te (β) ^{131}I	γ (81.2%), β	8 d	0.28, 0.36, 0.64	2.4 mm (0.8 mm)	S clusters
^{90}Y	^{90}Sr/^{90}Y-generator	β	2.7 d	2.28	12 mm (2.8 mm)	L clusters
^{67}Cu	^{64}Ni(α, p)^{67}Cu	β, γ	2.6 d	0.19	2.2 mm (0.7 mm)	S clusters
^{225}Ac	^{225}Ac-generaor	α	10 d	5.83, 5.79, 5.73	40–80 μm	Single cells, S clusters
^{223}Ra	^{227}Ac-generaor	α, γ	11.4 d	5.7–7.5	40–80 μm	Single cells, S clusters
^{211}At	Accelerator	α	7.2 h	5.87	40–80 μm	Single cells, S clusters
^{111}In	^{111}Cd (p, n) ^{111}In	Auger, γ	67 h	0.42	0.02–20 μm	Single cells

Note: [a]Rd: Radionuclide

[b]Radiation tumor tissue penetration maximum and mean range.

[c]Small, intermediate and large clusters correspond approximately to the intervals 10^4–10^6, 10^6–10^8 and 10^8–10^{10} tumor cells per clusters, respectively.[106]

L: large; I: intermediate; S: small.

Table 4. Characteristics of nanotargeted nuclear imaging modalities.[23–26,57]

Modality	Image probe (amount of probe)	Sensitivity	Spatial resolution	Depth active NP design
SPECT	^{99m}Tc, ^{111}In, ^{123}I, etc. loaded or labeled nanocarriers (ng)	10^{-10}–10^{-11} (M)	0.5–1 mm	No limit, surface bioconjugation or after loading antibodies, peptides or proteins.
PET	^{18}F, ^{64}Cu, etc. loaded or labeled nanocarriers (ng)	10^{-11}–10^{-12} (M)	1–2 mm	No limit, surface bioconjugation or after loading antibodies, peptides or proteins.

Note: SPECT: single photon emission computed tomography; PET: positron emission tomography.

most sensitive with subnanomolar amounts of molecular probes and the highest tissue penetration range (Table 4).

For targeted radionuclide internal radiotherapy applications, high and low energy between 0.1 and 2.2 MeV of β-emitters are ideal radioisotopes for the treatment of small to large clusters of tumor cells.[17,26,106] The maximum tissue penetration range (1–10 mm),[17,106] and cross-fire effects of β-particles with energy range between 0.1 and 2.2 MeV can kill tumor cells in close proximity to neovasculature.[17,24,106] Alpha-emitters hold great promise as therapeutics for small cancer lesions and micrometastatic cancers due to the high linear energy transfer (LET, 80 keV/μm) and short-range energy depositions with tissue penetration range of 50–100 μm. The mAb labeled with α-emitters have been demonstrated to have high specific killing effects and minimal normal-tissue damage in a tumor-bearing animal model.[17,54,106] Auger electrons have an energy of <30 keV and subcellular radiation penetration pathlength of 0.02–10 μm. Thus, auger electron emitters can exert their radiotoxic effects on cells when they are internalized into the cytoplasm.[98–100]

^{123}I-ITdu-mediated nano-irradiation of DNA induces efficiently death in HL60 leukemia cells and in doxorubicin, β- or γ-radiation-resistant cell lines have been examined. The experimental findings provide evidence that ultraselective nano-irradiation of DNA through auger electron-carrying metabolic substrates offers an extremely effective strategy for inducing cell death and breaking resistance to more conventional types of irradiation or

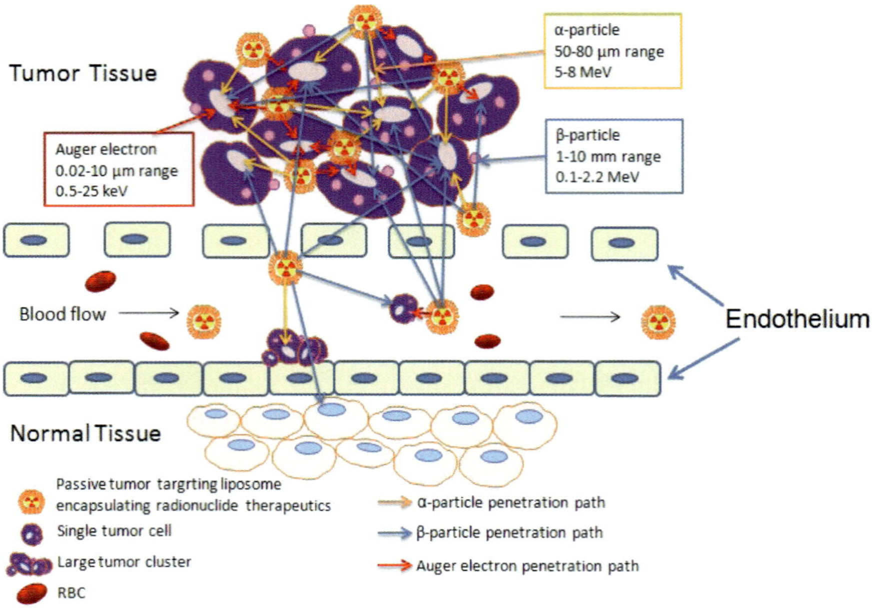

Fig. 3. Schematic illustration of tumor tissue penetration range of internal radiotherapy by NPs with auger electron (0.5–25 keV, 0.02–10 μm), α-emitters (5–8 MeV, 50–80 μm range) and β-emitters (0.1–2.2 MeV, 1–10 mm range) passively and actively nanotargeted radionuclide therapeutics. SPECT and PET diagnostic nuclear imaging for γ and positron emitters with energy emission range from 0.1 to 0.4 MeV, and tissue penetration depth range from 500 μm to 1.5 m, and the sensitivity of those imaging emitters range from 10^{-10} to 10^{-12} moles as shown in Tables 2 and 4.

chemotherapy.[107] The schematic illustration of tumor tissue penetration range of radiation emitters for passively and actively nanotargeted radionuclide therapy is shown in Fig. 3.[17,26,106]

5. Passive Nanotargeting Delivery of Radionuclides for Nuclear Imaging and Radiotherapy

Typically, nanotargeted radionuclides have a two-component architecture for passively targeted imaging and radiotherapeutics, e.g. a PEGylated nanoliposome loaded with radionuclide payloads for nuclear imaging or radiotherapeutics.[23,24,26,35] The research and applications of selected passively nanotargeted nuclear imaging agents and radiotherapeutics are summarized in Tables 5 and 6. The history and progress of the preclinical development of liposome-targeted treatments for cancer before 2000 were described in detail.[108] Optimal radiolabeled liposomes for tumor imaging have been

Table 5. Research on the selected passively nanotargeted radiopharmaceuticals for nuclear imaging applications.

NPs	Radionuclides	Imaging	Applications	References
Liposomes	^{99m}Tc, ^{111}In, ^{67}Ga	Gamma imaging	Diagnostics of tumor, infection, inflammation and lymphoscintigraphy	23, 27, 31, 32, 86, 108–110
Liposomes	^{111}In	Gamma/ SPECT imaging	Clinical biodistribution, pharmacokinetics and imaging studies of breast, head and neck, glioma and lung cancer patients	84, 109
Liposomes	^{188}Re	SPECT/ CT Imaging	MicroSPECT/CT imaging ,biodistribution and pharmacokinetics of ^{188}Re-liposome in mice tumor and ascites model	87, 89, 111–113
Liposomes	^{18}F	PET imaging	Liposomal tracking *in vivo* with ^{18}F-liposome-PET imaging	36, 81, 114, 115
Liposomes	^{64}Cu	PET imaging	Passive targeted delivery and imaging with bioconjugated ^{64}Cu-BAT- PEG-liposome	116
Liposomes	^{64}Cu	PET imaging	^{64}Cu-loaded liposomes as PET imaging agents	37
Liposomes	^{64}Cu	PET imaging	PET imaging of the stability of Cu-64 labeled dipalmitoyl and distearoyl lipids in liposomes.	79
Liposomes	^{111}In, ^{177}Lu	Gamma/ SPECT imaging	Gamma imaging of tumor targeting for C26 and HT29/luc animal models	117, 118
Polymeric glycol chitosan	^{131}I	Gamma imaging	Study the importance of radionuclide imaging for optimization of highly selective tumor targeting polymeric NPs with or without drug.	46

(*Continued*)

Table 5. (*Continued*)

NPs	Radionuclides	Imaging	Applications	References
Polymer	^{64}Cu	PET imaging	Strategies for optimized radiolabeling of NPs for *in vivo* PET Imaging.	45
IO	^{99m}Tc	SPECT/ MRI imaging	^{99m}Tc-bisphosphonate–IO NP conjugates for dual-modality biomedical imaging	119
IO	^{99m}Tc	SPECT/ MRI imaging	^{99m}Tc-labeled superparamagnetic IO NPs for multimodality SPECT/ MRI of sentinel lymph nodes	43
IO	^{64}Cu	PET/MRI imaging	*In vivo* evaluation of ^{64}Cu-labeled magnetic SPIO NPs as a dual-modality PET/ MR imaging agent	41
IO	^{68}Ga	PET/ MRI/ optical imaging	Superparamagnetic NPs: feasibility of multimodal detection using 3T MRI, small animal PET and fluorescence imaging	42
QDs	^{64}Cu	PET imaging	PET imaging and biodistribution of silicon QDs in mice	120
Carbon nanotubes	^{86}Y	PET imaging	PET imaging of soluble Y-86-labeled carbon nanotubes in mice	121
Gold nanoshells	^{64}Cu	PET imaging	*In vivo* PET imaging and biodistribution of radiolabeled gold nanoshells in rats with tumor xenografts	55
Latex	^{68}Ga	PET imaging	Latex NPs for multimodal imaging and detection *in vivo*	56

Table 6. Research on the selected passively nanotargeted radiopharmaceuticals for radiotherapy applications.

NPs	Radionuclides	Radiotherapy	Applications	References
Liposomes	^{131}I, ^{90}Y, ^{188}Re, ^{67}Cu	Radiotherapeutics	An analytical dosimetric study for the use of radionuclide–liposome conjugates in internal radiotherapy.	97
Liposomes	^{111}In, ^{125}I	Radiotherapeutics	^{111}In-surface/^{125}I-encapsulation, doubly radiolabeled liposomes.	122
Liposomes	^{186}Re	Radiotherapeutics	Intraoperative ^{186}Re-liposome radionuclide therapy in a head and neck squamous cell carcinoma xenograft positive surgical margin model.	123–125
Liposomes	^{111}In, ^{188}Re	Radiotherapeutics	Imaging, biodistribution, pharmacokinetics, dosimetric and comparative therapeutic efficacy studies of ^{111}In/^{188}Re-liposome on C26, HT-29, LS174T, solid, metastatic and ascites tumor-bearing animal models.	87, 89, 91, 92, 97, 101, 111, 117, 118, 126–128
Liposomes	^{225}Ac	Radiotherapeutics	Targeted α-particles emitters of ^{225}Ac-generators encapsulated in liposomes as therapeutic agents for micrometastases cancer.	75, 129–131
Liposomes	^{10}B	Radiotherapeutics	^{10}B-liposomes nanotargeted therapeutics for BNCT	132, 133
Liposomes	^{111}In, ^{188}Re	Co-delivery of radiochemotherapeutics	Imaging, biodistribution, pharmacokinetics, therapeutic efficacy and dosimetric studies of ^{111}In/^{188}Re-VNB/DXR-liposome on C26 and HT-29 tumor/ascites-bearing animal models.	90, 101–104, 118, 134, 135

Note: VNB: Vinorelbine; DXR: Doxorubicin.

established.[86] The clinical development for targeted therapy of cancer by passively targeted liposomes as vehicles have been summarized.[109] In addition, the potential areas for future development of liposome-targeted strategies have also been considered.[109]

Drug and radionuclides formulated within the liposome can occur in one of the three potential compartments: water-soluble agents are located in the central aqueous core of the liposome; lipid-soluble agents are carried in the liposome membrane; peptides and proteins tend to bind to the surface of the lipid bilayer.[108]

5.1. *Nuclear imaging*

Table 5 illustrates the selected type of passively targeted NPs (liposome, polymer, IO, QDs, carbon nanotubes, gold nanoshells and latex), the types of radionuclides imaging modalities (gamma imaging, SPECT, and PET) and the major nuclear imaging applications. Till date, nanotargeted liposome-encapsulated radionuclides is the frontier runner in both preclinical and clinical studies for medical applications.[108,109] The major characteristics of nanotargeted nuclear imaging modalities such as γ imaging, SPECT and PET are listed in Table 4. Liposomes are self-assembling colloidal particles composed of a spherical bilayer of small phospholipid vesicles which is spontaneously formed when water is added to a dried lipid mixture[108] (Fig. 2).

Delivery of ^{99m}Tc, ^{111}In and ^{67}Ga radionuclides by liposomes for γ-imaging and monitoring drug treatment have been reviewed and reported for preclinical and clinical studies.[23,27,31,32,86,108–110] A systemic study of optimal liposome formulation and encapsulation of radionuclides was also reported.[86,108] A novel amphiphilic probes for ^{18}F-radiolabeling-performed liposomes and determination of liposomal trafficking by PET was developed.[114] Liposomes encapsulating positron emitters of ^{18}F and ^{64}Cu were applicable for diagnostic imaging and real-time liposomal tracking *in vivo*.[36,37,79,81,114–116]

Wang *et al.* demonstrated an intravenous (*i.v.*) administration of ^{111}In-liposome by conjugating ^{111}In-oxine to DTPA/PEG-liposome followed by whole-body scintigraphy. Images revealed that the tumor clearly accumulated ^{111}In-liposome up to 48 h post-injection (p.i.).[117] In addition to the diagnostic imaging of ^{111}In-liposome, Lee *et al.* demonstrated the bifunctional imaging and bimodality therapeutic efficacy of radiochemotherapeutics of ^{111}In-VNB-liposomes in HT-29/luc mouse xenografts.[118] Passively nanotargeted radiopharmaceuticals of ^{188}Re-liposome was carried out for nuclear imaging of solid tumor of C26 bearing in BALB/c mouse model and

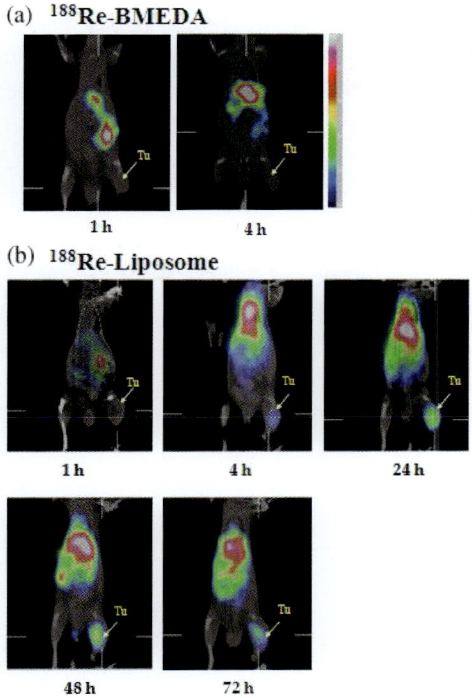

Time (h)	microSPECT (SUV)	Biodistribution (% ID/g)
1	1.94±0.10	1.87±0.22
4	2.72±0.27	2.10±0.21
24	2.81±0.26	3.62±0.73
48	2.17±0.14	2.49±0.35

Data were expressed as mean±standard error, n=2 and 5 for microSPECT imaging and biodistribution, respectively.

Fig. 4. Passively nanotargeted radiopharmaceuticals of ^{188}Re-liposome was carried out for nuclear imaging of nanotargeted solid tumor of C26 bearing in BALB/c mice animal model at 1, 4, 24, 48, 72 h (b), and compared with the control (a). The micro-SPECT imaging SUV of the ^{188}Re-liposome uptake on the solid tumor was also correlated with the biodistribution data (%ID/g) as shown in the right table (reprinted with permission from Ref. 111).

compared with the control as shown in the Fig. 4.[111] The micro-SPECT imaging standard uptake value (SUV) of the ^{188}Re-liposome on the solid tumor was also correlated with the biodistribution data (%ID/g).[111] The SPECT/CT imaging of ^{188}Re-liposome targeting on LS-174T human colorectal tumor bearing nude mouse model were also proved recently.[89]

SPECT/CT imaging of passive tumor targeting, pharmacokinetics and excretion of ^{188}Re-liposomes were studied in C26 murine colon carcinoma solid tumor-bearing mice, C26 colon carcinoma ascites-bearing mice, C26 colon carcinoma lung metastatic mice model and LS-174T human colon carcinoma solid tumor-bearing mice. Elimination half-lives ($T_{1/2\lambda z}$) of ^{188}Re-liposome are 14.8–19.7 h, C_{max} = 38.3–64.2% ID/g, Cl = 0.07–0.12 mL/h, AUC $_{(0\to\infty)}$ = 820–1390 h% ID/mL and MRT $_{(0\to\infty)}$ = 17.4–28.7 h, respectively, in these tumor-bearing mouse models. The pharmacokinetic results demonstrated the longer retention time (RT) in the blood circulation of PEGylated nanoliposome therapeutics loaded with ^{188}Re. For excretion studies, the total excreted fraction of the ^{188}Re-liposome was about 54–87%. The ^{188}Re-liposome was mostly excreted via feces suggesting the importance of hepatobiliary excretion for these radiolabeled NPs.[87,111,89–92] Gamma and SPECT imaging of ^{131}I and ^{64}Cu radiolabeled polymeric NPs were studied.[45,46] PET , SPECT/magnetic resonance imaging (MRI) dual-modality and SPECT/MRI/optical trimodality imaging of tumor targeting of ^{99m}Tc-, ^{64}Cu- and ^{68}Ga-radiolabeled IO NPs were demonstrated.[41–43,119] PET imaging of ^{64}Cu- and ^{86}Y-radiolabeled silicon QDs and carbon nanotubes were performed in mice.[120,121] PET imaging of ^{64}Cu-labeled gold nanoshells and ^{68}Ga-labeled latex passive NPs targeting mice tumor have also been demonstrated.[55,56]

5.2. *Radiotherapy*

Table 6 lists the selected passively nanotargeted radiopharmaceuticals for radiotherapy applications, which includes the types of NPs and radionuclides, and the major studied applications in solid, metastatic, ascites tumor-bearing animal models. An analytical dosimetric study for the use of ^{131}I, ^{90}Y, ^{188}Re and ^{67}Cu radionuclide-labeled liposome for internal radiotherapy has been reported, and the analysis suggested that the optimal liposome system for radiotherapy differs from the delivery system of chemotherapy.[97] In previous clinical-targeting tumor-imaging studies,[84] the results of the effective targeting of solid tumors in patients with advanced local cancers by radiolabeled PEGylated liposomes support the possible delivery of β-emitting radionuclide-loaded PEGylated liposome for the treatment of solid tumors, particularly those liposomes in head and neck patients. Radiotherapy with ^{111}In and ^{125}I doubly radiolabeled liposomes and ^{186}Re-liposomes have been studied in xenograft model.[122,123]

Bao *et al.* have developed a method of labeling liposomes with radionuclides using *N*,*N*-bis(2-Mercaptoethyl)-*N*′*N*′-diethylethylenediamine (BMEDA) to direct or after-load ^{99m}Tc or ^{186}Re into liposomes.[72,74] In addition to therapy via *i.v.* administration, the intratumoral and intraoperational therapy were also investigated for the potential use of ^{186}Re-liposomes.[123–125] High-resolution SPECT/CT images revealed the intratumoral distribution of therapeutic liposomes and indicated the potential use of ^{186}Re-liposomes for intratumoral therapy.[123,125] Intraoperative passive nanotargeted ^{186}Re-liposome therapy showed an excellent tumor suppression and minimal side-effect profile in the head and neck squamous cell carcinoma xenograft positive surgical margin model.[123]

Biodistribution, pharmacokinetics and nuclear imaging of passively nanotargeted radiotherapeutics of ^{111}In/^{188}Re-liposome on C26 and HT-29 colon carcinoma-bearing animal models have been extensively studied by our group.[87,111,117] In addition, biodistribution, pharmacokinetics, nuclear imaging and comparative therapeutic efficiency studies of ^{188}Re-liposome were also evaluated in LS-174T human colon carcinoma-bearing mice model,[89] C26 lung metastatic tumor model[92] and C26 ascites metastatic tumor model.[91] Therapeutic efficacy of ^{188}Re-liposome in a C26 murine colon carcinoma solid tumor model was studied recently.[126] Dosimetric evaluation of nanotargeted ^{188}Re-liposome with the MIRDOSE3 and Organ Level Internal Dose Assessment/Exponential Modeling (OLINDA/EXM) programs was compared.[127] Comparison of the therapeutic efficiency of ^{188}Re-liposome with liposome doxorubicin for murine breast tumor model was also reported.[128] ^{111}In has a γ-ray with 171 keV energy for nuclear imaging and an auger electron with 0.42 MeV energy in the nanometer tissue penetration range with specific single tumor cell or small tumor cluster killing effect (Table 3 and Fig. 3). ^{188}Re has a γ-ray with 155 keV energy for nuclear imaging and a high-energy β-emitter with 2.12 MeV energy for killing large tumor clusters. Both radionuclides can be used in bifunctional nuclear imaging and internal radiotherapeutic applications. The long-circulating PEGylated liposomes radiolabeled with ^{188}Re (^{188}Re-liposomes) showed a higher uptake in the tumor as compared with ^{188}Re-BMEDA alone. Passively nanotargeted ^{188}Re-liposomes were found to have a 7.1-fold higher tumor-to-muscle ratio as compared with *i.v.* administered unencapsulated ^{188}Re-BMEDA in a C26 murine colon carcinoma solid tumor animal model.[111] To evaluate the dose-dependent therapeutic efficacy of ^{188}Re-liposome after single dose, tumor growth inhibition and survival ratio were tested in C26 tumor-bearing BALB/c mice[126] and LS-174 tumor-bearing nude mice.[89] The therapeutic studies in both animal models demonstrated better survival time and therapeutic efficacy for mice that received ^{188}Re-liposome (100% MTD; 37 MBq) and ^{188}Re-liposome (80% MTD; 29.6 MBq).[89,126] To demonstrate the therapeutic efficacy of ^{188}Re-liposome, comparison of radiotherapy (^{188}Re-liposome) and chemotherapy (5-FU) efficacy was performed. Therapeutic studies indicated better survival time and therapeutic efficacy for mice that received ^{188}Re-liposome (80% MTD; 29.6 MBq) than mice that received 5-FU (80%MTD; 144 mg/kg).[89,91,92,126] Improvement of biodistribution and therapeutic index via increase of polyethylene glycol (PEG) from 0.9% to 6% on passively nanotargeted ^{111}In-liposome in an HT-29/luc xenografted mouse model was observed.[136] Figure 5 reveals the comparative therapeutic efficiency studies on passively nanotargeted radiopharmaceuticals of ^{188}Re-liposome with

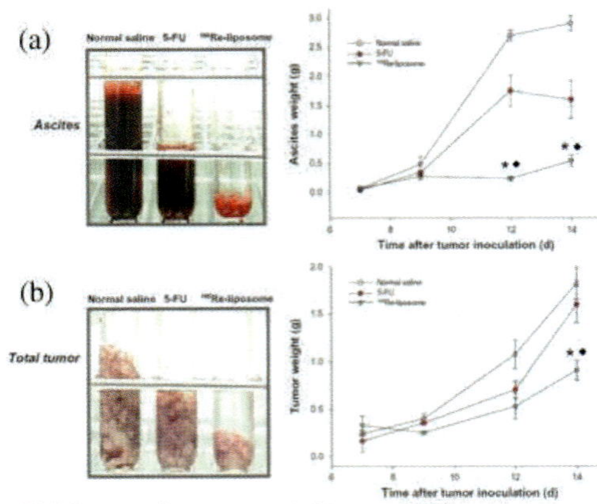

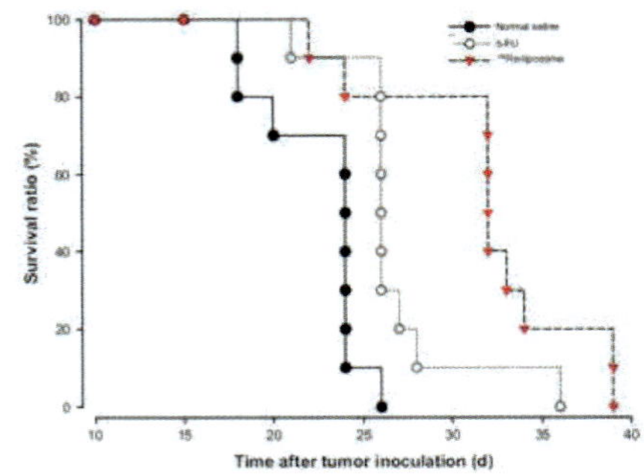

(d) Comparative therapeutic efficacy of different treatments at 7 days after intraperitoneal C26 peritoneal metastatic tumor inoculation

Group	Median survival time (d)	% ILS[a]	Significance P value
^{188}Re-liposome (29.6 MBq)	32.8	34.6	0.0006[b], 0.03[c]
5-FU (144 mg/kg)	26.7	9.6	0.001[b]
Normal saline (control)	24.3	–	–

Notes: [a]% increase in life span (% ILS) is expressed as (T/C−1) × 100%, where T is the median survival time of treated mice and C is the median survival time of control mice; [b]P values versus control were determined by use of Log-rank test (significant variation, $P < 0.01$); [c]P values vs 5-FU were determined by use of Log-rank test (significant variation, $P < 0.05$).

Fig. 5. Comparative therapeutic efficiency studies on passively nanotargeted radiopharmaceuticals of ^{188}Re-liposome with chemotherapeutics of 5-FU for treatment of murine colon C26 ascites animal model. The total ascites (a) and tumors (b) were sampled in the tubes and weighed at 0, 2, 4 and 7 days after administration of ^{188}Re-liposomes (29.6 MBq), 5-FU (144 mg/kg) and normal saline. (c) Kaplan–Meier survival curves for mice bearing C26 peritoneal metastatic tumor are shown. (d) Comparative therapeutic efficacy of different treatments is shown (reprinted with permission from Ref. 91).

chemotherapeutics of 5-FU for treatment of murine colon C26 ascites animal model. The results show both therapeutic efficiency and Kaplan–Meier survival curves for mice bearing C26 peritoneal metastatic tumor and ascites model. The therapeutic efficiency of nanotargeted radiotherapeutics of ^{188}Re-liposome demonstrates better than chemotherapeutics of 5-FU.[91] Targeting α-particle emitters are promising therapeutics for micrometastatic tumors. Enhanced loading of ^{225}Ac and retention of three α-particle-emitting daughters of ^{225}Ac by passively targeted liposomes have been demonstrated.[75,129–131]

Boron neutron capture therapy (BNCT) is a binary approach to cancer therapy involving the nuclear reaction that occurs when ^{10}B is irradiated with thermal neutrons to yield high LET of α-particles and lithium nuclei (2.4 MeV). These particles have a short range ($<10\,\mu m$) and deposit their energy within single cells. The efficacy and successful treatment of tumors by BNCT depend on the selective delivery of relatively high amounts of ^{10}B to tumors. There are three important parameters for development of boron compounds: (i) achieving tumor concentration in the range of 20–35 μg $^{10}B/g$; (ii) reaching a tumor/normal tissue ratio greater than 3–5; and (iii) illustrating sufficiently low toxicity. Application of passive stealth liposome-entrapped ^{10}B delivery systems has been studied for BNCT in animal models.[132,133] The results of the study on ^{10}B-PEG-liposome through *i.v.* injection suggested that passively targeted delivery of sodium mercaptoundecahydrododecaborate can increase the retention of ^{10}B by tumor cells, causing the suppression of tumor growth *in vivo* for BNCT.[133] A high level of ^{10}B concentration (22 ppm) was observed in tumor tissues at 24 h after the administration of boron liposomes, and the tumor was significantly suppressed.[132]

5.3. *Co-delivery of radionuclides and chemotherapeutics*

Concomitant chemotherapy and radiotherapy has been found to improve treatment outcome in a range of solid tumors. PEGylated liposome-encapsulated doxorubicin and cisplatin have shown to be the potential target drugs for tumors, showing increase in therapeutic efficacy and reduction in toxicity.[137] Trimodal cancer therapy combining anti-angiogenesis, chemotherapy and radiotherapy achieves beneficial effects when used as a clinical antitumor strategy.[138] Image-guided and passive nanocarrier-based polymeric nanomedicine for radiotherapy holds significant potential for improving the treatment of advanced solid tumors.[139]

Biodistribution, pharmacokinetics, nuclear imaging and comparative therapeutic efficacies were investigated for nanotargeted bifunctional co-delivery radiochemotherapeutics of $^{111}In/^{188}Re$-(vinorelbine/doxorubicin, VNB/DXR)-liposomes on colorectal carcinoma of HT-29 and C26 tumor and ascites-bearing animal models.[90,101–104,118,134,135] In addition to the diagnostic imaging of $^{111}In/^{188}Re$-liposome, the additive therapeutic efficacy was observed for the comparative co-delivery radiochemotherapeutics of specific killing auger electron emitters of ^{111}In-VNB-liposomes on HT-29/luc mouse xenografts.[118,135] Figure 6 indicates passively nanotargeted (^{111}In/VNB)-liposomes (VNB:vinorelbine), multimodality imaging and co-delivery radiochemotherapeutic efficacy studies on C26/tk-luc BALB/c tumor animal model. In (A) bioluminescence imaging (BLI) of C26/tk-luc tumor-bearing BALB/c mice after various treatments is shown. In (B) therapeutic

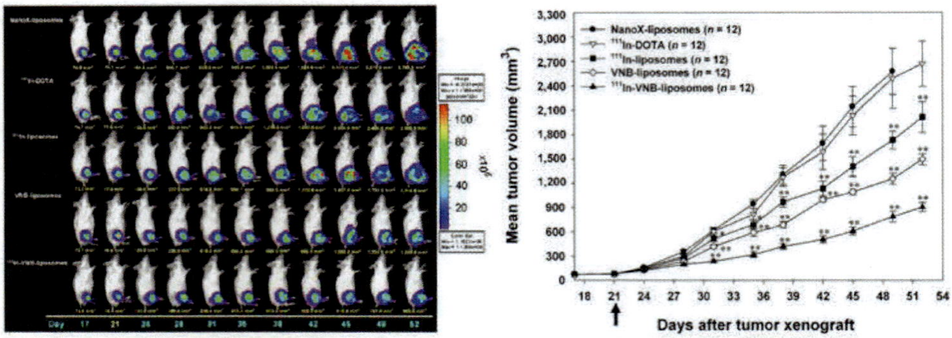

(a) (b)

Treatment	Tumor growth inhibition: Mean growth inhibition rate (%)	Expected growth inhibition rate (%)	Combination index	P	Survival time: Median survival time (d)	Increase in life span (%)*	P
^{111}In-vinorelbine liposomes	69	67	1.06	<0.01	35	+46	0.0045
Vinorelbine liposomes	51	–	–	<0.01	31	+29	0.0136
^{111}In-liposomes	33	–	–	<0.01	31	+29	0.0083
^{111}In-DOTA	3	–	–	NS	28	+17	NS
NanoX liposomes	–	–	–	–	24	–	–

*% increase in life span was expressed as (T/C-1) × 100%, where T is median survival time of treated mice, and C is median survival time of control mice.
NS = not significant.
Tumor growth tracing was measured twice per week using digital caliper. *P* values of survival time were determined by log-rank test.

(c)

Fig. 6. Passively nanotargeted (^{111}In/VNB)-liposomes, multimodality imaging and co-delivery radiochemotherapeutic efficacy studies on C26/tk-luc BALB/c tumor animal model are shown. (a) Bioluminescence (BLI) imaging of C26/tk-luc tumor-bearing BALB/c mice after various treatments is shown. (b) Therapeutic responses of C26/tk-luc tumor-bearing BALB/c mice after various treatments are depicted. Greatest tumor growth suppression was found in combination therapy using (^{111}In/VNB)-liposomes. (c) Tumor growth inhibition and survival time of C26/tk-luc tumor-bearing BALB/c mice (n = 12) after *i.v.* administration of different treatments are illustrated. VNB:vinorelbine, additive effect was demonstrated for (^{111}In/VNB)-liposomes (reprinted with permission from Ref. 101).

responses of CT-26/tk-luc tumor-bearing BALB/c mice after various treatments are depicted. In (C) additive effect found in tumor growth inhibition and survival time of C26/tk-luc tumor-bearing BALB/c mice animal model (n = 12) in combination therapy using (^{111}In/VNB)-liposomes[101] is illustrated. ^{188}Re-DXR-liposomes could provide a beneficial and promising strategy for the co-delivery of passively nanotargeted bimodality radiochemotherapeutics in the treatment of solid tumor and ascites.[103,110] The experimental results pointed to the potential benefit of the co-delivery of nanoliposome radiochemotherapeutics for adjuvant cancer treatment in oncology applications.[104]

Figure 7 reveals passively nanotargeted radiopharmaceuticals of radio-/radiochemotherapeutics of ^{188}Re-(DXR)-liposome for therapeutic efficacy studies on C26 colon solid tumor-bearing BALB/c animal model.[103] The results indicate that the synergistic effect is successfully demonstrated both in tumor growth inhibition (combinational index = 1.145>1) and survival time. Evaluation of pharmacokinetics of ^{111}In-VNB-liposome on C26/tk-luc after intraperitoneal (*i.p.*) and *i.v.* administration in a tumor/ascites mouse model was done and was compared, and the results indicated that the *i.p.* was a better approach than *i.v.* injection in the treatment of *i.p.* malignant tumor/ascites model.[102] Previous theoretical dosimetric studies have addressed the potential use of therapeutic nanoliposomes for the treatment of tumors via *i.v.*

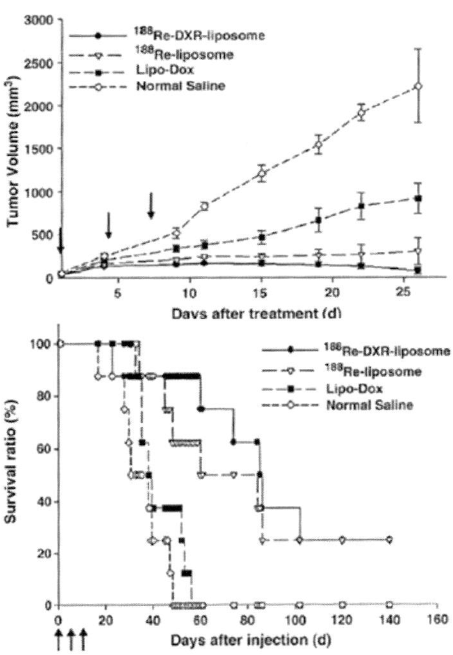

Therapeutic efficacy of radiochemotherapeutic combination ^{188}Re-DXR-liposomes on C26 colon tumor-bearing BALB/c mice

Treatment modality	Tumor growth inhibition			Survival		
	MGI[a]	Expected[b]	CI[c]	Median survival time (days)	*P* value[d]	Lifespan (%)[e]
NS				31		
Lipo-Dox	0.413			38	.104	+23
^{188}Re-Liposomes	0.134			60	.0027	+94
^{188}Re-DXR -Liposomes	0.048	0.055	1.145	74	.0009	+139

Fig. 7. Passively nanotargeted radiopharmaceuticals of radio-/radiochemiotherapeutics of ^{188}Re-(DXR)-liposome for therapeutic efficacy studies on C26 colon solid tumor-bearing BALB/c mice animal model-tumor growth and survival curve are shown (left figure). Tumor growth volume (mm^3) versus time (days) for BALB/c mice implanted with C26 murine colon tumors after administration of ^{188}Re-DXR-liposome (22.2 MBq of ^{188}Re and 2 mg/kg DXR) (●), ^{188}Re-liposome (22.2 MBq of ^{188}Re) (∇), Lipo-Dox (2 mg/kg DXR) (■) and normal saline (◊) by triple *i.v.* injection are illustrated and the duration time being 3 days. The arrow indicated an injection. The therapeutic efficacy is summarized in the right table. The synergistic effect is demonstrated for the co-delivery combination therapy with CI>1(1.145) (reprinted with permission from Ref. 103).

injection.[97,140,141] The comparative dosimetric evaluation of nanotargeted ^{188}Re-(DXR)-liposome derived from the biodistribution indicated that the delivery radiation doses were safe and feasible for further clinical translation research from bench to bedside.[90] The results for major organs doses for the ^{188}Re-(DXR)-liposome revealed that similar doses were received by spleen and liver, but a lower dose was given to kidney, compared with ^{111}In-DTPA-octreotide therapy. Lower doses were also received by total body and liver, compared with ^{111}In-DTPA-human epidermal growth factor radiotherapeutics (0.19 and 0.76 mGy/MBq, respectively). The absorbed doses for spleen, liver, kidney and red marrow in these studies are much lower than those from ^{90}Y-1,4,7,10-tetraazacyclododecane-*N*,*N*′,*N*″,*N*‴-tetraacetic acid tyrosine octreotide therapy.[90] The tumor growth inhibition and therapeutic efficacy studies of passively nanotargeted radionuclides of ^{111}In-(VNB)-liposome on HT-29/luc tumor-bearing in SCID mice animal model[118] and ^{188}Re-(DXR)-liposome on CT-26 solid tumor on BALB/c mice animal model were illustrated.[103] The synergistic therapeutic efficacy was also demonstrated in the co-delivery of nanotargeted radiochemotherapeutics of ^{188}Re-DXR-liposome.[103]

6. Active Nanotargeting Delivery of Radionuclides for Nuclear Imaging and Radiotherapy

Typically, nanotargeted radionuclides have a three-component architecture for active nuclear imaging and therapeutics, such as PEGylated nanoliposome surface bioconjugated with bioactive antibody, peptide, aptimer or ligand-like molecule, and encapsulated or bioconjugated radionuclide payloads for tumor-targeted nuclear imaging or radiotherapeutics.[23–26] In addition, tumor-specific receptor targeting of nanocarriers could provide for high antitumor therapeutic activity and imaging efficacy with low adverse side effects on healthy organs for practically any type of anticancer/imaging drug delivery systems.[34,57–59]

6.1. *Nuclear imaging*

Table 7 summarizes the research on the selected actively nanotargeted radiopharmaceuticals for nuclear imaging applications, which includes the type of NPs and radionuclides, nuclear imaging modalities and major applications. The ability to modify the surface of nanocarriers permits the improvement in the pharmacokinetics, bioavailability, toxicity and customization of nanocarrier formulations for particular actively nanotargeted tumor imaging[5,105,142]

Table 7. Research on the selected actively nanotargeted radiopharmaceuticals for nuclear imaging applications.

NPs	Radionuclides	Imaging	Applications	References
Immunoliposome	^{111}In	Gamma imaging	^{111}In-liposome-2C5(mAb) nucleosome-specific monoclonal 2C5 targeting delivery vehicles for tumor visualization of murine Lewis lung carcinoma and human HT-29 tumors.	64–66
Perfluorocarbon NPs	^{111}In	Gamma imaging	Imaging of targeted tumor angiogenesis of $\alpha_v\beta_3$-integrin in Vx-2 rabbit tumors.	67, 143
CNTs	^{111}In	Gamma or SPECT imaging	Multifunctional targeted delivery and imaging with. functionalized and bioconjugated ^{111}In DOTA-CNT-rituximab nanoconstructs.	53
CNTs	^{64}Cu	PET imaging	Efficient targeting of integrin positive tumor with RGD-SWCN.	144
CNTs	^{89}Zr	PET imaging	PET imaging of LS174T tumor animal model with SWCNT of 89ZrDFO-SWCNT-E4G10.	54
QDs	^{125m}Te	SPECT/CT	Studies on the fate of mAb-targeted Cd ^{125m}Te/ZnS NPs *in vivo*.	145
QDs	^{64}Cu	Bifunctional PET/NIRF imaging	Dual-functional targeted delivery with amine functionalized ^{64}Cu-DOTA-QD-RGD for tumor angiogenesis PET/NIRF imaging.	48
QDs	^{64}Cu	Bifunctional PET/NIRF imaging	Dual-functional targeted delivery with amine functionalized ^{64}Cu-DOTA-QD-VEGF for tumor angiogenesis PET/NIRF imaging.	49
QDs	^{18}F	Bifunctional PET/optical imaging	^{18}F-labeled phospholipids QD micelles for *in vivo* multimodal imaging.	50

(*Continued*)

Table 7. (*Continued*)

NPs	Radionuclides	Imaging	Applications	References
QDs	^{68}Ga	PET/Optical imaging	QDs modified by encapsulation of ligands with a long alkyl chain to affect multispecific and multimodal imaging.	52
QDs	^{64}Cu	PET imaging	Particle size, surface coating and PEGylation influence the biodistribution of QDs in living mice.	51
IO	^{111}In.	Gamma imaging	Development of tumor-targeting bioprobes of mAb-In-111-IO NPs for alternative magnetic field cancer therapy.	146
IO	^{125}I.	Gamma imaging	A novel type of dual-modality ^{125}I-IO -mAb molecular probe for MR and nuclear imaging of tumor.	147
IO	^{64}Cu	Bifunctional PET/ MRI imaging	PET/MRI dual-modality tumor angiogenesis imaging with ^{64}Cu-DOTA/NOTA-IO-RGD nanoconstructs.	38, 39, 148
IO	^{124}I	PET/MRI imaging	PET/MRI dual-modality tumor imaging	149
IO	^{18}F	Trimodel MRI/PET-CT/optical imaging	^{18}F-labeled IO for *in vivo* PET-CT imaging.	40
Polymer	^{99m}Tc	Scintigraphic images of tumor targeting	Targeting tumor angiogenesis: comparison of ^{99m}Tc–peptide and ^{99m}Tc–polymer–peptide conjugates.	44
Dendrimers	^{76}Br	RGD directed-dendrimers PET imaging	^{76}Br-labeled RGD-directed dendritic nanoprobes for PET imaging of angiogenesis.	47

Note: mAb: monoclonal antibody; CNT: carbon nanotubes; QD: quantum dots; IO: iron oxide.

Enhanced tumor accumulation and visualization by γ-scintigraphy with ^{111}In-labeled nucleosome-specific mAb 2C5 bioconjugated immunoliposome has been studied, and the results indicated better and faster imaging in various tumor-bearing mice.[64–66] Pharmaceutical lipid-based nanocarriers modified with mAb 2C5 could represent a new system for tumor-specific delivery of soluble, insoluble and radionuclide pharmaceuticals.[66] ^{111}In perfluorocarbon NPs-targeted $\alpha_v\beta_3$-integrin have been developed and studied for the detection of rabbit Vx-2 tumor angiogenesis. The circulatory half-life was estimated to be 5 h. The mean tumor uptake was fourfold higher than the nontargeted control. The specificity activity (mCi/mL) ^{111}In/NP may affect the tumor-to-muscle ratio in patients. The tumor-to-muscle uptake ratio for the nanotargeted ^{111}In/NP = 10 to ^{111}In/NP = 1 were 6.3 ± 0.2 to 5.1 ± 0.1, respectively. These data suggest that $\alpha_v\beta_3$-targeted ^{111}In perfluorocarbon NPs may provide a clinically useful tool for detecting angiogenesis in nascent tumors.[67,143] ^{111}In radiolabeled soluble functionalized multifunctional drug delivery platforms of active targeting with rituximab mAb bioconjugated on single-wall carbon nanotubes (SWCNT) (^{111}In-DOTA-CNT-rituximab) have been developed, and the selectivity of targeting disseminated human lymphoma were evaluated *in vitro* and *in vivo*.[53] The results of the ability to target tumor specifically with prototype-radiolabeled or fluorescent-labeled, antibody-appended carbon nanotube constructs are encouraging and suggest further investigation of carbon nanotubes as a novel radionuclide delivery platform.[53] PET imaging of ^{89}Zr- and ^{64}Cu-radiolabeled active SWCNT targeting angiogenesis has been reported.[54,144] The active nanotargeting radiopharmaceuticals for nuclear imaging and radiotherapy has been demonstrated. PET imaging of nanotargeted vascular of LS174T tumor animal model with SWCNT of ^{89}ZrDFO-SWCNT-E4G10 nanoconstructs has been investigated (E4G10:vasculature targeting antibody).[54]

The fate of mAb-targeted Cd^{125m}Te/ZnS NPs *in vivo* was studied with SPECT/CT imaging.[145] The development of a dual-function PET/near-infrared fluorescence (NIRF) molecular probe for the accurate assessment of pharmacokinetics and tumor-targeting efficacy of U87MG human glioblastoma tumor-bearing mice has been reported.[48] The amine-functionalized surface of QD bioconjugated with arginine-glycine-aspartic acid (RGD) peptides and 1,4,7,10-tetraazacyclododecane-1,4,7,10-tetraacetic acid (DOTA) for ^{64}Cu-radiolabeled ^{64}Cu-DOTA-QD-RGD nanoconstructs with 90 RGD/QD to target angiogenesis for application in integrin-$\alpha_v\beta_3$ PET/NIRF imaging was also illustrated.[48] This dual-function nuclear/optical *in vivo* molecular probe revealed a quantitative targeting ability in deep tumor lesions.[48]

Dual modality optical and PET imaging of vascular endothelial growth factor receptor (VEGFR) on tumor vasculature using QDs of ^{64}Cu-radiolabeled ^{64}Cu-DOTA-QD-VEGF was also investigated.[49] The U87MG tumor uptake of active nanotargeted ^{64}Cu-DOTA-QD-VEGF (1.52 ± 0.6% injected dose/gram [%ID/g], 2.81 ± 0.3%ID/g, 3.84 ± 0.4%ID/g, and 4.16 ± 0.5%ID/g at 1, 4, 16 and 24 h, respectively, p.i.) was 1%ID/g higher than that of passively targeted ^{64}Cu-DOTA-QD.[49] ^{18}F-labeled phospholipids QD micelles for *in vivo* PET and optical fluorescence imaging from cells to whole body have been designed and studied.[50] PET and PET/optical imaging of peptide-coated ^{64}Cu-labeled QDs 2 or 12 nm in diameter and ^{68}Ga-radiolabeled QDs were also reported.[51,52]

Research on the γ-imaging of ^{111}In- and ^{125}I-radiolabeled active nanotargeted IO of IO-mAb have been investigated.[146,147] Development of a bifunctional nanotargeted IO molecular probe for PET and MRI of tumor integrin-$\alpha_v\beta_3$ expression was reported; this bifunctional ^{64}Cu-DOTA-IO-RGD nanotargeted molecular imaging approach may allow for early tumor detection and may provide insight into the molecular mechanisms of cancer.[38,39,148] Radiolabeled ^{124}I-IO PET/MRI duality modality tumor imaging was also reported.[149] The synthesis and *in vivo* characterization of an ^{18}F-labeled trimodal (MRI/PET-CT/optical) IO for tumor imaging, the facile conjugation chemistry provides a simple platform for rapid and efficient IO labeling.[40] Tumor targeting angiogenesis and comparison of ^{99m}Tc-labeled peptide and ^{99m}Tc-labeled polymer–peptide nanocarrier conjugates were investigated.[44] Specific targeting of the $\alpha_v\beta_3$ integrin and nonspecific vascular permeability are both significant, but active specific targeting is more important than EPR of the carrier molecule. Nonspecific vascular permeability appears to be major factor in reducing tumor-to-normal tissue localization ratio for the peptide molecules.[44] Biodegradable ^{76}Br-labeled dendritic bioconjugated RGD bifunctional nanoprobes for the noninvasive PET imaging of angiogenesis were reported.[47]

6.2. *Radiotherapy*

Table 8 summarizes the selected actively nanotargeted radiopharmaceuticals for radiotherapy applications, which includes the developed NPs and used radionuclides, and the major radiotherapeutic applications. Dosimetric model for *i.p.* targeted liposomal radioimmunotherapy of ovarian cancer micrometastases and liposome-mediated radiotherapeutics-comparative dosimetric

Table 8. Research on the selected actively nanotargeted radiopharmaceuticals for radiotherapy applications.

NPs	Radionuclides	Radiotherapy	Applications	References
Immunoliposome	^{188}Re	Radiotherapeutics	Dosimetric model for *i.p.* targeted liposomal radioimmunotherapy of ovarian cancer micrometastases.	141
Immunoliposome	^{90}Y, ^{188}Re, *et al.*	Radiotherapeutics	Liposome-mediated radiotherapeutics-comparative dosimetric study for various radionuclides, liposome systems and a targeting antibody.	94
Immunoliposomes	^{90}Y	Radiotherapeutics	Targeted anti-angiogenesis of $\alpha_v\beta_3$-integrin or VEGFR2 anti-FLK-1 therapy with nanotargeted therapeutics of ^{90}Y-DTPA-liposome-IA (integrin antagonist) or ^{90}Y-DTPA-liposome-mAb.	63
Immunoliposomes	^{225}Ac	Radiotherapeutics	Targeted α-particles emitters of ^{225}Ac-generators encapsulated in liposomes as therapeutic agents for micrometastases cancer.	75, 129, 130
Immunoliposomes and folate-dendrimers	^{10}B	Radiotherapeutics	^{10}B-immunoliposomes-anti-EGFR and ^{10}B-PAMAM dendrimers anti-folate nanotargeted therapeutics for BNCT.	62,68
Streptavidin NPs	^{111}In	Radiotherapeutics	^{111}In-labeled three-component streptavidin (^{111}In-MORF/tat/trastuzumab) NPs for auger electron-induced antisense-mediated cytotoxicity of tumor cells.	69
IO	^{131}I	Radiotherapeutics	Using anti-VEGF mAb and magnetic NPs as double-targeting vector for the radioimmunotherapy of liver cancer.	150
IO	^{111}In	AMF therapy	Development of tumor targeting bioprobes (^{111}In-mAb IO) for alternating magnetic field cancer therapy.	146
Carbon nanotubes	^{225}Ac	Radiotherapeutics	Active nanotargeting radiopharmaceuticals of ^{225}AcDOTA-SWCNT-E4G10 nanoconstructs for radiotherapy.	54

Note: mAb: monoclonal antibody.

study for various radionuclides (^{188}Re, ^{90}Y, etc.), and immunoliposomes have been reported.[87,141] Targeted angiogenesis $\alpha_v\beta_3$ and VEGFR2 with three-component actively nanotargeted radionuclides of ^{90}Y-liposome (integrin antagonist) and ^{90}Y-liposome-anti-Flk-1 (mAb) have been reported in murine melanoma K1735-M2 and colon C26 animal models; the results demonstrated that ^{90}Y-liposome-anti-Flk-1 (mAb) was significantly more efficacious than conventional radioimmunotherapy in the mouse melanoma model.[63] Enhanced targeting, loading and retention of ^{225}Ac and three α-particle-emitting daughters of ^{225}Ac by actively nanotargeted immunoliposomes have also been illustrated.[75,129,130] The efficacy and successful treatment of tumors by BNCT depend on the selective delivery of relatively high amounts of ^{10}B to tumors. Application of active folate-receptor-targeted PAMAM-dendrimers, and active cetuximab immunoliposome-entrapped ^{10}B delivery systems have been studied for BNCT applications in animal models.[62,68]

Significant radiation-induced antisense-mediated cytotoxicity of tumor cells *in vitro* was achieved using an auger electron-emitting antisense antiR1α messenger RNA antisense morpholino (MORF) oligomer administered as a member of a three-component streptavidin-delivery NP (^{111}In-MORF/tat/trastuzumab).[69] Development of tumor-targeting bioprobes (^{111}In-mAb IO) for alternating magnetic field cancer, and usage of ^{131}I-radiolabelled anti-VEGF mAb and magnetic NPs of IO as double-targeting vector for the radioimmunotherapy of liver cancer have been reported.[146,150] Active nanotargeting radiopharmaceuticals of ^{225}AcDOTA-SWCNT-E4G10 nanoconstructs for α-emitters radiotherapy was investigated.[54] The active nanotargeting radiopharmaceuticals of α-emitters ^{225}Ac-DOTA-SWCNT-E4G10 nanoconstructs (E4G10:vasculature targeting antibody) for radiotherapy of nanotargeted vascular of LS174T tumor animal model has been demonstrated.[54]

Major challenges that have to be addressed by drug-delivery nanocarriers in cancer therapy are the low drug bioavailability of therapeutics within cancer cells and the high toxicities at normal organs due to the low tumor targeting or localization. The combination of molecular targeting of bioconjugated NPs or immunoliposomes can provide targeted cell internalization and intracellular drug release to improve anticancer therapeutic efficacy and to reduce toxicity.[151] Active receptor nanotargeted polymers, dendrimers and liposomes employed for targeting tumor-specific receptors can prevent serious adverse side effects on healthy organs. In addition, the internalization and intracellular distribution of nanocarriers in cancer cells indicated that tumor-specific receptor active targeting of nanocarriers could provide high antitumor therapeutic activity and imaging efficacy with low adverse side effects on normal tissues.[58]

7. Multimodality Nanotargeting Delivery of Radionuclides for Nuclear Imaging and Radiotherapy

NPs have advantages for cancer nuclear imaging and radiotherapy, which, due to many functionalities, can be added to the surface and interior of the nanocarriers. The ultimate goal in the design and preparation of multifunctional and multimodality NPs in drug delivery is the creation of combined diagnostics and therapeutics (or theranostics), and combined radiochemotherapeutics for the targeted diagnosis and treatment of cancer.[61,152] Recent advances in the field of nanotechnology and nanomedicine indeed offer the promise of better diagnostic and therapeutic options.[13–15] Newer generation of NPs have been designed and synthesized to target specific types of cell and molecule via affinity ligands from phage or small molecules or involving antibodies or peptides for nanotargeted radionuclide or drug concurrent delivery.

Synergistically integrated NPs with multifunctional and multimodality novel core platform for cancer nuclear imaging and radiotherapeutics have been developed.[151] Important multifunctions include imaging (single or multimodality), therapy (single drug or combination of two or more drugs) and targeting (one or more ligands) with multivalent. For example, binary NPs with two functions could be developed for simultaneous molecular imaging and targeted therapy, and ternary NPs with three functions could be designed for simultaneous imaging, therapy and targeting, targeted dual-modality imaging or targeted dual-modality therapy. Some typical and potential NPs for nuclear imaging and radiotherapeutics are illustrated as follows: (i) radionuclide (e.g. $^{111}In/^{188}Re/^{64}Cu$)-labeled passively nanotargeted multimodel nanoliposomes.[90,104,135] or actively nanotargeted multifunctional and multimodal immunoliposomes[64–66,153–155]; (ii) radionuclide (e.g. $^{18}F/^{64}Cu$)-labeled IO magnetic NPs for multimodal and multivalent MRI-PET-optical imaging agents and therapeutics[38–40]; (iii) radionuclide (e.g. $^{18}F/^{64}Cu$)-labeled QDs for multifunctional and multimodal imaging and therapeutics[48–50,156,157]; and (iv) silica NPs as a platform for multimodality imaging agents and therapeutics.[158,159] The simultaneous attainment of preferential localization and avoidance of the sequential biological barriers, such as RES system uptake, have been studied with a multifunctional multistage delivery system of mesoporous silicon particles for imaging and therapeutic applications.[160] Development of MDR is one of the most challenging aspects of cancer chemotherapy. Bimodality co-delivery chemotherapeutics in nano-emulsion formulations has shown to be very effective in enhancing the cytotoxicity in wild-type and resistant tumor cells.[161] Theranostic NPs contain diagnostic and therapeutic functions in one integrated system, enabling diagnosis, therapy and monitoring of therapeutic response at the same time. Tumor-targeting multifunctional NPs for theranosis, a new paradigm for cancer therapy, and nanotheranostics for personal medicine have been reported recently.[162,163]

8. Nanotoxicology and Radiotoxicology

8.1. *Nanotoxicology*

Although nanocarriers have provided some new breakthroughs for cancer diagnosis and therapy, the potential adverse human health effects, including nanotoxicology, resulting from exposure to NPs should also be a concern.[6,164,165] Research shows that NPs can stimulate and/or suppress the immune response, and that their compatibility with the immune system is largely determined by their surface chemistry. Modifying these factors can significantly reduce the immunotoxicity of NPs and make them useful platforms for drug delivery.[60] The biodistribution and movements of NPs through tissues, the phagocytosis and endocytosis of NPs would all likely affect the potential toxicity of NPs. The practical strategies for identifying and controlling interferences in common evaluation methods and the implications for regulation of NP immunotoxicity has been discussed and suggested[166]; the standardization of NP-tuned methods through international "round robin" interlaboratory testing was proposed.[166] Toxicity of nanocarrier systems involves physiological, physicochemical and molecular considerations. Nanocarrier systems may induce cytotoxicity and/or genotoxicity.[167] To minimize the risks posed by nanomaterials, there are two basic avenues. One is to develop new highly biocompatible nanomaterials with low toxicity. The other is surface modification of NPs with biocompatible chemicals. Many great efforts are being made to develop NPs that are satisfactory for clinical applications, but nanoregulation is still undergoing major changes to encompass environmental, health and safety issues.[60,164–167]

QDs larger than the renal filtration threshold quickly accumulate in the RES system following *i.v.* administration. Great concern has been raised over the use of QDs in living cells and animals due to their chemical composition of toxic heavy-metal atoms (e.g. Cd, Hg, Pb and As).[164] The nanotoxicology is challenging to understand and master. As a starting point, an earnest approach must acknowledge this complexity to make nanomedicine work in a safe manner. From the current literature, it is clear that no single methodology is likely to meet all of the important challenges. Rather, nanotoxicology is best studied through an interdisciplinary effort involving pharmacological risk assessment, materials science and engineering, selected animal models, as well as insight on drug delivery vehicles and the inner cellular and biochemical workings of the immune response to foreign materials.[168]

8.2. *Radiotoxicology*

In patients, radiolabeled PEGylated liposomes have demonstrated effective targeting of solid tumors by nuclear imaging.[84] There were no important adverse reactions attributable to the liposome infusion, and repeated hematological and biochemical profiles performed at day 10 showed no significant changes.[84] Radiation dosimetric assessments are required by the US FDA during review of clinical trials for approval of new diagnostic and therapeutic radiopharmaceuticals. The amount of radiation dose that is absorbed within different organs and the body as whole from radionuclide administration should be calculated by standardized methods described by the International Committee on Radiation Protection (ICRP) and Medical Internal Radiation Dose committee. One of the new standardized methods for internal dose assessment is OLINDA/EXM.[169] Generally, dosimetric safety assessments can be obtained from biodistribution and pharmacokinetic data. In the radiation assessments, the most critical organs are the organs which are the most sensitive (e.g. liver, red marrow, etc.) to accept radiation damage to develop cancer, which in the end may limit the radiopharmaceuticals administration dose. For most diagnostic imaging radiopharmaceuticals, the effective dose values are in the range 2–10 milli Sievert (mSv). Different dose limit ranges expressed as effective doses have been defined by ICRP.[170]

The radiation dosimetric analysis of passively nanotargeted radiotherapeutics of ^{188}Re-liposome and radiochemotherapeutics of ^{188}Re-DXR-liposome with OLINDA/EXM and MIRDOSE3 software for system-targeted radionuclide therapy has been established. The results showed that red marrow was to be the critical organ in determining the maximally tolerated absorbed doses, and it was promising and beneficial to carry out further preclinical and clinical investigations.[90,127] Absorbed dose calculations provide a scientific basis for evaluating the biological effects associated with administrated radiopharmaceuticals. In internal radionuclide cancer therapy, radiation dosimetry, supports treatment planning, dose–response analyses, predications of therapy effectiveness and safety.[171] An analytical dosimetric study for the use of radionuclide (^{67}Cu, ^{131}I, ^{188}Re and ^{90}Y)-liposome in internal radiotherapy has been reported.[97] Unlike the case with radioimmunotherapy, the dose-limiting organ is likely to be the liver, and strategies intended to reduce RES accumulation are needed to further improve such a tumor-targeting approach.[97]

To evaluate therapeutic strategies of ^{188}Re-liposome in clinical usage, it is essential to use animal models to reflect the important safety aspects before clinical application. Extended acute toxicity study of ^{188}Re-liposome was performed in rats which were administered by *i.v.* injection as a single dose.

During the 28-day study period, none of the rats died and no clinical sign was observed, and only male rats which were administered 185 MBq of ^{188}Re-liposome displayed a slight weight loss compared with the control rats. Although the WBC counts of both high-dose and medium-dose groups reduced 7 days p.i., they recovered to a normal range before termination. The information served as a safety reference for human clinical trials of ^{188}Re-liposome.[172] Compared with the radiation-absorbed dose estimates for ^{111}In- and ^{90}Y-ibritumomab tiuxetan, the radiation absorbed per unit administered activity (mGy/MBq) for nanotargeted ^{188}Re-(DXR)-liposome were much lower in the major organs.[112,171]

9. Clinical Research and Regulatory Issues

9.1. *Clinical research*

Clinical research with passive liposome radiopharmaceuticals demonstrated that a wide variety of tumors could be detected with good sensitivity and specificity. Clinical studies using radiolabeled liposomes for tumor diagnostic imaging of cancer and inflammation from 1979 to 2001 have been reported.[109,173] Conventional ^{111}In-based liposome (Vescan) preclinical and clinical performance and evaluation of lessons (Phases II–III study) learned from the formulation and process development has been discussed and summarized.[174,175] ^{99m}Tc has been performed clinical trial for gamma imaging.[173] ^{111}In has carried out clinical biodistribution, pharmacokinetics, gamma and SPECT imaging studies of breast, head and neck, glioma and lung cancer patients.[84,173] Effective targeting of solid tumors of breast (5.3 ± 2.6%ID/kg for a tumor volume of 234.7 ± 101.4 cm^3), head and neck (highest uptake of 33 ± 15.8%ID/kg for a tumor volume of 36.2 ± 18 cm^3), lung (18.3 ± 5.7%ID/kg for a tumor volume of 114.5 ± 42 cm^3), brain and cervix were also observed with gamma camera and SPECT imaging.[84]

Yet, despite these favorable clinical nuclear imaging results, no liposome radiopharmaceutical has been approved for clinical applications. Some of the reasons for this can be attributed to the requirement of more complex liposome-manufacturing procedures and the adoption of other competing imaging procedures. The best liposome formulations for radiopharmaceuticals have particle sizes in the 100 nm range and contain cholesterol and either a neutral surface charge or PEG surface coating as these features provide for prolonged circulation times. After-loading and surface chelation methods with preformed liposomes are the most efficient and practical radiolabeling methods. Of the 14 clinical studies outlined in the clinical review article,[173] seven studies focused on tumor imaging, six studies reported on inflammation/infection imaging and one report described an early sentinel lymph node imaging study. Continued research of liposome radiopharmaceutical design based on a better understanding of liposome and tumor biology, enhanced radionuclide-labeled capacity and stability and improved scale-up radiolabeled methodologies is warranted.[173]

9.2. *Regulatory issues*

Nanomaterials can have different chemical, physical or biological properties than their conventional materials in many drugs regulated by US FDA. Understanding nanotechnology remains the top FDA priority. All radiopharmaceuticals as well as drug products produced for clinical uses must follow cGMP, and QC tests guidance.[176–178] The product must be performed to assure acceptable quality (QC tests) prior to administration for human subjects.[179,180] Manufacturing of a liposomal radiopharmaceuticals will normally consist of two production steps, a nonradioactive process producing nanoliposomes product[176] and a radioactive process producing the final radiopharmaceutical product for human injection. Specifications for each part are necessary with different QC tests. Some of the nanoliposome recommended specifications include (i) physiochemical specifications (e.g. particle size and surface charge, pH and *in vitro* stability, phospholipid concentration, radiochemical purity and radiochemical identity, radiochemical impurities, specific activity, etc.) and (ii) biological specifications (endotoxin, sterility test). Generally, radiochemical purity is ≥90% which often can be evaluated by high-performance liquid chromatography (HPLC) or radio-thin layer chromatography analysis. Radiochemical impurities can arise from an incomplete loading or chelation during the labeling and loading of the preformed liposomes. The specifications of free radionuclide as well as other radionuclide impurities ensure that the activity imaged by the SPECT or PET reflects the biodistribution of the desired radionuclide. The radiochemical identity of a radiopharmaceutical must also be established by demonstrating similar chromatographic behavior between the radiolabeled form of the radionuclide and the non-radiolabeled form. For radiochemical identification of a liposomal radiopharmaceutical, the RT of the radiolabeled liposomal form could be compared with the RT of a standard non-radiolabeled liposome by use of analytical HPLC and size exclusion chromatography.[181]

All drug products or radiopharmaceuticals for clinical use must follow cGMP and QC specifications.[176–180] A new approach for obtaining early human pharmacokinetic and pharmacodynamics information of a diagnostic or therapeutic agent candidate is microdosing or human Phase 0 trials.[182] The term microdosing is defined as <1/100th of the dose calculated to yield a pharmacological effect of the drug candidate based on primary *in vitro* and *in vivo* data and administered at a maximum dose of ≤100 μg.[182] The FDA guidelines for an exploratory IND published in 2006[183] has given the ability to conduct a toxicology human trial such as outlined in the European Medicines Agency guidelines[182] and ensures the possibility of conducting clinical trials early according to the microdosing concept.[182,183] Institute of Nuclear Energy Research in Taiwan has carried out the world's first eIND trial for nanotargeted therapeutic radiopharmaceuticals of ^{188}Re-liposome for cancer patients since July 2012.

10. Conclusions and Future Prospective

Recent advances in the field of nanotechnology in biomedical research and applications offer the promise of better disease diagnostic and therapeutic options. Biology, medicine and synthetic scientists are making strides in developing nanoconstructs that can be used as core platforms for attaching different functionalities by surface conjugating or after-loading of various NPs for the purpose of disease molecular imaging and targeted drug delivery. Nanotechnology-based combinational diagnostic and therapeutic drug delivery platform can achieve the advantage of integration of drug synergistic effect, overcome MDR and reduction of toxicity. As compared with conventional targeted radionuclide therapy or radioimmunotherapy, the use of nanocarriers can allow for specific multivalent attachment of targeted molecules of antibodies, peptides or ligands to the surface of nanocarriers. Nanotargeted radiopharmaceuticals can deliver a high payload of radionuclides, chemotherapeutics and/or imaging agents to achieve multifunctional and multimodality targeting to disease cells and to increase the efficacy and safety of the new nanomedicine.

The major challenges and opportunities of the advancement and translation of the nanotargeted drugs or radiopharmaceuticals for personal diagnostics and therapeutics are the optimization of the nanopreparation, process scale-up production of the NP compositions, the simultaneous attainment of preferential targeting location and avoidance of sequential biological barriers, reducing immunotoxicity and chronic toxicity. Enhancement of the clinical trial studies are required to translate those advanced new technologies from the bench to the bedside.

Several passively nanotargeted radiolabeled nanocarriers have been successfully employed to image and treat tumor models both preclinically and clinically. Future studies should be designed to optimize these novel approaches and to combine targeted delivery, potent radionuclides, new biomarker or imaging agents, combination of chemotherapeutics and/or radiosensitizing agents. We have demonstrated that passively nanotargeted radiopharmaceuticals of ^{111}In-liposome and ^{188}Re-liposome and co-delivery of radiochemotherapeutics and simultaneous multifunctional imaging was an advantageous characteristics of nanotargeted radionuclides for cancer nuclear imaging and internal radiotherapy from 2006 to 2012. The eIND clinical study of passively nanotargeted ^{188}Re-liposome has been carrying out since July 2012. A good multi-discipline and multi-institute collaboration among the academia, research institutes and industry combing with an integrated "bench-to-clinic" translational approach would accelerate the progress in research of nanotargeted radiopharmaceuticals toward clinical applications for the improvement of the quality life and healthcare of the patients.

For the purpose of the benefit and useful reference source of the frontiers of nanobiomedical research and development in the future, the update statues of the art and the progress of the major research progress of the passive and active NPs, nanotargeted radionuclides, nanotargeted principle and strategies, multimodality and multifunctionality nanotargeting, radiotoxicology and nanotoxicology and clinical research and regulatory issues of the nanotargeted radionuclides for nuclear imaging and internal radiotherapy have been briefly overviewed and discussed in this chapter. We hope that the reader will find the useful and stimulating benefit of this chapter in the research and applications of novel nanotechnologies for the advancement of human disease healthcare worldwide.

Acknowledgments

Authors would like to thank Liang-Cheng, Chen who provided material that was used in this article, and Szu-Yen, Liu and Cheng-Hui, Chuang for preparation of figures and index.

References

1. K. Strebhardt and A. Ullrich, Paul Ehrlich's magic bullet concept: 100 years of progress. *Nat. Rev. Cancer* **8**(6), 473–480 (2008).
2. D. Hanahan and R. A. Weinberg, Hallmarks of cancer: the next generation. *Cell* **144**(5), 646–674 (2011).
3. A. Gabizon, H. Shmeeda and Y. Barenholz, Pharmacokinetics of pegylated liposomal Doxorubicin: review of animal and human studies. *Clin. Pharmacokinet.* **42**(5), 419–436 (2003).

4. T. M. Allen and P. R. Cullis, Drug delivery systems: entering the mainstream. *Science* **303**(5665), 1818–1822 (2004).
5. M. Ferrari, Cancer nanotechnology: opportunities and challenges. *Nat. Rev. Cancer* **5**(3), 161–171 (2005).
6. Y. Liu, H. Miyoshi and M. Nakamura, Nanomedicine for drug delivery and imaging: a promising avenue for cancer therapy and diagnosis using targeted functional nanoparticles. *Int. J. Cancer* **120**(12), 2527–2537 (2007).
7. W. E. Bawarski, *et al.*, Emerging nanopharmaceuticals *Nanomedicine* **4**(4), 273–282 (2008).
8. M. E. Davis, Z. G. Chen and D. M. Shin, Nanoparticle therapeutics: an emerging treatment modality for cancer. *Nat. Rev. Drug Discov.* 7(9), 771–782 (2008).
9. T. Lammers, W. E. Hennink and G. Storm, Tumour-targeted nanomedicines: principles and practice. *Br. J. Cancer* **99**(3), 392–397 (2008).
10. W. R. Sanhai, *et al.*, Seven challenges for nanomedicine. *Nat. Nanotechnol.* **3**(5), 242–244 (2008).
11. R. Duncan and R. Gaspar, Nanomedicine(s) under the microscope. *Mol. Pharm.* **8**(6), 2101–2141 (2011).
12. A. Schroeder, *et al.*, Treating metastatic cancer with nanotechnology. *Nat. Rev. Cancer* **12**(1), 39–50 (2011).
13. J. H. Grossman and S. E. McNeil, Nanotechnology in cancer medicine. *Phys. Today* **65**: 38–42 (2012).
14. Q. Sun, M. Radosz and Y. Shen, Challenges in design of translational nanocarriers. *J. Control Release* **164**(2), 156–169 (2012).
15. A. Wei, J. G. Mehtala and A. K. Patri, Challenges and opportunities in the advancement of nanomedicines. *J. Control Release* **164**(2), 236–246 (2012).
16. I. S. Vizirianakis and D. G. Fatouros, Personalized nanomedicine: paving the way to the practical clinical utility of genomics and nanotechnology advancements. *Adv. Drug. Deliv. Rev.* **64**(13), 1359–1362 (2012).
17. D. E. Milenic, E. D. Brady and M. W. Brechbiel, Antibody-targeted radiation cancer therapy. *Nat. Rev. Drug Discov.* **3**(6), 488–499 (2004).
18. R. M. Macklis and B. Pohlman, Radioimmunotherapy for non-Hodgkin's lymphoma: a review for radiation oncologists. *Int. J. Radiat. Oncol. Biol. Phys.* **66**(3), 833–841 (2006).
19. J. Y. Wong, Systemic targeted radionuclide therapy: potential new areas. *Int J. Radiat. Oncol. Biol. Phys.* **66**(Suppl 2), S74–S82 (2006).
20. A. J. Davies, Radioimmunotherapy for B-cell lymphoma: ^{90}Y- ibritumomab tiuxetan and ^{131}I- tositumomab. *Oncogene* **26**(25), 3614–3628 (2007).
21. H. A. Jacene, *et al.*, Comparison of ^{90}Y-ibritumomab tiuxetan and ^{131}I-tositumomab in clinical practice. *J. Nucl. Med.* **48**(11), 1767–1776 (2007).
22. S. H. Britz-Cunningham and S. J. Adelstein, Molecular targeting with radionuclides: state of the science. *J. Nucl. Med.* **44**(12), 1945–1961 (2003).
23. M. Hamoudeh, *et al.*, Radionuclides delivery systems for nuclear imaging and radiotherapy of cancer. *Adv. Drug Deliv. Rev.* **60**(12), 1329–1346 (2008).
24. A. Mitra, *et al.*, Nanocarriers for nuclear imaging and radiotherapy of cancer. *Curr. Pharm. Des.* **12**(36), 4729–4749 (2006).

25. H. Hong, *et al.*, Molecular imaging and therapy of cancer with radiolabeled nanoparticles. *Nano Today* **4**(5), 399–413 (2009).
26. G. Ting, C. H. Chang and H. E. Wang, Cancer nanotargeted radiopharmaceuticals for tumor imaging and therapy. *Anticancer Res.* **29**(10), 4107–4118 (2009).
27. I. Ogihara, S. Kojima and M. Jay, Differential uptake of gallium-67-labeled liposomes between tumors and inflammatory lesions in rats. *J. Nucl. Med.* **27**(8), 1300–1307 (1986).
28. A. L. Klibanov, *et al.*, Amphipathic polyethyleneglycols effectively prolong the circulation time of liposomes. *FEBS Lett.* **268**(1), 235–237 (1990).
29. D. Papahadjopoulos, *et al.*, Sterically stabilized liposomes: improvements in pharmacokinetics and antitumor therapeutic efficacy. *Proc. Natl. Acad. Sci. USA* **88**(24), 11460–11464 (1991).
30. D. D. Lasic, Doxorubicin in sterically stabilized liposomes. *Nature* **380**(6574), 561–562 (1996).
31. W. T. Phillips, Delivery of gamma-imaging agents by liposomes. *Adv. Drug Deliv. Rev.* **37**(1–3), 13–32 (1999).
32. O. C. Boerman, *et al.*, Radiolabeled liposomes for scintigraphic imaging. *Prog. Lipid Res.* **39**(5), 461–475 (2000).
33. A. Bao, *et al.*, A novel liposome radiolabeling method using ^{99m}Tc-"SNS/S" complexes: *in vitro* and *in vivo* evaluation. *J. Pharm. Sci.* **92**(9), 1893–1904 (2003).
34. S. Sofou, Surface-active liposomes for targeted cancer therapy. *Nanomed.* **2**(5), 711–724 (2007).
35. W. T. Phillips, B. Goins and A. Bao, Radioactive liposomes. *Wiley Interdiscipl. Rev. Nanomed. Nanobiotechnol.* **1**: 69–83 (2009).
36. N. Oku, *et al.*, PET imaging of brain cancer with positron emitter-labeled liposomes. *Int. J. Pharm.* **403**(1–2), 170–177 (2011).
37. A. L. Petersen, *et al.*, ^{64}Cu loaded liposomes as positron emission tomography imaging agents. *Biomaterials* **32**(9), 2334–2341 (2011).
38. B. R. Jarrett, *et al.*, Synthesis of ^{64}Cu-labeled magnetic nanoparticles for multimodal imaging. *Bioconjug. Chem.* **19**(7), 1496–1504 (2008).
39. H. Y. Lee, *et al.*, PET/MRI dual-modality tumor imaging using arginine-glycine-aspartic (RGD)-conjugated radiolabeled iron oxide nanoparticles. *J. Nucl. Med.* **49**(8), 1371–1379 (2008).
40. N. K. Devaraj, *et al.*, ^{18}F labeled nanoparticles for *in vivo* PET-CT imaging. *Bioconjug. Chem.* **20**(2), 397–401 (2009).
41. C. Glaus, *et al.*, *In vivo* evaluation of ^{64}Cu-labeled magnetic nanoparticles as a dual-modality PET/MR imaging agent. *Bioconjug. Chem.* **21**(4), 715–722 (2010).
42. L. Stelter, *et al.*, Modification of aminosilanized superparamagnetic nanoparticles: feasibility of multimodal detection using 3T MRI, small animal PET, and fluorescence imaging. *Mol. Imaging Biol.* **12**(1), 25–34 (2010).
43. R. Madru, *et al.*, ^{99m}Tc-labeled superparamagnetic iron oxide nanoparticles for multimodality SPECT/MRI of sentinel lymph nodes. *J. Nucl. Med.* **53**(3), 459–463 (2012).
44. B. R. Line, *et al.*, Targeting tumor angiogenesis: comparison of peptide and polymer-peptide conjugates. *J. Nucl. Med.* **46**(9), 1552–1560 (2005).
45. G. Sun, *et al.*, Strategies for Optimized Radiolabeling of Nanoparticles for *in vivo* PET Imaging. *Adv. Mater.* **19**, 3157–3162 (2007).

46. B. S. Lee, *et al.*, Tumor targeting efficiency of bare nanoparticles does not mean the efficacy of loaded anticancer drugs: importance of radionuclide imaging for optimization of highly selective tumor targeting polymeric nanoparticles with or without drug. *J. Control Release* **147**(2), 253–260 (2010).
47. A. Almutairi, *et al.*, Biodegradable dendritic positron-emitting nanoprobes for the noninvasive imaging of angiogenesis. *Proc. Natl. Acad. Sci. USA* **106**(3), 685–690 (2009).
48. W. Cai, *et al.*, Dual-function probe for PET and near-infrared fluorescence imaging of tumor vasculature. *J. Nucl. Med.* **48**(11), 1862–1870 (2007).
49. K. Chen, *et al.*, Dual-modality optical and positron emission tomography imaging of vascular endothelial growth factor receptor on tumor vasculature using quantum dots. *Eur. J. Nucl. Med. Mol. Imaging* **35**(12), 2235–2244 (2008).
50. F. Duconge, *et al.*, Fluorine-18-labeled phospholipid quantum dot micelles for *in vivo* multimodal imaging from whole body to cellular scales. *Bioconjug. Chem.* **19**(9), 1921–1926 (2008).
51. M. L. Schipper, *et al.*, Particle size, surface coating, and PEGylation influence the biodistribution of quantum dots in living mice. *Small* **5**(1), 126–134 (2009).
52. Y. K. Lee, *et al.*, Nanoparticles modified by encapsulation of ligands with a long alkyl chain to affect multispecific and multimodal imaging. *J. Nucl. Med.* **53**(9), 1462–1470 (2012).
53. M. R. McDevitt, *et al.*, Tumor targeting with antibody-functionalized, radiolabeled carbon nanotubes. *J. Nucl. Med.* **48**(7), 1180–1189 (2007).
54. A. Ruggiero, *et al.*, Imaging and treating tumor vasculature with targeted radiolabeled carbon nanotubes. *Int. J. Nanomedicine* **5**, 783–802 (2010).
55. H. Xie, *et al.*, *In vivo* PET imaging and biodistribution of radiolabeled gold nanoshells in rats with tumor xenografts. *Int. J. Pharm.* **395**(1–2), 324–330 (2010).
56. R. Cartier, *et al.*, Latex nanoparticles for multimodal imaging and detection *in vivo*. *Nanotechnology* **18**, 195102 (2007).
57. K. Kairemo, *et al.*, Nanoparticles in Cancer. *Curr. Radiopharmaceuticals* **1**(1), 30–36 (2008).
58. M. Saad, *et al.*, Receptor targeted polymers, dendrimers, liposomes: which nanocarrier is the most efficient for tumor-specific treatment and imaging? *J. Control Release* **130**(2), 107–114 (2008).
59. M. Shokeen, N. M. Fettig and R. Rossin, Synthesis, *in vitro* and *in vivo* evaluation of radiolabeled nanoparticles. *Q. J. Nucl. Med. Mol. Imaging* **52**(3), 267–277 (2008).
60. M. A. Dobrovolskaia and S. E. McNeil, Immunological properties of engineered nanomaterials. *Nat. Nanotechnol.* **2**(8), 469–478 (2007).
61. G. Lucignani, Nanoparticles for concurrent multimodality imaging and therapy: the dawn of new theragnostic synergies. *Eur. J. Nucl. Med. Mol. Imaging* **36**(5), 869–874 (2009).
62. S. Shukla, *et al.*, Synthesis and biological evaluation of folate receptor-targeted boronated PAMAM dendrimers as potential agents for neutron capture therapy. *Bioconjug. Chem.* **14**(1), 158–167 (2003).
63. L. Li, *et al.*, A novel antiangiogenesis therapy using an integrin antagonist or anti-Flk-1 antibody coated ^{90}Y-labeled nanoparticles. *Int. J. Radiat. Oncol. Biol. Phys.* **58**(4), 1215–1227 (2004).

64. T. A. Elbayoumi and V. P. Torchilin, Enhanced accumulation of long-circulating liposomes modified with the nucleosome-specific monoclonal antibody 2C5 in various tumours in mice: gamma-imaging studies. *Eur. J. Nucl. Med. Mol. Imaging* **33**(10), 1196–1205 (2006).
65. S. Erdogan, A. Roby and V. P. Torchilin, Enhanced tumor visualization by gamma-scintigraphy with ^{111}In-labeled polychelating-polymer-containing immunoliposomes. *Mol. Pharm.* **3**(5), 525–530 (2006).
66. T. A. Elbayoumi, *et al.*, Antinucleosome antibody-modified liposomes and lipid-core micelles for tumor-targeted delivery of therapeutic and diagnostic agents. *J. Liposome Res.* **17**(1), 1–14 (2007).
67. G. Hu, *et al.*, Imaging of Vx-2 rabbit tumors with alpha(nu)beta3-integrin-targeted ^{111}In nanoparticles. *Int. J. Cancer* **120**(9), 1951–1957 (2007).
68. X. Pan, *et al.*, Synthesis of cetuximab-immunoliposomes via a cholesterol-based membrane anchor for targeting of EGFR. *Bioconjug. Chem.* **18**(1), 101–108 (2007).
69. X. Liu, *et al.*, Auger radiation-induced, antisense-mediated cytotoxicity of tumor cells using a 3-component streptavidin-delivery nanoparticle with ^{111}In. *J. Nucl. Med.* **50**(4), 582–590 (2009).
70. S. Taurin, H. Nehoff and K. Greish, Anticancer nanomedicine and tumor vascular permeability; Where is the missing link? *J. Control Release* **164**(3), 265–275 (2012).
71. J. W. Seo, *et al.*, Liposomal Cu-64 labeling method using bifunctional chelators: poly(ethylene glycol) spacer and chelator effects. *Bioconjug. Chem.* **21**(7), 1206–1215 (2010).
72. A. Bao, *et al.*, ^{186}Re-liposome labeling using ^{186}Re-SNS/S complexes: *in vitro* stability, imaging, and biodistribution in rats. *J. Nucl. Med.* **44**(12), 1992–1999 (2003).
73. W. T. Phillips, *et al.*, A simple method for producing a technetium-99m-labeled liposome which is stable *in vivo. Int. J. Rad. Appl. Instrum. B* **19**(5), 539–547 (1992).
74. A. Bao, *et al.*, Direct ^{99m}Tc labeling of pegylated liposomal doxorubicin (Doxil) for pharmacokinetic and non-invasive imaging studies. *J. Pharmacol. Exp. Ther.* **308**(2), 419–425 (2004).
75. M. Y. Chang, J. Seideman and S. Sofou, Enhanced loading efficiency and retention of 225Ac in rigid liposomes for potential targeted therapy of micrometastases. *Bioconjug. Chem.* **19**(6), 1274–1282 (2008).
76. N. Mirahmadi, *et al.*, ^{99m}Tc-HMPAO-labeled liposomes: an investigation into the effects of some formulation factors on labeling efficiency and *in vitro* stability. *Nucl. Med. Biol.* **35**(3), 387–392 (2008).
77. Q. F. Ahkong and C. Tilcock, Attachment of ^{99m}Tc to lipid vesicles containing the lipophilic chelate dipalmitoylphosphatidylethanolamine-DTTA. *Int. J. Rad. Appl. Instrum. B* **19**(8), 831–840 (1992).
78. P. Laverman, *et al.*, A novel method to label liposomes with ^{99m}Tc by the hydrazino nicotinyl derivative. *J. Nucl. Med.* **40**(1), 192–197 (1999).
79. J. W. Seo, *et al.*, Positron emission tomography imaging of the stability of Cu-64 labeled dipalmitoyl and distearoyl lipids in liposomes. *J. Control Release* **151**(1), 28–34 (2011).
80. C. Tilcock, Q. F. Ahkong and D. Fisher, 99mTc-labeling of lipid vesicles containing the lipophilic chelator PE-DTTA: effect of tin-to-chelate ratio, chelate content and surface polymer on labeling efficiency and biodistribution behavior. *Nucl. Med. Biol.* **21**(1), 89–96 (1994).

81. J. Marik, *et al.*, Long-circulating liposomes radiolabeled with [^{18}F]fluorodipalmitin ([18F]FDP). *Nucl. Med. Biol.* **34**(2), 165–171 (2007).
82. V. P. Torchilin, Targeted pharmaceutical nanocarriers for cancer therapy and imaging. *AAPS J.* **9**(2), E128–147 (2007).
83. C. Underwood, *et al.*, Intravenous technetium-99m labelled PEG-liposomes in horses: a safety and biodistribution study. *Equine Vet. J.* **44**(2), 196–202 (2012).
84. K. J. Harrington, *et al.*, Effective targeting of solid tumors in patients with locally advanced cancers by radiolabeled pegylated liposomes. *Clin. Cancer Res.* 7(2), 243–254 (2001).
85. P. Laverman, O. C. Boerman and G. Storm, Radiolabeling of liposomes for scintigraphic imaging. *Methods Enzymol.* **373**, 234–248 (2003).
86. I. Ogihara-Umeda, *et al.*, Optimal radiolabeled liposomes for tumor imaging. *J. Nucl. Med.* **37**(2), 326–332 (1996).
87. L. C. Chen, *et al.*, Biodistribution, pharmacokinetics and imaging of ^{188}Re-BMEDA-labeled pegylated liposomes after intraperitoneal injection in a C26 colon carcinoma ascites mouse model. *Nucl. Med. Biol.* **34**(4), 415–423 (2007).
88. A. Soundararajan, *et al.*, Chemoradionuclide therapy with ^{186}Re-labeled liposomal doxorubicin in combination with radiofrequency ablation for effective treatment of head and neck cancer in a nude rat tumor xenograft model. *Radiology* **261**(3), 813–823 (2011).
89. C. W. Hsu, *et al.*, Comparative therapeutic efficacy of rhenium-188 radiolabeled-liposome and 5-fluorouracil in LS-174T human colon carcinoma solid tumor xenografts. *Cancer Biother. Radiopharm.* **27**(8), 481–489 (2012).
90. C. H. Chang, *et al.*, Comparative dosimetric evaluation of nanotargeted ^{188}Re-(DXR)-liposome for internal radiotherapy. *Cancer Biother. Radiopharm.* **23**, 749–758 (2008).
91. C. C. Tsai, *et al.*, Biodistribution and pharmacokinetics of ^{188}Re-liposomes and their comparative therapeutic efficacy with 5-fluorouracil in C26 colonic peritoneal carcinomatosis mice. *Int. J. Nanomedicine* **6**, 2607–2619 (2011).
92. L. C. Chen, *et al.*, Pharmacokinetics, dosimetry and comparative efficacy of ^{188}Re-liposome and 5-FU in a CT26-luc lung-metastatic mice model. *Nucl. Med. Biol.* **39**(1), 35–43 (2012).
93. G. Arora, *et al.*, PLGA nanoparticles for peptide receptor radionuclide therapy of neuroendocrine tumors: a novel approach towards reduction of renal radiation dose. *PLoS One* 7(3), e34019 (2012).
94. D. Emfietzoglou, *et al.*, Liposome-mediated radiotherapeutics within avascular tumor spheroids: comparative dosimetry study for various radionuclides, liposome systems, and a targeting antibody. *J. Nucl. Med.* **46**(1), 89–97 (2005).
95. G. Ting, *et al.*, Nanotargeted radionuclides for cancer nuclear imaging and internal radiotherapy. *J. Biomed. Biotechnol.* (2010).
96. L. Gedda, *et al.*, Experimental radionuclide therapy of HER2-expressing xenografts using two-step targeting nuclisome particles. *J. Nucl. Med.* **53**(3), 480–487 (2012).
97. D. Emfietzoglou, K. Kostarelos and G. Sgouros, An analytic dosimetry study for the use of radionuclide-liposome conjugates in internal radiotherapy. *J. Nucl. Med.* **42**(3), 499–504 (2001).
98. R. M. Reilly, *et al.*, ^{111}In-labeled EGF is selectively radiotoxic to human breast cancer cells overexpressing EGFR. *J. Nucl. Med.* **41**(3), 429–438 (2000).
99. P. Chen, *et al.*, Antitumor effects and normal tissue toxicity of ^{111}In-labeled epidermal growth factor administered to athymic mice bearing epidermal growth factor receptor-positive human breast cancer xenografts. *J. Nucl. Med.* **44**(9), 1469–1478 (2003).

100. R. M. Reilly, *et al.*, Preclinical pharmacokinetic, biodistribution, toxicology, and dosimetry studies of ^{111}In-DTPA-human epidermal growth factor: an auger electron-emitting radiotherapeutic agent for epidermal growth factor receptor-positive breast cancer. *J. Nucl. Med.* **47**(6), 1023–1031 (2006).
101. T. H. Chow, *et al.*, Therapeutic efficacy evaluation of ^{111}In-labeled PEGylated liposomal vinorelbine in murine colon carcinoma with multimodalities of molecular imaging. *J. Nucl. Med.* **50**(12), 2073–2081 (2009).
102. Y. Y. Lin, *et al.*, Evaluation of pharmacokinetics of ^{111}In-labeled VNB-PEGylated liposomes after intraperitoneal and intravenous administration in a tumor/ascites mouse model. *Cancer Biother. Radiopharm.* **24**(4), 453–460 (2009).
103. Y. J. Chang, *et al.*, Therapeutic efficacy and microSPECT/CT imaging of ^{188}Re-DXR-liposome in a C26 murine colon carcinoma solid tumor model. *Nucl. Med. Biol.* **37**(1), 95–104 (2010).
104. L. C. Chen, *et al.*, Pharmacokinetics, micro-SPECT/CT imaging and therapeutic efficacy of ^{188}Re-DXR-liposome in C26 colon carcinoma ascites mice model. *Nucl. Med. Biol.* **35**(8), 883–893 (2008).
105. Y. Liu and M. J. Welch, Nanoparticles labeled with positron emitting nuclides: advantages, methods, and applications. *Bioconjug. Chem.* **23**(4), 671–682 (2012).
106. J. Carlsson, *et al.*, Tumour therapy with radionuclides: assessment of progress and problems. *Radiother. Oncol.* **66**(2), 107–117 (2003).
107. S. N. Reske, *et al.*, ^{123}I-ITdU-mediated nanoirradiation of DNA efficiently induces cell kill in HL60 leukemia cells and in doxorubicin-, beta-, or gamma-radiation-resistant cell lines. *J. Nucl. Med.* **48**(6), 1000–1007 (2007).
108. K. J. Harrington, C. R. Lewanski and J. S. Stewart, Liposomes as vehicles for targeted therapy of cancer. Part 1: preclinical development. *Clin. Oncol. (R. Coll. Radiol.)* **12**(1), 2–15 (2000).
109. K. J. Harrington, C. R. Lewanski and J. S. Stewart, Liposomes as vehicles for targeted therapy of cancer. Part 2: clinical development. *Clin. Oncol. (R. Coll. Radiol.)* **12**(1), 16–24 (2000).
110. M. M. Kleiter, *et al.*, A tracer dose of technetium-99m-labeled liposomes can estimate the effect of hyperthermia on intratumoral doxil extravasation. *Clin. Cancer Res.* **12**(22), 6800–6807 (2006).
111. Y. J. Chang, *et al.*, Biodistribution, pharmacokinetics and microSPECT/CT imaging of ^{188}Re-BMEDA-liposome in a C26 murine colon carcinoma solid tumor animal model. *Anticancer Res.* **27**(4B), 2217–2225 (2007).
112. M. H. Chen, *et al.*, MicroSPECT/CT imaging and pharmacokinetics of ^{188}Re-(DXR)-liposome in human colorectal adenocarcinoma-bearing mice. *Anticancer Res.* **30**(1), 65–72 (2010).
113. F. Y. Huang, *et al.*, Imaging, autoradiography, and biodistribution of ^{188}Re-labeled PEGylated nanoliposome in orthotopic glioma bearing rat model. *Cancer Biother. Radiopharm.* **26**(6), 717–725 (2011).
114. T. Urakami, *et al.*, Novel amphiphilic probes for [^{18}F]-radiolabeling preformed liposomes and determination of liposomal trafficking by positron emission tomography. *J. Med. Chem.* **50**(26), 6454–6457 (2007).
115. N. Oku, Delivery of contrast agents for positron emission tomography imaging by liposomes. *Adv. Drug Deliv. Rev.* **37**(1–3), 53–61 (1999).
116. J. W. Seo, *et al.*, A novel method to label preformed liposomes with ^{64}Cu for positron emission tomography (PET) imaging. *Bioconjug. Chem.* **19**(12), 2577–2584 (2008).

117. H. E. Wang, *et al.*, Internal radiotherapy and dosimetric study for ^{111}In/^{177}Lu-pegylated liposomes conjugates in tumor-bearing mice. *Nucl. Instrum. Meth. A* **569**, 533–537 (2006).
118. W. C. Lee, *et al.*, Therapeutic efficacy evaluation of ^{111}In-VNB-liposome on human colorectal adenocarcinoma HT-29/luc mouse xenografts. *Nucl. Instrum. Meth. A* **569**, 497–504 (2006).
119. R. Torres Martin de Rosales, *et al.*, ^{99m}Tc-bisphosphonate-iron oxide nanoparticle conjugates for dual-modality biomedical imaging. *Bioconjug. Chem.* **22**(3), 455–465 (2011).
120. C. Tu, *et al.*, PET imaging and biodistribution of silicon quantum dots in mice. *ACS Med. Chem. Lett.* **2**(4), 285–288 (2011).
121. M. R. McDevitt, *et al.*, PET imaging of soluble yttrium-86-labeled carbon nanotubes in mice. *PLoS One* **2**(9), e907 (2007).
122. M. Mougin-Degraef, *et al.*, Doubly radiolabeled liposomes for pretargeted radioimmunotherapy. *Int. J. Pharm.* **344**(1–2), 110–117 (2007).
123. S. X. Wang, *et al.*, Intraoperative ^{186}Re-liposome radionuclide therapy in a head and neck squamous cell carcinoma xenograft positive surgical margin model. *Clin. Cancer Res.* **14**(12), 3975–3983 (2008).
124. W. T. Phillips, R. Klipper and B. Goins, Novel method of greatly enhanced delivery of liposomes to lymph nodes. *J. Pharmacol. Exp. Ther.* **295**(1), 309–313 (2000).
125. C. L. Zavaleta, *et al.*, Imaging of ^{186}Re-liposome therapy in ovarian cancer xenograft model of peritoneal carcinomatosis. *J. Drug Target* **16**(7), 626–637 (2008).
126. Y. J. Chang, *et al.*, Therapeutic efficacy of ^{188}Re-liposome in a C26 murine colon carcinoma solid tumor model. *Invest. New Drugs* **31**, 801–811 (2013).
127. C. H. Chang, *et al.*, Dosimetric evaluation of nanotargeted ^{188}Re-liposome with the MIRDOSE3 and OLINDA/EXM programs. *Ann. Nucl. Med.* **26**, 419–425 (2012).
128. C. M. Liu, *et al.*, Comparison of the therapeutic efficacy of ^{188}Rhenium-liposomes and liposomal doxorubicin in a 4T1 murine orthotopic breast cancer model. *Oncol. Rep.* **27**(3), 678–684 (2012).
129. S. Sofou, *et al.*, Engineered liposomes for potential alpha-particle therapy of metastatic cancer. *J. Nucl. Med.* **45**(2), 253–260 (2004).
130. S. Sofou, *et al.*, Enhanced retention of the alpha-particle-emitting daughters of Actinium-225 by liposome carriers. *Bioconjug. Chem.* **18**(6), 2061–2067 (2007).
131. G. Henriksen, *et al.*, Sterically stabilized liposomes as a carrier for alpha-emitting radium and actinium radionuclides. *Nucl. Med. Biol.* **31**(4), 441–449 (2004).
132. H. Nakamura, *et al.*, Development of boron nanocapsules for neutron capture therapy. *Appl. Radiat. Isot.* **67**, S84–87 (2009).
133. H. Yanagie, *et al.*, Application of boron-entrapped stealth liposomes to inhibition of growth of tumour cells in the *in vivo* boron neutron-capture therapy model. *Biomed. Pharmacother.* **60**(1), 43–50 (2006).
134. Y. Y. Lin, *et al.*, Pharmacokinetics and dosimetry of ^{111}In/^{188}Re-labeled PEGylated liposomal drugs in two colon carcinoma-bearing mouse models. *Cancer Biother. Radiopharm.* **26**(3), 373–380 (2011).
135. T. H. Chow, *et al.*, Diagnostic and therapeutic evaluation of ^{111}In-vinorelbine-liposomes in a human colorectal carcinoma HT-29/luc-bearing animal model. *Nucl. Med. Biol.* **35**(5), 623–634 (2008).

136. T. H. Chow, *et al.*, Improvement of biodistribution and therapeutic index via increase of polyethylene glycol on drug-carrying liposomes in an HT-29/luc xenografted mouse model. *Anticancer Res.* **29**(6), 2111–2120 (2009).
137. K. J. Harrington, *et al.*, Pegylated liposome-encapsulated doxorubicin and cisplatin enhance the effect of radiotherapy in a tumor xenograft model. *Clin. Cancer Res.* **6**(12), 4939–4949 (2000).
138. P. E. Huber, *et al.*, Trimodal cancer treatment: beneficial effects of combined antiangiogenesis, radiation, and chemotherapy. *Cancer Res.* **65**(9), 3643–3655 (2005).
139. T. Lammers, *et al.*, Image-guided and passively tumour-targeted polymeric nanomedicines for radiochemotherapy. *Br. J. Cancer* **99**(6), 900–910 (2008).
140. K. Kostarelos and D. Emfietzoglou, Tissue dosimetry of liposome-radionuclide complexes for internal radiotherapy: toward liposome-targeted therapeutic radiopharmaceuticals. *Anticancer Res.* **20**(5A), 3339–3345 (2000).
141. A. M. Syme, *et al.*, Dosimetric model for intraperitoneal targeted liposomal radioimmunotherapy of ovarian cancer micrometastases. *Phys. Med. Biol.* **48**(10), 1305–1320 (2003).
142. J. Huwyler, J. Drewe and S. Krahenbuhl, Tumor targeting using liposomal antineoplastic drugs. *Int. J. Nanomedicine* **3**(1), 21–29 (2008).
143. T. D. Tran, *et al.*, Clinical applications of perfluorocarbon nanoparticles for molecular imaging and targeted therapeutics. *Int. J. Nanomedicine* **2**(4), 515–526 (2007).
144. Z. Liu, *et al.*, *In vivo* biodistribution and highly efficient tumour targeting of carbon nanotubes in mice. *Nat. Nanotechnol.* **2**(1), 47–52 (2007).
145. S. J. Kennel, *et al.*, The fate of MAb-targeted ^{125m}CdTe/ZnS nanoparticles *in vivo*. *Nucl. Med. Biol.* **35**(4), 501–514 (2008).
146. S. J. DeNardo, *et al.*, Development of tumor targeting bioprobes ^{111}In-chimeric L6 monoclonal antibody nanoparticles) for alternating magnetic field cancer therapy. *Clin. Cancer Res.* **11**(19 Pt 2), 7087s–7092s (2005).
147. S. Liu, *et al.*, A novel type of dual-modality molecular probe for MR and nuclear imaging of tumor: preparation, characterization and *in vivo* application. *Mol. Pharm.* **6**(4), 1074–1082 (2009).
148. X. Yang, *et al.*, cRGD-functionalized, DOX-conjugated, and ^{64}Cu-labeled superparamagnetic iron oxide nanoparticles for targeted anticancer drug delivery and PET/MR imaging. *Biomaterials* **32**(17), 4151–4160 (2011).
149. J. S. Choi, *et al.*, A hybrid nanoparticle probe for dual-modality positron emission tomography and magnetic resonance imaging. *Angew Chem. Int. Ed. Engl.* **47**(33), 6259–6262 (2008).
150. J. Chen, *et al.*, Using anti-VEGF McAb and magnetic nanoparticles as double-targeting vector for the radioimmunotherapy of liver cancer. *Cancer Lett.* **231**(2), 169–175 (2006).
151. J. Cheon and J. H. Lee, Synergistically integrated nanoparticles as multimodal probes for nanobiotechnology. *Acc. Chem. Res.* **41**(12), 1630–1640 (2008).
152. J. W. Park, C. C. Benz and F. J. Martin, Future directions of liposome- and immunoliposome-based cancer therapeutics. *Semin. Oncol.* **31**(6 Suppl 13), 196–205 (2004).
153. T. A. Elbayoumi and V. P. Torchilin, Enhanced cytotoxicity of monoclonal anticancer antibody 2C5-modified doxorubicin-loaded PEGylated liposomes against various tumor cell lines. *Eur. J. Pharm. Sci.* **32**(3), 159–168 (2007).

154. N. Kamaly, *et al.*, Folate receptor targeted bimodal liposomes for tumor magnetic resonance imaging. *Bioconjug. Chem.* **20**(4), 648–655 (2009).
155. A. N. Lukyanov, *et al.*, Tumor-targeted liposomes: doxorubicin-loaded long-circulating liposomes modified with anti-cancer antibody. *J. Control Release* **100**(1), 135–144 (2004).
156. P. Diagaradjane, *et al.*, Imaging epidermal growth factor receptor expression *in vivo*: pharmacokinetic and biodistribution characterization of a bioconjugated quantum dot nanoprobe. *Clin. Cancer Res.* **14**(3), 731–741 (2008).
157. A. M. Smith, *et al.*, Bioconjugated quantum dots for *in vivo* molecular and cellular imaging. *Adv. Drug Deliv. Rev.* **60**(11), 1226–1240 (2008).
158. R. Koole, *et al.*, Paramagnetic lipid-coated silica nanoparticles with a fluorescent quantum dot core: a new contrast agent platform for multimodality imaging. *Bioconjug. Chem.* **19**(12), 2471–2479 (2008).
159. M. M. van Schooneveld, *et al.*, Improved biocompatibility and pharmacokinetics of silica nanoparticles by means of a lipid coating: a multimodality investigation. *Nano Lett.* **8**(8), 2517–2525 (2008).
160. E. Tasciotti, *et al.*, Mesoporous silicon particles as a multistage delivery system for imaging and therapeutic applications. *Nat. Nanotechnol.* **3**(3), 151–157 (2008).
161. S. Ganta and M. Amiji, Coadministration of Paclitaxel and curcumin in nanoemulsion formulations to overcome multidrug resistance in tumor cells. *Mol. Pharm.* **6**(3), 928–939 (2009).
162. S. Mura and P. Couvreur, Nanotheranostics for personalized medicine. *Adv. Drug Deliv. Rev.* **64**(13), 1394–1416 (2012).
163. J. H. Ryu, *et al.*, Tumor-targeting multi-functional nanoparticles for theragnosis: new paradigm for cancer therapy. *Adv. Drug Deliv. Rev.* **64**(13), 1447–1458 (2012).
164. M. C. Garnett and P. Kallinteri, Nanomedicines and nanotoxicology: some physiological principles. *Occup. Med. (Lond)* **56**(5), 307–311 (2006).
165. A. Nel, *et al.*, Toxic potential of materials at the nanolevel. *Science* **311**(5761), 622–627 (2006).
166. M. A. Dobrovolskaia, D. R. Germolec and J. L. Weaver, Evaluation of nanoparticle immunotoxicity. *Nat. Nanotechnol.* **4**(7), 411–414 (2009).
167. K. R. Vega-Villa, *et al.*, Clinical toxicities of nanocarrier systems. *Adv. Drug Deliv. Rev.* **60**(8), 929–938 (2008).
168. V. J. Thomas and P. Dan, Nanotoxicity and the importance of being earnest. Preface on theme issue on "Nanotoxicity: from bench to bedside." *Adv. Drug Deliv. Rev.* **64**(15), 1661–1662 (2012).
169. M. G. Stabin, R. B. Sparks and E. Crowe, OLINDA/EXM: the second-generation personal computer software for internal dose assessment in nuclear medicine. *J. Nucl. Med.* **46**(6), 1023–1027 (2005).
170. Protection, I.C.o.R. (ed.) *Radiological Protection in Biomedical Research*, Vol. 22., ICRP (1992).
171. D. R. Fisher, S. Shen and R. F. Meredith, MIRD dose estimate report No. 20: radiation absorbed-dose estimates for ^{111}In and ^{90}Y-ibritumomab tiuxetan. *J. Nucl. Med.* **50**(4), 644–652 (2009).
172. C. M. Liu, *et al.*, Extended acute toxicity study of ^{188}Re-liposome in rats. *J. Appl. Toxicol.* (2012).

173. B. A. Goins, Radiolabeled lipid nanoparticles for diagnostic imaging. *Expert Opin. Med. Diagn.* **2**, 853–873 (2008).
174. C. A. Presant, A. F. Turner and R. T. Proffitt, Potential for improvement in clinical decision-making: tumor imaging with In-111 labeled liposomes results of a phase ii-iii study. *J. Liposome Res.* **4**(2), 985–1008 (1994).
175. G. M. Jensen and T. H. Bunch, Conventional liposome performance and evaluation: lessons from the development of Vescan. *J. Liposome Res.* **17**(3–4), 121–137 (2007).
176. U.S. FDA's Guidance for Industry, Liposome Drug Products. Food Drug Administration (2002). Available from http://www.fda.gov.
177. U.S. FDA's Guidance, PET Drugs — Current Good Manufacturing Practice (CGMP). Food Drug Administration (2009). Available from http://www.fda.gov.
178. U.S. FDA's Guidance on Nanotechnology: Fact Sheet. Food Drug Administration (2012). Available from http://www.fda.gov.
179. U.S. FDA's Guidance for Industry, Developing Medical Imaging Drug and Biological Products, Part 1: Conducting Safety Assessments. Food Drug Administration (2004). Available from http://www.fda.gov.
180. U.S. FDA's Guidance for Industry, Nonclinical Evaluation of Late Radiation Toxicity of Therapeutic Radiopharmaceuticals. Food Drug Administration (2011). Available from http://www.fda.gov.
181. A. L. Petersen, *et al.*, Liposome imaging agents in personalized medicine. *Adv. Drug Deliv. Rev.* **64**(13), 1417–1435 (2012).
182. Position paper on non-clinical safety studies to support clinical trials with a single micro-dose. European Medicines Agency (EMA) (2004). Available from http://www.ema.europa.eu/ema.
183. U.S. FDA's Guidance for Industry, Investigators, and Reviewers, Exploratory IND Studies. Food Drug Administration (2006). Available from http://www.fda.gov.

Chapter 5

Nanoengineering at the Gas–Liquid Interface: Gas-Filled Microbubbles as Tools for Molecular Imaging and Drug Delivery

*Alexander L Klibanov**

Department of Medicine (Cardiovascular Division), Cardiovascular Research Center, Cardiovascular Imaging Center, Department of Radiology, Department of Biomedical Engineering, University of Virginia, P.O. Box 801394, Charlottesville VA 22908, USA
**sasha@virginia.edu; SKLIB1@gmail.com*

Abstract

Gas-filled microbubbles are becoming increasingly popular in the clinic as ultrasound contrast blood pool agents, with hundreds of thousands of patients treated annually. These particles are several micrometers in size, entrapping a water-insoluble gas. They are stabilized with a thin (monomolecular, typically ~2 nm thick) shell, made of lipids, surfactants, proteins and/or polymers. With the proper choice of the shell material, they can be stable for months and years following preparation, or available as precursors in sterile packaging for bedside use activation. Lately, these particles are being investigated as molecular imaging ultrasound contrast materials: targeting ligands are placed on the shell to achieve selective adhesion to the vascular endothelium markers of disease. Detection of microbubbles by ultrasound provides excellent sensitivity: each particle has sub-picogram mass and can be imaged in real time, several centimeter deep in the soft tissues of the body. Microbubbles can be also used as drug and gene delivery enhancers, by localized deposition of ultrasound energy and ultrasound-triggered payload release in the target vasculature. Overall, nanobiotechnology can successfully solve the complex tasks of microbubble preparation, stabilization, targeting and drug loading to prepare contrast agents that will find widespread use in the clinic and research.

1. Introduction

Targeted imaging has been in clinical and preclinical use for several decades.[1] After targeting ligand molecule is attached to a label moiety, the conjugate is administered in the body and its binding to the disease marker receptor is monitored by an imaging system. A wide variety of labels can be applied: fluorescent dyes and radioisotopes, as well as materials detectable by X-ray, MRI and ultrasound. Microparticles and nanoparticles are commonly investigated as contrast agents and drug delivery vehicles. Therefore, a design where a particle carries on its surface a molecule of interest, e.g. a ligand specific to the disease marker receptor, is quite common. In this chapter, we will discuss the nanoengineering techniques of an unusual type of particle, microbubble, which contains a gas core, as most suitable for ultrasound imaging and triggered drug delivery.

2. Microbubble: Contradictory Requirements to Particle Stabilization

For contrast ultrasound imaging, a wide variety of particles has been proposed initially, solid or liquid, but only one of the designs so far has reached practical application as a contrast agent in the clinic: gas-filled microbubble. Ultrasound is a pressure wave; an ultrasound transducer imaging probe sends series of short sound pulses into the tissue (typically, 1–15 MHz frequency). The reflected echo monitoring makes it possible to obtain the images of the anatomical structures and biological tissues. Sound reflection is based on the mismatch of acoustic impedance (i.e. the product of the density of the media and the speed of sound in it) for the adjacent tissues, and in some instances allows delineation of tissues, organs and areas of disease. Micrometer-sized gas bubbles in the aqueous environment are very effective scatterers of sound: unlike the surrounding tissues, which mostly consist of incompressible water, gas is extremely compressible. Therefore, when a gas bubble is placed in the medical ultrasound field, it rapidly compresses and expands, so the gas–liquid interface becomes a "speaker," and each bubble becomes a secondary point source of scattered sound. This scattered sound can be efficiently detected by the imaging probe, and bubble position in the body is determined with millimeter or sub-millimeter precision (if megahertz or multi-megahertz frequency is applied for imaging, respectively). To enhance bubble detection and reduce the background signal from the tissue surrounding the bubble, several approaches are applied. Modern imaging detection schemes monitor sound not just at a fundamental frequency[2] but also in harmonic or subharmonic

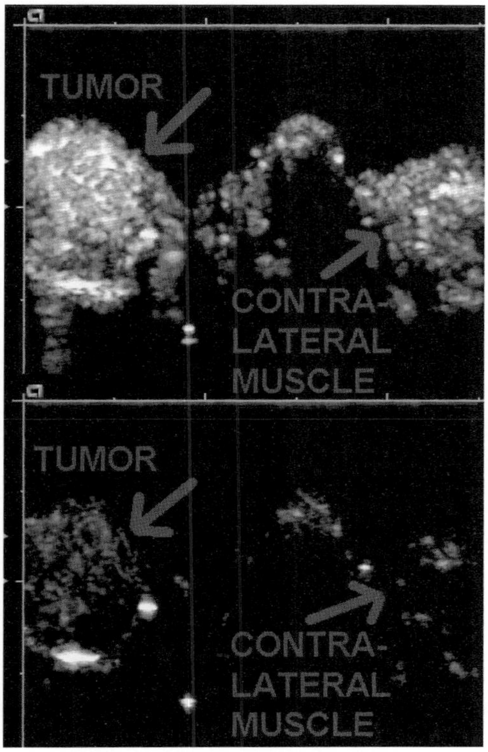

Fig. 1. Molecular ultrasound imaging of microbubbles that are decorated with an RGD peptide targeted to $\alpha_v\beta_3$ integrin expressed on the surface of endothelium in the tumor vasculature. Experimental conditions: 20 million microbubbles carrying cyclic RGD peptide injected intravenously in the C57BL/6 mouse carrying in the hind leg a subcutaneous tumor (murine adenocarcinoma, MC38 cells generously provided by J. Schlom, NIH). Imaging was performed with 15L8 probe of Siemens Sequoia 512 ultrasound system (7 MHz, Cadence CPS mode, image depth 2 cm), 30 s following *i.v.* bolus (top frame), to demonstrate perfusion by microbubbles in the tumor vasculature (left) as well as filling of normal vasculature in the contralateral leg (right). Bottom frame provides contrast ultrasound image 5 min after microbubble injection, so that the majority of circulating microbubbles have cleared from the bloodstream in the normal vasculature (bottom frame, right) but targeted bubbles are still adherent at the tumor vasculature (bottom frame, left).

modes: tissue and bubbles have distinct frequency responses. Even better are multipulse techniques,[3] where several ultrasound pulses are sent in a rapid succession, with varying polarity and intensity, and the processed combination of the echo responses clearly distinguishes a bubble and completely suppresses the signal from the surrounding non-target tissues. Figure 1 provides an example of the successful use for this technique (Cadence CPS) for molecular ultrasound imaging in a murine model: the signal of the tissue itself is limited to the speckles from bones and underlying heating pad.

A critical factor in the success of contrast ultrasound imaging is the compressibility of the gas bubble: lack of particle compressibility results in poor echo response. Therefore, the bubble with perfect acoustic response would be the one where gas–liquid interface is clean and uncoated, i.e. where there is no shell to constrain the vibration of the bubble. However, other issues come into play, among them the most important is the bubble stability. Gas microbubbles without a shell were the ones proposed initially,[4,5] where air was dispersed in saline or in X-ray contrast media solution.[6] Not only were these bubbles unstable and disappeared as soon as they floated to the top of the solution but even, if they were quickly administered in the bloodstream, they were unable to pass through microvasculature of the lung, and hence unusable for *i.v.* administration of contrast. Therefore, middle ground needs to be found between the stability of the bubble and compressibility of the shell. This chapter addresses the approaches to find such optimum and apply optimized particles for biomedical imaging and drug delivery.

3. Bubble Preparation and Stabilization by Nanometer-Thick Shell Deposition

3.1. *Microbubble preparation techniques*

The simplest methods of bubble preparation applied since 1980s is the kinematic mixing of air (or other gas) with aqueous media by application of shear. The hand-held apparatus consists of two half-filled medical syringes that contain gas and injectable liquid medium, connected by a stopcock valve. By moving the plungers of the syringes, gas and liquid are transferred through a narrow orifice of the valve, and bubbles (with a wide range of sizes) are created in an enclosed sterile setting, by the bedside.[7] There is a wide variety of "whip cream" devices: from a kitchen blender to laboratory high-shear mixers, sonicators and amalgamators. The major advantage of the latter approach is that the bubbles are prepared immediately before use, in an enclosed sterile sealed medical-grade vial, with an inexpensive (yet Food and Drug Administration [FDA]-approved) dentistry benchtop shaker.[8] An alternative, dry precursor formulation, implies having a dry lyophilized cake of a water-soluble matrix, such as a carbohydrate[9] or polyethylene glycol (PEG)[10,11] with the desired gas headspace in the vial. Formulation may contain stabilizers (e.g. lipids and surfactants) as well. Dry formulations have outstanding stability on storage. Sterile water is added when needed, and the matrix immediately dissolves, leaving behind gas bubbles in the aqueous phase. All these approaches as described above are easy to implement; a significant disadvantage is the wide size distribution of the resulting bubble product, from submicron to many micrometers. Sieve filter application to the (flexible) bubbles is not feasible to retain and filter out large particles (they will squeeze through the filter pores). Therefore, flotation techniques have been proposed, based on the size dependence of the limiting speed of the bubble movement in the gravity field governed by Navier–Stokes equation and viscous drag.[12–14] Lately, microfluidic approaches, which can prepare bubbles with an extremely tight size distribution of particles, are actively investigated, such as inkjet "printing"[15] of microbubbles or flow-through gas–liquid mixing microfluidic chambers.[16] These approaches will find practical use as soon as their productivity is improved and desired microbubble size distribution is obtained.[17]

3.2. *Microbubble shell as a stabilizer*

Examples of microbubble contrast agents manufactured in accordance with nanoengineering principles are provided in Table 1. Table 1 shows that the most practically useful design of gas microbubbles for ultrasound contrast is to coat the bubbles with the shell and select a proper (water insoluble) gas. Self-assembly of these particles is the most useful manufacturing approach, where the molecules used for the shell construction are present in the aqueous media during the formation of the bubbles in the aqueous environment, so that the newly formed gas–liquid interface, with its high surface energy, is immediately coated by properly oriented stabilizer molecules, to minimize the free energy of the system.[17] Most popular surfactants used are lipids,[18] e.g. phospholipids,[8,10] preferably ones that have higher phase transition temperature: if stabilizer molecules are in liquid crystal state, above their melting point, the shell is not the best stabilizer, and bubbles will likely fuse on collision. If the phase transition temperature of the stabilizer molecules is high, then at room temperature or even at 37°C, the shell molecules possess high cohesive energy, membranes are not prone to fusion and resulting bubbles can be stable in the aqueous environment for long periods of time. The issue of bubble stability due to gas dissolution was investigated by Epstein and Plesset decades ago[19]; they found that Laplace pressure inside the bubble, equaling 2 × (surface tension)/radius, caused rapid loss of air from bubbles by dissolution: non-shelled air bubble with the size of several micrometers will lose its air core within seconds. Gas is squeezed out of the bubble by

Table 1. Microbubbles approved for clinical use or evaluated in animal research testing: composition and design.

	Name (Manufacturer)	Shell composition	Gas core	Agent design	Status, References
1	Albunex, (MBI, Mallinckrodt)	Denatured albumin	air	Aqueous dispersion	Approved, not selling, 23
2	Levovist (Schering)	Palmitic acid	air	Dry galactose precursor	Approved, not selling, 9
3	Optison (GE)	Denatured albumin	C_3F_8	Aqueous dispersion	Approved, in clinic, 24
4	Definity (DMP115, Lantheus)	DPPC, DPPA, mPEG5000-DPPE	C_3F_8	Aqueous precursor	Approved, in clinic, 8
5	Sonazoid (GE)	Hydrogenated PS	C_4F_{10}	Dry PEG precursor	Approved, in clinic, 25
6	Sonovue (Bracco)	DSPC + DPPG	SF_6	Dry PEG precursor	Approved, in clinic, 11
7	Imagent (Alliance Pharma)	DMPC	C_6F_{14}+ N_2	Dry PEG precursor	Approved, not selling, 21
8	MP1950 (Mallinckrodt)	DSPC + PEG stearate	C_4F_{10}	Aqueous dispersion	Research agent: animal trials, 26
9	Imagify (AI-700, Acusphere)	PLGA + phosphatidylcholine	C_4F_{10}	Dry PEG precursor	Clinical trials, 27
10	Cardiosphere/PB127 (Point Biomedical)	PLGA + albumin	N_2	Dry powder precursor	Clinical trials: development stopped, 22
11	BR38 (Bracco)	DSPC + PEG-DSPE	C_4F_{10}+ N_2	Dry PEG precursor	Research agent: animal trials, 10
12	BR55 (Bracco)	DSPC + PEG-DSPE + VEGFR2-binding peptide-PEG-DSPE	C_4F_{10}+ N_2	Dry PEG precursor	Exploratory clinical trial, 28

surface tension and dissolves in the surrounding medium. Surface tension could be reduced somewhat, by detergents in the aqueous medium coming to the interface, but it will not abolish Laplace pressure or gas loss. If a "solid" shell is used to stabilize the bubble, such as a thin shell of denatured albumin,[20] Laplace pressure is counteracted by the shell and bubbles can be stable for days and months, e.g. as Albunex microbubble preparation. To achieve this for lipid bubbles, most popular microbubble shell phospholipids used are dipalmitoyl phosphatidylcholine (DPPC)[8] and especially distearoyl phosphatidylcholine (DSPC),[10,11] with thin-shell DMPC microbubbles being less stable.[21] However, air bubbles are not very well stabilized by lipid shells, a thicker biocompatible polymer shell (e.g. PLGA) is required, with tens of nanometer thickness.[22]

Despite being beneficial for unimpeded bubble vibration, the stabilizer shell with a thickness of several nanometer (lipid monolayer or albumin-coated) cannot serve as a good barrier for entrapped gas transfer: air rapidly leaves the core and bubble collapses if the bubbles are placed in air-depleted aqueous phase; one such example was provided when first-generation *i.v.* microbubble contrast agents with entrapped air (i.e. ~80% nitrogen), such as Albunex (albumin shell) or Levovist (palmitic acid shell) rapidly lost ultrasound contrast signal in the bloodstream when administered intravenously to experimental animals[23] or patients breathing pure oxygen, who had little dissolved nitrogen in blood. Therefore, insoluble gases, such as SF_6 and perfluorocarbon gases, have become standard: their solubility in water and blood are orders of magnitude less than of air or even helium. So, microbubbles with that gas core can stay in the aqueous phase for months and years in storage. Thin shell is still not an efficient barrier to gas exchange in circulation, so perfluorocarbon carbon gases are lost by the bubbles within a fraction of an hour after *i.v.* injection (mostly exhaled via lungs[24]). Lately, this approach has been augmented by another component in the shell composition: in addition to phospholipids, a thin layer of a water-insoluble polymer, PLGA, is added to improve circulation of perfluorocarbon bubbles even further; the resulting microbubble is now in clinical trials.[27]

So, overall design of intravascular ultrasound contrast agent is currently as follows: it is a microparticle with a fluorinated gas core, with the size of ~1–3 μm, stabilized by a thin (e.g. monomolecular phospholipids) shell and prepared by self-assembly stabilization. Thin shell allows almost unimpeded compression and expansion of the particles in the ultrasound field and allows excellent detection sensitivity: single particles with the mass under a picogram can be observed in real time.[26,29] These particles freely circulate in the bloodstream and therefore can be used for molecular imaging, by simply decorating the surface of the bubbles with the selective ligands, such as monoclonal antibodies against the specific markers present on vascular endothelium in the areas of disease. In the next section, we will discuss the optimal designs for targeted microbubble particles.

4. Attachment of Targeting Ligands to Microbubble Shell

A significant experience has been accumulated over the past decades on the use of targeting ligands, such as antibodies, peptides or peptide mimetics, hormones and carbohydrates for the purposes of molecular imaging or drug delivery. Targeting ligands are routinely attached to the surface of particles: liposomes evaluated for drug delivery,[30] red blood cells,[31] quantum dots[32] and MRI contrast agents.[33] Situation with a microbubble is somewhat different — its shell is just several nanometers thick, and in order to provide targeting in the conditions of fast flow, the shell has to be robust and not break in the presence of shear forces, e.g. when bubble is attached to its target on the vessel wall and subjected to viscous drag created by the flow of blood.[34]

Attachment of targeting ligands to the bubble shell can be covalent or noncovalent: targeting ligands can be simply added to the bubble preparation aqueous phase, so that when the new gas phase is generated, some of these targeting molecules are positioned at the gas–water interface and thus embedded in the shell.[35] This approach can be used for protein ligands (e.g. avidin as above), as well as for lipids capable of selective interactions: phosphatidylserine (PS) is a marker for phagocytosis, so that phagocytic cells that have access to the bloodstream, e.g. Kupffer cells in the liver,[36] or neutrophils adherent to the vessel wall in the areas of ischemia-reperfusion injury or acute inflammation[37, 38] will recognize PS and internalize microbubbles by phagocytosis. Sonazoid microbubbles with the shell entirely made of PS are approved for clinical use in Japan for imaging defects (i.e. tumor nodes) in normal liver parenchyma; the latter tissue will have functional Kupffer cells and take up Sonazoid microbubbles inside those cells.[25]

A universal and simple approach, yet not suitable for clinical application, is the use of avidin–biotin coupling: premade bubbles that carry biotin on their shell are incubated with streptavidin and then with biotinylated antibody.[39] There are numerous examples of this approach, because a wide variety of ligands is already available in biotinylated form or can be biotinylated using proper kits. Streptavidin can also be placed on premanufactured bubbles, so that biotinylated ligand can be added to the bubbles and the conjugate formed in a single step. However, streptavidin is not a human protein and to avoid undesired immune response to a foreign protein, this approach will not be considered for use in humans: for clinical applications, the ligands will be attached to the bubble shell by a covalent link.

Covalent attachment to the surface of microbubbles took place starting with the initial studies of microbubble targeting.[40,41] Briefly, microbubbles with carboxylic functional groups were first activated with water-soluble carbodiimide and *N*-hydroxysulfosuccinimide in acidic pH, followed by addition of the monoclonal antibody at a more alkaline pH, and formation of amide bond was achieved. Nearly 10^5 antibody molecules per bubble were attached this way, as needed for microbubble targeted to receptor antigen-expressing endothelial cells. To achieve this antibody load per bubble, large excess of added antibody had to be applied. Problem with this chemistry is the low coupling yield at low concentration of reagents: over 90% of the added antibody did not attach to the bubbles and was lost, due to the rapid hydrolysis of active *N*-hydroxysulfosuccinimide ester of the carboxyl on the bubble surface, as compared with its slow reaction with the primary aminogroups of the antibody at low concentration. Single-stage carbodiimide crosslinking at neutral pH[42] is not applicable here: protein's own carboxyls and amino groups would be first to react, resulting in the formation of protein aggregates, possibly also partially attached to the bubble shell.

An alternative reaction, coupling of a maleimide residue on the bubble shell with the thiol on the ligand, may be more useful: maleimide hydrolysis is much slower than its reaction with thiol. Therefore, thiolated ligands, such as single chain vascular endothelial growth factor (VEGF),[43] or E-selectin binding peptide[44] were covalently attached to the bubbles with much better coupling yield and reasonable surface density to ensure efficient targeting.

The most reliable approach, attachment of targeting ligand to the lipid anchor, followed by the conjugate purification and microbubble preparation, is most feasible for smaller molecules, such as peptides.[28] In this case, peptide synthesis and lipid attachment can be performed with high yield and purity, in an FDA-inspected established environment, and the resulting ligand–lipid complex can be then mixed with other lipids used for microbubble preparation. For larger protein ligands, which may be more sensitive to (sometimes) harsh conditions of microbubble manufacturing, attachment to preformulated bubbles might be investigated as the first feasible option.

4.1. *Microbubble shell external coating: perceived benefits of an extended spacer arm for the ligand*

Many microbubble coating designs provide an extra level of stabilizer coating on top of a phospholipids monolayer, a grafted polymer brush, usually PEG.[8,10]

It is attached to the phospholipid monolayer via an embedded lipid anchor (Fig. 2), traditionally PEG stearate[41] and/or PEG-phosphatidylethanolamine.[8,45] This PEG brush layer extends from the surface of the lipid shells by ~10–20 nm.[46] It should reduce already low chances of microbubble fusion on storage in the aqueous medium, by repelling adjacent bubbles. This brush also inhibits nonspecific adhesion of microbubbles in the vasculature, because it somewhat reduces complement C3 deposition.[47] However, if a small ligand molecule (e.g. biotin) is directly attached to the microbubble surface[45] and does not extend into the surrounding aqueous medium, its interaction with a receptor-coated surface will be inhibited. Therefore, it may make sense to place the targeting ligand on the tip of a long PEG spacer arm, so that the targeting ligand will not be buried in the PEG brush and will be able to interact with the receptor-coated surface, e.g. avidin-coated bead[45] or vascular endothelium, e.g. overexpressing P- or E-selectin in the area of inflammation.[48,49] There is still a scenario when hiding the ligand in the PEG brush could be beneficial: if it is desirable that targeting should take place only in a certain area of the vasculature and not in the rest of the body. To achieve such selectivity, acoustic

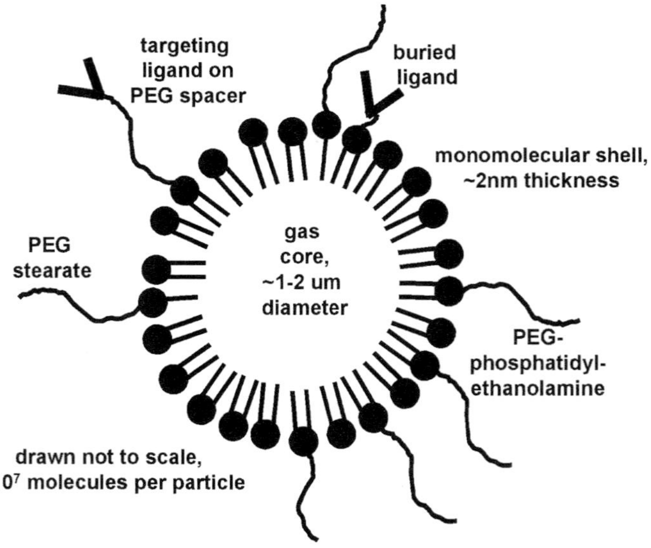

Fig. 2. Targeted microbubble design: a combination of phospholipid monolayer shell with PEG stearate (left) or PEG-phosphatidylethanolamine (right). Grafted PEG brush is used to stabilize the microbubble shell to reduce microbubble fusion, minimize complement activation and reduce adhesion to normal vasculature. Extended PEG polymer spacer arm (top left) is used to provide elongated flexible spacer arm to maximize the chance of ligand–receptor interaction. Binding of the ligand that is connected to the shell via a short spacer is inhibited by the brush (top right).

radiation force can be applied in an area of interest, so when the bubbles with "hidden" ligand circulate, they do not interact with the receptor antigen, unless acoustic radiation force brings the bubble in close proximity to the target on the vessel wall. PEG brush barrier is a dynamic/stochastic structure, which could be defeated by the radiation force, and ligand–receptor interaction would then take place where receptor is present.[50] Acoustic irradiation can be then stopped, and bubbles that did not adhere firmly to the vessel wall via a ligand–receptor interaction would be removed by the blood flow, leaving behind the bubbles only in the areas of receptor expression.

5. Targeted Gene Delivery with Microbubble Constructs

Ultrasound field can penetrate deep inside soft tissues in the body. It can be tightly focused, with the focal point as small as several millimeters. Ultrasound passes through soft tissue without significant energy deposition; ultrasound energy conversion to vibration of the gas–liquid interface takes place in the presence of compressible microbubbles. This technique was used to deliver circulating materials into cells: microbubble vibration can create pores in cell membrane and vessel wall, via sonoporation. Microbubbles can serve as drug delivery containers that release their payload in response to ultrasound action.[51] Transfection was one of the first techniques tested in this setting: microbubbles carrying positively charged lipids in the shell attach nucleic acid[52] and release it in the ultrasound-treated tissues, but only minimally in the non-insonated tissues or areas with insufficient acoustic pressure. This phenomenon has been successfully demonstrated for plasmid delivery and transfection of model genes, encoding β-galactosidase, luciferase, green and red fluorescent proteins, as well as transfection of therapeutic genes, such as VEGF165[53] for therapeutic angiogenesis, or thymidine kinase[54] for ganciclovir tumor therapy, or ultrasound-triggered siRNA delivery to inhibit expression of epidermal growth factor receptor in tumor therapy.[55] Particle design in all of these studies is standard (Fig. 3): a phospholipid monomolecular layer shell surrounding a perfluorocarbon gas core, where some of the lipids (e.g. DSTAP) in the shell carry a positive electrical charge, for electrostatic attachment of nucleic acid. In addition, the shell is decorated by a grafted brush of PEG, in the form of PEG-stearate[56] or PEG-PE[57]: this may help repel nucleases and minimize degradation of the immobilized nucleic acid.[57] Instead of direct attachment of nucleic acid to the bubble shell, genetic material can be packaged in a lipoplex particle[58] or an uptake-deficient virus,[59] which is then attached to the bubble. More recent improvements in the particle design (Fig. 3, top right) include placement of targeting ligands on the

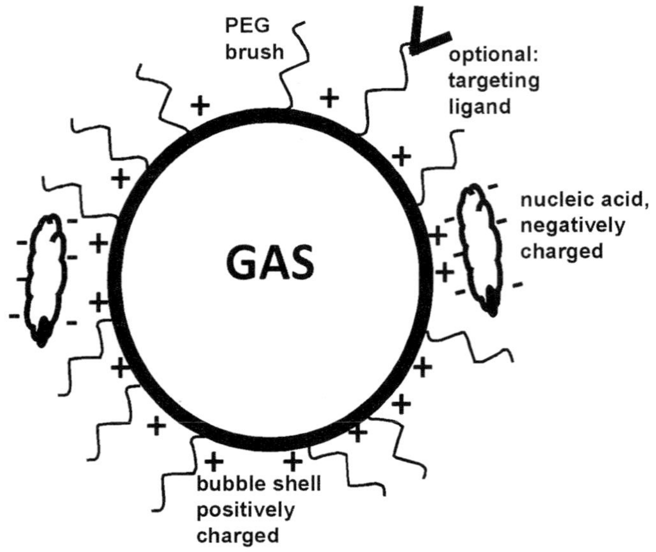

Fig. 3. Microbubble design for ultrasound-triggered nucleic acid delivery. Positively charged lipids of microbubble shell to which nucleic acid can attach via electrostatic interaction (left and right), combined with PEG brush coating and (optionally) with the targeting ligands to achieve active adhesion at the disease site (top right).

bubble: targeted microbubbles that carry nucleic acid are also decorated with a targeting ligand, such as a monoclonal antibody against MAdCAM-1,[60] or a P-selectin ligand for endothelium targeting.[61] Microbubbles–plasmid complex can be targeted to the tissues and cells where transfection is needed. Following *i.v.* bolus injection and 10 min dwell time, after circulating gene delivery complex is cleared from the bloodstream, only targeted bubbles remain attached to the activated endothelium target. Unfocused ultrasound can be then applied to achieve transfection — this will allow targeted transfection without the need to use focused ultrasound and without obtaining information about precise location of the lesions.

6. Ultrasound-Triggered Drug Delivery with Microbubble-Based Systems

Triggered release of drugs in the desired area of the body by an external stimulus, e.g. by mechanical activation of microbubble constructs with the action of ultrasound has been suggested more than a decade ago,[51] but has not yet reached clinical use. There are several complications with this approach: first, microbubbles stay in the bloodstream for only a fraction of an hour, then gas is lost and complexes no longer respond to ultrasound action; residual particle material will stay in the bloodstream until eventually picked up by phagocytes in the body. Therefore, significant improvement of therapeutic index is achievable only if the drug uptake and action in the target tissue is greatly enhanced by ultrasound application. Drug toxicity has to be low; preferably, drug should be nontoxic and inactive unless ultrasound is applied, just like a plasmid DNA, which is incapable of getting in the target cells on its own. For anticancer drugs, such as doxorubicin, drug effects on normal tissues may not always be critical: critical cardiotoxicity will be avoided, and drug–bubble complex accumulation in the liver will not be a major cause for concern (doxorubicin liposomes accumulate in the liver without major detriment to patients' condition), as long as enough drug will be delivered to the tumor cells to achieve tumor therapy.

The dose of a drug that a microbubble-based system can deliver will depend on several factors. First, it is the nature of the drug: if the drug is hydrophobic, such as paclitaxel, it can be embedded within the lipid monolayer shell — but only a small amount (Fig. 4(a)): if a typical bubble has ~20.10^6 lipid molecules in its coat, and, e.g. 25% of it will be the drug, then a billion bubbles will carry ~8 nanomoles of the drug, i.e. typically, several micrograms, which is too low to be useful for a clinical dose for most of the drugs as a human dose. Therefore, a hybrid microbubble formulation[62] has been suggested that will have a thicker shell (Fig. 4(b)) constructed of a drug carrier oil layer (e.g. glycerol trioleate) carrying the drug, stabilized with a lipid monolayer on the outside. If the oil layer is thick enough, several orders of magnitude more drug can be delivered in this scenario, directly related to the volume of the carrier.

For hydrophilic (or moderately hydrophilic) drug, placing it inside the monolayer-coated bubble should not be feasible — water will penetrate the lipid monolayer to dissolve the drug, which might not be useful for the stabilization of the gas–liquid interface; in this case, liposomes can help. First, liposomes (that can entrap the drug inside their internal aqueous space) were used to decorate the surface of microbubble (Fig. 4(c)). Liposomes consist of a lipid bilayer, which could be easily modified with biotin, and biotinylated lipids could be used to anchor biotinylated bubbles to biotinylated liposomes via a streptavidin linker.[63,64] Then, covalent attachment of bubbles with liposomes was successfully applied to prepare drug–liposome–microbubble pendant.[65] The best way to characterize the drug load of a pendant bubble–liposome conjugate is to compute the internal liposome volume available for payload[63]; for geometry considerations, it is proportional to the size of

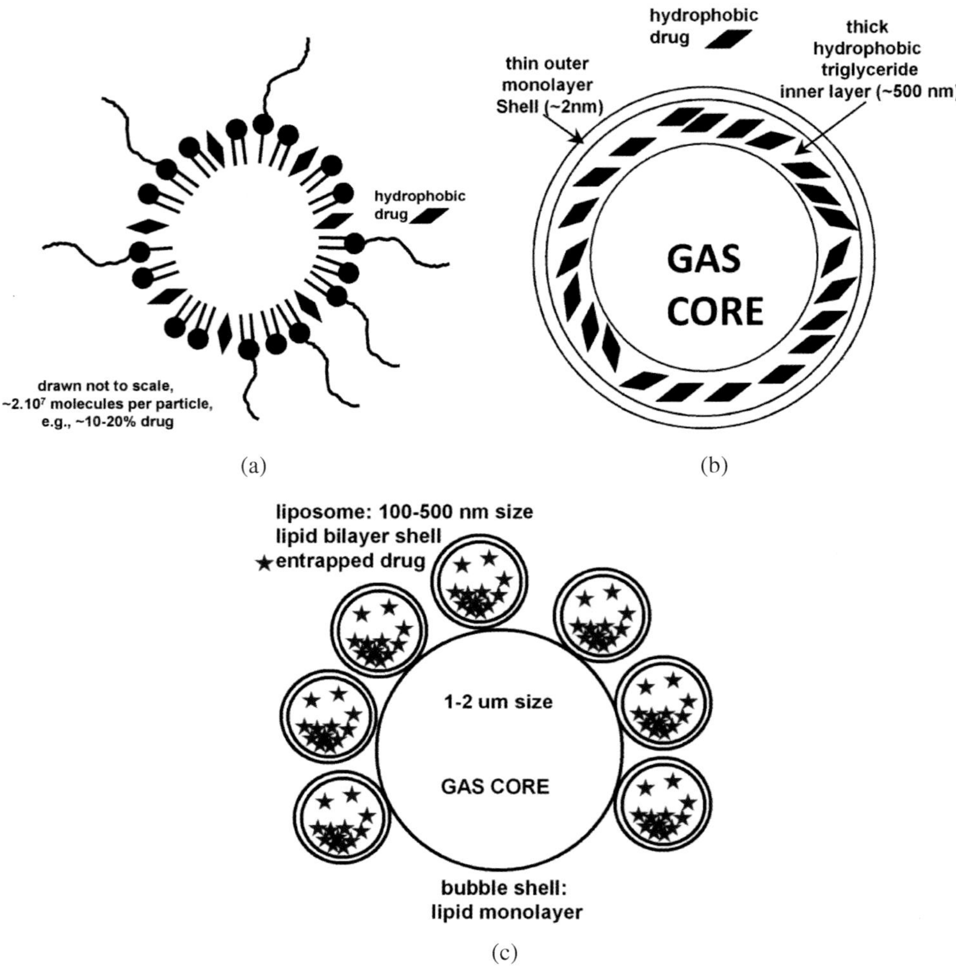

Fig. 4. Design of targeted microbubbles for ultrasound-triggered drug delivery. Hydrophobic drug can be embedded in the shell (a) or dissolved in the thick oil layer (b). Hydrophilic or hydrophobic drug can be entrapped in liposomes that decorate the bubble surface (c).

liposomes attached to the bubble surface: up to several cubic micrometer per bubble can be loaded, which translates to tens of microliters payload volume per billion bubbles, which should be sufficient for therapeutic action of potent cytotoxic agents (Fig. 4(c)). The most efficient use of payload volume is to have the drug precipitate and form crystals inside the liposomes — exactly as is done for doxorubicin salts for therapeutic applications of long-circulating liposomes.[66] Recently, paclitaxel-loaded liposome was conjugated with microbubbles; that hydrophobic drug formulation was shown to provide therapeutic drug dose high enough to achieve inhibition of tumor growth in animal model.[67]

It has been shown that ultrasound action on a microbubble in proximity to a giant liposome may cause content leakage from that liposome.[68] If a bubble is attached to smaller liposome, fractional content leakage is also reported, for model dyes,[63] drugs[64] and even therapeutic enzymes, such as thrombin[63] — the latter approach could lead to the ability to induce targeted clotting in the tumor vasculature by ultrasound activation of the complex.

Next step in the improvement of this approach is to place microbubble inside the liposome structure, so that the bubble is co-entrapped with the drug inside the internal aqueous core of a submicron liposome, termed "bubble-liposome"[69] or in a large multimicron diameter liposome.[70] This way, the ultrasound-responsive gas particle is located inside the drug-loaded "capsule"; its volume will significantly expand during the rarefaction phase of the ultrasound pressure wave, so we may expect complete (not fractional) content release on ultrasound treatment due to rupture of the outer membrane surrounding the complex.

7. Conclusions and Outlook

The issues of ultrasound contrast microbubble preparation, stabilization and targeting ligand attachment have been successfully solved with the aid of biomedical nanoengineering approaches. Several of these ultrasound contrast agents are already approved for clinical use worldwide; more are evaluated at the level of preclinical research and clinical studies. Targeted microbubble contrast agents give ultrasound the ability to obtain imaging information on the molecular level: nanoengineering allows placement of molecular targeting ligands on the surface of microbubbles. Ultrasound-triggered drug and gene delivery particles based on microbubbles are under preclinical investigation. Their design will also benefit from the nanoengineering techniques described in this chapter.

Acknowledgments

This work was supported in part via current NIH R21/33 CA102880 and earlier R01EB002185, R43/44 EB007857 and a research grant from Philips Research North America; research would not have been possible without input, advice and help of students, fellows, colleagues, collaborators and friends at UVA and other academic and industrial institutions in United States, Europe, Japan and China.

References

1. R. Weissleder and U. Mahmood, *Radiology* **219**, 316–33 (2001).
2. D. L. Miller, *J. Acoust. Soc. Am.* **104**, 2498–2505 (1998).
3. P. Phillips and E. Gardner, *Eur. Radiol.* **14**(Suppl 8), P4–10 (2004).
4. R. Gramiak and P. M. Shah, *Invest. Radiol.* **3**, 356–366 (1968).
5. R. Gramiak and P. M. Shah, *Radiology* **100**, 415–418 (1971).
6. H. Mudra, *et al.*, *Z. Kardiol.* **80**, 367–372 (1991).
7. J. Ophir and K. J. Parker, *Ultrasound Med. Biol.* **15**, 319–333 (1989).
8. T. A. Fritz, E. C. Unger and G. Sutherland, D. Sahn, *Invest. Radiol.* **32**, 735–740 (1997).
9. R. Schurmann and R. Schlief, *Radiol. Med.* **87**, 15–23 (1994).
10. M. Schneider, *et al.*, *Invest. Radiol.* **46**, 486–494 (2011).
11. M. Schneider, *et al.*, *Invest. Radiol.* **30**, 451–457 (1995).
12. S. Kvåle, H. A. Jakobsen, O. Asbjørnsen and T. Omtveit, *Separ. Tech.* **6**, 219–226 (1996).
13. J. A. Feshitan, C. C. Chen, J. J. Kwan and M. A. Borden, *J. Colloid Interface Sci.* **329**, 316–324 (2009).
14. K. E. Morgan, *et al.*, *IEEE Trans Ultrason Ferroelectr. Freq. Control* **47**, 1494–1509 (2000).
15. M. R. Bohmer, J. A. Steenbakkers and C. Chlon, *Colloid.s Surf. B Biointerfaces* **79**, 47–52 (2010).
16. E. Talu, *et al.*, *Langmuir* **24**, 1745–1749 (2008).
17. A. H. Dhanaliwala, J. L. Chen, S. Wang and J. A. Hossack, *Microfluid. Nanofluid.* **14**, 457–467 (2013).
18. J. S. D'Arrigo, R. H. Simon and S. Y. Ho, *J. Neuroimaging* **1**, 134–139 (1991).
19. P. Epstein and M. Plesset, *J. Chem. Phys.* **18**, 1505–1510 (1950).
20. M. W. Grinstaff and K. S. Suslick, *Proc. Natl. Acad. Sci. USA* **88**, 7708–7710 (1991).
21. T. J. Pelura, *Acad. Radiol.* **5** (Suppl 1), S69–S71; discussion S72–S74 (1998).
22. A. P. Patrianakos and M. I. Hamilos, *Curr. Opin. Investig. Drugs* **8**, 248–255 (2007).
23. J. H. Wible, Jr., *et al.*, *J. Am. Soc. Echocardiogr.* **9**, 442–451 (1996).
24. J. C. Hutter, H. M. Luu, P. M. Mehlhaff, A. L. Killam and H. C. Dittrich, *J. Ultrasound Med.* **18**, 1–11 (1999).
25. K. Sugimoto, *et al.*, *J. Ultrasound Med.* **31**, 529–538 (2012).
26. A. L. Klibanov, *et al.*, *Invest. Radiol.* **39**, 187–195 (2004).
27. R. Senior, *Exp. Rev. Cardiovasc. Ther.* **5**, 413–421 (2007).
28. A. H. Myrset, *et al.*, *Ultrasound Med. Biol.* **37**, 136–150 (2011).
29. A. L. Klibanov, *et al.*, *Acad. Radiol.* **9** (Suppl 2), S279–S281 (2002).
30. J. W. Park, *et al.*, *Clin. Cancer Res.* **8**, 1172–1181 (2002).
31. V. R. Muzykantov, *Exp. Opin. Drug Deliv.* **7**, 403–427 (2010).
32. X. Gao, *et al.*, *Curr. Opin. Biotechnol.* **16**, 63–72 (2005).
33. J. P. Earls and D. A. Bluemke, *Magn. Reson. Imaging Clin. N. Am.* **7**, 255–273 (1999).
34. J. J. Rychak, J. R. Lindner, K. Ley and A. L. Klibanov, *J. Control Release* **114**, 288–299 (2006).
35. G. Korpanty, P. A. Grayburn, R. V. Shohet and R. A. Brekken, *Ultrasound Med. Biol.* **31**, 1279–1283 (2005).
36. F. Forsberg, *et al.*, *Radiology* **222**, 824–829 (2002).

37. J. P. Christiansen, H. Leong-Poi, A. L. Klibanov, S. Kaul and J. R. Lindner, *Circulation.* **105**, 1764–1767 (2002).
38. J. R. Lindner, *et al.*, *Circulation* **102**, 2745–2750 (2000).
39. J. R. Lindner, *et al.*, *Circulation* **104**, 2107–2112 (2001).
40. A. L. Klibanov, *et al.*, *Acta. Radiol. Suppl.* **412**, 113–120 (1997).
41. F. S. Villanueva, *et al.*, *Circulation* **98**, 1–5 (1998).
42. G. T. Hermanson, *Bioconjugate Techniques.* San Diego: Academic Press (2008).
43. C. R. Anderson, *et al.*, *Invest. Radiol.* **45**, 579–585 (2010).
44. R. Pillai, *et al.*, *Bioconjug. Chem.* (2010).
45. D. H. Kim, A. L. Klibanov and D. Needham, *Langmuir* **16**, 2808–2817 (2000).
46. C. Jeppesen, *et al.*, *Science* **293**, 465–468 (2001).
47. N. G. Fisher, *et al.*, *J. Am. Coll. Cardiol.* **40**, 811–819 (2002).
48. A. S. Ham, A. L. Klibanov and M. B. Lawrence, *Langmuir* **25**, 10038–10044 (2009).
49. J. J. Rychak, *et al.*, *Mol. Pharm.* **3**, 516–524 (2006).
50. M. A. Borden, H. Zhang, R. J. Gillies, P. A. Dayton and K. W. Ferrara, *Biomaterials* **29**, 597–606 (2008).
51. E. C. Unger, E. Hersh, M. Vannan and T. O. Matsunaga, T. McCreery, *Prog. Cardiovasc. Dis.* **44**, 45–54 (2001).
52. E. C. Unger, E. Hersh, M. Vannan and T. McCreery, *Echocardiography* **18**, 355–361 (2001).
53. H. Leong-Poi, *et al.*, *Circ. Res.* **101**, 295–303 (2007).
54. A. R. Carson, *et al.*, *Ultrasound Med. Biol.* **37**, 393–402 (2011).
55. A. R. Carson *et al.*, *Cancer Res.* **72**, 6191–6199 (2012).
56. J. P. Christiansen, B. A. French, A. L. Klibanov, S. Kaul and J. R. Lindner, *Ultrasound Med. Biol.* **29**, 1759–1767 (2003).
57. D. S. Wang, *et al.*, *Radiology* **264**, 721–732 (2012).
58. I. Lentacker, *et al.*, *Mol. Pharm.* **6**, 457–467 (2009).
59. S. L. Taylor, A. A. Rahim, N. L. Bush, J. C. Bamber and C. D. Porter, *J. Gene Med.* **9**, 77–87 (2007).
60. J. L. Tlaxca *et al.*, *J. Control Release* **165**, 216–225 (2013).
61. A. Xie, *et al.*, *JACC Cardiovasc. Imaging* **5**, 1253–1262 (2012).
62. M. S. Tartis, *et al.*, *Ultrasound Med. Biol.* **32**, 1771–1780 (2006).
63. A. L. Klibanov, T. I. Shevchenko, B. I. Raju, R. Seip and C. T. Chin, *J. Control Release* **148**, 13–17 (2010).
64. I. Lentacker, B. Geers, J. Demeester, S. C. De Smedt and N. N. Sanders, *J. Control Release* **148**, e113–e114 (2010).
65. B. Geers, *et al.*, *J. Control Release* **152**, 249–256 (2011).
66. A. Fritze, F. Hens, A. Kimpfler, R. Schubert and R. Peschka-Suss, *Biochim. Biophys. Acta* **1758**, 1633–1640 (2006).
67. F. Yan, *et al.*, *Ultrasound Med. Biol.* **37**, 768–779 (2013).
68. P. Marmottant and S. Hilgenfeldt, *Nature* **423**, 153–156 (2003).
69. R. Suzuki, *et al.*, *J. Control Release* **117**, 130–136 (2007).
70. S. Ibsen, *et al.*, *J. Control Release* **155**, 358–366 (2011).

Chapter 6

Nanotechnology for Multimodality Imaging: Applications in Disease Detection and Treatment Guidance

Jinzi Zheng[*,§], *Raquel De Souza*[*,†]
David Jaffray[*,‡,§] *and Christine Allen*[*,†,¶]

[*]*STTARR Innovation Center, Princess Margaret Hospital, University Health Network, Toronto, ON, Canada*
[†]*Leslie Dan Faculty of Pharmacy, University of Toronto, Toronto, ON, Canada*
[‡]*Department of Radiation Physics, Radiation Medicine Program, Princess Margaret Hospital, University Health Network, Toronto, ON, Canada*
[§]*Techna Institute for the Advancement of Technology for Health, University Health Network, Toronto, ON, Canada*
[¶]*cj.allen@utoronto.ca*

1. Introduction

In clinical practice, specific imaging modalities are used routinely at different stages of disease diagnosis and treatment. For example, combinations of anatomical and functional imaging techniques such as computed tomography (CT), magnetic resonance (MR) imaging, positron emission tomography (PET), single photon emission computed tomography (SPECT), ultrasound (US) and optical imaging are routinely employed for disease diagnosis, staging, treatment planning and follow-up. These imaging systems are also being found increasingly in treatment rooms to guide delivery of a range of interventions including surgery, radiotherapy, thermal ablation and photodynamic therapy with the goal of improving the accuracy of these procedures.[1–9] The combined use of multiple imaging modalities can greatly benefit a number of medical procedures given that they provide complementary information on the anatomy, physiology and function of the biological system under investigation. For example, each modality relies on a different signal generating mechanism, and as such it can be challenging to identify common structures for spatial reference. Accurate definition of the anatomical location and boundary of the biological target is essential for therapy planning and delivery. From this perspective, signal generating agents that exhibit identical pharmacokinetics, biodistribution and disease targeting ability are considered advantageous. Their co-localization across multiple imaging modalities has the potential to yield significant improvements in the confidence associated with identification of diseased sites and their boundaries. In order to meet this need, multimodal nanotechnologies have been engineered to include two or more signal generating moieties within a single construct. These multimodal nanosystems have been explored pre-clinically for a wide-range of applications wherein the performance related properties and functional requirements for each application are unique. Specifically, multimodal agents can (1) increase the accuracy of image registration which will benefit patient alignment before and during an intervention; (2) bridge spatial and sensitivity scales thereby improving both the sensitivity and specificity of disease detection and characterization; (3) improve confidence in assessment of disease and normal tissue boundaries through signal confirmation obtained from more than one imaging modality. This chapter reviews and discusses many of the multimodal nanotechnologies that have been put forth for applications ranging from cancer imaging to *in vivo* cell-tracking. The chapter does not include an introduction to the signal generation mechanisms of the commonly used imaging techniques or the general requirements for successful engineering of multimodal nano-agents. A previously published introductory chapter on these topics is available for the interested reader.[10]

2. Cancer Imaging

An ongoing challenge in applied cancer research is the successful development of diagnostic tools with high sensitivity and specificity that are able to accurately detect the presence and anatomical location of neoplastic cells as well as to characterize their abnormal nature at a range of scales. In this context, much effort has been placed on the development of nanoparticles as diagnostic imaging agents. Neoplastic malignancies progress from molecular to micro and then to macro scales, as such the need for agents that can bridge these spatial ranges through multimodal imaging are of growing interest.

Overall, there are five rationales for employing nanoparticles in place of traditional small molecules as a new class of agents with the potential to lead to improvements in cancer diagnosis: (1) the critical size of nanoparticles and their tunable surface characteristics result in pharmacokinetics profiles that enable applications requiring longitudinal and/or repeated imaging sessions; (2) their extended pharmacokinetics allow for increased tumor tissue targeting, resulting in a greater target-to-background signal ratio in imaging applications; (3) the high payload of imaging agents that nanoparticles carry can be exploited for amplification of imaging signal, in particular when used in conjunction with a lower sensitivity imaging technique; (4) the versatility of nanoparticles, allows for differential compositions of their core and shell compartments resulting in their ability to transport molecules that differ significantly in terms of physico-chemical properties. From this perspective they are the ideal platform for use in multimodality imaging given that each imaging technique has different requirements, in terms of molecular structure and optimal concentration of its signal modifying moiety; (5) the suitability of nanoparticles for surface modification with peptide sequences and antibodies makes them an ideal vehicle for actively targeted delivery to specific cell populations.

Cancer detection is the most explored application for multimodal imaging agents. Nanoparticles, due to their macromolecular nature, can preferentially accumulate at tumor sites via the enhanced permeation and retention (EPR) effect, first described by Matsumura and Maeda in 1986.[11] EPR requires particles to have a prolonged vascular residence time (i.e. maintain high plasma area-under-the-curve — AUC for > 6 h in mice and rats[11–13]). It has been agreed that the degree of macromolecule accumulation in tumors is directly proportional to the blood AUC (or exposure) and inversely proportional to the rate of urinary clearance.[14–16]. Once the prerequisite of high exposure has been achieved, the transport of macromolecules, such as nanoparticles, into tumor tissues is further affected by the tumor vascular pore size (up to 400 nm)[17]. However, their subsequent intratumoral retention is a function of the particle's diffusion and transport in the tumor interstitial space,[18] the speed of the tumor venous return (usually slower than normal tissue[19,20]), as well as the presence of a poor lymphatic drainage system.[19,20] Altogether, macromolecules and nanoparticles not only preferentially accumulate in tumors via enhanced vascular permeability, but they are also preferentially retained there (for multiple hours to days). Conversely, low-molecular-weight agents are distributed systemically following administration, rapidly cleared from the circulating blood via renal clearance and their tumor accumulation is transient (on the order of minutes). Their small size allows them to readily return to the circulating blood system following extravasation into the tumor interstitial space[13,21] unless trapped by biochemical processes (e.g. fluorodeoxyglucose, FDG). The ability of the EPR effect to significantly increase tumor accumulation versus distribution to normal tissues can result in appreciable increases in the target-to-background signal ratio during image acquisition. EPR is the hallmark of nanoparticle-based delivery of diagnostic and therapeutic agents to tumors.[16]

Table 1 summarizes reports of multimodal nano-agents that have shown promise for cancer imaging in preclinical animal models (i.e. includes publications between 2008 and 2013). A wide range of nanoparticles with different sizes, shapes and surface chemistries have been explored for this purpose, including iron oxide particles,[22–28] cobalt ferrite particles,[29] chitosan nanoparticles,[30] gold nanorods,[31] upconversion nanophosphors,[32] silica nanoparticles,[33] dendrimers,[34] liposomes[35,36]] and polymer micelles.[37] These nanoparticles can either inherently provide positive or negative signal enhancement (i.e. iron oxide is an MR T_2 contrast agent) or they can be engineered to carry imageable moieties (i.e. fluorophore, Gd-chelate). These nanosystems can be further optimized for targeted imaging through the addition of selected ligands and peptide sequences[23,26,27,34,37,38] or inclusion of cleavable polymers that can provide a more elevated signal under specific physiological conditions (e.g. high concentration of matrix metalloproteinases, MMPs[22]).

For example in 2011, Zhang *et al.*[37] reported, the development and characterization of a dual SPECT and optical nanoparticle consisting of a PEG-coated, core-cross-linked polymeric micelle (CCPM, hydrodynamic diameter of 24 ± 8.9 nm)[39] targeted to the receptor tyrosine kinase EphB4, overexpressed in a number of tumors,[40–43] through the conjugation of the TNYL-RAW peptide. ^{111}In-SPECT and Cy7-fluorescence molecular tomography (FMT) demonstrated that this targeted nanoparticle selectively accumulates in EphB4-positive tumors and its binding to EphB4 is specific (i.e. the signal is greatly diminished when co-administered with an excess of cold TNYL-RAW peptide, Figs. 1(A) and 1(B)). Most interestingly, the tumor penetration of these micelles is also greatly improved by the incorporation of the targeting peptide sequence (Figs. 1(C) and 1(D)). The presence of both the gamma emitter ^{111}In and the NIR probe Cy7 allows for seamless bridging of quantitative whole body imaging using SPECT and high resolution *ex vivo* tissue and cellular imaging using fluorescence. Dual labeling strategies such as these are robust demonstrations of the benefit of combining complementary imaging techniques for assessment and validation of new targeted diagnostics and therapeutics.

Table 1. Multimodal nanoparticle systems designed for cancer detection and tumor imaging.

	Nanosystem composition	Imaging molecules and targeting moieties per nanoparticle	Fluorescence ex/em, MR relaxation rate constants (r_1, r_2) and radioisotope $t_{1/2}$	Tumor accumulation	Administration dose and route	Tumor model(s)	Imaging systems used
Hwang *et al.*, 2010 (Ref. 29)	^{67}Ga-labeled and AS1411 aptamer targeted magnetic cobalt ferrite nanoparticle with rhodamine B isothiocyanate embedded within its PEGylated silica shell Hydrodynamic diameter: 68.1 nm	^{67}Ga: 2.06 MBq/pmol nanoparticle AS1411 aptamer: 1.1×10^4/nanoparticle	^{67}Ga $t_{1/2}$: 78.2 h Rhodamine B: 556/570 nm r_2 = 9.04 $mM^{-1} s^{-1}$	Qualitative assessment	Intravenous injection (*i.v.*), dose not specified	Balb/c nude mice bearing subcutaneous bilateral C6 tumors	γ-Scintigraphy Imaging: ON 410 γ-camera equipped with a pinhole collimator (Ohio Nuclear) MRI: 1.5 T (GE)
Luo *et al.*, 2011 (Ref. 31)	Mesoporous silica-coated gold nanorods loaded with ICG Nanorod length: 44.44 ± 4.7 nm Nanorod width: 15.10 ± 1.7 nm Silica shell thickness: ~13 nm	N/A	ICG fluorescence: 785/810 nm	CT: ~16 HU increase (15 min *p.i.*) at tumor site compared to normal soft tissue Fluorescence: qualitative assessment	200 μL of 1.5 mg/mL of Au@SiO_2 administered intratumorally or subcutaneously at the chest level	Female nude mice bearing subcutaneous MGC803 gastric cancer tumors	CT: ONCOR impression OptiVue (Siemens) Fluorescence: Lumina XR (Caliper Life Sciences)

(*Continued*)

Table 1. (*Continued*)

	Nanosystem composition	Imaging molecules and targeting moieties per nanoparticle	Fluorescence ex/em, MR relaxation rate constants (r_1, r_2) and radioisotope $t_{1/2}$	Tumor accumulation	Administration dose and route	Tumor model(s)	Imaging systems used
Nam *et al.*, 2010 (Ref. 30)	Hydrophobically modified glycol chitosan nanoparticles with Cy5.5 and DOTA-Gd Hydrodynamic diameter: 350 ± 42 nm	Cy5.5: 0.5 ± 0.1/ nanoparticle Gd(III): 6.28 wt (%)	Cy5.5 Fluorescence: 675/680–720 nm	Fluorescence: Tumor signal was 3–7 times greater than normal tissue (*ex vivo*, at 24 h *p.i.*). MRI: Relative signal enhancement in tumor of 231% and 146% in muscle	5 mg/kg (3.3 μmol) administered *i.v.*	Nude mice bearing pectoral subcutaneous SCC7 squamous carcinoma tumors	Fluorescence: explore Optix (GE) MRI: 1.5 T Sonata (Siemens)
Lee *et al.*, 2008 (Ref. 27)	RGD-conjugated polyaspartic acid-coated iron oxide nanoparticles labeled with ^{64}Cu Core diameter: 5 nm Hydrodynamic diameter: 45 ± 10 nm	RGD: 35/ nanoparticle DOTA: 30/ nanoparticle ^{64}Cu: ~185 GBq/g of iron	^{64}Cu $t_{1/2}$: 12.7h $r_2 = 105.5$ $mM^{-1} s^{-1}$ $r_2^* = 165.5$ $mM^{-1} s^{-1}$	PET: 10.1 ± 2.1 %ID/g at 4h *p.i.* MRI: qualitative assessment	3.7 MBq of ^{64}Cu (300 μg of iron) *i.v.* per mouse	Female nude mice bearing subcutaneous U78MG glioblastoma tumors	PET: micro PET R4 (Siemens) MRI: 3.0 T Systems Revision 12.0 M5 (GE Healthcare)

(*Continued*)

Table 1. (*Continued*)

	Nanosystem composition	Imaging molecules and targeting moieties per nanoparticle	Fluorescence ex/em, MR relaxation rate constants (r_1, r_2) and radioisotope $t_{1/2}$	Tumor accumulation	Administration dose and route	Tumor model(s)	Imaging systems used
Lee *et al.*, 2012 (Ref. 28)	PEGylated Fe_3O_4/TaO_x core/shell nanoparticles with rhodamine-B isothiocyanate Core diameter: 8 nm Hydrodynamic diameter: 21 nm	N/A	$r_2 = 81.156$ $mM^{-1} s^{-1}$	CT: 45 HU increase in tumor X-ray attenuation at 24 h *p.i.* MRI: qualitative assessment	840 mg/kg *i.v.*	Rats bearing subcutaneous mammary adenocarcinoma (MAT B III) tumors	CT: Brilliance 64 (Philips) MRI: 3 T Trio Trim (Siemens)
Tan *et al.*, 2011 (Ref. 25)	Poly(lactide)-tocopheryl polyethylene glycol succinate (PLA-TPGS) nanoparticles co-encapsulating iron oxides and quantum dots Hydrodynamic diameter: 325.8 ± 5.2 nm with a PDI of 0.204 ± 0.065	Iron oxide (Fe 54): 60.00 ± 14.14 % encapsulation efficiency Quantum dots (Cd 111): 45.00 ± 7.07 % encapsulation efficiency	Fluorescence: 530/652 nm	MRI: 10% signal reduction in tumor (6 h *p.i.*) Fluorescence: 152.8% intensity increase in tumor (6 h, *ex vivo*)	6.0 mg Fe/kg equivalent to 1.5 mg Cd/kg *i.v.*	Female SCID mice bearing subcutaneous MCF-7 tumors	MRI: 7 T Clinscan (Bruker) Fluorescence: IVIS 100 (Xenogen)

(*Continued*)

Table 1. (*Continued*)

	Nanosystem composition	Imaging molecules and targeting moieties per nanoparticle	Fluorescence ex/em, MR relaxation rate constants (r_1, r_2) and radioisotope $t_{1/2}$	Tumor accumulation	Administration dose and route	Tumor model(s)	Imaging systems used
Bardhan *et al.*, 2010 (Ref. 26)	PEGylated gold nanoparticles with iron oxides in the core, ICG in the silica shell and conjugated to anti-HER2 antibodies Core diameter: 120 nm Hydrodynamic diameter: 148 nm	Anti-HER2 antibodies: 300–350/nanocomplex	Fluorescence : 785/830 nm ICG: 650 nM in ~ 9×10^9 particles/mL r_2 = 390 $mM^{-1} s^{-1}$	Fluorescence: 71.5% greater signal in BT-474-AZ (Her2+) than in MDA-MB-231 (Her2-) tumors (4 h *p.i.*) MRI: BT-474-AZ tumors 50.5 % darker than MDA-MB-231 tumors (24 h *p.i.*)	10 μL/g body weight in a 9×10^9 particles/mL solution (equivalent to 5 μg of ICG/kg of body weight and ≤ 0.1 μg/g of antibody)	Female nude mice bearing BT-474-AZ (HER2+) or MDA-MB-231 (HER2-) tumors in the mammary fat pad	Fluorescence: PhotonMax 512 CCD camera and a 28-mm Nikkor lens (Nikon) MRI: 9.4 T Avance Biospec (Bruker)

(*Continued*)

Table 1. (*Continued*)

	Nanosystem composition	Imaging molecules and targeting moieties per nanoparticle	Fluorescence ex/em, MR relaxation rate constants (r_1, r_2) and radioisotope $t_{1/2}$	Tumor accumulation	Administration dose and route	Tumor model(s)	Imaging systems used
Yan *et al.*, 2012 (Ref. 34)	PEGylated G5 PAMAM dendrimer labeled with Cy5.5, rhodamine B, DOTA-Gd and conjugated with angiopep-2 and c[RGDyK] Core diameter (dendrimer only): 7.1 nm Hydrodynamic diameter: 15.6 nm with a PDI of 0.224	Molar ratio of angiopep-2 / c[RGDyK] / DOTA-Gd / dendrimer is 5/6/54/1 Gd: 9.6%	Cy5.5: 750 / 800 nm Rhodamine B: 543/ 560 nm $r_1 = 7.1 \pm 0.2$ $mM^{-1}\,s^{-1}$	MRI: 0.36 ± 0.05 %ID/g at the tumor site and a 1.41 T/N ratio (24 h *p.i.*) Fluorescence: signal ratio of tumor to surrounding tissue was 1.49 (24 h *p.i.*)	0.05 mmol/kg [Gd^{3+}] In 0.25 mL PBS *i.v.*	U87MG glioblastoma tumors inoculated into the striatum of male Balb/c nude mice	MRI: 4.7 T Biospec 47/30 (Bruker) Fluorescence: Kodak Multispectral Imaging System (Carestream)

(*Continued*)

Table 1. (*Continued*)

	Nanosystem composition	Imaging molecules and targeting moieties per nanoparticle	Fluorescence ex/em, MR relaxation rate constants (r_1, r_2) and radioisotope $t_{1/2}$	Tumor accumulation	Administration dose and route	Tumor model(s)	Imaging systems used
Lee *et al.*, 2013 (Ref. 32)	^{124}I-labeled, PEGylated Er^{3+}/Yb^{3+} co-doped $NaGdF_4$ upconversion nanophosphors conjugated to $(cRGDyk)_2$ peptide Core diameter: 24 ± 1.5 nm Hydrodynamic diameter: 32 ± 9 nm	^{124}I: 0.88 GBq/g nanoparticle Ratio of UCNP to $(cRGDyk)_2$-to PEG: 1:200:800	^{124}I $t_{1/2}$: 4.2 d	MRI: signal-to-noise ratio increase of 1.8 ± 0.1 in tumors (4 h *p.i.*) PET: tumor uptake of 2.8 ± 0.8 %ID/g (4.5 h *p.i.*)	1.66 MBq/2 mg of nanoparticles *i.v.*	Balb/c nude mice bearing subcutaneous U87MG tumors	MRI: 3 T Magnetom Trio (Siemens) PET: Inveon microPET (Siemens)
Benezra *et al.*, 2011 (Ref. 33)	^{124}I- cRGDY-PEGylated Cy5 encapsulating core-shell silica nanoparticles Hydrodynamic diameter: ~7 nm	cRGDY peptide: 6-7/ nanoparticle Cy5: 2/ nanoparticle	^{124}I $t_{1/2}$: 4.2 d Fluorescence: 650/710 nm	PET: Accumulation of 3.6%ID/g in M21 (α_V+) tumors versus 0.7 %ID/g in M21L (α_V-) tumors (4 h *p.i.*)	200 μCi *i.v.* for tumor imaging	Male nude mice bearing subcutaneous M21 (α_V+) or M21L (α_V-) xenografts	PET: Focus 120 (Siemens) Fluorescence: Maestro (Perkin Elmer)

(*Continued*)

Table 1. (*Continued*)

	Nanosystem composition	Imaging molecules and targeting moieties per nanoparticle	Fluorescence ex/em, MR relaxation rate constants (r_1, r_2) and radioisotope $t_{1/2}$	Tumor accumulation	Administration dose and route	Tumor model(s)	Imaging systems used
Zhang *et al.*, 2011 (Ref. 37)	^{111}In-labeled Cy7core-crosslinked polymeric micelles conjugated to TNYL-RAW peptide <u>Hydrodynamic diameter:</u> 23 nm	<u>TNYL-RAW:</u> 60/micelle <u>Cy7:</u> 21/micelle <u>DTPA:</u> 19/micelle <u>^{111}In:</u> 180 MBq/mol	<u>Fluorescence:</u> 755/781 nm	<u>SPECT:</u> 2.87 %ID/g in PC-3M (EphB4+) tumors versus 1.37 %ID/g in A549 (EphB4-) tumors (24 h, *p.i.*) <u>Fluorescence:</u> qualitative assessment	1 × 10^4 nanoparticles (9.0 MBq) *i.v.*	Male nude mice bearing subcutaneous PC-3M (EphB4+) or A549 (EphB4-) prostate tumors	<u>SPECT:</u> × SPECT-CT (Gamma Medica) <u>Fluorescence:</u> FMT (VisEn)
Chou *et al.*, 2010 (Ref. 23)	Anti-Her2 antibody conjugated FePt nanoparticles <u>Diameter:</u> 12.80 ± 1.76 nm	N/A	N/A	<u>MRI:</u> relative signal intensity reduced by 51% at tumor site (24 h *p.i.*) <u>CT:</u> relative signal enhancement of 138% at tumor site (24 h *p.i.*)	28 mM Fe in 100 μL *i.v.*	Male C3H/HeN mice bearing subcutaneous MBT2 tumors	<u>CT:</u> Skyscan 1076 X-ray (Skyscan) <u>MRI:</u> 1.5 T (scanner type not reported)

(*Continued*)

Table 1. (*Continued*)

	Nanosystem composition	Imaging molecules and targeting moieties per nanoparticle	Fluorescence ex/em, MR relaxation rate constants (r_1, r_2) and radioisotope $t_{1/2}$	Tumor accumulation	Administration dose and route	Tumor model(s)	Imaging systems used
Cha *et al.*, 2011 (Ref. 22)	Cy5.5-MMP substrate and PEG conjugated iron oxide nanoparticles with silica coating Core-shell diameter: 19 nm Hydrodynamic diameter: 43.1 ± 6.3 nm	Average distance between Cy5.5 molecules: 14.21 nm (beyond FRET-based quenching)	Fluorescence: 670/700 nm	Fluorescence: 3–4-fold greater signal intensity at tumor site compared to normal tissue (*in* vivo, 12 h *p.i.*) and 5–10 fold at 24 h *p.i.* (*ex vivo*) MRI: 34.2% signal reduction in the tumor region (12h *p.i.*)	0.05 mL with 0.15 mg Fe/mL *i.v.*	Balb/c mice bearing subcutaneous SCC7 xenografts	MRI: 3 T Tim Trio (Siemens) Fluorescence: explore Optix (GE)

(*Continued*)

Table 1. (*Continued*)

	Nanosystem composition	Imaging molecules and targeting moieties per nanoparticle	Fluorescence ex/em, MR relaxation rate constants (r_1, r_2) and radioisotope $t_{1/2}$	Tumor accumulation	Administration dose and route	Tumor model(s)	Imaging systems used
Nahrendorf *et al.*, 2010 (Ref. 24)	18F-dextranated core-shell cross-linked iron oxide (CLIO)-VT680 or 64Cu-CLIO-VT680 Hydrodynamic diameter: 30 nm	VT680: 5/nanoparticle Fe: 2064/nanoparticle ^{18}F: 1.47 GBq/mg Fe ^{64}Cu: 1.28 GBq/mg Fe	^{18}F $t_{1/2}$: 109.8 min ^{64}Cu $t_{1/2}$: 12.7 h Fluorescence: 680/700 nm Detection thresholds: 0.025 μg Fe/mL (^{18}F) and 0.1 μg Fe/mL (^{64}Cu) by PET; 1.25 μg Fe/mL by FMT	Correlation between PET and FMT signal: r^2 ³ 0.99	Not reported	Balb/c mice bearing subcutaneous CT26 colon cancer tumors	PET/CT: Inveon (Siemens) Fluorescence: FMT-2500 (VisEn)

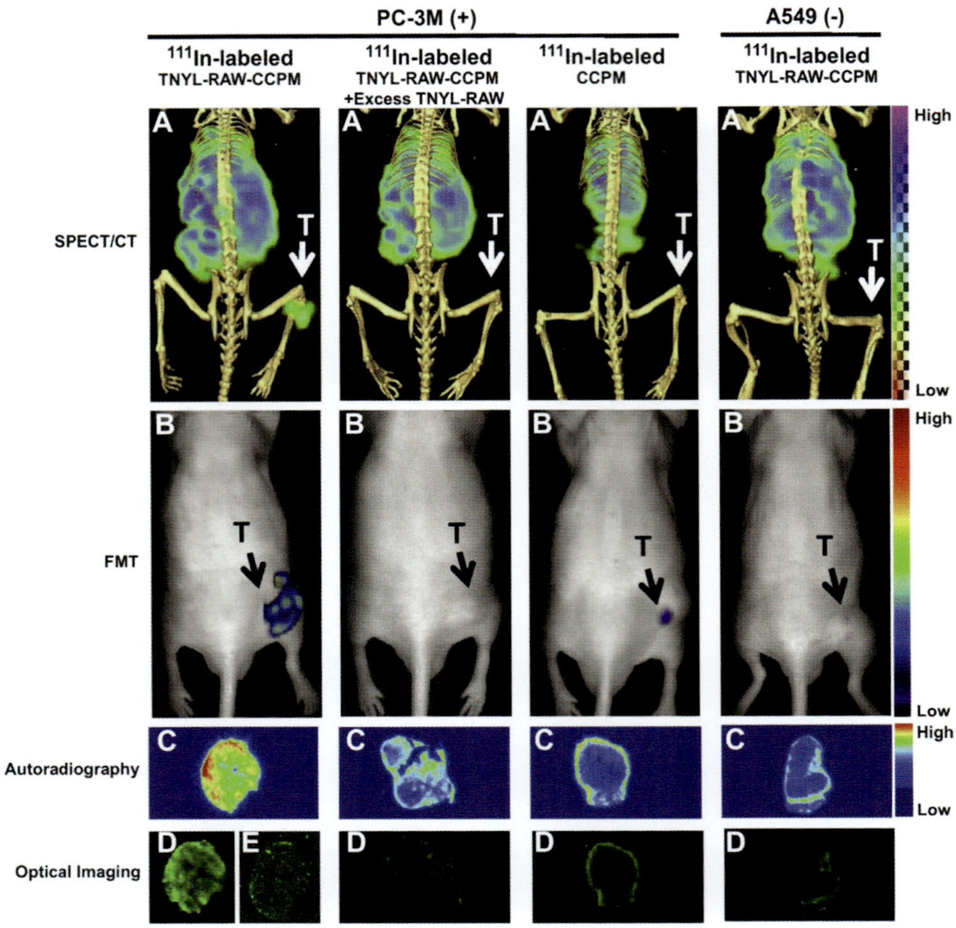

Fig. 1. Dual SPECT/CT and near-infrared fluorescence optical imaging of mice bearing a PC-3M or A549 tumor after intravenous (*i.v.*) administration of 111In-labeled TNYL-RAW-CCPM or CCPM. The mice bearing a PC-3M tumor received 111In-labeled TNYL-RAW-CCPM, 111In-labeled TNYL-RAW-CCPM plus excess cold TNYL-RAW peptide, or plain CCPM. The mice bearing an A549 tumor received 111In-labeled TNYL-RAW-CCPM. (A) Representative SPECT/CT images. (B) Representative fluorescence molecular tomographic images. (C) Representative autoradiographs of excised tumors. (D) Fluorescence images of the same slides in autoradiographic studies scanned at 800 nm. (E) Fluorescence image of the same slide in autoradiographic studies scanned at 520 nm after immunostaining. The section was stained with rabbit antihuman EphB4 antibody and Alexa Flour 488-conjugated goat anti-rabbit immunoglobulin, G. Arrows indicate tumors.

Figure reproduced with permission from Zhang *et al.*[37]

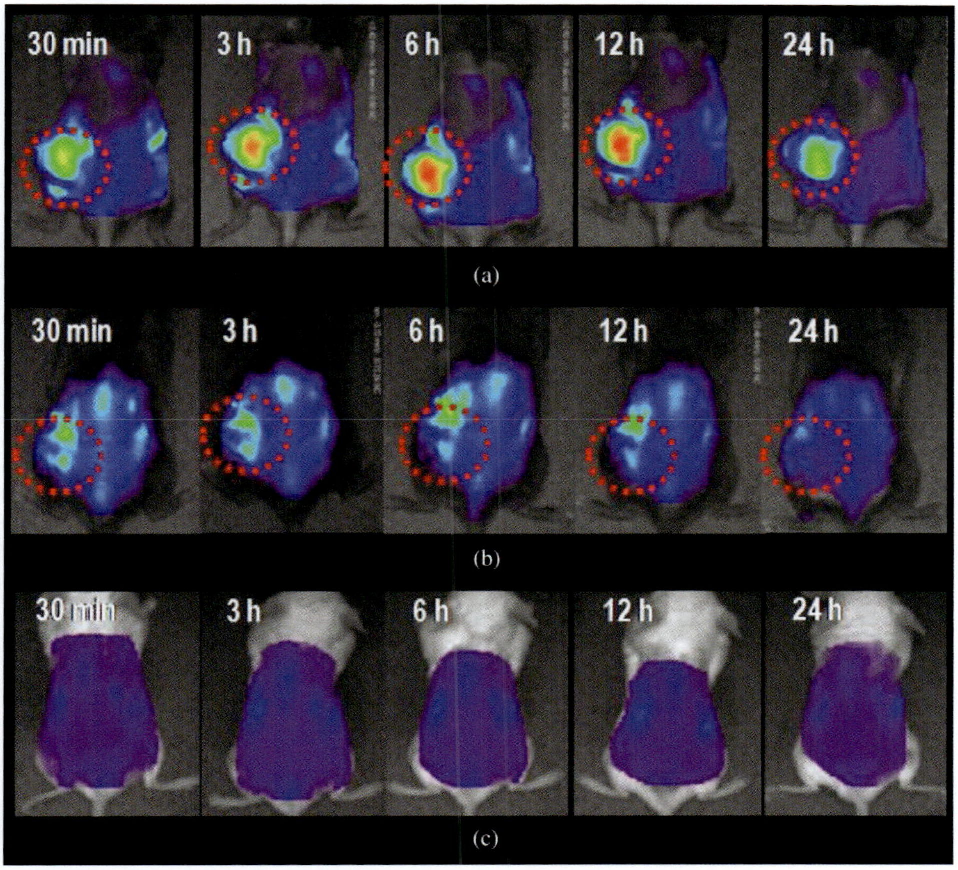

Fig 2. (Color online) NIRF tomographic images of tumor-bearing mice after injection of PCM-CS (A) without inhibitor, (B) with inhibitor and (C) normal mice after injection of PCM-CS (blue: low intensity, red: high intensity).

Figure reproduced with permission from Cha *et al.*[22]

Nanoparticles for multimodality imaging can also be designed to inform on the state of the tumor microenvironment[22] or to modulate tumor physiology.[34] Cha *et al.*[22] described the development of an MMP-responsive iron oxide-based MR/near infrared (NIR) fluorescence imaging probe which showed successful reporting of MMP-rich tumor environments. Its specificity was further validated by a decrease in signal measured following co-administration with MMP-2 inhibitors (Fig. 2). Similarly, Yan *et al.*[34] engineered an "active" nanoparticle which not only allowed for dual modality imaging using MR and NIR fluorescence, but it also targeted tumor vasculature and permeated the blood-brain barrier (BBB) due to the presence of surface conjugated cyclic RDGyK peptides and Angiopep-2 peptides, respectively. These "two-order" targeted nanoparticles are designed to first target the $\alpha_V\beta_3$ integrins on the tumor vasculature in the brain. The resulting increase in the local concentration of the nanoparticles then facilitates the association of Angiopep-2 peptides with low-density lipoprotein receptor-related protein (LRP) receptors, ultimately allowing for effective crossing of the BBB. The selectivity of this tumor targeting strategy is attributed to the high expression of both $\alpha_V\beta_3$ integrins and LRP receptors in brain cancer. This technology represents a powerful advancement as a significant number of brain tumors (i.e. 10% of glioblastoma multiforme,GBM and 30% of anaplastic astrocytoma, AA[44]) are not detectable by conventional Gd contrast enhanced MR imaging due to the uncompromised BBB.[34]

Another interesting example of the use of multimodality imaging probes is for validation of new imaging techniques against those which are established and proven. Nahrendorf *et al.*[24] performed *in vitro*, *in vivo* and *ex vivo* correlation of PET and FMT using an $^{18}F/^{64}Cu$ and VT680 co-labeled iron oxide core and dextran shell nanoparticle in CT26 colon tumors. Successful spatial coregistration and signal correlation, as well as similar detection thresholds validated the use of FMT as a potential surrogate imaging modality to be used for cost-effective development of new tumor targeted PET tracers.

Overall, the availability of a diverse selection of multimodality agents for cancer imaging will continue to support improved design and optimization of new diagnostic and therapeutic nanoplatforms. Innovation in this exciting space can further enable integrated imaging at the cell, animal and human scales in terms of both anatomy and function. Ultimately, comprehensive cancer profiling will be effectively achieved non-invasively.

3. Sentinel Lymph Node Imaging

Lymph node status is a significant prognostic factor for survival in a number of cancers including melanoma, vulvar and breast.[45–47] Sentinel Lymph Node (SLN) is the first node that drains the primary tumor and is at the highest risk for tumor metastasis. SLN mapping, biopsy and surgical resection are important steps in disease staging and management. The identification and removal of SLN and additional metastatic lymph nodes can greatly benefit from the use of an imaging agent. Current standard-of-care SLN mapping consists of the intraoperative use of ^{99m}Tc-sulfur colloids (SC), detected using a gamma probe (a Geiger–Muller tube or scintillation counter) or gamma camera and isosulfan blue dye detected visually by surgeons.[48] However, these methods are limited by poor spatial resolution and limited depth visualization ($\leq$ 2 mm from surface), respectively.[48] More recently, the use of the clinically approved indocyanine green (ICG) dye administered either pre-bound to human serum albumin (HSA) or alone has been explored in conjunction with a NIR fluorescence detection system.[47, 49–51] NIR fluorescence imaging of ICG enables a higher depth of penetration in comparison to isosulfan blue and results in significant improvements in detection sensitivity.[47] However, one of the major drawbacks associated with these clinically available SLN mapping agents is that they do not allow for 3D preoperative planning. To address this, much effort has been put forward by the research community to engineer nanoparticle-based multimodality imaging agents which can be detected using radioactivity, fluorescence or ultrasound by 3D imaging in the preoperative and intraoperative settings.[48, 52–58] Table 2 outlines the key characteristics of these nano-agents as well as their imaging performance in healthy and tumor-bearing animals.

It is generally agreed that the interstitial transport of agents to lymph nodes following an intradermal or subcutaneous injection can occur through (1) passive drainage via afferent lymphatics (Fig. 3) and/or (2) macrophage and/or dendritic cell-mediated migration.[52,59,60] Physico-chemical properties such as particle size, morphology and surface characteristics are known to determine the mode and kinetics of transport of nano-agents from their injection site to lymph nodes (i.e. which in turn determines the detection specificity and imaging time window). In addition, the ability of each nanoparticle platform to effectively load or chelate imaging moieties impacts their detection sensitivity. Although currently available reports on the use of multimodal nanoparticles for SLN mapping (Table 2) have focused on lymph node targeting following intradermal and subcutaneous injections, malignant lymph nodes can also be effectively detected by nanoparticles post *i.v.* administration through exploitation of the EPR effect.

MR is the preferred 3D preoperative anatomical imaging modality for visualization and localization of lymph nodes. The MR contrast enhancing moiety can consist of either a T_1 shortening gadolinium chelate (i.e. Gd-DTTA,[55] Gd-DTPA[56]) or a T_2 shortening iron oxide nanoparticle functionalized with a biocompatible coating such as PLGA,[58] PEG,[57] silica[52] or serum albumin.[54] For intraoperative surgical guidance, a number of signal generating moieties have been explored including fluorescence dyes,[48,52,55,56] US microbubbles,[58] as well as single photon and positron emitters. The variety of intraoperative agents is complemented by the availability of clinically approved hand-held devices specifically designed for detection of fluorescence, ultrasound and radioactivity signals in the operating room. Clinically

Table 2. Multimodal nanoparticle systems designed for SLN imaging.

	Nanosystem composition	Imaging molecules per nanoparticle	Fluorescence ex/em, MR relaxation rate constants (r_1, r_2) and radioisotope $t_{1/2}$	Administration dose and route	Animal model(s)	Imaging systems used	Detection sensitivity
Bradbury *et al.*, 2013 (Ref. 48)	Fluorescent core-shell silica nanoparticles (Cornell dots) with cRGDY-PEG labeled with ^{124}I. Diameter: ~6–7 nm	cRGDY: 6-7 peptides Cy5.5: 2 molecules ^{124}I: 6-7 atoms	Cy5.5: 675/694 nm ^{124}I $t_{1/2}$: 4.18 d	0.125–0.05 mCi/kg administered as a 4-quadrant, subdermal injection near the tumor site	Spontaneous melanoma miniswine model	PET: Focus 120 microPET (Siemens) Fluorescence: ArteMIS™ handheld fluorescence camera (O2view)	
Koyama *et al.*, 2007 (Ref. 56)	G6 dendrimer covalently attached to 1B4M-DTPA-Gd and 1B4M-Cy5.5. Hydrodynamic diameter: ~9 nm	Gd^{3+}: 172 atoms Cy5.5: 2 molecules	Cy5.5: 675/694 nm	30 mM Gd/25 μL G6-Cy5.5 (750 nmol Gd/9.3 nmol Cy5.5) injected into the mammary fat pad intradermally and/or subdermally	Female *nu/nu* mice	MRI: 3.0T Signal Excite (General Electric) Fuorescence: Maestro 2D Imager (Perkin Elmer)	MRI: 25 μL of 30 mM G6-Cy5.5 (750 nmol of Gd) Fluorescence: 10uL of 10 mM G6-Cy5.5 (1.2 nmol Cy5.5)

(Continued)

Table 2. (*Continued*)

	Nanosystem composition	Imaging molecules per nanoparticle	Fluorescence ex/em, MR relaxation rate constants (r_1, r_2) and radioisotope $t_{1/2}$	Administration dose and route	Animal model(s)	Imaging systems used	Detection sensitivity
Choi *et al.*, 2008 (Ref. 54)	^{124}I-labeled MnMEIO nanoparticle coated with albumin. Core diameter: 15 nm_ Hydrodynamic diameter: 32 nm		^{124}I $t_{1/2}$: 4.18 d r_2 = 321.6 $mM^{-1}\,s^{-1}$	229 μg/kg (110 μCi of ^{124}I) administered subcutaneously into the right forepaw of rats	Male Sprague-Dawley rats	PET: R4 Rodent Model (Concord) MRI: 1.5 T (scanner type not reported)	PET: 4 μCi/mL
Niu *et al.*, 2013 (Ref. 58)	Fe_3O_4 nanoparticle with a perfluorocarbon gas and doxorubicin core and a PLGA shell. Hydrodynamic diameter: 800–900 nm		r_2 = 0.9893 $mM^{-1}\,s^{-1}$	2 mL at a [Fe_3O_4] of 185.45 μg/mL administered into the foot pad followed by 5 min massage of injection site	Rabbits bearing intramuscular VX-2 tumors	Ultrasound: iU22 S9-3 probe (Philips) MRI: 3.0 T Achieva scanner(Philips)	

(*Continued*)

Table 2. (*Continued*)

	Nanosystem composition	Imaging molecules per nanoparticle	Fluorescence ex/em, MR relaxation rate constants (r_1, r_2) and radioisotope $t_{1/2}$	Administration dose and route	Animal model(s)	Imaging systems used	Detection sensitivity
Bumb *et al.*, 2011 (Ref. 52)	Silica-coated iron oxide nanoparticles (SCION) with Cy5.5 or AlexaFluor555. Hydrodynamic diameter: 18–20 nm	Fe: 4.46 mmol/g SCION	Cy5.5: 675/694 nm AlexaFluor555: 555/565 nm r_1 = 1.34 $mM^{-1} s^{-1}$ r_2 = 87.78 $mM^{-1} s^{-1}$	10 μL at a [Fe] of 40 mM administered intradermally into the foot pad (Cy5.5); 10 μL at a [Fe] of 30 mM injected intracutaneously (AlexaFluor555)	Female *nu/nu* mice (Cy5.5), B6(Cg)-*Tyr*$^{c-2J}$/J C57BL/6 or CD11c-EYFP mice (AlexaFluor555)	MRI: 3.0 T Achieva scanner (Philips) Fluorescence: Maestro 2D Imager (Perkin Elmer)	Fluorescence: 2 mM [Fe] SCION(Cy5.5)
Madru *et al.*, 2011 (Ref. 57)	Superparamagnetic iron oxide nanoparticles labeled with ^{99m}Tc Core diameter: 11 ± 2 nm Hydrodynamic diameter: 18 ± 3 nm		^{99m}Tc $t_{1/2}$: 6.01 h	0.05–0.07 mL of ~ 0.2 mg of Fe (25–50 MBq), administered subcutaneously in the right hind paw.	Male white Wistar rats	SPECT: scintillation camera (GE) MRI: 3.0 T Achieva scanner (Philips)	

(*Continued*)

Table 2. (*Continued*)

	Nanosystem composition	Imaging molecules per nanoparticle	Fluorescence ex/em, MR relaxation rate constants (r_1, r_2) and radioisotope $t_{1/2}$	Administration dose and route	Animal model(s)	Imaging systems used	Detection sensitivity
Huang *et al.*, 2012 (Ref. 55)	Mesoporous silica-based nanoprobe (MSN) with ZW800, Gd^{3+} and ^{64}Cu. Core diameter: 60 nm_ Hydrodynamic diameter: 76.8 ± 8.3 nm	ZW800: 30 μg/mg Gd^{3+}: 16 μg/mg ^{64}Cu: 0.2 mCi/mg	ZW800: 770/795 nm ^{64}Cu $t_{1/2}$: 12.7 h	50 μL (2 mg/mL, 20 μCi of ^{64}Cu) administered subcutaneously in the sole of the foot.	Female BALB/c mice bearing subcutaneous 4T1 cells, a model of metastatic breast cancer	Fluorescence: Maestro 2D Imager (Perkin Elmer) MRI: 7.05 T Biospin (Bruker) PET: Inveon microPET (Siemens)	
Cheng *et al.*, 2011 (Ref. 53)	USPIO adsorbed on a NaYF4-based nanoparticle with a PEGylated gold shell (MFNP). Core diameter: 160 nm Hydrodynamic diameter: ~230 nm		MFNP: 980/660 nm r_2 = 352.8 mM^{-1} s^{-1}	15 μL of 2.5 mg/mL solution administered intracutaneously in the rear paw	Female nu/nu mice bearing subcutaneous KB tumors	UCL: Maestro EX CRi (Perkin Elmer) MRI: 3.0 T clinical Biospin (Bruker)	UCL: 6.4 μg/L

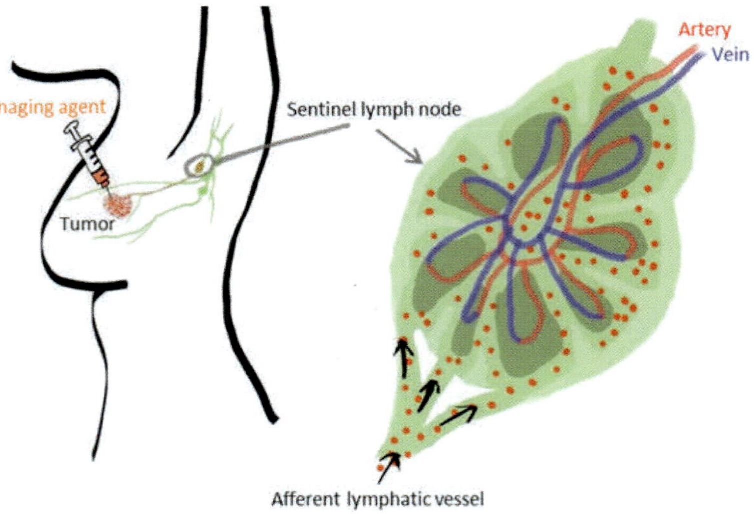

Fig. 3. Illustration of the interstitial transport and passive drainage of imaging agents to the SLN via afferent lymphatic vessels. Drawing by Lia Zheng.

available (i.e. approved for use in humans or currently in clinical trials) intraoperative detection systems include FLARE™ and mini-FLARE™ (The FLARE™ Foundation, Boston, MA, USA), ArteMIS™ (O2view, the Netherlands), Fluobeam™ (Fluoptics, Grenoble, France), SPY® and PINPOINT® (NOVADAQ, Mississauga, Canada) for fluorescence imaging, and Crystal Gamma Probes (Nuclear Fields Corp, Des Plaines, IL, USA), Node Seeker (IntraMedical Imaging, Los Angeles, CA, USA), Navigator GPS™ (Dynasil Products) and C-Trak® (Care Wise Medical Products, Morgan Hill, CA, USA) for SPECT and PET radioisotope detection.

The clear discrepancy between the large number of clinically available intraoperative detection systems specifically designed for guidance during SLN resection and the limited number of agents approved for human use for the same purpose highlights the need and opportunity for innovation and translation of new agents. Bradbury *et al.*[48] obtained approval from the US Food and Drug Administration (FDA) in 2011 for their Investigational New Drug Application (IND) on ultrasmall silica inorganic nanoparticles (diameter = 6–7 nm),[48] also known as Cornell or C dots,[33,61] surface modified with the cyclic arginine-glycine-aspartic acid-tyrosine (cRDGY) targeting peptides and labeled with ^{124}I, for PET imaging, and Cy5.5 (emission peak at 694 nm), for optical imaging. Their first-in-human trial entitled "PET imaging of patients with melanoma and malignant brain tumors using a ^{124}I-labeled cRGDY silica nanomolecular particle tracer: a microdosing study" is currently underway and is aimed at characterization of the pharmacokinetics, biodistribution and metabolic stability of the agent. Unlike other nanosystems proposed for use in multimodality imaging of SLN (Table 2), the ^{124}I-cRGDY-PEG-dots are detectable using PET and optical imaging. Both of these imaging techniques have high detection sensitivity, needing only tracer quantities of the ^{124}I-cRGDY-PEG-dots to be administered (i.e. 3.4–6.7 nmol into a patient[62]). As a result, Bradbury *et al.* successfully acquired FDA approval of their IND by presenting ^{124}I-cRGDY-PEG-dots as a "nanomolecular particle tracer". The group was also able to address the lack of anatomical imaging capability of the agent by using a PET/CT system for PET data acquisition. In this way, the corresponding CT data set may not only be used for PET attenuation correction, but the fused PET/CT data set allows for 3D anatomical localization of the ^{124}I-cRGDY-PEG-dots.

Despite the widespread use of agents designed for PET, SPECT imaging or gamma probe detection, a major drawback is that they are labeled with radioisotopes which require special handling and proper shielding both at the time of administration and surgical resection. Furthermore, patients are subjected to radioactivity, and their exposure can become significant particularly when a longer lived radioisotope such as ^{124}I (physical half-life of 4.18 days) is employed. Although ^{124}I has been used for thyroid imaging, its biological half-life may be significantly extended when attached to a nanoparticle compared to when it is administered as a small molecule. As such, it is expected that a patient receiving ^{124}I-cRGDY-PEG-dots would be subjected to higher total body radiation exposure relative to the dose associated with administration of ^{124}I alone. In fact, Benezra *et al.* estimated using the OLINDA dosimetry program that the effective dose of ^{124}I-cRGDY-PEG-dots for a 70 kg human would be 0.60 cSv/mCi.[33] This is approximately double the effective dose of ^{124}I when the thyroid uptake is 0%.[63] These considerations have steered new investigations toward exploration of non-radioactive imaging agent combinations, such as MR and fluorescence probes.[52,56] These have shown to allow for 3D preoperative imaging and localization of lymph nodes as well as high sensitivity intraoperative image-guided resection.

4. Imaging Atherosclerosis and Thrombosis

Atherosclerosis is a systemic disease characterized by inflammatory and proliferative responses of the arterial vessel wall.[64] The clinical consequences of atherosclerosis include myocardial infarction and ischemic stroke, and are often caused by a thromboembolic event at the site or downstream from an atherosclerotic plaque.[64] Research efforts in this disease area have therefore been focused on the detection and localization of atherosclerotic plaques as well as on the timely identification of thrombus formation.

The current gold-standard non-invasive techniques for imaging atherosclerosis and thrombosis include US, CT, MRI and FDG-PET.[64] Specifically, US is suitable for evaluation of the carotid artery stenosis and carotid intima-media thickness (CIMIT). It has a relatively high spatial resolution of 0.15 mm and the low cost associated with US systems' results in its widespread clinical availability and adoption. However, image acquisition using US is highly user dependent and suffers from limited field of view (FOV). As well, its 3D plaque assessment capabilities still need to be clinically validated. Coronary CT provides rapid measurement of the coronary lumen and wall thickness, but its relatively low spatial resolution (~ 0.4 mm) limits its effectiveness for detection of plaque burden. MRI has primarily been used for characterization of carotid atherosclerotic plaques due to its in-plane resolution of 0.25 × 0.25 mm. The lipid components in the plaque, which are used as evaluation criteria for unstable plaques, can be visualized using T_1, T_2 and proton density-weighted sequences. As a result, MRI has the ability to identify fibrous cap status (i.e. intact, thick or ruptured) and therefore has the potential to recognize vulnerable plaque before an ischemic event.[65,66] Nuclear imaging techniques have also been explored for functional characterization of atherosclerosis. In particular, [^{18}F]FDG-PET has been employed to image metabolically active macrophages and inflammatory cells found in vulnerable plaques.[67,68] Reports have also demonstrated the benefits of FDG-PET imaging as a means to detect early signs of response following lipid-modulating and anti-inflammatory treatment (i.e. decreased FDG uptake).[64] Overall, there is a need to obtain both anatomical and functional information of the disease state in order to improve diagnosis, therapy planning and follow up. As such, efforts have been put forward to engineer multimodality imaging agents (i.e. MR/optical and MR/PET) suitable for improved visualization of the anatomy and function of atherosclerosis and thrombosis[69–72] (Table 3).

Animal models of vascular inflammation employed for preclinical proof-of-principle investigations include (1) apo$E^{-/-}$ mice that exhibit a marked increase in plasma cholesterol levels and develop atherosclerotic lesions resembling those seen in humans, (2) mice with deep vein thrombosis induced through topical application of $FeCl_3$,[73,74] (3) rats subjected to carotid crush injury (i.e. carotid artery crushed with tweezers for 10 s) which causes

Table 3. Multimodal nanoparticle systems designed for imaging atherosclerosis and thrombosis.

	Nanosystem composition	Imaging molecules and targeting moieties per nanoparticle	Fluorescence ex/em, MR relaxation rate constants (r_1, r_2) and radioisotope $t_{1/2}$	Administration dose and route	Animal model	Imaging systems used
Ripplinger *et al.*, 2012 (Ref. 72)	1. CLIO-AF555: Crosslinked dextran-coated iron oxide nanoparticles conjugated to AlexaFluor555 Core diameter: 5 nm (Ref. 70) 2. CLIO-VT680: Crosslinked dextran-coated iron oxide nanoparticles conjugated to VivoTag680 Core diameter: 5 nm (Ref. 70)	CLIO-AF555: 3.6 mg Fe/mL and 1.8×10^{-4} M AF555 CLIO-VT680: 4.7 mg Fe/mL and 1.0×10^{-4} M VT680	AlexaFluor555: 555/565 nm VT680: 670/688 nm	CLIO-AF555: 10 mg Fe/kg *i.v.* CLIO-VT680: 10 mg Fe/kg *i.v.*	DVT induced in the distal femoral and proximal saphenous veins by $FeCl_3$ topically applied onto veins of C57Bl/6 male mice, and jugular vein thrombi induced using a similar procedure.	FMT/CT: FMT 2500 (PerkinElmer) and Inveon CT (Siemens)

(*Continued*)

Table 3. (*Continued*)

	Nanosystem composition	Imaging molecules and targeting moieties per nanoparticle	Fluorescence ex/em, MR relaxation rate constants (r_1, r_2) and radioisotope $t_{1/2}$	Administration dose and route	Animal model	Imaging systems used
Jarrett *et al.*, 2010 (Ref. 69)	1. Dextran sulfate coated iron oxide nanoparticles labeled with ^{64}Cu (^{64}Cu-CLIO) Core diameter: 7.9 nm (Ref. 100) Hydrodynamic diameter: 26.6 ± 7.3 nm (Ref. 100) 2. ^{64}Cu/Gd/TAMRA-labeled maleylated bovine serum albumin	Not reported	^{64}Cu-CLIO: r_1 = 14.46 mM^{-1}s^{-1} r_2 = 72.55 mM^{-1}s^{-1} (Ref. 100)	^{64}Cu-CLIO: 18–20 mg Fe/kg equivalent to 20–25 μCi of ^{64}Cu/g *i.v.* (mice) ^{64}Cu/Gd/TAMRA-BSA: 400–800 μCi administered *i.v.* to rats	1. Vascular inflammation with plaques induced in ApoE$^{-/-}$ mice by right carotid artery ligation followed by high cholesterol diet 2. Rats with mechanical injury of vascular inflammation induced by (i) carotid crush and (ii) implantation of copper cuff	MRI: 7T Biospec (Bruker) PET: custom-built microPETII [102]

(*Continued*)

Table 3. (*Continued*)

	Nanosystem composition	Imaging molecules and targeting moieties per nanoparticle	Fluorescence ex/em, MR relaxation rate constants (r_1, r_2) and radioisotope $t_{1/2}$	Administration dose and route	Animal model	Imaging systems used
McCarthy *et al.*, 2009 (Ref. 70)	Crosslinked iron oxide (CLIO) nanoparticles with: Cy7, Cy7-GPRPP, Cy7-GPSPP, VT680, VT680-FXIII, VT680-CXIII Core diameter: 5 nm	CLIO-VT680: 3 dye molecules CLIO-Cy7: 2 dye molecules CLIO-Cy7-GPRPP: 10–12 peptides, 2 dye molecules CLIO-Cy7-GPSPP: 10–12 peptides, 2 dye molecules CLIO-VT680-FXIII: 10–12 peptides, 3 dye molecules CLIO-VT680-CXIII: 10–12 peptides, 3 dye molecules	VT680: 660/680 nm Cy7: 762/800 nm	7.5 mg Fe/kg, 200 μL volume, *i.v.*	Thrombi induced by $FeCl_3$ topically applied onto the jugular vein of C57Bl/6 mice	MRI: 4.7 T Pharmascan (Bruker) Fluorescence: OV-100 (Olympus)
Nahrendorf *et al.*, 2008 (Ref. 71)	Monocrystalline iron oxide, dextran-coated nanoparticles with VT680 and ^{64}Cu (^{64}Cu-TNP). Target: plaque-associated macrophages. Hydrodynamic diameter: 20 nm	~5 VT680 molecules ~1 mCi of ^{64}Cu per 0.1 mg Fe	^{64}Cu $t_{1/2}$: 12.7 h VT680: detected at 680/700 nm $r_1 = 29\,mM^{-1}s^{-1}$ $r_2 = 60\,mM^{-1}s^{-1}$	PET/CT: 1.5 mg/kg *i.v.* (~300 μCi of ^{64}Cu and 1.5 mg/kg Fe) MRI: 30 mg/kg Fe *i.v.*	$ApoE^{-/-}$ mice with atheromata in their aortic root following high cholesterol diet	PET: FLEX X-PET/X-O (GE) MRI: 7 T horizontal-bore Pharmascan (Bruker) Fluorescence: OV-100 (Olympus)

inflammation in the endothelial layer through recruitment of macrophages,[69] and (4) rats implanted with a silicone-copper cuff around their right common carotid artery which induces chronic vascular inflammation up to 42 days post-implantation.[69,75]

Nahrendorf *et al.*,[71] Jarrett *et al.*[69] and Ripplinger *et al.*[72] reported the use of iron oxide nanoparticles with high inherent phagocytic avidity to passively target macrophages found at sites of atherosclerosis. McCarthy *et al.*[70] and Jarrett *et al.*[69] also used ligand-mediated active targeting strategies to direct the nanoagents to fibrin,[70] factor XIIIa (FXIIIa)[70] or the macrophage scavenger receptor A type I (SRA).[69] The T_2-shortening MR nanoparticles consist of a 3–7 nm iron oxide core surrounded by a dextran coating (final hydrodynamic diameter of 20–30 nm),[76,77] while the T_1-shortening MR agent was a Gd-DOTA bovine serum albumin particle.[69,78] The iron oxide particles are further labeled with fluorescence probes such as Cy7 (λ_{ex} = 747 nm, λ_{em} = 776 nm), VT680 (λ_{ex} = 670 nm, λ_{em} = 688 nm) or AlexaFluor555 (λ_{ex} = 555 nm, λ_{em} = 565 nm) for optical imaging. The PET imaging radioisotope used was ^{64}Cu (Refs. 69 and 71) (physical half-life of 12.7 h), which is well-matched to the longer circulation lifetime of nanoparticles in the body compared to small molecules and antibodies. Nahrendorf *et al.*[71] demonstrated the benefit of having a PET reporter in addition to an MR contrast agent given that PET is able to detect at least 50 times lower concentrations of the same nanoparticles compared to T_2*-weighted MR imaging. Combined MR/PET agents will likely be the ideal candidates for successful clinical translation, as PET allows for high sensitivity and yet non-invasive 3D whole body imaging. In contrast, fluorescence optical imaging, even in the NIR wavelength range has limited penetration depth, and is best used for endoscopic and intraoperative applications. Moreover, with the advent of the new clinical hybrid PET/MR scanners (i.e. Biograph mMR by Siemens) and PET/MR integrated platforms (i.e. two separate scanners lined up at opposite ends of a patient turntable — Ingenuity TF by Philips) simultaneous or sequential PET and MR image acquisition and registration are now possible. This ultimately enables full implementation of multimodality imaging through hardware, software and agent integration.

5. *In Vivo* Cell-Tracking

Cell-based therapies play an important role in tissue repair and immunomodulation. They have shown promise in treating neurological, cardiovascular and neoplastic diseases.[79–81] However, a major limitation in the development of new and improved cell-based therapies is the lack of appropriate evaluation tools to monitor cell fate and performance *in vivo*. Several cell-labeling strategies have been explored for non-invasive *in vivo* tracking of stem cells,[82,83] immune cells[84] and pancreatic islet cells[85] using nanoparticles designed for multimodality imaging. The majority of these employ direct cell-labeling, which relies on cells to either passively incorporate or phagocytose the imaging probe, or actively introduce it into cells in culture through the use of TAT-peptide, electroporation or microinjection mediated techniques.[86] Indirect cell-labeling techniques are also available, but these have limited potential for clinical translation as they introduce a reporter gene, rather than an imaging agent, into the target cells to produce imageable signals (i.e. fluorescent or bioluminescent proteins).[87–90]

The main concerns with current direct cell-labeling techniques are decreased cell viability, modification of cellular function (i.e. loss of stem cell pluripotency) and unsuitability for longitudinal tracking due to dilution of imaging probe as cells proliferate. Currently available cell-labeling nanoplatforms have all shown to be relatively inert and do not affect cell viability *in vitro* [82–85,91,92] (Table 4). These nanoparticle systems are based on iron oxide, gold, cobalt-ferrite, mesoporous silica nanoparticles (MSNs) and single-walled carbon nanotubes (SWNT) modified to include fluorescence tags (i.e. rhodamine B isothiocyanate,[92] AlexaFluor680,[84] FITC and ZW800[91]), T_1-weighted MR signal enhancing Gd^{3+} (i.e. DTDTPA-Gd,[85] Gd^{3+} ion loading into pore channels of MSN[91]), and the positron emitter ^{64}Cu for PET imaging (^{64}Cu-DOTA[91]). However, probe dilution following cell proliferation is an unresolved challenge that will need to be addressed. Likely, the short term solution to this problem will consist of performing comprehensive characterization of the imaging window (i.e. length of time after transplantation during which cells can be tracked) and improving detection sensitivity.

An interesting implementation of cell-tracking which does not rely on either direct or indirect cell-labeling is the co-encapsulation of a trimodality Au-DTDTPA-Gd nanoparticle with human cadaveric pancreatic islet cells inside an alginate-protamine sulfate-alginate (APSA) microcapsule (Fig. 4) suitable for *in vivo* delivery, immunoprotection and non-invasive imaging of pancreatic islet cell transplantation for treatment of type I diabetes.[85] This elegant approach circumvents the need to incorporate exogenous nanoparticles inside cells and the large size of the microcapsule ($542 \pm 19\,\mu m$) allows for a high nanoparticle to cell ratio to be achieved (i.e. 1 ng of gold and 87.7 pg of Gd^{3+} to 1 cell per capsule) thereby increasing detection sensitivity and the potential viability of the transplanted cells. *In vitro* tests demonstrated that one day after encapsulation, the viability of cells inside APSA was

Table 4. Multimodal nanoparticle systems designed for *in vivo* cell-tracking.

	Nanosystem composition	Retention time of nanoparticles in cells	Biocompatibility and impact on cell function	Fluorescence ex/em, MR relaxation rate constants (r_1, r_2) and radioisotope $t_{1/2}$	Administration dose and route	Animal model(s)	Imaging modalities	Detection sensitivity
Cheng *et al.*, 2013 (Ref. 82)	UCNP-IONPs coated with a PEGylated gold outer shell Core diameter: 180 nm (UCNP) Hydrodynamic diameter: 240 nm (UCNP-IONP + gold + PEG)	Signal present in cultured cells up to 10 days post-labeling	Non-cytotoxic to mMSCs up to 0.5 mg/mL and minimal oxidative stress to cells; no impact on differentiation of labeled stem cells	UCL: 980/540 nm $r_2 = 320.1$ $mM^{-1}s^{-1}$	*i.p.* injection of 10^6 labeled mMSCs	Wound model induced by skin peeling on both sides of abdomen in mice	UCL: Maestro spectroscopic imaging unit MRI: 3 T clinical Biospin (Bruker)	~100 labeled mMSCs injected s.c. into mice are detectable by UCL
Huang *et al.*, 2013 (Ref. 91)	HA coated mesoporous silica nanoparticles (HA-MSNs) labeled with FITC, ZW800, Gd^{3+} and ^{64}Cu Core diameter: 60 nm (MSNs + dyes)	Signal present up to 3 weeks post-labeling; no change in fluorescence up to 7 days post-labeling with HA-coated particles	Non-cytotoxic to mMSCs up to 1 mg/mL; no impact on cell proliferation or intracellular protein expression	^{64}Cu $t_{1/2}$: 12.7 h ZW800: 770/795 nm FITC: 490/525 nm	*i.v.* injection of 10^6 labeled mMSCs	Orthotopic U87MG glioblastoma mouse model	Fluorescence: Maestro spectroscopic imaging unit MRI: 7.05 T Biospin (Bruker) PET: Inveon (Siemens)	Not reported

(*Continued*)

Table 4. (*Continued*)

	Nanosystem composition	Retention time of nanoparticles in cells	Biocompatibility and impact on cell function	Fluorescence ex/em, MR relaxation rate constants (r_1, r_2) and radioisotope $t_{1/2}$	Administration dose and route	Animal model(s)	Imaging modalities	Detection sensitivity
Wang *et al.*, 2012 (Ref. 83)	Protamine conjugated, PEG-coated SWNT-PEG-PRO. Nanotube length: 150 nm Nanotube diameter: 1.2 nm	Raman signal present in cultured cells 10 days post-labeling, and detectable by MRI up to 2 weeks post-labeling; cells could be *tracked in vivo* for up to 7 days	Non-cytotoxic to hMSCs up to 20 nM; minimal oxidative stress to cells; no impact on differentiation of stem cells at the 10 nM	Raman excitation: 785 nm Raman G-band peak: 1590 cm^{-1} $r_2 = 482.6$ $mM^{-1}s^{-1}$	2×10^6 labeled hMSCs *s.c.* injected	*Nu/nu* mice	Raman: LabRam HR (Horiba) MRI: 3 T clinical MRI scanner equipped with small animal coil PAT: Home-made acoustic-resolution photoacoustic microscopy system	~500 labeled cells injected *s.c.* in mice detectable by Raman spectroscopy

(*Continued*)

Table 4. (*Continued*)

	Nanosystem composition	Retention time of nanoparticles in cells	Biocompatibility and impact on cell function	Fluorescence ex/em, MR relaxation rate constants (r_1, r_2) and radioisotope $t_{1/2}$	Administration dose and route	Animal model(s)	Imaging modalities	Detection sensitivity
Arifin *et al.*, 2011 (Ref. 85)	Gold nanoparticles functionalized with DTDTPA: gadolinium chelates (GG) co-encapsulated with human islets into alginate microcapsules GG nanoparticle diameter: 2.0–2.5 nm Microcapsule diameter: 542 ± 19 μm	1 human islet per capsule	Viability of encapsulated βTC-6 insulinoma cells not affected by microcapsules; encapsulated human islets continued to secrete insulin for > 15 days in culture	Not reported	6000 capsules containing 7.5×10^6 mouse βTC-6 insulinoma cells administered *i.p.*	C57BL/6J mice with streptozotocin-induced diabetes	MRI: 9.4 T horizontal bore magnet (Bruker Biospin) CT: X-SPECT (Gamma Medica) US: 40 MHz Vevo 660 (VisualSonics)	Single microcapsule detectable via US; 100 microcapsules detectable by CT

(*Continued*)

Table 4. (*Continued*)

	Nanosystem composition	Retention time of nanoparticles in cells	Biocompatibility and impact on cell function	Fluorescence ex/em, MR relaxation rate constants (r_1, r_2) and radioisotope $t_{1/2}$	Administration dose and route	Animal model(s)	Imaging modalities	Detection sensitivity
Thorek *et al.*, 2010 (Ref. 84)	SPIO nanoparticles with a cross-linked dextran shell labeled with NIR815 Core diameter: 5.6 nm [101] Hydrodynamic diameter: 53.5 nm	MRI signal in the spleen returned to baseline 2 weeks post-injection of labeled cells; NIR815 fluorescence was detectable up to 15 days	Non-cytotoxic to B cells;[101] no impact on expression of surface proteins involved in B cell activation	NIR815: 710–760/800 nm $r_2 = 82$ $mM^{-1}s^{-1}$	*i.v.* injection of 20×10^6 NIR815-labeled B cells with/without nanoparticles	C57BL/6 mice	MRI: 9.4 T (Varian) Fluorescence: Maestro spectroscopic imaging unit	Not reported

(*Continued*)

Table 4. (*Continued*)

	Nanosystem composition	Retention time of nanoparticles in cells	Biocompatibility and impact on cell function	Fluorescence ex/em, MR relaxation rate constants (r_1, r_2) and radioisotope $t_{1/2}$	Administration dose and route	Animal model(s)	Imaging modalities	Detection sensitivity
Hwang *et al.*, 2009 (Ref. 92)	Cobalt-ferrite magnetic particles with rhodamine-containing silica shell, PEG, amine, luciferase protein, ^{68}Ga. Core diameter (silica coated magnetic nanoparticles): ~50 nm Hydrodynamic diameter: 68.1 nm	Not reported	Not reported	Rhodamine: 555/578 nm ^{68}Ga $t_{1/2}$: 68 min $r_2 = 64$ $mM^{-1}s^{-1}$	*i.m.* or *s.c.* injection of 2×10^6 labeled C6 cells	BALB/c nude mice	Fluorescence/BLO: IVIS-200 (Xenogen) PET: microPET R4 scanner MRI: Bruker 7T micro-MRI	Not reported

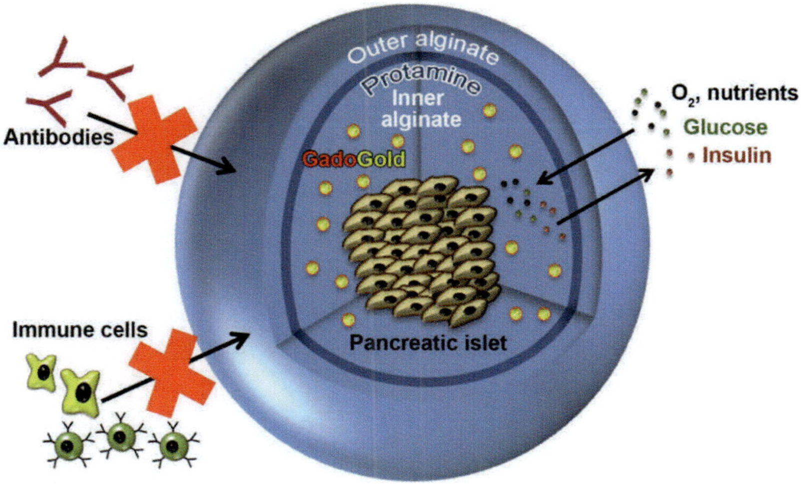

(A)

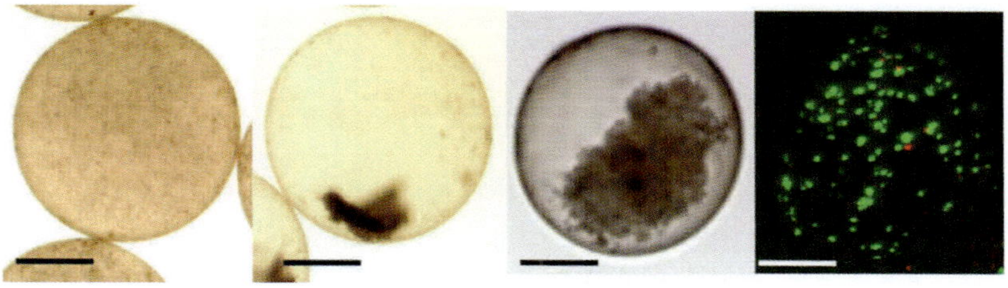

(B) (C) (D) (E)

Fig. 4. (A) Three-dimensional scheme of APSA microcapsule containing GG (APSA 1 GG). The semipermeable microcapsule allows diffusion of oxygen, nutrients, glucose and insulin, while passage of immune cells and antibodies is blocked. (B) and (C) Light microscopic images of APSA 1 GG microcapsule 1 h (B) and 2 days (C) after synthesis. (D) Human pancreatic islet inside APSA 1 GG microcapsule. (E) Fluorescence microscopic image shows viability staining of mouse insulinoma b TC-6 cells inside APSA 1 GG microcapsule. Green (fluorescein diacetate stain) = live cells, red (propidium iodide stain) = dead cells. Scale bars = 200 μm.

Figure reproduced with permission from Arifin *et al.*[85]

maintained in the presence and absence of the Au-DTDTPA-Gd nanoparticles. However, the levels of C-peptide secretion were decreased for the APSA encapsulated human islets compared to naked islets in culture over a 15-day period. The authors attributed this decrease in C-peptide secretion to the reduced rate of protein diffusion through APSA microcapsules into the culture media. *In vivo*, diabetic mice treated with APSA capsules containing βTC-6 insulinoma cells and Au-DTDTPA-Gd nanoparticles achieved blood glucose levels similar to those measured for healthy mice seven days after implantation and they remained normoglycemic for six weeks. The implanted cell capsules were also successfully imaged using MR, CT and US. MR proved to be the imaging modality with the highest detection sensitivity for the Au-DTDTPA-Gd nanoparticles. However, no quantitative measurement of the time dependent capsule migration or degradation *in vivo* was performed, possibly due to non-linearities associated with MR imaging as well as non-Gd dependent factors that contribute to relaxivity changes *in vivo* (i.e. degradation and declustering of APSA over time).

Overall, the use of multimodality imaging probes to track cells *in vivo* following implantation has shown promising initial feasibility and relatively high inertness. However, more in depth investigations need to be performed in order to quantitatively correlate the imaging signal to cell numbers over time, post-implantation. These cell-tracking methods can be further complemented by existing functional imaging techniques (i.e. FDG-PET) to assess cellular viability in addition to spatial localization and homing following transplantation.

6. Image-Guided Cancer Therapies

A major goal in cancer research is the development of therapeutic strategies that are effective at curing or controlling the disease, while being minimally invasive and having negligible side effects. Efforts are continuously being applied in the development of novel therapeutic platforms to improve the therapeutic ratio. Imaging has been increasingly exploited for assessment of the *in vivo* performance of therapeutic systems. Over the last five years, the number of reports of new nanosystems suitable for both imaging and therapy has surged to 257 (pubmed search nano + imaging + therapy, April 2007 to April 2013) from 55 (pubmed search nano + imaging + therapy, April 2002 to April 2007). In addition, at the 2010 Image-Guided Drug Delivery (IGDD) Summit held by the National Cancer Institute (NCI) and the FDA, it was concluded that the full implementation of IGDD requires (1) a drug that can be imaged, localized and activated at the tumor site, and (2) quantitative imaging techniques that can monitor the drug delivery process at appropriate spatial and temporal resolutions. A major challenge that currently hinders successful implementation of IGDD is the lack of integrated image-based platforms specifically designed for real-time quantification of the drug or drug carrier concentration at the target site immediately before and after therapy in order to provide feedback on the treatment location and dose delivered.

To date, examples of multimodality imaging nanoparticle-based platforms used for therapy guidance have been limited to four therapeutic interventions: IGDD,[93–95] image-guided surgery,[34] image-guided radiation therapy[35,36] and image-guided photo-thermal therapy (PTT)[53,96,97] (Table 5). It is interesting to note that despite the limited clinical translatability of optical imaging due to a variety of drawbacks including poor depth of penetration, background tissue autofluorescence (particularly when not operating in the NIR wavelength range), inaccurate 3D localization of the fluorescence source within the body and lack of robust quantification tools, all but one of the published reports to date include optical imaging as one of the multimodal components. In some cases, researchers take advantage of the inherent fluorescence property of the therapeutic agent used (i.e. doxorubicin,[95] lanthanide-doped rare earth nanocrystal[53]) while in other cases fluorescent labeling (i.e. IR-780[96]) is deemed advantageous as it allows for high resolution *ex vivo* investigation and confirmation of the cellular and subcellular localization of nanoparticle delivery and targeting.[34,93,94,96,97] In image-guided surgical applications, the optical imaging component becomes clinically translatable and valuable.[34]

MR imaging is also omnipresent in all published examples of multimodal agents for image-guided therapy applications. This is expected as MR has the ability to provide 3D imaging of relatively large FOVs with high inherent anatomical tissue contrast, relatively high sensitivity and spatial resolution. It is these features that make MR an ideal imaging modality for treatment planning. Furthermore, two different types of MR contrast generating moieties exist (Gd^{3+} for T_1-weighted imaging and iron oxide for T_2-weighted imaging) which provides versatility in terms of the choice of molecule when designing a new nanoplatform.

The majority of reports that illustrate examples of multimodality imaging nanosystems designed for PTT and drug delivery contain limited justification for employment of more than one imaging technique. Although the complementary nature of MR and optical imaging is always brought forward, the authors do not make convincing arguments that both modalities are necessary for their application of choice. In all cases, similar information is provided by each separate imaging system and MR always demonstrates superiority over 2D fluorescence optical imaging both in terms of image quality and clinical translatability. For image-guided radiation therapy applications, Zheng and Hoisak *et al.*[35,36,98,99] successfully demonstrated the ability of a CT/MR liposome system to provide spatially co-localized signal enhancement in the two imaging modalities for improved image-registration during the radiotherapy (RT) planning process (Fig. 5). In this particular example, the same contrast-enhancement information is needed from both imaging modalities in order to allow for accurate image registration. Specifically, the

Table 5. Multimodal nanoparticle systems designed for image-guided cancer therapies.

	Nanosystem Composition	Imaging molecules and targeting moieties per nanoparticle	Fluorescence ex/em, MR relaxation rate constants (r_1, r_2) and radioisotope $t_{1/2}$	Intervention/application	Administration dose and route	Animal model(s)	Imaging modalities
Yang *et al.*, 2012 (Ref. 97)	Reduced grapheme oxide-iron oxide nanoparticles with PEG (RGO-IONP-PEG) labeled with Cy5 Hydrodynamic diameter (RGO-IONP-PEG): 50 nm	Not reported	Cy5: 595/680 nm PAT excitation: 650 nm $r_2 = 108.1\ mM^{-1}s^{-1}$	Image-guided photothermal therapy	RGO-IONP-PEG: 2 mg/mL (200 μL of 2mg/mL) *i.v.*	Female Balb/c mice bearing *s.c.* 4T1 tumors	Fluorescence: CRi Maestro (Perkin Elmer) PAT: Custom-built setup MRI: 3 T Biospin (Bruker)
Yan *et al.*, 2012 (Ref. 34)	PAMAM dendrimer labeled with Cy5.5/rhodamine/Gd^{3+} and conjugated to angiopep-2 and c[RGDyK] peptides Hydrodynamic diameter: 15.6 nm	Molar ratio of angiopep-2/c[RGDyK]/Gd^{3+}-DOTA/dendrimer: 5/6/54/1	Cy5.5: 750/800 nm Rhodamine: 543 / 560 nm $r_1 = 7.1–7.4\ mM^{-1}s^{-1}$	Image-guided surgical resection	0.05 mmol/kg [Gd^{3+}] In 0.25 mL PBS *i.v.*	Orthotopic U87MG glioblastoma tumors induced in male Balb/c nude mice	MRI: Biospec 47/30 (Bruker) Fluorescence: Multispectral Imaging System (Kodak)

(*Continued*)

Table 5. (*Continued*)

	Nanosystem Composition	Imaging molecules and targeting moieties per nanoparticle	Fluorescence ex/em, MR relaxation rate constants (r_1, r_2) and radioisotope $t_{1/2}$	Intervention/application	Administration dose and route	Animal model(s)	Imaging modalities
Harrison *et al.*, 2012 (Ref. 94)	PGMA nanoparticles with SPIO and rhodamine B conjugated with polyethylenimine <u>Hydrodynamic diameter</u>: 157 ± 38 nm	Not reported	Rhodamine B: 540/625 nm $r_2 = 340\,mM^{-1}s^{-1}$	Image-guided drug delivery to the CNS	0.5 μg nanoparticles in a 0.5 μL volume injected intravitreally	Partial optic nerve transection induced in female PVG Hooded rats	<u>MRI:</u> 4.6 T AVANCE (Bruker) <u>Fluorescence:</u> Leitz Diaplan (Nikon), BX-51 fluorescence microscope
Cheng *et al.*, 2012 (Ref. 53)	Upconversion Y:Fe:Au nanoparticle with IONP coated with a PEGylated gold layer <u>Hydrodynamic diameter:</u> ~220 nm	Not reported	UCL: 850/640–670 nm	Nanoparticles for image-guided, magnetically targeted photothermal therapy	<u>MFNP-PEG</u>: 50 mg/kg in 200 μL volume *i.v.*	Balb/c nude mice bearing s.c. 4T1 tumors	<u>MRI:</u> 3T Biospin (Bruker) <u>UCL:</u> Maestro EX (Perkin Elmer)

(*Continued*)

Table 5. (*Continued*)

	Nanosystem Composition	Imaging molecules and targeting moieties per nanoparticle	Fluorescence ex/em, MR relaxation rate constants (r_1, r_2) and radioisotope $t_{1/2}$	Intervention/application	Administration dose and route	Animal model(s)	Imaging modalities
Gianella *et al.*, 2011 (Ref. 93)	Oleic acid coated ION crystals with rhodamine lipid, Cy7-PEG and prednisolone acetate valerate (PAV) Hydrodynamic diameter: ~50 nm	Not reported	Cy7 excitation: 710–760/810–875 nm $r_1 = 0.618\ mM^{-1}s^{-1}$ (inside LS174T tumor cells)	Image-guided drug delivery	Nanoemulsion dose equivalent to 30 mg FeO/kg and 10 mg of PAV/kg *i.v.*	Male Swiss nude mice bearing s.c. LS174T tumors	MRI: 3 T Achieva (Philips) Fluorescence: IVIS-200 (Xenogen)
Xu *et al.*, 2011 (Ref. 95)	UCNPs and IONPs coated with PS_{16}-*b*-PAA_{10} polymer and Saquaraine (UC-IO@Polymer-SQ) or doxorubicin (UC-IO@Polymer-DOX) Core diameter; (UCNPs and IONPs): 6-8 nm Hydrodynamic diameter: 200 nm	UC-IO@Polymer: 5.6% w/w UCNPs, 8% w/w IONPs DOX: 8% w/w	UC-IO@Polymer-SQ: 660/710 nm UCL emission band of UC-IO@Polymer-SQ/DOX: 500–560 nm $r_2 = 84\ mM^{-1}s^{-1}$	Magnetic targeted drug delivery and cell-tracking	Not reported	Athymic Balb/c nude mice bearing *s.c.* 4T1 tumors	Fluorescence/UCL: Maestro EX (Perkin Elmer) MRI: 3 T clinical scanner equipped with small animal coil

(*Continued*)

Table 5. (*Continued*)

	Nanosystem Composition	Imaging molecules and targeting moieties per nanoparticle	Fluorescence ex/em, MR relaxation rate constants (r_1, r_2) and radioisotope $t_{1/2}$	Intervention/application	Administration dose and route	Animal model(s)	Imaging modalities
Peng *et al.*, 2011 (Ref. 96)	IR-780 iodide-loaded and ^{188}Re labeled micelles Hydrodynamic diameter: 155–203 nm	6414 ± 641 IR-780 molecules per micelle	Fluorescence: 710–760/810–875 nm ^{188}Re $t_{1/2}$: 17.0 h	Image-guided photothermal therapy and targeted drug delivery	^{188}Re-labeled IR-780 micelles: i.v. injection (22 MBq, 1.25 mg/kg)	Female Balb/c athymic nude mice bearing *s.c.* HCT-116 human colon cancer xenografts	Fluorescence: IVIS (Xenogen) SPECT/CT: X-SPECT (GE)
Zheng *et al.*, 2006, 2007 (Ref. 35 and 36)	Liposome co-encapsulating iohexol and gadoteridol Hydrodynamic diameter: 74.4 ± 3.3 nm[36]	1.3×10^6 iodine and 1.3×10^5 gadolinium per liposome	X-ray attenuation: 37.8 ± 0.5 HU/(mg of iodine and 0.2 mg of gadolinium) $r_1 = 1.23 \pm 0.02\, mM^{-1}s^{-1}$ $r_2 = 1.46 \pm 0.02\, mM^{-1}s^{-1}$	Image-guided radiation therapy	340 mg/kg of iodine and 19 mg/kg of gadolinium *i.v.*	New Zealand White rabbits	CT: Discovery ST (GE) MR: 1.5 T Signa TwinSpeed (GE)

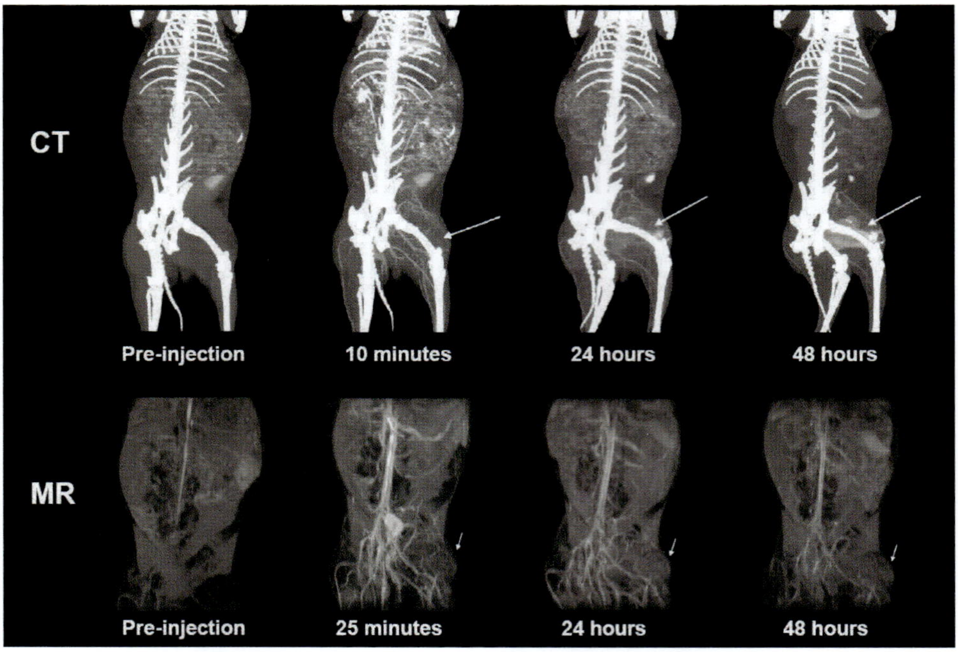

(a)

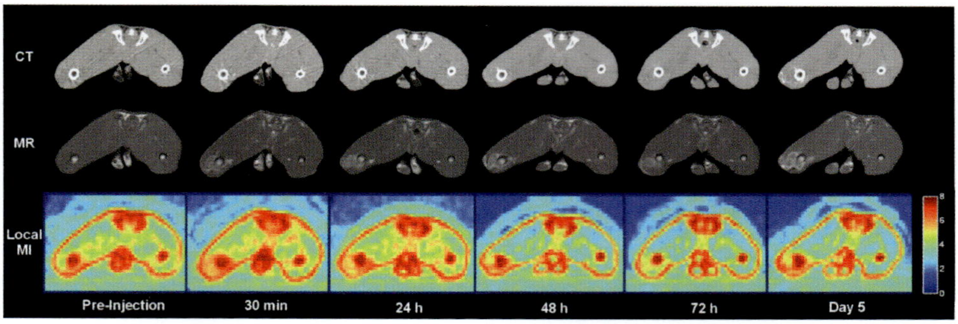

(b)

Fig. 5. (a) Three-dimensional maximum intensity projections (anterior view) of a VX2 tumor bearing male New Zealand White rabbit obtained at the indicated times before and post-administration of the liposomal iohexol and gadoteridol. The arrows point to the location of the tumor (left thigh, between hip and knee).

Figure reproduced with permission from Zheng *et al.*[98]

(b) Registered axial CT and MR images of VX2 tumor-bearing New Zealand White rabbit, and map of local mutual information (MI) (lower row). Images shown are from acquisitions at pre-injection of contrast agent to day 5, post contrast administration. Each pixel of the local MI map represents a 16×16 pixel region-of-interest (ROI) of the registered MR and CT image pairs. Pixel intensity represents the mutual information between the ROIs. Note the increase in mutual information within the tumor (left side of images) at 30 min, 24 h and 48 h post-injection of the contrast agent.

Figure reproduced with permission from Hoisak *et al.*[99]

registration of 3D CT and MR data sets are routinely performed in the clinic to obtain high contrast soft tissue visualization and tumor margin definition (offered by MR), along with a 3D tissue X-ray attenuation map needed for treatment dose planning and geometrically accurate bony structure visualization (provided by CT) required for accurate patient positioning on the treatment bed using radiotherapy delivery systems equipped with CT-based guidance (i.e. CBCT-guided linear accelerators — Elekta Synergy). Furthermore, the persistence of the signal provided in both imaging modalities following a single injection of the nanosystem is particularly attractive for longitudinal imaging procedures (i.e. pre- and intraoperative imaging for surgical guidance lasting hours, RT planning and treatment guidance over multiple treatment fractions lasting days).

In all image-guided therapy applications investigated, the extended biological half-life of nanoparticles, a prerequisite for EPR-mediated tumor accumulation, is further exploited for longitudinal and repeated imaging. In fact, long circulating nanoparticles only require a single administration to be performed for pre-, intra- and post-treatment imaging. Furthermore, there is an increasing interest in the design and development of new activatable and environmentally sensitive (i.e. temperature, pH, MMPs) nanoparticles for cancer therapy. These systems in particular can benefit from non-invasive imaging as a means to conformally guide treatment delivery (i.e. application of external heat sources to activate thermo-responsive carriers) and to evaluate treatment performance (i.e. chemodosimetry).

7. Conclusion

The world of imaging has seen an explosive increase in the development of innovative nanoplatforms aimed at providing persistent and co-localized signal enhancements across multiple imaging modalities. The ultimate goal is to seamlessly bridge spatial, temporal and sensitivity scales through the use of a common agent to improve and benefit a variety of clinical applications including disease diagnosis and characterization, treatment planning, intraoperative guidance and evaluation of therapeutic response.

This chapter discussed a wide range of nano-imaging agents categorized by their intended application. It is likely that a large number of nanosystems will continue to be developed in the preclinical setting, each with properties that best suit use in specific research applications. However, once the first multimodal nanosystem receives regulatory approval for use in humans this will most probably be tried in many clinical applications that could benefit from the use of multiple imaging modalities. Concurrently, research efforts will be applied to personalize the delivery and targeting of nano-agents. For example, novel strategies will be optimized to activate and deactivate different components of the nanoplatform following either endogenous biological cues (i.e. temperature, pH, MMP levels) or external triggers (i.e. irradiation, heat, magnetic field). Ultimately, the next generation of nanotechnology for multimodality imaging will likely be a "smart" modular platform which will enable the implementation of personalized nanomedicine.

References

1. M. Uematsu *et al.*, Daily positioning accuracy of frameless stereotactic radiation therapy with a fusion of computed tomography and linear accelerator (focal) unit: Evaluation of z-axis with a z-marker. *Radiother. Oncol.* **50**(3), 337–339 (1999).
2. D. A. Jaffray *et al.*, Flat-panel cone-beam computed tomography for image-guided radiation therapy. *Int. J. Radiat. Oncol. Biol. Phys.* **53**(5), 1337–1349 (2002).
3. E. R. McVeigh *et al.*, Real-time interactive MRI-guided cardiac surgery: aortic valve replacement using a direct apical approach. *Magn. Reson. Med.* **56**(5), 958–964 (2006).
4. M. Hashizume, MRI-guided laparoscopic and robotic surgery for malignancies. *Int. J. Clin. Oncol.* **12**(2), 94–98 (2007).
5. O. M. Rygh *et al.,* Intraoperative navigated 3-dimensional ultrasound angiography in tumor surgery. *Surg. Neurol.* **66**(6), 581–592 (2006); discussion 592.
6. D. Lindner *et al.*, Application of intraoperative 3D ultrasound during navigated tumor resection. *Minim. Invasive Neurosurg.* **49**(4), 197–202 (2006).
7. C. P. Hsu *et al.*, Liver tumor gross margin identification and ablation monitoring during liver radiofrequency treatment. *J. Vasc. Interv. Radiol.* **16**(11), 1473–1478 (2005).
8. M. P. Fried, F. A. Jolesz and P. R. Morrison, Image guidance with laser applications. *Otolaryngol. Clin. North Am.* **29**(6), 1063–1078 (1996).
9. P. Haigron *et al.*, Image-guided therapy: Evolution and breakthrough. *IEEE Eng. Med. Biol. Mag.* **29**(1), 100–104.
10. J. Zheng, D. A. Jaffray and C. Allen, Nanosystems for Multimodality *In Vivo* Imaging, in *Multifunctional Pharmaceutical Nanocarriers*, V. P. Torchilin (Ed.): Springer Science+Business Media, LLC (2008), pp. 409–430.
11. Y. Matsumura and H. Maeda, A new concept for macromolecular therapeutics in cancer chemotherapy: Mechanism of tumoritropic accumulation of proteins and the antitumor agent smancs. *Cancer Res.* **46**(12 Pt 1), 6387–6392 (1986).
12. Y. Noguchi *et al.*, Early phase tumor accumulation of macromolecules: A great difference in clearance rate between tumor and normal tissues. *Jpn. J. Cancer Res.* **89**(3), 307–314 (1998).
13. H. Maeda and Y. Matsumura, Tumoritropic and lymphotropic principles of macromolecular drugs. *Crit. Rev. Ther. Drug Carrier Syst.* **6**(3), 193–210 (1989).
14. L. W. Seymour *et al.*, Tumour tropism and anti-cancer efficacy of polymer-based doxorubicin prodrugs in the treatment of subcutaneous murine B16F10 melanoma. *Br. J. Cancer* **70**(4), 636–641 (1994).
15. Y. Takakura and M. Hashida, Macromolecular carrier systems for targeted drug delivery: Pharmacokinetic considerations on biodistribution. *Pharm. Res.* **13**(6), 820–831 (1996).

16. H. Maeda *et al.*, Tumor vascular permeability and the EPR effect in macromolecular therapeutics: A review. *J. Control. Release* **65**(1&2), 271–284 (2000).
17. F. Yuan *et al.*, Vascular permeability in a human tumor xenograft: Molecular size dependence and cutoff size. *Cancer Res.* **55**(17), 3752–3756 (1995).
18. F. Yuan *et al.*, Microvascular permeability and interstitial penetration of sterically stabilized (stealth) liposomes in a human tumor xenograft. *Cancer Res.* **54**(13), 3352–3356 (1994).
19. A. K. Iyer *et al.*, Exploiting the enhanced permeability and retention effect for tumor targeting. *Drug Discov. Today* **11**(17&18), 812–818 (2006).
20. S. A. Skinner, P. J. Tutton and P. E. O'Brien, Microvascular architecture of experimental colon tumors in the rat. *Cancer Res.* **50**(8), 2411–2417 (1990).
21. L. W. Seymour *et al.*, Influence of molecular weight on passive tumour accumulation of a soluble macromolecular drug carrier. *Eur. J. Cancer* **31A**(5), 766–770 (1995).
22. E. J. Cha *et al.*, Development of MRI/NIRF 'activatable' multimodal imaging probe based on iron oxide nanoparticles. *J. Control. Release* **155**(2), 152–158.
23. S. W. Chou *et al.*, *In vitro* and *in vivo* studies of FePt nanoparticles for dual modal CT/MRI molecular imaging. *J. Am. Chem. Soc.* **132**(38), 13270–13278.
24. M. Nahrendorf *et al.*, Hybrid PET-optical imaging using targeted probes. *Proc. Natl. Acad. Sci. USA* **107**(17), 7910–7915.
25. Y. F. Tan *et al.*, Multimodal tumor imaging by iron oxides and quantum dots formulated in poly (lactic acid)-D-alpha-tocopheryl polyethylene glycol 1000 succinate nanoparticles. *Biomaterials* **32**(11), 2969–2978.
26. R. Bardhan *et al.*, Tracking of multimodal therapeutic nanocomplexes targeting breast cancer *in Vivo. Nano Lett.*
27. H. Y. Lee *et al.*, PET/MRI dual-modality tumor imaging using arginine-glycine-aspartic (RGD)-conjugated radiolabeled iron oxide nanoparticles. *J. Nucl. Med.* **49**(8), 1371–1379 (2008).
28. N. Lee *et al.*, Multifunctional Fe_3O_4/TaO(x) core/shell nanoparticles for simultaneous magnetic resonance imaging and X-ray computed tomography. *J. Am. Chem. Soc.* **134**(25), 10309–10312.
29. W. Hwang do *et al.*, A nucleolin-targeted multimodal nanoparticle imaging probe for tracking cancer cells using an aptamer. *J. Nucl. Med.* **51**(1), 98–105.
30. T. Nam *et al.*, Tumor targeting chitosan nanoparticles for dual-modality optical/MR cancer imaging. *Bioconjug Chem.* **21**(4), 578–582.
31. T. Luo *et al.*, Mesoporous silica-coated gold nanorods with embedded indocyanine green for dual mode X-ray CT and NIR fluorescence imaging. *Opt. Express* **19**(18), 17030–17039.
32. J. Lee *et al.*, RGD peptide-conjugated multimodal NaGdF4:Yb3+/Er3+ nanophosphors for upconversion luminescence, MR, and PET imaging of tumor angiogenesis. *J. Nucl. Med.* **54**(1), 96–103.
33. M. Benezra *et al.*, Multimodal silica nanoparticles are effective cancer-targeted probes in a model of human melanoma. *J. Clin. Invest.* **121**(7), 2768–2780.
34. H. Yan *et al.*, Two-order targeted brain tumor imaging by using an optical/paramagnetic nanoprobe across the blood brain barrier. *ACS Nano* **6**(1), 410–420.
35. J. Zheng *et al.*, *In vivo* performance of a liposomal vascular contrast agent for CT and MR-based image guidance applications. *Pharm. Res.* **24**(6), 1193–1201 (2007).

36. J. Zheng *et al.*, Multimodal contrast agent for combined computed tomography and magnetic resonance imaging applications. *Invest Radiol.* **41**(3), 339–348 (2006).
37. R. Zhang *et al.*, Peptide-conjugated polymeric micellar nanoparticles for Dual SPECT and optical imaging of EphB4 receptors in prostate cancer xenografts. *Biomaterials* **32**(25), 5872–5879 .
38. E. Hwang, I. I. Smolyaninov and C. C. Davis, Surface plasmon polariton enhanced fluorescence from quantum dots on nanostructured metal surfaces. *Nano Lett.* **10**(3), 813–820.
39. Z. Yang *et al.*, Long-circulating near-infrared fluorescence core-cross-linked polymeric micelles: Synthesis, characterization, and dual nuclear/optical imaging. *Biomacromolecules* **8**(11), 3422–3428 (2007).
40. R. Erber *et al.*, EphB4 controls blood vascular morphogenesis during postnatal angiogenesis. *EMBO J.* **25**(3), 628–641 (2006).
41. S. R. Kumar *et al.*, Preferential induction of EphB4 over EphB2 and its implication in colorectal cancer progression. *Cancer Res.* **69**(9), 3736–3745 (2009).
42. S. R. Kumar *et al.*, Receptor tyrosine kinase EphB4 is a survival factor in breast cancer. *Am. J. Pathol.* **169**(1), 279–293 (2006).
43. G. Xia *et al.*, EphB4 receptor tyrosine kinase is expressed in bladder cancer and provides signals for cell survival. *Oncogene* **25**(5), 769–780 (2006).
44. J. N. Scott *et al.*, How often are nonenhancing supratentorial gliomas malignant? A population study. *Neurology* **59**(6), 947–949 (2002).
45. D. M. Gilmore *et al.*, Effective Low-dose Escalation of Indocyanine Green for Near-infrared Fluorescent SLN Mapping in Melanoma. *Ann. Surg. Oncol.*
46. B. E. Schaafsma *et al.*, Near-infrared fluorescence sentinel lymph node biopsy in vulvar cancer: A randomised comparison of lymphatic tracers. *BJOG* **120**(6), 758–764.
47. B. E. Schaafsma *et al.*, Clinical trial of combined radio- and fluorescence-guided sentinel lymph node biopsy in breast cancer. *Br. J. Surg.*
48. M. S. Bradbury *et al.*, Clinically-translated silica nanoparticles as dual-modality cancer-targeted probes for image-guided surgery and interventions. *Integr. Biol.* (*Camb*) **5**(1), 74–86.
49. O. R. Brouwer *et al.*, Comparing the hybrid fluorescent-radioactive tracer indocyanine green-99mTc-nanocolloid with 99mTc-nanocolloid for sentinel node identification: A validation study using lymphoscintigraphy and SPECT/CT. *J. Nucl. Med.* **53**(7), 1034–1040.
50. O. R. Brouwer *et al.*, Feasibility of sentinel node biopsy in head and neck melanoma using a hybrid radioactive and fluorescent tracer. *Ann. Surg. Oncol.* **19**(6), 1988–1994.
51. T. Buckle *et al.*, A self-assembled multimodal complex for combined pre- and intraoperative imaging of the sentinel lymph node. *Nanotechnology* **21**(35), 355101.
52. A. Bumb *et al.*, Trafficking of a dual-modality magnetic resonance and fluorescence imaging superparamagnetic iron oxide-based nanoprobe to lymph nodes. *Mol. Imaging Biol.* **13**(6), 1163–1172.
53. L. Cheng *et al.*, Multifunctional nanoparticles for upconversion luminescence/MR multimodal imaging and magnetically targeted photothermal therapy. *Biomaterials* **33**(7), 2215–2222.

54. J. S. Choi *et al.*, A hybrid nanoparticle probe for dual-modality positron emission tomography and magnetic resonance imaging. *Angew. Chem. Int. Ed. Engl.* **47**(33), 6259–6262 (2008).
55. X. Huang *et al.*, Long-term multimodal imaging of tumor draining sentinel lymph nodes using mesoporous silica-based nanoprobes. *Biomaterials* **33**(17), 4370–4378.
56. Y. Koyama *et al.*, A dendrimer-based nanosized contrast agent dual-labeled for magnetic resonance and optical fluorescence imaging to localize the sentinel lymph node in mice. *J. Magn. Reson. Imaging* **25**(4), 866–871 (2007).
57. R. Madru *et al.*, 99mTc-labeled superparamagnetic iron oxide nanoparticles for multimodality SPECT/MRI of sentinel lymph nodes. *J. Nucl. Med.* **53**(3), 459–463.
58. C. Niu *et al.*, Doxorubicin loaded superparamagnetic PLGA-iron oxide multifunctional microbubbles for dual-mode US/MR imaging and therapy of metastasis in lymph nodes. *Biomaterials* **34**(9), 2307–2317.
59. P. Elamanchili *et al.*, Characterization of poly(D,L-lactic-co-glycolic acid) based nanoparticulate system for enhanced delivery of antigens to dendritic cells. *Vaccine* **22**(19), 2406–2412 (2004).
60. V. Manolova *et al.*, Nanoparticles target distinct dendritic cell populations according to their size. *Eur. J. Immunol.* **38**(5), 1404–1413 (2008).
61. A. A. Burns *et al.*, Fluorescent silica nanoparticles with efficient urinary excretion for nanomedicine. *Nano Lett.* **9**(1), 442–448 (2009).
62. *PET Imaging of Patients With Melanoma and Malignant Brain Tumors Using an 124I-labeled cRGDY Silica Nanomolecular Particle Tracer: A Microdosing Study (NCT01266096)*. 2010 [cited.
63. L. Johansson *et al.*, Effective dose from radiopharmaceuticals. *Eur. J. Nucl. Med.* **19**(11), 933–938 (1992).
64. J. C. Tardif *et al.*, Imaging biomarkers in atherosclerosis trials. *Circ. Cardiovasc. Imaging* **4**(3), 319–333.
65. L. M. Mitsumori *et al.*, *In vivo* accuracy of multisequence MR imaging for identifying unstable fibrous caps in advanced human carotid plaques. *J. Magn. Reson. Imaging* **17**(4), 410–420 (2003).
66. C. Yuan *et al.*, Identification of fibrous cap rupture with magnetic resonance imaging is highly associated with recent transient ischemic attack or stroke. *Circulation* **105**(2), 181–185 (2002).
67. J. H. Rudd *et al.*, Imaging atherosclerotic plaque inflammation with [18F]-fluorodeoxyglucose positron emission tomography. *Circulation* **105**(23), 2708–2711 (2002).
68. A. Tawakol *et al.*, *In vivo* 18F-fluorodeoxyglucose positron emission tomography imaging provides a noninvasive measure of carotid plaque inflammation in patients. *J. Am. Coll. Cardiol.* **48**(9), 1818–1824 (2006).
69. B. R. Jarrett *et al.*, *In vivo* mapping of vascular inflammation using multimodal imaging. *PLoS One* **5**(10), e13254.
70. J. R. McCarthy *et al.*, Multimodal nanoagents for the detection of intravascular thrombi. *Bioconjug. Chem.* **20**(6), 1251–1255 (2009).
71. M. Nahrendorf *et al.*, Nanoparticle PET-CT imaging of macrophages in inflammatory atherosclerosis. *Circulation* **117**(3), 379–387 (2008).

72. C. M. Ripplinger *et al.*, Inflammation modulates murine venous thrombosis resolution *in vivo*: assessment by multimodal fluorescence molecular imaging. *Arterioscler. Thromb. Vasc. Biol.* **32**(11), 2616–2624.
73. F. A. Jaffer *et al.*, *In vivo* imaging of thrombin activity in experimental thrombi with thrombin-sensitive near-infrared molecular probe. *Arterioscler. Thromb. Vasc. Biol.* **22**(11), 1929–1935 (2002).
74. F. A. Jaffer *et al.*, Molecular imaging of factor XIIIa activity in thrombosis using a novel, near-infrared fluorescent contrast agent that covalently links to thrombi. *Circulation* **110**(2), 170–176 (2004).
75. W. Volker *et al.*, Copper-induced inflammatory reactions of rat carotid arteries mimic restenosis/arteriosclerosis-like neointima formation. *Atherosclerosis* **130**(1&2), 29–36 (1997).
76. B. R. Jarrett *et al.*, Size-controlled synthesis of dextran sulfate coated iron oxide nanoparticles for magnetic resonance imaging. *Nanotechnology* **18**(3), 035603 (2007).
77. R. Weissleder *et al.*, Cell-specific targeting of nanoparticles by multivalent attachment of small molecules. *Nat. Biotechnol.* **23**(11), 1418–1423 (2005).
78. B. Gustafsson, S. Youens and A. Y. Louie, Development of contrast agents targeted to macrophage scavenger receptors for MRI of vascular inflammation. *Bioconjug. Chem.* **17**(2), 538–547 (2006).
79. S. U. Kim and J. de Vellis, Stem cell-based cell therapy in neurological diseases: A review. *J. Neurosci. Res.* **87**(10), 2183–2200 (2009).
80. G. M. Mantia-Smaldone and C. S. Chu, A Review of Dendritic Cell Therapy for Cancer: Progress and Challenges. *BioDrugs*.
81. J. Ye and Y. Yeghiazarians, Cardiac Stem Cell Therapy: Review of the Native Cardiac Progenitor Cells and Future Direction. *J. Cardiovasc. Pharmacol.*
82. L. Cheng *et al.*, Multifunctional upconversion nanoparticles for dual-modal imaging-guided stem cell therapy under remote magnetic control. *Adv. Funct. Mater.* **23**, 272–280 (2013).
83. C. Wang *et al.*, Prostamine functionalized single-walled carbon nanotubes for stem cell labeling and *In Vivo* Raman/Magnetic Resonance/Photoscoustic Triple-Modal Imaging. *Adv. Funct. Mater.* **22**, 2363–2375 (2012).
84. D. L. Thorek *et al.*, *In vivo*, multimodal imaging of B cell distribution and response to antibody immunotherapy in mice. *PLoS One* **5**(5), e10655.
85. D. R. Arifin *et al.*, Trimodal gadolinium-gold microcapsules containing pancreatic islet cells restore normoglycemia in diabetic mice and can be tracked by using US, CT, and positive-contrast MR imaging. *Radiology* **260**(3), 790–798.
86. Hild, W.A., M. Breunig and A. Goepferich, Quantum dots — nano-sized probes for the exploration of cellular and intracellular targeting. *Eur. J. Pharm. Biopharm.* **68**(2), 153–168 (2008).
87. R. G. Blasberg and J. Gelovani, Molecular-genetic imaging: A nuclear medicine-based perspective. *Mol. Imaging* **1**(3), 280–300 (2002).
88. V. Ponomarev *et al.*, A novel triple-modality reporter gene for whole-body fluorescent, bioluminescent, and nuclear noninvasive imaging. *Eur. J. Nucl. Med. Mol. Imaging* **31**(5), 740–751 (2004).
89. B. A. Tannous *et al.*, Metabolic biotinylation of cell surface receptors for *in vivo* imaging. *Nat. Methods* **3**(5), 391–396 (2006).

90. R. Weissleder *et al.*, *In vivo* magnetic resonance imaging of transgene expression. *Nat. Med.* **6**(3), 351–355 (2000).
91. X. Huang *et al.*, Mesenchymal stem cell-based cell engineering with multifunctional mesoporous silica nanoparticles for tumor delivery. *Biomaterials* **34**(7), 1772–1780.
92. W. Hwang do *et al.*, Development of a quadruple imaging modality by using nanoparticles. *Chemistry* **15**(37), 9387–9393 (2009).
93. A. Gianella *et al.*, Multifunctional nanoemulsion platform for imaging guided therapy evaluated in experimental cancer. *ACS Nano* **5**(6), 4422–4433.
94. J. Harrison *et al.*, *In vivo* imaging and biodistribution of multimodal polymeric nanoparticles delivered to the optic nerve. *Small* **8**(10), 1579–1589.
95. H. Xu *et al.*, Polymer encapsulated upconversion nanoparticle/iron oxide nanocomposites for multimodal imaging and magnetic targeted drug delivery. *Biomaterials* **32**(35), 9364–9373.
96. C. L. Peng *et al.*, Multimodal image-guided photothermal therapy mediated by 188Re-labeled micelles containing a cyanine-type photosensitizer. *ACS Nano* **5**(7), 5594–5607.
97. K. Yang *et al.*, Multimodal imaging guided photothermal therapy using functionalized graphene nanosheets anchored with magnetic nanoparticles. *Adv Mater.* **24**(14), 1868–1872.
98. J. Zheng *et al.*, Longitudinal vascular imaging using a novel nano-encapsulated CT and MR contrast agent. *Proc. of SPIE* **6511,** 6511D1–6511D9 (2007).
99. J. Hoisak *et al.*, Improved CT and MR image registration with the introduction of a dual-modality contrast agent: performance assessment using quantitative and information theoretic methods. *Proc. of SPIE* **6914,** 69140Z (2008).
100. B. R. Jarrett *et al.*, Synthesis of 64Cu-labeled magnetic nanoparticles for multimodal imaging. *Bioconjug. Chem.* **19**(7), 1496–1504 (2008).
101. D. L. Thorek and A. Tsourkas, Size, charge and concentration dependent uptake of iron oxide particles by non-phagocytic cells. *Biomaterials* **29**(26), 3583–3590 (2008).
102. Y–C. Tai *et al.*, MicroPET II: design, development and initial performance of an improved microPET scanner for small-animal imaging. *Physics in Medicine and Biology* **48**, 1519–1537 (2003).

Chapter 7

Theranostic Nanopreparations for Medicine

Anthony J. McGoron

Department of Biomedical Engineering, Florida International University
10555 West Flagler St, Miami, FL, 33174, USA
mcgorona@fiu.edu

1. Introduction: Overview of Current Trends in Theranostics

Theranostics is a term that was reportedly introduced around the year 2000 by the now dissolved Pharmanetics Life Science Company to reflect the newly emerging trend of combining therapy with diagnostics.[1] However, the origin of commercialized "theranostics" might be the FDA approval in 1998 of Herceptin® for the treatment of breast cancer and HercepTest® for the diagnosis of the overexpression of Human Epidermal Growth Factor Receptor 2 (HER-2). This example fits the definition of theranostic as a system or process that integrates a diagnostic to test for the presence of molecular targets for which a specific therapy can be applied. The approach could provide improved therapy selection on the bases of specific molecular features of the disease (so-called "personalized medicine"), for greater prediction of efficacy and potential adverse effects and monitoring of the response to therapy.

Several companies have emerged with derivatives of "theranostics" (or "theragnostics" as is also used) in their name. The earliest appearance of the term in PubMed indexed journal was in the year 2000 in an obscure journal.[2] The term was used in 54 papers over the next decade. The journal *Theranostics* was started in 2011 (www.thno.org). Strictly speaking, the technologies that are commonly referred to as "theranostics" as reviewed in this chapter are not by nature "diagnostic" since the use of the formulations intended as theranostics would be used after an initial "diagnosis" has already been made, but would more precisely be used to either confirm a diagnosis during the treatment phase or to guide the therapy and ensure that the therapy is working as intended throughout the course of the therapy, and to generally predict treatment outcomes. The field of image-guided therapy, which is generally used to describe any use of imaging to direct or monitor the course of therapy, for example surgery, radiation, chemotherapy, immunotherapy and others, is more general and includes as a subset so-called theranostics. The imaging used for image guided therapy can be any of the typical methods, nuclear (PET and SPECT), MRI (and fMRI, spectroscopy), CT, ultrasound and the newly emerging area of medical optical imaging. This chapter reviews the history, as well as the state-of-the-art of theranostics, expounding on its advantages and disadvantages, and makes an attempt to predict the future direction and potential for increased clinical applications. This review is comprehensive, but not exhaustive. Due to the large and wide body of knowledge and activity in the field, this review cannot, and is not intended, to be inclusive of all of the exciting work that is being done in this area. Many of the technologies integrated with theranostics are reviewed in greater detail in other chapters of this volume. Apologies in advance for having not included references and contributors that are, nevertheless, important to the development and progress of the field. In fact, new work is added to the field every day.

2. Medical Imaging Applied to Cancer

Medical Imaging, as a contributor to the diagnosis of disease, has a long and rich history and has been reviewed by numerous authors. The introduction of the Anger camera (or gamma camera) in 1957, computed tomography (CT) in 1967, and X-ray CT in 1972 represent mile-stones in the application of medical imaging to diagnosing human disease and by extension, providing invaluable information in the decision making for the management of the patient's therapy. The continued advances in the technologies of computers and computer processing have solidified non-invasive medical imaging into standard patient care. The discovery of radioisotopes and the subsequent efforts in radiochemistry allowed for imaging of disease at the molecular level, rather than the more traditional anatomical imaging available by X-ray CT. Nuclear medicine Positron Emission Tomography (PET) and Single Photon Emission Tomography (SPECT) have increased our understanding of cellular molecular processes and allow *in vivo* and non-invasive (except for the injection of the nuclear tracer) diagnosis, assessment and management of disease. The introduction of Magnetic Resonance Imaging (MRI) has also become a standard in diagnosis of many diseases. While MRI is traditionally considered an anatomical, rather than a molecular imaging modality, functional MRI (fMRI) and nuclear magnetic resonance spectroscopy (NMRS) do provide molecular imaging. Ultrasound, due to its relative low cost, portability, safety and ability for imaging in real time also has an important place in the area of theranostics. Optical imaging is the newcomer to *in vivo* imaging, and has not yet found significant utility in clinical practice in humans. It nevertheless has great potential for *in vivo* molecular imaging. Below we will briefly present the advantages, disadvantages and applications of each modality.

X-ray CT imaging is most typically considered an anatomical imaging modality. Therefore, it lacks the richness in information compared to truly molecular imaging methods which report cellular and tissue function as well as anatomy. CT is commonly applied toward cancer diagnosis, but in some organs, for example liver, the contrast between cancer and normal tissue is poor without the use of an intravenous (*i.v.*) contrast enhancement agent, such as iodine,[3] usually administered in anionic form. It offers high spatial resolution (0.4 mm) and temporal resolution (seconds). If the attenuation (absorption) of the tumor tissue is not sufficiently different from the surrounding tissue, then the tumor will not be visible. A typical diagnostic CT exposes the patients to 10–100 mSv radiation dose, depending on the extent of the body being imaged.[4] While there is no consensus that exposure in this range increases the risk to cancer, repeated exposures should be avoided. Additionally, due to the limitations and identified toxicities and allergies of iodine,[5] and strong thyroid uptake, few drug formulations have been proposed that combine imaging and therapy in CT. Finally, CT is ill-suited to track changes in tissue metabolism so the detection of the effects of treatment may be delayed compared to using more classical molecular imaging, such as with 18F-FDG PET. That is, metabolic response to therapy often precedes anatomical changes.

Nuclear medicine (PET and SPECT) would appear to be ideal molecular imaging modalities and therefore ideal for theranostic applications, but it too has limitations. The spatial resolution of current PET and SPECT cameras are of the order of 5–7 mm, much poorer than some other imaging methods. Temporal resolution of PET and SPECT are also less than X-ray CT (as much as or even more than 5 min per body position). Nuclear Medicine imaging is further hampered by the fact that, in order to keep the exposure to ionizing radiation as low as possible, short half-lived isotopes are required which often negatively impact drug design. Molecules used for nuclear medicine imaging using short half-life isotopes must be labeled immediately *prior* to delivery to the patient, which requires special expertise. So-called "cold kits" have been developed so that the isotope need only be

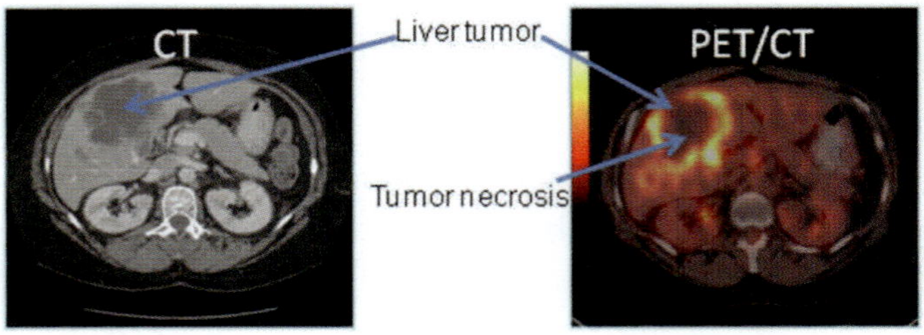

Fig. 1. (left) A CT image of a patient with liver tumor, (right) PET/CT image of the same patient but in this image necrotic core within the tumor is visible.

added to the kit for simple and fast labeling. Many kits have been developed and nuclear medicine remains the most used molecular imaging method with nearly 17 million nuclear medicine procedures done every year in the US[6] using more than 50 different radiopharmaceuticals (over 20 of them based on Tc-99m).[7] Nevertheless, alternatives to nuclear medicine imaging are desired. The main advantage of PET over CT is that PET reports function, while CT is limited to anatomy (Fig. 1). The hybrid PET/CT provides the best of both modalities.

MRI, functional MRI (fMRI) and NMRS represent another set of technologies that could find a place in the area of theranostics. The high resolution and relative safety of MRI has resulted in its expanding use. However, its poor sensitivity requires long imaging times. MRI is being investigated as a molecular imaging method through the use of MRI contrast agents (e.g. gadolinium), though the toxicity of the contrast agents and poor sensitivity has greatly limited their potential use for molecular imaging, particularly for patients with kidney disease.[8,9] It is used in areas of cancer, but with more applications toward diagnosing soft tissue damage and to neural applications. fMRI is not generally used to diagnose cancer because it is limited to measuring changes in blood flow and NMRS, while having much potential, has much too poor sensitivity for clinical diagnostic applications currently. NMRS remains mostly a research tool.

Ultrasound is low cost, safe, portable and provides images in real-time. Its resolution is somewhat limiting, but can be improved by new crystal design and image processing algorithms (0.3 mm at 5 MHz). It has a problem penetrating gas filled regions (lungs) and hard tissue (bone). Ultrasound imaging, except for Doppler flow imaging, is generally considered as an anatomical imaging modality, but can also be functional imaging with the introduction of echo contrast agents, typically echo enhancing bubbles in the micron diameter range. To refer to ultrasound as "molecular" imaging may not be appropriate. A major limitation of ultrasound as a molecular imaging modality is that the ultrasound reflectors must be large compared to the ultrasound wavelength and thus do not escape the vasculature and enter cells. A new method combining ultrasound with optical imaging probes, photo-acoustic imaging, has gained increased attention but its potential role in theranostics is a question yet to be answered. Photo-acoustic imaging does not require endogenous tracers and could potentially deliver real-time scans with very high resolution. The main drawback is a limited imaging depth.[10]

As discussed previously, *in vivo* tumor molecular imaging requires high contrast between tumor and normal surrounding tissues. Currently, the conventional imaging modalities such as MRI, nuclear imaging, CT and US molecularly image tissues with exogenous agents. MRI and CT techniques use substantial (~millimolar) amounts of the contrast agents while both nuclear and optical techniques require only small amounts (~ nanomolar) of the contrast agents. Near-infrared (NIR) fluorescent contrast agents have a potential to generate a strong signal (at least 10^{23} events/s) in comparison to the signal generated by the radioactive tracers ($< 10^4$ events/s) in nuclear imaging. In addition, the use of non-ionizing NIR radiation in optical techniques holds a promise for safe human imaging, with emphasis on quality patient care, in comparison to the use of ionizing gamma radiation in nuclear techniques. NIR light between the wavelengths of 700–900 nm propagates through tissues due to minimal DNA and water absorption[11] and preferential scattering, thus providing a unique approach for molecularly-based *in vivo* imaging.[12] An example of a multifunctional optical/nuclear imaging probe is presented in Fig. 2.[13] With the development of tumor-specific contrast agents, the technology has a potential for clinical translation for various non-invasive imaging applications.

Advantages of optical imaging are potentially high spatial and temporal resolution, high sensitivity, stable imaging molecules (compared to radio-isotopes) allowing for greater flexibility in chemical synthesis times, ability to incorporate into nanoparticle formulations, real-time imaging, low cost and portability. However, optical imaging suffers from poor penetration into, and scatter by, tissue and is limited to only millimeters in depth. This limitation of optical imaging has severely restricted its application in humans. While resolution in optical imaging is degraded due to the high degree of scattering by tissue, computer processing approaches have been developed to improve target location and identification.[14] As stated previously, photo-acoustic imaging is a relatively new technology focus that improves on the depth penetration and spatial resolution of traditional optical imaging.

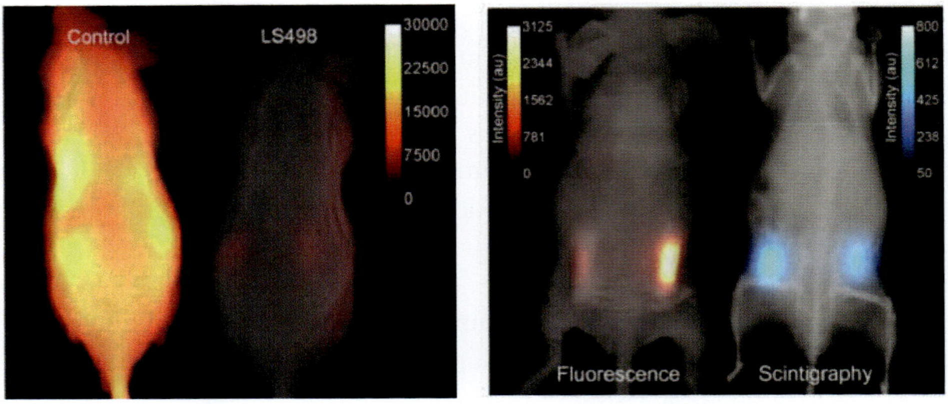

(a) (b)

Fig. 2. Imaging of multifunctional molecular probe in mice. (a) *In vivo* distribution of LS498 (right) and non-quenched control analog (left) in mice at 30 min after *i.v.* injection of the imaging agents. Fluorescence of the activatable probe was efficiently quenched by greater than 10-fold relative to the non-quenched control. (b) Multimodal fluorescence (left) and scintigraphic (right) imaging of a mouse with subcutaneously implanted tubes containing 64Cu-LS498 with BSA (left side) or caspase-3 (right side) 2 h after implantation. Tubes were implanted subcutaneously (about 1 mm) below the surface of the skin (reprinted with permission from Ref. 13).

3. Typical Therapies in Cancer

Having reviewed the available imaging ("nostics") component of theranostics, we now turn our attention to the therapeutic ("thera") component. Most approaches to cancer therapy include one or more of the following: surgery, chemotherapy and radiotherapy. More often than not, multiple therapy approaches are used to achieve maximum outcomes. For example, surgery alone is not able to treat distant metastases, and therefore chemotherapy must be used. Radiation (radiotherapy) is the treatment of choice for localized solid tumors, particularly in cases in which surgery is not an option. In most cases, chemotherapy is used, either alone, or in combination with surgery or radiotherapy in the attempt to ensure that distal metastases and residual cancer is treated. Often multiple chemotherapies are employed. In fact, few patients receive only a single chemotherapy through the course of their treatment.

Many metastatic cancers (e.g. advanced ovarian cancer) often spread to the peritoneal surfaces of the abdomen. Cytoreduction surgery (debulking) refers to the removal of part of a tumor that cannot be completely excised and increases the effectiveness of radiotherapy or chemotherapy. It is time-consuming with high potential for missing small cancerous nodules, but can be greatly facilitated by image guidance.[15–18] The combination of imaging agent with chemotherapy, particularly delivered by nanoparticles, has the potential of improving the outcomes of these late stage cancers.[15,16,18]

Adjuvant therapies refer to approaches that are meant to compliment the primary therapy. The goal is that the combinational therapy will be more effective than the sum of the individual therapies (synergism). For example, hyperthermia or hormone therapy may be used to enhance the response to the primary chemotherapies[19–21] and clinical trials for both are ongoing [www.clinicaltrials.gov]. Concentrating on chemotherapy, or other drug/protein/gene therapies, the delivery of the compound to the site-of-action is of major concern. A longstanding problem with traditional chemotherapy is the lack of targeting and the debilitating effects of off-site toxicities, particularly to bone marrow and the heart. Cardio-toxicity has been a major concern and has limited the total dose of many anticancer drug therapies.[22–25] Nevertheless, the search for a "magic bullet" to target a drug directly to the site-of-action, with controlled and triggered release, has the potential for improving clinical outcomes by more directly controlling the disease and limiting toxicities which will greatly improve the quality of life for cancer patients. Therefore, the response to the application of adjuvant therapies can be uniquely monitored by imaging. This has become the focus of many research groups and drug/biotech companies.

3.1. *Hyperthermia treatment of cancer combined with chemotherapy*

Hyperthermia for cancer therapy is not new.[26] However, technological advances have led to a renewed interest[27] with several methods currently under study clinically [www.cancer.gov]. A temperature of 42°C is lethal to cancer cells but not normal cells, probably because of protein denaturation, poor DNA synthesis and repair due to the comparatively lower oxygen level and pH of cancer cells[27] and likely other mechanisms as well. For example, hyperthermia may irreversibly damage cancer cells due to the inhibition of oxidative metabolism resulting in lower cellular pH. In addition, hyperthermia may promote cell death by inducing cell membrane and cytoskeleton damage, and enhanced apoptosis. Nevertheless, hyperthermia alone has not found a place in routine clinical practice, probably because of the difficulty in selectively and locally applying the heat and to non-invasively monitor the heat delivery. The synergistic effect of hyperthermia and chemotherapy (for example, adriamycin) has long been demonstrated *in vitro* and *in vivo*.[28–32]

The efficacy of hyperthermia combined with radiotherapy and/or chemotherapy has been shown clinically.[29,33,34] In combination with doxorubicin, a temperature of 43°C for 30 min reduced cell cancer survival in culture.[29] Doxorubicin uptake by cancer cell culture was increased following 15 min at 45.5°C,[30] which may be a mechanism of increased sensitization to chemotherapy. Cell killing was significant after 3 min of heating, but heating longer than 30 min[32] showed little or no enhancement. Hyperthermia can be induced in a number of ways, including whole body, regional or local, intraluminal or interstitial.[35] Direct heating of tissue can be accomplished by interstitial therapy, delivered directly to tumor tissue, which can be accomplished using laser fibers, ultrasound or microwave. But none is selective to the tumor cells. Besides, none of the procedures have come into general clinical use primarily because the specific and controlled heating of the tumors is difficult. Clinical trials are being conducted (www.cancer.gov). For example, liposomal doxorubicin, which prolongs plasma residence time, combined with hyperthermia has been shown to improve efficacy and reduce normal tissue toxicity.[20,36–38]

3.2. *Photothermotherapy*

Light in the NIR portion of the electromagnetic spectrum, particularly in the range 700–900 nm (referred to as the therapeutic window), avoids both DNA and water absorption and thus transmits deeper into tissue than visible light.[11] This characteristic allows for *in vivo* imaging of tumors that have accumulated targeted drugs that absorb and fluoresce in the NIR.[12] Indocyanine Green (ICG) is a tricarbocyanine type dye with NIR absorbing properties (peak absorption around 800 nm) that is approved as a diagnostic aid for blood determination of blood volume, cardiac output, liver function, visualization of retinal and choroidal vasculature and for photodynamic therapy (PDT). The main advantage is that its maximum absorption lies near the isobestic point of hemoglobin and oxyhemoglobin, it is confined to the vascular compartment because of binding to plasma proteins, and has very low toxicity. A recent review of the literature exploring the use of ICG in surgery attests to a continued interest.[39]

Some dyes that absorb energy in the NIR region can also release heat following exposure to the appropriate wavelength light, and thus kill or irreversibly damage cancer cells. With the dye incorporated into the drug delivery vehicle, therapy is image-guided because the drug is not activated until it has reached its intended target. ICG's ability to generate heat upon exposure to NIR illumination has been known for years,[40,41] but it was determined that a single treatment would not destroy all of the cells in a treated tumor. This may be addressed by optimizing the treatment parameters, employing multiple treatments, and/or combining heating with other modalities such as chemotherapy. Figure 3 shows the hyperthermia effect of two different NIR absorbing dyes and Fig. 4 demonstrates the synergistic effect between hyperthermia and chemotherapy.[314,326] Using phantom studies to model the treatment parameters, it was found that a 808 nm laser with a 1 cm beam radius and a power of 0.85 W for 12.5 min raised the temperature of a gel containing 0.27% ICG located 1 cm deep into chicken breast more than 6°C.[42] The same group later performed additional phantom studies and modeling and concluded that the optimal concentration and power combination is 0.083% and 0.97 W, respectively for a 12.5 min exposure.[43] However, drug delivery and selectivity continues to be obstacles, because 13 min continuous laser exposure does not seem practical.

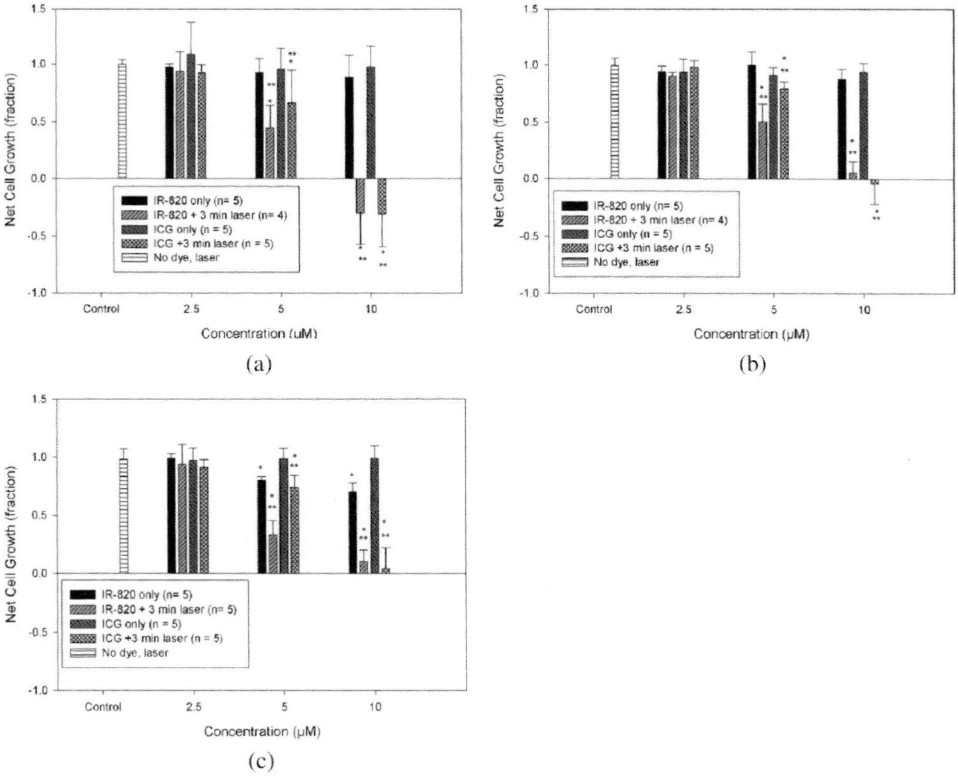

Fig. 3. IR820 and ICG cytotoxicity and hyperthermia effects in SKOV-3 (a), Dx5 (b), and MES-SA (c) cells. Net cell growth was normalized to a control group that was not exposed to either dye or laser. *$p < 0.05$ versus the control group. **$p < 0.05$ versus the same concentration of dye without laser.

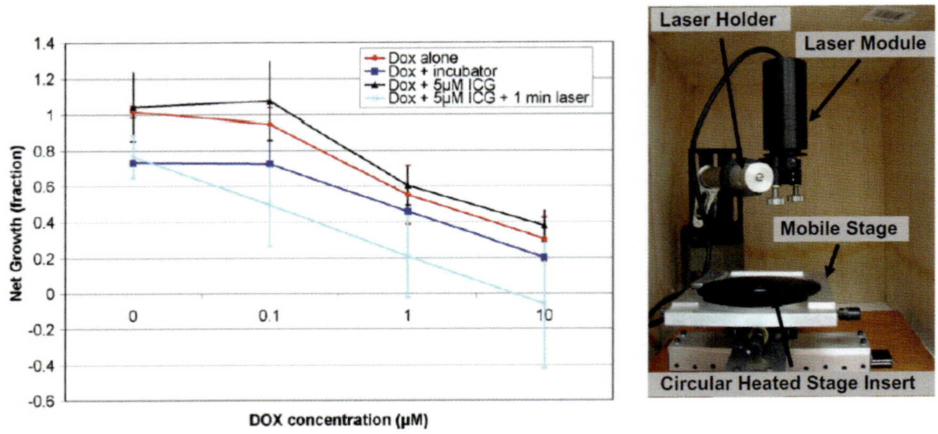

Fig. 4. Demonstration of the synergistic effect of NIR dye/laser induced hyperthermia with doxorubicin chemotherapy in SKOV-3 ovarian cancer cells. The image on the right is the laser system used to excite the dye loaded into the cells.

The remainder of this review will summarize the efforts toward directing drugs to their intended target while using imaging to monitor the drugs distribution and the response of the disease to the therapy.

4. Drug Delivery Methods

Having introduced the clinical need and current state of standard care of cancer patients, we now examine the approaches for drug delivery, i.e. administration and distribution of the drug therapy.

The choice of drug administration route is influenced by numerous factors, such as safety, required speed of drug action, the target for the drug and the possible need to avoid the first-pass hepatic clearance or metabolism. Drug administration is most typically enteral (gastrointestinal, either oral or rectal), parenteral (other than gastrointestinal, e.g. *i.v.*, subcutaneous (s.c.), intramuscular (i.m.), inhalation, direct placement on tissue) or transdermal (across the skin). Other forms include ophthalmic, inhalation, sublingual and others. The route of delivery chosen depends on the characteristics of the disease, the characteristics of the drug (and the drug vehicle) and the speed at which the therapeutic response is required. The safest mode of drug delivery is in the form of a topical solution or via a patch on the surface of the patient, but this is limited to small molecules due to the diffusion barrier of the skin. The most direct drug delivery, but perhaps the least safe, is directly on the target tissue. Direct application to tissues requires surgery and may not affect distant metastases in the case of cancer therapy. Both approaches are in commercial application for various indications. However, these approaches have not been combined with imaging to produce a theranostic. Direct injection of drugs to tumors has been investigated in pre-clinical studies, but this approach does not appear practical for clinical use as it requires surgery, results in an uneven distribution of the drug within the tumor, and also does not address distant metastases. The most common approach, and seemingly the most practical for drug delivery in cancer that will also reach distant metastases, is *i.v.* administration. *I.v.* administration will provide for a wider distribution of the drug and fast bioavailability. The limitation is that the drug is then able to interact more easily with non-target tissue and may be eliminated by the kidneys and liver. If the drug is in particle form, it is likely to be scavenged by the reticulo-endothelial system. This chapter will concentrate further on intravascularly administered nanopreparations.

Many of the drugs that have been found to be active against cancer are poorly soluble and/or unstable in water (or plasma) which presents a significant challenge for intravascular delivery. Such highly hydrophobic compounds are often opsonized by plasma proteins and removed by the liver or macgrophages of the reticuloendothelial system (RES). In addition, many compounds intended for neurological diseases are poorly transported across the blood–brain barrier. Polymer nanoparticle drug delivery is particularly attractive for the delivery of poorly soluble drugs because they can serve as a carrier that is soluble in water, can protect the active drugs from degradation, provide for controlled and prolonged release and increase the circulation time and thus bioavailability of the drugs, which can decrease their administration frequency, all for improved therapeutic effect. In addition to the critical physico-chemical and biocompatible requirements of candidate monomers for drug delivery, cost and availability are also important. Practical drug delivery systems are in fact often polymers, and more often copolymers to combine the advantages of different monomers. The choice of polymer systems must be closely matched to the drug to be delivered. In the case of water soluble drugs, such as proteins (for example insulin), hydrophilic polymers or liposomes, which contain a lipid exterior but an aqueous interior, could be chosen. The aqueous interior of liposomes helps to maintain the integrity of the drug while the lipid exterior allows the drug to cross membrane barriers.

We now turn our attention to attempts at designing drug delivery vehicles that can carry high payload of drug, deliver them to their intended site, avoid removal from the circulation before they reach their target, enter the target cells, and release their payload at the specific site for their therapeutic action. Design considerations include biocompatible, biodegradable, size, charge, solubility, surface chemistry, etc. For the practical design of a drug carrier, these factors must be optimized for the drugs being delivered, as well as the intended tissue and cellular target. Systemically administered nanoparticles have been shown to passively accumulate in a number of tumors. The reason for this seems to be the high leakiness of tumor blood vessels and the general lack of a lymphatic system in most tumors, which combine to generate the enhanced permeation and retention effect (EPR), and possibly due to a higher endocytosis of tumor-associated cells in the tumor's blood vessels.[44–46] Small (> ~30 but <~150 nm), neutral charge and hydrophilic coating seem to be common prerequisites for successful vascular delivery of drugs for cancer. Very small particles can be easily removed by the kidney and larger particles removed by the RES and liver Kupffer cells. Positively charged and hydrophobic particles tend to be opsonized in the plasma for recognition by resident macrophages and experience very fast plasma clearance. But at the cell, a positive charge tends to improve cellular transport due to the negatively charge cell membrane. The particular chemical composition of the drug carrier material must be matched to the drugs being delivered, and surface decoration is important for specific and active targeting. We will concentrate on drug carriers small enough to extravasate into the tissue and cross cell membranes (i.e. nanoparticles and macromolecules). The properties of various materials that have been investigated for drug delivery applications have been reviewed extensively and, therefore, will be only briefly described here.

4.1. *Organic nanoparticles*

It is not possible to thoroughly review all of the different polymers used for drug delivery, and even less practical to include the many groups that are active in the field. Complicating the issue is that most formulations are of copolymers and not necessarily of single polymers. We will attempt to review some of the more current literature, concentrating when possible on those formulations in clinical trial or commercially available. Natural and synthetic polymers are being studied extensively as drug carriers for cancer therapy. Table 1 lists some of the various formulations reported. The list is not meant to be exhaustive. Lists of nanoparticle drug formulations in clinical use (fully approved or in clinical trials) are available elsewhere[47–50] as well as in this volume.

Natural hydrophilic polymers are intrinsically biodegradable and biocompatible and are therefore attractive for drug delivery applications. They typically possess various functional groups allowing chemical modification for a

Table 1. Natural and synthetic polymers used for nanoparticle formulations.

Natural Polymers	Synthetic Polymers
Proteins	Poly(methyl methacrylate)[85,87,120]
Gelatin[53,88,96,121–123]	Poly(vinyl alcohol) (PVA) [124]
Albumin[53,78,92,125–128]	Poly(acrylic acid) [89,129]
Collagen[51,130,131]	Polyacrylamide[132]
Gums	Poly(ethylene glycol) [122]
Acacia[60,133,134]	Poly(methacrylic acid)[87]
Polysaccharides	Polylactides (PLA) [77,116,135,136]
Starch[54,137]	Polyglycolides (PGA) [138–139]
Dextran (primarily as a coating for stabilization or conjugate) [54,95,96,141–147]	Poly(lactide-co-glycolides) (PLGA) [71,72,148–152]
Hyaluronic acid[153–156]	Polyanhydrides[79,80]
Alginates[53,54,95,96]	Polyorthoesters[157,158]
Chitosan [53,54,95,96,114,115,141,159–163]	Polycyanoacrylates[164]
	Polycaprolactone[117]

wide-array of nanoparticle formulations and for conjugation of targeting and imaging moieties. Naturally derived polymer colloid systems for drug delivery have a long history.[51] Early on, controlling stability (particle aggregation) and size were challenges and efforts continue to deal with identifying the optimal formulation parameters, such as polymer and polymer ratios, temperature, pH, types and concentrations of solvents, surfactants and stabilizers, mixing and stirring rates. Of the natural polymers, chitosan has been widely investigated. It is obtained by the deacetylation of chitin, from the exoskeleton of crustaceans, and is commercially available. Its properties (i.e. charge, particle formation and potential to aggregate) are mainly determined by its MW and degree of deacetylation. Chitosan is soluble in most organic acidic solutions at pH below 6.5.[52–54] At or above physiological pH precipitation occurs. However, it can be glycolsilated for stability at physiological pH. Chitosan can be made more hydrophobic, and therefore more stable, by the covalent attachment of hydrophobic moieties. Chitosan is being investigated as a drug delivery vehicle for chemotherapy, proteins and siRNA for vascular/systemic applications. However, it binds quickly to plasma proteins, inducing clotting mechanisms, and so its approved biomedical use has been in bandages and other hemostatic agents.[55] It is capable of complexing with DNA through electrostatic interactions making it potentially useful for gene delivery.[56,57] It can bind to epithelial membranes making it attractive for applications requiring delivery across mucosal epithelia (oral or nasal administration routes) for efficient delivery of siRNA and proteins such as insulin and vaccines[54,56,58–60] or other applications in which epithelial transport is required.

The most attractive and commonly used biodegradable polymers to date are polyesters such as poly(lactic acid) (PLA), poly(glycolic acid) (PGA) and their copolymer poly(lactic-co-glycolic acid) (PLGA). Polyesters are synthetic polymers produced primarily by the reaction of dibasic acids with dihydric alcohols. PLA, one of the polymers in the co-polymer PLGA is found in a clinical application of drug delivery for cancer.[127] Nanoxel® docetaxel-loaded methoxy-poly(ethylene glycol)-block-poly(D,L-lactide) (mPEG-PDLLA) micelle[61] is a product in clinical trials being evaluated as a Radio-Sensitizer for locally advanced head and neck carcinoma.[62] Genexol-PM™ is another PEG-PLA polymeric micelle with paclitaxel in clinical trials for advanced malignancies.[63,64]

Possibly the most investigated synthetic polymer for nanoparticle drug delivery is the co-polymer PLGA. Introduced in 1970s for applications in resorbable sutures, it has long been considered biodegradable, biocompatible, with a low toxicity profile. Recent reports on PLGA as a nanoparticle drug carrier are too numerous to cite but several reviews are available.[65–69] Numerous commercial sustained drug release products are available that use PLGA,[70] mostly for oral, IM or SC delivery. A disadvantage of PLGA, and all polymer drug carriers is that they are limited by their solubilities for the drugs that can be efficiently loaded, their biphasic release profile, fast initial release followed by very slow extended release, and very slow degradation. Of course, these characteristics are dependent on the polymer ratio in co-polymer formulations (e.g. PLA and PGA in PLGA). Therefore, many different formulations of PLGA have been reported and reviewed. Our own group has developed a PLGA nanoparticle formulation that can encapsulate both chemotherapy drug (doxorubicin, DOX) and NIR imaging agent (ICG or IR820).[71,72,140,379] As stated previously, slow release and poor control of the drug release from the nanocarrier continues to present a problem for practical clinical applications. Further, DOX and ICG have different solubility characteristics which will affect their loading and release. The release of DOX from PLGA nanoparticles is much slower than the release of ICG. Numerous hydrophobic drugs have been successfully incorporated into PLGA.[69,73–75] Bind Biosciences is conducting a phase I trial of BIND-014, PEG-PLGA nanoparticles that contain the chemotherapeutic agent docetaxel that targets prostate-specific membrane antigen (PSMA). However, most FDA approved PLGA based drug delivery systems are based on microparticles injected im or sc to create a slow releasing depot, for example Lupron Depot® for the treatment of advanced prostate cancer.[70,76,77] Three recent reviews describe nanoparticle formulations undergoing clinical trials.[49,50,78]

Polyanhydrides are an important class of biodegradable polymers which are closely related to the polyesters. Polyanhydrides have been approved by the FDA as a surgical patch implant (Gliadel®) for delivery of Carmustine to the brain. Based on their favorable toxicity profile they have also been investigated as nanoparticle drug carriers.[79–81] Other hydrophilic polymers have been used to generate nanoparticles for drug delivery, for example many incorporate methacrylates and acrylates.[82–85,85–88] Poly(acrylic acids) have been used as coatings for inorganic nanoparticles (gold) to improve colloidal stability[89] and for temperature and pH-sensitive, controlled delivery of hydrophilic and hydrophobic drugs due to their swelling behavior in aqueous environments.[90,91] Most synthetic polymers tend to have a net negative charge. Highly charged particles tend to be cleared quickly from the vasculature. Other synthetic polymers such as Poly(ethylene glycol) (PEG) have been used extensively for surface modification of nanoparticles to render various nanoparticles stealth.[92] PEG induces a steric stabilization of the nanoparticles by creating a hydrophilic shell around the nanoparticle and has been used to stabilize many different organic and inorganic nanoparticle formulations. It masks nanoparticle charge and provide "stealth" properties. Doxil™ is a pegylated liposome loaded with doxorubicin for the treatment of breast cancer and Kaposi's sarcoma.[93,94] Doxil™ has been shown to increase plasma residence time and reduce cardiotoxicity compared to administration of free doxorubicin.[22]

Hydrogels are another very interesting class of materials with great potential for drug delivery. They are three-dimensional polymer networks of water-insoluble polymer chains that exhibit a semisolid morphology and are superabsorbent (containing over 99% water).[90,95–98] Hydrogel particles, flexible due to their high water content, are highly porous permitting loading of drugs into the gel matrix and subsequent drug release, which is dependent on the drug's molecular diffusion coefficient. Numerous natural and synthetic polymers have been formulated into hydrogels with different characteristics for drug delivery.

Most marketed nanoparticle based drug delivery formulations are based on liposomes (e.g. Doxil®, DaunoXome® and Myocet® doxorubicin loaded nanoparticles for treatment of multiple myeloma, refractory breast and ovarian cancer and Kaposi's sarcoma) or albumin (e.g. Abraxane®, albumin-bound paclitaxel nanoparticles for the treatment of metastatic breast cancer). Liposomes are particles made of natural phospholipids with a bi-lipid membrane and an aqueous core, most suitable for delivering hydrophilic drugs.

Liposomes are formed typically by extrusion to around 100 nm, with a very narrow size distribution, the size determined primarily by the extrusion filter pore as well as thermodynamic stability. Liposomes by nature are very biocompatible, but tend to be less stable than polymer nanoparticles. A table of liposome formulations available on the market for clinical use can be found elsewhere.[47] One can also find details of ongoing clinical trials using nanoparticles specifically in cancer at www.cancer.gov/clinicaltrials.

Polymersomes are from a relatively new class of nano (and micro) particles created through self-assembly of amphiphilic synthetic block copolymers. They are similar to liposomes, nanoparticles formed from naturally occurring lipids, but with a polymer instead of a phospholipid bilayer.[99–102] Both liposomes and polymersomes have an aqueous core and therefore are able to carry hydrophilic drugs. However, polymersomes are reported to exhibit increased stability and reduced permeability compared to liposomes. The use of synthetic polymers also gives designers greater flexibility in modifying the characteristics of the polymersome's membrane for better control of permeability, release rates, stability and other properties. Another very interesting class of nanoparticles is a micelle, which contains a hydrophilic head region in contact with a physiological aqueous environment, but a hydrophobic tail in the micelle center.[47,103–105,105–117] These nanoparticles are very efficient in delivering poorly water soluble drugs. Examples of micelles approved for clinical use are reviewed elsewhere.

Many of the polymers being evaluated for nanoparticle drug delivery applications were first developed as scaffolds for tissue engineering. However, the physico-chemical properties are different between tissue and nanoparticle drug delivery applications. In particular the mechanical properties and degradation requirements are different. Our own group is evaluating a new class of polymers which have been modified from a polymer originally designed for tissue engineering (poly-glycerol-dodecanedioic, PGD[118,119]). The new polymers include poly-glycerol-malic acid-dodecanedioic (PGMD) and poly-glycerol-citric acid-dodecanedioic (PGCD) for their potential in nanoparticle and microparticle drug delivery.[379] Modifications were made to allow for a wider range of sizes, glass transition temperatures, degradation rates, hydrophilicity and hydrophobicity etc.

4.2. *Inorganic nanoparticles*

Inorganic nanoparticles are very attractive for theranostic applications. They are often formulated with an inorganic core and an organic shell as the outer most part of the particle. Depending on the inorganic core composition and structure, the particles may be "tuned" for various applications, such as detection by imaging (absorbance, fluorescence, MRI and spectroscopy, X-ray CT), magnetic susceptibility (for magnetic control), heat generation (for hyperthermia or triggered drug release), drug delivery, etc. The ability to control size, size distribution and shape is in general greater than for polymer particles. The inorganic material confers stability and also provides functional groups for conjugation of drugs, targeting agents (antibodies, aptamers, peptides small molecules), and additional imaging agents for multifunctional applications. Inorganic nanoparticles developed for medical applications have a long history and can generally be made smaller than polymer nanoparticles. Nevertheless, the safety of inorganic nanoparticles continues to be a concern and the source of much debate.[165,166]

Due to their ability to shorten T2 relaxation times, the use of iron oxide nanoparticles, primarily in the form of magnetite (Fe_3O_4) or maghemite (Fe_2O_3), was first proposed around 1986 as contrast for MRI of the liver, spleen and bone marrow[167,168] and was approved by the FDA and commercialized in 1996 (Feridex®) as a tissue-specific MRI contrast agent. Super paramagnetic iron-oxide (SPIO) nanoparticles continue to attract considerable attention.[48,137,161,169–175] However, in 2008 the company AMAG Pharmaceuticals ceased manufacturing and marketing Feridex®(www.amagpharma.com). Another SPIO product developed by AMAG Pharmaceuticals for lymph node imaging (Combidex®) has also been discontinued. Bayer Healthcare's (healthcare.bayer.com) SPIO Resovist® for liver imaging was abandoned in 2009.

In addition to MRI imaging applications, iron oxide nanoparticles have also been evaluated for hyperthermia, either by magnetic or radiofrequency (RF) excitation.[176] Hyperthermia applications are reviewed in greater detail below. AMAG's Feraheme™ (Ferumoxytol), a SPIO being approved as an iron supplement,[177] is also being tested as a MRI contrast agent.[175] Despite the known toxicity of the widely used gadolinium (Gd) MRI contrast agent in patients with kidney disease,[8,9] there does not appear to be a FDA approved SPIO for MRI on the market at this time. Perhaps because they do not outperform current Gd based MRI contrast agents or due to manufacturing difficulties. To this end Osborne proposed a simple, rapid and reproducible method for the synthesis of dextran-coated iron oxide nanoparticles for MRI by rapid microwave-assisted synthesis.[171]

Gold, due to its inertness and surface functionalization (ability to bind amine and thiol groups) has attracted tremendous interest for drug delivery, imaging and theranostics and various forms are in clinical trials.[134,146,178,178–184] A strong broad size-dependent localized Surface Plasmon Resonance (SPR) absorption is observed in the spectra of many metallic nanoparticles, in particular those of gold and silver.[176,185] This phenomenon can be employed for imaging (fluorescence and Surface Plasmon Raman Spectroscopy) as well as photothermal heating of gold nanoparticles.[178,180,186–188] However, the potential clinical applications of photothermal heating are limited due to the restricted penetration of light in tissue as described above. Gold nanoshells can also produce heat in the presence of RF radiation.[176] Taking advantage of a unique optical property of gold nanoparticles Lukianova-Hleb demonstrated a novel cell-level transient phenomenon, the plasmonic nanobubble (PNB). When triggered by the short pulsed optical heating of gold NPs the PNB has a threshold sensitive to multiple variables, including NP size and aggregation state, to generate a super-heated region close to the PNB.[189] Gold can serve as an effective contrast agent for X-ray micro-radiology and micro-tomography as demonstrated in animal models.[190] The interest in nanoparticles as X-ray contrast agents is largely due to the known toxicities of iodine contrast media, especially in patients allergic to iodine, and iodine's unfavorable biokinetics.[5,191] Gold can also be used for combined imaging and hyperthermia, via RF field induced heat generation hyperthermia, and chemotherapy delivery.[180]

Single walled carbon nanotubes (SWCNT) are being investigated primarily for their potential in sensors, but are also interesting as possible theranostic agents including drug delivery, imaging and photothermal heating.[18,176,178,192–194] They are basically hollow cylindrical molecules of carbon atoms, open or capped at the ends, with high aspect ratios of as small as 1 nm in diameter to as much as several cm in length. In general, the particles are non-toxic and hydrophobic and can be made water soluble by surface modifications reducing the tendency to agglomerate. Surface modification can also be exploited for targeting.[193] They possess strong optical absorbance in the NIR which has been employed for optical imaging and photothermal ablation.[18,176,178] SWCNTs have a low fluorescent quantum yield relative to fluorescent dyes and gold and concerns persist regarding their long-term toxicities, particularly if inhaled.[195] SWCNTs are at a much earlier stage in terms of their consideration as theranostic agents than polymer materials or other inorganic materials. Nanodiamonds represent another exciting class of carbon based vehicles for drug delivery and optical imaging.[196–199] They have been shown to be highly biocompatibility, with relatively simple synthesis and allow for the surface-mediated binding of a wide variety of bioactive molecules. High binding affinity may result in slow drug release however.

Mesoporous silica (silicon dioxide) and ceramic nanoparticles are very attractive for use as a nano-drug carrier due to their tunable porosity, large pore volume, large surface area to volume ratio (high loading capacity), biocompatibility, biodegradability, relatively easy synthesis in very controllable size and size distribution, can be synthesized very small, are highly stable and their surface can be easily biofunctionalized.[155,182,200–208]. Based on these characteristics, they are perhaps the most practical inorganic material for nanoparticle based combined drug delivery and imaging as reflected in the large number of recent papers. Combinations of inorganic or inorganic and organic materials for theranostic applications are also very interesting. Superparamagnetic iron oxide nanocrystals were encapsulated inside mesostructured silica spheres which were labeled with fluorescent dye molecules and then coated with hydrophilic groups to prevent aggregation.[209] The Gold–Silica combination provides the high drug carrying capacity of silica with the optical properties of gold.[178] Nanospectra (www.nanospectra.com) is commercializing a silica core/gold shell nanoparticle for NIR heat activated ablation of tumors that continues to be studied in mice[188] but also in an ongoing clinical trial.

4.3. *Examples of nanoparticles responsive to external stimulus*

A significant problem associated with nanoparticle drug delivery is ensuring that the drug is not released prematurely (e.g. in the plasma or interstitial space if the target is intracellular) but is released at an appropriate rate once at the intended site (e.g. into the cell cytoplasm). The triggered release of drugs from nanoparticles has recently been reviewed.[126] To achieve this goal, nanoparticles have been formulated to allow for triggering of drug release due to externally applied temperature,[210,211] ultrasound,[212,213] intracellular pH,[102,110,214] or intracellular enzymes.[208,215,216] Wong[121] proposed a multistage nanoparticle that will reduce in size upon reaching the tumor microenvironment allowing for smaller drug carriers to then move more easily through the tumor. Given the fact that the ideal physico-chemical properties are different when the carrier is in the plasma, the interstitial space, and the intracellular space, a NP with multiple stages, each with a characteristic optimized to a particular stage of the path between administration and target is a very attractive approach.

4.4. *Nanoparticles for hyperthermia*

Compared to conventional forms of inducing hyperthermia for cancer therapy (i.e. whole-body or regional) localized hyperthermia with an energy absorbing (heat generating) agent has the advantage of a shorter time to target temperature, better control, and less impact on normal tissue which

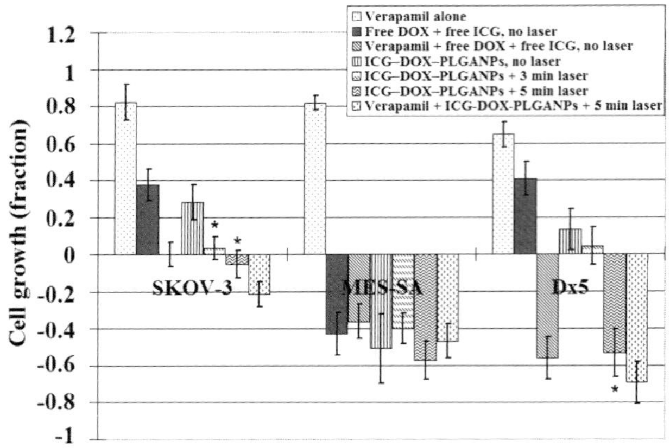

Fig. 5. Cytotoxicity of ICG-DOX-PLGANPs when excited by NIR laser. $n = 3$ experiments, 4 wells per treatment. The concentration of ICGDOX- PLGANPs was around 0.25 mg/ml, which contains 10 μM DOX and 6.2 μM ICG. $*P<0.05$ (by paired t-test) between laser-treated and non-treated cells of ICG-DOX-PLGANPs, indicating significant cytotoxicity difference due to hyperthermia.

does not contain the agent. Targeted local hyperthermia can be achieved by using chemical or metallic nanoparticles that convert electromagnetic energy to heat. Nanoparticles have been developed for localized photothermotherapy using NIR active dyes, gold and carbon nanotubes, magnetic fluid hyperthermia using iron nanoparticles, and RF ablation using gold or carbon nanotubes, and have shown promise in cells and animals.[18,169,187,217–231] These techniques are either non-invasive or minimally invasive. Figure 5 shows the effects of PLGA NPs loaded with DOX and ICG in a drug sensitive uterine cancer cell line (MES-SA) and its multidrug resistant derivative Dx5.[72]

Magnetic-particle hyperthermia is challenged by the efficiency of coupling of the field to the particles and the large amount of particles required for sufficient heating[219,232] relative to nanoparticles required for RF and photothermotherapy. RF heating has a disadvantage that capacitive RF fields can cause excessive heating in fat layers.[233,234] A limitation of photothermotherapy is that NIR light has less tissue penetration than a localized magnetic field used to heat magnetic nanoparticles[221,235] or RF fields to heat metal nanoparticles.[221,236,237] Another limitation of photothermal dyes is that they generally produce less heat when exposed to the NIR light than do gold nanoparticles.[187,225] But they have the advantage that they can easily be co-delivered with chemotherapy drugs (e.g. DOX as our group has shown) and can also be imaged using low cost optical cameras for image guided therapy. Combining metal nanoparticles in polymers (such as PLGA) with chemotherapy would be difficult because there would be little room left in the carrier for the chemotherapy drug; and, there is persistent apprehension regarding the long term effects of systemic administration of metal nanoparticles due to the still unresolved potential for toxicities. This is less of a concern for PLGA nanoparticles loaded with dye and DOX because they are completely biodegradable.

5. Targeting (HER-2, Transferrin, PSMA, Folate, CD44, Peptides, Etc.)

Active cell-targeted drugs can be obtained by attaching antigen-specific monoclonal antibodies or small proteins and molecules (folate, epidermal growth factor, PSMA, transferrin, CD44, interleukin –4 binding peptides, etc) to liposomes and other nanoparticles.[128,148,153,154,160,182,202,238–243] To be therapeutically effective, an encapsulated drug must be released either on or near the cell surface or inside the cell. It may be necessary that the complex be transported into the cell before the toxic portion of the drug is released from the delivery vehicle. The competing mechanisms that may be required to achieve these processes are described in an excellent review.[244] For instance, the choice and binding location of ligand (antibodies, antibody fragments or other tumor specific markers) on the drug complex may optimize target recognition but may not be optimal for prolonged vascular residence time. Alternatively, the vascular residence time may be optimized at the expense of allowing sufficient transport across the endothelial layer, or the rate at which the toxic element of the complex escapes the vehicle. In such a case, the toxic portion of the drug (chemotherapy agent) may be released systemically prematurely, rather than at or inside the target cell. Therefore, it is necessary to incorporate image guidance capabilities into the drug to verify that the drug has reached its target before the drug is activated.

Researchers have found that many aggressive cancers, among them ovarian, breast, oral and brain cancers, have high levels of folate receptors.[242,245,246] The receptors, located on the surfaces of cancer cells, are highly specific for binding folate. Such cells often express more folate receptors in secondary (metastasized) sites formed once the cancer becomes advanced throughout the body. Folate-mediated targeting in cancer and the coupling chemistry of folate and peptide to nanoparticles has been thoroughly reviewed.[247–251] Folate has been labeled to nanoparticles of albumin,[128] silica,[238] adenovirus,[239] dendrimers,[252] liposomes[253] and others. Liposome-entrapped doxorubicin pegylated to folate successfully targeted epithelial cancer cells.[253] Magnetite nanoparticles were targeted to BT-20 human breast cancer cells via the folate receptor.[254] More recently poly(3-hydroxybutyrate-co-3-hydroxyoctanoate) was used as a drug carrier modified with folic acid targeting the folate receptor.[255] Folate has also been successfully labeled for imaging with Tc-99m by various chelation techniques to actively target human KB epithelial carcinoma cells *in vitro* and *in vivo*.[256,257] Several other strategies to target cancer cells with PEG-liposomes attached to markers have been described. Biotin-targeting strepavidin,[258] monoclonal antibodies,[259–261] cytokines[262–264] and peptides[265] have been applied in liposome delivery to cancer. Liposome-encapsulated doxorubicin targeting of the CD44 cell-surface marker that binds to the carbohydrate hyaluronan was demonstrated.[266] Hyaluronic acid-ceramide based self-assembled nanoparticles have also been developed for *i.v.* docetaxel delivery.[153]

Human epidermal growth factor 2 (HER-2) is over expressed in 20%–30% of ovarian and breast cancers.[267] It plays an essential role in proliferation and antiapoptosis mechanisms in HER-2-positive breast cancer.[268] Tumor targeting using anti-HER-2 immunoliposomes loaded with doxorubicin has been successful in a mouse tumor model.[267] The HER-2 monoclonal antibody to antigen receptors on cancer cells can be conjugated to PLGA nanoparticles for targeted delivery.[240] Our group has demonstrated that HER-2 targeting using DOX loaded PLGA particles improves cellular uptake by HER-2 positive cells.[140,148] Figure 6 shows the effects of DOX and ICG loaded PLGA NPs with and without laser exposure on three different cancer cell lines, one of which (SKOV-3) expresses the HER-2 receptor. Jie[159] employed a novel peptide sequence-conjugated to magnetic nanoparticles for imaging in HER-2 positive ovarian cancer cells (SKOV-3) in mice. Chitosan-doxorubicin nanoparticles conjugated to Trastuzumab differentiate HER-2 positive from HER-2 negative cells *in vitro*.[163] Peptides have become particularly important for theranostic applications due to their ability to facilitate transport of particles and molecules (drugs or imaging agents) across the blood–brain barrier

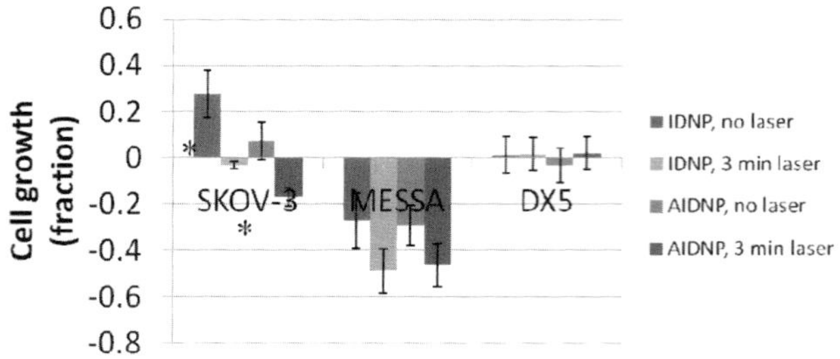

Fig. 6. Cytotoxicity of ICG+DOX PLGA NPs (IDNP) and HER-2 conjugated ICG+DOX PLGA NPs (AIDNP) excited by NIR laser. $*P < 0.05$ (by paired t-test) between laser treated and non-treated cells, indicating significant cytotoxicity difference due to hyperthermia.

(BBB). Yan demonstrated that tumor vasculature targeted cyclic (RGDyK) peptides and BBB-permeable Angiopep-2 peptides labeled on PAMAM-G5 dendrimer scan be used for targeted MRI imaging (by Gd3+-DTPA, Magnevist®) and NIR optical fluorescence imaging (by Cy5.5).[269]

6. Nanoparticles for Image-Guided Drug Delivery and Therapy

6.1. *General review*

The ideal theranostic NP will have the following features: (1) selectively and rapidly accumulate in diseased tissue; (2) deliver an effective therapeutic; (3) biodegrade with non-toxic byproducts and (4) report biochemical and morphological characteristics of the target tissue with high spatial and temporal resolution and high sensitivity. Up to this point we have reviewed the fundamental imaging modalities and the various nanoparticle drug delivery formulations, as well as the classic therapeutics. We now provide examples of image-guided drug delivery in which the nanoparticles serve the dual function of "thera" and "nostics". Of particular interested are those nanoparticles that can deliver multiple drugs simultaneously and/or that offer multiple imaging modalities. Since all therapies and imaging modalities have their strengths and weaknesses (outlined above), multimodal/multifunctional nanoparticles have the greatest potential for successful translation to humans. Figure 6 is a chematic of a multifunctional theranostics nanoparticle. Table 2 compares the imaging modalities for image guided therapy.

6.2. *Nuclear*

PET and SPECT imaging are often integrated with CT, though the CT image is typically not of diagnostic quality, intended instead to provide an

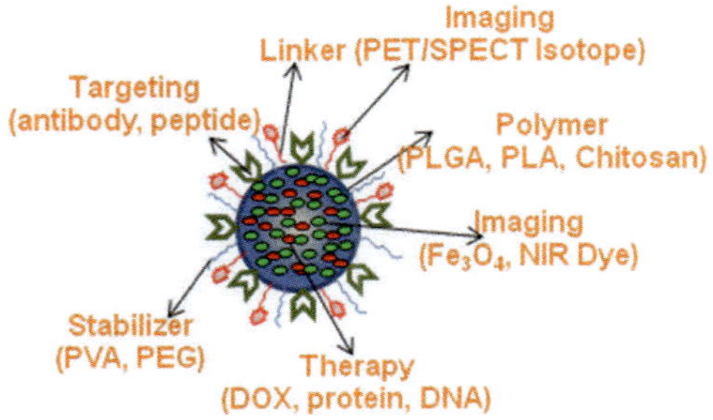

Fig. 7. Schematic illustration of polymer nanoparticle for image guided therapy and theranostics.

Table 2. Comparison of the characteristics of various clinical medical imaging modalities (adapted from Ref. 270.).

Modality	Δ (mm)	Δ (time)	Cost	Contrast agent	Comments
CT	0.4	L	$$	Au NP, Ag NP, Iodine	Spatial resolution and contrast resolution are inversely related.
SPECT	7	H	$$	125I, 131I, 99mTc, 111In	Spatial resolution degrades with distance from the detector.
PET	5	H	$$	18F, 64Cu, 89Zr, 68Ga, 124I	Spatial resolution improves closer to the center of the field
MRI (1.5T)	1.0	H	$$$	Iron Oxide NP, Gd^{3+}-	Resolution improves at higher magnetic field
Ultrasound (5MHz)	0.3	L	$	Requires microspheres	Limited by the wavelength of sound
Optical Computed Tomography	0.015	L	$	Quantum Dots, Au, Fluorescent Dye	Highly dependent on the system and light wavelength. Very low tissue penetration. Not in general clinical practice.[271,272]
Near-Infrared Fluorescence Optical Imaging	0.03	L	$	Quantum Dots, Au, Fluorescent Dye	Theoretical limit. Dominated by the scatter and absorption of light by tissue. Low tissue penetration. Not in general clinical practice.[273]

anatomical reference for the nuclear (PET or SPECT) image and provide an attenuation map for attenuation correction, which allows for true quantitative functional imaging. The key to nuclear imaging with nanoparticles is an efficient labeling scheme in which the nanoparticle vehicle containing drug can be manufactured in advance in a lyophilized form such that the radiolabel can be applied quickly, with high purity and yield. Since its discovery in the 1950s, technetium-99m (99mTc) has become the most widely used radioisotope for medical imaging. 99mTc has become the workhorse of nuclear imaging because of its favorable physical characteristics (140 keV and 6 h half-life), the introduction of the 99mTc/99Mo generator in 1958, and the development of radio-labeling techniques. Due to the high sensitivity and relatively low attenuation and scatter (at least relative to optical and ultrasound) nuclear is perhaps the most practical *in vivo* molecular imaging modality for human use. Liposomes, formed by extrusion along with 99mTc mebrofenin (Choletec®, Bracco Diagnostic Inc) which becomes encapsulated in the liposome, can be partially polymerized for added stability to be used for image-guided drug delivery.[274] The loaded liposomes were tagged with 50 nm gold particles to allow heating of the particles upon exposure to NIR light to trigger drug payload release. While the motivation was to use the radioisotope as a convenient way to examine drug leakage from the liposomes before and after triggering release, the approach could be used for theranostic SPECT and to confirm drug release via imaging. But, the approach would not be very practical since it requires loading the liposomes with the 6 h half-life isotope at the time of liposome generation. This would not be practical in a typical Nuclear Medicine lab in a hospital. A more practical approach would require conjugation of the radiolabel at the time of liposome drug injection. Organogermanium nanoparticles (60 to 80 nm) were labeled with 99mTc for imaging the spleen.[275] Labeling of the nanoparticles with 99mTc was carried out by the reduction of the 99mTc-pertechnetate ions by stannous chloride ($SnCl_2$) as has also been done by others.[276] Other methods have also been reported.[277] 99mTc-labelled gold nanoparticles[278] and PLGA nanoparticles[279] have been developed for sentinel lymph node (SLN) detection. Numerous kits have been developed for the labeling of molecules with 99mTc by the SnCl2 method,[280] which has the potential for being adapted to the fast labeling of nanoparticles.[281,282]

Another attractive isotope for theranostic SPECT imaging is 186Re because of its 137 keV gamma emission for imaging and 1.07 MeV energy beta emission with a tissue penetration depth range of 2–4 mm for therapy. Using the well-established ammonium sulfate gradient method both 186Re-BMEDA and doxorubicin were loaded into liposomes, combining imaging with therapy.[283] The product was tested in a human head and neck cancer xenograft mouse model. The relatively long half-life (3.78 days) is convenient for radiochemistry and allows for repeated imaging but results in a higher radiation dose compared to a shorter half-life isotope such as 99mTc. The gamma emissions will also contribute to a radiation dose far from the target tissue. 188Re, a pure beta emitter with a half-life of 17 h and a penetration range into tissue of 11 mm can be loaded into liposomes similarly as 186Re for therapy, though it cannot be imaged. The labeling of doxorubicin loaded liposomes with 99mTc and 111In-oxine has been reported.[93,284] The labeling method described by Soundararajan could be used to directly load therapeutic 186Re/188Re into the commercial product Doxil® without the requirement for the pre-labeling modification of the liposomes.[279] Again, the key to a successful product is the fast and efficient labeling of the NPs with the radioisotope.

The first application of theranostics as defined in this chapter, i.e. combining imaging and therapy, may be the use of radioiodine (particularly, 131I as NaI⁻), which is still in wide use. 131I decays with a half-life of about eight days with both beta and gamma emissions. The gamma emission is useful for scintigraphic imaging and the beta particles are useful for therapy, specifically ablation of the thyroid or treatment of cancers with active sodium iodide symporters as in differentiated thyroid cancer. The poor quality gamma emission limits 131I to planar imaging. 123I with a half-life or 13 h is used commonly for SPECT imaging. However, 123I is not useful for therapy and is more costly than 131I.

The iodination of molecules and materials is well-established and the chemistry is relatively simple. Liu[285] labeled polyacrylamide nanoparticles with 125I for imaging its distribution to the lungs. Like many preclinical applications of the radiolabeling of nanoparticles, it was used as a convenient method for high sensitivity measurement of the distribution of the nanoparticles. However, the success points to the potential for a clinical application. Chrastina[286] developed a rapid method for radiolabeling of pyrrolidone-capped 12 nm diameter silver nanoparticles with 125I in order to track the *in vivo* uptake of the nanoparticles into tissue after systemic administration. However, 125I is a poor choice for imaging because of its low gamma energy and long half-life (59 days). It is useful primarily for in *in vitro* biological assays (radio-ligand binding). 124I, with a half-life of 4.18 days is perhaps more attractive for imaging since it can be used in PET, which makes it more amenable for estimation of quantitative dosimetry for iodine therapy with 131I. Moderately long-lived PET radioisotopes like 124I are a good choice for immunotherapy and nanoparticles because of the long time required for cell uptake, washout from normal tissue and blood clearance. 124I is not a pure beta positron emitter and the non-511keV gamma components can complicate coincidence imaging. Improved energy discrimination instrumentation will allow for more routine use of 124I for PET imaging. A disadvantage of using iodine labeled molecules and particles in general is the potential for deiodination and the potential for released iodine to reach the thyroid.

PET, compared to SPECT, has the advantage of electronic collimation, which allows for much higher sensitivity, and many PET isotopes are also found in many biochemicals and therefore an isotopic substitution is possible. The use of radioisotopes labeled to nanoparticles for imaging and therapy have been extensively reviewed elsewhere.[181,287,288] The workhorse PET isotope is 18F (fluorine), primarily as 18F-fluoro-deoxy-glucose (FDG), used to measure the glucose metabolism of tissue, which is high in most tumors. The chemical binding of 18F is challenging and so few reports of 18F labeled nanoparticles exist. Most attempts at incorporating 18F into nanoparticles have been encapsulation or by weak ionic interactions. Devaraj demonstrated that "click" chemistry could be used to modify iron oxide nanoparticles with 18F for combined nuclear and MRI.[289] Lipid was labeled with 18F and incorporated into microbubble shells and used for quantitative biodistribution measurements in rats.[290] However, the other positron emitters besides 18F with greater than 90% β^+ decay (15O, 11C, 13N, 82Rb) have radioactive half-lives too short for practical labeling of NPs.

64Cu (in the form of $^{64}CuCl_2$) with a half-life of 12.7 h is a PET isotope that is more amenable for nanoparticle labeling than is 18F[291]. 64Cu and Cy5.5 (for dual PET/Fluorescence) were labeled by sarcophagine "cage-like" scaffold (actually a macromolecule rather than a nanoparticle) by a hetero-functional chelator, BaAn(Boc)Sar, targeted by a $c(RGDyK)_2$ peptide and evaluated *in vivo* in a mouse model.[291] The goal was that the PET tracer could be used to identify the tumor and the fluorescent tracer could direct surgery. It is conceivable that a therapeutic small molecule could be added to the construct and the entire synthesis accomplished via a dedicated synthesis module. Gold nanoshells have been labeled with 64Cu for combined PET imaging and thermotherapy.[183] The nanoshells were introduced directly into tumor tissue, which is not the most practical method for anticancer drug administration. In this case the radioisotope was used primarily as a way to determine if the nanoshells escaped the tumor tissue and the focus was on using the imaging to evaluate the transport of the nanoshells.

Like the 99mTc/99Mo generator that allowed 99mTc SPECT to become dominant in nuclear imaging, the PET isotope 68Ga is also easily available via a 68Ga/68Ge generator.[292] Unlike the 99mTc/99mMo generator with only a one to two week useful life due to the 66 h half-life of the parent 99Mo, current 68Ga/68Ge generators have a useful life of more than a year due to the 271 day half-life of 68Ge. A disadvantage of 68Ga relative to 99mTc is its much shorter half-life (68 min compared to 6 h), which limits the time available for radiolabeling of the active molecule/nanoparticle. However, the worldwide shortage of 99Mo [293,294] and the reduced production 99mTc/99Mo generators has caused alarm and reinforces the need for alternatives, such as 68Ga for routine clinical radiopharmaceuticals that can be produced on site in a nuclear medicine laboratory. The key to highly stable 68Ga-nanoparticle complexes is the presence of bifunctional chelators (such as EDTA, DTPA, DOTA, NOTA and TETA), which are molecules that bind strongly to metals possessing a metal binding moiety function and also a chemically reactive functional group that can bind to the nanoparticle or targeting molecule.[295] Of these chelators, NOTA binds with 68Ga with highest stability and radiosynthesis can be done at room temperature.[292] Relatively few examples exist for 68Ga labeled nanoparticles. However, it has been found that 68Ga binds tightly to chitosan for efficient nanoparticle labeling with high yield[296] but our experience is that chelation is required for *i.v.* administration because of the strong competition of transferrin for the 68Ga isotope (unpublished data from our group).

Examples of less common PET isotopes used for nanoparticle imaging are 76Br and 86Zr.[143,297,298] 76Br and 86Zr are PET isotopes with 16.2 and 3.3 day half-lives, respectively. There are very few reports of these isotopes labeled to NPs. Their relatively poor commercial availability limits their potential applications at this time.

6.3. *X-ray/CT*

X-ray imaging (including mammography and fluoroscopy) represents one of the oldest and most used forms of medical imaging. Its contrast is derived from the differences in the absorption of different tissues (or contrast agents), hard tissue absorbing greater than soft tissue. It is typically considered a modality for imaging anatomy rather than functional imaging. X-ray computed tomography (or simply CT) converts a set of 2D X-ray imaging into a 3D imaging similar to PET and SPECT. Advantages of CT imaging are its high temporal and spatial resolution. The disadvantage is the high radiation dose to the patient and the fact that different soft tissue organs have similar attenuation and therefore are difficult to differentiate. The addition of contrast agents, such as iodine, enhances the contrast in particular organs. In fact, most CT contrast agents are based on small iodinated molecules. They tend, however, to have short plasma residence time and poor tissue targeting capabilities. In addition, iodine can be toxic at high concentrations, especially to the thyroid.[5] But, like other compounds that are potentially toxic at high concentrations when administered systemically, targeting, for example in nanoparticles, can reduce unintended toxicities to non-target tissue. In an effort to reduce the systemic toxicity of iodine and allow for targeting, iodinated copolymeric nanoparticles (~30 nm) were developed for X-ray CT imaging.[83] Enhanced visibility of the blood pool was observed for 30 min following injection with delayed enhancement of lymph nodes, liver and spleen due to nanoparticle uptake by the RES. Another example is a liposome (~100 nm) encapsulated iodixanol contrast agent developed and explored for microCT imaging of lung cancer in mice.[299]

Dendrimer-stabilized silver (Ag) and gold (Au) nanoparticles have been used as contrast agents for X-ray CT imaging. The attenuation of Ag dendrimer-stabilized nanoparticles was found to be size-dependent, the 16 nm size showing X-ray attenuation intensity close to that of a clinically used iodine-based contrast agent (Omnipaque) at the same molar concentration of the active element (Ag versus iodine).[191] The Ag nanoparticles may also have longer residence time and less toxicity than free iodine. Gold also has higher X-ray absorption than iodine which can result in better contrast with a lower radiation dose. Nanoparticles are removed from the blood more slowly than are iodine agents permitting longer imaging times.[300] Gold nanoparticles (1.9 nm) were injected via tail vein into mice and imaged with a standard mammography unit. As expected for such small particles, elimination was primarily renal. Organs such as kidneys and tumors were reported to be observable with high spatial resolution. Blood vessels smaller than 100 μm in diameter could be identified and thus regions with increased vascularization and angiogenesis could be distinguished.[300] PEGylated dendrimer-entrapped gold nanoparticles for blood pool and tumor imaging have been investigated.[301] Chien[190] found that bare gold nanoparticles injected with heparin provided sufficient contrast to allow for the *in vivo* detection of small capillaries and proposed their use in *in vivo* imaging using synchrotron microradiology, microtomography and high resolution X-ray microscopy. Au NPs (~20 nm) stabilized by gum Arabic were evaluated as an X-ray CT contrast agent.[302] Anti-epidermal growth factor receptor conjugated gold nanoparticles (~30 nm) were evaluated in mice to treat head and neck cancer.[303] Small tumors undetectable through classical CT became visible by the molecularly-targeted gold nanoparticles. Though not approved for clinical use, gold nanoparticles for X-ray contrast imaging AurovistTM are commercially available from nanoprobes.com and intended for microCT imaging.

Pratx[304] described a dual-molecular/anatomical imaging modality, X-ray luminescence CT, which is based on the X-ray excitation and optical detection of NIR-emitting Gd_2O_2S:Eu phosphor nanoparticles. Other less common systems have also been proposed. For example, PEGylated Yb_2O_3:Er nanoparticles appear suitable for both X-ray CT imaging and upconversion luminescence imaging.[305] They reported that the X-ray absorption is greater than the commercial X-ray contrast agent Iobitridol. Still another approach consists of polymer-coated Bi_2S_3 nanoparticle.[306] They reported excellent stability at high concentrations, absorption five-fold better than iodine and *in vivo* circulation times > 2 h.

6.4. *MRI*

Like nuclear imaging (PET and SPECT) and CT, MRI is also practical for *in vivo* imaging in humans because of its deep penetration range. While high cost and poor temporal resolution are disadvantages, it does have spatial resolution comparable to CT. Gadolinium (Gd) is the most common MRI contrast agent and Gd based nanoparticles have been developed.[307–309] Pegylated poly-(L-glutamic acid) conjugates containing the photosensitizer mesochlorin, and conjugated to the MRI contrast agent Gd(III) by the DOTA chelator, were synthesized and tested in mice for minimally invasive image guided cancer treatment.[310] However, some patients have allergic reactions to Gd and renal toxicity of Gd has been reported.[8,9]

A large number of inorganic nanoparticles are under development for MRI imaging.[168] The earliest examples of nanoparticles applied to *in vivo* imaging are in MRI with iron oxide particles, which allow MRI to be closer to a functional (molecular) imaging modality then traditional MRI. Core-shell nanoparticles (~190 nm) were formulated by integrating PLGA, docetaxel and hydrophobic superparamagnetic iron oxide nanocrystals with prostate stem cell antigen antibody conjugated for targeting.[137] The polymer component allows for more efficient drug delivery. Peptide modified (for HER-2 targeting) PEGylated chitosan magnetic (iron oxide) nanoparticles (~270 nm) were tested for tumor imaging.[159] Polyethylene glycol-grafted polyethylenimine functionalized with a neuroblastoma cell-specific ligand was synthesized to deliver siRNA into neuroblastoma cells to induce apoptosis and inhibit cell growth.[174] The polymer was complexed with SPIO nanoparticles for MR imaging.

6.5. *Optical imaging*

Despite the obvious limitations of poor tissue penetration and high scatter (relative to nuclear, CT and MRI) optical imaging continues to gather intense interest in the research community because it is relatively inexpensive, portable, offers the potential for high sensitivity as well as spatial and temporal resolution. Optical imaging might find its widest niche as a tool for intraoperative guidance for surgical resection of cancer.[17,269,311,312] The majority of applications for *in vivo* optical imaging employ NIR fluorescent dyes, but gold nanoparticles,[180,188,313] carbon nanotubes[18,192] and quantum dots[180] have also gathered much interest. The NIR light source can induce fluorescence for optical imaging but can also be used to excite molecules or nanoparticles for PDT[110,313] and or photothermal (hyperthermia) therapy.[72,151,188,228,314]

Admittedly, much needs to be done to move optical imaging from the preclinical research into the clinic.[99,315] For that transition to occur, in addition to ensuring safety and efficacy, optical imaging system performance standards need to be developed so that the image system's sensitivity and can be measured and documented similar to how MRI, PET and SPECT systems undergo regular quality control.[316]

For optical imaging, cyanine dyes are an important family of synthetic molecules belonging to the polymethine group since many fluoresce in the NIR spectrum. As described above, ICG is a cyanine dye used in various medical diagnostics. It binds tightly to plasma and becomes confined to the vascular system and is excreted exclusively by hepatobiliary clearance. Heptamethine[317–320] and phthalocyanines[99,321] dyes are a subclass of the indocyanine dye family. Naphthalocyanine is a derivative of phthalocyanine that may be used in photodynamic cancer treatment.[322,323] Of the fluorescent dyes used for NIR optical imaging the cyanine dye ICG has the longest history and still garners much interest because it has been FDA approved for many years for systemic delivery. However, the quality that makes it a good tracer for indicator dilution studies, i.e. plasma binding and fast plasma clearance, make it a less than ideal candidate for *in vivo* NIR tissue imaging.[324,325] It also suffers from poor stability in aqueous solutions. Other dyes in the cyanine dye class have better physical and chemical characteristics than ICG[324,326–328] and attempts are ongoing to generate new cyanine dye analogs.[320,329] To improve its plasma residence, stability and tissue targeting, ICG has been loaded into NP and liposomes.[71,72,145,152,239,330,331] Saxena *et al.*[332] enhanced ICG stability by polymer encapsulation. Our group optimized the encapsulation of DOX and ICG into PLGA nanoparticles and also showed greater stability of the ICG.[71,140] Mordon *et al.*[333] delivered ICG to blood vessels by encapsulating it into heat sensitive liposomes. ICG has also been proposed for breast cancer imaging.[107,152,324,334–336] A hand-held fluorescence imaging device for detection of breast cancer is under development.[337,338]

Numerous NIR fluorescent dyes are commercially available, such as IR820,[326,339–341] the Cy dyes (Cy5.5, Cy7, Cy7.5),[154,269,273] IRDye800,[329,342] (as well as CF® dyes from Biotium, IRDye® dyes from Licor, QDot® from Invitrogen, Alexa Fluor dyes from Invitrogen, various cyanine dye (Cy) conjugates from GE Lifesciences and Lumiprobe). Cyanine dyes have been modified by numerous groups to improve on their physical and chemical characteristics and either encapsulated into or conjugated onto nanoparticles or conjugated to peptides or small molecules for targeting and/or therapy.[273,329]

Fan[311] described the use of poorly water-soluble chlorophyll from the leaf of Chimonanthus salicifolius encapsulated into liposomes (~22 nm) for NIR mouse axillary SLN mapping. The nanocomposite suspension exhibited brighter NIR fluorescence than the chlorophyll alone. McCann[203] compared avidin conjugated silica nanoparticles containing rhodamine (Av-SiR700) to similar compounds based on cyanine dyes (Av-Cy5.5 and Av-Alexa Fluor 680) in a mouse model of ovarian cancer and found better tumor to background ratio than Av-Cy5.5 and better stability than Av-Alexa Fluor 680. However, the absorption/emission wavelengths of the fluorophores are not optimal for *in vivo* imaging.

PDT is a procedure in which a photo-sensitive dye, after becoming localized in a tissue, is exposed to an appropriate wavelength of light to excite the dye to generate oxygen free radicals which are toxic to cells due to oxidative damage to cell membranes and DNA. The most widely used PDT compound is photofrin.[343] But, PDT has disadvantages, including prolonged cutaneous photosensitization and thus potential damage to non-target cells and fast elimination from the plasma. Encapsulation or incorporation into nanoparticles can address both of these disadvantages. For example, Reddy[344] developed a multifunctional polymeric nanoparticle consisting of a surface-localized tumor vasculature targeting peptide and encapsulated PDT and MRI agent (iron oxide) for brain tumor imaging.

6.6. *Multimodal*

The most recent developments in image-guided drug delivery that is made more practical with nanoparticles are not only the simultaneous delivery of imaging agent and therapeutic agent (or even multiple therapeutic agents or therapeutic modalities) but also the combining of different imaging modalities. As described above, each imaging modality has its advantages and disadvantages. By combining imaging modalities one can maximize outcomes by exploiting the advantages of each modality. PET/CT hybrid systems are standard in clinical nuclear medicine labs and imaging centers, but the CT part is typically not for diagnostic purposes, used instead to provide an anatomical reference for the PET molecular/functional imaging and to provide a 3D radiation absorption map for which to apply attenuation correction to the PET image. The low cost and small size of optical systems makes them especially attractive for including in a hybrid system, particularly for intraoperative surgical guidance. However, such hybrid imaging systems are not yet generally available for clinical use.

Nanoparticle drug delivery systems are particularly suitable for simultaneously delivering multiple imaging tracers, along with therapeutic drugs, and are also convenient for attaching a targeting moiety. There are numerous examples in the literature of such formulations. For example optical with nuclear.[291,312,342,345–348] optical with MRI,[161,349,350] nuclear with MRI,[277,351,352] MRI with CT,[353] optical with CT,[354,355] nuclear with CT,[356] optical with MRI and CT,[308] MRI with photoacoustic and Raman[309] have all been reported in the literature. Many of these references are for nanoparticle formulations that also include a therapeutic component. As multimodal tracers become more available and are shown to be effective, the clinical multimodal imaging system may begin to appear as well.

6.7. *Image guided surgery*

Hand-held gamma probes have been used for years in radio-immuno-guided surgery[357] and in identifying SNLs for surgical resection.[358–360] The advantage of a gamma probe is that the radiation detection undergoes low scatter and attenuation through tissue and therefore radiation from deep tissue can be easily detected. A disadvantage is that the need for collimation and shielding makes the probe heavy, and reduces sensitivity and spatial resolution making it very difficult to precisely determine the location of the radiation source. To overcome some of these disadvantages a new hand-held probe gamma-camera system which generates 2D images for image guided surgery (Sentinella®, Fig. 8) is commercially available.[361] On the other hand, optical imaging has poor penetration and high scattering, but the probe can be smaller and lighter and the origin of the fluorescence signal can be much more precisely identified. Nanoparticles have been found to be useful for SLN mapping using various imaging modalities.[10,311,362–366] and have been extensively reviewed.[367,368] Fugimaki[369] investigated ICG image-guided surgery for head and neck cancer in a mouse model. Examples of nanoparticle formulations used for intraoperative imaging guidance, primarily in SLN identification and peritoneal metastatic ovarian tumor identification, and brain tumor margin identification are numerous.[10,15,16,269,278,279,307,309,345,352,362–364,366,370–373] The commercial da Vinci® robotic surgery system (Fig. 9) available in many hospitals has been modified for NIR fluorescence image-guided therapy for SNL mapping and resection[374] and partial nephrectomy.[375] Other systems have been developed for NIR fluorescence image-guided glioma resection[376] and *in vivo* brain tumor[377] and epileptic brain lesion demarcation using diffuse reflectance optical spectroscopy.[378] However, these last two examples did not use exogenous fluorophores, but used the intrinsic properties of the tissue.

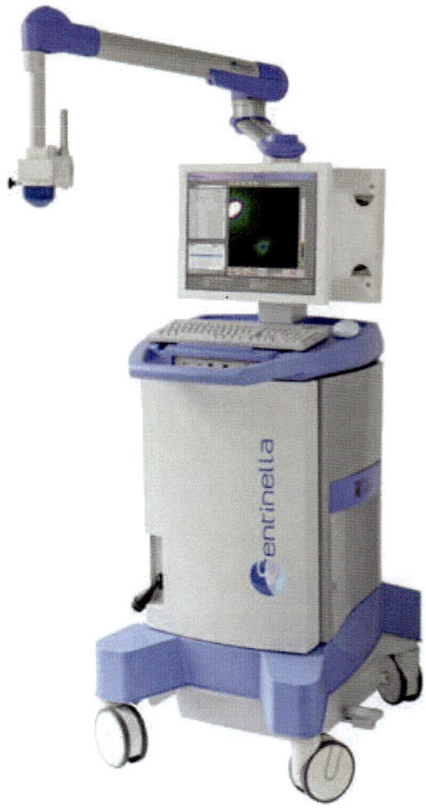

Fig. 8. Sentinella® gamma camera for intraoperative guidance of SLN detection and resection www.aplimedsupplies.com

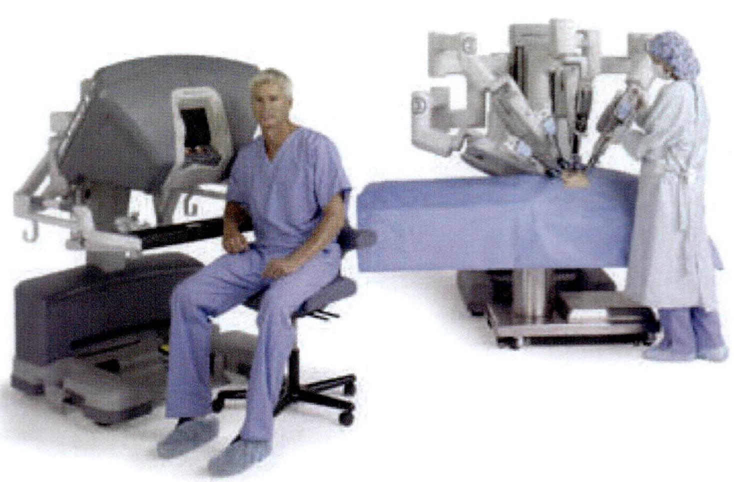

Fig. 9. The *da vinci*® surgical system (http://davincisurgery.com/).

7. Summary, Obstacles, and Future Directions

There are numerous challenges to bringing nanoparticle based targeted theranostics to regular clinical use, and their commercial success as drug targeting has been met with many pitfalls.[127] A primary feature of nanoparticle drug delivery is their non-specific targeting to tumors by the enhanced permeation and retention effect (EPR). However, not all tumors have leaky vessels and thus the dependence on EPR is not universal. Targeting is not ideal because of the occurrence of mutations and small differences in receptor specificity. Other obstacles include the potential immune reaction to antibodies, poor blood flow to avascular tumors, poor diffusion of nanoparticles in interstitial space, different requirement for vascular transport, interstitial transport, cell membrane transport and cellular localization, the requirement for carrier release from endosomes once inside the cells and the subsequent uncontrolled release of drug from the carrier. There is a potential for clumping of particles in lyophilized storage, the uneven uptake of nanoparticles by cancer cells even of the same type, and the concern for the potential long-term effects of nanoparticles. Safety, regulatory hurdles, and the development of cold kits for radionuclide labeling are additional obstacles to nanoparticle based imaging/drug delivery systems to becoming a reality. The scale up for commercial production and the reluctance of commercial investment complicate the issue even further. Despite these obstacles, just based on the sheer volume of related journal publications appearing every day, and the fact that there have been some clinical and commercial successes to date, the area of nanoparticle based theranostics development remains strong with continued high potential for breakthroughs. But success will require the contributions of many different types of scientists and engineers as well as entrepreneurs willing to take risks.

References

1. I. Gilham, Drug Discovery World (DDW) Web site. http://www.ddw-online.com/personalised-medicine/p148484-theranostics-an-emerging-tool-in-drug-discovery-and-commercialisation-fall-02.html. Published 2002. Accessed 06/10 (2014).
2. *Med. Device Technol.* **11**(8), 33 (2000).
3. K. T. Bae, *Radiology* **256**(1), 32–61 (2010).
4. E. C. Lin, *Mayo Clin. Proc.* **85**(12), 1142–1146 (2010).
5. J. Singh and A. Daftary, *J. Nucl. Med. Technol.* **36**(2), 69–74 (2008).
6. D. Delbeke and G. M. Segall, *J. Nucl. Med.* **52**(Suppl 2), 24S–28S (2011).
7. E. B. Silberstein, *J. Nucl. Med.* **46**(5), 13 (2005).
8. M. E. Bartolini, J. Pekar, D. R. Chettle *et al.*, *Magn. Reson. Imaging* **21**(5), 541–544 (2003).

9. M. A. Perazella, *Clin. J. Am. Soc. Nephrol.* **4**(2), 461–469 (2009).
10. W. J. Akers, C. Kim, M. Berezin *et al.*, *ACS Nano* **5**(1), 173–182 (2011).
11. C. R. Simpson, M. Kohl, M. Essenpreis and M. Cope, *Phys. Med. Biol.* **43**(9), 2465–2478 (1998).
12. R. Weissleder, *Radiology* **212**(3), 609–614 (1999).
13. H. Lee, W. J. Akers, P. P. Cheney *et al.*, *J. Biomed. Opt.* **14**(4), 040507 (2009).
14. B. Zhu, M. J. Eppstein, E. M. Sevick-Muraca and A. Godavarty, *Opt. Express.* **15**(18), 11285 (2007).
15. Y. L. Colson, R. Liu, E. B. Southard *et al.*, *Biomaterials* **32**(3), 832–840 (2011).
16. M. E. Werner, S. Karve, R. Sukumar *et al.*, *Biomaterials* **32**(33), 8548–8554 (2011).
17. G. M. van Dam, G. Themelis, L. M. Crane *et al.*, *Nat. Med.* **17**(10), 1315–1319 (2011).
18. N. H. Levi-Polyachenko, E. J. Merkel, B. T. Jones, D. L. Carroll and J. H. Stewart, 4th., *Mol. Pharm.* **6**(4), 1092–1099 (2009).
19. F. K. Storm, *Radiol. Clin. North Am.* **27**(3), 621–627 (1989).
20. P. Liu, L. X. Xu and A. Zhang, *Conf. Proc. IEEE Eng. Med. Biol. Soc.* **1**(43), 54–57 (2006).
21. A. Brufsky, *Semin. Oncol.* **33**(2 Suppl 7), S13–S17 (2006).
22. T. Safra, F. Muggia, S. Jeffers *et al.*, *Ann. Oncol.* **11**(8), 1029–1033 (2000).
23. G. Batist, G. Ramakrishnan, C. S. Rao *et al.*, *J. Clin. Oncol.* **19**(5), 1444–1454 (2001).
24. J. W. Park, *Breast Cancer Res.* **4**(3), 95–99 (2002).
25. G. Sharma, S. Anabousi, C. Ehrhardt and M. N. Ravi Kumar, *J. Drug Target* **14**(5), 301–310 (2006).
26. O. S. Selawry, M. N. Goldstein and C. T. Mc, *Cancer Res.* **17**(8), 785–791 (1957).
27. J. van der Zee, *Ann. Oncol.* **13**(8), 1173–1184 (2002).
28. G. M. Hahn, J. Braun and, I. Har-Kedar, *Proc. Natl. Acad. Sci. USA* **72**(3), 937–940 (1975).
29. M. Urano, J. Begley and R. Reynolds, *Int. J. Hyperthermia* **10**(6), 817–826 (1994).
30. G. C. Rice and G. M. Hahn, *Cancer Chemother. Pharmacol.* **20**(3), 183–187 (1987).
31. R. Supino, L. Bardella, N. Gibelli, G. Cairo and L. Schiaffonati, *Tumori* **73**(2), 109–116 (1987).
32. T. S. Herman, C. C. Sweets, D. M. White and E. W. Gerner, *J. Natl. Cancer Inst.* **68**(3), 487–491 (1982).
33. K. Engin, D. B. Leeper, L. Tupchong and F. M. Waterman, *Clin. Cancer Res.* **1**(2), 139–145 (1995).
34. B. Hildebrandt and P. Wust, *Cancer Treat Res.* **134**, 185–193 (2007).
35. C. Christophi, A. Winkworth, V. Muralihdaran and P. Evans, *Surg. Oncol.* 7(1&2), 83–90 (1998).
36. V. E. Kouloulias, C. E. Dardoufas, J. R. Kouvaris *et al.*, *Clin. Cancer Res.* **8**(2), 374–382 (2002).
37. J. Dvorak, Z. Zoul, B. Melichar *et al.*, *J. Chemother.* **16(**Suppl 5), 34–36 (2004).
38. Z. Vujaskovic, D. W. Kim, E. Jones *et al.*, *Int. J. Hyperthermia* **26**(5), 514–521 (2010).
39. J. T. Alander, I. Kaartinen, A. Laakso *et al.*, *Int. J. Biomed. Imaging* **2012**, 940585 (2012).
40. W. R. Chen, R. L. Adams, K. E. Bartels and R. E. Nordquist, *Cancer Lett.* **94**(2), 125–131 (1995).
41. W. R. Chen, R. L. Adams, S. Heaton, D. T. Dickey, K. E. Bartels and R. E. Nordquist, *Cancer Lett.* **88**(1), 15–29 (1995).

42. V. G. Liu, T. M. Cowan, S. W. Jeong, S. L. Jacques, E. C. Lemley and W. R. Chen, *Lasers Med. Sci.* **17**(4), 272–279 (2002).
43. C. A. Bailey, T. M. Cowan, V. G. Liu, E. C. Lemley and W. R. Chen, *J. Biomed. Opt.* **9**(3), 648–654 (2004).
44. F. Brasseur, P. Couvreur, B. Kante *et al., Eur. J. Cancer* **16**(11), 1441–1445 (1980).
45. H. Maeda, *Adv. Enzyme Regul.* **41**, 189–207 (2001).
46. V. Torchilin, *Adv. Drug Deliv. Rev.* **63**(3), 131–135 (2011).
47. C. Riggio, E. Pagni, V. Raffa and A. Cuschieri, *J. Nanomaterials* **1** (2011).
48. J. V. Jokerst and S. S. Gambhir, *Acc. Chem. Res.* **44**(10), 1050–1060 (2011).
49. A. Z. Wang, R. Langer and O. C. Farokhzad, *Ann. Rev. Med.* **63**, 185 (2012).
50. E. Frohlich, *Int. J. Nanomedicine* 7, 5577–5591 (2012).
51. M. S. El-Samaligy and P. Rohdewald, *J. Pharm. Pharmacol.* **35**(8), 537–539 (1983).
52. L. Illum, *Pharm. Res.* **15**(9), 1326–1331 (1998).
53. A. K. Sailaja, P. Amareshwar and P. Chakravarty, *Int. J. Pharm. Pharm. Sci.* **3**, 45 (2011).
54. T. A. Sonia and C. P. Sharma, *Adv. Polym. Sci.* **243**, 23 (2011).
55. P. Baldrick, *Regul. Toxicol. Pharmacol.* **56**(3), 290–299 (2010).
56. K. A. Howard, U. L. Rahbek, X. Liu *et al.*, *Mol. Ther.* **14**(4), 476–484 (2006).
57. H. Dai, X. Jiang, G. C. Tan *et al.*, *Int. J. Nanomedicine* **1**(4), 507–522 (2006).
58. S. A. Agnihotri, N. N. Mallikarjuna and T. M. Aminabhavi, *J. Control. Release* **100**(1), 5–28 (2004).
59. I. M. van der Lubben, G. Kersten, M. M. Fretz, C. Beuvery, J. Coos Verhoef and H. E. Junginger, *Vaccine* **21**(13&14), 1400–1408 (2003).
60. M. R. Avadi, A. M. Sadeghi, N. Mohamadpour Dounighi, R. Dinarvand, F. Atyabi and M. Rafiee-Tehrani, *ISRN Pharm.* **2011**, 860109 (2011).
61. S. W. Lee, M. H. Yun, Jeong *et al.*, *J. Control. Release* **155**(2), 262–271 (2011).
62. A. Nigam, A. Sharma and S. K. Singh, *J. Cancer Therapy* **3**, 44 (2012).
63. T. Y. Kim, D. W. Kim, J. Y. Chung *et al.*, *Clin. Cancer Res.* **10**(11), 3708–3716 (2004).
64. J. L. Lee, J. H. Ahn, S. H. Park *et al.*, *Invest. New Drugs* **30**(5), 1984–1990 (2012).
65. H. K. Makadia and S. J. Siegel, *Polymers (Basel)* **3**(3), 1377–1397 (2011).
66. J. M. Lu, X. Wang, C. Marin-Muller *et al.*, *Exp. Rev. Mol. Diagn.* **9**(4), 325–341 (2009).
67. S. Parveen and S. K. Sahoo, *J. Drug Target* **16**(2), 108–123 (2008).
68. M. Figueiredo and R. Esenaliev, *J. Drug Deliv.* **2012,** 767839 (2012).
69. R. Dinarvand, N. Sepehri, S. Manoochehri, H. Rouhani and F. Atyabi, *Int. J. Nanomedicine* **6**, 877–895 (2011).
70. Y. S. Rhee, C. W. Park, P. P. DeLuca and H. M. Mansour, *Pharm. Technol.* **1** (2010).
71. R. Manchanda, A. Fernandez-Fernandez, A. Nagesetti and A. J. McGoron, *Colloids and Surfaces B: Biointerfaces* **75**(1), 260–267 (2010).
72. Y. Tang, T. Lei, R. Manchanda *et al.*, *Pharm. Res.* **27**(10), 2242–2253 (2010).
73. A. Kumari, S. K. Yadav and S. C. Yadav, *Colloids Surf B Biointerfaces* **75**(1), 1–18 (2010).
74. F. Danhier, E. Ansorena, J. M. Silva, R. Coco, A. Le Breton and V. Preat, *J. Control. Release* **161**(2), 505–522 (2012).
75. A. Fernandez-Fernandez, R. Manchanda and A. J. McGoron, *Appl. Biochem. Biotechnol.* **165**(7&8), 1628–1651 (2011).
76. M. Chaubal, *Drug Deliv. Technol.* **2**(5), 34–35 (2002).
77. S. Hirsjärvi, Preparation and characterization of poly(lactic acid) nanoparticles for pharmaceutical use, University of Helsinki (2008).

78. C. Sheridan, *Nat. Biotechnol.* **30**(6), 471–473 (2012).
79. N. Kumar, R. S. Langer and A. J. Domb, *Adv. Drug Deliv. Rev.* **54**(7), 889–910 (2002).
80. W. C. Lee and I. M. Chu, *J. Biomed. Mater. Res. B Appl. Biomater.* **84**(1), 138–146 (2008).
81. P. Ojer, A. L. de Cerain, P. Areses, I. Penuelas and J. M. Irache, *Pharm. Res.* **29**(9), 2615–2627 (2012).
82. Y. Ma, H. Gao, W. Gu *et al.*, *Eur. J. Pharm. Sci.* **45**(1&2), 65–72 (2012).
83. H. Aviv, S. Bartling, F. Kieslling and S. Margel, *Biomaterials* **30**(29), 5610–5616 (2009).
84. C. W. Evans, M. J. Latter, D. Ho *et al.*, *New J. Chem.* **36**, 1457 (2012).
85. N. Jaulin, M. Appel, C. Passirani, G. Barratt and D. Labarre, *J. Drug Target* **8**(3), 165–172 (2000).
86. C. L. Peng, H. M. Tsai and S. J. Yang *et al.*, *Nanotechnology* **22**(26), 265608-4484/22/26/265608. Epub 2011 May 17 (2011).
87. B. Sibeko, Y. E. Choonara, L. C. du Toit *et al.*, *J. Drug Deliv.* **2012**, 579629 (2012).
88. J. Kreuter, *Pharm. Acta. Helv.* **53**(2), 33–39 (1978).
89. H. Jans, K. Jans, L. Lagae, G. Borghs, G. Maes and Q. Huo, *Nanotechnology* **21**(45), 455702–4484/21/45/455702. Epub 2010 Oct 14 (2010).
90. G. Deepa, A. K. Thulasidasan, R. J. Anto, J. J. Pillai and G. S. Kumar, *Int. J. Nanomedicine* 7, 4077–4088 (2012).
91. W. Xiong, W. Wang, Y. Wang *et al.*, *Colloids Surf B Biointerfaces* **84**(2), 447–453 (2011).
92. A. Chrastina, K. A. Massey and J. E. Schnitzer, *Wiley Interdiscip. Rev. Nanomed. Nanobiotechnol.* **3**(4), 421–437 (2011).
93. P. Laverman, M. G. Carstens, O. C. Boerman *et al.*, *J. Pharmacol. Exp. Ther.* **298**(2), 607–612 (2001).
94. D. W. Northfelt, B. J. Dezube, J. A. Thommes *et al.*, *J. Clin. Oncol.* **16**(7), 2445–2451 (1998).
95. M. Hamidi, A. Azadi and P. Rafiei, *Adv. Drug Deliv. Rev.* **60**(15), 1638–1649 (2008).
96. C. Gonçalves, P. Pereira and M. Gama, *Materials* **3**, 1420 (2010).
97. D. Gao, H. Xu, M. A. Philbert and R. Kopelman, *Nano Lett.* **8**(10), 3320–3324, (2008).
98. G. Nie, H. J. Hah, G. Kim *et al.*, *Small* **8**(6), 884–891 (2012).
99. J. Merian, J. Gravier, F. Navarro and I. Texier, *Molecules* **17**(5), 5564–5591 (2012).
100. W. Y. Ayen and N. Kumar, *Pharm. Res.* **29**, 2522 (2012).
101. K. K. Upadhyay, A. N. Bhatt, E. Castro *et al.*, *Macromol. Biosci.* **10**(5), 503–512 (2010).
102. W. Chen, F. Meng, R. Cheng and Z. Zhong, *J. Control. Release* **142**(1), 40–46 (2010).
103. H. S. Yoo and T. G. Park, *J. Control. Release* **96**(2), 273–283 (2004).
104. Z. Yang, S. Zheng, W. J. Harrison *et al.*, *Biomacromolecules* **8**(11), 3422–3428 (2007).
105. C. Zhang, G. Qu, Y. Sun *et al.*, *Biomaterials* **29**(9), 1233–1241 (2008).
106. Z. Gao, A. M. Kennedy, D. A. Christensen and N. Y. Rapoport, *Ultrasonics* **48**(4), 260–270 (2008).
107. V. B. Rodriguez, S. M. Henry, A. S. Hoffman, P. S. Stayton, X. Li and S. H. Pun, *J. Biomed. Opt.* **13**(1), 014025 (2008).
108. H. Zhao and L. Y. Yung, *Int. J. Pharm.* **349**(1&2), 256–268 (2008).

109. T. H. Kim, C. W. Mount, W. R. Gombotz and S. H. Pun, *Biomaterials* **31**(28), 7386–7397 (2010).
110. H. Koo, H. Lee, S. Lee *et al.*, *Chem. Commun. (Camb)* **46**(31), 5668–5670 (2010).
111. T. M. Sun, J. Z. Du, Y. D. Yao *et al.*, *ACS Nano* **5**(2), 1483–1494 (2011).
112. C. Zhu, S. Jung, S. Luo *et al.*, *Biomaterials* **31**(8), 2408–2416 (2010).
113. Q. Liu, R. Li, Z. Zhu *et al.*, *Int. J. Pharm.* **430**(1&2), 350–358 (2012).
114. P. T. Ha, M. H. Le, T. M. N. Hoang and T. T. H. Le, *Adv. Nat. Sci. Nanosci. Nanotechnol.* **3**, (2012).
115. W. Cui, X. Lu, K. Cui, J. Wu, Y. Wei and Q. Lu, *Nanotechnology* **22**(6), 065702-4484/22/6/065702. Epub 2011, 7 January (2011).
116. R. Tong, V. J. Coyle, L. Tang, A. M. Barger, T. M. Fan and J. Cheng, *Microsc. Res. Tech.* **73**(9), 901–909 (2010).
117. J. Cao, K. M. Xiu, K. Zhu, Y. W. Chen and X. L. Luo, *J. Biomed. Mater. Res A* **100**(8), 2079–2087 (2012).
118. F. Migneco, Y. C. Huang, R. K. Birla and S. J. Hollister, *Biomaterials* **30**(33), 6479–6484 (2009).
119. D. G. Barrett and M. N. Yousaf, *Soft Matter.* **6**, 5026 (2010).
120. C. Passirani, G. Barratt, J. P. Devissaguet and D. Labarre, *Pharm. Res.* **15**(7), 1046–1050 (1998).
121. C. Wong, T. Stylianopoulos, J. Cui *et al.*, *Proc. Natl. Acad. Sci. USA* **108**(6), 2426–2431 (2011).
122. J. Madan, N. Dhiman, S. Sardana, R. Aneja, R. Chandra and A. Katyal, *Anticancer Drugs* **22**(6), 543–555 (2011).
123. J. J. Marty, R. C. Oppenheim and P. Speiser, *Pharm. Acta. Helv.* **53**(1), 17–23 (1978).
124. J. K. Li, N. Wang and X. S. Wu, *J. Control. Release* **56**(1&3), 117–126 (1998).
125. V. Roy, B. R. LaPlant, G. G. Gross, C. L. Bane and F. M. Palmieri, North Central Cancer Treatment Group. *Ann. Oncol.* **20**(3), 449–453 (2009).
126. F. Danhier, O. Feron and V. Preat, *J. Control. Release* **148**(2), 135–146 (2010).
127. T. Lammers, F. Kiessling, W. E. Hennink and G. Storm, *J. Control. Release* **161**(2), 175–187 (2012).
128. Q. S. Tang, D. Z. Chen, W. Q. Xue *et al.*, *Int. J. Nanomedicine* **6**, 3077–3085 (2011).
129. R. M. Molnar, M. Bodnar, J. F. Hartmann and J. Borbely, *Colloid. Polym. Sci.* **287**, 739 (2009).
130. J. Wang, I. L. Lee, W. S. Lim *et al.*, *Int. J. Pharm.* **279**(1&2), 115–126 (2004).
131. M. Papi, V. Palmieri, G. Maulucci *et al.*, *J. Nanopart. Res.* (2011).
132. M. Kuruppuarachchi, H. Savoie, A. Lowry, C. Alonso and R. W. Boyle, *Mol. Pharm.* **8**(3), 920–931 (2011).
133. R. Hedayati, M. Jahanshahib and H. Attara, *J. Chem. Technol. Biotechnol.* (2012).
134. R. Kannan, A. Zambre, N. Chanda *et al.*, *Wiley Interdiscip. Rev. Nanomed. Nanobiotechnol.* **4**(1), 42–51 (2012).
135. F. Rancan, D. Papakostas and S. Hadam *et al.*, *Pharm. Res.* **26**(8), 2027–2036 (2009).
136. Q. Y. Tan, M. L. Xu, J. Y. Wu, H. F. Yin and J. Q. Zhang, *Pharmazie* **67**(4), 311–318 (2012).
137. X. Gao, Y. Luo, Y. Wang *et al.*, *Int. J. Nanomedicine* **7**, 4037–4051 (2012).
138. Y. C. Wang, M. C. Lin, D. M. Wang and H. J. Hsieh, *Biomaterials* **24**(6), 1047–1057 (2003).

139. S. Hurrell and R. E. Cameron, *J. Mater. Sci. Mater. Med.* **12**(9), 817–820 (2001).
140. Srinivasan, S. R. Manchanda, T. Lei, A. Nagesetti, Fernandez-Fernandez, A. J. McGoron, *Journal of Photobiology, B: Biology* **136C**, 81–90 (2014).
141. S. Bisht and A. Maitra, *Wiley Interdiscip. Rev. Nanomed. Nanobiotechnol.* **1**(4), 415–425 (2009).
142. R. A. Kroll, M. A. Pagel, L. L. Muldoon, S. Roman-Goldstein, S. A. Fiamengo and E. A. Neuwelt, *Neurosurgery* **43**(4), 879–886 (1998).
143. E. J. Keliher, J. Yoo, M. Nahrendorf *et al., Bioconjug. Chem.* **22**(12), 2383–2389 (2011).
144. D. Pouliquen, R. Perdrisot, A. Ermias, S. Akoka, P. Jallet and J. J. Le Jeune, *Magn. Reson. Imaging* 7(6), 619–627 (1989).
145. D. C. Turner, D. Moshkelani, C. S. Shemesh, D. Luc and H. Zhang, *Pharm. Res.* **29**(8), 2092–2103 (2012).
146. M. K. Yu, J. Park and S. Jon, *Theranostics* **2**(1), 3–44 (2012).
147. M. Zhao, D. A. Beauregard, L. Loizou, B. Davletov and K. M. Brindle, *Nat. Med.* 7(11), 1241–1244 (2001).
148. T. Lei, S. Srinivasan, Y. Tang *et al., Nanomedicine: Nanotechnology, Biology and Medicine.* 7, 324–332 (2011).
149. L. A. Guzman, V. Labhasetwar and C. Song *et al., Circulation* **94**(6), 1441–1448 (1996).
150. L. Bondioli, L. Costantino, A. Ballestrazzi *et al., Biomaterials* **31**(12), 3395–3403 (2010).
151. R. H. Patel, A. S. Wadajkar, N. L. Patel, V. C. Kavuri, K. T. Nguyen and H. Liu, *J. Biomed. Opt.* **17**(4), 046003 (2012).
152. C. Zheng, M. Zheng, P. Gong *et al., Biomaterials* **33**(22), 5603–5609 (2012).
153. H. J. Cho, H. Y. Yoon, H. Koo *et al., Biomaterials* **32**(29), 7181–7190 (2011).
154. K. Y. Choi, H. Chung, K. H. Min *et al., Biomaterials* **31**(1), 106–114 (2010).
155. Q. He, M. Ma, C. Wei and J. Shi, *Biomaterials* **33**(17), 4392–4402 (2012).
156. G. Jiang, K. Park, J. Kim *et al., Biopolymers* **89**(7), 635–642 (2008).
157. J. Heller, R. V. Sparer, G. M. Zentner, Poly(ortho esters), In M. Chaisin and R. Langer R (Eds.) *Biodegradable polymers as drug delivery systems* : New York, Dekker (1991), pp. 121–162.
158. K. E. Uhrich, S. M. Cannizzaro, R. S. Langer and K. M. Shakesheff, *Chem. Rev.* **99**(11), 3181–3198 (1999).
159. L. Y. Jie, L. L. Cai, L. J. Wang *et al., Int. J. Nanomedicine* 7, 3981–3989 (2012).
160. J. H. Kim, S. M. Bae, M. H. Na *et al., J. Control. Release* **157**(3), 493–499 (2012).
161. C. M. Lee, H. J. Jeong, S. J. Cheong *et al., Pharm. Res.* 27(4), 712–721 (2010).
162. A. P. Ranjan, K. Zeglam, A. Mukerjee, S. Thamake and J. K. Vishwanatha, *Nanotechnology* **22**(29), 295104-4484/22/29/295104. Epub 2011, 21 June 2011.
163. P. Yousefpour, F. Atyabi, E. Vasheghani-Farahani, A. A. Movahedi and R. Dinarvand, *Int. J. Nanomedicine* **6**, 1977–1990 (2011).
164. P. Calvo, B. Gouritin, H. Chacun *et al., Pharm. Res.* **18**(8), 1157–1166 (2001).
165. M. Auffan, J. Rose, J. Y. Bottero, G. V. Lowry, J. P. Jolivet and M. R. Wiesner, *Nat. Nanotechnol.* **4**(10), 634–641 (2009).
166. S. J. Soenen, P. Rivera-Gil, M. J. Montenegro, W. J. Parak, S. C. De Smedta and K. Braeckmans, *Nano Today* **6,** 446 (2011).

167. M. H. Mendonca Dias and P. C. Lauterbur, *Magn. Reson. Med.* **3**(2), 328–330 (1986).
168. H. B. Na, I. C. Song and T. Hyeon, *Adv. Mater.* **21,** 2133 (2009).
169. K. Kaaki, K. Herve-Aubert, M. Chiper *et al., Langmuir* **28**(2), 1496–1505 (2012).
170. M. M. Yallapu, S. F. Othman, E. T. Curtis, B. K. Gupta, M. Jaggi and S. C. Chauhan, *Biomaterials* **32**(7), 1890–1905 (2011).
171. E. A. Osborne, T. M. Atkins, D. A. Gilbert, S. M. Kauzlarich, K. Liu and A. Y. Louie, *Nanotechnology* **23**(21), 215602-4484/23/21/215602. Epub 2012 May 3 (2012).
172. Y. J. Wang, *Quant. Imaging Med. Surg.* **1,** 35 (2011).
173. J. Huang, X. Zhong, L. Wang, L. Yang and H. Mao, *Theranostics* **2**(1), 86–102 (2012).
174. M. Shen, F. Gong, P. Pang *et al., Int. J. Nanomedicine* 7, 3319–3332 (2012).
175. A. J. Cole, V. C. Yang and A. E. David, *Trends Biotechnol.* **29**(7), 323–332 (2011).
176. H. C. Huang, S. Barua, G. Sharma, S. K. Dey and K. Rege, *J. Control. Release* (2011).
177. M. Lu, M. H. Cohen, D. Rieves and R. Pazdur, *Am. J. Hematol.* **85**(5), 315–319 (2010).
178. J. K. Young, E. R. Figueroa and R. A. Drezek, *Ann. Biomed. Eng.* **40**(2), 438–459 (2012).
179. S. Jain, D. G. Hirst and J. M. O'Sullivan, *Br. J. Radiol.* **85**(1010), 101–113 (2012).
180. E. S. Glazer and S. A. Curley, *Cancer* **116**(13), 3285–3293 (2010).
181. Y. Liu and M. J. Welch, *Bioconjug. Chem.* **23**(4), 671–682 (2012).
182. H. Liu, T. Liu, X. Wu *et al., Adv. Mater.* **24**(6), 755–761 (2012).
183. H. Xie, B. Goins, A. Bao, Z. J. Wang, W. T. Phillips, *Int. J. Nanomedicine* 7, 2227–2238 (2012).
184. Y. Cheng, J. D. Meyers, A. M. Broome, M. E. Kenney, J. P. Basilion and C. Burda, *J. Am. Chem. Soc.* **133**(8), 2583–2591 (2011).
185. L. M. Liz-Marzan, *Langmuir* **22**(1), 32–41 (2006).
186. S. C. Boca and S. Astilean, *Nanotechnology* **21**(23), 235601-4484/21/23/235601. Epub 2010 May 13 (2010).
187. von G. Maltzahn, J. H. Park, A. Agrawal *et al., Cancer Res.* **69**(9), 3892–3900 (2009).
188. E. S. Day, P. A. Thompson, L. Zhang *et al., J. Neurooncol.* **104**(1), 55–63 (2011).
189. E. Y. Lukianova-Hleb, X. Ren, P. E. Constantinou *et al., PLoS One* 7(4), e34537 (2012).
190. C. C. Chien, H. H. Chen, S. F. Lai *et al., J. Nanobiotechnology* **10,** 10-3155-10-10 (2012).
191. H. Liu, H. Wang, R. Guo, X. Cao and J. Zhao, *Polym. Chem.* **1**, 1677 (2010).
192. A. M. Elhissi, W. Ahmed, I. U. Hassan, V. R. Dhanak, A . D'Emanuele, *J. Drug Deliv.* **2012,** 837327 (2012).
193. B. Kateb, V. Yamamoto, D. Alizadeh *et al., Methods Mol. Biol.* **651,** 307–317 (2010).
194. N. Daum, C. Tscheka, A. Neumeyer, M. Schneider, *Wiley Interdiscip. Rev. Nanomed. Nanobiotechnol.* **4**(1), 52–65 (2012).
195. N. Kobayashi, M. Naya, K. Mizuno, K. Yamamoto, M. Ema, J. Nakanishi, *Inhal. Toxicol.* **23**(13), 814–828 (2011).
196. B. Zhang, Y. Li, C. Y. Fang *et al., Small* **5**(23), 2716–2721 (2009).
197. L. K. Moore, M. Gatica, E. K. Chow and D. Ho, *Disruptive Sci. Technol.* **1**(1), 54 (2012).
198. R. Lam, D. Ho, *Exp. Opin Drug Deliv.* **6**(9), 883–895 (2009).
199. A. Bianco, K. Kostarelos and M. Prato, *Exp. Opin Drug Deliv.* **5**(3), 331–342 (2008).
200. Y. Liu, Y. Mi, J. Zhao and S. S. Feng, *Int. J. Pharm.* **421**(2), 370–378 (2011).
201. R. E. Serda, B. Godin, E. Blanco, C. Chiappini and M. Ferrari, *Biochim. Biophys. Acta.* **1810**(3), 317–329 (2011).

202. D. P. Ferris, J. Lu, C. Gothard *et al.*, *Small* 7(13), 1816–1826 (2011).
203. T. E. McCann, N. Kosaka, Y. Koide *et al.*, *Bioconjug. Chem.* **22**(12), 2531–2538 (2011).
204. F. Muhammad, M. Guo, W. Qi *et al.*, *J. Am. Chem. Soc.* **133**(23), 8778–8781 (2011).
205. O. Mykhaylyk, T. Sobisch, I. Almstatter *et al.*, *Pharm. Res.* **29**(5), 1344–1365 (2012).
206. Z. Wang, S. Zong, J. Yang, J. Li, Y. Cui, *Biosens. Bioelectron.* **26**(6), 2883–2889 (2011).
207. C. Wang, L. Cheng, Z. Liu, *Biomaterials* **32**(4), 1110–1120 (2011).
208. J. M. Rosenholm, C. Sahlgren, M. Linden, *Curr. Drug Targets* **12**(8), 1166–1186 (2011).
209. M. Liong, J. Lu, M. Kovochich *et al.*, *ACS Nano* **2**(5), 889–896 (2008).
210. W. Zhang, K. Gilstrap, L. Wu *et al.*, *ACS Nano* **4**(11), 6747–6759 (2010).
211. G. Wu, A. Mikhailovsky, H. A. Khant, C. Fu, W. Chiu and J. A. Zasadzinski, *J. Am. Chem. Soc.* **130**(26), 8175–8177 (2008).
212. M. C. Cochran, J. R. Eisenbrey, M. C. Soulen *et al.*, *Acad. Radiol.* **18**(11), 1341–1348 (2011).
213. C. Y. Lin, J. R. Li, H. C. Tseng, M. F. Wu and W. L. Lin, *Nanomedicine* **8**(6), 900–907 (2012).
214. E. K. Lim, Y. M. Huh, J. Yang, K. Lee, J. S. Suh and S. Haam, *Adv. Mater.* **23**(21), 2436–2442 (2011).
215. D. Putnam and J. Kopecek, *Bioconjug. Chem.* **6**(4), 483–492 (1995).
216. J. Kopecek, *Biomaterials* **5**(1), 19–25 (1984).
217. M. Salloum, R. Ma and L. Zhu., *Int. J. Hyperthermia* **25**(4), 309–321 (2009).
218. T. Kikumori, T. Kobayashi, M. Sawaki and T. Imai, *Breast Cancer Res. Treat* **113**(3), 435–441 (2009).
219. J. P. Fortin, F. Gazeau and C. Wilhelm, *Eur. Biophys. J.* **37**(2), 223–228 (2008).
220. A. Ito, H. Honda and T. Kobayashi, *Cancer Immunol. Immunother.* **55**(3), 320–328 (2006).
221. E. S. Day, J. G. Morton and J. L. West, *J. Biomech. Eng.* **131**(7), 074001 (2009).
222. D. P. O'Neal, L. R. Hirsch, N. J. Halas, J. D. Payne and J. L. West, *Cancer Lett.* **209**(2), 171–176 (2004).
223. G. S. Terentyuk, G. N. Maslyakova, L. V. Suleymanova *et al.*, *J. Biomed. Opt.* **14**(2), 021016 (2009).
224. J. L. Au, D. Zheng, F. Zhou, Z. Li and X. Li, *ACS Nano* **2**(8), 1645–1652 (2008).
225. X. Huang, I. H. El-Sayed, W. Qian and M. A. El-Sayed, *J. Am. Chem. Soc.* **128**(6), 2115–2120 (2006).
226. L. R. Hirsch, R. J. Stafford, J. A. Bankson *et al.*, *Proc. Natl. Acad. Sci. USA* **100**(23), 13549–13554 (2003).
227. F. Zhou, D. Xing, Z. Ou, B. Wu, D. E. Resasco and W. R. Chen, *J. Biomed. Opt.* **14**(2), 021009 (2009).
228. W. R. Chen, R. L. Adams, A. K. Higgins, K. E. Bartels and R. E. Nordquist, *Cancer Lett.* **98**(2), 169–173 (1996).
229. S. C. Gnyawali, Y. Chen, F. Wu *et al.*, *Med. Biol. Eng. Comput.* **46**(2), 159–168 (2008).
230. P. Cherukuri, E. S. Glazer and S. A. Curley, *Adv. Drug Deliv. Rev.* **62**(3), 339–345 (2009).
231. K. W. Hu, T. M. Liu, K. Y. Chung *et al.*, *J. Am. Chem. Soc.* **131**(40), 14186–14187 (2009).
232. R. Hergt and S. Dutz, *J. Magn. Magn. Mat.* **311**(1), 187–192 (2007).

233. H. Kato, M. Hiraoka, T. Nakajima and T. Ishida, *Int. J. Hyperthermia* **1**(1), 15–28 (1985).
234. H. Saito, K. Mitobe, A. Ito *et al., Cancer Sci.* **99**(4), 805–809 (2008).
235. S. E. Barry, *Int. J. Hyperthermia* **24**(6), 451–466 (2008).
236. S. A. Curley, P. Cherukuri, K. Briggs *et al., J. Exp. Ther. Oncol.* 7(4), 313–326 (2008).
237. C. J. Gannon, C. R. Patra, R. Bhattacharya, P. Mukherjee and S. A. Curley, *J. Nanobiotechnology* **6,** 2 (2008).
238. H. Yang, C. Lou, M. Xu, C. Wu, H. Miyoshi and Y. Liu., *Int. J. Nanomedicine* **6,** 2023–2032 (2011).
239. L. Shan, S. Cui, C. Du *et al., Biomaterials* **33**(1), 146–162 (2012).
240. M. N. Koopaei, R. Dinarvand, M. Amini *et al., Int. J. Nanomedicine* **6,** 1903–1912 (2011).
241. T. R. Daniels, E. Bernabeu, J. A. Rodriguez *et al., Biochim. Biophys. Acta.* **1820**(3), 291–317 (2012).
242. R. Mahato, W. Tai and K. Cheng, *Adv. Drug Deliv. Rev.* **63**(8), 659–670 (2011).
243. L. Cattel, M. Ceruti and F. Dosio, *J. Chemother.* **16(**Suppl 4), 94–97 (2004).
244. E. Mastrobattista, G. A. Koning and G. Storm, *Adv. Drug Deliv. Rev.* **40**(1&2), 103–127 (1999).
245. J. D. Byrne, T. Betancourt and L. Brannon-Peppas, *Adv. Drug Deliv. Rev.* **60**(15), 1615–1626 (2008).
246. S. A. Kularatne, P. S. Low, Targeting of nanoparticles: Folate receptor. In S. R. Grobmyer and B. M. Moudgil, (Eds.) *Cancer nanotechnology: Methods and protocols.* New York: Springer (2010), pp. 249–265.
247. P. S. Low and A. C. Antony, *Adv. Drug Deliv. Rev.* **56**(8), 1055–1058 (2004).
248. Y. Lu, E. Sega, C. P. Leamon and P. S. Low, *Adv. Drug Deliv. Rev.* **56**(8), 1161–1176 (2004).
249. J. A. Reddy, V. M. Allagadda and C. P. Leamon, *Curr. Pharm. Biotechnol.* **6**(2), 131–150 (2005).
250. J. A. Reddy and P. S. Low, *Crit. Rev. Ther. Drug Carrier Syst.* **15**(6), 587–627 (1998).
251. S. Franzen, *Exp. Opin Drug Deliv.* **8**(3), 281–298 (2011).
252. I. J. Majoros, C. R. Williams, A. Becker and J. R. Baker, Jr., *Wiley Interdiscip. Rev. Nanomed. Nanobiotechnol.* **1**(5), 502–510 (2009).
253. R. J. Lee and P. S. Low, *Biochim. Biophys. Acta.* **1233**(2), 134–144 (1995).
254. Y. Zhang, N. Kohler and M. Zhang, *Biomaterials* **23**(7), 1553–1561 (2002).
255. C. Zhang, L. Zhao, Y. Dong, X. Zhang, J. Lin and Z. Chen, *Eur. J. Pharm. Biopharm.* **76**(1), 10–16.
256. C. J. Mathias, D. Hubers, P. S. Low and M. A. Green, *Bioconjug. Chem.* **11**(2), 253–257 (2000).
257. W. Guo, G. H. Hinkle and R. J. Lee, *J. Nucl. Med.* **40**(9), 1563–1569 (1999).
258. H. C. Loughrey, A. Ferraretto, A. M. Cannon *et al., FEBS Lett.* **332**(1&2), 183–188 (1993).
259. T. M. Allen, I. Ahmad, D. E. Lopes de Menezes and E. H. Moase, *Biochem. Soc. Trans.* **23**(4), 1073–1079 (1995).
260. T. M. Allen, E. Brandeis, C. B. Hansen, G. Y. Kao and S. Zalipsky, *Biochim. Biophys. Acta.* **1237**(2), 99–108 (1995).
261. C. B. Hansen, G. Y. Kao, E. H. Moase, S. Zalipsky and T. M. Allen, *Biochim. Biophys. Acta.* **1239**(2), 133–144 (1995).

262. Y. Yuyama, M. Tsujimoto, Y. Fujimoto and N. Oku, *Cancer Lett.* **155**(1), 71–77 (2000).
263. E. Kedar, E. Braun, Y. Rutkowski, N. Emanuel and Y. Barenholz, *J. Immunother. Emphasis Tumor Immunol.* **16**(2), 115–124 (1994).
264. E. Kedar, Y. Rutkowski, E. Braun, N. Emanuel and Y. Barenholz, *J. Immunother. Emphasis Tumor Immunol.* **16**(1), 47–59 (1994).
265. H. H. Guan, W. Budzynski, R. R. Koganty *et al., Bioconjug. Chem.* **9**(4), 451–458 (1998).
266. R. E. Eliaz and F. C. Szoka, Jr., *Cancer Res.* **61**(6), 2592–2601 (2001).
267. J. W. Park, D. B. Kirpotin, K. Hong *et al., J. Control. Release* **74**(1–3), 95–113 (2001).
268. W. Tai, R. Mahato and K. Cheng, *J. Control. Release* **146**(3), 264–275 (2010).
269. H. Yan, L. Wang, J. Wang *et al., ACS Nano* **6**(1), 410–420 (2012).
270. J. T. Bushberg, J. A. Seibert, E. M. Leidholt and J. M. Boone, (Eds.), *The essential physics of medical imaging*. Philadelphia: Lippincott Williams & Wilkins publishers (2002).
271. J. G. Fujimoto, *Nat. Biotechnol.* **21**(11), 1361–1367 (2003).
272. P. H. Tomlins and R. K. Wang, *J. Phys. D, Appl. Phys.* **38,** 2519 (2005).
273. C. Quek and K. W. Leong, *Nanomaterials* **2,** 92 (2012).
274. G. Qin, Z. Li, R. Xia *et al., Nanotechnology* **22**(15), 155605-4484/22/15/155605. Epub 2011 Mar 10 (2011).
275. S. H. Park, H. J. Gwon and J. Shin, *J. Label Compd. Radiopharm.* **49**, 1163 (2006).
276. K. K. Halder, B. Mandal, M. C. Debnath, H. Bera, L. K. Ghosh and B. K. Gupta, *J. Drug Target.* **16**(4), 311–320 (2008).
277. R. Torres Martin de Rosales, R. Tavare, A. Glaria, G. Varma, A. Protti and P. J. Blower, *Bioconjug. Chem.* **22**(3), 455–465 (2011).
278. B. E. Ocampo-Garcia, M. Ramirez Fde, G. Ferro-Flores *et al., Nucl. Med. Biol.* **38**(1), 1–11 (2011).
279. S. Subramanian, P. Dandekar, R. Jain *et al., Cancer Biother. Radiopharm.* **25**(6), 637–644 (2010).
280. W. A. Volkert and S. Jurisson, *Topics in Current Chemistry* **176**, 123 (1996).
281. B. Ocampo-Garcia, G. Ferro-Flores, E. Morales-Avila and M. Ramirez Fde, *Nucl. Med. Commun.* **32**(11), 1095–1104 (2011).
282. W. T. Phillips, A. S. Rudolph, B. Goins, J. H. Timmons, R. Klipper and R. Blumhardt, *Int. J. Rad. Appl. Instrum B* **19**(5), 539–547 (1992).
283. A. Soundararajan, A. Bao, W. T. Phillips, R. Perez, 3rd and B. A. Goins, *Nucl. Med. Biol.* **36**(5), 515–524 (2009).
284. A. Bao, B. Goins, R. Klipper, G. Negrete and W. T. Phillips, *J. Pharmacol. Exp. Ther.* **308**(2), 419–425 (2004).
285. S. Liu, B. Jia, R. Qiao *et al., Mol. Pharm.* **6**(4), 1074–1082 (2009).
286. A. Chrastina and J. E. Schnitzer, *Int. J. Nanomedicine* **5**, 653–659 (2010).
287. G. Ting, C. H. Chang and H. E. Wang, *Anticancer Res.* **29**(10), 4107–4118 (2009).
288. M. Hamoudeh, M. A. Kamleh, R. Diab and H. Fessi, *Adv. Drug Deliv. Rev.* **60**(12), 1329–1346 (2008).
289. N. K. Devaraj, E. J. Keliher, G. M. Thurber, M. Nahrendorf and R. Weissleder, *Bioconjug. Chem.* **20**(2), 397–401 (2009).
290. M. S. Tartis, D. E. Kruse, H. Zheng *et al., J. Control. Release* **131**(3), 160–166 (2008).
291. S. Liu, D. Li, C. W. Huang *et al., Mol. Imaging Biol.* **14**(6), 718–724 (2012).
292. F. Roesch and P. J. Riss, *Curr. Top Med. Chem.* **10**(16), 1633–1668 (2010).

293. M. Lyra, P. Charalambatou, E. Roussou, S. Fytros and I. Baka, *Hell J. Nucl. Med.* **14**(1), 49–55 (2011).
294. P. Gould, *Nature* **460**(7253), 312–313 (2009).
295. M. W. Brechbiel and *Q. J. Nucl. Med. Mol. Imaging* **52**(2), 166–173 (2008).
296. A. Pourjavadi, M. Akhlaghi and A. R. Jalilian, *Nukleonika* **56**(1), 35 (2011).
297. A. Almutairi, R. Rossin, M. Shokeen *et al., Proc. Natl. Acad. Sci. USA* **106**(3), 685–690 (2009).
298. A. Ruggiero, C. H. Villa, J. P. Holland *et al., Int. J. Nanomedicine* **5**, 783–802 (2010).
299. C. T. Badea, K. K. Athreya, G. Espinosa *et al., PLoS One* 7(4), e34496 (2012).
300. J. F. Hainfeld, D. N. Slatkin, T. M. Focella and H. M. Smilowitz, *Br. J. Radiol.* **79**(939), 248–253 (2006).
301. C. Peng, L. Zheng, Q. Chen *et al., Biomaterials* **33**(4), 1107–1119 (2012).
302. E. Boote, G. Fent, V. Kattumuri *et al., Acad. Radiol.* **17**(4), 410–417 (2010).
303. T. Reuveni, M. Motiei, Z. Romman, A. Popovtzer and R. Popovtzer, *Int. J. Nanomedicine* **6,** 2859–2864 (2011).
304. G. Pratx, C. M. Carpenter, C. Sun, R. P. Rao and L. Xing, *Opt. Lett.* **35**(20), 3345–3347 (2010).
305. Z. Liu, Z. Li, J. Liu *et al., Biomaterials* **33**(28), 6748–6757 (2012).
306. O. Rabin, J. Manuel Perez, J. Grimm, G. Wojtkiewicz and R. Weissleder, *Nat. Mater.* **5**(2), 118–122 (2006).
307. H. Kobayashi, S. Kawamoto, M. Bernardo, M. W. Brechbiel, M. V. Knopp and P. L. Choyke, *J. Control. Release.* **111**(3), 343–351 (2006).
308. A. Xia, M. Chen, Y. Gao, D. Wu, W. Feng and F. Li, *Biomaterials* **33**(21), 5394–5405 (2012).
309. M. F. Kircher, A. de la Zerda, J. V. Jokerst *et al., Nat. Med.* **18**(5), 829–834 (2012).
310. A. Vaidya, Y. Sun, Y. Feng, L. Emerson, E. K. Jeong and Z. R. Lu, *Pharm. Res.* **25**(9), 2002–2011 (2008).
311. L. Fan, Q. Wu and M. Chu, *Int. J. Nanomedicine.* 7, 3071–3080 (2012).
312. Y. H. Jeon, Y. H. Kim, K. Choi *et al., Mol. Imaging Biol.* **12**(2), 155–162 (2010).
313. Y. Cheng, J. D. Meyers, A. M. Broome, M. E. Kenney, J. P. Basilion and C. Burda, *J. Am. Chem. Soc.* **133**(8), 2583–2591 (2011).
314. Y. Tang and A. J. McGoron, *J. Photochem. Photobiol. B, Biol.* **97**(3), 138–144 (2009).
315. W. Scheuer, van G. M. Dam, M. Dobosz, M. Schwaiger and V. Ntziachristos, *Sci. Trans. Med.* **4**(134), 134ps11 (2012).
316. B. Zhu, I. C. Tan, J. C. Rasmussen and E. M. Sevick-Muraca, *Technol. Cancer Res. Treat.* **11**(1), 95–104 (2012).
317. X. Peng, F. Song, E. Lu *et al., J. Am. Chem. Soc.* **127**(12), 4170–4171 (2005).
318. J. L. Gragg, *Synthesis of near-infrared heptamethine cyanine dyes,* PhD Chemistry, Georgia State University (2010).
319. X. Yang, C. Shi, R. Tong *et al., Clin. Cancer Res.* **16**(10), 2833–2844 (2010).
320. H. Lee, J. C. Mason and S. Achilefu, *J. Org. Chem.* **73**(2), 723–725 (2008).
321. X. Peng and D. Draney, *LabPlus International,* April/May:1-24-2013 (2004).
322. L. Song, H. Li, U. Sunar *et al., Int. J. Nanomedicine* **2**(4), 767–774 (2007).
323. M. Shopova, D. Woehrle, V. Mantareva and S. Mueller, *J. Biomed. Opt.* **4**(3), 276–285 (1999).
324. B. Ebert, B. Riefke, U. Sukowski and K. Licha, *J. Biomed. Opt.* **16**(6), 066003 (2011).

325. V. Saxena, M. Sadoqi and J. Shao, *J. Pharm. Sci.* **92**(10), 2090–2097 (2003).
326. A. Fernandez-Fernandez, R. Manchanda, T. Lei *et al., Mol. Imaging* 1–10 (2011).
327. G. Sun, M. Y. Berezin, J. Fan *et al., Nanoscale* **2**(4), 548–558 (2010).
328. Z. Zhang, M. Y. Berezin, J. L. Kao, A. d'Avignon, M. Bai and S. Achilefu, *Angew. Chem. Int. Ed. Engl.* **47**(19), 3584–3587 (2008).
329. M. Y. Berezin, K. Guo, W. Akers *et al., Biochemistry* **50**(13), 2691–2700 (2011).
330. V. Saxena, M. Sadoqi and J. Shao, *Int. J. Pharm.* **308**(1&2), 200–204 (2006).
331. H. S. Park, J. E. Lee, M. Y. Cho *et al., Nanotechnology* **22**(46), 465603-4484/22/46/465603. Epub 2011 Oct 27 (2011).
332. V. Saxena, M. Sadoqi and J. Shao, *J. Photochem. Photobiol. B* **74**(1), 29–38 (2004).
333. S. Mordon, T. Desmettre, J. M. Devoisselle and S. Soulie, *Lasers Surg. Med.* **20**(2), 131–141 (1997).
334. E. M. Sevick-Muraca, *Annu. Rev. Med.* **63**, 217–231 (2012).
335. C. Hirche, D. Murawa, Z. M. S. Kneir and M. Hiinerbein, *Breast Cancer Res. Treat.* **121,** 373 (2010).
336. M. Hutteman, J. S. Mieog, van J. der. R. Vorst *et al., Breast Cancer Res. Treat.* **127**(1), 163–170 (2011).
337. S. J. Erickson, S. L. Martinez, J. Gonzalez, L. Caldera and A. Godavarty, *Biomed. Opt. Exp.* **1**(1), 126–134 (2010).
338. J. Ge, S. J. Erickson and A. Godavarty, *Appl. Opt.* **49**(23), 4343–4354 (2010).
339. Q. Meng, M. Yu, B. Gu *et al., J. Drug Target.* **18**(6), 438–446 (2010).
340. A. Masotti, P. Vicennati, F. Boschi, L. Calderan, A. Sbarbati and G. Ortaggi, *Bioconjug. Chem.* **19**(5), 983–987 (2008).
341. S. I. Prajapati, C. O. Martinez, A. N. Bahadur *et al., Mol. Imaging* **8**(1), 45–54 (2009).
342. C. Li, W. Wang, Q. Wu *et al., Nucl. Med. Biol.* **33**(3), 349–358 (2006).
343. Z. Huang, *Technol. Cancer Res. Treat.* **4**(3), 283–293 (2005).
344. G. R. Reddy, M. S. Bhojani, P. McConville *et al., Clin. Cancer Res.* **12**(22), 6677–6686 (2006).
345. L. Sampath, S. Kwon, M. A. Hall, R. E. Price and E. M. Sevick-Muraca, *Transl. Oncol.* **3**(5), 307–317 (2010).
346. M. Nahrendorf, E. Keliher, B. Marinelli *et al., Proc. Natl. Acad. Sci. USA* **107**(17), 7910–7915 (2010).
347. A. A. Burns, J. Vider, H. Ow *et al., Nano Lett.* **9**(1), 442–448 (2009).
348. J. Culver, W. Akers and S. Achilefu, *J. Nucl. Med.* **49**(2), 169–172 (2008).
349. S. Santra, C. Kaittanis, J. Grimm and J. M. Perez, *Small* **5**(16), 1862–1868 (2009).
350. M. Kumar, M. Yigit, G. Dai, A. Moore and Z. Medarova, *Cancer Res.* **70**(19), 7553–7561 (2010).
351. R. Cartier, L. Kaufner, B. R. Paulke *et al., Nanotechnology* **18**(195102), 12-1 (2007).
352. R. Madru, P. Kjellman, F. Olsson *et al., J. Nucl. Med.* **53**(3), 459–463 (2012).
353. N. Lee, H. R. Cho, M. H. Oh *et al., J. Am. Chem. Soc.* **134**(25), 10309–10312 (2012).
354. A. Zhang, Y. Tu, S. Qin *et al., J. Colloid. Interface Sci.* **372**(1), 239–244 (2012).
355. T. Luo, P. Huang, G. Gao *et al., Opt. Exp.* **19**(18), 17030–17039 (2011).
356. J. M. Criscione, L. W. Dobrucki, Z. W. Zhuang *et al., Bioconjug. Chem.* **22**(9), 1784–1792 (2011).
357. G. Roncari, A. Benevento, L. Bianchi *et al., J. Nucl. Biol. Med.* **37**(1), 21–25 (1993).

358. S. A. Gulec, F. L. Moffat, R. G. Carroll and D. N. Krag, *Q. J. Nucl. Med.* **41**(3), 251–261 (1997).
359. K. Tsugawa, M. Noguchi, K. Miwa *et al., Breast Cancer* 7(1), 87–94 (2000).
360. S. Usmani, H. A. Khan, F. abu Huda *et al., Hell. J. Nucl. Med.* **13**(1), 30–34 (2010).
361. I. Stoffels, T. Poeppel, C. Boy *et al., J. Eur. Acad. Dermatol. Venereol.* **26**(3), 308–313 (2012).
362. D. Pan, X. Cai, C. Yalaz *et al., ACS Nano* **6**(2), 1260–1267 (2012).
363. Y. Sun, M. Yu, S. Liang *et al., Biomaterials* **32**(11), 2999–3007 (2011).
364. M. H. Oh, N. Lee, H. Kim *et al., J. Am. Chem. Soc.* **133**(14), 5508–5515 (2011).
365. E. M. Sevick-Muraca and J. C. Rasmussen, *J. Biomed. Opt.* **13**(4), 041303 (2008).
366. O. Khullar, J. V. Frangioni, M. Grinstaff and Y. L. Colson, *Semin. Thorac. Cardiovasc. Surg.* **21**(4), 309–315 (2009).
367. G. Ravizzini, B. Turkbey, T. Barrett, H. Kobayashi and P. L. Choyke, *Wiley Interdiscip. Rev. Nanomed. Nanobiotechnol.* **1**(6), 610–623 (2009).
368. R. Jain, P. Dandekar and V. Patravale, *J. Control. Release* **138**(2), 90–102 (2009).
369. M. Fujimaki, J. Yokoyama, S. Ohba *et al., Head Neck Oncol.* **4**(2), 50 (2012).
370. S. X. Wang, A. Bao, S. J. Herrera *et al., Clin. Cancer Res.* **14**(12), 3975–3983 (2008).
371. C. P. Parungo, S. Ohnishi, S. W. Kim *et al., J. Thorac. Cardiovasc. Surg.* **129**(4), 844–850 (2005).
372. C. G. Hadjipanayis, H. Jiang, D. W. Roberts and L. Yang, *Semin. Oncol.* **38**(1), 109–118 (2011).
373. R. Trehin, J. L. Figueiredo, M. J. Pittet, R. Weissleder, L. Josephson and U. Mahmood, *Neoplasia* **8**(4), 302–311 (2006).
374. R. W. Holloway, R. A. Bravo, J. A. Rakowski *et al., Gynecol. Oncol.* **126**(1), 25–29 (2012).
375. S. Tobis, J. Knopf, C. Silvers *et al., J. Urol.* **186**(1), 47–52 (2011).
376. H. Liao, M. Noguchi, T. Maruyama *et al., Med. Image Anal.* **16**(3), 754–766 (2012).
377. W. C. Lin, S. A. Toms, M. Johnson, E. D. Jansen and A. Mahadevan-Jansen, *Photochem. Photobiol.* **73**(4), 396–402 (2001).
378. W. C. Lin, D. I. Sandberg, S. Bhatia, M. Johnson, G. Morrison and J. Ragheb, *J. Biomed. Opt.* **14**(1), 014028 (2009).
379. T. Lei, R. Manchanda, A. Fernandez-Fernandez, Y. C. Huang, D. Wright, A. J. McGoron, *RSC Advances* **4**, 17959–17968 (2014).

Chapter 8

Fluorescent Silica Nanoparticles for Medical Imaging

Jeremy Tharkur[‡], *Maria Alejandra Ricaurte*[†,§] *and Swadeshmukul Santra*[*,¶]

[*] *NanoScience Technology Center, Advanced Materials Processing and Analysis Center, University of Central Florida Orlando, FL 32826, USA*

[†] *sanofi pasteur VaxDesign Campus 2501 Discovery Dr., STE 300, Orlando, FL 32826, USA*

[‡] *jeremyt@knights.ucf.edu*

[§] *Maria.Ricaurte@sanofipasteur.com*

[¶] *Swadeshmukul.Santra@ucf.edu*

1. Introduction

The study of silica nanoparticles (SiNPs) has widely expanded biomedical imaging through fluorescent systems capable of qualitative and quantitative identification. Over the past couple decades, applications of SiNPs in bioimaging and biosensing have led to increased medical understanding about bacteria, mammalian and cancerous cells.[1–5] The robust silica technology is optically transparent allowing for excitation and emission of encapsulated organic and inorganic dyes and agents without interference from the silica shell. In addition, the copious hydroxyl groups that surround the surface of the SiNP are easy to modify which enables further surface modification, chemical stability and improved biocompatibility.[1] In contrast to other widely used nanomaterials, SiNPs have a well-defined tunable structure, surface chemistry and surface area which all contribute to establishing this system as an ideal biocompatible agent for both biosensing and probing.[6]

During the silica growth process, the fluorescent dye (such as fluorescein isothiocyanate (FITC), tris(bipyridine)ruthenium(II) dichloride (RuBpy), rhodamine, coumarin or thioflavin T)[7–10] is embedded within the silica matrix. As the SiNP grows around the dye, it creates a protective shell shielding it from background environment fluorescence quenchers (substances that reduce fluorescence emission).[11] However, the ability to manipulate the fluorescence and alter the optical properties between completely quenched or completely unquenched is how the fluorescent silica nanoparticle (FSNP) finds its use.[12,13]

These nanoparticles (NPs) are capable of conjugation to multiple dyes (internal and external labeling) whereby a reference and sensor dye are covalently bound to the matrix and outer shell of the core/shell architecture, respectively.[14] This dual label enables the FSNP for labeling when the external dye is quenched and only the internal dye is observable. In addition, the conjugation of quenchers to the surface of the FSNP results in the absence of fluorescence. Once this bound particle is removed *via* intracellular molecules, the fluorescence restores.[15] Fluorescence restoration gives the FSNP activatable properties which help for real-time tracking.

Different cell types respond to FSNPs depending on their shape and mode of interaction. Most approaches to prokaryotic labeling involve extracellular labeling due to their reduced cell wall pore sizes. However, cell wall modification or degradation has proven helpful in assisting the intracellular labeling at the cost of cell viability.[1,7,16–26] As for eukaryotic labeling, larger cells require more direct mechanisms. Some cells will uptake the particle through phagocytosis (immune cells and cancer cells).[2,27–45] Other smaller molecules (proteins, antibodies, cofactors) require specialized conjugation to the SiNP.[38,46–61] Even the addition of magnetic NP conjugation to the FSNP[3,4,62] greatly assists in the recognition and tracking of different organic molecules.

The following chapter explains the significance of FSNPs as a reliable foundation for bioimaging and biosensing. A brief overview of common synthesis techniques, along with several examples regarding the specific functionality of the NP for various applications, is provided in order to show some of the revolutionary uses for FSNPs that have impacted medical imaging over the years. Currently, FSNP studies are carried out *in vitro,* however, human application is the future goal.

2. Biomedical Overview

FSNPs have a broad spectrum of applications that have been exploited for their efficient biodetection capabilities. The usefulness of FSNPs arises from their multifunctional capability. Design, structure, surface, labeling, core and shell can all be manipulated to meet a number of desired medically relevant diagnostic needs.[63] Figure 1 shows how FSNPs are synthesized and the color variation they can be designed for. Figure 1(a) demonstrates how the silica outer shell grows around the dye-labeled precursor, making the fairly monodispersed FSNPs visible through transmission electron microscopy in Fig. 1(b). Figure 1(c) shows the variety of fluorescent organic dyes that can be utilized. This section will address the many uses of FSNP starting from small biochemical molecules and advancing through large molecules, cells, tissues and body systems.

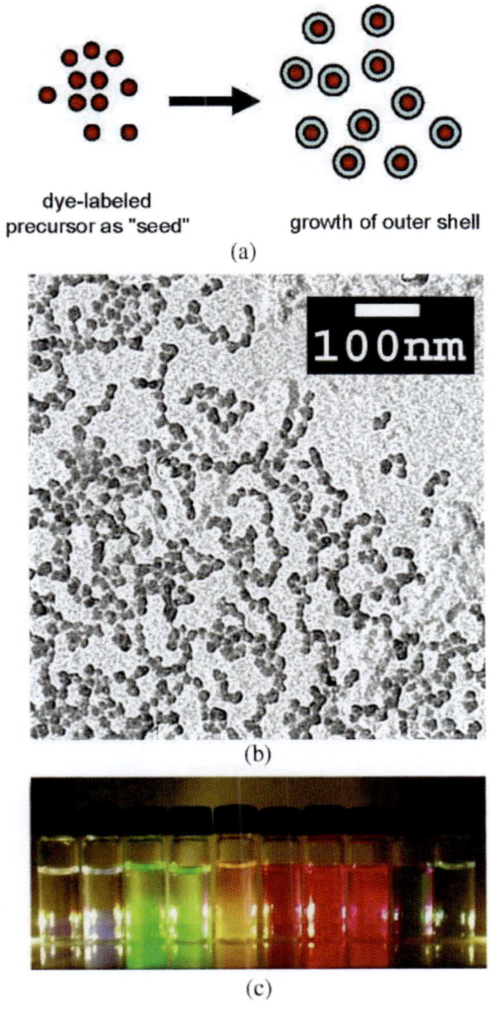

Fig. 1. (a) A scheme showing the synthesis of the core-shell architecture.[110] (b) Transmission electron microscopy image 30 nm particles. (c) A spectrum of FSNPs with different organic dyes.

2.1. *Sensors for adenosine-5′-triphosphate*

Adenosine-5′-triphosphate's (ATP) importance as the major chemical energy source for living cells has been the driving motivation behind developing a selective biomarker capable of tracking and sensing ATP concentrations for measuring cell viability and damage. Over the years, FSNP technology has enabled researchers to indicate for ATP through both chem/bioluminescence reactions[64] and DNA/RNA aptamers.[65]

Indicator dyes can be complexed with metal cations to detect for ATP. When these complexes are conjugated to FSNPs, the differences in fluorescence dye emission intensity can be used to detect and measure ATP. After the indicator dye interacts with the phosphate groups of ATP, the fluorescence changes are compared against the original FSNP fluorescence. Though this approach has its advantages, complications regarding ATP selectivity from the dye arise because of interference from the complexed metal ion.[64] For that reason, aptamers have moved to the forefront of ATP detection strategies.

Aptamers are single-stranded nucleic acids (DNA/RNA) that undergo conformational changes after binding to a specific target. Since aptamers are strands of DNA chosen from sequences that code for specific molecules, they have been employed to selectively probe for a number of small molecules and cell types.[44] In addition to being DNA selective, aptamers can utilize more robust, convincing fluorescence detection through the structure-switching/fluorescence–dequenching mechanism.[65] After target binding and conformation changing, the resulting shape creates distance changes between the fluorescent molecules and quenching molecules. The change in distance modifies the energy transfer between the molecules which is detected both optically and electrically.

2.2. *Sensors for hydrogen peroxide*

Hydrogen peroxide (H_2O_2) is produced by many biochemical reactions that take place in both prokaryotic and eukaryotic cells. Depending on the organic reaction, H_2O_2 acts as either a powerful oxidizing or reducing agent. The need for sensitive detection of H_2O_2 arose from the oxidative stress and pre-apoptotic damage that result from H_2O_2 production and lead to cell death.[66] While the common methods for H_2O_2 detection are based on enzyme affinity, the potential loss of enzyme activity mixed with the financial and labor expenses have forced researches to explore other means. FSNPs sustain the fluorescence of the doped organic dye from environmental photobleaching by encapsulating the dye within the silica. Photobleaching (loss of fluorescence) of the organic dye can only be achieved by passivating the silica shell. For example, when potassium iodide and H_2O_2 undergo redox reactions, the produced iodine is capable of reaching the organic dye and quench its fluorescence.[66] The degree of quenching serves as the indication for the concentration of H_2O_2 in the sample.

2.3. *Sensors for proteins*

Almost every biochemical reaction is dependent on the assistance of proteins and enzymes for operation.[67] Other processes, such as host defenses, rely on specific degradation enzymes to defend against invading pathogens and infections. Lysozyme, one of the host defense enzymes that fights against bacterial infections, cleaves the acetyl groups of the bacteria cell walls, whereas its concentration levels are studied in medicine to assess infections. Additionally, many aptamers, antibodies and signaling peptides are selective to protein biomarkers, which can be helpful in protein tracking for inflammation or metabolism.[47,53,68] Signaling peptides, like polyelectrolytes (CPE)[53] are fluorescent molecules that quench when interacting with target proteins.

Combining CPEs with other fluorescent dyes allows for target detection because of a process call fluorescence resonance energy transfer (FRET). FRET involves using the emission energy from a primary fluorophore to excite a secondary fluorophore. This method includes two differently labeled dyes doped within the silica matrix. Both the emission and excitation wavelengths exist within a short nanometer distance of each other. This method results in NPs that are capable of emitting double wavelengths through a single wavelength excitation and thus have the capacity to detect double targets.[69]

FRET allows for the energy emitted from the CPE fluorescence to excite another fluorescent molecule. When CPEs become conjugated to FSNPs, the combined vehicle becomes an effective detector for a wide range of proteins and enzymes like lysozyme because quenching of the CPE prevents excitation of the FRET dye as well. If the FRET dye quenches without CPE quenching, then target detection was not achieved and the CPE fluorescence emission intensity will still be visible.

2.4. *Sensors for specialized cells*

Expanding studies on FSNPs in medical diagnostics focus on measuring concentrations and activity of specific immune cells to help diagnose certain diseases. The immune system contains a vast array of cell types all working in tandem to protect against foreign cell invasion. At the heart of the immune system are the lymph nodes, which serve as the source for immune cells. Since deviations in lymph node structure and activity have been linked to many cancers tracking these immune cells as they circulate through the bloodstream using FSNPs could lead progressive cancer detection and therapy strategies.[70]

Monocytes are cells of the innate immune system that differentiate into macrophages and dendritic cells. Their cell count is used to assess a variety of inflammation associated diseases. Macrophages are white blood cells with phagocytic operations capable of ingesting other particles and have been associated with plaque formation in vessel walls.[49,71] Dendritic cells (DC) are a class of antigen presenting cell (APC) that bridge the gap between antigens and the adaptive immune response.[72] T-lymphocytes (T-cells) are a special group of cell-mediated immunity cells.[73] They are cytotoxic, capable of remembering antigens, regulate other T-cells, and help other immune cells protect against foreign antigens. $CD4^+$ T-cells are a special class of T-lymphocytes that assist white blood cells. Physicians use their cell count for HIV/AIDS screening.[74]

The ability of FSNPs to intracellulary detect specific cells, due to their reliable fluorescence properties and easily functionable outer shell, has enabled researchers to detect immune cell concentrations. Radioactive fluorescent dyes, specialized macrophage factors, manipulated shapes and pore sizes, inorganic fluorescent NP doping rather than conventional organic dye doping and antigen conjugation are just samples of applied functions for FSNPs for cells recognition and measurements.[70–74]

2.5. *Sensors for blood and plasma*

Vascular diseases contribute to some of the leading causes of death worldwide.[49] For this reason, the FSNP technology has been coupled with magnetic resonance imaging (MRI) for effective blood and plasma measuring. Devices capable of monitoring blood and plasma are useful for evaluating vascular efficiency. Many sensors are bio-toxic due to their composition. The toxicity arises from the body's inability to process and excrete the particle efficiently causing accumulation. Tracking the elimination efficiency of foreign substances in the blood by the reticuloendothelial system (RES) (mainly composed of the lymph nodes, liver and spleen)[75] provides reliable information about organ health, because accumulation of foreign substances is an indication of poor organ health.

MRI can be coupled with magnetic FSNPs for improved visualization. Plasma volume (PV) comprises about 25% of extracellular fluid and contains several vascular receptors necessary for monitoring cardiovascular function and organ perfusion.[60] Typical organic tracer dyes for PV are unreliable medical imagers because of their unstable fluorescence characteristics and complicated mechanism of clearance. However, when these dyes are attached to FSNPs they provide straightforward, single measurements of PV within 30 min and have minimal adverse affects to biological systems.[60]

2.6. *Sensors for stem cells*

Stem cell labeling and tracking with FSNPs has led to many advances in stem cell research. The attractiveness of stem cells comes from their applications in cell therapy as undifferentiated cells. Due to their magnetic functionality[4] and specialized dyes,[3] FSNPs can characterize both the origin and fate of stem cells after cellular uptake of the NP. The coating of FSNPs improves the attractiveness of the NP for stem cell uptake; the less cytotoxic the probe, the longer the tracking and imaging potential. This approach can also be applied to stem cell transplantation therapy to assess desired cell differentiation and any tissue/organ improvements or defects.[3]

3. Synthesis Methods

3.1. *Overview*

Compared to gold NPs and quantum dots, silicon dioxide is not inherently equipped with strong florescence usable for sensitive imaging applications. There are two advised techniques to create dye-doped SiNP: The reverse microemulsion method[2,28,29] and the Stober's sol–gel method.[76,77] SiNPs are made fluorescent through a process known as dye-doping,[78] which involves incorporating fluorescent dyes inside the matrix of the NP. In order to obtain maximum fluorescence, it is important to find the optimal amount of dye to be incorporated. Figure 2 shows the importance of dye concentration inside the SiNP, and how optimizing the loading efficiency regulates the overall fluorescence emission. Figures 2(a) and 2(c) are both schematic representations showing inappropriate amounts of dye additions and the effects that they have on fluorescence. Monitoring and testing the optimal amount of dye required is an essential step for obtaining maximum fluorescence emission. Utilizing dyes with large absorption coefficients is one approach to improving fluorescence emission, assuming they do not experience photobleaching during imaging.

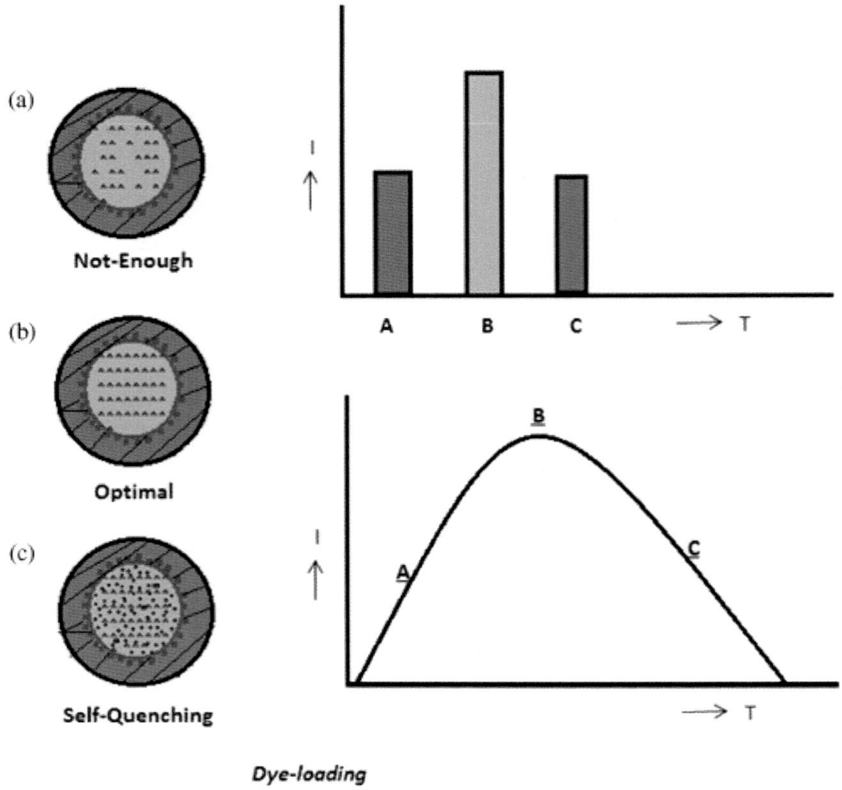

Fig. 2. (a) and (c) Representations of the negative effects that having not–enough or too much dye respectively have on the duration of the intensity of the fluorescence of the NPs. (b) Represents the optimal dye-loading and shows the maximum duration and strength of the fluorescent property of the NPs.

3.2. *Reverse microemulsion (water-in-oil) method*

The reverse microemulsion synthesis (water-in-oil) (w/o) method is one of the commonly used protocols to synthesize inorganic and organic dye-doped SiNP which serve as the base for medical imaging. This procedure has two main components, oil (hydrophobic bulk phase) and water (hydrophilic droplets). The appropriate surfactant is added to the mixture to serve as the interphase between the hydrophilic and hydrophobic phases and form the w/o microemulsion. Each nanosized water droplet is stabilized within the oil *via* the surfactant coat and houses a single nanoreactor that undergoes spontaneous, rapid collisions, fusions and separations. The mixing of the water contents within the water droplets generates condensation, precipitation and hydrolysis reactions. These series of events are a direct result from the thermodynamic stabilization provided by the surfactant interphase of oil and water, and facilitate the growth and nucleation of the nanoreactors. Controlling the NP size involves adding different water-to-surfactant molar ratios, which varies the dynamic properties of the reverse microemulsion.[79] Utilizing dyes with large absorption coefficients for improved fluorescence, assuming that these high-quantum-yield organic dyes will not experience photobleaching during imaging will allow for tunable control of the resulting fluorescence of the SiNP.

3.2.1. *Biosensing functionality using the reverse microemulsion (water-in-oil) method*

A new method to produce amine modified FSNPs as non-viral vectors for gene delivery has been reported. In this technique 2, 5-bis(5-tert-butyl-2-benzoxazolyl) thiopene and the w/o microemulsion method were used. The product was stable, highly monosdispersed, with green fluorescent protein gene surface modification to create a FSNP–DNA complex with promising imaging and therapy applications.[2] FSNPs can be made bi-functional by using water soluble Fe_3O_4 magnetic NPs and CdTe quantum dots and incorporating them into the silica shell by the reverse microemulsion. The result is an FSNP with high luminescence and high magnetism which can be used as a marker in medical imaging and manipulated by an external magnetic field.[80]

3.3. *Stober's method*

The Stober method easily synthesizes bulk amounts of FSNPs. In this procedure, a stable alcohol is formed by the based-catalyzed hydrolysis and condensation of various alkoxysilane compounds in a mixture of ammonia, ethanol and water. First, the amine groups of the silane coupling reagent and the thioisocyanate groups of the dye molecules are covalently attached. After, the solution is gently sealed and stirred in darkness for 24 h. In the second step, the dye-coupling reagent, a silane compound, H_2O, ammonium hydroxide and anhydrous ethanol are mixed in solution and then added to ethanol. This solution is stirred for 3 h at 40°C to form the FSNP. Adding more silane and aging this formula for three more hours can increase the shell thickness to create more stable SiNPs.[79] To purify the final sample, the collected particles are centrifuged and washed with ethanol several times. The result is a group of hybrid NPs with diameter lengths in the tens to hundreds nanometers range. Fairly monodispersed dye-doped SiNP can be synthesized with a few modifications to the traditional Stober's method.

3.3.1. *Organic dye loading*

Organic dyes are generally hydrophobic, so incorporating then into the hydrophilic silica matrix requires a few extra steps.[81] First an amine containing silane compound is reacted with an amine reactive organic dye (*e.g.* FITC), creating a thiourea linkage. Then, the conjugate of the amine reactive dye (FITC) and the amine containing silane is combined with the alkoxysilane compound (TEOS) and allowed to hydrolyze and condense, *via* the Stober's method, creating FITC doped SiNPs (the FSNP). Water dispersibility will be somewhat affected in the resulting FSNP, since some of the dye will be covalently attached to the surface of the particles. To solve this impediment and the possible photobleaching of the surface dye particles, an extra coat of pure silica is applied around the already synthesize FSNP.[82] Figure 3 shows how the SiNP protects and maintains the fluorescence of encapsulated dyes compared to unbound dyes. Figure 3(a) illustrates the rapid decline of fluorescence intensity strength of bare dyes. Figures 3(b) and 3(c) demonstrate how the intensity stabilizes after adding a protective shell. Figure 3(d) shows how the core-shell FSNP architecture protects the dye from photobleaching and thus maintains the intensity of the fluorescence for the longest time.

3.3.2. *Homogenous versus heterogeneous nucleation*

There are two known types of nucleations involved in the decomposition of the silane compound to the monomers, homogenous and heterogeneous. In the homogenous nucleation, the diameter of the particle grows as the silicon substrate concentration increases. In heterogeneous nucleation, the diameter of the NPs decreases in relation to the concentration of the fluorescent dye additive.[83] The kind of nucleation depends on the properties of the precursors and the additives used in the synthesis.

3.3.3. *Stober method limitations*

FSNPs produced by the Stober method are also called C-dots. Even though C-dots have shown to be promising for their superior brightness and quantum efficacy, their physical properties do not allow them to cross the cell membrane. Fuller and his team chose to use polyethyleneimine (PEI) to coat the C-dot particles, since it has shown to be an effective DNA delivery tool. The PEI coating did not affect toxicity or the visual appearance of the particles (Fig. 4).

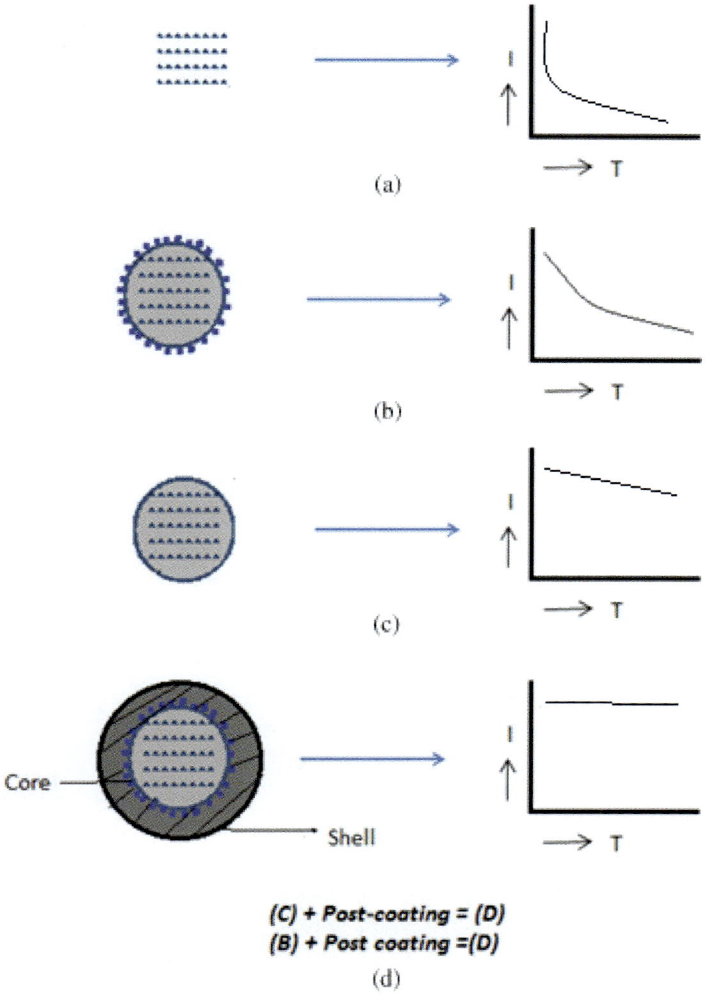

Fig. 3. Represents the significance and advantages that the silica shell has over the fluoresce property.

To ensure the efficacy of the PEI coating a side-to-side study was done with PEI coated C-dots and uncoated C-dots. Figure 5 shows the comparison. Figures 5(a)–5(d) show the PEI coated C-dots and Figs. 5(f)–5(i) show the uncoated C-dots. The scanning confocal laser microscopy (SCLM) images clearly demonstrate how the PEI-coated C-dots are distributed throughout both the cytoplasm and the nucleus of the cell, and thus can be use for intracellular detection and labeling.

Fig. 4. SEM micrographs of coated and uncoated C-dots.[111] No significant morphology differences were notices after coating C-dots with PEI.

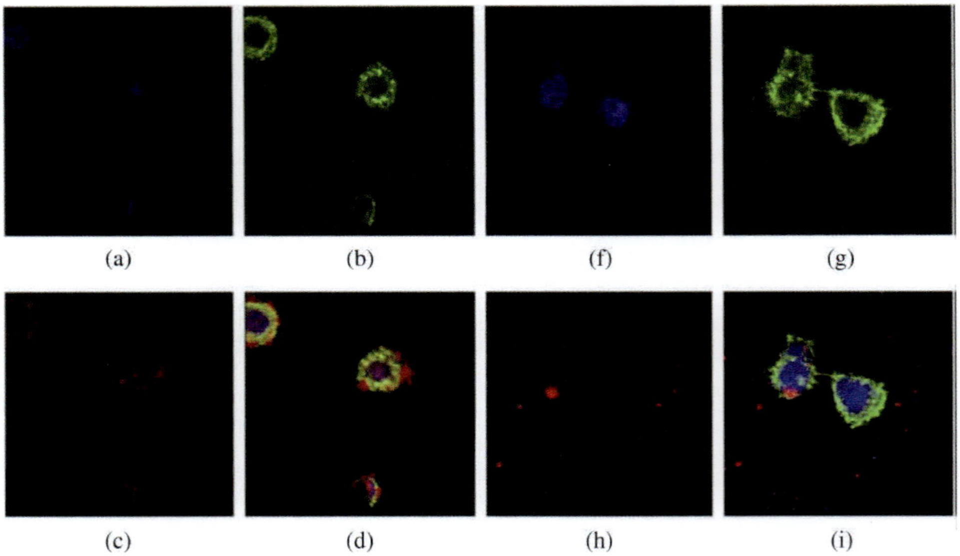

Fig. 5. Nucleus — blue (TROPO-3); endosomes — green (lysosensor — Green); C-dots — red[111].

3.3.4. *Modifications to the Stober method*

The Stober method can be modified to adjust to different research purposes. Wacker and his coworkers reported a controlled synthesis of FSNPs inside microfluidic droplets. In this approach, the classic Stober method was followed to yield a fluorescent silicon alkoxide precursor solution. This solution was then reacted with an aqueous ethanol–ammonia hydrolyzing mixture inside the microfluidic droplets, which were isolated using a microfluidic chip. This novel chip device can be utilized to experiment with fluid amounts below the pico-liter range. Synthesis in droplets drastically improves the uniformity of the NP sizes. The precise control of reagents concentration, reduces the NP size's standard deviation to 3%, when compared to the traditional methods.[84]

4. Particle Functionalizing

SiNP encapsulation creates a protective layer around the dye molecules protecting them against microbial attack, photobleaching due to reactive oxygen species (ROS) penetration, and is resistant to swelling due to changes in solvent polarity. Dye molecules encapsulated by the silica matrix[85] are blocked from interaction with any ROS or acidic environmental compounds, all of which could change and negatively affect the favorable florescence characteristics of the FSNP. The optical transparency of the silica matrix allows for the spectral properties of the encapsulated dye molecules to remain stable, maintaining a superior photostability compared to bare dyes.[85] Since FSNPs have reduced toxicity,[48,60] seemingly unlimited biocompatible applications, and increased magnetic properties,[4,6,37,49,75] they have become a promising and growing tool for bioimaging.

Surface functionalization and bioconjugation is highly desirable for bioimaging because FSNPs appropriately modified surfaces can vary depending on the accuracy of the imaging target. But for any bioimaging purpose the NPs should have the appropriate functional groups that enable the particle to recognize the specific target image. There are many ways to bioconjugate FSNPs such as surface coating, bioconjugation using avidin–biotin binding,[86] aminated particles,[24–26] or carboxylated particles.[87] The following section will explain various common applications for FSNPs made possible from the functioning of the FSNP.

4.1. *Surface coating*

The nascent FSNP is a bare SiNP doped with a fluorescent dye or inorganic fluorescent molecule and is absent of any tracking capabilities. Tracking is enabled through additional coating with a variety of molecules for broad range detection of cellular environments. For example, pH sensing is achieved through surface modification with carboxyl groups.[87] Carbohydrate macromolecules serve as the primary structures for sugars and glycoproteins, and are detected by non-specific carbohydrate–carbohydrate interaction.[61,88] FSNPs conjugated to specific binding proteins, lectins,[89] allow for sensing and tracking of the many different types of carbohydrates present on cells and circulating systemically.

4.2. *Fluorescence manipulation*

Designing FSNPs with co-localization of multiple fluorophores has increased both particle brightness and tracking because encapsulated dye molecules are decoupled from each other which prevents potential "blinking" (random switching between "ON" and "OFF" fluorescence that results from continuous excitation for extended periods of time).[90] In addition, FSNPs are a reliable solution to one of the greatest drawbacks to organic dyes, photobleaching. While some dyes can be toxic, non-specific or hard to clear systematically[60] one thing they all share is poor photostability. Some of the widely used fluorescent dyes (FITC, RuBpy, rhodamine, coumarin, thioflavin T)[7–10] dope the silica NP *via* encapsulation within the core of during synthesis.[2,29] Each dye is unique in fluorescence properties. Some are quenched by interactions with metal compounds while others from environmental oxidative stress molecules.[66,91]

Since quenching is the most noticeable change in dye fluorescence, most signaling events take advantage of quenching to dictate whether or not the dye has arrived at the desired destination.[10] The advantage to FSNPs is that the silica shell protects fluorescent dyes from non-specific, environmental photobleaching and traps the dyes securely to prevent leaking. In addition, the increased hydrophilicity associated with FSNPs allows for the organic dyes to remain stable in aqueous mediums.[9]

Figure 6(a) is schematic of the core/shell architecture with the reference dye (TRITC) encapsulated in the core coated by the sensor-dye (FITC) rich shell. Figures 6(b) and 6(c) are characterizations of the resulting FSNPs *via* scanning electron microscopy (SEM) and *via* spectrofluorometry respectively. Figure 7 demonstrates the architectural advantage to FSNPs. This comparison was done at intensities below saturation, measuring photobleaching between bare dye, FSNP and Q-dots. While the free dye is at least one order of magnitude dimmer that FSNPs and Q-dots, the differences in brightness intensity between FSNPs and Q-dots is very small.

Aggregation-caused quenching (ACQ) is a fluorescence quenching mechanism whereby dye fluorescence reduction results from aggregation of the dye particles.[92] However, some dyes, like tetraphenylethene (TPE) fluorescence from the opposite phenomenon, aggregation-induced emission

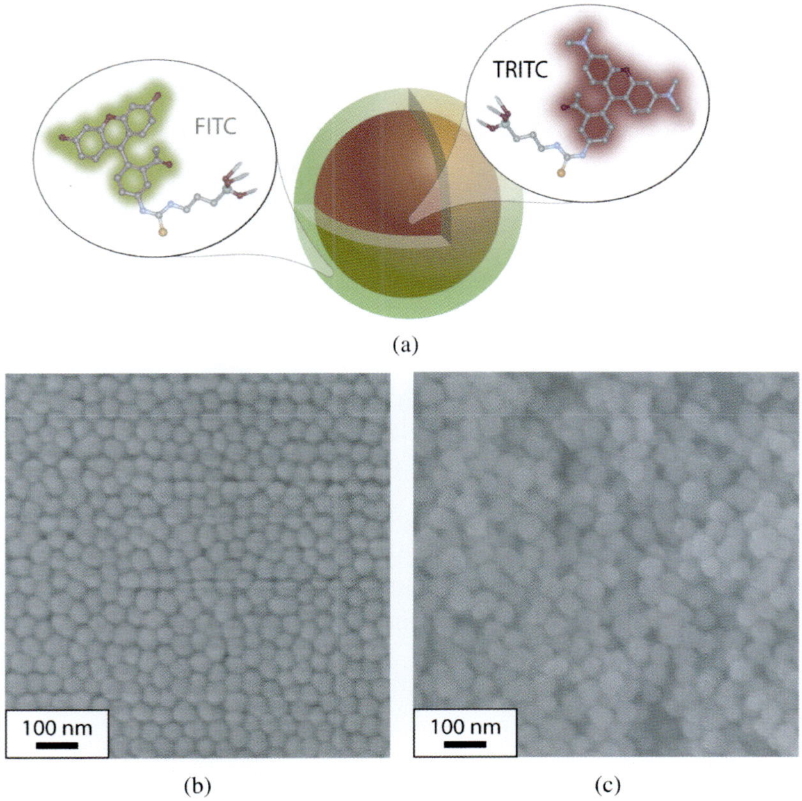

Fig. 6. (a) Schematic representation of core shell architecture[111]. (b) SEM images of 50 nm core. (c) SEM images of 70 nm core/shell.

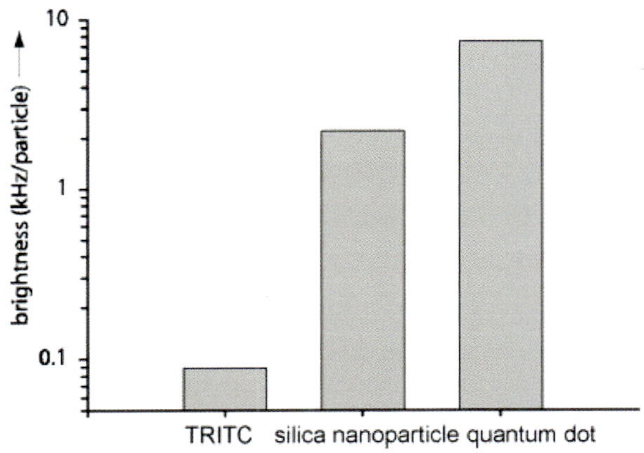

Fig. 7. Brightness comparison between TRITC, SNPs and Q-dots. The values were obtained using FCS measurements of the count/rate particle intensities.[112]

(AIE).[93] AIE results from dye particle aggregation that enhances the fluorescence properties of the dye, thus increasing the fluorescence intensity. Contrary to typical FSNPs that are doped with fluorescent dyes, AIE fluorescent dyes are conjugated to the outer shell in order to achieve better aggregation sensitivity. This type of sensing mechanism is very useful for target detection where intracellular molecules facilitate particle dissociation from the FSNP and cause aggregation.

4.3. *Nucleic acid functionalizing*

Aptamer conjugation to FSNPs is achieved through a process called biotinylation, whereby a biotin molecule is attached to the end of the aptamer and serve as half of the bridge that links the DNA to the FSNP. The advantage to using biotin lies in its strong binding to avidin.[44] Biotin is a ligand of avidin. Covalent conjugation through EDC chemistry immobilizes the avidin structure to the outer FSNP surface. Once the biotinylated aptamer binds with avidin, the resulting product is an FSNP specialized for nucleic acids.[86]

4.4. *Magnetic functionalizing*

Functionalized magnetic FSNPs (MFSNPs) expanded the range of applications for FSNPs by giving them multimodal sensing capabilities.[1] Both superparamagnetic magnetite (Fe_3O_4) and uniform magnetic (Fe_2O_3) NPs (MNP) enable for MRI imaging when conjugated to FSNPs.[41] The combination of the doped fluorescent dye and MRI image give the particle more sensitive, reliable detection for targets. These NPs are typically synthesized by co-precipitation under inert gas environments to prevent oxidation,[3] however other procedures have been established that can produce high quality MNPs such as the solvothermal[62] and gelatin growth methods.[94] The final products are usually covalently linked to the FSNP but since their metallic composition can interfere with fluorescence emission of the encapsulated organic dye, many techniques have been developed that allow for dissociation of the MNP from the FSNP to enhance or stimulate the FSNP's fluorescence.[2] In addition to better imaging capabilities, MNP-FSNPs are capable of separating bound targets through simple magnetic separation. When applied to biological systems, MNP-FSNPs are reliable detectors and isolators both intracellularly and extracellulary.

5. FSNP for Pathogenic Bacteria Detection

Depending on the functionality of the NP and the method of analysis, the ability for FSNPs to reliably detect different species of bacteria was one of the leading advancements to solidify FSNPs as a useful, targetable biotechnology over the past 10 years.[1] As a result, identification of infectious pathogens has become more rapid and sensitive. Using bioconjugated FSNPs, the timespan for accurate detection of bacteria has been reduced to a matter of minutes.[25] The advantageous functionality of FSNPs allows for bacteria detection though various ligand attachments.[7,16–24,26] These attachments can be interchanged to be specific to certain types of bacteria regardless of the diversity of the sample population.

Many different detection methods and analytical techniques have been developed from FSNPs. Their applications have expanded across many types of bacteria species. The following is an overview of different methods of FSNPs to detect and/or defend against different. Each particle, method and technique can be modified depending on the bacteria in question.

5.1. *Escherichia coli*

Escherichia coli (*E. coli*) are gram negative, rod-shaped bacteria capable of surviving in a variety of niches. Some strains of *E. coli* are highly pathogenic and lethal, while others exist relatively harmless and colonize the intestinal tract of animals. *E. coli* detection has become a topic of interest over the years because of its impact on food safety.[25] *E. coli* contamination in undercooked meats, unpasteurized milk and contaminated water[17] contributes to thousands of foodborne illnesses, hospitalizations and deaths worldwide.[21]

5.1.1. *FSNP advantages for selective targeting of Escherichia coli*

RuBpy-doped SiNP are one type of FSNPs that are made using a w/o microemulsion method. The RuBpy dye serves as the fluorescent chemical which illuminates the SiNP. When these NPs are modified to have an amine reactive or a carboxyl reactive surface, they can be attached to almost any type of ligand. Antibodies (Abs) that are specifically designed to bind to *E. coli*, like those that select against the *E. coli* O-antigen,[24–26] are covalently linked to the functionalized FSNP surface. In order to prevent non-specific binding (binding to particles and organic matter other than *E. coli*) Abs bound FSNPs are reacted with bovine serum albumin (BSA) (which can also be detected and confirmed using FSNPs[52]) and washed in phosphate buffer saline (PBS). The BSA will stabilize the Abs to only react with the *E. coli*, and the PBS will wash away any unbound Abs and BSA. The final product is an FSNP with a diameter of approximately 60 nm that is specific to only *E. coli*. The efficiency of these NPs has led to 20 min detection of *E. coli*.

5.1.2. *Detection of E. coli using antibody–antigen recognition*

When the Abs bound FSNPs are incubated with the *E. coli* they bind to the antigens on the cell surface. Since RuBPy is bound to the SiNP, its fluorescence stability is maintained longer and brighter than RuBpy bound to Abs without the SiNP. As a result, fluorescence imaging of the bacteria cells can be carried out to obtain first hand visual verification without compromising the particle's fluorescence integrity. To obtain definitive proof, alternative methods of characterization such as SEM[24] can be used to image the surface of *E. coli* cells and show Abs bound FSNPs on the cell wall. From there, the sample of bacteria can be diluted to minimum contamination concentrations (1 colony forming unit (CFU)), and spectrofluorometer analysis of the cell will be sufficient to detect the fluorescent signal of the FSNP bound to the cell surface of that single cell.[25]

One of the best techniques for rapid, quantifiable detection (of not only *E. coli* but other bacteria as well) is flow cytometry (FCM).[17,19] FCM uses laser excitation on the FSNP to detect fluorescence emission on the bacteria cells caused by the Abs bound FSNP. While the typical single dye Abs bound FSNPs serve as appropriate detection tools for FCM, another way to increase the sensitivity includes introducing a second fluorophore for the FCM to detect.

Alternate organic dyes that are specialized for bacteria can be used to discriminate the cell from the NP. When these dyes are used in tandem with the FSNP, colocalization of both the cell and NP is achieved. On the other hand, ligandmagnetic NPs (LMNPs)[17] can selectively attach to *E. coli* by using mannose residues as the linking agent, and serve as secondary markers on the cell surface. When the LMNPs are bound, the cells can be extracted using a magnet and then labeled with the Abs bound FSNP. This method ensures that only *E. coli* cells are being labeled, however it can be applied to detect many other bacteria such as *Mycobacterium tuberculosis*,[16] *Shigella flexneri*,[18] and *Campylobacter jejuni*.[95]

5.2. *Salmonella typhimurium*

Salmonella typhimurium (*S. typhimurium*) are gram negative, rod-shaped bacteria that are associated with many foodborne illnesses around the world. There are an estimated 1.4 million cases of infections in the United States each year.[23] As the number of infections grows, so does the importance for reliable detection methods of *S. typhimurium*. Common bacteria detection methods like enzyme-linked immunosorbent assay (ELIZA) or polymerase chain reaction (PCR) have been used to address the problem. However each technique is very time consuming and error prone. For that reason, functionalized FSNPs capable of detecting *S. typhimurium* quickly and reliably proves to be the most time efficient and equally reliable method of detection.

5.2.1. *Salmonella typhimurium specific oligonucleotide bioconjugation to FSNP*

Binding to DNA sequences requires modifying the synthesized with a linking molecule, like avidin.[86] Avidin serves as the bridge between the FSNP and the attached targeting molecule. Oligonucleotides are short DNA sequences that are specifically designed for target organisms. They can be designed to capture, target or indicate complementary DNA strands of the target organism.[23] The most convenient way to attach DNA is for it to have a terminal biotin label. The biotin contains several types of functional groups that enable it to be modified in numerous ways but the most convenient attachment is through biotin–avidin affinity.[23] The end result is achieved through a chemical interaction between the avidin linkage on the FSNP and the terminal biotin on the DNA.

5.2.2. *Salmonella typhimurium detection using specific oligonucleotide bioconjugated FSNPs*

A method known as sandwich-type DNA analysis, which uses pretreated 96-well plates, is employed to allow for complementary binding of free, single stranded DNA to the oligonucleotides bound to the FSNPs. In order to extract the DNA from the bacteria, most labs utilize DNA extraction kits that involve step-by-step instructions for lysing bacteria cells and collecting the fresh DNA. Since the FSNPs are fluorescent, after the *S. typhimurium* DNA binds to the attached oligonucleotides, the particles are collected and the DNA sequences are verified for *S. typhimurium*.[23]

5.3. *Listeria monocytogenes*

Listeria monocytogenes (*L. monocytogenes*) are gram positive, rod-shaped bacteria that cause listeriosis infections. As a pathogen, *L. monocytogenes* is a very common foodborne pathogen that commonly colonizes cheese, meats and vegetables.[22] As a disease, *L. monocytogenes* claims more lives than salmonella related cases because people with compromised immune systems are the ones who usually contract the disease. The conventional methods for detecting *L. monocytogenes,* like latex agglutination testing, PCR and ELIZA, are very laborious, often requiring days of sample preparation before reliable confirmations can be concluded. Even the less time sensitive methods like FCM are held back because of the time it takes to harvest enough *L. monocytogenes* sample.[22]

5.3.1. *FITC–IgG conjugated FSNP*

Conjugating immunoglobulin G (IgG) is very similar to conjugating any other antibody. The IgG is an antibody that specifically recognizes bacteria; therefore this type of particle can be employed to detect a wide variety of bacteria species. The only necessity is the right functional group on the surface of the FSNP. FITC doped FSNPs can be amine activated and linked with glutaraldehyde.[55] The glutaraldehyde will crosslink the FSNP to the IgG. Once the IgG is bound to the FSNP it can be combined with other immunological methods to specifically detect bacteria.

5.3.2. *Detection of listeria monocytogenes using FITC–IgG conjugated FSNP*

The sandwich-type immunoassay technique can be modified depending on the application. Since IgG is a broad antibody, there needs to be another antibody that is species specific in order to have reliable detection. These antibodies are immobilized in the plates so that when the *L. monocytogenes* are added, they bind to and hold the bacteria.[22] In order to evaluate the efficiency of the labeling against multiple bacteria species, the plates are washed so that any unbound bacteria do not randomly bind to the FITC–IgG conjugated FSNP. The captured bacteria are then labeled with the FITC–IgG conjugated FSNP which will glow in every spot where there is bacteria. This method has been used to reliably detect other bacteria species such as gram positives like *Streptococcus pneumonia* and *Staphylococcus aureus* and gram negatives like *Mycobacterium tuberculosis*[20] and the previously mentioned *L. monocytogenes.*

5.4. *Therapeutic bactericidal advantages to FSNP*

FSNPs are capable of fluorescing in almost any color depending on the organic dye being encapsulated in the silica matrix (Fig. 1). In addition to RuBPy, FITC is another commonly utilized organic dye for bioimaging. When SiNP encapsulates FITC, the resulting FSNPs protects the FITC fluorescence from any quenching that may occur from the surrounding environment.[7] When these FSNPs are bound to bactericidal agents, like zinc oxide (ZnO), they become lethal to bacteria cells. The ZnO surrounds the FSNP (like a coat), however since the FITC is embedded within the silica, the ZnO does not interact with the fluorescence characteristics of the detection. Cell staining is achieved when the ZnO coating interacts with the cell wall allowing for a temporary particle–bacteria attachment. When the particle has incubated long enough with the *E. coli*, the cell membrane begins to degrade from the toxic ZnO. As a result, the particle enters the cell. Using the second color detection method, a dye that only stains dead bacteria cells can be used to colocalize the particle with only dead cells and definitively conclude that the particle is toxic.[7]

6. Cancer Studies

6.1. *Overview*

Cancer diseases are the leading cause of death for patients below 85 years of age in USA.[29] Human cancer comprises a class of over 200 diseases (pancreatic cancer being the most debilitating[96]) in which malignant cell growth progresses exponentially and uncontrollably through the body's organ systems. Chemical mutation, disease infections, UV radiation, inherited genetic defects and teratogens are among the known causes of cancer.[97] Successful cancer treatment relies heavily an early diagnosis to prevent metastasis. Conjugated NPs with combined magnetic and fluorescent imaging properties could increase early detection rates. Over the last few years FSNPs have shown to be a promising tool for cancer detection.[2]

6.2. *Active targeting*

To ensure the specific accumulation of the NPs within the cancer cells, conjugation with targeting ligands is necessary. The interactive forces between the conjugated ligands and the functional groups of the diseased cells contributes to the high affinity and specificity of the targeting process.[2,86] The cell surface is covered in specific glucoproteins which can be targeted using lectin.

Lectin is a sugar-binding protein capable of recognizing and binding to glucoproteins that occur on the surface of cells.[89] The lectin–carbohydrate interactions have demonstrated active targeting through receptor–antigen attachment on the plasma membrane.

Many cancer cells overexpress specialized receptors which researchers have been able to take advantage for particle functionalization and cell targeting.[27,98] FSNP surface functionalization with folic acid, through amine reactive binding, increases the bioconjugation capability and the aqueous dispersibility of the NPs for intracellular.[99] Folate receptors active in the brain, ovary, kidney, breast and lungs[31,34,37] will uptake folate-conjugated NPs when possible through endocytosis (demonstrated through SCLM studies). Antibody-mediated tumor targeting through conjugation of monoclonal antibodies to FSNPs can also be useful since antigenic moieties are not significantly expressed in normal tissue.[2]

6.3. *Passive targeting*

Passive targeting engages an indirect approach for labeling tumors that centers around the environmental conditions presented by the tumor. One method, the enhanced permeation and retention (EPR) effect, involves the accumulation of NPs (diameter <100 nm) at tumors for targeting.[2] The specialized SiNPs avoid the RES, which enable them to circulate efficiently and arrange direct local delivery of the anticancer drug to tumors; introduction into the body sometimes involves injections or surgery. The drug remains inactive while conjugated to the FSNP until reaching the tumor. Enzymes and/or pH differences present in the tumor environment hydrolyze the linkages between the drug and FSNP to activate drug release.[2,29,99]

6.4. *Cancer detection and imaging with fluorescent silica nanoparticles*

Due to their size and properties, dye-doped SiNP are extremely specific and can detect even individual cancer cells.[40,100] Existing non-invasive diagnostic imaging techniques are effective for visualization of tumors macroscopically, but are not specialized or sensitive enough for microscopic visualization. In cancer, failing to detect a few malignant cells could be the difference between life and death for the patient. In the past decade, FSNPs have demonstrated reliable labeling of human oral carcinoma,[99] lung carcinoma[101–103] human leukemia,[38,102,104,105] HepG Liver cancer,[32,106] colon cancer,[37,107,108] bladder cancer,[39] gastric cancer,[33] and breast cancer[34] cells.

6.4.1. *Breast cancer*

Coupling FSNPS with specific recognition molecules produces both *in vitro* and *ex vivo* images of breast cancer cells. When conjugated to arginine–glycine–aspartic acid peptide (RGD-peptide), *in vitro*, FSNPs exhibit high target binding to $\alpha v\beta 3$ integrin in MDA-MB-231 human breast malignant tumors. Further studies demonstrate that by intravenously injecting RGD-peptide-labeled FSNPs, the ex vivo image of cancerous cells reached a maximum fluorescence after only 1 h. This rapid and extremely accurate labeling of cancer cells, demonstrates promising therapy.[34]

6.4.2. *Colon and lung cancer*

Iron oxide NPs that contain an organic dye and a protective silica coat enables for MRI scanning for detection and characterization of colon cancer cells.[108] Silica coating decreases the heavy metal cytotoxicity of the NPs,[31] and it also prevents dye photobleaching when they are successfully uptaken by cells. Conjugation to antibodies has allowed targeting of cancerous cells that can be observed using CLSM.[107] It is important to mention that the antibody-conjugated MFSNP attached to the target cells would migrate towards the external magnetic field.[37] A similar approach for detecting lung cancer cells utilizes these bifunctional nanocomposites with anticarcinoembryonic antigen (CEA) antibodies.[103]

6.4.3. *Leukemia*

Promising results have been found in tests for leukemia cells involving magnetic and fluorescent NP applications. Estevez *et al.* demonstrate these applications with a new generation of aptamers (specifically cell-SELEX).[38] This aptamer recognizes live, intact cells and shows high affinity for cancer cells. With this in mind, the researchers paired these aptamers with a dual-NP system; consisting of MNPs and FSNPs. As a result, enhanced selectivity and sensitivity enabled the complex for acute leukemia cell recognition due to the combined MRI and fluorescence imaging.[38]

6.4.4. *Gastric cancer*

Double immunofluorescence staining is a reliable method for diagnosis of MGC-803 gastric cancerous cells by SCLM. This approach requires incubation of the malignant cells in a mixture of anti-CEA (green) antibody-conjugated and anti-CK19 (red) antibody-conjugated FSNPs, followed by subsequent analysis with a SCLM. The cell lines show successful double labels (green and red signals from the FSNPS), and are distinguished through antigen–antibody recognition without cross reactivity. In relation with the classic double immunofluorescence staining, the results confirm that this technique has a potential advantage in photostability.[33]

6.5. *Comprehensive cancer theranostics with modified FSNPs*

Cheng *et al.* reported the development of theranostic compounds made from tri-functionalized mesoporous silica NPs (MSNPs). MSNPs are a type of FSNP. Theranostics involves specialized cancer therapy. Each person develops cancer differently, and therefore each case must be handled uniquely. MSNPs are functionalized in an organized sequence — first, with contrast agents for imaging, second, with drug payloads for therapeutic treatment and third, with biomolecular ligands for specific particle delivery. Utilizing specific contrast agents, drugs and biological ligands, these NPs have demonstrated highly accurate targeting and high therapeutic effect with low collateral damage, showing to be a promising biomedical tool for this cancer theranostics.[36]

Preliminary animal testing for the theranostic applications is the first step towards understanding the interactions of specialized MSNPs in mammalian specimens. The technique of colocalizing multiple fluorophores within a single particle was used to investigate the various intracellular pH values of rat basophilic leukemia mast cells.[109] It was initially found that the FSNPs were not internalized by the cells (Fig. 8(a)) but rather adhered to the surface. Phorbol-12,12-dibutyrate was added during incubation to increase the cell's endocytotic activity, and FSNP uptake followed[11] (Fig. 8(b)). Figure 9 shows the CSLM images of the FSNPs in the rat basophilic leukemia mast cell. Once the uptake impediment was resolved, the FSNPs with reference (Fig. 9(a)) and sensor (Fig. 9(b)) fluorophores were internalized in the mast cells and found to reside in different pH ranges (Fig. 9(d)) in different intracellular compartments. Overlay of reference and sensor dyes (Fig. 9(c)) show the sensitivity of each dye for a specific pH range. Altogether, this data demonstrates the feasibility for FSNPs to detect specific areas of mammalian cancer cells.

7. Conclusion

Existing optical imaging and diagnostic non-invasive imaging techniques such as computed tomography (CT), MRI, ultrasound (US), single photon

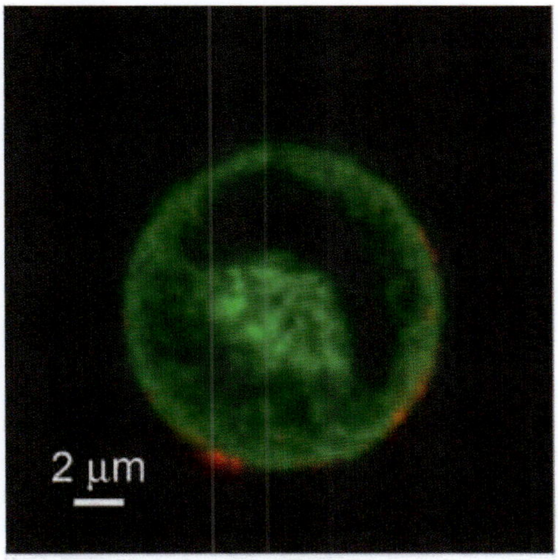

(a)

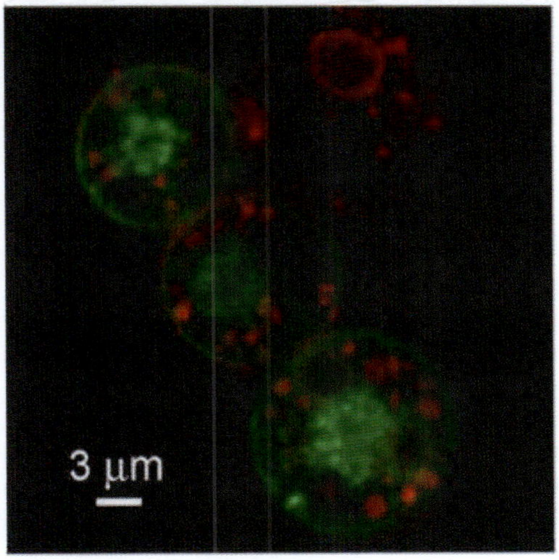

(b)

Fig. 8. (left). Confocal fluorescence microscopy images of 70 nm TRICT SiNPs (red) and Alexa Fluor 488-cholera toxin B (green).[11]

emission CT (SPECT), and positron emission tomography (PET) are effective for macroscopic visualization and diagnosis. Sensitive medical imaging is essential for accurate and early diagnosis of diseases. FSNPs have shown promising results towards the development of sophisticated and extremely accurate *in vivo* medical imaging methods capable of microscopic detection of a single abnormal cell. They are an excellent tool to better understand several diseases, and by extension, have expanded medical diagnostics through

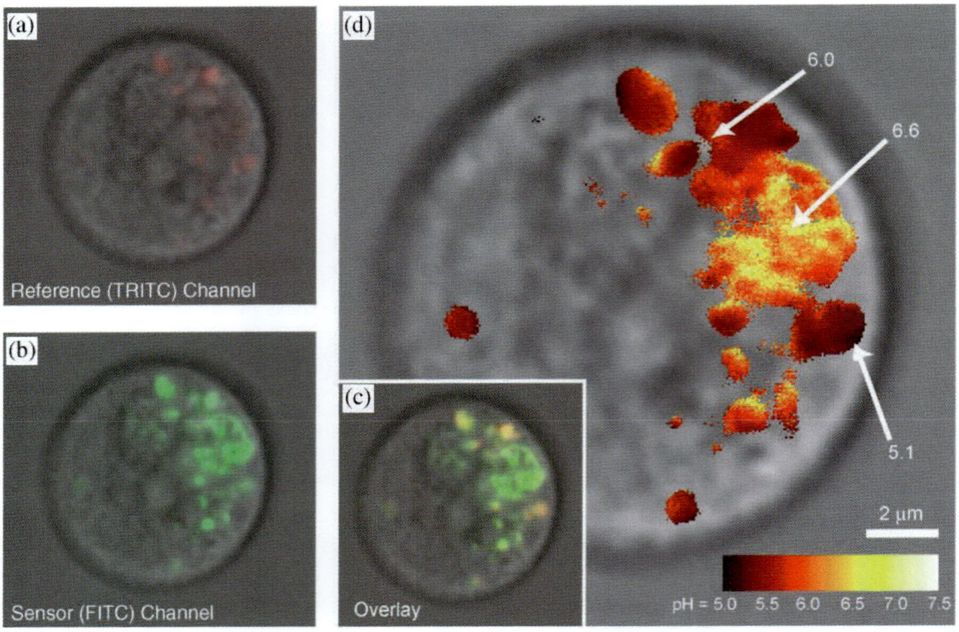

Fig. 9. (right) Confocal fluorescence microscopy images of pH sensors in RBL mast cells. (a) Reference dye channel, (b) sensor dye channel, (c) overlaid images, (d) false-color ratiometric imaging of pH in various intracellular compartment.[111]

novel medical approaches for targeting, detection, diagnosis and treatments. The robust SiNP technology offers an easy-to-modify surface,[24–26,85] improved biocompatibility,[1] highly accurate sensing capacity,[47,64,65] reduced photo bleaching over time and multiple functionalities.

Over the last couple decades, novel bioconjugation and functionalization methods have been used to produce FSNPs with diverse imaging abilities. These particles are successful at demonstrating intercellular fluctuations in chemical energy levels,[44,64,65] cell damage byproduct levels,[66] as well as enzyme-protein concentration levels,[47,53,68] allowing for detailed diagnosis and accurate predictions of the target cell's activity and state. Besides their sensory abilities, FSNPs are capable of treating diseased tissues. By loading the particles with specific drugs and functionalizing them accordingly, the drug release proceeds to the detected target.[36] Magnetic functionalization has expanded the range of applications of FSNPs by adding secondary sensing capabilities.[1] Both uniform magnetic NPs (Fe_2O_3) and superparamagnetic magnetite (Fe_3O_4), enable MRI imaging when conjugated to FSNPs.[41] The main advantage of these multimodal NPs, is their targeting capabilities through using an external magnetic field.

To provide a fair perspective, it is important to clarify that, even though FSNP-based optical contrast agents have a bright future in medical imaging, several aspects mentioned in this chapter need further investigation and understanding. The long-term goal for FSNPs as useful NPs is to make them capable of detecting and treating diseased cells without interfering or affecting the surrounding healthy cells and physiological processes. New applications for FSNPs are developed frequently, and this robust technology may hold the key to preventing many of the modern diseases and disorders that inflict millions of individuals around world.

Acknowledgments

The writers wish to acknowledge Dr. Jason Fuller, Dr. Andre Burns, and Dr. Hooisweng Ow for granting us permission to include their figures. They also acknowledge the Nanoscience Technology Center of University of Central Florida for their resources. This work has been partially supported by the National Science Foundation grant, CBET-1159500 (SS, AJG).

References

1. P. Tallury *et al.*, Nanobioimaging and sensing of infectious diseases. *Adv. Drug Delivery Rev.* **62**(4–5), 424–437 (2010).
2. S. Santra, *Fluorescent silica nanoparticles for cancer imaging.* Methods in molecular biology (Clifton, N.J.), **624**, 151–162 (2010).
3. K.-S. Park *et al.*, Characterization, in vitro cytotoxicity assessment, and in vivo visualization of multimodal, RITC-labeled, silica-coated magnetic nanoparticles for labeling human cord blood-derived mesenchymal stem cells. *Nanomed.-Nanotechnol. Biol. Med.* **6**(2), 263–276 (2010).
4. L. Wang *et al.*, Biodegradable magnetic-fluorescent magnetite/poly(DL-lactic acid-co-alpha,beta-malic acid) composite nanoparticles for stem cell labeling. *Biomaterials* **31**(13), 3502–3511 (2010).
5. Y. Sun *et al.*, Synthesis, Modification and Cell Image of Fluorescent Silica Nanoparticles. *Acta Chimica Sinica* **69**(8), 967–972 (2011).
6. J. L. Vivero-Escoto, R. C. Huxford-Phillips and W. B. Lin, Silica-based nanoprobes for biomedical imaging and theranostic applications. *Chem. Soc. Rev.* **41**(7), 2673–2685 (2012).
7. H. Wang *et al.*, Fluorescent dye encapsulated ZnO particles with cell-specific toxicity for potential use in biomedical applications. *J. Mater. Sci. — Mater. Med.* **20**(1), 11–22 (2009).
8. X. Gao *et al.*, Synthesis and characterization of functionalized rhodamine B-doped silica nanoparticles. *Opt. Mater.* **31**(11), 1715–1719 (2009).
9. E. Herz *et al.*, Relative quantum yield measurements of coumarin encapsulated in core-shell silica nanoparticles. *J. Fluorescence* **20**(1), 67–72 (2010).

10. B. Soptei *et al.*, Preparation and characterization of Thioflavin T doped silica nanoparticles. *Periodica Polytechnica — Chem. Eng.* **53**(2), 49–54 (2009).
11. A. S. Burns, P. Zedayko, T. B. Baird, and U. Wiesner, Core/shell fluorescent silica nanoparticles for chemical sensing: towards single-particle laboratories. *Small* **2**(6), 723–726 (2006).
12. E. Herz *et al.*, Large stokes-shift fluorescent silica nanoparticles with enhanced emission over free dye for single excitation multiplexing. *Macromol. Rapid Commun.* **30**(22), 1907–1910 (2009).
13. A. A. Burns *et al.*, Fluorescent silica nanoparticles, novel architectures with enhanced functionality. *Abs. Papers Am. Chem. Soc.* **230**, U1187–U1187 (2005).
14. G. Alberto *et al.*, Molecular engineering of hybrid dye-silica fluorescent nanoparticles, influence of the dye structure on the distribution of fluorophores and consequent photoemission brightness. *Chem. Mater.* **24**(14), 2792–2801 (2012).
15. A. A. Ansari *et al.*, Synthesis of optically active silica-coated NdF3 core-shell nanoparticles. *Spectrochimica Acta Part a — Mol. Biomol. Spectroscopy* **86**, 432–436.
16. A. Ekrami *et al.*, Validity of bioconjugated silica nanoparticles in comparison with direct smear, culture, and polymerase chain reaction for detection of Mycobacterium tuberculosis in sputum specimens. *Int. J. Nanomed.* **6**, 2729–2735 (2011).
17. X. He *et al.*, Rapid and ultrasensitive E. coli O157,H7 quantitation by combination of ligandmagnetic nanoparticles enrichment with fluorescent nanoparticles based two-color flow cytometry. *Analyst* **136**(20), 4183–4191 (2011).
18. Y. Li, and W. Xu, Highly sensitive detection of Shigella flexneri using fluorescent silica nanoparticles. *New Microbiologica* **32**(4), 377–383 (2009).
19. D. Qin *et al.*, Using fluorescent nanoparticles and SYBR Green I based two-color flow cytometry to determine Mycobacterium tuberculosis avoiding false positives. *Biosensors Bioelectron.* **24**(4), 626–631 (2008).
20. D. Qin *et al.*, Fluorescent nanoparticle-based indirect immunofluorescence microscopy for detection of Mycobacterium tuberculosis. *J. Biomed. Biotechnol.* (2007)
21. S. K. Rastogi *et al.*, Ag colloids and Ag clusters over EDAPTMS-coated silica nanoparticles, synthesis, characterization, and antibacterial activity against Escherichia coli. *Nanomed.-Nanotechnol. Biol. Med.* 7(3), 305–314 (2011).
22. Z. Wang *et al.*, Sensitive immunoassay of Listeria monocytogenes with highly fluorescent bioconjugated silica nanoparticles probe. *J. Microbiol. Meth.* **83**(2), 179–184 (2010).
23. Z. Wang *et al.*, Sensitive detection of Salmonella with fluorescent bioconjugated nanoparticles probe. *Food Chem.* **125**(2), 779–784 (2011).
24. W. Zhao, L. Wang, and W. Tan, Fluorescent nanoparticle for bacteria and DNA detection. In Bio-Applications of Nanoparticles, W. C. W. Chan (Ed) 129–135 (2007).
25. X. J. Zhao *et al.*, A rapid bioassay for single bacterial cell quantitation using bioconjugated nanoparticles. *Proceedings of the National Academy of Sciences of the United States of America*, **101**(42), 15027–15032 (2004).
26. L.-X. Zhou, *et al.*, Detection of E. coli O157, H7 by Fluorescent silica nanoparticles and SYBR green I based two-color microscopic imaging technique. *Chemical Journal of Chinese Universities-Chinese*, **32**(10), 2274–2279 (2011).
27. N. R. Jabir *et al.*, Nanotechnology-based approaches in anticancer research. *International Journal of Nanomedicine.* 7, 4391–4408.

28. X. X. He *et al.*, A novel fluorescent label based on organic dye-doped silica nanoparticles for HepG liver cancer cell recognition. *Journal of Nanoscience and Nanotechnology*, **4**(6), 585–589 (2004).
29. S. Santra *et al.*, Fluorescent nanoparticle probes for cancer imaging. *Technology in Cancer Research & Treatment*, **4**(6), 593–602 (2005).
30. J. K. Herr *et al.*, Aptamer-conjugated nanoparticles for selective collection and detection of cancer cells. *Analytical Chemistry*, **78**(9), 2918–2924 (2006).
31. H. Yoo, B. S. Nagaralli, and D. S. Chung, ANYL 164-Folate-PEG conjugated fluorescent silica nanoparticles for labeling cancer cells. *Abstracts of Papers of the American Chemical Society*, **232,** (2006).
32. J. Peng *et al.*, Identification of live liver cancer cells in a mixed cell system using galactose-conjugated fluorescent nanoparticles. *Talanta*, **71**(2), 833–840 (2007).
33. X. He *et al.*, FSiNPs mediated improved double immunofluorescence staining for gastric cancer cells imaging. *Talanta*, **76**(5), 1199–1206 (2008).
34. P. Wu *et al.*, Imaging breast cancer cells and tissues using peptide-labeled fluorescent silica nanoparticles. *Journal of Nanoscience and Nanotechnology*, **8**(5), 2483–2487 (2008).
35. M. C. Estevez *et al.*, Highly fluorescent dye-doped silica nanoparticles increase flow cytometry sensitivity for cancer cell monitoring. *Nano Research*, **2**(6), 448–461 (2009).
36. S.-H. Cheng *et al.*, Tri-functionalization of mesoporous silica nanoparticles for comprehensive cancer theranostics-the trio of imaging, targeting and therapy. *Journal of Materials Chemistry*, **20**(29), 6149–6157 (2010).
37. Y.-S. Cho *et al.*, Cetuximab-conjugated magneto-fluorescent silica nanoparticles for *in vivo* colon cancer targeting and imaging. *Cancer Lett.* **299**(1), 63–71 (2010).
38. M. C. Estevez *et al.*, Nanoparticle-aptamer conjugates for cancer cell targeting and detection. *Meth. Molecular Bio.* **624**, 235–248 (2010).
39. H. Guo *et al.*, Singlet oxygen-induced apoptosis of cancer cells using upconversion fluorescent nanoparticles as a carrier of photosensitizer. *Nanomed-Nanotechnol Biol. Med.* **6**(3), 486–495 (2010).
40. S. Jiang, M. K. Gnanasammandhan and Y. Zhang, Optical imaging-guided cancer therapy with fluorescent nanoparticles. *J. Roy. Soc. Interface* 7(42), 3–18 (2010).
41. C. Mi *et al.*, Multifunctional nanocomposites of superparamagnetic (Fe(3)O(4)) and NIR-responsive rare earth-doped up-conversion fluorescent (NaYF(4): Yb, Er) nanoparticles and their applications in biolabeling and fluorescent imaging of cancer cells. *Nanoscale* **2**(7), 1141–1148 (2010).
42. E. Besic Gyenge *et al.*, Uptake and fate of surface modified silica nanoparticles in head and neck squamous cell carcinoma. *J. Nanobiotechnol.* **9**, 32 (2011).
43. E. B. Gyenge *et al.*, Uptake and fate of surface modified silica nanoparticles in head and neck squamous cell carcinoma. *J. Nanobiotechnol.* **9** (2011).
44. C. D. Medley *et al.*, Aptamer-conjugated nanoparticles for cancer cell detection. *Anal. Chem.* **83**(3), 727–734 (2011).
45. Y. Mi *et al.*, Herceptin functionalized polyhedral oligomeric silsesquioxane — conjugated oligomers — silica/iron oxide nanoparticles for tumor cell sorting and detection. *Biomaterials* **32**(32), 8226–8233 (2011).
46. Y. Xu, and Q. Li, Multiple fluorescent labeling of silica nanoparticles with lanthanide chelates for highly sensitive time-resolved Immunofluorometric assays. *Clin. Chem.* **53**(8), 1503–1510 (2007).

47. H. Wu *et al.*, Dye-doped silica nanoparticle labels/protein microarray for detection of protein biomarkers. *Analyst* **133**(11), 1550–1555 (2008).
48. A. A. Burns *et al.*, Fluorescent Silica Nanoparticles with Efficient Urinary Excretion for Nanomedicine. *Nano Lett.* **9**(1), 442–448 (2009).
49. K. Lee *et al.*, Magnetic resonance imaging of macrophage activity in atherosclerotic plaques of apolipoprotein E-deficient mice by using polyethylene glycolated magnetic fluorescent silica-coated nanoparticles. *Current Appl. Phys.* **9**, S15–S18 (2009).
50. M. Lu, D. Wu and N. Guo, Novel functionalized ternary copolymer fluorescent nanoparticles, synthesis, fluorescent characteristics and protein immobilization. *J. Mater. Sci.-Mater. Med*, **20**(2), 563–572 (2009).
51. H. Shi *et al.*, Protein analysis based on molecular beacon probes and biofunctionalized nanoparticles. *Sci. China-Chem.* **53**(4), 704–719 (2010).
52. Y. Tian *et al.*, Silica-Coated CaF(2): Eu(3+) Nanoparticles functionalized with oxalic acid for bio-conjugation to BSA proteins. *Chin. J. Chem.* **28**(6), 921–927 (2010).
53. Y. Wang, K.-Y. Pu and B. Liu, Anionic conjugated polymer with aptamer-functionalized silica nanoparticle for label-free naked-eye detection of lysozyme in protein mixtures. *Langmuir* **26**(12), 10025–10030 (2010).
54. M. R. R. de Planque *et al.*, Electrophysiological characterization of membrane disruption by nanoparticles. *Acs Nano* **5**(5), 3599–3606 (2011).
55. N. Kim *et al.*, Development of an immobilized-antigen immunofluorescence glass slide system that exploits fluorescent silica nanoparticles. *Sensors Actuators B-Chem.* **160**(1), 563–570 (2011).
56. K. G. Lee *et al.*, Synthesis of Stable Silica-Dye Hybrid Nanomaterial as DNA Carrier. *J. Nanosci. Nanotechnol.* **11**(1), 686–690 (2011).
57. X. Li *et al.*, The packaging of siRNA within the mesoporous structure of silica nanoparticles. *Biomaterials* **32**(35), 9546–9556 (2011).
58. A. Liu *et al.*, Development of highly fluorescent silica nanoparticles chemically doped with organic dye for sensitive DNA microarray detection. *Anal. Bioanal. Chem.* **401**(6), 2003–2011 (2011).
59. Q. Yang *et al.*, Quantum dot-based immunochromatography test strip for rapid, quantitative and sensitive detection of alpha fetoprotein. *Biosensors Bioelectron*, **30**(1), 145–150 (2011).
60. C. Eisner *et al.*, Measurement of plasma volume using fluorescent silica-based nanoparticles. *J. Appl. Physiol.* **112**(4), 681–687 (2012).
61. J. Zhao *et al.*, Carbohydrate-coated fluorescent silica nanoparticles as probes for the galactose/3-sulfogalactose carbohydrate-carbohydrate interaction using model systems and cellular binding studies. *Bioconjugate Chem*, **23**(6), 1166–1173 (2012).
62. C.-W. Lai *et al.*, Iridium-complex-functionalized Fe(3)O(4)/SiO(2) core/shell nanoparticles, A facile three-in-one system in magnetic resonance imaging, luminescence imaging, and photodynamic therapy. *Small* **4**(2), 218–224 (2008).
63. A. Schulz, and C. McDonagh, Intracellular sensing and cell diagnostics using fluorescent silica nanoparticles. *Soft Matter* **8**(9), 2579–2585 (2012).
64. A. J. Moro *et al.*, Surface-functionalized fluorescent silica nanoparticles for the detection of ATP. *Chem. Commun.* **47**(21), 6066–6068 (2011).
65. Y. Wang, Y. Wang and B. Liu, Fluorescent detection of ATP based on signaling DNA aptamer attached silica nanoparticles. *Nanotechnology* **19**(41), (2008).

66. F. Luo *et al.*, A non-enzyme hydrogen peroxide sensor based on core/shell silica nanoparticles using synchronous fluorescence spectroscopy. *Microchimica Acta* **165**(1–2), 23–28 (2009).
67. M. Qhobosheane *et al.*, Biochemically functionalized silica nanoparticles. *Analyst* **126**, 1274–1278 (2001).
68. X.-X. He *et al.*, Study on the target interaction between mitochondria and signal peptide functionalized silica nanoparticles. *Chem. J. Chin. Univ.-Chin.* **30**(12), 2376–2380 (2009).
69. Y.-Q. Fei *et al.*, Synthesis of fluorescent nanoparticle sensors for simultaneous determination of double targets. *Chem. J. Chin. Univ.-Chin.* **29**(1), 83–85 (2008).
70. Y. H. Jeon *et al.*, *In vivo* imaging of sentinel nodes using fluorescent silica nanoparticles in living mice. *Mol. Imaging Biol.* **12**(2), 155–162 (2010).
71. E. W. Choi *et al.*, Effects of GM-CSF gene transfer using silica-nanoparticles as a vehicle on white blood cell production in dogs. *Experimental Hematolo.* **36**(7), 807–815 (2008).
72. A. M. Mumin *et al.*, Dendritic cell internalization of foam-structured fluorescent mesoporous silica nanoparticles. *J. Colloid Interface Sci* **353**(1), 156–162 (2011).
73. M. Bottini *et al.*, Quantum dot-doped silica nanoparticles as probes for targeting of T-lymphocytes. *Int. J. Nanomed.* **2**(2), 227–233 (2007).
74. H. Yun *et al.*, Simultaneous counting of two subsets of leukocytes using fluorescent silica nanoparticles in a sheathless microchip flow cytometer. *Lab on a Chip* **10**(23), 3243–3254 (2010).
75. K. Lee *et al.*, Blood Half-Life Study of Magnetic Fluorescent Silica Nanoparticles by Using MR T2 Relaxation Measurement: Long Circulating SPIO. *Journal of the Korean Physical Society*, **53**(5), 2535–2539 (2008).
76. W. Stober, A. Fink and E. Bohn, Controlled growth of monodisperse silica spheres in micron size range. *J. Colloid Interface Sci.* **26**, 62–67 (1968).
77. M. Qhobosheane, P. Zhang, W. H. Ta, Assembly of silica nanoparticles for two-dimensional nanomaterials. *J Nanos. Nanotechnol.* **4**, 635–640 (2004).
78. X. Song *et al.*, Synthesis of Fluorescent Silica Nanoparticles and Their Applications as Fluorescence Probes. *Journal of Fluorescence*, **21**(3), 1205–1212 (2011).
79. S. Zhao *et al.*, Controllable preparation and formation mechanism of monodispersed silica particles with binary sizes. *J. Colloid Interface Sci.* **388**, 40–46.
80. G. Wang *et al.*, The synthesis of magnetic and fluorescent bi-functional silica composite nanoparticles via reverse microemulsion method. *J. Fluorescence* **19**(6), 939–946 (2009).
81. X. Mu *et al.*, A facile and general approach for the synthesis of fluorescent silica nanoparticles doped with inert dyes. *Chin. Sci. Bull.* **56**(31), 3242–3246 (2011).
82. N. Zhang *et al.*, Synthesis, characterizations of dye-doped silica nanoparticles and their application in labeling cells. *Colloids Surfaces B-Biointerfaces* **89**, 133–138.
83. F. May *et al.*, Synthesis and Characterization of Photoswitchable Fluorescent SiO2 Nanoparticles. *Chemis-A Eur. J.* **18**(3), 814–821 (2012).
84. J. B. Wacker *et al.*, Controlled synthesis of fluorescent silica nanoparticles inside microfluidic droplets. *Lab on a Chip* **12**(17), 3111–31116 (2012).
85. S. Santra *et al.*, Fluorescence lifetime measurements to determine the core-shell nanostructure of FITC-doped silica nanoparticles: An optical approach to evaluate nanoparticle photostability. *J. Luminescence* **117**(1), 75–82 (2006).

86. Z.-Z. Chen *et al.*, Covalent conjugation of avidin with dye-doped silica nanopaticles and preparation of high density avidin nanoparticles as photostable bioprobes. *Biosensors Bioelectron.* **37**(1), 75–81 (2012).
87. L. Xue *et al.*, Carboxylate-modified squaraine dye doped silica fluorescent pH nanosensors. *Nanotechnology* **21**(21), (2010).
88. X. Wang, O. Ramstrom and M. Yan, Dye-doped silica nanoparticles as efficient labels for glycans. *Chem. Commun.* **47**(14), 4261–4263 (2011).
89. Y. Wang *et al.*, Photo- and Biophysical Studies of Lectin-Conjugated Fluorescent Nanoparticles Reduced Sensitivity in High Density Assays. *J. Phys. Chem. B* **114**(45), 14487–14494 (2010).
90. W. W. Yu *et al.*, Water-soluble quantum dots for biomedical applications. *Biochemi. Biophys. Res. Commun.* **348**(3), 781–786 (2006).
91. T.-W. Sung and Y.-L. Lo, Highly sensitive and selective sensor based on silica-coated CdSe/ZnS nanoparticles for Cu2+ ion detection. *Sensors Actuators B-Chem.* **165**(1), 119–125.
92. F. Mahtab *et al.*, Covalent Immobilization of Aggregation-Induced Emission Luminogens in Silica Nanoparticles Through Click Reaction. *Small* 7(10), 1448–1455 (2011).
93. M. Faisal *et al.*, Fabrication of Fluorescent Silica Nanoparticles Hybridized with AIE Luminogens and Exploration of Their Applications as Nanobiosensors in Intracellular Imaging. *Chem.-A Eur. J.* **16**(14), 4266–4272 (2010).
94. B. Perlstein *et al.*, Synthesis and Characterization of Functionalized Magnetic Maghemite Nanoparticles With Fluorescent Probe Capabilities for Biological Applications. *J. Biomed. Mater. Res. Part B-Appl. Biomat.* **92B**(2), 353–360 (2010).
95. W. Tansub *et al.*, Synthesis of Antibodies-Conjugated Fluorescent Dye-Doped Silica Nanoparticles for a Rapid Single Step Detection of Campylobacter jejuni in Live Poultry. *J. Nanomater.*
96. I. R. Rajiv Kumar, Y. Tymish Ohulchanskyy, N. Lalit Goswami, C. Adela Bonoiu, J. Earl Bergey, M. Kenneth Tramposch, An. Maitra, and Paras N. Prasad, Covalently Dye-Linked, Surface-Controlled, and Bioconjugated Organically Modified Silica Nanoparticles as Targeted Probes for Optical Imaging. *ACS NANO* **2**(3), 449–456 (2008).
97. L. S. G. B N Ames, and W C Willett, The causes and prevention of cancer. *PNAS* **92**(12), 5258–5265 (1995).
98. M. Yan *et al.*, Fluorescence immunosensor based on p-acid-encapsulated silica nanoparticles for tumor marker detection. *Analyst* **137**(12), 2834–2839.
99. S. Santra *et al.*, Folate conjugated fluorescent silica nanoparticles for labeling neoplastic cells. *J. Nanosci. Nanotechnol* **5**(6), 899–904 (2005).
100. H.-C. Huang *et al.*, Inorganic nanoparticles for cancer imaging and therapy. *J. Controlled Release* **155**(3), 344–357 (2011).
101. S. Santra *et al.* Chem Commun, TAT conjugated, FITC doped silica nanoparticles for bioimaging applications. *Chem Commun.* 2810–2811 (2004).
102. S. Santra *et al.*, Functional dye-doped silica nanoparticles for bioimaging, diagnosis and therapeutics. *Food Byproducts Process* **83**, 136–140 (2005).
103. J. Ma *et al.*, Synthesis of magnetic and fluorescent bifunctional nanocomposites and their applications in detection of lung cancer cells in humans. *Talanta* **81**(4–5), 1162–1169 (2010).

104. S. Santra, P. Zhang, K. M. Wang, R. Tapec and W. H. Tan, Conjugation of biomolecules with luminophore-doped silica nanoparticles for photostable biomarkers. *Anal. Chem.* **73**, 4988–4993 (2001).
105. S. Santra, K. M. Wang, R. Tapec and W. H. Tan, Development of novel dye-doped silica nanoparticles for biomarker application. *J. Biomed* **Opt**(6), 160–166 (2001).
106. X. X. He, J. H. Duan, K. M. Wang, W. H. Tan, X. Lin and C. M. He, A novel fluorescent label based on organic dye-doped silica nanoparticles for HepG liver cancer cell recognition. *J. Nanosci. Nanotechnol.* **4**, 585–589 (2004).
107. T. Y. Yoon, K. N. Y., E. Kim, J. S. Kim, B. G. Kim, S. H. Yun, B. H. Sohn, M. H. Cho and J. K., S. B. Lee, Park Specific targeting, cell sorting, and bioimaging with smart magnetic silica core–shell nanomaterials. *Small* (2), 209–215 (2006).
108. T. J. Yoon, J.S.K., B. G. Kim, K. N. Yu, M. H. Cho and J. K. Lee, Multifunctional nanoparticles possessing a "magnetic motor effect" for drug or gene delivery. *Angew. Chem. Int. Ed. Engl.* (44), 1068–1071 (2005).
109. M. Srivastava *et al.*, Single particle tracking of fluorescent silica nanoparticles bound to IgE receptors on RBL mast cells. *Biophys. J.* **82**(1), 498A–498A (2002).
110. H. Ow *et al.*, Bright and Stable Core-Shell Fluorecent Silica Nanoparticles. *Nano Lett,* **5**(1), 113–117 (2005).
111. J. E. Fuller *et al.*, Intracellular delivery of core-shell fluorescent silica nanoparticles. *Biomaterials* **29**(10), 1526–1532 (2008).
112. H. Ow *et al.*, Bright and stable core-shell fluorescent silica nanoparticles. *Nano Lett.* **5**(1), 113–117 (2005).

Chapter 9

Engineered Gold Nanoparticles as Contrast Agents for Biological Medical Imaging

*Anna M. Sitarski, Sanjeev Kandpal, Jeremy Grant, Andrew Doolittle, Matthew Newsom and Michael D. Mason**

Department of Chemical and Biological Engineering
The University of Maine, Orono, Maine 04469-5737, USA
**mmason@umche.maine.edu*

1. Introduction

Over the past three decades interest in exploiting the unusual properties of metal nanoparticles has exploded. Starting in 1990 the number of annual research publications on noble metal nanoparticles grew at a nearly exponential rate from less than 100 to more than 5,000 only 15 years later.[1–3] Since that time the numbers of publications that relate to specific nanoparticle applications continued to grow as the physical properties of specific systems were better understood and their potential applications envisioned. As a consequence of this effort, patent applications which make use of noble metal nanoparticles have been steadily increasing, with many now finding their way into industrial processes and commercial and products. Potential applications may include: printable inks for flexible electronics, biomedical assays, drug delivery, colorants and paints, solar cells, liquid crystal displays and chemical catalysis to name a few.[4–10]

To date, the largest industrial applications of metal nanoparticles have been as supported catalysts in high volume chemical reactors. The most common being: Rh, Ru, Pt, Pd, V and Fe, used in hydrogenation, dehydrogenation and selective oxidation reactions. Here the extremely high surface-to-volume ratio of packed finely divided (nanoparticulate) metals, as well as additional nanoscale effects, result in significant enhancement in chemical reaction rates and selectivity for desired processes.[11] Interestingly, even the most chemically inert bulk metals (Au) can exhibit efficient and selective catalytic activity when reduced to the nanoscale.[12] Unfortunately, the widespread transition of metal nanoparticles from the laboratory to use as catalysts at the industrial scale has been hindered somewhat by their high initial cost and limited supply.[10] Growing concerns about the potential for health and environmental risks have also slowed their application to industrial products and processes. Nonetheless, supported metal nanoparticles, including the noble metals, are now being used in a number of commercial products including high-end automotive catalytic converters.[13]

Like many other materials, the physical properties of noble metal nanoparticles deviate significantly from those of bulk. While some aspects of this size dependence were well understood by scientists as early as Michael Faraday[14] and even modeled as light-particle interactions by Gustav Mie,[15] the full extent to which dimensional confinement (size and geometry) impact physical properties was not understood until well after the advent of quantum mechanics. In fact, it has only been in the past 25 years that direct correlations between size and function are possible, largely due to the availability of electron microscopy. Today we recognize that many of the physical properties that we observe for metals at the macroscale: light scattering and absorption, electrical and thermal conductivity, heat capacity, malleability, etc., all arise from the relatively high mobility of electrons. By restricting the motion of these electrons, the observed physical properties can be modified. Hence the promise of the tunability of properties through nanoparticle size and shape control.

Beyond confirming our understanding of nanoparticle physics, high resolution imaging tools have made it possible to rapidly optimize chemical reaction conditions for a specific particle size or geometry, or find new geometries whose existence was previously obscured by ensemble averaging. As a result, a large number of distinct metal nanoparticle geometries have now been investigated. By varying the base metal, size and shape, nanoparticles can now be formed that interact with light spanning many orders of magnitude (from X-rays to microwaves) along the electromagnetic spectrum.[16] Typical gold nanoparticle geometries, which have shown promise for use in commercial, industrial and biological applications, including imaging, are shown in Fig. 1. Those imaging modalities to which they have been applied effectively are also listed. Other interesting geometries have included: prisms,[17] tetrapods,[18] gold decorated silica,[19] and cubes,[20] to name a few. While many complicated structures such as these have been identified as potential image contrasting agents, their synthetic complexity, cost and relative irreproducibility make them less likely candidates for most practical applications. Nonetheless, significant research is underway on these and other more complicated systems.

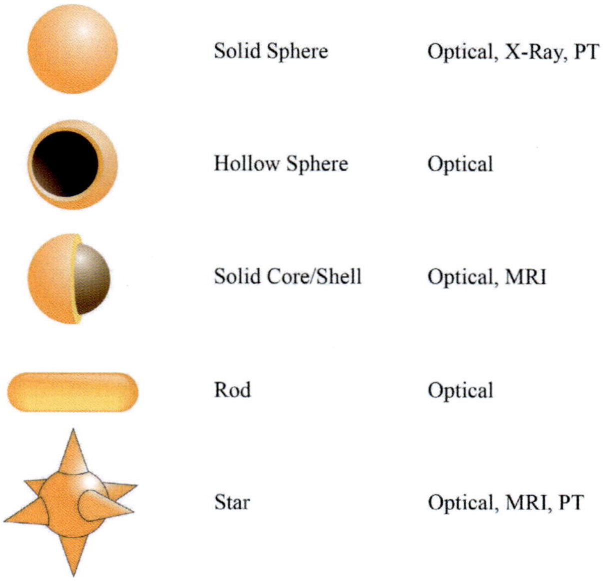

Fig. 1. Most common nanoparticle geometries investigated for use as contrast agents for imaging in biological or medical applications, and the respective imaging modalities.

While their physical properties hold great promise, metal nanoparticles generally hold some risks when considering their use in biological applications. Many metals are known to be toxic in the bulk, as metals, or as dissolved metal ions. Much of the early interest in gold nanoparticles as biological agents was based on the extreme inertness of bulk gold. Unfortunately, when reduced to the nanoscale, the toxicity of many metals, thought to be inert, increases dramatically due to their high surface activity. Some experimentally determinations of nanoparticle toxicities (metals and metal oxides) are listed in Table 1. For example, the noble (coinage) metals Pt, Pd, Ag, have all been shown to be cytotoxic. Though not fully inert, gold is an exception in that its chemical activity is far lower than that of the other noble metals, which are generally lower than all other groups of metals, making it the bestpossible candidate for biological applications involving true metals (as opposed to metal oxides).

In many ways the interest in using gold nanoparticles for biological applications is a result of a fortunate confluence of the relevant physical and chemical characteristics. First, gold nanoparticles can readily be formed using single pot room temperature aqueous redox chemistry.[23] Some size tunability can be achieved through simple modification of temperature, reducing agents and relative abundance of reactants. Shape control is also easily attained with the

Particle	Organism	Toxicity
C_{60}	Human dermal fibroblasts	0.00002 mg/ml
Gold nanorods	Human embryonic kidney	0.75 mg/ml
Nano copper	Living mice	0.413 mg/ml
Silver Spheres **	Living mice	>2 mg/ml
Gold Spheres*	Living mice	3.2 mg/ml
ZnO	*E. coli* bacterium	0.0211 mg/ml
CuO	*E. coli* bacterium	0.0645 mg/ml
Al_2O_3	*E. coli* bacterium	0.3261 mg/ml
La_2O_3	*E. coli* bacterium	0.4569 mg/ml
Fe_2O_3	*E. coli* bacterium	0.6383 mg/ml
SnO_2	*E. coli* bacterium	1.0456 mg/ml
TiO_2	*E. coli* bacterium	1.1048 mg/ml
30 nm Ni spheres	Zebrafish Embryos	0.328 mg/ml
60 nm Ni spheres	Zebrafish Embryos	0.361 mg/ml
100 nm Ni spheres	Zebrafish Embryos	0.221 mg/ml

Fig. 2. Comparison of observed toxicity of gold nanoparticles with other relevant nanoparticles systems.[2,20,21]

addition of common organic water-soluble surfactants. Second, the size tunability, achieved through chemical control, results in a commensurate change in light absorption and scattering properties that span the entire visible spectrum (which is why the system was found to be interesting in the earliest observations[14]). Third, in this same size range and using simple redox chemistry, gold nanoparticles can remain discrete and stable as a colloid for long periods of time due to electrostatic stabilization from residual reducing agent (Fig. 1). This stability made many of our earliest experiments possible. In contrast, other metals require higher temperature, are not amenable to aqueous reduction, are not as readily size tunable, and are unstable against aggregation and sedimentation.

Perhaps the single most significant characteristic of gold lies in the strength to which covalent sulfur–gold bonds can be formed. Energetically, these bonds are among the most stable for metals.[24] The combination of facile, and nearly irreversible, sulfhydryl chemistry, and the aqueous stability of the as-prepared nanoparticles provides for a sensible method whereby the surface of gold nanoparticles can be easily modified. In fact, this allows for nearly endless chemical and biochemical possibilities while preserving the underlying physical properties of the metal particle. Some of these are illustrated in (Fig. 3). In the context of biological applications, including medical imaging, the gold–thiol bond makes it possible to: (1) sterically stabilize nanoparticles against aggregation even in biological fluids,[25] (2) add fillers

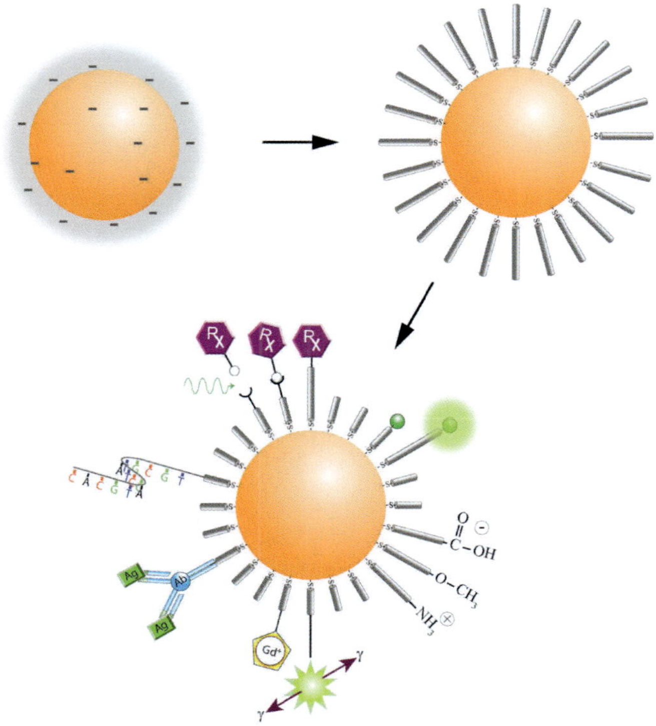

Fig. 3. Cartoon depiction illustrating the broad range of surface modifications made possible by the strong thiol–gold bond (S). These modifications are largely responsible for the utility of gold nanoparticles relative to other metal nanoparticle systems.

and spacers for use in metal enhanced fluorescence (MEF),[26] (3) modify cellular uptake through specific surface terminal groups, (4) target specific structures through DNA or antigen recognition, (5) deliver and release drugs and (6) make use of specific markers or radiotracers.[27] It is this range of possibilities that has stimulated the explosive growth in research and development of gold nanoparticle based imaging and therapeutic applications.

In this chapter, we review a specific class of gold nanoparticle applications; namely medical and diagnostic imaging. In the following sections, a number of established and emerging imaging modalities, enhanced through the use of gold nanoparticles, are described in some detail. Specific attention is given to describing the physical nature of the contrast mechanism exploited to generate diagnostic images. Additionally, connections between nanoparticle parameters (size and shape), and contrast or signal elements with respect to the surrounding tissue are presented. The extent to which nanoparticle surface modification plays a specific role in the specific contrast mechanism is also described. Finally, the likely future directions of each imaging modality, including remaining challenges, are discussed.

2. Optical Imaging

The excitement around noble metal nanoparticles originates primarily from the strong interaction with visible radiation. As early as 1850 Michael Faraday noted a strong visible change in the color of light transmitted through colloidal gold solutions with the addition of salt. Based on his understanding of electromagnetic radiation as well as phenomena such as sedimentation he recognized that there must exist a link between particle size, or aggregation state and the observed color of scattered or transmitted light.[14] While the underlying physics was not elucidated until much later, Faraday observed similar effects from a broad range of metals, though the specific colors differed significantly. He asserted that the observed optical properties must be a result of the specific metal and the extent to which the metal was divided, *i.e.* particle size.

More than 100 years ago, Gustav Mie developed a functional description of the scattering of electromagnetic radiation by colloidal spheres.[15] This general description made it possible to calculate fields both outside and inside spherical particles of arbitrary size, enabling the prediction of scattering and extinction cross-sections as well as any related angular dependence. Of relevance here was the realization that there exist material specific particle sizes, for which resonance conditions give rise to stronger than typical optical activity. Around the same time Lord Rayleigh developed a similar formalism specific to particles smaller than the wavelength of the incident radiation. The resulting functions are, in fact, a limit of the Mie description and are of specific interest in the area of nanotechnology, describing the size, wavelength, material and environmental dependence of both scattering and absorption of metal colloids:

$$\sigma_{\text{abs}} = -\frac{8\pi^2}{\lambda} R^3 \operatorname{Im}\left[\frac{\varepsilon_p - \varepsilon_m}{\varepsilon_p + 2\varepsilon_m}\right], \tag{1}$$

$$\sigma_{\text{sca}} = -\frac{128\pi^5}{\lambda^4} R^6 \left[\frac{\varepsilon_p - \varepsilon_m}{\varepsilon_p + 2\varepsilon_m}\right]^2. \tag{2}$$

Here R is the particle radius, λ is the wavelength of incident radiation, and ε_p and ε_m are the complex dielectric coefficients for the particle and surrounding medium, respectively. Due to the R^3 versus R^6 dependence, respectively, absorption tends to dominate for the very smallest particles (molecules). At increased particle sizes (35–100 nm for gold), scattering quickly begins to dominate. For metals an additional resonance arises due to the relative dielectric constants of the particles and the surrounding medium. Though indicated as constants in the above equations, both ε_p and ε_m are wavelength dependent and ε_p demonstrates its own additional size dependence.[28] Given that metals frequently have negative dielectric constants, at specific wavelengths where $\varepsilon_p = -2\varepsilon_m$, peaks in the absorption and scattering spectra are observed. This "resonance" condition results from coherent oscillation of conduction-band electrons giving rise to a size-dependent shift in the absorption and scattering maxima as demonstrated by the visible color shifts in the colloidal solutions in Fig. 4. This is referred to as a *surface plasmon resonance*, and is described in detail elsewhere.[1] It is the size dependence of this phenomenon that is responsible for the excitement with regard to the tunability of noble metal nanoparticles. In general, optical absorption is said to red-shift with an increase in particle radius (Fig. 5).

While Eqs. (1) and (2) specifically address spherical particles, they also provide valuable insight into the expected behavior of other geometries. For example, rod-shaped structures are characterized by three axes, suggesting the presence of three distinct "size" dependent resonances. Indeed, this behavior is routinely observed (Fig. 5), and even for more complicated structures can be treated with reasonable accuracy as a superposition of the apparent "sizes" present. In terms of technology, this implies that gold nanoparticles can be engineered to simultaneously present multiple absorption or transmission bands. As a result a number of nanoparticle geometries, each with different specified optical properties, have been synthesized including: rods,[29] prisms,[30] shells,[31] and stars,[32] to name a few.

Fig. 4. A series of colloidal gold particles with particle size increasing from left to right. The left most solution is dissolved gold ions in solution before chemical reduction is initiated.

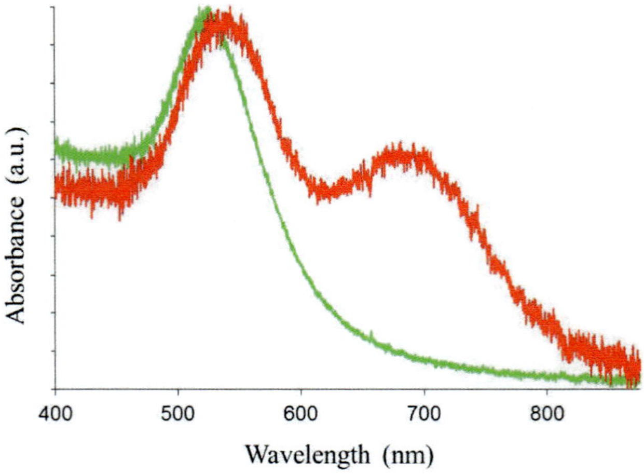

Fig. 5. Absorption spectra of as-prepared (a) 15 nm gold spheres and (b) gold rods with dimensions of approximately 15 nm × 15 nm × 100 nm.

In principle, gold nanoparticles can be implemented as contrast agents for biological or medical imaging in the same way as any other dye, stain or indicator. As such, virtually all instrumental geometries, including absorption and scattering, have been investigated. The relatively high extinction coefficient (ε) of a single gold nanoparticle is dominated by scattering (Eqs. (1) and (2).[33] In staining applications, the physical size of the gold nanoparticle will ultimately limit the achievable labeling density and access to constrained microenvironments. This is different from organic dyes where individual extinction coefficients are almost entirely due to absorption, though much smaller, but can reach higher effective concentrations and more widespread labeling. As such, gold nanoparticles may have minimal advantage over conventional dyes for brightfield (widefield) transmission imaging, where image contrast is dominated by differential absorption of sample features.

When image contrast is determined by light scattering, as in Darkfield Imaging, the performance of individual gold nanoparticles easily eclipses that of conventional dyes.[29, 34] The scattering cross-section of a single gold particle is many orders greater than that of any dye or fluorophore, according to Eq. (2). Additionally, because photons are not absorbed in the scattering process, signal rates are not limited by excited state dynamics as is the case with all absorption mechanisms. Further, metal nanoparticles do not exhibit photobleaching or intermittency, allowing for stable long-term imaging. In fact, individual nanoparticles can be readily viewed at video rates on glass, in live cells, or in relatively thin tissue sections using conventional transmission geometry Darkfield Microscopy, as illustrated in (Fig. 6). Unlike absorption

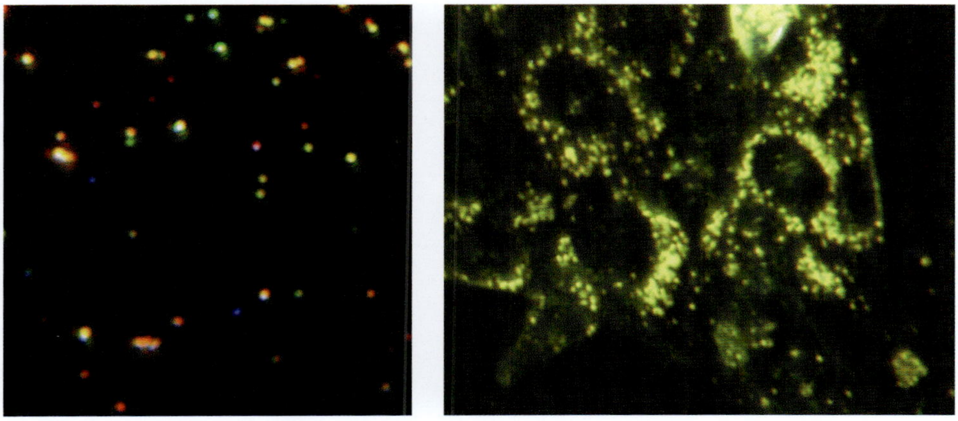

Fig. 6. (a) Darkfield scattering image of individual as prepared gold nanoparticles dispersed on glass. (b) Similar scattering image of gold nanoparticles in a single NIH3T3 cell following 24 h exposure. Here cell uptake is easily observed.

methods only those photons that arise from scattering, a process locally dominated by the gold nanoparticles, are collected and imaged.

Many examples of nanoparticle enhanced darkfield imaging now exist in the literature. Several early cancer studies made use of anti-epidermal growth factor (EGF) physisorbed to the surface of spherical gold nanoparticles to target the overexpressed receptor on the cancer cell surface.[35] Additional stability against localized aggregation and spectral shifting has been improved through the use of antibodies covalently bound through polymer linkages to the nanoparticle surface.[25] Through careful selection of nanoparticle geometry additional gains in labeling efficiency can be made. For example, rod-shaped particles have been shown to exhibit improved internalization through peptide mediated endocytosis, when compared to gold spheres.[36] It seems certain that future approaches will involve the intelligent selection of both particle geometry and surface targeting species to optimize cell labeling and internalization.

While the labeling strategy itself may not change, transmission imaging is not suitable for opaque or relatively thick samples. For these samples reflectance geometries are employed. In general, except at the highest labeling densities, the background reflectance is much higher than that of the individual nanoparticle. As such, poor contrast is observed using conventional reflection (epi-illumination) optics. Though this can be somewhat improved with the use of reflection-type darkfield optics, confocal approaches are now more widely implemented. The in plane selectivity of confocal optics generally yield improved signal-to-background and contrast.[37] A sample confocal

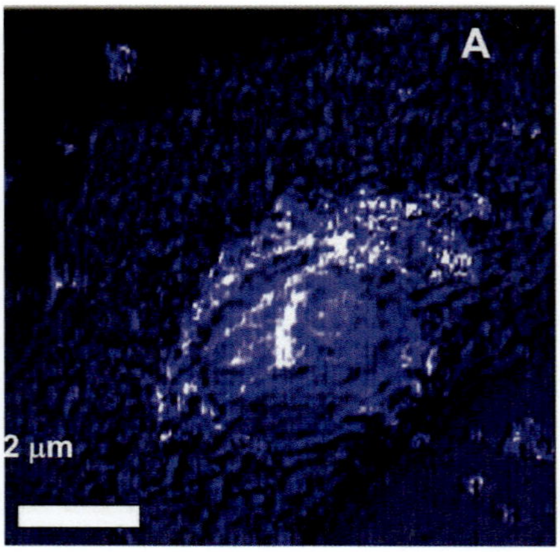

Fig. 7. Confocal reflectance image of a single NIH3T3 cell. Reflectance at the laser wavelength is matched to the peak in the gold nanoparticle scattering cross-section. Bright areas indicate regions of increased nanoparticle density.

reflectance image of gold nanoparticles localized inside of a single cell is shown in (Fig. 7).

The development of lower cost laser sources and dramatically improved detectors over the last three decades has brought about the application of highly sensitive, but extremely weak, vibrational methods such as Raman Spectroscopy. In general terms Raman Spectroscopy is an inelastic scattering method whereby the oscillatory field of incident photons induces a dipolar oscillation within a particle (or molecule). This oscillation can couple energetically to the vibrational motion of the molecule. The resulting dipole radiates at a frequency equal to that of the incident photon plus or minus that of the molecular vibration (Fig. 8). The observed (small) shifts in energy of the scattered photon provide a "fingerprint" of the underlying molecular structure.[38] Unfortunately, Raman signals are remarkably weak, owing to the relatively low probability of this process. Over the same time period, researchers discovered a dramatic Raman enhancement effect, when analytes were adsorbed onto roughened metal surfaces.[38,39] This effect has since been termed "surface enhanced Raman spectroscopy" (SERS), and the roughened surface has been extended to include colloidal suspensions of metal nanoparticles.[40]

Today, metal nanoparticles with pre-selected surface bound organic molecules are constructed to produce strong SERS spectral signatures, making it possible to label a broad range of biological structures.[41] One recent successful approach has been to functionalize a Raman active core with a large number of Raman active molecules. The resulting structure is then capped with a

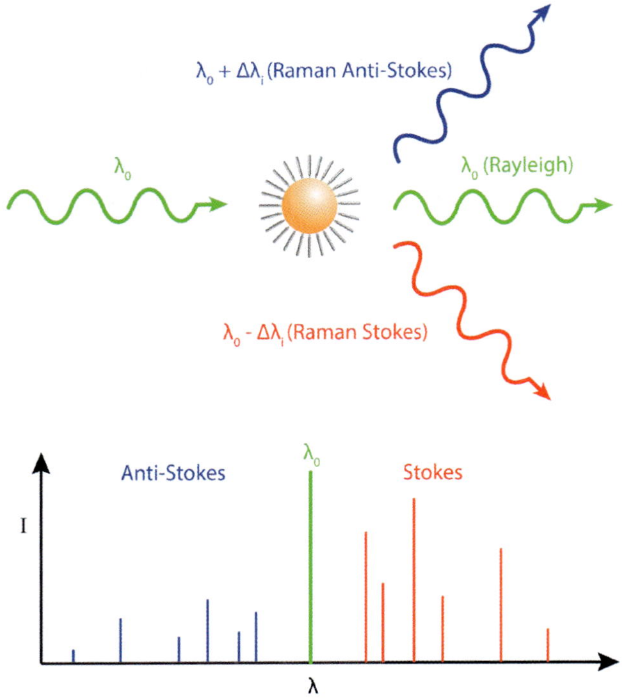

Fig. 8. Depiction of elastic Rayleigh Scattering and inelastic Raman Scattering.

layer of protective silica[29] or cross-linked polymer,[42] rendering the particle and the subsequent Raman signal, insensitive to fluctuations in the local chemical environment. The most recent enhancements based on this technology involve the further modification of the gold nanoparticle core to provide the largest possible SERS enhancement. Though not yet widely implemented, gold "nanostar" geometries have demonstrated additional enhancements above 1000X, producing Raman signals exceeding those of organic molecular fluorophores imaged under identical conditions.[43]

These particles could overcome some of the intrinsic limitation found with other commercial probes.[29,34,44] Unlike their fluorescent counterparts, these probes last indefinitely, and show no intermittency or photobleaching. Using even the most rigorous fluorescence methods, no more than five distinct probes can be observed simultaneously, yet many biological processes involve the interplay of a large number of molecules and structures. Due to their extremely narrow line widths (~20 cm^{-1}), many times narrower than those of fluorescent molecules, and distinct vibrational spectra, it becomes possible to distinguish many tags simultaneously using spectral deconvolution. This approach, now referred to as spectral multiplexing (SM) has been

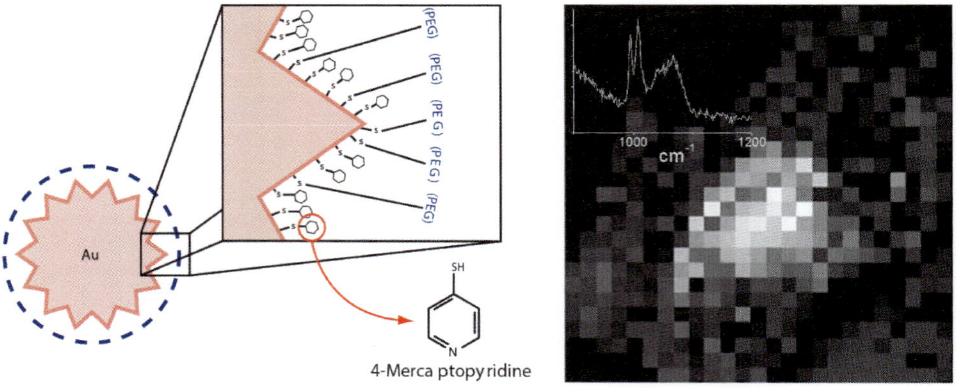

Fig. 9. (a) Illustration of SERS active gold nanoparticle geometry and surface modification. (b) Raman spectral image of single cell (from Fig. 7), with nanoparticles localized based on known particle spectral signature.

demonstrated for large molecular libraries,[45] but is only now being considered in biological applications. For example, SM has been used successfully to positively identify breast cancer cells using, simultaneously, a number of biomarkers (ergB-2, ki-67). While individual markers are not themselves conclusive, multiple unique markers used simultaneously represent a preponderance of evidence and can increase the accuracy of diagnosis.[46–48]

The use of gold nanoparticles as scattering based contrast agents in medical imaging and biotechnology is steadily increasing, however, their use is still dwarfed by fluorescence based imaging techniques. In fact, these two approaches have been viewed as mutually exclusive due to the observed quenching of fluorescence in the vicinity of metal surfaces (image dipole effect).[49] More recently it has been shown that by carefully controlling the distance from the fluorophore and the metal surface fluorescence emission can be dramatically enhanced relative to the free fluorophore in solution, or on glass.[50] This increase in fluorescence, now referred to as MEF, is thought to result from modification of the radiative decay rate (lifetime) as well as other photophysical limitations such as photobleaching and spectral selectivity.[51] MEF has significant promise for single molecule/single particle methods, where sensitive detection and localization requires optimal photon emission before bleaching, and has been challenging for all but the very best organic molecular fluorophores. Experiments requiring prolonged imaging or repeated exposure would also benefit from MEF, relative to traditional fluors. Though still limited in its use for *in vivo* imaging MEF is finding widespread application for a variety of plate and sandwich biochemical arrays.[52]

3. X-ray

Since its inception, X-ray imaging has become an incredibly valuable tool for dental and medical professionals allowing them to effectively visualize and diagnose a wide variety of conditions. For nearly 70 years X-ray imaging was limited to 2D images, analogous to standard optical photography, but where images are obtained from transmitted rather than scattered photons. Some four decades ago a new imaging method, computer assisted tomography (CAT or CT), was developed. In CT imaging, a series of typical 2D X-ray transmission images are obtained from different angles. Computer algorithms are then implemented which translate the series of 2D images into a reconstructed representation of the 3D structure. This method revolutionized X-ray imaging, and it continues to be improved upon today.[53]

Images produced by X-rays are heavily dependent on the fact that, as an X-ray passes through a medium, it *attenuates.* Attenuation, in general terms, is the gradual loss of intensity as a wave passes through a given medium. This is described by the familiar *Beer-Lambert law* which states: $I = I_0 e^{-ax}$ where I represents the measured intensity, I_0 the incident intensity, a the attenuation coefficient and x represents the path length. While the attenuation coefficient is typically treated phenomenologically, it is equivalent in mathematical form to the extinction coefficient (ε), used in optical (UV-Vis) absorption spectroscopy. As with ε, the attenuation coefficient, a, can be better described quantitatively using the (atomic) cross-section, σ_{total}, given as the sum of all underlying physical processes that cause attenuation:

$$\sigma_{\text{total}} = \sigma_{\text{scatt}} + \sigma_{\text{PE}} + \sigma_{\text{pair}} + \sigma_{\text{nuc}},$$

where the individual terms are due to scattering (elastic and inelastic), photoelectric absorption, electron–positron pair production and nuclear (photo and field) effects, respectively. For most elements the first two terms dominate all other processes and can be directly correlated to the number of electrons present, and hence, the atomic number, as indicated in Fig. 10.

Based on the atomic number and abundance of elements present, observed attenuation values vary depending on tissue type and can range from −90 to +90 Hounsfield units[a] (HU) for soft tissues and up to +1000 HU for bone. This large difference in X-ray attenuation coefficients means that bone and soft tissues are readily distinguished from each other in X-ray

[a] Hounsfield units are given by $\text{HU} = 1000 * \frac{a_y - a_{\text{water}}}{a_{\text{water}}}$ where a_y is the attenuation coefficient for material y and a_{water} is the attenuation coefficient for distilled water at STP.

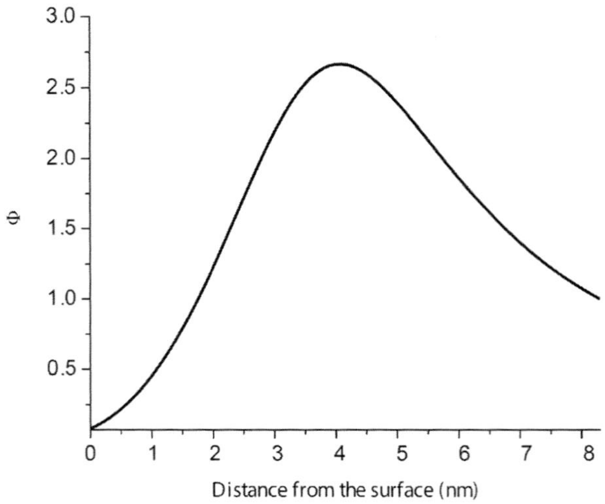

Fig. 10. Plot illustrating MEF intensity as a function of the fluorophore distance from the particle surface.

images. As a consequence, this has lead radiologists to classify common tissue types with their own effective attenuation coefficients. In order to differentiate between soft tissues, with more similar attenuation coefficients, either longer exposure times or additional chemical *contrast agents* are required. Clearly the latter is preferred due to the potential adverse effects of ionizing X-ray radiation.

In designing or selecting a contrast agent suitable for image enhancement, higher atomic number elements are generally more desirable as they will result in the greatest X-ray attenuation (Fig. 11). Commercially available contrast agents are primarily iodine based, with barium sulfate also seeing use, but primarily in gastrointestinal imaging.[54] Though widely used, these contrast agents are not without their pitfalls. For example, iodine-based contrast agents are cleared rapidly from the body, requiring careful timing to ensure optimal enhancement. Injection duration, clearance rate and volume of contrast agent all play a role in determining the time before *peak enhancement* of the target area occurs and technicians can image the area obtaining the best quality image. In addition to rapid clearing, there is a risk for renal toxicity (nephrotoxic), and in some cases allergic reaction to the iodinated contrast agent.[55]

Gold nanoparticles are being heavily investigated as a potential alternative to iodine-based CT contrast agents. Not only does elemental gold have a significantly higher attenuation coefficient than iodine (Fig. 11), but gold is viewed as inert and therefore non-toxic (biocompatible). Adding to their

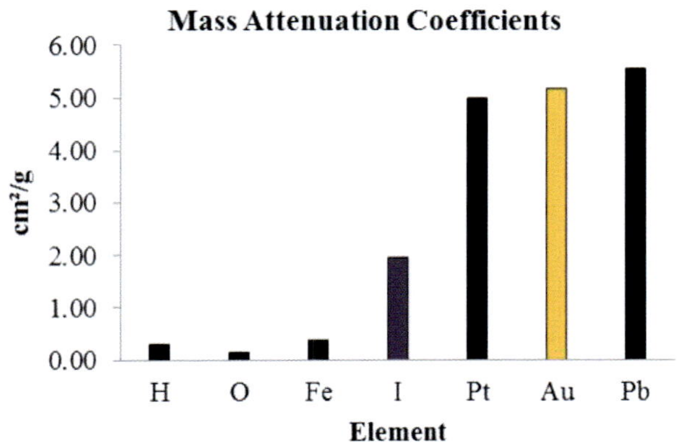

Fig. 11. Comparison of mass attenuation coefficients for a range of elements including those most commonly used as contrast agents.

appeal is recent data that shows that gold nanoparticles can be engineering that are cleared from the body at a significantly lower rate relative to iodine-based agents. Together, these characteristics suggest that gold nanoparticle-based agents could produce better contrast (per mass of material), and do so for longer periods of time.

Gold nanoparticles are now being investigated for specific use in diagnostic imaging of cancer, making use of two additional characteristics. During tumor growth blood vessels develop rapidly (angiogenesis), but as a consequence are poorly formed and highly permeable. Appropriately sized nanoparticles, engineering to circulate through the bloodstream, can diffuse through the vessels, becoming fixed within the tumor tissue. This "passive" targeting approach has been used successfully by a number of groups to effectively identify healthy from cancerous tissue using CT imaging. Second, the gold surface allows for facile attachment of cancer targeting species (antibodies).[25]

4. Photothermal Imaging

Visible light is strongly absorbed by gold nanoparticles due to coherent oscillation of conduction band electrons in strong resonance with light in the visible part of the electromagnetic spectrum. The type of metal, its dielectric properties relative to that of the surroundings, and the shape and size of the nanoparticle all influence this resonance condition, as described by Eqs. (1) and (2). This list of parameters allows for tunability of gold nanoparticles with regard to light interaction. As with any absorption event, the net energy of the particle must increase. Intuitively, one can expect that in the absence of illumination, the particle will eventually give up this excess energy back into

Tissue	CT Number (HU)
Bone	+1000
Liver	40-60
White Matter	-20 to -30
Grey Matter	-37 to -45
Blood	40
Muscle	10-40
Kidney	30
CSF	15
Water	0
Fat	-50 to -100
Air	-1000

Fig. 12. Table of effective (clinical) attenuation coefficients for various tissue types.

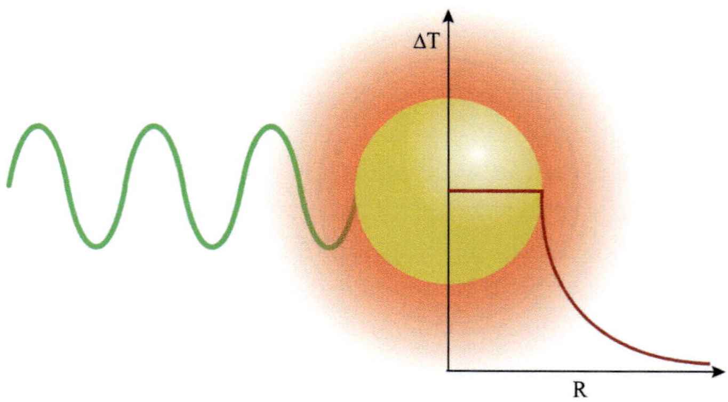

Fig. 13. Cartoon depiction of optically induced heating in the vicinity of a gold nanoparticle.

its surroundings. Depending on the extent to which the particle surface is coupled to its surrounding, and the magnitude of the energy, this process can involve visible or thermal wavelength photon emission or even ionization. In the case of metal particles of high relative surface area in aqueous media (water) illuminated with visible radiation, this transfer of energy results in heating of the media directly surrounding the nanoparticle, referred to as a photothermal effect as shown in Fig. 13.[56]

The photothermal properties of gold nanoparticles were studied mechanistically in detail using femtosecond transient absorption spectroscopy by Link and El-Sayed.[57] Their studies showed that photo-excitation of nanoparticles results in the formation of a *heated* electron cloud, which cools down by exchanging energy with the nanoparticle lattice. The nanoparticle lattice loses energy to the surrounding material through phonon–phonon interaction. This conversion happens within 100 ps. This rapid energy conversion and dissipation gives rise to the observed heating of the local environment.[58] This process of photothermal conversion is made exceptionally efficient in metal nanoparticles by strong resonant (surface plasmon resonance) absorption. As a result a number of imaging methods based on this phenomenon have emerged in recent years.

Photoacoustic tomography (PAT) is a non-invasive photothermal imaging technique, capable of resolution equivalent to that of modern ultrasound.[59] PAT makes use of the strong wavelength tunable absorption characteristics of gold nanoparticle constructs targeted to specific tissues through surface bound bioactive molecules. Once localized, pulsed laser light at a wavelength matching the peak absorption (plasmon resonance) of the particle is used to illuminate the sample. During each pulse, absorbed light is converted at the nanoparticle surface to heat which quickly dissipates into the surroundings when the pulse ends. Following the pulsed illumination, transient variations in temperature result in the generation of acoustic signals as a function of time and position, which are measured[60] and converted into image features.

PAT with finite element reconstruction has been used successfully to image millimeter-sized nanoparticle tagged features *in vitro* with high contrast resolution,[61] where local abundance data is quantitative.[60,61] In addition to standard gold spheres, more complicated nanoparticle geometries, such as hollow nanoshells,[31] nanocages,[62] and gold-speckled silica nanoparticles,[21] appear to show added efficiency as PAT contrast agents[60,63] and have been used for *in vivo* imaging of cerebral vasculature in mice.[63,64] While a relatively inexpensive and promising technique, PAT may ultimately be limited by the available peak absorption energy of the engineered nanoparticles, and the penetration of light at those wavelengths through biological tissues.[2]

Optical coherence tomography (OCT) is a related photothermal imaging method also making use of pulsed laser excitation. In this technique, a focused beam from a laser diode, at a wavelength matching the peak absorption of the contrast agent, is modulated at a specific frequency, while irradiating a solution/tissue containing contrast agent. Because of their large tunable optical cross-sections, the gold nanoparticles can efficiently absorb the incident light and convert it into heat. As described above, this produces small-scale localized temperature oscillations at the modulation frequency due to the photothermal properties of metal nanoparticles. These temperature oscillations then result in an effective optical path length due to temperature

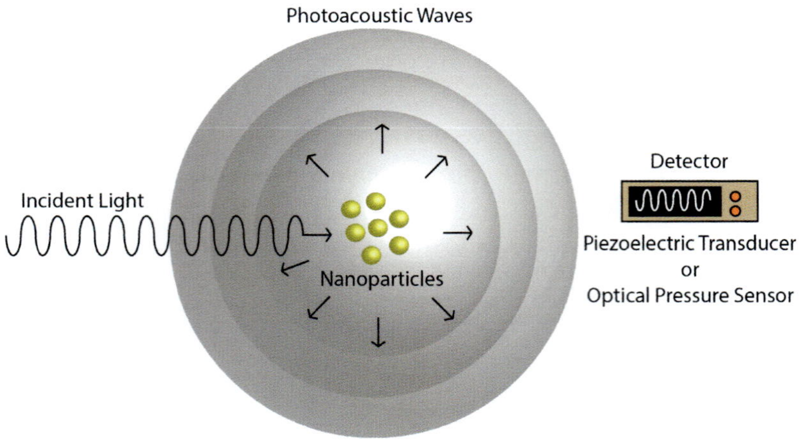

Fig. 14. Illustration of radially propagating acoustic waves generated by periodic illumination and subsequent local heating around a gold nanoparticle.

induced changes in refractive index of the surrounding tissue. These changes are detected by a phase-sensitive, swept source OCT system. The OCT system uses a double-buffered Fourier domain mode locked (FDML)[65] laser operating at a specific center wavelength and sweep rate. High contrast is observed between regions containing gold nanoparticles and regions without gold nanoparticles. This technique offers good signal-to-noise performance even at high operating speeds. The inherently high optical resolution of OCT permits the imaging of features such as tissue architectural morphology as well as some cellular features.

5. Magnetic Resonance Imaging

Magnetic resonance imaging (MRI) is a technique which employs nuclear magnetic resonance (NMR)[66] to non-invasively visualize physiological structures in 3D with effective resolution approaching 1 mm^3.[67] Recent advances in MRI techniques now make it possible to obtain near real-time MRI images, with temporal and effective spatial resolution on the order of 30 ms and 8 mm^3, respectively.[65] This makes MRI a valuable medical diagnostic tool, capable of visualizing and monitoring immediate structure and function relationships,[67] as well as longer term disease progression.[67] MRI scanners are similar in function to CT scanners in that they can image structures within the body with reasonably high spatial resolution, but far exceed the capabilities of CT scanning for resolving contrast, especially in biological tissues. This high contrast resolution allows for detection of tiny differences between otherwise very similar tissues. Additionally, because MRI does not use ionizing radiation, like X-rays in CT scans, it is believed to pose fewer negative health implications.[68]

Today there are numerous variants to traditional MRI techniques, though they all make use of similar principles. Generally, MRI scanners employ powerful magnets to induce a magnetic field around, and through, the patient. For atoms with net nuclear spin, this field causes an alignment of the nuclear spins either with (parallel) or against (antiparallel) the external magnetic field (B_0). The difference in these two populations, given by Boltzmann statistics,[69] is a result of the energy difference between these two states; ($\Delta E = h\gamma B_0$), where h is Planck's constant, γ is the gyromagnetic ratio for a given nucleus and B_0 is the magnetic field strength, and gives rise to a net magnetization parallel (longitudinal) to the external field. If the aligned nuclei are illuminated with radiofrequency (RF) photons with energy equal to this difference, the number of the higher energy antiparallel spins increases. As the two populations become equivalent, their spins cancel each other out, and the net longitudinal magnetization of the system vanishes. When the incident photons are removed, the populations revert to their equilibrium distribution, and radiofrequency photons are emitted. These photons, and the timescale over which they are emitted, give rise to the useful MRI signal. While a number of biologically relevant nuclei have been used (^{3}He, ^{13}C, ^{19}F, ^{17}O, ^{23}Na, ^{31}P), ^{1}H MRI is by far the most common due to the relative abundance of hydrogen in the water and fatty tissues of most organisms. Even though hydrogen is abundant, only 0.001% of hydrogen is detected making for very poor sensitivity. This is practically overcome by repeating the process many times (periodic RF pulse sequences), increasing the external magnetic field strength and by using contrast agents to enhance the signal by altering the local magnetic field,[69] or increasing the sensitivity of the local nuclei to the field.[70]

In most modern MRI instruments, image contrast is a function of the time required for excited nuclei to return to their equilibrium distribution following pulsed RF illumination. This dynamic process is referred to as relaxation, and is made up of two directional components; longitudinal and transverse, each of which has its own unique time constant T_1 and T_2, respectively. As indicated in Fig. 15, T_1 is the time constant describing the decay in measured RF photons (intensity) during the longitudinal relaxation of nuclear spins back to the equilibrium magnetization value along the applied magnetic field. During absorption of the RF photon, a net magnetization perpendicular to the applied magnetic field is generated. The time constant T_2 describes the gradual decrease in the transverse magnetization as the induced antiparallel spins revert to their equilibrium distribution. In general, T_2 is always shorter

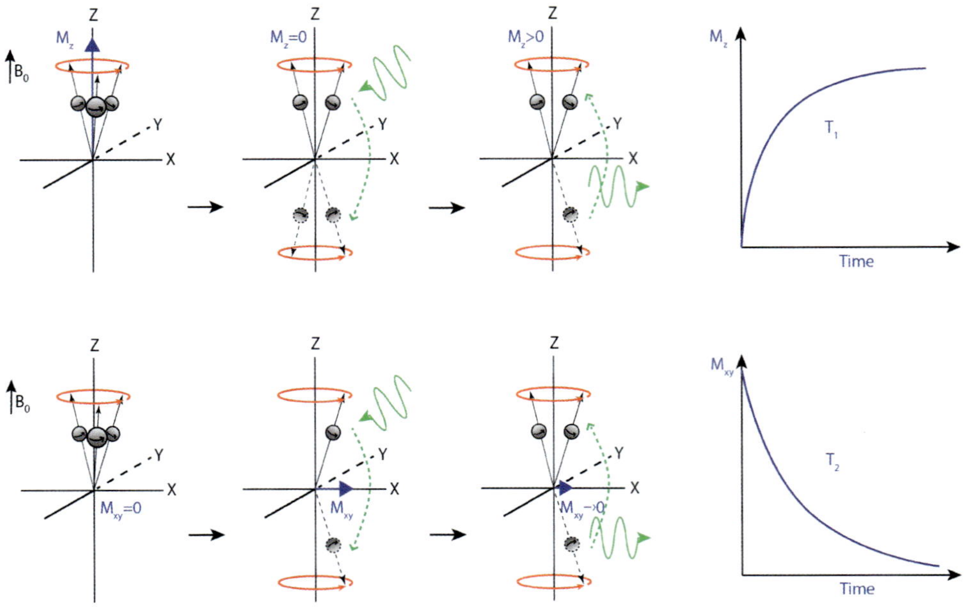

Fig. 15. Depiction of T_1 (a) and T_2 (b) nuclear relaxation processes.

than T_1, irrespective of field strength. The difference between T_1 and T_2, however, can change, and is a strong function of the nuclei and their surroundings.[70]

For a given external field strength, B_0, the intensity of emitted RF photons following each applied RF pulse can be expressed as:

$$I = M\left\{1 - e^{-T_R/T_1}\right\}e^{-T_E/T_2},$$

where T_R and T_E represent the time between successive RF pulses, and the time between the RF pulse and the intensity measurement, respectively,[66] and M is the bulk magnetization. Both T_R and T_E are free parameters that can be used by the MRI technician to affect image contrast by increasing or decreasing relative brightness. For fixed values of T_R, T_E and T_2, smaller values of T_1 produce larger intensity values (brightness) relative to those of neighboring tissue.[71] For fixed values of T_R, T_E and T_1, shorter T_2 values decrease the intensity, resulting in darker images.[71,72] By selecting the relative magnitudes of T_R and T_E, image contrast can be made more sensitive to variations in T_1 or T_2, respectively. This is referred to as "weighting", and is routinely accomplished for T_1 and T_2, both of which are then used to calculate ratiometric contrast such as T_1/T_2.[69]

There are a number of relevant issues that must be considered when selecting or engineering an improved MRI contrast agent. Most importantly, these agents must affect either T_1 or T_2, or both, in their immediate vicinity. The extent of this effect must be a function of the concentration of the contrast agent and is often expressed as:

$$T_i = \frac{T_{i,0}}{1 + T_{i,0}\gamma_i C},$$

where $T_{i,0}$ describes relaxation in the absence of the contrast agent, C is concentration of the agent, and r_i is the relaxivity $((\mathrm{mM}\cdot\mathrm{s})^{-1})$.[69] As this equation suggests, contrast agents tend to decrease both T_1 and T_2. The relaxivities, r_1 and r_2, establish the effectiveness of a contrast agent, whereas the ratio r_2/r_1 indicates for which contrast method (T_1 or T_2) the largest enhancement will be observed. As such, potential contrast agents are most often compared based on their relaxivities (Fig. 15). In addition to magnetic properties, other important chemical and biological properties must be considered when selecting contrast agents. These include, but are not limited to: delivery vector, chemical composition and possible toxicity (reactive groups or metal ions), solubility (hydrophobic or hydrophilic), accessible biodistribution (may include size and functionality), as well as handling and storage limitations.[69]

Today the most widely used contrast agent is the paramagnetic complex DTPA:Gd (diethylenetriamine–pentaaceticacid). Although free gadolinium ions (Gd^{3+}) are considered mildly toxic, a relatively high r_1 relaxivity (~3 $\mathrm{mM}^{-1}\ \mathrm{s}^{-1}$) makes it a reasonable choice for T_1-weighted contrast enhancement.[67] Unfortunately, gadolinium based contrast agents typically have short circulation times as they are quickly cleared from the body. As a result, the contrast agents are often administered in saturation doses that are not completely filtered by the kidneys or liver in healthy patients.[71,73] At these high levels, patients can safely receive only a single dose, and are required to remain motionless for up to 90 min. About 13% of all adults and 47% of patients over the age of 70 in the US have chronic kidney disease (CKD).[74] Many patients that are in need of scans have severe health conditions like CKD, and cannot safely excrete Gd.[74]

The toxicity issues of Gd^{3+} ions have been partially remediated through complexation into cyclic or linear chelates.[70] Unfortunately, even in a complexed state, Gd^{3+} stability can be easily influenced by pH, local ionic concentrations, temperature and surrounding ligands. Metals common in the body, like Zn and Cu, can displace complexed Gd^{3+} ions. These ions have been

Contrast agent	Core Material	r_1/mM^{-1}s^{-1}	r_2/mM^{-1}s^{-1}	Ref.
[illegible]	[illegible]	[illegible]	[illegible]	[illegible]
[illegible]	[illegible]	[illegible]	[illegible]	[illegible]
[illegible]	[illegible]	[illegible]	[illegible]	[illegible]
[illegible]	[illegible]	[illegible]	[illegible]	[illegible]
[illegible]	[illegible]	[illegible]	[illegible]	[illegible]
[illegible]	[illegible]	[illegible]	[illegible]	[illegible]
[illegible]	[illegible]	[illegible]	[illegible]	[illegible]

Fig. 16. Table of typical contrast agents and their relaxivities.

shown to deposit on the liver and in bones where they can irreversibly bind to serum proteins.[75] In recent years, a number of strategies have focused on designing Gd based contrast agents that target specific biological structures, reducing non-specific binding and toxicity to others areas in the body, though with only limited success.[75]

Due to the recurring risks of Gd toxicity, rapid clearance rates, and a continued need for increases in image resolution, significant effort is being dedicated to finding alternative contrast agents. Recent advances in nanotechnology have shown that nanoparticles can function as effective MRI contrast agents, and at relatively low concentrations can produce a large MRI contrast due to their high surface to volume ratio and relatively high relaxivities.[72] Typically nanoparticles are implemented as part of a complex where the nanoparticle is a magnetic core surrounded by a protective polymeric spacer/linker for attachment to biologically active targeting molecules directed towards specific structures or chemistries. These targeting molecules act as reactive probes and may be proteins, enzymes, antibodies or oligonucleotides that have an affinity for binding to a targeted site such as a cancer cell, particular protein, or nucleic acid sequence.

The most widely investigated nanoparticles are based upon a 1.5–15 nm superparamagnetic iron oxide (Fe_2O_3 or γ-Fe_2O_3) core.[69] These particles may be single or multiple domain, but show long range ordering and have magnetic moments ranging from ~4 to nearly 25,000 μ_B, respectively.[69] While iron oxide nanoparticles have r_1 values of generally less than 10 mM^{-1}s^{-1}, they can have r_2 values in excess of 400 mM^{-1}s^{-1}, making them excellent candidates for use as T_2 contrast agents. Iron oxide is considerably less toxic than Gd, and nanoparticles of this material can be processed more readily in the kidneys and liver.[74] Additional biocompatibility or functionality is easily attained through the use of surface coatings or direct surface attachment. Due to the relatively large magnetic moment of this system, this coating does not appreciably affect their ability to decrease T_2 in neighboring atoms (*i.e.* protons in water).[76] Iron oxide particles have been used to increase T_2 contrast at effective concentration of micromolar and even nanomolar ranges.[69]

Unfortunately, iron oxide based nanoparticles are not without their own issues. Interestingly, if not carefully administered, these particles can have too dramatic an effect on T_2, dramatically decreasing the image intensity in regions of elevated local concentration. Diagnostically these parts of the image have been mistakenly identified as a pathogenic condition.[46] Another concern are so-called "blooming effects" where a distortion of the magnetic field on the neighboring normal tissues can cause obscure images and ruin the background around lesions.[46] Other practical issues, such as poor shelf-life and biostability (against aggregation), have severely limited the widespread application of this promising system. As a result, Gd based T_1 contrast agents are still the most widely used in the clinical setting, despite their toxicity and stability concerns.[46]

Recently, multilayer approaches have been investigated that may take advantage of the magnetic properties of iron oxide (or more toxic Mn or Co doped iron systems),[77] but have an inert shell amenable to a broad range of surface chemistries to improve stability and biocompatibility.[72] Gold coatings are now being investigated for this purpose, where the noble metal is being used not for its magnetic properties, but rather as a highly desirable surface. As mentioned earlier, the gold coating can improve stability in aqueous dispersions through electrostatic stabilization and can be easily and irreversibly modified with thiol-terminal molecules to improve stability or add further functionality.[72] Hence, the gold coating provides a route whereby T_2 Fe_2O_3 based contrast agents can be modified to include surface bound T_1 contrast agents. In addition to its own interesting properties, namely efficient scattering and absorption at a surface plasmon resonance,[67,78] surface modification opens a pathway for a whole series of multimodal probes (X-ray, MRI, PET, PAT, etc.).[79]

Beyond thin surface coatings, solid gold nanoparticles are also of great interest as MRI contrast agents. In addition to the aforementioned biocompatibility, shelf-life,[80] improved circulation times, and facile surface chemistry, gold nanoparticles have been shown to exhibit unique magnetic properties by enhancing T_1 relaxation. In comparison to the widely used commercial contrast agent DTPA:Gd, gold nanoparticles can have surprisingly high relaxivity in T_1 (r_1 = 585 Mm^{-1} s^{-1}) as compared to 3.0 mM^{-1} s^{-1} for DTPA:Gd.[81] This large relaxivity may have more to do with the modified gold–thiol surface than the bulk of the gold particle.[71,82]

Magnetomotive imaging methods have emerged in the past few years, which make use of the magnetic response of some nanoparticles such as iron oxide (Fe_2O_3), as well as those with dopants such as Mn and Co.[68,69] Images are obtained by observing nanoparticle motion induced in response to a magnetic field gradient using secondary processes not related to nuclear spin response, as in MRI. The first examples of this method used linear magnetic field gradients and image contrast obtained using OCT.[55] The most recent advancement in this area, referred to as "*Gyromagnetic Imaging*", makes use of a periodic rotating magnetic field to induce particle motion, and a standard Darkfield microscope equipped for polarized light excitation and collection.[83]

Unlike the other methods described in this section, magnetomotive contrast agents must possess the desired physical characteristics for two distinct imaging methods. As such, a number of hybrid nanomaterials are being explored for this application which shows large magnetic response, but also exhibits strong scattering contrast. To this end there has been a focus on superparamagnetic and noble metal nanoparticles due to their extensive uses, biocompatibility, and well known properties.[81] Of the many possible candidates, gold coated iron oxide cores have emerged as the most promising, due to the large magnetic moment and the strong wavelength dependent scattering at the gold surface plasmon resonance, as well as previously mentioned biocompatibility of gold.[83]

Gold nanostars grown onto superparamagnetic iron oxide cores represent a specific hybrid nanoparticle geometry now used in the gyromagnetic imaging of cells and tissues.[84] In modest magnetic fields (80G), these particles are easily made to rotate with rotational frequencies (ω) between 1 and 10 Hz.[84] For sizes above 100 nm, the highly anisotropic structure of the nanostar results in strong (plasmonic) polarized scattering at wavelengths extending into the NIR, as illustrated in Fig. 17.[81] Contrast for gyromagnetic

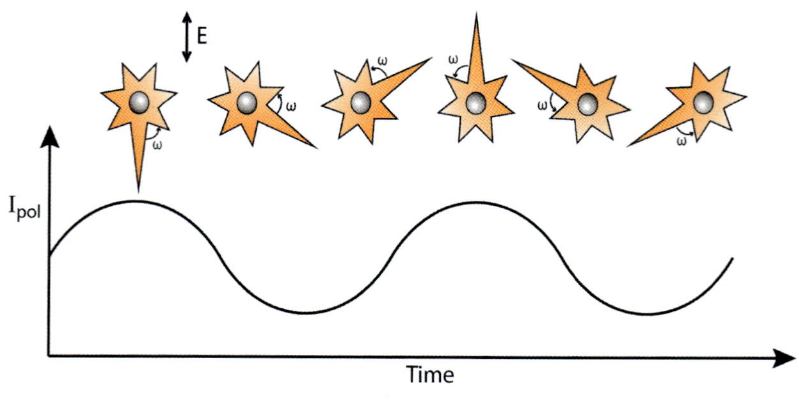

Fig. 17. Illustration of hybrid iron core/gold nanostar shell rotating in response to an applied magnetic field (a). (b) The resulting polarized scattered light signal.

imaging is obtained by synchronizing the scattered (polarized) light detection system with the rotational frequency of the field, allowing for the use of frequency domain methods, which typically show dramatic improvement in signal-to-noise.[81,84]

Unlike other scattering methods such as conventional darkfield microscopy, the additional frequency domain information made available by this method makes it possible to image individual nanoparticles even when surrounded by tissues and biological structures which tend to dominate other scattering signals. Hence, magnetomotive imaging holds significant potential for biological or medical applications, especially for surface or sectioned samples.

6. Conclusion

In recent years nanotechnology has increasingly made its way into the clinical and laboratory setting. Of the myriad of possible nanoparticle compositions, the noble metals, more specifically gold, have moved to the front due to their favorable signal generating properties as well as their stability and biocompatibility. In this chapter, we reviewed a number of emerging and evolving nanoparticle based medical and diagnostic imaging applications. In many ways, gold as a strong thiol-reactive surface is more useful in these applications than gold as a contrast agent. That being said, the flexibility and tunability of gold nanoparticles seems likely to ensure their study and implementation for many years to come.

References

1. M. A. E. -S. Susie Eustis, Why gold nanoparticles are more precious than pretty gold: Noble metal surface plasmon resonance and its enhancement of the radiative and nonradiative properties of nanocrystals of different shapes. *Chem. Soc. Rev.* **35**, 209–217 (2006).
2. S. Hainfeld and F. Smilowitz, Gold nanoparticles: a new X-ray contrast agent. *The British J. Radiol.* **79**(939), 5 (2006).
3. L. Jain and El-Sayed, Calculated absorption and scattering properties of gold nanoparticles of different size, shape, and composition: applications in biological imaging and biomedicine. *J. Phys. Chem. B.* **110**(14), 11 (2006).
4. S. B. Fuller, E. Wilhelm, J, and J. J. M., Ink-jet printed nanoparticle microelectromechanical systems. *J. Microelectromech. Syst.* **11**(1), (2002).
5. D. J. Maxwell, J. R. Taylor and S. Nie, Self-assembled nanoparticle probes for recognition and detection of biomolecules. *J. Am. Chem. Soc.* **124**(32), 9606–9612 (2002).
6. O. Salata, Applications of nanoparticles in biology and medicine. *J. Nanobiotechnol.* **2**, 3–8 (2004).

7. P. T. Bishop, The use of gold mercaptides for decorative precious metal applications. *Gold Bull.* **35**(3), 89–98 (2002).
8. M. Law, *et al.*, Nanowire dye-sensitized solar cells. *Nature Lett.* **4**, 455–459 (2005).
9. Y. Shiraishi, *et al.*, Frequency modulation response of a liquid-crystal electro-optic device doped with nanoparticles. *Appl. Phys. Lett.* **81**(15), 2845–2847 (2002).
10. M. Haruta, Gold Rush. *Nature* **437**, 1098–1099 (2005).
11. N. Raveendran Shiju and V. V. G., Recent developments in catalysis using nanostructured materials. *Appl. Catalysis A: General* **356**, 1–17 (2009).
12. M. A. Stefan Gilb, U. Heiz, Monodispersed clusterassembled materials. *Mater. Today*, **9**, 48–49 (2006).
13. M. D. Hughes, *et al.*, Tunable gold catalysts for selective hydrocarbon oxidation under mild conditions. *Nature Lett.* **437**, 1132–1135 (2005).
14. M. Faraday, The Bakerian lecture: experimental relations of gold (and other metals) to light. *Philosoph Trans Roy. Soc. Lon.* **147**, 145–181 (1857).
15. G. Mie, Beiträge zur optik trüber medien, speziell kolloidaler metallösungen. *Annalen der Physik* **25**(4), 377–445 (1908).
16. M. C. Daniel, D. A., Gold nanoparticles: assmebly, supramolecular chemistry, quantum-sized-related properties, and applications towards biology, catalysis, and nanotechnology. *Chem. Rev.* **104**(1), 293–346 (2004).
17. T. H. Ha, H. J. Koo and B. H. Chung, Shape-controlled syntheses of gold nanoprisms and nanorods influenced by specific adsorption of halide ions. *J. Phys. Chem. C* **111**(3), 1123–1130 (2007).
18. C. H. Kuo, and M. H. Huang, Synthesis of branched gold nanocrystals by a seeding growth approach. *Langmuir* **21**(5), 2012–2016 (2005).
19. A. M. Boies, *et al.*, Gas-phase production of gold-decorated silica nanoparticles. *Nanotechnology* **22**(31), (2011).
20. A. Miranda, *et al.*, Synthesis of gold nanocubes in aqueous solution with remarkable shape-selectivity. *J. Porphyrins Phthalocyan.* **15**(5–6), 441–448 (2011).
21. L. Nastassja, V. C. and R. Drezek, Cytotoxicity of nanoparticles. *Small* **4**(1), 26–49 (2008).
22. M. Pattwat, W. B., C. Thammacharoen, S. Ekgasit and T. Kaewamatawong, An evaluation of acute toxicity of colloidal silver nanoparticles. *Toxicology* **73**(11), 1417–1423 (2011).
23. B. V. Enustun, J. T., Coagulation of colloidal gold. *J. Am. Chem. Soc.* **85**(21), 3317–3328 (1963).
24. L. H. Dubois, R. G. N., Synthesis, structure, and properties of model organic-surfaces. *Ann. Rev. Phys. Chem.* **43**, 437–463 (1992).
25. G. C. Wolfgang Eck, A. Sigdel, G. Ritter, L. J. Old, L. Tang, M. F. Brennan and M. D. M. Peter J. Allen, PEGylated gold nanoparticles conjugated to monoclonal F19 antibodies as targeted labeling agents for human pancreatic carcinoma tissue. *ACS Nano* **11**, 2263–2272 (2008).
26. Kyung A Kang, J. W., J. B. Jasinski and S. Achilefu, Fluorescence manipulation by gold nanoparticles: from complete quenching to extensive enhancement. *J. Nanobiotechnol.* **9**(16) (2011).
27. V. Jesse, T. L. Jokerst, N. Richard Zare and S. S. Gambhir, Nanoparticle PEGylation for imaging and therapy. *Nanomedicine* **6**(4), 715–728 (2011).

28. M. V. Uwe Kreibig, *Optical Properties of Metal Clusters*, ed pp. 552. Vol. 25, Berlin: Springer. (1995).
29. S. P. Mulvaney, *et al.*, Glass-coated, analyte-tagged nanoparticles: a new tagging system based on detection with surface- enhanced raman scattering. *Langmuir* **19**, 4784–4790 (2003).
30. J. E. Millstone, *et al.*, Colloidal gold and silver triangular nanoprisms. *Small* **5**(6), 646–664 (2009).
31. R. D. Averitt, S. L. Westcott and N. J. Halas, Ultrafast optical properties of gold nanoshells. *J. Opti Soc. Am. B-Opt. Phys.* **16**(10), 1814–1823 (1999).
32. P. S. Kumar, *et al.*, High-yield synthesis and optical response of gold nanostars. *Nanotechnology* **19**(1) (2008).
33. R. H. Doremus, Optical properties of small gold nanoparticles. *J. Chem. Phys.*, **40**(8), 2389–2396 (1963).
34. J. Zheng, C. Zhang and R. Dickson, M., Higly fluorescent, water-soluble, size- tunable gold quantum dots. *Phys. Rev. Lett.* **93**, p. 077402(1)–077402(4) (2004).
35. J. Huang and El-Sayed, Gold nanoparticles: interesting optical properties and recent applications in cancer diagnostics and therapy. *Nanomed.* **2**(5), p. 12 (2007).
36. R. S. Nripen Chanda, Kattesh, V. Katti, R. Kannan, Gastrin releasing protein receptor specific gold nanorods: breast and prostate tumor avid nanovectors for molecular imaging. *Nano Lett.* **9**(5), p. 1798–1805 (2009).
37. K. T. Yong, *et al.*, Quantum rod bioconjugates as targeted probes for confocal and two-photon fluorescence imaging of cancer cells. *Nano Lett.* **7**(3), 761–765 (2007).
38. M. G. Albrecht, J. A. C., Anomalously intense Raman spectra of pyridine at a silver electrode. *J. Am. Chem Soc*, **99**, 5215–5219 (1977).
39. A. Campion, P. K., Surface-enhanced Raman scattering. *Chem. Soc. Rev.* **27**, 241–250 (1998).
40. J. Alan Creighton, C. G. B. and M. G. Albrecht, Plasma resonance enhancement of Raman scattering by pyridine adsorbed on silver or gold sol particles of size comparable to the excitation wavelength. *J. Chem. Soc. Faraday Trans. 2: Mol. Chem. Phys.* **75**, 790–798 (1979).
41. Amy L. Oldenburg, M. N. H., D. A. Zweifel, A. Wei and S. A. Boppart, Plasmon-resonant gold nanorods as low backscattering albedo contrast agents for optical coherence tomography. *Optics Express*, **14**(15), 6724–6738 (2006).
42. X. Qian, *et al.*, *In vivo* tumor targeting and spectroscopic detection with surface-enhanced Raman nanoparticle tags. *Nature Biotechnol.* **26**, 83–90 (2008).
43. E. S. Allgeyer, A. P. Michael Browne and M. D. Mason, Optical signal comparison of single fluorescent molecules and Raman active gold nanostars. *Nano Lett*, **9**(11), 3816–3819 (2009).
44. W. E. Doering and S. Nie, Spectroscopic tags using dye-embedded nanoparticles and surface-enhanced Raman scattering. *Anal. Chem.* **75**(22), 6171–6176 (2003).
45. H. Fenniri, *et al.*, Barcoded resins: a new concept for polymer-supported combinatorial library self-deconvolution. *J. Am. Chem. Soc.* **123**, 8151–8152 (2001).
46. G. von Maltzmahn, *et al.*, SERS-coded gold nanorods as a multifunctinal platform for densely multiplexed near-infrared imaging and photothermal heating. *Adv. Mater*, **21**, 3175–3180 (2009).
47. H. N. Wang and T. Vo-Dinh, Multiplex detection of breast cancer biomarkers using plasmonic molecular sentinal nanoprobes. *Nanotechnology* **20**(6), 065101 (2009).

48. K. K. Maiti, *et al.*, Multiplex cancer cell detection by SERS nanotags with cyanine and triphenylmethine Raman reporters. *Chem. Commun.* **47**, 3514–3515 (2011).
49. R. R. Chance, A. H. M., A. Prock and R. Silbey, Fluorescence and energy transfer near interfaces: The complete and quantitative description of the Eu+3/mirror systems. *J. Chem. Phys.* **63**, 7 (1975).
50. M. D. M. Krishanu Ray and R. D. Grober, Quantum yields of photoacid generation in 193-nm chemically amplified resists by fluorescence imaging spectroscopy. *Chem. Mater.* **16**, 5726–5730 (2004).
51. A. N. Joel Gersten, Spectroscopic properties of molecules interacting with small dielectric particles. *J. Chem. Phys.* **75**(3), 1139–1152 (1981).
52. Chris D. Geddes, I. G., J. Malicka, Z. Gryczynski and J. R. Lakowicz, Metal-enhanced fluorescence: potential applications in HTS. *Combinatorial Chem. High Throughput Screening* **6**, 109–117 (2003).
53. L. W. Goldman, Principles of CT and CT technology. *J. Nuclear Med.* **35**(3), 115–128 (2007).
54. K. L. Watkin, M. A. M., Multi-modal contrast agents: a first step. *Acad. Radiol*, **9**(2), (2002).
55. I. Goldenberga, *et al.*, Oral acetylcysteine as an adjunct to saline hydration for the prevention of contrast-induced nephropathy following coronary angiography: A rondomized conrolled trial and review of the current literature. *Eur. Heart J.* **25**(212–218), (2004).
56. A.O. : Govorov, H. H. R., Generating heat with metal nanoparticles. *Nano Today* **2**(1), 30–38 (2007).
57. S. Link, M. A. E.-S., Spectral properties and relaxation dynamics of surface plasmon electronic oscillations in gold and silver nanodots and nanorods. *J. Phys. Chem. B* **103**, 8410–8426 (1999).
58. V. K. Pustovalov, Theoretical study of heating of spherical nanoparticle in media by short laser pulses. *Chem. Phys.* **308**(1–2), 103–108 (2005).
59. C. G. A. Hoelen, d. M. F. F. M., R. Pongers and A. Dekker, Three-dimensional photoacoustic imaging of blood vessels in optical tissue. *Opt. Lett.* **23,** (1998).
60. J. A. Copland, E. M., V. L. Popov, N. Kotov, N. Mamedova, M. Motamedi and A. A. Oraevsky, Bioconjugated gold nanoparticles as a molecular based contrast agent: implications for imaging of deep tumors using optoacoustic tomography. *Mol. Imaging Biol* **6**, 341 (2004).
61. W. C. Yuan Z, H. Zhao and H. Jiang, Imaging of small nanoparticle-containing objects by finite-element-based photoacoustic tomography. *Opt. Lett.* **30,** (2005).
62. X. Yang, S. S. E., Z. Y. Li, Y. Xia and L. V. Wang, Photoacoustic tomography of a rat cerebral cortex *in vivo* with Au nanocages as an optical contrast agent. *Nano Lett.* **7,** (2007).
63. S. C. B. Parvesh Sharma, N. Bengtsson, Q. Zhang, G. A. Walter, S. R. Grobmyer, S. Santra, H. Jiang, E. W. Scott and B. M. Moudgil, Gold-speckled multimodal nanoparticles for noninvasive bioimaging. *Chem. Mater.* **20**(19), 6087–6094 (2008).
64. Y. Wang, X. X., X. Wang, G. Ku, K. Gill, D. P. O'Neil, G. Stoica and L. V. Wang, Photoacoustic tomography of a nanoshell contrast agent in the *in vivo* rat brain. *Nano Lett.* **4** (2004).
65. S. A. Huettel, A. W. Song and G. McCarthy, Functional magnetic resonance imaging, Sinauer Associates (2009).

66. M. M. Britton, Magnetic resonance imaging of chemistry. *Chem. Soc. Rev.* **39**, 4036–4043 (2010).
67. C. Alric, *et al.*, Gadolinium chelate coated gold nanoparticles as contrast agents for both X-ray conmputed tomography and magnetic resonance imaging. *J. Am. Chem. Soc.* **130**(18), 5908–5915 (2008).
68. J. H. L. Hyon Bin Na, K. An, Y. Il Park, M. Park, I. S. Lee, D.-H. Nam, S. T. Kim, S.-H. Kim, S.-W. Kim, K.-H. Lim, K.-S. Kim, S.-O. Kim and T. Hyeon, Development of a T1 contrast agent for magnetic resonance imaging using MnO nanoparticles *Angewandte Chemie* **46**, 1–6 (2007).
69. T. H. Nohyun Lee, Designed synthesis of uniformly sized iron oxide nanoparticles for efficient magnetic resonance imaging contrast agents. *Chem. Soc. Rev.* **41**, 2575–2589 (2012).
70. J. J. D. Wen-Yen Huang, Multimodality and nanoparticles in medical imaging. *Dalton Trans.* **40**, 6087 (2011).
71. S. R. Pierre-Jean Debouttière, Francis Vocanson, Claire Billotey, Olivier Beuf, and Y. L. A. Favre-Réguillon, S. Pellet-Rostaing, R. Lamartine, P. Perriat and O. Tillement, Design of gold nanoparticles for magnetic resonance imaging. *Adv. Funct. Mater.* **16**, 2330–2339 (2006).
72. K. M. Krishnan, Biomedical nanomagnetics: a spin through possibilities in imaging, diagnostics, and therapy. *IEEE Trans. Magnet.* **46**(7), 2523–2558 (2010).
73. C. S. Natalia Abrikossova and M. Ahren, Torbjorn Bengtsson, Kajsa Uvdal, Effects of gadolinium oxide nanoparticles on the oxidative burst from human neutrophil granulocytes. *Nanotechnology* **23**, 10 (2012).
74. Emine U. Saritas, P. W. G., Laura R. Croft, Justin J. Konkle, Kuan Lu, Bo Zheng and Steven M. Conolly, Magnetic particle imaging (MPI) for NMR and MRI researchers. *J. Magnetic Resonance* (2013).
75. B. Sharma, Walter, Santra, Moudgil, Nanoparticles for bioimaging. *Adva Colloid Interface Sci.* **123–126**, 14 (2006).
76. A. Timothy J. B. Larson, J. Aaron, K. Sokolov, Hybrid plasmonic magnetic nanoparticles as molecular specific agents for MRI/optical imaging and photothermal therapy of cancer cells. *Nanotechnology*, **18**, 8 (2007).
77. I. C. S. Hyon Bin Na, and T. Hyeon, Inorganic nanoparticles for MRI contrast agents. *Adv. Mater.* **21**, 2133–2148 (2009).
78. V. Elena M. I. B. Shevchenko, M. V. Kovalenko, D. V. Talapin, K. Rachel Smith, S. Aloni, W. Heiss and A. Paul Alivisatos, Gold/iron oxide core/hollow-shell nanoparticles. *Adv. Mater.* **20**, 4323–4329 (2008).
79. A. Louie, Multimodality imaging probes: design and challenges. *Chem. Rev.* **110**, 3146–3195 (2010).
80. M. F. Warsi, and V. Chechik, Strategies for increasing relaxivity of gold nanoparticles based MRI contrast agents. *Phys. Chem. Chem. Phys.* **13**(20), 9812–9817 (2011).
81. C. S. Yang Lu, M.-J. Hu, Y.-J. Xu , L. Yu, L.-P. Wen, Y. Zhao, W.-P. Xu and S.-H. Yu, Magnetic alloy nanorings loaded with gold nanoparticles: synthesis and applications as multimodal imaging contrast agents. *Adv. Funct. Mater.* **20**, 3701–3706 (2010).
82. G. Murphy, S. Sisco, A. Goldsmith, Baxter, Gold nanoparticles in biology: beyond toxicity to cellular imaging. *Acc. Chem. Res.* **41**(12), 9 (2008).

83. H.-M. S. Qingshan Wei, P. Alexei Leonov, J. A. H. Dongmyung, O. Quy, K. O. R. Kenneth, W. Alexander, Gyromagnetic imaging: dynamic optical contrast using gold nanostars with magnetic cores. *J. Am. Chem. Soc.* **131**, 9728–9734 (2009).
84. Q. W. Hyon-Min Song, K. Quy Ong and W. Alexander, Plasmon-resonant nanoparticles and nanostars with magnetic cores: synthesis and magnetomotive imaging. *ACS Nano* **4**(9), 5163–5173 (2010).

Chapter 10

Photoluminescent Nanoparticles for Optical Imaging in Biology and Medicine

*María Teresa Fernández-Argüelles, Rosario Pereiro, José M. Costa-Fernández and Alfredo Sanz-Medel**

Department of Physical and Analytical Chemistry, Faculty of Chemistry, University of Oviedo, Julian Clavería, 8,33006 Oviedo, Spain
**Corresponding author. asm@uniovi.es*

Abstract

Photoluminescence imaging techniques have become essential tools in biomedical studies because of their exceptional analytical performance. Their capabilities for time-resolved and kinetic measurements have enabled live imaging as well. In this line, one of the fastest moving and most exciting research activities in the area of optical bioimaging is the use of luminescent nanoparticles as tags for biomolecules. The unique optical properties of nanomaterials make them most attractive as *in vivo* and *in vitro* luminophores in a wide variety of biomedical investigations, where traditional fluorescent organic molecules fall short (*e.g.* in providing long-term stability and multiplex capabilities).

In this chapter, we come across recent advances on luminescent imaging techniques based on the use of different nanoparticles (NPs) (including carbon dots, dye-doped nanomaterials, quantum dots (QDs), gold NPs and rare-earth doped NPs).

The conjugation of fluorescence imaging with the benefits of nanotechnology has just begun to be realized and many new medical applications are likely to arise in the near future, as our ability to manipulate and understand the behavior of these materials improves. In this chapter, we aim at offering the reader a general outlook of reported bioanalytical strategies and the potential and limitations of such nanoparticles for biomedical optical imaging.

1. Introduction

Biotechnologies for high-throughput screening and medical diagnosis are increasingly demanded, being optical imaging techniques one of the most active areas. In fact, *in vitro* and *in vivo* optical imaging is becoming increasingly important for diagnosis, prognosis and monitoring of diseases. In particular, photoluminescent optical imaging provides a powerful tool for studying the disease-related cell and tissue functions. Photoluminescent imaging can potentially provide a great amount of scientific information: photoluminescent labels can achieve a high specificity and sensitivity, enabling to obtain critical biological and medical information (*e.g.* on the environment of the tagged biomolecules, on their location, and on possible interactions with their environment). As a matter of fact, recent advances in both labeling and detection imaging technologies are drastically pushing photoluminescence techniques impact in biomedicine.

Luminescent imaging techniques for biological and medical applications often require a proper labeling of the targeted biochemical compounds for detection. Photoluminescent tagging of biomolecules, a commonly used approach in biomedicine for immunoassays, cellular labeling and tissue imaging, has largely relied on conventional organic fluorophores and fluorescent proteins. The present great success of photoluminescence spectrometry in the field of biomolecules analysis derives from several well-recognized advantages, including[1]:

(a) High sensitivity of the detection (today many applications of fluorescence are focusing on even single molecule detection).
(b) Fast signal acquisition: each individual fluorescent label can potentially provide 107–108 photons per second for detection.
(c) Multi-colour dyes can be used for multiplexed assays.
(d) The luminescent signal is localized (unlike using some enzyme-linked amplification schemes).
(e) The labeling process can be straightforward provided that appropriate functional groups are available on the target biomolecule (analyte).

Of course, there are also some important limitations of photoluminescence for bioanalytics, most of them derived from the organic fluorescent labeling reagents commonly used. Apart from the labeling reaction intrinsic difficulties, many of the used fluorophor reagents suffer from photobleaching problems, pH sensitivity, reduced quantum yields (QYs) or short photoluminescence lifetimes.[1]

In order to solve such reported problems limiting the applicability of conventional organic dyes, scientists continue developing new fluorochromes appropriate for bioimaging applications. Researchers are searching for materials with improved fluorescence QYs, higher chemical and photostability, lower aggregation tendencies and low cytotoxicity.

In this context, fluorescence-based biochemical imaging research has quite recently leveraged the benefits of advances in nanotechnology. Significant progress in the ability to fabricate high-quality luminescent nanoparticles (NPs) and the understanding of their outstanding and size dependent physical and chemical features have stimulated their use in bioimaging. In particular, engineered photoluminescent colloidal NPs could overcome some of the limitations of more conventional organic dyes and, so, offer a significant thrust in bioimaging new developments.[2]

The present availability of photoluminescent NPs with a variety of shapes, sizes and compositions are about to change drastically the bioanalytical photoluminescence landscape. The seminal publications[3,4] describing the first biological uses of NPs-fluorophores foresaw that their unique properties could overcome many of the significant chemical and spectral limitations of classical molecular dye fluorophores. Actually, photoluminescent nanocrystals may offer substantial advantages, as compared to traditional dye labels for imaging techniques use and, so, to enable improved bioanalytical applications.

Depending on the composition and structure of the nanomaterial, NPs properties of interest to biologists include high luminescent QYs, high molar extinction coefficients (sometimes so high as ~10–100 times that of organic dyes), broad absorption with narrow, symmetric photoluminescence spectra spanning the UV to near-infrared, large effective Stokes shifts, high resistance to photobleaching and exceptional resistance to photo- and chemical degradation. Moreover, compared with molecular dyes, three features in particular stand out: NPs unparalleled ability to size-tune fluorescent emission as a function of core size (*e.g.* for binary semiconductor materials), idem of chemical composition, and their potential for signal amplifications resulting in higher detection sensitivities.[5]

On the other hand, NPs should fulfill some basic requirements to be used as labels in imaging applications: their photoluminescent emission should not be interfered by the biological matrix; in addition, NPs should be soluble in biological buffers, biocompatible and non-cytotoxic. Moreover, NPs bioconjugation to eventual recognition reagents (*e.g.* antibodies) has to be possible and effective to enable fluorescent labeling of the bioanalytes.[6] Unfortunately, so far all those requirements are hardly achieved and synthesizing practical NPs-based reagents is still a challenging task in research laboratories.

This chapter summarizes reported advances and limitations in the use of NPs as optical labels for optical imaging applications. Some representative imaging applications that have been successfully demonstrated using different nanomaterials will be here revised. Finally, although NPs offer potentially invaluable societal benefits such as drug targeting and *in vivo* biomedical imaging, they may also be risky to human health and the environment under certain conditions.[7] Therefore, a section about citotoxicity of the main NPs used in biomedical imaging is included.

2. Basic Physical Properties, Synthesis and Water Solubilization Approaches of Luminescent Nanoparticles

Optical imaging reagents can be classified into endogenous and exogenous agents. Endogenous agents typically use an enzyme mediated process inside the body to produce visible light when a given substrate is degraded (*e.g.* a fluorescent protein). Unfortunately, those reporter proteins often emit photoluminescence in the spectral region of 500 nm, which overlaps with absorption of the oxyhaemoglobin (HbO_2) and haemoglobin (Hb) present in blood, and with the autofluorescence of many tissues. (see Fig. 1).

Otherwise, the use of exogenous optical contrast agents added as reporters are a very common tool used nowadays in biology and medicine, ranging

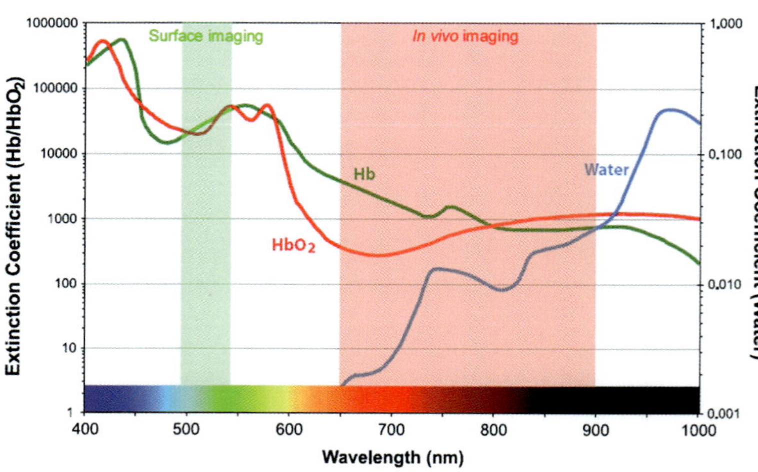

Fig. 1. Definition of the optical window for *in vivo* imaging. "Reprinted with permission from Kobayashi *et al.* Chem. Rev. 2010, 110, 2620–2640. Copyright 2010 American Chemical Society."

from colorimetric dyes to sensitive fluorescent probes. NP-based exogenous optical contrast agents such as gold nanoclusters (AuNCs), quantum dots (QDs), rare earth NPs, carbon dots (Cdots), and dye loaded NPs are recent incorporations to the list of exogenous contrast agents due to their excellent features on the way to overcome limitations found in traditional optical contrast agents. Moreover, actual knowledge on NPs surface functionalization makes possible to ensure solubility in biological buffers and biocompatibility and allows further simple and efficient bioconjugation to recognition elements.[8]

This section deals with an overview on the most common NPs used nowadays as optical contrast agents in biology and medicine.

2.1. *Gold nanocrystals*

Gold and silver NPs are commonly used as spectral absorptiometric contrast agents, whose intensity is a function of the size, shape and dielectric constant of the medium. However, fluorescent AuNCs synthesis is a relatively new field, still under development. It is noteworthy that metallic "nanoclusters" (NCs) used in fluorescence methodologies is increasing noticeably in the last few years.

AuNCs can be used for direct or indirect sensing, since their photoluminescence can be altered by interactions with other substances, such as metal ions (*e.g.* copper,[9] mercury[10]) or biomolecules like insulin.[11] While the synthesis of comparatively bigger NPs of Au is a well-known field in nanotechnology today, AuNCs synthesis is a relatively new field still in its early stages: the isolation and purification of AuNCs has been a rather recent scientific achievement.[12]

Stable fluorescent AuNCs can be obtained through a biomineralization one-pot synthesis in which incorporation of proteins with gold atom precursors takes place. This process is also known as "protein/peptide-directed," or "protein-stabilized" synthesis, and it is carried out in mild conditions, absence of strong reducing agents and aqueous solution. The synthesis of AuNCs strongly depends on the protein conformation.[13] Some proteins already employed to carry out the synthesis of AuNCs include BSA,[14] transferrin,[15] lactoferrin[16] or trypsin.[17] Additionally, it was observed that the selected amino acids play an important role in the nanocluster synthesis (as consequence of the differences on the reducing strength and their binding interactions with gold), and different sequence peptides are being investigated in order to obtain the desired shape and size of the nanocluster.[18] A highlight of these water-soluble AuNCs is how readily they can be used in bioapplications derived from their immediate biocompatibility.

Another key feature is that these AuNCs are very sensitive to their environment. In this sense, the effect of the capping/stabilizing ligand has been demonstrated to have a profound effect on their photoluminescence.[19] Although their optical properties strongly depend on their capping molecule, in general terms it can be said that protein-stabilized AuNCs emit strong photoluminescence in the visible spectrum, attributed to molecular-like transitions between HOMO and LUMO energies, as a consequence of their reduced size, with lifetimes in the order of a few nanoseconds.[20]

2.2. *Quantum dots*

Semiconductor QDs are spherical, luminescent nanocrystals with diameters on the order of 2–10 nm showing size-dependent optical properties.[4] Due to quantum confinement effects, they present narrow size-tunable emission, high QYs, and high resistance to photobleaching or to chemical degradation. For instance, a CdSe/ZnS core-shell QDs is 20 times brighter and about 100 times more stable to photobleaching than rhodamine 6G.[3] Furthermore, QDs of different sizes can be excited at a single wavelength and emit at multiple wavelengths, allowing their application in multiplexed assays.[21] With all these features, QDs have rapidly emerged as potential new fluorescent probes for the imaging of biological samples as far as they may overcome many of the chemical and spectral limitations of the more traditional molecular fluorophores.[22] However, biological uses of QDs have been slowed down by the difficulties to render such NPs into biocompatible probes.[23]

Cadmium selenide QDs are one of the most studied and understood semiconductor nanocrystals. Although first procedures for the synthesis of QDs gave rise to large size particle distributions, in 1993 Murray *et al.*[24] developed a new route for obtaining semiconductors with a high degree of monodispersity (particle size distribution < 5%) by using organic media and organometallic precursors. Since then, a wide variety of QDs synthesis approaches have been developed. The highest quality QDs are those obtained at high temperature in a coordinating solvent (as trioctylphosphine oxide, trioctylphosphine and hexadecylamine).[24,25] In addition, it is known that by overcoating QDs with higher band gap inorganic semiconductor materials the photoluminescence QYs and, especially, the chemical stability and photostability (due to the passivation of the non-radiative surface recombination sites) can be increased. Consequently core-shell type QDs, such as (CdSe)

ZnS QDs, have been widely used in both optoelectronic and biological applications.[26]

However, the surfactant-coated core-shell nanocrystals obtained have the drawback of only being soluble in organic, non-polar solvents (as toluene or chloroform) due to the hydrophobic alkyl chains of the ligands that extend from the NP surface.[27] In order to use QDs in biological applications they have to be transferred and stabilized into aqueous media. The design of robust and versatile surface chemistries for that purpose has been investigated during the last years, aiming at achieving solubilization and stabilization of QDs in biological buffers, keeping their original colloidal, photophysical and size properties and providing adequate reactive groups for further bioconjugations.[23] To that goal, three general phase-transfer approaches (that attach polar functional groups to the surface of the QD) have been developed (see Fig. 2).

A general method is the ligand exchange of the NPs original surface hydrophobic ligands (see Fig. 2) with bifunctional molecules which often

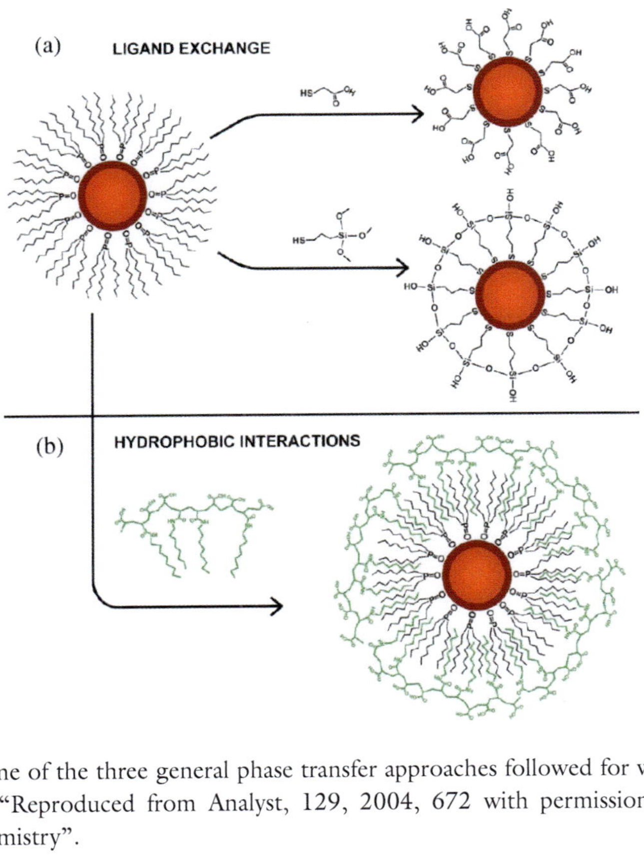

Fig. 2. Scheme of the three general phase transfer approaches followed for water solubilization of QDs. "Reproduced from Analyst, 129, 2004, 672 with permission of The Royal Society of Chemistry".

contain a thiol (–SH) functional group on one end (that binds to the ZnS shell) and a polar functional group on the other end, such as carboxyl (–COOH) or sulfonic (–SO_3H) acids, that extend away from the QD surface and provides the requires solubility in aqueous and polar media.[3] The use of bidentate ligands as dithiothreitol[28] and dihydrolipoic acid (DHLA),[29] to generate more stable coatings due to the multiple anchoring points has also been proposed.

A second approach consists on the surface silanization of the QDs through the generation of a silica shell around the NP surface.[4] A substitution of the hydrophobic ligands takes place when a molecule as mercaptopropyl-trimethoxysilane is added. The thiol groups react with the ZnS shell, and three methoxysilane groups can react with each other to form siloxane bonds generating a highly cross-linked protective shell around the NP (Fig. 2).

A third mode to transfer the non-polar nanoparticles into an aqueous media is based on the use of amphiphilic polymers. In this method, hydrophobic alkyl chains interact with the alkyl chains of QD ligands, and the hydrophilic groups, disposed to the external part, confer the required water-solubility.[30,31]

Of course, once the QDs are made soluble in a water medium, they can be conjugated to biological probes such as nucleic acids, proteins, antibodies or small molecules in order to be employed in bioanalytical applications.

2.3. *Rare earth nanoparticles*

Upconverting NPs (UCNPs), also known as upconverting nanophosphors, are NPs in which rare earth elements are embedded in a crystalline host matrix. Conversely to conventional fluorescence, which is based on single photon excitation, the rare earths present in these type of NPs absorb two or more low energy photons, *i.e.* infrared (IR) or near-IR (NIR) photons, giving rise to metastable and long-lived energy states, which upconvert to emit in the visible (VIS) spectrum (higher energy emission photons), a process known as antistokes emission. The upconversion mechanism has been explained as either sequential excitation of the same atom or as excitation of two centers and subsequent energy transfer.[32]

UCNPs possess several features that can be very useful in different types of applications, including lasers, solar cells, waveguides and display devices. However, there is a special interest in the use of these types of NPs for biomedical applications due to the benefits that UCNPs can offer, particularly that they can be excited in the IR or NIR region and emit in the VIS-NIR. The "optical transmission window" of biological tissues in the NIR region allows for deeper light penetration while the autofluorescence of tissues or living animals is minimal when irradiated with an IR source (in other words, the background signal is much lower allowing increased image contrast).[33,34]

Consequently, handicaps frequently encountered in conventional bioimaging such as tissue autofluorescence, short penetration depth, DNA damage or cell death upon long-term exposures to high energy radiation are reduced when using an IR excitation source.[35] Another feature that makes UCNPs attractive for biomedical applications is their low toxicity, especially when they are compared to Cd-based semiconductor QDs.[36] Moreover, due to their inherent high photon conversion efficiency and non-blinking emission, UCNPs allow reliable single-molecule imaging that challenges conventional staining agents.[37]

UCNPs consist of a nanocrystalline host and a dopant added at low concentrations.[38] Important requirements for photon upconversion, such as long lifetimes of the excited states and a ladder-like arrangement of the energy levels with similar spacing, are achieved for certain ions of the d and f elements. The dopant provides luminescent centers, and the host lattice with its crystal structure provides a matrix to bring these centers into optimal position.[39] Selection of the host lattice strongly affects to the upconversion process since it determines the distance between the dopant ions, their position, their coordination number and the types of anion that surround the dopant ions. Although numerous suitable hosts have been doped with transition metal or actinide ions, high upconversion efficiencies at room temperature have been observed for lanthanide-doped solids, typically of a LaF_3, YF_3, Y_2O_3, $LaPO_4$ or $NaYF_4$ as host lattice, and trivalent rare earth ions, such as Yb^{3+}, Er^{3+} and Tm^{3+}, as dopant centers.[40–42]

Long lifetimes of the excited states of the dopants are required. In some lanthanides, energy gaps between several subsequent energy levels are similar. Therefore, each absorption step requires the same photon energy, and excited lanthanide ions may sequentially absorb a second photon and reach a higher excited state. Some common dopants found in UCNPs include Er^{3+}, Tm^{3+} or Ho^{3+}, which possess an energy level structure that is suitable for this type of sequential excitation.[43]

However, forbidden transitions limit the absorption of light from the excitation source even if high concentrations of dopants are present in the UCNPs, and this is curtailing their applications. To overcome this constraint the material can be co-doped with strongly absorbing ions, known as "sensitizers", which ensure an energy transfer to the activator. The most common sensitizer employed for 980 nm excitation is Yb^{3+}, which has been added to systems based on Er^{3+}, Tm^{3+}, Ho^{3+} and Pr^{3+}.[39]

As with QDs the UCNPs obtained with hydrophobic organic ligands, such as oleic acid, have to be modified in order to be used in biological applications. There are several strategies described in the literature to transfer hydrophobic UCNPs into aqueous media and to provide functional groups as – COOH, $-NH_2$ or –SH, to their surfaces (that will be further employed to conjugate the desired biomolecules). Such general approaches are those described in more detail in the previous section for QDs and include: ligand exchange with bifunctional molecules, encapsulation with SiO_2, or amphiphilic polymer coating. In any case, the need to develop new and simpler methods for synthesizing appropriate hydrophilic and functionalized UCNPs is still pending.

2.4. *Carbon dots*

The observed limitations of the metal- or semiconductor-based NPs have prompted the search for alternative fluorescent NPs. For instance, the high content in heavy metals of semiconductor QDs has prompted health and environmental concerns in the community, as its applications might cause serious health risks particularly for *in vivo* applications. Therefore, research on the development and use of nanomaterials composed of non-toxic elements, as contrast agents for optical imaging *in vivo*, has become an increasingly important and urgent demand.[44]

In this sense, an interesting finding showed that small carbon NPs properly passivated with organic biomolecules were able to emit a strong fluorescence emission. Such nanosized pure carbon particles, named Cdots, carbogenic dots[45] or carbon NPs,[46] show a bright, colorful and stable photoluminescence, without blinking phenomena (both in solution and solid state) and being readily soluble in water. Cdots exhibit very high two-photon absorption cross-sections, enabling fluorescence imaging with one- and two-photon excitations on the same platform.[47] Mechanistically, the carbon-based photoluminescence has been attributed to passivated defects on the carbon particle surface acting as excitation energy traps.[44]

There are many different synthetic routes to obtain fluorescent Cdots.[48] In general terms, their synthesis can be operationally divided into the following two steps: first, synthesis of raw non-fluorescent Cdots, that can be carried out through top-down or bottom-up strategies; and second, passivation and functionalization operations, necessary to make Cdots fluorescent, such as oxidation of the surface carbons to carboxylic acid groups with nitric acid;[49] doping the oxidized Cdots with an inorganic salt[50] or capping the NP with different organic molecules *via* covalent linkage or chemical adsorption, such as polyethylene glycol (PEG) derivatives (with appropriate functional groups that can be used afterwards for bioconjugation purposes), or poly(propionylethyleneimine-co-ethyleneimine).[47] Even though these NPs exhibit photoluminescence with QYs lower than those of QDs, advantages in terms of biocompatibility and toxicity have been reported in cell lines and mice, thus suggesting that Cdots represent an emerging new class of QD-like fluorescent nanomaterials for both *in vitro* and *in vivo* imaging applications.[51,52]

2.5. *Dye loaded nanoparticles*

Although QDs present several outstanding optoelectronic properties that make them very attractive for imaging applications, issues related to their inherent toxicity encourage researchers to pay attention to other alternatives. In this context, the construction of new fluorescent nanotracers can also rely on the encapsulation or binding of low molecular weight fluorophores on nanoscale structures such as silica NPs, polymer-based NPs or lipid-based NPs. In general terms, the matrix in which the fluorophores are encapsulated plays a protecting role, preventing direct chemical contact between the dye and the biological fluids, and thus, avoiding problems related to aggregation or photobleaching. Moreover, the inclusion of several dyes in a single nanostructure increases their local concentration on the targeted site leading to potentially sensitive nanoprobes for fluorescence imaging.

Silica is commonly used for the development of dye loaded NPs because the active dye can be encapsulated inside the pores of the matrix or covalently attached to the silane groups in the core or the shell of the silica particles. Additionally, the synthesis protocols of silica NPs are well understood, and they are easy to carry out in the laboratory, obtaining silica NPs with good monodispersity, robust, transparent and mechanically stable.[2]

Polymer-based NP matrices may be another promising non-toxic alternative to inorganic NPs to encapsulate or bind active luminescent dyes. The polymer-matrix can be more versatile than the silica matrix in its ability to tune the hydrophobicity/hydrophilicity and modulate the analyte transport into the NP.[53] Additionally, this matrix can be designed to reduce cellular interferences, such as protein binding,[54] and differential sequestration by intracellular organelles.[55]

Alternatively, lipid-based NPs are based on the use of lipids and phospholipids that can be extracted and purified from natural products (*e.g.* animal or vegetable oils) or synthetically produced. Their main advantage is that they present very low intrinsic cytotoxicity, which makes them exceptional fluorescent nanocargos in clinical applications (including drug delivery).[56] Moreover, their lipidic nature might also improve their affinity for numerous tumours over-expressing lipoprotein receptors. However, lipid-based nanostructures may appear more difficult to tailor than polymer building blocks, which can be very finely tuned for each specific application.

3. Uses of Nanoparticles in Biology and Medicine

3.1. *Cellular labeling*

3.1.1. *Gold nanoclusters*

Fluorescent AuNCs with diameters between 16 and 55 nm were evaluated by Ren *et al.*[57] in 2008 for imaging of human epithelial cervical cancer cells (HeLa cells). After labeling the HeLa cells with AuNCs, they were illuminated with an intense light source, giving rise to a photobleaching effect of the cellular autofluorescence, but detecting the stable fluorescence emitted by the AuNCs (attached to the cell membrane or inside the cells). In this work, authors demonstrated a good behavior of antiphotobleaching of AuNCs as fluorescent probes in cell imaging. The AuNCs used showed a stable fluorescence emission centerd at 610 nm but with low QYs.

Trying to improve the photoluminescence QY, in 2010 Chen *et al.*[58] reported the use of gold nanocubes and nanorods with high Qy for imaging cell lines of human liver cancer and human embryo kidney. Results showed that QY obtained for gold nanocubes was about 200 times higher than the obtained for gold nanorods. Of course, these high Qy NPs are a leap forward in the search of useful fluorescent AuNCs.

Protein stabilized AuNCs constitute a new platform that allows performing an easy conjugation to receptor-targeting biomolecules for targeted imaging applications. Moreover, their strong emission in the NIR region can be very useful for tissue penetration in cell imaging. For instance, BSA stabilized AuNCs emitting in the NIR were conjugated to folic acid through anime linkages as an imaging probe for detection of folate receptors on oral carcinoma cells.[59] Another publication demonstrated that transferrin stabilized AuNCs can enter lung tumour cells and maintain strong red fluorescence.[15] Results showed that AuNCs could offer a viable method for selectively detecting and imaging cancer cells since only a minor reduction in the Qy was detected after their bioconjugation.

3.1.2. *Quantum dots*

Once the problem of water solubilization was overcome by Nie[3] and Alivisatos's[4] groups in 1998, the exploration of QDs in bioimaging has increased exponentially. Since then, QDs have been used as fluorescent imaging labels for the study of live cell single-molecule dynamics, monitoring of intracellular protein–protein interactions, disease detection in deeper tissues, detection of tumour cells based on selective binding of bioconjugated QDs to known cancer biomarkers and much more.[60]

QDs have been evaluated for *in vitro* cellular imaging, biomolecular tracking in cells and tissue staining. For instance, QDs linked to IgG and streptavidin were used to label different types of targets, including breast cancer marker Her2, actin, microtubule fibers or nuclear antigens, at different cell locations (surface, intracellular or inside the nucleus).[30] More recent investigations make use of molecules that are active recognition elements, such as peptides, antibodies or aptamers conjugated to the QDs to perform selective detection of different tumour biomarkers, actin filaments or specific cells through fluorescent imaging.[61]

One practical limitation on the use of QDs is the need of a method for robust and efficient cellular delivery of QDs in different parts of the cells, since different factors affect the QDs uptake into cells. Particularly, many researchers have concentrated their efforts on achieving appropriate delivery of the QDs to the cells cytosol, which is more complex than extracellular targeting. Several intracellular labeling approaches have been evaluated, including passive uptake, receptor-mediated and non-specific endocytosis, liposome mediated intracellular delivery, electroporation or micro-injection.[62] However, since QD material and surface ligand chemistry has found to be determinant for the final uptake of the NPs, further work is needed and is being carried out in order to avoid QD aggregation or deposition in endosomes or lysosomes.[63]

Authors have reported QDs use as intracellular fluorescent nanosensors. In this sense, it is known that fluorescence emission of QDs is affected by pH, since acidic pHs give rise to a decrease on the photoluminescence emission. Based on this principle, CdSe/ZnSe/ZnS QDs capped with mercaptoacetic acid ligands have been employed for the detection of changes of pH within SKOV-3 human ovarian cancer cells. After QDs were uptaken by endocytosis within the lysosomes, where the pH is rather low, addition of chloroquine produced an increase on the pH, and an enhancement of the photoluminescence intensity of QDs was detected.[64]

3.1.3. *Rare earth nanoparticles*

Fluorescence imaging is based on single-photon excitation, where a high energy source is employed. However, as said above, it is possible to take advantage of alternative "upconversion fluorescence" imaging, in which nanocrystals are excited in the IR or NIR region (*i.e.* low energy photons) and emit in the VIS-NIR. In this mode, fluorescence imaging with IR excitation could be achieved and so UCNPs stand up as a promising non-toxic alternative for the development of highly sensitive clinical imaging applications.[35,36,40]

Imaging techniques based on upconversion phenomena have been widely used for high-resolution imaging of cellular specimens. The first publication by Zijlmans *et al.*[65] in 1999 showed the advantages of these NPs for imaging applications. They used Y_2O_2S:Yb/Tm UCNPs to study the distribution of prostate-specific antigen (PSA) in human prostate tissue using standard immunohistological techniques. This work demonstrated that the non-specific tissue autofluorescence observed in conventional imaging was eliminated under NIR excitation, and there was not photobleaching after continuous exposure to the excitation source.

Since then, several publications have reported the use of UCNPs for target imaging of cancer cells making use of recognition elements such as antibodies or peptides. For instance, $NaYF_4$:Yb/Er UCNPs were conjugated to an antibody to carry out a highly specific staining of HeLa cells with antigen expressed on the cell membrane.[66] In other work, Y_2O_3:Er UCNPs modified with cyclic arginine–glycine–asparatic acid peptide were used for specifically imaging of cancer cells with elevated integrin avb3 expression, which can be very useful to evaluate the treatment efficacy in integrin-positive patients.[67]

3.1.4. *Carbon dots*

As mentioned previously, there is an increasing interest about searching alternative low toxic nanomaterials.[44] In this line, nanosized Cdots, though exhibiting lower QYs than QDs, offer better biocompatibility and toxicity in cell lines and mice. In fact, Cdots applicability for *in vitro* and *in vivo* imaging applications has been already demonstrated.[51,52]

One of the first studies showing the potential of Cdots for bioimaging was developed by Sun *et al.*[44] (where the Cdots cell uptake was monitored by confocal microscopy) and Cdots passivated with PEG were compared to CdSe/ZnS QDs demonstrating a very bright green emission. Moreover, in order to assess Cdots toxicity *in vitro*, some studies on human breast cancer cells and on human colorectal adenocarcinoma HT-29 were carried out. Results showed that cells were slightly affected by these NPs in terms of proliferation, mortality and viability. Cell viability of nitric acid passivated Cdots was also investigated in ascites carcinoma cells, concluding that a polymer passivation was not necessary and that no cytotoxicity effects were detected at the evaluated concentrations.[47]

Although applications of Cdots for cellular imaging are still scarce, the advantageous features shown by these NPs suggest that important advances in their use for imaging are expected in the near future.

3.1.5. *Dye-loaded nanoparticles*

The encapsulation of organic fluorophores into silica NPs has been evaluated for cancer cells imaging. One of the first reports was developed for leukaemia cell recognition with fluorescence microscopy. In this work, a metallorganic luminophore was loaded inside silica NPs and their surface was bioconjugated to an antibody, as the active recognition element.[68] In other approaches both, the fluorescent dye and the recognition element have been bound to the NP surface. For instance, organically modified silica NPs were synthesised covalently incorporating the fluorophore rhodamine-B and surface-functionalized with a variety of active groups (*e.g.* carboxyl groups). The carboxyl groups on the surface were then used to ensure conjugation with various bioactive molecules such as transferrin, as well as monoclonal antibodies such as anti-claudin 4 and anti-mesothelin, for targeted delivery to pancreatic cancer cell lines.[69] The simplicity of surface functionalization and incorporation of a variety of biotargeting molecules makes these fluorescent silica NPs good candidates as efficient probes for optical bioimaging.

Silica dye loaded NPs were also employed as nanosencers to detect small molecules in live cells. For instance, intracellular dissolved oxygen changes in concentration due to cell respiration were monitored in rat C6 glioma cells using several dyes acting as indicator and reference respectively, to develop a ratiometric based method.[70] A ratiometric method was also developed for detection of intracellular Zn^{2+} in HeLa cells making use of a reference dye embedded into the core of the SiO_2 particle and a Zn^{2+} sensitive fluorophore on the shell.[71]

Some authors have taken advantages of the combination of silica nanocomposites with QDs in order to overcome some of the limitations of the semiconductor NPs, such as cytotoxicity or colloidal stability problems. The silica NPs were functionalized with amino and phosphonate groups in order to carry out a bioconjugation with a mouse anti-human CD3 antibody *via* biotin–neutravidin interactions. The resultant nanocomposites, which presented a bright green fluorescence at 531 nm and were visible by conventional fluorescence microscope, were taken up by Jurkat T-cells through a specific CD3 receptormediated endocytosis. Controls with HeLa cells, which lack CD3 receptor, demonstrated the specificity of this intracellular nanoprobe.[72]

Even more advantageous features can be added to such silica-based nanocomposites *via* the combination with fluorescent QDs and magnetic NPs, enabling simultaneous labeling/imaging and cell sorting/separation. For instance, a bifunctional nanocomposite system consisting of Fe_2O_3 magnetic NPs and CdSe QDs coated with a thin silica shell have been used to carry out a specific labeling of live cell membranes (HepG2 human liver cancer cells, NIH-3T3 mouse fibroblast cells and 4T1 mouse breast cancer cells).[73]

The use of hybrid materials in which silica-coated polystyrene nanospheres contained QDs and Fe_3O_4 NPs in order to provide fluorescence and magnetic properties, respectively, has been reported as well. After bioconjugation to anti-epidermal growth factor receptor (anti-EGFR) antibodies, this nanocomposite can specifically label cells expressing the human EGFR on their surface, and their magnetic properties could be used for separating the targeted cells, since all the labeled cells could be collected quickly using a small magnet.[74]

There are also reports on dye-loaded NPs based on the use of indicator dyes acting as both, recognition element and optical transducer. For example, there are many approaches in the literature to develop nanosensors for calcium intracellular sensing. For instance, polyacrylamide-based NPs loaded with Ca^{2+}-sensitive Crimson dye were introduced to alveolar macrophages recovered from rat lung lavages. After induction of a slow increase in the Ca^{2+} concentration in the cell, an increase in the phagocyted Ca^{2+}-NP fluorescence was recorded as a function of the stimulation of the organelle.[75]

3.2. *In vivo and deep tissue imaging*

3.2.1. *Gold nanoclusters*

As said before, colloidal AuNCs show weaker fluorescence than that observed with alternative NPs, but they exhibit some special features that render them attractive for bioimaging applications. In any case, AuNCs weakest point is still the comparatively low emission QYs pending to be enhanced.[76] In this sense, the great potential of fluorescent gold nanorods for biological imaging was already showed in 2005 by Cheng *et al.*[77] They employed gold nanorods as two-photon luminescence imaging agents of proteins; *in vivo* imaging was performed in mouse ear blood vessels, observing a two-photon luminescence around 60 times brighter than that obtained using the traditional rhodamine-based imaging agent.

Also, the use of AuNCs with bluish green fluorescence using histidine, both as reductant and capping agent, was reported for an *in vivo* fluorescence imaging application. The low wavelength (480 nm) of such AuNCs gave rise to a poor tissue penetration, leading it to be useful only for superficial structures or cell imaging. Therefore, NIR dyes were conjugated to improve the tissue penetration needed for *in vivo* imaging. Such nanoassembly showed low cytotoxicity and high affinity to tumour at the cellular level, and its biodistribution in normal nude mice and MCF-7 tumour-bearing mice was investigated.[78]

Very small AuNCs (1.4 nm) coated with mercaptoundecanoic acid (MUA) were evaluated for *in vivo* imaging of a mouse leg, showing an induced tumour. Since the photoluminescence cannot travel through thick tissue, as it is not visible, an incision in the observed tissue must be carried out. Then, the emission from the MUA-AuNCs is visible at shallow locations underneath the surface in the tumour under UV illumination.[79]

Fluorescent AuNCs are promising systems. However, they still need to be further explored in several aspects (*e.g.* determination of AuNC local structure, enhancement of QYs, etc.). Such knowledge and better understanding on how to tune their optical properties will be critical for further exploitation of their potential for imaging applications.[80]

3.2.2. *Quantum dots*

QDs show a promising future in cancer diagnosis and therapy.[81] Cancer cells possess specific surface receptors, such as Her2 (human epidermal growth factor receptor 2); ER (estrogen receptor); EGFR; mTOR (mammalian target of rapamycin), etc. They can be used as targets which can be recognized by specific antibodies, peptides or proteins bioconjugated to the QDs. In this sense, the first *in vivo* application of QDs for cancer imaging was carried out in 2002 by Åkerman *et al.*[82] by conjugating QDs to three different peptides with affinity for tumour cells, which allowed their guidance to tumour vasculatures in mice.

The practical application of QDs in biomedical imaging can be widened by the synthesis of QDs emitting in the NIR region.[83] For instance, bright NIR QDs have been employed for *in vivo* imaging of the lymphatic and cardiovascular systems of mammals allowing a high sensitive real-time observation.[84,85] In fact, in 2004 Kim *et al.*[86] carried out a cancer surgery in pigs under complete image guidance in real time after NIR QDs intradermal injection, minimizing incision and dissection inaccuracies, and permitting real time confirmation of complete resection (see Fig. 3). Continuous developments in this area are taking place[31] and other applications of QDs for *in vivo* fluorescent imaging of proteins for cancer diagnosis include early detection of primary tumours such as prostate,[87] pancreatic[88] or liver cancer.[89]

Another relevant feature of QDs is that different sized QDs can be excited at a single wavelength but emitting photoluminescence at multiple wavelengths. Thus, QDs are especially appropriate to carry out multiplexed imaging. In this direction, a multiplexed immunohistological staining was performed to obtain a molecular mapping of tumour heterogeneity on human prostate cancer.[90] In this work, four protein biomarkers (E-cadherin, high-molecular-weight cytokeratin, p63 and α-methylacyl CoA racemase) were conjugated to four QDs of different emission wavelengths, thus targeting diverse tumour areas with different colors. This tumour heterogeneity has been observed at the molecular, cellular and architectural levels, allowing direct visualization of their structural transitions, an achievement not possible with traditional organic dyes.

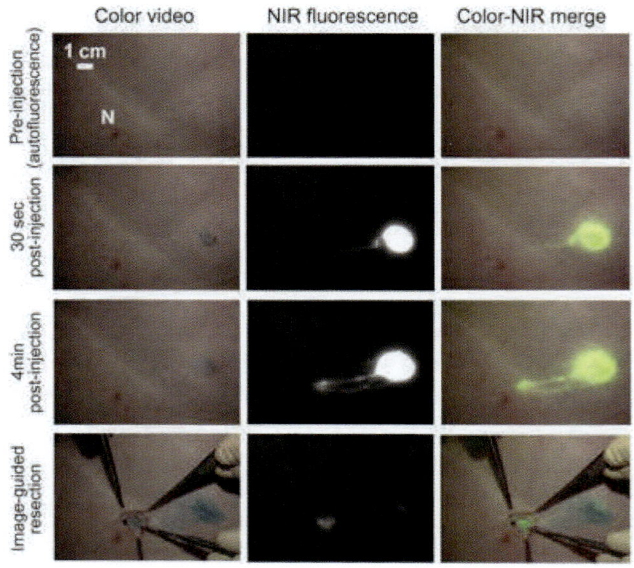

Fig. 3. Images of the surgical field in a pig injected intradermally with NIR QDs. From top to bottom: before injection (autofluorescence), 30 s after injection, 4 min after injection and during image-guided resection. From left to right: color image, NIR fluorescence and color-NIR merge images. "Reprinted by permission from Macmillan Publishers Ltd: [Nature Biotechnology] 22(1), 93–97, copyright 2003".

QDs have gained recognition as very versatile nanomaterials for fluorescence imaging *in vitro* and *in vivo*, and particularly for multiplexing studies. Nowadays there is a large variety of strategies for surface derivatization, and they present many advantages for long-term, high sensitivity and multiconstrast imaging of molecular dynamics in biological samples. Nevertheless, there are still certain disadvantages that need to be overcome before QDs full potential in this field can be achieved (*e.g.* increased size of the nanoparticles after their surface modification by bioconjugation that can hinder their access into cells for *in vivo* imaging).

3.2.3. *Rare earth nanoparticles*

The use of UCNPs for *in vivo* imaging has made a huge progress within the last few years, and has attracted a great interest. The first application of UCNPs in live organisms was carried out by Austin *et al.*,[33] who inoculated UCNPS in worms and carried out the imaging of their digestive system. They observed the distribution of the UCNPs in the intestines, and demonstrated that the UCNPS were biocompatible, since the worms did not exhibit unusual behavior in feeding.

UCNPs have been recently evaluated for optical imaging of blood vessels[91] and lymphatic nodes in mice[36] and they have been used as carriers of porphyrin for photodynamic therapy.[92] One of the most relevant applications described so far was developed by Chatterjee *et al.*[35] In such early paper, the authors demonstrated the great potential of UCNPs, when compared to green emitting QDs, for imaging of deep tissues in rats, which might be a prelude of the use of UCNPs in bigger animals. In 2013, UCNPs were used to perform optical imaging in human skin with a suppression of background due to the excitation light back-scattering and biological tissue autofluorescence to the level of the electronic noise.[93]

UCNPs conjugated with molecular recognition elements and certain drugs are good candidates for *in vivo* drug delivery. Some of the more important limitations of photodynamic therapy (PDT) are those caused by reduced tissue penetration of the light from the excitation source, and poor tumour-selectivity of the photosensitizer. In this vein, UCNPs possess the optical properties that can solve problems related to tissue penetration, and additionally, the appropriate size of the NPs allows anchoring several molecules of photosensitizer to the nanocrystal surface. In this sense, folate-modified amphiphilic chitosan was employed to immobilize zinc(II) phthalocyanine close enough to allow resonance energy transfer from UCNPs to the photosensitizer that acts as energy acceptor.[94] Selective tumour imaging of small animals was performed, and reactive oxygen species (ROS) were generated in cancer cells under a 1-cm tissue with higher efficiency when the excitation source was operating at 980 nm. As a consequence, this type of UCNPs conjugated to photosensitizers possesses a great therapeutic efficacy for PDT with tumour inhibition ratio up to 50%, which is much higher that the inhibition ratio obtained with conventional visible light-activated PDT (a value of 18%). Thus, multifunctionalized UCNPs are promising PDT agents for deep-seated tumour imaging and treatment (see Fig. 4).

3.2.4. *Carbon dots*

Cdots imaging in mice has been described through subcutaneous, interdermal (to track the migration through lymph vessels) and intravenous (for whole body circulation) injections.[52] A relatively slow diffusion and low NPs accumulation in the liver were observed and the intravenously injected Cdots were excreted *via* urine. Additionally, Cdots remained strongly fluorescent *in vivo*, which constitutes an important advantage for future *in vivo* imaging applications.

3.2.5. *Dye loaded nanoparticles*

In the last decade, with the arrival of small animal imaging devices, dye loaded NPs have been explored as fluorescent nanoprobes as well. One of the first architecture proposed for the design of fluorescent nanoprobes for *in vivo* imaging was based on the use of linear poly(lysine) chains bearing peptide–fluorophore moieties on pending amino groups. After peptide–fluorophore ratio was optimized, the spatial proximity of the fluorophores induced self-quenching and no emission was observed. After injection in the animal through vein tail, the presence of the enzyme gave rise to a peptide cleavage, and the fluorescence emission was recovered.[95]

Several publications have focused on the evaluation of the surface functionality of the NPs for *in vivo* applications. For instance, polymer NPs loaded

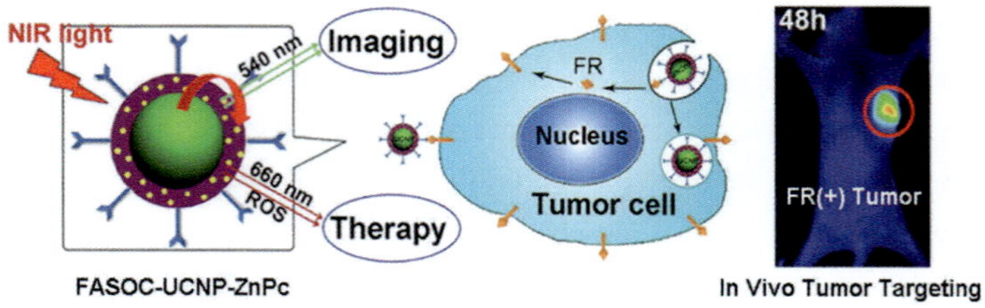

Fig. 4. *In vivo* tumour imaging and PDT based on UCNPs. "Reprinted with permission from ACS Nano 7 (2013) 676–688. Copyright (2013) American Chemical Society".

with the fluorescent dye Cy5 were evaluated as drug delivery systems, and were efficiently uptaken by the reticulo-endothelial system when they were functionalized with a PEG stealth coating.[96] Besides the possibility of transportation of both hydrophilic and hydrophobic molecules, PEG shells preserved the fluorophores stability, enhanced resistance to enzymatic oxidation and prolonged *in vivo* residence time.[97]

As stated at the end of Sect 2.5, the use of lipid-based NPs loaded with fluorescent dyes is also being studied due to their intrinsic low cytotoxicity. Passive tumour uptake of lipid-based NPs has been demonstrated in a variety of tumour models implanted sub-cutaneous in mice using fluorescence imaging, and for sentinel lymph node mapping.[98,99] In general terms, lipid-based NPs seem to display stronger affinity for lymphatic channels and lymph nodes than polymer NPs, and to several tumours that over-express lipoprotein receptors such as breast and prostate cancer or atherosclerotic lesions.[100]

3.3. *Nanoparticles and FRET for optical imaging*

Fluorescence (or Förster) resonance energy transfer (FRET) is a spectroscopic process by which a donor molecule non-radiatively transfers energy over a relatively long distance (1–10 nm) to an acceptor molecule. It is well known that the efficiency of this process is highly sensitive to the distance between the donor–acceptor pair. Hence, FRET is an effective tool for investigating a variety of biological phenomena involving changes in molecular proximity or conformation. FRET limitations often found when using traditional fluorophores, including photobleaching, spectral cross talk or direct acceptor excitation, are overcome by using the appropriate NPs. Most of the FRET-based applications so far make use of gold NPs and QDs. However, some applications with other types of fluorescent NPs are also appearing and will be discussed.

3.3.1. *Gold nanoparticles*

AuNCs are used in FRET-based biosensors because their molar extinction coefficient is high enough to make them excellent fluorescent acceptor materials.[101] However, even if are many applications in the FRET literature using AuNCs as energy acceptors, only few works developed further such methodologies for *in vitro* or *in vivo* imaging applications.

As example of such application, probing DNA damage caused by intracellular hydroxyl radicals with AuNCs in a FRET-based system was reported.[102] AuNCs, acting as acceptors, were conjugated to DNA oligomers containing a fluorescent dye (energy donor) at the other terminal position. The presence of hydroxyl radicals induced strand breakage in the single-stranded DNAs, so restoring the fluorescence of the energy donor. Fluorescence confocal microscopy imaging of macrophages and HepG2 revealed that the probe is cell-permeable and sensitive to intracellular hydroxyl radicals.

3.3.2. *Quantum dots*

One of the most frequently employed strategies regarding QDs as fluorescent labels is their use as energy donor molecules in FRET due to QDs advantageous spectroscopic properties.[103] Particularly important in this application is the possibility to control the emission by changing the NPs core size or composition, as this facilitates the spectral overlap between donor and acceptor. The large Stokes shift and wide absorption spectra are also favorable features. Moreover, QDs large surface area enables binding several proximal acceptors (proteins, peptides or oligonucleotides) to one NP. It is not surprising then that QDs have been used in many different FRET-based assays to detect proteins and genes. Some of these assays have been already used for imaging applications.

Bawendi *et al.*[104] demonstrated the possibility to carry out a dual conjugation of water-soluble QDs functionalized with a rhodamine dye as energy acceptor and with different proteins or peptides. Bioconjugation was *via* covalent bonding or *via* His-tag to provide the desired target capabilities. These multifunctional NPs were evaluated in HeLa cells to monitor the cellular receptor's local chemical environment.

The detection and quantitative measurement of enzyme/substrate interactions within living cells are of critical importance to understand cell physiology. Some of these enzymes are able to degrade specific peptide sequences located within proteins. This sequence-specific recognition and cleavage has been successfully applied to design NP FRET-based systems, in which both, donor and acceptor, are conjugated to the peptide. For instance, the activity of MMP-7, a protease involved in the degradation of extracellular matrix during cancer cell metastasis, could be measured as the fluorescence of the QD (acting as donor) is quenched by TAMRA dyes, but the presence of MMP-7 causes cleavage of the peptide, restoring the intrinsic photoluminescence of the QDs.[105]

In a different type of work, Farokhzad *et al.*[106] functionalized QDs with the A10 RNA aptamer that recognizes the extracellular domain of the prostate-specific membrane antigen (PSMA). They developed a system capable of differential uptake and imaging of prostate cancer cells that express the PSMA protein. The authors also intercalated doxorubicin (Dox), a widely used drug with fluorescent properties, in the double-stranded stem of the A10 aptamer to produce a QD-Apt(Dox) conjugate with reversible self-quenching properties based on a Bi-FRET mechanism: FRET between QD and Dox, and FRET between Dox and aptamer (see Fig. 5).

3.3.3. *Rare earth and dye-loaded nanoparticles*

UCNPs, due to their intense visible emission when they are excited under NIR light, are starting to be assayed for FRET imaging applications. An example is the use of UCNPs of the $NaYF_4$ type modified with carboxylic groups and conjugated to streptavidin, to develop a FRET-based nanoprobe for DNA imaging by attachment of a biotinylated oligonucleotide.[107]

As said before, recent advances in the development of nanoprobes for molecular imaging move towards NIR wavelengths, due to its advantages for *in vivo* applications. These types of dye loaded nanoparticles for FRET detection are made of diblock copolymers that encapsulate both, donor and acceptor dyes, generating a large Stokes shift (>100 nm) and able to reduce the background autofluorescence observed in live animals.[108]

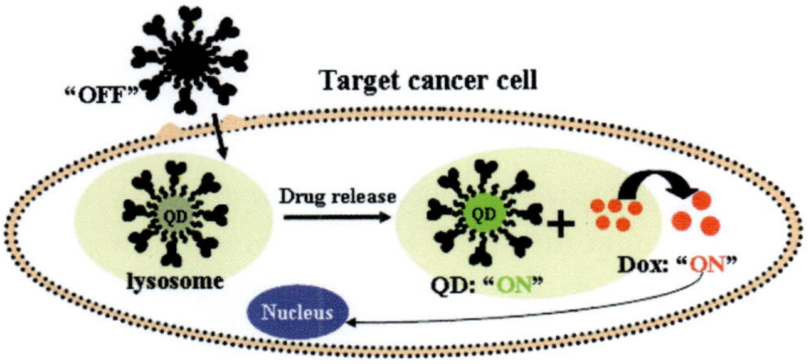

Fig. 5. Schematic illustration of QD-Apt(Dox) Bi-FRET system. (a) In the first step, the CdSe/ZnS QD are functionalized with the A10 PSMA aptamer. The intercalation of Dox within the A10 PSMA aptamer on the surface of QDs results in the formation of the QD-Apt(Dox) and quenching of both QD and Dox fluorescence through a Bi-FRET mechanism (the fluorescence of the QD is quenched by Dox while simultaneously the fluorescence of Dox is quenched by intercalation within the A10 PSMA aptamer) resulting in the "OFF" state. (b) Schematic illustration of specific uptake of QD-Apt(Dox) conjugates into target cancer cell through PSMA mediated endocytosis. The release of Dox from the QD-Apt(Dox) conjugates induces the recovery of fluorescence from both QD and Dox ("ON" state), thereby sensing the intracellular delivery of Dox and enabling the synchronous fluorescent localization and killing of cancer cells. "Reprinted with permission from Nano Lett. 10, 2007, 3065. Copyright 2007 American Chemical Society".

Interestingly, FRET systems can be also used to help and track drug delivery systems. For instance, polymeric micelles used with copolymers fluorescently labeled were assembled in solution around hydrophobic fluorescent probes to form FRET pairs. Real-time tracking of the release of hydrophobic probes, located in the core of the polymeric micelle, into the extracellular space was accomplished by FRET imaging.[109] Other approach, using FRET in the drug delivery action, used chitosan NPs loaded with Cy5 and Dox. After the NPs were uptaken by cells, the drug release was induced as a consequence of the chitosan breakdown in the acidic endosome, and accumulation of Dox in the nuclei was followed by FRET measurements.[110]

Dye loaded silica NPs have also been used for controlled release of anticancer chemotherapeutics.[111] Atto-647 (acting as energy acceptor) and a pH-sensitive linker were used to attach Dox (used as energy donor). When the drug was encapsulated inside the silica NPs, two emission peaks were measured as consequence of a FRET phenomenon. When the release of Dox takes place, the intensities of both peaks decrease. Thus, a real time monitoring of the Dox concentration can be followed by measuring FRET changes.

4. Potential Toxicity Associated with Long-term Exposure of Live Cells to Particular Luminescent Nanoparticles

As shown in previous sections, advances in nanotechnology are paving the way for new biomedical applications of nanosized materials. Particularly, inorganic NPs are playing an increasingly important role in biological applications and medical diagnosis by enabling non-invasive and long-term imaging of the whole body or of the complex cellular environment. The *in vitro* labeling of cells with NPs is becoming more and more common, but concerns about the possible negative effects of the NPs on the cultured cells or on the living organisms are also increasing. In fact, the transformation of nanotechnological improvements into the field of biomedicine (and particularly in routine optical bioimaging) is made difficult by several factors, one of the most important being the current incomplete knowledge of interactions between NPs and cells. To date, many ambiguous findings have been reported in the literature, mostly based on experiments with cultured cells.[112]

The research on the cytotoxic properties of different types of nanomaterials has significantly increased in the last few years, and there are reviews on this topic.[113] Some recent findings, including the intracellular degradability of NPs or the close correlation between intracellular localization and local NPs concentration and their cytotoxic effect,[114–116] have provided novel insights in the present understanding of NPs effects on cells. A brief overview of important studies on cytotoxic effects of inorganic NPs commonly used in cellular labeling, is presented here.

4.1. *Gold nanoparticles*

As mentioned before, AuNCs have been widely used in the development of biomedical applications. As gold is considered an inert element, the established view is that AuNCs should also be non-toxic (that is, disruptive to cellular integrity or viability) and non-genotoxic.[117] However, if AuNCs are sufficiently small (less than 2 nm)[118] or can be conjugated with specific nuclear receptors,[119] the particles can indeed enter the nucleus and interact directly with DNA and/or chromosomes and induce citotoxicity.

Though the general perception views nanosized gold colloids as relatively inert towards biological systems, a closer analysis of results from recent studies aimed at evaluating the real effect of AuNCs on cells reveals a more complex picture. While the chemical composition of NPs plays an important role, further biocompatibility determining aspects have been unveiled. The majority of assays concerning AuNCs toxicity were performed *in vitro* and reveal a considerable dependency of toxic effects on size, zeta potential and surface functionalization.[120] *In vivo* studies on this subject are still scarce. Based on the existing data, it can be assumed that a dosage of under $< 400\,\mu$g Au/kg showed no unwanted effects. However, when higher amounts were applied, toxicity depended on route of administration and particle size.

A critical point to be considered when exposing biological specimens to nanoparticles is the NP size. For AuNCs it was observed that, as expected, the smaller the particles (the higher the surface area to volume ratio) the larger their interaction with cellular or intracellular components. Also, their capability to reach intracellular locations which cannot be reached by larger sized NPs (*e.g.* the nucleus) increases. Therefore, smaller sized particles are often considered more dangerous.[121] AuNCs show a size-dependent cellular uptake with an optimal size of approximately 50 nm.[122,123] In any case, toxic effects exerted by NPs are linked to their cellular uptake levels, and therefore size influence on toxicity may not be so straightforward.

On the other hand, conventional NPs coating agents (*e.g.* cetyltrimethylammonium bromide) typically used during the NP synthesis for water-solubilization and stabilization may easily desorb from the AuNC surface, so inducing cell death (irrespective of the actual effects of the AuNCs themselves). Using polymer-coated particles, the toxic effects of the AuNCs were greatly reduced.[114,124]

Recently, attention has been directed to standardization of nanotoxicity tests, in view of the many papers describing a limited number of effects of a single NP type on one particular cell type for a specific incubation protocol.[125] In that line, a detailed approach has been proposed to assess cell–AuNC interactions[126]: by using multiple cell types and different NP incubation concentrations and times, a deeper view of how NPs can interfere with normal cellular homeostasis has been achieved. A concentration of 10 nM of phorbol myristate acetate-coated 4 nm diameter AuNCs was found not to lead to any significant effects with respect to cell viability, ROS induction, cell morphology, or functionality.

That final non-cytotoxic concentration of 10 nM that was defined is approximately 10-fold lower than the concentration that was obtained used in common cell viability assays (100 nM), further emphasizing the importance of including multiple parameters for a meaningful evaluation of the toxic effects of nanosized materials on cultured cells. As multiple sensitive cell types have been tested, such 10 nM concentration seems to be a good reference point for any safe cell labeling studies employing AuNCs with similar physicochemical properties.

Kang *et al.*[127] showed that NP size may play an important role to induce DNA damage as well. AuNCs of 100 and 200 nm size resulted in elevated levels of single-strand breaks (DNA lesions) at concentrations of 25, 50 and 100 μg/mL, whereas 4 and 15 nm NPs did not have an effect at such concentrations. Overall, it appears that AuNC concentrations of 50 μg/mL and above might have the potential to induce DNA damage.

4.2. *Semiconductor quantum dots*

Translation of QDs applications toward routine biomedical applications appears to be hindered by their questioned toxicity. In fact, the main drawback of QDs for *in vivo* imaging applications appears to be the inherent chemical cytotoxicity expected from their constitutive individual ions (*i.e.* Cd^{2+}, Se^{2+} and Te^{2+}).

Toxicity is certainly one of the most pressing drawbacks of QD probes since they were first introduced for biological applications. This concern mainly comes from the fact that QD cores are often synthesized using metals, such as Cd, Se, Zn, Te or Pb, known to be toxic at sufficiently high concentrations. Actually, such concerns are originating a perception that these nanomaterials are a "global health risk". Therefore, in despite of the high potential in bioanalysis of QDs, the toxicity risk consideration is somewhat dampening further developments in biological applications. On the other hand, determining the potential toxic effects of QDs is a highly complex challenge as a result of the different core compositions that can be synthesized, the different shells and surface coatings available, etc., each of which may seriously impact final toxicity. Therefore, long-term health risks of exposure to QDs would be intimately related to final synthesis conditions affecting QDs stability in biological media.

Neither all QDs are the same in toxicity terms nor all engineered QDs should be considered a uniform group of nanomaterials. In fact, each individual type of QD exhibits its own unique physicochemical properties, which in turn determines its eventual potential toxicity. A revision of the literature regarding QDs toxicity reveals many discrepancies that can be attributed to several factors: the lack of standardized toxicological studies, the variety of QD dosage/exposure concentrations used and the widely varying physicochemical properties of individual QDs (QD size, charge, concentration, coating materials, etc.).[7] The need for a reliable assessment of QDs potential long-term cytotoxicity has gathered much attention during the last decade.[128,129] Although mainly limited to cadmium-based nanomaterials (CdS, CdSe or CdTe cores, with or without a ZnS shell), studies so far carried out have shown that QDs become toxic to cells when present at micromolar concentrations in the growth medium, and that toxicity is directly related to the accessibility of the core surface cadmium atoms to the surrounding medium.

Although release of Cd^{2+} is generally considered as the main cause of QD-induced cellular toxicity,[130] the use of stabilizing coatings, passivating shell layers and the development of cadmium-free QDs are all being investigated in order to reduce any Cd^{2+} release and thereby promote further biomedical use of QDs.[131] Despite Cd^{2+} release being the primary cause of toxicity, other effects such as ROS production may not be neglected as they can lead to various toxicological effects different from those induced by Cd^{2+} release. Of course, such additional factors may also play an important role in the toxic effects of cadmium free QDs.[132]

QDs may be systemically distributed and may accumulate in organs and tissues. Such processes are, not surprisingly, highly variable for different QDs because of the wide variation in the QD physicochemical properties. QD size, charge, concentration and surface coating contribute to not only the potential toxicity of a given QD but also to their absorption, distribution, metabolism and excretion characteristics. Several *in vitro* studies have shown QDs to be incorporated *via* endocytic mechanisms by a variety of cell types (*e.g.* mammalian and dictyostelium discoideum cells) were observed to incorporate avidin and DHLA conjugated CdSe/ZnS QDs *via* endocytosis,[133] and rat primary hepatocytes were observed to incorporate mercaptoacetic acid–CdSe QDs.[128] In one instance, QD size was shown to be a determining factor in subcellular distribution. Lovric *et al.*[134] observed 5.2 nm cationic CdTe QDs to localize throughout the cytoplasm of N9 cells (murine microglial cell line) but not in the nucleus. In contrast, 2.2-nm cationic CdTe QDs were observed to localize in the nuclear compartment within the same time frame.

Regarding genotoxicity of QDs, different *in vitro* studies on the capability of QDs to induce oxidative damage to DNA have been recently conducted. As an example, an *in vitro* study reported that QDs made of (CdSe)ZnS induced DNA damage under photoactivation.[135] In this study, DNA strand breaks were not observed when the QDs were tested in the dark; thus, the mechanism of DNA damage was not related to the release of Cd^{2+}. A detailed mechanism for QD-induced oxidative damage to DNA was proposed: after QDs photoactivation, the NPs undergo a non-radiative energy transfer to molecular oxygen which induces the formation of reactive oxygen species, that should attack DNA. On the basis of other reported *in vitro* studies, it appears that (CdSe)ZnS QDs-induced DNA damage is mediated by either photogenerated ROS or by surface oxide-generated ROS. Conversely, cadmium ions do not appear to have a significant role in inducing oxidative damage to DNA when QDs are synthesized with a protective shell.

Finally, some studies have reported that the use of QDs made of less toxic metals tend to exhibit reduced or no toxicity at all, even without a ZnS-overcoating shell.[136]

4.3. *Carbon Dots, rare earth and dye-loaded nanoparticles: are they less toxic nanomaterials?*

Considerable efforts have been devoted to the development of heavy-metal-free novel photoluminescence nanoprobes in the past several years.[47] In this context, carbon NPs, or Cdots, as a class of "zero-dimensional" carbon nanomaterials, possess some of the major advantageous features of semiconductor QDs (*e.g.* high photostability, tunable emission and large two-photon excitation cross-sections) while they are more environment friendly and could be much safer for biological use.[51,52] As a result, studies on synthesis of Cdots as well as their applications in bioimaging have appeared recently.

As a previous step to their use for biological studies, Tao *et al.*[137] studied the *in vitro* cytotoxicity of three different Cdots on human kidney embryonic 293T cell line. Varied concentrations of Cdots made from different original materials were added to the cells cultured in 96 well-plates and incubated for 24 h. Subsequently, a standard assay was performed to assess the cell viabilities (after various Cdots treatments). No significant reduction in cell viability was observed for cells treated with Cdots even at very high concentrations (up to 0.5 mg/mL) demonstrating that the studied Cdots were not toxic at least from *in vitro* experiments.

The same authors systematically investigated the *in vivo* toxicology of Cdots to female mice over Three months of exposure.[137] Results did not show any obvious sign of toxic side effect for Cdot-injected mice at the dose of 20 mg/kg within 90 days. Neither death nor significant body weight drop was noted in the treatment group. However, even if such pioneer studies showed that Cdots were not apparently toxic to cell cultures or treated mice, further systematic investigation on their pharmacokinetics, biodistribution and toxicology are still required.

UCNPs have also been proposed as a promising new class of luminescent labels for bioimaging applications.[138] Moreover, upconverting photoluminescence could lead to predictive models with potential clinical applications.[139] Only a few studies of citotoxicity of UCNPs have been performed so far. However, some recent investigations have revealed the comparatively low toxicity of such NPs.[140,141] As an example, the use of PEG-conjugated phospholipids to coat UCNPs seems to offer a valuable strategy that not only render the NPs dispersible in aqueous solution (in view of the biocompatible, hydrophilic, non-immunogenic and non-antigenic features of PEG), but also the PEGylated UCNPs showed low toxicity.[142]

Reported *in vitro* cytotoxicity studies suggest that UCNPs have no or low toxicity when used within a certain range of concentration and within a limited time period of incubation. However, up to date, only a few long-term toxicological reports of UCNPs using animal models are available. Xiong *et al.*[143] studied the long-term *in vivo* biodistribution and toxicity of polyacrylic acid (PAA)-coated NaYF4 upconversion nanophosphors. *In vitro* cytotoxicity experimental results showed that PAA-UCNPs have no significant effects on the proliferation of KB cells. Furthermore, biodistribution results showed that most of the PAA-UCNPs in the liver and spleen were cleared from the body of mice in a very slow manner. *In vivo* toxicity studies indicated that mice intravenously injected with 15 mg/kg of PAA-UCNPs survived for 115 days without any evident toxic effects. These studies provide preliminary information to further use PAA-UCNPs for long-term *in vivo* imaging.

As already described, the loading of NPs with organic fluorophores to design dedicated *in vivo* fluorescent nanoprobes has been explored for imaging purposes.[144] In such context, dye-loaded polymer NPs offer an interesting alternative for imaging purposes (a large and diverse group of reporters with one or more analytical reagents incorporated into the polymeric matrix are possible). When using polymer-dye encapsulation, the appropriate choice of the matrix polymer is crucial to achieve low-cytotoxicity dye-loaded nanomaterials. Biologically inert polymer matrices can be designed to both minimize molecular/cellular interferences (*e.g.* protein binding), while protecting at the same time the intracellular environment from the potential cytotoxicity of the embedded analytical reagents.[145]

Encapsulation of fluorescent dyes, either by the entrapment using molecular affinity (*i.e.* a lipophilic dye will present a strong affinity with a lipophilic matrix) or by covalent conjugation to the material constituting the particle core (*e.g.* a hydrophobic polymer chain), are some of the described approaches followed to design such nanoprobes. Many different possibilities can be envisaged, and thus it is not simple to provide a general rule describing the general toxicity of such fluorescent tags. An important point to underline, however, is the intrinsic very low cytotoxicity of dye-loaded lipid-based NPs (with IC50 typical values as high as 0.1 to 1 mg/mL) reported for clinical applications.[99] As an example, near infrared encapsulated dyes based on oil-in-water nanoemulsions display very low cytotoxicity (IC50 ≈ 300 µg/mL of lipids), a high *in vivo* tolerance dose (> 150 mg lipids/kg) and they distribute in liver in 30 min following their intravenous injection in healthy animals.[99]

As said before, organic fluorophores encapsulation into silica NPs has been carried out as well for the development of both intracellular nanosensors and cell targeting systems. Studies of genotoxicity and cytotoxicity of these nanomaterials have shown no significant toxic effects at the molecular and cellular levels below a concentration of 0.1 mg/mL, indicating that they can be promising labeling reagents for biomedical applications.[146]

Acrylates and methacrylates are also used frequently for dye-encapsulation on the development of NP sensors. Although limited so far, studies assessing the toxicity of methacrylate and acrylate polymeric NPs tend to indicate the very low or non-toxic nature of such NPs. However, the inherent cytotoxicity and genotoxicity of their monomeric compounds has been widely reported.[147] Once again, assessment of the potential toxicity of these polymeric materials is a consideration of scale: cell viability assays by flow cytometry show a concentration dependant decrease in the viability of lymphocytes (up to 80%) when exposed to methacrylate monomer concentrations in the millimolar range.[148] At such high concentrations, methacrylate and acrylate monomers can also potentially exhibit DNA damage. Nevertheless, it is unlikely that such high concentrations might be reached when used as a biomolecule tag for imaging purposes.

Clearly there is an urgent need today to address issues associated with NPs toxicity to living cells, tissues and animals. It constitutes a pressing problem that the scientific community interested in NPs and biology should continue to explore to get better insights and understanding. NPs toxicity problem is really complex as the wide variety of NPs, the different possibilities of their functionalization, the environmental conditions and many other parameters affect the actual toxic effects of the used NPs. Moreover, the number and variety of NPs being developed for biomedical applications is steadily increasing. Present nanotoxicology studies indicate that there are several parameters which may be influenced during the NP production process and so should be carefully controlled (*e.g.* size and shape, composition purity, chemistry and NP's surface charge).[125]

5. Concluding Remarks and Future Outlook

The emergence and fast development of luminescent engineered NPs is challenging the capabilities of more traditional imaging reagents. Long-term imaging of cell biomarkers using conventional organic dyes is hindered by photobleaching and limited to a relatively few colors. The advent of photoluminescent NPs might solve many of the technical problems of such organic dyes. What is more, after NPs appropriate functionalization, they could penetrate endothelial barriers to reach intracellular organs to enable improved imaging performance.

Additionally to the different types of NPs revised along this chapter, recent advances in nanotechnology have given rise to new fluorescent metal NCs. The attractive set of features exhibited by the NCs, such as ultrasmall size, good biocompatibility and excellent photostability, also make them very promising fluorescent labels for bioimaging applications.

Although traditional fluorescent dyes continue to be used, engineered NPs could serve as a platform to synthesise advanced multifunctional probes for multiplexed imaging. Direct one- and two-photon fluorescence or indirect FRET to probe protein–protein/DNA interactions and track protein receptors in the cytosol or on the membrane of live cells are imaging active research areas where NPs-based approaches are offering important advances.

However, NPS-based labeling reagents still require further improvements in order to enable robust and practical applications. For example, the chemical synthesis of inorganic nanomaterials in solution often yields a large distribution of particle sizes and shapes, which may lead to heterogeneous nanomaterials with not well-defined optical properties. Of course, improving the colloidal stability of the nanocrystals in aqueous buffers and in biological media is another must at this point. Moreover, present conjugation strategies are still suffering from inefficiency, crossreactivity and strong dependence on the conditions and reagents used. In addition, water solubility, surface functionalization and chemical purity can all affect the eventual performance and stability of such nanomaterial based reagents in bioimaging.

Actual research is focused on improved surface coating strategies and/or the consolidation of the existing water-solubilization techniques into a platform able to provide NPs stable enough over a wide range of biologically-relevant conditions (*e.g.* lower or higher pHs, salty media, counterions excess, etc.). Efforts to improve such surface coatings should be also oriented to ensure biocompatibility of the label nanomaterials and to allow intracellular delivery of the NPs (often required for many medical imaging studies).

On the way to obtain better nanomaterials for bioimaging, additionally to search novel synthetic routes, proper analytical characterization and purification of the obtained nanomaterials is urgently needed. Nanomaterials purity, narrowing their size/shape/conjugation distributions, etc. may demand appropriate separation techniques.[149]

Moreover, NPs conjugation to biomolecules is crucial today to achieve the required biological recognition and activity of the resulting NP-bioconjugates. Development of more efficient and easy to implement conjugation strategies and the bioconjugates separation and characterization is still needed but largely underexplored.

Regarding the toxicological effects of nanoparticles, a key aspect if NPs-based imaging technologies are eventually implemented for *in vivo* studies, it is an aspect still unclear: the question whether inorganic NPs are safe to be used for biomedical purposes is open to debate. More and more data are becoming available regarding NPs toxicity. However, a lot of further research effort is still required in order to advance and clarify our knowledge in this important field. The wide variety of cell labeling conditions (*e.g.* incubation time, cell types, NP concentrations to be used, etc.) and the varied physico-chemical properties of NPs (size, surface charge, core composition, surface functionalization, etc.) have arisen a substantial amount of disparate data and conclusions on the topic so far. In fact, the safety of nanomaterials for biomedical applications and (cultured) cells exposure effects to NPs is still a pending problem. Standardization of incubation conditions, careful analytical characterization of NPs and bioconjugates in their biologically relevant environment and large scale comparative studies could be a first step aiming at in increasing our understanding in this area, particularly for *in vivo* imaging applications.

Acknowledgments

Financial support from project CTQ2010-16636 of the Spanish Ministry of Science and Innovation (MICINN) and the European FEDER program co-financing is gratefully acknowledged.

References

1. A. Waggoner, *Curr. Opin. Chem. Biol.* **10**, 62 (2006).
2. P. Sharma, S. Brown, G. Walter, S. Santra and B. Moudgil, *Adv. Colloid Interfac.* **123**, 471 (2006).
3. W. C. W. Chan and S. Nie, *Science* **281**, 2016 (1998).
4. M. Bruchez, M. Moronne, P. Gin, S. Weiss and A. P. Alivisatos, *Science* **281**, 2013 (1998).
5. A. M. Coto-García, E. Sotelo-González, M. T. Fernández-Argüelles, R. Pereiro, J. M. Costa-Fernández and A. Sanz-Medel, *Anal. Bioanal. Chem.* **399**, 29 (2011).
6. J. L. West and N. J. Halas, *Ann. Rev. Biomed. Eng.* **5**, 285 (2003).
7. R. Hardman, *Environ. Health Perspectives* **114**, 165 (2006).
8. R. Gill, M. Zayats and I. Willner, *Angew. Chem. Int. Ed.* **47**, 7602 (2008).
9. C. V. Durgadas, C. P. Sharma and K. Sreenivasan, *Analyst* **136**, 933 (2011).
10. J. P. Xie, Y. G. Zheng and J. Y. Ying, *Chem. Commun.* **46**, 961 (2010).
11. C.-L. Liu, H.-T. Wu, Y.-H. Hsiao, C.-W. Lai, C.-W. Shih, Y.-K. Peng, K.-C. Tang, H.-W. Chang, Y.-C. Chien, J.-K. Hsiao, J.-T. Cheng and P.-T. Chou, *Angew. Chem. Int. Ed.* **50**, 7056 (2011).
12. J. Akola, M. Walter, R. L. Whetten, H. Hakkinen and H. Gronbeck, *J. Am. Chem. Soc.* **130**, 3756 (2008).
13. N. Goswami, R. Saha and S. K. Pal, *J. Nanopart. Res.* **13**, 5485 (2011).
14. D. Hu, Z. Sheng, P. Gong, P. Zhang and L. Cai, *Analyst* **135**, 1411 (2010).
15. X. L. Guével, N. Daum and M. Schneider, *Nanotechnology* **22**, 275103 (2011).
16. P. L. Xavier, K. Chaudhari, P. K. Verma, S. K. Pal and T. Pradeep, *Nanoscale* **2**, 2769 (2010).
17. H. Kawasaki, K. Yoshimura, K. Hamaguchi and R. Arakawa, *Anal. Sci.* **27**, 591 (2011).
18. Z. Zhong, S. Patskovsky, P. Bouvrette, H. T. Luong and A. Gedanken, *J. Phys. Chem. B*, **108**, 4046 (2004).
19. D. M. Chevrier, A. Chatt and P. Zhang, *J. Nanophoton.* **3**, 1 (2012).
20. M. Zhu, C. M. Aikens, F. J. Hollander, G. C. Schatz and R. Jin, *J. Am. Chem. Soc.* **130**, 5883 (2008).
21. M. K. Wagner, F. Li, J. Li, X. Li and X. C. Le, *Anal. Bioanal. Chem.* **397**, 3213 (2010).
22. S. G. Penn, L. Hey and M. J. Natan, *Current Opinion Chem. Biol.* **7**, 609 (2003).
23. F. Pinaud, X. Michalet, L. A. Bentolila, J. M. Tsay, S. Doose, J. J. Li, G. Lyer and S. Weiss, *Biomaterials* **27**, 1679 (2006).
24. C. B. Murray, D. J. Norris and M. G. Bawendi, *J. Am. Chem. Soc.* **115**, 8706 (1993).
25. Z. A. Peng and X. Peng, *J. Am. Chem. Soc.* **123**, 183 (2001).
26. X. Ji, J. Zheng, J. Xu, V. K. Rastogi, T. Cheng, J. J. DeFrank and R. M. Leblanc, *J. Phys. Chem. B* **109**, 3793 (2005).

27. A. M. Smith and S. Nie, *Analyst* **129**, 672 (2004).
28. S. Pathak, S. K. Choi, N. Arnheim and M. E. Thompson, *J. Am. Chem. Soc.* **123**, 4103 (2001).
29. H. Mattoussi, J. M. Mauro, E. R. Goldman, G. P. Anderson, V. C. Sundar, F. V. Mikulec and M. G. Bawendi, *J. Am. Chem. Soc.* **122**, 12142 (2000).
30. X. Y. Wu, H. J. Liu, J. Q. Liu, K. N. Haley, J. A. Treadway, J. P. Larson, N. F. Ge, F. Peale and M. P. Bruchez, *Nat. Biotechnol.* **21** (2003) 41.
31. B. Dubertret, P. Skourides, D. J. Norris, V. Noireaux, A. H. Brivanlou and A. Libchaber, *Science* **298**, 1759 (2002).
32. S. F. Lim, R. Riehn, C. Tung, W. S. Ryu, R. Zhuo, J. Dalland and R. H. Austin, *Nanotechnology* **20**, 405701 (2009).
33. S. F. Lim, R. Riehn, W. S. Ryu, N. Khanarian, C. K. Tung, D. Tank and R. H. Austin, *Nano Lett.* **6**, 169 (2006).
34. M. Nyk, R. Kumar, T. Y. Ohulchanskyy, E. J. Bergey, and P. N. Prasad, *Nano Lett.* **8**, 3834 (2008).
35. D. K. Chatterjee, A. J. Rufaihah and Y. Zhang, *Biomaterials* **29**, 937 (2008).
36. H. Kobayashi, N. Kosaka, M. Ogawa, N. Y. Morgan, P. D. Smith, C. B. Murray, X. Ye, J. Collins, G. A. Kumar, H. Belld and P. L. Choyke, *J. Mater. Chem.* **19**, 6481 (2009).
37. F. Wang, D. Banerjee, Y. Liu, X. Chen and X. Liu, *Analyst* **135**, 1839 (2010).
38. A. Aebischer, M. Hostettler, J. Hauser, K. Krämer, T. Weber, H.-U. Güdel and H.-B. Bürgi, *Angew. Chem. Int. Ed.* **45**, 2802 (2006).
39. M. Haase and H. Schäfer, *Angew. Chem. Int. Ed.,* **50**, 5808 (2011).
40. H. Wang and T. Nann, *ACS Nano* **3**, 3804 (2009).
41. Wang, L. and Li, Y. *Chem. Commun.* 2557 (2006).
42. L. Wang, Y. Zhang, and Y. Zhu, *Nano Res.* **3**, 317 (2010).
43. F. Wang and X. Liu, *Chem. Soc. Rev.* **38**, 976 (2009).
44. Y. Sun, B. Zhou, Y. Lin, W. Wang, K. A. S. Fernando, P. Pathak, M. J. Meziani, B. A. Harruff, X. Wang, H. Wang, P. Luo, H. Yang, M. E. Kose, B. Chen, L. M. Veca and S. Xie, *J. Am. Chem. Soc.* **128**, 7756 (2006).
45. A. B. Bourlinos, A. Stassinopoulos, D. Anglos, R. Zboril, M. Karakassides and E.P. Giannelis, *Small* **4**, 455 (2008).
46. S. L. Hu, K. Y. Niu, J. Sun, J. Yang, N. Q. Zhao and X. W. Du, *J. Mater. Chem.* **19**, 484 (2009).
47. L. Cao, X. Wang, M. J. Meziani, F. Lu, H. Wang, P. G. Luo, Y. Lin, Y, B. A. Harruff, L. M. Veca, D. Murray, S. Y. Xie, and Y. P. Sun, *J. Am. Chem. Soc.* **129**, 11318 (2007).
48. J. C. G. Esteves da Silva and H. M. R. Goncalves, *Trends Anal. Chem.* **30**, 1327 (2011).
49. X .Y. Xu, R. Ray, Y. L. Gu, H. J. Ploehn, L. Gearheart, K. Raker and W. A. Scrivens, *J. Am. Chem. Soc.* **126**, 12736 (2004).
50. Y. P. Sun, X. Wang, F. S. Lu, L. Cao, M. J. Meziani, J. G. Luo, L. R. Gu and L. M. Veca, *J. Phys. Chem. C* **112**, 18295 (2008).
51. S. Yang, X. Wang, H. Wang, F. Lu, P. G. Luo, L. Cao, M. J. Meziani, J. Liu, Y. Liu, M. Chen, Y. Huang and Y. Sun, *J. Phys. Chem. C* **113**, 18110 (2009).
52. S. Yang, L. Cao, P. G. Luo, F. Lu, X. Wang, H. Wang, M. J. Meziani, Y. Liu, G. Qi and Y. Sun, *J. Am. Chem. Soc.* **131**, 11308 (2009).
53. M. J. Ruedas-Rama, J. D. Walters, A. Orte and E. A. H. Hall, *Anal. Chim. Acta* **751**, 1 (2012).

54. Y. Cao, Y. -E. L. Koo, S. M. Koo and R. Kopelman, *Photochem. Photobiol.* **81**, 1489 (2005).
55. H. A. Clark, M. Hoyer, M. A. Philbert and R. Kopelman, *Anal. Chem.* **71**, 4831 (1999).
56. W. Weyenberg, P. Filev, D. Van den Plas, J. Vandervoort, K. De Smet, P. Sollie, and A. Ludwiga, *Int. J. Pharm.* **337**, 291 (2007).
57. H. He, C. Xie and J. Ren, *Anal. Chem.* **80**, 5951 (2008).
58. X. Wu, T. Ming, X. Wang, P. Wang, J. Wang and J. Chen, *ACS Nano* **4**, 113 (2010).
59. J. Xie, Y. Zheng and J. Y. Ying, *J. Am. Chem. Soc.* **131**, 888 (2009).
60. A. M. Smith, G. Ruan, M. N. Rhyner and S. Nie, *Ann. Biomed. Eng.* **34**, 3 (2006).
61. C. Wang, X. Gao, X. Su, *Anal. Bioanal. Chem.* **397**, 1397 (2010).
62. J. B. Delehanty, C. E. Bradburne, K. Boeneman, K. Susumu, D. Farrell, B. C. Mei, J. B. Blanco-Canosa, G. Dawson, P. E. Dawson, H. Mattoussi and I. L. Medintz, *Integ. Biol.* 2, 265 (2010).
63. Q A. M. Smith, H. Duan, A. M. Mohs and S. Nie, *Adv. Drug. Deliver. Rev.* **60**, 1226 (2008).
64. Y. S. Liu, Y. Sun, P. T. Vernier, C. H. Liang, S. Y. C. Chong and M. A. Gundersen, *J. Phys. Chem. C* **111**, 2872 (2007).
65. H. Zijlmans, J. Bonnet, J. Burton, K. Kardos, T. Vail, R. S. Niedbala and H. J. Tanke, *Anal. Biochem.* **267**, 30 (1999).
66. M. Wang, C. C. Mi, W. X. Wang, C. H. Liu, Y. F. Wu, Z. R. Xu, C. B. Mao and S. K. Xu, *ACS Nano* **3**, 1580 (2009).
67. T. Zako, H. Nagata, N. Terada, A. Utsumi, M. Sakono, M. Yohda, H. Ueda, K. Soga and M. Maeda, *Biochem. Biophys. Res. Commun.* **381**, 54 (2009).
68. S. Santra, P. Zhang, K. Wang, R. Tapec and W. Tan, *Anal. Chem.* **73**, 4988 (2001).
69. R. Kumar, I. Roy, T. Y. Ohulchanskyy, L. N. Goswami, A. C. Bonoiu, E. J. Bergey, K. M. Tramposch, A. Maitra and P. N. Prasad, *ACS Nano.* **2**, 449 (2008).
70. H. Xu, J. W. Aylott, R. Kopelman, T. J. Miller and M. A. Philbert, *Anal. Chem.* **73**, 4124 (2001).
71. C. He, W. Zhu, Y. Xu, Y. Zhong, J. Zhou and X. Qian, *J. Mater. Chem.* **20**, 10755 (2010).
72. F. D. A. Massimo Bottini, A. Magrini, F. Cerignoli, Y. Arimura, M. I. Dawson, E. Bergamaschi, N. Rosato, A. Bergamaschi and T. Mustelin, *Int. J. Nanomed.* **2**, 227 (2007).
73. S. T. Selvan, P. K. Patra, C. Y. Ang and J. Y. Ying, *Angew. Chem. Int. Ed.* **46**, 2448 (2007).
74. Y. Wu, M. Chu, B. Shi and Z. Li, *Appl. Biochem. Biotechnol.* **163**, 813 (2011).
75. H. A. Clark, M. Hoyer, S. Parus, M. A. Philbert and R. Kopelman, *Microchim. Acta* **131**, 121 (1999).
76. C. J. Lin, T. Yang, C. Lee, S. H. Huang, R. A. Sperling, M. Zanella, J. K. Li, J. Shen, H. Wang, H. Yeh, W. J. Parak and W. H. Chang, *ACS Nano.* **3**, 395 (2009).
77. H. Wang, T. B. Huff, D. A. Zweifel, W. He, P. S. Low, A. Wei and J. Cheng, *PNAS* **102**, 15752 (2005).
78. H. Chen, B. Li, C. Wang, X. Zhang, Z. Cheng, X. Dai, R. Zhu and Y. Gu, *Nanotechnology* **24**, 055704 (2013).
79. F. Lai, C. Chien, W. Chen, H. Chen, Y. Chen, C. Wang, Y. Hwu, C. S. Yang, C. Y. Chen, K. S. Liang, C. Petibois, H. Tan, E. Tok and G. Margaritondo, *Biotechnol, Adv.* **31**, 362 (2013).

80. D. M. Chevrier, A. Chatt and P. Zhang, *J. Nanophotonics,* **6**, 064504 (2012).
81. C. W. Peng and Y. Li, *J. Nanomater* 676839 (2010).
82. M. E. Åkerman, W. C. W. Chan, P. Laakkonen, S. N. Bhatia and E. Ruoslahti (2002) *Proc. Natl. Acad. Sci. USA* **99**, 12617 (2002).
83. R. G. Aswathy, Y. Yoshida, T. Maekawa and D. S. Kumar, *Anal. Bioanal. Chem.* **397**, 1417 (2010).
84. H. Kobayashi, Y. Hama, Y. Koyama, T. Barrett, C. A. S. Regino, Y. Urano and P. L. Choyke, *Nano Lett.* 7, 1711 (2007).
85. M. Hikage, K. Gonda, M. Takeda, T. Kamei, M. Kobayashi, M. Kumasaka, M. Watanabe, S. Satomi and N. Ohuchi, *Nanotechnology* **21**, 185103 (2010).
86. S. Kim, Y. T. Lim, E. G. Soltesz, A. M. De Grand, J. Lee, A. Nakayama, J. A. Parker, T. Mihaljevic, R. G. Laurence, D. M. Dor, L. H. Cohn, M. G. Bawendi and J. V. Frangioni, *Nat. Biotechnol.* **22**, 93 (2004).
87. X. Gao, Y. Cui, R. M. Levenson, L. W. K. Chung and S. Nie, *Nat. Biotechnol.* **22**, 969 (2004).
88. L. Yang, H. Mao, Y. A. Wang, Z. Cao, X. Peng, X. Wang, H. Duan, C. Ni, Q. Yuan, G. Adams, M. W. Smith, W. C. Wood, X. Gao and S. Nie, *Small.* **5**, 235 (2009).
89. X. Yu, L. Chen, K. Li, Y. Li, S. Luo, J. Liu, L. Deng, D. Pang and Q. Wang, *J. Biomed. Opt.* **12**, 014008 (2007).
90. J. Liu, S. K. Lau, V. A. Varma, R. A. Moffitt, M. Caldwel, T. Liu, A. N. Young, J. A. Petros, A. O. Osunkoya, T. Krogstad, B. Leyland-Jones, M. D. Wang and S. Nie, *ACS Nano.* **4**, 2755 (2010).
91. S. A. Hilderbrand, F. Shao, C. Salthouse, U. Mahmood and R. Weissleder, *Chem. Commum.* 4188 (2009).
92. B. Ungun, R. K. Prud'homme, S. J. Budijono, J. Shan, S. F. Lim, Y. Ju and R. Austin, *Opt. Express* **17**, 80 (2009).
93. Z. Song, Y. G. Anissimov, J. Zhao, A. V. Nechaev, A. Nadort, D. Jin, T. W. Prow, M. S. Roberts and A. V. Zvyagin, *J. Biomed. Opt.* **18**, 061215 (2013).
94. S. Cui, D. Yin, Y. Chen, Y. Di, H. Chen, Y. Ma, S. Achilefu and Y. Gu, *ACS Nano.* 7, 676 (2013).
95. C. H. Tung, U. Mahmood, S. Bredow and R. Weissleder, *Cancer Res.* **60**, 4953 (2000).
96. P. Couvreur and C. Vauthier, *Pharm. Res.* **23**, 1417 (2006).
97. A. Almutairi, W. J. Akers, M. Y. Berezin, S. Achilefu and J. M. Fréchet, *J Mol. Pharm.* **5**, 1103 (2008).
98. I. Texier, M. Goutayer, A. Da Silva, L. Guyon, N. Djaker, V. Josserand, E. Neumann, J. Bibette and F. Vinet, *J. Biomed. Opt.* **14**, 054005 (2009).
99. F. P. Navarro, M. Berger, S. Guillermet, V. Josserand, L. Guyon, E. Neumann, J. L. R. Coll and I. Texier, *J. Biomed. Nanotechnol.* **8**, 730 (2012).
100. J. Mérian, J. Gravier, F. Navarro and I. Tessier, *Molecules* **17**, 5564 (2012).
101. L. Li, J. Song, S. Li and C. Fan, *J. Phys. D Appl. Phys.* **42**, 203001 (2009).
102. B. Tang, N. Zhang, Z. Chen, K. Xu, L. Zhuo, L. An and G. Yang, *Chem. Eur. J.* **14**, 522 (2008).
103. I. L. Mendintz and H. Mattoussi, *Phys. Chem. Chem. Phys.* **11**, 17 (2009).
104. W. Liu, M. Howarth, A. B. Greytak, Y. Zheng, D. G. Nocera, A. Y. Ting, and M. G. Bawendi, *J. Am. Chem. Soc.* **130**, 1274 (2008).
105. Y. P. Kim, Y. H. Oh, E. Oh and H.-S. Kim, *Biochip J.* **1**, 228 (2007).

106. V. Bagalkot, L. Zhang, E. Levy-Nissenbaum, S. Jon, P. W. Kantoff, R. Langer and O. C. Farokhzad, *Nano Lett.* 7, 3065 (2007).
107. Z. Chen, H. Che, H. Hu, M. Yu, F. Li, Q. Zhang, Z. Zhou, T. Yi and C. Huang, *J. Am. Chem. Soc.* **130**, 3023 (2008).
108. A. Wagh, S. Y. Qian and B. Law, *Bioconjugate Chem.* **23**, 981 (2012).
109. H. Chen, S. Kim, L. Li, S. Wang, K. Park and J. X. Cheng, *Proc. Natl. Acad. Sci. USA* **105**, 6596 (2008).
110. K. J. Chen, Y. L. Chiu, Y. M. Chen, Y. C. Ho and H. W. Sung, *Biomaterials* **32**, 2586 (2011).
111. C. H. Lee, S. H. Cheng, I. P. Huang, J. S. Souris, C. S. Yang, C. Y. Mou and L. W. Lo, *Angew. Chem. Int. Ed. Engl.* **49**, 8214 (2010).
112. S. J. Soenen and M. De Cuyper, *Nanomedicine* **5**, 1261 (2010).
113. M. A. Maurer-Jones, K. C. Bantz, S. A. Love, B. J. Marquis and C. L. Haynes, *Nanomedicine* **4**, 219 (2009).
114. S. J. Soenen, U. Himmelreich, N. Nuytten, T. R. Pisanic, A. Ferrari and M. De Cuyper, *Small* **6**, 2136 (2010).
115. A. D. Lehmann, W. J. Parak, F. Zhang, Z. Ali, C. Rocker, G. U. Nienhaus, P. Gehr and B. Rothen-Rutishauser, *Small* **6**, 753 (2010).
116. C. Brandenberger, C. Muhlfeld, Z. Ali, A. G. Lenz, O. Schmid, W. J. Parak, P. Gehr and B. Rothen-Rutishauser, *Small* **6**, 1669 (2010).
117. J. Ponti, R. Colognato, F. Franchini, S. Gioria, F. Simonelli, K. Abbas, C. Uboldi, C. J. Kirkpatrick, U. Holzwarth and F. Rossi, *Nanotoxicology* **3**, 296 (2009).
118. M. Tsoli, H. Kuhn, W. Brandau, H. Esche and G. Schmid, *Small* **1**, 841 (2005).
119. B. Kang, M. A. Mackey and M. A. El-Sayed, *J. Am. Chem. Soc.* **132**, 1517 (2010).
120. U. Taylor, A. Barchanski, W. Garrels, S. Klein, W. Kues, S. Barcikowski and D. Rath, *Nano-Biotechnol. Biomed. Diag. Res. Adv. Experim. Med. Biol.* **733**, 125 (2012).
121. U. Simon and W. Jahnen-Dechent, *Nanomedicine* **3**, 601 (2008).
122. W. C. W. Chan, B. D. Chithrani and A. A. Ghazani, *Nano Lett.* **6**, 662 (2006).
123. B. D. Chithrani and W. C. W. Chan, *Nano Lett.* 7, 1542 (2007).
124. C. Y. Chen, Y. Qiu, Y. Liu, L. M. Wang, L. G. Xu, R. Bai, Y. L. Ji, X. C. Wu, Y. L. Zhao and Y. F. Li, *Biomaterials* **31**, 7606 (2010).
125. S. J. Soenen, P. Rivera Gil, J. M. Montenegro, W. J. Parak, De Smedt and K. Braeckmans, *Nano Today* **6**, 446 (2011).
126. C-A J. Lin, T-Y Yang, C-H Lee, S. H. Huang, R. A. Sperling, M. Zanella, J. K. Li, J-L Shen, H-H Wang, H-I Yeh, W. J. Parak and W. H. Chang, *ACS Nano* **3**, 395 (2009).
127. J. S. Kang, Y. N. Yum, J. H. Kim, H. Song, J. Jeong, Y. T. Lim, B. H. Chung and S. N. Park, *Biomol. Ther.* **17**, 92 (2009).
128. A. M. Derfus, W. C. W. Chan and S. N. Bhatia, *Nano Lett.* **4**, 11 (2004).
129. C. Kirchner, T. Liedl, S. Kudera, T. Pellegrino, A. Munoz Javier, H. E. Gaub, S. Stolzle, N. Fertig and W. J. Parak, *Nano Lett.* **5**, 331 (2005).
130. K. G. Li, J. T. Chen, S. S. Bai, X. Wen, S. Y. Song, Q. Yu, J. Li and Y. Q. Wang, *Toxicol. In Vitro* **23**, 1007 (2009).
131. M. J. D. Clift, M. S. P. Boyles, D. M. Brown and V. Stone, *Nanotoxicology* **4**, 139 (2010).
132. M. Tang, M. Wang, T. Xing, J. Zeng, H. Wang and D. Y. Ruan, *Biomaterials* **29**, 4383 (2008).

133. J. K. Jaiswal, H. Mattoussi, J. M. Mauro and S. M. Simon, *Nat. Biotechnol.* **21**, 47 (2003).
134. J. Lovric, H. S. Bazzi, Y. Cuie, G. R. A. Fortin, F. M. Winnik and D. Maysinger, *J. Mol. Med.* **83**, 377 (2005).
135. A. Anas, H. Akita, H. Harashima, T. Itoh, M. Ishikawa and V. Biju, *J. Phys. Chem. B* **112**, 10005 (2008).
136. S. J. Cho, D. Maysinger, M. Jain, B. Roder, S. Hackbarth and F. M. Winnik, *Langmuir* **23**, 1974 (2007).
137. H. Tao, K. Yang, Z. Ma, J. Wan, Y. Zhang, Z. Kang and Z. Liu, *Small* **8**, 281 (2012).
138. J. Zhou, Y. Sun, X. Du, L. Xiong, H. Hu and F. Li, *Biomaterials* **31**, 3287 (2010).
139. L. Xiong, T. Yang, Y. Yang, C. Xu and F. Li, *Biomaterials* **31**, 7078 (2010).
140. D. K. Chatterjee, M. K. Gnanasammandhan and Y. Zhang, *Small* **6**, 2781 (2010).
141. M. Haase and H. Schafer, *Angew. Chem. Int. Ed.* **50**, 5808 (2011).
142. G. Tian, Z. Gu, L. Zhou, W. Yin, X. Liu, L. Yan, S. Jin, W. Ren, G. Xing, S. Li and Y. Zhao, *Adv. Mater.* **24**, 1226 (2012).
143. L. Xiong, T. Yang, Y. Yang, C. Xu and F. Li, *Biomaterials* **31**, 7078 (2010).
144. K. Licha and C. Olbrich, *Adv. Drug Deliv. Rev.* **57**, 1087 (2005).
145. J. R. Henderson, D. A. Fulton, C. J. McNeil and P. Manning, *Biosens. Bioelectron.* **24**, 3608 (2009).
146. Y. Jin, S. Kannan, M. Wu and J. X. Zhao, *Chem. Res. Toxicol.* **20**, 1126 (2007).
147. H. Schweikl, K. A. Hiller, A. Eckhardt, C. Bolay, G. Spagnuolo, T. Stempfl and G. Schmalz, *Biomaterials* **29**, 1377 (2008).
148. T. Poplawski, E. Pawlowska, M. Wisniewska-Jarosinska, D. Ksiazek, K. Wozniak, J. Szczepanska and J. Blasiak, *Chem.-Biol. Interact.* **180**, 69 (2009).
149. S. K. Ratanathanawongs Williams, J. R. Runyon and A. A. Ashames, *Anal. Chem.* **83**, 634 (2011).

Chapter 11

Metal Particles for Photoacoustic Imaging

Dipanjan Pan

Assistant Professor in Medicine, Research,
Division of Cardiology, Department of Medicine,
Washington University School of Medicine,
Campus Box 8215, 660 Euclid Ave, St Louis, MO 63108
dipanjan@wustl.edu

Abstract

Photoacoustic imaging (PAI) is a hybrid, non-ionizing modality, offers excellent spatial resolution and high soft tissue contrast. In this technique, a signal is generated based on the optical absorption of a material, which is sensitive to physiological parameters. Myriad advancement in this area has shown that this technique has the ability to provide both functional and molecular imaging *in vivo*. This chapter recapitulates various strategies to produce exogenous contrast for photoacoustic (PA) application emphasizing metal-based approaches. Various angles are discussed with regards to PA signal amplification *in vivo*, their association with homing agents for targeting biological epitopes and factors that govern *in vivo* distribution. Overall, this chapter brings out a brief description of exogenous contrast agents used for PAI.

1. Introduction

Research has been developing immensely in the area of advanced imaging since conception of nuclear medicine then and expanded to almost all clinically relevant modalities. Despite the preliminary success of detecting molecular changes with these techniques (MRI, CT, US, PET), new modalities are emerging to make them more powerful. Photoacoustic imaging (PAI) is one such technique that is showing high potential for its satisfactory spatial resolution and high soft tissue contrast.[1–6] The novelty behind these methods encompass a range of optical and ultrasonic methods that converge to perform a powerful, hybrid and non-ionizing imaging modality that are sensitive to optical absorption of biological tissue proteins such as hemoglobin and circulating erythrocytes.

In a typical photoacoustics (PAs) system, the tissue is irradiated with a short-pulsed laser beam. The applied laser beam is non-ionizing and within the acceptable level of ANSI limit. Once irradiated, protein absorbs the optical energy to result in a thermoelastic expansion and radiates PA waves. A clinical, wide-band ultrasonic transducer (5–50 MHz) receives (only) the PA waves and an image is formed without the usual acoustic speckle associated with interactions of transmitted and returning acoustic waves. PAT is sensitive to physiological parameters (e.g. concentration and oxygenation of hemoglobin) and therefore it can provide functional and molecular imaging information *in vivo*. Based on these endogenous contrasts, PAT has long been used for imaging and quantifying the overall levels of vascularization and oxygen saturation in tumors[1–5] (Fig. 1). PAT becomes particularly useful for detecting changes in association with angiogenesis and hypoxia accompanying malignant cancers.[7,8]

Myriad advances have been made in terms of developing exogenous contrast agents for PA molecular imaging application. As discussed earlier, in PAT, once the tissue is irradiated with laser, protein absorbs the optical energy to result in a thermoelastic expansion. So contrast enhancement in PA is highly-dependent on the optical properties of the exogenous agents. Several optically active materials have been suggested as a potential PA contrast agents, e.g. small molecule dyes, gold nanoparticles, single walled carbon nanotubes (SWNT), copper nanoparticles etc.[2,9,10] The prerequisite features to provide strong PA contrast enhancement beyond the endogenous blood signal include defined *in vivo* properties and biocompatibility, stability during circulatory transit and imaging, and tolerance to sterilization processing. These agents can be classified in two broad categories: small molecule-based and nanoparticulate based, Fig. 1. The examples of small molecular agents include near infrared dyes (e.g. cyanine dyes) and other optically active agents (e.g. porphyrins). The examples of nanoparticulate contrast agents include several different classes of gold nanoparticles of varying sizes, morphology (spherical versus rod, cages); carbon nanotubes (SWNTs), hybrid nanoparticles (gold coated SWNTs) and copper nanoparticles.

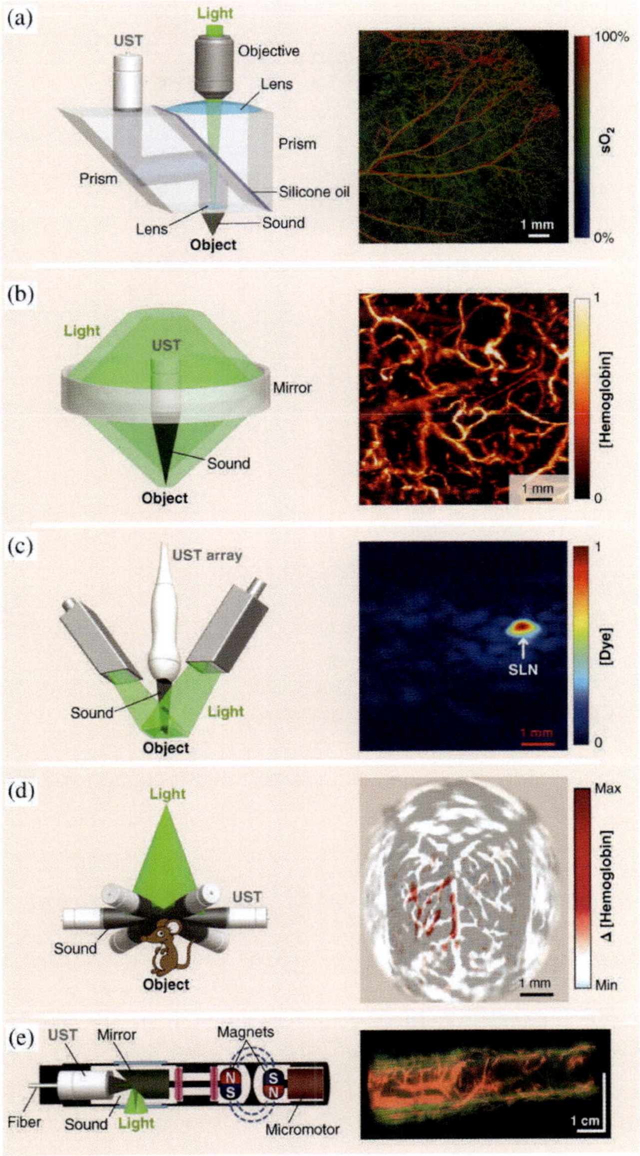

Fig. 1. Capabilities of PAT: (a) OR-PAM of sO2 in a mouse ear. (b) AR-PAM of normalized total hemoglobin concentration, [hemoglobin], in a human palm. (c) Linear-array PACT of normalized Methylene Blue concentration, [dye], in a rat sentinel lymph node (SLN). (d) Circular-array PACT of cerebral hemodynamic changes, Δ[hemoglobin], in response to one-sided whisker stimulation in a rat. (e) PAE of a rabbit esophagus and adjacent internal organs, including the trachea and lung. Ultrasonic transducer (UST). Reproduced with permission from Ref. 2.

Optical-resolution PAM: OR-PAM; acoustic-resolution PAM: AR-PAM; photoacoustic computed tomography: PACT; photoacoustic endoscopy (PAE).

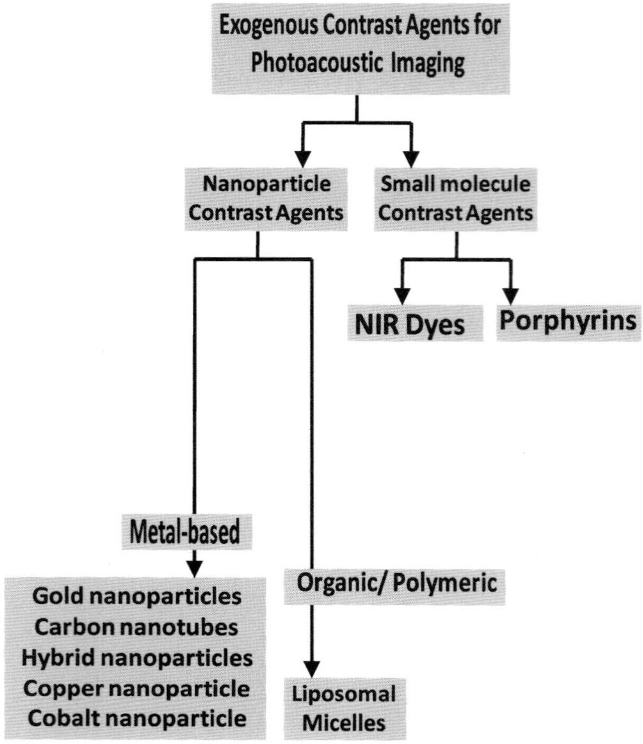

Fig. 2. Classification of exogenous contrast agents for PAI.

1.1. *Exogenous contrast agents based on metallic nanoparticles*

1.2. *Gold nanoparticles*

Gold nanoparticles exhibit a phenomenon known as localized surface plasmon resonance (LSPR) due to their inherent abilities to absorb and scatter light at specific wavelengths. Gold's SPR can be broadly tuned by changing their morphology, shape and surface coating. Owing to these properties, gold is an attractive choice as an optical agent. Nanoparticles of gold are easily excitable in the near-infrared (NIR) range ($\lambda = 650$–900 nm). This range is within the "optical transmission window" of biological tissues, which allows for deeper light penetration, lower autofluorescence, and reduced light scattering. In addition to their strong optical absorption, another major advantage of gold particles is their resistance to photo-bleaching. Small molecule fluorophores (excitable in the NIR range using single- or two-photon excitation) are often suffered by the intrinsic loss of optical properties due to photo-bleaching.[11–13] Various possibilities with gold nanoparticles has been explored, which are discussed in detail below.

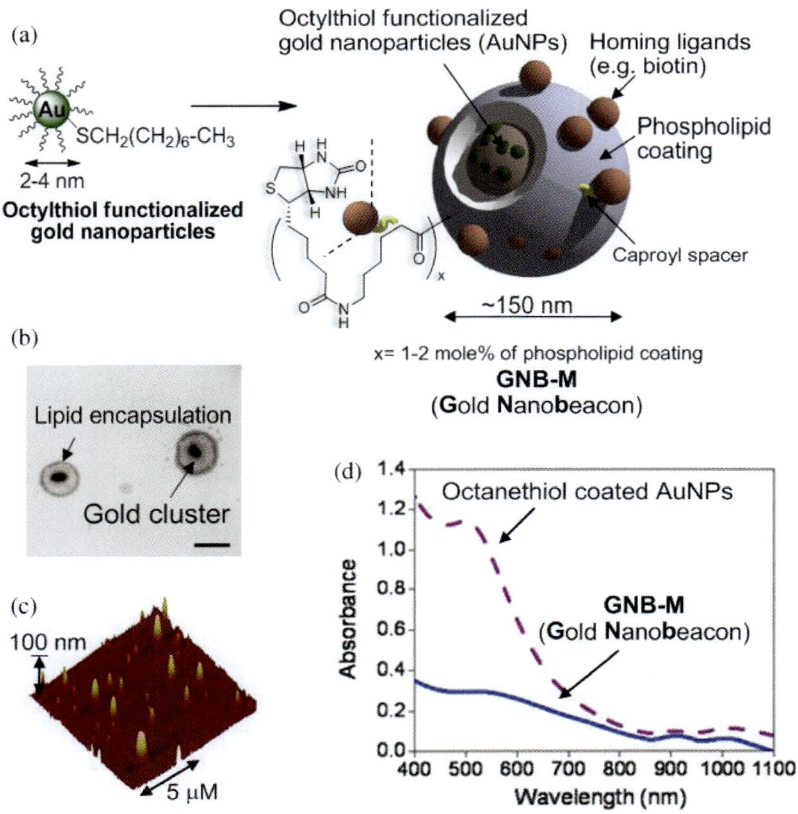

Fig. 3. (Color online) Gold nanobeacons: (a) Preparation of gold nanobeacons from octanethiol-functionalized gold nanoparticles (AuNPs). $x = 1–2$ mol% phospholipid coating. (b) TEM image of gold nanobeacons (drop deposited over nickel grid, 1% uranyl acetate; scale bar: 100 nm). (c) AFM image of gold nanobeacons. Average height H_{av}/nm = (10151) nm. (d) UV/Vis spectroscopic profile. Solid blue line: gold nanobeacons; purple dashed line: octanethiol-coated AuNPs. Spectra are not normalized. Reproduced with permission from Ref. 14.

Gold nanobeacons: Gold nanobeacons (GNB) were developed by entrapping multiple copies of small gold nanoparticles (spherical, 2–4 nm) within a larger colloidal particle, encapsulated with a biocompatible amphiphilic agent.[14–17] It was presumed that the entrapment of high payload of small gold particles will greatly amplify the signal for each binding event, effectively similar to a larger single gold particle, without exceeding the renal elimination threshold size (<10 nm).

The first generation gold nanobeacon (GNB_1) were synthesized by suspending octanethiol-coated spherical gold nanoparticles (AuNPs, 2 w/v%,) in vegetable oil (20 vol%) matrix followed by homogenization of the mixture with phospholipid surfactants mixture (2 w/v%), Fig. 3. The phospholipids mixture was comprised of phosphatidylcholine (PC, 91 mol%) and cholesterol (8 mol%) and included biotin-caproyl-PE (1 mol%) or a phospholipid anchored homing ligand (0.3 to 1.0 mole %) at the equimolar expense of PC dependent on the biological target under question. The GNB_1 particles were 154 ± 10 nm with polydispersity and zeta potential of 0.08 ± 0.03 and −47 ± 7 mV, respectively (Brookhaven Instrument Co.). Gold content, determined by ICP-MS, was 1080 μg/g of the 20% colloid suspension. UV-vis spectroscopy confirmed the absorbance at ~520 nm and in the near-infrared (NIR) window (~900 nm). These particles exhibited remarkable stability as studied by observing their particle size and zeta potential over time. These parameters only varied <5% over 100 days when stored at 4°C under inert atmosphere in sealed serum vials.

These agents were intended for molecular imaging application. However, high contrast in the presence of blood is a requisite for GNB_1 for imaging intraluminal microthrombus associated with ruptured atherosclerotic plaques. So, the signal sensitivity of GNB was compared with blood *in vitro*. The PA signal from a 20% GNB suspension mixed with blood (rat blood; 1:1) within Tygon® tubing (I.D. 250 μm, O.D. 500 μm) was 15 times higher than the pure blood.[14] The signal sensitivity was monitored over the entire 740–820 nm NIR window, demonstrating the average PA signal enhancement by GNB over blood more than 10-fold (Fig. 4). This confirmed the probability of using these agents for enhanced blood pool imaging. However, it was realized that the inherent contrast from endogenous agents (i.e. hemoglobin) alone or amplified by other exogenous agents (i.e. methylene blue), is usually adequate for such clinical need.

For targeting fibrin in the thrombus, GNB_1 and the control nanobeacons (i.e. presenting no gold) were pre-targeted using a high avidity antifibrin antibody. Nanoparticles were homed to artificially prepared acellular fibrin clot phantoms with a three-step pre-targeting approach by the classic avidin-biotin approach. A well-characterized biotinylated anti-human fibrin-specific monoclonal antibody (NIB5F3) was used.[18] The clots were made at the bottom of the low density polyethylene (LDPE) tubes (~1 cc volume, I. D. ~6 mm) and treated with biotinylated-GNB and control (targeted, no gold nanobeacon). Cross-sectional PAT images of the LDPE tubes are shown in Figs. 4(e) and 4(f). A curved array PAT system[19] was used to scan these phantoms with an 800 nm wavelength laser. As prominent from these images, clots targeted with biotinylated GNB_1 resulted in very high contrast in comparison to the control clot.

The cross-sectional PAT images were obtained using a PA breast scanner system,[20] of the same control and targeted plasma clots, respectively,

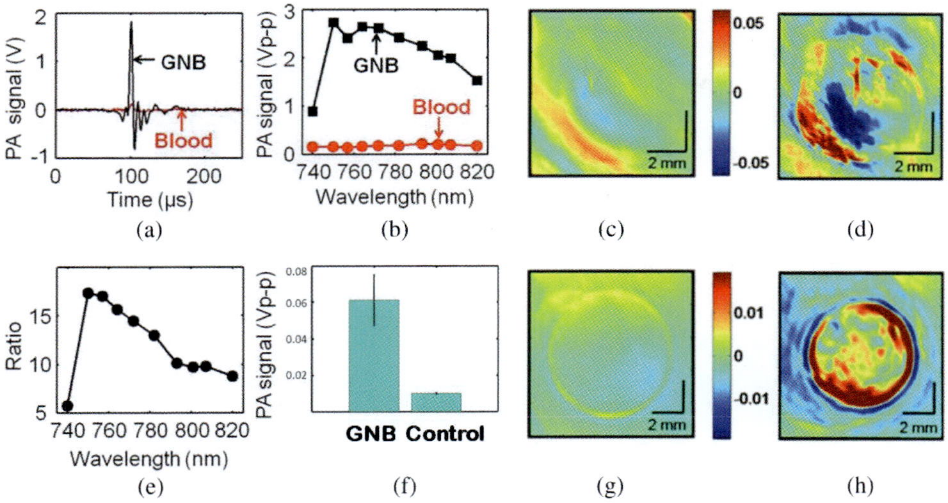

Fig. 4. (a) PA signals generated from a Tygon® tube (I.D. 250 μm, O.D. 500 μm) filled with GNB and rat blood. The excitation light is of 764 nm wavelength. (b) PA spectrum of GNB and blood over a 740–820 nm range of NIR wavelengths. (c) Ratio of the peak-to-peak PA signal amplitudes generated from GNB to those of blood. (d) PA signal to noise ratio for the GNB and the control (no gold). Cross-sectional PA image of a low density polyethylene (LDPE) tube (~1cc volume, I.D. ~6 mm) filled with plasma clot: (e) control, (f) targeted with GNB using a curved array PA system ($\lambda = 800$ nm). (g) Control, (h) targeted with GNB using a PA breast scanner system ($\lambda = 532$ nm). Reproduced with permission from Ref. 14.

Figs. 4(g) and 4(h).[14] Similar to the previous experiments, the targeted plasma clot was clearly distinguishable (Fig. 4(f)) in the PAT image, whereas the control clot image was not appreciable (Fig. 4(e)). Analytical support of the targeted clots was made by using inductively couple plasmon resonance (ICP-MS) spectrometry. The total gold content of the clots with targeted GNB_1, non-targeted GNB_1 and targeted non-metallic nanobeacons were found to be 47 μg/g, not detected (ND, <0.02 μg/g) and ND respectively, by ICP-MS. This experiment, for the first time, demonstrated that PAI can be applied for specifically targeting biological epitopes.

Gold nanobeacons with concentrating nanorods: Optical properties of gold can further be tuned toward NIR absorption by manipulating their shape and morphology. It is well-known that elongated or rod-shaped gold nanoparticles exhibits two surface plasmon (SP) bands corresponding to the transverse and longitudinal SP bands in the visible ($\lambda = 520$ nm) and the NIR ($\lambda = 900$ nm) regions, respectively.[11–13] This distinct optical property has been exploited for fluorescence, light scattering and two-photon luminescence imaging due to its intrinsic large extinction coefficient of longitudinal band.

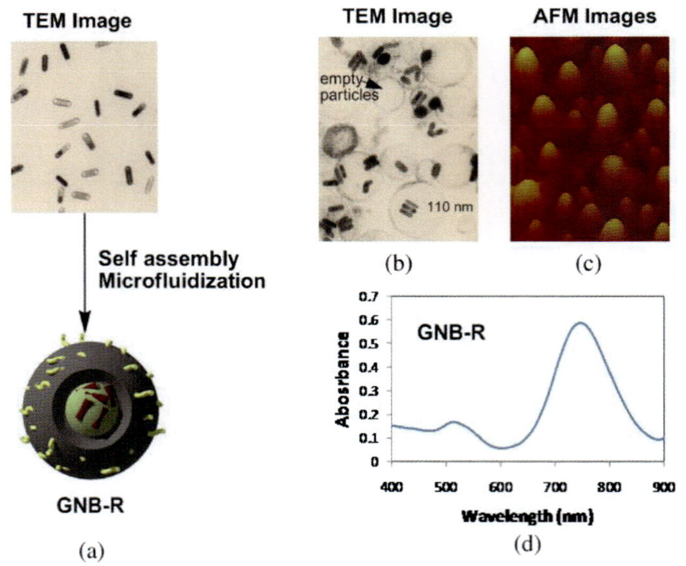

Fig. 5. (a) TEM image of gold nanorods drop deposited over Ni-grid; (b) TEM images of GNB_R drop deposited over Ni-grid; (c) AFM images of GNB_R drop deposited over glass; (d) UV-vis spectrum of gold nanorods showing LSPR and TSPR bands at 750 nm and 530 nm, respectively. Reproduced with permission from Ref. 16.

It is well-known that the longitudinal SP bands are very sensitive to the aggregation of gold nanorods and the peak points of the SP bands predominantly depend on the anisotropic shape of each gold nanorod the aggregation of rods generates isotropic (random) coupling of SP oscillations, and produces in dramatic changes of the longitudinal SP bands.

The self-assembled gold nanobeacons (rods) [GNB_R] was based on the self-assembly of phospholipids in aqueous media to entrap multiple copies (i.e. hundreds) of gold atoms.[16] Parent gold nanorods (polymer coated GNR; transmission electron microscope axial diameter: 25 ± 5 nm, length: 80 ± 8 nm; lLSPR: 750 nm, lTSPR: 530 nm), Fig. 5(a) were dispersed in chloroform and premixed with vegetable oil (20% v/v almond oil) core matrix before encapsulated within phospholipids coating.

ICP-MS confirmed the encapsulation of ~127 gold metal atoms (8.12 μg of gold/g of 20% colloidal nanobeacon), markedly less than GNB_1. These particles had a nominal hydrodynamic diameter of 129 ± 7 nm as measured by dynamic light scattering (DLS). The polydispersity and zeta potential were measured as 0.06 ± 0.02 and −41 ± 12 mV (Brookhaven Instrument Co.), respectively. The dehydrated state diameter (D_{av}) and height parameters (H_{av}) of the GNB_R were 105 ± 28 nm and 60 ± 24 nm, as measured by transmission electron microscopy (TEM) and atomic force microscopy measurements (AFM), respectively (Figs. 5(b) and 5(c)). The encapsulation efficiency of the gold nanorods was low per GNB_R, presumably due to the bigger size of the gold nanorods. This also provided with less *in vitro* contrast (approximately three-fold with respect to whole blood). Moreover, the bigger size of gold nanorods would likely preclude bioelimination through a renal route in non-rodent mammals. Due to these obvious reasons, subsequent studies from the group mainly utilized spherical gold nanoparticles.

Sentinel Lymphnode Imaging: In the clinic today, breast cancer patients need to undergo invasive sentinel lymph node (SLNB) biopsy to exclude metastatic disease. However, the technique has reported been tricky and the recovery of lymph nodes (LNs) is less adequate than expected and often the tissue is missed.[21,22] The technique has frequently been associated with post biopsy complications e.g. local seroma formation, lymphedema, nerve injury and numbness of arms.[23] It is therefore crucial to develop non-invasive imaging methods for axillary staging for breast cancer patients in combination with minimally invasive percutaneous fine-needle biopsy (FNAB). PAI may have the potential to offer a marked advancement over invasive exploration and resection when combined with suitably designed contrast agent. It is hypothesized that contrast agents injected subcutaneously will migrate into lymphatics and travel to the sentinel LNs and can be detected at depths of 4 cm to 5 cm. In a series of studies, Pan *et al.* demonstrated that the lymphatic transport of the nanopartculate contrast agents was highly governed by size and mass.

Sentinel lymph node imaging was tested first in a rodent model.[15] A sagittal maximum amplitude projection (MAP) PA image of the axillary area showed with limited contrast that within 5 min GNB_l accumulated in the SLN. However, inorder to achieve a more robust sensitivity, further approaches were tested. Substantially a larger polymeric PA agent (GNB_P) with more gold per particle than GNB_l was synthesized.

GNB_P used a di-block co-polymer (polystyrene-b-polyacrylic acid) PS-*b*-PAA (Refs. 24–27) (Mn × 10^3: 0.8-*b*-29.3 PDI = 1.18, 0.0033 mmoles) to encapsulate octanethiol coated AuNPs (2w/v%). Polysorbate (sorbitan monolaureate (5 vol%) was used as a co-surfactant and microfluidized with PS-*b*-PAA dispersion (0.5 vol%) yielding a 10% colloidal suspension of nanoparticles (Fig. 6). GNB_P (289 ± 24 nm) had a narrow distribution (polydispersity index: 0.15 ± 0.04). The gold content of GNB_l (6120 metal atoms/particle) was enhanced to reach 71,493 gold metal atoms per GNB_P. GNB_Pproduced significantly higher PA signal *in vitro* than GNB_l (Ref. 15) but was unable to provide any signal enhancement at all *in vivo*, Fig. 6(d) So, despite a ten-fold increase in

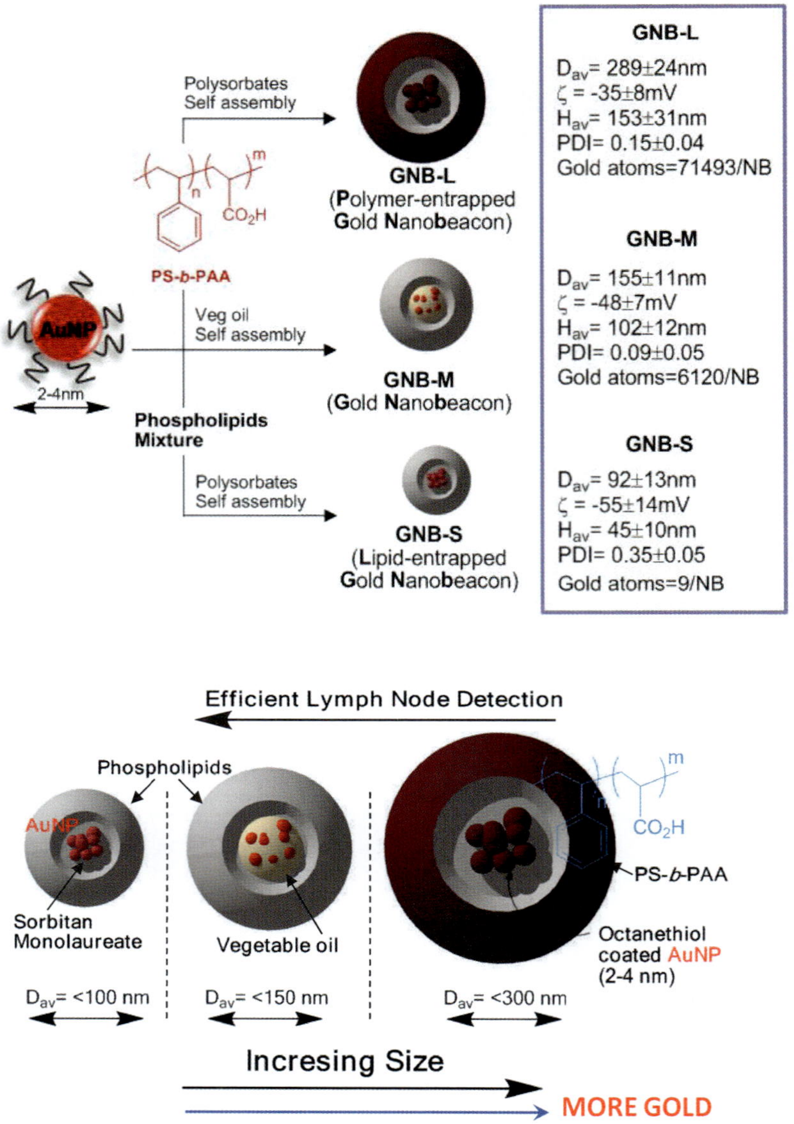

Fig. 6. Top: Synthesis of GNBs. (D_{av} = Number averaged (DLS); ζ = electrophoretic (zeta) potential; H_{av} = average height (AFM); PDI = polydispersity index (DLS)). Bottom: a general scheme showing the effect of particle size in facile lymphatic distribution. Reproduced with permission from Ref. 15.

gold content GNB_P was ineffective for *in vivo* LN imaging. It was presumed that the larger size (and or mass) of GNB_P was the reason for the poor migration and uptake of the particles by the lymphatic system. To support this hypothesis, a third smaller gold nanobeacon (GNB_S), was produced and examined.[15]

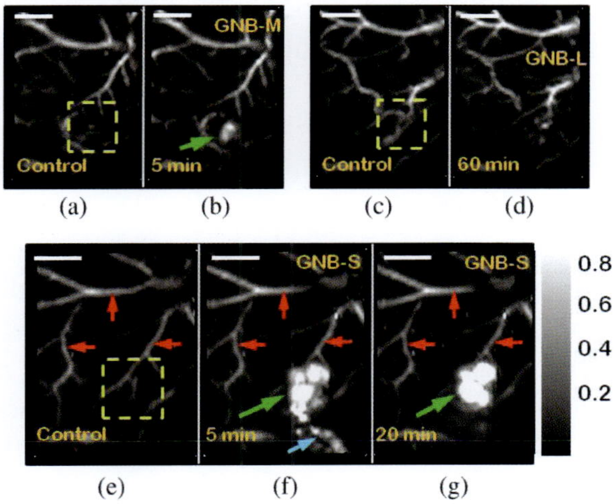

Fig. 7. *In vivo* non-invasive PAI of sentinel lymph nodes in rat (λ =767 nm). (A–G) Scale bar is 5 mm. 150 mL of nanobeacons were injected intradermally in all the cases. GNB_l: (a). Control PA image. (b) 5 min post-injection image of GNB_l (5 mM). GNB_P: (c). Control PA image. (d). Lymph node is not visible in a 60 min post-injection image of GNB_P (680 nM). GNB_S: (e). Sagittal maximum amplitude projection (MAP)[28] pre-injection control image: Bright parts represent optical absorption from blood vessels, marked with red arrows. (f). PA image (MAP) acquired 5 min after GNB_S injection (10 nM). SLNs are clearly visible, marked with green arrow. Lymphatic vessel is also visible, marked with blue arrow. (g). 20 min post-injection PA image. Reproduced with permission from Ref. 15.

GNB_S was prepared similar as GNB_l by retuning the core matrix to polysorbates (Fig. 6). Hydrodynamic particle size for GNB_S was 92 ± 12 nm with a polydispersity index of 0.35 ± 0.05. The zeta potential (ζ) was −35 ± 8 nm, confirming appropriate particle encapsulation. Gold metal atom (9 gold metal atoms per GNB_S) content per particle was determined to be much less in comparison to GNB_l and GNB_P.

The results with GNBS was dramatically improved. In the rat sentinel lymphnode model, the microvasculature at baseline was again clearly visible (marked with red arrows) while the lymph node, devoid of optical absorbers, was not seen. Figures 7(f) and 7(g) are two MAP images of the same axillary area 5 min and 20 min after the GNB-S injection, respectively, which show the SLN as a bright spot (green arrow), dramatically more robust than observed with other GNB particles. GNB_S was clearly tracked in transit within the lymphatic ductile and accumulated in the sentinel lymphnode at 5 min post injection. Within 10 min post GNB injection, the nanobeacons completed migration through the lymphatic ductile and with persistent and robust signal from the lymph node was appreciated over the next hour. The contrast of the sentinel LN to the surrounding blood vessel was calculated to be 9:1 (ratio of the peak-to-peak PA signal amplitude obtained from blood vessel and SLN) after GNB_S injection.[15]

In summary, the smaller GNB_S, with less gold particles per nanoparticle, were more effective in detecting LN in comparison to the larger GNB contrast agents.

Integrin-specific PAI of angiogenesis: Angiogenesis, the growth of new capillary blood vessels in the body, is an important natural process in the body used for healing and reproduction. Angiogenesis is an essential microanatomical component of tumor and cardiovascular progression that may be effective as a quantitative indicator of tumor or cardiovascular plaque progression or response to therapy.[29–31] Abnormal blood vessel growth, either excessive or insufficient, is now recognized as a "common denominator" underlying many deadly and debilitating conditions, including cancer, skin diseases, age-related blindness, diabetic ulcers, cardiovascular disease, stroke and many others. The list of diseases that have angiogenesis as an underlying mechanism grows longer every year. The $\alpha_v\beta_3$-integrin, a heterodimeric transmembrane glycoprotein, is expressed by numerous cell types, including endothelial cells,[32] macrophages,[33] platelets,[34] lymphocytes,[35] smooth muscle cells[36] and tumor cells.[37,38] Fortunately, the steric constraint of perfluorocarbon nanoparticles to the vasculature precludes significant interaction with non-endothelial integrin-expressing cells, which greatly enhances neovascular target specificity.[39] Although PAT generates high-resolution images of red blood cells in the microvasculature,[40–46] it is unable to discriminate immature neovasculature and mature microvessels. The adjunctive of an angiogenesis-specific molecular PAI agent not only resolved red-blood cell containing angiogenic vessels from similar but more mature microvessels, but also revealed the presence of neovascular sprouts forming and bridging between neovessels.

PA contrast agents targeting angiogenesis have been designed around integrin-targeted indocyanine green (ICG)-fluorescent-peptide conjugates.[47–51] Unfortunately these agents may easily diffuse from the neovascular vasculature, where numerous cell types including macrophages, platelets, lymphocytes, smooth muscle cells and tumor cells express $\alpha_v\beta_3$⁻ integrin. GNB_1 precluded microscopically detectable extravasation beyond the "leaky" neovasculature of the Matrigel™ plug. This observation was further supported by other reports using copolymer and nanoemulsion particles.[39,52] Such steric constraint with $\alpha_v\beta_3$-⁻GNB_1 nanoparticles greatly enhanced neovascular homing specificity, which will be essential for quantitative biomedical applications.

$\alpha_v\beta_3$-targeted GNB_1 was selected for study[17] for vascular constraint properties due to particle size (160 nm), as opposed to smaller nanobeacons.

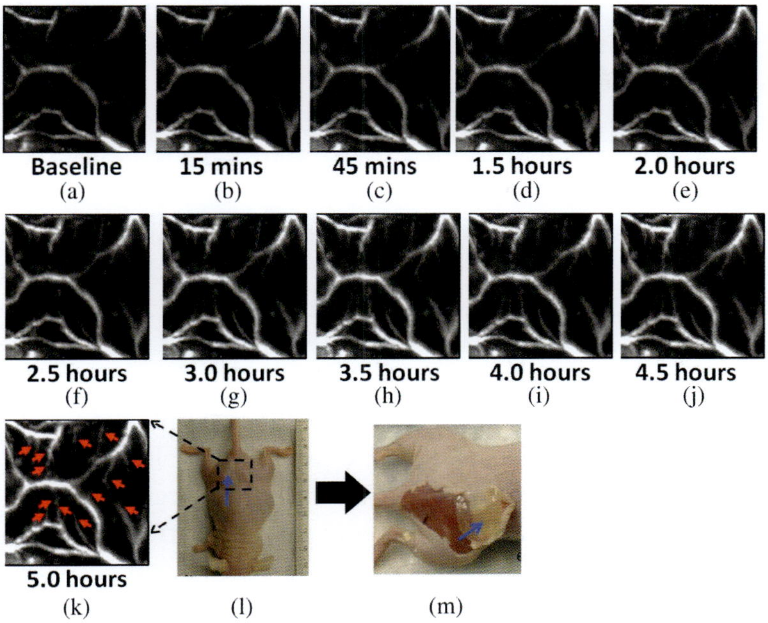

Fig. 8. (Color online) 0.75 mL of Matrigel™ was implanted subcutaneously in nude mouse. The mouse was imaged photoacoustically 8–20 days after Matrigel™ implantation. (a) Photoacoustic (PA) maximum amplitude projection (MAP) image of the dotted area. This is a control image. After the control image was taken targeted gold nanobeacons ($\alpha_v\beta_3$-GNB-M) were injected intravenously using the tail vein. In a time course study (B–K), PA images were acquired with an interval of approximately 1/2 h up to 5 h. (g) Three hour post-injection PA image. Red arrows point to the angiogenic sprout (not visible in A). (K) Five hour post-injection PA image. For all PA images $\lambda = 767$ nm, scale bar = 5 mm. (L) Digital photograph of a mouse implanted with Matrigel™ plug. Blue arrow points to the plug. The black dotted area was imaged. The smallest tick: 1 mm. (M) Digital photograph of the sacrificed mouse after all the image acquisition was over. The skin was removed to show the Matrigel™ plug (blue arrow). Reproduced with permission from Ref. 17.

$\alpha_v\beta_3$-targeted GNB_1 was produced by microfluidization as discussed before by incorporating an $\alpha_v\beta_3$-peptidomimetic antagonist. [53–55]

For the angiogenesis imaging study, a Matrigel™ plug model of angiogenesis was used (BD Biosciences, San Jose, California). Matrigel™ (750 μl) enriched with fibroblast growth factor-2 (500 nm/mL; Sigma Aldrich, St. Louis, MO) and heparin (64 U/mL) was implanted subcutaneously along the flank of mice. PAT imaging was performed prior to treatment and serially imaged over 5 h post-injection on days 16 and 17 post Matrigel™ implantation, Fig. 8. Animals were injected with: (1) $\alpha_v\beta_3$-gold nanobeacons

($\alpha_v\beta_3$-GNB$_1$, $n = 6$) $\alpha_v\beta_3$-nanobeacons without gold ($\alpha_v\beta_3$-NB) followed by $\alpha_v\beta_3$-GNB$_1$ (competitive blockage, $n = 2$) non-targeted gold nanobeacons (NT-GNB$_1$, $n = 3$) or saline ($n = 2$).

For the first time, non-invasive PAT imaging of angiogenesis was demonstrated (10 MHz). PA signal changes in the Matrigel™ plug were monitored serially over 5 h or more (Fig. 7). $\alpha_v\beta_3$-GNB$_1$ produced a 600% increase in signal in a Matrigel™ plug mouse model relative to the inherent pretreatment contrast from hemoglobin. Interestingly $\alpha_v\beta_3$-GNB$_1$ also allowed visualization of numerous angiogenic sprouts and bridges that were undetectable before contrast injection. Competitive inhibition of $\alpha_v\beta_3$-GNB$_1$ with $\alpha_v\beta_3$-NB blocked contrast enhancement to pretreatment levels. These developing, nascent neovessels were presumed immature and incomplete to support any further blood flow. PA images in the saline control animals showed no change in vascular anatomy over the same time course. Non-targeted GNB was seen to be passively accumulating in the tortuous neovascularity, but provided low (less than half) contrast enhancement of the targeted agent.

This experiment was also supported histologically. Microscopic studies further established that the vascular-constrained, rhodamine-labeled $\alpha_v\beta_3$-GNB$_1$ targeted specifically to immature neovasculature (PECAM$^+$, Tie-2$^-$) along the immediate tumor periphery, but not to nearby mature microvasculature (PECAM$^+$, Tie-2$^+$), Fig. 9. With this experiment, it was further demonstrated that PAT sensitively discriminated angiogenesis and microvasculature in presence of $\alpha_v\beta_3$-GNB$_{1.}$

Gold nanocages as PA contrast agents: A general problem arises when gold nanoparticles LSPR peaks need to be turned on to the NIR region to obtain maximum penetration depth *in vivo*. Such unmet need is due to the (i) soft tissue and blood attenuation of light is low[56] and (ii) traditional methods of keeping nanoparticles in their conventional spherical shape does not allow it to do so. Yonan Xia *et al.* embark upon this problem by developing a novel class of nanoparticles called gold nanocages (AuNC). These particles were specified by their hollow interiors, single-crystal ultrathin structure and porous walls.[57] Xia *et al.* demonstrated that these AuNCs exhibited wavelength plasticity and could be manipulated to any wavelength in the NIR region based on their modifications of the nanocage sizes and thickness. Additionally, due to their novel properties which included high absorption, low cytotoxicity and their ability to be easily conjugated to tumor-specific ligands, these AuNCs provided as an ideal contrast agents for PA imaging.[58–60]

The synthesis of the gold nanocages followed a galvanic replacement reaction using Ag solid as a template and $HAuCl_4$ as a reducing agent.

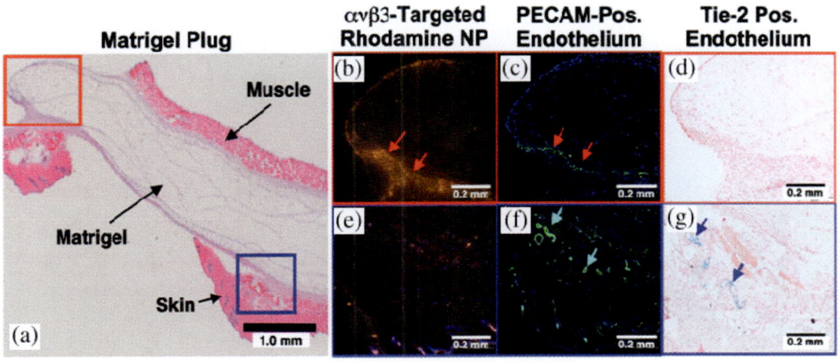

Fig. 9. (Color online) Microscopic examination of FGF Matrigel subcutaneous explant from FVB/N-TgN(TIE2LacZ)182Sato mice following injection (IV) of $\alpha_v\beta_3$-targeted rhodamine labelled GNB-M nanoparticles. These transgenic mice carry a beta-galactosidase reporter gene under the control of the murine Tek (Tie2) promoter. LacZ is expressed specifically in vascular endothelial cells in embryonic and adult mice. (a) H&E staining of the excised implant providing spatial orientation of the matrix with respect to skin and muscle. The red box region is further examined in figures b, c and d. The blue box region is studied in more detail in panels e, f and g. Panels B and E depict the accumulation of $\alpha_v\beta_3$-targeted rhodamine nanoparticles in the red and blue tissue regions, respectively. Note the brilliant and dense accumulation of NPs in panel B (red arrows) and little to no accumulation of particles in the panel E. Region. Panels C and F depict the staining of vascular endothelium for PECAM (CD34) in the red (c) and blue (f) regions of the matrigel plug. There was dense vascularity in both locations (Red arrows in C and turquoise arrows in F. Panels D and G depict the LacZ signal for beta-galactosidase under Tie2 promoter control. In panel D, no LacZ signal was appreciated, reflecting a paucity of mature microvessels. In contradistinction, there is strong LacZ signal in panel G. These results indicated that the $\alpha_v\beta_3$-GNB nanoparticles were specifically targeted to angiogenic endothelial cells (PECAM + /Tie-2 -) and not to more mature microvessels, which were PECAM +, Tie-2 +. The data corroborated that PAT imaging with $\alpha_v\beta_3$-targeted GNB specifically distinguished and enhanced the angiogenic neovasculature from new, but more matured and differentiated microvessels. Reproduced with permission from Ref. 17.

Electrochemical potential difference of two metals helped to form the growth of Au atoms.

$$3Ag\ (s) + AuCl_4^-(aq) \text{ --> } Au\ (s) + 3Ag^+\ (aq) + 4Cl^-\ (aq)$$

Ag nanocubes were synthesized using $AgNO_3$ or CF_3COOAg as a precursor to Ag, and a polyol reduction by using ethylene glycol as a solvent. A reducing agent further a capping agent, poly(vinyl pyrrolidone) (PVP) further assisted the process. Subsequently, the addition of catalytic amounts of NaHS and or Cl^- ions imparted rapid nucleation of single-crystal seeds while simultaneously, due to oxidative etching, discarded twinned seeds and

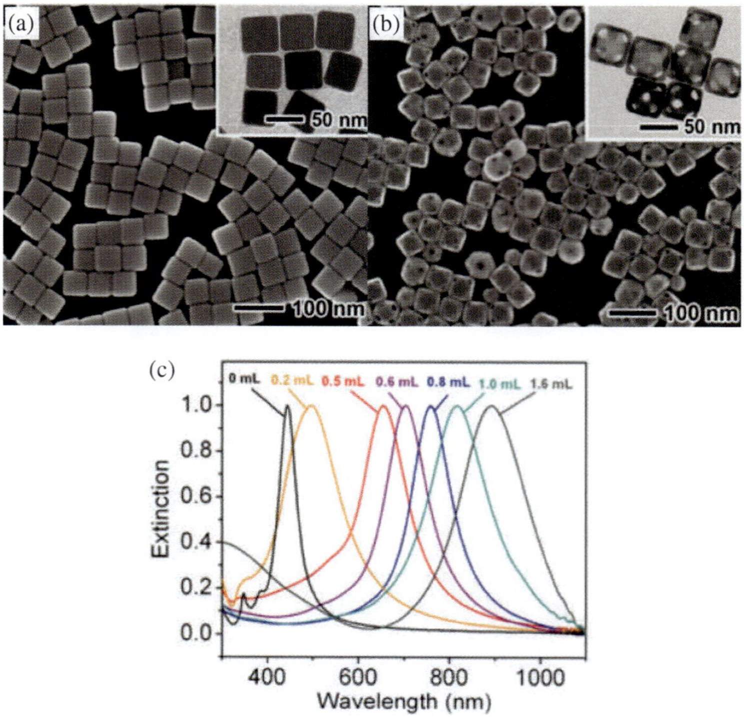

Fig. 10. SEM images of (a) Ag nanocubes and (b) AuNCs. The inset shows the corresponding TEM images of the same sample. (c) UV-vis spectra of the samples obtained by titrating Ag nanocubes with different volumes of 0.1 mM $HAuCl_4$ solution.

ultimately formed single-crystal Ag nanocubes. Titration of these Ag nanocubes with $HAuCl_4$ was followed at 100°C in order to ensure maximal growth of thin layered Au on the surface of the Ag crystals and to evade precipitation of AgCl. Interestingly, the formation of AuNCs can be distinctively observed as the extinction spectra of the titration of Ag nanocubes changes. With the addition of more $HAuCl_4$ epitaxial growth of AuNCs were promoted and the LSPR peaks shifted from the visible to the NIR region. Characterization of these particles was subsequently performed by obtaining SEM and TEM images as shown in Fig. 10.

Imaging of cerebral cortex via AuNCs: Cerebral cortex of small animals have been imaged with PA techniques.[61–63] Combining the properties of the leaky vasculature and dysfunctional lymphatic system of a tumor, the small sizes of the nanoparticles, via enhanced permeability and retention (EPR) effect, AuNCs are able to accumulate at the site of a tumor, which is highly effective in brain tumor detection. Another property that adds to the arsenal of properties of these particles is the denial of access of these particles to healthy vessels

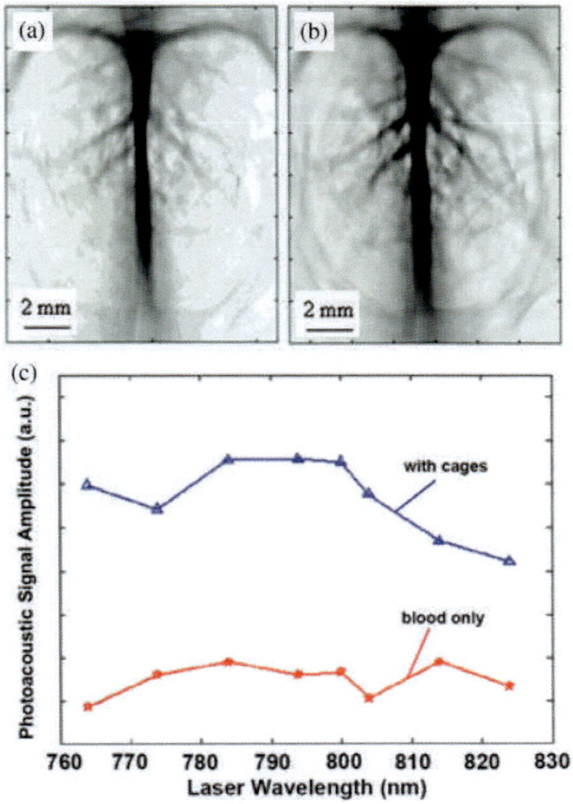

Fig. 11. (a) The measured PA signal amplitude generated with and without AuNCs in rat blood at several wavelengths. Non-invasive PA imaging of a rat's cerebral cortex, (b) before the injection of AuNCs and (c) about 2 h after the final injection of nanocages, which is the peak enhancement point. Reproduced with permission from (65), copyright 2007 American Chemical Society.

in the blood brain barrier, thus allowing accurate imaging of brain tumors with enhanced efficacy and contrast without ambiguities of differentiating healthy and tumorigenic cells. Wang *et al.*[64] initially studied the imaging of the vasculature of a rat brain using gold nanoshells with >100 nm diameters, but Xia's method of utilizing smaller sized AuNCs (<50 nm) and its higher optical properties exhibited much better feasibility towards PA imaging of a rat's cerebral cortex.[65] Figure 11 shows the comparative results of an *ex vivo* PA study produced from pure rat blood and a mixture of AuNC-rat blood, and an *in vivo* study of injections of AuNCs in the cerebral cortex of a rat model.

Sentinel lymph node mapping via AuNCs: As discussed previously, sentinel lymphnode imaging is becoming an alternative way to non-invasively

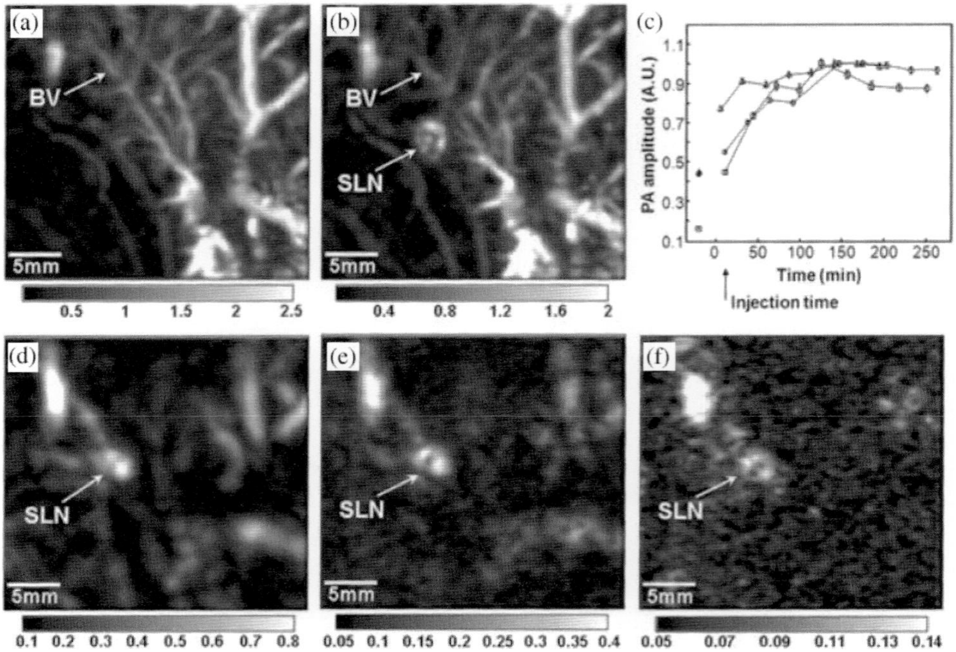

Fig. 12. PA images of the axillary region of a rat taken (a) before and (b) 28 min after the injection of AuNCs. (c) The changes of PA signal amplitude as a function of the post-injection time. After the injection, PA signals increased with time, which means gradual accumulations of the nanocages. (d–f) Depth capability of SLN mapping with AuNCs. The PA images were acquired after the injection of nanocages for: (d) 126 min with a total imaging depth of 10 mm by placing a layer of chicken breast tissue on the axillary region; (e) 165 min with a total imaging depth of 21 mm by adding another layer of chicken breast tissue; and (f) 226 min with a total imaging depth of 33 mm by using three layers of chicken breast tissue. The bars represent the optical absorption. BV, blood vessel. SLN, sentinel lymph node. Reproduced with permission from (72), copyright 2009 American Chemical Society.

stage breast cancer.[66–71] Similar to the GNB system, Xia *et al.* also demonstrated that AuNCs can be employed for sentinel lymph nodes mapping,[72] giving high spatial resolution, low cost, non-toxicity and prominent imaging depth (33 mm), Fig. 12.

Imaging of melanoma with AuNCs: AuNC was conjugated to [Nle^4,D-Phe^7]-*α*-melanocyte-stimulating hormone ([Nle^4,D-Phe^7]-*α*-MSH), Ref. 73 for binding *α*-MSH receptors, over expressed on melanomas. PA signals were increased significantly with time as shown in Figs. 13 and 14 for the targeted [Nle^4,D-Phe^7]-*α*-MSH-AuNCs than the non-targeted AuNC, giving almost a 300% enhancement in signal. These results indicated that the uptake of the AuNCs can be actively targeted to melanoma cells.

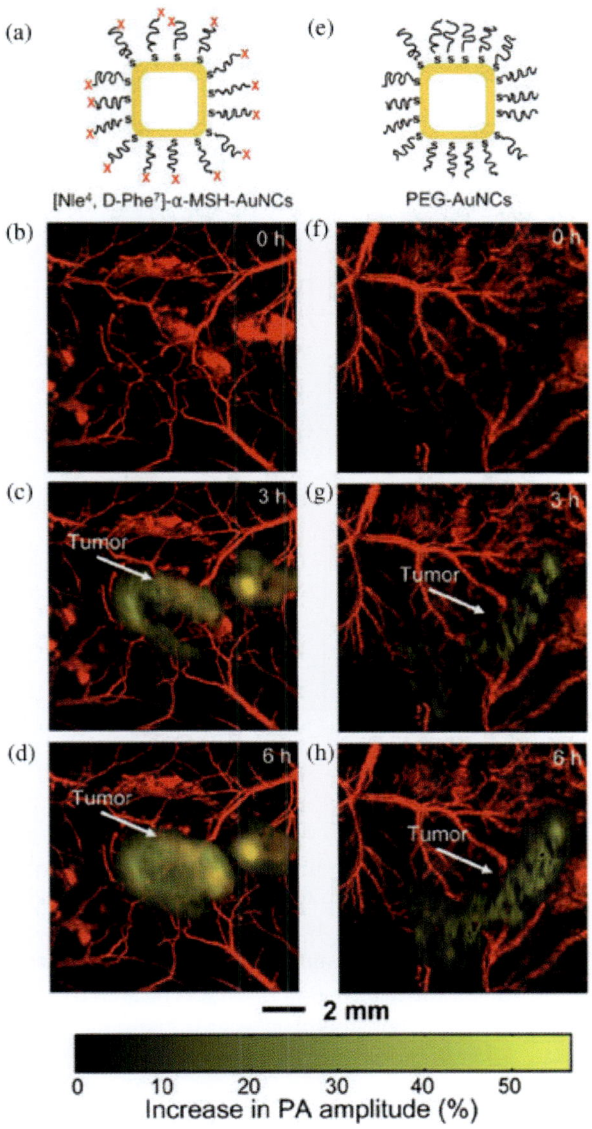

Fig. 13. *In vivo* non-invasive PA time-course coronal MAP images of B16 melanomas using [Nle^4, D-Phe^7]-a-MSH- and PEG-AuNCs. (a, e) a schematic of the [Nle^4, D-Phe^7]-a-MSH- and PEG-AuNCs. Time-course PA images of the B16 melanomas after intravenous (i.v.) injection with 100 μl of 10 nM (b–d) [Nle^4, D-Phe^7]-a-MSH- and (f–h) PEG-AuNCs through the tail vein. The background vasculature images were obtained using the PA microscope at 570 nm (ultrasonic frequency = 50 MHz), and the melanoma images were obtained using the PA macroscope at 778 nm (ultrasonic frequency = 10 MHz). reproduced with permission from (74), copyright 2010 American Chemical Society.

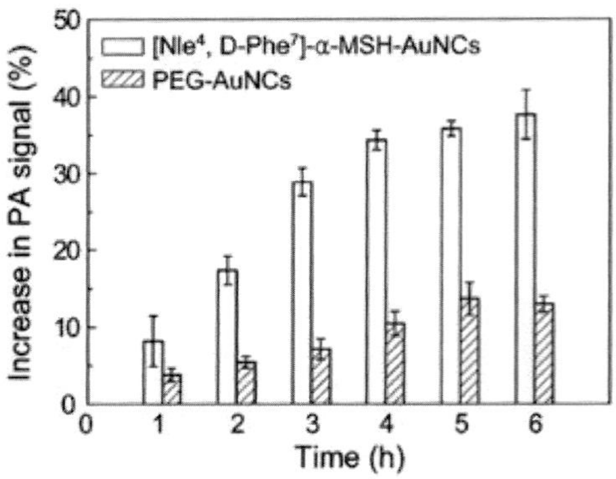

(a)

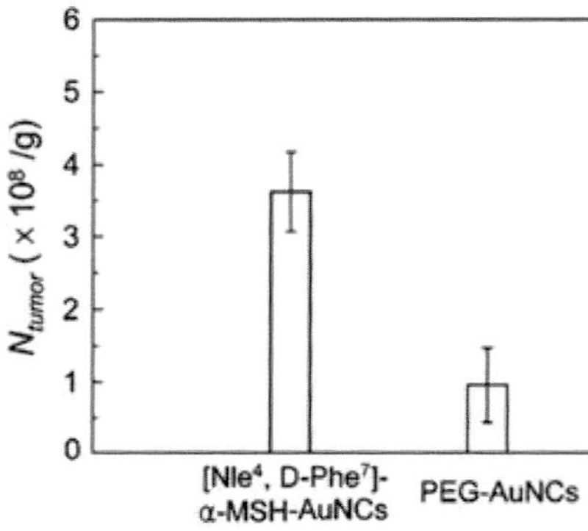

(b)

Fig. 14. (a) Increase of PA amplitude in the melanoma tumors after *i.v.* injection of [Nle^4, D-Phe^7]-a-MSH- and PEG-AuNCs (n = 4 mice for each group), respectively, for different periods of time. The PA signals increased up to 38 ± 6 % for [Nle^4, D-Phe^7]-a-MSH-AuNCs while the maximum signal increase only reached 13 ± 2% for PEG-AuNCs at a post-injection time of 6 h ($p < 0.0001$). (b) The average number of AuNCs accumulated in the melanomas dissected at 6 h post injection for the two types of AuNCs as measured by ICP-MS. Here N_{tumor} denotes the number of AuNCs per unit tumor mass (g). Reproduced with permission from (73), copyright 2010 American Chemical Society.

2.2. *Carbon nanotubes for advanced contrast agents*

There has been a growing interest in carbon nanotubes (CNT) over the past years due to their ability to optically absorb within wide ranges of wavelengths. Owing to their unique optical properties within the NIR range, they can act as ideal contrast agents for PA imaging. Many studies were devoted to CNTs to explore their potential application for PA imaging. Zharov *et al.*[74] initially demonstrated CNTs ability to act as an exogenous contrast agents for PA. Many other studies followed including targeting peptide conjugated CNT to tumor vasculatures in mice.[75] Modifications of CNT chemistry and coating for enhancement of contrast[76–78] were thoroughly studied followed by their use to map SLNs and identification of tumor cells.[78–83] CNTs also found applications in therapeutic hyperthermia and demonstrated its ability to eliminate tumor and bacterial cells.[84–92]

2.3. Carbon nanotube hybrids

An unmet need regarding CNTs is their relatively low NIR absorption coefficient compared to gold nanoparticles.[91] As a result of this, very high local concentrations of CNTs are needed for sensitive PA imaging.[77,78] Usually single-walled nanotubes (SWNT) present a very high surface area (~1300 m^2/g) allowing optimal interaction with optical radiation. To augment this, a hybrid SWNT (core) and gold nanoparticles (layer) can be prepared (SWNT-Au)[78], Fig. 15b. The hybrid SWNT-Au presented the advantages and potential for bioconjugation and biocompatibility of gold while delineating high optical NIR absorbance. Compared to the SWNT without the gold layer, the plasmon responses of the SWNT-Au hybrids exhibited greatly enhanced absorbance and minimal cytotoxicity.

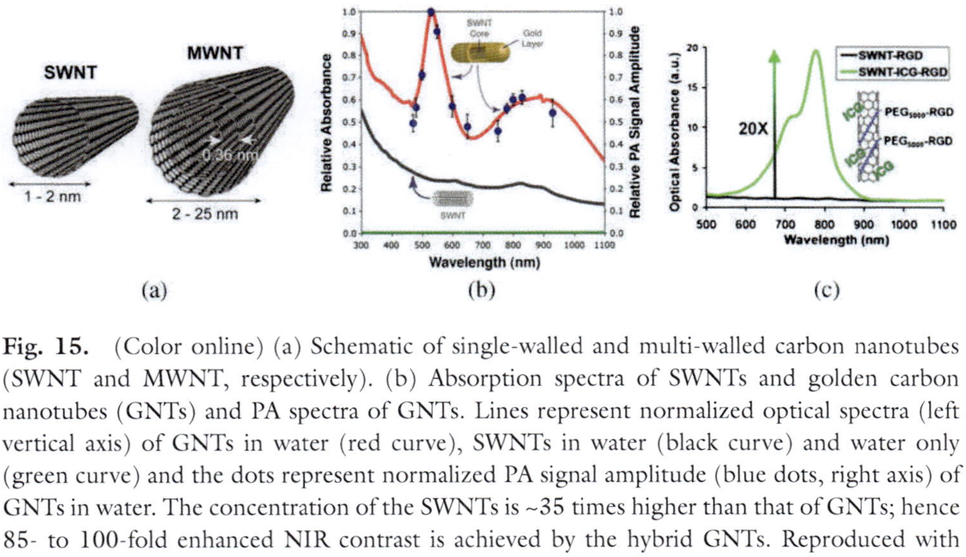

Fig. 15. (Color online) (a) Schematic of single-walled and multi-walled carbon nanotubes (SWNT and MWNT, respectively). (b) Absorption spectra of SWNTs and golden carbon nanotubes (GNTs) and PA spectra of GNTs. Lines represent normalized optical spectra (left vertical axis) of GNTs in water (red curve), SWNTs in water (black curve) and water only (green curve) and the dots represent normalized PA signal amplitude (blue dots, right axis) of GNTs in water. The concentration of the SWNTs is ~35 times higher than that of GNTs; hence 85- to 100-fold enhanced NIR contrast is achieved by the hybrid GNTs. Reproduced with permission from Kim *et al.* (78). (c) Optical absorption spectrum of SWNT-RGD (black curve) and indocyanine green-enhanced SWNT-RDG (SWNT-ICG-RGD, green curve). The optical absorbance spectrum of plain SWNT-RGD is relatively flat with slight gradual absorption decrease as the wavelength increase. However, by attaching a large number of ICG molecules to the SWNT surface, a 20-fold increase in optical absorption results at 780 nm. Reproduced with permission from de la Zerda *et al.*[76]

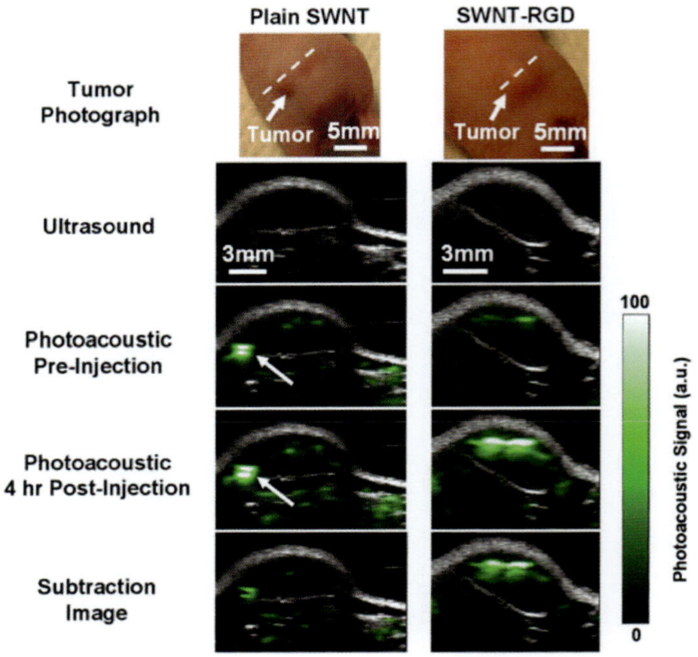

Fig. 16. (Color online) Single-walled carbon nanotube (SWNT) argine-glyice-aspartic acid (SWNT-RGD) tumor targeting in living mice. Ultrasound (gray) and PA (green) images of a vertical slice (white dotted line) through the tumors of mice injected with SWNT-RGD (right column) and control plain SWNTs (left column). Subtraction images were calculated as 4 h post-injection minus pre-injection to remove tissue background signal from the PA image. Mice injected with SWNT-RGD showed an averaged 7-fold PA signal increase in the tumor over mice injected with control untargeted SWNTs. The high PA signal in the mouse injected with plain SWNTs (indicated by the white arrow) is not seen in the subtraction image, suggesting that it is due to a large blood vessel and not SWNTs. Reproduced with permission from de la Zerda *et al.* (94)

As an alternate approach, the low absorption properties of SWNT in the NIR region could be modified by presenting small molecule dyes on the high surface area of the nanotubes. Liu *et al.*[93] was able to demonstrate this by loading indocyanine green (ICG) molecules via pi-pi stacking bonds onto SWNTs through its hydrophobic benzene rings. This hybrid SWNT with ICG exhibited a 20-fold increase in optical absorption, Fig. 15c.

Applications of carbon nanotubes geared towards targetedPAI: For targeting vascular integrins $\alpha_v\beta_3$, SWNTs have been conjugated with arginylglycylaspartic acid (RGD peptide) via PL-PEG$_{5400}$.[94] SWNT-RGD were injected in mice bearing U87 gliobastoma xenograft tumors and showed a ~7-fold increase in PA signal in the tumor compared to that of the untargeted SWNTs, Fig. 16.

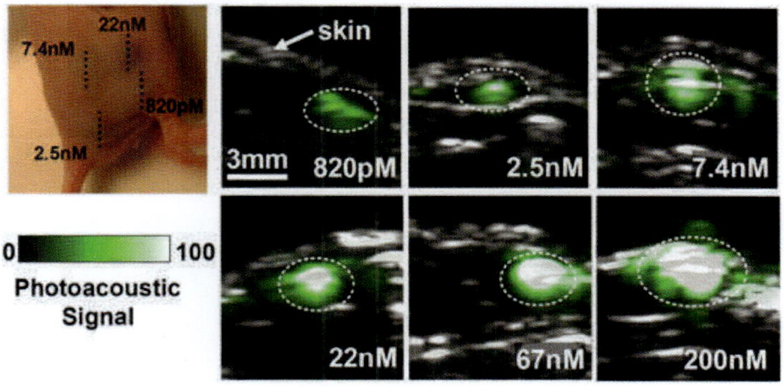

Fig. 17. (Color online) PA detection of single-walled carbon nanotube indocyanine green (SWNT-ICG) in living mice. Vertical slices of ultrasound images (gray) and PA images (green) of mice injected subcutaneously with SWNT-ICG-RGD at concentrations of 0.82–200 nm (dotted black line). The white dotted lines on the images illustrate the approximate edges of each inclusion. Quantitative analysis of the images estimated that 170 pm of SWNT-ICG-RGD gives the equivalent PA signal as the tissue background. Reproduced with permission from de la Zerda *et al.* (76)

To further enhance SWNTs' sensitivity and signal, small molecule optical dyes, (i.e. ICG) were conjugated. *In vitro* testing of SWNT-ICG-RGD molecules for PA imaging have shown increased stability in serum. Results showed increase in sensitivity by over 300-fold compared to SWNT-RGD which suggested the higher optical activity and ability of the ICG molecules to shift the laser excitation wavelength to 780 nm from 690 nm. Their ability to target $\alpha_v\beta_3$ integrinw in angiogenic tumors were also demonstrated *in vivo*. However, despite the photo resistive nature of the native SWNT molecule, photo bleaching was observed in SWNT-ICG-RGD, Fig. 17.[76]

2.4. *Hybrid magnetic-gold nanoparticles*

One major problem in the clinic is the absence of immediate tools to detect the rapid and metastatic pervasion of cancer. This is typically achieved via an assay that detects levels of circulating tumor cells (CTCs). However, such assay has its precincts due to its limited sensitivity, sampling errors and high sample volume requirement (patient blood volume). Development of methods to assay and characterize CTCs is emerging as an important area of Nanomedicine research. A multiplex targeting strategy was shown to synthesize a conjugated nanoparticle consisting of a magnetic (MNP) and gold nanoparticle (GNP), which have the NIR properties for detection of two colors at laser wavelengths of 639 nm and 900 nm, Figs. 18A and 18B. These

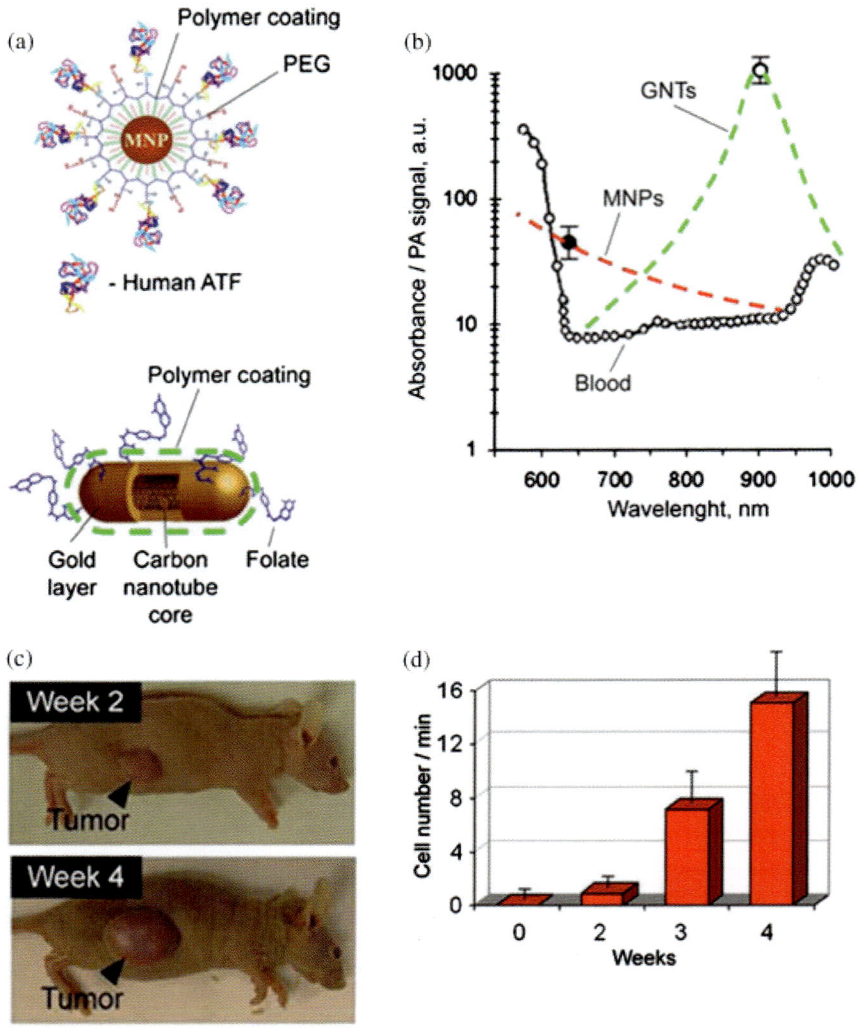

Fig. 18. *In vivo* multiplex two-color PA detection of CTCs. (a) The 10 nm magnetic NPs (MNPs) coated with amphiphilic triblock polymers, polyethylene glycol (PEG) and the amino-terminal fragment of urokinase plasminogen activator (ATF). The 12 × 98 nm GNTs coated with PEG and folic acid. (b) PA spectra of ~70 μm veins in mouse ear (open circles). Absorption spectra of the MNPs and GNTs (dashed curves) normalized to PA signals from CTC labeled with MNPs (black cricle) and GNTs (open circle). (c) The size of the primary breast cancer xenografts at different time stages of tumor development. (d) Average rate of CTCs in mouse ear vein. Reproduced with permission from Galanzha *et al.* (77)

conjugated nanoparticles were specifically synthesized to evade normal blood cells, target urokinase plasminogen activator (ATF) and folate receptors.[77] PA signals indicated that the optimal NP concentration and solution volume were 10^9 NPs/ml for GNTs and 10–20 μl, respectively. Also, there were no detectable PA signals in mouse ear microvessels above blood background, which suggested that PA signals were able to differentiate between bound and unbound NPs, supporting its specificity. In order to detect CTCs in a primary tumor, PA flow cytometry (PAFC) was used to measure the progressive increase in the number of CTCs over a course of four weeks, and showed a high correlation with respect to the progression of the tumor (Figs. 18C and 18D). This '*in vivo* blood cancer test,' may become an applicable and potential measure for early PA diagnosis of primary tumors and prevention of metastasis.

3. Small Molecule-Based and Other Exogenous PA Agents

3.1. *NIR cyanine dyes for real-time sentinel lymph node imaging*

In clinic, SLN techniques use methylene blue dye or radioactive colloid preoperatively for the localization of the SLNs.[95–97] However, despite its widely used practice, radioactive colloids leads to very low lateral and negligible vertical resolution. This gives low precision and localization, which can accrue to potentially harmful effects when resecting the SLN during surgical operations. On the other hand, methylene blue stain gives only a sample of raw data that must be approximated for the localization and precision of SLN, having potential adverse effects. Additionally, allergic reactions are frequent with the blue stain.[98]

In order to address these problems a sub 20 nm "soft" polymeric nanoparticle has been synthesized by Pan *et al.*[99] with the aim to directly address safety issues with safety and facile lymphatic transport Fig. 19. Additionally, with the supplementation of real time imaging via PA imaging, detection and resection of SLN would be more efficient and would potentially decrease post-operative complications.

The one pot synthesis of the self-assembled nanoparticle involved mixing of polysterene-*b*-poly(acrylic acid) (PS_8-*b*-PAA_{400}, Mnx10^3, 0.8-b-33.0, polydispersity index: PDI = 1.18, 0.00033 mmol, 0.5 mole%) with a cosurfactant (sorbitan monoleate) to restrict the particle diameter within 20 nm. A near infrared cyanine dye (ADS832WS, λ_{ex} = 824 nm, 1.90×10^5 L mol^{-1} cm^{-1}) was incorporated as a means to act as a surrogate photoacoustic payload as shown in Fig. 19.

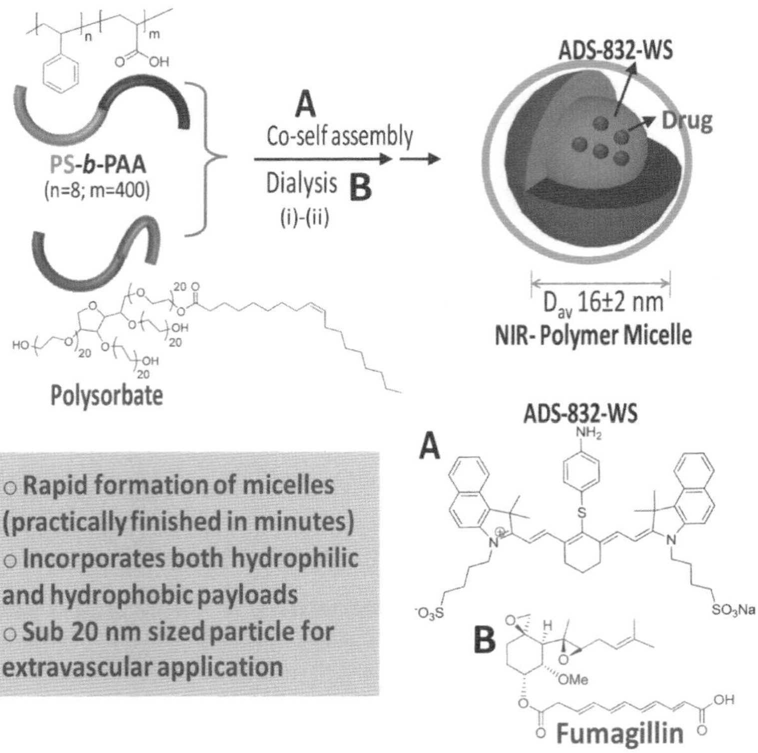

Fig. 19. Synthesis and characterization of theranostic polymeric micelles: (i) co-self assembly of amphiphilic diblock copolymer and sorbitan monooleate, ADS-832-WS (a) and/or fumagillin (b), sonication, 25°C, 1 min; (ii) dialysis 10 kDa cellulosic membrane, nanopure water (0.2 μM). Reproduced with permission from Pan *et al.* (99)

The hydrodynamic particle diameter (Da_v) were determined by light scattering method as 16 ± 2 nm with a polydispersity of 0.021. The anhydrous particle diameter was 12 ± 0.4 nm based on TEM images (uranyl acetate) and presented a height of 15 ± 0.4 nm based on AFM images. Due to the incorporation of the NIR dye, the particle was highly fluorescent and had confirmed absorbencies in the NIR range (750–860 nm). Additionally, these particles consisted of long shelf-life stability (~40 days at ambient temperatures). The loading efficiency of the NIR dye was determined to be ~97%, Fig. 20.

Intradermal injection of the nanoparticles in rats showed PA signal enhancement of SLN ~510% (Fig. 21) in comparison to the background. The signal enhancement was retained for 30 min at the SLNs and within 110 min, was almost completely drained. Such results were indicative the tiny nanoparticles maintained their structural integrity and migrated out of the lymph

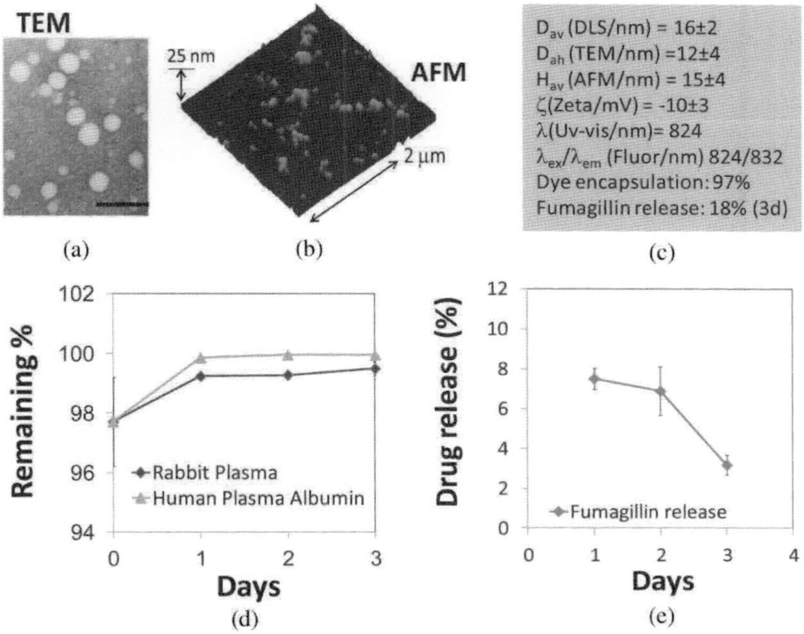

Fig. 20. (a) TEM image of the polymeric micelles drop desposited over nickel brid, scale bar = 100 nm; (b) AFM image of the micelles drop deposited over glass; (c) physio-chemical characterization table for a NIR-polymer micelle; (d) dissolution of ADS-832-WS from micelle when incubated against rabbit plasma and human plasma albumin; (e) time-dependent release of fumagillin from micelle. Reproduced with permission from Pan *et al.* (99)

nodes into the circulation system. Their tendency for not being entrapped within the lymph nodes can be attributed to their miniature sizes.

3.2. *Copper as a contrast agent in PA imaging of SLN*

Interestingly, Pan *et al.*[100] has demonstrated for the first time how copper could be utilized as an exogenous PA contrast agents. Cu-neodecanoate complexes were self-assembled with polysorbates (sorbitan mono-9-octadecenoate poly(oxy-1,2-ethanediyl)) a phospholipids surfactants to produce (Fig. 22) particles with hydrodynamic diameter 86 ± 0.6 nm (D_h) and low polydispersity of 0.21 ± 0.02.

Negative zeta potential of −12 ± 07 mV indicated a complete encapsulation of the copper-polysorbate core with the surfactant. Cu-particles exhibited a PA signal that was significantly higher than blood. Copper particles exhibited a strong absorption in the NIR region compared to a weak one displayed by hemoglobin. Figure 23 shows a baseline image of the rat before

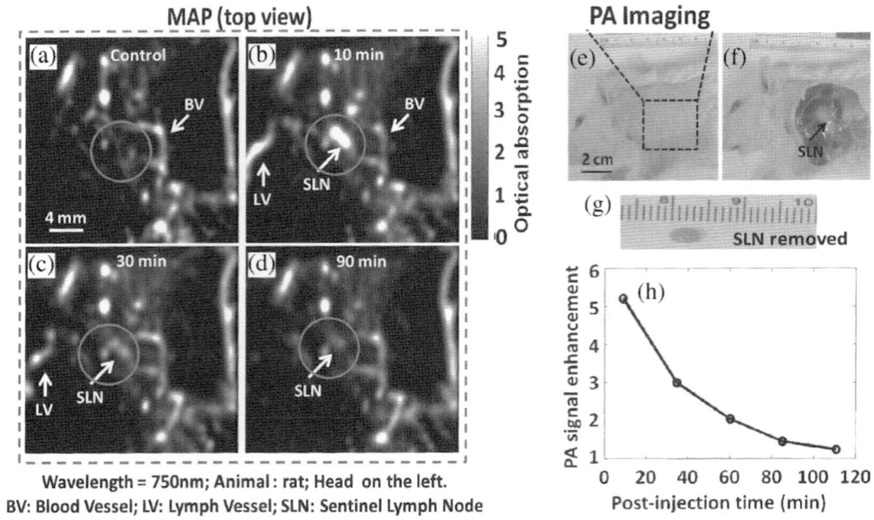

Fig. 21. Non-invasive *in vivo* PA images of SLN in rat: For all PA images, the laser was tuned to 750 nm wavelength. (a) Control PA image acquired before polymeric nanoparticle injection. Bright parts represent optical absorption, here, from blood vessels (BV). (b) PA image (MAP) acquired immediately after the polymeric nanoparticle injection. (c) 35 min post-injection PA image. (d) 110 min post-injection PA image Blood vessel (BV), lymph vessel (LV) and sentinel lymph node (SLN) are marked with arrows, and the SLN is visible in (b)–(d), however, invisible in (a). (e) Photograph of the rat after the hair was removed from the scanning region before taking the PA images. The scanning region is marked with a black dotted square. (f) Photograph of the rat with the skin removed after PA imaging. (g) Excised lymph node. Smallest tick: 1 mm. (h) PA signal enhancement in the SLN after the injection of polymeric nanoparticle as a function of post-injection time. For (a)–(d): FOV = 25 mm × 24 mm, step size along the x direction = 0.2 mm, step size along the *y* direction = 0.4 mm, total scan time = ~23 min. No signal averaging was used. Reproduced with permission from Pan *et al.* (99)

contrast administration, Fig. 23b. Rats were injected with copper nanoparticles in their forepaw and after 60 min a significant enhancement in contrast was seen indicating an increase in uptake of the nanoparticles by the draining of the SLN (Fig. 23d).

3.3. *Porphysome nanovesicles for multimodal biophotonic contrast agents*

Nanoparticles based on inorganic molecules have been widely studied,[101–104] however, perhaps due to their questionable long term safety,[105–108] they have not been able to be translated into the clinic. On the other hand, organic

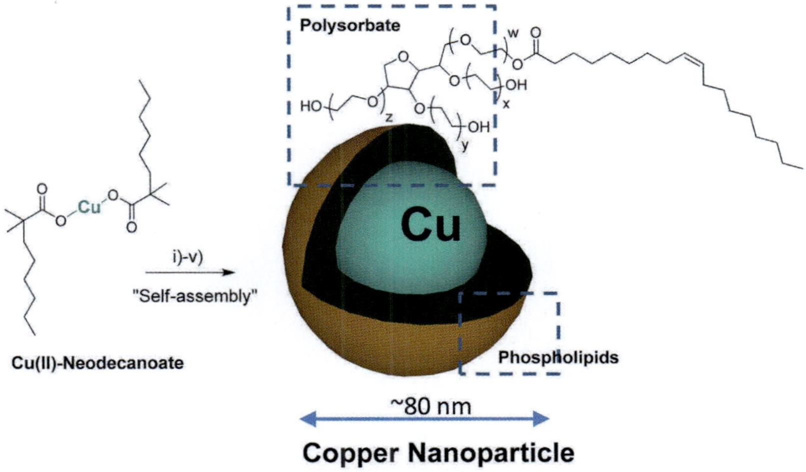

Physico-Chemical Characterization Table
Hydrodynamic diameter (D_h)= 86 ± 06 nm
Polydispersity index (PDI)= 0.21 ± 0.02
Zeta potential (ζ)= -12 ± 07 mV
Particle height (H_{av}) = 60 ± 14 nm
UV-vis spectroscopy = λ 603-746 nm
EDX = C 22 wt%, O 6 wt%, Cu 11 wt%
ICP-OES= 16.99 mg/L of Cu (20% colloidal suspension)

Fig. 22. Synthesis and physiochemical characterization of self-assembled nanoparticles of copper neodecanoate. Schematic describing the preparation of copper-enriched nanoparticles: (i) suspension of copper neodecanoate (1) in sorbitan sesquioleate, vigorously vortex and mixing, filter using cotton bed, vortex; (ii) dissolve phospholipids in anhydrous chloroform and preparation of phospholipid thin film by slow evaporation of solvent at 45°C under reduced pressure; (iii) resuspension of the thin film in water (0.2 μM); (iv) self-assembly by high pressure homogenization at 4°C, 20,000 psi (141 MPa), 4 min; (v) dialysis (cellulosic membrane, MWCO 20 k); in box, characterization table for nanoparticles of copper neodecanoate. Reproduced with permission from Pan *et al.* (100)

nanoparticles have been shown to have more biocompatibility.[109] Lovell *et al.*[110] synthesized a self-assembled nanoparticle comprising of porphyrin-lipid conjugates (porphysome). Porphysomes were also coated with polyethylene glycol (PEG)-lipids to improve its *in vivo* pharmacokinetic properties (Fig. 24). These particles were spherical (100 nm) with two lipid bilayers of higher-density material separated by 2 nm gaps. The nanoparticle showed two distinct absorption peaks at 400 nm and 680 nm. By inserting metal ions into the porphyrin structure, they were able to demonstrate the shifts in optical density bands at 440 nm and 670 nm. This suggested that the porphyrin molecules within the lipid layer are easily accessible for further chelation with metals.

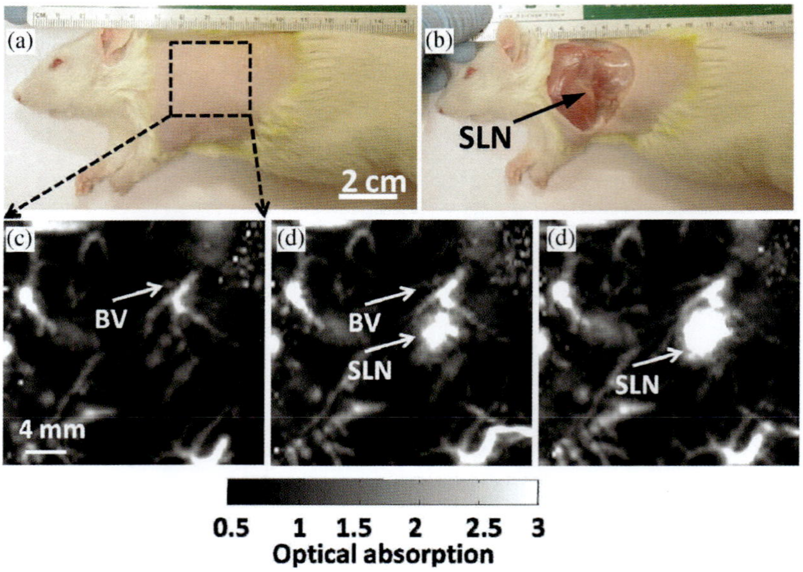

Fig. 23. Non-invasive *in vivo* PA images (maximum intensity projections, MIP) of the SLN in a rat. (a) Photograph of the rat with region of interest depilated before scanning. The scanning region is delineated by a black dotted square. (b) Baseline PA image acquired before NanoCuN injection. Bright regions represent inherent optical absorption from a blood vessel (BV). (c) Pa image (MIP) acquired almost immediately after NanoCuN administration. (d) PA image 60 min post injection showing a marked signal enhancement corresponding to increased NanoCuN uptake by the node: Blood vessel (BV) and sentinel lymph node (SLN) are marked with arrows. The SLN is visible in both (c) and (d), however not apparent in (b). (e) Photograph of the rat with the skin excised after PA imaging. For (b)–(d), FOV = 25 mm × 24 mm, step size along the x direction = 0.2 mm, step size along the y direction = 0.4 mm, total scan time = ~ 23 min. No signal averaging was used. Reproduced with permission from Pan *et al.* (100).

Porphysomes were further tested for its photochemical properties by exposing it to laser irradiation — thermal energy was released which was similar in efficacy to that of gold nanorods. PA amplitudes of porphysomes were much higher than methylene blue a routinely used contrast agents in the clinic for SLN detection (Figs. 25(a), 25(b) and 25(c)).

To further show that porphysomes could be used concurrently with PA tomography, they were injected intradermally in rats. Figure 25(d) shows that the local lymphatic network was clearly visible and detectable within 15 min post-injection. It further suggested the nature of the porphysomes as draining to the lymph vessels and nodes.

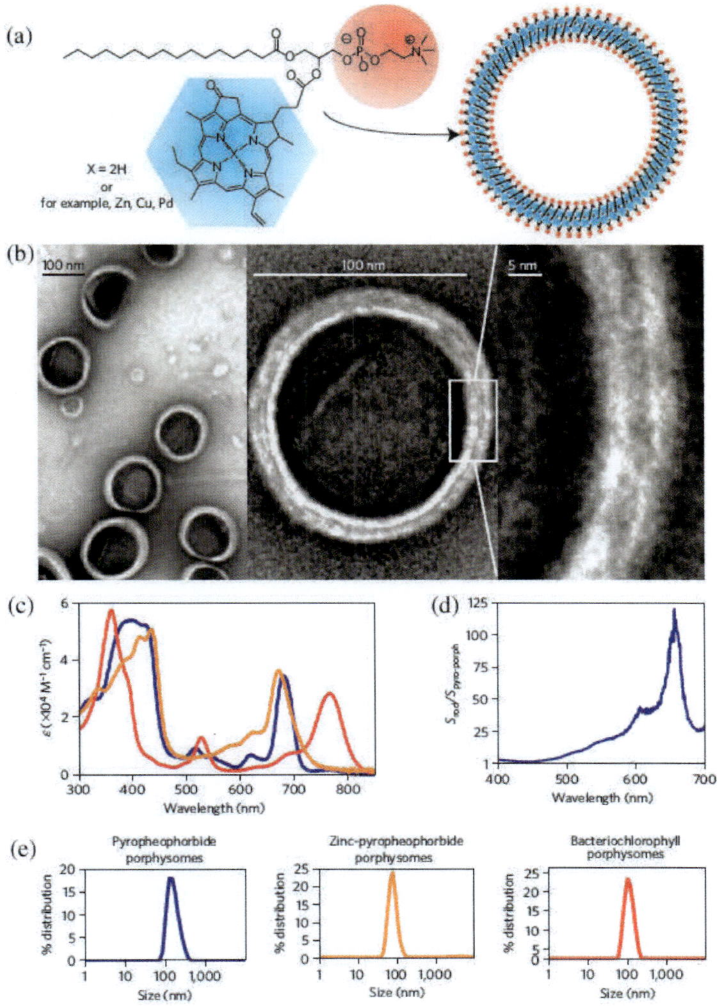

Fig. 24. (Color online) Porphysomes are optically active nanovesicles formed from porphyrin bilayers. (a) Schematic representation of a pyropheophorbide–lipid porphysome. The phospholipid headgroup (red) and porphyrin (blue) are highlighted in the subunit (left) and assembled nanovesicle (right). (b) Electron micrographs of negatively stained porphysomes (5% PEG-lipid, 95% pyropheophorbide-lipid). (c) Absorbance of the porphyrin–lipid subunits incorporated in porphysomes formed from pyropheophorbide (blue), zinc pyropheophorbide (orange) and bacteriochlorophyll (red) in PBS. (d) Resonance light scattering spectra ratio between gold nanorods and pyropheophorbide porphysomes. Nanorod and porphysome concentration was adjusted to have equal-optical density at 680 nm. (e) Dynamic light scattering size profiles of indicated porphysomes recorded in PBS. Reproduced with permission from Lovell *et al.* (110)

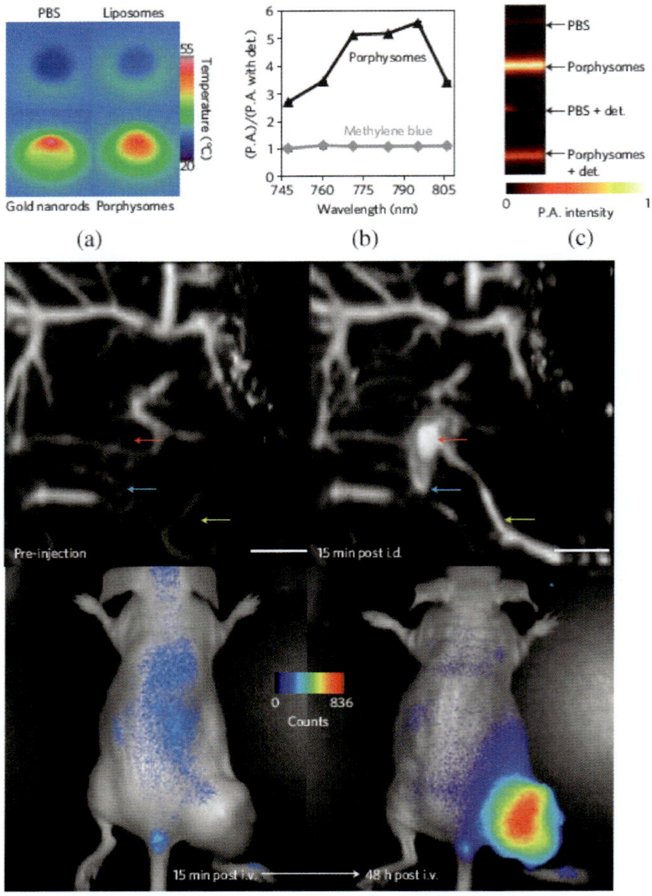

(d)

Fig. 25. (Color online) Multimodal optical utility of porphysomes. (a) Photothermal transduction. Solutions were irradiated with a 673 nm laser and imaged with a thermal camera. (b) Ratio of photoacoustic amplitudes measured for porphysomes and methylene blue _0.5% Triton X-100 (mean _ s.e.m. from 10 measurements). det., detergent. (c) PA images of tubing containing porphysomes and PBS measured _0.5% Triton X-100. (d) Dual modality for PA contrast and activatable fluorescence. Top, lymphatic mapping. Rats were imaged using PA tomography before and after intradermal (i.d.) injection of porphysomes (2.3 pmol). Secondary lymph vessels (cyan), lymph node (red), inflowing lymph vessel (yellow) and 5 mm scale bar are indicated. Bottom, fluorescence activation after *i.v.* injection of porphysomes (7.5 pmol) in a KB xenograft-bearing mouse. Reproduced with permission from Lovell *et al.* (110)

3.4. *Cobalt-Gold hybrid nanoparticles for MR-PAI*

A unique hybrid metallic nanoparticle of cobalt and gold has been reported very recently for MRI and photoacoustic tomography (PAT) detection at the picomolar level sensitivity.[111] The nanoconstruct is composed of ferromagnetic (Co) particles coated with gold (Au) for biocompatibility and optical absorption over NIR range. This dual-modality probe has been found to be very useful for the detection of trace amounts of nanoparticles in biological tissues, in which MRI provides volume detection and PA contributes edge detection.

4. Conclusion

PAI offers unprecedented potential for *in vivo* biological imaging. PAT is a comparatively lower cost methodology and its versatile properties have been well-demonstrated to assess critical early biosignatures of diseases (e.g. angiogenesis). The prototype PAT scanners are putting their initial steps into the clinics with the aim to augment ultrasound based techniques. For most of the preclinical work, the instrumentation typically utilizes a laboratory robotic scanner with an acoustic receiver frequency of 10 MHz. The clinical applicability garnered from the usage of PAT with approved dyes (e.g. methylene blue) are significant, and can be further improved by extending the concurrent usages of ultrasound to assess tissue vasculature that cannot be done simply with the formers alone. Based on the nanoparticle-based research, the additional inclusion of safer nanoparticles opens up possibilities of non invasive sentinel lymph node imaging with increased detection sensitivity and rapid acquisition. One of the critical unmet needs is the detection of lumen stenosis in high-risk stroke patients, which cannot be perceived with current ultrasound techniques. Microthrombus formed over ruptured atherosclerotic plaque has been targeted with nanoparticles (e.g. GNB) for PA to augment the capabilities of US for sensitive detection. Various nanoparticles (e.g. SWNT, GNB, AuNC) have been bioconjugated with homing agents for targeting biological epitopes. PAI has been used successfully to image nascent biological signatures e.g. integrins. The prominent optical properties in metals (e.g. gold, copper etc.) have been exploited cleverly for sensitive PA detection. Finally, organic dyes, which might possibly contain less cytotoxicity than any other contrast agents, can also be used as ideal components of PAT imaging. The plethora of potentialities and opportunities of PAT towards clinical utility have garnered remarkable interest in this area. The demand for its powerful implementation can only be anticipated and expected for those wanting to break the shackles of traditional medical diagnosis and treatment.

References

1. X. D. Wang, P. Y. Pang, G. Ku, X. Y. Xie, G. Stoica and L. H. V. Wang, *Nat. Biotechnol.* **21**, 803–806 (2003).
2. L. V. Wang and S. Hu, *Science* **335**, 1458–1462 (2012).
3. X. D. Wang, X. Y. Xie, G. Ku and L. H. V. Wang, *J. Biomed. Opt.* **11**, 024015 (2006).
4. G. F. Lungu, M. L. Li, X. Y. Xie, L. H. V. Wang and G. Stoica, *Int. J. Oncol.* **30**, 45–54 (2007).
5. G. Ku, X. D. Wang, G. Stoica and L. H. V. Wang, *Phys. Med. Biol.* **49**, 1329–1338 (2004).
6. G. Ku, X. D. Wang, X. Y. Xie, G. Stoica and L. H. V. Wang, *Appl. Opt.* **44**, 770–775 (2005).
7. B. P. Schneider and K. D. Miller, *J. Clin. Oncol.* **23**, 1782–1790 (2005).
8. P. Vaupel, A. Mayer, S. Briest and M. Hockel, *Oxygen Transport to Tissue* XXVI **566**, 33–342 (2005).
9. A. Agarwal, S. W. Huang, M. O'Donnell, K. C. Day, M. Day, N. Kotov and S. Ashkenazi, *J. Appl. Phys.* **102**, 064701–064704 (2007).
10. A. D. Zerda, C. Zavaleta, S. Keren, S. Vaithilingam, S. Bodapati, Z. Liu, J. Levi and B. R. Smith, O T-J Ma, Oralkan, Z. Cheng, X. Chen, H. Dai, B. T. Khuri-Yakub and S. S. Gambhir, *Nat. Nanotech.* **3**, 557–562 (2008).
11. H. Kojima and Y. Zasshi, Development of near infrared fluorescent probes for near infrared imaging. **128**, 1653 (2008).
12. V. Ntziachristos, C. Bremer and R. Weissleder, Fluorescence imaging with nearinfrared light: New technological advances that enable *in vivo* molecular imaging. *Eur. Radiol.* **13**, 195 (2003).
13. B. C. Rostro-Kohanloo, L. R. Bickford, C. M. Payne, E. S. Day, L. J. Anderson, M. Zhong, S. Lee, K. M. Mayer, T. Zal, L. Adam, C. P. Dinney, R. A. Drezek, J. L. West and J. L. Hafner, The stabilization and targeting of surfactant-synthesized gold nanorods. *Nanotechnology* **20**, 434005 (2009).
14. D. Pan, M. Pramanik, A. Senpan, X. Yang, M. J. Scott, K. H. Song, P. J. Gaffney, S. A. Wickline, L. V. Wang and G. M. Lanza, Molecular Photo Acoustic Imaging (PAI) with ligand-directed gold nanobeacons. *Angew. Chem. Int. Ed.* **48**, 4170–4173 (2009).
15. D. Pan, M. Pramanik, A. Senpan, S. Ghosh, S. A. Wickline, L. V. Wang and G. M. Lanza, Near infrared photoacoustic detection of sentinel lymphnodes with gold nanobeacons. *Biomaterials* **31**, 4088–4093 (2010).
16. D. Pan, M. Pramanik, A. Senpan, S. A. Wickline, L. V. Wang and G. M. Lanza, A facile synthesis of novel self-assembled gold nanorods designed for near-infrared imaging. *J. Nanosci. Nanotech.* **10**, 8118–8123 (2010).
17. D. Pan, M. Pramanik, A. Senpan, A. Stacy, H. Zhang, S. A. Wickline, L. V. Wang and G. M. Lanza, Molecular photoacoustic imaging of angiogenesis with integrin targeted gold nanobeacons. *FASEB J.* (published online) (2010).
18. S. Raut and P. J. Gaffney, *J. Thromb. Haemostasis* **76**, 56–64 (1996).
19. K. H. Song and L. H. V. Wang, *J. Biomed. Opt.* **12**, 060503 (2007).
20. M. Pramanik, G. Ku, C. H. Li and L. H. V. Wang, *Med. Phys.* **35**, 2218–2223 (2008).
21. D. Krag, D. Weaver, T. Ashikaga, F. Moffat, V. S. Klimberg, C. Shriver The sentinel node in breast cancer — A multicenter validation study. *New Engl. J. Med.* **339**, 941–946 (1998).

22. K. M. McMasters, T. M. Tuttle, D. J. Carlson, C. M. Brown, R. D. Noyes and R. L. Glaser, Sentinel lymph node biopsy for breast cancer: A suitable alternative to routine axillary dissection in multi-institutional practice when optimal technique is used. *J. Clin. Oncol.* **18**, 2560–2566 (2000).
23. S. L. Chen, D. M. Iddings, R. P. Scheri and A. J. Bilchik, Lymphatic mapping and sentinel node analysis. *CA Cancer J. Clin.* **56**, 292–309 (2006).
24. G. Laruelle, J. Francios and L. Billon, Self-assembly in aqueous media of amphiphilic poly acrylic acid based di-block copolymers synthesized by direct nitroxide-mediated polymerization. *Macromol Rapid Comm.* **25**, 1839–1844 (2004).
25. H. Y. Huang, T. Kowalewski, E. E. Remsen, R. Gertzmann and K. L. Wooley, Hydrogel-coated glassy nanospheres: A novel method for the synthesis of shell cross-linked knedels. *J. Am. Chem. Soc.* **119**, 11653–11659 (1997).
26. J. Wu and A. Eisenberg, Proton diffusion across membranes of vesicles of poly (styrene-b-acrylic acid) diblock copolymers. *J. Am. Chem. Soc.* **128**, 2880–2884 (2006).
27. J. Xu, G. Sun, R. Rossin, A. Hagooly, Z. Li, K. Fukukawa, Labeling of polymer nanostructures for medical imaging: Importance of charge density, spacer length, and crosslinking extents. *Macromolecules* **40**, 2971–2973 (2007).
28. H. F. Zhang, K. Maslov, G. Stoica and L. H. V. Wang, *Nat. Biotechnol.* **24**, 848–851 (2006).
29. P. C. Brooks, R. A. F. Clark and D. A. Cheresh, Requirement of vascular integrin avb3 for angiogenesis. *Science* **264**, 569–571 (1994).
30. P. Winter, A. Neubauer, S. Caruthers, T. Harris, J. Robertson, T. Williams, A. Schmieder, G. Hu, J. Allen, E. Lacy, S. Wickline and G. Lanza, Endothelial alpha(nu)beta(3)-integrin targeted fumagillin nanoparticles inhibit angiogenesis in atherosclerosis. *Arterioscler. Thromb. Vasc. Biol.* **26**, 2103–2109 (2006).
31. P. M. Winter, A. H. Schmieder, S. D. Caruthers, J. L. Keene, H. Zhang, S. A. Wickline and G. M. Lanza, Minute dosages of alpha(nu)beta3-targeted fumagillin nanoparticles impair Vx-2 tumor angiogenesis and development in rabbits. *FASEB J.* **22**, 2758–2767 (2008).
32. D. A. Cheresh, Integrins in thrombosis, wound healing and cancer. *Biochem. Soc. Trans.* **19**, 835–838 (1991).
33. M. Friedlander, C. L. Theesfeld, M. Sugita, M. Fruttiger, M. A. Thomas, S. Chang and D. A. Cheresh, Involvement of integrins anb3 and anb5 in ocular neovascular diseases. *Proc. Natl Acad. Sci. USA* **93**, 9764–9769 (1996).
34. M. De Nichilo and G. Burns, Granulocytemacrophage and macrophage colonystimulating factors differentially regulate a v integrin expression on cultured human macrophages. *Proc. Natl Acad. Sci. USA* **90**, 2517–2521 (1993).
35. O. Helluin, C. Chan, G. Vilaire, S. Mousa, W. F. DeGrado and J. S. Bennett, The activation state of avb 3 regulates platelet and lymphocyte adhesion to intact and thrombin-cleaved osteopontin. *J. Biol. Chem.* **275**, 18337–18343 (2000).
36. H. Itoh, P. Nelson, L. Mureebe, A. Horowitz and K. Kent, The role of integrins in saphenous vein vascular smooth muscle cell migration. *J. Vasc. Surg.* **25**, 1061–1069 (1997).
37. F. Carreiras, Y. Denoux, C. Staedel, M. Lehman, F. Sichel and P. Gauduchon, Expression and localization of a v integrins and their ligand vitronectin in normal ovarian epithelium and in ovarian carcinoma. *Gynecol. Oncol.* **62**, 260–267 (1996).

38. T. Kageshita, C. V. Hamby, S. Hirai, T. Kimura, T. Ono and S. Ferrone, Differential clinical significance of a(v)B(3) expression in primary lesions of acral lentiginous melanoma and of other melanoma histotypes. *Int. J. Cancer.* **89**, 153–159 (2000).
39. G. Hu, M. Lijowski, H. Zhang, Imaging of Vx-2 rabbit tumors with avb3-integrintargeted 111In nanoparticles. *Int. J. Cancer* **120**, 1951–1957 (2007).
40. X. D. Wang, Y. I. Pang, G. Ku, X. Y. Xie, G. Stoica and L. V. Wang, Noninvasive laser-induced photoacoustic tomography for structural and functional *in vivo* imaging of the brain. *Nat. Biotech.* **21**, 803–806 (2003).
41. R. I. Siphanto, K. K. Thumma, R. G. Kolkman, T. G. van Leeuwen, F. F. de Mul, J. W. van Neck, L. N. van Adrichem and W. Steenbergen, Serial noninvasive photoacoustic imaging of neovascularization in tumor angiogenesis. *Optic. Express* **13**, 89–95 (2005).
42. K. Maslov, H. F. Zhang, S. Hu and L. V. Wang, Optical-resolution photoacoustic microscopy for *in vivo* imaging of single capillaries. *Opt. Lett.* **33**, 929–931 (2008).
43. Y. Lao, D. Xing, S. Yang and L. Xiang, Noninvasive photoacoustic imaging of the developing vasculature during early tumor growth. *Phys. Med. Biol.* **53**, 4203–4212 (2008).
44. E. Z. Zhang, J. G. Laufer, R. B. Pedley and P. C. Beard, *In vivo* high-resolution 3D photoacoustic imaging of superficial vascular anatomy. *Phys. Med. Biol.* **54**, 1035–1046 (2009).
45. J. Jose, S. Manohar, R. G. Kolkman, W. Steenbergen and T. G. van Leeuwen, Imaging of tumor vasculature using twente photoacoustic systems. *J. Biophotonics* **2**, 701–717 (2009).
46. S. Hu, S. Oladipupo, J. Yao, A. C. Santeford, K. Maslov, J. Kovalski, J. M. Arbeit and L. V. Wang, Optical resolution photoacoustic microscopy of angiogenesis in a transgenic mouse model. *Proc. SPIE* **7564**, 756406 (2010).
47. L. V. Wang, X. Xie, J. T. Oh, M. N. Li, G. Ku, S. Ke, S. Similache, C. Li and G. Stoica, Combined photoacoustic and molecular fluorescence imaging *in vivo. IEEE Eng. Med. Biol. Proc.* 7, 190–192 (2005).
48. X. Xie, J. T. Oh, M. L. Li, G. Ku, X. Wang, S. Ke, C. Li, S. Similache, G. Stoica and L. V. Wang, Photoacoustic tomography and molecular fluorescence imaging: Dual modality imaging of small animal brains *in vivo. Proc. SPIE* **5697**, 107–110 (2005).
49. A De la Zerda, C. Zavaleta, S. Keren, S. Vaithilingam, S. Bodapati, Z. Liu, J. Levi, B. R. Smith, T. J. Ma, O. Oralkan, Z. Cheng, X. Chen, H. Dai, B. T. Khuri-Yakub and S. S. Gambhi, Carbon nanotubes as photoacoustic molecular imaging agents in living mice. *Nat. Nanotechnol.* **3**, 557–562 (2008).
50. X. Xie, M. L. Li, J. T. Oh, G. Ku, W. Wang, C. Li, S. Similache, G. F. Lungu, G. Stoica and L. V. Wang, Photoacoustic molecular imaging of small animals *in vivo. Proc. SPIE* **6086**, 608606 (2006).
51. L. Xiang, Y. Yuan, D. Xing, Z. Ou, S. Yang and F. Zhou, Photoacoustic molecular imaging with antibody-functionalized single-walled carbon nanotubes for early diagnosis of tumor. *J. Biomed. Opt.* **14**, 021008 (2009).
52. R. Weissleder, A. Jr. Bogdanov, C. H. Tung and H. J. Weinmann, Size optimization of synthetic graft copolymers for *in vivo* angiogenesis imaging. *Bioconjug. Chem.* **12**, 213–219 (2001).
53. US patent 6,511,648.
54. D. F. Meoli, M. M. Sadeghi, S. Krassilnikova, B. N. Bourke, F. J. Giordano, D. P. Dione, H. Su, D. S. Edwards, S. Liu, T. D. Harris, J. A. Madri, B. L. Zaret and A. J. Sinusas,

Noninvasive imaging of myocardial angiogenesis following experimental myocardial infarction. *J. Clin. Invest.* **113**, 1684–1691 (2004).
55. A. Miyauchi, J. Alvarez, E. M. Greenfield, A. Teti, M. Grano, S. Colucci, A. Zambonin-Zallone, F. P. Ross, S. L. Teitelbaum, D. Cheresh, Recognition of osteopontin and related peptides by an alpha v beta 3 integrin stimulates immediate cell signals in osteoclasts. *J. Biol. Chem.* **266**, 20369–20374 (1991).
56. R. A. Weissleder, Clearer vision for *in vivo* imaging. *Nat. Biotechnol.* **19**, 316–317 (2001).
57. J. Y. Chen, B. Wiley, Z. Y. Li, D. Campbell, F. Saeki, H. Cang, L. Au, J. Lee, X. Li and Y. Xia, Gold nanocages: Engineering their structure for biomedical applications. *Adv. Mater.* **17**, 2255–2261 (2005).
58. S. E. Skrabalak, J. Chen, L. Au, X. Lu, X. Li and Y. Xia, Gold nanocages for biomedical applications. *Adv. Mater.* **19**, 3177–3184 (2007).
59. S. E. Skrabalak, J. Chen, Y. Sun, X. Lu, L. Au, C. Cobley and Y. Xia, Gold nanocages: Synthesis, properties, and applications. *Accounts Chem. Res.* **41**, 1587–1595 (2008).
60. J. Chen, M. Yang, Q. Zhang, E. C. Cho, C. M. Cobley, C. Claus, C. Kim, L. V. Wang, M. J. Welch and Y. Xia, Gold nanocages: A novel class of multifunctional nanomaterials for theranostic applications. *Adv. Funct. Mater.* **20**, 3684–3694 (2010).
61. C. G. A. Hoelen, F. F. M. de Mul, R. Pongers and A. Dekker, Three dimensional photoacoustic imaging of blood vessels in tissue. *Opt. Lett.* **23**, 648–650 (1998).
62. R. A. Kruger, D. R. Reinecke and G. A. Kruger, Thermoacoustic computed tomography technical considerations. *Med. Phys.* **26**, 1832–1837 (1999).
63. M. Xu and L. V. Wang, Photoacoustic imaging in biomedicine. *Rev. Sci. Instrum.* 77, 1–22 (2006).
64. Y. Wang, X. Xie, X. Wang, G. Ku, K. L. Gill, D. P. O'Neal, G. Stoica, L. V. Wang, Photoacoustic tomography of a nanoshell contrast agent in the *in vivo* rat brain. *Nano Lett.* **4**, 1689–1692 (2004).
65. X. Yang, S. Skrabalak, Z. Li, Y. Xia and L. Wang, Photoacoustic tomography of a rat cerebral cortex *in vivo* with Au nanocages as an optical contrast agent. *Nano Lett.* 7, 3798–3802 (2007).
66. D. L. Morton, D. R. Wen, J. H. Wong, J. S. Economou, L. A. Cagle, F. K. Storm, L. J. Foshag and A. J. Cochran, Intraoperative lymphatic mapping and selective lymphadenectomy: Technical details of a new procedure for clinical stage I melanoma. *Arch. Surg.* **127**, 392–399 (1992).
67. D. N. Krag, D. L. Weaver, J. C. Alex and J. T. Fairbank, Surgical resection and radiolocalization of the sentinel lymph node in breast cancer using a gamma probe. *Surg. Oncol.* **2**, 335–340 (1993).
68. Y. Koyama, V. S. Talanov, M. Bernardo,Y. Hama, C. A. Regino, M. W. Brechbiel, P. L. Choyke and H. A. Kobayashi, Dendrimer based nanosized contrast agent dual labeled for magnetic resonance and optical fluorescence imaging to localize the sentinel lymph node in mice. *J. Magn. Reson. Imag.* **25**, 866–871 (2007).
69. M. Nakajima, M. Takeda, M. Kobayashi, S. Suzuki and N. Ohuchi, Nano sized fluorescent particles as new tracers for sentinel node detection: Experimental model for decision of appropriate size and wavelength. *Cancer Sci.* **96**, 353–356.
70. S. C. Michel, T. M. Keller, J. M. Frohlich, D. Fink, R. Caduff, B. Seifert, B. Marincek and R. A. Kubik-Huch, Preoperative breast cancer staging: MR imaging of the axilla with

Ul trasmall superparamagnetic iron oxide enhancement. *Radiology* **225**, 527–536 (2002).
71. M. G. Torchia, R. Nason, R. Danzinger, J. M. Lewis, J. A. Thliveris, Interstitial MR lymphangiography for the detection of sentinel lymph nodes. *J. Surg. Oncol.* **78**, 151–156 (2001).
72. K. H. Song, C. Kim, C. M. Cobley, Y. Xia and L. V. Wang, Near infrared gold nanocages as a new class of tracers for photoacoustic sentinel lymph node mapping on a rat model. *Nano Lett.* **9**, 183–188 (2009).
73. C. Kim, E. C. Cho, J. Chen, K. H. Song, L. Au, C. Favazza, Q. Zhang, C. M. Cobley, F. Gao, Y. Xia and L. V. Wang, *In vivo* molecular photoacoustic tomography of melanomas targeted by bioconjugated gold nanocages. *ACS Nano* **4**, 4559–4564 (2010).
74. V. P. Zharov, E. I. Galanzha, E. V. Shashkov, J. W. Kim, N. G. Khlebtsov and V. V. Tuchin, Photoacoustic flow cytometry: Principle and application for real time detection of circulating single nanoparticles, pathogens, and contrast dyes *in vivo*. *J. Biomed. Opt.* **12**, 051503 (2007).
75. de la Zerda A *et al.*, Carbon nanotubes as photoacoustic molecular imaging agents in living mice. *Nat. Nanotechnol.* **3**, 557–562 (2008).
76. A. de la Zerda, Ultrahigh sensitivity carbon nanotube agents for photoacoustic molecular imaging in living mice. *Nano Lett.* **10**, 2168–2172 (2007).
77. E. I. Galanzha, E. V. Shashkov, T. Kelly, J. W. Kim, L. Yang and V. P. Zharov, *In vivo* magnetic enrichment and multiplex photoacoustic detection of circulating tumour cells. *Nat. Nanotechnol.* **4**, 855–860 (2009).
78. J. W. Kim, E. I. Galanzha, E. V. Shashkov, H. M. Moon and V. P. Zharov, Golden carbon nanotubes as multimodal photoacoustic and photothermal high contrast molecular agents. *Nat. Nanotechnol.* **4**, 688–694 (2009).
79. E. I. Galanzha, E. V. Shashkov, M. S. Kokoshka, J. A. Myhill, Zharov. *In vivo* noninvasive detection of metastatic cells in vasculature and sentinel lymph nodes by photoacoustic cytomery. *Laser Surg. Med.* **S20**, 81–82 (2008).
80. E. I. Galanzha, M. S. Kokoska, E. V. Shashkov, J. W. Kim, V. V. Tuchin and V. P. Zharov, *In vivo* fiber based multicolor photoacoustic detection and photothermal purging of metastasis in sentinel lymph nodes targeted by nanoparticles. *J. Biophotonics* **2**, 528–539 (2009).
81. S. L. N. Wang and M. Pramanik, *In vivo* carbon nanotube enhanced non invasive photoacoustic mapping of the sentinel lymph node. *Phys. Med. Biol.* **54**, 3291–3301.
82. M. Pramanik, M. Swierczewska, D. Green, B. Sitharaman and L. V. Wang, Single walled carbon nanotubes as a multimodal thermoacoustic and photoacoustic contrast agent. *J. Biomed. Opt.* **14**, 034018 (2009).
83. E. I. Galanzha, J. W. Kim and V. P. Zharov, Integrated nanotechnology based photoacoustic and photothermal flow cytometry platform for *in vivo* detection and killing of circulating cancer stem cells. *J. Biophoton* **2**, 725–735 (2009).
84. V. P. Zharov, V. Galitovsky and M. Viegas, Photothermal detection of local thermal effects during selective nanophotothermolysis. *Appl. Phys. Lett.* **83**, 4897–4899 (2003).
85. V. P. Zharov, E. N. Galitovskaya, C. Jonson and T. Kelly, Synergistic enhancement of selective nanophotothermolysis with gold nanoclusters: Potential for cancer therapy. *Laser Surg. Med.* **37**, 219–226 (2005).
86. V. P. Zharov, R. R. Letfullin and E. N. Galitovskaya, Microbubbles overlapping mode for laser killing of cancer cells with absorbing nanoparticle clusters. *J. Phys. D Appl. Phys.* **38**, 2571–258 (2005).

87. V. P. Zharov, J. W. Kim, M. Everts and D. T. Curiel, Self assembling nanoclusters in living system application for integrated photothermal nanodiagnostics and nanotherapy. *Nanomedicine* **1**, 326–345 (2005).
88. V. P. Zharov, K. E. Mercer, E. N. Galitovskaya and M .S. Smeltzer, Photothermal nanotherapeutics and nanodiagnostics for selective killing of bacteria targeted with gold nanoparticles. *Biophys. J.* **90**, 619–628 (2006).
89. J. W. Kim, E. I. Galanzha, E. V. Shashkov, N. Kotagiri and V. P. Zharov, Photothermal antimicrobial nanotherapy and nanodiagnostics with self assembling carbon nanotube clusters. *Lasers Surg. Med.* **39**, 622–634 (2007).
90. B. Khlebtsov, V. Zharov, A. Melnikov, V. Tuchin and N. Khlebtsov, Optical amplification of photothermal therapy with gold nanoparticles and nanoclusters. *Nanotechnology* **17**, 5167–5179 (2006).
91. L. R. Hirsch, R. J. Stafford, J. A. Bankson, Nanoshell mediated thermal therapy of tumors under magnetic resonance guidance. *Proc. Natl. Acad. Sci. USA* **100**, 13549–13554 (2003).
92. N. W. Kam, M. O'Connell, J. A. Wisdom and H. Dai, Carbon nanotubes as multifunctional biological transporters and near infrared agents for selective cancer cell destruction. *Proc. Natl. Acad. Sci. USA* **102**, 11600–11605 (2005).
93. Z. Liu, X. Sun, N. Nakayama-Ratchford and H. Dai, Chemistry on watersoluble carbon nanotubes for drug loading and delivery. *ACS Nano* **1**, 50–56 (2007).
94. Z. Liu, *In vivo* biodistribution and highly efficient tumour targeting of carbon nanotubes in mice. *Nat. Nanotechnol.* **2**, 47–52 (2007).
95. B. Cady, M. D. Stone and J. G. Schuler, The new era in breast cancer. Invasion, size, and nodal involvement dramatically decreasing as a result of mammographic screening. *Arch. Surg.* **131,** 301 (1996).
96. B. G. Haffty, B. Ward and P. Pathare, Reappraisal of the role of axillary lymph node dissection in the conservative treatment of breast cancer. *J. Clin. Oncol.* **15**, 691 (1997).
97. U. Veronesi, G. Paganelli and G. Viale, A randomized comparison of sentinel-node biopsy with routine axillary dissection in breast cancer. *N. Engl. J. Med.* **349**, 546 (2003).
98. D. N. Krag, S. J. Anderson, T. B. Julian, A. M. Brown, S. P. Harlow, J. P. Costantino, T. Ashikaga, D. L. Weaver, E. P. Mamounas, L. M. Jalovec, T. G. Frazier, R. D. Noyes, A. Robidoux, H. M. C. Scarth and N. Wolmark, Sentinel-lymph-node resection compared with conventional axillary-lymph-node dissection in clinically node-negative patients with breast cancer: Overall survival findings from the NSABP B-32 randomised phase 3 trial. *Lancet Oncol.* **11**, 927 (2010).
99. D. Pan, X. Cai, B. Kim, A. L. Stacy, L. V. Wang and G. M. Lanza, Rapid synthesis of near infrared polymeric micelles for real-time sentinel lymph node imaging. *Adv. Healthcare Mater.* **1**, 582–589 (2012).
100. D. Pan, X. Cai, C. Yalaz, A. Senpan, K. Omanakuttan, S. A. Wickline, L. V. Wang and G. M. Lanza, Photoacoustic sentinel lymph node imaging with self-assembled copper neodecanoate nanoparticles. *ACS Nano* **6**, 1260–1267 (2012).
101. W. C. W. Chan and S. Nie, Quantum dot bioconjugates for ultrasensitive nonisotopic detection. *Science* **281**, 2016–2018 (1998).
102. J. J. Storhoff, A. D. Lucas, V. Garimella, Y. P. Bao, and U. R. Muller, Homogeneous detection of unamplified genomic DNA sequences based on colorimetric scatter of gold nanoparticle probes. *Nat. Biotechnol.* **22**, 883–887 (2004).
103. A. A. Oraevsky, Laser optoacoustic tomography for medical diagnostics: Principles. *Proc. SPIE* **2676**, 22–31 (1996).

104. S. Lal, S. E. Clare, and N. J. Halas, Nanoshell-enabled photothermal cancer thereapy: Impending clinical impact. *Accounts Chem. Res.* **41**, 1842–1851 (2008).
105. P. Ghosh, G. Han, M. De, C. K. Kim, and V. M. Rotello, Gold nanoparticles in delivery applications. *Adv. Drug. Deliv. Rev.* **60**, 1307–1315 (2008).
106. N. Lewinski, V. Colvin and R. Drezek, Cytotoxicity of nanoparticles. *Small.* **4**, 26–49 (2008).
107. T. Nel, a. Xia, L. Madler and N. Li, Toxic potential of materials at the nanolevel. *Science* **311**, 622–627 (2006).
108. D. Peer *et al.*, Nanocarriers as an emerging platform for cancer therapy. *Nat. Nanotech.* **2**, 751–760.
109. D. Peer *et al.*, Nanocarriers as an emerging platform for cancer therapy. *Nature Nanotech.* **2**, 751–760 (2007).
110. J. F. Lovell *et al.*, Porphysome nanovesicles generated by porphyrin bilayers for use as multimodal biophotonic contrast agents. *Nat. Mater.* **10**, 324–332 (2011).
111. L. S. Bouchard *et al.*, Picomolar sensitivity MRI and photoacoustic imaging of cobalt nanoparticles. *Proc. Natl. Acad. Sci USA* **106**, 4085–4089 (2009).

Chapter 12

Nanoelectrodes for Biomedical Applications

Haibin Shi and Joanne I. Yeh*,†,‡*

**Department of Structural Biology,*
University of Pittsburgh
3501 5th Avenue, BST 1043, Pittsburgh, PA, USA 15260

†Department of Bioengineering,
University of Pittsburgh
3501 5th Avenue, BST 1043, Pittsburgh, PA, USA 15260
‡jiyeh@pitt.edu

Abstract

Biosensors are analytical devices that are comprised of a biological component that specifically recognizes the analyte and a physiochemical component that transduces the signal produced when the biological interacts with a target analyte. An electrochemical biosensor, a focus of this chapter, is a type of biosensor based on the production or consumption of electrons to detect analyte. In electrochemical biosensors, individual electrodes or electrode platforms serve as the physiochemical component and the nature and dimensions of the electrode have broad influence on the overall biosensor. In this chapter, we discuss how geometrical and chemical nature of the electrodes affects the overall performance of biosensors, focusing primarily on nanoelectrodes. Nanoelectrodes are a kind of electrode with a critical dimension in the one to hundreds of nanometer range and include individual nanoelectrodes, nanoelectrode ensembles (NEEs) and nanoelectrode arrays (NEAs). Carbon nanotubes (CNTs), carbon nanofibers, metallic nanowires (NWs), metal oxide nanowires (NWs) and magnetic nanoparticles have been employed to fabricate nanoelectrodes. In this review, applications of single nanoelectrodes, NEAs and ensembles are briefly evaluated, with an emphasis on *in vivo* intracellular analysis of single living cells. Considerable benefits are gained when multidisciplinary methods are appropriately merged, exemplified by the fabrication of microchips and incorporation of wireless technology to further evolve the platforms. Diverse biosensors have been produced using principles of coordinated biosensing, an approach integrating molecular and functional properties of the biological component with spatial characteristics of the physiochemical component. Coordinated biosensing is based on exploiting the innate sensitivity and specificity of biological systems (e.g. proteins, nucleic acids), harnessing the innate specificity and sensitivities of proteins to ultrasensitive detection of diverse analytes. We describe how structural and functional data of proteins can be applied to preserve biochemical activity and to align the signal transducing units in each component of a biosensor. Clinically-relevant biosensors possessing enhanced sensitivities and exhibiting robust performance capabilities are now highly feasible and cost-effective to produce. Consequently, nanoelectrodes have immense potential in the development of efficient, specific, sensitive, and intelligent biosensors. These biosensors will soon become a new class of low cost, robust, reliable, easy-to-use, and ultrasensitive diagnostics, which may spur a dramatic increase in the number of point-of-care (POC) diagnostics, enabling more personalized medical care.

1. Introduction

Biosensors are analytical devices that utilize biological molecules to detect analytes. An electrochemical biosensor, a focus of this chapter, is a type of biosensor based on the production or consumption of electrons to detect analytes. Typically, redox enzymes catalyze the rapid production/consumption of electrons, coupling the binding of a target analyte to measurable changes in electronic signals. When the potential of the working electrode is fixed, the concentration of the analyte is proportional to the current (rate of electron flow). Electrochemical biosensors have a long history since Dr. Leland C. Clark introduced the first enzyme electrode with immobilized glucose oxidase (GOx) in 1962.[2] Now, they have a significant impact in everyday life that are applied to characterize and reveal the biological processes proceeding, and determine a variety of *in situ*, *in vivo*, and *in vitro* biomedical measurements, such as blood glucose monitoring, drug discovery, drug analysis and whole blood analyzers.[3–6]

The electrode is a key component of the electrochemical biosensor, which measures the change in current, voltage, impedance or conductance resulting from a biochemical reaction.[7–9] Recently, significant advances in the nanotechnology have produced a large number of new nanomaterials with unique physical and chemical properties such as adsorption, surface modification, catalytic and electric features. And, such preferable properties of nanomaterials are enabling a novel way in the development and application of nanoelectrodes for construction of innovative electrochemical biosensors over the past decade.[7–13] Nanoelectrodes are with critical dimension in the range from one nanometer to hundreds of nanometers. The nanoelectrodes bring to biosensors the improved analytical performance, the possibility for construction of biosensor arrays for high throughput parallel measurements and the possibility for integration of biosensors with microfluidics and nanofluidic for construction of lab-on-a-chip devices that would be used as point-of-care (POC) diagnostic devices.[7–9] Moreover, nanoelectrodes based on nanostructured materials or nanomaterials have proposed a method to enhance biocompatibility of biosensors and a possible solution to the problem of fouling of implantable biosensors for *in vivo* real-time sensing.

This review summarizes recent research highlights involving a few illustrative examples such as use of individual nanoelectrodes on single cell monitoring, utilization of nanoelectrode arrays (NEAs) and nanoelectrode ensembles (NEEs) for fabrication of microfluidic devices, and the development of coordinated biosensors for the biological and medical applications.

2. Advantages of Nanoelectrodes

Nanoelectrodes can be categorized as individual nanoelectrodes, NEAs and NEEs).[7–9] Individual nanoelectrodes produced from carbon fibers and metals wires (e.g. platinum, gold) have ultra-small electrode tips from a few tens to several hundreds of nanometers, which enable the biosensor access to microenvironments (e.g. living cells) not accessible to conventional macroelectrodes.[7–9] Thus, individual nanoelectrodes enable the application of analytical capabilities to *in vivo* monitor single cell secretion and even detect single molecule. NEAs and NEEs are assembled from carbon nanotubes (CNTs), carbon fiber nanoelectrodes (CFNEs), metallic nanowires (NWs), metal oxide NWs, and magnetic nanoparticles.[7] NEAs consist of periodic arrangement of individual nanoelectrodes, whereas NEEs are more random collections of nanoelectrodes. NEAs and NEEs can be facilely prepared through the template synthesis methods on a small size platform, which can be easily used in fabrication of microchip devices. These microchip devices are emerging as powerful tools (e.g. POC devices) for the direct detection of biological and chemical species, including determining low concentrations of biochemicals, proteins and viruses.[14,15]

The nanoelectrodes have displayed unique characteristics, which are not expected for conventional macroelectrodes (millimeter to centimeter diameter), such as favorable Faradic-to-capacitive current (signal-to-noise, S/N) ratios, low double-layer charging currents, significantly enhanced detection limits, short response times, high mass-transport rates and fast electron transfer rates.[16,17] The S/N ratios are determined by the relationship between Faradaic current (I_F) and background current (double-layer capacitive current, I_C). I_F occurs at the electrode during the redox reaction of the analyte. I_C is the major component of background noise predominantly caused by a double-layer charging current at the electrode — solution interface in voltammetry measurements. The I_F is proportional to the total geometric area (A_{geom}; e.g. nanodisc plus insulator area) of the NEE (or NEA), whereas I_C is proportional to the sum of the areas of the electrode elements in the NEE (or NEA) (A_{act}, active area). Therefore, the S/N ratios at NEEs (or NEAs) and conventional electrodes with the same geometric area are related by Eq. (1):[17]

$$(I_F/I_C)_{NEE} = (I_F/I_C)_{conv}/f \tag{1}$$

The f is named fractional electrode area that is the ratio of the active area to the geometric area (Eq. (2)).

$$f = A_{act}/A_{geom} \tag{2}$$

The typical f values for NEEs (or NEAs) are between 10^{-3} and 10^{-2}, therefore, the S/N ratio for NEE (or NEA) is significantly higher than for conventional electrodes (2–3 orders of magnitude) with the same geometric area.[17] Such an improvement in the S/N current ratios results in the detection limits at nanoelectrodes can be 2–3 orders of magnitude lower than for conventional electrode.[7–9] Recently, the immobilization of gold nanoparticles (AuNPs) onto the surface of NEEs to increase the active area (A_{act}) of NEEs has been described.[18] And the analysis showed that the presence of AuNPs reflects in an approximately ten-times increase in the active area of the NEEs, which results in the further improved sensitivity.[19]

Moreover, the mass transport rate to and from the nanoelectrode increases and leads to a steady-state voltammograms as the electrode size decreases. Models and experiments predict substantially higher mass transfer rates for nanoelectrodes due to radial (non-planar) compared to measurements using bulk electrodes that operate via planar diffusion. As a consequence of the increase in mass transport rates and the reduced I_C, nanoelectrodes will also exhibit excellent S/N ratio in comparison to conventional macroelectrodes. The significantly improved S/N ratio of nanoelectrodes makes them particularly suitable for direct determination of electroactive species at low concentrations.[17] Enhanced detection capabilities for biological events on nanoelectrode platforms and their suitability for biological and medical diagnostics applications have been demonstrated.[7–9]

3. Nanoelectrode for Single Cell Monitoring

Studying biology in living cells is one of the most challenging tasks in biological and medical fields. Advantages of individual nanoelectrodes are the ultra-small dimensions, increased S/N ratios, low double-layer charging currents and fast mass transport rates. Therefore, individual nanoelectrodes provide a novel tool for *in vivo* detecting intracellular components or real-time monitoring of biological phenomena in a single living cell because the diameter of a cell is approximately 10–15 μm. Nanoelectrode can penetrate the cell membrane with minimal cell damage when the diameter of nanoelectrode is ≤400 nm.[8]

CFNE produced with ultra-small tips (diameter of 100–300 nm) has been used for monitoring the secretion — dopamine (DA) from a single cell with high spatial and temporal resolution.[20–22] DA is one of the most important neurotransmitters, playing an important physiological role as an extracellular chemical messenger in the functioning of the central nervous, renal and hormonal systems. It is known that the loss of DA may result in serious diseases, such as Parkinson's.[23,24] Compared to traditional microelectrodes, enhanced sensitivities and resolution of detection for DA levels by using CFNE have been reported.[20–22] Furthermore, a microchip capillary electrophoresis (CE) with end-column amperometric detection based on an integrated CFNE has been developed.[25] The use of microchip-based systems for analysis of cellular secretion is a rapidly growing research that includes detection of single cell quantal exocytosis, analysis of intracellular contents and monitoring cell confluency.[14] The integration of CFNE as working electrodes into a microchip platform resulted in better reproducibility and markedly improved sensitivity and selectivity. Compared to traditional microelectrodes (1.7×10^{-7} M detection limit for DA), one to two orders of magnitude enhancement in sensitivity (5.9×10^{-8} M detection limit for DA) is obtained using miniaturized platforms.[25] The CFNE-based microchip CE system has been successfully applied to the determination of DA in cultured rat pheochromocytoma (PC12) cells (shown as Fig. 1), and the results were in good agreement with that reported in the literature.[25] Communication between cells by releasing specific biochemical or chemical messengers via exocytosis (such as neurotransmitters, hormones and pheromones) plays crucial roles in biological process.[26] Consequently, single cells and even single subcellular structures can be directly sampled on microchip platforms enabling quantitative and kinetic detection of biologically important compounds at higher sensitivity and resolution by nanoelectrodes compared to traditional electrode formats.

Gold individual nanoelectrode with a very sharp tip (diameter of 30 nm) has been used to monitor biological phenomena — photosynthesis in a living

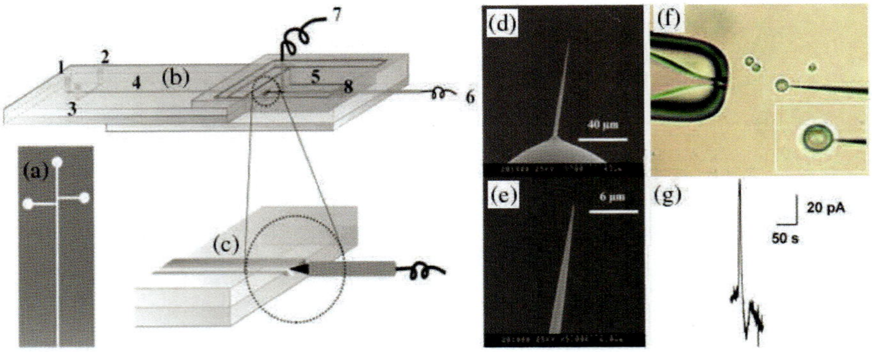

Fig. 1. Amperometric monitoring of DA release from PC12 single cells by a microchip integrated with CFNE. (a) Layout of the chip; (b) Configuration of electrophoresis system with an integrated electrochemical detector: (1) sample reservoir, (2) buffer reservoir, (3) sample waste reservoir, (4) separation channel, (5) detection cell, (6) working electrode, (7) reference electrode and (8) epoxy; (c) Enlarged view of the outlet of separation channel; (d) SEM images of CFNE; (e) SEM image of CFNE tip; (f) Photographs of actual nanoelectrode-cell arrangement; (g) Electrochemical signal of DA detected by CFNE.[22,25]

algal cell.[27] Abundant solar energy is stored and converted into chemical bond energy by photosynthesis in vascular plants, algae and some bacteria. However, only a small fraction of solar energy is converted to a form of energy that can be readily exploited. Therefore, gold nanoelectrode has been employed to investigate the high energy electrons (e^-) generated during photosynthesis by directly extracting electrons from the photosynthetic electron transport chain. A living algal cell (from *Chlamydomonas reinhardtii*) was trapped and maintained in an open, microfluidic substrate. The ultrasharp gold nanoelectrode was penetrated into the algal cell and photosynthetic electrons (1.2 pA at 6000 mA/m^2) were directly extracted by applying an overvoltage.[27] This result may represent an initial step in collecting "high energy e^-" in steps of the photosynthetic electron transport chain and prior to the downstream processes associated with energy loss.

Some metal oxide nanomaterials (such as copper oxide (CuO),[28,29] Zinc oxide (ZnO),[30–32] etc.) have attracted considerable attention for fabrication of nanobiosensors because of their unique advantages including the high surface-to-volume ratio, non-toxicity, chemical stability, electrochemical activity and high electron communication features.[7] ZnO nanoelectrodes including nanorods, NWs, and nanotubes have been used for *in vitro* detection and diagnosis of different biological and chemical components, such as cholesterol[31,32] and calcium ion (Ca^{2+}).[30] Due to the small size of ZnO nanorods (diameter from 70 nm to 120 nm, length from 700 nm to 1000 nm), they have been grown on a silver-coated tip of a borosilicate glass capillary

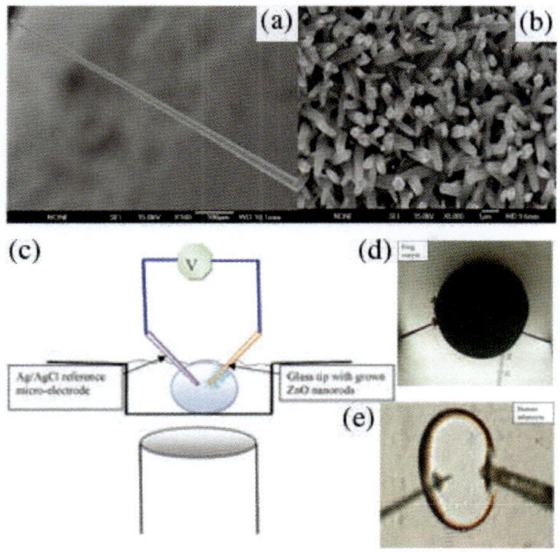

Fig. 2. Functionalized ZnO nanorods based electrochemical sensor for intracellular analysis. (a) and (b): Scanning electron microscopy (SEM) images of the ZnO nanorods grown on Ag-coated glass capillaries; (c) Schematic diagram illustrating the intracellular measurement setup; (d) and (e) Microscope images of a single human fat cell (d) and a single frog egg cell (e) during measurements with a ZnO nanorods coated probe as a working electrode and with an Ag/AgCl reference microelectrode.[36]

(700 nm in diameter) to create intracellular biosensors for monitoring *in vivo* biological process within single cells[33–36] (shown as Fig. 2). The intracellular pH measurements have been performed in a single human fat cell using ZnO nanorods as a working electrode with a Ag/AgCl reference microelectrode.[33] Furthermore, after covering polymeric membrane containing Ca^{2+}-selective ionophores or magnesium ion (Mg^{2+})-selective ionophores, ZnO nanorods exhibit a Ca^{2+}-dependent or Mg^{2+}-dependent electrochemical potential difference versus an Ag/AgCl reference electrode over a large concentration range of 100 nM–10 mM for Ca^{2+} and 500 nM–100 mM for Mg^{2+}.[34,35] The Ca^{2+} and Mg^{2+} are the most abundant intracellular divalent cations in living cells, where they play major biological roles in regulating metabolism, enzyme activity, neuronal activity, muscle contraction, cell death and development. The measurements of Ca^{2+} and Mg^{2+} concentrations using ZnO nanorods sensor in human fat cells or in frog egg cells (shown as Fig. 2) were consistent with values of Ca^{2+} and Mg^{2+} concentrations reported in the literature. The ZnO nanorods have also been functionalized with GOx for selectively measuring intracellular glucose.[36] The GOx-coated ZnO nanorods have exhibited a glucose-dependent electrochemical potential difference versus an Ag/AgCl reference microelectrode over the concentration range of interest (0.5–1000 μM). The measured glucose concentration in human fat cells or frog egg cells using ZnO-nanorod sensor has also been consistent

with values of glucose concentration reported in the literature. This kind of nanoelectrode devices demonstrates a simple technique to measure intracellular biochemical species in specific locations within single living cells.

Nanobiosensors based on individual nanoelectrodes and metal oxide nanomaterials have shown great promise for detection, diagnosis, and classification of different biochemical species within single cells in order to understand cellular function and may offer enormous potential to cell biology. However, the research is still at a very early stage, and many efforts are necessary to obtain reliable instrumentation for intracellular measurements.

4. Gold Nanoelectrodes and Nickel NWs

Gold NEEs and NEAs are a widely-used nanoelectrode platform, which can be facilely prepared through the template synthesis methods and other approaches.[7–9,37,38] Recently, three-dimensional (3D) structured NEEs have aroused considerable interest for electrochemical studies and application due to their unique configurations and large surface areas.[7–9,37] 3D structured NEEs are typically produced by etching upper layers of the template membrane from a flat two-dimensional (2D) array, creating additional active surfaces and chemistries for detection. The immobilization of AuNPs onto the surface of 2D NEEs has also used to increase the active area and result in the further improved sensitivity.[19] An apparent advantage of gold NEEs is an enhanced signal-to-background current ratio, leading to enhanced detection limits.[7–9] This makes gold NEEs very attractive for fabrication of electrochemical sensors for use in clinical diagnostic applications. Gold NEEs and NEAs have been used to directly detect diverse electroactive species of pharmacological, medical and healthy interests such as iodide,[39] cytochrome c,[40] phenothiazines,[41] levodopa,[42] organophosphorous pesticides,[43] daunorubicin (DNR),[44] arsenic,[45] human serum chorionic gonadotrophin[46] and DNA.[47] Compare with conventional gold disc electrode, the detection limits of above electrochemical sensors have been enhanced 10^2–10^3 times. For example, a nanoscale approach using 3D gold NEEs for quick and reliable detection of DNA has been reported with a detection limit of 1 pM.[48,49] Furthermore, gold NEEs have been demonstrated to promote both sensitivities and detection limits when used to produce enzyme biosensors after the immobilization of redox-active proteins (horseradish peroxidase (HRP),[50] GOx,[51,52] hemoglobin (Hb),[53] tyrosinase[54] at electrode surface. At present, development of biosensors for reliable early detection of cancer from biological fluids via minimally invasive methods is a high priority. Biosensor approaches utilizing gold NEEs have been reported for the sensitive detection of the cancer biomarkers. For example, the detection limits of biosensors on gold NEEs to prostate-specific antigen (PSA, prostate cancer biomarker),[55] human endothelial receptor 2 (HER2, breast cancer biomarker),[56] ovarian cancer antigen-125 (CA 125, ovarian cancer biomarker)[57] and Mucin-16 (ovarian cancer biomarker)[58] has been promoted to 1 pM,[55] 0.04 ng/μL,[56] 0.5 U/mL[57] and 5×10^{-4} U/mL,[58] respectively. These results indicate that gold NEE is a good immobilization matrix, providing a large well-defined surface area with the capacity to modify the nanowire surfaces by linking to biomolecules, where these biomolecules maintain their activity and functional states.

Miniaturization of the entire experimental setup is a key requirement for the development of POC testing devices. The POC devices based on MEMS-based microchip have bring a number of advantages including less sample consumption, better portability, faster detection speed and lower chip cost.[59] Gold nanoelectrodes have displayed their advantages due to the easy synthesis and assembly in microscale or nanoscale platform. A nanobiosensor with proposed architecture integrated with gold NWs as working electrode has been set up for estimating total blood cholesterol.[60] Cholesterol oxidase and cholesterol esterase were immobilized on the gold nanowire and an enhanced sensitivity of 69 nA/mM was achieved. Along with the advent of assembling gold NWs on micro-fluidic platforms and their compatibility with standard MEMS/microfabrication technology, the proposed architecture can be integrated with an array of other sensors/systems on a lab-on-chip format.[60]

The microchip CE combined with electrochemical detection has received increased attention due to the sensitive detection, the high efficiency separation, and the possibility of the integration with a miniaturized system. A miniaturized total analysis system-microchip CE have been set up by integrating gold NEEs, [61–63] shown as Fig. 3. The fabricated microchip has been used for

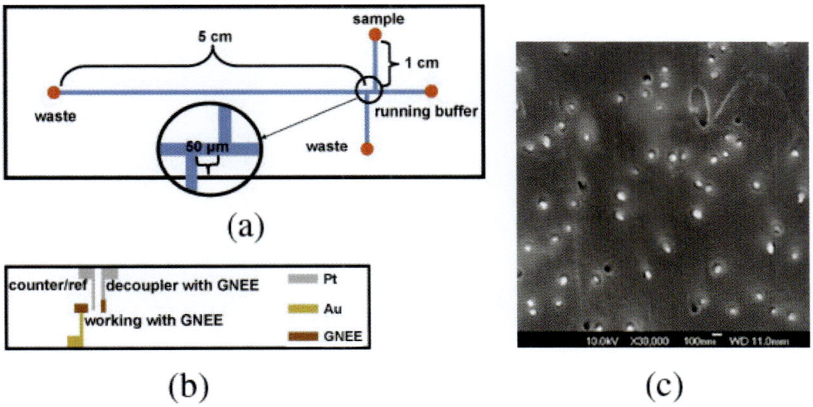

Fig. 3. (a) Overall microchannel configuration of the microchip CE embedded with gold NEE; (b) Amplified gold NEE working electrode and different decoupler electrodes; (c) SEM image of the gold NEE film.[61,62]

detection of in-column urea/urease reactions with a satisfactory detection limit of 14.8 and 62.8 μM, for detection of standard NH_4^+ solution and the urea/urease reaction product of NH_4^+, respectively. Furthermore, this microchip CE has been improved by featuring embedded gold NEE working and decoupler electrodes.[62,63] The function of gold NEE decoupler electrode was to reduce both the amplitude of the charge current and the baseline drift at higher separation voltages. Therefore, the measured baseline current drift for the microchip equipped the proposed gold NEE decoupler electrode was decreased around three times and the calculated signal response was approximately five times higher, which resulted in an enhanced detection limit of 1 μM toward DA[62] and 100 nM toward mercaptopurine,[63] respectively.

Recently, a dual channel configuration microchip CE has been developed to avoid the interference from applied potential used in the CE separation,[64] shown as Fig. 4A. The amperometric signals can be measured without any potential shift and interference from the applied CE potential. The improved performance of this dual channel CE has resulted in a high detectability of 0.1 pA for analysis of DA, which corresponds to a 1.8 nM concentration.[64] And a novel format of gold nanoelectrodes–gold nanoband electrodes have been fabricated on SU-8 substrates with a 100-nm-width gold layer and introduced into the dual-channel microchip CE to be an electrochemical sensor[65] (shown as Figs. 4B and 4C). Due to the nano-sized width of the gold nanoband electrodes, the noise of the amperometric detection was further significantly reduced and a high separation resolution was achieved for monitoring the analytes. The detection sensitivity of this microchip CE was improved by high S/N ratio with a low detection limit of 2.09 nM toward p-aminophenol.[65]

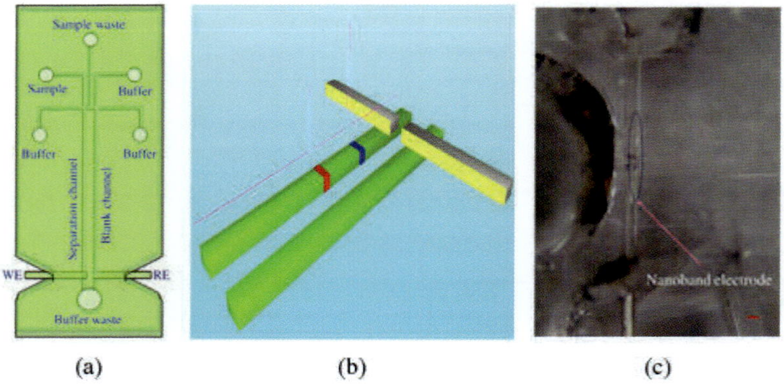

Fig. 4. (a) The schematic of the dual-channel CE microchip with nanoband electrodes; (b) The alignment sketch of the nanoband electrodes mounted at the exit of the channels as WE and RE, respectively; (c) CCD image of the nanoband electrodes inserted into PDMS microchip.[65]

Consequently, these results indicate that the proposed micro devices with embedded gold NEEs provide a feasible solution for performing high resolution, high speed and low cost detection in lab-on-chip and micro total analysis systems applications. It is anticipated that future economic and portable POC devices will rely on these microchip analysis systems.

Metallic NWs have offered new possibilities for use in electrochemical biosensors because of their easy preparation, stability, and large active surfaces allowing greatly enhanced sensitivity. Nickel (Ni) NWs are becoming a new simple easy-to-use, and disposable enzyme-free nanoelectrodes for fast and reliable total carbohydrate sensing in real analytical domains.[66,67] Ni NWs and nickel–copper (Ni–Cu) NWs have been synthesized on disposable screen printed platforms by using a simple, easy, and low-cost template electroplating method.[67] Both the Ni NWs and Ni–Cu NWs exhibited excellent analytical performance for detection of sugars, such as low detection limits (0.06 mM for glucose and 0.04 mM for fructose) and fast analytical time (a rate of 1 sample per 30 s). The Ni NWs have been further coated with platinum (Pt) particles and both Ni NWs and Pt/Ni NWs have been developed to an enzyme-free electrochemical sensor for detection of glutamate.[68] The detection limits of Ni NWs and Pt/Ni NWs-based sensors were 68 μM and 83 μM, respectively and the sensitivity of Ni NWs has been found to be approximately 6 to 9 times higher than the reported glutamate sensors. Therefore, the Ni NWs have showed significant promise for fabricating enzyme-free sensor platform. Especially, the metallic NWs based disposable screen-printed platforms are easily connected to miniaturized instrumentation (e.g. microfluidics) and represent a novel candidate that could be used as POC testing for biological and medical analysis.

5. CNTs NEA and CFNE Array

CNTs are an important class of nanomaterials due to their high degree of chemical, electrochemical and thermal stability and exceptional mechanical properties.[7–9] CNTs including single walled CNTs (SWCNTs) and multi-walled CNTs (MWCNTs) are ideal electrode materials as they provide a compatible microenvironment for enzyme, helping enzyme to retain its bioactivity and directing electron transfer between enzyme's active site and electrode, which has produced highly sensitive biosensors.[69,70] A vertically aligned CNTs NEA can be easily synthesized through simple but effective methods (e.g. chemical vapor deposition (CVD), photolithography), which also enables the density of nanoelectrodes controllable for diverse purposes.[70–73] The array of nanoelectrodes can produce a much higher current than a single nanoelectrode, which can improve the signal to noise ratio. Consequently,

CNTs NEA is an excellent ultra sensitive electrochemical component for chemical and biological sensing. As an example of utilizing CNTs NEAs, the detection of trace concentrations of species, such as lead (II), at ppb level has been successfully achieved.[74] NEA consisting of probe biomolecules immobilized on CNTs tips has attracted much attention for the development of nanobiosensors.[7–9] A number of different biomolecules (enzyme, DNA, etc.) have been immobilized on CNTs via COOH groups at the nanotube ends.[75] The resulting biosensors can promote both electron transfer rates (1500 s^{-1})[76] and detection limits (1 attomole to target DNA).[77–80] All of these results indicate that the CNT NEA is a promising platform due to its biocompatibility to fabricate biosensor with good sensitivity.

The high ordered CNT NEA provides a possibility for fabrication of multiplexed sensors monitoring multiple analytes at once.[81,82] DNA oligonucleotide served as a molecular link to address the site-specific delivery of a large number of different proteins to spatially distinct regions of a CNT NEA. The sequence specificity of DNA hybridization ensured that each enzyme carrying a single-stranded DNA (ssDNA) oligo tag would bind exclusively to the nanoelectrode region modified with the complementary single strand. Different enzymes were simultaneously addressed onto a CNT NEA to form a multiplexed amperometric biosensor capable of monitoring several substrates in real time, shown as Fig. 5. Moreover, the reversibility of DNA hybridization allows removing inactive enzyme from the CNT NEA and replacing it with fresh enzyme to renew and extend the lifetime of the

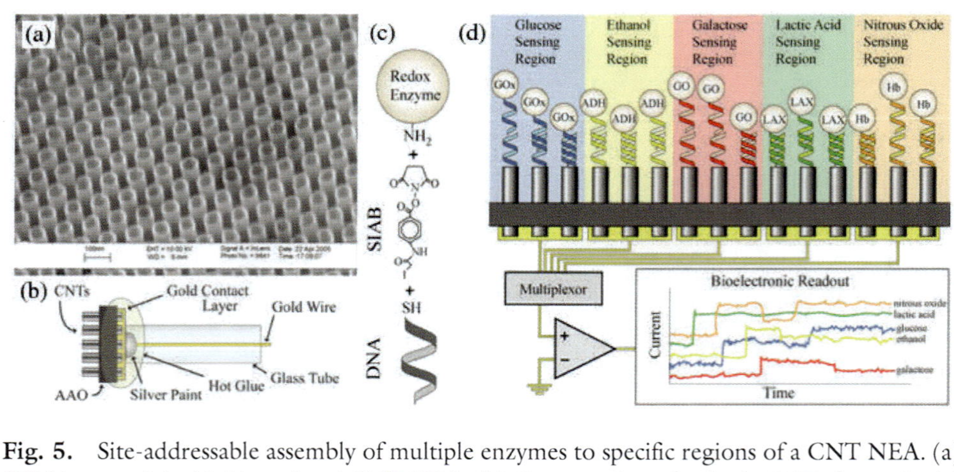

Fig. 5. Site-addressable assembly of multiple enzymes to specific regions of a CNT NEA. (a) SEM image of the highly ordered CNT NEA; (b) Construction of a single CNT electrode; (c) The enzyme–ssDNA assembly; (d) Parallel hybridization of the five enzymes to distinct regions of a CNT NEA modified with differing ssDNA "addresses". Electrocatalytic currents originating from the enzymes indicate the corresponding substrate concentrations.[81]

system.[82] Overall, this DNA–CNT NEA linking strategy offers a relatively simple and programmable means of directing and controlling the nanoscale assembly of individual biomolecules and nanoelectronic elements while preserving high levels of biofunctionality and providing a stable physical link and an efficient electrical contact. Furthermore, this method is suitable for any redox enzyme-based bioelectronics device, and can vastly facilitate the controllable and site-addressable assembly, disassembly and replacement of fragile bio-components by virtue of sequence-specific DNA hybridization. Such CNT NEA will have the capability for performing a high level of multiplexed sensing or analysis, which will have tremendous functionality while using only small amounts of sensor material, and thus could provide a universal platform for cost-effective analysis.

Individual CFNE has shown promise as neurochemical recording electrodes due to their high sensitivity, rapid electron transfer kinetics, biocompatibility and manufacturing versatility.[20–22,25] Recently, a patterned and vertically aligned CFNE array has been synthesized by controlling electrode size, spacing, and array size, which provides high stability and improved S/N ratio.[83,84] The CFNEs have also demonstrated its capability of both stimulating and monitoring electrophysiological signals from brain tissues *in vitro* and monitoring dynamic information of neuroplasticity.[85] For an example, the vertically aligned CFNEs have been found to be effective for the simultaneous detection of DA and serotonin (5-HT) in the presence of excess ascorbic acid (AA) with the detection limits of 50 nM for DA and 250 nM for 5-HT, respectively.[86] Moreover, the electrochemical detection of DNA targets,[87] ricin[88] and cardiac troponin-I[89] by using vertically aligned CFNE array has been reported, providing a simple, reliable and low-cost approach in developing practical biosensors for applications like pathogen detection, early cancer diagnosis and environmental monitoring. Especially, the CFNE array has been integrated with the Wireless Instantaneous Neurotransmitter Concentration Sensor System (WINCS) for neurochemical analysis.[90] The sensor devices of patterned CFNE arrays (diameter of ~100 nm) were prepared on a 4-inch silicon wafer using standard silicon processing technology, shown as Fig. 6. These devices are insulated with silicon dioxide, with only the tips of the CFNEs exposed for sensing. Each device consists of 3 × 3 electrode pads (200 μm square), containing CFNEs spaced at 1 μm intervals and allowing for simultaneous neurochemical detection in 9 locations. Therefore, the sensor device provides high spatial resolution over a larger area of interrogation. As shown in Figs. 6D and 6E, the CFNEs showed a clear DA oxidation current upon DA injection over a range of concentrations (0.5 to 2.5 μM), occurring at 25 sec after the start of the data acquisition and

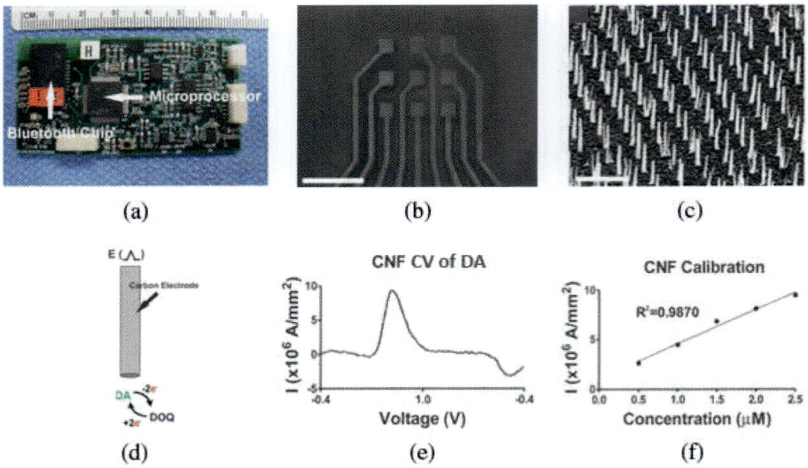

Fig. 6. WINCS device for DA detection. (a) Printed circuit board indicating the FSCV microcontroller and the Bluetooth module for wireless communication with a data-acquisition computer; (b) SEM image of CFNE-based 3 × 3 sensor pad array; (c) High-resolution SEM image of CFNEs on a sensor pad; (d) Schematic of carbon electrode and DA redox reaction; (e) Cyclic voltammogram at CFNE for DA injection; (f) Measured current densities at varied DA concentrations.[90]

ending at 60 sec. And the DA detection performance by CFNE arrays was comparable to that of traditional carbon fiber microelectrodes, demonstrating that CFNE arrays can be utilized as an alternative carbon electrode for neurochemical monitoring. The relative ease by which CFNE arrays can be replicated in novel electrode designs is expected to facilitate the detection of neurochemical release over a large area of interrogation, yet with high spatial resolution. The CFNEs may potentially serve as a precise, informative, and biocompatible device for monitoring of biological activity at the single-cell level and even inside the cell.

6. Coordinated biosensor

As discussed above, the electrochemical biosensors have received great attention due to their unique properties toward developing sensitive, simple, inexpensive, and rapid diagnostic devices. However, the full realization of the potential of biosensors has not yet to be reached because the rates of reactions and sensitivities of the biomolecules used as detectors are generally significantly lower when used *in vitro* than when they are in the endogenous, biological environment. This limitation is largely due to the random nature in which the biomolecule (e.g. redox enzyme) acting as detector is situated with respect to the sensing component of the biosensor (e.g. an electrode), or which in turn affects the signal transduction. For an example, enzymes are mostly globular, with their active sites buried at the center or at oligomeric (multimeric) interfaces. When fabricate enzyme biosensors, the electrons produced from a reaction cannot efficiently get out to the surface to be used or are not directed to the electrode and lost to solution if redox enzyme is not optimally positioned with the electrode. This results that the electron transfer rate constants in electrode-contacted enzymes are far lower than those between enzymes and their natural electron acceptors. Consequently, the markedly decreased electron transfer rates cause the electronic signals too weak to be detectable, making these enzyme-based biosensors impractical and limited in their sensitivities. This leads biosensors that are functional in the lab may not be of use in the practical field (such clinic field) because of less sensitivity and narrow dynamic range, which makes the translation from bench to bedside outside the research setting more difficult. Therefore, control of the structure and relative orientation of the biomolecule is therefore an important consideration for the design and performance of an electrochemical biosensor.

To address these issues, an approach of "coordinated biosensor" based on exploiting the innate sensitivity and specificity of nature biomolecules (e.g. proteins, nucleic acids), harnessing the innate specificity and sensitivities of these biomolecules has been developed to detect diverse specific targets.[91–93] The biomolecule (e.g. redox enzyme) is precisely and modularly linked to a nanoelectrode platform (e.g. gold nanoparticles, gold NEEs, CNTs) through a "nano-biolinker" composed of metallizing DNA, peptide and peptide nucleic acid (PNA). The nano-biolinker directs electron transfer efficiently from enzyme to electrode surface, resulting in enhanced sensitivities and activities by several orders of magnitude. A coordinated nanobiosensor by using metallizing peptides to serve as conduits of electronic signals that bridge between a redox enzyme and a CNT NEA has first been developed.[92,93] Synthetic peptide-based biosensors have been developed in recent years according to different strategies.[94] In the coordinated biosensor, a peptide comprised of a leucine zipper motif was mutated to bind divalent metals, conferring conductivity into the peptide. A thiolate at the end of peptide reacted with the sulfenic acid of the NADH peroxidase (Npx) enzyme to form a peptide-enzyme bioassembly and this bioassembly was immobilized to the tips of CNT electrodes. The formation of a CNT NEA–Npx bioassembly electrode is shown in Fig. 7. The metallized peptide provided a bridge of directing the electrons generated at the enzyme center to the electrode. Also, the peptide served as a spacer between the protein and the electrode surface,

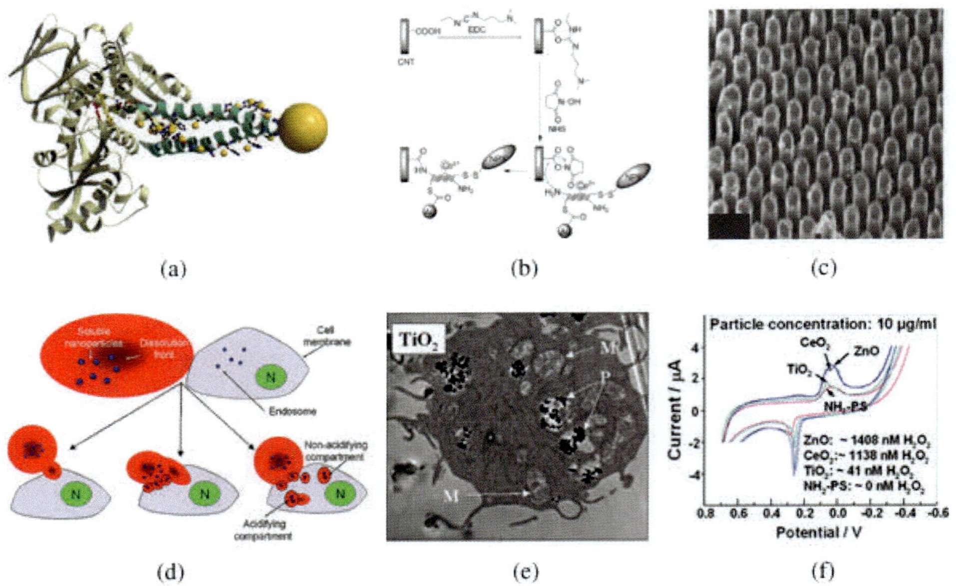

Fig. 7. (a) Model of the Npx-peptide bioassembly; (b) Fabrication process of CNT NEA–Npx bioassembly electrode; (c) SEM image of Npx bioassembly immobilized on tips of CNT electrode; (d) Schematic of nanoparticles induced toxicity in cells, leading to the generation of reactive oxygen species; (e) Use of electron microscopy to determine the uptake and subcellular localization of nanoparticles; (f) Cyclic voltammograms of individual NP demonstrate a clear redox signal for ZnO, CeO_2 and TiO_2 by using the coordinated biosensor.[91,92,95 and 96]

helping to prevent surface denaturation. Therefore, the fabricated CNT NEA–Npx bioassembly nanobiosensor had a markedly improved electron transfer rate of 3000 s^{-1},[91] which was two orders of magnitude higher than the endogenous electron transfer rate from NADH to the native enzyme (27 s^{-1}). Moreover, the sensitivity of this coordinated biosensor to H_2O_2 was enhanced to 1 pM that has been successfully used for evaluation of the safety of engineered nanomaterials. The safety of engineered nanomaterials is of great importance to the nanotechnology industry that could impact the implementation and acceptance of new nanomaterials in society. However, the emerging toxicity of nanomaterials (such as TiO_2, CeO_2 and ZnO) may cause most oxidative injury (reactive oxygen species and oxidative stress) within cells that can impair or destroy cell proteins, lipids and DNA, leading to the deterioration in cell function linking diverse diseases including cancer, asthma and cardiovascular ailments (shown as Fig. 7). This nanobiosensor has been successfully used to detect H_2O_2 production from these nanoparticles, which could offer more information for the study in oxidative stress and cytotoxicity.[95,96] These results suggest that measuring oxidative stress could eventually screen the thousands of new nanoparticles made every year. Overall, the results highlight the unique electrical and reaction properties of CNT NEA and the potential for coordinated linkage of biological components using biological linkers (such as metalized peptides, PNA, DNA, etc.) to produce highly sensitive biological nanosensors.

Early cancer diagnosis is for patient survival and successful prognosis of the disease. When cancer is diagnosed in its earliest stages, treatment for the disease is much more effective and most likely to succeed. As the cancer progresses, treatment methods change and their efficacies decrease. Most cancers present a number of aberrant proteins or increased production of other proteins in blood, urine and other body fluids, which can be used as cancer biomarkers. However, early diagnosis of cancer is difficult because of the low initial concentrations of these biomarkers in blood samples, generally in the ng mL^{-1} range or even lower. A coordinated biosensor utilizing metallizing PNA has been further developed to fabricate electrochemical biosensors for ultrasensitive detection of prostate cancer biomarker — PSA.[97] Just like DNA and RNA, PNA forms duplexes by Watson Crick hybridization. X-ray crystallography studies have shown that PNA duplexes adopt a helical structure with a larger displacement from the helical axis and smaller twist than DNA duplexes.[98] It is suggested that this geometrically restricted structure contributes orientation and formation of well-ordered nanostructures on the electrode. As with pseudo-peptide backbone, PNA is easily modified with biological and chemical elements, such as thiol groups,[99] amino groups,[100] biotin[101] and metal chelating groups,[102] which make PNA particularly useful for creating biosensors. In this PSA coordinated biosensor, the lysine groups situated at the both ends of the PNA are used for covalently linking PSA-antibody and gold nanoelectrode on microfluidic chip respectively. The microfluidic biosensor chip is built with an embedded three-electrode configuration with controlled inter-electrode distance in a nanoliter-scale sample volume (shown as Fig. 8). The on-chip three-electrode cell consists of a 5 × 5 array of gold working nanoelectrodes with radii between 60 and 120 nm, a Ag/AgCl reference electrode, and a gold counter electrode. A major improvement is that this chip-based platform incorporating a complete electrochemical cell which can detect specific biomolecular signals in nanoliter volumes and in a flow-cell environment. Compared to other approaches, a distinguishing feature of coordinated biosensing method for sensor fabrication is that the PSA-antibody is precisely linked to a gold nanoelectrode through a PNA nano-biolinker, shown in Fig. 8e. The stable helical structure of PNA makes it serve as a spacer between the molecule and the electrode surface, helping to prevent surface denaturation. The other advantage of the

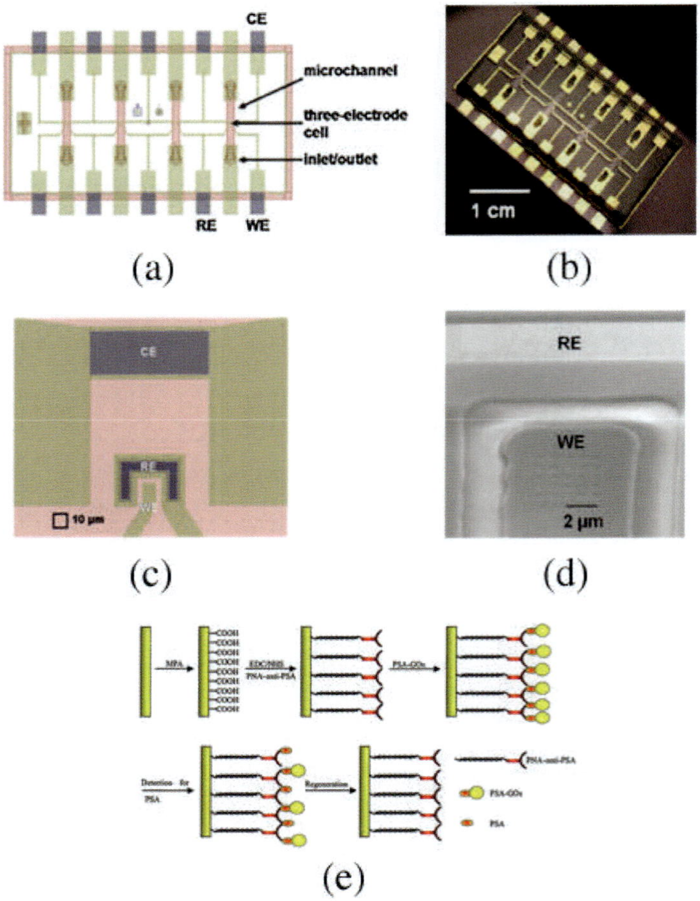

Fig. 8. (a) Layout of the microchip; (b) Microchip with microchannel assembled on the glass substrate; (c) Layout of the three-microelectrode cell; (d) SEM image of gold NEAs; (e) Schematic diagram of the assembly of coordinated nanobiosensor on gold nanoelectrode.[97]

PNA linker is to provide a "nano-bridge" by directing the electrons generated at the enzyme to the electrode, and thus enhances the electron transfer rate. The electron transfer properties of PNAs show that the electron transfer rates of PNA are from 0.02 to 2000 s^{-1} depending on the sequence and length of the PNA.[103,104] In this PSA coordinated biosensor, a strategy to incorporate metal in PNA duplexes to enhance electron transfer rate has been developed. The metal-containing PNA duplexes are formed by hydrogen bonding in nucleobase pairs and coordination bonding in the metal complexes (e.g. $Cu^{+/2+}$), which act as alternative base pairs.[105] This strategy affords excellent control over the type and position of metal ions incorporated in the PNA duplexes. Such an addition of metal in the PNA significantly improves the electron transfer rate (5.4 s^{-1}), which is two orders higher than 7 mer single stranded PNA (e.g. Cys-TTTGTTT, 0.076 s^{-1}).[103,104] Therefore, the detection limit is approximately 10 pg/mL (~270 fM), which corresponds to roughly 30,000 copies of PSA in the microchannel test volume. This should be readily adaptable to cancer biomarkers having very low levels. These experimental results demonstrate a promising approach for the development of ultrasensitive bio-signal transducers in a chip-based miniaturized analytical platform, which may be of importance to the development of POC devices of the future. The system is being functionalized with other linker assemblies for the detection of various biomolecular targets with high sensitivity.

The efficient electrical communication between biomolecules (e.g. redox proteins) and electrode surfaces is always a critical problem in fabrication of biosensor. A biomimetic, modular coordinated biosensor strategy has been reported that have very high accuracy, quick response time, compact in size, modular for facile integration into handheld devices and economical to produce.

7. Conclusion

In the past years a wide range of biosensors based on nanoelectrodes has been developed to identify and detect diseases associated proteins, nucleic acid sequences, key biochemicals in metabolic pathways, which can provide highly detailed knowledge of patients' conditions, disease pathways, or the presence of contamination to biomedical researchers and healthcare professionals. As highlighted in this review, nanoelectrodes offer significant advantages including increased S/N ratio, improved sensitivity, high mass-transport rate and fast electron transfer rate. The nanoscale dimensions of nanoelectrodes lead to facile integration into microchips, providing a feasible solution for intracellular analysis in single living cells. Integrating the nanobiosensors with wireless technology has enabled the biosensors to be interrogated remotely. Molecular knowledge of the 3D atomic structures and functions of proteins and nucleic acids coupled to precise linkage of the biomolecules to nanoelectrodes is the basis of coordinated nanobiosensors. Utilizing geometric and topological constraints to coordinate electron transfer centers to direct efficient signal transduction between redox proteins and electrode surfaces through biolinkers enhances sensitivities and activities by several orders of magnitude. Therefore, nanoelectrodes have already been applied for ultrasensitive detection methodologies and show great potential for further development as efficient, specific, sensitive and intelligent devices. These devices would be novel powerful detection platforms that have potential to significantly impact analysis of biological activity and diagnosis of diseases, such as cancers and metabolic disorders, genetic disorders screening and new drug discovery.

According to the Frost & Sullivan (http://www.transportation.frost.com), the biosensors market is categorized as a growth market and expected to grow from $6.72 billion in 2009 to $14.42 billion in 2016. However, the goal of low-cost, robust, reliable, easy-to-use, ultrasensitive multiplexed clinical diagnostic nanobiosensor devices are yet to be truly realized. Integration of diverse expertise, in materials fabrication, molecular biology, biochemistry, surface, and analytical chemistries through collaborative efforts can be tremendously productive. The transition of nanobiosensors functional in the lab to commercial devices will open up new avenues for the development of innovative POC diagnostic devices that allow for personalized medicine, enabling real-time monitoring of the user and their environment (such as regular health checkup and detection of hazardous substances causing health problems) and direct coupling of this sensor data to new response mechanisms (such as targeted drug release or immediate diagnosis). The major breakthroughs based on nanoelectrode technology will occur in biology and medicine in the near future, with enormous implications for science and society.

Acknowledgment

This work was supported.

References

1. N. J. Ronkainen, H. B. Halsall and W. R. Heineman, Electrochemical biosensors. *Chem. Soc. Rev.* **39**(5), 1747–1763 (2010). Epub 2010/04/27, doi: 10.1039/b714449k, PubMed PMID: 20419217.
2. L. C. Clark, Jr. and C. Lyons, Electrode systems for continuous monitoring in cardiovascular surgery. *Ann. New York Acad. Sci.* **102**, 29–45 (1962). Epub 1962/10/31, PubMed PMID: 14021529.
3. J. Wang, Electrochemical biosensors: towards point-of-care cancer diagnostics. *Biosens. Bioelectron.* **21**(10), 1887–1892 (2006). Epub 2005/12/07, doi: 10.1016/j.bios.2005.10.027, PubMed PMID: 16330202.
4. J. Wang, Electrochemical glucose biosensors. *Chem. Rev.* **108**(2), 814–825 (2008). Epub 2007/12/25, doi: 10.1021/cr068123a, PubMed PMID: 18154363.
5. P. D'Orazio, Biosensors in clinical chemistry — 2011 update. *Clin. Chim. Acta* **412**(19–20), 1749–1761 (2011). Epub 2011/07/07, doi: 10.1016/j.cca.2011.06.025, PubMed PMID: 21729694.
6. P. Arora, A. Sindhu, N. Dilbaghi and A. Chaudhury, Biosensors as innovative tools for the detection of food borne pathogens. *Biosens. Bioelectron.* **28**(1), 1–12 (2011). Epub 2011/07/19, doi: 10.1016/j.bios.2011.06.002, PubMed PMID: 21763122.

7. D. Krapf, Nonergodicity in nanoscale electrodes. *Phys. Chem. Chem. Phys.* **15**(2), 459–465 (2013). Epub 2012/10/17, doi: 10.1039/c2cp42838e, PubMed PMID: 23070026.
8. H. Shi and J. I. Yeh, Part I: Recent developments in nanoelectrodes for biological measurements. *Nanomedicine (Lond)* **2**(5), 587–598 (2007). Epub 2007/11/03, doi: 10.2217/17435889.2.5.587, PubMed PMID: 17976022.
9. H. Shi, T. Xia, A. E. Nel and J. I. Yeh, Part II: Coordinated biosensors — development of enhanced nanobiosensors for biological and medical applications. *Nanomedicine (Lond)* **2**(5), 599–614 (2007). Epub 2007/11/03, doi: 10.2217/17435889.2.5.599, PubMed PMID: 17976023.
10. D. Wei, M. J. A. Bailey, P. Andrew and T. Ryhanen, Electrochemical biosensors at the nanoscale. *Lab Chip* **9**(15), 2123–2131 (2009). PubMed PMID: ISI:000268033900002.
11. T. Vo-Dinh and Y. Zhang, Single-cell monitoring using fiberoptic nanosensors. *Wiley Interdiscip. Rev. Nanomed. Nanobiotechnol.* **3**(1), 79–85 (2011). doi: 10.1002/wnan.112, PubMed PMID: ISI:000285789700007.
12. L. M. Bellan, D. Wu and R. S. Langer, Current trends in nanobiosensor technology. *Wiley Interdiscip. Rev. Nanomed. Nanobiotechnol.* **3**(3), 229–246 (2011). Epub 2011/03/11, doi: 10.1002/wnan.136, PubMed PMID: 21391305.
13. M. Campas, D. Garibo and B. Prieto-Simon, Novel nanobiotechnological concepts in electrochemical biosensors for the analysis of toxins. *Analyst* **137**(5), 1055–1067 (2012). Epub 2012/01/12, doi: 10.1039/c2an15736e, PubMed PMID: 22234474.
14. A. S. Johnson, A. Selimovic and R. S. Martin, Microchip-based electrochemical detection for monitoring cellular systems. *Anal. Bioanal. Chem.* (2013). Epub 2013/01/24, doi: 10.1007/s00216-012-6682-3, PubMed PMID: 23340999.
15. S. K. Yadav, P. Chandra, R. N. Goyal and Y. B. Shim, A review on determination of steroids in biological samples exploiting nanobio-electroanalytical methods. *Anal. Chim. Acta* **762**, 14–24 (2013). Epub 2013/01/19. doi: 10.1016/j.aca.2012.11.037, PubMed PMID: 23327941.
16. T. Matsue, Development of biosensing devices and systems using micro/nanoelectrodes. *Bull. Chem. Soc. Jpn.* **85**(5), 545–557 (2012). doi: 10.1246/bcsj.20110249, PubMed PMID: ISI:000305168800002.
17. M. Ongaro and P. Ugo, Bioelectroanalysis with nanoelectrode ensembles and arrays. *Anal. Bioanal. Chem.* (2012). Epub 2012/11/29, doi: 10.1007/s00216-012-6552-z, PubMed PMID: 23187824.
18. M. Silvestrini and P. Ugo, Ensembles of nanoelectrodes modified with gold nanoparticles: characterization and application to DNA-hybridization detection. *Anal. Bioanal. Chem.* **405**(2–3), 995–1005 (2013). Epub 2012/09/08, doi: 10.1007/s00216-012-6354-3, PubMed PMID: 22955672.
19. M. Silvestrini and P. Ugo, Ensembles of nanoelectrodes modified with gold nanoparticles: characterization and application to DNA-hybridization detection. *Anal. Bioanal. Chem.* (2012). Epub 2012/09/08, doi: 10.1007/s00216-012-6354-3, PubMed PMID: 22955672.
20. W. H. Huang, D. W. Pang, H. Tong, Z. L. Wang and J. K. Cheng, A method for the fabrication of low-noise carbon fiber nanoelectrodes. *Anal. Chem.* **73**(5), 1048–1052 (2001). Epub 2001/04/06, PubMed PMID: 11289416.
21. R. S. Chen, W. H. Huang, H. Tong, Z. L. Wang and J. K. Cheng, Carbon fiber nanoelectrodes modified by single-walled carbon nanotubes. *Anal. Chem.* **75**(22), 6341–6345 (2003). Epub 2003/11/18, doi: 10.1021/ac0340556, PubMed PMID: 14616019.

22. W. Z. Wu, W. H. Huang, W. Wang, Z. L. Wang, J. K. Cheng, T. Xu *et al.*, Monitoring dopamine release from single living vesicles with nanoelectrodes. *J. Am. Chem. Soc.* **127**(25), 8914–8915 (2005). Epub 2005/06/23, doi: 10.1021/ja050385r, PubMed PMID: 15969544.
23. J. D. Elsworth and R. H. Roth, Dopamine synthesis, uptake, metabolism, and receptors: relevance to gene therapy of Parkinson's disease. *Exp. Neurol.* **144**(1), 4–9 (1997). Epub 1997/03/01, doi: S0014-4886(96)96379-7 [pii]10.1006/exnr.1996.6379, PubMed PMID: 9126143.
24. M. Greer and C. M. Williams, Dopamine metabolism in Parkinson's disease. *Neurology* **13**, 73–76 (1963). Epub 1963/01/01, PubMed PMID: 13950225.
25. H. Cheng, W. H. Huang, R. S. Chen, Z. L. Wang and J. K. Cheng, Carbon fiber nanoelectrodes applied to microchip electrophoresis amperometric detection of neurotransmitter dopamine in rat pheochromocytoma (PC12) cells. *Electrophoresis* **28**(10), 1579–1586 (2007). Epub 2007/04/21, doi: 10.1002/elps.200600603, PubMed PMID: 17447239.
26. W. Wang, S. H. Zhang, L. M. Li, Z. L. Wang, J. K. Cheng and W. H. Huang, Monitoring of vesicular exocytosis from single cells using micrometer and nanometer-sized electrochemical sensors. *Anal. Bioanal. Chem.* **394**(1), 17–32 (2009). doi: 10.1007/s00216-009-2703-2, PubMed PMID: ISI:000265035700006.
27. W. Ryu, S. J. Bai, J. S. Park, Z. Huang, J. Moseley, T. Fabian *et al.*, Direct extraction of photosynthetic electrons from single algal cells by nanoprobing system. *Nano Lett.* **10**(4), 1137–1143 (2010). Epub 2010/03/06, doi: 10.1021/nl903141j, PubMed PMID: 20201533.
28. D. Xiao, Z. J. Zhuang, X. D. Su, H. Y. Yuan, Q. Sun and M. M. F. Choi, An improved sensitivity non-enzymatic glucose sensor based on a CuO nanowire modified Cu electrode. *Analyst* **133**(1), 126–132 (2008). PubMed PMID: ISI:000251684300025.
29. L. Wang, G. F. Wang, Y. Wei, W. Zhang, X. J. Zhang and B. Fang, Enzyme-free amperometric sensing of glucose using Cu–CuO nanowire composites. *Microchim. Acta* **168**(1–2), 87–92 (2010). PubMed PMID: ISI:000274098300013.
30. M. H. Asif, O. Nur, M. Willander, M. Yakovleva and B. Danielsson, Studies on calcium ion selectivity of ZnO nanowire sensors using ionophore membrane coatings. *Res. Lett. Nanotechnol.* **2008**, 1–4 (2008).
31. P. R. Solanki, A. Kaushik, A. A. Ansari and B. D. Malhotra, Nanostructured zinc oxide platform for cholesterol sensor. *Appl. Phys. Lett.* **94**(14), (2009). PubMed PMID: ISI:000265083700072.
32. M. Q. Israr, J. R. Sadaf, O. Nur, M. Willander, S. Salman and B. Danielsson, Chemically fashioned ZnO nanowalls and their potential application for potentiometric cholesterol biosensor. *Appl. Phys. Lett.* **98**(25), (2011). PubMed PMID: ISI:000292039900075.
33. S. M. Al-Hilli, M. Willander, A. Ost and P. Stralfors, ZnO nanorods as an intracellular sensor for pH measurements. *J. Appl. Phys.* **102**(8), (2007). PubMed PMID: ISI:000250589300114.
34. M. H. Asif, A. Fulati, O. Nur, M. Willander, C. Brannmark, P. Stralfors *et al.*, Functionalized zinc oxide nanorod with ionophore-membrane coating as an intracellular Ca(2+) selective sensor. *Appl. Phys. Lett.* **95**(2), (2009). doi: 10.1063/1.3176441, PubMed PMID: ISI:000268089200113.
35. M. H. Asif, S. M. Ali, O. Nur, M. Willander, U. H. Englund and F. Elinder, Functionalized ZnO nanorod-based selective magnesium ion sensor for intracellular measurements. *Biosens. Bioelectron.* **26**(3), 1118–1123 (2010). Epub 2010/09/18, doi: S0956-5663(10)00521-X [pii]10.1016/j.bios.2010.08.017, PubMed PMID: 20846846.

36. M. H. Asif, S. M. Ali, O. Nur, M. Willander, C. Brannmark, P. Stralfors *et al.*, Functionalised ZnO-nanorod-based selective electrochemical sensor for intracellular glucose. *Biosens. Bioelectron.* **25**(10), 2205–2211 (2010). Epub 2010/03/23, doi: S0956-5663(10)00095-3[pii]10.1016/j.bios.2010.02.025, PubMed PMID: 20303253.
37. Y. B. Mollamahalle, M. Ghorbani and A. Dolati, Electrodeposition of long gold nanotubes in polycarbonate templates as highly sensitive 3D nanoelectrode ensembles. *Electrochim. Acta* **75**, 157–163 (2012). doi: 10.1016/j.electacta.2012.04.119, PubMed PMID: ISI:000306884100021.
38. L. M. Moretto, M. Tormen, M. De Leo, A. Carpentiero and P. Ugo, Polycarbonate-based ordered arrays of electrochemical nanoelectrodes obtained by e-beam lithography. *Nanotechnology* **22**(18), 185305 (2011). Epub 2011/03/24, doi: S0957-4484(11)73544-9 [pii]10.1088/0957-4484/22/18/185305, PubMed PMID: 21427473.
39. F. C. Pereira, L. M. Moretto, M. De Leo, M. V. Zanoni and P. Ugo, Gold nanoelectrode ensembles for direct trace electroanalysis of iodide. *Anal. Chim. Acta* **575**(1), 16–24 (2006). Epub 2007/08/29, doi: S0003-2670(06)01143-3 [pii]10.1016/j.aca.2006.05.056, PubMed PMID: 17723566.
40. P. Ugo, N. Pepe, L. M. Moretto and M. Battagliarin, Direct voltammetry of cytochrome c at trace concentrations with nanoelectrode ensembles. *J. Electroanal. Chem.* **560**, 51–58 (2003). doi: 10.1016/j.jelechemn.2003.06.007, PubMed PMID: ISI:000186955700007.
41. B. Brunetti, P. Ugo, L. M. Moretto and C. R. Martin, Electrochemistry of phenothiazine and methylviologen biosensor electron-transfer mediators at nanoelectrode ensembles. *J. Electroanal. Chem.* **491**(1–2), 166–174 (2000). PubMed PMID: ISI:000165773500020.
42. J. A. A. Ho, S. Viswanathan, W. C. Liao, C. C. Huang and W. L. Hsu, Rapid analysis of L-dopa in urine samples using gold nanoelectrode ensembles. *Talanta* **74**(2), 229–234 (2007). doi: 10.1016/j.talanta.2007.05.056, PubMed PMID: ISI:000251629800009.
43. Y. H. Yang, X. X. Lu, H. P. Bai, Q. Ruan, M. H. Yang, G. M. Yang *et al.*, Direct determination of pesticides in vegetable samples using gold nanoelectrode ensembles. *Int. J. Environ. Anal. Chem.* **88**(11), 813–824 (2008). doi: 10.1080/03067310801902500, PubMed PMID: ISI:000258312500005.
44. L. X. Cao, P. S. Yan, K. N. Sun and D. W. Wirk, Tailor-made gold brush nanoelectrode ensembles modified with L-cysteine for the detection of daunorubicine. *Electrochim. Acta* **53**(28), 8144–8148 (2008). doi: 10.1016/j.electacta.2008.06.016, PubMed PMID: ISI:000259835300006.
45. A. Mardegan, P. Scopece, F. Lamberti, M. Meneghetti, L. M. Moretto and P. Ugo, Electroanalysis of trace inorganic arsenic with gold nanoelectrode ensembles. *Electroanalysis* **24**(4), 798–806 (2012). doi: 10.1002/elan.201100555, PubMed PMID: ISI:000302160000012.
46. G. M. Yang, X. Y. Yang, C. Y. Yang and Y. H. Yang, A reagentless amperometric immunosensor for human chorionic gonadotrophin based on a gold nanotube arrays electrode. *Colloid Surf. A Physicochem. Eng. Asp.* **389**(1–3), 195–200 (2011). PubMed PMID: ISI:000296937900028.
47. M. Silvestrini, L. Fruk and P. Ugo (2013). Functionalized ensembles of nanoelectrodes as affinity biosensors for DNA hybridization detection. *Biosens. Bioelectron.* **40**(1),

265–270 (2013). Epub 2012/08/18, doi: S0956-5663(12)00483-6 [pii]10.1016/j.bios.2012.07.041, PubMed PMID: 22898659.

48. S. O. Kelley, R. Gasparac, B. J. Taft, M. A. Lapierre-Devlin, A. D. Lazareck and J. M. Xu, Ultrasensitive electrocatalytic DNA detection at two- and three-dimensional nanoelectrodes. *J. Am. Chem. Soc.* **126**(39), 12270–12271 (2004). doi: 10.1021/ja0458221, PubMed PMID: ISI:000224219900039.
49. S. O. Kelley, M. A. Lapierre-Devlin, C. L. Asher, B. J. Taft, R. Gasparac and M. A. Roberts, Amplified electrocatalysis at DNA-modified nanowires. *Nano Lett.* **5**(6), 1051–1055 (2005). doi: 10.1021/nl050483a, PubMed PMID: ISI:000229729900010.
50. M. Delvaux, A. Walcarius and S. Demoustier-Champagne, Electrocatalytic H2O2 amperometric detection using gold nanotube electrode ensembles. *Anal. Chim. Acta* **525**(2), 221–230 (2004). doi: 10.1016/j.aca.2004.08.054, PubMed PMID: ISI:000224709900009.
51. M. Delvaux, S. Demoustier-Champagne and A. Walcarius, Flow injection amperometric detection at enzyme-modified gold nanoelectrodes. *Electroanaysis* **16**(3), 190–198 (2004). doi: 10.1002/elan.200402778, PubMed PMID: ISI:000220047100004.
52. J. C. Claussen, M. M. Wickner, T. S. Fisher and D. M. Porterfield, Transforming the fabrication and biofunctionalization of gold nanoelectrode arrays into versatile electrochemical glucose biosensors. *ACS Appl. Mater. Interfaces* **3**(5), 1765–1770 (2011). Epub 2011/04/27, doi: 10.1021/am200299h, PubMed PMID: 21517070.
53. Y. He, M. Yang, F. Qu, Y. Li, G. Shen and R. Yu, Direct electrochemistry of hemoglobin in gold nanowire array. *Biosens. Bioelectron.* **23**(3), 414–420 (2007). doi: 10.1016/j.bios.2007.05.003, PubMed PMID: ISI:000250612700015.
54. A. Pinho, S. Viswanathan, S. Ribeiro, M. B. P. P. Oliveira and C. Delerue-Matos, Electroanalysis of urinary L-dopa using tyrosinase immobilized on gold nanoelectrode ensembles. *J. Appl. Electrochem.* **42**(3), 131–137 (2012). doi: 10.1007/s10800-012-0379-3, PubMed PMID: ISI:000301302500001.
55. S. O. Kelley and M. A. Roberts, Ultrasensitive detection of enzymatic activity with nanowire electrodes. *J. Am. Chem. Soc.* **129**(37), 11356–+ (2007). doi: 10.1021/ja074546y, PubMed PMID: ISI:000249464900030.
56. P. Ugo, S. P. Mucelli, M. Zamuner, M. Tormen and G. Stanta, Nanoelectrode ensembles as recognition platform for electrochemical immunosensors. *Biosens. Bioelectron.* **23**(12), 1900–1903 (2008). doi: 10.1016/j.bios.2008.02.027, PubMed PMID: ISI:000256736300025.
57. S. Viswanathan, C. Rani, S. Ribeiro and C. Delerue-Matos, Molecular imprinted nanoelectrodes for ultra sensitive detection of ovarian cancer marker. *Biosens. Bioelectron.* **33**(1), 179–183 (2012). Epub 2012/01/24, doi: S0956-5663(11)00871-2 [pii]10.1016/j.bios.2011.12.049, PubMed PMID: 22265879.
58. S. Viswanathan, C. Rani and C. Delerue-Matos, Ultrasensitive detection of ovarian cancer marker using immunoliposomes and gold nanoelectrodes. *Anal. Chim. Acta* **726**, 79–84 (2012). Epub 2012/05/01, doi: S0003-2670(12)00442-4 [pii]10.1016/j.aca.2012.03.025, PubMed PMID: 22541017.
59. J. L. Kong, X. L. Xu, S. Zhang and H. Chen, Integration of electrochemistry in micro-total analysis systems for biochemical assays: recent developments. *Talanta* **80**(1), 8–18 (2009). PubMed PMID: ISI:000271055700002.

60. S. Bhansali, S. Aravamudhan, A. Kumar and S. Mohapatra, Sensitive estimation of total cholesterol in blood using Au nanowires based micro-fluidic platform. *Biosens. Bioelectron.* **22**(9–10), 2289–2294 (2007). doi: 10.1016/j.bios.2006.11.027, PubMed PMID: ISI:000245541600067.
61. K. T. Liao, C. M. Chen, H. J. Huang and C. H. Lin, Poly(methyl methacrylate) microchip device integrated with gold nanoelectrode ensemble for in-column biochemical reaction and electrochemical detection. *J. Chromatogr. A* **1165**(1–2), 213–218 (2007). Epub2007/08/19,doi:S0021-9673(07)01315-5[pii]10.1016/j.chroma.2007.07.063, PubMed PMID: 17692860.
62. C. M. Chen, G. L. Chang and C. H. Lin, Performance evaluation of a capillary electrophoresis electrochemical chip integrated with gold nanoelectrode ensemble working and decoupler electrodes. *J. Chromatogr. A* **1194**(2), 231–236 (2008). Epub 2008/05/20, doi: 10.1016/j.chroma.2008.04.056, PubMed PMID: 18485353.
63. C. M. Chen, Y. H. Ho, S. M. Wu, G. L. Chang and C. H. Lin, A new method for CE-EC determination of mercaptopurine (MP) in a PMMA biochip with on-chip gold nano-electrode ensemble (GNEE) working and decouple electrodes. *J. Nanosci. Nanotechnol.* **9**(2), 718–722 (2009). Epub 2009/05/16, PubMed PMID: 19441378.
64. C. Chen and J. H. Hahn, Dual-channel method for interference-free in-channel amperometric detection in microchip capillary electrophoresis. *Anal. Chem.* **79**(18), 7182–7716 (2007). Epub 2007/08/22, doi: 10.1021/ac070721h, PubMed PMID: 17708674.
65. J. H. Hahn, C. Chen and W. Teng, Nanoband electrode for high-performance in-channel amperometric detection in dual-channel microchip capillary electrophoresis. *Electrophoresis* **32**(8), 838–843 (2011). doi: 10.1002/elps.201000661, PubMed PMID: ISI:000289352900006.
66. M. Garcia and A Escarpa, Disposable electrochemical detectors based on nickel nanowires for carbohydrate sensing. *Biosens. Bioelectron.* **26**(5), 2527–2533 (2011). PubMed PMID: ISI:000286904400119.
67. M. Garcia and A. Escarpa, A class-selective and reliable electrochemical monosaccharide index in honeys, as determined using nickel and nickel–copper nanowires. *Anal. Bioanal. Chem.* **402**(2), 945–953 (2012). doi: 10.1007/s00216-011-5453-x, PubMed PMID: ISI:000298645300036.
68. M. Jamal, M. Hasan, A. Mathewson and K. M. Razeeb, Disposable sensor based on enzyme-free Ni nanowire array electrode to detect glutamate. *Biosens. Bioelectron.* **40**(1), 213–218 (2013). Epub 2012/08/11, doi: 10.1016/j.bios.2012.07.024, PubMed PMID: 22878082.
69. A. K. Wanekaya, Applications of nanoscale carbon-based materials in heavy metal sensing and detection. *Analyst* **136**(21), 4383–4391 (2011). PubMed PMID: ISI:000295727900001.
70. Z. G. Zhu, L. Garcia-Gancedo, A. J. Flewitt, H. Q. Xie, F. Moussy and W. I. Milne, A critical review of glucose biosensors based on carbon nanomaterials: carbon nanotubes and graphene. *Sensors* **12**(5), 5996–6022 (2012). doi: Doi 10.3390/S120505996, PubMed PMID: ISI:000304539200049.
71. L. T. Qu, Y. Zhao, Y. Hu, H. Zhang, Y. Li, W. Guo *et al.*, Controlled removal of individual carbon nanotubes from vertically aligned arrays for advanced nanoelectrodes.

J. Mater. Chem. **20**(18), 3595–3599 (2010). doi: 10.1039/c0jm00071j, PubMed PMID: ISI:000277086200009.

72. G. Z. Sun, Y. X. Huang, L. X. Zheng, Z. Y. Zhan, Y. N. Zhang, J. H. L. Pang *et al.*, Ultra-sensitive and wide-dynamic-range sensors based on dense arrays of carbon nanotube tips. *Nanoscale* **3**(11), 4854–4858 (2011). PubMed PMID: ISI:000296659000053.
73. A. Gholizadeh, S. Shahrokhian, A. Iraji Zad, S. Mohajerzadeh, M. Vosoughi, S. Darbari *et al.*, Fabrication of sensitive glutamate biosensor based on vertically aligned CNT nanoelectrode array and investigating the effect of CNTs density on the electrode performance. *Anal. Chem.* **84**(14), 5932–5938 (2012). Epub 2012/06/30, doi: 10.1021/ac300463x, PubMed PMID: 22742619.
74. Y. H. Lin, Y. Tu, W. Yantasee and Z. F. Ren, Carbon nanotubes based nanoelectrode arrays: fabrication, evaluation, and application in voltammetric analysis. *Electroanalysis* **17**(1), 79–84 (2005). doi: 10.1002/elan.200403122, PubMed PMID: ISI:000226791400010.
75. A. Gholizadeh, S. Shahrokhian, A. I. Zad, S. Mohajerzadeh, M. Vosoughi, S. Darbari *et al.*, Mediator-less highly sensitive voltammetric detection of glutamate using glutamate dehydrogenase/vertically aligned CNTs grown on silicon substrate. *Biosens. Bioelectron.* **31**(1), 110–115 (2012). doi: 10.1016/j.bios.2011.10.002, PubMed PMID: ISI:000300468400018.
76. G. D. Withey, A. D. Lazareck, M. B. Tzolov, A. Yin, P. Aich, J. I. Yeh *et al.*, Ultra-high redox enzyme signal transduction using highly ordered carbon nanotube array electrodes. *Biosens. Bioelectron.* **21**(8), 1560–1565 (2006). doi: 10.1016/j.bios.2005.07.014, PubMed PMID: ISI:000235492900020.
77. J. Li, H. T. Ng, A. Cassell, W. Fan, H. Chen, Q. Ye *et al.*, Carbon nanotube nanoelectrode array for ultrasensitive DNA detection. *Nano Lett.* **3**(5), 597–602 (2003). doi: 10.1021/nl0340677, PubMed PMID: ISI:000183040100007.
78. J. Li, J. Koehne, A. M. Cassell, H. Chen, Q. Ye, H. T. Ng *et al.*, The fabrication and electrochemical characterization of carbon nanotube nanoelectrode arrays. *J. Mater. Chem.* **14**(4), 676–684 (2004). doi: 10.1039/b311728f, PubMed PMID: ISI:000220224100033.
79. J. Li, J. Koehne, H. Chen, A. M. Cassell, Q. Ye, H. T. Ng *et al.*, Ultrasensitive label-free DNA analysis using an electronic chip based on carbon nanotube nanoelectrode arrays. *Nanotechnology*, **14**(12), 1239–1245 (2003). PubMed PMID: ISI:000187944500003.
80. J. E. Koehne, J. Li, A. M. Cassell, H. Chen, Q. Ye, J. Han *et al.*, System optimization for the development of ultrasensitive electronic biosensors using carbon nanotube nanoelectrode arrays. *Mech. Chem. Biosyst.* **1**(1), 69–80 (2004). Epub 2006/06/21, PubMed PMID: 16783947.
81. G. D. Withey, J. H. Kim and J. Xu, Wiring efficiency of a metallizable DNA linker for site-addressable nanobioelectronic assembly. *Nanotechnology* **18**(42), 424025 (2007). doi: 10.1088/0957-4484/18/42/424025, PubMed PMID: ISI:000249735800026.
82. G. D. Withey, J. H. Kim and J. Xu, DNA-programmable multiplexing for scalable, renewable redox protein bio-nanoelectronics. *Bioelectrochemistry* **74**(1), 111–117 (2008). Epub 2008/06/27, doi: S1567-5394(08)00092-3 [pii]10.1016/j.bioelechem.2008.05.004, PubMed PMID: 18579451.
83. P. U. Arumugam, S. Siddiqui, H. Chen, J. Li and M. Meyyappan, Characterization of carbon nanofiber electrode arrays using electrochemical impedance spectroscopy: effect of scaling down electrode size. *ACS Nano* **4**(2), 955–961 (2010). PubMed PMID: ISI:000274635800047.

84. A. V. Melechko, R. C. Pearce, D. K. Hensley, M. L. Simpson and T. E. McKnight, Challenges in process integration of catalytic DC plasma synthesis of vertically aligned carbon nanofibres. *J. Phys. D Appl. Phys.* **44**(17), (2011). doi: 10.1088/0022-3727/44/17/174008, PubMed PMID: ISI:000289512700008.
85. Z. Yu, T. E. McKnight, M. N. Ericson, A. V. Melechko, M. L. Simpson and B. Morrison 3rd, Vertically aligned carbon nanofiber as nano-neuron interface for monitoring neural function. *Nanomedicine* **8**(4), 419–423 (2012). Epub 2012/03/13, doi: 10.1016/j.nano.2012.02.011, PubMed PMID: 22406183.
86. E. Rand, A. Periyakaruppan, Z. Tanaka, D. A. Zhang, M. P. Marsh, R. J. Andrews *et al.*, A carbon nanofiber based biosensor for simultaneous detection of dopamine and serotonin in the presence of ascorbic acid. *Biosens. Bioelectron.* **42C**, 434–438 (2012). Epub 2012/12/12, doi: 10.1016/j.bios.2012.10.080, PubMed PMID: 23228495.
87. P. U. Arumugam, E. Yu, R. Riviere and M. Meyyappan, Vertically aligned carbon nanofiber electrode arrays for nucleic acid detection. *Chem. Phys. Lett.* **499**(4–6), 241–216(2010).doi:10.1016/j.cplett.2010.09.048,PubMedPMID:ISI:000283244700011.
88. A. Periyakaruppan, P. U. Arumugam, M. Meyyappan, J. E. Koehne, Detection of ricin using a carbon nanofiber based biosensor. *Biosens. Bioelectron.* **28**(1), 428–433 (2011). Epub 2011/08/20, doi: S0956-5663(11)00509-4 [pii]10.1016/j.bios.2011.07.061, PubMed PMID: 21852102.
89. A. Periyakaruppan, R. P. Gandhiraman, M. Meyyappan and J. E. Koehne, Label-free detection of cardiac troponin-i using carbon nanofiber based nanoelectrode arrays. *Anal. Chem.* (2013). Epub 2013/02/07, doi: 10.1021/ac302801z, PubMed PMID: 23384128.
90. J. E. Koehne, M. Marsh, A. Boakye, B. Douglas, I. Y. Kim, S. Y. Chang *et al.*, Carbon nanofiber electrode array for electrochemical detection of dopamine using fast scan cyclic voltammetry. *Analyst* **136**(9), 1802–1805 (2011). PubMed PMID: ISI:000289365100003.
91. J. I. Yeh, M. B. Zimmt and A. L. Zimmerman, Nanowiring of a redox enzyme by metallized peptides. *Biosens. Bioelectron.* **21**(6), 973–978 (2005). doi: 10.1016/j.bios.2005.02.017, PubMed PMID: ISI:000233346300019.
92. J. I. Yeh, A. Lazareck, J. H. Kim, J. Xu and S. Du, Peptide nanowires for coordination and signal transduction of peroxidase biosensors to carbon nanotube electrode arrays. *Biosens. Bioelectron.* **23**(4), 568–574 (2007). Epub 2007/09/21, doi: 10.1016/j.bios.2007.06.019, PubMed PMID: 17881214; PubMed Central PMCID: PMC3158592.
93. J. I. Yeh, Coordinated biosensors: integrated systems for ultrasensitive detection of biomarkers. In *Nanomaterials for Biosensors*, C. Kumar (Ed.), Weinheim: Wiley-VCH (2007), pp. 259–277.
94. S. Pavan and F. Berti, Short peptides as biosensor transducers. *Anal. Bioanal. Chem.* **402**(10), 3055–3570 (2012). Epub 2011/12/16, doi: 10.1007/s00216-011-5589-8, PubMed PMID: 22169951.
95. T. Xia, M. Kovochich, J. Brant, M. Hotze, J. Sempf, T. Oberley *et al.*, Comparison of the abilities of ambient and manufactured nanoparticles to induce cellular toxicity according to an oxidative stress paradigm. *Nano Lett.* **6**(8), 1794–1807 (2006). Epub 2006/08/10, doi: 10.1021/nl061025k, PubMed PMID: 16895376.
96. T. Xia, M. Kovochich, M. Liong, L. Madler, B. Gilbert, H. Shi *et al.*, Comparison of the mechanism of toxicity of zinc oxide and cerium oxide nanoparticles based on dissolution and oxidative stress properties. *ACS Nano* **2**(10), 2121–2134 (2008). Epub

2009/02/12, doi: 10.1021/nn800511k10.1021/nn800511k [pii], PubMed PMID: 19206459.

97. N. Triroj, P. Jaroenapibal, H. Shi, J. I. Yeh, R. Beresford, Microfluidic chip-based nanoelectrode array as miniaturized biochemical sensing platform for prostate-specific antigen detection. *Biosens. Bioelectron.* **26**(6), 2927–2933 (2011). Epub 2010/12/31, doi: S0956-5663(10)00805-5 [pii]10.1016/j.bios.2010.11.039, PubMed PMID: 21190835.
98. J. I. Yeh, E. Pohl, D. Truan, W. He, G. M. Sheldrick, S. Du *et al.*, The crystal structure of non-modified and bipyridine-modified PNA duplexes. *Chemistry* **16**(39), 11867–11875 (2010). Epub 2010/09/23, doi: 10.1002/chem.201000392, PubMed PMID: 20859960; PubMed Central PMCID: PMC3194003.
99. T. H. Degefa and J. Kwak, Electrochemical impedance sensing of DNA at PNA self assembled monolayer. *J. Electroanal. Chem.* **612**(1), 37–41 (2008). doi: 10.1016/j.jelechem.2007.09.004, PubMed PMID: ISI:000252481200006.
100. S. Reisberg, L. A. Dang, Q. A. Nguyen, B. Piro, V. Noel, P. E. Nielsen *et al.*, Label-free DNA electrochemical sensor based on a PNA-functionalized conductive polymer. *Talanta* **76**(1), 206–210 (2008). doi: 10.1016/j.talanta.2008.02.044, PubMed PMID: ISI:000256934200034.
101. C. Y. Yao, T. Y. Zhu, J. Tang, R. Wu, Q. H. Chen, M. Chen *et al.*, Hybridization assay of hepatitis B virus by QCM peptide nucleic acid biosensor. *Biosens. Bioelectron.* **23**(6), 879–885 (2008). doi: 10.1016/j.bios.2007.09.003, PubMed PMID: ISI:000252790000018.
102. R. M. Franzini, R. M. Watson, G. K. Patra, R. M. Breece, D. L. Tierney, M. P. Hendrich *et al.*, Metal binding to bipyridine-modified PNA. *Inorg. Chem.* **45**(24), 9798–9811 (2006). PubMed PMID: AN 2006:1160111.
103. A. Paul, R. M. Watson, P. Lund, Y. J. Xing, K. Burke, Y. F. He *et al.*, Charge transfer through single-stranded peptide nucleic acid composed of thymine nucleotides. *J. Phys. Chem. C* **112**(18), 7233–7240 (2008). doi: 10.1021/Jp711764q, PubMed PMID: ISI:000255486800019.
104. A. Paul, S. Bezer, R. Venkatramani, L. Kocsis, E. Wierzbinski, A. Balaeff *et al.*, Role of nucleobase energetics and nucleobase interactions in single-stranded peptide nucleic acid charge transfer. *J. Am. Chem. Soc.* **131**(18), 6498–6507 (2009). Epub 2009/04/23, doi: 10.1021/ja9000163, PubMed PMID: 19382798.
105. H. Shi and J. I. Yeh, University of Pittsburgh, Assignee. Biosensors and Related Methods, #8198039. United States 2012.

Chapter 13

Nanosensors for Biomedicine

Mary K. Balaconis and Heather A. Clark*[†,‡]

[*] *Bioengineering Department, Northeastern University*
[†] *Department of Pharmaceutical Sciences, Northeastern University*
360 Huntington Avenue, Boston, MA 02115
[‡] *h.clark@neu.edu*

Abstract

Fluorescent molecular probes are highly sensitive and enable non-invasive monitoring of biological phenomena. However, unprotected from their environment, these probes have the potential to experience interference from biomolecules and may induce cytotoxicity. As a result, fluorescent sensor development has turned toward the use of nano-based systems that can enhance the functionality of traditional molecular probes and have unique sensing properties of their own. Fluorescent nanosensors are sufficiently small to interact at biological scales and are capable of functionalization and complex sensing mechanisms difficult to achieve with molecular probes. Currently, fluorescent nanosensors can detect important biological molecules ranging from reactive species to DNA polymorphisms. This chapter discusses the advantages and disadvantages of fluorescent nanosensors and reviews the major platforms, including carbon nanotubes, quantum dots, hydrogels, liposomes, dendrimers and optodes. The advantages and disadvantages of each platform are provided along with examples that highlight sensing designs and applications relevant to monitoring biological phenomena. The chapter concludes with design considerations for current and future fluorescent nanonsensors based on specific biological applications.

1. Introduction

Fluorescence as a transduction mechanism has benefits that include high sensitivity down to the single molecular level, non-invasive sensing and relatively inexpensive equipment. The advent of fluorescence sensing began in the 1970s and 1980s with the development of voltage-sensitive[1] and calcium-sensitive dyes,[2,3] respectively.[4] The class of fluorescent sensors has primarily consisted of environmentally-sensitive dyes and fluorescent proteins. In the presence of the target analyte, these probes may undergo inherent fluorescent changes upon analyte binding or undergo mechanisms such as Förster resonance energy transfer (FRET).[4,5] For example, Fura-2, a well-known fluorescent probe, undergoes a ratiometric change in fluorescence excitation at 340 and 380 nm upon binding calcium.[3] Fura-2 provides ratiometric fluorescent measurements which eliminate the effects of uneven dye loading and instrument variability.[3] However, tuning and synthesis of new molecular probes with these characteristics is difficult due to complicated organic synthesis and purification steps.[4]

Drug delivery systems, imaging, and sensing fields are all investigating the use of nanomaterials to improve biomedical products.[6] The diversity and unique properties of these materials have the potential to ameliorate the shortcomings of traditional molecular probes. For example, liposomes,[7] dendrimers,[8] and hydrogel[9] nanoparticles can all act as encapsulation vehicles for molecular probes, and other sensing moieties and protect sensing components from the biological environment. Carbon nanotubes (CNTs) and quantum dots (QDs) also have unique fluorescent properties such as resistance to photobleaching for longer sensing lifetimes.[10–12] Nanosensors effectively increase the size of molecular probes while interacting at biological scales. However, nanosensors can still enter subcellular compartments such as dendrites,[8] they cause minimum cellular perturbation,[13] and they may be easily implanted *in vivo* through injection techniques.[14] The high surface area to volume ratio of nanosensors also enables fast sensor response times for detection of rapid cellular events such as ion channel fluxes.[14,15] Finally, nanoparticles can be functionalized with biocompatible coatings for minimizing sensor cytotoxicity and bioaccumulation.[16,17] Given the advantages of nanoparticle-based systems, research into nanosensor design has lead to sensors for a range of biomolecules that include ions and small molecules.

In this chapter, fluorescent nanosensors for biological sensing are reviewed. Nanosensors are defined here as nanoparticle systems within the size range of 1–1,000 nm and use fluorescence as their only transduction mechanism for sensing. The chapter begins with a discussion of the advantages of fluorescent nanosensors over traditional molecular probes with examples highlighting the unique sensing capabilities of nanosensors. Then, a review of specific nanosensor platforms that includes QDs, CNTs, liposomes, hydrogels, dendrimers and optodes is provided. For each platform, a general review of the sensor platform is discussed including advantages and disadvantages of the design along with specific examples and their use in biological applications. Other nanosensor materials, such as gold and silica, are not reviewed here and the reader is directed to the cited reviews for more information on these sensor materials.[18–20] Multimodule nanosensors that combine two signal transduction mechanisms (*i.e.* combined fluorescence and magnetic-based nanosensors) have also been excluded from this review. The chapter ends with an analysis of design considerations for nanosensors that are important for biological sensing.

2. Molecular Probes Versus Nanosensors

Fluorescent molecular probes are an essential tool in biomedical research for monitoring intracellular and *in vivo* phenomena. The advantages of these probes are non-invasive measurements and fast response.[20] For example, calcium-sensitive probes have been used for applications ranging from determining the calcium distribution in smooth muscle cells[21] to the calcium dynamics of neurons *in vivo*.[22] Calcium green was loaded into pyramidal neurons of a rat and used to determine calcium dynamics upon whisker stimulation.[22] From this study, increases in calcium concentrations were correlated to the incidence of sodium action potentials. Calcium-sensitive probes are the most prominent, but molecular probes for other biologically relevant analytes such as pH and oxygen are also available.[23]

Molecular probes have become a research staple, but their long-term monitoring is prohibitive. Since molecular probes are small molecules, they freely diffuse throughout the intracellular and extracellular environments and the bloodstream. Unprotected from biological environments, they may interact with proteins and other biomolecules that cause interference with probe measurements.[24,25] Interactions between proteins and intracellular pH probes have been shown to cause a fluorescent spectral shift and alter ratiometric measurements.[25] Furthermore, cytotoxicity from direct interaction between probes and the biological environment requires that low dye concentrations be used.[26] Intracellularly, probe leaking from cells during experiments and organelle sequestration can also severely limit monitoring times and confound fluorescent results, respectively.[8,27] For example, sodium-sensitive CoroNa Green leaked from neurons, causing a loss of over 50% of its initial fluorescence within 1 min of loading.[8] Incorporation of traditional molecular probes with nanoparticle-based platforms may ameliorate these issues and extend probe lifetime.

Nanoparticle-based sensors offer many advantages over traditional molecular probes as has been reviewed previously.[20,110] (Insert reference 20 and 110.) Similar to probes, nanoparticles are still negligible in size compared to the total cell volume. For example, a 20–600 nm particle occupies only 1–1,000 part per million (PPM) of the total cell volume causing minimum physical perturbation to the intracellular environment.[13] However, encapsulation or conjugation of probes to nanoparticles effectively increases probe size to prevent probe leakage out of cells.[5] In contrast to free CoroNa Green mentioned previously, CoroNa Green encapsulated in dendrimers showed stable fluorescent signal or no leakage for at least 1 h when loaded into neurons.[8] For CNTs, their small diffusion constant even allowed spatial resolution of adenosine triphosphate (ATP) in cells.[28]

Nanoparticles also act as encapsulation vehicles for molecular probes and other sensing components protecting these moieties from harsh biological environments. In the presence of 0.01% albumin, the pH fluorescent indicator, 5-(and 6-)carboxynaphthofluorescein (CNF), experienced a shift in ratiometric emission of nearly 90% correlating to a pH shift of one unit; however, isolation of CNF in hydrogel particles in the presence of albumin only resulted in a shift in fluorescence equivalent to a hundredth of a pH unit.[9] Isolation further minimizes exposure of cells to probe cytotoxicity[9] and in the case of hydrophobic nanoparticles expands the range of sensing components to hydrophobic moieties.[29] Finally, nanoparticles not only improve probe lifetime, but can also improve sensor fluorescent properties. QDs and CNTs are unique in that they are resistant to photobleaching.[10–12] Under constant illumination, CNTs exhibited no photobleaching threshold, suggesting theoretical infinite sensor lifetime.[10] Since nanoparticles can also encapsulate several fluorescent indicators, ratiometric sensing is possible without synthesis of new dyes.[9]

Apart from improving upon molecular probes, nanoparticles are commonly surface modified for targeting and for imparting biocompatibility. For example, initial toxicity studies showed adverse effects to bare CNTs;[30,31] however, with increased surface functionalization, bioaccumulation of these particles was limited.[32] All of these examples highlight that nanoparticle-based systems offer many advantages as a sensing platform. The following review provides further detail on major classes of nanosensors that include QDs, CNTs, hydrogels, liposomes, optodes and dendrimers.

3. Types of Fluorescent Nanosensors

3.1. *Quantum dot-based nanosensors*

QDs are colloidal semiconducting nanocrystals with fluorescent properties desirable for biological sensing applications.[11,33] Their fluorescence derives from excitation of valence electrons above their band gap energy and generation of electron–hole pairs.[11] Recombination of an electron with its hole results in an emission of a photon. When QDs are only several nanometers in diameter, they exhibit size dependent emission properties. QDs have unique fluorescent properties that include resistance to photobleaching[11,12] and high quantum yields[34] compared to organic fluorophores. The broad excitation spectrum with size tunable emission of QDs facilitates simultaneous measurements of multiple biological phenomena using different size QDs[33] phenomena. In this section, specific examples of QD sensors for ions, small molecules and biomolecules are covered and the reader is directed to several reviews that include QD-based sensors for further examples.[33,35]

QD-based sensors function through different energy transfer mechanisms, such as radiationless resonance energy transfer,[36] photo-induced electron transfer,[37] and FRET.[38] However, the majority of QD sensor designs involve FRET-based mechanisms. FRET is the non-radiative transfer of energy between two fluorescent species where an excited donor molecule transfers energy to an acceptor molecule.[39] The transfer of energy results in a decrease in donor fluorescence with a corresponding increase in acceptor fluorescence. For biological applications, FRET is efficient at distances between 2 and 9 nm with 50% FRET efficiency typically occurring between 2 and 6 nm.[40] Because of these distance requirements, FRET is sensitive to binding events on the molecular and nanoscale. For FRET-based sensing, QDs provide not only an alternative but distinct advantages over the use of organic fluorophores. QDs have size tunable spectral properties capable of adjusting their emission spectra to overlap closely with the excitation spectrum of the acceptor molecule while also accommodating multiple acceptor molecules on their surface for increased FRET efficiency.[41]

QD-based nanosensors have been developed for a range of important physiological ions that include sodium,[42] potassium,[43] chloride,[37] pH,[44,45] and calcium[46] (Table 1). For these sensors, simple environmentally-sensitive fluorophores[44,45] to DNAzymes[47] attached to QDs have served as ion-sensitive

Table 1. Analytes detected with QD-based nanosensors.

	Analytes detected
Ions	Sodium,[42] Potassium,[43] Chloride,[37] pH,[44,45] Calcium,[46] Carbonate,[48] Zinc,[36,49] Manganese,[36] Copper,[47,50] Lead[47]
Small molecules	Glucose,[51] Acetylsalicylic Acid,[52] Folic Acid,[53] Maltose[38]
Biomolecules	Enzymes,[44,46,54,55] DNA,[56–61] RNA,[62,63] Peptides,[62,63] microRNA,[64] Nucleic Acids,[65] Immunoassays[66]

recognition elements. Wang and coworkers designed chloride-sensitive QDs for measuring intracellular fluxes in epithelial cells.[37] Cadmium selenide/zinc sulfide (CdSe/ZnS) QDs were conjugated with 1-(2-mercapto-ethyl)-3-phenyl-thiourea and a linker. 1-(2-mercapto-ethyl)-3-phenyl-thiourea is an anion-sensitive molecule and upon binding to chloride causes quenching of QD fluorescence through photo-induced electron transfer. The chloride-sensitive QDs were calibrated at physiologically relevant chloride concentrations and then loaded into CF-PAC cells, a cystic fibrosis human pancreatic epithelial cell line, using liposomal carriers. When cells were exposed to a chloride channel inhibitor, glibenclamide, chloride-sensitive QDs fluorescently detected changes in intracellular chloride concentrations with greater sensitivity and signal changes than commonly used chloride fluorescent indicators.

QD-based nanosensors have also been designed for detecting small molecules such as glucose,[51] acetylsalicylic acid,[52] folic acid,[53] and maltose[38] (Table 1). For example, CdSe–ZnS QDs were used in a competitive binding scheme to quantitatively monitor changes in maltose concentrations.[38] Maltose binding proteins (MBPs) and a β-cyclodextrin-QSY9, a non-emitting acceptor, were coated on the QDs (Fig. 1(a)). β-cyclodextrin-QSY9 attached to the maltose binding site on the MBPs and quenched QD photoluminescence. When maltose was present, maltose displaced β-CD-QSY9 and QD photoluminescence was restored in a concentration dependent manner with an apparent dissociation constant of 6.9 μM. Additionally, a two-step FRET QD sensor based on the same MPB competitive sensing mechanism was also developed.[38] Since FRET is distance dependent, QD size poses limitations because of increased distance between donor and acceptor resulting in decreased FRET efficiency. In the two-step mechanism, MBP and β-cyclodextrin were labeled with Cy3 and Cy3.5, respectively (Fig. 1(b)). QDs acted as the donor and Cy3 as the acceptor in the first FRET step. Cy3 then underwent FRET again with Cy3.5 as the acceptor. When maltose was present, it displaced β-cyclodextrin-Cy3.5 and Cy3 emission increased. The presence of

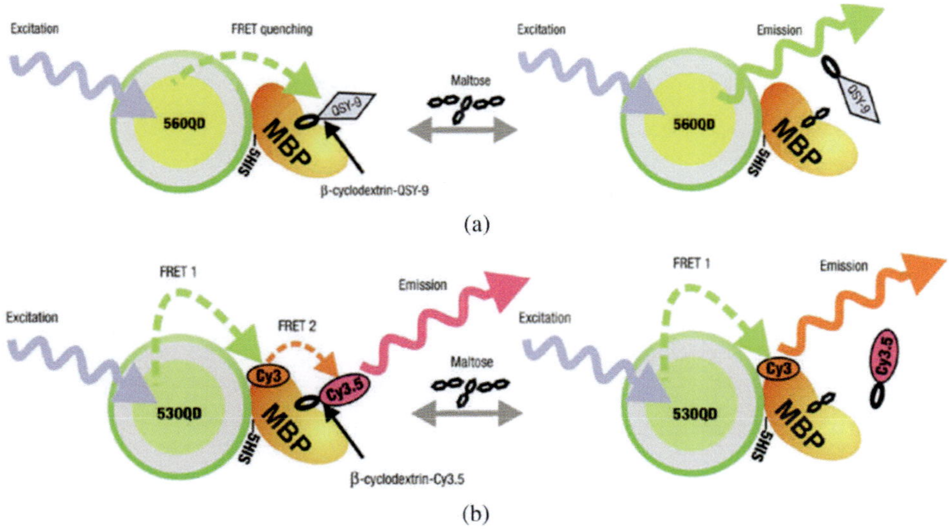

Fig. 1. Two QD-based sensing mechanisms for the detection of maltose. (a) MBP-β-cyclodextrin-QSY9 complex bound to QDs quench QD emission. When maltose is added to the system, maltose displaces β-cyclodextrin-QSY9 and QD emission is restored. (b) Two-step FRET sensing mechanism for maltose. QD fluorescence undergoes FRET 1 with MBP-Cy3 which then undergoes FRET 2 with β-cyclodextrin-Cy3.5 ultimately resulting in Cy3.5 emission. When maltose is added to the system, β-cyclodextrin-Cy3.5 is displaced and Cy3 emission is restored. Figures modified from Ref. 38. Reprinted with permission from Ref. 38. Copyright 2003 Nature Publishing Group.

the relay station was estimated to increase the FRET energy of the sensor from 6% to 20%, but when no relay station was involved, the QD-based sensor was unsuccessful at sensing maltose.

The major class of QD-based sensors, however, is for the detection and monitoring the activity of biomolecules including enzymes[44,46,54,55] (Table 1). QD-based sensors for enzymatic activity can detect single molecules and determine enzymatic rate kinetics. Zhang and coworkers demonstrated these capabilities with their renin-sensitive QD sensor.[54] QDs were streptavidin-coated with biotinylated Cy5 labeled peptide. When no renin was present, QD fluorescence was quenched by Cy5 through FRET-based mechanisms. As renin was introduced into the system, the enzyme cleaved the peptide sequence, which resulted in loss of FRET efficiency between Cy5 and QD and thus a decrease in Cy5 fluorescence. Total internal reflectance fluorescence (TIRF) imaging techniques were used to detect renin activity at the single molecule level. The sensors had a limit of detection of 25 pM and at least a two-fold increase in sensitivity over other fluorescent detection methods. Additionally, the Michaelis–Menten kinetic parameters of renin using the

QD-based system matched those reported in the literature. The high signal to noise ratio of single molecule detection and high FRET efficiency of the QD–Cy5 system were possible reasons for the improved sensitivity.

QD-based sensors have also been used for the detection of other biomolecules that include ribonucleic acid (RNA)-peptide interactions,[62,63] micro RNAs,[64] nucleic acid hybridization,[65] deoxyribonucleic acid (DNA),[56–61] and immunoassay based systems[66] (Table 1). Similar to enzymatic approaches, QDs can detect low concentrations of DNA sequences without requiring amplification or separation techniques.[61] The three component system by Wang and coworkers consisted of CdSe–ZnS QDs conjugated with streptavidin, a reporter probe tagged with a Cy5 fluorescent reporter, and a capture probe conjugated with biotin (Fig. 2).[61] When the target DNA sequence was

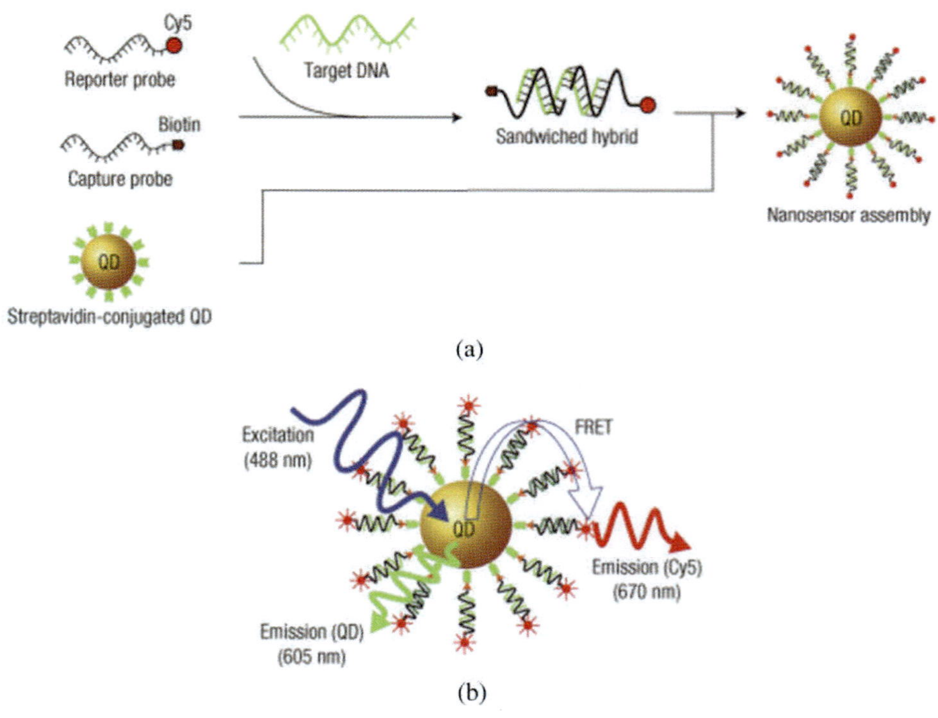

Fig. 2. Schematic for sensitive detection of DNA sequences. (a) In the presence of the target DNA, a sandwich hybrid is formed between a reporter probe tagged with a Cy5 fluorescent reporter, a capture probe conjugated with biotin and the target DNA. The hybrid then binds to the CdSe–ZnS QDs conjugated with streptavidin. (b) Upon the sandwich hybrid binding to the QDs through biotin–streptavidin binding, FRET occurs between the QD and Cy5. Figure modified from Ref. 61. Reprinted with permission from Ref. 61. Copyright 2005 Nature Publishing Group.

present, the reporter and capture probe formed a sandwich hybrid with the target DNA sequence. The hybrid then bound to the QDs through biotin–streptavidin binding and FRET occured between the QD and Cy5. The QD platform concentrated the target sequence because multiple hybrids with fluorescent reporters can bind to a single QD. Using this sensing mechanism, the detection of the target sequence occurred even with less than 50 targets present in the sample and had a detection limit 100 fold greater than commonly used molecular beacons.

QD-based sensors monitor a diverse set of analytes from small ions to DNA sequences, but properties of QDs raise concerns about sensor reproducibility and toxicity issues. QDs derive their fluorescence from generation of electron–hole pairs; however, surface defects create "traps" for excited electrons and holes to fill resulting in random fluorescent "blinking."[11,67] This fluorescent-off state of QDs can last between 1 ms to 1 s.[68] If the biological phenomenon occurs on a faster time scale, fluorescent blinking may affect sensor measurements. The composition of QDs has also raises concerns about their biocompatibility and toxicity, especially since they have been shown to cause DNA damage[69] and remain within the body for up to 2 years.[70] QDs can be surface functionalized with biocompatible coatings.[71] Furthermore, toxicity for *in vivo* applications may be reduced if QDs with a hydrodynamic diameter less than 5.5 nm are used, since they are excreted through renal clearance.[72]

3.2. *Carbon nanotube-based nanosensors*

CNTs are single layers of graphene rolled into a one-dimensional cylindrical shape.[73] They are made at various lengths and with diameters of 1 nm that can result in high aspect ratios of 28,000,000 to 1 (length to diameter).[73] CNTs have unique and versatile mechanical, electrical and thermal properties. For example, simply changing how the CNTs are rolled results in either semiconducting or metallic properties.[73] Similar to QDs, CNTs do not derive their fluorescent properties from organic fluorophores, but rather CNT fluorescence originates from the van Hove singularities characteristically found on one-dimensional electronic materials.[74–76] Light absorption across these singularities results in fluorescence emission that is tunable based on nanotube chirality.[75] CNTs emit in the near infrared (NIR) range of 820–1600 nm which allows for greater tissue penetration and is outside the range of tissue autofluorescence.[74] Photobleaching studies have also concluded that CNTs are more photostable than QDs.[10] CNTs can be functionalized and their fluorescence is especially sensitive to surface adsorption with single molecule detection limits.[74] Lastly, CNTs can be monitored either by intensity or wavelength changes that allows for greater multiplexing and specific analyte spectrum fingerprinting.[77] In this section, CNT-based sensors designed for ions, small molecules, reactive species, and DNA are covered along with specific examples (Table 2); however, for further discussion of these sensors, the reader is directed to several reviews written about this sensor class.[73,74,78,79]

CNT-based sensors have been developed for the detection of ions that include silver,[80] iron,[81] mercury,[82,83] lead,[84] and pH.[85,86] Several of these sensor mechanisms rely on divalent ion interactions with nucleobases. For example, mercury ions exhibit a high affinity for thymine–thymine base mismatches in DNA.[87] Due to π-stacking interactions, single stranded DNA (ssDNA) non-covalently adsorbs onto CNTs.[82] Wang and coworkers used these properties to develop CNT-based nanosensors for detection of mercury ions at nanomolar concentrations.[82] The molecular probe 6-carboxyfluorescein was attached to ssDNA and its fluorescence was quenched when the DNA adsorbed onto the surface of the single walled CNT (SWCNT). The ssDNA released from the CNT upon incorporation of mercury ions into the system due to mercury ions interacting with the thymine-dense ssDNA and forming a double helical structure. Once detached, the fluorescence of the fluorescein-based fluorophore was restored. The detection limit of these sensors ranged from 14.5 nM to 8 μM and they exhibited slight interference from other divalent ions. Another CNT-based mercury sensor used similar principles that involved thymine-dense ssDNA and an intercalating fluorophore that both adsorbed onto the surface of SWCNTs.[83] Upon addition of another thymine-dense ssDNA strand and mercury, a double stranded DNA (dsDNA) complex formed resulting in restoration of fluorescence. These sensors also had a nanomolar limit of detection, which was lower than previously seen. The selectivity of this sensing mechanism may be changed by altering the nucleobases present such as using cytosine-dense DNA for detection of silver ions.[80]

Table 2. Analytes detected with CNT-based nanosensors.

	Analytes detected
Ions	Iron,[81] Mercury,[82,83] Lead,[84] pH,[85,86] Silver[80]
Small molecules	Glucose,[89–92] ATP,[28,93,94] Glycan[95]
Reactive species	Hydrogen Peroxide,[96–98] Nitric Oxide[99,100]
Biomolecules	DNA,[101–104] Amino Acids[80]

In all previously discussed sensor designs, the sensing moieties are non-covalently adsorbed onto the surface; however, for molecular dyes, physical adsorption is not sufficient to maintain the CNT-dye interaction.[88] To prevent dye leaching from pH-sensitive CNTs, Zhao and coworkers covalently linked the self-ratiometric pH-sensitive dye, 6,8-dihydroxy-1,3-pyrenedisulfonic acid disodium salt to multi-walled CNTs (MWCNTs).[86] An additional zinc phthalocyanine coating was also added to further prevent dye leaching and interaction with other biomolecules. The sensors had a dynamic range from pH 5.6 to 8.3 that is relevant for physiological measurements. Jo and coworkers also used a FRET-based sensing mechanism for monitoring reversible changes in pH by linking CNTs to pyrene with the pH-sensitive polymeric chain, polysulfonamide.[85] When the pH is above the pK_a of polysulfonamide, the linker extends away from the CNT and pyrene fluoresces; however as the pH drops below the pK_a, the polymer linker recoils and the fluorescence of pyrene is quenched by the CNT. The emission intensity rapidly dropped after pH = 7.2 and had an on–off reversible response.

CNTs have also been developed for the detection of small molecules especially glucose.[89–92] CNT-based glucose sensors were first developed by adsorbing the enzyme, glucose oxidase (GOx) and ferricyanide, a CNT fluorescence quencher, onto the CNT surface (Fig. 3).[90] Upon glucose addition, GOx

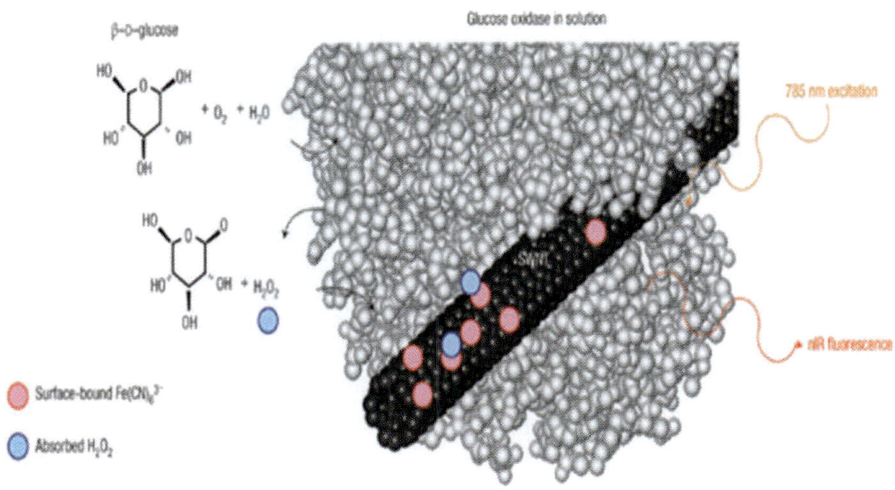

Fig. 3. Schematic for CNT-based sensor for glucose detection. Ferricyanide quenches CNT fluorescence. Upon glucose addition, GOx oxidizes glucose and produces hydrogen peroxide as a side product. Hydrogen peroxide then reduces ferricyanide, resulting in an increase in CNT fluorescence. Figure modified from Ref. 90. Reprinted with permission from Ref. 90. Copyright 2004 Nature Publishing Group.

oxidizes glucose to D-glucono-1,5-lactone and hydrogen peroxide. Hydrogen peroxide reduces ferricyanide, increasing the fluorescence intensity of the CNTs. The derivatized-CNTs were visualized in blood samples and could detect glucose at physiological levels. Popular glucose detection schemes that are used in other sensing platforms such as those that utilize boronic acids[91] or concanavalin A,[92] have also been coupled with the inherent fluorescence of CNTs to yield glucose sensors. Sensors based on these mechanisms do not require cofactor regeneration which is an advantage over GOx-based sensors.

In addition to glucose, CNT sensors have also been designed for ATP using enzymes[28] and aptamers.[93,94] Strano and coworkers labeled CNTs with poly(ethylene glycol) (PEG) and luciferase, an ATP-selective enzyme.[28] In the presence of D-luciferin and magnesium, luciferase reacts with ATP, resulting in light-emitting luminescent oxyluciferin that ultimately quenches the fluorescence of the CNTs. The sensors detected ATP at a concentration of 240 nM and distinguished temporal and spatial differences in ATP in HeLa cells; however, the sensors were not reversible due to inhibition of luciferase by oxyluciferin making measurements of real-time changes in cellular ATP impossible. A different ATP-sensing mechanism using CNTs improved the detection limit of the sensor previously mentioned. A 6-carboxy-fluorescein (FAM) labeled anti-ATP DNA aptamer non-covalently interacts with CNTs, resulting in a quenching of the fluorophore.[93] Upon ATP addition and interaction with the aptamer, the aptamer undergoes a conformational change altering the distance between FAM and the CNT that results in FAM fluorescence recovery. The sensors exhibited a lower detection limit for ATP of 4.5 nM and had a linear response within the range of 10–800 nM. Additionally, the sensors could detect the presence or absence of ATP in cell lysate.

More complex sensing mechanisms for CNTs have been employed to monitor glycan–lectin binding with the long–term goal of characterizing protein glycosylation.[95] SWCNTs were wrapped in chitosan gel and then functionalized with *N*α, *N*α-bis(carboxymethyl)-L-lysine-nickel complex (NTA–Ni^{2+}). The functionalization enables binding of His-tagged lectin. The proximity of nickel to the SWCNT causes quenching of nanotube fluorescence; however, upon glycan binding, fluorescence intensity is recovered because of the now increased distance between nickel and the nanotube. The detection and binding kinetics of fucose to PA-IIL lectin and *N*-acetylglucosamine to GafD lectin exhibited dissociation constants of 106 and 19 μM for each pair respectively and selectivity of the sensors for their respective glycan. Though the percent change in signal of the CNTs was small (<5%), this system does carry advantages over other glycan assays because it does not require large amounts of material, no labeling is necessary, and it can be multiplexed with different lectins.

CNT-based sensors have also been developed for reactive oxygen species (ROS) and reactive nitrogen species (RNS). These species play an important role in the life cycle of cells, but organic fluorescent dyes used to monitor reactive species are typically irreversible.[96] The detection of hydrogen peroxide has been used in several designs for the indirect detection of glucose where GOx oxidizes glucose with hydrogen peroxide as a byproduct.[89,90] CNTs embedded in collagen thin films extend hydrogen peroxide detection to the single molecule level.[96–98] When hydrogen peroxide contacts the CNT surface, CNT fluorescence is quenched in a reversible process. With the SWCNT-collagen array, hydrogen peroxide was measured from individual A431 human epidermal carcinoma cells and NIH-3T3 murine fibroblast cells.[98] The cells were stimulated with epidermal growth factor (EGF), which initiated hydrogen peroxide formation upon EGF activation of EGF receptors. The CNT array distinguished differences in hydrogen peroxide production between the two cell types with A431 cells exhibiting a higher quenching rate. These results correlate to the fact that A431 cells had a density of epidermal growth factor receptors that was 10 times higher than was found on 3T3 cells. The CNT array was also used to study the generation of hydrogen peroxide in human umbilical vein endothelial cells after exposure to vascular endothelial growth factor A (VEGF-A) and europium(III) hydroxide nanorods that both have pro-angiogenic properties.[96] The array had the spatial resolution to determine that hydrogen peroxide was produced near the cellular membrane after VEGF-A stimulation but was absent after nanorod stimulation, indicating that these stimulants induce angiogenesis through two independent signaling pathways.

Nitric oxide (NO) is also a major cell signaling molecule that is desirable but difficult to detect. Strano and coworkers developed NO selective sensors by wrapping SWCNTs with 3,4-diaminophenyl-functionalized dextran.[99] The amine moieties present in the surface coating dope CNTs with electrons which are then transferred to NO, resulting in bleaching of the CNTs. These CNT sensors were reversible and selective for NO over other reactive oxygen and nitrogen species with a lower detection limit of 70 nM. The NO-sensitive CNT sensors were placed in dialysis tubing and implanted into an asphyxiated mouse. Upon injection of NO near the dialysis tubing, the fluorescence intensity of the sensors was bleached, indicating that changes in fluorescence of the CNTs can be visualized through the skin. Later work on NO detection showed that wrapping CNTs in specific DNA sequences created sensors that selectively detected NO at the single molecule level.[100] The development of CNT-based sensors for reactive species has made progress in the reversible monitoring and single molecule detection of these highly biologically important species.

As has been highlighted earlier in this section, DNA has the capability to wrap around CNTs because of π-stacking interactions.[82] This property has not only been exploited for the detection of divalent ions, but also for the detection of DNA hybridization[101,102] and the presence of DNA polymorphisms.[103,104] Strano and coworkers first demonstrated the detection of DNA hybridization on the surface of SWCNTs through SWCNTs' NIR fluorescence.[101] In this work, ssDNA was absorbed onto the surface of SWCNTs and hybridization of the complimentary strand of DNA resulted in a hypochromic shift in the NIR fluorescence with a detection limit of 6 nM; however, the kinetics of this reaction was 13 h compared to less than 2 min for free DNA. Later studies concluded that the preadsorption of DNA onto the SWCNTs was energetically unfavorable and caused slow hybridization times.[102] SWCNTs with bound ssDNA are also sensitive enough to detect single nucleotide polymorphisms in complimentary DNA upon binding.[104] The presence of polymorphisms resulted in a lower emission energy difference than the full complement DNA sequence. Unfortunately, when this work was expanded to include other DNA sequences, this phenomenon was not observed because different DNA sequences have different coating efficiencies onto CNTs. The authors concluded that for the detection of single nucleotide polymorphisms, submonolayer coverage of the ssDNA sequence is necessary.

CNT attributes such as NIR fluorescence and resistance to photobleaching make them a desirable sensing platform; however, some CNTs are structurally similar to asbestos and not biodegradable raising concerns about their potential toxicity.[105] Injection of MWCNTs with lengths greater than 20 μm in mice caused the formation of granuloma lesions.[30] In other animal studies, MWCNTs induced mesothelioma, but metal impurities may have confounded these results.[31] Other studies in mice have shown that surface functionalized MWCNTs were excreted with limited bioaccumulation.[32] Future studies are necessary for more conclusive results regarding the ultimate toxicity risks of CNTs and thus their full potential as *in vivo* monitoring sensors.

3.3. *Hydrogel-based nanosensors*

Hydrogels are hydrophilic polymer networks that form an insoluble gel when crosslinked and are highly water absorbent.[106,107] Because of their high water retention rates, hydrogels have similar characteristics to the extracellular matrix imparting them with excellent biocompatibility and they are commonly used for drug delivery and tissue engineering applications.[106,107] From a sensing perspective, hydrogels are beneficial not only because of their biocompatibility, but their hydrophilicity also allows the diffusion of water soluble analytes into the sensor for detection.[23] Furthermore, the encapsulation of biological molecules, such as enzymes, expands the range of analytes and processes that can be monitored.[108,109]

Hydrogel-based fluorescent nanosensors have been mainly developed for the detection of ions and small molecules. Photonic explorer for biomedical use with biologically localized embedding (PEBBLEs) composed of polyacrylamide (PAA) hydrogel constitute a large class of hydrogel-based fluorescent nanosensors.[20,110,111] PAA nanoparticles serve as a neutral hydrophilic platform in which indicator dyes, reference dyes and recognition elements are encapsulated.[20] Currently, PEBBLEs have been designed to detect ions, radicals, and small molecules that include calcium,[9,27] magnesium,[112] hydroxide radicals (.OH),[113] and glucose[108] (Table 3). An overview with several examples of these sensors will be provided here, but more extensive reviews of this class of sensors are available.[20,110,111] Additionally, PEBBLE sensors based on optode-sensing technology will be covered in the optode-based fluorescent nanosensor section of this chapter.

Kopelman and coworkers first introduced PEBBLE sensors for pH, oxygen and calcium.[9] Fluorescent indicators for sensing and a reference dye were entrapped within an acrylamide matrix through microemulsion polymerization techniques. PEBBLEs 20–200 nm in diameter were produced and exhibited 1 ms response times, full signal reversibility, cell viability and minimum protein biofouling.[9,27] When introduced into single cells, calcium-sensitive PEBBLEs with Calcium Green and sulforhodamine 101 monitored the release of calcium from the membrane permeability transition of mitochondria.[27] Ratiometric PEBBLEs for measuring magnesium ions within living cells have also been developed.[112] These PEBBLEs were the first magnesium nanosensors that had a high selectivity for magnesium versus their divalent counterpart, calcium, at intracellular levels. When injected into cells, the magnesium-sensitive

Table 3. Analytes detected with hydrogel-based nanosensors.

	Analytes detected
Ions	Calcium,[9,27] Magnesium,[112] pH,[9] Inorganic Phosphate,[114] Zinc,[115] Iron,[116] Copper[117]
Small molecules	Glucose,[108,118] Oxygen[9,119]
Reactive species	.OH,[113] Hydrogen Peroxide,[109] ROS[109]

PEBBLEs responded to changes in intracellular magnesium levels upon ion channel opening.

Other hydrogel-based sensing platforms not based on PAA have also been developed. Glucose nanosensors were designed that take advantage of the ability of hydrogels to swell.[118] Poly(*N*-isopropylacrylamide) (NIPAM) nanoparticles were copolymerized with fluorescein and rhodamine derivatives that undergo FRET when within close proximity. They were covalently bound with different phenylboronic acid derivative receptor molecules that have an affinity for glucose. Upon binding to glucose, the boronic acids become negatively charged which increases water uptake and induces hydrogel swelling. Swelling separates the distance between fluorophores that results in a change in the fluorescence emission. These nanosensors were 250 nm in diameter, showed a linear response to glucose and the glucose response could be tailored at different pHs by interchanging boronic acids with different pK_a.

The high water retention of hydrogels also makes them suitable for biomolecule encapsulation of enzymes[108,109] and proteins.[114] Nanoparticle encapsulation of bioactive molecules can impact their biological activity[109,114] and large biomolecules can inhibit hydrogel polymerization.[109] However, several groups have successfully fabricated hydrogel-based sensors with enzymes.[108,109] To measure oxidative stress within macrophages, Pishko and coworkers encapsulated horseradish peroxidase (HRP) in a PEG hydrogel through reverse emulsion photopolymerization.[109] In these studies, HRP was covalently attached to the polymer PEG backbone, preventing its leaching from the hydrogel spheres. The enzyme remained active and catalyzed the reaction between hydrogen peroxide and Amplex Red that produced a red fluorescent product, resorufin. The HRP hydrogel spheres also sensed changes in ROS production induced by both exogenous and endogenous sources of oxidative stress while contained in phagocytotic vacuoles within macrophages. However, in this case, the reactant, Amplex Red, was not coencapsulated with HRP within the hydrogel nanoparticle. The PEBBLE hydrogel platform has also been exploited for the incorporation of enzymes. GOx, a ruthenium-based oxygen-sensitive fluorescent dye, and Oregon Green reference dye were all encapsulated within the PAA matrix.[108] In the presence of GOx, glucose is reduced to gluconic acid and results in a decrease in oxygen levels that can be measured using the oxygen-sensitive dye. The glucose-sensitive PEBBLEs were reversible and had a dynamic range between 0.3 and 8 mM; however, the local oxygen depletion caused non-linearity in the sensor response at high concentrations.

Lastly, proteins have been incorporated into hydrogel particles because of the hydrogel's ability to protect the protein from degradative enzymes.[114,117]

Inorganic phosphate sensing proteins (FLIPPi) were encapsulated in PAA hydrogels.[114] FLIPPi functions through a FRET-based sensing mechanism with a three-component molecule: with a cyan fluorescent protein and yellow fluorescent protein sandwiching a phosphate binding protein. When phosphate is bound to FLIPPi, the FRET efficiency between the two fluorescent reporters varies, resulting in changing fluorescent intensities. Monomer concentrations and cross-linking concentrations were adjusted to yield optimal protein loading efficiency and protein activity within the particle. In the presence of proteases, the PAA matrix slowed the degradation of FLIPPi activity compared to free FLIPPi.

As presented here, hydrogel-based particles have many advantages as a sensing platform due to their biocompatibility and high water retention.[107] However, this platform also presents itself with several drawbacks. As shown in the FLIPPi example, reducing pore size can cause inactivation of bioactive sensing molecules.[114] Additionally, though hydrogels may limit biofouling by proteins, ions can still permeate the matrix and interact with the sensing molecules or quench the fluorescent signal.[23]

3.4. *Liposome-based nanosensors*

Liposomes have been researched as nanocarriers for pharmaceuticals with several Food and Drug Administration (FDA) approved liposomal systems now currently available.[120] Liposomes are spherical lipid bilayers formed using amphiphilic molecules that self-assemble. Upon liposome formation, entrapped molecules, such as fluorescent indicators, maintain their solution properties.[5] Liposomes are either unilamellar (one bilayer) or multilamellar (concentric bilayers) and range from 100 nm up to 5 μm.[120] The use of liposomes as a vehicle for sensing provides several advantages over other sensing platforms. Liposomes are biocompatible and have been extensively surface-modified for increased circulation time *in vivo* and for targeting.[120] Liposomes readily enter into cells and have even been used as a delivery vehicle for entry of other nanosensors into cells.[37] Because of their unique bilayer construction, liposomes entrap hydrophilic and hydrophobic agents in the core or lipid membrane, respectively, expanding the range of recognition elements and fluorescent indicators that can be used.[120] The entrapment also isolates the sensing components from the surrounding media, minimizing interference from other biomolecules. Furthermore, these sensing elements can be incorporated non-covalently, minimizing the need for specific functional groups for attachment.[5]

Table 4. Analytes detected with liposome-based nanosensors.

	Analytes detected
Ions	pH,[121,123] Calcium[124]
Small molecules	Glucose,[125] Oxygen [7,121,126]

Liposome-based nanosensors have been developed for oxygen,[7,121,126] pH,[121–123] calcium,[124] and glucose[125] (Table 4). Dimyristoylphospatidylcholine, cholesterol and dihexadecyl phosphate are used to form unilamellar liposomes approximately 70 nm in diameter for several of these lipsome-based nanosensors for the detection of oxygen,[7,121] pH,[121,122] and calcium.[124] For oxygen sensing, a ruthenium-based oxygen-sensitive dye was encapsulated at an efficiency of 1% and the dye-loaded liposomes remained stable for 8 days at room temperature.[7] The oxygen-sensitive liposomes experienced a two-fold increase in fluorescence when the oxygen concentration changed due to the oxidation of glucose by GOx. Similarly, pH-sensitive liposomes containing a fluorescein derivative attached to the phospholipids fluorescently detected the corresponding changes in pH as a result of the reaction between glucose and GOx.[121] Covalently linking the dye to the phospholipid membrane prevented leaching. As an alternative to fluorescein labeled phospholipids, Ostafin and coworkers developed pH-sensitive liposomes with a calcium phosphate shell to prevent dye leakage and interactions between biomolecules and the dual wavelength fluorescent indicator, 5-(and-6)-carboxy–seminaphthorhodafluor-1 (cSNARF-1).[123] When in the presence of human plasma, the fluorescence changes of free cSNARF-1 could not be used to determine the pH of the plasma, but the pH could be determined when cSNARF-1 was protected in liposomes from interfering molecules in liposomes.

Calcium and glucose-sensitive lipsome-based nanosensors have more complex sensing mechanisms than simple inclusion of an environmentally sensitive fluorescent indicator.[124,125] For calcium detection, a calcium ion-binding protein, calmodulin, was labeled with the fluorescent indicator, Alexa-488, and embedded in the lipid bilayer of liposomes.[124] Calmodulin undergoes a conformational change upon binding calcium, which results in a decrease in the fluorescence intensity of Alexa-488. The embedded calmodulin–Alexa-488 complex had a similar dynamic range to the free protein-dye complex, but extended the lifetime of the protein from 2 to 24 h. Additionally, the sensors were reversible when increasing the temperature or upon addition of a chelating compound and exhibited selectivity for calcium over magnesium.

Liposomes were also used as a platform in the competitive binding mechanism between boronic acids, glucose, and the fluorescent indicator, alizarin red S (ARS) (Fig. 4).[125] For the liposome, 4-(4-dodecyloxybiphenyl-4-yloxy) butyl trimethylammonium bromide was used as the amphiphile to form the lipid bilayer. This amphiphile contains a cationic quaternary ammonium salt that electrostatically couples to the negatively charged ARS. ARS forms a boronate ester with phenylboronic acid resulting in high fluorescence intensity. Upon glucose addition, the fluorescence intensity decreases as the ARS-phenylboronic acid complex is replaced by a glucose-phenylboronic acid complex. The use of the vesicle improved the sensitivity of the system for glucose between seven and eight-fold compared to components in free solution.

The extension of liposomes from drug delivery vehicles to nanosensor platform has been established, but concerns are raised over liposome

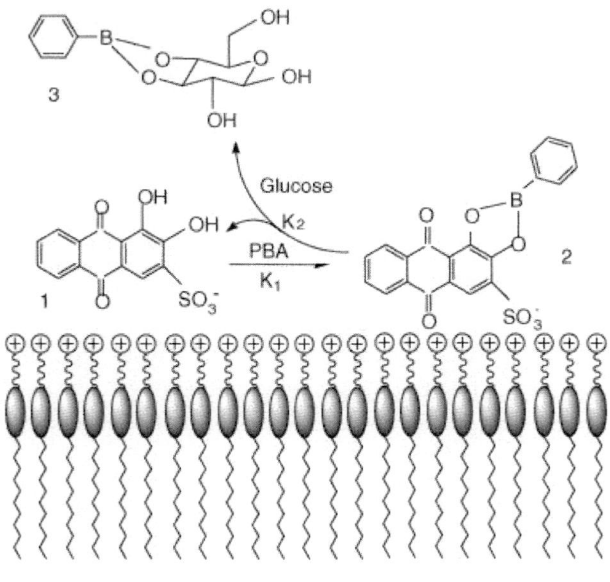

Fig. 4. Liposome-based glucose nanosensor. ARS is electrostatically coupled to quaternary ammonium salts located on the liposome surface and forms a complex with phenylboronic acid, resulting in high fluorescence intensity. When glucose is added to the system, glucose displaces ARS and forms a complex with phenylboronic acid. This displacement of ARS results in a decreased fluorescence intensity of the sensor. Reprinted from Sensors and Actuators B: Chemical, 119/2, Qiusheng Wang, Guangquan Li, Wenyi Xiao, Haixia Qi, Guowen Li, Glucose-responsive vesicular sensor based on boronic acid–glucose recognition in the ARS/PBA/DBBTAB covesicles, 695–700, Copyright (2003) with permission from Elsevier.

instability that results in dye leakage and, thus, limits their ability for long-term monitoring. In storage, liposomes undergo degradation from hydrolysis and oxidation.[121] Dye leakage is shown to have a temperature dependence, with oxygen-sensitive liposomes losing fluorescence signal faster at 37°C than at 25°C due to liposome instability at these temperatures.[121] Lipsomal stability is also influenced by membrane composition, encapsulated material, and pH, resulting in multiple parameters that hinder long-term monitoring by liposomal-based sensors.[121]

3.5. *Optode-based nanosensors*

Another class of polymeric sensors is based on the sensing mechanism employed rather than the sensing material. Optodes are the optical equivalent to ion-selective electrodes.[127] They are highly plasticized lipophilic membranes or particles in which recognition elements and fluorescent indicators are embedded.[127,128] They function through the bulk extraction of a particular analyte into the polymeric matrix by the recognition element. This extraction causes a concentration change within the matrix that results in optical changes of the sensors. For example, an ionophore extracts the cation of interest into the sensor (Fig. 5). This additional positive charge causes deprotonation of a pH-sensitive chromoionophore resulting in a decrease in fluorescence intensity. An ionophore, chromoionophore, and an additive to maintain charge neutrality are typical components in ion-selective optode sensing mechanisms. By simply substituting different ionophores, other ions of interest can be detected. Additionally, alternative sensing mechanisms have also been designed for the detection of small molecules such as glucose.[29] By altering the ratio of sensing components, the dynamic range and sensitivity can be tuned to within the desired physiological ranges.[16]

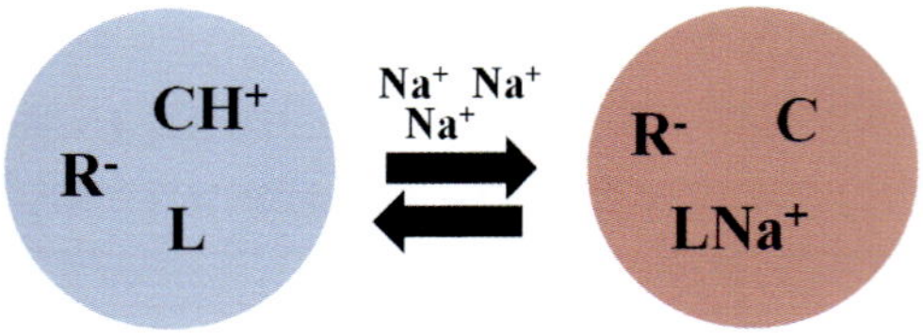

Fig. 5. Schematic of optode-based sensing mechanism for sodium. The three main sensing components are the sodium ionophore (L), pH-dependent chromoionophore (C), and a negatively charged additive (R) to maintain charge neutrality. When sodium is introduced, L extracts sodium into the sensor and C deprotonates in order to maintain charge neutrality. Deprotonation of C results in a change in fluorescence emission of the sensors.

Similar to other sensing platforms, the hydrophobic particle isolates sensing components from the surrounding biological media, protecting the sensor from interferents.[9] Biocompatible coatings, such as PEG, have been incorporated onto the surface of optode-based nanosensors to prevent sensor biofouling and reduce toxicity.[16] Further improvements to the plasticized poly vinyl chloride (PVC) particle are also being made to improve biocompatibility. For example, sensors composed of biodegradable polycaprolactone and a biocompatible citrate-based plasticizer have recently been developed[129] and plasticizer free sensors[130] have also been designed to minimize toxicity concerns over plasticized PVC in medical applications.[131,132]

The range of analytes detectable with optode-based nanosensors include sodium,[16,129,133,134] potassium,[135–137] and chloride[138] along with small molecules such as glucose[29] and histamine[139] with applications ranging from single cell analysis to *in vivo* monitoring (Table 5). This range of applications for these sensors is highlighted by the work of Clark and coworkers on the development of sodium-sensitive nanosensors.[15,16,140] Initially tuned for monitoring in the intracellular space, sodium-sensitive nanosensors were capable of capturing sodium fluxes from ion channels in cardiac myocytes.[15] The changing fluorescent signal of the sensors was correlated to voltage-gated sodium channel openings. By altering the sensing component ratio, these same nanosensors were then tuned to respond to sodium at extracellular physiological levels.[140] A mouse was coinjected with sodium-sensitive nanosensors and sodium solutions at different concentrations (Fig. 6(a)). The sodium calibration generated *in vivo* was similar to *in vitro* calibrations. When coinjected with 0 mM sodium and 150 mM sodium into the backs of mice, the sodium-sensitive nanosensors were able to fluorescently track changes in sodium as the concentration at the injection site equilibrated back to physiological levels (Fig. 6(b)).

Optode-based nanosensors have also been expanded to include the detection of the small molecules, glucose[29] and histamine.[139] For histamine detection, an ammonium ionophore having an affinity for amine-containing molecules was used to extract histamine into the sensor.[139] The extraction causes a pH change that results in a change in fluorescence of the sensors.

Table 5. Analytes detected with optode-based nanosensors.

	Analytes detected
Ions	Sodium,[16,129,133,134] Potassium,[135–137] Chloride[138]
Small molecules	Glucose,[29] Histamine[139]

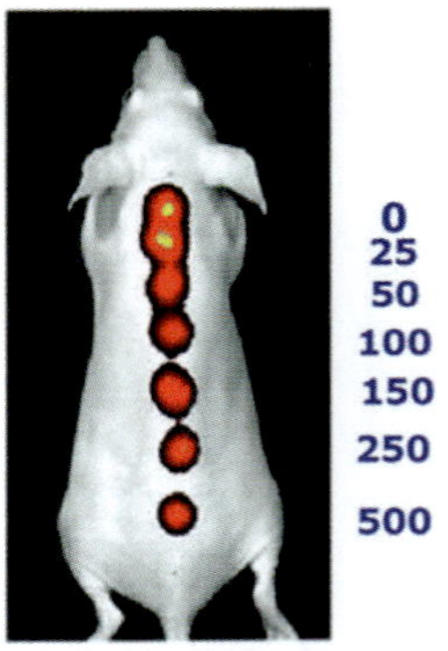

(a)

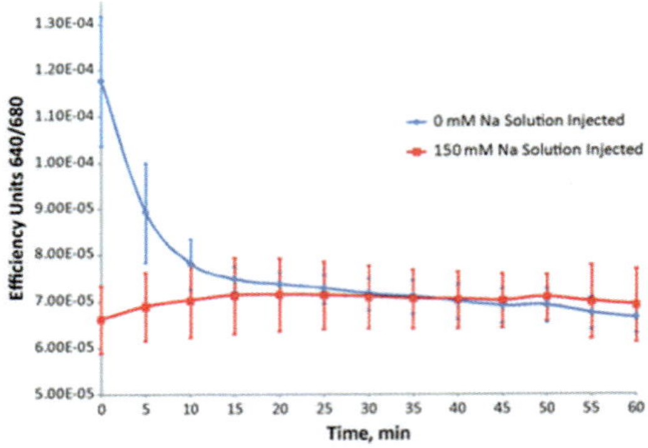

(b)

Fig. 6. *In vivo* response of sodium-sensitive optode-based nanosensors. (a) Sodium-sensitive nanosensors were coinjected along with different millimolar concentrations of sodium solutions along the back of a mouse. Fluorescence intensity of the sensors decreased with increasing sodium concentration. (b) Sodium-sensitive nanosensors were coinjected with either 0 mM or 150 mM sodium-sensitive along the backs of mice. The fluorescence intensity of the sensors responded to changes in the local sodium concentration *in vivo*. Figures modified from Ref. 140. [Dubach, J. M.; Lim, E.; Zhang, N.; Francis, K. P.; Clark, H., *Integr Biol (Camb), In vivo* sodium concentration continuously monitored with fluorescent sensors 2011, *3*(2), 142–8.] — Reproduced by permission of The Royal Society of Chemistry (http://pubs.rsc.org/en/Content/ArticleLanding/2011/IB/c0ib00020e).

The nanosensors were able to track systemic histamine changes in mice after an intraperitonal histamine injection. The design of glucose nanosensors diverges from the typical ion-selective optode sensing scheme, but instead is based on the competitive binding scheme between a boronic acid recognition molecule, the fluorescent indicator, alizarin and glucose as previously described in the liposome-based section.[29,125] When no glucose is present, boronic acid is bound to alizarin yielding a highly fluorescent species; however, as glucose is added to the system, the boronic acid extracts glucose into the sensor, displacing alizarin which results in a loss of fluorescence intensity. These sensors were injected into the subcutaneous space along the backs of mice. Upon administration of a glucose solution to mice, the sensors tracked changes in glucose concentration that correlated with a commercially-available blood glucose meter.

These examples highlight the scope of optode-based nanosensors for biological applications. The expansion of this sensor platform, though, to other analytes is limited to the availability of ionophores or recognition molecules with high specificity for the analyte of interest.[128] Efforts have also been undertaken to improve their biocompatibility for *in vitro* and *in vivo*

Table 6. Analytes detected with dendrimer-based nanosensors.

	Analytes detected
Ions	Cobalt,[143–145] Nickel,[145,146] Iron,[145,146] Zinc,[146] Lead,[145,146] Magnesium,[145] Sodium,[8] Potassium,[147] Chloride,[147] pH[142,147,148]
Small molecules	Amino Alcohols[149–151]

applications.[16,129,130] *In vitro*, sodium-sensitive nanosensors functionalized with PEG have not shown signs of cellular toxicity,[16] but no extensive biodistribution and histological studies have been conducted to assess potential *in vivo* toxicity.

3.6. *Dendrimer-based nanosensors*

Dendrimers are highly structured and branched macromolecules or nanostructures with tightly controlled fabrication resulting in monodisperse samples.[141] They can be biocompatible[141] and functionalized with fluorophores[142] or they can non-covalently trap fluorophores within their structural cavities.[8] Similar to other nanosensors discussed in this chapter, dendrimer-based nanosensors provide improvements to molecular probes, such as preventing dye leakage out of cells.[8]

Dendrimers have been designed to detect metal cations,[143–146] the chirality of amino alcohols,[149–151] and other important physiological parameters such as sodium,[8] potassium,[147] chloride,[147] and pH[142,147,148] (Table 6). Several different designs have been used for pH sensing. For example, Vinogradov and coworkers developed two different porphyrin-based dendrimers sensitive to changes in pH.[148] By encapsulating the porphyrins in dendrimers, the water-solubility of the porphyrins improved and their p*K*s were tuned to the physiological pH range by adjusting the peripheral charges on the surface of the dendrimers. The two dendrimers were spectrally independent and were capable of distinguishing proton gradients across liposomal membranes when each dendrimer was spatially separated by the membrane. Albertazzi and coworkers also designed pH-sensitive dendrimer sensors that are ratiometric, tunable and targeted to different cellular compartments based on dendrimer surface charge and size.[142] Three pH-sensitive fluorophores, Oregon Green 488, 5(6)-carboxyfluorescein, and 7-hydroxy-4-methyl-3-coumarinylacetic acid, were covalently attached to separate dendrimers along with a pH-insensitive fluorophore, Rhodamine RedX, to make three separate pH sensors that operate in different ranges. Cationic fluorescein-based dendrimer sensors were electroporated into HeLa cells and targeted to lysosomes and the plasma membrane. The pH difference between these compartments was clearly shown in the ratio of the fluorescein to rhodamine fluorescence of the two compartments. In later work by Albertazzi, pH-sensitive dendrimer-PEG sensors were introduced into the extracellular space of the brain in anesthetized mice and imaged using two-photon fluorescence imaging.[147] The fluorescence of the dendrimers remained stable in the extracellular matrix for hours and changes in fluorescence were visualized when the mouse was exposed to hypercapnia conditions that acidified the blood.

Another unique application of dendrimer-based sensors is their use as sodium sensors in neurons.[8] Lamy and coworkers developed sodium-sensitive poly (amidoamine) dendrimers (nanoprobes) that entrapped the sodium sensitive dye, CoroNa Green. These nanoprobes do not rapidly leach out of the cell which occurs with free dyes and they are small enough (~7 nm) to easily diffuse into small cellular compartments of neurons, such as dendrites. The sodium-sensitive nanoprobes had a linear response to sodium in the physiological range and did not experience interference from potassium, pH, or calcium. Additionally, the probes experienced little change in their fluorescence over an *in vitro* storage period of more than a month. When electroporated into live neurons in acute neocortical brain slices, the nanoprobes diffused throughout the dendrites and axons and detected sodium transients caused by action potentials.

Dendrimer-based nanosensors provide an exciting opportunity to create highly controllable and monodisperse sensors. However, high costs are associated with the scale-up of dendrimer manufacturing.[141] From a toxicity perspective, both *in vitro* and *in vivo* biocompatibility of dendrimers is dependent upon their surface charge with cationic functional groups exhibiting greater toxicity than anionic or neutral groups.[17,141,152,153] To further address toxicity concerns, biodegradable dendrimers have also been developed.[154]

4. Design Considerations for Fluorescent Nanosensors

From the current review of fluorescent nanosensors, a list of basic design considerations for nanosensors in biomedical applications. As in any engineering design, each of these criteria is weighted according to its necessity in the desired application. For example, fluorescent nanosensors with emission wavelengths in the NIR provide greater signal to noise over shorter emitting nanosensors *in vivo*. Since sensor design many on the application, no "one size fits all" nanosensor exists and trade-offs are made between criteria.

Throughout this section, these criteria will be highlighted along with examples. Table 7 provides a summary of these considerations.

4.1. *Fluorescent wavelength considerations*

Wavelength selection for fluorescent nanosensors is important for imparting nanosensors with sufficient signal to noise ratios in biological samples. Biological samples naturally fluoresce below 550 nm[155] and this autofluorescence contributes to background noise and masks the fluorescence of nanosensors. Autofluorescence is mainly attributed to mitochondria and lysosomes on the cellular level and from the extracellular matrix on the *in vivo* level.[156] For *in vitro* cell culture work, nanosensors that excite at wavelengths greater than 550 nm are recommended to minimize the effects of autofluorescence; however, if not, then fluorescent nanosensors with sufficient fluorescent quantum efficiency should be selected.[27] For *in vivo* fluorescent detection and monitoring, stricter criteria of indicators fluorescing in the NIR (700–1000 nm)[157] is optimal in order to achieve a high signal to noise ratio above autofluorescence and tissue absorbance. The three basic fluorescent indicators presented in this chapter (molecular probes, QDs and CNTs) are available in the NIR range. QDs and CNTs are more versatile in this respect

Table 7. Summary of design considerations.

Design consideration	Recommended guidelines
Fluorescent wavelength	*In vitro*: Sensors with emission > 550 nm *In vivo*: Sensors with emission 700–1000 nm Quantitative measurements: Spectrally independent emission
Photostability	QDs and CNTs: high frequency measurements
Size	Intracellular: < 100 nm *In vivo*: ~1 μm or non-spherical geometries
Response characteristics	Response time: Faster than biological phenomenon Dynamic range: Capture physiological concentrations Sensitivity: Capture physiological changes Reversibility: Continuous monitoring Limited interference from other biomolecules
Biocompatibility	Biocompatible surface functionalization Biodegradable components or size control for renal clearance

because their fluorescence can be tuned to the NIR[158] or are inherently NIR fluorescent,[74] respectively. Lastly, if quantitative measurements are desired, then dual emission or two independent fluorescent indicators may be used. Selection of these indicators is mainly dependent upon the emission wavelengths being spectrally independent or resolvable. The size tunability and narrow emission spectra of QDs make this platform desirable for incorporation as a reference.[11]

4.2. *Photostability*

Photobleaching can also be a major problem for fluorescent nanosensors depending upon if measurements or images are collected at high frequencies or with or without laser illumination. Capturing the dynamics of ion channel fluxes or action potentials requires near constant excitation and QDs and CNTs offer a distinct advantage because of their resistance to photobleaching.[10–12] However in some applications, constant imaging is not required because important fluxes in analyte concentration occur at slower time scales. For example, a Medtronic Minimed Guardian RT glucose monitor for diabetic care acquires readings every 10 s and provides the patient with an average glucose reading every 5 min.[159] For the optical equivalent of these measurements, the lifetime of organic fluorophores would be extended over longer monitoring times.

4.3. *Nanosensor size*

In this chapter, nanosensors were presented that ranged in size between 1 and 1,000 nm. The desired size range for nanosensors is highly dependent on their application. For intracellular measurements, particles less than 100 nm are desirable for minimal cellular perturbation.[160] As an example, PEBBLE sensors typically in the range of 20–600 nm only occupy 1-1,000 PPB of the total mammalian cell volume.[13] Sensor size is also critical for monitoring in small intracellular areas, such as dendrites. This design feature is highlighted in the work of Lamy, where 7 nm dendrimer-based sodium sensors diffused into dendrites and measured fast sodium dynamics.[8] However, for constant *in vivo* monitoring, it may be desirable that sensors remain in a defined location for constant monitoring. Diffusion and cellular uptake at the injection site has been shown with quantum dots.[161] In this case, larger nanosensors, closer to a micron, or different sensor geometries may hinder diffusion and cellular uptake, prolonging sensor monitoring *in vivo*.[162]

4.4. *Nanosensor response characteristics*

Typical response characteristics reported with nanosensor development are: response time, dynamic range, sensitivity, and reversibility. The most basic requirement for sensor response time is that it must be less than the duration of the biological phenomenon being monitored. Rapidly occurring phenomena like action potentials or short-lived species like reactive oxygen species (half life of hydrogen peroxide is 1 ms)[163] require nanosensors with near instantaneous response in order to capture their kinetics. For example, to monitor sodium flux from ion channels in cardiac myocytes, optode-based sodium nanosensors were required to have response times less than a millisecond to capture action potential dynamics.[15] Longitudinal measurements *in vivo* such as glucose measurements, however, do not require millisecond response times since measurements may be taken on the order of seconds to minutes.[159] Therefore, sensors with slower response kinetics are still applicable in this case.

Sensor dynamic range and sensitivity are also important characteristics when monitoring biological analytes. Biological species can have low concentrations, fluctuate in a narrow concentration ranges, or both. These restrictions make it imperative that fluorescent nanosensors have dynamic ranges and sensitivities that are able to detect and monitor the species of interest's fluctuations. Several examples of CNT-based nanosensors were provided that described sensitivities down to the single molecular level.[96–98,100] With respect to dynamic range and sensitivity, optode-based nanosensors have advantages because their response can be adjusted by tailoring the ratio of sensor components.[16] However similar to molecular probes, the sensitivity of these recognition elements is ultimately governed by the chemistry of the recognition molecule itself and further improvements would require the synthesis of new recognition elements.[128] Additionally, if continuous monitoring both *in vivo* and *in vitro* is desired, fluorescent nanosensors must be reversible over many cycles without regeneration. This requires selection of appropriate recognition molecules and sensor platform to maximize stability. Lastly, in complex biological media, it is important to recognize any potential sensor interferents. In the case of sodium sensing, potassium can be a major interferent for intracellular measurements where its concentration is over 10 times greater than that of sodium.[164] The impact of interference is highly dependent on the interferent's physiologically relevant concentration and whether this interference causes a significant shift in sensor response.

4.5. *Biocompatibility of fluorescent nanosensors*

Whether *in vivo* or *in vitro*, biocompatibility of fluorescent nanosensors is important to minimize cytotoxicity, inflammation, and an immune responses. These reactions may not only affect nanosensor measurements, but may also lead to cell or animal death. Several approaches are used to improve nanosensor biocompatibility, including the addition of surface coatings. For example, PEG is a popular biocompatible coating that has been used as a surface coating for optodes[16] and dendrimers.[17] PEGylated dendrimers demonstrated limited hemolytic activity and cytotoxicity, whereas their cationic counterparts exhibited hemolysis and cytotoxicity.[17] Additionally, when tested *in vivo*, PEGylated dendrimers did not show signs of toxicity either through mortality or abnormal blood chemistry. Size selection can also influence long term toxicity of nanosensor use. *In vivo* work has shown that QDs less, than 5.5 nm were removed from the body through the renal system whereas larger particles accumulated in the liver, kidneys, and spleen.[72] Thus, size can be exploited for QDs, because concerns have been raised about their heavy metal composition. Another approach for minimizing toxicity *in vivo* is to design fluorescent nanosensors composed of biodegradable components, such as polycaprolactone.[165] For example, sodium optode-based nanosensors have recently been designed out of polycaprolactone, an FDA approved polymer, and they degraded in the presence of lipases at physiological concentrations.[129,165]

5. Summary and Conclusion

Fluorescent nanosensors provide unprecedented access to both *in vitro* and *in vivo* spaces while providing longer monitoring times previously not possible with traditional molecular probes. Coupled with advantages of fluorescence imaging, such as highly-sensitive and non-invasive measurements, nanosensors provide additional advantages that include rapid response times,[15] resistance to photobleaching,[10–12] size dependent tunable emission,[11] and single molecule detection.[96–98,100] Nanosensors can also be fabricated with biocompatible and biodegradable materials and coatings that reduce cytotoxicity and long-term toxicity.[17,129] This chapter has illustrated the use of nanosensors for detecting and monitoring single molecules,[96–98,100] DNA polymorphisms,[103,104] action potentials in dendrites,[8] and small animal sodium and glucose fluxes.[29,140] Currently, no "one size fits all" nanosensor platform exists that can be applied to all applications and there is always a trade-off when choosing specific nanosensor platforms over another. Fundamental drawbacks to fluorescence, such as tissue penetration *in vivo,* also exist and limit *in vivo* monitoring capabilities.[157] Despite these disadvantages, the understanding of biological phenomena is still only realizable if the tools for monitoring them are accessible. As presented here, the present and future range of the diversity of nanosensors provides an expanding toolbox for researchers for monitoring their biological phenomenon of choice.

Acknowledgments

This work was supported by NIH Grant 5RO1GM084366 and DARPA Young Investigator Award D12AP00246. We also thank Christopher G. Skipwith for his technical guidance in preparing this chapter.

References

1. A. Waggoner, Optical probes of membrane potential. *J. Membr. Biol.* **27**(4), 317–334 (1976).
2. R. Y. Tsien, New calcium indicators and buffers with high selectivity against magnesium and protons: design, synthesis and properties of prototype structures. *Biochemistry* **19**(11), 2396–2404 (1980).
3. G. Grynkiewicz, M. Poenie and R. Y. Tsien, A new generation of Ca2+ indicators with greatly improved fluorescence properties. *J. Biol. Chem.* **260**(6), 3440–3450 (1985).
4. E. A. Lemke and C. Schultz, Principles for designing fluorescent sensors and reporters. *Nat. Chem. Biol.* 7(8), 480–483 (2011).
5. L. Basabe-Desmonts, D. N. Reinhoudt and M. Crego-Calama, Design of fluorescent materials for chemical sensing. *Chem. Soc. Rev.* **36**(6), 993–1017 (2007).
6. V. Wagner, A. Dullaart, A. K. Bock and A. Zweck, The emerging nanomedicine landscape. *Nat. Biotechnol.* **24**(10), 1211–1217 (2006).
7. K. P. McNamara and Z. Rosenzweig, Dye-encapsulating liposomes as fluorescence-based oxygen nanosensors. *Anal. Chem.* **70**(22), 4853–4859 (1998).
8. C. M. Lamy, O. Sallin, C. Loussert and J. Y. Chatton, Sodium sensing in neurons with a dendrimer-based nanoprobe. *ACS Nano.* **6**(2), 1176–1187 (2012).
9. H. A. Clark, M. Hoyer, M. A. Philbert and R. Kopelman, Optical nanosensors for chemical analysis inside single living cells. 1. Fabrication, characterization and methods for intracellular delivery of PEBBLE sensors. *Anal. Chem.* **71**(21), 4831–4836 (1999).
10. D. A. Heller, S. Baik, T. E. Eurell and M. S. Strano, Single-walled carbon nanotube spectroscopy in live cells: towards long-term labels and optical sensors. *Adv. Mater.* **17**(23), 2793–2799 (2005).
11. X. Michalet, F. F. Pinaud, L. A. Bentolila, J. M. Tsay, S. Doose, J. J. Li, G. Sundaresan, A. M. Wu, S. S. Gambhir and S. Weiss, Quantum dots for live cells, in vivo imaging and diagnostics. *Science* **307**(5709), 538–544 (2005).
12. X. Wu, H. Liu, J. Liu, K. N. Haley, J. A. Treadway, J. P. Larson, N. Ge, F. Peale and M. P. Bruchez, Immunofluorescent labeling of cancer marker Her2 and other cellular targets with semiconductor quantum dots. *Nat. Biotechnol.* **21**(1), 41–46 (2003).

13. S. M. Buck, H. Xu, M. Brasuel, M. A. Philbert and R. Kopelman, Nanoscale probes encapsulated by biologically localized embedding (PEBBLEs) for ion sensing and imaging in live cells. *Talanta.* **63**(1), 41–59 (2004).
14. K. J. Cash and H. A. Clark, Nanosensors and nanomaterials for monitoring glucose in diabetes. *Trends. Mol. Med.* **16**(12), 584–593 (2010).
15. J. M. Dubach, S. Das, A. Rosenzweig and H. A. Clark, Visualizing sodium dynamics in isolated cardiomyocytes using fluorescent nanosensors. *Proc. Natl. Acad. Sci. USA* **106**(38), 16145–16150 (2009).
16. J. M. Dubach, D. I. Harjes and H. A. Clark, Fluorescent ion-selective nanosensors for intracellular analysis with improved lifetime and size. *Nano Lett.* 7(6), 1827–1831 (2007).
17. H. T. Chen, M. F. Neerman, A. R. Parrish and E. E. Simanek, Cytotoxicity, hemolysis and acute in vivo toxicity of dendrimers based on melamine, candidate vehicles for drug delivery. *J. Am. Chem. Soc.* **126**(32), 10044–10048 (2004).
18. R. A. Sperling, P. Rivera Gil, F. Zhang, M. Zanella and W. J. Parak, Biological applications of gold nanoparticles. *Chem. Soc. Rev.* **37**(9), 1896–908 (2008).
19. E. Boisselier and D. Astruc, Gold nanoparticles in nanomedicine: preparations, imaging, diagnostics, therapies and toxicity. *Chem. Soc. Rev.* **38**(6), 1759–1782 (2009).
20. Y. E. Lee, R. Smith and R. Kopelman, Nanoparticle PEBBLE sensors in live cells and in vivo. *Annu. Rev. Anal. Chem. (Palo Alto Calif).* **2** 57–76 (2009).
21. D. A. Williams, K. E. Fogarty, R. Y. Tsien and F. S. Fay, Calcium gradients in single smooth muscle cells revealed by the digital imaging microscope using Fura-2. *Nature* **318**(6046), 558–561 (1985).
22. K. Svoboda, W. Denk, D. Kleinfeld and D. W. Tank, In vivo dendritic calcium dynamics in neocortical pyramidal neurons. *Nature* **385**(6612), 161–165 (1997).
23. O. S. Wolfbeis, Materials for fluorescence-based optical chemical sensors. *J. Mater. Chem.* **15**(27–28), 2657–2669 (2005).
24. B. Morelle, J. M. Salmon, J. Vigo and P. Viallet, Are intracellular ionic concentrations accessible using fluorescent probes? The example of Mag-indo-1. *Cell. Biol. Toxicol.* **10**(5–6), 339–344 (1994).
25. M. L. Graber, D. C. DiLillo, B. L. Friedman and E. Pastoriza-Munoz, Characteristics of fluoroprobes for measuring intracellular pH. *Anal. Biochem.* **156**(1), 202–212 (1986).
26. I. Johnson, Fluorescent probes for living cells. *Histochem. J.* **30**(3), 123–140 (1998).
27. H. A. Clark, R. Kopelman, R. Tjalkens and M. A. Philbert, Optical nanosensors for chemical analysis inside single living cells. 2. Sensors for pH and calcium and the intracellular application of PEBBLE sensors. *Anal. Chem.* **71**(21), 4837–4843 (1999).
28. J. H. Kim, J. H. Ahn, P. W. Barone, H. Jin, J. Zhang, D. A. Heller and M. S. Strano, A Luciferase/Single-Walled Carbon Nanotube Conjugate for Near-Infrared Fluorescent Detection of Cellular ATP. *Ange. Chem.* **122**(8), 1498–1501 (2010).
29. K. Billingsley, M. K. Balaconis, J. M. Dubach, N. Zhang, E. Lim, K. P. Francis and H. A. Clark, Fluorescent nano-optodes for glucose detection. *Anal. Chem.* **82**(9), 3707–3713 (2010).
30. C. A. Poland, R. Duffin, I. Kinloch, A. Maynard, W. A. Wallace, A. Seaton, V. Stone, S. Brown, W. Macnee and K. Donaldson, Carbon nanotubes introduced into the abdominal cavity of mice show asbestos-like pathogenicity in a pilot study. *Nat. Nanotechnol.* **3**(7), 423–428 (2008).

31. A. Takagi, A. Hirose, T. Nishimura, N. Fukumori, A. Ogata, N. Ohashi, S. Kitajima and J. Kanno, Induction of mesothelioma in p53+/- mouse by intraperitoneal application of multi-wall carbon nanotube. *J. Toxicol. Sci.* **33**(1), 105–116 (2008).
32. L. Lacerda, A. Soundararajan, R. Singh, G. Pastorin, K. T. Al-Jamal, J. Turton, P. Frederik, M. A. Herrero, S. L. A. Bao, D. Emfietzoglou, S. Mather, W. T. Phillips, M. Prato, A. Bianco, B. Goins and K. Kostarelos, Dynamic Imaging of functionalized multi-walled carbon nanotube systemic circulation and urinary excretion. *Adv. Mater.* **20**(2), 225–230 (2008).
33. I. L. Medintz, H. T. Uyeda, E. R. Goldman and H. Mattoussi, Quantum dot bioconjugates for imaging, labelling and sensing. *Nat. Mater.* **4**(6), 435–446 (2005).
34. B. O. Dabbousi, J. RodriguezViejo, F. V. Mikulec, J. R. Heine, H. Mattoussi, R. Ober, K. F. Jensen and M. G. Bawendi, (CdSe)ZnS core-shell quantum dots: synthesis and characterization of a size series of highly luminescent nanocrystallites. *J. Phys. Chem. B.* **101**(46), 9463–9475 (1997).
35. R. Freeman and I. Willner, Optical molecular sensing with semiconductor quantum dots (QDs). *Chem .Soc. Rev.* **41**(10), 4067–4085 (2012).
36. M. J. Ruedas-Rama and E. A. Hall, Multiplexed energy transfer mechanisms in a dual-function quantum dot for zinc and manganese. *Analyst* **134**(1), 159–169 (2009).
37. Y. Wang, H. Mao and L. B. Wong, Dynamic [Cl(-)](i) measurement with chloride sensing quantum dots nanosensor in epithelial cells. *Nanotechnology* **21**(5), 055101 (2010).
38. I. L. Medintz, A. R. Clapp, H. Mattoussi, E. R. Goldman, B. Fisher and J. M. Mauro, Self-assembled nanoscale biosensors based on quantum dot FRET donors. *Nat. Mater.* **2**(9), 630–638 (2003).
39. A. R. Clapp, I. L. Medintz and H. Mattoussi, Forster resonance energy transfer investigations using quantum-dot fluorophores. *Chemphyschem.* 7(1), 47–57 (2006).
40. J. R. Lakowicz, Kluwer Academic Pub *Principles of Fluorescence Spectroscopy* (1999).
41. A. R. Clapp, I. L. Medintz, J. M. Mauro, B. R. Fisher, M. G. Bawendi and H. Mattoussi, Fluorescence resonance energy transfer between quantum dot donors and dye-labeled protein acceptors. *J. Am. Chem. Soc.* **126**(1), 301–310 (2004).
42. Y. Wang, H. Mao and L. Wong, Na+(-)sensing quantum dots for cell-based screening of intracellular Na+ concentrations ([Na+]i). *Talanta* **85**(1), 694–700 (2011).
43. C. Y. Chen, C. T. Cheng, C. W. Lai, P. W. Wu, K. C. Wu, P. T. Chou, Y. H. Chou and H. T. Chiu, Potassium ion recognition by 15-crown-5 functionalized CdSe/ZnS quantum dots in H2O. *Chem. Commun. (Camb).* (3), 263–265 (2006).
44. M. Suzuki, Y. Husimi, H. Komatsu, K. Suzuki and K. T. Douglas, Quantum dot FRET biosensors that respond to pH, to proteolytic or nucleolytic cleavage, to DNA synthesis, or to a multiplexing combination. *J. Am. Chem. Soc.* **130**(17), 5720–5725 (2008).
45. P. T. Snee, R. C. Somers, G. Nair, J. P. Zimmer, M. G. Bawendi and D. G. Nocera, A ratiometric CdSe/ZnS nanocrystal pH sensor. *J. Am. Chem. Soc.* **128**(41), 13320–13321 (2006).
46. D. E. Prasuhn, A. Feltz, J. B. Blanco-Canosa, K. Susumu, M. H. Stewart, B. C. Mei, A. V. Yakovlev, C. Loukov, J. M. Mallet, M. Oheim, P. E. Dawson and I. L. Medintz, Quantum dot peptide biosensors for monitoring caspase 3 proteolysis and calcium ions. *ACS Nano* **4**(9), 5487–5497 (2010).
47. C. S. Wu, M. K. Khaing Oo and X. Fan, Highly sensitive multiplexed heavy metal detection using quantum-dot-labeled DNAzymes. *ACS Nano* **4**(10), 5897–5904 (2010).

48. C. Han, Z. Cui, Z. Zou, D. Sabahaiti Tian and H. Li, Urea-type ligand-modified CdSe quantum dots as a fluorescence "turn-on" sensor for CO3(2-) anions. *Photochem. Photobiol. Sci.* **9**(9), 1269–1273 (2010).
49. M. J. Ruedas-Rama and E. A. Hall, Azamacrocycle activated quantum dot for zinc ion detection. *Anal. Chem.* **80**(21), 8260–8268 (2008).
50. M. Koneswaran and R. Narayanaswamy, L-Cysteine-capped ZnS quantum dots based fluorescence sensor for Cu2+ ion. *Sens. Actuator B-Chem.* **139**(1), 104–109 (2009).
51. L. Cao, J. Ye, L. Tong and B. Tang, A new route to the considerable enhancement of glucose oxidase (GOx) activity: the simple assembly of a complex from CdTe quantum dots and GOx and its glucose sensing. *Chemistry* **14**(31), 9633–9640 (2008).
52. M. Algarra, B. B. Campos, F. R. Aguiar, J. E. Rodriguez-Borges and J. da Silva, Novel beta-cyclodextrin modified CdTe quantum dots as fluorescence nanosensor for acetylsalicylic acid and metabolites. *Mater. Sci. Eng. C-Mater. Biol. Appl.* **32**(4), 799–803 (2012).
53. M. Geszke-Moritz, G. Clavier, J. Lulek and R. Schneider, Copper- or manganese-doped ZnS quantum dots as fluorescent probes for detecting folic acid in aqueous media. *J. Lumines.* **132**(4), 987–991 (2012).
54. Y. Long, L. F. Zhang, Y. Zhang and C. Y. Zhang, Single quantum dot based nanosensor for renin assay. *Anal. Chem.* **84**(20), 8846–8852 (2012).
55. T. Qiu, D. Zhao, G. Zhou, Y. Liang, Z. He, Z. Liu, X. Peng and L. Zhou, A positively charged QDs-based FRET probe for micrococcal nuclease detection. *Analyst* **135**(9), 2394–2399 (2010).
56. C. Y. Zhang and L. W. Johnson, Quantum dot-based fluorescence resonance energy transfer with improved FRET efficiency in capillary flows. *Anal. Chem.* **78**(15), 5532–5537 (2006).
57. I. L. Medintz, L. Berti, T. Pons, A. F. Grimes, D. S. English, A. Alessandrini, P. Facci, H. Mattoussi, A reactive peptidic linker for self-assembling hybrid quantum dot-DNA bioconjugates. *Nano Lett.* 7(6), 1741–1748 (2007).
58. S. E. Diltemiz, R. Say, S. Buyuktiryaki, D. Hur, A. Denizli and A. Ersoz, Quantum dot nanocrystals having guanosine imprinted nanoshell for DNA recognition. *Talanta* **75**(4), 890–896 (2008).
59. C. Y. Zhang and J. Hu, Single quantum dot-based nanosensor for multiple DNA detection. *Anal. Chem.* **82**(5), 1921–1927 (2010).
60. X. Wang, X. Lou, Y. Wang, Q. Guo, Z. Fang, X. Zhong, H. Mao, Q. Jin, L. Wu, H. Zhao and J. Zhao, QDs-DNA nanosensor for the detection of hepatitis B virus DNA and the single-base mutants. *Biosens. Bioelectron.* **25**(8), 1934–1940 (2010).
61. C. Y. Zhang, H. C. Yeh, M. T. Kuroki and T. H. Wang, Single-quantum-dot-based DNA nanosensor. *Nat. Mater.* **4**(11), 826–831 (2005).
62. C. Y. Zhang and L. W. Johnson, Quantifying RNA-peptide interaction by single-quantum dot-based nanosensor: an approach for drug screening. *Anal. Chem.* **79**(20), 7775–7781 (2007).
63. C. Y. Zhang and L. W. Johnson, Quantum-dot-based nanosensor for RRE IIB RNA-Rev peptide interaction assay. *J. Am. Chem. Soc.* **128**(16), 5324–5325 (2006).
64. Y. Zhang and C. Y. Zhang, Sensitive detection of microRNA with isothermal amplification and a single-quantum-dot-based nanosensor. *Anal. Chem.* **84**(1), 224–231 (2012).
65. W. R. Algar and U. J. Krull, Multiplexed interfacial transduction of nucleic acid hybridization using a single color of immobilized quantum dot donor and two acceptors in fluorescence resonance energy transfer. *Anal. Chem.* **82**(1), 400–405 (2010).

66. F. Long, C. Gu, A. Z. Gu and H. Shi, Quantum dot/carrier-protein/haptens conjugate as a detection nanobioprobe for FRET-based immunoassay of small analytes with all-fiber microfluidic biosensing platform. *Anal. Chem.* **84**(8), 3646–3453 (2012).
67. M. Nirmal, B. O. Dabbousi, M. G. Bawendi, J. J. Macklin, J. K. Trautman, T. D. Harris and L. E. Brus, Fluorescence intermittency in single cadmium selenide nanocrystals. *Nature* **383**(6603), 802–804 (1996).
68. J. M. Klostranec and W. C. W. Chan, Quantum dots in biological and biomedical research: recent progress and present challenges. *Adv. Mater.* **18**(15), 1953–1964 (2006).
69. M. Green and E. Howman, Semiconductor quantum dots and free radical induced DNA nicking. *Chem. Commun. (Camb).* (1), 121–123 (2005).
70. B. Ballou, L. A. Ernst, S. Andreko, T. Harper, J. A. Fitzpatrick, A. S. Waggoner and M. P. Bruchez, Sentinel lymph node imaging using quantum dots in mouse tumor models. *Bioconjug. Chem.* **18**(2), 389–396 (2007).
71. W. W. Yu, E. Chang, J. C. Falkner, J. Zhang, A. M. Al-Somali, C. M. Sayes, J. Johns, R. Drezek and V. L. Colvin, Forming biocompatible and nonaggregated nanocrystals in water using amphiphilic polymers. *J. Am. Chem. Soc.* **129**(10), 2871–2879 (2007).
72. H. S. Choi, W. Liu, P. Misra, E. Tanaka, J. P. Zimmer, B. Itty Ipe, M. G. Bawendi and J. V. Frangioni, Renal clearance of quantum dots. *Nat. Biotechnol.* **25**(10), 1165–1170 (2007).
73. M. S. Strano, A. A. Boghossian, W. J. Kim, P. W. Barone, E. S. Jeng, D. A. Heller, N. Nair, H. Jin, R. Sharma and C. Y. Lee, The chemistry of single-walled nanotubes. *MRS Bull.* **34**(12), 950–961 (2009).
74. A. A. Boghossian, J. Zhang, P. W. Barone, N. F. Reuel, J. H. Kim, D. A. Heller, J. H. Ahn, A. J. Hilmer, A. Rwei, J. R. Arkalgud, C. T. Zhang and M. S. Strano, Near-infrared fluorescent sensors based on single-walled carbon nanotubes for life sciences applications. *ChemSusChem.* **4**(7), 848–863 (2011).
75. S. M. Bachilo, M. S. Strano, C. Kittrell, R. H. Hauge, R. E. Smalley and R. B. Weisman, Structure-assigned optical spectra of single-walled carbon nanotubes. *Science* **298**(5602), 2361–2366 (2002).
76. M. J. O'Connell, S. M. Bachilo, C. B. Huffman, V. C. Moore, M. S. Strano, E. H. Haroz, K. L. Rialon, P. J. Boul, W. H. Noon, C. Kittrell, J. Ma, R. H. Hauge, R. B. Weisman and R. E. Smalley, Band gap fluorescence from individual single-walled carbon nanotubes. *Science* **297**(5581), 593–596 (2002).
77. D. A. Heller, H. Jin, B. M. Martinez, D. Patel, B. M. Miller, T. K. Yeung, P. V. Jena, C. Hobartner, T. Ha, S. K. Silverman and M. S. Strano, Multimodal optical sensing and analyte specificity using single-walled carbon nanotubes. *Nat. Nanotechnol.* **4**(2), 114–120 (2009).
78. P. W. Barone and M. S. Strano, Single walled carbon nanotubes as reporters for the optical detection of glucose. *J. Diabetes .Sci .Technol.* **3**(2), 242–252 (2009).
79. L. M. Uusitalo and N. Hempel, Recent advances in intracellular and in vivo ROS sensing: focus on nanoparticle and nanotube applications. *Int. J. Mol. Sci.* **13**(9), 10660–10679 (2012).
80. C. Zhao, K. Qu, Y. Song, C. Xu, J. Ren and X. Qu, A reusable DNA single-walled carbon-nanotube-based fluorescent sensor for highly sensitive and selective detection of Ag+ and cysteine in aqueous solutions. *Chemistry* **16**(27), 8147–8154 (2010).
81. L. Jing, C. Liang, X. Shi, S. Ye and Y. Xian, Fluorescent probe for Fe(III) based on pyrene grafted multiwalled carbon nanotubes by click reaction. *Analyst* **137**(7), 1718–1722 (2012).

82. L. Zhang, T. Li, B. Li, J. E. Li and E. Wang, Carbon nanotube-DNA hybrid fluorescent sensor for sensitive and selective detection of mercury(II) ion. *Chem. Commun. (Camb).* **46**(9), 1476–1478 (2010).
83. L. Q. Guo, N. Yin, D. D. Nie, J. R. Gan, M. J. Li, F. F. Fu and G. N. Chen, Label-free fluorescent sensor for mercury(II) ion by using carbon nanotubes to reduce background signal. *Analyst* **136**(8), 1632–1636 (2011).
84. J. Yao J. Li, J. Owens and W. Zhong, Combing DNAzyme with single-walled carbon nanotubes for detection of Pb(II) in water. *Analyst* **136**(4), 764–768 (2011).
85. E. S. Cho, S. W. Hong and W. H. Jo, A New pH Sensor Using the Fluorescence Quenching of Carbon Nanotubes. *Macromol. Rapid Commun.* **29**(22), 1798–1803 (2008).
86. L. P. Zhao, T. Nakayama, H. Tomimoto, Y. Shingaya and Q. Huang, Functionalization of carbon nanotubes with a pH-responsive molecule to produce a pH sensor. *Nanotechnology* **20**(32) (2009).
87. Y. Miyake, H. Togashi, M. Tashiro, H. Yamaguchi, S. Oda, M. Kudo, Y. Tanaka, Y. Kondo, R. Sawa, T. Fujimoto, T. Machinami and A. Ono, MercuryII-mediated formation of thymine-HgII-thymine base pairs in DNA duplexes. *J. Am. Chem. Soc.* **128**(7), 2172–2173 (2006).
88. N. Nakayama-Ratchford, S. Bangsaruntip, X. Sun, K. Welsher and H. Dai, Noncovalent functionalization of carbon nanotubes by fluorescein-polyethylene glycol: supramolecular conjugates with pH-dependent absorbance and fluorescence. *J. Am. Chem. Soc.* **129** (9), 2448–2449 (2007).
89. V. A. Karachevtsev, A. Y. Glamazda, V. S. Leontiev, O. S. Lytvyn and U. Dettlaff-Weglikowska, Glucose sensing based on NIR fluorescence of DNA-wrapped single-walled carbon nanotubes. *Chem. Phys. Lett.* **435**(1–3), 104–108 (2007).
90. P. W. Barone, S. Baik, D. A. Heller and M. S. Strano, Near-infrared optical sensors based on single-walled carbon nanotubes. *Nat. Mater.* **4**(1), 86–92 (2005).
91. K. Yum, J. H. Ahn, T. P. McNicholas, P. W. Barone, B. Mu, J. H. Kim, R. M. Jain and M. S. Strano, Boronic acid library for selective, reversible near-infrared fluorescence quenching of surfactant suspended single-walled carbon nanotubes in response to glucose. *ACS Nano* **6**(1), 819–830 (2012).
92. P. W. Barone and M. S. Strano, Reversible control of carbon nanotube aggregation for a glucose affinity sensor. *Angew. Chem. Int .Ed. Engl.* **45**(48), 8138–8141 (2006).
93. L. Zhang, H. Wei, J. Li, T. Li, D. Li, Y. Li and E. Wang, A carbon nanotubes based ATP apta-sensing platform and its application in cellular assay. *Biosens. Bioelectron.* **25**(8), 1897–1901 (2010).
94. S. J. Zhen, L. Q. Chen, S. J. Xiao, Y. F. Li, P. P. Hu, L. Zhan, L. Peng, E. Q. Song and C. Z. Huang, Carbon nanotubes as a low background signal platform for a molecular aptamer beacon on the basis of long-range resonance energy transfer. *Anal. Chem.* **82** (20), 8432–8437 (2010).
95. N. F. Reuel, J. H. Ahn, J. H. Kim, J. Zhang, A. A. Boghossian, L. K. Mahal and M. S. Strano, Transduction of glycan-lectin binding using near-infrared fluorescent single-walled carbon nanotubes for glycan profiling. *J. Am. Chem. Soc.* **133**(44), 17923–17933 (2011).
96. J. H. Kim, C. R. Patra, J. R. Arkalgud, A. A. Boghossian, J. Zhang, J. H. Han, N. F. Reuel, J. H. Ahn, D. Mukhopadhyay and M. S. Strano, Single-molecule detection of

H(2)O(2) mediating angiogenic redox signaling on fluorescent single-walled carbon nanotube array. *ACS Nano* **5**(10), 7848–7857 (2011).

97. H. Jin, D. A. Heller, J. H. Kim and M. S. Strano, Stochastic analysis of stepwise fluorescence quenching reactions on single-walled carbon nanotubes: single molecule sensors. *Nano Lett.* **8**(12), 4299–4304 (2008).
98. H. Jin, D. A. Heller, M. Kalbacova, J. H. Kim, J. Zhang, A. A. Boghossian, N. Maheshri and M. S. Strano, Detection of single-molecule H2O2 signalling from epidermal growth factor receptor using fluorescent single-walled carbon nanotubes. *Nat. Nanotechnol.* **5**(4), 302–309 (2010).
99. J. H. Kim, D. A. Heller, H. Jin, P. W. Barone, C. Song, J. Zhang, L. J. Trudel, G. N. Wogan, S. R. Tannenbaum and M. S. Strano, The rational design of nitric oxide selectivity in single-walled carbon nanotube near-infrared fluorescence sensors for biological detection. *Nat. Chem.* **1**(6), 473–481 (2009).
100. J. Zhang, A. A. Boghossian, P. W. Barone, A. Rwei, J. H. Kim, D. Lin, D. A. Heller, A. J. Hilmer, N. Nair, N. F. Reuel and M. S. Strano, Single molecule detection of nitric oxide enabled by d(AT)15 DNA adsorbed to near infrared fluorescent single-walled carbon nanotubes. *J. Am. Chem. Soc.* **133**(3), 567–581 (2011).
101. E. S. Jeng, A. E. Moll, A. C. Roy, J. B. Gastala and M. S. Strano, Detection of DNA hybridization using the near-infrared band-gap fluorescence of single-walled carbon nanotubes. *Nano Lett.* **6**(3), 371–375 (2006).
102. E. S. Jeng, P. W. Barone, J. D. Nelson and M. S. Strano, Hybridization kinetics and thermodynamics of DNA adsorbed to individually dispersed single-walled carbon nanotubes. *Small* **3**(9), 1602–1609 (2007).
103. D. A. Heller, E. S. Jeng, T. K. Yeung, B. M. Martinez, A. E. Moll, J. B. Gastala and M. S. Strano, Optical detection of DNA conformational polymorphism on single-walled carbon nanotubes. *Science* **311**(5760), 508–511 (2006).
104. E. S. Jeng, J. D. Nelson, K. L. Prather and M. S. Strano, Detection of a single nucleotide polymorphism using single-walled carbon-nanotube near-infrared fluorescence. *Small* **6**(1), 40–43 (2010).
105. K. Donaldson, R. Aitken, L. Tran, V. Stone, R. Duffin, G. Forrest and A. Alexander, Carbon nanotubes: a review of their properties in relation to pulmonary toxicology and workplace safety. *Toxicol. Sci.* **92**(1), 5–22 (2006).
106. K. Deligkaris, T. S. Tadele W. Olthuis, and A. van den Berg, Hydrogel-based devices for biomedical applications. *Sens. Actuator B-Chem.* **147**(2), 765–774 (2010).
107. T. R. Hoare and D. S. Kohane, Hydrogels in drug delivery: Progress and challenges. *Polymer* **49**(8), 1993–2007 (2008).
108. H. Xu, J. W. Aylott and R. Kopelman, Fluorescent nano-PEBBLE sensors designed for intracellular glucose imaging. *Analyst* **127**(11), 1471–1477 (2002).
109. S. H. Kim, B. Kim, V. K. Yadavalli and M. V. Pishko, Encapsulation of enzymes within polymer spheres to create optical nanosensors for oxidative stress. *Anal. Chem.* 77(21), 6828–6833 (2005).
110. Y. E. L. Koo, R. Agayan, M. A. Philbert, A. Rehemtulla, B. D. Ross and R. Kopelman, Photonic explorers based on multifunctional nanoplatforms: in vitro and in vivo biomedical applications. In *New Approaches in Biomedical Spectroscopy*, K. Kneipp, R. Aroca H. Kneipp, and E. Wen trupByrne, (Eds), Vol. 963. Washington: Amer Chemical Soc, pp. 200–218 (2007).

111. S. M. Buck, Y. E. Koo, E. Park, H. Xu, M. A. Philbert, M. A. Brasuel and R. Kopelman, Optochemical nanosensor PEBBLEs: photonic explorers for bioanalysis with biologically localized embedding. *Curr. Opin. Chem. Biol.* **8**(5), 540–546 (2004).
112. E. J. Park, M. Brasuel, C. Behrend, M. A. Philbert and R. Kopelman, Ratiometric optical PEBBLE nanosensors for real-time magnesium ion concentrations inside viable cells. *Anal. Chem.* **75**(15), 3784–3791 (2003).
113. M. King and R. Kopelman, Development of a hydroxyl radical ratiometric nanoprobe. *Sens. Actuator B-Chem.* **90**(1–3), 76–81 (2003).
114. H. Sun, A. M. Scharff-Poulsen, H. Gu, I. Jakobsen, J. M. Kossmann, W. B. Frommer and K. Almdal, Phosphate sensing by fluorescent reporter proteins embedded in polyacrylamide nanoparticles. *ACS Nano* **2**(1), 19–24 (2008).
115. J. P. Sumner, J. W. Aylott, E. Monson and R. Kopelman A fluorescent PEBBLE nanosensor for intracellular free zinc. *Analyst* **127**(1), 11–16 (2002).
116. J. P. Sumner and R. Kopelman, Alexa Fluor 488 as an iron sensing molecule and its application in PEBBLE nanosensors. *Analyst* **130**(4), 528–533 (2005).
117. J. P. Sumner, N. M. Westerberg, A. K. Stoddard, C. A. Fierke and R. Kopelman, Cu+- and Cu2+-sensitive PEBBLE fluorescent nanosensors using DsRed as the recognition element. *Sens. Actuator B-Chem.* **113**(2), 760–767 (2006).
118. G. Zenkl and I. Klimant, Fluorescent acrylamide nanoparticles for boronic acid based sugar sensing — from probes to sensors. *Microchim. Acta.* **166**(1–2), 123–131 (2009).
119. Y. Cao, Y. E. Lee Koo and R. Kopelman, Poly(decyl methacrylate)-based fluorescent PEBBLE swarm nanosensors for measuring dissolved oxygen in biosamples. *Analyst* **129**(8), 745–750 (2004).
120. V. P. Torchilin, Recent advances with liposomes as pharmaceutical carriers. *Nat. Rev. Drug Discov.* **4**(2), 145–160 (2005).
121. K. P. McNamara, N. Rosenzweig and Z. Rosenzweig, Liposome-based optochemical nanosensors. *Mikrochim. Acta.* **131**(1–2), 57–64 (1999).
122. T. Nguyen, K. P. McNamara and Z. Rosenzweig, Optochemical sensing by immobilizing fluorophore-encapsulating liposomes in sol-gel thin films. *Anal. Chim. Acta.* **400**, 45–54 (1999).
123. Y. C. Chen, A. Ostafin and H. Mizukami, Synthesis and characterization of pH sensitive carboxySNARF-1 nanoreactors. *Nanotechnology* **21**(21), 215503 (2010).
124. T. Nguyen and Z. Rosenzweig, Calcium ion fluorescence detection using liposomes containing Alexa-labeled calmodulin. *Anal. Bioanal. Chem.* **374**(1), 69–74 (2002).
125. Q. S. Wang, G. Q. Li, W. Y. Mao, H. X. Qi and G. W. Li, Glucose-responsive vesicular sensor based on boronic acid-glucose recognition in the ARS/PBA/DBBTAB covesicles. *Sens. Actuator B-Chem.* **119**(2), 695–700 (2006).
126. Z. Cheng and C. A. Aspinwall, Nanometre-sized molecular oxygen sensors prepared from polymer stabilized phospholipid vesicles. *Analyst* **131**(2), 236–243 (2006).
127. E. Bakker and P. Buhlmann, E. Pretsch, Carrier-Based Ion-Selective Electrodes and Bulk Optodes. 1. General Characteristics. *Chem. Rev.* **97**(8), 3083–3132 (1997).
128. R. D. Johnson and L. G. Bachas, Ionophore-based ion-selective potentiometric and optical sensors. *Anal. Bioanal. Chem.* **376**(3), 328–341 (2003).
129. M. K. Balaconis and H. A. Clark, Biodegradable optode-based nanosensors for in vivo monitoring. *Anal. Chem.* **84**(13), 5787–5793 (2012).

130. S. Peper, I. Tsagkatakis and E. Bakker, Cross-linked dodecyl acrylate microspheres: novel matrices for plasticizer-free optical ion sensing. *Anal. Chim. Acta.* **442**(1), 25–33 (2001).
131. J. A. Tickner, T. Schettler, T. Guidotti, M. McCally and M. Rossi, Health risks posed by use of Di-2-ethylhexyl phthalate (DEHP) in PVC medical devices: a critical review. *Am. J. Ind. Med.* **39**(1), 100–111 (2001).
132. M. Rahman and C. S. Brazel, The plasticizer market: an assessment of traditional plasticizers and research trends to meet new challenges. *Prog. Polym. Sci.* **29**(12), 1223–1248 (2004).
133. J. M. Dubach, D. I. Harjes and H. A. Clark, Ion-selective nano-optodes incorporating quantum dots. *J. Am. Chem. Soc.* **129**(27), 8418–8419 (2007).
134. M. Brasuel, R. Kopelman, I. Kasman, T. J. Miller and M. A. Philbert, Ion concentrations in live cells from highly selective correlation fluorescent nano-sensors for sodium. *In Proc. IEEE. Sensors, 2002.* Vol. 1, pp. 288–292 (2002).
135. M. Brasuel, R. Kopelman, T. J. Miller, R. Tjalkens and M. A. Philbert, Fluorescent nanosensors for intracellular chemical analysis: decyl methacrylate liquid polymer matrix and ion-exchange-based potassium PEBBLE sensors with real-time application to viable rat C6 glioma cells. *Anal. Chem.* **73**(10), 2221–2228 (2001).
136. M. J. Ruedas-Rama and E. A. Hall, K+-selective nanospheres: maximising response range and minimising response time. *Analyst* **131**(12), 1282–1291 (2006).
137. H. A. Clark, M. Hoyer, S. Parus, M. A. Philbert and M. Kopelman, Optochemical nanosensors and subcellular applications in living cells. *Mikrochim. Acta.* **131**(1–2), 121–128 (1999).
138. M. G. Brasuel, T. J. Miller, R. Kopelman and M. A. Philbert, Liquid polymer nano-PEBBLEs for Cl- analysis and biological applications. *Analyst* **128**(10), 1262–1267 (2003).
139. K. J. Cash and H. A. Clark, In vivo histamine optical nanosensors. *Sensors (Basel).* **12**(9), 11922–11932 (2012).
140. J. M. Dubach, E. Lim, N. Zhang, K. P. Francis and H. Clark, In vivo sodium concentration continuously monitored with fluorescent sensors. *Integr. Biol. (Camb).* **3**(2), 142–148 (2011).
141. C. C. Lee, J. A. MacKay, J. M. Frechet and F. C. Szoka, Designing dendrimers for biological applications. *Nat. Biotechnol.* **23**(12), 1517–1526 (2005).
142. L. Albertazzi, B. Storti, L. Marchetti and F. Beltram, Delivery and subcellular targeting of dendrimer-based fluorescent pH sensors in living cells. *J. Am. Chem. Soc.* **132**(51), 18158–18167 (2010).
143. F. Vogtle, S. Gestermann, C. Kauffmann, P. Ceroni, V. Vicinelli and V. Balzani, Coordination of Co2+ ions in the interior of poly(propylene amine) dendrimers containing fluorescent dansyl units in the periphery. *J. Am. Chem. Soc.* **122**(42), 10398–10404 (2000).
144. V. Balzani, P. Ceroni, S. Gestermann, C. Kauffmann, M. Gorka and F. Vogtle, Dendrimers as fluorescent sensors with signal amplification. *Chem. Commun.* (10), 853–854 (2000).
145. I. Grabchev, J. M. Chovelon, V. Bojinov and G. Ivanova, Poly(amidoamine) dendrimers peripherally modified with 4-ethylamino-1,8-naphthalimide. Synthesis and photophysical properties. *Tetrahedron* **59**(48), 9591–9598 (2003).

146. I. Grabchev, J. M. Chovelon and X. H. Qian, A polyamidoamine dendrimer with peripheral 1,8-naphthalimide groups capable of acting as a PET fluorescent sensor for metal cations. *New J. Chem.* **27**(2), 337–340 (2003).
147. L. Albertazzi, M. Brondi, G. M. Pavan, S. S. Sato, G. Signore, B. Storti, G. M. Ratto and F. Beltram, Dendrimer-based fluorescent indicators: in vitro and in vivo applications. *PLoS ONE.* **6**(12), e28450 (2011).
148. O. Finikova, A. Galkin, V. Rozhkov, M. Cordero, C. Hagerhall and S. Vinogradov, Porphyrin and tetrabenzoporphyrin dendrimers: tunable membrane-impermeable fluorescent pH nanosensors. *J. Am. Chem. So.* **125**(16), 4882–4893 (2003).
149. V. J. Pugh, Q. S. Hu and L. Pu, The first dendrimer-based enantioselective fluorescent sensor for the recognition of chiral amino alcohols. *Angew. Chem. Int. Edit.* **39**(20), 3638–3641 (2000).
150. L. Z. Gong, Q. S. Hu and L. Pu, Optically active dendrimers with a binaphthyl core and phenylene dendrons: light harvesting and enantioselective fluorescent sensing. *J. Org. Chem.* **66**(7), 2358–2367 (2001).
151. L. Pu, Synthesis and study of binaphthyl-based chiral dendrimers. *J. Photochem. Photobiol. A-Chem.* **155**(1–3), 47–55 (2003).
152. N. Malik, R. Wiwattanapatapee, R. Klopsch, K. Lorenz, H. Frey, J. W. Weener, E. W. Meijer, W. Paulus and R. Duncan, Dendrimers: relationship between structure and biocompatibility in vitro and preliminary studies on the biodistribution of 125I-labelled polyamidoamine dendrimers in vivo. *J. Control. Release* **65**(1–2), 133–148 (2000).
153. M. F. Neerman, W. Zhang, A. R. Parrish and E. E. Simanek, In vitro and in vivo evaluation of a melamine dendrimer as a vehicle for drug delivery. *Int. J. Pharm.* **281** (1–2), 129–132 (2004).
154. M. W. Grinstaff, Biodendrimers: new polymeric biomaterials for tissue engineering. *Chemistry* **8**(13), 2839–2846 (2002).
155. A. Mokhir, A. Kiel, D. P. Herten and R. Kraemer, Fluorescent sensor for Cu2+ with a tunable emission wavelength. *Inorg. Chem.* **44**(16), 5661–5666 (2005).
156. M. Monici, Cell and tissue autofluorescence research and diagnostic applications. *Biotechnol. Annu. Rev.* **11**, 227–256 (2005).
157. J. V. Frangioni, In vivo near-infrared fluorescence imaging. *Curr. Opin. Chem. Biol.* 7(5), 626–634 (2003).
158. S. Kim, Y. T. Lim, E. G. Soltesz, A. M. De Grand, J. Lee, A. Nakayama, J. A. Parker, T. Mihaljevic, R. G. Laurence, D. M. Dor, L. H. Cohn, M. G. Bawendi and J. V. Frangioni, Near-infrared fluorescent type II quantum dots for sentinel lymph node mapping. *Nat. Biotechnol.* **22**(1), 93–97 (2004).
159. H. G. Piper, J. L. Alexander, A. Shukla, F. Pigula, J. M. Costello, P. C. Laussen, T. Jaksic and M. S. Agus, Real-time continuous glucose monitoring in pediatric patients during and after cardiac surgery. *Pediatrics* **118**(3), 1176–1184 (2006).
160. J. W. Aylott, Optical nanosensors — an enabling technology for intracellular measurements. *Analyst* **128**(4), 309–312 (2003).
161. N. V. Gopee, D. W. Roberts, P. Webb, C. R. Cozart, P. H. Siitonen, A. R. Warbritton, W. W. Yu, V. L. Colvin, N. J. Walker and P. C. Howard, Migration of intradermally injected quantum dots to sentinel organs in mice. *Toxicol. Sci.* **98**(1), 249–257 (2007).
162. G. Ozaydin-Ince, J. M. Dubach, K. K. Gleason and H. A. Clark, Microworm optode sensors limit particle diffusion to enable in vivo measurements. *Proc. Natl. Acad. Sci. USA.* **108**(7), 2656–2661 (2011).

163. B. D'Autreaux and M. B. Toledano, ROS as signalling molecules: mechanisms that generate specificity in ROS homeostasis. *Nat. Rev. Mol. Cell. Biol.* **8**(10), 813–824 (2007).
164. R. L. Bijlani, and Manjunatha, *Understanding Medical Physiology: A Textbook for Medical Students.* Jp Medical Pub: (2010).
165. M. A. Woodruff and D. W. Hutmacher, The return of a forgotten polymer-Polycaprolactone in the 21st century. *Prog. Polym. Sci.* **35**(10), 1217–1256 (2010).

Chapter 14

In Vivo Photoacoustic Detection of Circulating Cells and Nanoparticles

Jin-Woo Kim

Bio/Nano Technology Laboratory, Institute for Nanoscience and Engineering, University of Arkansas, Fayetteville, AR, USA; Department of Biological and Agricultural Engineering, University of Arkansas, Fayetteville, AR, USA
jwkim@uark.edu

*Ekaterina I. Galanzha and Vladimir P. Zharov**

Phillips Classic Laser and Nanomedicine Laboratories, University of Arkansas for Medical Sciences, Little Rock, AR, USA; Arkansas Nanomedicine Center, University of Arkansas for Medical Sciences, Little Rock, AR, USA
ZharovVladimirP@uams.edu

1. Introduction

The vascular dissemination of pathological features is a main cause of progression of many life-threatening diseases.[1,2] The migration of tumor cells through the vascular circulation is a mandatory step in the metastasis of solid tumors, which accounts for nine out of every 10 cancer-related deaths.[3] Also, bacteria reaching the circulation is a major problem worldwide due to the possibility of septic shock and the development of systemic infection. But, the treatment of bacteremia is increasingly hampered by the rapid emergence of antibiotic-resistant strains.[4] Furthermore, arterial and pulmonary thromboembolisms due to circulation of blood clots often lead to severe cardiovascular and cerebrovascular events. Therefore, the detection and enumeration of circulating pathological features, such as circulating tumor cells (CTCs), circulating bacterial cells (CBCs) and clots, as well as cancer biomarkers such as microparticles and exosomes, which contain important microRNA signatures and other abnormal cells like malaria parasites in biological fluids, are clinically significant for many fatal diseases. Unfortunately, even in a disease setting, circulating abnormal cells are rare events that are easily concealed by the overwhelming background material in whole blood. To date, most detection methods estimate circulating abnormal cells in a small volume (5–20 mL) of blood drawn from patients.[1,2,5–16] The small sample volume limits the ability of *ex vivo* techniques to detect exceedingly rare populations. Particularly, low cell counts in small volumes acquired at fixed time points hinders the promise for real-time diagnosis of metastatic diseases at earlier stages.[1,2,14] Moreover, the use of small blood volumes limits the collection of abnormal features for later analysis to investigate their role in disease progression.

Recent development of nanomedicine that uses nanoparticles (NPs) with unique biophysicochemical properties offers exciting opportunities for therapeutics and diagnostics (theranostics) of cancer and infections, as well as "smart" drug delivery and regenerative medicine.[1,2,11,17–22] The ability to manipulate matters at the nanoscale enables noble properties that are desirable and exploitable. This allows engineering NPs and NP clusters that target specific cells, such as tumors or other pathological cells and permit *in vivo* monitoring and destruction of the respective cell targets. Various NPs with different shapes and compositions have been proven effective as theranostic contrast nanoagents.[11,17–22] Recently, tunable near-infrared (NIR)-responsive plasmonic NPs, including gold nanoshells (GNSs), gold nanorods (GNRs) and golden carbon nanotubes (GNTs) among many others,[11,13,18,23] have attracted attention for minimally invasive imaging and therapy owing to their high NIR absorption (*e.g.* ~700–900 nm) in the window of optical transparency of most biological tissues as well as high efficiency conversion of absorbed energy into thermal and acoustic energy.[1,2,4,11,13–16] Also, a variety of approaches have been reported to design and self-assemble NPs and their nanocomposites with desirable size, shape and functionality for medical theranostic applications.[1,17,24,25] However, despite excellent promises of NPs and their composites, many questions remain about their safety and risk-benefit characteristics of their usage for human. Particularly, their reliable and reproducible *in vivo* pharmacokinetic data, including blood circulation and clearance rate, organ biodistribution and accumulation and their cell- or organ-targeting efficacy, must be available before realizing the many envisioned biomedical applications of NPs. However, assessing *in vivo* pharmacokinetic parameters is a formidable challenge as it is not trivial to monitor their behaviors *in vivo* and there is no clinically relevant technology available yet for accurate, real-time and quantitative assessments of NPs' pharmacokinetics *in vivo*.

A prerequisite to achieve the enormous theranostic potential of circulating abnormal features, including pathogenic cells and biomolecules and nanoagents, is the development of effective and efficient detection platforms capable of interrogating the entire circulation *in vivo* and in real time. Recent advances in laser-based photothermal (PT) and photoacoustic (PA) techniques demonstrated a great potential in various *in vivo* biomedical diagnostic, therapeutic and theranostic (integration of diagnosis and therapy) applications.[1,2,4,11,13–23] Of note is the *in vivo*, real-time monitoring of circulating abnormal cells, biomolecules and NPs by integrating PT, PA and flow cytometry (PTFC/PAFC, respectively) was first demonstrated in 2004 by Zharov *et al.*[26] These technical platforms showed high potential to overcome the aforementioned theranostic challenges to the circulating abnormal features.[1,2,4,11,13–16] Moreover, an innovative *in vivo* non-invasive multicolor and multimodal concept using multimodal NIR-responsive NP-based contrast agents and multicolor laser-based *in vivo* PAFC/PTFC to detect, capture and purge rare circulating tumor and other pathological features in the peripheral circulation were recently introduced by our team (Fig. 1),[1,2,4,11,13–16] demonstrating a broad range of clinical applications including early disease theranostics, evaluation of disease progression and *in vivo* pharmacokinetic assessments. Some groups also proposed similar approaches to achieve real-time detection of CTCs using fluorescence-labeled biomarkers with some degree of success.[27,28] Although more work is deemed to realize the PAFC/PTFC's promise, the *in vivo* platforms, in particular nanotechnology-based theranostic (nanotheranostic) technologies, carry enormous clinical potential for the therapy and management of life-threatening diseases, as well as *in vivo* characterizations of abnormal features.

This chapter discusses the latest PA technological platforms concerning *in vivo* real-time detection and characterization of CTCs, CBCs, clots and other abnormal features in circulation. Particular focus is put into the recent advances in *in vivo* PA nanotheranostic approaches using multispectral high-pulse-repetition-rate lasers, tunable plasmonic nanoprobes with ultrasharp (up to ~1 nm) PA spectral resonances, positive and negative PA contrasts, *in vivo* magnetic enrichment, time-of-flight cell velocity measurement, PA feedback to control of nanotherapy efficiency and further integration of

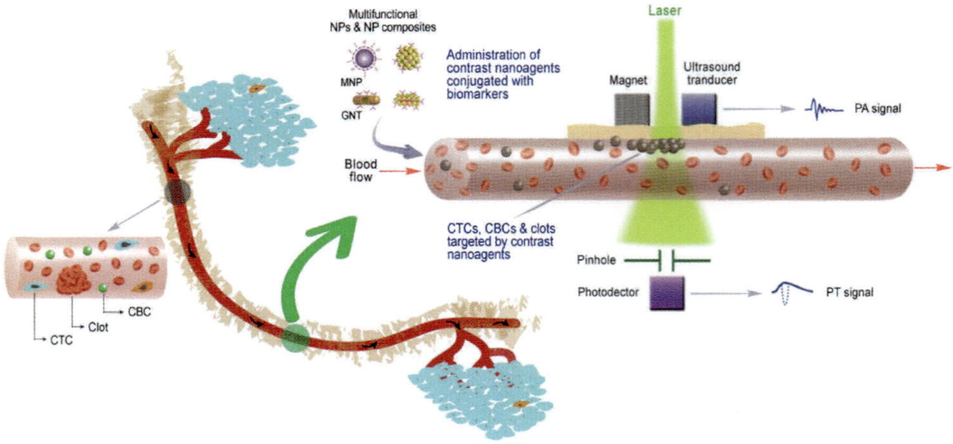

Fig. 1. *In vivo* multimodal multicolor nanotheranostic platform integrating magnetic enrichment, PA detection, PT therapy and real-time PA monitoring of PT therapeutic efficacy of circulating pathological features, including CTCs, CECs, CCSCs, CBCs, emboli, and other biomolecules (*i.e.* DNA, RNA, exosomes and microparticles, *etc.*). Adapted with permission from Ref. 1. Copyright 2013 American Chemical Society.

other modalities, such as fluorescent and Raman spectral methods. It also emphasizes a critical need for *in vivo* real-time multifunctional nanotheranostic platforms comprised of multifunctional contrast nanoagents to realize a breakthrough for early detection and characterization of harmful diseases disseminated through the circulation and their successful translation to clinical practice.

2. Recent Advances in *In Vivo* PA Diagnostics

2.1. *PA nanotheranostic platform for in vivo studies*

The initial demonstration of *in vivo* real-time PAFC detection platform in 2004 by Zharov *et al.*[26] opened a new possibility to realize theranostic potential of circulating abnormal features. It was realized by uniquely combining FC, which is one of the powerful analytical tools in cell biology and molecular disease diagnosis,[29,30] and laser-induced PT and PA effects in circulating objects in flows. Laser energy absorbed by the objects is transformed mostly into heat and then into amplified sound, which makes their PA "listening" as an alternative to the fluorescent "seeing".[31] PA methods usually do not require conventional labeling and are not sensitive to light scattering. They may use either endogenous biomolecules with natural PA contrast properties, such as hemoglobin, melanin, cytochromes or carotinoids or exogenous contrast agents.[11] Such exogenous agents include small molecule optical dyes, such as evans blue, indocyanin green (ICG), lymphazurin and methylen blue, as well as synthetic NPs such as gold, magnetic and carbon-based NPs.

In 2009, a novel multicolor PAFC/PTFC using novel multimodal NPs as diagnostic and therapeutic agents was reported, demonstrating the first *in vivo* real-time imaging and therapy of CTCs and circulating cancer stem-like cells (CCSCs).[1,2,11,13–16] Other theranostic targets were also explored, including bacteria (*i.e.* CBCs), micro/Nps and theromboemboli.[1,2,4,11] In these studies, prospective capability of *in vivo* multispectral PAFC/PTFC were demonstrated for ultra-sensitive non-invasive molecular quantitative detection, intravascular magnetic enrichment and targeted laser ablation of pathological cells in blood and lymph circulation. The multicolor multi-parameter high-speed PAFC was capable of clinically relevant molecular detection and enumeration of cells of interest among highly heterogeneous population of blood cells *in vivo* that has never been possible before. Integration of PA and PT approaches using the same technical platform allowed robust and relatively inexpensive combination of diagnosis and targeted therapy of rare abnormal cells directly in blood circulation *in vivo*. The potential advantages of PAFC in comparison with other *in vitro* approaches include (1) analyzing mobile cells and other abnormal objects, such as NPs, in moving fluids, (2) fast testing (the rate up to 10^6 and 10^{10} cells per second in small and large, such as human jugular vein vessel, respectively) of the largest blood volume (up to whole blood volume, *i.e.* ~50–100 times more than volume of conventional samples) at unrestricted speeds and (3) unprecedented sensitivity threshold at one pathological cell, such as a CTC, among a billion normal blood cells and millions of white blood cells (WBCs) in blood and lymph flow.[1,2] The advantages of PAFC over existing *in vivo* technologies also include (1) time-resolved and deep detection of specific acoustic signals from individual metastatic cells (depth up to a few cm), (2) multicolor detection, (3) use of functionalized NPs as unique NIR-absorbing low toxic super-contrast PA molecular agents and (4) non-invasiveness for normal tissue due to a relatively low level of laser energy, well within safety parameters of laser use for humans.[1,2] Here we summarize the unique features and recent advances of the *in vivo* PAFC nanotheranostic technical platform for circulating cells and other abnormal features, such as NPs.

2.1.1. PAFC schematics

The underlying principle of *in vivo* PAFC is based on the irradiation of selected vessels using short laser pulses followed by time-resolved detection of laser-induced acoustic waves (referred to as PA signals) with an ultrasound transducer gently attached to the skin near examined vessels (Fig. 1).[1,2] The physical mechanism of PAFC is based upon the PA phenomenon associated with fast (picosecond scale), non-radiative relaxation of absorbed laser energy into heat and then through the transient thermoelastic expansion of the energy into thermoacoustic waves.[32–38] PAFC combines high sensitivity and spectral specificity of optical methods with the high spatial resolution and depth penetration of ultrasound techniques.

Laser radiation can be delivered to biotissue either by using a microscope schematic with a customized condenser to create the desired linear beam shapes ranging from 5 × 50 μm to 25 × 1,000 μm, or a fiber with a miniature tip and cylindrical optics.[1,2] PAFC molecular specificity is provided either by label-free intrinsic absorption spectroscopic contrast (*e.g.* hemoglobin or melanin) or by strongly absorbing, low-toxicity, bio-functionalized contrast nanoagents, such as NPs. Preclinical feasibility studies of PAFC *in vivo* involved nude mouse tumor models of melanoma and breast cancer. The non-invasive PA detection of cells in blood and lymphatic systems was carried out using (1) the mouse ear and (2) the skinfold chamber. A typical setup was equipped with five high-pulse-repetition-rate nanosecond lasers with the following parameters[2]: (1) 532-nm wavelength, 100-μJ pulse energy, 5-ns pulse width and repetitions rate up to 100 kHz, (2) 671 nm, 35 μJ, 25 ns and 100 kHz, respectively, (3) 820 nm, 75 μJ, 8 ns and 30 kHz, respectively, (4) 904 nm, 5 μJ, 15 ns, 10 kHz (diode laser), respectively and (5) 1064 nm, 100 μJ, 10 ns and 1–750 kHz, respectively. The PA signals had a bipolar shape with a typical duration of 0.1–1 μs. PA signals were detected by ultrasound transducers (*e.g.* unfocused: model 6528101, 3.5 MHz, 5.5 mm in diameter, Imasonic Inc and focused: model V316-SM, 20 MHz, focal length 12.5 mm, Panametrics). To collect PA signals, the PAFC setups were equipped with a high-speed, analog-to-digital converter boards (*e.g.* National Instruments Corp.). In analogy to conventional flow cytometry, final PA data were presented as PA signal traces, which were analyzed using customized software.

The use of a single technique limits the range of detectable cells with different optical properties. Recently, advances of real-time integration of PAFC and fluorescent FC (FFC), termed PAFFC,[39] were demonstrated using pulsed and continuous wave (CW) lasers as traditional sources for the generation of PA and fluorescent signals, respectively. Conventional PA and fluorescence techniques use preferential positive imaging contrasts when signals from strongly absorbing or fluorescent cells are above absorption or autofluorescence background, respectively. In PAFFC, conventional positive contrast mode was combined with the new negative PA and fluorescent contrast modes, in particular, for PA detection of cells with lower absorption than red blood cells (RBCs) as in case of platelets or WBCs. PAFC was also integrated with PT techniques, *i.e.* PTFC and PT therapy.[1,2] In the PT thermal lens diagnostic mode, a laser-induced refractive heterogeneity causes defocusing of a collinear He–Ne laser probe beam and hence a reduction in the beam's intensity at its center, as detected by a photodiode with a pinhole (referred to as PT signals). PT signals from single cells in a linear mode (*i.e.* without notable cell photodamage) represent a standard positive peak associated with rapid (*i.e.* pico to nanosecond scale) cell heating and a slower, microsecond scale tail corresponding to cell cooling. PA and PT methods beneficially supplement each other and in combination, provide a very powerful diagnostic and therapeutic tool. For example, non-invasive PA diagnostics can be integrated with PT killing of CTCs and CBCs using more powerful laser pulses triggered by PA signals from these cells.[1,2]

2.1.2. *In vivo* labeling

In vivo PAFC allows cell detection without labeling,[1,2,39,40] for example by using the positive and negative PA contrast of RBCs and WBCs versus the negative and positive fluorescent contrast of the same cells. Intrinsic chromophores and pigments, such as hemoglobin, melanin, cytochromes or carotenoids, can generate PA signals. Also, cells with a low endogenous absorption can be labeled directly in the bloodstream through *i.v.* injection of strongly absorbing functionalized NPs.[1,2,11] *In vivo* labeling procedure using mouse models usually takes from 10–20 min to 1 h depending upon the properties of cells and NPs. Labeling specificity is defined by the proper selection of molecular markers that are highly expressed in target cells, but almost absent in normal blood and endothelial cells (*e.g.* folates in CTCs).[14,27] Frequent collisions between NPs and sabnormal cells (*e.g.* CTCs) in partly turbulent blood flow is related to high labeling efficiency. Much higher PA signals are obtained from targeted cells with a typical NP number that ranges from 5×10^2 to 5×10^3 NPs per CTC cell, than the PA/PT signal background from RBCs, unbound NPs with typical numbers of 1–10 in the detected volume, or NPs non-specifically bound to normal blood cells. NP clustering around naturally densely packed cell markers would lead to a significant linear (5–50 fold) and non-linear (100–300 fold) nanobubble-based enhancement in PA/PT signal amplitude and a red-shift effect in the absorption of coupled NPs in clusters,[18] both of which serve as indicators of successful cell targeting.

2.1.3. Detection threshold

The detection sensitivity of PAFC *ex vivo* was evaluated with two types of NPs including GNTs[13] and magnetic NPs (MNPs).[14] GNTs consist of hollow shortened CNTs surrounded by thin gold layers, averaging 11 nm × 98 nm in diameter and length, respectively. They exhibit high water solubility and biocompatibility, low cytotoxicity (due to the protective layer of gold) and spectrally tunable high plasmon absorption in the NIR range of 700–1000 nm. The minimal number of NPs in the detected (*i.e.* irradiated) volume that produced readable PA signals in the background of mouse blood in the 120-μm microscopic slide was estimated to be 120 ± 14 for MNPs at 639 nm and 7 ± 2 for GNTs at 900 nm under laser fluence of 100 mJ/cm^2 in non-linear mode at both wavelengths and 35 for GNTs and 600 for MNPs at laser fluence of 20 mJ/cm.[2,14]

The detection threshold sensitivity of PAFC *in vivo* was estimated as follows using GNTs.[13] The 10-μL GNT suspension (~10^{11}/mL) was injected *i.v.* into mice with total blood volume of ~2 mL. Assuming ideal conditions with no loss of GNTs during injection and in circulation, the average GNT concentration in the blood pool should be 5 × 10^8 GNTs/mL. The irradiated (detected) volume in blood vessel of mouse ear was limited by laser beam size and the diameter of the vessels. Assuming a linear beam shape with a width of 15 μm and a vessel diameter of 60 μm in mouse ear, the detected volume was estimated as 4.2×10^{-8} cm^3. These parameters correspond to the estimated maximum number of GNTs (around 21) in the detected volume. Considering the threshold sensitivity as 7 GNTs at laser fluence 100 mJ/cm^2, it is expected that the signal-to-noise ratio (SNR) *in vivo* in blood vessel is ~3 (*i.e.* 21/7). However, it was experimentally determined that the SNR is ~7. This discrepancy can be explained by uncontrollable formation of small GNT aggregates in flow, which provided a stronger PA signal as compared to suspensions of single GNTs. Injection of high concentrations of each unconjugated NP alone (i.v. ~10^{11} NPs/50 μL) in mice revealed their fast clearance during 15–20 min, with no PA detection at low concentrations (~10^9 NPs/50 μL).

Comparison of PA data *in vivo* and *in vitro* revealed that the threshold sensitivity of PAFC *in vivo* could be estimated as ~1 CTC/mL.[1,2,11,41] The small blood volume (2 mL) in mice primarily limits the threshold sensitivity rather than the PAFC parameters. Hence, in humans, larger blood volumes (~5 L) circulating through the peripheral vasculature should increase the sensitivity. According to modeling, a sensitivity of 1 CTC/10 mL (*i.e.* one order better than existing assays) can be achieved by monitoring 200–300 μm vessel over ~1 h and a sensitivity of 1 CTC/100 mL by examining 1–2 mm peripheral blood vessels.

2.1.4. *In vitro* verification of PAFC capacity

To verify target specificity of PAFC technique, the MDA-MB-231 cells in PBS and mouse blood were treated *in vitro* with conjugated GNT-folate, MNP–ATF-uPA (amino-terminal fragment of the urokinse plasminogen activator), MNP–CD44 and GNT–CD44 complexes, as well as unconjugated MNPs and GNTs. A double molecular NP probe for targeting bulk CTCs was developed by mixing folate-conjugated GNTs with Ab-conjugated magnetic NPs.[2,15,16] The GNTs had a maximum absorption at 900 nm and the folate was specific to CTC folate receptors. The 10-nm MNPs (Ocean NanoTech) had an optimum PA contrast in a blood background at 639 nm and weak absorption at 900 nm and the Ab conjugate was specific to the protease-activated receptor. Both types of receptors (folate receptor and protease-activated receptor) are highly expressed in human breast tumors and are either not expressed or expressed at low levels in normal blood, lymph and endothelial cells. Excellent duplex targeting efficiency of breast cancer cells ($96 \pm 2.1\%$) compared with non-specific and/or unconjugated NP binding (5–9%) was verified with fluorescence and PA techniques *in vitro*.[15,16]

The presence of NPs in individual cancer cells was validated by PAFC with increases of PA signals from cells with MNPs at 639 nm and cells with GNTs at 900 nm and by fluorescent imaging of targeted cells (Figs. 2(a) and 2(b)).[15,16] Counting of labeled cells was performed by automatic scanning of a microscopic slide on the microscopic stage at a fixed position of focused laser beam with diameter of ~10 μm and a laser fluence level of 15–100 mJ/cm^2. PAFC of CD44+ cells among bulk cancer cells demonstrated that readable PA signals were obtained from 70% to 80% of cancer cells. The fluorescent imaging confirmed the presence of conjugated NPs in these cells (Fig. 2(a)) and, thus, verified that the PA signals were associated with CD44+ cancer cells.[15,16] In addition, the percentage of CD44+ cells was verified by data of conventional FC (Fig. 2(c)). The cancer cells incubated with the unconjugated NPs at the same concentration as conjugated NPs yielded the level of detectable signals $\leq 0.2\%$, demonstrating negligible influence of non-specific binding at the adjusted low pulse energy of 15 mJ/cm^2. These findings suggested that (1) the PA signals from GNT–CD44 or MNP–CD44 conjugates were associated with binding of Abs to CD44+ breast cancer cells, (2) the cells overexpressed CD44 might be selected by adjusted laser energies, (3) measurement of PA signals as a function of the number of NPs revealed that the sensitivity limit is only 35 GNTs/cell and (4) the influence of background signals from unbound or non-specifically bound NPs could be alleviated by appropriate reduction of the laser fluencies, which are low enough to provide signals only from cells of interest with highest NP concentration.

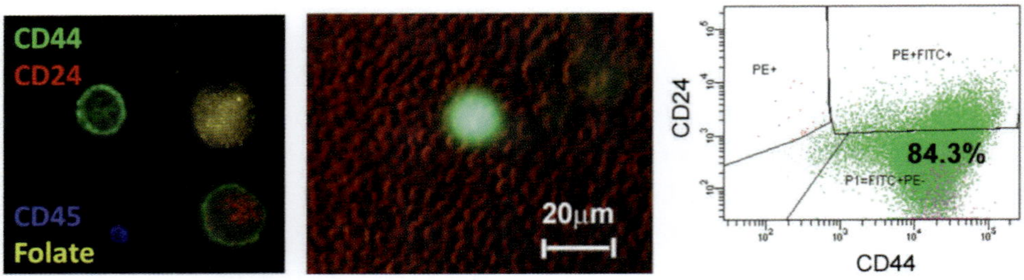

(a) (b) (c)

Fig. 2. (a) High resolution fluorescent imaging distinguishing different subpopulations of breast cancer cells (CD44+/CD24− cell is green and red; Folate+ cell is yellow and CD44+/CD24+ cell is green) and leukocyte (blue). (b) Integrated fluorescent and transmission image of mouse blood spiked with breast cancer cells, which were prior labeled with FITC. (c) Conventional flow cytometry of CD44+/CD24− cells among bulk breast cancer cells (MDA-MB-231 cell line). Adapted with permission from Ref. 16. Copyright 2013 Wiley.

The attachment of a magnetic tip to the top cover of the slides with MNP-CD44 alone led to concentration of MNPs near the magnet.[15,16] This was visualized even by the naked eyes as a dark spot and verified by fluorescent microscopy, which showed increases in fluorescent intensity from magnet-induced MNP clusters compared to the intact area before magnet action. The PA signal from the area with high local concentration of MNPs after magnet action was significantly higher (10–20 fold) compared to that from intact sample. Similar effects were obtained from single cancer cells labeled with MNP–CD44 before and after magnet exposure. Specifically, the magnet action led to notable enhancement of local fluorescence gradient within one cell compared to relatively homogenous spatial fluorescent light distribution before magnet action (Figs. 3(a) and 3(b)). This suggests magnet-induced clustering of MNPs within single cell, yielding significant (*i.e.* 6.6 fold in this particular experiment) enhancement of PA signals from these cells (Figs. 3(c) and 3(d)). The appearance of the local intracellular zones with dense MNPs is likely due to the accumulation of moving (under magnet action) MNPs near cellular membranes as the mechanical obstacles.

Attaching a magnet to the testing tube captured 10-nm MNP-labeled cells in phosphate buffer saline (PBS) at a broad range of flow velocity (0.1–10 cm/s) and this was accompanied by strong PA signals from the area under the magnet that were stronger than those signals outside the magnet corresponding to rare free cells and unbound MNPs.[14] To examine the application of this novel tool for targeted therapy of bulk CTCs and CCSCs, PA technique was integrated with PT therapy based on laser-induced microbubbles around overheated GNTs as a main mechanism of cell damage.[13–16]

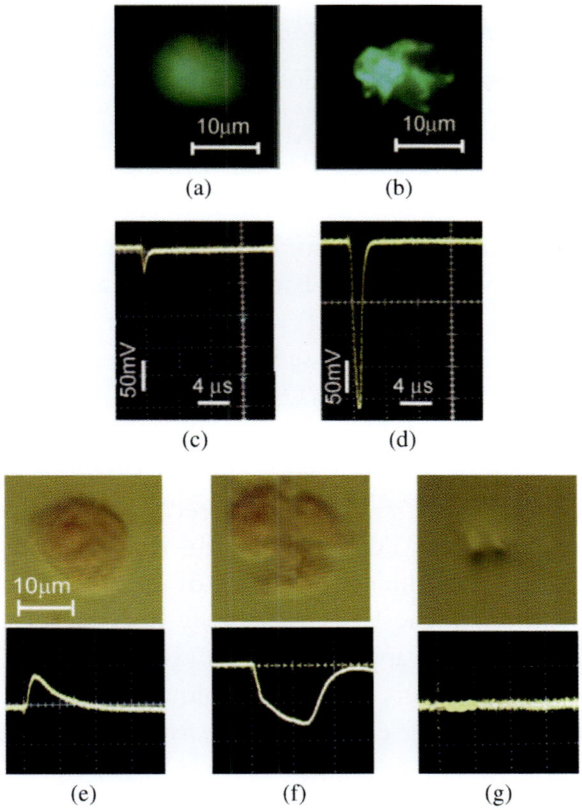

Fig. 3. Magnet-induced nanoparticle clustering and signal amplification in breast cancer CD44+ cells targeted by MNP-CD44 conjugates: (a) and (b) Fluorescent images of the single breast cancer cell targeted by MNP-CD44-FITC conjugates before (a) and after (b) the magnetic action during 3 min; (c) and (d) PT signal amplitudes from the same cell in (a) and (b) before (c) and after (d) magnetic action. Laser parameters: wavelength, 639 nm; beam diameter of 15 mm; fluence, 50 mJ/cm^2. Laser ablation of MDA-MB-231 breast cancer cells labeled by GNT-folate conjugates; (e) linear PT response from live cell; (f) nonlinear PT response from damaged cell; (g) the disappearance of PT signal after cell disintegration by high energy laser pulse as feedback control of therapeutic efficacy. Adapted with permission from Ref. 16. Copyright 2009 Wiley.

We applied the PT technique to demonstrate ablation of the folate+ and CD44+ breast cancer cells labeled by GNT–folate and GNT–CD44 conjugates, correspondingly. A low laser pulse energy (20 mJ/cm^2) at 850 nm produced a linear PT responses *in vitro* from single breast cancer cells labeled with GNTs with characteristic unipolar positive pattern associated with fast cell heating and slower cooling (Fig. 3(e), bottom). However, at increased pulse energy from 20 to 100 mJ/cm^2, a negative peak was observed (Figs. 3(f) and 3(g), top), which is indicative of microbubble formation, with subsequent cell disintegration (Fig. 3(f), bottom).[2]

2.2. *In vivo* PAFC applications

2.2.1. *In vivo* label-free PA detection of melanoma CTCs

Melanoma is an optimal target for PAFC that can enable routine, label-free, *in vivo* clinical assessment of CTCs for earlier detection of this most aggressive cancer with increasing incidence rates.[1,2] The overexpressed melanin could be used as an intrinsic high contrast PA NIR agent, which provided PA signals above the blood absorption background. Real-time PA counting of metastatic melanoma CTCs (B16F10) was performed in 50 μm diameter vessels in the ear, 200 μm abdominal vessels and 0.9 mm aorta of tumor-bearing mice (Figs. 4(a) and 4(b)).[41] CTCs could be readily detected with PAFC weeks before any evidence of detectable metastasis appears in the tissue samples using conventional techniques (Fig. 4(c)). The PAFC's sensitivity threshold was found ~1 CTC/mL. This unprecedented threshold sensitivity indicates high promise of PAFC as a powerful research tool to study CTC behaviors and their role in the development of metastasis at an early cancer stage. As aforementioned, the PAFC sensitivity has potential to be substantially improved for human applications, *i.e.* 100-fold (~1 CTC/100 mL), which cannot be achieved with existing techniques.

Ten-time increase of laser energy fluence from 60 to 600 J/cm^2 at 820-nm wavelength yielded about six-time increase in PA contrast of the CTCs

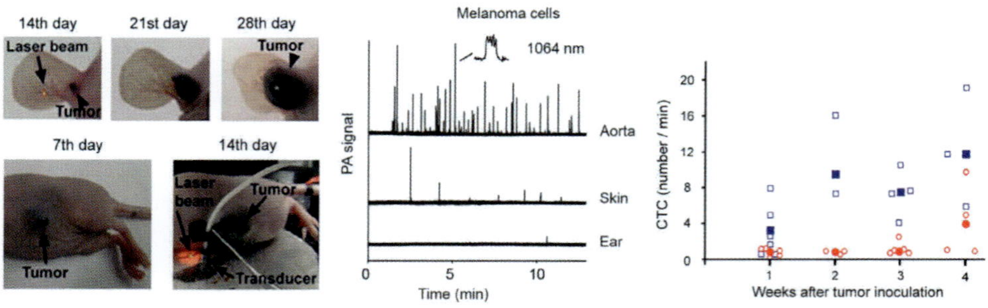

(a) (b) (c)

Fig. 4. *In vivo* label-free PA detection of melanoma CTCs. (a) Melanoma tumor growth in the mouse ear (top) and skin (bottom). (b) Average melanoma CTC rates in the ear and abdominal skin vessels, as well as the aorta, in B16F10 tumor-bearing nude mice 1 week after tumor development. (c) Change in the CTC count in the vessels of the abdominal skin as a function of time after B16F10 tumor cell inoculation in the ear (red hollow circle) and skin (blue hollow square). The dark red circle and blue square indicate averaged data. Laser parameters: wavelength, 904 nm; pulse energy fluence, 100 mJ/cm^2; pulse rate, 10 kHz). Adapted with permission from Ref. 2. Copyright 2012 Elsevier.

above the RBC background.[41] This phenomenon was associated with laser-induced nanobubbles around overheated strongly absorbing melanin nanoclusters in melanoma cells, which served as a non-linear PA signal amplifier compared to linear PA signals from RBCs with homogenous hemoglobin distribution (*i.e.* with no nanobubble formation). Also, the rates of CTCs have been gradually decreased from 12 to 1–2 CTC/Min over 1 h monitoring period, demonstrating effective PT purging of circulating melanoma cells. This was also related with the generation of nanobubbles, which served as melanoma cell killers with dominant mechanical action.[16] Further study could determine whether this new PA/PT theranostics are effective enough to be used alone, or whether it should be used in combination with other therapies such as chemotherapy or radiation therapy.

The *in vivo* label-free PA technology may have a tremendous clinical significance because it does not require time-consuming labeling procedures. It also allows highly sensitive CTC detection *in vivo* at CTC concentrations much below the sensitivity of other methods. Potential clinical applications may include[1,2]: (1) blood screening for early CTCs before the progression of metastasis; (2) testing for cancer recurrence; (3) personalized assessment of the therapeutic intervention and its efficiency (*e.g.* surgery, chemo or radiation) through real-time CTC counting and (4) potential for metastasis inhibition, if not prevention by a well-timed therapy. The following further improvements would facilitate its successful translation to clinical practices:[1,2] (1) reducing background signals from blood through changes in its oxygenation, osmolarity and hematocrit, within physiological norms[41]; (2) assessing CTCs in deep large vessels (*e.g.* jugular vein) with a focused ultrasound transducer;[41,42] (3) increasing PA contrast by drug-induced activation of melanin synthesis in melanoma cells[41,43]; (4) melanogenesis activation in melanoma and even non-melanoma (*e.g.* breast cancer) cells *via* transfection with tyrosinase-activating plasmids[41,43]; (5) using melanin NPs as new PA contrast agents[44] for targeting of melanoma and other non-pigmented CTCs and (6) targeting melanoma CTCs *in vivo* by MNP conjugated with specific antibody and magnetic CTC enrichment.[44]

2.2.2. *In vivo* magnetic enrichment and multiplex PA detection of CTCs and CCSCs

A noble multiplex, multicolor and multimodal nanotheranostic approach was reported in 2009[14] by uniquely combining magnetic enrichment of CTCs directly in bloodstream with duplex targeting dual color detection strategy (Fig. 5) in order to overcome the limits of *ex vivo* CTC detection methods

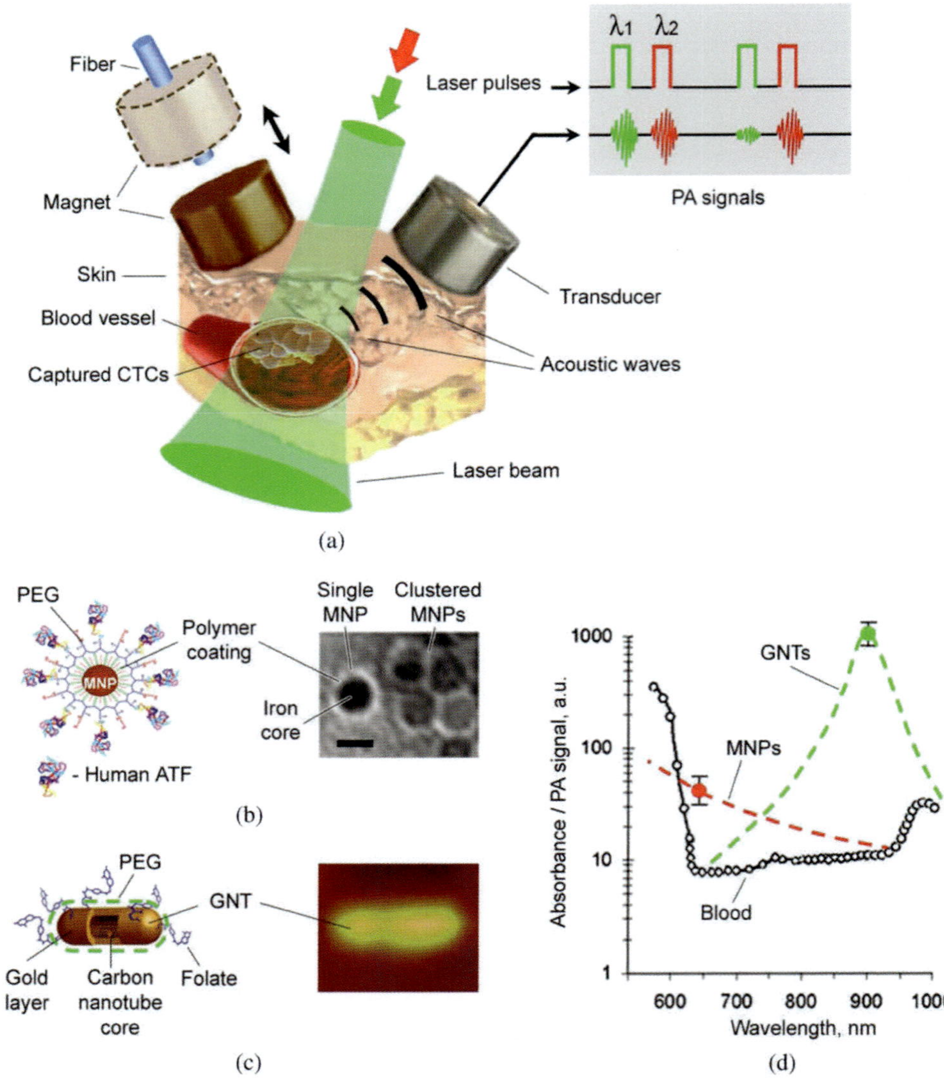

Fig. 5. *In vivo* magnetic enrichment and two-color PA detection of CTCs. (a) Schematic with laser beam outside magnet or fiber-based laser delivery through the hole in magnet (dashed lines). (b) The 10 nm MNPs coated with amphiphilic triblock polymers, polyethylene glycol (PEG) and the amino-terminal fragment (ATF). (c) The 12 × 98-nm GNTs coated with PEG and folate. (d) PA spectra of ~70 μm veins in mouse ear (open circles). The average standard deviation for each wavelength is 18%. Absorption spectra of the MNPs and GNTs (dashed curves) normalized to PA signals from CTC labeled with MNPs (red circle) and GNTs (green circle). Adapted with permission from Ref. 14. Copyright 2009 Nature Publishing Group.

(*i.e.* small blood sample volume) and increase specificity of *in vivo* CTC detection. Highly NIR responsive GNTs (Fig. 5(c)) with an absorption maximum at 900 nm and a minimum at 639 nm (Fig. 5(d)) were coated with polyethylene glycol (PEG) and conjugated with folates.[14] As the second NPs, 10-nm MNPs coated with PEG and amphiphilic triblock polymers were conjugated with the ATF-uPAR (Fig. 5(b)). MNPs have an absorption *in* broad NIR range; nevertheless, absorption spectrum is different from that of GNTs (Fig. 5(d)). Injection of cancer cells (*i.v.* 10^5) previously labeled *in vitro* with conjugated MNPs revealed readable PA signals from mimic CTCs with ~40- to 60-min clearance. Cell injection alone followed by *i.v.* injection of NP cocktail (20:80 conjugated GNT:conjugated MNP) revealed, using two-color (639/900 nm) PAFC, effective labeling within ~5 min after NP injection. The rate of detectable PA signals from CTCs that were double-labeled *in vivo* had clearance rates similar to cells labeled in vitro.[14] To detect CTCs originating from a primary tumor, 5 $\times 10^6$ MDA-MB-231 cells were inoculated subcutaneously into nude mice. At 2, 3 and 4 weeks of tumor development (Fig. 6(a)), a cocktail of the conjugated NPs (GNT–folate + MNP–ATF) was injected *i.v.* into the circulation. Two-color PA detection of CTCs at 20 min after injection (to allow clearance of most free NPs) showed that the ratio of the numbers of CTCs in mouse ear to those in abdominal vessels (CTCs per minute) increased from $(0.9 \pm 0.3)/(6 \pm 2.1)$ at 2 weeks to $(15.1 \pm 2.7)/(47 \pm 6.4)$ at 4 weeks (Fig. 6(b)). These data approximately correlated with the stage of the primary tumor progression. Attaching a magnet with a field strength of 0.39 Tesla at week 1 led to immediate increases in both PA signal amplitude and rate and changed the character of the PA signal from infrequent flashes to a continuous increase of permanent PA signals up to 88-fold within 1.5 h (Figs. 6(d) and 6(e)), indicating successful magnetic CTC capturing.[14] Applying this method to clinical use, the patients may potentially carry a magnet attached to selected peripheral vessels (*e.g.* in wrist area) for trapping of CTCs, followed by a quick PA detection of the trapped CTCs and, if necessary, local PT treatment or removal of CTCs by using syringe-based systems for further molecular analysis.

In vivo targeting and detection of breast CTCs with a stem-like phenotype, which are naturally shed from the parent tumor on mice models, were performed with two color and dual modal strategy using functionalized GNTs and MNPs.[14] GNTs and MNPs conjugated with folic acid and antibodies to CD44 were selected for the detection of stem-like CTCs. For identification of bulk CTCs, we used markers described in the previous section. At week 4 of the tumor inoculation (Fig. 6(a)), when the metastatic disease was well recognized by metastasis in the liver, GNT–Folate and GNT–CD44

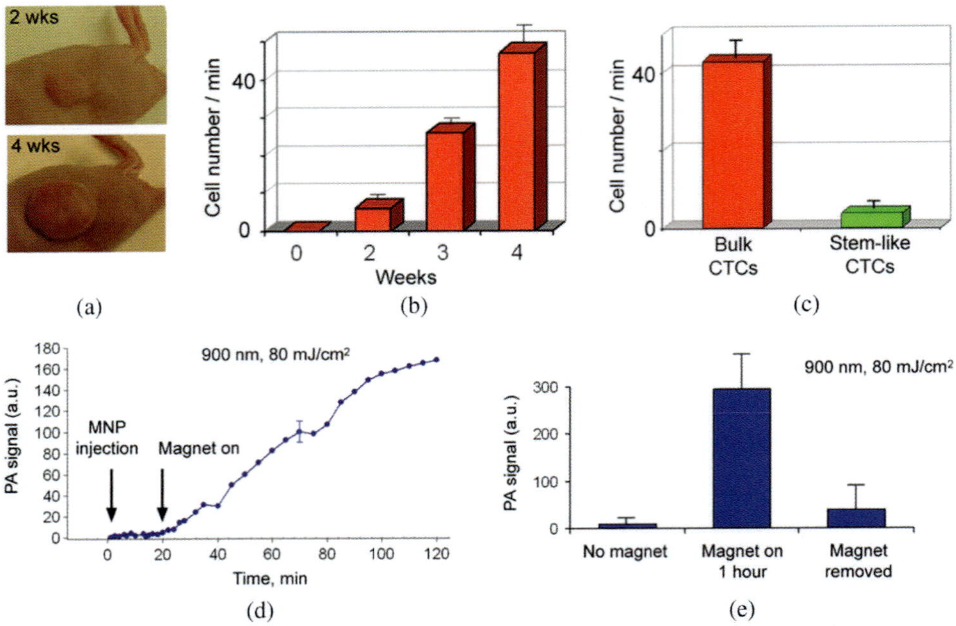

Fig. 6. PA detection and magnetic enrichment of CTCs in tumor-bearing mice. (a) The size of the primary breast cancer xenografts at different stages of tumor development. (b) The average rate of CTCs in 200 μm abdominal skin vein over a period of several weeks. (c) Average rate of bulk CTCs and stem-like CTCs in the abdominal skin blood vessels at week 4 of tumor development. (d) PA signals from CTCs in abdominal skin vessels obtained with fibre schematics at week 1 of tumor development before and after magnet action. The average standard deviation for each wavelength is 24%. (e) PA signals from CTCs in abdominal skin vessels before, during (3 min) and after magnetic action at week 2 of tumor development. The error bars in b, c and e represent standard error (n = 3). Figures were adapted with permission from: (a), (c), Ref. 16, copyright 2009 Wiley; (b), (d), (e), Ref. 14, copyright 2009 Nature Publishing Group.

were separately injected *i.v.* through mouse tail vein at concentrations of 10^9/mL in 100 μL PBS (n = 3). To allow effective labeling of CTCs in bloodstream, PA monitoring of blood vessels began at 20 min after injection.[26,39,41] The average mean rate of PA signals, which is equivalent to the cell rate (*i.e.* one PA signal corresponds to one cell or cell cluster crossing laser beam) after introduction of GNT–Folate was 42.9 ± 6.5 cells/min. After subsequent injection of GNT–CD44, the rate of flashing PA signals increased to 46.7 ± 6.8 cells/min (Fig. 6(c)). It implies that 3.8 ± 0.6 cells/min or 8.8% of all detectable CTCs can be exclusively associated with stem-like subpopulation of CTCs. This study demonstrated for the first time that the multifunctional PAFC nanotheranostic platform has the potential for ultrasensitive PA molecular detection of stem-like CTCs *in vivo*. Additional studies are needed to develop a fully validated platform, including further exploration of the role of the folate receptor and addition of CD24 and CD45 markers to increase detection and specificity of stem-like CTCs and to exclude possible false positive signals from leukocytes, respectively.

2.2.3. *In vivo* PA detection of CTCs in lymph flow and sentinel lymph nodes

The lymphatic test has been paid much less attention compared to the blood test, in spite of the recent discoveries that indicate the key role of lymphatic function in the development of many specific diseases such as tumor metastasis, infections or immunity disorders.[1,2] Many questions of lymphatic function remain to be answered. Unlike the blood vasculature, lymph vessels are colorless, with relatively low pressures and low cell concentrations. This makes lymph sampling impractical due to very low sample yields (only a few microliters at a time) and long-term cannulation.

Furthermore, blood and lymph vessels are common pathways for cell dissemination from one location to another through numerous anatomical interconnections between lymph vessels, lymph nodes and blood circulatory system.[45] The close interrelations of blood and lymph systems can be illustrated by simplified schematic of dissemination of CTCs in cancer.[1,45] Logically, both lymph and blood pathways should be examined simultaneously to understand disease progression. However, cells (*e.g.* CTCs) in blood and lymph have been studied separately until now, with a particular focus on the role of blood cells. This askew perspective cannot provide a comprehensive understanding of the mechanisms of disease development. Consequently, we currently do not have the clinical capability to intervene and stop development of many diseases at early stage when well-timed treatment is most effective.

Recently, an integrated blood and lymph PAFC was demonstrated in combination with PA lymphography and PA cytometry of SLNs.[1,2,45] This PAFC platform provided simultaneous real-time quantitative monitoring of blood and lymph CTCs *in vivo* and defined *in vivo* cross-correlations between lymph CTCs, blood CTCs, size of primary tumor and SLN and distant metastases.[45] The unprecedented sensitivity and specificity allows for the study of CTC pathways at the earliest possible stages of metastatic disease. Specifically, in a preclinical mouse melanoma model, we found that early metastatic cells in latent metastatic disease (~1 week after tumor inoculation) can disseminate equally by blood and lymph pathways (Table 1). However, in a few cases, metastatic cells appeared in lymph vessels in week 1 without any

Table 1. Correlation between primary tumor aize, metastasis in SLN and number of tumor cells in lymph and blood flow of melanoma-bearing mice.[1]

	Tumor size, (mm^2)	Rate of lymph CTCs (cell/min)	Rate of blood CTCs (cell/min)	Number of PA signals associated with metastasis in SLNs	Histology
1 week	1.0 ± 0.2	0.26 ± 0.05	0.85 ± 0.03	493	NO
2 weeks	3.6 ± 0.5	2.13 ± 0.30	1.07 ± 0.05	3,188	YES

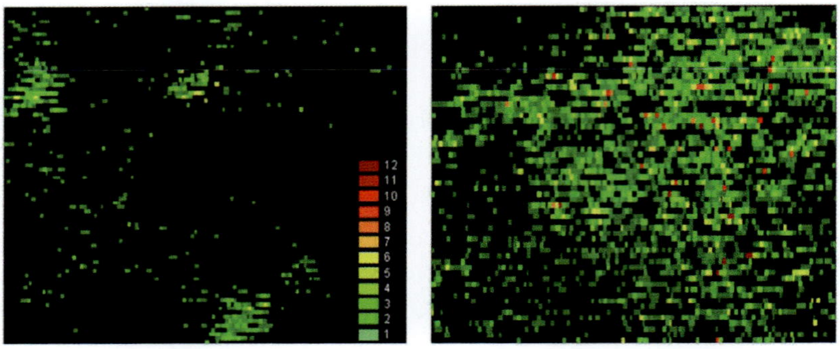

Fig. 7. PA scanning cytometry of melanoma metastasis in the sentinel lymph node (1.6 × 3 mm) at a single cancer cell level using the tumor-bearing mouse model at weeks one (a) and two (b) of tumor development. Red pseudo-color peaks indicate PA signals with maximum amplitudes. Each angle spot is associated with single metastatic cells. Adapted with permission from Ref. 15. Copyright 2009 Wiley.

cells detected in the blood vessels and vice versa. On average, over the course of two weeks, an increase in the size of a primary tumor by ~3.5 times is associated with an increase in the quantity of lymphatic CTCs by ~10 times, with only a small increase in blood CTCs. The number of PA signals (*i.e.* number of micrometastases and/or clusters of metastatic cells)[15] from the SLNs increased by 6.5-fold (Fig. 7). These data showed that PAFC could clarify dynamic interrelationships of blood and lymph CTCs *in vivo* and their impact on metastasis development.

Also, PA detection of melanoma and breast cancer micrometastasis in sentinel lymph nodes (SLNs) was demonstrated using mouse models.[13] At 5 min after the injection of GNT–folate conjugates in the mouse ear, strong PA signals above the background, with a contrast of 10–15, appeared in the SLNs, indicating the delivery of the CTCs through the lymph vessels to the SLNs and the process of micrometastasis. Subsequent application of the therapeutic laser pulse, with enhanced energy fluence, led to a decrease of these strong signals to a background level, which implied laser-induced PT eradication of tumor cells. However, when the experiment was repeated with unconjugated GNTs, low PA signals with a contrast of ~3 above the background was observed, suggesting random distribution of GNTs in the SLN volume. To further validate the data, *ex vivo* experiments mimicking the lymph node micrometastasis were carried out using conjugated GNT–folate and fluorescent dye to visualize tumor cells.[15] The results supported the feasibility of the PAFC-PT technique for the *in vivo* detection and killing of metastatic cells in SLNs.

This *in vivo* PAFC platform, in combination with microarrays, would enable assessments of the biological properties of lymph disseminated tumor cells (DTCs), *e.g.* molecular profiles and viability, compared to those in the primary tumor, regional and distant metastasis and blood CTCs during metastasis progression, with a focus on identifying the "so-called" tumor initiating cells among the bulk lymph DTCs.[2] The expanded knowledge of lymphatic-related metastasis may catalyze a paradigm shift in the diagnostic clinical oncology, from conventional assessment of early metastasis in sentinel lymph nodes toward lymph DTC testing. Considering the safe nature of the novel *in vivo* lymph cancer tests as supplementary (or in some cases as alternative) to conventional blood tests, we anticipate a quick translation of this new technology for use in humans.

2.2.4. *In vivo* PA detection of CTCs cerebrospinal fluid

The spreading of cancer, such as leukemia, lymphoma, breast cancer and melanoma, into the central nervous system through hematogenous spread, direct release from the tumor itself, or by migration along perineural or perivascular spaces leads to neurological symptoms, such as neoplastic meningitis and rapid mortality.[2] Particularly, cerebrospinal fluid (CSF) tumor cells may serve as a marker of disease progression and early-stage brain metastasis in breast cancer. The CSF is a colorless body fluid with a total volume of 135–150 mL in adults and contains $1.5–5 \times 10^5$ mononuclear cells. CSF is produced at the rate of 0.2–0.7 mL/min and circulates under pressure up to 132 Pa (~1 mmHg) through the ventricular system around and inside the brain and the spinal cord. The CSF flow has pulsatile "forward–backward" characteristics that correlate with cardiac cycles. Forward velocity in the spinal canal ranges from 10 mm/s in the craniocervical junction to 1 mm/s in the lumbar part of the canal; the velocity of backward flow is approximately two times slower than that of forward flow. The CSF function is bathing the central nervous system and bridging the vascular and lymphatic systems. Currently available tools to detect CTC dissemination into the CSF including cytology, *in vitro* flow cytometry and others[46,47] are limited by small sample volume and suffer from a lack of sensitivity. Many of these problems could be resolved by increasing the sensitivity and specificity of CSF examination and by the analysis of a relatively large volume of circulating CSF *in vivo* using non-invasive PAFC and functionalized NPs as PA contrast agents.

GNR–folate conjugates with maximum absorption near 670 nm were *i.v.* injected in tumor-bearing mice 10 weeks after innoculation.[2] At day 1, irradiation of a nanosecond laser pulse at 671 nm through a 300-μm fiber of CSF in the neck area of nude mice revealed rare PA signals with rate of a few signals per hour, which were associated with targeted breast tumor cells flowing though the CSF. Using the methodology described in Secs. 2.2.1–2.2.3, the assessment of lymph nodes and the brain confirmed the presence of micrometastasis as potential source of tumor cells in the CSF. The sensitivity was estimated to be approximately a few CTCs among a million leukocytes over a 4-h period. This approach may provide a semi-automated PA molecular analysis that vastly improves the sensitivity, reliability, objectivity and accuracy of detecting CSF tumor cells compared to CSF cytology. Due to high depth assessment (up to few cm), fiber-based PAFC could be applied to different sites of the spinal canal (*e.g.* cervical, lumbar) and ventricles using a procedure that is well established for lumbar puncture. If clinically successful, this tool may provide a tremendous leap forward in new scientific areas such as *in vivo* CSF cell testing including cell trafficking, cell-to-cell interactions in the CSF and the route of CSF circulation especially in the terminal end of the spinal canal.

2.2.5. *In vivo PA detection of CBCs*

Bacterial infections are a significant cause of morbidity and mortality worldwide.[4] In particular, multidrug resistant pathogens in the bloodstream are increasingly prevalent problems that complicate the care of many patients. The critical steps in the development of bacterial infections include their penetration into the blood system, interactions with blood cells flowing in the circulatory system or with endothelial cells and further translocations in the host organisms. Little is known about CBC kinetics in the blood. This includes their clearance and adherence rates, which might be very important for understanding the transition from the bacteremic stage to the tissue invasive stage and for development of an effective therapy. Previously, we demonstrated the capability of the PAFC/PTFC platform to detect and kill, *in vitro*,

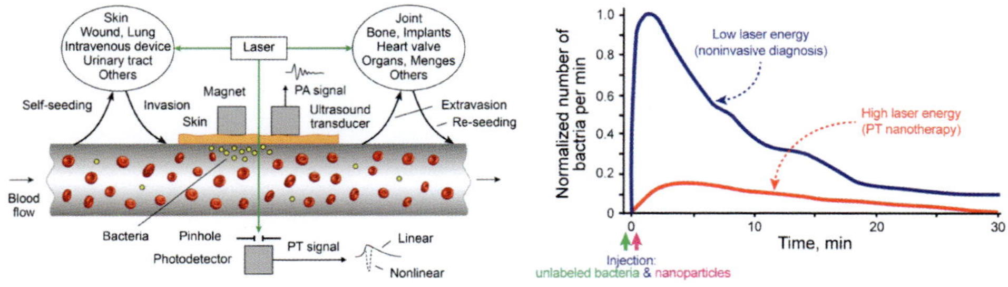

Fig. 8. PA molecular diagnosis and PT targeted eradication of circulating *S. aureus* in the blood of the mouse model with real-time PA monitoring of PT nanotherapeutic efficacy. (a) Schematic; (b) experimental data. Adapted with permission from Ref. 4. Copyright 2012 Galanzha *et al.*

Staphylococcus aureus and *Escherichia coli* labeled with gold NPs and carbon nanotubes (CNTs).[48–50] Recently, we extended this platform to include *in vivo* magnetic enrichment, multiplex PA detection and PT eradication of circulating *S. aureus* on the basis of the *in vivo* nanotheranostic platform developed for CTCs (Fig. 8(a)).[1,2,4,14] Bacteria were targeted directly in the bloodstream through *i.v.* injection of silica-coated MNPs (siMNP) and GNRs functionalized with antibodies specific for *S. aureus* protein A and lipoproteins, respectively. Both antibodies are highly expressed in *S. aureus* and absent in mammalian cells. GNRs had a maximum absorption near 820 nm. After successful two-color PA detection of targeted CBCs at low energy fluence (50 mJ/cm^2) of lasers at 671 and 820 nm, mice were subjected to PT therapy by 1 h laser exposure of a 300 μm abdominal blood vessel with a laser fluence of 0.8 J/cm^2 at 820 nm, coinciding with an absorption spectrum of GNRs. PT nanotherapy led to a significant decrease in the PA signal rate (Fig. 8(b), red curve) compared to the control group (Fig. 8(b), blue curve) at a laser fluence shown to be safe for blood cells. To confirm this therapeutic effect, mice were euthanized and the blood was examined for the presence of viable bacteria. Blood from mice in the control and PA diagnostic groups showed similar bacterial growth, while the number of bacteria from the PT therapeutic group was reduced by 10-fold. This study implied that the PAFC/PTFC nanotheranostic technical platform may offer unique advantages compared to existing diagnostic and therapeutic approaches: (1) ultrahigh sensitivity (0.5 CFU/mL); (2) physical, PT-based destruction of bacteria; thus retaining its therapeutic efficacy irrespective of antibiotic resistance; (3) integration of multiplex PA molecular detection and PT targeted elimination of CBCs with real-time PA monitoring of therapeutic efficacy and (4) high spectral specificity based on distinct spectral properties of NPs.[1,2,4]

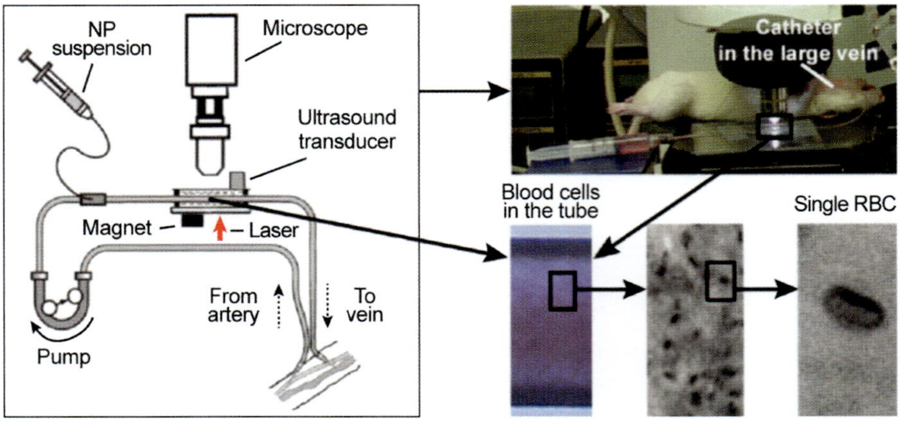

Fig. 9. Principle and animal model for *in vivo* PA detection, molecular targeting, and PT purging of infected blood with an extracorporeal schematic. Catheters are placed in a large artery or vein (e.g. jugular) to create a bypass. MNPs conjugated with Ab against staphylococcal protein A (Spa), were injected into a tube. The distance between the injection site and detection points could vary by changing the tube length. Cells labeled in flow were captured by the magnet. Laser irradiation of the area near the magnet generates PA signals, which were detected with an ultrasound transducer attached to the tube. Simultaneously, laser irradiation at higher energy allowed PT killing of targeted bacteria. Conventional transmission imaging made it possible to simultaneously control the positions of the laser beam, magnet, and transducer. High-speed imaging also allowed visualization of individual moving cells at the single-cell level as shown in the images (right, bottom) obtained at different magnifications (4×, 20×, and 100×, respectively). Thus, the extracorporeal (bypass) schematic could provide continuous PA monitoring of blood flow in external tubes, and permit efficient capture of abnormal objects (e.g. bacteria or its toxins targeted by MNPs directly in the extracorporeal flow). Magnetic capture of both abnormal objects and unbound magnetic nanoparticles prevents them from being transported further in the systemic circulation.

Because several gold and magnetic nanoparticles were already approved for pilot study on humans and clinical potential and safety of PA technology has been successfully demonstrated in several pilot trials, early laser-based detection and treatment of CBCs could be feasible in a 0.5–1.5 mm hand vein at 1–2 mm depths, which is within well-documented capacity of PA technology to assess much deeper (3 cm) and larger (1 cm) human blood vessels.[4,13,14,41] As an alternative, the bypass schematics (Fig. 9) could be useful for purging of infected blood during sepsis crisis with simultaneous magnetic removal of CBCs and MNPs from circulation.

2.2.6. *In vivo* PA detection of circulating blood clots

When a blood vessel is injured, the normal physiological response of the body is clot (thrombus) formation to prevent blood loss. Alternatively, in the absence of vessel injury, a pathological condition called thromboembolism may lead to formation of circulating blood clots that can block vessels at distant locations, in particular, in peripheral veins (venous thromboembolism), lungs (pulmonary embolism), brain (embolic stroke), heart (myocardial infarction), kidney or gastrointestinal tract. Thromboembolism is a significant cause of morbidity and mortality, especially in adults. However, little progress has been made in the development of methods for real-time detection and identification of circulating clots. Most emboli are not detected until a vessel obstruction manifests as a serious clinical event.

Commonly used *ex vivo* methods of detecting clots are time-consuming, suffer from poor sensitivity due to the small volumes of blood samples and are limited by discrete time-point sampling with difficult access to clinically relevant sites. Most *in vivo* methods are only able to detect fixed or slowly moving large clots. Pulse Doppler ultrasound is a promising technique for the detection of circulating clots, but this technique cannot detect small clots or assess clot composition and may be affected by artifacts. To overcome these limitations, we proposed an *in vivo* PAFC for real-time detection of clots of different compositions (Fig. 10(a))[40,51] using a combination of positive and negative contrast modes. Laser irradiation of blood vessels in normal vessels creates a constant PA background determined by the absorption of RBCs randomly distributed in the irradiated volume. Depending on the size of vessels, hematocrit (Ht) and PAFC spatial resolution, the number of RBCs in the detected volume can vary from one or a few RBCs in a capillary to thousands in larger vessels. When a RBC-rich red clot passes through the irradiated blood volume, a transient increase in the local absorption, which is associated with a high concentration of hemoglobin (Hb), results in a sharp positive PA peak (Fig. 10(b), left). Red clots can be detected when they have a higher local absorption than the normal RBC background in the detected volume. When a weakly absorbing, platelet rich white clot passes through the laser-irradiated vessel volume, a transient decrease in the local absorption results in a sharp negative PA signal (Fig. 10(b), middle). A mixed clot with both RBC-rich (*i.e.* high-absorbing) and platelet-rich (*i.e.* low-absorbing) local zones produces a pattern of positive and negative PA signals (Fig. 10(b), right).

This phenomenological model was demonstrated in preclinical studies using the mouse model of myocardial infarction created by total ligation of left coronary artery.[40] The readable transient PA signals were observed with different patterns of negative, positive and combined contrasts (Fig. 10(c)) associated with white, red and mixed clots, respectively, compared to no signals in normal control mice. The concentration and size of clots were measured with a threshold of a few clots with diameter of >20 μm in the entire

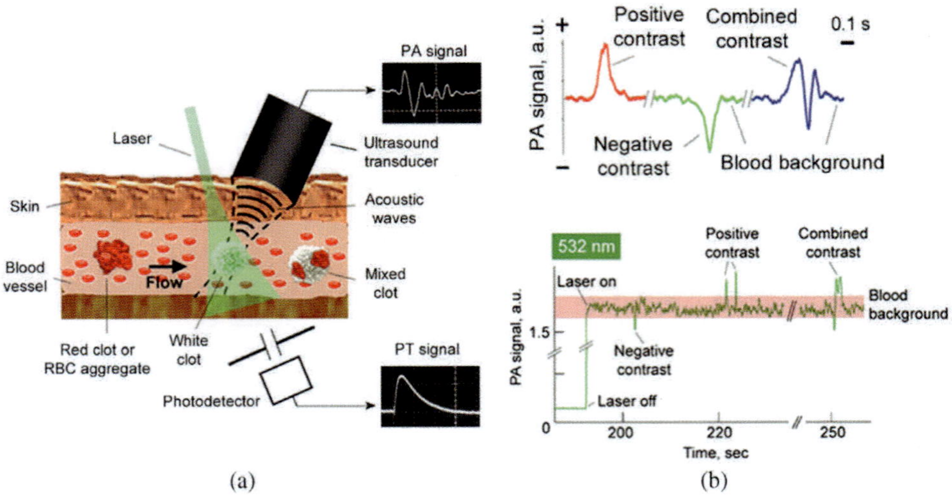

Fig. 10. The principles of fluctuation positive and negative contrast PA flow cytometry. (a) Schematic. (b) Example of PA positive, negative, and combined contrasts from circulating clots of different compositions. (c) Monitoring of clot formation during and after surgery modeled by small skin incision. Adapted with permission from Ref. 2. Copyright 2012 Elsevier.

circulation (*i.e.* 1–3 clots/mL in mice). This PAFC-based diagnostic platform can be used in real time defining risk factors for cardiovascular diseases, as well as for prognosis and potential prevention of stroke by using a well-timed therapy or for a clot count as a marker of therapy efficacy. Besides label-free negative contrast PA detection of clots, they can be potentially targeted by NPs (as CTCs and CBCs) and then destroyed by laser irradiation.

Moreover, NPs with enhanced PA/PT contrast properties were integrated with ultrasound contrast agent (*i.e.* 0.5–2 μm microbubbles) to enhance the treatment of emboli. Microbubbles were prepared according to standard manufacturer's procedure with modifications by adding absorbing NPs, *i.e.* CNTs, in various concentrations. PA/PT signals from microbubbles with absorbing agents were significantly higher (10–50 fold) compared to signals from microbubbles or NPs alone. In particular, for intact microbubbles, the threshold for laser-induced evaporation at 850 nm was in the range of 2–8 J/cm^2, while the presence of CNTs reduced this threshold to 10–30 mJ/cm^2 compared to 100–200 mJ/cm^2 for CNTs in water at the similar concentration. Gradual increase of laser energy above evaporation threshold led to sudden non-linear (10–30 fold) amplification of both PA and PT signals compared to NP solution alone, accompanied by microbubble degradation at high laser energy. After *i.v.* injection of 100-μL microbubble solution in concentration range of 10^{10}–10^{11}/mL to mouse circulatory system, PAFC revealed quick appearance of many microbubbles with typical clearance of 3–6 min. In developed model of mouse thrombosis, thrombi with microbubbles could be visualized *in vivo* in blood vessels. Laser irradiation of microbubbles and thrombi *in vitro* and *in vivo* using safe for human doses (*e.g.* 50–100 mJ/cm^2 at 1064 nm) led to disappearance of both microbubbles and thrombi, suggesting effective destruction and clearance of thrombi.

2.2.7. *In vivo* PA time-of-flight velocity measurement of single cells and NPs

Existing optical techniques for *in vivo* measurement of blood flow velocity are not quite applicable for determination of velocity of individual cells or NPs.[2] PAFC may solve this problem by using time-of-flight technique with single or multiple laser beams (Fig. 11(a)). In the one-beam scheme, an object crossing a laser beam created transient PA signals with a width t_L, defined as $t_L \approx (D + d)/V_F$ where D is the diameter of object, d is the width of the linear laser beam and V_F is the velocity of the object. Thus, V_F can be measured for objects and laser beam of known sizes. In the three-beam scheme, velocity of an object is determined by the time it takes for the object to travel between multiple laser beams (Fig. 11(a), bottom). In this case, an object in flow produces multiple consequent trains of PA signals. Object velocity can be estimated by using the time intervals between the trains and the distances between the beams. This measurement is independent of the object size. These approaches were demonstrated *in vivo* using an animal (mouse) model by estimating velocity of GNRs, melanin NPs, RBCs, WBCs and CTCs in the broad range of flow velocity from 0.1 to 20 cm/s.[52] In particular, monitoring of PA signal widths in mouse ear microvessels from WBCs labeled with GNRs revealed the double maxima in peak-width histograms (Fig. 11(b)) associated with the fast moving portion of WBCs in central flow and slowly rolling WBCs. In general, the peak-width analysis in the time-of-flight mode provided important information on objects' velocities, cell aggregation manifested by the appearance of wider peaks with complex shapes and on rolling effects that result in slower cell velocity. However, width histograms alone were not sufficient to distinguish a large aggregate of cells with high velocity and an individual rolling cell moving slowly because both would produce peaks of similar widths. We determined that these cells could be identified with a two-parameter plot representing PA peak amplitude and peak widths obtained from time-of-flight measurements. Fig. 11(c) illustrates verification of this approach for label-free detection of melanoma B16F10 cells. Three

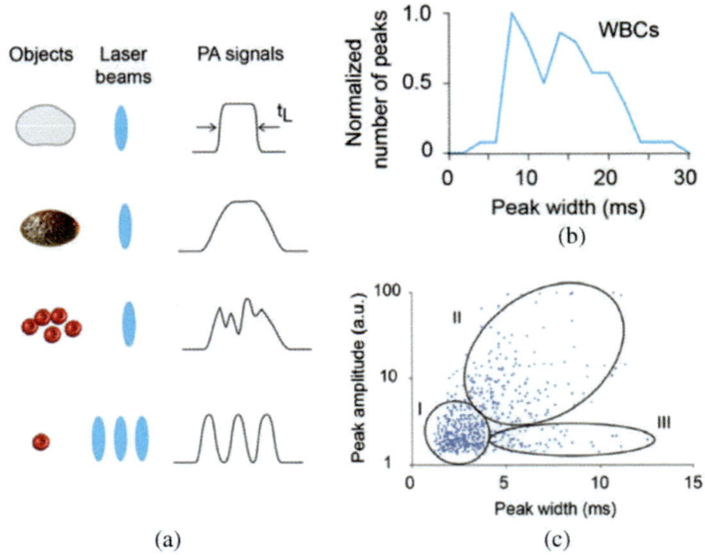

Fig. 11. *In vivo* PA time-of-flight velocity measurements of cells and nanoparticles. (a) Shapes of peaks in PA trace for particles of different sizes and various beam geometries. (b) Peak widths distributions in the WBCs in the mouse circulatory system targeted by GNRs with antibodies specific for the CD45 receptor. Laser parameters: wavelength: 820 nm, pulse width, 8 ns; pulse repetition rate: 10 kHz; laser pulse energy: 20 μJ. (c) Scatter plot of the height and width of peaks for label-free PA detection of circulating B16F10 cells in an arteriole of the mouse ear. Area I indicates signals from individual cells. Area II includes aggregates of several cells, and Area III indicates possible "rolling" cells. Adapted with permission from Ref. 2. Copyright 2012 Elsevier.

areas of interest were selected for this scatter plot: (I) the area with the highest density of events, corresponding to fast-moving individual cells; (II) aggregates of cells that were larger and had more melanin than individual cells and (III) peaks with PA signal too small to consider these objects as aggregates. However, large widths in III suggested that these cells were moving more slowly than the other cells. Therefore, multi-parameter analysis of PA data may provide a criterion for distinguishing aggregates and individual rolling cells.[52]

2.2.8. *In vivo* PA detection of circulating contrast agents for their pharmacokinetic assessments

The rapidly growing application of *in vivo* imaging modalities (*e.g.* optical, MRI or ultrasound) and nanotechnology-based probes has placed new demands on monitoring the clearance rate of various contrast agents including NPs, drug carriers (*e.g.* liposomes), dyes and microbubbles.[2] No clinically relevant method has been developed for rapid and ultrasensitive study of the pharmacokinetics of these agents including their clearance rate in different vessels. As most contrast agents have intrinsic optical absorption, PAFC is an almost ideal tool for real-time, label-free monitoring of their pharmacokinetics. We demonstrated PAFC's capability to monitor the circulation lifetime of NPs (*e.g.* GNRs, GNTs, MNPs), dyes (ICG, methylene blue [MB] and trypan blue [TB], evans blue and lymphazurine), microbubbles, normal blood cells and CTCs (melanoma, breast cancer, scamous carcinoma, pancreatic) using a laser with different wavelengths.[14,41,42,52–56] Using PAFC, circulating objects can be detected when PA signals from them are above the blood background associated preferentially with absorption on RBCs (Fig. 12(a)). After injection of contrast agents, PAFC provides two typical signal trace patterns (Fig. 12(a), middle and bottom) as an increase in the baseline level above blood backgrounds and strong fluctuations above of baseline. The first pattern is associated with homogenous random distribution of dye molecules, individual microbubbles, or NPs in the detection volume, while the second

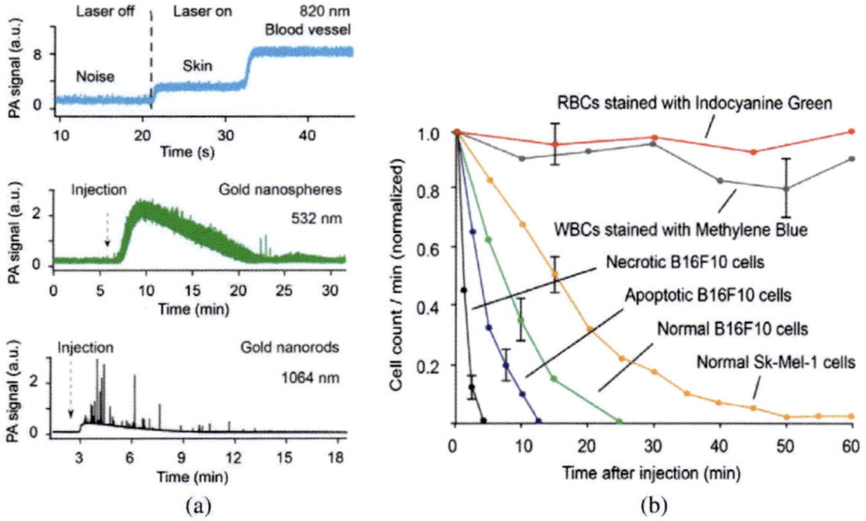

Fig. 12. PA monitoring of NP and cell clearance rates. (a) *In vivo* PA signal levels from the mouse ear vein and surrounding skin, compared to baseline noise, when the laser is off (top). Laser parameters: wavelength: 820 nm; fluence, 0.2 J/cm^2; pulse rate: 10 kHz. PA monitoring of the clearance rate of 30-nm spherical gold NPs in a blood vessel of the mouse ear (middle). Laser parameters: 532 nm/cm^2, 9 kHz. PA monitoring of GNR clearance in a blood vessel of the mouse ear (bottom). Laser parameters: 1064 nm, 0.1 J/cm^2, 9 kHz. (b) PA monitoring of the clearance rates of melanoma cells at different functional states and metastatic activity and normal blood cells labeled with contrast dyes. Adapted with permission from Ref. 2. Copyright 2012 Elsevier.

pattern is related with the presence of their aggregates with a more strongly localized absorption. The clearance rate for most NPs, depending on their protective layer properties, was in the range of 20–40 min and in rare cases a few hours, while liposomes demonstrated a long-term circulation of up to few days.[39] Many NPs demonstrated aggregation in blood flow especially during injection procedures. The described approach should be useful to the routine evaluation of the possible influence of the natural properties of NPs, the protective materials and the coating procedures on their clearing. The approach *in vivo* allows minimization to the animal number compared to *ex vivo* methods with periodical blood sampling procedures.

I.v. injection of normal blood cells and tumor cells labeled with various contrast agents and having functional states revealed their different clearance rates (Fig. 12(b)): 1–3 min for necrotic cells, 5–15 min for apoptotic cells, 30–60 for high metastatic tumor cells, 1–2 h for cancer cells with lower metastatic activity and 3–5 days and 1–3 days for normal RBCs and WBCs, respectively. These data showed that the immune responses to abnormal and more metastatic cells were stronger than those to normal blood cells.

PA monitoring of the clearance of dyes (ICG, MB and TB) revealed increased dynamic PA signal fluctuations, which can be associated with interactions of dyes with circulating blood cells and plasma proteins.[56] In a study of the injection of TBs, which is broadly used for viability tests *in vitro*, we revealed rare notable PA signals *in vivo* which can be associated with rare, dead cells uptaking TB directly in bloodstream. Thus, in addition to previously demonstrated detection of circulating normal and apoptotic cells,[57] the potential to use PAFC for detection of circulating rare dead cells was demonstrated. This is important for many applications including studies of cell metabolism in normal and pathological states or response to various therapies. PAFC can provide also detection circulating microbubbles labeled with NPs (*e.g.* GNTs) as an ultrasound and PT contrast agents for imaging and nanothrombosis.[11] The typical clearance rate of microbubbles was in the range of 5–10 min.

2.2.9. Integrated PA and fluorescent FC

In vivo PAFC and FFC systems were previously used separately using preferentially positive contrast and a pulsed and continuous wave laser sources, respectively. Recently we introduced a multimodal *in vivo* flow cytometry (PAFFC) which integrates both techniques and can use positive and negative contrasts and pulsed lasers.[39] The supplementary nature of PA and fluorescence methods provided dramatic increases in the detectable range of

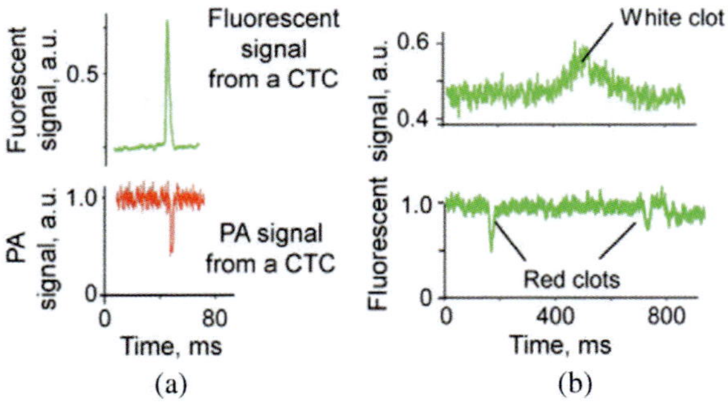

Fig. 13. Positive and negative contrasts in PA and fluorescence flow cytometry. (a) Negative PA, and positive fluorescent contrast signals from circulating C8161-GFP cells in 40-μm ear artery of the mouse. (b) *In vivo* fluorescent monitoring of white and red clots with positive and negative contrast, respectively. Laser excitation wavelength: 488 nm; intensity: 80 W/cm^2. Adapted with permission from Ref. 2. Copyright 2012 Elsevier.

absorbing and fluorescent contrast agents. These included liposomes loaded with Alexa-660 dye, RBCs labeled with ICG, B16F10 melanoma cells co-expressing melanin and green fluorescent protein (GFP), C8161-GFP melanoma cells targeted by functionalized MNPs, MTLn3 adenocarcinoma cells expressing novel NIR iRFP protein and quantum dot (QD)-CNT conjugates. The use of pulsed laser provided time-resolved discrimination of objects with long fluorescence lifetime (*e.g.* QDs) from shorter autofluorescence background (*e.g.* from blood plasma). Simultaneous PA and fluorescent detection of CTC-GFP (*e.g.* melanoma C8161-GFP) in tumor-bearing mice injection of MNPs functionalized with antibody specific to melanoma receptors was useful for verification of CTC targeting by NPs directly in the bloodstream.[39] Our data showed that novel FC platform using negative contrast can provide label-free detection of low absorbing or weakly fluorescent cells in strong background blood absorption and in an autofluorescence background, respectively. Fig. 13 illustrates detection of low-pigmented melanoma C8161-GFP cells, which simultaneously provided positive fluorescence and negative PA signals. These data demonstrated that negative contrast, label-free PAFC has the potential to detect single low absorbing cells with sizes down to 10–15 μm in 30–50 μm blood vessels (*e.g.* non-pigmented CTCs and likely WBCs). FFC also demonstrated the capability to use label-free detection of red clots with weekly fluorescent RBCs providing negative contrast in blood plasma autofluorescence and white clot with fluorescent WBCs, which provide positive contrasts above fluorescence background. Thus, the combination of the positive and negative contrasts can be used for PA label-free enumeration of RBC and WBC aggregates, respectively. On the contrary, in

FFC these cells and its aggregates produce label-free negative and positive contrast respectively. This can provide synergy in identification of these cells using integrated PAFFC.

3. Summary and Outlook

Recent technological advances in the PAFC platform showed its excellent promise for *in vivo* real-time detection of rare circulating cells and other abnormal features in biological fluids, such as blood, lymph and cerebrospinal fluid, with the overwhelming background material (*i.e.* billions of normal blood cells and millions of white blood cells in lymph) that has been never possible before with the existing techniques. *In vivo* PAFC allows for the interrogation of the entire (~5 L) circulation, which is expected to increase detection sensitivity by at least two if not three order of magnitude over current *ex vivo* approaches and offers a number of novel applications that may include label-free detection of various objects with intrinsic absorption or targeted by functionalized PA probes (*e.g.* NPs) in a variety of vessels such as arterioles, capillaries, veins and afferent and efferent lymphatics in assorted locations including ear, skin and deep organs for many disease models (*e.g.* cancer, infections or cardiovascular immune system disorders). Also, it would allow realizing real-time and quantitative assessments of pharmacokinetics of nano-objects *in vivo*. Furthermore, this platform is capable of integrating magnetic enrichment and PT therapy with a potential for multiplex targeting of diverse biomarkers using multicolor NPs. The PAFC nanotheranostic approach is also clinically relevant given that low toxicity NPs are used as contrast agents and diagnostic lasers are not only used at safe fluence levels, but also penetrate deeply in biological tissues (*e.g.* up to 3 cm through skin in NIR region, that is human jugular vein located at 1.5–2.0 cm).

Despite recent progress, there still is much room to improve for the realization of the *in vivo* nanotechnology-based PAFC theranostic platform and its translation into clinical practice. Continued exploration of contrast nanoagents must find spectrally distinct, non-toxic NPs that avoid opsonization and liver/spleen accumulation. Unique complexes of NPs, *e.g.* "customizable" NP nanocomposites, which respond to multiple modalities as well as deliver therapeutic agents or local hyperthermia are a particularly fertile area of exploration.[1,17] Engineering the sizes and structural configurations of nanocomposites offers promise to tailor plasmonic NPs.[17] Geometric factors and compositions affect the plasmonic properties, such as extinction and scattering spectra, of nanoscale materials. Thus, the assembly of plasmonic NPs of various sizes and shapes into desired patterns and geometries could achieve enhanced optical response over several wavelength bands. With nanocomposites of finely tuned plasmonic spectra, multispectral, multi-color imaging could be realized. In addition, combining the useful functionalities of different NPs in a single nanostructure would open another significant possibility of multimodal theranostics, for example, using a novel anisotropic mono-functionalization strategy to place different number of biological linkers, such as DNA or RNA, on a NP at specific angles to each other and to synthesize NP building blocks with their well-defined arrangements in all three dimension srecently reported by Kim *et al.*[17,24,25] The successful merger of desirable functionalities of various NPs in a controlled manner should enable us to realize highly sensitive, multiplex, multicolor nanoscale nanotheranostics for circulating tumor cells and pathological biomolecules with multiple complementary modalities beyond the inherent limitations of individual existing technologies.[17] Furthermore, ultrasharp PA/PT resonances in plasmonic NPs with a spectral width down to 0.8 nm (*i.e.* 100-fold narrower than linear absorption spectra) was recently discovered.[58,59] This would offer a promising way to overcome the challenges related to the relatively broad absorption bands of PA/PT contrast agents (*e.g.* 60–150 nm for GNRs or GNSs), which limit their multicolor capacity (*i.e.* restrict them to only two simultaneous colors).[1,2,11,14,58,59] Separately, the success of *in vivo* PAFC nanotheranostics for circulating abnormal features will hinge upon the specificity. More specific biomarkers are needed to overcome not only high background noise in highly complex *in vivo* system but also inter- and intra-patient heterogeneity. Also, given the heterogeneous nature of circulating pathogenic cells, a single marker will not be sufficient to provide a complete account of those cells. The development of multifunctional contrast agents would help overcome this hurdle by enabling multicolor and multimodal capability.

Although challenges are ahead, the presented innovative *in vivo* PAFC technology would enable a paradigm shift in medicine from early diagnostics and treatment of life-threatening diseases, including cancers and infections, to effective clinical management of individual patients, ultimately decreasing mortality. It would also solve very complex and largely unexplored areas of medicine related to *in vivo* theranostics of infections and stem, dendritic and metastatic cells in different functional states (*e.g.* apoptosis) in lymph and blood flow at the single cell level. As the technology progresses, appealing new directions would arise that could transform the field of nanotheranostics beyond those described in this chapter.

References

1. J.-W. Kim, E. I. Galanzha, D. Zaharoff, R. J. Griffin and V. P. Zharov, Nanotheranostics of circulating tumor cells, infections and other pathological features *in vivo*. *Mol. Pharmaceutics* **10**, 813–830 (2013).
2. E. I. Galanzha and V. P. Zharov, Photoacoustic flow cytometry. *Methods* **57**, 280–296 (2012).
3. C. L. Chaffer and R. A. Weinberg, A perspective on cancer cell metastasis. *Science* **331**, 1559–1564 (2011).
4. E. I. Galanzha, E. V. Shashkov, M. Sarimollaoglu, K. E. Beenken, A. G. Basnakian, M. E. Shirtliff, J.-W. Kim, M. S. Smeltzer and V. P. Zharov, *In vivo* magnetic enrichment, photoacoustic diagnosis and photothermal purging of infected blood using multifunctional gold and magnetic nanoparticles. *PLoS One* 7, e45557 (2012).
5. K. Pantel and C. Alix-Panabieres, Circulating tumour cells in cancer patients: challenges and perspectives. *Trends Mol. Med.* **16**, 398–406 (2010).
6. M. Yu, S. Stott, M. Toner, S. Maheswaran and D. A. Haber, Circulating tumor cells: approaches to isolation and characterization. *J. Cell Biol.* **192**, 373–382 (2011).
7. C. Alix-Panabieres, H. Schwarzenbach and K. Pantel, Circulating tumor cells and circulating tumor DNA. *Ann. Rev. Med.* **63**, 199–215 (2012).
8. P. Paterlini-Brechot and N. L. Benali, Circulating tumor cells (CTC) detection: clinical impact and future directions. *Cancer Lett.* **253**, 180–204 (2007).
9. S. Maheswaran and D. A. Haber, Circulating tumor cells: a window into cancer biology and metastasis. *Curr. Opin. Genet. Dev.* **20**, 96–99 (2010).
10. X. Zhe, M. L. Cher and R. D. Bonfil, Circulating tumor cells: finding the needle in the haystack. *Am. J. Cancer Res.* **1**, 740–751 (2011).
11. A. de la Zerda, J.-W. Kim, E. I. Galanzha, S. S. Gambhir and V. P. Zharov, Advanced contrast nanoagents for photoacoustic molecular imaging, cytometry, blood test and photothermal theranostics. *Contrast Media Mol. Imaging* **6**, 346–369 (2011).
12. A. D. Hughes and M. R. King, Nanobiotechnology for the capture and manipulation of circulating tumor cells. *WIREs Nanomed. Nanobiotechnol.* **4**, 291–309 (2012).
13. J.-W. Kim, E. I. Galanzha, E. V. Shashkov, H.-M. Moon and V. P. Zharov, Golden carbon nanotubes as multimodal photoacoustic and photothermal high-contrast molecular agents. *Nat. Nanotechnol.* **4**, 688–694 (2009).
14. E. I. Galanzha, E. V. Shashkov, T. Kelly, J.-W. Kim, L. Yang and V. P. Zharov *In vivo* magnetic enrichment and multiplex photoacoustic detection of circulating tumour cells. *Nat. Nanotechnol.* **4**, 855–860 (2009).
15. E. I. Galanzha, M. S. Kokoska, E. V. Shashkov, J.-W. Kim, V. V. Tuchin and V. P. Zharov, *In vivo* fiber-based multicolor photoacoustic detection and photothermal purging of metastasis in sentinel lymph nodes targeted by nanoparticles. *J. Biophotonics* **2**, 528–539 (2009).
16. E. I. Galanzha, J.-W. Kim and V. P. Zharov, Nanotechnology-based molecular photoacoustic and photothermal flow cytometry platform for *in-vivo* detection and killing of circulating cancer stem cells. *J. Biophotonics* **2**, 725–735 (2009).
17. J.-W. Kim and R. Deaton, Molecular self-assembly of multifunctional nanoparticle composites with arbitrary shapes and functions: challenges and strategies. *Part Part Syst. Charact.* **30**, 117–132 (2009).

18. V. P. Zharov, J.-W. Kim, D. T. Curiel and M. Everts, Self-assembling nanoclusters in living systems: application for integrated photothermal nanodiagnostics and nanotherapy. *Nanomedicine* **1**, 326–345 (2005).
19. S. Nie, Y. Xing, G. J. Kim and J.-W. Simons, Nanotechnology applications in cancer. *Ann. Rev. Biomed. Eng.* **9**, 257–288 (2007).
20. M. De, P. S. Ghosh and V. M. Rotello, Applications of Nanoparticles in Biology. *Adv. Mater.* **20**, 4225–4241 (2008).
21. S. E. Lee and L. P. Lee, Biomolecular plasmonics for quantitative biology and nanomedicine. *Curr. Opin. Biotechnol.* **21**, 489–497 (2012).
22. D. A. Giljohann, D. S. Seferos, W. L. Daniel, M. D. Massich, P. C. Patel and C. A. Mirkin, Gold nanoparticles for biology and medicine. *Angew. Chem. Int. Ed. Engl.* **49**, 3280–3294 (2010).
23. S. Lal, S. E. Clare and N. J. Halas, Nanoshell-enabled photothermal cancer therapy: impending clinical impact. *Acc. Chem. Res.* **41**, 1842–1851 (2008).
24. J.-W. Kim, J. H. Kim and R. Deaton, DNA-linked nanoparticle building blocks for programmable matter. *Angew. Chem. Int. Ed. Engl.* **50**, 9185–9190 (2011).
25. J.-W. Kim, J. H. Kim and R. Deaton, Programmable construction of nanostructures: assembly of nanostructures with various nanocomponents. *IEEE Nanotechnol.* **6**(1), 19–23 (2012).
26. V. P. Zharov, E. I. Galanzha and V. V. Tuchin, Photothermal imaging of moving cells in lymph and blood flow *in vivo*, *Proc. SPIE.* **5320,** 185, doi:10.1117/12.532107 (2004).
27. W. He, H. Wang, L. C. Hartmann, J. X. Cheng and P. S. Low, *In vivo* quantitation of rare circulating tumor cells by multiphoton intravital flow cytometry. *Proc. Natl. Acad. Sci. USA.* **104**, 11760–11765 (2007).
28. I. Georgakoudi, N. Solban, J. Novak, W. L. Rice, X. Wei, T. Hasan and C. P. Lin, *In vivo* flow cytometry: a new method for enumerating circulating cancer cells. *Cancer. Res.* **64**, 5044–5047 (2004).
29. H. M. Shapiro, *Practical Flow Cytometry*, 4th edn. New York: Wiley (2003).
30. U. Sack, A. Tárnok and G. Rothe (Eds.) *Cellular diagnostics: basic principles, methods and clinical applications of flow cytometry.* Karger, Basel, Freiburg, Paris (2008).
31. V. V. Tuchin, E. I. Galanzha and V. P. Zharov, *In vivo* photothermal and photoacoustic flow cytometry in *Advanced Optical Flow Cytometry: Methods and Disease Diagnoses*, V. V. Tuchin, (Ed.). KGaA, Weinheim, Germany Wiley-VCH Verlag GmbH & Co. doi: 10.1002/9783527634286.ch17 (2011).
32. A. G. Bell, On the production and reproduction of sound by light. *Am. J. Sci.* **20**, 305–324 (1880).
33. V. P. Zharov, V. S. Letokhov, *Laser Optoacoustic Spectroscopy.* New York: Springer-Verlag (1986).
34. A. A. Oraevsky and A. A. Karabutov, Optoacoustic tomography, in *Handbook of Biomedical Photonics,* (*Ed.*). T. Vo-Dinh (Florida: CRC Press 2003).
35. L. V. Wang, Multiscalephotoacoustic microscopy and computed tomography. *Nat. Photon.* **3**, 503–509 (2009).
36. D. Rasansky, M. Distel, C. Vinegoni, R. Ma, N. Perrimon, R. W. Köster, V. Ntziachristos, Multispectral opto-acoustic tomography of deep-seated fluorescent proteins *in vivo. Nat. Photon.* **3,** 412–417 (2009).

37. S. Y. Emelianov, P. C. Li, M. O'Donnell, Photoacoustics for molecular imaging and therapy. *Phys. Today* **62**(5), 34–39 (2009).
38. P. C. Beard, Biomedical photoacoustic imaging. *Interface Focus* **1**, 602–631 (2011).
39. D. A. Nedosekin, M. Sarimollaoglu, E. I. Galanzha, R. Sawant, V. P. Torchilin, V. V. Verkhusha, J. Ma, M. H. Frank, A. S. Biris and V. P. Zharov, Synergy of photoacoustic and fluorescence flow cytometry of circulating cells with negative and positive contrasts. *J. Biophoton.* doi: 10.1002/jbio.201200047 (*in press*) (2013).
40. E. I. Galanzha, M. Sarimollaoglu, D. A. Nedosekin, S. G. Keyrouz, J. L. Mehta and V. P. Zharov, *In vivo* flow cytometry of circulating clots using negative photothermal and photoacoustic contrasts. *Cytometry A* **79**, 814–824 (2011).
41. E. I. Galanzha, E. V. Shashkov, P. M. Spring, J. Y. Suen and V. P. Zharov, *In vivo*, non-invasive, label-free detection and eradication of circulating metastatic melanoma cells using two-color photoacoustic flow cytometry with a diode laser. *Cancer Res.* **69**, 7926–7934 (2009).
42. V. P. Zharov, E. I. Galanzha, E. V. Shashkov, N. G. Khlebtsov and V. V. Tuchin, *In vivo* photoacoustic flow cytometry for monitoring of circulating single cancer cells and contrast agents. *Opt. Lett.* **31**, 3623–3625 (2006).
43. D. A. Nedosekin, E. I. Galanzha, S. Ayyadevara, R. J. Shmookler Reis and V. P. Zharov, Photothermal confocal spectromicroscopy of multiple cellular chromophores and fluorophores. *Biophys. J.* **102**, 672–681 (2012).
44. D. A. Nedosekin, M. Sarimollaoglu, J. H. Ye, E. I. Galanzha and V. P. Zharov, *In vivo* ultra-fast photoacoustic flow cytometry of circulating human melanoma cells using near-infrared high-pulse rate lasers. *Cytometry, Part A* **79**, 825–833 (2011).
45. E. I. Galanzha, Blood and lymph circulating cells: well-known systems, well-forgotten interdependence. *J. Blood Lymph.* **1**, 1–2 (2011).
46. C. L. Weston, M. J. Glantz and J. R. Connor, Detection of cancer cells in the cerebrospinal fluid: current methods and future directions. *Fluids Barriers CNS* **8**, 14, doi: 10.1186/2045-8118-8-14 (2011).
47. A. S. Patel, J. E. Allen, D. T. Dicker, K. L. Peters, J. M. Sheehan, M. J. Glantz, W. S. El-Deiry, Identification and enumeration of circulating tumor cells in the cerebrospinal fluid of breast cancer patients with central nervous system metastases. *Oncotarget* **2**, 752–760 (2011).
48. V. P. Zharov, E. I. Galanzha, E. V. Shashkov, J.-W. Kim, N. G. Khlebtsov and V. V. Tuchin, Photoacoustic flow cytometry: principle and application for real-time detection of circulating single nanoparticles, pathogens and contrast dyes *in vivo*. *J. Biomed. Opt.* **12**, 051503 (2007).
49. J.-W. Kim, E. V. Shashkov, E. I. Galanzha, N. Kotagiri and V. P. Zharov, Photothermal antimicrobial nanotherapy and nanodiagnostics with self-assembling carbon nanotube clusters. *Lasers Surg. Med.* **39**, 622–634 (2007).
50. V. P. Zharov, K. E. Mercer, E. N. Galitovskaya and M. S. Smeltzer, Photothermal nanotherapeutics and nanodiagnostics for selective killing of bacteria targeted with gold nanoparticles. *Biophys. J.* **90**, 619–627 (2006).
51. E. I. Galanzha and V. P. Zharov, *In vivo* photoacoustic and photothermal cytometry for monitoring multiple blood rheology parameters. *Cytometry A* **79**, 746–757 (2011).
52. M. Sarimollaoglu, D. A. Nedosekin, Y. Simanovsky, E. I. Galanzha and V. P. Zharov, *In vivo* photoacoustic time-of-flight velocity measurement of single cells and nanoparticles.*Opt. Lett.* **36**, 4086–4088 (2011).

53. V. P. Zharov, E. I. Galanzha, E. V. Shashkov, J.-W. Kim, N. G. Khlebtsov and V. V. Tuchin, Photoacoustic flow cytometry: principle and application for real-time detection of circulating single nanoparticles, pathogens and contrast dyes *in vivo*. *J. Biomed. Opt.* **12**, 051503 (2007).
54. A. S. Biris, E. I. Galanzha, Z. Li, M. Mahmood, Y. Xu and V. P. Zharov, *In vivo* raman flow cytometry for real-time detection of carbon nanotube kinetics in lymph, blood and tissues. *J. Biomed. Opt.* **14**, 021006 (2009).
55. D. A. Nedosekin, M. Sarimollaoglu, E. V. Shashkov, E. I. Galanzha and V. P. Zharov, Ultra-fast photoacoustic flow cytometry with a 0.5 MHz pulse repetition rate nanosecond laser. *Opt. Exp.* **18**, 8605–8620 (2010).
56. M. Proskurnin, T. V. Zhidkova, D. S. Volkov, M. Sarimollaoglu, E. I. Galanzha, D. Mock and V. P. Zharov, *In vivo* multispectral photoacoustic and photothermal flow cytometry with multicolor dyes: a potential for real-time assessment of circulation, dye-cell interaction and blood volume.*Cytometry A* **79**A, 834–884 (2011).
57. E. I. Galanzha, E. V. Shashkov, V. V. Tuchin and V. P. Zharov, *In vivo* multispectral, multiparameter, photoacoustic lymph flow cytometry with natural cell focusing, label-free detection and multicolor nanoparticle probes. *Cytometry A* **73**A, 884–894 (2008).
58. V. P. Zharov, Ultrasharp non-linear photothermal and photoacoustic resonances and holes beyond the spectral limit. *Nat. Photon.* 5, 110–116 (2011).
59. J. Shao, R. J. Griffin, E. I. Galanzha, J.-W. Kim, N. Koonce, J. Webber, T. Mustafa, A. Biris, D. A. Nedosekin and V. P. Zharov, Photothermal nanodrugs: potential of TNF-gold nanospheres for cancer theranostics. *Sci. Rep.* **3**, 1293. doi:10.1038/srep01293 (2013).

Chapter 15

Raman Micro-Spectroscopy as a Non-Invasive Tool to Follow the Intracellular Fate of Nanoparticles

Tatyana Chernenko[*,§], Max Diem[†,¶] and Vladimir Torchilin[‡,‖]*

[*] Department of Pathology and Laboratory Medicine
University of California— Davis, Sacramento, CA 95817, USA
[†] Department of Chemistry and Chemical Biology
Northeastern University, Massachusetts 02115, USA
[‡] Department of Pharmaceutical Sciences, School Pharmacy
Northeastern University, Massachusetts 02115, USA
[§] taty.chernenko@gmail.com
[¶] m.diem@neu.edu
[‖] v.torchilin@neu.edu

1. Introduction

Nanotechnology is on the verge of integration into everyday therapeutic strategies, since nanoscale intervention capabilities are vital for eradication of disease at its molecular core. Pharmacological aspects of nanomedicine are concerned with improving the therapeutic effect of existing chemotherapeutics with specialized delivery systems, or vehicles. Such systems are needed to solubilize hydrophobic agents, and/or increase the concentration of the therapeutic payload at the tumor interstitial. This occurs if the delivery system can avoid recognition by the immune system, bypass excretion from the body, and remain circulating within the blood for sustained amount of time to allow population of the tissue of interest. The physical characteristics of the designed nanocarrier thus must be optimized with respect to the toxicity levels. Once the nanoparticle has reached the tissue, it must traverse the cell and subsequently avoid intracellular degradation, in order to deliver the drug to the site-of-action.[1–5]

Hence, optimization of the materials implemented for nanocarrier preparation is of vital consequence. Various materials have been under scrutiny for their integration into the nanopharmaceutics. Such materials include inorganics — such as gold or silver colloids — used as beacons, as well as various organic structures, such as phospholipids. The phospholipid structures are being exploited for their biocompatible nature as building blocks of various nanostructures such as nanospheres, nanocapsules, micelles and liposomes. For this reason they have been applied to serve as drug carriers, allowing for an enhanced delivery of loaded drugs compared to free drugs. Nano-encapsulation of drugs also prevents premature degradation and excretion as well as circumvents non-specific toxicity.

Efficacy and specificity of various nanodelivery systems can be further enhanced with various modifications with cationic compounds, which augment their internalization and may even increase affinity for specific organelles, depending on the modification type. Drug-loaded pharmaceutical nanocarriers can be delivered into diseased sites by passive targeting (spontaneous accumulation of nanocarriers in the areas with affected vasculature) or by active targeting (via site-specific ligands attached to the surface of drug-loaded nanocarriers). Subsequent level of targeting requires cellular internalization of nanocarriers and their specific association with certain individual cell organelles. Among those organelles of interest, mitochondria may be targeted for initiating apoptosis in target cells, to a chemotherapeutic advantage.

Enhancement of the internalization of the carriers was achieved by modifying their surface with various cationic molecules.[6–10] One such cationic modification was with a TAT-peptide (TATp). This peptide is a small moiety of 11-mer peptide sequence taken from an Human Immunodeficiency virus (HIV), which uses this protein sequence to aid in cellular invasion. Such positively charged peptide sequences are termed "cell-penetrating-peptides," and we have exploited their properties in order to augment liposome internalization. Other cationic modifications, which are less controversial, are small cationic molecules, which can be covalently attached to the phospholipid head. We have analyzed three such cationic systems: 1,2-dioleoyl-3-trimethylammonium-propane (DOTAP)-, stearyl amine (SA)- and stearyltriphenylphosphonium (STPP)-modified liposomes. Out of the three cationic modification units, STPP was shown to accumulate specifically within the mitochondria-rich regions inside cells.[11,12]

Accurate visualization of the nano-assemblies and their fate *in situ* would yield information as to the cellular behaviors and responses, which is of consequence in optimization of carriers for targeted delivery of a chemotherapeutic. However, the widely implemented tools in cellular imaging include fluorescence microscopy, a modality which is confounded with potential barriers, such as the use of exogenous labels and tags, which are used to illuminate systems of interest, inadvertently potentially compromising the specimen. In order to circumvent these problems, we employ vibrational (Raman) micro-spectroscopy coupled with optical microscopy to provide a non-invasive and non-destructive method for cellular imaging. The technique is based upon the identification of molecular vibrations that are characteristic of different functional groups contained within molecules.

The control over intracellular distribution of pharmaceutical nanocarriers requires effective and non-perturbing methods of their visualization inside cells. Raman, micro-spectroscopy offers such a capability, allowing for the identification of molecular fingerprints of the functional groups within any biological species. For cells, such fingerprints arise from the functional groups contained within proteins, carbohydrates, nucleic acids, lipids and phospholipids, among others, which constitute basic building blocks of the cell. Vibrational spectra may be obtained by illuminating the specimen either with broadband infrared radiation and monitoring its absorption, or with narrowband visible light from a laser and detecting scattered photons. The latter approach detects inelastically scattered photons and is known as the Raman effect.

Raman micro-spectroscopy affords minimal sample preparation, high sensitivity to small intracellular fluctuations as well as high spatial resolution. Raman spectra from within the cell reflect the biochemical composition found within the laser focal volume, which is determined by the diffraction limit of the applied laser light. Using the spectral parameters of a cell's components, it is possible to image sub-cellular organelles at the resolution of conventional microscopy.

Cellular spectra exhibit features originating from peptides, DNA/RNA, lipids, etc., within the lower wavenumber spectral region, termed "fingerprint region," (750–1800 cm^{-1}) as well as the C–H stretching region (2800–3100 cm^{-1}) of the spectrum.[11–13] Different cells have distinct functions within the body. Subsequently, their intracellular environment must be biochemically unique to perform the inherent functions. Thus by using vibrational spectroscopy, we are able to detect those changes, since a spectrum mirrors cellular biochemistry in form of vibrational bands which shift or increase/decrease in intensity.[14,16–20] Raman micro-spectroscopy can classify diverse cellular types within the body based on their inherent biochemistry. This chemical information is manifest in the entire spectral region, which is a vast superposition of lipids, phospholipids, proteins, nucleic acids and many other functional groups.

An introduction of an exogenous moiety, in the form of a nanostructure, into the cell contributes a changes in the spectral signatures of the chemical micro-environments, which are distinguishable with Raman micro-spectroscopy.[21–23] However, the phospholipid-based biocompatible nanoparticle employed — such as liposomes or micelles, which are comprised of phospholipids — would exhibit vibrational peaks overlapping with signal originating from various intracellular compartments. Hence, in order to increase the sensitivity to the phospholipid-based nanoparticle system within the intracellular environment, we exploit the properties of isotopic (deuterium) labeling to produce vibrational signals that do not occur in the undeuterated compounds.

Subsequently, biochemical contrast is enhanced by employing various image reconstruction algorithms which are based on spectral unmixing or clustering methods. Since any one particular Raman cellular map is comprised of more than 10,000 spectra, visual analysis of the entire data-cube is rendered virtually impossible without the implementation of statistical methods of data analysis. Such mathematical methods using the entire spectrum, as opposed to visualizing the data based on intensity of a certain peak (as is done in fluorescence imaging) or spectral region, are referred to as "multivariate" methods, such as Vertex Component Analysis (VCA).[25,26]

2. Materials and Methods

2.1. *Raman data acquisition*

The cellular images presented in this paper were recorded using a CRM200 confocal Raman microscope (WITec GmbH, Ulm, Germany) employing 488 nm radiation at ~5 mW power at sample). Excitation is provided by a solid state laser (Spectral Physics model "Cyan"). A water immersion objective (60x/NA 1.00, WD 2.0 mm), is used to focus the radiation onto the sample, which results in a diffraction limited excitation spot. The exciting laser radiation was coupled into a Zeiss microscope through a wavelength-specific single-mode optical fiber. The incident laser beam was collimated via an achromatic lens and passed a holographic band-pass filter before being focused onto the sample through the microscope objective.

The sample was located on a piezo-electrically driven microscope scanning stage with an x,y resolution of ca. 3 nm and a repeatability of ±5 nm and z resolution of ca. 0.3 nm and ±2 nm repeatability. The sample was scanned through the laser focus in a raster pattern at a constant stage speed of fractions of a micrometer per second. Data were acquired corresponding to a step size of 500 nm at 250 ms integration time. The continuous motion prevents sample degradation at the focal point of the laser beam.

Raman backscattered radiation was collected through the microscope objective and passed through a holographic edge filter to block Rayleigh scattering and reflected laser light before being focused into a multimode optical fiber. The 50 μm diameter multimode output fiber provided the optical apertures for the confocal measurement. The light emerging from the output optical fiber was dispersed by a 30 cm focal length, f/4 CzernyTurner monochromator, incorporating interchangeable gratings (1800/mm, blazed at 500 nm and 600/mm, blazed at 500 nm). The light was finally detected by a back-illuminated deep-depletion, 1024 × 128 pixel charge-coupled device camera operating at −82°C. The analysis methods for the hyperspectral datasets will be discussed below.

2.2. *Data analysis: Raman micro-spectroscopy for biochemical functional cellular maps*

An optimal imaging modality, which offers unbiased and uncompromised view of the biological system of interested, is an ideal candidate for analysis. These requirements are fulfilled with the use of a Raman micro-spectroscopy. Vibrational spectroscopy has the capability to image the sample based on its inherent biochemical fingerprint, consequently bypassing extensive staining procedures. The technique is based on the identification of molecular vibrations that are characteristic of distinct functional groups comprised within all molecules. For cells and tissues, these vibrational fingerprints arise from the functional groups of proteins, nucleic acids, lipids, phospholipids and carbohydrates — the basic building blocks of mammalian cells. Such a biochemical map can provide information regarding a samples' chemical composition, protein conformation and various sub-cellular metabolic activities, which are all reflected in the spectral features of cellular compartments.

In Raman spectroscopy a very small volume, or a voxel, is sampled at one point in time, ca. $0.3 \times 0.3 \times 1.3\ \mu m^3$ (or 10^{-12} g) in size. The spectral information thus acquired within the diffraction-limited volume, reflects the biochemical composition of the particular volume sampled. Our particular instrument setup also has the potential to image a full volume of a specimen, rendering a three-dimensional biochemical map of the sample. This is achieved with a creation of pinholes with the use of optical fibers focusing on the sample and subsequently the detector. The setup thus allows us to interrogate the sample on any focal plane, analyzing intracellular and intercellular responses to any given micro-environment.

Within the aforementioned sampling volume, however, minute biochemical variations are overwhelming with information regarding protein, lipids, phospholipids, as well as other biological entities within the intracellular environment. For example, if a voxel spectrum recorded in the nucleus of a cell is scrutinized, the observed spectrum describes contributions due to nucleic acids and proteins; if the voxel is in the perinuclear region, the protein signal is overlaid with signals originating from membrane-rich organelles, such as the endoplasmic reticulum, Golgi or mitochondria. Due to the complex biochemical information contained within one Raman spectrum, interpixel variations within the sample are difficult to discriminate without the use of multivariate methods of analysis.

Work with phospholipid-based nanostructures required two different human cell lines, human HeLa cells (cell line CCL-2, ATCC, Manassas, VA) and an MCF-7 cell line (cell line HTB-22, ATCC, Manassas, VA). Both were grown in 75 cm^2 culture flasks (Fisher Scientific) with 7 ml of growth media (either Dulbecco's Modified Eagle's medium (DMEM, ATCC) for Human HeLa or Eagle's Minimum Essentials Medium (MEM, ATCC), for MCF-7 line) and supplemented with 10% fetal bovine serum (FBS, ATCC). Cells were incubated at 37°C and 5% CO_2. Cells were grown on CaF_2 windows to avoid background signals from the regular glass slides.

Prior to Raman biochemical analysis, cells were fixed in 10% formalin (Sigma-Aldrich, St. Louis, MO) and subsequently submerged into the phosphate-buffered saline. Fixation of the cellular samples was carried out primarily due to required data acquisition times for hyperspectral datasets (with a 0.25 s/pixel dwell time and upward of 10,000 pixels/cell, yields an estimated 30 min to 1 h data accumulation time/cell).

An example of three extracted spectra from three intracellular regions is portrayed in Fig. 1, namely those of the nucleus, nucleolus, as well as a point within the cytoplasm. The most prominent spectral bands in such a biological spectrum are the C–H stretching modes of saturated and unsaturated aliphatic chains (2800–3020 cm^{-1}). The remaining peaks in the 500 cm^{-1} to 1800 cm^{-1}, or fingerprint region, are of smaller intensity.

The cellular image in Fig. 1 shows the topography associated with the higher and lower cellular densities when comparing spectral intensities of the C–H stretching region from each spectrum collected from the entire cell. Here, brighter yellow areas denote larger intensities. This method of data representation is better known as univariate mapping or imaging, since it uses a single intensity variant from a recorded spectrum, whether it is peak height, peak height ratio, peak area or peak area ratio, to describe the abundance of a chemical moiety and hence the biomolecule. This is due to the fact that the

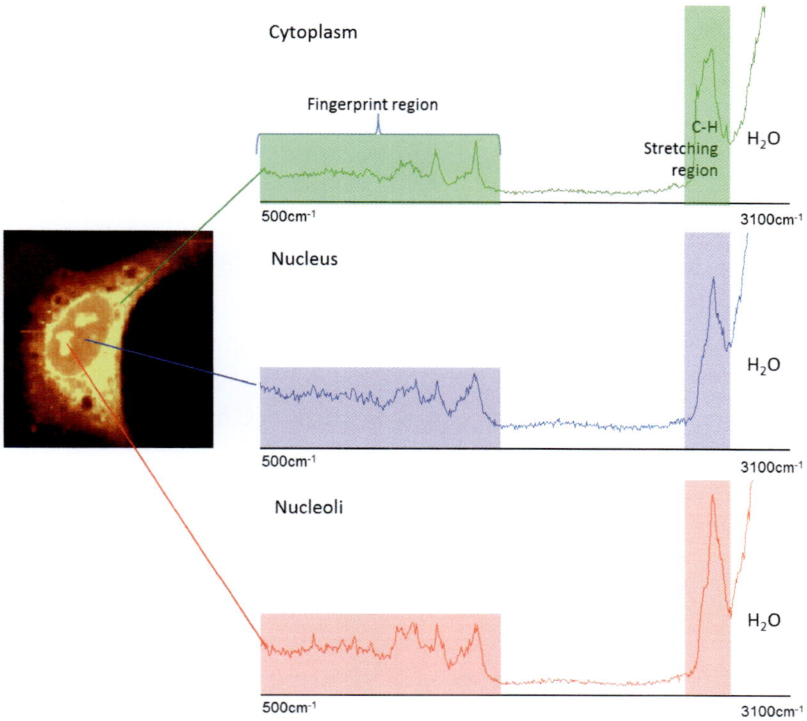

Fig. 1. Representative univariate intensity image of a HeLa cell, which has been fixed and imaged *in situ*. The image was generated by plotting the integrated band intensities of the C–H stretching region (2800–3020 cm⁻1) versus the coordinates spectra were recorded from. The three spectra displayed were randomly selected from pixels inside the nucleolus, nucleus and cytoplasm. Two spectral regions are highlighted, the fingerprint region and the C–H stretching region.

peak intensity of a vibrational band can be directly related to the concentration of the entity giving rise to that particular vibration. Thus differences due to sample density, protein or water content are mostly reflected when calculating univariate maps using the C–H stretching region of the Raman spectrum.

Gross biochemical changes may thus become manifest in the spectral signatures within the acquired cellular dataset. Visual analysis of the three spectra shows distinct variations within the fingerprint region as well as within the C–H stretching region between the three biochemically-distinct sub-cellular environments. Notice, however, that in the cellular image that the perinuclear region exhibits undistinguishable intensities from the intensities of the nucleoli, which underscores the limitation of univariate imaging. Thus it would be beneficial to use statistical means of data decomposition, which can provide chemical contrast between all the different cellular organelles or environments.

2.3. *Multivariate data analysis: VCA*

All cellular Raman images were analyzed using multivariate color decomposition — VCA.[21] It decomposes the spectrum of each pixel into contributions from the most dissimilar spectra of the dataset, which are considered the "pure component" or "endmember" spectra. Each endmember spectrum has a corresponding abundance or intensity variation associated with a region within the cellular architecture. In other words, each Raman spectrum can be represented as a vector defined by intensity measurements of channels or frequencies. The Raman-collected spectra discussed here are represented by VCA in a 1024-dimensional space, since there are 1024 frequency points, which comprise the spectrum. VCA then proceeds with a step-by-step reduction of the dimensionality of the system via a simplex-method, until we get a predetermined geometrical shape, such as a tetrahedron, which would yield us four spectral endmembers. Thus, from an n-simplex, VCA reduces the problem to $(n - 1)$-simplex after a specific amount of iterations.

If more than one mixed spectrum is considered and the spectra are identical, they will have identical vectors with the same orientation. Spectra, which have same composition but different total intensities, will have different length superimposed vectors. If we plot a multitude of data points collected at two specified frequencies, we get a two-dimensional plane with discrete spectra as endpoints of a line. If three component spectra are needed to describe the material, they will be the corners of a triangle in a two-dimensional space, etc. These endpoint spectra represent the closest to being pure component spectra and are termed "endmembers".

The rendered endmembers and their abundance vectors are scaled and converted to monochrome intensity images of the specimen of interest. Ultimately three such abundance maps are overlaid and converted to a red, green and blue (RGB) map, showing the intensity of each of the rendered endmembers. The intensities depict the extent of spectral contribution of the endmember spectrum to each of the pixels within the dataset. An overlay of these RGB color maps yield a pseudo-color map with mixed colors indicating the intensity values of each of the three endmembers.

Biochemical contrast may thus be achieved with the aid of VCA, as is shown in Fig. 2. Figure 2(b) is an image of the cell shown previously in Fig. 1, subsequently analyzed using the VCA algorithm. The multivariate spectral unmixing method easily discerns the nucleus from nucleoli as well as signal from the perinuclear environment and the rest of the cytoplasm. Nucleus and nucleoli are shown in gradations of dark and bright orange,

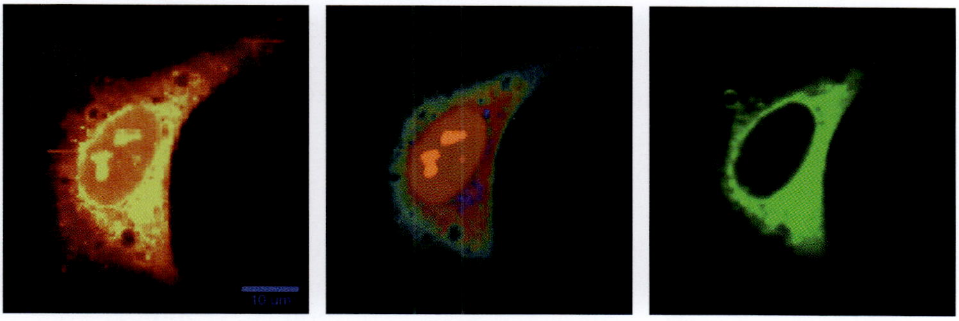

(a) (b) (c)

Fig. 2. Univariate image of a HeLa cell, same as in Fig. 1. (b) VCA-analyzed image of the cell, showing nucleus (brown), nucleolus (bright brown/orange), as well as gradation of the cytoplasm. Mitochonrdia aggregations are seen in gradations of blue and dark brown in the perinuclear region. (c) is the same cell stained with MitoTracker® post Raman acquisition.

while the cytoplasm is pseudo-colored purple, brown, blue and green. The pseudo-colored dark purple and dark brown gradations within the cytoplasm are correlating with the high mitochondrial content of the perinuclear region. This is further corroborated with the MitoTracker Green® stain for mitochondrial content, shown in Fig. 2(c).

3. Non-Invasive Monitoring of Intracellular Liposomal Dynamics

Traditional ways of drug administration result in the molecule's entry via an endocytic pathway, being trapped in the endosome, and ultimately degraded inside the lysosome. In order to circumvent the degradation pathway diverse nanocarrier systems have been developed to encapsulate and deliver the vital drug or DNA to its intracellular destination by use of a transduction mechanism. Several nanoparticulates, such as nanospheres, nanocapsules, micelles, liposomes and dendrimers have been applied to serve as drug carriers, allowing for an enhanced delivery of loaded drugs compared to free drugs. Nanoencapsulation of drugs also prevents premature degradation and excretion as well as circumvents non-specific toxicity.[27–30]

Increased interest in the intracellular behaviors, dynamics and side-effects of the various biocompatible nanodelivery systems has enabled a slew of research to be conducted and optimized for visualization of these nanostructures. Here, we are reporting on implementation of Raman microspectral imaging as well as isotopic labeling to monitor and record dynamics of biocompatible nanoparticulates.

Due to an inherent drawback of Raman spectroscopy as a weak effect, we have exploited the properties of deuterium labeling, which is non-destructive to the labeled molecule and non-invasive to the cell. For liposomal preparation we have utilized a deuterated phospholipid, which has covalently labeled fatty acid side chains. This molecular structure exhibits strong carbon-deuterium (C–D) vibrational intensities in the 2050–2300 cm^{-1} spectral region, which is commonly barren of any peak intensities from biological samples. The internalized nanoparticles form a dense material, allowing the detection of their aggregations even in the small focal volume of ca. 300 nm × 300 nm × 1 μm.

Liposomes were prepared from 1,2-Dipalmitoyl-*d62-sn*-glycero-3-phosphocholine DPPC-*d*62 and cholesterol (85:15 mol%) by the lipid film hydration method. (DPPC-*d*62), Purified TATp was received from Tufts University Core Facility (Boston, MA). For preparing TATp-liposomes, TATp-PEG3400-PE was added to the lipid film. The hydration and sizing of TATp-liposomes were done as described above. TATp-PEG3400-PE was prepared by reacting TATp with p-nitrophenyl carbonyl-PEG3400-DOPE (pNP-PEG-PE). Briefly, pNPPEG-PE film was hydrated with a citrate buffer, pH 5.0 (10 mM sodium citrate, 150 mM sodium chloride), at 0.95 mM concentration. TATp was dissolved in a borate buffer, pH 9.2 (100 mM sodium borate, 140 mM NaCl), and mixed at 1.7 molar excess of pNP-PEG-PE. The mixture was incubated with stirring overnight at room temperature, followed by further incubation for a day at 4°C. During the incubation, TATp gets attached to the distal ends of pNP-PEG-PE yielding loose micelles of TATp-PEG-PE. TATp-PEG-PE was purified by dialysis against PBS, pH 8.0 (dialysis tubing cut-off size of 12,000 Da) for a day and then against water for two days at 4°C. TATp-PEG-PE was dried under a freeze-dryer and dissolved in chloroform for storage at −80°C until use. 1,2-dioleoyl-3-trimethylammonium-propane (DOTAP) were obtained from Avanti Polar Lipids, Inc (Alabaster, AL). The dried lipid film prepared from the chloroform solution of DPPC-*d*62 and cholesterol was hydrated and vortexed with HEPES buffered saline (HBS), pH 7.4, at final lipid concentration of 5 mg/ml. The resulting liposomes were sized through double stacked 200 nm pore size polycarbonate membranes (Nucleopore) using a hand-held extruder (Avanti). When required, 1.5mol % of stearyl amine (SA), or stearyltriphenylphosphonium (STPP) (kindly provided by Prof. R. Hanson, Northeastern University) or 1,2-dioleoyl-3-trimethylammonium-propane (DOTAP) (Sigma) was added during liposome preparation.

Liposome size and size distribution were determined by Coulter N4 MD Submicron Particle Size Analyzer (Beckman-Coulter, Fullerton, CA). The zeta-potential values of the various liposomal formulations were measured at

25°C in water using the Zeta Plus Particle Analyzer (Brookhaven Instruments Corp, Santa Barbara, CA).

Dynamics of the internalized liposomes were assessed after performing a time-course experiment on human breast adenocarcinoma MCF-7 cells. This particular cell line has inherent lipid inclusions scattered throughout the cell body, appearing as bright areas in the Raman maps, due to high C–H stretching intensities.[22] Such inclusions provide very distinct spectral variations in the C–H stretching region of a Raman spectrum that can be associated to the nature of lipid inclusions, due to their highly-rich content of CH_3 and CH_2 moieties.

The MCF-7 cell culture was incubated with the growth media containing liposomal suspension for three, six or nine hours. The primary time point of analysis was taken as the initial time at which all cells within the cell culture show significant liposomal inclusions within the cell body.[7,22,27,31] There were studies conducted previously, which showed a saturation behavior of nanoparticles, showing linear dependencies at low concentrations and reaching a plateau at high dosages.[32] As we have tried to avoid a saturation of the cell cultures with the nanoparticles in order to detect their uptake and translocation kinetics, the appropriate time points were chosen as starting points for each of the nanostructures analyzed.

The data represented here has been analyzed using the VCA algorithm. The utility of the mathematical algorithm stems from the fact that it can achieve superior biochemical contrast, deconvolve and unmix the abundant information collected from the acquired data hypercube. This mathematical data manipulation would ultimately yield manageable classes of spectra, characterized based on their chemical composition, due to the principle that biological samples are comprised of "pure" component spectra (i.e. DNA, lipids, proteins, etc.). Hence, the VCA algorithm decomposes or unmixed all the spectra from the entire dataset into the most dissimilar biochemical contributions, termed "endmembers," which are represented by a monochrome intensity or an abundance map. When overlaying any three of such monochrome intensity maps, each one is assigned a red, green or blue intensity scale to produce a composite false color RGB image, or a functional biochemical map, of the cell. The principle sub-cellular regions detected by this method of analysis are those related to internalized liposomes, which are assigned a red channel, the nuclei and cell body proteins (blue channel), and various membrane-rich organelles such as the mitochondria, ER and Golgi, or vesicular compartments (green channel).

Initial penetration of the liposomes can be observed at the periphery of the cell, as can be seen from the bright red inclusions, describing the

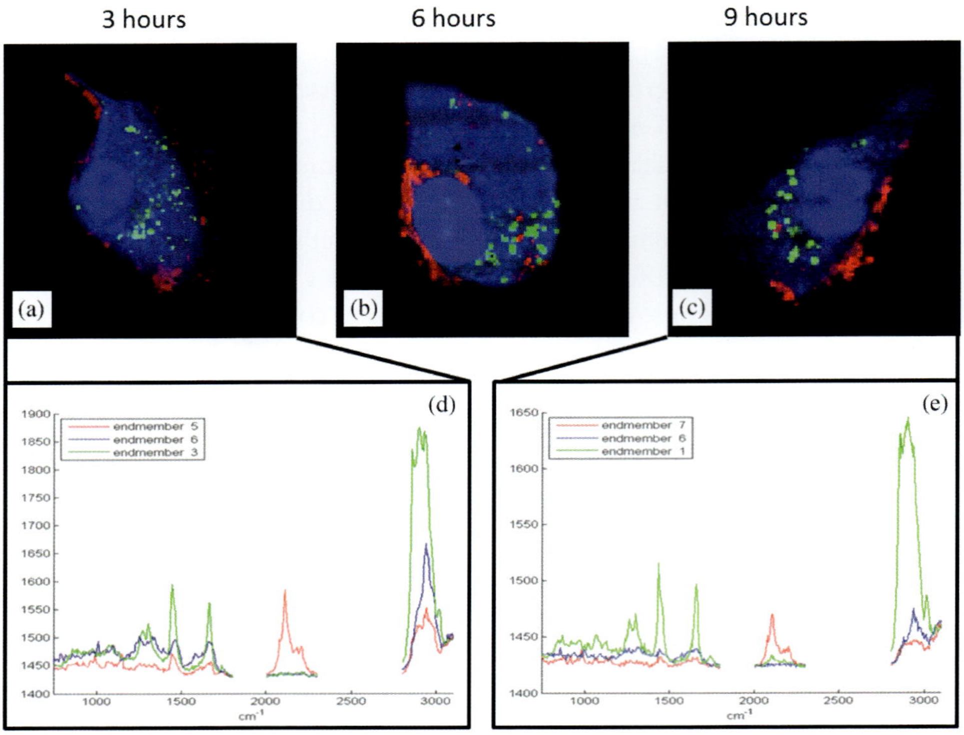

Fig. 3. Intracellular dynamics of TATp-modified liposomes. VCA-analyzed MCF-7 cells with three different incubation times of cell culture with liposomes. The rendered images represent the cell body in blue, high lipid content inclusions in green, deuterated liposomes in red. The corresponding abundance spectra for the 3 h and the 9 h incubation points are shown in the images d and e.

internalized liposomes in Fig. 3. Internalization of the nanoparticles is easily corroborated with the use of a depth scanning capability of the Raman instrumentation, which underscores the nanoparticlulate penetration into the cell, rather than precipitating on the cell surface.[22] TATp seems to enhance liposomal penetration into the cytosol drastically as well as increase the initial accumulations of the cationic liposomes by ca. a factor of 10, when compared with non-cationic liposomes, from our previous studies. The quantification of the liposomes within the cytosol is rendered difficult due to the confocal nature of the imaging system utilized, since a portion or a slice of a cell is assessed at a time, representative of ca. 20% of the cell.

Following a six to nine hours incubation of the liposomes with the cell culture reveals a plateau for nanoparticle internalization into the cells, saturating roughly 10% of the cell (data taken from our previous studies, which may be found at Matthaeus *et al.*, *Mol. Pharm.* 2008). After saturation is reached, the signal from the nanoparticles starts to be observed in the naturally occurring lipid inclusions, which are inherent to the cell line. This can be seen from Figs. 3(b), 3(c) and 3(e). The spectra reveal that with time the deuterated phospholipids from the liposomal complex exchange and integrate with inherent sub-cellular phospholipid inclusions within the cell. This is revealed from the spectra, or rather the c–d stretching region within the spectrum, in Fig. 3(e) when compared with Fig. 3(d).

The three spectra in Fig. 3(d) show representative spectra for the three distinct chemical microenvironments of a cell incubated for 3 h with the liposome-containing media. This image reveals the high c–d content, seen from the intense peaks within the 2050–2300 cm^{-1} spectral range (red trace). The spectra are depicting here the early internalization patterns of freshly engulfed liposomes. This is evident from a lack of superimposed cellular signatures with signatures pertaining to liposomal aggregates. It also shows no c–d stretching vibrations within the nucleus or any compartment within the cytosol, as can be seen from a lack of any c–d peaks within the same spectral region in the green and blue spectra. However, as we analyze a 9 h incubation period, we get a glimpse into the intracellular response to engulfed nanostructures. As can be seen from Fig. 3(e), the green trace shows distinct peaks in the 2050–2300 cm^{-1} spectral region.

Little is known about the integration of the liposomes with the cellular compartments and membranes. This phenomenon may be diffusion controlled or via cellular trafficking mechanisms. However, a model study was reported that provided evidence to a slow transfer of phospholipids from diverse nanoparticle structures with membranous vesicles of the cell.[33,34] This phenomenon is visible from the spectra in Fig. 3. As there is no concrete information on the dynamics of the exogenous phospholipid inclusions within a cell, Raman spectroscopy offers a window into that phenomenon. By scrutinizing the spectra of the inherent intracellular lipid inclusions, we are able to detect the dynamic interchange of the nanoparticle system, via the integration of the deuterated phospholipids of the liposomes with sub-cellular compartments.

4. Studies of Degradation of Micelles-Based Nanoparticle System as Followed by Raman Micro-Spectroscopy

Another phospholipid-based nanoparticle system heavily studied in pharmaceutical sciences is based on a micelle system. Polymeric micelles are self-assembly nanoparticulates, which effectively solubilize and transport poorly soluble chemotherapeutics. Their small size also allows for the extravasation into the tumor interstitium, thus enhancing drug bioavailability.[35,36] We have, therefore, tried to elucidate the degradation patterns of a pure micelle-based system, enhanced with cell-penetrating-peptide TAT-modification. It is important to note that the cell line for the following experiments was switched from MCF-7 to a cervical cancer cell line (HeLa). This alteration was mainly done for unbiased elucidation to the degradation patterns of the nanoparticulates.

For the preparation of the micelle composition, PEG–PE and TATp-PEG_{1000}-PE was prepared in chloroform at a molar ratio of 97.5:2.5. Chloroform was removed by rotary evaporation and freeze-drying. The film was hydrated with HEPES-buffered saline (HBS), pH 7.4 at room temperature and vortexed for 5 min. Micelle size was measured using a Coulter N4 Plus Submicron Particle Analyzer (Coulter Corporation, Miami, FL, USA).

As we have previously validated, the phospholipid-based nanoparticles, namely liposomes, showed unequivocal patterns of degradation and subsequent integration of the phospholipid molecules within the inherent cellular structures. Hence a cell line heavily expressing various phospholipid inclusions is not an ideal candidate to study fine biochemical variation due to intracellular micelle degradation. HeLa cells do not exhibit any intracellular inclusions, which renders an ideal canvas for such an experiment.

A time-course experiment was performed using the TATp-micelles to ascertain the intracellular dynamics. The cell culture was incubated with the micelle-containing medium for an hour, a time point, which was established as a starting point based on the saturation and presence of the nanoparticles within each cell of the culture. As can be seen in Fig. 4, three cells for each time point are shown as representatives for the culture. The micelles, which are pseudo-colored red, are seen penetrating the periphery of the cells. The nucleus and the cell body are shown in blue hues, which the membrane-rich organelles including the mitochondria, endoplasmic reticulum and Golgi, are shown in green.

In order to detect the cellular behaviors and responses toward the exogenous micelle system, the cell cultures with the nanoparticle-containing media were gently washed after an hour of incubation. Subsequently, the cell cultures were submerged into fresh media for distinct periods of time, fixed, and consequently imaged. Migration of the micelles toward the perinuclear region is clearly seen from 3 h of incubation with fresh medium, to a 6 h incubation period. The mass migration of the TATp-micelles toward the nucleus is the consequence of the HIV-peptide modification, which helps the virus infiltrate not only the cellular membrane, but also proceed toward the nucleus, where the retrovirus is known to use cellular machinery in order to replicate. The 3 h and 6 h degradation periods allowed the

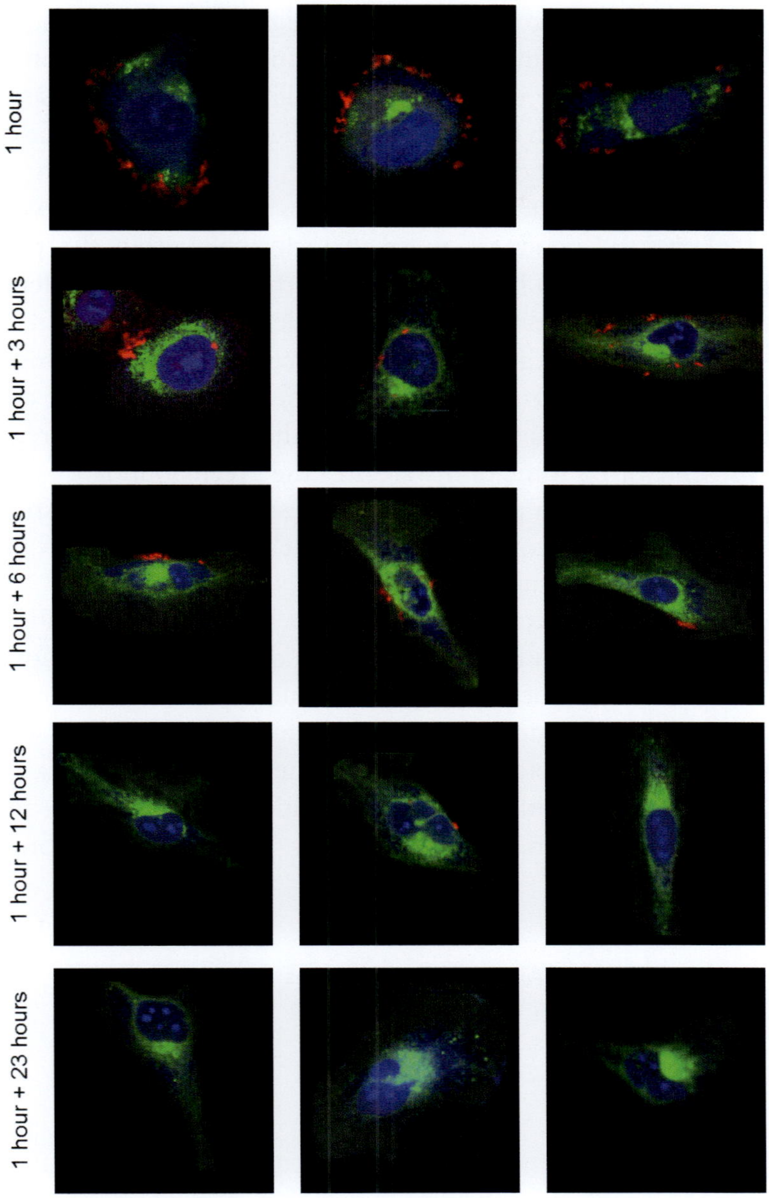

Fig. 4. Time-course degradation experiment of TATp-modified micelles. Top row: 1 h incubation with D-PE TATp-modified micelles. Second row: 1 h incubation with micellese followed by a 3 h incubation with fresh medium. Third row: 1 h incubation with micelles followed by a 6 h incubation with fresh medium. Fourth row: 1 h incubation with micells followed by a 12 h incubation with fresh medium. Bottom row: 1 h incubation with micelles, followed by a 23 h incubation in fresh medium. The VCA algorithm produced three distinct spectra which correspond to cell body and nucleus (blue), membrane-rich organelles (green) and internalized micelles (red).

nanoparticles to co-localize with the regions high in ER and Golgi, which are seen as the network of thin green fibers radiating from the perinuclear area and extending toward the edges of the cell (Fig. 4). This is consistent with previous investigations that observed biodegradable nanoparticles being sorted into Golgi-associated late endosomes.[37]

The initially large intracellular aggregates of micelles observed at the periphery of the cell condense to small inclusions within the perinuclear region after the 3 h and 6 h of incubation with fresh medium. Hence, most of the internalized nanoparticles have been degraded and cleared from the cell via the lysosomal pathway, which is discernible from the intracellular inclusions exhibiting C–D stretching intensities. However, close analysis of the cellular spectra reveal a small intensity in the C–D stretching region sporadically present within some of the areas of the green abundance map of the cells after 12 h incubation period. This suggests not only a digestion of the nanoparticles, but also incorporation of the deuterated lipids into the membrane-rich sub-cellular compartments.

Micelle nanoparticles completely disappear by a 23 h degradation period (Fig. 4), as indicated by only two major VCA-derived contrast maps present, one for the nucleus and the cell body (blue hues), while the other is for ER- and Golgi-rich regions (green hues). This time period shows a completed digestion of the phospholipid-based nanostructures, which is seen via a decrease of micelle-specific signal with increasing incubation time, where the amount of internalized micelles decreases exponentially during degradation. Raman

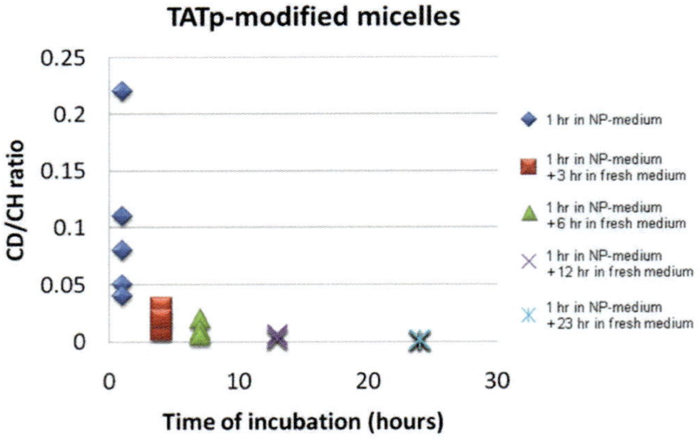

Fig. 5. Graph depicting an exponential decrease in micelle signal of the internalized nanoparticles as the incubation time is increased from 3 h to 23 h in fresh medium. Each point on the graph represents CD/CH ratio analyzed for each cell.

micro-spectroscopy thus has the potential to demonstrate the biochemical behaviors of the cells, deriving their metabolic responses from the changes in the peak heights and ratios in the C–H stretching intensities and the fingerprint region, in relation to the C–D stretching intensities of the introduced nanosystems. A tabulated representation of this phenomenon is shown in Fig. 5.

5. Raman Imaging and Analysis of Targeted Liposome-Based Nanoparticle Systems

A chemotherapeutic nanocarrier can yield a more effective and specific system if its outermost layer was modified with cationic compounds, which enhance internalization. Also, depending on the type of the modification molecule implemented, affinity and specificity to the cell or organelle may also be fine-tuned. There do exist major disadvantages to a cationic nanosystem, such as non-specificity on a sub-cellular level as well as inherent toxicity, however in lieu of these events research has been underway to develop a safe and a side-effect-free carrier.

We have analyzed three distinct cationic moieties for their enhanced internalization kinetics as well as validate and characterize mitochondria targeting capabilities. The three cationic systems analyzed include 1,2-dioleoyl-3-trimethylammonium-propane (DOTAP)-, stearyl amine (SA)-, and stearyltriphenylphosphonium (STPP)-modified liposomes. Out of the three cationic modification units, STPP was shown to have a high affinity for the mitochondria.[11,12] The phospholipids utilized for the bulk liposomal nano-assemblies were the same as analyzed in our previous work, namely deuterated phospholipids. As was mentioned previously, deuteration dramatically increases the sensitivity and specificity of Raman micro-spectral imaging analysis.

Just as for the micelle experimentation, a HeLa cell line has been chosen, as it proved ideal for detection and assessment of minute biochemical variations due to intracellular dynamics of deuterated systems. The time points analyzed were chosen based on previous experimentation with non-modified liposomal systems as well as established work with nanoparticles varying in cationic composition.[7,27,31] The cationic nanoparticles have been previously reported to penetrate the cellular periphery within minutes. Longer incubation times have been found to result in mass internalization of the nano-assemblies as well as their subsequent translocation within the cytosol. Hence, after a 1 h incubation of the HeLa cell cultures with our liposomal nanoparticles, consistent presence of the liposomes were observed within every cell analyzed. Shorter incubation times resulted in nanoparticle aggregations in fewer cells.

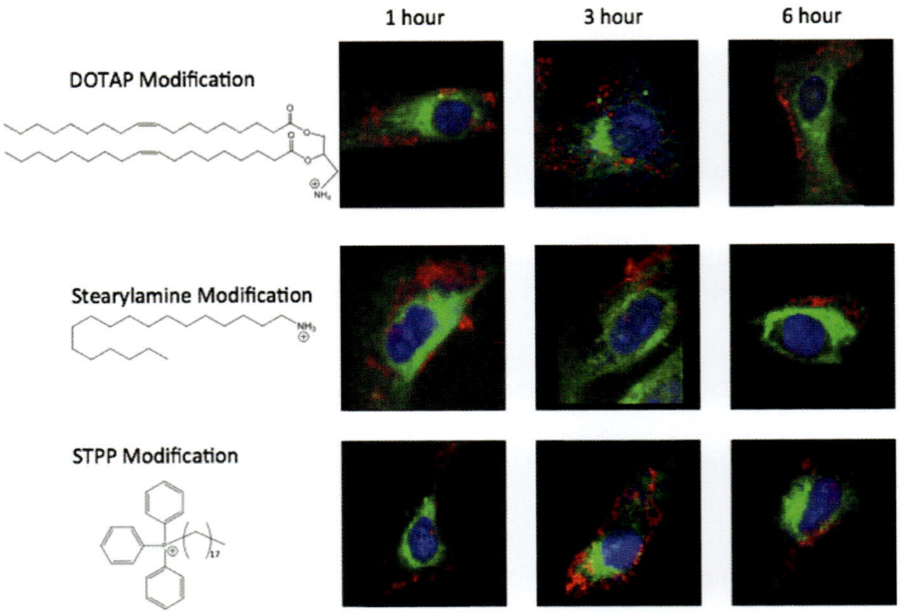

Fig. 6. Raman images of HeLa cells incubated with DOTAP-modified liposomes (top row), SAmodified liposomes (middle row) and STPP-modified liposomes (bottom row), structures of which are shown on the left. Cell cultures were incubated for 1 h, 3 h and 6 h, with DOTAP-, SA- or STPPmodified liposomes. Generated VCA-abundance images. The colors are representative of nuclei and cell body (blue), mitochondria-rich regions (green) and internalized liposomes (red).[38]

As degradation of the modified liposomes was not the priority, but rather their internalization and translocation patterns, we have analyzed three distinct incubation periods of the HeLa cell cultures with each of the three cationic liposomes (Fig. 6). Representative cells were taken to demonstrate the biochemical imaging capabilities of our Raman system coupled to multivariate analyses methods, namely VCA. Compared with our previous results with non-modified neutral liposomes, all three cationic nanoparticle systems demonstrated greatly enhanced internalization kinetics. While the neutral liposomes were detected after a 12 h incubation period with the cell culture, the influx of the cationic nanostructures was observed within an hour of incubation, which is comparable with internalization patterns of TATp-modified systems. Intracellular concentration of the cationic composites was also vastly greater within a shorter time frame when compared with the non-modified liposomes. Due to our 500 nm spatial resolution, we were able to observe dynamic changes of the internalized liposomes, whereas non-modified liposomes were sequestered within the cytosol in small quantities, inclusions of which range from 1–3 μm, the mass of intracellular aggregates of cationic liposomes is vastly greater, as can be seen in Fig. 6.

VCA-generated functional maps of the cells incubated with the three different cationic liposomes for an hour did not demonstrate any distinctions between liposomal internalization patterns, as can be seen in Fig. 6. Their sub-cellular presence was not seen to follow any specific segregation and compartmentalization patterns, but rather aggregated within the cellular periphery, with a few distinct pixels of liposome-characteristic signal within the perinuclear region. This can be seen from the bright red inclusions within the cytosol of all three cells presented in Fig. 6.

Distinctions between the sub-cellular behaviors of the three cationic liposomes within the rich cellular information become apparent after a 3 h incubation period. The 3 h and the 6 h incubation times for the cells with the three different liposomal solutions did not result in an increase in their internalization. Also, as with the 1 h incubation period, the majority of the internalized DOTAP- and SA-modified liposomes did not seem to adhere to any specific pattern of compartmentalization within the cell body. Most of the signal detected for these two aforementioned cationic systems was seen to aggregate at the periphery of the cell, with a few distinct pixels of liposome-characteristic signal within the perinuclear region. However, sub-cellular preferences of the STPP-modified liposomes become apparent solely after application of a multivariate method of spectral analysis.

After a 3 h incubation period of the cell culture with the STPP-modified liposomes, the spectra revealed not only the distinct accumulations of the nanoparticles within the cellular periphery, but also their mitochondrial affinity. The bright red inclusions within the cytosol exhibit the signal of the aggregated and sequestered nanoparticles. This is evident from the signal superposition pertaining to STPP-modified liposomes with the biochemical compartment encapsulating them. The green abundances within the image, on the other hand, possesses chemical properties pertaining to mitochondrial aggregations, as we have previously established,[20] while also showing a distinct C–D stretching vibrational peak (spectra are not shown and can be found at Chernenko *et al.*, *Mol. Pharm.* 2012). This phenomenon is due to the STPP moiety on the liposomes, which demonstrates their clear affinity for the mitochonrdia-rich perinuclear regions. We have corroborated this finding with subsequent use of Mitotracker® dye, which is known to label mitochondria. An overlay of the green intensity map generated from our VCA analysis with the fluorescence image of the mitochondrial aggregations, shows an exact correlation between the two perinuclear regions. Hence, VCA-generated spectral information has the capabilities to demonstrate the internalization kinetics as well as demonstrate the sorting of the liposomes into distinct cellular compartments.

This result demonstrates the sensitivity of the micro-spectroscopy coupled with statistical means of hyperspectral analysis. VCA is capable of distinguishing the liposomes co-localized with distinct sub-cellular organelles. Whether liposomes are fully or partially intact is still under investigation, however previous studies have reported seeing a fusion of the liposomal monomers with the sub-cellular organelles upon their degradation.[39,40]

6. Conclusion

Raman micro-spectroscopy in synergy with multivariate methods of analysis thus offers an advantage over diverse staining techniques such as fluorescence microscopy. In combination with optical microscopy, Raman spectroscopic images display a biochemical snapshot of a cell, along with any exogenous systems therein. Multivariate analyses almost go hand-in-hand with the spectroscopic imaging techniques, as they allow for unmixing of a vast amount of biochemical information reaped from the spectral data hypercube. Over the years we have shown the utility of the spectroscopic techniques within the biological sciences as well as the pharmaceutical sciences, where either the system itself is of vast interest or the chemotherapeutic carrier within the organism.

References

1. L. E. van Vlerken, S. Yadav, M. M Amiji and J.-M., Multi-Functional nanocarriers to overcome tumor drug resistance. *Cancer Treat. Rev.* **34**, 592–602 (2008).
2. L. E. Vlerken, Z. Duan, S. R. Little, M. V. Seiden and M. M. Amiji, Augmentation of therapeutic efficacy in drug-resistant tumor models using ceramide coadministration in temporal-controlled polymer-blend nanoparticle delivery systems. *AAPS J.* **12**, 171–180 (2010).
3. V. Torchilin, Multifunctional and stimuli-sensitive pharmaceutical nanocarriers. *Euro. J. Pharm. Biopharm.* **71**, 431–444 (2009).
4. Clinical Relevance of Mol Mechanisms_Links 1999. **1999**, 1–21.
5. PEG-Gelatin NPs for Intracellular Delivery_Amiji 2002. **2002**, 1–7.
6. L. Wasungu and D. Hoekstra, Cationic lipids, lipoplexes and intracellular delivery of genes. *J. Control. Release* **116**, 255–264 (2006).
7. C. Lonez, M. Vandenbranden and J.-M. Ruysschaert, Cationic liposomal lipids: from gene carriers to cell signaling. *Prog. Lipid Res.* **47**, 340–347 (2008).
8. M. Ross, G. Kelso, F. Blaikie, A. James, H. Cocheme, A. Filipovska, T. Da Ros, T. Hurd, R. Smith and M. Murphy, Lipophilic triphenylphosphonium cations as tools in mitochondrial bioenergetics and free radical biology. *Biochemistry* **70**, 222–230 (2005).
9. V. Torchilin and T. Levchenko, TATp-Liposomes:_novel intracellular drug carrier. *Curr. Protein Pept. Sci.* **4**, 133–140 (2003).

10. V. Torchilin, R Rammohan, V. Weissig and T. Levchenko, TAT peptide on the surface of liposomes affords their efficient intracellular delivery even at low temperature and in the presence of metabolic inhibitors. *Proc. Natl. Acad. Sci.* **98**, 8786–8791 (2001).
11. R. Smith, C. Porteous, A. Gane and M. Murphy, Delivery of bioactive molecules to mitochondria in vivo. *Proc. Natl. Acad. Sci.* **100**, 5407–5412 (2003).
12. S. M. Ojovan, D. A. Knorre, O. V. Markova, E. A. Smirnova, L. E. Bakeeva and F. F. Severin, Accumulation of dodecyltriphenylphosphonium in mitochondria induces their swelling and ROS-dependent growth inhibition in yeast. *J. Bioenerg. Biomembr.* **43**, 175–180 (2011).
13. Z. Movasaghi, S. Rehman and I. U. Rehman, Raman spectroscopy of biological tissues. *Appl. Spectrosc. Rev.* **42**, 493–541 (2007).
14. D. Naumann, Ft-Infrared and Ft-Raman spectroscopy in biomedical research. *Appl. Spectrosc. Rev.* **36**, 239–298 (2001).
15. S. Wartewig and R. Neubert, Pharmaceutical applications of mid-IR and Raman spectroscopy. *Adv. Drug Deliv. Rev.* **57**, 1144–1170 (2005).
16. E. O. Potma and X. S. Xie, Detection of single lipid bilayers with Coherent Anti-Stokes Raman Scattering (CARS) microscopy. *J. Raman Spectrosc.* **34**, 642–650 (2003).
17. C. Krafft, B. Dietzek and J. Popp, Raman and CARS microspectroscopy of cells and tissues. *Analyst* **134**, 1046 (2009).
18. N. Stone, C. Kendall, J. Smith, P. Crow and H. Barr, Raman spectroscopy for identification of epithelial cancers. *Faraday Disc.* **126**, 141 (2004).
19. H. G. Schulze, S. O. Konorov, N. J. Caron, J. M. Piret, M. W. Blades and R. F. B. Turner, Assessing differentiation status of human embryonic stem cells noninvasively using Raman microspectroscopy. *Anal. Chem.* **82**, 5020–5027 (2010).
20. C. Matthäus, T. Chernenko, J. A. Newmark, C. M. Warner and M. Diem, Label-Free detection of mitochondrial distribution in cells by nonresonant Raman microspectroscopy. *Biophys. J.* **93**, 668–673 (2007).
21. A. A. van Apeldoorn, H.-J. van Manen, J. M. Bezemer, J. D. de Bruijn, C. A van Blitterswijk and C. Otto, Raman imaging of PLGA microsphere degradation inside macrophages. *J. Am. Chem. Soc.* **126**, 13226–13227 (2004).
22. C. Matthäus, A. Kale, T. Chernenko, V. Torchilin and M. Diem, New ways of imaging uptake and intracellular fate of liposomal drug carrier systems inside individual cells, based on Raman microscopy. *Mol. Pharm.* **5**, 287–293 (2008).
23. B. G. Saar, L. R. Contreras-Rojas, X. S. Xie and R. H. Guy, Imaging drug delivery to skin with stimulated Raman scattering microscopy. *Mol. Pharm.* **8**, 969–975 (2011) .
24. T. Chernenko, C. Matthäus, L. Milane, L. Quintero, M. Amiji and M. Diem, Label-Free Raman spectral imaging of intracellular delivery and degradation of polymeric nanoparticle systems. *ACS Nano* **3**, 3552–3559 (2009).
25. N. Keshava, A survey of spectral unmixing algorithms. *Lincoln Lab. J.* 1–24 (2003).
26. J. M. P. Nascimento and J. M. B. Dias, Vertex component analysis: a fast algorithm to unmix hyperspectral data. *IEEE Trans. Geosci. Remote Sensing* **43**, 898–910 (2005).
27. V. Torchilin, Recent approaches to intracellular delivery of drugs and DNA and organelle targeting. *Annu. Rev. Biomed. Eng.* **8**, 343–375 (2006).
28. V. P. Torchilin, Drug tarteting. Eur. J. Pharm. Sci. 11 Suppl. S81–S91 (2000).
29. V. P. Torchilin, Micellar nanocarriers: pharmaceutical perspectives. *Pharm. Res.* **24**, 1–16 (2006).
30. P. Couvreur and C. Vauthier, Nanotechnology: intelligent design to treat complex disease. *Pharm. Res.* **23**, 1417–1450 (2006).

31. Y. Aramaki, Cationic liposomes induce macrophage apoptosis through mitochondrial pathway. *Arch. Biochem. Biophys.* **392**, 245–250 (2001).
32. P. Nativo, I. A. Prior and M. Brust, Uptake and intracellular fate of surface-modified gold nanoparticles. *ACS Nano* **2**, 1639–1644 (2008).
33. Palmerini, C. A.; Cametti, C.; Sennato, S.; Gaudino, D.; Carlini, E.; Bordi, F.; Arienti, G. Role of Cholesterol, DOTAP, and DPPC in Prostasome/Spermatozoa Interaction and Fusion. *J. Membr. Biol.* **211**, 185–190 (2006).
34. J. Kristl, Interactions of solid lipid nanoparticles with model membranes and leukocytes studied by EPR. *Int. J. Pharm.* **256**, 133–140 (2003).
35. A. Lukyanov, Micelles from polyethylene glycol/phosphatidylethanolamine conjugates for tumor drug delivery. *J. Control. Release* **91**, 97–102 (2003).
36. V. Torchilin, Lipid-Core micelles for targeted drug delivery. *Curr. Drug Deliv.* **2**, 319–327 (2005).
37. J. Panyam and V. Labhasetwar, Targeting intracellular targets. *Curr. Drug Deliv.* **1**, 235–247 (2004).
38. T. Chernenko, R. R. Sawant, M. Miljkovic, L. Quintero, M. Diem and V. Torchilin, Raman microscopy for noninvasive imaging of pharmaceutical nanocarriers: intracellular distribution of cationic liposomes of different composition. *Mol. Pharm.* **9**, 930–936 (2012).
39. M. Miljković, T. Chernenko, M. J. Romeo, B. Bird, C. Matthäus and M. Diem, Label-Free imaging of human cells: algorithms for image reconstruction of Raman hyperspectral datasets. *Analyst* **135**, 2002 (2010).
40. C. Krafft, T. Knetschke, R. H. W. Funk and R. Salzer, Studies on stress-induced changes at the subcellular level by Raman microspectroscopic mapping. *Anal. Chem.* **78**, 4424–4429 (2006).

Chapter 16

Clinical Experience with Nanoparticles in Imaging

Patrick D. Sutphin[*,†] *and Mukesh Harisinghani*[‡,§]

[*]*Division of Vascular Imaging and Intervention*
Department of Radiology, Massachusetts General Hospital
55 Fruit St, Boston, Massachusetts, US
[‡]*psutphin@partners.org*
[†]*Division of Abdominal Imaging and Intervention*
Department of Radiology, Massachusetts General Hospital
55 Fruit St, Boston, Massachusetts, US
[§]*mharisinghani@mgh.harvard.edu*

Abstract

Imaging with nanoparticles is an emerging technology which holds the promise of increased diagnostic accuracy resulting in improved therapeutic selection, monitoring and surgical planning ultimately leading to better patient outcomes. An important advance in nanoparticle imaging is the development of ultrasmall superparamagnetic iron oxide (USPIO) nanoparticles as a novel magnetic resonance imaging (MRI) contrast agent. USPIO nanoparticles have a long blood pool half-life and cross the capillary endothelium into the interstitial space and are transported to the lymph node and internalized by macrophages. Malignant lymph nodes fail to retain USPIO nanoparticles as tumor tissue disrupts the normal lymph node architecture and replaces macrophages. Accumulation of USPIO nanoparticles decreases T2 relaxation times resulting in signal intensity loss, negative enhancement, on T2-weighted images in normal lymph nodes; whereas, the signal intensity of malignant lymph nodes remains unchanged on post-USPIO nanoparticle images. This method of lymph node imaging is known as lymphotropic nanoparticle MRI (LNMRI) and has been shown by numerous studies to offer greater accuracy in the detection of malignant lymph nodes than by anatomic evaluation.

1. Introduction

Cross-sectional imaging has revolutionized the practice of medicine with the accurate detection of disease and ability to provide regular surveillance through non-invasive means. Computed tomography (CT), ultrasound and magnetic resonance imaging (MRI) take advantage of differences in tissue characteristics such as density, in the case of CT, to generate anatomic cross-sectional images for diagnostic purposes. Of the three modalities, CT has been the most routinely adopted with a virtual explosion in the number of CT examinations performed per year over the last three decades with an estimated 3 million studies performed in 1980 to 70 million performed in 2006.[1,2] The clinical preference for CT over ultrasound and MRI is likely secondary to the relative speed with which CT examinations can be performed as well as the uniformity of the exam and ease of use.

The wide-spread inclusion of CT in Oncology guidelines has markedly altered the way cancers are staged. For example, in the pre-CT era, Hodgkin's disease was initially staged with clinical examination and a number of studies including PA and lateral chest radiographs, tomography of the mediastinum and hila, skeletal surveys, liver/spleen nuclear imaging, bone scans, lymphangiography and intravenous (*i.v.*) urograms. In addition, given the limited sensitivity of the physical exam and conventional X-rays, staging laparotomy was often performed because of the high rate of treatment failure of 30%–50% secondary to radiographically occult abdominal disease in patients with Stages I and II disease treated with radiation alone.[3] Improved sensitivity for detection of both thoracic and intra-abdominal disease with CT along with evolving treatment regimens have essentially made staging laparotomy irrelevant to the modern-day Hodgkin's patient.[3,4] In addition, the battery of imaging exams routinely performed on Hodgkin's disease patients was supplanted by CT which is widely-available, non-invasive and highly reproducible. Ultrasound and MRI are also incorporated into cancer staging but are largely used as adjuncts to CT to answer specific diagnostic questions based on patient symptomatology such as spine MRI and endoscopic ultrasound or testicular ultrasound.[3]

Despite enormous advances in anatomic staging with imaging, limitations remain in accurately defining the extent of disease. One such limitation is the ability to accurately characterize lymph nodes as benign or malignant based on imaging findings. Lymph node size is the most widely used criterion for judging whether a lymph node is benign or malignant, yet no universally accepted size exists for this distinction. In addition, as numerous studies have demonstrated not all metastatic lymph nodes are enlarged (Fig. 1),[5] thus any

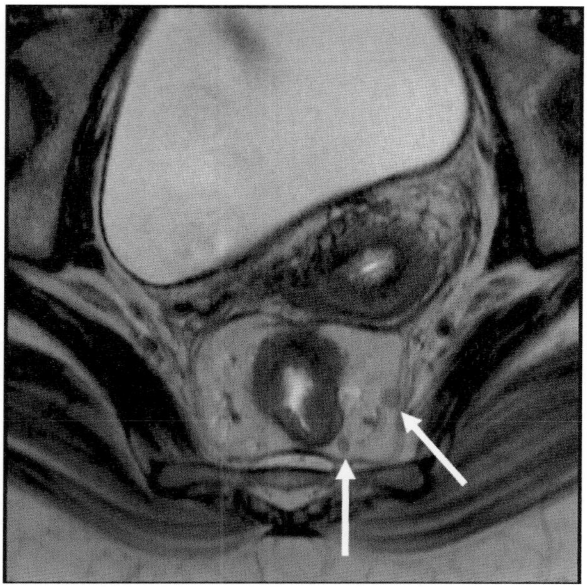

Fig. 1. Normal-sized malignant lymph nodes. Axial T2-weighted MR image in a patient with rectal cancer reveals two normal-sized perirectal lymph nodes (white arrows). Based on size criterion the lymph nodes were characterized as benign, but were found to be malignant on pathologic evaluation.

size threshold would reduce the sensitivity accordingly. Similarly, lymph nodes may be enlarged for reasons other than cancer (Fig. 2). Inaccurate detection of malignant lymph nodes has direct clinical impact on patients leading to either the under treatment or overtreatment of disease with suboptimal therapies resulting in significant costs both to the affected patients and our healthcare system. In the case of prostate cancer, malignant pelvic lymph nodes have such important implications in patient management that pelvic lymph node dissection remains the gold standard for lymph node staging, despite the associated risks and morbidity associated with the invasive procedure. The addition of pelvic lymph node dissection to radical prostatectomy is based on risk of regional lymph node invasion and the current National Comprehensive Cancer Network Practice Guidelines (2011) recommend pelvic lymph node dissection in patients with a 2% risk of lymph node invasion.[6] Even in the context of lymph node dissection there is controversy with how much lymph node dissection is enough, with many advocating for extended lymph node dissection.[7,8]

Unfortunately, routine imaging modalities have insufficient sensitivity for the detection of malignant lymph nodes. A recent meta-analysis of CT and

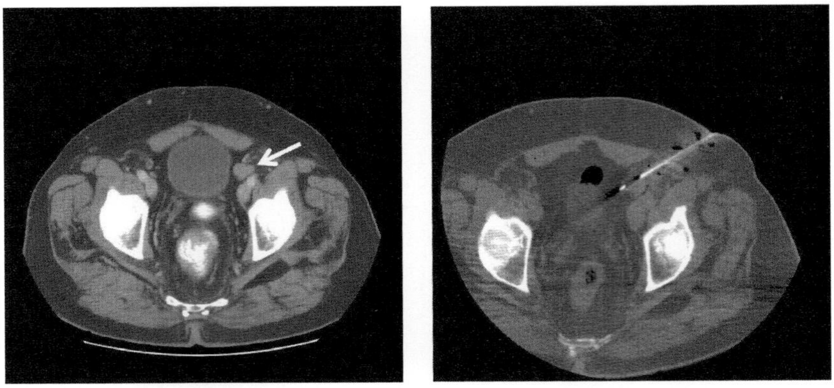

(a) (b)

Fig. 2. 80 year old man with history of prostate cancer and colon cancer with an enlarged left external iliac lymph node. (A) Routine surveillance CT exam revealed an enlarged 1.8 cm left external iliac lymph node (white arrow). The node was interpreted as malignant based on size criterion, but CT-guided biopsy, (B) demonstrated a reactive lymph node.

MRI in pelvic lymph node staging in the context of prostate cancer found a sensitivity of 42% and 39% by CT and MRI, respectively for the detection of malignant lymph nodes.[9]

The limitations of CT and MRI in the detection of metastatic lymph nodes are widely recognized and significant effort is underway to improve the diagnostic accuracy for detection of lymph node disease. Many of the imaging techniques under development are leveraging concepts from molecular imaging or nanotechnology in conjunction with CT or MRI to enhance the accuracy of malignant disease detection. ^{18}Fluorodeoxyglucose Positron Emission Tomography (^{18}FDG-PET) has been extensively studied in lymph node staging. A recent meta-analysis of ^{18}FDG-PET in colorectal cancer found a pooled sensitivity of 42.9% and a specificity of 87.9% for the detection of metastatic lymph nodes with a global diagnostic accuracy of 70.6%. PET has limited spatial resolution below 1 cm, and accordingly subcentimeter lymph nodes in the studies evaluated were often falsely classified as benign, reducing sensitivity for detection.[10–12] An additional source of lymph node misclassification is secondary to masking of lymph node uptake and confusion of lymph node uptake with renally excreted^{18} FDG in the urinary system, particularly in the ureters which course along the lymphatic pathways of the retroperitoneum and pelvis. Finally,^{18} FDG-PET in the detection of neoplastic disease is dependent on the transport and retention of glucose molecules and analogs such as^{18} FDG into neoplastic cells as a marker of tumor metabolic activity. The spectrum of tumor^{18} FDG-avidity varies widely with some tumor types such as lymphoma which demonstrates high levels of[18] FDG uptake, whereas some tumor types such as prostate cancer demonstrate low-sensitivity on [18]FDG-PET.[13,14] As such, a number of non-[18] FDG radiotracers are currently under development which will target alternative pathways for imaging which include DNA synthesis and cell surface markers/receptors to increase detection sensitivity in tumors that are not dependent on glucose metabolism.[15] Even with the addition of new radiotracers the sensitivity for the detection of tumor less than 5 mm with PET in metastatic lymph nodes is poor as demonstrated in a rabbit model of lymph node metastasis where 0 of 7 lymph nodes with metastatic disease less than 5 mm were detected.[16]

An alternate imaging strategy to lymph node staging, which will be discussed in detail, involves the application of nanotechnology in imaging with the use of magnetic nanoparticles in MR imaging to evaluate lymph nodes known as lymphotropic nanoparticle enhanced MRI (LNMRI).

2. History and Mechanism of LNMRI

2.1. *History*

It was recognized early after the development of MRI that iron overload, as in hemochromatosis, alters and enhances T2 relaxation times resulting in significant signal loss in the liver, spleen and pancreas. Fortuitously, in the context of hepatic tissue signal loss, hepatic lesions were found to be well-demonstrated by MRI in patients with hemochromatosis.[17] Moreover, both experimental animal models and human clinical observations revealed that liver tissue signal loss due to iron overload could be induced by excess iron in the diet or through blood transfusion.[18] These two observations sparked the idea of administering superparamagnetic ferrite particles as a novel MR contrast agent. Saini *et al.* tested superparamagnetic particles as a contrast agent by administering ferrite particles (0.5 to 1.0 μm) intravenously into animal models and demonstrated a marked decrease in T2 relaxation times in the liver, spleen and bone marrow with no changes in tissue signal intensity observed in the lung, kidney, or muscle.[19] Electron microscopy confirmed sequestration of the ferrite particles in the hepatic reticuloendothelial system.[19]

Subsequent experiments with the superparamagnetic ferrite particles in animal models of liver cancer showed localization of the ferrite particles to the hepatic reticuloendothelial cells and not hepatocytes or tumor tissue. Accordingly, the ferrite particles enhanced signal differences between the liver and tumor, improving detection of hepatic lesions in animal models of liver cancer.[20] Similarly, ferrite particles were found to increase the conspicuity of splenic metastases in a rat model of splenic metastasis with a 43-fold increase in tumor to spleen contrast-to-noise ratio.[21]

The initial superparamagnetic iron oxide particles were heterogeneous in size with particles ranging in size from 30 to 1000 nm and localized to the mononuclear phagocytic system of the liver and spleen. It was hypothesized that the pharmacokinetics and biodistribution of the nanoparticles could be modified by reducing the size of the particles through size fractionation. After size fractionation of AMI-25 with gel chromatography 70% of the particles were smaller than 10 nm. Initial studies of these ultrasmall superparamagnetic iron oxide (USPIO) particles demonstrated a significantly longer blood pool time with half-life of 81 min versus 6 min for AMI-25 in a rat model. In addition, the biodistribution of the particles was altered with approximately half the distribution to the liver and spleen seen with the unfractionated AMI-25. Lymph node uptake was also demonstrated with intravenously administered USPIO which was only previously appreciated with AMI-25 administered interstitially.[22]

2.2. *Mechanism of lymph node targeting*

Following the *i.v.* administration of USPIO nanoparticles, the nanoparticles reach the lymph nodes in two distinct pathways. The first pathway is through direct transcapillary transport through the high endothelial venules within individual lymph nodes. The second pathway is through the extravasation of the nanoparticles into the interstitial space with subsequent uptake into the lymph vessels and transport to the lymph nodes. In the lymph node parenchyma the nanoparticles are engulfed by the phagocytic cells of the mononuclear phagocyte system (MPS) leading to the retention and accumulation of the USPIO nanoparticles in the lymph nodes. The accumulation of the USPIO nanoparticles alters the T2 signal characteristics of the lymph node which are detectable as signal loss on MRI. If the MPS is disrupted or replaced as with tumor cells, the lymph nodes fail to retain nanoparticles and the T2 signal characteristics remain unchanged.[5,23,24] The nanoparticle structure and the route of the nanoparticles to the lymph nodes following *i.v.* administration is depicted in Fig. 3.

In LNMRI pre-contrast images are obtained and post-USPIO images are obtained in order to better identify signal loss in USPIO nanoparticle containing lymph nodes. Accumulation of USPIO nanoparticles in lymph nodes is a slow process, thus post-contrast images are obtained 24 h to 36 h following *i.v.* administration of nanoparticles. Direct comparison is then made of the lymph nodes on pre- and post-contrast images. As shown in Fig. 4(A),

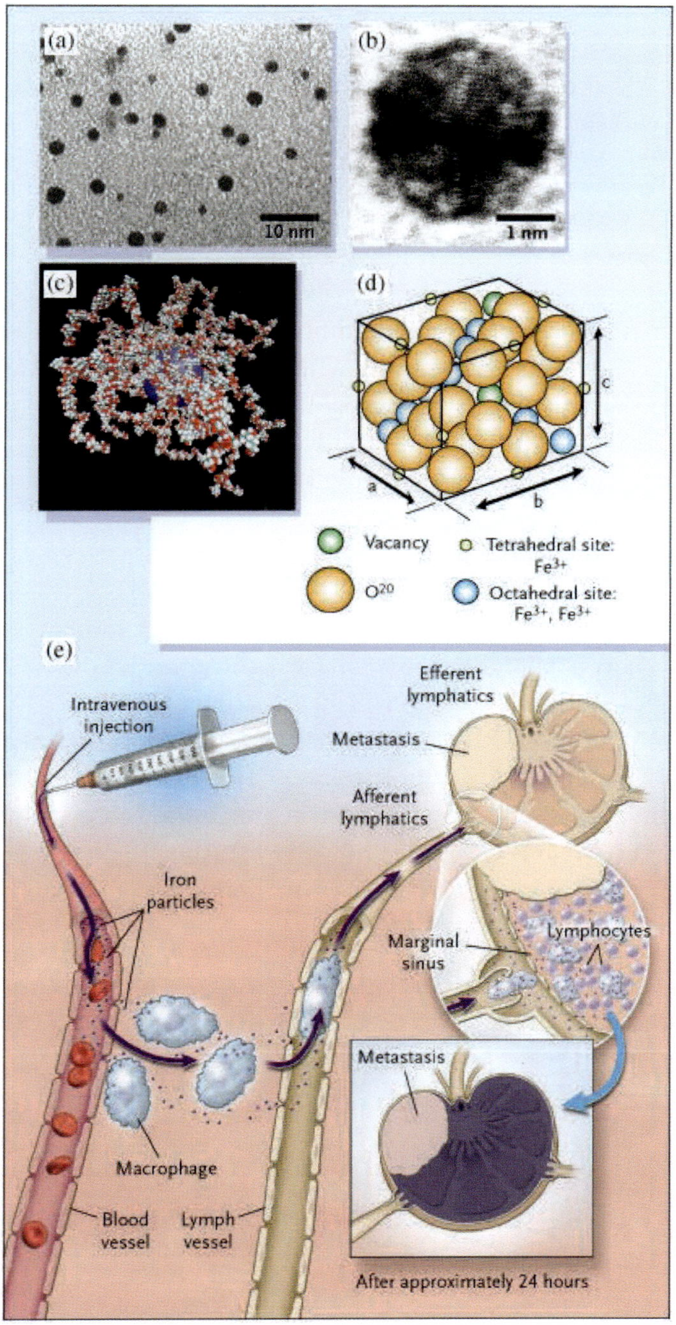

Fig. 3. Electron micrograph of hexagonal lymphotropic superparamagnetic nanoparticles (Panels A and B), molecular model of surface-bound 10-kD Dextrans and packing of iron oxide crystals (Panels C and D), and mechanism of action of lymphotropic superparamagnetic nanoparticles. The model lymphotropic superparamagnetic nanoparticles shown here measure 2 nm to 3 nm on average (Panels A and B). The mean overall particle size of 10-kD dextran is 28 nm (Panels C and D). In Panel E, the systemically injected long-circulating particles gain

Fig. 3. (*Continued*) access to the interstitium and are drained through lymphatic vessels. Disturbances in lymph flow or in nodal architecture caused by metastases lead to abnormal patterns of accumulation of lymphotropic superparamagnetic nanoparticles, which are detectable by MRI. From Ref. 5. Copyright © 2003 Massachusetts Medical Society. Reprinted with permission.

pre-contrast axial T2*-weighted images from a prostate cancer patient demonstrate an enlarged hyperintense left external iliac lymph node. Axial T2* images obtained one day after the administration of ferumoxtran-10 show homogeneous signal loss compatible with a normal lymph node, Fig. 4(B). In contrast, malignant lymph nodes maintain signal intensity on post-USPIO images as seen in Fig. 5, in which the lymph node is hyperintense on both pre- and post-contrast images reflecting loss of normal lymph node architecture. A subsequent biopsy confirmed malignancy. Characterization of lymph nodes as either benign or malignant is based on the pattern of signal loss, negative enhancement, observed on post-contrast images.

Typical imaging sequences in LNMRI include gradient echo (GRE) T1-weighted, fast spin-echo (FSE) T2-weighted and GRE T2*-weighted sequences. These sequences are performed before and 24 h to 36 h following administration of the USPIO nanoparticles. In addition, to mapping lymph nodes USPIO nanoparticles demonstrate T1 shortening in the vascular space early after administration and can be used to map the vasculature. The use of a 3D T1-weighted GRE can be combined with the lymph node anatomy to create a lymphovascular map for surgical planning and correlation (Fig. 6).[25]

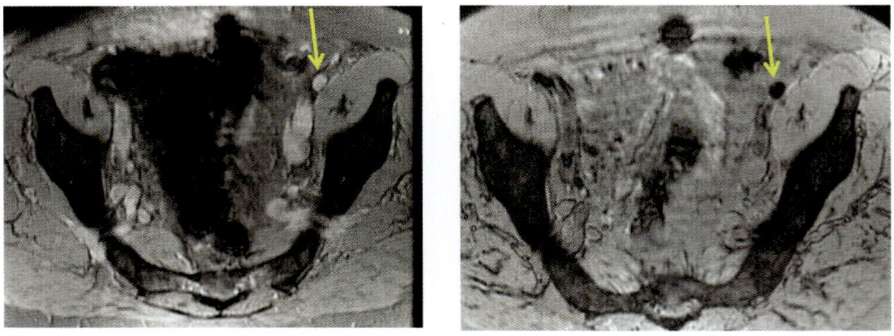

(a) (b)

Fig. 4. LNMRI in a prostate cancer patient with a benign enlarged left external iliac lymph node. (A) T2*-weighted gradient-echo MR image shows hyperintense enlarged left external iliac node (yellow arrow) in a patient with prostate cancer which was characterized as metastatic on the basis of size criterion. (B) T2*-weighted image obtained 24 hours after ferumoxtran-10 administration demonstrates homogenous decrease in the signal intensity (yellow arrow) compatible indicating benignity.

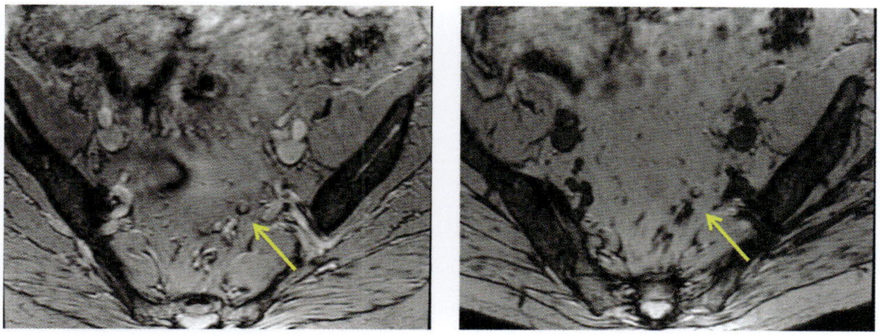

(a) (b)

Fig. 5. Lymphotropic MRI in a patient with bladder cancer with small malignant pelvic lymph node. (A) T2*-weighted gradient-echo MR image shows hyperintense subcentimeter pelvic lymph node (yellow arrow) in a patient with bladder cancer which was characterized as benign on the basis of size criterion. (B) T2*-weighted image obtained 24 h after ferumoxtran-10 administration demonstrates unchanged signal intensity (yellow arrow) indicating malignancy.

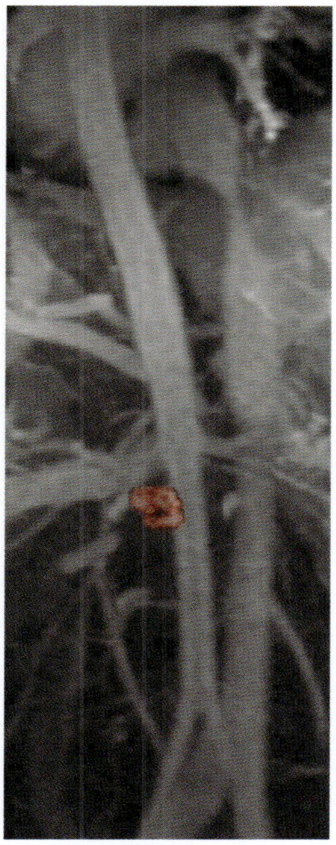

Fig. 6. Three-dimensional lymphovascular map for surgical planning. Three-dimensional lymphovascular maps can be created by combining a T1-weighted 3D sequence with the T2*-weighted lymph node images for improved surgical planning.

T1-weighted sequences are also important for identifying if a lymph node has a fatty hilum which can mimic a metastatic deposit on T2-weighted images.[25,26]

The critical sequences for lymph node characterization are the T2*-weighted sequences. On the pre-contrast images the lymph nodes are hyperintense on T2-weighted sequences, whereas 24 h after the administration of USPIO nanoparticles there is T2 signal loss in normal lymph nodes which accumulate the USPIO nanoparticles. Malignant lymph nodes, due to lack of the MPS, are unable to retain the nanoparticles and thus remain T2 hyperintense.[5,25] Lymph nodes are interpreted based on established guidelines for characterization of lymph nodes based on the pattern of negative enhancement with USPIO nanoparticles as seen in Fig. 7.[26–28]

The LNMRI imaging protocol is lengthy and inconvenient for patients as they are required to be present for two consecutive days of imaging, pre-contrast and 24 h post-contrast administration. The lengthy protocol may reduce patient compliance if there is widespread adoption of the technique. In addition, if feasible to perform a single post-contrast MRI there would be

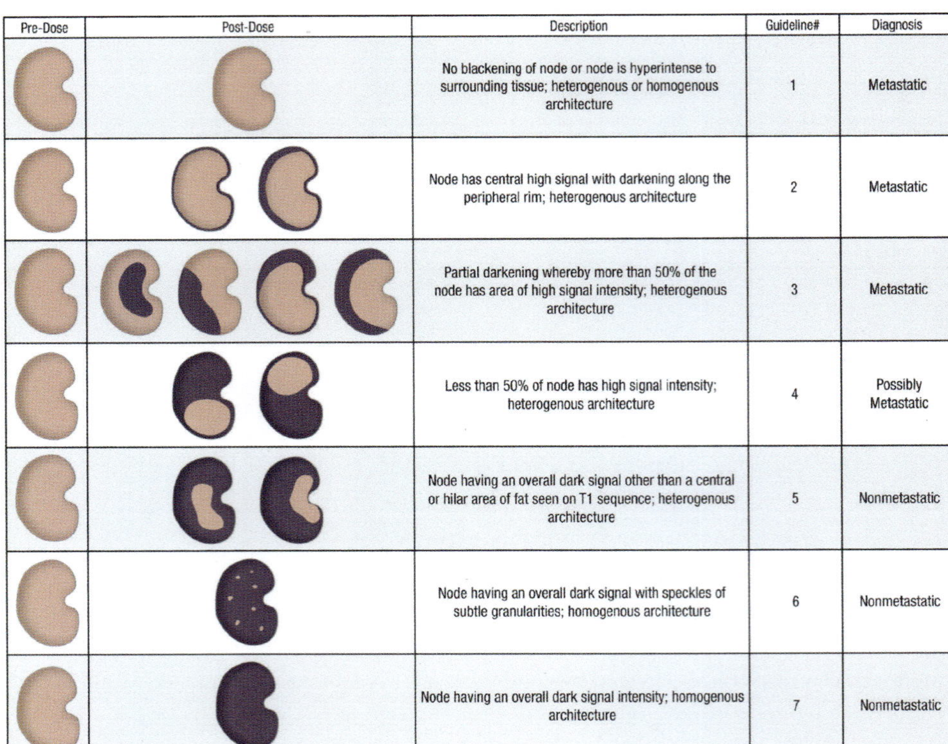

Fig. 7. Example of guidelines used for nodal characterization on lymphotropic nanoparticle-enhanced MRI for characterization of benign versus malignant lymph nodes. The chart was adapted from Harisinghani *et al.*[28]

a significant cost reduction. A study was performed to determine if results are equivalent in interpreting only post-contrast images versus pre- and post-contrast images in the characterization of malignant lymph nodes. A total of 77 patients were enrolled with a variety of malignancies including bladder, breast, endometrial, prostate, rectal and renal cancer. Two independent blinded reviewers, of differing levels of experience, first evaluated lymph nodes on post-contrast T2* images using a six-point scale. In a second session, the blinded reviewers evaluated the lymph nodes using both the pre- and post-contrast images using the same six-point scale. The study found a statistically significant difference for the inexperienced reviewer, but no such difference for the experienced reviewer. Thus, in the context of a radiologist experienced in LNMRI, it may be feasible to evaluate malignant lymph nodes based solely on post-contrast images.[28] Additional studies have found similar accuracy with the use of only post-USPIO images and meta-analysis of LNMRI studies did not show a significant difference between post-contrast only and paired MRI.[29–31]

3. LNMRI in Clinical Practice

The basic structure of the lymphotropic USPIO nanoparticle used clinically is an iron crystalline core of 4.3 nm to 6.0 nm covered by a low-molecular-weight dextran. This basic USPIO nanoparticle has been referred to in the literature by a number of different trade names including Ferumoxtran-10 also known as Combidex, Advanced Magnetics, Cambridge, MA, USA; BMS 180549, Bristol-Meyers Squibb Diagnostics, Princeton, NJ, USA; Sinerem, Guerbet Laboratories, Aulnay-sous-Bois, France; AMI-7227 and AMI-227.[25] Whenever possible the USPIO agent will be referred to as originally described in the respective studies.

The first USPIO agent administered intravenously to humans was a dextran-coated superparamagnetic iron oxide nanoparticle BMS 180549. The study was performed as a phase-I clinical trial in five healthy human subjects given five different doses ranging from 0.3 to 1.7 mg Fe/kg. MRI with T1, T2 and proton density-weighted images were performed at 1, 4 and 24 h post administration of the contrast agent. Preliminary results demonstrated a drop in signal intensity in head and neck lymph nodes 24 h after the administration of 1.7 mgFe/kg.[32]

Although, the vast majority of LNMRI studies have been performed with ferumoxtran, it is important to note that it never gained FDA approval. Future studies are likely to involve ferumoxytol, a third-generation carboxymethyl dextran bolus-injectable USPIO. Ferumoxytol contains a crystalline iron oxide core which measures 6.8 +/- 0.4 nm and was designed with a different coating to minimize immunogenicity and allow for bolus administration.[33] A blood pool half-life of 10 to 14 h and T1 shortening also make ferumoxytol useful as an MR angiographic contrast agent.[34] In 2009, ferumoxytol was approved for the treatment of iron-deficiency anemia in adult patients with chronic kidney disease [35]. The use of ferumoxytol in LNMRI is off-label but clinical trials demonstrated an excellent safety profile and a small study of prostate cancer patients demonstrated similar performance in the detection of malignant lymph nodes.[33,35]

3.1. *Head and neck cancer*

Lymph node staging is a critical determinant of prognosis and treatment selection in patients with head and neck cancer. The presence of a single regional metastatic cervical lymph node can decrease patient survival up to 50% correcting for tumor type, location and size.[29,36] Despite the superficial location of lymph nodes in the neck, assessment for metastatic lymph nodes through physical examination and palpation has had limited success. Imaging with CT and MRI are routinely used to evaluate for metastatic lymph nodes but depend on both size and morphologic features to characterize lymph nodes as either benign or malignant. The reported sensitivity and specificity for CT with a 10 mm size threshold for malignancy is 88% and 39%, respectively and 56% and 84% when the size threshold is increased to 15 mm.[37] Additional morphologic features such as nodal necrosis may add specificity, which has a reported specificity between 95% and 100%.[38]

Anzai *et al.* extended their findings from the phase I clinical trial with USPIO nanoparticles to test the ability of USPIO-enhanced MRI to differentiate benign and malignant lymph nodes. MRI was performed in 12 patients with suspected extracranial head and neck cancer pre- and post-administration of BMS 180549 of 1.7 mg Fe/kg. One of the patients suspected of having head and neck cancer underwent lymph node dissection which demonstrated an infected branchial cleft cyst with associated inflammatory lymph nodes and was excluded from the study. Of the remaining 11 patients who underwent lymph node dissection a total of 426 lymph nodes were harvested. Almost, 91 of the lymph nodes could be confidently correlated with USPIO enhanced MRI. Of the 91 lymph nodes 42 were classified as malignant and 49 benign by histopathologic evaluation. Lymph nodes were classified malignant by MRI if the lymph nodes had persistent high-signal intensity on post-contrast images and benign if the lymph nodes exhibited homogeneous signal loss on post-contrast images. Of the 42 malignant lymph nodes, 40 demonstrated high signal intensity or a mixture of high and low signal intensity and 41 of the 49 benign lymph nodes had homogeneous drop in signal intensity on post-contrast images. The study had a sensitivity of 95% and specificity of 84% in the detection of malignant lymph nodes with LNMRI. The authors speculated that the specificity may have been reduced secondary to underdosing of contrast agent as four of the eight false-positive nodes were noted in a patient whose serum iron level at 24 h was much lower than the remaining 11 patients.[39]

Another study evaluated 30 patients with biopsy-proven squamous cell carcinoma of the head and neck with LNMRI. In this study, surgical planning was modified based on findings from pre-operative LNMRI. Surgical resection and lymph node dissection provided 1029 lymph nodes for histopathologic evaluation, 69 of which were classified as malignant. LNMRI was performed in patients before and 24 h to 36 h after the administration of 2.6 mg Fe/kg Sinerem as part of a multicenter phase III trial within days of undergoing surgery. The surgical plan was based on physical exam, endoscopic exam and ultrasound with modifications based on LNMRI findings if necessary. Three of the 30 patients were no longer considered for surgical resection based on MRI findings due to extensive tumor infiltration with unresectable lymph node metastases. Evaluation on a node-by-node basis demonstrated a sensitivity of 86%, specificity of 100%, positive predictive value (PPV) of 100% and a negative predictive value (NPV) of 99%. In a patient-by-patient comparison there was complete concordance between MRI and histopathologic findings. Of the 27 patients, 26 were found to have at least one malignant lymph node. Importantly, the surgical plan was modified based on MRI findings in seven of the 27 patients. The most common scenario (five patients) was identification of additional lymph node metastases on the contralateral side resulting in limited or radical dissection on the contralateral side.[40]

Sigal *et al.* studied 81 previously untreated patients with squamous cell carcinoma of the head and neck with LNMRI as part of a multicenter phase III clinical trial of Sinerem. In the study, they compared LNMRI to pre-contrast MRI in classifying lymph nodes as either benign or malignant with histopathologic examination as the gold standard. Pre- and post-Sinerem MRI was performed within 15 days of surgery. Each patient was administered 2.6 mg Fe/kg Sinerem and post-contrast MRI was performed 24 h to 36 h later. On pre-contrast MRI, a lymph node was considered metastatic if its short-axis diameter was greater than or equal to 10 mm. On post-contrast MRI lymph nodes were categorized as benign if there was homogeneous reduction in signal intensity over the entire lymph node on T2- or T2*-weighted images, the lymph node was otherwise classified as metastatic.

Due to the difficulty in mapping lymph nodes harvested from lymph node dissection to imaging, comparison was made by lymph node groups. The authors designated a lymph node group as metastatic if at least one lymph node in that group was metastatic on MRI and non-metastatic when no metastatic nodes were identified in that group. A node-by-node analysis was also performed for individual lymph nodes with one axial dimension greater than or equal to 10 mm on histopathology. Lymph node dissection across the 81 patients yielded 3092 lymph nodes in 517 lymph node groups, 101 groups contained one or more metastases. The sensitivity for detection of metastasis in lymph node groups was ≥ 59% and the specificity of ≥ 81%, PPV of ≥ 51%, and NPV of ≥ 90%. The sensitivity for detection of metastasis in individual lymph nodes was ≥ 88% with a specificity of ≥ 81%.[29]

This study also examined the efficacy of malignant lymph node detection in 28 patients classified as N0 based on physical examination and routine traditional imaging. Lymph node dissection yielded 184 lymph node groups 19 of which were deemed metastatic over 12 of the 28 patients. In the context of N0 patients LNMRI performed no better than non-contrast MRI.[29]

Finally, the authors evaluated the efficacy of post-Sinerem images alone in the detection of malignant lymph nodes and found that Sinerem images alone performed as well as the combination of pre-contrast and post-contrast images, suggesting that in clinical practice the time and cost associated with LNMRI may be reduced by only evaluating post-contrast images only.[29]

An additional study of 20 patients with head and neck cancer similarly evaluated the use of USPIO-enhanced MRI in the detection of malignant lymph nodes. MRI was performed prior to and 24 h to 36 h following *i.v.* administration of 2.6 mg Fe/kg ferumoxtran-10. Surgery was performed either the same day after post-contrast imaging or the day after. Selected lymph nodes were removed based on MRI findings. All 63 lymph nodes were surgically removed with 36 characterized as benign, 25 malignant and two inflammatory on histopathology. Lymph nodes were categorized as malignant on non-contrast MRI if the short-axis diameter was ≥ 10 mm. Non-contrast MRI demonstrated a sensitivity of 64%, specificity of 78.9%, PPV of 75% and NPV of 76.9% which was improved with ferumoxtran-10-enhanced MRI to a sensitivity of 96%, specificity of 78.9%, PPV of 75% and NPV of 96.8%. One important note should be made of the methodology in interpreting the results. In this study, lymph nodes were prospectively selected prior to surgery by both the radiologist and surgeon and MRI images were available intraoperatively to ensure selection of the appropriate lymph nodes, thus a direct node-by-node mapping was performed without the inclusion of additional lymph nodes identified at surgery.[41] This methodology is thus likely to overestimate the sensitivity of lymphotropic MRI as compared to the study of Sigal *et al.*[29,41]

3.2. *Pelvic malignancies*

The initial evaluation of USPIO-enhanced MRI in pelvic malignances was performed as part of a multi-institutional phase II trial of Combidex in patients with known pelvic malignancy. In a small preliminary study of six patients a total of 15 lymph nodes were correlated with histopathology either from lymph node biopsy or from lymph node dissection. Of the 15 lymph nodes, 9 were malignant and six of the lymph nodes were benign by histopathology. Eight of the nine malignant lymph nodes demonstrated no evidence of signal drop following administration of 1.7 mg Fe/kg Combidex, the remaining lymph node was found to have a micrometastasis and in retrospect had heterogeneous decrease in signal, different from that observed in the benign lymph nodes. Three of the six benign lymph nodes exhibited signal loss on post-USPIO images, while no change in signal intensity was found in the remaining three lymph nodes which were found to have inflammatory changes.[42]

A subsequent study, performed as part of a phase II and III clinical trial of Combidex, included two prostate cancer patients from the phase II study described above. A total of 11 patients were involved in the study with primary pelvic or abdominal malignancy with a total of 27 lymph nodes available for histopathologic correlation within 15 days of MR evaluation. Nearly, 11of the lymph nodes were benign and 16 were malignant on histopathologic evaluation. Of the 11 benign lymph nodes, 9 were observed to have a homogenous drop in signal intensity on post-Combidex images. Two of the benign lymph nodes had unchanged signal intensity and were found to have inflammatory changes on pathologic evaluation. Nearly, 15 of the 16 malignant lymph nodes demonstrated unchanged T2 signal on post-Combidex images, while one with a micrometastasise exhibited heterogeneous signal decrease. Overall, the results of LNMRI were quite encouraging with a sensitivity of 93.8% and specificity of 88.2%.[43] This work was further extended as part of the phase II/III clinical trial of Combidex with a total of 19 patients and 49 lymph nodes with short axis greater than 1 cm, mean size 1.4 cm, evaluated. Of the 49 lymph nodes evaluated 20 were benign and 29 were malignant. In this study, all 20 of the benign lymph nodes displayed homogenous signal loss on post-contrast imaging. In contrast, 27 of the 29 malignant lymph nodes retained signal intensity on post-contrast imaging. The remaining two malignant lymph nodes were described as having a heterogeneous loss of signal intensity on post-contrast images. The sensitivity and specificity of the study was 93.1% and 100% respectively.[44] These results were very encouraging for the characterization of lymph nodes of 1 cm or greater in the short axis.

A similar study also performed as part of a phase II trial of AMI-227 included 30 patients with known urologic or pelvic malignancy and with the suspicion of metastatic lymph node involvement. AMI-227 was administered at dose of 1.7 mg Fe/kg intravenously. The authors found that each of the 27 histologically-proven malignant lymph nodes retained signal intensity on post AMI-227 images, while 16 of the 21 benign lymph nodes demonstrated the expected signal loss on post-AMI-227 images. The sensitivity and specificity of the study were 100% and 80%, respectively.[45] The authors speculated that the low dose of 1.7 mg Fe/kg was responsible for the lack of signal drop in five of 21 normal lymph nodes and supported this by a study which demonstrated increased signal loss in normal lymph nodes at the 2.6 mg Fe/kg dose compared to the 1.7 mg Fe/kg dose with no effect on the signal intensity of malignant lymph nodes.[46,47]

In the largest early evaluation of USPIO-enhanced MRI for lymph node characterization 80 patients with resectable prostate cancer, T-stage T1-3, underwent USPIO-enhanced MRI prior to surgical treatment. Nearly, 75 of the patients underwent radical prostatectomy, 60 open and 15 laparoscopic and the remaining five patients had CT-guided biopsy of lymph node without resection. MRI was performed on a 2-days protocol on a 1.5 T MRI before and 24 h after injection with 2.6 mg Fe/kg Combidex/Sinerem. Lymph nodes were evaluated prospectively with both conventional MRI and LNMRI. Lymph nodes were classified as malignant on conventional MRI if the short axis diameter was elongated and exceeded 10 mm or was rounded and exceeded 8 mm. The criteria for malignancy on LNMRI was a decrease in signal intensity less than 30% on post-contrast T2-weighted images, heterogeneous signal or focal high signal intensity consistent with a focal defect, and nodes with a central area of hyperintensity with a peripheral decrease in signal intensity. A total of 334 lymph nodes were evaluated both histopathologically and with LNMRI. Almost, 63 (18.9%) of these lymph nodes from 33 patients were classified histopathologically as malignant. Conventional MRI classified 18 lymph nodes as malignant, sensitivity 35.4%, while LNMRI correctly classified 57 lymph nodes, sensitivity 90.5%, as malignant. LNMRI did, however, identify all patients with lymph node malignancy with a sensitivity of 100% compared to sensitivity of 45.4% with conventional MRI. Importantly, sensitivity for detection of malignant lymph nodes with LNMRI dropped dramatically in lymph nodes with a short-axis less than 5 mm, sensitivity of LNMRI for lymph nodes between 5–10 mm was 96.4% compared to 41.1% for lymph nodes less than 5 mm in the short-axis.[5]

LNMRI was further evaluated in a prospective multicenter study in patients with prostate cancer in the Netherlands. Nearly, 375 consecutive prostate cancer patients with a greater than 5% risk of lymph node disease based on nomograms were imaged with both CT and LNMRI prior to either pelvic lymph node dissection for CT-guided fine needle aspiration across 11 hospitals in the Netherlands. CT and LNMRI were performed within one week of each other and each within 8 weeks of pelvic lymph node dissection. All CT scans were performed with a multidetector CT with both oral and *i.v.* contrast. LNMRI was performed on a 1.5 T magnet 24 h to 36 h following *i.v.* infusion of ferumoxtran-10. Lymph nodes were characterized as malignant if the short axis exceeded 10 mm for an oval lymph node or 8 mm for a round lymph node while lymph nodes on LNMRI were classified as malignant if the lymph node demonstrated homogeneous or focal high T2 signal intensity. A number of 361 patients underwent pelvic lymph node dissection and 14 patients underwent CT-guided lymph node fine needle aspiration. Similar to prior studies, LNMRI demonstrated superior sensitivity relative to size-based detection of malignant lymph nodes with a sensitivity of 82% versus 34%, respectively. In addition, to improved sensitivity LNMRI also demonstrated a superior NPV of 96% compared to CT which had an NPV of 88%. Interestingly, the authors of the study discovered a discrepancy between LNMRI performed at experienced hospitals versus less-experienced hospitals. The sensitivity and NPV for malignant lymph nodes was 90% and 98% at the more experienced hospitals compared to 40% and 91% at the less-experienced hospitals. These results raise the obvious question of whether this technique can be widely adapted to the community setting or whether it should remain in the domain of the specialized center. The high NPV of LNMRI of 96% is consistent with less than a 4% chance of lymph node disease and suggests that in patients with a negative LNMRI evaluation with pelvic lymph node dissection can be omitted.[48]

A recent prospective study in 75 patients sought to determine the diagnostic accuracy of LNMRI in the detection of lymph node metastases in normal-sized lymph nodes in patients with prostate or bladder cancer with extended pelvic lymph node dissection as the reference standard. Patients with lymph nodes greater than 10 mm in the short axis for bladder cancer and ≥8 mm in the short axis for prostate cancer were excluded from the study. All MRI was performed on a 3T MR unit with pre- and post-contrast imaging performed 24 h to 36 h after *i.v.* administration of 2.6 mg Fe/kg Ferumoxtran-10. A first pre-operative reading was performed by two reviewers and discussed with the surgeon to ensure removal of suspicious lymph nodes. A second post-operative reading was performed by three reviewers blinded to the results. Results from the first reading demonstrated a sensitivity of 55%, specificity of 85.5%, and NPV of 83.9% and the second reading had a sensitivity of 58.3%, specificity of 83%, and NPV of 84.4%. Relative to prior studies a reduced sensitivity and NPV was observed which is likely secondary to the exclusion of patients with enlarged lymph nodes and the high number of lymph nodes, an average of 39 per patient, from the extended pelvic lymph node dissection. Despite the relatively low sensitivity, there was accurate detection of metastases 5 mm and greater, with the majority of missed metastases less than or equal to 3 mm in size. Given the number of missed metastases 3 mm and smaller, the authors felt that LNMRI is not an adequate substitute for pelvic lymph node dissection, but could provide important information by directing the removal of suspicious lymph nodes not included in standard lymph node dissection.[49]

Standard of care in prostate cancer is to evaluate for risk of lymph node involvement through the use of nomograms or formulas based on serum prostate-specific antigen (PSA) levels and Gleason score (GS). A recent study compared the use of the nodal risk formula (NRF) to LNMRI in predicting the presence of lymph node involvement. NRF: Pmet = (2/3 * PSA) + [(GS-6) X 10], where Pmet represents the probability of lymph node metastasis. The NRF score has been used to stratify patients as either low nodal risk, < 15% risk, from those with a high nodal risk (15%) for consideration of whole pelvic radiotherapy. The patient cohort is identical to that of Heesakkers *et al.*[48] with a total of 375 patients evaluated. Both NRF and LNMRI were found to be independent predictors of lymph node metastasis. NRF was found to have a sensitivity of 79% and a specificity of 38% while LNMRI had a sensitivity of 82% and a specificity of 93%. The probability of lymph node metastasis with a negative LNMRI was 4% with no additional information provided by the NRF. The authors thus concluded that LNMRI is better than the NRF for lymph node risk assessment and that use of the NRF alone can lead to overtreatment of pelvic lymph nodes.[50]

3.3. *Breast cancer*

In the context of breast cancer, axillary lymph node status is a critical determinant of patient prognosis and thus treatment planning. Axillary lymph node dissection delivers the most sensitive and specific technique as each lymph node is specifically examined microscopically. Approximately 40%–70%

people who undergo lymph node dissection are negative for metastatic lymph nodes,[51] thus a substantial fraction of patients are needlessly subjected to the risks and morbidity of lymph node dissection which include nerve damage leading to paresthesias, seroma formation and lymphedema. The overtreatment of these patients has been recognized and has led to the investigation of less invasive means to assess lymph node status such as sentinel lymph node biopsy, ^{18}FDG-PET/CT, and novel MRI techniques to reduce the number of patients needing axillary lymph node dissection.[52–54]

Performed as part of a multicenter phase III study, 21 patients with breast cancer from two sites were evaluated with LNMRI. MRI was performed before and 24 h to 36 h after *i.v.* administration of 2.6 mg Fe/kg contrast agent. The patients then underwent axillary dissection according to protocol with no changes in surgery based on imaging findings unless a suspicious lymph node was identified outside of the anticipated surgical field. A total of 12 of the 21 patients were removed from the analysis, two patients had a minor adverse event which limited the total administered dose to less than 80%. Two patients did not undergo lymph node dissection and technical problems were noted in eight patients which resulted in unsatisfactory image quality which was discarded without interpretation. Axillary lymph node dissection identified 52 lymph nodes over nine patients, that could be definitively correlated with imaging. The authors of the study compared non-contrast MRI, post-contrast MRI, and the combination of both in the classification of lymph nodes as benign or malignant. Pre-contrast images had a sensitivity of 63% and a specificity of 86% and post-contrast images alone were 56% sensitive and 96% specific. The most accurate result was when lymph node classification was based on both pre- and post-contrast imaging with a sensitivity of 81% and specificity of 92%. While the combination of pre-and post-contrast images resulted in the greatest accuracy the NPV remained only 82% which the authors felt was not adequate to supplant axillary lymph node dissection from routine practice.[51]

A similar study of USPIO-enhanced MRI in malignant axillary lymph node detection was performed in 20 patients by Michel *et al.* Patients were imaged 24 h to 36 h post-administration of 2.6 mg Fe/kg Combidex. Two of the patients were found to have extensive tumor burden and lymphadenopathy and it was felt that these patients would benefit from neoadjuvant chemotherapy, and were thus excluded from the study. Data from axillary lymph node dissection is thus available from 18 patients, 11 of which were positive for lymph node disease and the remaining seven patients were negative. USPIO-enhanced MRI detected malignancy at the patient level with a sensitivity of

82%, specificity of 100%, PPV of 100% and NPV of 78% by two independent readers. On a node-by-node analysis sensitivity by both readers was 83% and 73% and specificity was 96% and 97%, respectively. As patients were not imaged prior to USPIO administration contrast images were not available for comparison of loss of signal intensity in the lymph nodes. The NPV of 78% again remains insufficient for LNMRI to replace lymph node dissection, but did change management in two patients and may be useful in the preoperative surgical planning.[55]

In a study of 33 patients with stage II or III breast cancer Harada *et al.* performed MRI of the axilla prior to and 24 h to 36 h after USPIO administration of 2.6 mg Fe/kg Ferumoxtran-10. Histopathologic examination of dissected lymph nodes was then compared with analysis by conventional MRI, combined pre- and USPIO-enhanced images, and USPIO-enhanced MRI alone. Lymph node dissection performed in the 33 patients yielded 622 lymph nodes (503 metastatic and 119 non-metastatic). Lymph nodes were classified as malignant by conventional MRI by two size criteria > 5 mm or > 10 mm and by morphology. For the size criteria of > 5 mm the sensitivity was 59.1% and the specificity was 86.7% and for the > 10 mm threshold the sensitivity was 15.7% and 99.2% specificity. Analysis of morphology (replacement of central fatty hilum or irregular margins) by conventional MRI resulted in a sensitivity of 36.5% and specificity of 94.1%. Analysis of both the pre- and post-USPIO images demonstrated a sensitivity of 86.4%, specificity of 97.5%, PPV of 91.1% and NPV of 96.1%. Finally, evaluation of post-USPIO images alone had a sensitivity of 84.7%, specificity of 96.8%. Patient-based results of evaluation of post-USPIO images alone demonstrated 100% sensitivity and 80% specificity, which was identical to results of analysis of the combined pre- and post-USPIO images. These results demonstrate an improved detection of metastatic lymph nodes with USPIO-enhanced MRI compared to conventional MRI. In addition, on the patient based analysis post-USPIO images were equivalent to analysis of pre- and post-contrast images, suggesting a streamlined protocol with post-USPIO images alone may be sufficient for analysis.[30]

Another interesting application of USPIO-enhanced MRI in breast cancer is mapping of lymph nodes for inclusion in radiation planning. In the context of breast cancer, radiation is targeted to axillary lymph nodes with the intent to eradicate any tumor cells harbored in the lymph nodes of the axilla. Treatment planning to the axilla is typically performed using CT, but the radiation treatment fields are largely based on bony anatomy alone, as axillary lymph nodes are often poorly demonstrated on CT. MacDonald *et al.* set out to better delineate the axillary lymphatics for more precise delivery of radiation.

A number of 23 patients underwent USPIO-enhanced MRI and results were registered with the reference treatment planning CT. The study found that standard radiation fields encompassed 81% of all lymph nodes (82% malignant and 79% benign) and were covered by the 45-Gy isodose line. They found that 8% of lymph nodes received < 50% of the prescribed dose. In addition, lymph node regions were also contoured according to Radiation Therapy Oncology Group (RTOG) guidelines. Nodal volumes contoured without a margin had 86% of the lymph nodes within the contoured volumes and this increased to 99% when a 5 mm expansion was included. The results of this study validate the RTOG contouring guidelines, but also introduce the concept of LNMRI in the mapping axillary lymph nodes to maximize individual patient benefit.[56]

3.4. *Safety*

Establishing therapeutic efficacy which can be delivered in a safe manner is the goal of clinical trials. To this end, ferumoxtran-10 has been extensively studied in phase I to III clinical trials with a meticulous record of adverse events experienced by patients. Bernd *et al.* provide a comprehensive analysis of the safety and tolerability of ferumoxtran-10 with data pooled from 37 clinical trials encompassing a total of 1777 patients. Of the 1777 patients, 1663 received ferumoxtran-10 (1527 patients and 136 healthy volunteers). An additional 75 patients received placebo and 39 patients were enrolled in the studies but ultimately did not receive the contrast agent.[57]

Pregnant women and children were excluded from the clinical trials, but other patients traditionally considered high risk, such as patients with history of allergy, elderly patients, as well as patients with cardiac disease and renal or hepatic failure were included in the trials, thus allowing establishment of safety and tolerability in a broad patient population. In all, 830 adverse events were reported in 386 patients (23.2%). The majority of adverse events were mild (56.2%), followed by moderate adverse events (30.2%), and severe adverse events (11.6%). Information regarding the severity of reaction was not available in 2% of the patients. The most common adverse event was back pain experienced by 3.4% of patients followed by headache, 2.9%; pruritus, 2.5%; urticaria, 2.5%; and nausea, 2.3%.

The vast majority, 86.3%, of back pain was considered mild to moderate in severity. A typical presentation for back pain, nearly 50%, was the onset of back pain within 5 min after initiation of infusion and approximately 80% within 30 min of initiation of infusion. The general recommendation is to stop the infusion after the onset of back pain and to monitor the patient for resolution of back pain. Once the back pain has resolved the infusion can be resumed at a slower rate with the patient under constant observation. Back pain was also reported with larger SPIO particles such as AMI-25 and it is thought to be secondary to the colloid effect in the microvasculature of the paraspinal muscle.[58]

Patients with a history of hypersensitivity had a significantly higher rate of adverse events relative to patients that did not. Among the patients studied, 361 (21.7%) reported a history of allergy and had a higher rate of multiple adverse events relative to patients without allergy history including headache 4.4% to 2.4%, pruritus 5.8% to 1.6%, urticaria 4.4 to 1.8%, flushing 3.6% to 1.4% and hot flush 1.15% to 0.2% all of which were statistically significant. Interestingly, no difference was observed in the rate of adverse events between ferumoxtran-10 and placebo in patients with history of allergy.[57]

Serious adverse events, defined as life-threatening, resulting in death, requiring hospitalization or resulting in incapacity or disability, were reported in 44 patients (2.6%). Only seven of the 76 reported serious adverse events were considered to be related to ferumoxtran-10 administration. Events linked to ferumoxtran-10 treatment included two cases of hypotension and one case of death, chest pain, dyspnea, skin rash and oxygen desaturation.

Nearly, 12 deaths were reported in the clinical trials. Only one of the deaths was directly attributable to ferumoxtran-10. The patient was a 70 year old man with multiple competing comorbidities including poor general health, metastatic cancer, chemotherapy-related cardiac failure or cardiomyopathy, and history of allergy to iodinated contrast agents. Approximately 4 min after the bolus injection of undiluted ferumoxtran-10 the patient went into anaphylactic shock with flushing, dyspnea, asthenia, urinary incontinence, coma and pulselessness. The remainder of deaths occurred 7 days to up to 536 days following ferumoxtran-10 administration with causes of death attributable to stage IV lung cancer, cerebrovascular accident and septic shock among other maladies. It should be noted that the risk of serious adverse event was higher in patients who received an undiluted bolus injection than a diluted infusion. As the risk of serious adverse event appeared higher after undiluted bolus injection, this method of delivery was abandoned early in development and subsequent clinical trials were designed with a diluted infusion of ferumoxtran-10.[57]

3.5. *Interpretation pitfalls*

Interpretation of LNMRI can at times be difficult as correct interpretation of lymph node status may require careful review of multiple imaging sequences with attention to subtle signal characteristics over a potentially large number of lymph nodes. As one might expect, given the complexity of the imaging findings experience with the imaging modality is a factor in the accuracy of interpretation particularly if only post-USPIO images are interpreted.[28,48,59]

In addition to experience, understanding the basics of lymph node architecture may be useful in characterizing unexpected signal characteristics. Normal lymph node parenchyma is comprised of the cortex and medulla. The lymph node hilum contains the efferent lymphatic vessel as well as the lymph node vasculature, artery and vein. Lymph enters the lymph node through the afferent vessels which enter through the fibrous capsule of the lymph node at the periphery and into the subcapsular and marginal sinuses. The lymph then flows through the reticular meshwork of the lymph node and then exits the lymph node through the efferent lymphatic vessel at the hilum. The macrophages, which phagocystose the USPIO nanoparticles, are distributed in the subcapsular sinus and in the medullary sinuses. Consistent with lymph node architecture, high resolution MRI with a 9.4 T unit and histologic comparison of individual rat lymph nodes revealed accumulation of USPIO nanoparticles in the medullary sinuses with nanoparticle sparing in the lymphocyte-containing follicles.[47,60] At lower magnetic field strengths such as 1.5T and 3.0T used in routine clinical practice the anatomic distribution in the lymph node is not appreciated due to reduced spatial resolution and volume averaging, thus normal lymph nodes appear to have a homogenous reduction in signal intensity.[60]

Early in the development of LNMRI a high fraction of benign lymph nodes were classified as malignant. For example, in one study 24% of normal lymph nodes did not demonstrate the characteristic signal intensity loss expected in normal lymph nodes. It was suggested at the time that this may be related to an insufficient contrast dose thus the standard dose of contrast agent was increased from 1.7 mg Fe/kg to 2.6 mg Fe/kg.[45,47,61] This observation has important implications in interpreting images particularly in patients who received a fraction of the dose either secondary to contrast extravasation or adverse event. Interpretation of such studies could lead to the classification of benign lymph nodes as falsely positive. One strategy employed by Stets *et al.* was to exclude patients who received less than 80% of the prescribed dose.[51]

Additional etiologies of false classification of benign lymph nodes as malignant include processes which lead to the destruction of the normal lymph node architecture by obstructing the transport of lymph to lymph nodes or processes that displace macrophages in the medullary sinuses and subcapsular sinuses. Benign etiologies which have been described to lead to false-positive results include granulomatosis, fibrosis, sinus histocytosis, lipomatosis and focal fatty infiltration.[26,49,62] In the case of fat, the T1-weighted sequence can be important for identifying a fatty hilum.[25] Reactive hyperplasia has also been reported to lead to false-positive results. It may result in diminished uptake of contrast material. Absence of uptake of nanoparticles has been described at the periphery in lymph nodes has been described as a pattern in hyperplastic nodes leading to a halo on T2*-weighted images.

As with other imaging studies, artifact from motion can degrade the quality of images and prevent accurate interpretation. Patient motion may come from gross patient movement or from physiologic motion related to the cardiac cycle, respiration or bowel motility. Cardiac and respiratory motion can interfere with image acquisition of the thorax and axilla and bowel motility tends to interfere with evaluation of abdominal and pelvic lymph nodes. Image acquisition can be performed with breath holds, respiratory gating and/or cardiac gating to reduce motion artifact.[63] Bowel motility can be reduced with the administration of 1.0 mg of the spasmolytic agent glucagon.[49,64]

Finally, microscopic metastases are a major source of false-negatives in LNMRI. Although lymph node metastases as small as 2 mm have been detected by LNMRI,[5] meticulous pathologic examination identified metastases < 1 mm in at least one study.[49] If the metastasis is confined to the cortex and does not significantly disrupt or replace the macrophages in the medullary sinuses, the uptake of USPIO nanoparticles will remain unchanged and the lesion will go undetected. Micrometastases that only partially disrupt the nodal architecture leading to reduced USPIO uptake in a portion of the node may be obscured by blooming artifact from USPIO uptake in the normal portion of the lymph node.[62,65]

4. Thinking Outside the Lymph Node

USPIO nanoparticles are internalized by macrophages which are abundant in normal lymph nodes, thus as reviewed in detail earlier, USPIO-enhanced MRI is a valuable technique for assessing lymph node status. Macrophages, however, are highly mobile and migrate outside of lymph nodes to sites of inflammation. USPIO-enhanced MRI can detect focal aggregation of macrophages through T2* signal intensity loss due to the T2-shortening from USPIO nanoparticle uptake, as in lymph nodes, and is being investigated as an imaging marker of inflammation. Although very early in clinical development USPIO-enhanced MRI has been applied to numerous inflammatory disorders including multiple sclerosis,[66,67] type-I diabetes,[68] carotid atherosclerosis, abdominal aortic aneurysms and myocardial infarction.

4.1. *Carotid disease, the vulnerable plaque*

Atherosclerotic disease in the carotid arteries has important implications in cerebral perfusion and stroke risk with atherosclerotic plaque instability representing a major contributor to the risk of ischemic complications. An accumulation of macrophages has been identified at autopsy studies at sites of plaque rupture and the degree of macrophage infiltration has been found to increase the risk of plaque rupture.[69,70] Thus, the application of USPIO-enhanced MRI for the detection of macrophages in atherosclerotic plaque may provide important insight into the risk of plaque rupture. Kooi *et al.* performed a study of 11 symptomatic patients scheduled for carotid endarterectomy. The patients had MRIs performed before, 24 h and 72 h after the administration of 2.6 mg Fe/kg Sinerem. Correlation of imaging findings and histopathologic analysis demonstrated that there was accumulation of USPIO in macrophages in ruptured and rupture prone atherosclerotic plaques which was detected as signal loss on MR images.[71] Similarly, Trivedi *et al.* evaluated eight patients scheduled for endarterectomy with Sinerem-enhanced MRI at 12, 24, 36, 48 and 72 h following contrast administration. Correlation with imaging and histologic findings demonstrated identification of macrophages on MRI.

The ATHEROMA (Atorvastatin Therapy: Effects on Reduction of Macrophage Activity) study used USPIO-enhanced MRI to investigate the effects of low-dose and high-dose atorvastatin on macrophage infiltration in carotid atheroma and in essence plaque vulnerability. A number of 20 patients were randomized to the low-dose atorvastatin group (10 mg daily) and 20 patients in the high-dose atorvastatin group (80 mg daily). MRI examinations were performed prior to initiation of statin therapy and after a 12 week course. In this study, authors found a significant reduction in macrophage infiltration with aggressive lipid lowering therapy with high-dose atorvastatin. Importantly, the authors emphasized the use of USPIO-enhanced MRI as a means to evaluate the efficacy of modulators of inflammation on carotid disease early in clinical development in a less-expensive timely fashion as a screening mechanism prior to the entrance to conventional large-scale clinical trials.[72]

4.2. *Aneurysm surveillance*

Current surveillance of abdominal aortic aneurysms (AAA) relies on the size of the aneurysm to predict risk of rupture. This, of course, assumes uniform structural integrity and strength of the aneurysmal sac amongst aneurysms. Risk assessment based on size alone is insensitive and it is estimated that approximately 20% of ruptured AAA are under 5.5 cm in diameter, a commonly used threshold for intervention. Most AAA arise in the setting of atherosclerosis, an inflammatory process, with the immune-mediated release and activation of proteolytic enzymes which lead to the irreversible degradation and remodeling of the extracellular matrix consequently weakening the aortic wall. At least two studies have evaluated the use of USPIO-enhanced MRI to detect macrophages as a marker of inflammation.[73,74] Richards *et al.* examined 29 asymptomatic patients with Sinerem-enhanced MRI with AAA ranging from 4.0 to 6.6 cm. In patients who underwent open AAA repair there was good concordance between imaging findings and histologic findings. Interestingly, the authors discovered that patients with mural uptake of USPIO had a three-fold higher growth rate than those with none or non-specific uptake despite a similar size distribution.[73] Although the study did not address risk of rupture directly, the findings are nevertheless intriguing and further investigation is warranted to determine if UPSIO-enhanced MRI will improve risk stratification compared to size alone.

4.3. *Myocardial infarction*

Acute myocardial infarction is known to induce an inflammatory response with the mobilization and infiltration of immune cells to the infarcted region to clear necrotic debris and enable remodeling of the infarcted myocardium. Excessive post-myocardial infraction inflammation, however, may have detrimental effects on tissue remodeling and contribute to the development of heart failure. Thus, the inflammatory response and its mediators have been implicated as potential targets for the development of novel therapeutics to improve infarction outcomes. Two studies recently evaluated the use of USPIO-enhanced MRI in patients following acute ST-elevation myocardial infarction (STEMI). The first study by Alam *et al.* examined 16 patients following acute STEMI. Patients underwent three sequential MRI studies within five days of admission. A pre-contrast MRI was performed and repeat studies were performed at 24 h and 48 hours following the administration of ferumoxytol. USPIO-nanoparticles were found to localize in the infarct and peri-infarct zone and to a lesser extent myocardium remote from the infarct. A second study of 14 patients following acute-STEMI had similar findings with USPIO uptake observed in the infarct and peri-infarct regions as well as the presumably healthy remote myocardium.[75] While neither of the studies have direct histopathologic correlation, the distribution of USPIO uptake is consistent with cellular inflammation seen in other models of myocardial infarction. USPIO-enhanced MRI provides direct non-invasive visualization of the post-MI inflammatory response which may be useful in risk stratification in left ventricular remodeling and evaluating therapies directed at improving left ventricular remodeling.

5. Future Directions

The initial superparamagnetic iron oxide particles studied had a short blood half-life, 6 min for AMI-25, and quickly accumulated in the liver and spleen.[23] The size fractionation of SPIO particles into USPIO nanoparticles resulted in a significantly increased blood half-life and altered biodistribution. USPIO nanoparticles were found to enter the interstitial space and eventually the lymphatic system. The marked increase in the blood half-life and the ability of the nanoparticles to cross the capillary endothelium sparked the concept that the distribution of the nanoparticles could be more finely controlled and even targeted to specific tissues or cells.

Weissleder *et al.* was the first to demonstrate the selective targeting of USPIO nanoparticles by targeting the asialoglycoprotein receptor on hepatocytes.[76,77] USPIO nanoparticles in the study were modified through the conjugation of arabinogalactan (AG-USPIO), a galactose-terminated polysaccharide and ligand for the asialoglycoprotein receptor. Selective targeting of the nanoparticles to hepatocytes was shown through electron microscopy *in vitro* and by the *in vivo* targeting of the rat liver by MRI. The diagnostic utility of targeted nanoparticles was further validated with the detection of liver tumors in a rat model of liver cancer. Uptake of AG-USPIO nanoparticles in normal hepatocytes resulted in decreased liver signal intensity relative to tumor tissue which did not accumulate the modified nanoparticles.[77]

Selective targeting of the hepatocyte with AG-USPIO nanoparticles was an important proof of concept that USPIO nanoparticles could be modified to target receptors specific to tissue types and possibly pathology. This sparked a tremendous interest in conjugating additional tissue-specific and disease-specific targeting ligands to USPIO nanoparticles to selectively image specific tissues and disease processes. Numerous ligands have been conjugated to USPIO nanoparticles for the purpose of pathology-specific imaging and evaluated in preclinical studies *in vitro* and in animal studies. Experimental ligand-USPIO conjugates include transferrin-USPIO, breast cancer;[78] folate-USPIO, breast cancer;[79] antiGPC3-USPIO, hepatocellular carcinoma;[80] and anti-CXCR4-USPIO, pancreatic cancer[81] to list just a few of the experimental USPIO conjugates and their respective targets.

6. Conclusions and Outlook

Imaging with nanoparticles is a developing concept which will one day allow the interrogation of pathophysiology for the improved detection and surveillance of disease relative to routine anatomic imaging modalities. In this chapter we reviewed the most widely studied, technologically mature, and only nanoparticle in clinical use, the USPIO nanoparticle. USPIO nanoparticles have been studied extensively in the setting of LNMRI and have demonstrated superior diagnostic performance compared to conventional MRI in lymph node staging in patients with cancer.[31] Despite the high diagnostic accuracy of LNMRI, with the limitations of resolution of MRI at approximately 1 mm microscopic metastases remained undetected in a number of studies. Since false-negative results may have devastating clinical consequences it is unlikely that the adoption of LNMRI will replace conventional methods of lymph node staging such as pelvic lymph node dissection in patients with pelvic malignancies or sentinel lymph node biopsy in patients with breast cancer. LNMRI, however, may play an important adjunctive role in stratifying patient risk for lymph node metastasis as well as in surgical planning by detecting malignant lymph nodes outside the routine surgical dissection field. The precise role of LNMRI in oncology will require further investigation on a disease-specific basis.

USPIO-enhanced MRI has expanded beyond the lymph node and is currently being evaluated in a number of inflammatory disorders. The most thoroughly studied is the application of USPIO-enhanced MRI in the interrogation of macrophages in atherosclerotic disease.

Finally, USPIO nanoparticles are modifiable through the conjugation of ligands to the coating of the monocrystaline iron core allowing for the selective targeting of the nanoparticles to specific tissue types and disease processes. Numerous examples of selectively-targeted nanoparticles exist as investigative tools in preclinical development, but adaptation to clinical use in humans remains distant. Nevertheless, these experimental nanoparticles offer a glimpse into the future of imaging.

Acknowledgments

References

1. D. J. Brenner and E. J. Hall, Computed Tomography — An Increasing Source of Radiation Exposure. *New Engl. J. Med.* **357**, 2277–2284 (2007).

2. R. Smith-Bindman, Radiation Dose Associated With Common Computed Tomography Examinations and the Associated Lifetime Attributable Risk of Cancer Radiation Dose From CT and Risk of Cancer. *Arch. Intern. Med.* **169**, 2078 (2009).
3. M. K.Gospodarowicz, Hodgkin's Lymphoma—Patients Assessment and Staging. *Cancer J.* **15**, 138–142 (2009).
4. S. J.Vinnicombe and R. H Reznek, Computerised tomography in the staging of Hodgkin's disease and non-Hodgkin's lymphoma. *Eur. J. Nucl. Med. Mol. Imaging* **30 Suppl 1**, S42–S55 (2003).
5. M. G Harisinghani *et al.*, Noninvasive Detection of Clinically Occult Lymph-Node Metastases in Prostate Cancer. *New Engl. J. Med.* **348**, 2491–2499 (2003).
6. F. Abdollah *et al.,* National Comprehensive Cancer Network practice guidelines 2011: Need for more accurate recommendations for pelvic lymph node dissection in prostate cancer. *J. Urol.* **188**, 423–428 (2012).
7. S. Joniau *et al.*, Mapping of Pelvic Lymph Node Metastases in Prostate Cancer. *Eur. Urol.* doi:10.1016/j.eururo.2012.06.057 (2012).
8. A. Briganti, N. Suardi, A. Gallina, F. Abdollah and F. Montorsi, Pelvic Lymph Node Dissection in Prostate Cancer: The Mystery Is Taking Shape. *Eur. Urol.* (2012). doi:10.1016/j.eururo.2012.08.029 .
9. A. M. Hövels *et al.*, The diagnostic accuracy of CT and MRI in the staging of pelvic lymph nodes in patients with prostate cancer: a meta-analysis. *Clin. Radiol.* **63**, 387–395 (2008).
10. I. Kantorová *et al.*, Routine (18)F-FDG PET preoperative staging of colorectal cancer: comparison with conventional staging and its impact on treatment decision making. *J. Nucl. Med.* **44**, 1784–1788 (2003).
11. U. Tateishi *et al.*, Non-enhanced CT versus contrast-enhanced CT in integrated PET/CT studies for nodal staging of rectal cancer. *Eur. J. Nucl. Med. Mol. Imaging* **34**, 1627–1634 (2007).
12. Y.-Y. Lu *et al.*, A systematic review and meta-analysis of pretherapeutic lymph node staging of colorectal cancer by 18F-FDG PET or PET/CT. *Nucl. Med. Commun.* **33**, 1127–1133 (2012).
13. L. M. Freeman and M. D. Blaufox, Letter from the Editors: low-sensitivity FDG-PET studies. *Semin. Nucl. Med.* **42**, 219–220 (2012).
14. S. T. Lee, N. Lawrentschuk and A. M. Scott, PET in prostate and bladder tumors. *Semin. Nucl. Med.* **42**, 231–246 (2012).
15. S. L. Rice, C. A. Roney, P. Daumar and J. S.Lewis, The next generation of positron emission tomography radiopharmaceuticals in oncology. *Semin. Nucl. Med.* **41**, 265–282 (2011).
16. S. H. Choi *et al.*, Lymph Node Metastasis: Ultrasmall Superparamagnetic Iron Oxide-enhanced MR Imaging versus PET/CT in a Rabbit Model. *Radiology* **242**, 137–143 (2006).
17. A. W. Leung, R. E. Steiner and I. R. Young, NMR imaging of the liver in two cases of iron overload. *J. Comput. Assist. Tomogr.* **8**, 446–449 (1984).
18. D. D. Stark *et al.*, Magnetic resonance imaging and spectroscopy of hepatic iron overload. *Radiology* **154**, 137–142 (1985).
19. S. Saini *et al.*, Ferrite particles: a superparamagnetic MR contrast agent for the reticuloendothelial system. *Radiology* **162**, 211–216 (1987).
20. S. Saini *et al.*, Ferrite particles: a superparamagnetic MR contrast agent for enhanced detection of liver carcinoma. *Radiology* **162**, 217–222 (1987).

21. R. Weissleder *et al.*, MR imaging of splenic metastases: ferrite-enhanced detection in rats. *AJR Am. J. Roentgenol.* **149**, 723–726 (1987).
22. R. Weissleder *et al.*, Experimental lymph node metastases: enhanced detection with MR lymphography. *Radiology* **171**, 835–839 (1989).
23. R. Weissleder *et al.*, Ultrasmall superparamagnetic iron oxide: characterization of a new class of contrast agents for MR imaging. *Radiology* **175**, 489–493 (1990).
24. R. Weissleder *et al.*, Ultrasmall superparamagnetic iron oxide: an intravenous contrast agent for assessing lymph nodes with MR imaging. *Radiology* **175**, 494–498 (1990).
25. M. G. Harisinghani *et al.*, MR Lymphangiography: Imaging Strategies to Optimize the Imaging of Lymph Nodes with Ferumoxtran-10. *Radiographics* **24**, 867–878 (2004).
26. M. Saksena *et al.*, Comparison of lymphotropic nanoparticle-enhanced MRI sequences in patients with various primary cancers. *AJR Am. J. Roentgenol.* **187**, W582–W588 (2006).
27. Y. Anzai *et al.*, Evaluation of Neck and Body Metastases to Nodes with Ferumoxtran 10-enhanced MR Imaging: Phase III Safety and Efficacy Study. *Radiology* **228**, 777–788 (2003).
28. M. G. Harisinghani, Ferumoxtran-10-Enhanced MR Lymphangiography: Does Contrast-Enhanced Imaging Alone Suffice for Accurate Lymph Node Characterization? *Amer. J. Roentgenol.* **186**, 144–148 (2006).
29. R. Sigal *et al.*, Lymph node metastases from head and neck squamous cell carcinoma: MR imaging with ultrasmall superparamagnetic iron oxide particles (Sinerem MR) -- results of a phase-III multicenter clinical trial. *Eur. Radiol.* **12**, 1104–1113 (2002).
30. T. Harada, N. Tanigawa, M. Matsuki, T. Nohara and I. Narabayashi, Evaluation of lymph node metastases of breast cancer using ultrasmall superparamagnetic iron oxide-enhanced magnetic resonance imaging. *Eur. J. Radiol.* **63**, 401–407 (2007).
31. L. Wu, Y. Cao, C. Liao, J. Huang and F. Gao, Diagnostic performance of USPIO-enhanced MRI for lymph-node metastases in different body regions: A meta-analysis. *Eur. J. Radiol.* **80**, 582–589 (2011).
32. Y. Anzai, S. McLachlan, M. Morris, R. Saxton and R. B Lufkin,. Dextran-coated superparamagnetic iron oxide, an MR contrast agent for assessing lymph nodes in the head and neck. *AJNR Am. J. Neuroradiol.* **15**, 87–94 (1994).
33. M. Harisinghani, R. W. Ross, A. R. Guimaraes and R. Weissleder, Utility of a new bolus-injectable nanoparticle for clinical cancer staging. *Neoplasia* **9**, 1160–1165 (2007).
34. M. Sigovan, W. Gasper, H. F. Alley, C. D. Owens and D. Saloner, USPIO-enhanced MR angiography of arteriovenous fistulas in patients with renal failure. *Radiology* **265**, 584–590 (2012).
35. M. Lu, M. H. Cohen, D. Rieves and R. Pazdur, FDA report: Ferumoxytol for intravenous iron therapy in adult patients with chronic kidney disease. *Am. J. Hematol.* **85**, 315–319 (2010).
36. C. F. Thompson *et al.*, Diagnostic value of sentinel lymph node biopsy in head and neck cancer: a meta-analysis. *Eur. Archives of Oto-Rhino-Laryngology* (2012). doi:10.1007/s00405-012-2320-0
37. H. D. Curtin *et al.*, Comparison of CT and MR imaging in staging of neck metastases. *Radiology* **207**, 123–130 (1998).
38. J. K. Hoang, J. Vanka, B. J. Ludwig and C. M. Glastonbury, Evaluation of Cervical Lymph Nodes in Head and Neck Cancer With CT and MRI: Tips, Traps, and a Systematic Approach. *Amer. J. Roentgenol.* **200**, W17–W25 (2012).

39. Y. Anzai *et al.*, Initial clinical experience with dextran-coated superparamagnetic iron oxide for detection of lymph node metastases in patients with head and neck cancer. *Radiology* **192**, 709–715 (1994).
40. M. G. Mack, J. O. Balzer, R. Straub, K. Eichler and T. J. Vogl, Superparamagnetic iron oxide-enhanced MR imaging of head and neck lymph nodes. *Radiology* **222**, 239–244 (2002).
41. L. Curvo-Semedo *et al.*, USPIO-enhanced magnetic resonance imaging for nodal staging in patients with head and neck cancer. *J. Magn. Reson. Imaging* **24**, 123–131 (2006).
42. M. G. Harisinghani, S. Saini, G. J. Slater, M. D. Schnall and M. D. Rifkin, MR imaging of pelvic lymph nodes in primary pelvic carcinoma with ultrasmall superparamagnetic iron oxide (Combidex): preliminary observations. *J. Magn. Reson. Imaging* 7, 161–163 (1997).
43. M. G. Harisinghani, S. Saini, P. F. Hahn, R. Weissleder and P. R. Mueller, MR imaging of lymph nodes in patients with primary abdominal and pelvic malignancies using ultrasmall superparamagnetic iron oxide (Combidex). *Acad. Radiol.* **5 Suppl 1**, S167–S169, discussion S183–184 (1998).
44. M. G. Harisinghani *et al.*, MR lymphangiography using ultrasmall superparamagnetic iron oxide in patients with primary abdominal and pelvic malignancies: radiographic-pathologic correlation. *AJR Am. J. Roentgenol.* **172**, 1347–1351 (1999).
45. M. F. Bellin *et al.*, Lymph node metastases: safety and effectiveness of MR imaging with ultrasmall superparamagnetic iron oxide particles--initial clinical experience. *Radiology* **207**, 799–808 (1998).
46. P. A. Hudgins, Y. Anzai, M. R. Morris and M. A. Lucas, Ferumoxtran-10, a superparamagnetic iron oxide as a magnetic resonance enhancement agent for imaging lymph nodes: a phase 2 dose study. *AJNR Am. J. Neuroradiol.* **23**, 649–656 (2002).
47. M. F. Bellin, L. Lebleu and J. B. Meric, Evaluation of retroperitoneal and pelvic lymph node metastases with MRI and MR lymphangiography. *Abdom. Imaging* **28**, 155–163 (2003).
48. R. A. M. Heesakkers *et al.*, MRI with a lymph-node-specific contrast agent as an alternative to CT scan and lymph-node dissection in patients with prostate cancer: a prospective multicohort study. *Lancet Oncol.* **9**, 850–856 (2008).
49. M. Triantafyllou *et al.*, Ultrasmall superparamagnetic particles of iron oxide allow for the detection of metastases in normal sized pelvic lymph nodes of patients with bladder and/ or prostate cancer. *Eur. J. Cancer* (2012). doi:10.1016/j.ejca.2012.09.034
50. W. M. L. L. G. Deserno *et al.*, Comparison of Nodal Risk Formula and MR Lymphography for Predicting Lymph Node Involvement in Prostate Cancer. *Int. J. Radiation Oncol. Biol. Phys.* **81**, 8–15 (2011).
51. C. Stets, *et al.* Axillary lymph node metastases: a statistical analysis of various parameters in MRI with USPIO. *J Magn Reson Imaging* **16**, 60–68 (2002).
52. U. Veronesi *et al.*, Sentinel-node biopsy to avoid axillary dissection in breast cancer with clinically negative lymph-nodes. *Lancet* **349**, 1864–1867 (1997).
53. T. W. Stadnik *et al.*, Breast imaging. Preoperative breast cancer staging: comparison of USPIO-enhanced MR imaging and 18F-fluorodeoxyglucose (FDC) positron emission tomography (PET) imaging for axillary lymph node staging--initial findings. *Eur. Radiol.* **16**, 2153–2160 (2006).
54. D. Groheux, M. Espié, S. Giacchetti and E. Hindié, Performance of FDG PET/CT in the Clinical Management of Breast Cancer. *Radiology* **266**, 388–405 (2013).

55. S. C. A. Michel *et al.*, Preoperative Breast Cancer Staging: MR Imaging of the Axilla with Ultrasmall Superparamagnetic Iron Oxide Enhancement. *Radiology* **225**, 527–536 (2002).
56. S. M. MacDonald *et al.*, Nanoparticle-enhanced MRI to evaluate radiation delivery to the regional lymphatics for patients with breast cancer. *Int. J. Radiat. Oncol. Biol. Phys.* **77**, 1098–1104 (2010).
57. H. Bernd, E. De Kerviler, S. Gaillard and B. Bonnemain, Safety and tolerability of ultrasmall superparamagnetic iron oxide contrast agent: comprehensive analysis of a clinical development program. *Invest. Radiol.* **44**, 336–342 (2009).
58. Y. Anzai, Superparamagnetic iron oxide nanoparticles: nodal metastases and beyond. *Top. Magn. Reson. Imaging* **15**, 103–111 (2004).
59. M. Harisinghani, Nanoparticle-enhanced MRI: are we there yet? *The Lancet Oncology* **9**, 814–815 (2008).
60. A. S. Lee, R. Weissleder, T. J. Brady and J. Wittenberg, Lymph nodes: microstructural anatomy at MR imaging. *Radiology* **178**, 519–522 (1991).
61. M. Torabi, S. L. Aquino and M. G. Harisinghani, Current concepts in lymph node imaging. *J. Nucl. Med.* **45**, 1509–1518 (2004).
62. P. Narayanan, T. Iyngkaran, S. A. Sohaib, R. H. Reznek and A. G. Rockall, Pearls and Pitfalls of MR Lymphography in Gynecologic Malignancy1. *Radiographics* **29**, 1057–1069 (2009).
63. H. K. Pannu, K. P. Wang, T. L. Borman and D. A. Bluemke, MR imaging of mediastinal lymph nodes: evaluation using a superparamagnetic contrast agent. *J. Magn. Reson. Imaging* **12**, 899–904 (2000).
64. K. Rafat Zand *et al.*, Artifacts and pitfalls in MR imaging of the pelvis. *J. Magn. Reson. Imaging* **26**, 480–497 (2007).
65. J. Lei, H. D. Xue, Z. Li, S. Li and Z. Y. Jin, Possible pathological basis for false diagnoses of lymph nodes by USPIO-enhanced MRI in rabbits. *J. Magn. Reson. Imaging* **31**, 1428–1434 (2010).
66. M. M. Vellinga *et al.*, Pluriformity of inflammation in multiple sclerosis shown by ultra-small iron oxide particle enhancement. *Brain* **131**, 800–807 (2008).
67. T. Tourdias *et al.*, Assessment of disease activity in multiple sclerosis phenotypes with combined gadolinium- and superparamagnetic iron oxide-enhanced MR imaging. *Radiology* **264**, 225–233 (2012).
68. J. L. Gaglia *et al.*, Noninvasive imaging of pancreatic islet inflammation in type 1A diabetes patients. *J. Clin. Invest.* **121**, 442–445 (2011).
69. R. Ross, Atherosclerosis--an inflammatory disease. *N. Engl. J. Med.* **340**, 115–126 (1999).
70. N. Nighoghossian, L. Derex and P. Douek, The vulnerable carotid artery plaque: current imaging methods and new perspectives. *Stroke* **36**, 2764–2772 (2005).
71. M. E. Kooi, Accumulation of Ultrasmall Superparamagnetic Particles of Iron Oxide in Human Atherosclerotic Plaques Can Be Detected by In Vivo Magnetic Resonance Imaging. *Circulation* **107**, 2453–2458 (2003).
72. T. Y. Tang *et al.*, The ATHEROMA (Atorvastatin Therapy: Effects on Reduction of Macrophage Activity) Study. *J. Am. College of Cardiol.* **53**, 2039–2050 (2009).
73. J. M. J. Richards *et al.*, Abdominal Aortic Aneurysm Growth Predicted by Uptake of Ultrasmall Superparamagnetic Particles of Iron Oxide: A Pilot Study. *Circulation: Cardiovasc. Imaging* **4**, 274–281 (2011).

74. U. Sadat *et al.*, Ultrasmall superparamagnetic iron oxide-enhanced magnetic resonance imaging of abdominal aortic aneurysms--a feasibility study. *Eur. J. Vasc. Endovasc. Surg.* **41**, 167–174 (2011).
75. A. Yilmaz *et al.*, Imaging of myocardial infarction using ultrasmall superparamagnetic iron oxide nanoparticles: a human study using a multi-parametric cardiovascular magnetic resonance imaging approach. *Eur. Heart J.* **34**, 462–475 (2013).
76. R. Weissleder, P. Reimer, A. S. Lee, J. Wittenberg and T. J. Brady, MR receptor imaging: ultrasmall iron oxide particles targeted to asialoglycoprotein receptors. *AJR Am. J. Roentgenol.* **155**, 1161–1167 (1990).
77. P. Reimer, R. Weissleder, A. S. Lee, J. Wittenberg and T. J. Brady, Receptor imaging: application to MR imaging of liver cancer. *Radiology* **177**, 729–734 (1990).
78. M. Kresse *et al.* Targeting of ultrasmall superparamagnetic iron oxide (USPIO) particles to tumor cells in vivo by using transferrin receptor pathways. *Magn. Reson. Med.* **40**, 236–242 (1998).
79. R. Meier *et al.*, Breast cancers: MR imaging of folate-receptor expression with the folate-specific nanoparticle P1133. *Radiology* **255**, 527–535 (2010).
80. Y. Li, Z. Chen, F. Li, J. Wang and Z. Zhang, Preparation and in vitro studies of MRI-specific superparamagnetic iron oxide antiGPC3 probe for hepatocellular carcinoma. *Int. J. Nanomedicine* 7, 4593–4611 (2012).
81. Y. He *et al.*, Anti-CXCR4 monoclonal antibody conjugated to ultrasmall superparamagnetic iron oxide nanoparticles in an application of MR molecular imaging of pancreatic cancer cell lines. *Acta. Radiol.* **53**, 1049–1058 (2012).

Index

Vol. 3

Frontiers in Nanobiomedical Research

HANDBOOK OF NANOBIOMEDICAL RESEARCH

Fundamentals, Applications and Recent Developments

❷ Applications in Therapy

Frontiers in Nanobiomedical Research

ISSN: 2251-3965

Series Editors: Martin L. Yarmush *(Harvard Medical School, USA)*
Donglu Shi *(University of Cincinnati, USA)*

Published

Vol. 1: Handbook of Immunological Properties of Engineered Nanomaterials
edited by Marina A. Dobrovolskaia and Scott E. McNeil (SAIC-Frederick, Inc., USA)

Vol. 2: Tissue Regeneration: Where Nano-Structure Meets Biology
edited by Qing Liu (3D Biotek, USA and Tongji University, China) and Hongjun Wang (Stevens Institute of Technology, USA)

Vol. 3: Handbook of Nanobiomedical Research: Fundamentals, Applications and Recent Developments (In 4 Volumes)
edited by Vladimir Torchilin (Northeastern University, USA)

Forthcoming titles

Cancer Therapeutics and Imaging: Molecular and Cellular Engineering and Nanobiomedicine
edited by Kaushal Rege (Arizona State University, USA)

Nano Vaccines
edited by Balaji Narasimhan (Iowa State University, USA)

Nano Pharmaceuticals
edited by Rajesh N. Dave (New Jersey Institute of Technology, USA)

Thermal Aspects in Nanobiomedical Systems and Devices
by Dong Cai (Boston College, USA)

Nano Mechanochemistry in Biology
edited by Jeffrey Ruberti (Northeastern University, USA)

Nanomaterial Probes of Biological Processes and Systems
edited by David Mast (University of Cincinnati, USA)

Handbook of Biomaterials
edited by Donglu Shi (University of Cincinnati, USA) and Xuejun Wen (Clemson University, USA)

HANDBOOK OF NANOBIOMEDICAL RESEARCH

Fundamentals, Applications and Recent Developments

❷ Applications in Therapy

editor

Vladimir Torchilin

Northeastern University, USA

Vol. 3

Frontiers in Nanobiomedical Research

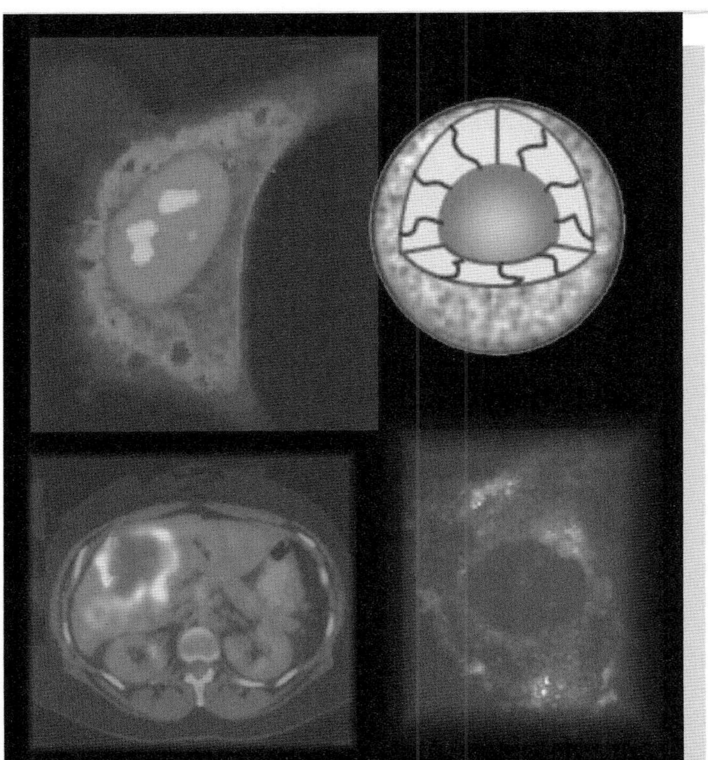

World Scientific

NEW JERSEY • LONDON • SINGAPORE • BEIJING • SHANGHAI • HONG KONG • TAIPEI • CHENNAI

Published by

World Scientific Publishing Co. Pte. Ltd.
5 Toh Tuck Link, Singapore 596224
USA office: 27 Warren Street, Suite 401-402, Hackensack, NJ 07601
UK office: 57 Shelton Street, Covent Garden, London WC2H 9HE

Library of Congress Cataloging-in-Publication Data
Handbook of nanobiomedical research : fundamentals, applications, and recent developments / editor, Vladimir Torchilin.
p. ; cm. -- (Frontiers in nanobiomedical research ; vol. 3)
Includes bibliographical references and index.
ISBN 978-9814520645 (set : alk. paper) -- ISBN 978-9814520676 (volume 1 : alk. paper) --
ISBN 978-9814520683 (volume 2 : alk. paper) -- ISBN 978-9814520690 (volume 3 : alk. paper) --
ISBN 978-9814520706 (volume 4 : alk. paper)
I. Torchilin, V. P., editor. II. Series: Frontiers in nanobiomedical research ; v. 3. 2251-3965
[DNLM: 1. Nanomedicine. 2. Nanostructures. 3. Nanotechnology. QT 36.5]
R857.N34
610.28--dc23

2014024090

British Library Cataloguing-in-Publication Data
A catalogue record for this book is available from the British Library.

Typeset by Stallion Press
Email: enquiries@stallionpress.com

Printed in Singapore

Contents

Chapter 1

The Application of Nanomedicine to Cardiovascular Diseases

Kevin M. Bardon, Olivier Kister and Jason R. McCarthy

Center for Systems Biology, Harvard Medical School and Massachusetts General Hospital, 149 13th St., 6th Floor, Charlestown, MA 02129, USA
jason_mccarthy@hms.harvard.edu

Abstract

Cardiovascular disease (CVD) remains to be the leading cause of death in the developed world. In contradistinction to other diseases, including cancers, whose mortality rates have recently decreased, mortality from CVD has steadily increased. Thus, new technologies are needed to better address this unmet need. Nanomedicine possesses the potential to fill this void, as nano-agents can be generated targeting specific receptors for the detection and treatment of disease. This chapter will thus focus on the most poignant recent advances in cardiovascular nanomedicine, particularly its use in atherosclerosis, thrombosis, myocardial infarction, aortic aneurysms, and heart transplant rejection.

1. Introduction

Cardiovascular disease (CVD) is an overarching term used to describe a group of diseases that affect the heart and vasculature. Typically, when people discuss CVD, they are actually referring to atherosclerotic vascular disease and its sequelae, including myocardial infarction, stroke, and aortic aneurysms. CVD has been amongst the leading causes of death in the developed world for over a decade. In fact, 24.6% of deaths in the US in 2009 were caused by diseases of the heart.[1] Whereas the death rates from other diseases, particularly cancer, continue to decrease year-over-year, the persistence of CVD as the leading cause of death signals a need for improved resources and technologies for the diagnosis and treatment of this disorder.

Over the past decade, nanotechnology has demonstrated explosive growth, due in part to governmental efforts to foster the field through funding and outreach. Owing to this, nanomedicine, or the application of these technologies to the improvement of human health, has made significant progress into surpassing many of the hurdles presented to clinicians and researchers for a number of diseases, including cancer and CVD, although clinical realization of many of these advances are yet to be achieved. Interestingly, the vast majority of nanomedical researches have focused on cancers. This is somewhat surprising, as tumor targeting necessitates traversing numerous biological barriers to reach the site of action. Vascular targets, as are seen in CVD, are readily accessible to agents administered intravenously. Due in part to this fact, recent interest in the field of cardiovascular nanomedicine has certainly increased.

This chapter will focus on the application of nanotechnologies to the imaging and therapy of CVD, including atherosclerosis, thrombosis, myocardial infarction, aortic aneurysms, and heart transplant rejection. Due to the breadth of the discipline, an all-encompassing review would be exceedingly difficult to compile. Thus, this chapter will focus on the most poignant recent advances in the field.

2. Atherosclerosis

Atherosclerotic vascular disease is a process that occurs over decades, and is often asymptomatic until the onset of clinically significant sequelae.[2] In its earliest stages, leukocytes, which are normally resistant to attachment to the endothelium of blood vessels, adhere and translocate to the intima. This initial step is prompted by inflammation, and subsequent upregulation of receptors, including vascular cell adhesion molecule-1 (VCAM-1). These monocytes undergo morphological changes, becoming macrophages, which feature several scavenger receptors capable of internalizing modified lipoproteins, such as oxidized low density lipoproteins, eventually taking on a foam cell phenotype. These resident cells are responsible for the release of a number of cytokines and proteases, promoting disease progression and eventual plaque rupture.

Atherosclerotic lesions, in particular those that are inflamed and prone to rupture, present a plenitude of potential imaging and therapeutic targets to which nanomaterials can be addressed. This section will thus focus on recent advancements in these areas.

2.1. *Nanomaterials for the detection of atherosclerotic lesions*

2.1.1. *Probes of phagocytic activity*

Several strategies have been devised for the localization of nanomaterials to atherosclerotic plaques. In its simplest form, targeting is achieved via the non-specific uptake of nanomaterials by inflamed lesions with increased endothelial permeability, and subsequent uptake by phagocytic cells. This strategy has been used extensively to image macrophages, one of the most prevalent cell types within lesions, and indicative of inflamed, rupture-prone lesions.

Cross-linked dextran-coated iron oxide nanoparticles have been used extensively to study macrophages due to the avidity of this cell type for these particles, with demonstrated applications in cancer, diabetes, and CVD. Their ability to home in on atherosclerotic lesions has been demonstrated with both fluorescence and magnetic resonance imaging.[3] In 2008, Nahrendorf *et al.* increased the sensitivity of detection by incorporating a positron emission tomography (PET) radiotracer, (^{64}Cu) via surface modification of the nanoparticle. The resulting agent was initially examined *in vitro*, and the sensitivity of detection was analyzed by both MR and nuclear imaging. In agar phantoms, the lowest threshold of detectability by MR imaging was 5 μg Fe/mL using T_2-weighted MRI, while it was 0.1 μg Fe/mL for PET-X-ray computed tomographic (CT) imaging, a full 50-fold lower. When injected into atherosclerotic, apolipoprotein E deficient (apoE$^{-/-}$) mice, particle accumulation was readily observed in the aortic root, arch, and the carotid artery by PET-CT. This localization was subsequently correlated *ex vivo.* Interestingly, the authors also analyzed the cell-specific uptake of the agent by digesting the aortae of treated animals to form a single cell suspension, which was then labeled with a cocktail of fluorescent antibodies and subject to flow cytometric analysis. Interestingly, 74% of the observed activity was localization to macrophages, with significantly lower amounts in other cells such as neutrophils, lymphocytes, smooth muscle cells, and endothelial cells.

We have used these magnetofluorescent iron oxide particles, in conjunction with a fluorogenic probe for matrix metalloproteinases (MMPs), proteases release by inflammatory macrophages, to detect the inflammatory-modulating therapeutic properties of pioglitazone (PIO), a peroxisome proliferator — activated receptor γ agonist, within murine atherosclerotic lesions.[4] In this study (Fig. 1), serial intravital fluorescence microscopy was performed at baseline (26 weeks, after 16 weeks of high cholesterol diet) and eight weeks later (34 weeks, after randomization and dosing with PIO). In the control cohort not receiving the drug, both the signal from the particles phagocytosed by the macrophages, as well as from the MMP sensor increased versus baseline

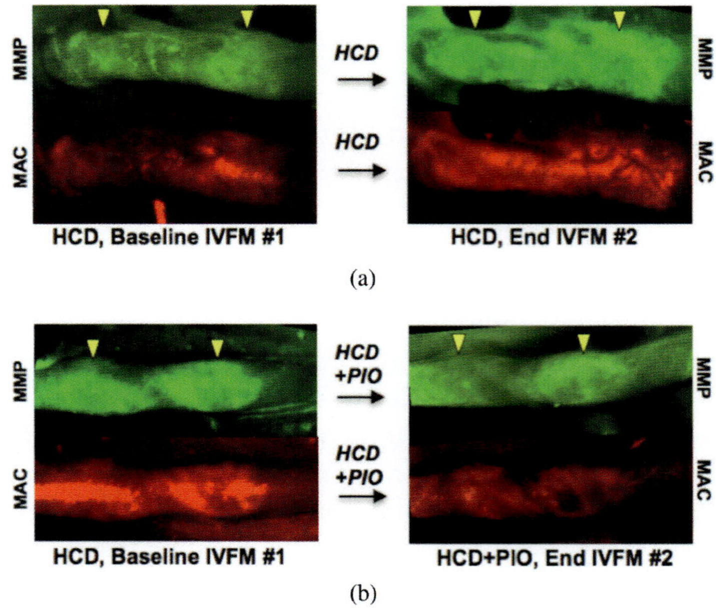

Fig. 1. (Color online) Longitudinal *in vivo* tracking and quantification of multicomponent plaque inflammation in apoE$^{-/-}$ mice. The same right carotid plaque from each animal underwent serial IVFM. (B) In HCD- treated mice, plaque MMP and Mac phagocytic activities (green and red, respectively) increased significantly over eight weeks. (C) In contrast, PIO treatment reduced plaque MMP and Mac activities. Representative images processed and windowed identically. Reproduced with permission from Ref. 4.

(1.4 fold for both macrophages and MMPs). For the PIO treated group, the signals from both probes showed decreases (0.8 for both macrophages MMPs). Histologically, sections of the excised carotid arteries for the treated group showed a reduced presence of MMP-9 and macrophages, while immunoblotting corroborated these observations. Interestingly, we found that PIO treatment increased the plaque collagen content by 72%, which may be attributed to a decrease in interstitial collagenases. Overall, these studies demonstrate the ability of macrophage targeted nanomaterials to detect inflammatory cells within atherosclerotic lesions in murine models of the disease.

Iron oxides have also been used to examine atherosclerotic lesions clinically. Kooi *et al.* demonstrated uptake of ultra-small superparamagnetic particles of iron oxide (USPIO, Sinerem) in 11 patients scheduled for carotid endarterectomy, correlating the results *ex vivo* using histological and electron microscopic analysis.[5] This site-specific uptake was further used by Tang *et al.* in 2009 to study the effect of low and high dose atorvastatin on carotid plaque inflammation. About 47 patients were enrolled in the study, with each showing baseline plaque MR signal in post-contrast images after injection of USPIO. The participants were split into two cohorts, with one receiving 10 mg and the other 80 mg of atorvastatin everyday. While the low dose group did not show appreciable MR signal changes as compared to baseline at 6th and 12th weeks of treatment, the high dose group showed significant decrease in inflammation, further demonstrating the utility of iron oxides in the monitoring of therapeutic interventions.

This methodology has also been used by Fayad and co-workers for the delivery of a nanoparticulate CT contrast agent. In their initial report, the authors generated this novel iodinated contrast agent by milling down ethyl-3,5-bis(acetylamino)-2,4,6-triiodobenzoate, an esterified version of the CT contrast agent diatrizoic acid, with a biocompatible surfactant, resulting in 259 nm particles.[6] This agent was subsequently injected into atherosclerotic New Zealand White (NZW) rabbits, which were subject to CT imaging. As compared to rabbits receiving a conventional iodinated contrast agent or non-diseased rabbits, the nanoparticulate agent readily localized and identified multiple lesions in the atherosclerotic aortae. The co-localization of the agent with macrophages was further confirmed histologically. The authors followed up this initial report with an examination of whether the intensity of the signal observed in CT images could be correlated with inflammatory activity, as determined by ^{18}F-fluorodeoxyglucose (FDG) PET-CT.[7] In this study, atherosclerotic NZW rabbits were subject to pre- and post-contrast imaging after injection of the CT agent. One week later, the same rabbits received ^{18}F-FDG and underwent PET–CT imaging. As compared to the control rabbits, the atherosclerotic rabbits demonstrated significantly higher signal for both agents, with a strong correlation between inflammation and the CT signal for the nanoagent.

Whereas most investigations utilizing passive targeting focus on detecting inflammatory macrophages, Zhang *et al.* used semipermanent perfluorocarbon-core nanoparticles to probe vascular barrier disruption in advanced lesions.[8] The author's hypothesis was that the larger particles, approximately 250 nm in diameter, would not traverse the endothelial layer in acute disease, yet would be found in the lesions of animals bearing extensive endothelial dysfunction. To test this, the particles were injected into NZW rabbits on 3–14 months of high cholesterol diet, and allowed to circulate for 6 h prior to sacrifice and excision of the aortae. Histologically, the fluorescently-labeled particles showed minimal uptake in rabbits on three months of cholesterol feeding, yet demonstrated extensive endothelial penetration in both the 9 and 12 month animals. As these particles contain a perfluorocarbon, they are also applicable to ^{19}F imaging, allowing for the quantification of the number of nanoparticles per cubic centimeter within the lesions. This measurement ultimately demonstrated that rabbits on 7–14 months of diet, took up 10-fold more particles than animals on only three months of high cholesterol diet. Importantly, the particles used in this study have been proven safe in clinical trials for use in humans, and may thus find utility in the examination of a number of diseases, including atherosclerosis.

2.1.2. *Targeted probes*

Nanomaterials have also been generated which are targeted to specific receptors via affinity ligands, or which make use of the inherent targeting of lipoproteins to atherosclerotic lesions. The earliest examples of these agents include magnetofluorescent iron oxides targeted to VCAM-1 via anti-VCAM-1 antibodies.[9] Although not ultimately tested in atherosclerotic animals, these particles demonstrated avid *in vivo* binding to activated endothelium in a TNF-α stimulated model of vascular inflammation. Kelly *et al.* improved upon this initial report by performing phage screening to identify *de novo* peptide targeting sequences for the VCAM-1 receptor. In their first report, they identified a cyclically constrained peptide, CVHSPNKKC, which, when attached to a fluorescent dye or magnetofluorescent nanoparticle, demonstrated avid binding to atherosclerotic lesions in apoE$^{-/-}$ mice.[10] In their subsequent work, the authors identified a linear sequence with increased affinity for VCAM-1 via *in vivo* phage display.[11,12] This sequence, VHPKQHR, when appended to magnetofluorescent nanoparticles via a GGSKGC linker sequence, was capable of readily identifying atherosclerotic lesions within diseased mice by both fluorescence and MR imaging. Importantly, the authors also demonstrated that these particles could be used to monitor therapeutic efficacy. Atherosclerotic mice on high cholesterol diet were divided into two cohorts, with one receiving atorvastatin to reduce serum cholesterol and inflammation. After eight weeks of therapy, the mice were injected with the nanoagent and examined by MR imaging, with the treated animals demonstrating a significant reduction in signal change intimating decreased uptake of the agent and less inflammation. These results were correlated histologically, with the plaques in the treated mice displaying minimal particle fluorescence.

Pan *et al.* used the same targeting sequence to localize their lipid-coated perfluorocarbon nanoparticles to the activated endothelium.[13] To incorporate the peptide, the authors conjugated the VHPKQHR peptide to a linker sequence derived from melittin, which inserts itself into lipid monolayers. The VCAM-1 targeting potential of the particles was assayed *in vivo* in apoE$^{-/-}$ mice by ^{19}F magnetic resonance spectroscopy. After 2 h of circulation, the mice were sacrificed and the aortae excised. As compared to controls, the targeted particles demonstrated a 4.1-fold increase in accumulation, a result that was further corroborated by histology.

Other cell adhesion molecules have also been targeted, including selectins. Jacobin-Valat *et al.* have developed pegylated dextran-coated iron oxide nanoparticles targeted to P-selectin via an antihuman P-selectin antibody (VH10).[14] P-selectin is overexpressed on the surface of activated platelets, which adhere to the luminal surface of atherosclerotic endothelium, as well as the activated endothelium, itself. *In vitro*, the authors demonstrated that the nanoprobe bound to both activated platelets and activated endothelial cells by flow cytometry and fluorescence microscopy, respectively. *In vivo*, the particles demonstrated significant accumulation in the aortae of atherosclerotic mice, as determined by MR imaging. Kang *et al.* have developed magneto-fluorescent nanomaterials for the detection of E-selectin.[15] These cross-linked dextran-coated particles were modified with an antihuman E-selectin antibody, and assayed *in vitro* for their binding to activated endothelial cells.

Integrins, in particular $\alpha_v\beta_3$ integrin, are indicative of neovascularization. With respect to the progression of atherosclerosis, plaque angiogenesis is associated with plaques that are more prone to rupture.[16] Over the past decade, Lanza and Wickline have imaged this receptor extensively using lipid-coated perfluorocarbon nanoparticles.[17–21] In their earliest work, these gadolinium-loaded particles were modified with an Arg-Gly-Asp (RGD) mimetic, the ligand sequence for the integrin.[19] When injected into atherosclerotic rabbits, the particles were readily observed within atherosclerotic lesions by MR imaging as a 47% increase in contrast. More recently, the authors used similar particles to characterize neovascularization in a rat model of atherosclerosis, and to subsequently monitor angiogenesis in a therapeutic model using a cardioprotective therapy, benfluorex. The model rats, the James C. Russell: LA corpulent rat strain develop spontaneous atherosclerosis due to obesity and insulin resistance, a phenotype approximating metabolic syndrome. At 8th and 16th weeks after the introduction of the drug into their diet, the treated or control (lean) rats demonstrated MR enhancement similar to that observed at day 0, whereas the control (obese) rats showed greater than a two-fold increase. Histological staining for von Willebrand factor, which is indicative of endothelialization, further corroborated these results.

A number of strategies have also been employed to specifically target macrophages. Lipinski *et al.* have modified gadolinium-containing lipid-based nanoparticles with antibodies specific to macrophage scavenger receptor B (CD36).[22] The cellular uptake of these particles was initially tested *in vitro* using macrophages derived from human THP-1 monocytes. Whereas the control agents showed minimal uptake, the targeted agent demonstrated significant accumulation within the cells. The particles were then tested in cadaveric human aortae. The targeted particle or control were incubated with the tissue for 24 h, washed, and examined by MR imaging. As compared to the pre-contrast images, the targeted particles exhibited a 52.5% increase in contrast-to-noise, whereas the control particles showed a 17.2% increase. Immunofluorescent confocal microscopy was used to further demonstrate the co-localization of the targeted agent with macrophages.

Dr. Angelique Louie and co-workers have developed superparamagnetic nanoparticles to enable the imaging of macrophage scavenger receptor A (SR-A), via the sulfation of dextran-coated iron oxides.[23] *In vitro*, the particles were incubated with P388D1 murine macrophages, which were subsequently lysed and examined by T_2 MR imaging, showing a 3.5-fold decrease in signal, as compared to the parent, non-sulfated particle. This uptake could be abrogated by including an excess of a competitor, namely dextran sulfate, in the assay. *In vivo*, in atherosclerotic apo$E^{-/-}$ mice, the SR-A-targeted agent demonstrated superiority to the non-sulfated particle at both 4 h and 24 h after injection, revealing a four-fold increase in contrast ratio at the earlier time point.

Stabilin-2 is a receptor that has been identified in activated macrophages, and is responsible for endocytosis and engulfment of apoptotic cells. It had previously been shown to be present in a number of tissues, yet had not been identified in atherosclerotic lesions. Lee *et al.* have used phage display to identify a peptide sequence, CRTLTVRKC, which binds stabilin-2, and used a nanoparticulate conjugate of that peptide to show that stabilin-2 is expressed on macrophages, endothelial cells, and smooth muscle cells within atherosclerotic plaques.[24] Initial studies used immunofluorescent staining of atherosclerotic endarterectomy sections to co-localize stabilin-2 with alpha-smooth muscle actin-, CD68-, and CD31-positive cells. Using the same sections, the authors subsequently illustrated the binding of the dye-labeled peptide. The peptide was then appended to the surface of hydrophobically-modified glycol chitosan (HGC) nanoparticles, and modified with Cy5.5. When injected into apo$E^{-/-}$ mice, the particles became enriched within atherosclerotic plaques, as compared to particles bearing a control sequence.

In advanced atherosclerotic lesions, there is a significant number of apoptotic and necrotic cells. While all cells contain phosphatidylserine, it is intracellular until such time that the cells begin dying. Annexin-A5 had been used extensively to identify these cells in a variety of scenarios. van Tilburg *et al.*

have conjugated Annexin-A5 to micelles containing gadolinium for MR imaging and a dye for fluorescence detection.[25] When injected into atherosclerotic mice, the post-contrast MR images revealed significant uptake of the agent in aortic lesions, as compared to control micelles or wild-type control mice. These results were correlated *ex vivo* by fluorescence reflectance imaging (FRI). Histological examination of the tissues revealed co-localization of the targeted agent with macrophages and apoptotic cells within the intima.

Low density lipoproteins (LDL) are responsible for the trafficking of cholesterol and fatty acids. In instances of endothelial dysfunction, as is seen in atherosclerosis, these particles deposit their contents within lesions. As such, this innate targeting may be useful for the imaging of atherosclerotic plaques. Interestingly, due to the size of its lipoprotein (apoB-100, 534 kD), these particles cannot be synthesized; instead, they must be isolated and modified. Yamakoshi *et al.* synthesized a DOTA gadolinium chelate, bearing a single oleic acid for intercalation into LDL.[26] Preliminary experiments were performed in atherosclerotic LDL receptor knockout mice, where the nanoagent or a small molecule MR imaging agent were administered and allowed to circulate for 24 h, at which time the mice were sacrificed and the aortae dried and subjected to inductively coupled plasma mass spectrometry (ICP-MS) to quantify gadolinium content. Whereas animals injected with the nanoprobe showed 0.20 ± 0.03 ppm gadolinium, there was no accumulation observed in the control group. More recent studies demonstrated, by MR imaging, that injection of the LDL particle-based imaging probe showed significant signal enhancement within atherosclerotic lesions of apoE$^{-/-}$ mice.[27]

Fayad and co-workers have extensively studied the development of imaging agents from HDL and its mimics.[28–31] Initial work involved the isolation and dilapidation of normal human HDL, followed by incorporation of gadolinium modified phospholipids or fluorescently labeled phospholipids.[32] These particles demonstrated accumulation within lesions, as determined by MR imaging 24, 48, 72, 96 h after intravenous (IV) injection. Several years later, these researchers created a number of synthetic agents with nanocrystal cores (i.e. iron oxide, gold or quantum dot). These syntheses involved the phospholipid coating of the hydrophobic core, followed by the incorporation of apoA-I. A number of phospoholipids were used, dependent upon the application, including those that were gadolinium- or dye-labeled.[30] After *in vitro* characterization of nano-agent uptake by relevant cells, the particles were injected into atherosclerotic apoE$^{-/-}$ mice and imaged by MR imaging, with particle accumulation noted 24 h after injection (Fig. 2).

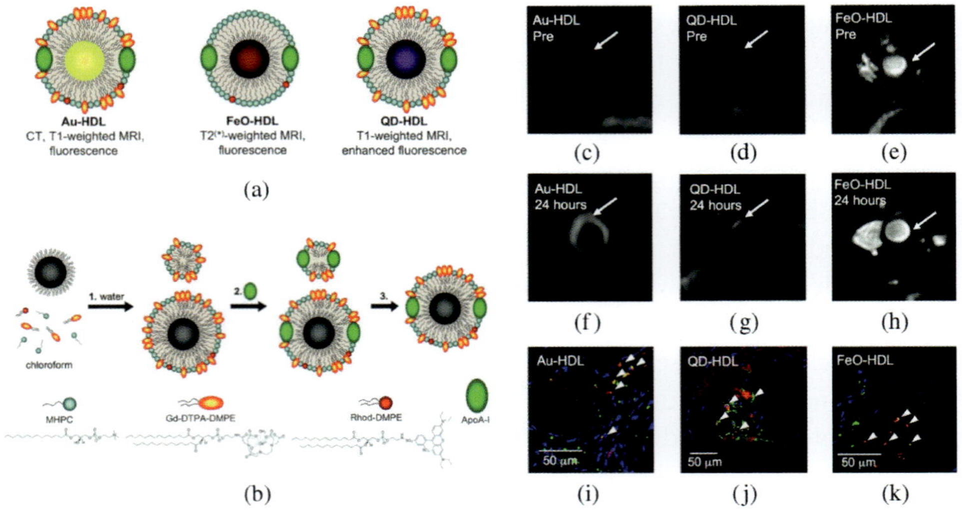

Fig. 2. (Color online) Nanocrystal core high density lipoprotein. (a) Schematic depiction of the different agents in this study. (b) Summary of the synthesis procedure of the agents where 1, the phospholipids and nanocrystal in chloroform are added to water, 2, apoA-I is added and 3, the "empty" particles are removed. Multimodality imaging of atherosclerosis using nanocrystal high density lipoprotein. T_1-weighted MR images of the aorta of apoE KO mice pre- [(c) and (d)] and 24 h post-injection [(f) and (g)] with Au-HDL or QD-HDL. Arrows indicate areas enhanced in the post images. (e) and (H) T_2^*-weighted images of an apoE KO mouse pre- and 24 h post-injection with FeO-HDL. [(i),(j),(k)] Confocal microscopy images of aortic sections of mice injected with nanocrystal HDL. Red is nanocrystal HDL, macrophages are green, and nuclei are blue. Reproduced with permission from Ref. 30.

2.2. *Nanoparticle-based drug delivery*

While there has been an exceedingly large amount of research accomplished for the detection and characterization of atherosclerosis using nanomaterials, disproportionately fewer drug delivery strategies have been devised and reported.

Using the $\alpha_v\beta_3$ integrin-targeted perfluorocarbon nanoparticles described above, Winter *et al.* were able to deliver an antiangiogenic drug, fumagillin, to atherosclerotic lesions.[20] The authors incorporated the lipophilic drug into the lipid monolayer of the particles by replacing some of the lecithin in the outer shell. Atherosclerotic NZW rabbits were injected with the particles at day 0, on which a baseline MR image was acquired. One week later, the rabbits were reinjected with the agent or controls, and were reimaged. For all animals treated with the targeted particles bearing the drug, the MR signal decreased drastically between imaging sessions, whereas the group that received the targeted agent without drug showed a net overall increase in signal. Histological examination of the aortae of the rabbits revealed a difference in the number of PECAM-positive microvessels per slice between the control (73 ± 28) and treated animals (24 ± 5). These particles were later investigated to determine the effect of a prolonged antiangiogenesis regimen.[21] Concomitantly, the authors also studied the synergism between the drug loaded particles and statin therapy. In this study, the rabbits received the drug loaded delivery vehicle at week 0, and the status of the $\alpha_v\beta_3$ integrin receptor was examined weekly with the targeted, but non-drug loaded particles. In all cases, the particles decreased angiogenesis for up to four weeks, at which time the MR signal from the control animals matched that of the treated animals. For the synergistic study, the animals were dosed at 0 and 4 weeks with the drug-loaded particles. Whereas atorvastatin alone did not affect the MR signal, when used in conjunction with the drug loaded particles, a pattern similar to rabbits receiving just the particles was observed during fourth week. Yet, when the rabbits received a second dose of the drug-loaded particles, a more durable response was observed in 8 weeks, where the signal from the atherosclerotic lesions was only half of that observed for the controls.

Lobatto *et al.* have focused on decreasing inflammation within atherosclerotic lesions via the liposomal delivery of the glucocorticoid steroid prednisolone phosphate, as this should affect plaque regression and stabilization.[33] The 103 nm liposomes were formulated to include gadolinium chelates, allowing for detection by T_1 MR imaging, and rhodamine, for *ex vivo* analysis of particle localization. When administered to atherosclerotic NZW rabbits, the particles demonstrated accumulation within aortae 48 h after injection. *Ex vivo* fluorescence imaging showed focal accumulation of the particles at points along the aorta, which were further confirmed by confocal fluorescence microscopy. The therapeutic efficacy of the particles was examined using ^{18}F-FDG, which is capable of identifying inflammation. Within seven days of treatment, PET–CT imaging revealed significant decreases in the level of inflammation within the aortae of the treated rabbits, as compared to animals receiving control, non-therapeutic particles. Immunohistochemical analyses were performed to examine whether the decreased ^{18}F-FDG was due to decreased macrophage burden within lesions. As was hypothesized, the number of macrophages within the histological sections correlated well with the signals observed by PET–CT ($R = 0.78$). Importantly, this study demonstrated that an anti-inflammatory effect could be accomplished within atherosclerotic lesions within a short timescale.

We have investigated the utility of macrophage-targeted photodynamic therapy for the focal ablation of these inflammatory cells within atherosclerotic lesions.[34] Using the macrophage avid cross-linked dextran-coated magnetofluorescent nanoparticles described above, we have developed a light-activated therapeutic nanoparticle via the conjugation of a photosensitizer to the particle surface. *In vitro*, these particles were readily internalized by RAW264.7 murine macrophages, demonstrating a 1900-fold increase in phototoxicity as compared to the gold standard, chlorine$_6$. When injected into atherosclerotic apoE$^{-/-}$ mice, the particle readily localized to lesions, as determined by intravital fluorescence microscopy (IVFM) after surgical exposure of the carotid artery. The mice receiving treatment were then irradiated with a 650 nm laser to induce the therapeutic effect, and their surgical incisions were closed. One week later, the animals were reinjected with the agent and reimaged by IVFM. In the treated group there was negligible uptake of the agent, intimating focal macrophage ablation, which was subsequently confirmed histologically.

3. Thrombosis

Thrombosis is the formation of a blood clot within a vessel. Normally, this is a protective event aimed at preventing blood loss (hemostasis). Yet, under certain conditions, pathological thrombi form, occluding blood flow and causing a number of sequelae, including myocardial infarction, stroke, and pulmonary embolism. Arterial clots have the potential for more immediate and devastating consequences, as ischemia may lead to tissue death, especially that of the heart or brain. Venous thromboses, including deep vein thrombosis and pulmonary embolism, have the potential for long term complications (e.g. post-thrombotic syndrome (PTS), difficulty breathing, chest pain on inspiration, and palpitations) in addition to the risk of death.

Clinically, a number of gold standard techniques are used to diagnose thrombosis. D-dimer, a soluble fibrin degradation product that is formed during fibrinolysis, can be detected using a blood test. Importantly, this assay has a high negative predictive value, with the absence of D-dimer indicative of a decreased likelihood of thrombus. Depending upon the symptoms with which a patient presents, a number of different imaging modalities may be used for diagnosis, including Doppler ultrasound, CT, and MR imaging. Given the reliability of these methodologies, what need is there for the development of molecular imaging agents for thrombosis? Answers to this question fall into three main categories, each of which will be further discussed below:

- For research purposes. Molecular imaging allows for further insight into the biology and pathology related to the disease. This may include monitoring the influx of cells into thrombus or the investigation of the therapeutic benefits of drugs in relevant animal models.

- To provide biological information about clot composition and age in the clinical setting concomitant with the gold standard imaging regime. This would allow for a more informed determination of the optimal interventional strategy for each patient.
- As platforms for the targeted delivery of therapeutics. Nanoparticulate agents for the detection of intravascular thrombi may also serve as excellent delivery vehicles for targeted thrombolysis or other therapeutic interventions.

3.1. *Nanomaterials for the detection and monitoring of thrombosis*

Thrombi are composed of a finite number of components, including fibrin, platelets, and coagulation factors. Upon vessel damage, platelets initially form a hemostatic plug, preventing blood loss, which is subsequently strengthened by the deposition and cross-linking of insoluble strands of fibrin by coagulation factors. As such, a number of molecular imaging agents have been developed to monitor this process. These have included dye or radionuclide-labeled peptides and antibodies, and more recently, nanoparticles.

The earliest examples of these nano-agents came from the laboratories of Drs. Wickline and Lanza at the Washington University Medical School. These lipid-encapsulated perfluorocarbon nanoparticles were loaded with gadolinium chelates, allowing for their detection by magnetic resonance (MR) imaging. In the initial report, the authors used a three-step targeting strategy, facilitated by the incorporation of biotin within the outer lipid monolayer of the particles, and biotinylated antifibrin antibodies.[35] To assess the binding efficiency and detectability by MR imaging, plasma clots were generated *in vitro*, and were incubated with the antibodies, followed by washing and subsequent incubation with avidin. Lastly, a sandwich complex was formed by incubation with the biotinylated particles. As compared to control clots treated with just phosphate buffered saline, the surface of the treated clots demonstrated significant signal enhancement in T_1-weighted images, as would be expected for bound agents. Binding was further correlated by scanning electron microscopy (SEM), demonstrating particle accumulation within the fibrin strands.

In a subsequent report, Lanza and Wickline directly modified the surface of the perfluorocarbon nanoparticles with the antifibrin antibody to enable facile *in vivo* targeting.[36] These particles were again tested *in vitro* to ensure adequate binding to fibrin clots, as determined by MR imaging and SEM.

They were then tested *in vivo* in a canine thrombus model, in which thrombi were formed by the introduction of nylon monofilament bearing cotton strands soaked with thrombin into the external jugular vein. After clot formation, the thrombi were isolated via snare enclosures and the targeted agent was introduced into that segment of vessel. After incubation for 1 h, the clots were reintroduced into circulation and imaged. As compared to the contralateral control thrombus that did not receive the probe, the vein bearing the thrombus demonstrated significant T_1-enhancement due to the binding of the probe. This methodology was further tested in human endarterectomy specimens with the fibrin-targeted particles displaying significant signal enhancement.

Drs. Lanza and Wickline have also developed a novel lipid encapsulated nanoagent with an organically soluble divalent copper core as a potent T_1w MR contrast agent (NanoQ).[37] A surfactant mixture comprised of phosphatidylcholine and biotin-caproyl-phosphatidylethanolamine was mixed with copper(II) oleate suspended in sorbitan sesquioleate, which was subsequently subjected to high-pressure homogenization to form 210 nm particles bearing 18,000 biotins for conjugation to the antifibrin antibody-based targeting ligand. The resulting particles demonstrated ionic relaxivities comparable to commercially available gadolinium-based contrast agents, whereas their relaxivities on a per particle basis were massive ($r1 = 66{,}000 \pm 2200$ mM^{-1} s^{-1} at 1.5 T). Similar to the *in vitro* clot binding experiments performed above, the NanoQ demonstrated significant binding, as compared to the controls. Due to the high relaxivity and sensitivity of these agents, the authors believe that they may be capable of the *in vivo* detection of extremely small microthrombi.

We have designed and synthesized two nano-agents targeted via peptide affinity ligands to different components of thrombi, namely fibrin and the coagulation factor, activated factor XIII (FXIII).[38] Importantly, fibrin is omnipresent during thrombogenesis, whereas FXIII is a hallmark of acute thrombi, as it is only present during the early stages of clot formation. Thus, this tandem of agents may allow for the investigation of the biological age of the thrombus, informing decisions as to the optimal therapeutic strategy. The fibrin targeting sequence was derived from N-terminus of the α-chain of fibrin. Other investigators searching for therapeutic moieties had previously found that the sequence Gly-Pro-Arg was capable of inhibiting fibrin and thrombin clotting.[39] Subsequent iterations found the sequence Gly-Pro-Arg-Pro-Pro (GPRPP) possessed significant affinity for fibrin.[39] As such, radiolabeled derivatives have been used to scintigraphically image pulmonary emboli in swine. To enable conjugation to a nano-agent, we appended the peptide with a C-terminal conjugation sequence (GGSKGC). For the detection of the coagulation factor, the sequence, GNQEQVSPLTLLKC, which is derived from the N-terminus of α_2-antiplasmin, was used. α_2-antiplasmin is a plasmin inhibitor that is cross-linked into clots by FXIII, making them more resistant to fibrinolysis. We had previously demonstrated the ability of a fluorescent derivative of this peptide to localize and cross-link into clots.[40,41] For this study, both peptides were conjugated to cross-linked dextran-coated iron oxide nanoparticles, each bearing a different fluorophores. For the fibrin probe (CLIO-GPRPP), a dye in the 750 nm channel was chosen (Cy7), whereas a 680 nm channel dye (VivoTag 680, VT680) was used for the FXIII probe (CLIO-FXIII). This allowed for simultaneous acquisition of the localization of both agents using a number of fluorescence imaging systems after concurrent injection. Initial *in vitro* studies with clots derived from fresh frozen plasma demonstrated avid binding by both probes, as investigated by FRI. Importantly, as the core particle had been widely used as a T_2 contrast agent, we also demonstrated *in vitro* clot binding by MR imaging. The particles were later injected into mice bearing jugular thrombi, which were resected and examined by FRI. Targeted nano-agent accumulation was significant in all cases, as compared to particles bearing control peptide sequences.

The fibrin-targeted magnetofluorescent nanoparticle has also demonstrated efficacy in the intravital detection of stent microthrombi using an experimental catheter-based optical frequency domain imaging (OFDI)/near-infrared fluorescence (NIRF) imaging system.[42] OFDI, also known as frequency-domain optical coherence tomography allows for the rapid acquisition of the structural features of a vessel wall in three dimensions at microscopic resolution. When combined with NIRF imaging, complementary cellular and molecular information can be acquired. In the initial experiments with this system, a stent coated with CLIO-GPRPP-labeled clots was deployed within the iliac artery of a rabbit. The catheter based system was then used to simultaneously acquire a three-dimensional image of the artery. As can be seen in Fig. 3, the vessel could be readily reconstructed with the stent struts clearly visible in the OFDI image. Importantly, the location of the nano-agent-labeled clots became evident in the OFDI/NIRF merged image. The clinical potential of this imaging system is currently under investigation.

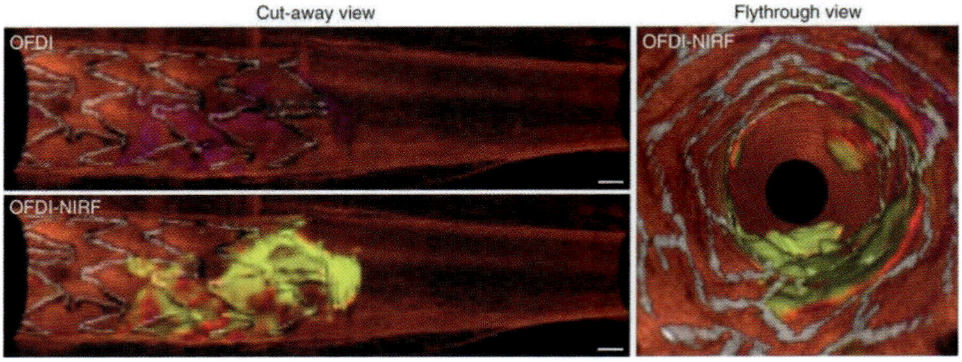

Fig. 3. (Color online) Dual-modality OFDI-NIRF images of an iliac artery of a rabbit with an implanted NIR-fluorescent fibrin-coated stent, attained *in vivo*. Three-dimensional rendering of the stented right iliac artery of a living rabbit. Structural components were segmented and color-coded in OFDI images for clear visualization. Red, artery wall; white, stent; purple, thrombus; yellow, NIR fluorescent fibrin. Scale bars, 500 μm. Reproduced with permission from Ref. 42.

3.2. *Nanoparticle-based therapeutic interventions*

To ensure optimal outcomes after pathological thrombosis, it is exceedingly important to reestablish vascular patency. This can be accomplished via the administration of exogenous plasminogen activators (PAs), such as recombinant tissue plasminogen activator (tPA) or streptokinase (SK). Circulating plasminogen interacts with these PAs, which cleaves the zymogen, forming plasmin, which in turn degrades fibrin clots. Unfortunately, these PAs suffer from suboptimal efficacy[43,44] and side-effects including intracerebral hemorrhage, which can result in death or neurological deficit.[45] Many of the impediments listed above may be circumvented by the site specific localization of the PAs to the thrombus, decreasing the required dose while increasing therapeutic efficacy. As well, preventing the thrombolytic from crossing the blood brain barrier may decrease the incidence of ICH, increasing the safety profile of the drugs.

Although the diagnostic agents described in the previous section may not have any immediate utility in the clinical assessment of thrombosis, their multivalent nature may enable them to serve as platforms for the targeted delivery of therapeutic moieties. For example, the perfluorocarbon and iron oxide nanoparticles described above have been used for the localization of thrombolytics[46–49] and anticoagulants[50] to thrombi.

Marsh *et al.* generated SK-modified particles via the covalent coupling of succinimidyl acetylthioacetate (SATA)-functionalized SK with the maleimide-functionalized particle surface, resulting in approximately 20 SK per 240 nm particle.[48] As opposed to the MR-active particles above, these derivatives did not contain gadolinium chelates. Instead, the authors made use of the intrinsic acoustic reflectivity of the nanoparticles to detect their *in vitro* binding to plasma clots, and their subsequent dissolution via ultrasound imaging. As expected, clots treated with the nano-agent in the presence of plasminogen demonstrated significant lysis, whereas the control particles without SK, or the active particles in the absence of plasminogen showed minimal degradation.

The authors have expanded upon this original work by covalently conjugating urokinase to their fibrin-targeted perfluorocarbon nanoparticles, resulting in up to 400 PAs and 40 antibodies per particle. When the *in vitro* fibrinolytic efficacy of these particles were compared to the previously-described SK-modified particles, with comparable numbers of targeting ligands and PAs, the urokinase-modified agents proved superior at 120 min, resulting in significantly less clot present. The urokinase-modified particles were ultimately investigated *in vivo* in a canine model of femoral artery injury. The targeted or control particles were injected into the contralateral femoral vein in the respective cohorts and were allowed to circulate for 2 h prior to sacrifice and resection of the thrombosed vessel. Histological examination of the tissue demonstrated that the targeted particles homed to the clot, whereas the irrelevantly-targeted particles displayed no binding.

Wickline and Lanza have also used the lipid-encapsulated perfluorocarbon nanoparticle platform for the delivery of anticoagulants to thrombi, specifically Phe[D]-Pro-Arg-Chloromethylketone (PPACK), which inhibits thrombin, a key mediator in thrombogenesis. These particles used a perfluoro 15-crown-5 ether core, as opposed to the perfluoro-octylbromide core used above, and incorporated 1.5 mole percent 1,2-distearoyl-sn-glycero-3-phosphoethanolamine-N-(carboxy[polyethylene glycol]-2000), which provided a carboxylic acid moiety to which the PPACK could be conjugated via amide bond formation. Analysis of this conjugation step revealed 13,650 PPACK coupled to each particle. *In vitro* thrombin inhibition was investigated using a chromogenic assay, Chromozym TH, and demonstrated no decrease in activity due to conjugation to the particle surface, as compared to free PPACK. The particles were subsequently tested *in vivo* using a photochemical injury model of the carotid arteries of C57Bl/6 mice. Heparin, PPACK, the PPACK modified particles, control particles, or saline were administered 10 min prior to thrombus induction and the time to occlusion was measured using an ultrasonic flow probe. As compared to PPACK or the controls, the occlusion time was two-fold longer. Nanoparticle accumulation was quantified *ex vivo* using ^{19}F MR imaging of excised arteries, with the injured arteries retaining almost eight-fold more agent than uninjured. Histological examination of the treated clots showed significant fibrin, with sparse clusters of platelets, while the controls were more tightly packed with significantly greater platelet accumulation. While this initial study demonstrated reasonable efficacy, additional study must be performed to investigate a number of systemic effects.

We have used the FXIII-targeted magnetofluorescent nanoparticle platform described above for the targeted delivery of tPA to intravascular thrombi.[46,47] This strategy was chosen due to the presence of FXIII in acute, thus more lysable thrombi. tPA possesses a solitary free cysteine residue within its structure, which served as an attachment point to the particle. CLIO–FXIII was initially modified with a discrete polyethylene glycol terminated with a maleimide, which was then reacted with excess tPA. The clot binding potential of the particle was assayed at each step of the build, after the enzymatic activity of the tPA was inhibited by pre-incubation with PPACK to prevent lysis. As was anticipated, the targeted particle demonstrated significant affinity for clots, as compared to the control particles. Enzymatic efficacy of the tPA conjugates was analyzed via chromogenic amidolytic assay and D-dimer formation. Whereas the thrombolytic particles demonstrated diminished amidolytic activity as compared to free tPA, possibly due to steric constraints incurred by the tPA, when the activity was matched, the fibrinolytic potential was comparable to the free drug. The *in vivo* targeting ability of particles was subsequently tested using a ferric chloride induced injury model. Almost, 60 min after thrombus induction in the femoral vein, the targeted or control particles were injected into the respective cohorts of mice, and were allowed to circulate for 90 additional minutes, at which time the localization of the particles was examined by intravital fluorescence microscopy. As can be seen in Fig. 4, the targeted particle demonstrated significant accumulation within the thrombus, whereas the control agent showed minimal accumulation. The thrombolytic potential of the agent was finally assayed in a murine model of pulmonary embolism. As compared to the free drug, the targeted agent showed comparable therapeutic efficacy. Although the nano-agent did not display superiority in the treatment of thrombosis, its benefit may lie in an improved safety profile, the study of which is currently underway.

Several other nanomaterial-based strategies have been used for thrombolytic therapies. The most basic have involved the use of iron oxide nanoparticles for the magnetic targeting of thrombi. Bi *et al.* conjugated urokinase to the surface of aminated dextran-coated iron oxides via a glutaraldehyde linker, and assessed its thrombolytic potential in a rat arteriovenous shunt

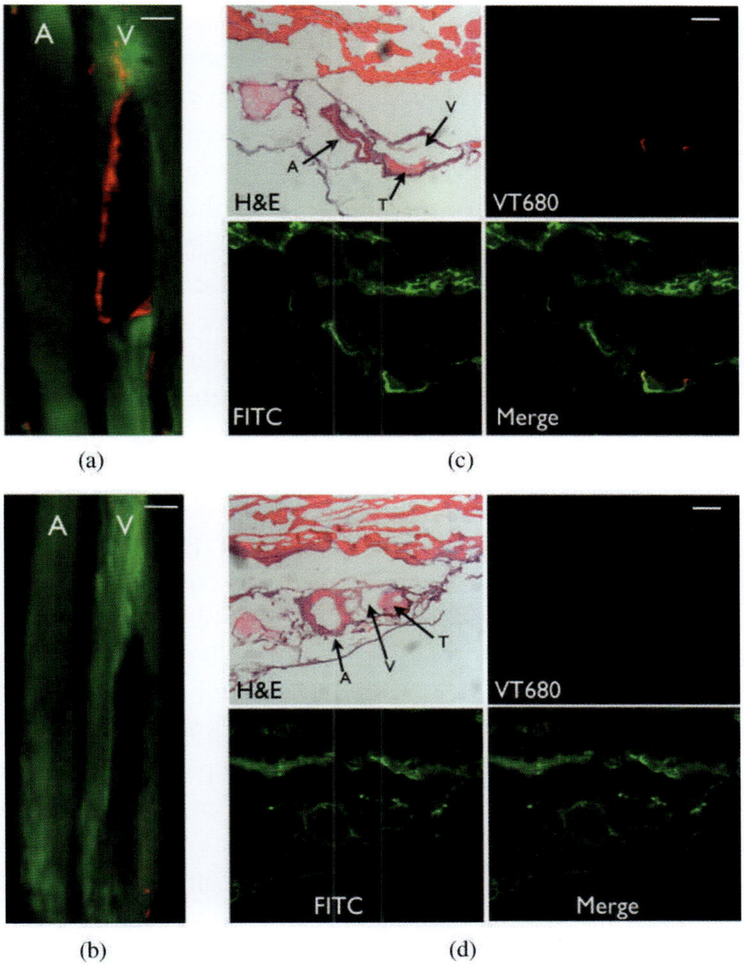

Fig. 4. (Color online) *In vivo* examination of nanoagent binding to arterial and venous thrombi. Representative intravital fluorescence microscopy images of the localization to intravascular thrombi of targeted (a) and control (b) agents demonstrating significant binding of the FXIIIa-targeted agent to the periphery of the thrombus. Histological examination of targeted (c) and control (d) agent localization (scale bars = 100 μm). (a–d) Green fluorescence can be attributed to fluorescein-labeled dextran, which is used to delineate the vasculature, while red fluorescence is attributable to the fluorophore on the nanoparticle. Reproduced with permission from Ref. 47.

model.[51] As the particles are superparamagnetic, an external magnetic field induces magnetism and traps the particles within the field. If the magnet is placed above the thrombus, this will effectively localize the particles to the site of action. The authors tested the therapeutic potential of the urokinase-modified particles in the presence and absence of an external magnetic field, and contrasted it with the free drug and saline, with the magnetically targeted particles displaying superior lytic potential. The systemic consequences of magnetically targeted thrombolysis were also assessed. About 30 min after injection, the magnetically targeted group demonstrated significantly lower levels of circulating drug, and shorter tail bleeding times than the control groups. Additionally, the percentage of residual circulating fibrinogen, a marker of the non-specific cleavage by the circulating fibrinolytic, was comparable between the magnetically targeted group and rats receiving saline, whereas those receiving the free drug or the drug modified particles without magnetic targeting showed significant decreases (up to 28%). These data intimate the likelihood that targeted strategies for the lysis of intravascular thrombi may significantly decrease detrimental systemic effects.

In the same year as the work above, Ma *et al.* also reported on the synthesis of a magnetically targeted thrombolytic.[51] The authors synthesized polyacrylic acid-coated iron oxides, to which was conjugated tPA via amide bond formation. *In vitro* analyses revealed that the resulting particles retained their enzymatic activity after functionalization. The nanothrombolytic was then tested *in vivo* in a rat embolic model, in which a pre-formed thrombus was embolized into the right ilial artery, followed 5 min later by injection of tPA, the tPA-modified particle, or the unmodified particle, each under magnetic guidance. Ultimately, the only group that showed reperfusion was the tPA-modified particle.

One exceedingly novel strategy for the site specific delivery of thrombolytics was reported by Dr. Donald Ingber and co-workers.[52] In thrombosed vessels, sheer stress is increased by one- or two-orders of magnitude. This phenomenon is responsible for the local activation of platelets and their adherence to the vessel surface. Using this model, the authors designed nanotherapeutics based upon the aggregation of multiple nanoparticles that remain as microparticles in the blood, yet which break apart under high sheer stress. Initially, the microparticles were formed by spray-drying concentrated solutions of poly(lactic-co-glycolic acid) (PLGA) nanoparticles, and were assayed to determine the mechanisms by which the particles would dissociate. Using a rheometer, the investigators applied sheer stress to the fluorescently-labeled particles and found that at 100 dyne/cm^2, there was an 8–12-fold increase in free nanoparticle. These particles were subsequently subjected to analysis in a microfluidic device designed to mimic vascular narrowing with 90% lumen obstruction, resulting in a 16-fold increase in the number of free nanoparticles. The microparticles were next coated with tPA using biotin-streptavidin chemistry, resulting in 5×10^5 tPA per particle. The particles were tested in mice bearing ferric chloride-induced thrombi of the mesenteric artery and imaged by IVFM. When injected intravenously, the particles accumulated within the region of clot formation, and rapidly lysed the clot within 5 min. The particles also significantly increased the time to reocclusion. The sheer stress activated nanoagent was subsequently investigated in a murine model of pulmonary embolism (PE), in which fluorescently labeled clots were injected through the jugular vein, resulting in their lodging in the peripheral vessels of the lung. If the agent was injected immediately after embolization, therapeutic intervention resulted in a 60% reduction of clot area and clot number, while injection 30 min after embolization resulted in a 30% decrease. Ultimately, the authors tested the agent in a model of massive PE, one in which larger clots were injected and accumulated in the main pulmonary arteries. In this case, the mice were infused with the tPA-functionalized particles or the carrier fluid for 45 min, with all control animals dying within 1 h of embolism. In contrast, 80% of the mice in the treatment group survived without obvious symptoms of respiratory distress. Overall, this novel strategy resulted in a therapeutic moiety that required 1% of the dose to elicit the identical response as the free drug, potentiating an increase in overall efficacy and safety.

4. Myocardial Infarction

Myocardial infarction (MI), popularly referred to as "heart attack", occurs roughly 1.5 million times annually in the US alone and carries a mortality rate of about 33%. Acute MI results from severe and/or enduring ischemia of the cardiomyocytes (CMs) and causes cardiac muscle death. Most MIs are caused by acute coronary artery thrombosis, and most thrombi originate from a disrupted atherosclerotic plaque. Prolonged ischemia results in insufficient oxygen delivery, energy production and noxious waste removal leading to a functional lack of contractility; while these early effects may be reversible if reperfusion is quickly achieved, severe ischemia lasting at least 20 to 40 min causes irreversible damage.[53]

Diagnostic and therapeutic advances over the latter part of the 20th century have contributed to a dramatic decrease in mortality from MI; these advances include: coronary angiography imaging; aspirin prophylaxis; pharmaceutical treatments like statins; improved arrhythmia control; the establishment of coronary care units (CCUs) in major hospitals; angioplasty and endovascular stents; thrombolysis treatments; and coronary artery bypass surgery. The standard clinical lab evaluation of MI is based on the analysis of blood samples for leaked macromolecules indicative of damaged cell membranes, with the most important of these markers being the cardiac troponins (cTnT and cTnI) and the myocardial isoform of creatine kinase (CK-MB); these markers possess high specificity and sensitivity for myocardial damage, with elevated cardiac troponin levels considered the gold standard for MI diagnosis.[54]

Although it is typically not possible to grossly observe MIs less than 12 h old, the infarct becomes identifiable between 12 and 24 h as a reddish blue discoloration.[53] On the cellular level, this corresponds to the usual healing time course in which neutrophilic infiltration begins 12–24 h post-MI, followed by phagocytic macrophages; these recruited leukocytes secrete a variety of enzymes- matrix MMPs and cathepsin B protease, in particular- to aid in the healing process and remove cellular debris. Current understanding of MI continues to be enriched yet still falls short in many ways, thus nanomedicine presents an attractive means to overcome these shortcomings.

4.1. *Nanoparticles in the diagnosis and treatment of acute myocardial infarction*

It is critically important to correctly identify and treat acute MI as early as possible, as doing so correlates strongly with reduced mortality and improved healing. Current generation troponin assays fail to detect the increased levels in many patients presenting with acute coronary syndrome (ACS) in the absence of a corroborating ST-interval elevation on electrocardiogram, possibly leading to misdiagnosis of unstable angina and the missed opportunity for intervention. Ultrasensitive cardiac troponin assays are currently in development, many of which exploit the advantages of nanoscale elements to achieve enhanced detection. One such assay, from the Mirkin research group at Northwestern University, was built on a nanoparticle-based bio-barcode platform and allowed for earlier detection in sub-acute circumstances.[55] Abbott, Roche, Siemens and Nanosphere are all engaged in next-generation, nanotechnology-based cardiac troponin assay development; this significant investment by highly respected corporations underscores the viability of applied nanoscience to this problem and the inherent need for improved assays.[54,56]

Another clinically relevant class of biomarkers is the reactive oxygen and nitrogen species (ROS/RNS) which arises as a response to acute tissue injury, as in MI. These compounds are mostly generated by myeloperoxidase (MPO) as part of the $MPO/H_2O_2/Cl^-$ system, and may modify critical cellular components to exacerbate the initial insult; specifically, the $MPO/H_2O_2/Cl^-$

system has been shown to alter the crucial N-terminal cysteine residue of tissue inhibitor metalloproteinase-1 (TIMP-1), allowing for dysregulation of MMPs. Earlier methods of ROS/RNS monitoring, including absorbance spectrophotometry, ESR assays, MRI and chemiluminescence assays all suffered from drawbacks such as unfavorable washout kinetics or a lack of specificity. Panizzi *et al.* developed a dye-conjugated nanoparticle system to overcome these limitations.[57] This sensor localized to the infarcted zone (Fig. 5), as confirmed by later tissue section analysis, and the novel fluorogenic dye allowed for sensitive and specific detection of hypochlorous acid/anion (HOCl/OCl^-) and peroxynitrite ($ONOO^-$), two clinically important ROS/RNS. Furthermore, the biocompatibility of the nanoagent and its favorable kinetics enable less-invasive fluorescence imaging studies to be performed to identify areas of ischemic injury in order to map areas of acute MI.

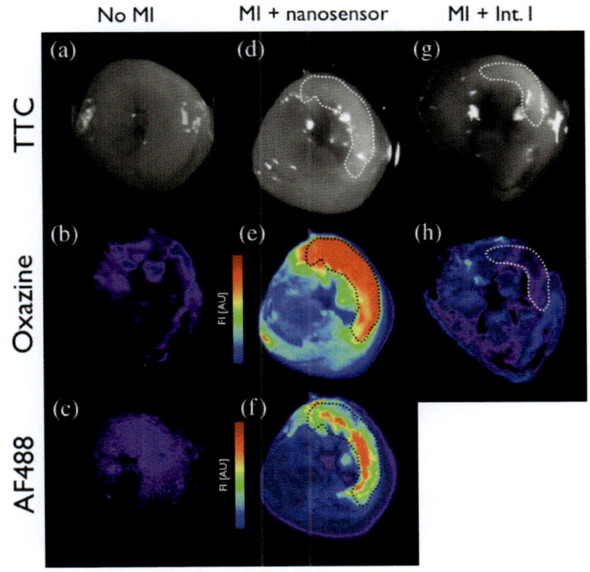

Fig. 5. *Ex vivo* comparison of ROS/RNS sensor or Intermediate I injected mouse tissue and fluorescence signal associated with oxazine release, accumulation, and probe washout kinetics. FRI of control (a–c), ROS/RNS sensor (d–f), and Intermediate I (g) and(h); injected myocardial rings are shown. White light images of TTC- stained heart slices are shown [(a),(d) and (g)] with infarcted region highlighted by the dashed line throughout. Accumulation of the activated oxazine probe is detectable in the Cy5 channel [(b), (e) and (h)], while fluorescence associated with the covalent attached AF488-labeled nanoparticles is also shown [(c) and (f)]. Reproduced with permission from Ref. 57.

Nanoscale systems are being investigated for the treatment of acute MI, as well. Leuschner *et al.* reported the use of lipidoid nanoparticles to deliver siRNA to the specific inflammatory monocyte subset via the chemokine receptor 2 (CCR2), as CCR2 acts to recruit the inflammatory monocytes to the site of injury, therapeutic targeting of this receptor could limit harmful inflammation while maintaining the activity of non-inflammatory immune cells.[58] Previous experiments attempting this aim relied on small molecules, microparticles and neutralizing antibodies but all fell short in terms of cellular specificity and efficacy. These lipidoid agents delivering siRNA showed precise cell targeting, nearly quantitative siRNA entrapment, efficient cell delivery, and effective mRNA knockdown, resulting in a marked reduction in the number of monocytes and macrophages in the hearts of mice that had been treated. This study established the feasibility for lipidoid nanoparticle-based siRNA delivery to attenuate the production of specific target proteins *in vivo* to great clinical effect.

Similarly, Gray *et al.* transcended the earlier work with small molecule inhibitors and microparticles from other groups by developing *N*-acetylglucosamine (GlcNAc)-decorated polyketal nanoparticles to reduce infarct size and improve acute cardiac function.[59] These agents were loaded with the p38 inhibitor SB239063, selectively bound and internalized by CM and shown to significantly decrease infarct size in the ischemia reperfusion (IR) animal model, with infarct size an established indicator of CM damage. Building on these findings, this same nanoparticle system could be used to encapsulate a wide variety of compounds — from siRNA to proteins — to serve as the vehicle for the intracellular delivery to CMs.

The influence of sugar moieties on cellular uptake was also exploited to good effect by Tranter *et al.*[60] In this study, a novel series of glycopolymer-based delivery vehicles able to complex with polynucleotides to form nanoparticles (termed polyplexes) were investigated. These polyplexes were found to possess virus-like efficiency in delivering oligonucleotides to the nucleus of primary cells, with several advantages over standard adenovirus systems, including more control over effects, more freedom over payload, and no risk of genetic modification due to insertion and inflammatory signaling. In effectively knocking down nuclear factor kappa-light-chain-enhancer of B cells (NF-kB) signaling in CMs with excellent delivery efficiency and low toxicity both *in vitro* and *in vivo*, these nanoparticulate polyplexes demonstrated their utility as inhibitors of the transcriptional activation of NF-kB and further associated damage from MI; ongoing work will focus on interrogating these polyplexes as delivery vehicles for numerous polynucleotide types, including transcription factor decoys, transgenes, antisense oligos, and other oligodeoxynucleotides.

4.2. *The utility of nanoparticles post-myocardial infarction*

Surviving acute MI is only the beginning — nearly 75% of these survivors have one or more complications, including contractile dysfunction, arrhythmias, myocardial ruptures and pericarditis.[53] The infarcted area must undergo a healing and remodeling process, and the newly-formed scar must be viewed as a weak point in normal cardiac function; the loss of functional contractile tissue post-MI, especially with repeated infarcts, greatly increases the risk of catastrophic heart failure.[56] Nanomedicine is playing a pivotal role in improving outcomes for survivors of acute MI, both before progression to scar formation and after when fibrotic areas must be managed. The efforts of researchers are primarily focused on improved imaging of the infarcted area and treatments designed to return the damaged area to as normal function as possible.

In 2007, Sosnovik *et al.* established the feasibility and value of applying fluorescence molecular tomography (FMT) to the imaging of MI via their magnetofluorescent nanoparticle system.[61] Using a murine model of post-MI myocardial macrophage infiltration and combining FMT with gadolinium-enhanced MR imaging, these researchers demonstrated the sensitive detection of nanomolar concentrations of fluorophores within the heart, allowing for multispectral imaging simultaneously and non-invasively. MR imaging of the infarcted mice showed left ventricular dilatation with thinning and akinesis of the anterior, lateral, and inferolateral walls, while accumulation of the iron oxide particles permitted clear imaging of the damaged areas of myocardium via both MR imaging and FMT.

Nahrendorf *et al.* advanced the utility of FMT even further by demonstrating its capacity for spatiotemporal assessment of phagocytic and protease activity post-MI.[62] As inefficient infarct healing leads to infarct expansion and heart failure and this healing process is highly dependent upon temporal regulation of enzymes secreted by invading leukocytes, these researchers tailored the magnetofluorescent nanoparticle system to enable imaging and localization in specific neutrophils and macrophages. FMT imaging in the mouse MI model dosed with nanoparticles showed robust fluorescence in the heart region four days post-MI (Fig. 6), indicating normal leukocyte infiltration to the infarcted area, whereas the murine model for impaired inflammatory response receiving the same nano-agents showed significantly decreased fluorescent signal, corresponding to a decrease in specific leukocytes and subsequent decrease in released proteases, specifically cathepsin B. Using FMT in this manner allowed for the non-invasive detection of impaired recruitment of leukocytes, visualization of decreased local protease activity and the ability to monitor the healing process post-MI.

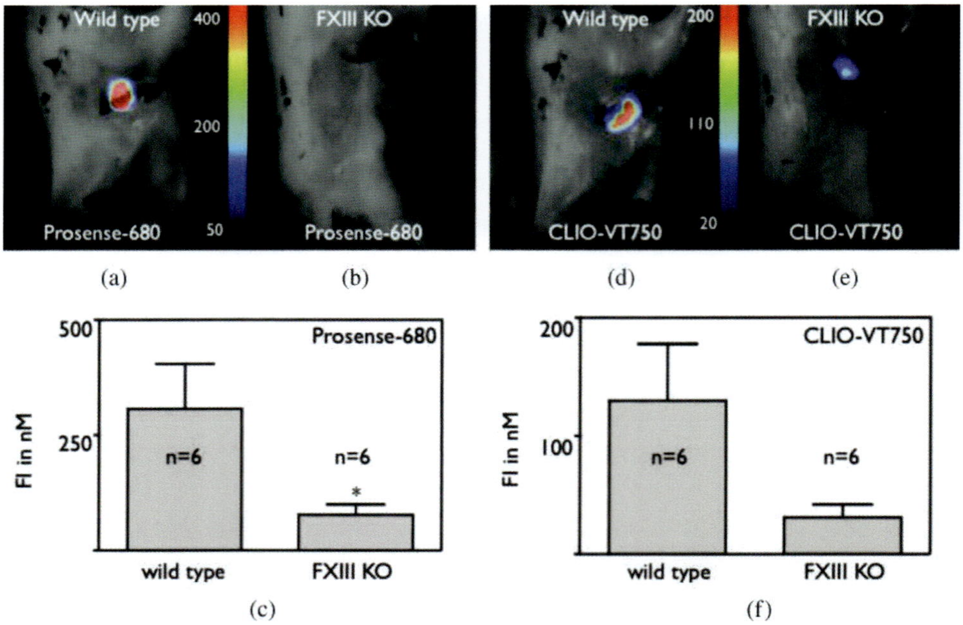

Fig. 6. Dual channel FMT detects impaired protease activity and leukocyte recruitment in FXIII$^{-/-}$ mice. (A) and (B), (D) and (E), Maximum intensity projection of FMT signal in wild-type and FXIII$^{-/-}$ mice (both genotypes windowed identically). (C) and (F), In FXIII$^{-/-}$ mice, fluorescence signal was significantly diminished in both channels, consistent with decreased protease activity and leukocyte recruitment. $^{*}P<0.05$. Reproduced with permission from Ref. 62.

In 2009, Sosnovik *et al.* reported on an improved annexin-labeled iron oxide nanoparticle system to image CM apoptosis within the first 4–6 h of ischemic onset.[63] Prior studies had established annexin-V, a cellular protein, as a useful means to visualize cell death, although these earlier works were limited by the spatial resolution of the applied imaging platform (SPECT, or single-photon emission computed tomography) and by its inability to distinguish between CM apoptosis and necrosis. The authors built upon this earlier work, modified the nanoparticle system and applied it to delayed enhancement (DE) MRI with a novel gadolinium chelate to clearly differentiate between apoptosis and necrosis in CMs; this dual contrast, MR imaging approach presents the opportunity for new insights into CM death after ischemia.

Applications of the powerful advantages of nanoparticles were detailed by Yilmaz *et al.* in 2012.[64,65] This group conducted two prospective phase III trials to compare infarct imaging using SPIO/USPIO (Super Paramagnetic Iron Oxide/Ultrasmall SPIO nanoparticles) and respective T_2/T_2^*-weighted cardiac MRI (CMR) techniques permitted a more detailed characterization of infarct pathology. The first trial, using the SPIO ferucarbotran (Resovist), did not achieve this aim; the second trial, using the USPIO ferumoxytol was able to substantially add detail. With this 30 nm nanoparticle system, it was demonstrated for the first time in humans that the use of this type of contrast agent could lead to signal changes in the infarct area when different CMR pulse sequences were employed. This enhanced characterization may be clinically valuable for several reasons, including: (1) infiltrating macrophages are of both diagnostic and therapeutic interest; (2) USPIOs may be used in patients with contraindications to standard gadolinium-based contrast agents (such as those with renal failure); (3) USPIOs may distinguish between acute MI and chronic myocardial fibrosis; and (4) USPIOs are capable of surface functionalization to allow for more targeted molecular imaging.

The treatment and management of post-MI patients has greatly improved in recent times, with much of the success attributable to advances in nanomedicine. Formerly, approaches to treating the infarcted heart involved the supply of growth factors, cytokines, drugs and other biomolecules via numerous delivery methods, including direct injection, microparticle injection or biogel application to the left ventricle, although significant disadvantages limited the efficacy of these treatments. An alternative strategy transplanted functional biomaterials to the scar tissue site; however, this carried the risks and costs associated with surgery. Dvir *et al.* reported in 2011 on a nanoparticulate system which largely overcame these earlier limitations.[66] Because angiotensin-1 (AT1) receptors are known to be overexpressed in the infarcted heart, nanoagents decorated with AT1 were developed and shown to accumulate primarily at the site of infarction. The targeted nanoparticles may be injected intravenously, circulated in the body for an extended time and allowed to cross endothelia. Ongoing research into decreasing the particle size, adding a therapeutic payload and optimizing the targeting peptide density could further enhance this promising system.

Gene therapy offers the potential to therapeutically correct for a missing or altered gene product, although the outcome of gene therapy *in vivo* is highly dependent upon the effective delivery of the gene of interest to the target site; in the context of MI, gene therapy could allow for more complete revascularization or angiogenesis after acute MI. Direct delivery of the gene for vascular endothelial growth factor (VEGF) to the myocardium using non-viral systems failed to induce significant levels of angiogenesis and myocardial viability due to low transfection efficiencies. Viral vectors are known to improve gene delivery efficiency and endurance, although host immune response may result in clearance of transduced cells, immunogenicity or interference with tissue-specific promoter activity. Paul *et al.* circumvented these challenges by employing a Baculovirus (Bac) gene delivery platform, as Bac is considered non-pathogenic and non-toxic.[67] The transduction efficiency was further improved via the use of non-viral endosomolytic Tat/plasmid DNA nanoparticles to mediate cell attachment. The system was evaluated using the angiopoietin-1 (Ang-1) angiogenic transgene to induce myocardial angiogenesis in an infarcted rat heart model. Results showed that the Bac-nanoparticle-mediated Ang-1 gene therapy clearly inhibited the progression of the impairment in cardiac performance three weeks after MI; this improvement in cardiac function may be attributable to the reduction in the infarct size and the increase in blood vessel density that was observed. Additional studies suggest that a combination of intramyocardial delivery of both Ang-1 and VEGF using Bac nanoagents might further improve upon these current results.

Zhang *et al.* also sought to exploit the important function of VEGF in repairing the heart after acute MI by designing a nanoparticle-based gene delivery system.[68] Stable magnetic nanobeads complexed with Adeno virus were constructed, inject intravenously and, under the guidance of an applied magnetic field, targeted to ischemically damaged hearts; administration of the hVEGF-coded complexes in a MI rat model resulted in hVEGF detection in the heart and improved right ventricle performance one week after MI. While concerns over immunogenicity and CD8+ T cell activation resulting from the use of the adenovirus vector need to be addressed, this study highlights the potential for magnetically-controlled gene delivery to improve post-MI outcomes.

Cristallini *et al.* likewise sought to improve post-MI outcome via a nanomolecular approach, albeit with a different therapeutic target — the cardiac extracellular matrix (ECM).[69] Because the cardiac ECM plays a fundamental role in healing after MI, and this ECM is largely controlled by the MMP proteolytic system, these researchers synthesized molecularly imprinted polymer nanoparticles (MIPNPs) capable of selectively modulating the balance of MMPs and TIMPs in the ECM. More extensive studies into the efficacy of MIPNPs as a viable clinical approach are necessary, but this work established the proof of concept for MIPNPs and further underscored the capacity for nanotechnology to address the clinical problems and improve the patient outcomes following MI.

5. Aortic Aneurysms

Aortic aneurysms (AAs) are vascular disorders which appear predominantly in the elderly population and are defined by the dilatation of the aorta to a diameter of 1.5 times normal.[70] If untreated, they can expand and eventually rupture, resulting in death rates as high as 90%.[71] AAs are subdivided anatomically into thoracic aortic aneurysms (TAAs) and abdominal aortic aneurysms (AAAs), each with differing underlying pathogenesis.[72] Whereas TAAs display loss of smooth muscle cells, the degradation of elastic fibers, and the amassing of proteoglycans, AAAs typically demonstrate severe atherosclerotic lesions possessing significant inflammatory cells, resulting in the remodeling of the elastic media. As AAAs are the most common subtype of AA, this section will only focus on the imaging techniques developed to diagnose and monitor this condition.

For most patients, AAA is asymptomatic, until such time that the aneurysm ruptures. Interestingly, the presence of AAA can be easily detected by non-invasive ultrasound imaging. In worst-case scenarios, surgical intervention is necessary in order to prevent rupture. Over the past few years, researchers have begun to focus on the development of new diagnostic and therapeutic moieties to study the rate of AAA development and to provide personalized treatment for individual patients.

Turner *et al.* have evaluated the use of USPIOs as a marker for the detection of macrophage infiltration in AAA via MR imaging.[73] The purpose of this study was to demonstrate the efficiency of USPIO in detecting the development of AAA by following the progression of the acute inflammatory process. Initially, AAA was induced in mice by a continuous infusion of angiotensin II (ang-II). Once AAA was detected, the USPIO contrast agent ferumoxtran-10 was administered via the tail vein, and the mice were imaged. After imaging, the aortae were harvested for *ex vivo* analyses including histology, iron quantification, and gene expression. As a result of these experiments, the authors showed that significant inflammation is caused by macrophage migration, which can be detected and examined using USPIO as an imaging marker. Unfortunately, *in vivo* imaging of USPIO uptake in mouse aortic aneurysms proved somewhat difficult due to the accumulation of the contrast agents in surrounding tissues such as the liver. Further investigation with larger cohorts is still needed to complete the analysis, and despite the encouraging results reported therein, the ability to detect early inflammation before AAA development using USPIO labeling is still limited.

Nchimi *et al.* have reported the magnetic resonance imaging of iron phagocytosis in intraluminal thrombi of AAA in humans.[74] The aim of the study was to determine if the phagocytic activities of specific cell types could be visualized by MR imaging using superparamagnetic iron oxide (SPIO) nanoparticles, known for their affinity for leukocytes. Patients that were enrolled in the study were referred for MR analysis of their AAA prior to surgical repair. Pre- and post-contrast MR images were acquired with patients undergoing imaging at 1 h and 24 h after receiving an intravenous infusion of ferumoxide SPIO. After surgery, the authors were able to analyze the tissues and correlate the imaging data with histology, zymography, and reverse transcriptase polymerase chain reaction data. The authors observed a strong relationship between the thrombus signal and the level of proteins such as CD68, CD66b and enzymes such as MMP-9, which plays a role in inflammation in AAA. Importantly, SPIO uptake was mainly localized at the luminal interface of the thrombus in high-risk AAAs in correlation with the abundance of leukocytes, the key mediators of proteolytic activity, and thus the source of aneurysm rupture.

Nahrendorf *et al.* have also used macrophage-targeted nanoparticles for the detection of AAA in a murine model, yet attempted to increase the sensitivity of the system by labeling the particle with ^{18}F for PET-CT.[75] The purpose of the study was to detect inflammatory aneurysms before rupture. The nanoparticle size and coating were firstly optimized to target macrophages. The particles were also co-labeled with near infrared fluorophores to ease the validation of agent location using optical imaging techniques. In addition, the authors used a recently developed PET isotope labeling strategy based on inverse electron demand Diels Alder "click" chemistry, that was found to be advantageous for the rapid labeling of nanoparticles with the ^{18}F radiotracer.[76] AAAs were induced via administration of angiotensin II over a prolonged period. Using flow cytometry, the authors initially showed a huge increase in the number of macrophages (18.6 fold) present in aneurysms, as compared to the aortic wall of control mice. When the ^{18}F-derivatized nanoparticles were subsequently injected into the AAA-bearing animals, the particles were avidly taken up by phagocytic cells as was clearly observed in the PET–CT (Fig. 7). Overall, these findings suggest that the use of macrophage-targeted imaging may be suitable to evaluate the activity of the disease in nascent AAs.

Using the knowledge gleaned from earlier studies reporting on the inflammatory process and angiogenesis in AAA, Kitagawa *et al.* have demonstrated the targeted imaging of the disorder in murine models, by using Arg-Gly-Asp (RGD)-conjugated human ferritin nanoparticles (HFn) in experimental abdominal aortic aneurysm disease.[77] The nanoparticles were developed for the targeting of $\alpha_v\beta_3$ integrin. AAAs were first induced in apo $E^{-/-}$ mice, by a continuous angiotensin II infusion via subcutaneous implanted osmotic mini-pumps. The targeted (RGD-HFn) and non-targeted (HFn) nanoparticles were chemically labeled with the Cy5.5 and injected intravenously for near-infrared fluorescence imaging. The FMT signal of RGD-HFn

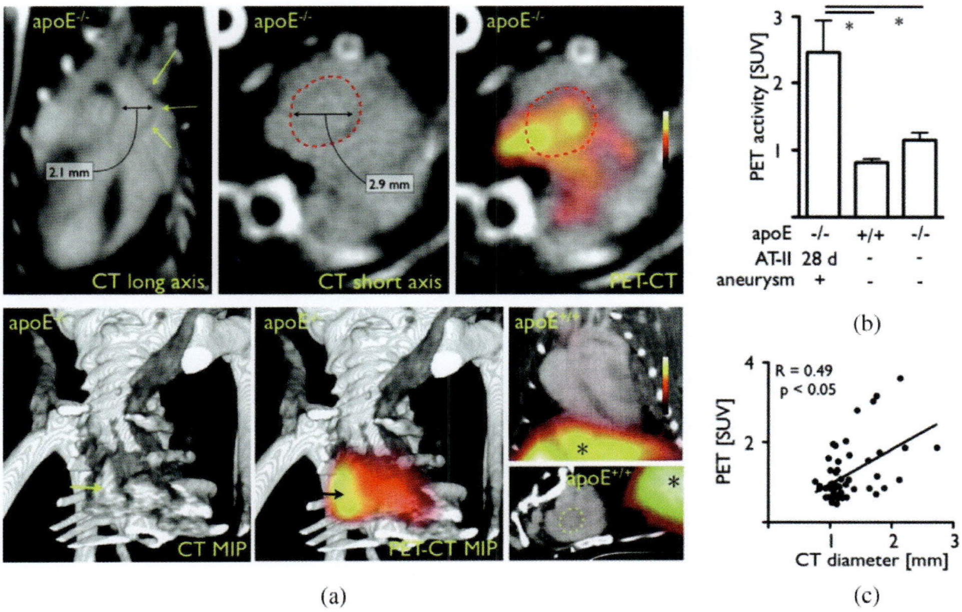

Fig. 7. (Color online) PET–CT imaging in mice with AAs. A, Representative PET-CT images. Yellow arrows and dotted lines outline the aneurys- matic aorta. MIP indicates maximum intensity projection. B, *In vivo* PET signal derived from the aneurysmatic vessel wall at the site of the aneurysm (left) wild-type controls (middle), and apoE$^{-/-}$ mice that did not receive AT-II (right bar). Data are presented as mean SEM, $^{*}P < 0.01$. C, A correlation of the PET signal with aortic diameter (measured by CT) after AT-II delivery in apoE$^{-/-}$ mice. Reproduced with permission from Ref. 75.

was significantly higher than HFn in AAAs as compared to non-diseased areas, both *in situ* and *ex vivo*. Histologically, the authors showed that the targeted nanoparticles readily co-localized with both macrophages and with areas of increased angiogenesis. While the authors have certainly developed a promising imaging method for vascular inflammation and angiogenesis in AAA, the using FMT in humans for the detection and quantification of deeper AAAs is limited due to fluorescence attenuation with depth. One solution would be the incorporation of other imaging agents in HFn and the combination of FMT with other clinically available imaging techniques (e.g. MRI, PET, SPECT, CT).

In one final, promising study on humans with AAAs, Richards *et al.* have shown, by MRI, differential USPIO uptake in aneurysms can be readily associated with high rates of aneurysm expansion.[78] In particular, this method could differentiate between patients who need earlier surgical intervention and those with only periluminal uptake who need continuous monitoring and medical treatment. Whereas this was just a small pilot study, the results clearly demonstrate that nanoparticulate agents are capable of having potential impact in informing decisions as to the proper course of intervention.

6. Nanomaterials for the Detection and Prevention of Heart Transplant Rejection

Heart transplantation is the last treatment option for patients with severe cardiac failure or severe coronary artery disease. Implanting an organ from a recently deceased donor is the most common procedure (allograft), which may significantly increase the quality of life for patients. However, the transplantation procedure is far from perfect. The main problem encountered is the acute rejection of the organ, characterized by an immune system attack on the donor tissue, and often leading to rapid graft loss. The standard diagnostic tool for detecting rejection is morphological analysis of endomyocardial biopsies obtained from the right ventricle with access through the right jugular vein. However, this procedure is invasive and risky for the patient. Thus, researchers have investigated a number of alternative methods for the non-invasive detection and treatment of acute rejection.

Penno *et al.* have reported the use of USPIO nanoparticles (AMI-227, a 35 nm dextran-coated iron oxide) as an agent for MR imaging in acute cardiac transplant rejection.[79] The purpose of this study was to demonstrate that the nanoparticles were taken up by macrophages in a rejecting transplanted organ with a consequent reduction in T_2^*. AMI-227 was injected at low and high doses in two groups of rats that received either a syngeneic or allogeneic heart transplantations. Unfortunately, no difference in the change in T_2^* was found between the rejecting and non-rejecting grafts. This initial study concluded that, in a rat model, measurements of T_2^* after myocardial macrophage uptake of AMI-227, in a clinical MR 1.5T scanner, were not useful for the diagnostic of acute rejection.

Christen *et al.* have gone beyond this initial report, reporting on a dual imaging technique based on a fluorescent protease sensor (ProSense-680) and a magnetofluorescent phagocytosis marker (CLIO-VT750) for the study of macrophage function in heart transplant rejection.[80] The purpose of their study was to develop a non-invasive method for the detection of parenchymal rejection by providing an *in vivo* three-dimensional functional map of macrophages. The authors transplanted allografts (BALB/c) and isografts (B/6) into B/6 recipient mice. During rejection, they observed significantly higher phagocytic uptake of the magnetofluorescent probe by MR imaging and higher protease activity by FMT in allografts than in isografts (Fig. 8). They

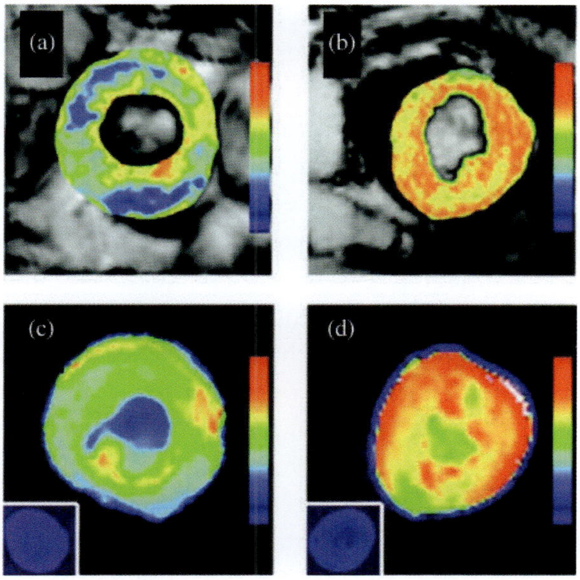

Fig. 8. T_2^*-weighted MRI after phagocytosis sensor injection resulted in negative signal enhancement in the grafted heart, with the signal represented as CNR. On seventh day, the contrast was stronger in the allograft (b) than in the isograft (a). FRI after phagocytosis sensor injection validates *in vivo* imaging. (c) The isograft exhibited a considerably higher signal than the native heart (inset shows native heart of same animal), pointing to ongoing ischemia/reperfusion injury. (d) The signal in the allograft was higher than in the native heart (inset shows native heart of same animal) and higher than in the isograft (c). Reproduced with permission from Ref. 80.

also managed to correlate the degree of rejection with the intensity of the *in vivo* imaging of macrophage response.

Lourenço-Filho *et al.* have demonstrated the efficacy of a cholesterol-rich nanoemulsion (LDE) associated with paclitaxel in the decrease of heart transplant vascular disease in a rabbit graft model.[81] The authors showed that LDE is concentrated in grafted hearts almost four-fold greater than in native hearts. Treatment of animals with LDE-paclitaxel reduced the degree of stenosis in the coronary arteries of the grafted hearts by roughly 50% and resulted in a seven-fold reduction in the macrophage infiltrate. The luminal area in the group receiving LDE-paclitaxel treatment was three-fold larger than that seen in the control group that were only treated with saline. This new treatment approach improved the post-transplantation injury to grafted hearts. Nevertheless, the association of LDE with paclitaxel is still to be tested in clinical trials to ensure limited toxicity to patients undergoing heart transplantation.

Guo *et al.* have developed a targeted polymeric nanocarrier for the T-cell-based early diagnosis and immune-suppressive therapy of allogeneic heart transplantation in a rat model.[82] This theranostic nanosystem is based on a poly(ethylene glycol)-grafted polyethyleneimine (PEG-g-PEI) complex functionalized with the CD3 single-chain antibody (scAbCD3) coating the surface of SPION. Polyplexes were subsequently formed with pDNA for gene therapy. The authors showed that they were able to monitor, by magnetic resonance and optical imaging, the *in vivo* uptake of the targeted nanocarrier by the T-cells gathered in hearts of transplanted rats and the efficient suppression of the acute immune rejection when the pDNA DGKα was successfully delivered. These encouraging results opened a novel approach for the diagnosis and treatment of acute rejection due to the efficiency of the targeted-theranostic nanosystem in the real-time monitoring of primary T-cells and the inhibition of the immune response after heart transplantation.

Transplantation has become the standard procedure for end-stage heart failure. However, the major issue with this technique is the acute rejection caused by the immune response to the donor tissue. Usually, morphological analysis of graft biopsies is done to detect rejection. Unfortunately, it is extremely invasive, which may involve complications and cause discomfort and anxiety to the patients. Thus, investigators have studied a number of different nanomaterial-based techniques to enable more efficient diagnosis of acute rejection. Although demonstrating promising results, most of these methods are still far from clinically relevant.

7. Conclusions and Outlook

This chapter has focused on the applicability of nanomaterials to CVD and its sequelae, including atherosclerosis, thrombosis, and myocardial infarction. Whereas conventional methodologies have been extensively studied for this class of diseases, the use of nanoagents for the detection and treatment of these conditions may allow for more specifically targeted interventions, increasing efficacy while decreasing deleterious side effects. While a number of nanomedicine-based agents have already made it to clinic for the treatment of cancer, this has yet to be realized for CVD. Given the rapid pace of the research in this field, clinical translation may be realized in the near future.

Acknowledgment

This work was supported in part by funds from the NIH NHLBI under contract number HHSN268201000044C.

References

1. M. Heron, *Natl. Vital. Stat. Rep.* **61**, 1 (2012).
2. P. Libby, *Nature* **420**, 868 (2002).
3. J. R. McCarthy and R. Weissleder, *Adv. Drug Deliv. Rev.* **60**, 1241 (2008).
4. K. Chang, S. A. Francis, E. Aikawa, J. L. Figueiredo, R. H. Kohler, J. R. McCarthy, R. Weissleder, J. Plutzky and F. A. Jaffer, *Arterioscler. Thromb. Vasc. Biol.* **30**, 1933 (2010).
5. M. E. Kooi, V. C. Cappendijk, K. B. Cleutjens, A. G. Kessels, P. J. Kitslaar, M. Borgers, P. M. Frederik, M. J. Daemen and J. M. van Engelshoven, *Circulation* **107**, 2453 (2003).
6. F. Hyafil, J. C. Cornily, J. E. Feig, R. Gordon, E. Vucic, V. Amirbekian, E. A. Fisher, V. Fuster, L. J. Feldman and Z. A. Fayad, *Nat. Med.* **13**, 636 (2007).
7. F. Hyafil, J. C. Cornily, J. H. Rudd, J. Machac, L. J. Feldman and Z. A. Fayad, *J. Nucl. Med.* **50**, 959 (2009).
8. H. Zhang, L. Zhang, J. Myerson, K. Bibee, M. Scott, J. Allen, G. Sicard, G. Lanza and S. Wickline, *PLoS One* **6**, e26385 (2011).
9. A. Tsourkas, V. R. Shinde-Patil, K. A. Kelly, P. Patel, A. Wolley, J. R. Allport and R. Weissleder, *Bioconjug. Chem.* **16**, 576 (2005).
10. K. A. Kelly, J. R. Allport, A. Tsourkas, V. R. Shinde-Patil, L. Josephson and R. Weissleder, *Circ. Res.* **96**, 327 (2005).
11. K. A. Kelly, M. Nahrendorf, A. M. Yu, F. Reynolds and R. Weissleder, *Mol. Imaging Biol.* **8**, 201 (2006).
12. M. Nahrendorf, F. A. Jaffer, K. A. Kelly, D. E. Sosnovik, E. Aikawa, P. Libby and R. Weissleder, *Circulation* **114**, 1504 (2006).
13. H. Pan, J. W. Myerson, L. Hu, J. N. Marsh, K. Hou, M. J. Scott, J. S. Allen, G. Hu, S. San Roman, G. M. Lanza, R. D. Schreiber, P. H. Schlesinger and S. A. Wickline, *FASEB J.* **27**, 255 (2013).
14. M. J. Jacobin-Valat, K. Deramchia, S. Mornet, C. E. Hagemeyer, S. Bonetto, R. Robert, M. Biran, P. Massot, S. Miraux, S. Sanchez, A. K. Bouzier-Sore, J. M. Franconi, E. Duguet and G. Clofent-Sanchez, *NMR Biomed.* **24**, 413 (2011).
15. H. W. Kang, L. Josephson, A. Petrovsky, R. Weissleder and A. Bogdanov, *Bioconjug. Chem.* **13**, 122 (2002).
16. O. J. de Boer, A. C. van der Wal, P. Teeling and A. E. Becker, *Cardiovasc. Res.* **41**, 443 (1999).
17. K. Cai, S. D. Caruthers, W. Huang, T. A. Williams, H. Zhang, S. A. Wickline, G. M. Lanza and P. M. Winter, *JACC Cardiovasc. Imaging* **3**, 824 (2010).
18. A. M. Neubauer, H. Sim, P. M. Winter, S. D. Caruthers, T. A. Williams, J. D. Robertson, D. Sept, G. M. Lanza and S. A. Wickline, *Magn. Reson. Med.* **60**, 1353 (2008).
19. P. M. Winter, A. M. Morawski, S. D. Caruthers, R. W. Fuhrhop, H. Zhang, T. A. Williams, J. S. Allen, E. K. Lacy, J. D. Robertson, G. M. Lanza and S. A. Wickline, *Circulation* **108**, 2270 (2003).
20. P. M. Winter, A. M. Neubauer, S. D. Caruthers, T. D. Harris, J. D. Robertson, T. A. Williams, A. H. Schmieder, G. Hu, J. S. Allen, E. K. Lacy, H. Zhang, S. A. Wickline and G. M. Lanza, *Arterioscler. Thromb. Vasc. Biol.* **26**, 2103 (2006).
21. P. M. Winter, S. D. Caruthers, H. Zhang, T. A. Williams, S. A. Wickline and G. M. Lanza, *JACC Cardiovasc. Imaging* **1**, 624 (2008).

22. M. J. Lipinski, J. C. Frias, V. Amirbekian, K. C. Briley-Saebo, V. Mani, D. Samber, A. Abbate, J. G. Aguinaldo, D. Massey, V. Fuster, G. W. Vetrovec and Z. A. Fayad, *JACC Cardiovasc. Imaging* **2**, 637 (2009).
23. C. Tu, T. S. Ng, H. K. Sohi, H. A. Palko, A. House, R. E. Jacobs and A. Y. Louie, *Biomaterials* **32**, 7209 (2011).
24. G. Y. Lee, J. H. Kim, G. T. Oh, B. H. Lee, I. C. Kwon and I. S. Kim, *J. Control Release* **155**, 211 (2011).
25. G. A. van Tilborg, E. Vucic, G. J. Strijkers, D. P. Cormode, V. Mani, T. Skajaa, C. P. Reutelingsperger, Z. A. Fayad, W. J. Mulder and K. Nicolay, *Bioconjug. Chem.* **21**, 1794 (2010).
26. Y. Yamakoshi, H. Qiao, A. N. Lowell, M. Woods, B. Paulose, Y. Nakao, H. Zhang, T. Liu, S. Lund-Katz and R. Zhou, *Chem. Commun.* (*Camb*) **47**, 8835 (2011).
27. A. N. Lowell, H. Qiao, T. Liu, T. Ishikawa, H. Zhang, S. Oriana, M. Wang, E. Ricciotti, G. A. FitzGerald, R. Zhou and Y. Yamakoshi, *Bioconjug. Chem.* **23**, 2313 (2012).
28. W. Chen, E. Vucic, E. Leupold, W. J. Mulder, D. P. Cormode, K. C. Briley-Saebo, A. Barazza, E. A. Fisher, M. Dathe and Z. A. Fayad, *Contrast Media Mol. Imaging* **3**, 233 (2008).
29. D. P. Cormode, K. C. Briley-Saebo, W. J. Mulder, J. G. Aguinaldo, A. Barazza, Y. Ma, E. A. Fisher and Z. A. Fayad, *Small* **4**, 1437 (2008).
30. D. P. Cormode, T. Skajaa, M. M. van Schooneveld, R. Koole, P. Jarzyna, M. E. Lobatto, C. Calcagno, A. Barazza, R. E. Gordon, P. Zanzonico, E. A. Fisher, Z. A. Fayad and W. J. Mulder, *Nano Lett.* **8**, 3715 (2008).
31. T. Skajaa, D. P. Cormode, P. A. Jarzyna, A. Delshad, C. Blachford, A. Barazza, E. A. Fisher, R. E. Gordon, Z. A. Fayad and W. J. Mulder, *Biomaterials* **32**, 206 (2011).
32. J. C. Frias, K. J. Williams, E. A. Fisher and Z. A. Fayad, *J. Am. Chem. Soc.* **126**, 16316 (2004).
33. M. E. Lobatto, Z. A. Fayad, S. Silvera, E. Vucic, C. Calcagno, V. Mani, S. D. Dickson, K. Nicolay, M. Banciu, R. M. Schiffelers, J. M. Metselaar, L. van Bloois, H. S. Wu, J. T. Fallon, J. H. Rudd, V. Fuster, E. A. Fisher, G. Storm and W. J. Mulder, *Mol. Pharm.* **7**, 2020 (2010).
34. J. R. McCarthy, E. Korngold, R. Weissleder and F. A. Jaffer, *Small* **6**, 2041 (2010).
35. X. Yu, S. K. Song, J. Chen, M. J. Scott, R. J. Fuhrhop, C. S. Hall, P. J. Gaffney, S. A. Wickline and G. M. Lanza, *Magn. Reson. Med.* **44**, 867 (2000).
36. S. Flacke, S. Fischer, M. J. Scott, R. J. Fuhrhop, J. S. Allen, M. McLean, P. Winter, G. A. Sicard, P. J. Gaffney, S. A. Wickline and G. M. Lanza, *Circulation* **104**, 1280 (2001).
37. D. Pan, S. D. Caruthers, A. Senpan, C. Yalaz, A. J. Stacy, G. Hu, J. N. Marsh, P. J. Gaffney, S. A. Wickline and G. M. Lanza, *J. Am. Chem. Soc.* **133**, 9168 (2011).
38. J. R. McCarthy, P. Patel, I. Botnaru, P. Haghayeghi, R. Weissleder and F. A. Jaffer, *Bioconjug. Chem.* **20**, 1251 (2009).
39. A. P. Laudano and R. F. Doolittle, *Proc. Natl. Acad. Sci. USA* **75**, 3085 (1978).
40. C. H. Tung, N. H. Ho, Q. Zeng, Y. Tang, F. A. Jaffer, G. L. Reed and R. Weissleder, *ChemBioChem* **4**, 897 (2003).
41. F. A. Jaffer, C. H. Tung, J. J. Wykrzykowska, N. H. Ho, A. K. Houng, G. L. Reed and R. Weissleder, *Circulation* **110**, 170 (2004).
42. H. Yoo, J. W. Kim, M. Shishkov, E. Namati, T. Morse, R. Shubochkin, J. R. McCarthy, V. Ntziachristos, B. E. Bouma, F. A. Jaffer and G. J. Tearney, *Nat. Med.* **17**, 1680 (2011).

43. *N. Engl. J. Med.* **333**, 1581 (1995).
44. H. V. Anderson and J. T. Willerson, *N. Engl. J. Med.* **329**, 703 (1993).
45. P. W. Armstrong and D. Collen, *Circulation* **103**, 2862 (2001).
46. S. S. Erdem, I. Y. Sazonova, T. Hara, F. A. Jaffer and J. R. McCarthy, *Methods Enzymol.* **508**, 191 (2012).
47. J. R. McCarthy, I. Y. Sazonova, S. S. Erdem, T. Hara, B. D. Thompson, P. Patel, I. Botnaru, C. P. Lin, G. L. Reed, R. Weissleder and F. A. Jaffer, *Nanomedicine (Lond)*, 7, 1017 (2012).
48. J. N. Marsh, A. Senpan, G. Hu, M. J. Scott, P. J. Gaffney, S. A. Wickline and G. M. Lanza, *Nanomed.* **2**, 533 (2007).
49. J. N. Marsh, G. Hu, M. J. Scott, H. Zhang, M. J. Goette, P. J. Gaffney, S. D. Caruthers, S. A. Wickline, D. Abendschein and G. M. Lanza, *Nanomedicine (Lond)* **6**, 605 (2011).
50. J. Myerson, L. He, G. Lanza, D. Tollefsen and S. Wickline, *J. Thromb. Haemost.* **9**, 1292 (2011).
51. F. Bi, J. Zhang, Y. Su, Y. C. Tang and J. N. Liu, *Biomaterials* **30**, 5125 (2009).
52. N. Korin, M. Kanapathipillai, B. D. Matthews, M. Crescente, A. Brill, T. Mammoto, K. Ghosh, S. Jurek, S. A. Bencherif, D. Bhatta, A. U. Coskun, C. L. Feldman, D. D. Wagner and D. E. Ingber, *Science* **337**, 738 (2012).
53. V. Kumar, A. K. Abbas, N. Fausto and R. N. Mitchell (eds.), *Robbins Basic Pathology* (Saunders, an imprint of Elsevier, Philadelphia, 2007), pp. 390–397.
54. T. Reichlin, W. Hochholzer, S. Bassetti, S. Steuer, C. Stelzig, S. Hartwiger, S. Biedert, N. Schaub, C. Buerge, M. Potocki, M. Noveanu, T. Breidthardt, R. Twerenbold, K. Winkler, R. Bingisser and C. Mueller, *N. Engl. J. Med.* **361**, 858 (2009).
55. E. D. Goluch, J. M. Nam, D. G. Georganopoulou, T. N. Chiesl, K. A. Shaikh, K. S. Ryu, A. E. Barron, C. A. Mirkin and C. Liu, *Lab Chip.* **6**, 1293 (2006).
56. D. B. Buxton, *Nanomedicine (Lond)* 7, 173 (2012).
57. P. Panizzi, M. Nahrendorf, M. Wildgruber, P. Waterman, J. L. Figueiredo, E. Aikawa, J. McCarthy, R. Weissleder and S. A. Hilderbrand, *J. Am. Chem. Soc.* **131**, 15739 (2009).
58. F. Leuschner, P. Dutta, R. Gorbatov, T. I. Novobrantseva, J. S. Donahoe, G. Courties, K. M. Lee, J. I. Kim, J. F. Markmann, B. Marinelli, P. Panizzi, W. W. Lee, Y. Iwamoto, S. Milstein, H. Epstein-Barash, W. Cantley, J. Wong, V. Cortez-Retamozo, A. Newton, K. Love, P. Libby, M. J. Pittet, F. K. Swirski, V. Koteliansky, R. Langer, R. Weissleder, D. G. Anderson and M. Nahrendorf, *Nat. Biotechnol.* **29**, 1005 (2011).
59. W. D. Gray, P. Che, M. Brown, X. Ning, N. Murthy and M. E. Davis, *J. Cardiovasc. Transl. Res.* **4**, 631 (2011).
60. M. Tranter, Y. Liu, S. He, J. Gulick, X. Ren, J. Robbins, W. K. Jones and T. M. Reineke, *Mol. Ther.* **20**, 601 (2012).
61. D. E. Sosnovik, M. Nahrendorf, N. Deliolanis, M. Novikov, E. Aikawa, L. Josephson, A. Rosenzweig, R. Weissleder and V. Ntziachristos, *Circulation* **115**, 1384 (2007).
62. M. Nahrendorf, D. E. Sosnovik, P. Waterman, F. K. Swirski, A. N. Pande, E. Aikawa, J. L. Figueiredo, M. J. Pittet and R. Weissleder, *Circ. Res.* **100**, 1218 (2007).
63. D. E. Sosnovik, E. Garanger, E. Aikawa, M. Nahrendorf, J. L. Figuiredo, G. Dai, F. Reynolds, A. Rosenzweig, R. Weissleder and L. Josephson, *Circ. Cardiovasc. Imaging* **2**, 460 (2009).
64. A. Yilmaz, M. A. Dengler, H. van der Kuip, H. Yildiz, S. Rosch, S. Klumpp, K. Klingel, R. Kandolf, X. Helluy, K. H. Hiller, P. M. Jakob and U. Sechtem, *Eur. Heart J.* (2012).

65. A. Yilmaz, S. Rosch, H. Yildiz, S. Klumpp and U. Sechtem, *Circulation* **126**, 1932 (2012).
66. T. Dvir, M. Bauer, A. Schroeder, J. H. Tsui, D. G. Anderson, R. Langer, R. Liao and D. S. Kohane, *Nano Lett.* **11**, 4411 (2011).
67. A. Paul, Z. M. Binsalamah, A. A. Khan, S. Abbasia, C. B. Elias, D. Shum-Tim and S. Prakash, *Biomaterials* **32**, 8304 (2011).
68. Y. Zhang, W. Li, L. Ou, W. Wang, E. Delyagina, C. Lux, H. Sorg, K. Riehemann, G. Steinhoff and N. Ma, *PLoS One* **7**, e39490 (2012).
69. C. Cristallini, M. Gagliardi, N. Barbani, D. Giannessi and G. D. Guerra, *J. Mater. Sci. Mater. Med.* **23**, 205 (2012).
70. K. W. Johnston, R. B. Rutherford, M. D. Tilson, D. M. Shah, L. Hollier and J. C. Stanley, *J. Vasc. Surg.* **13**, 452 (1991).
71. H. C. Kochanek, J. Xu, S. Murphy, A. M. Minino and H. C. Kung, *Natl. Vital. Stat. Rep.* **59**, 1 (2011).
72. T. S. Absi, T. M. r. Sundt, W. S. Tung, M. Moon, J. K. Lee, R. R. J. Damiano and R. W. Thompson, *J. Thorac. Cardiovasc. Surg.* **126**, 344 (2003).
73. G. H. Turner, A. R. Olzinski, R. E. Bernard, K. Aravindhan, R. J. Boyle, M. J. Newman, S. D. Gardner, R. N. Willette, P. J. Gough and B. M. Jucker, *J. Magn. Reson. Imaging* **30**, 455 (2009).
74. A. Nchimi, O. Defawe, D. Brisbois, T. K. Broussaud, J. O. Defraigne, P. Magotteaux, B. Massart, J. M. Serfaty, X. Houard, J. B. Michel and N. Sakalihasan, *Radiology* **254**, 973 (2010).
75. M. Nahrendorf, E. Keliher, B. Marinelli, F. Leuschner, C. S. Robbins, R. E. Gerszten, M. J. Pittet, F. K. Swirski and R. Weissleder, *Arterioscler. Thromb. Vasc. Biol.* **31**, 750 (2011).
76. N. K. Devaraj, E. J. Keliher, G. M. Thurber, M. Nahrendorf and R. Weissleder, *Bioconjug. Chem.* **20**, 397 (2009).
77. T. Kitagawa, H. Kosuge, M. Uchida, M. M. Dua, Y. Iida, R. L. Dalman, T. Douglas and M. V. McConnell, *Mol. Imaging Biol.* **14**, 315 (2012).
78. J. M. Richards, S. I. Semple, T. J. MacGillivray, C. Gray, J. P. Langrish, M. Williams, M. Dweck, W. Wallace, G. McKillop, R. T. Chalmers, O. J. Garden and D. E. Newby, *Circ. Cardiovasc. Imaging* **4**, 274 (2011).
79. E. Penno, C. Johnsson, L. Johansson and H. Ahlström, *Acta Radiol.* **47**, 264 (2006).
80. T. Christen, M. Nahrendorf, M. Wildgruber, F. K. Swirski, E. Aikawa, P. Waterman, K. Shimizu, R. Weissleder and P. Libby, *Circulation* **119**, 1925 (2009).
81. D. D. Lourenco-Filho, R. C. Maranhao, C. A. Mendez-Contreras, E. R. Tavares, F. R. Freitasand N. A. Stolf, *J. Thorac. Cardiovasc. Surg.* **141**, 1522 (2011).
82. Y. Guo, W. Chen, W. Wang, J. Shen, R. Guo, F. Gong, S. Lin, D. Cheng, G. Chen and X. Shuai, *ACS Nano*, DOI: 10.1021/nn3037573 (2012).

Chapter 2

Nanomedicines for Restenosis Therapy

J. E. Tengood, I. Fishbein, R. J. Levy and M. Chorny

Department of Pediatrics, The Children's Hospital of Philadelphia, 3401 Civic Center Blvd, Philadelphia, PA 19104, US
chorny@email.chop.edu

Abstract

Injury-triggered renarrowing (restenosis) of arteries treated with angioplasty to relieve atherosclerotic obstruction presents a formidable challenge due to a lack of therapies combining adequate efficacy and safety. Drug eluting stents (DES) introduced as a platform for site-specific pharmacotherapy have substantially reduced the incidence of restenosis in coronary artery disease patients with non-complex lesions. However, the finite drug payload and inability to adjust the drug dose and release kinetics to complex settings presented by real-world patients are significant limitations of the DES technology. Thus, restenosis remains a serious medical condition posing an unmet need for novel treatment strategies with significantly improved therapeutic performance. A versatile design and adjustable properties of nanocarriers formulated with different classes of therapeutic agents make them a highly promising experimental approach for preventing the disease while minimizing untoward effects, and can provide the basis for the development of new, clinically viable antirestenotic therapies. This chapter discusses restenosis and its pathological pathways as a target for nanocarrier-based treatment modalities and provides an overiew of nanoparticle formulations evaluated to date in terms of their design, safety and performance.

1. Arterial Restenosis: Definitions and Pathophysiology

1.1. *Atherosclerosis and its regional manifestations*

Atherosclerosis, the principle cause of myocardial and cerebral infarction, is a maladaptive response to insults to the endothelium and smooth muscle cells (SMC) of the wall of an artery, triggered primarily by the accumulation of modified lipoproteins in the sub-endothelial space of large and medium-size arteries. Atherosclerosis begins with the formation of non-occluding fatty streaks that progress to fibrous lesions (accompanied by inflammation) and further mature in many cases to advanced and complicated lesions in which either plaque itself, or plaque rupture-induced thrombus may occlude the artery.[1,2] The atherosclerotic process underlies a group of diseases characterized by an arterial obstruction and accounts for 653,000 deaths as the contributing cause in 2003,[3] thus being responsible for nearly three-fourths of all deaths from cardiovascular disease (CVD).[4]

1.2. *Coronary artery disease*

Coronary artery disease (CAD), i.e. clinically consequential coronary atherosclerosis, results in luminal narrowing, reduced compliance of the vessel wall and either a gradual reduction or a dramatic and sudden loss of blood supply to the myocardium.[5] As of 2006, CAD is the single largest killer of Americans, causing 1 in every 6 deaths in the US.[6,7] As the incidence of death caused by CAD and heart disease overall has decreased steadily over the past 10 years,[8] the need for repeated intervention for CAD has progressively increased over this time period.[9]

1.3. *Interventional treatment of atherosclerotic lesions*

Percutaneous coronary intervention (PCI), including percutaneous transluminal coronary angioplasty (PTCA) has improved the outcomes for the majority of patients with CAD, compared to coronary artery bypass graft surgery (CABG), and has become the most common revascularization procedure for CAD.[10] In the course of this procedure a balloon catheter is advanced through the circulatory system to site of arterial obstruction, and the balloon is inflated for a few seconds to both physically dilate the obstructed artery and compress the blockage against the arterial wall. Following balloon angioplasty, a stent, an expandable metallic scaffold, is used to maintain patency of the blood vessel lumen, and serves as a mechanical support. The major drawback of PCI is a high rate of restenosis (defined as more than 50% loss of initial gain in the blood vessel diameter, determined by follow-up angiogram within 3–6 months following PCI)[11,12] triggered by the arterial wall trauma caused by this procedure, and the resultant need for repeated intervention.[13–15]

After stent placement, activated leukocytes invade the arterial wall as soon as 15 min post-implantation, following deposition of platelets and fibrinogen at the injury site. Also, in the time frame of 1–6 weeks post-implantation, proliferation and accumulation of SMC occur in the medial and intimal arterial wall layers, triggered by the direct mechanical stimuli from the over-expanded stent, and by the plethora of cytokines and growth factors released by the activated platelets and leukocytes. Concurrently with SMC proliferation, centripetal migration of SMC from the media to intima commences, forming the neointima. Thus, while stent deployment reduces the occurrence of restenosis from 30–40% to 20% when compared to PTCA alone, it also contributes to some of the pathways involved in the restenosis pathophysiology.[16] It is important to note that the patterns of restenosis caused by balloon angioplasty and by placement of a metallic stent are governed by considerably different mechanisms, with vascular remodeling being more prominent in the case of balloon angioplasty; however, both result in a similar pathological outcome.[17–19] The complex nature of the processes contributing to restenosis also makes it challenging to identify appropriate animal disease models (a prerequisite for developing effective antirestenotic treatment strategies), as the response to arterial injury in animals may vary considerably from that in humans, due to underlying differences in anatomy.[20,21]

An antirestenotic approach that could be uniformly applicable and equally effective in all patients may not be feasible in clinical practice. When choosing the appropriate treatment for each patient, it is important to consider the individual's risk factors. For example, a need for target vessel revascularization due to restenosis amounted to 28% of patients with insulin-dependent diabetes compared to 16% in patients without diabetes, in a study of 954 patients undergoing PCI.[22] A similar restenosis rate was seen in a later study in diabetic patients.[23] The higher risk for restenosis in patients with diabetes might be explained by endothelial dysfunction (which is also observed in patients with hypertension)[24–26] and increased growth factor and cytokine production, platelet aggregability and thrombogenicity.[27,28] Also, a trend has been reported, demonstrating the risk for restenosis following balloon angioplasty increasing with age.[29] Thus, an optimal treatment modality should be based on a flexible therapeutic approach, enabling adjustment and optimization for different categories of patients, as well as the type and disease status of the treated blood vessel.

1.4. *Contributing mechanisms and stages of restenosis*

1.4.1. *Mechanical trauma*

Current evidence suggests that restenosis is a response to trauma and injury inflicted during angioplasty, beginning with thrombosis and inflammation in the first week, followed by cellular proliferation and extracellular matrix production over the next six months and ultimately resulting in lumen narrowing.[10,30] The lumen loss can be separated into distinct phases: 1) early loss due to elastic recoil within 15 min of balloon inflation and 2) late loss resulting from the combined effects of neointimal hyperplasia initiated by platelet adhesion and aggregation as well as SMC proliferation,[10,15,31–34] and negative remodeling owing to compression of the outer layer of the blood vessel by newly formed collagen fibers, similar to scar contraction.[35,36] By providing a rigid scaffold for the vessel, endovascular stents reduce the extent of lumen loss due to acute recoil and negative remodeling, but in many cases also greatly increase neointimal hyperplasia.[10,37]

1.4.2. Endothelial denudation

Vascular injury in the course of balloon angioplasty leads to endothelial denudation, causing extensive endothelial layer disruption. Leukocyte recruitment and infiltration occur at sites of vascular injury where the lining endothelial cells have been denuded, and platelets and fibrin have been deposited.[38] In addition to the loss of direct physical shielding of sub-endothelial space from blood elements, the loss of nitric oxide (NO) production by damaged endothelium plays crucial role in the rapid ingress of platelets and leukocytes. Endothelial cells are also responsible for the production of other antithrombotic factors such as prostacyclin and bradykinin,[10] which are known to synergize with NO in regulating vascular tone by inhibiting smooth muscle contraction and platelet aggregation. The disruption of the continuous endothelial lining by acute angioplasty-related trauma, thus, leads to parietal thrombus formation (typically in the absence of major lumen-occluding thrombi). This non-obstructing parietal thrombus, however, creates a provisional matrix for SMC proliferation and migration into the site, facilitating formation of neointimal hyperplasia at the later stages.[13,39] These steps are set into motion by an elaborate interplay of events including activation of platelets and consequent release of growth factors, primarily platelet derived growth factor (PDGF) and basic fibroblast growth factor (bFGF), pro-inflammatory cytokines, and the coagulation-fibrinolysis system.[13,28]

1.4.3. Inflammation

Platelet, neutrophil and monocyte recruitment brought on by endothelial denudation is a key player in inflammatory response initiation after vascular injury.[15,38] In addition to promoting the accumulation of leukocytes at sites of vascular injury, the binding of platelets to neutrophils amplifies the inflammatory response by inducing neutrophil activation, upregulating cell adhesion molecule expression, and generating signals that promote integrin activation and chemokine synthesis.[38] Neutrophil-platelet and monocyte-platelet aggregates have been identified in the peripheral blood of patients with CAD and may be used as markers of disease activity.[40,41] Additionally, elevation of plasma C-reactive protein (CRP) is commonly used as a marker of inflammation and is associated with the progression of CVD,[42] where high CRP levels predicted the development of restenosis.[15]

Central to the recruitment of leukocytes to areas of vascular injury are also chemokines, a group of chemoattractant cytokines produced by a variety of somatic cells, including SMCs, endothelial cells and leukocytes. One such chemokine, interleukin IL-8, plays a critical role in the recruitment of leukocytes to areas of vascular injury.[38] Another chemokine, monocyte chemoattractant protein (MCP)-1, participates in the recruitment of monocytes.[43] In the arterial wall the recruited monocytes mature to macrophages, which are thought to influence vascular repair by producing a variety of mediators, including members of the interleukin family (IL-1, IL-6, IL-12), tumor necrosis factor, MCP-1 and growth factors, including PDGF, bFGF and heparin-binding epidermal growth factor.[44] PDGF plays the most significant role, contributing to SMC migration and attraction to the site of injury through its action on the PDGF-receptor β (PDGFR β), and has been considered as an important target for antirestenotic therapies.[45,46] The bFGF, also a major contributor to SMC activation and proliferation, has been investigated as another molecular target for preventing restenosis.[47]

1.4.4. Increased generation of reactive oxygen species

Although neutrophils are not known to secrete growth factors, they contribute to tissue injury through the release of reactive oxygen species (ROS) and proteases.[44] Direct injury of the vessel as well as the infiltration of neutrophils and monocytes/macrophages result in additional injury to vascular tissue causing the loss of antioxidant mechanisms and enhanced production of superoxide and other ROS. Evidence has also been presented that ROS contribute to the initial vascular SMC (VSMC) apoptosis (discussed below), as well as remodeling, proliferation and migration of medial VSMCs or adventitial myofibroblasts.[48] Subsequent proliferation of neointimal cells has also been linked to ROS.[48] ROS production in arterial tissue is remarkably increased immediately after balloon injury, and remains elevated compared to uninjured arteries for up to 10 min.[49]

1.4.5. SMC apoptosis

Apoptotic death of vascular cells is a prominent feature of blood vessel remodeling that occurs during normal development and fibro-proliferative disorders of the vessel wall.[50] Additionally, changes in flow affect wall tension and cell-matrix interactions, and it may be these factors that alter the survival characteristics of vascular cells.[50] Consequently, VSMC apoptosis has been shown to occur when wall tension is diminished[51] or enhanced[52] or when the expression of matrix components[53] or matrix metalloproteinases[54] (MMPs) is altered. All these processes are known to take place during or following balloon angioplasty. Several studies demonstrated that balloon injury of blood vessels induces two waves of VSMC apoptosis. The first wave is a rapid burst of apoptosis in the media occurring within hours of the injury, resulting in a marked decrease in vessel wall cellularity. Initial studies in the rat carotid model of vascular injury showed that balloon denudation leads to a rapid and relatively synchronous induction of medial VSMC apoptosis, where apoptotic marker expression peaked at 1 h after injury but was no longer evident by 4 h after injury.[55] Early-onset apoptosis in medial VSMCs could exacerbate neointima lesion formation at later time points by provoking a greater wound healing response to overcome the cellular deficit.[50] As cells undergo apoptosis they release cytokines, which could enhance the proliferative response after traumatic balloon injury.[56] The second wave of apoptosis occurs days to weeks after injury at much lower frequencies.[50] However, VSMC accumulation in the neointima of injured rat carotid arteries reaches a maximal level at two weeks after injury, and cellular proliferation continues for up to 12 weeks.[57]

2. Currently used Approaches for Restenosis Prevention and Treatment

As of 2012, CABG is frequently not the first line therapy for CAD patients.[10,58] More commonly, a PCI such as stent implantation or a PTCA, also known as balloon angioplasty (plaque compression) is carried out,[10,15] with about 2 million procedures performed worldwide each year. In up to 40% of the cases, entailing balloon angioplasty as a sole intervention, the arterial damage and subsequent neointimal tissue proliferation triggered by PTCA lead to clinically significant restenosis.[59,60] To decrease the incidence of arterial renarrowing, balloon angioplasty is often accompanied by an atherectomy where the plaque is scraped, brushed and vacuumed out, as well as stent implantation intended to prevent acute recoil and negative remodeling. Although more effective than conventional balloon angioplasty, these modalities do not provide an ultimate, universally applicable solution for the prevention of restenosis, as discussed below.[61]

2.1. *Drug eluting stent deployment*

The implantation of a metallic stent, serving as mechanical support for the vessel and preventing its elastic recoil, typically reduces restenosis frequency to 17–30%, when compared to balloon angioplasty alone[62–64] (the incidence of restenosis is higher with stents that have thicker stent struts).[65] While the implantation of stents has considerably improved the outcome of coronary angioplasty procedures and also reduced re-intervention rates,[66] showing similar results to CABG with lower costs, at one-year follow up appointments[67] restenosis, developing following stent deployment (namely, in-stent restenosis) remains a problem frequently encountered in everyday clinical practice.[68]

When clinical advancements were thought to be exhausted through variations of stent geometry, the next degree of device sophistication evolved with transforming the stent into a drug- delivery platform.[69] As a new development in pharmacological interventions that have potential to aid in preventing restenosis, drug eluting stent (DES) technology enabling local drug delivery to the injured blood vessel has become a well-researched field[70–76] providing the basis for the most widely used restenosis prevention therapy to date.[68] The use of the DES platform for delivery of antiproliferative and antiinflammatory small molecule therapeutic agents as well as for modulation of genes with plasmid DNA and RNA has been explored. Clinically used DES release antiproliferative agents directly into the vessel wall providing high local drug concentration at the site of the lesion.[77–79] As of 2005, DES are used in 85% of coronary interventions.[80] A partial listing of FDA approved DES includes: Sirolimus (rapamycin)-eluting stents (SESs; Cypher®, Cordis Corporation, Bridgewater, NJ, USA), paclitaxel-eluting stents (PESs; Taxus®, Boston Scientific, Maple Grove, MN, USA), everolimus-eluting stents (Xience V®, Abbott Cardiovascular Systems, Santa Clara, CA, USA) and zotarolimus-eluting stents (Endeavor®, Medtronic Vascular, Santa Rosa, CA, USA).[58]

DES have been shown to decrease the incidence of restenosis to about 10% or less in patients with non-complex lesions[81]; however, their use has

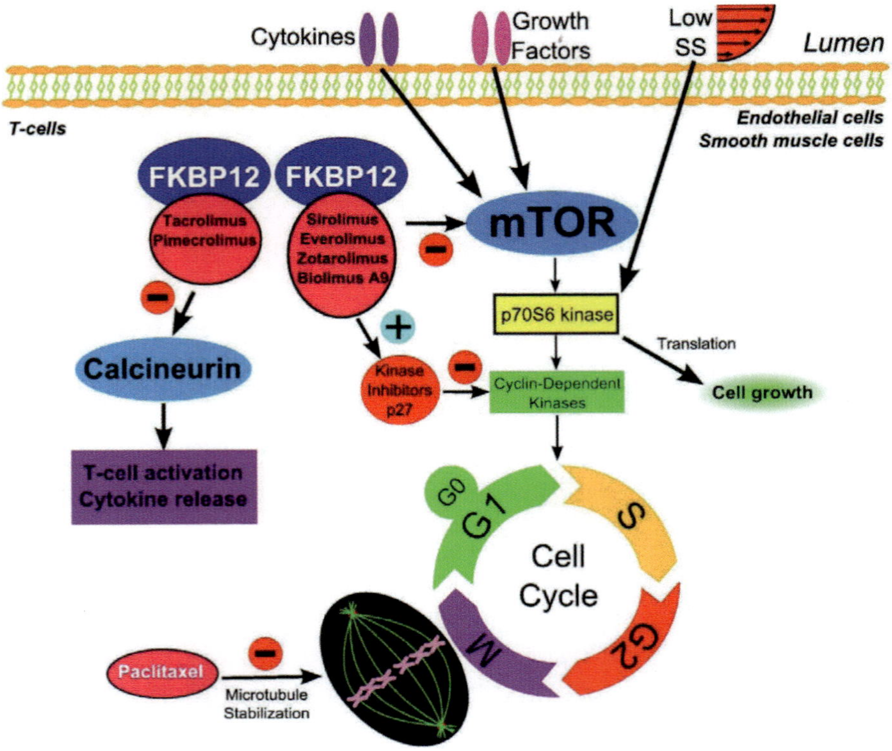

Fig. 1. Pathophysiological pathway of neo-intimal hyperplasia and primary mechanisms of action of pharmacological agents: limus family compounds and paclitaxel. Tissue injury during stent implantation initiates an inflammatory response including T-cell cycle progression and activation, and cytokine release. Growth factors (e.g. PDGF and fibroblast growth factor) and cytokines (e.g. IL-2) upregulate protein kinases of the mTOR pathway leading to augmented cell proliferation and growth. Similarly, flow-generated low shear stress stimulates proliferation and growth of endothelial and SMC through a common pathway. The "sirolimus analogs-FKBP12" and "tacrolimus/pimecrolimus-FKBP12" complexes inhibit separate target molecules, i.e. mTOR and calcineurin, respectively. Paclitaxel is a microtubule stabilizing agent which primarily impacts the formation of the mitotic spindle, and leads to cell apoptosis or death. FKBP, FK binding protein; *G*, growth; *M*, mitosis; mTOR, mammalian target of rapamycin; *S*, synthesis; SS, shear stress. Reproduced with permission from (M.I. Papafaklis *et al.*, Pharmacol. Ther. **134** (1), (43–58, 2012).

been associated with delayed endothelialization[82,83] and increased thrombogenicity,[84,85] introducing a new complication known as late in-stent thrombosis[86] underlying the overall suboptimal results.[87,88] Late in-stent thrombosis has been linked to fatal events, and the risk can persist up to 30 months after DES implantation,[82,89] often times presenting when antiplatelet therapy is discontinued.[90–93] Late thrombosis has been reported to occur in 0.6% of patients following DES implantation in the coronary artery and is associated with increased mortality, non-fatal myocardial infarction and the need for target-vessel revascularization.[85,94] The literature reports late in-stent thrombosis up to six years after DES implantation.[95–97] Additionally, off-label use of DES in peripheral arteries has been reported with an even higher incidence of late in-stent thrombosis.[98] Although the exact pathophysiology of late in-stent thrombosis remains unclear, autopsy studies have shown incomplete endothelialization and persistent fibrin deposition in DES compared to bare metal stents (BMSs).[99]

Overall, in-stent restenosis following DES implantation (namely, DES restenosis) is physiologically and histologically different from the development following BMS implantation, specifically when assessing temporal presentation, morphological patterns, underlying mechanisms, tissue composition and response to treatment.[68] According to various *in vitro* and *in vivo* experimental studies the efficiency of a specific DES is not determined exclusively by the type of drug used and its pharmacokinetic properties, but is rather multifactorial and also depends on parameters such as the total amount of drug loaded on the stent, the release kinetics and duration, the chemistry of polymeric coating used as the drug depot, stent strut material and other variables in the manufacturing technology.[100–102] In contrast to their efficacy demonstrated for CAD patients with non-complex lesions, DES placement in peripheral arteries has been largely ineffective in preventing vessel renarrowing,[103,104] which was particularly evident when applied to longer lesions (greater than 10 cm).[104]

Important shortcomings limiting the effectiveness of the DES approach include poorly adjusted drug release kinetics and the inability to replenish the drug payload once its finite amount has been exhausted.[105] In human patients restenosis has been shown to progress in many cases beyond six months,[106,107] at which point therapeutically adequate drug levels in the arterial wall may not be maintained by the DES. Additionally, DES is manufactured with only one therapeutic agent, when in reality, a combination of therapeutic agents might be necessary to fully prevent restenosis in certain patients. Generally, the DES technology lacks the flexibility required for achieving optimal results in different patient populations and in different clinical settings. Notably, certain categories of higher risk patients, such as those with diabetes mellitus and renal failure, tend to develop DES restenosis at considerably higher rates.[108] DES restenosis is particularly difficult to manage clinically, and is often treated by repeat stenting ("stent sandwich" technique) in the absence of more effective treatment modalities.[68,72] However, repeat stenting can be associated with an increased risk for in-stent thrombosis due to the overall higher drug exposure. In the cases when DES fails to prevent neointimal formation, further treatment options are limited due to the fact that the neointima formed despite the effect of antiproliferative agents is generally less susceptible to pharmacological control.

However, the so-called second-generation DES fabricated with thinner struts, and releasing everolimus or zotarolimus, may be more effective than first-generation stents, which contained either controlled release paclitaxel or sirolimus, using a relatively larger stent platform. Although direct comparisons are not readily available in patients similarly treated with antiplatelet drugs, a number of meta-analyses and large registries have shown that the risk of both restenosis and late in-stent thrombosis is lower with the second-generation DES.[109] Furthermore, in a study analyzing 94,384 consecutive stent deployments, second-generation DES were associated with a 38% lower risk of restenosis, a 43% lower risk of in-stent thrombosis, and a 23% lower risk of death against first-generation stents.[110]

2.2. *Systemic pharmacologic treatments*

Due to the limitations of the stent-based therapies, many alternative pharmacologic strategies have been investigated for the prevention of restenosis.[79] The extensive involvement of ROS in the restenosis pathophysiology and their detrimental effect on NO crucial to preventing vasoconstriction, activation of macrophages and resulting inflammatory response[111] prompted investigating the antirestenotic activity of the antioxidant probucol (a lipid lowering agent, which has been shown to inhibit SMC proliferation)[112] with promising results obtained in the animal disease models (rats, rabbits and pigs).[113–115] Probucol has also shown efficacy in humans, lowering the incidence of restenosis and need for repeat angioplasty, but only when patients received the drug as a pre-treatment to angioplasty.[116–119] Although these results were positive, the risks of reducing high-density lipoprotein levels and prolonging the QT interval in the heart's electrical cycle have prevented this drug from being used clinically.[79] In an effort to avoid these risks, the probucol molecule was modified to provide a more metabolically stable analogue, resulting in the potent antioxidant and antiinflammatory drug succinobucol (AGI-1067) that does not prolong the QT interval.[120] In clinical trials, a reduction in restenosis achieved with AGI-1067 was however not found to be statistically significant.[121]

Due to the role played by platelet activation in the neointima formation process, antiplatelet agents have also been the focus of pharmaceutical development for the reduction of restenosis.[79,122] Although these treatments were able to prevent the formation of thrombi at the site of PTCA and reduce immediate complications of angioplasty, only cilostazol (selective phosphodiesterase type III blocker and antiplatelet agent approved for use by the Food and Drug Administration (FDA) in 1998) has shown significant efficacy in preventing restenosis.[79] Through increasing VSMC relaxation and mildly lowering lipid levels cilostazol was able to reduce neointimal hyperplasia and restenosis rates,[123,124] particularly when applied in combination with DES.[125–130]

Because restenosis is a quasi-neoplastic process, caused by proliferation of SMCs,[131,132] the use of antineoplastic agents has been one of the most promising clinical and experimental avenues for restenosis prevention.[133] Neoplastic agents such as cytarabine[133,134] (interferes with the synthesis of DNA, killing proliferating cells), doxorubicin[133,135] (intercalates DNA), vincristine[133] (mitotic inhibitor), and paclitaxel[136,137] (mitotic inhibitor) have all been explored as either systemic treatments, local treatments in the form of a drug-eluting stent or both.

Other pharmacological agents such as tranilast (an allergy medication that has potent effects on VSMC proliferation and intimal hyperplasia)[138], trapidil[139], and pioglitazone[140] (a thiazolidinedione drug, clinically used to treat type 2 diabetes, but also exhibiting antithrombogenic activity) have been investigated in the context of restenosis prevention, but have not shown significant results in clinical trials.[79] Overall, promising results observed with systemic treatments in preclinical studies have rarely translated to clinically meaningful efficacy in humans, leaving a clinical need for effective and safe therapies capable of reducing the risk of restenosis in patients undergoing angioplasty.[79]

2.3. *Alternative therapies*

The sub-optimal results observed with systemic pharmacological treatments and DESs have led researchers to investigate other approaches.[141,142] These include two modifications of the balloon angioplasty: drug-eluting balloon angioplasty and cutting-balloon angioplasty. The drug-eluting balloon was developed to avoid uneven drug distribution and the inflammatory effects resulting from polymeric coating materials used to provide controlled drug release in DES.[77,142,143] This method would also allow for the drug concentration to be the highest at the time of injury, when the neointimal formation cascade is triggered.[142] In a randomized, double-blind, multicenter pilot study a balloon catheter coated with paclitaxel significantly reduced the incidence of restenosis in comparison to an uncoated balloon catheter in coronary angioplasty.[70] However, escape of the drug into the bloodstream was observed prior to the balloon placement at the lesion site. While the results from drug coated balloon studies are still awaited and the understanding of the mechanism of action of balloon drug delivery constantly grows, a new generation of drug coated balloons is currently under development in order to provide more effective drug delivery, minimize toxicity due to increased drug doses on longer balloons, prevent heterogeneous drug uptake from different tissue types and distal microembolization due to fragmentation of the coating.[102] Cutting balloons were developed for stenoses that are resistant to conventional balloons,[144,145] and were shown to reduce the amount of elastic recoil and vessel wall trauma, resulting in less inflammation and neointima formation.[146] The cutting balloon consists of a balloon catheter with three to four blades designed to create discrete longitudinal incisions in the atherosclerotic lesion during balloon inflation.[58,147] Results with the cutting balloon, although preliminary, have been promising, with some studies indicating that a higher efficacy is achievable when used in combination with a low-pressure balloon angioplasty.[148]

Additionally, the antiproliferative effect of radiation therapy[149] has also been explored in the context of in-stent restenosis treatment.[150,151] This method, known as vascular brachytherapy, where antiproliferative γ (iridium-192) or β (phosphorus-32) irradiation is applied locally to the target lesions via dedicated catheters,[58] has been shown to reduce neointima by more than 50% (Refs. 152 and 153) however not without the risk of serious side effects, such as late-stent thrombosis.[150]

Biodegradable stents have been developed in an attempt to prevent late thrombosis, a major concern for patients who receive a metal stent and require long term antiplatelet therapy.[154] Following its complete degradation a biodegradable stent is expected to leave behind the healed vessel, allowing restoration of vasoreactivity and reducing the risk of late thrombosis, thereby obviating the need for prolonged antiplatelet therapy.[143] The mechanical strength of the material and its ability to expand with sufficient radial force, degradation rates and products, as well as biocompatibility are important considerations when designing a bioabsorbable stent. Poly-L-lactic acid (PLA), polyglycolic acid (PGA), poly-D,L-lactide/glycolide (PLGA) and polycaprolactone (PCL) have all been proposed as potentially suitable materials for bio-absorbable stents.[155]

3. Nanomedicines as Experimental Antirestenotic Therapies: Rationale and Challenges

Currently used strategies for preventing arterial reobstruction are associated with high incidence of serious complications, including the poorly manageable instent restenosis and the potentially life-threatening thrombosis observed with DES,[61,67,156] which creates an unmet need for safer and more effective antirestenotic strategies. Nanoparticulate delivery systems, which, depending on the targeted pharmacological pathway and the choice of therapeutic agent, can potentially be applied as either local or systemic therapies with minimal untoward effects providing a well-matched experimental approach capable of addressing the limitations of the existing treatment modalities.

The term "nanomedicine" has previously been defined as the versatile "application of nanotechnology for treatment, diagnosis, monitoring, and control of biologic systems".[157] In the context of restenosis prevention/treatment therapies, liposomes, micelles and polymer-based nanoparticles, as well as some unique nanocarrier formulations, can all be considered as examples of nanomedicines. The advantages of a nanoparticulate delivery system include a flexible design that can be adapted to provide bio-distribution and release kinetics optimal for the particular therapeutic agent and/or pathological pathway, as well as achieving controlled release and extended retention at a target site.[61] Furthermore, nanomedicines can be designed to provide rapid and effective uptake of a delivered therapeutic agent, a small-molecule drug or a gene, by a target cell. In this fashion, controlled release of kinetics, extended retention at a target site and increased cellular uptake achievable with nanocarrier formulations can address the solubility, stability and bio-availability issues often observed with pharmaceuticals or genetic material.[61,158,159] The versatility of nanomedicines and available techniques for their production allow for rational formulation design based on a particular mode of action, including ability to provide longevity in circulation and adjustable rates of degradation and drug release, where release kinetics is dictated by the structure and composition of the carrier, chemical and physical properties of the therapeutic agent and the drug-carrier binding mode (i.e. drug covalently attached, entrapped or adsorbed to nanoparticle surface).

Due to the complex nature of the restenotic process with its multiple pathophysiological components, the route of administration, drug dose and duration of the desired response best-suited for acting effectively on the targeted pathway should be considered when designing a nanoparticle-based delivery system.[160] The cell type and the molecular pathway being targeted will determine whether a local or systemic delivery approach should be chosen for a specific nanocarrier formulation. As an example, growth inhibition of SMCs that make up the intima of the injured artery requires a local effect making local delivery the preferable route of administration. Alternatively, certain types of systemically administered formulations designed with capacity for homing and site-specific action on injured arteries can be considered as discussed below. If the goal is to target systemic processes contributing to

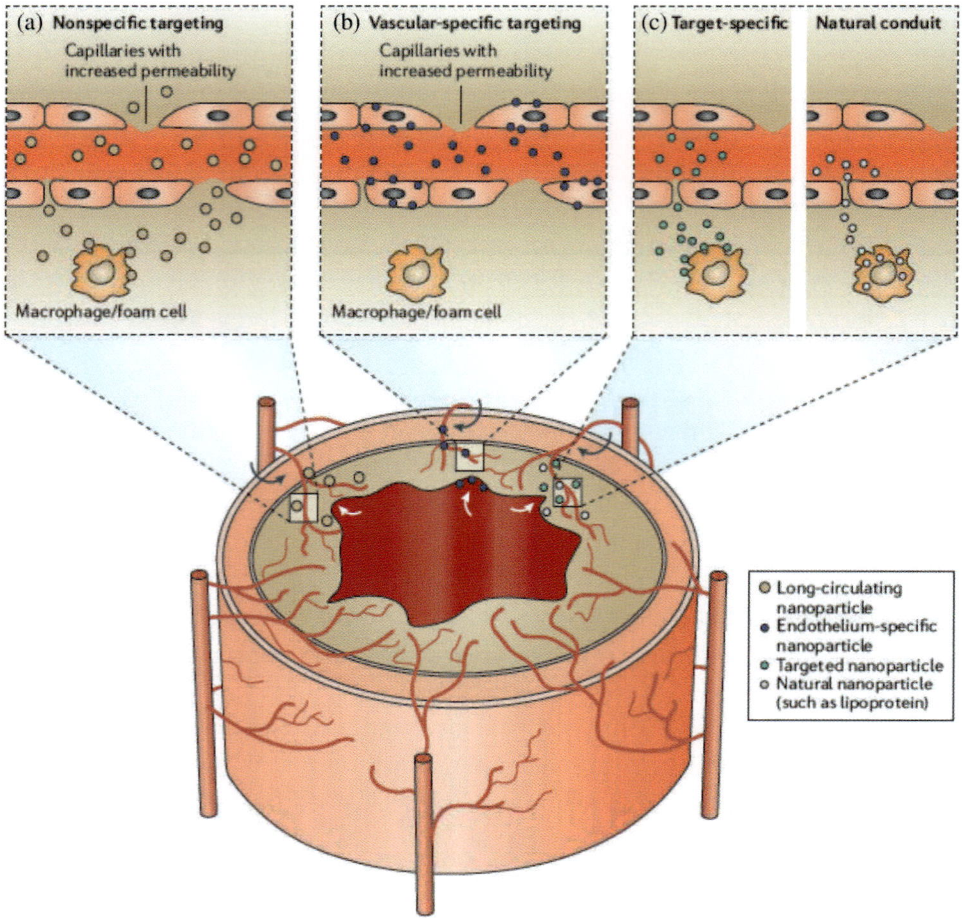

Fig. 2. Principles of nanoparticle targeting to vasculature. Non-specific targeting and specific targeting of the vasculature (parts *a* and *b*, respectively), and specific targeting of vessel wall components (part *c*) with either synthetic nanoparticles or via interaction through a natural conduit. Depending on the targeting principle applied, the cellular distribution of nanoparticles in the plaque will vary considerably. Reproduced with permission from (M.E. Lobatto *et al.*, Nat. Rev. Drug Discov. **10** (11), (835–852, 2011).

restenosis, such as those associated with the activation and recruitment of circulating inflammatory cells, the required systemic effect prompts using systemic delivery for best therapeutic performance.

When feasible, systemic delivery offers a number of advantages, including the ability to treat multiple lesions with a single treatment, as well as convenient re-administration when multiple treatments are required for optimal efficiency. Notably, choosing the right delivery mode based on the pharmacological target is of essential importance as demonstrated by numerous failures of systemically administered therapies intended to control local processes within the arterial wall, emphasizing the difficulty of achieving therapeutically adequate local drug levels at the site of interest without causing significant systemic adverse effects. In practice, therapeutically adequate drug levels at the site of arterial injury are more readily achievable with local delivery, as prior to the introduction of nanocarrier formulations was demonstrated using periadventitial application of antiproliferative drugs formulated in polymeric matrices. Perivascularly implanted polymeric matrices were shown to be compatible with chemically labile agents and provided high local and low systemic levels of a drug[161] as well as readily controllable release rates.[162,163] In these studies implants were fabricated from ethylene vinyl acetate (EVA), a nontoxic, biocompatible material that can incorporate both lipophilic and hydrophilic molecules, while maintaining excellent mechanical properties. More

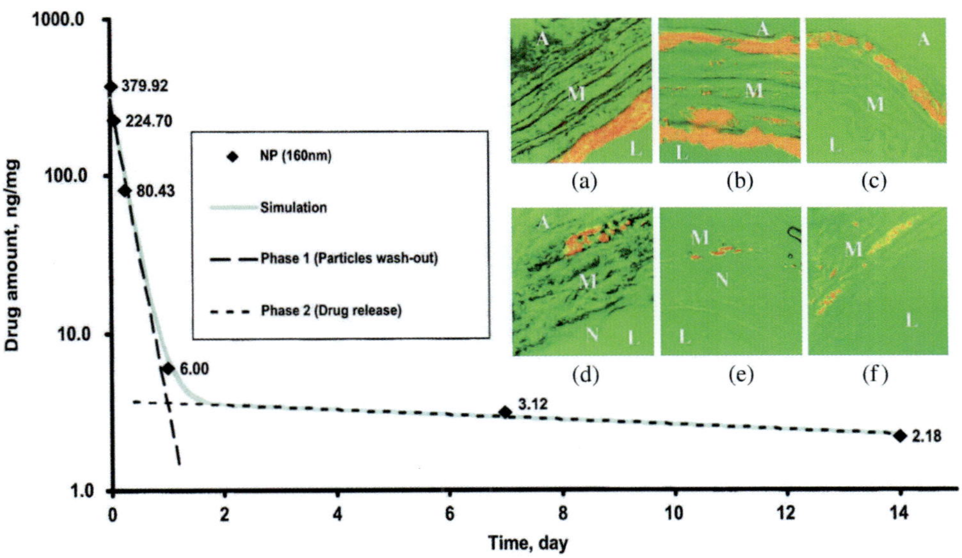

Fig. 3. (Color online) Distribution and retention of intraluminally administered nanoparticles in the arterial tissue. Line graph showing arterial levels of tyrphostin AG-1295 formulated in PLA-based nanoparticles (average size: 160 nm) at different time points after local delivery in a rat carotid model. Inset: Confocal images of rat carotid arteries after local delivery of similarly prepared nanoparticles loaded with a fluorescent compound (Nile Red) used as a model of a lipophilic small-molecule therapeutic agent. The arteries were harvested 5 min (a), 90 min (b), 1 day (c), 7 days (d), and 14 days (e) after 15 min of intraluminal delivery and 6 h after delivery (f). Discrete, granular, fluorescent foci of particle aggregates are clearly distinguished in *f*. *L*, *M*, *N*, and *A* indicate the lumen, media, neointima, and adventitia, respectively. Reproduced from Ref. 170.

recently developed drug-loaded nanoparticles suitable for intraluminal delivery are primarily distributed in the intima and innermost media of the vessel, and thus can effectively target lumen-localized processes.[164]

The deep penetration into the vessel wall typically seen with locally applied nanoparticles helps to create a depot, from which the drug is released at a rate which can be optimized based on the time course of the targeted pathological processes.[164–169] Fishbein *et al.* showed protracted release of a small-molecule PDGFRβ-specific tyrosine kinase inhibitor, AG-1295, formulated in PLA-based nanoparticles after intraluminal delivery in a rat carotid model.[170] The size of nanoparticles is an important determinant of the arterial wall delivery efficiency and subsequent retention as was shown in the same study where smaller sized nanocarriers (90 nm) accumulated and remained associated with the arterial tissue at higher levels compared to larger particles (160 nm). In experiments conducted in a rabbit model, Westedt *et al.* observed that nanoparticles with average sizes of 100 nm and 200 nm were able to penetrate to the inner layers of the vessel wall, while 514 nm particles accumulated predominantly at the luminal surface.[171] This suggests that selective action on specific cell populations localized in different layers within the arterial wall can potentially be accomplished through adjusting the formulation variables controlling the size of the drug carrier particles. Notably, smaller-sized nanoparticles having a high surface area to volume ratio are more effectively modified with targeting ligands,[166,167,172] which may be desirable in some instances for improving their arterial binding and retention.

Therapeutic Agents. Due to the multifaceted nature of restenosis, there are multiple levels at which therapeutic agents formulated in nanocarriers can interfere with the mechanisms contributing to the pathology. Small-molecule pharmaceuticals, biotherapeutics, including gene vectors and proteins, and cells can potentially be used for preventing the progression of various stages in the restenotic process.

3.1. *Small-molecule pharmaceuticals*

Since 1980, more than 150 clinical studies have investigated the use of antiproliferative, antihyperlipidemic, antioxidant, antiplatelet, anti-inflammatory, immunosuppressant, antithrombotic, calcium antagonists and antimigratory pharmacological agents for prevention of restenosis.[168,169,173–175] In some studies the use of these therapeutic agents was investigated as a pretreatment prior to the angioplasty procedure in order to act on the earliest events contributing to restenosis.[168,169] Pharmacological treatments aimed at all phases of the restenotic process (antithrombotic, antiplatelet, antiproliferative, anti-migration, anti-inflammatory) have shown promise in animal studies; however these results have not translated to clinically viable therapies.[168,169] It is believed that the poor translation from animal studies to clinical practice might be due to a discrepancy in dosing between animals and humans, as well as the inability to provide controlled administration of these drugs in patients for an adequate period of time. In other words, it is the administration mode inadequate for acting effectively on the relevant pathological events rather than the drug itself that has failed.[164,168,169,176] The key processes in the restenosis development where effectiveness of small-molecule therapeutic agents formulated in nanocarriers has been experimentally demonstrated (often in stark contrast to free drugs shown ineffective in prior studies) include inflammation, intimal cell proliferation, platelet deposition and activation.

Inflammation plays an essential role in the initiation and progression of restenosis. The vascular injury caused by angioplasty initiates the inflammatory response that involves recruitment of macrophages, T-cells and a small number of B-cells to the site of injury.[177] Analysis of balloon-injured arteries demonstrated that neutrophils are present in abundance within hours of balloon injury, followed by prolonged macrophage accumulation.[178–180] Monocyte-derived macrophages, which are key players in inflammation and eventual activation of SMCs after balloon angioplasty, accumulate in the vessel wall and stimulate SMCs by secretion of cytokines, metalloproteinases, and growth factors. Clinical evidence shows that macrophage infiltration predicts restenosis after coronary angioplasty, and that late lumen loss after coronary angioplasty is associated with the activation status of circulating monocytes before treatment.[160,181] This has provided the rationale for examining the antirestenotic efficiency of anti-inflammatory agents such as prednisone, dexamethasone, as well as nitrogen-containing bisphosphonates (BP) that inhibit monocytes by blocking the mevalonate pathway (e.g. alendronate) as a systemic approach aimed at reducing macrophage accumulation at the injury site.[180,182] BP are drugs clinically used for treating bone disorders, where they act via osteoclast depletion.[183,184] Macrophages share the same hematopoietic progenitor cells with osteoclasts and have similar capacity for phagocytosis. Notably, due to their strong negative charge-free BP they are membrane-impermeable and are not taken up by monocytes to any significant extent. However, their formulation in properly designed nanocarriers, liposomes or polymeric nanoparticles, readily phagocytosed by these cells is the key to effectively blocking macrophage infiltration and achieving inhibition of restenosis in animal disease models.

The hypercellularity of restenotic lesions versus hypocellularity of atherosclerotic plaque points at the significant role of SMC proliferation in restenosis.[168,169] This proliferation occurring in two phases (early medial SMCs and adventitial myofibroblast proliferation over 2–6 days post-injury and late intimal SMC proliferation over 2–4 weeks post-injury)[168,169] has prompted the use of antiproliferative and antineoplastic molecules for restenosis prevention therapies. Pharmaceuticals, explored for this purpose, that have previously been used for cancer treatment or otherwise include cytarabine, doxorubicin, vincristine (Oncovin), tranilast (Rizaben), paclitaxel, U-86, tyrphostins AG-1295 and AGL-2043, actinomycin D, cytochalasin D, 17-beta-estradiol, mycophenolic acid and angiopeptin. Notably, tyrphostins evaluated for their antirestenotic effects are tyrosine kinase blockers specific for the PDGF receptor β expressed by VSMCs. By blocking the activation of the signal transduction cascade initiated by PDGF, these agents effectively and selectively inhibit VSMC proliferation and migration, and thus represent a unique class of cell type-specific antirestenotic compounds capable of providing a therapeutically relevant effect without causing delayed vessel re-endothelialization and healing.[16,164,168,169,185–187]

Preclinical studies have demonstrated that systemically administered immunosuppressant sirolimus (rapamycin), a potent inhibitor of cytokine and growth factor-mediated cell-proliferation reduces neointimal proliferation after vascular injury.[58,177,188,189] Several clinical trials confirmed the benefit of oral sirolimus in reducing in-stent restenosis after BMS implantation,[58,190,191] making it a possible effective and cost-saving alternative to DES implantation. However, the long-term results of the OSIRIS trial[58,192] reported by Kufner in 2009, showed an attenuated benefit of oral sirolimus after four years. Statins, clinically used to treat hypercholesterolemia, also exert antiproliferative effects on VSMCs.[174,193] Specifically, simvastatin has been shown to reduce SMC proliferation and neointimal formation in a rat carotid stenting model. Statin drugs have also been shown to increase NO stability, thereby improving endothelial cell function.[174]

3.2. *Strategies aimed at endothelium recovery*

The trigger for inflammation and consequent restenosis is the disruption of the endothelial layer, whose structural and functional integrity is fundamental for the vessel homeostasis and effective damage repair.[194,195] It has been shown that the development of neointimal hyperplasia was most prominent in denuded areas that had not re-endothelialized after seven days.[196]

Consequently, facilitated re-endothelialization has been the focus of a number of experimental restenosis prevention therapies. Attempts to accelerate endothelium recovery following balloon angioplasty have been made using different strategies that include accelerating endothelial or endothelial progenitor cell proliferation (growth factors,[197] heparin,[198] estradiol[199]) as well as delivery of endothelial or endothelial progenitor cells, either via catheters or by seeding on the stent surface.[195,200–203]

Mesenchymal stem cells (MSCs), non-hematopoietic multipotent stem-like cells capable of differentiating into both mesenchymal and non-mesenchymal lineages, have also been investigated in the context of cell therapy for restenosis prevention.[204,205] The rationale for exploring MSC delivery as a therapeutic approach is based on their ability to differentiate into functional cell types able to repair the diseased or injured tissue in which they are localized,[204] as well as their ability to control function of both *B* and *T* cells.[206,207] Additionally MSCs produce NO, which promotes restoration of functional endothelium and assists with vascular remodeling.[204]

3.3. *Gene delivery*

Gene therapy within the vasculature has focused on a host of targets, with the prevention of restenosis following balloon angioplasty typically directed towards the same concepts as drug delivery: antiproliferative,[208] antimigratory,[209] antifibrotic,[210] antioxidative,[211] antithrombotic,[212] anti-inflammatory,[213] etc. In particular, gene therapies able to block SMC migration and proliferation have shown promise as a therapeutic approach against restenosis.[214,215]

Proliferation of SMC can be limited via a cytotoxic approach (e.g. by inducing cell death via the Herpes Simplex Virus thymidine kinase gene (HSV-tk) and the nucleoside analogue ganciclovir)[216] or a cytostatic approach (inhibiting progression of the cell cycle via overexpression of a constitutively active mutant form of Rb and CKI protein p21).[217] Due to the redundant nature of pathways involved in restenosis, it is believed that genes exhibiting pleiotropic actions may represent a more effective therapeutic means than those acting on distinct mechanisms of lesion formation. A well studied pleiotropic candidate employed in studies of vascular injury is NO synthase (NOS), which catalyzes the synthesis of NO, an important player in vascular maintenance that acts by inhibiting platelet aggregation and SMC proliferation and promoting endothelium regeneration.

4. Nanocarrier Formulations for Restenosis Prevention: Design and Applications

4.1. *Nucleic acid delivery*

Restenosis lends itself to the application of gene therapy due to the following features: 1) the time of injury is known in advance, 2) the location of pathology is restricted to the site of injury and is thus well defined spatially, 3) the local environment of pathology is accessible at the onset of the pathological process.[214,218] When selecting a therapeutic transgene, there are a number of factors that should be considered, including whether the protein encoded by the transgene acts intracellularly or extracellularly and if the restriction of expression to specific cell types is critical. Non-viral delivery of plasmid DNA and antisense RNA are favored for their low toxicity, but is typically inefficient in the absence of complexing agents.[164,219,220] (Recombinant viral vectors, such as adenoviruses enable efficient gene transfer, but carry the risk of inflammatory and immune responses).[164,220]

To avoid potential risks associated with viral gene transfer, Muhs *et al.* developed a gene transfer approach where cationic lipid-DNA complexes (lipoplexes) were applied to achieve overexpression of inducible NOS (iNOS) for inhibiting stent-induced neointimal lesion formation in a porcine model.[221] In this study, the infiltrator device was used to deliver varying doses of the iNOS lipoplexes to both carotid and femoral arteries. A progressive, dose-dependent decrease in surviving medial cells was observed with up to 50% loss of medial cells in the group treated with the highest dose. Immunohistochemical analysis showed a significant increase in the oxidative stress marker nitrotyrosine in the arteries of the animals treated with the highest dose of iNOS lipoplexes. In the group treated with a low dose of iNOS lipoplexes, intravascular ultrasound analysis of stented femoral artery segments performed four weeks post treatment showed a reduction of in-stent plaque area and plaque ratio in comparison with contralateral control arteries. A clinical trial was performed by von der Leyen *et al.* to determine toxicity of these lipoplexes.[222] Patients received up to three applications (depending on the length of the lesion) of varying doses, with each application at the same dosing level. Overall, it could be concluded that the lipoplex-mediated iNOS gene transfer was well tolerated and appears to be safe.[222] However, the ability to achieve therapeutically adequate gene transfer levels with this approach still needs be experimentally shown.

Brito *et al.* examined therapeutic efficacy of endothelial NOS (eNOS) locally overexpressed using non-viral gene delivery with a lipoplex formulation administered from stents in a rabbit iliac stenting model.[223] The authors utilized poly(beta-amino ester) (PbAE) pre-complexed plasmid DNA encapsulated in lipoplexes, which were applied onto stainless steel stents with a gelatin coating. Transfection with the GFP reporter protein was confirmed 24 h after implantation of the lipoplex-coated stents at both mRNA and protein expression levels.[223] Despite the positive preliminary results with *in vivo* gene transfer, one of the issues with this approach was the rapid dissolution of the gelatin coating after stent implantation upon contact with blood. This could potentially be prevented by using a cross-linked form of gelatin or adding biodegradable polymer (e.g. PLGA) to prevent the rapid dissolution and allow for sustained delivery *in vivo*.

Complexes of siRNA with cationic polymers (polyplexes) have also been explored for restenosis prevention.[224] Non-toxic low molecular weight polyethyleneimine (PEI1.8, 1.8 kDa) was covalently conjugated to deoxycholic acid (DA) via carbodiimide-mediated coupling reactions. The resulting DA-modified PEI1.8 conjugate (DA-PEI1.8) formed self-assembled nanostructured complexes with siRNA targeted against MMP-2. SMCs were transfected with siRNA polyplexes formed at different polymer to siRNA weight ratios and varying particle incubation times. Results showed that the DA modification enhanced cellular uptake of the polyplexes, with increased siRNA delivery and a gradual decrease in MMP-2 gene expression observed at higher polymer to siRNA weight ratios. The siRNA transduction and consequent decrease in MMP-2 expression coincided with a reduction in *in vitro* SMC migration in a scratch-wound assay, suggesting that DA-PEI1.8/MMP-2 siRNA polyplexes could provide a potential strategy for the suppression of neointima formation following vascular injury through the prevention of SMC migration by enabling local delivery of MMP-2 siRNA following balloon angioplasty.[224]

Lawrie *et al.* investigated gene transfer to vascular cells using ultrasound exposure and microbubble echocontrast agents, and was able to achieve transgene expression levels *in vitro* approximately 300-fold higher than those with naked plasmid DNA alone.[225] The transfection efficiency could be further increased to approximately 3000-fold over naked DNA by combining this approach with a polyplex transfection reagent.[225] However, the challenge of successfully translating these findings to *in vivo* setting still remains to be addressed.

4.2. *Albumin-based formulations*

Since albumin is an endogenous protein, properly designed nanoparticles composed of human albumin are biodegradable and exhibit low toxicity and antigenicity.[226,227] Markovsky *et al.* developed a novel desolvation-based method for the preparation of additive- and cross-linker-free albumin nanoparticles via electrostatic interaction of the protein with an acidic BP drug (alendronate).[228] These particles showed rapid release of alendronate (95% released after 1 h), which is a prerequisite for its strong inhibitory action on monocytes and macrophages. In agreement with this finding, a significant inhibition of RAW264 macrophage growth was observed with alendronate-albumin nanoparticles *in vitro*. In a rat model, where nanoparticles were administered via tail vein injection, restenosis was found to be strongly reduced as evidenced by neointima to media ratio significantly lower compared to blank NPs and free alendronate.[228]

Kolodgie *et al.* studied the antirestenotic effect of nanosuspensions prepared by high-pressure homogenization of paclitaxel in the presence of human serum albumin.[229] The nanosuspension was systemically administered at varying doses in a rabbit bilateral iliac stented-artery model.[230] Systemic delivery of paclitaxel formulations produced a dose-dependent reduction of neointimal growth after 28 days, however this reduction was accompanied by incomplete healing.[230] A single dose at the highest concentration administered at the time of stenting failed to sustain neointimal growth suppression after 90 days. However, by adding a repeat intravenous dose of a lower concentration on the 28th day, persistent reduction of neointimal formation after 90 days was achieved.

These nanosuspension formulations were further explored by Margolis *et al.* in a clinical trial designed to establish the safety and optimal systemic dose for reducing in-stent restenosis in humans.[137] Patients received varying doses intravenously after stenting of a single lesion greater than or equal to 3 mm in diameter. Moderate neutropenia, moderate sensory neuropathy, and mild to moderate, reversible alopecia occurred only at the highest doses. No major adverse cardiac events were reported for 2 months, however, after 6 months, restenosis was observed in several cases in the lowest and highest dose groups. Additional studies are necessary to explore the potential utility of this therapy more extensively.[137]

4.3. *Micelles*

Polymeric micelles produced by self-assembly of block co-polymers have gained increasing popularity since they were first introduced in the late 1980s,[231,232] and have recently been investigated as drug carriers for the purpose of preventing restenosis. In one example, doxorubicin-loaded polymeric micelles obtained by self-assembly of polyethyleneglycol and polyaspartic acid[135,233] were studied in a single injury model and a double injury model (repeated balloon injury two weeks after the first injury) in rats.[135] This micelle formulation suppressed the vascular lesion formation, after balloon injury, significantly more effectively than free doxorubicin by inhibiting SMC proliferation without enhancing apoptosis. Furthermore, measurement of tissue concentrations of doxorubicin confirmed the greater effectiveness of the micelle formulation in delivering the agent to the balloon-injured artery.[135] The authors suggest that the formulation with its small size (40 nm) could take advantage of the enhanced permeability and retention in the arterial wall resulting from endothelial denudation, leading to its increased arterial uptake after systemic administration.[61]

Chen *et al.* developed a nanoscale drug-entrapment strategy for loading sirolimus in the form of Pluronic block copolymer micelles on stents for local drug delivery.[234] When these stents were incubated with cultured rat aortic SMC for three days, cell viability was significantly reduced compared to incubation with sirolimus-coated stents. In a rabbit model, a transverse aortotomy was made and sirolimus-loaded micelles coated onto stents were deployed into the rabbit infrarenal abdominal aorta. During the sixth week after stenting, the intimal response to stent implantation was greatly reduced by stent-bound micellar sirolimus. In the group treated with the unloaded (control) stent, a marked neointima formation was observed with a large number of RAM11-positive macrophages around the stent struts. However, the presence of CD31-positive cells, an indicator of vessel re-endothelialization, was more prominent in arteries implanted with the control stent, in agreement with the *in vitro* data where sirolimus suppressed proliferation of both SMC and endothelial cells.[234]

In a study by Kim *et al.*, a layer-by-layer fabrication technique was utilized to coat paclitaxel in a micellar formulation on non-thrombogenic heparinized metallic substrate[235] First, heparin was stably immobilized on the cobalt-chromium alloy surface, and then a multilayer composed of anionic PLGA-graft-hyaluronic acid (HA-g-PLGA) micelles encapsulating paclitaxel, heparin, and poly-L-lysine was built up in a layer-by-layer fashion. The multilayer coating composed of releasable heparin and paclitaxel effectively inhibited the proliferation of human coronary artery SMC *in vitro*.[235]

4.4. *Liposomes*

Liposomes representing lipid vesicles with a bilayered membrane composed of natural or synthetic amphiphilic lipid molecules[236] can be used to favorably alter pharmacokinetics of different classes of drug molecules, a property that can be utilized to achieve effective therapeutic drug action and prevent side effects.[180] Liposomal formulations are designed to protect their cargo, and depending on their composition, size, and surface charge, can be rapidly phagocytosed by blood monocytes.

A series of studies investigated liposomal delivery of the BP drugs to achieve systemic depletion of monocytes and decreasing the ingress of macrophages in the arterial wall. The use of liposomes as a delivery system for BPs was first introduced by van Rooijen,[237] who used liposomes to deliver clodronate to phagocytic cells *in vivo*. This approach, termed liposome-mediated macrophage "suicide" technique, has been used extensively as a means to eliminate blood monocytes and macrophages in different body compartments and to study their role in a range of pathological and immunological conditions.[180] Danenberg *et al.* demonstrated that partial systemic inactivation and transient depletion of monocytes and macrophages by liposomal BPs (pamidronate, clodronate, ISA or alendronate) reduces neointimal hyperplasia and restenosis in several animal models[182,238,239] with liposomal alendronate formulation exhibiting the most potent activity.[239] Researchers examined blood monocyte, tissue macrophage, and the total white blood cell (WBC) count before and after treatment with liposomal BP in the rabbit model of balloon angioplasty with and without stenting.[182,239] The WBC count increased slightly two days after surgery, with no significant differences between controls and the liposomal clodronate and liposomal alendronate groups,[239] whereas the number of monocytes at one and two days after balloon injury and stenting was significantly lower in BP-treated animals. By six days post-procedure, WBC and monocyte counts returned to baseline levels. However, the transient inactivation of monocytes after systemic administration of liposomal BP was shown to result in reduced numbers of tissue macrophages in the injured artery.[182] Thus, the outcome of the systemically administered therapy was manifested as a local antirestenotic effect on the injured artery.

In a comprehensive study of the bioactivity of liposomal BP the inhibitory effect of the formulation on the target cells (monocyte/macrophage), and its potential toxicity to non-target cells *in vitro* was assessed.[240] Liposomal alendronate was found to be twice as potent as liposomal clodronate in RAW 264, J774 and h-monocyte cell cultures.[240] An improved liposomal alendronate formulation was prepared by a modified thin lipid film hydration technique followed by extrusion, resulting in smaller sized vesicles, narrower size distribution and low drug to lipid ratio in comparison to the reverse phase evaporation method.[241] The new formulation exhibited increased stability (2.5 years) as determined by the insignificant changes in the vesicle size, no drug leakage, extensive lipid and drug stability, *in vitro* bioactivity (macrophage inhibition), and *in vivo* depletion of circulating monocytes and inhibition of restenosis.[241]

The ultimate effect of this treatment strategy on proliferating SMCs, which comprise the major component of human restenotic lesions, is indirectly mediated via the inhibitory effect of the liposomal formulation on macrophages. *In vitro* studies demonstrated that macrophages respond with greater sensitivity to the liposomal alendronate formulation than SMCs.[240] Epstein-Barash *et al.* showed that the BP-loaded liposomes affected SMC growth by reducing cytokine secretion due to monocyte depletion. The impact of monocyte depletion by liposomal BP on vascular repair of stented blood vessels was even more significant.[240] These results support the major role of monocytes in neointimal formation after stent injury and establish monocyte/macrophage depletion as an important target in its modulation.[180,240] Based on the results of a clinical trial, the liposomal alendronate formulation has been proposed as an adjunct therapy in high-risk patients.[180]

4.5. *Polymeric nanoparticles*

Polymeric nanoparticles are solid, colloidal particles that range in size from 10–1000 nm and are made from synthetic or natural polymeric materials.[242,243] They are a convenient and effective drug delivery system,[233,243] in which the compound of interest can be dissolved or entrapped in the polymeric matrix, adsorbed onto the surface, or chemically bound to the polymer.[243–246] Biodegradable nanoparticles made of preformed polymers have been most widely prepared using a family of the "polymer precipitation" methods, including emulsification/solvent evaporation, salting-out technique, emulsification/solvent diffusion process, and nanoprecipitation.[245] These fabrication methods allow for the production of both nanospheres with a solid core and nanocapsules possessing a polymeric rate-limiting membrane and enabling encapsulation of lipophilic and hydrophilic drugs.[245,247] The biodegradability of the resultant nanoparticles, drug encapsulation yield, and release kinetics are controlled by polymer chemistry and molecular weight, as well as by the specific mode of drug incorporation in the particle. In addition to synthetic biodegradable polymers (polyesters, polyalkylcyanoacrylates, polyanhydrides, etc.), nanoparticles for therapeutic applications can also be formulated using biopolymers (gelatin, albumin, casein, polysaccharides, lectins, etc.). Natural polymers as particle-forming materials are characterized by good bio-compatibility, however they may not be ideal for providing controlled release of small-molecule therapeutic agents. In a number of studies this limitation was addressed by applying these materials in combination with synthetic polymers to create composite nanoparticles that exhibit controlled drug release kinetics. In one example, the anti-inflammatory agent, prednisolone acetate, was loaded into chitosan/PLA particles that were shown to release the drug at a constant rate over a period of 11 days, making this formulation an interesting candidate for treating in-stent restenosis.[248]

The ultimate choice of the polymer and the drug encapsulation strategy for the nanocarrier formulation depend on the chemistry of a therapeutic agent and a specific therapeutic application, and can be readily adjusted to provide optimal performance and acceptable bio-compatibility.

4.5.1. Biodegradable polyester-based nanoparticles

The bio-degradable aliphatic polyesters most widely applied for producing nanoparticle formulations for biomedical applications are polylactide (PLA), polyglycolide (PGA), and their copolymer poly(lactic-co-glycolic acid) (PLGA), where the ratio of PLA and PGA can be adjusted to control the degradation rate of the nanoparticles. Klugherz *et al.* showed promising results with locally delivered PLGA nanoparticles loaded with probucol in a rabbit iliac model as an approach for achieving high arterial intramural drug levels while providing sustained intramural retention.[249]

Nanoparticles loaded with the PDGFRβ-specific tyrphostin AG-1295 were shown to have an inhibitory effect on the activation, migration, and proliferation of porcine and human SMCs *in vitro*[187] and resulted in a 50% inhibition of neointimal formation when delivered via an infusion catheter in a balloon injury pig model.[187] Banai *et al.* also reported PLA-based nanoparticles formulated with another PDGFRβ-specific tyrphostin, AGL-2043, and administered intraluminally to balloon-injured rat carotid and stented pig coronary arteries. This study demonstrated that catheter-based local delivery of tyrphostin-loaded PLA nanoparticles can effectively reduce neointima formation in stented coronary arteries extending the potential application of this approach and making it a viable alternative to DES that is not limited by stent design or type of injury.[250]

Cohen-Sela *et al.* explored the use of PLGA nanoparticles formulated with alendronate as a nanocarrier for depleting monocytes and macrophages in a hypercholesterolaemic rabbit model.[251] Following subcutaneous delivery, the reduction in restenosis observed in the animals injected with alendronate-loaded nanoparticles correlated with depletion of macrophages.[251] In a study performed by this group with PLGA based nanoparticles formulated with another amine-containing BP, ISA, their intravenous administration resulted in a significant attenuation of restenosis in a rat model 14 days after carotid artery injury, in comparison to control groups treated with free ISA, buffer or blank nanoparticles.[252] However, the antirestenotic effect was not observed at a longer time point (30 days post injury). PLGA nanoparticles reported in a different study and formulated with an amphiphilic small-molecule antiproliferative agent, mithramycin, using the nanoprecipitation technique failed to inhibit restenosis in the rat carotid model presumably due to less potent depletion of circulating monocytes compared to BP, and the lack of arterial targeting.[253]

Paclitaxel-loaded PLGA nanoparticles prepared by the oil-in-water emulsion/solvent evaporation protocol and surface modified using the cationic surfactant didoceylmethylammonium bromide (DMAB) to enhance arterial retention[254] effected a dose-dependent inhibition of intimal proliferation in balloon injured rabbit carotid arteries at 28 days post treatment. The results demonstrated complete absence of restenosis achievable at a 30 mg/mL nanoparticle dose, suggesting that surface modified paclitaxel-loaded nanoparticles can provide an effective means for inhibiting the proliferative response to vascular injury.[254]

Westedt *et al.* demonstrated a 50% reduction in neointimal area four week post-treatment with paclitaxel-loaded nanoparticles administered locally to balloon injured rabbit iliac arteries using a porous balloon catheter, compared to control vessels treated with blank nanoparticles.[255] In another study, sirolimus encapsulated in PLA nanoparticles[256] effectively prevented the proliferation of SMC *in vitro*, potentially making this formulation an interesting candidate for restenosis prevention and warranting further investigation *in vivo*.

5. Targeted Delivery Strategies

Implementation of targeted therapies for restenosis prevention has the potential to markedly increase efficacy and safety compared to non-targeted therapies by reducing distribution to peripheral organs and allowing for delivery of a therapeutically adequate drug dose without causing toxic effects at non-target sites. Local delivery as a strategy for spatially restricting the drug to the site of interest is applicable in most clinical scenarios through the use of devices, such as catheters, to localize the therapeutic agent to the site of arterial injury. Alternatively, achieving arterial localization following systemic delivery may require the use of strategies that enable control over the biodistribution of nanocarriers either by passive or active guidance mechanisms. Passive targeting takes advantage of size, charge, natural forces or mechanisms that route carriers to specific sites or organs. Active targeting refers to a range of strategies based on affinity interactions or physical forces (e.g. magnetic targeting).[257] Research into the delivery and targeting of therapeutic agents involves the identification of precise targets (cells and receptors) related to specific clinical conditions and development of the appropriate carriers.[157]

5.1. *Nanoparticle surface modification for affinity targeting*

Surface modification of nanoparticles in order to increase the affinity for the arterial wall and improve site-specificity and retention is one of the most extensively investigated approaches for developing nanocarrier-based antirestenotic therapies.[164] The arterial wall uptake of surface-modified PLGA nanoparticles loaded with an antiproliferative agent, U-86983, was evaluated in a dog femoral artery model.[258] Nanoparticles coated with a cationic surface-active compound, didodecyldimethylammonium bromide (DMAB), demonstrated seven–ten-fold greater arterial drug levels compared to unmodified nanoparticles. The enhanced arterial uptake of the DMAB surface-modified nanoparticles appears to be governed by the alteration in the nanoparticle surface charge.[258]

Perfluorocarbon doxorubicin- or paclitaxel-loaded nanoparticles targeted to tissue factor (using a polyclonal antibody specific to the extracellular domain of recombinant porcine tissue factor) were fabricated by Lanza *et al.* and investigated in cultured VSMC *in vitro*.[259] Targeting of the nanoparticles to VSMC surface epitopes significantly increased the antiproliferative effect, particularly for paclitaxel. Another *in vitro* study by Nah *et al.*, focused on characterization of polyplexes formed with poly(ethylene glycol)-graft-poly(L-lysine) conjugated to arterial wall-binding targeting peptide (AWBP). The application of these polyplexes prepared with a luciferase reporter plasmid to bovine aortic endothelial cells resulted in transfection efficiencies 150–180 times higher than those of control polyplexes formed with polylysine or polylysine-PEG.[260]

In another example a modification with collagen IV-targeting peptide was applied by Chan *et al.* in order to modify the bio-distribution and achieve arterial localization of 60 nm lipid-polymeric nanoparticles.[261,262] Notably, the maximal tolerated dose of paclitaxel formulated in these targeted nanocarriers was found to be 3.5-fold higher than that of free paclitaxel. When targeted paclitaxel-loaded nanoparticles were delivered to the carotid artery in a rat model, a significant reduction in neointima-to-media ratio was observed after two weeks in comparison to animals treated with free paclitaxel or non-targeted nanoparticles.

5.2. *Magnetically targeted delivery to stented arteries*

Magnetic targeting for drug delivery was first explored in a study by Widder *et al.*, where this concept was applied in order to reach the microcirculation of the targeted tissue and deliver the active ingredient on a cellular or subcellular level.[263] This method is now among the most intensively explored targeting approaches because of its potential to reduce systemic drug exposure and make a local therapeutic effect achievable at significantly lower drug doses. Chorny *et al.* investigated the novel two-source magnetic targeting scheme for guiding therapeutic agents to injured arteries implanted with a magnetizable stent.[264] In this approach a stent is used as the physical platform for targeted delivery due to its ability to focus the magnetic force in its vicinity and attract agents configured in magnetically responsive particles to the site of stent implantation.

Notably, this targeted delivery approach is potentially applicable to deep blood vessels and is not restricted to superficial regions, does not require the use of permanently magnetic materials (thus obviating safety concerns), and can be realized using magnetic field intensities and sources that are currently widely used in the clinic (e.g. MRI scanners and magnetic navigation systems). The feasibility of this strategy was shown in a rat model of carotid stenting, where targeted PLA-based magnetic nanoparticles (MNPs) loaded

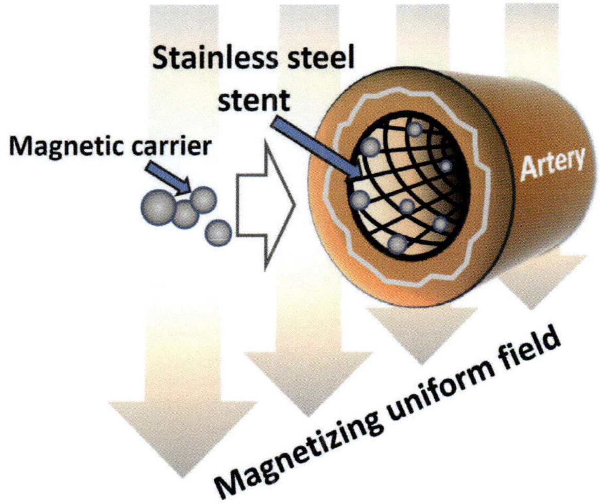

Fig. 4. Schematically shown stent-targeted delivery of magnetic carriers using a two-source guidance strategy. The uniform field magnetizes the carrier particles and the stent, creating a particle capture region spatially restricted to the stented arterial segment. Reproduced from Ref. 278.

with paclitaxel effectively inhibited neointima formation compared with non-magnetically treated control animals. *In vitro* experiments also showed that magnetic treatment of cultured arterial SMC with paclitaxel-loaded MNPs caused significant inhibition of cell growth, which was not observed under non-magnetic conditions.[264]

Magnetically targeted drug delivery was also explored by Räthel *et al.* where rapamycin was adsorbed onto superparamagnetic nanoparticles, and the antiproliferative activity under magnetic conditions was demonstrated in cultured primary porcine vascular cells (SMC and endothelial cells).[265] The particles incorporated into lipid based microbubbles were magnetically

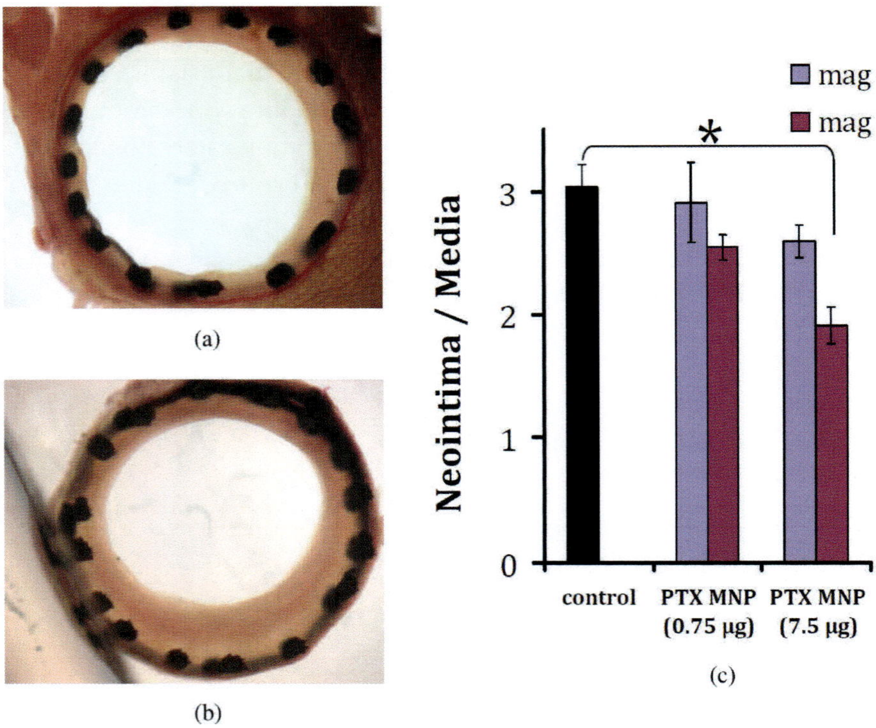

Fig. 5. The antirestenotic effect of paclitaxel-loaded PLA-based MNPs targeted to stented arteries using a two-source magnetic guidance strategy in a rat carotid stenting model. Stented carotid segments of animals treated with MNPs at paclitaxel doses of 7.5 μg and 0.75 μg under magnetic versus non-magnetic conditions were harvested 14 days post surgery. The control group were stented animals untreated with MNPs. A representative Verhoeff–van Gieson-stained section of an artery treated with 7.5 μg paclitaxel under magnetic conditions (a) is shown in comparison with a "no treatment" control (b). Original magnification 100×. Morphometric results expressed as neointima/media ratios (c) are shown as a function of the magnetic field exposure and paclitaxel dose ($n \geq 6$). Adapted from Ref. 264.

guided to nickel plated stents *in vitro* in the presence of high hydrodymanic forces using the magnetizing effect of neodymium-based permanent magnets. This work demonstrates that drugs with established antirestenotic efficacy, such as rapamycin, can be incorporated into magnetically responsive nanoparticle-loaded microbubbles and delivered to magnetizable stents under flow if the blood vessel is accessible to externally applied magnetic fields.

Magnetically driven gene transfer mediated by iron oxide-loaded biodegradable nanoparticles formulated with plasmid DNA was studied using a green fluorescent protein (GFP) reporter plasmid in cultured arterial SMC and endothelial cells using a high-gradient magnetic field.[266] SMC growth inhibition after MNP-mediated adiponectin plasmid transfection was shown as an example of a therapeutically relevant end point with negligible transfection observed in control cells in the absence of a magnetic field. In a different study Chorny *et al.* formulated affinity complexes of adenoviral vectors with PLA-based MNPs and demonstrated their ability to substantially increase transgene expression in vascular cells in culture.[267] Fluorescently labeled nanoparticle-adenovirus complexes exhibited rapid cell binding kinetics in both SMC and endothelial cells following a brief exposure to a high gradient magnetic field. Gene transfer levels significantly greater than those with either free vector or non-magnetic control treatment were achievable without compromising the cell viability or growth kinetics.[267] Importantly, in contrast to free adenovirus, the cell uptake of these complexes was shown to be independent from the Coxsackie adenovirus receptor (CAR), which is poorly expressed by arterial cells. This is an important finding since the effectiveness of gene transfer to injured arteries with free adenovirus is limited by the CAR expression levels[268] and therefore requires the use of large, potentially toxic vector doses.

Kempe *et al.* studied the use of MNPs coated with immobilized tissue plasminogen activator (tPA) for the treatment of in-stent thrombosis.[269] The hemolytic activity of the nanoparticles in blood was negligible, and intravascular injection of the nanoparticles in a stented brachial artery in a porcine model showed no short-term adverse effects. *In vitro* evaluation in a flow-through model proved that the nanoparticles were captured efficiently to the surface of a ferromagnetic coiled wire at the fluid velocities typical for human arteries.[269] Preliminary results obtained with the tPA-nanoparticle conjugates in a pig model suggested that under magnetic guidance this formulation can potentially be used for treatment of in-stent thrombosis in coronary arteries.

As previously discussed, rapid restoration of a continuous and functional endothelial layer can be essential for preventing thrombosis as well as inhibiting the progression of vessel narrowing, making magnetically guided endothelial cell delivery a particularly attractive experimental treatment modality for restenosis prevention. However, the application of endothelial cells for restoring endothelium after denudation in the course of angioplasty is limited by poor delivery efficiency and retention of viable cells at the injury site. Magnetically driven cell delivery may provide a strategy potentially capable of overcoming these limitations as was first demonstrated *in vitro* using magnetically responsive endothelial cells guided to stents.[264,270,271] Hofmann *et al.*[272] used MNPs associated with lentiviral vectors to combine cell transduction with magnetically driven positioning of endothelial cells (a concept first explored by Consigny *et al.*).[273] The amount of nanoparticles internalized by cells progressively increased with the dose. Interestingly, the authors reported strong levels of transduction with MNP-lentivirus complexes observed even when the incubation was carried out under hypothermic conditions (4°C), which normally inhibit all energy-dependent routes of endocytosis. Using strong permanent magnets placed next to targeted blood vessels cells loaded with magnetic complexes were successfully captured and positioned at the target site in the mouse model.[272] Kyrtatos *et al.* used an externally applied magnetic device for enhancing endothelial progenitor cell delivery to the site of carotid artery injury in a rat model.[274] Following angioplasty, MNP-loaded cell engraftment to rat carotid arteries was 5.4-fold higher under the presence of a magnet, compared to non-magnetic conditions.[274] While organ distribution of the magnetically targeted versus non-targeted cells was not addressed in this study, the authors emphasize two potential limitations of their one-source approach relying entirely on an externally applied magnetic field: 1) extensive cell capture in non-target regions is likely to occur in agreement with a mathematical model developed by Grief *et al.*[275] stating the impossibility of magnetically targeting interior regions while avoiding more superficially localized surrounding tissue that inevitable experiences a stronger magnetic exposure; 2) in the absence of additional mechanisms contributing to cell retention, captured cells will be rapidly redistributed from the target site upon the removal of the magnetic field source.

Pislaru *et al.* explored a magnetic cell delivery strategy where porcine endothelial cells were loaded with superparamagnetic iron oxide microspheres and delivered to premagnetized stents implanted in porcine coronary and femoral arterties.[270,271] The delivered cells were found attached to the stent struts and were also distributed within the adjacent denuded vessel wall at 24 h post-implantation, suggesting that magnetic forces can be used to rapidly place endothelial cells at the site of a magnetized intravascular stent. The translation of this approach may be limited by the use of a permanently magnetized

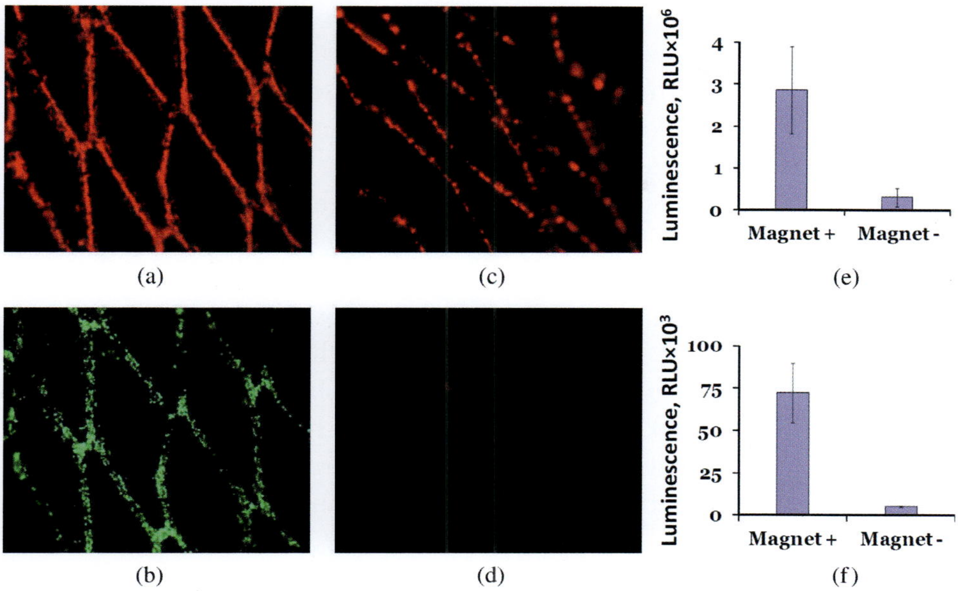

Fig. 6. (Color online) Magnetic targeting of endothelial cells to stents. Endothelial cells loaded with fluorescent labeled PLA-based MNPs captured *in vitro* onto a 304-grade stainless-steel stent as evidenced by the red fluorescence of MNPs (a) and Calcein green staining of live cells (b). (c) MNP-loaded cells transthoracically injected into the left ventricular cavity and captured onto a 304-grade stainless-steel stent deployed in the rat carotid artery following a 5 min exposure to a magnetic field of 1,000 G for 5 min. (d) Control rats treated identically without a magnetic exposure. *In vivo* magnetic cell delivery using a two-source magnetic guidance strategy in a rat carotid stenting model under stop-flow conditions and uninterrupted blood-flow conditions (e and f, respectively). In the magnetic group (Mag+), the injection was carried out with animals placed in a magnetic field of 1,000 G, and the field was maintained for a total of 5 min after delivery. In control rats (Mag- group) no magnetic field was applied. MNP-loaded endothelial cells transduced with a firefly luciferase reporter were used in all animal groups, and the animals were imaged 48 h after delivery by local perivascular administration of luciferin admixed in a pluronic gel. Adapted from Ref. 274.

implant presenting a safety concern, and may also provide suboptimal results, as the field generated by a magnetized implant decays quickly, which greatly reduces the effective capture distance compared to the two-source targeting strategy.[276] Stents made of certain types of stainless steel exhibiting negligible magnetic remanence (e.g. 304) can provide a platform for magnetically targeted cell therapy as a part of a two-source configuration. This method enables effective magnetic capture due to a combination of a deep-penetrating magnetizing field with high magnetic field gradients at a target site. Polyak *et al.* applied this approach to magnetically responsive endothelial cells loaded with biodegradable, iron oxide-loaded nanoparticles to address delayed endothelial healing.[277] The effectiveness of this approach was demonstrated in the rat carotid stenting model both with cells delivered locally under stop-flow conditions and following aortic arch delivery in the presence of blood flow where significantly greater cell localization to the stented arterial region was observed 48 h post treatment in comparison to non-magnetic control animals.

Chorny *et al.* further optimized the bio-degradable MNP formulation used for generating magnetically responsive endothelial cells for stent-targeted delivery and extended its use to provide adequate magnetic cell properties in combination with magnetically enhanced gene transfer without compromising cell viability and function.[278] Cell loading achieved in the absence of toxic effects simultaneously with magnetically driven transduction using a brief exposure to a high-gradient field can significantly improve the bio-compatibility and performance of stent-targeted cell delivery as an experimental restenosis prevention strategy and warrants further investigation in animal disease models.

6. Conclusion

Arterial restenosis following angioplasty is a serious complication caused by interplay of multiple and complex pathophysiological mechanisms. Its effective management requires the development of novel and flexible treatment modalities, easily adjusted for achieving optimal results in different patient populations and clinical settings. The use of nanocarriers with versatile design and properties enabling controlled delivery of therapeutic agents aimed at various pathological pathways within the restenotic process represents a highly promising experimental approach for preventing the disease while minimizing untoward effects, and can provide the basis for developing new, safe and effective therapies.

7. Acknowledgements

The authors would like to acknowledge financial support from American Heart Association Scientist Development grants (I.F. and M.C.), The W.W. Smith Charitable Trust (M.C.), Foerderer Foundation Grant (J.E.T.), U.S. National Heart, Lung, and Blood Institute grants HL 094816 (R.J.L.), HL 111118 (M.C.), T32 HL 007915 (J.E.T.), and donated research funds to the Children's Hospital of Philadelphia including the William J. Rashkind Endowment, Erin's Fund, and The Kibel Foundation (R.J.L.).

References

1. R.Ross, . The pathogenesis of atherosclerosis — an update, *N. Engl. J. Med.* **314**(8), 488–500 (1986).

2. R.Ross, The pathogenesis of atherosclerosis: A perspective for the 1990s, *Nature* **362**(6423), 801–809 (1993).
3. T. Thom *et al.*, Heart disease and stroke statistics–2006 update: a report from the American Heart Association Statistics Committee and Stroke Statistics Subcommittee, *Circulation* **113**(6), 85–151 (2006).
4. D. S. Celermajer *et al.*, Cardiovascular disease in the developing world: prevalences, patterns, and the potential of early disease detection, *J. Am. Coll. Cardiol.* (2012).
5. L. E. Perkins, Preclinical models of restenosis and their application in the evaluation of drug-eluting stent systems, *Vet. Pathol.* **47**(1), 58–76 (2010).
6. D. Lloyd-Jones *et al.*, Executive summary: heart disease and stroke statistics — 2010 update: a report from the American Heart Association, *Circulation* **121**(7), 948–954 (2010).
7. V. L. Roger *et al.*, Heart disease and stroke statistics–2011 update: a report from the American Heart Association, *Circulation* **123**(4), e18–e209 (2011).
8. A. M. Minino, Death in the United States, 2009. *NCHS Data Brief* (64), 1–8 (2011).
9. K. M. Kenney *et al.*, Percutaneous coronary intervention outcomes in a low-volume center: survival, stent thrombosis, and repeat revascularization, *Circ. Cardiovasc. Qual. Outcomes* **2**(6), 671–677 (2009).
10. W. S. Weintraub, The pathophysiology and burden of restenosis, *Am. J. Cardiol.* **100**(5A), 3K–9K (2007).
11. D. E. Cutlip *et al.*, Clinical end points in coronary stent trials: a case for standardized definitions, *Circulation* **115**(17), 2344–2351 (2007).
12. B. L. van der Hoeven *et al.*, Sirolimus-eluting stents versus bare-metal stents in patients with ST-segment elevation myocardial infarction: 9-month angiographic and intravascular ultrasound results and 12-month clinical outcome results from the MISSION! Intervention Study, *J. Am. Coll. Cardiol.* **51**(6), 618–626 (2008).
13. W. R. Agema *et al.*, Genetic aspects of restenosis after percutaneous coronary interventions: towards more tailored therapy, *Eur. Heart J.* **22**(22), 2058–2074 (2001).
14. R. Mehran and G. D. Dangas. Off-label use of drug-eluting stents: assessing the risk, *Nat. Clin. Pract. Cardiovasc. Med.* **4**(11), 594–595 (2007).
15. J. W. Jukema *et al.*, Restenosis after PCI. Part 1: pathophysiology and risk factors, *Nat. Rev. Cardiol.* **9**(1), 53–62 (2012).
16. A. Levitzki, PDGF receptor kinase inhibitors for the treatment of restenosis, *Cardiovasc. Res.* **65**(3), 581–586 (2005).
17. R. Hoffmann *et al.*, Patterns and mechanisms of in-stent restenosis. A serial intravascular ultrasound study, *Circulation* **94**(6), 1247–1254 (1996).
18. M. Nakatani *et al.*, Mechanisms of restenosis after coronary intervention: difference between plain old balloon angioplasty and stenting, *Cardiovasc. Pathol.* **12**(1), 40–48 (2003).
19. N. A.Scott, Restenosis following implantation of bare metal coronary stents: pathophysiology and pathways involved in the vascular response to injury, *Adv. Drug Deliv. Rev.* **58**(3), 358–376 (2006).
20. S. M. Schwartz *et al.*, The intima. Soil for atherosclerosis and restenosis, *Circ. Res.* 77(3), 445–465 (1995).
21. M. W. Majesky *et al.*, The adventitia: a dynamic interface containing resident progenitor cells, *Arterioscler. Thromb. Vasc. Biol.* **31**(7), 1530–1539 (2011).
22. A. Abizaid *et al.*, The influence of diabetes mellitus on acute and late clinical outcomes following coronary stent implantation, *J. Am. Coll. Cardiol.* **32**(3), 584–589 (1998).

23. J. Gilbert *et al.*, Meta-analysis of the effect of diabetes on restenosis rates among patients receiving coronary angioplasty stenting, *Diabetes Care* **27**(4), 990–994 (2004).
24. D. E. Cutlip *et al.*, Clinical restenosis after coronary stenting: perspectives from multicenter clinical trials, *J. Am. Coll. Cardiol.* **40**(12), 2082–2089 (2002).
25. W. R. Agema *et al.*, Current PTCA practice and clinical outcomes in The Netherlands: The real world in the pre-drug-eluting stent era, *Eur. Heart J.* **25**(13), 1163–1170 (2004).
26. M. Singh *et al.*, Clinical and angiographic predictors of restenosis after percutaneous coronary intervention: insights from the Prevention of Restenosis With Tranilast and Its Outcomes (PRESTO) trial, *Circulation* **109**(22), 2727–2731 (2004).
27. D. Aronson *et al.*, Potential mechanisms promoting restenosis in diabetic patients, *J. Am. Coll. Cardiol.* **27**(3), 528–535 (1996).
28. M. S. Lee *et al.*, Molecular and cellular basis of restenosis after percutaneous coronary intervention: the intertwining roles of platelets, leukocytes, and the coagulation-fibrinolysis system, *J. Pathol.* **203**(4), 861–870 (2004).
29. W. S. Weintraub *et al.*, Can restenosis after coronary angioplasty be predicted from clinical variables? *J. Am. Coll. Cardiol.* **21**(1), 6–14 (1993).
30. B. Bhargava *et al.*, New approaches to preventing restenosis, *Br. Med. J.* **327**(7409), 274–279 (2003).
31. W. Casscells *et al.*, Mechanisms of restenosis, *Tex. Heart Inst. J.* **21**(1), 68–77 (1994).
32. R. Pakala *et al.*, Effect of serotonin, thromboxane A2, and specific receptor antagonists on vascular smooth muscle cell proliferation, *Circulation* **96**(7), 2280–2286 (1997).
33. M. A. Costa and D. I. Simon, Molecular basis of restenosis and drug-eluting stents, *Circulation* **111**(17), 2257–2273 (2005).
34. T. Inoue *et al.*, Vascular inflammation and repair: implications for re-endothelialization, restenosis, and stent thrombosis, *JACC Cardiovasc. Interv.* **4**(10), 1057–1066 (2011).
35. R. Riessen *et al.*, Distribution of hyaluronan during extracellular matrix remodeling in human restenotic arteries and balloon-injured rat carotid arteries, *Circulation* **93**(6), 1141–1147 (1996).
36. V. Rajagopal and S. G. Rockson. Coronary restenosis: a review of mechanisms and management, *Am. J. Med.* **115**(7), 547–553 (2003).
37. U. Sigwart *et al.*, Intravascular stents to prevent occlusion and restenosis after transluminal angioplasty, *N. Engl. J. Med.* **316**(12), 701–706 (1987).
38. F. G. Welt and C. Rogers. Inflammation and restenosis in the stent era, *Arterioscler. Thromb. Vasc. Biol.* **22**(11), 1769–1776 (2002).
39. P. S. Monraats *et al.*, Genetic predictive factors in restenosis, *Pathol. Biol. (Paris)* **52**(4), 186–195 (2004).
40. I. Ott *et al.*, Increased neutrophil-platelet adhesion in patients with unstable angina, *Circulation* **94**(6), 1239–1246 (1996).
41. M. I. Furman *et al.*, Increased platelet reactivity and circulating monocyte-platelet aggregates in patients with stable coronary artery disease, *J. Am. Coll. Cardiol.* **31**(2), 352–358 (1998).
42. J. Danesh *et al.*, Low grade inflammation and coronary heart disease: prospective study and updated meta-analyses, *Br. Med. J.* **321**(7255), 199–204 (2000).
43. B. J. Rollins, Monocyte chemoattractant protein 1: a potential regulator of monocyte recruitment in inflammatory disease, *Mol. Med.* Today **2**(5), 198–204 (1996).
44. P. Libby *et al.*, A cascade model for restenosis. A special case of atherosclerosis progression, *Circulation* **86**(6 Suppl), III 47–52 (1992).

45. G. A. Ferns *et al.*, Inhibition of neointimal smooth muscle accumulation after angioplasty by an antibody to PDGF, *Science* **253**(5024), 1129–1132 (1991).
46. R. D. Kenagy *et al.*, Primate smooth muscle cell migration from aortic explants is mediated by endogenous platelet-derived growth factor and basic fibroblast growth factor acting through matrix metalloproteinases 2 and 9, *Circulation* **96**(10), 3555–3560 (1997).
47. V. Lindner and M. A. Reidy, Proliferation of smooth muscle cells after vascular injury is inhibited by an antibody against basic fibroblast growth factor, *Proc. Natl. Acad. Sci. USA* **88**(9), 3739–3743 (1991).
48. K. K. Griendling and G. A. FitzGerald, Oxidative stress and cardiovascular injury: Part II: animal and human studies, *Circulation* **108**(17), 2034–2040 (2003).
49. H. P. Souza *et al.*, Vascular oxidant stress early after balloon injury: evidence for increased NAD(P)H oxidoreductase activity, *Free Radic. Biol. Med.* **28**(8), 1232–1242 (2000).
50. K. Walsh *et al.*, Vascular cell apoptosis in remodeling, restenosis, and plaque rupture, *Circ. Res.* **87**(3), 184–188 (2000).
51. I. M. Bayer *et al.*, Atrophic remodeling of the artery-cuffed artery, *Arterioscler. Thromb. Vasc. Biol.* **19**(6), 1499–1505 (1999).
52. M. Lauth *et al.*, Elevated perfusion pressure upregulates endothelin-1 and endothelin B receptor expression in the rabbit carotid artery, *Hypertension* **35**(2), 648–654 (2000).
53. P. L. Jones *et al.*, Regulation of tenascin-C, a vascular smooth muscle cell survival factor that interacts with the alpha v beta 3 integrin to promote epidermal growth factor receptor phosphorylation and growth, *J. Cell Biology* **139**(1), 279–293 (1997).
54. A. H. Baker *et al.*, Divergent effects of tissue inhibitor of metalloproteinase-1, -2, or -3 overexpression on rat vascular smooth muscle cell invasion, proliferation, and death *in vitro*. TIMP-3 promotes apoptosis, *J. Clin. Invest* **101**(6), 1478–1487 (1998).
55. H. Perlman *et al.*, Evidence for the rapid onset of apoptosis in medial smooth muscle cells after balloon injury, *Circulation* **95**(4), 981–987 (1997).
56. K. Miwa *et al.*, Caspase 1-independent IL-1beta release and inflammation induced by the apoptosis inducer Fas ligand, *Nat. Med.* **4**(11), 1287–1292 (1998).
57. A. W. Clowes *et al.*, Kinetics of cellular proliferation after arterial injury. I. Smooth muscle growth in the absence of endothelium, *Lab. Invest.* **49**(3), 327–333 (1983).
58. J. W. Jukema *et al.*, Restenosis after PCI. Part 2: prevention and therapy, *Nat. Rev. Cardiol.* **9**(2), 79–90 (2012).
59. S. Elezi *et al.*, Clinical and angiographic follow-up after balloon angioplasty with provisional stenting for coronary in-stent restenosis, *Catheter. Cardiovasc. Interv.* **48**(2), 151–156 (1999).
60. G. D. Dangas *et al.*, In-stent restenosis in the drug-eluting stent era, *J. Am. Coll. Cardiol.* **56**(23), 1897–1907 (2010).
61. L. Brito and M. Amiji, Nanoparticulate carriers for the treatment of coronary restenosis, *Int. J. Nanomedicine* **2**(2), 143–161 (2007).
62. A. Kastrati *et al.*, Influence of lesion length on restenosis after coronary stent placement, *Am. J. Cardiol.* **83**(12), 1617–1622 (1999).
63. A. Kastrati *et al.*, Restenosis after coronary placement of various stent types, *Am. J. Cardiol.* **87**(1), 34–39 (2001).
64. M. A. Ferguson *et al.*, A history of prior bare metal stent restenosis is associated with major adverse cardiac events in subsequent bare metal stenting of de novo coronary lesions, *Cardiology* **116**(3), 190–193 (2010).

65. J. Pache *et al.*, Intracoronary stenting and angiographic results: strut thickness effect on restenosis outcome (ISAR-STEREO-2) trial, *J. Am. Coll. Cardiol.* **41**(8), 1283–1288 (2003).
66. D. M. Martin and F. J. Boyle, Drug-eluting stents for coronary artery disease: A review, *Med. Eng. Phys.* **33**(2), 148–163 (2011).
67. P. W. Serruys *et al.*, Comparison of coronary-artery bypass surgery and stenting for the treatment of multivessel disease, *N. Engl. J. Med.* **344**(15), 1117–1124 (2001).
68. F. Alfonso, Treatment of drug-eluting stent restenosis the new pilgrimage: Quo vadis? *J. Am. Coll. Cardiol.* **55**(24), 2717–2720 (2010).
69. B. Godin *et al.*, Emerging applications of nanomedicine for the diagnosis and treatment of cardiovascular diseases, *Trends Pharmacol. Sci.* **31**(5), 199–205 (2010).
70. B. Scheller *et al.*, Treatment of coronary in-stent restenosis with a paclitaxel-coated balloon catheter, *N. Engl. J. Med.* **355**(20), 2113–2124 (2006).
71. P. W. Serruys *et al.*, Coronary-artery stents, *N. Engl. J. Med.* **354**(5), 483–495 (2006).
72. A. Aminian *et al.*, Treatment of drug-eluting stent restenosis: an emerging challenge, *Catheter. Cardiovasc. Interv.* **74**(1), 108–116 (2009).
73. M. Unverdorben *et al.*, Paclitaxel-coated balloon catheter versus paclitaxel-coated stent for the treatment of coronary in-stent restenosis, *Circulation* **119**(23), 2986–2994 (2009).
74. Y. Ozaki *et al.*, A quantitative coronary angiography-matched comparison between a prospective randomised multicentre cutting balloon angioplasty and bare metal stent trial (REDUCE III) and the Rapamycin-Eluting Stent Evaluation At Rotterdam Cardiology Hospital (RESEARCH) study, *EuroIntervention* **6**(3), 400–406 (2010).
75. S. J. Park *et al.*, Comparison of plain balloon and cutting balloon angioplasty for the treatment of restenosis with drug-eluting stents vs bare metal stents, *Circ. J.* **74**(9), 1837–1845 (2010).
76. J. M. Stolker *et al.*, Predicting restenosis of drug-eluting stents placed in real-world clinical practice: derivation and validation of a risk model from the EVENT registry, *Circ. Cardiovasc. Interv.* **3**(4), 327–334 (2010).
77. J. W. Moses *et al.*, Sirolimus-eluting stents versus standard stents in patients with stenosis in a native coronary artery, *N. Engl. J. Med.* **349**(14), 1315–1323 (2003).
78. G. W. Stone *et al.*, One-year clinical results with the slow-release, polymer-based, paclitaxel-eluting TAXUS stent: the TAXUS-IV trial, *Circulation* **109**(16), 1942–1947 (2004).
79. J. S. Douglas, Jr., Pharmacologic approaches to restenosis prevention, *Am. J. Cardiol.* **100**(5A), 10K–16K (2007).
80. D. E. Kandzari *et al.*, Frequency, predictors, and outcomes of drug-eluting stent utilization in patients with high-risk non-ST-segment elevation acute coronary syndromes, *Am. J. Cardiol.* **96**(6), 750–755 (2005).
81. M. Degertekin *et al.*, Sirolimus-eluting stent for treatment of complex in-stent restenosis: the first clinical experience, *J. Am. Coll. Cardiol.* **41**(2), 184–189 (2003).
82. A. T. Ong *et al.*, Late angiographic stent thrombosis (LAST) events with drug-eluting stents, *J. Am. Coll. Cardiol.* **45**(12), 2088–2092 (2005).
83. A. V. Finn *et al.*, Vascular responses to drug eluting stents: Importance of delayed healing, *Arterioscler. Thromb. Vasc. Biol.* **27**(7), 1500–1510 (2007).
84. G. W. Stone *et al.*, Safety and efficacy of sirolimus- and paclitaxel-eluting coronary stents, *N. Engl. J. Med.* **356**(10), 998–1008 (2007).

85. S. Windecker and B. Meier. Late coronary stent thrombosis, *Circulation* **116**(17), 1952–1965 (2007).
86. R. Tung *et al.*, Narrative review: Drug-eluting stents for the management of restenosis: a critical appraisal of the evidence, *Ann. Intern. Med.* **144**(12), 913–919 (2006).
87. S. H. Duda *et al.*, Sirolimus-eluting versus bare nitinol stent for obstructive superficial femoral artery disease: the SIROCCO II trial, *J. Vasc. Interv. Radiol.* **16**(3), 331–338 (2005).
88. S. H. Duda *et al.*, Drug-eluting and bare nitinol stents for the treatment of atherosclerotic lesions in the superficial femoral artery: Long-term results from the SIROCCO trial, *J. Endovasc. Ther.* **13**(6), 701–710 (2006).
89. A. E. Rodriguez *et al.*, Coronary stent thrombosis in the current drug-eluting stent era: Insights from the ERACI III trial, *J. Am. Coll. Cardiol.* **47**(1), 205–207 (2006).
90. F. Liistro and A. Colombo, Late acute thrombosis after paclitaxel eluting stent implantation, *Heart* **86**(3), 262–264 (2001).
91. M. Cassin *et al.*, Late coronary thrombosis in paclitaxel-eluting stents. Case reports, *J. Cardiovasc. Med. (Hagerstown)* **8**(4), 284–288 (2007).
92. L. M. de Abreu *et al.*, Late coronary thrombosis secondary to implantation of paclitaxel-eluting stent without restenosis of conventional stents, *Arq. Bras. Cardiol.* **88**(5), e111–114 (2007).
93. D. Trabattoni *et al.*, Stent thrombosis after sirolimus- and paclitaxel-eluting stent implantation in daily clinical practice: analysis of a single center registry, *Catheter. Cardiovasc. Interv.* **70**(3), 415–421 (2007).
94. J. L. Chen *et al.*, Incidence of thrombosis after implantation of drug-eluting stents in patients with coronary artery disease, *Chin. Med. J. (Engl)* **121**(21), 2144–2147 (2008).
96. S. A. Kim and S. Y. Lim. A case of stent thrombosis occurred at 5 years after sirolimus-eluting stent implantation, *Chonnam Med. J.* **47**(2), 124–126 (2011).
95. T. Nagrani *et al.*, In-stent thrombosis after 68 months of implantation inspite of continuous dual antiplatelet therapy: A case report. *Cases J.* **3**: 68 (2010).
97. R. Rayoo *et al.*, Histological confirmation of hypersensitivity as a contributor to very-late coronary stent thrombosis, *Int. J. Cardiol.* **157**(2), e29–30 (2012).
98. S. S. Naidu, Three years since the FDA advisory panel on drug-eluting stents: What have we learned about off-label use and stent thrombosis? *J. Invasive Cardiol.* **22**(1), 20–21 (2010).
99. F. Zhang *et al.*, Super late stent thrombosis occurred at 8years after drug-eluting stent implantation, *Int. J. Cardiol.* (2011).
100. M. Roque *et al.*, Effect of p27 deficiency and rapamycin on intimal hyperplasia: *In vivo* and *in vitro* studies using a p27 knockout mouse model, *Lab. Invest.* **81**(6), 895–903 (2001).
101. T. Suzuki *et al.*, Stent-based delivery of sirolimus reduces neointimal formation in a porcine coronary model, *Circulation* **104**(10), 1188–1193 (2001).
102. D. Karnabatidis *et al.*, Below-the-knee drug-eluting stents and drug-coated balloons, *Expert Rev. Med. Devices* **9**(1), 85–94 (2012).
103. E. P. Strecker *et al.*, Femoropopliteal artery stent placement: Evaluation of long-term success, *Radiology* **205**(2), 375–383 (1997).
104. H. A. Deutschmann *et al.*, Placement of Hemobahn stent-grafts in femoropopliteal arteries: Early experience and midterm results in 18 patients, *J. Vasc. Interv. Radiol.* **12**(8), 943–950 (2001).

105. H. M. van Beusekom *et al.*, Coronary stent healing, endothelialisation and the role of co-medication, *Neth. Heart J.* **15**(11), 395–396 (2007).
106. L. Machan, Clinical experience and applications of drug-eluting stents in the noncoronary vasculature, bile duct and esophagus, *Adv. Drug Deliv. Rev.* **58**(3), 447–462 (2006).
107. P. R. Umashankar *et al.*, Effect of blood flow on drug release from DES: an experimental study, *Int. J. Cardiol.* **131**(3), 415–417 (2009).
108. D. R. Holmes Jr. *et al.*, Analysis of 1-year clinical outcomes in the SIRIUS trial: A randomized trial of a sirolimus-eluting stent versus a standard stent in patients at high risk for coronary restenosis, *Circulation* **109**(5), 634–640 (2004).
109. T. Palmerini *et al.*, Stent thrombosis with drug-eluting and bare-metal stents: Evidence from a comprehensive network meta-analysis, *Lancet* **379**(9824), 1393–1402 (2012).
110. G. Sarno *et al.*, Lower risk of stent thrombosis and restenosis with unrestricted use of 'new-generation' drug-eluting stents: A report from the nationwide Swedish Coronary Angiography and Angioplasty Registry (SCAAR), *Eur. Heart J.* **33**(5), 606–613 (2012).
111. J. C. Tardif *et al.*, Prevention of restenosis with antioxidants: mechanisms and implications, *Am. J. Cardiovasc. Drugs* **2**(5), 323–334 (2002).
112. K. Miyauchi *et al.*, Effect of probucol on smooth muscle cell proliferation and dedifferentiation after vascular injury in rabbits: Possible role of PDGF, *Cardiovasc. Drugs Ther.* **12**(3), 251–260 (1998).
113. C. L. Jackson and K. S. Pettersson. Effects of probucol on rat carotid artery responses to balloon catheter injury, *Atherosclerosis* **154**(2), 407–414 (2001).
114. A. K. Lau *et al.*, Probucol promotes functional reendothelialization in balloon-injured rabbit aortas, *Circulation* **107**(15), 2031–2036 (2003).
115. T. Yokoyama *et al.*, Effect of probucol on neointimal thickening in a stent porcine restenosis model, *Jpn. Heart. J.* **45**(2), 305–313 (2004).
116. J. H. O'Keefe, Jr. *et al.*, Lovastatin plus probucol for prevention of restenosis after percutaneous transluminal coronary angioplasty, *Am. J. Cardiol.* 77(8), 649–652 (1996).
117. K. Watanabe *et al.*, Preventive effects of probucol on restenosis after percutaneous transluminal coronary angioplasty, *Am. Heart J.* **132**(1 Pt 1), 23–29 (1996).
118. J. C. Tardif *et al.*, Probucol and multivitamins in the prevention of restenosis after coronary angioplasty. Multivitamins and Probucol Study Group, *N. Engl. J. Med.* **337**(6), 365–372 (1997).
119. H. Yokoi *et al.*, Effectiveness of an antioxidant in preventing restenosis after percutaneous transluminal coronary angioplasty: The Probucol Angioplasty Restenosis Trial, *J. Am. Coll. Cardiol.* **30**(4), 855–862 (1997).
120. C. L. Sundell *et al.*, AGI-1067: a multifunctional phenolic antioxidant, lipid modulator, anti-inflammatory and antiatherosclerotic agent, *J. Pharmacol. Exp. Ther.* **305**(3), 1116–1123 (2003).
121. J. C. Tardif *et al.*, Effects of the antioxidant succinobucol (AGI-1067) on human atherosclerosis in a randomized clinical trial, *Atherosclerosis* **197**(1), 480–486 (2008).
122. J. M. Maraganore. The arterial thrombotic process and emerging drugs for its control, *Tex. Heart Inst. J.* **20**(1), 43–47 (1993).
123. E. Tsuchikane *et al.*, Impact of cilostazol on restenosis after percutaneous coronary balloon angioplasty, *Circulation* **100**(1), 21–26 (1999).
124. J. S. Douglas, Jr. *et al.*, Coronary stent restenosis in patients treated with cilostazol, *Circulation* **112**(18), 2826–2832 (2005).

125. Y. Kubota *et al.*, Pharmacologic treatment of intimal hyperplasia after metallic stent placement in the peripheral arteries. An experimental study, *Invest. Radiol.* **30**(9), 532–537 (1995).
126. S. W. Lee *et al.*, Comparison of triple versus dual antiplatelet therapy after drug-eluting stent implantation (from the DECLARE-Long trial), *Am. J. Cardiol.* **100**(7), 1103–1108 (2007).
127. S. W. Lee *et al.*, Drug-eluting stenting followed by cilostazol treatment reduces late restenosis in patients with diabetes mellitus the DECLARE-DIABETES Trial (A Randomized Comparison of Triple Antiplatelet Therapy with Dual Antiplatelet Therapy After Drug-Eluting Stent Implantation in Diabetic Patients), *J. Am. Coll. Cardiol.* **51**(12), 1181–1187 (2008).
128. D. L. Jennings and J. S. Kalus. Addition of cilostazol to aspirin and a thienopyridine for prevention of restenosis after coronary artery stenting: A meta-analysis. *J. Clin. Pharmacol.* **50**(4), 415–421 (2010).
129. A. K. Kamal *et al.*, Cilostazol versus aspirin for secondary prevention of vascular events after stroke of arterial origin, *Cochrane Database Syst. Rev.* (1), CD008076 (2011).
130. S. W. Lee *et al.*, A randomized, double-blind, multicenter comparison study of triple antiplatelet therapy with dual antiplatelet therapy to reduce restenosis after drug-eluting stent implantation in long coronary lesions: results from the DECLARE-LONG II (Drug-Eluting Stenting Followed by Cilostazol Treatment Reduces Late Restenosis in Patients with Long Coronary Lesions) trial, *J. Am. Coll. Cardiol.* **57**(11), 1264–1270 (2011).
131. M. D. Waterfield *et al.*, Platelet-derived growth factor is structurally related to the putative transforming protein p28sis of simian sarcoma virus, *Nature* **304**(5921), 35–39 (1983).
132. D. W. Muller *et al.*, Colchicine and antineoplastic therapy for the prevention of restenosis after percutaneous coronary interventions, *J. Am. Coll. Cardiol.* **17**(6 Suppl B), 126B–131B (1991).
133. R. Voisard *et al.*, The in-vitro effect of antineoplastic agents on proliferative activity and cytoskeletal components of plaque-derived smooth-muscle cells from human coronary arteries, *Coron. Artery Dis.* **4**(10), 935–942 (1993).
134. A. H. Cragg *et al.*, Effect of antineoplastic agents on smooth muscle cell proliferation *in vitro*: implications for prevention of restenosis after transluminal angioplasty, *J. Vasc. Interv. Radiol.* **3**(2), 273–277 (1992).
135. T. Uwatoku *et al.*, Application of nanoparticle technology for the prevention of restenosis after balloon injury in rats, *Circ. Res.* **92**(7), e62–69 (2003).
136. K. J. Gruchalla and J. J. Nawarskas, The paclitaxel-eluting stent in percutaneous coronary intervention: part I: background and clinical comparison to bare metal stents, *Cardiol. Rev.* **14**(2), 88–98 (2006).
137. J. Margolis *et al.*, Systemic nanoparticle paclitaxel (nab-paclitaxel) for in-stent restenosis I (SNAPIST-I): a first-in-human safety and dose-finding study, *Clin. Cardiol.* **30**(4), 165–170 (2007).
138. A. Takahashi *et al.*, Tranilast inhibits vascular smooth muscle cell growth and intimal hyperplasia by induction of p21(wafl/cipl/sdil) and p53, *Circ. Res.* **84**(5), 543–550 (1999).
139. S. Okamoto *et al.*, Effects of trapidil (triazolopyrimidine), a platelet-derived growth factor antagonist, in preventing restenosis after percutaneous transluminal coronary angioplasty, *Am. Heart J.* **123**(6), 1439–1444 (1992).

140. T. Takagi *et al.*, Pioglitazone reduces neointimal tissue proliferation after coronary stent implantation in patients with type 2 diabetes mellitus: An intravascular ultrasound scanning study, *Am. Heart J.* **146**(2), E5 (2003).
141. G. W. Stone *et al.*, Comparison of a polymer-based paclitaxel-eluting stent with a bare metal stent in patients with complex coronary artery disease: A randomized controlled trial, *JAMA* **294**(10), 1215–1223 (2005).
142. A. De Labriolle *et al.*, Paclitaxel-eluting balloon: from bench to bed, *Catheter. Cardiovasc. Interv.* **73**(5), 643–652 (2009).
143. R. Virmani *et al.*, Drug-eluting stents: caution and concerns for long-term outcome, *Coron. Artery Dis.* **15**(6), 313–318 (2004).
144. P. Barath *et al.*, Cutting balloon: A novel approach to percutaneous angioplasty, *Am. J. Cardiol.* **68**(11), 1249–1252 (1991).
145. C. Engelke *et al.*, Using 6-mm Cutting Balloon angioplasty in patients with resistant peripheral artery stenosis: Preliminary results, *Am. J. Roentgenol.* **179**(3), 619–623 (2002).
146. P. Dick *et al.*, Conventional balloon angioplasty versus peripheral cutting balloon angioplasty for treatment of femoropopliteal artery in-stent restenosis: Initial experience, *Radiology* **248**(1), 297–302 (2008).
147. L. Mauri *et al.*, Cutting balloon angioplasty for the prevention of restenosis: Results of the Cutting Balloon Global Randomized Trial, *Am. J. Cardiol.* **90**(10), 1079–1083 (2002).
148. L. Bergersen *et al.*, Randomized trial of cutting balloon compared with high-pressure angioplasty for the treatment of resistant pulmonary artery stenosis, *Circulation* **124**(22), 2388–2396 (2011).
149. R. Nath and K. B. Roberts. Vascular irradiation for the prevention of restenosis after angioplasty: a new application for radiotherapy, *Int. J. Radiat. Oncol. Biol. Phys.* **36**(4), 977–979 (1996).
150. M. B. Leon *et al.*, Localized intracoronary gamma-radiation therapy to inhibit the recurrence of restenosis after stenting, *N. Engl. J. Med.* **344**(4), 250–256 (2001).
151. R. Waksman *et al.*, Use of localised intracoronary beta radiation in treatment of in-stent restenosis: the INHIBIT randomised controlled trial, *Lancet* **359**(9306), 551–557 (2002).
152. P. S. Teirstein, *et al.*, Catheter-based radiotherapy to inhibit restenosis after coronary stenting, *N. Engl. J. Med.* **336**(24), 1697–1703 (1997).
153. R. Waksman *et al.*, Intracoronary gamma-radiation therapy after angioplasty inhibits recurrence in patients with in-stent restenosis. *Circulation* **101**(18), 2165–2171 (2000).
154. R. Waksman, Update on bioabsorbable stents: from bench to clinical, *J. Interv. Cardiol.* **19**(5), 414–421 (2006).
155. R. S. Stack *et al.*, Interventional cardiac catheterization at Duke Medical Center, *Am. J. Cardiol.* **62**(10 Pt 2), 3F–24F (1988).
156. G. Lanza *et al.*, Nanomedicine opportunities in cardiology, *Ann. NY Acad. Sci.* **1080**: 451–465 (2006).
157. S. M. Moghimi *et al.*, Nanomedicine: Current status and future prospects, *FASEB J.* **19**(3), 311–330 (2005).
158. L. Zhang *et al.*, Nanoparticles in medicine: Therapeutic applications and developments. *Clin. Pharmacol. Ther.* **83**(5), 761–769 (2008).

159. O. C. Farokhzad and R. Langer, Impact of nanotechnology on drug delivery, *ACS Nano* **3**(1), 16–20 (2009).
160. E. Cohen-Sela *et al.*, Nanoparticles for the Treatment of Restenosis, in *Nanoparticles for Pharmaceutical Applications*. American Scientific Publishers (2007), pp. 123–134.
161. G. Golomb *et al.*, Controlled delivery of a tyrphostin inhibits intimal hyperplasia in a rat carotid artery injury model, *Atherosclerosis* **125**(2), 171–182 (1996).
162. E. R. Edelman *et al.*, C-myc in vasculoproliferative disease, *Circ. Res.* **76**(2), 176–182 (1995).
163. M. A. Lovich *et al.*, Drug clearance and arterial uptake after local perivascular delivery to the rat carotid artery, *J. Am. Coll. Cardiol.* **29**(7), 1645–1650 (1997).
164. M. Chorny *et al.*, Drug delivery systems for the treatment of restenosis, *Crit. Rev. Ther. Drug Carrier Syst.* **17**(3), 249–284 (2000).
165. R. L. Wilensky *et al.*, Direct intraarterial wall injection of microparticles via a catheter: a potential drug delivery strategy following angioplasty, *Am. Heart J.* **122**(4 Pt 1), 1136–1140 (1991).
166. J. J. Rome *et al.*, Anatomic barriers influence the distribution of *in vivo* gene transfer into the arterial wall. Modeling with microscopic tracer particles and verification with a recombinant adenoviral vector, *Arterioscler. Thromb.* **14**(1), 148–161 (1994).
167. T. K. Nasser *et al.*, Microparticle deposition in periarterial microvasculature and intramural dissections after porous balloon delivery into atherosclerotic vessels: quantitation and localization by confocal scanning laser microscopy, *Am. Heart J.* **131**(5), 892–898 (1996).
168. I. Fishbein *et al.*, Treatment of restenosis by controlled-release delivery systems of tyrphostins, *Drug Develop. Res.* **50**(3&4), 487–496 (2000).
169. I. Fishbein *et al.*, Nanoparticulate delivery system of a tyrphostin for the treatment of restenosis, *J. Control Release* **65**(1&2), 221–229 (2000).
170. I. Fishbein *et al.*, Formulation and delivery mode affect disposition and activity of tyrphostin-loaded nanoparticles in the rat carotid model, *Arterioscler. Thromb. Vasc. Biol.* **21**(9), 1434–1439 (2001).
171. U. Westedt *et al.*, Deposition of nanoparticles in the arterial vessel by porous balloon catheters: localization by confocal laser scanning microscopy and transmission electron microscopy, *AAPS Pharm Sci.* **4**(4), E41 (2002).
172. H. Suh *et al.*, Regulation of smooth muscle cell proliferation using paclitaxel-loaded poly(ethylene oxide)-poly(lactide/glycolide) nanospheres, *J. Biomed. Mater. Res.* **42**(2), 331–338 (1998).
173. W. B. Batchelor *et al.*, A comparison of ionic versus nonionic contrast medium during primary percutaneous transluminal coronary angioplasty for acute myocardial infarction (GUSTO IIb). Global Use of Strategies to Open Occluded Coronary Arteries in Acute Coronary Syndromes, *Am. J. Cardiol.* **85**(6), 692–697 (2000).
174. A. M. Lefer *et al.*, Vascular effects of HMG CoA-reductase inhibitors (statins) unrelated to cholesterol lowering: new concepts for cardiovascular disease, *Cardiovasc. Res.* **49**(2), 281–287 (2001).
175. U. Speck *et al.*, Local drug delivery — the early Berlin experience: single drug administration versus sustained release, *EuroIntervention* **7 (Suppl K)**: K17–22 (2011).
176. J. Lefkovits and E. J. Topol. Pharmacological approaches for the prevention of restenosis after percutaneous coronary intervention, *Prog. Cardiovasc. Dis.* **40**(2), 141–158 (1997).

177. A. Farbet *et al.*, Morphological predictors of restenosis after coronary stenting in humans, *Circulation* **105**(25), 2974–2980 (2002).
178. F. G. Welt *et al.*, Neutrophil, not macrophage, infiltration precedes neointimal thickening in balloon-injured arteries, *Arterioscler. Thromb. Vasc. Biol.* **20**(12), 2553–2558 (2000).
179. K. Toutouzas *et al.*, Inflammation and restenosis after percutaneous coronary interventions, Eur. *Heart J.* **25**(19), 1679–1687 (2004).
180. D. Gutman and G. Golomb, Liposomal alendronate for the treatment of restenosis, *J. Control Release* (2011).
181. A. Pietersma *et al.*, Late lumen loss after coronary angioplasty is associated with the activation status of circulating phagocytes before treatment, *Circulation* **91**(5), 1320–1325 (1995).
182. H. D. Danenberg *et al.*, Macrophage depletion by clodronate-containing liposomes reduces neointimal formation after balloon injury in rats and rabbits, *Circulation* **106**(5), 599–605 (2002).
183. R. G. Russell *et al.*, The pharmacology of bisphosphonates and new insights into their mechanisms of action, *J. Bone Miner. Res.* **14 (Suppl 2)**: 53–65 (1999).
184. J. E. Dunford *et al.*, Structure-activity relationships for inhibition of farnesyl diphosphate synthase *in vitro* and inhibition of bone resorption *in vivo* by nitrogen-containing bisphosphonates, *J. Pharmacol. Exp. Ther.* **296**(2), 235–242 (2001).
185. G. E. Bilder *et al.*, Tyrphostins inhibit PDGF-induced DNA synthesis and associated early events in smooth muscle cells, *Am. J. Physiol.* **260** (4 Pt 1), C721–730 (1991).
186. A. Levitzki and A. Gazit. Tyrosine kinase inhibition: an approach to drug development, *Science* **267**(5205), 1782–1788 (1995).
187. S. Banai *et al.*, PDGF-receptor tyrosine kinase blocker AG1295 selectively attenuates smooth muscle cell growth *in vitro* and reduces neointimal formation after balloon angioplasty in swine, *Circulation* **97**(19), 1960–1969 (1998).
188. R. Gallo *et al.*, Inhibition of intimal thickening after balloon angioplasty in porcine coronary arteries by targeting regulators of the cell cycle, *Circulation* **99**(16), 2164–2170 (1999).
189. S. O. Marx and A. R. Marks. Bench to bedside: the development of rapamycin and its application to stent restenosis, *Circulation* **104**(8), 852–855 (2001).
190. J. Hausleiter *et al.*, Randomized, double-blind, placebo-controlled trial of oral sirolimus for restenosis prevention in patients with in-stent restenosis: The Oral Sirolimus to Inhibit Recurrent In-stent Stenosis (OSIRIS) trial, *Circulation* **110**(7), 790–795 (2004).
191. A. E. Rodriguez *et al.*, Percutaneous coronary intervention with oral sirolimus and bare metal stents has comparable safety and efficacy to treatment with drug eluting stents, but with significant cost saving: long-term follow-up results from the randomised, controlled ORAR III (Oral Rapamycin in A Rgentina) study, *EuroIntervention* **5**(2), 255–264 (2009).
192. S. Kufner *et al.*, Long-term risk of adverse outcomes and new malignancies in patients treated with oral sirolimus for prevention of restenosis, *JACC Cardiovasc. Interv.* **2**(11), 1142–1148 (2009).
193. S. Bellosta *et al.*, Direct vascular effects of HMG-CoA reductase inhibitors, *Atherosclerosis* **137(Suppl),** S101–109 (1998).

194. A. Lerman and E. Eeckhout. Coronary endothelial dysfunction following sirolimus-eluting stent placement: Should we worry about it? *Eur. Heart J.* **27**(2), 125–126 (2006).
195. D. Versari, *et al.*, The importance of reendothelialization after arterial injury, *Curr. Pharm. Des.* **13**(17), 1811–1824 (2007).
196. C. C. Haudenschild and S. M. Schwartz, Endothelial regeneration. II. Restitution of endothelial continuity, *Lab. Invest.* **41**(5), 407–418 (1979).
197. T. Asahara *et al.*, Local delivery of vascular endothelial growth factor accelerates reendothelialization and attenuates intimal hyperplasia in balloon-injured rat carotid artery, *Circulation* **91**(11), 2793–2801 (1995).
198. M. Oberhoff *et al.*, Local and systemic delivery of low molecular weight heparin stimulates the reendothelialization after balloon angioplasty, *Cardiovasc. Res.* **38**(3), 751–762 (1998).
199. K. Krasinski *et al.*, Estradiol accelerates functional endothelial recovery after arterial injury, *Circulation* **95**(7), 1768–1772 (1997).
200. M. Herring *et al.*, A single-staged technique for seeding vascular grafts with autogenous endothelium, *Surgery* **84**(4), 498–504 (1978).
201. V. Lindner *et al.*, Basic fibroblast growth factor stimulates endothelial regrowth and proliferation in denuded arteries, *J. Clin. Invest.* **85**(6), 2004–2008 (1990).
202. N. Kipshidze *et al.*, Role of the endothelium in modulating neointimal formation: Vasculoprotective approaches to attenuate restenosis after percutaneous coronary interventions, *J. Am. Coll. Cardiol.* **44**(4), 733–739 (2004).
203. A. T. Ong *et al.*, How to accelerate the endothelialization of stents, *Arch. Mal. Coeur. Vaiss.* **98**(2), 123–126 (2005).
204. A. Forte *et al.*, Mesenchymal stem cells effectively reduce surgically induced stenosis in rat carotids, *J. Cell Physiol.* **217**(3), 789–799 (2008).
205. A. Forte *et al.*, Mesenchymal stem cells: a good candidate for restenosis therapy? *Curr. Vasc. Pharmacol.* 7(3), 381–393 (2009).
206. I. Rasmusson *et al.*, Mesenchymal stem cells inhibit the formation of cytotoxic T lymphocytes, but not activated cytotoxic T lymphocytes or natural killer cells, *Transplantation* **76**(8), 1208–1213 (2003).
207. A. Corcione *et al.*, Human mesenchymal stem cells modulate B-cell functions, *Blood* **107**(1), 367–372 (2006).
208. S. M. Sanz-Gonzalez *et al.*, Increased p53 gene dosage reduces neointimal thickening induced by mechanical injury but has no effect on native atherosclerosis, *Cardiovasc. Res.* **75**(4), 803–812 (2007).
209. T. W. Johnson *et al.*, Stent-based delivery of tissue inhibitor of metalloproteinase-3 adenovirus inhibits neointimal formation in porcine coronary arteries, *Arterioscler. Thromb. Vasc. Biol.* **25**(4), 754–759 (2005).
210. S. Miyagawa *et al.*, Myocardial regeneration therapy for heart failure: hepatocyte growth factor enhances the effect of cellular cardiomyoplasty, *Circulation* **105**(21), 2556–2561 (2002).
211. M. O. Laukkanen *et al.*, Adenovirus-mediated extracellular superoxide dismutase gene therapy reduces neointima formation in balloon-denuded rabbit aorta, *Circulation* **106**(15), 1999–2003 (2002).

212. P. Zoldhelyi *et al.*, Thromboresistance of balloon-injured porcine carotid arteries after local gene transfer of human tissue factor pathway inhibitor, *Circulation* **101**(3), 289–295 (2000).
213. K. Egashira *et al.*, Local delivery of anti-monocyte chemoattractant protein-1 by gene-eluting stents attenuates in-stent stenosis in rabbits and monkeys, *Arterioscler. Thromb. Vasc. Biol.* **27**(12), 2563–2568 (2007).
214. C. E. Appleby and P. A. Kingston, "Gene therapy for restenosis — what now, what next? *Curr. Gene Ther.* **4**(2), 153–182 (2004).
215. C. Lande *et al.*, Innovative erythrocyte-based carriers for gene delivery in porcine vascular smooth muscle cells: basis for local therapy to prevent restenosis, *Cardiovasc. Hematol. Disord. Drug Targets* (2012).
216. R. J. Guzman *et al.*, *In vivo* suppression of injury-induced vascular smooth muscle cell accumulation using adenovirus-mediated transfer of the herpes simplex virus thymidine kinase gene, *Proc. Natl. Acad. Sci. USA* **91**(22), 10732–10736 (1994).
217. Z. Y. Yang *et al.*, Role of the p21 cyclin-dependent kinase inhibitor in limiting intimal cell proliferation in response to arterial injury, *Proc. Natl. Acad. Sci. USA* **93**(15), 7905–7910 (1996).
218. P. Libby, Gene therapy of restenosis: promise and perils, *Circ. Res.* **82**(3), 404–406 (1998).
219. N. Sarkar *et al.*, Nonsurgical direct delivery of plasmid DNA into rat heart: time course, dose response, and the influence of different promoters on gene expression, *J. Cardiovasc. Pharmacol.* **39**(2), 215–224 (2002).
220. H. Cao *et al.*, Gene therapy: light is finally in the tunnel, *Protein Cell* **2**(12), 973–989 (2011).
221. A. Muhs *et al.*, Preclinical evaluation of inducible nitric oxide synthase lipoplex gene therapy for inhibition of stent-induced vascular neointimal lesion formation, *Hum. Gene Ther.* **14**(4), 375–383 (2003).
222. H. E. von der Leyen *et al.*, A prospective, single-blind, multicenter, dose escalation study of intracoronary iNOS lipoplex (CAR-MP583) gene therapy for the prevention of restenosis in patients with de novo or restenotic coronary artery lesion (REGENT I extension), *Hum. Gene Ther.* **22**(8), 951–958 (2011).
223. L. A. Brito *et al.*, *In vitro* and *in vivo* studies of local arterial gene delivery and transfection using lipopolyplexes-embedded stents, *J. Biomed. Mater. Res. A* **93**(1), 325–336 (2010).
224. D. Kim *et al.*, Facial amphipathic deoxycholic acid-modified polyethyleneimine for efficient MMP-2 siRNA delivery in vascular smooth muscle cells, *Eur. J. Pharm. Biopharm.* **81**(1), 14–23 (2012).
225. A. Lawrie *et al.*, Microbubble-enhanced ultrasound for vascular gene delivery, *Gene Ther.* 7(23), 2023–2027 (2000).
226. C. Christiansen *et al.*, Lack of an immune response to Albunex, a new ultrasound contrast agent based on air-filled albumin microspheres, *Int. Arch. Allergy Immunol.* **104**(4), 372–378 (1994).
227. V. Schafer *et al.*, Phagocytosis and degradation of human serum albumin microspheres and nanoparticles in human macrophages, *J. Microencapsul.* **11**(3), 261–269 (1994).
228. E. Markovsky *et al.*, Additive-free albumin nanoparticles of alendronate for attenuating inflammation through monocyte inhibition, *Nanomedicine* (*Lond*) **2**(4), 545–553 (2007).

229. N. K. Ibrahim *et al.*, Phase I and pharmacokinetic study of ABI-007, a Cremophor-free, protein-stabilized, nanoparticle formulation of paclitaxel, *Clin. Cancer Res.* **8**(5), 1038–1044 (2002).
230. F. D. Kolodgie *et al.*, Sustained reduction of in-stent neointimal growth with the use of a novel systemic nanoparticle paclitaxel, *Circulation* **106**(10), 1195–1198 (2002).
231. M. Yokoyama *et al.*, Characterization and anticancer activity of the micelle-forming polymeric anticancer drug adriamycin-conjugated poly(ethylene glycol)-poly(aspartic acid) block copolymer, *Cancer Res.* **50**(6), 1693–1700 (1990).
232. A. V. Kabanov *et al.*, Pluronic block copolymers as modulators of drug efflux transporter activity in the blood-brain barrier, *Adv. Drug Deliv. Rev.* **55**(1), 151–164 (2003).
233. T. Nakanishi *et al.*, Development of the polymer micelle carrier system for doxorubicin, *J. Control Release* **74**(1–3), 295–302 (2001).
234. M. C. Chen *et al.*, A nanoscale drug-entrapment strategy for hydrogel-based systems for the delivery of poorly soluble drugs, *Biomaterials* **30**(11), 2102–2111 (2009).
235. T. G. Kim *et al.*, Controlled release of paclitaxel from heparinized metal stent fabricated by layer-by-layer assembly of polylysine and hyaluronic acid-g-poly(lactic-co-glycolic acid) micelles encapsulating paclitaxel, *Biomacromolecules* **10**(6), 1532–1539 (2009).
236. V. P. Torchilin, Liposomes as targetable drug carriers, *Crit. Rev. Ther. Drug Carrier Syst.* **2**(1), 65–115 (1985).
237. N. van Rooijen and R. van Nieuwmegen Elimination of phagocytic cells in the spleen after intravenous injection of liposome-encapsulated dichloromethylene diphosphonate. An enzyme-histochemical study, *Cell Tissue Res.* **238**(2), 355–358 (1984).
238. H. D. Danenberg *et al.*, Systemic depletion of macrophages by liposomal bisphosphonates reduces neointimal formation following balloon-injury in the rat carotid artery, *J. Cardiovasc. Pharmacol.* **42**(5), 671–679 (2003).
239. H. D. Danenberg *et al.*, Liposomal alendronate inhibits systemic innate immunity and reduces in-stent neointimal hyperplasia in rabbits, *Circulation* **108**(22), 2798–2804 (2003).
240. H. Epstein-Barash *et al.*, Physicochemical parameters affecting liposomal bisphosphonates bioactivity for restenosis therapy: internalization, cell inhibition, activation of cytokines and complement, and mechanism of cell death, *J. Control Release* **146**(2), 182–195 (2010).
241. H. Epstein *et al.*, Preparation of alendronate liposomes for enhanced stability and bioactivity: *in vitro* and *in vivo* characterization, *Aaps J.* **10**(4), 505–515 (2008).
242. D. N. Patel and S. R. Bailey. Nanotechnology in cardiovascular medicine, Catheter. *Cardiovasc. Interv.* **69**(5), 643–654 (2007).
243. G. McDowell *et al.*, Nanotechnology for the treatment of coronary in stent restenosis: a clinical perspective, *Vasc. Cell* **3**(1), 8 (2011).
244. V. P. Torchilin, Structure and design of polymeric surfactant-based drug delivery systems, *J. Control Release* **73**(2–3), 137–172 (2001).
245. M. Chorny *et al.*, Biodegradable nanoparticles as drug delivery systems for parenteral administration, in *Tissue Engineering and Novel Delivery Systems.* CRC Press 393–422 (2003).
246. V. P. Torchilin, Targeted polymeric micelles for delivery of poorly soluble drugs, *Cell Mol. Life Sci.* **61**(19-20), 2549–2559 (2004).
247. E. Allemann *et al.*, *In vitro* extended-release properties of drug-loaded poly(DL-lactic acid) nanoparticles produced by a salting-out procedure, *Pharm. Res.* **10**(12), 1732–1737 (1993).

248. C. Sarisozen *et al.*, Optimization of prednisolone acetate-loaded chitosan microspheres using a 2(3) factorial design for preventing restenosis, *Drug Delivery* **17**(3), 178–186 (2010).
249. B. D. Klugherz *et al.*, Sustained Intramural Retention and Regional Redistribution Following Local Vascular Delivery of Polylactic-Coglycolic Acid and Liposomal Nanoparticulate Formulations Containing Probucol, *J. Cardiovasc. Pharmacol. Ther.* **4**(3), 167–174 (1999).
250. S. Banai *et al.*, Locally delivered nanoencapsulated tyrphostin (AGL-2043) reduces neointima formation in balloon-injured rat carotid and stented porcine coronary arteries, *Biomaterials* **26**(4), 451–461 (2005).
251. E. Cohen-Sela *et al.*, Alendronate-loaded nanoparticles deplete monocytes and attenuate restenosis, *J. Control Release* **113**(1), 23–30 (2006).
252. E. Cohen-Sela *et al.*, Nanospheres of a bisphosphonate attenuate intimal hyperplasia, *J. Nanosci. Nanotechnol.* **6**(9&10), 3226–3234 (2006).
253. E. Cohen-Sela *et al.*, Single and double emulsion manufacturing techniques of an amphiphilic drug in PLGA nanoparticles: Formulations of mithramycin and bioactivity, *J. Pharm. Sci.* **98**(4), 1452–1462 (2009).
254. L. Mei *et al.*, Modified paclitaxel-loaded nanoparticles for inhibition of hyperplasia in a rabbit arterial balloon injury model, *Pharm. Res.* **24**(5), 955–962 (2007).
255. U. Westedt *et al.*, Poly(vinyl alcohol)-graft-poly(lactide-co-glycolide) nanoparticles for local delivery of paclitaxel for restenosis treatment, *J. Control Release* **119**(1), 41–51 (2007).
256. F. Luderer *et al.*, Biodegradable sirolimus-loaded poly(lactide) nanoparticles as drug delivery system for the prevention of in-stent restenosis in coronary stent application, *J. Biomater. Appl.* **25**(8), 851–875 (2011).
257. J. C. Roth *et al.*, Cell vehicle targeting strategies. *Gene Ther.* **15**(10), 716–729 (2008).
258. V. Labhasetwar *et al.*, Arterial uptake of biodegradable nanoparticles: Effect of surface modifications, *J. Pharm. Sci.* **87**(10), 1229–1234 (1998).
259. G. M. Lanza *et al.*, Targeted antiproliferative drug delivery to vascular smooth muscle cells with a magnetic resonance imaging nanoparticle contrast agent: Implications for rational therapy of restenosis, *Circulation* **106**(22), 2842–2847 (2002).
260. J. W. Nah *et al.*, Artery wall binding peptide-poly(ethylene glycol)-grafted-poly(L-lysine)-based gene delivery to artery wall cells, *J. Control Release* **78**(1–3), 273–284 (2002).
261. J. M. Chan *et al.*, Spatiotemporal controlled delivery of nanoparticles to injured vasculature, *Proc. Natl. Acad. Sci. USA* **107**(5), 2213–2218 (2010).
262. J. M. Chan *et al.*, *In vivo* prevention of arterial restenosis with paclitaxel-encapsulated targeted lipid-polymeric nanoparticles, *Proc. Natl. Acad. Sci. USA* **108**(48), 19347–19352 (2011).
263. K. J. Widder *et al.*, Magnetically responsive microspheres and other carriers for the biophysical targeting of antitumor agents, *Adv. Pharmacol. Chemother.* **16**: 213–271 (1979).
264. M. Chorny *et al.*, Targeting stents with local delivery of paclitaxel-loaded magnetic nanoparticles using uniform fields, *Proc. Natl. Acad. Sci. USA* **107**(18), 8346–8351 (2010).

265. T. Rathel *et al.*, Magnetic stents retain nanoparticle-bound antirestenotic drugs transported by lipid microbubbles, *Pharm. Res.* **29**(5), 1295–1307 (2012).
266. M. Chorny *et al.*, Magnetically driven plasmid DNA delivery with biodegradable polymeric nanoparticles, *FASEB J.* **21**(10), 2510–2519 (2007).
267. M. Chorny *et al.*, Magnetically responsive biodegradable nanoparticles enhance adenoviral gene transfer in cultured smooth muscle and endothelial cells, *Mol. Pharm.* **6**(5), 1380–1387 (2009).
268. M. Chorny *et al.*, Adenoviral gene vector tethering to nanoparticle surfaces results in receptor-independent cell entry and increased transgene expression, *Mol. Ther.* **14**(3), 382–391 (2006).
269. M. Kempe *et al.*, The use of magnetite nanoparticles for implant-assisted magnetic drug targeting in thrombolytic therapy, *Biomaterials* **31**(36), 9499–9510 (2010).
270. S. V. Pislaru *et al.*, Magnetic forces enable rapid endothelialization of synthetic vascular grafts, *Circulation* **114**(1 Suppl), I314–318 (2006).
271. S. V. Pislaru *et al.*, Magnetically targeted endothelial cell localization in stented vessels, *J. Am. Coll. Cardiol.* **48**(9), 1839–1845 (2006).
272. A. Hofmann *et al.*, Combined targeting of lentiviral vectors and positioning of transduced cells by magnetic nanoparticles, *Proc. Natl. Acad. Sci. USA* **106**(1), 44–49 (2009).
273. P. M. Consigny *et al.*, Use of endothelial cells containing superparamagnetic microspheres to improve endothelial cell delivery to arterial surfaces after angioplasty, *J. Vasc. Interv. Radiol.* **10**(2 Pt 1), 155–163 (1999).
274. P. G. Kyrtatos *et al.*, Magnetic tagging increases delivery of circulating progenitors in vascular injury, *JACC Cardiovasc. Interv.* **2**(8), 794–802 (2009).
275. A. D. Grief and G. Richardson, Mathematical modelling of magnetically targeted drug delivery, *J. Magn. Magn. Mat.* **293**(1), 455–463 (2005).
276. B. B. Yellen *et al.*, Targeted drug delivery to magnetic implants for therapeutic applications, *J. Magn. Magn. Mat.* **293**(1), 647–654 (2005).
277. B. Polyak *et al.*, High field gradient targeting of magnetic nanoparticle-loaded endothelial cells to the surfaces of steel stents, *Proc. Natl. Acad. Sci. USA* **105**(2), 698–703 (2008).
278. M. Chorny *et al.*, Formulation and *in vitro* characterization of composite biodegradable magnetic nanoparticles for magnetically guided cell delivery, *Pharm. Res.* **29**(5), 1232–1241 (2012).

Chapter 3

Nanopreparations for Cancer Treatment and Diagnostics

Jayant Khandare, Shashwat Banerjee† and Tamara Minko‡,§*

**Maharashtra Institute of Pharmacy, MIT Campus, Pune, 411038, India*
†*Actorius Innovation and Research, 100 NCL Innovation Park Pune-411008, India*
‡*Department of Pharmaceutics, Rutgers, the State University of New Jersey, Piscataway, NJ, USA*
§*minko@rci.rutgers.edu*

Abstract

A number of nanopreparations for cancer treatment have been recently developed. Several advantages of nanoscale-based cancer therapeutics determine a continuous interest to these preparations. Formulation of conventional anticancer drugs as nanopreparations substantially changes their properties and allows for improving their water solubility, stability, anticancer activity, pharmacokinetics and release profile. It also permits the inclusion of several active components with different functionalities in one delivery system. Modern nanopreparations are capable of delivering hydro- and lipophilic anticancer drugs, nucleic acids, peptides, proteins and almost any other class of therapeutic agents. Such multicomponent multifunctional nanopreparations possess newly enhanced capabilities that exceed the potentials of each individual component applied separately. In addition to traditional chemotherapy, nanopreparations can be designed to be used in several other treatment protocols. They also can be employed as imaging agents to detect primary tumors and metastases. In addition, nanotechnology approach for anticancer formulations allows for the targeting of chemotherapeutic drugs and other active substances (nucleic acids, peptides, etc.) specifically to cancer cells therefore enhancing their anticancer activity and limiting severe adverse side effects upon healthy organs and tissues. The present work discusses examples of a few novel approaches to the formulations of nanopreparations for cancer treatment and imaging.

1. Introduction

Nanomedicine, as medical application of nanotechnology, has high potential to address important problems in cancer therapy and diagnostics, primarily by delivery of drug molecules with high pay load at the targeted site.[1] Delivery of anticancer drugs and imaging agents by nanoscale-based carriers allows for enhancing their uptake specifically by cancer cells, overcoming or suppressing drug resistance and limit adverse side effects upon healthy organs and tissues.[2–18] Polymeric and inorganic nanoparticles, liposomes, micelles, self-assemblies, lipid emulsions and prodrug conjugates have been recently defined as "nanopreparations."[19] This chapter presents an overview of a few nanopreparations implicated in cancer therapy and diagnostics. Particularly, the role of liposomes and polymeric prodrug conjugates of poly(ethylene glycol) (PEG) and hyperbranched polymers (e.g.dendrimers and polyglycerols) are highlighted.

Engineered anticancer nanopreparations with optimal physicochemical traits have offered substantial therapeutic advantages over their free non-bound counterparts. These nanotechnology-based medicines have been tested and used in oncology for more than three decades (Fig. 1). Additionally,

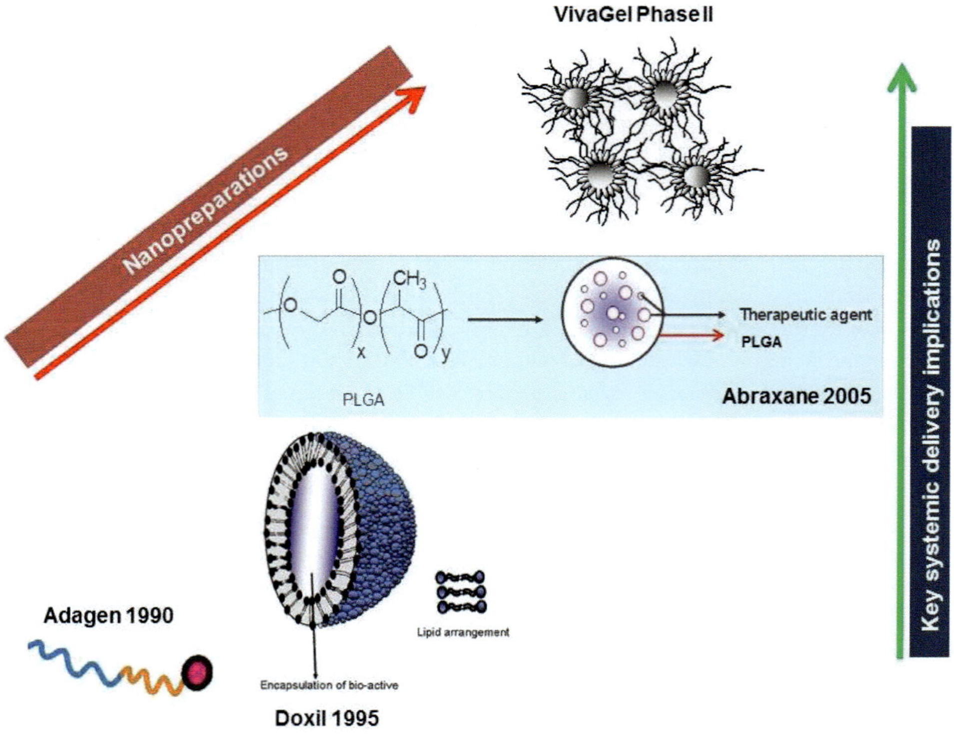

Fig. 1. Systemic delivery of key nanopreparations in clinic. VivaGel is currently in phase 3 clinical trials for the treatment of bacterial vaginosis.

many new nanopreparations are being currently evaluated for non-cancer therapeutics and tested in clinical trials.

At the cross roads of interdisciplinary sciences, the nanopreparations are continuously envisioned with specific functional roles including but not limiting to (i) enhancing the therapeutic activity, (ii) achieving long circulation time, (iii) increasing specificity at the cellular level and (iv) providing for a cancer specific tissue targeting. In general, nanotechnology has offered "tune ability" of biological relevant materials with improved drug delivery capacity due to controlled matter on the nanoscale.[20] The envisioned goals in nanomedicine are yet to be entirely achieved, particularly, for systemic delivery applications in oncological disorders. Therefore, the *Holy Grail* in nanomedicine is in designing advanced macromolecular nanopreparations and to translate them from laboratory to clinic. Nanopreparations for highly specific medical needs are at the nanometric scale. In this, the classical molecular bottom-up approach (1–100 nm) is joined by a top-down approach.[20] Bottom-up approach involves interactions between molecules or colloidal particles in assembling the discrete nanoscale structures in two and three dimensions. On the other hand, top-down approach involves lithographic methods to pattern nanoscale architectures. Designing advanced nanosized preparation is achieved by bringing together a cluster of functional and biocompatible multicomponents. Furthermore, the nanopreparations are envisioned to increase the cellular permeation, with high-resolution imaging ability, and to decrease systemic toxicity. More recently, carbon allotropes (e.g. carbon nanotubes (CNTs) and graphene) have emerged as delivery vehicles for bioactives (Fig. 2). Conversely, multicomponents of a nanopreparation would combine the specific functionality of each component to perform several tasks in parallel.[4,6,9,21] Such multicomponent nanosystems would offer great potential with enhanced efficiency and versatility for numerous applications in treating diseases or as diagnostics. However, the design of a multifunctional system with multiple components is often technically challenging due to the lack of robust chemical methodologies. Typically, the number of approaches that have been introduced to fabricate the nanopreparations consists of either encapsulation or by covalent attachment of bioactive molecules.

2. Liposomal Nanopreparations

Liposome-like structures were first described by Watkins *et al.* who studied diffusion of univalent ions across the lamellae of swollen phospholipids.[22] Later, such lipid assemblies containing microscopic phospholipid bubbles with a bilayered membrane structures (Fig. 3) were termed as "liposomes."[23]

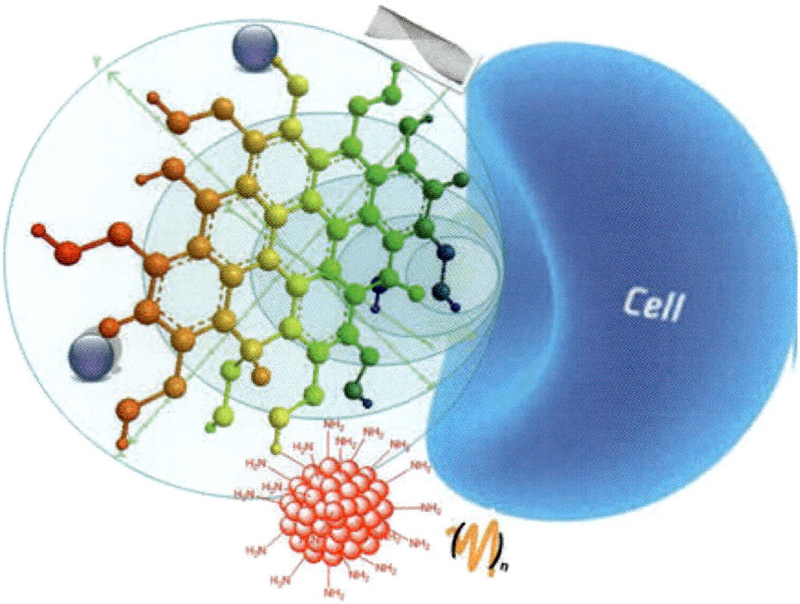

Fig. 2. Schematic illustration of nanopreparations for cancer treatment.

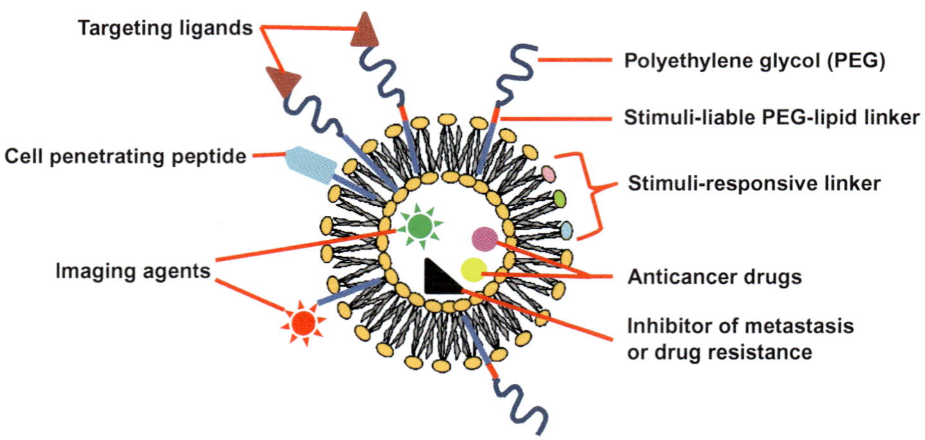

Fig. 3. Liposome nanopreparation with PEG polymeric chains and stimuli-labile PEG-lipid linker. Modified from Ref. 25.

Among various nanopreparations, liposomes represent a relatively simple form of nanovectors. Liposome-encapsulated doxorubicin (Doxil) was approved 18 years ago and is used for the treatment of Kaposi's sarcoma, breast cancer and refractory ovarian cancer. Liposomes accumulate in cancer neovasculature, thereby predominantly increasing the drug concentration at tumor sites.[24] Liposomal nanopreparations are now continuously implicated for clinical use, among other nanodelivery systems due to enhanced efficacy and decreased systemic toxicity.[23,25] Nonetheless, liposomal formulations are constantly under optimization for improved *in vivo* stability, high drug loading and stimuli-sensitive targeted delivery of the drug at the tumor site.[25]

Recently, Rangger *et al.* reported liposomal nanopreparation with cyclic arginyl–glycyl–aspartic acid (RGD) peptide targeting sequence (RLPs) using a combination of 1-palmitoyl-2-oleoyl-*sn*-glycero-3-phosphocholine, cholesterol, diethylenetriaminepentaacetic acid-derivatized lipids for radiolabeling, a PEG building block and a lipid-based RGD building block.[26] The authors studied the influence of PEGylation and RGD loading of different liposomal nanoparticles in relation to their targeting traits, and specific to image tumor-induced angiogenesis.

In addition to systemic route of delivery, liposomes were also used for a local topical delivery of different drugs and imaging agents, including transdermal and inhalation routes.[6,7,27–34] Several liposomal or other types of lipid formulations that can be used for pulmonary delivery were developed in the laboratory of Dr. Minko[5–7,18] These formulations can be dispensed by intratracheal injection or by inhalation and were used for the delivery of different drugs, antisense oligonucleotides (ASO) and small interfering RNA (siRNA). The preparations were extensively characterized *in vitro* and tested in animal models of lung cancer and idiopathic pulmonary fibrosis. A special nose-only exposure chamber was developed for inhalation of small laboratory animals and used for testing of these liposomal formulations (Figs. 4a–c). Animal testing on orthotopic mouse model of human lung cancer showed high antitumor efficacy of the proposed formulations (Fig. 4d). The suppression of tumor growth after inhalation treatment with the complex liposomal delivery system was documented by bioluminescence and ultrasound systems. It can be seen from Fig. 4d that doxorubicin (DOX) delivered by *i.v.* or inhalation partially decreased tumor growth (Fig. 4d, curve 2). Liposomal form of DOX was more effective after *i.v.* injection (Fig. 4d, curve 3) and especially after the inhalation delivery (Fig. 4d, curve 4). However, only inhalation treatment with liposomal systems containing a combination of DOX with ASO targeted to MRP1 and BCL2 mRNA were able to suppress tumor growth. These ASO were used to suppress multidrug resistance (MRP1) and antiapoptotic cellular defense (BCL2) in order to enhance the ability of the anticancer drug to kill tumor cells more efficiently. The "cocktail" of two systems (Fig. 4d, curve 5) each containing DOX and one type of ASO and one complex system containing the anticancer drugs and both types of ASO (Fig. 4d, curve 6) were able

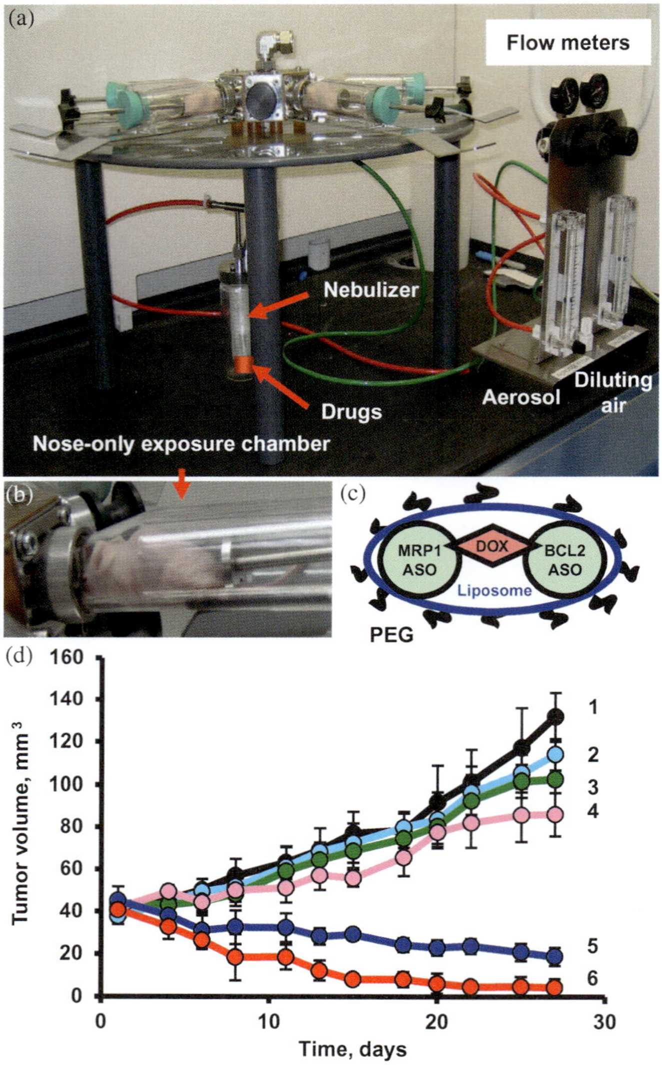

Fig. 4. Inhalation treatment of mice with orthotopic human lung tumor by DOX combined with inhibitors of pump and non-pump cellular resistance significantly decreases tumor size. Anorthotopic lung tumor model was created by intratracheal instillation of human A549 lung cancer cells in nude mice. (a) Installation for inhalation treatment. (b) A mouse inside the nose-only exposure chamber. (c) PEGylated liposomal drug delivery system containing anticancer drug (DOX) and suppressors of pump (ASO targeted to MRP1 mRNA) and non-pump (ASO targeted to BCL2 mRNA) resistance. (d) Changes in tumor volume after beginning of the treatment. The mice were treated on days 0, 3, 7, 11, 14, 17, 21 and 24. Means ± SD are shown. 1-Untreated tumor; 2-DOX (*i.v.*); 3-Lip-DOX (*i.v.*); 4-Lip-DOX (Inhal.); 5-cocktail: Lip-DOX-MRP1 ASO + Lip-DOX-BCL2 ASO (Inhal.); 6-complex delivery system: Lip-DOX-MRP1-BCL2 ASOs. Modified from Ref. 6.

to significantly inhibit the growth of lung tumors. It can be seen that the complex DDS containing DOX, MRP1 ASO and BCL2 ASO was more effective than the cocktail approach. It should also be stressed that inhalation delivery of anticancer drugs in combination with cell death enhancers limits their penetration to the systemic circulation and therefore prevented adverse side effects of this highly toxic system on healthy organs.[6] However, local delivery of such system does not prevent the damage of healthy lung cells. In order to avoid the toxicity to healthy lung cells, additional active targeting can be employed. Dr. Minko pioneered in the use of peptide similar to luteinizing-hormone-releasing hormone (LHRH) as cancer targeting moiety for many different types of nanoparticles for treatment of different cancers.[2–4,8,11,15–18] It was found that LHRH receptors are over expressed in plasma membrane of many types of cancer cells. In contrast, they do not express in healthy visceral organs and their expression in healthy tissues from reproductive organs is significantly less when compared with cancerous tissues. An inclusion of LHRH peptide as a cancer targeting moiety in lipid nanoparticles led to their preferential accumulation in lung cancer cells after inhalation leaving healthy lung tissues intact.[18]

3. Polymeric Nanopreparations

The utility of polymers in medicine has brought a paradigm change in delivery ability of many drug molecules and other bioactive compounds.[35–37] Polymeric drug preparations that are used for treatment and diagnostics can be commonly termed as "polymer therapeutics." Translation of biopolymers from laboratory to clinic has been very enduring, though the primary focus was to deliver anticancer drugs and other biologically active substances. In this context, the objectives of designing polymeric nanopreparation is primarily driven to enhance the efficacy of the drug by (i) increasing the aqueous solubility (relevant for many drugs with low aqueous solubility), (ii) improving physiological stability against degrading enzymes or reducing the uptake by reticulo-endothelial system (RES) and (iii) targeting a drug to its specific site of action (in case of anticancer therapeutics — targeting to cancer cells).[38–40] Nanopreparations have immense potential in treating metastatic cancer with increased capacity to carry molecular payloads to the major sites of metastasis, e.g. lungs, liver and lymph nodes. They are currently used for targeting specific cells in cancer and extended for early detection of cancer.

The nanopreparations that entered the clinical trials were primarily designed with the utility of PEG (as PEG–prodrug conjugates), which were mostly used for the delivery of anticancer drugs, namely, camptothecin

(CPT), SN38 and irinotecan (Table 1). However, the first PEG-based preparations were designed for the delivery of asparaginase, and subsequently were also evaluated for treatment of other than cancer diseases, e.g. infectious diseases (e.g. PEG-interferons), and age-related diseases, including macular degeneration and arthritis.[39]

The perspective of newer forms of polymer nanotherapeutics in clinical development continues to grow.[20] Interestingly, the early developments of the linear polymers (e.g. PEG, HPMA, etc.) in the form of longer polymer chains, was then followed by the introduction of defined polymeric topologies, especially in the form of highly branched dendritic architectures. For

Table 1. Various PEG prodrug preparations in clinic or in development. Modified from Ref. 39.

Brand	Composition	Therapeutic use	Route of administration	Stage
PEG–protein conjugates				
Oncaspar®	PEG-asparaginase	Acute lymphocytic leukemia	*i.v./i.m.*	Market
Adagen®	PEG-adenosine deaminase	Severe combined immune deficiency syndrome	*i.m.*	Market
Somavert®	PEG-HGH antagonist	Acromegaly	*s.c.*	Market
PEGIntron®	PEG-interferon alpha 2b hepatitis C	Hepatitis C	*s.c.*	Market
NeulastaTM®	PEG-rhGCSFchemotherapy	Chemotherapy-induced neutropenia	*s.c.*	Market
Pegasys®	PEG-interferon alpha 2a hepatitis C	Hepatitis C	*s.c.*	Market
CimziaTM®	PEG-anti-TNF Fab	Rheumatoid arthritis, Crohn's disease	*s.c.*	Market
Mircera®	PEG-EPO	Anemia associated with chronic kidney disease	*i.v./s.c.*	Market
Puricase®	PEG-uricase	Gout	*i.v.*	Market
Macugen®	PEG-aptamer	Age-related macular degeneration	*intravitreal*	Market
PEG–drug conjugates				
NKTR-102	PEG-irinotecan	Cancer— metastatic breast	*i.v.*	Phase II
PEG-SN38	Multiarm PEG-camptothecin derivative	Cancer — various	*i.v.*	Phase II
NKTR-118	PEG-naloxone	Opioid-induced constipation	oral	Phase II

s.c., subcutaneous; *i.v.*, intravenous; *i.m.*, intramuscular.

example, in milieu of newer nanopreparations, the hyperbranched polymers (e.g. PAMAM dendrimers or polyglycerols dendrimers) have been considered as unique platforms due to their nanosize, monodisperisty and degree of branching, facilitating the conjugation (or even complexation) of drugs and solubilizing agents.[41] The dendrimers and nanopreparations with other architectures are evaluated for the following traits: (i) biocompatibility, (ii) non-immunogenicity and (ii) targeting to drug action side. Various nanopreparations, including quantum dots, dendrimers, polymer–drug conjugates, polymer protein conjugates, polyplexes and polymeric micelles, have been compared schematically with respect to their size and general utility (Fig. 5).[15,20]

3.1. *Nanopreparations for passive targeting*

It is known that macromolecules passively can accumulate in solid tumors.[38,42,43] Initially, this phenomenon was not well understood for several years, and later proved to be a potential biological target in achieving tumor-selective drug delivery. The existence of the enhanced permeability and retention (EPR) effect was first experimentally confirmed in laboratories of Maeda and Jain for

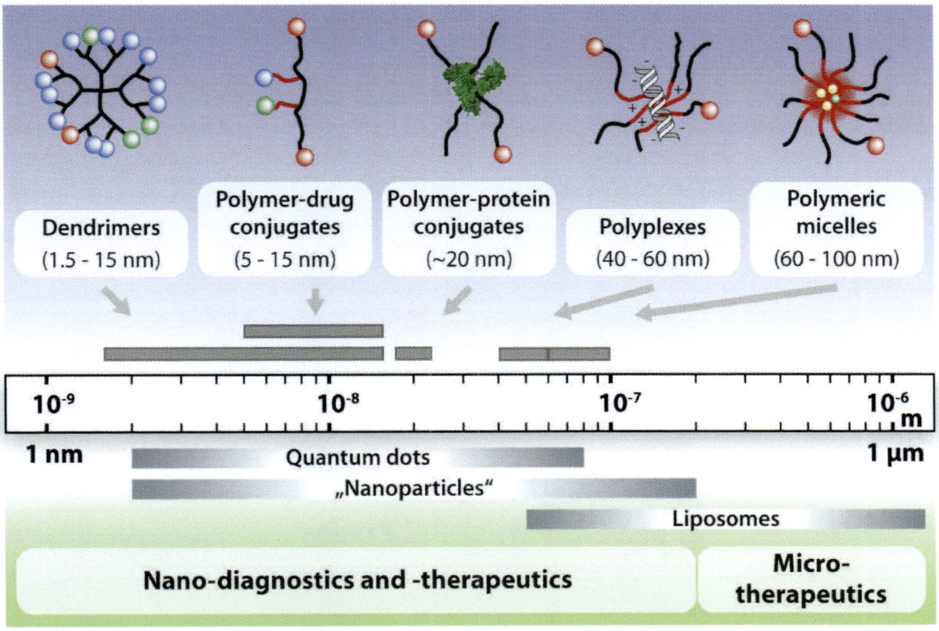

Fig. 5. Various nanopreparation architectures with comparative size used for diagnostics and cancer therapeutics. Modified from Refs 10, 15 and 20.

macromolecular drug delivery systems.[44–46] The EPR effect creates the basis for so-called "passive" tumor targeting of macromolecular anticancer drugs. The EPR effect leads to an increased "passive" accumulation of macromolecules in the tumor tissue and is predominantly governed by the hyperpermeability of tumor vasculature.[20] Such enhanced permeability allows selective extravasation of macromolecules into the tumor, while poor lymphatic drainage increases the retention of macromolecules in the tumors. This leads to an increased retention of macromolecules in the tumor and inflamed area of body.[47–50] The EPR effect now is embarked as one of the prudent anticancer prodrug drug delivery strategies.

Nanopreparations that utilize passive tumor targeting vary in physicochemical traits to carry small drug molecules or biomacromolecules such as proteins or nucleic acids (ASO or siRNA).[5,6,12–14,51–53] Figure 6 represents the major types of nanopreparations for the treatment of metastatic cancer with different strategies from the point-of-view of the cell.[54] The ability of

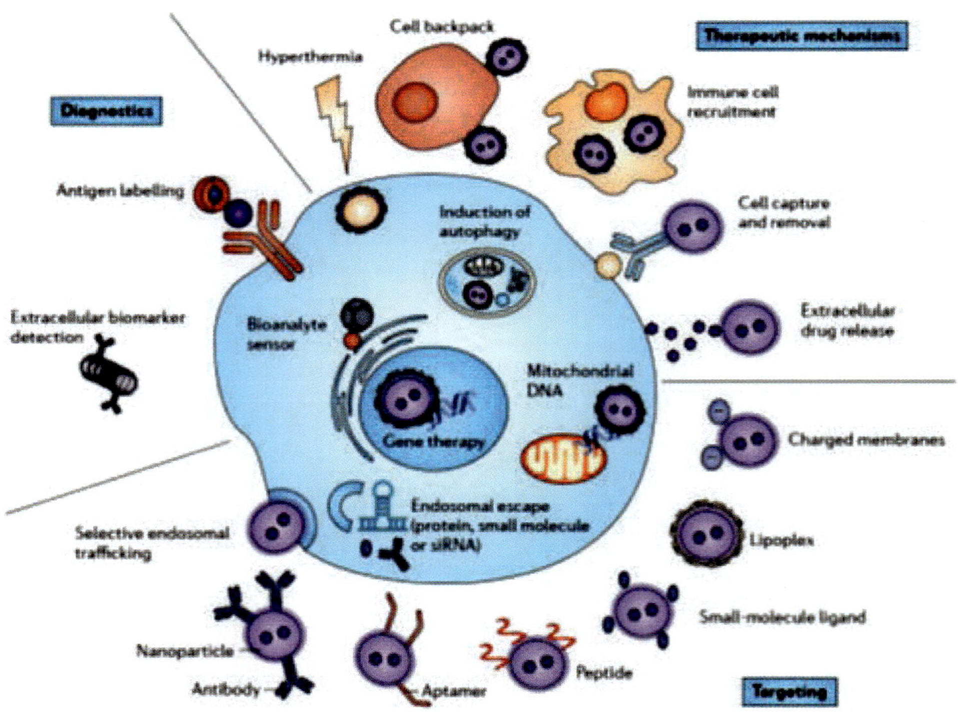

Fig. 6. Nanopreparations strategies from the point-of-view of the cell. The schematic shows the multiple strategies to achieve targeting, utility of the over expressed receptors for possible therapeutic mechanisms to attain the interactions with cancer cell. Reproduced with permission from Ref. 54.

nanoparticles to target cancer cells (also referred to as secondary targeting) and to influence their uptake into specific cellular compartments (referred to as tertiary targeting) is now well defined. Overall, passive targeting increases the concentration of the drug in the tumor and therefore "passively" forces the polymeric drug to enter the cells by means of the concentration gradient between the intracellular and extracellular spaces. However, passive targeting is not very efficient method to achieve the final objective in cancer therapy. The more efficient way to provide the targeting is by "active targeting".[9,38]

3.2. *Active cancer targeting*

The internalization of the nanopreparations into the cancer cells is efficiently achieved by simple diffusion for low molecular weight drugs or receptor-mediated endocytosis for high molecular weight preparations.[4] Large molecules enter the cell through membrane-bound vesicles by the process called as "invagination" and/or the plasma membrane process termed as "endocytosis" (Figs. 7 and 8). This process has two mechanistic features: (i) phagocytosis (the uptake of large particles) and (ii) pinocytosis (the uptake of fluid and solutes).[39] Phagocytosis is an active and highly regulated process which involves specific cell-surface receptors and signaling cascades mediated by

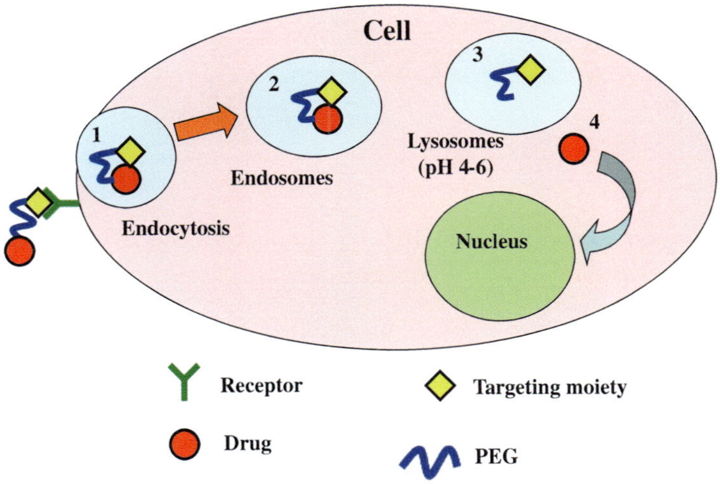

Fig. 7. Receptor-mediated endocytosis. (**1**) Polymer-conjugated drug is internalized by tumor cells through receptor-mediated endocytosis following ligand–receptor docking; (**2**) transport of DDS in membrane-limited organelles; (**3**) fusion with lysosomes; (**4**) the drug will usually be released intracellularly on exposure to lysosomal enzymes or lower pH (pH 6.5–<4.0). Modified from Refs. 38 and 39.

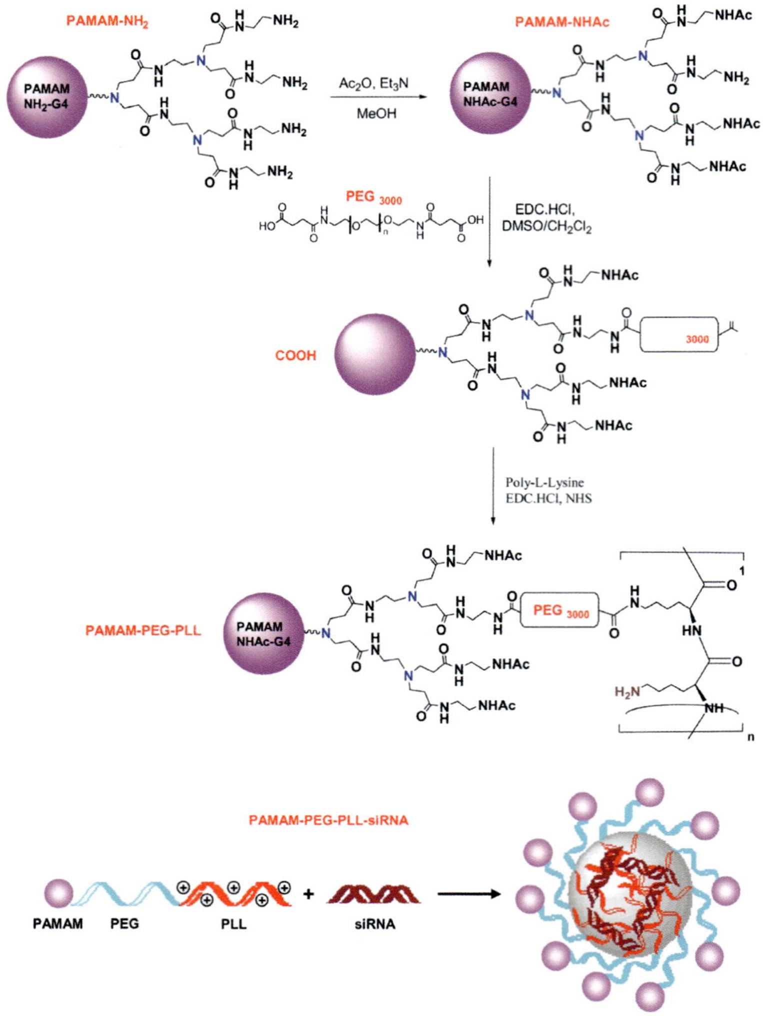

Fig. 8. Synthesis of triblock PAMAM–PEG–PLL nanocarrier. Reproduced with permission from Ref. 13.

Rho-family GTPases.[55] When the drug is chemically conjugated to the polymer, it is either hydrolyzed by an acid-sensitive linker or by enzymatic process and drug molecule is further trapped by the tumor for longer period of time.

Active targeting of nanocarriers to cancer cells can be achieved by incorporating of special targeting moieties specific to cancer cells. Such targeting enhances the internalization of the carrier and entire system by cancer cells and limits their accumulation in healthy cells and tissues. This limits adverse side effects on healthy tissues and substantially facilitates their penetration inside cancer cells. Internalization of cancer-targeted systems in most cases is achieved by active receptor-mediated endocytosis, which is significantly more effective when compared with "simple" endocytosis. Such shift in internalization mechanisms substantially increases the ability of relatively large (when compared with small drug molecules) nanopreparations to cross the plasma membrane of cancer cells and finally enhance their specific toxicity. For details, mechanisms of this phenomena and different approaches to active tumor targeting, the reader is referred to the corresponding comprehensive review.[9]

Nanopreparations of next generation have been proposed for delivery of drugs, genes and other bioactive substances. Many of these preparations remain to be highly specific in delivering active components. Minko *et al.* developed *in vivo* gene knock down delivery system to eliminate the following limitations of naked nucleic acids: (i) low cellular uptake, (ii) decreased endosomal escape, (iii) major liver and renal clearance, (iv) enzymatic degradation in the blood and extracellular environment and (v) inefficient gene silencing.[2–4,6,12–14,16–18,38] The authors reported design, synthesis and evaluation of a triblock multifunctional nanopreparation consisting of PAMAM–PEG–poly-l-lysine (PLL) triblock polymer (Fig. 8). The entities in the triblock offered three functionalities, namely (i) PLL supports cationic primary amine groups for electrostatic interactions with negatively charged siRNA, (ii) PAMAM dendrimer imparts tertiary amine groups for proton sponge effect and (iii) PEG deliberates nuclease stability with blood serum.

Haag group reported size-dependent delivery of polyglycerols using non-malignant normal endothelia cells (HUVEC).[56] The intracellular localization and cellular uptake of HPG–dye conjugates showed localization in perinuclear region using different human cells (Fig. 9). Upon staining of the cytoskeleton component actin (green), HPG–dye (red) was primarily localized inside of the cell after incubation for 4 h both in the epidermal (A431) and lung (A549) tumor cell lines, as well as in the non-malignant normal endothelia cells (HUVEC). The majority of HPG–dye conjugate were colocalized with the late endosome marker after incubation of 4 h, indicating the uptake and metabolism of HPG–475 dye conjugate by endocytosis.

An original engineering approach for the preparation of cancer-targeted nanopreparations for the delivery of siRNA was recently proposed in the laboratory of Dr Minko.[16] The complex multifunctional nanopreparations were fabricated in four distinct steps (Fig. 10a). Initially, complexes of poly(propyleneimine) (PPI) generation five dendrimers were formed with siRNA. In the step 2, the resulting complexes were caged with dimethyl-3-3′-dithiobispropionimidate-HCl (DTBP). Such caging prevented degradation

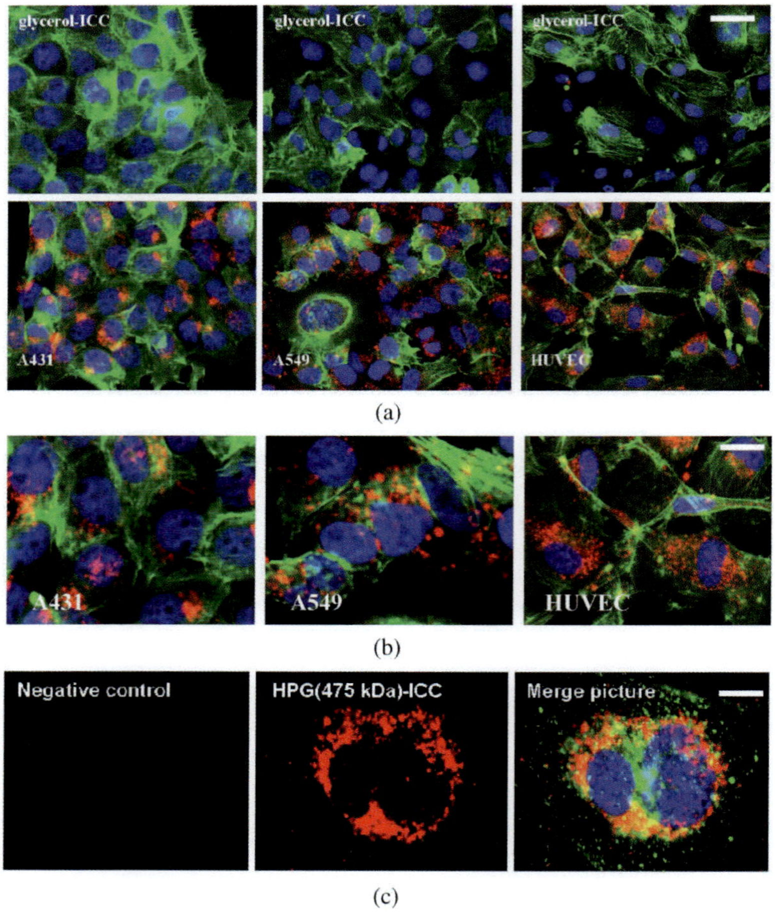

Fig. 9. Cellular uptake of dendritic polyglycerol. (a) Different human cancer cell lines were incubated with ICC-conjugated glycerol or PG-475 kDa for 4 h (red). Actin was stained with Phalloidin conjugated with Alexa488 (green). Nuclear staining with DAPI (blue). (b) Confocal microscopy of HUVEC cells cultured with PG (475 kDa)-ICC (red) for 4 h. Double staining with a late endosome marker (green). **(c)** Colocalization (yellow) in the merge picture. Optical sectioning in the Z-axis. Section of the nucleus level. Reproduced with permission from Ref. 56.

of nanoparticles and siRNA under the harsh action of biological fluids during their journey to the site of action increasing the stability of nucleic acids and entire particles. Later, the resulting caged nanoparticles were covered with PEG conjugated with a cancer targeting moiety (LHRH peptide). Resulting complex multifunctional nanopreparations demonstrated a very favorable organ distribution of dendrimer (Fig. 10b) and delivered siRNA (Fig. 10c) leading to the preferential accumulation of siRNA in tumor.

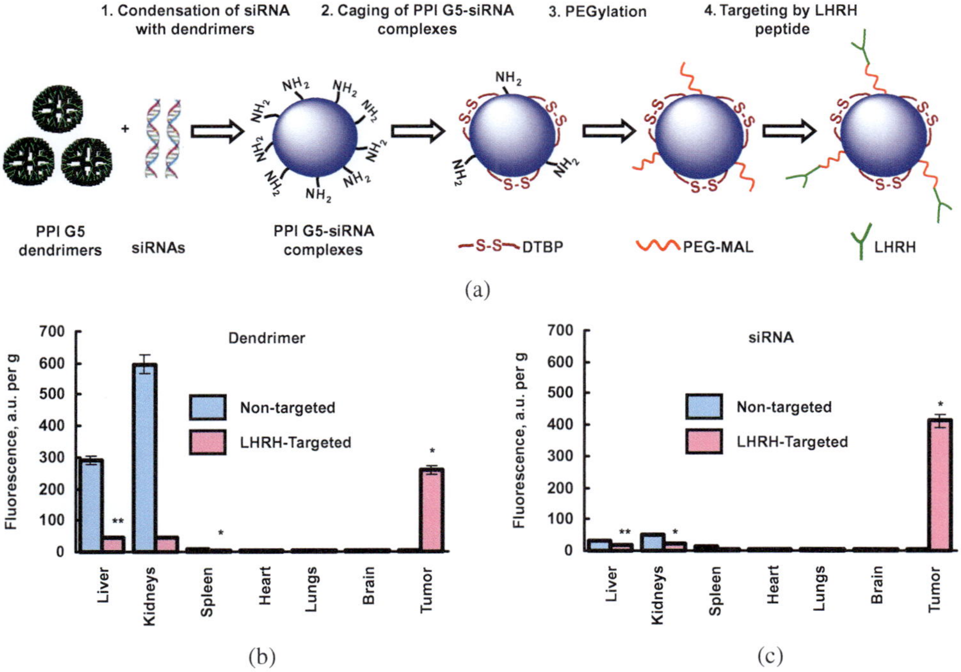

Fig. 10. Engineering approach for the preparation of tumor-targeted, stable nanoparticles for the siRNA delivery. **(a)** Synthesis of dendrimer-based tumor-targeted complex multifunctional delivery system. **(b, c)** Organ and tumor content of non-targeted and LHRH-targeted PPI dendrimers and siRNA. Human A549 cancer cells were transfected with luciferase and injected subcutaneously into the flanks of nude mice. Dendrimers and siRNA were labeled with near-infrared Cy5.5 dye and DY-547 (siGLO red), respectively. PPI–siRNA dendrimers were injected into the mice. The distribution of labeled dendrimers and siRNA was analyzed in live anesthetized animals 72 h after injection using IVIS Xenogen imaging system. Tumor and organs were excised and their bioluminescence and fluorescence were registered and processed by the imaging system. The bottom panel demonstrates average concentration per gram of organ weight of labeled dendrimers **(b)** or siRNA **(c)**. Means ± SD are shown. $^*P < 0.05$ when compared with non-targeted complex. Modified from Ref. 16.

4. Carbon Nanopreparations

Nanotechnology is now a well-established multidisciplinary field primarily involving different fields of engineering, biology, physics and chemistry. The primary focus of this field has been on designing nanovectors for targeted imaging and therapeutics. Nanopreparations are now a more studied discipline of science involving the application of nanosize within the size range of 1–1000 nm that can facilitate the understanding of disease pathophysiology, nano-imaging, nanomedicines and nanodiagnostics.[57] Furthermore, such nanovectors have emerged in bimolecular identification from tissue and serum samples.[24] This chapter will cover few novel recent nanovectors for targeting cancer cells through their contrast imaging and thermal properties. The fate of these nanovectors in biology is implicated for their utility, with safety and biocompatibility.

Carbon nanovectors, such as carbon nanotubes (CNTs), and more recently graphene, have generated much interest in their potential medical applications.[58] Although CNTs have been widely investigated for biosensing, drug delivery and molecular imaging applications, graphene and its oxidation form, graphene oxide (GO), have been relatively less explored in the biomedical field.[59,60] As produced, CNTs and graphenes lack dispersion ability in aqueous medium, which is considered as the foremost critical aspect for any biological applications. Therefore, for any type of biological application, the CNT and graphene surface needs to be chemically modified. Different strategies have been developed to make CNTs and graphenes compatible with the biological milieu. The two main methodologies are based on the non-covalent coating of the surfaces with amphiphilic molecules (for example, lipids and polymers), and the covalent functionalization of the surfaces by grafting various chemical groups directly onto the backbone (Figs. 11 and 12).[61,62]

4.1. *Carbon nanotube*

CNTs have attracted particular attention as delivery vehicles of biologically relevant molecules due to their unique physical, chemical and physiological properties. CNTs are seamless cylindrical structures containing single or multiple concentric graphene sheets. One of the key benefits of CNTs implications in biomedical applications is that they are internalized easily by cancer cells. CNTs have been investigated for application in cancer imaging. For instance, a multicomponent fluorescent CNT nanosystem (Fe_3O_4–CNT–PEG–FITC) encompassing covalently conjugated magnetite nanoparticles (Fe_3O_4), PEG and fluorophore fluorescein isothiocyanate (FITC) to the CNT surface (Fig. 13a) was used for the delivery of fluorescence dye (FITC) into the cytoplasm.[63] This multicomponent CNT demonstrated enhanced delivery of FITC inside the cells when compared with free non-bound dye (Fig. 13b). The internalized CNT-based nanosystem was localized mostly in the cytoplasm and at perinuclear region. In contrast, free FITC showed weak fluorescence and was mainly present in the cell cytoplasm. Furthermore, multiwalled CNTs can be used for photothermal ablation therapy owing to their release of vibrational energy upon near-infrared light exposure.[64] However, potential toxicity associated with CNTs must be addressed before clinical translation.

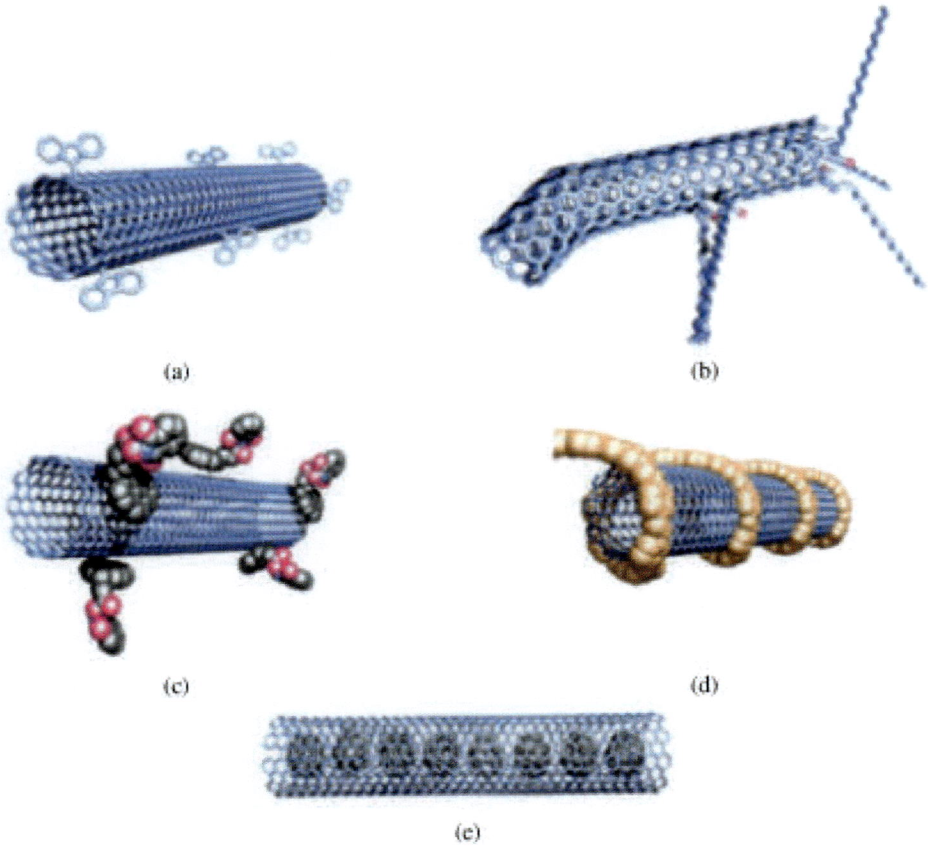

Fig. 11. Strategies for chemically functionalizing SWNTs. (a) Covalent sidewall chemistry. (b) Covalent chemistry at defects or open ends. (c) Non-covalent surfactant encapsulation. (d) Non-covalent polymer wrapping. (e) Molecular insertion into the SWNT interior. Reproduced with permission from Ref. 61.

4.2. *Graphene*

Graphene, due its unique 2D shape, together with fascinating physical and chemical properties, has aroused substantial attention in biomedicine. Graphene is a form of carbon consisting of single-atom-thick sheets of hexagonally arranged sp^2-bonded carbon atoms. Functionalized graphene has been used for cell and tumor imaging, as a novel carrier for drug and gene delivery, as well as photothermal therapy of cancer. More recently, inorganic nanoparticles have been grown/conjugated on the surface of graphene to obtain graphene-based nanocomposites for multimodal bioimaging, and imaging guided cancer therapy.[65] For example, a multicomponent graphene-based nanostructured

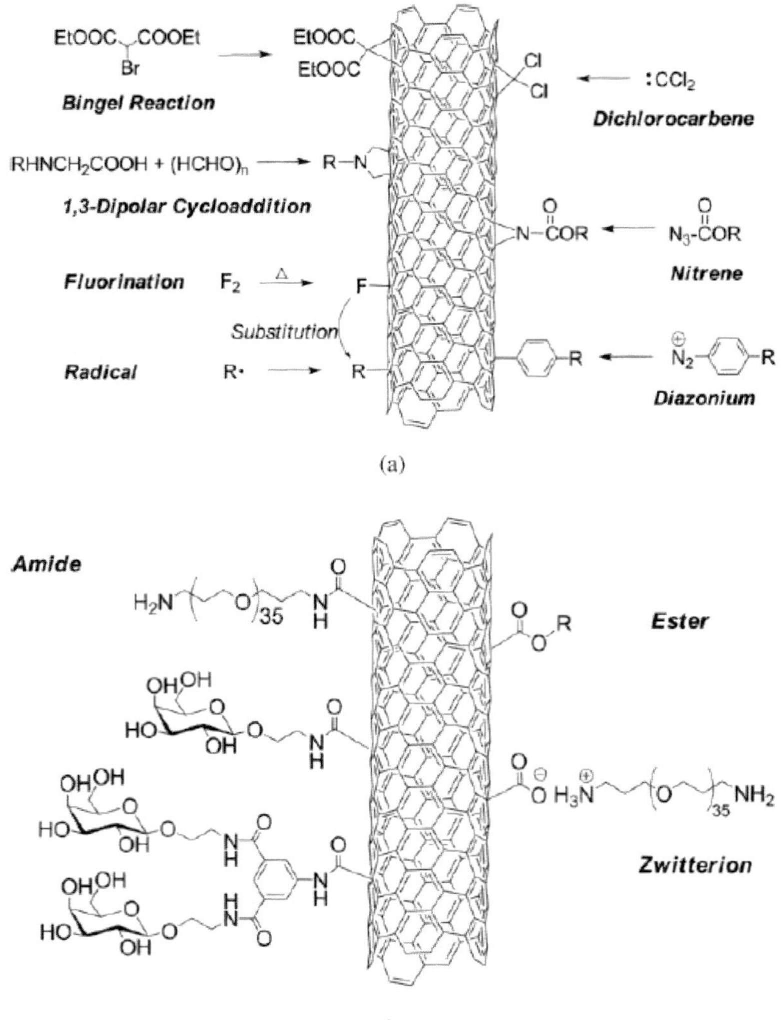

(b)

Fig. 12. Modifications of nanotubes. (a) Covalent addition reactions on the sidewall of carbon nanotubes. (b) Reactions of targeting carboxylic acids (derived from nanotube surface defects). Reproduced with permission from Ref. 62.

system (GO–G4–Fe–Cy) consisting of covalently attached Fe_3O_4 nanoparticles, PAMAM–G4–NH_2 (G4) dendrimer and flurophore cyanine 5 on the surface exhibited successful uptake by MCF-7 breast cancer cells (Fig. 14). The system showed bright and stable fluorescent signal with varied cellular distribution kinetics compared to free cyanine 5.0.[21] Despite these exciting results demonstrated by graphene in biomedicine, there may still be a long way

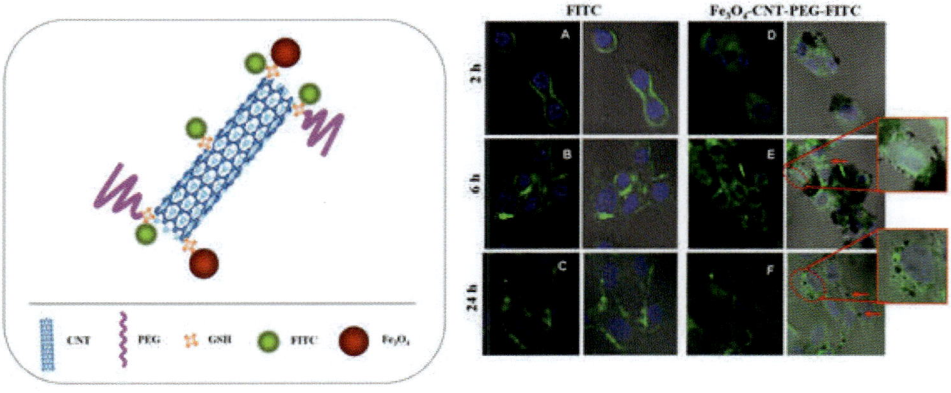

(a) (b)

Fig. 13. Multicomponent nanotube-based delivery system. **(a)** Schematic illustration of multicomponent CNT-based nanosystem for cellular imaging; **(b)** confocal laser scanning microscopic images of MCF7 cells incubated with Fe_3O_4–CNT–PEG–FITC nanosystem for 2, 6 and 24 h. **(a)** Left: Merged image of the nuclei stained with blue DAPI and FITC or Fe_3O_4–PEG–FITC–CNT. Right: Merged image of the nuclei stained with blue DAPI, FITC or Fe_3O_4–PEG–FITC–CNT and differential interference contrast (DIC). Reproduced with permission from Ref. 63.

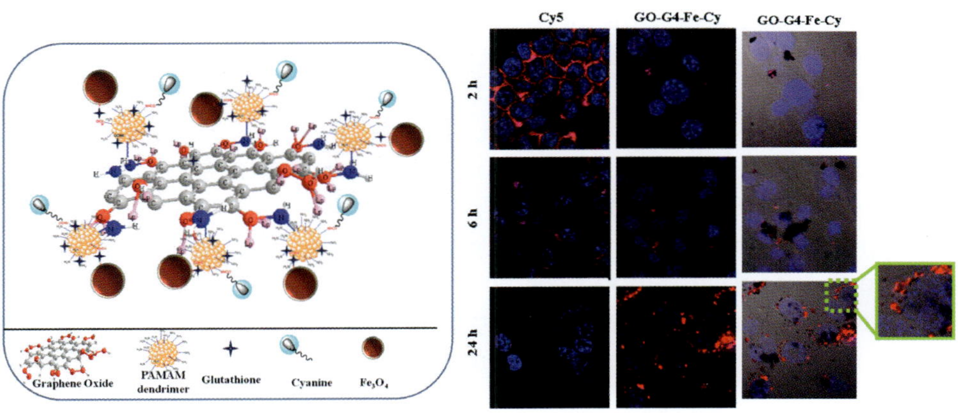

(a) (b)

Fig. 14. Graphene-based delivery system. (a) Schematic illustration of multicomponent magneto-dendritic GO nanosystem for cellular imaging; (b) confocal laser scanning microscopic images of MCF-7 breast cancer cells and the internalization kinetics comparison of cells incubated with free Cy, and GO–G4–Fe–Cy systems for 2, 6 and 24 h. Right panel: Merged image of the nuclei stained with blue DAPI and GO–G4–Fe–Cy and DIC. Reproduced with permission from Ref. 21.

to go before it can be really applied in a clinic. The potential toxicity in biological systems of graphene and its derivatives has generated growing debate in recent years. The risk–benefit balance for these materials will ultimately determine their clinical fate.

5. Iron Oxide Nanoparticle

Iron oxide nanoparticles have the potential to significantly impact and revolutionize current clinical diagnostic and therapeutic techniques. Iron oxide nanoparticles such as magnetite have been widely used for *in vivo* biomedical applications, including magnetic resonance imaging (MRI) contrast enhancement, tissue-specific release of therapeutic agents, hyperthermia and magnetic field-assisted radionuclide therapy. Iron oxide nanoparticles have received particular attention owing to their proven biocompatibility and biodegradability. Iron from degraded nanoparticles is used in the body's natural iron stores such as hemoglobin in red blood cells. These nanoparticles develop superparamagnetism at the nanometerscale, as each particle becomes a single magnetic domain that is free to rotate at room temperature.

After synthesis, unmodified iron oxide nanoparticles require surface coating for *in vivo* applications. These surface coatings is used to (i) prevent nanoparticle agglomeration and avoid opsonization, (ii) provide chemical handles for the conjugation of drug molecules, specific binding sites or other functional groups and (iii) retards non-specific cell interactions. The coating strategies can roughly be divided into two major groups: (i) coating with organic shells, such as polymers (PEG, dextran, chitosan, PEI and phospholipids) and surfactant and/or (ii) coating with inorganic components, such as silica, carbon, precious metals (e.g. Ag, Au) or oxides.[66]

Early clinical trials of epirubicin-loaded colloidal iron oxide nanocarriers directed toward solid tumors have demonstrated successful accumulation in the target site in about half the patients in the study.[67,68] These magnetic iron oxide nanocarriers were also shown to be well tolerated by patients. Unfortunately, several problems have been identified with this technique. First, the difficulty in scaling up from animal models due to limited field penetration of commercial magnets. Second, the magnetic nanocarriers can cause embolization of the blood vessels in the region once they are accumulated in the target sites.

Another important application of iron oxide nanoparticle is isolation of circulating tumor cell (CTC). The techniques rely on targeting moiety bound to the iron oxide nanoparticles for capturing CTCs (Fig. 15). For example, a biofunctionalized MDNS designed by combining multiple components such

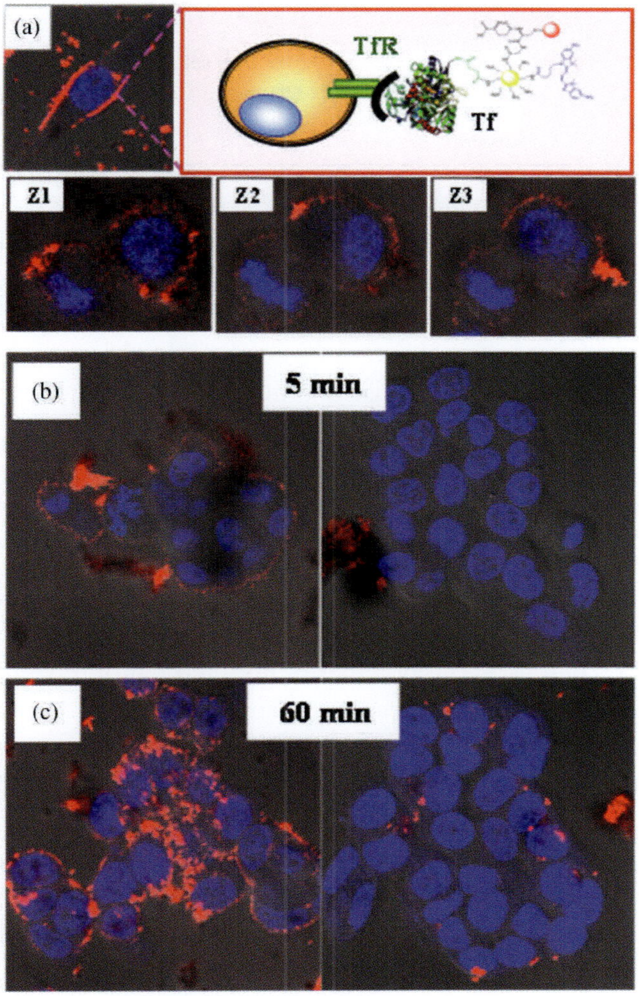

Fig. 15. Cellular uptake of a biofunctionalized magneto-dendritic nanosystem (MDNS). Images are simultaneously taken in Cy5.5 (MDNS), DAPI (nucleus) and brightfield (cell outline) channels to compare the interaction of TfR+ HCT116 cells with MDNS and a Tf-lacking variant of MDNS (Tf–). **(a)** Magnified image of a cell after 5 min of incubation shows the localization of MDNS particles on the cell membrane (left) and confirms the hypothesis of specific interaction of Tf with the TfRs (right). **(b)** Rapidity and specificity of the MDNS–cell interaction are confirmed by the images of Tf+ MDNS particles attached to the HCT116 cells after just 5 min of exposure. In contrast, there are hardly any Tf-lacking particles present on the cell surface. **(c)** Target specificity of MDNS interaction remains even after long exposure. After 60 min of exposure to both types of particles, only the Tf+ MDNS are present in large numbers on the cells. Reproduced with permission from Ref. 69.

as transferrin (Tf), iron oxide (Fe_3O_4) nanoparticles, cyanine 5 NHS (Cy5) dye, fourth generation (G4) dendrimers and glutathione (GSH) was able to capture transferrin receptor (TfR) overexpressed cancer cells from an artificial CTC-like suspension.[69] The MDNS platform exhibited (i) rapid capture (~5 min) of TfR-overexpressing (TfR+) cancer cells at clinically relevant concentrations (~1 CTC per 105 blood cells), (ii) magnetic isolation of the captured cells owing to the presence of Fe_3O_4 nanoparticles, (iii) high-resolution fluorescence imaging of the isolated cells due to the presence of Cy5 and (iv) enhanced aqueous dispersibility (and hence bioavailability) of the nanoparticles resulting from the presence of dendrimers. The Tf-based MDNS technology showed promise for future development into a CTC capture platform targeted to TfR+ cancer cells.

6. Future Direction

Many new chemical entities fail in preclinical studies and during clinical phases due to their low solubility, stability and high toxicity. Numerous newly developed molecules exhibit short half-life and limited specific accumulation in cancer cells. Therefore, the polymeric-modified nanopreparations are considered promising approach in treating many diseases, e.g. cancer, cardiovascular diseases, bacterial infections, viral infections. Nanopreparations offer unique versatility to enhance the therapeutic benefit with reduction in untoward side effects and reduction in drug. They also offer to increase solubility of drugs, improve body residence and site-specific cellular uptake; to enhance accumulation in malignant tissues by active or passive targeting; to eliminate undesirable body interactions and adverse side effects and to allow further biodegradation.

References

1. K. Bourzac, *Nature* **S58** (2012).
2. S. S. Dharap and T. Minko, *Pharm. Res.* **889** (2003).
3. S. S. Dharap, B. Qiu, G. C. Williams, P. Sinko, S. Stein and T. Minko, *J. Control Release* **61** (2003).
4. S. S. Dharap, Y. Wang, P. Chandna, J. J. Khandare, B. Qiu, S. Gunaseelan, P. J. Sinko, S. Stein, A. Farmanfarmaian and T. Minko, *Proc. Natl. Acad. Sci. USA* **12962** (2005).
5. O. B. Garbuzenko, M. Saad, S. Betigeri, M. Zhang, A. A. Vetcher, V. A. Soldatenkov, D. C. Reimer, V. P. Pozharov and T. Minko, *Pharm. Res.* **382** (2009).
6. O. B. Garbuzenko, M. Saad, V. P. Pozharov, K. R. Reuhl, G. Mainelis and T. Minko, *Proc. Natl. Acad. Sci. USA* **10737** (2010).
7. V. Ivanova, O. B. Garbuzenko, K. R. Reuhl, D. C. Reimer, V. P. Pozharov and T. Minko, *Eur. J.Pharm. Biopharm.* **335** (2012).

8. J. J. Khandare, P. Chandna, Y. Wang, V. P. Pozharov and T. Minko, *J. Pharmacol. Exp. Ther.* **929** (2006).
9. T. Minko, Receptor mediated delivery systems for cancer therapeutic, In *Advances in Delivery Science and Technology*, R. Siegel, S. Juergen and R. M (Eds.), New York: CRS Press (2012), p. 329.
10. T. Minko, J. J. Khandare, A. A. Vetcher, V. A. Soldatenkov, O. B. Garbuzenko, M. Saad and V. P. Pozharov, Multifunctional nanotherapeutics for cancer (Chapter 10), In *Multifunctional Pharmaceutical Nanocarriers*, V. P. Torchilin (Eds.), New York: Springer (2008), p.
11. T. Minko, M. L. Patil, M. Zhang, J. J. Khandare, M. Saad, P. Chandna and O. Taratula, *Methods Mol. Biol.* **281** (2010).
12. M. L. Patil, M. Zhang, S. Betigeri, O. Taratula, H. He and T. Minko, *Bioconjug. Chem* **1396** (2008).
13. M. L. Patil, M. Zhang and T. Minko, *ACS Nano.* **1877** (2011).
14. M. L. Patil, M. Zhang, O. Taratula, O. B. Garbuzenko, H. He and T. Minko, *Biomacromolecules* **258**(2009).
15. M. Saad, O. B. Garbuzenko, E. Ber, P. Chandna, J. J. Khandare, V. P. Pozharov and T. Minko, *J. Control Release* **107** (2008).
16. O. Taratula, O. B. Garbuzenko, P. Kirkpatrick, I. Pandya, R. Savla, V. P. Pozharov, H. He and T. Minko, *J. Control Release* **284** (2009).
17. O. Taratula, O. Garbuzenko, R. Savla, Y. A. Wang, H. He and T. Minko, *Curr. Drug Deliv.* **59** (2011).
18. O. Taratula, A. Kuzmov, M. Shah, O. B. Garbuzenko and T. Minko, *J. Control Release* (2013).
19. L. Zhu and V. P. Torchilin, *Integr. Biol.* (Camb), **96** (2013).
20. J. Khandare, M. Calderon, N. M. Dagia and R. Haag, *Chem. Soc. Rev.* **2824** (2012).
21. P. S. Wate, S. S. Banerjee, A. Jalota-Badhwar, R. R. Mascarenhas, K. R. Zope, J. Khandare and R. D. Misra, *Nanotechnology* **415101** (2012).
22. A. D. Bangham, M. M. Standish and J. C. Watkins, *J. Mol. Biol.* **238** (1965).
23. V. P. Torchilin, *Nat. Rev. Drug Discov.* **145** (2005).
24. M. Ferrari, *Nat. Rev. Cancer* **161** (2005).
25. F. Perche and V. P. Torchilin, *J. Drug Deliv.* **705265** (2013).
26. C. Rangger, A. Helbok, E. von Guggenberg, J. Sosabowski, T. Radolf, R. Prassl, F. Andreae, G. C. Thurner, R. Haubner and C. Decristoforo, *Int. J. Nanomedicine* **5889** (2012).
27. M. Carboni, A. M. Falchi, S. Lampis, C. Sinico, M. L. Manca, J. Schmidt, Y. Talmon, S. Murgia and M. Monduzzi, *Adv. Healthc. Mater.* **692** (2013).
28. A. Gillet, P. Compere, F. Lecomte, P. Hubert, E. Ducat, B. Evrard and G. Piel, *Int. J. Pharm.* **223** (2011).
29. N. O. Knudsen, L. Jorgensen, J. Hansen, C. Vermehren, S. Frokjaer and C. Foged, *Int. J. Pharm.* **478** (2011).
30. H. Zhang, X. Ran, C. L. Hu, L. P. Qin, Y. Lu and C. Peng, *PLoS One* **e31157** (2012).
31. M. M. Gaspar, O. Gobbo and C. Ehrhardt, *J. Liposome Res.* **55** (2010).
32. A. Gibbons, N. G. McElvaney and S. A. Cryan, *AAPS PharmSciTech.* **1411** (2010).
33. A. Gibbons, D. Padilla-Carlin, C. Kelly, A. J. Hickey, C. Taggart, N. G. McElvaney and S. A. Cryan, *Pharm Res.* **2233** (2011).

34. W. Wijagkanalan, S. Kawakami, Y. Higuchi, F. Yamashita and M. Hashida, *J. Control Release* **42** (2011).
35. R. Duncan, *Nat. Rev. Drug Discov.* **347** (2003).
36. R. Duncan, *Nat. Rev. Cancer* **688** (2006).
37. J. M. Harris and R. B. Chess, *Nat. Rev. Drug Discov.* **214** (2003).
38. J. Khandare and T. Minko, *Prog. Polym. Sci.* **359** (2006).
39. S. S. Banerjee, N. Aher, R. Patil and J. Khandare, *J. Drug Deliv.* **103973** (2012).
40. S. K. Pawar, A. J. Badhwar, F. Kharas, J. J. Khandare and P. R. Vavia, *Int. J. Pharm.* **183** (2012).
41. J. Khandare and S. Kumar, Biodegradable dendrimers and dendritic polymers, In *Handbook of Biodegradable Polymers*, A. Lendlein and A. Sisson (Eds.),Weinheim: Wiley-VCH Verlag GmbH & Co. KGaA(2011), p. 237.
42. A. David, P. Kopeckova, T. Minko, A. Rubinstein and J. Kopecek, *Eur. J. Cancer* **148** (2004).
43. T. Minko, P. Kopeckova and J. Kopecek, *Int. J. Cancer* **108** (2000).
44. L. W. Seymour, Y. Miyamoto, H. Maeda, M. Brereton, J. Strohalm, K. Ulbrich and R. Duncan, *Eur. J. Cancer* **766** (1995).
45. R. K. Jain, *Cancer Res.* **3039** (1987).
46. Y. Matsumura and H. Maeda, *Cancer Res.* **6387** (1986).
47. M. Belting, S. Sandgren and A. Wittrup, *Adv. Drug Deliv. Rev.* **505** (2005).
48. M. Bolla, L. Collette, L. Blank, P. Warde, J. B. Dubois, R. O. Mirimanoff, G. Storme, J. Bernier, A. Kuten, C. Sternberg, J. Mattelaer, J. Lopez Torecilla, J. R. Pfeffer, C. Lino Cutajar, A. Zurlo and M. Pierart, *Lancet* **103** (2002).
49. K. Greish, J. Fang, T. Inutsuka, A. Nagamitsu and H. Maeda, *Clin. Pharmacokinet.* **1089** (2003).
50. E. A. Goun, T. H. Pillow, L. R. Jones, J. B. Rothbard and P. A. Wender, *Chembiochem.* **1497** (2006).
51. M. E. Davis, J. E. Zuckerman, C. H. Choi, D. Seligson, A. Tolcher, C. A. Alabi, Y. Yen, J. D. Heidel and A. Ribas, *Nature* **1067** (2010).
52. S. D. Li, S. Chono and L. Huang, *Mol. Ther.* **942** (2008).
53. L. Ma, J. Young, H. Prabhala, E. Pan, P. Mestdagh, D. Muth, J. Teruya-Feldstein, F. Reinhardt, T. T. Onder, S. Valastyan, F. Westermann, F. Speleman, J. Vandesompele and R. A. Weinberg, *Nat. Cell Biol.* **247** (2010).
54. A. Schroeder, D. Heller, M. Winslow, J. Dahlman, J. Pratt, R. Langer, T. Jacks and D. Anderson, *Nature Reviews, Cancer* **39** (2012).
55. Y. Wakai, J. Matsui, K. Koizumi, S. Tsunoda, H. Makimoto, I. Ohizumi, K. Taniguchi, S. Kaiho, H. Saito, N. Utoguchi, Y. Tsutsumi, S. Nakagawa, Y. Ohsugi and T. Mayumi, *Jpn J. Cancer Res.* **1319** (2000).
56. S. Reichert, P. Welker, M. Calderon, J. Khandare, D. Mangoldt, K. Licha, R. K. Kainthan, D. E. Brooks and R. Haag, *Small.* **820** (2011).
57. R. Savic, L. Luo, A. Eisenberg and D. Maysinger, *Science* **615** (2003).
58. K. Kostarelos, *Nat. Biotechnol.* **774** (2008).
59. A. K. Geim and K. S. Novoselov, *Nat. Mater.* **183** (2007).
60. S. Park and R. S. Ruoff, *Nat. Nanotechnol.* **217** (2009).
61. A. Hirsch, *Angew Chem. Int. Ed. Engl.* **1853** (2002).
62. F. Lu, L. Gu, M. Meziani, X. Wang, P. Luo, L. Veca and Y. Sun, *Adv. Mater.* **139** (2009).

63. J. J. Khandare, A. Jalota-Badhwar, S. D. Satavalekar, S. G. Bhansali, N. D. Aher, F. Kharas and S. S. Banerjee, *Nanoscale* **837** (2012).
64. A. Burke, X. Ding, R. Singh, R. A. Kraft, N. Levi-Polyachenko, M. N. Rylander, C. Szot, C. Buchanan, J. Whitney, J. Fisher, H. C. Hatcher, R. D'Agostino, Jr N. D. Kock, P. M. Ajayan, D. L. Carroll, S. Akman, F. M. Torti and S. V. Torti, *Proc. Natl. Acad. Sci. USA* **12897** (2009).
65. K. Yang, Y. Li, X. Tan, R. Peng and Z. Liu, *Small* (2012).
66. A. H. Lu, E. L. Salabas and F. Schuth, *Angew. Chem. Int. Ed. Engl.* **1222** (2007).
67. A. S. Lubbe, C. Bergemann, H. Riess, F. Schriever, P. Reichardt, K. Possinger, M. Matthias, B. Dorken, F. Herrmann, R. Gurtler, P. Hohenberger, N. Haas, R. Sohr, B. Sander, A. J. Lemke, D. Ohlendorf, W. Huhnt and D. Huhn, *Cancer Res.* **4686** (1996).
68. A. S. Lubbe, C. Alexiou and C. Bergemann, *J. Surg. Res.* **200** (2001).
69. S. S. Banerjee, A. Jalota-Badhwar, S. D. Satavalekar, S. G. Bhansali, N. D. Aher, R. R. Mascarenhas, D. Paul, S. Sharma and J. J. Khandare, *Adv. Healthc. Mater.* **800** (2013).

Chapter 4

Nanoparticles in the Gastrointestinal Tract

Abraham Rubinstein

*The Hebrew University of Jerusalem, Faculty of Medicine,
The School of Pharmacy Institute for Drug Research
P. O. Box 12065, Jerusalem 91120, Israel*

1. Introduction

The possibility of manipulating material at the scale of individual atoms has attracted the imagination of engineers, metallurgists and life science professionals, since 1959, when the physicist, Nobel laureate Richard Feynman, first suggested to miniaturize materials to the nanoscale, followed by the works of Norio Taniguchi, in 1974, who described his research in semiconductor processes of brittle materials as nanotechnology.[1] It was the electronics industry that first showed interest in downsizing of materials to the order of magnitude of several atoms in order to develop tools to create smaller and faster electronic devices on silicon chips. A typical example is the electron beam lithography that was developed by IBM in the early 1970s to create nanostructures and devices as small as 40–70 nm. Key inventions in the field of nanotechnology were the scanning tunneling microscope (1982) and the atomic force microscope (1986).

In general, nanotechnology deals with particles that, due to their nanoscale size, possess unexpected physical, electrical or chemical properties compared with the same materials with larger dimensions. The changes are due to the profound increase in the surface area. A typical example is silver nanoparticles that exhibit strong bactericidal properties and, therefore, can be embedded in filters for water purification.[2] Quantum effect may cause a change in the transparency, color, electrical conductivity and magnetic properties of materials.[3] When the size of titanium dioxide particles (widely used as a white pigment in pharmaceutics) is reduced to the nanosize range (less than 300 nm), the particles acquire new photochemical properties and can block UV-light while transferring visible light. This enables nanoparticles of titanium dioxide to serve in sunscreen creams. In addition, because nanoparticles of titanium dioxide become hydrophilic upon exposure to ultraviolet light, they can be used as antifogging and self-cleaning coating on glass windows.[4]

Broadly, nanomaterials are a result of fascinating technologies. These include films that are just one atom or a few atoms thick, organic nanofibers, biopolymers, nanoparticles of metals and metal oxides, amorphous carbon (carbon black, also known as synthetic soot), fullerenes (spherical hollow pentagon or hexagon constructs of carbon atoms), dendrimers (spherical, highly branched polymers) and quantum dots (nanocrystals of semiconductor material). Innovative nanomaterials that have been discovered and synthesized in the past 30 years are already used in electronics, metallurgy, agriculture, textiles, coatings, medicine and cosmetics. Fullerenes are chemically reactive and are used in catalysis, construct materials, lubricants, solar cells, proton exchange membranes in fuel cells and also drug delivery.[5,6] Ceramic nanoparticles such as zirconia (nano ZrO2) are poor thermal conductors and, therefore, used as thermal barrier coatings in jet turbines and diesel engines to allow operation at higher temperatures. Quantum dots, metallic nanocrystals with specific fluorescence, magnetic or electrical properties, are increasingly mentioned as potential imaging agents in medicine.[7] One of the most interesting and well-studied nanoparticles is the cylindrical shaped carbon nanotube. Single-walled or multiwalled, they are characterized by extreme mechanical strength (100 times the strength of steel whilst being only 1/6th its density).[8]

In medicine, nanotechnology continuously invades the discipline of drug delivery at an exponential rate. Because high-throughput technologies are commonly used by pharma companies nowadays, more new drug entities with water-solubility problems enter the market and pose dissolution and absorption challenges for the pharmaceutical formulators. The method of choice for improving the bioavailability of these molecules is nanotechnology. Interestingly, as nanotechnology becomes more available, drug manufacturers are revisiting shelved molecules that were declined in the past due to solubility problems after being synthesized.[9] The route of choice for administering drug-nanoparticles to the body is the systemic route (i.e. intravenously). However, increasing amounts of pre-clinical and clinical studies are employing nanotechnology methods for oral administration of drugs. Some products, such as Renagel® (Genzyme), a nanoparticle product of Sevelamer (a copolymer of epychlorohydrin and allylamine) for adsorbing duodenal phosphate in hyperphosphatemia and Triglide®, (SkyPharma), nanocrystals of fenofibrate for improving its bioavailability after oral ingestion, are already on the market for more than 10 years now.

This chapter reviews the medical and technological aspects, as well as biological constraints of nanotechnology in the convoluted constituency of oral delivery.

2. Historical Perspective

The observation that the intestinal epithelium is permeable to nanoparticles was investigated intensively in the 1960s when the intestinal uptake of lipid droplets was explored. A typical study is the one involving latex particles in the order of magnitude of 200 nm that were demonstrated to cross the rat intestinal epithelium, accumulate in the lymph vessels of the mucosa, ending in the hepatic sinusoids within 2–4 days.[10] In 1994, Hillery *et al.* studied the passage of much smaller polystyrene particles (60 nm) in the rat intestine and found that about 10% of the administered dose was recovered from the entire intestine, with a significant higher amount taken up through the lymphoid compared with non-lymphoid tissues. About 60% of the uptake occurred through the Peyer's patches, despite the fact that their total surface area is significantly lesser in relation to the surface area of the intestinal epithelium. The largest amount of particle uptake occurred in the lymphoid sections of the colon.[11] Numerous theories tried to elucidate the mechanism of particle uptake by the intestinal mucosa, including endocytosis by enterocytes, persorption,[12] migration through tight junctions,[13] uptake by intestinal macrophages[14,15] and uptake through the gut-associated lymphoid tissues (GALT).[16–19] Yet, the implication of intestinal uptake of particulate matter for drug delivery is controversial due to the difficulty in quantifying the magnitude of the uptake and its variability.[20]

In a focused review, Florence summarized the phenomena of nanoparticulates translocation from the intestinal mucosa after oral administration, by claiming that this is a normal occurrence. Particulate uptake does take place, not only via the M-cells in the Peyer's patches and the GALT, but also via the normal intestinal enterocytes. Factors affecting uptake include particle size, surface charge and surface hydrophobicity. It can be enhanced by conjugating surface ligands. Typical examples are the covalent attachment of lectins[21,22] or invasion molecules. Although it has been shown that a small fraction of the ingested dose of nanoparticles can be absorbed, the routine administration of drug molecules that are not normally absorbed from the gut is not yet demonstrated clinically.[23]

3. Current Reasoning for Oral Delivery of Drugs in Nano Carriers

Grossly, nanoparticles are designed to be administered orally for the following medical reasons: (a) increasing the oral bioavailability of poorly water-soluble, (b) oral immunization, and (c) localized therapy in the GI tract.

3.1. *Increasing oral bioavailability by downsizing drug particles*

The obvious means of improving the bioavailability of poorly water-soluble drug substances is by reducing their particle size. This was exemplified, clinically, 50 years ago by griseofulvin[26] and later in the digoxin (used for the treatment of cardiac arrhythmia) case, in which it was demonstrated that alterations in the *in vitro* dissolution rate of digoxin from various commercial tablets were correlated with digoxin bioavailability in man after oral administration. The variations were observed even after the administered digoxin tablets completely disintegrated.[27] It was concluded that bioavailability could be predicted by the drug powder dissolution rates[28,29] and that the latter could be accelerated by comminuting the drug particles to predetermined limits[30] so that a sufficient amount of drug powder is dissolved and absorbed in the appropriate intestinal absorption window.[27] In a cross-over study the oral bioavailability of three digoxin tablets made of different particle size was

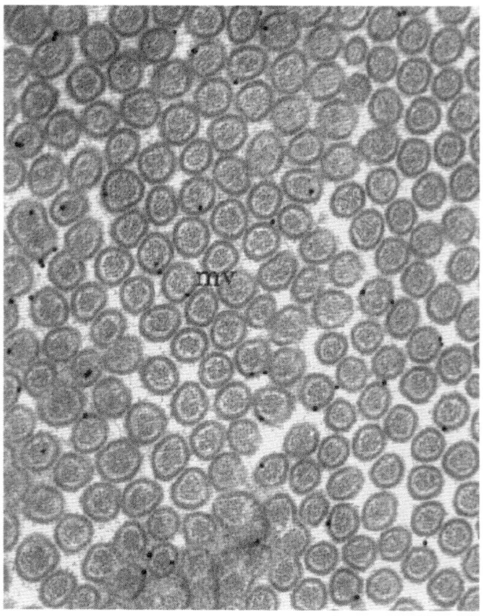

Fig. 1. Cross-section, immunostained, rat duodenal microvilli, 10 min after oral administration of insulin labeled with gold particles (black dots), demonstrating their presence adjacent to and inside the microvilli (× 50,000) (from Ref. 24 with permission).

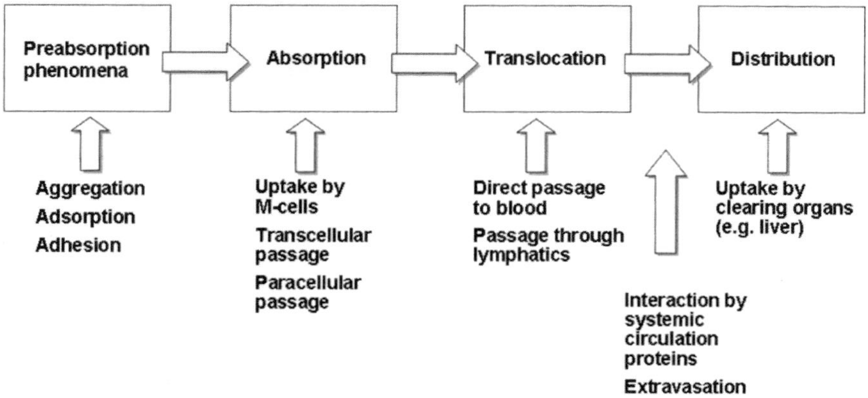

Fig. 2. Processing of nanoparticles by the intestinal mucosa after luminal administration (from Ref. 25 with permission).

measured in healthy volunteers, using an alcoholic solution of the drug as a reference standard. The bioavailability in tablets with a particle sizes of 7–13 microns was 78–97% of that of digoxin in solution, while the tablet containing particles size of 102 microns showed a profound lower bioavailability (39%). Interestingly, it was observed that there was a lower limit for particle size reduction beyond which it was impossible to improve bioavailability.[31] This discrepancy could be attributed to augmented interfacial interactions (e.g. increased agglomeration) that interfere with hydration of the very minute particles.

Particle down-sizing of poorly water-soluble drugs

The incorporation of high throughput screening methods in the drug industry resulted in increasing amounts of new molecules with poor water-solubility properties.[9,32] This poses a serious challenge to the pharmaceutical formulator as water-solubility is directly correlated with drug oral bioavailability. If salt making is not possible and the molecule is not non-ionizable, poorly water-soluble molecules (belonging to the biopharmaceutical classification system class II) may be solubilized by micellization. However, this requires the use of solubilizing agents that could be toxic, as was found for Cremophor EL which was used to solubilize paclitaxel in parenteral formulations.[33] It seems that particle comminution to the nanorange is the safest way to obtain an increase in the drug particles' dissolution and hence (unless they are subjected to intestinal metabolic or permeation-related interference; i.e. biopharmaceutical classification system class IV) their intestinal absorption. As will be discussed below, due to surface phenomena the down-sizing is not a straightforward solution however, it is a reasonable tactic for increasing dissolution rate. The velocity at which poorly water-soluble drugs interact with the gastrointestinal (GI) fluids is crucial due to the restricted intestinal absorption window lasting several hours under harsh and altering biological conditions.[34]

The correlation between the dissolution rate of poorly water-soluble drugs and its particle surface area is clearly described by the Noyes Whitney equation:

$$\frac{dX}{dt} = D \cdot A\left(C_S - C_t\right)$$

where, dX/dt is the dissolution rate of X amount of a drug; D is the macroscopic diffusion constant of the drug (composed of microscopic constants, depending on the drug molecule characteristics and the nature of the dissolution medium); C_s is the solubility of the drug (i.e. its saturation concentration in the medium) and C_t is the drug concentration at time t in the tested system. As will be discussed below (Sec. 5.1), down-sizing particles to the nano-range can increase their C_s and, thus, the concentration gradient of the Noyes Whitney equation. Recent examples include nifedipine[35] and naproxen. The use of nanoparticles not only increases the bioavailability and the exposure time of a drug, but can also reduce subject variability.[34]

Drawbacks in down-sizing drug particles

Occasionally, the reduction of particle size of hydrophobic substances does not increase the rate of drug dissolution because of interference with the ability of the dissolution medium to moisten the surface of the particles.[36] Particle comminution, in these cases, increases the resistant surface area. In other words, the effective surface area available for dissolution is lowered. Surface wetting is a crucial step in the dissolution process. Luner and VanDer Kamp found that the composition of the micellar phase is the most important parameter for particle surface wettening.[37]

3.2. *Formulating drugs into nanoparticulate vehicles for increased intestinal uptake*

Although the extent of nanoparticle uptake by the intestinal mucosa is low, mucosal uptake of plain surface nanoparticles into the GI lymphatic system can reach an order of magnitude of 2–3% of the orally administered dose. This value can even be increased when recognition moieties are attached to

Table 1. Factors enhancing GALT uptake of particles (from Ref. 23 with permission).

Parameter	Uptake
Stability of particles in the GI fluids	
Reducing particle dimensions to the low nanorange	⇧
Surface charge	⇩
Surface hydrophobicity	⇧
Surface grafted recognition moieties	⇧

the nanoparticles' surface.[23] It has already been mentioned that Hillery *et al.* found that 60 nm polystyrene nanoparticles were uptaken, primarily, by the lymphoid compartment of the rat intestine, with 2/3 of the uptake occurring in the Peyer's patches, despite the fact that their total surface area is significantly lesser in relation to the entire surface area of the intestinal epithelium.[11] The magnitude of particulate matter uptake remains a crucial issue if translocation of nanoparticles across the mucosal barrier of the intestine is considered seriously. Although it is clear that particulate matter can pass not only injured epithelium but also healthy regions and that this transport is size-dependent[38] and can be enhanced by conjugating targeting moieties (see below), the prospect of the medical use of drugs entrapped in nanoparticles depends on the ability to accomplish epithelial transport of quantities in the order of magnitude of milligrams by the intestinal GALT (the only tissue that was shown to translocate nanoparticles in significant quantities). The factors affecting GALT uptake of particles are summarized in Table 1.

Oral delivery of protein drugs by nanoparticles — insulin as a model drug

Being highly potent (i.e. active in pico-nanogram doses), but suffering from limited stability in the lumen of the GI tract and poor intestinal absorption, protein and peptide drugs are the most obvious drug candidates for nanotechnology.[39] Due to its potential market, and the fact that it can be absorbed in the intestine if protected by a protease inhibitor,[40] insulin is most frequently used as a protein drug model in oral delivery of nanoparticles.[41] Damge *et al.* prepared nanocapsules made of polyalkylcyanoacrylate with a mean size of 220 nm, loaded with insulin. After oral administration to normal or diabetic rats, the blood glucose levels were reduced by 50–60% in a dose-dependent and prolonged manner.[42] Larger nanoparticles are also able to deliver insulin after oral administration. Sarmento *et al.* loaded insulin into

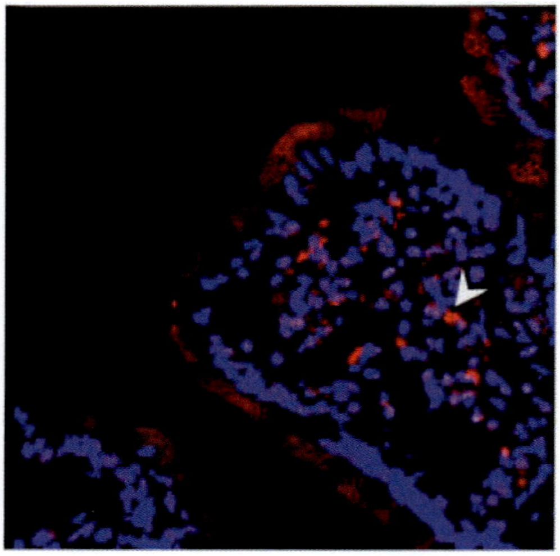

Fig. 3. (Color online) Confocal microphotograph of Cy3-labeled mucoadhesive, enteric-coating nanoparticles (red) in a duodenal microvillus of the rat after oral administration, showing adhered, as well as internalized nanoparticles. The nanoparticles (250 nm) were made of pentetic acid conjugated to poly gamma-glutamic acid (as a protease inhibitor with paracellular permeation enhancer properties) and chitosan and loaded with insulin (from Ref. 45 with permission).

750 nm mucoadhesive nanoparticles made of an alginate chitosan complex (designed to be resistant to the gastric fluids) and found a 40% reduction in blood glucose levels; a four-fold increased effect compared to oral administration of insulin solution.[43] Reviewing the numerous studies conducted in this field, it appears that average relative bioavailability values (commonly compared to subcutaneous administration) of 5–7% are rarely surpassed when nanotechnology is used for oral administration of insulin.[44] To increase this limit, functional adjuvants should be incorporated into the nanoparticles, such as protease inhibitors.[45]

3.3. *Nanoparticles for oral immunization*

Some 20 years ago, Maloy *et al.* demonstrated that oral administration of ovalbumin entrapped in poly (lactide-co-glycolide) (PLG) microparticles induced cell-mediated immunity by stimulating cytotoxic T-lymphocytes.[46] Indeed, microparticles possess considerable potential for mucosal vaccination, via the oral route, against a wide range of pathogens, particularly against those which initially infect mucosal sites. Challacombe *et al.* loaded poly (lactic-co-glycolic acid) (PLGA) microparticles with ovalbumin and administered them orally to rats. They found that the nanoparticles were taken up by antigen-presenting cells in the rat Peyer's patches followed by a slow degradation. The entrapped antigen was released slowly and induced a prolonged immune response at both the mucosal and systemic levels.[47] O'Hagan *et al.* also showed that PLG microparticles were able to induce comparable immune responses to cholera toxin, the most potent oral adjuvant available.[48] It was concluded that in mice, oral immunization with PLG microparticles is a potent mucosal and systemic vehicle for the entrapment of antigens.[49]

Although nanoparticles have been suggested as potent adjuvants for recombinant vaccines in small animal models, studies that were conducted in larger animal models and in human clinical trials have been largely disappointing. It should be realized that the ability of particulate carriers to function as effective adjuvants following mucosal administration depends on their uptake into the specialized mucosal associated lymphoid tissue (MALT).[50] Such uptake can be improved by the attachment of recognition moieties toward distinct carbohydrate receptors on the intestinal M-cells to the nanoparticle surface. PLGA nanoparticles were loaded with surface antigen of the hepatitis B virus (HBsAg) together with trehalose and magnesium hydroxide as stabilizers. *Ulex europaeus* 1 lectin was anchored to the nanoparticle surface as an M-cell recognition moiety. The immune stimulating activity was tested in Balb/c mice after oral administration and compared to non-lectinized formulations, as well as a conventional alum-based formulation of the antigen that was injected intramuscularly. The immune response was found to be superior to the response obtained after oral administration of non-lectinized nanoparticles and was comparable to the intramuscularly administered formulations.[51]

To overcome stability problems associated with direct oral administration (i.e. delivery without a supporting platform such as an enteric coating capsule) poly lactic acid nanoparticles of hepatitis B surface antigen were copolymerized with PEG. After oral administration to mice, an improved immune response was obtained compared to the response obtained with the non-PEGylated nanoparticles.[52]

PEGylated PLGA-based nanoparticles displaying RGD molecules at their surface were designed to target human M cells. The transport of RGD-labeled nanoparticles increased significantly in co-cultures, due to interactions between the RGD ligand and the beta(1) integrins detected at the apical surface of co-cultures. *In vivo* studies demonstrated that nanoparticles decorated with RGD particularly concentrated in M cells. Ovalbumin that was loaded into the RGD-labeled nanoparticles induced an IgG response, attesting antigen ability to elicit an immune response after oral delivery to mice.[53]

Because vaccines alone are not sufficiently taken up after mucosal administration, they need to be co-administered with penetration enhancers, adjuvants or encapsulated in particles. Chitosan easily forms microparticles and nanoparticles which encapsulate large amounts of antigens such as ovalbumin, diphtheria toxoid or tetanus toxoid. It has been shown that ovalbumin loaded chitosan microparticles were taken up by the Peyer's patches of the GALT. Additionally, after co-administering chitosan with antigens in nasal-vaccination studies, a strong enhancement of both mucosal and systemic immune responses was observed. It was concluded that chitosan nanoparticles, powders and solutions are promising candidates for mucosal vaccine delivery.[19]

DNA vaccine against infectious haematopoietic necrosis virus that affects rainbow trout was incorporated (by incorporating the virus gene plasmid) into PLGA nanoparticles to examine the possibility to immunize the fish in its natural maritime environment. It was found that after feeding the fish, the nanoparticles were detected in the epithelial cells of the lower intestine within 96 h. Low levels of gene expression and antivirus neutralizing antibodies were identified in both fed fish and fish incubated with the PLGA nanoparticles. The fish survival was increased at 6 weeks, but not at 10 weeks post-vaccination. It was concluded that immune response with an orally delivered DNA vaccine could be possible after improving the particulate matter formulation.[54] This can be done by adding a PEG component to the nanoparticles. For example, amino-pegylated poly(methyl vinyl ether-co-maleic anhydride) nanoparticles were prepared and loaded with genomic salmon testes DNA. Apart from possessing high mucoadhesive properties and enterocyte penetrability, the particles provided sufficient protection to the entrapped DNA.[55]

3.4. *Localized therapy in the GI tract by nanoparticles*

IBD

One of the most explored disciplines in medical nanotechnology inside the GI tract is inflammatory bowel disease (IBD). The question whether inflamed mucosal tissues can be targeted by nanoparticles is discussed in numerous studies. It is believed that a mucosal tissue accumulation in the inflamed intestine can be obtained by reducing the dimensions of particulate drug delivery

Table 2. Some typical examples of nanoparticle delivery systems tested *in vivo* for mucosal targeting of inflamed intestinal tissues.

Particle type	Drug	Reference
Polystyrene	None	58
PLGA	Rolipram	57
Anionic liposomes	Superoxide dismutase, tempamine, catalase	61
Silica	5-Aminosalicylic acid	62
PLA coated with PVA, embedded in crosslinked alginate and chitosan hydrogel	Lys-Pro-Val (tripeptide)	63
Termed thioketal	TNFα-siRNA	64
Gelatin-B encapsulated in PCL microcapsules	TNFα-siRNA	65
Anionic liposomes decorated with antitransferrin receptor	None	66
Cationic polymethacrylate (Eudragit RL)	Clodronate	67

Abbreviations: **Eudragit RL**, cationic polymethacrylate; **PCL**, poly(epsilon-caprolactone); **PLA**, polylactic acid; **PLGA**, poly (lactic co glycolic acid); **PVA**, polyvinyl alcohol.

systems.[56] This was first demonstrated by Lamprecht *et al.* who showed that charged polystyrene particles[57] and later, PLG nanoparticles[58] adhered preferentially (more than three-fold compared with non-inflamed regions) to the inflamed epithelium in the rat intestine. Anchoring recognition ligands to the nanoparticle surface can improve this preferential mucoadhesion. Still, it seems that the materials comprising the nanoparticles are important for the entrapped drug release kinetics rather than targeting efficacy. Table 2 summarizes typical examples for this medical approach. For a comprehensive review and a detailed list of particulate drug delivery systems developed for topical therapy in IBD, see Ref. 56. As will be discussed later (Sec. 4) the mucus gel lining of the epithelial tissues discriminates nanoparticles according to their size, charge and functional groups, and therefore plays a crucial role in targeting particulate delivery systems to the inflamed tissues in IBD. Moreover, because mucus is constantly renewed, particles that adhere first to this hydrogel layer are removed first. Once in the vicinity of the mucosa, small particles (100 nm–1 μm) will adhere to the epithelium, while larger particles (e.g. 10 μm) will be uptaken by the GALT. In the design process one should take into consideration the overproduction of mucus in IBD, which leads to a thicker mucus layer especially in the ulcerated areas.[59,60]

The 1998 approval of infliximab (Remicade®), a chimeric immunoglobulin IgG1 monoclonal antibody (mAb) against tumor necrosis factor alpha (TNFα) for the treatment of Crohn's disease, triggered intensive research activity to explore novel biological drugs for the treatment of IBD. Typical examples include the IgG4 humanized monoclonal antibody, CDP 571 (Humicade®), Certolizumab pegol® (CDP-870) a PEGilated Fab' fragment of humanized anti-TNFα mAb and Adalimumab (D2E7, Humira®), a fully human anti-TNFα antibody, which was approved by the FDA for the treatment of Crohn's disease. However, the use of infliximab is associated with typical adverse effects, such as the formation of antibodies against the murine fraction of the chimeric antibody.[68] A possible solution for this problem could be the formulation of the biological drugs into nanoparticles if targeted successfully to the site of inflammation. So far, the clinical outcome of nanotechnology for targeting intestinal inflammation is premature and the bench to bedside transfer of mucosal delivery systems of potent, modern biological drugs is, as yet, wishful thinking.[69]

Intestinal radio protection

An amifostine metabolite, a cytoprotective agent used to prevent chemo- and radiotherapy of cancer was loaded into PLGA nanoparticles. Its protective effect against intestinal injury following whole body gamma irradiation was assessed in mice. A significant radioprotection after whole body gamma irradiation in mice was obtained after oral administration of the nanoparticles (compared with the effect obtained after oral administration of the drug in a solution) 1 h prior to the radiation.[70]

4. Relevant Biological and Physiological Constraints Relevant to Oral Delivery of Nanoparticles

The performance of nanoparticles in the GI tract is subjected to complicated events that, in many cases, are not understood completely.[71] The most relevant physiological parameters that should be taken into consideration, when designing nanoparticulate delivery systems for the oral route, are reviewed here briefly. For a more comprehensive review see Ref. 72.

Because of its large surface area, the small intestine is the primary organ in which drug absorption occurs after oral ingestion, due to its huge surface (200 m^2), resulted by the presence of millions of villi. They house a dynamic, self-renewing population of epithelial cells that include absorptive cells (enterocytes), secretory cells and endocrine cells. Out of the three segments of the small bowel; duodenum, jejunum and ileum, the former is considered to be the best absorptive organ and its epithelium is of the leaky type.[73] Mucosal secretory cells as well as adjacent functional organs (pancreas, gall-bladder) secret proteolytic enzymes into the lumen of the gut. The serine proteases, trypsin and chymotrypsin, are the two major pancreatic proteases secreted into the duodenum. Pancreatic lipase is also secreted into the lumen to assist in the digestion of fatty acids by their lipase-mediated hydrolysis, after which micellization of the lipophilic compounds is performed by duodenal bile salts. The micelles formed in the duodenum enable the absorption of hydrophobic drugs such as steroids. They pose, however, a serious constraint for the stability of drug delivery carriers such as liposomes and emulsions. Unlike the small intestine epithelium, the colonic columnar epithelium does not have villi projections. Its surface area is lined with numerous depressions (crypts) and its absorbing surface area per unit length is limited. In terms of drug absorption, the small surface area is compensated by the long residence time of colonic contents. Another constraint, which minimizes drug absorption in the colon, is the extensive luminal metabolism caused by its active bacterial flora.[74]

Although the surface area of the rectum is small (200–400 cm^2, with columnar epithelial cells without villi), it is an interesting absorption cavity because of its venous and lymphatic drainage. The rectal upper hemorrhoidal vein is drained to the portal system, and the middle and lower hemorrhoidal veins are drained to the inferior vena cava. Therefore, drugs that are absorbed in the proximal rectum are more susceptible to first-pass metabolism than drugs absorbed in the lower part of the rectum.[75] Because of the extensive lymphatic drainage in the organ, first pass of rectally administered drugs may also be avoided if lymphatic absorption is achieved.

GI motility and its role in dosage forms transit and spreading

A major problem associated with orally administered nanoparticles is the uncertainty of location at any point in time following ingestion. There are two modes of motility patterns in the stomach and consequently in the small intestine. The digestive (fed) pattern consists of continuous motor activity, characterized by a constant emptying of chyme from the stomach into the duodenum. The interdigestive (fasted) pattern (the "migrating motor complex", or MMC) is organized into alternating cycles of activity and quiescence and can be roughly subdivided into three phases: I (basal), II (pre-burst), and III (burst).[76] The MMC pattern of the stomach can be disrupted by either the presence of food or infusion of nutrient into the small intestine. The duration of this disruption depends on meal composition. It is also longer for solid meals than for liquids.

Gastric emptying of dosage forms

While large, non-disintegrating solids (e.g. tablets) of 13 mm and above can be retained in the fed stomach for several hours, there is no exact cutoff size for smaller particles and their emptying from a fed stomach. Particles smaller than 1 mm disperse in the liquid content of the stomach and empty as a fluid.[77] Gastric emptying patterns of liquids depend on their volume and viscosity. For small volumes of 150 ml or less the emptying pattern is controlled by the MMC cycle. Thus, little or no liquid empties in phase I. The onset of discharge occurs in phase II and usually finishes before phase III. Following oral administration with a small volume of water, a dosage form such as a gelatin capsule containing pellets will disintegrate and release its content. Dispersion of the capsule content will occur due to the quiescent nature of the stomach in phase I. The small volume of co-administered water will empty during phase II without any particles.[78] Even with a large volume of liquid, less than 5% of the ingested particles will be emptied from the stomach together with the liquid. Graham *et al.* found that in the empty stomach of human volunteers tablets and particulate matter do not differ in their dispersion and concluded that formulation of a drug in a microencapsulated multiple-unit dosage form does not guarantee wide dispersion or absence of high local concentration of drug.[79] A possible explanation for this observation is the entrapment of particles in mucus clots whether in the fasted stomach[78] or the fasted small intestine and the ability of these particle-loaded clots to travel over relatively large intestinal distances.[80]

Intestinal transit of liquids and solids in fed and fasted states

Transit time of dosage forms in the small intestine determine the exposure time to mucosal tissue and the extent of drug absorption and the derived bioavailability of orally administered drug molecules. The mean small bowel transit time for both liquids and solids regardless of particle size is similar, namely around 3–4 h in both the fasted and fed state.[81] Common pharmaceutical excipients could alter small intestinal transit time.[82]

Intestinal spreading of solid dosage forms

Spreading of particulate dosage forms is highly affected by the rate of gastric emptying in the fasted state. Once the particles have left the stomach there is little, if any, additional spreading.[78,80] Hardy *et al.* found that following oral administration of non-disintegrating capsules and multipellet dosage forms, no distribution of the latter occurred in the stomach or along the length of the small bowel in fasted volunteers. Both preparations moved together through the stomach and small intestine and reached the colon approximately 4 h after administration. It was only after they reached the colon that they dispersed and moved at a slower rate.[83]

Motility of the colon and transit of liquids and solids

All three functions of the colon: water reabsorption, residual carbohydrate fermentation, and storage and propulsion of fecal material in its distal portion are time consuming and require a slow exposure of the colonic mucosa to the luminal content. The resulting contractions are less organized than in the small intestine. They are less propulsive, and results in to-and-fro movements.[84] The mean colonic transit time measured by radiopaque markers and confirmed by gamma scintigraphy could be as high as 40 h. Different materials behave differently in the colon. However, under normal conditions, transit through the ascending colon is similar for solids and small volumes (10 ml) of liquids.[85]

Luminal pH and transverse pH gradient

The pH of the resting stomach is acidic (average value of 1.5). Immediately after meal the stomach pH increases sharply to 5–6 and decreases again slowly when gastric acid secretion over-rides the buffering capacity of the food. This decline lasts 2–4 h, until returning to the basal levels of the empty stomach. In fasted humans the duodenal pH is around 6.4.[86] The pH of the colon lumen is 5.5 in its proximal part due to acidification of the colonic contents by bacterial fermentation products (short chain fatty acids).

In addition to the longitudinal pH gradient, there is a transverse pH gradient of at least 1 pH unit (e.g. from 7.1 to 6.1, Ref. 87) starting with the lumen to the absorbing surface.[88] This transverse pH gradient should be taken into consideration when designing enteric coated nanoparticles that are supposed to possess pH resistant properties.

Boundary layers

Boundary layers in the GI tract contribute to the effect of intestinal fluid hydrodynamics on drug absorption by both diffusional and carrier-mediated processes. In a well-defined isolated *in situ* model such as perfused intestine of the rat, a good estimate of the gut wall permeability, which is the vector of convective diffusive mass transfer, passive diffusion and carrier-mediated transport, can be accomplished.[89] Two major physical obstacles create epithelial boundary layers in the intestine: (a) the hydrodynamic gradient during fluid flow, which causes transverse changes from the center of the lumen to the gut wall (chyme flow is rapid in the center and slower at the periphery, assuming a laminar, non-turbulent flow,[90] and (b) the physical nature of the fluid layers (e.g. increase in the viscosity of mucus adjacent to the enterocyte membrane due to the cell membrane glycocalix, or transverse changes in mucin content) and their interactions with epithelial components.[91] When analyzing a transverse migration of a drug molecule toward the epithelium, a two compartment model should be considered: diffusion in the aqueous part and diffusion in the mucus part which can be looked upon as the unstirred layer, resistant to diffusion.[92]

GI mucus and turnover rates

Mucus is a hydrogel, made of approx 5% mucin (a heavily glycosylated protein) in water. The GI mucus that covers the entire surface of the GI tract also contains proteins, lipids, and cellular debris, entrapped in the hydrogel matrix.[93,94] Gastric mucus is saturated with bicarbonate and serves as a thick barrier against hydrogen ions and pepsin, being a major reason, together with the small surface area of the gastric epithelium, for the inability of the stomach, to be an absorptive organ.[95] The average thickness of the mucus gel that covers the gastric and intestinal epithelium is a consequence of the steady state between the mucus secretion and erosion rates and varies in the various organs of the gut.

Similar to gastric mucus the function of the mucus gel layering the intestinal epithelium is to protect the epithelium underneath against mechanical, chemical and enzymatic erosion and to serve as a barrier against diffusion of macromolecules (toxins), particles and bacteria[96,97] (Fig. 4).

Mechanical damage is avoided due to the unique construct of the mucus gel made of firmly bound layers, adjacent to the GI epithelium and loosely bound layers on top. This loosely bound mucus interacts immediately with mucoadhesive polymers and, eventually, determines their fate after oral administration. Thus, residence time of mucoadhesive drug-delivery in the vicinity of the mucosa of the GI tract is rate limited by mucus-turnover rather than by tightness of the adhesive polymer to the mucus lining. Lehr *et al.* calculated the turnover time of mucus in the intestine of the perfused rat preparation (a common animal model in the study of bioadhesive drug platforms) and found it to range between 47 and 270 min.[99] The variability in mucus turnover presents another complexity in the design of mucoadhesive drug platforms.

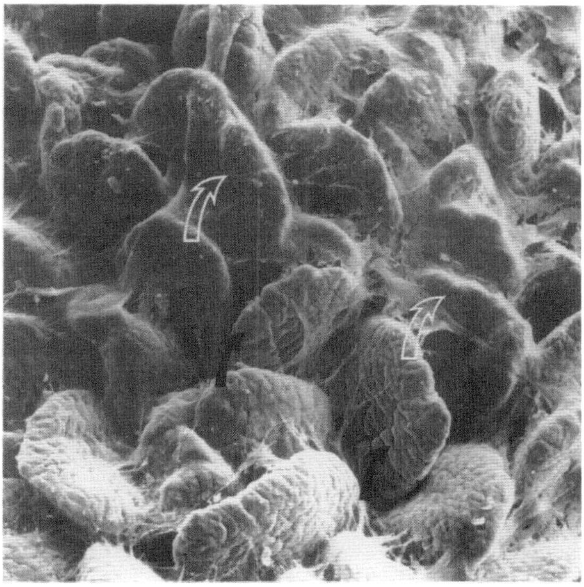

Fig. 4. Scanning electron micrograph (× 100) of villi of the rat jejunum covered with a thin mucus layer (open arrows). Disrupted mucus region is marked by a solid arrow (from Ref. 98 with permission).

Mucus thickness varies throughout the human GI tract. Typical values that were measured in the rat are shown in Fig. 5.

Diffusional path in mucus layers and possible drug interactions

The thickness of the mucus layer in the human stomach is about 570 μm,[101] while in the human colon the thickness is 100–150 μm.[102] Ryu and Grim suggested that there is little mixing between the villi and the unstirred layer over the tip of the villi of the canine jejunum, which is about 500–1000 μm thick. Therefore, for a solute to reach the lateral surfaces of the villi, an additional barrier of as much as 800 μm needs to be traversed.[103] It was shown that native mucus gel from rat small intestine reduced the diffusion coefficients of ^3H-water, urea, benzoic acid, antipyrine, aminopyrine, alpha-methylglucoside and L-phenylalanine by 37–53% compared with buffer solution.[104] For large molecular weight substances, such as proteins, or microparticulates, mucin can offer considerable resistance to diffusion.[105]

Apart from being a diffusional barrier, mucus can also interact with drugs to decrease their bioavailability, as has been shown with tetracycline, phenylbutazone and warfarin. On the other hand, studies in rats showed that binding

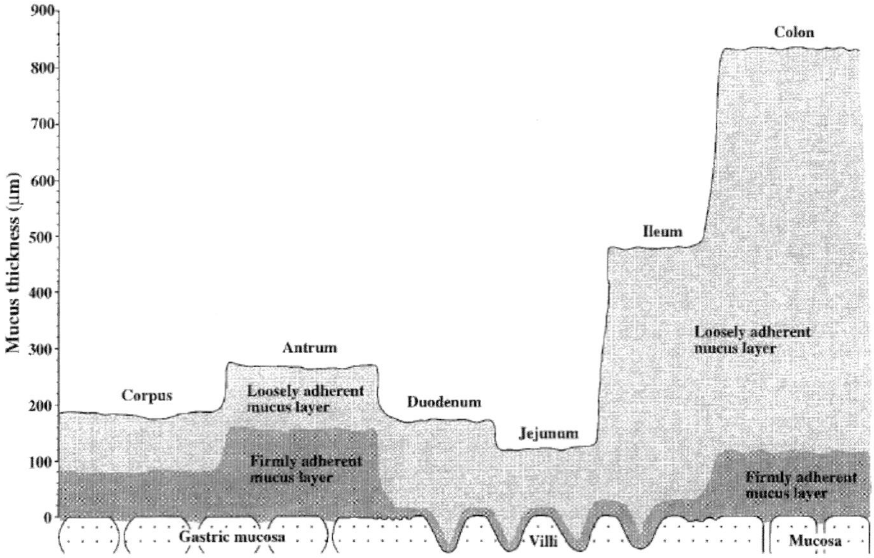

Fig. 5. The thicknesses of the loosely adherent and firmly adherent mucus in the rat GI tract (from Ref. 100 with permission).

of some water-soluble drugs to intestinal mucus was essential for their absorption and that damage to the mucus significantly reduced absorption.[106] The acidic mucus is essential for lipid absorption and could be important for the diffusion of lipophilic drugs. It was shown that the acidic microclimate is an essential determinant in fatty acid uptake after micellization.[107] The presence of a low-pH compartment facilitates the dissociation of mixed micelles made of taurocholate and oleic acid. The rate of fatty acid diffusion in the mucin layer was estimated to be 400% of that in a buffer solution.[108]

Nanoparticle transport in mucus

Because mucus is highly viscous (10^3–10^4 times higher than water viscosity) and due to its constant turnover along the alimentary canal, it is generally impermeable to hydrophilic macromolecules. However, there are exceptions. Viruses as large as 30 nm (e.g. Norwalk and papilloma virus) diffuse freely in cervical mucus.[109] It is assumed that the average pore size in the mucus gel matrix is 100 nm (Ref. 110). Thus, the development of mucus-penetrating nanoparticle formulations appears to be limited to the dimensions of small viruses (see Sec. 5.4 for recent formulative resolutions).

Mucus traps digested particles by polyvalent adhesive interactions with their surfaces. Mucus is negatively charged due to the presence of carboxyl or sulfate groups on the mucin proteoglycans.[111,112] In addition, it contains hydrophobic domains. The fibrous nature of mucin allows multiple accesses to different types of particles' surfaces, both hydrophobic and charged, causing an efficient system of particle removal from the GI tract. The hydrophobic interactions enable the mucus to immobilize pathogens such as *E. Coli*, *P. Aeruginosa* and *Y. Enterocolitica.*[113] When considering mucus-nanoparticle interaction during the design of a particulate delivery system, one should take into account that mucoadhesion may become saturated.[114]

Nanoparticle uptake in the GI tract

Probing macromolecules from maternal milk by mucosal M cells in the intestine of newborns is well-established. Likewise, it is established that particles can pass into the lymphatic system of the adults.[23] It is believed that particulate matter is transported across the intestinal epithelium by three routes: (a) M cells (Fig. 6), (b) normal enterocytes, (c) tight junctions between the enterocytes (paracellular route) (Table 3). The M cells translocate particles such as bacteria, viruses and foreign particulate matter by endocytosis. Florence *et al.* showed that 6%–7% of an orally delivered dose of 50 nm

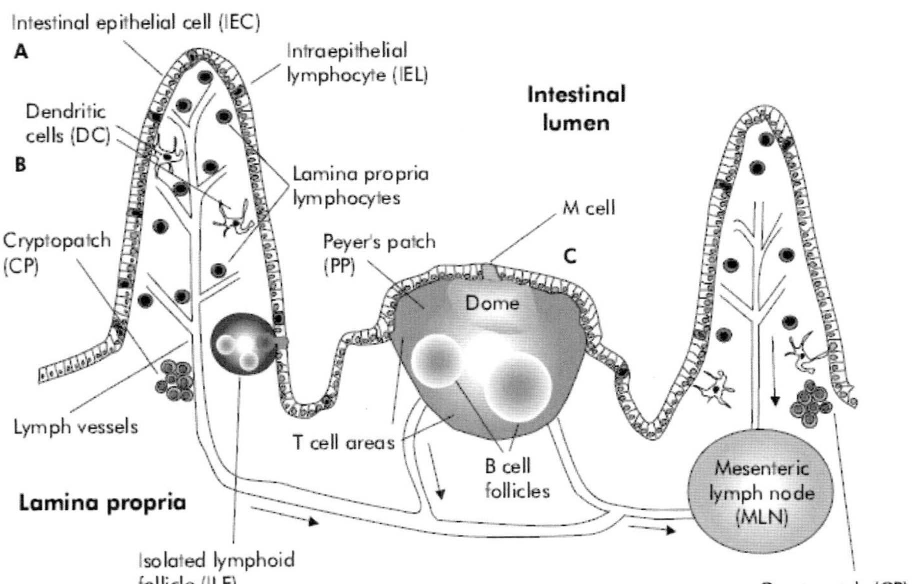

Fig. 6. Illustration of the lymphoid elements of the gut associated lymphatic (GALT) system involved in nanoparticle uptake. The nanoparticles can be taken up by intestinal epithelial cells (A), dendritic cells (DC), and M cells. The lymphatic drainage of PP underneath the M cells and villus lamina propria is drained into the mesenteric lymph node (MLN). Direction of lymph flow is indicated by arrows. (reproduced with permission from Ref. 117).

Table 3. Suggested modes of nanoparticle uptake by various GI organelles and their relation to nanoparticle dimensions.[16]

Uptake site	Mode of uptake	Nanoparticle dimensions
Villus tips	Persorption	1–150 microns
Intestinal macrophages	Phagocytosis	1 micron
Enterocytes	Endocytosis	< 200 nm
M cells	Paracellular	< 10 nm

particles are found later in the liver.[115] The surface properties of the particles are crucial for the transport process. When polystyrene particles were coated with hydrophilic poloxamers their intestinal uptake was reduced. Size is the most important factor. It seems that particles in the range of 3–10 microns are sequestered within the Peyer's patches and do not migrate into the mesenteric lymph system.[23] Kreuter *et al.* found that 4% of an orally administered single dose of polymethyl methacrylate nanoparticles was absorbed and 0.15% of the dose was found in the lung, spleen and liver.[116]

5. Common Technologies for Producing Nanoparticles for the Oral Route

5.1. *Down sizing particles to the nano range*

The most apparent method to increase water solubility of poorly water-soluble drugs is increasing their surface area by reducing their particle size. In time, powder micronization to the range of 1–10 microns was found to be insufficient to overcome bioavailability problems of biopharmaceutical classification system class II molecules and further reduction to the nano-size range was required. Comminuting particles to the order of magnitude of single nanometers, while maintaining their crystalline properties (that is, not transforming them to an amorphous state) lead to the creation of nanocrystals. Their major physical-chemical property is the change in their saturation solubility (C_s), which leads to a larger solubility difference in the Noyes Whitney equation (see Sec. 3 above) and, hence, to a larger value of the dissolution rate. This increase in the concentration gradient is caused by the increase in the dissolution pressure over the nanoparticle surface.[118] Increased C_s can also be obtained by transforming the physical form of the nanoparticle to the amorphous state. A typical example is chloramphenicol palmitate, in which the amorphous particles are 10-fold more soluble that the crystalline ones.[119]

Table 4. Some marketed oral products containing powders that were down-sized to the nanorange.

Product name	Drug	Medical indication	Dosage form
Rapamune	Sirolimus (rapamycin)	Immunosuppressant	Tablets, suspension
Emend	Aprepitant,	Emesis	Pellets in hard gelatin capsules
Tricor	Fenofibrate	Hypercholesterolemia	Tablets
Megace ES	Megestrol acetate*	Appetite loss	Suspension

*Well-absorbed in the GI tract, however, highly variable drug.

In principle, downsizing particles to the nano-range is performed by milling (top down), e.g. Elan's pearl mill technology,[34] precipitation of drug solutions (bottom up)[120] and homogenization, which introduce very strong shear forces on a dispersion of the comminuted particles in a collision chamber.[118] The nanoparticles thus formed can be dried but can also be kept in a suspension form (nanosuspensions) if aimed at intravenous (IV) administration.[121] Some oral commercial products containing nanoparticles are shown in Table 4.

5.2. *Formulating drugs into polymeric nanoparticles*

By reducing particle size to the nanoscale a single aspect is being resolved in the oral bioavailability of drugs, that is — dissolution rates in GI fluids (and the derived passive diffusion through boundary layers and enterocytes membrane). To sophisticate the performance of nanoparticles aimed at oral delivery they should be engineerable to enable their (a) targeting to selected regions (b) local protection against harsh biological conditions, and (c) improved mucoadhesion. For that purpose a variety of polymeric vehicles are mixed with the drug molecules to form different types of nano delivery systems. This includes micelles[122] liposomes[123,124] (the first injectable nano formulation was Doxil®, Ref. 125) and nanocapsules or nanomatrices.[126] The use of polymers in nanotechnology allows anchoring recognition and bioadhesive molecules to the surface of the nanoparticles.[127] They also allow the attachment of fluorescent dyes for imaging purposes in theranostics.[69]

The basic procedure for the preparation of polymeric nanoparticles involves dispersing the active ingredient(s) with a mixture of the functional surface active agents (typically, Pluronic F68, polyvinyl alcohol, Tween 20 or Tween 80) and the selected polymer [typically, poly(lactide), poly(lactide-co-glycolide), poly(isobutylcyanoacrylate), poly(n-butylcyanoacrylate) or polysaccharides such chitosan or alginate] in water or organic solvent. The

Table 5. Some recent examples of enhancement of oral bioavailability (relative bioavailability) of poorly absorbed drugs by their incorporation into polymeric nanoparticle formulations as tested in the rat.

Drug	Nanoparticle formulation	Bioavailability increase	Reference
Methotrexate	Solid lipid	7-fold	130
Candesartan cilexetil	Solid lipid	12-fold	131
Docetaxel	PLGA (in Eudragit L microcapsules)	156-fold	132
Paclitaxel	PEGylated poly(anhydride)	9–70-fold	128
γ-Tocotrienol	Solid lipid	3-fold	133

formed emulsion (or double emulsion) is then homogenized to final, nanosize droplets and the external phase is removed by evaporation or precipitating the droplets by salting out. Alternatively, *in situ* polymerization is performed to obtain suspended nanoparticles. Finally, drying is performed by lyophilization or spray drying.[126]

Recent reports have demonstrated that the oral bioavailability of poorly absorbed drugs was increased (in rat studies) by formulating them into polymeric nanoparticles (Table 5). This increase could be related to the interaction between the nanoparticles and the intestinal mucosa, leading to a possible effect on efflux pumps and prolonging the residence time of the encapsulated drug in the vicinity of the intestinal wall, due to mucoadhesion.[128] Appropriate formulation could enhance the bioavailability of highly water-soluble (but poorly absorbed, biopharmaceutical classification system class III) drugs. A phospholipid complex of the water-soluble salvianolic acid B and Poloxamer F68 was prepared in the form of nanoparticles and tested in the perfused intestine rat model. It was found that the nanoparticles enabled intestinal lymphatic transport of the active probe, which increased the bioavailability primarily when perfused in the colon.[129]

5.3. *Mucoadhesion of nanoparticles and its improvement by recognition ligands*

The concept of adhesive, drug-loaded polymers, that adhere to epithelial tissues of the GI tract after oral ingestion, as a means of retarding intestinal transit and, thus, prolonging drug absorption was introduced, separately, in the 1980s by Robinson[134] and Peppas.[135] A typical polymer that was suggested to be most suitable for the purpose of mucoadhesion is polycarbophil, a loosely cross-linked polyacrylic acid with divinyl glycol. However, in human studies performed by Davis *et al.*, the effect of mucoadhesion on particle transit was found to be negligible, most probably due to mucus turnover.[135] (see also Sec. 4 above). In addition to the synthetic acrylic polymers, a variety of natural polysaccharide and proteinaceous polymers can serve as mucoadhesive platforms, such as gelatin, alginates, pectin[136] and most commonly, chitosan derivatives.[137]

Although mucoadhesion technology is questionable for large particles (e.g. 1 mm, beads[78,138]), it is believed that reducing particulate matter to the nanoscale and grafting recognition molecules onto their surfaces could lead to adhesiveness in the GI tract if formulated properly.[139] Given a successful overcoming of the viscous mucus gel barrier that covers the entire surface of the GI epithelium, nanoparticles could be used for a variety of medical applications, such as absorption enhancement of poorly water-soluble drugs, absorption enhancement of peptide and protein drugs, mucosal immunization and local treatment of regional inflammation. For this reason, it is highly important to carefully select the appropriate biological model for challenging the nanoparticles' performance. Cell cultures are a poor model as they do not provide information about mucus turnover rates.[114] Mathiowitz *et al.* prepared biodegradable adhesive nanoparticles made of poly(fumaric-co-sebacic) anhydride. The mucoadhesive properties and the small dimensions enabled them to cross the intestinal epithelium and pass the follicle-associated epithelium covering the lymphoid tissue of Peyer's patches. A proof of concept was demonstrated in the rat using dicumarol, insulin and plasmid DNA.[140] The spray-dried polyanhydride microspheres showed excellent mucoadhesive properties in everted intestinal sac preparations of the rat colon. As much as 65% of the initial dose adhered to the mucosa after a 30 min incubation.[141]

Improving mucoadhesion by surface recognition ligands

The passage of nanoparticles across the intestinal epithelium depends on their size[142] and surface structure.[143,144] Attempts to improve mucoadhesion by incorporating recognition ligands to the nanoparticles' surface, either by conjugation or by using coating techniques, were reported more than 20 years ago. For example, the uptake of polystyrene particles was enhanced in the rabbit Peyer's patches by decorating the particles with rabbit M cell-specific monoclonal antibody.[145]

Negatively charged nanoparticles made of poly(methyl vinyl ether-co-maleic anhydride) were prepared and coated with mannosamine (targets the lectin concanavalin A). Mucoadhesion was quantified and bioadhesion kinetics was calculated and compared to non-mannosylated (control) nanoparticles, after oral administration to rats. The adherence of the mannosylated nanoparticles was found to be two-fold stronger and longer, with a strong uptake by the Peyer's patches, compared with the control nanoparticles.[146] Tomato lectins were conjugated to the surface of polystyrene nanoparticles that, in turn, were administered via oral gavage to rats for five days. It was found that 23% of the dose was uptaken systemically compared to less than 0.5% of lectin-containing nanoparticles blocked with N-acetylchitotetraose (a 50-fold increase across the intestine). Also, intestinal uptake of the lectin-containing nanoparticles was 15 times higher through the villi than GALT tissues.[147]

The poorly water-soluble drug, bufalin, was loaded into lipid nanoparticles that were further conjugated with wheat germ agglutinin. A significantly higher adhesion between WGA-grafted lipid nanoparticles and the intestinal mucosa was found compared with that of nanoparticles that were not grafted with the wheat germ agglutinin. The oral bioavailability of bufalin was 2.7-fold higher compared with bufalin suspensions when administered to fasted rats.[148] Still, furnishing nanoparticles with targeting ligands does not necessarily improve mucoadhesion. For example, nanoparticles made of poly(methyl vinyl ether-co-maleic anhydride) were coated with either lectin from *Sambucus nigra* or bovine serum albumin (control) and administered orally to rats. Mucoadhesion was quantified by measuring the affinity of the nanoparticles, the intensity of the adherence, the adhesion time and the elimination rate of the adhered nanoparticles. As expected, the lectin-containing nanoparticles adhered for a longer period of time than the control particles. However, it took longer for the former to accumulate in the tested intestinal tissues. Moreover, in the stomach, the mucoadhesion intensity was profoundly higher for the control nanoparticles compared with the targeted nanoparticles.[149] Chen *et al.* incorporated modified *Ulex europaeus* agglutinin I into polymerized liposomes and improved their intestinal uptake by approx 11% compared to agglutinin-free liposomes.[150]

It should be realized that in the lumen of the GI tract, surface recognition ligands are exposed to premature interaction with carbohydrates, bacteria and proteins soaked in the mucus gel matrix.[94,151] Therefore, the design of targetable nanoparticles should comprise a profound understanding of the biological barriers confronting drug platforms after oral administration.

5.4. *Mucus penetrating nanoparticles*

It is well-established that PEG increases nanoparticle stability in biological fluids after oral administration. For example, PLA nanoparticles containing[125] I-radiolabeled tetanus toxoid were coated with PEG and tested in rats following oral ingestion. PEG prevented the degradation of the polymer and assisted in retaining the association of tetanus toxoid to the nanoparticles in simulated gastric fluids. Oral administration to rats revealed an improved uptake of the antigen into their blood and lymph fluid.[152] In 2007, Hanes *et al.* suggested to use high density PEG as a penetrating agent to assist the passage of nanoparticles through native mucus gels. PEG was suggested because it was realized that a protective polymer should possess strong hydrophilicity and neutral charge concurrently. These two characteristics are owned by PEG. Although early reports indicated that PEG possess mucoadhesive properties,[153] it was hypothesized, employing polystyrene nanoparticles, that modifying the nanoparticle surface with a high density, low molecular weight (2 kDa) PEG, would lead to a product with improved mucus penetration properties. Using a high-speed silicon-intensified target camera that allowed quantification of transport rates of individual particles of various sizes, Lai *et al.* measured the diffusion rates of 200 nm and 500 nm nanoparticles (larger than the mucus mesh-pore size, which is 10–200 nm). It was observed that the diffusion rate through native cervical mucus was four-fold (for the 200 nm particles) and six-fold (for the 500 nm particles) lower than the diffusion rate of the same particles in water, when coated with PEG. Surprisingly, the diffusion rate of smaller (100 nm) particles but identically coated, was found to be 200-fold lower in mucus than in water. Uncoated particles in the size range of 100–500 nm diffused 2,400-fold to 40,000-fold lower in mucus than in water. The fraction of the 100 nm particles that was immobilized by the mucus was much larger than that of the 200–500 nm particles. This surprising observation led the authors to suggest that large nanoparticles, if properly coated with PEG, can penetrate human mucus and, thus, may serve for mucosal drug delivery.[154] Employing a similar strategy, this group suggested the use of single-site-functionalized PEG dendrons to prepare gene vectors that penetrate human mucus barriers. They used PAMAM S-S dendrimers that contained a high density of primary amino groups on their surface together with a cleavable disulfide bond in their core. PEGylating the terminal amino groups, cleaving the disulfide bond in the core, and coupling various cationic polymers to the free sulfhydryl group, lead to the creation of nanostructures containing cationic polymer cores, conjugated with PEG modified dendrons that could be designed for gene-delivery applications. PEG dendron-coated gene vectors assembled in this way, with highly PEGylated PEI/DNA, showed much larger particle size and/ or incomplete particle assembly, displaying a near-neutral surface charge, with a dense PEG coating that did not reduce the ability of cationic polymers to efficiently compact the plasmid DNA.[155]

6. Toxicological Aspects

With the development of nanotechnology over the past 20 years, it became obvious that this discipline requires attention to safety issues. There are two major toxicology aspects associated with the handling and use of nanoparticles in the GI tract. The first is the risk of inhaling raw materials while processing the formulation components. The other is related to body distribution of the formulation components after being absorbed through the GI tract. Nanomaterials can induce pulmonary inflammation, oxidative stress (due to inhalation of surface active nanoparticles) and distal organ involvement.

The protective role of the mucus lining and its role in the entrapment of nanoparticles have already been discussed.[112] In terms of immunological risks it has been shown that intestinal absorption of endotoxin (LPS) could not be enhanced by chitosan nanoparticles due to electrostatic repulsion by mucus).[156] Moreover, in contrast to injected nanoparticles that circulate for days in the body and end up in the liver (about 70%), most of the ingested nanoparticles (about 90%) leave the GI tract with the feces. Still, the risk of immune perturbation by nanoparticles exists and is not completely understood. Although the reticuloendothelial system sequesters nanoparticles, the possible association with systemic protein may change their antigenicity and initiate autoimmune or primary and secondary immune responses.[157] Han *et al.* found that copper-loaded chitosan nanoparticles increased villi height, decreased crypt depth and increased the activity of trypsin, amylase lipase, sucrase and lactase after oral administration, compared to $CuSO_4$ aqueous solutions.[158] Although polystyrene nanoparticles are thought to be biocompatible, their chronic and acute oral administration impaired iron uptake in the chicken intestine loop model. The increased iron transport was caused by the nanoparticle disruption of the mucosal cells of the intestine. On the other hand, iron absorption in chickens that were acutely exposed to carboxylated particles was lowered. It was noted that chronic exposure to the nanoparticles caused remodeling of the intestinal villi, which increased the surface area available for iron absorption.[159]

Another toxicology aspect of nanotechnology is the ingredients used to produce nanoparticles. For example, Cremophor EL which, in contrast to earlier reports is not an inert vehicle. Its use has been associated with severe anaphylactoid hypersensitivity reactions, hyperlipidaemia, abnormal lipoprotein patterns, aggregation of erythrocytes and peripheral neuropathy. Its pharmacokinetics is dose-independent and may affect the disposition of drugs by changing the unbound drug concentration through micellar encapsulation. In addition, it has been shown that the compound can modify the toxicity profile of certain anticancer agents administered concomitantly.[33]

7. Future Prospects

In the past 25 years numerous studies have been reported in which poorly water-soluble or poorly absorbed drugs were loaded into a vast variety of polymeric vehicles to form orally administered nanoparticles. Yet, the only nano-technology that reached clinical phase and medical markets is the ordinary comminution technique in which particles are ground to the nanoscale in order to increase dissolution rates (Table 4). To some extent oral vaccination (e.g. Dukoral® that contains killed whole cells of *Vibrio cholerae* O1) may be classified as oral delivery of nanoparticles. However, it is time for innovation in the area of orally delivered nanoparticles.

Guan *et al.* have reported self-folding microstructures made by a soft-lithographic technique that can fold its micro-arms deep into the mucus layer, allowing the device to resist the surface erosion of the mucus layer covering the intestinal epithelium. The authors also suggested that the self-folding microstructures could grasp themselves into the intestinal villi, thus residing longer at the mucosal surface. The particulate microstructures consisted of three polymers, PLGA, chitosan, and poly(ethylene glycol methacrylate) and poly(ethylene glycol dimethacrylate).[160] A combination of a film and nanoparticles was suggested by Cui *et al.* It consists of nanoparticles made of carboxylated chitosan grafted in bilaminated films made of the chitosan-ethylenediaminetetraacetic acid hydrogel layer and ethylcellulose layer.[161] This approach of grafting nanoparticles into hydrogel platforms was also adopted by Schoener *et al.* who entrapped polymethyl methacrylate nanoparticles in poly(methacrylic-grafted-ethylene glycol) films for the purpose of shuttling doxorubicin locally in the colon.[162] Khazanov and Rubinstein suggested using grafted fluorescent nanoparticles, tagged with relevant antibodies for the *in vivo* detection of gastric cancer biomarkers in the lumen of the stomach.[163] At a later stage they entrapped the nanoparticles in a superporous hydrogel[164] to achieve a pseudo-zero order release of the antibody tagged fluorescent nanoparticles for diagnostic purposes (data not shown).

These few examples demonstrate that nanotechnology bears interesting possibilities and medical opportunities in the oral route of nanoparticulates. Innovative thinking will lead to novel prospects in drug treatment, targeting and imaging in the GI tract.

References

1. D. M. S. Eigler and E. K. Positioning single atoms with a scanning tunneling microscope. *Nature* **344**, 524–526 (1990).

2. Z. Sheng and Y. Liu, Effects of silver nanoparticles on wastewater biofilms. *Water Res.* **45**(18), 6039–6050 (2011).
3. M. Cavallini, M. Facchini, C. Albonetti and F. Biscarini, Single molecule magnets: from thin films to nano-patterns. *Phys. Chem. Chem. Phys.* **10**(6), 784–793 (2008).
4. K. Schilling, B. Bradford, D. Castelli, E. Dufour, J. F. Nash, W. Pape, S. Schulte, I. Tooley, J. van den Bosch and F. Schellauf, Human safety review of "nano" titanium dioxide and zinc oxide. *Photochem. Photobiol.* **9**(4), 495–509 (2010).
5. G. B. Adams, O. F. Sankey, J. B. Page, M. O'Keeffe and D. A. Drabold, Energetics of large fullerenes: balls, tubes, and capsules. *Science* **256**(5065), 1792–1795 (1992).
6. A. Montellano, T. Da Ros, A. Bianco and M. Prato, Fullerene C(6)(0) as a multifunctional system for drug and gene delivery. *Nanoscale* **3**(10), 4035–4041 (2011).
7. B. Dubertret, P. Skourides, D. J. Norris, V. Noireaux, A. H. Brivanlou and A. Libchaber, *In vivo* imaging of quantum dots encapsulated in phospholipid micelles. *Science* **298**(5599), 1759–1762 (2002).
8. M. F. De Volder, S. H. Tawfick, R. H. Baughman and A. J. Hart, Carbon nanotubes: present and future commercial applications. *Science* **339**(6119), 535–539 (2013).
9. R. Bawa, Nanoparticle-based therapeutics in humans: a survey. *Nanotechnol. Law Business* **5**(2), 135–155 (2008).
10. E. Sanders and C. T. Ashworth, A study of particulate intestinal absorption and hepatocellular uptake. Use of polystyrene latex particles. *Exp. Cell Res.* **22**, 137–145 (1961).
11. A. M. Hillery, P. U. Jani and A. T. Florence, Comparative, quantitative study of lymphoid and non-lymphoid uptake of 60 nm polystyrene particles. *J. Drug Target* **2**(2), 151–156 (1994).
12. G. Volkheimer, Passage of particles through the wall of the gastrointestinal tract. *Environ. Health Perspect* **9**, 215–225 (1974).
13. M. Aprahamian, C. Michel, W. Humbert, J. P. Devissaguet and C. Damge, Transmucosal passage of polyalkylcyanoacrylate nanocapsules as a new drug carrier in the small intestine. *Biol. Cell* **61**(1–2), 69–76 (1987).
14. C. L. Wells, M. A. Maddaus, S. L. Erlandsen and R. L. Simmons, Evidence for the phagocytic transport of intestinal particles in dogs and rats. *Infect Immun.* **56**(1), 278–282 (1988).
15. C. L. Wells, M. A. Maddaus and R. L. Simmons, Proposed mechanisms for the translocation of intestinal bacteria. *Rev. Infect Dis.* **10**(5), 958–979 (1988).
16. D. T. O'Hagan, The intestinal uptake of particles and the implications for drug and antigen delivery. *J. Anat.* **189**, 477–482 (1996).
17. M. E. LeFevre, R. Olivo, J. W. Vanderhoff and D. D. Joel, Accumulation of latex in Peyer's patches and its subsequent appearance in villi and mesenteric lymph nodes. *Proc. Soc. Exp. Biol. Med.* **159**(2), 298–302 (1978).
18. M. E. LeFevre and D. D. Joel, Distribution of label after intragastric administration of 7Be-labeled carbon to weanling and aged mice. *Proc. Soc. Exp. Biol. Med.* **182**(1), 112–119 (1986).
19. I. M. van der Lubben, J. C. Verhoef, G. Borchard and H. E. Junginger, Chitosan for mucosal vaccination. *Adv. Drug Deliv. Rev.* **52**(2), 139–144 (2001).
20. G. M. Hodges, E. A. Carr, R. A. Hazzard, C. O'Reilly and K. E. Carr, A commentary on morphological and quantitative aspects of microparticle translocation across the gastrointestinal mucosa. *J. Drug Target* **3**(1), 57–60 (1995).

21. J. F. Woodley, Lectins for gastrointestinal targeting — 15 years on. *J. Drug Target* 7(5), 325–333 (2000).
22. F. Gabor, E. Bogner, A. Weissenboeck and M. Wirth, The lectin-cell interaction and its implications to intestinal lectin-mediated drug delivery. *Adv. Drug Deliv. Rev.* **56**(4), 459–480 (2004).
23. A. T. Florence, The oral absorption of micro- and nanoparticulates: neither exeptional, nor unusual. *Pharm. Res.* **14**(3), 259–266 (1997).
24. E. Ziv and M. Bendayan, Intestinal absorption of peptides through the enterocytes. *Microsc. Res. Tech.* **49**(4), 346–352 (2000).
25. A. T. Florence, Issues in oral nanoparticle drug carrier uptake and targeting. *J. Drug Target* **12**(2), 65–70 (2004).
26. R. M. Atkinson, C. Bedford, K. J. Child and E. G. Tomich, Effect of particle size on blood griseofulvin-levels in man. *Nature* **193**, 588–589 (1962).
27. J. G. Wagner, Appraisal of digoxin bioavailability and pharmacokinetics in relation to cardiac therapy. *Am. Heart J.* **88**(2), 133–138 (1974).
28. E. J. Fraser, R. H. Leach and J. W. Poston, Bioavailability of digoxin. *Lancet* **2**(7776), 541 (1972).
29. J. Lindenbaum, V. P. Butler, Jr., J. E. Murphy and R. M. Cresswell, Correlation of digoxin-tablet dissolution-rate with biological availability. *Lancet* **1**(7814), 1215–1217 (1973).
30. T. R. Shaw, J. E. Carless, M. R. Howard and K. Raymond, Particle size and absorption of digoxin. *Lancet* **2**(7822), 209–210 (1973).
31. A. J. Jounela, P. J. Pentikainen and A. Sothmann, Effect of particle size on the bioavailability of digoxin. *Eur. J. Clin. Pharmacol.* **8**(5), 365–370 (1975).
32. D. F. Veber, S. R. Johnson, H. Y. Cheng, B. R. Smith, K. W. Ward and K. D. Kopple, Molecular properties that influence the oral bioavailability of drug candidates. *J. Med. Chem.* **45**(12), 2615–2623 (2002).
33. H. Gelderblom, J. Verweij, K. Nooter and A. Sparreboom, Cremophor EL: the drawbacks and advantages of vehicle selection for drug formulation. *Eur. J. Cancer* **37**(13), 1590–1598 (2001).
34. E. Merisko-Liversidge, G. G. Liversidge and E. R. Cooper, Nanosizing: a formulation approach for poorly-water-soluble compounds. *Eur. J. Pharm. Sci.* **18**(2), 113–120 (2003).
35. J. Hecq, M. Deleers, D. Fanara, H. Vranckx and K. Amighi, Preparation and characterization of nanocrystals for solubility and dissolution rate enhancement of nifedipine. *Int. J. Pharm.* **299**(1–2), 167–177 (2005).
36. B. C. Lippold and A. Ohm, Correlation between wettability and dissolution rate of pharmaceutical powders. *Int. J. Pharm.* **28**, 67–74 (1986).
37. P. E. Luner and D. Vander Kamp, Wetting behavior of bile salt-lipid dispersions and dissolution media patterned after intestinal fluids. *J. Pharm. Sci.* **90**(3), 348–359 (2001).
38. C. He, L. Yin, C. Tang and C. Yin, Size-dependent absorption mechanism of polymeric nanoparticles for oral delivery of protein drugs. *Biomaterials* **33**(33), 8569–8578 (2012).
39. S. H. Bakhru, S. Furtado, A. P. Morello and E. Mathiowitz, Oral delivery of proteins by biodegradable nanoparticles. *Adv. Drug Deliv. Rev.* **65**(6), 811–821 (2013).
40. E. Danforth, Jr. and R. O. Moore, Intestinal absorption of insulin in the rat. *Endocrinology* **65**(1), 118–123 (1959).

41. G. P. Carino and E. Mathiowitz, Oral insulin delivery. *Adv. Drug Deliv. Rev.* **35**(2–3), 249–257 (1999).
42. C. Damge, C. Michel, M. Aprahamian and P. Couvreur, New approach for oral administration of insulin with polyalkylcyanoacrylate nanocapsules as drug carrier. *Diabetes* **37**(2), 246–251 (1988).
43. B. Sarmento, A. Ribeiro, F. Veiga, P. Sampaio, R. Neufeld and D. Ferreira, Alginate/chitosan nanoparticles are effective for oral insulin delivery. *Pharm. Res.* **24**(12), 2198–2206 (2007).
44. K. Chaturvedi, K. Ganguly, M. N. Nadagouda and T. M. Aminabhavi, Polymeric hydrogels for oral insulin delivery. *J. Control. Release* **165**(2), 129–138 (2013).
45. F. Y. Su, K. J. Lin, K. Sonaje, S. P. Wey, T. C. Yen, Y. C. Ho, N. Panda, E. Y. Chuang, B. Maiti and H. W. Sung, Protease inhibition and absorption enhancement by functional nanoparticles for effective oral insulin delivery. *Biomaterials* **33**(9), 2801–2811 (2012).
46. K. J. Maloy, A. M. Donachie, D. T. O'Hagan and A. M. Mowat, Induction of mucosal and systemic immune responses by immunization with ovalbumin entrapped in poly(lactide-co-glycolide) microparticles. *Immunology* **81**(4), 661–667 (1994).
47. S. J. Challacombe, D. Rahman, H. Jeffery, S. S. Davis and D. T. O'Hagan, Enhanced secretory IgA and systemic IgG antibody responses after oral immunization with biodegradable microparticles containing antigen. *Immunology* **76**(1), 164–168 (1992).
48. D. T. O'Hagan, J. P. McGee, J. Holmgren, A. M. Mowat, A. M. Donachie, K. H. Mills, W. Gaisford, D. Rahman and S. J. Challacombe, Biodegradable microparticles for oral immunization. *Vaccine* **11**(2), 149–154 (1993).
49. D. T. O'Hagan, In: *Novel Delivery Systems for Oral Vaccines*, D. T. O'Hagan (Ed.), Boca Raton: CRC Press (1994), p. 175.
50. D. T. O'Hagan, Recent developments in vaccine delivery systems. *Curr. Drug Targets Infect Disord.* **1**(3), 273–286 (2001).
51. P. N. Gupta, K. Khatri, A. K. Goyal, N. Mishra and S. P. Vyas, M-cell targeted biodegradable PLGA nanoparticles for oral immunization against hepatitis B. *J. Drug Target* **15**(10), 701–713 (2007).
52. A. K. Jain, A. K. Goyal, N. Mishra, B. Vaidya, S. Mangal and S. P. Vyas, PEG-PLA-PEG block copolymeric nanoparticles for oral immunization against hepatitis B. *Int. J. Pharm.* **387**(1–2), 253–262 (2009).
53. M. Garinot, V. Fievez, V. Pourcelle, F. Stoffelbach, A. des Rieux, L. Plapied, I. Theate, H. Freichels, C. Jerome, J. Marchand-Brynaert, Y. J. Schneider and V. Preat, PEGylated PLGA-based nanoparticles targeting M cells for oral vaccination. *J. Control. Release* **120**(3), 195–204 (2007).
54. M. Adomako, S. St-Hilaire, Y. Zheng, J. Eley, R. D. Marcum, W. Sealey, B. C. Donahower, S. Lapatra and P. P. Sheridan, Oral DNA vaccination of rainbow trout, Oncorhynchus mykiss (Walbaum), against infectious haematopoietic necrosis virus using PLGA [Poly(D,L-Lactic-Co-Glycolic Acid)] nanoparticles. *J. Fish Dis.* **35**(3), 203–214 (2012).
55. K. Yoncheva, M. N. Centelles and J. M. Irache, Development of bioadhesive amino-pegylated poly(anhydride) nanoparticles designed for oral DNA delivery. *J. Microencapsul.* **25**(2), 82–89 (2008).
56. E.-M. Collnot, H. Ali and C. M. L. Lehr, Nano- and microparticulate drug carriers for targeting of the inflamed intestinal mucosa. *J. Control. Release* **161**(2), 235–246 (2012).

57. A. Lamprecht, U. Schafer and C. M. Lehr, Size-dependent bioadhesion of micro- and nanoparticulate carriers to the inflamed colonic mucosa. *Pharm. Res.* **18**(6), 788–793 (2001).
58. A. Lamprecht, N. Ubrich, H. Yamamoto, U. Schafer, H. Takeuchi, P. Maincent, Y. Kawashima and C. M. Lehr, Biodegradable nanoparticles for targeted drug delivery in treatment of inflammatory bowel disease. *J. Pharmacol. Exp. Ther.* **299**(2), 775–781 (2001).
59. A. M. Dvorak and G. R. Dickersin, Crohn's disease: transmission electron microscopic studies. I. Barrier function. Possible changes related to alterations of cell coat, mucous coat, epithelial cells, and Paneth cells. *Hum. Pathol.* **11**(5 Suppl), 561–571 (1980).
60. A. Rubinstein and B. Tirosh, Mucus gel thickness and turnover in the gastrointestinal tract of the rat, responses to cholinergic stimulus and implication for mucoadhesion. *Pharm. Res.* **11**(6), 794–799 (1994).
61. T. T. Jubeh, M. Nadler-Milbauer, Y. Barenholz and A. Rubinstein, Local treatment of experimental colitis in the rat by negatively charged liposomes of catalase, TMN and SOD. *J. Drug Target* **14**(3), 155–163 (2006).
62. B. Moulari, D. Pertuit, Y. Pellequer and A. Lamprecht, The targeting of surface modified silica nanoparticles to inflamed tissue in experimental colitis. *Biomaterials* **29**(34), 4554–4560 (2008).
63. H. Laroui, G. Dalmasso, H. T. Nguyen, Y. Yan, S. V. Sitaraman and D. Merlin, Drug-loaded nanoparticles targeted to the colon with polysaccharide hydrogel reduce colitis in a mouse model. *Gastroenterology* **138**(3), 843–853 (2010).
64. D. S. Wilson, G. Dalmasso, L. Wang, S. V. Sitaraman, D. Merlin and N. Murthy, Orally delivered thioketal nanoparticles loaded with TNF-alpha-siRNA target inflammation and inhibit gene expression in the intestines. *Nat. Mater.* **9**(11), 923–928 (2010).
65. C. Kriegel and M. Amiji, Oral TNF-alpha gene silencing using a polymeric microsphere-based delivery system for the treatment of inflammatory bowel disease. *J. Control. Release* **150**(1), 77–86 (2011).
66. E. Harel, A. Rubinstein, A. Nissan, E. Khazanov, M. Nadler Milbauer, Y. Barenholz and B. Tirosh, Enhanced transferrin receptor expression by proinflammatory cytokines in enterocytes as a means for local delivery of drugs to inflamed gut mucosa. *PLoS One* **6**(9), e24202 (2011).
67. W. Niebel, K. Walkenbach, A. Beduneau, Y. Pellequer and A. Lamprecht, Nanoparticle-based clodronate delivery mitigates murine experimental colitis. *J. Control. Release* **160**(3), 659–665 (2012).
68. E. Abutbul and A. Rubinstein, In *Biodrug Delivery Systems: Fundamentals, Applications and Clinical Development,* M. P. Morishita, K. (Ed.), Vol. 194, New York: Informa Healthcare (2009), pp. 138–171.
69. E. Elinav and D. Peer, Harnessing nanomedicine for mucosal theranostics--a silver bullet at last? *ACS Nano* 7(4), 2883–2890 (2013).
70. S. Pamujula, V. Kishore, B. Rider, K. C. Agrawal and T. K. Mandal, Radioprotection in mice following oral administration of WR-1065/PLGA nanoparticles. *Int. J. Radiat. Biol.* **84**(11), 900–908 (2008).
71. D. Y. Stainier, No organ left behind: tales of gut development and evolution. *Science* **307**(5717), 1902–1904 (2005).
72. A. Rubinstein, In *Gastrointestinal anatomy, physiology and permeation pathways,* E. Touitou and B. W. Barry (Eds.), *Enhancement in Drug Delivery*: Boca Raton: CRC (2006), pp. 3–37.

73. H. Kurz, In *Pharmacology of Intestinal Absorption: Gastrointestinal Absorption of Drugs*, W. Forth and W. Rummel (Eds.), Vol. 1, Oxford: Pergamon Press (1975), pp. 245–296.
74. R. R. Scheline, Metabolism of foreign compounds by gastrointestinal microorganisms. *Pharmacol. Rev.* **25**(4), 451–523 (1973).
75. A. G. de Boer, E. J. van Hoogdalem and D. D. Breimer, Rate-controlled rectal peptide drug absorption enhancement. *Adv. Drug Deliv. Rev.* **8**(2–3), 237–251 (1992).
76. C. F. Code and J. A. Marlett, The interdigestive myo-electric complex of the stomach and small bowel of dogs. *J. Physiol.* **246**, 289–309 (1975).
77. A. J. Coupe, S. S. Davis, D. F. Evans and I. R. Wilsing, Do pellets empty from the stomach with food? *Int. J. Pharm.* **92**, 167–175 (1993).
78. P. Gruber, A. Rubinstein, H. K. V. Li, P. Bass and J. R. Robinson, Gastric emptying of nondigestible solids in the fasted dog. *J. Pharm. Sci.* **76**(2), 117–122 (1987).
79. D. Y. Graham, J. L. Smith and A. A. Bouvet, What happens to tablets and capsules in the stomach: endoscopic comparison of disintegration and dispersion characteristics of two microencapsulated potassium formulations. *J. Pharm. Sci.* **79**(5), 420–424 (1990).
80. B. Tirosh and A. Rubinstein, Migration of adhesive and non-adhesive particles in the rat intestine under altered mucus secretion conditions. *J. Pharm. Sci.* **87**(4), 453–456 (1998).
81. S. S. Davis, J. G. Hardy and J. W. Fara, Transit of pharmaceutical dosage forms through the small intestine. *Gut* **27**(8), 886–892 (1986).
82. D. A. Adkin, S. S. Davis, R. A. Sparrow, P. D. Huckle, A. J. Phillips and I. R. Wilding, The effect of pharmaceutical exipients on small intestinal transit. *Br. J. Clin. Pharmacol.* **39**(4), 381–387 (1995).
83. J. G. Hardy, C. G. Wilson and E. Wood, Drug delivery to the proximal colon. *J. Pharm. Pharmacol.* **37**, 874–877 (1985).
84. S. K. Sarna, Physiology and pathophysiology of colonic motor activity (1). *Dig Dis. Sci.* **36**(6), 827–862 (1991).
85. M. Proano, M. Camilleri, S. F. Phillips, G. M. Thomforde, M. L. Brown and R. L. Tucker, Unprepared human colon does not discriminate between solids and liquids. *Am. J. Physiol.* **260**, G13–G16 (1991).
86. J. B. Dressman, R. R. Berardi, L. C. Dermentzoglou, T. L. Russell, S. P. Schmaltz, J. L. Barnett and K. M. Jarvenpaa, Upper gastrointestinal (GI) pH in young, healthy men and women. *Pharm. Res.* 7, 756–761 (1990).
87. G. T. McEwan, B. Schousboe and E. Skadhauge, Direct measurement of mucosal surface pH of pig jejunum *in vivo. Zentralbl Veterinarmed A* **37**(6), 439–444 (1990).
88. Y. F. Shiau, P. Fernandez, M. J. Jackson, S. McMonagle, Mechanisms maintaining a low-pH microclimate in the intestine. *Am. J. Physiol.* **248**(6 Pt 1), G608–G617 (1985).
89. J. H. Kou, D. Fleisher and G. L. Amidon, Calculation of the aqueous diffusion layer resistance for absorption in a tube: application to intestinal membrane permeability determination. *Pharm. Res.* **8**(3), 298–305 (1991).
90. M. D. Levitt, C. A. Fetzer, J. M. Kneip, J. H. Bond and D. G. Levitt, Quantitative assessment of luminal stirring in the perfused small intestine of the rat. *Am. J. Physiol.* **252** (3 Pt 1), G325–G332 (1987).
91. K. W. Smithson and D. B. Millar, Intestinal diffusion barrier: unstirred water layer or membrane surface mucous coat? *Science* **214**, 1241–1244 (1981).

92. D. Winne, H. Goig and U. Muller, Closed rat Jejunum segment *in situ*: role of pre-epithelial diffusion resistance (unstirred layer) in absorption process and model analysis. *Naunyn Schmiedebergs Arch. Pharmacol.* **335**(2), 204–215 (1987).
93. S. K. Lai, Y. Y. Wang, D. Wirtz and J. Hanes, Micro- and macrorheology of mucus. *Adv. Drug Deliv. Rev.* **61**(2), 86–100 (2009).
94. L. M. Ensign, R. Cone and J. Hanes, Oral drug delivery with polymeric nanoparticles: the gastrointestinal mucus barriers. *Adv. Drug Deliv. Rev.* **64**(6), 557–570 (2012).
95. E. Engel, P. H. Guth, Y. Nishizaki and J. D. Kaunitz, Barrier function of the gastric mucus gel. *Am. J. Physiol.* **269**(6 Pt 1), G994–G999 (1995).
96. A. Allen and G. Flemstrom, Gastroduodenal mucus bicarbonate barrier: protection against acid and pepsin. *Am. J. Physiol. Cell Physiol.* **288**(1), C1–C19 (2005).
97. J. T. Lamont, In *Neuro-Immuno Physiology of the Gastrointestinal Mucosa, Implications for Inflammatory Diseases*, R. H. Stead, M. H. Perdue, H. Cook, D. W. Powel and K. E. Barrett (Eds.), Vol. 664, New York: New York Academy of Sciences (1992), pp. 190–201.
98. P. Sherman, N. Fleming, J. Forstner, N. Roomi, G. Forstner, Bacteria and the mucus blanket in experimental small bowel bacterial overgrowth. *Am. J. Pathol.* **126**(3), 527–534 (1987).
99. C.-M. Lehr, F. G. P. Poelma, H. E. Junginger and J. J. Tukker, An estimate of turnover time of intestinal mucus gel layer in the rat in situ loop. *Int. J. Pharm.* **70**, 235–240 (1991).
100. C. Atuma, V. Strugala, A. Allen and L. Holm, The adherent gastrointestinal mucus gel layer: thickness and physical state *in vivo*. *Am. J. Physiol. Gastrointest Liver Physiol.* **280**(5), G922–G929 (2001).
101. M. Bickel and G. L. Kauffman, Jr., Gastric gel mucus thickness: effect of distention, 16,16-dimethyl prostaglandin e2, and carbenoxolone. *Gastroenterology* **80**(4), 770–775 (1981).
102. R. D. Pullan, G. A. O. Thomas, M. Rhodes, R. G. Newcombe, G. T. Williams, A. Allen and J. Rhodes, Thickness of adherent mucus gel on colonic mucosa in humans and its relevance to colitis. *Gut* **35**, 353–359 (1994).
103. K. H. Ryu and E. Grim, Unstirred water layer in canine jejunum. *Am. J. Physiol.* **242**(4), G364–G369 (1982).
104. D. Winne and W. Verheyen, Diffusion coefficient in native mucus gel of rat small intestine. *J. Pharm. Pharmacol.* **42**(7), 517–519 (1990).
105. G. Flemstrom, A. Hallgren, O. Nylander, L. Engstrand, E. Wilander and A. Allen, Adherent surface mucus gel restricts diffusion of macromolecules in rat duodenum *in vivo*. *Am. J. Physiol.* **277**(2 Pt 1), G375–382 (1999).
106. J. Nakamura, K. Shima, T. Kimura, S. Muranishi and H. Sezaki, Role of intestinal mucus in the absorption of quinine and water-soluble dyes from the rat small intestine. *Chem. Pharm. Bull (Tokyo)*. **26**(3), 857–863 (1978).
107. Y. F. Shiau, Mechanism of intestinal fatty acid uptake in the rat: the role of an acidic microclimate. *J. Physiol.* **421**, 463–474 (1990).
108. Y. F. Shiau, R. J. Kelemen and M. A. Reed, Acidic mucin layer facilitates micelle dissociation and fatty acid diffusion. *Am. J. Physiol.* **259**(4 Pt 1), G671–G675 (1990).
109. S. S. Olmsted, J. L. Padgett, A. I. Yudin, K. J. Whaley, T. R. Moench and R. A. Cone, Diffusion of macromolecules and virus-like particles in human cervical mucus. *Biophys. J.* **81**(4), 1930–1937 (2001).

110. S. K. Lai, Y. Y. Wang and J. Hanes, Mucus-penetrating nanoparticles for drug and gene delivery to mucosal tissues. *Adv. Drug Deliv. Rev.* **61**(2), 158–171 (2009).
111. D. A. Norris and P. J. Sinko, Effect of size, surface charge, and hydrophobicity on the translocation of polystyrene microspheres through gastrointestinal mucin. *J. Appl. Polym. Sci.* **63**, 1481–1492 (1997).
112. R. A. Cone, Barrier properties of mucus. *Adv. Drug Deliv. Rev.* **61**(2), 75–85 (2009).
113. U. S. Sajjan, M. Corey, M. A. Karmali and J. F. Forstner, Binding of Pseudomonas cepacia to normal human intestinal mucin and respiratory mucin from patients with cystic fibrosis. *J. Clin. Invest.* **89**(2), 648–656 (1992).
114. C. Durrer, J. M. Irache, F. Puisieux, D. Duchene and G. Ponchel, Mucoadhesion of latexes. II. Adsorption isotherms and desorption studies. *Pharm. Res.* **11**(5), 680–683 (1994).
115. P. Jani, G. W. Halbert, J. Langridge and A. T. Florence, Nanoparticle uptake by the rat gastrointestinal mucosa: quantitation and particle size dependency. *J. Pharm. Pharmacol.* **42**(12), 821–826 (1990).
116. M. Nefzger, J. Kreuter, R. Voges, E. Liehl and R. Czok, Distribution and elimination of polymethyl methacrylate nanoparticles after peroral administration to rats. *J. Pharm. Sci.* **73**(9), 1309–1311 (1984).
117. T. W. Spahn and T. Kucharzik, Modulating the intestinal immune system: the role of lymphotoxin and GALT organs. *Gut* **53**(3), 456–465 (2004).
118. J. U. Junghanns and R. H. Muller, Nanocrystal technology, drug delivery and clinical applications. *Int. J. Nanomedicine* **3**(3), 295–309 (2008).
119. B. C. Hancock and M. Parks, What is the true solubility advantage for amorphous pharmaceuticals? *Pharm. Res.* **17**(4), 397–404 (2000).
120. V. K. Tzitzios, A. Bakandritsos, V. Georgakilas, G. Basina, N. Boukos, A. B. Bourlinos, D. Niarchos and D. Petridis, Large-scale synthesis, size control, and anisotropic growth of gamma-Fe2O3 nanoparticles: organosols and hydrosols. *J. Nanosci. Nanotechnol.* 7(8), 2753–2757.
121. R. H. Muller, C. Jacobs and O. Kayser, Nanosuspensions as particulate drug formulations in therapy. Rationale for development and what we can expect for the future. *Adv. Drug Deliv. Rev.* **47**(1), 3–19 (2001).
122. G. Gaucher, P. Satturwar, M. C. Jones, A. Furtos and J. C. Leroux, Polymeric micelles for oral drug delivery. *Eur. J. Pharm. Biopharm.* **76**(2), 147–158 (2010).
123. J. A. Rogers and K. E. Anderson, The potential of liposomes in oral drug delivery. *Crit. Rev. Ther. Drug Carrier Syst.* **15**(5), 421–480 (1998).
124. V. P. Torchilin, Recent advances with liposomes as pharmaceutical carriers. *Nat. Rev. Drug Discov.* **4**(2), 145–160 (2005).
125. Y. Barenholz, In *Handbook of Harnessing Biomaterials in Nanomedicine: Preparation, Toxicity and Applications*, D. Peer (Ed.), Singapore: Pan Stanford (2012), pp. 335–398.
126. C. Vauthier and K. Bouchemal, Methods for the preparation and manufacture of polymeric nanoparticles. *Pharm. Res.* **26**(5), 1025–1058 (2009).
127. R. R. Sawant, R. M. Sawant, A. A. Kale and V. P. Torchilin, The architecture of ligand attachment to nanocarriers controls their specific interaction with target cells. *J. Drug Target* **16**(7), 596–600 (2008).

128. V. Zabaleta, G. Ponchel, H. Salman, M. Agueros, C. Vauthier and J. M. Irache, Oral administration of paclitaxel with pegylated poly(anhydride) nanoparticles: permeability and pharmacokinetic study. *Eur. J. Pharm. Biopharm.* **81**(3), 514–523 (2012).
129. Q. Peng, Z. R. Zhang, X. Sun, J. Zuo, D. Zhao and T. Gong, Mechanisms of phospholipid complex loaded nanoparticles enhancing the oral bioavailability. *Mol. Pharm.* 7(2), 565–675 (2010).
130. R. Paliwal, S. Rai, B. Vaidya, K. Khatri, A. K. Goyal, N. Mishra, A. Mehta and S. P. Vyas, Effect of lipid core material on characteristics of solid lipid nanoparticles designed for oral lymphatic delivery. *Nanomedicine* **5**(2), 184–191 (2009).
131. Z. Zhang, F. Gao, H. Bu, J. Xiao and Y. Li, Solid lipid nanoparticles loading candesartan cilexetil enhance oral bioavailability: in vitro characteristics and absorption mechanism in rats. *Nanomedicine* **8**(5), 740–747 (2011).
132. T. Nassar, S. Attili-Qadri, O. Harush-Frenkel, S. Farber, S. Lecht, P. Lazarovici and S. Benita, High plasma levels and effective lymphatic uptake of docetaxel in an orally available nanotransporter formulation. *Cancer Res.* **71**(8), 3018–3028 (2011).
133. B. S. Abuasal, C. Lucas, B. Peyton, A. Alayoubi, S. Nazzal, P. W. Sylvester and A. Kaddoumi, Enhancement of intestinal permeability utilizing solid lipid nanoparticles increases gamma-tocotrienol oral bioavailability. *Lipids* **47**(5), 461–469 (2012).
134. H. S. Ch'ng, H. Park, P. Kelly and J. R. Robinson, Bioadhesive polymers as platforms for oral controlled drug delivery II: synthesis and evaluation of some swelling, water-insoluble bioadhesive polymers. *J. Pharm. Sci.* **74**, 399–405 (1985).
135. R. Gurney, J. M. Meyer and N. A. Peppas, Bioadhesive intraoral release system: design, testing and analysis. *Biomaterials* **5**, 336–340 (1984).
136. J. D. Smart, I. W. Kellaway and H. E. C. Worthington, An in vitro investigation of mucosa adhesive materials for use in controlled drug delivery. *J. Pharm. Pharmacol.* **36**(5), 295–299 (1984).
137. M. N. V. Ravi Kumar, R. A. A. Muzzarelli, C. Muzzarelli, H. Sashiwa and A. J. Domb, Chitosan chemistry and pharmaceutical perspectives. *Chem. Rev.* **104**, 6017–6084 (2004).
138. R. Khosla and S. S. Davis, The effect of polycarbophil on the gastric emptying of pellets. *J. Pharm. Pharmacol.* **39**, 47–49 (1987).
139. E. Mathiowitz, D. E. Chickering III and C. M. Lehr, *Bioadhesive Drug Delivery Systems, Fundamentals, Novel Approaches and Development.* New York: Marcel Dekker (1999).
140. E. Mathiowitz, J. S. Jacob, Y. S. Jong, G. P. Carino, D. E. Chickering, P. Chaturvedi, C. A. Santos, K. Vijayaraghavan, S. Montgomery, M. Bassett and C. Morrell, Biologically erodable microspheres as potential oral drug delivery systems. *Nature* **386**, 410–414 (1997).
141. D. Chickering, J. Jacob and E. Mathiowitz, Poly(fumaric-co-sebacic) microspheres as oral drug delivery systems. *Biotechnol. Bioeng.* **52**(1), 96–101 (1996).
142. M. P. Desai, V. Labhasetwar, G. L. Amidon and R. J. Levy, Gastrointestinal uptake of biodegradable microparticles effect of particle size. *Pharm. Res.* **13**(12), 1838–1845 (1996).
143. P. Arbos, M. A. Campanero, M. A. Arangoa, M. J. Renedo and J. M. Irache, Influence of the surface characteristics of PVM/MA nanoparticles on their bioadhesive properties. *J. Control. Release* **89**(1), 19–30 (2003).

144. J. Buske, C. Konig, S. Bassarab, A. Lamprecht, S. Muhlau and K. G. Wagner, Influence of PEG in PEG-PLGA microspheres on particle properties and protein release. *Eur. J. Pharm. Biopharm.* (2012).
145. J. Pappo, T. H. Ermak and H. J. Steger, Monoclonal antibody-directed targeting of fluorescent polystyrene microspheres to Peyer's patch M cells. *Immunology* **73**(3), 277–280 (1991).
146. H. H. Salman, C. Gamazo, M. A. Campanero and J. M. Irache, Bioadhesive mannosylated nanoparticles for oral drug delivery. *J. Nanosci. Nanotechnol.* **6**(9–10), 3203–3209 (2006).
147. N. Hussain, P. U. Jani and A. T. Florence, Enhanced oral uptake of tomato lectin-conjugated nanoparticles in the rat. *Pharm. Res.* **14**(5), 613–618 (1997).
148. Y. Liu, P. Wang, C. Sun, J. Zhao, Y. Du, F. Shi and N. Feng, Bioadhesion and enhanced bioavailability by wheat germ agglutinin-grafted lipid nanoparticles for oral delivery of poorly water-soluble drug bufalin. *Int. J. Pharm.* **419**(1–2), 260–265 (2011).
149. P. Arbos, M. A. Arangoa, M. A. Campanero and J. M. Irache, Quantification of the bioadhesive properties of protein-coated PVM/MA nanoparticles. *Int. J. Pharm.* **242** (1–2), 129–136 (2002).
150. H. Chen, V. P. Torchilin and R. Langer, Lectin-bearing polymerized liposomes as potential oral vaccine carriers. *Pharm. Res.* **13**(9), 1378–1383 (1996).
151. J. M. Irache, C. Durrer, D. Duchene and G. Ponchel, Bioadhesion of lectin-latex conjugates to rat intestinal mucosa. *Pharm. Res.* **13**(11), 1716–1179 (1996).
152. M. Tobio, A. Sanchez, A. Vila, I. I. Soriano, C. Evora, J. L. Vila-Jato and M. J. Alonso, The role of PEG on the stability in digestive fluids and *in vivo* fate of PEG-PLA nanoparticles following oral administration. *Colloids Surf B Biointerfaces* **18**(3–4), 315–323 (2000).
153. L. Serra, J. Domenech and N. A. Peppas, Design of poly(ethylene glycol)-tethered copolymers as novel mucoadhesive drug delivery systems. *Eur. J. Pharm. Biopharm.* **63**(1), 11–18 (2006).
154. S. K. Lai, D. E. O'Hanlon, S. Harrold, S. T. Man, Y. Y. Wang, R. Cone and J. Hanes, Rapid transport of large polymeric nanoparticles in fresh undiluted human mucus. *Proc. Natl. Acad. Sci. USA* **104**(5), 1482–1487 (2007).
155. A. J. Kim, N. J. Boylan, J. S. Suk, M. Hwangbo, T. Yu, B. S. Schuster, L. Cebotaru, W. G. Lesniak, J. S. Oh, P. Adstamongkonkul, A. Y. Choi, R. M. Kannan and J. Hanes, Use of single-site-functionalized PEG dendrons to prepare gene vectors that penetrate human mucus barriers. *Angew. Chem. Int. Ed. Engl.* **52**(14), 3985–3988 (2013).
156. K. Sonaje, K. J. Lin, M. T. Tseng, S. P. Wey, F. Y. Su, E. Y. Chuang, C. W. Hsu, C. T. Chen and H. W. Sung, Effects of chitosan-nanoparticle-mediated tight junction opening on the oral absorption of endotoxins. *Biomaterials* **32**(33), 8712–8721 (2011).
157. A. Nel, T. Xia, L. Madler and N. Li, Toxic potential of materials at the nanolevel. *Science* **311**(5761), 622–627 (2006).
158. X. Y. Han, W. L. Du, Q. C. Huang, Z. R. Xu and Y. Z. Wang, Changes in small intestinal morphology and digestive enzyme activity with oral administration of copper-loaded chitosan nanoparticles in rats. *Biol. Trace Elem. Res.* **145**(3), 355–360 (2011).
159. G. J. Mahler, M. B. Esch, E. Tako, T. L. Southard, S. D. Archer, R. P. Glahn and M. L. Shuler, Oral exposure to polystyrene nanoparticles affects iron absorption. *Nat. Nanotechnol.* 7(4), 264–271 (2012).

160. J. Guan, H. He, L. J. Lee and D. J. Hansford, Fabrication of particulate reservoir-containing, capsulelike, and self-folding polymer microstructures for drug delivery. *Small* **3**(3), 412–418 (2007).
161. F. Cui, C. He, L. Yin, F. Qian, M. He, C. Tang and C. Yin, Nanoparticles incorporated in bilaminated films: a smart drug delivery system for oral formulations. *Biomacromolecules* **8**(9), 2845–2850 (2007).
162. C. A. Schoener, H. N. Hutson and N. A. Peppas, pH-responsive hydrogels with dispersed hydrophobic nanoparticles for the oral delivery of chemotherapeutics. *J. Biomed. Mater. Res. A* **101**(8), 2229–2236 (2013).
163. E. Khazanov, E. Yavin, A. Pascal, A. Nissan, Y. Kohl, M. Reimann-Zawadzki and A. Rubinstein, Detecting a secreted gastric cancer biomarker molecule by targeted nanoparticles for real-time diagnostics. *Pharm. Res.* **29**(4), 983–993 (2012).
164. J. Chen, H. Park and K. Park, Syntesis of superporous hydrogels: hydrogels with fast swelling and superabsorbent properties. *J. Biomed. Mater. Res. Part A* **44**, 53–62 (1999).

Chapter 5

Nanopreparations for Oral Administration

D. Hubbard[*,‡], *D. J. Brayden*[§] *and H. Ghandehari*[*,†,‡]

[*]*Department of Bioengineering,*
University of Utah
Sorenson Molecular Biotechnology Building,
36 S. Wasatch Dr., Salt Lake City, UT, 84112, USA
[†]*Pharmaceutics and Pharmaceutical Chemistry,*
University of Utah
Sorenson Molecular Biotechnology Building,
36 S. Wasatch Dr., Salt Lake City, UT, 84112, USA
[‡]*Utah Center for Nanomedicine,*
Nano Institute of Utah, University of Utah
Sorenson Molecular Biotechnology Building,
36 S. Wasatch Dr., Salt Lake City, UT, 84112, USA
[§]*School of Veterinary Medicine, University College Dublin,*
Room 214, Veterinary Sciences Centre,
University College Dublin, Belfield, Dublin 4, Ireland
hamil.ghandehari@pharm.utah.edu

Abstract

Oral delivery has distinct advantages over parenteral administration. These include ease of administration and improved patient compliance. Physiological barriers to oral drug delivery include low pH environment of the stomach, high enzymatic activity of the small intestine, and poor permeability of large molecular weight and hydrophilic compounds across the gastrointestinal epithelial layer. These barriers reduce stability and absorption of drug molecules to the systemic circulation. In order to overcome these barriers, nanopreparations have been investigated for their potential in oral delivery. In this chapter, selected nanoscale formulations investigated for oral drug delivery are discussed.

1. Introduction

Oral administration of pharmaceuticals has long been the preferred route compared to parenteral injections. The benefits of the oral route include flexible dosing, higher patient compliance and reduced health care costs.[1] Many small molecules and biologicals have limited oral bioavailability due to low solubility, limited stability in the GI (gastrointestinal) tract as well as the low and variable permeation.[2,3] These challenges form the basis of the design of nanopreparations for oral delivery.[4] Oral nanopreparations have been investigated for improving the bioavailability (F) of various drugs such as heparin,[5] enalaprilat,[6] tobramycin,[7] and antitubercular drugs.[8] Nanopreparations may enhance solubility, increase stability and absorption, and control spatiotemporal release.[2] Already, several nanopreparations have been approved by regulatory agencies for use in the clinic as oral dosages (e.g. Élan's Nanocrystal ® technology[9]). A variety of other systems are currently under investigation (see Table 3).

The long standing need for the oral delivery of proteins and peptides has been recognized.[10] Nanopreparations have been viewed as possible formulations to improve oral bioavailability of such molecules.[11,12] For example in the case of insulin, the human small intestine mesenteric blood delivers nutrients and drugs into the hepatic portal vein which makes oral delivery of insulin more physiologically relevant given that the liver is its primary target when normally released from pancreatic beta cells.

Many active small molecular weight drugs are discovered which lack oral permeability and solubility.[13] Poorly soluble drugs may often have low or variable permeability making dosing difficult to predict. Nanopreparations have been investigated for improving pharmacokinetics of poorly soluble drugs and poorly permeable drugs.[9] Compared to intravenous (*i.v.*) administration, in oral delivery a delayed time to maximum plasma concentration (t_{max}) and a lower peak plasma concentration (C_{max}) are observed. These attributes can facilitate a more flexible dosing regimen as well as decreased risk of infection from *i.v.* lines in chronic patients.

Oral delivery of poorly soluble and permeable drugs has been facilitated by use of penetration enhancers[14,15] and microparticles.[16–19] Permeation enhancers have been used to improve transport across the epithelial barrier, while microparticles have been used to protect drug molecules from the harsh gastrointestinal environment. Depending on structure and composition, nanoscale system can potentially carry out both of these important functions aiding in oral drug delivery.[11,20]

2. Physiological Barriers to Nanopreparations for Oral Delivery

Clinical success for nanopreparations has been difficult due to the inherent physiological environment of the gastrointestinal tract. The intestines are designed to breakdown and absorb proteins, carbohydrates and lipids into component formats.[3]

Orally delivered drugs have to survive the stomach acidic environment, the renin and pepsin enzymes, and the pancreatic enzymatic activity of the small intestine. In addition to these biological barriers, the drug has to be able to traverse across the intestinal epithelium.[3,21]

Many drug compounds are hydrophobic in nature and hence poorly soluble in aqueous environments. Many of the drugs of biopharmaceutics classification systems (BCS) class II or class IV category require significant formulation in order to enhance their solubility. Nanopreparations can be used to solvate these drugs and to create greater drug exposure to the absorption site of the targeted organ.[24,25] The dimension and shape of the delivery system can also have a significant effect on dissolution rate in the intestinal tract and subsequent oral bioavailability.[2,26–28]

An important note for oral nanopreparations is that the stomach secretions can reach a pH level as low as 1. This pH can denature or hydrolyze many exposed proteins or other acid labile linkers, thus making the stomach milieu a formidable chemical barrier to oral formulations.[3] On a related note, stomach pH can rise as high as 7 in the fed state.[29] This can influence release from enteric polymer-coated delivery systems, such as Eudragit ®, which are often used in order to protect acid or pepsin susceptible materials in solid dosage forms from degrading in the stomach. Another consideration is that the pH level of the small intestine ranges from 6.6 ± 0.5 to 7.5 ± 0.4.[3],[30]

Particle transit time varies based on region and size.[31] Other physiological factors also affect transit time. The transit time from the stomach to the duodenum depends on the type of content in the stomach and the amount of time since previous food intake. During fasting, the stomach to duodenum transit time for saline is only 12 min, but with a large meal can take as long as 4 h.[32] The carbohydrate portions usually arrive, in the small intestine, first while the fatty portions arrive last. Liquids can by-pass the solid portions of the meal and enter the duodenum rapidly.[33]

Therefore, the time for an oral nanopreparation to reach the small intestine can be drastically affected by the contents of the stomach, and the body's state of fasting. This can be critical, if for example, in cases where an oral drug is stable for only 2 h. But the transit time to the small intestine takes up to 4 h. Generally, like most nutrients and drug molecules, nanopreparations are not absorbed in the stomach region, but more distally in the small intestine or colon. Some nanopreparations may not even reach the absorptive surface in an intact form.

The actual flow speed of chyme through the small intestine is approximately 1–4 cm/min.[33] Bile salts, help bicarbonate to neutralize acid in fluid emerging from the stomach into the duodenum, but their primary role is to emulsify fatty contents. They are actively reabsorbed via the ileal bile acid transporter and returned to the liver for reuse.[32] Bile salts can affect oral dosage forms by changing the properties of nanopreparations that are formed from micelles or surfactants. Pancreatic juice also delivers proteases, lipases and carbohydrases, essential for further digestion in the small intestinal tract.[22] These enzymes can cause degradation or denaturation of proteins, DNA and other labels or linkers that potentially have role in stabilizing an oral nanopreparation. (See Table 1 for common intestinal enzymes.)

Table 1. Major enzymes of the gastrointestinal tract.[22,23]

Enzyme	Site of action	Substrate	Products of action
Pepsinogen:pepsin	Stomach	Proteins	Large peptides
Pancreatic α-amylase	Small intestine	Starch	Disaccharides
Trypsinogen:trypsin	Small intestine	Proteins	Large peptides
Chymotrypsin	Small intestine	Proteins	Large peptides
Elastase	Small intestine	Elastin	Large peptides
Carboxypeptidases	Small intestine	Large peptides	Oligopeptides
Aminopeptidases	Small intestine	Large peptides	Oligopeptides
Lipase	Small intestine	Triglycerides	Monoglycerides
Nucleases	Small intestine	Nucleic acids	Nucleotides
Enterokinase	Small intestine	Trypsinogen	Trypsin
Disaccharidases	Small intestine[a]	Trypsinogen	Trypsin
Peptidases	Small intestine[a]	Oligopeptides	Amino acids
Nucleosidases	Small intestine[a]	Nucleosides	Nucleosidases, phosphoric acid
Nucleotidases	Small intestine[a]	Nucleotides	Sugars, purines, pyrimidines
Diverse bacterial enzymes	Large intestine	Various	Vitamin K, Various

[a]Intracellular.

Mucous layer interactions can also interfere with absorption.[34] A thick layer of mucus is secreted from the inner surface of the stomach protecting it from its acidic contents. Thinner layers coat the small and large intestine lubricating bolus food transport and protecting the delicate microvilli. This mucous can block nanoparticles from penetrating the intestinal wall, or provide a high turnover reservoir for temporary attachment and release.

Monoglycerols, fatty acids, and glycerol from food diffuses through the epithelial membrane into the local cells where they are reassembled to triglycerides. These triglycerides are formed into small droplets of approximately 150 μm in diameter called chylomicrons. The chylomicrons are then transported by the golgi complex to the basolateral side of the cell and are exocytosed. The chylomicrons enter the central lacteal flow and are carried from the lymphatic system to vena cava to be injected back into the systemic circulation.[22] This route through the lymphatic system can potentially be exploited for lipid based nanoparticle oral absorption.[35]

A formidable challenge to oral delivery is the first pass metabolism effect in the liver. The intestinal blood flow drains in to the hepatic portal vein and then enters the liver. Here a host of metabolic enzymes and phagocytic cells reside that can sequester nanopreparations and their cargoes.

Intact nanopreparations are generally limited to <5% entering systemic circulation when orally delivered.[36] Variables operating on the physiological environment such as state of fasting, disease and enzymatic content of the intestines can affect the oral absorption of these systems. Nevertheless, nanopreparations can potentially be conferred with characteristics that increase gastrointestinal regional residence time, enable transepithelial transport, and optimize pharmacokinetic parameters.[37]

2.1. *Tight junctions*

Epithelial cells along the surface of the intestine are joined by intercellular tight junctions. These junctions between cells are formed by a variety of proteins including claudins and occludins. Unless transported by another transcellular mechanism, these proteins form a "belt" around the epithelial cell, sealing off the majority of hydrophilic molecules from passage into the blood stream.[38] Nanopreparations of ~1–3 nm[39,40] or smaller may penetrate through tight junctions without modifying intestinal physiology. Many nanopreparations with a larger diameter open tight junctions, facilitating their transport.[11,41] Endocytic mechanisms, such as clathrin and caveolin mediated pathways contribute significantly to nanoparticle uptake.[42,43]

Tight junctions have a range of pore sizes in different intestinal regions.[44] The molecular regulation of these junctions is also becoming better understood. It has been postulated that these junctions can open and close the intracellular space in a time-dependent manner.[45,46] While the tight junctional areas only account for 0.1–0.01% of the surface area of the intestinal tract (approximately 200–2000 cm^2), their potential impact on oral nanopreparations should not be overlooked.[3,40,47] Modern research places more emphasis on penetration enhancers that modulate tight junctions with reduced adverse effects.[44]

The permeability of nanopreparations may be significantly increased by the use of targeted protein penetration enhancers for tight junction proteins.

2.2. *Peyer's patches*

Peyers patches are small areas of the intestinal tract that are unique to the gastrointestinal environment. These patches are usually elongated in shape and measure a few centimeters in diameter.[48] They are mostly found in the ileum and are generally absent from the jejunum. Similar patches are also found in the colon, but are referred to as lymphoepithelial or lymphoglandular complexes. These patches penetrate from the mucosa to the submucosa. They are part of the mucosa associated lymphatic system (MALT) and are composed of a surface layer of M-cells (micro-fold cells) with sublayers of other immune cells which can elicit an immune response based on antigens passed to them through the M-cells.[49] The uptake of poly(lactic-co-glycolic acid) (PLGA) nanoparticles with an average size of about 100 nm was enhanced in Peyers patches using the *in situ* loop model with biopsy punches of the patch area for separation.[31] Electron micrographs of rat ileum showed that PLA nanoparticles were endocytosed to a high extent in Peyer's patch tissue (Fig. 1). Peyer's patches have also been shown to express β_1-integrins and CD9 cell adhesion molecules. The β_1-integrin expressing cells showed enhanced uptake of 200 nm polystyrene beads coated with adsorbed pluronic-arginine-glycine-aspartic acid (RGD) peptide.[50] This interaction of nanopreparation with the Peyer's patches makes a plausible route for oral delivery, especially for vaccines, which can be presented to the mucosal immune system via M-cells.

2.3. *Vitamin B_{12} transport*

Vitamin B_{12} (also called cobalamin) is the largest water soluble essential nutrient that is absorbed intact by the intestinal tract. This molecule has a

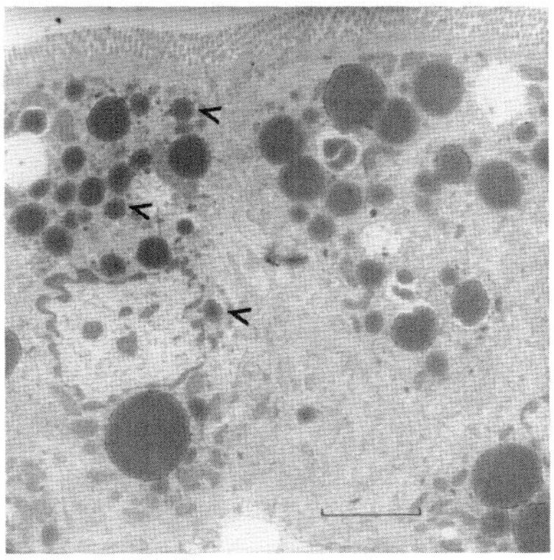

Fig. 1. Electron micrograph of rat ileum Peyer's patch after *in situ* instillation of ferritin loaded PLA nanoparticles for 30 min (10 mg/mL). Image displays uptake of over 30 particles inside two different epithelial cells (black arrows). Scale Bar = 3.1 μm. Image courtesy, Ref. 51.

molecular weight of 1357 Da and is taken up in combination with intrinsic factor by intrinsic factor binding protein (or hemopoietic factor) receptor. Some methods of drug delivery using nanopreparations have focused on vitamin B_{12} targeting.[52–56] The difficulty of proving vitamin B_{12} mediated receptor-dependent particle uptake is a major issue. Problems for this type of targeting include complicated chemistry of covalent attachment and poor translation from rats to larger models. Bile released from the liver inhibits vitamin B_{12} binding to intrinsic factor and must generally be avoided if this route is targeted.[57] The Vitamin B_{12} route is also limited due the fact that only 1–2 μg of vitamin B_{12} can be absorbed into the body per day.[58,59]

3. Types of Nanopreparations

The general principles governing the physiochemical properties of drug molecules that influence oral absorption have been an active area of research.[60–65] For nanopreparations, many factors can influence their absorption into the systemic circulation. These include particle composition, pKa of the material components, molecular weight, particle aggregation, particle size, pH, inherent stability of the particle, and surface charge.[3] Physiological parameters can also vary based on state of anesthesia, fasting or disease. These physiological parameters modify the permeability of the membrane, surface area of the epithelium and mucous layer properties.[32]

The single most formidable barrier for nanopreparations to enter the systemic circulation is the permeability of the intestinal membrane. Major types of penetration enhancers include surfactants, chelating agents, bile salts, cationic and anionic polymers, acylcarnitines, and fatty acids.[3] Generally, penetration enhancers work by physiological alteration of the epithelium to exploit transport mechanisms.[66]

Exploited transport mechanisms can be sub-grouped into methods that disrupt the structural integrity of the epithelial layer, decrease the mucous viscosity, open tight junctions or increase membrane fluidity.[3] Penetration enhancement principles assume a sufficient quantity of therapeutic agent and penetration enhancers at the site of absorption.[67] This formulation goal can be difficult to accomplish due to variance in intestinal motility, gastric emptying and dilution factors in the gastrointestinal tract. Approximately, 9 L of fluid enters the small intestine each day, of which only about 2 L is food and liquid ingested, giving a dilution factor > 4 fold.[32] Nanoparticles can increase effectiveness of penetration enhancers by carrying them to the intestinal barrier and releasing them in a site specific manner to aid in particle absorption. Penetration enhancers, which have been reformulated into nanopreparations, aid bioavailability by increasing the local delivery of drug and enhancer. Makhlof *et al.* showed that a solution of co-delivered penetration enhancer spermine, and nanoparticles was not as effective as nanoparticles encapsulating spermine in delivery of salmon calcitonin to rats.[68] This provides rationale for encapsulating penetration enhancers such as chitosan and glycerides into nanoparticles alongside the therapeutic cargoes.[8,41,69–72]

3.1. *Chitosan nanoparticles*

Chitosan has been widely researched for uses as both a nanopreparation and as a permeation enhancer in oral delivery.[73–75] Chitosan is made from deacetylated chitin. Chitosan is a linear polysaccharide made of varying ratios of N-acetyl-D-glucosamine and D-glucosamine groups. It has an extremely high oral LD_{50} in mice of 16 g/kg orally (as non-toxic as sugar and salt).[76] Primary amines and hydroxyl groups on chitosan can be modified to attach functional moieties. Chitosan has a positive charge at pH < 6.5 due to the fact that it has free amino groups with a pKa of 6.1. Native chitosan has limited solubility at pH >7.[77] Lack of charge and solubility of chitosan at neutral and basic pH levels causes it to be less effective as a penetration enhancer in the small intestine where pH can rise above 6.5. Modifications were thus developed to increase chitosan's solubility in the small intestine. Thiolated chitosan, quaternization of the primary amine with methyl groups (trimethylchitosan), PEG (poly (ethylene glycol)) chitosan, N-succinyl chitosan and carboxymethyl chitosan derivatives have all been synthesized for oral delivery.[78–81] Some of these have been used in oral nanopreparations with varying effects.

Chitosan nanoparticles have shown to cause a decrease in transepithelial electrical resistance (TEER) and increase in FITC-dextran permeability in Caco-2 models.[84,85] The mechanism by which they open tight junctions appears to be through F-actin and ZO-1 protein depolymerization.[86] The penetration enhancing effect of chitosan solutions appears to be dependent on the molecular weight and degree of deacetylation of the chitosan. The best results for absorption enhancement have been observed for chitosans having 65% deacetylation and a >170 kDa molecular weight in rat-size small intestine. This information can be useful for the preparation of nanoparticles as well.[87] Chitosan nanoparticles have exhibited particular advantages over free chitosan solutions in that they are endo- and transcytosed and better protect cargo drugs.[88]

Substantial research has been conducted on the oral delivery of insulin using chitosan as a nanocarrier.[89–92] Many of these have utilized the mild ionic gelation techniques that can maintain insulin's three-dimensional structure. In one example, nanopreparations were synthesized by ionotropic gelation in the size range of 250–400 nm using 88.9% deacetylated chitosan.[70] When dosed to diabetic rats, hypoglycemic blood glucose levels were maintained for >15 h. The relative oral bioavailability of the insulin was as high as 14.9%.[70]

In order to increase the bioavailability of insulin, modified chitosan has been studied. The basis for modification is to maintain the charged nature of the amine groups in order to increase the penetration enhancing effects at neutral and basic pH. For this purpose quaternized chitosan nanoparticles have been developed.[93]

Trimethyl- and dimethyl chitosan showed enhanced insulin delivery in rats over chitosan solutions and unmodified chitosan nanoparticles during colonic delivery.[93]

These were estimated to improve bioavailability based on the increased solubility at neutral and basic pH. The nanoparticle delivery showed highest effect when free chitosan solutions and chitosan nanoparticles were admixed prior to delivery.

Active targeting of chitosan using peptides has also been investigated for goblet cell delivery. Trimethylchitosan nanoparticles with goblet cell targeting peptides (CSK) showed improvement over trimethylchitosan insulin delivery.[94] In Caco-2/HT29-MTX co-cultures, transepithelial transport was increased using CSK peptide-conjugated nanoparticles. Similar nanoparticles loaded with insulin showed increased relative bioavailability compared to subcutaneous injection in rats for targeted versus non-targeted nanoparticles (5.66% versus 3.69% respectively).[94] This study provided the rationale for active targeting of chitosan nanopreparations in oral delivery.

Table 2. Structural variations of chitosan.

Name	Description	Structure	Reference
Chitin	– Most abundant polymer next to cellulose – Charged at pH <6.5	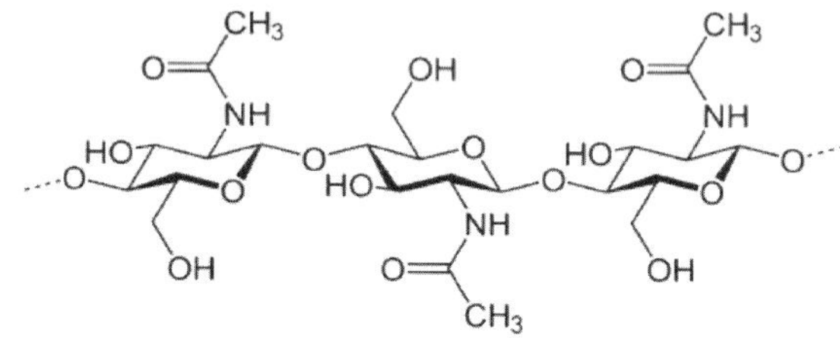	74
Chitosan	– Varying degrees of acetylation – Charged at pH <6.5 – Increases permeability of intestinal mucosa	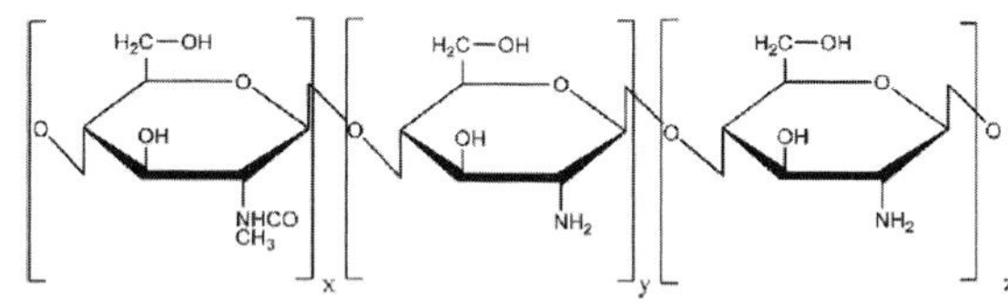	82
Trimethylchitosan	– Positively charged at acidic & basic pH – Retains permeation enhancing effect on intestinal mucosa	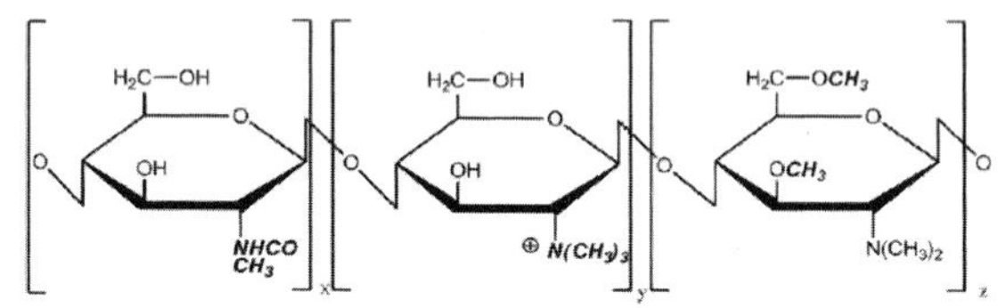	82

For further structures used in nanopreparations see Ref. 83.

Combinations of nanopreparations have been studied for improved oral delivery of therapeutic cargoes with chitosan. An example is alginate: A biodegradable anionic polymer composed of β-D-mannuronic acid and α-L-glucuronic acid.

The anionic properties of alginate make it ideal for complexation with chitosan as this increases their stability. Combinations of chitosan and alginate also aid in oral insulin delivery by facilitating the complexation of the protein and nanoparticle.[74]

Nanoparticles formulated by ionotropic pre-gelation of an alginate core followed by chitosan complexation with the surface were ~750 nm in diameter and obtained an insulin loading capacity of 9.9%.[41] When dosed orally to diabetic rats they showed a significant hypoglycemic effect from 8–14 h with decreased levels lasting for 18 h. The duration of hypoglycemia was longer than previous studies with chitosan nanoparticles.[11,70] Insulin solution mixed with the nanoparticles showed only a minor decrease in blood glucose levels, highlighting the importance of encapsulation in the nanoparticles.[41] The relative bioavailability of insulin in this nanoparticle formulation was 7% versus subcutaneous insulin.[41]

Some of the most promising results *in vivo* involved the encapsulation of insulin in chitosan, facilitated by the incorporation of glutamic acid residues and zinc (a known complexing agent for insulin).[95] Chitosan-γ-glutamic acid nanoparticles of diameter ~197 nm showed a loading efficiency of insulin of 57%. They reduced glycaemia by 60% for >10 h in streptozotocin induced diabetic rat models.[11] The benefit of these chitosan-glutamate nanopreparations was improved solubility at a range of pH values. Unfortunately these systems degraded *in vivo* in the neutral-basic pH of the intestine causing premature release and degradation of the protein.[11]

In order to overcome this problem Sonaje *et al.* lyophilized the nanoparticles and inserted them into enteric capsules. This approach increased the oral bioavailability of insulin to ~20%, however conclusive evidence in large animals and man is necessary.[96–98]

Inhibition of intestinal proteolytic enzymes has been explored for enhancing oral delivery of proteins.[99] Chitosan alone is insufficient to inhibit trypsin and carboxypeptidases (Table 1).[100] Further development of chitosan anti-enzyme technology led to the conjugation of ethylenediaminetetraacetic acid (EDTA) to chitosan, but these conjugates also were not effective against trypsin, chymotrypsin and elastase.[99] Competitive enzyme inhibitors (i.e. papain, chymostatin) were used in conjunction with EDTA-chitosan to slow degradation of encapsulated insulin. An amount of 40–60% of insulin remained after 4.5 h when including inhibitors compared to 90% degradation without.[101] This combination highlights the potential role of enzyme inhibitors in oral nano drug delivery.

Chitosan-DNA complexes have been studied for the purpose of improving transfection efficiency via the oral route. Attempts have been made to optimize transfection efficiency based on charge ratio, chitosan molecular weight, endosomes escaping peptide conjugation and DNA concentration.[102] These studies have resulted in formation of stable DNA-chitosan nanoparticles of approximately 100–500 nm in size (102 kDa chitosan). However, they produced expression levels 200 fold less than Lipofectamine on Cos-1 cells. Surprisingly, when instilled in an upper rabbit small intestine, chitosan showed better transfection than lipid based DNA carriers.

Chitosan nanoparticles have also been studied for oral antihypersensitivity gene therapy. In one of these systems, cDNA for peanut antigen was encapsulated in monolithic high molecular weight chitosan (390 kDa) nanoparticles. The particle size was assessed by TEM and appeared on the order of 150–300 nm in size. The testing of anaphylactic shock scoring after oral administration of chitosan-DNA complexes, carrying the peanut antigen, showed that this nanopreparation was able to reduce hypersensitivity response to peanut antigen *in vivo*.[103]

Lack of clinical translation of these systems may be related to low transfection efficiency due to mucous entrapment of the chitosan nanoparticles occurring in the intestine. Uptake of the chitosan nanoparticles and subsequent release of the DNA may give higher transfection rates than pre-endocytosis release. The mucous adhesion of the chitosan may increase local concentrations of the target gene at the epithelial surface, but limit uptake of the nanoparticles themselves, thus limiting transfection.[104] Improved transfection efficiency will probably be needed to merit further development and translation.

In spite of significant efforts for developing oral chitosan based delivery systems they have not yet reached clinical trials.[74,104] A major concern may be the toxicity of chitosan. *In vitro* blood coagulation studies have shown that 2% solution of unmodified chitosan (126 kDa) causes a 40% decrease in coagulation time with heparinized blood, whole blood, and defibrinated blood.[105] Varying degrees of quaternization have showed modulated *in vitro* toxicity, bringing to light the importance of evaluation of novel chitosan derivatives. This effect is illustrated by the study in MCF-7 cells where increasing charge density (quaternization) exhibited a decreasing IC_{50} at an exposure time of 6 h.[106] Future clinical development may require extensive toxicity evaluation of chitosan which could be a possible reason for lack of translation.[107] Also the ability of chitosan to increase oral bioavailability may not be sufficient. Difficulties with scale up of production may be another source of concern. Despite these problems, the ease of synthesis and penetration enhancing effects of chitosan have prompted continued research of the material. Novel modifications, such as thiol-protected chitosan hold promise for future improvement.

3.2. *Poly (Alkyl Cyanoacrylates)*

Poly (alkyl cyanoacrylates), PCA have been extensively used for drug delivery.[109,110] These polymeric systems are made from the common acrylate monomer system with an additional cyanate group. The use of PCA in drug delivery is based on their ability to encapsulate small hydrophobic drugs and also proteins.[111] PCA was the first polymer nanoparticle to be used for oral delivery of peptides.[111] This landmark study using insulin set the stage for many other nanoparticle-peptide oral drug delivery studies.[112] PCA has been shown to improve the oral bioavailability of small molecular weight drugs.[113,114] A large number of studies have been done on the delivery of proteins and drugs using PCA demonstrating its versatility as a drug delivery agent.[115,116] PCA is considered bioerodable due to the hydrolysis of its side chain ester moieties producing the corresponding alkyl alcohol and poly(cyano acrylic acid).[117–119] In the gut, pancreatic esterases are able to degrade PCA nanoparticles and release their cargoes. This hydrolysis proceeds at variable time rates depending on the side alkyl chain length. PCA above molecular weight 10,000 Da is probably not excreted from the body.[120]

PCA nanoparticles can be synthesized in acidic water solutions with strong mixing and added surfactants for stabilization.[115] Polymerization begins immediately upon the addition of the monomer to any solution containing nucleophilic groups (i.e. OH). The rate of reaction can be slowed by lowering the pH of aqueous polymerization solutions. The size of the nanoparticles can be adjusted between 40 nm and 200 nm based on the concentration of surfactants used in pure PCA nanoparticle synthesis. The combination of different polymers (chitosan, dextran) can change the size of the end product as well as the zeta potential. For *in vivo* work all products need to be maintained sterile during polymerization. Heat, gas, filtration or radiative sterilizations are not appropriate for PCA nanoparticles.[115] Nanocapsules (nanoparticles with hollow cores) can be formed using PCA. These can be either water or oil-cored to carry cargoes.

Isobutylcyanoacrylate (IBCA) nanocapsules have been used for oral delivery of insulin in streptozotocin diabetic rats (see Fig. 2 for structural differences of PCA monomers). Results obtained showed hypoglycemia

Table 3. General overview of nanopreparations for oral delivery.[108]

	Physiochemical characteristics	Cargo	Therapeutic applications
Micelles	~20–50 nm Dynamic structure Hydrophilic corona Hydrophobic core	Encapsulated in hydrophobic core Mainly poorly soluble drugs	Solubilization Inhibition of Pgp efflux
PCA	40–200 nm Static structure Non-degradable backbone	Entrapped in core Hydrophilic or hydrophobic Small or large MW	Protection of drugs Localized release
Chitosan	>100 nm Static structure Variable surface properties	Entrapped in core Hydrophilic or hydrophobic Small or large MW	Protection of drugs Localized release Gene Delivery
Dendrimers	1–10nm Static structure Variable surface properties	Surface attachment/entrapped space Hydrophilic or hydrophobic Small or large MW	Solubilization Localized release Protection of drugs
Liposomes	>50 nm Dynamic structure	Entrapped in core or hydrophobic bilayer Hydrophilic or hydrophobic Small or large MW	Protection of drugs Localized release Solubilization
Solid Lipid Nanoparticles	>100 nm Static structure Hydrophobic/crystalline interior	Entrapped in core Hydrophilic or hydrophobic	Solubilization Localized release
PLA/PGA	>50 nm Static structure Customizable degradation rate	Entrapped in core Hydrophilic Small or large MW	Protection of drugs Localized release
Nanosized Formulations	>100 nm Static structure Stabilized by surfactant	Small MW Hydrophobic	Increased solubility

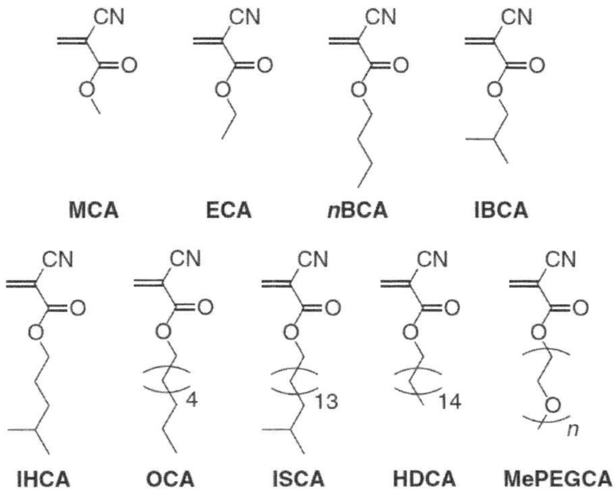

Fig. 2. Chemical structures of common PCA derivatives: Methyl cyanoacrylate (MCA), ethyl cyanoacrylate (ECA), n-butyl cyanoacrylate (nBCA), isobutyl cyanoacrylate (IBCA), isohexyl cyanoacrylate (IHCA), octyl cyanoacrylate (OCA), isostearyl cyanoacrylate (ISCA), hexadecyl cyanoacrylate (HDCA), and methoxypoly(ethylene glycol) cyanoacrylate (MePEGCA). Image courtesy of Ref. 120.

of 65% lasting for up to 18 days after administration. Delivery to the ileum showed the greatest hypoglycemic effect using the IBCA nanoparticles.[121] This demonstrates the ability of nanoparticles of IBCA to deliver proteins orally, but other results have shown lack of efficacy. Nanoparticles of IBCA entrapping calcitonin underwent slow degradation of the peptide in simulated intestinal fluids, but *in vivo* absorption was not significantly increased.[122] Initial studies by Lowe *et al.* showed that rapid release of calcitonin occurred and little protection from intestinal proteins was afforded by IBCA nanoparticles.[122]

Nanoparticles using PCA have also increased the oral bioavailability of small molecular weight drugs. Hexylcyanoacrylate (HCA) nanoparticles have shown to increase the oral bioavailability of vincamine in rabbits.[114] The oral bioavailability of vincamine in HCA nanoparticles was increased from 25% to 40% when drug was entrapped. Also the oral bioavailability of mitoxantrone was increased by incorporation in poly(butyl cyanoacrylate), (BCA) nanoparticles.[123] This result could be explained by the local release of the drug next to the epithelial barrier following the mucoadhesion.

Poly(isohexyl cyanoacrylate) (PIHCA) nanoparticle formulations have been investigated for the delivery of cyclosporine A. Nanoparticle delivery could potentially decrease acute kidney damage caused by this drug in humans. An *in vivo* study of PIHCA nanoparticles in mice showed a 13-fold increase in bioavailability of the nanoparticle over an emulsion of cyclosporine A.[124] The study also showed a greatly reduced liver AUC as well as reduced kidney exposure to the drug. This study demonstrates the further ability of these systems to avoid reticuloendothelial uptake.

BCA loaded with calcitonin showed a 45% decrease in calcium levels compared to control *i.v.* injection following oral administration to rats. Absorption enhancers such as deoxycholic acid and sodium lauryl sulfate were also used to enhance permeability.[125]

Specific properties of PCA nanoparticles can be modified in order to modulate interactions with mucous proteins. This can optimize absorption enhancing properties of nanoparticles and cargo release. Size and surface characteristics are the major properties that affect mucous interactions.[126–128] Other properties, including surface modification with other polymers such as large MW PEG and chitosan, can increase mucoadhesion.[129]

Toxicity of PCA has been noted for shorter chain cyanoacrylates (methyl cyanoacrylate) in humans when used as a wound closure agent, and subsequently was replaced by longer chain cyanoacrylate adhesives.[110] It has been reported that the degradation products of PCA could potentially be converted to formaldehyde through the inverse Knoevenagel reaction. This led to several reports attributing toxicity of nanoparticles to this mechanism.[130–132] It has been postulated that this mechanism is too slow to compete with much more rapid metabolic pathways that degrade PCA.[110,131,133] The results of an *i.v.* phase III clinical trial of PIHCA nanoparticles for the purpose of doxorubicin delivery are of relevance (Livatag ®, BioaAlliance Pharma, France). This trial was later suspended based on severe pulmonary adverse events.[112] This could have been due to aggregation of the nanoparticles but the exact mechanism is unknown. Toxicity of these systems needs to be further addressed for *i.v.* and oral dosing.[112]

3.3. *Dendrimers*

Dendrimers were first described in 1978 by Fritz Vögtle and his group at University of Bonn.[134] Since then a variety of these branched structures have been developed.[135] Dendrimers are characterized by their hyper-branched structure resulting from repeating branches off of a central core atom. This branching results in a tree like structure, that forms a spherical or near spherical macromolecule with multiple sites for attachment at the surface (see Fig. 3).[136] Also the internal spaces remaining between the branched structure create loci for hydrophobic encapsulation of other drugs or nanoparticles in the core.[137] This multifunctional surface and encapsulating core,

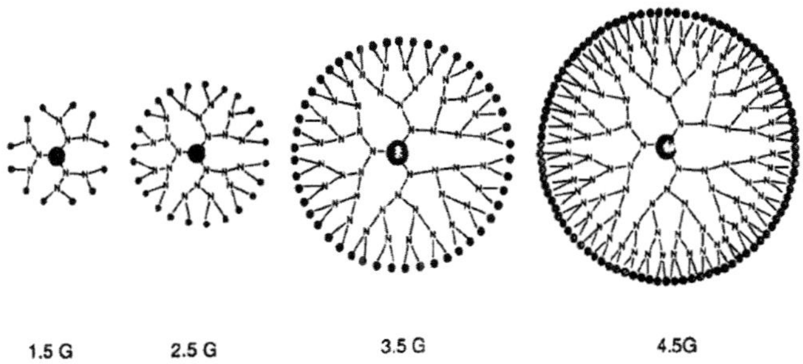

Fig. 3. Branching PAMAM dendrimers enable attachment of moieties to multiple surface groups or the encapsulation of molecules in the interior spaces. Reprinted with permission from M. F. Ottaviani, S. Bossmann, N. J. Turro, and D. A. Tomalia, Characterization of starburst dendrimers by the EPR technique. 1. Copper complexes in water solution, *J. Amer. Chem. Soc.* 116(2), 661–671 (1994). Copyright 1994, American Chemical Society.

combined with low polydispersity and prolonged plasma half-life make these polymers suited for drug delivery.[138]

3.3.1. Poly(amido amine)

Poly(amido amine), in short PAMAM, dendrimers have been evaluated for oral drug delivery.[136] These dendrimers are synthesized through repeated Michael addition to the free terminal amine groups.[139] This synthesis results in alternating amine terminated or carboxyl terminated surfaces. Full generation amine terminated (G1.0, G2.0, etc.) or half-generation carboxyl terminated (G0.5, G1.5, etc.) dendrimers allow for modification of the surface charge and modulated toxicity profiles.[140] PAMAM dendrimers were first shown to cross the gut mucosa in an everted intestinal sac model.[141] Altering the termini can change the bioavailability of these dendrimer conjugates. These characteristics give favorable qualities for PAMAM dendrimer drug delivery.

The potential of PAMAM dendrimers to permeate across the intestinal epithelial barrier was first shown by Wiwattanapatapee *et al.*[141] Using everted rat gut sacs it was observed that serosal transfer rate of anionic dendrimers was high (2.3–2.7 microL/mg protein/h) for G3 and G4.[141] Transepithelial transport was observed to be dependent on a variety of factors including size, surface charge, concentration and incubation time.[142]

PAMAM dendrimers have shown toxicity both *in vitro*[136] and *in vivo.*[140] Caco-2 cell culture studies and CD-1 mice experiments showed that toxicity was generation, size, surface charge and incubation time dependent.[140,143] Toxicity for PAMAM dendrimer surface groups ranks in the order of hydroxyl-terminated<carboxyl-terminated<amine-terminated systems.[144] Carboxyl terminated, G3.5 and G4.5 PAMAM dendrimers were cytotoxic using an LDH release assay from Caco-2 cells at concentrations higher than 10.0 mM whereas amine terminated dendrimers, G3.0 and G4.0, were cytotoxic at much lower concentrations (1.0mM). Concentrations of 0.01mM G4.0 were found to be non-cytotoxic by LDH assay and visual inspection in Caco-2 cells.[145]

Orally dosed PAMAM dendrimers showed similar toxicity to *i.v.* injections of the same in CD-1 mice. At toxic doses, disseminated intravascular coagulation-like effect and fibrin degradation products with resultant intravascular coagulation and hemorrhage were observed. It was noted that the amine terminated dendrimers were more toxic than carboxyl terminated in CD-1 mice.[140] PAMAM dendrimer surface groups have been modified to reduce toxicity. PAMAM dendrimers with lauroyl chains attached to the surface have shown decreased toxicity and increased permeability across Caco-2 cell monolayers.[146] Other surface modifications such as PEGylation, acetylation and FITC labeling have been observed to reduce toxicity and improve transport across Caco-2 monolayers.[147–149]

Anionic dendrimers have been observed to open tight junctions and increase mannitol transport across Caco-2 cell monolayers. In the size range studied, larger dendrimers appeared to have a greater effect. Higher permeability of mannitol across Caco-2 cell monolayers was noted for G2.5, G3.5, and G4.5 dendrimers, compared to smaller G-0.5, G0.5 and G1.5 dendrimers.[143]

Fluorescently labeled G2 dendrimers showed increased permeation with increasing times ranging from 90–150 min in Caco-2 cell culture. It was also shown that the transport was energy-dependent with decreased permeability at 4°C compared to 37°C.[150] A portion of transport may be dependent on endocytic mechanisms. PAMAM dendrimers have been shown to be transported across Caco-2 cell monolayers via paracellular and transcellular routes.[42,43,145,148,150]

The ability of PAMAM dendrimers to permeate the epithelial barrier and enter systemic circulation was recently investigated.[151] It was shown that 106,000 Da G6.5 dendrimers were able to be absorbed orally in CD-1 mice. 1 mg/kg of radiolabeled G6.5 dendrimers were dosed by oral gavage to mice, and animals were sacrificed at 4 h. 9.4% of the macromolecular dose was found to be absorbed into the plasma after exclusion of unbound radiolabel by size exclusion chromatography.[151] This *in vivo* study shows promise for enhancement of bioavailability of highly potent drugs using PAMAM dendrimers.

The potential of PAMAM dendrimers to increase transepithelial flux has been harnessed for the delivery of propanolol.[146] Lauroyl modified dendrimers with 2–6 moles of lauroyl per dendrimer were attached to G3.0. The study showed increased transport of propanolol when lauroyl chains were attached compared to dendrimer and propanolol alone. The endocytosis inhibitor cyclosporin A did not decrease transport, but reduced temperature (4 °C) did. This phenomenon may be based on the energy dependent endocytosis of PAMAM dendrimers. More studies are needed to evaluate the stability of dendrimer-lauroyl chain conjugates.

PAMAM dendrimers have also been shown to increase the transepithelial transport of naproxen when covalently conjugated.[152] The stability of the conjugated system to G0 dendrimers was dependent on the bond type with the amide bond being more stable than the ester. Using liver homogenate and plasma it was found that the amide conjugate was stable for up to 48 h. Further, attachment of lauroyl chains increased permeability across Caco-2 cells.[153]

PAMAM dendrimers have shown increased transepithelial transport of anticancer compounds. Kolhatkar *et al.* complexed SN38 (7-Ethyl-10-hydroxy-camptothecin) with PAMAM G4.0. SN-38 is a highly potent topoisomerase poison which has poor water solubility and dose limiting toxicity. The prodrug of SN38, CPT-11, is used in the clinic, but is 1000 fold less active than SN-38. Complexation or conjugation with dendrimers can potentially reduce toxicity and improve oral bioavailability. Complexation with dendrimers led to a 100 fold increase in transepithelial transport across Caco-2 cell monolayers, however the complex was not stable.[154] Attempts to remedy this problem by covalent attachment were made.[155] Stability of the conjugates of SN38 with carboxyl-terminated G3.5 using glycine and β-alanine linkers demonstrated that the β-alanine linkers are more stable. Transepithelial transport across Caco-2 cells was increased for both conjugates. Studies performed with these conjugates found reduced cytotoxicity compared to the free drug for both.[156] Further studies with PAMAM dendrimers for delivery of small molecular weight drugs has been reviewed previously.[136,144]

PAMAM dendrimers have potential to improve the bioavailability of poorly absorbed drugs. Toxicity of these polymers must be monitored in order to avoid systemic effects and local irritation. In addition linker chemistries need to be developed that are stable in the GI tract, during transport across the intestinal barrier and in the blood stream and allow release at the target site. Another barrier to translation of PAMAM dendrimers is challenge with scale up of the synthesis, which takes several reaction steps to produce.[135]

3.3.2. Polylysine dendrimers

Polylysine dendrimers have been synthesized for improved stability and reduced synthetic steps (reduced cost) over PAMAM dendrimers.[135] These dendrimers have also been studied for oral absorption. Florence *et al.* synthesized and characterized polylysine dendrimers originating from glycine cores with lipid termini. Dendrimers with 2.5 nm diameter were chosen for dosing to rats at 14 mg/kg and 28 mg/kg. A maximum of 3% of the tritiated dendrimer was found in the blood at 6 h.[157] This uptake was not in excess of that for 50 nm latex beads. More studies are needed to investigate in detail the stability of polylysine radiolabel conjugates in the GI tract, during transport and in the systemic circulation.

3.4. *Poly(lactic acid) and Poly(glycolic acid) nanoparticles*

Poly(lactic acid) (PLA) and poly(glycolic acid) (PGA) have been studied as early as 1971 for drug delivery purposes.[17] PLA and PGA have the unique advantages of being both degradable and resorbable. The two monomers can be copolymerized (poly(lactic-co-glycolic acid)) at varying feed ratios to customize the crystalinity and hydrolysis rate.[158] Both nano-and micro-drug delivery has been studied using PLGA particulate systems. The biodistribution of the PLGA particles can be influenced based on the size of the particles. Particles that are > 250–300 nm are predominantly taken up by the spleen, while particles that are several microns in diameter are taken up by the lung.[159] PLGA nanoparticles have been shown to increase intestinal tissue uptake of certain drugs in the *in situ* loop model systems validating their use for oral drug delivery.[31] Uptake into the intestinal tissue of the free nanoparticles was especially high for 100 nm PLGA particles in the ileum of anesthetized rats. It was noted that uptake was higher in Peyers patches compared to villous tissue.

Increased absorption has also been observed using PLGA nanoparticles for oral delivery of drugs. PLGA nanoparticles were used for the purpose of administering antituberculosis drugs isoniazid, rifampicin and pyrazinamide orally.[160] These nanoparticles had a size of 186–290 nm and

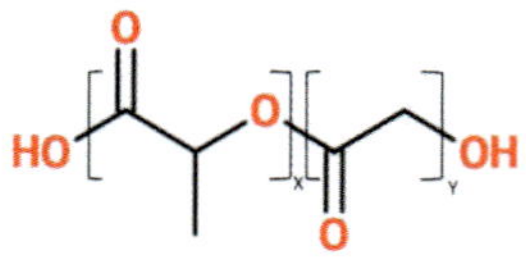

Fig. 4. Structure of poly(lactic-co-glycolic acid) (*x*=lactic acid, *y*=glycolic acid).

showed an increased mean residence time (in the blood-plasma compartment) and bioavailability in guinea pigs. The levels of drug in the plasma were sustained for 7–12 days when encapsulated, while the drug alone was only detectable for 24 h. Guinea pigs infected with mycobacterium tuberculosis showed no signs of infection following administration of 5 oral doses of antibacterial loaded PLGA nanoparticles, compared to 46 doses of the free drug. This study shows promise for reducing dose frequency for the antituberculosis drugs.[160] Improvements on this study were performed by attaching wheat germ agglutinin to the PLGA nanoparticles. Bioavailability and residence time were increased from 4–6 days to 6–7 days for rifampicin, and from 8–9 days to 13–14 days for pyrazinamide and isoniazid.[161]

PLA-pluronic-PLA triblock copolymers have been used for effective oral insulin delivery.[162] These nanopreparations were synthesized due to their vesicular nature and theoretical biocompatibility. The ability of these polymer combinations to encapsulate insulin was demonstrated. The release profile for 56 nm PLA-pluronic-PLA vesicles showed a biphasic pattern, with initial burst release and subsequent sustained release over a 4 h time period. Induction of hypoglycemia in rats was demonstrated by gavaging 50 IU/kg dose in vesicles which yielded a sustained effect for over >18.5 h.[163]

PLGA nanoparticles have been covalently linked to RGD peptides for the purpose of increasing *M* cell uptake.[164] These nanopreparations showed increased uptake in human *M* cell co-cultures. *In vivo* ovalbumin-loaded PLGA nanoparticles showed significant concentration in mice *M* cells and were able to elicit an immune response.

Immune cell targeting is one of the goals of PLGA nanoparticle oral delivery. Previous work on oral vaccination against virulent strains of E. Coli have been carried out in humans using microspheres. This has lead to vaccine efficacy of 30% in early studies.[165] Later, Katz *et al.* used PGA microspheres carrying CS6 antigen from enterotoxigenic E. coli, in order to induce immunity, but there was no apparent difference in response to the encapsulated antigen compared to the unencapsulated.[166] While these were microparticles, their equivocal results may be the chief reason why PLGA particle systems have not been continued in further clinical trials for vaccine delivery. Other reasons why these systems appear to have been excluded from further clinical testing in oral delivery may be due to instability, lack of uptake from the intestine, instability of the antigen during synthesis and low antigen loading.[167,168]

3.5. *Polymeric micelles*

Polymeric micelles represent a colloidal dispersion ranging from 5 nm to 100 nm in diameter. They are generally composed of amphiphilic materials, which spontaneously self-assemble to form a hydrophobic core and hydrophilic corona in aqueous environments. This formation occurs when the concentration of amphiphilic material rises above the critical micelle concentration (CMC) and when the temperature of the system remains below the CMT (critical micelle temperature). The association of the amphiphilic molecules is based on the decrease of free energy of the system due to the loss of hydrophobic interactions with water molecules, and the reestablishment of the hydrogen bond network in water.[169] Micelles are different from liposomes in that they lack an aqueous core, but instead maintain a hydrophobic core where poorly soluble cargoes (drugs) can be encapsulated. In the intestinal tract bile salts are believed to naturally form micelles aiding in fat digestion and transport.[170–172] Micelles have been used in oral drug delivery and enhancement in C_{max}, AUC_{0-24} and T_{max} of poorly soluble drugs *in vivo*.[173,174] Polymeric micelles generally do not affect TEER.[175]

Pluronic (poloxamer) micelles have shown potential for oral drug delivery.[176] Pluronics are triblock copolymers made of a central hydrophobic chain of poly(propylene oxide) (PO) with two hydrophilic chains of poly(ethylene oxide) (EO) on the outside. These can be tailored for specific hydrophilic/hydrophobic balance by customizing the length of the hydrophilic EO and hydrophobic PO portions of the polymer. Pluronics spontaneously form micelles above a critical micelle concentration in the range of 10–50 nm.[177] Pluronic F127 of average compositions EO_{99} PO_{67} EO_{99}, showed increased oral delivery of genistein in rats compared to genistein powder.[178] The plasma AUC was three-fold higher at a dose of 4 mg/kg genistein in anesthetized rats ($p > 0.5$). The micelle particle size was on average 27 nm. Pluronics inhibit P-gp efflux showing potential to aid in transepithelial transport.[174]

The ability of micelles to improve bioavailability of small molecule drugs (BCS class II) were exhibited using monomethyletherpoly(oxyethylene glycol$_{750}$)-poly(caprolactone-co-trimethylene carbonate) (mmePEG$_{750}$ P(CL-co-TMC)) micelles and risperidone.[179] Micelles, 20 nm in diameter (10% weight polymer solution), were dosed to rats exhibiting a two-fold increase in risperidone bioavailability compared to risperidone solution. T_{max} was extended by 3 h to match that of the radiolabeled micelle constituents. This correlation between drug and micelle concentration maxima suggests that the drug could remain encapsulated during transport. Drug half-life was extended 1.3 times showing a sustained release effect.

The increased transepithelial transport of micelle encapsulated drugs remains controversial.[175] Some evidence suggests that increasing concentration of micelles decreases transport of micellar components across the membrane. ^{14}C labeled mmePEG$_{750}$ P(CL-co-TMC) micelles showed decreased permeability across Caco-2 cells with increasing concentration of polymer.[179] This could be due to passage of unimers and lack of passage of micelles, leading to saturated transport of unimers at or above the CMC. Opposing this report, polyoxyethylenecetylether hydroxypropylcellulose (HM-HPC) mixed with poloxomer 85 caused a three-fold increase in permeability of cyclosporin A across Caco-2 cells.[180] It was hypothesized that increased solubility may be the primary reason for increased transport. In spite of opposing *in vitro* studies, transport of micellar components appears to be significant *in vivo*. Micellar components were observed to reach a bioavailability as high as 40% upon oral administration of mmePEG$_{750}$ P(CL-co-TMC) micelles to rats. These results validate the oral absorption of micellar components *in vivo*.[179]

D-α-tocopheryl poly(ethylene glycol) succinate (TPGS), poloxomers and a variety of other common micelle components have shown inhibitory effects for *P*-glycoprotein (Pgp) efflux pumps.[175] This has been used as rationale for oral drug delivery using polymeric micelles. The mechanism by which TPGS inhibits Pgp efflux appears to be through allosteric modulation of the Pgp protein, while that of poloxomer appears to be through membrane fluidization and decreased ATPase activity.[181,182] Increased transepithelial transport due to Pgp efflux is most pronounced at concentrations below the CMC, which is likely due to higher presence of the unencapsulated drug in solution.[183,184] The exact mechanism of Pgp efflux inhibition by these micellar components is still not fully understood.[175]

The effects on Pgp efflux may be observed in an oral administration study of paclitaxel in micelles of poly(ethylene glycol) poly(2-(4-vinylbenzyloxy)-N,N-diethylnicotinamide) (PEG-b-P(VBODENA)).[185] Upon intraduodenal administration in rats of the paclitaxel micelle formulation (drug loading of 37.4 wt%, 105–120 nm diameter), bioavailability was 12.4% compared to a previously reported oral bioavailability of Taxol R of 6.5%.[186] The ability of the micelle formulation to increase the bioavailability may be dually related to the increase in solubility of paclitaxel and the inhibition of Pgp efflux.

For BCS class II drugs poor solubility is a significant concern that may be surmountable using micellar compositions. The physiochemical structure of micelles with the hydrophobic core and hydrophilic corona make them well suited for formulation of such drugs. Increases in solubility have been observed to be as high as 15,000 fold for paclitaxel.[185] The ease of synthesis of these systems is also well suited for manufacture, in many cases requiring only simple mixing of components. Micelles for oral delivery have not yet been extensively investigated *in vivo*, likely due to interaction of bile salts and other food components with these systems leading to low stability.[175] The bioavailability of drugs in micellar formulations may still be too low or too variable to merit clinical development.[108]

3.6. *Liposomes*

Oral delivery of liposomes has been investigated for the purpose of drug and protein delivery.[167] Liposomes consist of an inner aqueous core surrounded by a membrane. Membrane components can be composed of naturally derived phospholipids such as egg phosphatidylethanolamine or dioleoylphosphatidylethanolamine (DOPE), phosphotidyl choline or phosphotidyl inositol.[3] Oral liposomes have special requirements beyond *i.v.* liposomes due to the harsh gastrointestinal environment. Liposomes generally need to maintain stability in the gastrointestinal milieu and release their drug at the site of absorption. Liposomal systems can be unstable in the presence of bile salts, phospholipids, lipases, acidic environments and other physiochemical and enzymatic barriers native to the gastrointestinal environment. Especially in the case of liposomes carrying small hydrophobic species that can easily precipitate in aqueous environments, the understanding of lipid metabolism and colloidal changes due to additional endogenous lipids can be important.[187]

In spite of these difficulties, liposomes for oral delivery have shown some promising results. The oral delivery of insulin using bile salt sodium glycocholate and soybean phosphatidylcholine as liposome forming agents was evaluated in rats.[188]

These agents have both penetration enhancing and enzyme inhibiting effects. Sodium glycocholate liposomes showed an 5.7% and 11.4% insulin bioavailability in non-diabetic and diabetic rats, respectively. Hypoglycemic levels as low as 63% of control were achieved. Oral dosing of controls involving admixed liposomes and insulin (no encapsulated insulin) were done. No significant increase in blood insulin levels or hypoglycemic effect was observed for control admixed solutions.[188] This confirms the importance of the liposome as a protective carrier for insulin. Also of note is that liposomes of size 2 μm showed no significant hypoglycemic effect or insulin bioavailability, confirming the importance of nanoparticle size.[188] This recent study highlights the positive effects of liposomes on oral drug delivery and represents a relatively high bioavailability compared to other studies with chitosan[69,96,189,190] and liposomes.[191] Mechanistic studies support the absorption of the liposome with cargo as opposed to free insulin solution. Higher bioavailability could be due to the better protective effect against proteolysis in the GIT.

PEGylated liposomes were also studied for the oral delivery of human epidermal growth factor in rats.[192] Dipalmitoylphosphatidylcholine (DPPC) liposomes carrying recombinant epidermal growth factor showed a 2.5 fold increase in AUC using liposomal formulations over growth factor solution. Gastric healing of ulcers was modulated using these liposomal carriers.

Studies with PEGylated liposomes have been performed for oral vaccination purposes.[193] PEGylated liposomes were observed to cause an effective mucosal immune response. The effect of lipid ratio on antigenicity was analyzed. It was observed that a higher relative amount of phospholipid dose (12.5 μmol versus 5 μmol) for encapsulation of the antigen protein led to a higher (1.7 fold) IgG response. On the other hand, it was found that an IgA response was reduced by increasing the phospholipid dose.

Despite these results, most attempts to orally deliver vaccines using liposomes have ended in failure.[194] For example Orasomes were developed for oral delivery of drugs and also for targeting *M*-cells for vaccination purposes. These nanopreparations were initially promising based on the stable cross linking lipid components making them resistant to detergent disruption.[168] *In vivo* studies demonstrated increased uptake in M-cells through the use of UEA-1 (Ulex europaeus 1) mouse *M*-cell targeting protein,[195] but these systems did not advance to clinical trials.[196] The reasons why these systems have been unsuccessful is probably multifaceted. Insufficient loading, irreproducible manufacturing, difficulties in scale up and low uptake by *M* cells may be causes of the lack of translation.[167]

While we still await a successful clinical outcome from oral liposomal delivery of drugs or proteins, future research should be focused on increasing permeability and absorption into systemic circulation. Higher loading and longer release profiles may also be critical for increased drug delivery to the site of action. Stability is also of importance and related to both of these aims. New work in the area of archaeosomes (liposomes made from lipids extracted from archaebacterias) could help achieve these goals, as they are based on non-saponifiable lipids. Archaeosomes have shown 3.5-fold increased absorption of Tm_{99} DTPA in rats over conventional liposomes.[197] Other combination strategies using polymer coated liposomes could also aid in stability, release and drug loading goals.

3.7. *Solid lipid nanoparticles*

Solid lipid nanoparticles as drug delivery systems have been actively studied for many years.[198] These systems are similar to liposomes, but use solid lipids instead of liquid, create a homogenous hydrophobic core, and can maintain stability for sustained periods of time *in vivo*.[199] These nanopreparations are generally composed of glyceride molecules manufactured easily through high pressure liquid homogenization or phase inversion temperature shifts.[200] These lipids are usually of varying chain length because single size lipids may form highly crystalline particles and expel drugs from the interior regions. The heterogeneity of the lipid chains has found to be conducive to drug loading.[198] Generally lipids such as caprylic acid or capric acid can be used for these systems with PEG and small amounts of lecithin is added for stability. These compounds are currently used in oral, topical and parenteral administrations and are generally considered safe.[199–201]

Oral administration of solid stearic acid nanoparticles coated with poloxamer 188 (mean diameter 196.8 nm, ξ-potential −69mV) loaded with camptothecin showed increased absorption *in vivo* compared to camptothecin solution.[202] Interestingly the solid lipid nanoparticle formulation displayed a double peak, presumably from the initial burst release and subsequent sustained release from the particle core. Delivery of all-trans retinoic acid (ATRA) has been investigated by oral gavage in male rats.[203] It was found that solid lipid nanoparticles increased ATRA oral bioavailability approximately five-fold over ATRA solution. These studies provide the rationale for oral solid lipid nanopreparations for drug delivery.

Cyclosporin A has also been dosed orally in solid lipid nanoparticle formulations.[204–206] Microemulsions of cyclosporin have shown nephrotoxicity and variable bioavailability.[207] Attempts have been made to overcome these problems using nanoparticles loaded with 20% cyclosporin. The lipid Imwitor was used with Tagat S (PEG-30 glyceryl stearate) and sodium cholate to stabilize the nanoparticles. The solid lipid nanoparticle formulation was found to eliminate the initial peak in cyclosporin blood concentration and exhibited sustained blood concentrations.[206]

Treatment of rats with insulin loaded solid cetyl palmitate lipid nanoparticles showed a hypoglycemic effect.[208] These nanoparticles prepared by solvent emulsification-evaporation were 350 nm in diameter and had a 43% loading efficiency. Poloxamer 407 was used to stabilize the nanoparticles. After oral administration to rats, a hypoglycemic effect or area above the curve (AAC) was 484 versus 260 for insulin loaded nanoparticles, and insulin alone respectively. The hypoglycemic effect lasted for 24 h and was significantly lower than insulin control.

Solid lipid nanoparticles formulated from glyceryl monostearate were used to encapsulate vinpocetine.[209] These nanoparticles were studied for drug release and particle stability. Vinpocetine which is normally highly insoluble, was stable in these nanoparticles for up to a year. Tween 80 and poly(oxyethylene) hydrogenated castor oil were used to stabilize the nanoparticles, but were also found to enhance the oral bioavailability of the nanoparticles. The nanoparticle formulation increased the bioavailability of vinpocetine significantly. For a more exhaustive analysis of types of therapeutic agents included into solid lipid nanoparticles readers are referred to a relevant review.[199]

Mucosolvan™ (ambroxol) is the only solid lipid nanoparticle product that has reached the market to date for treatment of acute and chronic bronchopulmonary disease. (Boehringer–Ingelheim, Germany). Other products in this area, such as Rifamsolin™ that was in phase I clinical trials in 2005, appear to have been abandoned (AlphaRX now UMeWorld, Canada).[210] Reasons for the lack of further development of some solid lipid nanoparticle formulations appears to be instability in the gastrointestinal tract.[211] Other problems associated with these systems could be the lack of sufficient improvement of absorption to warrant clinical success. Future research into the use of solid lipid nanoparticle could focus on less degradable lipid formulations (such as the previously mentioned archaeosomes) and increased penetration enhancement.

4. Nanosizing of Standard Formulations

A large portion of both current drugs on the market and investigational pharmaceutical agents have low solubility in water.[13,206] Nanopreparations have been able to increase the solubility of hydrophobic drugs by increasing the surface area exposure of the solid formulation. Reducing the size of solid pharmaceutical ingredients can increase the dissolution rate and saturation solubility of drug. This can be of critical importance for low solubility drugs. The increase in saturation solubility of the drug creates a greater local concentration gradient across the intestinal epithelium aiding in absorption.[212] This has been shown to increase the oral bioavailability of several formulations (in addition to creating novel intellectual property for drugs coming off of patent). The theoretical basis for this occurrence revolves around the Nernst–Brunner equation:[213]

$$\frac{dw}{dt} = \frac{D}{h} \cdot S(C_s - C_t),$$

where:

dw/dt: Dissolution rate (mg/s)
S: Effective surface area of the solid drug (cm^2)
Cs - Ct: Concentration gradient (mg/mL)

Cs: Saturated concentration (mg/mL)
Ct: Concentration of the solute at time t (mg/mL)
h: Effective diffusion layer thickness (cm)
D: Diffusion coefficient (cm/s).

Dissolution rate is positively affected by increasing the surface area based on this equation. This can be achieved through reducing the size to nanodimensions. It has been observed that nanosizing can have a dramatic effect on the oral bioavailability of the drug *in vivo*. For example in an oral dosing of naproxen in rats the [214] AUC_{0-240} of the nanoparticle suspension was 125% that of the free solution.[214] Indeed, one of the first reports to analyze the effect of nanosizing a formulation on oral bioavailability showed a 50–80% increase in bioavailability of phenylbenzoylurea derivative HO-221 by decreasing the particle size from 17 μm to 450 nm in rats and dogs.[215,216]

The demonstration of clinical successes using this solubilization for small molecules is the greatest of any oral nanopreparation category to date. A large quantity of pharmaceuticals have already been approved using the nanosizing technology (see Table 4). Tricor sales alone have exceeded $1 billion, soundly placing this paradigm into mainstream pharmaceutical industry.[217] The nano formulation showed reduced variability between fed and fasted states for the delivery of fenofibrate and allowed reduction in the dosage level. A variety of other products have progressed to clinical trials for oral delivery using this technology.[37]

Élan Pharmaceutical Technologies (now Alkermes, Ireland) has trademarked the methods used for creating nanosized pharmaceuticals as Nanocrystal™ while SkyPharma uses the term Dissocubes™.[24,26] These nanoparticles are produced in a top down approach, where larger particles are

Table 4. Marketed oral nanopreparations.[12,212]

Product name	Drug	Technology	Company
Rapamune ®	Sirolimus	Nanosuspensions	Wyeth Pharmaceuticals/Élan
Megace ES	Megestrol acetate	Nanosuspensions	Par Pharmaceuticals/Élan
R Emend ®	Aprepitant	Nanosuspensions	Merck/Élan
Tricor ®	Fenofibrate	Nanosuspensions	Abbott Laboratories/Élan
Triglide ®	Fenofibrate	Nanosuspensions	SkyePharma/Sciele
Avinza ®	Morphine sulfate	Nanosuspensions	King Pharmaceuticals
Focalin XR™	Dexmethyl-phenidate HCl	Nanosuspensions	Novartis
Ritalin LA ®	Methylphenidate HCl	Nanosuspensions	Novartis
Zanaflex ®	Tizanidine HCl	Nanosuspensions	Acorda

milled or homogenized to smaller particle sizes. Resins are usually 500 μm beads and milling can be completed within a few hours. Difficulties of nanosizing formulations include stabilization of the nanoparticles after synthesis. Stabilization is necessary because of the high surface energy created by the small size. If this surface energy is not tailored to the right range, Ostwald ripening or agglomeration can occur. The optimal ratio of stabilizer is usually 20:1 to 2:1 and can include cellulosics, pluronics, polysorbates and povidones. Combinations of ionic and non-ionic stabilizers are often used to decrease aggregation and Ostwald ripening. In spite of additional stabilizers the dispersion can be made into tablets, capsules and fast melts adding additional versatility to this approach.[2,26]

This approach is limited to those drugs that are dissolution rate limited since nanosizing itself does little to increase penetration of the intestinal epithelium. Other problems that still exist for nanosizing technology include the empirical method of screening excipient stabilizers for each drug and lack of miniaturized processes supporting discovery. Future work into novel mechanisms of synthesis, such as spray freezing into liquid or evaporative precipitation into aqueous solution could aid in reducing particle size below 100 nm. Patents for this technology will soon end, opening the door for other companies to take advantage of nanosizing drug formulations for insoluble molecules.[217] Many small molecule drugs have yet to be tested for nanosizing effects, but the benefits may be well worth the cost.[2]

5. Conclusions

The oral route remains the preferred route for dosing of pharmaceuticals. Decreased stringency for production as well as increased patient compliance are the driving forces behind manufacture of oral formulations. Nanoscale constructs aid in the delivery of poorly soluble and poorly permeable drugs. Stemming from past research in biomaterials and drug delivery many current nanopreparations have been created from biodegradable and biocompatible materials. These have shown the ability to encapsulate and protect drugs for oral delivery. Enhanced absorption has been observed using these formulations, partly due to protection of cargo and partly due to penetration enhancing properties. The most successful class of nanopreparations for oral delivery have been solubility-enhancing nanoparticles for active pharmaceuticals. With more than nine current oral nanopreparations of this type already on the market, it is likely that additional products will be commercialized in the near future.

Novel materials have been slower to reach the clinic, likely due to the high development cost and reluctance by industry to use new agents when established approved excipients are available. Instability issues and lack of penetration enhancement may also oppose translation.[107] Micelles appear to be poorly absorbed by the intestinal epithelium and suffer from instability and synthesis issues.[218] Solid nanoparticles, such as chitosan and PCA systems appear to be more stable and easier to synthesize for oral delivery than micelles, but have not shown high enough translocation for clinical development to date. It is difficult to draw conclusions on what system would be optimal for oral delivery because of variability in size, loading, charge, animal model and study methods.[219] Also the ability to successfully deliver cargoes in humans is not easy to extrapolate from small animal work and *in vitro* studies. In spite of these obstacles a general consensus has been reached that diameters of <200 nm is best for oral nanopreparation delivery.[108]

The almost one hundred[220] year old struggle to delivery of insulin orally still remains unconquered, with no marketed oral formulation available to date. Nanopreparations have shown improvement, but low bioavailability still remains a major issue. Highly variable bioavailability also is of considerable concern when drugs have a low therapeutic index. High doses are generally required to see effective absorption *in vivo* which may not be feasible in human trials. In spite of obstacles, recent results are optimistic for the field. Chiasma (Israel) and Roche (Switzerland) have entered agreement for the advancement of octreotide loaded oral particle formulation to phase III clinical trials. This is promising for the field of oral particle-protein delivery.[67,221]

Overcoming penetration and stability obstacles represents a first line of success. Other barriers still exist such as defining nanomedicines to regulatory authorities as well as explaining the risks to the public.[222] Regulatory agencies such as the Food and Drug Administration have convened in order to address these issues and define roadmaps.[223] Future challenges that need addressing for nanoparticle pharmaceuticals include: Biodistribution, toxicity assessment, mass transport, mathematical models, risk/benefit tables covering known nanoparticles and developing standard materials for comparison. Many of these studies are currently being undertaken by research institutions worldwide and will likely lead to the development of the field. Some have proposed delving into technical aspects of fabrication, and large scale production as well for creating feasible systems for pharmaceutical industry.[224] While important findings have been made, many formulation advances are required for oral nanopreparations of poorly permeable molecules to be translated.

References

1. G. Liu, E. Franssen, M. I. Fitch and E. Warner, Patient preferences for oral versus intravenous palliative chemotherapy, *J. Clin. Oncology* **15**(1), 110–115 (1997), ISSN 0732-183X, 1527-7755. URL http://jco.ascopubs.org/content/15/1/110.
2. F. Kesisoglou, S. Panmai and Y. Wu, Nanosizing — oral formulation development and biopharmaceutical evaluation, *Adv. Drug Deliv. Rev.* **59**(7), 631–644 (2007). ISSN 0169-409X. doi: 10.1016/j.addr.2007.05.003. URL http://www.sciencedirect.com/science/article/pii/S0169409X0700083X.
3. J. Renukuntla, A. D. Vadlapudi, A. Patel, S. H. Boddu and A. K. Mitra, Approaches for enhancing oral bioavailability of peptides and proteins, *Int. J. Pharm.* (2013). URL http://www.sciencedirect.com/science/article/pii/S0378517313001658.
4. R. Pandey, Z. Ahmad, S. Sharma and G. Khuller, Nano-encapsulation of azole antifungals: Potential applications to improve oral drug delivery. *Int. J. Pharm.* **301**(1–2), 268–276 (2005). ISSN 0378-5173. doi: 10.1016/j.ijpharm.2005.05.027. URL http://www.sciencedirect.com/science/article/pii/S0378517305003376.
5. Y. Jiao, N. Ubrich, M. Marchand-Arvier, C. Vigneron, M. Hoffman, T. Lecompte and P. Maincent, *In vitro* and *in vivo* evaluation of oral heparin-loaded polymeric nanoparticles in rabbits. *Circulation* **105**(2), 230–235 (2002).
6. P. Ahlin, J. Kristl, A. Kristl and F. Vrečer, Investigation of polymeric nanoparticles as carriers of enalaprilat for oral administration, *Int. J. Pharm.* **239**(1&2), 113–120, (2002).
7. R. Cavalli, A. Bargoni, V. Podio, E. Muntoni, G. Zara and M. Gasco, Duodenal administration of solid lipid nanoparticles loaded with different percentages of tobramycin, *J. Pharm. Sci.* **92**(5), 1085–1094 (2003).
8. R. Pandey, A. Zahoor, S. Sharma and G. Khuller, Nanoparticle encapsulated antitubercular drugs as a potential oral drug delivery system against murine tuberculosis, *Tuberculosis* **83**(6), 373–378 (2003). ISSN 1472-9792. doi: 10.1016/j.tube.2003.07.001. URL http://www.sciencedirect.com/science/article/pii/S1472979203000763.
9. J. A. H. Junghanns and R. H. Müller, Nanocrystal technology, drug delivery and clinical applications, *Int. J. Nanomed.* **3**(3), 295–310 (2008). ISSN 1176-9114. URL http://www.ncbi.nlm.nih.gov/pmc/articles/PMC2626933/PMID: 18990939 PMCID: PMC2626933.
10. W. Shen, Oral peptide and protein delivery: Unfulfilled promises? *Drug Discov. Today* **8**(14), 607–608 (2003). ISSN 1359-6446. doi: 10.1016/S1359-6446(03)02692-8. URL http://www.sciencedirect.com/science/article/pii/S1359644603026928.
11. Y. Lin, F. Mi, C. Chen, W. Chang, S. Peng, H. Liang and H. Sung, Preparation and characterization of nanoparticles shelled with chitosan for oral insulin delivery. *Biomacromolecules.* **8**(1), 146–152 (2007).
12. P. P. Desai, A. A. Date and V. B. Patravale, Overcoming poor oral bioavailability using nanoparticle formulations — opportunities and limitations, *Drug Discov. Today Technologies* **9**(2), e87–e95 (2012). ISSN 1740-6749. doi: 10.1016/j.ddtec.2011.12.001. URL http://www.sciencedirect.com/science/article/pii/S1740674911000345.
13. C. M. Keck and R. H. Müller, Drug nanocrystals of poorly soluble drugs produced by high pressure homogenization, *Eur. J. Pharm. Biopharm.* **62**(1), 3–16 (2006). ISSN

0939–6411. doi: 10.1016/j.ejpb.2005.05.009. URL http://www.sciencedirect.com/science/article/pii/S0939641105001797.

14. W. Saski, Effect of a nonionic surface-active polymer on passage of hydrocortisone across rat intestine *in vitro*, *J. Pharm. Sci.* **57**(5), 836–841 (1968). ISSN 1520–6017. doi: 10.1002/jps.2600570525. URL http://onlinelibrary.wiley.com/doi/10.1002/jps.2600570525/abstract.
15. M. Gibaldi and S. Feldman, Mechanisms of surfactant effects on drug absorption, *J. Pharm. Sci.* **59**(5), 579–589 (1970). ISSN 1520-6017. doi: 10.1002/jps.2600590502. URL http://onlinelibrary.wiley.com/doi/10.1002/jps.2600590502/abstract.
16. D. T. O'Hagan, D. Rahman, J. P. McGee, H. Jeffery, M. C. Davies, P. Williams, S. S. Davis and S. J. Challacombe, Biodegradable microparticles as controlled release antigen delivery systems, *Immunology* **73**(2), 239–242 (1991). ISSN 0019–2805. URL http://www.ncbi.nlm.nih.gov/pmc/articles/PMC1384472/PMID: 2071168 PMCID: PMC1384472.
17. L. Brannon-Peppas, Recent advances on the use of biodegradable microparticles and nanoparticles in controlled drug delivery, *Int. J. Pharm.* 116(1), 1–9 (Mar., 1995). ISSN 0378–5173. doi: 10.1016/0378–5173(94)00324-X. URL http://www.sciencedirect.com/science/article/pii/037851739400324X.
18. S. Cohen and H. Bernstein, Microparticulate Systems for the Delivery of Proteins and Vaccines. (CRC Press, Aug. 1996). ISBN 9780824797539.
19. S. A. Agnihotri, N. N. Mallikarjuna and T. M. Aminabhavi, Recent advances on chitosan-based micro-and nanoparticles in drug delivery, *J. Controlled Release* **100**(1), 5–28 (Nov., 2004). ISSN 0168–3659. doi: 10.1016/j.jconrel.2004.08.010. URL http://www.sciencedirect.com/science/article/pii/S0168365904003803.
20. K. Yamabe, Y. Kato, H. Onishi and Y. Machida, Potentiality of double liposomes containing salmon calcitonin as an oral dosage form, *J. Controlled Release* **89**(3), 429–436 (2003). ISSN 0168–3659. doi: 10.1016/S0168-3659(03)00160-3. URL http://www.sciencedirect.com/science/article/pii/S0168365903001603.
21. F. Antunes, F. Andrade, D. Ferreira, H. Morck Nielsen and B. Sarmento, Models to predict intestinal absorption of therapeutic peptides and proteins, *Current Drug Metabolism* **14**(1), 4–20, (2013). doi: 10.2174/138920013804545160.
22. D. Randall, W. Burggren and K. French, *Eckert Animal Physiology* (Macmillan, 2001). ISBN 9780716738633.
23. F. Gabor, C. Fillafer, L. Neutsch, G. Ratzinger and M. Wirth, Improving oral delivery, *Drug Deliv.* 345–398 (2010). URL http://www.springerlink.com/index/K657233V1M63N390.pdf.
24. R. Müller, C. Jacobs and O. Kayser, Nanosuspensions as particulate drug formula-tions in therapy: Rationale for development and what we can expect for the future, *Adv. Drug Deliv. Rev.* **47**(1), 3–19 (2001).
25. B. Rabinow, Pharmacokinetics of drugs administered in nanosuspension, *Discov. Medicine* **5**(25), 74–79 (2009). URL http://www.discoverymedicine.com/Barrett-Rabinow/2009/07/17/pharmacokinetics-of-drugs-administered-in-nanosuspension/.
26. E. Merisko-Liversidge, G. G. Liversidge and E. R. Cooper, Nanosizing: A formulation approach for poorly-water-soluble compounds, *Eur. J. Pharm. Sci.* **18**(2), 113–120 (2003). ISSN 0928–0987. doi: 10.1016/S0928–0987(02)00251-8. URL http://www.sciencedirect.com/science/article/pii/S0928098702002518.

27. M. Chittchang, A. K. Mitra and T. P. Johnston, Interplay of secondary structure and charge on the diffusion of a polypeptide through negatively charged aqueous pores, *Pharm. Res.* **24**(3), 502–511 (2007). URL http://link.springer.com/article/10.1007/s11095-006-9166-3.
28. M. Tomita, M. Shiga, M. Hayashi and S. Awazu, Enhancement of colonic drug absorption by the paracellular permeation route, *Pharm. Res.* **5**(6), 341–346 (1988). URL http://link.springer.com/article/10.1023/A% 3A1015999309353.
29. G. Van den Mooter, Colon drug delivery, *Exp. Op. Drug Deliv.* **3**(1), 111–125 (2006). ISSN 1742-5247, 1744-7593. doi: 10.1517/17425247.3.1.111. URL http://informahealthcare.com/doi/abs/10.1517/17425247.3.1.111.
30. D. F. Evans, G. Pye, R. Bramley, A. G. Clark, T. J. Dyson and J. D. Hardcastle, Measurement of gastrointestinal pH profiles in normal ambulant human subjects, *Gut.* **29**(8), 1035–1041 (1988). ISSN , 1468-3288. doi: 10.1136/gut.29.8.1035. URL http://gut.bmj.com/content/29/8/1035.
31. M. P. Desai, V. Labhasetwar, G. L. Amidon and R. J. Levy, Gastrointestinal uptake of biodegradable microparticles: Effect of particle size, *Pharm. Res.* **13** (12), 1838–1845 (1996). ISSN 0724-8741. doi: 10.1023/A:1016085108889. URL http://www.springerlink.com/content/xk034l82465150r2/abstract/.
32. D. U. Silverthorn, *Human Physiology: An Integrated Approach.* 4th ed. Pearson Benjamin Cummings (2007).
33. D. N. Granger, J. A. Barrowman and P. R. Kvietys, *Clinical Gastrointestinal Physiology.* Saunders (1985). URL http://www.getcited.org/pub/102382265.
34. S. K. Lai, D. E. O'Hanlon, S. Harrold, S. T. Man, Y. Wang, R. Cone and J. Hanes, Rapid transport of large polymeric nanoparticles in fresh undiluted human mucus, *Proc. Nat. Acad. Sci.* **104**(5), 1482–1487 (2007). ISSN 0027-8424, 1091-6490. doi: 10.1073/pnas.0608611104. URL http://www.pnas.org/content/104/5/1482.
35. R. A. Hazzard, G. M. Hodges, J. D. Scott, C. B. McGuinness and K. E. Carr, Early intestinal microparticle uptake in the rat, *J. Anatomy* **189**(2), 265–271 (1996). ISSN 0021-8782. URL http://www.ncbi.nlm.nih.gov/pmc/articles/PMC1167743/. PMID: 8886948 PMCID: PMC1167743.
36. E. Allémann, J. Leroux and R. Gurny, Polymeric nano-and microparticles for the oral delivery of peptides and peptidomimetics, *Adv. Drug Deliv. Rev.* **34**(2&3), 171–189 (1998). ISSN 0169-409X. doi: 10.1016/S0169-409X(98)00039-8. URL http://www.sciencedirect.com/science/article/B6T3R-3V7JG5N-5/2/a88eae800879f9aacf25e7d3f6fe5c29.
37. B. E. Rabinow, Nanosuspensions in drug delivery, *Nat. Rev. Drug Discov.* **3**(9), 785–796 (2004). ISSN 1474-1776. doi: 10.1038/nrd1494. URL http://www.nature.com/nrd/journal/v3/n9/full/nrd1494.html?type=access_denied.
38. N. Salamat-Miller and T. P. Johnston, Current strategies used to enhance the paracellular transport of therapeutic polypeptides across the intestinal epithelium, *Int. J. Pharm.* **294**(1&2), 201–216 (2005). ISSN 0378-5173. doi: 10.1016/j.ijpharm.2005.01.022. URL http://www.sciencedirect.com/science/article/pii/S0378517305000918.
39. N. Pantzar, S. Lundin, L. Wester and B. R. Westrom, Bidirectional small-intestinal permeability in the rat to some common marker molecules *in vitro*, *Scandinavian J. Gastroenterol.* **29**(8), 703–709 (1994). URL http://informahealthcare.com/doi/abs/10.3109/00365529409092497.

40. H. N. Nellans, Mechanisms of peptide and protein absorption: (1) paracellular intestinal transport: Modulation of absorption, *Adv. Drug Deliv. Rev.* 7(3), 339–364 (1991). ISSN 0169-409X. doi: 10.1016/0169-409X(91)90013-3. URL http://www.sciencedirect.com/science/article/pii/0169409X91900133.
41. B. Sarmento, A. Ribeiro, F. Veiga, P. Sampaio, R. Neufeld and D. Ferreira, Alginate/Chitosan nanoparticles are effective for oral insulin delivery, *Pharma. Res.* **24**(12), 2198–2206 (2007). ISSN 0724-8741, 1573-904X. doi: 10.1007/s11095-007-9367-4. URL http://link.springer.com/10.1007/s11095-007-9367-4.
42. D. S. Goldberg, H. Ghandehari and P. W. Swaan, Cellular entry of G3.5 poly(amido amine) dendrimers by clathrin-and dynamin-dependent endocytosis promotes tight junctional opening in intestinal epithelia, *Pharm. Res.* **27**(8), 1547–1557 (2010). ISSN 1573-904X. doi: 10.1007/s11095-010-0153-3. URL http://www.springerlink.com/content/gv981x24g9t8rr1j/?MUD=MP. PMID: 20411406.
43. K. M. Kitchens, R. B. Kolhatkar, P. W. Swaan and H. Ghandehari, Endocytosis inhibitors prevent poly(amidoamine) dendrimer internalization and permeability across Caco-2 cells, *Mol. Pharm.* **5**(2), 364–369 (2008). ISSN 1543-8384. doi: 10.1021/mp700089s. URL http://pubs.acs.org/doi/abs/10.1021/mp700089s. PMID: 18173246.
44. M. Kondoh, A. Takahashi and K. Yagi, Spiral progression in the development of absorption enhancers based on the biology of tight junctions, *Adv. Drug Deliv. Rev.* **64**(6), 515–522 (2012). ISSN 0169-409X. doi: 10.1016/j.addr.2011.07.004. URL http://www.sciencedirect.com/science/article/pii/S0169409X11001980.
45. P. Claude, Morphological factors influencing transepithelial permeability: A model for the resistance of the zonula occluden, *J. Membrane Biol.* **39**(2), 219–232 (1978). ISSN 0022-2631. doi: 10.1007/BF01870332. URL http://www.springerlink.com/content/m55un248u7688787/abstract/.
46. H. Chiba, M. Osanai, M. Murata, T. Kojima and N. Sawada, Transmembrane proteins of tight junctions, Biochimica et Biophysica Acta (BBA) — *Biomembranes* **1778**(3), 588–600 (2008). ISSN 0005-2736. doi: 10.1016/j.bbamem.2007.08.017. URL http://www.sciencedirect.com/science/article/pii/S0005273607003045.
47. J. Pappenheimer, Physiological regulation of transepithelial impedance in the intestinal mucosa of rats and hamsters, *J. Membrane Biol.* **100**(1), 137–148 (1987). ISSN 0022-2631. doi: 10.1007/BF02209146. URL http://www.springerlink.com/content/u74n12062503272q/abstract/.
48. D. J. Brayden, M. A. Jepson and A. W. Baird, Keynote review: Intestinal peyer's patch M cells and oral vaccine targeting, *Drug Discov. Today* **10**(17), 1145–1157 (2005). ISSN 1359-6446. doi: 10.1016/S1359-6446(05)03536-1. URL http://www.sciencedirect.com/science/article/pii/S1359644605035361.
49. K. A. Heel, R. D. McCauley, J. M. Papadimitriou and J. C. Hall, Review: Peyer's patches, *J. Gastroenterol. Hepatol.* **12**(2), 122–136 (1997). ISSN 1440-1746. doi: 10.1111/j.1440-1746.1997.tb00395.x. URL http://onlinelibrary.wiley.com/doi/10.1111/j.1440-1746.1997.tb00395.x/abstract.
50. E. Gullberg, R. V. Keita, S. Y. Salim, M. Andersson, K. D. Caldwell, J. D. Soderholm and P. Artursson, Identification of cell adhesion molecules in the human Follicle-Associated epithelium that improve nanoparticle uptake into the peyer's patches, *J. Pharm. Exp. Therapeut.* **319**(2), 632–639 (2006). ISSN , 1521-0103 (Online). doi: 10.1124/jpet.106.107847. URL http://jpet.aspetjournals.org/content/319/2/632.

51. S. McClean, E. Prosser, E. Meehan, D. O'Malley, N. Clarke, Z. Ramtoola and D. Brayden, Binding and uptake of biodegradable poly-dl-lactide micro-and nanoparticles in intestinal epithelia, *Eur. J. Pharm. Sci.* **6**(2), 153–163 (1998). ISSN 0928–0987. doi: 10.1016/S0928–0987(97)10007–0. URL http://www.sciencedirect.com/science/article/pii/S0928098797100070.
52. M. F. Francis, M. Cristea and F. M. Winnik, Exploiting the vitamin B12 pathway to enhance oral drug delivery via polymeric micelles, *Biomacromolecules* **6**(5), 2462–2467 (2005). ISSN 1525–7797. doi: 10.1021/bm0503165. URL http://dx.doi.org/10.1021/bm0503165.
53. G. J. Russell-Jones, S. W. Westwood and A. D. Habberfield, Vitamin B12 mediated oral delivery systems for granulocyte-colony stimulating factor and erythropoietin, *Bioconjug. Chem.* **6**(4), 459–465 (Aug., 1995). ISSN 1043–1802. URL http://www.ncbi.nlm.nih.gov/pubmed/7578366. PMID: 7578366.
54. A. Habberfield, K. Jensen-Pippo, L. Ralph, S. W. Westwood and G. J. Russell-Jones, Vitamin B12-mediated uptake of erythropoietin and granulocyte colony stimulating factor *in vitro* and *in vivo*, *Int. J. Pharm.* **145** (1–2), 1–8 (1996). ISSN 0378–5173. doi: 10.1016/S0378–5173(96)04690-X. URL http://www.sciencedirect.com/science/article/pii/S037851739604690X.
55. G. J. Russell-Jones, Use of vitamin B12 conjugates to deliver protein drugs by the oral route, Critical Reviews in Therapeutic Drug Carrier Systems. **15**(6), 557–586 (1998). ISSN 0743–4863. URL http://www.ncbi.nlm.nih.gov/pubmed/9883390. PMID: 9883390.
56. G. J. Russell-Jones, L. Arthur and H. Walker, Vitamin B12-mediated transport of nanoparticles across Caco-2 cells, *Int. J. Pharm.* **179**(2), 247–255 (1999). ISSN 0378–5173. URL http://www.ncbi.nlm.nih.gov/pubmed/10053217. PMID: 10053217.
57. B. Seetharam, M. Jimenez and D. H. Alpers, Effect of bile and bile acids on binding of intrinsic factor to cobalamin and intrinsic factor-cobalamin complex to ileal receptor, *Amer. J. Phys. -Gastrointestinal and Liver Physiol.* **245**(1), G72–G77 (1983). ISSN 0193–1857, 1522–1547. URL http://ajpgi.physiology.org/content/245/1/G72.
58. P. W. Swaan, Recent advances in intestinal macromolecular drug delivery via receptor-mediated transport pathways, *Pharm. Res.* **15**(6), 826–834, (1998). ISSN 0724–8741. doi: 10.1023/A:1011908128045. URL http://www.springerlink.com/content/l674787627252118/abstract/.
59. S. M. Clardy, D. G. Allis, T. J. Fairchild and R. P. Doyle, Vitamin B12 in drug delivery: Breaking through the barriers to a B12 bioconjugate pharmaceutical, *Exp. Opin. Drug Deliv.* **8**(1), 127–140 (2011). ISSN 1742–5247, 1744-7593. doi: 10.1517/17425247.2011.539200. URL http://informahealthcare.com/doi/abs/10.1517/17425247.2011.539200.
60. C. Lipinski, F. Lombardo, B. Dominy and P. Feeney, Experimental and computational approaches to estimate solubility and permeability in drug dis-covery and development settings, *Adv. Drug Deliv. Rev.* **23**(1–3), 3–25 (1997). URL http://www.scopus.com/inward/record.url?eid=2-s2. 0–0031024171&partnerID=40&md5=5229a5a941c1af07a367239e8b9033d7.
61. D. F. Veber, S. R. Johnson, H. Cheng, B. R. Smith, K. W. Ward and K. D. Kopple, Molecular properties that influence the oral bioavailability of drug candidates, *J. Med.*

Chem. 45(12), 2615–2623 (2002). ISSN 0022–2623. doi: 10.1021/jm020017n. URL http://dx.doi.org/10.1021/jm020017n.

62. F. Yoshida and J. G. Topliss, QSAR model for drug human oral bioavailability1, *J. Med. Chem.* **43**(13), 2575–2585 (2000). ISSN 0022–2623. doi: 10.1021/jm0000564. URL http://dx.doi.org/10.1021/jm0000564.
63. B. J. Aungst, Novel formulation strategies for improving oral bioavailability of drugs with poor membrane permeation or presystemic metabolism, *J. Pharm. Sci.* **82**(10), 979–987(1993). URL http://onlinelibrary.wiley.com/doi/10.1002/jps.2600821002/abstract.
64. T. Vasconcelos, B. Sarmento and P. Costa, Solid dispersions as strategy to improve oral bioavailability of poor water soluble drugs, *Drug Discovery Today.* 12(23-24), 1068–1075, (2007). URL http://www.sciencedirect.com/science/article/pii/S1359644607003753.
65. W. K. Sietsema, The absolute oral bioavailability of selected drugs, International Journal of Clinical Pharmacology, Therapy, and Toxicology. **27**(4), 179–211 (1989). ISSN 0174–4879. URL http://www.ncbi.nlm.nih.gov/pubmed/2654032. PMID: 2654032.
66. A. Leone-Bay, N. Santiago, D. Achan, K. Chaudhary, F. DeMorin, L. Falzarano, S. Haas, S. Kalbag and D. Kaplan, N-Acylated-α-Amino acids as novel oral delivery agents for proteins, *J. Med. Chem.* **38**(21), 4263–4269 (1995). ISSN 0022–2623. doi: 10.1021/jm00021a015. URL http://dx.doi.org/10.1021/jm00021a015.
67. S. Maher and D. J. Brayden, Overcoming poor permeability: translating permeation enhancers for oral peptide delivery, *Drug Discov. Today Technol.* **9**(2), e113–e119 (2012). ISSN 1740–6749. doi: 10.1016/j.ddtec.2011.11.006. URL http://www.sciencedirect.com/science/article/pii/S1740674911000266.
68. A. Makhlof, M. Werle, Y. Tozuka and H. Takeuchi, A mucoadhesive nanoparticulate system for the simultaneous delivery of macromolecules and permeation enhancers to the intestinal mucosa, *J. Controlled Release* **149**(1), 81–88 (2011). ISSN 0168–3659. doi: 10.1016/j.jconrel.2010.02.001. URL http://www.sciencedirect.com/science/article/pii/S0168365910001070.
69. L. Yin, J. Ding, C. He, L. Cui, C. Tang and C. Yin, Drug permeability and mucoadhesion properties of thiolated trimethyl chitosan nanoparticles in oral insulin delivery, *Biomaterials* **30**(29), 5691–5700 (2009).
70. Y. Pan, Y. Li, H. Zhao, J. Zheng, H. Xu, G. Wei, J. Hao and F. Cui, Bioadhesive polysaccharide in protein delivery system: Chitosan nanoparticles improve the intestinal absorption of insulin *in vivo*, *Int. J. Pharm.* **249**(1), 139–147 (2002).
71. Z. Ma, T. M. Lim and L. Lim, Pharmacological activity of peroral chitosan-insulin nanoparticles in diabetic rats. *Int, J. Pharm.* **293**(1&2), 271–280 (2005). ISSN 0378–5173. doi: 10.1016/j.ijpharm.2004.12.025. URL http://www.ncbi.nlm.nih.gov/pubmed/15778065. PMID: 15778065.
72. Z. Ma and L. Lim, Uptake of chitosan and associated insulin in Caco-2 cell monolayers: A comparison between chitosan molecules and chitosan nanoparticles. *Pharm. Res.* **20**(11), 1812–1819 (2003). ISSN 0724–8741. doi: 10.1023/B:PHAM.0000003379.76417.3e. URL http://www.springerlink.com/content/wp307875x8742575/abstract/.
73. R. Rosenthal, D. Günzel, C. Finger, S. M. Krug, J. F. Richter, J. Schulzke, M. Fromm and S. Amasheh, The effect of chitosan on transcellular and paracellular mechanisms in the

intestinal epithelial barrier, *Biomaterials* **33**(9), 2791–2800 (2012). ISSN 0142–9612. doi: 10.1016/j.biomaterials.2011.12.034. URL http://www.sciencedirect.com/science/article/pii/S0142961211015043.

74. A. Chaudhury and S. Das, Recent advancement of chitosan-based nanoparticles for oral controlled delivery of insulin and other therapeutic agents, *AAPS PharmSciTech.* **12**(1), 10–20 (2010). ISSN 1530–9932. doi: 10.1208/s12249–010–9561–2. URL http://www.ncbi.nlm.nih.gov/pmc/articles/PMC3066343/. PMID: 21153572 PMCID: PMC3066343.
75. C. Prego, M. García, D. Torres and M. J. Alonso, Transmucosal macromolecular drug delivery, *J. Controlled Release* **101**(1–3), 151–162 (2005). ISSN 0168–3659. doi: 10.1016/j.jconrel.2004.07.030. URL http://www.ncbi.nlm.nih.gov/pubmed/15588901. PMID: 15588901.
76. K. Arai, T. Kinumaki and T. Fujita, Toxicity of chitosan, *Bull. Tokai. Region. Fish. Res. Lab.* (56), 89–94 (1968).
77. M. Thanou, J. C. Verhoef, J. H. M. Verheijden and H. E. Junginger, Intestinal absorption of octreotide using trimethyl chitosan chloride: Studies in pigs, *Pharm. Res.* **18**(6), 823–828 (2001). ISSN 0724–8741, 1573–904X. doi: 10.1023/A:1011092613951. URL http://link.springer.com/article/10.1023/A% 3A1011092613951.
78. A. Bernkop-Schnürch, M. Hornof and T. Zoidl, Thiolated polymers—thiomers: synthesis and *in vitro* evaluation of chitosan–2-iminothiolane conjugates, *Int. J. Pharm.* **260**(2), 229–237 (2003). ISSN 0378–5173. doi: 10.1016/S0378–5173(03)00271–0. URL http://www.sciencedirect.com/science/article/pii/S0378517303002710.
79. M. Roldo, M. Hornof, P. Caliceti and A. Bernkop-Schnürch, Mucoadhesive thiolated chitosans as platforms for oral controlled drug delivery: Synthesis and *in vitro* evaluation, *Eur. J. Pharm. Biopharm.* **57**(1), 115–121 (2004). ISSN 0939–6411. doi: 10.1016/S0939-6411(03)00157–7. URL http://www.sciencedirect.com/science/article/pii/S0939641103001577.
80. M. D. Hornof, C. E. Kast and A. Bernkop-Schnürch, *In vitro* evaluation of the viscoelastic properties of chitosan–thioglycolic acid conjugates, *Eur. J. Pharm. Biopharm.* **55**(2), 185–190 (2003). ISSN 0939–6411. doi: 10.1016/S0939–6411(02)00162-5. URL http://www.sciencedirect.com/science/article/pii/S0939641102001625.
81. F. Andrade, F. Goycoolea, D. A. Chiappetta, J. das Neves, A. Sosnik and B. Sarmento, Chitosan-Grafted copolymers and Chitosan-Ligand conjugates as matrices for pulmonary drug delivery, *Int. J. Carbohydrate Chem.* **2011**, 1–14 (2011). ISSN 1687–9341, 1687–935X. doi: 10.1155/2011/865704. URL http://www.hindawi.com/journals/ijcc/2011/865704/.
82. M. Amidi, E. Mastrobattista, W. Jiskoot and W. E. Hennink, Chitosan-based delivery systems for protein therapeutics and antigens, *Adv. Drug Deliv. Rev.* **62**(1), 59–82 (2010). ISSN 0169–409X. doi: 10.1016/j.addr.2009.11.009. URL http://www.sciencedirect.com/science/article/pii/S0169409X09003494.
83. Y. Zambito. Nanoparticles based on chitosan derivatives. In Advances in Biomaterials Science and Biomedical Applications. In Tech, (2013). URL http://cdn.intechopen.com/pdfs/43738/InTech-Nanoparticles_based_on_chitosan_derivatives.pdf.
84. D. Vllasaliu, R. Exposito-Harris, A. Heras, L. Casettari, M. Garnett, L. Illum and S. Stolnik, Tight junction modulation by chitosan nanoparticles: Comparison with chitosan solution, *Int. J. Pharm.* **400**(1&2), 183–193 (2010).

85. L. Kudsiova and M. Lawrence, A comparison of the effect of chitosan and chitosan-coated vesicles on monolayer integrity and permeability across Caco-2 and 16HBE14o-cells, *J. Pharm. Sci.* **97**(9), 3998–4010 (2008). ISSN 1520-6017. doi: 10.1002/jps.21262. URL http://onlinelibrary.wiley.com/doi/10.1002/jps.21262/abstract.
86. P. Artursson, T. Lindmark, S. S. Davis and L. Illum, Effect of chitosan on the permeability of monolayers of intestinal epithelial cells (Caco-2), *Pharm. Res.* **11**(9), 1358–1361 (1994). ISSN 0724–8741. doi: 10.1023/A:1018967116988. URL http://www.springerlink.com/content/g80u530547197066/abstract/.
87. N. G. Schipper, K. M. Vårum and P. Artursson, Chitosans as absorption enhancers for poorly absorbable drugs. 1: Influence of molecular weight and degree of acetylation on drug transport across human intestinal epithelial (Caco-2) cells, *Pharm. Res.* **13**(11), 1686–1692 (1996). ISSN 0724–8741. PMID: 8956335.
88. G. Sandri, M. C. Bonferoni, S. Rossi, F. Ferrari, C. Boselli and C. Caramella, Insulin-loaded nanoparticles based on N-trimethyl chitosan: *In vitro* (Caco-2 model) and ex *vivo* (Excised rat jejunum, duodenum, and ileum) evaluation of penetration enhancement properties, *AAPS PharmSciTech.* **11**(1), 362–371 (2010). ISSN 1530–9932. doi: 10.1208/s12249-010-9390-3. URL http://www.ncbi.nlm.nih.gov/pmc/articles/PMC2850491/. PMID: 20232266 PMCID: PMC2850491.
89. C. Prego, D. Torres and M. J. Alonso, The potential of chitosan for the oral administration of peptides, *Exp. Opin. Drug Deliv.* **2**(5), 843–854 (2005). ISSN 1742–5247, 1744–7593. doi: 10.1517/17425247.2.5.843. URL http://informahealthcare.com/doi/abs/10.1517/17425247.2.5.843.
90. B. Sarmento, A. Ribeiro, F. Veiga, D. Ferreira and R. Neufeld, Oral bioavailability of insulin contained in polysaccharide nanoparticles, *Biomacromolecules* **8**(10), 3054–3060 (2007). ISSN 1525–7797. doi: 10.1021/bm0703923. URL http://dx.doi.org/10.1021/bm0703923.
91. S. Shu, X. Zhang, D. Teng, Z. Wang and C. Li, Polyelectrolyte nanoparticles based on water-soluble chitosan–poly(l-aspartic acid)–polyethylene glycol for controlled protein release, *Carbohydrate Res.* **344**(10), 1197–1204 (2009). ISSN 0008–6215. doi: 10.1016/j.carres.2009.04.018. URL http://www.sciencedirect.com/science/article/pii/S0008621509001724.
92. P. Mukhopadhyay, R. Mishra, D. Rana and P. P. Kundu, Strategies for effective oral insulin delivery with modified chitosan nanoparticles: A review, *Prog. Polymer Sci.* **37**(11), 1457–1475 (2012). ISSN 0079–6700. doi: 10.1016/j.progpolymsci.2012.04.004. URL http://www.sciencedirect.com/science/article/pii/S007967001200041X.
93. A. Bayat, F. A. Dorkoosh, A. R. Dehpour, L. Moezi, B. Larijani, H. E. Junginger and M. Rafiee-Tehrani, Nanoparticles of quaternized chitosan derivatives as a carrier for colon delivery of insulin: *Ex vivo* and *in vivo* studies, *Int. J. Pharm.* **356**(1&2), 259–266 (2008). ISSN 0378–5173. doi: 10.1016/j.ijpharm.2007.12.037. URL http://www.sciencedirect.com/science/article/pii/S037851730701068X.
94. Y. Jin, Y. Song, X. Zhu, D. Zhou, C. Chen, Z. Zhang and Y. Huang, Goblet cell-targeting nanoparticles for oral insulin delivery and the influence of mucus on insulin transport, *Biomaterials* **33**(5), 1573–1582 (2012). ISSN 0142-9612. doi: 10.1016/j.biomaterials.2011.10.075. URL http://www.sciencedirect.com/science/article/pii/S0142961211013081.

95. S. Emdin, G. Dodson, J. Cutfield and S. Cutfield, Role of zinc in insulin biosynthesis, *Diabetologia* **19**(3), 174–182 (1980). ISSN 0012–186X. doi: 10.1007/BF00275265. URL http://www.springerlink.com/content/l2345h846t38v21h/abstract/.
96. K. Sonaje, Y. Chen, H. Chen, S. Wey, J. Juang, H. Nguyen, C. Hsu, K. Lin and H. Sung, Enteric-coated capsules filled with freeze-dried chitosan/poly(γ-glutamic acid) nanoparticles for oral insulin delivery, *Biomaterials* **31**(12), 3384–3394 (2010). ISSN 0142–9612. doi: 10.1016/j.biomaterials.2010.01.042. URL http://www.sciencedirect.com/science/article/pii/S014296121000058X.
97. K. Sonaje, K. Lin, S. Wey, C. Lin, T. Yeh, H. Nguyen, C. Hsu, T. Yen, J. Juang and H. Sung, Biodistribution, pharmacodynamics and pharmacokinetics of insulin analogues in a rat model: Oral delivery using pH-Responsive nanoparticles vs. subcutaneous injection, *Biomaterials* **31**(26), 6849–6858 (2010). ISSN 0142-9612. doi: 10.1016/j.biomaterials.2010.05.042. URL http://www.sciencedirect.com/science/article/pii/S0142961210006605.
98. H. Sung, K. Sonaje, Z. Liao, L. Hsu and E. Chuang, pH-Responsive nanoparticles shelled with chitosan for oral delivery of insulin: From mechanism to therapeutic applications, *Acc. Chem. Res.* **45**(4), 619–629 (2012). ISSN0001–4842. doi: 10.1021/ar200234q. URL http://dx.doi.org/10.1021/ar200234q.
99. A. Bernkop-Schnürch and M. E. Krajicek, Mucoadhesive polymers as platforms for peroral peptide delivery and absorption: Synthesis and evaluation of different chitosan-EDTA conjugates, *J. Controlled Release* **50**(1–3), 215–223 (1998). ISSN 0168–3659. URL http://www.ncbi.nlm.nih.gov/pubmed/9685888. PMID: 9685888.
100. H. L. Luessen, B. J. de Leeuw, M. W. Langemeÿer, A. B. de Boer, J. C. Verhoef and H. E. Junginger, Mucoadhesive polymers in peroral peptide drug delivery. VI. carbomer and chitosan improve the intestinal absorption of the peptide drug buserelin *in vivo*, *Pharm. Res.* **13**(11), 1668–1672 (1996). ISSN 0724–8741. URL http://www.ncbi.nlm.nih.gov/pubmed/8956332. PMID: 8956332.
101. A. Bernkop-Schnürch, Ed., *Oral Delivery of Macromolecular Drugs*: New York, Springer (2009). ISBN 978–1–4419–0199–6. URL http://www.springerlink.com/content/978–1–4419–0199–6#section=99315&page=1.
102. F. C. MacLaughlin, R. J. Mumper, J. Wang, J. M. Tagliaferri, I. Gill, M. Hinchcliffe and A. P. Rolland, Chitosan and depolymerized chitosan oligomers as condensing carriers for *in vivo* plasmid delivery, *J. Controlled Release* **56**(1–3), 259–272 (1998). ISSN 0168–3659. doi: 10.1016/S0168–3659(98)00097–2. URL http: //www.sciencedirect.com/science/article/pii/S0168365998000972.
103. K. Roy, H. Q. Mao, S. K. Huang and K. W. Leong, Oral gene delivery with chitosan–DNA nanoparticles generates immunologic protection in a murine model of peanut allergy, *Nat. Med.* **5**(4), 387–391 (1999). ISSN 1078–8956. doi: 10.1038/7385. URL http://www.ncbi.nlm.nih.gov/pubmed/10202926. PMID: 10202926.
104. K. Bowman and K. Leong, Chitosan nanoparticles for oral drug and gene delivery, *Int. J. Nanomed.* **1**(2), 117 (2006).
105. S. B. Rao and C. P. Sharma, Use of chitosan as a biomaterial: Studies on its safety and hemostatic potential, *J. Biomed. Mat. Res.* **34**(1), 21–28 (1997). URL http://www.calidaddeinformacioncfr.com/biopiel/Use_of_chitosan_as_a_biomaterial_Studies_on_its_safety.pdf.

106. T. Kean, S. Roth and M. Thanou, Trimethylated chitosans as non-viral gene delivery vectors: Cytotoxicity and transfection efficiency, *J. Controlled Release* **103**(3), 643–653 (2005). ISSN 0168–3659. doi: 10.1016/j.jconrel.2005.01.001. URL http://www.sciencedirect.com/science/article/pii/S0168365905000076.
107. P. Baldrick, The safety of chitosan as a pharmaceutical excipient, *Regulat. Toxicol. Pharmacol.* **56**(3), 290–299 (2010). ISSN 0273–2300. doi: 10.1016/j.yrtph.2009.09.015. URL http://www.sciencedirect.com/science/article/pii/ S0273230009002001.
108. L. Plapied, N. Duhem, A. des Rieux and V. Préat, Fate of polymeric nanocarriers for oral drug delivery, *Current Opin. Colloid Int. Sci.* **16**(3), 228–237 (2011). URL http://www.sciencedirect.com/science/article/pii/S1359029411000033.
109. N. Al Khouri Fallouh, L. Roblot-Treupel, H. Fessi, J. Devissaguet and F. Puisieux, Development of a new process for the manufacture of polyisobutylcyanoacrylate nanocapsules, *Int. J. Pharm.* **28**(2–3), 125–132 (1986). ISSN 0378–5173. doi: 10.1016/0378–5173(86)90236-X. URL http://www.sciencedirect.com/science/article/pii/037851738690236X.
110. C. Vauthier, C. Dubernet, E. Fattal, H. Pinto-Alphandary and P. Couvreur, Poly(alkylcyanoacrylates) as biodegradable materials for biomedical applications, *Adv. Drug Deliv. Rev.* **55**(4), 519–548 (2003). ISSN 0169–409X. doi: 10.1016/S0169-409X(03)00041–3. URL http://www.sciencedirect.com/science/article/pii/S0169409X03000413.
111. C. Damgé, C. Michel, M. Aprahamian and P. Couvreur, New approach for oral administration of insulin with polyalkylcyanoacrylate nanocapsules as drug carrier, *Diabetes* **37**(2), 246–251 (1988). ISSN 0012–1797, 1939–327X. doi: 10.2337/diab.37.2.246. URL http://diabetes.diabetesjournals.org/content/37/2/246.
112. M. J. Alonso, N. S. Csaba, D. E. Thurston, A. Martinez and D. Fox, *Nanostructured Biomaterials for Overcoming Biological Barriers.* Royal Society of Chemistry (2012). ISBN 9781849733632.
113. P. Maincent, J. P. Devissaguet, R. LeVerge, P. A. Sado and P. Couveur, Preparation and *in vivo* studies of a new drug delivery system, *Appl. Biochem. Biotechnol.* **10**(1–3), 263–265 (1984). ISSN 0273–2289, 1559–0291. doi: 10.1007/BF02783760. URL http://link.springer.com/article/10.1007/BF02783760.
114. P. Maincent, R. Le Verge, P. Sado, P. Couvreur and J. P. Devissaguet, Disposition kinetics and oral bioavailability of vincamine-loaded polyalkyl cyanoacrylate nanoparticles, *J. Pharm. Sci.* **75**(10), 955–958 (1986). ISSN 1520–6017. doi: 10.1002/jps.2600751009. URL http://onlinelibrary.wiley.com/doi/10.1002/jps.2600751009/abstract.
115. C. Vauthier, D. Labarre and G. Ponchel, Design aspects of poly(alkylcyanoacrylate) nanoparticles for drug delivery, *J. Drug Targeting* **15**(10), 641–663 (2007). ISSN 1061–186X, 1029–2330. doi: 10.1080/10611860701603372. URL http://www.informapharmascience.com/doi/abs/10.1080/10611860701603372.
116. A. Graf, A. McDowell and T. Rades, Poly(alkycyanoacrylate) nanoparticles for enhanced delivery of therapeutics – is there real potential? *Exp. Opin. Drug Deliv.* **6**(4), 371–387 (2009). ISSN 1742–5247, 1744–7593. doi: 10.1517/17425240902870413. URL http://informahealthcare.com/doi/abs/10.1517/17425240902870413.
117. L. Vansnick, P. Couvreur, D. Christiaens-Leyh and M. Roland, Molecular weights of free and drug-loaded nanoparticles, *Pharm. Res.* **2**(1), 36–41 (1985). URL http://link.springer.com/article/10.1023/A%3A1016366022712.

118. V. Lenaerts, P. Couvreur, D. Christiaens-Leyh, E. Joiris, M. Roland, B. Rollman and P. Speiser, Degradation of poly (isobutyl cyanoacrylate) nanoparticles, *Biomaterials* **5**(2), 65–68 (1984). ISSN 0142–9612. doi: 10.1016/0142-9612(84)90002–4. URL http://www.sciencedirect.com/science/article/pii/0142961284900024.
119. K. Langer, E. Seegmüller, A. Zimmer and J. Kreuter, Characterisation of poly-butylcyanoarylate nanoparticles: I. quantification of PCBA polymer and dextrans, *Int. J. Pharm.* **110**(1), 21–27 (1994). ISSN 0378-5173. doi: 10.1016/0378–5173(94)90371–9. URL http://www.sciencedirect.com/science/article/pii/0378517394903719.
120. J. Nicolas and P. Couvreur, Synthesis of poly (alkyl cyanoacrylate)-based colloidal nanomedicines, Wiley Interdisciplinary Reviews: *Nanomed. Nanobiotechnol.* **1**(1), 111–127 (2009). URL http://onlinelibrary.wiley.com/doi/10.1002/wnan.15/full.
121. C. Michel, M. Aprahamian, L. Defontaine, P. Couvreur and C. Damgé, The effect of site of administration in the gastrointestinal tract on the absorption of insulin from nanocapsules in diabetic rats, *J. Pharm. Pharmacol.* **43**(1), 1–5 (1991). ISSN 2042–7158. doi: 10.1111/j.2042-7158.1991.tb05437.x. URL http://onlinelibrary.wiley.com/doi/10.1111/j.2042–7158.1991.tb05437.x/abstract.
122. P. Lowe and C. Temple, Calcitonin and insulin in isobutylcyanoacrylate nanocapsules: Protection against proteases and effect on intestinal absorption in rats, *J. Pharm. Pharmacol.* **46**(7), 547–552 (1994). ISSN 00223573.
123. P. H. Beck, J. Kreuter, W. E. G. Müller and W. Schatton, Improved peroral delivery of avarol with polybutylcyanoacrylate nanoparticles, *Eur. J. Pharm. Biopharm.* **40**(3), 134–137 (1994). URL http://cat.inist.fr/?aModele=afficheN&cpsidt=4100137.
124. S. Bonduelle, M. Carrier, C. Pimienta, J. Benoit and V. Lenaerts, Tissue concentration of nanoencapsulated radio-labelled cyclosporin following peroral delivery in mice or ophthalmic application in rabbits, *Eur. J. Pharm. Biopharm.* **42**(5), 313–319 (1996). ISSN 0939–6411. URL http://cat.inist.fr/?aModele=afficheN&cpsidt=3259960.
125. H. Vranckx, M. Demoustier and M. Deleers, A new nanocapsule formulation with hydrophilic core: Application to the oral administration of salmon calcitonin in rats, *Eur. J. Pharm. Biopharm.* **42**(5), 345–347 (1996). ISSN 09396411.
126. G. Ponchel and J. Irache, Specific and non-specific bioadhesive particulate systems for oral delivery to the gastrointestinal tract, *Adv. Drug Delivery Rev.* **34** (2&3), 191–219 (1998). ISSN 0169–409X. doi: 10.1016/S0169–409X(98)00040–4. URL http://www.sciencedirect.com/science/article/pii/S0169409X98000404.
127. A. T. Florence, Issues in oral nanoparticle drug carrier uptake and targeting, *J. Drug Target.* **12**(2), 65–70 (2004). URL http://informahealthcare.com/doi/ pdf/10.1080/10611860410001693706.
128. I. Bravo-Osuna, C. Vauthier, A. Farabollini, G. F. Palmieri and G. Ponchel, Mucoadhesion mechanism of chitosan and thiolated chitosan-poly(isobutyl cyanoacrylate) core-shell nanoparticles, *Biomaterials* **28**(13), 2233–2243 (2007). ISSN 0142-9612. doi: 10.1016/j.biomaterials.2007.01.005. URL http://www.sciencedirect.com/science/article/pii/S0142961207000117.
129. S. C. Yang, H. X. Ge, Y. Hu, X. Q. Jiang and C. Z. Yang, Formation of positively charged poly(butyl cyanoacrylate) nanoparticles stabilized with chitosan, *Colloid and Polymer Sci.* **278**(4), 285–292 (2000). ISSN 0303–402X, 1435–1536. doi: 10.1007/s003960050516. URL http://link.springer.com/article/10.1007/s003960050516.

130. C. W. R. Wade and F. Leonard, Degradation of poly(methyl 2-cyanoacrylates), *J. Biomed. Mat. Res.* **6**(3), 215–220 (1972). ISSN 1097–4636. doi: 10.1002/jbm.820060310. URL http://onlinelibrary.wiley.com/doi/10.1002/jbm.820060310/abstract.
131. W. R. Vezin and A. T. Florence, *In vitro* heterogeneous degradation of poly(n-alkyl α-cyanoacrylates), *J. Biomed. Mat. Res.* **14**(2), 93–106 (1980). ISSN 1097–4636. doi: 10.1002/jbm.820140202. URL http://onlinelibrary.wiley.com/doi/10.1002/jbm.820140202/abstract.
132. H. Vinters and H. Ho, Effects of isobutyl 2-cyanoacrylate polymer on cultured cells derived from murine cerebral microvessels, *Toxicol. in Vitro* **2**(1), 37–41 (1988). ISSN 08872333.
133. A. W. Cooper, P. J. Harris, G. K. Kumar and J. C. Tebby, Hydrolysis of alkyl cyanoacrylate and ethoxycarbonylpropenoylphosphonate polymers, *J. Polym. Sci.. Part A: Polym. Chem.* **27**(6), 1967–1974 (1989). ISSN 1099-0518. doi: 10.1002/pola.1989.080270617. URL http://onlinelibrary.wiley.com/doi/10.1002/pola.1989.080270617/abstract.
134. E. Buhleier, W. Wehner and F. VoGtle, "Cascade"-and "Nonskid-Chain-like" syntheses of molecular cavity topologies. *Synthesis* **1978**(02), 155–158 (1978). ISSN 0039–7881, 1437–210X. doi: 10.1055/s-1978-24702. URL https://www.thieme-connect.com/ejournals/abstract/10.1055/s-1978-24702.
135. V. Marx, Poised to branch out, *Nat. Biotechnol.* **26**(7), 729–732 (2008). ISSN 1087–0156. doi: 10.1038/nbt0708-729. URL http://usrexp-sandbox.nature.com/nbt/journal/v26/n7/full/nbt0708-729.html.
136. S. Sadekar and H. Ghandehari, Transepithelial transport and toxicity of PAMAM dendrimers: Implications for oral drug delivery, *Adv. Drug Deliv. Rev.* **64**(6), 571–588 (2012). ISSN 0169–409X. doi: 10.1016/j.addr.2011.09.010. URL http://www.sciencedirect.com/science/article/pii/S0169409X11002419.
137. W. Ke, Y. Zhao, R. Huang, C. Jiang and Y. Pei, Enhanced oral bioavailability of doxorubicin in a dendrimer drug delivery system, *J. Pharm. Sci.* **97**(6), 2208–2216 (2008). ISSN 1520–6017. doi: 10.1002/jps.21155. URL http://onlinelibrary.wiley.com/doi/10.1002/jps.21155/abstract.
138. R. Duncan, Polymer conjugates as anticancer nanomedicines, *Nat. Rev. Cancer* **6**(9), 688–701 (2006). ISSN 1474–175X. doi: 10.1038/nrc1958. URL http://www.nature.com/nrc/journal/v6/n9/full/nrc1958.html.
139. D. A. Tomalia, H. Baker, J. Dewald, M. Hall, G. Kallos, S. Martin, J. Roeck, J. Ryder and P. Smith, A new class of polymers: Starburst-dendritic macro-molecules, *Polym. J.* **17**(1), 117–132 (1985). ISSN 0032–3896. URL http://cat.inist.fr/?aModele=afficheN&cpsidt=9170924.
140. K. Greish, G. Thiagarajan, H. Herd, R. Price, H. Bauer, D. Hubbard, A. Burckle, S. Sadekar, T. Yu, A. Anwar, A. Ray and H. Ghandehari, Size and surface charge significantly influence the toxicity of silica and dendritic nanoparticles, *Nanotoxicology* **6**, 713–723 (2012). ISSN 1743–5404. doi: 10.3109/17435390.2011.604442. URL http://www.ncbi.nlm.nih.gov/pubmed/21793770. PMID: 21793770.
141. R. Wiwattanapatapee, B. Carreño-Gomez, N. Malik and R. Duncan, Anionic PAMAM dendrimers rapidly cross adult rat intestine in vitro: A potential oral delivery system? *Pharm. Res.* **17**(8), 991–998 (2000). ISSN 0724–8741. URL http://www.ncbi.nlm.nih.gov/pubmed/11028947. PMID: 11028947.

142. M. El-Sayed, M. Ginski, C. Rhodes and H. Ghandehari, Transepithelial transport of poly(amidoamine) dendrimers across Caco-2 cell monolayers *J. Controlled Release* **81**(3), 355–365 (2002).
143. M. El-Sayed, M. Ginski, C. A. Rhodes and H. Ghandehari, Influence of surface chemistry of poly(amido amine) dendrimers on Caco-2 cell monolayers *J. Bioactive Compat. Polym.* **18**(1), 7–22 (2003). doi: 10.1177/0883911503018001002. URL http://jbc.sagepub.com/content/18/1/7.abstract.
144. K. Kitchens, M. El-Sayed and H. Ghandehari, Transepithelial and endothelial transport of poly(amidoamine) dendrimers *Adv. Drug Deliv. Rev.* **57**(15), 2163–2176 (2005).
145. K. M. Kitchens, A. B. Foraker, R. B. Kolhatkar, P. W. Swaan and H. Ghandehari, Endocytosis and interaction of poly(amido amine) dendrimers with Caco-2 cells, *Pharm. Res.* **24**(11), 2138–2145 (2007). ISSN 0724–8741. doi: 10.1007/s11095-007–9415–0. URL http://www.ncbi.nlm.nih.gov/pubmed/17701324.PMID: 17701324.
146. R. Jevprasesphant, J. Penny, D. Attwood, N. B. McKeown and A. D'Emanuele, Engineering of dendrimer surfaces to enhance transepithelial transport and reduce cytotoxicity, *Pharm. Res.* **20**(10), 1543–1550 (2003). ISSN 0724–8741. URL http://www.ncbi.nlm.nih.gov/pubmed/14620505. PMID: 14620505.
147. D. M. Sweet, R. B. Kolhatkar, A. Ray, P. Swaan and H. Ghandehari, Transepithelial transport of PEGylated anionic poly(amido amine) dendrimers: Implications for oral drug delivery, *J. Controlled Release* **138**(1), 78–85 (2009). ISSN 1873–4995. doi: 10.1016/j.jconrel.2009.04.022. URL http://www.ncbi.nlm.nih.gov/pubmed/19393702. PMID: 19393702.
148. K. M. Kitchens, R. B. Kolhatkar, P. W. Swaan, N. D. Eddington and H. Ghandehari, Transport of poly(amido amine) dendrimers across Caco-2 cell monolayers: Influence of size, charge and fluorescent labeling, *Pharm. Res.* **23**(12), 2818–2826 (2006). ISSN 0724–8741. doi: 10.1007/s11095-006–9122–2. URL http://www.springerlink.com/content/vm6156776212pt54/fulltext.html.
149. R. B. Kolhatkar, K. M. Kitchens, P. W. Swaan and H. Ghandehari, Surface acetylation of poly(amido amine) (PAMAM) dendrimers decreases cytotoxicity while maintaining membrane permeability, *Bioconjug. Chem.* **18**(6), 2054–2060 (2007). doi: 10.1021/bc0603889. URL http://dx.doi.org/10.1021/bc0603889.
150. M. El-Sayed, C. A. Rhodes, M. Ginski and H. Ghandehari, Transport mechanism (s) of poly (amido amine) dendrimers across Caco-2 cell monolayers, *Int. J. Pharm.* **265**(1&2), 151–157 (2003).
151. G. Thiagarajan, S. Sadekar, K. Greish, A. Ray and H. Ghandehari, Evidence of oral translocation of anionic G6.5 dendrimers in mice, *Mol. Pharm.* **10**(3), 988–998 (2013). ISSN 1543–8384. doi: 10.1021/mp300436c. URL http://dx.doi.org/10.1021/mp300436c.
152. M. Najlah and A. D'Emanuele, Crossing cellular barriers using dendrimer nanotechnologies, *Curr. Opin. Pharmacol.* **6**(5), 522–527 (2006). ISSN 1471–4892. doi: 10.1016/j.coph.2006.05.004. URL http://www.sciencedirect.com/science/article/pii/S1471489206001275.
153. M. Najlah, S. Freeman, D. Attwood and A. D'Emanuele, *In vitro* evaluation of dendrimer prodrugs for oral drug delivery, *Int. J. Pharm.* **336**(1), 183–190 (2007). ISSN 0378–5173. doi: 10.1016/j.ijpharm.2006.11.047. URL http://www.sciencedirect.com/science/article/pii/S0378517306010064.

154. R. B. Kolhatkar, P. Swaan and H. Ghandehari, Potential oral delivery of 7-Ethyl-10-hydroxy-camptothecin (SN-38) using poly(amido amine) dendrimers, *Pharm. Res.* **25**(7), 1723–1729 (2008). ISSN 0724–8741. doi: 10.1007/s11095–008–9572–9. PMID: 18438703 PMCID: 2734974.
155. N. Vijayalakshmi, A. Ray, A. Malugin and H. Ghandehari, Carboxyl-terminated PAMAM-SN38 conjugates: Synthesis, characterization, and *in vitro* evaluation, *Bioconjug. Chem.* **21**(10), 1804–1810 (2010). ISSN 1520–4812. doi: 10.1021/bc100094z. URL http://www.ncbi.nlm.nih.gov/pubmed/20836544. PMID: 20836544.
156. D. S. Goldberg, N. Vijayalakshmi, P. W. Swaan and H. Ghandehari, G3.5 PAMAM dendrimers enhance transepithelial transport of SN38 while minimizing gastrointestinal toxicity, *J. Controlled Release* **150**(3), 318–325 (2011). ISSN 0168–3659. doi: 10.1016/j.jconrel.2010.11.022. URL http://www.sciencedirect.com/science/article/pii/S0168365910009508.
157. A. T. Florence, T. Sakthivel and I. Toth, Oral uptake and translocation of a polylysine dendrimer with a lipid surface, *J. Controlled Release* **65**(1&2), 253–259 (2000). ISSN 0168–3659. URL http://www.ncbi.nlm.nih.gov/pubmed/10699285. PMID: 10699285.
158. R. A. Miller, J. M. Brady and D. E. Cutright, Degradation rates of oral resorbable implants (polylactates and polyglycolates): Rate modification with changes in PLA/PGA copolymer ratios, *J. Biomed. Mat. Res.* **11**(5), 711–719 (1977). ISSN 1097–4636. doi: 10.1002/jbm.820110507. URL http://onlinelibrary.wiley.com/doi/10.1002/jbm.820110507/abstract.
159. P. Scholes, A. Coombes, L. Illum, S. Daviz, M. Vert and M. Davies, The preparation of sub-200 nm poly(lactide-co-glycolide) microspheres for site-specific drug delivery, *J. Controlled Release* **25**(1–2), 145–153 (1993). ISSN 0168-3659. doi: 10.1016/0168-3659(93)90103-C. URL http://www.sciencedirect.com/science/article/pii/016836599390103C.
160. A. Sharma, S. Sharma and G. K. Khuller, Lectin-functionalized poly (lactide-co-glycolide) nanoparticles as oral/aerosolized antitubercular drug carriers for treatment of tuberculosis, *The J. Antimicrob. Chemotherapy* **54**(4), 761–766 (2004). ISSN 0305–7453. doi: 10.1093/jac/dkh411. URL http://www.ncbi.nlm.nih.gov/pubmed/15329364. PMID: 15329364.
161. A. Sharma, R. Pandey, S. Sharma and G. Khuller, Chemotherapeutic efficacy of poly (dl-lactide-co-glycolide) nanoparticle encapsulated antitubercular drugs at sub-therapeutic dose against experimental tuberculosis, *Int. J. Antimicrobial Agents* **24**(6), 599–604 (2004). ISSN 0924–8579. doi: 10.1016/j.ijantimicag.2004.07.010. URL http://www.sciencedirect.com/science/article/pii/S0924857904003036.
162. X. Y. Xiong, Y. P. Li, Z. L. Li, C. L. Zhou, K. C. Tam, Z. Y. Liu and G. X. Xie, Vesicles from pluronic/poly(lactic acid) block copolymers as new carriers for oral insulin delivery, *J. Controlled Release* **120**(1&2), 11–17 (2007). ISSN 0168–3659. doi: 10.1016/j.jconrel.2007.04.004. URL http://www.sciencedirect.com/science/article/pii/S0168365907001794.
163. X. Y. Xiong, K. C. Tam and L. H. Gan, Synthesis and aggregation behavior of pluronic f127/poly (lactic acid) block copolymers in aqueous solutions, *Macromolecules* **36**(26), 9979–9985 (2003). URL http://pubs.acs.org/doi/abs/10.1021/ma035292d.
164. M. Garinot, V. Fiévez, V. Pourcelle, F. Stoffelbach, A. des Rieux, L. Plapied, I. Theate, H. Freichels, C. Jérome, J. Marchand-Brynaert, Y. Schneider and V. Préat, PEGylated

PLGA-based nanoparticles targeting M cells for oral vaccination, *J. Controlled Release* **120**(3), 195–204 (2007). ISSN 0168–3659. doi: 10.1016/j.jconrel.2007.04.021. URL http://www.sciencedirect.com/science/article/pii/S016836590700209X.

165. C. O. Tacket, R. H. Reid, E. C. Boedeker, G. Losonsky, J. P. Nataro, H. Bhagat and R. Edelman, Enteral immunization and challenge of volunteers given enterotoxigenic E. coli CFA/II encapsulated in biodegradable microspheres, *Vaccine* **12**(14), 1270–1274 (1994). ISSN 0264–410X. PMID: 7856290.
166. D. E. Katz, A. J. DeLorimier, M. K. Wolf, E. R. Hall, F. J. Cassels, J. E. van Hamont, R. L. Newcomer, M. A. Davachi, D. N. Taylor and C. E. McQueen, Oralimmunization of adult volunteers with microencapsulated enterotoxigenic escherichia coli (ETEC) CS6 antigen, *Vaccine* **21**(5&6), 341–346 (2003). ISSN 0264–410X. doi: 10.1016/S0264–410X(02)00613-8. URL http://www.sciencedirect.com/science/article/pii/S0264410X02006138.
167. E. G. Walsh, B. E. Adamczyk, K. B. Chalasani, S. Maher, E. B. O'Toole, J. S. Fox, T. W. Leonard and D. J. Brayden, Oral delivery of macromolecules: rationale underpinning gastrointestinal permeation enhancement technology (GIPET R), *Therap. Deliv.* **2**(12), 1595–1610 (2011). ISSN 2041–5990, 2041–6008. doi: 10.4155/tde.11.132. URL http://www.future-science.com/doi/abs/10.4155/tde.11.132?journalCode=tde.
168. D. J. Brayden and A. W. Baird, Microparticle vaccine approaches to stimulate mucosal immunization, *Microbes and Infection* **3**(10), 867–876 (2001). ISSN 1286-4579. doi: 10.1016/S1286-4579(01)01445-9. URL http://www.sciencedirect.com/science/article/pii/S1286457901014459.
169. V. P. Torchilin, Recent advances with liposomes as pharmaceutical carriers, *Nat. Rev. Drug Discov.* **4**(2), 145–160 (2005). ISSN 1474–1776. doi: 10.1038/nrd1632. URL http://www.nature.com/nrd/journal/v4/n2/abs/nrd1632.html.
170. W. C. N. Charman and V. J. Stella, *Lymphatic Transport of Drugs*: CRC Press (1992).
171. V. P. Torchilin, *Polymeric micelles for therapeutic applications in medicine, Polymer based nanostructures: Medical applications. RSC nanoscience and nanotechnology*. Cambridge: Royal Society of Chemistry (2010), pp. 261–299. URL http://gendocs.ru/docs/33/32534/conv_1/file1.pdf#page=278.
172. P. van Hasselt, G. Janssens, T. Slot, M. van der Ham, T. Minderhoud, M. Talelli, L. Akkermans, C. Rijcken and C. van Nostrum, The influence of bile acids on the oral bioavailability of vitamin k encapsulated in polymeric micelles, *J. Controlled Release* **133**(2), 161–168 (2009). ISSN 0168–3659. doi: 10.1016/j.jconrel.2008.09.089. URL http://www.sciencedirect.com/science/article/pii/S0168365908006238.
173. V. P. Sant, D. Smith and J. Leroux, Enhancement of oral bioavailability of poorly water-soluble drugs by poly(ethylene glycol)-block-poly(alkyl acrylate-co-methacrylic acid) self-assemblies, *J. Controlled Release* **104**(2), 289–300 (2005). ISSN 0168–3659. doi: 10.1016/j.jconrel.2005.02.010. URL http://www.sciencedirect.com/science/article/pii/S0168365905000763.
174. L. Bromberg, Polymeric micelles in oral chemotherapy, *J. Controlled Release* **128**(2), 99–112 (2008). ISSN 0168–3659. doi: 10.1016/j.jconrel.2008.01.018. URL http://www.sciencedirect.com/science/article/pii/S0168365908000679.
175. G. Gaucher, P. Satturwar, M. Jones, A. Furtos and J. Leroux, Polymeric micelles for oral drug delivery, *Eur. J. Pharm. Biopharm.* **76**(2), 147–158 (2010). ISSN 0939–6411.

doi: 10.1016/j.ejpb.2010.06.007. URL http://www.sciencedirect.com/science/article/pii/S0939641110001566.

176. V. Alakhov, G. Pietrzynski, K. Patel, A. Kabanov, L. Bromberg and T. A. Hatton, Pluronic block copolymers and pluronic poly(acrylic acid) microgels in oral delivery of megestrol acetate, *J. Pharm. Pharmacol.* **56**(10), 1233–1241 (2004). ISSN 2042-7158. doi: 10.1211/0022357044427. URL http://onlinelibrary. wiley.com/doi/10.1211/0022357044427/abstract.

177. K. R. Lange, *Surfactants.* Munich: Hanser Publishers (1999). URL http://elibrary.bsu.az/azad/new/1946.pdf.

178. S. H. Kwon, S. Y. Kim, K. W. Ha, M. J. Kang, J. S. Huh, Y. M. Kim, Y. M. Park, K. H. Kang, S. Lee and J. Y. Chang, Pharmaceutical evaluation of genistein-loaded pluronic micelles for oral delivery, *Archives of Pharmacal Research* **30**(9), 1138–1143 (2007). URL http://link.springer.com/article/10.1007/BF02980249.

179. F. Mathot, L. van Beijsterveldt, V. Préat, M. Brewster and A. Ariën, Intestinal uptake and biodistribution of novel polymeric micelles after oral administration, *J. Controlled Release* **111**(1&2), 47–55 (2006). ISSN 0168-3659. doi: 10.1016/j.jconrel.2005.11.012. URL http://www.sciencedirect.com/science/article/pii/S0168365905006553.

180. M. F. Francis, M. Cristea, Y. Yang and F. M. Winnik, Engineering polysaccharide-based polymeric micelles to enhance permeability of cyclosporin A across Caco-2 cells, *Pharm. Res.* **22**(2), 209–219 (2005). ISSN 0724-8741, 1573-904X. doi: 10.1007/s11095-004-1188-0. URL http://link.springer.com/article/10.1007/s11095-004-1188-0.

181. E. V. Batrakova, S. Li, Y. Li, V. Y. Alakhov and A. V. Kabanov, Effect of pluronic p85 on ATPase activity of drug efflux transporters, *Pharm. Res.* **21**(12), 2226–2233 (2004). URL http://link.springer.com/article/10.1007/s11095-004-7675-5.

182. E. Collnot, C. Baldes, U. F. Schaefer, K. J. Edgar, M. F. Wempe and C. Lehr, Vitamin TPGS P-Glycoprotein inhibition mechanism: Influence on conformational flexibility, intracellular ATP levels, and role of time and site of access, *Mol. Pharm.* **7**(3), 642–651 (2010). ISSN 1543-8384. doi: 10.1021/mp900191s. URL http://dx.doi.org/10.1021/mp900191s.

183. E. V. Batrakova, H. Han, V. Y. Alakhov, D. W. Miller and A. V. Kabanov, Effects of pluronic block copolymers on drug absorption in Caco-2 cell monolayers, *Pharm. Res.* **15**(6), 850–855 (1998). ISSN 0724-8741, 1573-904X. doi: 10.1023/A:1011964213024. URL http://link.springer.com/article/10.1023/A% 3A1011964213024.

184. E. V. Batrakova, S. Li, W. F. Elmquist, D. W. Miller, V. Y. Alakhov and A. V. Kabanov, Mechanism of sensitization of MDR cancer cells by pluronic block copolymers: Selective energy depletion, *British J. Cancer.* **85**(12), 1987–1997 (2001). ISSN 0007-0920. doi: 10.1054/b joc.2001.2165. URL http://www.ncbi.nlm.nih.gov/pmc/articles/PMC2364003/. PMID: 11747344 PMCID: PMC2364003.

185. S. C. Lee, K. M. Huh, J. Lee, Y. W. Cho, R. E. Galinsky and K. Park, Hydrotropic polymeric micelles for enhanced paclitaxel solubility: *In vitro* and *in vivo* characterization, *Biomacromolecules* **8**(1), 202–208 (2007). ISSN 1525-7797. doi: 10.1021/bm060307b. URL http://dx.doi.org/10.1021/bm060307b.

186. S. Peltier, J. Oger, F. Lagarce, W. Couet and J. Benoît, Enhanced oral paclitaxel bioavailability after administration of paclitaxel-loaded lipid nanocapsules, *Pharm. Res.*

23(6), 1243–1250 (2006). URL http://link.springer.com/article/10.1007/s11095-006-0022-2.

187. C. J. H. Porter, N. L. Trevaskis and W. N. Charman, Lipids and lipid-based formulations: Optimizing the oral delivery of lipophilic drugs, *Nat. Rev. Drug Discov.* **6**(3), 231–248 (2007). ISSN 1474–1776. doi: 10.1038/nrd2197. URL http://www.nature.com/nrd/journal/v6/n3/full/nrd2197.html.
188. M. Niu, Y. Lu, L. Hovgaard, P. Guan, Y. Tan, R. Lian, J. Qi and W. Wu, Hypoglycemic activity and oral bioavailability of insulin-loaded liposomes containing bile salts in rats: The effect of cholate type, particle size and administered dose, *Eur. J. Pharm. Biopharm.* **81**(2), 265–272 (2012). ISSN 0939–6411. doi: 10.1016/j.ejpb.2012.02.009. URL http://www.sciencedirect.com/science/article/pii/S0939641112000483.
189. Y. Lin, K. Sonaje, K. M. Lin, J. Juang, F. Mi, H. Yang and H. Sung, Multi-ion-crosslinked nanoparticles with pH-responsive characteristics for oral delivery of protein drugs, *J. Controlled Release* **132**(2), 141–149 (2008). ISSN 0168–3659. doi: 10.1016/j.jconrel.2008.08.020. URL http://www.sciencedirect.com/science/article/pii/S0168365908005191.
190. M. Li, W. Lu, J. Wang, X. Zhang, H. Zhang, X. Wang, C. Wu and Q. Zhang, Preparation and characterization of insulin nanoparticles employing chitosan and poly(methylmethacrylate/methylmethacrylic acid) copolymer, *J. Nanosci. Nanotechnol.* **6**(9&10), 2874–2886 (2006). doi: 10.1166/jnn.2006. 411.
191. F. Cui, K. Shi, L. Zhang, A. Tao and Y. Kawashima, Biodegradable nanoparticles loaded with insulin–phospholipid complex for oral delivery: Preparation, *in vitro* characterization and *in vivo* evaluation, *J. Controlled Release* **114**(2), 242–250, (2006). URL http://www.sciencedirect.com/science/article/pii/S0168365906002355.
192. H. Li, J. Song, J. Park and K. Han, Poly(ethylene glycol)-coated liposomes for oral delivery of recombinant human epidermal growth factor, *Int. J. Pharm.* **258**(1–2), 11–19 (2003). ISSN 0378–5173. doi: 10.1016/S0378-5173(03)00158-3. URL http://www.sciencedirect.com/science/article/pii/S0378517303001583.
193. S. Minato, K. Iwanaga, M. Kakemi, S. Yamashita and N. Oku, Application of poly(ethylene glycol) (PEG)-modified liposomes for oral vaccine: Effect of lipid dose on systemic and mucosal immunity, *J. Controlled Release* **89**(2), 189–197 (2003). ISSN 0168–3659. doi: 10.1016/S0168-3659(03)00093-2. URL http://www.sciencedirect.com/science/article/pii/S0168365903000932.
194. J. A. Rogers and K. E. Anderson, The potential of liposomes in oral drug delivery, *Critical Rev. Therapeutic Drug Carrier Syst.* **15**(5), 421–480 (1998). ISSN 0743–4863. URL http://www.ncbi.nlm.nih.gov/pubmed/9822867. PMID: 9822867.
195. H. Chen, V. Torchilin and R. Langer, Lectin-bearing polymerized liposomes as potential oral vaccine carriers, *Pharm. Res.* **13**(9), 1378–1383 (1996). URL http://link.springer.com/article/10.1023/A%3A1016030202104.
196. D. R. Serrano Lopez and A. Lalatsa, Peptide pills for brain diseases? reality and future perspectives, *Therapeutic Deliv.* **4**(4), 479–501 (2013). ISSN 2041–5990, 2041–6008. doi: 10.4155/tde.13.5. URL http://www.future-science.com/doi/abs/10.4155/tde.13.5.
197. M. J. Morilla, D. M. Gomez, P. Cabral, M. Cabrera, H. Balter, M. V. Tesoriero, L. Higa, D. Roncaglia and E. L. Romero, M cells prefer archaeosomes: an *in vitro/in vivo* snapshot upon oral gavage in rats, *Current Drug Deliv.* **8**(3), 320–329 (2011). URL http://www-biblio.inti.gov.ar/TRABinti/DefainMC.pdf.

198. R. H. Müller, K. Mader and S. Gohla, Solid lipid nanoparticles (SLN) for controlled drug delivery – a review of the state of the art, *Eur. J. Pharm. Biopharm.* **50**(1), 161–177 (2000). ISSN 0939-6411. doi: 10.1016/S0939–6411(00)00087-4. URL http://www.sciencedirect.com/science/article/pii/S0939641100000874.
199. P. Severino, T. Andreani, A. S. Macedo, J. F. Fangueiro, M. H. A. Santana, A. M. Silva and E. B. Souto, Current state-of-art and new trends on lipid nanoparticles (SLN and NLC) for oral drug delivery, *J. Drug Deliv.* **2012**, 1–10 (2012). ISSN 2090–3014, 2090–3022. doi: 10.1155/2012/750891. URL http://www.hindawi.com/journals/jdd/2012/750891/.
200. N. Huynh, C. Passirani, P. Saulnier and J. Benoit, Lipid nanocapsules: A new platform for nanomedicine, *Int. J. Pharm.* **379**(2), 201–209 (2009). ISSN 0378–5173. doi: 10.1016/j.ijpharm.2009.04.026. URL http://www.sciencedirect.com/science/article/pii/S0378517309002488.
201. B. Heurtault, P. Saulnier, B. Pech, J. Proust and J. Benoit, Physico-chemical stability of colloidal lipid particles, *Biomaterials* **24**(23), 4283–4300 (2003). URL http://www.sciencedirect.com/science/article/pii/S0142961203003314.
202. S. Yang, J. Zhu, Y. Lu, B. Liang and C. Yang, Body distribution of camptothecin solid lipid nanoparticles after oral administration, *Pharm. Res.* **16**(5), 751–757 (1999).
203. L. Hu, X. Tang and F. Cui, Solid lipid nanoparticles (SLNs) to improve oral bioavailability of poorly soluble drugs, *J. Pharm. Pharmacol.* **56**(12), 1527–1535 (2004). ISSN 2042–7158. doi: 10.1211/0022357044959. URL http://onlinelibrary.wiley.com/doi/10.1211/0022357044959/abstract.
204. S. Runge. Feste Lipid-Nanopartikel (SLN) als kolloidaler Arzneistofftrager für Cyclosporin A. Ph. D. thesis, University of Berline (1968).
205. L. Penkler, R. Müller, S. Runge and V. Ravelli. Pharmaceutical cyclosporin formulation with improved biopharmaceutical properties, improved physical quality and greater stability, and method for producing said formulation (1999). URL http://patentscope.wipo.int/search/en/WO1999056733.
206. R. Müller, S. Runge, V. Ravelli, W. Mehnert, A. Thünemann and E. Souto, Oral bioavailability of cyclosporine: Solid lipid nanoparticles (SLN R) versus drug nanocrystals, *Int. J. Pharm.* **317**(1), 82–89 (2006). ISSN 0378–5173. doi: 10.1016/j.ijpharm.2006.02.045. URL http://www. sciencedirect.com/science/article/pii/S037851730600192X.
207. R. Müller, S. Runge, V. Ravelli, A. Thünemann, W. Mehnert and E. Souto, Cyclosporine-loaded solid lipid nanoparticles (SLN R): drug–lipid physicochemical interactions and characterization of drug incorporation, *Eur. J. Pharm. Biopharm.* **68**(3), 535–544 (2008). ISSN 0939–6411. doi: 10.1016/j.ejpb.2007.07.006. URL http://www.sciencedirect.com/science/article/pii/S0939641107002603.
208. B. Sarmento, S. Martins, D. Ferreira and E. B. Souto, Oral insulin delivery by means of solid lipid nanoparticles, *Int. J. Nanomedicine* **2**(4), 743–749 (2007). ISSN 1176–9114. URL http://www.ncbi.nlm.nih.gov/pmc/articles/PMC2676823/. PMID: 18203440 PMCID: PMC2676823.
209. Y. Luo, D. Chen, L. Ren, X. Zhao and J. Qin, Solid lipid nanoparticles for enhancing vinpocetine's oral bioavailability, *J. Controlled Release* **114**(1), 53–59 (2006). ISSN 0168–3659. doi: 10.1016/j.jconrel.2006.05.010. URL http://www.sciencedirect.com/science/article/pii/S0168365906002203.

210. M. Uner and G. Yener, Importance of solid lipid nanoparticles (SLN) in various administration routes and future perspectives, *Int. J. Nanomedicine* **2**(3), 289–300 (2007). ISSN 1176–9114. URL http://www.ncbi.nlm.nih.gov/pmc/articles/PMC2676658/. PMID: 18019829 PMCID: PMC2676658.
211. V. Mathur, Y. Satrawala, M. S. Rajput, P. Kumar, P. Shrivastava and A. Vishvkarma, Solid lipid nanoparticles in cancer therapy, *Int. J. Drug Deliv.* **2**(3), (2011). URL http://www.arjournals.org/index.php/ijdd/ article/view/43.
212. R. Shegokar and R. Müller, Nanocrystals: Industrially feasible multifunctional formulation technology for poorly soluble actives, *Int. J. Pharm.* **399**(1&2), 129–139 (2010). ISSN 03785173. doi: 10.1016/j.ijpharm.2010.07.044.
213. T. GÜ LSÜ N, R. N. GÜ RSOY and L. Ö NER, Nanocrystal technology for oral delivery of poorly water-soluble drugs, *FABAD J. Pharm. Sci.* **34**, 55–65 (2009). URL http://www.fabad.org.tr/fabad.org/pdf/volum34/issue1/55-65.pdf.
214. G. G. Liversidge and P. Conzentino, Drug particle size reduction for decreasing gastric irritancy and enhancing absorption of naproxen in rats, *Int. J. Pharm.* **125**(2), 309–313 (1995). ISSN 0378-5173. doi: 10.1016/0378-5173(95)00148-C. URL http://www.sciencedirect.com/science/article/pii/037851739500148C.
215. S. Panmai and S. Deshpande, Development of nanoformulations: Selection of polymeric stabilizers based on adsorption isotherm, PMSE Preprints. **89**, 808–809 (2003).
216. F. Stenger, S. Mende, J. Schwedes and W. Peukert, Nanomilling in stirred media mills, *Chem. Eng. Sci.* **60**(16), 4557–4565 (2005). URL http://www. sciencedirect.com/science/article/pii/S000925090500182X.
217. R. Müller and C. Keck, Twenty years of drug nanocrystals: Where are we, and where do we go? *Eur. J. Pharm. Biopharm.* **80**(1), 1–3 (2012). ISSN 09396411. doi: 10.1016/j.ejpb.2011.09.012. URL http://linkinghub.elsevier.com/retrieve/pii/S0939641111002736.
218. C. Chia-Ming and N. Weiner, Gastrointestinal uptake of liposomes. i. *in vitro* and *in situ* studies, *Int. J. Pharm.* **37**(1–2), 75–85 (1987). ISSN 0378-5173. doi: 10.1016/0378-5173(87)90011-1. URL http://www.sciencedirect.com/science/article/pii/0378517387900111.
219. A. des Rieux, V. Fievez, I. Théate, J. Mast, V. Préat and Y. Schneider, An improved *in vitro* model of human intestinal follicle-associated epithelium to study nanoparticle transport by M cells, *Eur. J. Pharm. Sci.* **30**(5), 380–391 (2007). ISSN 0928-0987. doi: 10.1016/j.ejps.2006.12.006. URL http://www.sciencedirect.com/science/article/pii/S092809870700005X.
220. L. Rosenfeld, Insulin: Discovery and controversy. *Clin. Chem.* **48**(12), 2270–2288 (2002). ISSN 0009-9147, 1530-8561. URL http://www.clinchem.org/content/48/12/2270.
221. C. Harrison, Deal watch: Chiasma and Roche partner in oral peptide drug delivery. *Nat. Rev. Drug Discov.* **12**(4), 255–255 (2013). ISSN 1474-1776. doi: 10.1038/nrd3989. URL http://www.nature.com/nrd/journal/v12/n4/full/nrd3989.html.
222. M. Eaton, Nanomedicine: Industry-wise research, *Nat. Mat.* **6**(4), 251–253 (2007). ISSN 1476-1122. doi: 10.1038/nmat1879. URL http://www.nature.com/nmat/journal/v6/n4/abs/nmat1879.html.
223. W. R. Sanhai, J. H. Sakamoto, R. Canady and M. Ferrari, Seven challenges for nanomedicine, *Nat. Nanotechnol.* **3**(5), 242–244 (2008). ISSN 1748-3387. doi: 10.1038/

nnano.2008.114. URL http://www.nature.com/nnano/journal/v3/n5/abs/nnano.2008.114.html.

224. A. Puri, K. Loomis, B. Smith, J. Lee, A. Yavlovich, E. Heldman and R. Blumenthal, Lipid-based nanoparticles as pharmaceutical drug carriers: From concepts to clinic, *Critical Rev. Therapeutic Drug Carrier Syst.* **26**(6), 523–580 (2009). ISSN 0743-4863. URL http://www.ncbi.nlm.nih.gov/pmc/articles/PMC2885142/.PMID: 20402623 PMCID: PMC2885142.

Chapter 6

Nanopreparations for Central Nervous System Diseases

Leyuan Xu and Hu Yang*,†*

**Department of Biomedical Engineering,*
Virginia Commonwealth University,
Richmond, Virginia 23284, US
†Massey Cancer Center, Virginia Commonwealth University
P.O. Box 843067, Richmond, Virginia 23298, US
†hyang2@vcu.edu

Abstract

Therapeutic treatment effectiveness of many types of central nervous system (CNS) diseases relies on efficient entry of drugs into the brain. Numerous efforts have been attempted to develop nanostructured vehicles and formulations for efficient delivery of CNS drugs. In this chapter, we review advances of CNS drug delivery systems based on nanoparticles (NPs).

1. Introduction

The central nervous system (CNS) is protected by two major barriers, the blood–brain barrier (BBB) and the blood-cerebrospinal fluid barrier (BCSFB). In human the BBB is a system of vascular cellular structures with a total capillary length of approximately 600 km and a surface area of approximately 20 m^2.[1] The central function of the BBB is to maintain the overall homeostasis of the CNS. The most prominent feature of the BBB is tight junctions, responsible for regulating passive diffusion of ions and small non-charged solutes between the apical and basolateral plasma membrane domains and keeping osmotic and hydrostatic balance between the interstitial fluid (ISF) and plasma.[2,3]

1.1. *Transport mechanisms*

The transport of molecules across the BBB relies on two major routes: paracellular and transcellular pathways (Fig. 1). Hydrophilic molecules cross the BBB via the paracellular pathway. Hydrophobic molecules that possess molecular weights below 400 Da and do not bind with plasma proteins are able to cross the BBB following the transcellular pathway. Endogenous molecules such as glucose and transferrin cross the BBB following the transporter-mediated transcellular pathway. These transporters are expressed at both the apical and the basolateral membranes of the endothelium. Drugs structurally similar to endogenous substrates may enter the brain via this route.

Endocytosis pathway can be divided into adsorption- and receptor-mediated endocytosis. Adsorption-mediated endocytosis is initiated by the attraction of cationic substances to the negative charges on the plasma membrane; whereas receptor-mediated endocytosis is triggered by receptor-ligand binding. The substance is subsequently internalized and transported to the early endosome and then to the lysosome, and eventually transcytosed to the plasma membrane. Therefore, endocytosis is believed to be the only practical

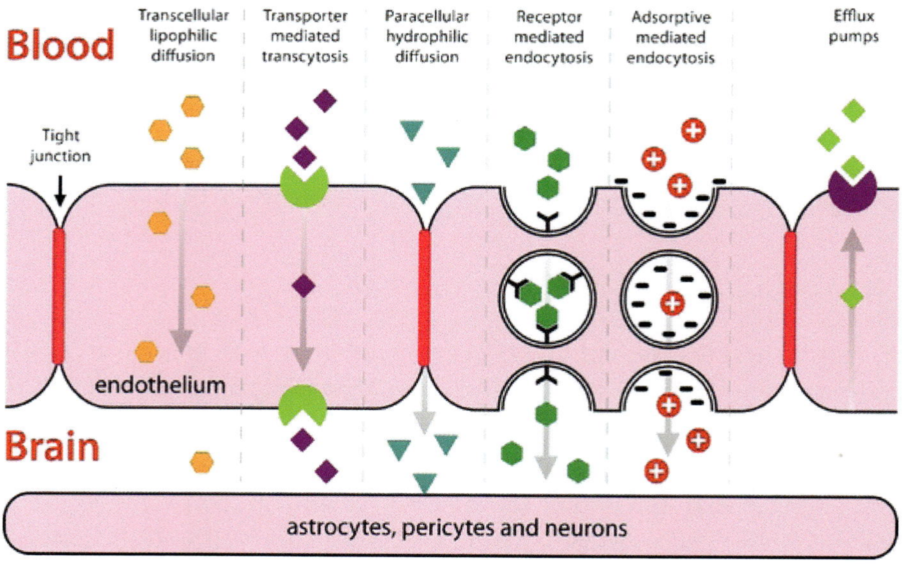

Fig. 1. Pathways across the BBB (reproduced from Ref. 4).

pathway for large-molecular–weight molecules such as antibodies, lipoproteins, proteins and nanoparticles (NPs) to cross the BBB. Efflux transporters such as P-glycoprotein, multiple drug resistance (MDR)-related protein and ATP-binding cassette (ABC) transporters at the BBB are important because they help remove harmful substances from the brain. However, the presence of efflux transporters raises another barrier for drug transport across the BBB and is believed to be a major cause for drug resistance.[4]

1.2. *Challenges to brain drug delivery*

CNS diseases such as stroke, trauma, infections, Alzheimer's disease, Parkinson's disease, epilepsy and brain tumors can significantly alter the BBB integrity and consequently drug permeability across the BBB. In some pathological conditions, the transport mechanism might be up- or down-regulated at the BBB.[4] Most large-molecular–weight agents such as recombinant proteins, monoclonal antibodies, and gene therapeutics are unable to permeate into the brain. Some small molecular weight CNS drugs may enter the brain in their free form via transcellular diffusion, but most cannot because of the lack of desired physicochemical properties. As a result, therapeutic agents may easily diffuse into normal tissues and cause serious side effects.[4] Thus, the major challenge for CNS disease therapy is to deliver drugs efficiently and specifically to the brain.[1]

2. Therapeutic NPs

NPs can facilitate drugs to cross the BBB. Typical NPs that have been actively investigated for CNS drug delivery are shown in Fig. 2.[5] Some materials of nanometric dimensions are in clinical trials.[6] NPs are capable of manipulating drug release rates and possessing brain-targeting specificity. NPs enter the endothelial cells on the BBB, most likely via adsorptive or receptor-mediated endocytosis. Transcytosis of nanoparticle to the basal side of the BBB is required for NPs to enter into the brain. Brain drug concentration depends on pharmacokinetic parameters and many physiological and biological factors, such as the plasma concentration-time curve, the extent of plasma protein binding, the permeability across the BBB, the efflux out of the brain by efflux transporters, the metabolic conversion by enzymes, the binding to membranes or intracellular sites in the brain, and the continual secretion and drainage of cerebrospinal fluid (CSF) and brain ISF.[4] For CNS drug delivery, ideal NPs are expected to be: inexpensive for manufacturing, non-toxic, non-thrombogenic, non-immunogenic, biodegradable, and biocompatible. From a

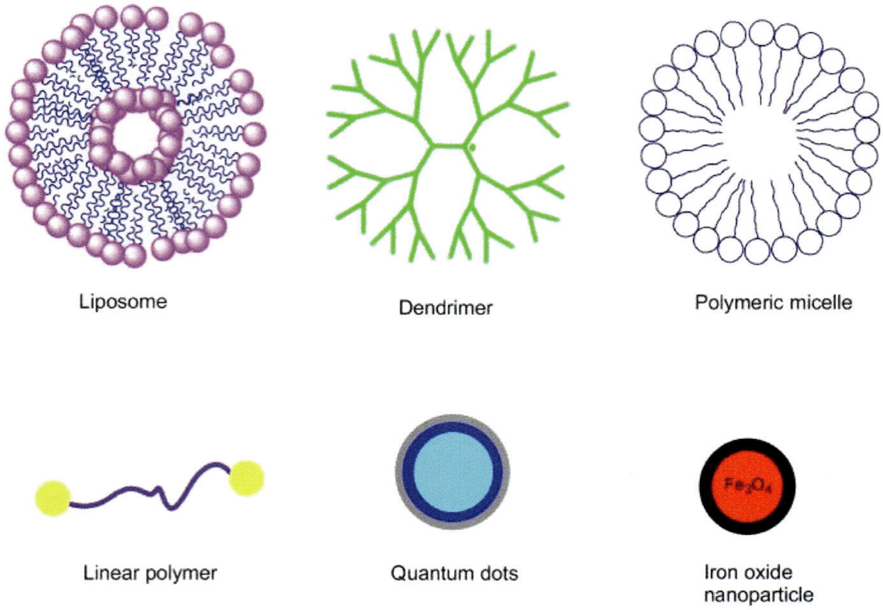

Fig. 2. Illustration of typical NPs used in brain drug delivery, imaging, and diagnosis (reproduced from Ref. 5).

pharmacokinetic standpoint, it is of interest to determine at what rate and to what extent drug uptake occurs.[4] The main advantages and disadvantages of nanocarriers for brain drug delivery are summarized in Table 1. The major *in vivo* studies for the applications of NPs in the treatment of CNS diseases are summarized in Table 2.

2.1. *Liposomes*

Liposomes are non-toxic, biocompatible and biodegradable carriers that are made up of lipid bilayers, typically phospholipid bilayers, as shown in Fig. 3. Liposomes can deliver hydrophilic, lipophilic and amphoteric drug molecules, which are entrapped either inside the carrier or on the micellar surface. However, liposomes have such problems as short clearance time and low transport rate. Liposomes with specificity may mitigate these disadvantages. For instance, liposomes surface modified with targeting ligands may facilitate encapsulated drug molecules to the target tissue or organ. Targeting ligands include monoclonal antibody (mAb) (e.g. mAb against transferrin receptor), cationized proteins (e.g. albumin), endogenous peptides or plasma proteins. Liposomal systems have been developed and evaluated for delivery of various

Table 1. Main advantages and disadvantages of nanosystems for drug delivery to the brain (adapted from Ref. 2).

Nano system	Advantages	Disadvantags
Ciclodextrins	• solubility enhancement • bioavailability enhancement • biotransformation reduction	• expensive
Diendrimers	• lower polydispersity index • surface with multiple functional groups • synthesized and designed for specific applications	• multi-step and tedious synthesis
Nanogels	• high loading capacity • low toxicity	unknown
Micells	• good drug solubilization properties • surface with multiple functional groups	• relatively narrow pH range stability
Liposomes	• solubilization properties for both oil-soluble and water-soluble drugs • biocompatible, biodegradable and nonimmunogenic • administered through most routes of administration, easy surface decoration,	• production cost is high • Sometimes phospholipid undergoes oxidation and hydrolysis
Lipid NPs	• easy surface decoration • hydrophilic and hydrophobic drugs • biocompatible • avoid organic solvent • easy to scale-up and sterilize • excipients of low cost • lipid NPs can be produced as physically stable aqueous or non-aqueous suspensions	• drug loading capacity limited by solubility of drug in lipid melt and drug expulsion after polymeric transition during storage
Inorganic NPs	• diagnostic improvement structurally robust	• potential toxicity • low efficiency
Polymeric NPs	• biocompatible • FDA approved materials • non-toxic • production under mild conditions • controlled release • targeted site over a period of time • targeting to non accessed tumors • potential for delivery of multiple agents at the same site	• expensive polymers (some) • surface modification • increased complexity of the nanoparticles • possible particle aggregation (physical handling difficult in liquid and dry forms)

Table 2. Representative studies using nanomaterials in neuro-oncology (adapted from Ref. 7).

	Delivery system	Model	Route	Aim and main conclusions	Ref.
NPs	Doxorubicin delivered by Polysorbate-80 coated PBCA NPs	Rat intracranial glioblastoma (101/8)	Intravenous (i.v.)	• Investigation of the antitumor capacity of Dox-loaded NPs. • Distribution of doxorubicin into the brain in vivo and therapy of brain tumors.	8
	Paclitaxel delivered by methoxy poly(ethylene glycol)-poly ε-caprolactone NPs	Mouse intracranial gliosarcoma (C6)	i.v.	• Investigation of the anti-tumor capacity of drug-loaded NPs. • NPs enhanced anti-glioblastoma activity of paclitaxel.	9
	Paclitaxel delivered by PEGylated poly(trimethylene carbonate) NPs	Mouse intracranial glioblastoma (U87)	i.v.	• NPs enhanced the accumulation of paclitaxel in tumor tissue. • Survival rate of glioma-bearing rats was significantly prolonged after anti-glioblastoma therapy of paclitaxel NPs.	10
	Paclitaxel delivered by magnetic NPs	Rat intracranial glioma	i.v.	• Antitumor capacity of paclitaxel-loaded super paramagnetic NPs. • Survival rate of glioma-bearing rats was significantly prolonged after therapy with NPs.	11
	pDNA (pORF-hTRAIL) delivered by cationic albumin conjugated PEG-NPs	Mouse intracranial gliosarcoma (C6)	i.v.	• Systemic administration of pDNA loaded NPs for gene therapy of glioma. • Tumor-specific NPs delivery with induction of apoptosis and significant retardation of tumor growth.	12

(Continued)

Table 2. (*Continued*)

	Delivery system	Model	Route	Aim and main conclusions	Ref.
superpara-magnetic	Dextran- and aminosilane-coated superparamagnetic iron oxide superparamagnetic	Rat intracranial glioblastoma (RG2)	Intratumoral (i.t.)	• Investigation of the feasibility and tolerability of thermotherapy using magnetic NPs on rats with GBM. • Thermotherapy with aminosilane-coated NPs led to prolongation of survival rate over dextran-coated particles.	13
	Aminosilane-coated superparamagnetic iron oxide NPs	Patients with GBM (under clinical trial)	i.t.	• Development of NP-based hyperthermia therapy for cancer treatment. • The clinical efficacy of intratumoural thermotherapy using magnetic NPs in combination with fractionated stereotactic radiotherapy was demonstrated.	14, 15
	PEGylated NPs of polyacrylamide (PAA) core; photosensitizers; MRI contrast agents and molecular targeting groups	Rat intracranial gliosarcoma (C6)	i.t.	• Investigation of photodynamic NPs for extracellular cancer therapy. • Improvement in residence time and tumor specific delivery, along with a significant increase in animal survival.	16
Nanoworms	Multifunctional iron oxide nanoworms: tumor-homing peptide (CGKRK), pro-apoptotic peptide ($_{D}$[KLAKLAK]$_2$)	Mouse intracranial glioblastoma (U87)	i.v.	• Investigation of anti-angiogenic therapy by using a theranostic platform. • Targeted GBM tumors, providing effective therapeutic activity with diagnostic function.	17

(*Continued*)

Table 2. (*Continued*)

	Delivery system	Model	Route	Aim and main conclusions	Ref.
Dendrimers	Methotrexate Polyamidoamine dendrimer conjugated to cetuximab	Rat intracranial glioblastoma (F98)	i.t.	• Investigation of the antitumor capacity of methotrexate-loaded dendrimers. • Specific tumors cells targeting and reduction of toxicity in normal brain tissue.	18
Microspheres	Mitoxantrone delivered by PLGA microspheres	Rat intracranial glioblastoma (RG2)	i.t.	• Mitoxantrone-loaded PLGA microspheres prevented glioma growth, minimizing side effects.	19
	Imatinib mesylate delivered by PLGA microspheres	Mouse intracranial glioblastoma (U87)	i.t.	• Imatinib-microspheres were shown to significantly reduce tumor growth within two weeks post injection.	20
Liposomes	Sodium borocaptate delivered by immunoliposomes (transferrin antibody)	Mouse intracranial glioblastoma (U87)	i.v.	• Investigations of therapeutic efficacy by boron neutron capture therapy. • Enhanced survival rate was reported.	21
	Sodium borocaptate delivered by Immunoliposomes (EGFR antibody)	Mouse intracranial-glioblastoma (U87)	i.v.	• Investigations of therapeutic efficacy by boron neutron capture therapy. • High levels of sodium borocaptate were delivered into brain tumors.	22
	Irinotecan-loaded liposome (co-administered with Doxil®)	Rat intracranial glioblastoma (U251 and U87)	i.t.	• Therapeutic efficacy of co-administration between doxorubicin and irinotecan-loaded liposomes in the treatment of glioblastoma tumors. • Co-administration showed potential for brain tumor therapy.	23

(*Continued*)

Table 2. (*Continued*)

	Delivery system	Model	Route	Aim and main conclusions	Ref.
	Non-PEGylated liposome encapsulated with both topoCED™ and gadoCED™	Rat intracranial glioblastoma (U87)	i.t.	• Therapeutic efficacy from co-administration of topoCED™ with gadoCED™. • Real-time imaging of the lipsome distribution, and a significant increase in the survival rate were reported.	24
	Doxorubicin delivered by PEGylated liposome (plaque-specific peptide AP-1 antibody)	Mouse intracranial glioblastoma (GBM8401-luc)	i.v.	• Investigation of applicability of repeated pulsed high-intensity focused ultrasound (HIFU) when treating GBM with liposomal doxorubicin. • Combination of repeated pulsed HIFU enhanced delivery of liposomal doxorubicin to the brain tumor and improved the antitumor effect.	25
	Co-delivery of Paclitaxel and pEGFP-hTRAIL gene by PEGylated Liposome-conjugated to Angiopep-2	Mouse intracranial glioma (U87)	i.v.	• Investigation of paclitaxel delivery by PEGylated liposomes modified with Angiopep-2 for the targeting and treatment of gliomas. • Dual targeting co-delivery system showed potential for glioma therapy in clinical application	26
Carbon nanotubes	MWNT coated with Pluronic PF-108	Mouse intracranial glioblastoma (GL261)	i.t.	• Evaluation of the uptake and toxicity of MWNT in a glioma model. • CNT injection led to a transient and self-limiting local inflammatory response.	27

(*Continued*)

Table 2. (*Continued*)

Delivery system	Model	Route	Aim and main conclusions	Ref.
CpG oligonucleotide coated SWNT (chemically functionalised with PL-PEG)	Mouse intracranial glioma (GL261 and GL261)	i.t.	• Enhanced CpG uptake by tumor-associated inflammatory cells. • CpG immunotherapy eradicated glioma and protected against tumor re-challenge.	28
Doxorubicin and PEGygylated MWNT-conjugated to Angiopep-2 (Dox and PEG were physically adsorbed to nanotubes)	Mouse intracranial glioma(C6)	i.v.	• Investigation of targeted (with Angiopep-2) PEGylated MWNT for targeting and treatment of gliomas. • Increased glioma targeting and improved survival reported.	29

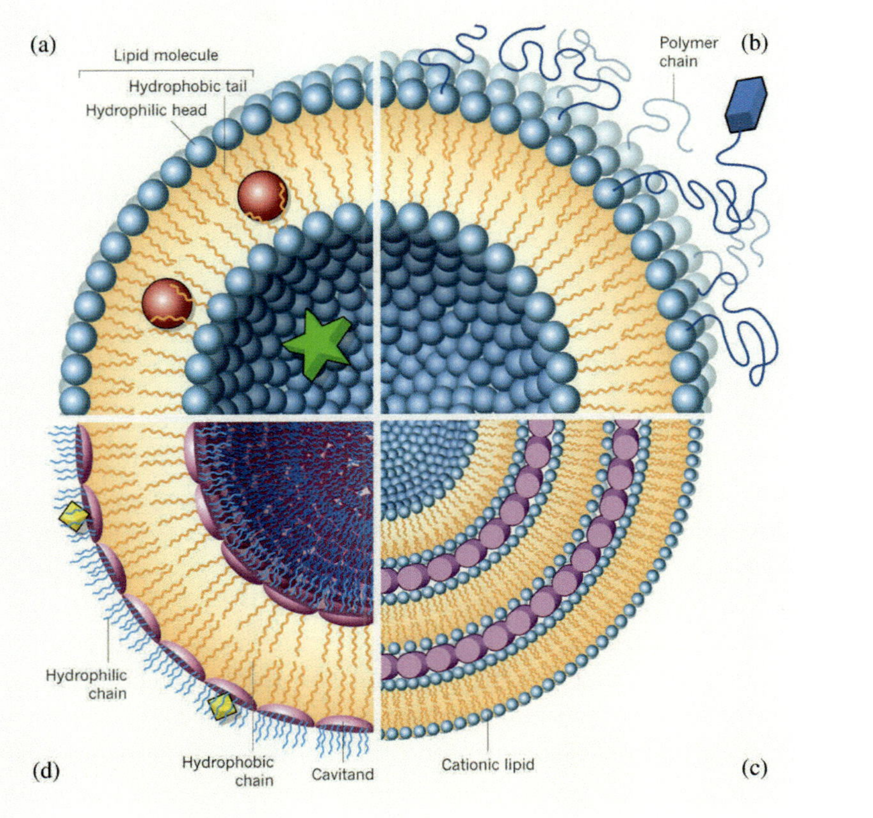

Fig. 3. The evolution of liposomes. (a) Simple liposomes are vesicles that have a shell consisting of a lipid bilayer. A liposome can trap hydrophobic guest molecules a few nanometres in diameter (red spheres) within the hydrophobic bilayer, and hydrophilic guests up to several hundred nanometres (green star) in its larger interior. (b) In "stealth", liposomes developed for drug-delivery applications, the lipid bilayer contains a small percentage of polymer lipids. Peptides (blue rectangle) that target specific biological targets may also be attached to the polymers. (c) Most cationic liposome–DNA complexes have an onion-like structure, with DNA (purple rods) sandwiched between cationic membranes. (d) Liposomes in which the bilayer assembles from cavitands — vase-shaped molecules — to which attached hydrophobic and hydrophilic chains. The cavitands can trap ångström-sized guest compounds (yellow diamonds) in their hydrophobic cavities. These vesicles can therefore encapsulate guest molecules of different sizes in the cavitands, the bilayer and the liposome's interior. (Reproduced from Ref. 32.)

CNS drugs, such as chemotherapeutic compounds, antiretrovirals, anti-epilepsy drugs and anti-ischemia drugs. The mechanism for liposomes coupled with brain-targeting ligand to enter the brain through absorption- or receptor-mediated transcytosis. Cationic liposomes have been shown to cross the BBB primarily through absorption-mediated transcytosis. Liposomes with sulphatide groups such as sulphate ester of galactocerebrosides were found to gain enhanced BBB permeability. Liposome structural parameters such as surface charge, site membrane lipid packing and extent of steric stabilization can be modulated to prolong liposome circulation. In addition, polymers such as polypropylene oxide and polyethylene glycol (PEG) as surface modifier help extend liposome circulation time, as a result of enhanced steric stability, reduced phagocytosis, and weakened interaction between serum opsonins.[5,30–32]

2.1.1. Liposome preparation methodologies

A common method for liposome preparation is oil-in-water (o/w) emulsification, in which the oil phase is dispersed as the droplets of submicron size and then stabilized by surfactants. Liposomes made by o/w are suitable for delivery of lipophilic drugs. Recently, saquinavir, an antiretroviral drug for HIV therapy, was evaluated for brain delivery in an oral formulation of flaxseed oil-based nano-emulsion with an average size of 100–200 nm in diameter.[33] The pharmacokinetic profile of saquinavir in mice was improved in multiple aspects: a three-fold increase in the saquinavir plasma concentration, a three-fold increase in the area-under-the curve (AUC), and a five-fold increase in the maximum brain drug concentration. In addition to enhanced BBB permeability, it has been shown that nano-emulsions of such size could cross other barriers such as the gastrointestinal tract following oral administration.

2.1.2. Liposome-based delivery systems

Antibody-coupled PEGylated liposomes, namely immunoliposomes, have been developed for receptor-mediated delivery of various therapeutics such as nucleic acids to the brain.[34–36] Feng *et al.* developed anti-EGFR antibody-carrying immunoliposomes to deliver sodium borocaptate (BSH) for boron capture neuron therapy.[22] This delivery system was able to selectively deliver a large amount of boron to glioma cells *in vitro* and *in vivo*. In contrast, only low levels of BSH in both tumor and normal tissues were achieved with the control liposome without anti-EGFR mAb.

Liposomes have also been studied to deliver agents that inhibit reactive oxygen species (ROS) to treat auto-immune demyelinative diseases such as multiple sclerosis (MS) or neurodegenerative diseases.[37] For instance, Kizelsztein *et al.* developed sterically stabilized nanoliposomes by mixing the liposome-forming lipid, cholesterol and the lipopolymer^{2000}PEG-DSPE to deliver tempamine (TMN), a stable radical with anti-oxidant and proapoptotic activities. The TMN-loaded liposomes penetrated into the brain causing three- to six-fold higher drug concentrations in mice with autoimmune encephalomyelitis than in normal mice. This formulation was able to extend the half-life of TMN in plasma up to 5 h in normal mice and more than 10 h in experimental autoimmune encephalomyelitis (EAE) mice, whereas free TMN was completely cleared from plasma within 1 h.

PEGylated liposomes have been explored to encapsulate vector/DNA plasmid complexes such as polyethylenimine/oligodeoxynucleotides (PEI/ODN).[38] As a consequence, the blood clearance of gene delivery systems was extended. Furthermore, the accumulation of ODN in the brain was significantly improved when transferrin receptor-specific antibody 8D3 was non-covalently bound to the PEGylated liposome through streptavidin–biotin binding.

Non-PEGylated liposomes were utilized by Grahn *et al.* for simultaneous topotecan convection enhanced delivery (topoCED) and gadodiamide convection enhanced delivery (gadoCED) for treatment of glioblastoma multiforme and real-time monitor of NP distribution.[24] The liposomes loaded with topotecan and gadodiamide were administered by convection enhanced delivery (CED) using an *in vivo* U87MG intracranial rodent xenograft mouse model. Desirable topotecan pharmacokinetics and co-convection of gadodiamide have been observed. In particular, a half-life of approximately 1 day in the brain and an AUC of 153.8 μg day/g at an administration dose of 10 μg (0.5 mg/ml) were achieved; whereas an AUC of 5.5 μg day/g resulted from administration of free topotecan at the same dose.[39]

2.2. *Polymeric NPs*

Synthetic and natural polymers with different structures (linear, branched, or dendritic) have attracted considerable attention for biomedical applications including CNS drug delivery. In general, polymers used for drug delivery possess biocompatibility, drug compatibility, suitable biodegradation kinetics and mechanical properties and ease of processing.[40] Examples include poly(alkyl-a-cyano acrylates), poly(amino aicds), poly(acrylamides), poly(amides), ploy(esters), poly(orthoesters) and poly(urethanes). Aliphatic poly(esters) such as polylactic acid (PLA), polyglycolic acid (PGA), and copolymer poly(lactic-co-glycolic acid) (PLGA) have spurred tremendous interest in biomedical applications because of their excellent biocompatibility and biodegradability. PLGA degrades in the body through hydrolysis and breaks down into non-toxic lactic acid and glycolic acid monomers, which can be efficiently eliminated from the body. By virtue of their safety, PLGA copolymers have been approved by the FDA for the use in drug delivery. PLGA are commonly denoted by the molar ratio of lactic acid (LA) to glycolic acid (GA) in the polymer. The degradation rate of PLGA depends on several factors: the molar ratio of LA to GA, molecular weight of copolymer, degree of crystallinity and glass transition temperature (*Tg*) of the polymer.[40]

2.2.1. PLGA NP preparation methodologies

Single- and double-emulsion-solvent evaporation methods are most used to prepare PLGA NPs (Fig. 4). The single-emulsion method is essentially o/w emulsification; whereas the double-emulsion process is water-in-oil-in-water (w/o/w) emulsification. The o/w emulsification method is ideal for hydrophobic lipophilic drug encapsulation, and the w/o/w emulsification method is more suitable for hydrophilic drug encapsulation. The solid/oil/water (s/o/w) emulsification method has been developed to utilize PLGA-based microspheres to deliver a high drug loading for hydrophilic peptides, such as insulin.[41]

To prepare PLGA NPs via o/w emulsification, PLGA is first dissolved in a water-immiscible, volatile organic solvent, commonly dichloromethane (DCM). Hydrophobic drug is then dissolved in the polymer-solvent solution to form a polymer-solvent-drug solution/dispersion. Emulsification is performed with sonication, homogenization, or stirring in a larger volume of water in the presence of an stabilizer, e.g., poly(vinyl alcohol) (PVA) or poloxamer 188 (Pluronic® F-68). The stabilizer-coated NPs are obtained after removal of solvent. After washing, the drug encapsulated PLGA NPs are collected following filtration, sieving, centrifugation and lyophilization.

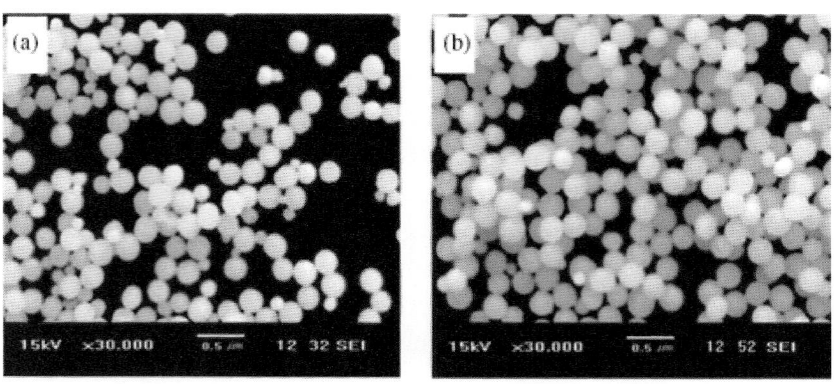

Fig. 4. Scanning electron microscopy (SEM) image of chitosan coated (0.5 mg/mL) (a) and uncoated (b) PLGA/PVA NPs (reproduced from Ref. 42).

To use w/o/w emulsification to encapsulate hydrophilic drugs, a small volume of drug-containing water is added into PLGA-DCM solution. The obtained polymer-solvent-drug-water is emulsified by the methods described above and then transferred to the emulsifier solution for a second emulsification. The resulting hydrophilic drug-encapsulated PLGA NPs are collected in the same way as used in o/w emulsification.

2.2.2. Roles of surfactants

PLGA NPs have been applied for a variety of biomedical applications including CNS drug delivery applications. It has been well-documented that non-ionic surfactants, e.g., polysorbate 80 (Tween® 80), can be used to coat NPs such as poly(butylcyanoacrylate) (PBCA) NPs to enhance brain targeting. Polysorbate 80-coated PBCA NPs can selectively adsorb certain plasma proteins, namely apolipoproteins *E* and *B*, which, in turn, promote receptor-mediated endocytosis of the particles at the BBB. However, binding of dalargin or loperamide to polysorbate 80-coated PBCA NPs may induce considerable analgesia in the CNS, counteracting the drug efficacy. Because of the toxicity and short duration of pharmacological effect, PBCA NPs may not be suitable drug carriers for chronic CNS diseases.[43]

Since surfactant coating has been considered to be essential for successful brain drug delivery, Gelperina *et al.* investigated surfactant-coated PLGA NPs for brain drug delivery. Both poloxamer 188 and polysorbate 80 coated PLGA NPs displayed high efficiency in getting drugs into the CNS following *i.v.* injection. The delivery efficiency and therapeutic efficacy are influenced by factors such as the nature of the core polymer, drug type, and stabilizer. PVA-stabilized NPs have been reported to be more efficient than those coated/stabilized with human serum albumin.[44] Tosi *et al.* developed NPs made of PLGA conjugated with a simil-opioid glycopeptide (g7) along with PLGA coupled with a near-infrared (NIR) probe (DY-675) to deliver Loperamide to the CNS. The mechanism for crossing the BBB of these drug carriers was believed to be adsorption-mediated endocytosis. As shown in Fig. 5, the NPs are able to sustain drug release and promote drug distribution.[45]

2.2.3. Targeted PLGA NPs

Solvent evaporation is a necessary step for PLGA particle preparation; however, it causes particle aggregation. Inclusion of polymer stabilizers such as PVA, poly(vinyl pyrrolidone) and polysorbate 80 helps prevent PLGA particles from forming aggregates. Because of concern about toxicity of certain

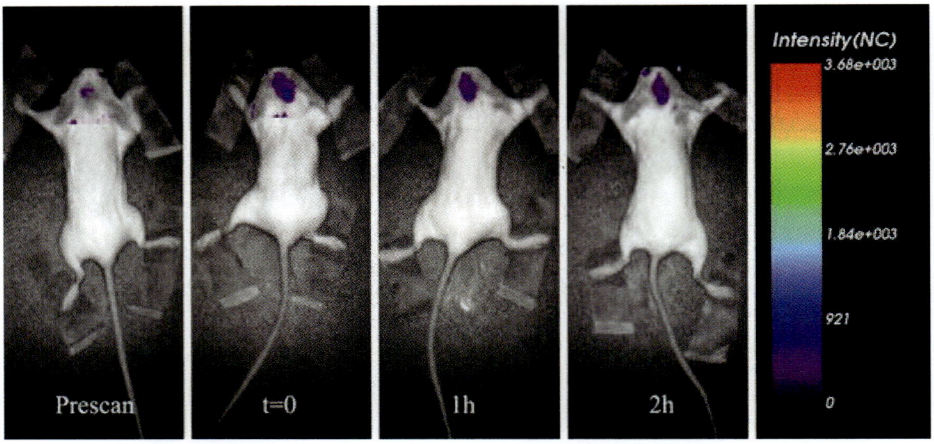

Fig. 5. Biodistribution of DY-g7-NPs in a bulb/C mouse after i.v. injection. Targeting of the probe to the brain area was monitored: fluorescence emission in regions of interest encompassing the brain were acquired at different time intervals post-injection and normalized (reproduced from Ref. 45).

types of stabilizers, complete removal of stabilizer from particles by thorough washing is deemed a necessary step.[46] Methods to avoid stabilizers in particle preparation have been developed. Chang *et al.* developed a new PLGA particle-based brain drug system without the use of stabilizer. They coated PLGA NPs with transferrin (Tf). Protein adsorption on the surface of the NPs was stable for at least 12 d at 37°C. It was found that Tf-coated PLGA NPs cross the BBB via an energy-dependent receptor mediated endocytosis following the caveolae pathway.[47]

Yu *et al.* developed a brain-targeted drug delivery system based on biodegradable PEG-PLGA polymersomes (POS) conjugated with lactoferrin (Lf). A self-assembly method was applied to prepare POS.[48] In particular, a mixture of methoxy-PEG-PLGA and maleimide-PEG-PLGA was dissolved in acetone and then mixed with pH 7.4 PBS under mild stirring at room temperature. Following removal of the organic solvent by dialysis, POS particles formed via self-assembly. Fluorescent probe such as 6-coumarin can be encapsulated during the self-assembly process. Lf was thiolated and then reacted with POS to form Lf-POS conjugates. The average sizes of the polymersomes and Lf-POS were approximately 90 nm and 120 nm, respectively. Pharmacokinetic profiles showed the optimal number of lactoferrin per polymersomes was 101. The brain tissue % ID/g of 6-coumarin-Lf101-POS was more than three-fold higher than that of 6-coumarin-POS. S14G-humanin peptides that were loaded into Lf-POS (SHN-Lf101-POS) provided a protective effect on neurons of rats treated with amyloid-β. (25–35). Administration of

SHN-Lfl01-POS significantly attenuated the expression of Bax and caspase-3 positive cells, indicating that Lfl01-POS is a promising drug delivery system for delivery of peptide drugs to the CNS.[48]

2.2.4. PEGylated PLGA NPs

Similar to any colloidal drug carrier that is not purposely designed to escape from mononuclear phagocyte system (MPS) uptake, PLGA NPs are rapidly cleared from the blood stream after vascular administration and preferentially accumulate in the liver and the spleen. Due to the lack of stealthiness, a significant portion of PLGA NPs are taken up by the MPS and the blood half-lives are generally around 2–3 min. Using PEG to modify PLGA NPs to obtain PEGylated PLGA immunonanoparticles is a viable approach to extending their blood half-lives and achieving sustained drug release. A variety of drugs including proteins, plasmid DNAs, and oligonucleotides can be loaded into PEGylated PLGA immunonanoparticles to gain long-acting efficacy. Optimization of PLGA NPs for brain-specific delivery and effective drug release in the brain remain to be investigated.

2.3. *Dendrimers*

Dendrimers have a highly branched, three-dimensional architecture at nanoscale with very low polydispersity and high functionality, comprising a central core, several internal braches and multiple reactive terminal groups on the surface. As dendrimer generation (G) increases, the number of branches and functional groups increases geometrically along with a linear increase in diameter as shown in Fig. 6. The presence of numerous surface groups allows for a high drug payload and multifunctionality. Widely acclaimed as the most versatile nanoscale building blocks, dendrimers have been extensively utilized for construction of nanodevices and nanomedicines. Among polymers with dendritic architecture, poly(amidoamine) (PAMAM) dendrimers are most investigated. The PAMAM dendrimers with a core (e.g. ethylene diamine) and an amidoamine repeat branching structure can be divided by full generations with amine surface groups and half-generations with carboxyl surface groups.

2.3.1. Drug-dendrimer coupling strategies

Hydrophobic anticancer drugs such as doxorubicin (DOX) and camptothecin (CPT) can be encapsulated into the interior of PAMAM dendrimers to gain

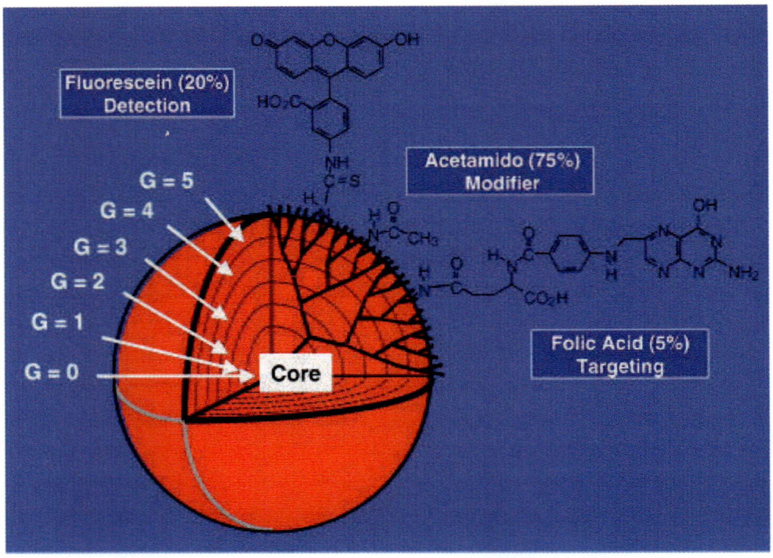

Fig. 6. Schematic presentation of dendrimers as nanoscaffold for the attachment of cell-specific ligands, modifiers, and fluorescence tags. Abbreviation: G, generation (reproduced from Ref. 49, Copyright 2013 Elsevier).

enhanced biologically and pharmacokinetically desirable properties including improved water solubility, prolonged blood circulation and reduced side effects during chemotherapy.[50–52] However, the amount of drugs that can be encapsulated by dendrimers is limited, and burst drug release is significant. Covalently coupling drug to the dendrimer surface can result in a high drug payload. Drug release can be controlled by a variety of linkages for drug-dendrimer coupling, such as acid labile *cis*-aconityl,[53,54] acyl hydrazone,[55] disulfide,[56] or triglycine.[57] For instance, DOX–PAMAM dendrimer conjugates with the *cis*-aconityl linkages are acid-sensitive. The acid-triggered release of DOX from the PAMAM dendrimer conjugates in tumor brings about a high concentration of DOX in tumor, resulting in a remarkable antitumor activity.[54] Using stimuli-sensitive linkage to connect drug to the dendrimer may minimize drug release during transport to the brain, hence reducing drug toxicity to normal tissues and organs.

Another interesting example is inclusion of the disulfide linkage in dendrimer-based nanocarrier for controlled intracellular drug release.[56] *N*-acetyl cysteine (NAC) conjugated PAMAM dendrimers via a disulfide linkage is sensitive to glutathione. Significant amounts of NAC were released from the conjugates in 1 h in response to intracellular glutathione. In the presence of bovine serum albumin (BSA), no NAC was released from the conjugates over

24 h, suggesting that the conjugates protect the drug from premature release in the blood circulation. As a result, the conjugates with disulfides were 16 times more potent than free NAC.[56]

2.3.2. Dendrimer transport mechanisms

The transport of PAMAM or surface-modified PAMAM dendrimers transport across the cell monolayer relies on primarily endocytosis-mediated cellular internalization. With the aid of dendrimers, therapeutics can cross various cell membranes or biological barriers such as the BBB.[5] To enhance receptor-mediated endocytosis and promote uptake by a specific brain lesion, dendrimers can be modified with brain targeting ligands, such as angiopep,[58,59] epidermal growth factor (EGF),[57] rabies virus glycoprotein (RVG29),[60] tamoxifen (TAM), transferrin (Tf)[50,61] and monoclonal antibody OX26.[62] For instance, PAMAM dendrimers modified by Tf and TAM significantly enhanced the ability of the carrier in crossing the BBB. Tf mediates the BBB transportation by transferrin receptor-mediated endocytosis; while TAM inhibits the expression of drug efflux proteins such as the ABC transporters on the BBB.[52] Another intriguing dendrimer delivery system utilized a combination of Tf and wheat germ agglutinin (WGA) as surface modifiers. The BBB transport of this system was enhanced synergistically by two different endocytosis mechanisms: Tf for receptor-medicated endocytosis and WGA for adsorption-mediated endocytosis.[50]

Angiopep is a peptide, which targets the low-density lipoprotein receptor-related protein-1 (LRP1). Angiopep modified PAMAM dendrimers can be internalized by brain capillary endothelial cells via receptor-mediated endocytosis and trafficked through clathrin- and caveolae-mediated energy dependent pathways.[59] Nonetheless, dendrimers are non-biodegradable and may accumulate in lysosomes. The precise mechanisms of dendrimer internalization and subsequent fate remain to be elucidated.

2.3.3. PAMAM dendrimer-EGF conjugates

Epidermal growth factor receptor (EGFR) is preferentially overexpressed in various types of human solid tumors, including cancers of the head and neck, lung, breast, colon, and brain. This provides a rationale for design of tumor-specific delivery systems by coupling ligands that bind to EGFR specifically. Yuan *et al.* developed a dendrimer-EGF delivery system, in which EGF is covalently conjugated to PAMAM dendrimer G4.0 through a triglycine spacer as shown in Fig. 7.[57] In particular, the carboxyl group of triglycine was

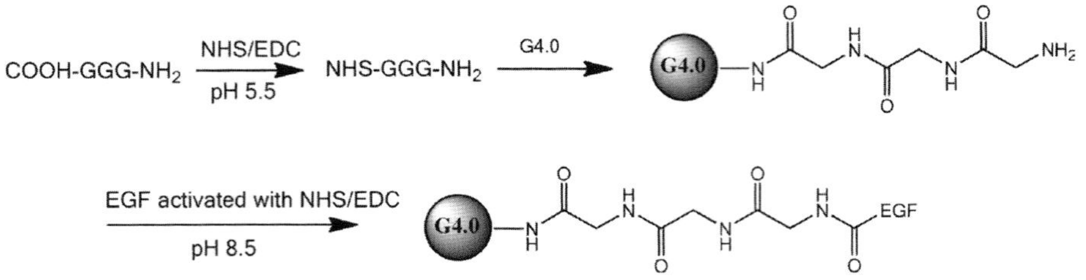

Fig. 7. Synthesis of dendrimer-EGF conjugates through triglycine spacer (GGG); (Reproduced from Ref. 57. Copyright 2013 Elsevier).

activated into ester by *N*-hydroxysuccinimide (NHS)/1-ethyl-3-[3-dimethylaminopropyl]carbodiimide hydrochloride (EDC) reaction. The resulting NHS-activated triglycine (NHS-GGG) was then reacted with the amino group of PAMAM dendrimer G4 in a bicarbonate buffer solution (pH 8.5) for 2 h. Similarly, the carboxyl group of human recombinant EGF was activated by NHS/EDC. The resultant EGF-NHS esters were then reacted with G4-GGG to form G4.0-GGG-EGF conjugates. They observed that dendrimer-EGF conjugates facilitated intracellular uptake in an EGFR-dependent manner without triggering EGFR signaling pathways in EGFR-overexpressing cells. Furthermore, they found that dendrimer-EGF conjugates improved siRNA delivery and gene knockdown efficiency. The enhanced knockdown efficiency of dendrimer-EGF conjugates compared to unmodified dendrimer or TransIT (a commercial keratinocyte transfection reagent) transfections indicates this delivery system is suitable for introduction of nucleic acids into cells via a receptor-targeted mechanism.[57]

2.3.4. PEGylated PAMAM dendrimers with bis-aryl hydrazone linkages

The "proton sponge" property conferred by amines in the dendrimer is crucial for successful gene delivery. However, surface modification of PAMAM dendrimers with PEG or other moieties has to utilize surface amine groups and inadvertently impairs dendrimer buffering capacity. To address this issue, Yuan *et al.* employed a new methodology to make PEGylated dendrimers by using a bis-aryl hydrazone (BAH) linker as shown in Fig. 8.[55] To use BAH to connect PEG to the dendrimer, PAMAM dendrimer G4.0 is activated with succinimidyl 4-hydrazinonicotinate acetone hydrazone (SANH), while monofunctional methoxy polyethylene glycol amine (PEG5000) is activated with succinimidyl 4-formylbenzoate (SFB). A subsequent coupling reaction between

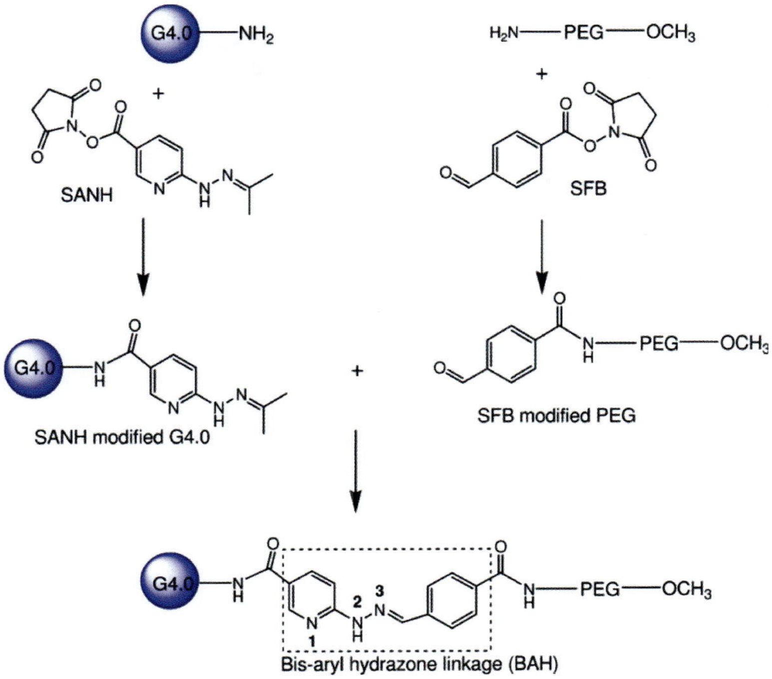

Fig. 8. Synthesis of PEGylated PAMAM dendrimers with BAH linkages (reproduced from Ref. 55, copyright 2013, American Chemical Society).

SFB-activated PEG and SANH-activated PAMAM forms a BAH linkage. The authors found that use of BAH linkage simultaneously enables a high degree of PEGylation and an increase in buffering capacity of the dendrimer vector. Higher transfection efficiency was also observed for this vector compared to unmodified dendrimer.[55]

2.3.5. PEGylated full generation PAMAM dendrimer-angiopep-2 conjugates

Tumor necrosis factor (TNF)-related apoptosis-inducing ligand (TRAIL) can selectively induce cell apoptosis of glioma without affecting normal cells. Huang *et al.* engineered a dendrimer-based system to deliver TRAIL plasmid to the brain.[58] As a ligand against LRP1, Angiopep-2 was conjugated to the PAMAM dendrimer G5 via bifunctional PEG. This PAMAM-PEG-angiopep conjugates were then complexed with the DNA to form PAMAM-PEG-Angiopep/DNA NPs. Both *in vitro* and *in vivo* biodistribution studies showed PAMAM-PEG-Angiopep/DNA NPs accumulated in the brain in

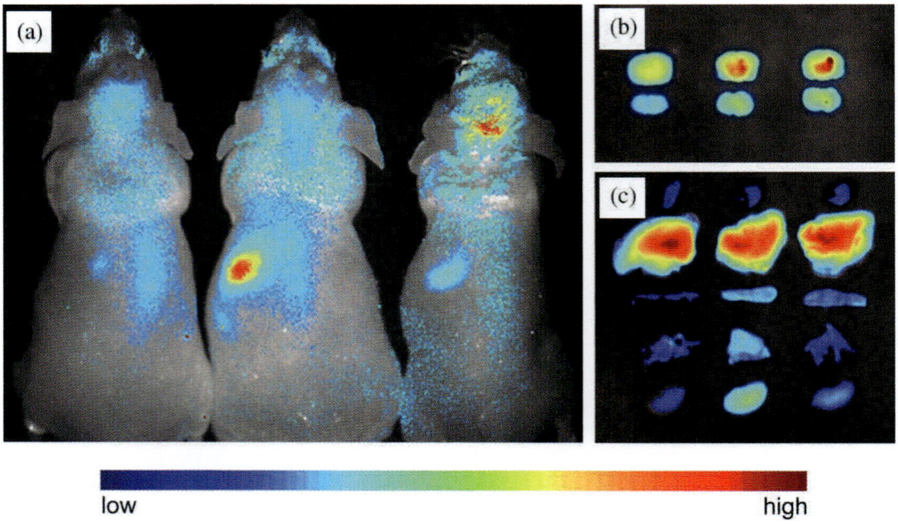

Fig. 9. *In vivo* (a) and *in vitro* (b) and (c), imaging of mice administrated with PAMAM/DNA NPs (Left), PAMAM-PEG/DNA NPs (Midst) or PAMAM-PEG-Angiopep/DNA NPs (Right). Images were taken 120 min after NPs systemic administration (reproduced from Ref. 58).

significant amounts and preferentially at the tumor site (Fig. 9). As a result, the median survival time for brain tumor bearing mice treated with PAMAM-PEG-Angiopep/pORF-TRAIL NPs was significantly longer than that of control groups. Notably, the NPs alone showed low cytotoxicity after *in vitro* transfection. Thus, angiopep-2 could be exploited as a specific ligand to facilitate NPs to permeate through the BBB, and PAMAM-PEG-Angiopep/DNA NPs can serve as a potential non-viral delivery system for gene therapy of glial tumor.

2.3.6. **PAMAM dendrimer-OX26 conjugates**

Due to the susceptibility of bioactive molecules, systemic administration (i.v. injection) of therapeutic NPs is commonly used to avoid the first-pass effect and enhance the drug bioavailability. However, for chronic CNS diseases, this route may lead to poor patient compliance. It is highly demanded to explore a non-invasive and safe administration route for CNS therapeutic NPs. Recently, Yuan *et al.* developed a new dendrimer-based delivery system, constituted of targeting ligand OX26 that was coupled to PAMAM dendrimers via a PEG spacer and opioid peptide DPDPE (Fig. 10).[62] They tested

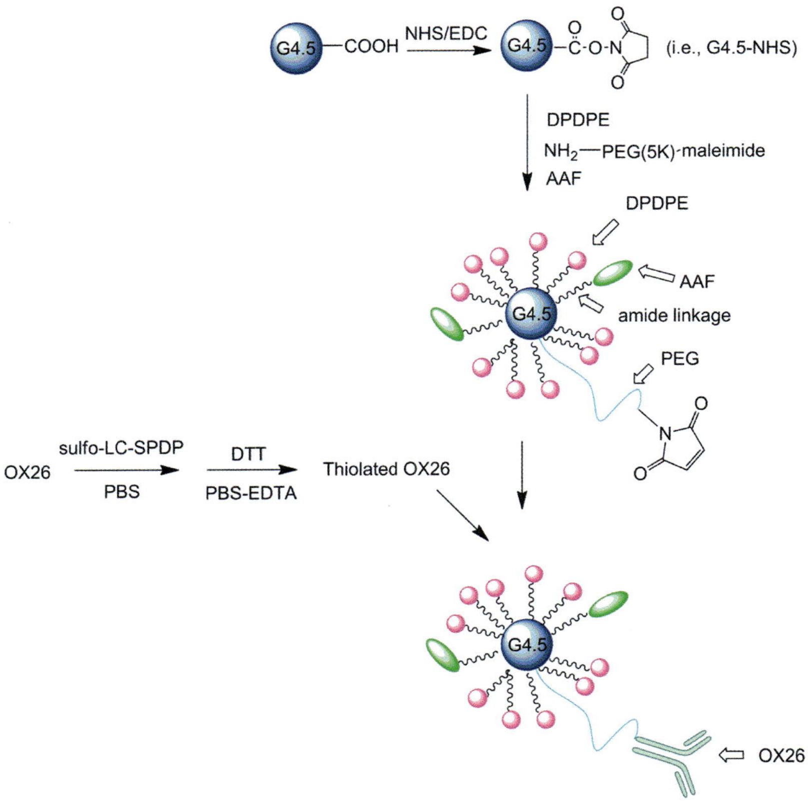

Fig. 10. Synthesis of PEG-G4.5 (AAF)-DPDPE and OX26-PEG-G4.5 (AAF)-DPDPE conjugates (reproduced from Ref. 62, copyright 2013 American Chemical Society).

tranbuccal delivery of this new system. Functionalized PAMAM dendrimer G4.5 displayed a high permeability across the buccal mucosa. Notably, co-administration of bile salt sodium glycodeoxycholate (NaGDC) with OX26-PEG-G4.5-DPDPE NPs could further increase the permeability of the carriers across the porcine buccal mucosal by multiple folds.[62] Given such high adaptability of PAMAM dendrimers, a wide range of CNS drugs including therapeutic drugs, peptides, proteins and genes can benefit from dendrimer-based delivery and buccal administration.

From the examples discussed above, dendrimers appear to be a very promising class of carrier for specific delivery of drugs and genetic materials to the CNS. The size, architecture and surface properties (e.g. charge) of dendrimer-based NPs affect their brain drug delivery efficiency. Therefore, these factors should be taken into account for the rational design of brain-targeted drug delivery systems in conjunction with toxicity, biocompatibility and biodistribution profiles.

3. Conclusions and Outlook

Nanocarriers can significantly enhance drug delivery to the brain. To develop efficient, safe and brain-targeted nanostructured delivery systems, we believe the following factors should be taken into account: an in-depth understanding of barrier architectures and pathological conditions, selection of NPs with adaptable structures and high capacity for drug loading and multifunctionalization, and safe routes for NP administration particularly for chronic CNS diseases. NPs such as liposomes, PLGA NPs and dendrimers as discussed above have shown promising structures and functions desirable for brain drug delivery and hence warrant further examination.

Acknowledgments

The National Science Foundation (CAREER Award CBET0954957) and Jeffress Memorial Trust (J-1043).

References

1. H. L. Wong, X. Y. Wu and R. Bendayan, Nanotechnological advances for the delivery of CNS therapeutics, *Adv. Drug Delivery Rev.* **64**, 686–700 (2012).
2. L. Martin-Banderas, M. A. Holgado, J. L. Venero, J.Alvarez-Fuentes and M. Fernandez-Arevalo, Nanostructures for drug delivery to the brain, *Curr. Med. Chem.* **18**, 5303–5321 (2011).
3. F. L. Cardoso, D. Brites and M. A.Brito, Looking at the blood-brain barrier: molecular anatomy and possible investigation approaches, *Brain Res. Rev.* **64**, 328–363 (2010).
4. R. I. van, S. Cakir-Tascioglu, W. E. Hennink, G. Storm R. M. Schiffelers and E. Mastrobattista, *In vivo* methods to study uptake of nanoparticles into the brain, *Pharm. Res.* **28**, 456–471 (2011).
5. H. Yang, Nanoparticle-mediated brain-specific drug delivery, imaging, and diagnosis, *Pharm. Res.* **27**, 1759–1771 (2010).
6. G. Caruso, M. Caffo, C. Alafaci, G. Raudino, D. Cafarella, S. Lucerna, F. M. Salpietro and F. Tomasello, Could nanoparticle systems have a role in the treatment of cerebral gliomas? *Nanomedicine* 7, 744–752 (2010).
7. A. Nunes, K. T. Al-Jamal and K. Kostarelos, Therapeutics, imaging and toxicity of nanomaterials in the central nervous system, *J. Control. Release* **161**, 290–306 (2012).
8. S. C. Steiniger, J. Kreuter, A. S. Khalansky, I. N. Skidan, A. I. Bobruskin, Z. S. Smirnova, S. E. Severin, R. Uhl, M. Kock K. D. Geiger and S. E. Gelperina, Chemotherapy of glioblastoma in rats using doxorubicin-loaded nanoparticles, *Int. J. Cancer* **109**, 759–767 (2004).

9. H. Xin, L. Chen, J. Gu, X. Ren, Z. Wei, J. Luo, Y. Chen, X. Jiang, X. Shaand X. Fang, Enhanced anti-glioblastoma efficacy by PTX-loaded PEGylated poly(varepsilon-caprolactone) nanoparticles: *In vitro* and *in vivo* evaluation, *Int. J. Pharm.* **402**, 238–247 (2010).
10. X. Jiang, H. Xin, X. Sha, J. Gu, Y. Jiang, K. Law, Y. Chen, L. Chen, X. Wang and X. Fang, PEGylated poly(trimethylene carbonate) nanoparticles loaded with paclitaxel for the treatment of advanced glioma: *In vitro* and *in vivo* evaluation, *Int. J. Pharm.* **420**, 385–394 (2011).
11. M. Zhao, C. Liang, A. Li, J. Chang, H. Wang, R. Yan, J. Zhang and J. Tai, Magnetic paclitaxel nanoparticles inhibit glioma growth and improve the survival of rats bearing glioma xenografts, *Anticancer Res.* **30**, 2217–2223 (2010).
12. W. Lu, Q. Sun, J. Wan, Z. She and X. G. Jiang, Cationic albumin-conjugated pegylated nanoparticles allow gene delivery into brain tumors via intravenous administration, *Cancer Res.* **66**, 11878–11887 (2006).
13. A. Jordan, R. Scholz, K. Maier-Hauff, F. K. van Landeghem, N. Waldoefner, U. Teichgraeber, J. Pinkernelle, H. Bruhn, F. Neumann, B. Thiesen, A. von Deimling and R. Felix, The effect of thermotherapy using magnetic nanoparticles on rat malignant glioma, *J. Neuro-oncol.* **78**, 7–14 (2006).
14. F. K. van Landeghem, K. Maier-Hauff, A. Jordan, K. T. Hoffmann, U. Gneveckow, R. Scholz, B. Thiesen, W. Bruckand A. von Deimling, Post-mortem studies in glioblastoma patients treated with thermotherapy using magnetic nanoparticles, *Biomaterials* **30**, 52–57 (2009).
15. K. Maier-Hauff, R. Rothe, R. Scholz, U. Gneveckow, P. Wust, B. Thiesen, A. Feussner, A. von Deimling, N. Waldoefner, R. Felix and A. Jordan, Intracranial thermotherapy using magnetic nanoparticles combined with external beam radiotherapy: Results of a feasibility study on patients with glioblastoma multiforme, *J. Neuro-oncol.* **81**, 53–60 (2007).
16. R. Kopelman, Y.-E. Lee Koo, M. Philbert, B. A. Moffat, G. Ramachandra Reddy, P. McConville, D. E. Hall, T. L. Chenevert, M. S. Bhojani, S. M. Buck, A. Rehemtulla and B. D. Ross, Multifunctional nanoparticle platforms for *in vivo* MRI enhancement and photodynamic therapy of a rat brain cancer, *J. Magn. Magn. Mat.* **293**, 404–410 (2005).
17. L. Agemy, D. Friedmann-Morvinski, V. R. Kotamraju, L. Roth, K. N. Sugahara, O. M. Girard, R. F. Mattrey, I. M. Verma and E. Ruoslahti, Targeted nanoparticle enhanced proapoptotic peptide as potential therapy for glioblastoma, *Proc. Nat. Acad. Sci. USA* **108**, 17450–17455 (2011).
18. G. Wu, R. F. Barth, W. Yang, S.;Kawabata, L. Zhang and K. Green-Church, Targeted delivery of methotrexate to epidermal growth factor receptor-positive brain tumors by means of cetuximab (IMC-C225) dendrimer bioconjugates, *Mol. Cancer Ther.* **5**, 52–59 (2006).
19. M. Yemisci, S. Bozdag, M. Cetin, F. Soylemezoglu, Y. Capan, T. Dalkara and I. Vural, Treatment of malignant gliomas with mitoxantrone-loaded poly (lactide-co-glycolide) microspheres, *Neurosurg.* **59**, 1296–1302; discussion 1302–1303 (2006).
20. O. Benny, L. G. Menon, G. Ariel, E. Goren, S. K. Kim, C. Stewman, P. M. Black, R. S. Carroll and M. Machluf, Local delivery of poly lactic-co-glycolic acid microspheres containing imatinib mesylate inhibits intracranial xenograft glioma growth, *Clin. Cancer Res.* **15**, 1222–1231 (2009).
21. A. Doi, S. Kawabata, K. Iida, K. Yokoyama, Y. Kajimoto, T. Kuroiwa, T. Shirakawa, M. Kirihata, S. Kasaoka, K. Maruyama, H. Kumada, Y. Sakurai, S. Masunaga, K. Ono and

S. Miyatake, Tumor-specific targeting of sodium borocaptate (BSH) to malignant glioma by transferrin-PEG liposomes: A modality for boron neutron capture therapy, *J. Neuro-oncol.* **87**, 287–294 (2008).

22. B. Feng, K. Tomizawa, H. Michiue, S. Miyatake, X. J. Han, A. Fujimura, M. Seno, M. Kirihata and H. Matsui, Delivery of sodium borocaptate to glioma cells using immunoliposome conjugated with anti-EGFR antibodies by ZZ-His, *Biomaterials* **30**, 1746–1755 (2009).
23. M. T. Krauze, C. O. Noble, T. Kawaguchi, D. Drummond, D. B. Kirpotin, Y. Yamashita, E. Kullberg, J. Forsayeth, J. W. Park and K. S. Bankiewicz, Convection-enhanced delivery of nanoliposomal CPT-11 (irinotecan) and PEGylated liposomal doxorubicin (Doxil) in rodent intracranial brain tumor xenografts, *Neuro. Oncol.* **9**, 393–403 (2007).
24. A. Y. Grahn, K. S. Bankiewicz, M. Dugich-Djordjevic, J. R. Bringas, P. Hadaczek, G. A. Johnson, S. Eastman and M. Luz, Non-PEGylated liposomes for convection-enhanced delivery of topotecan and gadodiamide in malignant glioma: Initial experience, *J. Neuro-oncol.* **95**, 185–197 (2009).
25. F. Y. Yang, M. C. Teng, M. Lu, H. F. Liang, Y. R. Lee, C. C. Yen, M. L. Liang and T. T. Wong, Treating glioblastoma multiforme with selective high-dose liposomal doxorubicin chemotherapy induced by repeated focused ultrasound, *Int. J. Nanomed.* **7**, 965–974 (2012).
26. X. Sun, Z. Pang, H. Ye, B. Qiu, L. Guo, J. Li, J. Ren, Y. Qian, Q. Zhang, J. Chen and X. Jiang, Co-delivery of pEGFP-hTRAIL and paclitaxel to brain glioma mediated by an angiopep-conjugated liposome, *Biomaterials* **33**, 916–924 (2012).
27. M. VanHandel, D. Alizadeh, L. Zhang, B. Kateb, M. Bronikowski, H. Manohara and B. Badie, Selective uptake of multi-walled carbon nanotubes by tumor macrophages in a murine glioma model, *J. Neuroimmunol.* **208**, 3–9 (2009).
28. D. Zhao, D. Alizadeh, L. Zhang, W. Liu, O. Farrukh, E. Manuel, D. J. Diamond and B. Badie, Carbon nanotubes enhance CpG uptake and potentiate antiglioma immunity, *Clin. Cancer Res.* **17**, 771–782 (2011).
29. J. Ren, S. Shen, D. Wang, Z. Xi, L. Guo, Z. Pang, Y. Qian, X. Sun and X. Jiang, The targeted delivery of anticancer drugs to brain glioma by PEGylated oxidized multi-walled carbon nanotubes modified with angiopep-2, *Biomaterials* **33**, 3324–3333 (2012).
30. M. I. Alam, S. Beg, A. Samad, S. Baboota, K. Kohli, J. Ali, A. Ahuja and M. Akbar, Strategy for effective brain drug delivery, *Eur. J. Pharm. Sci.* **40**, 385–403 (2010).
31. H. L. Wong, X. Y. Wu and R. Bendayan, Nanotechnological advances for the delivery of CNS therapeutics, *Adv. Drug Delivery Rev.* **64**, 686–700 (2012).
32. C. R. Safinya and K. K. Ewert, Materials chemistry: Liposomes derived from molecular vases, *Nature* **489**, 372–374 (2012).
33. T. K. Vyas, A. Shahiwala and M. M. Amiji, Improved oral bioavailability and brain transport of Saquinavir upon administration in novel nanoemulsion formulations, *Int. J. Pharm.* **347**, 93–101 (2008).
34. W. M. Pardridge, Vector-mediated drug delivery to the brain, *Adv. Drug Delivery Rev.* **36**, 299–321 (1999).
35. R. J. Boado, Y. Zhang, Y. Zhang and W. M. Pardridge, Humanization of anti-human insulin receptor antibody for drug targeting across the human blood–brain barrier, *Biotechnol. Bioeng.* **96**, 381–391 (2007).

36. J. C. Olivier, R. Huertas, H. J. Lee, F. Calon and W. M. Pardridge, Synthesis of pegylated immunonanoparticles, *Pharm. Res.* **19**, 1137–1143 (2002).
37. P. Kizelsztein, H. Ovadia, O. Garbuzenko, A. Sigal and Y. Barenholz, Pegylated nanoliposomes remote-loaded with the antioxidant tempamine ameliorate experimental autoimmune encephalomyelitis, *J. Neuroimmunol.* **213**, 20–5 (2009).
38. Y. T. Ko, R. Bhattacharya and U. Bickel, Liposome encapsulated polyethylenimine/ODN polyplexes for brain targeting, *J. Controlled Release* **133**, 230–237 (2009).
39. E. Afergan, H. Epstein, R. Dahan, N. Koroukhov, K. Rohekar, H. D. Danenberg and G. Golomb, Delivery of serotonin to the brain by monocytes following phagocytosis of liposomes. *J. Controlled Release* **132**, 84–90 (2008).
40. S. Acharya and S. K. Sahoo, PLGA nanoparticles containing various anticancer agents and tumour delivery by EPR effect, *Adv. Drug Delivery Rev.* **63**, 170–83 (2011).
41. J. M. Lu, X. Wang, C. Marin-Muller, H. Wang, P. H. Lin, Q. Yao and C. Chen, Current advances in research and clinical applications of PLGA-based nanotechnology, *Exp. Rev. Mol. Diag.* **9**, 325–341 (2009).
42. B. S. Kim, C. S. Kim and K. M. Lee, The intracellular uptake ability of chitosan-coated Poly (D,L-lactide-co-glycolide) nanoparticles, *Arch. Pharm. Res.* **31**, 1050–1004 (2008).
43. J. C. Olivier, Drug transport to brain with targeted nanoparticles. *Neurorx.* **2**, 108–119 (2005).
44. S. Gelperina, O. Maksimenko, A. Khalansky, L. Vanchugova, E. Shipulo, K. Abbasova, R. Berdiev, S. Wohlfart, N. Chepurnova and J. Kreuter, Drug delivery to the brain using surfactant-coated poly(lactide-co-glycolide) nanoparticles: Influence of the formulation parameters. *Eur. J. Pharm. Biopharm.V* **74**, 157–163 (2010).
45. G. Tosi, L. Bondioli, B. Ruozi, L. Badiali, G. M. Severini, S. Biffi, A. De Vita, B. Bortot, D. Dolcetta, F. Forni and M. A. Vandelli, NIR-labeled nanoparticles engineered for brain targeting: *in vivo* optical imaging application and fluorescent microscopy evidences, *J. Neural Trans.* **118**, 145–153 (2011).
46. J.-C. Olivier, L. Fenart, R. Chauvet, C. Pariat, R. Cecchelli and W. Couet, Indirect Evidence that Drug Brain Targeting Using Polysorbate 80-Coated Polybutylcyanoacrylate Nanoparticles Is Related to Toxicity, *Pharm. Res.* **16**, 1836–1842 (1999).
47. J. Chang, Y. Jallouli, M. Kroubi, X. B. Yuan, W. Feng, C. S. Kang, P. Y. Pu and D. Betbeder, Characterization of endocytosis of transferrin-coated PLGA nanoparticles by the blood-brain barrier, *Int. J. Pharm.* **379**, 285–292 (2009).
48. Y. Yu, Z. Pang, W. Lu, Q. Yin, H. Gao and X. Jiang, Self-assembled polymersomes conjugated with lactoferrin as novel drug carrier for brain delivery, *Pharm. Res.* **29**, 83–96 (2012).
49. S. Svenson and D. A. Tomalia, Dendrimers in biomedical applications--reflections on the field. *Adv. Drug Delivery Rev.* **57**, 2106–2129 (2005).
50. H. He, Y. Li, X. R. Jia, J. Du, X. Ying, W. L. Lu, J. N. Lou and Y. Wei, PEGylated Poly(amidoamine) dendrimer-based dual-targeting carrier for treating brain tumors, *Biomaterials* **32**, 478–487 (2011).
51. M. T. Morgan, Y. Nakanishi, D. J. Kroll, A. P. Griset, M. A. Carnahan, M. Wathier, N. H. Oberlies, G. Manikumar, M. C. Wani and M. W. Grinstaff, Dendrimer-encapsulated camptothecins: Increased solubility, cellular uptake, and cellular retention affords enhanced anticancer activity *in vitro*, *Cancer Res.* **66**, 11913–11921 (2006).

52. Y. Li, H. He, X. Jia, W. L. Lu, J. Lou and Y. Wei, A dual-targeting nanocarrier based on poly(amidoamine) dendrimers conjugated with transferrin and tamoxifen for treating brain gliomas, *Biomaterials* **33**, 3899–3908 (2012).
53. S. Zhu, M. Hong, L. Zhang, G. Tang, Y. Jiang and Y. Pei, PEGylated PAMAM dendrimer-doxorubicin conjugates: *In vitro* evaluation and *in vivo* tumor accumulation, *Pharma. Res.* **27**, 161–174 (2010).
54. S. Zhu, M. Hong, G. Tang, L. Qian, J. Lin, Y. Jiang,and Y. Pei, Partly PEGylated polyamidoamine dendrimer for tumor-selective targeting of doxorubicin: the effects of PEGylation degree and drug conjugation style, *Biomaterials* **31**, 1360–1371 (2010).
55. Q. Yuan, W. A. Yeudall,and H. Yang, PEGylated polyamidoamine dendrimers with bis-aryl hydrazone linkages for enhanced gene delivery. *Biomacromolecules* **11**, 1940–1947 (2010).
56. Y. E. Kurtoglu, R. S. Navath, B. Wang, S. Kannan, R. Romero and R. M. Kannan, Poly(amidoamine) dendrimer-drug conjugates with disulfide linkages for intracellular drug delivery, *Biomaterials* **30**, 2112–2121 (2009).
57. Q. Yuan, E. Lee, W. A. Yeudalland H. Yang, Dendrimer-triglycine-EGF nanoparticles for tumor imaging and targeted nucleic acid and drug delivery, *Oral Oncol.* **46**, 698–704 (2010).
58. S. Huang, J. Li, L. Han, S. Liu, H. Ma, R. Huangand C. Jiang, Dual targeting effect of Angiopep-2-modified, DNA-loaded nanoparticles for glioma, *Biomaterials* **32**, 6832–6838 (2011).
59. W. Ke, K. Shao, R. Huang, L. Han, Y. Liu, J. Li, Y. Kuang, L. Ye, J. Lou and C. Jiang, Gene delivery targeted to the brain using an Angiopep-conjugated polyethyleneglycol-modified polyamidoamine dendrimer, *Biomaterials* **30**, 6976–6985 (**2009**).
60. Y. Liu, R. Huang, L. Han, W. Ke, K. Shao, L. Ye, J. Louand C. Jiang, Brain-targeting gene delivery and cellular internalization mechanisms for modified rabies virus glycoprotein RVG29 nanoparticles, *Biomaterials* **30**, 4195–4202 (2009).
61. R. Q. Huang, Y. H. Qu, W. L. Ke, J. H. Zhu, Y. Y. Peiand C. Jiang, Efficient gene delivery targeted to the brain using a transferrin-conjugated polyethyleneglycol-modified polyamidoamine dendrimer. *FASEB J.* **21**, 1117–1125 (**2007**).
62. Q. Yuan, Y. Fu, W. J. Kao, D. Janigro and H. Yang, Transbuccal Delivery of CNS Therapeutic Nanoparticles: Synthesis, Characterization, and *In Vitro* Permeation Studies, *ACS Chem. Neurosci.* **2**, 676–683 (2011).

Chapter 7

Nanoparticles for Dermal and Transdermal Delivery: Permeation Pathways and Applications

Marianna Foldvari, Marjan Gharagozloo and Christine Li

School of Pharmacy and Waterloo Institute of Nanotechnology, University of Waterloo
foldvari@uwaterloo.ca

Abstract

Nanoparticles may prove more effective in delivering drugs across the skin compared to traditional topical dosage forms. A wide range of different nanoparticle designs have been evaluated for both dermal and transdermal delivery. The balance of two factors, penetration and safety, is to be considered when developing new, more effective delivery systems. The properties of nanoscale materials as building blocks for dermal/transdermal drug delivery nanoparticles determine how the nanoparticles interact with the skin and what will be the mechanism of drug delivery.

1. Introduction

In recent years, significant attention has been devoted to the beneficial use and potential toxicity of nanomaterials. The skin is the largest organ of the body and with an average surface area of $1.8\,m^2$ it provides for significant exposure to these nanomaterials. There are several new questions arising within the field of dermal and transdermal drug delivery related to the use of nanoscale materials. Will nanosizing active ingredients improve skin penetration? Is size the most important parameter in skin permeation for nanoparticles? Will nanosizing increase toxicity?

Although, the primary function of skin is to defend the body against foreign substances,[1] it has also been extensively studied for the transdermal delivery of therapeutics.[2] The skin is an excellent barrier to most substances, but it may be permeable to some degree depending on the chemical properties (*e.g.* lipid solubility) of the compound in question. The mechanisms of permeation have been the subject of extensive study; however, the fine details and multiple pathways of the delivery process across the skin still need to be elucidated. The skin is composed of three main layers: epidermis, dermis and subcutaneous tissue (Fig. 1). In the epidermis, the stratum corneum (a non-viable, 10–15 μm thick layer) can be distinguished from the viable epidermis (thickness: 200 μm). The stratum corneum is made up of corneocytes (bricks) and intercellular spaces filled with lipids (mortar), the former being a hydrophilic domain and the latter hydrophobic. Absorption may occur through (i) the intracellular route, which is more typical of hydrophilic substances, (ii) the intercellular route, which is more typical of lipophilic substances or (iii) the appendages (hair follicles, sweat ducts). The main barrier function of the skin is attributed to the stratum corneum, due to the compact stacking of corneocytes and the hydrophobic nature of the intercellular lipid ("brick and mortar" model).[3–5] In humans, the intercellular lipids, which make up 8–10% of the total stratum corneum weight[6] consist of fatty acids (mainly C22 and C24, 15–25%), long chain ceramides (CER1-CER8) (mostly C24–C26,

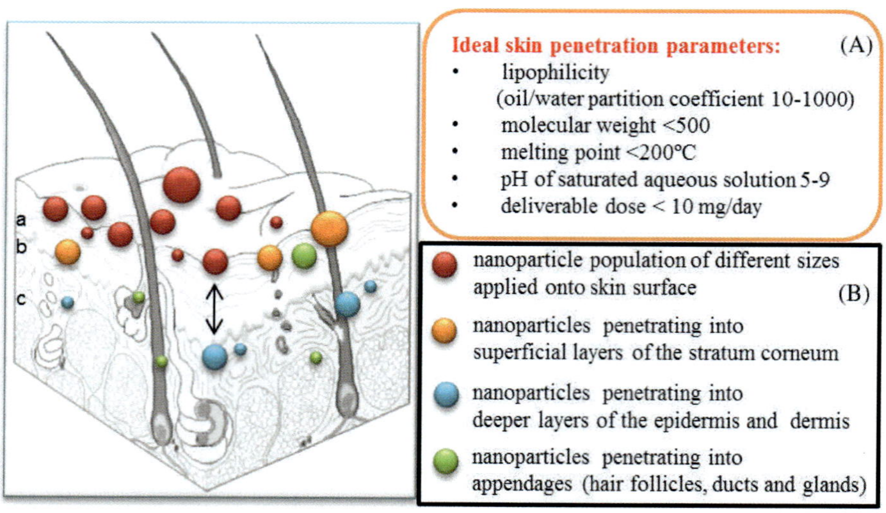

Fig. 1. Schematic picture of skin — (a) stratum corneum, (b) viable epidermis, (c) dermis — and considerations for optimum skin permeability parameters for small chemical entities (A) and general size representation of nanoparticles capable of reaching to some depth within the skin illustrated on the skin schematic drawing (B).

35–50%, and CER1 and CER4 with C30–C32), cholesterol (15–25%) and cholesterol sulfate (5–10%).[7–9]

2. Skin as a Treatment Site and A Route of Entry into Systemic Circulation

The topical application of drugs can be performed for two distinct purposes, either as a localized treatment for dermatological conditions or for systemic delivery. There are specific set of strategies for either of these two applications, although the first hurdle in both cases is to cross the stratum corneum barrier.

In healthy, uncompromised skin the physical barrier properties of the skin prevail and prevent most external substances to permeate into the body. However, there may be compounds or materials that relatively easily penetrate the skin due to ideal physicochemical properties. Unfortunately, these are typically substances with considerable toxicities and their skin uptake is unwanted. The properties of most drugs fall outside the optimum range of permeability (Fig. 1) and hence require some type of an enhancer to be therapeutically useful. Permeation and penetration enhancers are substances that temporarily increase the permeability of the skin and facilitate drug delivery. The main mechanism of permeation/penetration enhancers is related to alteration of the properties of the stratum corneum lipids.

Models which help explain the formation and molecular organization of the stratum corneum lipids include the *Landmann model* (1986)[10] and *Norlen's membrane folding model* (2001).[11] The Landmann model showed that lamellar granules produced by granular cells in the upper viable epidermis fuse with the cell membranes and extrude into the extracellular spaces to form stacks of bilayer sheets. Other more recent studies indicate that the structure of lipids that are being extruded during barrier lipid formation may be in other polymorphic states such as cubic phase which is a bicontinuous structure where amphipatic molecules are arranged in a porous connected bilayer structure in three dimensions.[11] The final intercellular lipid structure is built from two stacked lipid bilayers composed of different ceramide molecules (CER1-CER8) in an interdigitated arrangement with a periodicity of approximately 13 nm, also called the Landmann unit.[12,13]

3. How Nanotechnology is Defined in Dermal and Transdermal delivery

Nanotechnology is also having an impact on drug delivery through the skin. The interest is motivated by two factors, whether nanoscale particles can penetrate into the skin and could be a source of toxicity and whether nanoscaling delivery systems would enhance dermal/transdermal delivery. The application of the "nano" definition in dermatology and transdermal drug delivery context is fairly complex. The particle size requirement is more loosely considered and often particles under 1,000 nm may be presented by research groups as "nanoparticles."

Health Canada defines nanotechnology as "the application of scientific knowledge to manipulate and control matter in the nanoscale to make use of size- and structure-dependent properties and phenomena distinct from those associated with individual atoms or molecules or with bulk materials." The term "nanoscale" is defined in the range 1–100 nm *inclusive*'. Similar definitions exist by the Food and Drug Administration (FDA) (approximately 1 nm to 100 nm), the European Union-European Medicines Agency (EU-EMA) (nanometer scale ranges from the atomic level at around 0.2 nm (2Å) up to around 100 nm) and the International Organization for Standardization (ISO) (at least one dimension of size range from approximately 1 nm to 100 nm) with regards to the size range of nanomaterials. However, the term "approximately" or "typically" is used to describe the 1 nm to 100 nm size range to reflect that nanorelated properties may be associated with an internal structural property or behavior.

Importantly all agencies point out that the definition also includes "a material that has structure in the 1–100 nm range (*e.g.* porous and fibrous materials; ISO) or "physical or chemical properties or biological effects, that are attributable to its dimension(s), even if these dimensions fall outside the nanoscale range, up to 1 μm" (FDA). Health Canada's Working Definition of Nanomaterial (http://www.hc-sc.gc.ca/sr-sr/pubs/nano/pol-eng.php) also "considers any manufactured substance or product and any component material, ingredient, device, or structure to be nanomaterial if: (a) it is at or within the nanoscale in at least one external dimension, or has internal or surface structure at the nanoscale, or (b) it is *smaller or larger than the nanoscale in all dimensions* and exhibits one or more *nanoscale properties/ phenomena*.

In dermal and transdermal delivery these nanodefinitions provide more grounds for considerations and even deeper questions about the molecular level interactions of chemical penetration enhancers (CPE), drug delivery systems and physical technologies (ballistic, microneedles, ultrasound, electroporation, etc.) with the skin. For example, CPEs or formulations containing particles in the micrometer size range capable of extracting lipids from the stratum corneum can open "pores" of less than 100 nm.

4. Nanoparticle Types Applied onto Skin — Do they permeate and/or Do they Enhance Delivery?

Similarly to small molecules, the potential routes of entry of nanoparticles into the skin could include intercellular, intracellular pathways or through the appendages (*e.g.* mainly trans-follicular). While in case of small molecules, the penetrating ability depends primarily on their molecular weight, o/w partition coefficient and lipophilicity (Fig. 1) and proceeds by diffusion according to Fick's law, nanoparticle penetration into skin may depend on more complex set of parameters.

Large molecules typically are not able to penetrate into the skin and the limit in this regard may be around 500–1000 Da[14] or 5–7 nm to be able to diffuse through the lipidic bilayers of the SC[15,16] or smaller than 36 nm to occupy the aqueous pores within the stratum corneum.[17–19]

Generally, the intracellular penetration route is unlikely to occur with nanoparticles as it would require active transportation and/or partitioning of relatively large particulate matter from cell-to-cell. In case of sweat glands and pilosebaceous units outward secretion of sweat and sebum interferes with permeation, however, in many cases deposition into hair follicles can be demonstrated for liposomes and polymeric types of particles[20–22] making these potential delivery systems for diseases of the appendages. Systemic delivery through the trans-follicular route is not considered a significant opportunity.

The intercellular route of penetration proceeds via the "lipid channels" which are lipids arranged in a highly organized stacked bilayer structure. Many studies have shown that only formulations or enhancers that are able to compromise this organized barrier can facilitate drug delivery. Perturbation of the "lipid channels" may range from slight disorder to complete extraction of the lipids and typically result in correspondingly increased levels of toxicity. Therefore, it is difficult to achieve just the right balance of permeation enhancement-toxicity.

A recent theoretical paper by Watkinson *et al.* discussed the unlikelihood of any nanoparticles permeating into or through the skin based on passive diffusion mechanism,[23] considering the exponential relationship between molecular weight of a compound and permeability coefficient or flux.[24,25] This has been largely confirmed in many studies with a wide range of nanoparticles under conditions that included intact healthy human skin and compatible vehicles.[26–30] However, there are still possibilities for drug delivery by nanocarriers based on other mechanisms of interaction or facilitative action (flexing or stretching the skin).

Nanoparticles can be constructed from a wide variety of materials resulting in many different shapes, sizes, surface features, internal structures and charge. Usually, the combination of physicochemical and structural parameters of a nanoparticle affects its penetration through skin. The set of most important parameters is unique to the type of nanomaterial (Fig. 2).

It appears that from the point of view of intra-cutaneous penetration, nanoparticles could be placed into two major categories, solid particles and structured particles. Solid particles include metal, metal oxide, crystalline, certain polymeric, ceramic and silica particles that have a uniformly homogeneous inside without significant ability to deform or fragment. Structured particles, on the other hand, include particles that are prepared or self-assembled from man-made or (bio)materials such as lipids, peptides, carbohydrates, polymers and carbon and have the capability to deform, fragment, dis- and re-assemble and potentially influence the molecular organization of the skin barrier.

The growing body of literature indicates that variations in physical parameters beyond these two major categories, such as size, surface features and shape have a lesser influence on uptake into the skin.

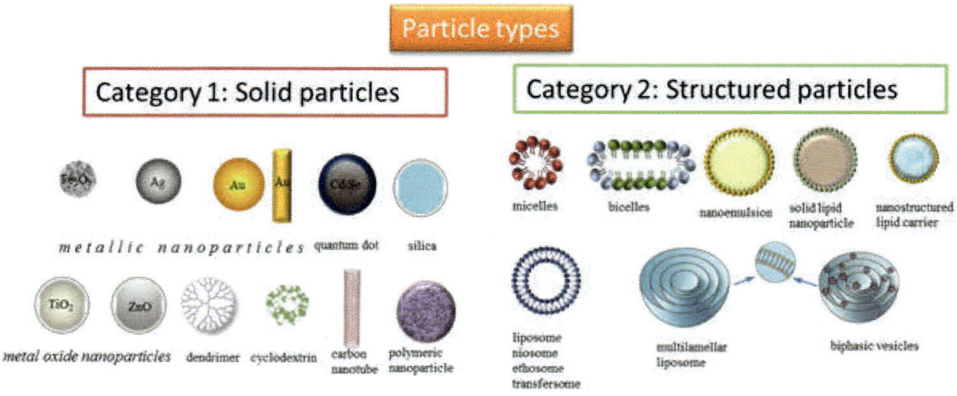

Influence of individual particle properties on skin uptake:	Category 1	Category 2
Size	some effect	limited effect
Surface area	no effect	no effect
Shape	some effect	no effect
Surface charge	medium effect	medium effect
Surface coatings	some effect	some effect
Structure	no significant effect	significant effect

Fig. 2. Nanoparticle types and their relevant properties in skin interactions.

The most effective dermatological treatments match the drug delivery system to the target region of the skin. This maybe a simple approach when for example the target site is on the skin surface as in the case of camouflage or cosmetic preparations, protective films, sunscreens, antifungal or antibacterial formulations. Stratum corneum is the target site for emollients, moisturizers and keratolytics. Topical corticosteroids, anti-inflammatory agents, anticancer drugs, antioxidants, need to reach the viable epidermis and local anesthetics, antihistamines and collagen-blockers/stimulating agents must reach the dermis. Antiperspirants, exfoliants, depilatory agents and antimicrobials target the appendages. Traditional dosage forms, such as emulsions, suspensions, creams and ointments have achieved reasonably good rates of drug delivery, however, their effect reached the maximum and small modifications will not be enough to achieve higher efficiency or more specific or more targeted performance.

4.1. *Solid particles and their applications*

Gold nanoparticles (AuNPs) are versatile materials, typically within the size range of 0.8 nm to 200 nm. The widespread use of AuNP in biological applications has been due to their simple synthesis methods, ease of surface modification with peptides, DNA and antibodies, and unique physicochemical properties such as excellent absorbance and scattering of light. AuNPs have multiple potential therapeutic and diagnostic applications.[31]

When AuNP with size at the lower end of the scale (mean size of 5 nm) was examined in mice for the delivery of a model protein antigen (ovalbumin),[32] Au-NPs were found to cross the stratum corneum layer of the skin and induced temporary openings which facilitated antigen delivery into the viable epidermis and dermis resulting in a measurable immune response.[32]

Penetration after topical application on intact skin is dependent on size,[33,34] but importantly also on the surface properties of AuNPs and potentially on the dispersion medium.[35,36] AuNPs with hydrophobic character even if their size is more than two times greater (6 nm vs 15 nm) penetrated better into deeper layers of the stratum corneum.[35] The main mechanism of AuNP interaction was likely through hydrophobic binding to skin lipids leading to limited distribution within the stratum corneum layers.[34] Larger AuNPs showed even lower levels of penetration into the stratum corneum.[30] In case of spherical siRNA-gold nanoparticle conjugates based on a 13 nm gold core, nanoparticles were detected in the epidermis and dermis of mouse skin *in vivo* within three hours after application in Aquaphore ointment base.[36] Aquaphore ointment is an occlusive vehicle containing petrolatum, mineral oil and ceresin (a beeswax substitute), which may be responsible for the penetration enhancement of siRNA–AuNPs.

Silver nanoparticles (AgNP) are known for their bactericidal properties. The antibacterial potency is dependent on particle size and dispersion status. Results of an electron microscopic study on the penetration of silver nanoparticles through intact and damaged human skin revealed that silver nanoparticles with a mean size of 25 nm penetrated into the stratum corneum and upper layers of the epidermis of damaged skin whereas minimal uptake was seen in intact skin.[37] Another study also confirmed minimal penetration of silver nanoparticles into skin and implied safety.[38] However, the most common side effect associated with prolonged dermal exposure to AgNP is argyria, a permanent bluish-gray discoloration of the skin.[31] This permanent discoloration is not physically harmful but remains an inherent cosmetic problem. AgNP have been used in several biomedical products, including wound or burn dressings, catheters, and bone cement based on the molecular mechanism of silver ions known to cause damage to bacterial DNA, proteins, enzymes, as well as the bacterial cell wall.[39] Dressings and bandages coated with AgNP are used for prevention of sepsis in skin wounds like burns and ulcers.[40]

Magnetic nanoparticles are particles generally made up of iron or its derivatives and are magnetic in nature. They are generally below 100 nm and have many important therapeutic and diagnostic applications in nanomedicine. [41,42] Stabilized iron oxide nanoparticles with hydrophilic surface properties and size around 5–7 nm (measured by transmission electron microscopy [TEM]) were shown to penetrate mostly into the stratum corneum layers, stratum corneum-stratum granulosum boundary and into hair follicles.[37] The study used human abdominal full-thickness skin and was carried out in vertical diffusion cells with PBS as receptor fluid for up to 24 h.[37]

Titanium dioxides and zinc oxides are groups of nanomaterials extensively utilized in sunscreen and sunblock formulations with the intention that the particles stay on the skin surface during use. The size of these particles range between 20–200 nm and they maybe coated or uncoated.[43] Several research groups have shown that these nanoparticles are only found in the outermost layer of the stratum corneum in both intact skin and skin exposed to solar radiation.[26,27,43–47] Some experiments showed that nanoparticles were also deposited in the openings of pilosebaceous follicles.[27] In a recent study TiO2 nanoparticles were applied to hairless rat skin once a day for 56 days in a water-in-oil emulsion which showed that the particles were only located in the stratum corneum layer of epidermis and hair follicles and no particles were detected in the viable epidermis.[27,44,46]

Quantum dots (QDs) are semiconducting nanocrystals with a metal core and additional shell coating. While the core is typically smaller than 10 nm , the overall particle size ranges between 20–40 nm. Recent studies conducted on human skin showed conflicting results regarding QDs' penetration into deeper skin layers and more experiments are needed before reaching a conclusion.[28,48,49] QD penetration of skin depends on several factors such as the type, shape, size, surface charge, however, without flexing the skin most types of QDs primarily remain on the surface of the skin or in the uppermost stratum corneum bilayers and sometimes near hair follicles.[50,51] Biodegradable polymers such as poly(D,L-lactic-co-glycolic acid) (PLGA) have been used to formulate QDs in the form of nanoparticles.[52] The surface coatings by polyethylene glycol or carboxyl groups can greatly influence the depth of skin penetration of the NP.[53] QDs may be used as an alternative to organic fluorophores in some fluorescent probes to image and study biological processes and the penetration of NP in different skin layers.[53] Research has shown that these particles were often able to penetrate the stratum corneum and reach the epidermis and sometimes the dermis.[48,50,51,54] However, a few factors contributed to variations in results among different studies. First, shape plays a role in determining the nanomaterial's penetration potential. By examining only neutral QDs, spherical nanomaterials had the greatest penetration, followed by ellipsoid, and then nail-shaped ones.[50,51,54] Last, most of the contributions used porcine[51,54] and rat skin,[50,55] and the results may differ when one conducts the same experiments with human skin.

Silica-based nanomaterials, including colloidal, sol–gel and mesoporous silica nanoparticles are emerging as a new and promising class of nanoparticles that offer several attractive features for diagnostics and targeted drug delivery.[56] Silica-based nanomaterials, owing to their unique advantages such as small and uniform pore size, large surface area and non-toxic nature, have possible applications in drug delivery systems.[39] The penetration of silica nanoparticles into the skin appears to be size dependent, particles greater than 75 nm do not penetrate, even after skin barrier disruption.[57]

Carbon nanomaterials include carbon nanotubes (CNTs) and spherical fullerenes (buckyballs).[58] A fullerene-nuclear localization peptide conjugate could only penetrate into porcine skin if the skin was continuously flexed for 60–90 min, the uptake was attributed to passive penetration and explained by the relatively small size of individual fullerenes.[59] Other recent studies showed that fullerenes can permeate deep into the stratum corneum and epidermis but did not reach the dermis.[60,61] The dispersing solvent played a significant role in penetration, as significant penetration was shown when nanotubes were dispersed in toluene, cyclohexane or chloroform, but not when dosed in mineral oil.[60] CNTs have recently been investigated for potential topical application and transdermal delivery of antineoplastic drug molecules including doxorubicin (hydrophilic and weak basic drug) and indomethacin (as an example of hydrophobic and weak acidic drug).[62]

Polymeric nanoparticles can be prepared from natural and synthetic polymers for dermatological applications. Both natural and synthetic polymers are widely used in nanoparticle formulations. The widely used synthetic polymers in skin delivery are polyalkylcyanoacylates, poly-lactic acid (PLA), poly-ε-caprolactone (PCL), poly-glycolic acid (PGA) or their co-polymers as poly-lactic-co-glycolic acid (PLGA).[63–65] PLGA nanoparticles can also successfully deliver encapsulated drug into the skin.[66] These capsules are usually oily particles coated with polymers and may carry positive or negative charge. Mannosylated polyethyleneimine nanoparticles (DermaVir)[67] with (average size 100 nm) were shown to be effective dendritic cell targeting of plasmid DNA encoding a combination of antigens in HIV-infected individuals,[68,69] after topical application to skin that was scrubbed to remove the stratum corneum layers.

The ability of polymeric nanoparticles, mainly carboxylate styrene and PLGA particles, to penetrate the stratum corneum was the focus of several studies.[63,70] Campbell *et al.*, using detailed confocal microscopic analysis, showed that fluorescent polystyrene nanoparticles in the size range of 20–200 nm, applied in aqueous suspension, did not penetrate beyond the most superficial layers of stratum corneum.[71] Alvarez-Roman *et al.* applied fluorescent styrene particles of sizes 20 nm and 200 nm on skin using water as the dispensing vehicle for 2 h.[63] Results showed that both particles failed to cross the stratum corneum and were only found on skin surface, although the 20 nm particles reached deeper in hair follicles compared to the 200 nm ones.[63] In other studies, styrene particles, mostly those that are smaller, were shown to penetrate into hair follicles using skin samples only when subjected to massage.[70,72] Another experiment using inflamed skin showed that polymeric nanoparticles of neutral, positive and negative charges all accumulated in pilosebaceous units.[73] It has been also shown that PCL nanoparticles containing sunscreen agents were able to permeate only into the stratum corneum.[64] These experiments illustrated that the rigid polymeric particles are unable to penetrate the stratum corneum but do take on the trans-follicular route of penetration.[70,72,73] They also showed that experimental set-up and application tools have significant impact on experimental outcome.

Dendrimers used in targeted drug delivery are usually 10 nm to 100 nm. Dendrimers form by synthesizing repeated branching around a central core which produces larger and larger particles as the number of generations or layers increase. Poly(amidoamine) (PAMAM) dendrimers are multivalent and can be their surface functional groups can be easily controlled and tailored for specific purposes.[74] It has been recently shown that size, surface charge and hydrophobicity play a role in skin interactions.[75] Smaller PAMAM dendrimers (generation 2 (G2)) penetrate the skin layers more efficiently than the larger ones (G4).[75] PAMAM dendrimers were identified as potential skin permeation enhancers for 5-fluorouracil (5-FU),[76,77] NSAIDs (ketoprofen-PAMAM dendrimer complex and diflunisal-PAMAM dendrimer complex) in *in vitro* permeation studies with excised rat skin.[78] Dendrimers increased skin permeation by increasing the skin partitioning of 5-FU. Chauhan *et al.*[79] reported that steady-state flux and C(max) values were significantly higher with G4-NH2 and G4-OH dendrimer formulations of ibuprofen. PAMAM dendrimers form host-guest complexes with 8-methoxypsoralene (8-MOP), the photosensitizer for PUVA therapy.[80] Interestingly, PAMAM type dendrimers were also shown to act as penetration retardants.[81]

Cyclodextrins (CDs) are cyclic oligosaccharides composed of cyclically linked glucopyranose units with a hydrophilic outer surface and a lipophilic core. The α-, β- and γ-CDs are the most common natural CDs, consisting of 6–8 glucose units.[82] Cyclodextrin can form complexes with other agents such as chitosan,[83] silica particles,[84] PLGA[85] or soybean oil,[86] to enhance the performance of other topical drug delivery systems. CDs have a wide range of applications for dermal delivery of drugs in animal models and humans. Various studies reported that CDs increase the solubility, stability and skin permeation of poor water-soluble drugs like curcumin,[87] ferulic acid[88] and isotretinoin.[82] CDs are also able to enhance the dermal delivery of NSAIDs, such as flurbiprofen,[85] indomethacin,[89] zaltoprofen[90] and piroxicam.[91] A recently published study showed that cyclodextrin could also enhance the stability of the drug formulations such as betamethasone-cyclodextrin complexes encapsulated in liposomes while they did not improve its penetration.[92]

4.2. *Structured particles*

Micelles are the simplest self-assembled systems formed from amphipatic materials. Solubilization of drugs by surfactants enhances thermodynamic properties and can create supersaturated formulations for enhanced drug delivery. Pluronic-based chitosan-conjugated micelles (50–60 nm), in the form of a nanogel, were developed as a transdermal delivery vehicle for hydrophilic proteins. *In vitro* permeation studies using human cadaver skin demonstrated successful transdermal delivery of FITC-BSA (67 kDa) and FITC-Insulin (6 kDa) with cumulative amount absorbed in the order of about 10 μg/cm^{2}in 24 h.[93] The authors speculated that the mechanism of delivery by chitosan-conjugated pluronic surfactant carriers might be due to the electrostatic interaction between chitosan and the negatively charged stratum corneum and to the disturbance of keratin structure or tight junctions in the stratum corneum and epidermis.[93]

A new type of discoidal shape micelles, named bicelles, were developed using two types of lipids, dimyristoylphosphocholine and dihexanoylphosphocholine.[94,95] Bicelles, 15–50 nm in diameter, have the ability to adhere to the skin surface and partly interact with skin lipids and reinforce the lamellar order within the stratum corneum as shown by SAXS studies.[95] These properties could make them suitable candidates as vehicles for skin surface drug reservoirs.

Sub-micron emulsions and nanoemulsions are thermodynamically stable formulations composed of emulsified oil phase in an aqueous external phase.[96] The droplet size is usually under 1 μm and typically smaller than 150 nm, respectively. Several small molecular weight drugs were shown to benefit from formulating them as nanoemulsions. Transdermal delivery of celecoxib[97] and aceclofenac,[98] selective cyclo-oxygenase 2 (COX-2) inhibitors, was enhanced when formulated as a nanoemulsion with an average particle size of 16 nm, using a combination of Sefsol-218 and Triacetin (oil phase), Cremophor-EL or Tween 80 (surfactants) and Transcutol-P (permeation enhancer) in water or Carbopol-940 gel.[97] Both formulations were shown to enhance the bioavailability of celecoxib by about three-fold compared to the oral capsule in a rat model. The authors observed disruption of lipid bilayers in the stratum corneum which may be the reason for the increase in transdermal delivery. Kumar *et al.* used nanoemulsion as vehicle for transdermal delivery of amlodipine.[99] Microemulsion formulation prepared from oleic acid, Tween 80, propylene glycol, and water and ethanol enhanced progesterone transdermal flux compared to the saturated drug solution in 14% aqueous propylene glycol (control).[100] *In vitro* transdermal delivery of carvedilol, a poorly soluble cardiovascular drug was enhanced by a nanoemulsion prepared from oleic acid:isopropyl myristate, Tween 80, Transcutol P and water.[101]

Percutaneous delivery of ropinirole was facilitated by a specific nanoemulsion composition comprised of capryol 90, Tween 20, Carbitol and water[102] and of ketoprofen by a nanoemulsion system with benzyl alcohol, ethanol, non-ionic solubilizers and water *in vitro* through mouse skins.[103] In a recent study, Schwartz *et al.* showed that the transdermal delivery of flufenamic acid solubilized in a nanoemulsion made from a surfactant, glycerol distearate and a neutral oil was enhanced and the cumulative amount in 24 h was 120 $\mu g/cm^2$, whereas from a bicellar system it was retarded.[104]

In most, if not all of these cases, transdermal permeation enhancement was due to being able to use micro- or nanoemulsion technology to incorporate penetration enhancers or create a thermodynamically stable supersaturated system. Particle size is not as relevant in case of emulsions as interaction with skin is related to the emulsion components (surfactants, solvents) diffusing into the stratum corneum and inducing disorder.

Certain compositions appear to localize the drug within the skin without promoting transdermal delivery. For example, a nanoemulsion with droplet size of 70–80 nm was used to improve the topical antifungal efficacy of nystatin by localizing the drug within the skin without systemic absorption.[105] To localize paclitaxel (PCL) deeper skin layers, viable epidermis and dermis while minimizing systemic absorption, a nanoemulsion was formulated for psoriasis treatment.[106] In another study, enhanced penetration of prednicarbate from a positively charged nanoemulsion, containing phytosphingosine to attribute the positive charge, compared to a negatively charged nanoemulsion was recommended for topical use for the treatment of atopic dermatitis.[107]

The mechanism of skin penetration may include release of drug from the emulsion droplets on the surface of the skin,[108] partial penetration of emulsion components, especially surfactants, together with the drug.

Liposomes, uni- or multilamellar vesicles made up of lipid bilayers enclose an aqueous compartments, were first applied in dermal drug delivery by Mezei and his group. They are typically composed of combinations of phospholipids and cholesterol and may be negatively or positively charged.

A wide variety of liposomal formulations have been described in the past 33 years for dermal and transdermal delivery. The early liposome compositions encapsulating small molecules (local anesthetic agents, corticosteroids, prostaglandin, small peptides)[109–119] are continuing to improve by a number of next generation technologies that are suitable for not only low molecular weight dermal drug delivery[120–123] but also for macromolecules such as human epidermal growth factor, low molecular weight heparin and plasmid DNA.[124–126]

Since in most cases, traditional liposomes do not penetrate deep into the skin, but rather remain in the stratum corneum and viable epidermis, newer strategies have been developed to enhance the ability of skin penetration in liposomes.

Niosomes are similar vesicular particles as liposomes, except they are composed of mixtures of non-ionic surfactants, cholesterol and may contain small amounts of phospholipids. Niosomes are flexible particles that may carry a charge.[127] Niosome formulations[128] and amphoteric and cationic liposome compositions[129] have been designed to target the hair follicle route, with some success shown in certain experiments.[130]

Ethosomes were developed to increase the elasticity of the lipid vesicle, which in turn improves skin penetration. They contain around 20–45% of ethanol and usually carry a negative charge with fluid lipid bilayers.[131] Ethanol is an efficient permeation enhancer, increases vesicle flexibility, lipid fluidity and cell membrane permeability.[132] A recent comparative study on ethosomes and liposomes with particle size lower than 200 nm demonstrated that the ethosomal system was a highly effective carrier for tacrolimus delivery to deeper layers of rat skin. In contrast to liposomes, ethosomes are less rigid due to the effects of ethanol which might give the vesicles soft flexible characteristics to more easily penetrate into deeper layers of the skin.[133]

Transfersomes are designed to provide vesicle with elasticity and deformability through supplementation of surfactants. They are considered ultradeformable vesicles that usually carry a negative charge.[134] Transfersomes are able to squeeze through channels smaller than their size, allowing them to spontaneously penetrate intact skin, even with sizes up to 200–300 nm.[135,136] Examples of these types of particles under investigation are anesthetics[137] and anti-inflammatory drugs.[138] Transfersomes can increase skin penetration and permeation small and high molecular weight proteins (glucocorticoids, ketoprofen, albumin and insulin).[139,140]

As mentioned previously, the size of lipid-based nanoparticles generally range above 100 nm, which would indicate that the penetration through the stratum corneum might be unfavorable. The majority of contributions in fact showed that these types of particles may penetrate into the outermost layer of stratum corneum but are mostly unable to permeate deeper,[141,142] except for ethosomes and transfersomes. Ethosomes have been shown to permeate the stratum corneum in some experiments[143] and to improve the release and action of encapsulated drugs.[144,145] Transfersomes, too, has shown to penetrate into skin under non-occluded conditions[110,146,147] as intact vesicles were found in the stratum corneum[147] and some at the boundary of stratum corneum and stratum granulosum.[141]

Solid lipid nanoparticles (SLN) are sub-micron size range carriers (50–1000 nm) composed of a single lipid that is solid at body temperature with crystalline lattice matrices. They are coated with a stabilizing surfactant and thus may carry a positive or negative charge[148] and mostly designed for extended drug delivery from the skin surface. Compared to other types of nanoparticles such as polymeric nanoparticles, emulsions and liposomes, SLN are physically more stable, relatively easier to make and show good encapsulation efficiency. A few of the drugs formulated with SLN are clotrimazole, betamethasone, podophyllotoxin, isotretinoin, vitamin E.[149] In case of griseofulvin SLN (developed for dermatophytosis) the authors found nanoparticles in the skin layers and attributed the penetration of nanoparticles to their nanosize.[150]

Nanostructured lipid carriers (NLC) are made up of both solid and liquid lipids, whose mixture is in general solid at body temperature. As with SLN, they are covered with a stabilizing surfactant and may carry a charge,[148] however, their matrix structure can take on different organizations: an irregular crystalline lattice, an amorphous lipid blend, or a solid lipid matrix containing liquid lipid components.[151] NLC particle size also ranges between 50 nm and 1000 nm and are used in cosmetics and dermatology and often mixed into solutions or cream bases.[148] Topical application of lutein as an antioxidant and photoprotectant was evaluated in both SLN and NLC for dermal delivery 150–350 nm.[152] *In vitro* studies using pig ear skin showed that both SLN and NLC localized lutein within the skin and no systemic absorption was detected. Importantly, physicochemical stability of lutein could be also improved as lutein in both nanocarriers was protected against UV degradation. Highest protection was by SLN (0.06% degradation) followed by NLC (6–8%), nanoemulsion (14%) and lutein powder suspended in corn oil (50%). Resveratrol was formulated as a SLN and NLC made from Compritol 888ATO, Myglyol, Poloxamer188 and Tween80 for dermal applications as an antioxidant.[153] *In vitro* studies showed that NLC delivered resveratrol deeper into the epidermis.

It is evident that both SLN and NLC could be suitable dermal delivery systems, but not because of their small particle size but due to their optimized composition. Mechanistic studies indicated that these formulations formed an occlusive film on the skin surface and were mostly found in the outermost layers of the stratum corneum.[151,154,155] In certain cases vehicles, in which SLN and NLC was dispersed, facilitated deposition of encapsulated drugs in the hair follicles.[156]

Peptide-based nanoparticles is a new group of delivery systems which have only recently been evaluated as potential transdermal carriers. For example, one of the cell penetrating peptides (CPPs), Tat peptide enhanced the skin permeation of celecoxib nanoparticles. A peptide containing 11 arginine amino acids (R11) was used to deliver Spantide II and ketoprofen into the deeper skin layers and thus resulted in reduction in inflammation associated with allergic contact dermatitis.[157] Patlolla *et al.* studied the use of cell penetrating peptides for dermal delivery.[158] TAT peptide was used to coat nanolipid crystal nanoparticles made from Compritol, Miglyol and DOGS-NTA-Ni lipids by hot melt homogenization. Confocal microscopy indicated the penetration of fluorescent dye (DID-oil) into rat skin *in vitro* to a depth of 120 μm, mainly distributing in the hair follicles and epidermis.[158]

Several studies have demonstrated the potential therapeutic application of peptide nanoparticles in animal models. Potential application of peptide nanoparticles in siRNA delivery has been reported recently in the treatment of atopic dermatitis. Topical application of RelA siRNA conjugated to AT1002, a synthetic oligomer capable of reversibly opening tight junctions, on to an atopic dermatitis mouse model reduced ear thickness, the severity of skin disease at the site of application, and cutaneous cytokine and serum IgE levels.[159] A similar peptide-siRNA conjugate against microphthalmia-associated transcription factor (MITF) was used to reduce facial melisma, a common disorder of cutaneous hyperpigmentation.[160]

Mixed lipid-based systems with novel structures include biphasic vesicles recently developed for dermal macromolecule delivery,[161–165] mesophasic systems[166] and gemini nanoparticles.[167–170] Biphasic vesicles are a multicompartmental lipid vesicle system for enhanced encapsulation and cutaneous delivery of drugs formed from a submicron/nano emulsion system and phospholipid bilayers.[163] In addition, the flexible components include viscosity increasing agents and emollients and permeation enhancers, which can be added to the emulsion system or the bilayer forming lipids. The resulting vesicles can be considered a structurally organized form of synergistic mixture of permeation enhancers (Fig. 3). Delivery of proteins, plasmid DNA and various small drugs was enhanced by biphasic vesicles after topical application onto intact skin. The mechanism of uptake and drug delivery is related to molecular organization switching (MOS), a novel nanopathway, characterized by lamellar-to-cubic polymorphic phase change within the stratum corneum intercellular lipid structure.

Gemini nanoparticles are built from a new class of surfactants, dicationic N,N-bis(dimethylalkyl)-α,ω-alkene-diammonium compounds as promising building blocks for non-viral gene delivery systems.[172–176] Gemini surfactants (two interlinked monomeric surfactants) are amphiphilic molecules, composed of at least two hydrophobic tails and two hydrophilic head groups linked by a spacer group. This basic design serves as a template for a number of possible structures.[177,178] An interesting structural feature of gemini nanoparticles is that, more than one type of polymorphic phase (lamellar,

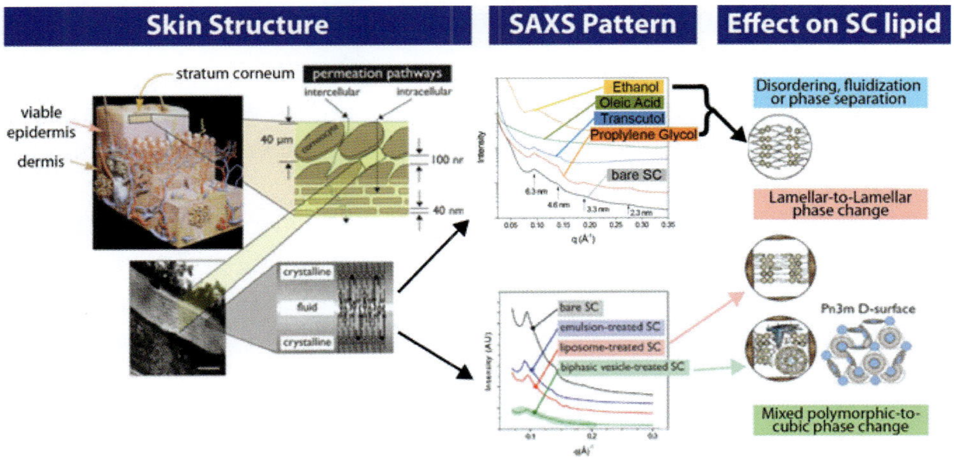

Fig. 3 Mechanism of interaction of biphasic vesicles with human skin. Dermal and transdermal delivery is limited by the outermost layer of the skin called the stratum corneum. Chemical permeation enhancers and drug delivery systems interact with the intercellular lipids in the stratum corneum. This interaction modifies the structural order of the lipids which are originally arranged in multiple bilayer stacks (left column) [electron micrograph and bilayer model reproduced from Ref. 16 with permission]. This modification, fluidization, disorder or rearrangement of the lipids can be monitored by SAXS/WAXS (middle column). Conventional permeation enhancers such as ethanol, oleic acid, propylene glycol and a marketed enhancer, Transcutol, causes disordering of the lipids, but do not disrupt the bilayer configuration (right column). Among the lipid-based delivery systems, liposomes and a sub-micron emulsion also have a disordering effect, whereas biphasic vesicles appear to cause rearrangement of the organization of the stratum corneum lipids into a Pn3m cubic phase configuration (SAXS pattern-middle column, model-right column). This cubic phase could be an intercellular permeation nanopathway that may explain the increased delivery of IFN α by biphasic vesicles.[165] Figure and legend reproduced from Ref. 171 with permission.

hexagonal, cubic phases) can coexist within the particles depending on the gemini surfactant–DNA charge ratio, for example hexagonal-cubic and lamellar-cubic mixed polymorphic structures.[179] This polymorphic flexibility is thought to be responsible for the increased transfection ability and dermal delivery.[167,178,180] In a another recent study reversed hexagonal (HII) mesophase, formed from glycerol mono-oleate and tricaprylin (9:1 weight ratio and 18% water), as solubilization reservoir for diclofenac was used to enhance transdermal delivery.[181] *In vitro* delivery through porcine ear skin was enhanced in the presence of an additional component in the mesophas, penetratin, a cell penetrating peptide.[181]

5. Most Promising Directions and Clinical Advancement of Nanoparticles

Nanotechnology-based formulations first made their way into the market as cosmetic, cosmeceutical and over-the-counter products with significantly improved performance over traditional formulations.

Liposomal formulations of sunscreens, for example Actinica (Spirig Pharma Ltd., Australia) was developed with a combination of six sun filters (but no zinc oxide or titanium dioxide particles) for people with a high risk of skin cancer with an impressively low 0.01% UVA transmission (Australian requirement is less than 10% transmission). A significant benefit of reduced skin irritation was achieved when benzoyl peroxide was incorporated into porous microspheres (NeoBenz® Micro, Intendis, Bayer Group). Silver nanoparticles are employed in antibacterial dressings containing nanocrystalline silver, where silver is incorporated within the dressing rather than being applied as separate silver compounds. SilvaSorb® silver nanoparticle gel or dressing (Acrymed) is available as antimicrobial for scrapes , cuts and infections. Another example of silver dressing are Acticoat-7 (Smith and Newphew, UK), which utilizes a nanocrystalline Silcryst technology with an increased surface area to volume ratio.

Glymesason is a dermal ointment of dexamethasone-β-cyclodextrin, marketed by Fujinaga/Japan.

Dendrimer based nanoparticle formulation by StarPharma (Melbourne, Australia), Vivagel®, is a product for the prevention of recurrent bacterial vaginosis, HIV and HSV infections.[182,183] SPL7013 is active in VivaGel™ and contains a specifically designed polyanionic surface facilitating binding to viruses and blocking viral attachment to cells thereby preventing infection.[184,185]

Estrasorb® (Novavax) estradiol hemihydrate micellar nanoparticles[186] and Elestin® (BioSante) (estradiol gel with calcium phosphate nanoparticles) for menopausal symptoms were developed. Micellar nanoparticle estradiol emulsion (MNPEE) is FDA approved for the transdermal delivery of estrogen, used for treatment of menopause related vasomotor symptoms.[187]

Several clinical trials were conducted for dermal technologies involving nanosized particles, for example, a nano all-transretinoic acid and hydroquinon combination[188] and a nanoemulsion of adapalene and clindamycin for acne vulgaris.[189] Peptide nanoparticle therapeutics that have reached clinical trials are mainly used for topical use toward scarring, psoriasis, pain, wrinkles and excessive sweating, although none of them has been approved by FDA yet.

A topically applied botulinum toxin type A (BoNT-A), RT001 (Revance Therapeutics Inc, USA), is currently being studied for the treatment of lateral canthal lines and hyperhidrosis.[190] RT001 uses a TransMTS technology, a purified 150 kDa BoNT-A molecule combined with a positively charged lysine oligopeptide and TAT peptide domains, which facilitates percutaneous delivery without altering the function of BoNT-A. [191] Recently, published clinical trial reported that RT001 was a safe and well-tolerated treatment for improvement of lateral canthal lines. AZX-100 (Capstone Therapeutics, USA) is a phosphopeptide analog of heat shock protein (HSP20) that has been attached to a protein transduction domain to allow penetration into cells. AZX100 relaxes smooth muscles after topical application. It has been evaluated in Phase II trials for prevention of dermal/keloid scarring.[192,193]

DermaVir, developed by Genetic Immunity, is a topical polymeric nanoparticle system for HIV vaccination. In Phase II clinical trials it showed excellent safety and immunogenicity.[68]

Biphasic vesicle technology (Biphasix™) is a non-invasive topical delivery system for IFNα. In one of several Phase II studies, 41 women with cytologically confirmed, HPV-induced low-grade squamous intraepithelial lesions (LSIL, PapIIw to Pap IIID) of the cervix were studied across four clinical sites. The preclinical development of Biphasix™-IFNα indicated excellent dermal safety and product stability. Clinical trial results in patients with LSIL/CIN1 indicated that 46.7% of the women in the treated per-protocol population had their abnormal Pap smears revert to normal during the 12 week period, compared with only 15.8% of the untreated women. Topical Interferon Alpha-2b for treatment of patients with cervical intraepithelial neoplasia grade 1 or 2 lesions (CIN 1 or CIN 2, respectively) is entering Phase III clinical trials.

6. Conclusions and Outlook

This article has focused on nanotechnology trends in dermal and transdermal drug delivery. Nanoparticles built from a wide variety of materials offer novel benefits in dermatology and systemic drug delivery. The technologies for these new products constitute a translation of their capabilities to the clinic.

Acknowledgments

This paper was supported by grants from the Canadian Institutes of Health Research (CIHR) and the Natural Sciences and Engineering Research Council of Canada (NSERC). NSERC Undergraduate Student Research

Award for C. Li is highly appreciated. The generous support of the Canada Research Chairs Program, the Canada Foundation for Innovation and the Ontario Research Fund for M. Foldvari is gratefully acknowledged.

References

1. P. M. Elias, Stratum corneum defensive functions: an integrated view. *J. Invest. Dermatol.* **125**, 183–200 (2005).
2. J. Hadgraft and M. E. Lane, Skin permeation: the years of enlightenment. *Int. J. Pharm.* **305**, 2–12 (2005).
3. P. M. Elias, Epidermal lipids, barrier function, and desquamation. *J. Invest. Dermatol.* **80**, 44s–49s (1983).
4. W. Abraham and D. T. Downing, Lamellar structures formed by stratum corneum lipids *in vitro*: a deuterium nuclear magnetic resonance (NMR) study. *Pharm. Res.* **9**, 1415–1421 (1992).
5. K. Tojo, Random brick model for drug transport across stratum corneum. *J. Pharm. Sci.* **76**, 889–891 (1987).
6. N. Y. Schurer, G. Plewig and P. M. Elias, Stratum corneum lipid function. *Dermatologica* **183**, 77–94 (1991).
7. M. A. Lampe, M. L. Williams and P. M. Elias, Human epidermal lipids: characterization and modulations during differentiation. *J. Lipid Res.* **24**, 131–140 (1983).
8. P. W. Wertz, Lipids and barrier function of the skin. *Acta Derm Venereol. Suppl. (Stockh)* **208**, 7–11 (2000).
9. J. A. Bouwstra, G. S. Gooris, F. E. Dubbelaar and M. Ponec, Phase behavior of lipid mixtures based on human ceramides: coexistence of crystalline and liquid phases. *J. Lipid Res.* **42**, 1759–1770 (2001).
10. L. Landmann, Epidermal permeability barrier: transformation of lamellar granule-disks into intercellular sheets by a membrane-fusion process, a freeze-fracture study. *J. Invest. Dermatol.* **87**, 202–209 (1986).
11. L. Norlen, Skin barrier formation: the membrane folding model. *J. Invest. Dermatol.* **117**, 823–829 (2001).
12. K. J. Robson, M. E. Stewart, S. Michelsen, N. D. Lazo and D. T. Downing, 6-Hydroxy-4-sphingenine in human epidermal ceramides. *J. Lipid Res.* **35**, 2060–2068 (1994).
13. J. A. Bouwstra, P. L. Honeywell-Nguyen, G. S. Gooris and M. Ponec, Structure of the skin barrier and its modulation by vesicular formulations. *Prog. Lipid Res.* **42**, 1–36 (2003).
14. J. D. Bos and M. M. Meinardi, The 500 Dalton rule for the skin penetration of chemical compounds and drugs. *Exp. Dermatol.* **9**, 165–169 (2000).
15. M. E. Johnson, D. Blankschtein and R. Langer, Evaluation of solute permeation through the stratum corneum: lateral bilayer diffusion as the primary transport mechanism. *J. Pharm. Sci.* **86**, 1162–1172 (1997).
16. J. A. Bouwstra and M. Ponec, The skin barrier in healthy and diseased state. *Biochim. Biophys. Acta* **1758**, 2080–2095 (2006).
17. G. Cevc, Lipid vesicles and other colloids as drug carriers on the skin. *Adv. Drug Deliv. Rev.* **56**, 675–711 (2004).

18. H. Tang, S. Mitragotri, D. Blankschtein and R. Langer, Theoretical description of transdermal transport of hydrophilic permeants: application to low-frequency sonophoresis. *J. Pharm. Sci.* **90**, 545–568 (2001).
19. R. O. Potts and M. L. Francoeur, The influence of stratum corneum morphology on water permeability. *J. Invest. Dermatol.* **96**, 495–499 (1991).
20. R. Toll *et al.*, Penetration profile of microspheres in follicular targeting of terminal hair follicles. *J. Invest. Dermatol.* **123**, 168–176 (2004).
21. S. Dokka, S. R. Cooper, S. Kelly, G. E. Hardee and J. G. Karras, Dermal delivery of topically applied oligonucleotides via follicular transport in mouse skin. *J. Invest. Dermatol.* **124**, 971–975 (2005).
22. B. Illel, H. Schaefer, J. Wepierre and O. Doucet, Follicles play an important role in percutaneous absorption. *J. Pharm. Sci.* **80**, 424–427 (1991).
23. A. C. Watkinson, A. L. Bunge, J. Hadgraft and M. E. Lane, Nanoparticles do not penetrate human skin-a theoretical perspective. *Pharm. Res.* **30**, 1943–1946 (2013).
24. R. O. Potts and R. H. Guy, A predictive algorithm for skin permeability: the effects of molecular size and hydrogen bond activity. *Pharm. Res.* **12**, 1628–1633 (1995).
25. B. M. Magnusson, Y. G. Anissimov, S. E. Cross and M. S. Roberts, Molecular size as the main determinant of solute maximum flux across the skin. *J. Invest. Dermatol.* **122**, 993–999 (2004).
26. A. V. Zvyagin *et al.*, Imaging of zinc oxide nanoparticle penetration in human skin *in vitro* and *in vivo*. *J. Biomed. Opt.* **13**, 064 031 (2008).
27. P. Filipe *et al.*, Stratum corneum is an effective barrier to TiO2 and ZnO nanoparticle percutaneous absorption. *Skin Pharmacol. Physiol.* **22**, 266–275 (2009).
28. T. Gratieri *et al.*, Penetration of quantum dot particles through human skin. *J. Biomed. Nanotechnol.* **6**, 586–595 (2010).
29. F. Larese Filon *et al.*, Human skin penetration of gold nanoparticles through intact and damaged skin. *Nanotoxicology* **5**, 493–501 (2011).
30. D. C. Liu *et al.*, The human stratum corneum prevents small gold nanoparticle penetration and their potential toxic metabolic consequences. *J. Nanomater.* **2012**, 8 (2012).
31. A. M. Schrand *et al.*, Metal-based nanoparticles and their toxicity assessment. *Wiley Interdiscip. Rev. Nanomed. Nanobiotechnol.* **2**, 544–568 (2010).
32. Y. Huang, *et al.*, Co-administration of protein drugs with gold nanoparticles to enable percutaneous delivery. *Biomaterials* **31**, 9086–9091 (2010).
33. G. Sonavane *et al.*, *In vitro* permeation of gold nanoparticles through rat skin and rat intestine: effect of particle size. *Colloids Surf. B. Biointerfaces* **65**, 1–10 (2008).
34. D. C. Liu *et al.*, The human stratum corneum prevents small gold nanoparticle penetration and their potential toxic metabolic consequences. *J. Nanomater.* (2012).
35. H. I. Labouta, L. K. el-Khordagui, T. Kraus and M. Schneider, Mechanism and determinants of nanoparticle penetration through human skin. *Nanoscale* **3**, 4989–4999 (2011).
36. D. Zheng *et al.*, Topical delivery of siRNA-based spherical nucleic acid nanoparticle conjugates for gene regulation. *Proc. Natl. Acad. Sci. USA* **109**, 11975–11980 (2012).
37. B. Baroli *et al.*, Penetration of metallic nanoparticles in human full-thickness skin. *J. Invest. Dermatol.* **127**, 1701–1712 (2007).
38. T. W. Prow *et al.*, Nanoparticles and microparticles for skin drug delivery. *Adv. Drug Deliv. Rev.* **63**, 470–491 (2011).

39. R. Saraceno, A. Chiricozzi, M. Gabellini and S. Chimenti, Emerging applications of nanomedicine in dermatology. *Skin Res. Technol.* **19**, e13–19 (2013).
40. R. Bhattacharya and P. Mukherjee, Biological properties of "naked" metal nanoparticles. *Adv. Drug Deliv. Rev.* **60**, 1289–1306 (2008).
41. Royal Society and Royal Academy of Engineering on Nanotechnology and Nanoscience. Available at: http:// www.nanotec.org.uk/finalReport.htm (22 November 2008).
42. S. K. Jain *et al.*, Solid lipid nanoparticles bearing flurbiprofen for transdermal delivery. *Drug Deliv.* **12**, 207–215 (2005).
43. G. J. Nohynek, J. Lademann, C. Ribaud and M. S. Roberts, Grey goo on the skin? Nanotechnology, cosmetic and sunscreen safety. *Crit. Rev. Toxicol.* **37**, 251–277 (2007).
44. S. E. Cross *et al.*, Human skin penetration of sunscreen nanoparticles: *in-vitro* assessment of a novel micronized zinc oxide formulation. *Skin Pharmacol. Physiol.* **20**, 148–154 (2007).
45. C. Miquel-Jeanjean *et al.*, Penetration study of formulated nanosized titanium dioxide in models of damaged and sun-irradiated skins. *Photochem. Photobiol.* **88**, 1513–1521 (2012).
46. K. Adachi, N. Yamada, Y. Yoshida and O. Yamamoto, Subchronic exposure of titanium dioxide nanoparticles to hairless rat skin. *Exp. Dermatol.* **22**, 278–283 (2013).
47. N. Sadrieh *et al.*, Lack of significant dermal penetration of titanium dioxide from sunscreen formulations containing nano- and submicron-size TiO2 particles. *Toxicol. Sci.* **115**, 156–166 (2010).
48. A. Gartner, L. Collin and G. Lalli, Nucleofection of primary neurons. *Methods Enzymol.* **406**, 374–388 (2006).
49. T. W. Prow *et al.*, Quantum dot penetration into viable human skin. *Nanotoxicology* **6**, 173–185 (2012).
50. L. W. Zhang and N. A. Monteiro-Riviere, Assessment of quantum dot penetration into intact, tape-stripped, abraded and flexed rat skin. *Skin Pharmacol. Physiol.* **21**, 166–180 (2008).
51. L. W. Zhang, W. W. Yu, V. L. Colvin and N. A. Monteiro-Riviere, Biological interactions of quantum dot nanoparticles in skin and in human epidermal keratinocytes. *Toxicol. Appl. Pharmacol.* **228**, 200–211 (2008).
52. P. W. Lee *et al.*, Multifunctional core-shell polymeric nanoparticles for transdermal DNA delivery and epidermal Langerhans cells tracking. *Biomaterials* **31**, 2425–2434 (2010).
53. L. W. Zhang and N. A. Monteiro-Riviere, Use of confocal microscopy for nanoparticle drug delivery through skin. *J. Biomed. Opt.* **18**, 061214–061214 (2013).
54. J. P. Ryman-Rasmussen, J. E. Riviere and N. A. Monteiro-Riviere, Penetration of intact skin by quantum dots with diverse physicochemical properties. *Toxicol. Sci.* **91**, 159–165 (2006).
55. V. Kulvietis, E. Zurauskas and R. Rotomskis, Distribution of polyethylene glycol coated quantum dots in mice skin. *Exp. Dermatol.* **22**, 157–159 (2013).
56. B. Fadeel and A. E. Garcia-Bennett, Better safe than sorry: understanding the toxicological properties of inorganic nanoparticles manufactured for biomedical applications. *Adv. Drug Deliv. Rev.* **62**, 362–374 (2010).
57. F. Rancan *et al.*, Skin penetration and cellular uptake of amorphous silica nanoparticles with variable size, surface functionalization, and colloidal stability. *ACS Nano* **6**, 6829–6842 (2012).

58. M. Foldvari and M. Bagonluri, Carbon nanotubes as functional excipients for nanomedicines: I. pharmaceutical properties. *Nanomedicine* **4**, 173–182 (2008).
59. J. G. Rouse, J. Yang, J. P. Ryman-Rasmussen, A. R. Barron and N. A. Monteiro-Riviere, Effects of mechanical flexion on the penetration of fullerene amino acid-derivatized peptide nanoparticles through skin. *Nano lett.* 7, 155–160 (2007).
60. X. R. Xia, N. A. Monteiro-Riviere and J. E. Riviere, Skin penetration and kinetics of pristine fullerenes (C60) topically exposed in industrial organic solvents. *Toxicol. Appl. Pharmacol.* **242**, 29–37 (2010).
61. S. Kato, H. Aoshima, Y. Saitoh and N. Miwa, Biological safety of LipoFullerene composed of squalane and fullerene-C60 upon mutagenesis, photocytotoxicity, and permeability into the human skin tissue. *Basic Clin. Pharmacol. Toxicol.* **104**, 483–487 (2009).
62. I. T. Degim, D. J. Burgess and F. Papadimitrakopoulos, Carbon nanotubes for transdermal drug delivery. *J. Microencapsul.* **27**, 669–681 (2010).
63. R. Alvarez-Roman, A. Naik, Y. N. Kalia, R. H. Guy and H. Fessi, Skin penetration and distribution of polymeric nanoparticles. *J. Control. Release* **99**, 53–62 (2004).
64. R. Alvarez-Roman, A. Naik, Y. N. Kalia, R. H. Guy and H. Fessi, Enhancement of topical delivery from biodegradable nanoparticles. *Pharm. Res.* **21**, 1818–1825 (2004).
65. F. Rancan *et al.*, Investigation of polylactic acid (PLA) nanoparticles as drug delivery systems for local dermatotherapy. *Pharm. Res.* **26**, 2027–2036 (2009).
66. J. Luengo *et al.*, Influence of nanoencapsulation on human skin transport of flufenamic acid. *Skin Pharmacol. Physiol.* **19**, 190–197 (2006).
67. J. Lisziewicz *et al.*, DermaVir: a novel topical vaccine for HIV/AIDS. *J. Invest. Dermatol.* **124**, 160–169 (2005).
68. J. Lisziewicz *et al.*, Single dermavir immunization: dose-dependent expansion of precursor/memory T cells against all HIV antigens in HIV-1 infected individuals. *PLoS One* 7, e35416 (2012).
69. J. Lisziewicz, L. Kelly and F. Lori, Topical dermavir vaccine targeting dendritic cells. *Curr. Drug. Deliv.* **3**, 83–88 (2006).
70. J. Lademann *et al.*, Nanoparticles — An efficient carrier for drug delivery into the hair follicles. *Eur. J. Pharm. Biopharm.* **66**, 159–164 (2007).
71. C. S. Campbell, L. R. Contreras-Rojas, M. B. Delgado-Charro and R. H. Guy, Objective assessment of nanoparticle disposition in mammalian skin after topical exposure. *J. Control. Release* **162**, 201–207 (2012).
72. A. Vogt *et al.*, 40 nm, but not 750 or 1,500 nm, nanoparticles enter epidermal CD1a+ cells after transcutaneous application on human skin. *J. Invest. Dermatol* **126**, 1316–1322 (2006).
73. M. M. Abdel-Mottaleb, B. Moulari, A. Beduneau, Y. Pellequer and A. Lamprecht, Surface-charge-dependent nanoparticles accumulation in inflamed skin. *J. Pharm. Sci.* **101**, 4231–4239 (2012).
74. J. H. Myung, K. A. Gajjar, J. Saric, D. T. Eddington and S. Hong, Dendrimer-mediated multivalent binding for the enhanced capture of tumor cells. *Angew. Chem. Int. Ed. Engl.* **50**, 11769–11772 (2011).
75. Y. Yang *et al.*, Effect of size, surface charge, and hydrophobicity of poly(amidoamine) dendrimers on their skin penetration. *Biomacromolecules* **13**, 2154–2162 (2012).
76. V. V. Venuganti and O. P. Perumal, Poly(amidoamine) dendrimers as skin penetration enhancers: influence of charge, generation, and concentration. *J. Pharm. Sci.* **98**, 2345–2356 (2009).

77. V. V. Venuganti, P. Sahdev, M. Hildreth, X. Guan and O. Perumal, Structure-skin permeability relationship of dendrimers. *Pharm. Res.* **28**, 2246–2260 (2011).
78. Y. Cheng *et al.*, Transdermal delivery of nonsteroidal anti-inflammatory drugs mediated by polyamidoamine (PAMAM) dendrimers. *J. Pharm. Sci.* **96**, 595–602 (2007).
79. A. S. Chauhan *et al.*, Dendrimer-mediated transdermal delivery: enhanced bioavailability of indomethacin. *J Control. Release* **90**, 335–343 (2003).
80. K. Borowska, S. Wolowiec, K. Glowniak, E. Sieniawska and S. Radej, Transdermal delivery of 8-methoxypsoralene mediated by polyamidoamine dendrimer G2.5 and G3.5 — *In vitro* and *in vivo* study. *Int. J. Pharm.* **436**, 764–770 (2012).
81. H. R. Moghimi, R. Varshochian, F. Kobarfard and M. Erfan, Reduction of percutaneous absorption of toxic chemicals by dendrimers. *Cutan. Ocul. Toxicol.* **29**, 34–40 (2010).
82. S. Fireman, O. Toledano, K. Neimann, N. Loboda and N. Dayan, A look at emerging delivery systems for topical drug products. *Dermatol. Ther.* **24**, 477–488 (2011).
83. H. Ammar, H. Salama, S. El-Nahhas and H. Elmotasem, Design and evaluation of chitosan films for transdermal delivery of glimepiride. *Curr. Drug deliv.* **5**, 290–298 (2008).
84. S. Hbaieb, R. Kalfat and Y. Chevalier, Loading antifungal drugs onto silica particles grafted with cyclodextrins by means of inclusion complex formation at the solid surface. *Int. J. Pharm.* (2012).
85. E. Vega *et al.*, Flurbiprofen PLGA-PEG nanospheres: role of hydroxy-β- cyclodextrin on *ex vivo* human skin permeation and *in vivo* topical anti-inflammatory efficacy. C*olloids Surf. B. Biointerfaces.* (2013).
86. M. Hamoudi, J. Saunier, C. Gueutin, E. Fattal and A. Bochot, Beads made of α-cyclodextrin and soybean oil: the drying method influences bead properties and drug release. *Drug Dev. Ind. Pharm.* 1–9 (2012).
87. R. Ghanghoria, P. Kesharwani, H. B. Agashe and N. Jain, Transdermal delivery of cyclodextrin-solubilized curcumin. *Drug Deliv. Transl. Res.* 1–14 (2013).
88. D. Monti *et al.*, Permeation and distribution of ferulic acid and its α-cyclodextrin complex from different formulations in hairless rat skin. *AAPS Pharm Sci Tech.* **12**, 514–520 (2011).
89. S. Jambhekar, R. Casella and T. Maher, The physicochemical characteristics and bioavailability of indomethacin from beta-cyclodextrin, hydroxyethyl-beta-cyclodextrin, and hydroxypropyl-beta-cyclodextrin complexes. *Int. J. Pharm.* **270**, 149–166 (2004).
90. J. S. Baek *et al.*, Enhanced transdermal drug delivery of zaltoprofen using a novel formulation. *Int. J. Pharm.* (2013).
91. A. Doliwa, S. Santoyo and P. Ygartua, Transdermal iontophoresis and skin retention of piroxicam from gels containing piroxicam: Hydroxypropyl-β-cyclodextrin complexes. *Drug Dev. Ind. Pharm.* **27**, 751–758 (2001).
92. S. C. Raposo, S. D. Simões, A. J. Almeida and H. M. Ribeiro, Advanced systems for glucocorticoids' dermal delivery. *Expert Opin. Drug Deliv.* **10**, 857–877 (2013).
93. W. I. Choi *et al.*, Efficient skin permeation of soluble proteins via flexible and functional nano-carrier. *J Control. Release* **157**, 272–278 (2012).
94. G. Rodriguez *et al.*, Bicellar systems to modify the phase behaviour of skin stratum corneum lipids. *Phys. Chem. Chem. Phys.* **14**, 14523–14533 (2012).
95. G. Rodriguez, *et al.*, A unique bicellar nanosystem combining two effects on stratum corneum lipids. *Mol. Pharm.* **9**, 482–491 (2012).

96. M. Changez, M. Varshney, J. Chander and A. K. Dinda, Effect of the composition of lecithin/n-propanol/isopropyl myristate/water microemulsions on barrier properties of mice skin for transdermal permeation of tetracaine hydrochloride: *in vitro*. *Colloids Surf. B. Biointerfaces* **50**, 18–25 (2006).
97. F. Shakeel, S. Baboota, A. Ahuja, J. Ali and S. Shafiq, Skin permeation mechanism and bioavailability enhancement of celecoxib from transdermally applied nanoemulsion. *J. Nanobiotechnol.* **6**, 8 (2008).
98. F. Shakeel *et al.*, Nanoemulsions as vehicles for transdermal delivery of aceclofenac. *AAPS Pharm. Sci. Tech.* **8**, E104 (2007).
99. D. Kumar, M. Aqil, M. Rizwan, Y. Sultana and M. Ali, Investigation of a nanoemulsion as vehicle for transdermal delivery of amlodipine. *Pharmazie* **64**, 80–85 (2009).
100. G. M. El Maghraby, Occlusive and non-occlusive application of microemulsion for transdermal delivery of progesterone: mechanistic studies. *Sci. Pharm.* **80**, 765–778 (2012).
101. N. Dixit, K. Kohli and S. Baboota, Nanoemulsion system for the transdermal delivery of a poorly soluble cardiovascular drug. *PDA J. Pharm. Sci. Technol.* **62**, 46–55 (2008).
102. A. Azeem, F. J. Ahmad, R. K. Khar and S. Talegaonkar, Nanocarrier for the transdermal delivery of an antiparkinsonian drug. *AAPS Pharm. Sci. Tech.* **10**, 1093–1103 (2009).
103. B. S. Kim, M. Won, K. M. Lee and C. S. Kim, *In vitro* permeation studies of nanoemulsions containing ketoprofen as a model drug. *Drug Deliv.* **15**, 465–469 (2008).
104. L. Rubio *et al.*, Bicellar systems as new delivery strategy for topical application of flufenamic acid. *Int. J. Pharm.* **444**, 60–69 (2013).
105. F. Fernandez-Campos, B. Clares Naveros, O. Lopez Serrano, C. Alonso Merino and A.C. Calpena Campmany, Evaluation of novel nystatin nanoemulsion for skin candidosis infections. *Mycoses* **56**, 70–81 (2013).
106. S. Khandavilli and R. Panchagnula Nanoemulsions as versatile formulations for paclitaxel delivery: peroral and dermal delivery studies in rats. *J. Invest. Dermatol.* **127**, 154–162 (2007).
107. Y. Baspinar and H. H. Borchert, Penetration and release studies of positively and negatively charged nanoemulsions — Is there a benefit of the positive charge? *Int. J. Pharm.***430**, 247–252 (2012).
108. López-Quintela MA, Quibén J, Rivas J. Use of microemulsions in the production of nanostruc — tured materials. In *Industrial applications of microemulsions*, C. Solans, H. Kunieda (Eds.). New York: Marcel Dekker, Inc. (1997), pp. 247–265.
109. M. Foldvari, C. Oguejiofor, S. Afridi, T. Kudel and T. Wilson, Liposome encapsulated prostaglandin E1 in erectile dysfunction: correlation between *in vitro* delivery through foreskin and efficacy in patients. *Urology* **52**, 838–843 (1998).
110. G. Cevc, D. Gebauer, J. Stieber, A. Schatzlein and G. Blume, Ultraflexible vesicles, Transfersomes, have an extremely low pore penetration resistance and transport therapeutic amounts of insulin across the intact mammalian skin. *Biochim. Biophys. Acta* **1368**, 201–215 (1998).
111. D. Fleisher *et al.*, Topical delivery of growth hormone releasing peptide using liposomal systems: an *in vitro* study using hairless mouse skin. *Life Sci.* **57**, 1293–1297 (1995).
112. M. Foldvari, *In vitro* cutaneous and percutaneous delivery and *in vivo* efficacy of tetracaine from liposomal and conventional vehicles. *Pharm. Res.* **11**, 1593–1598 (1994).
113. E. Touitou *et al.*, Dyphylline liposomes for delivery to the skin. *J. Pharm. Sci.* **81**, 131–134 (1992).

114. I. Abraham, J. C. Hilchie and M. Mezei, Pharmacokinetic profile of intravenous liposomal triamcinolone acetonide in the rabbit. *J. Pharm. Sci.* **72**, 1412–1415 (1983).
115. M. Mezei and V. Gulasekharam, Liposomes — A selective drug delivery system for the topical route of administration: gel dosage form. *J. Pharm. Pharmacol.* **34**, 473–474 (1982).
116. M. Mezei and V. Gulasekharam, Liposomes — A selective drug delivery system for the topical route of administration. Lotion dosage form. *Life Sci.* **26**, 1473–1477 (1980).
117. J. Lasch and W. Wohlrab, Liposome-bound cortisol: a new approach to cutaneous therapy. *Biomed. Biochim. Acta* **45**, 1295–1299 (1986).
118. A. Gesztes and M. Mezei, Topical anesthesia of the skin by liposome-encapsulated tetracaine. *Anesth. Analg.* **67**, 1079–1081 (1988).
119. H. C. Korting, H. Zienicke, M. Schafer-Korting and O. Braun-Falco, Liposome encapsulation improves efficacy of betamethasone dipropionate in atopic eczema but not in psoriasis vulgaris. *Eur. J. Clin. Pharmacol.* **39**, 349–351 (1990).
120. I. Scognamiglio *et al.*, Nanocarriers for topical administration of resveratrol: a comparative study. *Int. J. Pharm.* **440**, 179–187 (2013).
121. M. A. El-Nabarawi, E. R. Bendas, R. T. El Rehem and M. Y. Abary, Transdermal drug delivery of paroxetine through lipid-vesicular formulation to augment its bioavailability. *Int. J. Pharm.* **443**, 307–317 (2013).
122. S. Duangjit, P. Opanasopit, T. Rojanarata and T. Ngawhirunpat, Evaluation of meloxicam-loaded cationic transfersomes as transdermal drug delivery carriers. *AAPS Pharm. Sci. Tech.* **14**, 133–140 (2013).
123. C. Balaguer-Fernandez *et al.*, Elastic vesicles of sumatriptan succinate for transdermal administration: characterization and *in vitro* permeation studies. *J. Liposome Res.* **21**, 55–59 (2011).
124. S. O. Jeon *et al.*, Enhanced percutaneous delivery of recombinant human epidermal growth factor employing nano-liposome system. *J. Microencapsul.* **29**, 234–241 (2012).
125. Y. K. Song, S. Y. Hyun, H. T. Kim, C. K. Kim and J. M. Oh, Transdermal delivery of low molecular weight heparin loaded in flexible liposomes with bioavailability enhancement: comparison with ethosomes. *J. Microencapsul.* **28**, 151–158 (2011).
126. J. Li *et al.*, Gene therapy for psoriasis in the K14-VEGF transgenic mouse model by topical transdermal delivery of interleukin-4 using ultradeformable cationic liposome. *J. Gene Med.* **12**, 481–490 (2010).
127. M. J. Choi and H. I. Maibach, Liposomes and niosomes as topical drug delivery systems. *Skin Pharmacol. Physiol.* **18**, 209–219 (2005).
128. S. N. Ciotti and N. Weiner, Follicular liposomal delivery systems. *J. Liposome Res.* **12**, 143–148 (2002).
129. S. Jung *et al.*, Innovative liposomes as a transfollicular drug delivery system: penetration into porcine hair follicles. *J. Invest. Dermatol.* **126**, 1728–1732 (2006).
130. S. Jung *et al.*, Strategy of topical vaccination with nanoparticles. *J. Biomed. Opt.* **14**, 021001 (2009).
131. M. M. Elsayed, O. Y. Abdallah, V. F. Naggar and N. M. Khalafallah, Lipid vesicles for skin delivery of drugs: reviewing three decades of research. *Int. J. Pharm.* **332**, 1–16 (2007).
132. B. Godin and E. Touitou, Ethosomes: new prospects in transdermal delivery. *Crit. Rev. Ther. Drug Carrier Syst.* **20** (2003).

133. G. Li *et al.*, Preparation and *in vitro* evaluation of tacrolimus-loaded ethosomes. *Scientific World Journal* **2012**, 874–053 (2012).
134. G. Cevc, Transfersomes, liposomes and other lipid suspensions on the skin: permeation enhancement, vesicle penetration, and transdermal drug delivery. *Crit. Rev. Ther. Drug Carrier Syst.* **13**, 257–388 (1996).
135. G. Cevc and G. Blume, Lipid vesicles penetrate into intact skin owing to the transdermal osmotic gradients and hydration force. *Biochim. Biophys. Acta* **1104**, 226–232 (1992).
136. G. Cevc and D. Gebauer, Hydration-driven transport of deformable lipid vesicles through fine pores and the skin barrier. *Biophys. J.* **84**, 1010–1024 (2003).
137. A. Taddio, H. K. Soin, S. Schuh, G. Koren and D. Scolnik, Liposomal lidocaine to improve procedural success rates and reduce procedural pain among children: a randomized controlled trial. *CMAJ* **172**, 1691–1695 (2005).
138. D. Paolino, G. Lucania, D. Mardente, F. Alhaique and M. Fresta, Ethosomes for skin delivery of ammonium glycyrrhizinate: *in vitro* percutaneous permeation through human skin and in vivo anti-inflammatory activity on human volunteers. *J. Control. Release* **106**, 99–110 (2005).
139. H. C. Korting and M. Schäfer-Korting, Carriers in the topical treatment of skin disease. *Handb. Exp. Pharmacol.* **197**, 435–468 (2010).
140. J. Malakar, S. O. Sen, A. K. Nayak and K. K. Sen, Formulation, optimization and evaluation of transferosomal gel for transdermal insulin delivery. *Saudi Pharm. J.* **20**, 355–363 (2012).
141. P. L. Honeywell-Nguyen, G. S. Gooris and J. A. Bouwstra, Quantitative assessment of the transport of elastic and rigid vesicle components and a model drug from these vesicle formulations into human skin in vivo. *J. Invest. Dermatol.* **123**, 902–910 (2004).
142. D. D. Verma, S. Verma, G. Blume and A. Fahr, Liposomes increase skin penetration of entrapped and non-entrapped hydrophilic substances into human skin: a skin penetration and confocal laser scanning microscopy study. *Eur. J. Pharm. Biopharm.* **55**, 271–277 (2003).
143. S. S. Bodade, K. S. Shaikh, M. S. Kamble and P. D. Chaudhari, A study on ethosomes as mode for transdermal delivery of an antidiabetic drug. *Drug Deliv.* **20**, 40–46 (2013).
144. D. Ainbinder, D. Paolino, M. Fresta and E. Touitou, Drug delivery applications with ethosomes. *J. Biomed. Nanotechnol.* **6**, 558–568 (2010).
145. R. Rakesh and K. R. Anoop, Formulation and optimization of nano-sized ethosomes for enhanced transdermal delivery of cromolyn sodium. *J. Pharm. Bioallied Sci.* **4**, 333–340 (2012).
146. A. Ahad *et al.*, Formulation and optimization of nanotransfersomes using experimental design technique for accentuated transdermal delivery of valsartan. *Nanomedicine* **8**, 237–249 (2012).
147. G. Cevc, A. Schatzlein and H. Richardsen, Ultradeformable lipid vesicles can penetrate the skin and other semi-permeable barriers unfragmented. Evidence from double label CLSM experiments and direct size measurements. *Biochim. Biophys. Acta* **1564**, 21–30 (2002).
148. M. Schafer-Korting, W. Mehnert and H. C. Korting, Lipid nanoparticles for improved topical application of drugs for skin diseases. *Adv. Drug. Deliv. Rev.* **59**, 427–443 (2007).
149. M. Auffan *et al.*, Is there a Trojan-horse effect during magnetic nanoparticles and metalloid cocontamination of human dermal fibroblasts? *Environ. Sci. Technol.* **46**, 10789–10796 (2012).

150. D. C. Liu, M. O. Meyers, C. B. Hill and W. A. Jr. Loe, Laparoscopic splenectomy in children with hematological disorders: preliminary experience at the Children's Hospital of New Orleans. *Am. Surg.* **66**, 1168–1170 (2000).
151. S. A. Wissing and R. H. Muller, Solid lipid nanoparticles as carrier for sunscreens: *in vitro* release and in vivo skin penetration. *J. Control. Release* **81**, 225–233 (2002).
152. K. Mitri, R. Shegokar, S. Gohla, C. Anselmi and R. H. Muller, Lipid nanocarriers for dermal delivery of lutein: preparation, characterization, stability and performance. *Int. J. Pharm.* **414**, 267–275 (2011).
153. D. C. Liu, J. Rodriguez, C. B. Hill and W. A. Jr. Loe, Transanal mucosectomy in the treatment of Hirschsprung's disease. *J. Pediatr. Surg.* **35**, 235–238 (2000).
154. T. Ohrt, L. H. Thorell, I. Sjodin and G. D'Elia, Are dysfunctional attitudes in depressive disorder trait or state dependent? *Acta Psychiatr. Scand.* **97**, 419–422 (1998).
155. L. B. Jensen, K. Petersson, and H. M. Nielsen, *In vitro* penetration properties of solid lipid nanoparticles in intact and barrier-impaired skin. *Eur. J. Pharm. Biopharm.* **79**, 68–75 (2011).
156. Y. K. Lin, S. A. Al-Suwayeh, Y. L. Leu, F. M. Shen and J. Y. Fang, Squalene-containing nanostructured lipid carriers promote percutaneous absorption and hair follicle targeting of diphencyprone for treating alopecia areata. *Pharm. Res.* **30**, 435–446 (2013).
157. S. A. Nasrollahi, C. Taghibiglou, E. Azizi and E. S. Farboud, Cell-penetrating peptides as a novel transdermal drug delivery system. *Chem. Biol. Drug Des.* **80**, 639–646 (2012).
158. R. R. Patlolla, P. R. Desai, K. Belay and M. S. Singh, Translocation of cell penetrating peptide engrafted nanoparticles across skin layers. *Biomaterials* **31**, 5598–5607 (2010).
159. T. Uchida, T. Kanazawa, M. Kawai, Y. Takashima and H. Okada, Therapeutic effects on atopic dermatitis by anti-RelA short interfering RNA combined with functional peptides Tat and AT1002. *J. Pharmacol. Exp. Ther.* **338**, 443–450 (2011).
160. X. Yi *et al.*, MITF-siRNA formulation is a safe and effective therapy for human melasma. *Mol. Ther.* **19**, 362–371 (2010).
161. V. Alcon *et al.*, Mucosal delivery of bacterial antigens and CpG oligonucleotides formulated in biphasic lipid vesicles in pigs. *Aaps J.* **7**, E566–E571 (2005).
162. S. Babiuk, M. E. Baca-Estrada, R. Pontarollo and M. Foldvari, Topical delivery of plasmid DNA using biphasic lipid vesicles (Biphasix). *J. Pharm. Pharmacol.* **54**, 1609–1614 (2002).
163. M. Foldvari, Biphasic vesicles: a novel topical drug delivery system. *J Biomed. Nanotechnol.* **6**, 543–557 (2010).
164. M. Foldvari *et al.*, Biphasic vesicles for topical delivery of interferon alpha in human volunteers and treatment of patients with human papillomavirus infections. *Curr. Drug Deliv.* **8**, 307–319 (2011).
165. M. Foldvari *et al.*, Topical delivery of interferon alpha by biphasic vesicles: evidence for a novel nanopathway across the stratum corneum. *Mol. Pharm.* **7**, 751–762 (2010).
166. M. Cohen-Avrahami, D. Libster, A. Aserin and N. Garti, Penetratin-induced transdermal delivery from H(II) mesophases of sodium diclofenac. *J Control. Release* **159**, 419–428 (2012).
167. I. Badea *et al.*, In vivo cutaneous interferon-gamma gene delivery using novel dicationic (gemini) surfactant-plasmid complexes. *J. Gene Med.* **7**, 1200–1214 (2005).
168. I. Badea, C. Virtanen, R. E. Verrall, A. Rosenberg and M. Foldvari, Effect of topical interferon-gamma gene therapy using gemini nanoparticles on pathophysiological markers of cutaneous scleroderma in Tsk/+ mice. *Gene Ther.* (2011).

169. I. Badea, S. Wettig, R. Verrall and M. Foldvari, Topical non-invasive gene delivery using gemini nanoparticles in interferon-gamma-deficient mice. *Eur. J. Pharm. Biopharm.* **65**, 414–422 (2007).
170. M. Donkuru, S. D. Wettig, R. E. Verrall, I. Badea and M. Foldvari, Designing pH-sensitive gemini nanoparticles for non-viral gene delivery into keratinocytes. *J. Mater. Chem.* **22**, 6232–6244 (2012).
171. J. Torin Huzil, S. Sivaloganathan, M. Kohandel and M. Foldvari, Drug delivery through the skin: molecular simulations of barrier lipids to design more effective noninvasive dermal and transdermal delivery systems for small molecules, biologics, and cosmetics. *Wiley Interdiscip. Rev. Nanomed. Nanobiotechnol.* (2011).
172. S. Karaborni *et al.*, Simulating the self-assembly of gemini (dimeric) surfactants. *Science* **266**, 254–256 (1994).
173. F. M. Menger and J. S. Keiper, Gemini surfactants. *Angew. Chem. Int. Ed. Engl.* **39**, 1906–1920 (2000).
174. L. Karlsson, M. C. van Eijk and O. Soderman, Compaction of DNA by gemini surfactants: effects of surfactant architecture. *J. Colloid Interface Sci.* **252**, 290–296 (2002).
175. A. J. Kirby *et al.*, Gemini surfactants: new synthetic vectors for gene transfection. *Angew. Chem. Int. Ed. Engl.* **42**, 1448–1457 (2003).
176. C. Bombelli *et al.*, Role of the spacer of cationic gemini amphiphiles in the condensation of DNA. *Langmuir* **21**, 10271–10274 (2005).
177. A. J. Kirby *et al.*, Gemini surfactants: new synthetic vectors for gene transfection. *Angew. Chem. Int. Ed. Engl.* **42**, 1448–1457 (2003).
178. S. D. Wettig, R. E. Verrall and M. Foldvari, Gemini surfactants: A new family of building blocks for non-viral gene delivery systems. *Curr. Gene Ther.* **8**, 9–23 (2008).
179. M. Foldvari, I. Badea, S. Wettig, R. Verrall and M. Bagonluri, Structural characterization of novel gemini non-viral DNA delivery systems for cutaneous gene therapy. *J. Exp. Nanosci.* **1**, 165–176 (2006).
180. S. D. Wettig, I. Badea, M. Donkuru, R. E. Verrall and M. Foldvari, Structural and transfection properties of amine-substituted gemini surfactant-based nanoparticles. *J. Gene Med.* **9**, 649–658 (2007).
181. M. Cohen-Avrahami, D. Libster, A. Aserin and N. Garti, Penetratin-induced transdermal delivery from HII mesophases of sodium diclofenac. *J. Control. Release* **159**, 419–428 (2012).
182. S. Mignani, S. El Kazzouli, M. Bousmina and J. P. Majoral, Expand classical drug administration ways by emerging routes using dendrimer drug delivery systems: a concise overview. *Adv. Drug Deliv. Rev* (2013).
183. C. F. Price *et al.*, SPL7013 Gel (VivaGel®) retains potent HIV-1 and HSV-2 inhibitory activity following vaginal administration in humans. *PloS one* **6**, e24095 (2011).
184. R. Rupp, S. L. Rosenthal and L. R. Stanberry, VivaGel™(SPL7013 Gel): a candidate dendrimer–microbicide for the prevention of HIV and HSV infection. *Int. J. Nanomed.* **2**, 561 (2007).
185. I. McGowan *et al.*, Phase 1 randomized trial of the vaginal safety and acceptability of SPL7013 gel (VivaGel®) in sexually active young women (MTN-004). *AIDS* **25**, 1057 (2011).
186. R. W. Lee, D. B. Shenoy and R. Sheel, Micellar nanoparticles: applications for topical and passive transdermal drug delivery. In *Handbook of Non-Invasive Drug Delivery Systems*, Burlington, MA: Elsevier Inc. (2010), pp. 37–58.

187. P. Valenzuela and J. A. Simon, Nanoparticle delivery for transdermal HRT. *Nanomedicine* **8** (Suppl 1), S83–89 (2012).
188. K. Sato *et al.*, A clinical trial of topical bleaching treatment with nanoscale tretinoin particles and hydroquinone for hyperpigmented skin lesions. *Dermatol. Surg.* **33**, 937–944 (2007).
189. S. Prasad *et al.*, Efficacy and safety of a nano-emulsion gel formulation of adapalene 0.1% and clindamycin 1% combination in acne vulgaris: a randomized, open label, active-controlled, multicentric, phase IV clinical trial. *Indian J. Dermatol. Venereol. Leprol.* **78**, 459–467 (2012).
190. R. G. Glogau, Topically applied botulinum toxin type A for the treatment of primary axillary hyperhidrosis: results of a randomized, blinded, vehicle-controlled study. *Dermatol. Surg.* **33**, S76–80 (2007).
191. R. Glogau *et al.*, Results of a randomized, double-blind, placebo-controlled study to evaluate the efficacy and safety of a botulinum toxin type A topical gel for the treatment of moderate-to-severe lateral canthal lines. *J. Drugs Dermatol.* **11**, 38–45 (2012).
192. E. Koren and V. P. Torchilin, Cell-penetrating peptides: breaking through to the other side. *Trends Mol. Med.* **18**, 385–393 (2012).
193. L. C. Papadopoulou and A. S. Tsiftsoglou, The potential role of cell penetrating peptides in the intracellular delivery of proteins for therapy of erythroid related disorders. *Pharmaceuticals* **6**, 32–53 (2013).

Chapter 8

Lysosomes and Nanotherapeutics: Diseases, Treatments, and Side Effects

Rachel L. Manthe[*] *and Silvia Muro*[*,†]

[*]*Fischell Department of Bioengineering,*
University of Maryland, College Park, USA
[†]*Institute for Bioscience and Biotechnology Research,*
University of Maryland, College Park, USA
muro@umd.edu

Abstract

Endolysosomal-related organelles, such as lysosomes, phagosomes, autophagosomes, and others, are essential to numerous cell processes; hence, their defective function associates with a number of diseases. For instance, lysosomal storage disorders are a group of about 50 genetic diseases characterized by lysosomal accumulation of undigested substrates, resulting in damage to peripheral organs and the central nervous system. Lysosomal aberrations are also implicated in autophagy-mediated disease progression, such as in cancer, neurodegenerative conditions (Alzheimer's, Parkinson's, and Huntington's diseases), auto-immune defects, and inflammatory diseases. Although several treatment modalities help alleviate symptoms and improve quality of life for patients affected by these conditions, these therapies are still suboptimal in normalizing lysosomal-related functions. In order to overcome this caveat, strategies employing nanoscale drug delivery systems are being investigated. This chapter aims to review the features of lysosome-related diseases, the limitations of current treatments, and the potential utility of drug carriers to improve these strategies. A basic review of drug delivery vehicles is provided, along with considerations on the importance of proper carrier design and evaluation of the potential lysosomal toxicity associated with drug delivery systems.

1. Nanocarriers for Drug Delivery

Nanotechnology has had a profound impact on the medical field, including delivery of therapeutic agents such as small chemicals and protein- and nucleic acid-based therapeutics.[1,2] Nanoscale (<1 μm) drug delivery systems (referred to herein as nanocarriers) can be fabricated from a wide array of materials including synthetic, semi-synthetic, and biological products to form structures with unique compositions and architectures.[1,2] To improve the efficacy of therapeutic agents, nanocarriers must be able to overcome the barriers that limit the efficacy of drugs in solution, such as issues with poor drug solubility and stability, rapid degradation and/or clearance from the circulation, systemic toxicity, poor accumulation in disease target sites, difficulty in maintaining drug concentrations within therapeutic windows, and subcellular transport to the molecular targets requiring therapeutic intervention[3,4] (Fig. 1). These factors can be controlled by the physicochemical properties of the nanocarrier, which can be modified by additional functionalization with polymers and biological compounds.[5,6]

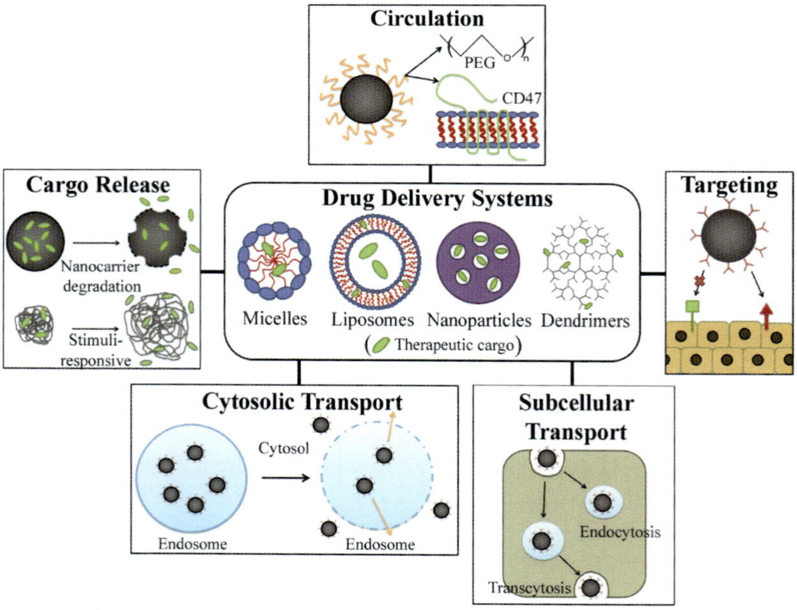

Fig. 1. Structure and function of drug delivery systems. Various types of drug delivery systems currently exist and provide enhanced therapeutic efficacy compared to drugs in solution, due to their ability to prolong circulation, bind to specific sites and gain access to therapeutic targets of interest through enhanced transcellular, intracellular, and cytosolic transport, while also providing control over cargo release.

1.1. *Nanocarrier types and properties*

Liposomes are the oldest and probably most investigated drug delivery system.[7] Based on natural phospholipids found in cellular membranes, liposomes self-assemble in aqueous solution into bilayered spherical vesicles with a diameter ranging from 50 nm to several μm.[8,9] Their main advantage derives from the amphiphilic nature of phospholipids, allowing for hydrophilic molecules to be entrapped in the liposome inner aqueous lumen or onto its surface, while hydrophobic molecules partition into the bilayer.[5] Liposomes also have low toxicity and can be functionalized to improve their circulation time and to provide site-specific targeting.[6,8] They have modest drug-loading capacity and can tolerate a low degree of the aforementioned modifications, which affects their stability and clearance.[10,11]

Closely related to liposomes are polymersomes and polymeric micelles. Polymersomes are analogous to liposomes, but are composed of amphiphilic block copolymers as opposed to phospholipids.[8,12] They also form bilayered vesicles of a comparable size range and can be loaded with hydrophilic and hydrophobic agents.[8] While polymersomes are more rigid and less permeable to drugs than liposomes,[8] they allow for a higher degree of functionalization, chemical stability, and structural control.[8,12] Amphiphilic block copolymers can also self-assemble into micelles (a few nm in diameter), which possess only an inner hydrophobic core and an outer hydrophilic layer.[4,8] They can be functionalized for targeting, have low toxicity and high loading capacity for water-insoluble drugs, and offer advantages regarding their simplicity of preparation, controlled release profile, and prolonged blood circulation time.[4,8]

Nanoparticles fabricated from biodegradable polymers (synthetic, natural, or hybrid) have become one of the most popular drug delivery systems due to their versatility as therapeutic carriers.[4,13,14] They can exist in a variety of shapes and sizes (10–1000 nm) as hollow or solid structures with varying degrees of porosity.[8,15] A therapeutic agent can be encapsulated, adsorbed onto the particle surface, or dispersed throughout the polymer matrix.[4,5] Control of drug release can be achieved via diffusion, surface erosion, or bulk degradation of the nanoparticle over a period of time.[8,15]

Dendrimers consist of highly branched polymers radiating from a central core.[8] Sequential and symmetrical growth of branched layers (i.e. generations) creates a spherical cage with small cavities for drug entrappment.[5,8] Based on the generation, dendrimer properties such as size, shape, and surface chemistry can be controlled to regulate delivery properties.[16,17] Dendrimers have a high surface-to-volume ratio and allow for multiple functional groups, amenable for high density attachment of therapeutic, diagnostic, and targeting moieties.[5,16,17] Their small size (only a few nm in diameter) permits enhanced permeability through cellular layers, which is highly advantageous for delivery across the gastrointestinal tract in oral delivery.[17] However, they require a laborious preparation and suffer from somewhat poor stability and potential toxicity.[8]

With regard to materials, some of the most commonly used biocompatible synthetic polymers are polyanhydrides, polyorthoesters, polycaprolactone, polylactides, polyglycolides, poly(lactide-co-glycolides), poly(methyl methacrylate), poly(acrylic acid), poly(vinyl alcohol) and poly(ethylene glycol), which is commonly referred to as PEG.[16,18] Poly(lactic-co-glycolic acid) (PLGA) is one of the most widely investigated polymers for nanocarriers because of its FDA-approval (owing to its good biocompatibility)[19] and controllable rate of degradation (hence, controlled drug release) via modification of particle size and the ratio of lactic to glycolic acid.[15] PEG is also widely used in drug delivery systems because of its ability to solubilize hydrophobic drugs and its relatively low interaction with components of the immune system, thereby allowing for prolonged blood circulation.[15,20] Since some synthetic polymers have somewhat poor batch-to-batch variability and residual chemicals from their fabrication may pose safety concerns,[8] natural polymers such as chitosan, gelatin, albumin, natural gum, and sodium alginate are also utilized.[18] A number of polymers are also sensitive to biologically-relevant stimuli such as changes in pH, which results in swelling of the drug delivery system with concomitant drug release.[21,22] These polymers include chitosan, polyethylenimine, poly(acrylic acid), poly(L-lysine), and others.[21,22] They are especially useful in oral delivery and endolysosomal escape,[21–23] which is discussed in more detail in Sec. 1.4.

Other nanoscale systems are based on inorganic materials, such as silica nanoparticles, metal nanoparticles, quantum dots, and carbon nanotubes.[4] While often used in diagnostic applications, these carriers are also advantageous for drug delivery due to their small size (<50 nm) and ease of functionalization with targeting moieties, drugs, etc.[4] However, they are still under investigation in regards to their safety. Some of these systems also possess inherent imaging capability, allowing for real-time monitoring of biodistribution and treatment efficacy (theranostics).[24–26] One relevant example is that of 10–100 nm superparamagnetic iron-oxide nanoparticles,[4,24,26] which have become increasingly popular for drug delivery applications because they are relatively biocompatible, biodegradable, can serve as an MRI contrast agent,[4,24] and their distribution can be manipulated by applying an external magnetic field over the desired area of accumulation.[24]

1.2. *Blood circulation and clearance*

One of the major challenges in designing drug delivery systems is how to prevent or limit their rapid clearance from the circulation following intravenous (*i.v.*) administration.[20] Prolonged blood circulation is necessary for nanocarrier accumulation at therapeutic levels in body sites that require passive targeting (e.g. by the enhanced permeability and retention (EPR) effect),[27] as well as for applications in vascular drug delivery.[20] As with any foreign object, nanocarrier clearance is initially mediated via the opsonization process, encompassing adsorption of plasma proteins called opsonins onto the nanocarrier surface.[20] These opsonins, including complement proteins, various immunoglobulins, von Willebrand factor, and several others, induce receptor-mediated phagocytosis by cells of the reticuloendothelial system (RES), including monocytes and neutrophils in the blood, Kupffer cells in the liver, and macrophages in the spleen, liver, lungs, and lymph nodes.[20,28] In addition, nanocarriers of a certain size (≥200 nm) can be mechanically filtered from the body via fenestrations in the spleen and liver.[8]

The physicochemical properties of nanocarriers, including their size, surface charge, chemical composition, etc., directly impact opsonization and subsequent clearance by the RES.[20,28,29] Differences in the extent of opsonization and rate of RES clearance have been observed for nanocarriers with varying properties as they attract different combinations of opsonins.[20] Yet, how a certain combination of opsonins or their conformation on the carrier surface contribute to clearance is not well understood.[29] In general, larger nanocarriers (≥ 200 nm) activate the complement system better than smaller counterparts and are captured by the spleen and liver.[5,8] Alternatively, smaller particles (10–100 nm) can avoid the RES and mechanical clearance, but their larger surface area-to-volume ratio promotes aggregation, which changes their clearance kinetics.[5] Nanocarriers with a positive charge and hydrophobic surface also have a higher clearance rate than an electrically neutral or negatively-charged hydrophilic surface.[28,29]

A good strategy to prevent rapid clearance of nanocarriers is to limit opsonization and immune system recognition via adsorption or covalent attachment of PEG to the nanocarrier surface, a process referred to as PEGylation.[5,29] PEG is a hydrophilic polymer that can exist in linear or branched structures[28,29] and, when added to a nanocarrier surface, forms a hydrophilic layer in a mushroom or brush configuration that helps minimize opsonin adsorption and complement activation via steric hindrance.[5,28,29] As a result, nanocarriers demonstrate prolonged blood circulation extending up to a few days.[9,11] This outcome depends on the molecular mass, chain length, and structure (linear or branched) of PEG, thereby requiring PEGylation to be optimized for each specific application.[29]

Despite these advantageous properties, in some cases PEG can interfere with nanocarrier interactions with cellular membranes, targeting, endosomal escape, and can also cause certain immune reactions.[28,30] Some approaches have been devised to overcome this, including cleavage of PEG from the nanocarrier surface after tissue extravasation or cellular uptake via pH and/or temperature change, enzymatic reactions, or ultrasonic stimulation.[28] Additionally, other "stealth" polymers, including polyoxazolines, poly(amino acids), polybetaines, polyglycerols and polysaccharides, have been used with promising results.[28] Also, incorporation of CD47 derivatives on a nanocarrier surface has been explored to evade phagocytosis.[31] CD47 is a transmembrane protein found on red blood cells and acts as a marker of the "self": it avoids phagocytosis by leukocytes by binding to a leukocyte-surface receptor, leadings to a signaling cascade that impairs cytoskeletal reorganization into engulfment structures.[31]

1.3. *Targeting strategies*

Regardless if therapeutic intervention is required at one localized site or at multiple sites throughout various organ systems in the body, targeting can enhance the efficacy of drug delivery systems by maximizing their bioadhesion.[1,3] This can be achieved in a non-specific manner by targeting common elements (e.g. negative charge of the cell surface, affinity to ubiquitous macromolecules) or more specifically by taking advantage of physiological features unique to the diseased tissue, including increased vascular permeability (passive targeting) and specific cell/tissue markers that are overexpressed during the pathological state (active targeting).[3,32] Specific targeting also helps maintain therapeutic levels of a drug at the diseased site(s) while minimizing potential side effects in healthy tissues.[1,3]

The most common passive targeting strategy is the EPR effect exploited in cancer therapy, which takes advantage of the leaky vasculature of tumor areas to enhance nanocarrier accumulation within the cancerous tissue.[27] Nanocarriers that rely on the EPR effect are approved clinically for the treatment of several cancers.[4] A representative example is Doxil, a PEGylated liposome that encapsulates the chemotherapeutic agent doxorubicin, used in the treatment of ovarian cancer, breast cancer, and Kaposi's sarcoma.[4] The EPR phenomenon also occurs in other diseases where inflammatory processes have disrupted the permeability barrier of the vascular endothelium, such as in infection, infarction, rheumatoid arthritis, etc.[4,20] Similar to the EPR effect, nanocarriers also accumulate within organs characterized by discontinuous blood vessels that do not present a barrier to diffusion of compounds, such as the liver and spleen.[3] By the first pass phenomenon, drug delivery systems can also be passively targeted to organs directly downstream of the administration site via normal blood flow.[3]

In some disorders where multiple organ systems are affected, non-specific active targeting can enhance ubiquitous delivery of nanocarriers throughout the body.[33,34] This can be achieved by relying on non-specific interactions of nanocarriers toward general biological features, such as in the case of positively-charged moieties in HIV Tat peptide, RGD counterparts, antennapedia, and other sequences that can provide affinity to the negatively-charged plasmalemma of cells.[3,35] Coating nanocarriers with cationic polymers, such as chitosan, serves a similar purpose.[21]

Specific active targeting is typically employed when localized treatment is required.[3] This is achieved by addressing nanocarriers to specific markers of diseased sites (including extracellular or cell surface determinants) using affinity moieties such as proteins or peptides, antibodies or their fragments, sugars, or aptamers.[3,36,37] Targeting moieties can be coupled to nanocarriers via hydrophobic or electrostatic interactions, chemical conjugation, etc.[3] This approach can also be designed to target specific cell surface receptors involved in endocytic transport into or across cells as discussed below.[3,38]

1.4. *Transport into cells and across cellular barriers*

In most cases, the targets for therapeutic action are intracellular. Hence, once drug carriers reach the diseased tissues they must gain access to particular intracellular compartments to act on the molecular component(s) responsible for a given pathology.[1,3,37,39] Most commonly, this requires drug delivery to the cell cytosol for action in this location or in order to gain access to other organelles.[1,3] Physical mechanisms such as ultrasound or electroporation can be used to temporarily increase the permeability of the plasma membrane for entry into the cytosol, although these techniques are only applicable locally and/or are relatively toxic.[3,40] Alternatively, some toxins or positively-charged cell-penetrating peptides that achieve cell surface binding via forming pores or by electrostatic interactions, respectively, can facilitate translocation through the phospholipid bilayer.[35,41] This route is typically restricted to molecular conjugates rather than larger nanocarriers and also lacks site-specificity since it mainly exploits general plasmalemma features.[3] In order to improve these aspects, endocytosis is often used.[3,42] This is a route where substances and objects bound to the cell surface are engulfed by the plasma membrane into membrane-enveloped vesicles that pinch off into the cytosol.[38] Because binding to the cell surface can be achieved non-specifically via electrostatic interaction or specifically via affinity moieties that recognize particular receptors, both of these modalities can be exploited.[3] Different endocytic pathways can be induced, depending on the treatment requirements.[3,38] Uptake of particulate ligands via phagocytosis or macropinocytosis-mediated uptake of fluid into micrometer-sized vesicles is amenable for drug delivery systems with a variety of sizes and shapes.[3,38] Yet, these pathways associate with specialized cells of the immune system, thereby preventing delivery to other cell types.[3,38] Clathrin- or caveolar-mediated endocytosis operates in most cells, but they associate with smaller endocytic vesicles (~100 nm or ~70 nm, respectively).[3,38] There are a few other pathways different from these ones, including cell adhesion molecule (CAM)-mediated endocytosis induced by binding to ICAM-1,[43] as well as that of GPI-anchored proteins, IL-2β receptor, proteoglycans, and bound growth factors (e.g. FGF2),[44] although they are less understood.

The route of entry within the cell often determines the intracellular fate of the internalized material.[3] In some instances, clathrin- or CAM-mediated endocytosis can deliver contents back to the cell surface via recycling endosomes.[45–47] Instead, caveolar pathways may also lead to retrograde transport to the Golgi apparatus or the endoplasmic reticulum.[3] Most commonly though, materials internalized by endocytosis traffic to lysosomal compartments.[3] This is beneficial for treatment of lysosomal storage disorders[48] (discussed in Secs. 4.1 and 4.2) and can be used for drug release from pH-sensitive carriers.[22] Nevertheless, lysosomal destination may also result in drug entrapment within this compartment, impairing access to other subcellular locations and leading to premature drug degradation due to the acidic pH and hydrolytic enzymes in lysosomes.[49] In these cases, endosomal escape with cytosolic release is required.[3] This has been pursued through a variety of means, including the use of cationic polymers that disrupt the endosomal membrane via osmotic swelling (i.e. the proton sponge effect)[22,23] or fusogenic peptides (e.g. GALA peptides, hemagglutinin) that form pores in endolysosomal membranes.[3,41]

Targeting of pathways associated with transport across cellular barriers is necessary for nanocarrier accumulation in tissues and can be achieved by endocytosis in certain instances.[34] Prominent examples of such cellular barriers are the endothelial lining in the blood-brain interface and the epithelial lining of the gastrointestinal tract, where tight junctions between neighboring cells prevent free transport of substances between the blood and tissue.[50,51] To overcome these permeability barriers, paracellular and transcellular pathways are utilized.[34] The paracellular route involves transport between neighboring cells via transient disruption of tight junctions, which can be induced by certain nanocarriers (e.g. small dendrimers) or by application of hyperosmotic solutions, vasoactive agents, ultrasound, solvents, or stabilizers.[17,52–54] Since this route is often associated with leakage of substances between the compartments separated by the barrier, the transcellular route is used as an alternative.[34] This involves nanocarrier binding to specific receptors located on the apical surface of cells that constitute the barrier, followed by endocytosis, traffic across the cell body, and final exocytosis from the abluminal cell surface.[34] This process is called transcytosis and can be induced by targeting certain receptors associated with clathrin-, caveolar-, or CAM-mediated endocytosis.[55–57]

2. Lysosomes and Lysosomal Storage Disorders

Constituting approximately 5% of the intracellular volume of mammalian cells, lysosomes are dynamic membrane-bound organelles that play a crucial role in a number of cellular processes required to maintain homeostasis.[58] Therefore, any lysosomal aberration could potentially lead to disease. The most well-known class of diseases caused by dysfunctional lysosomes are the lysosomal storage disorders (LSDs), a group of about 50 different diseases characterized by lysosomal accumulation of non-degraded metabolites.[49] Although individually rare, the combined prevalence of LSDs is estimated to be 1 in every 2000 live births, with some LSDs occurring at a much higher frequency in certain populations, such as the prevalence of Gaucher disease in the Ashkenazi Jewish population (1 in 855 births).[59]

2.1. *Lysosome structure and function*

Lysosomes are heterogeneous in size and morphology, and often contain electron-dense deposits and membranous whorls.[58,60] More than 50 acid hydrolases have been described within the lumen of this compartment, whose pH is kept between pH 4.5–5.0 by mediation of vacuolar H^+-ATPases.[49,61] Along with activator molecules, these hydrolases are responsible for the degradation of different substrates (proteins, lipids, sugars, nucleic acids) taken up from the extracellular environment via endocytosis.[62] Lysosomes also offer protection against foreign substances and pathogens that invade the cell, and contribute to the disposal of intracellular components that are no longer needed through a process called autophagy, where degradation of intracellular macromolecular assemblies render monomeric units that may be returned to the cytosol and recycled as new building blocks.[49,58,62] In addition, lysosomes play an important role in removing plasmalemma lipids and proteins, and repairing this structure after injury (e.g. due to mechanical stress in muscle cells).[63,64] For instance, plasmalemma disruption leads to increased intracellular calcium, which triggers targeted fusion of the lysosome to the plasmalemma for addition of a new phospholipid membrane to repair the damaged area.[63,64]

With regard to lysosomal hydrolases, these enzymes enter the rough endoplasmic reticulum (ER) during their synthesis[48,58] (Fig. 2). Oligosaccharides containing mannose residues are then added to these enzymes, followed by transport to the Golgi apparatus where phosphorylation of the mannose residues yields enzymes tagged with mannose-6-phosphate (M6P).[65,66] Binding of M6P residues to their respective receptors in the *trans*-Golgi network (TGN) mediates intracellular trafficking of the enzymes to lysosomes, where

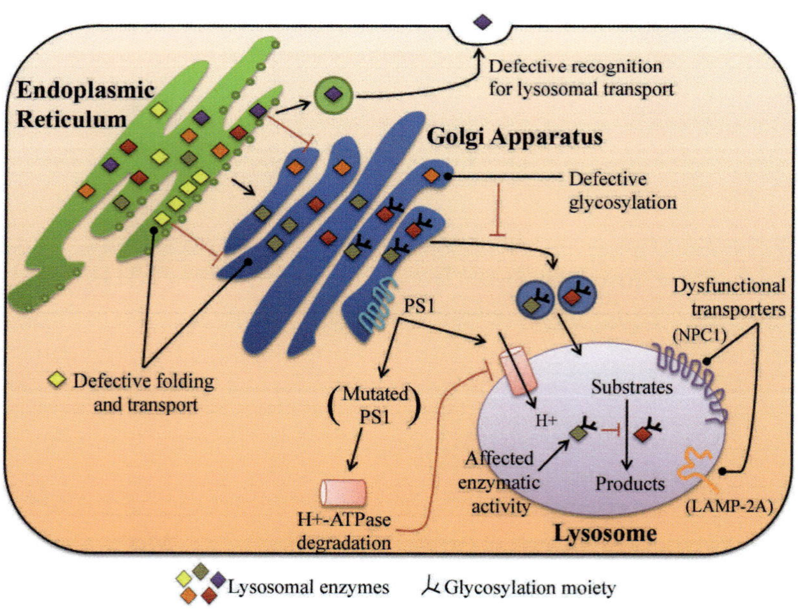

Fig. 2. Overview of some aberrations identified in LSDs. Genetic mutations associated with LSDs cause dysfunctions in upstream compartments, lysosomal membrane components, and lysosomal enzymes, all of which affect lysosome function. Defects include incorrect folding of lysosomal enzymes in the ER, dysfunctional trafficking due to defective recognition or defective glycosylation, insufficient lysosomal acidification, and dysfunctional integral membrane proteins of the lysosome. Consequently, LSDs are characterized by aberrant accumulation of undigested substrates in the lysosome.

they dissociate from M6P receptors due to the acidic luminal pH in this compartment.[65,66] Delivery occurs via multiple fusion and fission cycles with late endosomes, with M6P receptors being recycled back to the TGN.[58] In addition, a fraction of the lysosomal enzymes are secreted to the extracellular milieu via exocytosis.[66,67] Secreted enzymes that contain M6P bind to M6P receptors on the surface of nearby cells, leading to clathrin-mediated endocytosis with transport to lysosomes.[66,67] Some secreted enzymes contain mannose instead of M6P and can be taken up by mannose receptors on monocytes and macrophages.[67] Last, lysosomes also contain highly glycosylated integral proteins in their membrane, including LAMP-1, LAMP-2, tetraspanin CD63 or LAMP-3, LIMP-2, etc.[62,68] These proteins assist in maintaining lysosome stability and integrity, transport of soluble metabolites out of lysosomes, and chaperone-mediated autophagy and macroautophagy (discussed in Sec. 3.1).[49,68] LIMP-2 also acts as a receptor for sorting several lysosomal proteins in an M6P-independent manner.[58]

Mammalian lysosomes share many similarities with yeast vacuoles, likely reflecting their evolutionary origin.[69] Lysosomes also closely resemble other organelles of the biosynthetic, endocytic, autophagic, and secretory pathways, especially endosomes and autophagosomes,[70,71] although they are generally smaller than these compartments, have a higher buoyant density, and have a higher proton and calcium content.[71] Compared to endosomes, lysosomes also contain greater levels of mature acid hydrolases, are more enriched with LAMPs, and lack M6P receptors.[70] Most or all of these features are shared with several other lysosome-related organelles specific to certain cell types, including melanosomes, lytic granules, major histocompatibility complex (MHC) class II compartments, platelet-dense granules, basophil granules, and neutrophil azurophil granules.[70] However, these compartments serve a storage and secretion (not degradative) role.[70] Regardless, abnormalities in both lysosomes and lysosome-related organelles have been found to occur in human genetic diseases.[49,70]

2.2. *Lysosomal storage disorders*

LSDs consist of about 50 different diseases affecting humans and several animal species.[49] Most of these disorders are caused by dysfunctional lysosomal enzymes, resulting in lysosomal accumulation of unmetabolized substrates, including lipids, glycoproteins, and glycosaminoglycans[49] (Fig. 2). However, several LSDs are caused by defective integral membrane proteins, such as LAMP-2A in Danon disease and NPC1 in Niemann-Pick type C (NPC) disease.[49] Also in a number of LSDs, the primary defects affect proteins in the

ER, Golgi, and endosomal pathways involved in the synthesis, modification, and trafficking of lysosomal proteins.[49,72,73] This is the case in mucolipidosis (ML) type II (I Cell disease) and ML type III (pseudo-Hurler polydystrophy), which cause lysosomal enzymes to be secreted from the cell.[49] Interestingly, no LSD is caused by defective vacuolar H^+-ATPases, likely due to lethal effects during development,[72] yet certain LSDs associate with a more neutral lysosomal pH[74,75] that affects hydrolase and vesicle maturation, intracellular trafficking, recycling of M6P receptors back to the Golgi, secretion, signaling, and other functions.[34,76,77]

All LSDs are monogenic diseases inherited as autosomal recessive traits, with the exception of Fabry disease, mucopolysaccharidosis (MPS) type II (Hunter disease), and Danon disease.[73,78] In some cases, although only one gene is defective, the enzyme it encodes may modify the activity of several other enzymes,[49] leading to multiple enzyme deficiencies such as the case of multiple sulphatase deficiency (MSD).[79] The prevalence and location of gene mutations frequently differ in patients with the same LSD, and the same mutation can lead to varying degrees of severity in different patients.[49] Depending on the age when clinical symptoms begin to develop, LSDs are classified as infantile, juvenile, or adult forms.[49] Since lysosomes are essential components of almost all cell types, LSDs associate with a wide range of phenotypic symptoms in multiple tissues and organs.[49] Infantile forms are the most severe and involve dysfunction of the central nervous system (CNS).[49] CNS involvement is not limited to newborns though, as late onset of LSDs characterized by neurological symptoms are also seen in juvenile and adult forms.[49,80] Overall, at least half of all LSDs are associated with severe CNS disorders (seizures, dementia, ataxia, brainstem dysfunction, vision loss, movement disorders, etc.) that result in early death.[34,49] Less severe LSDs manifest in infancy and progress slowly into adulthood, mainly causing peripheral disease, including damage to the liver, spleen, striated muscle, kidneys, lungs, bone, immune system, etc.[34,49,59] Inflammation caused by oxidative stress and leukocyte activation from biochemical and metabolic stresses often causes the initial pathology to further spread and increase in severity.[34,81] Moreover, multiple epigenetic and environmental factors contribute further to the phenotypic variability of LSDs.[49,59]

2.3. *Clinical and experimental treatments*

A number of strategies have been proposed to treat LSDs, including organ/cell transplantation, gene therapy, small molecule therapies, lysosome exocytosis, and enzyme replacement therapy[34,48] (Table 1). Organ or cell

Table 1. Treatment modalities for LSDs.

Treatment	Goal	LSD examples	Bottlenecks	Refs.
Organ transplantation	Replacement of damaged organ(s) (e.g. liver and kidneys)	Fabry and Cystinosis	Limited to treatment of peripheral organs; graft rejection; lack of suitable donors	48, 49, 97
Cell therapy	Transplantation of normal cells from a healthy donor	Sandhoff; GM1 gangliosidosis; Tay-Sachs; Pompe; Gaucher; Krabbe; Hurler; others	Limited to treatment of peripheral organs; graft rejection; lack of suitable donors	48, 81, 82
Gene therapy	Use of viral vectors to deliver exogenous lysosomal gene sequences to restore functional enzyme activity	Sanfilippo; Pompe; Fabry; Sandhoff; NPC; Gaucher; MPS type VII; other MPS types	Limited to treatment of peripheral organs; safety concerns (potential for immunogenicity and random insertion into host gene)	81, 84–86
Substrate reduction therapy	Block synthesis of accumulating substrates	Gaucher; Fabry; Tay-Sachs; NPC; Cystinosis; others	Lack of complete inhibitor specificity; potential side effects of depleting substrate	73, 96, 97
Use of chemical chaperones	Assist in correct folding of lysosomal enzymes	Gaucher; Fabry; GM1 gangliosidosis	Requires knowledge of the particular mutation(s); limited to LSDs caused by incorrect enzyme folding	99–101

(Continued)

Table 1. (*Continued*)

Treatment	Goal	LSD examples	Bottlenecks	Refs.
Use of cytosolic molecules	Reduce substrate storage by increasing levels of cytosolic molecule(s) involved in vesicle trafficking	NPC	Limitations associated with use of gene therapy or protein transduction to increase levels of the cytosolic molecule(s)	48, 102
Enhanced lysosome exocytosis	Lysosome exocytosis to reduce substrate accumulation	NPC; Wolman; Batten; Fabry; Farber; several others	Loss of lysosomes with functional enzymes; disruption of autophagic processing	73, 103–106
Enzyme replacement therapy	Replacement of defective lysosomal enzyme(s) with recombinant enzyme(s)	Hurler; Hunter; Sanfilippo type A; MPS type VI and VII; Fabry, Gaucher type I-III; Pompe; Krabbe; Sandhoff; others	Rapid clearance from the circulation; inactivation by proteases; reduced access through tissue and cellular barriers; immunogenicity	48, 73, 81, 87–95

transplantation have been clinically implemented to alleviate phenotypic symptoms in several LSDs, such as Gaucher, Pompe and Krabbe diseases, MPS type I (Hurler syndrome), and others.[48,81] Organ transplantation, commonly of the liver and kidneys, is performed in severe cases of organ failure.[49] Similarly, cell therapy involves transplantation of bone marrow or mesenchymal stem cells obtained from a healthy donor.[82] Healthy transplanted cells or organs can secrete normal or wild-type lysosomal enzymes, which can be endocytosed and transported to lysosomes in neighboring diseased cells for repair of the defective enzyme activity (cross correction).[48,83] These options alleviate LSD symptoms in organs of the RES, but are limited in scope due to lack of suitable donors and graft rejection.[82]

Gene therapy is a treatment option with great promise in terms of therapeutic efficacy.[81,84] Various viral vectors (e.g. adenovirus, adeno-associated virus, retrovirus, and lentivirus) have been utilized for delivery of wild-type cDNA sequences into diseased cells to restore functional enzymes.[81,84,85] This strategy is currently approved for human clinical trials for MPS type III (Sanfilippo syndrome) and Pompe disease, and is being investigated in animal models for Fabry disease and other types of MPS.[81,86] The efficacy of this approach is greater in peripheral organs, especially those of the RES, while CNS treatment remains elusive and intracerebral administration is being explored.[34] The overall safety of gene therapy (e.g. immunogenicity, random insertion in the host genome) is also being investigated.[34]

Enzyme replacement therapy (ERT) is the most widely used clinical therapy for LSDs with products currently FDA-approved for the treatment of Hurler syndrome, Hunter disease, Maroteaux-Lamy syndrome (MPS type VI), Fabry disease, Gaucher type I disease, and Pompe disease.[81] In this approach, wild-type recombinant lysosomal enzymes are intravenously infused in patients.[48] ERT relies on the cross correction effect and the fact that relatively low amounts of the "correcting enzyme" (≤10–20% of endogenous levels) are needed for therapeutic effects.[83] ERT has proven valuable with regard to treatment of peripheral tissues, but several limitations reduce its efficacy.[48,73] Enzyme delivery is suboptimal due to rapid clearance from the circulation, inactivation by proteases, reduced access across cellular barriers, e.g. the blood–brain barrier (BBB), and immunogenicity with production of antibodies against recombinant enzymes.[33,87] Several approaches have focused on overcoming these caveats, including optimization of enzyme glycosylation, modification of said residues, or enhancement of mannose or M6P receptor expression to improve enzyme targeting.[67,88–90]

Enzyme targeting via glycosylation-independent mechanisms is another option. This is achieved by designing fusion proteins that contain the enzyme fused to an affinity peptide for recognition by cell surface receptors.[48] Examples of this strategy have shown to improve delivery of β-glucuronidase (MPS type VII), α-L-iduronidase (Hurler syndrome), and α-glucosidase (Pompe disease) through use of insulin-like growth factor II (ILF-II) peptides to target M6P receptors, receptor-associated protein (RAP) that binds to low-density lipoprotein receptors (LDLRs), or others, all of which result in clathrin-dependent endocytosis and lysosomal transport.[34,91,92] Chimeras containing protein transduction domains (e.g. HIV Tat) have also been used in the case of β-glucuronidase (MPS type VII), glucocerebrosidase (Gaucher disease), and galactocerebrosidase (Krabbe disease) to improve charge-mediated binding to the plasmalemma and internalization via passive uptake within endocytic vesicles and/or direct penetration through the plasmalemma.[93–95]

Small molecule therapies have also been explored, including substrate reduction, chemical chaperones, and cytosolic molecules.[34] In substrate reduction therapy, inhibitory molecules reduce the biosynthesis of accumulating substrates, such as *N*-butyldeoxynojirimycin (Miglustat or Zavesca®)[49,73] used in Gaucher, Fabry, Tay-Sachs, and NPC diseases.[34,96] This molecule crosses the BBB and reduces neurological symptoms.[73] Cysteamine is another substrate inhibitor successfully used to treat Cystinosis.[97] However, lack of complete specificity and long-term effects of impairing anabolic routes needs to be fully investigated. Certain accumulated metabolites can be removed using small molecules such as cyclodextrins, which are small sugar molecules with a ring-link structure capable of removing excess cholesterol and glycosphingolipids in NPC disease.[98] In addition, some of these inhibitors act as chemical chaperones that assist in lysosomal protein folding.[99] This is helpful in the case of mutations that cause incorrect enzyme folding (not affecting the catalytic site) and facilitates transport into lysosomes since misfolded enzymes often accumulate in upstream compartments.[99] Examples in clinical trials include *N*-nonyldeoxynojirimycin for Gaucher disease, 1-deoxygalactonojirimycin for Fabry disease, and *N*-octyl-4-epi-β-valienamine for GM1 gangliosidosis.[99–101]

Other approaches have focused on lysosomal associated transport. When substrate accumulation is due to defective lysosomal transport, storage can be reduced by increasing the intracellular levels of molecules controlling intracellular vesicle trafficking.[34] This is the case of Rab9 overexpression, which reduces cholesterol and sphingolipid storage in NPC disease.[102] However, increasing intracellular levels of cytosolic molecules requires gene therapy or protein transduction and, therefore, is bound to the limitations of these methods. Alternatively, lysosome exocytosis may help by releasing non-degraded metabolites extracelluarly. This can be mediated by cyclodextrin, which induces calcium-dependent lysosome exocytosis similar to the events of plasma membrane repair discussed in Sec. 2.1.[103] Cholesterol expulsion from NPC cells via exosomes has also been shown.[104] Lysosomal exocytosis has also been documented in MSD and MPS type IIIA via overexpression of transcription factor EB (TFEB),[105] or by using δ-tocopherol, as shown for NPC and Wolman diseases.[106] Moreover, δ-tocopherol has been shown to reduce lipid accumulation in LSD cell models of Batten, Fabry, Farber, Sanfilippo type B, Tay-Sachs, and Niemann-Pick disease type A.[106]

Although certainly valuable, most these approaches are suboptimal, particularly for neurological symptoms. Hence, there is a compelling need for approaches to more effectively treat these diseases.

3. Lysosomal-Related Organelles and their Role in Disease

In most LSDs, lysosomal accumulation of substrates causes secondary accumulation in components of the autophagic and endolysosomal pathways.[71,107] Some of these systems, particularly autophagy components, are also impaired in non-LSD diseases, including neurodegenerative disorders, inflammatory diseases, and a variety of cancers.[107–112]

3.1. *Mechanisms of autophagy and connection to lysosomal storage disorders*

Autophagy is the process by which cells degrade intracellular proteins and organelles. The two main types of autophagy important to this discussion are macroautophagy and chaperone-mediated autophagy (CMA)[71,73,113] (Fig. 3). In macroautophagy, the elements to be digested are sequestered from the cytosol into double-membrane-bound vesicles called autophagosomes.[71] Autophagosomes fuse with early and late endosomes, and finally fuse with mature lysosomes (autolysosomes) for degradation of the entrapped materials.[71] This is the most common form of autophagy and is responsible for degradation of cellular components such as mitochondria, RNA, ER fragments, soluble proteins, carbohydrates, and lipids.[108] Macroautophagy (herein referred to as autophagy) is controlled by autophagy-related genes (ATGs), with the beclin-1/phosphatidylinositol 3-kinase (PI3K) type III complex and mammalian target of rapamycin (mTOR) kinase being two major protein regulators.[108]

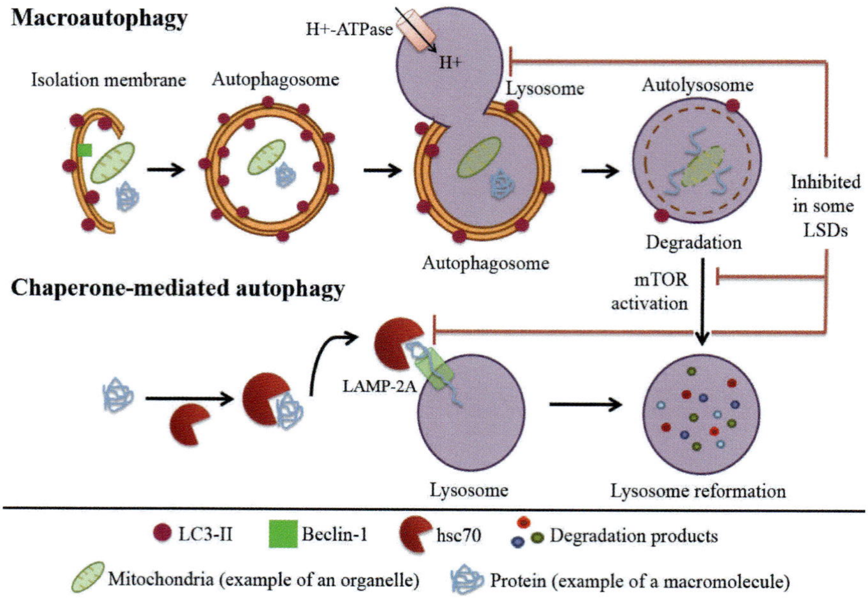

Fig. 3. Mechanisms of autophagy. Macroautophagy involves bulk sequestration of cytosolic materials into autophagosomes decorated with LC3-II, followed by fusion with lysosomes, degradation of the entrapped materials with release into the cytosol (not depicted), and regeneration of a mature lysosome via mTOR activation. Conversely, chaperone-mediated autophagy selectively targets proteins with a KFERQ motif for chaperone-mediated transport via hsc70 to lysosomes and translocates the protein across the lysosomal membrane upon interaction with LAMP-2A. Some LSDs display dysfunctions in these pathways, including impaired autophagosome-lysosome fusion, inhibition of mTOR activation, and inhibition of protein translocation.

Autophagy is a ubiquitous process and contributes to the maintenance of cellular homeostasis and disease prevention.[107,108] TFEB (transcription factor EB), involved in lysosomal biogenesis and function, also controls autophagosomes by regulation of the coordinated lysosomal expression and regulation (CLEAR) gene network.[114–116] Under cellular stress, mTOR-dependent phosphorylation of TFEB is reduced, causing TFEB to travel to the nucleus for transcriptional induction of CLEAR genes.[113,114,117] In several cell culture models of LSDs, including MPS type II, MPS type IIIA and MSD, TFEB localizes primarily to the nucleus, indicating that the autophagic pathway is activated in these diseases.[115]

Several LSDs are also known to impair autophagy.[118,119] Lysosomal storage can delay the rate at which autophagic vacuoles are processed or affect autophagosome-lysosome fusion, causing retention of materials in autophagosomes, as shown in Pompe disease, ML type IV, MPS type IIIA, and

MSD.[107,118,119] In cell culture models of MPS type IIIA and MSD, this effect associates with abnormal distribution of SNARE proteins in cholesterol-rich regions of the lysosomal membrane, which locks the SNAREs in place and inhibits their capacity to assist in autophagosome-lysosome fusion.[120] Autophagosome accumulation in LSDs can also be caused by further induction of autophagy, as observed in NPC disease, Batten disease, and GM1 gangliosidosis.[118,119] Under normal physiological conditions, autolysosomes are transient organelles: reformation of lysosomes occurs quickly after degradation of the entrapped material is complete,[71] which is largely mediated by mTOR activity.[121] Unfortunately, mTOR is also autophagy-dependent and exists in a non-active state until degradation of the products in autolysosomes is complete.[121] Since LSDs are characterized by failure of autolysosome clearance, mTOR cannot be reactivated and lysosome reformation is impaired.[121]

With regard to CMA, proteins containing a pentapeptide (KFERQ) motif are recognized by the constitutively expressed form of heat shock cognate protein of 70 kDa (hsc70).[122] Subsequently, these captured proteins are delivered to lysosomes with help from co-chaperones that aid in unfolding the protein for translocation across the lysosomal membrane upon interaction with LAMP-2A.[113,123] Improper CMA has also been identified in several LSDs, including Danon disease, ML type IV, and Galactosialidosis, with accumulation of autophagic vacuoles in the cytosol of cells.[124–126]

3.2. *Impaired autophagy in other diseases*

Neurodegenerative diseases such as Alzheimer's disease (AD), Huntington's disease (HD), and Parkinson's disease (PD) affect millions of people worldwide in the middle to late stages of life.[127] Patients progressively lose cognitive and motor abilities due to neuronal death resulting from protein accumulation in autophagic vacuoles, which is similar to the neurological effects of LSDs[110,111,128] (Table 2).

AD is the most common neurodegenerative disorder and is characterized by intraneuronal deposits of neurofibrillary tangles and extracellular senile plaques of β-amyloid (Aβ) deposits.[110] Dysfunction of the autophagic pathway in AD is caused by defective lysosomal proteolysis,[128] leading to the accumulation of undigested substrates in lysosomes and autophagic vacuoles within dystrophic axons and dendrites.[110] The cause of this dysfunction is mutated presenilin-1 (PS1), which affects proper glycosylation and transport of vacuolar H^+-ATPases, resulting in poor lysosome acidification and, hence, poor activation of lysosomal hydrolases.[110] Severe autophagy failure in AD also associates with overexpression of mutated amyloid precursor protein

Table 2. Elements defective in the autophagic-lysosomal pathway in non-LSDs.

Disease	Defective component(s)	Refs.
Alzheimer's disease	Defective lysosomal proteolysis due to a mutation in PS1 that reduces lysosomal acidification. Slowed autophagic flux due to overexpression of mutated amyloid precursor protein.	110, 128
Huntington's disease	Prevention of normal degradation by proteasome action due to the large size of mutant huntingtin. Inability of autophagosomes to sequester cytosolic materials for degradation.	109, 128, 129
Parkinson's disease	Altered CMA activity due to high binding affinity of mutated α-synuclein with LAMP-2A and interaction of defective forms of ubiquitin-terminal esterase L1 with LAMP-2A. Impaired lysosomal acidification, decreased hydrolase activity, and impaired removal of autophagic vacuoles due to defective lysosomal type 5 P-type ATPase.	113, 130, 133, 134
Crohn's disease	Excessive inflammatory response and impaired bacterial clearance due to mutations in three autophagic genes: ATG16L1, NOD2, IRGM.	108, 138
Cystic Fibrosis	Down-regulation of beclin-1 with dysfunction of p62/SQSTM1 that normally aids in autophagy degradation, and inactivation of the beclin-1/PI3K type III complex required for autophagy initiation.	108, 139
Cancer	Transcriptional activation of autophagy inducers due to p53 localization in the nucleus, loss of normal beclin-1 activity, inactivation of other autophagy-related genes, dysfunctions in p62/SQSTM1, alterations in lysosomal trafficking and localization, increased expression of cathepsins, and altered localization of heat-shock protein 70.	112, 143–146

(APP),[110] causing ubiquitinated proteins and Aβ peptides to accumulate within the cell and a reduction in autophagic flux which, combined with other oxidative damage impairs the endolysosomal and autophagic pathways.[110]

Lysosomal dysfunction and impaired autophagy are also found in HD, an autosomal dominant disorder due to a CAG trinucleotide repeat expansion (>35 repeats) at the N-terminus of the protein huntingtin (htt).[109,128] Mutant htt activates autophagy and the endolysosomal pathway with increased tubulation of endosomal membranes, lysosome number, and protease levels.[129]

Lysosomes secrete proteases to activate caspase-3 for cleavage of N-terminal fragments of htt.[129] With wild-type htt, these cleaved fragments are degraded by the proteasome; however, the larger expanded fragments of mutant htt cannot be degraded by this mechanism, leading to compensation via autophagocytosis.[129] Consequently, this results in slow autophagosome turnover and accumulation of these vesicles in the cytosol and perinuclear regions.[109,129]

Lastly, PD is caused by the aggregation of α-synuclein, a cytosolic protein located primarily in neural tissue, whose mutations or increased expression lead to pathology.[117,130,131] In the healthy brain, α -synuclein is degraded by the ubiquitin-proteasome system and CMA.[117,132] However, CMA cannot degrade mutated forms of α-synuclein due to their high affinity to LAMP-2A, which blocks uptake and degradation by lysosomes.[130] Defective forms of ubiquitin-terminal esterase L1 are also associated with altered CMA activity due to aberrant interaction with LAMP-2A in PD patients.[113,133] Blockage leads to storage of α-synuclein and other materials in Lewy bodies, as well as the accumulation of large secondary lysosomes containing lipofuscin in the brainstem and cortex.[133] In an atypical form of PD (Kufor–Rakeb syndrome), a mutation in the gene that encodes for lysosomal type 5 P-type ATPase also affects autophagic and lysosomal function.[134]

As mentioned previously, the CNS pathology of these neurodegenerative diseases is very similar to some LSDs, highlighting common mechanisms of pathology.[110,128] AD-like neurofibrillary tangles and accumulation of autophagic vacuoles occurs in MPS type IIB and NPC disease, with NPC patients also presenting Aβ deposits.[110,135] Many PD patients are carriers of Gaucher's disease mutations and both diseases associate with astrogliosis and Lewy body-like compartments containing α-synuclein.[136,137] Moreover, a number of LSDs including Tay–Sachs, GM1 gangliosidosis, NPC disease, and several types of MPS have been linked to PD based on phenotypic similarities.[136]

Impaired autophagy is also linked to inflammatory and auto-immune disorders, including pulmonary disorders, metabolic diseases such as obesity and diabetes, systemic lupus erythematosus, and others.[108] This is the case for Crohn's disease, which is caused by excessive inflammatory response and is exacerbated by impaired intracellular bacterial clearance due to defective autophagic activity.[108] Three genes associated with autophagy (ATG16L1, NOD2, and IRGM) have been identified as contributors to Crohn's disease.[138] Cystic Fibrosis (CF), a genetic disorder caused by mutations in the gene encoding the Cystic Fibrosis transmembrane conductance regulator (CFTR), also exhibits impaired autophagy with accumulation of protein aggregates in lung epithelial cells, which leads to increased inflammation.[139]

Cancer is another disease group where impaired autophagy is observed.[112] Under normal conditions, autophagy aids in tumor suppression by preventing the accumulation of damaged organelles/proteins and regulating cell growth.[140] Yet, in response to high metabolic demand and hypoxic conditions of established tumors, autophagy can be activated and allows tolerance of these adverse conditions for cancer cell survival and proliferation.[141] Depending on its cellular localization, the tumor suppressor protein p53 can induce or inhibit autophagy.[112] Autophagy is inhibited when p53 is localized in the cytoplasm and is induced by transcriptional activation when p53 is located in the nucleus.[112] The initiator of autophagy, beclin-1, also acts as a tumor suppressant[112] and loss of beclin-1 activity is found in 40–75% of human breast, prostrate, and ovarian cancers.[142]

Substantial cross-talk between autophagy and apoptosis exists so that one mechanism may persist if another is disabled.[112] This is possible because the two pathways share a number of cellular mediators, such as the anti-apoptotic beclin-2 family of proteins and the contribution of p62/SQSTM1 to enhanced TRAIL-mediated apoptosis, which plays a large role in cell death of multiple tumor types.[112] Indeed, autophagy plays an important role in apoptosis, especially when in the presence of other apoptotic signals.[112] Lysosomes also play a role in the induction of apoptosis,[143] e.g. disruption of the lysosomal membrane can lead to cytosol release of lysosomal hydrolases (including cathepsins) with caspase activation and mitochondrial outer-membrane permeabilization.[143,144] Alternatively, other lysosomal alterations can actually facilitate cancer progression.[143] For instance, alterations in lysosomal trafficking and increased expression of aspartate cathepsin, as well as cathepsin B, D, and L, are associated with tumor invasion and metastasis.[143,145] Moreover, translocation of hsp70 to the inner leaflet of endolysosomal membranes in cancer cells prevents lysosomal membrane permeabilization and, hence, cell death.[143,146]

3.3. *Treatment for autophagy-related diseases*

Given that impaired autophagy is a prominent contributor to these diseases, regulation of the autophagic-lysosomal pathway represents a potential treatment option. In the context of LSDs, the lack of fusion between autophagosomes and lysosomes presents an obstacle for ERT.[107] As discussed, defective fusion of these compartments may be due to inhibition of SNARE proteins in regions of endolysosomal membranes with an excess of cholesterol.[120] Therefore, one potential method to restore autophagosome-lysosome fusion is to use cyclodextrins for cholesterol reduction, as shown for MSD and MPS type IIIA.[120] Another method to restore normal autophagic function and potentially repair cellular defects found in LSDs is by activation of TFEB and the CLEAR network.[73] For instance, overexpression of TFEB enhances lysosomal clearance of glycosaminoglycans in LSD animal models, as well as mutant htt accumulated in a cell culture model of HD.[115] Alternatively, inactivation of an essential component in autophagosome formation (ATG7) improved recombinant acid α-glucosidase trafficking and normalized glycogen storage in the skeletal muscle of mice with Pompe disease.[147]

With regard to neurodegenerative diseases, rapamycin (and its analogs) is commonly used to induce autophagy since it is the natural inhibitor of mTOR.[107] Rapamycin-induced autophagy results in degradation of α-synuclein and mutant htt fragments in cell culture models of PD and HD, respectively.[148,149] Through clearance of damaged mitochondria, rapamycin-induced autophagy also offers protection against apoptosis.[150] Besides rapamycin, molecules that act independently of the mTOR pathway, such as lithium (also tested for treatment of Batten disease),[151] carbamazepine, valproic acid, and trehalose have shown promising results for the treatment of HD and PD.[107,117] Last, deletion of the protease inhibitor, cystatin B, is able to raise cathepsin D, B, and L levels back to normal in mice with AD.[152] Consequently, there is significant clearance of Aβ, ubiquitinated proteins, and other substrates in autophagosomes, which improves cognitive abilities such as learning and memory functions.[152]

Rapamycin and other mTOR inhibitors have also been used to induce autophagy in cancer cells, associated with sensitization of tumor cells to DNA-damaging agents[153] and tumor regression.[154] However, rapamycin and its analogues have only been successful in clinical treatment of renal cell carcinoma, neuroendocrine carcinoma, and lymphoma[155] since this agent only inhibits one complex of mTOR (i.e. mTORC1).[112] A number of other drugs are also capable of inducing autophagy in cancer cells including chloroquine and hydroxychloroquine, which enhance efficacy of chemotherapeutics in breast, colorectal, lung, prostate, and other forms of cancer.[112] Other therapies that target lysosomal abnormalities in cancer cells include those focused on preventing the secretion of cathepsins,[143] or regulating the expression/activity or subcellular localization of hsp70.[143,146,156]

Importantly, most autophagy regulators lack specificity and the best treatment is one that attempts to constantly maintain autophagic activity at proper levels, as opposed to shifting its activity to one extreme or the other.[112] Nevertheless, combination therapies involving autophagy regulators could offer invaluable therapeutic benefit for the treatment of these disorders.

4. Lysosomal Delivery and Effects Mediated by Nanocarriers

To overcome the limitations posed by current LSD treatment modalities, nanocarrier systems are being explored. During the 1970s and early 1980s, liposomes were utilized to improve ERT.[157–162] However, with the development of a greater variety of drug delivery systems in recent years,[1,2] new opportunities exist for treatment of LSDs.

4.1. *Nanocarriers for enzyme replacement therapy of lysosomal storage disorders*

As discussed in Sec. 2.3, ERT is a promising treatment option for patients with LSDs and is clinically available for a number of these diseases.[81] However, delivery of "naked" recombinant enzymes poses several limitations, including limited circulation, lack of protection from premature degradation, immunogenicity with production of antibodies, suboptimal targeting, lack of transport across cellular barriers such as the BBB, limited regulation over the subjacent endocytic uptake and lysosomal transport mechanism, and lack of control over enzyme release rate in lysosomes.[34] Unfortunately, the use of fusion proteins improves only aspects related to targeting and uptake.[48] Overall, these drawbacks make ERT a limited and expensive treatment, averaging ~$50,000–$165,000 per patient each year.[163]

To overcome these obstacles, recombinant lysosomal enzymes can be coupled to drug delivery carriers[34] (Table 3). Enzyme encapsulation can protect from degradation and offer control over the release rate.[34] Addition of "stealth" polymers such as PEG to the nanocarrier surface can help protect against degradation by proteases, immunological responses, and prolong systemic circulation.[28,29] Nanocarrier coating with multiple copies of affinity moieties can provide enhanced targeting via enhanced avidity due to multivalency,[164] also allowing selection over the mechanism of transport across cellular barriers and into cells.[34] Enzyme-carrying systems could also be co-loaded with small cell-permeable compounds so that lysosomal degradation of nanocarriers will release these cargoes within the cell for combined therapeutic interventions (e.g. using recombinant enzymes and small molecule treatments).[34]

Liposomes were the first drug delivery systems investigated in the context of LSD treatments. Initial studies conducted during the 1970s and early 1980s tested the potential of liposomes for enzyme delivery of β-fructofuranosidase, α-mannosidase, β-glucuronidase, β-glucosidase, and β-galactosidase.[157–162] *In vivo* performance of each formulation depends heavily on the

Table 3. Advantages of using nanocarriers in ERT for LSDs.

Challenges of ERT for LSDs	Nanocarrier benefits
Rapid clearance from the circulation	PEG or other stealth strategies; modification of nanocarrier physicochemical properties
Loss of enzyme activity by proteases or generation of blocking antibodies	Nanocarriers provide protection by encapsulation or by posing steric hindrance to proteases and the immune system
Glycosylation defects of lysosomal enzymes	Use of glycosylation-independent mechanisms for enzyme targeting
Preferential accumulation in the RES	Carrier functionalization for active targeting
Inadequate transport across tissue and cellular barriers	Carrier functionalization to enhance transport
Gaining cell entry	Targeting cell surface receptors involved in various forms of endocytosis
Need frequent administrations due to inefficient delivery and lack of sustained release	Nanocarriers provide controlled release of cargo by modulation of nanocarrier composition

physicochemical properties of the liposome, which can be influenced by the phospholipid composition, entrapped lysosomal enzyme, preparation method, etc.[158–160,162] For example, liposomal formulations showed enhanced uptake of β-glucuronidase (MPS type VII) in the liver,[160] where negatively-charged liposomes accumulated in lysosomes more efficiently and positively-charged liposomes showed longer enzyme activity (11 days versus 2 days).[160] The addition of PEG to liposomal formulations enhanced both enzyme delivery and prolonged activity in lysosomes.[165] In cases where lysosomal delivery is required to treat immune cells (e.g. in Batten disease),[166,167] liposomes can be coated with autologous heat-aggregated IgG or apolipoprotein *E* to enhance accumulation in lysosomes of polymorphonucleocytes.[166] In a similar technique, phagocytes were targeted by liposomes coated with heat-aggregated isologous IgM.[167] More recently, liposomes modified with lysosomotrophic agents (e.g. octadecyl-rhodamine B (Rh)) and loaded with velaglucerase alfa (VPRIV™) for ERT of Gaucher disease type I were shown to improve uptake by monocytes and fibroblasts in culture, accumulation in lysosomes, and enzymatic activity.[168–170]

Apart from liposomes, limited work has been published on the use of other drug delivery systems for LSD enzyme replacement. At the micro-scale, calcium alginate microspheres loaded with glucocerebrosidase have been tested in cell culture to mimic local implantation in Gaucher disease type I.[171] This would achieve prolonged enzyme release in areas of the body that are difficult to access from the circulation, as current ERT by *i.v.* administration is sub-optimal at alleviating bone pathologies characteristic of this disease.[171] Immobilization of lysosomal enzymes on magnetic microparticles has also been reported.[172]

At the sub-micrometer scale, several nanocarrier modalities have been used for endocytic uptake in cells other than those of the immune system. With regard to inorganic particles, QDs with surface-immobilized enzymes resulted in lysosomal transport with sufficient enzyme delivery to restore normal substrate levels in a cell culture model.[173] Gold nanoparticles (13 nm) coated with cell-penetrating peptides (penetratin and Tat) and lysosomal sorting proteins (L1 and L2) showed targeted localization into lysosomes in cell cultures and have been proposed to be applicable for delivery of lysosomal enzymes.[174] As per polymer-based particles, polyelectrolyte-based complexes (<200 nm) of trimethyl chitosan displayed pH-dependent dissociation of α-galactosidase with enzyme release within the lysosomal compartment in a cell culture model.[175]

Overall, these approaches have shown enhanced enzyme circulation and accumulation in RES organs *in vivo*, or improved enzyme stability with lysosomal transport and activity in cell cultures. In addition, a series of studies using targeted polymer nanocarriers have shown active targeting of lysosomal enzymes with transport across cellular barriers and into lysosomes in non-immune cells (the main targets for most LSDs) in cell cultures and animal models, as described below.

Long after the liposomal studies of the 1970s and 1980s, the first strategy that attempted delivery of recombinant lysosomal enzymes using nanocarriers was published in 2004 and focused on targeting intercellular adhesion molecule 1 (ICAM-1) using model polymer nanoparticles (~200 nm) coated with anti-ICAM antibodies.[176] ICAM-1 is a transmembrane glycoprotein involved in inflammation[177,178] and was selected for targeting because of several reasons. ICAM-1 is expressed on most cell types in the body, which is amenable for targeting both peripheral organs and the CNS.[177] It is particularly overexpressed during pathology, including inflammation associated with LSDs, which can provide preferential targeting to most affected sites.[177] ICAM-1 is abundant on endothelial cells lining the vasculature, in direct contact with lysosomal enzymes injected in circulation, and is uniquely involved in mediating transcellular extravasation of leukocytes from the bloodstream to tissues without opening of the cell junctions, which may facilitate transport across cellular barriers.[177,178] In addition, ICAM-1 had been found to mediate endocytosis with transport to lysosomes via CAM-mediated endocytosis[43] (Fig. 4). This is a unique mechanism different from classical macropinocytosis, phagocytosis, and clathrin- or caveolar-mediated endocytosis, and has been shown in all cell types tested to date, including endothelial, epithelial, mesothelial, fibroblasts, skeletal muscle, neurons, cells of the RES, etc.[177] Further, it provides uptake of drug carriers within an ample range of sizes (few hundred nm to a few μm) in non-immune cells.[179,180] This is possible due to specific ICAM-1-mediated activation of the sodium-proton exchanger protein NHE1 and the sphingomyelin/ceramide pathway, which regulate the physicochemical properties of the cell surface and reorganization of the cytoskeleton into actin stress fibers, facilitating engulfment and endocytosis of micron-sized objects targeted to ICAM-1.[46,179,180]

Several studies used model polystyrene nanocarriers or biocompatible PLGA counterparts to improve delivery of recombinant lysosomal enzymes

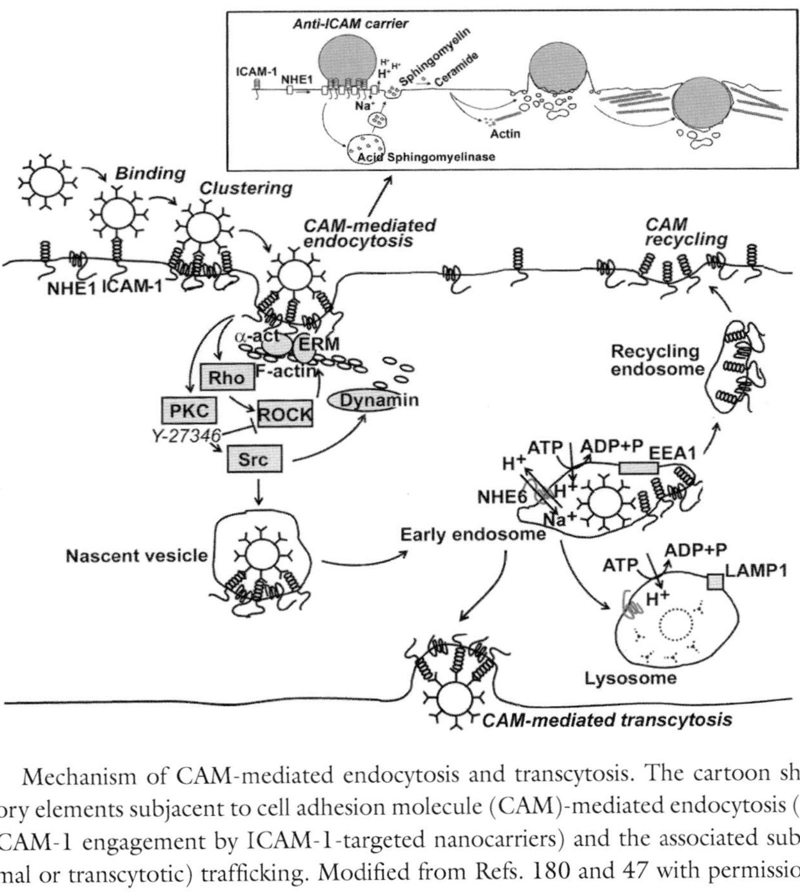

Fig. 4. Mechanism of CAM-mediated endocytosis and transcytosis. The cartoon shows the regulatory elements subjacent to cell adhesion molecule (CAM)-mediated endocytosis (induced upon ICAM-1 engagement by ICAM-1-targeted nanocarriers) and the associated sub-cellular (lysosomal or transcytotic) trafficking. Modified from Refs. 180 and 47 with permission.

by targeting ICAM-1, including acid sphingomyelinase (ASM) deficient in types A and B Niemann–Pick disease,[179,181–184] α-galactosidase A deficient in Fabry disease,[55,185] and α-glucosidase deficient in Pompe disease.[186] This strategy reduced the circulation time of lysosomal enzymes (preferable to avoid immunogenicity), while providing enhanced enzyme accumulation in most LSD target organs after *i.v.* administration in mouse models, as compared to free enzyme.[181,183–186] For instance, ICAM-1-targeted nanocarriers achieved ~200-fold and ~15-fold enhanced enzyme accumulation in the lung and liver (major targets for type B NPC disease), ~5-fold and ~1.5-fold increases in the heart and kidneys (major targets for Fabry disease), and ~10-fold and ~6-fold increases in the heart and skeletal muscle (major targets for Pompe disease), respectively.[181,185,186] Additionally, ICAM-1-targeted nanocarriers tested in cell cultures improved uptake, lysosomal transport, and lysosomal degradation of the intended substrate compared to free enzymes.[179,182,185,186] The rate of CAM-mediated endocytosis and lysosomal transport can be controlled by modulation of several design parameters, such as carrier size and shape,[179] the molecular epitope to which carriers bind,[187] and pharmacological agents that impact the cell machinery involved in regulation of endocytic transport.[188,189]

Importantly, it has recently been found that ICAM-1-targeted nanocarriers can transport lysosomal enzymes across cellular barriers, such as the case of the BBB *in vivo*.[190] Transport occurs via endocytosis without opening of the cell junctions or leakage of other substances (e.g. serum albumin) into the brain.[55,190] As a result, this strategy provided markedly enhanced accumulation of lysosomal enzymes compared to free counterparts in the brain of mice after a single *i.v.* injection: ~7-fold increase for acid sphingomyelinase, ~4-fold increase for α-galactosidase and ~6-fold increase for α-glucosidase.[185,186] In addition, CAM-mediated transcytosis was also shown to efficiently and safely transport α-galactosidase across a cell model of the gastrointestinal epithelium,[55] thereby providing an avenue to design future oral delivery applications of lysosomal ERT.

4.2. *Nanocarriers used in other lysosomal treatment modalities*

Apart from ERT, nanocarrier-mediated gene therapy and cell therapy have also been studied for LSD treatment, although in a very limited capacity.[191–193] As discussed in Sec. 2.3, gene therapy represents a highly promising treatment modality because it has the capacity to correct the genetic defect(s) that cause LSDs.[81,86] Somatic gene therapy via viral vectors has been preferred as viral systems perform advantageously in this application and are actually approved for human clinical trials of several LSDs, including Pompe disease and Sanfilippo syndrome.[84–86] However, non-viral vectors also present interesting features, e.g. they present less risk for gene insertion errors and immune responses, and are less expensive to manufacture.[193,194] Furthermore, while non-viral systems may be less efficacious than viral approaches, they still can achieve sufficient expression for a therapeutic response in LSDs since only a relatively low amount of the deficient enzyme is needed to alleviate phenotypic symptoms due to the cross correction effect.[193]

As with the delivery of many other biological agents, nanocarriers for nucleic acid-based therapeutics can be created from a number of different biological and synthetic polymers, and can be further functionalized with stealth polymers, targeting moieties, cell-penetrating peptides, etc.[194] Cationic polymers are used to allow interaction with negatively-charged nucleic acid and charge neutralization renders nucleic acid condensation into compact structures for cell internalization and reduced degradation by nucleases.[194] In fact, formulations based on cationic lipids are the most widely used delivery system for nucleic acids.[194] Both liposomes and solid lipid nanoparticles have been utilized to deliver plasmid DNA encoding for lysosomal enzymes, including hexosaminidases deficient in Tay–Sachs and Sandhoff disease,[193] α-galactosidase A deficient in Fabry disease,[191] and β-galactosidase deficient in GM1 gangliosidosis.[195] For example, *i.v.* administration of cationic liposomes loaded with plasmid cDNAs for either sub-unit of hexosaminidase was tested in Sandhoff disease mice.[193] This resulted in 10–35% hexosaminidase expression in the peripheral organs of mice compared to normal levels, and a reduction of GA2 and GM2 substrates by 10% and 50% after 3 days, respectively. Further, by day 7, these levels reached 60% GA2 reduction and 70% GM2 reduction. Unfortunately, the efficacy of this treatment was found to be limited due to RES-mediated uptake of liposomal aggregates, and this system could only be used for LSDs lacking CNS pathology, as hexosaminidase expression was not observed in the brain.

In an alternative approach, recombinant fibroblast cells engineered via non-viral gene therapy to express and secrete β-glucuronidase have been developed and tested in MPS type VII-deficient mice.[192,196] Since the main problem with cell transplantation is the potential of an immunological reaction leading to graft rejection, the non-autologous cells were encapsulated in alginate microcapsules prior to *in vivo* administration.[192,196] When injected directly into the lateral ventricles of the mouse brain, β-glucuronidase delivery from the microencapsulated cells was achieved throughout most of the CNS and was sufficient to reverse histological pathology and behavioral abnormalities.[196] Similarly, intraperitoneal (*i.p.*) implantation of the microcapsules resulted in enhanced β-glucuronidase activity (66% of normal levels by two weeks) and a significant reduction in lysosomal accumulation of undegraded glycosaminoglycans over the 8-week study, especially within the liver and spleen.[192]

4.3. *Lysosomal side effects induced by nanocarriers*

As described in previous sections, most nanocarriers enter cells via the endocytic pathway and are often sorted to lysosomes.[42] Therefore, it is important to consider the potential detrimental effects that nanocarriers present to lysosome structure and function, as well as the direct consequences on the autophagic pathway (Table 4). Overall, better understanding of how nanomaterials impact lysosome function will help advance the field of nanotechnology and improve the therapeutic efficacy of drug delivery systems.

In order to escape the lysosomal compartment and gain access to the cytosol, which is required for many therapeutic applications, a number of nanocarrier-based strategies rely on lysosomal membrane permeabilization (LMP), including cationic polystyrene nanoparticles,[197] cationic PAMAM dendrimers,[198] PLGA nanoparticles,[199] metal nanoparticles,[200] carbon nanotubes,[201] and many others.[202] This may result in oxidative stress and release of lysosomal hydrolases including cathepsins.[143] Subsequently, this leads to permeabilization of the outer mitochondrial membrane, inducing further ROS production and eventually resulting in cell apoptosis.[143] In the case of cationic nanocarriers, protonation of the surface amine groups neutralizes the acidic pH of lysosomes and ruptures the lysosomal membrane due to increasing osmolarity.[197] This mechanism is called the proton sponge effect and is the most common cause of nanoparticle-induced LMP.[202] For non-cationic nanocarriers, however, generation of ROS is a likely mechanism underlying LMP, as demonstrated in cell culture models.[203] Nanocarrier functionalization with certain protein- and peptide-based agents derived from bacteria or viruses results in formation of pores in the lysosomal membrane or lysis of this membrane via interaction with the phospholipid bilayer, with similar outcomes.[41]

Alkalization of lysosomes alone (i.e. neutralization of lysosomal pH) can also predispose cells to LMP.[204] For example, pH-dependent dissolution of toxic Zn^{2+} ions from zinc oxide nanoparticles decreased lysosomal stability, resulting in macrophage cell death and significant lung injury in treated rats.[204] In addition to LMP, alkalization of lysosomes can also lead to impairment of lysosomal degradation by disrupting the vacuolar ATPase proton pump, as observed with gold nanoparticles.[205] Ultimately, the gold

Table 4. Lysosomal and autophagic side effects induced by nanocarriers.

Side effect	Mechanism	Nanocarrier types	Refs.
Lysosomal membrane permeabilization	Proton sponge effect	Cationic nanocarriers, such as chitosan nanoparticles, PAMAM dendrimers and cationic polystyrene nanoparticles, or formulations containing poly(ethylenimine)	41, 143, 197, 198, 202
Lysosomal membrane permeabilization	ROS generation	Titanium oxide nanoparticles, fullerenes, carbon nanotubes, cationic polystyrene nanospheres	203
Lysosomal membrane permeabilization	Creation of pores in the lysosomal membrane or membrane lysis via interaction with the phospholipid bilayer	Nanocarriers functionalized with fusogenic proteins or peptides derived from bacteria or viruses, such as GALA and hemagglutinin	41
Lysosomal membrane permeabilization	Alkalization (i.e. neutralization of pH)	Zinc oxide and gold nanoparticles	204, 205
Autophagy impairment	Lysosomal accumulation of undigested nanocarriers resulting in reduced autophagic flux with autophagosome accumulation	Gold, polystyrene and titanium dioxide nanoparticles	197, 202, 205, 207–209
Autophagy induction	Autophagy induction due to ROS generation and oxidative stress or as a defensive response	Gold nanoparticles, fullerenes, dendrimers, quantum dots, magnetic nanoparticles, and neodymium oxides	211–219
NLRP3 inflammasome formation and release of pro-inflammatory IL-1β	Cathepsin leakage and ROS generative from damaged mitochondria following nanocarrier-induced LMP	Amino-functionalized polystyrene nanoparticles, silver, silica and titanium dioxide nanoparticles	220–222

nanoparticles could not be degraded, which hindered normal autophagic processing.[205] Similar to the outcome of alkalization, a number of nanoparticles contain lysosomotropic drugs, which are known to accumulate in lysosomes, inhibit the activity of several lysosomal enzymes, disrupt fusion events and induce autophagy,[206] which is discussed in more detail below.

Without even compromising lysosomal pH, autophagy dysfunction can result from lysosomal accumulation of undigested nanocarriers, which is similar to the cause of LSDs and neurodegenerative diseases.[202,207] Due to the non-degradable nature of some nanocarriers, including gold nanoparticles, polystyrene nanoparticles, and titanium dioxide nanoparticles, they cannot be degraded by lysosomes and accumulate within this compartment.[197,205,208,209] As discussed, lysosomal storage reduces autophagic flux with subsequent autophagosome accumulation.[202] However, the major concern of this outcome is that autophagy dysfunction results in cellular toxicity and cell death associated with the pathologies discussed in Sec. 3.2.[202] Prolonged storage and high accumulation of single-walled carbon nanohorns also destabilizes the lysosomal membrane, initiating downstream processes detrimental to cell survival.[210]

Alternatively, a variety of nanoparticles including gold nanoparticles, fullerenes, dendrimers, QDs, and neodymium oxides are autophagy activators.[211–215] There are several plausible mechanisms as to how nanoparticles initiate the autophagic pathway. First, autophagy could be induced due to oxidative stress following nanoparticle exposure, as observed for fullerenes in HeLa cells.[215] Another likely possibility is that autophagy activation acts as a defense mechanism against the foreign nanomaterials.[202] Supportive evidence for this theory is provided by numerous reports showing visible nanoparticles in the double-membrane of autophagosomes.[214,216] Recent studies have also demonstrated that nanocarriers undergo ubiquitination, which is a necessary process for transport to autophagosomes by p62.[217–219] Overall, induction of autophagy can also lead to the accumulation of autophagosomes, resulting in detrimental consequences including disease onset and progression.

Last, nanoparticle-induced LMP in human macrophages has led to formation of the NLRP3 inflammasome and release of pro-inflammatory IL-1β.[220–222] This process is initially mediated by cathepsin leakage into the cytosol and subsequent ROS production from damaged mitochondria following LMP.[222] Ultimately, NLRP3 inflammasome complexes assemble and regulate maturation of IL-1β through caspase-1 activation and cleavage of the pro-form of IL-1β.[223] Following maturation, IL-1β is released into the extracellular environment to perform its immunological functions.[222] Inflammasome formation via this mechanism has been observed for a number of nanocarrier systems including amino-functionalized polystyrene nanoparticles, silver nanoparticles, silica nanoparticles, and titanium dioxide nanoparticles.[220–222]

5. Conclusions and Future Outlook

In summary, lysosomes and lysosomal-like organelles are critical components of mammalian cells and function to maintain cellular homeostasis. Caused by the accumulation of undigested substrates, LSDs are characterized by widespread dysfunction in both peripheral organs and the CNS. While treatment for LSD patients has remarkably improved due to biotechnological therapies, including organ/cell transplantation, substrate reduction therapy, gene therapy, small molecule therapy and ERT, significant limitations reduce the effectiveness of these strategies and warrant further development of more advanced treatment options. The recent implementation of nanocarriers in these treatment modalities has shown to be a promising avenue. The most advantageous features of drug delivery systems for this application include controlled circulation time, effective targeting and cellular uptake, transcellular and intracellular trafficking and preservation of the therapeutic cargo with controlled release within the lysosome. Nevertheless, further research into the effectiveness of these systems, as well as their potential to disrupt lysosome function, is necessary before they can transition to clinical application. Last, the effect of lysosomal aberrations on impaired autophagic-lysosomal processing implicated in several prominent human diseases (including neurodegenerative disorders, inflammatory diseases, and cancer) was highlighted. The underlying mechanisms shared amongst these disease groups, especially LSDs and neurodegenerative diseases, present a unique opportunity to explore new treatment strategies involving therapeutic combinations to address lysosomal dysfunction and autophagy regulation.

6. Acknowledgments

The authors would like to acknowledge financial support from the NSF Graduate Research Fellowship Program (DGE-0750616) and the Flagship Fellowship Program of the University of Maryland, College Park (RLM), as well as the NIH (R01 HL098416) (SM).

References

1. R. Langer, Drug delivery and targeting, *Nature* **392**, 5–10 (1998).
2. V. P. Torchilin, Multifunctional nanocarriers, *Adv. Drug Deliv. Rev.* **58**, 1532–1555 (2006).
3. S. Muro, Challenges in design and characterization of ligand-targeted drug delivery systems, *J. Control Release* **164**, 125–137 (2012).
4. S. Parveen, R. Misra and S. K. Sahoo, Nanoparticles: a boon to drug delivery, therapeutics, diagnostics and imaging, *Nanomedicine* **8**, 147–166 (2012).

5. S. M. Moghimi, A. C. Hunter and J. C. Murray, Nanomedicine: current status and future prospects, *FASEB J.* **19**, 311–330 (2005).
6. S. K. Sahoo, S. Parveen and J. J. Panda, The present and future of nanotechnology in human health care, *Nanomedicine* **3**, 20–31 (2007).
7. A. D. Bangham, M. M. Standish and J. C. Watkins, Diffusion of univalent ions across the lamellae of swollen phospholipids, *J. Mol. Biol.* **13**, 238–252 (1965).
8. M. Beija, R. Salvayre, N. Lauth-de Viguerie and J. D. Marty, Colloidal systems for drug delivery: from design to therapy, *Trends Biotechnol.* **30**, 485–496 (2012).
9. T. Musacchio and V. P. Torchilin, Recent developments in lipid-based pharmaceutical nanocarriers, *Front. Biosci.* **16**, 1388–1412 (2011).
10. S. Betigeri, M. Zhang, O. Garbuzenko and T. Minko, Non-viral systemic delivery of siRNA or antisense oligonucleotides targeted to Jun N-terminal kinase 1 prevents cellular hypoxic damage, *Drug. Deliv. Transl. Res.* **1**, 13–24 (2011).
11. S. M. Moghimi and J. Szebeni, Stealth liposomes and long circulating nanoparticles: critical issues in pharmacokinetics, opsonization and protein-binding properties, *Prog. Lipid Res.* **42**, 463–478 (2003).
12. D. E. Discher and A. Eisenberg, Polymer vesicles, *Science* **297**, 967–973 (2002).
13. R. Duncan and M. J. Vicent, Polymer therapeutics-prospects for 21st century: The end of the beginning, *Adv. Drug Deliv. Rev.* **65**, 60–70 (2013).
14. A. V. Kabanov and T. Okano, Challenges in polymer therapeutics: state of the art and prospects of polymer drugs, *Adv. Exp. Med. Biol.* **519**, 1–27 (2003).
15. J. Panyam and V. Labhasetwar, Biodegradable nanoparticles for drug and gene delivery to cells and tissue, *Adv. Drug Deliv. Rev.* **55**, 329–347 (2003).
16. H. Ghandehari, Materials for advanced drug delivery in the 21st century: a focus area for Advanced Drug Delivery Reviews, *Adv. Drug Deliv. Rev.* **60**, 956 (2008).
17. K. M. Kitchens, M. E. El-Sayed and H. Ghandehari, Transepithelial and endothelial transport of poly (amidoamine) dendrimers, *Adv. Drug Deliv. Rev.* **57**, 2163–2176 (2005).
18. O. Kayser, A. Lemke and N. Hernandez-Trejo, The impact of nanobiotechnology on the development of new drug delivery systems, *Curr. Pharm. Biotechnol.* **6**, 3–5 (2005).
19. R. C. Mundargi, V. R. Babu, V. Rangaswamy, P. Patel and T. M. Aminabhavi, Nano/micro technologies for delivering macromolecular therapeutics using poly(D,L-lactide-co-glycolide) and its derivatives, *J. Control Release* **125**, 193–209 (2008).
20. S. M. Moghimi, A. C. Hunter and J. C. Murray, Long-circulating and target-specific nanoparticles: theory to practice, *Pharmacol. Rev.* **53**, 283–318 (2001).
21. S. A. Agnihotri, N. N. Mallikarjuna and T. M. Aminabhavi, Recent advances on chitosan-based micro- and nanoparticles in drug delivery, *J. Control Release* **100**, 5–28 (2004).
22. D. Schmaljohann, Thermo- and pH-responsive polymers in drug delivery, *Adv. Drug Deliv. Rev.* **58**, 1655–1670 (2006).
23. N. D. Sonawane, F. C. Szoka, Jr. and A. S. Verkman, Chloride accumulation and swelling in endosomes enhances DNA transfer by polyamine-DNA polyplexes, *J. Biol. Chem.* **278**, 44826–44831 (2003).
24. T. K. Jain *et al.*, Magnetic nanoparticles with dual functional properties: drug delivery and magnetic resonance imaging, *Biomaterials* **29**, 4012–4021 (2008).
25. C. E. Probst, P. Zrazhevskiy, V. Bagalkot and X. Gao, Quantum dots as a platform for nanoparticle drug delivery vehicle design, *Adv. Drug Deliv. Rev.* **65**, 703–718 (2013).

26. V. I. Shubayev, T. R. Pisanic, 2nd and S. Jin, Magnetic nanoparticles for theragnostics, *Adv. Drug Deliv. Rev.* **61**, 467–477 (2009).
27. H. Maeda, J. Wu, T. Sawa, Y. Matsumura and K. Hori, Tumor vascular permeability and the EPR effect in macromolecular therapeutics: a review, *J. Control Release* **65**, 271–284 (2000).
28. Z. Amoozgar and Y. Yeo, Recent advances in stealth coating of nanoparticle drug delivery systems, *Wiley Interdiscip. Rev. Nanomed. Nanobiotechnol.* **4**, 219–233 (2012).
29. F. Alexis, E. Pridgen, L. K. Molnar and O. C. Farokhzad, Factors affecting the clearance and biodistribution of polymeric nanoparticles, *Mol. Pharm.* **5**, 505–515 (2008).
30. H. Du, P. Chandaroy and S. W. Hui, Grafted poly-(ethylene glycol) on lipid surfaces inhibits protein adsorption and cell adhesion, *Biochim. Biophys. Acta* **1326**, 236–248 (1997).
31. S. J. Stachelek *et al.*, The effect of CD47 modified polymer surfaces on inflammatory cell attachment and activation, *Biomaterials* **32**, 4317–4326 (2011).
32. J. K. Vasir and V. Labhasetwar, Targeted drug delivery in cancer therapy, *Technol. Cancer Res. Treat* **4**, 363–374 (2005).
33. J. Hsu and S. Muro, Nanomedicine and drug delivery strategies for treatment of genetic diseases. In *Human Genetic Disease* D. Plaseska-Karanfilska (Ed.), In Tech (2011), Chapter 14: pp: 241–246.
34. S.Muro, Strategies for delivery of therapeutics into the central nervous system for treatment of lysosomal storage disorders, *Drug Deliv. Trans. Res.* **2**, 169–186 (2012).
35. J. S. Suk *et al.*, Gene delivery to differentiated neurotypic cells with RGD and HIV Tat peptide functionalized polymeric nanoparticles, *Biomaterials* **27**, 5143–5150 (2006).
36. A. R. Hilgenbrink and P. S. Low, Folate receptor-mediated drug targeting: from therapeutics to diagnostics, *J. Pharm. Sci.* **94**, 2135–2146 (2005).
37. W. M. Pardridge, Biopharmaceutical drug targeting to the brain, *J. Drug Target* **18**, 157–167 (2010).
38. A. T. Jones, Gateways and tools for drug delivery: endocytic pathways and the cellular dynamics of cell penetrating peptides, *Int. J. Pharm.* **354**, 34–38 (2008).
39. A. Chrastina, K. A. Massey and J. E. Schnitzer, Overcoming in vivo barriers to targeted nanodelivery, *Wiley Interdiscip, Rev. Nanomed. Nanobiotechnol.* **3**, 421–437 (2011).
40. C. Trollet, D. Scherman and P. Bigey, Delivery of DNA into muscle for treating systemic diseases: advantages and challenges, *Methods Mol. Biol.* **423**, 199–214 (2008).
41. A. K. Varkouhi, M. Scholte, G. Storm and H. J. Haisma, Endosomal escape pathways for delivery of biologicals, *J. Control Release* **151**, 220–228 (2011).
42. G. Sahay, D. Y. Alakhova and A. V. Kabanov, Endocytosis of nanomedicines, *J. Control Release* **145**, 182–195 (2010).
43. S. Muro *et al.*, A novel endocytic pathway induced by clustering endothelial ICAM-1 or PECAM-1, *J. Cell. Sci.* **116**, 1599–1609 (2003).
44. R. V. Stan, Endocytosis pathways in endothelium: how many? *Am. J. Physiol. Lung Cell. Mol. Physiol.* **290**, L806–808 (2006).
45. A. Dautry-Varsat, Receptor-mediated endocytosis: the intracellular journey of transferrin and its receptor, *Biochimie* **68**, 375–381 (1986).
46. S. Muro, C. Gajewski, M. Koval and V. R. Muzykantov, ICAM-1 recycling in endothelial cells: a novel pathway for sustained intracellular delivery and prolonged effects of drugs, *Blood* **105**, 650–658 (2005).
47. B. Ding, T. Dziubla, V. V. Shuvaev, S. Muro and V. R. Muzykantov, Advanced drug delivery systems that target the vascular endothelium, *Mol Interventions* **6**, 98–112 (2006)

48. S. Muro, New biotechnological and nanomedicine strategies for treatment of lysosomal storage disorders, *Wiley Interdiscip. Rev. Nanomed. Nanobiotechnol.* **2**, 189–204 (2010).
49. A. H. Futerman and G. van Meer, The cell biology of lysosomal storage disorders, *Nat. Rev. Mol. Cell Biol.* **5**, 554–565 (2004).
50. E. Dejana, Endothelial cell–cell junctions: happy together, *Nat. Rev. Mol. Cell Biol.* **5**, 261–270 (2004).
51. T. Stevens, J. G. Garcia, D. M. Shasby, J. Bhattacharya and A. B. Malik, Mechanisms regulating endothelial cell barrier function, *Am. J. Physiol. Lung Cell Mol. Physiol.* **279**, L419–422 (2000).
52. M. N. Azmin *et al.*, The effect of non-ionic surfactant vesicle (niosome) entrapment on the absorption and distribution of methotrexate in mice, *J. Pharm. Pharmacol.* **37**, 237–242 (1985).
53. J. P. Hanig, J. M. Morrison, Jr. and S. Krop, Ethanol enhancement of blood-brain barrier permeability to catecholamines in chicks, *Eur. J. Pharmacol.* **18**, 79–82 (1972).
54. D. Kobiler, S. Lustig, Y. Gozes, D. Ben-Nathan and Y. Akov, Sodium dodecylsulphate induces a breach in the blood-brain barrier and enables a West Nile virus variant to penetrate into mouse brain, *Brain Res.* **496**, 314–316 (1989).
55. R. Ghaffarian, T. Bhowmick and S. Muro, Transport of nanocarriers across gastrointestinal epithelial cells by a new transcellular route induced by targeting ICAM-1, *J. Control Release* **163**, 25–33 (2012).
56. P. Oh *et al.*, Live dynamic imaging of caveolae pumping targeted antibody rapidly and specifically across endothelium in the lung. *Nat. Biotechnol.* **25**, 327–337 (2007).
57. W. M. Pardridge, Blood-brain barrier delivery, *Drug Discov. Today* **12**, 54–61 (2007).
58. J. P. Luzio, P. R. Pryor and N. A. Bright, Lysosomes: fusion and function, *Nat. Rev. Mol. Cell Biol.* **8**, 622–632 (2007).
59. P. J. Meikle, J. J. Hopwood, A. E. Clague and W. F. Carey, Prevalence of lysosomal storage disorders, *JAMA* **281**, 249–254 (1999).
60. E. Holtzman, *Lysosomes* In Cellular Organelles, P. Siekevitz (Series Ed.), New York: Plenum Press (1989), pp. 1–455.
61. I. Mellman, R. Fuchs and A. Helenius, Acidification of the endocytic and exocytic pathways, *Annu. Rev. Biochem.* **55**, 663–700 (1986).
62. D. Sabatini and M. B Adesnik, The biogenesis of membranes and organelles. In *The Metabolic and Molecular Bases of Inherited Disease*, Eds. C. Scriver, A. Beaudet, W. Sly, D. Valle, B. Childs *et al.*, New York: McGraw-Hill (2001), Chapter 16.
63. N. W. Andrews, Regulated secretion of conventional lysosomes, *Trends Cell Biol.* **10**, 316–321 (2000).
64. P. L. McNeil and T. Kirchhausen, An emergency response team for membrane repair, *Nat. Rev. Mol. Cell Biol.* **6**, 499–505 (2005).
65. S. Kornfeld, M. L. Reitman, A. Varki, D. Goldberg and C. A. Gabel, Steps in the phosphorylation of the high mannose oligosaccharides of lysosomal enzymes, *Ciba Found. Symp.* 138–156 (1982).
66. M. G. Rosenfeld, G. Kreibich, D. Popov, K. Kato and D. D. Sabatini, Biosynthesis of lysosomal hydrolases: their synthesis in bound polysomes and the role of co- and post-translational processing in determining their subcellular distribution, *J. Cell Biol.* **93**, 135–143 (1982).
67. E. F. Neufeld, The uptake of enzymes into lysosomes: an overview, *Birth Defects Orig. Artic. Ser.* **16**, 77–84 (1980).

68. E. L. Eskelinen, Y. Tanaka and P. Saftig, At the acidic edge: emerging functions for lysosomal membrane proteins, *Trends Cell Biol.* **13**, 137–145 (2003).
69. S. C. Li and P. M. Kane, The yeast lysosome-like vacuole: endpoint and crossroads, *Biochim. Biophys. Acta* **1793**, 650–663 (2009).
70. E. C. Dell'Angelica, C. Mullins, S. Caplan and J. S. Bonifacino, Lysosome-related organelles, *FASEB J.* **14**, 1265–1278 (2000).
71. F. M. Platt, B. Boland and A. C. van der Spoel, The cell biology of disease: Lysosomal storage disorders: The cellular impact of lysosomal dysfunction, *J. Cell Biol.* **199**, 723–734 (2012).
72. G. M. Mancini, A. C. Havelaar and F. W. Verheijen, Lysosomal transport disorders, *J. Inherit. Metab. Dis.* **23**, 278–292 (2000).
73. M. L. Schultz, L. Tecedor, M. Chang and B. L. Davidson, Clarifying lysosomal storage diseases, *Trends Neurosci.* **34**, 401–410 (2011).
74. D. A. Pearce, T. Ferea, S. A. Nosel, B. Das and F. Sherman, Action of BTN1, the yeast orthologue of the gene mutated in Batten disease, *Nat. Genet.* **22**, 55–58 (1999).
75. M. K. Raychowdhury *et al.*, Molecular pathophysiology of mucolipidosis type IV: pH dysregulation of the mucolipin-1 cation channel, *Hum. Mol. Genet.* **13**, 617–627 (2004).
76. D. Mijaljica, M. Prescott and R. J. Devenish, V-ATPase engagement in autophagic processes, *Autophagy* **7**, 666–668 (2011).
77. J. A. Sobota, N. Back, B. A. Eipper and R. E. Mains, Inhibitors of the V0 subunit of the vacuolar H+-ATPase prevent segregation of lysosomal- and secretory-pathway proteins, *J. Cell Sci.* **122**, 3542–3553 (2009).
78. A. Kloska, A. Tylki-Szymanska and G. Wegrzyn, Lysosomal storage diseases--an overview, *Postepy Biochem.* **57**, 128–132 (2011).
79. B. Schmidt, T. Selmer, A. Ingendoh and K. von Figura, A novel amino acid modification in sulfatases that is defective in multiple sulfatase deficiency, *Cell* **82**, 271–278 (1995).
80. C. Verity, A. M. Winstone, L. Stellitano, R. Will and A. Nicoll, The epidemiology of progressive intellectual and neurological deterioration in childhood, *Arch. Dis. Child* **95**, 361–364 (2010).
81. S. S. Seregin and A. Amalfitano, Gene therapy for lysosomal storage diseases: progress, challenges and future prospects, *Curr. Pharm. Des.* **17**, 2558–2574 (2011).
82. J. J. Malatack, D. M. Consolini and E. Bayever, The status of hematopoietic stem cell transplantation in lysosomal storage disease, *Pediatr. Neurol.* **29**, 391–403 (2003).
83. R. W. Barton and E. F. Neufeld, The Hurler corrective factor. Purification and some properties, *J. Biol. Chem.* **246**, 7773–7779 (1971).
84. M. S. Sands and B. L. Davidson, Gene therapy for lysosomal storage diseases, *Mol. Ther.* **13**, 839–849 (2006).
85. N. M. Ellinwood, C. H. Vite and M. E. Haskins, Gene therapy for lysosomal storage diseases: the lessons and promise of animal models, *J. Gene Med.* **6**, 481–506 (2004).
86. A. Gritti, Gene therapy for lysosomal storage disorders, *Expert Opin. Biol. Ther.* **11**, 1153–1167 (2011).
87. U. Matzner *et al.*, Non-inhibitory antibodies impede lysosomal storage reduction during enzyme replacement therapy of a lysosomal storage disease, *J. Mol. Med. (Berl)* **86**, 433–442 (2008).
88. R. Dhami and E. H. Schuchman, Mannose 6-phosphate receptor-mediated uptake is defective in acid sphingomyelinase-deficient macrophages: implications for Niemann-Pick disease enzyme replacement therapy, *J. Biol. Chem.* **279**, 1526–1532 (2004).

89. F. S. Furbish, C. J. Steer, N. L. Krett and J. A. Barranger, Uptake and distribution of placental glucocerebrosidase in rat hepatic cells and effects of sequential deglycosylation, *Biochim. Biophys. Acta* **673**, 425–434 (1981).
90. Y. Zhu, X. Li, E. H. Schuchman, R. J. Desnick and S. H. Cheng, Dexamethasone-mediated up-regulation of the mannose receptor improves the delivery of recombinant glucocerebrosidase to Gaucher macrophages, *J. Pharmacol. Exp. Ther.* **308**, 705–711 (2004).
91. J. H. LeBowitz *et al.*, Glycosylation-independent targeting enhances enzyme delivery to lysosomes and decreases storage in mucopolysaccharidosis type VII mice, *Proc. Natl. Acad. Sci. USA* **101**, 3083–3088 (2004).
92. W. S. Prince *et al.*, Lipoprotein receptor binding, cellular uptake, and lysosomal delivery of fusions between the receptor-associated protein (RAP) and alpha-L-iduronidase or acid alpha-glucosidase, *J. Biol. Chem.* **279**, 35037–35046 (2004).
93. K. O. Lee *et al.*, Improved intracellular delivery of glucocerebrosidase mediated by the HIV-1 TAT protein transduction domain, *Biochem. Biophys. Res. Commun.* **337**, 701–707 (2005).
94. K. O. Orii *et al.*, Defining the pathway for Tat-mediated delivery of beta-glucuronidase in cultured cells and MPS VII mice, *Mol. Ther.* **12**, 345–352 (2005).
95. X. Y. Zhang, A. Dinh, J. Cronin, S. C. Li and J. Reiser, Cellular uptake and lysosomal delivery of galactocerebrosidase tagged with the HIV Tat protein transduction domain, *J. Neurochem.* **104**, 1055–1064 (2008).
96. F. M. Platt *et al.*, Prevention of lysosomal storage in Tay-Sachs mice treated with N-butyldeoxynojirimycin, *Science* **276**, 428–431 (1997).
97. W. A. Gahl, J. G. Thoene and J. A. Schneider, Cystinosis, *N. Engl. J. Med.* **347**, 111–121 (2002).
98. F. Camargo *et al.*, Cyclodextrins in the treatment of a mouse model of Niemann-Pick C disease, *Life Sci.* **70**, 131–142 (2001).
99. G. H. Yam, C. Zuber and J. Roth, A synthetic chaperone corrects the trafficking defect and disease phenotype in a protein misfolding disorder, *FASEB J.* **19**, 12–18 (2005).
100. A. R. Sawkar *et al.*, Gaucher disease-associated glucocerebrosidases show mutation-dependent chemical chaperoning profiles, *Chem. Biol.* **12**, 1235–1244 (2005).
101. Y. Suzuki, Chemical chaperone therapy for GM1-gangliosidosis, *Cell Mol. Life Sci.* **65**, 351–353 (2008).
102. K. Narita *et al.*, Protein transduction of Rab9 in Niemann-Pick C cells reduces cholesterol storage, *FASEB J.* **19**, 1558–1560 (2005).
103. F. W. Chen, C. Li and Y. A. Ioannou, Cyclodextrin induces calcium-dependent lysosomal exocytosis, *PLoS One* **5**, e15054 (2010).
104. K. Strauss *et al.*, Exosome secretion ameliorates lysosomal storage of cholesterol in Niemann-Pick type C disease, *J. Biol. Chem.* **285**, 26279–26288 (2010).
105. D. L. Medina *et al.*, Transcriptional activation of lysosomal exocytosis promotes cellular clearance, *Dev. Cell* **21**, 421–430 (2011).
106. M. Xu *et al.*, Delta-Tocopherol reduces lipid accumulation in Niemann-Pick type C1 and Wolman cholesterol storage disorders, *J. Biol. Chem.* **287**, 39349–39360 (2012).
107. A. P. Lieberman *et al.*, Autophagy in lysosomal storage disorders, *Autophagy* **8,** 719–730 (2012).
108. B. Levine, N. Mizushima and H. W. Virgin, Autophagy in immunity and inflammation, *Nature* **469**, 323–335 (2011).

109. M. Martinez-Vicente *et al.*, Cargo recognition failure is responsible for inefficient autophagy in Huntington's disease, *Nat. Neurosci.* **13**, 567–576 (2010).
110. R. A. Nixon and D. S. Yang, Autophagy failure in Alzheimer's disease-locating the primary defect, *Neurobiol. Dis.* **43**, 38–45 (2011).
111. A. M. Sanchez-Perez, B. Claramonte-Clausell, J. V. Sanchez-Andres and M. T. Herrero, Parkinson's disease and autophagy, *Parkinsons Dis.* **2012**, 429524 (2012).
112. Z. J. Yang, C. E. Chee, S. Huang and F. A. Sinicrope, The role of autophagy in cancer: therapeutic implications, *Mol. Cancer Ther.* **10**, 1533–1541 (2011).
113. E. Bejarano and A. M. Cuervo, Chaperone-mediated autophagy, *Proc. Am. Thorac. Soc.* 7, 29–39 (2010).
114. M. Palmieri *et al.*, Characterization of the CLEAR network reveals an integrated control of cellular clearance pathways, *Hum. Mol. Genet.* **20**, 3852–3866 (2011).
115. M. Sardiello *et al.*, A gene network regulating lysosomal biogenesis and function, *Science* **325**, 473–477 (2009).
116. C. Settembre *et al.*, TFEB links autophagy to lysosomal biogenesis, *Science* **332**, 1429–1433 (2011).
117. T. Pan, S. Kondo, W. Le and J. Jankovic, The role of autophagy-lysosome pathway in neurodegeneration associated with Parkinson's disease, *Brain* **131**, 1969–1978 (2008).
118. A. Ballabio, Disease pathogenesis explained by basic science: lysosomal storage diseases as autophagocytic disorders, *Int. J. Clin. Pharmacol. Ther.* **47** (**Suppl 1**), S34–38 (2009).
119. C. Settembre *et al.*, A block of autophagy in lysosomal storage disorders, *Hum. Mol. Genet.* **17,** 119–129 (2008).
120. A. Fraldi *et al.*, Lysosomal fusion and SNARE function are impaired by cholesterol accumulation in lysosomal storage disorders, *EMBO J.* **29**, 3607–3620 (2010).
121. L. Yu *et al.*, Termination of autophagy and reformation of lysosomes regulated by mTOR, *Nature* **465**, 942–946 (2010).
122. J. F. Dice *et al.*, A selective pathway for degradation of cytosolic proteins by lysosomes, *Semin. Cell Biol.* **1**, 449–455 (1990).
123. A. M. Cuervo and J. F. Dice, Regulation of lamp2a levels in the lysosomal membrane, *Traffic* **1**, 570–583 (2000).
124. A. M. Cuervo, L. Mann, E. J. Bonten, A. d'Azzo and J. F. Dice, Cathepsin A regulates chaperone-mediated autophagy through cleavage of the lysosomal receptor, *EMBO J.* **22**, 47–59 (2003).
125. I. Nishino *et al.*, Primary LAMP-2 deficiency causes X-linked vacuolar cardiomyopathy and myopathy (Danon disease), *Nature* **406**, 906–910 (2000).
126. B. Venugopal *et al.*, Chaperone-mediated autophagy is defective in mucolipidosis type IV, *J. Cell Physiol.* **219**, 344–353 (2009).
127. S. B. Prusiner, Shattuck lecture-neurodegenerative diseases and prions, *N. Engl. J. Med.* **344**, 1516–1526 (2001).
128. L. Zhang, R. Sheng and Z. Qin, The lysosome and neurodegenerative diseases, *Acta Biochim. Biophys. Sin (Shanghai)* **41**, 437–445 (2009).
129. K. B. Kegel *et al.*, Huntingtin expression stimulates endosomal-lysosomal activity, endosome tubulation, and autophagy, *J. Neurosci.* **20**, 7268–7278 (2000).
130. A. M. Cuervo, L. Stefanis, R. Fredenburg, P. T. Lansbury and D. Sulzer, Impaired degradation of mutant alpha-synuclein by chaperone-mediated autophagy, *Science* **305**, 1292–1295 (2004).

131. M. H. Polymeropoulos *et al.*, Mutation in the alpha-synuclein gene identified in families with Parkinson's disease, *Science* **276**, 2045–2047 (1997).
132. M. Martinez-Vicente and A. M. Cuervo, Autophagy and neurodegeneration: when the cleaning crew goes on strike, *Lancet Neurol.* **6**, 352–361 (2007).
133. H. J. Lee, F. Khoshaghideh, S. Patel and S. J. Lee, Clearance of alpha-synuclein oligomeric intermediates via the lysosomal degradation pathway, *J. Neurosci.* **24**, 1888–1896 (2004).
134. B. Dehay *et al.*, Loss of P-type ATPase ATP13A2/PARK9 function induces general lysosomal deficiency and leads to Parkinson disease neurodegeneration, *Proc. Natl. Acad. Sci. USA* **109**, 9611–9616 (2012).
135. K. Ohmi *et al.*, Sanfilippo syndrome type B, a lysosomal storage disease, is also a tauopathy, *Proc. Natl. Acad. Sci. USA* **106**, 8332–8337 (2009).
136. T. Shachar *et al.*, Lysosomal storage disorders and Parkinson's disease: Gaucher disease and beyond, *Mov. Disord.* **26**, 1593–1604 (2011).
137. K. Wong *et al.*, Neuropathology provides clues to the pathophysiology of Gaucher disease, *Mol. Genet. Metab.* **82**, 192–207 (2004).
138. J. C. Barrett *et al.*, Genome-wide association defines more than 30 distinct susceptibility loci for Crohn's disease, *Nat. Genet.* **40**, 955–962 (2008).
139. A. Luciani *et al.*, Defective CFTR induces aggresome formation and lung inflammation in cystic fibrosis through ROS-mediated autophagy inhibition, *Nat. Cell Biol.* **12**, 863–875 (2010).
140. R. Mathew *et al.*, Autophagy suppresses tumorigenesis through elimination of p62, *Cell* **137**, 1062–1075 (2009).
141. K. Degenhardt *et al.*, Autophagy promotes tumor cell survival and restricts necrosis, inflammation, and tumorigenesis, *Cancer Cell* **10**, 51–64 (2006).
142. X. Qu *et al.*, Promotion of tumorigenesis by heterozygous disruption of the beclin 1 autophagy gene, *J. Clin. Invest.* **112**, 1809–1820 (2003).
143. G. Kroemer and M. Jaattela, Lysosomes and autophagy in cell death control, *Nat. Rev. Cancer* **5**, 886–897 (2005).
144. P. Boya *et al.*, Lysosomal membrane permeabilization induces cell death in a mitochondrion-dependent fashion, *J. Exp. Med.* **197**, 1323–1334 (2003).
145. J. A. Joyce and D. Hanahan, Multiple roles for cysteine cathepsins in cancer, *Cell Cycle* **3**, 1516–1619 (2004).
146. J. Nylandsted *et al.*, Heat shock protein 70 promotes cell survival by inhibiting lysosomal membrane permeabilization, *J. Exp. Med.* **200**, 425–435 (2004).
147. N. Raben *et al.*, Suppression of autophagy permits successful enzyme replacement therapy in a lysosomal storage disorder-murine Pompe disease, *Autophagy* **6**, 1078–1089 (2010).
148. J. L. Webb, B. Ravikumar, J. Atkins, J. N. Skepper and D. C. Rubinsztein, Alpha-Synuclein is degraded by both autophagy and the proteasome, *J. Biol. Chem.* **278**, 25009–25013 (2003).
149. B. Ravikumar *et al.*, Inhibition of mTOR induces autophagy and reduces toxicity of polyglutamine expansions in fly and mouse models of Huntington disease, *Nat. Genet.* **36**, 585–595 (2004).
150. B. Ravikumar, Z. Berger, C. Vacher, C. J. O'Kane and D. C. Rubinsztein, Rapamycin pre-treatment protects against apoptosis, *Hum. Mol. Genet.* **15**, 1209–1216 (2006).

151. J. W. Chang, H. Choi, S. L. Cotman and Y. K. Jung, Lithium rescues the impaired autophagy process in CbCln3(Deltaex7/8/Deltaex7/8) cerebellar cells and reduces neuronal vulnerability to cell death via IMPase inhibition, *J. Neurochem.* **116**, 659–668 (2011).
152. D. S. Yang *et al.*, Therapeutic effects of remediating autophagy failure in a mouse model of Alzheimer disease by enhancing lysosomal proteolysis, *Autophagy* 7, 788–789 (2011).
153. I. Beuvink *et al.*, The mTOR inhibitor RAD001 sensitizes tumor cells to DNA-damaged induced apoptosis through inhibition of p21 translation, *Cell* **120**, 747–759 (2005).
154. J. P. Dutcher, Mammalian target of rapamycin inhibition, *Clin. Cancer Res.* **10**, 6382S–6387S (2004).
155. F. Meric-Bernstam and A. M. Gonzalez-Angulo, Targeting the mTOR signaling network for cancer therapy, *J. Clin. Oncol.* **27**, 2278–2287 (2009).
156. E. Schmitt *et al.*, Chemosensitization by a non-apoptogenic heat shock protein 70-binding apoptosis-inducing factor mutant, *Cancer Res.* **63**, 8233–8240 (2003).
157. I. P. Braidman and G. Gregoriadis, Rapid partial purification of placental glucocerebroside beta-glucosidase and its entrapment in liposomes, *Biochem. J.* **164**, 439–445 (1977).
158. G. Gregoriadis and B. E. Ryman, Lysosomal localization of β-fructofuranosidase-containing liposomes injected into rats, *Biochem. J.* **129**, 123–133 (1972).
159. H. Patel and B. E. Ryman, Alpha-mannosidase in zinc-deficient rats. Possibility of liposomal therapy in mannosidosis, *Biochem. Soc. Trans.* **2**, 1014–1017 (1974).
160. L. D. Steger and R. J. Desnick, Enzyme therapy. VI: Comparative in vivo fates and effects on lysosomal integrity of enzyme entrapped in negatively and positively charged liposomes, *Biochim. Biophys. Acta* **464**, 530–546 (1977).
161. G. Takada, H. Onodera and K. Tada, Delivery of fungal beta-galactosidase to rat brain by means of liposomes, *Tohoku J. Exp. Med.* **136**, 219–229 (1982).
162. F. Umezawa, Y. Eto, T. Tokoro, F. Ito and K. Maekawa, Enzyme replacement with liposomes containing beta-galactosidase from Charonia lumpas in murine globoid cell leukodystrophy (twitcher), *Biochem. Biophys. Res. Commun.* **127**, 663–667 (1985).
163. J. Schmitz, L. W. Poll and S. vom Dahl, Therapy of adult Gaucher disease, *Haematologica* **92**, 148–152 (2007).
164. S. Muro *et al.*, Endothelial targeting of high-affinity multivalent polymer nanocarriers directed to intercellular adhesion molecule 1, *J. Pharmacol. Exp. Ther.* **317**, 1161–1169 (2006).
165. S. Mumtaz and B. K. Bachhawat, Enhanced intracellular stability and efficacy of PEG modified dextranase in the treatment of a model storage disorder, *Biochim. Biophys. Acta* **1199**, 175–182 (1994).
166. N. H. Ansari, Q. He, J. D. Cook, J. Wen and S. K. Srivastava, Delivery of liposome-sequestered hydrophobic proteins to lysosomes of normal and Batten disease cells, *J. Neurosci. Res.* **47**, 341–347 (1997).
167. G. Weissmann *et al.*, A general method for the introduction of enzymes, by means of immunoglobulin-coated liposomes, into lysosomes of deficient cells, *Proc. Natl. Acad. Sci. USA* **72**, 88–92 (1975).
168. A. Koshkaryev, R. Thekkedath, C. Pagano, I. Meerovich and V. P. Torchilin, Targeting of lysosomes by liposomes modified with octadecyl-rhodamine B, *J. Drug Target* **19**, 606–614 (2011).

169. I. Meerovich, A. Koshkaryev, R. Thekkedath and V. P. Torchilin, Screening and optimization of ligand conjugates for lysosomal targeting, *Bioconjug. Chem.* **22**, 2271–2282 (2011).
170. R. Thekkedath, A. Koshkaryev and V. P. Torchilin, Lysosome-targeted octadecyl-rhodamine B-liposomes enhance lysosomal accumulation of glucocerebrosidase in Gaucher's cells in vitro, *Nanomedicine (Lond)* **8**, 1055–1065 (2013).
171. C. C. Barrias, M. Lamghari, P. L. Granja, M. C. Sa Miranda and M. A. Barbosa, Biological evaluation of calcium alginate microspheres as a vehicle for the localized delivery of a therapeutic enzyme, *J. Biomed. Mater. Res. A* **74**, 545–552 (2005).
172. J. L. Corchero *et al.*, Enzymatic characterization of highly stable human alpha-galactosidase A displayed on magnetic particles, *Biochem. Eng. J.* **67**, 20–27 (2012).
173. S. Sarrazin, B. Wilson, W. S. Sly, Y. Tor and J. D. Esko, Guanidinylated neomycin mediates heparan sulfate-dependent transport of active enzymes to lysosomes, *Mol. Ther.* **18**, 1268–1274 (2010).
174. C. D. Dekiwadia, A. C. Lawrie and J. V. Fecondo, Peptide-mediated cell penetration and targeted delivery of gold nanoparticles into lysosomes, *J. Pept. Sci.* **18**, 527–534 (2012).
175. M. I. Giannotti, O. Esteban, M. Oliva, M. F. Garcia-Parajo and F. Sanz, pH-responsive polysaccharide-based polyelectrolyte complexes as nanocarriers for lysosomal delivery of therapeutic proteins, *Biomacromolecules* **12**, 2524–2533 (2011).
176. S. Muro, E. H. Schuchman and V. Muzykantov, Immunotargeting to ICAM-1 provides binding, internalization and lysosomal delivery of acid sphingomyelinase, *Am. J. Hum. Genet.* **75**, PN268 (2004).
177. S. Muro, Intercellular adhesion molecule 1 and vascular cell adhesion molecule 1. In *Endothelial Biomedicine*, W. Aird (Ed.), New York: Cambridge University Press (2007), pp. 1058–1070.
178. R. Rothlein, M. L. Dustin, S. D. Marlin and T. A. Springer, A human intercellular adhesion molecule (ICAM-1) distinct from LFA-1, *J. Immunol.* **137**, 1270–1274 (1986).
179. S. Muro *et al.*, Control of endothelial targeting and intracellular delivery of therapeutic enzymes by modulating the size and shape of ICAM-1-targeted carriers, *Mol. Ther.* **16**, 1450–1458 (2008).
180. D. Serrano, T. Bhowmick, R. Chadha, C. Garnacho and S. Muro, Intercellular adhesion molecule 1 engagement modulates sphingomyelinase and ceramide, supporting uptake of drug carriers by the vascular endothelium, *Arterioscler. Thromb. Vasc. Biol.* **32**, 1178–1185 (2012).
181. C. Garnacho *et al.*, Delivery of acid sphingomyelinase in normal and niemann-pick disease mice using intercellular adhesion molecule-1-targeted polymer nanocarriers, *J. Pharmacol. Exp. Ther.* **325**, 400–408 (2008).
182. S. Muro, E. H. Schuchman and V. R. Muzykantov, Lysosomal enzyme delivery by ICAM-1-targeted nanocarriers bypassing glycosylation- and clathrin-dependent endocytosis, *Mol. Ther.* **13**, 135–141 (2006).
183. I. T. Papademetriou, C. Garnacho, E. H. Schuchman and S. Muro, In vivo performance of polymer nanocarriers dually-targeted to epitopes of the same or different receptors, *Biomaterials* **34**, 3459–3466 (2013).
184. J. Papademetriou *et al.*, Comparative binding, endocytosis, and biodistribution of antibodies and antibody-coated carriers for targeted delivery of lysosomal enzymes to ICAM-1 versus transferrin receptor, *J. Inherit. Metab. Dis.* **36**, 467–477 (2013).

185. J. Hsu *et al.*, Enhanced endothelial delivery and biochemical effects of alpha-galactosidase by ICAM-1-targeted nanocarriers for Fabry disease, *J. Control Release* **149**, 323–331 (2011).
186. J. Hsu, L. Northrup, T. Bhowmick and S. Muro, Enhanced delivery of alpha-glucosidase for Pompe disease by ICAM-1-targeted nanocarriers: comparative performance of a strategy for three distinct lysosomal storage disorders, *Nanomedicine* **8**, 731–739 (2012).
187. C. Garnacho, S. M. Albelda, V. R. Muzykantov and S. Muro, Differential intra-endothelial delivery of polymer nanocarriers targeted to distinct PECAM-1 epitopes, *J. Control Release* **130**, 226–233 (2008).
188. S. Muro *et al.*, Slow intracellular trafficking of catalase nanoparticles targeted to ICAM-1 protects endothelial cells from oxidative stress, *Am. J. Physiol. Cell. Physiol.* **285**, C1339–1347 (2003).
189. S. Muro *et al.*, Control of intracellular trafficking of ICAM-1-targeted nanocarriers by endothelial Na+/H+ exchanger proteins, *Am. J. Physiol. Lung Cell Mol. Physiol.* **290**, L809–817 (2006).
190. J. Hsu, T. Bhowmick, C. Garnacho and S. Muro, Design of ICAM-1-targeting strategies for brain delivery of lysosomal therapies, *Mol. Genet. Metab.* **102**, S3 (2011).
191. A. P. R. de Garibay, D. Delgado, A. del Pozo-Rodriguez, M. A. Solinis and A. R. Gascon, Multicomponent nanoparticles as nonviral vectors for the treatment of Fabry disease by gene therapy, *Drug Des. Dev. Ther.* **6**, 303–310 (2012).
192. C. J. Ross, L. Bastedo, S. A. Maier, M. S. Sands and P. L. Chang, Treatment of a lysosomal storage disease, mucopolysaccharidosis VII, with microencapsulated recombinant cells, *Hum. Gene Ther.* **11**, 2117–2127 (2000).
193. A. Yamaguchi *et al.*, Plasmid-based gene transfer ameliorates visceral storage in a mouse model of Sandhoff disease, *J. Mol. Med.* (*Berl*) **81**, 185–193 (2003).
194. J. J. Green, Rita Schaffer lecture: nanoparticles for intracellular nucleic acid delivery, *Ann. Biomed. Eng.* **40**, 1408–1418 (2012).
195. T. Ochiya *et al.*, Evaluation of cationic liposome suitable for gene transfer into pregnant animals, *Biochem. Biophys. Res. Commun.* **258**, 358–365 (1999).
196. C. J. Ross, M. Ralph and P. L. Chang, Somatic gene therapy for a neurodegenerative disease using microencapsulated recombinant cells, *Exp. Neurol.* **166**, 276–286 (2000).
197. T. Xia, M. Kovochich, M. Liong, J. I. Zink and A. E. Nel, Cationic polystyrene nanosphere toxicity depends on cell-specific endocytic and mitochondrial injury pathways, *ACS Nano* **2**, 85–96 (2008).
198. T. P. Thomas *et al.*, Cationic poly(amidoamine) dendrimer induces lysosomal apoptotic pathway at therapeutically relevant concentrations, *Biomacromolecules* **10**, 3207–3214 (2009).
199. J. Panyam, W. Z. Zhou, S. Prabha, S. K. Sahoo and V. Labhasetwar, Rapid endo-lysosomal escape of poly(DL-lactide-co-glycolide) nanoparticles: implications for drug and gene delivery, *FASEB J.* **16**, 1217–1226 (2002).
200. S. Hussain *et al.*, Carbon black and titanium dioxide nanoparticles elicit distinct apoptotic pathways in bronchial epithelial cells, *Part Fibre Toxicol.* 7, 10 (2010).
201. Z. Yang *et al.*, Pharmacological and toxicological target organelles and safe use of single-walled carbon nanotubes as drug carriers in treating Alzheimer disease, *Nanomedicine* **6**, 427–441 (2010).

202. S. T. Stern, P. P. Adiseshaiah and R. M Crist, Autophagy and lysosomal dysfunction as emerging mechanisms of nanomaterial toxicity, *Part Fibre Toxicol*, **9**, 20 (2012).
203. N. Li, T. Xia and A. E. Nel, The role of oxidative stress in ambient particulate matter-induced lung diseases and its implications in the toxicity of engineered nanoparticles, *Free Radic. Biol. Med.* **44**, 1689–1699 (2008).
204. W. S. Cho *et al.*, Progressive severe lung injury by zinc oxide nanoparticles; the role of Zn2+ dissolution inside lysosomes, *Part Fibre Toxicol.* **8**, 27 (2011).
205. X. Ma *et al.*, Gold nanoparticles induce autophagosome accumulation through size-dependent nanoparticle uptake and lysosome impairment, *ACS Nano* **5**, 8629–8639 (2011).
206. P. Schneider, T. A. Korolenko and U. Busch, A review of drug-induced lysosomal disorders of the liver in man and laboratory animals, *Microsc. Res. Tech.* **36**, 253–275 (1997).
207. E. Frohlich *et al.*, Action of polystyrene nanoparticles of different sizes on lysosomal function and integrity, *Part Fibre Toxicol* **9**, 26 (2012).
208. P. Nativo, I. A. Prior and M. Brust, Uptake and intracellular fate of surface-modified gold nanoparticles, *ACS Nano* **2**, 1639–1644 (2008).
209. R. C. Stearns, J. D. Paulauskis and J. J. Godleski, Endocytosis of ultrafine particles by A549 cells, *Am. J. Respir. Cell Mol. Biol.* **24**, 108–115 (2001).
210. Y. Tahara *et al.*, Lysosomal membrane destabilization induced by high accumulation of single-walled carbon nanohorns in murine macrophage RAW 264.7, *Biomaterials* **33**, 2762–2769 (2012).
211. Y. Chen, L. Yang, C. Feng and L. P. Wen, Nano neodymium oxide induces massive vacuolization and autophagic cell death in non-small cell lung cancer NCI-H460 cells, *Biochem. Biophys. Res. Commun.* **337**, 52–60 (2005).
212. C. Li *et al.*, PAMAM nanoparticles promote acute lung injury by inducing autophagic cell death through the Akt-TSC2-mTOR signaling pathway, *J. Mol. Cell Biol.* **1**, 37–45 (2009).
213. J. J. Li, D. Hartono, C. N. Ong, B. H. Bay and L. Y. Yung, Autophagy and oxidative stress associated with gold nanoparticles, *Biomaterials* **31**, 5996–6003 (2010).
214. O. Seleverstov *et al.*, Quantum dots for human mesenchymal stem cells labeling. A size-dependent autophagy activation, *Nano Lett.* **6**, 2826–2832 (2006).
215. Q. Zhang *et al.*, Autophagy-mediated chemosensitization in cancer cells by fullerene C60 nanocrystal, *Autophagy* **5**, 1107–1117 (2009).
216. T. Yokoyama *et al.*, EGFR-targeted hybrid plasmonic magnetic nanoparticles synergistically induce autophagy and apoptosis in non-small cell lung cancer cells, *PLoS One* **6**, e25507 (2011).
217. L. Calzolai, F. Franchini, D. Gilliland and F. Rossi, Protein-nanoparticle interaction: identification of the ubiquitin-gold nanoparticle interaction site, *Nano Lett.* **10**, 3101–3105 (2010).
218. M. Chen and A. von Mikecz, Formation of nucleoplasmic protein aggregates impairs nuclear function in response to SiO2 nanoparticles, *Exp. Cell Res.* **305**, 51–62 (2005).
219. A. von Mikecz, M. Chen, T. Rockel and A. Scharf, The nuclear ubiquitin-proteasome system: visualization of proteasomes, protein aggregates, and proteolysis in the cell nucleus, *Methods Mol. Biol.* **463**, 191–202 (2008).

220. O. Lunov *et al.*, Amino-functionalized polystyrene nanoparticles activate the NLRP3 inflammasome in human macrophages, *ACS Nano* **5**, 9648–9657 (2011).
221. M. Winter *et al.*, Activation of the inflammasome by amorphous silica and TiO2 nanoparticles in murine dendritic cells, *Nanotoxicology* **5**, 326–340 (2011).
222. E. J. Yang, S. Kim, J. S. Kim and I. H. Choi, Inflammasome formation and IL-1beta release by human blood monocytes in response to silver nanoparticles, *Biomaterials* **33**, 6858–6867 (2012).
223. C. Jin and R. A. Flavell, Molecular mechanism of NLRP3 inflammasome activation, *J. Clin. Immunol.* **30**, 628–631 (2010).

Chapter 9

Nanostructured Biomaterials for Inhibiting Cancer Cell Functions

*Lijuan Zhang and Thomas J. Webster**

Department of Chemical Engineering, Northeastern University, 360 Huntington Avenue, Boston, MA 02115, USA
**th.webster@neu.edu*

Abstract

While numerous studies have been conducted to determine the role of nanoparticles on decreasing cancer cell functions, the impact of material nanotopographies on healthy and carcinoma lung and breast cell functions [such as adhesion, proliferation, apoptosis, morphology and vascular endothelial growth factor (VEGF) secretion] has not been systematically studied. Recent studies have suggested that cancer cells can recognize and respond differently to nanotopographies compared to flat topographies. However, many of these studies have altered both material topography and chemistry making it difficult to ascertain the influence of one property compared to the other. Studies presented in this chapter created materials with similar chemistry but altered nanoscale topographies allowing for the accurate assessment of the role of nanometer features on cancer cells. In summary, this chapter will report on nanotopography-mediated cancer and healthy cell functions in relationship to surface properties and protein adsorption, depicting some possible clues behind topography-mediated cell functions important for regenerating healthy tissue and keeping cancerous tissue from returning after lung or breast tumor resection.

1. Introduction: Summary of Relevant Studies

Great attention has been paid to exploring the cell-material interface in biomaterial science, as many studies have revealed that the physiochemical properties of an implant has a large impact on regulating cell behavior and functions.[1–9] In particular, many studies have shown the effects of substrate topography or roughness,[10–12] surface chemistry,[13] surface charge[14] and surface free energy[13] on cell morphology, adhesion, migration, proliferation, and gene expression. Recent studies have demonstrated that nanotopographies (as small as 23 nm) inhibited both lung cancer and breast cancer cell functions and enhanced lung healthy and breast healthy cell proliferation. Also, changing roughness is easier and is more likely to be cleared by the FDA faster than proposing new, FDA-unapproved, chemistries. Therefore, topography is of specific interest because of the great potential in the application of ideal anticancer implants. However, the effect of nanoscale topography on cancer and normal cell functions has not been throughly understood until date. Especially, the infinite number of potential combinations of cell types, biomaterial compositions, and topographical patterns have increased the difficulty in extricating the precise interaction mechanisms between implants and cell responses.[15] For example, in addition to the myriad of different topographical cues intentionally presented to cells, all materials contain topographies. In addition, the functional, behavioral, and morphologic differences between cell types imply that distinct cell types may respond differently to nanotopographical cues.[16]

A possible mechanism suggests that substrate nanotopographical discontinuities lead to altered surface energetics and preferential protein adsorption, and subsequent protein patterning.[15] Patterned protein deposition has been observed on substrates modified to have topographical cues and such proteins were found to be effective in patterning cells.[17–19]. Particularly, in a recent study, a 23 nm PLGA surface topography had patterns similar in scale as proteins, which could have an important effect on directing protein deposition due to the suitable "dock-in" sites. Moreover, protein adsorption and subsequent bioactivity are highly related to cell responses including adhesion, proliferation, morphology, and apoptosis. For example, it was observed that avidin adsorption on polystyrene (PS) increased with surface roughness on nanometer-size topographical patterns most likely caused by the increase of the effective surface area.[20] Another study revealed that significantly greater quantities of vitronectin adsorbed on nanophase ceramics compared to conventional ceramics, which, subsequently, may have contributed to the observed enhanced adhesion of osteoblasts (bone forming cells).[21] Similarly, it was also found that more fibronectin (a protein which promotes osteoblast adhesion) adsorption occurred on self-assembled chemistries called rosette nanotubes (RNT) when coated on hydrogels than uncoated hydrogels and osteoblast adhesion and proliferation were subsequently greater.[22]

Here, we will emphasize a series of studies in which PLGA surface properties (including surface topography or roughness, surface chemistry, surface charge, and surface energy) were changed systematically by creating different nanosize surface features to subsequently change protein adsorption events leading to altered lung and breast cancer responses. Therefore, studying the amount of protein adsorbed on nanotopographies could be an important clue to uncover the mechanisms of the novel topography-mediated cell responses which will be reported here.

In addition, it is claimed that if the topographical effect was assessed using substrates of the same chemistry, the process of cell sensing and adapting to topographies (such as focal adhesion, protein secretion, gene expression, cell-matrix signal pathways, etc.) may play an important role in determining the fate of cells.[23] Especially, isotropic nanotopographies could direct more cell functions with relatively less dramatic changes in cell shape. Specifically, it was found that a separation of ≥ 73 nm between cell adhesive gold nanodots coated with a cyclic RGDfK peptide resulted in limited MC3T3-osteoblast attachment and spreading, and dramatically reduced the formation of focal adhesion and actin stress fibers.[24] Another study by Matthew *et al.* [25] indicated that 13 nm high islands increased fibroblast functions compared to planar surfaces through gene up-regulation, notably in the areas of cell signaling, proliferation, cytoskeleton, and production of extracellular matrix proteins. Also, it has been proposed that the cell nucleus can turn the mechanical cues arising from nanotopographically induced morphological changes into gene changes.[26]. Moreover, during the cell sensing and adaptation process, a topographical pattern determines which cell functions are regulated and the scale affects the degree to which that function is regulated.

With this in mind, studies presented in this chapter were designed to explore the correlation between the various surface properties and cell behaviors. Specifically, the impact and correlation between surface topography and surface chemistry on cell functions will be discussed. Moreover, the surface free energy of various substrates will be emphasized and its influence on cell functions evaluated. Last, the adsorption of cell adhesion proteins (fibronectin and vitronectin) will be emphasized with a strong relationship between protein adsorption and cell responses.

2. The Influence of Surface Topography and Roughness of PLGA on Cancer Cells

2.1. *Creation of nanoscale PLGA surfaces*

AFM images of different surface topographies of nanopatterned PLGA surfaces which have been shown to decrease lung and breast cancer cell functions are shown in Fig. 1. The nomenclature refers to the size of PS beads used to create molds in which soluble PLGA was poured into. Thus, for PLGA0, PLGA23, PLGA300 and PLGA400, pattern sizes corresponded to the diameters of the PS beads used and PLGA23 showed pattern sizes less than 100 nm. Topographical results obtained from AFM confirmed the evolution of surface pattern sizes increasing with the diameters of the PS beads. Quantitative roughness values (R_{rms}) and real geometric surface areas of the nanopatterned PLGA surfaces are listed in Table 1. Specifically, as expected, PLGA0 had a much smaller R_{rms} value than PLGA23, PLGA300 and PLGA400, but PLGA0 still had some nanoscale roughness and, thus, was called nanosmooth PLGA. In addition, as the pattern sizes increased from 23 nm to 300 nm and 400 nm, the surface roughness values (R_{rms}) increased from 3.73 nm to 10.78 nm and 24.11 nm. Although the diameters of the patterns increased largely from the nanoscale (23 nm) to the submicron scale (300 nm and 400 nm), the relatively small change of roughness values (R_{rms}) could be attributed to the very slight alteration in vertical heights of the patterns when the horizontal diameters of the patterns were greatly changed. The surface geometric area values also showed that PLGA0, PLGA23, PLGA300 and PLGA400 had very similar surface area values (100 μm^2, 100.1 μm^2, 100.6 μm^2 and 102.3 μm^2) and further confirmed the very small alteration in vertical heights of the patterns when using largely different diameters of PS beads.

Studies have also coated such novel PLGA tissue engineering surfaces with alginate (Alg) and chitosan (Chi) in a further attempt to understand the relationship between changing surface topography surface chemistry on cancer cells. This was accomplished via a layer-by-layer (LBL) deposition process. For the LBL deposited PLGA surfaces, the nanopattern remained on PLGA23/Alg and PGA23/Chi, but the deposition of alginate and chitosan monolayers changed the surface roughness of the PLGA0 and PLGA23. Specifically, the deposition of the alginate monolayer largely increased the surface roughness values of PLGA0 and PLGA23. However, the further deposition of the chitosan monolayer slightly decreased the surface roughness values. Thus, the surface roughness measurement revealed the following trend as: PLGA0<PLGA0/Chi<PLGA0/Alg and PLGA23<PLGA23/

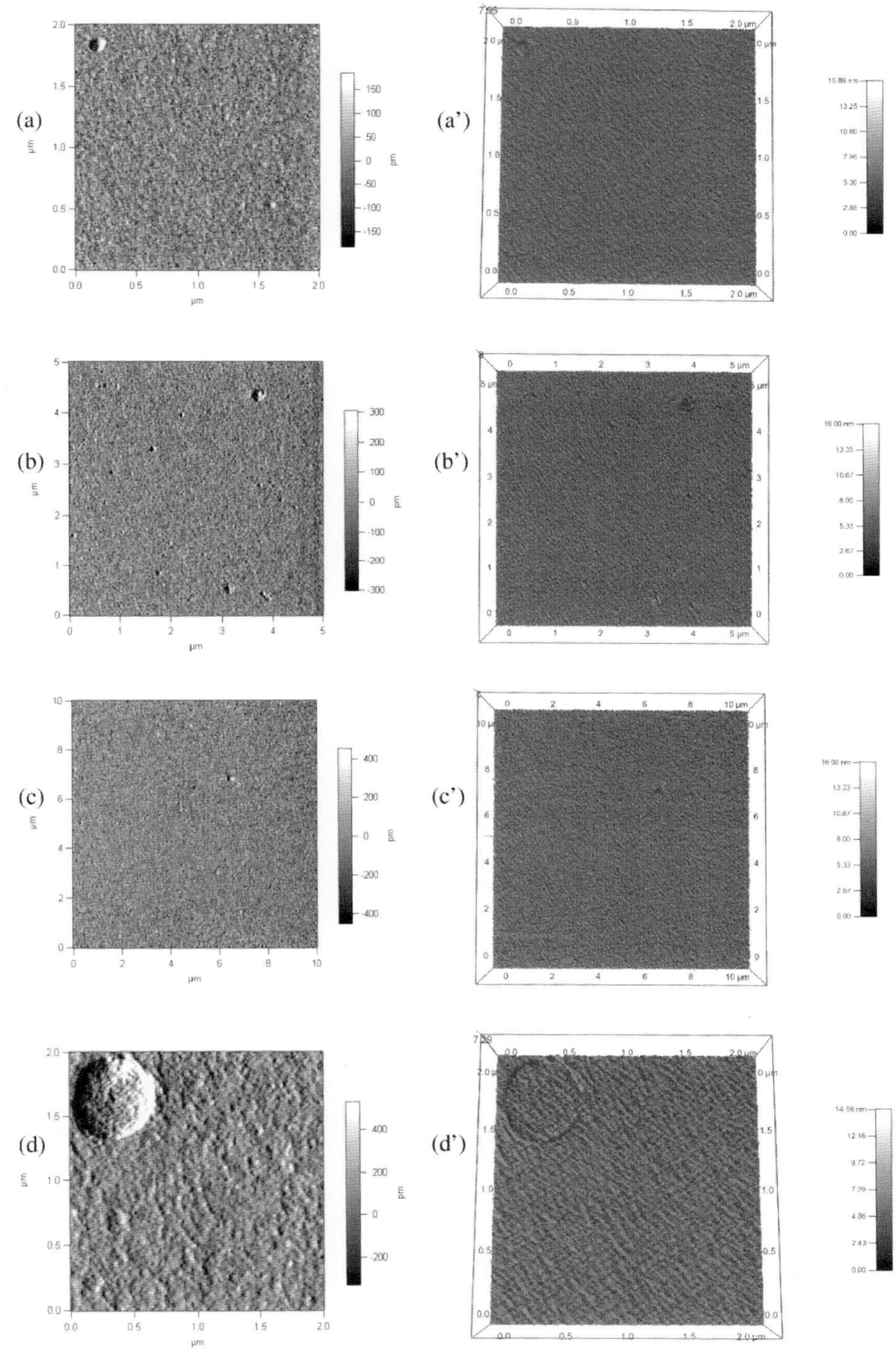

Fig. 1. 2 μm × 2 μm, 5 μm × 5 μm, and 10 μm × 10 μm 2D and 3D AFM images of PLGA0 (a, b, c and a', b' c'), PLGA23 (d, e, f and d', e', f'), PLGA300 (g, h, i and g', h', i'), PLGA400 (j, k, l and j', k' l'), PLGA0/Alg (m, n, o and m', n' o'), PLGA0/Chi (p, q, r and p', q', r'), PLGA23/Alg (s, t, u and s', t', u') and PLGA23/Chi (v, w, x and v', w', x').

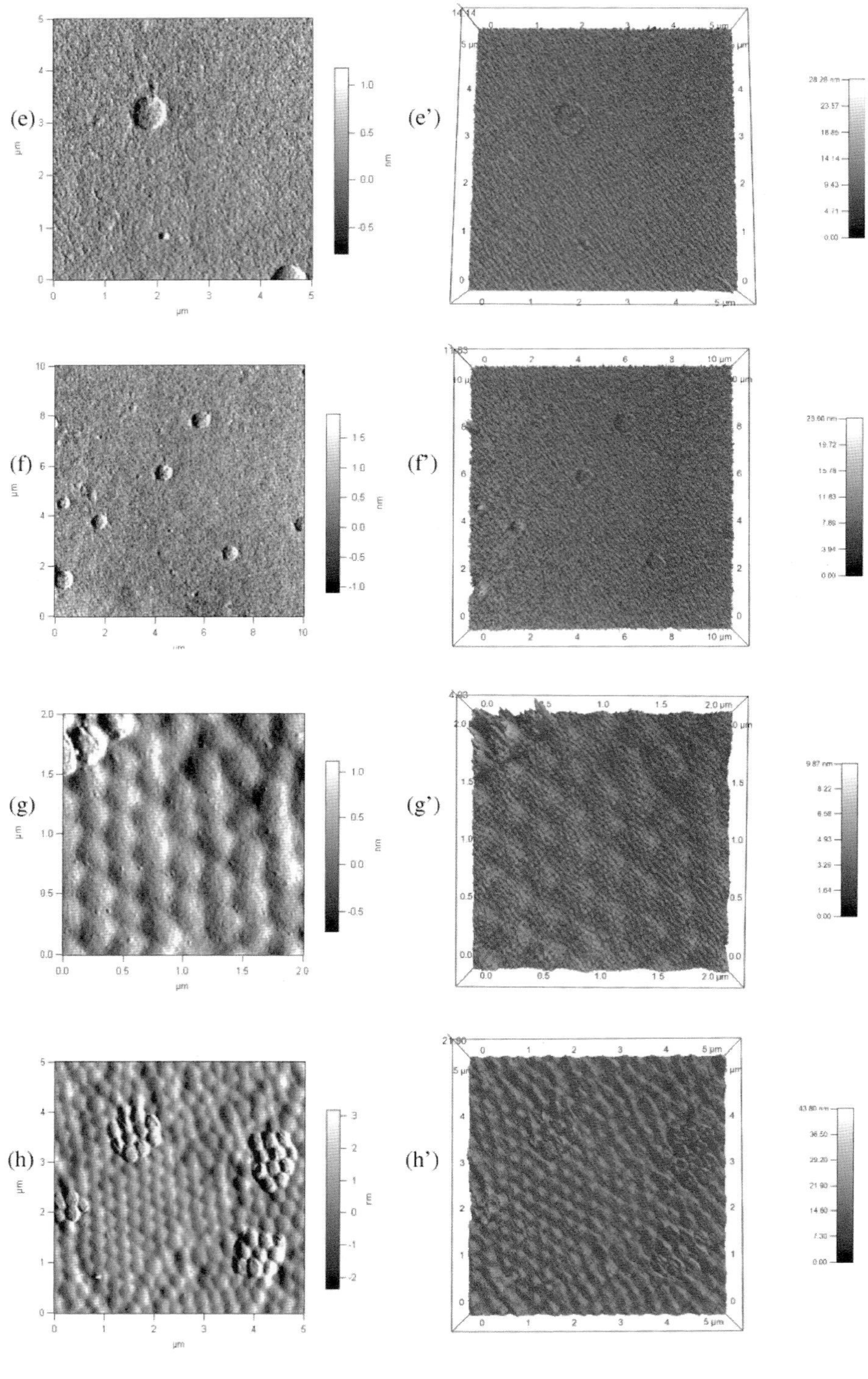

Fig. 1. *(Continued)*

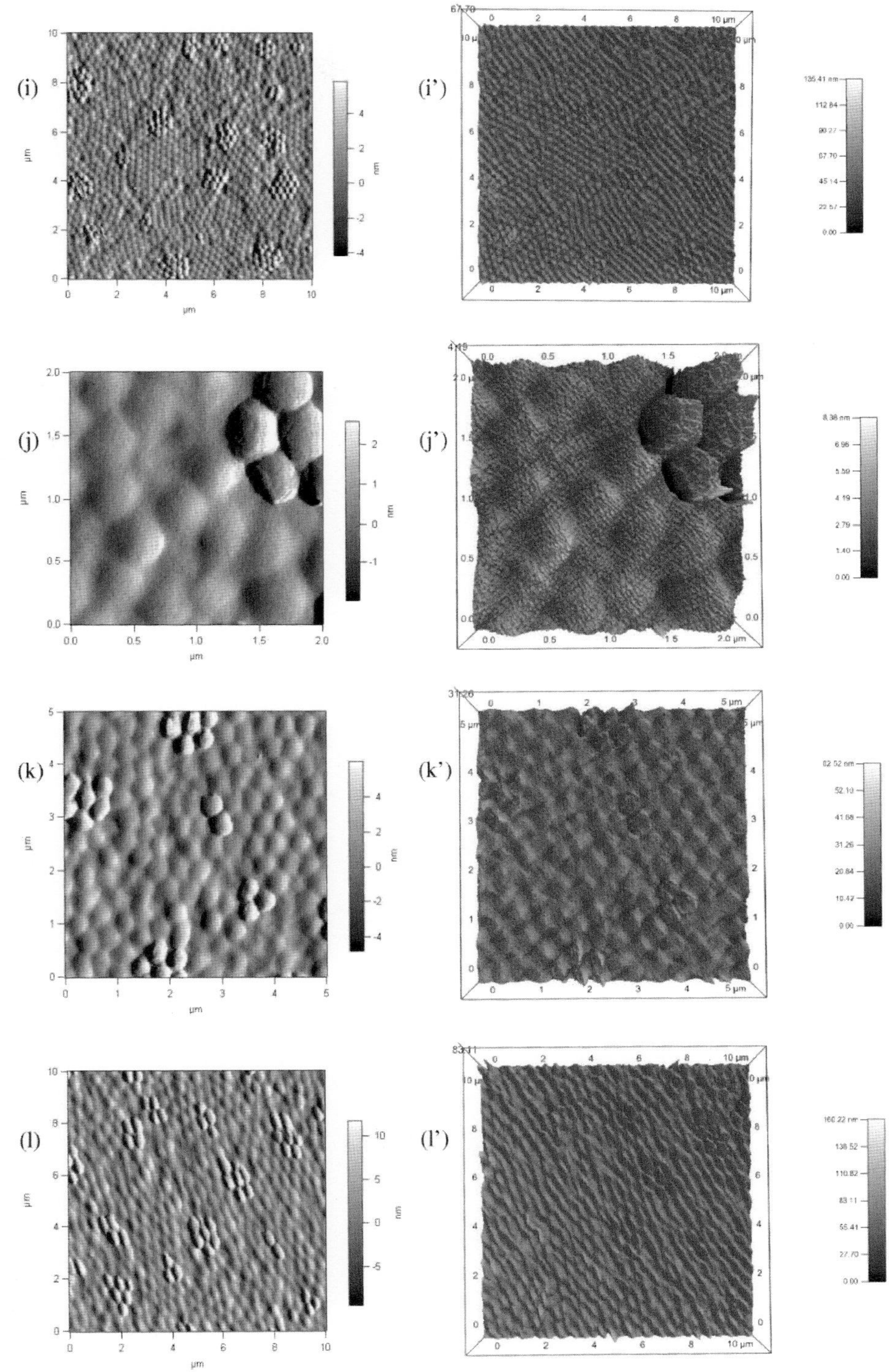

Fig. 1. *(Continued)*

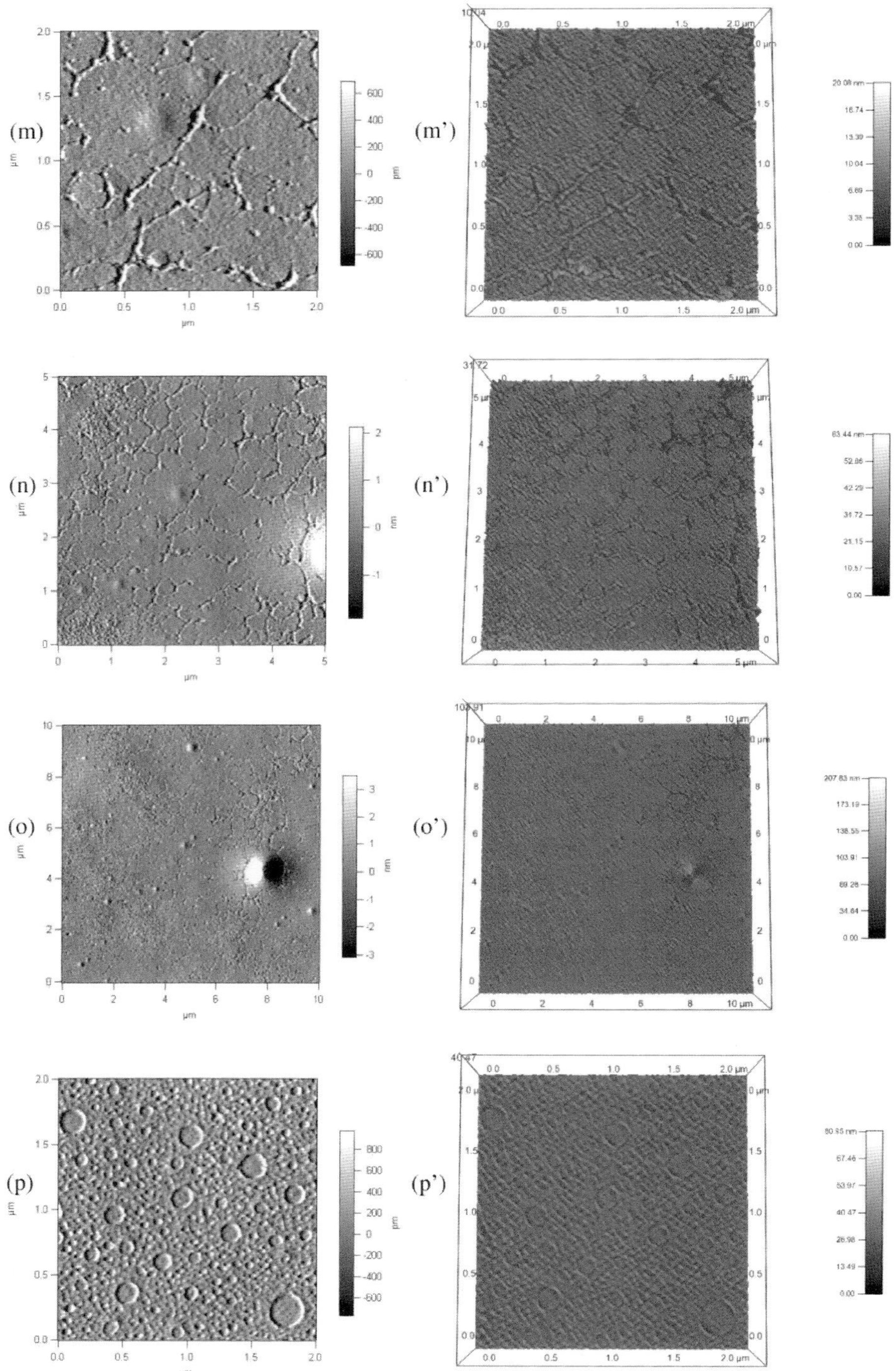

Fig. 1. *(Continued)*

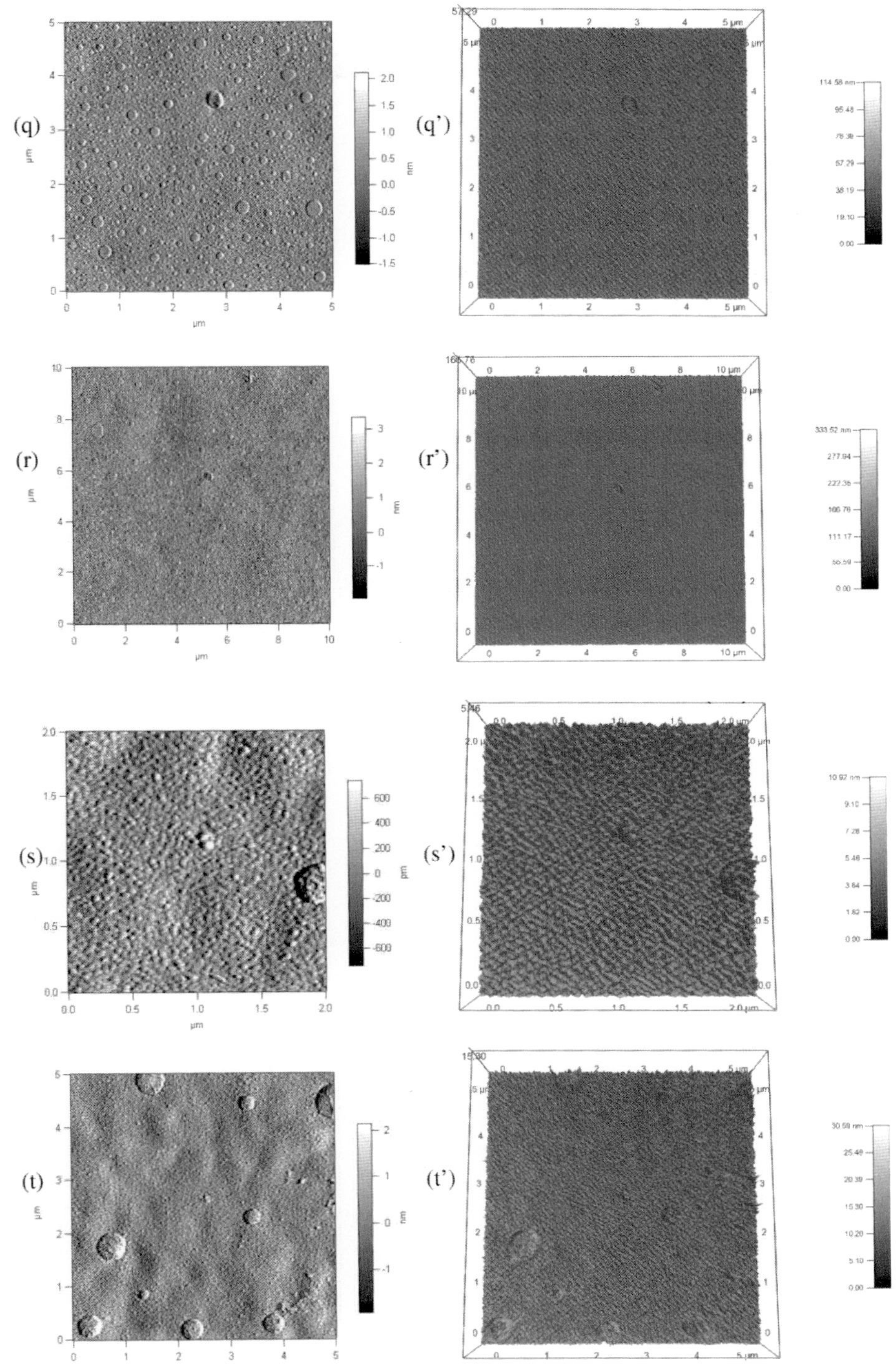

Fig. 1. *(Continued)*

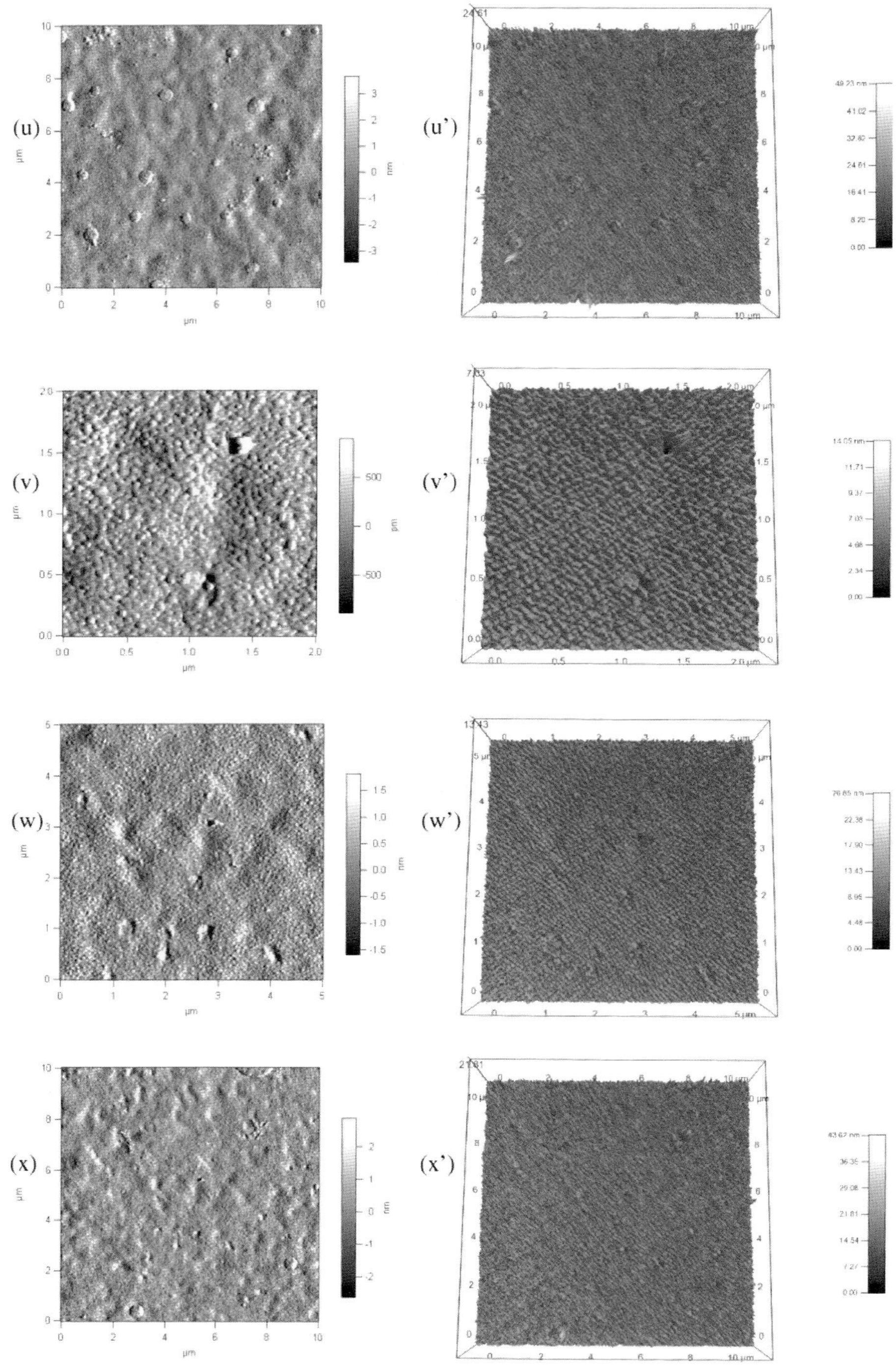

Fig. 1. (*Continued*)

Table 1. Surface roughness (R_{rms}) and local geometric area (S_{geo}) of all the samples measured by AFM (10×10 m^2 scan area).

Sample	R_{rms} (nm)	S_{geo} (μm^2)
PLGA0	0.54	100.0
PLGA23	3.73	100.1
PLGA300	10.78	100.6
PLGA400	24.11	102.3
PLGA0/Alg	5.38	100.3
PLGA0/Chi	3.31	100.1
PLGA23/Alg	7.41	100.3
PLGA23/Chi	5.01	100.2

Chi<PLGA23/Alg. As expected, the deposition of alginate and chitosan monolayers altered the surface roughness at the same time as they changed surface chemistry. For PLGA0, the LBL deposition of alginate on the nanosmooth PLGA surface increased the surface roughness from 0.54 nm to 5.38 nm, but the deposition of another monolayer of chitosan decreased the surface roughness to 3.31 nm. For PLGA23, although the LBL deposition changed the surface roughness values from 3.73 nm for PLGA23 to 7.41 nm and 5.01 nm for PLGA23/Alg and PLGA23/Chi, the tightly arrayed surface pattern was still observed from the AFM images of PLGA23/Alg and PLGA23/Chi. Again, the surface area values of the LBL deposited PLGA and non-modified PLGA were very similar, but a slight decrease was demonstrated from PLGA0/Alg to PLGA0/Chi, from PLGA23/Alg to PGA23/Chi. This result also agreed with the fact that the deposition of another monolayer of chitosan decreased the surface roughness of PLGA0 and PLGA23. In this study, the deposition of the alginate and chitosan monolayer not only changed the surface chemistry, but it also altered the surface roughness.

2.2. *The influence of nanoscale PLGA topographies on surface wettability and surface free energy*

Of course, it has long been speculated and demonstrated that creating nanoscale surface features on substrates can influence surface energy since more of that material is exposed to influence interactions with aqueous

Table 2. Contact angles measured by water and glycerol on nanopatterned substrates.

Substrate	Contact angles measured by different liquids (°)	
	θ (H_2O)	θ (Glycerol)
PLGA0	98.1 ± 0.4	94.9 ± 1.8
PLGA23	93.8 ± 1.8	94.5 ± 2.3
PLGA300	91.6 ± 1.6	92.4 ± 2.6
PLGA400	96.6 ± 0.5	101.6 ± 2.9

solutions. As shown in Table 2, although water contact angle measurements revealed slightly different results, the difference among the aforementioned PLGA0, PLGA23, PLGA300 and PLGA400 was statistical. All the nanosmooth and nanopatterned PLGA surfaces were hydrophobic (θ_{H2O} > 90°). As nanopatterns were present on the PLGA, the surfaces became more hydrophilic as contact angles decreased. Clearly, surface wettability can be influenced by surface topography, but the observed impact of topography on controlling PLGA surface wettability was slight while keeping the same surface chemistry. Among all the PLGA substrates of interest here, PLGA300 and PLGA400 revealed the highest surface free energy (17.9 mN/m and 18.3 mN/m, respectively), PLGA23 had the second highest (16.5 mN/m), and PLGA0 demonstrated the lowest surface free energy (13.6 mN/m) (Fig. 2). Generally, nanopatterned PLGA surfaces exhibited higher surface free energy than nanosmooth surfaces. Considering that all the nanosmooth and nanopatterned PLGA had the same surface chemistry, this increase could be attributed solely to the influence of surface nanotopographies.

It is also important to discuss the change in surface energy for the PLGA nanopatterned samples coating with alginate and chitosan. The contact angle results on alginate and chitosan deposited PLGA substrates are shown in Table 3. Clearly, the deposition of alginate and chitosan on PLGA nanosurfaces decreased water contact angles compared to pure PLGA surfaces, indicating an increase in hydrophilicity. Moreover, the increase in hydrophilicity on nanopatterned surfaces (PLGA23/Alg and PLGA23/Chi) was more significant than that on nanosmooth surfaces (PLGA0/Alg and PLGA0/Chi). This result is obvious since both alginate and chitosan are more hydrophibic than PLGA, resulting in increased water wettability than

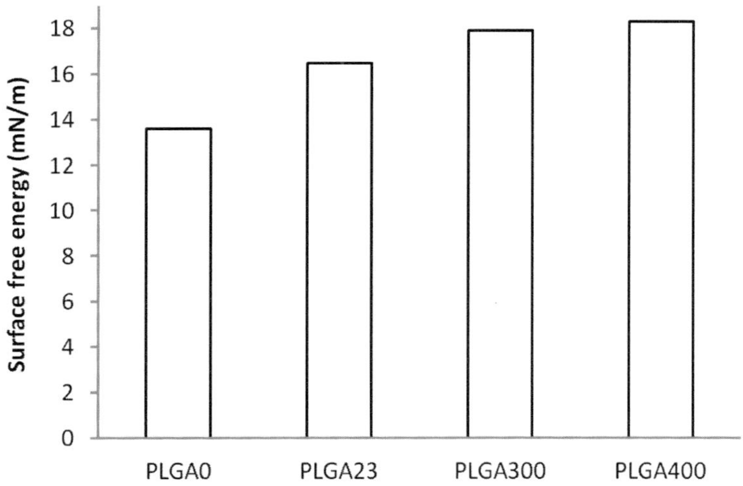

Fig. 2. Calculated surface free energy (using contact angle results of water and glycerol) of PLGA0, PLGA23, PLGA300 and PLGA400.

Table 3. Contact angles measured by water and formamide on LBL deposited substrates.

	Contact angles measured by different liquids (°)	
Substrate	θ (H_2O)	θ (Formamide)
PLGA0	98.1 ± 0.4	65.4 ± 2.6
PLGA0/Alg	80.0 ± 0.9	56.5 ± 2.9
PLGA0/Chi	87.4 ± 0.6	59.3 ± 1.8
PLGA23	93.8 ± 1.8	59.4 ± 2.9
PLGA23/Alg	76.3 ± 2.2	61.3 ± 2.9
PLGA23/Chi	68.5 ± 1.4	45.9 ± 3.6

pure PLGA surfaces. Here, the deposition of alginate and chitosan changed both surface chemistry and surface roughness, both of which could contribute to the alteration of surface wettability.

Surface free energy results of alginate and chitosan deposited PLGA nanopatterned substrates are shown in Fig. 3. Clearly, the deposition of alginate and chitosan on PLGA surfaces decreased surface free energy. Among all the samples, PLGA23/Alg had the lowest surface free energy (31.7 mN/m), PLGA0/Alg revealed the second lowest (36.6 mN/m), and PLGA23

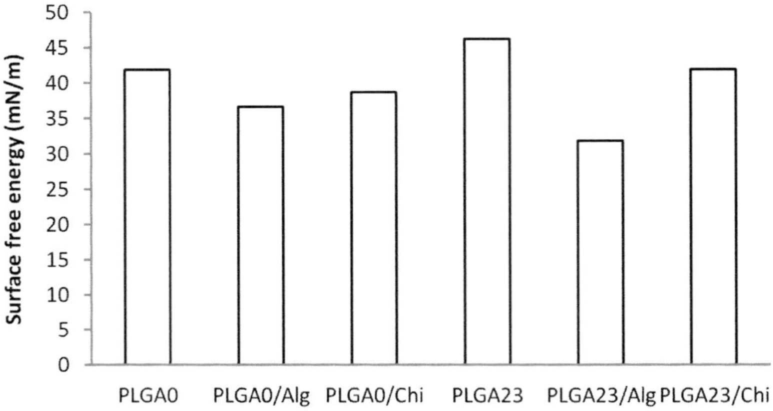

Fig. 3. Calculated surface free energy (using contact angle results of water and formamide) of PLGA0, PLGA0/Alg, PLGA0/Chi, PLGA23, PLGA23/Alg and PLGA23/Chi.

exhibited the highest surface free energy (46.5 mN/m). However, compared to chitosan, alginate exhibited a more significant influence on decreasing surface free energy. As expected, the alteration of surface free energy was related to the surface wettability of various substrates. The increased surface wettability on the alginate and chitosan deposited surfaces can be attributed to the decreased surface free energy. Moreover, surface topography and terminal chemistries (here, PLGA, alginate and chitosan) on the substrate surfaces were known to play a substantial influence on surface free energy.[29,30] Here, the alteration of surface chemistry and surface roughness can have a collective impact on surface free energy.

2.3. *The Influence of PLGA nanotopographies on protein adsorption*

Due to the change in surface energetics, it can be expected that protein adsorption will be different in nanopatterned versus nanosmooth surfaces. Indeed, studies have shown different adsorption trends for fibronectin and vitronectin on the aforementioned PLGA nanopatterns (Fig. 4). Specifically, PLGA23 showed 20% more fibronectin adsorption compared to PLGA0. However, no significant difference was demonstrated between PLGA0, PLGA300 and PLGA400. Vitronectin adsorption revealed a significant difference between nanosmooth and nanopatterned PLGA surfaces. PLGA23, PGA300 and PLGA400 all showed significantly higher vitronectin adsorption than that on PLGA0, 13%, 30% and 26%, respectively. Among the

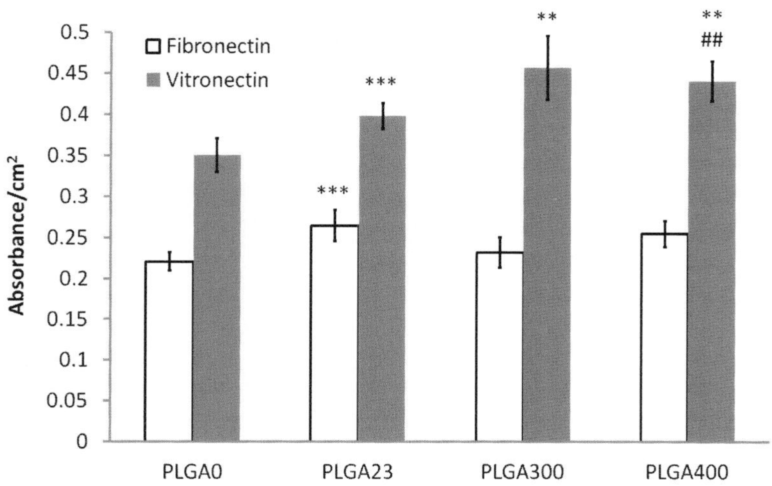

Fig. 4. Fibronectin and vitronectin adsorption per substrate size (expressed by light absorbance) on PLGA0, PLGA23, PLGA300 and PLGA400 measured by the enzyme-linked immunosorbent assay (ELISA). Data are expressed as mean ± SEM ($N = 3$). $**p < 0.05$ and $***p < 0.1$ compared to PLGA0; $\#\#p < 0.05$ compared to PLGA23.

nanopatterned PLGA surfaces, vitronectin adsorption on PGA300 was slightly higher than that on PLGA23 and PLGA300, although the difference was not statistically significant.

Results also revealed decreased vitronectin adsorption but increased fibronectin adsorption on alginate and chitosan deposited PLGA nanosurfaces (i.e. PLGA0/Alg. PLGA0/Chi, PLGA23/Alg and PLGA23/Chi) compared to untreated pure PLGA surfaces (i.e. PLGA0 and PLGA23; Fig. 5). Specifically, PLGA0/Alg, PLGA0/Chi, PLGA23/Alg and PLGA23/Chi all exhibited significantly lower vitronectin adsorption than that on PLGA0 and PLGA23. As a result, although PLGA23 had higher vitronectin adsorption than PLGA0, PLGA23/Alg demonstrated 36% lower vitronectin adsorption compared to PLGA0. However, fibronectin adsorption results exhibited the opposite trend as alginate and chitosan deposited surfaces had increased fibronectin adsorption compared to pure PLGA surfaces. PLGA0/Alg and PLGA0/Chi both showed significantly higher fibronectin adsorption than that on PLGA0, 31% and 63% more, respectively. However, the increase in fibronectin adsorption on the nanopatterned surfaces was not as significant as the increase on nanosmooth substrates. PLGA23/Chi showed 12% more vitronectin adsorption compared to PLGA23. No significant difference was observed between PLGA23 and PLGA23/Alg. The results indicated that surface chemistry played an important role on protein adsorption, which may

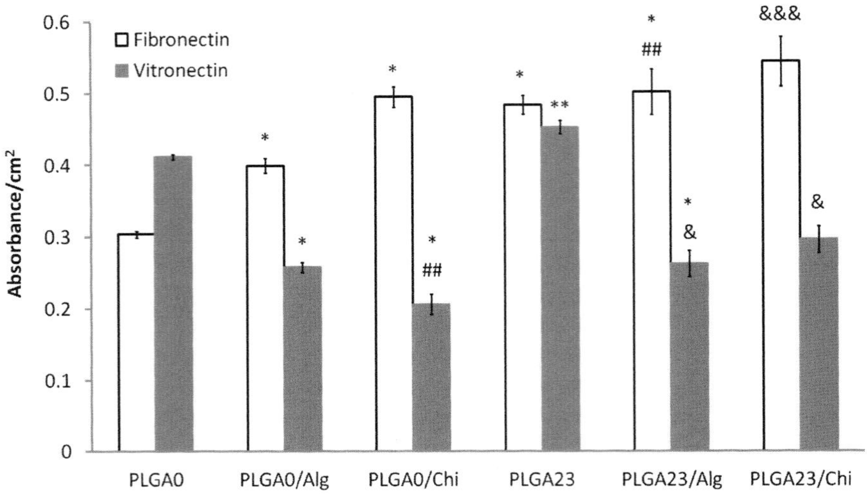

Fig. 5. Fibronectin and vitronectin adsorption per substrate size (expressed by light absorbance) on PLGA0, PLGA0/Alg, PLGA0/Chi, PLGA23, PLGA23/Alg and PLGA23/Chi measured by the ELISA. Data are expressed as mean ± SEM ($N = 3$). *$p < 0.01$ and **$p < 0.05$ compared to PLGA0; ## $p < 0.05$ compared to PLGA0/Alg; & $p < 0.01$ and &&& $p < 0.1$ compared to PLGA23.

have partially masked the influence of nanotopography on the nanopatterned surfaces. Such results are critically important since both vitronectin and fibronectin have been shown to influence cancer cell functions (such as adhesion and migration).

2.4. *The impact of PLGA surface nanopatterns on cancer cell functions*

It is known that proteins immediately get adsorbed on biomaterial surfaces, once a biomaterial is in contact with a protein containing solution and cells subsequently recognize this initial layer of proteins to initiate integrin-mediated adhesion.[31] Fibronectin and vitronectin have been recognized as two major adhesion proteins to mediate cell-biomaterial adhesion. A study by Lafrenie *et al.* found that lung cancer (A549) cells were deficient in the vitronectin receptor $\alpha v\beta 3$ but did express a fibronectin receptor. However, another study showed that A549 cells did express integrin $\alpha v\beta 3$.[32] In this way, it is reasonable to expect that A549 cell adhesion had a significant dependence on fibronectin but possibly a limited dependence on vitronectin. For the presently emphasized PLGA nanopatterned surfaces, as mentioned, results have shown that PLGA23, PLGA300 and PLGA400 had significantly higher vitronectin adsorption than nanosmooth PLGA, whereas only PLGA23 showed significantly higher fibronectin adsorption compared to PLGA0. This result suggested that both fibronectin and vitronectin preferentially and competitively adsorbed onto various PLGA surfaces. The reason for the observed increase in select protein adsorption resulted from the alterations in surface topography from a conventional to a nanopatterned surface.

As also previously mentioned, greater surface wettability was shown on nanopatterned PLGA surfaces and this increase could be linked to greater protein adsorption. A study by Kooten *et al.*[33] found that an assembled fibronectin matrix was higher on the polystyrene substrates with higher wettability exposed to plasma treatment to increase oxygen surface incorporation, and subsequent endothelial cell proliferation increased with increasing oxygen incorporation. In addition, as the nanometer surface roughness increased, there was a greater charge density at the surface which can absorb greater quantities of proteins.[31]

Of course, it is important to emphasize how all this change in surface topography, surface energy, and protein adsorption relate to breast and lung cancer cell responses. Results demonstrated that although the 23 nm surface features on PLGA increased lung cancer cell responses, the highest amount of apoptosis (programmed cell death) and vascular endothelial growth factor (VEGF), a factor which leads to increased vascularization of tumors, was observed on those surfaces. Thus, the optimal surface features on PLGA tissue engineering materials that may be used to regenerate healthy lung tissue and keep cancer from returning after a lung tumor is removed, is 23 nm. Trends on breast cancer cell functions will be reported next, but as can be seen from Table 4, the same exact trends of high apoptosis and lowest amount of VEGF synthesis was observed on the 23 nm PLGA samples. Importantly, these results imply that breast implants used for breast cancer patients should possess 23 nm surface features, as those features would inhibit the return of cancer after breast tumor resection. Such results are unprecedented and highlight that medical device surfaces with 23 nm surface features may be optimal to selective decrease in cancer cell functions for two much different cancers (lung and breast).

However, similar to the lung cancer cell density, 23 nm surface features on PLGA did not reduce breast cancer cell density the most as a significantly higher number of breast cancer cells (MCF-7) were observed on PLGA23 compared to PLGA0. MCF-7 cells were shown to have a greater affinity and bound well to both fibronectin and vitronectin,[34,35] and, thus, MCF-7 cell adhesion could be related to the adsorption of both fibronectin and vitronectin. This increase could have been due to the greater protein adsorption of both fibronectin and vitronectin on PLGA23 compared to PLGA0. However, although PLGA300 and PLGA400 had greater vitronectin adsorption than PLGA0, the breast cancer cell adhesion density was not promoted on PLGA300 and PLGA400. The possible reason could be that the conformation and biological activities of vitronectin coated on nanopatterned PLGA was not favorable to breast cancer cell adhesion which remains to be tested.

Importantly, a study by Underwood *et al.*[36] provided evidence that antibodies specific to a cell attachment region of fibronectin bound preferentially to the conformation of fibronectin on the more hydrophilic tissue culture polystyrene compared to untreated hydrophobic polystyrene. In contrast, antibodies specific to the cell-binding region in vitronectin displayed slightly increased binding to vitronectin coated on hydrophobic polystyrene. Here, the nanopatterned PLGA was more hydrophilic compared to nanosmooth PLGA, which may limit the conformation and bioactivity of adsorbed vitronectin and, thus, breast cancer cell adhesion was not promoted on PLGA300 and PLGA400.

Clearly, different cell functions (including adhesion, proliferation, apoptosis and VEGF secretion) showed varied cell responses to PLGA surface nanotopography. Both lung cancer and breast cancer cells exhibited decreased cell adhesion and proliferation density on PLGA0 and PLGA400 compared to PLGA23 and PLGA300. However, all four cell lines (lung cancer cell, lung normal cell, breast cancer cell and breast normal cell) had higher cell proliferation density on PLGA23 compared to PLGA0. As discussed above, cell adhesion and proliferation was related to protein adsorption and surface free energy influenced by surface topography. PLGA23 had nanoscale surface roughness comparable to the dimensions of proteins, which was favorable for the protein conformation and bioactivity due to the suitable "dock-in" sites. Moreover, the increased surface free energy of PLGA23 can promote the dispersion of cell aggregates and cell adhesivity to the material surface, and thus high cell adhesion and proliferation was observed on 23 nm topographical surfaces.

But as mentioned, most importantly, although PLGA23 showed higher cell adhesion and proliferation density, a higher percentage of cell apoptosis and decreased VEGF secretion was discovered on PLGA23 for both cancer cells after further investigation of critical cell bioactivity indicators. For both lung cancer and breast cancer cells, a greater percentage of cancer cell apoptosis occurred on PLGA23. Moreover, the secretion of VEGF, an essential growth factor for cancer growth and development, also revealed that cancer cells on PLGA23 secreted the least VEGF. Clearly, the lack of VEGF could significantly inhibit the growth of cancer cells and thus limit the formation of tumors. It is known that nanotopography can alter gene expression profiles of various cell types[15] and influence integrin-mediated signaling cascades.[23] The proposed model showing how polymer surface nanotopography directed cancer cell fates is presented schematically in Fig. 6. Cell interactions with the ECM are mediated by integrins that control cell gene expression, apoptosis and growth factor synthesis in a synergistic manner. The predicted size of surface occupancy by the head of an integrin heterodimer is about 10 nm in diameter.[37,38] This suggested that a 23 nm tightly arrayed semispherical

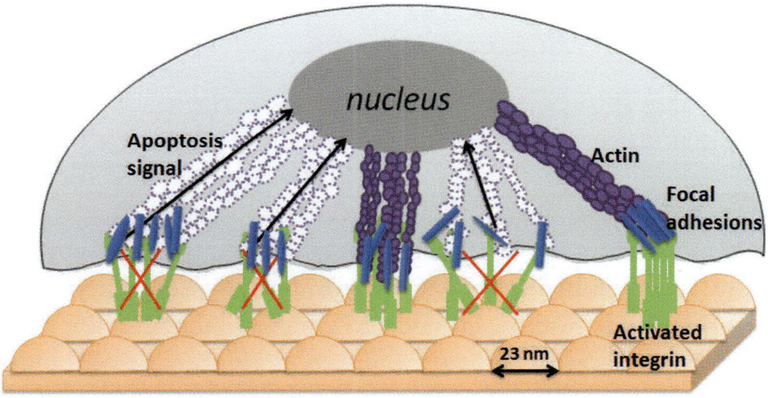

(a)

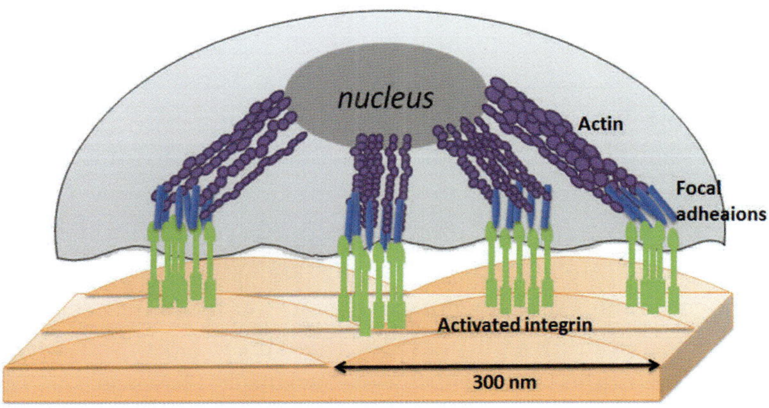

(b)

Fig. 6. Schematic presentation of nanopatterns which direct cancer cell apoptosis: (a) 23 nm patterns severely limit intergrin clustering and activation, resulting in dramatically reduced cell spreading, proliferation and eventually cell apoptosis; (b) 300 nm surface patterns support clustering and activation of integrins, resulting in assembly of actin stress fibers and signaling to the nucleus.

pattern would restrict the clustering of integrins and the following formation of focal adhesion and actin stress fibers through disordering the assembly of intergins by continuous protrusions. Therefore, 23 nm patterns severely limit intergrin activation and cell-biomaterial communication, resulting in dramatically reduced cell spreading, proliferation and migration and eventually cell apoptosis. However, 300 nm or 400 nm surface patterns had relatively much larger sizes compared to integrins, and, thus, clustering of integrins could be allowed to form on the pattern surface or spacing of patterns. The activated integrins could effectively promote the formation of focal adhesions and actin stress fibers, resulting in cell spreading and proliferation.

2.5. *The impact of nanopatterns and LBL monolayers on cell functions*

Results have also showed that the adsorption of the cell adhesive proteins fibronectin and vitronectin were significantly influenced by the alteration in surface chemistry due to alginate and chitosan adsorption. Altered surface properties (surface chemistry, surface wettability, surface charge and surface free energy) and competitive adsorption of fibronectin and vitronectin on substrate surfaces could be responsible for the altered protein adsorption. Specifically, the deposition of alginate and chitosan increased the hydrophilicity of substrate surfaces. This is expected as both alginate and chitosan are hydrophilic and, thus, LBL deposition on PLGA surfaces increased hydrophilicity. Fibronectin showed a strong affinity to more hydrophilic surfaces as more fibronectin adsorbed on the alginate and chitosan deposited surfaces. Some studies also discovered that hydrophilic substrates absorbed more fibronectin molecules than hydrophobic substrates. For example, Sousa *et al.*[39] found that fibronectin displayed higher affinity for the pure titanium with a titanium oxide layer, the more hydrophilic one, and lower for the TiO_2 sputtered on Si substrate, the more hydrophobic one. Another study also showed that the amounts of serum proteins, including fibronectin, adsorbed from serum containing media were greater on the more hydrophilic plasma-treated polytetrafluoroethylene surfaces than those adsorbed to the more hydrophobic unmodified polytetrafluoroethylene.[40] However, no consistent conclusion has been established since other studies also revealed increased fibronectin binding on more hydrophobic surfaces.[41,42] For instance, increased fibronectin binding and more rapid adsorption kinetics were observed on hydrophobic silanized-silica in contrast to hydrophilic silica.[42] Moreover, the increased fibronectin adsorption could be attributed to other surface properties in addition to surface wettability. The fibronectin adsorption result clearly indicated that positively charged chitosan-terminated films adsorbed higher amount of fibronectin adsorption in comparison to negatively charged alginated-terminated surfaces. This observation can be well explained by the strong electrostatic attraction between negatively charged proteins (i.e. fibronectin and vitronectin) and positively charged chitosan at or near the substrate surfaces. A similar observation has also been reported by Wittmer *et al.*[43] on LBL (poly (L-lysine) and dextran sulfate) assembly multilayer films.

However, vitronectin adsorption exhibited the opposite trend in comparison to fibronectin adsorption and less vitronectin adsorbed on alginate- and chitosan-terminated substrates. Here, vitronectin exhibited a preference for more hydrophobic surfaces. Similarly, Toromanov *et al.*[44] reported that vitronectin is adsorbed in increasing amount as the hydrophobicity of the substrate increases (the fraction of -OH groups decreases). In addition, electrostatic interactions could play an important role in vitronectin adsorption. Because vitronectin bears a net negative charge, the anionic alginate-terminated surfaces tended to adsorbed less protein than the cationic chitosan-terminated surfaces. However, more vitronectin adsorption was still observed on PLGA0/Alg in comparison to PLGA0/Chi. Complex competitive adsorption of fibronectin with vitronectin may be responsible for the observed results. Fibronectin has been shown to adsorb more than vitronectin onto same substrates under competitive condition.[45] Here, it is also observed that more fibronectin adsorbed on the alginate- and chitosan-terminated surfaces, leading to the binding sites for vitronectin to decrease and less vitronectin to adsorb onto these surfaces under the presence of competitive proteins (i.e. fibronectin and vitronectin).

Lung carcinoma cells exhibited higher cell proliferation on alginate- and chitosan-terminated substrates compared to pure unmodified PLGA. This result can be well explained by the increased fibronectin adsorption on these surfaces. Since lung cancer cells express fibronectin receptor but are deficient in vitronectin receptor $\alpha v\beta 3$,[46] the increased fibronectin adsorption could induce more lung cancer cell adhesion and subsequent cell proliferation. Moreover, fibronectin could be reorganized in an extracellular matrix-like structure on more hydrophilic surfaces by cells,[47] and extensive rearranging of fibronectin molecules can happen on hydrophilic surfaces, whereas, on hydrophobic substrates, the rearrangement is restricted due to a higher binding force between the protein and surface.[48] Thus, high lung cancer cell proliferation was observed on more hydrophilic surfaces with more fibronectin adsorption in this study.

However, lung cancer cells showed the lowest cell proliferation only on alginate-terminated surfaces (PLGA0/Alg and PLGA23/Alg). This observation can be well explained by the total least protein adsorption of fibronectin and vitronectin on PLGA0/Alg and PLGA23/Alg. Since MCF-7 express both fibronectin and vitronectin receptors,[49] cell proliferation can be affected by the adsorption of both fibronectin and vitronectin. On average, alginate-terminated surfaces had the least total protein adsorption including fibronectin and vitronecin (PLGA0/Alg<PLGA0/Chi<PLGA0, PLGA23/Alg<PLGA23/Chi<PLGA23), and decreased protein adsoprtion on alginate-terminated substrates had potential to decrease breast cancer cell adhesion and further less live cells were observed on these surfaces. In addition, the lower surface free energy of alginate-terminated substrates could limit the "wetting" of cell aggregates on their surfaces and induce decreased cancer cell aggregate spreading and subsequently cell proliferation.[28]

Clearly, through the deposition of alginate and chitosan on PLGA surfaces, surface properties and corresponding cell functions were significantly changed. Both positive and negative impacts were observed on cancer and healthy cells. Lung cancer cell proliferation increased on alginate- and chitosan-terminated surfaces, while breast cancer cell proliferation was significantly decreased on alginate-terminated surfaces. The possible reason was related to surface protein (fibronectin and vitronectin) adsorption as discussed above. However, lung healthy cell proliferation exhibited little correlation with surface chemistry but a significant correlation with surface nanopatterns. Lung healthy cell proliferation was significantly increased on nanopatterned substrates (PLGA23, PLGA23/Alg and PLGA23/Chi) no matter what surface chemistry compared to substrates without nanopatterns (PLGA0, PLGA0/Alg and PLGA0/Chi). However, for breast healthy cells, alginate- and chitosan-terminated substrates revealed significantly lower cell proliferation compared to corresponding pure PLGA substrates. Moreover, chitosan exhibited more profound inhibition on breast healthy cell proliferation in comparison to alginate. The mechanism(s) of these phenomena need to be clarified in future studies.

More importantly, further investigation on VEGF secretion showed that alginate had a negative effect on both lung cancer and breast cancer cells. On alginate-terminated surfaces (PLGA0/Alg and PLGA23/Alg), lung cancer and breast cancer cells had significantly lower VEGF concentrations compared to any other substrates. To the best of our knowledge in 2012, this is the first report concerning decreased VEGF secretion of lung cancer and breast cancer cells on alginate-terminated surfaces in comparison to PLGA and chitosan. Surface free energy results revealed that alginate decreased substrate surface free energy as PLGA23/Alg had lowest surface free energy and PLGA0/Alg the second lowest. The lower surface energy of alginate-terminated substrates could have potential to predict "wetting" of cancer cell aggregates on their surfaces and this potential could inhibit cancer cell spreading, cell proliferation and eventually decrease cancer bioactivity and less VEGF secretion. A study by Yang *et al.*[28] provided evidence that there was a strong correlation between cell responses and the surface free energy and found that nanocrystalline diamond which was treated with NH_3 plasma had the highest free energy and exhibited the best osteoblast responses, whereas the worst osteoblast responses was found on the lowest surface free energy diamond surface which was treated by H_2 plasma. Moreover, cancer cells on PLGA23/Alg revealed slightly lower VEGF concentration than PLGA0/Alg. This result indicated that surface nanopattern and surface chemistry simultaneously inhibited cancer cell VEGF secretion, but surface chemistry still played a leading role. Inhibition of VEGF has been recognized as a novel approach for cancer therapy. However, in some cases where chemistry change needs to be avoided, surface nanotopography could be an ideal candidate to inhibit cancer cell functions since only physical changes introduced and no any pharmacy or chemistry used.

3. Conclusions

This chapter summarizes the surface properties and cell functions on nanopatterned or LBL (alginate or chitosan) PLGA coatings. Surface topography and roughness, surface free energy and protein adsorption were believed to play profound roles on regulating cell functions. It was concluded that cell adhesive protein adsorption can affect cancer cell adhesion and subsequent cell proliferation. Moreover, it was believed that surface free energy was related to cell spreading and cell proliferation. However, competitive roles were present to cells when positive or negative impacts were induced by different factors. In addition, on alginate- and chitosan-terminated substrates, surface roughness, surface chemistry and surface charged all had significant influences on cell functions and especially, alginate-terminated nanopatterned substrates exhibited optimal cancer cell behaviors. Clearly, these results provided important guidance and cues for designing custom biomaterials to achieve specific goals.

References

1. G. E. Park, *et al.*, *Biomaterials* **26**(16), 3075–3082 (2005).
2. G. A. Dunn and A. F. Brown, *J. Cell Sci.* **83**, 313–340 (1986).
3. J. Lu *et al.*, *Acta Biomater.* **4**(1), 192–201 (2008).
4. M. Amaral *et al.*, *J. Biomed. Mat. Res. Part A* **87A**(1), 91–99 (2008).

5. K. M. Woo *et al.*, *Biomaterials* **28**(2), 335–343 (2007).
6. P. A. Tran *et al.*, *J. Mat. Chem.* **19**(18), 2653–2659 (2009).
7. G. Vazquez and A. Pellon, *Aesthet. Plastic Surg.* **31**(4), 330–336 (2007).
8. T. V. Slenters *et al.*, *J. Mat. Chem.* **18**(44), 5359–5362 (2008).
9. D. E. MacDonald *et al.*, *Biomaterials* **25**(16), 3135–46 (2004).
10. S. Tsukiyama *et al.*, *Jpn. J. Appl. Phys.* **47**(2), 1429–1434 (2008).
11. A. Ranjan and T. J. Webster, *Nanotechnology* **20**(30), - (2009).
12. D. Khang, *et al.*, *Int. J. Nanomed.* **1**(1), 65–72 (2006).
13. C. A. Scotchford *et al. J. Biomed. Mat. Res.* 41(3), 431–442 (1998).
14. T. H. Chung *et al.*, *Biomaterials* **28**(19), 2959–2966 (2007).
15. C. J. Bettinger, R. Langer and J. T. Borenstein, *Angew. Chem. -Int. Ed.* **48**(30), 5406–5415 (2009).
16. D. Hoffman-Kim, J. A. Mitchel and R. V. Bellamkonda, *Ann. Rev. Biomed. Eng.* **12**, 203–231 (2010).
17. Q. Tang *et al., Synthet. Metals* **147**(1–3), 247–252 (2004).
18. J. Zemla *et al.*, *Biomacromol.* **10**(8), 2101–2109 (2009).
19. J. L. Tan *et al.*, *Tissue Eng.* **10**(5&6), 865–872 (2004).
20. E. M. Blanco, M. A. Horton and P. Mesquida, *Langmuir* **24**(6), 2284–2287 (2008).
21. T. J. Webster *et al.*, *J. Biomed. Mat. Res.* **51**(3), 475–483 (2000).
22. L. Zhang *et al.*, *Biomaterials* **30**(7), 1309–1320 (2009).
23. J. Y. Lim and H. J. Donahue, *Tissue Eng.* **13**(8), 1879–1891 (2007).
24. M. Arnold *et al.*, *Chemphyschem.* **5**(3), 383–388 (2004).
25. M. J. Dalby *et al.*, *Exp. Cell Res.* **276**(1), 1–9 (2002).
26. M. J. Dalby, *Med. Eng. Phys.* **27**(9), 730–742 (2005).
27. G. Lamour *et al.*, *Biomaterials* **31**(14), 3762–3771 (2010).
28. L. Yang *et al.*, *J. Mat. Chem.* **22**(1), 205–214 (2012).
29. W. C. Clem *et al.*, *Biomaterials* **29**(24&25), 3461–3468 (2008).
30. T. Lechleitner *et al.*, *Biomaterials* **29**(32), 4275–4284 (2008).
31. D. C. Miller, K. M. Haberstroh and T. J. Webster, *J. Biomed. Mat. Res. Part A* **73A**(4), 476–484 (2005).
32. O. Heikkila *et al.*, *J. Gen. Vir.*. **90**, 197–204 (2009).
33. T. G. van Kooten, H. T. Spijker and H. J. Busscher, *Biomaterials* **25**(10), 1735–1747 (2004).
34. R. D. Blumenthal, H. J. Hansen and D. M. Goldenberg, *Cancer Res.* **65**(19), 8809–8817 (2005).
35. S. Hiscox *et al.*, *Clin. Exp. Met.* **24**(3), 157–167 (2007).
36. P. A. Underwood, J. G. Steele and B. A. Dalton, *J. Cell Sci.* **104**, 793–803 (1993).
37. J. Takagi *et al.*, *Cell* **110**(5), 599–611 (2002).
38. J. Park *et al.*, *Nano Lett.* 7(6), 1686–1691 (2007).
39. S. R. Sousa, P. Moradas-Ferreira and M. A. Barbosa, *J. Mat. Sci. -Mat. Med.* **16**(12), 1173–1178 (2005).
40. A. Dekker *et al.*, *Biomaterials* **12**(2), 130–138 (1991).
41. A. J. Campillo-Fernandez *et al.*, *Tissue Eng. Part A* **15**(6), 1331–1341 (2009).
42. G. K. Iwamoto *et al.*, *J. Colloid Interface Sci.* **106**(2), 459–464 (1985).
43. C. R. Wittmer *et al.*, *Biomaterials* **28**(5), 851–860 (2007).
44. G. Toromanov *et al.*, *Polymer* **51**(11), 2329–2336 (2010).

45. C. C. Barrias *et al.*, *Biomaterials* **30**(3), 307–316 (2009).
46. R. M. Lafrenie *et al.*, *Cancer Res.* **52**(8), 2202–2208 (1992).
47. G. Altankov and T. Groth, *J. Mat. Sci. -Mat. Med.* **5**(9&10), 732–737 (1994).
48. L. Renner *et al.*, *Langmuir* **21**(10), 4571–4577 (2005).
49. A. Nista *et al.*, *Int. J. Cancer* **72**(1), 133–141 (1997).

Chapter 10

Nanomedicine in Otorhinolaryngology

Carl Philpott

Department of Medicine, University of East Anglia, Norwich Medical School, Norwich, NR4 7TJ, UK
C. Philpott@uea.ac.uk

1. Introduction

Nanotechnology as applied to Otorhinolaryngology is, to date, relatively virgin territory but may provide the next generation of breakthrough treatments. The specialty has witnessed great developments in the visualization and access to otorhinolaryngological pathology with, for example, advances in endoscopy and neuronavigation and trans-oral robotic surgery, but this is very much at the macroscopic end of the spectrum and therefore nanomedicine opens up a whole new world of possibilities to be explored.[1–3] In this chapter, I will look at the early areas of promise in our specialty as well as future directions for research and development. The evidence cited here has been collated using standard online medical reference databases PubMed, Google Scholar, ISI Web of Science and Science Direct with search terms "Nanotechnology, Nanomedicine" in combination with "Otolaryngology, ENT, Rhinology, Otology, Head and Neck Surgery, Laryngology" in turn. Any significant abstracts were further scanned for relevance. In addition, personal knowledge and colleague recommendations were used. The recent archives of relevant journals *e.g.* "Nanomedicine" were hand searched for further articles.

2. Otology

The cochlear implant has been a revolution for both congenital and adult-onset hearing loss offering an option to sufferers of profound deafness to return to the auditory world around them. However, this is an invasive surgical option and due to the costs involved is limited to selected patients. However, if a therapeutic agent could be delivered nanomedically to the inner ear, this may forgo the need for surgery. Using a guinea pig model, researchers in Hannover have been able to demonstrate that hyperbranched polylysine nanoparticles injected into the scala tympani of the cochlea were detectable in the basilar membrane and stria vascularis after 48 h with a confocal laser scanning microscope.[4] Whilst this is only preliminary work that has been replicated,[5] it shows that transportation of therapeutic agents to the site of pathology may become eminently more feasible with further research. Vestibular disorders could similarly be targeted in this manner which would prove to be a great advance for patients where therapeutic options are currently limited in pharmacological terms or more radical approaches that involve other sacrifices *e.g.* hearing with intratympanic gentamicin, risks to adjacent structures in surgical approaches such as endolymphatic sac decompression. Recent literature shows that this is a developing area with several centres showing varying degrees of success in transporting nanoparticles to the vestibulocochlear apparatus[6–8] but recognizing that biotoxicity remains to be determined before these developments translate to the bedside.[9] In the short-term, these novel delivery methods may help with existing surgical interventions, by providing neuronal protection.[10]

3. Rhinology

Epistaxis poses a common problem for otorhinolaryngologists both in an acute setting and in the elective setting of endoscopic sinus surgery. Controlling this bleeding without the need for packing or agents with potential systemic side-effects would have widespread benefit, as the heavy intra-operative bleeding seen in inflammatory sino-nasal disease during endoscopic sinus surgery could be easily controlled to keep the field free of blood and reduce the operative time spent combating bleeding. In 2006, Ellis-Behnke *et al.* undertook research to utilize a self-assembling peptide to treat brain injuries.[11] What they discovered as a fascinating sideline was that the peptides (N-hydroxysulfosuccinimide) created a barrier of nanofibres that achieves a rapid and complete haemostasis which seems viable in any setting from cerebral to overt vascular bleeding.[12] This novel therapy arrests bleeding without the use of pressure, cauterization, vasoconstriction, coagulation, or cross-linked adhesives and achieves this in about 10 s. The self-assembling solution is non-toxic and non-immunogenic and the breakdown products are amino acids, which are tissue building blocks that can be used to repair the site of injury. At this time, this research has only been conducted in animal models and will require verification in human subjects.

Allergic fungal rhinosinusitis is a disease entity, which can be challenging to treat and control due to its inherent propensity to recur after surgery unless medical treatment is unrelenting. AFRS has been a matter of controversy regarding its aetiology and pathogenesis[13] and there is evidence to support the use of antifungals therapy in controlling relapses[14–17] but also evidence to the contrary that standard modality amphotericin B is not significantly more effective than other topical agents in combating chronic rhinosinusitis (CRS).[18] This may however be a delivery issue to be overcome. Furthermore, there would clearly also be a useful application in cases of invasive fungal sinusitis that require systemic therapy. A recent study by Nahar *et al.* in the last year looked at the development of a nanoparticulate carrier of amphotericin B to allow controlled delivery and reduced toxicity.[19,20] Despite being a potent antifungal agent, the problems with amphotericin B include its pronounced side effects (*e.g.* chills, fever, nausea, haemolytic toxicity and nephrotoxicity). The gelatin nanoparticles were found to deliver the amphotericin B without causing significant haematological or nephrotoxicity.

Similarly, nanoparticle delivery of itraconazole has also been established *in vitro*.[21] Itraconazole in traditional formulations has the potential to cause hepatotoxicity. Ensuring sustained drug delivery whilst avoiding the systemic side-effects may enable effective antifungal nanomedical treatment. Mosqueira *et al.*[22] looked at the nanocapsules for delivering the antifungal albaconazole using atomic force microscopy and found that they deform on a mica surface. The suggestion from their work is that the nanocapsules would be able to cross intracellular spaces.

In the wider field of CRS with nasal polyps, the use of nanosilver which has antifungal, antibacterial and antiviral properties has been shown to inhibit the activation of epithelial cells derived from human nasal polyps.[23] This *in vitro* study will require *in vivo* confirmation to advance this possibility and once again it will be the final human clinical studies that determine the viability of deploying this technology in clinical practice.

4. Laryngology/Head & Neck

Nanofibrous scaffold fabrication offers great potential in regenerative medicine by means of synthetic, biomimetic extracellular matrix analogs.[24] One of the goals of regenerative medicine is the successful creation of "spare parts for surgery": the Holy Grail for tissue engineers. Although almost any surgeon would find uses for extra cartilage, bone or muscle; it is perhaps in laryngology and head and neck surgery where transplant and augmentation would have the most impact; in fact this already has started to become a reality.[25] One of the difficulties in tissue engineering is the creation of suitable nanoscale extracellular matrices (ECM). The ECM is composed of structural proteins arranged to allow a specific function and these tissues are organized over multiple levels of size. The matrix acts as a local reservoir for growth factors, anchorage and as a mechanotransduction element transmitting internal and external stresses to the cells within it. Several methods have been developed including the hallmark of nanotech, self-assembly, to allow creation of scaffolds on which cells may be coaxed into tissue.[24] Examples of this include a thermally induced phase separation (TIPS) process in which a solution of biodegradable polymer is cast into a porous scaffold, resulting in a nanofibrous pore-wall structure with the fibres having a diameter 50–500 nm.[26] By placing molecules such as growth factors into these scaffolds, tissue regeneration can be controlled for the requirements of the site of reconstruction.

Moving back to the theme of enhanced drug delivery, nanoparticle photo-ablation provides the opportunity for nanotransportation to allow targeting of tumor cells by circulating prodrugs that require photoactivation.[27] The principle here is to use metal nanodots which would be activated by photons of light and heat the dots and subsequently their environs.[28] In some scenarios highly energetic oxygen molecules may be produced from the dots and will react with and destroy adjacent organic molecules. The major advantage of this technique is that it is highly more focused than traditional chemotherapy that leaves its mark on other cell populations with a high turnover. The key focus for researchers in this area will be to determine the exact structure of the nanodots and their bioavailability in order to produce the optimum effect in the target tissue. Also known as *Q*-dots, these nanoparticles may well replace current radiolabelling techniques for identifying active disease processes such as metastatic disease.[29]

Targeting malignant cells selectively may also be possible by virtue of their structural differences from normal cells. Malignant transformation of cells involves a loss of normal cytoskeletal architecture and therefore rigidity.[30] Researchers in Oxford have shown that deformability is reliably increased in malignant cells and can be detected by an "optical stretcher" during flow cytometry (Fig. 2).[31] Malignant cells are much "floppier" than healthy ones, therefore by using photons (which have mass) to bombard the cells with a laser can cause them to deform; watching them as they fall back to their original shape. The measureable compliance of an abnormal cell could be a great help to cytological diagnosis, adding a number to the qualitative assessment. A further trial of this "mechanical phenotyping" in oral cancer cells[32] showed that the normal and abnormal cells could be clearly distinguished to a p value of < 0.01. What is more, the cells are not stained or destroyed during the process and the optical stretcher can mechanically sort the cells into two populations based on their properties, allowing further genomic, proteomic etc. testing.

5. Future Directions

5.1. *Chronic otitis media*

Chronic otitis media with effusion, known as glue ear in children, is a common problem in young primary school-aged children. Biofilm formation in the middle ear is now believed to be partly responsible for this chronic source of pathology[33–36] and being targeted for treatment.[37] However, in the nanomedical domain recent animal studies by Lensing *et al.* have looked at using nanoporous silica layers as a host for molecules which can then be released over a defined time period. In their rabbit model, there were clear microbiological and histological differences between the control group and the treated group, both of which were inoculated with *P. aeruginosa*. The nanoporous silica layer (coated Bioverit(R)II implants) were loaded with ciprofloxacin (or without in the control group) and implanted into the middle ears of the rabbits. This model of delivery system showed convincing efficacy in this *in vivo* experiment and heralds the opportunity for the work to be replicated in a human model.[38]

5.2. *Head and neck cancer*

Diagnostic imaging and nanopharmaceuticals may enable imaging at the molecular level with subsequent targeting of specific cancer cells. Following this it may become possible to monitor for the presence of metastatic cells in a patient's blood to determine the effectiveness of the intervention and to monitor for recurrence without the reliance on more macroscopic imaging and visualization techniques. Beyond this there is the opportunity to study the genetic aspects of disease at the molecular level directly.

Perhaps the most advanced area in medical nanotechnology is cancer theranostics. The portmanteau of diagnosis and therapeutics is meant to convey the simultaneous action of tissue-specific or even cell-specific technologies that, like Feynman's swallowed "mechanical surgeon," both diagnose and treat the malignant cell. In fact with the advent of capsule endoscopy, such a concept may not be that far away[39] and in fact some very recent work has shown promising findings for personalized polymer-based immunotherapy for head and neck squamous cell carcinomas.[40] Theranostics seeks to enable molecular profiling of cancers to determine patient-specific chemotherapy regimens that may avoid unnecessary toxicity from unselective agents. Warenius *et al.* in Southampton looked at 18 human cancer cell lines including squamous cell carcinoma of the larynx and found variability in the cell survival curves for the four chemotherapeutic agents tested with cisplatin and etoposide being the most effective.[41] Gold nanoparticles have already shown promise in the remit of detection and are far more effective than existing techniques such as ELISA[27,42] and by attaching antibodies to gold nanoparticles which subsequently photoactivate, selective binding to tumor cells has been possible.[43] Magnetic drug targeting is another nanomedical approach whereby the toxic effects of chemotherapeutic agents such as mitoxantrone on human cancer cell lines can be enhanced when bound to magnetic nanoparticles.[44] Silicon-based nanowires have also been shown able to electrically detect tumor marker proteins and give a measured response.[45] Similarly microfabricated silicon cantilevers which when bound to the complementary molecule show a deflection that is detectable by an optical beam.[46]

5.3. *Olfactory disorders*

Anosmia currently represents a sometimes challenging problem for otorhinolaryngologists and can be a considerable source of frustration for both the doctor and the patient, with many patients' problem being neglected. Repair of central nervous system tissue after trauma using nanotechnologic molecular cell assembly with nanofibres has already been shown to be possible. As discussed above the option of cochlear implantation exists for patients with profound bilateral sensorineural hearing loss, but there is currently no equivalent option for anosmic patients. Nanotechnologic advances have however been achieved in the development of the electronic nose in mimicking the human sense of olfaction, although to date the application for this has been on an industrial level or for diagnosis of infections.[47] Through appropriate development of the apparatus at this smaller scale, it may be possible to achieve the olfactory equivalent of the cochlear implant and restore olfactory sensation to these often-maligned patients. The devices are currently hand-held and would need to be scaled appropriately, as well as the consideration of how to stimulate the neuroepithelium in the olfactory cleft in an effective manner to produce a meaningful and desirable response.

It may be, however, that making a sophisticated device is a heavy-handed approach to a simpler problem and like the example of transmission of drugs nanomedically to the inner ear, an equivalent step exists for the olfactory apparatus. Recent work at Auburn University in the USA has found that zinc metal nanoparticles stimulated rat olfactory neuroepithelium to enhance the odorant response in a dose-dependent manner.[48] This is an interesting development as initial hopes for zinc compound remedies for hyposmia[49,50] and URTIs were conversely found to have the opposite effect[51,52] and have been withdrawn after FDA recommendations that patients should not use them,[53] however there remains controversy about the exact role of zinc compounds in causing or treating olfactory disorders.[54] $ZnSO_4$ irrigation is a standard method of olfactory epithelium ablation in olfactory animal research. As we have explored in this article, nanoparticles have the potential of differing properties to their parental compounds and Viswaprakash *et al.*, have suggested that the zinc nanoparticles are closely related to the interface between the guanine nucleotide binding protein and the receptor proteins and hence have an involvement in making the initial signal transfer in the presence of olfactory stimulation thus leading us to suspect that enhancement of impaired olfaction may be possible at this level. This proposal represents a modification of the theories surrounding the role of zinc ions proposed by Turin in 1996 (Ref. 55) as well as highlighting recent proposals by Brookes *et al.* and Gane *et al.* on odour recognition.[56,57]

It is also possible that akin to the theranostic example above, identifying the best choice of agent for the individual case of anosmia is a distinct possibility. Shiga *et al.* have proposed that administrating thallium-201 nasally and following this with single photon emission computed tomography, X-ray computed tomography and magnetic resonance imaging might be useful in demonstrating the progression or lack of the thallium into the olfactory bulbs and hence selection for appropriate trials of medical treatment.[58]

Aside from isolated olfactory disorders, neurodegenerative disorders have a significant link with olfaction as anosmia may be the first indication of impending motor dysfunction in Alzheimer's and Parkinson's diseases.[59–62] With recent proposals that Parkinson's may be detected earlier using biomarkers and olfactory loss,[63,64] it may be that nanomedicine has a role to play both in the detection and treatment of these disabling diseases which have a significant healthcare burden despite current drug therapeutics.[65,66]

5.4. *Chronic rhinosinusitis*

Two of the current theories around the pathogenesis of CRS include biofilm formation in cases without polyps and staphylococcal superantigens in cases with polyps; in either case, *S. aureus* is commonly yielded from sinus aspirates in CRS patients. Recently, Ndieyira *et al.* have demonstrated the ability to detect the interaction between MRSA and vancomycin through the deformation of cantilever arrays. Thus, it may become possible to carefully examine the interaction between coagulase negative *S. aureus* and antibiotics perhaps providing more targeted treatment. Beyond this there has also been recent work by Das *et al.* to suggest that surface enhanced laser desorption/ionization time-of-flight mass spectrometry (SELDI-TOF-MS) proteomic profiling of serum samples may be used to distinguish CRS subtypes which in turn may allow for a blood test to highlight a severe subtype.[67] With a heterogenous disease such as CRS, patient profiling to determine careful endotypes and phenotypes will be a major advance in helping to avoid a "monotherapy" approach for all cases regardless of response and is a stated research priority of the recent iteration of the European Position Paper on Rhinosinusitis and Nasal Polyps.[68]

6. Conclusions

The future treatment of otorhinolaryngological diseases could be revolutionized by advances in nanomedicine and nanotechnology, and treatment in many areas of the specialty may become very focused on the disease in hand. For instance, being able to diagnose and treat a patient with head and neck cancer that is specific to that individual and likely to be free of the significant side-effects of current medical and surgical interventions will be a major advance, not least for patients' quality of life. Other potential benefits will be seen in the nose and the larynx to name but a few. However, whilst these new technologies provide exciting potential for revolutionizing patient care in our specialty and beyond, there will always need to be an element of caution to ensure that we abide by Hippocrates and ensure no harm is done to our patients in the quest of greater good. Overall, nanotoxicology is an area where caution will need to continue to be excised in order to prevent any iatrogenic disease in the future and this concern is already being expressed by researchers in the field.[69] Nonetheless, nanomedicine yields great promise for many aspects of the field of otorhinolaryngology.

References

1. R. T. Sataloff and C. Wei, Nanomedicine: Important new concepts for otolaryngology. *Ear Nose Throat J.* **86**(9), 528–531 (2007).
2. C. Wei, N. Liu, P. Xu, M. Heller, D. A. Tomalia, D. T. Haynie *et al.*, From bench to bedside: Successful translational nanomedicine: highlights of the Third Annual Meeting of the American Academy of Nanomedicine. *Nanomedicine (London)* **3**(4), 322–331 (2007).
3. L. Balogh, The future of nanomedicine and the future of Nanomedicine: NBM. *Nanomedicine (London)* **5**, 1 (2009).
4. V. Cheper, M. Wolf, M. Scholl, Z. Kadlecova, T. Perrier, H. A. Klok *et al.*, Potential novel drug carriers for inner ear treatment: Hyperbranched polylysine and lipid nanocapsules. *Nanomedicine (London)* **4**(6), 623–635 (2009).
5. W. Zhang, Y. Zhang, M. Lobler, K. P. Schmitz, A. Ahmad, I. Pyykko *et al.*, Nuclear entry of hyperbranched polylysine nanoparticles into cochlear cells. *Int. J. Nanomedicine* **6**, 535–546 (2011).
6. C. O. Pritz, M. Bitsche, W. Salvenmoser, J. Dudas, A. Schrott-Fischer and R. Glueckert, Endocytic trafficking of silica nanoparticles in a cell line derived from the organ of Corti. *Nanomedicine (London)*. **8**(2), 239–252 (2013).
7. Y. Zhang, W. Zhang, A. H. Johnston, T. A. Newman, I. Pyykko and J. Zou, Targeted delivery of Tet1 peptide functionalized polymersomes to the rat cochlear nerve. *Int. J. Nanomedicine* 7, 1015–1022 (2012).
8. E. V. Surovtseva, A. H. Johnston, W. Zhang, Y. Zhang, A. Kim, M. Murakoshi *et al.*, Prestin binding peptides as ligands for targeted polymersome mediated drug delivery to outer hair cells in the inner ear. *Int. J. Pharm.* **424**(1&2), 121–127 (2012).
9. C. O. Pritz, J. Dudas, H. Rask-Andersen, A. Schrott-Fischer and R. Glueckert, Nanomedicine strategies for drug delivery to the ear. *Nanomedicine (London)* **8**(7), 1155–1172 (2013).
10. H. Meyer, T. Stover, F. Fouchet, G. Bastiat, P. Saulnier, W. Baumer *et al.*, Lipidic nanocapsule drug delivery: Neuronal protection for cochlear implant optimization. *Int. J. Nanomedicine* 7, 2449–2464 (2012).
11. J. Guo, H. Su, Y. Zeng, Y. X. Liang, W. M. Wong, R. G. Ellis-Behnke *et al.*, Reknitting the injured spinal cord by self-assembling peptide nanofiber scaffold. *Nanomedicine* **3**(4), 311–321 (2007).
12. R. G. Ellis-Behnke, Y. X. Liang, D. K. Tay, P. W. Kau, G. E. Schneider, S. Zhang *et al.*, Nano hemostat solution: Immediate hemostasis at the nanoscale. *Nanomedicine* **2**(4), 207–215 (2006).
13. W. J. Fokkens, C. van Drunen, C. Georgalas and F. Ebbens, Role of fungi in pathogenesis of chronic rhinosinusitis: The hypothesis rejected. *Curr. Opin. Otolaryngol. Head Neck Surg.* **20**(1),19–23 (2012).
14. K. O. Chan, K. A. Genoway and A. R. Javer, Effectiveness of itraconazole in the management of refractory allergic fungal rhinosinusitis. *J. Otolaryngol. Head Neck Surg. Le Journal d'oto-rhino-laryngologie et de chirurgie cervico-faciale* **37**(6), 870–874 (2008).
15. J. U. Ponikau, D. A. Sherris, A.Weaver and H. Kita, Treatment of chronic rhinosinusitis with intranasal amphotericin B: A randomized, placebo-controlled, double-blind pilot trial. *J. Allergy Clin. Immunol.* **115**(1), 125–131 (2005).

16. E. B. Kern, D. Sherris, A. M. Stergiou, L. M. Katz, L. C. Rosenblatt and J. Ponikau, Diagnosis and treatment of chronic rhinosinusitis: Focus on intranasal Amphotericin B. *Ther. Clin. Risk Manag.* **3**(2), 319–325 (2007).
17. J. U. Ponikau, D. A. Sherris, H. Kita and E. B. Kern, Intranasal antifungal treatment in 51 patients with chronic rhinosinusitis. *J. Allergy Clin. Immunol.* **110**(6), 862–866 (2002).
18. F. A. Ebbens, G. K. Scadding, L. Badia, P. W. Hellings, M. Jorissen, J. Mullol *et al.*, Amphotericin B nasal lavages: Not a solution for patients with chronic rhinosinusitis. *J. Allergy Clin. Immunol.* **118**(5), 1149–1156 (2006).
19. M. Nahar and N. K. Jain, Preparation, characterization and evaluation of targeting potential of amphotericin B-loaded engineered PLGA nanoparticles. *Pharm. Res.* **26**(12), 2588–2598 (2009).
20. M. Nahar, D. Mishra, V. Dubey and N. K. Jain, Development, characterization, and toxicity evaluation of amphotericin B-loaded gelatin nanoparticles. *Nanomedicine (London)* **4**(3), 252–261 (2008).
21. S. Mukherjee, S. Ray and R. S. Thakur, Design and evaluation of itraconazole loaded solid lipid nanoparticulate system for improving the antifungal therapy. *Pak.J. Pharm. Sci.* **22**(2), 131–138 (2009).
22. V. C. F. Mosqueira, E. A. Leite, C. M. Barros, J. M. C. Vilela and M. S. Andrade, Polymeric Nanostructures for Drug Delivery: Characterization by Atomic Force Microscopy. *Microsc. Microanal.* **11**(SupplementS03), 36–39 (2005).
23. S. H. Shin and M. K. Ye, The effect of nano-silver on the activation of nasal polyp epithelial cells by Alternaria, Der P1 and staphylococcal enterotoxin B. *Int. Immunopharmacol.* **11**(11), 1691–1696 (2011).
24. P. A. Madurantakam, C. P. Cost, D. G. Simpson and G. L. Bowlin, Science of nanofibrous scaffold fabrication: Strategies for next generation tissue-engineering scaffolds. *Nanomedicine (London)* **4**(2), 193–206 (2009).
25. M. A. Birchall, Cell- and tissue-engineered organ replacements. *Br. J. Surg.* **96**(6), 565–566 (2009).
26. I. O. Smith, X. H. Liu, L. A. Smith and P. X. Ma, Nanostructured polymer scaffolds for tissue engineering and regenerative medicine. *Wiley Interdiscip. Rev.* **1**(2), 226–236 (2009).
27. D. Falcone, J. Li, A. Kale and G. B. Jones, Photoactivated enediynes as targeted antitumoral agents: Efficient routes to antibody and gold nanoparticle conjugates. *Bioorganic & Med. Chem. Lett.* **18**(3), 934–937 (2008).
28. M. Ratner and D. Ratner, *Nanotechnology: A Gentle Introduction to the Next Big Idea.* Upper Saddle River, NJ: Prentice Hall (2002).
29. K. J. Morrow, Jr., R. Bawa and C. Wei, Recent advances in basic and clinical nanomedicine. *The Medical Clinics of North America* **91**(5), 805–843 (2007).
30. A. Ben-Ze'ev, The cytoskeleton in cancer cells. *Biochim. Biophys. Acta.* **780**, 197–212 (1985).
31. J. Guck, S. Schinkinger, B. Lincoln, F. Wottawah, S. Ebert, M. Romeyke *et al.*, Optical Deformability as an Inherent Cell Marker for Testing Malignant Transformation and Metastatic Competence. **88**(5), 3689–3698 (2005).
32. T. W. Remmerbach, F. Wottawah, J. Dietrich, B. Lincoln, C. Wittekind and J. Guck, Oral Cancer Diagnosis by Mechanical Phenotyping. *Cancer Res.* **69**(5), 1728–1732 (2009).

33. H. Lampikoski, A. A. Aarnisalo, J. Jero and T. J. Kinnari, Mastoid biofilm in chronic otitis media. *Otol. Neurotol.: Official Publication of the American Otological Society, American Neurotology Society [and] European Academy of Otology and Neurotology* **33**(5), 785–788 (2012).
34. G. Saylam, E. C. Tatar, I. Tatar, A. Ozdek and H. Korkmaz, Association of adenoid surface biofilm formation and chronic otitis media with effusion. *Arch. Otolaryngol. Head Neck Surg.* **136**(6), 550–555 (2010).
35. H. Coates and S. Vijayasekaran, Bacterial biofilm in chronic suppurative otitis media. *Otolaryngol.Head Neck Surg.: Official Journal of American Academy of Otolaryngology-Head and Neck Surgery* **142**(5), 778 (2010). author reply 78
36. M. R. Lee, K. S. Pawlowski, A. Luong, A. D. Furze and P. S. Roland, Biofilm presence in humans with chronic suppurative otitis media. *Otolaryngol.Head Neck Surg.: Official Journal of American Academy of Otolaryngology-Head and Neck Surgery* **141**(5), 567–571 (2009).
37. M. Daniel, R. Chessman, S. Al-Zahid, B. Richards, C. Rahman, W. Ashraf *et al.*, Biofilm eradication with biodegradable modified-release antibiotic pellets: A potential treatment for glue ear. *Arch. Otolaryngol. Head Neck Surg.* **138**(10), 942–949 (2012).
38. R. Lensing, A. Bleich, A. Smoczek, S. Glage, N. Ehlert, T. Luessenhop *et al.*, Efficacy of nanoporous silica coatings on middle ear prostheses as a delivery system for antibiotics: An animal study in rabbits. *Acta Biomaterialia* **9**(1), 4815–4825 (2013).
39. A. Moglia, A. Pietrabissa and A. Cuschieri, Capsule endoscopy. *British Med. J.* **339**, b3420 (2009).
40. S. Prasad, V. Cody, J. K. Saucier-Sawyer, W. M. Saltzman, C. T. Sasaki, R. L. Edelson *et al.*, Polymer nanoparticles containing tumor lysates as antigen delivery vehicles for dendritic cell-based antitumor immunotherapy. *Nanomedicine* (2010).
41. H. M. Warenius, L. Seabra, L. Kyritsi, R. White, R. Dormer, S. Anandappa *et al.*, Theranostic proteomic profiling of cyclins, cyclin dependent kinases and Ras in human cancer cell lines is dependent on p. 53 mutational status. *Int. J. Oncol.* **32**(4), 895–907 (2008).
42. C. A. Mirkin, C. S. Thaxton and N. L. Rosi, Nanostructures in biodefense and molecular diagnostics. *Expert Rev. Mol. Diagn.* **4**(6), 749–751 (2004).
43. X. Huang, I. H. El-Sayed, W. Qian and M. A. El-Sayed, Cancer cell imaging and photothermal therapy in the near-infrared region by using gold nanorods. *J. Am. Chem. Soc.* **128**(6), 2115–2120 (2006).
44. S. Durr, S. Lyer, J. Mann, C. Janko, R. Tietze, E. Schreiber *et al.*, Real-time cell analysis of human cancer cell lines after chemotherapy with functionalized magnetic nanoparticles. *Anticancer Res.* **32**(5), 1983–1989 (2012).
45. G. Zheng, F. Patolsky, Y. Cui, W. U. Wang and C. M. Lieber, Multiplexed electrical detection of cancer markers with nanowire sensor arrays. *Nat. Biotechnol.* **23**(10), 1294–1301 (2005).
46. R. McKendry, J. Zhang, Y. Arntz, T. Strunz, M. Hegner, H. P. Lang *et al.*, Multiple label-free biodetection and quantitative DNA-binding assays on a nanomechanical cantilever array. *Proc. Natl. Acad. Sci. USA* **99**(15), 9783–9788 (2002).
47. M. E. Shykhon, D. W. Morgan, R. Dutta, E. L. Hines and J. W. Gardner Clinical evaluation of the electronic nose in the diagnosis of ear, nose and throat infection: A preliminary study. *J. Laryngol. Otol.* **118**(9), 706–709 (2004).

48. N. Viswaprakash, J. C. Dennis, L. Globa, O. Pustovyy, E. M. Josephson, P. Kanju *et al.*, Enhancement of odorant-induced responses in olfactory receptor neurons by zinc nanoparticles. *Chem. Senses* **34**(7), 547–557 (2009).
49. T. Aiba, M. Sugiura, J. Mori, K. Matsumoto, K. Tomiyama, F. Okuda *et al.*, Effect of zinc sulfate on sensorineural olfactory disorder. *Acta Otolaryngol. Suppl.* **538**, 202–204 (1998).
50. S. K. Williams, T. Gilbey and S. C. Barnett, Immunohistochemical studies of the cellular changes in the peripheral olfactory system after zinc sulfate nasal irrigation. *Neurochem. Res.* **29**(5), 891–901 (2004).
51. B. W. Jafek, M. R. Linschoten and B. W. Murrow, Anosmia after intranasal zinc gluconate use. *Am. J. Rhinol.* **18**(3), 137–141 (2004).
52. T. H. Alexander and T. M. Davidson, Intranasal zinc and anosmia: The zinc-induced anosmia syndrome. *Laryngoscope* **116**(2), 217–220 (2006).
53. Warnings on Three Zicam Intranasal Zinc Products. Available at: http://www.fda.gov/ForConsumers/ConsumerUpdates/ucm166931.htm.
54. B. Slotnick, A. Sanguino, S. Husband, G. Marquino and A. Silberberg, Olfaction and olfactory epithelium in mice treated with zinc gluconate. *Laryngoscope* **117**(4), 743–749 (2007).
55. L. Turin, A spectroscopic mechanism for primary olfactory reception. *Chem. Senses* **21**(6), 773–791 (1996).
56. J. C. Brookes, F. Hartoutsiou, A. P. Horsfield and A. M. Stoneham, Could humans recognize odor by phonon assisted tunneling? *Phys. Rev. Lett.* **98**(3), 038101 (2007).
57. S. Gane, D. Georganakis, K. Maniati, M. Vamvakias, N. Ragoussis, E. M. C. Skoulakis *et al.*, Molecular Vibration-Sensing Component in Human Olfaction. *PLoS ONE* **8**(1), (2013).
58. H. Shiga, J. Yamamoto and T. Miwa, Clinical diagnosis of the olfactory nerve transport function. *Yakugaku Zasshi: Journal of the Pharmaceutical Society of Japan* **132**(11), 1263–1266 (2012).
59. G. S. Solomon, Anosmia in Alzheimer disease. *Perceptual and Motor Skills* **79**(3 Pt 1), 1249–1250 (1994).
60. K. Imamura, S. Matumoto, N. Mabuchi, Y. Kobayashi, N. Okayasu and K. Watanabe, Relationship between the regional cerebral blood flow and the cognitive function and anosmia in patients with Parkinson disease and Alzheimer disease. *Brain Nerve = Shinkei kenkyu no shinpo* **61**(6), 683–690 (2009).
61. A. Siderowf, A. Newberg, K. L. Chou, M. Lloyd, A. Colcher, H. I. Hurtig *et al.*, [99mTc] TRODAT-1 SPECT imaging correlates with odor identification in early Parkinson disease. *Neurology* **64**(10), 1716–1720 (2005).
62. R. L. Doty, D. P. Perl, J. C. Steele, K. M. Chen, J. D. Pierce Jr., P. Reyes *et al.*, Olfactory dysfunction in three neurodegenerative diseases. *Geriatrics* **46**(Suppl 1), 47–51 (1991).
63. D. Berg, Marker for a preclinical diagnosis of Parkinson's disease as a basis for neuroprotection. *J. Neural Transm. Suppl.* **71**, 123–132 (2006).
64. D. Berg, Biomarkers for the early detection of Parkinson's and Alzheimer's disease. *Neurodegener. Dis.* **5**(3&4), 133–136 (2008).
65. T. Keranen, S. Kaakkola, K. Sotaniemi, V. Laulumaa, T. Haapaniemi, T. Jolma *et al.*, Economic burden and quality of life impairment increase with severity of PD. *Parkinsonism Relat. Disord.* **9**(3), 163–168 (2003).

66. R. Soumen, A. H. Johnston, S. T. Moin, J. Dudas, T. A. Newman, B. Hausott *et al.*, Activation of TrkB receptors by NGFbeta mimetic peptide conjugated polymersome nanoparticles. *Nanomedicine* **8**(3), 271–274 (2012).
67. S. Das, P. A. Maeso, A. M. Becker, J. D. Prosser, B. L. Adam and S. E. Kountakis, Proteomics blood testing to distinguish chronic rhinosinusitis subtypes. *Laryngoscope* **118**(12), 2231–2234 (2008).
68. W. J. Fokkens, V. J. Lund, J. Mullol, C. Bachert, I. Alobid, F. Baroody *et al.*, European Position Paper on Rhinosinusitis and Nasal Polyps 2012. *Rhinol. Supplement* **2012**(23), 3 p. preceding table of contents, 1–298.
69. S. Bi, J. Zhang and J. Cheng, Call from China for joint nanotech toxicity-testing effort. *Nature* **461**, 593 (2009).

Chapter 11

Nanomedicines for Ocular Administration

Yolanda Diebold

Ocular Surface Group, Institute for Applied Ophthalmo-Biology (IOBA), University of Valladolid Edificio IOBA, Campus Miguel Delibes, Paseo de Belén 17, 47011 Valladolid, Spain
yol@ioba.med.uva.es

Abstract

Nanotechnology-based medicines provide clear opportunities for broader therapeutic choices to treat ocular diseases. The eye is a good candidate organ for the delivery of nanotechnology-based therapeutics for several reasons. This chapter will present different eye diseases and nanomedicines that have the potential to cure or alleviate them. This overview cannot be exhaustive as the interest in this particular field has grown exponentially during the last two decades. Rather, this chapter includes recent reviews that provide comprehensive information for the interested reader and highlights some promising experiments or achievements that are likely to lead to novel and more efficacious ocular treatments in the future.

1. Introduction

Nanomedicine can be broadly defined as the application of nanobiotechnologies to different segments of medicine (Table 1).[1] One of the most exciting aspects of this emerging discipline involves the development of pharmaceuticals or medicines (nanomedicines) that utilize nanoscale-level technologies. Ocular diseases are a particular and promising example of what may be revolutionized in a near future by this novel approach to therapeutics.[2–4]

Table 1. "Nanomedicine" is the application of different nanotechnologies to the three main segments of medicine.

Diagnostic	Nanodiagnostic	Nanobiosensors Nanoparticle-size contrast agents Molecular diagnostic
Treatment	Nanopharmaceuticals or nanomedicines	Drug delivery systems Nanocoated stents Nanotechnology-based drugs Nanopumps Targeted drug delivery
Surgery	Nanorobotics	Nanorobots for vascular surgery Nanorobots for detection and destruction of cancer cells
	Nanosurgery	Nanolaser surgery Nanosensors implanted in catheters to provide real-time data during surgery
	Reconstructive surgery	Artificial tissues and organs Nanotechnology-based scaffolds for cell growth Tissue engineering

Source: Adapted from Ref. 1.

The threat of losing vision to potentially blinding diseases such as corneal opacity, glaucoma or degenerative diseases of the retina among others deserves research efforts that are frequently at the leading edge of scientific and technological knowledge. This explains why the eye is such a good organ to test nanomedicines as therapeutic options. Additionally, the morphological and physiological features of the eye readily lend themselves to the development of nanobased therapeutics.

For a better understanding of this chapter, a brief description of eye anatomy is first provided. The eyeball is comprised of three layers of tissue. The outermost is a fibrous layer that consists of the sclera and the main ocular surface structures, including the cornea, conjunctiva and limbus. The middle layer is the uvea that consists of the highly vascular choroid, ciliary body and iris. Finally, the innermost layer is the retina that consists of neural tissues. These three layers of the eyeball surround the lens and two fluid-filled compartments. One compartment is anterior to the lens and contains the aqueous humor. The other chamber is posterior to the lens and contains the gel-like vitreous body. These two compartments are of high relevance from a pharmacology point of view. Also, on the exposed surface of the eye, the pre-ocular tear film moisturizes the cornea and protects the eyeball from the outside environment.

When the eye is described, it is usually divided into the anterior and posterior segments. The iris is used to establish the division between the two segments. Anterior segment structures, located in front of the iris, include the eye lids, ocular surface, anterior chamber filled with aqueous humor, lens and uvea. Posterior segment structures, located behind the iris, include the vitreous body, retina, choroid and the optic nerve head.

Ocular physiology is amazingly complex and exquisitely regulated to maintain eyesight. It all happens in a small, closed organ located in the orbit of the skull. While well vascularized, the eye remains almost isolated from the outside environment and the systemic circulation by an important series of physiological barriers. This latter fact has a pivotal impact in ocular pharmacokinetics and influences the design of drug delivery systems to battle eye diseases. By virtue of the anatomical location of the eye and the transparency of the cornea, the eye can be visualized by non-invasive means, and it features a certain immune privilege.[5] Both of these characteristics are supportive for the trial of new nanotechnology-based drug delivery systems.

2. Routes for Ocular Nanomedicine/Drug Administration

Nanomedicines, as well as conventional ophthalmic drug formulations, can reach target eye tissues after administration in different sites and using different methods (Table 2). Topical application of a liquid form (eyedrops) is the preferred site and mode of delivery mainly because it is not invasive, and patients can easily comply with treatment. However, it is a very inefficient system. The normal pre-corneal space is less than 10 μl,[6] but the average

Table 2. Routes for drug administration to the eye.

Routes	Drug form	Anatomical place
Topical	Liquid (eyedrop)	Ocular surface and tear film
Intraocular injection	Liquid or suspension	Anterior chamber Vitreous body
	Solid depot	Vitreous body
Periocular injection	Liquid or suspension	Subconjunctival space Transeptal Sub-Tenon's space (anterior or posterior Retrobulbar space (inside muscle cone)
Implant (surgery)	Solid form	Subconjunctival space Suprachoroidal space Vitreous body

volume instilled using eyedrops is 40 μl. Excess volume is drained through the nasolacrimal duct and can reach systemic circulation. Another important disadvantage of topical delivery is the necessity of frequent instillations to locally reach therapeutic drug levels. Less than 5% of an instilled drug is absorbed into the eye.[7] Moreover, tear turnover rapidly washes away what has been instilled; thus for most drugs, the residence time is too short to be very effective. Nevertheless, this administration route is frequently used to treat chronic disorders, such as dry eye and glaucoma.

To reach the inner structures of the eye, intravitreal injection is the preferred administrative route. It consists of direct injection of the drug into the vitreous body through a small gauge needle. The needle has to traverse all three layers of the eyeball, which implies the risk of microbial contamination, retinal detachment, hemorrhages and fibrosis. In addition, intravitreal injection has to be done under local anesthesia, it requires a well-trained ophthalmologist and it is difficult for patients to accept periodic injections, thus resulting in poor compliance or patient dropout.

Drug delivery through implants is an interesting option that allows for a sustained drug release at the implantation site. Even though the intraocular placement is invasive and requires not only anesthesia but also an operating room, implants achieve effective drug levels for long periods of time, typically many months. Common eye locations to place implants are subconjunctival and suprachoroidal spaces and the vitreous body.

An ideal ocular therapy should be non-invasive, easily administered, effective with a single or few applications and capable of maintaining effective drug levels for as long as possible. Let us see how the above-mentioned limitations affect drug delivery and why nanomedicines are so interesting for eye therapy.

3. Ocular Pharmacokinetics

To understand why nanotechnology-based therapies are so appealing to develop ocular treatments, one has to understand the limitations that conventionally marketed ophthalmic formulations must overcome to reach eye structures. The eye as an organ has quite a unique anatomy, physiology and biochemistry. However, the very same physiological barriers and mechanisms that maintain homeostasis also reduce drug access to target ocular tissues in a disease situation.[8] With regard to the anterior part of the eye, the cornea along with the tear film constitute a formidable barrier for topically administered drugs. Mechanisms such as lacrimation and tear turnover wash drugs out in a matter of minutes after administration. This means that a conventional drug solution, instilled as eyedrops, remains in contact with the ocular surface for a very short period of time, typically 1–2 min, before it is eliminated by blinking and tear drainage. Moreover, cells of the corneal epithelium are tightly sealed with tight junctions that make it almost impermeable to hydrophilic substances. Under these circumstances, we can understand just how poor the efficacy is of topical forms of conventional drug formulations and why frequent instillations are needed to treat chronic diseases such as dry eye or glaucoma.

The conjunctival epithelium is also sealed with tight junctions; however, it is more permeable than the cornea to hydrophilic molecules, and it allows both passive and carrier-mediated transport of drugs.[9] The conjunctiva is always involved whenever topical application of drugs is used to treat the ocular anterior segment. Along with the sclera, which is even more permeable, the conjunctiva drains topically applied drugs throughout the transcleral pathway to the anterior chamber, thus reducing local drug absorption.

Regarding the intraocular structures, there are several other kinds of barriers for conventional drug formulations to overcome. For instance, several blood–ocular tissue barriers compartmentalize the eye and block foreign materials or substances from reaching the retina. Those barriers include the blood–aqueous and the blood–retina barriers. The blood–aqueous barrier is formed by the endothelial cells of iris blood vessels and the pigmented cell layer of ciliary epithelium. The blood–retina barrier is formed by the retinal capillaries and the retinal pigment epithelium. Both barriers prevent large hydrophilic molecules in the plasma from accessing the aqueous humor and retina. At the same time, they greatly reduce the access of drugs to inner parts of the eye. In addition, the vitreous body, a viscoelastic gel that fills the posterior compartment of the eye, forms an additional barrier,[10] although it is not as tight as the previously described ones. Nevertheless, it has a proteoglycan structure that reduces the mobility of large molecules, particularly those that are positively charged. Also, convection forces in the vitreous body generated under normal or pathophysiological situations affect drug diffusion in ways that depend on the characteristics of each drug.[11]

Transport of drugs applied topically as eyedrops to intraocular tissues poses paramount challenges. Although the conjunctiva is much more permeable than the cornea, and the conjunctival–scleral pathway favors the delivery of hydrophilic drugs to the posterior part of the eye, the anatomical distance is quite long, and unproductive drug absorbance that bypasses tissues may be a serious risk. However, promising results in this regard came early from Liaw *et al.*[12] They used poly(ethylene oxide)–poly(propylene oxide)–poly(ethylene oxide) (PEO–PPO–PEO) polymeric micelles to deliver by topical administration a model plasmid gene to the eyes of nude mice and rabbits. After two days, the reporter gene expression was detected in the conjunctiva, sclera, iris, vitreous body and retinal pigment epithelium, and there were no signs of tissue inflammation or damage. To achieve successful topical delivery for intraocular problems, we now know that there are many different promising strategies based upon the target cell or tissue, disease state and the properties of drug and nanomaterial used.[3] In fact, topical administration of pazopanib eyedrops, prepared as a suspension in the presence of carboxymethyl cellulose, has been demonstrated to alleviate retinal complications in a rat model of streptozotocin-induced diabetic retinopathy.[13] In a pilot clinical trial with 19 patients, topical eyedrops of 1.5% dexamethasone in cyclodextrin microparticles applied three to six times a day for four weeks were efficacious in treating patients with diabetic macular edema.[14] All of the patients had good tolerance of the formulation. After 4 weeks of treatment, 12 patients showed decreased central macular thickness, and 14 had improved visual acuity. These are but a few encouraging examples of how nanomedicines can benefit many patients in a near future.

4. Nanomedicines for the Ocular Anterior Segment

Topical drug administration is mainly used to treat local disorders of the ocular surface, including the cornea, conjunctiva, limbus and tear film. These include infectious diseases and non-infectious ones such as allergies, dry eye or other immune-based diseases. In addition, treatments for intraocular disorders, such as anterior uveitis or glaucoma, are often topically administered. Bearing in mind that less than 5% of drugs administered in that way reach the target, the help sought from the nanotechnology field is intended to prolong the residence time at the ocular surface, which is usually very low (less than 2 min),[15] to improve drug bioavailability. Different strategies are being explored with such a purpose.[16,17]

There are many nanoparticulate delivery platforms for ocular applications, and they have been extensively reviewed.[18,19] Briefly, one approach to nanoparticulate systems utilizes mucoadhesive polymers. Chitosan (CS) nanoparticles,[20,21] liposome–CS nanoparticles complexes[22] and hyaluronic acid (HA)–CS nanoparticles[23] have been studied *in vitro* and *in vivo* by our group with promising results. We reported that neither deleterious effects nor inflammation of corneal and conjunctival cells and tissues occurred using any of those nanosystems. Interestingly, we found unique tissue-related distribution patterns in rabbit cornea and conjunctiva after topical administration of albumin-loaded CS nanoparticles.[21] The conjunctival epithelium more avidly took up the nanoparticles than did the corneal epithelium. However, that tissue pattern was reverse when using HA–CS nanoparticles.[23] These results point out potential differences in endocytic mechanisms of different tissues, and consequently, the possibility of using selected nanosystems for tissue-targeted delivery. Also, HA–CS nanoparticles are capable of encapsulating macromolecules of both hydrophilic and hydrophobic nature, as well as plasmid DNA,[24] which makes them quite versatile to use for different drug formulations.

As inflammation is the main process underlying the above-mentioned diseases affecting the anterior segment of the eye, different drug carriers have been studied to deliver glucocorticoids and non-steroidal anti-inflammatory drugs (NSAIDs) to reduce or abolish the inflammation.[25,26] In a rabbit model, flurbiprofen-loaded nanoparticles have recently been shown to be more effective in reducing chemically induced inflammation of the cornea than commercially available flurbiprofen formulations.[27] CS-coated liposomes containing diclofenac sodium showed prolonged precorneal retention time and improved transcorneal penetration compared to uncoated liposomes or the control drug solution in rabbits.[28] The potential uses of nanosuspensions,[29] micro- and nanoemulsions, lipid nanoparticles, including solid lipid nanoparticles and nanostructured lipid carriers,[30] and drug nanocrystals[25] have been also recently reviewed.

To treat allergies, different nanomedicines are being explored. Azelastine hydrochloride-loaded Eudragit RL100 microspheres were recently investigated in a rat model of allergic conjunctivitis.[31] There was a more pronounced reduction in eosinophil count number in animals treated with loaded microspheres than with the marketed control formulation (Azelast®). Another interesting approach is based on CS in combination with disodium α-D-glucose 1-phosphate loaded with levocetirizine dihydrochloride, an anti-allergy drug.[32] This unique formulation is a thermosensitive hydrogel that showed rapid drug release and enhanced corneal penetration in an animal model of allergic conjunctivitis. The residence time was significantly prolonged, and it was more effective in reducing disease effects than the drug solution in conventional eyedrop form.

Although pre-clinical studies have rendered promising results, unfortunately, none of the nanomedicines described above has yet been approved and marketed. To the best of our knowledge, those studies have not yet reached clinical phases.

5. Nanomedicines for the Ocular Posterior Segment

Intraocular injections are the main way to administer nanomedicines to treat disorders affecting intraocular structures such as the choroid, the retina or the optic nerve. Diseases of these tissues are mainly associated with inflammation and degeneration, both inherited and non-inherited. As mentioned above, topical administration is also used for some chronic disorders such as glaucoma or intraocular inflammation.

5.1. *Glaucoma*

Glaucoma is a neurodegenerative disorder affecting the optic nerve and progressively causes apoptosis of retinal ganglion cells (RGCs). Axonal damage associated with glaucoma is considered to be due mainly to elevated intraocular pressure (IOP); however, other events, such as oxidative stress and genetic changes, are involved as well. Glaucoma causes irreversible blindness in over 60 million people worldwide.[33] Current treatments are either pharmacological or surgical, with the principal goal of lowering IOP. Other pharmacological treatments try to prevent oxidative damage of RGCs. Conventional approaches are demonstrably insufficient for vision restoration and just aim at delaying vision loss as much as possible.

Many efforts have been directed to efficiently deliver common IOP-reducing drugs, and interesting results have been obtained from preclinical studies. Because of its mucoadhesive properties, different CS-based formulations have been explored. HA-modified CS nanoparticles loaded with antiglaucomatous drugs, such as timolol maleate or dorzolamide hydrochloride, significantly reduced rabbit IOP compared to drug solution or CS nanoparticles alone, with no irritant effects.[34] Also, CS films loaded with timolol maleate were assayed in albino rabbits.[35] Loaded films remained implanted for at least two weeks, and lowered the IOP over a 10-week period of time. Also, a novel CS polyacrylic, acid-based, *in situ* gelling system loaded with timolol has been recently reported.[36] IOP in rabbits was elevated by betamethasone injection, and then the eyes were treated topically with the drug-loaded CS polyacrylic acid-based gelling system to lower the IOP. The timolol residence time was prolonged in the CS formulation, and the reduction of the elevated IOP was higher compared to other formulations without CS or the marketed formulation (Glucomol®). The good biological response and a considerably lower drainage of timolol into the systemic circulation as compared to conventional eyedrops, along with the absence of damage to ocular structures, make this system particularly interesting.

Finally, excessive scar formation sometimes follows glaucoma filtration surgery and is considered to be the main determinant of surgery failure in glaucoma patients.[37] The targeted delivery of CS nanoparticles loaded with an antifibrotic agent such as mitomycin C has been proposed to reduce this occurrence.

Liposomes have also been used to deliver antiglaucomatous drugs. For instance, Natarajan *et al.*[38] recently reported the loading of latanoprost into di-palmitoyl-phosphatidyl-choline liposomes. A single subconjunctival injection in rabbits lowered IOP for as much as 50 days with no toxicity or local inflammation. In contrast, topical administration of the same liposomal formulation did not induce any IOP reduction.

As mentioned above, the current concept of glaucoma treatment involves more than the IOP control. Thus, controlling vision loss will require other approaches such as avoiding oxidative stress, providing neuroprotection and locally delivering antiapoptotic agents.[39] These new approaches will utilize novel nanomedicines to confront the different challenges. As many of these approaches are using non-viral drug delivery systems for gene therapy, we return to this in later sections of this chapter. However, it is worth mentioning here the results obtained in rats in which superparamagnetic nanoparticles were used to induce local magnetic hyperthermia.[40] The induction of heat shock proteins by hyperthermia is a recently considered modality that can provide powerful protection of the optic nerve. Jeun *et al.* designed superparamagnetic $Mn0.5Zn0.5Fe_2O_4$ nanoparticles as agents to be delivered locally to the eyeball while avoiding undesirable side effects associated with whole body hyperthermia.[41] The engineered magnetic nanoparticles were intravitreally injected where they diffused to the retina without causing inflammation or tissue alteration. This innovative approach will require demonstration of localized and controlled heating in the optic nerve head and successful induction of heat shock proteins.

5.2. *Intraocular inflammation*

Intraocular inflammation, also called uveitis, is a potentially blinding disease in which the uveal tract is severely affected. It may be unilateral or bilateral, recurrent or chronic and infectious or non-infectious in origin. Recurring anterior segment uveitic diseases are usually treated with corticosteroids that can be administered topically in the less severe cases. In more severe cases, the treatment is provided by periocular or systemic injection. There are also a variety of immunomodulatory agents, such as cyclosporine A, tacrolimus or rapamycin, among others, that are lipophilic in nature, and consequently, the intraocular penetration is insufficient. Besides the limitations already mentioned for these drug administration routes, serious complications from corticosteroid exposure for long periods of time can occur, such as cataract and elevated IOP that can give rise to glaucoma. However, corticosteroids, which remain the main therapeutic option for the management of patients with uveitis, have experienced the most exciting nanotechnology-based improvements (Table 3).[42] A few of these drugs are already commercially available, and the others have provided promising results.

Ganciclovir is an antibiotic used to combat the retinitis induced by the herpesvirus cytomegalovirus, mainly in patients with the acquired immune deficiency syndrome (AIDS). This drug was initially encapsulated in liposomes or in microspheres that were administered by intravitreal injection in animal models. With promising results, the formulations evolved and an implantable device was developed that was well tolerated in rabbits[43] and in AIDS patients.[44] The current ganciclovir implant, Vitrasert® (Chiron Vision, Irvine, CA, USA), is non-biodegradable, and it needs to be surgically implanted into the patient's vitreous body. The implant is made of a ganciclovir tablet coated with polyvinyl alcohol and also partially coated with ethylene vinyl acetate. In phase III clinical trials, it has achieved higher levels of ganciclovir in the vitreous body and fewer side effects than the *i.v.* ganciclovir injection.[45]

Retisert®, marketed by Bausch & Lomb (Rochester, NY, USA) was the first intravitreal fluocinolone acetonide implant approved for the treatment of chronic non-infectious uveitis affecting the posterior segment of the eye. It is loaded with 0.59 mg of the drug and delivers sustained levels for about 30 months. It also needs to be surgically implanted. Clinical studies have shown its efficacy, including reduced uveitis recurrences, improved visual acuity and decreased need for adjunctive therapy.[46] However, it is not completely free of side effects, as increased IOP and cataract progression have been reported.[46]

Also, a biodegradable intrascleral implant for sustained intraocular delivery of betamethasone has been developed.[47] It is made of poly(Dl-lactide) and contains 25% betamethasone phosphate. It is disk shaped, 0.5 mm thick and 4 mm in diameter, and has been surgically implanted in rabbit scleras where it showed good biocompatibility.[47] A non-biodegradable implant has been further studied in rabbits, where it released the drug for about a month.[48] This non-biodegradable implant, made of ethylene vinyl acetate copolymer, had a similar disk shape and showed similar good results in normal rabbits. However, no preclinical studies with disease animal models or clinical studies have been reported so far using any of the intrascleral implant modalities.

Ozurdex® (Allergan, Inc, Irvine, CA, USA) is a biodegradable implant that lasts approximately six months and can be inserted in an office setting by intravitreal injection.[49] This implant delivers 700 μg of dexamethasone instead of fluocinolone acetonide, that is placed in a poly(lactic acid–coglycolic acid) matrix. It has been approved by the FDA for the treatment of macular edema following branch or central retinal vein occlusion, as well as for non-infectious posterior uveitis.[50,51] This implant has completed phase III clinical trials in

Table 3. Nanomedicines for intraocular inflammation.

Delivery approach	Drug	Administration route	References
Non-biodegradable implant	Ganciclovir (Vitrasert®)	Intravitreal (surgical implantation)	Chiron Vision/Bausch and Lomb (http://www.bausch.ca/en_US/ecp/pharma/product/vitrasert.aspx) 43–45
Non-biodegradable implant	Fluocinolone acetonide (Retisert®)	Intravitreal injection	Bausch and Lomb (http://www.bausch.com/en/ECP/Our-Products/Rx-Pharmaceuticals/Rx-Pharmaceuticals-ECP/Retisert) 46
Biodegradable and non-biodegradable implants	Betamethasone	Intrascleral (surgical implantation)	47, 48
Biodegradable implant	Dexamethasone (Ozurdex®)	Intravitreal injection	(http://www.ozurdex.com/) 49–52
PLGA microspheres	Dexamethasone	Intravitreal injection	53
Liposomes in HA gel	Vasoactive intestinal peptide	Intravitreal injection	54

patients with intermediate and posterior uveitis. It is effective and has fewer side effects than Retisert® or intravitreal triamcinolone injection, at least for one injection.[52]

Dexamethasone has also been formulated in microspheres prepared from poly(lactic acid–coglycolic acid) by our group.[53] In preclinical studies of endotoxin-induced uveitis in rabbits, a single intravitreal injection of dexamethasone-loaded microspheres seven days before the induction of uveitis reduced the inflammation after one day and completely abolished it in seven days. Importantly, it avoided the commonly associated complications of steroid treatment such as elevated IOP, cataract formation and macular edema.[53] Also, as human uveitis often recurs, a second injection of the endotoxin was delivered 30 days after the first one, simulating the disease recurrence. The initially administered, dexamethasone-loaded microspheres maintained a continuous drug release to the vitreous body, thus limiting the recurrence to a mild inflammatory reaction. Clinical studies with this nanoparticle are ongoing.

Also, a novel HA gel formulation of liposomes loaded with the neuropeptide vasoactive intestinal peptide (VIP) as an immunosuppressive agent[54] was tested in a rat model of endotoxin-induced uveitis. A single intravitreal injection of loaded gels was delivered seven days before the disease induction. The results showed that delivery of VIP was sustained in both ocular tissues and in regional lymph nodes, and the clinical scores and numbers of inflammatory cells infiltrating the eye were significantly reduced.

As mentioned before, the search for effective nanomedicines suitable for topical application is also of particular interest in this group of diseases, especially for the more severe chronic ones. However, some experts consider that topical immunomodulatory therapy will not replace systemic therapy, at least immediately.[55] Nevertheless, Pereira *et al.* tested the ability of topically applied gold nanoparticles to reduce intraocular inflammation induced by endotoxin injected into rat eyes.[56] They found that the gold nanoparticles, delivered topically 24 h prior to injection of the endotoxin, reduced both oxidative damage to the iris and activation of intracellular inflammatory pathways. These results open an interesting avenue to control inflammation by other means.

5.3. *Retinal degenerative diseases*

Retinal degenerative diseases collectively refer to the different entities that affect the retina, are associated with aging, infection, genetic alteration or inflammation and cause progressive visual impairment. The functional effects of retinal loss include impaired vision, loss of peripheral or central vision and total loss of vision. The main diseases included in this group are wet and dry age-related macular degeneration (AMD), macular edema secondary to retinal vein occlusion, diabetic retinopathy, retinitis pigmentosa and proliferative vitreoretinopathy.

The increasing incidence of retinal degenerative diseases associated with the aging of populations in developed countries is stimulating the search for more efficacious nanotechnology-based therapeutic approaches. There are several comprehensive reviews focused on different drug delivery strategies to specifically treat retinal diseases.[3,4,57,58] Table 4 summarizes the few systems that are currently in clinical trials or in the clinical practice for diabetic macular edema, dry and wet forms of AMD and retinitis pigmentosa. Mostly non-biodegradable implants loaded with different corticosteroids, such as triamcinolone or fluocinolone acetonides, have shown significant long-term improvement of retinal disease symptoms. However, these novel systems are not free of induced complications, such as increased IOP and accelerated cataract progression.[46]

Iluvien™ (Alimera Sciences, Inc., Alpharetta, GA, USA) is a recent novel injectable intravitreal insert that delivers fluocinolone acetonide in a low dose (190 μg) for up to three years.[59] It is a cylindrical polymide tube that contains the corticosteroid drug and is inserted into the vitreous body with a proprietary 25-gauge injector system through a minimally invasive procedure. It is marketed for diabetic macular edema treatment. Phase III clinical trials for the treatment of diabetic macular edema with this technology are ongoing.

The I-vation™ implant (Surmodics Inc., Eden Prairie, MN, USA) is a platform for intraocular drug delivery that consists of a titanium helical coil that is coated with the corticoid triamcinolone acetonide (925 μg) and two non-biodegradable polymers, poly(methyl methacrylate) and ethylene-vinyl acetate.[60] Even though it has been designed to deliver a variety of drugs, the main development has been to deliver triamcinolone intravitreally for at least two years. It recently completed the phase I trial that demonstrated it is safe and well tolerated (http://clinicaltrials.gov/ct2/show/NCT00692614?term=I-vation&rank=1). More than 70% of patients showed improved visual acuity;[61] however, no further clinical trials are ongoing.

NT-501 is a polymer implant containing human retinal epithelial cells that are genetically modified to secrete ciliary neurotrophic factor (CNTF).[62] This device (Neurotech Pharmaceuticals, Cumberland, RI, USA) achieves long-term and stable intraocular release of CNTF to the vitreous body. It consists of a sealed semipermeable, hollow fiber, membrane capsule surrounding a scaffold of six strands of polyethylene terephthalate, and can be

Table 4. Nanomedicines in clinical trials or in clinical use for retinal degenerative diseases.

Delivery approach	Drug	Administration route	Disease	References
Non-biodegradable implant	Fluocinolone acetonide (Iluvien™, formerly known as Medidur®)	Intravitreal injection (25-gauge needle)	Diabetic macular edema	Alimera Sciences Inc. (http://www.alimerasciences.com/) 59
Non-biodegradable implant	Triamcinolone acetonide (I-vation™)	Intravitreal injection (27-gauge needle)	Diabetic macular edema AMD	Surmodics Inc. (http://www.surmodics.com/medical-device/coating-technologies/drug-delivery-coatings-overview/durable-materials/) 60, 61
Non-biodegradable implant	hRPE cells secreting CNTF[a] (NT-501)	Intrascleral (surgical implantation)	Dry AMD Retinitis pigmentosa	Neurotech Pharmaceuticals (http://www.neurotechusa.com/ect-platform.html) 62–65
Liposomes Light-induced system	Verteporfin (Visudyne®)	*i.v.* injection Activation using a non-thermal red laser light	Choroidal neovascularization Wet AMD	http://visudyne.com/ 67, 68

[a]Human retinal pigment cells genetically engineered to secrete the growth factor CNTF.

loaded with the human retinal epithelial cells. The semipermeable hollow fiber membrane allows diffusion in two ways: (i) the CNTF diffuses out toward the vitreous body and (ii) nutrients in the vitreous body diffuse into support cell survival. One disadvantage of this technology is that it must be anchored in the vitreous body by surgical implantation through a scleral incision. Currently, there are several clinical trials underway with this technology.[62–64] So far, phase II studies in patients with atrophic AMD have shown significantly stabilized visual acuity accompanied by positive structural tissue changes.[63,64] In addition, NT-501 prevents secondary cone degeneration in patients with retinitis pigmentosa.[65] Neurotech has developed another similar device, NT-503, that encapsulates human cells modified to secrete VEGF receptor Fc-fusion protein.[66] It is also intended for intravitreal delivery, and it has shown constant release for up to a year in preclinical studies. A phase I clinical trial for neovascular AMD is underway.

Photodynamic therapy of the retina, based on light-activated drugs delivered using liposomes or micelles, is clinically available for treating neovascularization associated with wet AMD.[67] This technology achieves the occlusion of neovascular vessels after a photosensitizing drug, verteporfin, reaches the retinal target and is activated with a non-thermal laser light.[68] It is painless and the administration route is *i.v.* So far, however, this kind of therapy does not completely solve the problem, and neovascular vessels may reappear. In addition, patients with systemic liver disease or severe retinal or choroidal ischemia are not suitable candidates for photodynamic therapy. Exciting novel approaches using other light-activated delivery systems are underway in pre-clinical studies. Vectosomes are vesicles made of Herpes simplex structural protein that are bound to an antisense oligonucleotide (AS-ODN). When intravitreally injected vectosomes were hit by a laser pulse, the ODN was released and redistributed in different retinal layers.[69]

Another example of photodynamic therapy uses gold nanoparticles (2–3 nm) incorporated in liposomes loaded with calcein, which has been tested on a human retinal pigment epithelial cell line (ARPE-19).[70] UV light-induced heating of the gold nanoparticles, which act as energy collectors, sensitizes the liposome carrier and enables the release of calcein without affecting cell viability *in vitro*. Similarly, light-sensitive hydrogels with an incorporated chromophore can be activated by UV or visible light to release the content. Chromophores absorb the light and transfer the energy as heat to the hydrogel molecules, which then undergo a gel–sol phase transition.[71] There are two critical aspects necessary for the success of the photodynamic systems. The first is that the wavelength of the light used to activate the system must not cause any collateral damage to the retinal tissue. The second aspect is the speed of the delivery upon activation.

However, there are many interesting approaches still in pre-clinical study. For instance, albumin nanoparticles[72] loaded with ganciclovir, which has improved uptake by cytomegalovirus-infected human cells *in vitro*, were safe and well-tolerated after intravitreal injection in rats. Particles loaded with the ganciclovir ODN analog formivirsen were similarly safe and well tolerated. Interestingly, a single injection was sufficient to maintain drug presence in the vitreous and ciliary bodies for at least two weeks. Also, preclinical studies using a variety of different nanoparticulate system composed mainly of lipids have been recently reviewed because of improved biocompatibility with ocular structures.[73] These systems include solid lipid nanoparticles, nanostructured lipid carriers and lipid drug conjugates and have faster diffusion and better cell internalization compared to other systems, such as microparticles. This makes them promising carriers not only for conventional drugs but also for genetic material delivered to the retina.

Microneedles are also a promising nanotechnology-based medicine with an ocular application using the transcleral administration route.[74] Microneedles are micrometer-size needles that can be solid or hollow. While they were initially designed for minimally invasive transdermal drug delivery, they were soon tested in the eye. The effectiveness of pilocarpine delivery into rabbit eyes by coated microneedles was evaluated by measuring pupil constriction. The delivery method was successful and caused no adverse effects.[75] Hollow microneedles loaded with fluorescent nanospheres and microparticles were tested for transcleral delivery in human cadaver eyes.[74] In addition, microneedles have been reported to be painless by human volunteers.[76] *Ex vivo* experiments on animal and human cadaver eyes have been done for transcleral[75] and suprachoroidal[77] delivery. Also, *in vivo* suprachoroidal delivery of different marker molecules to the posterior structures in rabbit eyes has been achieved with no adverse effects.[78]

6. Gene Therapy for Ocular Diseases

Roughly, gene therapy consists of inserting foreign genetic material (a transgene) into host cells. Under the proper circumstances, the transduced cells will use the inserted genetic material to replace or enhance an altered gene or even to produce a novel protein that could not be made before. The elimination or reduction of a gain-of-function mutated allele is another form of gene therapy. In the eye, the purpose of gene therapy is to correct defects that are associated with gene-based ocular diseases.

Increasingly, we are aware of gene mutations responsible for the development of severe eye diseases. Among all eye diseases, retinal problems are the most studied in terms of gene therapy-based development of experimental treatments. So far, around 200 genes involved in retinal degenerations have been identified (https://sph.uth.edu/Retnet/sum-dis.htm last accessed January 2013). This fact, along with the threat that many of them pose to vision and the burden on the health system associated with their management, makes retinal degenerations quite attractive to attempt to correct at the genetic level. There are several recent reviews dealing with this particular topic.[79–81]

Application of gene therapy for ocular diseases started less than 20 years ago, and significant advances have been made due to methodological improvements in the delivery methods. There are currently 28 gene therapy clinical trials for ocular diseases ongoing (*http://www.wiley.com/legacy/wileychi/genmed/clinical/Last updated January 2013*). Most protocols for delivering genetic material to the eyes of both animal and human involve viral vectors, such as adenovirus, retrovirus and mainly, recombinant adeno-associated virus (AAV).[82] These vectors are often efficient carriers for the genetic material. Especially in the case of retinal degenerations, AAV-mediated gene transference to a variety of retinal cell types has rendered good results because they achieve long-term transgene expression with minimal host immune response.[83] Particularly, AAV vectors can mediate efficient gene replacement therapy in large animal models of inherited retinal diseases due to loss-of-function mutations, such as the RPE65 gene-deficient canine model of Leber's congenital amaurosis (LCA).[84] The dog model resembles very closely the human disease, which is an inherited autosomal recessive form of retinitis pigmentosa caused by mutations in anyone of the 14 genes identified so far. LCA is one of the most serious retinal diseases as it causes severe impairment of vision at birth and the first months of life. It represents around the 20% of all blind children and has no successful treatment. In dogs that are almost blind, the novel gene therapy strategy has achieved a stable improvement in visual function that was maintained in many animals for longer than three years.[85] AAV-mediated RPE65 gene transfer has also been applied in human clinical trials in which the gene was delivered by subretinal injection in LCA patients.[86] Although less impressive than in the dog model, the LCA patients showed functional improvements in light sensitivity and in visual acuity that persisted through at least 1.5 years in some patients. There were no safety concerns and the immune response levels were tolerable.[81,86]

Other severe ocular problems, such as the posterior uveitis and the effects of the retinal neovascularization accompanying AMD or posterior uveitis have also benefited from this viral vector-based technology. Autoimmune posterior uveitis is an inflammatory disease that can cause irreversible damage to photoreceptors due to abnormal immune responses against retinal antigens. In experimental models, it can be reduced by AAV-mediated gene therapy without undesirable systemic side effects.[87] Also, AAV-mediated intravitreal delivery of pigment epithelium-derived factor (PEDF) is in ongoing phase I clinical trials for neovascular AMD patients. In preclinical experiments, this factor was reported to induce regression of neovessels like those that develop in AMD patients.[88] A group of patients receiving the highest dose of virus showed some improvements in outcome measures, such as neovascular lesion size; however, inflammatory responses developed after viral vector administration.

This last example highlights the fact that viral vectors, while they can be efficient carriers for genetic material, are not free of limitations that are mostly related to tissue tropism, host immunity and potential persistence in the brain.[89] This is the main reason that has moved technologists to develop non-viral nanotechnology-based gene delivery tools. In the eye, there are different examples of experimental use of non-viral vectors, such as polymer-based nanoparticles, liposomes and dendrimers. Unfortunately, there are no ocular gene therapy clinical trials using nanoparticles or other non-viral polymer-based vectors yet.

There are recent and comprehensive reviews dealing with the application of different nanosystems, such as nanoparticles, dendrimers and many others to deliver genetic material to the different structures of the eye.[3,4,80,91–94] Some of these applications are for the topical delivery of genetic material to the ocular surface structures. For instance, Tong *et al.* used polymeric micelles to deliver the beta-galactosidase gene and the cornea-specific promoters keratin 12 and keratocan to the cornea of rabbit and mouse eyes.[95] With administration three times a day, the gene was expressed without any toxic effects.

Also, our group has recently developed a cationized gelatin, nanoparticle-based, topical gene-delivery nanomedicine for dry eye.[96] A plasmid encoding a modified MUC5AC protein (pMUC5AC) was loaded in the gelatin nanoparticles and showed efficient *in vitro* and *in vivo* transfection.[97] The most interesting results were obtained after instillation in experimental dry eye mice where MUC5AC expression, clinical signs, corneal fluorescein staining, tear production and inflammatory markers were evaluated.[97] pMUC5AC-NPs induced the expression of the modified MUC5A in the ocular surface tissues. This expression was associated with a reduction of the inflammation-related changes present in the diseased mice. Histopathologic evaluation of anterior segment structures confirmed the absence of adverse or other tissue alterations induced by the pMUC5AC-NPs.

Nanoparticle-based gene therapy for the retina has been extensively reviewed in recent years.[2–4,80,90,94] Glaucoma causes a blinding optic nerve neuropathy and is another example of severe ocular pathology in which the application of both viral and non-viral vector-mediated gene therapy is rendering promising results.[2,4,94] Targets can be different tissues involved in the pathogenesis of the disease, such as the trabecular meshwork, ciliary epithelium, Müller cells and RGCs. The delivered genetic material is usually intended for neuroprotection.[98,99] There are several reports showing that delivery of neurotrophic factor genes increases RGC survival in rodent models of glaucoma.[100–102] In those reports, different kinds of viral carriers were used to deliver the genetic material. One advance in development of non-viral nanocarriers that are potentially useful for glaucoma is intravenously administered, pegylated immunoliposomes carrying plasmid DNA under the control of an opsin promoter.[103] In rhesus monkeys, one injection enabled plasmid expression that was restricted to retina and lasted at least 14 days. Also, superparamagnetic nanoparticles safely induced heat shock protein as a neuroprotectant agent for rat RGCs.[104] Solid lipid nanoparticles containing methazolamide[105] and 6-*O*-lauryl-l-ascorbic acid (coagels) containing acetazolamide[106] are also showing interesting *in vivo* results. They provide improved ocular bioavailability of the assayed drugs compared to conventional solutions and have better therapeutic effectiveness in reducing elevated IOP.[105,106]

There are also genetic-based ocular therapies using genetic material other than DNA, such as AS-ODNs, small interfering RNAs (siRNAs) and aptamers. All of them are synthetic or natural molecules composed of a small number of nucleotides capable of blocking, interfering or modulating protein translation by different mechanisms. Therapeutic applications of all of these molecules in the eye have been extensively reviewed.[90,107–109] It is worth mentioning the frequent use in ophthalmology of the RNA aptamer Pegaptanib sodium.[113] It is a very selective RNA aptamer that exclusively inhibits the isoform 165 of vascular endothelial growth factor (VEGF). This VEGF isoform is primarily responsible for pathologic ocular neovascularization and vascular permeability. It was the first commercially available aptamer approved for therapeutic use in humans (Macugen®, Pfizer-EyeTech Pharmaceuticals Inc., New York, NY, USA)[114] and the first VEGF inhibitor available for the treatment of wet AMD and other ocular vascular diseases. Because of the excellent ocular and systemic safety profiles shown in clinical trials,[115] patients with diabetic macular edema, proliferative diabetic retinopathy and retinal vein occlusion generally benefit from the intravitreal application. A recent review[116] presented extensive evidence of its successful use in the clinical setting since its FDA approval in 2004.

There are two other marketed anti-VEGF agents used for wet AMD,[117] ranibizumab (Lucentis®, Genentech/Novartis, San Francisco, CA, USA) and bevacizumab (Avastin®, Genetech/Novartis), that is being used off-label.[118] It is important to note that neither agent is an ODN. Rather, they are humanized monoclonal antibody fragments produced by recombinant DNA technology.

AS-ODNs, aptamers and siRNAs can pass across the cornea through the paracellular pathway. However, because of their high molecular weight and a negative charge, they do not exhibit good intraocular permeation. Diffusion of these molecules across the sclera is being studied.[108] Also, different kinds of drug delivery systems, such as cationic lipids and polymers, dendrimers, liposomes, microparticles, nanoparticles and vectosomes are being explored *in vitro* and *in vivo* with variable success.[108,119,120]

Finally, another genetic-based therapeutic strategy based upon "naked" genetic material administration (i.e. without the help of a protective carrier) has received extensive attention.[121] Different chemical approaches have been used, with limited success due to the mostly inefficient entrance of the genetic material into the target cells and the short duration of transgene expression.[122] However, plasmid DNA compacted by polyethylene glycol (PEG)-substituted 30-mer lysine peptides (CK30PEG) has demonstrated suitable characteristic for high transfectivity with no toxicity in the eye.[123] Interestingly in a recent work, compacted DNA nanoparticles containing the Rds gene were administered to the retinas of a Rds-deficient mice, a model of retinitis pigmentosa.[124] The DNA nanoparticles were taken up by almost all of the photoreceptor cells into which the Rds gene was efficiently transferred. The transfection induced a significant structural and biochemical rescue of the disease without evidence of toxicity in the animals.

7. Conclusions

This chapter has presented an overview of different nanomedicines currently under study or marketed for ocular administration. The progress made in this field during the last 20 years has been extraordinary. There is no doubt about the future success in terms of achieving commercially available nanomedicines for some of the nanotechnology-based therapies reviewed in this chapter. I and many others believe that the key for progress in this challenge is the multidisciplinary approach to solve arising problems. This will enable a deeper understanding of the pathological mechanisms of blinding diseases and contribute to the advancement in pharmaceutical technology.

Acknowledgments

I would like to thank Margarita Calonge, MD, and Rosa M. Coco, MD, for valuable comments on the early drafts, and Britt Bromberg, PhD, ELS, of Xenofile Editing (http://www.xenofileediting.com) for assistance in the final editing of the chapter.

References

1. K. K. Jain, *Med. Princ. Pract.* **17**, 89–101 (2008).
2. D. W. Pita-Thomas and J. L. Goldberg, *Curr. Opin. Ophthalmol.* **24**,130–135 (2013).
3. U. B. Kompella, A. C. Amrite and R. P. Ravi, *Prog. Retin. Eye Res.* **XX**, 1–27 (2013).
4. M. A. Zarbin, C. Montemagno, J. F. Leary and R. Ritch, *Curr. Opin. Pharmacol.* **13**, 134–148 (2013).
5. J. W. Streilein, *Nat. Rev. Immunol.* **3**, 879–889 (2003).
6. S. Mishima, A. Gasset, S. D. Klyce Jr. and J. L. Baum, *Invest. Ophthalmol. Vis. Sci.* **5**, 264–276 (1966).
7. J. C. Keister, E. R. Cooper, P. J. Missel, J. C. Lang and D. F. Hager, *J. Pharm. Sci.* **80**, 50–53 (1991).
8. A. Urtti A, *Adv. Drug Deliv. Rev.* **58**, 1131–1135 (2006).
9. K. Hosoya, V. H. L. Lee and K.-J. Kim, *Eur. J. Pharm. Biopharm.* **60**, 227–240 (2005).
10. L. Pitkänen, M. Ruponen, J. Nieminen and A. Urtti, *Pharm. Res.* **20**, 576–583 (2003).
11. J. Park, P. M. Bungay, R. J. Lutz, J. J. Augsburger, R. W. Millard, A. S. Roy and R. K. Banerjee, *J. Control. Release* **105**, 279–295 (2005).
12. J. Liaw, S. F. Chang and F. C. Hsiao, *Gene Therap.* **8**, 999–1004 (2001).
13. A. Thakur, R. I. Scheinman, V. R. Rao and U. B. Kompella, *Microvasc. Res.* **82**, 346–350 (2011).
14. M. Tanito, K. Hara, Y. Takai, Y. Matsuoka, N. Nishimura, P. Jansook, T. Loftsson, E. Stefansson and A. Ohira, *Invest. Ophthalmo. Vis. Sci.* **52**, 7944–7948 (2011).
15. K. Jarvinen, T. Jarvinen and A. Urtti, *Adv. Drug Deliv. Rev.* **16**, 3–19 (1995).
16. I. P. Kaur and M. Kanwar, *Drug Dev. Ind. Pharm.* **28**, 473–93 (2002).
17. I. P. Kaur and R. Smitha, *Drug Dev. Ind. Pharm.* **28**, 353–69 (2002).
18. A. Sánchez and M. J. Alonso, In *Nanoparticulates as Drug Carriers*, V. Torchilin (Ed.), World Scientific-Imperial College Press, ISBN 1-86094-630-5, 2006) p. 1–30.
19. S. Liu, L. Jones and F. X. Gu, *Macromol. Biosci.* **12**, 608–620 (2012).
20. A. M. de Campos, Y. Diebold, E. L. S. Carvalho, A. Sánchez and M. J. Alonso, *Pharm. Res.* **21**, 803–810 (2004).
21. A. Enríquez-de-Salamanca, Y. Diebold, M. Calonge, C. García-Vázquez, S. Callejo, A. Vila and M. J. Alonso, *Invest. Ophthalmol. Vis. Sci.* **47**, 1416–1425 (2006).
22. Y. Diebold, M. Jarrín, V. Sáez, E. L. S. Carvalho, M. Orea, M. Calonge, B. Seijo and M. J. Alonso, *Biomaterials* **28**, 1553–1564 (2007).

23. L. Contreras-Ruiz, M. de la Fuente, C. García-Vázquez, V. Sáez, B. Seijo, M. J. Alonso, M. Calonge and Y. Diebold, *Cornea* **29**, 550–558 (2010).
24. M. de la Fuente, B. Seijo and M. J. Alonso, *Macromol. Biosci.* **8**, 441–450 (2008).
25. J. Araújo, E. Gonzalez, M. A. Egea, M. L. Garcia and E. Souto, *Nanomedicine* **5**, 394–401 (2009).
26. Y. Diebold, L. Contreras-Ruiz, I. Arranz-Valsero and L. García-Posadas, *Anti-Inflammatory & Anti-Allergy Agents in Medicinal Chemistry (formerly Current Medicinal Chemistry — Anti-Inflammatory and Anti-Allergy Agents)* **10**, 203–214 (2011).
27. E. Vega, M. A. Egea, O. Valls, M. Espina and M. L. García, *J. Pharm. Sci.* **95**, 2393–2405 (2006).
28. N. Li, C. Zhuang, N. Wang, X. Sun, S. Nie and W. Pan, *Int. J. Pharm.* **379**, 131–138 (2009).
29. M. A. Kassem, A. A. Abdel Rahman, M. M. Ghorab, M. B. Ahmed and R. M. Khalil, *Int. J. Pharm.* **349**, 126–133 (2007).
30. E. B. Souto, S. Doktorovova, E. Gonzalez-Mira, M. A. Egea and M. L. Garcia, *Curr. Eye Res.* **35**, 537–552 (2010).
31. U. A. Shinde, J. N. Shete, H. A. Nair and K. H. Singh, *J. Microencapsul.* **29**, 511–519 (2012).
32. X. Chen, X. Li, Y. Zhou, X. Wang, Y. Zhang, Y. Fan, Y. Huang and Y. Liu, *Biomater. Appl.* **27**, 391–402 (2012).
33. H. A. Quigley and A. T. Broman, *Br. J. Ophthalmol.* **90**, 262–267 (2006).
34. S. Wadhwa, R. Paliwal, S. R. Paliwal and S. P. Vyas, *J. Drug Target.* **18**, 292–302 (2010).
35. G. O. Fulgencio, F. A. Bretas Viana, R. R. Ribeiro, M. I. Yoshida, A. Gomes Feraco and A. da Silva Cunha-Júnior, *J. Ocul. Pharm. Therap.* **28**, 305–359 (2012).
36. S. Gupta and S. P. Vyas, *Sci. Pharm.* **78**, 959–976 (2010).
37. T. Shao, X. Li and J. Ge, *Diagn. Pathol.* **6**, 64 (2011).
38. J. V. Natarajan, S. Chattopadhyay, M. Ang, A. Darwitan, S. Foo, M. Zhen, M. Koo, T. T. Wong and S. S. Venkatraman, *PLoS One.* **6**, e24513 (2011).
39. S. Alqawlaq, J. T. Huzil, M. V. Ivanova and M. Foldvari, *Nanomedicine* **7**, 1067–1083 (2012).
40. M. Jeun *et al.*, *Biomaterials* **32**, 387–394 (2011).
41. J. M. Kim, K. H. Park, Y. J. Kim, H. J. Park and D. M. Kim, *Invest. Ophthalmol. Vis. Sci.* **47**, 4888–4894 (2006).
42. S. R. Taylor, H. Isa, L. Joshi and S. Lightman, *Ophthalmologica* **224**, 46–53 (2010).
43. T. J. Smith *et al.*, *Arch. Ophthalmol.* **110**, 255–258 (1992).
44. G. E. Sanborn *et al.*, *Arch. Ophthalmol.* **110**, 188–195 (1992).
45. D. C. Musch, D. F. Martin, J. F. Gordon, M. D. Davis and B. D. Kuppermann, *N. Engl. J. Med.* **337**, 83–90 (1997).
46. G. J. Jaffe, D. Martin, D. Callanan, P. A. Pearson, B. Levy and T. Comstock, *Ophthalmology* **113**, 1020–1027 (2006).
47. J. Okabe, H. Kimura, N. Kunou, K. Okabe, A. Kato and Y. Ogura, *Invest. Ophthalmol. Vis. Sci.* **44**, 740–744 (2003).
48. K. Okabe, H. Kimura, J. Okabe, A. Kato, N. Kunou and Y. Ogura, *Invest. Ophthalmol. Vis. Sci.* **44**, 2702–2707 (2003).
49. Ozurdex [package insert]. Irvine, CA: Allergan, Inc. (2009).

50. B. D. Kuppermann *et al.*, *Arch. Ophthalmol.* **125**, 309–317 (2007).
51. J. A. Haller *et al.*, *Ophthalmology* **117**, 1134–1146 (2010).
52. C. A. Arcinue, O. M. Cerón and C. S. Foster, *J. Ocul. Pharmacol. Ther.* **29**, 501–507 (2013).
53. E. Barcia, R. Herrero-Vanrell, A. Díez, C. Alvarez-Santiago, I. López and M. Calonge, *Exp. Eye Res.* **89**, 238–245 (2009).
54. L. Lajavardi, S. Camelo, F. Agnely, W. Luo, B. Goldenberg, M. C. Naud, F. Behar-Cohen, Y. de Kozak and A. Bochot, *J. Control. Release.* **139**, 22–30 (2009).
55. E. Bertelmann and U. Pleyer, *Ophthalmologica* **218**, 359–367 (2004).
56. D. V. Pereira *et al.*, *Invest. Ophthalmol. Vis. Sci.* **53**, 8036–8041 (2012).
57. K. G. Janoria, S. Gunda, S. H. Boddu and A. K. Mitra, *Expert Opin. Drug Deliv.* **4**, 3713–3788 (2007).
58. N. Kuno and S. Fuji, *Drugs Aging* **27**, 117–134 (2010).
59. F. E. Kane, J. Burdan, A. Cutino and K. E. Green, *Expert Opin. Drug Deliv.* **5**, 1039–1046 (2008).
60. P. J. Barnett, *31st Annual International Conference of the IEEE EMBS*, Minneapolis, Minnesota, USA, September (2009).
61. P. U. Dugel, D. Eliott, H. L. Cantrill, T. Mahmoud, R. Avery and S. R. Erickson, *Association for Research in Vision and Ophthalmology Annual Meeting*, Fort Lauderdale, FL, May (2009).
62. P. A. Sieving *et al.*, *Proc. Natl. Acad. Sci. USA.* **103**, 3896–3901 (2006).
63. K. Zhang *et al.*, *Proc. Natl. Acad. Sci. USA.* **108**, 6241–6245 (2011).
64. K. Kauper *et al.*, *Invest. Ophthalmol. Vis. Sci.* **53**, 7484–7491 (2012).
65. K. E. Talcott *et al.*, *Invest. Ophthalmol. Vis. Sci.* **52**, 2219–2226 (2011).
66. K. Kauper *et al.*, *Association for Research in Vision and Ophthalmology Annual Meeting*, Fort Lauderdale, FL, May (2012).
67. R. Z. Renno and J. W. Miller, *Adv. Drug Deliv. Rev.* **52**, 63–78 (2001).
68. R. Weinreb and E. Cotlier, *Surv. Ophthalmol.* **45**, 195–214 (2000).
69. N. Normand, F. Valamanesh, M. Savoldelli, F. Mascarelli, D. BenEzra, Y. Courtois and F. Behar-Cohen, *Mol. Vis.* **11**, 184–191 (2005).
70. L. Paasonen *et al.*, *J. Control. Release* **147**, 136–143 (2010).
71. L. A. Wells, F. Lasowski, S. D. Fitzpatrick and H. Sheardown, *Crit. Rev. Biomed. Eng.* **38**, 487–509 (2010).
72. J. M. Irache, M. Merodio, A. Arnedo, M. A. Campanero, M. Mirshahi and S. Espuelas, *Mini Rev. Med. Chem.* **5**, 293–305 (2005).
73. A. del Pozo-Rodríguez, D. Delgado, A. R. Gascón and M. A. Solinís, *J. Ocul. Pharmacol. Ther.* **29**, 173–188 (2013).
74. J. Jiang, J. S. Moore, H. F. Edelhauser and M. R. Prausnitz, *Pharm. Res.* **26**, 395–403 (2009).
75. J. Jiang, H. S. Gill, D. Ghate, B. E. McCarey, S. R. Patel, H. F. Edelhauser and M. R. Prausnitz, *Invest. Ophthalmol. Vis. Sci.* **48**, 4038–4043 (2007).
76. H. S. Gill, D. D. Denson, B. A. Burris and M. R. Prausnitz, *Clin. J. Pain* **24**, 585–594 (2008).
77. S. R. Patel, A. S. Lin, H. F. Edelhauser and M. R. Prausnitz, *Pharm. Res.* **28**, 166–176 (2011).
78. S. R. Patel, D. E. Berezovsky, B. E. McCarey, V. Zarnitsyn, H. F. Edelhauser and M. R. Prausnitz, *Invest. Ophthalmol. Vis. Sci.* **53**, 4433–4441 (2012).

79. D. C. Chung, V. Lee and A. M. Maguire, *Curr. Opin. Ophthalmol.* **20**, 377–381 (2009).
80. G. Thumann, *Curr. Genomics* **13**, 350–362 (2012).
81. P. Colella and A. Auricchio, *Hum. Gene Ther.* **23**, 796–807 (2012).
82. R. O. Snyder, *J. Gene Med.* **1**, 166, 175 (1999).
83. P. K. Buch, J. W. Bainbridge and R. R. Ali, *Gene Ther.* **15**, 849–857 (2008).
84. G. M. Acland, G. D. Aguirre, J. Bennett, T. S. Aleman, A. V. Cideciyan and J. Bennicelli, *Mol. Ther.* **12**, 1072–1082 (2005).
85. J. W. Bainbridge *et al.*, *N. Engl. J. Med.* **358**, 2231–2239 (2008).
86. F. Simonelli *et al.*, *Mol. Ther.* **18**, 643–650 (2010).
87. C. A. Broderick *et al.*, *Mol. Ther.* **12**, 369–373 (2005).
88. P. A. Campochiaro *et al.*, *Hum. Gene Ther.* **17**, 167–176 (2006).
89. J. Bennett, *Gene Ther.* **10**, 977–982 (2003).
90. Y. Diebold and M. Calonge, *Prog. Retin. Eye Res.* **29**, 596–609 (2010).
91. S. P. Kambhapati and R. M. Kannan, *J. Ocul. Pharm. Therap.* **29**, 1–15 (2013).
92. T. Borrás, *Exp. Eye Res.* **76**, 643–652 (2003).
93. S. M. Conley and M. I. Naash, *Prog. Retin. Eye Res.* **29**, 376–397 (2010).
94. A. M. Demetriades, *Curr. Opin. Ophthalmol.* **22**, 73–77 (2011).
95. Y. C. Tong, S. F. Chang, C. Y. Liu, W. W. Kao, C. H. Huang and J. Liaw, *J. Gene Med.* **9**, 956–966 (2007).
96. G. K. Zorzi, L. Contreras-Ruiz, J. E. Párraga, A. López-García, R. Romero Bello, Y. Diebold, B. Seijo and A. Sánchez, *Mol. Pharm.* **8**, 1783–1788 (2011).
97. L. Contreras-Ruiz, G. K. Zorzi, D. Hileeto, A. López-García, M. Calonge, B. Seijo, A. Sánchez and Y. Diebold, *Gene Ther.* doi: 10.1038/gt.2012.56 (2012).
98. S. Alqawlaq, J. T. Huzil, M. V. Ivanova and M. Foldvari, *Nanomedicine* **7**, 1067–1083 (2012).
99. A. M. Wilson and A. Di Polo, *Gene Ther.* **19**, 127–136 (2012).
100. A. Di Polo, L. J. Aigner, R. J. Dunn, G. M. Bray and A. J. Aguayo, *Proc. Natl. Acad. Sci. USA.* **95**, 3978–3983 (1998).
101. B. A. Van Adel, C. Kostic, N. Deglon, A. K. Ball and Y. Arsenijevic, *Hum. Gene Ther.* **14**, 103–115 (2003).
102. S. G. Leaver, Q. Cui, O. Bernard and A. R. Harney, *Eur. J. Neurosci.* **24**, 3323–3332 (2006).
103. Y. Zhang, F. Schlachetzki, J. Y. Li, R. J. Boado and W. M. Pardrige, *Mol. Vis.* **9**, 465–472 (2003).
104. M. Jeun *et al.*, *Biomaterials* **32**, 387–394 (2011).
105. R. Li, S. Jiang, D. Liu, X. Bi, F. Wang, Q. Zhang and Q. Xu, *J. Microencapsul.* **28**, 134–141 (2011).
106. L. I. Tártara, D. A. Quinteros, V. Saino, D. A. Allemandi and S. D. Palma, *J. Ocul. Pharmacol. Ther.* **28**, 102–109 (2012).
107. S. P. Henry, E. G. Marcusson, T. M. Vincent and N. M. Dean, *Trends Pharm. Sci.* 25, 523–527 (2004).
108. E. Fattal and A. Bochot, *Adv. Drug Dev. Rev.* **58**, 1203–1223 (2006).
109. E. Levy-Nissenbaum, A. F. Radovic-Moreno, A. Z. Wang, R. Langer and O. C. Farokhzad, *Trends Biotechnol.* **26**, 442–449 (2008).
110. D. G. Shcharbin, B. Klajnert and M. Bryszewska, *Biochemistry (Mosc).* **74**, 1070–1079 (2009).

111. S. S. Venkatraman, L. L. Ma, J. V. Natarajan and S. Chattopadhyay, *Front. Biosci. (Schol Ed)*. **2**, 801–814 (2010).
112. R. K. Khar, G. K. Jain, M. H. Warsi, N. Mallick, S Akhter, S. A. Pathan and F. J. Ahmad, *Indian J. Pharm. Sci.* **72**, 675–688 (2010).
113. E. W. Ng, D. T. Shima, P. Calias, E. T. Cunningham Jr, D. R. Guyer and A. P. Adamis, *Nat. Rev. Drug Discov.* **5**, 123–132 (2006).
114. S. A. Doggrell, *Expert Opin. Pharmacother.* **6**, 1421–1423 (2005).
115. M. Tolentino, *Surv. Ophthalmol.* **56**, 95–113 (2011).
116. E. T. Cunningham Jr. *et al.*, *Ophthalmology* **112**, 1747–1757 (2005).
117. The CATT (Comparison of Age-Related Macular Degeneration Treatments Trials) Research Group, *N. Engl. J. Med.* **364**, 1897–1908 (2011).
118. J. P. Hubschman, S. Reddy and S. D. Schwartz, *Clin. Ophthalmol.* **3**, 155–166 (2009).
119. O. Boussif, F. Lezoualc'h, M. A. Zanta, M. D. Mergny, D. Scherman, B. Demeneix and J. P. Behr, *Proc. Natl. Acad. Sci. USA.* **92**, 7297–7301 (1995).
120. I. Jaaskelainen, K. Lappalainen, P. Honkakoski and A. Urtti, *Methods Enzymol.* **387**, 210–230 (2004).
121. H. Herweijer and J. A. Wolff, *Gene Ther.* **10**, 453–458 (2003).
122. D. Luo and W. M. Saltzman, *Nat. Biotechnol.* **18**, 893–895 (2000).
123. R. Farjo, J. Skaggs, A. B. Quiambao, M. J. Cooper and M. I. Naash, *PLoS One* **1**, e38 (2006).
124. X. Cai, Z. Nash, S. M. Conley, S. J. Fliesler, M. J. Cooper and M. I. Naash, *PLoS One* **4**, e5290 (2009).

Chapter 12

Nano-antibiotics: Nanotechnology in Fighting Against Infectious Diseases

Jessica Kemp, Julius Edson*[†], *Young Jik Kwon*[*,†,‡,§]

[*]*Department of Pharmaceutical Sciences,*
University of California, Irvine, CA 92697, USA
[†]*Department of Chemical Engineering and Materials Science,*
University of California, Irvine, CA 92697, USA
[‡]*Department of Biomedical Engineering,*
University of California, Irvine, CA 92697, USA
[§]*Department of Molecular Biology and Biochemistry,*
University of California, Irvine, CA 92697, United States
youngjik.kwon@gmail.com

Abstract

Infectious diseases pose a worldwide threat due to the development of multiple drug resistance, demanding a rapidly increasing need for the production of new therapeutics. However, the cost, labor, and time required to develop, synthesize, and implement new antibiotics is ever increasing, and it has become necessary to investigate the utility of nanomaterials in treating infections. Many nanomaterials have intrinsic antimicrobial properties in addition to easily tunable physicochemical properties that allow for greater specificity and efficacy over traditional antibiotics. This book chapter summarizes the advantages of nanomaterials and nano-delivery systems over current therapeutics, the mechanisms of action, the current applications in industrial and clinical settings, and the future challenges to be met in the fight against infectious disease.

1. Introduction

Antibiotics were developed in the mid-20th century to impede the infectious diseases that were rampantly spreading throughout the world. After the emergence of penicillin, a plethora of therapeutics were developed that demonstrated even stronger potencies than the aforementioned against infections and disease. However, deadly microorganisms have evolved in various ways to form resistances against antibiotics, essentially nullifying the advances which had been made in the fight against disease. Although the progression of new drugs continues despite this downfall, it can be postulated that further resistances will continue to form, and a critical point will be reached where the rate of the resistance formation by microorganisms will exceed the rate of effective therapeutic development.[1] In order to circumvent this challenge, it is necessary to explore complementary materials to enhance the efficacy of current remedies that use alternative mechanisms to work against resistance formation.

Nano-antibiotics can be defined as nanomaterials that either show antimicrobial activity[1] or heighten the efficiency and safety of antibiotics administration.[2–4] Though the potential toxicity upon long-term exposure for nano-antibiotics is unknown, antimicrobial nanoparticles (NPs) may not cause direct or acute adverse effects, unlike many antimicrobial agents currently used in clinics. Production and preparation of antimicrobial NPs could be cheaper, compared with current antibiotics development and synthesis. Another advantage of nanomaterials is a long shelf life[5] and the ability to withstand harsh conditions such as high temperatures during sterilization. Antibiotics delivery using nanomaterials offers several advantages: (i) regulated and uniform distribution in target tissue, (ii) enhanced solubility, (iii) sustained and controlled release, (iv) superior patient compliance, (v) reduced side effects, and (vi) heightened cellular internalization.[6–8] This chapter seeks to elucidate the properties and mechanisms of antibacterial nanomaterials and antibiotics nano delivery systems, their advantages and disadvantages over existing therapeutics, and current industrial and clinical applications.

2. Antimicrobial Nanomaterials

Antibacterial NPs are typically composed of either metals or metal oxides, naturally occurring antibacterial substances, carbon-based nanomaterials or surfactant-based nanoemulsions (NEs).[1] Effective antimicrobial activities of various nanomaterials can be attributed to high surface area to volume ratios and unique chemico-physical properties.[5] Various proposed antimicrobial mechanisms of nanomaterials include (i) photocatalytic production of reactive oxygen species (ROS) that damage cellular and viral components, (ii) compromising the bacterial cell wall/membrane, (iii) interruption of energy transduction, and (iv) inhibition of enzyme activity and DNA synthesis (Fig. 1 and Table 1).[1,4,5,9,10,16,23]

2.1. *Inorganic nanomaterials*

2.1.1. Silver NPs

Since ancient times, the antibacterial properties of silver have been investigated and utilized since ancient times. Silver has been used for burn wound treatment, dental work, catheters, and bacterial infection control, in the forms of metallic silver, silver nitrate, and silver sulfadiazine.[30] The use of silver became less prominent after penicillin was introduced in the 1940s,[31] however the recent influx of antibiotics-resistant bacteria and the limited capability of antibiotics revived clinical use of silver such as wound dressings.[30] Silver nanoparticles (Ag NPs) have proven to be the most effective against bacteria, viruses, and other eukaryotic microorganisms amongst the many different types of metallic and metal oxide NPs.[14,32–34] Ag NPs work by attacking the respiratory chain and cell division that finally lead to cell death, while simultaneously

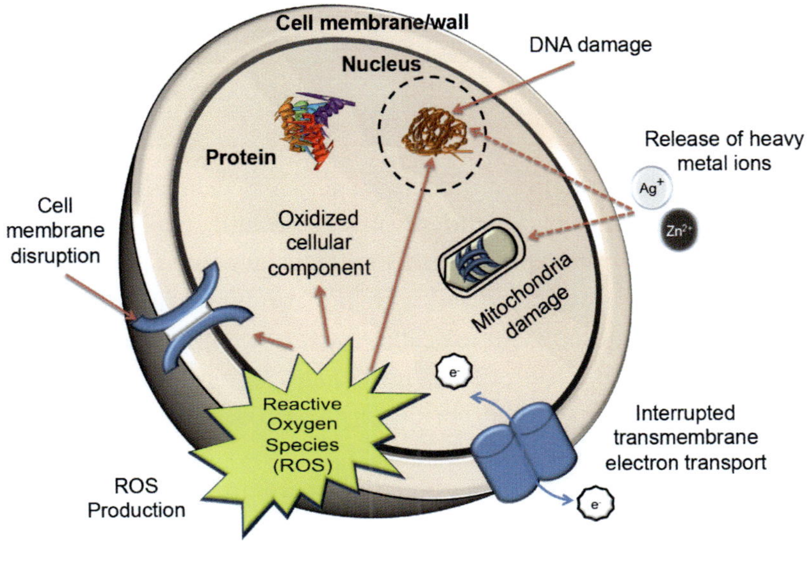

Fig. 1. Diagram depiction of various toxicity mechanisms through which NP-based antibiotics work against microorganisms.

Table1. Antibacterial mechanisms of certain nanomaterials and their applications adapted from Ref. 29.

Nanoparticle material	Antimicrobial mechanism	Industrial/clinical application	Reference
Ag	Silver ion release; cell membrane and DNA damage; disruption of electron transport	Surgical wound dressings, medical device coatings, water filters	1,10, 11–13
Au	Potent electrostatic interaction with cell membrane	Antifungal agent, used in conjunction with antibacterial agents for serious infection, photothermal therapy	14,15
ZnO	Zinc ion release, disruption of cell membrane, H_2O_2 production	Medical device coating, antibacterial creams, mouthwash	16,17,18
TiO_2	Production of ROS, DNA and cell wall destruction	Food sterilization, water and air purification systems	19, 20–22
Chitosan	Increased permeability and disruption of cell membrane, trace metal chelation, inactivation of enzymes	Biomedical microbiocide, water disinfectant, immobilizer of bacteria	1,23,24
C_{60}	Cell membrane disruption, increases activity of infiltrating neutrophil	Potential disinfectant	25–29

releasing Ag ions that enhance antibacterial activity.[11] The antimicrobial properties of Ag NPs is inversely dependent on size[12,13] and shape[10] of the particle. Enhanced and synergistic antimicrobial effects against Gram-positive and Gram-negative bacteria was achieved with the combined use of Ag NPs with antibiotics, such as penicillin G, amoxicillin, erythromycin, and vancomycin.[2,35,36] Wound dressings, coating for medical devices and surgical masks, impregnated textile fabrics, nanogels, and nanolotions are some of the various applications of Ag NPs.[37–40]

Though Ag NPs provide many benefits, there are coupled disadvantages to their use. Extended exposure to soluble silver-containing compounds may lead to an irreversible pigmentation in the skin (argyria) and the eyes (argyrosis), as well as other toxic effects, including organ damage (i.e. liver and kidney), irritation (i.e. eyes, skin, respiratory and intestinal tract), and changes in blood cell counts.[41] But in some studies, metallic silver appears to pose a minimal risk to health and Ag NPs are suggested to be non-toxic,[15,42] while some studies reported concentration-dependent adverse effects of Ag NPs on the mitochondrial activity.[43,44] Therefore, the employment of Ag NPs as favorable antimicrobial nanomaterials requires further investigation of their potential toxicity.

2.1.2. Gold NPs

Gold nanoparticles (Au NPs), nanorods (NRs), nanoshells (NSs), and nanocages (NCs) in the near-infrared (NIR) light-absorbing range have been used to treat bacterial infections by irradiation with focused laser pulses of the appropriate wavelengths.[33,45] The antimicrobial activity of gold seems to be attributed to the strong electrostatic attractions to the negatively charged bilayer of the cell membrane,[14,15] further supported by the observation that cationic particles were found to be moderately toxic while anionic particles were not.[46] Au NPs conjugated with antimicrobial agents and antibodies have been explored to obtain selective antimicrobial effects.[47] Many studies have reported strong antimicrobial effects against Gram-positive and Gram-negative bacteria, including antibiotic-resistant strains,[33,48] by gold/drug nanocomposites (e.g. Au NPs coated with antibiotics such as streptomycin, gentamicin and neomycin). Therefore, Au NPs are promising adjuvants for antibiotics therapy in treating serious bacterial infections with minimal adverse effects at a reduced antibiotics dosage.

2.1.3. Aluminum and copper NPs

Although antibacterial activity of Ag NPs has been proven to be most effective in general among the metallic NPs,[4,34,49] the antibacterial activity of NPs varies depending on the microbe.[50] Aluminum oxide (Al_2O_3) NPs display moderate inhibitory effects on microbial growth by the disruption of the cell wall but only at very high concentrations.[51] Additionally, though copper is a structural constituent of many enzymes in many living microorganisms, free ionic Cu^{2+} at high concentrations can generate toxic effects through the generation of ROS that disrupts the amino acid synthesis and DNA.[52] For example, Cu NPs have greater affinity to amines and carboxyl groups on the surface of *Bacillus subtilis* at a high density in comparison to Ag NPs, which results in superior antibacterial activity.[50,53] Aside from the selective microbial targeting, copper oxide (CuO) is cheaper than silver, easily miscible with polymers, and is relatively chemically and physically stable.[54]

2.1.4. Zinc oxide NPs

Some metal oxide NPs are not only stable under harsh processing conditions and have selective toxicity to bacteria but also exhibit minimal effects on human and animal cells.[55–57] One example is zinc oxide (ZnO) NPs which are non-toxic and biocompatible, have been used as drug carriers, cosmetics ingredients, and medical filling materials.[58,59] Low production cost, a white appearance and UV-blocking properties are some of the advantages ZnO NPs have over Au NPs.[17] In recent studies, ZnO NPs were found to have antibacterial activity against important food-borne pathogens, such as *E. coli O157:H7* and enterotoxigenic *E. coli*.[32,55,60] These studies lead to the suggestion that the application of ZnO NPs may be effective for preserving agricultural products and food. In terms of their mechanism, ZnO NPs are believed to destruct lipids and proteins of the bacterial cell membrane, resulting in a leakage of intracellular contents and ultimately the death of bacterial cells.[16,32] In addition, generation of hydrogen peroxide and Zn^{2+} ions were suggested to be key antibacterial mechanisms.[61]

2.1.5. Titanium dioxide NPs

Titanium dioxide (TiO_2) NPs are the most studied for photocatalytic antimicrobial activity among various NPs.[62] Ample amount of information over the last 20 years[20] demonstrated the strong bactericidal activity of TiO_2 upon receiving irradiation with near-UV light and UV-A. The necessary concentration for killing bacteria varies in the range of 100–1000 ppm, depending on the TiO_2 NPs size as well as the intensity and wavelength of the light source.[19] The production of ROS such as free hydroxyl radicals and peroxide is responsible for the photocatalytic antibacterial activity of TiO_2.[21] Hydroxyl radicals generated by photocatalytic TiO_2 are very effective oxidants with wide-ranging reactivity, and the microbial surface is the primary target of the initial oxidative attack by irradiated TiO_2 NPs. Damaged membrane structure hinders many crucial biological functions, such as semipermeability, respiration, and oxidative phosphorylation reactions.[19] Some unknown antimicrobial properties of TiO_2 NPs are indicated by irradiation-independent bacterial death.[63]

The potential for activation by visible light (e.g. sunlight) is an appealing feature of disinfection by TiO_2. In addition, metal doping (e.g. Ag/TiO_2) improved photocatalytic inactivation of bacteria and viruses, as well as the light absorbance of TiO_2.[20,64] Ag/(C,S)-TiO_2 NPs were shown to have strong light-independent antimicrobial activities against both *E. coli* and *B. subtilis* spores, by utilizing the combined bactericidal activity of Ag and TiO_2.[65]

TiO_2 is especially useful for water treatment because it is stable in water, non-toxic by ingestion, and cost-effective. The photocatalytic activity by UV-A and the potential to be activated by visible light, when doped with metals, makes TiO_2-mediated disinfection especially advantageous in developing countries.[1] The inactivation of various microorganisms that are resistant to dehydration and UV radiation, also makes TiO_2 a promising agent for improving process hygiene and product safety in food industries and cosmetics.[20,64] In addition, antibacterial effects of TiO_2 on *Lactobacillus acidophilus* would assist the use in orthodontic tools, such as toothbrushes, dental implants, and screws.[21]

2.2. *Organic nanomaterials*

2.2.1. Carbon-based materials: fullerene (C_{60}) and carbon nanotubes

C_{60}'s antimicrobial properties are of very recent findings, therefore there is limited knowledge.[66] Although native C_{60} are nearly insoluble by aqueous means, they can be dispersed in water by recently developed methods.[67] Numerous techniques for creating stable colloidal C_{60} aggregates (nC_{60}) in water are highlighted for their effective and broad antibacterial activity.[25,54]

The hypothesized antibacterial mechanism for nC_{60} includes photocatalytic ROS production in eukaryotic cells.[26–28] Some studies emphasize that antibacterial activity of nC_{60} to prokaryotic cells is mediated via lipid peroxidation in the cell membrane.[25,29] ROS-independent toxicity mechanism of nC_{60} is indicated by deficient protein oxidation as well as the light- and oxygen-independent antibacterial properties of nC_{60}.[28] Various C_{60} derivatives exhibit different antimicrobial mechanisms. For example, the antimicrobial activity of carboxyfullerene is facilitated by injection into the cell wall, followed by disruption of the cell membrane structure.[66] For alkylated C_{60}-bis(*N,N*-dimethylpyrrolidinium iodide) derivatives, the respiratory chain is affected and bacteria growth is effectively inhibited; comparable to the antimicrobial effects of vancomycin.[68] Fullerols, polyhydroxylated C_{60} [$C_{60}(OH)_n$], exhibit superb antimicrobial activity over a broad range of microorganisms with lower toxicity than nC_{60}.[69] $C_{60}(OH)_n$ can also be used as a drug carrier that bypasses the blood ocular barriers.[70] The aggregated form of C_{60} in water (fullerene water suspensions [FWS]) has unique properties, including antimicrobial activity different from those of bulk solid C_{60}.[54,71]

New observations suggest that the observed toxicity of nC_{60} in both human cells and microorganisms may be caused by solvent contaminants (e.g. tetrahydrofuran [THF]) used or generated during C_{60} preparation.[54,67,72] The suspensions of nC_{60} prepared by THF-independent methods were demonstrated to be non-toxic, particularly after γ-irradiation.[67,73,74] Additionally, nC_{60} prepared without the use of any polar organic solvent resulted in no acute or subacute toxicity in rodents. The livers of the rodents were also protected from damages by free-radicals in a dose-dependent manner.[75] Although the acute toxicity of $C_{60}(OH)_n$ is low, long retention in the body generates alarms about chronic toxic effects.[76–78]

Carbon nanotubes (CNTs), hollow cylindrical structures with graphene walls, have multiple unique properties (electrical, optical, thermal, etc.) that have prompted their use in several fields of research.[79–80] Single-walled nanotubes (SWNTs), a single pipe several nanometers wide in diameter, have been suggested to possess antimicrobial properties.[81] However, due to their low aqueous solubility, stabilization via surfactants or polymers are needed to significantly improve dispersion, or through functionalization of the surface with lysine using microwave irradiation.[79,82,83] A recent study demonstrates that the correlation between cytotoxicity in *E. coli* and surface functionalization relies upon aggregate morphology and dispersion more than individual physicochemical properties of the functional groups.[80]

Most carbon-based nanomaterials are generally cytotoxic, and SWNTs demonstrate the highest antimicrobial activity,[80,81,84] attributed to oxidative stress and membrane damage working synergistically.[81,85] The toxicity mechanisms for functionalized SWNTs and multi-walled CNTs (MWNTs) examined on *E. coli* demonstrate that MWNTs physically damage cell membrane through piercing, while SWNTs inhibit cell division of the bacteria by strong adhesion to the surface.[86] In order to fully utilize CNTs as antimicrobial agents, aggregation patterns, organic matter stabilization effects, and bioavailabilty must be accounted for.[79] It was suggested that the ability of SWNTs to pierce the cell membrane also relies on the dispersive properties and collision force.[87] This was demonstrated by rope-like CNT agglomerates having higher cytotoxicity than the dispersed CNTs at equal concentration.[88] To further elucidate the molecular mechanism behind CNT toxicity through oxidative stress, the generation of ROS and glutathione by PEGylated and uncoated SWNTs was investigated in neuronal P12 cells.[89] SWCNTs taken up by the PC12 cells caused upregulation of oxidoreductases, and ROS was generated via a Fenton reaction.[90] At higher levels of oxidative stress, membrane damage and mitochondria cytotoxicity may be attributed to depletion of glutathione levels. In contrast, SWCNT-PEGs (polyethylene glycol) migrated into PC12 cells and upregulated six genes involved in the lipid metabolic process. Fewer ROS were also generated, possibly as an effect of the surface PEGylation, resulting in less membrane damage and cytotoxicity. SWNTs have shown high efficacy against biofilm formation and biofouling of surfaces.[91,92] Surface functional groups are simple to adhere to SWNTs, making them attractive nanomaterials, along with their high chemical stability.[85] However, one disadvantage is the extreme cytotoxicity of CNTs in alveolar macrophage.[88,93] MWNTs are not eliminated rapidly in the lungs, and as a result, an upregulation of TNF-α causes inflammation and fibrosis in the lungs.[92] Instillation of SWNTs resulted in a temporary lung inflammation and injury.[94,95] ROS- and non-ROS mediated toxic effects are summarized in Fig. 2.

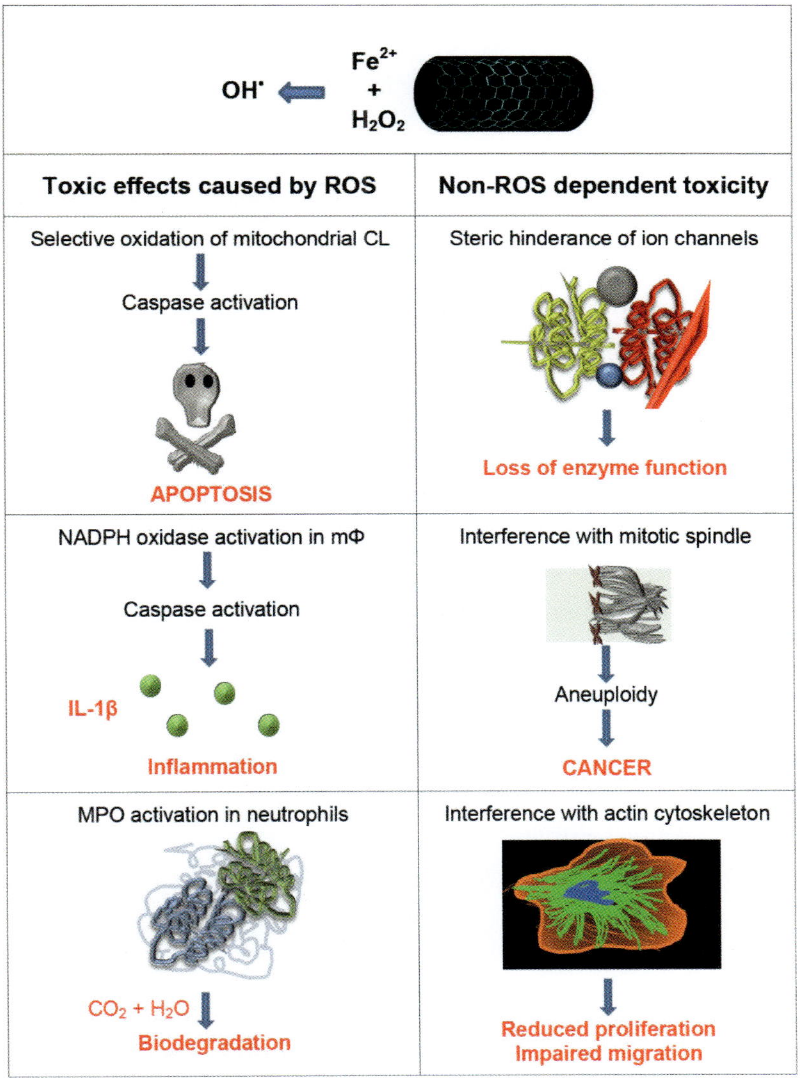

Fig. 2. Toxic effects of CNTs and C_{60}, through ROS production and non-ROS pathways. Adapted from Ref. 96.

2.2.2. Natural antimicrobial polymer: chitosan

Chitosan is a partially deacetylated chitin that has been discovered to possess a wide-spectrum of antibacterial activity.[97] Not until recently have the antimicrobial effects of chitosan against bacteria, viruses, and fungi surfaced, after the engineering of chitosan into nanomaterials.[23,24,97] Chitosan was found to be more effective for controlling fungal and viral infections than bacterial ones,[23] and the antimicrobial activity of chitosan had previously been considered greater for Gram-positive bacteria than Gram-negative.[24,98] A further clarified tendency has been shown by recent studies.[99,100] Chitosan's antimicrobial effect is very dependent on the molecular weight of chitosan being used for the NPs and intrinsic differences in target bacterial wall structure. Chitosan of a low molecular weight generated high antimicrobial effects on Gram-negative bacteria while the inverse is observed with chitosan of a high molecular weight on Gram-positive bacteria.[101] A study also showed chitosan's synergistic antimicrobial activity against drug resistant *P. aeruginosa* when used with sulfamethoxazole.[102] The clinical potential for such combinatory uses of chitosan in overcoming untreatable resistant infections could be vast. In a recent study, chitosan-capped Au NPs coupled with ampicillin exhibited a doubling in the increase in antimicrobial activity, compared with that of free ampicillin.[14]

Chitosan's antimicrobial mechanisms have been explained by various mechanisms. One is the binding to the negatively charged bacterial surface to cause clumping, increasing the permeability of the microbial wall, which eventually leads to a leakage of intracellular components.[24] In another proposed mechanism, chitosan chelates trace metals and thereby inhibits enzyme activities and the microbial growth.[103] It was also proposed that chitosan liberated from the fungal wall by host hydrolytic enzymes penetrates to the nucleus of fungi and inhibits RNA and protein syntheses.[23] The restricted solubility only in acidic media and precipitation in the culture medium prevent accurate investigation of the antibacterial activity of chitosan and its mechanisms.[97] Water-soluble derivatives of chitosan had better antimicrobial effects than native chitosan through the disruption of the outer and inner membranes of bacteria.[104]

Utilizing chitosan is a promising, cost-effective and technologically affordable sterilization method that is particularly useful in developing countries. Nanoscale chitosan could be used to disinfect microbes in membranes, sponges or surface coatings of water storage tanks.[24] It has several advantages over other disinfectants, such as a high antibacterial activity, a broad spectrum of activity, a high microbe-killing efficiency, and a low toxicity on mammalian cells.[23]

2.2.3. Synthetic anti-microbial linear polymers

Positively charged polymers, i.e. those with quaternary ammonium groups, are the most explored for anti-microbial purposes, including acrylic and methacrylic polymers, cationic conjugated polyelectrolytes, polysiloxanes, oxazolines, polymers with quaternary nitrogens on the main chain, and heterocyclic polymers.[105] Due to the bacterial cell wall negative charge, these antimicrobial polymers create destructive interactions with the bacterial membrane.[106] Polyisohexylcyanoacrylate NPs were shown to diffuse through the membrane of a human cell, and interact with the cell wall of intracellular bacterial parasites.[107] Non-quaternary polydialylamines, sulfonates, and phosphonates (latter two in slightly acidic conditions) display high biocidal efficacy due to strong electrostatic interactions between polymer and membrane.[105] The protonation of the phospholipids in the membrane results in a positively charged lipid membrane, facilitating the hydrogen bond network with the polymers and increasing the permeability of the membrane.[108] Copolymers containing ammonium or phosphonium salts were shown to cause cell membrane disruption and release of potassium ions.[109] Cyclic *N*-halamines are another large class of polymeric biocides characterized by a halogen covalently bonded to the nitrogen atoms of the polymers. Bacterial cell inactivation is caused by the slow release of an oxidative halogen by the nitrogen that targets thiol or amine groups on proteins.[106] Recently, a novel *N*-halamine epoxide coating was shown to be effective against *E. coli* and *S. aureus* bacteria.[110] Antimicrobial peptides found naturally are being mimicked through synthesis of polymers and peptides. Arylamide foldamers have shown strong bactericidal activity against *E. coli* due to significant changes in the permeability of the bacterial cell outer membrane through the binding of arylamide to the lipopolysaccharide groups.[111]

2.2.4. Dendrimers

Dendrimers are hyperbranched polymers with superior size control due to the layer-by-layer synthesis, high drug loading capacity through multiple branching points, and facile surface functionality.[112] Increased *in vivo* reactivity to micro-organisms may be attributed to the large surface area to size ratio of the branched structures. The high surface functionality allows for specific binding to various receptors, and hydrophobic and hydrophilic drugs can be incorporated either into the internal cavities or multivalent surfaces, respectively.[113–115] As previously mentioned, the quaternary ammonium moieties on the dendrimer facilitate antibacterial activity.[113–116] An amine-functionalized poly(propyleneimine) dendrimer has the capability of NO storage and release, creating a dual-modality of attack on bacteria.[117] Dendrimers may be functionalized with alcohol groups to interact with O antigens of the bacteria membrane, sugars to mimic the bacteria cell surface and interact with lectins, peptides to induce bacteria membrane recognition, carboxylic acid moieties to bind to Ca^{2+} and Mg^{2+} or charged amine groups to displace those ions.[118] It is important to consider, however, how the number of functional groups on the dendrimer will affect the ability to penetrate the cell wall to arrive at the plasma membrane. Polyamidoamine (PAMAM) is the first and most widely studied dendrimer; however, its cytoxicity due to the terminal amine groups limits clinical usage,[119] and [8] PAMAM dendrimers with carboxyl or hydroxyl termini have shown to be more biocompatible and easy to conjugate antimicrobial agents.[8,114]. Cationic dendrimers such as poly(propyleneimine) and carbosilane, and anionic Janus peptide dendrimers have also shown to have potent bactericidal activity.[120,121]

3. Nano-Antibiotics Delivery

The NPs have unique tunable properties that make for a viable platform for drug delivery and overcoming limitations such as drug resistance and intracellular infections.[113,122] NP-based antibiotic delivery enables increased solubility of hydrophobic drugs, improved pharmacokinetics, and optimization of dosage requirements.[7,122] Moreover, targeted delivery of NPs ensures lowering of systemic side effects, and simultaneous loading of synergistic therapies may increase efficacy and inhibit drug resistance as well.[113,122] Table 2 lists examples of infectious microorganisms which can be effectively targeted using a variety of NPs.

3.1. *Nitric oxide delivery*

Nitric oxide (NO) is a regulator for immune response to infection, and nanocarriers of NO have potential for being an alternative antimicrobial agent.[131] NO free radicals interact with ROS to form reactive nitrogen species (RNS) such as peroxynitrite, *S*-nitrosothiols, nitrogen dioxide, dinitrogen trioxide, and dinitrogen tetroxide (Fig. 3).[132] These RNS interact with proteins at various sites, modify lipids, and cause oxidative damage to DNA as well as several other DNA alterations.[133]. *S*-nitrosoglutathione (GSNO) contributes to the *S*-nitrosation of enzymatic proteins which contain sulfhydryl groups, which ultimately inhibits survival of microbes. GSNO is effectively generated from NO-releasing NPs in the presence of glutathione and shows improved antimicrobial properties.[134]

Table 2. NP-based antibiotic delivery increases therapeutic efficacy against microorganisms.

Target	NP type	NP composition	Antibiotic agent	Advantage over free antibiotic	Reference
S. aureus	Liposome	1,2-dipalmitoyl-phosphatidylcholine (DPPC), Cholesterol (Chol), dimethyl ammonium ethane carbamoyl-chol (DC-chol)	Benzyl penicillin	Decrease in drug concentration and exposure time	123
	Polymeric	Glycosylated polyacrylate	*N*-sec-butylthio β-lactam, Ciprofloxacin	Improved bioavailability	124
	Dendrimer	PAMAM	Silver salt	Prolonged circulation, high payload	125
Methicillin-resistant *S. aureus* (MRSA)	Liposome	Egg-PC(EPC), Diacetylphosphate (DCP), Chol	Vancomycin or Teicoplanin	Enhanced uptake by macrophages, enhanced intracellular antimicrobial effect	126
		Polyacrylate (PAA)	*N*-methylthiolated β-lactams	Enhanced solubilty of drug, anti-MRSA activity of PAA	127
E. coli	Liposome	Hydrogenated soybean-PC (HSPC), Chol, distearoyl phosphatidylglycerol (DSPG)	Amikacin	Prolonged residence in plasma and tissue, reduced renal clearance	128
	Solid lipid	Stearic acid (SA), soybean-PC(SPC), sodium taurocholate (STC)	Ciprofloxacin hydrochloride	Prolonged release of agent	129
	Dendrimer	PAMAM	Sulfamethoxazole, Nadifloxacin and Prulifloxacin	Sustained release of drug, increased solubility, enhanced penetration of cell membrane	115,130

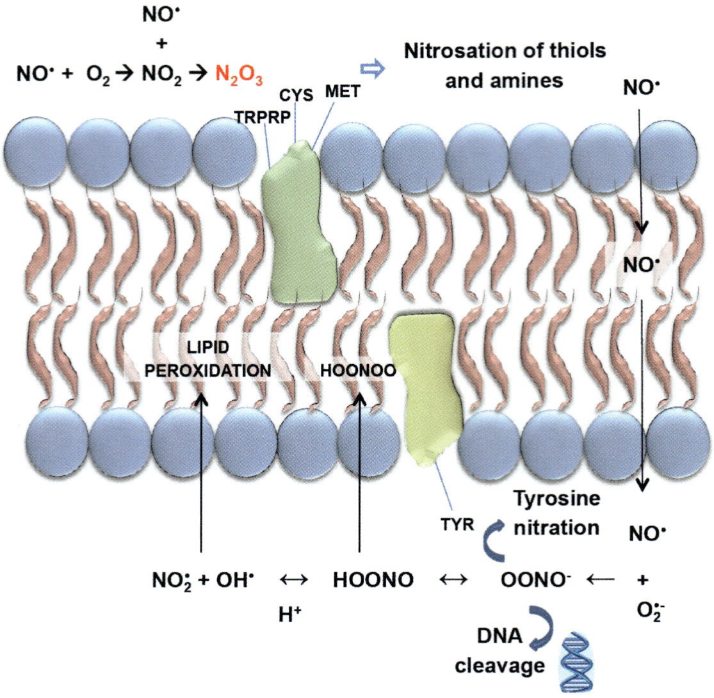

Fig. 3. NO free radical effects on species within, outside, and inside of the microorganism cell membrane.

Although there are multiple vehicles for NO storage and delivery, there are limiting factors associated with each. Carriers must prevent NO from oxidizing to NO_2 therefore it is crucial to have limited contact with the outside environment.[135] Dendrimers can store and release large amounts of NO, and are easily functionalizable; however, there is a lack of control over the rate of release.[132] Metal-containing materials such as quantum dots and nanoporous solid materials may cause acute and/or chronic toxicity, while biofriendly carriers, i.e. PLGA NPs, have high production cost.[136] Zeolites have a lattice structure capable of storing a large amount of NO and allowing for reversible constraint.[137] In aqueous conditions at physiological temperature and pH, spontaneous NO release is achieved using NP-based scaffolds such as silica NPs and biodegradable polymeric NPs.[136,138,139] NO-releasing silica NPs, non-toxic to mammalian cells, were shown to be effective against Gram-negative, Gram-positive bacteria, and fungi.[138,140] NO-releasing NPs were shown to reduce angiogenesis in MRSA abscesses through the increased production of IFN-γ, ultimately preventing bacterial propagation to other tissues.[141] NO-releasing NPs possess tunable properties such as size, surface charge, and hydrophilicity/phobicity making NPs superior to small molecular NO donors.[139] Recently, electrospun polymeric nanofibers were utilized to release NO under physiological pH. The structure of the fibrous material allows for support of cell attachment and growth, while NO release provides protection against infection and enhances wound healing by increasing and stimulating fibroblast migration and collagen deposition.[142,143]

3.2. *Liposomes*

Liposomes, nano- or microvesicles composed of an aqueous core surrounded by phospholipid bilayer, are the most commonly used vehicle for antimicrobial drug delivery.[3,8] Their outer lipid layer is mimicry of the cell membrane, creating an accessible mode of fusion to infectious microbes[113] and easy incorporation of hydrophobic drugs.[144] Likewise, the aqueous core allows for facile loading of hydrophilic antimicrobial therapeutics.[8] In order to effectively utilize liposomes in antimicrobial drug delivery, considerations must be taken for numerous variables such as stability, particle size, polydispersity, surface charge, drug loading, lipid properties, and scalability.[113] The mononuclear phagocytic system (MPS) clears liposomes from the blood quickly after administration,[7] and the incorporation of certain glycolipids has resulted in extended circulation time.[145–147] Conjugation of PEG with the liposome surface also improves circulation time, and tethering of targeting ligands to the surface allows for superior precise delivery of antimicrobial drugs.[113,148,149]

Bacterial cells and eukaryotes possess negatively charged surfaces allowing for a strong interaction with liposomes carrying positively charged molecules on the surface.[150] It was demonstrated that cationic liposomal encapsulations of meropenem or gentamicin were more effective than neutral or anionic formulations against Gram-positive and Gram-negative bacteria irrespective of the antibiotic.[151] Upon fusion of the liposome with the cell membrane, a significant dose of the antimicrobial agents is instantly unloaded into the bacteria, possibly outriding the efflux pumps and inhibiting the drug resistance of microbes.[113,152] The direct infusion and interaction between liposomes and bacterial cells have aided in the eradication of drug-resistant strains of bacteria.[150] *P. aeruginosa* resistance to antibiotics may be attributed to low permeability of the outer membrane, alterations in cell surface or efficient efflux systems.[153–155] Transmission electron microscopy (TEM), flow cytometry, lipid mixing assay, and immunocytochemistry were used to show the membrane deformation of bacteria caused by the interaction of aminoglycosides-loaded liposomes with the outer membrane of *P. aeruginosa*.[156] Polymyxin B is an antibacterial drug effective against *P. aeruginosa* infections;

however, its use is limited by the toxic side effects in the ears, kidneys, and neuromuscular blockade.[152] Incapsulation of polymyxin B into liposomes have drastically decreased unwanted side effects and increased antimicrobial activity.[157] In a murine model, the treatment of MRSA-induced pneumonia, PEGylated liposome encapsulation of vancomycin enhanced intracellular killing of MRSA by increased deposition in lung, liver and spleen and prolonged circulation.[158] Liposomal formulations of several drugs have proven to be more effective and stable than the free drug administration including ampicillin, penicillin, streptomycin, gentamicin, and ceftazidime.[123,159–161] Liposomal ciprofloxacin was cleared from the blood quickly and remained in the liver and spleen at least 3 h after the last administration, demonstrating its usefulness against systemic salmonella infection.[128] Radioactively labeled PEGylated liposomes have been used to effectively image infectious and inflammatory centers as well.[145]

3.3. *Solid lipid nanoparticles*

Solid lipid nanoparticles (SLNPs) utilize lipids as drug carriers, with therapeutics either homogeneously dispersed within the lipid matrix, loaded into the core of a lipid shell, or loaded onto the outside of a lipid sphere.[162] SLNPs combine the benefits of traditional NPs and liposomes, while circumventing several of their drawbacks.[163,164] For example, SLNPs have very minor toxicity, prevent the degradation of sensitive molecules, can easily encapsulate lipophilic molecules, have prolonged drug release, use of organic solvents can be avoided in preparation, and preparation process can be easily scaled up.[165] In the design of SLNPs, the fatty acid selected plays an important role in drug release rates along with preparative method.[166,167] Using SLNPs to improve bioavailability and precise delivery of antimicrobial drugs has been investigated through topical, oral, pulmonary, parenteral, and ocular administration.[168–171] Topical application of SLNPs provides a long residence time on the stratum corneum[172] through the formation of a dense hydrophobic film.[173] Water-insoluble azole antifungal drugs encapsulated in SLNPs had increased transdermal diffusion in several studies.[113,172–175] Additionally, antimicrobial drug delivery via oral administration can be realized in various SLNP formulations.[176] Poor intestinal absorption of certain drugs like tobramycin is caused by P-glycoproteins, an ATP-dependent efflux pump that actively exports the drugs.[171] Incorporation of tobramycin into an SLNP was one method proposed to improve intestinal absorption.[170] Enzymes in the gut degrade the lipids to form active mono- and diglycerides on the surface of SLNPs, from which micelles are formed. The drug dissolved in the lipid is taken up into the micelle, which then interact with bile salts to form mixed micelles. Subsequently, the drug is absorbed through the gastrointestinal (GI) tract.[165] Tobramycin-encapsulating SLNPs also demonstrate increased capability of crossing blood–brain barriers (BBB)[169] and greater bioavailability in the aqueous humor than that of standard eye drops.[171] Local delivery of SLNPs loaded with ciprofloxacin show promise for extended release in ocular and skin infections.[129]

Colloidal drug carriers injected intravenously are taken up by the MPS, necessitating the alteration of lipid surface or variation of administration route.[162,170] Inhalable SLNPs are advantageous over liposomes and polymeric NPs due to their stability, drug incorporation capability, and lower risk of organic solvent contamination.[164] Alveolar macrophages engulf the SLNPs, which are then transported to the lymphoid tissues.[172–176] Drug-loaded SLNPs were nebulized to infected guinea pigs on a weekly basis, and elimination of tubercle bacilli in the lungs and spleen was observed. The free drug administration required daily dosage to achieve the equivalent effects.[172,177]

Despite the multiple advantages of SLNPs, a need for improved drug-loading capacity was met through the development of nanostructured lipid carriers (NLCs). NLCs combine solid lipids with oils to produce particles which are solid at room temperature, in the form of imperfect nanostructured solid matrix, multiple O/F/W carrier or solid amorphous matrix.[165] Along with enhanced drug-loading capacity, NLCs have long-term stability due to the fixed network formed in highly concentrated solutions, whereas SLNPs undergo restructuring during storage to low energy modifications in the crystal lattice leading to drug expulsion.[178,179] In a comparative study of SLNPs and NLCs, the drug release profile of SLNPs was shown to change after three-month storage versus the NLCs, which remained constant.[180]

3.4. *Polymeric NPs*

Since the mid-1970s, biocompatible polymers have been intensively explored for controlled drug release, enhanced intracellular delivery, and reduced rapid clearance by the RES (reticuloendothelial system).[181] One of the first antimicrobial polymers to be evaluated against *E. coli* was the electrolyte salt polyhexamethylene biguanide chloride, where the bactericidal effect was proposed to be through attraction and adsorption to the cell surface, impairment of outer membrane, binding to inner membrane causing a K^+ loss, and bacteriostasis ultimately leading to complete membrane dysfunction and leakage of intracellular components.[182] Polymeric NP antibiotic drug delivery has several benefits including structural stability in biological and chemical environments, properties that can be finely tuned through manipulation of polymer lengths or surfactants, and versatility in surface modification.[3,7,113] NPs conjugated with lectin were selective toward carbohydrate receptors on *H. pylori* and efficiently released antimicrobial agents into the targeted bacteria.[183]

Linear polymers and amphiphilic block copolymers are the two main types of polymeric NPs that have been widely used for drug delivery. Linear polymer NPs are typically nanocapsules, utilizing a polymeric membrane to surround the solvated drug and control rate of release; or nanospheres, where drugs are distributed homogenously throughout the polymeric matrix.[124,184] Amphiphilic block copolymers are capable of self-assembly into micelles containing a drug-loaded hydrophobic core within a hydrophilic outer region.[185,186] Commonly used hydrophobic polymers include poly(lactic acid) (PLA), poly(cyanoacrylate)(PCA), poly(ε-carprolactone)(PCL) and poly(lactic-*co*-glycolic acid) (PLGA). PEG is widely used for the hydrophilic segment, often with targeting ligands attached to the termini.[113,187,188] Polymeric NPs demonstrate improved bioactivity relative to the free drug therapies. In a mouse model of *Salmonella typhimurium* infections, the poly(isohexyl cyanoacrylate) NP encapsulation of ampicillin had a 103-fold increase in efficacy when compared to free ampicillin.[189] Polyethylcyanoacrylate NPs have increased absorption in the GI tract[190] and the PEGylation of these also reduced phagocytosis, increased serum half-life, and portrayed mucoadhesive properties.[191] Polyacrylate NPs conjugated with *N*-thiolated β-lactam antibiotics showed improved bioactivity against MRSA compared to the free drug, and a PLA/PLGA-gentamicin NP demonstrated potent antimicrobial activity against intracellular *Brucella* infection.[127,192] Although polymeric NPs provide an effective means for treatment against bacteria, it is important to evaluate the biocompatibility of the materials used.

3.5. *Nanoemulsions (NEs)*

Recently, certain thermodynamically stable oil-in-water (O/W) micro- and NEs, either transparent or translucent, have demonstrated antimicrobial properties suggested to be through microbe membrane disruption.[193–198] NEs are often comprised of essential oils having antimicrobial effects which can be characterized according to three major active components: (i) phenols, (ii) aldehydes, and (iii) terpenes.[199] Phenolic compounds are suggested to interact with the membrane to cause leakage of cellular components, ion exchange, impairment of metabolism, and alteration of genetic material synthesis.[199–201] Emulsions allow for a significantly higher concentration of poorly soluble antimicrobial materials in conjunction with concentrations of aqueous phase bioactive molecules to remain stable. O/W NE-based anthrax and hepatitis B vaccines administered through the nasal cavity were found to be well tolerated in a murine model and caused no inflammation.[202,203] Important considerations in the design of NEs are the partition ratio between the organic phase and aqueous phase and the contributions of the emulsifier.[201]

Antimicrobial O/W microemulsions (MEs) consisting of Tween 29, pentanol and ethyl oleate were successful in killing *S. aureus*, resistant *P. aeruginosa*,[197] and *P. aeruginosa* biofilms.[198] BCTP, NEs of tri-*n*-butyl phosphate and soybean oil, emulsified with Triton X-56 were shown to inactivate spores of various *Bacillus* species and disrupt membranes of enveloped viruses and bacteria.[30,196] NB-401 is a mixture of BCTP and P10 liposomes which is an effective agent against planktonically grown bacteria, either as a biofilm or in the sputum.[198,204]

4. Promises and Challenges of Nano-Antibiotics

The use of NPs for the delivery of antimicrobial agents is a promising archetype for the engineering of effective therapeutics against pathogenic bacteria[30] and demonstrates several clinical advantages. First, NPs are easy to fabricate so that activation by environmental stimuli such as pH, magnetic field, heat or chemical interaction may enhance targeted delivery or prompt a biological sensor.[205,206] In one example, specific release of antibiotics in acidic environment was achieved using amoxicillin freeze-dried in conjunction with chitosan and polyvinyl pyrrolidone,[207] which has potential for use in treatment of abscess. Second, through the versatility of the physicochemical properties of NPs, they can be designed in a manner that helps to minimize unwanted side effects by improving solubility and stability of antimicrobial agents.[113,208,209] Manipulation of size and material has shown to be effective in transporting antimicrobial NPs across the blood–brain barrier,[210] and NPs can be formulated with a variety of shapes to further enhance specific properties (Fig. 4). Third, the resistance built in many pathogenic bacteria to traditional therapeutics can be superseded through the use of NP-based delivery systems.[30] Fourth, NPs provide a means to improve the pharmacokinetics of antimicrobial drugs by prolonged circulation and controlled release.[113] Multiple studies have confirmed that antimicrobial NPs exhibit greater efficacy against bacteria than the free antibiotic alone.[208] Antimicrobial NPs are also economical in the sense of their cost-effective preparation and decrease in administration frequency in addition to the capability of concurrent delivery of synergistic antimicrobial agents[5].[122]

As with any new technology, careful consideration of the drawbacks of nano-antibiotics must be taken into account prior to the attainment of clinical

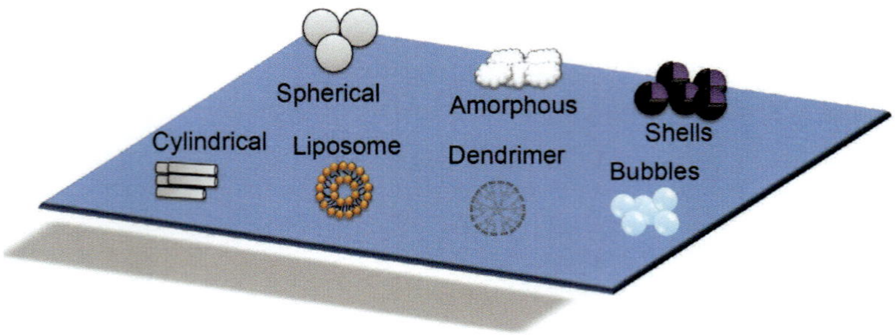

Fig. 4. NP formulations in different shapes.

employment. Thorough evaluation of the chemical and biological interactions of antimicrobial NPs is necessary to calibrate dosage, identify administration routes, and most importantly, assess the toxicity of NPs.[125,205,211] Intravenous or pulmonary administration of NPs can accumulate in organs such as the lung, liver, heart, spleen or brain[211–213] due to the enhanced cellular uptake and transcytosis across endothelial and epithelial cells.[23]

The oxidative stress caused by certain antimicrobial NPs may contribute to mitochondrial failure, glycoloysis, and enhanced ketogenesis, contributing to toxic effects in the liver, lungs or kidneys.[214,215] Additionally, NPs can adversely affect heart rate and the reproductive system.[216–218] There is a lack of standardized protocol for the administration of NP therapeutics due to their size-specific properties, therefore it is necessary to develop novel characterization techniques and biological media that will not be affected by NP properties.[212,214,219] The mechanisms behind NP toxicity at the molecular level may further be explored with the use of toxicogenomics, proteomics, and other new technologies.[213] A recent study further elucidated the genotoxic mechanism of a silver NP-based hydrogel by global gene expression analysis and cytokinesis-block micronucleus assay.[220] It was suggested that genotoxicity of the Ag NP–hydrogel is associated with the signaling proteins and regulatory pathways involved in anti-ROS response, DNA damage, apoptosis, and mitosis. The investigation of nephrotoxic effects such as widespread renal proximal tubule necrosis after administration of nanocopper particles revealed a significant alteration in renal gene expression profiles in male Wistar rats.[221] There were vast gene expression changes associated with the pathways of oxidative phosphorylation, valine, leucine and isoleucine degradation, glutathione metabolism, MAPK signaling, cell cycling, and complement and coagulation cascades. A genome-wide toxicity assay of *E. coli* confirmed the antibacterial mechanisms of amine-functionalized polystyrene nanomaterials to be through destabilization of the outer membrane and ROS production intracellularly and at the cell membrane.[222]

Certain NPs also have the potential to produce harmful effects in the environment. Ag, CuO and ZnO NPs have been shown to be toxic against the soil microbe *P. putida* KT2440, known to be a strong root colonizer and possess bioremediation potential.[223] In a study on a virus–bacteria binary system, treatment with Ag and ZnO NPs increased the transportation of bacteriophages into *E. coli* by two to six orders of magnitude, which could potentially facilitate drug-resistance in bacteria.[224] Nanomaterials used for water and waste treatment can spread multidrug resistance through the horizontal DNA transfer. Nano-alumina in water promoted the conjugative transfer of the RP4 plasmid from *E. coli* to *Salmonella* species by 197-fold.[225] Careful consideration must be taken into account during the production of nano-antibiotics and the use of nanomaterials in water treatment so as not to disrupt the balances in the environment.

5. Applications

5.1. *Treatment of drug-resistant microorganisms and biofilms*

Antimicrobial resistance to antibiotics is a major challenge in fighting infections, and may be remedied through the incorporation of multiple agents into a single NP for concomitant delivery.[113] For example, Au NPs in conjunction with various antibiotics showed enhanced activity against bacteria versus sole administration of the drugs.[226–228] Liposomes encapsulating vancomycin and teicoplanin demonstrated increased efficacy against MRSA, likewise polyacrylate and carbohydrate NPs conjugated with various antibiotics.[124,126,127,229]

Recurring infections are often caused by bacterial biofilms, aggregates of microorganisms that are irreversibly adhered to a surface and are non-responsive to antimicrobial agents.[123,230] Two major approaches can be used against the formation of bacterial biofilms: attack or defend. Recently, a smart polymer capable of defense and attack by switching between cationic and zwitterionic states demonstrated the killing, releasing, and defending against *E. coli*.[231] Under dry conditions, the *N,N*-dimethyl-2-morpholinone surface kills over 55.9% of *E. coli*, and in non-acidic aqueous conditions, the hydrolyzed carboxy betaine moiety allows release of dead bacteria and resistance of adhesion of new bacteria. Antibiotics have difficulty penetrating the matrix of biofilms, thus resulting in the survival of the bacteria on the interior of the biofilm. Through the utilization of carboxyl-functionalized superparamagnetic iron-oxide nanoparticles (SPIONs), there was an eightfold higher efficiency against resistant strain of staphylococci than free gentamicin. It was suggested that the SPIONs' antibacterial activity relies on the production of ROS and damage to the cell membrane via electrostatic interactions. The use of magnetic targeting produced 28% more dead bacteria on an established *Staphylococcus epidermidis* biofilm.[232] Antibiofilm activity can be found in antibiotic-encapsulating liposomes, NO-releasing silica NPs, MEs, and NEs.[123,233–239] Recently, a new *N*-halamine-based polyurethane surface showed potent antimicrobial effects against MRSA, *E. coli*, *S. aureus*, and vancomycin-resistant *Enterococcus faecium*, and was effective in the prevention of biofilm formation.[240]

5.2. *Targeted delivery and local administration*

Specific delivery of NPs to target tissues may be passive, relying on vasculature properties surrounding infection; active, utilizing receptor–ligand interaction; or extrinsic, using external guidance such as light, magnetic or thermal applications.[233,241–245] Acidic environments of infections may be exploited through functionalization of NPs to release drug payload under a lower pH. An *in vitro* study suggests that using surface charge-switching polymeric NPs to deliver acid-sensitive antibiotics may improve drug targeting and circulation.[246] Imidazole groups of poly-L-histidine on the surface of the encapsulated drug are protonated in acidic infection environments, allowing for the increase in binding to bacteria in order to deliver the drug payload. NPs loaded with antibiotics can enter the host cells to release a payload of the drug, effectively eliminating antibiotic-resistant microorganisms that survive intracellularly.[148,229,247] Mesoporous silica pH-gated NPs loaded with isoniazid, designed to release inside acidified endosomes of *M. tuberculosis*-infected macrophages, showed much higher efficacy than free isoniazid.[246] Improved targeted delivery to alveolar macrophages (AMs) through pulmonary administration has been achieved through conjugation of mannose to NP surfaces.[248–252]

The administration of antibiotics typically occurs orally or *i.v.*, and these routes often cause distribution of the agent to various organs and tissues, ultimately leading to systemic adverse effects. Local administration decreases the chance of drug alteration and loss, and helps to diminish unwanted side effects.[253,254] NPs can be engineered to selectively deliver drugs to nasal epithelia or the lung, and even specifically to the peripheral lung.[8,254,255] Employment of helper hydrophilic polymers such as chitosan and poly(vinyl alcohol) may help to achieve optimal size and surface properties to regulate deposition patterns in addition to controlling distribution properties by proper shape formulation.[256,257] SLNPs have been used for prolonged release of antibiotics delivered intravitreously to treat ocular infections, resulting in less irritation.[129,148,171,172] Surgical wound infections treated with tobramycin-loaded NPs had a significant reduction in *P. aeruginosa* counts for an extended period of time.[247] Sustained release of antibiotics has also been exploited through the use of biodegradable carriers such as PLA, PGA or poly(trimethylene carbonate) (PTMC) in orthopedic surgery for the treatment of bone infections.[258–262] D,L-Dilactide containing pefloxacin and nano-hydroxyapatite pellets were effective against MRSA-induced bone infections over a prolonged period and stimulated reconstruction of new bone.[263,264] Various types of bioceramics have been shown to effectively regenerate bone in addition to the local delivery of antibiotics.[265]

6. Conclusion

Antibiotics have been fighting against infectious diseases for over a half century; however, the growing resistance to therapeutics is a major threat. Although pharmaceutical companies have attempted to design more potent drugs, the prevalence of resistance may ultimately become even stronger and uncontrollable. Nano-antibiotics have the potential to overcome these resistances through their intrinsic mechanism of toxicity, co-delivery of synergistic drugs or through the capability of fine-tuning the physicochemical properties to enable specific targeting. Despite great promises, this emerging, highly demanded area of employing nanomaterials in treating infectious diseases requires better understanding of their interactions in biological systems and optimally engineered formulations.

References

1. G. Taubes, *Science* **321**, 356–361 (2008).
2. M. Rai, A. Yadav and A. Gade, *Biotechnol. Adv.* **27**, 76–83 (2009).
3. S. C. Abeylath and E. Turos, *Expert Opin. Drug Deliv.* **5**(9), 931–949 (2008).
4. J. S. Kim, E. Kuk, K. N. Yu, J. H. Kim, S. J. Park, H. J. Lee, S. H. Kim, Y. K. Park, Y. H. Park, C. Y. Hwang, Y. K. Kim, Y. S. Lee, D. H. Jeong and M. H. Cho, *Nanomedicine* **3**, 95–101 (2007).
5. E. Weir, A. Lawlor, A. Whelan and F. Regan, *Analyst* **133**, 835–845 (2008).
6. H. M. Mansour, Y. S. Rhee and X. Wu, *Int. J. Nanomedicine* **4**, 299–319 (2009).
7. N. S. Santos-Magalhães and V. C. Mosqueira, *Adv. Drug Deliv. Rev.* **62**, 560–575 (2010).
8. A. Sosnik, A. M. Carcaboso, R. J. Glisoni, M. A. Moretton and D. A. Chiappetta, *Adv. Drug Deliv. Rev.* **62**, 547–559 (2010).

9. A. J. Huh and Y. J. Kwon, *Journal of Controlled Release*, **156**(2), 128–145 (2011).
10. S. Pal, Y.K. Tak and J. M. Song, *Environ. Microbiol.* 27(6), 1712–1720 (2007).
11. H. J. Klasen, *Burns* **26**, 117–130 (2000).
12. F. Raimondi, G. G. Scherer, R. Kötz and A. Wokaun, *Angew. Chem. Int. Ed Engl.* **44**, 2190–2209 (2005).
13. I. Sondi, B. Salopek-Sondi, *J. Colloid Interface Sci.* **275** 177–182 (2004).
14. M. Chamundeeswari, S. S. Sobhana, J. P. Jacob, M. G. Kumar, M. P. Devi, T. P. Sastry and A. B. Mandal, *Biotechnol. Appl. Biochem.* **55**, 29–35 (2010).
15. H. J. Johnston, G. Hutchison, F. M. Christensen, S. Peters, S. Hankin and V. Stone, *Crit. Rev. Toxicol.* **40**(4), 328–346 (2010).
16. Z. Huang, X. Zheng, D. Yan, G. Yin, X. Liao, Y. Kang, Y. Yao, D. Huang and B. Hao, *Langmuir* **24**(8), 4140–4144 (2008).
17. R. Dastjerdi and M. Montazer, *Colloids Surf. B Biointerfaces* **79**(1), 5–18 (2010).
18. S. S. Uğur, M. Sarıışık, A. H. Aktaş, M. C. Uçar and E. Erden, *Nanoscale Res. Lett.* **5**, 1204–1210 (2010).
19. P. C. Maness, S. Smolinski, D. M. Blake, Z. Huang, E. J. Wolfrum and W. A. Jacoby, *Appl. Environ. Microbiol.* **65**(9), 4094–4098 (1999).
20. M. P. Reddy, A. Vengopal and M. Subrahmanyam, *Water Res.* **41**, 379–386 (2007).
21. J. Y. Choi, K. H. Kim, K. C. Choy, K. T. Oh and K. N. Kim, *J. Biomed. Mater. Res. B Appl. Biomater.* **80**(2), 353–359 (2007).
22. K. P. Kühn, I. F. Cahberny, K. Massholder, M. Stickler, V. W. Benz, H. Sonntag and L. Erdinger, *Chemosphere* **53**, 71–77 (2003).
23. E. I. Rabea, M. E. Badawy, C. V. Stevens, G. Smagghe and W. Steurbaut, *Biomacromolecules* **4**(6), 1457–1465 (2003).
24. 23. L. Qi, Z. Xu, X. Jiang, C. Hu and X. Zou, *Carbohydr. Res.* **339**, 2693–2700 (2004).
25. D. Y. Lyon, J. D. Fortner, C. M. Sayes, V. L. Colvin and J. B. Hughe, *Environ. Toxicol. Chem.* **24**(11), 2757–2762 (2005).
26. C. M. Sayes, A. M. Gobin, K. D. Ausman, J. Mendez, J. L. West and V. L. Colvin, *Biomaterials* **26**, 7587–7595 (2005).
27. Z. Markovic, B. Todorovic-Markovic, D. Kleut, N. Nikolic, S. Vranjes-Djuric, M. Misirkic, L. Vucicevic, K. Janjetovic, A. Isakovic, L. Harhaji, B. Babic-Stojic, M. Dramicanin and V. Trajkovic, *Biomaterials* **28**, 5437–5448 (2007).
28. D. Y. Lyon, L. Brunet, G. W. Hinkal, M. R. Wiesner and P. J. Alvarez, *Nano Lett.* **8**(5), 1539–1543 (2008).
29. J. Fang, D. Y. Lyon, M. R. Weisner and P. J. Alvarez, *Environ. Sci. Technol.* **41**(7), 2636–2642 (2007).
30. P. Dibrov, J. Dzioba, K. K. Gosink and C. C. Häse, *Antimicrob. Agents Chemother.* **46**(8), 2668–2670 (2002).
31. I. Chora and J. *Antimicrob. Chemother.* **59**, 587–590 (2007).
32. Y. Liu, L. He, A. Mustapha, H. Li, Z. Q. Hu and M. Lin, *J. Appl. Microbiol.* **107**, 1193–1201 (2009).
33. W. C. Huang, P. J. Tsai and Y. C. Chen, *Small* **5**(1), 51–56 (2009).
34. V. K. Sharma, R. A. Yngard and Y. Liu, *Adv. Colloid Interface Sci.* **145**(1–2), 83–96 (2008).
35. A. R. Shahverdi, A. Fakhimi, H. R. Shahverdi and S. Minaian, *Nanomedicine* **3**(2), 168–171 (2007).

36. A. M. Fayaz, K. Balaji, M. Girilal, R. Yadav, M. Tech, P.T. Kalaichelvan and R. Venketesan, *Nanomedicine* **6**, 103–109 (2010).
37. F. Furno, K. S. Morley, B. Wong, B. L. Sharp, P. L. Arnold, S. M. Howdle, R. Bayston, P.D. Brown, P. D. Winship and H. J. Reid, *J. Antimicrob. Chemother.* **54**, 1019–1024 (2004).
38. M. Ip, S. L. Lui, V. K. Poon, I. Lung and A. Burd, *J. Med. Microbiol.* **55**, 59–63 (2006).
39. D. J. Leaper, *Int. Wound J.* **3**(4), 282–294 (2006).
40. Y. Li, P. Leung, L. Yao, Q. W. Song and E. Newton, *J. Hosp. Infect.* **62**, 58–63 (2006).
41. P. L. Drake and K. J. Hazelwood, *Ann. Occup. Hyg.* **49** (7) 575–585 (2005).
42. G. Oberdörster, E. Oberdörster and J. Oberdörster, *Environ. Health Perspect.* **113**(7), 823–939 (2005).
43. L. Braydich-Stolle, S. Hussainn, J. J. Schlager and M. C. Hofmann, *Toxicol. Sci.* **88**, 412–419 (2005).
44. S. M. Hussain, K. L. Hess, J. M. Gearhart, K. T. Geiss and J. J. Schlager, *Toxicol. In Vitro* **19**, 975–983 (2005).
45. B. S. Sekhon and S. R. Kamboj, *Nanomedicine* **6**(5), 612–618 (2010).
46. C. M. Goodman, C. D. McCusker, T. Yilmaz and V. M. Rotello, *Bioconjug. Chem.* **15**, 897–900 (2004).
47. D. Pissuwan, C. H. Cortie, S. M. Valenzuela and M. B. Cortie, *Trends Biotechnol.* **28**(4), 207–213 (2010).
48. A. N. Grace and K. Pandian, *Colloids Surf.* A **297**, 63–70 (2007).
49. H. L. Pape, F. S. Serena, P. Contini, C. Devillers, A. Maftah and P. Leprat, *Carbon* **40**, 2947–2954 (2002).
50. J. P. Ruparelia, A. K. Chatterijee, S. P. Duttagupta and S. Mukherji, *Acta Biomater.* **4**, 707–716 (2008).
51. I. M. Sadiq, B. Chowdhury, N. Chandrasekaran and A. Mukherjee, *Nanomedicine* **5**, 282–286 (2009).
52. L. Esteban-Tejeda, F. Malpartida, A. Esteban-Cubillo, C. Pecharromán and J.S. Moya, *Nanotechnology* **20**(50), 505701 (2009).
53. K. Y. Yoon, J. Hoon Byeon, J. H. Park and J. Hwang, *Sci. Total Environ.* **373**(2–3), 572–575 (2007).
54. G. Ren, D. Hu, E. W. Cheng, M. A. Vargas-Reus, P. Reip and R. P. Allaker, *Int. J. Antimicrob. Agents* **33**(6), 587–590 (2009).
55. R. Brayner, R. Ferarri-Iliou, N. Brivois, S. Djediat, M. F. Benedetti and F. Fiévet, *Nano Lett.* **6**, 866–870 (2006).
56. K. M. Reddy, K. Feris, J. Bell, D. G. Wingett, C. Hanley and A. Punnoose, *Appl. Phys. Lett.* **90**, 2139021–2139023 (213902) (2007).
57. G. Fu, P. S. Vary and C. T. Lin, *Anatase. J. Phys. Chem. B* **109**, 8889–8898 (2005).
58. P. Jiang, J. J. Zhou, H. F. Fang, C. Y. Wang, Z. L. Wang and S. S. Xie, *Adv. Funct. Mater.* **17** 303–1310 (2007).
59. J. Zhou, N. Xu and Z. L. Wang, *Adv. Mater.* **18**, 2432–2435 (2006).
60. M. Roselli, A. Finamore, I. Garaguso, M. S. Britti and E. Mengheri, *J. Nutr.* **133**(12), 4077–4082 (2003).
61. J. Sawai, *J. Microbiol. Methods* **54**(2), 177–182 (2003).
62. S. Gelover, L. A. Gómez, K. Reyes and M. T. Leal, *Water Res.* **40**, 3274–3280 (2006).
63. L. K. Adams, D. Y. Lyon and P. J. Alvarez, *Water Res.* **40**(19), 3527–3532 (2006).

64. P. Muranyi, C. Schraml and J. Wunderlich, *J. Appl. Microbiol.* **108**(6), 1966–1973 (2010).
65. D. B. Hamal, J. A. Haggstrom, G. L. Marchin, M. A. Ikenberry, K. Hohn and K. J. Klabunde, *Langmuir* **26**(4), 2805–2810 (2010).
66. N. Tsao, T. Y. Luh, C. K. Chou, T. Y. Chang, J. J. Wu, C. C. Liu and H. Y. Lei, *J. Antimicrob. Chemother.* **49**, 641–649 (2002).
67. J. A. Brant, J. Labille, J. Y. Bottero and M. R. Wiesner *Langmuir* **22**, 3878–3885 (2006).
68. T. Mashino, D. Nishikawa, K. Takahashi, N. Usui, T. Yamori, M. Seki, T. Endo and M. Mochizuki, *Bioorg. Med. Chem. Lett.* **13**(24), 4395–4397 (2003).
69. H. Aoshima, K. Kokubo, S. Shirakawa, M. Ito, S. Yamana and T. Oshima, *Biocontrol Sci.* **14** (2), 69–72 (2009).
70. Z. Q. Ji, H. Sun, H. Wang, Q. Xie, Y. Liu and Z. Wang, *J. Nanopart. Res.* **8**, 53–63 (2006).
71. D. Lyon, L. Adams and J. Falkner, *Environ. Sci. Technol.* **40**, 4360–4366 (2006).
72. X. R. Xia, N. A. Monteiro-Riviere and J. E. Riviere, *Toxicol. Lett.* **197**(2), 128–134 (2010).
73. A. Isakovic, Z. Markovic, N. Nikolic, B. Todorovic-Markovic, S. Vranjes-Djuric, L. Harhaji, N. Raicevic, N. Romcevic, D. Vasiljevic-Radovic, M. Dramicanin and V. Trajkovic, *Biomaterials* **27**(29), 5049–5058 (2006).
74. G. L. Baker, A. Gupta, M. L. Clark, B. R. Valenzuela, L.M. Staska, S. J. Harbo, J. T. Pierce and J.A. Dill, *Toxicol. Sci.* **101**(1), 122–131 (2008).
75. N. Gharbi, M. Pressac, M. Hadchouel, H. Szwarc, S. R. Wilson and F. Moussa, *Nano Lett.* **5**(12), 2578–2585 (2005).
76. A. Isakovic, Z. Markovic, B. Todorovic-Markovic, N. Nikolic, S. Vranjes-Djuric, M. Mirkovic, M. Dramicanin, L. Harhaji, N. Raicevic, Z. Nikolic and V. Trajkovic, *Toxicol. Sci.* **91**(1), 173–183 (2006).
77. J. E. Roberts, A. R. Wielgus, W. K. Boyes, U. Andley and C. F. Chignell, *Toxicol. Appl. Pharmacol.* **228**(1), 49–58 (2007).
78. A. R. Wielgus, B. Zhao, C. F. Chignell, D. N. Hu and J. E. Roberts, *Toxicol. Appl. Pharmacol.* **242**(1), 79–90 (2010).
79. H. Hyung, J. D. Fortner, J. B. Hughes and J. H. Kim, *Environ. Sci. Technol.* **41**(1), 179–184 (2007).
80. L.M. Pasquini, S. M. Hashmi, T.J. Sommer, M. Elimelech, J. B. Zimmerman, *Environ. Sci. Technol.* 2012, **46**, 6297-6305.
81. S. Kang, M. Pinault, L. D. Pfefferle and M. Elimelech, *Langmuir* **23**, 8670–8673 (2007).
82. A. Amiria, H. Z. Zardinib, M. Shanbedia, M. Maghrebia, M. Baniadama, B. Tolueiniab, *Materials Letters* 72, 153–156 (2012).
83. F. Ahmed, C. M. Santos, R. Aileen, M. V. Vergara, M. Celeste, R. Tria, R. Advincula and D.F. Rodrigues, *Environ. Sci. Technol.* **46**, 1804–1810 (2012).
84. A. S. Brady-Estévez, T. H. Nguyen, L. Gutierrez and M. Elimelech, *Water Res.* **44**, 3773–3780 (2010).
85. S. Aslan, C. Z. Loebick, S. Kang, M. Elimelech, L.D. Pfefferle and P. R. Van Tassel, *Nanoscale* **2**(9), 1789–1794 (2010).
86. Y. Younga, H. Leeb, Y. Shena, S. Tsenga, C. Leea and Nyan-Hwa Taia, *Materials Chemistry and Physics* **134**, 1, 15 279–286 (2012).

87. S. Liu, A. Keong Ng, R. Xu, J.Wei, C. M. Tan, Y. Yanga and Y. Chen, *Nanoscale* **2**, 2744–2750 (2010).
88. P. Wick, P. Manser, L. K. Limbach, U. Dettlaff-Weglikowska, R. Krumeich, S. Roth, W. J. Stark and A. Bruinink, *Toxicol. Lett.* **168**, 121–131 (2007).
89. Y. Zhang, Y. Xu, Z. Li , T. Chen, S.M. Lantz, P.C. Howard, M. Paule, W. Slikker, Jr., F. Watanabe, T. Mustafa, A. S. Biris and S. Ali. *ACS Nano,* **5** (9), 7020–7033 (2011).
90. C. Thomas, M. Mackey, A. Diaz and D. Cox, *Redox Report* **14**(3), 102–108 (2009).
91. A. S. Brady-Estévez, S. Kang, and M. Elimelech, *Small* **4**(4), 481–484 (2008).
92. J. Muller *et al., Toxicol. Appl. Pharmacol.* **207**, 221–231 (2005).
93. G. Jia, H. Wang, L. Yan, X. Wang, R. Pei, T. Yan, T. Zhao and X. Guo, *Environ. Sci. Technol.* **39**(5), 1378–1383 (2005).
94. D. B. Warheit, B. R. Laurence, K. L. Reed, D.H. Roach, G.A. Reynolds and T.R. Webb, *Toxicol. Sci.* 77(1), 117–125 (2004).
95. C. W. Lam, J. T. James, R. McCluskey and R.L. Hunter, *Toxicol. Sci.* 77(1), 126–134 (2004).
96. A. Shvedovaa, A. Pietroiustic, B. Fadeeld and V.E. Kagane, *Toxicology and Applied Pharmacology* **261**(2), 121–133 (2012).
97. Y. C. Chung, H.L. Wang, Y.M. Chen and S. L. Li, *Bioresour.Technol.* **88**, 179–184 (2003).
98. T. M. Don, C. C. Chen, C. K. Lee, W. Y. Cheng and L. P. Cheung, *J.Biomater. Sci. Polym. Ed.* **16**(12), 1503–1519 (2005).
99. H. K. No, N. Y. Park, S. H. Lee and S. P. Meyers, *Int. J. Food Microbiol.* **74**(1), 65–72 (2002).
100. J. C. Fernandes, F. K. Tavaria, J. C. Soares, O. S. Ramos, M. João Monteiro, M.E. Pintado and F. Xavier Malcata, *Food Microbiol.* **25**, 922–928 (2008).
101. J. C. Fernandes, F. K. Tavaria, S. C. Fonseca, O. S. Ramos, M. E. Pintado and F.X. Malcata, *J. Microbiol. Biotechnol.* **20**(2), 311–318 (2010).
102. S. Tin, K. R. Sakharkar, C. S. Lim and M. K. Sakharkar, *Int. J. Biol. Sci.* **5**(2), 153–160 (2009).
103. R. G. Cureo, G. Osuji and A. Washington, *Biotechnol. Lett.* **13**, 441–444 (1991).
104. J. Y. Je and S. K. Kim, *J. Agric. Food Chem.* **54**(18), 6629–6633 (2006).
105. A. Mũnoz-Bonilla and M. Fernández-García; *Progress in Polymer Science* **37**, 281– 339 (2012).
106. E. Kenawy, F. Abdel-Hay, A. El-Shanshoury and M.H. El-Newehy. *J. Polym. Sci. Part A Polym. Chem.* **40**, 2384–93 (2002).
107. O. Balland, H. Pinto-Alphandary, A. Viron, E. Puvion, A. Andremont and P. Couvreur, *J. Antimicrob. Chemother.* **37**, 105–115 (1996).
108. A. K. Berkovicha, V. N. Orlovb and N. S. Melik Nubarova, *Polymer Science, Ser. A*, **51**, **6**, 648–657 (2009).
109. E. Kenawy and Y. Mahmoud, *Macromol. Biosci.* **3**(2), 107–116 (2003).
110. H. B. Kocer, I. Cerkez, S. D. Worley, R. M. Broughton and T. S. Huang, *ACS Appl. Mater. Interfaces* **3**, 2845–2850 (2011).
111. B. Mensa, Y. H. Kim, S. Choi, R. Scott, G. A. Caputo and W. F. DeGrado, *Antimicrobial Agents and Chemotherapy* **55**(11), 5043–5053 (2011).
112. S. Svenson, *Eur. J. Pharm. Biopharm.* **71**, 445–462 (2009).

113. L. Zhang, D. Pornpattananangkul, C. M. Hu, C. M. Huang, *Curr. Med. Chem.* 17 (2010) 585–594.
114. E. R. Gillies and J. M. Fréchet, *Drug Discov. Today* **10**(1), 35–43(2005).
115. Y. Cheng, H. Qu, M. Ma, Z. Xu, P. Xu, Y. Fang and T. Xu, *Eur. J. Med. Chem.* **42**, 1032–1038 (2007).
116. C. Z. Chen and S. L. Cooper, *Biomaterials* **23**(16), 3359–3368 (2002).
117. Y. Lu, B. Sun, C. Li and M. H. Schoenfisch, *Chem. Mater.* **23**, 4227–4233 (2011).
118. A. Castonguay, E. Ladd, T. G. M. van de Ven and A. Kakkar, *New J. Chem.*, **36**, 199–204 (2012).
119. H. K. Sajja, M. P. East, H. Mao, Y. A. Wang, S. Nie and L. Yang, *Curr. Drug Discov. Technol.* **6**(1), 43–51 (2009).
120. L. Ouyang, D. Yin and Y. Zhaoa, *Synlett* **23**, 1937–1940 (2012).
121. M. A. Mintzer, E. L. Dane, G. A. O'Toole and M. W. Grinstaff, *Mol. Pharmaceutics* **9**, 342–354 (2012).
122. L. Zhang, F.X. Gu, J. M. Chan, A. Z. Wang, R. S. Langer and O. C. Farokhzad, *Clin. Pharmacol. Ther.* **83**(5), 761–769 (2008).
123. H. J. Kim and M. N. Jones, *J. Liposome Res.* **14** (3–4), 123–139 (2004).
124. S. C. Abeylath, E. Turos and S. Dickey, D. V. *Limb Bioorg. Med. Chem.* **16**(5), 2412–2418 (2008).
125. S. S. Suri, H. Fenniri and B. Singh, *J. Occup. Med. Toxicol.* **2**, 16(2007).
126. C. O. Onyeji, C. H. Nightingale and M. N. Marangos, *Infection* **22**(5), 338–342 (1994).
127. E. Turos, J. Y. Shim, Y. Wang, K. Greenhalgh, G. S. Reddy, S. Dickey and D. V. Lim, *Bioorg. Med. Chem. Lett.* **17**(1), 53–56 (2007).
128. M. Magallanes, J. Dijkstra and J. Fierer, *Antimicrob. Agents Chemother.* **37**(11), 2293–2297 (1993).
129. D. Jain, R. Banerjee, *J. Biomed. Mater. Res. B Appl. Biomater.* **86**(1) 105–112 (2008).
130. M. Ma, Y. Cheng, Z. Xu, P. Xu, H. Qu, Y. Fang, T. Xu and L. Wen, *Eur. J. Med. Chem.* **42**, 93–98 (2007).
131. F. C. Fang, *Nat. Rev. Microbiol.* **2**, 820–832 (2004).
132. N. A. Stasko, T. H. Fischer and M. H. Schoenfisch, *Biomacromolecules* **9**, 834 (2008).
133. F .C. Fang, *J Clin Invest.* **99**(12), 2818–2825 (1997).
134. A. J. Friedman, K. Blecher, D. Schairer, C. Tuckman-Vernon, P. Nacharaju, D. Sanchez, P. Gialanella, L. R. Martinez, J. M. Friedman and J. D. Nosanchuk; *Nitric Oxide* **25**, 381–386 (2011).
135. M. L. Jones, J. G. Ganopolsky, A. Labbé, C. Wahl and S. Prakash, *Appl. Microbiol. Biotechnol.* **88**, 401–407 (2010).
136. A. B. Seabra and N. Duran, *J. Mater. Chem.* **20**, 1624–1637 (2010).
137. M. Mowbray, X. Tan, P. S. Wheatley, R.E. Morris and R.B. Weller, *J. Invest. Dermatol.* **128**, 352–360 (2009).
138. E .M. Hetrick, J. H. Shin, N. A. Stasko, C. B. Johnson, D. A. Wespe, E. Holmuhamedov and M. H. Schoenfisch, *ACS Nano* **2**(2) 235–246 (2008).
139. R. P. Bagwe, L. R. Hilliard and W. Tan, *Langmuir* **22**, 4357–4362 (2006).
140. N. Barraud, D. J. Hassett, S. H. Hwang, S. A. Rice, S. Kjelleberg and J. S. Webb, *J. Bacteriol.* **188**, 7344–7353 (2006).
141. G. Han, L. R. Martinez, M. R. Mihu, A. J. Friedman and J. M. Friedman, *PLoS One* **4**(11), 7804 (2009).

142. K. A. Wold, V. B. Damodaran, L. A. Suazo, R. A. Bowen and M. M. Reynolds; *ACS Appl. Mater. Interfaces* **4**, 3022-3030 (2012).
143. G. Han, L. N. Nguyen, C. Macherla, Y. Chi, J. M. Friedman, J. D. Nosanchuk and L. R. Martinez. *The American Journal of Pathology,* **180**(4), 1465–1473 (2012).
144. D. D. Lasic, *Trends Biotechnol.* **16**(7), 307–321 (1998).
145. I. A. Bakker-Woudenberg, *Int. J. Antimicrob. Agents* **19**, 299–311 (2002).
146. M. Moonis, I. Ahmad and B. K. Bachhawat, *Antimicrob. Chemother.* **33**(3), 571–583 (1994).
147. T. M. Allen, C. Hansen and J. Rutledge, *Biochem. Biophys. Acta* **981**(1), 27–35 (1989).
148. H. P. Alphandary, A. Andremont and P. Couvreur, *Int. J. Antimicrob. Agents* **13** 155–168 (2000).
149. K. Maruyama, S. J. Kennel and L. Huang, *Proc. Natl. Acad. Sci. U.S.A.* **87**, 5744–5748 (1990).
150. Z. Drulis-Kawa and A. Dorotkiewicz-Jach, *International Journal of Pharmaceutics* **387**, 187–198 (2010).
151. Z. Drulis-Kawa, J. Gubernator, A. Dorotkiewicz-Jach, W. Doroszkiewicz and A. Kozubek, *Int. J. Pharm.* **315**, 59–66 (2006).
152. A. Omri, Z. E. Suntres and P. N. Shek, *Biochem. Pharmacol.* **64**, 1407–1413 (2002).
153. R. E. Hancock and F. S. Brinkman, *Ann. Rev. Microbiol.* **56**, 17–38 (2002).
154. D. M. Livermore, *J. Antimicrob. Chemother.* 47, 247–250 (2001).
155. D. H. Kwon and C. D. Lu, *Antimicrob. Agents Chemother.* **50**, 1615–1622 (2006).
156. C. Mugabe, M. Halwani, A. O. Azghani, R. M. Lafrenie and A. Omri, *Antimicrob. Agents Chemother.* **50**(6), 2016–2022 (2006).
157. M. Alipour, M. Halwani, A. Omri and Z. E. Suntres, *Int. J. Pharm.* **355,** 293–298 (2008).
158. K. Muppidi, J. Wang, G. Betageri and A. S. Pumerantz *Antimicrobial Agents and Chemotherapy,* 4537–4542 (2011).
159. I. Schumacher and R. Margalit, *J. Pharm. Sci.* **86**(5), 635–641 (1997).
160. R. M. Fielding, R. O. Lewis and L. M. McDermott, *Pharm. Res.* **15**(11), 1775-1781 (1998).
161. I. A. Bakker-Woudenberg, M. T. ten Kate, L. E. Stearne-Cullene and M. C. Woodle, *J. Infect. Dis.* **171**(4), 938–947 (1995).
162. M. Üner and G. Yener, *International Journal of Nanomedicine* **2**(3), 289–300 (2007).
163. S. A. Wissing and R. H. Müller, *Int. J. Pharm.* **254**, 65–68 (2003).
164. R. H. Müller, K. Mäder and S. Gohla, *Eur. J. Pharm. Biopharm.* **50**(1), 161–177 (2000).
165. S. Das and A. Chaudhury, *AAPS PharmSciTech* **12**, 62–76 (2011).
166. S. Xiea, L. Zhua, Z. Donga, X. Wanga, Y. Wanga, X. Li and W. Zhou; *Colloids and Surfaces B: Biointerfaces* **83,** 382–387 (2011).
167. A. A. Pawara, D. Chenb and C. Venkataramana, *International Journal of Pharmaceutics* **430**, 228– 237 (2012).
168. R. Pandey and G.K. Khuller, *Tuberculosis* **85**, 227–234 (2005).
169. A. Bargoni, R. Cavalli, G.P. Zara, A. Fundarò, O. Caputo and M.R. Gasco, *Pharmacol. Res.* **43**(5), 497–502 (2001).
170. M. A. Videira, M. F. Botelho, A. C. Santos, L. F. Gouveia, J. J. de Lima and A. J. Almeida, *J. Drug Target* **10**(8), 607–613 (2002).

171. R. Cavalli, M.R. Gasco, P. Chetoni, S. Burgalassi and M.F. Saettone, *Int. J. Pharm.* **238**, 241–245 (2002).
172. E.B. Souto, S.A. Wissing, C.M. Barbosa and R.H. Müller, *Int. J. Pharm.* **278**, 71–77 (2004).
173. E.B. Souto and R.H. Müller, *Int. J. Cosmet. Sci.* **30**(3), 157–165 (2008).
174. A.K. Gupta and E.A. Cooper, *Mycopathologia* **166,** 353–367 (2008).
175. V. Sanna, E. Gavini, M. Cossu, G. Rassu and P. Giunchedi, *J. Pharm. Pharmacol.* **59**(8), 1057–1064 (2007).
176. C.W. Pouton, *Eur. J. Pharm. Sci.* **11**(Suppl. 2) S93–S98 (2000).
177. S. Gelperina, K. Kisich, M.D. Iseman and L. Heifets, *Am. J. Respir. Crit. Care Med.* 172 (12) 1487–1490 (2005).
178. M. Radtke, E.B. Souto and R.H. Müller *Pharm, Technol. Eur.* **17**(4), 45–50 (2005).
179. R. H. Müller, M. Radtke and S.A. Wissing, *Int. J. Pharm.* **242**, 121–128.
180. S. Das, W. K. Ng and R.B.H. Tan, *European Journal of Pharmaceutical Sciences* **47**, 139–151 (2012).
181. K.S. Soppimath, T.M. Aminabhavi, A.R. Kulkarni and W.E. Rudzinski, *J. Control. Release* **70**, 1–20. (2001).
182. L.Timofeeva and N. Kleshcheva, *Appl. Microbiol. Biotechnol.* **89**, 475–492 (2011).
183. R. B. Umamaheshwari and N.K. Jain, *J. Drug Target* **11**(7) 415–424 (2003).
184. C. Vauthier, C. Dubernet, E. Fattal, H. Pinto-Alphandary and P. Couvreur, *Adv. Drug Deliv. Rev.* **55**, 519–548 (2003).
185. D.E. Owens 111 and N.A. Peppasnt. *J. Pharm.* **307**, 93–102 (2006).
186. S. R. Croy and G. S. Kwon, *Curr. Pharm. Des.* **12,** 4669–4684 (2006).
187. J. Cheng, B. A. Teply, I. Sherifi, J. Sung, G. Luther, F.X. Gu, E. Levy-Nissenbaum, A.F. Radovic-Moreno, R. Langer, O.C. Farokhzad, *Biomaterials* **28**, 869–876 (2007).
188. F. Gu, L. Zhang, B.A. Teply, N. Mann, A. Wang, A.F. Radovic-Moreno, R. Langer, O.C. Farokhzad, *Proc. Natl. Acad. Sci. U.S.A.* **105**(7), 2586–2591 (2008).
189. E. Fettal, M. Youssef, P. Couvreur and A. Andremont, *Antimicrob. Agents Chemother.* **33**(9), 1540–1543 (1989).
190. G. Fontana, G. Pitarresi, V. Tomarchio, B. Carlisi and P.L. San Biagio, *Biomaterials* **19** 1009–1017 (1998).
191. G. Fontana, M. Licciardi, S. Mansueto, D. Schillaci and G. Giammona *Biomaterials* **22**, 2857–2865 (2001).
192. S. Prior, C. Gamazo, J. M. Irache, H. P. Merkle and B. Gander, *Int. J. Pharm.* **196**, 115–125 (2000).
193. T. Hamouda, M. M. Hayes, Z. Cao, R.Tonda, K. Johnson, D. C. Wright, J. Brisker and J. R. Baker, Jr. *Journal of Infectious Diseases* **180,** 1939–49 (1999).
194. M.J. Lawrencea and G.D. Reesb, *Adv. Drug Deliv. Rev.* **45,** 89–121 (2000).
195. T. Hamouda and J.R. Baker Jr., *J. Appl. Microbiol.* **89**, 397–403 (2000).
196. T. Hamouda, M.M. Hayes, Z. Cao, R. Tonda, K. Johnson, D.C. Wright, J. Brisker and J.R.Baker Jr., *J. Infect. Dis.* **180**, 1939–1949 (1999).
197. I.S. Al-Adham, E. Khalil, N.D. Al-Hmoud, M. Kierans and P.J. Collier, *J. Appl. Microbiol.* **89,** 32–39 (2000).
198. I.S. Al-Adham, N.D. Al-Hmoud, E. Khalil, M. Kierans and P.J. Collier, *Lett. Appl. Microbiol.* **36,** 97–100 (2003).

199. E. Ceylan and D. Fung, *Journal of Rapid Methods and Automation in Microbiology* **12,** 1–55 (2004).
200. S. Burt, *International Journal of Food Microbiology* **94,** 223–253 (2004).
201. F. Donsìb, M. Annunziataa, M. Vincensia and G. Ferrari; *Journal of Biotechnology* **159**, 342– 350 (2012).
202. P. E. Makidon, A. U. Bielinska, S. S. Nigaverkar, K. W. Janczak, J. Knowlton, A. J. Scott, N. Mank, Z. Cao, S. Rathinavelu, M. R. Beer, J. E. Wilkinson, L.P. Blanco, J. J. Landers and J. R. Baker Jr., *PLoS One* **3**(8), e2954 (2008).
203. A. U. Bielinska, K. W. Janczak, J. J. Landers, P. Makidon, L. E. Sower, J. W. Peterson and J. R. Baker Jr., *Infect. Immun.* **75**(8), 4020–4029 (2007).
204. J. J. LiPuma, S. Rathinavelu, B. K. Foster, J. C. Keoleian, P. E. Makidon, L. M. Kalikin and J. R. Baker Jr, *Antimicrob. Agents Chemother.* **53**(1), 249–255 (2009).
205. S. Sandhiya, S. A. Dkhar and A. Surendiran, *Clin. Pharmacol.* **23**(3), 263–269 (2009).
206. M. H. Dufresne, D. L. Garrec, V. Sant, J. C. Leroux and M. Ranger, *Int. J. Pharm.* **277**, 81–90 (2004).
207. M. V. Risbud, A. A. Hardikar, S. V. Bhat and R. R. Bhonde, *J. Control. Release* **68**(1), 23–30 (2000).
208. E. M. Hetrick, J. H. Shin, N. A. Stasko, C. B. Johnson, D. A. Wespe, E. Holmuhamedov and M. H. Schoenfisch, *ACS Nano,* **2**(2), 235–246 (2008).
209. S. M. Moghimi, A. C. Hunter and J. C. Murray, *FASEB J.* **19**(3), 311–330 (2005).
210. Y. L. Hu and J. Q. Gao, *Int. J. Pharm.* **394**(1–2) 115–121 (2010).
211. A. El-Ansary and S. Al-Daihan, *J. Toxicol.* 754810 (2009).
212. W. I. Hagens, A. G. Oomen, W. H. de Jong, F. R. Cassee and A. J. Sips, *Regul. Toxicol. Pharmacol.* **49**(3), 217–229 (2007).
213. A. Poma and M. L. Di Giorgio, *Curr. Genomics* **9**, 571–585 (2008).
214. W. H. De Jong and P. J. Borm, *Int. J. Nanomedicine* **3**(2), 133–149 (2008).
215. R. Lei, C. Wu, B. Yang, H. Ma, C. Shi, Q. Wang, Q. Wang, Y. Yuan and M. Liao, *Toxicol. Appl. Pharmacol.* **232**(2), 292–301 (2008).
216. D. C. Chalupa, P.E. Morrow, G. Oberdörster, M. J. Utell and M. W. Frampton, *Environ. Health Perspect.* **112**(8), 879–882 (2004).
217. S. Yoshida, K. Hiyoshi, S. Oshio, H. Takano and T. Ichinose, *Fertil. Steril.* **93**(5), 1695–1699 (2010).
218. T. Komatsu, M. Tabata, M. Kubo-Irie, T. Shimizu, K. Suzuki, Y. Nihei and K. Takeda, *Toxicol. In Vitro* **22**(8), 1825–1831 (2008).
219. A. Kroll, M. H. Pillukat, D. Hahn and J. Schnekenburger, *Eur. J. Pharm. Biopharm.* **72**, 370–377 (2009).
220. L. Xu, X. Li, T. Takemura, N. Hanagata, G. Wu and L. L. Chou, *Journal of Nanobiotechnology* 10:16 (2012).
221. M. Y. Liao and H. G. *Liu, Env. Toxic. Pharm.* **34**, 67–80 (2012).
222. A. Ivask, E. Suarez, T. Patel, D. Boren, Z. Ji, P. Holden, D. Telesca, R. Damoiseaux, K. A. Bradley and H. Godwin, *Environ. Sci. Technol.* **46**, 2398–2405 (2012).
223. P. Gajjar, B. Pettee, D. W. Britt, W. Huang, W. P. Johnson and A. J. Anderson, *Journal of Biological Engineering* 3–9 (2009).
224. J. You, Y. Zhang and Z. Hu, *Colloids and Surfaces B: Biointerfaces* **85**(2), 161–167 (2011).

225. Z. Qiua, Y.Yua, Z. Chena, M. Jina, D. Yanga, Z. Zhaoa, J. Wanga, Z. Shena, X. Wanga, D. Qianc, A. Huanga, B. Zhangc and J. Lia; *PNAS* **109**(13), 4944–4949 (2012).
226. A. M. Fayaz, K. Balaji, M. Girilal, R. Yadav, P. T. Kalaichelvan, R. Venketesan, *Nanomedine* **6**, 103–109 (2010).
227. A. De Souza, D. Mehta and R. W. Leavitt, *Curr. Sci.* **91,** 926–929 (2006).
228. R. T. Tom, V. Suryanarayanan, P. G. Reddy, S. Baskaran and T. Pradeep, *Langmuir* **20**(5) 1909–1914 (2004).
229. E. Turos, G. S. Reddy, K. Greenhalgh, P. Ramaraju, S. C. Abeylath, S. Jang, S. Dickey and D. V. Lim, *Bioorg. Med. Chem. Lett.* **17**, 3468–3472 (2007).
230. C. Walsh, *Nature* **406**, 775–781 (2000).
231. Z. Cao, L. Mi, J. Mendiola, J. Ella-Menye, L. Zhang, H. Xue and S. Jiang, *Angew. Chem. Int. Ed.* **51**, 2602–2605 (2012).
232. G. Subbiahdoss, S. Sharifia, D. W. Grijpma, S. Laurent, H. C. van der Mei, M. Mahmoudi and H. J. Busscher, *Acta Biomaterialia* **8**, 2047–2055 (2012).
233. L. R. Martinez, G. Han, M. Chacko, M. R. Mihu, M. Jacobson, P. Gialanella, A. J. Friedman, J. D. Nosanchuk and J. M. Friedman, *J. Invest. Dermatol.* **129**, 2463–2469 (2009).
234. R. Weller, R.J. Price, A. D. Ormerod, N. Benjamin and C. Leifert, *J. Appl. Microbiol.* **90**, 648–652 (2001).
235. J. Lellouche, E. Kahana, S. Elias, A. Gedanken and E. Banin, *Biomaterials* **30**, 5969–5978 (2009).
236. N. M. Sanderson and M. N. Jones, *J. Drug Target* **4**(3), 181–189 (1996).
237. P. C. Teixeira, G. M. Leite, R. J. Domingues, J. Silva, P. A. Gibbs and J. P. Ferreira, *Int. J. Food Microbiol.* **118**(1), 15–19 (2007).
238. D. M. Livermore, *J. Antimicrob. Chemother.* **47**, 247–250 (2001).
239. D. Pérez-Conesa, L. McLandsborough and J. Weiss, *J. Food Prot.* **69**(12), 2947–2954 (2006).
240. X. Sun, Z. Cao, N. Porteous and Y. Sun, *Acta Biomaterialia* **8**, 1498–1506 (2012).
241. K. Blecher, A. Nasir and A. Friedman, Virulence **2**(5), 395-401 (2011).
242. P. Gao, X. Nie, M. Zou, Y. Shi and G. Cheng, *The Journal of Antibiotics* **64**, 625–634 (2011).
243. K. Nagahama, M. Hashizume, H. Yamamoto, T. Ouchi and Y. Ohya, *Langmuir* **25**(17), 9734–9740 (2009).
244. M. Qiao, D. Chen, X. Ma and Y. Liu, *Int. J. Pharm.* **294**, 103–112 (2005).
245. K. T. Peng, C. F. Chen, I. M. Chu, Y. M. Li, W. H. Hsu, R. W. Hsu and P. J. Chang, *Biomaterials* **31**, 5227–5236 (2010).
246. A. F. Radovic-Moreno, T. K. Lu, V. A. Puscasu, C. J. Yoon, R. Langer and O. C. Farokhzad, *ACS Nano.* **6**(5), 4279–4287 (2012).
247. C. I. Price, J. W. Horton and C. R. Baxter, *J. Surg. Res.* **49**(2), 174–178 (1990).
248. S. Chono, T. Tanino, T. Seki and K. Morimoto, *J. Control. Release* **127**, 50–58 (2008).
249. X. He, H. Zhang, Y. Ma, W. Bai, Z. Zhang, K. Lu, Y. Ding, Y. Zhao and Z. Chai, *Nanotechnology* **21**(28), 285103 (2010).
250. J. M. Irache, H. H. Salman, C. Gamazo and S. Espuelas, *Expert Opin. Drug Deliv.* **5**(6), 703–724 (2008).
251. W. Wijagkanalan, S. Kawakami, M. Takenaga, R. Igarashi, F. Yamashita and M. Hashida, *J. Control. Release* **125**(2), 121–130 (2008).

252. D. L. Poelma, L. J. Zimmermann, H. H. Scholten, B. Lachmann and J. F. Van Iwaarden, *Am. J. Physiol. Lung Cell. Mol. Physiol.* **283**, L648–L654 (2002).
253. J. S. Patton, C. S. Fishburn and J. G. Weers, *Proc. Am. Thorac. Soc.* **1**(4), 338–344 (2004).
254. M. M. Bailey and C. J. Berkland, *Med. Res. Rev.* **29**(1), 196–212 (2009).
255. I. Roy and N. Vij, *Nanomedicine* **6**, 237–244 (2010).
256. F. Ungaroa, I. d'Angeloa, C. Colettab, R. d'Emmanuele di Villa Biancab, R. Sorrentinob, B. Perfettoc, M.A. Tufanoc, A. Miroa, M. I. La Rotondaa and F. Quagliaa; *Journal of Controlled Release*, **157**, **1**(10), 149–159 (2012).
257. A. Garcia, P. Mack, S. Williams, C. Fromen, T. Shen, J. Tully, J. Pillai, P. Kuehl, M. Napier, J. M. DeSimone and B. W. Maynor; *Journal of Drug Delivery*. Vol. 2012, 1–10 (2012).
258. K. Kanellakopoulou and E. J. Giamarellos-Bourboulis, *Drugs* **59**(6), 1223–1232 (2000).
259. J. K. Koort, E. Suokas, M. Veiranto, T. J. Mäkinen, J. Jalava, P. Törmälä and H. T. Aro, *J. Biomed. Mater. Res.* A 1:**78**(3), 532–540 (2006).
260. P. K. Naraharisetti, H. C. Guan Lee, Y. C. Fu, D. J. Lee and C. H. Wang, J. Biomed. *Mater. Res. B Appl. Biomater.* 77(2) 329–337 (2006).
261. L. C. Li, J. Deng and D. Stephens *Adv. Drug Deliv. Rev.* **54**, 963–986 (2002).
262. D. Neut, O. S. Kluin, B. J. Crielaard, H. C. van der Mei, H. J. Busscher, D. W. Grijpma, *Acta Orthop.* **80**(5), 514–519 (2009).
263. K. Kanellakopoulou, N. Galanakis, E. J. Giamarellos-Bourboulis, C. Rifiotis, K. Papakostas, A. Andreopoulos, E. Dounis, P. Karagianakos and H. Giamarellou, *J. Antimicrob. Chemother.* **46**(2), 311–314 (2000).
264. J. Jiang, Y. Li, T. Fang, J. Zhou, X. Li, Y. Wang and J. Dong. *Inflamm. Res.* **61**, 207–215 (2012).
265. W. T. Jia, X. Zhang, S. H. Luo, X. Liu, W. H. Huang, M. N. Rahaman, D. E. Day, D. Q. Zhang, Z. P. Xie and J. Q. Wang, *Acta Biomater.* **6**, 812–819 (2010).

G. Taubes, *Science* **321**, 356–361 (2008).

L.M. Pasquini, S. M. Hashmi, T.J. Sommer, M. Elimelech, J. B. Zimmerman, *Environ. Sci. Technol.* 2012, **46**, 6297-6305.

K. Xing, X. G. Chen, C.S. Liu, D. S. Cha and H. J. Park, *Int. J. Food Microbiol.* **132**, 127–133 (2009).

N. A. Stasko, T. H. Fischer and M. H. Schoenfisch, *Biomacromolecules* **9**, 834 (2008).

E .M. Hetrick, J. H. Shin, N. A. Stasko, C. B. Johnson, D. A. Wespe, E. Holmuhamedov and M. H. Schoenfisch, *ACS Nano* **2**(2) 235–246 (2008).

M. Magallanes, J. Dijkstra and J. Fierer, *Antimicrob. Agents Chemother.* **37**(11), 2293–2297 (1993).

Chapter 13

Nanoparticles as Adjuvants for Vaccination

Juan M. Irache[*]*, Ana I. Camacho, Carlos Gamazo*

Immunoadjuvant Unit, University of Navarra,
31080 — Pamplona (Spain)
[*]*jmirache@unav.es*

Abstract

In the following chapter, we will present the main classes of adjuvants for vaccination and immunotherapy. More particularly, the chapter focuses on the capabilities and potential of polymeric nanoparticles and the influence of their physicochemical and technological aspects on their effects as antigen delivery systems. Enhancement of adjuvant activity through the use of nanoparticulate delivery systems is particularly exciting. These carriers offer many advantages to antigens, including the possibility to deliver relatively large quantities of particle-associated antigen inside the antigen presenting cells (APCs) and, therefore, to prolong antigen release leading to extended antigen presentation compared with soluble antigen. Polymeric nanoparticles also offer the possibility to co-encapsulate the antigen with immunostimulant compounds and to enhance their interaction with appropriate immune cells through manipulation of their surface properties.

1. Introduction

Vaccines are one of the most effective preventative health tools available. They may be effective against infectious diseases, allergy, cancer, autoimmune diseases or drug dependencies.[1–3] From a general point of view, the goal of vaccination is to induce the appropriate immune response to the administered antigens with a long-term memory.

Many of the most effective vaccines designed to prevent infectious diseases are live attenuated variants of the target pathogen. They elicit strong cellular and antibody responses and often confer lifelong immunity with only one or two doses. However, they present some important drawbacks, including the possibility for this attenuated pathogen to revert to a virulent form and cause the disease. In addition, these vaccines may induce symptoms in immunocompromised individuals. Another limitation is that live attenuated vaccines usually need to be refrigerated to maintain their potency and efficacy. This fact represents an important practical problem for their shipment and distribution in developing countries or after natural disasters and emergency situations.

In the last decades, the research in this field has evolved toward non-living approaches, including subunit, toxoid, conjugate or DNA vaccines. All of them are based on the use of well-characterized and highly purified antigens, focusing on the protective aspects avoiding side effects. On the other side, the use of highly refined antigens, with often poorly immunogenic properties, require the incorporation of additional components to induce strong and prolonged protective immunity. These extra components, termed adjuvants (from "***adjuvare***," to help), provide the necessary "help" to enhance and even modulate the immunogenicity of the vaccine antigens.

What does it mean "modulate" the "immunogenicity"? Immunogenicity means, literally, generation of an immune response. Which one? There are many variables behind the term immunogenicity that need to be clarified when we intend to administer a vaccine. The classical paradigm established interplay between Th1/Th2 subsets; however, new subsets have been identified. Currently, we can assume the following arsenal of T-lymphocytes: (i) the Th1, Th2 and Th17 subsets of helper T cells; (ii) regulatory T cells (Tregs), divided into natural (nTreg) and inducible Tregs (iTregs) and (iii) cytotoxic T lymphocytes. It has been shown that the presence of signature cytokines during T-cell activation directs the resulting response.[4–6] Thus, Th1 cells are inflammatory cells that release interferon-gamma (IFN-γ) as a key signature and are mainly involved in inflammation and protective immunity against intracellular pathogens. Notably, a new Th1 subset is able to release both, pro- and anti-inflammatory cytokines, IFN-γ and IL-10 (interleukin-10). Th2 cells are primarily involved in B-cell help as they release B-cell growth factors like IL-4. Th17 cells may exacerbate inflammation and play a critical role in host defense against a variety of bacteria and fungi. These cells are characterized as preferential producers of IL-17A, IL-17F, IL-21 and IL-22.

On the other hand, there are several subsets of T-regulatory cells that suppress the function of effector T cells and are then prominent in both the maintenance of mucosal immunity and the control of inflammatory diseases.

Throughout this chapter, we will deal with the importance of the right activation of dendritic cells (DCs), as pivotal antigen presenting cells (APCs), to lead the appropriate immunogenicity. DCs regulate the activation and differentiation of these T-cell subsets through the pattern of cytokines secreted. For example, IL-12 (p70) promote Th1 cells, IL-10 favors Th2 or Treg responses and stimuli that induce DC to release TGF-β, IL-6 and IL-23 promote Th17 differentiation. Intestinal DCs are environmentally conditioned to induce Treg differentiation.

Additionally, any vaccine discussion must not forget the mucosa. Considering most human pathogens enter the body through a mucosal surface, the mucosal immune system is the largest and most complex part of our immune system. Through innate and acquired immunity, the mucosal immune system contains inductive and effector components along the vast epithelial surface area, ranging from the oral and nasal cavities to the respiratory, intestinal or genitourinary tracts. For instance, GALT (gut-associated lymphoid tissue) contains more than 10^6 lymphocytes/g tissue, the largest pool of immunocompetent cells in the body — B cells, plasma cells, subsets of CD4 T cells and even CD8 T cells. The relative concentration of CD4 T subsets in the mucosal-associated lymphoid tissue (MALT) influences the outcome of the response. Moreover, the activation of mucosal immunity induces both mucosal and systemic immune responses. In contrast, parenteral immunization usually fails to stimulate mucosal lymphatic tissues to generate protective immunity at these sites. Therefore, mucosal routes may be used as the most convenient choice for the route of vaccine administration, and constitute a great window for the new antigen delivery systems.

With these variables and opportunities in focus, we can now review briefly the concept of "adjuvants" and their benefits in vaccination.

2. Adjuvants

From a clinical point of view, adjuvants may be used for different purposes, including the following: (i) to enhance the immunity of antigens, increasing the generation of higher specific antibodies titers and/or cellular immune responses; (ii) to reduce the amount of antigens and so allow immunizations with low doses;[7–10] (iii) to reduce the number of administrations minimizing compliance issues and logistic challenges;[11] (iv) to improve the efficacy of vaccines in populations with reduced responsiveness because of age (both infants and elderly), pharmacological treatments, disease-induced immunodeficiencies or therapeutic interventions[12,13] and (v) to permit the delivery and uptake of antigens by the skin or mucosa.[14,15]

Adjuvants, under an immunological point of view, are compounds that increase the magnitude and quality of an adaptive response. The former quality is becoming increasingly important since, as it was mentioned above, it is critical to guide the type of adaptive response to produce the most effective forms of immunity in order to get protection.[16,17]

For instance, some adjuvants may be capable to induce in immune cells the more adequate response (e.g. T helper 1 [Th1] cell versus Th2 cell, CD8 versus CD4 T cells)[11,18] or to increase the generation of memory, especially T-cell memory.[19–21] In other cases, adjuvants may also increase the speed of the initial response, which may be critical in a pandemic outbreak of infection[19,22,23] and to alter the breadth, specificity or affinity of the elicited response.[23,24]

Ideally, adjuvants should be stable with long shelf life, biodegradable and cheap to produce.[25,26] In addition, there are marked differences in the efficacy of adjuvants depending on the administration route, e.g. between mucosal and parenteral routes. This lack of efficacy is particularly challenging when a mucosal route of vaccination is required. For example, mucosal immunization regimes are often compromised by antigen degradation and/or immune-tolerance phenomena.

In order to classify the adjuvants, we can attend to their "*modus operandi.*" Adjuvants may act in several ways. For example, adjuvants may function as delivery systems, modifying the antigens in such a way that prolong their residency or generating a particle more likely to be recognized and captured by the host immune system.[27–29] In other words, the adjuvant may allow the progressive release of the antigens, delay their clearance and improve their exposure to the immune system. As a result, these adjuvants increase the provision of antigen-loaded APC's for cognate naive T cells, promoting upregulation of co-stimulatory signals or major histocompatibility complex (MHC) expression, inducing cytokine release and thus enhancement of the magnitude and duration of the immune response.[30–32]

Another group of adjuvants includes those components displaying an innate ability to directly activate immune cells during antigen recognition and presentation.[33,34] These immunoenhancers or immunostimulants include pathogen invasion signals, natural compounds extracted from plants (i.e. saponins) and endogenous immunoactive compounds such as cytokines, chemokines and co-stimulatory molecules.

In spite of, on the last decades, a number of different compounds have been proposed as immunostimulants, the list of approved adjuvants for human use is still limited (Table 1). Nevertheless, there is a demand for safe adjuvants capable to induce the appropriate and adequate immune response.

Table 1. Examples of adjuvant types approved for human use.

Adjuvant	Description	Vaccine (Target)	Reference
Alum	• Aluminum phosphate or aluminum hydroxide	• Multiple vaccines • Worldwide	35,36
MPL	• 3-*O*-desacyl-4'-monophosphoryl lipid A • Detoxified derivative of the LPS isolated from the Gram-negative bacterium *Salmonella minnesota* R595 strain • TLR 4 agonist	• Supervax® (hepatitis B)	37
AS04	• Mixture of alum (1 mg/mL) and MPL (0.1 mg/mL)	• Cervarix® (human papillomavirus) • Fendrix® (hepatitis B)	38
Montanide ISA 51	• w/o emulsion • Mixture of mineral oil and mannide-monooleate	• CimaVax® EGF (non-small cell lung cancer)	39
MF59	• o/w emulsion • Squalene (2.5%), Tween 80 (0.25%) and Span 85 (0.25%) in 10 mM citrate (pH 6) • Mean size of approximately 165 nm	• Fluad® (influenza) • Focetria® (influenza)	40,41
AS03	• o/w emulsion • Squalene (2.5%), α-tocopherol and Tween® 80 (0.9%) in PBS (pH 6.8) • Mean size of approximately 150–155 nm	• Arepanrix® (influenza) • Pandemrix® (influenza)	42,43
AF03	• o/w emulsion • Squalene (2.5%), Montane 80, Eumulgin B1 PH	• Humenza® (influenza)	44
Virosomes	• Unilamellar phospholipid vesicles incorporating functional viral envelope glycoproteins (influenza virus hemagglutinin and neuraminidase) • Mean size of approximately 150 nm	• Inflexal® (influenza) • Epaxal® (hepatitis A) • Invivac® (influenza)	45

This demand is crucially needed to stimulate strong immune responses through mucosal administration.

2.1. *Aluminum salts (Alum)*

Aluminum-based compounds (principally, aluminum hydroxide or phosphate) have been used extensively as adjuvants. Nowadays, alum continue to monopolize human and veterinary vaccines in spite of comparative studies show that it is a weak adjuvant.[46] In fact, alum salts are relatively poor adjuvants in many situations, particularly to induce cellular immune responses.[47] Alum stimulate primarily Th2-type responses rather than Th1 or Th17 responses, which are necessary for protection against intracellular or extracellular pathogens, respectively.[48,49]

Aluminum salts form a depot at the injection site,[35] from which the antigen is slowly released.[50] In addition, the antigens absorbed on the aluminum salts would be exposed in a particulate multivalent form, making them more efficiently internalized by APCs.[51] More recently, it has been demonstrated that aluminum salts also activate APCs in a TLR-independent signaling, but dependent of the "inflammasome" complex, leading to the processing and release of pro-inflammatory cytokines such as IL-1β and IL-18.[36,52]

Apart from the inability of aluminum salts to induce Th1 or Th17 responses, the use of this adjuvant is restricted to the subcutaneous and intramuscular routes of administration.[53] In addition, some studies have reported local reactions (mainly granulomas) as well as IgE production, allergenicity and potential neurotoxicity after administration of vaccines using alum.[53,54] In addition, because alum-adjuvanted vaccines have to be kept in liquid form (freezing or lyophilization destroys the colloidal structure), they suffer from limited storage stability.[46]

2.2. *Microorganism-derived adjuvants*

The named "microorganism-derived adjuvants" are constituted by microbial components, known as "pathogen-associated molecular patterns" (PAMPs), which are molecular compounds shared by multiple invading pathogens that directly stimulate immune cells through interaction with receptors, including toll-like receptors (TLRs).[55] To date, 12 members of the TLR family have been identified, of which TLR1, TLR2, TLR4, TLR5 and TLR6 are located on the cell surface whereas TLR3, TLR7, TLR8 and TLR9 are localized in endosomal/lysosomal compartments of DCs.[30,56] Activation of these receptors plays a key part in DCs maturation and cytokine production. Among others, this group of immunopotentiators includes the lipopolysaccharide (LPS), murein, CpG motifs, flagellin or the cholera toxin B subunit.

The LPS, derived from the outer membrane of Gram-negative bacteria, is a major immunostimulant adjuvant. The principal structural element of LPS responsible for their adjuvant effect is lipid A. In low acid conditions, lipid A can be hydrolyzed to obtain 3-*O*-desacyl-4'-monophosphoryl lipid A (MPL), a compound which retains the adjuvant activity of lipid A with reduced toxicity.[57] MPL is a TLR4 agonist and acts on innate immune cells according to the same mechanism of LPS. It has been shown to stimulate the expression of co-stimulatory molecules and cytokine release, thereby improving qualitatively and quantitatively both antibody and cellular immune responses depending on the antigen considered.[58]

Another group of TLR agonists are those constituted by sequences of bacterial DNA enriched in CpG domains, that means, unmethylated ones.[59,60] These CpG motifs directly cause activation of immune cells (such as macrophages and DCs) through TLR9 yielding in upregulation of co-stimulatory molecules and pro-inflammatory cytokines (TNF-α, IL-1, IL-6 and IFN-γ) capable to promote a Th1-polarized immune response.[60]

In a similar way, polyriboinosinic–polyribocytidylic acid (poly I:C), a double-stranded polyinosinic:polycytidylic acid, is a synthetic analog of viral dsRNA that interacts with TLR3,[61,62] leading to increased production of type-I interferons.[63] When stabilized with poly-l-lysine and carboxymethylcellulose (poly ICLC), it shows higher stability and a greater activity.[61,64]

Other immunopotentiators include the synthetic low-molecular weight imidazoquinolines — imiquimod and resiquimod (R-848)[65] — and certain guanine nucleoside analogs[66] which bind to TLR7 and TLR8. These adjuvants induce cytokine production and upregulate co-stimulatory molecules and MHC I/II in DCs.[66] Imidazoquinolines have been demonstrated to augment T-cell responses against viral infections, such as HIV infections, and have increased antiviral protective immunity in mice.[67] In the particular case of resiquimod, it has been shown to be a potent topical adjuvant that significantly enhances cutaneous immune responses.[68]

2.3. *Emulsions and nanoemulsions*

Water-in-oil (w/o) emulsions have a long history in vaccine formulation. In the first half of the 20th century, Jules T. Freund described a mineral oil mixed with an aqueous suspension of killed *Mycobacterium*, which is now known as Freund's complete adjuvant (FCA).[69] The same w/o emulsion without mycobacteria is known as Freund's incomplete adjuvant (FIA). FIA is still applied in veterinary medicine but has been prohibited for use in humans because of its high reactogenicity.

More recently, in response to this safety concern, other w/o emulsions have been developed. Among the ones undergoing clinical trial assessment are Montanide w/o emulsions formulated as a mixture of a mineral oil with a mannide monooleate as surfactant.[70] Montanides have been proposed for the development of vaccines against prostate cancer,[71] ovarian cancer[72] and non-small-cell lung cancer.[73]

In addition to w/o emulsions, oil-in-water (o/w) emulsions have also been developed. Most of them are squalene-based adjuvants associated with different hydrophilic surfactants in order to form a fine emulsion with typically mean droplet sizes lower than 200 nm (i.e. MF59, AS03 and AF03). Overall, o/w emulsions appear to be more stable, less viscous and easier to administer with less resulting granulomas than the w/o emulsions.[74,75]

The mode of action of the emulsions involves innate inflammatory responses, APCs recruitment and activation, enhancement of antigen persistence at the injection site and presentation to immune-competent cells, in addition to elicitation of different patterns of cytokines.[76] However, the precise mechanism of action is still unclear and there are clear differences between both types of emulsions. Thus, FIA induces a predominantly Th2-biased response with some Th1 cellular response whereas, for instance, MF59 is a potent stimulator of both cellular (Th1) and antibody (Th2) immune responses.[77,78]

2.4. *Particulate adjuvants*

Particulate adjuvants act as effective antigen delivery systems, enhancing and/or facilitating the uptake of antigens by APCs, such as DCs or macrophages.[79] APCs have evolved to engulf microorganisms, and thus, it is not surprising that particulate adjuvants with sizes in the range of pathogens are efficiently captured and processed by these cells. Furthermore, particle-based antigen vehicles may serve as a depot for the controlled release of the loaded antigens, thereby increasing the availability of antigens to the immune cells. In this way, it has been reported that extended antigen release may enhance not only the level but also the quality of the elicited immune responses.[80,81]

In this heterogeneous group of adjuvants, some particulate-delivery systems are also capable to protect the integrity of antigens against degradation until delivered to the immune cells.[82,83] This is particularly important in mucosal (oral) vaccine formulations where antigens must be protected from the harsh acidic conditions of the stomach and enzymatic degradation in the gastrointestinal tract.[84,85] Some of them may be easily combined with the previously described immunopotentiators (i.e. microbial-derived adjuvants) in order to promote the most adequate immune response.[86,87] Furthermore, these particles can generate CD8 T-cell responses.[88] Last but not least, vaccines based on particulate adjuvants can be, in general, presented in the form of a dried powder, facilitating their storage, shipping and preservation.

From the different particulate platforms, two main groups can be distinguished: the (i) lipid- and (ii) the polymer-based particulates. In the former, numerous lipid-based technologies are currently being investigated for their potential as vaccine delivery systems. These include liposomes,[89–96] virosomes,[43] immune-stimulating complexes (ISCOMs),[97–99] cochleates,[100] non-ionic surfactant vesicles (NISVs)[101] and NISV incorporating bile salts (bilosomes).[102] In the latter, the group of polymer-based particulates includes dendrimers,[103] polymeric micelles[104–106] and polymeric nanoparticles.

3. Polymeric Nanoparticles

Polymeric nanoparticles are versatile devices for vaccine design due to the availability of a number of materials (i.e. polymers, macromolecules, lipids) and methods for nanoparticle synthesis. Common polymer compositions of nanoparticles include biodegradable or bioeliminable synthetic polymers [e.g. poly(esters), poly(anhydrides), poly(amino acids), poly(ethylene glycol)s (PEGs)] and natural polymers (chitosan, alginate, albumin, hyaluronic acid), copolymers and polymer blends.[107,108] On the other hand, the formulation process and the surface properties of the resulting nanoparticles play important roles on their efficacy as adjuvants for vaccination.[109,110] These factors are discussed in the next sections.

3.1. *Factors influencing the efficacy of nanoparticles as adjuvants*

Nanoparticle formulation parameters like composition, particle size, surface properties, load of antigen and the presence of co-stimulatory compounds may limit the route of immunization and are decisive for the presentation of antigen to immune cells as well as for controlling the overall elicited immune response.[79,111] Some of these parameters influence the distribution and behavior of nanoparticles *in vivo* and their uptake by the immunocompetent cells. For instance, the size and the surface properties of nanoparticles (including the presence of targeting ligands on their surfaces) would be major factors that determined the interaction and their capture by DCs, macrophages or M-cells.

3.1.1. The nature of the polymer: hydrophobicity

One of the main important factors influencing the behavior and the fate of particulates *in vivo* are the material used for its production. The choice of the polymer is guided by a number of factors such as biocompatibility, degradation rate, hydrophobicity/hydrophilicity, capability for co-incorporation of immunostimulatory agents, possibility to decorate the resulting particulates and route of administration. As a result, the resulting particulates will encompass the following properties: antigen loading, antigen release kinetics, stability and interaction with APCs.[112]

Particularly, hydrophobicity is thought to be a damage-associated molecular pattern that causes innate immune activation.[113] Particulate adjuvants with hydrophobic surfaces would facilitate the binding of IgM antibodies and other proteins, leading to the recruitment and activation of the complement component 1q (C1q) and subsequent activation of the classical pathway of the complement cascade. This leads to a better interaction with APCs.[113,114] In addition, artificial hydrophilic surfaces, such as those of pluronic-coated nanoparticles, activate the alternative, rather than the classical, pathway of the complement cascade.[79] This fact is in line with the observations of Moyano and coworkers who demonstrated a linear increase in immune activity with the increase of hydrophobicity of engineered gold nanoparticles.[115]

The hydrophobicity/hydrophilicity ratio as well as the charge of nanoparticulate surface can alter the microstructure of mucus and determine their behavior in the mucosa and, thus, their efficacy as oral vaccine delivery system. In general, hydrophobic surfaces facilitate the interaction of particulates with the mucus layer, minimizing the possibility of these carriers to arrive and interact with the APCs. On the contrary, a hydrophilic surface facilitates the transport of particulates through the mucus layer and the possibility to reach the surface of the epithelium and the interaction with immunocompetent cells.[116–118] In this context, one possible approach to render more hydrophilic the surface of nanoparticles is the conjugation of polymers and/or coating of nanoparticles with "hydrophilic" compounds such as PEG or dextrans.

3.1.2. Polymers as PRR agonists

In some cases, the polymer used to prepare the nanoparticles is capable to act as immune-stimulating agent "*per se*." Some of these materials show a certain ability to trigger innate immune receptors such as TLRs or nucleotide-binding domain leucine-rich repeat containing receptors (NLRs), among others. These receptors, known collectively as pattern recognition receptors (PRRs) are specific for conserved moieties associated with cellular damage or invading organisms, known as PAMPs (see Sec.2.2).[119,120]

This is the case for alginates, which have been reported to elicit cytokine production by monocytes through TLR2 and TLR4 activation.[121] Another interesting material is the copolymer of methyl vinyl ether and maleic anhydride (Gantrez® AN) which, in the form of nanoparticles, acts as agonist of various TLRs (TLR2 and TLR4), triggering a Th1-profile cytokine release

(IFN-γ and IL-12).[14,122] Another material widely employed as antigen-delivery system is chitosan. Chitosan would potently activate the NLRP3 inflammasome in DCs, whereas chitin does not.[123,124] In a similar way, nanoparticles from the copolymer of glutamic acid and l-phenylalanine (γ-PGE–Phe) would induce innate immune responses in DCs through a TLR4-dependent manner.[125] However, this effect would be restricted to copolymers displaying a high molecular weight (around 2,000 kDa).[126] Certain lipid constituents (i.e. glycolipids, lipopeptides or glycosyl-phosphatidylinositol) have also been described as potent TLR4 agonists.[127,128] Other TLRs–polymer agonists have been recently described, including a carbonyl-modified polyethylene[129] and lipomanann.[130]

3.1.3. Methods for antigen accommodation in the nanoparticles

The method used to incorporate the antigens in the particle delivery system may have a great influence not only in the encapsulation rate but also in the topographical disposition of the carried antigens and, hence, on the elicited immune responses. Several methods have been used to load antigens of interest into nanoparticles. These include encapsulation, surface chemical conjugation and surface physical adsorption. Entrapment or encapsulation of the antigens into the particulates could protect antigens from degradation and, more important, prolong the presentation of the antigen through a sustained release phenomenon. However, the release of antigens from nanoparticles is directly determined by their degradability. This aspect was put in evidence by Broaders and coworkers[131] by comparing ovalbumin (OVA)-loaded acetylated-dextran particulates with similar-sized nanoparticles consisting of poly(acrylamide), poly(lactic acid) (PLGA) and iron oxide. Two particles, acetylated-dextran and poly(acrylamide), with similar dissolution rates of 1.6 and 2.0 h, performed an order of magnitude better than PLGA and iron oxide nanoparticles with respect to MHC I presentation via bone-marrow-derived DCs.[131]

The physical adsorption of antigens onto particulates is simple and, in general, rapid. However, surface chemical binding presents several drawbacks: (i) it is likely to provide less protection to the antigens; (ii) the extra step of conjugation may prolong the preparative process of the antigen-delivery system; (iii) chemical reactions may hide critical epitopes; (iv) upon dilution and/or during the biodistribution of particulates in the body, some adsorbed antigens may be released. This last fact limits its use for oral administration.[132,133]

3.1.4. Correlation of particle size and immune stimulation

The size of antigens is an important factor for their efficient uptake by APCs and their arrival to lymphoid organs. Regarding the uptake of particulates, in general, it is assumed that nanoparticles of 20–200 nm are taken up via receptor-mediated endocytosis whereas larger particles are taken up via phagocytosis or micropinocytosis.[134] On the other hand, small particulates (20–200 nm) efficiently enter the lymphatic vessels and, thus, may rapidly reach the lymphoid organs (within hours after administration).[135] By contrast, particles larger than 200–500 nm do not enter directly in the lymph capillaries and need to be carried out into the lymphatic system by specialized cells such as DCs.[136,137] In this last case, the transport of these large particulates takes approximately 24 h to arrive in lymph nodes.[136] By a mucosal route of administration (i.e. oral or nasal), nanoparticles can be efficiently taken up by M-cells of the MALT. Jepson and coworkers[138] quantified this phenomenon in rabbits and calculated that about 10^5 particles of about 500 nm were taken up per each lymphoid follicle dome of the intestinal Peyer patch in 45 min.

More controversy exists in the effect of the size of particulates on the modulation of the immune response. Particulates of 20–200 nm are reported to trigger a CD4, CD8, Th1-type cellular immune response. In contrast, particles higher than 500 nm are associated with stimulation of an antibody immune response.[134] Recent studies with antigen-conjugated poly(styrene) particulates have confirmed that discrete sizes induce a cytotoxic T-cell response and that the optimal size for effective immune response would be in the viral range size (around 50 nm).[139,140] Similar results were found with poly(lactic acid) (PLA) particles. Immunization with PLA nanoparticles (200–600 nm) was associated with higher levels of IFN-γ production related to the Th1-type immune response. In contrast, immunization with PLA microparticles (2–8 μm) would promote a Th2-type immune response.[141]

Nevertheless, other reports are against that correlation. Thus, in an interesting work using bovine serum albumin (BSA) as a model antigen entrapped into poly(lactic-*co*-glycolic acid) (PLGA) particulates of different sizes (200, 500 and 1000 nm). Gutierro and collaborators[142] reported that the 1,000 nm-sized BSA-loaded PLGA particles generally elicited a higher serum IgG response than that obtained with the vaccination of 500 or 200 nm-sized particles, using parenteral or mucosal routes for the administration. In the same line, Cohen and coworkers[143] found that antigen-loaded poly(acrylamide) hydrogel particles of 3.5 μm or 35 nm direct a similar specific cytotoxic T-cell response.[143] Overall, it is difficult to predict the type of the immune response based just on the particle size. Other properties, such as surface or polymer nature, may have great influence in the outcome.[111] Table 2 summarizes some

Table 2. Influence of different physicochemical properties of particulates (polymer, size, method of antigen loading) on the elicited immune responses.

Polymer	Antigen	Particle size	Route of administration	Comments	Reference
Poly(styrene)	OVA/covalently bound to the surface	Between 20 and 200 nm	*i.d.*	• The strongest immune surface was found for 40 nm particulates.	139
Poly(styrene)	OVA/covalently bound to the surface	Between 20 and 2,000 nm	*i.d.*	• Small influence of size on the production of IgG1. • Small particulates (40–50 nm) induced higher IFN-γ levels. • Particulates 90–120 nm induced high levels of IL-4.	140
PLA	HBsAg/encapsulated	200–600/ 2,000–8,000 nm	*i.m.*	• Nanoparticles (200–600 nm) elicited lower antibody titers than microparticles (2–8 μm). • Nanoparticles (200–600 nm) induced higher levels of IFN-γ favoring Th1-type immune response. • Microparticles promoted IL-4 secretion, favored Th2-type immune response.	141
PLGA	BSA/encapsulated	200/500/1,000 nm	*s.c.*, oral, *i.n.*	• Particulates of 200 nm and 500 nm elicited similar immune responses. • Microparticulate elicited higher immune responses than nanoparticles.	142

(Continued)

Table 2. (*Continued*)

Polymer	Antigen	Particle size	Route of administration	Comments	Reference
Sulfo-butylated PVA-*g*-PLGA	Tetanus toxoid/adsorbed	100/500/1,000 nm	Oral, *i.n.*	• Antibody titers induced with small particulates were higher than with microparticles.	144
PLGA	Glycoprotein env. from HIV-1/adsorbed	110/800–900 nm	*i.n.* + *i.m.* boost	• Both types of particulates led to comparable immune responses in mice.	145
PLGA	NenB from *Neisseria meningitidis*/adsorbed	110/800–900 nm	*i.m.*, *i.p.*	• Both types of particulates led to comparable immune responses in mice.	145
Chitosan	OVA/adsorbed	Between 400 and 3,000 nm	*i.n.*, *i.p.*	• By the *i.n.* route, particles of 400 and 1,000 nm induced higher IgA levels than larger particles. • IgG levels were comparable when particulates were *i.n.* or *i.p.* administered.	146

i.d., intradermal; *i.m.*, intramuscular; *s.c.*, subcutaneous; *i.n.*, intranasal; *i.p.*, intraperitoneal.

reports about the influence of size and other physicochemical properties of particulates on the induction of immune responses.

3.1.5. Co-delivery of antigens and immunostimulants

The inherent adjuvanticity of particulates can be further enhanced or modulated by combination with immune response modulators (i.e. immunostimulants). Probably, the possibility of polymeric nanoparticles to deliver antigens and immunoenhancers in a concomitantly way to the same single APC would be one of the greatest benefits of these delivery systems.[147] Furthermore, the adequate selection of the immunostimulant for co-delivery would determine the preferential induction of a Th (T-helper lymphocytes) subset. In addition, this association would minimize the potential toxic effects of the immunostimulant, because of restricted systemic distribution and lower maximal available concentrations.[31,148,149]

The TLR9 ligand CpG has been tested in association with different types of particulate vaccine formulations increasing the immunogenicity of the encapsulated antigens.[150–154] PLGA nanoparticles loaded with *Leishmania* antigens and CpG sequences induced in mice showed a more pronounced and stronger Th1 immune response than formulations in which CpG motifs were physically mixed with either antigen-loaded nanoparticles or the antigen alone.[151] In another interesting work, *N*-trimethyl chitosan nanoparticles loaded with OVA and cross-linked with CpG sequences were prepared. After nasal administration, nanoparticles containing CpG induced a 10-fold higher IgG2a response (Th1) and enhanced the number of OVA-specific IFN-γ-producing T cells in the spleen than similar nanoparticles prepared in the absence of CpG.[152] In the same way, alginate-coated chitosan particles have also been used successfully through an intranasal route to co-deliver entrapped hepatitis B surface antigen and CpG sequences. These particulates produced Th1-biased antigen-specific circulating antibody.[153] Likewise, CpG co-encapsulated with tetanus toxoid in PLGA nanoparticles induced protective antibody titers characterized by strong T-cell responses, particularly of the Th1 type.[154]

Ligands for the targeting of TLR4 have also been widely employed. For example, MPL derived from the LPS of *Salmonella enterica* serovar Minesota showed an acceptable safety profile and great capacity to promote Th1 responses.[155,156] Sarti and coworkers[157] co-encapsulated the model antigen OVA and MPLA into PLGA nanoparticles. These nanoparticles, orally administered in mice, induced a stronger IgG immune response and significantly higher IgA titers than that induced by the nanoparticles in the absence of the LPS derivative.[157] In a similar way, PLGA nanoparticles containing MPL induced a significantly higher activation of DCs (measured as expression of MCH class II and CD86 molecules) than the same nanoparticles but in the absence of the immunostimulant.[158] In another interesting work, Bal and coworkers[159] evaluated the possibility to improve the immune response induced by OVA-loaded *N*-trimethyl chitosan nanoparticles by the co-encapsulation of LPS from *Escherichia coli*. After nasal vaccination, nanoparticles containing LPS elicited higher IgG, IgG1 and sIgA levels than control OVA nanoparticles. Similarly, by the intradermal route, LPS containing nanoparticles elicited higher IgG titers than control nanoparticles.[159]

Flagellin, a TLR5 agonist, is a potent immunostimulant. Stano *et al.*[160] covalently bound flagellin to OVA-loaded poly(propylene) nanoparticles and, when given intranasally, these nanoparticles induced cytotoxic T-lymphocytic responses in lung and spleen tissues, as well as high antibody levels in mucosal airways.[160] In a similar way, one single subcutaneous or oral dose of OVA loaded into poly(anhydride) nanoparticles coated with flagellin elicited higher IgG1 (Th2), IgG2a (Th1) and sIgA (mucosal) levels of specific antibodies than those induced by conventional OVA nanoparticles or OVA associated to alum.[161]

Ligands for the intracellular TLR3 or TLR7 have also been tested in association with particulate vaccine formulation.[161–164] For example, Heuking and Borchard[162] showed that the incorporation of a TLR7 agonistic moiety (9-benzyl-8-hydroxyadenine covalently linked to a PEG spacer) into carboxymethyl-*N,N,N*-trimethylchitosan nanoparticles containing a pDNA increased significantly the IL-8-related immunity. The TLR3 ligand poly(inosinic acid)–poly(cytidyl acid), or poly(I:C), a synthetic dsRNA analog, also potentiates the adjuvant effect of nanoparticles.[163,164] Thus, the co-encapsulation of OVA and poly(I:C) in polysaccharide nanoparticles triggered upregulation of inflammatory cytokines and bactericidal mediators.[164]

3.1.6. Targeting

Oral immunization using conventional antigen-delivery particulates is, in general, not able to elicit a sufficient systemic immune response compared to parenteral routes, and consequently, high and multiple oral doses are required. This fact has been related with a low capacity of conventional antigen-loaded nanoparticles to target lymphoid tissues within the gut after oral administration. In addition, these particulates are rapidly eliminated from the intestinal mucosa due to the continuous mucus turnover and intestinal peristaltism. Thus, the "decoration" of nanoparticulate surface may be an adequate strategy to improve the capabilities of these antigen-delivery systems to specifically reach and interact with the immune system of the gastrointestinal tract (GALT).

For targeting purposes, different types of molecules have been tested, including lectins,[165–168] vitamin B_{12} derivatives[169,170] and microorganism-derived adhesive factors.[87,171] Particularly interesting is the idea to mimic the bacteria and virus behavior concerning their strategies to adhere and invade the host through a mucosa. For this purpose, microorganisms have been developed using a number of different approaches, including the use of adhesins (bacterial flagella or fimbriae, among others) and glycoconjugates (i.e. mannose proteins). Following this biomimetic approach, the coating of poly(anhydride) nanoparticles with flagellin from *Salmonella enteritidis* yielded particles with a distribution profile within the gut similar to that observed for the wild microorganism.[87] Using OVA as model antigen, these "Salmonella-like" nanoparticles were able to induce a much more strong mucosal IgA response than control particles.[85]

Another interesting approach would be the coating of particulates with glycoconjugates enriched in mannose residues, which have been found to promote their interaction with the mucosal tissue of the gastrointestinal tract.[172] Recently, mannose coating nanoparticles have been proposed for oral vaccination. Using OVA as antigen model, these nanoparticles elicited a higher and more balanced specific antibody response (IgG1 and IgG2a) compared with conventional nanoparticles which elicited a typical Th2 response. In addition, mannose-coated nanoparticles were able to elicit a significant intestinal secretory IgA (s-IgA) for at least six weeks.[85]

3.2. *The polymer choice: immunization with polymeric nanoparticles*

PLA and PLGA are, so far, the major synthetic polymers for the preparation of polymeric biodegradable nanoparticles as vaccine formulations, where antigens are encapsulated. The well-documented biocompatibility and safety of these materials, together with their biodegradability and controlled release capacity, has already led to their FDA approval for a number of clinical applications in humans, and also to a number of marketed products.[173–176] However, the use of these polymers can be limited by acid hydrolytic degradation products harmful to the immunogenicity of the entrapped antigens on storage. Also, the organic solvents commonly used to prepare particles can be detrimental for the antigens.

In any case, a number of microbial antigens (e.g. hepatitis B, *Salmonella enterica* serovar *equi*, tetanus toxoid, diphtheria toxoid, etc.) have been successfully encapsulated in PLA and PLGA nanoparticles.[144,177,178] For example, hepatitis B therapeutic vaccines were designed and formulated by loading the hepatitis B core antigen (HBcAg) into PLGA nanoparticles (300 nm) with or without monophospholipid A (MPL) adjuvant.[179] A single subcutaneous immunization with HBcAg and MPL in PLGA nanoparticles induced stronger cellular immune response than those induced by the antigen alone or mixed with MPLA in a murine model.[179] PLGA nanospheres have also been used to co-encapsulate inactivated *Leishmania major* and CpG motifs. Mice subcutaneously immunized with these nanoparticles showed the highest IgG2a/IgG1 ratio, IFN-γ production and lowest IL-4 production compared to control groups, being correlated with protection in a model of infection.[151]

For vaccination through a mucosal route, PLA and PLGA nanoparticles usually elicit weak immune responses. This fact would be related with their hydrophobicity and, thus, their inability to cross the mucus layer and interact with the epithelium. In order to solve this drawback, PEGylated derivatives have been employed. Thus, PEG–PLA–PEG copolymers were used to prepare nanoparticles loaded with hepatitis B surface antigen (HBsAg).[180] Oral administration of these nanoparticles to animals significantly increased mucosal (sIgA) and cellular immune Th1 responses with respect to control nanoparticles.[180] In another report, PEGylated PLA nanoparticles were also found to enhance the transport of tetanus toxoid across the nasal mucosa of sensible rats.[181] More recently, PLGA–PEG nanoparticles containing a recombinant HBsAg were administered to the lungs of guinea pigs eliciting protective IgG and IgA levels.[181]

An alternative approach to the use of PEG and its derivatives is the attachment of the polyester PLGA to a hydrophilic backbone based on poly(vinyl alcohol) (PVA).[183,184] In particular, branched copolymers based on sulfobutylated-PVA resulted to be promising for the nasal delivery of tetanus toxoid, resulting in enhanced IgG and IgA antibody production in mice.[144]

Regarding the eventual damage of the antigens when loaded in these poly(ester) nanoparticles due to the acidic environment generated, a possible solution would be the replacement of the polyester nanoparticle matrix by an intimate blend composition incorporating polyoxyethylene derivatives such as poloxamers (Pluronic®) or poloxamines (Tetronic®) into the carriers.[185] These nanoparticles were found to transport pDNA across the nasal mucosa and to transfect the adequate cells, thus, resulting in significant systemic IgG antibody responses against the encoded protein. Moreover, the immunization studies showed that the DNA-loaded nanoparticles elicited a fast and strong response, significantly more pronounced than that corresponding to the naked pDNA for up to six weeks.[185]

Like PLGA or PLA, poly(ε-caprolactone) (PCL) is a biodegradable polyester that can be also used for the formulation of nanoparticles.[186] The degradation of PCL is slower than that of PLGA, making it more appropriate for long-term delivery systems.[187,188] Furthermore, PCL particles, unlike polylactides, do not generate an acidic environment that could negatively affect the antigenicity of the encapsulated antigens.[189] For example, PCL nanoparticles together with mucoadhesive polymers (alginate or chitosan) were used as a potential carrier for *Streptococcus equi* surface proteins. Serum IgG antibody levels of animals vaccinated with *S. equi* antigens encapsulated or adsorbed onto PCL particles were significantly higher than those caused by free antigens or even free antigens adjuvanted with cholera toxin B subunit.[190]

Another group of interesting polymers for vaccination purposes are the poly(anhydrides). In general, poly(anhydride) nanoparticles possess the ability to develop strong bioadhesive interactions with components of the mucosa, increasing the residence in close contact with the epithelium.[191–193] In addition, the surface of these poly(anhydride) nanoparticles can be easily decorated by simple incubation with excipients or ligands in order to modify their *in vivo* distribution.[87,194–196] Particularly, nanoparticles from the copolymer of methyl vinyl ether and maleic anhydride (Gantrez® AN) have demonstrated a very high capability to enhance both Th1 and Th2 markers (IgG2a and IgG1, respectively) after parenteral,[197] oral,[85] nasal[11] and ophthalmic[12] administration. For example, Vandamme and coworkers[198] evaluated the adjuvanticity of Gantrez® AN nanoparticles in an oral immunization experiment of pigs against F4+ enterotoxigenic *Escherichia coli*. Vaccination with this formulation improved protection against an experimental infection.[198] In another interesting work, these nanoparticles orally administered have been proved to protect a model of sensitized mice to OVA from anaphylactic shock.[199] More recently, poly(anhydride) nanoparticles loaded with allergenic proteins of rye-grass pollen and containing LPS as immunopotentiator were capable to decrease the mortality rate and the mMCP-1 plasma levels in sensitized mice compared with controls.[200]

From the group of polysaccharides, chitosan and their derivatives are the most popular macromolecules investigated for vaccine delivery. Chitosan is a biocompatible cationic polysaccharide consisting of *N*-acetylglucosamine and D-glucosamine units. It is characterized by mucoadhesiveness (i.e. due to ionic interactions with the negatively charged sialic acid groups in mucin) and the ability to open the tight junctions guarding the paracellular pathway, thus facilitating cellular permeability of biopharmaceutics.[111] In the last years, a number of derivatives have been proposed in order to improve its mucoadhesive and/or permeation-enhancing properties, including trimethylated chitosan, mono-*N*-carboxymethyl chitosan and thiolated chitosan derivatives (e.g. chitosan–cysteine conjugates or chitosan thioglycolic acid conjugates).[201]

Chitosan nanoparticles have shown potential for the delivery of a variety of antigens in different animal models, promoting satisfactory immune responses. Thus, significant systemic and mucosal immune responses after nasally administered chitosan nanoparticles have been demonstrated for tetanus toxoid, or antigens from influenza, *Bordetella pertussis* or *Corynebacterium diphteriae*.[202–204]

Furthermore, the high cationic charge of the chitosan backbone makes this polysaccharide as an ideal carrier for the mucosal delivery of polyanionic biomacromolecules (i.e. DNA). For example, plasmid DNA encoding respiratory syncytial virus (RSV) antigens was efficiently delivered and expressed following its intranasal administration in chitosan nanoparticles, resulting in efficient production of specific IgG and IgA antibodies, IFN-γ and also in the induction of cytotoxic T cells.[205] In a subsequent study, a significant reduction of the pulmonary virus load was also observed following an experimental RSV challenge.[206] Good results were also obtained against the major structural protein (VP1) of coxsackievirus (responsible for acute and chronic myocarditis) following the nasal administration of VP1 encoding plasmid DNA complexed with chitosan in comparison to naked plasmid.[207]

For oral vaccination studies, Roy and coworkers[208] have demonstrated the effectiveness of orally delivered chitosan–DNA nanoparticles in inducing protective immunity in the peanut allergy mouse model. More specifically, chitosan nanoparticles loaded with the gene for Ara h2 (main peanut allergen) allowed to obtain a significant reduction in the levels of serum IgE and plasma histamine, and the increase of specific mucosal IgA antibodies.[208] Other studies show a significant amelioration of OVA-induced food allergy symptoms when the TGF-β expressing DNA vector was orally administered after its encapsulation in chitosan nanoparticles.[209]

However, in spite of all the described advantageous properties, a major concern is the low solubility of chitosan at physiological pH and, therefore, the partial lost of its ability to enhance the permeability and absorption of biomacromolecules.[210] As a consequence, different derivatives have been proposed, such as the use of trimethylated chitosan nanoparticles.[211] Other interesting alternative is the use of chitosan together with other biodegradable polyelectrolyte polymers with opposite charge to obtain combined nanoparticles suitable for mucosal vaccinations. In this way, it has been demonstrated that alginate–chitosan nanoparticles is an effective system for subcutaneous and oral vaccination with the rHBsAg.[153,212]

Table 3 summarizes some studies related with the evaluation of the capacity of polymeric nanoparticles to induce immune and protective responses.

4. Conclusions

Subunit vaccines are proved to be the safest option for vaccination or immunotherapy but require additional components necessary to confer the adequate protective immunity. This led us to the full concept of vaccine formulation, enclosing not only antigens but also other chemical substances with adjuvant properties. Adjuvants development has been mainly based on empirical studies and, therefore, the mechanisms of action of adjuvants are still not completely known, precluding their extensive applicability. Furthermore, this erratic comprehension of their effects leads to several safety concerns that limit their use. The understanding of the immune mechanisms will support the rational design of adjuvants filling the still remaining gaps. However, achieving the appropriate effect depends not only on the type of adjuvants but also on the route of administration.[220,221] Upon this situation, novel strategies based on polymeric particulate delivery systems have merged with power since they put together immunological properties as well as technical and practical features. Polymeric particles not only act as inert transport of antigens but are also capable of interacting with the immune system, exhibiting real adjuvant properties. An additional advantage is that these particle delivery systems can increase retention of antigens on mucosal surface, protect them from proteases in the mucus gels and then, increase antigen uptake and immune responses; otherwise, relatively large doses of vaccine are required being impossible to determine exactly what dose actually crosses the mucosa.[28,27]

Furthermore, as it was reviewed in this chapter, antigens associated with particles mimic the particulate nature of pathogens. Indeed, dimensions of nanoparticulate vaccines are comparable to those of common pathogens, against which the immune system has evolved to react, and they are readily taken up by APCs. Thus, particles larger than 0.5 μm are internalized by APCs via phagocytosis, whereas uptake of soluble antigen or smaller particles is primarily mediated by endocytosis.[136,137,222] Internalization of particulate vaccines through phagocytosis into phagosomes has important consequences because phagosomes are known to be competent organelles for antigen cross-presentation, thereby allowing the induction of cytotoxic T-cell responses, a

Table 3. Non-exhaustive list of recent polymeric nanoparticles developments for vaccination.

Antigen	Carrier	Characteristics	Observations	Reference
Tetanus toxoid	Chitosan nanoparticles (CS-NP)	Size: 350 nm Antigen loading around 5%	• By the *i.n.* route, CS-NP elicited a high and long-lasting humoral immune response. • The IgA levels generated by CS-NP were higher than that obtained by controls.	203
Tetanus toxoid	Chitosan NP *N*-trimethyl chitosan NP (TMC-NP) mono-*N*-carboxymethyl chitosan NP (MCC-NP)	CS-NP: 400 nm TMC-NP: 300 nm MCC-NP: 90 nm	• By the *i.n.* route, CS-NP and TMC-NP resulted in higher levels of IgG than that obtained with MCC-NP.	213
HBsAG	Chitosan NP *N*-trimethyl chitosan NP (TMC-NP)	CS-NP: 220 nm TMC-NP: 195 nm	• Alum-based HBsAg-induced strong humoral but negligible mucosal immunity. • TMC-NP induced higher IG1 and IgG2a antibody titers than CS-NP.	211
HBsAG	Alginate-coated chitosan NP	210 nm	• By the *i.n.* route, NP elicited a potent IgA immune response. • The co-encapsulation of CpG motifs was necessary to yield Th1 responses.	153
HBsAG	PEG–PLA–PEG NP	220 nm	• NP produced better humoral (IgG) and sIgA responses than control (PLA-NP), and comparable to that produced by alum-vaccine. • The overall immune response was a Th1/Th2 mixed immune response.	214
HBsAG	PEG–PLA–PEG NP	110 nm	• Orally, NP exhibited effective levels of humoral immunity along with the mucosal (sIgA) and cellular immune response (Th1).	180

(Continued)

Table 3. (*Continued*)

Antigen	Carrier	Characteristics	Observations	Reference
Diphteria toxoid (DT)	PCL NP PLGA NP PLGA–PEC NP	270 nm Loading: 1.6%	• By the *i.n.* route, all NP induced a response higher than free DT. • PCL NP induced the highest serum IgG antibody titers. • By the *i.m.* route, again, PCL NP elicited the highest IgG antibody response.	215
Streptococcus equi antigen	Glycolchitosan-coated PLA NP	Size: 250–450 nm Loading: 5–11%	• By the *i.n.* route, balanced mucosal Th1/Th2 immune response was observed. • PLA NP coated with glycolchitosan offered better results than when coated with spermine, alginate or PVA.	216
OVA	Polypropylene sulfide NP	Flagellin co-encapsulated Size: 50 nm	• By the *i.n.* route, NP induced cytotoxic T-lymphocytic responses in lung and spleen, and humoral response in mucosal airways.	160
HIV-1-Tat	Eudragit L-coated PMMA NP	Size: 220 nm Loading up to 20%	• NP delivered and released efficiently antigen intracellularly. • NP elicited strong immune responses in mice.	217
pDNA encoding protein of *Schistosoma mansoni*	Chitosan NP	Size: 300 nm	• By the oral route, NP induced high levels of IL-10: this resulted in a significant reduction of liver pathology. • No complete protection in mice of infection challenge with *S. mansoni* worms.	218
Recombinant influenza hemagglutinin (rHA)	Poly-γ-glutamic acid and chitosan nanoparticles	Size: 900 nm	• Intranasal administration induced a high IgA response in lung and IgG response in serum but also an influenza virus-specific cell-mediated immune response. • NP-protected mice against challenge with a lethal dose influenza A H5N1 virus.	219

PMMA, poly(methylmethacrylate); *i.n.*, intranasal; *i.m.*, intramuscular.

feature hardly achievable when using soluble antigen.[223,224] Moreover, during transport to the lymph nodes, soluble antigens are susceptible to premature degradation by proteolytic enzymes. By contrast, association of antigens with carrier particles can protect against such degradation.

Other interesting features of particulate antigens include the possibility to deliver relatively large quantities of particle-associated antigen inside the APCs and, therefore, to prolong antigen release leading to extended antigen presentation compared with soluble antigen. To this respect, it still remains to be established whether continuous delivery of antigen involves a better induction of antibody responses.[27,225] Moreover, studies have described that particulate vaccines are able to induce long-term immune protection.[29] It is well known that protection against infectious diseases not only relies on the magnitude of the immune response, but more importantly on the quality and longevity of it.[226] The mechanism by which these nanoparticulated adjuvants may achieve long-term protection is unclear but several studies support its potential.[227,228]

Finally, of particular interest is to highlight the possibility of the co-delivery of either different antigens and different immunostimulatory components[27,159,229,230] known as "multivalent strategy" in order to achieve broad protection against different pathogens.[231–234]

Acknowledgments

The authors wish to thank the following agencies and foundations for their economic support: "Instituto de Salud Carlos III" (Project PI12/01358), "Ministerio Ciencia e Innovación" (PRI-PIBAR-2011–1377) and Foundations "Ramón Areces" and "Caja Navarra" in Spain.

References

1. A. Umar, B. K. Dunn and P. Greenwald, *Nat. Rev. Cancer* **12**, 835 (2012).
2. B. Linhart and R. Valenta, *Curr. Opin. Immunol.* **24**, 354 (2012).
3. G. J. Nabel, *N. Engl. J. Med.* **368**, 551 (2013).
4. D. Schenten and R. Medzhitov, *Adv. Immunol.* **109**, 87 (2011).
5. C. S. Ma, E. K. Deenick, M. Batten and S. G. Tangye, *J. Exp. Med.* **209**, 1241 (2012).
6. B. Jin, T. Sun, X. H. Yu, Y. X. Yang and A. E. Yeo, *Clin. Dev. Immunol.* **2012**, 836485 (2012).
7. A. Banzhoff, R. Gasparini, F. Laghi-Pasini, T. Staniscia, P. Durando, E. Montomoli, P. L. Capecchi, P. di Giovanni, L. Sticchi, L. C. Gentile, A. Hilbert, V. Brauer, S. Tilman and A. Podda, *PLoS ONE* **4**, e4384 (2009).
8. J. Boyle, D. Eastman, C. Millar, S. Camuglia, J. Cox, M. Pearse, J. Good and D. Drane, *Vaccine* **25**, 2541 (2007).

9. T. F. Schwarz, T. Horacek, M. Knuf, H. G. Damman, F. Roman, M. Drame, P. Gillard and W. Jilg, *Vaccine* **27**, 6284 (2009).
10. S. A. Halperin, S. Dobson, S. McNeil, J. M. Langley, B. Smith, R. McCall-Sani, D. Levitt, G. V. Nest, D. Gennevois and J. J. Eiden, *Vaccine* **24**, 20 (2006).
11. R. L. Coffman, A. Sher and R. A. Seder, *Immunity* **33**, 492 (2010).
12. J. Beran. *Expert Opin. Biol. Ther.* **8**, 235 (2008).
13. A. Podda, *Vaccine* **19**, 2673 (2001).
14. A. I. Camacho, J. de Souza, S. Sánchez-Gómez, M. Pardo-Ros, J. M. Irache and C. Gamazo, *Vaccine* **29**, 8222 (2011).
15. R. Da Costa Martins, C. Gamazo, M. Sánchez-Martínez, M. Barberán, I. Peñuelas and J. M. Irache, *J. Control. Release* **162**, 553 (2012).
16. R. T. Kenney and A. S. Cross, In *New Generation Vaccines*, M. M. Levine, G. Dougan, M. F. Good, M. A. Liu, G. J. Nabel, J. P. Nataro and R. Rappuoli (Eds.), New York: Informa Healthcare USA (2010), p. 250.
17. B. Pulendran, J. Powell and R. A. Flavell, In *New Generation Vaccines*, M. M. Levine, G. Dougan, M. F. Good, M. A. Liu, G. J. Nabel, J. P. Nataro and R. Rappuoli (Eds.), New York: Informa Healthcare USA (2010), p. 183.
18. M. Murillo, M. M. Goñi, J. M. Irache, M. A. Arangoa, J. M. Blasco and C. Gamazo, *J. Control. Release* **85**, 237 (2002).
19. G. Galli, K. Hancock, K. Hoschler, J. DeVos, M. Praus, M. Bardelli, C. Malzone, F. Castellino, C. Gentile, T. McNally, G. Del Giudice, A. Banzhoff, V. Brauer, E. Montomoli, M. Zambon, J. Katz, K. Nicholson and I. Stephenson, *Proc. Natl. Acad. Sci. USA* **106**, 7962 (2009).
20. I. Leroux-Roels, F. Roman, S. Forgus, C. Maes, F. De Boever, M. Drame, P. Gillard, R. van der Most, M. Van Mechelen, E. Hanon and G. Leroux-Roels, *Vaccine* **28**, 849 (2010).
21. P. Vandepapeliere, Y. Horsmans, P. Moris, M. Van Mechelen, M. Janssens, M. Koutsoukos, P. Van Belle, F. Clement, E. Hanon, M. Wettendorff, N. Garçon and G. Leroux-Roels, *Vaccine* **26**, 1375 (2008).
22. J. W. Huleatt, A. R. Jacobs, J. Tang, P. Desai, E. B. Kopp, Y. Huang, L. Song, V. Nakaar and T. J. Powell, *Vaccine* **25**, 763 (2007).
23. S. Khurana, W. Chearwae, F. Castellino, J. Manischewitz, L. R. King, A. Honorkiewicz, M. T. Rock, K. M. Edwards, G. Del Giudice, R. Rappuoli and H. Golding, *Sci. Transl. Med.* **2**, 15ra5 (2010).
24. L. Malherbe, L. Mark, N. Fazilleau, L. J. McHeyzer-Williams and M. G. McHeyzer-Williams, *Immunity* **28**, 698 (2008).
25. C. Palena, S. I. Abrams, J. Schlom and J. W. Hodge, *Adv. Cancer Res.* **95**, 115 (2006).
26. M. T. De Magistris, *Adv. Drug Deliv. Rev.* **58**, 52 (2006).
27. G. Kersten and H. Hirschberg, *Expert Rev. Vaccines* **3**, 453 (2004).
28. T. Storni, T. M. Kundig, G. Senti and P. Johansen, *Adv. Drug Del. Rev.* **57**, 333 (2005).
29. C. Ardavin, *Nat. Rev. Immunol.* **3**, 582 (2003).
30. S. De Koker, B. N. Lambrecht, M. A. Willart, Y. van Kooyk, J. Grooten, C. Vervaet, J. P. Remon and B. G. De Geest, *Chem. Soc. Rev.* **40**, 320 (2011).
31. M. L. De Temmerman, J. Rejman, J. Demeester, D. J. Irvine, B. Gander and S. C. De Smedt, *Drug Discov. Today* **16**, 569 (2011).
32. S. Jain, D. T. O'Hagan and M. Singh, *Expert Rev. Vaccines* **10**, 1731 (2011).

33. V. E. J. C. Schijns and W. G. J. Degen, *Clin. Pharmacol. Therap.* **82**, 750 (2007).
34. L. J. Peek, C. R. Middaugh and C. Berkland, *Adv. Drug Del. Rev.* **60**, 915 (2008).
35. V. E. Schijns and E. C. Lavelle, *Expert Rev. Vaccines* **10**, 539 (2011).
36. M. Kool, K. Fierens and B. N. Lambrecht, *J. Med. Microbiol.* **61**, 927 (2012).
37. F. Steinhagen, T. Kinjo, C. Bode and D. M. Klinman, *Vaccine* **29**, 3341 (2011).
38. N. Garçon, S. Morel, A. Didierlaurent, D. Descamps, M. Wettendorff and M. Van Mechelen, *BioDrugs* **25**, 217 (2011).
39. P. C. Rodríguez, G. Rodríguez, G. González and A. Lage, *MEDICC Rev.* **12**, 17 (2010).
40. D. T. O'Hagan, G. S. Ott, E. De Gregorio and A. Seubert, *Vaccine* **30**, 4341 (2012).
41. D. T. O'Hagan and M. Singh, In *Vaccine Adjuvants and Delivery Systems*, M. Singh (Ed.), Hoboken, New York: John Wiley & Sons (2007), p. 115.
42. C. B. Fox, *Molecules* **14**, 3286 (2009).
43. N. Garçon, D. W. Vaughn and A. M. Didierlaurent, *Exp. Rev. Vaccine* **11**, 349 (2012).
44. M. F. Klucker, F. Dalençon, P. Probeck and J. Haensler, *J. Pharm. Sci.* **101**, 4490 (2012).
45. C. Herzog, K. Hartmann, V. Künzi, O. Kürsteiner, R. Mischler, H. Lazar and R. Glück, *Vaccine* **27**, 4381 (2009).
46. E. De Gregorio, E. Tritto and R. Rappuoli, *Eur. J. Immunol.* **8**, 2068 (2008).
47. R. Schirmbeck, K. Melber, T. Mertens and J. Reimann, *J. Virol.* **68**, 1418 (1994).
48. M. J. McGeachy and S. J. McSorley, *J. Immunol.* **189**, 3285 (2012).
49. C. Biondo, G. Mancuso, C. Beninati, C. Iaria, O. Romeo, A. Cascio and G. Teti, *Eur. Rev. Med. Pharmacol. Sci.* **16**, 1506 (2012).
50. E. B. Lindblad, *Vaccine* **22**, 3658 (2004).
51. G. L. Morefield, A. Sokolovska, D. Jiang, H. Hogen Esch, J. P. Robinson and S. L. Hem, *Vaccine* **23**, 1588 (2005).
52. B. N. Lambrecht, M. Kool, M. A. Willart and H. Hammad, *Curr. Opin. Immunol.* **21**, 23 (2009).
53. S. Vecchi, S. Bufali, D. A. Skibinski, D. T. O'Hagan and M. Singh, *J. Pharm. Sci.* **101**, 17 (2012).
54. R. L. Hunter, *Vaccine* **20**, S7 (2002).
55. F. M. Audibert and L. D. Lise, *Immunol. Today* **14**, 281 (1993).
56. R. Klippstein and D. Pozo, *Nanomedicine* **6**, 523 (2010).
57. M. A. Tomai and A. G. Johnson, *J. Biol. Resp. Modifiers* **8**, 625 (1989).
58. A. Didierlaurent, I. S. More, L. Lockman, S. Giannini, M. Bisteau, H. Carlsen, A. Kielland, O. Vosters, N. Vanderheyde, F. Schiavetti, D. Larocque, M. van Mechelen and N. Garçon, *J. Immunol.* **183**, 6186 (2009).
59. G. J. Weiner, L. Hsin-Ming, J. E. Wooldridge, C. E. Dahle and A. M. Krieg, *Proc. Natl. Acad. Sci. USA* **94**, 10833 (1997).
60. D. Higgins, J. D. Marshall, P. Traquina, G. Van Nest and B. D. Livingston, *Expert Rev. Vaccines* **6**, 747 (2007).
61. H. Kato, O. Takeuchi, S. Sato, M. Yoneyama, M. Yamamoto, K. Matsui, S. Uematsu, A. Jung, T. Kawai, K. J. Ishii, O. Yamaguchi, K. Otsu, T. Tsujimura, C. S. Koh, C. Reis e Sousa, Y. Matsuura, T. Fujita, S. Akira, *Nature* **441**, 101 (2006).
62. S. Marshall-Clarke, J. Downes, I. Haga, A. G. Bowie, P. Borrow, J. L. Pennock, R. K. Grencis, P. Rothwell, *J. Biol. Chem.* **282**, 24759 (2007).

63. O. Schulz, S. Diebold, M. Chen, T. I. Näslund, M. A. Nolte, L. Alexopoulou, Y. T. Azuma, R. A. Flavell, P. Lilieström, C. Reis e Sousa, *Nature* **433**, 887 (2005).
64. C. H. Ewel, W. J. Urba, W. C. Kopp, J. W. Smith II, R. G. Steis, J. L. Rossio, D. L. Longo, M. J. Jones, W. G. Alvord, C. M. Pinsky, J. M. Beveridge, K. L. McNitt and S. P. Creekmore, *Cancer Res.* **52**, 3005 (1992).
65. M. Moser and O. Leo, *Vaccine* **28**, C2 (2010).
66. J. Lee, T. Chuang, V. Redecke, L. She, P. M. Pitha, D. A. Carson, E. Raz and H. B. Cottam, *Proc. Natl Acad. Sci. USA* **100**, 6646 (2003).
67. A. K. Zuber, A. Brave, G. Engstrom, B. Zuber, K. Ljungberg, M. Fredriksson, R. Benthin, M. G. Isaguliants, E. Sandström, J. Hinkula and B. Wahren, *Vaccine* **22**, 1791 (2004).
68. D. N. Sauder, M. H. Smith, T. Senta-McMillian, I. Soria and T. C. Meng, *Antimicrob. Agents Chemother.* **47**, 3846 (2003).
69. F. R. Vogel, C. Caillet, I. C. Kusters and J. Haensler, *Expert Rev. Vaccines* **8**, 483 (2009).
70. J. Aucouturier, S. Ascarateil and L. Dupuis, *Vaccine* **24**, 44 (2006).
71. S. Feyerabend, S. Stevanovic, C. Gouttefangeas, D. Wernet, J. Hennenlotter, J. Bedke, K. Dietz, S. Pascolo, M. Kuczyk, H. G. Rammensee and A. Stenzl, *Prostate* **69**, 917 (2009).
72. N. Leffers, A. J. Lambeck, M. J. Gooden, B. N. Hoogeboom, R. Wolf, I. E. Hamming, B. G. Hepkema, P. H. Willemse, B. H. Molmans, H. Hollema, J. W. Drijfhout, W. J. Sluiter, A. R. Valentijn, L. M. Fathers, J. Oostendorp, A. G. van der Zee, C. J. Melief, S. H. van der Burg, T. Daemen and H. W. Nijman, *Int. J. Cancer* **125**, 2104 (2009).
73. M. Barve, J. Bender, N. Senzer, C. Cunningham, F. A. Greco, D. McCune, R. Steis, H. Khong, D. Richards, J. Stephenson, P. Ganesa, J. Nemunaitis, G. Ishioka, B. Pappen, M. Nemunaitis, M. Morse, B. Mills, P. B. Maples, J. Sherman and J. J. Nemunaitis, *J. Clin. Oncol.* **26**, 4418 (2008).
74. A. Podda and G. Del Giudice, *Expert Rev. Vaccines* **2**, 197 (2003).
75. Y. Wu, R. D. Ellis, D. Shaffer, E. Fontes, E. M. Malkin, S. Mahanty, M. P. Fay, D. Narum, K. Rausch, A. P. Miles, J. Aebig, A. Orcutt, O. Muratova, G. Song, L. Lambert, D. Zhu, K. Miura, C. Long, A. Saul, L. H. Miller and A. P. Durbin, *PLoS One* **3**, e2636 (2008).
76. N. Garçon, P. Chomez and M. Van Mechelen, *Expert Rev. Vaccines* **6**, 723 (2007).
77. G. Ott, G. L. Barchfeld, D. Chernoff, R. Radhakrishnan, P. van Hoogevest and G. Van Nest, *Pharm. Biotechnol.* **6**, 277 (1995).
78. A. Seubert, E. Monaci, M. Pizza, D. T. O'Hagan and A. Wack, *J. Immunol.* **180**, 5402 (2008).
79. S. T. Reddy, A. J. van der Vlies, E. Simeoni, V. Angeli, G. J. Randolph, C. P. O'Neil, L. K. Lee, M. A. Swartz and J. A. Hubbell, *Nat. Biotech.* **25**, 1159 (2007).
80. A. C. Rice-Ficht, A. M. Arenas-Gamboa, M. M. Kahl-McDonagh and T. A. Ficht, *Curr. Opin. Microbiol.* **13**, 106 (2009).
81. B. San Román, J. M. Irache, S. Gómez, N. Tsapis, C. Gamazo and M. S. Espuelas, *Eur. J. Pharm. Biopharm.* **70**, 98 (2008).
82. S. Gómez, C. Gamazo, B. San Roman, M. Ferrer, M. L. Sanz, S. Espuelas and J. M. Irache, *Eur. J. Pharm. Biopharm.* **70**, 711 (2008).
83. B. Slütter, P. C. Soema, Z. Ding, R. Verheul, W. Hennink and W. Jiskoot, *J. Control. Release* **143**, 207 (2010).

84. D. T. O'Hagan and M. Singh, *Expert Rev. Vaccines* **2**, 269 (2003).
85. H. H. Salman, J. M. Irache and C. Gamazo, *Vaccine* **27**, 4784 (2009).
86. S. K. Mallapragada and B. Narasimhan, *Int. J. Pharm.* **364**, 265 (2008).
87. H. H. Salman, C. Gamazo, M. A. Campanero and J. M. Irache, *J. Control. Release* **106**, 1 (2005).
88. S. Jain, W. T. Yap and D. J. Irvine, *Biomacromolecules* **6**, 2590 (2005).
89. M. Mori, M. Nishida, N. Maekawa, H. Yamamura, Y. Tanaka, M. Kasai, M. Taneichi and T. Uchida, *Int. Arch. Allergy Immunol.* **136**, 83 (2005).
90. G. F. A. Kersten, A. M. van de Put, T. Teerlink, E. C. Beuvery and D. J. Crommelin, *Infect. Immun.* **56**, 1661 (1988).
91. C. Underwood and A. W. van Eps, *Vet. J.* **193**, 12 (2012).
92. A. Takagi, M. Matsui, S. Ohno, H. Duan, O. Moriya, N. Kobayashi, H. Oda, M. Mori, A. Kobayashi, M. Taneichi, T. Uchida and T. Akatsuka, *Clin. Vaccine Immunol.* **16**, 1383 (2009).
93. R. Ravindran, S. Bhowmick, A. Das and N. Ali, *BMC Microbiol.* **10**, 181 (2010).
94. K. K. Peachman, Q. Li, G. R. Matyas, S. B. Shivachandra, J. Lovchik, R. C. Lyons, C. R. Alving, V. B. Rao and M. Rao, *Clin. Vaccine Immunol.* **19**, 11 (2012).
95. V. Heravi Shargh, M. R. Jaafari, A. Khamesipour, S. A. Jalali, H. Firouzmand, A. Abbasi and A. Badiee, *Parasitol. Res.* **111**, 105 (2012).
96. M. Degregorio, G. T. Wurz, A. Gutierrez and M. Wolf, *Oncoimmunology* **1**, 1422 (2012).
97. M. Schnurr, M. Orban, N. C. Robson, A. Shin, H. Braley, D. Airey, J. Cebon, E. Maraskovsky and S. Endres, *J. Immunol.* **182**, 1253 (2009).
98. E. Maraskovsky, M. Schnurr, N. S. Wilson, N. C. Robson, J. Boyle and D. Drane, *Immunol. Cell Biol.* **87**, 371 (2009).
99. A. Schiött, K. Larsson, S. Manniche, S. Kalliomäki, A. V. Heydenreich, K. Dalsgaard and N. Kirkby, *Int. J. Pharm.* **414**, 312 (2011).
100. O. Perez, G. Bracho, M. Lastre, N. Mora, J. del Campo, D. Gil, C. Zayas, R. Acevedo, D. González, J. A. López, C. Taboada, C. Turtle and R. L. Solis, *Cell Biol.* **82**, 603 (2004).
101. J. F. Mann, V. A. Ferro, A. B. Mullen, L. Tetley, M. Mullen, K. C. Carter, J. Alexander and W. H. Stimson, *Vaccine* **22**, 2425 (2004).
102. M. Conacher, J. Alexander and J. M. Brewer, *Vaccine* **19**, 2965 (2001).
103. J. Tarradas, M. Monsó, M. Muñoz, R. Rosell, L. Fraile, M. T. Frías, M. Domingo, D. Andreu, F. Sobrino and L. Ganges, *Vaccine* **29**, 4422 (2011).
104. V. P. Torchilin, *Pharm. Res.* **24**, 1 (2007).
105. J. K. Eby, K. Y. Dane, C. P. O'Neil, S. Hirosue, M. A. Swartz and J. A. Hubbell, *Acta Biomater.* **8**, 3210 (2012).
106. Y. Yang, Y. Kuang, Y. Lin, W. Li, Z. Jiang, L. Xiao and M. Li, *Comp. Immunol. Microbiol. Infect. Dis.* **34**, 65 (2011).
107. A. des Rieux, V. Fievez, M. Garinot, Y. J. Schneider and V. Preat, *J. Control. Release* **116**, 1 (2006).
108. R. C. Mundargi, V. R. Babu, V. Rangaswamy, P. Patel and T. M. Aminabhavi, *J. Control. Release* **125**, 193 (2008).
109. Y. K. Katare, T. Muthukumaran and A. K. Panda, *Int. J. Pharm.* **301**, 149 (2005).
110. J. Wendorf, M. Singh, J. Chesko, J. Kazzaz, E. Soewanan, M. Ugozzoli and D. T. O'Hagan, *J. Pharm. Sci.* **95**, 2738 (2006).

111. S. Sharma, T. K. S. Mukkur, H. A. E. Benson and Y. Chen, *J. Pharm. Sci.* **98**, 812 (2009).
112. J. Panyam and V. Labhasetwar, *Adv. Drug Deliv. Rev.* **55**, 329 (2003).
113. S. Y. Seong and P. Matzinger, *Nat. Rev. Immunol.* **4**, 469 (2004).
114. B. Bottazzi, A. Doni, C. Garlanda and A. Mantovani, *Annu. Rev. Immunol.* **28**, 157 (2010).
115. D. F. Moyano, M. Goldsmith, D. J. Solfiell, D. Landesman-Milo, O. R. Miranda, D. Peer and V. M. Rotello, *J. Am. Chem. Soc.* **134**, 3965 (2012).
116. S. K. Lai, Y. Y. Wang and J. Hanes, *Adv. Drug Deliv. Rev.* **61**, 158 (2009).
117. P. Arbós, M. Wirth, M. A. Arangoa, F. Gabor and J. M. Irache, *J. Control. Release* **83**, 321 (2002).
118. M. Sánchez-Martínez, R. da Costa Martins, G. Quincoces, C. Gamazo , C. Caicedo, J. M. Irache and I. Peñuelas, *Rev. Esp. Med. Nucl. Imagen. Mol.* in press (doi:pii: S2253-654X(12)00243-0).
119. S. J. Green, *Clin. Pharmacol. Ther.* **83**, 813 (2008).
120. B. Guy, *Nat. Rev. Microbiol.* **5**, 505 (2007).
121. T. H. Flo, L. Ryan, E. Latz, O. Takeuchi, B. G. Monks, E. Lien, O. Halaas, S. Akira, G. Skjak-Braek, D. T. Golenbock and T. Espevik, *J. Biol. Chem.* **277**, 35489 (2002).
122. I. Tamayo, J. M. Irache, C. Mansilla, J. Ochoa-Reparaz, J. J. Lasarte and C. Gamazo, *Clin. Vaccine Immunol.* **17**, 1356 (2010).
123. H. Li, S. B. Willingham, J. P. Ting and F. Re, *J. Immunol.*, **181**, 17 (2008).
124. C. L. Bueter, C. K. Kee, V. A. K. Rathinam, G. J. Healy, C. H. Taron, C. A. Specht and S. M. Levitz, *J. Biol. Chem.* **286**, 35447 (2011).
125. T. Y. Lee, Y. H. Kim, S. W. Yoon, J. C. Choi, J. M. Yang, C. J. Kim, J. T. Schiller, M. H. Sung and H. Poo, *Cancer Immunol. Immunother.* **58**, 1781 (2009).
126. T. W. Kim, T. Y. Lee, H. C. Bae, J. H. Hahm, J. H. Kim, C. Park, T. H. Kang, C. J. Kim, M. H. Sung and H. Poo, *J. Immunol.* **179**, 775 (2007).
127. R. Xu and Z. H. Jiang, *Carbohydr. Res.* **343**, 7 (2008).
128. L. M. Wetzler, *Vaccine* **21**, S55 (2003).
129. R. Maitra, *PLoS One* **3**, e2438 (2008).
130. M. Hultqvist, *Annal. Rheum. Dis.* **69**, 1527 (2010).
131. K. E. Broaders, J. A. Cohen, T. T. Beaudette, E. M. Bachelder and J. M. Frechet, *Proc. Natl. Acad. Sci. USA* **106**, 5497 (2009).
132. J. M. Irache, M. Huici, M. Konecny, S. Espuelas, M. A. Campanero and P. Arbós, *Molecules* **10**, 126 (2005).
133. A. Albanese, P. S. Tang and W. C. W. Chan, *Ann. Rev. Biomed. Eng.* **14**, 1 (2012).
134. S. D. Xiang, A. Scholzen, G. Minigo, C. David, V. Apostolopoulos, P. L. Mottram and M. Plebanski, *Methods* **40**, 1 (2006).
135. L. V. Leak, *J. Cell Biol.* **50**, 300 (1971).
136. V. Manolova, A. Flace, M. Bauer, K. Schwarz, P. Saudan and M. F. Bachmann, *Eur. J. Immunol.* **38**, 1404 (2008).
137. H. Pflicke and M. Sixt, *J. Exp. Med.* **206**, 2925 (2009).
138. M. A. Jepson, N. L. Simmons, G. L. Hirst and B. H. Hirst, *Cell Tissue Res.* **273**, 127 (1993).
139. T. Fifis, A. Gamvrellis, B. Crimeen-Irwin, G. A. Pietersz, J. Li, P. L. Mottram, I. F. McKenzie and M. Plebanski, *J. Immunol.* **173**, 3148 (2004).

140. P. L. Mottram, D. Leong, B. Crimeen-Irwin, S. Gloster, S. D. Xiang, J. Meanger, R. Ghildyal, N. Vardaxis and M. Plebanski, *Mol. Pharm.* **4**, 73 (2007).
141. V. Kanchan and A. K. Panda, *Biomaterials* **28**, 5344 (2007).
142. I. Gutierro, R. M. Hernandez, M. Igartua, A. R. Gascón, J. L. Pedraz, *Vaccine* **21**, 67 (2002).
143. J. A. Cohen, T. T. Beaudette, W. W. Tseng, E. M. Bachelder, I. Mende, E. G. Engleman and J. M. Frechet, *Bioconjug. Chem.* **20**, 111 (2009).
144. T. Jung, W. Kamm, A. Breitenbach, K. D. Hungerer, E. Hundt and T. Kissel, *Pharm. Res.* **18**, 352 (2001).
145. J. Wendorf, J. Chesko, J. Kazzaz, M. Ugozzoli, M. Vajdy and D. T. O'Hagan, *Hum. Vaccine* **4**, 44 (2008).
146. T. Nagamoto, Y. Hattori, K. Takayama and Y. Maitani, *Pharm. Res.*, **21**, 671 (2004).
147. E. Schlosser, M. Mueller, S. Fischer, S. Basta, D. H. Busch, B. Gander and M. Groettrup, *Vaccine* **26**, 1626 (2008).
148. K. A. Woodrow, K. M. Bennett and D. D. Lo, *Annu. Rev. Biomed. Eng.* **14**, 17 (2012).
149. A. L. Blasius and B. Beutler, *Immunity* **32**, 305 (2010).
150. Y. R. Lee, Y. H. Lee, S. A. Im, I. H. Yang, G. W. Ahm, K. Kim and C. K. Lee, *Arch. Pharm. Res.* **33**, 1859 (2010).
151. M. Tafaghodi, A. Khamesipour and M. R. Jaafari, *Parasitol. Res.* **108**, 1265 (2011).
152. B. Slutter and W. Jiskoot, *J. Control. Release* **148**, 117 (2010).
153. O. Borges, A. Cordeiro-da-Silva, J. Tavares, N. Santarem, A. de Sousa, G. Borchard and J. E. Junginger, *Eur. J. Pharm. Biopharm.* **69**, 405 (2008).
154. M. Diwan, P. Elamanchili, M. Cao and J. Samuel, *Curr. Drug Deliv.* **1**, 405 (2004).
155. C. H. Cho, B. K. Lee, S. M. Kwak and J. D. Kim, *Yonsei Med. J.* **40**, 20 (1999).
156. P. Vandepapeliere, Y. Horsmans, P. Moris, M. van Mechelen, M. Janssens, M. Koutsoukos, P. Van Belle, F. Clement, E. Hanon, M. Wettendorff, N. Garçon and G. Leroux-Roels, *Vaccine* **26**, 1375 (2008).
157. F. Sarti, G. Perera, F. Hintzen, K. Kotti, V. Karageorgiou, O. Kammona, C. Kiparissides and A. Bernkop-Schnürch, *Biomaterials* **32**, 4052 (2011).
158. P. Elamanchili, M. Diwan, M. M. Cao and J. Samuel, *Vaccine* **22**, 2406 (2004).
159. S. M. Bal, B. Slütter, R. Verheul, J. A. Bouwstra and W. Jiskoot, *Eur. J. Pharm. Sci.* **45**, 475 (2012).
160. A. Stano, A. J. van der Vlies, M. M. Martino, M. A. Swartz, J. A. Hubbell and E. Simeoni, *Vaccine* **29**, 804 (2011).
161. D. N. Nguyen, K. P. Mahon, G. Chikh, P. Kim, H. Chung, A. P. Vicari, K. T. Love, M. Goldberg S. Chen, A. M. Krieg, J. Chen, R. Langer and D. G. Anderson, *Proc. Natl. Acad. Sci. USA* **109**, E797 (2012).
162. S. Heuking and G. Borchard, *J. Pharm. Sci.* **101**, 1166 (2012).
163. O. Schulz, S. S. Diebold, M. Chen, T. I. Näslund, M. A. Nolte, L. Alexopoulou, Y. T. Azuma, R. A. Flavell, P. Liljeström and C. Reis e Sousa, *Nature* **433**, 887 (2005).
164. G. Tincer, S. Yerlikaya, F. C. Yagci, T. Kahraman, O. M. Atanur, O. Erbatur and I. Gursel, *Biomaterials* **32**, 4275 (2011).
165. C. M. Lehr, J. A. Bouwstra, W. Kok, A. B. J. Noach, A. G. de Boer and H. E. Junginger, *Pharm Res.* **9**, 547 (1992).
166. J. M. Irache, C. Durrer, D. Duchene and G. Ponchel, *Biomaterials* **11**, 899 (1994).
167. C. Bies, C. M. Lehr and J. F. Woodley, *Adv. Drug Deliv. Rev.* **56**, 425 (2004).

168. Y. Mo and L. Y. Lim, *J. Control. Release* **108**, 244 (2005).
169. G. J. Russell-Jones, L. Arthur and H. Walker, *Int. J. Pharm.* **179**, 247 (1999).
170. H. H. Salman, C. Gamazo, P. C. de Smidt, G. Russell-Jones and J. M. Irache, *Pharm. Res.* **25**, 2859 (2008).
171. H. H. Salman, C. Gamazo, M. A. Campanero and J. M. Irache, *J. Nanosci. Nanotechnol.* **6**, 3203 (2006).
172. F. Dalle, T. Jouault, P. A. Trinel, J. Esnault, J. M. Mallet, P. d'Athis, D. Poulain and A. Bonnin, *Infect. Immun.* **71**, 7061 (2003).
173. N. Csaba, M. Garcia-Fuentes and M. J. Alonso, *Adv. Drug Deliv. Rev.* **61**, 140 (2009).
174. R. C. Mundargi, V. R. Babu, V. Rangaswamy, P. Patel and T. M. Aminabhavi, *J. Control. Release* **125**, 193 (2008).
175. F. Ungaro, I. d'Angelo, A. Miro, M. I. La Rotonda and F. Quaglia, *J. Pharm. Pharmacol.* **64**, 1217 (2012).
176. F. Danhier, E. Ansorena, J. M. Silva, R. Coco, A. Le Breton and V. Préat, *J. Control. Release* **161**, 505 (2012).
177. J. J. Moon, H. Suh, M. E. Polhemus, C. F. Ockenhouse, A. Yadava and D. J. Irvine, *PLoS One* 7, e31472 (2012).
178. D. M. Santos, M. W. Carneiro, T. R. de Moura, K. Fukutani, J. Clarencio, M. Soto, S. Espuelas, C. Brodskyn, A. Barral, M. Barral-Netto and C. I. de Oliveira, *Int. J. Nanomedicine* 7, 2115 (2012).
179. C. S. Chong, M. Cao, W. W. Wong, K. P. Fischer, W. R. Addison, G. S. Kwon, D. L. Tyrrell and J. Samuel, *J. Control. Release* **102**, 85 (2005).
180. A. K. Jain, A. K. Goyal, N. Mishra, B. Vaidya, S. Mangal and S. P. Vyas, *Int. J. Pharm.* **387**, 253 (2010).
181. A. Vila, A. Sanchez, C. Evora, I. Soriano, O. McCallion and M. J. Alonso, *Int. J. Pharm.* **292**, 43 (2005).
182. P. Muttil, C. Prego, L. Garcia-Contreras, B. Pulliam, J. K. Fallon, C. Wang, A. J. Hickey and D. Edwards, *AAPS J.* **12**, 330 (2010).
183. A. Breitenbach, Y. X. Li and T. Kissel, *J. Control. Release* **64**, 167 (2000).
184. L. A. Dailey, M. Wittmar and T. Kissel, *J. Control. Release* **101**, 137 (2005).
185. N. Csaba, A. Sanchez and M. J. Alonso, *J. Control. Release* **113**, 164 (2006).
186. J. M. Irache, I. Esparza, C. Gamazo, M. Agueros and S. Espuelas, *Vet. Parasitol.* **180**, 47 (2011).
187. A. Kumari, S. K. Yadav and S. C. Yadav, *Coll. Surf. B* **75**, 1 (2010).
188. V. R. Sinha, K. Bansal, R. Kaushik, R. Kumria and A. Trehan, *Int. J. Pharm.* **278**, 1 (2004).
189. S. R. Jameela, N. Suma, A. Misra, R. Raghuvanshi, S. Ganga and A. Jayakrishnan, *Curr. Sci.* **70**, 669 (1996).
190. H. F. Florindo, S. Pandit, L. Lacerda, L. M. D. Goncalves, H. O. Alpar and A. J. Almeida, *Biomaterials* **30**, 879 (2009).
191. D. E. Chickering, J. S. Jacob, T. A. Desai, M. Harrison, C. N. Morrell, P. Chaturvedi and E. Mathiowitz, *J. Control. Release* **48**, 35 (1997).
192. P. Arbos, M. A. Campanero, M. A. Arangoa, M. J. Renedo and J. M. Irache, *J. Control. Release* **89**, 19 (2003).
193. S. Gomez, C. Gamazo, B. San Roman, C. Vauthier, M. Ferrer and J. M. Irache, *J. Nanosci. Nanotechnol.* **6**, 3283 (2006).

194. H. H. Salman, C. Gamazo, M. Agüeros and J. M. Irache, *Vaccine* **25**, 8123 (2007).
195. K. Yoncheva, S. Gómez, M. A. Campanero, C. Gamazo and J. M. Irache, *Expert Op. Drug Del.* **2**, 205 (2005).
196. J. M. Irache, H. H. Salman, S. Gomez, S. Espuelas and C. Gamazo, *Front. Biosci.* **2**, 876 (2010).
197. J. Ochoa, J. M. Irache, I. Tamayo, A. Walz, V. G. DelVecchio and C. Gamazo, *Vaccine* **25**, 4410 (2007).
198. K. Vandamme, V. Melkebeek, E. Cox, J. P. Remon and C. Vervaet, *Vet. Immunol. Immunopathol.* **139**, 148 (2011).
199. S. Gomez, C. Gamazo, B. San Roman, M. Ferrer, M. L. Sanz and J. M. Irache, *Vaccine* **25**, 5263 (2007).
200. S. Gomez, C. Gamazo, B. San Roman, A. Grau, S. Espuelas, M. Ferrer, M. L. Sanz and J. M. Irache, *J. Immunol. Meth.* **348**, 1 (2009).
201. A. Bernkop-Schnürch, M. Hornof and D. Guggi, *Eur. J. Pharm. Biopharm.* **57**, 9–17 (2004).
202. A. Vila, A. Sanchez, M. Tobio, P. Calvo and M. J. Alonso, *J. Control. Release* **78**, 15 (2002).
203. A. Vila, A. Sanchez, K. Janes, I. Behrens, T. Kissel, J. L. Vila Jato and M. J. Alonso, *Eur. J. Pharm. Biopharm.* **57**, 123 (2004).
204. L. Illum, I. Jabbal-Gill, M. Hinchcliffe, A. N. Fisher and S. S. Davis, *Adv. Drug Del. Rev.* **51**, 81 (2001).
205. M. Kumar, A. K. Behera, R. F. Lockey, J. Zhang, G. Bhullar, C. P. De La Cruz, L. C. Chen, K. W. Leong, S. K. Huang and S. S. Mohapatra, *Hum. Gene Ther.* **13**, 1415 (2002).
206. M. Iqbal, W. Lin, I. Jabbal-Gill, S. S. Davis, M. W. Steward and L. Illum, *Vaccine* **21**, 1478 (2003).
207. W. Xu, Y. Shen, Z. Jiang, Y. Wang, Y. Chu and S. Xiong, *Vaccine* **22**, 3603 (2004).
208. K. Roy, H. Q. Mao, S. K. Huang and K. W. Leong, *Nat. Med.* **5**, 387 (1999).
209. F. Li, L. Wang, X. M. Jin, C. H. Yan, S. Jiang and X. M. Shen, *Immunobiology* **214**, 87 (2009).
210. S. A. Agnihotri, N. N. Mallikarjuna and T. M. Aminabhavi, *J. Control. Release* **100**, 5 (2004).
211. S. Mangal, D. Pawar, N. K. Garg, A. K. Jain, S. P. Vyas, D. S. Rao and K. S. Jaganathan, *Vaccine* **29**, 4953 (2011).
212. O. Borges, G. Borchard, J. C. Verhoef, A. De Sousa and H. E. Junginger, *Int. J. Pharm.* **299**, 155 (2005).
213. B. Sayin, S. Somavarapu, X. W. Li, M. Thanou, D. Sesardic, H. O. Alpar and S. Senel, *Int. J. Pharm.* **363**, 139 (2008).
214. A. K. Jain, A. K. Goyal, P. N. Gupta, K. Khatri, N. Mishra, A. Mehta, S. Mangal, S. P. Vyas, *J. Control. Release* **136**, 161 (2009).
215. J. Singh, S. Pandit, V. W. Bramwell and H. O. Alpar, *Methods* **38**, 96 (2006).
216. H. F. Florindo, S. Pandit, L. M. D. Gonçalves, H. O. Alpar and A. J. Almeida, *Vaccine* **27**, 1230 (2009).
217. R. Voltan, A. Castaldello, E. Brocca-Cofano, G. Altavilla, A. Caputo, M. Laus, K. Sparnacci, B. Ensoli, S. Spaccasassi, M. Ballestri and L. Tondelli, *Pharm. Res.* **24**, 1870 (2007).
218. C. R. Oliveira, C. M. Rezende, M. R. Silva, O. M. Borges, A. P. Pêgo and A. M. Goes, *Scientific World Journal* **2012**, 938457 (2012).

219. H. J. Moon, J. S. Lee, M. R. Talactac, M. Y. Chowdhury, J. H. Kim, M. E. Park, Y. K. Choi, M. H. Sung and C. J. Kim, *Vet. Microbiol.* **160**, 277 (2012).
220. V. Pavot, N. Rochereau, C. Genin, B. Verrier and S. Paul, *Vaccine* **30**, 142 (2012).
221. I. M. Belyakov and J. D. Ahlers, *J. Immunol.* **183**, 6883 (2009).
222. S. Burgdorf and C. Kurts, *Curr. Opin. Immunol.* **20**, 89 (2008).
223. H. Shen, A. L. Ackerman, V. Cody, A. Giodini, E. R. Hinson, P. Cresswell, R. L. Edelson, W. M. Saltzman and D. J. Hanlon, *Immunology* **117**, 78 (2006).
224. S. Chadwick, C. Kriegel and M. Amiji, *Adv. Drug Deliv. Rev.* **62**, 394 (2010).
225. R. K. Gupta, M. Singh and D. T. O'Hagan, *Adv. Drug Deliv. Rev.* **32**, 225 (1998).
226. R. Ahmed and D. Gray, *Science* **272**, 54 (1996).
227. S. L. Demento, W. Cui, J. M. Criscione, E. Stern, J. Tulipan, S. M. Kaech and T. M. Fahmy, *Biomaterials* **33**, 4957 (2012).
228. F. A. Sharp, D. Ruane, B. Claass, E. Creagh, J. Harris, P. Malvala, M. Singh, D. T. O'Hagan, V. Pétrilli, J. Tschopp, L. A. O'Neill and E. C. Lavelle, *Proc. Natl. Acad. Sci. USA* **106**, 870 (2009).
229. P. Malyala, J. Chesko, M. Ugozzoli, A. Goodsell, F. Zhou, M. Vajdy, D. T. O'Hagan and M. Singh, *J. Pharm. Sci.* **97**, 1155 (2008).
230. L. C. Agren, L. Ekman, B. Lowenadler and N. Y. Lycke, *J. Immunol.* **158**, 3936 (1997).
231. M. F. Kotturi, J. Botten, J. Sidney, H. H. Bui, L. Giancola, M. Maybeno, J. Babin, C. Oseroff, V. Pasquetto, J. A. Greenbaum, B. Peters, J. Ting, D. Do, L. Vang, J. Alexander, H. Grey, M. J. Buchmeier and A. Sette, *PLoS. Pathog.* **5**, e1000695 (2009).
232. F. Afrin, R. Rajesh, K. Anam, M. Gopinath, S. Pal and N. Ali, *Infect. Immun.* **70**, 6697 (2002).
233. S. Prasad, V. Cody, J. K. Saucier-Sawyer, T. R. Fadel, R. L. Edelson, M. A. Birchall and D. J. Hanlon, *Pharm. Res.* **29**, 2565 (2012).
234. M. A. Hall, S. D. Stroop, M. C. Hu, M. A. Walls, M. A. Reddish, D. S. Burt, G. H. Lowell and J. B. Dale, *Infect. Immun.* **72**, 2507 (2004).

Chapter 14

Nanopreparations for Gene Delivery

*Viola B. Morris and Vinod Labhasetwar**

Department of Biomedical Engineering, Lerner Research Institute, 9500 Euclid Avenue, Cleveland Clinic, Cleveland, OH 44195, USA
**labhasv@ccf.org*

Abstract

Over the past decade, significant progress has been made in the development of various non-viral methods of gene delivery. In this chapter, we have summarized various nanoformulations that are being investigated for gene delivery, their advantages and limitations and the future of gene therapy to treat various diseases.

1. Introduction

Gene therapy can be defined as the treatment of genetic diseases of both acquired and inherited type, through the insertion of genetic material like DNA into diseased cells in order to supplement or alter the missing or non-functional genes of the cell. It can be broadly classified into two types, somatic and germ line gene therapies. In somatic gene therapy, therapeutic genes are delivered into somatic cells at the tissue or cellular level to treat a specific individual. While in the case of germ line gene therapy, therapeutic genes are inserted into reproductive cells or possibly into embryos to correct genetic defects that could be inherited and passed on to later generations.[1]

Potential scope of gene therapy is considerably increasing with the progress of Human Genome Project. Thousands of diseases have been identified as resulting directly from abnormal genes including acquired diseases like cancer, neurodegenerative diseases, etc. and genetic disorders like cystic fibrosis, Huntington's disease, etc.

Delivery of gene or recombinant DNA to the nucleus of cells can be accomplished using molecular delivery vehicles called gene delivery vectors. Due to the natural ability of viruses to invade into cells, viral vectors that have been genetically altered are used as vectors for gene delivery. Few of the different types of viral vectors used for gene delivery are retrovirus, adenovirus, lentivirus, herpes simplex virus, pox virus and adeno-associated virus.[2] However, there are several risk factors associated with the use of viral vectors. For example, the risk of mutations to the host genome due to insertion of gene at the wrong location in the DNA. Another drawback of viral vectors is the overexpression of protein which could trigger immune reactions. Further, certain viral vectors do not work as efficiently *in vivo* as *in vitro* and their repeated administration could trigger immune response such as with adenoviral vector. Limited packing capacity, toxicity, short-lived nature and the possibility that they might cause disease are the other disadvantages with viral vectors.

The above mentioned disadvantages of viral vectors lead to the development of non-viral vectors which showed certain advantages compared to viral vectors such as their easy large scale production, low host immunogenicity, etc. Non-viral gene delivery includes direct injection of naked DNA, physical and chemical methods of gene delivery. In this chapter, we have focused on the recent advances in chemical methods, particularly the nanoparticle-based gene delivery methods (Fig. 1).

2. Chemical Methods of Gene Transfer

Chemical methods of gene transfer involve the use of a positively charged delivery vector in order to overcome the negatively charged DNA to cross the negatively charged cell membrane. DNA containing particles are taken up by cells by processes such as via endocytosis, macropinicytosis or phagocytosis. The DNA particles then escape from endosomes into the cytoplasm. General convention is that DNA dissociated from the complex into cytoplasm migrates into the nucleus but few studies have also reported that DNA dissociation from the complex takes place in the nucleus.[3–5] Commonly used chemical vectors are calcium phosphate, Diethylaminoethyl-Dextran (DEAE-Dextran), cationic lipids and neutral or cationic polymers. Compared to DEAE-Dextran or calcium phosphate method of gene transfection, the attractive features of lipid and polymer based gene delivery are the possibility of engineering these nanoparticles with

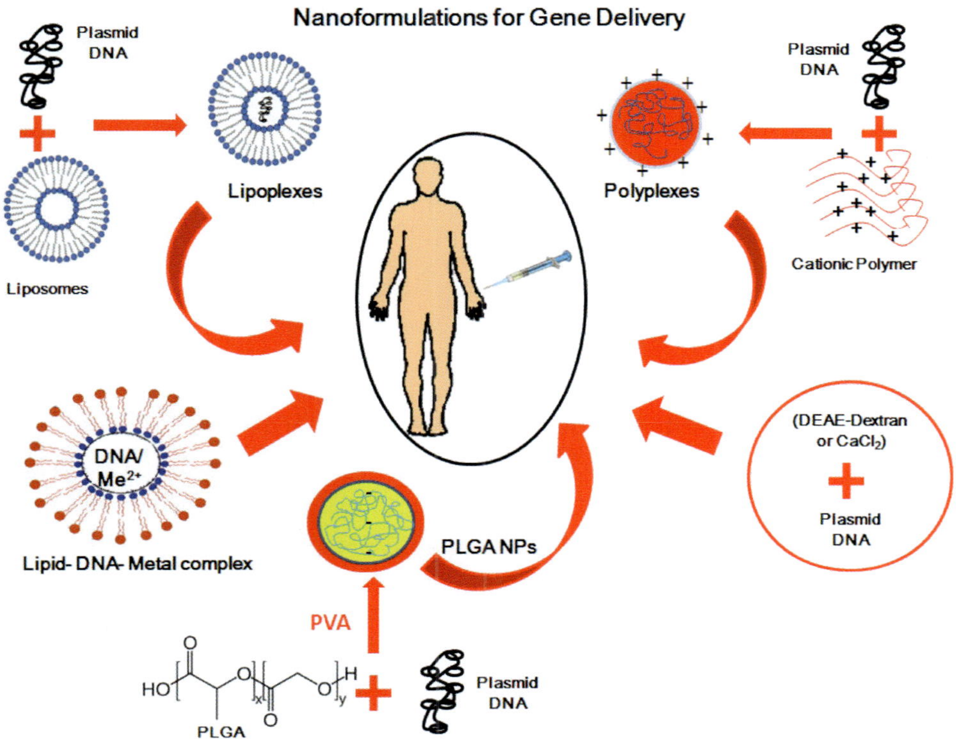

Fig. 1. Various nanoformulations for gene delivery.

functional moieties to improve in terms of the performance, specifically circulation time, targetability, stimuli sensitivity, cellular and nuclear penetration, etc.[6]

2.1. *DEAE-Dextran*

DEAE-Dextran is a polycationic derivative of carbohydrate dextran which was the first non-viral transfection method verified by Vaheri and Pagano in 1965.[5] Positively charged dextran derivative electrostatically interacts with the anionic phosphodiester backbone of DNA. The resultant cationic complex comes into contact with negatively charged cell membrane and is subsequently internalized by endocytosis. Primary disadvantages of this method include, limited range of cell types with which it works effectively as well as toxicity especially when applying supplemental chemical shocks. The most routinely used chemical agents are glycerol and dimethyl sulfoxide (DMSO).[7] Sodium butyrate and chloroquine are other chemical agents applied with DEAE-Dextran to improve transfection. About 10–20% of glycerol or DMSO is applied to the cells after adding DNA for a period of 30 s to few minutes. These treatments modify cell membrane structure to enhance uptake of DNA. Chloroquine is applied to cells along with DNA to inhibit DNA degradation by lysosomes.[8] Sodium butyrate treatment following the transfection results in the formation of more active chromatin structure of the newly transfected genes which results in high levels of transient expression.[9] Simplicity, low cost, and reproducibility are the advantages of this method.

2.2. *Calcium phosphate*

Gene delivery with calcium phosphate was first introduced by Graham and Van der Eb in 1973.[10] In this method, DNA is first mixed with calcium chloride and then co-precipitated by carefully adding the DNA/$CaCl_2$ solution into phosphate buffered saline followed by incubation at room temperature. This results in the formation of DNA-calcium phosphate co-precipitate in a period of ~20 min which adheres to the surface of cells and is taken up through endocytosis or phagocytosis. This method is widely used because of its simplicity, low cost, its applicability in wide variety of cell types and its ability for long-term gene expression. Efficiency of this method depends on various factors such as the cell type, amount of DNA, size and quality of the precipitate, buffer salt concentrations, temperature and pH. Highly transformed and adherent cell types such as human cervical carcinoma cells (HeLa), human osteosarcoma cell lines such as U2OS and SAOS2, human adenocarcinoma cells (AdAH), nasopharyngeal carcinoma hybrid cells NPC-KT, etc. are found to show 20–100% transfection efficiency. Like in the case of DEAE-Dextran, transfection efficiency can be improved in some cell lines by applying chemical shocks with DMSO or glycerol. Transfection can also be improved by using bis-hydoxyethyl amino ethane sulphonate (BES).[11] Sensitivity of this method upon slight changes as well as its low transfection efficiency compared to other transfection methods like cationic polymers and lipids are some of the limiting factors.

2.3. *Lipofection*

The use of cationic lipids for DNA Delivery is known as lipofection and was first introduced by Felgner and colleagues in 1987. Cationic liposomes for gene delivery are composed of a positively charged hydrophilic head group, hydrophobic hydrocarbon tails and a flexible linker group like ester or ether. The presence of positively charged head groups such as primary, secondary tertiary or quaternary head groups facilitates binding with phosphate groups of DNA by electrostatic interaction resulting in small quasi-stable particles called lipoplexes.[3] Hydrophobic head group act as an anchor, which consists of either two hydrocarbon chains or cholesterol. Of various linker groups used, glycerol is the commonly used linker between the head group and anchor group. Transfection efficiency of cationic liposomes depends on various factors such as the chemical structure of the cationic lipid, the charge ratio between cationic lipids and DNA, size and structure of lipoplexes, total amount of lipoplexes applied and cell types. In addition, the efficiency is dependent on the structure and proportion of neutral or helper lipids which facilitate the entrapped DNA to escape the endosomes by fusion of the liposome with the membrane. One of the most commonly used helper lipids is dioleoylphosphatidylethanolamine (DOPE). High concentration of DOPE has a strong tendency to form an inverse hexagonal phase, which gives rise to the formation of a non-bilayer lipid structure, and causes membrane perturbation and endosome destruction. However, some multivalent cationic lipopolyamines undergoing endosomal escape by proton sponge mechanism, briefly, the protonable amines absorbs the protons and slow down the acidification process inside the endosomes, prevents the endosome to lysosome transition.[12,13] Optimal lipid structure for better *in vivo* transfection is (i) cationic head group to anchoring hydrocarbon chain should be in a 1, 2-relationsip on the backbone, (ii) an ether bond as the linker group and (iii) paired oleyl chains as the hydrophobic anchor. The structure of lipoplexes depends on various factors including charge ratio, concentration of individual lipids and DNA, structure of each component lipids, salt concentration and preparation methods. The chemical structure of each component lipid determines the overall structure and physical properties of liposomes.

2.3.1. **Structure of Cationic Head Group**

As mentioned earlier, positive charge on the head group domain not only helps to bind the DNA but it also promotes interaction of charged components of the cell membrane and its uptake. Structure of hydrophobic head group is also an important factor as it determines the transfection efficiency and cytotoxicity of liposomes. Many cationic lipid compounds have been formulated using various types of head groups such as primary, secondary, tertiary, or quaternary ammonium salts and guanidine,[14] amidine[15] and cyclic guanidine.[16] In order to spread the cationic charge, compounds with heterocyclic rings like morpholine, piperazine, imidazole or pyridine rings[17] have also been attempted.

Heterocyclic rings not only improved the transfection efficiency, but also reduced the cytotoxicity. Polyvalent cationic lipids such as lipopolyamines and lipospermine were also used.[18] Even though multivalent cationic lipids are efficient in compacting DNA, they also caused increased cytotoxicity. In contrast, Wölk *et al.* reported that a multivalent cationic lipid of two lysine residues bound via tris(aminoethyl)amine spacer to malonic acid diamine backbone showed better transfection efficiency with low toxicity.[19]

2.3.2. **Structure of hydrophobic domain**

Hydrophobic hydrocarbon chain of cationic lipids is mainly composed of aliphatic chains or steroid domains. Aliphatic chains is typically either linear and saturated, or linear and mono-unsaturated. The structure, chemical bonds and position of these chains determine the phase transition temperature and fluidity of the bilayer, which in turn influences the stability and toxicity of the complex. From various studies, it was concluded that unsaturated C18:1 oleyl was the optimal aliphatic chain for good transfection. Because of the rigidity, biodegradability and fusion activity, cholesterol is the better replacement for the aliphatic chains. Vitamin D,[13] bile acids,[20] antibiotic,[21] lithocholic acid and cholestane[22] are the other commonly used steroid hydrophobic moieties.

2.3.3. **Structure of linker groups**

Commonly used linker groups used to connect between the hydrophobic hydrocarbon and cationic head group are ether, ester,[23] amide[24] carbamate group or redox sensitive disulphide groups.[25] A linker group should be chemically stable as well as biodegradable to render better transfection efficiency and low cytotoxicity. 3β-[N-(N',N'-dimethylaminoethane)-carbamoyl] cholesterol (DC-Chol) with carbamate linker group is the cationic lipid, which was first used in clinical trials because of its combined properties of transfection efficiency, stability and low toxicity.[26] For carbamate linker group, pH change in the endosome is expected to be the triggering factor for the breaking down of the bond. Similarly, lipids with disulfide linker group are stable in the circulation system, and will eventually break bonds and release DNA after entering the endosome in cells.

2.3.4. **Preparation of cationic liposomes**

After designing suitable lipid components, liposomes can be prepared by the conventional manufacturing method which involves four basic steps. The first step is dissolving each component lipids in a suitable volatile organic solvent mainly chloroform, ether or methanol. Second step involves drying down of lipids from organic solvents under reduced pressure. In the third step, the dried lipid film can then be dispersed in aqueous media by vortexing and sonicating properly. Finally, the resultant liposomes can be purified and used. One of the main disadvantages of this method is the presence of residual organic solvents, known as organic volatile impurities, which is used to dissolve lipids. These impurities affect the chemical structure of the entrapped substance and contribute to the toxicity. In order to avoid the use of volatile organic solvents, there are several methods that have been developed. Those include polyol dilution method, the bubble method and heating method. Polyol dilution method consist of a mixing of membrane components with water-soluble non-volatile organic solvents such as glycerin, propylene glycol, etc., instead of using volatile organic solvents.[27] In the bubble method, inert gas is bubbled into a coarse dispersion of non-hydrated lipids or non-ionic surfactant to produce stable liposome/noisome dispersions.[28] In the heating method, liposome components are hydrated in an aqueous medium followed by the heating of these components in the presence of 3% (v/v) glycerol up to 120°C. Since glycerol is a physiologically acceptable chemical having an ability to increase the stability of lipid vesicles, it does not need to be removed from the final liposomal product. These preformed liposomes can be used to interact electrostatically with plasmid DNA molecules.[29]

2.3.5. **Preparation of lipid- DNA–Metal complex**

Despite promising features of cationic liposomes as gene carriers in human gene therapy, they exhibit few major drawbacks which limit their extensive application. One of the disadvantages, is the instability of cationic lipoplexes in the presence of serum *in vivo*.[30] The cationic nature of the lipoplexes is also attributed to cytotoxicity and immunogenic response.[31] These factors lead to the development of self- assembled ternary system of neutral lipids-DNA and bivalent metal cations in water solutions (L-DNA-Me^{2+}). The neutral or zwitterionic lipids offer lower inherent cytotoxicity and longer circulation time. Bivalent metal cations such as Mn^{2+}, Ca^{2+}, Co^{2+}, Mg^{2+} or Fe^{2+}, common in biological cells, act as DNA condensing agents.[31–35] These metal ions also work as a bridge between the phosphate groups of DNA and zwitterionic lipid head groups.[36]

In the method reported by Pisani *et al.*, for the preparation of ternary complex, as a first step, lipid was dissolved in chloroform and the solvent was evaporated by a stream of nitrogen. To remove traces of the solvent, samples were incubated under vacuum for 5 h. Dried film was then dispersed in 20 mM HEPES buffer (pH-7.2) and incubated at 50°C for 2 h. During the hydration period, the mixture was vortexed several times and then stored at 4°C. In the second step, the triple complexes were prepared by mixing equal volumes of the solutions of lipid, DNA and metal at different relative molar ratios and incubated for few hours at room temperature.[36]

In another method, by Huang's group, an anionic lipid was used for the preparation of ternary complex. In this method, the anionic phospholipid, dioleoyl phosphatidic acid (DOPA) acts as a pre-coating reagent during the formation of DNA–metal complex. The anionic lipid coated DNA metal core was prepared by water-in-oil microemulsion method. In this method, $CaCl_2$ and siRNA were dispersed in cyclohexane/Igepal CO-520 (71/29 v/v) solution to form well dispersed water-in-oil reverse micro emulsion. The anionic lipid, DOPA, was dispersed in chloroform was added into 25 mM Na_2HPO_4 in a separate 15 ml oil phase. The two solutions were mixed for 20 min, 30 ml of absolute ethanol was then added to the microemulsion and the mixture was centrifuged at 12,000 g for at least 15 min to remove cyclohexane and surfactants. The resultant lipid coated DNA-metal core (CaP core) was then extensively washed by ethanol 2–3 times and the pellets were dissolved in chloroform and stored. The lipid coating rendered the DNA–metal complex soluble in chloroform and therefore prevented aggregation. The DOPA layer also acts as the inner leaflet lipid for the formation of bilayer lipid. The bilayer lipid can be prepared by mixing the CaP core with outer leaflet lipid such as 1,2-dioleoyl-3-trimethylammonium-propane chloride Salt (DOTAP)/cholesterol or 1,2-distearoryl-sn-glycero-3-phosphoethanolamine (DSPE)/cholesterol and DSPE-PEG-2000 or DSPE-PEG-anisamide in chloroform. Chloroform was then evaporated and residual lipid was dispersed in Tris-HCl buffer to form the bilayer lipid–DNA complex. Since PEG-ligand modified lipid can be used as the outer leaflet, it is not necessary to perform PEGylation of NPs by post insertion.[37]

2.4. *Biodegradable nanoparticles-mediated gene delivery*

Biodegradable polymers such as polycaprolatone,[38,39] polyglycolic acid (PGA), polylactic acid (PLA),[40] and poly(lactic *co*-glycolic acid) (PLGA)[41] have been extensively studied as delivery carriers of therapeutic agents such as DNA, drugs, enzymes, etc. These carriers can be used either in the form of nano/microparticles or in the form of three- dimensional scaffolds. Biodegradable ester bonds of these polymers undergo hydrolysis in the body forming biologically compatible and metabolizable moieties which will then be removed from the body by citric acid cycle. This degradation of the polymer matrix leads to slow and sustained release of DNA or other therapeutic agents entrapped into it. DNA entrapped in the nanoparticles or scaffolds is protected from extracellular and intracellular nuclease degradation. These nano/microparticles undergo effective intracellular uptake, endosomal escape and subsequent slow release of the encapsulated gene in the cytoplasm.[42,43] On the other hand, DNA delivered from the scaffold is taken up by the surrounding cells as it is released at the implant site and this avoids unspecific exposure of the DNA in other areas.

DNA release from nanoparticles depended on both size and its distribution, and DNA loading efficiency which in turn is influenced by the preparation methods.[44] Several methods such as spray drying, emulsification and diffusion, emulsion-solvent evaporation, phase separation or coacervation, nanoprecipitation, cross-flow filtration salting-out technique, etc., are used for the preparation of PLGA nanoparticles. Out of these methods, emulsion-solvent evaporation method is the most extensively used one for DNA loading. The formulation parameters that impact on the delivery and transfection of DNA loaded NPs include type and concentrations of surfactants used, volume of organic and aqueous phase, the concentration of drug and polymer, molecular weight of the polymer and power and duration of the energy applied. In the following section, we are describing these factors more in depth.

2.4.1. Materials for the formulation of DNA loaded PLGA NPs

Primary materials used for the preparation of PLGA NPs by emulsion solvent evaporation method are PLGA, appropriate solvent for PLGA based on its molecular weight and co-polymer ratio, plasmid DNA and Poly(vinylalcohol) (PVA) as emulsifier. Acetylated bovine serum albumin (Ac-BSA, nuclease free) provides stability to the primary emulsion, and stabilizers like cetyltrimethylammonium bromide (CTAB), didodecyldimethylammonium bromide (DMAB), etc., for the surface modification, are the other major components for the preparation of NPs.

PLGA: PLGA, the copolymer of poly (lactic acid) (PLA) and poly (glycolic acid) (PGA) is synthesized by the ring opening polymerization of corresponding cyclic dimmers using stannous octanoate or stannous chloride as the initiator. The copolymer ratio of lactide to glycoside influence gene loading and releasing properties.[45] Mehta *et al.* reported that the initial burst release can be controlled by increasing the encapsulation efficiency.[46] The encapsulation efficiency can be controlled by manipulating the precipitation of the polymer on the surface of the dispersed phase and to prevent DNA loss into the continuous phase.[47] Mehta's study also concludes that, higher the solubility, longer the solidification time which resulted in low encapsulation efficiency.[48] Glycolic acid provides more hydrophilicity to the co-polymer. Since hydrophilic PLGA is relatively less soluble in organic solvents, it enhances the stability of the primary emulsion and precipitates more quickly in the dispersed phase, and this enhances the encapsulation efficiency. Alternatively, encapsulation efficiency can also be increased by increasing the interaction between PLGA and the protein or DNA through hydrogen bonding and polar interactions.[49,50] Moreover, the increase in the hydrophilic property increases the degradation property of the polymer and thereby release rate of the NPs. A copolymer ratio of 50:50 is accepted to provide slow and sustained release of the gene.[51,52]

Encapsulation efficiency and DNA release from particles are also affected by the molecular weight of the polymer. In general, high molecular weight PLGA enhances DNA encapsulation efficiency. An increase in the molecular weight of PLGA from 12 to 143 kDa resulted in increase in the DNA loading efficiency as well as transfection efficiencies.[53] This is due to the higher viscosity and rapid solidification of the polymer that results in better entrapment of DNA. On the other hand, low molecular weight polymers are more soluble in the organic solvents and undergo slow solidification which resulted in more porous particles and burst release.[54,55] It also attributed to the smaller size of the particles which also provide more surface area for drug diffusion.

Solvents of PLGA: PLGA is soluble in a wide range of solvents like chlorinated solvents, tetrahydrofuran, benzene, acetone or ethyl acetate, depending on its molecular weight and co-polymer ratio. The solidification rate during the solvent evaporation method depends on the polymer and its solvent, therefore solubility is directly related to microparticle properties like drug incorporation and matrix porosity. Methylene chloride is more soluble in water than chloroform and benzene. Therefore, a rapid mass transfer occurs between the dispersed and continuous phases that leads to fast precipitation of the polymer which enhances loading efficiency.[47]

Surfactants: In emulsion-solvent evaporation method, stability of the emulsion can be enhanced by using emulsifying agents like BSA, PVA, polyvinyl pyrrolidone (PVP), Tween-80 or span-80. PVA and PVP are the most commonly used stabilizers. Generally, 1%–7% (w/v) PVA solution has been used as the stabilizer in which 4% PVA solution results in maximum DNA loading and protection of DNA from nuclease degradation.[56] As the concentration of stabilizers increases, emulsification property increases and thereby size of the particle decreases. This is due to the prevention of coalescence by the presence of precipitated surfactants on the surface of the o/w droplets.[56,57] Our earlier studies showed that DNA encapsulation efficiency improved with increase in PVA concentration.[58]

Our earlier studies suggested that, the fraction of the PVA that remains on the surface of the nanoparticles affects the physical and surface properties of nanoparticles, such as particle size, zeta potential, polydispersity index, loading efficiency, surface hydrophobicity and cellular uptake. Even though the particle size of the nanoparticles decreased with increase in concentration of PVA, the hydrophilicity of the nanoparticle also increased at high concentrations of PVA. Therefore, the cellular uptake of the nanoparticles reduces when there is a high content of residual PVA.[59]

In order to improve cellular uptake, we investigated the interaction of surface modified nanoparticles with an endothelial cell model membrane. We used three different types of surfactants such as dichained DMAB or single-chained CTAB and dodecyltrimethylammonium bromide [DTAB].[60] The results demonstrated that, compared to CTAB and DTAB, significant changes in surface pressure and more cellular uptake were observed with DMAB modified nanoparticles.[60] In other words, the molecular structure of the surfactants at the nanoparticle interface is affecting the NPs interaction with cell membrane. The factors affecting this enhanced uptake are (i) the electrostatic interactions between the positive surface charge of nanoparticles and the negative charge of the cell membranes, (ii) penetration of NPs into cell membrane which in turn depends on the cell membrane biophysical characteristics such as membrane tension, membrane fluidity/rigidity and adhesion strength/binding energy between the NPs and the cell membrane,[61] and (iii) cell type.

In agreement with this finding,[62] modified PLGA nanoparticles with DMAB formulated using emulsion solvent diffusion method and examined for their cellular uptake properties were also found to improve oral bioavailability of the drugs. In our recent study, we demonstrated greater cellular uptake of DMAB-modified NPs into cancer cells *in vitro* and greater intratumor retention *in vivo* in a murine model of prostate cancer. It also resulted in tumor regression and greater animal survival than controls (unpublished data).

BSA is another example of component used to provide stability to the primary emulsion, resulting better encapsulation of DNA in NPs.[51] BSA also facilitate the release of DNA.[58]

Primarily, small particle size and uniform particle size distribution are important to enhance the nanoparticle-mediated gene expression.[43] In our recent study, we used PLGA NPs of average diameter 280 nm to deliver p53 gene in mice model having human prostate tumor xenografts. Both local and systemic administration of NPs led to tumor regression and improved animal survival.[51,52] In a different study, biodegradable nanoparticles prepared using tribock copolymer of monomethoxypoly(ethylene glycol)-poly(lactic *co*-glycolic acid)-poly-l-lysine (mPEG-PLGA-PLL) was used for the delivery of vascular endothelial growth factor (VEGF). From our previous studies, we demonstrated sustained marker gene expression in rat bone osteotomy model even after 5 weeks of surgery.[63] We have investigated various factors affecting sustained release of gene and its expression and has been reported elsewhere.[41]

2.5. *Cationic polymer-mediated gene delivery*

Polyplex are the polyelectrolyte complexes of DNA and a cationic polymer formed by electrostatic interaction and self-assembly. These polyplexes are homogeneous and nanoparticulate in nature. Enormous numbers of polymers in the form of linear, branched or spherical configurations have been studied for *in vitro* and *in vivo* gene delivery. Those polymers include polyethyleneimine (PEI), poly L-lysine (PLL), poly-amidoamine, chitosan, cationic dextran, poly-propylamine dendrimers, and cationic peptides.

2.5.1. **Gene delivery using poly-L-lysine and its derivatives**

Poly-L-lysine (PLL) is the first cationic polymer used for gene delivery. The primary ε-amino group in the back bone of the PLL is responsible for electrostatic interaction with phosphate group of DNA.[64] PLL having molecular weight of greater than 3000 are suitable for gene delivery. However, high molecular weight PLL/DNA complex showed relatively high cytotoxicity and aggregation tendency.[65,66] There are several studies that have been conducted to reduce the cytotoxicity and enhance transfection efficiency of PLL. Out of those studies, the most significant was the conjugation with poly ethyleneglycol (PEG). PEG conjugation helps to prevent interparticular aggregation of the complex and provides stability in the presence of serum proteins.[67] PEG can be conjugated to PLL either by block co-polymerization or by graft polymerization methods. PLL-PEG block co-polymer formed polyplex of <100 nm with DNA and it showed low cytotoxicity and prolonged *in vivo* circulation compared to naked plasmid DNA.[68]

In a recent study, hybrid copolymer of PEG-bPLL-b-PLLeu (poly-L-leucine) micelles was found to be more efficient in terms of *in vitro* transfection, cytotoxicity, and bio compatibilities. It was interesting to note that *in vivo* transfection efficiency of these micelles was two to three times higher than that of PEI (25K).[69] In order to improve buffering capacity and thus to improve transfection efficiency, Boylan *et al.* inserted poly-L histidine between PEG and PLL. The triblock copolymer system thus developed showed ~20 fold *in vitro* gene transfer and ~3 fold *in vivo* gene transfer over PEG-PLL nanoparticles without poly-L histidine.[70]

2.5.2. **Gene delivery using PEI and its derivatives**

PEI has been the most extensively studied polymer for gene delivery since Behr first introduced it in 1995.[71] Both branched and linear PEI are suitable for gene delivery studies, out of which branched PEI having molecular weight of 25 kDa was considered as a gold standard in non-viral gene delivery. But due to its cytotoxicity, less toxic linear PEI of 22 kDa emerged as a premier PEI for gene delivery.[72] Branched PEI (BPEI) has primary:Secondary:Tertiary amine groups in the ratio of 1:1:1, on the other hand amino groups of linear PEI are mostly secondary amines except the terminal group. Higher transfection efficiency of linear PEI compared to branched may be due to the less stable complex with DNA which can dissociate more efficiently than BPEI/DNA. It has also been reported that LPEI/DNA complexes enter the nucleus more readily than branched PEI/DNA complex.[3]

Different strategies have been developed to reduce the cytotoxicity of the PEI without compromising the transfection efficiency. Few of those are conjugating branched PEI with polysaccharides, hydrophilic polymers like PEG, disulfide linkers, lipid moieties, etc. The presence of primary amino groups of BPEI makes it more susceptible to functional group modifications which make it more efficient in transfection and less toxic to cells.[73] PEG modification of PEI is one of the most broadly studied transfecting agents as it drastically reduces the surface charge and aggregation tendency in the presence of serum and therefore increases circulation half- life in bloodstream. Even though PEG modification reduces cell toxicity, transfection efficiency is still comparable or may sometimes be reduced. For example, in an attempt to reduce its cytotoxicity, some researchers have conjugated PEG with low molecular weight PEI. Although it exhibited reduced toxicity, transfection efficiency was also reduced compared to BPEI 25 kDa.[74,75]

Erbacher *et al.* created a brush corona layer of PEG on the surface of PEI polyplexes in order to increase blood circulation time. However, transfection efficiency was lower than the control PEI due to the charge masking properties of PEG group.[76] Transfection efficiency and cytotoxicity of PEI–PEG conjugates depends on the degree of PEG substitution and molecular weight of PEG group. For example, Park *et al.* synthesized biodegradable copolymers of PEI and PEG by varying molecular weight with PEG-diacrylate. From the *in vitro* studies in Hep G2 and MG64 cell lines, copolymer of PEI-*alt*-PEG having molecular weight of 258 Da was found to show greater gene transfection than PEI 25 kDa.[77]

Other than PEG conjugation, Kim *et al.* prepared a degradable PEI with acid labile imine linkers.[78] Due to the degradation of acid labile linkers, acid-labile PEI was less toxic compared to BPEI 25 kDa. In a recent study of synthesizing brush like copolymer of PEI, Wei *et al.* tried controlled living radical polymerization technique. They copolymerized PEI with modified poly(glycidyl methacrylate) (P(GMA)). Out of the various modifications, RGD peptide modified copolymer showed increased gene delivery efficiency to HeLa cells as compared to analogous polymers lacking RGD.[79]

PEI derivatives are also found to be promising for *in vivo* studies. β-Cyclodextrin conjugated low molecular weight PEI with an anticancer drug 5-fluoro-2'–deoxyuridine showed improved bioavailability and stronger ability to induce apoptosis compared to the free drug in the treatment of glioma.[80] A starshaped copolymer was developed by Li *et al.* by copolymerizing eight armed polyethylene glycol and low molecular weight (600 Da) PEI. It is then conjugated with a peptide which target selectively to fibroblast growth factor receptor (FGFR). The resultant vector exhibited fourfold higher luciferase expression in tumor bearing mice and tumor regression with CSK-R interferon plasmid.[81] Low molecular weight PEI of 10 kDa was conjugated to chondrotin sulphate (CS) via Michael addition method. Cytotoxicity of CS modified PEI was reduced without sacrificing transfection efficiency. They also observed that the cellular uptake of the product was primarily based on clatherin-dependent and CD44-mediated endocytosis.[82]

Jiang *et al.*[83] prepared a cationic polysaccharide with low molecular weight PEI (1200 Da) followed by coupling with folic acid for tumor targeted gene delivery and it showed stronger green fluorescent (GFP) and luciferase expression compared to non-targeted vector. In another approach, oligoalkylaminosiloxane conjugated BPEI (25 kDa) high level of gene expression and negligible cytotoxicity in KB cell lines compared to BPEI (25 kDa).[84] Its targeted derivative with PEG-folic acid showed high level of tumor accumulation in Balb/C mice bearing tumor.[85] Shi *et al.* prepared a self-assembled amphiphilic triblock copolymer based on polyethylene glycol-ε-caprolactone-polyethyleneimine (mPEG-PCL-g-PEI) and co-delivered doxorubicin with gene. High level of gene expression and cellular uptake was observed along with loading of doxorubicin.[86]

Glioblastoma multifore (GBM) of astrocytic origin is a good candidate for gene therapy because it is difficult to treat this tumor by conventional therapies. Zhan *et al.*[87] used RGD peptide to target integrin $\alpha_v\beta_3$ receptors which overexpressed in GBM with PEG-PEI. RGD-PEG-PEI formed 73-nm-sized nanoparticles with plasmid DNA and exhibited high level of gene delivery which leads to prolonged survival of intracranial gliobastoma-bearing nude mice.

As a new approach to gene delivery, Kim *et al.*[88] functionalized nanodiamonds (NDs) with PEI to transfect small interfering RNAs (siRNA). In order to elucidate the binding mechanism between ND, PEI and siRNA, they have performed multi-scale simulation and results were in good agreement with their experimental results. The model was able to predict loading trends and saturation limits for PEI and siRNA.

Combined system of polymers and liposomes have also been evaluated for gene delivery. It is believed that structure of DNA complexes in the combined system may be more effective for endosomal escape and to facilitate transportation to nucleus.[10] Combined system includes polyplexes of DNA with the polymer and then a coating of cationic liposomes,[89] anionic liposomes,[90] or amphiphilic polymers with or without helper lipids.[91] The system appears to provide better protection for DNA and more efficient in transfection than lipid–DNA complex.

2.5.3. Gene delivery using chitosan and its derivatives

Chitosan is a biodegradable and biocompatible natural polycations having minimum immunogenicity and toxicity.[92] Primary amino groups in the backbone of chitosan enable it to bind negatively charged DNA and protect it against nuclease degradation.[93] Chitosan–pDNA complexes has found to be stable and exhibited controlled release of pDNA, leading to a sustained gene expression. It can transfect in various kinds of cell lines such as human embryonic kidney cells (HEK293), human osteosarcoma cell lines (MG63),[94] human epithelial cervical cancer cells (HeLa), B16 melanoma cells,[95] etc. Transfection efficiency of chitosan depends on various factors like molecular weight, degree of deacetylation, pH of transfecting medium, the charge ratio of the complex,[96,97] cell type, etc. Even though *in vitro* transfection efficiency of chitosan varied according to cell lines, it exhibited successful gene expression in various *in vivo* studies like gene expression for osteoarthritis,[98] for intranasal DNA vaccination,[99] nasal mucosal immunization against hepatitis B,[100] for intramuscular immunization as well as corneal gene therapy.[101] Moreover, mucoadhesive property of chitosan makes it an excellent vector for oral gene delivery.[102] Other than chitosan, its derivatives were also found to be more efficient for gene delivery. For instant, colloidal stability, circulation time and cellular uptake were improved upon PEGylation of chitosan.[103] Poly-L arginine derivatized PEG-Chitosan showed enhanced *in vitro* and *in vivo* gene silencing.[104]

Zhang *et al.* copolymerized and synthesized PEG-Chitosan-PEI so as to improve hydrophilicity, bio compatibility and transfection efficiency.[105] The product showed efficient gene expression in the presence of serum than BPEI-25 kDa. Hydrophobic aminoacid grafted chitosan (AGC) was also demonstrated to be efficient by Layek *et al.* for improved cellular uptake and transfection in HEK293 cells compared to unmodified chitosan. This study proved that hydrophobic amino acid substitution enhances pDNA release at cytosolic pH.[106]

Quaternization of chitosan is another well known modification of chitosan, since it renders more solubility over a wide range of pH and reduces the aggregation tendency.[107] Quaternized chitosan found to exhibit superior transfection efficiency compared to oligomeric chitosan in DOS-1 cell lines.[108] Chitosan has also been modified with ligands such as antibodies, aptamers, peptides, vitamins and carbohydrates to provide specific targeting.[109] For example, conjugation of folic acid ligand to PEG-chitosan improved transfection efficiency in cells having folate receptors.[109,110] In another study, arginine or histidine modified trimethylated chitosan with PEG-FA showed enhanced transfection efficiency, colloidal stability, cellular uptake and nuclear localization.[111,112] Antisense oligodeoxynucleotide (asODNs) delivered by the folate conjugated hydroxypropyl chitosan (HPCS) was found to inhibit tumor growth by 35% compared to bare as ODNs.

In a recent study, Zhang *et al.*[113] designed and synthesized ternary polymeric nanoparticles using thiolated trimethyl chitosan (TTMC), siRNA and tripophosphate for oral delivery of siRNA. The ternary system was found to exhibit suitable structural stability and siRNA protection in the intestinal tract, good permeability across *ex vivo* rat ileum, high systemic biodistribution and superior cellular uptake and gene silencing efficiency.

3. Challenges in Gene Delivery and Targeting Approaches

Three main challenges in gene delivery using non-viral vectors are insufficient accumulation of the DNA nanoparticles at the target tissue, intracellular trafficking and nuclear uptake.[114] Targeted gene therapy offers several advantages such as localized delivery of gene in high concentration and reduced the potential side effects.[115] An ideal targeted gene delivery system should enable transport of gene from the injection site through the blood stream to distant target tissue. Once the vector reaches the site, it has to extravasate and internalize into the target tissue, and pass through cellular and nuclear barriers. High concentration of the delivered gene at the tissue would lead to enhanced cellular uptake and high level of gene expression.[114]

3.1. *Systemic targeting*

The first obstacle of a gene delivery vector is the systemic transportation. The cationic surface charge of the nanoparticles leads to the unspecific interactions with negatively charged cellular blood components and plasma proteins.[116] In general, targeting through systemic level can be accomplished by either passive or active method. In passive targeting, systemic half-life of the nanoparticles can be increased by masking the net positive charge of the DNA nanoparticles using surface modifiers such as PEG, poly hydroxypropylmethacrylamide (pHPMA) or PVP.[117] Such surface modifications inhibit non-specific interactions with blood components and enhances the circulation time, which leads to the accumulation of nanoparticles via leaky vasculature in pathological sites, like tumors, inflammations and infracted areas. The accumulated nanoparticles can be successfully retained in the dense internal connective tissue. This effect is called the enhanced permeation and retention (EPR) effect.[118] One of the drawbacks of this method is that, in many pathological situations, the integrity of vascular endothelium remains unaffected. In such situation, long circulation time may not make possible the EPR effect.

Nanoparticles accumulation at pathological sites can also be achieved using magnetic targeting. In this method, DNA nanoparticles of cationic polymer or lipid are transported into the cells with the help of a magnetic field. DNA nanoparticles are coated with super-paramagnetic iron oxide nanoparticles (SPION) to make it magnetic. It allows the concentration of nanoparticles on the cell surface and promotes its entry into the cells. This method can combine the advantages of both nanoparticles based transfection and physical methods such as electroporation and gene gun based gene delivery. The *in vitro* and *ex vivo* studies conducted using this method showed an enhancement in the efficiency.[119,120] Early *in vivo* studies into airway epithelium also showed better results.[121]

The most natural and advanced way to target nanoparticles into pathological sites is based on the attachment of specific ligands such as certain carbohydrates, proteins like growth factors or antibodies, and low-molecular-weight ligands like folic acid to the surface of the nanoparticles to recognize and bind the pathological cells.[114] These ligands are not only helping the nanoparticles to target at the specific tissue but also helps to facilitate receptor mediated cell internalization. For example, folate conjugated gene delivery vectors targets toward certain tumor cells such as lung, ovary, kidney, endometrium, breast, brain or colon, which overexpress the folate receptors and aid folate- receptor-mediated endocytosis.[115] Epidermal growth factor (EGF) receptor and ErbB2 are other tumor markers which over express in breast and prostate cancer.[117,122,123] Receptor mediated endocytosis also enables to cross the therapeutic agents through blood–brain barrier (BBB). Nanoparticles conjugated with ligands such as insulin, insulin-like growth factors or transferrin, binds specifically on BBB and mediate endocytosis.[124]

3.2. *Cellular targeting*

The first obstacle of a gene delivery vector in cellular level is plasma membrane. Receptor-mediated endocytosis enables the nanoparticles to cross this barrier. Other than receptor-mediated endocytosis cellular entry can be made possible via fluid phase endocytosis, macropinocytosis or phagocytosis.[125] Use of positively charged nanoparticles is one of the strategies to cross via these pathways due to the presence of negatively charged proteoglycans and glycoprotein on the cell surface. Cellular uptake can also be facilitated by protein transduction domain (PTD) mediated endocytosis. Nanoparticles having ligands such as TAT protein derived from HIV-1 virus,[126] the drosophila antennapedia transcription factor (ANTP)[127] and herpes simplex virus type-1 VP22 transcription factor aids PTD-mediated endocytosis.[128]

After the nanoparticles enter into the cytoplasm via endocytosis, in the form of endosome, the next barrier of delivery vector is endosomal escape. Usually the vesicular escape of the nanocomplex occurs via a pH-dependent process. Molecules having buffering capacity hinder the acidification of the endosome, causing the endosomal burst. Recently, Pittella *et al.* demonstrated the endosomal escape by destabilization of endosomal membrane using a polycation poly*N*-[*N*-(2-aminoethyl)-2-aminoethyl] aspartamide [PAsp(DET)] via selective cleavage of cis-aconitylamide in acidic endosome.[129]

3.3. *Nuclear targeting*

The final barrier of a non-viral gene delivery vector is the nucleus and its entry into nucleus enables transcription of mRNA and gene expression. The nuclear envelop is a double membrane that can be interrupted by proteins less than

40 kDa in molecular weight or 9–11 nm in diameter.[130] Nuclear membrane degradation during mitosis is one of the ways to enter into the nucleus. Other than mitosis the only way to enter into the nucleus is through the nuclear pore complex. Nanoparticles attached to proteins or peptides having nuclear localization signals (NLS) can help to overcome this barrier. TAT protein is an example of ligand having NLS, and TAT-mediated nuclear import is faster and is capable of importing large nanoparticles.

4. Conclusions

Despite development of several vectors, there may not be a single vector that would work in different disease conditions. This has to do with the difference in the pathophysiology of various diseases. Therefore, depending on the end use, a suitable non-viral vector for gene delivery can be chosen. For instant, cationic polymer based gene expression may be suitable for certain diseases like Huntington's disease, which requires only a short term but high level of gene expression. While certain other diseases like cancer and cystic fibrosis require vectors having sustained gene expression. The vector choice would also depend upon the stability and potency of the expressed protein. Despite several challenges, a growing number of studies in the field of non-viral gene expression vectors points that these could soon be used in treating different genetic diseases.

Acknowledgments

The study described here from our laboratory was supported by a grant 1R01EB003975 (to V.L.) from the National Institute of Biomedical Imaging and Bioengineering of the National Institutes of Health. We would like to thank Miss Shushi Kabu with her help in editing the chapter.

References

1. A. M. Hedgecoe, in *Encyclopedia of Applied Ethics*, C. Ruth (Ed) 2nd ed. San Diego: Academic Press (2012) p 407.
2. X. Zhang and W. T. Godbey, *Adv. Drug Del. Rev.* **58**, 515 (2006).
3. X. Gao, K.-S. Kim and D. Liu, *AAPS J.* **9**, E92 (2007).
4. D. T. Klink, S. Chao, M. C. Glick and T. F. Scanlin, *Mol. Ther.* **3**, 831 (2001).
5. H. Eliyahu, Y. Barenholz and A. J. Domb, *Molecules* **10**, 34 (2005).
6. L. Zhu and V. P. Torchilin, *Integr. Biol.* **5**, 96 (2013).
7. M. A. Lopata, D. W. Cleveland and B. Sollner-Webb, *Nucleic Acids Res.* **12**, 5707 (1984).

8. H. Luthman and G. Magnusson, *Nucleic Acids Res.* **11**, 1295 (1983).
9. C. M. Gorman, B. H. Howard and R. Reeves, *Nucleic Acids Res.* **11**, 7631 (1983).
10. D. Liu, E. Chia and H. Tian, in *Gene Delivery to Mammalian Cells*, W. Heiser (Ed.), Humana Press (2004) p 3.
11. R. E. Kingston, C. A. Chen and H. Okayama, *Current Protocols in Immunology*, Ed (John Wiley & Sons, Inc. (2001)).
12. D. C. Litzinger and L. Huang, *Biochim. Biophys. Acta* **1113**, 201 (1992).
13. T. Ren, G. Zhang, F. Liu and D. Liu, *Bioorg. Med. Chem. Lett.* **10**, 891 (2000).
14. V. Floch, S. Loisel, E. Guenin, A. C. Hervé, J. C. Clément, J. J. Yaouanc, H. des Abbayes and C. Férec, *J. Med. Chem.* **43**, 4617 (2000).
15. M. F. Lensink, C. Lonez, J.-M. Ruysschaert and M. Vandenbranden, *Langmuir* **25**, 5230 (2009).
16. M. Frederic, D. Scherman and G. Byk, *Tetrahedron Lett.* **41**, 675 (2000).
17. M. A. Ilies, W. A. Seitz, B. H. Johnson, E. L. Ezell, A. L. Miller, E. B. Thompson and A. T. Balaban, *J. Med. Chem.* **49**, 3872 (2006).
18. J. P. Behr, B. Demeneix, J. P. Loeffler and J. Perez-Mutul, *Proc. Natl. Acad. Sci. U.S.A.* **86**, 6982 (1989).
19. C. Wölk, M. Heinze, P. Kreideweiß, M. Dittrich, G. Brezesinski, A. Langner and B. Dobner, *Int. J. Pharm.* **409**, 46 (2011).
20. R. A. S. Randazzo, R. Bucki, P. A. Janmey and S. L. Diamond, *Bioorg. Med. Chem.* **17**, 3257 (2009).
21. A. Kichler, C. Leborgne, P. B. Savage and O. Danos, *J. Control. Release* **107**, 174 (2005).
22. T. Fujiwara, S. Hasegawa, N. Hirashima, M. Nakanishi, and T. Ohwada, *Biochim. Biophys. Acta* **1468**, 396 (2000).
23. R. Leventis and J. R. Silvius, *Biochim. Biophys. Acta* **1023**, 124 (1990).
24. J. H. Felgner, R. Kumar, C. N. Sridhar, C. J. Wheeler, Y. J. Tsai, R. Border, P. Ramsey, M. Martin and P. L. Felgner, *J. Biol. Chem.* **269**, 2550 (1994).
25. G. Byk, C. Dubertret, V. Escriou, M. Frederic, G. Jaslin, R. Rangara, B. Pitard, J. Crouzet, P. Wils, B. Schwartz and D. Scherman, *J. Med. Chem.* **41**, 224 (1998).
26. K. Son, F. Sorgi, X. Gao and L. Huang, In *Gene Therapy Protocols*, (Ed) (1996), p 329.
27. H. Kikuchi, H. Yamauchi and S. Hirota, *J. Liposome Res.* **4**, 71 (1994).
28. H. Talsma, M. J. van Steenbergen, J. C. Borchert and D. J. Crommelin, *J. Pharm. Sci.* **83**, 276 (1994).
29. Mozafari, *J. Aerosol. Med.* **17**, 100 (2004).
30. S. Chesnoy and L. Huang, *Annu. Rev. Biophys. Biomol. Struct.* **29**, 27 (2000).
31. G. Caracciolo, C. Sadun, R. Caminiti, M. Pisani, P. Bruni and O. Francescangeli, *Chem. Phys. Lett.* **397**, 138 (2004).
32. J. J. McManus, J. O. Radler and K. A. Dawson, *J. Phys. Chem. B* **107**, 9869 (2003).
33. V. G. Budker, A. A. Godovikov, L. P. Naumova and I. A. Slepneva, *Nucleic Acids Res.* **8**, 2499 (1980).
34. J. Marra and J. Israelachvili, *Biochemistry (Mosc).*, **24**, 4608 (1985).
35. J. J. McManus, J. O. Radler and K. A. Dawson, *J. Phys. Chem. B* **107**, 9869 (2003).
36. M. Pisani, P. Bruni, G. Caracciolo, R. Caminiti and O. Francescangeli, *J. Phys. Chem. B* **110**, 13203 (2006).
37. J. Li, Y. Yang and L. Huang, *J. Control. Release* **158**, 108 (2012).

38. H. Lboutounne, J.-F. Chaulet, C. Ploton, F. Falson and F. Pirot, *J. Control. Release* **82**, 319 (2002).
39. R. Arote, T.-H. Kim, Y.-K. Kim, S.-K. Hwang, H.-L. Jiang, H.-H. Song, J.-W. Nah, M.-H. Cho and C.-S. Cho, *Biomaterials* **28**, 735 (2007).
40. Y. Ataman-Önal, S. Munier, A. Ganée, C. Terrat, P.-Y. Durand, N. Battail, F. Martinon, R. Le Grand, M.-H. Charles, T. Delair and B. Verrier, *J. Control. Release* **112**, 175 (2006).
41. O. Jeon, H.-W. Lim, M. Lee, S. J. Song and B.-S. Kim, *J. Drug Target.* **15**, 190 (2007).
42. J. Panyam, W. Z. Zhou, S. Prabha, S. K. Sahoo and V. Labhasetwar, *FASEB J.* **16**, 1217 (2002).
43. J. Panyam and V. Labhasetwar, *Adv. Drug Del. Rev.* **64**, 61 (2012).
44. T. S. J. Kashi, S. Eskandarion, M. Esfandyari-Manesh, S. M. A. Marashi, N. Samadi, S. M. Fatemi, F. Atyabi, S. Eshraghi and R. Dinarvand *Int. J. Nanomedicine*, 7, 221 (2012).
45. J. S. Blum and W. M. Saltzman, *J. Control. Release* **129**, 66 (2008).
46. R. C. Mehta, R. Jeyanthi, S. Calls, B. C. Thanoo, K. W. Burton and P. P. DeLuca, *J. Control. Release* **29**, 375 (1994).
47. R. Bodmeier and J. W. McGinity, *Int. J. Pharm.* **43**, 179 (1988).
48. R. C. Mehta, B. C. Thanoo and P. P. Deluca, *J. Control. Release* **41**, 249 (1996).
49. P. Johansen, Y. Men, R. Audran, G. Corradin, H. Merkle and B. Gander, *Pharm. Res.* **15**, 1103 (1998).
50. E. Walter, D. Dreher, M. Kok, L. Thiele, S. G. Kiama, P. Gehr and H. P. Merkle, *J. Control. Release* **76**, 149 (2001).
51. B. Sharma, W. Ma, I. Adjei, J. Panyam, S. Dimitrijevic and V. Labhasetwar, *Drug Deliv. Transl. Res.* **1**, 43 (2011).
52. S. Prabha and V. Labhasetwar, *Mol. Pharmaceutics*, **1**, 211 (2004).
53. D. Wang, D. R. Robinson, G. S. Kwon and J. Samuel, *J. Control. Release* **57**, 9 (1999).
54. Y.-Y. Yang, T.-S. Chung and N. Ping Ng, *Biomaterials* **22**, 231 (2001).
55. S. Cohen, T. Yoshioka, M. Lucarelli, L. Hwang and R. Langer, *Pharm. Res.* **8**, 713 (1991).
56. Y. Capan, B. H. Woo, S. Gebrekidan, S. Ahmed and P. P. DeLuca, *J. Control. Release* **60**, 279 (1999).
57. Y.-Y. Hsu, T. Hao and M. L. Hedley, *J. Drug Target.* 7, 313 (1999).
58. S. Prabha and V. Labhasetwar, *Pharm. Res.* **21**, 354 (2004).
59. S. K. Sahoo, J. Panyam, S. Prabha and V. Labhasetwar, *J. Control. Release* **82**, 105 (2002).
60. C. Peetla and V. Labhasetwar, *Langmuir* **25**, 2369 (2009).
61. V. Labhasetwar, C. Song, W. Humphrey, R. Shebuski and R. J. Levy, *J. Pharm. Sci.* **87**, 1229 (1998).
62. A. R. Xu, M. F. Yao, G. K. Xu, J. Y. Ying, W. C. Ma, B. Li and Y. Jin, *Int. J. Nanomed.* 7, 3547 (2012).
63. M. Subrahmanyam, S. Kaneco, N. Alonso-Vante, V. Labhasetwar, J. Bonadio, S. A. Goldstein and R. J. Levy, *Colloids Surf. B Biointerfaces* **16**, 281 (1999).
64. D. Y. Kwoh, C. C. Coffin, C. P. Lollo, J. Jovenal, M. G. Banaszczyk, P. Mullen, A. Phillips, A. Amini, J. Fabrycki, R. M. Bartholomew, S. W. Brostoff and D. J. Carlo, *Biochim. Biophys. Acta* **1444**, 171 (1999).

65. Y. H. Choi, F. Liu, J.-S. Kim, Y. K. Choi, P. Jong Sang and S. W. Kim, *J. Control. Release* **54**, 39 (1998).
66. G. Liu, M. Molas, G. A. Grossmann, M. Pasumarthy, J. C. Perales, M. J. Cooper and R. W. Hanson, *J. Biol. Chem.* **276**, 34379 (2001).
67. M. Lee and S. W. Kim, *Pharm. Res.* **22**, 1 (2005).
68. S. Katayose and K. Kataoka, *Bioconjug. Chem.* **8**, 702 (1997).
69. J. Deng, N. Gao, Y. Wang, H. Yi, S. Fang, Y. Ma and L. Cai, *Biomacromolecules* **13**, 3795 (2012).
70. N. J. Boylan, A. J. Kim, J. S. Suk, P. Adstamongkonkul, B. W. Simons, S. K. Lai, M. J. Cooper and J. Hanes, *Biomaterials* **33**, 2361 (2012).
71. O. Boussif, F. Lezoualc'h, M. A. Zanta, M. D. Mergny, D. Scherman, B. Demeneix and J. P. Behr, *Proc. Natl. Acad. Sci. U S.A* **92**, 7297 (1995).
72. L. Wightman, R. Kircheis, V. Rössler, S. Carotta, R. Ruzicka, M. Kursa and E. Wagner, *J. Gene Med.* **3**, 362 (2001).
73. A. Kichler, *Gene Med.* **6**, S3 (2004).
74. C.-H. Ahn, S. Y. Chae, Y. H. Bae and S. W. Kim, *J. Control. Release* **80**, 273 (2002).
75. M. A. Gosselin, W. Guo and R. J. Lee, *Bioconjug. Chem.* **12**, 989 (2001).
76. P. Erbacher, T. Bettinger, P. Belguise-Valladier, S. Zou, J. L. Coll, J. P. Behr and J. S. Remy, *J. Gene Med.* **1**, 210 (1999).
77. M. R. Park, K. O. Han, I. K. Han, M. H. Cho, J. W. Nah, Y. J. Choi and C. S. Cho, *J. Control. Release* **105**, 367 (2005).
78. Y. H. Kim, J. H. Park, M. Lee, Y.-H. Kim, T. G. Park and S. W. Kim, *J. Control. Release* **103**, 209 (2005).
79. H. Wei, J. A. Pahang and S. H. Pun, *Biomacromolecules* **14**, 275 (2012).
80. X. Lu, Y. Ping, F. J. Xu, Z. H. Li, Q. Q. Wang, J. H. Chen, W. T. Yang and G. P. Tang, *Bioconjug. Chem.* **21**, 1855 (2010).
81. D. Li, Y. Ping, F. Xu, H. Yu, H. Pan, H. Huang, Q. Wang, G. Tang and J. Li, *Biomacromolecules* **11**, 2221 (2010).
82. Y. L. Lo, K. H. Sung, C. C. Chiu and L. F. Wang, *Mol. Pharmaceutics* **10**, 664.
83. Q. Jiang, P. Shi, C. Li, Q. Wang, F. Xu, W. Yang and G. Tang, *Macromol. Biosci.* **11**, 435 (2011).
84. V. B. Morris and C. P. Sharma, *Biomaterials* **31**, 8759 (2010).
85. V. B. Morris and C. P. Sharma, *Biomaterials* **32**, 3030 (2011).
86. S. Shi, X. C. Zhu, Q. F. Guo, Y. J. Wang, T. Zuo, F. Luo and Z. Y. Qian, *Int. J. Nanomedicine* 7, 1749 (2012).
87. C. Zhan, Q. Meng, Q. Li, L. Feng, J. Zhu and W. Lu, *Chem. Asian J.* 7, 91 (2012).
88. H. Kim, H. Bin Man, B. Saha, A. M. Kopacz, O. S. Lee, G. C. Schatz, D. Ho and W. K. Liu, *J. Phys. Chem. Lett.* **3**, 3791 (2012).
89. X. Gao and L. Huang, *Biochemistry (Mosc)* **35**, 1027 (1996).
90. R. J. Lee and L. Huang, *J. Biol. Chem.* **271**, 8481 (1996).
91. L. K. Lee, C. L. Williams, D. Devore and C. M. Roth, *Biomacromolecules* 7, 1502 (2006).
92. S. Mansouri, P. Lavigne, K. Corsi, M. Benderdour, E. Beaumont and J. C. Fernandes, *Eur. J. Pharm. Biopharm.* **57**, 1 (2004).
93. F. C. MacLaughlin, R. J. Mumper, J. Wang, J. M. Tagliaferri, I. Gill, M. Hinchcliffe and A. P. Rolland, *J. Control. Release* **56**, 259 (1998).

94. K. Corsi, F. Chellat, L. H. Yahia and J. C. Fernandes, *Biomaterials* **24**, 1255 (2003).
95. F. Langenbucher, *Eur. J. Pharm. Biopharm.* **53**, 1 (2002).
96. M. Huang, C.-W. Fong, E. Khor and L.-Y. Lim, *J. Control. Release* **106**, 391 (2005).
97. T. Sato, T. Ishii and Y. Okahata, *Biomaterials*, **22**, 2075 (2001).
98. X. Zhang, C. Yu, XuShi, C. Zhang, T. Tang and K. Dai, *Biochem. Biophys. Res. Commun.* **341**, 202 (2006).
99. E. A. McNeela and K. H. G. Mills, *Adv. Drug Del. Rev.* **51**, 43 (2001).
100. K. Khatri, A. K. Goyal, P. N. Gupta, N. Mishra and S. P. Vyas, *Int. J. Pharm.*, **354** 235 (2008).
101. E. A. Klausner, Z. Zhang, R. L. Chapman, R. F. Multack and M. V. Volin, *Biomaterials* **31**, 1814 (2010).
102. E. Ramos, J. Relucio and C. Torres-Villanueva, *Mar. Biotechnol.* 7, 89 (2005).
103. X. Jiang, H. Dai, K. W. Leong, S. H. Goh, H. Q. Mao and Y. Y. Yang, *J. Gene Med.* **8**, 477 (2006).
104. S. M. Noh, M. O. Park, G. Shim, S. E. Han, H. Y. Lee, J. H. Huh, M. S. Kim, J. J. Choi, K. Kim, I. C. Kwon, J.-S. Kim, K.-H. Baek and Y.-K. Oh, *J. Control. Release* **145**, 159 (2010).
105. W. Zhang, S. Pan, Y. Wen, X. Luo and X. Zhang, *J. Biomater. Sci. Polym. Ed.*, **21**, 741 (2010).
106. B. Layek and J. Singh, *Biomacromolecules* (2013).
107. O. Germershaus, S. Mao, J. Sitterberg, U. Bakowsky and T. Kissel, *J. Control. Release* **125**, 145 (2008).
108. M. Thanou, B. I. Florea, M. Geldof, H. E. Junginger and G. Borchard, *Biomaterials* **23**, 153 (2002).
109. V. B. Morris, C. K. S. Pillai and C. P. Sharma, *Polym. Int.* **60**, 1097 (2011).
110. Y. Zheng, Z. Cai, X. Song, B. Yu, Y. Bi, Q. Chen, D. Zhao, J. Xu and S. Hou, *Int. J. Pharm.* **382**, 262 (2009).
111. V. B. Morris and C. P. Sharma, *J. Colloid Interface Sci.* **348**, 360 (2010).
112. V. B. Morris and C. P. Sharma, *Int. J. Pharm.* **389**, 176 (2010).
113. J. Zhang, C. He, C. Tang and C. Yin, *Pharm. Res.* 1 (2013).
114. M. K. Aneja, J. P. Geiger, A. Himmel and C. Rudolph, *Expert Opin. Drug Deliv.* **6**, 567 (2009).
115. W. Xia and P. S. Low, *J. Med. Chem.* **53**, 6811 (2010).
116. M. Ogris, S. Brunner, S. Schuller, R. Kircheis and E. Wagner, *Gene Ther.* **6**, 595 (1999).
117. M. Ogris and E. Wagner, *Drug Discov. Today* 7, 479 (2002).
118. S. K. Kim, M. B. Foote and L. Huang, *Cancer Lett.*, DOI: 10.1016/j.bbr.2011.03.031 (2012).
119. S. W. Kamau, P. O. Hassa, B. Steitz, A. Petri-Fink, H. Hofmann, M. Hofmann-Amtenbrink, B. von Rechenberg and M. O. Hottiger, *Nucleic Acids Res.* **34**, e40 (2006).
120. F. Krotz, H. Y. Sohn, T. Gloe, C. Plank and U. Pohl, *J. Vasc. Res.* **40**, 425 (2003).
121. S. Xenariou, U. Griesenbach, S. Ferrari, P. Dean, R. K. Scheule, S. H. Cheng, D. M. Geddes, C. Plank and E. W. F. W. Alton, *Gene Ther.* **13**, 1545 (2006).
122. D. B. Agus, R. W. Akita, W. D. Fox, G. D. Lewis, B. Higgins, P. I. Pisacane, J. A. Lofgren, C. Tindell, D. P. Evans, K. Maiese, H. I. Scher and M. X. Sliwkowski, *Cancer Cell* **2**, 127 (2002).
123. S. T. Lo, A. Kumar, J. T. Hsieh and X. Sun, *Mol. Pharm.* **10**, 793 (2013).

124. A. Beduneau, P. Saulnier and J. P. Benoit, *Biomaterials* **28**, 4947 (2007).
125. S. D. Conner and S. L. Schmid, *Nature* **422**, 37 (2003).
126. A. D. Frankel and C. O. Pabo, *Cell* **55**, 1189 (1988).
127. A. Joliot, C. Pernelle, H. Deagostini-Bazin and A. Prochiantz, *Proc. Natl. Acad. Sci. USA*. **88**, 1864 (1991).
128. G. Elliott and P. O'Hare, *Cell* **88**, 223 (1997).
129. F. Pittella, M. Z. Zhang, Y. Lee, H. J. Kim, T. Tockary, K. Osada, T. Ishii, K. Miyata, N. Nishiyama and K. Kataoka, *Biomaterials* **32**, 3106 (2011).
130. S. R. Wente, *Science* **288**, 1374 (2000).

Chapter 15

Cancer Hyperthermia Using Nanomagnetic Particles and Induction of Immune Responses

Takeshi Kobayashi

School of Bioscience and Biotechnology, Chubu University, Matsumoto-cho 1200, Kasugai, Aichi 487-8501, Japan
kobatake@isc.chubu.ac.jp

Abstract

Magnetic nanoparticle-mediated hyperthermia has the potential to achieve heating a local tumor tissue to the desired temperature without damaging normal tissue. The technique consists of targeting magnetic nanoparticles to tumor tissue followed by application of an external alternating magnetic field that induces heat generation by the magnetic nanoparticles. Temperature in tumor tissue is increased to above 43°C, which causes necrosis of cancer cells but does not damage surrounding normal tissue. Magnetite has been extensively studied among available magnetic nanoparticles. Recent years have seen remarkable advances in magnetite nanoparticle-mediated hyperthermia; both functional magnetite nanoparticles and alternating magnetic field generators have been developed. By applying magnetite nanoparticle-mediated hyperthermia treatment, a lot of heat shock proteins (HSPs) are expressed within and around tumor tissue, which causes unexpected biological responses such as tumor-specific immune response. These results suggest that the hyperthermia is able to kill not only local tumors exposed to heat treatment but also tumors at distant sites, including metastatic cancer cells. Currently, some researchers are attempting to begin clinical trials, suggesting that time may have come for clinical applications. This chapter describes recent advances in magnetic nanoparticle-mediated hyperthermia.

1. Introduction

Hyperthermia, as a therapeutic approach, consists in rising the temperature of the whole body (whole body hyperthermia), part of the body (regional hyperthermia) or tissue (local hyperthermia), with increasing temperature applicable regarding deleterious effects of heat.[1,2] For whole body hyperthermia, temperature increases are limited to the range of 39 to 41°C, also called fever-like range. It is worth to discern fever from hyperthermia: both are defined as elevation in body temperature but, in contrast to fever, hyperthermia is not associated with a change in the set point or thermo effectors such as pyretogenous cytokines.[1,2] In whole body hyperthermia, the physiological response to thermal stress led at first to immediate adaptation of hemodynamic parameters such as elevation of heart rate, cardiac output and blood flow with vasodilatation. Hemodynamic changes can redistribute blood especially in solid tumors, which are often characterized by limited and deficient vascularization.The effect of whole body hyperthermia on pharmacokinetics of antineoplastic agents is not precisely known. Changes in metabolism, very sensitively unbalanced by heat, distribution volume and excretion, are difficult to predict.[1] Due to these facts, whole body hyperthermia is not clinically common.

With the regional or local hyperthermia modalities, temperatures ranging from 40 to 46°C are applicable. A very important practical consideration distinguishes moderate hyperthermia modality corresponding to temperature below 42°C, which can routinely be achieved in clinic.[1] A consensus landmark is the threshold temperature of 43°C.[3,4] Direct cell-killing effect can be observed for temperatures ranging from 41 to 47°C for human healthy cells with variation of a 10-fold factor in sensitivity with the cell type. Above the "breakpoint" temperature of 43°C, heat cytotoxicity is markedly higher.[1,4] The threshold energy required to trigger lethal injury appears comparable to that triggering protein denaturation, thus suggesting that the direct cytotoxic effect of hyperthermia is mainly based on cellular protein denaturation. Heat sensitivity can be increased at lower temperature in hypoxic and acidic microenvironment, as it is the case in a solid tumor for cancer cells as well as for infiltrated healthy cells. Hyperthermia is able to induce both necrosis and apoptosis, depending on temperature. Susceptibility to apoptosis seems important below the consensus temperature of 43°C, whereas necrosis seems more likely above.

A major technical problem with the currently available regional or local hyperthermia modalities is the difficulty of heating a local tumor tissue to the desired temperature without damaging normal tissue. High temperatures above 43°C can kill a great number of tumor cells, but normal tissues are also severely damaged under these conventional hyperthermia treatments. Therefore, the development of novel hyperthermia systems which can heat tumor tissue above 43°C and are capable of specifically targeting tumor tissue is required.

Magnetic nanoparticle-mediated hyperthermia is a largely experimental modality for hyperthermia application which has the potential to overcome these shortcomings.[5,6] This technique consists of targeting magnetic nanoparticles to tumor tissue, and then applying an external alternating magnetic field (AMF) to induce heat generation in the nanoparticles via hysteresis loss or Néel relaxational loss. Recent years have seen remarkable advances in magnetic nanoparticle-mediated hyperthermia; both tumor-targeted magnetic nanoparticles and AMF generators have been developed, and some of these are just entering into clinical trials.[7–12] This review covers recent advances in magnetic nanoparticle-mediated hyperthermia and discusses future directions.

2. Magnetic Nanoparticles for Hyperthermia

In order to increase tumor tissue above 43°C, inductive heating was at first proposed. This approach has been based on the heating through Joule effect of the eddy currents induced in a metallic conductor by AMF. This was applied to metallic elements used for other indications such as orthopedic elements and stents, and to dedicated material specially developed for this use known as thermoseeds.[13] Ceramics and glass, already clinically applied in bone surgery, can acquire through incorporation of metallic constituent add the capability of heating for further treatment of bone neoplastic processes. Implantable rods have also been developed for soft tissue indication. *In situ*, a thermoseed acts as a focal heat source, reaching rapidly very high temperature associated with a steep temperature gradient in surrounding tissue. Thus, homogeneous heating of the target volume technique requires the insertion of thermoseeds in adequate number and pattern; limit of this approach consists in the poor control of heat distribution with focal under heating and overheating.

Therefore, magnetic soft materials such as ferro- and ferri-magnetic iron oxides (maghemite(γ-Fe_2O_3) and magnetite(Fe_3O_4)) have been used to interact more specifically by magnetic induction with AMF.[14,15] The magnetization depends on the magnetic permeability of the material, which in turn depends on the temperature. Above a specific temperature corresponding to a threshold thermal energy overcoming spin lattice organization, namely the Curie point, the material loses its magnetic permeability. This annihilates magnetization and magnetic interactions, and thus stops heat dissipation. This occurs for common magnetic material at high temperature (for instance, magnetite has the Curie point at 850°C), but for alloys, the Curie point can be decreased to low temperature in relation with lattice composition. This led to the development of so-called self-regulating thermoseeds, which are composed of specific alloy having a Curie point precisely set at the maximum desired temperature in clinical applications.[16,17] For this purpose, barium is preferred due to its relative biocompatibility. An inevitable technical problem with hyperthermia is the difficulty of uniformly heating only the tumor region to the required temperature without damaging surrounding normal tissue. As a result, magnetic materials of very fine particles such as 10 nm are preferred to thermoseeds, since they can be injected *in vein* or directly into tumor tissue. It is not easy to prepare fine particles of the magnetic materials with low Curie point, and this restricts the magnetic materials with low Curie point only to the application for thermoseeds.

When considering pure magnetic interactions with AMF, the mechanisms of heating rely now on mesoscopic and microscopic magnetic considerations, which is best described by statistical mechanics with the magnetic domain theory.[18] One domain in a lattice is represented by a population of molecular dipoles with isotropic orientation (magnetic molecular field) of finite size enclosed by a bordering population of molecular dipoles with anisotropic orientation, the so-called domain walls.[16] The randomly patterned magnetic domains have a resultant zero magnetization at macroscopic level. One isolated domain, that is called a single-domain, is only possible below a minimal domain wall size. Above this critical size, domain walls divide the lattice in the so-called multi-domain structure. This accounts for an important distinction at mesoscopic level between multi-domain and single-domain magnetic structure of ferri- and ferromagnetic materials, which drastically changes magnetization process in static or dynamic conditions. In the case of ferri- and ferromagnetic materials, single-domain magnetic structure is observed for the particles of about 10 nm in diameter. Such a ferri- or ferromagnetic material appears thus to adopt a pseudo paramagnetic behavior that is called superparamagnetism. In clinical applications, this superparamagnetic behavior is desired to prevent adverse magnetic aggregation in the body, especially critical for the *i.v.* route.

A multi-domain magnetic microstructure is characterized by hysteresis losses. Magnetization process occurs through domain walls displacement so that domains preferentially oriented along the axis of magnetic induction grow at the expense of the other domains not preferentially oriented. On the other hand, a single-domain magnetic structure is characterized by Néel relaxation. Néel relaxation is based on the rotation of each individual molecular dipole belonging to a static lattice, which collectively relaxes into random orientation. Thermal energy arises from orbital friction and dissipation to the environment.

In general, magnetic characteristics depend on the particle size and on the methods used for the preparation of submicron magnetic particles. Multi-domain ferromagnetic characteristics change to single-domain ferromagnetic, and finally, to superparamagnetic characteristics as the particle size decreases.[19] Multi-domain ferromagnetic particles possess lower hysteresis loss than single-domain ferromagnetic particles. Therefore, single-domain ferromagnetic particles generate more heat when exposed to AMF as shown by Shinkai *et al.*[20] and Hergt *et al.*[21] Superparamagnetic particles have no hysteresis loss, and generate heat due to Néel relaxational loss in AMF as mentioned by Hergt *et al.*[21] Therefore, two types of loss mechanisms have been found to be of interest for hyperthermia: hysteresis loss and Néel relaxational loss.[21,22] Both loss types show a non-monotonic dependence of loss with particle size: i.e. there exist optimum particle sizes which are different for each loss mechanism. Hysteresis loss increases with decreasing particle size due to increasing remanence and coercivity until the Néel relaxation effects appear. There, in a narrow transition region to superparamagnetic behavior, remanence and coercivity decrease abruptly.[21,22]

Magnetic iron oxides, such as maghemite or magnetite, are the two ferrimagnetic materials accepted presently in clinical use. Mono-domain particles are called superparamagnetic iron oxides nanoparticles (SPIONs). Reproducible SPIONs synthesis methods are available,[23] but synthesis of highly controlled monodisperse mono-domain magnetic particles dispersion is still a challenge. Practically, this has strong repercussions on the heat dissipation capacities.[24,25] Due to their high surface energy, SPIONs agglomeration occurs rapidly and irreversibly, resulting in loss of superparamagnetism by dipole–dipole attraction and leading to a ferrimagnetic behavior characteristic of larger particles. SPIONs agglomeration is prevented by suitable coating, showing the importance of stabilizing agent or coating to control biological interaction.[26]

FDA has approved the clinical use of coated SPIONs as magnetic resonance imaging (MRI) contrast agent in products such as Endoderm™, Feridex® and Resovist®. For example, Resovist® is a complex of magnetic nanoparticles (diameter, 50 nm) surrounded by dextran, which is chemically bound to its surface, and is stable as a colloid and can be mixed with various solvents or serum without aggregating or precipitating. The literature[27] states that the core is maghemite, and Hergt *et al.*[21] agreed that the core of these particles is maghemite judging from analysis using X-ray diffraction. Resovist® was first applied clinically for detecting liver cancer, since Resovist® is taken up rapidly by the reticuloendothelial systems, such as the Kupffer cells in liver, and when compared to uptake by liver cancer cells. Kupffer cells exist in normal liver but are not present in liver cancer. Therefore, Resovist® has the potential to enhance relaxivity in a normal liver under MRI conditions. However, these products have no targeting ability against tumor tissues, and the application is limited only for detection of liver cancer.

Both multi-domain and single-domain magnetic microstructures can contribute to further development of complex pharmaceutical formulations designed for magnetically mediated hyperthermia. Injectable formulations carrying magnetic particles are interesting for magnetically mediated hyperthermic treatment of soft tissue tumor. The most important point in hyperthermia is heating uniformly only the tumor region to the required temperature without damaging surrounding normal tissue. As a result, some researchers have proposed the use of "intracellular" hyperthermia to provide a tumor-specific hyperthermia system, and submicron magnetic particles (typically, less than 100 nm in diameter) with a targeting ability against tumor tissues have been developed for this purpose. When magnetic particles with a targeting ability against tumor tissues have been prepared, they are injected *in vein* or into tumor tissues directly and intracellular hyperthermia becomes possible as shown in Fig. 1a.

Any submicron magnetic particles which can generate heat under AMF can theoretically be used for intracellular hyperthermia. However, the most important criterion is that the magnetic particles be non-toxic. Because of this requirement, magnetite and maghemite particles have been the focus of most studies. Maghemite is produced by the oxidation of magnetite above 300°C, and the steps required to produce magnetite are simpler than those required to produce maghemite. The heating properties of magnetite and maghemite are comparable for use in intracellular hyperthermia. Therefore, most studies of submicron magnetic particles for intracellular hyperthermia have focused on magnetite.

In 1979, Gordon *et al.* first proposed the concept of intracellular hyperthermia.[28] About 6 nm diameter magnetite particles suspended in a sucrose solution were injected into a tail vein, and the selective uptake of particles by a tumor was observed with light microscopy and electron microscopy.

In 1993, Jordan *et al.* developed 15 nm aminosilan-coated magnetite nanoparticles in order to enhance cellular uptake and prevent intracellular digestion of the particles.[29] As a result, intracellular hyperthermia showed a more potent cell killing effect than heating in a water bath. Thus, the use of

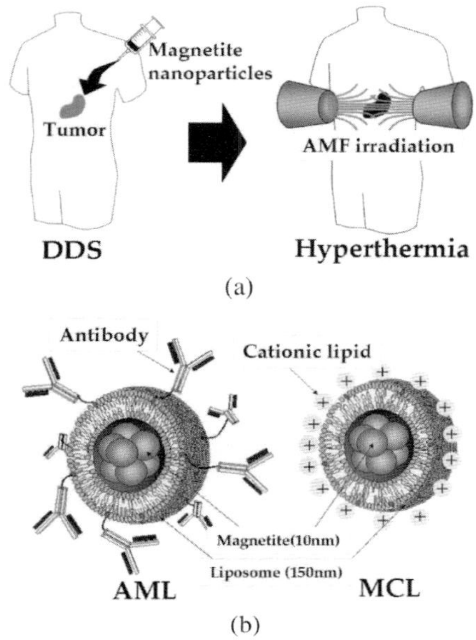

Fig. 1. Hyperthermia using magnetic nanoparticles. (a) Scheme of hyperthermia treatment. Magnetic nanoparticles are concentrated in tumor tissue by the drug delivery system (DDS). Then, the nanoparticles are irradiated with an AMF produced outside of the human body, resulting in tumor-specific intracellular hyperthermia. (b) Liposomal drugs containing magnetite nanoparticles for DDS. Left, antibody-conjugated magnetoliposome (AML); right, magnetite cationic liposome (MCL).

coated magnetite nanoparticles provided a promising approach for intracellular hyperthermia.

Melanogenesis, i.e. biosynthesis of melanin pigments in the presence of tyrosinase, uniquely occurs in melanocytic cells. Tyrosinase converts tyrosine to dopa and subsequently to dopaquinone which will be then autoxidized or oxidized by tyrosinase-related proteins to form melanin pigments. In this oxidation process, highly cytotoxic free radicals are produced, and tyrosine analogs that are tyrosinase substitutes are good candidates for melanoma-specific targeting and destroy them. Jimbow group has focused on the application of the melanogenesis substrate, *N*-propionyl-cysteaminylphenol (NPrCAP), since NPrCAP is selectively incorporated into melanoma cells and inhibits their growth by production of cytotoxic free radicals.[30] In fact, NPrCAP could selectively disintegrate follicular melanocytes after single or multiple *i.p.* administration to new-born or adult C57 black mice.[31] A high, specific uptake of *N*-acetyl-cysteaminylphenol (NAcCAP) was seen by melanoma cell lines compared to non-melanoma cells.[30] Also only melanoma cells showed the specific cytotoxicity of NPrCAP on the *in vitro* DNA synthesis

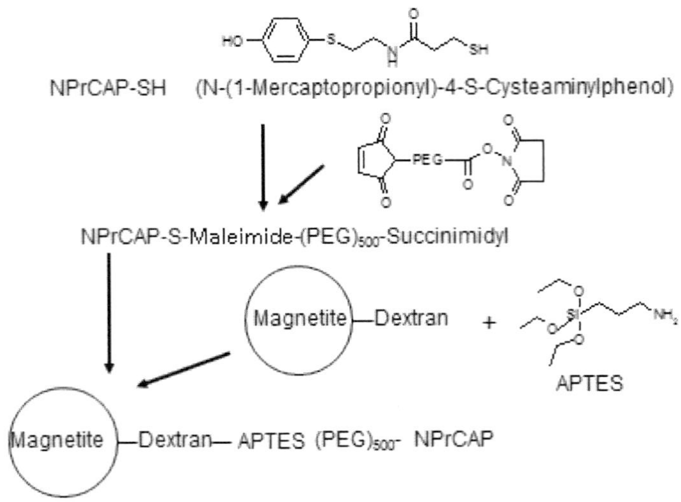

Fig. 2. Preparation method of NPrCAP/PEG/M. NPrCAP: *N*-(1-mercaptopropionyl)-4-cysteaminylphenol, (PEG)500: polymer of about 500 molecules of ethylene glycol, APTES: aminopropyl triethoxysilane.

and the *in vivo* growth inhibition studies. The *in vivo* lung metastasis experiment also showed the decreased number of lung melanoma colonies.[32]

In order to develop a novel anti-melanoma therapy using the unique biological properties of NPrCAP, Jimbow *et al.* further established a novel chemothermotherapy which was demonstrated in a B16 mouse melanoma system.[9,10] They constructed a particle with a conjugated NPrCAP on the surface of dextran-coated magnetite nanoparticles (NPrCAP/M) or NPrCAP/PEG/M which was made by conjugating polyethylene glycol (PEG) between NPrCAP and magnetite nanoparticles as shown in Fig. 2. NPrCAP/M or NPrCAP/PEG/M selectively leads to the disintegration of melanoma cells and generate heat upon exposure to AMF. Importantly, there was a significant melanoma growth inhibition of re-challenge melanoma transplants and increased life span of the host animals, i.e. 30–50% complete rejection.[9,10,33–35]

For DDS, liposomal coatings provide a promising approach. Kobayashi group used DDS techniques with liposomes to provide intracellular hyperthermia.[36] An accumulation of magnetite nanoparticles (with a diameter of 10 nm) in tumor cells can be enhanced by conferring a positive surface charge to the liposomal surface. This group has developed "magnetite cationic liposomes (MCLs)" with improved adsorption and accumulation properties as shown in Fig 1b (right).[36,37] MCLs, which have a positive surface charge,

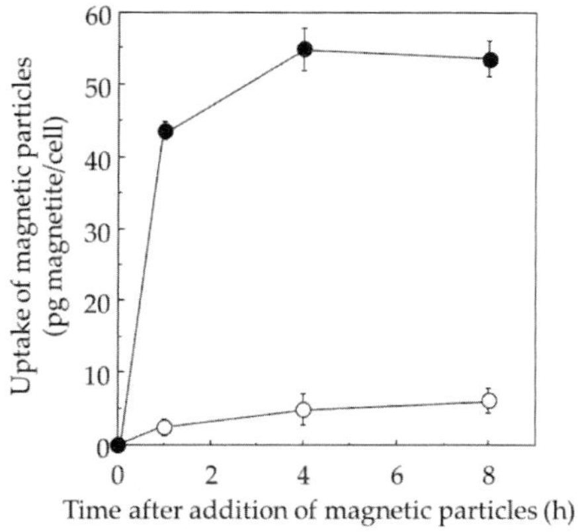

Fig. 3. Comparison of magnetite uptake between MCLs and MLs. Open and closed symbols indicate data for MLs and MCLs, respectively. The maximum MCL uptake (55 pg/cell) was achieved after 4 h, and was 10 times higher than that for MLs.

have a 10-fold higher affinity for glioma cells than neutrally charged magnetoliposomes (MLs) as shown in Fig. 3.

Furthermore, a significant development in intracellular hyperthermia occurred when Kobayashi group developed antibody-conjugated PEG derivatives containing magnetite[38] and antibody-conjugated liposomes containing magnetite nanoparticles as shown in Fig. 1B (left) (AMLs; core diameter, 10nm). They constructed AMLs using mouse G22 monoclonal antibodies (MAb) against human glioma cells,[39,40] mouse G250 MAb against human renal cell carcinomas[41,42] and humanized MAb against human epidermal growth factor receptor-2 (HER2) (Herceptin®),[43,44] MAb against human high molecular weight–melanoma associated antigen (HMW–MAA).[11]They demonstrated in an animal model that AMLs using G250 MAb against human renal cell carcinomas have the tumor-specific targeting ability for human renal cell carcinomas as shown in Fig. 4.[42]

Recently, Kobayashi group has also developed oligosaccharide-conjugated liposomes containing magnetite nanoparticles (oligosaccharides-conjugated magnetoliposomes, OMLs).[45] In the case of mannotriose-conjugated magnetoliposomes, macrophages specifically recognize OMLs via carbohydrate receptors such as the macrophage–mannose-receptor (CD206), and these can be used as a cellular vehicle for targeting macrophages.

When tumors are located in organs with a high blood flow, the temperature of the tumors heated with magnetic nanoparticles does not increase as much as desired, because heat is dissipated by the blood flow. Needle-type metal implants[46,47] have been developed for such situations. The needle-type

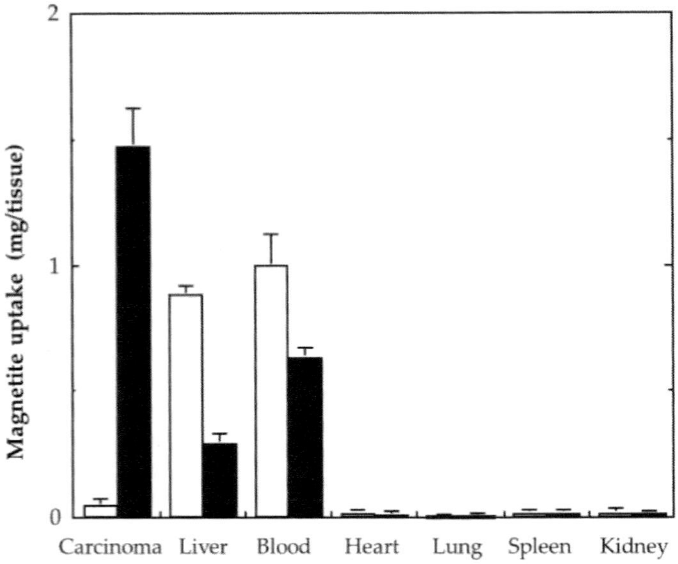

Fig. 4. AMLs for active targeting. A G250 antibody was covalently linked on the surface of MLs. Magnetite uptake of G250-AMLs for carcinomas and various organs 48 h after the injection (0.4 mL, net magnetite: 3 mg) is shown by open (MLs without antibody (control)) and closed (G250-AMLs) symbols, respectively. With G250-AMLs, the uptake in carcinomas was 1.5 mg per tissue, which corresponded to 50% of the total injected amount. This was approximately 27 times higher than the uptake of the control.

metal implants can generate heat and increased temperatures in organs with a high blood flow. However, the temperature of tumor tissues located at a distance from the implants does not increase above 43°C. In such cases, regrowth of the tumor can occur from tumor cells which were located at a distance from the implants.

Furthermore, the implants must be removed from the body after hyperthermia. Kobayashi group developed a magnetite needle,[48,49] in which magnetite nanoparticles are molded with carboxymethyl cellulose into a needle shape. It was possible to administer the needle within a few minutes, and the temperature rise was very rapid in organs with a high blood flow due to the very high magnetite concentration. These results suggest that magnetite needles could provide very simple and effective particulate heating mediators.

One of the characteristics of magnetite nanoparticles is the distribution of the particles within tumor tissues after repeated hyperthermia (RH)[50] as shown in Fig. 5. When MCLs were injected into tumor tissues, they remained at the injected site because of electrostatic interactions between the MCLs and the tumor cell membrane. When the first hyperthermic treatment was

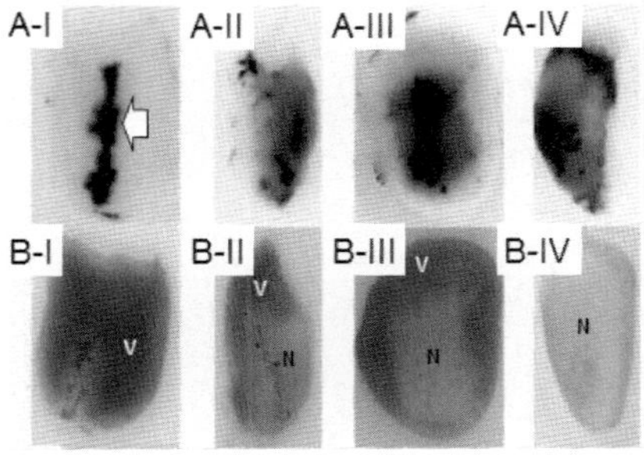

Fig. 5. Photographs of tumor specimens. Tumors were resected at 24 h after hyperthermia and were (A) paraffinized and (B) histologically stained with hematoxylin-eosin. I: without AMF irradiation, II: irradiated once for 30 min, III: irradiated twice for 30 min, IV: irradiated three times for 30 min. An arrow indicates MCLs. N and V in the photographs indicate necrotic tissues and viable tissues, respectively.

applied, the temperature of the MCLs increased above 43°C, and the tumor cells located near the MCLs were killed. In this necrotic area, the MCLs diffused and spread within the tumor. After repeated hyperthermic treatments, additional diffusion occurs, and the MCLs can expand into the entire tumor tissue. In this case, the removal of the particles is not necessary, because the particles are carried away by blood flow after several hyperthermic treatments.

Several types of magnetite nanoparticles developed for intracellular hyperthermia are listed in Table 1. The magnetite core sizes range from 5 nm to 15 nm. When used for hyperthermia, the effective absorption of the energy of AMF is an important property of magnetite nanoparticles. Chan *et al.* reported on modified dextran magnetite and its hyperthermic effect using several human carcinoma cell lines *in vitro*.[51] The specific absorption rate (SAR) (which indicates the heat evolution rate in hyperthermia) of the dextran magnetite particles was low because conventional dextran magnetite with a size of less than 10 nm behaves as a superparamagnetic particle rather than a ferromagnetic one due to their small size, so that the hysteresis loss was very low. Chan *et al.* controlled the oxygen concentration during the preparation of the dextran magnetite and selected particles of around 15 nm.[51]

Okawa *et al.*[52] synthesized four kinds of magnetite particles having average sizes of 7, 18, 40 and 80 nm, and investigated their heating ability when they were dispersed in an agar gel and exposed to AMF at 120 kHz. The particles which had an average diameter of 18 nm possessed the highest heating ability, although they exhibited narrow hysteresis loops when compared to particles having average diameters of 40 and 80 nm. This indicated that

Table 1. Magnetic nanoparticles for intracellular hyperthermia.

Name	Core size	Characters	References
Magnetite	6 nm	The first reported for intracellular hyperthermia	28
Aminosilane-coated magnetite	15 nm	Enhances the uptake by cancer cells, prevents intracellular digestion	29
Dextran magnetite	5–10 nm	Enhances the blood circulation time, stabilizes the colloidal solution	57
Magnetite cationic liposome (MCL)	10 nm	Enhances the uptake by cancer cells, stabilizes the colloidal solution	36, 37
Antibody-conjugated magnetoliposome (AML)	10 nm	Targets to tumor cells, stabilizes the colloidal solution	11, 38–42
Oligosaccharides-conjugated magnetoliposome (OML)	10 nm	Targets to macrophages	45
NPrCAP-conjugated magnetite nanoparticle (NPrCAP/M)	10 nm	Targets to melanoma cells, exerts chemotherapeutic effects	9, 10, 34

hysteresis loss did not contribute much to the heat rise generated by the 120 kHz alternating field, and the Néel relaxation loss contributed predominantly to the heat rise caused by the 18-nm sized particles.

Magnetic materials suitable for hyperthermia are also discussed by Atsumi *et al.* in terms of their magnetic properties, practical limitations in treatment conditions and of instrumentation.[53] The experimental results suggested that either ferromagnetic particles with a very low anisotropy constant, or superparamagnetic particles with a moderate anisotropy constant are suitable. Considering the magnetic and biocompatibility of the particles, magnetite SPIONs with a diameter of 11–13 nm were considered the most promising by this group.

3. Magnetic Field Applicators for Intracellular Hyperthermia

During the past decade or so, various magnetic particles possessing biocompatiblilty, injectability and high levels of accumulation in the target tumor have been developed for intracellular hyperthermia. After the particles have been selectively taken up by tumor cells, an external AMF is applied to tumor tissues. However, even if an AMF generator is not available, the temperature of the tumor tissues is preferentially elevated when radiofrequency capacitive heating is applied as reported by Kobayashi group.[54,55] Radiofrequency capacitive heating is popular in clinical hospital settings in Japan, and intracellular hyperthermia using radiofrequency capacitive heating is one of the choices available for clinical applications. In this case, magnetite particles in the tumor tissues may attract more electrical current than normal tissues, and the temperature difference between the tumor tissues and the non-tumor tissues can reach 2 to 3°C. However, the temperature of the non-tumor tissues located between the electrodes inevitably increases in the case of radiofrequency capacitive heating, and a selective heating system for tumor tissues is desirable. Therefore, the development of AMF generators is indispensable for precisely targeted intracellular hyperthermia.

The AMF generators have been developed by some companies and universities. In Germany, MagForce Nanotechnologies developed the MFH 300F 100 kHz magnetic field applicator for intracellular hyperthermia.[56] The magnetic applicator in the MFH 300F instrument is a "gap type" generator with a C-type toroidal core, and patients are placed inside the aperture. The width of the aperture is 70 cm and the gap size is adjustable from 21 to 45 cm. Using the MFH 300F, Gneveckow *et al.*[56] found a magnetic field intensity of up to 18 kA/m in a cylindrical treatment area 20 cm in diameter and an aperture height of up to 30 cm. The SAR can be controlled directly by the magnetic field strength during the treatment. In Japan, Mitsumori *et al.*[57] developed an AMF generator which is a cooled, gap-type generator with a C-type toroidal core. It generates a 100-kHz AMF. They reported that the magnetic field was uniform in the 10 cm × 10 cm area between the cores, and gap space is 25 cm.

In Japan, some solenoid-type magnetic applicators have also been developed. The Yamamoto Vinyter Co. (Osaka, Japan) and Tazawa *et al.* developed the inductive heating system TY-1, which consisted of four rolls of water-cooled hollow coils.[58] TY-1 generates a 500-kHz AMF. Furthermore, Tazawa *et al.* described a "portable" inductive heating system.[59] The device can use an alternating current (AC) of 200 V (maximum output 3 kW), and it is extremely small (40 × 40 × 40 cm long). In addition, the frequency can be adjusted continuously within the range of 100–500 kHz.

In 1996, the Dai-Ichi High Frequency Co. (Tokyo, Japan) and Kobayashi group developed a solenoid-type magnetic applicator.[36] AMF was generated by a horizontal coil (inner diameter, 7 cm; length, 7 cm) driven by a transistor inverter at a frequency of 118 kHz. In preclinical studies using rats, the animals were placed inside of the coil so that the region containing the subcutaneously transplanted glioma tumor was at the center, and exposure to AMF was found to produce strong therapeutic effects.[60] However, it is technically difficult to scale up the coil size in this solenoid-type applicator for clinical use because the very large coil which would be required to accommodate a human body may be accompanied by serious risks associated with the presence of the high voltage between the two ends of the solenoid. Therefore, a new device was recently developed, called a "Ferrite core-inserted solenoid type."[61] A near-lossless MnZn ferrite core, which has a high relative intrinsic permeability, typically 3000–4000 at low magnetic field strengths, and a sharp transition from the ferromagnetic to non-magnetic states was inserted into a vertical coil. An AMF was generated by a vertical coil (inner diameter, 7 cm; length, 9 cm) driven by a transistor inverter at a frequency of 360 kHz. The ferrite core inside a solenoid coil was designed to concentrate the magnetic field generated by solenoid coil, resulting in the emission of the magnetic field from the surface of the device. Using this magnetic applicator, Kawai *et al.* reported that in the case of a phantom with MCLs, a phantom 10 mm distant from the ferrite core-inserted solenoid could show an increase in temperature of 8°C within 5 min,[61] suggesting that the ferrite core-inserted solenoid was suitable for heating a target positioned outside of the coils.

The AMF generators cited here are listed in Table 2. These magnetic applicators generate AMF with several frequencies ranging from 100 to 500 kHz. Because the heating properties of magnetite nanoparticles are proportional to the frequencies of the magnetic applicator,[59] higher temperatures can be generated by magnetite nanoparticles by irradiation with a higher frequency. In our experience, however, higher frequencies, such as those over 400 kHz, caused non-specific heating due to eddy currents (unpublished results). This phenomenon was observed when a solenoid-type applicator was used, suggesting that the appropriate frequency used in solenoid-type applicators should be relatively low frequencies ranging from 100 to 200 kHz. The heat generated is proportional to the concentration of magnetite

Table 2. Magnetic field applications for intracellular hyperthermia.

Manufacturer	Type of applicator	Frequency	References
MagForce Nanotechnologies	Gap type	100 kHz	7, 56
Sumitomo Metal Industries	Gap type	100 kHz	57
Yamamoto Vinyter	Solenoid type	500 kHz	58
Komatsu Power Tron	Solenoid type (portable)	100–400 kHz	59
Dai-ichi High Frequency	Solenoid type	118 kHz	36
	Ferrite core-inserted solenoid type	360 kHz	61

nanoparticles, and also proportional to the 1.6 (obtained experimentally[62]) or 2.0 (estimated theoretically[63]) power of the magnetic field intensity. In addition, the particle size and the preparation methods used to make the submicron magnetic particles are strongly associated with their ability to generate heat as mentioned above. Thus, a number of parameters are important, and the relationship of the frequency used, the particle size used, the preparation methods used to make submicron magnetic particles, their concentration and the type of applicator used must be optimized for an effective therapy involving intracellular hyperthermia.

4. *In vivo* Experimental Results Using Intracellular Hyperthermia

The first *in vivo* experiment with intracellular hyperthermia was performed by Gordon *et al.*[28] in 1979. Their study was conducted on 26 Sprague Dawley rats bearing mammary tumors. Non-coated magnetite nanoparticles (100 mg) with a 6 nm diameter were injected into a vein. Microscopic tissue analysis revealed that the tumor cells had taken up the particles. Two days after injection, the rats were exposed to a 450 kHz AMF for 12 min. Tumor temperatures increased about 8°C. It was also reported that there were no side effects or toxic reactions to the particles, and histologic evidence for tumor necrosis was shown.

During the past decades from this first *in vivo* experiment by Gordon *et al.*,[28] many researchers have reported encouraging results using intracellular hyperthermia with magnetic nanoparticles, such as magnetite-core dextran nanoparticles[64,65] and aminosilane-coated magnetite.[66,67] One of the most systematic and thoroughgoing experiments using liposomal magnetite was conducted by Kobayashi and his colleagues. They demonstrated the efficacy of intracellular hyperthermia using magnetite nanoparticles covered with liposomes (using MCLs and AMLs as mentioned above) in animals with several types of tumors, including B16 mouse melanoma,[68] MM46 mouse mammary carcinoma,[69] PC3 and LNCaP human prostate cancer cells in athymic mice,[70] spontaneously occurring primary melanoma in transgenic mice,[71] T-9 rat glioma,[60,72] rat prostate cancer PLS10,[73] Os515 hamster osteosarcoma,[74] LM8 mouse osteosarcoma,[75] VX-7 squamous cell carcinomas in rabbit tongue[76] and human breast cancer BT474 (HER2-positive) cells in nude mice.[44] These results are shown in Table 3. In these therapeutic experiments, MCLs (net magnetite amount, 3 mg/tumor) were directly injected into solid tumors and the animals were irradiated several times (RH) for 30 min with a "solenoid-type" AMF of 118 kHz. The temperature of the tumor was

Table 3. *In vivo* experimental results of intracellular hyperthermia using MCLs.

Animal	Tumor (size)	Therapy protocol (RH: repeated hyperthermia)	Complete tumor regression	References
Mouse	B16 melanoma (5–6 mm)	2-T at 46°C for 30 min	90% (9/10)	68
	MM46 mammary carcinoma (7 mm)	3-T at 45°C for 30 min	100% (5/5)	69
	MM46 mammary carcinoma (15 mm)	2-T at 45°C for 30 min (RH), RH was conducted 1–6 times	100% (5/5)	69
	LMB osteosarcoma	3-T at 45°C for 30 min	100% (7/7)	75
Athymic mouse	PC-3 human prostate cancer (7 mm)	3-T at 46°C for 30 min (RH), RH was conducted 1–5 times	100% (5/5)	70
	LNCaP human prostate cancer (7 mm)	3-T at 46°C for 30 min (RH), RH was conducted 2–5 times	100% (5/5)	70
	BT474 human breast cancer	2-T at 45°C for 30 min	100% (5/5)	44
Transgenic mouse	Primary skin Melanoma (5–7 mm)	3-T at 45°C for 30 min (RH), RH was conducted 1–3 times	100% (5/5)	71
Rat	T-9 glioma (13 mm)	3-T at 43–44°C for 30 min	80% (4/5)	60
	PLS10 prostate cancer	3-T at 45°C for 30 min	100% (5/5)	73
Hamster	Os515 osteosarcoma (10 mm)	3-T at 42°C for 30 min	100% (4/4)	74
Rabbit	VX-7 squamous cell carcinoma (10 mm)	3-T at 43°C for 30 min	100% (4/4)	76

2-T: Irradiated 2 times, 3-T: Irradiated 3 times.

elevated rapidly by magnetic heating and reached the intended temperature (42–46°C). In contrast, the rectal temperature or temperatures in tumors lacking the MCLs did not increase. After the AMF irradiation, the tumor volume decreased markedly, and complete tumor regression was observed in 97% (63/65) of the animals in these experiments (Table 3). These results indicate that MCLs can be an effective tool for hyperthermia, and RH using magnetite nanoparticles is a promising approach for cancer therapy.

Since magnetite particles larger than 10 nm in diameter are ferromagnetic and they are moved by the action of a magnet, magnetic force-based targeting for drug and gene delivery has demonstrated the efficacy of this technique.[77] When tumor tissues are located at peripheral portions of the body, magnetic forces enhance the delivery of magnetite nanoparticles into tumor tissues. This was shown in a rat model by Kobayashi group.[78]

As shown in Fig. 3, MCLs have a 10-fold higher affinity for cells than neutrally charged MLs. Because MCLs may attach to most cells via electrostatic interaction, administration of MCLs has been limited to their direct injection into tumor tissues, which poses problems in treating impalpable lymph node metastases. Kobayashi group proposed the application of MLs in such cases.[79] The lymphatic system is an important pathway for metastasis and the lymphatic system is known to selectively clear particulates of up to hundreds of nanometers in diameter. MLs are 94 nm in size, and MLs injected into the submucosa or intramuscularly were delivered selectively to regional lymph nodes. Subsequently, AMF irradiation was found capable of generating induced hyperthermia in lymph node metastases. If an MAb against a tumor cell is available, the use of AMLs will be more effective for targeting tumor cells in lymph nodes.

5. Induction of Antitumor Immunity by Intracellular Hyperthermia

Kobayashi group[80] observed that tumor-specific hyperthermia using magnetite nanoparticles induced antitumor immunity. They demonstrated this event by transplanting a tumor to each femur of a rat. MCLs were directly injected into only one tumor and the hyperthermia treatment was applied. The temperature at the tumor increased to 45°C, whereas that at the other tumor did not increase (Fig. 6A). However, both the tumors disappeared completely after 28 days (Fig. 6B). Immunohistochemical assay revealed that NK cells and $CD8^+$ and $CD4^+$ T cells migrated in both tumors after the hyperthermia treatment.[80] These results suggest that therapeutic magnetite nanoparticles are potentially effective tools for hyperthermic treatment of tumors, because

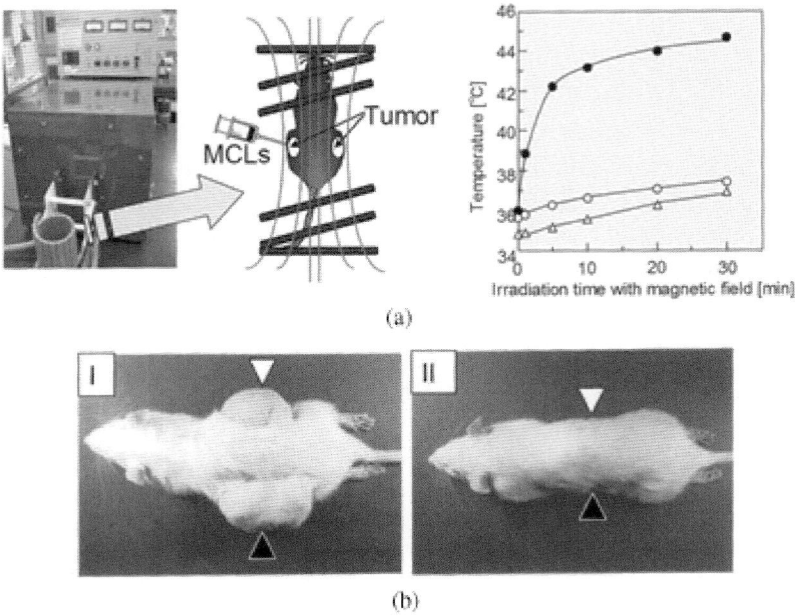

Fig. 6. Immune response induced by hyperthermia using MCLs. Rats bearing tumors on each side of the body were prepared. MCLs were injected into the left tumor only, and the rats were irradiated with an AMF apparatus shown in a (left). Temperature of the left tumor at which MCLs were injected (closed circles), increased specifically, whereas temperature of the right tumor (open circles) and rectum (open triangles) remained below 37°C (a, right). The tumor-specific hyperthermia treatment induced an antitumor immune response, and both tumors had disappeared on the 28th day after hyperthermia treatment. (I) Control rat without AMF irradiation; (II) rat with AMF irradiation. Open triangles in b, the side without MCLs; closed triangle in b, the side with MCLs.

in addition to the killing of tumor cells by heat, a host immune response is strongly induced. As a mechanism for recognition of tumor antigens by the host immune system, Kobayashi group has proposed HSP-mediated antitumor immunity.[81,82]

5.1. *Augmentation of tumor immunogenicity by intracellular hyperthermia*

The HSPs, which are abundant intracellular proteins, function as molecular chaperons that control the folding and prevent the aggregation of proteins.[83] Because expression of HSPs protects cells from heat-induced apoptosis,[84] HSP expression has been considered a complicating factor in hyperthermia. Because hyperthermia is a physical treatment, it is likely to have fewer side effects than chemotherapy. Consequently, an advantage of hyperthermia is the feasibility of frequent repeated treatment.[69] However, conventional hyperthermia systems involve treatment only once or twice per week, performed at an interval of more than 48 h, to prevent thermotolerancc[85] caused by HSP expression.

On the other hand, recent reports have shown the importance of HSPs in immune reactions, including HSP70, HSP90 and glucose-regulated protein 96 (gp96), and studies suggest that HSPs chaperon tumor antigens.[86–88] With regard to the mechanism of antitumor immunity induced by hyperthermia using MCLs, findings as shown in Fig. 6 suggest two possible mechanisms of antigen presentation via HSP70 expression during hyperthermia.[81,82,89]

One possible mechanism is heat-induced augmentation of tumor immunogenicity due to presentation of antigenic peptides via MHC class I antigens of tumor cells. Srivastava *et al.* proposed the following "relay line model" for tumor antigenic peptide transfer during antigen processing and presentation by HSPs as shown in Fig. 7.[90,91] (i) The tumor antigens are decomposed to the

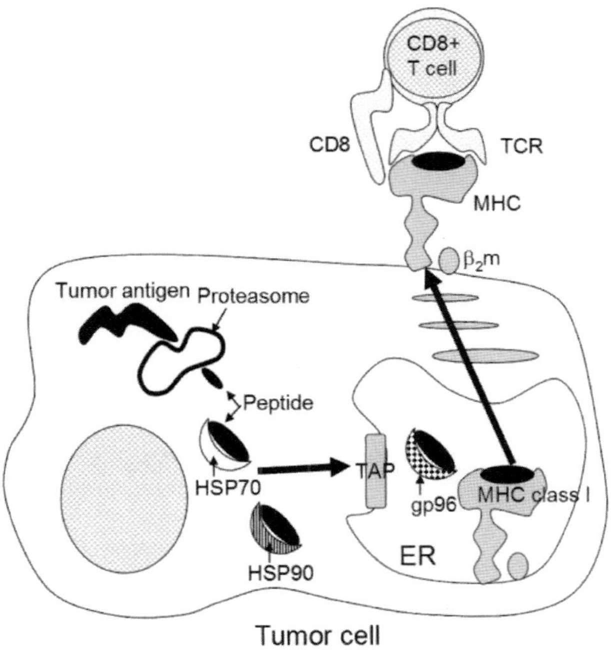

Fig. 7. Relay line model for tumor antigenic peptide transfer during antigen processing and presentation by HSPs. HSP family, including HSP70 and HSP90 in cytoplasm, and gp96 in endoplasmic reticulum (ER), is involved in peptide transfer to MHC class I molecule. CD8$^+$ T cells recognize tumor antigenic peptide displayed on MHC class I molecules and are activated.

peptides in the proteasome. (ii) The peptides are bound to HSP70 or HSP90, which carry them to the ER via the transporter associated with antigen processing (TAP). (iii) The peptides are transferred to gp96 in the lumen of the ER. (iv) In the terminal step, gp96 transfers the peptides to the MHC class I-β_2 microglobulin complexes. Wells *et al.* demonstrated that stably transferred B16 melanoma cells that constitutively expressed human HSP70 exhibited significantly increased levels of MHC class I antigens on their surface.[92,93] Kobayashi group has also shown that augmentation of MHC class I antigens on the tumor cell surface via HSP70 expression causes immune induction.[89]

The working hypothesis I is illustrated in Fig. 8.[81,82] In the study by Kobayashi group, HSP70 expression reached its maximum 24 h after heating, and the augmentation of MHC class I surface expression began 24 h after heating and reached its maximum 48 h after heating.[89] The expression of other immunologic mediators, such as intracellular adhesion molecule-1 (ICAM-1) and MHC class II, did not increase. In an *in vivo* experiment, growth of T-9 cells in immunocompetent syngenic rats (F344) was significantly inhibited by hyperthermia, with augmentation of MHC class I antigen surface expression, whereas growth of T-9 cells was not inhibited in nude rats,[89] suggesting that the effector cells were T lymphocytes. Furthermore, compared with lymphocytes from non-immunized rats or rats injected with non-heated T-9 cells, the splenic lymphocytes of rats injected with heated T-9 cells displayed specific cytotoxicity against T-9 cells.[89]

Okamoto *et al.*[94] reported that immunization of mice with heat-treated colon-26 cell extract, which was enriched in HSP70, elicited antitumor

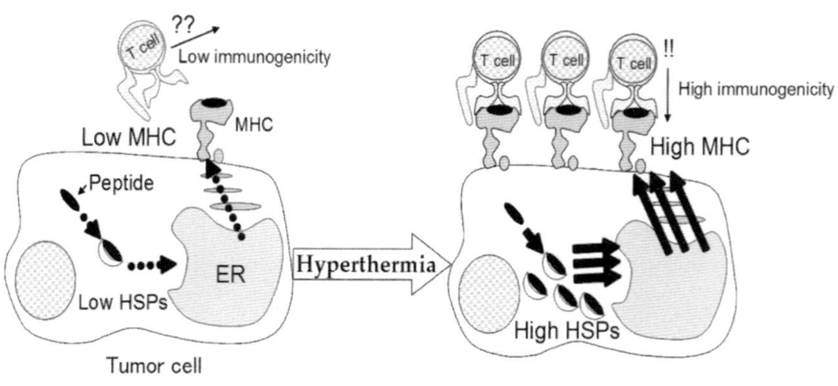

Fig. 8. Working hypothesis I of induction of antitumor immune response by hyperthermia. Augmentation of tumor immunogenicity by an increase of the number of MHC class I molecules on the surface of cancer cells via inducible HSP expression. $CD8^+$ T cells are activated by the recognition of tumor antigenic peptide displayed on MHC class I molecules.

immunity against subcutaneously injected colon-26 cells. Also, similar immunization enhanced the cytotoxic activity of mouse splenic lymphocytes against heat-treated colon-26 cells. Because cell extracts contained various antigens, HSP-independent antigens may be also involved in the antitumor immunity. However, these results suggest that HSP70 is an important modulator of tumor cell immunogenicity during hyperthermia, and that cytotoxic T-lymphocytes (CTLs) are the effector cells.

5.2. *Vaccine-like effect of HSP-peptide complex released from necrotic cells*

An alternative mechanism of recognition of tumor antigens by the host immune system in hyperthermia is cross-presentation of antigenic peptides by dedicated APCs, including DCs.[81,82] The working hypothesis II is illustrated in Fig. 9. HSP-mediated antitumor immunity may be caused by a vaccine-like effect of HSP–peptide complexes released from dying tumor cells. The released HSP–peptide complexes encounter APCs that express

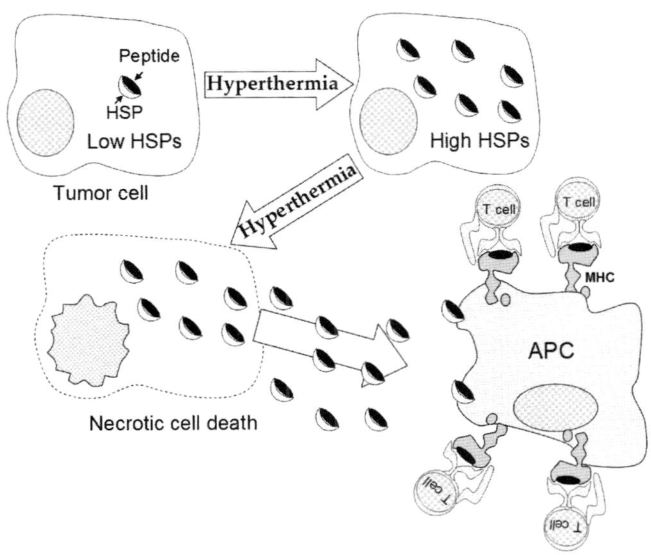

Fig. 9. Working hypothesis II of induction of antitumor immune response by hyperthermia. *In situ* vaccination by HSP–peptide complexes released from dying tumor cells via necrosis by hyperthermia treatment. HSP–peptide complexes encounter antigen-presenting cells (APCs) such as dendritic cells (DCs). Interaction of HSP–peptide complexes with the receptors of APCs leads to receptor-mediated endocytosis, processing of the antigenic peptide by the endogenous MHC class I pathway, and re-presentation on the cell surface to $CD8^+$ T cell.

receptors such as CD91,[95] CD40[96] and toll-like receptors 2/4.[97] Interaction of HSP–peptide complexes with these receptors leads to receptor-mediated endocytosis, processing of the antigenic peptide by the endogenous MHC class I pathway, and re-presentation on the cell surface to $CD8^+$ T-cell receptors (TCRs). Additionally, HSP70 functions as a direct activator of APCs, stimulating cytokine secretion from monocytes and inducing the maturation of DCs via CD14 and/or CD91 receptors.[98] This cytokine-like ability of HSP70 to stimulate the innate immune system is independent of the peptides they chaperone, suggesting that HSP70 is a natural adjuvant.[99,100]

Sato *et al.* proposed a similar mechanism of recognition of tumor antigens.[101] They used NPrCAP/M as a magnetic nanoparticle and applied AMF in C57BL/6 mice melanoma model. The hyperthermic treatment caused cytotoxic reaction as well as heat shock responses, leading to elicitation of antitumor immune response. The level of HSP70 was increased in the cell lysate and culture supernatant after intracellular hyperthermia. Melamona-specific $CD8^+$ T-cell response to DCs loaded with hyperthermia-treated tumor lysate was enhanced when compared with non-treated tumor lysate. When HSP70 was immuno-deleted from hyperthermia-treated tumor cell lysate, specific $CD8^+$ T cell response was abolished. From these results, they concluded that the antitumor immune response induced by hyperthermia using NPrCAP/M is derived from the release of HSP–peptide complex from degraded tumor cells.

Classical immunology theory is based on the distinction between "self and nonself." However, the self and non-self model is sometimes not very useful, particularly in cancer immunity because cancer cell is derived from normal cell that is "self," originally. Matzinger proposed the "Danger model," which suggests that the immune system is more concerned with damage than with foreignness.[102] In the "Danger model", HSPs are considered an important "Danger signal."In particular, when HSPs, which are stress indicators and normally present in cytoplasm, are released into the bloodstream, they can act as a signal to the immune system of an abnormal situation. Yanase *et al.* demonstrated that HSP70 expression following hyperthermia using MCLs induced antitumor immunity in rats with T-9 rat glioma.[80] Because MCLs are themselves heated in intracellular hyperthermia system, the distribution of magnetite nanoparticles within tumors is an important issue. When MCLs were repeatedly heated, the surrounding tumor tissues underwent necrosis, and magnetite nanoparticles subsequently expanded into the necrosis area within the tumor, resulting in wide distribution of magnetite nanoparticles. Thus, the entire tumor area was necrosed by repeated (three times) hyperthermia (with a 24-h interval over three days).[50]

The 24-h interval corresponded to the time when HSP70 expression in T-9 cells reached its maximum, and a large amount of HSP70 was detected in the tumor tissue. Thus, the intracellular hyperthermia system using MCLs overcame thermotolerance and induced necrotic cell death that correlated with HSP70 expression. A reason why necrotic cell death via HSP70 expression is effective for cancer therapy is that necrotic cell death may strongly induce a "Danger signal." Ito *et al.* observed that numerous and diverse kinds of immunocytes, such as $CD8^+$ and $CD4^+$ T cells, NK cells, macrophages and DCs, infiltrated into the necrotic area of tumors after hyperthermia treatment.[50,80] Todryk *et al.* observed infiltration of such cells into B16 melanoma nodules transfected with the HSP70 gene,[103] suggesting that HSP70 expression is a "Danger signal" for recruitment of immunocytes.

Next, Ito *et al.* purified the HSP70–peptide complexes from the tumor after intracellular hyperthermia, and found that immunization of rats with T-9 HSP70 strongly suppressed tumor growth.[50] HSP70–peptide complexes from liver were also purified, and their vaccine-like effects were examined, but no antitumor effects were observed. These results suggest that HSP70 in tumor cells chaperons some antigenic peptides after hyperthermia. Also, in order to investigate the vaccine-like effect of the tumor cells killed via hyperthermia-induced necrosis in rats, a tumor rejection assay was performed after hyperthermia treatment of implanted T-9 cells; tumor growth was strongly suppressed, and 50% of the rats were protected from challenge with T-9 cells.[50]

Recently, Ito *et al.* investigated the diversity of TCRs in tumor-infiltrating lymphocytes after hyperthermia using NPrCAP/M in mouse model.[104] NPrCAP/M was injected into B16 melanomas in mice, which were subjected to an AMF for hyperthermia treatment. They found that T cells were clonally expanded in B16 tumors after hyperthermia, in which the variable region of TCR β-chain 11^+ T cells were preferentially expanded. The DNA sequences of the complementarity-determining region (CDR)3 of the TCR β-chain in tumor-infiltrating lymphocytes revealed the presence of clonally expanded T cells. These results indicate that the T-cell response in B16 melanoma after hyperthermia is dominated by T cells directed toward a limited number of epitopes and that epitope-specific T cells frequently use a restricted TCR repertoire. In near future, artificial CTL will be constructed by referring the DNA sequences of the preferentially expanded CDR3, and the combination of TCR gene-engineered CTL with hyperthermia will be promising for patients with early metastatic tumor.

In 2002, a vaccine consisting of autologous tumor-derived gp96–peptide complexes (HSPPC-96, Oncophage; Antigenics, Inc, Woburn, MA) has entered clinical trials, and the feasibility of its use to treat metastatic melanoma patients has been demonstrated.[105] Because HSP–peptide complexes must be extracted from tumors in the body, surgery is needed in this therapeutic protocol. In contrast, in hyperthermia using MCLs, no surgery or extraction is necessary. Udono *et al.* reported that the vaccination effect of HSP70–peptide complexes was directly dependent on the dose.[106] The HSPs in the tumor can be regarded as an antigen source, and 1 g (approximately 10^8 cells) of tumor tissue may contain approximately 2 mg HSP70. This represents a much higher dose than that used for the clinical trials. In the hyperthermia system using MCLs,[50,80] the expression of HSP70 was also enhanced, and tissue lysis via necrosis was observed throughout the tumor. The hyperthermia system using MCLs can produce HSP–peptide complexes (possibly including HSPs such as HSP90 and gp 96), resulting in vaccination.

These results suggest that the hyperthermia system confers antitumor immunity via release of HSP70–peptide complexes during necrotic tumor cell death *in vivo*. This phenomenon, which may be called *in situ* vaccination, has important implications for development of novel antitumor therapies. A proposed scenario in which HSPs function during successive stages of an antitumor response after hyperthermia is summarized and illustrated in Fig. 10: (i) A poorly immunogenic tumor cell with a low concentration of

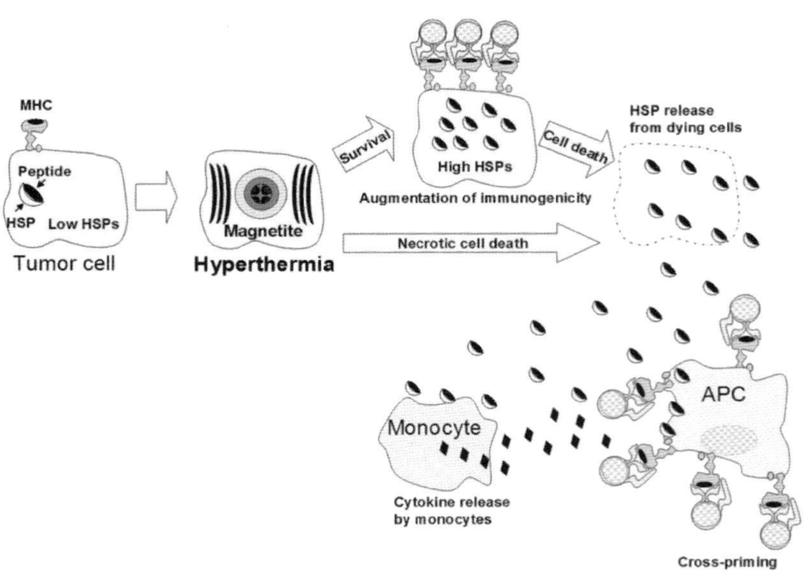

Fig. 10. Proposed scenario of the mechanism of induction of antitumor immune response by intracellular hyperthermia. In addition to the working hypothesis I and II, HSP–peptide complexes function as direct activator of APCs, and stimulate cytokine secretion from monocytes.

intracellular HSP–peptide complexes, decreased function of the endogenous antigen processing machinery, and a very low level of MHC class I–peptide complexes at the cell surface. (ii) A sublethal stress response induced by hyperthermia using MCLs results in increased levels of intracellular HSP–peptide complexes, enhanced processing of endogenous antigens and an increase in the density of MHC class I–peptide complexes at the cell surface. These tumor cells are then recognized directly by MHC class I-restricted CTLs. (iii) Dying tumor cells, which are killed by the CTLs or by lethal hyperthermia treatment, release their intracellular contents, including HSP–peptide complexes. (iv) The released HSPs and/or antigenic peptides activate neighboring monocytes to produce pro-inflammatory cytokines and recruit APCs. (v) The HSP–peptide complexes are taken up by DCs, and are in turn presented to T cells via MHC class I and/or II antigens (cross-priming).

As shown in Fig. 6, very strong antitumor immune response was induced by intracellular hyperthermia in T-9 rat glioma model.[80] The CTL activity for T-9 cells was approximately seven times higher in spleen cells of cured rats as compared with that from spleen cells of naïve rats, and all cured rats (5/5) rejected re-challenge test against T-9 cells. In the case of mouse B16 melanoma model,[68] the CTL activity for B16 cells was approximately two times higher in spleen cells of cured mice as compared with that from spleen cells of naïve mice, and 66% of the cured mice (4/6) rejected the melanoma cells. Similar results as shown in Fig. 6 were not observed in mouse B16 melanoma model. These differences seem reflection of immunogenicity of tumor cells tested.

Takahashi *et al.* developed a transgenic mouse, in which the *ret* oncogene was used under the control of the mouse metallothionein (MT)-I promoter.[107,108] This transgenic mouse showed hyperpigmented skin due to aberrant melanogenesis, and the melanocytic tumor progressed to melanoma accompanied by distant metastasis.[108] Ito *et al.* used this MT/*ret* transgenic mouse as a hereditary melanoma model of low immunogenicity.[109] Several melanoma nodules (4.5 on average) spontaneously developed in a MT/*ret* transgenic mouse in three month after birth. MCLs were injected in one melanoma nodule, and AMF irradiation was carried out for 30 min (the nodule temperature was maintained at 45°C) three times at 24-h intervals (RH). When this nodule began to grow, another RH treatment was applied, and the nodule finally disappeared in the hyperthermia group ($n = 5$). Interestingly, although only one nodule was subjected to intracellular hyperthermia, some of the other nodules (3/7 in a mouse) also disappeared completely.[109]

Kobayashi group is currently developing a novel cancer immunotherapy based on the mechanism of antitumor immune response via HSP expression

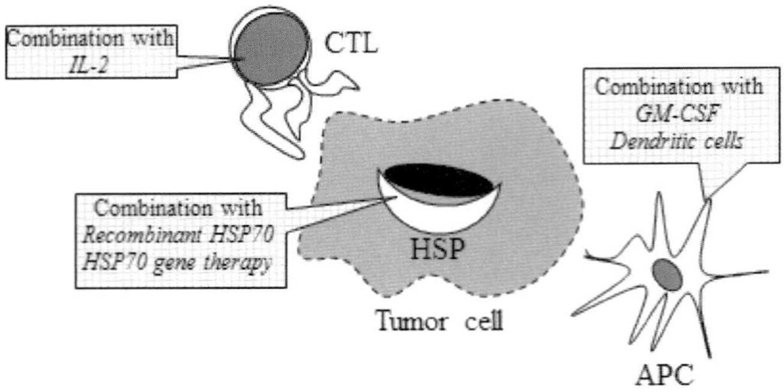

Fig. 11. Strategies for novel cancer immunotherapy based on the immune response induced by hyperthermia. Three key elements may be involved in this mechanism: (i) $CD8^+$ CTLs as an effector cell, (ii) APC as an antigen-processing cell and APC for the HSP–peptide complex released from necrotic cells, (iii) HSPs as a natural and powerful immunostimulant. In order to enhance the antitumor immunity induced by hyperthermia, one of the three elements was combined with the hyperthermia treatment.

as shown in Fig. 11. Three key elements may be involved in this mechanism: (i) CTL as an effector cell, (ii) APC as an antigen-processing cell and APC for HSP–peptide complex released from necrotic cells, (iii) HSPs as a natural and powerful immunostimulant. This strategy is based on combinations of hyperthermia using MCLs with injection of cytokines (IL-2 and GM-CSF),[110] heat-inducible TNF-α gene therapy,[111] injection of recombinant HSP70,[112] HSP70 gene therapy[113] and injection of immature DC therapy.[114,115] By applying combination of intracellular hyperthermia and one of the immunotherapy shown in Fig. 11, significantly improved results were obtained.

It is very important to study and establish combination effect of hyperthermia and immunotherapy based on the concept of heat-controlled necrosis with HSP expression.

6. Clinical Trials for Intracellular Hyperthermia

Magnetic nanoparticle-mediated hyperthermia has been a largely experimental modality for hyperthermia, because of limitations which must be addressed by the developments of both drugs (functional and practical magnetic nanoparticles) and apparatus (AMF generators).

For clinical applications, the toxic properties of the particles are an important issue to consider. Kobayashi group[116] investigated the toxicity resulting from the systemic administration of MCLs (90 mg, *i.p.*) in mice; none of the 10 MCL-treated mice died during the study. Transient accumulation of magnetite was observed in the liver and spleen, but magnetite nanoparticles were cleared from circulation by hepatic Kupffer cells and/or fixed macrophages in the spleen by the 30th day after administration.

The safety of any apparatus must also be guaranteed, as well as the safety of any magnetite agents, before clinical applications can be pursued. Because AMF generators are electric appliances which use high voltages and currents, insulators must be carefully placed where any contact with a patient's body will occur. Temperature monitoring systems are also important for the safe use of intracellular hyperthermia generated by AMF generators. Generally, temperature monitoring during AMF irradiation is performed by using fiber-optic thermometry probes which should not be affected by magnetic field exposure. In experimental animal models, the probes were positioned at tumor surfaces,[80] at the margin of tumors[117] and/or inserted into the center of the tumors.[118]

In 2005, Johannsen *et al.*[7] described the first clinical application of magnetite nanoparticle-mediated hyperthermia in locally recurrent prostate cancer, using the sophisticated MFH 300F apparatus which was mentioned above. In this study, the feasibility of hyperthermia was evaluated using the MFH 300F apparatus and aminosilane-coated magnetite nanoparticles. The particles were prepared by MagForce Nanotechnology Co. (Berlin, Germany), but the preparation method has not been published. The nanoparticles were injected transperineally into the prostate of a 67-year-old patient under transrectal ultrasound and fluoroscopy guidance. Treatments were performed using an AMF of 100 kHz. The temperature was measured in the first and last of six weekly hyperthermia sessions with 60 min durations using fiber-optic thermometry probes positioned in the prostate, urethra, rectum, perineum, scrotum and left ear. In addition, since the magnetite nanoparticles displayed a visible contrast at therapeutic concentrations on a CT scanner, and could be detected down to 0.01 g/l Fe in the MRI, a non-invasive temperature estimation system was used for the clinical hyperthermia treatment planning system.[119] Surprisingly, it was reported that the treatment was well tolerated without anesthesia by using a specially designed cooling device. In the first patient treated, the maximum and minimum intraprostatic temperatures were 48.5°C and 40.0°C during the first treatment, and 42.5°C and 39.4°C during the sixth treatment. These data mark a significant development in intracellular hyperthermia.

Based upon the animal experiments which have been done by Jimbow group at Sapporo Medical University, a preliminary human clinical trial has been carried out by employing NPrCAP/PEG/M plus AMF after they obtained the approval of their human clinical trial for a limited number of stage III and IV melanoma patients.[10] The therapeutic protocol followed the basically identical experimental schedule as that of the animal experiments. In the clinical trials, they evaluated the feasibility of using NPrCAP/PEG/M and also resultant antitumor immunity. Among two of four patients showed complete and partial responses to their treatment and have been able to carry out normal daily activities after the hyperthermia treatment. In one patient, four distant cutaneous metastasis sites were evaluated and either significant regression or shrinkage of all of these four melanoma lesions was seen. The patient was able to survive 30 months after several trials of hyperthermia treatment. The pathological and immunological specimens revealed dense aggregation of lymphocytes and macrophages at the site of the original melanoma lesion. Importantly, there was a trend to have an almost identical distribution of CD8+ T cells and MHC class I cells. Another patient had numerous lymph node metastasis, but sill has been surviving more than 32 months.

In October 2007, Takata *et al.* at Shinshu University in collaboration with Kobayashi group began a phase I clinical trial using AMLs conjugated with MAbs against human HMW-MAA in melanoma for four patients after they obtained the approval of their human clinical trial.[11] The magnetic field applicator (a ferrite core-inserted solenoid type) was applied with an AMF of 110 kHz as shown in Fig. 12. The safety of the applicator, as well as of the AMLs, was investigated before the clinical trial. The AMLs were injected into

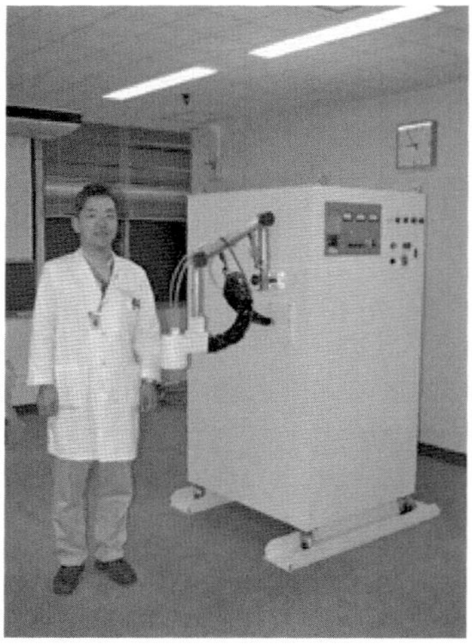

Fig. 12. Magnetic field applicator for clinical applications. This applicator is a ferrite core-inserted solenoid type developed by the Dai-Ichi High Frequency Co. For cooling purposes, a closed loop of tubes with circulating cold medium made of fluorocarbons is placed in the head of the applicator. An AMF with a frequency of 110 kHz and variable field strength are used.

melanoma nodules, and treatments were performed for 30 min, repeated three times at 24-h intervals. The temperatures at the surface of the melanoma nodules were monitored with a fiber-optic thermometer, and were maintained between 44 and 46°C. The patients did not suffer from pain derived from the hyperthermia treatments, and the adverse effects were not observed for all four patients.

Imai *et al.* at Nagoya University Hospital in collaboration with Kobayashi group began a phase I clinical trial using MCLs and the same AMF as shown in Fig. 12 from November 2009.[12] The group member is consisted from Department of Breast and Endocrine Surgery, Department of Otolaryngology, Department of Oral Surgery and Department of Orthopedic Surgery at Nagoya University Hospital. Patients with breast, head and neck, soft tissue malignant tumors refractory to conventional therapy were enrolled. Informed consent was obtained from all patients. The amount of administered MCLs to the tumor was increased in the three-stepwise increment. The number of patients of each step was three. Following direct injection of MCLs into the tumors, AMF was applied to the tumor for 30 min twice at an interval of 24 h as shown in Fig. 13. Output power of AMF was adjusted so that the temperature of the skin overlying the tumor was elevated to approximately 43°C.

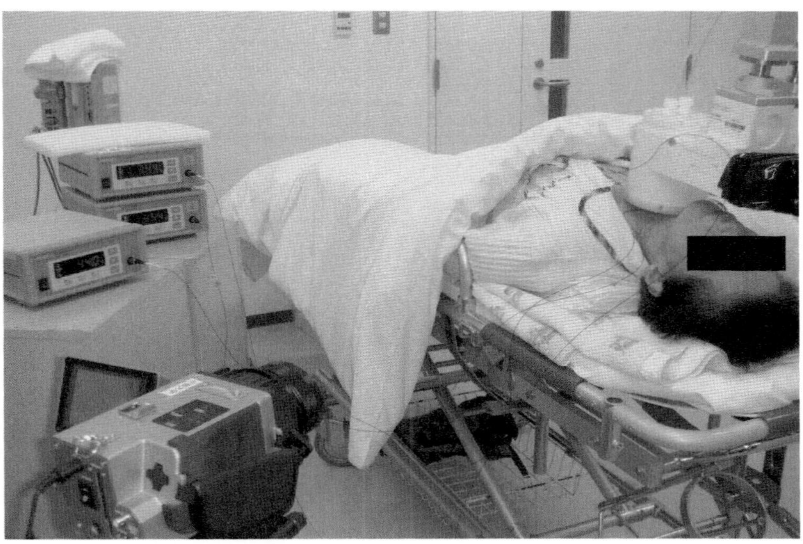

Fig. 13. Photograph during hyperthermia treatment for a patient with papillary thyroid cancer at Nagoya University Hospital. Fiber-optic thermometry probes are positioned at the surface of the papillary thyroid cancer, at the peripheral position of the cancer and the skin surface 3 cm apart from the cancer. Temperature at the surface of the papillary thyroid cancer was kept constant (approximately 43°C) manually by adjusting AC field strength.

Table 4. Summary of phase I clinical trial at Nagoya University Hospital.

Level	Type	Age/sex	Site	Pathological effect	Adverse event
1	Papillary thyroid ca.	79/F	anterior neck	1a	none
1	breast ca.	52/F	right side chest	2	none
1	Papillary thyroid ca.	45/F	left neck	0	none
2	tongue ca.	34/M	left neck	0	none
2	soft tissue sarcoma	32/F	right sole	1b	none
			right knee	1b	
			occipital	1a	
2	follicular thyroid ca.	55/M	right neck (center of the target)	3	none
			right neck (peripheral zone of the target)	1a	

Pathological effect 0: ineffective, 1a: minor effect, 1b: moderate effect, 2: considerable effect, 3: complete effect.

Two weeks after administration of MCLs, the treated tumor was excised under local anesthesia. Hyperthermic effect was examined pathologically. The result of this clinical trial is summarized in Table 4. Elevation of temperature of the injection site was demonstrated in all six patients of phase I trial. The temperature of adjacent site did not show significant change. No significant adverse effects were observed in level 1 and 2. As regard to pathological effect, necrosis in more than one-third of tumor cells was observed in four cases out of six cases.

7. Concluding Remarks

Many continuous efforts have been made to construct a safe and effective intracellular hyperthermia system. Finally, some researchers are opening the door to clinical trials, suggesting that the time has come to use intracellular hyperthermia. Other groups are now preparing to begin clinical studies. Once the door has opened, many potent combination therapies based on intracellular hyperthermia will also be available for clinical trials. The author hopes that intracellular hyperthermia will provide a novel effective therapy for cancer patients.

Acknowledgments

I thank Dr Masashige Shinkai, Dr Akira Ito, Professor Hiroyuki Honda of Nagoya University for their contribution to develop the intracellular hyperthermia system using MCLs and AMLs. Collaborations with Professor Toshiaki Saida (Department of Dermatology, Shinshu University), and Dr Tsuneo Imai and Dr Toyone Kikumori (Department of Breast and Endocrine Surgery, Nagoya University Hospital) are acknowledged. I also acknowledge Professor Kowichi Jimbow (Department of Dermatology, Sapporo Medical University) for many valuable comments.

References

1. R. W. Y. Hildebrandt, P. Wust, O. Ahlers, A. Dieing, G. Sreenivasa, *et al.*, *Critical Reviews in Oncology/Hematology* **43**, 33 (2002).
2. M. H. Falk and R. D. Issels, *Int. J. Hyperthermia* **17**, 1 (2001).
3. M. W. Dewhirst, B. L. Viglianti, M. Lora-Michiels, M. Hanson and P. J. Hoopes, *Int. J. Hyperthermia* **19**, 267 (2003).
4. S. B. Field and C. C. Morris, *Radiother. Oncol.* **1**, 179 (1983).
5. P. Moroz, S. K. Jones and B. N. Gray, *Int. J. Hyperthermia* **18**, 267 (2002).
6. A. Ito, M. Shinkai, H. Honda and T. Kobayashi, *J. Biosci. Bioeng.* **100**, 1 (2005).
7. M. Johannsen, U. Gneveckow, L. Eckelt, A. Feussner, N. Waldofner, *et al.*, *Int. J. Hyperthermia* **21**, 637 (2005).
8. G. Sreenivasa, J. Gellermann, B. Rau, J. Nadobny, P. Schlag, *et al.*, *Int. J. Rad. Oncol. Biol. Phys.* **55**, 407 (2003).
9. K. Jimbow, O. Y. Ishii, S. Ito, Y. Tamura, A. Ito, *et al.*, *J. Skin Cancer* accepted, (2013).
10. K. Jimbow, A. Yoneta, J. Kato, T. Takeda, M. Sato, *et al.*, *28th Annual Meeting of Jpn. Soc. for Thermal Medicine, Nagoya*, S3–S4, (2011).
11. M. Takata, K. Matsumoto, T. Saida and T. Kobayashi, *25th Annual Meeting of Jpn. Soc. for Thermal Medicine, Nagoya*, S3–S4, (2008).
12. T. Imai, T. Kikumori, M. Akiyama, K. Yokota, Y. Nishida, *et al.*, *28th Annual Meeting of Jpn. Soc. for Thermal Medicine, Nagoya*, S3–S6, (2011).
13. I. A. Brezovich and R. F. Meredith, *Radiol. Clin. North Am.* **27**, 589 (1989).
14. B. D. Cullity and C. D. Graham, *Introduction to Magnetic Materials*, Wiley-IEEE, (2008).
15. R. M. Cornell, In *The Iron Oxides: Structure, Properties, Reactions, Occurrence and Uses*, E. Weinheim (Ed.), VCH (1996).
16. R. D. Tucker, *J. Endourol.* **17**, 601 (2003).
17. S. Deger, K .Taymoorian, D. Boehmer, T. Schink, J. Roigas, *et al.*, *Eur. Urol.* **45**, 574 (2004).
18. A. E. Hubert and R. Schäfer, *Magnetic Domains*, Springer-Verlag (1998).
19. Q. A. Panhurst, J. Connelly, S. K. Jones and J. Dobson, *J. Phys. D: Appl. Phys.* **36**, R167 (2003).
20. M. Shinkai, M. Matsui and T. Kobayashi, *Jpn. J. Hyperthermic Oncol.* **10**, 168 (1994).
21. R. Hergt, S. Dutz, R. Mueller and M. Zeisberger, *J. Phys.: Condensed Mater.* **18**, S2919 (2006).
22. R. Hergt, W. Andrae, C. G. d'Ambly, I. Hilger, W. A. Kaiser, *et al.*, *IEEE Trans. Magnetics* **34**, 3745 (1998).
23. U. Schwertmann, *Iron Oxides in the Laboratory: Preparation and Characterization.* 2nd ed. Wiley-VCH (2000).

24. M. Ma, Y. Wu, J. Zhou, Y. Sun, Y. Zhang, *et al.*, *J. Magnet. Magnetic Mater.* **268**, 33 (2004).
25. R. Hergt, S. Dutz and M. Roder, *J. Phys.: Condensed Mater.* **20**, 385214 (2008).
26. A. Gupta and M. Gupta, *Biomaterials* **26**, 3995 (2005).
27. M. Hasegawa and S. Hokkoku, *US Patent 4* **101**, 435 (1978).
28. R. T. Gordon, J. R. Hines and D. Gordon, *Med Hypothesis* **5**, 83 (1979).
29. A. Jordan, P. Wust, H. Fähling, W. John, A. Hinz, *et al.*, *Int. J. Hyperthermia* **9**, 51 (1993).
30. P. D. Thomas, H. Kishi, H. Cao, M. Ota, T. Yamashita, S. Singh and K. Jimbow, *J. Invest. Dermatol.* **113**, 928 (1999).
31. S. Ito, T. Kato, K. Ishikawa, T. Kasuga and K. Jimbow, *Biochem. Pharmacol.* **36**, 2007 (1987).
32. F. Alena, T. Iwashina, A. Gili and K. Jimbow, *Cancer Res.* **54**, 2661 (1994).
33. M. Sato, T. Yamashita, M. Ohkura, Y. Osai, A. Sato, *et al.*, *J. Invest. Dermatol.* **129**, 2233 (2009).
34. K. Jimbow, Y. Tamura, A. Yoneta, T. Kamiya, I. Ono, *et al.*, *J. Biomater. Nanobiotechnol.* **3**, 140 (2012).
35. K. Jimbow, T. Takada, M. Sato, A. Sato, T. Kamiya, *et al.*, *Pigment Cell Melanoma Res.*, **21**, **243** (2008).
36. M. Shinkai, M. Yanase, H. Honda, T. Wakabayashi, J. Yoshida and T. Kobayashi, *Jpn. J. Cancer Res.* **87**, 1179 (1996).
37. M. Shinkai, M. Yanase, M. Suzuki, H. Honda, T. Wakabayashi, J. Yoshida and T. Kobayashi, *J. Magnet. Magnetic Mater.* **194**, 176 (1999).
38. M. Suzuki, M. Shinkai, M. Kamihira, T. Hanaichi and T. Kobayashi, *Biotechnol. Appl. Biochem.* **21**, 335 (1995).
39. M. Shinkai, M. Suzuki, S. Iijima and T. Kobayashi, *Biotechnol. Appl. Biochem.* **21**, 125 (1994).
40. M. Suzuki, H. Honda, T. Kobayashi, T. Wakabayashi, J. Yoshida and M. Takahashi, *Brain Tumor Pathol.* **13**, 127 (1996).
41. B. Le, M. Shinkai, T. Kitade, H. Honda, T. Kobayashi, *et al.*, *J. Chem. Eng. Jpn.* **34**, 66 (2001).
42. M. Shinkai, B. Le, H. Honda, K. Yoshikawa, T. Kobayashi, *et al.*, *Jpn. J. Cancer Res.* **92**, 1138 (2001).
43. A. Ito, Y. Kuga, H. Honda, H. Kikkawa, T. Kobayashi, *et al.*, *Cancer Lett.* **212**, 167 (2004).
44. T. Kikumori, T. Kobayashi, M. Sawaki and T. Imai, *Breast Cancer Res. Treat.* **113**, 435 (2009).
45. Y. Ikehara, T. Niwa, B. Le, S. Kabata, T. Kobayashi, *et al.*, *Cancer Res.* **66**, 8740 (2006).
46. H. Takahashi, R. Tanaka, T. Uzuka and I. Grinev, *Jpn. J. Hyperthermic Oncol.* **21**, 221 (2005).
47. R. Stauffer, T. C. Cetas and T. C. Fletcheretal, *IEEE Trans. Biomed. Eng.* BME, **31**, 76 (1984).
48. T. Ohno, T. Wakabayashi, A. Takemura, J. Yoshida, T. Kobayashi, *et al.*, *J. Neuro-Oncol.* **56**, 233 (2002).
49. M. Shinkai, K. Ueno, H. Honda and T. Kobayashi, *Jpn. J. Hyperthermia Oncol.* **18**, 191 (2002).

50. A. Ito, M. Shinkai, H. Honda, K. Yoshikawa, T. Kobayashi, *et al.*, *Cancer Immunol. Immunother.* **52**, 80 (2003).
51. D. Chan, D. B. Kirpotin and P. A. Bunn, *J. Magnet. Magnetic Mater.* **122**, 374 (1993).
52. K. Okawa, M. Sekine, M. Maeda, M. Tada and M. Abe, *J Appl Physics* **99**, 08H102 (2006).
53. T. Atsumi, B. Jeyadevan, Y. Sato and K. Tohji, *J. Magnet. Soc. Jpn.* **30**, 555 (2006).
54. M. Shinkai, K. Ueda, S. Ohtsu, H. Honda and T. Kobayashi, *Jpn. J. Cancer Res.* **90**, 699 (1999).
55. M. Shinkai, K. Ueda, S. Ohtsu, H. Honda, K. Kohri and T. Kobayashi, *Jpn. J. Cancer Res.* **93**, 103 (2002).
56. U. Gneveckow, A. Jordan, R. Scholz, V. Bruss, N. Waldofner, *et al.*, *Med. Phys.* **31**, 1444 (2004).
57. M. Mitsumori, M. Hiraoka, T. Shibata, Y. Okuno, Y. Nagata, *et al.*, *Hepatogastro.* **43**, 1431 (1996).
58. S. Wada, K. Tazawa, I. Furuta, H. Nagae, *Oral Dis.* **9**, 218 (2003).
59. K. Tazawa, S. Wada, M. Yatsuzuka, T. Saito, K. Tazawa, *et al.*, *Jpn. J. Hyperthermic Oncol.* **19**, 79 (2003).
60. M. Yanase, M. Shinkai, H. Honda, T. Wakabayashi, J. Yoshida and T. Kobayashi, *Jpn. J. Cancer Res.* **89**, 463 (1998).
61. N. Kawai, A. Ito, Y. Nakahara, H. Honda, T. Kobayashi, *et al.*, *Prostate* **66**, 718 (2006).
62. M. Shinkai, M. Matsui and T. Kobayashi, *Jpn. J. Hyperthermic Oncol.* **10**, 168 (1994).
63. R. Hergt, S. Dutz, R. Mueller and M. Zeisberger, *J. Phys.: Condens. Mater.* **18**, S2919 (2006).
64. M. Mitsumori, M. Hiraoka, T. Shibata, Y. Okuno, S. Masunaga, *et al.*, *Int. J. Hyperthermia* **10**, 785 (1994).
65. M. Mitsumori, T. Shibata, Y. Nagata, M. Hiraoka, M. Hasegawa, *et al.*, *Jpn. J. Hyperthermic Oncol.* **17**, 85 (2001).
66. A. Jordan, R. Scholz, K. Maier-Hauff, F. K. van Landeghem, N. Waldoefner, *et al.*, *J. Neurooncol.* **78**, 7 (2006).
67. M. Johannsen, B. Thiesen, A. Jordan, K. Taymoorian, U. Gneveckow, *et al.*, *Prostate* **64**, 283 (2005).
68. M. Suzuki, M. Shinkai, H. Honda and T. Kobayashi, *Melanoma Res.* **13**, 129 (2003).
69. A. Ito, K. Tanaka, H. Honda, S. Abe, S. Yamaguchi and T. Kobayashi, *J. Biosci. Bioeng.* **96**, 364 (2003).
70. N. Kawai, A. Ito, Y. Nakahara, M. Futakuchi, T. Kobayashi, *et al.*, *Prostate* **66**, 718 (2006).
71. A. Ito, Y. Nakahara, M. Fujioka, T. Kobayashi, T. Takeda, *et al.*, *Jpn. J. Hyperthermic Oncol.* **21**, 139 (2005).
72. M. Yanase, M. Shinkai, H. Honda, T. Wakabayashi, J. Yoshida and T. Kobayashi, *Jpn. J. Cancer Res.* **88**, 630 (1997).
73. N. Kawai, A. Ito, Y. Nakahara, M. Futakuchi, T. Kobayashi, *et al.*, *Prostate* **64**, 373 (2005).
74. F. Matsuoka, M. Shinkai, H. Honda, T. Kubo, T. Sugita and T. Kobayashi, *BioMag. Res. Technol.* **2**, 3 (2004).
75. Y. Shido, Y. Nishida, Y. Suzuki, T. Kobayashi and N. Ishiguro, *Orthopaedic Res.* **92-B**, 580 (2010).

76. H. Matsuno, I. Tohnai, K. Mitsudo, K. Hayashi, T. Kobayashi, *et al.*, *Jpn. J. Hyperthermic Oncol.* **17**, 141 (2001).
77. J. Dobson, *Nanomed.* **1**, 31 (2006).
78. M. Suzuki, M. Shinkai, M. Yanase, A. Ito, H. Honda and T. Kobayashi, *Jpn. J. Hyperthermic Oncol.* **15**, 79 (1999).
79. S. Hamaguchi, I. Tohnai, A. Ito, K. Mitsudo, T. Kobayashi, *et al.*, *Cancer Sci.* **94**, 834 (2003).
80. M. Yanase, M. Shinkai, H. Honda, T. Wakabayashi, J. Yoshida and T. Kobayashi, *Jpn. J. Cancer Res.* **89**, 775 (1998).
81. A. Ito, T. Kobayashi and H. Honda, *Jpn. J. Hyperthermic Oncol.* **21**, 1 (2005).
82. A. Ito, H. Honda and T. Kobayashi, *Cancer Immunol. Immunother.* **55**, 320 (2006).
83. S. Lindquist, *Ann. Rev. Biochem.* **55**, 1151 (1986).
84. D. D. Mosser, A. W. Caron, L. Bourget, A. B. Merlin, M. Y. Sherman, *et al.*, *Mol. Cell Biol.* **20**, 7146 (2000).
85. J. R. Subjeck, J. J. Sciandra and R. J. Johnson, *Br. J. Radiol.* **55**, 579 (1982).
86. A. Menoret and R. Chandawarkar, *Semin. Immunol.* **25**, 654 (1998).
87. P. K. Srivastava, A. Menoret, S. Basu, R. Binder and K. Quade, *Immunity* **8**, 657 (1998).
88. Y. Tamura and N. Sato, *Jpn. J. Hyperthermic Oncol.* **19**, 131 (2003).
89. A. Ito, M. Shinkai, H. Honda, T. Wakabayashi, J. Yoshida and T. Kobayashi, *Cancer Immunol. Immunother.* **50**, 515 (2001).
90. P. K. Srivastava and M. Heike, *Semin. Immunol.* **3**, 57 (1991).
91. P. K. Srivastava, H. Udono, N. F. Blachere and Z. Li, *Immunogenetics* **39**, 93 (1994).
92. A. D. Wells, S. K. Rai, M. S. Salvato, H. Band and M. Malkovsky, *Int. Immunol.* **10**, 609 (1998).
93. A. D. Wells and M. Malkovsky, *Immunol. Today* **21**, 129 (2000).
94. M. Okamoto, K. Tazawa, T. Kawagoshi, M. Maeda, T. Honda, *et al.*, *Int. J. Hyperthermia* **10**, 263 (2000).
95. S. Basu, K. J. Binder, T. Ramalingam and P. K. Srivastava, *Immunity.* **14**, 303 (2001).
96. T. Becker, F. U. Hartl and F. Wieland, *J. Cell Biol.* **158**, 1277 (2002).
97. A. Asea, M. Rehli, L. Kabingu, J. A. Boch, O. Bare, *et al.*, *J. Biol. Chem.* **277**, 15028 (2003).
98. A. Asea, S. K. Kraeft, L. A. Kurt-Jones, M. A. Stevenson, L. B. Chen, *et al.*, *Nat. Med.* **6**, 435 (2000).
99. P. K. Srivastava, *Nat. Rev. Immunol.* **2**, 185 (2002).
100. P. K. Srivastava, *Annu. Rev. Immunol.* **20**, 395 (2002).
101. A. Sato, Y. Tamura, N. Sato, T. Yamashita, T. Takeda, *et al.*, *Cancer Sci.* **101**, 1939 (2010).
102. P. Matzinger, *Science* **296**, 301 (2002).
103. S. Todryk, A. A. Melcher, N. Hardwick, E. Linardakis, A. Bateman, *et al.*, *J. Immunol.* **163**, 1398 (1999).
104. A. Ito, M. Yamaguchi, N. Okamoto, Y. Sanematsu, T. Kobayashi, *et al.*, *Nanomed.*, doi: 10.2217/nnm.12.142 (2013).
105. A. Testori, J. Richards, E. Whitman, G.B. Mann, J. Lutzky, *et al.*, *J. Clin Oncol.* **26**, 955 (2008).
106. H. Udono and P. K. Srivastava, *J. Immunol.* **152**, 5398 (1994).
107. M. Takahashi, G. M. Coope, *Mol. Cell Biol.* 7, 1378 (1987).

108. M. Kato, M. Takahashi, A. A. Akhand, W. Liu, Y. Dai, *et al.*, *Oncogene* **17**, 1885 (1998).
109. A. Ito, Y. Nakahara, M. Fujioka, T. Kobayashi, K. Takeda, *et al.*, *Jpn. J. Hyperthermic Oncol.* **21**, 139 (2005).
110. A. Ito, K. Tanaka, K. Kondo, M. Shinkai, T. Kobayashi, *et al.*, *Cancer Sci.* **94**, 308 (2003).
111. A. Ito, M. Shinkai, H. Honda and T. Kobayashi, *Cancer Gene Ther.* **8**, 649 (2001).
112. A. Ito, M. Matsuoka, H. Honda and T. Kobayashi, *Cancer Immunol. Immunother.* **53**, 26 (2004).
113. A. Ito, M. Matsuoka, H. Honda, T. Kobayashi, *Cancer Gene Ther.* **10**, 918 (2003).
114. K. Tanaka, A. Ito, T. Kobayashi, T. Kawamura, S. Shimada, *et al.*, *J. Biosci. Bioeng.* **100**, 112 (2005).
115. K. Tanaka, A. Ito, T. Kobayashi, T. Kawamura, S. Shimada, *et al.*, *Int. J. Cancer* **116**, 624 (2005).
116. A. Ito, Y. Nakahara, K. Tanaka, Y. Kuga, H. Honda and T. Kobayashi, *Jpn. J. Hyperthermic Oncol.* **19**, 151 (2003).
117. S. Wada, K. Tazawa, I. Furuta and H. Nagae, *Oral Dis.* **9**, 218 (2003).
118. G. Sreenivasa, J. Gellermann, B. Rau, J. Nadobny, P. Schlag, P. Deuflhard, *et al.*, *Int. J. Rad. Oncol. Biol. Phys.* **55**, 407 (2003).
119. P. Wust, B. Hildebrandt, G. Sreenivasa, B. Rau, J. Gellermann, *et al.*, *Lancet Oncol.* **3**, 487 (2002).

Chapter 16

Clinical Experience with Liposomal Doxorubicin

Giuseppe Visani and Alessandro Isidori*

Hematology and Hematopoietic Stem Cell Transplant Center, Marche Nord Hospital, Pesaro, Italy
**g.visani@ospedalimarchenord.it*

Abstract

Anthracyclines, including doxorubicin, are effective antineoplastic agents with a very broad anticancer spectrum that includes many solid and hematological tumors. However, the use of anthracyclines is limited by a dose-dependent cardiotoxicity and increasing survivors' susceptibility to treatment-related complications that can remarkably affect their quality of life. Surviving patients have an increased rate of heart failure, coronary artery disease and cerebrovascular accidents compared to the general population. Several preventive measures are currently being used, including limiting cumulative dose, altering anthracycline administration, using anthracyclines analogs and adding cardioprotectants to the regimen.

Liposomal anthracyclines were developed to reduce the cardiotoxicity of doxorubicin while preserving its antitumor efficacy. Among liposomal anthracyclines, the most commonly used in oncohematology is liposomal doxorubicin. The rationale used as the basis for the design of *i.v.* liposomal doxorubicin is that the liposomes cannot exit the circulation in healthy tissues, but can exit through leaky tumor-associated vessels. As a matter of fact, liposomal conjugation of doxorubicin results in preferential distribution of the drug in the tumor compared with normal tissue. Two liposomal formulations are currently available: (i) non-pegylated liposomal doxorubicin (NPLD) (Myocet®, Cephalon, USA) and (ii) pegylated liposomal doxorubicin (PLD) (Caelix®/Doxil®, Schering-Plough/Orto Biothech, USA). In this chapter, we will review the clinical activity of liposomal doxorubicin in oncohematological malignancies.

1. Introduction

Anthracyclines remain among the most widely prescribed and effective anticancer agents, but, unluckily, life-threatening cardiotoxicity is still a matter of fact that compromises their usefulness. The most important and most extensively used anthracycline is doxorubicin, an antitumor antibiotic originally isolated from a variant of the *fungus* producing daunorubicin. It was first introduced in the 1970s and, since that time, has become the most commonly used drug for the treatment of both hematological and solid tumors. Doxorubicin is composed of a fluorescent tetracyclic chromophore, adriamycinone (which gives the characteristic red color to the drug), linked to the aminosugar daunosamine via a glycosidic bond. The main metabolite doxorubicinol (adryamicinol) also has antiproliferative and antineoplastic effect.

The major drawback of conventional chemotherapeutic drugs and, thus, of doxorubicin is their non-specific distribution in the body. This means that chemotherapeutic agents have an effect on both cancer and normal cells and tissues, thereby limiting the dose attainable within the tumor and also resulting in excessive toxicities. A major complication of doxorubicin is its adverse cardiovascular effects. In both children and adults, the risk of clinical cardiotoxicity increases with the cumulative dose of doxorubicin.[1,2] In fact, approximately 2% of patients who have received a cumulative (lifetime) doxorubicin dose of 450–500 mg/m^2 will experience this condition. An approach to ameliorate doxorubicin-related toxicity is to use drug carriers, which engender a change in the pharmacological distribution of the drug, resulting in reduced drug levels in the heart.[3]

In the early 1980s, liposomes were found to be useful as anthracycline carriers, due to their ability of reducing toxicity while preserving powerful antitumor activity and, therefore, increasing the therapeutic index of anthracyclines.[3] Liposomes are formed spontaneously when amphiphilic lipids, such as phospholipids are dispersed in water. The liposome is a supramolecular assembly, whereby phospholipids form a closed bilayer, creating a vesicle with an entrapped water phase that is separated physically from the external medium.[3] Liposomes represent ideal drug delivery systems, as the microvasculature in tumors is typically discontinuous, having pore sizes (100–780 nm) large enough for liposomes to move from the blood compartment into the extravascular space surrounding the tumor cells.[4] Liposomes, having a size range of 100–200 nm, readily extravasate within the site of tumor providing a targeted drug delivery. Liposomal formulations of doxorubicin have been developed with the aim of improving the therapeutics index of doxorubicin by reducing cardiotoxicity. All the same, liposomal conjugation of doxorubicin results in preferential distribution of the drug in the tumor compared with normal tissue, maintaining its antitumor efficacy. Early preclinical studies in dogs[5] indicated that liposomal doxorubicin was less cardiotoxic than conventional doxorubicin. Furthermore, other studies indicated that cardiac muscle cells are less affected by liposomal daunorubicin.[6,7] Nevertheless, higher cumulative doses of liposomal daunorubicin (600–900 mg/m^2) are cardiotoxic.[8,9]

2. Pharmacology of Conventional Doxorubicin

The anthracyclines represent a broad family of antibiotics that exhibit activity in numerous tumors. The first anthracyclines, doxorubicin and daunorubicin, were isolated from *Streptomyces var. peucetius*. They were shown to be composed of a tetracyclic ring system with adjacent quinone–hydroquinone moieties, a short side chain with a carbonyl group and an aminosugar bound to the C-7 of the four-ring system. Doxorubicin and daunorubicin only differ in the side chain terminus ($-CH_2OH$ in doxorubicin versus $-CH_3$ in DNR). Second-generation anthracyclines, like epirubicin and idarubicin, were obtained after minor chemical modifications of doxorubicin or daunorubicin, respectively.

When injected by standard *i.v.* infusion, anthracyclines show a rapid distribution phase, a high distribution volume at steady state (~15 l/kg), a slow elimination phase (successive plasma half-lives of approximately 5 min, 1 h and 30 h). Anthracyclines are excreted mostly through bile, which imposes special care in patients with hepatic dysfunction.[10] In comparison with doxorubicin, epirubicin is characterized by a unique glucuronidation that accelerates its systemic body clearance and imposes administering epirubicin at doses 1.5 times higher than those of doxorubicin.[11] Doxorubicin undergoes hepatic metabolism and is predominantly excreted in the bile. The major metabolite of doxorubicin is doxorubicinol, which usually represents approximately 40–50% of the free doxorubicin concentrations.

Anthracyclines have long been known to kill tumor cells by inhibiting topoisomerase II.[12] Anthracyclines act by stabilizing a reaction intermediate, in which DNA strands are cut and covalently linked to topoisomerase II, eventually impeding DNA resealing. Anthracycline intercalation into DNA plays a role in this reaction; in fact, anthracycline rings that do not intercalate into DNA probably stabilize the complex between topoisomerase II and the DNA that it has nicked.[13] Anthracycline- and topoisomerase II-mediated DNA damage is followed by growth arrest in G1 and G2 and apoptosis.[14] This is usually, but not always, relayed by p53 and the consequent induction of the WAF1/CIP1 p21 gene product, a strong inhibitor of cyclin-dependent kinases that favor cell cycle progression through the G1 to S transition.[15]

Although topoisomerase II inhibition remains the most persuasive mechanism to explain the antitumor activity of anthracyclines, clinically relevant concentrations of anthracyclines were shown to induce apoptosis through additional mechanisms that were not bound to the topoisomerase II — p53 machinery. These mechanisms include, among others: (i) the activation of neutral sphingomyelinases, followed by ceramide formation and converse activation of cell death effectors (c-Jun N-terminal kinase) or downregulation of survival pathways (Akt/protein kinase B); (ii) mitochondrial dysfunction, followed by cytochrome *c* release and apoptosome formation; (iii) induction of lipid peroxidation and formation of malondialdehyde–DNA adducts, followed by the reduced activity of cyclin E- and cyclin B-associated kinase activities and growth arrest in both p53-proficient and p53-deficient cells; (iv) inhibition of the proteasome, followed by an accumulation of undegraded ubiquinated proteins which signal apoptosis. The mechanisms (i–iii) are triggered by reactive oxygen species (ROS) that are major by-products of anthracycline metabolism. ROS may also enable anthracyclines to damage and shorten telomeres, long sequences of base repeats that otherwise would delay cell senescence and apoptosis by preventing the degradation and legation of the end of chromosomes; however, anthracycline-induced telomere damage and dysfunction would be relayed to apoptosis through p53.[16]

Anthracycline treatment may be accompanied by the acquisition of a resistance phenotype through a combination of pharmacokinetic and pharmacodynamic mechanisms. On pharmacokinetic grounds, tumor resistance is caused by the reduced accumulation and/or an altered distribution of anthracyclines in tumor cells, usually mediated by overexpression of drug transporters that belong to the ATP-binding cassette family of proteins and are collectively referred to as ABC proteins (P-glycoprotein/Pgp, multidrug resistance protein 1/MRP1, breast cancer-resistance protein/BCRP). On pharmacodynamic grounds, tumor resistance may be caused by such diverse mechanisms as topoisomerase II mutation or redundancy, overexpression and preferred nuclear localization of proteasome α-type subunits (leading to an anomalous degradation of topoisomerase II), genetic deletion or loss-of-function mutations of p53, overexpression of ROS-detoxifying enzymes, overexpression of Bcl-2 (leading to a diminished cytochrome *c* release), etc. When taken in isolation, however, none of these factors would universally predict anthracycline-resistance in a given tumor or another.[15]

These limits of conventional doxorubicin stimulated the development of new formulations of the drug.

3. Pharmacology of Liposomal Doxorubicin

The mechanism thought to be responsible for the killing of tumor cells made by doxorubicin is obviously the same for liposomal doxorubicin. In fact, the liposomal encapsulation does not modify the mechanism of action of the native drug. Nevertheless, when considering the biological activity of liposomal formulations of doxorubicin, it is essential to keep in mind that (i) doxorubicin is released from the liposomes following *i.v.* administration, (ii) the distribution of doxorubicin loaded into liposomes is dependent, in part, on the biodistribution characteristics of the liposomes and (iii) the biological activity of liposomal doxorubicin is dependent on when, where and at what rate the drug is released.[3] As a consequence, differences in doxorubicin-mediated toxicity and efficacy are dependent on liposomal lipid composition, which in turn influences drug release properties and liposome biodistribution pattern.

Generally speaking, liposomal encapsulation results in an increase of the therapeutic index of the native molecules. In fact, there are several potential barriers to the delivery of a drug in its active form to solid tumors. Most small-molecule chemotherapeutics agents have a large volume of distribution on *i.v.* administration.[17] The result of this is always a narrow therapeutic index due to high toxicity in healthy tissues. Through encapsulation in liposomes, the volume of distribution is extensively reduced and the concentration of drug in tumor is significantly improved.[18] As a consequence, the quantity of the drug that can be successfully distributed to the tumor is increased and, thereafter, unwanted toxicities are reduced. The liposome, in fact, shields the drug from metabolism and inactivation in the plasma, and due to size limitations in the transport of large molecules or carriers across healthy endothelium, the drug accumulates to a reduced extent in healthy tissues.[18] However, leaky tumor-associated vessels have been shown to result in an increased extravasation of large carriers and, in combination with an impaired lymphatics supporting the tumor area, an accumulation of liposomal doxorubicin in the tumor.[18] This effect is often referred to as the enhanced permeability and retention effect ("EPR" phenomenon).[19]

To be effective as a carrier, liposome must be able to efficiently balance stability in the circulation with the ability to make the drug bioavailable at the tumor. In one hand, instantaneous release of doxorubicin from liposomal carrier following *i.v.* administration will result in a drug with substantially no differences from the unencapsulated form, whereas on the other hand, complete drug retention without any release will result in an agent that is neither toxic nor efficacious.

Two liposomal formulations are currently available: NPLD (Myocet®, Cephalon, USA) and PLD (Caelix®/Doxil®, Schering-Plough/Orto Biothech, USA).

NPLD is a doxorubicin citrate complex encapsulated in a liposome, an aqueous dispersion of egg phosphtidylcholine and cholesterol (55:45 molar ratio), sterile, non-pyrogenous that traps doxorubicin by pulling it into the interior of the vesicle (TLC D-99 model) by the generation of an electropotential across the liposome membrane. This mechanism for remote loading involves the generation of a pH gradient between the inside and the extraliposomal buffer. At the end of the encapsulation process, the ratio of doxorubicin:lipid in NPLD is approximately 1:4 and the pH is in the range of 5.5–6.5. The maximum activity of doxorubicin and other anthracyclines manifests itself in the S-phase of the cellular cycle.

The encapsulation of a cytostatic agent within a macromolecular vector such as a liposome reduces drastically its distribution volume, diminishing its diffusion in the organism and thus the toxicity for healthy tissues while increasing the concentration within the neoplastic tissue.[20] In ideal conditions, the drug can be transported in the circulatory system within the liposome's aqueous space and arrive to the site of the tumor in active form. Encapsulation in a liposome means the drug is protected from inactivation while in the blood stream and, furthermore, its diffusion through the healthy endothelium is limited, while it can diffuse through tumoral endothelium which would present discontinuities. Thus, NPLD should preferentially direct doxorubicin away from sites of potential toxicity, but leave the tumor exposed.

NPLD is a "conventional liposome"-encapsulated formulation of doxorubicin, with altered pharmacokinetics and biodistribution with respect to both native and pegylated doxorubicin. This results in a higher area under the curve (AUC), a smaller volume of distribution (V_D) and a preferential distribution to liver, spleen and lymphatics, when compared with conventional doxorubicin.[21] Pre-clinical studies comparing equal doses of liposome-encapsulated doxorubicin and conventional doxorubicin showed that the use of NPLD resulted in a significantly lower cardiac and gastrointestinal (GI) toxicity, with similar antitumor efficacy.[21] In the clinical setting, following the administration of liposomal doxorubicin to patients with breast cancer, the area under the plasma concentration–time curve from time zero to infinity ($AUC\infty$) and maximum plasma concentration (C_{max}) values for total doxorubicin (encapsulated and unencapsulated) were approximately 20- and 10-fold higher than those following the administration of conventional doxorubicin.[21] The authors observed that the plasma AUC is not a good index of drug exposure, because most of the doxorubicin is encapsulated and is, therefore, not bioavailable or active.[21] In patients with breast cancer, the clearance of total doxorubicin was approximately ninefold lower in patients treated with NPLD with respect to those treated with conventional doxorubicin; furthermore, the terminal half-life was shorter for NPLD.

The PLD (Caelix®/Doxil®, Schering-Plough/Orto Biothech, USA) is composed of hydrogenated soya phosphatidylcholine, cholesterol and poly-ethylene glycol (PEG)-modified phosphatidylethanolamine (55:40:5 molar ratio); doxorubicin is confined in liposomes that have been sterically stabilized by grafting PEG onto the surface ("Stealth Liposome"). PLD has a circulation half-life of ≈73.9 h, whereas doxorubicin has a half-life of <10 min. Prolonged circulation is primarily determined by the clearance of the pegylated liposomal carrier, and facilitates greater uptake of PLD by tumor tissue. The hydrophilic coating of the PLD formulation reduces interaction between various circulating plasma components and the liposome surface, thereby preventing reticuloendothelial system-mediated uptake of the circulating liposomes. This results in a slower plasma clearance (CL) and a smaller volume of distribution (V_D) of PLD with respect to conventional doxorubicin, and therefore in a longer circulation time of PLD in the bloodstream after *i.v.* administration. Mean plasma clearance values were low after administration of PLD 10–60 mg/m^2 (0.003 [range 0.008–0.15]L/h/m^2) or 20 mg/m^2 (0.041 L7h/m^2). Conventional doxorubicin has a much higher rate of plasma clearance (24–35 L/h/m^2) compared to PLD.[22] The volume of distribution of PLD (over the dose range of 10–60 mg/m^2) was ≈2 L/m^2, and approximated the plasma value; this is significantly lower than that seen for conventional doxorubicin (700–1,100 L/m^2).[22] The mean area under the plasma–concentration–time curve (AUC) for doxorubicin is large (590 μg h/mL) following administration of PLD 20 mg/m^2 in patients with Kaposi's sarcoma, with a corresponding peak plasma concentration (C_{max}) of the drug of 8.34 μg/mL (mean value).[22] In contrast, AUC and C_{max} values were lower with conventional doxorubicin 60 mg/m^2 (≈2 μg h/mL and ≈ 5 μg/mL).

As a consequence, PLD accumulates selectively in metastatic breast carcinoma tissue, resulting in 10-fold higher intracellular drug concentrations compared with adjacent normal tissue.[23] Pegylated liposomal encapsulation also reduces plasma levels of free doxorubicin and may reduce drug delivery to normal tissue, which may reduce toxicity. Biodistribution animal studies indicate superior accumulation of PLD into various implanted mouse–human tumors, with an augmentation of liposomal drug tumor levels compared with free drugs.[22] The extended circulation time of pegylated liposomes and their ability to extravasate through the leaky vasculature of tumors results in the increased delivery of PLD to the tumor site in patients with cancer.

The PLD has been approved for clinical use in a variety of cancer types due to its antitumor efficacy. However, the safety profile needs further improvement, as long as hand–foot syndrome remains a dose-limiting toxic effect of Doxil, maybe due to the considerable amount of drug being delivered to the skin owing to long circulation of the drug in the bloodstream.

4. Clinical Activity

4.1. *Non-pegylated liposomal doxorubicin*

The NPLD (Myocet®) is a liposome-encapsulated formulation of doxorubicin, which differs from PLD, as well as from unencapsulated, conventional doxorubicin, resulting in an alteration in the pharmacokinetics and biodistribution. The active drug is loaded into liposomes in the hospital pharmacy just before *in vivo* administration. The maximum single dose is 75 mg/m^2, with a recommended schedule every three weeks. Maximum cumulative dose, associated with 5% of cardiac toxicity, is approximately 785 mg/m^2, and the drug's dose-limiting toxicities are neutropenia and stomatitis.

Pre-clinical studies comparing equal doses of liposome-encapsulated doxorubicin and conventional doxorubicin showed that the use of NPLD resulted in a significantly lower cardiac and GI toxicity, with similar antitumor efficacy.[24] NPLD is approved for the treatment of metastatic breast cancer (MBC) in Europe.

The most common non-hematological side effect of NPLD reported in all clinical trials is alopecia (Table 1), occurring in approximately 90% of

Table 1. Adverse effect profile of conventional and liposomal doxorubicin.

Toxicity	Doxorubicin	PLD	NPLD
Infusion reaction	–	+[a]	++[a]
Vescicant effect	+	–/+	++
Nausea/vomiting	++	+	+
Grade III–IV neutropenia	+++	+++	+++
Stomatitis/mucositis	+	+++	+
Palmar/plantar erythrodysesthesia	–	+++	–
Cardiotoxicity	+++	+	+
Alopecia	+++	++	+++
Maximum tolerated dose	60 mg/m^2	50 mg/m^2	80 mg/m^2
Maximum cumulative dose	450 mg/m^2	undefined >650 mg/m^2	undefined >900 mg/m^2

[a]Drip-rate and pre-medication dependent.

PLD, pegylated liposomal doxorubicin; NPLD, non-pegylated liposomal doxorubicin; –, no relevant toxicity; +, toxicity occurring in <25% of patients; ++, toxicity occurring in 26–50% of patients; +++, toxicity occurring in >50% of patients.

patients. The incidence of stomatitis/mucositis was significantly lower in the liposomal doxorubicin plus cyclophosphamide group than the conventional doxorubicin plus cyclophosphamide group in one trial ($P = 0.008$).[25] However, there was no significant difference between the same groups in terms of grade III–IV GI toxicity.[25] In another study,[26] grade III stomatitis/mucositis was experienced by significantly more liposomal doxorubicin plus cyclophosphamide recipients than epirubicin plus cyclophosphamide recipients, while injection site toxicity (all grades) was experienced by significantly fewer liposomal doxorubicin plus cyclophosphamide recipients.[26] No other significant between-group differences in adverse events were reported in any of the trials. Remarkably, severe palmar–plantar erythrodysesthesia (PPE), which is a dose-limiting toxicity of PLD, was not observed in any patients receiving NPLD.[21]

Grade III–IV neutropenia, on the other hand, was the most common hematological side effect, reported in all trials in both solid and hematological cancers, occurring in approximately 50–91% of patients. Nevertheless, the relative dose intensity reported in all the trials was >90%, and treatment delays were uncommon.

4.1.1. NPLD in breast cancer

The NPLD was designed to reduce the cardiotoxicity of doxorubicin while preserving its antitumor efficacy. In the early 2000, the safety and efficacy of NPLD was tested in cancer patients, after the demonstration of a significantly lower cardiac and GI toxicity, with similar antitumor efficacy showed by NPLD in pre-clinical studies.

The first studies were conduced in breast cancer patients. These phase III, randomized, multicenter trials were designed to test the hypothesis that NPLD, alone or in combination with other drugs, would result in significantly less cardiac toxicity than the same dose and schedule of conventional doxorubicin and other drugs, while providing comparable antitumor efficacy in first-line treatment of MBC.

Batist *et al.*[25] randomized 297 patients with MBC and no prior chemotherapy for metastatic disease were randomized to receive either 60 mg/m^2 of Myocet (M) or conventional doxorubicin (A), in combination with 600 mg/m^2 of cyclophosphamide (C), every three weeks until disease progression or unacceptable toxicity. Antitumor efficacy of MC versus AC was comparable: objective response rates (RRs), 43% versus 43%; median time to progression (TTP), 5.1% versus 5.5 months; median time to treatment failure (TTF), 4.6 versus 4.4 months and median survival, 19 versus 16 months.[25]

Cardiotoxicity was a primary end point parameter in all treated patients.[25] Cardiac toxicity, sufficient for removal of a patient from study, was defined as a decrease in resting left ventricular ejection fraction (LVEF) of ≥ 20 ejection fraction (EF) units from baseline to a final value of less than 50%, or a decrease of ≥10 EF units from baseline to a final value of less than 50%, or clinical evidence of congestive heart failure (CHF). LVEFs were assessed using serial multigated blood-pool imaging (MUGA) scans, which have been shown to be a reliable and serially reproducible method of evaluating cardiac function in patients receiving anthracycline therapy.

To ensure accuracy and objectivity, each center was required to have its equipment and methodology used for MUGA scans reviewed and certified by a cardiologist at the Core Laboratory at Yale University before enrolling patients. During the trial, all MUGA scans were sent to the Core Laboratory at Yale, where they were read by the same cardiologist blinded to the patient's treatment.[25] To minimize the risk of CHF, all scans were read in real time and results provided to the site before the next scheduled dose of anthracycline therapy. CHF was determined on the basis of a treatment-blinded review of records from patients for whom the investigator had made a diagnosis of CHF, as well as patients who had a LVEF of ≤ 30%. LVEF ≤ 30% was selected as the cutoff because these patients are at significant risk for CHF.[25] The blinded review was conducted by a second cardiologist at Yale University noted for his expertise in doxorubicin-induced cardiotoxicity. All MUGA scan data were interpreted and LVEF values were estimated at a core laboratory on a blinded basis. Nine patients (6%) treated with MC developed protocol-defined cardiotoxicity compared with 33 patients (21%) treated with AC (log-rank $P = 0.0001$). Five cases of CHF, all in the AC arm (log-rank $P = 0.02$), were observed after cumulative lifetime doses ranging from 360 to 480 mg/m^2. Four of the five patients with CHF were anthracycline-naive before this study; one patient had 240 mg/m^2 of prior adjuvant doxorubicin. All other patients with cardiotoxicity had an asymptomatic decrease in LVEF of ≥ 10 EF units from baseline to a final value less than 50%. The estimated (Kaplan–Meier) median cumulative lifetime dose of doxorubicin at the first occurrence of protocol-defined cardiac toxicity was more than 2,220 mg/m^2 for the MC arm versus 480 mg/m^2 for the AC arm. The hazard ratio of 4.8 shows that patients treated with MC were 80% less likely to develop cardiotoxicity relative to patients treated with AC.

Similarly, there was a highly significant difference in the time to onset of cardiotoxicity when measured from the start of protocol therapy. The estimated median onset of protocol-defined cardiotoxicity was more than 22 months for MC versus 10 months for AC (log-rank $P = 0.0003$). There was a gradual increase in the median change from baseline LVEF to the last post-treatment LVEF among patients treated with either regimen, but this was more pronounced in the AC-treated group.[25]

In the subset of patients with recognized risk factors for cardiac toxicity, the hazard ratio was increased to 16, indicating that these patients were more than 90% less likely to develop cardiac toxicity with MC relative to AC. About 4% of MC-treated patients developed a protocol-defined cardiac event versus 22% of AC-treated patients.[25] The median lifetime cumulative dose of doxorubicin at onset was 480 mg/m^2 for AC versus more than 2,220 mg/m^2 for MC ($P = 0.0001$). Four of the five patients with CHF were in this subgroup with increased risk of cardiac toxicity.[25]

In conclusion, the improved therapeutic index for NPLD predicted by the preclinical data and indicated by the phase I/II clinical trials was confirmed in this phase III, randomized, multicenter trial. Statistically, significantly fewer patients treated with NPLD in combination with cyclophosphamide experienced cardiac toxicity defined by reductions in LVEF or clinical CHF.[25]

Another interesting study was carried on by Harris *et al.*[27] About 224 patients with MBC and no prior therapy for metastatic disease were randomized to receive either TLC D-99 (75 mg/m^2) or doxorubicin (75 mg/m^2) every three weeks, in the absence of disease progression or unacceptable toxicity. The primary efficacy endpoint was RR. The primary safety endpoint was cardiotoxicity.

Electrocardiograms and MUGA scans were done at baseline, and after reaching a lifetime cumulative doxorubicin dose of 300, 400, 500 mg/m^2, and before each subsequent dose, at off-study, and at three-month follow-up. All MUGA scan data were transferred electronically to a core laboratory for blinded interpretation and estimation of LVEF. Before the first interim analysis, endomyocardial biopsies were performed at a selected number of institutions after lifetime cumulative doxorubicin dose of 425 mg/m^2. All patients whose LVEF declined by greater than 10% to a value of greater than or equal to 50%, or by greater than 6% to a value of less than 50% were to have cardiac biopsies regardless of lifetime cumulative doses.[27] All biopsies were read by a core pathologist, and the results were scored according to the Billingham scale. The purpose of biopsies was to validate the results of MUGA scans. After the first interim analysis, it was determined that MUGA scans were adequate to monitor cardiac function and cardiac biopsies were discontinued.[27] All MUGA scan data were sent to an independent central laboratory for estimation and interpretation of LVEF values without knowledge of treatment arm.

Cardiac events, sufficient for removal of a patient from study, were more than twice as frequent in doxorubicin-treated patients than TLC D-99–treated patients (29%versus 13%, log-rank $P = 0.0001$).[27] With the increasing lifetime cumulative dose of doxorubicin and TLC D-99, there was a gradual increase in the median change from baseline LVEF to the first post-treatment LVEF among patients treated with either agent, but this was more pronounced in the doxorubicin group.[27] A Kaplan–Meier estimate of the probability of the first onset of a cardiac event as related to the lifetime cumulative dose of doxorubicin or TLC D-99 showed that risk of cardiotoxicity was much higher with doxorubicin treatment than TLC D-99 (HR = 3.56) ($P = 0.0001$).[27] Two patients (2%) on TLC D-99 developed clinical CHF. One patient, after 13 cycles of TLC D-99 and a cumulative dose of 1,110 mg/m^2, had a decrease of 14 EF units in her LVEF to 46% and was taken off-study. Two months after the last dose, she presented with shortness of breath and bilateral pleural effusions and was hospitalized for CHF. Another patient, with prior adjuvant doxorubicin dose of 290 mg/m^2 and prior chest wall irradiation, received five cycles of TLC D-99 for a total lifetime doxorubicin dose of 785 mg/m^2, and went off-study for PD. Four months after the last dose, a MUGA scan showed a LVEF of 46% (a 16-point decrease from baseline). Later, the patient received five cycles of mitomycin plus mitoxantrone, and 11 months after the last study treatment, she received a diagnosis of CHF.[27] Nine patients (8%) on doxorubicin developed clinical CHF at lifetime doses of 525–765 mg/m^2. Three patients had CHF within 30 days of the last dose of study treatment, including one who died of CHF after 585 mg/m^2. All nine cases were attributed to study drug treatment.[27]

Before the first interim analysis, 51 patients were treated at 8 participating institutions performing endomyocardial biopsies. Of those, 36 patients qualified for the procedure, and all 36 patients had cardiac biopsies. All biopsies were read by a core pathologist, blinded to treatment assignment and the results were scored according to the Billingham scale. There was a significant difference between the two treatment groups favoring TLC D-99 and the number of patients who had a score of greater than or equal to 2.5 (26% versus 71%; $P = 0.02$).[27]

Regarding efficacy, the overall response rate (ORR) was 26% in both treatment groups. The median TTP was 2.9 months on TLC D-99 versus 3.1 months on doxorubicin. Median survival was 16 versus 20 months with a non-significant trend in favor of doxorubicin ($P = 0.09$).[27]

In another randomized study, Chan *et al.* tried to ascertain the efficacy and tolerability of NPLD and epirubicin combined with cyclophosphamide in the first-line treatment of patients with MBC.[26] About 160 anthracycline-naïve MBC patients were randomized to receive NPLD (M; 75 mg/m^2) or epirubicin (E; 75 mg/m^2) in combination with cyclophosphamide (C; 600 mg/m^2), every three weeks for up to eight cycles. Cardiotoxicity was low in both treatment groups: nine patients on MC and eight on EC had asymptomatic LVEF reductions at comparable cumulative doses. For MC, there were two cases at 100–299 mg/m^2, four at 300–399 mg/m^2 and three at 500–599 mg/m^2; for EC, there was one case at 200–299 mg/m^2, four at 300–399 mg/m^2 and three at 400–499 mg/m^2. There was no clinical evidence of CHF in any patient. ORRs were 46% and 39% for MC and EC treatment, respectively ($P = 0.42$). MC was superior to EC with respect to median TTF (5.7 versus 4.4 months; $P = 0.01$) and median time to disease progression (7.7 versus 5.6 months; $P = 0.02$). Median survival times were 18.3 and 16.0 months for MC and EC, respectively ($P = 0.504$). Unsurprisingly, the results from this study[26] suggested that, at equimolar doses, in combination with cyclophosphamide, NPLD has modest but significant advantages over epirubicin for some efficacy end points and a non-significant trend toward improvement in others.

Some comments to better interpret the results arising from these studies are necessary. First, in all trials, treatment with NPLD was administered every three weeks and continued until disease progression or the development of unacceptable toxicity. Second, baseline patient characteristics (ECOG performance status (PS), median age at enrolment, prior radiotherapy, prior hormonal therapy, adjuvant chemotherapy) were well balanced between treatment groups; in one trial, significantly more patients receiving conventional than liposomal doxorubicin had a positive progesterone receptor status (50 versus 39%; $P = 0.02$).[27] In one study,[26] all patients were anthracycline naive; in the other trials,[25,27] prior adjuvant doxorubicin therapy was permitted if the cumulative dose was ≤ 300 mg/m^2. Third, the primary endpoint in all trials was the objective tumor RR (complete response (CR) plus partial response (PR) rate), and all the trials were designed to demonstrate the non-inferiority of liposomal doxorubicin versus conventional doxorubicin in an intention-to-treat population. Looking at the results, two studies[25,27] clearly demonstrated reduced cardiotoxicity of NPLD when compared to standard doxorubicin and significantly less mucositis, with a superimposable antitumor efficacy of the two drugs. The third study showed a trend of higher efficacy for NPLD with respect to epirubicin, with equal cardiotoxicity.[26] Additionally, in small phase I/II and II trials in patients with advanced or MBC, NPLD showed antitumor activity in combination with trastuzumab,[28] paclitaxel and trastuzumab,[29] gemcitabine and docetaxel,[30] or cyclophosphamide and fluorouracil.[31]

4.1.2. NPLD in non-Hodgkin lymphoma

The first study testing the safety and efficacy of NPLD in non-Hodgkin lymphoma (NHL) (AIDS-related) was conducted by Alexandra Levine and coworkers in 2004.[32] The study enrolled 24 patients with newly diagnosed AIDS-related NHL (median age: 43 years). About 67% of patients had a high or high–intermediate International Prognostic Index (IPI) scores at diagnosis. Serum LDH was high in 17 patients (71%). A total of 21 patients (88%) had extranodal disease, and 12 patients (50%) had 2 or more sites of extranodal involvement.

The primary objective of the study was to evaluate the safety and efficacy of NPLD when substituted for doxorubicin in the CHOP in patients with newly diagnosed AIDS-related non-Hodgkin's lymphoma (AIDS-NHL). NPLD at doses of 40, 50, 60 and 80 mg/m^2 was given with fixed doses of cyclophosphamide (750 mg/m^2), vincristine (1.4 mg/m^2) and prednisone (40 mg/m^2) every 21 days.[32] The maximum tolerated dose (MTD) of Myocet was defined as the dose at which less than half of the patients on a cohort experienced a dose-limiting toxicity.[32] No dose escalations were allowed in individual patients. The MTD of NPLD was not reached at the 80 mg/m^2 dose. As a high CR rate was seen at all dose levels, a dose of 50 mg/m^2 of NPLD was chosen for the phase II portion of the trial, which was expanded to a total of 24 patients, including 10 on the initial phase I dose-escalation portion.[32] All patients received highly active antiretroviral therapy (HAART) while on chemotherapy.

The results of the study were encouraging. No dose-limiting toxicities were observed at any level, with myelosuppression being the most frequent toxicity (grade IV neutropenia in 75% of patients, neutropenic fever in 12% and neutropenic sepsis in 8%).[32] ORR was 88%, with a CR rate of 75%, and a PR rate of 13%. The median duration of CR was 15.6 months (range, 1.7 to 43.5 months).

Moreover, 16 out of 24 AIDS-related NHL patients were evaluated for multidrug resistance (MDR-1) expression on their histological samples. The authors demonstrated that this NPLD-based regimen was equally effective in both MDR-1-positive and MDR-1-negative cases. The authors postulated that the efficacy of this regimen was eventually related to the ability of NPLD to overcome excessive drug efflux due to P-gp (MDR-1) overexpression. The MDR-1 expression did not correlate with response in this study, suggesting that NPLD might evade this resistance mechanism.[32]

In terms of potential cardiac toxicity, LVEF was obtained at baseline and at study termination in 19 patients. No patient developed a decline in LVEF to below normal (45%). A total of 14 patients had no significant change in LVEF over time, while 5 patients (26%) experienced a 10% or greater decline, though their LVEF values still remained within normal range. None of these five patients had any signs or symptoms of cardiac dysfunction, and the values returned to baseline levels in the two patients who had follow-up cardiac studies performed, three months after completion of chemotherapy.

Starting from that background, the same group of the Keck School of Medicine designed in 2006 a phase I-II trial to evaluate the safety of the same regimen (NPLD, cyclophosphamide, vincristine and prednisone every 21 days) in the treatment of newly diagnosed aggressive NHL patients.[33]

A total of 47 patients (median age: 55 years) were enrolled in the study. The vast majority of the patients had diffuse large B-cell NHL (37/47). Liposomal doxorubicin at doses of $40\,mg/m^2$, $50\,mg/m^2$, $60\,mg/m^2$ and $80\,mg/m^2$ was given with fixed doses of cyclophosphamide ($750\,mg/m^2$), vincristine ($1.4\,mg/m^2$) and prednisone ($40\,mg/m^2$). Chemotherapy cycles were repeated every 21 days.[33]

No dose-limiting toxicities were observed at any level. Reversible grade III/IV neutropenia was the most common toxicity (95.8%). Most non-hematologic side effects were nausea, vomiting, fatigue and fever and were primarily grade I/II in severity. Stomatitis of mild or moderate severity was reported in 23% of patients.[33]

Two out of 47 patients (4%) developed clinically silent cardiac toxicity, with a decline in LVEF of 20% each. None of the patients presented any symptoms of cardiotoxicity. The decline in LVEF occurred in one patient after a cumulative NPLD dose of $240\,mg/m^2$ and in the other one after a cumulative NPLD dose of $640\,mg/m^2$. In both patients, the declines returned to baseline two months and six months from discontinuation of NPLD, respectively.[33]

Complete remissions were documented in 31 of 46 evaluable patients (67.4%) and partial remissions in 7 (15.2%), for an overall major RR of 82.6%. Responses were attained after a median of four cycles of therapy. Patients with T-cell lymphomas fared poorly with this regimen, with none of them achieving a CR lasting more than six months.[33]

The median duration of complete remission was 27.7 months (range, 2.4 months to 59.8 months). Reported median follow-up was 3.1 years. Median survival was not reached at the time of publication. The two-year estimated overall survival (OS) probability was 65% (95% CI, 50–77%), with a three-year estimated OS probability of 59% (95% CI, 44–73%). A total of 19 patients had died at the time of publication.[33]

Regarding the relationship between MDR-1 expression and outcome, MDR-1-related *p*-glycoprotein expression was assessed in lymphoma tissues from 27 patients. Eight patients (30%) were MDR-1-positive at diagnosis. No difference in CR rates were observed when comparing patients whose tumors expressed MDR-1 versus those who did not (63% in MDR-1-positive and 74% in the MDR-1-negative lymphomas, $P = 0.66$), indicating that NPLD might overcome MDR-1 *in vivo*.[33]

Rigacci and collaborators[34] designed a prospective study to assess the efficacy and safety of the combination of NPLD with cyclophosphamide, vincristine, prednisone and rituximab (R-COMP) in patients with aggressive non-Hodgkin's B-cell lymphomas and concurrent cardiac disease or pretreated with anthracyclines.

A total of 21 patients were selected for the presence of cardiac comorbidity and/or previous treatment with anthracycline-based regimens. NPLD at a dose of 50 mg/m^2 was administered in association with cyclophosphamide (750 mg/m^2), vincristine (1.4 mg/m^2), prednisone (40 mg/m^2) and rituximab (375 mg/m^2) every 21 days for 4 to 6 cycles unless progression or unacceptable toxicity occurred. CR rate was 76%, whereas PR rate was 14%, with an ORR of 90%. Two patients (10%) did not respond to therapy. After a median follow-up of 13 months (range 2–36 months), 2/16 CR patients relapsed, with a cumulative disease-free survival (DFS) rate of 78%.[34]

Regarding toxicities, the authors observed only a CHF in one patient, among a total of 115 chemotherapy cycles administered. LVEF was evaluated after the third cycle and at the end of treatment in all but one patient, who developed CHF with LVEF 20% after the first cycle of R-COMP.[34] Between the 20 evaluable patients, median LVEF was 60% (range 38–74%) after three cycles and 60% (range 40–69%) at the end of the treatment. There was no significant difference between LVEF at baseline, after the third cycle, and at the end of study in patients with cardiac comorbidity (52, 58 and 60%, respectively), in those without cardiac diseases (62, 61 and 60%, respectively) and in those previously treated with anthracyclines. Patients were evaluated every six months after the end of the study. Among the 15 patients who were followed-up for at least 12 months after the end of therapy, none of them presented any cardiac dysfunction or significant decrease of LVEF.[34]

The authors stated that rituximab plus NPLD-based regimen was well tolerated and highly effective in inducing clinical responses in this group of patients at high risk for cardiac toxicity or previously treated with anthracyclines. The tolerability profile was favorable, with a low incidence of cardiac events (1/21 patients).[34]

Visani and coworkers[35] conducted a pilot study to assess the toxicity and efficacy of the combination of NPLD with cyclophosphamide, vincristine, prednisone and rituximab (R-COMP) in frail and elderly patients with aggressive non-Hodgkin's B-cell lymphomas.

The NPLD at a dose of 50 mg/m^2 was administered in association with cyclophosphamide (750 mg/m^2), vincristine (1.4 mg/m^2), prednisone (40 mg/m^2) and rituximab (375 mg/m^2) every 21 days for 4 to 6 cycles unless progression or unacceptable toxicity occurred. A total of 20 frail patients (median age: 73 years), as defined by Balducci and Extermann[36] with diffuse large B-cell or grade IIIb follicular lymphoma, either at diagnosis (15 patients) or relapsed (5 patients), were prospectively enrolled.[35]

About 13 out of 20 NHL patients (65%) had a CR and an additional 5 patients (25%) achieved a PR, with an ORR of 90%. Notably, the median age of the study population (73 years) was the oldest ever reported in the literature treated with this combination.[35] As a matter of fact, this population consisted of particularly elderly and frail patients with advanced disease at the time of study entry. These patients had indeed an extremely high probability of being unable to complete one of the standard anthracyclines-based regimens or to suffer from cardiotoxicity.[35]

Therapy was well tolerated. Grade III or IV neutropenia occurred in 26% of cycles and febrile neutropenia in 5%. There was no significant difference between LVEF at baseline, after the third cycle, and at the end of study in 16/19 patients. Notably, two patients presented a CHF (NYHA 3) after one and three cycles of R-COMP, respectively, and were shifted to receive an anthracycline-free regimen while in CR after R-COMP. In these two patients, the authors observed a 20% decrease of the LVEF which partially recovered after medical therapy. The cumulative percentage of cardiovascular complications was 15%, slightly higher than that reported for elderly patients treated with R-CHOP 14 or 21. However, this patient population consisted only of frail patients having a poor WHO PS and important comorbidities. Indeed, it is not possible to compare these patients to those previously reported in other series and any conclusions might not be drawn from this small trial regarding cardiotoxicity. This study demonstrated for the first time the safety and efficacy of the R-COMP 21 regimen in frail and elderly patients with aggressive NHL. Cardiotoxicity was low and a promising RR was observed in this setting patients.[35]

Dell'Olio *et al.*[37] reported on 80 elderly patients (median age 70.9 years) affected by newly diagnosed poor-risk diffuse large B-cell lymphoma (DLBCL) treated with the same R-COMP 21 regimen. Patients had to be more than 55 years of age, with a LVEF of at least 30%, with an IPI between 2 and 4 and with a disease stage II–IV according to the Ann–Arbor staging system. The presence of a comorbidity, even cardiovascular, was not an exclusion criterion. A total of 63 out of 80 patients (79%) had a CR, 3 a CRu and

11 additional patients achieved a PR for an ORR of 96%. The Fisher exact test showed disease stage and IPI being associated with RR ($P = 0.003$ and 0.05, respectively).[37] With a median follow-up of 31 months (range 8–42) as of December 2009, DFS was 77.5% and only 15/80 (18.7%) enrolled patients died. No therapy-related cardiac event was observed during the study period. Interestingly, the authors observed a slight, not significant increase in LVEF over time in patients with a basal LVEF of <50% or with cardiac comorbidities. However, this is the only study reporting an increasing of LVEF after treatment with NPLD, and the small sample size of the study may overestimate the real benefit of the drug on cardiac tissue.[37]

Another study regarding the safety and the efficacy of the R-COMP combination was recently published.[38] A total of 75 elderly patients with DLBCL were studied. Only patients with LVEF > or = 50% were allowed. R-COMP regimen was administered every three weeks for three cycles, followed by additional five cycles in case of CR orPR.[38]

A total of 75 patients with a median age of 72 years were registered, and 72 were evaluable for response assessment. About 56% of patients had high or high–intermediate IPI score. Median LVEF at baseline was 61%. About 38 patients had history of abnormal cardiovascular conditions. The ORR was 71%, with a CR rate of 57%. After a median follow-up of 33 months, the three-year OS, failure-free survival and progression-free survival (PFS) rates were 72%, 39% and 69%, respectively. Neutropenia (54%) was the most frequent grade III–IV adverse event; 21% of patients experienced cardiac adverse events, graded as III–IV in 4% of the cases.[38]

Corazzelli *et al.*[39] published about biweekly administration of R-COMP (R-COMP 14) in elderly patients with poor-risk DLBCL and moderate to high life-threatening cardiopathy. In this regimen, an equal dose of NPLD was substituted for doxorubicin within the R-CHOP-14 schedule. Even if the study was designed for patients with a life-threatening cardiopathy, as a matter of fact, patients with a history of CHF or with a LVEF <45% were excluded from the study. Nevertheless, 65% of patients were receiving a beta-blocker and/or an angiotensin converting enzyme (ACE) inhibitor or an angiotensin receptor blocker (ARB), and, among patients with high impact heart-related conditions all of them were under active treatment with beta-blocker and/or ACE inhibitors or ARBs.[40,41] It is well known and demonstrated that both beta-blockers and ACE inhibitors or ARBs have a cardioprotective effect, mostly due to their ability in inhibiting the formation of reactive oxygen-free radicals. Besides, when looking at the results on the safety of this regimen, it is important to remind that NPLD was at least co-adjuvated by beta-blockers, ACE inhibitors and ARBs in preventing cardiovascular events. Nevertheless, the cumulative incidence of failures due to early and late cardiac events was 17%, higher than reported in other series.[34,35,37,38] Other toxicities were acceptable, with grade III–IV hematological toxicities occurring only in approximately 20% of the patients. In this respect, all patients received primary prophylaxis for febrile neutropenia with conventional G-CSF from day 6 to day 10, every cycle. The scheduled six courses of R-COMP-14 were completed by only 67% of the patients, including three patients who were considered to have completed their treatment after five courses. The ORR was 73% and CR rate was 68%. The four-year TTF was 49% without a significant advantage for patients younger than 70 years. As a whole, these dose-dense regimen R-COMP-14 performed worst with respect to the standard R-COMP-21 regimen.

Isidori *et al.*[42] evaluated the toxicity and efficacy of NPLD when substituted for conventional doxorubicin in the CHOP regimen in 129 consecutive elderly patients with newly diagnosed aggressive B-cell NHL. Patients were split into two groups according to the Multidimensional Geriatric Assessment (MDGA). A total of 39 patients with an Activities of Daily Living (ADL) = 6 were addressed to receive dose-dense R-COMP every two weeks (14), whereas 90 patients with an ADL <6 were addressed to receive R-COMP every three weeks (21). The median age of patients was 74 years (range: 65–89). At baseline 94/129 (73%) patients had stage III–IV disease, and 60/129 (46%) patients had an intermediate or high risk IPI score. Median PS was WHO 1 (range 0–3) and median number of comorbidities was 2 (range 1–6). The median LVEF before starting chemotherapy was 59% (range 40–73). A total of 715 cycles of chemotherapy were administered (234 R-COMP 14 and 481 R-COMP 21). All these variables were comparable between the two groups (14 versus 21). Toxicity was mainly hematological in both groups. Grade III/IV neutropenia occurred in 11% and 22% of cycles in the R-COMP 14 and 21 groups, respectively, with an incidence of febrile neutropenia of 3% and 8%, respectively. The relative dose intensity for the regimens was 93% for the R-COMP 14 group and 90% for the R-COMP 21 group, respectively. Regarding cardiotoxicity, only 8/129 patients presented a grade II–IV WHO toxicity.[42]

All patients were evaluable for response. The ORR was 89% (CR 73%, PR 16%). With a median follow-up of 24 months (range 2–20) as of January 2012, 87/129 patients (67%) are alive and disease free, whereas 32/129 (25%) are dead and 8/129 (6%) are alive with active disease. Log rank analysis showed high risk IPI (4/5; $P = 0.02$) to be a variable predictive for shorter event-free survival (EFS), whereas again high risk IPI (4/5; $P = 0.04$), age >70 years ($P = 0.009$), advanced-stage disease (stage III–IV; $P = 0.01$) and

PS 2–3 ($P = 0.003$) were all predictive of shorter OS. The Cox proportional hazards regression model identified age >70 years ($P = 0.03$) and PS 2–3 ($P = 0.02$) as the variables with a negative impact on OS. Finally, no statistically significant difference in terms of response was observed within the two groups (14 versus 21).[42]

Up to now, this is the largest series of elderly aggressive NHL patients treated with NPLD instead of conventional doxorubicin reported in literature. The data by Isidori *et al.* suggest that NPLD induces a high rate of long-lasting CR in a population of elderly patients with relevant comorbidities. The overlapping rate of CR between the dose-dense R-COMP 14 group and the standard R-COMP 21 group suggest that the use of NPLD outruns the impact of a dose-dense regimen in a substantial proportion of very elderly patients, representing a therapeutic opportunity for a category of patients not suitable for a dose-dense treatment.[42]

In summary, NPLD seems to be a promising drug for a potentially curative treatment in poor-risk elderly patients with aggressive NHL. In fact, irrespectively of dose intensification, patients with comorbidity, underlying heart disease and/or poor PS due to the presence of several comorbidities are often precluded "*a priori*" from a curative treatment. These conditions should, by themselves, contraindicate the use of anthracyclines, which remain the mainstay of treatment for aggressive NHL. The biological properties of liposomes, which increase the therapeutic index of doxorubicin reducing side effects, allow the "magic" transformation of a "big bad wolf" into a "magic bullet," with curative potential even in a subset of "almost untouchables" elderly patients with NHL.

4.2. *Pegylated liposomal doxorubicin*

The PLD (Caelix®/Doxil®, Schering-Plough/Orto Biothech, USA) consists of an aqueous suspension of single lamellar lipid vesicles with an approximate mean size of 90 nm. PLD is a liposomal formulation with a distinct pharmacokinetic profile, characterized by an extended circulation time and a reduced volume of distribution. PLD has been approved for clinical use in a variety of cancer types due to its antitumor efficacy and exclusive safety profile: in 1995, for AIDS-related Kaposi sarcoma; in 1999, for recurrent ovarian carcinoma by FDA in the US; in 2003, for MBC patients at increased cardiac risk, by the European Medicines Agency (EMA) (Table 2). Studies of PLD suggest that a dose of 45 to 50 mg/m^2 every four weeks is well tolerated, with little nausea or vomiting, mild myelosuppression, minimal alopecia and very little cardiotoxicity.

Table 2. Approved dosage and schedule for pegylated liposomal doxorubicin (PLD) in patients with various cancer in the European Union (EU) and the US.

Cancer	Treatment schedule	Region	Approved indication by region and prescribing information
AIDS-related Kaposi's sarcoma	PLD 20 mg/m^2 every two or three weeks	EU	For treatment of AIDS-KS in patients with low CD4 + cell counts (< 200 cells/mL) and extensive mucocutaneous or visceral disease. PLD may be used as first-line systemic chemotherapy, or as second-line chemotherapy with disease that has progressed with, or in patients intolerant to, prior combination systemic chemotherapy comprising at least two of the following agents: a vinca alkaloid, bleomycin and conventional doxorubicin (or other anthracycline). The initial dose of PLD should be administered over 30 min. at an initial rate of ≤ 1 mg/min to minimize the risk of infusion reactions. Treatment intervals <10 days should be avoided, as PLD accumulation and increased toxicity is possible. Treatment should be continued for two to three months to achieve a therapeutic response. PLD should not be used to treat AIDS-KS that may be treated effectively with local therapy or systemic interferon-α.
AIDS-related Kaposi's sarcoma	PLD 20 mg/m^2 every three weeks	US	For the treatment of AIDS-KS in patients after failure of prior systemic chemotherapy or intolerance to such therapy. The initial dose of PLD should be administered over 30 min at an initial rate of ≤1 mg/min to minimize the risk of infusion reactions; if no reaction is observed, subsequent infusions may be administered over a 1-h period. Treatment should be continued for as long as patients respond satisfactory and tolerate treatment.
Metastatic Breast Cancer	PLD 50 mg/m^2 every four weeks	EU	As monotherapy in patients with metastatic breast cancer, where there is an increased cardiac risk. The initial dose of PLD should be administered at a rate of ≤1 mg/min to minimize the risk of infusion reactions; if no reaction is observed, subsequent infusions may be administered over a 1-h period. Treatment should be repeated as long as treatment is tolerated and the disease does not progress.

(*Continued*)

Table 2. (*Continued*)

Cancer	Treatment schedule	Region	Approved indication by region and prescribing information
Multiple myeloma	PLD 30 mg/m^2 (day 4) in association with bortezomib 1.3 mg/m^2 (days 1, 4, 8 and 11) every three weeks	EU	In combination with bortezomib for the treatment of progressive multiple myeloma in patients who have received at least one prior therapy and who have already undergone or are unsuitable for bone marrow transplantation. The initial dose of PLD should be administered at a rate of ≤ 1mg/min to minimize the risk of infusion reactions; if no reaction is observed, subsequent infusions may be administered over a 1-h period. Treatment should be repeated as long as treatment is tolerated and the response is satisfactory. Botezomib is administered as an *i.v.* bolus. Doses of bortezomib should be at least 72 h apart. PLD should be administered on day 4, immediately after the bortezomib infusion.
Multiple myeloma	PLD 30 mg/m^2 (day 4) in association with bortezomib 1.3 mg/m^2 (days 1, 4, 8 and 11) every three weeks	US	In combination with bortezomib for the treatment of patients with multiple myeloma who have not previously received bortezomib and have received ≥1 prior therapy. Patients may be treated with this regimen for up to eight cycles, until disease progression or the occurrence of unacceptable toxicity. Botezomib is administered as an *i.v.* bolus. PLD should be administered on day 4, immediately after the bortezomib infusion.
Ovarian cancer	PLD 50 mg/m^2 every four weeks	EU	For advanced ovarian cancer in patients who have failed a first-line platinum-based chemotherapy regimen. The initial dose of PLD should be administered at a rate of ≤1mg/min to minimize the risk of infusion reactions; if no reaction is observed, subsequent infusions may be administered over a 1-h period. Treatment should be repeated as long as treatment is tolerated and the disease does not progress.

(*Continued*)

Table 2. (*Continued*)

Cancer	Treatment schedule	Region	Approved indication by region and prescribing information
Ovarian cancer	PLD 50 mg/m^2 every four weeks	US	For the treatment of patients with ovarian cancer whose disease has progressed or recurred after platinum-based chemotherapy. The initial dose of PLD should be administered at a rate of ≤ 1mg/min to minimize the risk of infusion reactions; if no reaction is observed, subsequent infusions may be administered over a 1-h period. Treatment should be repeated as long as treatment is tolerated and the disease does not progress. A minimum of four cycles is recommended.

Note: PLD is administered as an *i.v.* infusion.

Sources:

- □ EMA. Summary of product characteristics: Caelix (doxorubicin hydrochloride in a pegylated liposomal formulation) 2 mg/mL concentrate for infusion [online]. Available from URL: http://www.ema.europa.eu/docs/en_GB/document_library/EPAR_-_Product_Information/human/000089/WC500020180.pdf [Accessed 21 Jan 2013].
- □ Alza Corporation. Doxil_ (doxorubicin HCl liposome injection) for *i.v.* infusion: US prescribing information [online]. Available from URL: http://www.doxiline.com [Accessed 21 Nov 2011].

The PLD has a toxicity profile extremely different from that of both free doxorubicin and NPLD (Table 1). In general, the most commonly reported treatment-related adverse events with PLD comprise hematological (specially, neutropenia) and GI toxicities (stomatitis and mucositis), as well as skin and subcutaneous tissues disorders (PPE or hand–foot syndrome, alopecia and cutaneous rash). Stomatitis usually appears after the first cycle, increases in frequency and severity with higher doses and is a dose limiting toxicity. The reduction of 20% of the dose of PLD is, most of the times, sufficient to prevent/reduce the severity of mucositis in subsequent cycles. PPE or hand–foot syndrome, which is characterized by severe dermal lesions on the soles of the feet and the palms of the hand, is seen in patients receiving long-term continuous infusion of free doxorubicin.[43] This particular toxicity is a consequence of the toxic effect of doxorubicin on the rapidly dividing keratinocytes. As a consequence of pegylation, it is possible that the liposome drug-delivery results in a slow-release mechanism of doxorubicin into the tissues that mimics the effects seen with continuous infusion chemotherapy. Alternatively, several studies have shown that long-circulating liposomes accumulate to a limited degree in skin,[44] where slow release in the near vicinity of keratinocytes can give rise to their toxicity. As a matter of fact, PPE occurs after two or more cycles of treatment and increases in risk of incidence with frequently repeated treatments. PPE was frequently reported by PLD recipients in clinical trials, occurring in ≈ 50% of patients with MBC or ovarian cancer receiving PLD 50 mg/m^2 monotherapy and in 16% of patients with multiple myeloma receiving PLD 30 mg/m^2 plus bortezomib.[21] Severe (grade III or IV) PPE was reported in ≈20% of patients with MBC or ovarian cancer receiving PLD monotherapy. Other manifestations of skin toxicity include follicular rashes, pigmentation spots and radiation recall reactions. Other minor non-hematological toxicities include nausea, vomiting and acute infusion reactions, approximately occurring in <10% of the patients treated with PLD.

4.2.1. PLD in Kaposi sarcoma

The PLD started its clinical development on AIDS-related Kaposi's sarcoma. PLD has a preferential tropism for skin lesions of Kaposi sarcoma due to EPR, and its concentration in Kaposi sarcoma's skin is 53 times higher with respect to normal skin. Therefore, PLD has an increased selectivity with respect to standard doxorubicin and, as a consequence, could be administered at lower dosages than is required for free doxorubicin.[45] The therapeutic efficacy of PLD in the treatment of AIDS-related Kaposi's sarcoma is well established and has been extensively reviewed previously.[46] Thus, this subchapter briefly discusses randomized, multinational phase III trials of PLD in patients with AIDS-related Kaposi's sarcoma.[47,48] PLD was compared with the commonly used regimens of bleomycin plus vincristine and standard doxorubicin,[47] or bleomycin plus vincristine.[48] These trials were conducted before[47,48] the introduction of HAART. Antiretrovirals (*albeit* not HAART) were administered to a proportion of patients in the two trials conducted before HAART was introduced.[47,48] Monotherapy with PLD 20 mg/m^2 every two to three weeks was effective in the treatment of AIDS-related Kaposi's sarcoma, resulting in ORRs of 46–59%.[47,48] PLD was significantly more effective than bleomycin plus vincristine[48] and bleomycin plus vincristine and standard doxorubicin[47] in terms of ORRs, and was not significantly different to paclitaxel in terms of PFS (primary efficacy endpoint) and ORR. In addition, PLD was at least as effective as these comparator regimens for other RR endpoints.[47,48] The median duration of response did not differ significantly between PLD and bleomycin plus vincristine (142 versus 123 days)[48] or bleomycin plus vincristine and standard doxorubicin (90.0 versus 92.0 days).[47]

In a small phase II study,[49] testing the efficacy of PLD in combination with HAART (with protease inhibitors, non-nucleoside reverse transcriptase inhibitors, or both) for the treatment of patients with moderate to advanced AIDS-related Kaposi's sarcoma, PLD was given at 20 mg/m^2 every three weeks for a median number of 11 cycles. Results were extremely encouraging in favor of combined treatment, with RRs at 48 weeks significantly (P = 0.003) higher in PLD plus HAART recipients with respect to HAART alone (76% versus 20%; ITT population).[49]

4.2.2. PLD in breast cancer

The first randomized, multicenter, phase III study was designed to demonstrate that efficacy in terms of PFS of Caelix® [pegylated liposomal doxorubicin HCl (PLD)] was non-inferior to doxorubicin with significantly less cardiotoxicity in first-line treatment of women with MBC.[50] Women (n = 509) with MBC and normal cardiac function were randomized to receive either PLD 50 mg/m^2 (every four weeks) or doxorubicin 60 mg/m^2 (every three weeks). Cardiac event rates were based on reductions in LVEF as a function of cumulative anthracycline dose.

In this open-label, multicenter trial, patients were randomized in a 1:1 ratio by an independent central third party according to a computer-generated randomization program. Patients received either PLD [50 mg/m^2 *i.v.* infusion for up to 60 min every four weeks] or doxorubicin [60 mg/m^2 *i.v.* infusion for 60 min every three weeks]. Patients were prospectively stratified, basing on three criteria to balance major prognostic risk factors between treatment groups: prior adjuvant anthracycline exposure; presence of bone metastases as only site of disease; presence of at least one cardiac risk factor.[50] Cardiac risk factors were defined as prior mediastinal irradiation, age ≥65 years, history of heart disease (previous myocardial infarction, arrhythmia or angina, not requiring treatment) or had hypertension, or diabetes requiring medical treatment.[50]

Regarding cardiotoxicity, MUGA scans were performed to measure LVEF before onset of treatment, after 300 mg/m^2 cumulative anthracycline exposure, and after every additional 100 mg/m^2 of PLD and every 120 mg/m^2 of doxorubicin. Compliance with the protocol on performing MUGA evaluations was high. Of the 283 patients who reached doses ≥300 mg/m^2 cumulative anthracyclines, all but 20 patients (9 PLD, 11 doxorubicin) had a baseline MUGA evaluation and at least one follow-up MUGA evaluation during treatment.[50]

Overall, 339 patients (152 PLD and 187 doxorubicin) had electronic MUGA scan data for evaluation of cardiotoxicity (baseline and at least one scan during treatment) and were included in the analysis. Patients in the PLD arm had a median cumulative anthracycline dose of 398 mg/m^2 (including prior anthracycline exposure). Patients in the conventional doxorubicin arm had a median cumulative anthracycline dose of 421 mg/m^2 (including prior anthracycline exposure). About 58 patients (10 PLD, 48 doxorubicin) met the protocol-defined LVEF criteria for cardiotoxicity during treatment and/or follow-up. The risk of developing cardiotoxicity was significantly higher for patients receiving doxorubicin than for those receiving PLD ($P < 0.001$, HR = 3.16 for comparison of cumulative anthracycline dose at the first, protocol-specified, cardiac event).[50] The increase in risk of developing cardiotoxicity on doxorubicin versus PLD was observed in all subgroups analyzed, including those at high risk for developing CHF. In the subgroup that received prior adjuvant anthracycline therapy, the risk of developing cardiotoxicity was sevenfold higher with doxorubicin than with PLD. None of the 10 PLD-treated patients, who had cardiotoxicity by LVEF criteria, developed clinical signs or symptoms of CHF, whereas 10 of 48 doxorubicin-treated patients, who had cardiotoxicity by LVEF criteria, developed signs or symptoms of CHF. Two patients in each group developed clinical CHF, but did not have a corresponding decrease in LVEF. As expected with doxorubicin, the mean percentage change from baseline in LVEF was positively correlated with the increase in cumulative anthracycline dose. However, with PLD, only a 2–3% mean decrease in LVEF was observed as the cumulative anthracycline dose increased. At cumulative doses at or above 450 mg/m^2, a sevenfold greater mean percentage decrease in LVEF was observed with doxorubicin versus PLD (−17.2% versus −2.3%; mean percentage change from baseline in LVEF in doxorubicin-treated and PLD-treated patients, respectively).[50]

Regarding efficacy, PLD and doxorubicin were comparable with respect to PFS (6.9 versus 7.8 months, respectively) and to OS (21 and 22 months for PLD and doxorubicin, respectively. PPE (48% versus 2%), stomatitis (22% versus 15%) and mucositis (23% versus 13%) were more often associated with PLD than doxorubicin. In conclusion, in first-line therapy for MBC, PLD provides comparable efficacy to doxorubicin, with significantly reduced cardiotoxicity.[50]

Another multicentric, randomized trial was published on the same year.[51] This trial was designed to compare the efficacy of PLD with that of a common salvage regimen (comparator) in patients with taxane-refractory advanced breast cancer. Following failure of a first- or second-line taxane-containing regimen for metastatic disease, 301 women were randomly assigned to receive PLD (50 mg/m^2 every 28 days); or comparator-vinorelbine (30 mg/m^2 weekly) or mitomycin C (10 mg/m^2 day 1 and every

28 days) plus vinblastine (5 mg/m^2 day 1, day 14, day 28 and day 42) every six to eight weeks.[51]. Changes in LVEF values were only assessed in patients receiving PLD. Cardiac toxicity was defined as either a decrease of ≥ 15 points from baseline or a ≥ 5-point decrease from baseline with a level below the lower limit of normal for the institution. About 22 patients developed LVEF changes consistent with cardiac toxicity. However, decreases in LVEF did not correlate with cumulative anthracycline dose and none of these patients developed clinical CHF. The majority (n = 14) discontinued due to progressive disease. There were four patients who discontinued treatment due to cardiac toxicity (LVEF decrease), three patients who discontinued due to non-cardiac adverse events and one patient who discontinued due to noncompliance. PFS and OS were similar for PLD and comparator.[51]

The efficacy of PLD as maintenance therapy after first-line induction chemotherapy with docetaxel and doxorubicin in patients with MBC was also evaluated in a phase III study.[51] Patients who had a response or stable disease after first-line induction chemotherapy with doxorubicin 75 mg/m^2 every three weeks for three cycles then docetaxel 100 mg/m^2 every three weeks for three cycles were randomized to maintenance therapy with low dose PLD (40 mg/m^2 every 28 days for six cycles; n = 78) or to observation (n = 77).[52] Patients had a median age of 56 years, good PS (>90% ECOG 0/1), and most of them (>60%) had visceral metastases. The PLD maintenance group and the observation group were comparable in terms of patients who had received prior chemotherapy, hormonal therapy or adjuvant therapy with an anthracycline. In the intent-to-treat population, PLD was proven to be effective as maintenance therapy, resulting in a significantly (P = 0.0002) longer median TTP (primary endpoint) relative to observation (8.4 versus 5.1 months; HR 0.54 [95% CI 0.39, 0.76]).[52]

4.2.3. PLD in ovarian cancer

The therapeutic efficacy of PLD in the treatment of ovarian cancer is continuously debated, and an exhaustive discussion of this topic is beyond the goal of this chapter. The therapeutic efficacy of *i.v.* PLD as second-line or salvage chemotherapy in women (median age 56–63 years) with progressive ovarian cancer has been evaluated in randomized, multicenter phase III trials.[53–57] These trials, that will be briefly discussed in this section, evaluated the efficacy of PLD as monotherapy as compared with that of gemcitabine[53–55] and topotecan[54] or evaluated the efficacy of combining PLD plus carboplatin compared with standard carboplatin plus paclitaxel[57] or PLD plus trabectedin compared with PLD alone.[56]

An improved efficacy and safety profile was seen when PLD was compared with topotecan in recurrent ovarian cancer in the seminal phase III study of Gordon *et al.*[54] A total of 474 patients with epithelial ovarian carcinoma that recurred after or did not respond to first-line, platinum-based chemotherapy were treated with either PLD (239) or topotecan (235). Patients with measurable disease were randomized to receive either PLD 50 mg/m^2 every four weeks or topotecan 1.5 mg/m^2/d for five consecutive days every three weeks. The overall PFS rates were similar between the two arms ($P = 0.095$). Data analyzed in platinum-sensitive patients demonstrated a statistically significant benefit from PLD for PFS ($P = 0.037$), with medians of 28.9 for PLD versus 23.3 weeks for topotecan. For OS, PLD was significantly superior to topotecan in platinum-sensitive patients ($P = 0.008$), with a median of 108 weeks versus 71.1 weeks. However, median OS times in all patients, according to intention-to-treat analysis, did not significantly differ between PLD recipients (60 weeks) and topotecan recipients (56.7 weeks). Nevertheless, the survival advantage reached statistical significance in the overall intent-to-treat patient population in an extended follow-up analysis.[58] As a result, PLD has become the new standard therapy for recurrent ovarian cancer, and most of the subsequent studies testing novel agents had PLD as the comparator.

Ferrandina and coworker[53] designed a phase III randomized multicenter trial to compare gemcitabine (1,000 mg/m^2 on days 1, 8 and 15 every 28 days) with PLD (40 mg/m^2 every 28 days) in ovarian cancer patients who experienced treatment failure with only one platinum/paclitaxel regimen and who experienced recurrence or progression within 12 months after completion of primary treatment. About 153 patients were randomly assigned to PLD ($n = 76$) or gemcitabine ($n = 77$). The ORR was 16% in the PLD arm compared with 29% in the gemcitabine arm ($P = 0.056$). No statistically significant difference in TTP curves according to treatment allocation was documented ($P = 0.411$). However, there was a trend in terms of longer OS in the PLD arm compared with the gemcitabine arm, with a borderline P-value ($P = 0.048$). Grade III or IV PPE was documented in a higher proportion of PLD patients (6%) versus gemcitabine patients (0%; $P = 0.061$). The authors concluded that gemcitabine does not provide an advantage compared with PLD in terms of TTP in ovarian cancer patients who experience recurrence within 12 months after primary.[53]

Mutch and coworkers[55] randomly assigned 195 patients with platinum-resistant (defined as progressive disease within six months of platinum-based therapy in the primary setting) taxane-pretreated ovarian cancer to receive either gemcitabine 1,000 mg/m^2 (days 1 and 8; every 21 days) or PLD 50 mg/m^2 (day 1; every 28 days) until disease progression or unacceptable toxicity. Optional cross-over therapy was allowed at disease progression or at withdrawal because of toxicity. Primary end point was PFS. No significant between-group differences were seen in terms of the reported primary end-points of median PFS (3.6 months gemcitabine group versus 3.1 months PLD group) and ORR was 6.1% versus 8.3%; and in the subset of patients with measurable disease, ORR was 9.2% versus 11.7%, respectively. None of the other efficacy end points showed a statistically significant difference between treatment groups (median OS 12.7 versus 13.5 months).[55] The PLD group experienced significantly more hand–foot syndrome and mucositis; the gemcitabine group experienced significantly more constipation, nausea/vomiting, fatigue and neutropenia but not febrile neutropenia. The authors concluded that, although the study was not designed and powered as an equivalency study, gemcitabine and PLD seem to have an equivalent therapeutic index in platinum-resistant taxane-pretreated patients with ovarian cancer.[55]

Pujade-Lauraine *et al.*[57] designed a randomized, multicenter, phase III non-inferiority trial enrolling 976 patients to test the efficacy and safety of the combination of PLD 30 mg/m^2 with carboplatin AUC 5 every four weeks compared with standard carboplatin AUC 5 plus paclitaxel 175 mg/m^2 every three weeks, for at least six cycles, in patients with platinum-sensitive relapsed/recurrent ovarian cancer.

In these patients, PLD plus carboplatin was significantly ($P = 0.005$) more effective than the standard carboplatin plus paclitaxel regimen in terms of median PFS (primary endpoint) [HR 0.82; 95% CI 0.72, 0.94] after non-inferiority had first been established ($P < 0.001$). Furthermore, the combination of PLD plus carboplatin remained significantly ($P < 0.01$) more effective irrespective of a number of baseline predictive factors, including time since platinum-based therapy (6–12 or >12 months), measurable disease (yes or no) and CA-125 levels (<100 or >100 U/mL).[57] Regarding toxicity, overall severe non-hematologic toxicity (36.8% versus 28.4%; $P < 0.01$) leading to early discontinuation (15% versus 6%; $P < 0.001$) occurred more frequently in the carboplatin plus paclitaxel arm. The authors concluded that the combination of PLD plus carboplatin is superior in terms of PFS and has a better therapeutic index over standard carboplatin plus paclitaxel.[57]

Finally, co-administration of randomly assigned PLD 30 mg/m^2 and trabectedin 1.1 mg/m^2 was associated with a significantly ($P = 0.019$) longer median PFS than PLD 50 mg/m^2 monotherapy in 672 patients with recurrent ovarian cancer after failure of first-line, platinum-based chemotherapy (primary endpoint).[56] This corresponded to a 21% relative reduction in the risk of disease progression or death (HR 0.79; 95% CI 0.65, 0.96) in the combination therapy arm. When patients were stratified according to platinum sensitivity, combination therapy with PLD plus trabectedin was significantly ($P = 0.017$) more effective than PLD monotherapy in patients with platinum-sensitive disease (median PFS 9.2 versus 7.5 months; HR 0.73; 95% CI 0.56, 0.95), although this effect was not observed in those with platinum-resistant disease (median PFS 4.0 versus 3.7 months).[56] There was no significant difference in the duration of median OS between the PLD plus trabectedin and PLD alone treatment groups (Table III) [HR 0.86; 95% CI 0.72, 1.02], according to a final analysis of survival data determined after 522 deaths had occurred[56]; the pre-identified number of patient deaths required for this analysis was 520.[56]

4.2.4. PLD in NHL

The PLD has been substituted for conventional doxorubicin in the CHOP regimen in a number of trials. In phase II studies in elderly patients with DLBCL, ORR of approximately 65% was achieved: 50% CR and 15% PR, an estimated one-year OS of 55%, and an estimated two-year EFS of 45%. Neutropenia was the only grade III–IV toxicity observed.[59,60]

Martino *et al.*[59] tested in a prospective phase II study the feasibility of a regimen including PLD 30 mg/m^2 on day 1 plus vincristine 2 mg on day 1, cyclophosphamide 750 mg/m^2 on day 1 and prednisone 60 mg/m^2 on days 1–5 (CCOP) in patients with DLBCL above the age of 60 years. Treatment cycles were scheduled to be given every 21 days for a total of six (or eight in case of PR after six). A total of 33 patients, with a median age of 74 years, were enrolled in the study. A total of 187 cycles of CCOP were given, with a median of six cycles per patient (range 2–8).[59] About 13 patients (39%) received fewer than six cycles of CCOP. Four patients died during therapy without confirmation of disease progression. Two further patients did not complete therapy, one due to oral mucositis (which recurred after switching to CHOP chemotherapy) and the other's loss from follow-up, and seven patients progressed during CCOP and were withdrawn from the study. About 67 cycles (36%) were delayed in 22 patients (67%), mostly due to neutropenia. Only two patients developed febrile neutropenia, and G-CSF was used prophylactically in these two patients and in a third one who developed recurrent grade IV neutropenia. There were no other relevant toxicities. The LVEF was assessed in 16 patients both before and after treatment, without any relevant changes in any case.[59] About 16 patients (49%) achieved a CR and five (39%) a PR following CCOP, for an OR rate of 64%. The only variable that was significantly associated with a lower CR rate was having a DLBCL that had transformed from a prior low grade lymphoma (0/5 versus 16/28 in *de novo* cases, $P = 0.04$).[59] The median follow-up for all 33 cases was 13 months (range, 1–32) and for the 17 survivors 20 months (range, 10–32). The estimated one-year overall and EFSs were 55% (95% CI, 38–72) and 45% (95%CI, 28–62), respectively. The authors suggested that CCOP appeared to be an interesting treatment for elderly patients with DLBCL.[59]

Visani *et al.*[60] treated 13 NHL patients, either "frail" front line (five patients) or heavily pretreated (refractory/relapsed, eight patients) with a combination of PLD 40 mg/m^2 on day 1, vincristine 2 mg on day 1, cyclophosphamide 750 mg/m^2 on day 1 and prednisone 60 mg/m^2 on days 1–5 (CDOP schedule). Therapy was administered every three weeks, for a total of six cycles, adding Rituximab on day 15 of every cycle.[60] Three patients had a very poor PS (grade IV according to the WHO scale). A total of 12 patients (92%) completed the six CDOP/Rituximab courses, all of them being evaluable for response. Seven patients (53%) achieved a CR, 4(31%) a PR, leading to an ORR of 84%.[60] The treatment was very well-tolerated even by "frail" patients. No major toxicity (WHO grade III/IV) occurred. The administration of PLD was not related to severe side effects, with only two patients showing mild oral mucositis and another one a neurological toxicity (grade I). No cardiac toxicity was observed.[60]

Cutaneous T-cell lymphoma (CTCL) is a specific niche in which PLD has proven to be very active at low doses in a similar manner to Kaposi sarcoma. A retrospective, multicenter study[61] was performed evaluating 34 patients (31 male). A total of 27 patients received PLD 20 mg/m^2, 5 patients received PLD 203–0 mg/m^2, and 2 patients received PLD 40 mg/m^2. PLD was administered *i.v.* every two weeks in six patients, every two to three weeks in four patients, and every four weeks in 23 patients.[61] About 34 patients received at least one cycle of PLD. Disease was classified as mycosis fungoides in 28 patients, mycosis fungoides with follicular mucinosis in two patients, small or medium-sized pleomorphic CTCL in two patients, Sèzary syndrome in one patient, and CD30 positive CTCL in one patient. Fifteen patients achieved either a CR or a CRu, and 15 patients achieved a PR, resulting in an ORR of 88.2%. Two patients dropped out: one patient after a single PLD infusion because of Grade III capillary leakage syndrome and one patient after two cycles because of a suicide attempt that was not related to treatment or to CTCL. All other patients received at least four cycles of PLD. The OS was 17.8 months +/− 10.5

months ($n = 33$ patients), EFS was 12.0 months +/− 9.5 months and DFS was 13.3 +/− 10.5 months ($n = 16$ patients). Adverse effects were seen in 14 of 34 patients (41.2%); they were temporary and generally mild. Only six patients had Grade III or IV adverse effects. This multicenter study provided evidence of high efficacy of PLD monotherapy in a subset of patients with recurrent, unresponsive or rapidly progressive disease.[61]

Recently, PLD was investigated by the EORTC in a multicenter trial involving in a clearly defined patient population with advanced MF.[62] To be eligible for the trial, the patients had to had stage IIB, IVA, or IVB MF, refractory or recurrent after at least two previous systemic therapies. Patients were intended to receive a maximum of six cycles of PLD 20 mg/m^2 on days 1 and 15, every 28 days (one cycle). The primary end point was RR.[62] As a whole, 49 patients received a median number of five chemotherapy cycles. No grade III/IV hematological toxicities were reported. Grade III or IV non-hematological toxicities included cardiac symptom (2%), allergy/hypersensitivity (2%), constitutional symptom (4%), hand and foot reaction (2%), other dermatologic toxicity (6%), other GI toxicity (4%), infection (4%), pulmonary embolism (2%) and cardiac ischemia (2%).[62] ORR was 40.8% (20/49 patients); CR rate was 6.1%, whereas PR rate was 34.7%. A 50% or greater reduction of cutaneous manifestations was observed in 26 (60.5%) of 43 assessable patients. Two early deaths were reported, resulting from related cardiovascular toxicity and disease progression. Median TTP and median duration of response were 7.4 and six months, respectively.[62] The authors concluded that PLD has an acceptable safety profile in patients with advanced MF, together with promising efficacy.[62]

Finally, PLD was recently tested in patients with AIDS-related lymphoma.[63] Levine and Tulpule coordinated a prospective, multi-institutional phase II trial, employing PLD 40 mg/m^2, cyclophosphamide 750 mg/m^2, vincristine 2 mg and rituximab 375 mg/m^2 given on day 1, coupled with prednisone 100 mg orally on days 1 through 5 (DR-COP regimen). Chemotherapy was associated with concomitant antiretroviral therapy.[63] In 40 evaluable patients, median CD4 cells was 114/μL, and median HIV-1 viral load (VL) was 25,000 copies/mL. High or intermediate/high aaIPI was present in 28%. ORR was 67.5% and CR rate was 47.5% (95% CI, 31.5 to 63.9).[63] Of 19 complete responders, 84% had extranodal disease, 47% had CD4 <100/μL, and 47% had VL > 50,000 copies/mL; one relapsed. With a median follow-up of 25.5 months, 62% (95% CI, 44 to 75) of patients are alive. A total of 16 patients (40%) experienced 22 infections, with grade IV in only two (5%).[63] No patient died as a result of infection during treatment; one had opportunistic infection.[63] The authors concluded that profound immunodeficiency and high HIV-1 VL do not preclude attainment of CR after DR-COP with highly active antiretroviral therapy. Furthermore, the regimen was well tolerated, and the use of rituximab was not associated with death as a result of infection during treatment. PLD seems to be promising in patients with AIDS-related NHL. It would be interesting to randomize patients with this disease to receive PLD or NPLD, in order to eventually establish which liposomal formulation is better for the treatment of this patient population.

4.2.5. PLD in multiple myeloma

Indeed, PLD has been shown to have equal efficacy and a better safety profile in the treatment of multiple myeloma, when compared with conventional doxorubicin combinations. A review of all the studies published on multiple myeloma has already been done elsewhere,[64] and is not the objective of this section.

This paragraph reports on the pivotal, randomized, open-label, multinational phase III trial that evaluated PLD in combination with bortezomib versus bortezomib alone for the treatment of relapsed or refractory multiple myeloma.[65] A total of 646 patients were randomly assigned to receive either *i.v.* bortezomib 1.3 mg/m^2 on days 1, 4, 8 and 11 of an every 21-days cycle, or the same bortezomib regimen with PLD 30 mg/m^2 on day 4.

The primary efficacy endpoint in the phase III trial was TTP in the ITT population. Combination therapy with PLD plus bortezomib was significantly more effective than bortezomib alone in terms of the primary endpoint; median TTP was increased from 6.5 months for bortezomib to 9.3 months with the PLD + bortezomib combination (P = 0.000004). The improvement in TTP in the combination group corresponded to a 45% reduction in the risk of disease progression compared with the monotherapy group (HR 1.82; 95% CI 1.41, 2.35). Among the other efficacy endpoints, median PFS, 15-month OS and duration of response also significantly favored the combination over bortezomib alone. The 15-month survival rate for PLD + bortezomib was 76% compared with 65% for bortezomib alone (P = 0.03). The complete plus PR rate was 41% for bortezomib and 44% for PLD + bortezomib, a difference that was not statistically significant. Median duration of response was increased from 7.0 to 10.2 months (P = 0.0008) with PLD + bortezomib. While there was no significant between-group difference in the number of deaths at the cut-off date for the planned interim analysis (28 April 2006), an updated analysis (28 November 2006) showed a significant (P < 0.05) survival advantage for the combination compared with bortezomib alone (mortality 18% versus 25%; HR 1.41 [95% CI 1.002, 1.97]).[21,65]

Grade III/IV adverse events were more frequent in the combination group (80% versus 64%), with safety profiles consistent with the known toxicities of the two agents. An increased incidence in the combination group was seen for grade III/IV neutropenia, thrombocytopenia, asthenia, fatigue, diarrhea and hand–foot syndrome.

More recently, the subgroup analysis of the trial produced other interesting results. The first paper made from the same investigators of the "DOXIL-MMY-3001 Study" was aimed to determine (i) the efficacy of PLD plus bortezomib versus bortezomib alone in patients with multiple myeloma who had failed on prior thalidomide/lenalidomide (immunomodulatory drug [IMiD]) treatment and (ii) the efficacy and safety profile of PLD plus bortezomib in IMiD-exposed and IMiD-naive patients.[66] In the IMiD-exposed subgroup, PLD plus bortezomib maintained the higher median TTP compared with bortezomib alone (270 days versus 205 days; HR, 1.62; 95% CI, 1.08–2.41 [$P = 0.018$]), similar to the results in the total study population reported previously.[65,66] A similar advantage for combination therapy compared with bortezomib alone was observed among IMiD-naive patients. Within the PLD plus bortezomib arm, the median TTP was comparable in both IMiD-naive and IMiD-exposed patients (295 days and 270 days, respectively). There was no statistically significant difference in the heterogeneity test between subgroups (IMiD-naive versus IMiD-exposed; $P = 0.446$).[66] Furthermore, in the PLD plus bortezomib combination arm, the objective RR (CRs plus PRs) was similar both for IMiD-exposed patients (48%) and for IMiD-naive patients (47%). Likewise, in patients who achieved a response, the response duration was comparable for IMiD-naive patients and IMiD-exposed patients in the combination treatment group and lasted a median of 310 days and 319 days, respectively. The incidence of grade III/IV adverse events was similar with PLD plus bortezomib regardless of prior IMiD exposure. Overall, seven patients, including three IMiD-naive patients (two who received PLD plus bortezomib and one who received bortezomib alone) and four IMiD-exposed patients (two who received PLD plus bortezomib and two who received bortezomib alone), had drug-related AEs that resulted in death.[66] The authors concluded that the subgroup analysis demonstrated that neither the efficacy nor the safety of the PLD plus bortezomib combination or bortezomib alone in patients with recurrent or refractory multiple myeloma is affected by prior exposure to IMiD therapy. In addition, the therapeutic advantage of a higher TTP observed with PLD plus bortezomib compared with bortezomib alone was maintained in this patient population regardless of prior IMiD exposure.

In another paper, from the same pool of 646 relapsed/refractory myeloma patients were extrapolated 193 patients with renal insufficiency (creatinine clearance [CrCl] < 60 mL/min) who were randomized in the same phase III trial comparing bortezomib +/– PLD.[67] The RR (49% versus 42%) and median time to disease progression (331 days versus 199 days) were comparable or slightly better for patients with renal insufficiency treated with PLD/bortezomib compared with patients treated with bortezomib alone. There was a steady, clinically meaningful improvement in renal function for patients with renal insufficiency in both treatment arms. However, patients with impaired renal function were at a slightly increased risk of a drug-related serious adverse event (28% versus 19% for CrCl < 60 and > or = 60 mL/min, respectively.[67]

The last subgroup analysis was performed in high-risk patients.[68] Four high- and low-risk subgroup categories were identified, including age ≥ 65, ≥ 2 previous therapies, International Staging System stage II/III, and disease refractory to last previous therapy.[68] Compared with bortezomib alone, PLD plus bortezomib significantly prolonged the TTP and duration of response in all of these subgroups. PLD plus bortezomib was well tolerated in all subgroups, and had a safety profile that was not affected by response to previous therapy. Treatment of relapsed/refractory myeloma with the combination of PLD plus bortezomib provides better outcomes over bortezomib alone, even in the presence of high-risk prognostic factors. These results suggest that PLD plus bortezomib may represent an additional standard of care for this population of patients with multiple myeloma.[68]

5. Future Perspective and Conclusions

Cancer patients undergoing chemotherapy have an increased risk of developing cardiovascular complications; improving survival means that long-term survivors will, possibly, face late cardiac toxicity, with subclinical or clinical pictures of heart failure and coronary artery disease. The anthracyclines are most frequently implicated; however, recent pathogenetical acquisitions have markedly changed the perception of anthracycline cardiotoxicity. From a "cumulative dose" era, we have shifted to a more prudential approach, considering a persistent biological damage due to anthracyclines, with a particular emphasis on late cardiac toxicity. Thus, the design of liposomal anthracyclines meets the objective of reducing the toxicity, and couples profitably with strategies intended to implement cardioprotection. The niches of fragile patients are now enlarging, due to new therapeutic ways to approach elderly and very elderly individuals; this point, as well as a clear view of the importance to grant a safe life to the increasing number of long survivors, should help clinicians to overcome the cost issues related to liposomal anthracyclines and, in the long run, contribute to reduce and possibly minimize the frequency of late cardiac toxicity. Focusing on future research, even if the clinically available liposomal formulations of anthracyclines have improved the therapeutic index of the naive drugs, the third generation of liposomes, the so-called ligand-targeted liposomes (LTLs) are extensively studied. We expect to get advantage from the combination of long circulation and active intracellular drug delivery offered by LTLs, to further improve, in the future, the therapeutic index of anticancer drugs.

References

1. D. D. Von Hoff, M. Layard, P. Basa, *et al.*, Risk factors for doxorubicin-induced congestive heart failure. *Ann. Int. Med.* **91**, 7710–717 (1979).
2. S. M. Swain, F. S. Whaley and M. S. Ewer., Congestive heart failure in patients treated with doxorubicin: a retrospective analysis of three trials. *Cancer* **97**, 2869–2879 (2003).
3. S. A. Abraham, D. N. Waterhouse, L. D. Mayer, *et al.*, The liposomal formulation of doxorubicin. *Methods Enzymol.* **391**, 71–97 (2005).
4. S. K. Hobbs, W. L. Monsky, F. Yuan, *et al.*, Regulation of transport pathways in tumor vessels: role of tumor type and microenvironment. *Proc. Natl. Acad, Sci. USA* **95**, 4607–4612 (1998).
5. C. E. Swenson, W. R. Perkins, P. Roberts, *et al.*, Liposome technology and the development of Myocet™ (liposomal doxorubicin citrate). *Breast* (2), 1–7 (2001).
6. J. F. Money-Kyrle, F. Bates, J. Ready, *et al.*, Liposomal daunorubicin in advanced Kaposi's sarcoma: a phase II study. *Clin Oncol (R Coll Radiol)* **5**(6), 367–371 (1993).
7. P. S. Gill, J. Wernz, D. T. Scadden, *et al.*, Randomized phase III trial of liposomal daunorubicin versus doxorubicin, bleomycin, and vincristine in AIDS-related Kaposi's sarcoma. *J. Clin. Oncol.* **14**(8), 2353–2364 (1996).
8. A. Fassas, R. Buffels, A. Anagnostopoulos, *et al.*, Safety and early efficacy assessment of liposomal daunorubicin (DaunoXome) in adults with refractory or relapsed acute myeloblastic leukaemia: a phase I-II study. *Br. J. Haematol.* **116**(2), 308–315 (2002).
9. K. J. O'Byrne, A. L. Thomas, R. A. Sharma, *et al.*, A phase I dose-escalating study of Daunoxome, liposomal daunorubicin, in metastatic breast cancer. *Brit. J. Cancer* **87**, 15–20 (2002).
10. J. Robert and L. Gianni, Pharmacokinetics and metabolism of anthracyclines. *Cancer Surv.* **17**, 219–252 (1993).
11. F. Innocenti, *et al.*, Epirubicin glucuronidation is catalyzed by human UDP glucuronosyltransferase 2B7. *Drug Metab. Dispos.* **29**, 686–692 (2001).
12. D. A. Gewirtz, Critical evaluation of the mechanism of action proposed for the antitumor effects of the anthracycline antibiotics doxorubicin and daunorubicin. *Biochem. Pharmacol.* **57**, 727–741 (1999).
13. P. Menna, S. Recalcati, G. Cairo, G. Minotti, An introduction to the metabolic determinants of anthracycline cardiotoxicity. *Cardiovasc. Toxicol.* 7(2), 80–85 (2007).
14. O. Tacar, P. Sriamornsak, C. R. Dass, Doxorubicin: an update on anticancer molecular action, toxicity and novel drug delivery systems. *J. Pharm. Pharmacol.* **65**(2), 157–170 (2013).

15. G. Minotti, Anthracyclines. In *Encyclopedic Reference of Molecular Pharmacology*, S. Offermanns and W. Rosenthal (Eds.), Springer (2008).
16. G. Minotti, P. Menna, E. Salvatorelli, *et al.*, Anthracyclines: molecular advances and pharmacologic developments in antitumor activity and cardiotoxicity. *Pharmacol. Rev.* **56**(2), 185–229 (2004).
17. B. A. Chabner and D. L. Longo, *Cancer Chemotherapy and Biotherapy: Principles and Practice*. Philadelphia: Lippincott-Raven (1996).
18. D. C. Drummond, O. Meyer, K. Hong, *et al.*, Optimizing liposomes for delivery of chemotherapeutic agents to solid tumors. *Parmacol. Rev.* **51**(4), 691–743 (1999).
19. Y. Matsamura and H. Maeda, A new concept for macromolecular therapeutics in cancer chemotherapy: mechanism of tumoritropic accumulation of proteins and the antitumor agent SMANCS. *Cancer Res.* **46**, 6387–6392 (1986).
20. D. D. Lasic, *Liposomes: From Physics to Applications*. Amsterdam: Elsevier (1993) p. 575.
21. T. M. Dando and G. M. Keating, Liposomal doxorubicin: a review of its use in metastatic breast cancer and potential in non-Hodgkin lymphoma. *Am. J. Cancer* **4**(3), 193–206 (2005).
22. S. T. Duggan and G. M. Keating. Pegylated liposomal doxorubicin: a review of its use in metastatic breast cancer, ovarian cancer, multiple myeloma and AIDS-related Kaposi's sarcoma. *Drugs* **71**(18), 2531–2558 (2011).
23. A. Gabizon, Liposome circulation time and tumor targeting: implications for cancer chemotherapy. *Adv. Drug Deliv. Rev.* **16**, 285–294 (1995).
24. K. Lotfi, A. L. Zackrisson and C. Peterson, Comparison of idarubicin and daunorubicin regarding intracellular uptake, induction of apoptosis, and resistance. *Cancer Lett.* **178**(2), 141–149 (2002).
25. G. Batist, G. Ramakrishnan, C. S. Rao, *et al.*, Reduced cardiotoxicity and preserved antitumor efficacy of liposome-encapsulated doxorubicin and cyclophosphamide compared with conventional doxorubicin and cyclophosphamide in a randomized, multicenter trial of metastatic breast cancer. *J. Clin. Oncol.* **19**, 1444–1454 (2001).
26. S. Chan, N. Davidson, E. Juozaityte, *et al.*, Phase III trial of liposomal doxorubicin and cyclophosphamide compared with epirubicin and cyclophosphamide as first-line therapy for metastatic breast cancer. *Ann. Oncol.* **15**(10), 1527–1534 (2004).
27. L. Harris, G. Batist, R. Belt, *et al.*, TLC D-99 Study Group. Liposome-encapsulated doxorubicin compared with conventional doxorubicin in a randomized multicenter trial as first-line therapy of metastatic breast carcinoma. *Cancer* **94**(1), 25–36 (2002).
28. M. Theodoulou, G. Batist, S. Campos, *et al.*, Phase I study of nonpegylated liposomal doxorubicin plus trastuzumab in patients with HER2-positive breast cancer. *Clin. Breast Cancer* **9**(2), 101–107 (2009).
29. J. Cortes, S. Di Cosimo, M. A. Climent, *et al.*, Spanish Breast Cancer Cooperative Group SOLTI. Nonpegylated liposomal doxorubicin (TLC-D99), paclitaxel, and trastuzumab in HER-2-overexpressing breast cancer: a multicenter phase I/II study. *Clin. Cancer Res.* **15**(1), 307–314 (2009).
30. P. Schmid, J. Krocker, C. O. Schulz, *et al.*, Primary chemotherapy with gemcitabine, liposomal doxorubicin and docetaxel in patients with locally advanced breast cancer: results of a phase I trial. *Anticancer Drugs* **16**(1), 21–29 (2005).
31. V. Valero, A. U. Buzdar, R. L. Theriault, *et al.*, Phase II trial of liposome-encapsulated doxorubicin, cyclophosphamide, and fluorouracil as first-line therapy in patients with metastatic breast cancer. *J. Clin. Oncol.* **17**(5), 1425–1434 (1999).

32. A. M. Levine, A. Tulpule, B. Espina, *et al.*, Liposome-encapsulated doxorubicin in combination with standard agents (cyclophosphamide, vincristine, prednisone) in patients with newly diagnosed AIDS-related non-Hodgkin's lymphoma: results of therapy and correlates of response. *J. Clin. Oncol.* **22**(13), 2662–2670 (2004).
33. A. Tulpule, B. M. Espina, N. Berman, *et al.*, Phase I/II trial of nonpegylated liposomal doxorubicin, cyclophosphamide, vincristine, and prednisone in the treatment of newly diagnosed aggressive non-Hodgkin's lymphoma. *Clin. Lymphoma Myeloma* 7(1), 59–64 (2006).
34. L. Rigacci, S. Mappa, L. Nassi, *et al.*, Liposome-encapsulated doxorubicin in combination with cyclophosphamide, vincristine, prednisone and rituximab in patients with lymphoma and concurrent cardiac diseases or pre-treated with anthracyclines. *Hematol. Oncol.* **25**(4), 198–203 (2007).
35. G. Visani, F. Ferrara, F. Alesiani, *et al.*, R-COMP 21 for frail elderly patients with aggressive B-cell non-Hodgkin lymphoma: a pilot study. *Leuk. Lymphoma* **49**(6), 1081–1086 (2008).
36. L. Balducci and M. Extermann, Management of cancer in the older person: a practical approach. *Oncologist* **5**(3), 224–237 (2000).
37. M. Dell'Olio, R. P. Scalzulli, G. Sanpaolo, *et al.*, Non-pegylated liposomal doxorubicin (Myocet®) in patients with poor-risk aggressive B-cell non-Hodgkin lymphoma. *Leuk. Lymphoma.* **52**(7), 1222–1229 (2011).
38. S. Luminari, A. Montanini, D. Caballero, *et al.*, Nonpegylated liposomal doxorubicin (MyocetTM) combination (R-COMP) chemotherapy in elderly patients with diffuse large B-cell lymphoma (DLBCL): results from the phase II EUR018 trial. *Ann. Oncol.* **21**(7), 1492–1499 (2010).
39. G. Corazzelli, F. Frigeri, M. Arcamone, *et al.*, Biweekly rituximab, cyclophosphamide, vincristine, non-pegylated liposome-encapsulated doxorubicin and prednisone (R-COMP-14) in elderly patients with poor-risk diffuse large B-cell lymphoma and moderate to high 'life threat' impact cardiopathy. *Br. J. Haematol.* **154**(5), 579–589 (2011).
40. N. Noguchi, N. Nishino and E. Niki, Antioxidant action of the antihypertensive drug, carvedilol, against lipid peroxidation. *Biochem. Pharmacol.* **59**, 1069–1076 (2000).
41. H. Matsui, I. Morishima, Y. Numaguchi, *et al.*, Protective effects of carvedilol against doxorubicin-induced cardiomyopathy in rats. *Life Sciences* **65**, 1265–1274 (1999).
42. A. Isidori, F. Ferrara, F. Merli, *et al.*, Non-pegylated liposomal doxorubicin outruns the impact of a dose-dense regimen in very elderly patients with aggressive b-cell non hodgkin lymphoma: an Italian multicentric study on 129 patients. *Haematologica* **97**(1), 95 (abs236) (2012).
43. J. J. Lokich and C. Moore, Chemotherapy-associated palmar-plantar erythrodysesthesia syndrome. *Ann. Intern. Med.* **101**(6), 798–799 (1984).
44. A. Gabizon, H. Shmeeda, Y. Barenholz, Pharmacokinetics of pegylated liposomal doxorubicin: review of animal and human studies. *Clin. Pharmacokinet.* **42**, 419–436 (2003).
45. D. W. Northfelt, F. J. Martin, P. Working, *et al.*, Doxorubicin encapsulated in liposomes containing surface-bound polyethylene glycol: pharmacokinetics, tumor localization, and safety in patients with AIDS-related Kaposi's sarcoma. *J. Clin. Pharmacol.* **36**, 55–63 (1996).
46. M. Sharpe, S. E. Easthope, G. M. Keating, *et al.*, Polyethylene glycol-liposomal doxorubicin: a review of its use in the management of solid and haematological malignancies and AIDS-related Kaposi's sarcoma. *Drugs* **62**(14), 2089–2126 (2002).

47. D. W. Northfelt, B. J. Dezube, J. A. Thommes, *et al.*, Pegylated liposomal doxorubicin versus doxorubicin, bleomycin, and vincristine in the treatment of AIDS-related Kaposi's sarcoma: results of a randomized phase III clinical trial. *J. Clin. Oncol.* **16**(7), 2445–2451 (1998).
48. S. Stewart, H. Jablonowski, F. D. Goebel, *et al.*, Randomized comparative trial of pegylated liposomal doxorubicin versus bleomycin and vincristine in the treatment of AIDS-related Kaposi's sarcoma. International Pegylated Liposomal Doxorubicin Study Group. *J. Clin. Oncol.* **16**(2), 683–691 (1998).
49. L. Martin-Carbonero, A. Barrios, P. Saballs, *et al.*, Pegylated liposomal doxorubicin plus highly active antiretroviral therapy versus highly active antiretroviral therapy alone in HIV patients with Kaposi's sarcoma. *AIDS* **18**(12), 1737–1740 (2004).
50. M. E. O'Brien, N. Wigler, M. Inbar, *et al.*, CAELYX Breast Cancer Study Group. Reduced cardiotoxicity and comparable efficacy in a phase III trial of pegylated liposomal doxorubicin HCl (CAELYX/Doxil) versus conventional doxorubicin for first-line treatment of metastatic breast cancer. *Ann. Oncol.* **15**(3), 440–449 (2004).
51. A. M. Keller, R. G. Mennel, V. A. Georgoulias, *et al.*, Randomized phase III trial of pegylated liposomal doxorubicin versus vinorelbine or mitomycin C plus vinblastine in women with taxane-refractory advanced breast cancer. *J. Clin. Oncol.* **22**(19), 3893–3901 (2004).
52. E. Alba, M. Ruiz-Borrego, M. Margeli, *et al.*, Maintenance treatment with pegylated liposomal doxorubicin versus observation following induction chemotherapy for metastatic breast cancer: GEICAM 2001–01 study. *Breast Cancer Res. Treat.* **122**(1), 169–176 (2010).
53. G. Ferrandina, M. Ludovisi, D. Lorusso, *et al.*, Phase III trial of gemcitabine compared with pegylated liposomal doxorubicin in progressive or recurrent ovarian cancer. *J. Clin. Oncol.* **26**(6), 890–896 (2008).
54. A. N. Gordon, J. T. Fleagle, D. Guthrie, *et al.*, Recurrent epithelial ovarian carcinoma: a randomized phase III study of pegylated liposomal doxorubicin versus topotecan. *J. Clin. Oncol.* **19**(14), 3312–3322 (2001).
55. D. G. Mutch, M. Orlando, T. Goss, *et al.* Randomized phase III trial of gemcitabine compared with pegylated liposomal doxorubicin in patients with platinum-resistant ovarian cancer. *J. Clin. Oncol.* **25**(19), 2811–2818 (2007).
56. B. J. Monk, T. J. Herzog, S. B. Kaye, *et al.*, Trabectedin plus pegylated liposomal doxorubicin in recurrent ovarian cancer. *J. Clin. Oncol.* **28**(19), 3107–3114 (2010).
57. E. Pujade-Lauraine, U. Wagner, E. Aavall-Lundqvist, *et al.*, Pegylated liposomal doxorubicin and carboplatin compared with paclitaxel and carboplatin for patients with platinum-sensitive ovarian cancer in late relapse. *J. Clin. Oncol.* **28**(20), 3323–3329 (2010).
58. A. N. Gordon, M. Tonda, S. Sun, *et al.*, Long-term survival advantage for women treated with pegylated liposomal doxorubicin compared with topotecan in a phase 3 randomized study of recurrent and refractory epithelial ovarian cancer. *Gynecol. Oncol.* **95**(1), 1–8 (2004).
59. R. Martino, G. Perea, M. D. Caballero, *et al.*, Cyclophosphamide, pegylated liposomal doxorubicin (Caelyx), vincristine and prednisone (CCOP) in elderly patients with diffuse large B-cell lymphoma: results from a prospective phase II study. *Haematologica* **87**, 822–827 (2002).
60. G. Visani, B. Guiducci, F. D'Adamo, *et al.*, Cyclophosphamide, pegylated liposomal doxorubicin, vincristine and prednisone (CDOP) plus rituximab is effective and well tolerated in poor performance status elderly patients with non-Hodgkin's lymphoma. *Leuk. Lymphoma* **46**(3), 477–479 (2005).

61. U. Wollina, R. Dummer, N. H. Brockmeyer, *et al.*, Multicenter study of pegylated liposomal doxorubicin in patients with cutaneous T-cell lymphoma. *Cancer* **98**, 993–1001 (2003).
62. R. Dummer, P. Quaglino, J. C. Becker, *et al.*, Prospective international multicenter phase II trial of intravenous pegylated liposomal doxorubicin monochemotherapy in patients with stage IIB, IVA, or IVB advanced mycosis fungoides: final results from EORTC 21012. *J. Clin. Oncol.* **30**(33), 4091–4097 (2012).
63. A. M. Levine, A. Noy, J. Y. Lee, *et al.*, Pegylated liposomal doxorubicin, rituximab, cyclophosphamide, vincristine, and prednisone in AIDS-related lymphoma: AIDS Malignancy Consortium Study 047. *J. Clin. Oncol.* **31**(1), 58–64 (2013).
64. G. L. Plosker, Pegylated liposomal doxorubicin: a review of its use in the treatment of relapsed or refractory multiple myeloma. *Drugs* **68**(17), 2535–2551 (2008).
65. R. Z. Orlowski, A. Nagler, P. Sonneveld, *et al.*, Randomized phase III study of pegylated liposomal doxorubicin plus ortezomib compared with bortezomib alone in relapsed or refractory multiple myeloma: combination therapy improves time to progression. *J. Clin. Oncol.* **25**(25), 3892–3901 (2007).
66. P. Sonneveld, R. Hajek, A. Nagler, *et al.*, Combined pegylated liposomal doxorubicin and bortezomib is highly effective in patients with recurrent or refractory multiple myeloma who received prior thalidomide/lenalidomide therapy. *Cancer* **112**(7), 1529–1537 (2008).
67. J. Blade, P. Sonneveld, M. J. F. San, *et al.*, Pegylated liposomal doxorubicin plus bortezomib in relapsed or refractory multiple myeloma: efficacy and safety in patients with renal function impairment. *Clin. Lymphoma Myeloma* **8**(6), 352–355 (2008).
68. L. Blade, P. Sonneveld, J. F. San Miguel, *et al.*, Efficacy and safety of pegylated liposomal doxorubicin in combination with bortezomib for multiple myeloma: effects of adverse prognostic factors on outcome. *Clin. Lymphoma Myeloma Leuk.* **11**(1), 44–49 (2011).

Chapter 17

Clinical Experience with Nanoparticle Albumin Bound (Nab) Paclitaxel (Abraxane®)

Prakash Vishnu[*] *and Vivek Roy*[†,‡]

[*]*Floyd & Delores Jones Cancer Institute, Virginia Mason Medical Center, Seattle, WA 98101, USA*
[†]*Division of Hematology Oncology, Mayo Clinic, Jacksonville, FL, USA*
[‡]*roy.vivek@mayo.edu*

Abstract

Paclitaxel is a highly active chemotherapeutic agent used in the treatment of a number of malignancies. Nanoparticle albumin bound (*nab*)-paclitaxel is a solvent free, albumin-bound 130-nanometer particle formulation of paclitaxel (ABI-007; Abraxane®, Abraxis Bioscience). This preparation avoids toxicity of polyoxyethylated castor oil and ethanol vehicle used in solvent-based paclitaxel formulation. Solvent free formulation also allows higher doses of paclitaxel to be administered over a shorter infusion time without the need for special infusion sets or pre-medications. In addition to better safety and side-effect profile compared to conventional paclitaxel, *nab*-paclitaxel is associated with higher response rates and improved survival in metastatic breast cancer (MBC) patients receiving it as second line therapy. *Nab*-paclitaxel has been approved in the US and many other countries for treatment of anthracycline treated breast cancer. Studies have also evaluated its role in combination with other agents for first-line treatment of MBC and other cancers. It was approved in the US in 2012 for use in combination with carboplatin for the initial treatment of patients with locally advanced or metastatic non-small cell lung cancer (NSCLC) who are not candidates for curative surgery or radiation therapy. It was approved in 2013 for the first-line treatment of patients with metastatic adenocarcinoma of the pancreas, in combination with gemcitabine.

1. Introduction

Taxanes (e.g paclitaxel, docetaxel) are among the most active and widely used chemotherapy agents for a number of commonly occurring cancers including lung and breast cancer.[1–3] Taxanes are cell-cycle specific agents that stabilize cellular microtubules by binding to β-microtubule chain and promoting tubulin polymerization. This inhibits critical cellular mechanisms of cellular motility, intracellular transport and mitosis leading to apoptotic cell death. Taxanes also block antiapoptotic effects of BCL-2 gene family proteins, induce *TP*53 gene activation resulting in mitotic arrest and cell death.[4] Despite their efficacy there are notable drawbacks of taxanes. Being hydrophobic, paclitaxel has to be formulated for clinical use with a solvent which introduces the risk of hypersensitivity reaction, peripheral neuropathy and also impairs drug delivery.[5,6] A solvent-free 130-*n*anometer *a*lbumin-*b*ound paclitaxel (*nab*-paclitaxel, Abraxane®, Abraxis Bioscience) was approved for clinical use in the USA by U.S. Food and Drug Administration (FDA) in January 2005 and in Europe by European Medicines Agency (EMEA) in January 2008 for use in patients with metastatic breast cancer (MBC) previously treated with an anthracycline containing chemotherapy regimen.

In this chapter, we provide an overview of clinical pharmacology, safety and efficacy profile of *nab*-paclitaxel, and evaluate its use in treatment of breast cancer and other malignancies.

2. Side-Effects and Drawbacks of Solvent-Based Taxanes

Paclitaxel is a naturally occurring diterpinoid product extracted from the bark of pacific yew tree. It was first approved for clinical use in the USA in 1992. Being highly hydrophobic, Cremaphor EL (CrEL), a poly-oxy-ethylated castor oil mixed 1:1 with dehydrated ethanol is used to solubilize it for *i.v.* administration. Cremaphor is associated with adverse effects including hypersensitivity reactions, neuropathy and impaired drug delivery, limiting the clinical effectiveness.[5,6] CrEL based paclitaxel formulation requires the use of premedications and prolonged infusion time to prevent hypersensitivity reactions and the use of special infusion set to avoid exposure to di(2-ethylhexyl) phthalate (DHEP), which may leach from standard polyvinyl chloride sets. Even with these precautions, sometimes severe hypersensitivity reactions can still occur.[6] CrEL also alters the pharmacodynamic properties of paclitaxel leading to drug retension in plasma compartment, non-linear pharmacokinetics and reduced systemic clearance causing increase risk of systemic toxicity.[5] This may also partly explain why doses beyond the standard of care do not lead to improved efficacy,[7] while more frequent dosing (such as weekly administration) is associated with improved efficacy.[8]

To overcome these drawbacks of solvent-based paclitaxel, the nanoparticle albumin platform was developed which harnesses the natural properties of albumin to increase drug delivery to the tumor and eliminates the toxicity associated with solvents.

3. Nanomedicine and *Nab*-Technology

Nanomedicine utilizes materials and devices at extremely small scale in the diagnosis and treatment of diseases.[9] Albumin has a number of features that make it desirable as a drug delivery system. It is a natural carrier of endogenous hydrophobic molecules such as vitamins, hormones and other water-insoluble plasma substances and it plays an important role in endothelial transcytosis of protein-bound and unbound plasma constituents. Its binding to a cell-surface 60 kDa glycoprotein (gp60) receptor (albondin) on the endothelial cell membrane leads to the activation of caveolin-1 in the membrane vesicles and receptor mediated internalization of the albumin–drug complex. The secreted protein acidic and rich in cysteine (SPARC) is a matricellular glycoprotein that is produced by tumor and/or neighboring stroma, which binds to the albumin–drug complex with the resultant release of the drug in the vicinity of tumor cells.[9,10] Because of the potential of improved drug delivery and favorable therapeutic ratio, *nab* formulations of many water-insoluble compounds are being developed. *Nab*-paclitaxel is the most developed of these compounds and the only one approved for clinical use.

4. Clinical Experience with *Nab*-Paclitaxel in Breast Cancer

Nab-paclitaxel was demonstrated to have better intratumoral concentration and antitumor activity in preclinical models of multiple tumor types.[10,11] In mice xenograft tumor models, *nab*-paclitaxel was associated with higher LD_{50} (lethal dose, 50%) values and maximum tolerated dose (MTD) than CrEL-paclitaxel. Transfer across cell membrane and intratumor drug concentration was also higher for *nab*-paclitaxel, an effect that was abolished by inhibition of endothelial gp60 receptor and caveolar-mediated transport.[11] Zhou *et al* reported similar antitumor activity of *nab*-paclitaxel in hepatocelluar carcinoma (HCC) cell lines.[12] In HCC cell lines, the effective IC_{50} dose of *nab*-paclitaxel was 15-fold lower than solvent-based paclitaxel or docetaxel alone, and ~450-fold less compared to doxorubicin. SPARC is a type of caveolin-1 that has a sequence homology with gp60 and binds to albumin which has been suggested as a possible mechanism of better antitumor activity of *nab*-paclitaxel because of enhanced uptake and intratumoral accumulation of *nab*-paclitaxel[10] although mechanisms independent of SPARC may also exist.[13] SPARC is known to be overexpressed in several tumor types including breast cancer and may be associated with a poor prognosis. SPARC may be able to serve as a bio-marker for *nab*-paclitaxel effectiveness. In preclinical breast cancer models[10] as well as in a clinical study[14] in head and neck cancer, SPARC expression and its interaction with albumin was associated with enhanced uptake and intratumoral accumulation of *nab*-paclitaxel. Serum and tumor biomarkers (caveolin-1 and SPARC) are being assessed as predictors of response to *nab*-paclitaxel in a large Phase III study (NCT00785291) in MBC.

4.1. *Phase I studies*

Different dose schedules of *nab*-paclitaxel have been evaluated in Phase I and pharmacokinetics studies. A 3-weekly schedule was evaluated in 19 patients with advanced solid tumors.[15] Patients received *nab*-paclitaxel as a 30 min infusion given every 3 weeks without premedication using doses from 135 to 375 mg/m^2. Treatment was well tolerated without infusion reactions and only mild non-cumulative hematologic toxicity. Dose-limiting toxicity of sensory neuropathy (3 patients), stomatitis (2 patients) and superficial keratopathy (2 patients) occurred in 3 of 6 patients receiving the highest dose level of 375 mg/m^2. The MTD was determined to be 300 mg/m^2 every 3 weeks which is substantially higher than the CrEL-paclitaxel typically used in clinical practice. Unlike the non-linear kinetics of solvent-based paclitaxel, whole blood paclitaxel concentrations and area under the curve (AUC) values increase linearly over the dose range of 135–300 mg/m^2 of *nab*-paclitaxel.

Weekly dosing schedule was studied by Nyman *et al* in another Phase I trial[16] in 39 patients with advanced non-hematological malignancies. *Nab*-paclitaxel, 80–200 mg/m^2 was infused without premedication over 30 min once a week for 3 weeks every month. One third of patients received more than 6 cycles. After enrollment of the first cohort, patients were registered in two separate cohorts of "lightly" or "heavily" pretreated patients based on prior chemotherapy exposure. MTDs for these two cohorts were 150 mg/m^2 and 100 mg/m^2; dose-limiting toxicity were grade 3 peripheral neuropathy and grade 4 neutropenia. Linear pharmacokinetics was again noted and there were no dose-dependant changes in plasma clearance. Encouraging evidence of activity was noted including partial responses in CrEL-paclitaxel treated patients.

An interesting study by Gardner *et al.*[17] compared pharmacokinetics of *nab*-paclitaxel and CrEL-paclitaxel in the same patient using a randomized cross over study design. *Nab*-paclitaxel (260 mg/m^2 as a 30-min infusion) or CrEL-paclitaxel (175 mg/m^2 as a 3 h infusion) was given and crossed over to the alternative regimen after one cycle. Pharmacokinetic studies showed that the unbound free fraction and exposure to paclitaxel was higher with *nab*-paclitaxel.

Taxanes are commonly combined with cisplatin or carboplatin in clinical practice because of the mechanistic synergy and largely non-overlapping toxicity. To study the feasibility of *nab*-paclitaxel and platinum combination, a Phase I study evaluated three different dosing schedules of *nab*-paclitaxel in combination with carboplatin[18]. Forty-one patients with advanced heavily pre-treated solid tumors received *nab*-paclitaxel and carboplatin AUC of 6 on day 1. Group A received *nab*-paclitaxel at doses ranging from 220 to 340 mg/m^2 on day 1 every 21 days; group B received *nab*-paclitaxel at 100 or 125 mg/m^2 on days 1, 8, and 15 every 28 days; and group C received *nab*-paclitaxel 25 or 150 mg/m^2 on days 1 and 8 every 21 days. MTD of *nab*-paclitaxel in combination with carboplatin was 300, 100 and 125 mg/m^2 in groups A, B and C, respectively with myelosuppression as the primary dose limiting toxicity in all the groups.

4.2. *Phase II studies*

Encouraging safety data, favorable pharmacokinetics and evidence of activity led to several Phase II studies to evaluate antitumor efficacy of various dosing schedules and combinations. A multicenter Phase II study evaluated *nab*-paclitaxel for first line or subsequent therapy in 63 patients with MBC. Overall response rate (RR) of 48% (64% in chemotherapy naive patients), median time to progression (TTP) of 26.6 weeks and median overall survival (OS) of 63.6 weeks was noted. The main toxicities were neutropenia and sensory neuropathy. Grade 4 neutropenia occurred in 24% and grade 3 sensory neuropathy in 11%. No patient experienced grade 4 neuropathy. Hypersensitivity reactions were not seen despite lack of premedication.[19] *Nab*-paclitaxel has clinical activity in a subset of patients who have progressed after prior solvent-based paclitaxel treatment. In a phase II study patients with MBC who had progressed on paclitaxel or relapsed within 12 months of taxane therapy patients were treated with 100 mg/m^2 or 125 mg/m^2

nab-paclitaxel weekly for 3 weeks every 28 days. Objective responses were seen in 14–16% of patients and an additional 12–21% of patients had stable disease for 16 or more weeks.[20]

Mirtsching *et al.* studied weekly *nab*-paclitaxel as a first-line therapy of MBC, as monotherapy and in combination with trastuzumab.[21] *Nab*-paclitaxel (125 mg/m^2) was administered by 30 min *i.v.* infusion weekly for 3 of 4 weeks. Patients whose tumors overexpressed HER2 also received trastuzumab. 72 patients were enrolled; 22 patients had HER2 + breast cancer. The ORR was 42.2%; 5 patients had a CR and 22 patients had a PR. Additionally, 17 patients experienced SD, providing a disease control rate (CR + PR + SD) of 68.8%. Patients with HER2+ disease had an ORR of 52.4% compared to 38.1% in the HER2 negative population. Median progression free survival (PFS) was 14.5 months and survival rates at 1 year and 2 years were 69% and 62%, respectively. The most commonly observed toxicities were pain (64%), fatigue (58%), sensory neuropathy (54%), infection (46%), nausea (38%), alopecia (33%) and anemia (33%).

In a randomized multicenter Phase II trial 300 previously untreated patients with MBC received 300 mg/m^2 *nab*-paclitaxel every 3 weeks (n = 76), or 150 mg/m^2 every week (n = 74), or 100 mg/m^2 every week (n = 76) or docetaxel 100 mg/m^2 every 3 weeks (n = 74).[22] Weekly 150 mg/m^2 *nab*-paclitaxel was associated with significantly longer PFS than docetaxel (12.9 versus 7.5 months). Overall RR was also better with 150 mg/m^2 and 100 mg/m^2 per week regimen of *nab*-paclitaxel (49% and 45%, respectively) compared to 35% with docetaxel, but this did not reach statistical significance. RR and PFS were similar in the 3-weekly *nab*-paclitaxel and docetaxel cohorts. *Nab*-paclitaxel had a more favorable toxicity profile than docetaxel with lower incidence of grade 3/4 fatigue, neutropenia and febrile neutropenia in all *nab*-paclitaxel arms. All treatment groups experienced similar rates and severity of peripheral neuropathy but this resolved more rapidly after treatment withdrawal with *nab*-paclitaxel compared with patients who received docetaxel. *Nab*-paclitaxel was shown to be safe and effective in patients over the age of 65 in a post-hoc analysis of the four armed study that compared weekly *nab*-paclitaxel compared with 3-weekly *nab*-paclitaxel, solvent-based paclitaxel or docetaxel.[23]

Nab-paclitaxel in combination with gemcitabine was studied by Roy *et al.*[24] in a multicenter Phase II study in patients with MBC. In this open-labe trial, 50 patients with previously untreated MBC were treated with 125 mg/m^2 of *nab*-paclitaxel and 1000 mg/m^2 of gemcitabine on days 1 and 8 of a 21 days cycle until disease progression. A large majority of patients (80%) had visceral involvement and 60% of patients had at least 3 metastatic sites. ORR was 50% (8% CR and 42% PR). Median PFS was 7.9 months. PFS and OS at 6 months was 60% and 92% respectively. Neutropenia was the most common toxicity, grade 3 and 4 neutropenia occurring in 43% and 12%, respectively). Grade 3 neuropathy was noted in only 4 patients (8%), and no patient developed grade 4 neuropathy.

Based on the demonstrated superiority of dose-dense over conventional schedule for adjuvant treatment of breast cancer,[25] a pilot study evaluated sequential adjuvant dose-dense (every-2-week) doxorubicin (A) plus cyclophosphamide (C) followed by dose-dense *nab*-paclitaxel for early-stage breast cancer.[26] Women with high risk breast cancer (T_{1-3} N_{1-2}, or T_2 N_0 disease) were enrolled to receive four cycles of dose-dense A (60 mg/m^2) plus C (600 mg/m^2) with peg-filgrastim, followed by dose-dense *nab*-paclitaxel (260 mg/m^2) with peg-filgrastim given as needed. 30 patients received four cycles of dose-dense AC and 29 women proceeded to *nab*-paclitaxel therapy. Twenty-seven received all 4 doses (mean cumulative dose, 959); one discontinued *nab*-paclitaxel after 2 doses due to unacceptable AEs. Of 29 patients, 34% received peg-filgrastim during *nab*-paclitaxel therapy and 31% had a treatment delay, mainly due to hematologic toxicity. Four (14%) patients had a grade 3 *nab*-paclitaxel related neuropathy (no grade 4 neuropathy was seen) which improved after cessation of treatment. Kaplan–Meier probability estimates of the proportion of patients having grade 1 or less neuropathy at the end of treatment, 2, and 8 months after treatment were 59, 79, and 97%. The authors concluded that administering adjuvant dose-dense AC followed by 260 mg/m2 dose-dense *nab*-paclitaxel was feasible with manageable toxicity.

Sequential administration of *nab*-paclitaxel and anthracycline was evaluated in a neoadjuvant setting for locally advanced breast cancer (LABC) in a Phase II study by Robidoux *et al.*[27] Sixty-six treatment naïve patients with LABC irrespective of hormone receptor or HER2 status were treated with 100 mg/m^2 *nab*-paclitaxel weekly for 12 weeks followed by 5-fluorouracil, epirubicin and cyclophosphamide (FEC) every 3 weeks for 4 cycles. Trastuzumab was given in HER2-positive (HER2+) patients. Sixty-three patients completed 4 cycles of *nab*-paclitaxel following which 58 completed 4 cycles FEC. Seventeen of 19 HER2+ patients received trastuzumab. Pathologic complete response (pCR) was the primary objective of the study and it occurred in 29% (19 of 65) of patients. The pCR rate for HER2 positive subset of patients was 58% (11 of 19). Both *nab*-paclitaxel and FEC were well tolerated. The most significant toxicity were grade 2/3 neuropathy (16%) with *nab*-paclitaxel and grade 3/4 febrile neutropenia (7%) with FEC.

Other investigators have explored the possibility of 3 or 4 drug combinations with gemcitabine, carboplatin, bevacizumab or trastuzumab in early or advanced stage breast cancer. The reported studies show the feasibility of such combinations and encouraging evidence of efficacy.[23,28–33]

4.3. *Phase III studies*

A large (n = 460), multicenter, multi-national Phase III trial compared 260 mg/m^2 every 3-weeks *nab*-paclitaxel to CrEL-paclitaxel 175 mg/m^2 in patients with MBC.[34] The majority of patients were previously treated (59%), had more than 3 sites of disease (76%), and had visceral disease (79%). Actual paclitaxel dose delivered was almost 50% higher in the *nab*-paclitaxel group and associated with significantly higher RR (33% versus 19%, p = 0.001). *Nab*-paclitaxel as first line or subsequent therapy was associated with RR of 42% and 27% compared to 27% and 13%, respectively with CrEL-paclitaxel. *Nab*-paclitaxel was also associated with longer TTP (23 versus 16.9 weeks; hazard ratio [HR] = 0.75; p = 0.006). The overall survival of all patients was similar in the *nab*-paclitaxel and CrEL-paclitaxel groups (65 versus 55.7 weeks; p = 0.374); however, *nab*-paclitaxel as second-line or later therapy was associated with significantly longer survival than CrEL-paclitaxel (56.4 versus 46.7 weeks; HR = 0.73; p = 0.024). *Nab*-paclitaxel had lower incidence of neutropenia but higher likelihood of sensory neuropathy (10% versus 2%) which improved to grade 1–2 in a median of 22 days. Severe hypersensitivity reactions were not seen with *nab*-paclitaxel despite the lack of routine prophylactic premedications and shorter administration time.

Superior results with *nab*-paclitaxel were also seen in a multicenter randomized study from China in patients with MBC. The study compared *nab*-paclitaxel 260 mg/m^2 over 30 min every 3 weeks with no premedication to CrEL-paclitaxel 175 mg/m^2 over 3 hr every 3 weeks with standard premedication[35]. Improved ORR (54% versus 29%; p < 0.001), median TTP (7.6 months versus 6.2 months; p = 0.078), median PFS (7.6 months versus 6.2 months; p = 0.118) were seen in the *nab*-paclitaxel group.

Taken together, these studies provide strong evidence of better toxicity profile and improved efficacy of *nab*-paclitaxel[34], including responses in patients previously treated with solvent-based paclitaxel.[20] In particular, the incidence of any grade 3 and 4 neutropenia was similar or less in patients receiving *nab*-paclitaxel compared with those receiving paclitaxel, despite an increase in the dose of paclitaxel. Grade 3 sensory neuropathy was higher with *nab*-paclitaxel but manageable and quickly improved after cessation of therapy. Interestingly, responses were seen with *nab*-paclitaxel in some patients whose disease had progressed on solvent-based taxane. Prolongation of survival was in patients with MBC receiving *nab*-paclitaxel as second line or greater therapy.[34]

Because of the benefits of *nab*-paclitaxel seen in breast cancer patients, it has been tested in variety of other cancers, including non-small cell lung cancer (NSCLC), pancreatic cancer and melanoma at different stages of disease, as monotherapy and in combination with other cytotoxic chemotherapy and/or biologic agents.

5. Tolerability and Economic Considerations

The favorable characteristics of *nab*-paclitaxel have to be balanced against the higher cost of the medication. In the phase III trial by Gradidhar *et al.* trial[34] that compared *nab*-paclitaxel to CrEL-paclitaxel, the tolerability of the two preparations including Quality of Life were similar despite the higher dose of paclitaxel delivered in the *nab*-paclitaxel group (85.13 versus 57.02 mg/m^2/week). The near absence of hypersensitivity reactions even without corticosteroid or antihistamine premedication is a major difference between *nab*-paclitaxel and CrEL. *Nab*-paclitaxel may be safely used in patients who have previously experienced hypersensitivity reaction to CrEL.[36–38] The spectrum of toxicities of *nab*- and solvent-based paclitaxel are similar but there are some notable differences; neutropenia is less frequent with *nab*-paclitaxel whereas sensory neuropathy and gastrointestinal symptoms tend to be more frequent with *nab*-paclitaxel but sensory neuropathy resolves rapidly after treatment withdrawal. *Nab*-paclitaxel is cleared in the body mainly by non-renal mechanisms. However, caution should be exercised when using nab-paclitaxel in patients with renal dysfunction since it has not been studied in patients with renal compromise. Patients with serum creatinine >2 mg/dL were excluded from clinical trials. *Nab*-paclitaxel should be used with caution in patients with hepatic dysfunction, myelosuppression or preexisting neuropathy. Specific recommendations for dose modifications for various degrees of organ dysfunction are available in the prescribing information from the manufacturer.[39]

Although *nab*-paclitaxel is more expensive than solvent based taxanes, its better toxicity profile, faster infusion time and lack of need for premedications are significant cost savings compared to solvent based product. An economic analysis from the United Kingdom National Health Service (NHS) perspective was performed based on the resource utilization data of the Phase II randomized study comparing 3 different doses of *nab*-paclitaxel to standard 3-weekly docetaxel as first line therapy in patients with MBC. The total cost of therapy (medication cost, drug administration, monitoring, supportive care drugs and hospitalization due to toxicity) in the four arms was compared. Treatment costs in 100 mg/m^2 once weekly and 300 mg/m^2 every 3 weeks *nab*-paclitaxel arms were comparable to docetaxel arm while the *nab*-paclitaxel 150 mg/m^2 weekly arm had significantly higher overall costs.[40] In a similar economic analysis of treatment of MBC in the Canadian setting, the incremental cost per Quality-adjusted Life Year (gained) was more favorable for *nab*-paclitaxel than docetaxel.[41]

Thus, given its safety profile and better efficacy with comparable costs, *nab*-paclitaxel could be considered as a reasonable alternative to docetaxel as first-line chemotherapy in MBC. No such data exist for direct comparison of cost of treatment with *nab*-paclitaxel versus solvent based paclitaxel.

6. *Nab*-Paclitaxel for Cancers Other Than Breast

The large majority of clinical studies of nab-paclitaxel have been in the treatment of breast cancer. To take advantage of favorable toxicity profile and the potential for improved efficacy, it has also been investigated in a variety of other solid tumors including breast, lung, melanoma and pancreas. *Nab*-paclitaxel monotherapy was evaluated in a Phase II trial in 20 patients with advanced pancreatic cancer that had progressed on gemcitabine. Patients received 100 mg/m^2 *nab*-paclitaxel once a week for 3 weeks, every 28 days. Clinical benefit (1 partial response and 6 stable disease) was seen in 7 of 19 evaluable patients (37%). Median overall survival was 7.3 months.[42] SPARC expression was shown to correlate with higher RR and longer PFS in patients with metastatic pancreatic carcinoma treated with *nab*-paclitaxel and gemcitabine.[43] In a mouse xenograft model of human pancreatic cancer, intratumoral gemcitabine concentration was 2.8 fold higher when it was administered with *nab*-paclitaxel than as monotherapy.[44] In a Phase I/II clinical trial for patients with previously untreated pancreatic carcinoma gemcitabine 1000 mg/m^2 plus 125 mg/m^2 *nab*-paclitaxel weekly for 3 weeks every 28 days was determined to be the MTD. RR of 48% and median OS of 12.2 months and 48% one year overall survival was noted. SPARC expression in the stroma but not in the tumor correlated with survival.[44] Based on the promising activity of this combination a randomized, multinational Phase III trial was performed (NCT00844649). Patients with metastatic pancreatic cancer (n = 861) were randomized to gemcitabine 1000 mg/m^2 plus 125 mg/m^2 *nab*-paclitaxel on days 1, 8 and 15 for every 28 days or gemcitabine 1000 mg/m^2 weekly for 7 weeks in cycle 1 followed by the same dose on days 1, 8 and 15 for every 28 days. *Nab*-paclitaxel and gemcitabine combination was well tolerated and associated with statistically significant improvement in overall RR (23% versus 7%), median PFS (5.5 m versus 3.7 m) and overall survival (8.5 m versus 6.7 m).[45] Based on these data, the FDA in the USA approved it for the first-line treatment of patients with metastatic adenocarcinoma of the pancreas, in combination with gemcitabine in 2013.

A Phase II study[46] of 3-weekly 260 mg/m^2 *nab*-paclitaxel in advanced NSCLC found acceptable tolerability and RR of 16%, disease control rate of 49%, median TTP of 6 months and median survival of 11 months. A Phase I/II study[47] evaluated weekly *nab*-paclitaxel in patients with NSCLC, three dose levels (100 mg/m^2, 125 mg/m^2, and 150 mg/m^2) weekly for 3 weeks every 28 days were evaluated. Two DLTs (grade III sensory neuropathy and neutropenic fever) occurred at the highest dose level leading to the selection of 125 mg/m^2 as Phase II dose. Forty patients were treated at this dose. Overall RR of 30%, PFS of 5 months and median survival of 11 months was observed. *Nab*-paclitaxel and carboplatin in combination with bevacizumab was shown to be well tolerated and showed encouraging evidence of activity in a Phase II trial as first line treatment of patients with advanced non-small cell lung carcinoma.[48]

A large Phase III study[49] involving 1,052 patients compared *nab*-paclitaxel (100 mg/m^2 weekly) versus CrEL-paclitaxel (200 mg/m^2 every 3 weeks) in combination with carboplatin (AUC 6 mg × min/mL) every 3 weeks as a first line treatment for stage IIIB or IV NSCLC. The objective RR as determined by independent radiologic review for *nab*-paclitaxel/carboplatin was 33% versus 25% for paclitaxel/carboplatin. The *nab*-paclitaxel/carboplatin combination was particularly effective for patients with squamous cell carcinoma, with an objective RR of 41% versus 24% in the paclitaxel/carboplatin arm. Sensory neuropathy, myalgia and neutropenia occurred in significantly more patients on paclitaxel while thrombocytopenia and anemia were significantly more likely in patients who received *nab*-paclitaxel. Based on these data, the Food and Drug Administration of USA (FDA) approved *nab*-paclitaxel (Abraxane®; Abraxis Bioscience) for patients with locally advanced or metastatic NSCLC in October 2012.

Based on evidence of activity of paclitaxel in metastatic melanoma and bladder cancer, *nab*-paclitaxel was evaluated for these indications. In a Phase II trial by North Central Cancer Treatment Group (NCCTG) combination of *nab*-paclitaxel (100 mg/m^2) and carboplatin (AUC 2) on days 1, 8 and 15 every 28 days was studied in chemotherapy naïve (41) or previously treated (35) patients. The treatment was well tolerated and led to an objective RR of 25.6% chemotherapy naïve and 9% in previously treated cohort. Median PFS was 4.5 months and 4.1 months, and overall survival was 11.1 and 10.9 months, respectively.[50] Another Phase II multicenter study treated 74 patients with chemotherapy naïve (37) or previously treated (37) metastatic melanoma with single agent *nab*-paclitaxel weekly for 3 of 4 weeks. Chemotherapy naïve patients received 150 mg/m^2 whereas previously treated patients were treated with 100 mg/m^2. *Nab*-Paclitaxel was found to be well tolerated and active in both cohorts with RR of 2.7% and 21.6%, PFS of 3.5 months and 4.5 months, and median survival of 12.1 months and 9.6 months, respectively.[51] *Nab*-paclitaxel in combination with carboplatin and gemcitabine was evaluated as neoadjuvant therapy in patients with locally advanced carcinoma of the bladder. The combination was feasible with manageable toxicity, mainly myelosuppression. Twenty-two of the 27 patients were evaluable for response 6 of whom achieved a pathologic CR and 5 additional patients had only residual carcinoma *in situ*.[52] Promising evidence of activity has been reported in ovarian, fallopian tube and primary peritoneal malignancies.[53,54] Further studies are needed to precisely define the role of *nab*-paclitaxel monotherapy or combinations in these and other malignancies.

References

1. NCCN Clinical Practice Guidelines in Oncology/Breast Cancer v 3.2012. 2012; http://www.nccn.org/professionals/physician_gls/PDF/breast.pdf. Accessed 11/29/2012.
2. C. G. Azzoli, S. Baker Jr. and S. Temin *et al.*, American Society of Clinical Oncology clinical practice guideline update on chemotherapy for stage IV non-small-cell lung cancer, *J. Clin. Oncol.* **27**(36), 6251–6266 (2009).
3. V. Kataja and M. Castiglione, Primary breast cancer: ESMO clinical recommendations for diagnosis, treatment and follow-up, *Ann. Oncol.* **20**(Suppl 4), 10–14 (2009).
4. C. Dumontet and B. I. Sikic, Mechanisms of action of and resistance to antitubulin agents: microtubule dynamics, drug transport, and cell death, *J. Clin. Oncol.* **17**(3), 1061–1070 (1999).
5. A. Sparreboom, L. van Zuylen and E. Brouwer *et al.*, Cremophor EL-mediated alteration of paclitaxel distribution in human blood: clinical pharmacokinetic implications, *Cancer Res.* **59**(7), 1454–1457 (1999).
6. A. J. ten Tije, J. Verweij, W. J. Loos and A. Sparreboom, Pharmacological effects of formulation vehicles: implications for cancer chemotherapy, *Clinical pharmacokinetics* **42**(7), 665–685 (2003).
7. E. P. Winer, D. A. Berry and S. Woolf *et al.*, Failure of higher-dose paclitaxel to improve outcome in patients with metastatic breast cancer: cancer and leukemia group B trial 9342, *J. Clin. Oncol.* **22**(11), 2061–2068 (2004).
8. A. D. Seidman, "Will weekly work"? Seems to be so, *J. Clin. Oncol.* **23**(25), 5873–5874 (2005).
9. B. Y. Kim, J. T. Rutka and W. C. Chan, Nanomedicine, *N. Engl. J. Med.* **363**(25), 2434–2443 (2010).
10. N. P. Desai, V. Trieu, L. Y. Hwang, R. Wu P. Soon-Shiong and W. J. Gradishar, Improved effectiveness of nanoparticle albumin-bound (nab) paclitaxel versus polysorbate-based

docetaxel in multiple xenografts as a function of HER2 and SPARC status, *Anticancer Drugs* **19**(9), 899–909 (2008).
11. N. Desai, V. Trieu and Z. Yao *et al.*, Increased antitumor activity, intratumor paclitaxel concentrations, and endothelial cell transport of cremophor-free, albumin-bound paclitaxel, ABI-007, compared with cremophor-based paclitaxel, *Clin. Cancer Res.* **12**(4), 1317–1324 (2006).
12. Q. Zhou, A. K. Ching and W. K. Leung *et al.*, Novel therapeutic potential in targeting microtubules by nanoparticle albumin-bound paclitaxel in hepatocellular carcinoma, *Int. J. Oncol.* **38**(3), 721–731 (2011).
13. H. Shao, H. Tang and O. E. Salavaggione *et al.*, Improved response to nab-paclitaxel compared with cremophor-solubilized paclitaxel is independent of secreted protein acidic and rich in cysteine expression in non-small cell lung cancer, *J. Thorac. Oncol.* **6**(6), 998–1005 (2011).
14. N. Desai, V. Trieu, B. Damascelli and P. Soon-Shiong, SPARC Expression Correlates with Tumor Response to Albumin-Bound Paclitaxel in Head and Neck Cancer Patients, *Translational oncology* **2**(2), 59–64 (2009).
15. N. K. Ibrahim, N. Desai and S. Legha. *et al.*, Phase I and pharmacokinetic study of ABI-007, a Cremophor-free, protein-stabilized, nanoparticle formulation of paclitaxel, *Clin. Cancer Res.* **8**(5), 1038–1044 (2002).
16. D. W. Nyman, K. J. Campbell and E. Hersh *et al.*, Phase I and pharmacokinetics trial of ABI-007, a novel nanoparticle formulation of paclitaxel in patients with advanced nonhematologic malignancies, *J. Clin. Oncol.* **23**(31), 7785–7793 (2005).
17. E. R. Gardner, W. L. Dahut and C. D. Scripture *et al.*, Randomized crossover pharmacokinetic study of solvent-based paclitaxel and nab-paclitaxel, *Clin. Cancer Res.* **14**(13), 4200–4205 (2008).
18. T. E. Stinchcombe, M. A. Socinski and C. M. Walko *et al.*, Phase I and pharmacokinetic trial of carboplatin and albumin-bound paclitaxel, ABI-007 (Abraxane) on three treatment schedules in patients with solid tumors, *Cancer Chemother. Pharmacol.* **60**(5), 759–766 (2007).
19. N. K. Ibrahim, B. Samuels and Page R *et al.*, Multicenter phase II trial of ABI-007, an albumin-bound paclitaxel, in women with metastatic breast cancer, *J. Clin. Oncol.* **23**(25), 6019–6026 (2005).
20. J. L. Blum, M. A. Savin and G. Edelman *et al.*, Phase II study of weekly albumin-bound paclitaxel for patients with metastatic breast cancer heavily pretreated with taxanes, *Clin. Breast Cancer* **7**(11), 850–856 (2007).
21. B. Mirtsching, T. Cosgriff, G. Harker, M. Keaton, T. Chidiac and M. Min, A Phase II Study of Weekly Nanoparticle Albumin-Bound Paclitaxel With or Without Trastuzumab in Metastatic Breast Cancer, *Clin. Breast Cancer* 1–8 (2011).
22. W. J. Gradishar, D. Krasnojon and S. Cheporov *et al.*, Significantly longer progression-free survival with nab-paclitaxel compared with docetaxel as first-line therapy for metastatic breast cancer, *J. Clin. Oncol.* **27**(22), 3611–3619 (2009).
23. M. Aapro, S. Tjulandin, P. Bhar and W. Gradishar. Weekly nab-paclitaxel is safe and effective in ≥65 years old patients with metastatic breast cancer: a post-hoc analysis, *Breast* **20**(5), 468–474 (2011).
24. V. Roy, B. R. LaPlant, G. G. Gross, C. L. Bane and F. M. Palmieri, Phase II trial of weekly nab (nanoparticle albumin-bound)-paclitaxel (nab-paclitaxel) (Abraxane) in combination

with gemcitabine in patients with metastatic breast cancer (N0531), *Ann. Oncol.* **20**(3), 449–453 (2009).
25. M. L. Citron, D. A. Berry and C. Cirrincione *et al.*, Randomized trial of dose-dense versus conventionally scheduled and sequential versus concurrent combination chemotherapy as postoperative adjuvant treatment of node-positive primary breast cancer: first report of Intergroup Trial C9741/Cancer and Leukemia Group B Trial 9741, *J. Clin. Oncol.* **21**(8), 1431–1439 (2003).
26. N. Robert, L. Krekow, C. Stokoe, A. Clawson, J. Iglesias and J. O'Shaughnessy. Adjuvant dose-dense doxorubicin plus cyclophosphamide followed by dose-dense nab-paclitaxel is safe in women with early-stage breast cancer: a pilot study, *Breast Cancer Res. Treat.* **125**(1), 115–120 (2011).
27. A. Robidoux, A. U. Buzdar and E. Quinaux *et al.*, A Phase II Neoadjuvant Trial of Sequential Nanoparticle Albumin-Bound Paclitaxel Followed by 5-Fluorouracil/ Epirubicin/Cyclophosphamide in Locally Advanced Breast Cancer, *Clin. Breast Cancer.* **10**(1), 81–86 (2010).
28. C. Lobo, G. Lopes and O. Baez *et al.*, Final results of a phase II study of nab-paclitaxel, bevacizumab, and gemcitabine as first-line therapy for patients with HER2-negative metastatic breast cancer, *Breast Cancer Res. Treat* **123**(2), 427–435 (2010).
29. H. L. McArthur, H. Rugo and B. Nulsen *et al.*, A feasibility study of bevacizumab plus dose-dense doxorubicin-cyclophosphamide (AC) followed by nanoparticle albumin-bound paclitaxel in early-stage breast cancer, *Clin. Cancer Res.* **17**(10), 3398–3407 (2011).
30. D. W. Northfelt, A. C. Dueck and T. P. Flynn *et al.*, Phase II trial combining nab-paclitaxel (NP), gemcitabine (G), and bevacizumab (B) in patients (pts) with metastatic breast cancer (MBC): NCCTG N0735, *J. Clin. Oncol.* **29**(suppl), abstr 1126 (2011).
31. I. B. Paz, S. Lau and C. Garberoglio, *et al.*, Nab-paclitaxel and carboplatin with or without trastuzumab (trast) as part of neoadjuvant chemotherapy (NCT) in patients (pts) with stage II-III breast cancer (BC), *J. Clin. Oncol.* **26**(suppl), abstr 567 (2008).
32. A. D. Seidman, A. K. Conlin and A. Bach *et al.*, Phase II study of weekly nanoparticle albumin bound (nab)paclitaxel with carboplatin and trastuzumab as 1st-line therapy for HER2-positive metastatic breast cancer (MBC), *J. Clin. Oncol.* **26**(suppl), abstr 1047 (2008).
33. D. A. Yardley, E. Raefsky and R. Castillo *et al.*, Phase II Study of Neoadjuvant Weekly nab-Paclitaxel and Carboplatin, With Bevacizumab and Trastuzumab, As Treatment For Women With Locally Advanced HER2(+) Breast Cancer, *Clin. Breast Cancer* **11**(5), 297–305 (2011).
34. W. J. Gradishar, S. Tjulandin and N. Davidson *et al.*, Phase III trial of nanoparticle albumin-bound paclitaxel compared with polyethylated castor oil-based paclitaxel in women with breast cancer, *J. Clin. Oncol.* **23**(31), 7794–7803 (2005).
35. Z. Z Guan, Q. L. Li and F. Feng *et al.*, Superior efficacy of a Cremophor-free albumin-bound paclitaxel compared with solvent-based paclitaxel in Chinese patients with metastatic breast cancer. *Asia Pac, J. Clin. Oncol.* **5**(3), 165–174 (2009).
36. A. N. Fader and P. G. Rose. Abraxane for the treatment of gynecologic cancer patients with severe hypersensitivity reactions to paclitaxel, *Int. J. Gynecol. Cancer* **19**(7), 1281–1283 (2009).
37. J. P. Micha, B. H. Goldstein, C. L. Birk, M. A. Rettenmaier and J. V. Brown III3, Abraxane in the treatment of ovarian cancer: the absence of hypersensitivity reactions, *Gynecol Oncol.* **100**(2), 437–438 (2006).

38. J. Timoney, D. McGuffy and D. DeRitis *et al.*, Incidence of hypersensitivity reactions (HSRs) to albumin-bound paclitaxel (ABI) in patients (pts) with a history of HSR to cremophor-based paclitaxel (P), *J. Clin. Oncol.* **29**s(abstract): abstract 2566 (2011).
39. ABRAXANE Healthcare Professional Prescribing Information. 2012; http://www.abraxane.com/mbc/hcp/. Accessed 111/29/2012.
40. G. Dranitsaris, R. Coleman and W. Gradishar, nab-Paclitaxel weekly or every 3 weeks compared to standard docetaxel as first-line therapy in patients with metastatic breast cancer: an economic analysis of a prospective randomized trial, *Breast Cancer Res. Treat.* **119**(3), 717–724 (2010).
41. G. Dranitsaris, W. Cottrell, B. Spirovski and S. Hopkins, Economic analysis of albumin-bound paclitaxel for the treatment of metastatic breast cancer, *J. Oncol. Pharm. Pract.* **15**(2), 67–78 (2009).
42. P. J. Hosein, G. D. Lopes Jr. and C. M. Gomez *et al.*, A phase II trial of nab-paclitaxel (NP) in patients with advanced pancreatic cancer (PC) who have progressed on gemcitabine (G)-based therapy, *J. Clin. Oncol.* **28**(15s), abstr 4120 (2010).
43. D. D. Von Hoff, R. K. Ramanathan and M. J. Borad *et al.*, SPARC correlation with response to gemcitabine (G) plus nab-paclitaxel (nab-P) in patients with advanced metastatic pancreatic cancer: A phase I/II study, *J. Clin. Oncol.* **27**(15s), abstr 4525 (2009).
44. D. D. Von Hoff, R. K. Ramanathan and M. J. Borad *et al.*, Gemcitabine Plus nab-Paclitaxel Is an Active Regimen in Patients With Advanced Pancreatic Cancer: a Phase I/II Trial, *J. Clin. Oncol.* **29**(34), 4548 (2011).
45. D. D. Von Hoff, T. J. Ervin and F. P. Arena *et al.*, Randomized phase III study of weekly nab-paclitaxel plus gemcitabine versus gemcitabine alone in patients with metastatic adenocarcinoma of the pancreas (MPACT), *J. Clin. Oncol.* **30**(34s), abstr LBA 148 (2013).
46. M. R. Green, G. M. Manikhas and S. Orlov *et al.*, Abraxane, a novel Cremophor-free, albumin-bound particle form of paclitaxel for the treatment of advanced non-small-cell lung cancer, *Ann. Oncol.* **17**(8), 1263–1268 (2006).
47. N. A. Rizvi, G. J. Riely and C. G. Azzoli *et al.*, Phase I/II trial of weekly intravenous 130-nm albumin-bound paclitaxel as initial chemotherapy in patients with stage IV non-small-cell lung cancer, *J. Clin. Oncol.* **26**(4), 639–643 (2008).
48. C. Reynolds, D. Barrera and R. Jotte *et al.*, Phase II trial of nanoparticle albumin-bound paclitaxel, carboplatin, and bevacizumab in first-line patients with advanced nonsquamous non-small cell lung cancer, *J. Thorac. Oncol.* **4**(12), 1537–1543 (2009).
49. M. A. Socinski, I. Bondarenko and N. A. Karaseva *et al.*, Weekly nab-Paclitaxel in Combination With Carboplatin Versus Solvent-Based Paclitaxel Plus Carboplatin as First-Line Therapy in Patients With Advanced Non–Small-Cell Lung Cancer: Final Results of a Phase III Trial, *J. Clin. Oncol.* 2012 **30**(17), 2055–2062 (2012).
50. L. A. Kottschade, V. J. Suman and T. Amatruda III *et al.*, A phase II trial of nab-paclitaxel (ABI-007) and carboplatin in patients with unresectable stage IV melanoma: a North Central Cancer Treatment Group Study, N057E(1), *Cancer* **117**(8), 1704–1710 (2011).
51. E. M. Hersh, S. J. O'Day and A. Ribas *et al.*, A phase 2 clinical trial of nab-paclitaxel in previously treated and chemotherapy-naive patients with metastatic melanoma, *Cancer* **116**(1), 155–163 (2010).
52. D. C. Smith, P. Grivas and S. Daignault *et al.*, A phase II trial of neoadjuvant ABI-007, carboplatin, and gemcitabine (ACG) in patients with locally advanced carcinoma of the bladder, *J. Clin. Oncol.* **29**(Suppl 7), abstr 244 (2011).

53. R. L. Coleman, W. E. Brady and D. S. McMeekin *et al.*, A phase II evaluation of nanoparticle, albumin-bound (nab) paclitaxel in the treatment of recurrent or persistent platinum-resistant ovarian, fallopian tube, or primary peritoneal cancer: a Gynecologic Oncology Group study, *Gynecol. Oncol.* **122**(1), 111–115 (2011).
54. M. G. Teneriello, P. C. Tseng and M. Crozier *et al.*, Phase II evaluation of nanoparticle albumin-bound paclitaxel in platinum-sensitive patients with recurrent ovarian, peritoneal, or fallopian tube cancer, *J. Clin. Oncol.* **27**(9), 1426–1431 (2009).

Index